HANDBOOK OF

Media for Environmental Microbiology

SECOND EDITION

HANDBOOK OF

Media for Environmental Microbiology

SECOND EDITION

By

RONALD M. ATLAS

CRC Press is an imprint of the
Taylor & Francis Group, an informa business
A TAYLOR & FRANCIS BOOK

First published 2005 by Taylor & Francis

Published 2019 by CRC Press
Taylor & Francis Group
6000 Broken Sound Parkway NW, Suite 300
Boca Raton, FL 33487-2742

First issued in paperback 2019

ISBN 13: 978-0-367-45418-0 (pbk)
ISBN 13: 978-0-8493-3560-0 (hbk)

Visit the Taylor & Francis Web site at
http://www.taylorandfrancis.com

and the CRC Press Web site at
http://www.crcpress.com

Library of Congress Card Number 2004065561

Library of Congress Cataloging-in-Publication Data

Atlas, Ronald M., 1946-
Handbook of media for environmental microbiology / Ronald M. Atlas.--2nd ed.
p. cm.
Includes index.
ISBN 0-8493-3560-4 (alk. paper)
1. Microbiology--Cultures and culture media--Handbooks, manuals, etc. 2. Sanitary microbiology --Handbooks, manuals, etc. 3. Water--Microbiology--Handbooks, manuals, etc. 4. Microbial ecology --Handbooks, manuals, etc. I. Title.

QR66.3.A848 2005
579'.028--dc22 2004065561

About the Author:

Ronald M. Atlas, Ph.D., is dean of the Graduate School, professor of Biology, professor of Public Health, and codirector of the Center for the Deterrence of Biowarfare and Bioterrorism at the University of Louisville. He received his B.S. from the State University of New York at Stony Brook in 1968 and his M.S. and Ph.D. from Rutgers in 1970 and 1972, respectively. After spending a year at the Jet Propulsion Laboratory in Pasadena, California, he joined the faculty at the University of Louisville. He also has served as an Adjunct Professor at the University of Puerto Rico, External Examiner at the National University of Singapore, External Examiner at the Chinese University in Hong Kong, and Extraordinary Professor of Microbiology at the University of Pretoria. Dr. Atlas has received a number of honors including: The University of Louisville Excellence in Research Award, Johnson and Johnson Fellowship for Biology, the American Society for Microbiology (ASM) Award in Applied and Environmental Microbiology, the ASM Founders Award, and the Edmund Youde Lectureship Award in Hong Kong.

He has taught a variety of courses in microbiology at the University of Louisville and has authored several textbooks in general microbiology and microbial ecology. He has written nearly 300 research papers and authored or edited more than 20 books. He has conducted studies on the fate of oil in the sea. As part of these studies, he has extensively characterized marine bacterial populations and examined the diversity of microorganisms. He pioneered the field of bioremediation for marine oil spills. He also has conducted studies on the application of molecular techniques to environmental problems. His studies have included the development of "suicide vectors" for the containment of genetically engineered microorganisms and the use of gene probes and the polymerase chain reaction for environmental monitoring, including the detection of pathogens and indicator bacteria for water quality monitoring.

He served as president of the American Society for Microbiology. Additionally, he has served on the National Institutes of Health (NIH) Recombinant Advisory Committee (RAC), as well as on various advisory boards for the Environmental Protection Agency (EPA), Federal Bureau of Investigation, and Department of Homeland Security. He has been chairperson of the Environmental Committee and the Task Force on Biological Weapons of the Public and Scientific Affairs Board of the American Society for Microbiology. He has been a national lecturer for Sigma Xi, an American Society for Microbiology Foundation lecturer, and an Australian Society for Microbiology national lecturer. He has served on the editorial boards of *Applied and Environmental Microbiology, Advances in Microbial Ecology, BioScience, Biotechniques*, *Journal of Environmental Science, Environmental Microbiology, Journal of Industrial Microbiology,* and *Biosecurity and Bioterrorism*. He is editor of *Critical Reviews in Microbiology.*

Contents

Introduction

Overview

The second edition of the *Handbook of Media for Environmental Microbiology* includes descriptions of nearly 2000 media that are used for environmental microbial analyses, e.g., of water quality; for the isolation of microorganisms from soils, waters, and other environmental samples; and for the cultivation and maintenance of environmentally relevant microorganisms, including plant pathogens. It provides the formulations for the wide range of media needed to cultivate the ever expanding diversity of microorganisms that are culurable.

Organization

The media described in the second edition of the *Handbook of Media for Environmental Microbiology* are organized alphabetically. Synonyms for media are listed. The description of each medium includes its name(s), composition, instructions for preparation, commercial sources, safety cautions where needed, and uses.

Names of Media

Media often have numerous names. For the most part the second edition of the *Handbook of Media for Environmental Microbiology* retains the original names assigned in the literature. In some cases media with identical compositions produced by different companies have different names. For example Trypticase™ Soy Agar produced as a BBL product of BD Diagnostic Systems, Tryptone Soy Agar produced by Oxoid Unipath, and Tryptic Soy Agar produced as a Difco product of BD Diagnostic Systems have identical compositions. Many media also are known by acronyms. TSA, for example, is the common acronym for Trypticase Soy Agar.

Trademarks

The names of some media, components of media, and other terms are registered trademarks. The trademarked items referred to in the second edition of the *Handbook of Media for Environmental Microbiology* are listed below.

American Type Culture Collection® and ATCC® are trademarks of the American Type Culture Collection.

Bacto®, BiTek®, and Difco® are trademarks of Difco Laboratories (registered trademarks owned by Becton Dickinson and Company).

Oxoid® and Lab–Lemco® are trademarks of Unipath Ltd.

Acidase®, BBL®, Biosate®, CTA Medium®, DTA Medium®, DCLS Agar®, Desoxycholate®, Desoxycholate Agar®, Desoxycholate Citrate Agar®, Enterococcosel®, Eugonagar®, Eugonbroth®, GC-Lect®, Gelysate®, IsoVitaleX®, Mycobactosel®, Mycophil®, Mycosel®, Myosate®, Phytone®, Polypeptone®, Selenite-F Enrichment®, Thiotone®, Trichosel®, Trypticase®, TSA II®, and TSI Agar® are trademarks of Becton Dickinson and Co.

Composition of Media

Media for the cultivation of microorganisms contain the substances necessary to support the growth of microorganisms. Due to the diversity of microorganisms and their diverse metabolic pathways, there are numerous media. Even slight differences in the composition of a medium can result in dramatically different growth characteristics of microorganisms.

When methods for culturing microorganisms were first developed in the 19th century, largely by Robert Koch and his colleagues, animal and plant tissues were principally used as sources of nutrients used to support microbial growth. One of the major discoveries of Fanny Hesse in Koch's laboratory was that agar could be used to form solidified culture media on which microorganisms would grow. Extracts of plants and animal tissues were prepared as broths or mixed with agar to form a variety of culture media. Virtually any plant, animal, or animal organ was considered for use in preparing media. Infusions were prepared from beef heart, calf brains, and beef liver, as a few examples. These classic infusions still form the primary components of many media that are widely used.

The composition section of each medium describes the ingredients that make up the medium, their amounts, and the pH. It lists those ingredients in order of decreasing amount. Solids are listed first showing the weights to be added, followed by liquids showing the volumes to be included in the medium.

The composition uses generic terms where these are applicable. For example, pancreatic digest of casein is marketed by various manufacturers as trypticase, tryptone, and other commercial product names. While there may well be differences between these products, such differences are undefined. Variations also occur between batches of products produced as digests of animal tissues.

Media for the cultivation of microorganisms have a source of carbon for incorporation into biomass. For autotrophs, the carbon source most often is carbon dioxide, which may be supplied as bicarbonate within the medium. Carbohydrates, such as glucose, or other organic compounds, such as acetate, various lipids, proteins, hydrocarbons, and other organic compounds, are included in media as sources of carbon for heterotrophs. These carbon sources may also serve as the supply of energy. Other compounds, such as ammonium ions, nitrite ions, elemental sulfur, and reduced iron, may be used as the sources of energy for the cultivation of autotrophs. Nitrogen also is required for microbial growth. It may be supplied as inorganic nitrogen compounds for the cultivation of some microorganisms but more commonly is supplied as proteins, peptones, or amino acids. Phosphates and metals, such as magnesium and iron, are also necessary components of microbiological media. Phosphates may also serve as buffers to maintain the pH of the medium within the growth tolerance limits of the microorganism being cultivated. Various additional growth factors may also be included in the media.

Agars

Agar is the most common solidifying agent used in microbiological media. Agar is a polysaccharide extract from marine algae. It melts at 84°C and solidifies at 38°C. Agar concentrations of 15.0g/L typically are used to form solid media. Lower concentrations of 7.5–10.0g/L are used to produce soft agars or semisolid media. Below are some agars used as solidifying agents in various media.

Agar Bacteriological (Agar No. 1)
An agar with low calcium and magnesium. Available from Oxoid Unipath.

Agar, Bacto
A purified agar with reduced pigmented compounds, salts, and extraneous matter. Available from BD Diagnostic Systems.

Agar, BiTek™
Agar prepared as a special technical grade. Available from BD Diagnostic Systems.

Agar, Flake
A technical-grade agar. Available from BD Diagnostic Systems.

Agar, Grade A
A select-grade agar containing minerals. Available from BD Diagnostic Systems.

Agar, Granulated
A high-grade granulated agar that has been filtered, decolorized, and purified. Available from BD Diagnostic Systems.

Agar, Purified
A very high-grade agar that has been filtered, decolorized, and purified by washing and extraction of refined agars. It has reduced mineral content. Available from BD Diagnostic Systems.

Agar, Technical (Agar No. 3)
A technical-grade agar. Available from BD Diagnostic Systems and Oxoid Unipath.

Agarose
A low-sulfate neutral gelling fraction of agar that is a complex galactose polysaccharide of near neutral charge.

Ionagar
A purified agar. Available from Oxoid Unipath.

Noble Agar
An agar that has been extensively washed and is essentially free of impurities. Available from BD Diagnostic Systems.

Purified Agar
An agar that has been extensively washed and extracted with water and organic solvent. Available from BD Diagnostic Systems and Oxoid Unipath.

Peptones

Many complex media, that is, media in which not all the specific chemical components are known, contain peptones as the source of nitrogen. Peptones are hydrolyzed proteins formed by enzymatic or acidic digestion. Casein most often is used as the protein substrate for forming peptones, but other substances, such as soybean meal, also are commonly employed. Below is a list of some of the peptones that are used as ingredients in various media.

Acidase™ Peptone
A hydrochloric acid hydrolysate of casein. It has a nitrogen content of 8% and is deficient in cystine and tryptophan. Available from BD Diagnostic Systems.

Bacto Casitone
A pancreatic digest of casein. Available from BD Diagnostic Systems.

Bacto Peptamin
A peptic digest of animal tissues. Available from BD Diagnostic Systems.

Bacto Peptone
An enzymatic digest of animal tissues. It has a high concentration of low molecular weight peptones and amino acids. Available from BD Diagnostic Systems.

Bacto Proteose Peptone
An enzymatic digest of animal tissues. It has a high concentration of high molecular weight peptones. Available from BD Diagnostic Systems.

Bacto Soytone
A enzymatic hydrolysate of soybean meal. Available from BD Diagnostic Systems.

Bacto Triplets
An enzymatic hydrolysate containing numerous peptides, including those of higher molecular weights. Available from BD Diagnostic Systems.

Bacto Tryptone
A pancreatic digest of casein. Available from BD Diagnostic Systems.

Biosate™ Peptone
A hydrolysate of plant and animal proteins. Available from BD Diagnostic Systems.

Casein Hydrolysate
A hydrolysate of casein prepared with hydrochloric acid digestion under pressure and neutralized with sodium hydroxide. It contains total nitrogen of 7.6% and NaCl of 28.3%. Available from Oxoid Unipath.

Gelatone
A pancreatic digest of gelatin. Available from BD Diagnostic Systems.

Gelysate™ Peptone
A pancreatic digest of gelatin deficient in cystine and tryptophan and which has a low carbohydrate content. Available from Oxoid Unipath.

Lactoalbumin Hydrolysate
A pancreatic digest of lactoalbumin, a milk whey protein. It has high levels of amino acids. It contains total nitrogen of 11.9% and NaCl of 1.4%. Available from BD Diagnostic Systems and Oxoid Unipath.

Liver Digest Neutralized
A papaic digest of liver that contains total nitrogen of 11.0% and NaCl of 1.6%. Available from Oxoid Unipath.

Mycological Peptone
A peptone that contains total nitrogen of 9.5% and NaCl of 1.1%. Available from Oxoid Unipath.

Myosate™ Peptone
A pancreatic digest of heart muscle. Available from BD Diagnostic Systems.

Neopeptone
An enzymatic digest of protein. Available from from BD Diagnostic Systems.

Peptone Bacteriological Neutralized
A mixed pancreatic and papaic digest of animal tissues. It contains total nitrogen of 14.0% and NaCl of 1.6%. Available from BD Diagnostic Systems and Oxoid Unipath.

Peptone P
A peptic digest of fresh meat that has a high sulfur content and contains total nitrogen of 11.12% and NaCl of 9.3%. Available from BD Diagnostic Systems and Oxoid Unipath.

Peptonized Milk
A pancreatic digest of high-grade skim milk powder. It has a high carbohydrate and calcium concentration. It contains total nitrogen of 5.3% and NaCl of 1.6%. Available from Oxoid Unipath.

Phytone™ Peptone
A papaic digest of soybean meal. It has a high vitamin and a high carbohydrate content. Available from BD Diagnostic Systems.

Polypeptone™ Peptone
A mixture of peptones composed of equal parts of pancreatic digest of casein and peptic digest of animal tissue. Available from BD Diagnostic Systems.

Proteose Peptone
A specialized peptone prepared from a mixture of peptones that contains a wide variety of high molecular weight peptides. It contains total nitrogen of 12.7% and NaCl of 8.0%. Available from BD Diagnostic Systems and Oxoid Unipath.

Proteose Peptone No. 2
An enzymatic digest of animal tissues with a high concentration of high molecular weight peptones. Available from BD Diagnostic Systems.

Proteose Peptone No. 3
An enzymatic digest of animal tissues. It has a high concentration of high molecular weight peptones. Available from BD Diagnostic Systems.

Soya Peptone
A papaic digest of soybean meal with a high carbohydrate concentration. It contains total nitrogen of 8.7% and NaCl of 0.4%. Available from Oxoid Unipath.

Soytone
A papaic digest of soybean meal. Available from BD Diagnostic Systems.

Special Peptone
A mixture of peptones, including meat, plant, and yeast digests. It contains a wide variety of peptides, nucleotides, and minerals. It contains total nitrogen of 11.7% and NaCl of 3.5%. Available from Oxoid Unipath.

Thiotone™ E Peptone
An enzymatic digest of animal tissue. Available from BD Diagnostic Systems.

Trypticase™ Peptone
A pancreatic digest of casein. It has a very low carbohydrate content and a relatively high tryptophan content. Available from BD Diagnostic Systems.

Tryptone
A pancreatic digest of casein. It contains total nitrogen of 12.7% and NaCl of 0.4%. Available from Oxoid Unipath.

Tryptone T
A pancreatic digest of casein with lower levels of calcium, magnesium, and iron than tryptone. It contains total nitrogen of 11.7% and NaCl of 4.9%. Available from BD Diagnostic Systems and Oxoid Unipath.

Tryptose
An enzymatic hydrolysate containing high molecular weight peptides. It contains total nitrogen of 12.2% and NaCl of 5.7%. Available from BD Diagnostic Systems and Oxoid Unipath.

Meat and Plant Extracts

Meat and plant infusions are aqueous extracts that are commonly used as sources of nutrients for the cultivation of microorganisms. Such infusions contain amino acids and low molecular weight peptides, carbohydrates, vitamins, minerals, and trace metals. Extracts of animal tissues contain relatively high concentrations of water-soluble protein components and glycogen. Extracts of plant tissues contain relatively high concentrations of carbohydrates.

With regard to infusions, many media list as an ingredient infusion from beef heart or another animal tissue. This ingredient is prepared by boiling a given amount of the animal tissue (e.g., 500.0g), and then using the liquid or, more commonly, drying the broth and using the solids from the infusion. The actual weight of the dry solids extracted from the hot water used to create the infusion varies, and so the ingredient typically is simply listed as 500.0g beef heart infusion, although the actual weight of solids recovered from the infusion and used in the medium is far less.

Below is a list of some of the meat and plant extracts that are used as ingredients in various media.

Bacto Beef
A desiccated powder of lean beef. Available from BD Diagnostic Systems.

Bacto Beef Extract
An extract of beef (paste). Available from BD Diagnostic Systems.

Bacto Beef Extract Desiccated
An extract of desiccated beef. Available from BD Diagnostic Systems.

Bacto Liver
A desiccated powder of beef liver. Available from BD Diagnostic Systems.

Lab-Lemco
A meat extract powder. Available from Oxoid Unipath.

Liver Desiccated
Dehydrated ox livers. Available from Oxoid Unipath.

Malt Extract
A water-soluble extract from germinated grain dried by low-temperature evaporation. It has a high carbohydrate content. It contains total nitrogen of 1.1% and NaCl of 0.1%.

Growth Factors

Many microorganisms have specific growth factor requirements that must be included in media for their successful cultivation. Vitamins, amino acids, fatty acids, trace metals, and blood components often must be added to media. In some cases, specific defined components are used to meet the growth factor requirements. Incorporation of growth factors are used to enrich, that is, to increase the numbers of particular species of microorganisms. Most often, mixtures of growth factors are used in microbiological media. Acid hydrolysates of casein commonly are used as sources of amino acids. Extracts of yeast cells also are employed as sources of amino acids and vitamins for the cultivation of microorganisms. Many media, particularly those employed in the clinical laboratory, contain blood or blood components that serve as essential nutrients for fastidious microorganisms. X factor (heme) and V factor (nicotinamide adenine dinucleotide) often are supplied by adding hemoglobin, IsoVitaleX, and/or Supplement VX.

Selective Components

Many media contain selective components that inhibit the growth of nontarget microorganisms. Selective media are especially useful in the isolation of specific microorganisms from mixed populations. In many media for the study of microorganisms in nature, compounds are included in the media as sole sources of carbon or nitrogen so that only a few types of microorganisms can grow. Selective toxic compounds are also frequently used to select for the cultivation of particular microbial species. The isolation of a pathogen from a stool specimen, for example, where there is a high abundance of nonpathogenic normal microbiota, requires selective media. Often, antimicrobics or other selectively toxic compounds are incorporated into media to suppress the growth of the background microbiota while permitting the cultivation of the target organism of interest. Bile salts, selenite, tetrathionate, tellurite, azide, phenylethanol, sodium lauryl sulfate, high sodium chloride concentrations, and various dyes—such as eosin, Crystal Violet, and Methylene Blue—are used as selective toxic chemicals. Antimicrobial agents used to suppress specific types of microorganisms include ampicillin, chloramphenicol, colistin, cycloheximide, gentamicin, kanamycin, nalidixic acid, sulfadiazine, and vancomycin. Various combinations of antimicrobics are effective in suppressing classes of microorganisms, such as enteric bacteria.

Differential Components

The differentiation of many microorganisms is based upon the production of acid from various carbohydrates and other carbon sources or the decarboxylation of amino acids. Some media include indicators, particularly of pH, that permit the visual detection of changes in pH resulting from such metabolic reactions.

Below is a list of some commonly used pH indicators.

pH Indicator	pH Range	Acid Color	Alkaline Color
m-Cresol Purple	0.5 – 2.5	Red	Yellow
Thymol Blue	1.2 – 2.8	Red	Yellow
Bromphenol Blue	3.0 – 4.6	Yellow	Blue
Bromcresol Green	3.8 – 5.4	Yellow	Blue
Chlorcresol Green	4.0 – 5.6	Yellow	Blue
Methyl Red	4.2 – 6.3	Red	Yellow
Chlorphenol Red	5.0 – 6.6	Yellow	Red
Bromcresol Purple	5.2 – 6.8	Yellow	Purple
Bromothymol Blue	6.0 – 7.6	Yellow	Blue
Phenol Red	6.8 – 8.4	Yellow	Red
Cresol Red	7.2 – 8.8	Yellow	Red
m-Cresol Purple	7.4 – 9.0	Yellow	Purple
Thymol Blue	8.0 – 9.6	Yellow	Blue
Cresolphthalein	8.2 – 9.8	Colorless	Red
Phenolphthalein	8.3 – 10.0	Colorless	Red

pH Buffers

Maintaining the pH of media usually is accomplished by the inclusion of suitable buffers. Because microorganisms grow optimally only within certain limits of a pH range, the pH generally is maintained within a few tenths of a pH unit. For the phosphage buffers, the pH is established by using varying volumes of equimolar concentrations of Na_2HPO_4 and NaH_2PO_4.

pH	Na_2HPO_4 (mL)	NaH_2PO_4 (mL)
5.4	3.0	97.0
5.6	5.0	95.0
5.8	7.8	92.2
6.0	12.0	88.0
6.2	18.5	81.5
6.4	26.5	73.5
6.6	37.5	62.5
6.8	50.0	50.0
7.0	61.1	38.9
7.2	71.5	28.5
7.4	80.4	19.6
7.6	86.8	13.2
7.8	91.4	8.6
8.0	94.5	5.5

Preparation of Media

The ingredients in a medium are usually dissolved, and the medium is then sterilized. When agar is used as a solidifying agent, the medium must be heated gently, usually to boiling, to dissolve the agar. In some cases where interactions of components, such as metals, would cause precipitates, solutions must be prepared and occasionally sterilized separately before mixing the various solutions to prepare the complete medium. The pH often is adjusted prior to sterilization, but in some cases sterile acid or base is used to adjust the pH of the medium following sterilization. Many media are sterilized by exposure to elevated temperatures. The most common method is to autoclave the medium. Different sterilization procedures are employed when heat-labile compounds are included in the formulation of the medium.

Tyndallization

Exposure to steam at 100°C for 30 min will kill vegetative bacterial cells but not endospores. Such exposure can be achieved using flowing steam in an Arnold sterilizer. By allowing the medium to cool and incubate under conditions where endospore germination will occur and by repeating the 100°C–30 min exposure on 3 successive days, the medium can be sterilized because all the endospores will have germinated and the heat exposure will have killed all the vegetative cells. This process of repetitive exposure to 100°C is called tyndallization, after its discoverer, John Tyndall.

Inspissation

Inspissation is a heat exposure method that is employed with high-protein materials, such as egg-containing media, that cannot withstand the high temperatures used in autoclaving. This process causes coagulation of the protein without greatly altering its chemical properties. Using an Arnold sterilizer or a specialized inspissator, the medium is exposed to 75°–80°C for 2 hr on each of 3 successive days. Inspissation using an autoclave employs exposure to 85°–90°C for 10 min achieved by having a mixture of air and steam in the chamber, followed by a 15 min exposure during which the temperature is raised to 121°C using only steam under pressure in the chamber; the temperature then is slowly lowered to less than 60°C.

Autoclaving

Autoclaving uses exposure to steam, generally under pressure, to kill microorganisms. Exposure for 15 min to steam at 15 psi—121°C is most commonly used. Such exposure kills vegetative bacterial cells and bacterial endospores. Media containing carbohydrates often are sterilized at 116°–118°C in order to prevent the decomposition of the carbohydrate and the formation of toxic compounds that would inhibit microbial growth. Below is a list of pressure–temperature relationships.

Pressure—psi	Temperature—°C
0	100.0
1	101.9
2	103.6
3	105.3
4	106.9
5	108.4
6	109.8
7	111.3
8	112.6
9	113.9
10	115.2
11	116.4
12	117.6
13	118.8
14	119.9
15	121.0
16	122.0
17	123.0
18	124.0
19	125.0
20	126.0
21	126.9
22	127.8
23	128.7
24	129.6
25	130.4

Filtration

Filtration is commonly used to sterilize media containing heat-labile compounds. Liquid media are passed through sintered glass or membranes, typically made of cellulose acetate or nitrocellulose, with small pore sizes. A membrane with a pore size of 0.2mm will trap bacterial cells and, therefore, sometimes is called a bacteriological filter. By preventing the passage of microorganisms, filtration renders fluids free of bacteria and eukaryotic microorganisms, that is, free of living organisms, and hence sterile. Many carbohydrate solutions, antibiotic solutions, and vitamin solutions are filter sterilized and added to media that have been cooled to temperatures below 50°C.

Caution about Hazardous Components

Some media contain components that are toxic or carcinogenic. Appropriate safety precautions must be taken when using media with such components. Basic fuchsin and acid fuchsin are carcinogens, and caution must be used in handling media with these compounds to avoid dangerous exposure that could lead to the development of malignancies. Thallium salts, sodium azide, sodium biselenite, and cyanide are among the toxic components found in some media. These compounds are poisonous, and steps must be taken to avoid ingestion, inhalation, or skin contact. Azides also react with many metals, especially copper, to form explosive metal azides. The disposal of azides must avoid contact with copper or achieve sufficient dilution to avoid the formation of such hazardous explosive compounds. Media with sulfur-containing compounds may result in the formation of hydrogen sulfide, which is a toxic gas. Care must be used to ensure proper ventilation. Media with human blood or human blood components must be handled with great caution to avoid exposure to human immunodeficiency virus and other pathogens that contaminate some blood supplies.

Uses of Media

The *Handbook of Media for Environmental Microbiology, Second Edition* contains the media used for the testing of waters and wastewaters recommended by the USEPA for the standard methods examination of water and food. It also includes the formulations of media used to cultivate the microorganisms of environmental significance that are held in the major culture collections of the world.

References

Below is a list of references that can be consulted for further information about media used for the isolation, cultivation, and differentiation of microorganisms.

Archaea: A Laboratory Manual. 1995. F.T. Robb, A.R. Place, K.R. Sowers, H.J. Schreier, C. DasSarma, and E.M. Fleischmann. Cold Spring Harbor Laboratory Press, Cold Spring Harbor, New York.

Archaea: A Laboratory Manual: Halophiles. 1995. DasSarma, S. and E.M. Fleischmann. Cold Spring Harbor Laboratory Press. Cold Spring Harbor, New York.

Archaea: A Laboratory Manual: Methanogens. 1995. Sowers, K.R. and H.J. Schreier. Cold Spring Harbor Laboratory Press. Cold Spring Harbor, New York.

Archaea: A Laboratory Manual: Thermophiles. 1995. Robb, F.T. and A.R. Place. Cold Spring Harbor Laboratory Press, Cold Spring Harbor, New York.

Difco & BBL Manual: Dehydrated Culture Media and Reagents for Microbiology. 2003. Becton, Dickinson and Co., Sparks, Maryland.

The Oxoid Manual. 1990. E.Y. Bridson, ed. Unipath Ltd. Basingstoke, Hampshire, England.

The Prokaryotes: An Evolving Electronic Resource for the Microbiological Community. 1999. M. Dworkin, Springer-Verlag Inc. New York.

Standard Methods for the Examination of Water and Wastewater. 1998. Clesceri, L.S., A.E. Greenberg, and A.D. Eaton, American Public Health Association Publications, Washington D.C.

Web Resources

Below is a list of Web sites that provide information about media and microbial cultures.

American Type Culture Collection (ATCC), a global biological resource in the United States with many services. http://www.atcc.org/catalogs/catalogs.html

Belgian Coordinated Collections of Microorganisms of the Laboratorium Voor Microbiologie at the University of Gent (LMG). http://www.belspo.be/bccm/db/media.htm

Czechoslovokian Collection of Microorganisms. (CCM). http://www.sci.immuni.cz/ccm

Finish Culture Collection, Valtion Teknillinen Tutkimuskeskus (VTT). http://www.inf.vtt.fi

German Collection of Microorganisms (DSMZ). http://www.gbf-braunschweig.de

Japanese Collection of Microorganisms and Microbial Cultures. http://www.jcm.riken.go.jp

Netherlands Centraalbureau voor Schimmelcultures (CBS). http://www.cbs.know.know.nl/databases.

Russian Specialized Collection of Alkanotrophs. http://www.ecology. psu.ru/iegmcol

Spanish Collection of Microorganisms (Colección Española de Cultivos Tipo Catalogo de Cepas). http://www.cect.org/english/index.htm

United Kingdom National Collection of Yeast Cultures. http://www.ifr.bbsrc.ac.uk/ncyc

United Kingdom National Culture Collection Microbiological Resources. http://www.ukncc.co.uk

United States Food and Drug Administration Bacteriological Analytical Manual. http://vm.cfsan.fda.gov/~embam/bam-toc.html.

World Federation of Culture Collections. http://www. wfcc.info

A 1 Broth

Composition per liter:

Pancreatic digest of casein 20.0g
Lactose 5.0g
NaCl 5.0g
Salicin 0.5g
Triton™ X-100 .... 1.0mL
pH 6.9 ± 0.1 at 25°C

Source: This medium is available as a premixed powder from BD Diagnostic Systems.

Preparation of Medium: Add components to distilled/deionized water and bring volume to 1.0L. Mix thoroughly. Gently heat and bring to boiling. Distribute into test tubes containing an inverted Durham tube. Autoclave for 10 min at 15 psi pressure–121°C.

Use: For the detection of fecal coliforms in treated wastewater, and seawater by a most-probable-number (MPN) method. Multiple dilutions of samples (3, 5, or 10 replicates per dilution) are added to tubes containing A 1 broth. After incubation, test tubes with gas accumulation in the Durham tubes are scored positive and those with no gas as negative. A MPN table is consulted to determine the most probable number of fecal coliforms.

Acanthamoeba Medium

Composition per liter:

Proteose peptone 15.0g
Glucose 15.0g
KH_2PO_4 .... 0.3g
L-Methionine 14.9mg
Thiamine 1.0mg
Biotin 0.2mg
Vitamin B_{12} 1.0μg
Salt solution 1.0mL
pH 5.5 ± 0.2 at 25°C

Salt Solution:

Composition per 100.0mL:

$MgSO_4 \cdot 7H_2O$ 2.46g
$CaCl_2 \cdot 2H_2O$ 0.15g
$FeCl_3$ 0.02g

Preparation of Salt Solution: Add components to distilled/deionized water and bring volume to 100.0mL. Mix thoroughly.

Preparation of Medium: Add components to distilled/deionized water and bring volume to 1.0L. Mix thoroughly. Adjust pH to 5.5. Filter through Whatman filter paper to remove particles. Distribute into screw-capped tubes or flasks. Autoclave for 15 min at 15 psi pressure–121°C.

Use: For the cultivation of *Acanthamoeba* species.

ACC Medium

Composition per liter:

Proteose peptone 20.0g
Agar 12.0g
Glycerol 1.5g
K_2SO_4 .... 1.5g
$MgSO_4 \cdot 7H_2O$ 1.5g
Antibiotic solution 10.0mL
pH 7.2 ± 0.2 at 25°C.

Antibiotic Solution:

Composition per 10.0mL:

Cycloheximide 0.075g
Ampicillin 0.05g
Chloramphenicol 0.0125g

Preparation of Antibiotic Solution: Add components to distilled/deionized water and bring volume to 10.0mL. Mix thoroughly. Filter sterilize.

Preparation of Medium: Add components, except antibiotic solution, to distilled/deionized water and bring volume to 990.0mL. Mix thoroughly. Gently heat and bring to boiling. Autoclave for 15 min at 15 psi pressure–121°C. Cool to 45°–50°C. Aseptically add sterile antibiotic solution. Mix thoroughly. Pour into sterile Petri dishes or distribute into sterile tubes.

Use: For the selective isolation and cultivation of fluorescent *Pseudomonas* species.

Acetamide Agar

Composition per liter:

Agar 15.0g
Acetamide 10.0g
NaCl 5.0g
K_2HPO_4 .... 1.0g
$NH_4H_2PO_4$ 1.0g
$MgSO_4 \cdot 7H_2O$ 0.2g
Bromothymol Blue 0.08g
pH 6.9 ± 0.2 at 25°C

Preparation of Medium: Add components to distilled/deionized water and bring volume to 1.0L. Mix thoroughly. Gently heat and bring to boiling. Adjust pH. Distribute into tubes or flasks. Autoclave for 15 min at 15 psi pressure–121°C. Cool tubes in a slanted position to produce a long slant.

Use: For the differentiation of nonfermentative Gram-negative bacteria, especially *Pseudomonas aeruginosa.* Can be used as a confirmatory test for water analysis. Bacteria that deamidate acetamide turn the medium blue.

Acetamide Agar

Composition per liter:

Agar 15.0g
Acetamide 10.0g
NaCl 5.0g
K_2HPO_4 .... 1.39g
KH_2PO_4 .... 0.73g
$MgSO_4 \cdot 7H_2O$ 0.5g
Phenol Red 0.012g
pH 6.9 ± 0.2 at 25°C

Source: This medium is available as a premixed powder from BD Diagnostic Systems.

Preparation of Medium: Add components to distilled/deionized water and bring volume to 1.0L. Mix thoroughly. Gently heat and bring to boiling. Adjust pH. Distribute into tubes or flasks. Autoclave for 15 min at 15 psi pressure–121°C. Cool tubes in a slanted position to produce a long slant.

Use: For the differentiation of nonfermentative Gram-negative bacteria, especially *Pseudomonas aeruginosa.* Can be used as a confirmatory test for water analysis. Bacteria that deamidate acetamide turn the medium blue.

Acetamide Broth

Composition per liter:

Acetamide 10.0g
NaCl 5.0g
K_2HPO_4 .... 1.39g
KH_2PO_4 .... 0.73g

$MgSO_4 \cdot 7H_2O$ 0.5g
Phenol Red 0.012g
pH 6.9 ± 0.2 at 25°C

Preparation of Medium: Add components to distilled/deionized water and bring volume to 1.0L. Mix thoroughly. Adjust pH. Autoclave for 15 min at 15 psi pressure–121°C.

Use: For the differentiation of nonfermentative Gram-negative bacteria, especially *Pseudomonas aeruginosa.* Can be used as a confirmatory test for water analysis. Bacteria that deamidate acetamide turn the broth purplish red.

Acetate Agar

Composition per liter:
Agar 15.0g
Yeast extract 2.0g
Sodium acetate 1.0g
Pancreatic digest of casein 1.0g
pH 7.4 ± 0.2 at 25°C

Preparation of Medium: Add components to distilled/deionized water and bring volume to 1.0L. Mix thoroughly. Gently heat and bring to boiling. Distribute into tubes or flasks. Autoclave for 15 min at 15 psi pressure–121°C. Pour into sterile Petri dishes or leave in tubes.

Use: For the cultivation and maintenance of *Caryophanon latum.*

Acetate Differential Agar (Sodium Acetate Agar) (Simmons' Citrate Agar, Modified)

Composition per liter:
Agar 20.0g
NaCl 5.0g
Sodium acetate 2.0g
$(NH_4)H_2PO_4$ 1.0g
K_2HPO_4 .... 1.0g
$MgSO_4 \cdot 7H_2O$ 0.2g
Bromothymol Blue 0.08g
pH 6.8 ± 0.2 at 25°C

Source: This medium is available as a premixed powder from BD Diagnostic Systems.

Preparation of Medium: Add components to cold distilled/deionized water and bring volume to 1.0L. Mix thoroughly. Gently heat and bring to boiling. Distribute into tubes to produce a 1 cm butt and 30 cm slant. Autoclave for 15 min at 15 psi pressure–121°C. Cool tubes in a slanted position.

Use: For the differentiation of *Shigella* species from *Escherichia coli* and also for the differentiation of nonfermenting Gram-negative bacteria. Bacteria that can utilize acetate as the sole carbon source turn the medium blue.

Acetivibrio cellulolyticus Medium

Composition per 1170.0mL:
Cellobiose or cellulose (MN 300, Whatman CF II, Kleenex tissue paper, or HCl-treated cotton) 3.0g
$NaHCO_3$ 2.0g
L-Cysteine·HCl 0.25g
$Na_2S \cdot 9H_2O$ 0.25g
$FeSO_4 \cdot 7H_2O$ 0.02g
Resazurin 0.001g
Mineral solution 1 .... 75.0mL
Mineral solution 2 .... 75.0mL
Cellobiose solution 50.0mL
Trace elements solution 10.0mL
Vitamin solution 10.0mL
Reducing agent solution 10.0mL
pH 7.2 ± 0.2 at 25°C

Mineral Solution 1:
Composition per liter:
K_2HPO_4 .... 3.9g

Preparation of Mineral Solution 1: Add K_2HPO_4 to distilled/deionized water and bring volume to 1.0L. Mix thoroughly.

Mineral Solution 2:
Composition per liter:
$(NH_4)_2SO_4$ 6.0g
K_2HPO_4 .... 2.4g
$MgSO_4 \cdot 7H_2O$ 1.2g
$CaCl_2 \cdot 2H_2O$ 0.72g
NaCl 0.59g

Preparation of Mineral Solution 2: Add components to distilled/deionized water and bring volume to 1.0L. Mix thoroughly.

Cellobiose Solution:
Composition per 50.0mL:
D-Cellobiose 5.0g

Preparation of Cellobiose Solution: Add cellobiose to distilled/deionized water and bring volume to 50.0mL. Mix thoroughly. Sparge under 100% N_2 gas for 3 min. Filter sterilize. Store under N_2 gas.

Trace Elements Solution:
Composition per liter:
$MgSO_4 \cdot 7H_2O$ 3.0g
Nitrilotriacetic acid 1.5g
$CaCl_2 \cdot 2H_2O$ 1.0g
NaCl 1.0g
$MnSO_4 \cdot 2H_2O$ 0.5g
$CoSO_4 \cdot 7H_2O$ 0.18g
$ZnSO_4 \cdot 7H_2O$ 0.18g
$FeSO_4 \cdot 7H_2O$ 0.1g
$NiCl_2 \cdot 6H_2O$ 0.025g
$KAl(SO_4)_2 \cdot 12H_2O$ 0.02g
$CuSO_4 \cdot 5H_2O$ 0.01g
H_3BO_3 .... 0.01g
$Na_2MoO_4 \cdot 2H_2O$ 0.01g
$Na_2SeO_3 \cdot 5H_2O$ 0.3mg

Preparation of Trace Elements Solution: Add nitrilotriacetic acid to 500.0mL of distilled/deionized water. Dissolve by adjusting pH to 6.5 with KOH. Add distilled/deionized water to 1.0L. Add remaining components. Mix thoroughly. Adjust pH to 7.0 with 1*N* KOH.

Vitamin Solution:
Composition per liter:
Pyridoxine·HCl 10.0mg
Calcium DL-pantothenate 5.0mg
Lipoic acid 5.0mg
Nicotinic acid 5.0mg
p-Aminobenzoic acid 5.0mg
Riboflavin 5.0mg
Thiamine·HCl 5.0mg
Biotin 2.0mg
Folic acid 2.0mg
Vitamin B_{12} 0.1mg

Preparation of Vitamin Solution: Add components to distilled/deionized water and bring volume to 1.0L. Mix thoroughly.

Reducing Agent Solution:
Composition per 110.0mL:

L-Cysteine·HCl·H_2O 2.5g
$Na_2S \cdot 9H_2O$ 2.5g

Preparation of Reducing Agent Solution: Add 110.0mL of distilled/deionized water to a 250.0mL flask. Boil under N_2 gas for 1 min. Cool to room temperature. Add L-cysteine·HCl·H_2O and dissolve. Adjust to pH 9 with 5*N* NaOH. Add washed $Na_2S \cdot 9H_2O$ and dissolve. Distribute under N_2 gas in 10.0mL volumes into tubes. Autoclave for 10 min at 15 psi pressure–121°C.

Preparation of Medium: Add components, except cellobiose solution and reducing agent solution, to distilled/deionized water and bring volume to 940.0mL. Gently heat and bring to boiling. Continue boiling for 3 min. Cool to room temperature under 80% N_2 + 20% CO_2. Adjust pH to 7.6 by gassing. Distribute anaerobically under 80% N_2 + 20% CO_2. Autoclave for 15 min at 15 psi pressure–121°C. After autoclaving, the pH of the medium will be 7.2. Prior to inoculation of cultures, aseptically and anaerobically add 0.1mL of sterile reducing agent solution and 0.5mL of sterile cellobiose solution to each tube containing 9.4mL of sterile basal medium.

Use: For the cultivation and maintenance of *Acetivibrio cellulolyticus,* an anaerobic, cellulolytic bacterium.

Acetivibrio Desulfovibrio Medium (LMG Medium 105)

Composition per liter:

Solution A 869.0mL
Solution C 100.0mL
Solution D 10.0mL
Solution E 10.0mL
Solution F 10.0mL
Solution B 1.0mL

pH 7.7 ± 0.2 at 25°C

Solution A:
Composition per 869.0mL:

Na_2SO_4 .... 3.0g
NaCl 1.0g
KCl 0.5g
$MgCl_2 \cdot 6H_2O$ 0.4g
NH_4Cl 0.3g
KH_2PO_4 .... 0.2g
$CaCl_2 \cdot 2H_2O$ 0.15g

Preparation of Solution A: Add components to distilled/deionized water and bring volume to 869.0mL. Mix thoroughly. Prepare and autoclave part A under 80% N_2 + 20% CO_2. Autoclave for 15 min at 15 psi pressure–121°C. Cool to room temperature.

Solution B:
Composition per liter:

$FeCl_2 \cdot 4H_2O$ 1.5g
$CoCl_2 \cdot 6H_2O$ 0.19g
$MnCl_2 \cdot 4H_2O$ 0.1g
$ZnCl_2$ 0.07g
H_3BO_3 .... 0.06g
$Na_2MoO_4 \cdot 2H_2O$ 0.04g
$NiCl_2 \cdot 6H_2O$ 0.02g
$CuCl_2 \cdot 2H_2O$ 0.02g
HCl, 25% 10.0mL

Preparation of Solution B: Add the $FeCl_2 \cdot 4H_2O$ to the HCl. Add distilled/deionized water and bring volume to 1.0L. Add remaining components. Mix thoroughly. Autoclave under 100% N_2 for 15 min at 15 psi pressure–121°C. Cool to room temperature.

Solution C:
Composition per 100.0mL:

$NaHCO_3$ 5.0g

Preparation of Solution C: Add the $NaHCO_3$ to distilled/deionized water and bring volume to 100.0mL. Mix thoroughly. Filter sterilize. Gas with 80% N_2 + 20% CO_2 to remove residual O_2.

Solution D:
Composition per 10.0mL:

Sodium butyrate 0.7g
Sodium caproate 0.3g
Sodium octanoate 0.15g

Preparation of Solution D: Add components to distilled/deionized water and bring volume to 10.0mL. Mix thoroughly. Autoclave under 100% N_2 for 15 min at 15 psi pressure–121°C. Cool to room temperature.

Solution E:
Composition per 10.0mL:

Yeast extract 1.0g
Thiamine·HCl 100.0μg
p-Aminobenzoic acid 40.0μg
D(+)-Biotin 10.0μg

Preparation of Solution E: Add components to distilled/deionized water and bring volume to 10.0mL. Mix thoroughly. Autoclave under 100% N_2 for 15 min at 15 psi pressure–121°C. Cool to room temperature.

Solution F:
Composition per 10.0mL:

$Na_2S \cdot 9H_2O$ 0.4g

Preparation of Solution F: Add $Na_2S \cdot 9H_2O$ to distilled/deionized water and bring volume to 10.0mL. Mix thoroughly. Autoclave under 100% N_2 for 15 min at 15 psi pressure–121°C. Cool to room temperature.

Preparation of Medium: To 869.0mL of sterile cooled Part A, aseptically add the remaining sterile solutions in the following order: solution B, solution C, solution D, solution E, and solution F. Mix thoroughly. Adjust pH to 7.7. Anaerobically distribute under 90% N_2 + 10% CO_2 into sterile tubes or flasks.

Use: For the cultivation of *Acetivibrio ethanolgignens* and *Desulfovibrio sapovorans* which reduces sulfate.

Acetobacter diazotrophicus Agar

Composition per liter:

Glucose 50.0g
$CaCO_3$ 30.0g
Agar 25.0g
Yeast extract 10.0g

pH 5.5 ± 0.2 at 25°C

Preparation of Medium: Add components to distilled/deionized water and bring volume to 1.0L. Mix thoroughly to evenly distribute $CaCO_3$. Bring pH to 5.5. Gently heat and bring to boiling. Distribute into tubes or flasks. Autoclave for 15 min at 15 psi pressure–121°C. Cool rapidly to 50°–55°C. Pour into sterile Petri dishes or leave in tubes.

Use: For the cultivation and maintenance of *Acetobacter diazotrophicus,* a nitrogen fixing endophyte.

Acetobacterium dehalogenans Medium (DSMZ Medium 787)

Composition per liter:

$NaHCO_3$	10.0g
Yeast extract	2.0g
NH_4Cl	1.0g
K_2HPO_4	0.45g
KH_2PO_4	0.33g
$MgSO_4 \cdot 7H_2O$	0.1g
Resazurin	1.0mg
$NaHCO_3$ solution	30.0mL
Na_2CO_3 solution	20.0mL
Trace elements solution	20.0mL
Vitamin solution	20.0mL
Na-syringate solution	10.0mL
Cysteine solution	10.0mL

pH 7.4 ± 0.2 at 25°C

$NaHCO_3$ Solution:
Composition per 100.0mL:

$NaHCO_3$	10.0g

Preparation of $NaHCO_3$ Solution: Add $NaHCO_3$ to distilled/deionized water and bring volume to 100.0mL. Mix thoroughly. Sparge with 80% N_2 + 20% CO_2. Filter sterilize.

Cysteine Solution:
Composition per 10.0mL:

L-Cysteine·HCl·H_2O	0.3g

Preparation of Cysteine Solution: Add L-cysteine·HCl·H_2O to distilled/deionized water and bring volume to 10.0mL. Mix thoroughly. Sparge with 100% N_2. Autoclave for 15 min at 15 psi pressure–121°C. Cool to 25°C.

Na-syringate Solution:
Composition per 20.0mL:

Na-syringate	1.2g

Preparation of Na-syringate Solution: Add Na-syringate to distilled/deionized water and bring volume to 20.0mL. Mix thoroughly. Sparge with 100% N_2. Filter sterilize.

Trace Elements Solution:
Composition per liter:

$MgSO_4 \cdot 7H_2O$	3.0g
Nitrilotriacetic acid	1.5g
NaCl	1.0g
$MnSO_4 \cdot 2H_2O$	0.5g
$CoSO_4 \cdot 7H_2O$	0.18g
$ZnSO_4 \cdot 7H_2O$	0.18g
$CaCl_2 \cdot 2H_2O$	0.1g
$FeSO_4 \cdot 7H_2O$	0.1g
$NiCl_2 \cdot 6H_2O$	0.025g
$KAl(SO_4)_2 \cdot 12H_2O$	0.02g
H_3BO_3	0.01g
$Na_2MoO_4 \cdot 4H_2O$	0.01g
$CuSO_4 \cdot 5H_2O$	0.01g
$Na_2SeO_3 \cdot 5H_2O$	0.3mg

Preparation of Trace Elements Solution: Add nitrilotriacetic acid to 500.0mL of distilled/deionized water. Dissolve by adjusting pH to 6.5 with KOH. Add remaining components. Add distilled/deionized water to 1.0L. Mix thoroughly.

Vitamin Solution:
Composition per liter:

Pyridoxine-HCl	10.0mg
Thiamine-HCl·$2H_2O$	5.0mg
Riboflavin	5.0mg
Nicotinic acid	5.0mg
D-Ca-pantothenate	5.0mg
p-Aminobenzoic acid	5.0mg
Lipoic acid	5.0mg
Biotin	2.0mg
Folic acid	2.0mg
Vitamin B_{12}	0.1mg

Preparation of Vitamin Solution: Add components to distilled/deionized water and bring volume to 1.0L. Mix thoroughly. Sparge with 80% N_2 + 20% CO_2. Filter sterilize.

Preparation of Medium: Prepare and dispense medium under 80% N_2 + 20% CO_2 gas atmosphere. Add components, except $NaHCO_3$ solution, Na_2CO_3 solution, vitamin solution, Na-syringate solution, and cysteine solution, to distilled/deionized water and bring volume to 900.0mL. Mix thoroughly. Gently heat and bring to boiling. Boil for 5 min. Cool while sparging with 80% N_2 + 20% CO_2. Distribute 9.0mL aliquots into serum bottles. Autoclave under 80% N_2 + 20% CO_2 for 15 min at 15 psi pressure–121°C. Aseptically and anaerobically add approximately 0.20mL sterile Na_2CO_3 solution to each 9.0mL of medium so that pH is adjusted to 7.4. For every 9.0mL of medium inject 1.0mL $NaHCO_3$ solution, 0.15mL Na-syringate solution, 0.2mL vitamin solution, and 0.17mL cysteine solution.

Use: For the cultivation of *Acetobacterium dehalogenans,* which dehalogenates methyl chloride.

Acetobacterium tundrae Medium (DSMZ Medium 900)

Composition per liter:

Yeast extract	1.0g
$MgCl_2 \cdot 6H_2O$	0.52g
KCl	0.33g
NH_4Cl	0.33g
KH_2PO_4	0.33g
$CaCl_2 \cdot 2H_2O$	0.3g
Resazurin	0.5mg
Fructose solution	30.0mL
$NaHCO_3$ solution	20.0mL
L-Cysteine solution	10.0mL
$Na_2S \cdot 9H_2O$ solution	10.0mL
Trace mineral solution SL-10	1.0mL
Seven vitamin solution	1.0mL

pH 7.0 ± 0.2 at 25°C

L-Cysteine Solution:
Composition per 10.0mL:

L-Cysteine·HCl·H_2O	0.3g

Preparation of L-Cysteine Solution: Add L-cysteine·HCl·H_2O to distilled/deionized water and bring volume to 10.0mL. Mix thoroughly. Sparge with 100% N_2. Autoclave for 15 min at 15 psi pressure–121°C.

$NaHCO_3$ Solution:
Composition per 20.0mL:

$NaHCO_3$	5.0g

Preparation of $NaHCO_3$ Solution: Add $NaHCO_3$ to distilled/deionized water and bring volume to 20.0mL. Mix thoroughly. Sparge with 80% N_2 + 20% CO_2. Filter sterilize.

$Na_2S \cdot 9H_2O$ Solution:
Composition per 10mL:

$Na_2S \cdot 9H_2O$	0.3g

Preparation of $Na_2S·9H_2O$ Solution: Add $Na_2S·9H_2O$ to distilled/deionized water and bring volume to 10.0mL. Mix thoroughly. Autoclave under 100% N_2 for 15 min at 15 psi pressure–121°C. Cool to room temperature.

Fructose Solution:
Composition per 30.0mL:
Glucose 5.0g

Preparation of Fructose Solution: Add glucose to distilled/deionized water and bring volume to 30.0mL. Mix thoroughly. Sparge with 100% N_2. Filter sterilize.

Seven Vitamin Solution:
Composition per liter:
Pyridoxine hydrochloride 300.0mg
Thiamine-HCl·$2H_2O$ 200.0mg
Nicotinic acid 200.0mg
Vitamin B_{12} 100.0mg
Calcium pantothenate 100.0mg
p-Aminobenzoic acid 80.0mg
D(+)-Biotin 20.0mg

Preparation of Seven Vitamin Solution: Add components to distilled/deionized water and bring volume to 1.0L. Sparge with 100% N_2. Mix thoroughly. Filter sterilize.

Trace Elements Solution SL-10:
Composition per liter:
$FeCl_2·4H_2O$ 1.5g
$CoCl_2·6H_2O$ 190.0mg
$MnCl_2·4H_2O$ 100.0mg
$ZnCl_2$ 70.0mg
$Na_2MoO_4·2H_2O$ 36.0mg
$NiCl_2·6H_2O$ 24.0mg
H_3BO_3 .... 6.0mg
$CuCl_2·2H_2O$ 2.0mg
HCl (25% solution) 10.0mL

Preparation of Trace Elements Solution SL-10: Add $FeCl_2·4H_2O$ to 10.0mL of HCl solution. Mix thoroughly. Add distilled/deionized water and bring volume to 1.0L. Add remaining components. Mix thoroughly. Sparge with 80% N_2 + 20% CO_2. Autoclave for 15 min at 15 psi pressure–121°C.

Preparation of Medium: Prepare and dispense medium under 80% N_2 + 20% CO_2 gas atmosphere. Add components, except seven vitamin solution, $NaHCO_3$ solution, fructose solution, L-Cysteine solution, and $Na_2S·9H_2O$ solution, to distilled/deionized water and bring volume to 929.0mL. Mix thoroughly. Sparge with 80% N_2 + 20% CO_2. Distribute into anaerobe tubes or bottles. Autoclave for 15 min at 15 psi pressure–121°C. Aseptically and anaerobically add per liter of medium, 1.0mL seven vitamin solution, 20.0mL $NaHCO_3$ 30.0mL fructose solution, 10.0mL L-Cysteine-HCl·H_2O solution, and 10.0mL $Na_2S·9H_2O$ solution. Mix thoroughly. The final pH should be 7.0.

Use: For the cultivation of *Acetobacterium tundrae,* a psychrophilic, acetigenic bacterium from tundra soils.

Acetogen Medium

Composition per 421.8mL:
$NaHCO_3$ 2.4g
NH_4Cl 0.2g
Yeast extract 0.2g
Stock salts solution #1 .... 40.0mL
Potassium phosphate buffer 20.0mL
Clarified rumen fluid 20.0mL
Stock salts solution #2 .... 4.0mL
Trace minerals 4.0mL
Vitamin solution 4.0mL
Reducing agent 4.0mL
Tungstate solution 0.4mL
Resazurin (0.1% solution) 0.4mL

Potassium Phosphate Buffer:
Composition per 830.0mL:
K_2HPO_4 .... 15.68g
KH_2PO_4 .... 4.72g

Preparation of Potassium Phosphate Buffer: Dissolve K_2HPO_4 in 600.0mL of distilled/deionized water and KH_2PO_4 in 230.0mL of distilled/deionized water. Mix the two solutions together and use.

Stock Salts Solution #1:
Composition per liter:
KCl 1.6g
NaCl 1.4g
$MgSO_4·7H_2O$ 0.2g

Preparation of Stock Salts Solution #1: Add components to distilled/deionized water and bring volume to 1.0L. Mix thoroughly.

Stock Salts Solution #2:
Composition per liter:
$CaCl_2·2H_2O$ 0.1g

Preparation of Stock Salts Solution #2: Add components to distilled/deionized water and bring volume to 1.0L. Mix thoroughly.

Trace Minerals:
Composition per liter:
Nitrilotriacetic acid 1.5g
$MgSO_4·7H_2O$ 3.0g
$MnSO_4·H_2O$ 0.5g
NaCl 1.0g
$NiCl_2·6H_2O$ 0.1g
$FeSO_4·7H_2O$ 0.1g
$CoCl_2·6H_2O$ 0.1g
$CaCl_2$ 0.1g
$ZnSO_4·7H_2O$ 0.1g
$Na_2SeO_3·5H_2O$ 0.01g
$CuSO_4·5H_2O$ 0.01g
$AlK(SO_4)_2·12H_2O$ 0.01g
H_3BO_3 .... 0.01g
$Na_2MoO_4·2H_2O$ 0.01g

Preparation of Trace Minerals: Add nitrilotriacetic acid to 500.0mL of distilled/deionized water. Dissolve by adjusting pH to 6.5 with KOH. Bring volume to 1.0L with distilled/deionized water. Add remaining components. Mix thoroughly.

Vitamin Solution:
Composition per liter:
Pyridoxine·HCl 10.0mg
Ascorbic acid 5.0mg
Calcium pantothenate 5.0mg
Choline chloride 5.0mg
Lipoic acid 5.0mg
i-Inositol 5.0mg
Niacinamide 5.0mg
Nicotinic acid 5.0mg
p-Aminobenzoic acid 5.0mg
Pyridoxal·HCl 5.0mg
Riboflavin 5.0mg
Thiamine·HCl 5.0mg
Biotin 2.0mg
Folic acid 2.0mg
Vitamin B_{12} 0.1mg

Preparation of Vitamin Solution: Add components to distilled/deionized water and bring volume to 1.0L. Mix thoroughly. Store frozen.

Tungstate Solution:
Composition per liter:
$Na_2WO_4 \cdot 2H_2O$ 99.0mg

Preparation of Tungstate Solution: Add components to distilled/deionized water and bring volume to 1.0L. Mix thoroughly.

Reducing Agent:
Composition per 110.0mL:
L-Cysteine·HCl·H_2O 2.5g
$Na_2S \cdot 9H_2O$ 2.5g

Preparation of Reducing Agent: Add 110.0mL of distilled/deionized water to a 250.0mL round-bottomed flask. Boil under N_2 gas for 1 min. Cool to room temperature. Add L-cysteine·HCl and dissolve. Adjust to pH 9 with 5*N* NaOH. Add washed $Na_2S \cdot 9H_2O$ and dissolve. Distribute in amounts needed into tubes or flasks. Autoclave for 10 min at 15 psi pressure–121°C.

Preparation of Medium: Add components, except $NaHCO_3$ and reducing agent, to distilled/deionized water and bring volume to 417.8mL. Mix thoroughly. Gently heat and bring to boiling under 80% N_2 + 20% CO_2. Cool to 45°–50°C. Add $NaHCO_3$ and reducing agent. Distribute into tubes or flasks under 80% N_2 + 20% CO_2. Autoclave for 15 min at 15 psi pressure–121°C. After inoculation, exchange headspace with 80% H_2 + 20% CO_2.

Use: For the cultivation and maintenance of acetogenic anaerobes such as some *Clostridium* species.

Acetohalobium Medium

Composition per 1025.0mL:
NaCl 150.0g
$MgCl_2 \cdot 6H_2O$ 4.0g
$CaCl_2 \cdot 2H_2O$ 0.33g
KCl 0.33g
KH_2PO_4 .... 0.33g
NH_4Cl 0.33g
Resazurin 1.0mg
Trace elements solution 10.0mL
Vitamin solution 10.0mL
Trimethylamine·HCl solution 10.0mL
Yeast extract solution 5.0mL
$NaHCO_3$ solution 5.0mL
$Na_2S \cdot 9H_2O$ solution 5.0mL
pH 7.6 ± 0.2 at 25°C

Trace Elements Solution:
Composition per liter:
$MgSO_4 \cdot 7H_2O$ 3.0g
Nitrilotriacetic acid .... 1.5 g
$CaCl_2 \cdot 2H_2O$ 1.0g
NaCl 1.0g
$MnSO_4 \cdot 2H_2O$ 0.5g
$CoSO_4 \cdot 7H_2O$ 0.18g
$ZnSO_4 \cdot 7H_2O$ 0.18g
$FeSO_4 \cdot 7H_2O$ 0.1g
$NiCl_2 \cdot 6H_2O$ 0.025g
$KAl(SO_4)_2 \cdot 12H_2O$ 0.02g
$CuSO_4 \cdot 5H_2O$ 0.01g
H_3BO_3 .... 0.01g
$Na_2MoO_4 \cdot 2H_2O$ 0.01g
$Na_2SeO_3 \cdot 5H_2O$ 0.3mg

Preparation of Trace Elements Solution: Add nitrilotriacetic acid to 500.0mL of distilled/deionized water. Dissolve by adjusting pH to 6.5 with KOH. Add distilled/deionized water to 1.0L. Add remaining components. Mix thoroughly. Adjust pH to 7.0 with KOH.

Vitamin Solution:
Composition per liter:
Pyridoxine·HCl 10.0mg
Calcium DL-pantothenate 5.0mg
Lipoic acid 5.0mg
Nicotinic acid 5.0mg
p-Aminobenzoic acid 5.0mg
Riboflavin 5.0mg
Thiamine·HCl 5.0mg
Biotin 2.0mg
Folic acid 2.0mg
Vitamin B_{12} 0.1mg

Preparation of Vitamin Solution: Add components to distilled/deionized water and bring volume to 1.0L. Mix thoroughly. Filter sterilize.

Trimethylamine·HCl Solution:
Composition per 10.0mL:
Trimethylamine·HCl 2.4g

Preparation of Trimethylamine·HCl Solution: Add trimethylamine·HCl to distilled/deionized water and bring volume to 10.0mL. Mix thoroughly. Sparge under 100% N_2 gas for 3 min. Autoclave for 15 min at 15 psi pressure–121°C. Store under N_2 gas. One may use 4.5g of glycine betaine in place of trimethylamine·HCl.

Yeast Extract Solution:
Composition per 5.0mL:
Yeast extract 0.05g

Preparation of Yeast Extract Solution: Add yeast extract to distilled/deionized water and bring volume to 5.0mL. Mix thoroughly. Sparge under 100% N_2 gas for 3 min. Autoclave for 15 min at 15 psi pressure–121°C. Store under N_2 gas.

$NaHCO_3$ Solution:
Composition per 5.0mL:
$NaHCO_3$ 0.5g

Preparation of $NaHCO_3$ Solution: Add $NaHCO_3$ to distilled/deionized water and bring volume to 5.0mL. Mix thoroughly. Sparge under 100% N_2 gas for 3 min. Autoclave for 15 min at 15 psi pressure–121°C. Store under N_2 gas.

$Na_2S \cdot 9H_2O$ Solution:
Composition per 5.0mL:
$Na_2S \cdot 9H_2O$ 0.5g

Preparation of $Na_2S \cdot 9H_2O$ Solution: Add $Na_2S \cdot 9H_2O$ solution to distilled/deionized water and bring volume to 5.0mL. Mix thoroughly. Sparge under 100% N_2 gas for 3 min. Autoclave for 15 min at 15 psi pressure–121°C. Store under N_2 gas.

Preparation of Medium: Add components, except vitamin solution, trimethylamine·HCl solution, yeast extract solution, $NaHCO_3$ solution, and $Na_2S \cdot 9H_2O$ solution, to distilled/deionized water and bring volume to 1.0L. Adjust pH to 7.2–7.4 with NaOH. Mix thoroughly. Gently heat and bring to boiling. Boil for a few minutes. Allow to cool to room temperature under 100% N_2. Distribute into tubes or flasks under 100% N_2. Autoclave for 15 min at 15 psi pressure–121°C. Cool to room temperature. Before inoculation, aseptically and anaerobically add vitamin solution, trimethylamine·HCl solution, yeast extract solution, $NaHCO_3$ solution, and $Na_2S·9H_2O$ solution. If necessary, adjust pH 7.6 to with sterile anaerobic Na_2CO_3 solution.

Use: For the cultivation and maintenance of *Acetohalobium arabaticum,* which is capable of chemolithotrophic growth on $H_2 + CO_2$ or CO, of methylotrophic growth on trimethylamine, and of organotrophic growth on betaine, lactate, etc.

Acetylglucosamine Medium (*N*-Acetylglucosamine Medium)

Composition per liter:

N-Acetylglucosamine	20.0g
Beef extract	10.0g
Peptone	10.0g
Yeast extract	5.0g
K_2HPO_4	2.0g
Triammonium citrate	2.0g
$MgSO_4·7H_2O$	0.2g
$MnSO_4·4H_2O$	0.05g
Tween™ 80	1.0mL

pH 6.2 ± 0.4 at 25°C

Preparation of Medium: Add components to distilled/deionized water and bring volume to 1.0L. Mix thoroughly. Adjust pH to 6.2. Distribute into tubes or flasks. Autoclave for 15 min at 15 psi pressure–121°C.

Use: For the cultivation of bacteria that can utilize *N*-acetylglucosamine.

Acid Glucose Salts Medium

Composition per liter:

Glucose	5.0g
$MgSO_4·7H_2O$	0.5g
$(NH_4)_2SO_4$	0.15g
KH_2PO_4	0.1g
KCl	50.0mg
$Ca(NO_3)_2$	10.0mg

pH 3.0 ± 0.2 at 25°C

Preparation of Medium: Add components to distilled/deionized water and bring volume to 1.0L. Mix thoroughly. Distribute into tubes or flasks. Autoclave for 15 min at 15 psi pressure–121°C.

Use: For the cultivation of *Thiobacillus organoparus.*

Acid Rhodospirillaceae Medium

Composition per 1050.0 mL:

Ammonium acetate	1.5g
KH_2PO_4	0.5g
$MgSO_4.7H_2O$	0.4g
NaCl	0.4g
NH_4Cl	0.4g
Disodium succinate	0.25g
Yeast extract	0.2g
$CaCl_2·2H_2O$	0.05g
Ferric citrate solution	5.0mL
Trace elements solution SL-6	1.0mL
Vitamin B_{12} solution	0.4mL
Neutralized sulfide solution	variable

pH 5.7 ± 0.2 at 25°C

Ferric Citrate Solution:
Composition per 10.0mL:

Ferric citrate	10.0mg

Preparation of Ferric Citrate Solution: Add ferric citrate to distilled/deionized water and bring volume to 10.0mL. Mix thoroughly. Sparge under 100% N_2 gas for 3 min. Autoclave for 15 min at 15 psi pressure–121°C. Store under N_2 gas.

Trace Elements Solution SL-6:
Composition per liter:

$MnCl_2·4H_2O$	0.5g
H_3BO_3	0.3g
$CoCl_2·6H_2O$	0.2g
$ZnSO_4·7H_2O$	0.1g
$Na_2MoO_4·2H_2O$	0.03g
$NiCl_2·6H_2O$	0.02g
$CuCl_2·2H_2O$	0.01g

Preparation of Trace Elements Solution SL-6: Add components to distilled/deionized water and bring volume to 1.0L. Mix thoroughly.

Vitamin B_{12} Solution:
Composition per 100.0mL:

Vitamin B_{12}	10.0mg

Preparation of Vitamin B_{12} Solution: Add Vitamin B_{12} to distilled/deionized water and bring volume to 100.0mL. Mix thoroughly. Sparge under 100% N_2 gas for 3 min. Autoclave for 15 min at 15 psi pressure–121°C. Store under N_2 gas.

Neutralized Sulfide Solution:
Composition per 100.0mL:

$Na_2S·9H_2O$	1.5g

Preparation of Neutralized Sulfide Solution: Add $Na_2S·9H_2O$ to distilled/deionized water in a 250.0mL screw-capped bottle fitted with a butyl rubber septum and bring volume to 100.0mL. Add a magnetic stir bar. Mix thoroughly. Sparge under 100% N_2 gas for 3 min. Autoclave for 15 min at 15 psi pressure–121°C. Cool to room temperature. Adjust pH to about 7.3 with sterile 2*M* H_2SO_4. Do not open the bottle to add H_2SO_4; use a sterile syringe. Stir the solution continuously to avoid precipitation of elemental sulfur. The final solution should be clear and yellow in color.

Preparation of Medium: Add components, except neutralized sulfide solution, to distilled/deionized water and bring volume to 1050.0mL. Mix thoroughly. Gently heat and bring to boiling. Boil for 3–4 min under a stream of 100% N_2. Distribute 45.0mL of the prepared medium into 50.0mL screw-capped tubes that have been flushed with 100% N_2. Autoclave for 15 min at 15 psi pressure–121°C. Cool to room temperature. Before inoculation, aseptically and anaerobically add 0.25–0.50mL of neutralized sulfide solution.

Use: For the cultivation and maintenance of members of the family Rhodospillaceae, including *Rhodomicrobium vannielii* and *Rhodopseudomonas acidophila.*

Acid Tomato Broth

Composition per liter:

Glucose	10.0g
Peptone	10.0g
Yeast extract	5.0g
$MgSO_4·7H_2O$	0.2g
$MnSO_4·4H_2O$	0.05g
Tomato juice	250.0mL
L-Cysteine solution	0.5mL

L-Cysteine Solution:
Composition per 10.0mL:

L-Cysteine	0.1g

Preparation of L-Cysteine Solution: Add 0.1g of L-cysteine to distilled/deionized water and bring volume to 10.0mL. Mix thoroughly. Filter sterilize.

Preparation of Medium: Add components, except L-cysteine solution, to distilled/deionized water and bring volume to 999.5mL. Mix thoroughly. Adjust pH to 4.8. Autoclave for 15 min at 15 psi pressure–121°C. Aseptically add 0.5mL of sterile L-cysteine solution. Mix thoroughly. Aseptically distribute into sterile tubes or flasks.

Use: For the cultivation of a variety of fungi.

Acidaminobacter Medium

Composition per liter:

$NaHCO_3$	2.0g
Glycine	1.5g
NaCl	1.2g
KCl	0.4g
$MgCl_2 \cdot 6H_2O$	0.4g
$Na_2S \cdot 9H_2O$	0.3g
KH_2PO_4	0.2g
Na_2SO_4	0.2g
Yeast extract	0.2g
$CaCl_2 \cdot 2H_2O$	0.15g
Resazurin	1.0mg
$Na_2SeO_3 \cdot 5H_2O$	30.0μg
Vitamin solution	10.0mL
$NaHCO_3$ solution	5.0mL
$Na_2S \cdot 9H_2O$ solution	5.0mL
Trace elements solution SL-10	1.0mL

pH 7.4 ± 0.2 at 25°C

Trace Elements Solution SL-10:
Composition per liter:

$FeCl_2 \cdot 4H_2O$	1.5g
$CoCl_2 \cdot 6H_2O$	190.0mg
$MnCl_2 \cdot 4H_2O$	100.0mg
$ZnCl_2$	70.0mg
$Na_2MoO_4 \cdot 2H_2O$	36.0mg
$NiCl_2 \cdot 6H_2O$	24.0mg
H_3BO_3	6.0mg
$CuCl_2 \cdot 2H_2O$	2.0mg
HCl (25% solution)	10.0mL

Preparation of Trace Elements Solution SL-10: Add $FeCl_2 \cdot 4H_2O$ to 10.0mL of HCl solution. Mix thoroughly. Add distilled/deionized water and bring volume to 1.0L. Add remaining components. Mix thoroughly.

Vitamin Solution:
Composition per liter:

Pyridoxine·HCl	10.0mg
Calcium DL-pantothenate	5.0mg
Lipoic acid	5.0mg
Nicotinic acid	5.0mg
p-Aminobenzoic acid	5.0mg
Riboflavin	5.0mg
Thiamine·HCl	5.0mg
Biotin	2.0mg
Folic acid	2.0mg
Vitamin B_{12}	0.1mg

Preparation of Vitamin Solution: Add components to distilled/deionized water and bring volume to 1.0L. Mix thoroughly. Filter sterilize.

$NaHCO_3$ Solution:
Composition per 5.0mL:

$NaHCO_3$	0.5g

Preparation of $NaHCO_3$ Solution: Add $NaHCO_3$ to distilled/deionized water and bring volume to 5.0mL. Mix thoroughly. Sparge under 100% N_2 gas for 3 min. Autoclave for 15 min at 15 psi pressure–121°C. Store under N_2 gas.

$Na_2S \cdot 9H_2O$ Solution:
Composition per 5.0mL:

$Na_2S \cdot 9H_2O$	0.5g

Preparation of $Na_2S \cdot 9H_2O$ Solution: Add $Na_2S \cdot 9H_2O$ solution to distilled/deionized water and bring volume to 5.0mL. Mix thoroughly. Sparge under 100% N_2 gas for 3 min. Autoclave for 15 min at 15 psi pressure–121°C. Store under N_2 gas.

Preparation of Medium: Add components, except vitamin solution, $NaHCO_3$ solution, and $Na_2S \cdot 9H_2O$ solution, to distilled/deionized water and bring volume to 1.0L. Mix thoroughly. Gently heat and bring to boiling. Boil for a few minutes. Allow to cool to room temperature under 100% N_2. Distribute into tubes or flasks under 100% N_2. Autoclave for 15 min at 15 psi pressure–121°C. Cool to room temperature. Before inoculation, aseptically and anaerobically add vitamin solution, $NaHCO_3$ solution, and $Na_2S \cdot 9H_2O$ solution. Mix thoroughly. Check that final pH is 7.4.

Use: For the cultivation and maintenance of *Acidaminobacter hydrogenoformans,* a fermentative bacterium from estuarine mud.

Acidaminococcus fermentans Medium

Composition per liter:

Casamino acids	10.0g
Glucose	5.0g
Pancreatic digest of casein	5.0g
Yeast extract	5.0g
Sodium glutamate	4.0g
KH_2PO_4	2.0g
Arginine	1.0g
Glycine	1.0g
L-Cysteine·HCl	0.5g
DL-Tryptophan	0.1g
Tween™ 80	0.5mL

pH 7.0 ± 0.2 at 25°C

Preparation of Medium: Add components to distilled/deionized water and bring volume to 1.0L. Mix thoroughly. Adjust pH to 7.0. Distribute into tubes or flasks. Autoclave for 15 min at 15 psi pressure–121°C.

Use: For the cultivation and maintenance of *Acidaminococcus fermentans* from rumen.

Acidaminococcus Medium VR

Composition per liter:

Acid-hydrolyzed casein (vitamin and salt free)	20.0g
Glucose	5.0g
L-Cysteine·HCl·H_2O	0.35g
DL-Tryptophan	0.1g
Guanine	0.01g
Uracil	0.01g
Hypoxanthine	0.01g
Pyridoxal	1.0mg
Calcium pantothenate	1.0mg
Thiamine	50.0μg
Niacin	50.0μg
Riboflavin	50.0μg
p-Aminobenzoic acid	10.0μg
Biotin	2.0μg

Folic acid 1.0μg
Vitamin B_{12} 1.0μg
VR salts A 30.0mL
VR salts B 4.0mL
pH 7.0 ± 0.2 at 25°C

VR Salts A:
Composition per 500.0mL:
Na_2HPO_4 .... 37.5g
KH_2PO_4 .... 12.5g

Preparation of VR Salts A: Add components to distilled/deionized water and bring volume to 500.0mL. Mix thoroughly.

VR Salts B:
Composition per liter:
$MgSO_4 \cdot 7H_2O$ 24.0g
$CaCl_2 \cdot 2H_2O$ 0.5g
$FeSO_4 \cdot 7H_2O$ 0.5g
$ZnSO_4 \cdot 7H_2O$ 0.25g
$MnSO_4 \cdot H_2O$ 0.25g
$CoCl_2 \cdot 6H_2O$ 0.25g
$VSO_4 \cdot 7H_2O$ 0.25g
$Na_2MoO_4 \cdot 2H_2O$ 0.25g
$CuSO_4 \cdot 5H_2O$ 0.125g

Preparation of VR Salts B: Add components to distilled/deionized water and bring volume to 700.0mL. Add 2.0mL of concentrated HCl and heat until dissolved. Add 5.0g of nitrilotriacetic acid to 300.0mL distilled/deionized water. Adjust pH with 10*N* NaOH to 7.0. Stir vigorously. Slowly add the nitrilotriacetic acid solution to the larger volume of salt solutions while stirring until dissolved. Add distilled/deionized water and bring volume to 1.0L. Filter through paper. Store in a cool place.

Preparation of Medium: Filter sterilize vitamins as separate solution. Add aseptically to sterile basal medium. If necessary, adjust pH with solid K_2CO_3 to 7.0. Prepare and distribute medium anaerobically using Hungate techniques with 100% N_2 gas.

Use: For the cultivation and maintenance of *Acidaminococcus fermentans from rumen.*

Acidianus brierleyi Medium

Composition per liter:
Sulfur flowers 10.0g
$(NH_4)_2SO_4$ 3.0g
$K_2HPO_4 \cdot 3H_2O$ 0.5g
$MgSO_4 \cdot 7H_2O$ 0.5g
KCl 0.1g
$Ca(NO_3)_2$ 0.01g
Yeast extract solution 10.0mL
pH 1.5–2.5 at 25°C

Yeast Extract Solution:
Composition per 10.0mL:
Yeast extract 0.2g

Preparation of Yeast Extract Solution: Add yeast extract to distilled/deionized water and bring volume to 10.0mL. Mix thoroughly. Autoclave for 15 min at 15 psi pressure–121°C.

Preparation of Medium: Add components, except sulfur flowers and yeast extract solution, to distilled/deionized water and bring volume to 990.0mL. Mix thoroughly. Gently heat and bring to boiling. Autoclave for 15 min at 15 psi pressure–121°C. Sulfur flowers are sterilized separately by steaming for 3 hr on 3 consecutive days. Aseptically combine the basal solution, sterile sulfur flowers, and sterile yeast extract solution. Adjust pH to 1.5–2.5 with 6*N* H_2SO_4.

Use: For the cultivation and maintenance of *Acidianus brierleyi,* a facultatively anaerobic thermoacidophilic archaean.

Acidianus infernus Medium

Composition per liter:
$(NH_4)_2SO_4$ 1.3g
Yeast extract 1.0g
Sulfur flowers 1.0g
KH_2PO_4 .... 0.28g
$MgSO_4 \cdot 7H_2O$ 0.25g
$CaCl_2 \cdot 2H_2O$ 0.07g
$FeCl_3 \cdot 6H_2O$ 0.02g
$Na_2B_4O_7 \cdot 10H_2O$ 4.5mg
$MnCl_2 \cdot 4H_2O$ 1.8mg
$ZnSO_4 \cdot 7H_2O$ 0.22mg
$CuCl_2 \cdot 2H_2O$ 0.05mg
$Na_2MoO_4 \cdot 2H_2O$ 0.03mg
$VOSO_4 \cdot 2H_2O$ 0.03mg
$CoSO_4$ 0.01mg
pH 2.5 ± 0.2 at 25°C

Preparation of Medium: Add components to distilled/deionized water and bring volume to 1.0L. Mix thoroughly. Adjust pH to 2.5 with 10*N* H_2SO_4. Distribute into tubes or flasks. Autoclave for 15 min at 15 psi pressure–121°C.

Use: For the aerobic cultivation and maintenance of *Acidianus infernus, Acidianus brierleyi, Acidianus infernus,* and *Desulfurolobus ambivalens.*

Acidic Rhodospirillaceae Medium

Composition per 1006.0mL:
Disodium succinate 1.0g
KH_2PO_4 .... 0.5g
$MgSO_4 \cdot 7H_2O$ 0.4g
NaCl 0.4g
NH_4Cl 0.4g
Yeast extract 0.2g
$CaCl_2 \cdot H_2O$ 50.0mg
Ferric citrate solution 5.0mL
Trace elements solution 1.0mL

Ferric Citrate Solution:
Composition per 100.0mL:
Ferric citrate 0.1g

Preparation of Ferric Citrate Solution: Add ferric citrate to distilled/deionized water and bring volume to 100.0mL. Mix thoroughly. Filter sterilize.

Trace Elements Solution:
Composition per liter:
H_3BO_3 .... 0.3g
$CoCl_2 \cdot 6H_2O$ 0.2g
$ZnSO_4 \cdot 7H_2O$ 0.1g
$MnCl_2 \cdot 4H_2O$ 0.03g
$Na_2MoO_4 \cdot 2H_2O$ 0.03g
$NiCl_2 \cdot 6H_2O$ 0.02g
$CuCl_2 \cdot 2H_2O$ 0.01g

Preparation of Trace Elements Solution: Add components to distilled/deionized water and bring volume to 1.0L. Mix thoroughly. Filter sterilize.

Preparation of Medium: Add components, except ferric citrate solution and trace elements solution, to distilled/deionized water and bring volume to 1.0L. Mix thoroughly. Autoclave for 15 min at 15 psi pressure–121°C. Aseptically add 5.0mL of sterile ferric citrate solution and 1.0mL of sterile trace elements solution. Mix thoroughly. Aseptically distribute into sterile tubes or flasks.

Use: For the cultivation of *Rhodopseudomonas acidophila.*

Acidic Tomato Medium for *Leuconostoc*

Composition per liter:

Agar	15.0g
Glucose	10.0g
Peptone	10.0g
Yeast extract	5.0g
$MgSO_4 \cdot 7H_2O$	0.2g
$MnSO_4 \cdot 4H_2O$	0.05g
Tomato juice	250.0mL

pH 4.8 ± 0.2 at 25°C

Preparation of Medium: Add solid components to 750.0mL of distilled/deionized water. Add tomato juice. Mix well and warm gently until dissolved. Autoclave for 15 min at 15 psi pressure–121°C. Pour into sterile Petri dishes.

Use: For the cultivation and maintenance of *Leuconostoc oenos* and other *Leuconostoc* species.

Acidimicrobium Medium

Composition per liter:

$MgSO_4 \cdot 7H_2O$	0.5g
$(NH_4)_2SO_4$	0.4g
K_2HPO_4	0.2g
KCl	0.1g
$FeSO_4 \cdot 7H_2O$	10.0mg
Yeast extract solution	20.0mL

pH 2.0 ± 0.2 at 25°C

Yeast Extract Solution:

Composition per 20.0mL:

Yeast extract	10.0g

Preparation of Yeast Extract Solution: Add yeast extract to distilled/deionized water and bring volume to 20.0mL. Mix thoroughly. Autoclave for 15 min at 15 psi pressure–121°C.

Preparation of Medium: Add components, except yeast extract solution, to distilled/deionized water and bring volume to 980.0mL. Mix thoroughly. Adjust pH to 2.0 with H_2SO_4. Autoclave for 15 min at 15 psi pressure–121°C. Aseptically add 20.0mL of sterile yeast extract solution. Mix thoroughly. Aseptically distribute into sterile tubes or flasks.

Use: For the heterotrophic cultivation of *Sulfobacillus acidophilus.*

Acidimicrobium Medium

Composition per liter:

$FeSO_4 \cdot 7H_2O$	13.9g
$MgSO_4 \cdot 7H_2O$	0.5g
$(NH_4)_2SO_4$	0.4g
K_2HPO_4	0.2g
KCl	0.1g

pH 1.7 ± 0.2 at 25°C

Preparation of Medium: Add components to distilled/deionized water and bring volume to 1.0L. Mix thoroughly. Adjust pH to 1.7 with H_2SO_4. Distribute into tubes or flasks. Autoclave for 15 min at 15 psi pressure–121°C.

Use: For the autotrophic cultivation of *Sulfobacillus acidophilus.*

Acidiphilium Medium

Composition per liter:

$(NH_4)_2SO_4$	2.0g
K_2HPO_4	0.5g
$MgSO_4 \cdot 7H_2O$	0.5g
KCl	0.1g
Glucose solution	10.0mL
Yeast extract solution	10.0mL

pH 3.0 ± 0.2 at 25°C

Glucose Solution:

Composition per 10.0mL:

D-Glucose	1.0g

Preparation of Glucose Solution: Add glucose to distilled/deionized water and bring volume to 10.0mL. Mix thoroughly. Filter sterilize.

Yeast Extract Solution:

Composition per 10.0mL:

Yeast extract	0.3g

Preparation of Yeast Extract Solution: Add yeast extract to distilled/deionized water and bring volume to 10.0mL. Mix thoroughly. Autoclave for 15 min at 15 psi pressure–121°C.

Preparation of Medium: Add components, except glucose solution and yeast extract solution, to distilled/deionized water and bring volume to 1080.0mL. Mix thoroughly. Gently heat and bring to boiling. Adjust pH to 3.0 using 1*N* H_2SO_4. Autoclave for 15 min at 15 psi pressure–121°C. Cool to room temperature. Before inoculation, aseptically add glucose solution and yeast extract solution. Mix thoroughly.

Use: For the cultivation and maintenance of *Acidiphilium cryptum.*

Acidobacterium Medium

Composition per liter:

$(NH_4)_2SO_4$	2.0g
Glucose	1.0g
K_2HPO_4	0.5g
$MgSO_4 \cdot 7H_2O$	0.5g
KCl	0.1g
Yeast extract	0.1g

pH 3.5 ± 0.2 at 25°C

Preparation of Medium: Add components to distilled/deionized water and bring volume to 1.0L. Mix thoroughly. Adjust pH to 3.5 with H_2SO_4. Distribute into tubes or flasks. Autoclave for 15 min at 15 psi pressure–121°C.

Use: For the cultivation of *Acidobacterium capsulatum.*

Acidolobus aceticus Medium (DSMZ Medium 901)

Composition per 1055mL:

Sulfur, powdered	10.0g
NH_4Cl	0.33g
KCl	0.33g
KH_2PO_4	0.33g
$MgCl_2 \cdot 6H_2O$	0.33g
$CaCl_2 \cdot 2H_2O$	0.33g
Resazurin	0.5mg
Yeast extract solution	30.0mL
$Na_2S \cdot 9H_2O$ solution	15.0mL
Vitamin solution	10.0mL
Trace mineral solution SL-10	1.0mL

pH 3.5-3.8 at 25°C

Trace Elements Solution SL-10:

Composition per liter:

$FeCl_2 \cdot 4H_2O$	1.5g
$CoCl_2 \cdot 6H_2O$	190.0mg
$MnCl_2 \cdot 4H_2O$	100.0mg
$ZnCl_2$	70.0mg
$Na_2MoO_4 \cdot 2H_2O$	36.0mg
$NiCl_2 \cdot 6H_2O$	24.0mg
H_3BO_3	6.0mg
$CuCl_2 \cdot 2H_2O$	2.0mg
HCl (25% solution)	10.0mL

Preparation of Trace Elements Solution SL-10: Add $FeCl_2 \cdot 4H_2O$ to 10.0mL of HCl solution. Mix thoroughly. Add distilled/deionized water and bring volume to 1.0L. Add remaining components. Mix thoroughly. Sparge with 80% N_2 + 20% CO_2. Filter sterilize.

Vitamin Solution:

Composition per liter:

Pyridoxine-HCl	10.0mg
Thiamine-HCl·$2H_2O$	5.0mg
Riboflavin	5.0mg
Nicotinic acid	5.0mg
D-Ca-pantothenate	5.0mg
p-Aminobenzoic acid	5.0mg
Lipoic acid	5.0mg
Biotin	2.0mg
Folic acid	2.0mg
Vitamin B_{12}	0.1mg

Preparation of Vitamin Solution: Add components to distilled/deionized water and bring volume to 1.0L. Mix thoroughly. Sparge with 80% H_2 + 20% CO_2. Filter sterilize.

$Na_2S \cdot 9H_2O$ Solution:

Composition per 20.0mL:

$Na_2S \cdot 9H_2O$	0.6g

Preparation of $Na_2S \cdot 9H_2O$ Solution: Add $Na_2S \cdot 9H_2O$ to distilled/deionized water and bring volume to 20.0mL. Mix thoroughly. Autoclave under 100% N_2 for 15 min at 15 psi pressure–121°C. Cool to room temperature. Neturalize to pH 7.0 with HCl.

Yeast Extract Solution:

Composition per 30.0mL:

Yeast extract	3.0g

Preparation of Yeast Extract Solution: Add yeast extract to distilled/deionized water and bring volume to 30.0mL. Mix thoroughly. Sparge with 100% N_2. Autoclave under 100% N_2 for 15 min at 15 psi pressure–121°C. Cool to room temperature.

Preparation of Medium: Prepare and dispense medium under 100% CO_2. Add components, except vitamin solution, $Na_2S \cdot 9H_2O$ solution, and yeast extract solution, to distilled/deionized water and bring volume to 1.0L. Mix thoroughly. Adjust pH to 3.5 with H_2SO_4. Distribute to anaerobe tubes or bottles. Heat to 90°C for 1 h on each of 3 successive days. Aseptically and anaerobically add per liter of medium, 10.0mL sterile vitamin solution, 15.0mL of sterile $Na_2S \cdot 9H_2O$ solution and 30.0mL sterile yeast extract solution. Mix thoroughly. The final pH should be 3.5-3.8.

Use: For the cultivation of *Acidilobus aceticus,* a novel anaerobic thermoacidophilic archaeon from continental hot vents in Kamchatka.

Acidomonas Agar

Composition per liter:

Solution A	500.0mL
Solution B	500.0mL

Solution A:

Composition per 500.0mL:

Glucose	10.0g
Peptone	5.0g
Malt extract	3.0g
Yeast extract	3.0g

Preparation of Solution A: Add components to distilled/deionized water and bring volume to 1.0L. Mix thoroughly. Adjust pH to 4.0. Autoclave for 15 min at 15 psi pressure–121°C. Cool to 50°–55°C.

Solution B:

Composition per 500.0mL:

Agar	20.0g

Preparation of Solution B: Add 20.0g of agar to distilled/deionized water and bring volume to 500.0mL. Mix thoroughly. Gently heat and bring to boiling. Autoclave for 15 min at 15 psi pressure–121°C. Cool to 50°–55°C.

Preparation of Medium: Aseptically mix 500.0mL of solution A with 500.0mL of solution B. Pour into sterile Petri dishes or leave in tubes.

Use: For the cultivation and maintenance of *Acidomonas methanolica,* an acidophilic facultatively methylotrophic bacterium.

Acidomonas methanolica Agar

Composition per liter:

Solution A	500.0mL
Solution B	500.0mL

pH 4.0 ± 0.2 at 25°C

Solution A:

Composition per 500.0mL:

Glucose	20.0g
Yeast extract	5.0g
$(NH_4)_2SO_4$	3.0g
KH_2PO_4	1.0g
$MgSO_4 \cdot 7H_2O$	0.7g
NaCl	0.5g
$Ca(NO_3)_2 \cdot 4H_2O$	0.4g
$K_2HPO_4 \cdot 3H_2O$	0.16g

Preparation of Solution A: Add components to distilled/deionized water and bring volume to 500.0mL. Mix thoroughly. Autoclave for 15 min at 15 psi pressure–121°C. Cool to 50°–55°C.

Solution B:

Composition per 500.0mL:

Agar	20.0g

Preparation of Solution B: Add agar to distilled/deionized water and bring volume to 500.0mL. Mix thoroughly. Gently heat and bring to boiling. Autoclave for 15 min at 15 psi pressure–121°C. Cool to 50°–55°C.

Preparation of Medium: Aseptically mix 500.0mL of solution A and 500.0mL of solution B. Mix thoroughly. Aseptically adjust pH to 4.0. Pour into sterile Petri dishes or distribute into sterile tubes.

Use: For the cultivation and maintenance of *Acidomonas methanolica,* an acidophilic facultatively methylotrophic bacterium.

Acidophilic *Bacillus stearothermophilus* Agar

Composition per liter:

Part B 600.0mL
Part A 400.0mL

pH 5.0 ± 0.2 at 25°C

Part A:

Composition per 400.0mL:

Soluble starch 10.0g
Pancreatic digest of casein 5.0g
Yeast extract 5.0g
KH_2PO_4 ....1.0g
$CaCl_2 \cdot 2H_2O$ 0.5g
$MnCl_2 \cdot 4H_2O$ 0.5g

Preparation of Part A: Add components to distilled/deionized water and bring volume to 400.0mL. Mix thoroughly. Gently heat and bring to boiling. Adjust pH to 4.7. Autoclave for 15 min at 15 psi pressure–121°C. Cool to 50°C.

Part B:

Composition per 600.0mL:

Agar 20.0g

Preparation of Part B: Add agar to distilled/deionized water and bring volume to 600.0mL. Autoclave for 15 min at 15 psi pressure–121°C. Cool to 50°C.

Preparation of Medium: Aseptically combine solution A and solution B. Mix thoroughly. Adjust pH to 5.0. Pour into sterile Petri dishes.

Use: For the cultivation and maintenance of *Bacillus stearothermophilus* and other acidophilic *Bacillus* species.

Acidophilic *Bacillus stearothermophilus* Broth

Composition per liter:

Soluble starch 10.0g
Pancreatic digest of casein 5.0g
Yeast extract 5.0g
KH_2PO_4 ....1.0g
$CaCl_2 \cdot 2H_2O$ 0.5g
$MnCl_2 \cdot 4H_2O$ 0.5g

pH 5.0 ± 0.2 at 25°C

Preparation of Medium: Dissolve all components in 1.0L of distilled/deionized water. Mix thoroughly. Gently heat and bring to boiling. Adjust to pH 5.0. Autoclave for 15 min at 15 psi pressure–121°C.

Use: For the cultivation and maintenance of *Bacillus stearothermophilus* and other acidophilic *Bacillus* species.

Acidophilium Agar

Composition per liter:

Solution A 500.0mL
Solution B 500.0mL

pH 3.5 ± 0.2 at 25°C

Solution A:

Composition per 500.0mL:

Agar 12.0g

Preparation of Solution A: Add agar to distilled/deionized water and bring volume to 500.0mL. Mix thoroughly. Gently heat and bring to boiling. Autoclave for 15 min at 15 psi pressure–121°C. Cool to 50°C.

Solution B:

Composition per 500.0mL:

Mannitol 1.0g
$MgSO_4 \cdot 7H_2O$ 0.5g
$(NH_4)_2SO_4$ 0.1g
Tryptone soya broth 0.1g
KCl 50.0mg
KH_2PO_4 ....50.0mg
$Ca(NO_3)_2$ 10.0mg

Preparation of Solution B: Add components to distilled/deionized water and bring volume to 1.0L. Mix thoroughly. Bring pH to 3.5 Autoclave for 15 min at 15 psi pressure–121°C. Cool to 50°C.

Preparation of Medium: Aseptically combine 500.0mL of solution A with 500.0mL of solution B. Mix thoroughly. Pour into sterile Petri dishes or aseptically distribute into sterile tubes.

Use: For the cultivation and maintenance of *Acidiphilium cryptum* and other *Acidiphilium* species.

Activated Carbon Medium (DSMZ Medium 811)

Composition per liter:

Agar 15.0g
$Na_2HPO_4 \cdot 12H_2O$ 9.0g
Activated carbon 5.0g
KH_2PO_4 ....1.5g
NH_4Cl 1.5g
$MgSO_4 \cdot 7H_2O$ 0.2g
$CaCl_2 \cdot 2H_2O$ 20.0mg
NH_4-Fe-III-Citrate 1.2mg
Trace elements solution TS2 ....1.0mL

pH 7.5 ± 0.1 at 25°C

Trace Elements Solution TS2:

Composition per liter:

$Na_2MoO_4 \cdot 4H_2O$ 900.0mg
H_3BO_3 ....300.0mg
$CoCl_2 \cdot 6H_2O$ 200.0mg
$ZnSO_4 \cdot 7H_2O$ 100.0mg
$MnCl_2 \cdot 4H_2O$ 30.0mg
$NiCl_2 \cdot 6H_2O$ 20.0mg
Na_2SeO_3 ....20.0mg
$CuCl_2 \cdot 2H_2O$ 10.0mg

Preparation of Solution A: Add components to distilled/deionized water and bring volume to 1.0L. Mix thoroughly.

Preparation of Medium: Add components, except activated carbon, to distilled/deionized water and bring volume to 900.0mL. Mix thoroughly. After all components are dissolved add activated carbon. Bring volume to 1.0L with distilled/deionized water. Adjust pH to 7.5. Distribute into tubes or flasks. Autoclave for 15 min at 15 psi pressure–121°C. Pour into Petri dishes or leave in tubes.

Use: For the cultivation of *Streptomyces thermoautotrophicus*.

Aeromonas Differential Agar (Dextrin Fuchsin Sulfite Agar)

Composition per liter:

Dextrin 15.0g
Agar 13.0g
Pancreatic digest of casein 10.0g
Na_2HPO_4 ....7.75g
NaCl 5.0g
Beef extract 3.0g
Na_2SO_3 ....1.6g
Acid Fuchsin solution 50.0mL

pH 7.5 ± 0.2 at 25°C

Acid Fuchsin Solution:
Composition per 50.0mL:
Acid Fuchsin 0.25g
Aquesou dioxan, 5% 50.0mL

Preparation of Acid Fuchsin Solution: Add Acid Fuchsin to 50.0mL of 5% aqueous dioxan. Mix well to dissolve.

Caution: Acid Fuchsin is a potential carcinogen and care must be taken to avoid inhalation of the powdered dye and contamination of the skin.

Preparation of Medium: Add components to distilled/deionized water and bring volume to 1.0L. Mix thoroughly. Gently heat while stirring and bring to boiling. Distribute into tubes or flasks. Autoclave for 15 min at 15 psi pressure–121°C. Pour into sterile Petri dishes or leave in tubes.

Use: For the isolation and differentiation of *Aeromonas* species from other Gram-negative rods such as *Pseudomonas* and Enterobacteriaceae. Specimens with low numbers of *Aeromonas* may first be enriched by growth in starch broth for 4–9 days. After 24 hrs of growth on this agar, colonies are sprayed with Nadi reagent (1% solution of *N,N,N′,N′*-tetramethyl-*p*-phenylene-diammonium dichloride). A positive Nadi reaction (dextrin degradation) is indicated by a purple color at the periphery of the colony. Dextrin fermentation is also indicated by red colonies. *Aeromonas* species appear as large, convex, dark red colonies with a purple periphery.

Aeromonas hydrophila Medium

Composition per liter:
Inositol 10.0g
Pancreatic digest of casein 10.0g
L-Ornithine·HCl 5.0g
Proteose peptone 5.0g
Agar 3.0g
Yeast extract 3.0g
Mannitol 1.0g
Ferric ammonium citrate 0.5g
$Na_2S_2O_3 \cdot 5H_2O$ 0.4g
Bromcresol Purple 0.02g
pH 6.7 ± 0.2 at 25°C

Preparation of Medium: Add components to distilled/deionized water and bring volume to 1.0L. Mix thoroughly. Gently heat until dissolved. Adjust pH to 6.7. Distribute into tubes in 5.0mL volumes. Autoclave for 12 min at 15 psi pressure–121°C.

Use: For the isolation and cultivation of *Aeromonas hydrophila.*

Aeromonas Medium (Ryan's *Aeromonas* Medium)

Composition per liter:
Agar 12.5g
$Na_2S_2O_3$ 10.67g
Proteose peptone 5.0g
NaCl 5.0g
Xylose 3.75g
L-Lysine·HCl 3.5g
Yeast extract 3.0g
Sorbitol 3.0g
Bile salts No. 3 .... 3.0g
Inositol 2.5g
L-Arginine·HCl 2.0g
Lactose 1.5g
Ferric ammonium citrate 0.8g
Bromothymol Blue 0.04g
Thymol Blue 0.04g
pH 8.0 ± 0.1 at 25°C

Source: This medium is available as a dehydrated powder from Oxoid Unipath.

Preparation of Medium: Add components to distilled/deionized water and bring volume to 1.0L. Mix thoroughly. Gently heat and bring to boiling. Do not autoclave. Cool to 50°C and aseptically add 5.0mg of ampicillin. Pour into sterile Petri dishes.

Use: For the isolation and selective differentiation of *Aeromonas hydrophila* and other *Aeromonas* species. *Aeromonas* species appear as small (0.5–1.5mm), dark green colonies with darker centers.

Agrobacterium Agar

Composition per liter:
Agar 15.0g
Mannitol 8.0g
NaCl 5.0g
Yeast extract 5.0g
$(NH_4)_2SO_4$ 2.0g
Casamino acids 0.5g
pH 6.6 ± 0.2 at 25°C

Preparation of Medium: Add components to distilled/deionized water and bring volume to 1.0L. Mix thoroughly. Adjust pH to 6.6. Gently heat and bring to boiling. Distribute into tubes or flasks. Autoclave for 15 min at 15 psi pressure–121°C. Pour into sterile Petri dishes or leave in tubes.

Use: For the cultivation and maintenance of *Agrobacterium rhizogenes* and *Agrobacterium tumefaciens.*

Agrobacterium Agar

Composition per liter:
Agar 15.0g
Glucose 10.0g
Yeast extract 10.0g
$(NH_4)_2SO_4$ 1.0g
KH_2PO_4 .... 0.25g

Preparation of Medium: Add components to tap water and bring volume to 1.0L. Mix thoroughly. Gently heat and bring to boiling. Distribute into tubes or flasks. Autoclave for 15 min at 15 psi pressure–121°C. Pour into sterile Petri dishes or leave in tubes.

Use: For the cultivation and maintenance of *Agrobacterium azotophilum, Agrobacterium radiobacter, Agrobacterium rhizogenes, Agrobacterium rubi, Agrobacterium tumefaciens,* and *Agrobacterium vitis.*

Agrobacterium Agar with Biotin

Composition per liter:
Agar 15.0g
Mannitol 8.0g
NaCl 5.0g
Yeast extract 5.0g
$(NH_4)_2SO_4$ 2.0g
Casamino acids 0.5g
Biotin solution 0.1mL
pH 6.6 ± 0.2 at 25°C

Biotin Solution:
Composition per 10.0mL:
Biotin 0.2mg

Preparation of Biotin Solution: Add biotin to distilled/deionized water and bring volume to 10.0mL. Mix thoroughly. Filter sterilize.

Preparation of Medium: Add components, except biotin solution, to distilled/deionized water and bring volume to 999.9mL. Mix thoroughly. Gently heat and bring to boiling. Adjust pH to 6.6. Autoclave for 15 min at 15 psi pressure–121°C. Cool to 50°–55°C. Aseptically add 0.1mL of sterile biotin solution. Mix thoroughly. Pour into sterile Petri dishes or distribute into sterile tubes.

Use: For the cultivation and maintenance of slow-growing *Agrobacterium rhizogenes* and *Agrobacterium tumefaciens*.

Agrobacterium Mannitol Medium

Composition per liter:

Mannitol	10.0g
L-Glutamate	2.0g
KH_2PO_4	0.5g
Yeast extract	0.3g
$MgSO_4 \cdot 7H_2O$	0.2g
NaCl	0.2g

pH 7.0 ± 0.2 at 25°C

Preparation of Medium: Add components to distilled/deionized water and bring volume to 1.0L. Mix thoroughly. Adjust pH to 7.0. Autoclave for 15 min at 15 psi pressure–121°C.

Use: For the cultivation of *Agrobacterium rhizogenes*.

Agrobacterium Medium

Composition per liter:

Agar	20.0g
Mannitol	10.0g
$NaNO_3$	4.0g
$MgCl_2$	2.0g
Calcium propionate	1.2g
$Mg_3(PO_4)_2$	0.2g
$MgSO_4$	0.1g
$MgCO_3$	0.075g
$NaHCO_3$	0.075g
Supplement	100.0mL

pH 7.1 ± 0.2 at 25°C

Supplement:
Composition per 100.0mL:

Berberine	0.275g
Cycloheximide	0.2g
Bacitracin	0.1g
Na_2SeO_3	0.1g
Penicillin G	0.06g
Streptomycin sulfate	0.03g
Tyrothricin	1.0mg

Preparation of Supplement: Add components to distilled/deionized water and bring volume to 100.0mL. Mix thoroughly. Filter sterilize.

Preparation of Medium: Add components, except supplement, to distilled/deionized water and bring volume to 900.0mL. Mix thoroughly. Gently heat and bring to boiling. Autoclave for 15 min at 15 psi pressure–121°C. Cool to 45°–50°C. Aseptically add 100.0mL of sterile supplement. Mix thoroughly. Pour into sterile Petri dishes or distribute into sterile tubes.

Use: For the selective isolation and cultivation of *Agrobacterium* species.

Agrobacterium Medium D1

Composition per liter:

Agar	15.0g
Mannitol	15.0g
LiCl	6.0g
$NaNO_3$	5.0g
K_2HPO_4	2.0g
$MgSO_4 \cdot 7H_2O$	0.2g
Bromothymol Blue	0.1g
$Ca(NO_3)_2 \cdot 4H_2O$	0.02g

pH 7.2 ± 0.2 at 25°C

Preparation of Medium: Add components to distilled/deionized water and bring volume to 1.0L. Mix thoroughly. Gently heat and bring to boiling. Distribute into tubes or flasks. Autoclave for 15 min at 15 psi pressure–121°C. Cool to 45°–50°C. Adjust pH to 7.2. Pour into sterile Petri dishes or leave in tubes.

Use: For the selective isolation and cultivation of *Agrobacterium* species.

Agrobacterium tumefaciens Modified Roy and Sasser Medium for Grapevine Strains

Composition per 1020mL:

Agar	15.0g
Adoniitol	4.0g
H_3BO_3	1.0g
K_2HPO_4	0.9g
KH_2PO_4	0.7g
NaCl	0.2g
$MgSO_4 \cdot 7H_2O$	0.2g
Yeast extract	0.14g
Cycloheximide solution	10.0mL
Triphenyl tetrazolium chloride solution	1.0mL
D-Cycloserine solution	1.0mL
Trimethoprin solution	1.0mL

Cycloheximide Solution:
Composition per 10.0mL:

Cycloheximide	0.025g

Preparation of Cycloheximide Solution: Add cycloheximide to distilled/deionized water and bring volume to 10.0mL. Mix thoroughly. Filter sterilize.

Triphenyl Tetrazolium Chloride Solution:
Composition per 10.0mL:

Triphenyl tetrazolium chloride	0.8g

Preparation of Triphenyl Tetrazolium Chloride Solution: Add triphenyl tetrazolium chloride to distilled/deionized water and bring volume to 10.0mL. Mix thoroughly. Filter sterilize.

D-Cycloserine Solution:
Composition per 10.0mL:

D-Cycloserine	0.2g

Preparation of D-Cycloserine Solution: Add D-cycloserine to distilled/deionized water and bring volume to 10.0mL. Mix thoroughly. Filter sterilize.

Trimethoprim Solution:
Composition per 10.0mL:

Trimethoprim	0.025g

Preparation of Trimethoprim Solution: Add trimethoprim to distilled/deionized water and bring volume to 10.0mL. Add a drop of dilute HCl. Mix thoroughly. Gently heat while mixing until dissolved. Filter sterilize.

Preparation of Medium: Add components, except cycloheximide solution and tetrazolium chloride solution, D-cycloserine solution, and trimethoprin solution to distilled/deionized water and bring volume to 1.0L. Mix thoroughly. Adjust pH to 7.2. Distribute 100.0mL into flasks. Gently heat and bring to boiling. Autoclave for 15 min at 15 psi pressure–121°C. Cool to 50°C. Aseptically add per 100.0mL of medium, 0.1 mL sterile tetrazolium chloride solution, 0.1mL sterile D-cycloserine solution, 0.1mL sterile trimethoprin solution, and 1.0mL cycloheximide solution. Mix thoroughly. Aseptically pour into sterile Petri dishes.

Use: For the selective cultivation of *Agrobacterium tumefaciens* biovar 3.

Agrobacterium tumefaciens Selective Medium

Composition per 1020mL:

Agar	15.0g
L(−)Arabitol	3.04g
K_2HPO_4	1.04g
KH_2PO_4	0.54g
Sodium taurocholate	0.29g
$MgSO_4 \cdot 7H_2O$	0.25g
NH_4NO_3	0.16g
Cycloheximide solution	10.0mL
Selenite solution	10.0mL
Crystal Violet (0.1% solution)	2.0mL

Selenite Solution:
Composition per 10.0mL:

NaOH	0.5g
$Na_2SeO_3 \cdot 5H_2O$	0.1g

Preparation of Selenite Solution: Add components to distilled/deionized water and bring volume to 1.0L. Mix thoroughly. Filter sterilize.

Cycloheximide Solution:
Composition per 10.0mL:

Cycloheximide	0.02g

Preparation of Cycloheximide Solution: Add cycloheximide to distilled/deionized water and bring volume to 10.0mL. Mix thoroughly. Filter sterilize.

Preparation of Medium: Add components, except cycloheximide solution and selenite solution, to distilled/deionized water and bring volume to 1.0L. Mix thoroughly. Distribute 100.0mL into flasks. Gently heat and bring to boiling. Autoclave for 15 min at 15 psi pressure–121°C. Cool to 50°C. Aseptically add per 100.0mL of medium, 1.0mL sterile selenite solution and 1.0mL cycloheximide solution. Mix thoroughly. Aseptically pour into sterile Petri dishes.

Use: For the selective cultivation of *Agrobacterium tumefaciens* biovar 1.

Agrobacterium tumefaciens Selective Medium

Composition per 1020mL:

Agar	15.0g
Erythritol	3.05g
K_2HPO_4	1.04g
KH_2PO_4	0.54g
Sodium taurocholate	0.29g
$MgSO_4 \cdot 7H_2O$	0.25g
NH_4NO_3	0.16g
Cycloheximide solution	10.0mL
Selenite solution	10.0mL
Malachite Green (0.1% solution)	5.0mL
Yeast extract (1% solution)	1.0mL

Selenite Solution:
Composition per 10.0mL:

NaOH	0.5g
$Na_2SeO_3 \cdot 5H_2O$	0.1g

Preparation of Selenite Solution: Add components to distilled/deionized water and bring volume to 1.0L. Mix thoroughly. Filter sterilize.

Cycloheximide Solution:
Composition per 10.0mL:

Cycloheximide	0.02g

Preparation of Cycloheximide Solution: Add cycloheximide to distilled/deionized water and bring volume to 10.0mL. Mix thoroughly. Filter sterilize.

Preparation of Medium: Add components, except cycloheximide solution and selenite solution, to distilled/deionized water and bring volume to 1.0L. Mix thoroughly. Distribute 100.0mL into flasks. Gently heat and bring to boiling. Autoclave for 15 min at 15 psi pressure–121°C. Cool to 50°C. Aseptically add per 100.0mL of medium, 1.0mL sterile selenite solution and 1.0mL cycloheximide solution. Mix thoroughly. Aseptically pour into sterile Petri dishes.

Use: For the selective cultivation of *Agrobacterium tumefaciens* biovar 2.

AKI Medium

Composition per liter:

Peptone	15.0g
NaCl	5.0g
Yeast extract	4.0g
Sodium bicarbonate solution	30.0mL

pH 7.2 ± 0.2 at 25°C

Sodium Bicarbonate Solution:
Composition per 100.0mL:

$NaHCO_3$	10.0g

Preparation of Sodium Bicarbonate Solution: Add sodium bicarbonate to distilled/deionized water and bring volume to 100.0mL. Mix thoroughly. Filter sterilize. Use freshly prepared solution.

Preparation of Medium: Add components, except sodium bicarbonate solution, to distilled/deionized water and bring volume to 970.0mL. Mix thoroughly. Autoclave for 15 min at 15 psi pressure–121°C. Cool to 45°–50°C. Aseptically add sterile sodium bicarbonate solution. Mix thoroughly. Aseptically distribute into sterile tubes or flasks. Prepare medium freshly.

Use: For the cultivation of *Vibrio cholerae* and other *Vibrio* species.

Alcaligenes N5 Medium

Composition per liter:

Sodium succinate·$2H_2O$	5.0g
KH_2PO_4	0.75g
NH_4Cl	0.67g
K_2HPO_4	0.61g
$MgSO_4 \cdot 7H_2O$	0.2g
$CaCl_2 \cdot 2H_2O$	0.03g
$MnCl_2 \cdot 4H_2O$	3.0mg
$FeCl_3$	2.4mg
$Na_2MoO_4 \cdot 2H_2O$	1.0mg

Preparation of Medium: Add components to distilled/deionized water and bring volume to 1.0L. Mix thoroughly. Gently heat while stirring and bring to boiling. Distribute into tubes or flasks. Autoclave for 15 min at 15 psi pressure–121°C.

Use: For the cultivation and maintenance of *Alcaligenes faecalis.*

Alcaligenes NA YE Medium (*Alcaligenes* Nutrient Agar Yeast Extract Medium)

Composition per liter:

Agar	15.0g
Pancreatic digest of gelatin	5.0g
Yeast extract	5.0g
Beef extract	3.0g

pH 7.0 ± 0.2 at 25°C

Preparation of Medium: Add components to distilled/deionized water and bring volume to 1.0L. Mix thoroughly. Gently heat while stirring and bring to boiling. Distribute into tubes or flasks. Autoclave for 15 min at 15 psi pressure–121°C. Pour into sterile Petri dishes or leave in tubes.

Use: For the cultivation and maintenance of *Alcaligenes* species.

Alcaligenes NB YE Agar (*Alcaligenes* Nutrient Broth Yeast Extract Agar)

Composition per liter:

Agar	15.0g
Pancreatic digest of gelatin	5.0g
Yeast extract	5.0g
Beef extract	3.0g

Preparation of Medium: Add components to distilled/deionized water and bring volume to 1.0L. Mix thoroughly. Gently heat while stirring and bring to boiling. Distribute into tubes or flasks. Autoclave for 15 min at 15 psi pressure–121°C. Pour into sterile Petri dishes or leave in tubes.

Use: For the cultivation and maintenance of *Alcaligenes faecalis.*

Alcaliphilic *Amphibacillus* Strains Medium (DSMZ Medium 931)

Composition per liter:

Na_2CO_3	63.6g
$NaHCO_3$	50.4g
KH_2PO_4	0.2g
$MgCl_2$	0.1g
NH_4Cl	0.5g
KCl	0.2g
Resazurin	0.01g
Sucrose solution	50.0mL
$Na_2S \cdot 9H_2O$ solution	10.0mL
Yeast extract solution	10.0mL
Vitamin solution	10.0mL
Trace elements	1.0mL

pH 9.5-10.0 at 25°C

Sucrose Solution:
Composition per 50.0mL:

Sucrose	5.0g

Preparation of Sucrose Solution: Add sucrose to distilled/deionized water and bring volume to 50.0mL. Mix thoroughly. Sparge with 100% N_2. Autoclave for 15 min at 15 psi pressure–121°C. Cool to room temperature.

Yeast Extract Solution:
Composition per 10.0mL:

Yeast extract	0.2g

Preparation of Yeast Extract Solution: Add yeast extract to distilled/deionized water and bring volume to 10.0mL. Mix thoroughly. Sparge with 100% N_2. Autoclave under 100% N_2 for 15 min at 15 psi pressure–121°C. Cool to room temperature.

$Na_2S \cdot 9H_2O$ Solution:
Composition per 10mL:

$Na_2S \cdot 9H_2O$	0.7g

Preparation of $Na_2S \cdot 9H_2O$ Solution: Add $Na_2S \cdot 9H_2O$ to distilled/deionized water and bring volume to 10.0mL. Mix thoroughly. Autoclave under 100% N_2 for 15 min at 15 psi pressure–121°C. Cool to room temperature.

Trace Elements Solution:
Composition per liter:

$MgSO_4 \cdot 7H_2O$	3.0g
Nitrilotriacetic acid	1.5g
NaCl	1.0g
$MnSO_4 \cdot 2H_2O$	0.5g
$CoSO_4 \cdot 7H_2O$	0.18g
$ZnSO_4 \cdot 7H_2O$	0.18g
$CaCl_2 \cdot 2H_2O$	0.1g
$FeSO_4 \cdot 7H_2O$	0.1g
$NiCl_2 \cdot 6H_2O$	0.025g
$KAl(SO_4)_2 \cdot 12H_2O$	0.02g
H_3BO_3	0.01g
$Na_2MoO_4 \cdot 4H_2O$	0.01g
$CuSO_4 \cdot 5H_2O$	0.01g
$Na_2SeO_3 \cdot 5H_2O$	0.3mg

Preparation of Trace Elements Solution: Add nitrilotriacetic acid to 500.0mL of distilled/deionized water. Dissolve by adjusting pH to 6.5 with KOH. Add remaining components. Add distilled/deionized water to 1.0L. Mix thoroughly.

Vitamin Solution:
Composition per liter:

Pyridoxine-HCl	10.0mg
Thiamine-HCl·2H_2O	5.0mg
Riboflavin	5.0mg
Nicotinic acid	5.0mg
D-Ca-pantothenate	5.0mg
p-Aminobenzoic acid	5.0mg
Lipoic acid	5.0mg
Biotin	2.0mg
Folic acid	2.0mg
Vitamin B_{12}	0.1mg

Preparation of Vitamin Solution: Add components to distilled/deionized water and bring volume to 1.0L. Mix thoroughly. Sparge with 80% H_2 + 20% CO_2. Filter sterilize.

Preparation of Medium: Prepare and dispense medium under 100% N_2 gas atmosphere. Add components, except $NaHCO_3$, NH_4Cl, Na_2CO_3, sucrose solution, $Na_2S \cdot 9H_2O$ solution, yeast extract solution, and vitamin solution, to distilled/deionized water and bring volume to 920.0mL. Mix thoroughly. Gently heat and bring to boiling. Boil for 5 min. Cool to room temperature while sparging with 100% N_2. Add solid $NaHCO_3$, NH_4Cl, and Na_2CO_3. Mix thoroughly. Distribute into anaerobe tubes or bottles. Autoclave for 15 min at 15 psi pressure–121°C. Aseptically and anaerobically add per liter of medium 50.0mL sucrose solution, 10.0mL yeast extract solution, 10.0mL $Na_2S \cdot 9H_2O$ solution, and 10.0mL vitamin solution. The final pH should be 9.5-10.0.

Use: For the cultivation of *Amphibacillus fermentum* and *Amphibacillus tropicus.*

Alginate Utilization Medium

Composition per liter:

Solution B	500.0mL
Solution A	400.0mL
Solution C	100.0mL

Solution A:

Composition per 400.0mL:

Marine salts	38.0g

Preparation of Solution A: Add marine salts to distilled/deionized water and bring volume to 400.0mL. Mix thoroughly. Autoclave for 15 min at 15 psi pressure–121°C.

Solution B:

Composition per 500.0mL:

Agar	20.0g
Sodium alginate	10.0g

Preparation of Solution B: Add components to distilled/deionized water and bring volume to 500.0mL. Mix thoroughly. Autoclave for 15 min at 15 psi pressure–121°C.

Solution C:

Composition per 100.0mL:

Tris·HCl buffer	0.067g
$NaNO_3$	0.047g
Ferric EDTA	66.5mg
Sodium glycerophosphate	6.67mg
Thiamine·HCl	67.0μg
Vitamin B_{12}	1.3μg
Biotin	0.67μg

Preparation of Solution C: Add components to distilled/deionized water and bring volume to 100.0mL. Mix thoroughly. Filter sterilize.

Preparation of Medium: Aseptically combine solutions A, B, and C. For liquid medium, omit agar from solution B.

Use: For the cultivation of microorganisms that can utilize alginate as a carbon source. Growth on alginate (production of alginase) is a diagnostic test used in the differentiation of *Vibrio* species.

Alicyclobacillus acidoterrestris Agar

Composition per 1001.0mL:

Solution A	500.0mL
Solution C	500.0mL
Solution B	1.0mL

pH 4.0 ± 0.2 at 25°C

Solution A:

Composition per 500.0mL:

Glucose	5.0g
KH_2PO_4	3.0g
Yeast extract	2.0g
$MgSO_4·7H_2O$	0.5g
$CaCl_2·7H_2O$	0.25g
$(NH_4)_2SO_4$	0.2g

Preparation of Solution A: Add components to distilled/deionized water and bring volume to 500.0L. Mix thoroughly. Adjust pH to 4.0. Distribute into tubes or flasks. Autoclave for 15 min at 15 psi pressure–121°C. Cool to 50°–55°C.

Solution B:

Composition per liter:

$MnCl_2·4H_2O$	0.5g
H_3BO_3	0.3g
$CoCl_2·6H_2O$	0.2g
$ZnSO_4·7H_2O$	0.1g
$Na_2MoO_4·2H_2O$	0.03g
$NiCl_2·6H_2O$	0.02g
$CuCl_2·2H_2O$	0.01g

Preparation of Solution B: Add components to distilled/deionized water and bring volume to 1.0L. Mix thoroughly. Filter sterilize. Warm to 50°–55°C.

Solution C:

Composition per 500.0mL:

Agar	15.0g

Preparation of Solution C: Add agar to distilled/deionized water and bring volume to 500.0L. Mix thoroughly. Gently heat and bring to boiling. Autoclave for 15 min at 15 psi pressure–121°C. Cool to 50°–55°C.

Preparation of Medium: Aseptically combine 500.0mL of solution A, 1.0mL of solution B, and 500.0mL of solution C. Mix thoroughly. Pour into sterile Petri dishes or distribute into sterile tubes.

Use: For the cultivation and maintenance of *Alicyclobacillus acidoterrestris.*

Alicyclobacillus Agar (DSMZ Medium 402)

Composition per liter:

Glucose	5.0g
KH_2PO_4	3.0g
Yeast extract	2.0g
$MgSO_4·7H_2O$	0.5g
$CaCl_2·2H_2O$	0.25g
$(NH_4)_2 SO_4$	0.2g
Agar solution	500.0mL
Trace Elements Solution SL-6	1.0mL

pH 4.0 ± 0.2 at 25°C

Trace Elements Solution SL-6:

Composition per liter:

$MnCl_2·4H_2O$	0.5g
H_3BO_3	0.3g
$CoCl_2·6H_2O$	0.2g
$ZnSO_4·7H_2O$	0.1g
$Na_2MoO_4·2H_2O$	0.03g
$NiCl_2·6H_2O$	0.02g
$CuCl_2·2H_2O$	0.01g

Preparation of Trace Elements Solution SL-6: Add components to distilled/deionized water and bring volume to 1.0L. Mix thoroughly. Autoclave for 15 min at 15 psi pressure–121°C. Cool to room temperature.

Agar Solution:

Composition per 500mL:

Agar	15.0g

Preparation of Agar Solution: Add agar to distilled/deionized water and bring volume to 500.0mL. Mix thoroughly. Gently heat and bring to boiling. Autoclave for 15 min at 15 psi pressure–121°C. Cool to 45°C.

Preparation of Medium: Add components, except trace elements solution SL-6 and agar solution, to distilled/deionized water and bring volume to 499.0mL. Mix thoroughly. Adjust pH to 4.0. Autoclave for 15 min at 15 psi pressure–121°C. Cool to 45°C. Aseptically add 1.0mL of sterile trace elements solution SL-6 and 500.0mL agar solution. Mix thoroughly. Pour into sterile Petri dishes or aseptically distribute into sterile tubes or flasks.

Use: For the cultivation and maintenance of *Alicyclobacillus* spp., *Bacillus* sp., *and Bacillus naganoensis.*

Alicyclobacillus cycloheptanicus Agar (LMG Medium 174)

Composition per 1001.0mL:

Solution A	500.0mL
Agar solution	500.0mL
Trace elements solution SL-6	1.0mL

pH 4.0 ± 0.2 at 25°C

Solution A:

Composition per 500.0mL:

Yeast extract	5.0g
Glucose	5.0g
K_2HPO_4	3.0g
$MgSO_4 \cdot 7H_2O$	0.5g
$CaCl_2 \cdot 2H_2O$	0.25g
$(NH_4)_2SO_4$	0.2g

Preparation of Solution A: Add components to distilled/deionized water and bring volume to 500.0mL. Mix thoroughly. Adjust to pH 4.0. Autoclave for 15 min at 15 psi pressure–121°C. Cool to 45°-50°C.

Trace Elements Solution SL-6:

Composition per liter:

H_3BO_3	0.3g
$CoCl_2 \cdot 6H_2O$	0.2g
$ZnSO_4 \cdot 7H_2O$	0.1g
$MnCl_2 \cdot 4H_2O$	0.03g
$Na_2MoO_4 \cdot H_2O$	0.03g
$NiCl_2 \cdot 6H_2O$	0.02g
$CuCl_2 \cdot 2H_2O$	0.01g

Preparation of Trace Elements Solution SL-6: Add components to distilled/deionized water and bring volume to 1.0L. Mix thoroughly. Adjust pH to 3.4. Filter sterilize.

Agar Solution:

Composition per 500.0mL:

Agar	30.0g

Preparation of Agar Solution: Add agar to distilled/deionized water and bring volume to 500.0mL. Mix thoroughly. Gently heat and bring to boiling. Autoclave for 15 min at 15 psi pressure–121°C. Cool to 45°-50°C.

Preparation of Medium: Aseptically combine 500.0mL solution A, 500.0mL sterile agar solution and 1.0mL sterile trace elements solution SL-6. Mix thoroughly. Aseptically pour into sterile Petri dishes or distribute into sterile tubes.

Use: For the cultivation of *Alicyclobacillus cycloheptanicus.*

Alkaline *Bacillus* Medium

Composition per liter:

Agar	15.0g
Peptone	10.0g
Glucose	10.0g
Yeast extract	5.0g
K_2HPO_4	1.0g
Na_2CO_3 solution	100.0mL

pH 8.5–11.0 at 25°C

Na_2CO_3 Solution:

Composition per 100.0mL:

Na_2CO_3	10.0g

Preparation of Na_2CO_3 Solution: Add Na_2CO_3 to distilled/deionized water and bring volume to 100.0mL. Mix thoroughly. Filter sterilize.

Preparation of Medium: Add components, except Na_2CO_3 solution, to distilled/deionized water and bring volume to 900.0mL. Gently heat while stirring and bring to boiling. Autoclave for 15 min at 10 psi pressure–115°C. Cool to 45°–50°C. Aseptically add sterile Na_2CO_3 solution. Mix thoroughly. Pour into sterile Petri dishes or distribute into sterile tubes.

Use: For the cultivation and maintenance of alkalophilic microorganisms such as *Bacillus alcalophilus*, *Bacillus circulans*, and other *Bacillus* species.

Alkaline Cellulose Agar

Composition per liter:

Solution A	900.0mL
Solution B	100.0mL

Solution A:

Composition per 900.0mL:

Agar	15.0g
Cellulose powder MN 300	15.0g
NH_4NO_3	2.0g
K_2HPO_4	1.0g
Peptone	1.0g
Yeast extract	0.5g
$CaCl_2$	0.4g
$MgSO_4 \cdot 7H_2O$	0.4g

Preparation of Solution A: Add components to distilled/deionized water and bring volume to 900.0mL. Mix thoroughly. Adjust pH to 7.0 with 1*N* HCl. Gently heat and bring to boiling. Autoclave for 15 min at 15 psi pressure–121°C. Cool to 50°–55°C.

Solution B:

Composition per 100.0mL:

Na_2CO_3	0.5g

Preparation of Solution B: Add 0.5g of Na_2CO_3 to distilled/deionized water and bring volume to 100.0mL. Mix thoroughly. Adjust pH to 9.4 with 6% $NaHCO_3$ solution. Autoclave for 15 min at 15 psi pressure–121°C. Cool to 50°–55°C.

Preparation of Medium: Aseptically combine 900.0mL of solution A with 100.0mL of solution B. Mix thoroughly. Pour into sterile Petri dishes or leave in tubes.

Use: For the cultivation and maintenance of cellulose-utilizing bacteria.

Alkaline Nutrient Agar

Composition per liter:

Agar	15.0g
Peptone	5.0g
Beef extract	3.0g

pH 9.5–10.0 at 25°C

Preparation of Medium: Add components to distilled/deionized water and bring volume to 1.0L. Mix thoroughly. Gently heat and bring to boiling. Distribute into tubes or flasks. Autoclave for 15 min at 15 psi pressure–121°C. Aseptically adjust to pH 9.5-10.0 with sterile 9% Na_2CO_3 solution. Pour into sterile Petri dishes or leave in tubes.

Use: For the cultivation and maintenance of *Bacillus alcalophilus* and *Bacillus* species.

Alkaline Nutrient Agar

Composition per liter:

Agar 15.0g
Peptone 5.0g
NaCl 5.0g
Yeast extract 2.0g
Beef extract 1.0g
Sodium sesquicarbonate solution 100.0mL

pH 9.7 ± 0.2 at 25°C

Sodium Sesquicarbonate Solution:

Composition per 100.0mL:

Na_2CO_3, anhydrous 10.6g
$NaHCO_3$ 8.42g

Preparation of Sodium Sesquicarbonate Solution: Add components to distilled/deionized water and bring volume to 100.0mL. Mix thoroughly. Filter sterilize. Warm to 50°–55°C.

Preparation of Medium: Add components, except sodium sesquicarbonate solution, to distilled/deionized water and bring volume to 900.0mL. Mix thoroughly. Gently heat and bring to boiling. Autoclave for 15 min at 15 psi pressure–121°C. Cool to 50°–55°C. Aseptically add sterile sodium sesquicarbonate solution. Mix thoroughly. Adjust pH to 9.7. Pour into sterile Petri dishes or distribute into sterile tubes.

Use: For the cultivation of alkaliniphilic bacteria, including *Bacillus alcalophilus, Bacillus cohnii,* and other *Bacillus* species.

Alkaline Peptone Agar

Composition per liter:

NaCl 20.0g
Agar 15.0g
Peptone 10.0g

pH 8.5 ± 0.2 at 25°C

Preparation of Medium: Add components to distilled/deionized water and bring volume to 1.0L. Mix thoroughly. Gently heat and bring to boiling. Adjust pH to 8.5. Distribute into tubes. Autoclave for 15 min at 15 psi pressure–121°C. Allow tubes to cool in a slanted position.

Use: For the cultivation of *Vibrio cholerae* and other *Vibrio* species.

Alkaline Peptone Water

Composition per liter:

Peptone 10.0g
NaCl 5.0g

pH 8.4 ± 0.2 at 25°C

Preparation of Medium: Add components to distilled/deionized water and bring volume to 1.0L. Mix thoroughly. Adjust pH to 8.4. Distribute into tubes or flasks. Autoclave for 20 min at 15 psi pressure–121°C.

Use: For the cultivation of a variety of alkalophilic microorganisms.

Alkaline Polypectate Agar

Composition per liter:

Agar 16.0g
Na_2CO_3 .... 10.0g
Peptone 6.0g
Sodium polypectate 5.0g
Yeast extract 3.0g
K_2HPO_4 .... 1.0g
$MgSO_4 \cdot 7H_2O$ 0.2g
$MnSO_4$ 40.0mg

pH 10.0 ± 0.2 at 25°C

Preparation of Medium: Add components to distilled/deionized water and bring volume to 1.0L. Mix thoroughly. Bring pH to 10.0. Gently heat and bring to boiling. Distribute into tubes or flasks. Autoclave for 15 min at 15 psi pressure–121°C. Pour into sterile Petri dishes or leave in tubes.

Use: For the cultivation of pectinolytic bacteria.

Alkaline Starch Agar

Composition per liter:

Starch 20.0g
Agar 16.0g
Na_2CO_3 .... 10.0g
Peptone 6.0g
Yeast extract 3.0g
K_2HPO_4 .... 1.0g
$MgSO_4 \cdot 7H_2O$ 0.2g
$MnSO_4$ 40.0mg

pH 9.7 ± 0.2 at 25°C

Preparation of Medium: Add components to distilled/deionized water and bring volume to 1.0L. Mix thoroughly. Bring pH to 9.7. Gently heat and bring to boiling. Distribute into tubes or flasks. Autoclave for 15 min at 15 psi pressure–121°C. Pour into sterile Petri dishes or leave in tubes.

Use: For the cultivation of alkiliniphilic starch-utilizing bacteria.

Alkaliphilic Methanogen Medium

Composition per liter:

NaCl 15.0g
$NaHCO_3$ 10.0g
Methanol 5.0g
Na_2CO_3 .... 4.0g
$Na_2S \cdot 9H_2O$ 1.0g
NH_4Cl 0.5g
Yeast extract 0.5g
KH_2PO_4 .... 0.3g
$NiCl_2 \cdot 6H_2O$ 2.0mg
Resazurin 0.5mg
Wolfe's mineral solution 10.0mL
Selenite-tungstate solution 1.0mL

pH 9.2–9.4 at 25°C

Wolfe's Mineral Solution:

Composition per liter:

$MgSO_4 \cdot 7H_2O$ 3.0g
Nitrilotriacetic acid 1.5g
NaCl 1.0g
$MnSO_4 \cdot 2H_2O$ 0.5g
$CoCl_2 \cdot 6H_2O$ 0.1g
$ZnSO_4 \cdot 7H_2O$ 0.1g
$CaCl_2 \cdot 2H_2O$ 0.1g
$FeSO_4 \cdot 7H_2O$ 0.1g
$NiCl_2 \cdot 6H_2O$ 0.025g
$KAl(SO_4)_2 \cdot 12H_2O$ 0.02g
$CuSO_4 \cdot 5H_2O$ 0.01g
H_3BO_3 .... 0.01g
$Na_2MoO_4 \cdot 2H_2O$ 0.01g
$Na_2SeO_3 \cdot 5H_2O$ 0.3mg

Preparation of Wolfe's Mineral Solution: Add nitrilotriacetic acid to 500.0mL of distilled/deionized water. Adjust pH to 6.5 with KOH. Add remaining components. Add distilled/deionized water to 1.0L. Adjust pH to 6.8.

Selenite-Tungstate Solution:
Composition per liter:
NaOH 0.5g
$Na_2WO_4 \cdot 2H_2O$ 4.0mg
$Na_2SeO_3 \cdot 5H_2O$ 3.0mg

Preparation of Selenite-Tungstate Solution: Add components to distilled/deionized water and bring volume to 1.0L. Mix thoroughly. Sparge with 100% N_2. Autoclave for 15 min at 15 psi pressure–121°C.

Preparation of Medium: Prepare and dispense medium under 100% N_2. Add components, except $NaHCO_3$ and $Na_2S \cdot 9H_2O$, to distilled/deionized water and bring volume to 1.0L. Mix thoroughly. Gently heat and bring to boiling. Continue boiling for 5 min. Cool to room temperature while sparging with 100% N_2. Add $NaHCO_3$ and $Na_2S \cdot 9H_2O$. Mix thoroughly. Anaerobically distribute into tubes. Autoclave for 15 min at 15 psi pressure–121°C. Adjust pH to 9.2–9.4.

Use: For the cultivation of *Methanohalophilus zhilinae.*

Alkaliphilic Spirochete Medium

Composition per 1011.0mL:
Na_2CO_3 .... 10.0g
NaCl 10.0g
NH_4Cl 1.0g
K_2HPO_4 .... 0.2g
KCl 0.2g
Yeast extract 0.5g
$NaHCO_3$ solution 50.0mL
Sucrose solution 20.0mL
$Na_2S \cdot 9H_2O$ solution 10.0mL
Wolfe's vitamin solution 10.0mL
Trace elements solution SL-6 .... 1.0mL
pH 9.7 ± 0.2 at 25°C

$NaHCO_3$ Solution:
Composition per 50.0mL:
$NaHCO_3$ 15.0g

Preparation of $NaHCO_3$ Solution: Add $NaHCO_3$ to distilled/deionized water and bring volume to 50.0mL. Mix thoroughly. Filter sterilize. Sparge with 100% N_2.

Sucrose Solution:
Composition per 20.0mL:
Sucrose 5.0g

Preparation of Sucrose Solution: Add sucrose to distilled/deionized water and bring volume to 20.0mL. Mix thoroughly. Sparge with 100% N_2. Autoclave for 15 min at 15 psi pressure–121°C.

$Na_2S \cdot 9H_2O$ Solution:
Composition per 10.0mL:
$Na_2S \cdot 9H_2O$ 1.0g

Preparation of $Na_2S \cdot 9H_2O$ Solution: Add $Na_2S \cdot 9H_2O$ to distilled/deionized water and bring volume to 10.0mL. Mix thoroughly. Sparge with 100% N_2. Autoclave for 15 min at 15 psi pressure–121°C. Before use, neutralize to pH 7.0 with sterile HCl.

Wolfe's Vitamin Solution:
Composition per liter:
Pyridoxine·HCl 10.0mg
p-Aminobenzoic acid 5.0mg
Lipoic acid 5.0mg
Nicotinic acid 5.0mg
Riboflavin 5.0mg
Thiamine·HCl 5.0mg
Calcium DL-pantothenate 5.0mg
Biotin 2.0mg
Folic acid 2.0mg
Vitamin B_{12} 0.1mg

Preparation of Wolfe's Vitamin Solution: Add components to distilled/deionized water and bring volume to 1.0L. Mix thoroughly. Filter sterilize. Sparge with 100% N_2.

Trace Elements Solution SL-6:
Composition per liter:
$MnCl_2 \cdot 4H_2O$ 0.5g
H_3BO_3 .... 0.3g
$CoCl_2 \cdot 6H_2O$ 0.2g
$ZnSO_4 \cdot 7H_2O$ 0.1g
$Na_2MoO_4 \cdot 2H_2O$ 0.03g
$NiCl_2 \cdot 6H_2O$ 0.02g
$CuCl_2 \cdot 2H_2O$ 0.01g

Preparation of Trace Elements Solution SL-6: Add components to distilled/deionized water and bring volume to 1.0L. Mix thoroughly.

Preparation of Medium: Prepare and dispense medium under 100% N_2. Add components, except $NaHCO_3$ solution, sucrose solution, $Na_2S \cdot 9H_2O$ solution, and Wolfe's vitamin solution, to distilled/deionized water and bring volume to 910.0mL. Mix thoroughly. Adjust pH to 9.7 with 6*N* NaOH (about 15.0mL). Gently heat and bring to boiling. Cool to room temperature while sparging with 100% N_2. Autoclave for 15 min at 15 psi pressure–121°C. Aseptically and anaerobically add 50.0mL of sterile $NaHCO_3$ solution, 20.0mL of sterile sucrose solution, 10.0mL of sterile $Na_2S \cdot 9H_2O$ solution, and 10.0mL of sterile Wolfe's vitamin solution. Mix thoroughly. Aseptically distribute into sterile tubes or flasks.

Use: For the cultivation of *Spirochaeta africana, Spirochaeta alkalica,* and *Spirochaeta asiatica.*

Alkaliphilic Sulfur Respiring Strains Medium (DSMZ Medium 925)

Composition per liter:
Mineral base 997.0mL
KSCN solution 10.0mL
Trace elements solution 2.0mL
Magnesium chloride solution 1.0mL
pH 10.0 ± 0.2 at 25°C

Mineral Base:
Composition per liter:
Na_2CO_3 .... 20.0g
$NaHCO_3$ 10.0g
NaCl 5.0g
K_2HPO_4 .... 1.0g

Preparation of Mineral Base: Add components to distilled/deionized water and bring volume to 1.0L. Mix thoroughly. Sparge with 100% N_2. Autoclave for 20 min at 110°C. The pH should be about 10.0.

Trace Elements Solution:
Composition per liter:
H_3BO_3 .... 300.0mg
$CoCl_2 \cdot 6H_2O$ 200.0mg
$ZnSO_4 \cdot 7H_2O$ 100.0mg
$MnCl_2 \cdot 4H_2O$ 30.0mg
$Na_2MoO_4 \cdot 4H_2O$ 30.0mg
$NiCl_2 \cdot 6H_2O$ 20.0mg
$CuCl_2 \cdot 2H_2O$ 10.0mg

EDTA 5.0mg
$FeSO_4·7H_2O$ 2.0mg

Preparation of Trace Elements Solution: Add components to distilled/deionized water and bring volume to 1.0L. Adjust pH to 3.0 with HCl. Mix thoroughly. Sparge with 100% N_2. Autoclave for 15 min at 15 psi pressure–121°C.

Magnesium Chloride Solution:
Composition per 10.0mL:
$MgCl_2·6H_2O$ 2.0g

Preparation of Magnesium Chloride Solution: Add $MgCl_2·6H_2O$ to distilled/deionized water and bring volume to 10.0mL. Mix thoroughly. A white colloid will dissolve after mixing. Autoclave for 15 min at 15 psi pressure–121°C.

KSCN Solution:
Composition per 10mL:
KSCN 1.5g

Preparation of KSCN Solution: Add KSCN to distilled/deionized water and bring volume to 10.0mL. Mix thoroughly. Autoclave for 15 min at 15 psi pressure–121°C. Cool to room temperature.

Preparation of Medium: Aseptically add 10.0mL sterile KSCN solution, 2.0mL sterile trace elements solution, and 1.0mL sterile magnesium chloride solution to 987.0mL sterile mineral base. Aseptically distribute to sterile tubes, flasks, or bottles.

Use: For the cultivation of *Thialkalivibrio paradoxus* and *Thialkalivibrio thiocyanoxidans.*

Alkaliphilic *Thermococcus* Medium (DSMZ Medium 926)

Composition per 1082.0mL:
Base solution 1000.0mL
Glycine solution 50.0mL
Yeast extract solution 20.0mL
Polysulfide solution 12.0mL

Base Solution:
Composition per 2000.0mL:
NaCl 27.7g
$MgSO_4·7H_2O$ 7.0g
$MgCl_2·6H_2O$ 5.5g
K_2HPO_4 .... 1.0g
KCl 0.65g
$NaHCO_3$ 0.32g
NaBr 0.1g
H_3BO_3 .... 0.03g
KI 15.0mg
$CaCl_2·2H_2O$ 0.05mg
Trace elements solution 20.0mL

Preparation of Base Solution: Sparge 2.0L of distilled/deionized water with 100% N_2 to remove O_2. Add components to 2000.0mL of O_2-free distilled/deionized water. Mix thoroughly. Sparge with 100% N_2. Autoclave for 15 min at 15 psi pressure–121°C. Cool to room temperature. Do not adjust the pH.

Trace Elements Solution:
Composition per liter:
$MgSO_4·7H_2O$ 3.0g
Nitrilotriacetic acid 1.5g
NaCl 1.0g
$MnSO_4·2H_2O$ 0.5g
$CoSO_4·7H_2O$ 0.18g
$ZnSO_4·7H_2O$ 0.18g
$CaCl_2·2H_2O$ 0.1g
$FeSO_4·7H_2O$ 0.1g
$NiCl_2·6H_2O$ 0.025g
$KAl(SO_4)_2·12H_2O$ 0.02g
H_3BO_3 .... 0.01g
$Na_2MoO_4·4H_2O$ 0.01g
$CuSO_4·5H_2O$ 0.01g
$Na_2SeO_3·5H_2O$ 0.3mg

Preparation of Trace Elements Solution: Add nitrilotriacetic acid to 500.0mL of distilled/deionized water. Dissolve by adjusting pH to 6.5 with KOH. Add remaining components. Add distilled/deionized water to 1.0L. Mix thoroughly.

Polysulfide Solution:
Composition per 100.0mL:
$Na_2S·9H_2O$ 1.2g
Sulfur 0.16g

Preparation of Polysulfide Solution: Sparge 100.0mL distilled/deionized water with 100% N_2. Add $Na_2S·9H_2O$. Mix thoroughly. Add sulfur. The solution will be dark yellow. Sparge with 100% N_2. Autoclave for 15 min at 15 psi pressure–121°C. Cool to room temperature.

Yeast Extract Solution:
Composition per 100.0mL:
Yeast extract 10.0g

Preparation of Yeast Extract Solution: Add yeast extract to distilled/deionized water and bring volume to 100.0mL. Mix thoroughly. Sparge with 100% N_2. Autoclave for 15 min at 15 psi pressure–121°C. Cool to room temperature.

Glycine Solution:
Composition per 100.0mL:
Glycine 15.0g

Preparation of Glycine Solution: Add glycine to distilled/deionized water and bring volume to 100.0mL. Mix thoroughly. Sparge with 100% N_2. Autoclave for 15 min at 15 psi pressure–121°C. Cool to room temperature.

Preparation of Medium: Aseptically and anaeriobically add 50.0mL sterile glycine solution, 20.0mL sterile yeast extract solution, and 12.0mL sterile polysulfide solution to 1.0L sterile base solution. Mix thoroughly. Aseptically and anaerobically distribute to sterile tubes or bottles. Do not adjust pH.

Use: For the cultivation of *Thermococcus alcaliphilus.*

Alkalophile Medium

Composition per liter:
Agar 15.0g
Peptone 5.0g
NaCl 5.0g
Yeast extract 2.0g
Beef extract 1.0g
Sodium sesquicarbonate solution 15.0mL
pH 9.5 ± 0.2 at 25°C

Sodium Sesquicarbonate Solution:
Composition per 100.0mL:
Sodium sesquicarbonate 9.0g

Preparation of Sodium Sesquicarbonate Solution: Add sodium sesquicarbonate to distilled/deionized water and bring volume to 100.0L. Mix thoroughly. Filter sterilize.

Preparation of Medium: Add components, except sodium sesquicarbonate solution, to distilled/deionized water and bring volume to 985.0L. Mix thoroughly. Gently heat and bring to boiling. Autoclave for 15 min at 15 psi pressure–121°C. Cool to 50°–55°C. Aseptically add 15.0mL of filter-sterilized sodium sesquicarbonate solution to adjust pH to 9.5. Mix thoroughly. Pour into sterile Petri dishes or distribute into sterile tubes.

Use: For the cultivation of *Bacillus alcalophilus, Bacillus circulans, Bacillus submarinus*, and other *Bacillus* species.

Alkalophilic Halophile Agar

Composition per liter:

Solution A	500.0mL
Solution B	500.0mL

pH 9.5 ± 1.0 at 25°C

Solution A:

Composition per 500.0mL:

NaCl	200.0g
$Na_2CO_3 \cdot 10H_2O$	50.0g

Preparation of Solution A: Add components to distilled/deionized water and bring volume to 500.0L. Mix thoroughly. Autoclave for 15 min at 15 psi pressure–121°C. Cool to 50°–55°C.

Solution B:

Composition per liter:

Agar	20.0g
Yeast extract	10.0g
Casamino acids	7.5g
Trisodium citrate	3.0g
KCl	2.0g
$MgSO_4 \cdot 7H_2O$	1.0g
$FeSO_4 \cdot 7H_2O$	50.0mg
$MnCl_2 \cdot 4H_2O$	0.36mg

Preparation of Solution B: Add components to distilled/deionized water and bring volume to 500.0L. Mix thoroughly. Autoclave for 15 min at 15 psi pressure–121°C. Cool to 50°–55°C.

Preparation of Medium: Aseptically combine 500.0mL of solution A with 500.0mL of solution B. Mix thoroughly. Bring pH to 9.5. Pour into sterile Petri dishes or distribute into sterile tubes.

Use: For the cultivation and maintenance of *Natronobacterium gregoryi, Natronobacterium magadii, Natronobacterium pharaonis, Natronobacterium vacuolata,* and *Natronococcus occultus.*

Allen and Arnon Medium with Nitrate

Composition per 1000.25mL:

Noble agar	10.0g
KNO_3	0.253g
$NaNO_3$	0.212g
Solution A	25.0mL
Solution B	6.25mL

Solution A:

Composition per 2.0L:

$MgSO_4 \cdot 7H_2O$ (4% solution)	500.0mL
$CaCl_2 \cdot 2H_2O$ (1.2% solution)	500.0mL
NaCl (3.8% solution)	500.0mL
Microelements stock solution	500.0mL

Preparation of Solution A: Prepare individual solutions and combine.

Microelements Stock Solution:

Composition per 1090.0mL:

H_3BO_3	572.0mg
$MnCl_2 \cdot 4H_2O$	360.0mg
$ZnSO_4 \cdot 7H_2O$	44.0mg
MoO_3	36.0mg
$CuSO_4 \cdot 5H_2O$	15.8mg
$CoCl_2 \cdot 6H_2O$	8.0mg
NH_4VO_3	4.6mg
A & A FeEDTA solution	160.0mL

Preparation of Microelements Stock Solution: Add components to distilled/deionized water and bring volume to 1090.0mL. Mix well.

A & A FeEDTA Solution:

Composition per 550.0mL:

Disodium EDTA·$2H_2O$	20.4g
$FeSO_4 \cdot 7H_2O$	13.7g
KOH	5.2g

Preparation of A & A FeEDTA Solution: Dissolve 5.2g of KOH in 186.0mL of distilled/deionized water. Add 20.4g of disodium EDTA·2H2O. Add 13.7g of FeSO4·7H2O to 364.0mL of distilled/deionized water. Combine the EDTA solution with the FeSO4 solution. Sparge solution with filtered air until color changes. pH about 3.5.

Solution B:

Composition per 500.0mL:

K_2HPO_4	28.0g

Preparation of Solution B: Add K_2HPO_4 to distilled/deionized water and bring volume to 500.0mL.

Preparation of Medium: Add agar, KNO_3, and $NaNO_3$ to distilled/deionized water and bring volume to 969.0mL. Mix thoroughly. Gently heat and bring to boiling. Add 25.0mL of solution A. Autoclave for 15 min at 15 psi pressure–121°C. Add 6.25mL of solution B aseptically after sterilization.

Use: For the cultivation and maintenance of *Anabaena* species and *Nostoc* species.

AMB Agar

Composition per liter:

Agar	15.0g
Starch, soluble	5.0g
Pancreatic digest of casein	2.5g
$MgSO_4 \cdot 7H_2O$	0.5g
K_2HPO_4	0.25g

Preparation of Medium: Add components to distilled/deionized water and bring volume to 1.0L. Mix thoroughly. Gently heat and bring to boiling. Distribute into tubes or flasks. Autoclave for 15 min at 15 psi pressure–121°C. Pour into sterile Petri dishes or leave in tubes.

Use: For the cultivation of myxobacteria.

AMB Broth

Composition per liter:

Starch, soluble	5.0g
Pancreatic digest of casein	2.5g
$MgSO_4 \cdot 7H_2O$	0.5g
K_2HPO_4	0.25g

Preparation of Medium: Add components to distilled/deionized water and bring volume to 1.0L. Mix thoroughly. Distribute into tubes or flasks. Autoclave for 15 min at 15 psi pressure–121°C.

Use: For the cultivation of myxobacteria.

Amino-butyric Acid Medium

Composition per liter:

Agar	15.0g
DL-Amino-butyric acid	10.0g
K_2HPO_4	7.0g
Glucose	5.0g
KH_2PO_4	3.0g
$(NH_4)_2SO_4$	1.0g
$MgSO_4 \cdot 7H_2O$	0.5g

pH 7.0 ± 0.2 at 25°C

Preparation of Medium: Add components to distilled/deionized water and bring volume to 1.0L. Mix thoroughly. Gently heat and bring to boiling. Autoclave for 15 min at 15 psi pressure–121°C. Pour into sterile Petri dishes.

Use: For the cultivation and maintenance of *Serratia marcescens* and other microorganisms that can utilize amino-butyric acid as a carbon source.

Ammonifex Medium

Composition per 1010.0mL:

NaCl	1.0g
$Na_2S \cdot 9H_2O$	0.5g
$K_2HPO_4 \cdot 3H_2O$	0.3g
KH_2PO_4	0.22g
$(NH_4)_2SO_4$	0.22g
$NaHCO_3$	0.2g
$MgSO_4 \cdot 7H_2O$	0.09g
$CaCl_2 \cdot 2H_2O$	0.06g
$FeSO_4 \cdot 7H_2O$	2.0mg
$NiCl_2 \cdot 6H_2O$	0.2mg
Resazurin	0.5mg
KNO_3 solution	10.0mL
Selenite-tungstate solution	3.0mL
Wolfe's mineral solution	1.0mL

pH 5.4 ± 0.2 at 25°C

Selenite-Tungstate Solution:

Composition per liter:

NaOH	0.5g
$Na_2WO_4 \cdot 2H_2O$	4.0mg
$Na_2SeO_3 \cdot 5H_2O$	3.0mg

Preparation of Selenite-Tungstate Solution: Add components to distilled/deionized water and bring volume to 1.0L. Mix thoroughly. Sparge with 100% N_2. Autoclave for 15 min at 15 psi pressure–121°C.

Wolfe's Mineral Solution:

Composition per liter:

$MgSO_4 \cdot 7H_2O$	3.0g
Nitrilotriacetic acid	1.5g
NaCl	1.0g
$MnSO_4 \cdot 2H_2O$	0.5g
$CoCl_2 \cdot 6H_2O$	0.1g
$ZnSO_4 \cdot 7H_2O$	0.1g
$CaCl_2 \cdot 2H_2O$	0.1g
$FeSO_4 \cdot 7H_2O$	0.1g
$NiCl_2 \cdot 6H_2O$	0.025g
$KAl(SO_4)_2 \cdot 12H_2O$	0.02g
$CuSO_4 \cdot 5H_2O$	0.01g
H_3BO_3	0.01g
$Na_2MoO_4 \cdot 2H_2O$	0.01g
$Na_2SeO_3 \cdot 5H_2O$	0.3mg

Preparation of Wolfe's Mineral Solution: Add nitrilotriacetic acid to 500.0mL of distilled/deionized water. Adjust pH to 6.5 with KOH. Add remaining components. Add distilled/deionized water to 1.0L. Adjust pH to 6.8.

KNO_3 Solution:

Composition per 100.0mL:

KNO_3	10.0g

Preparation of KNO_3 Solution: Add KNO_3 to distilled/deionized water and bring volume to 100.0mL. Mix thoroughly. Sparge with 100% N_2. Autoclave for 15 min at 15 psi pressure–121°C.

Preparation of Medium: Prepare and dispense medium under 100% N_2. Add components, except $Na_2S \cdot 9H_2O$ and KNO_3 solution, to distilled/deionized water and bring volume to 1.0L. Mix thoroughly. Add $Na_2S \cdot 9H_2O$. Mix thoroughly. Adjust pH to 6.5 with 20% HCl. Sparge with 100% N_2 for 30 min. Dispense anaerobically into tubes in 10.0mL aliquots. Evacuate headspace under vacuum. Pressurize tubes to 200 kPa with 80% H_2 + 20% CO_2 gas mixture. Autoclave for 15 min at 15 psi pressure–121°C. Adjust pH to 5.4. Prior to use, aseptically and anaerobically add 0.1mL of sterile KNO_3 to each tube.

Use: For the cultivation of *Ammonifex* species that form ammonium from nitrate during chemolithoautotrophic growth of this extremely thermophilic bacterium.

AMS Agar (Ammonium Mineral Salts Agar)

Composition per liter:

Agar	15.0g
$MgSO_4 \cdot 7H_2O$	1.0g
K_2HPO_4	0.7g
KH_2PO_4	0.54g
NH_4Cl	0.5g
$CaCl_2 \cdot 2H_2O$	0.2g
$FeSO_4 \cdot 7H_2O$	4.0mg
H_3BO_4	0.3mg
$CoCl_2 \cdot 6H_2O$	0.2mg
$ZnSO_4 \cdot 7H_2O$	0.1mg
$Na_2MoO_4 \cdot 2H_2O$	0.06mg
$MnCl_2 \cdot 4H_2O$	0.03mg
$NiCl_2 \cdot 6H_2O$	0.02mg
$CuCl_2 \cdot 2H_2O$	0.01mg

pH 6.8 ± 0.2 at 25°C

Preparation of Medium: Add components to distilled/deionized water and bring volume to 1.0L. Mix thoroughly. Gently heat and bring to boiling. Autoclave for 15 min at 15 psi pressure–121°C. Add sterile methanol to a concentration of 0.5% aseptically to cooled basal medium.

Use: For the cultivation and maintenance of bacteria that can utilize methanol as a carbon source, such as *Methylobacterium* species, *Methylomonas* species, and *Methylophilus* species.

AMS Agar without Methanol (Ammonium Mineral Salts Agar without Methanol)

Composition per liter:

Agar	15.0g
$MgSO_4 \cdot 7H_2O$	1.0g
K_2HPO_4	0.7g
KH_2PO_4	0.54g
NH_4Cl	0.5g
$CaCl_2 \cdot 2H_2O$	0.2g
$FeSO_4 \cdot 7H_2O$	4.0mg
H_3BO_4	0.3mg
$CoCl_2 \cdot 6H_2O$	0.2mg
$ZnSO_4 \cdot 7H_2O$	0.1mg
$Na_2MoO_4 \cdot 2H_2O$	0.06mg
$MnCl_2 \cdot 4H_2O$	0.03mg

$NiCl_2 \cdot 6H_2O$ 0.02mg
$CuCl_2 \cdot 2H_2O$ 0.01mg

pH 6.8 ± 0.2 at 25°C

Preparation of Medium: Add components to distilled/deionized water and bring volume to 1.0L. Mix thoroughly. Gently heat and bring to boiling. Autoclave for 15 min at 15 psi pressure–121°C.

Use: For the cultivation and maintenance of *Methylosinus trichosporium* and other methane-oxidizing bacteria. Cultures are grown under an atmosphere of 50% methane.

AN1 Medium

Composition per liter:

Pancreatic digest of casein 10.0g
Sulfur, powdered 8.0g
NaCl 2.5g
K_2HPO_4 .... 1.5g
Disodium thioglycolate solution 1.0g
Resazurin 1.0mg

pH 7.3 ± 0.2 at 25°C

Disodium Thioglycolate Solution:
Composition per 10.0mL:

Disodium thioglycolate 1.0g

Preparation of Disodium Thioglycolate Solution: Add disodium thioglycolate to distilled/deionized water and bring volume to 10.0mL. Mix thoroughly. Filter sterilize.

Preparation of Medium: Add components, except powdered sulfur and disodium thioglycolate solution, to distilled/deionized water and bring volume to 990.0mL. Mix thoroughly. Gently heat and bring to boiling. Allow to cool under 100% N_2. Autoclave for 15 min at 15 psi pressure–121°C. Powdered sulfur is sterilized separately by steaming for 3 hr on 3 consecutive days. Aseptically and anaerobically combine the basal solution, sterile powdered sulfur, and sterile disodium thioglycolate solution under 100% N_2.

Use: For the cultivation of *Thermococcus* species.

Anacker and Ordal Medium

Composition per liter:

Agar 10.0g
Pancreatic digest of casein 0.5g
Yeast extract 0.5g
Sodium acetate 0.2g
Beef extract 0.2g

pH 7.3 ± 0.1 at 25°C

Preparation of Medium: Add components to distilled/deionized water and bring volume to 1.0L. Mix thoroughly. Gently heat and bring to boiling. Distribute into tubes or flasks. Autoclave for 15 min at 15 psi pressure–121°C. Pour into sterile Petri dishes or leave in tubes.

Use: For the cultivation and maintenance of *Flexibacter columnaris*.

Anacker and Ordal Medium, Enriched

Composition per liter:

Agar 10.0g
Pancreatic digest of casein 5.0g
Yeast extract 0.5g
Sodium acetate 0.2g
Beef extract 0.2g

pH 7.3 ± 0.1 at 25°C

Preparation of Medium: Add components to distilled/deionized water and bring volume to 1.0L. Mix thoroughly. Gently heat and bring to boiling. Distribute into tubes or flasks. Autoclave for 15 min at 15 psi pressure–121°C. Pour into sterile Petri dishes or leave in tubes.

Use: For the cultivation and maintenance of *Flexibacter psychrophilus*.

Anaerobaculum thermoterrenum Medium (DSMZ Medium 104a)

Composition per liter:

Yeast extract 10.0g
NaCl 8.0g
Trypticase peptone 5.0g
Peptone 5.0g
Beef extract 5.0g
Glucose 5.0g
K_2HPO_4 .... 2.0g
L-Cysteine·HCl 0.5g
Resazurin 1.0mg
$Na_2S \cdot 9H_2O$ solution 50.0mL
Salt solution 40.0mL
Glucose solution 10.0mL

pH 7.0± 0.2 at 25°C

Salt Solution:
Composition per liter:

$NaHCO_3$ 10.0g
NaCl 2.0g
K_2HPO_4 .... 1.0g
KH_2PO_4 .... 1.0g
$MgSO_4 \cdot 7H_2O$ 0.5g
$CaCl_2 \cdot 2H_2O$ 0.25g

Preparation of Salt Solution: Add components to distilled/deionized water and bring volume to 1.0L. Mix thoroughly.

$Na_2S \cdot 9H_2O$ Solution:
Composition per 100.0mL:

$Na_2S \cdot 9H_2O$ 5.0g

Preparation of $Na_2S \cdot 9H_2O$ Solution: Add $Na_2S \cdot 9H_2O$ to distilled/deionized water and bring volume to 100.0mL. Sparge with N_2. Autoclave for 15 min at 15 psi pressure–121°C. Cool to 25°C. Store anaerobically.

Glucose Solution:
Composition per 100.0mL:

D-Glucose 10.0g

Preparation of Glucose Solution: Add D-glucose to distilled/deionized water and bring volume to 100.0mL. Mix thoroughly. Sparge with N_2. Autoclave for 15 min at 15 psi pressure–121°C. Cool to 25°C. Store anaerobically.

Preparation of Medium: Add components, except L-cysteine, glucose solution, and $Na_2S \cdot 9H_2$ solution, to distilled/deionized water and bring volume to 940.0mL. Mix thoroughly. Gently heat and bring to boiling. Sparge with CO_2. Add L-cysteine. Autoclave for 15 min at 15 psi pressure–121°C. Cool to 25°C. Aseptically add 50.0mL sterile $Na_2S \cdot 9H_2O$ solution and 10.0mL sterile glucose solution. Mix thoroughly. Adjust pH to 7.0 with 8*N* NaOH. Distribute into sterile tubes or flasks under anaerobic N_2.

Use: For the cultivation and maintenance of *Anaerobaculum thermoterrenum*.

Anaerobic Agar

Composition per liter:

Pancreatic digest of casein 20.0g
Agar 15.0g
Yeast extract 15.0g

NaCl 5.0g
Sodium thioglycolate 2.0g
Sodium formaldehyde sulfoxylate 1.0g
pH 7.2 ± 0.2 at 25°C

Preparation of Medium: Add components to distilled/deionized water and bring volume to 1.0L. Mix thoroughly. Gently heat and bring to boiling. Adjust pH to 7.2. Distribute into tubes until medium is 3 inches deep. Autoclave for 20 min at 15 psi pressure–121°C.

Use: For the anaerobic cultivation of *Bacillus* species, especially *Bacillus larvae*, *Bacillus popilliae*, and *Bacillus lentimorbus*.

Anaerobic Cellulolytic Medium

Composition per liter:
NH_4Cl 1.0g
Cellobiose 1.0g
Yeast extract 1.0g
$MgSO_4$ 0.5g
KCl 0.5g
L-Cysteine·HCl·H_2O 0.5g
K_2HPO_4 .... 0.4g
Resazurin 1.0mg
Wolfe's mineral solution 20.0mL
Na_2CO_3 solution 10.0mL
$Na_2S·9H_2O$ solution 10.0mL
pH 6.9 ± 0.1 at 25°C

Wolfe's Mineral Solution:
Composition per liter
$MgSO_4·7H_2O$ 3.0g
Nitrilotriacetic acid 1.5g
NaCl 1.0g
$MnSO_4·H_2O$ 0.5g
$FeSO_4·7H_2O$ 0.1g
$CoCl_2·6H_2O$ 0.1g
$CaCl_2$ 0.1g
$ZnSO_4·7H_2O$ 0.1g
$CuSO_4·5H_2O$ 0.01g
$AlK(SO_4)_2·12H_2O$ 0.01g
H_3BO_3 .... 0.01g
$Na_2MoO_4·2H_2O$ 0.01g

Preparation of Wolfe's Mineral Solution: Add nitrilotriacetic acid to approximately 500.0mL of distilled/deionized water and adjust pH to 6.5 with KOH to dissolve. Bring volume to 1.0L with distilled/deionized water. Add other compounds. Mix thoroughly.

Na_2CO_3 Solution:
Composition per 100.0mL:
Na_2CO_3 .... 10.0g

Preparation of Na_2CO_3 Solution: Add Na_2CO_3 to distilled/deionized water and bring volume to 100.0mL. Mix thoroughly. Filter sterilize.

$Na_2S·9H_2O$ Solution:
Composition per 100.0mL:
$Na_2S·9H_2O$ 15.0g

Preparation of $Na_2S·9H_2O$ Solution: Add $Na_2S·9H_2O$ to distilled/deionized water and bring volume to 100.0mL. Mix thoroughly. Filter sterilize.

Preparation of Medium: Add components, except Na_2CO_3 solution and $Na_2S·9H_2O$ solution, to distilled/deionized water and bring volume to 980.0mL. Boil medium under 80% N_2 + 10% CO_2 + 10% H_2 until medium is colorless. Cool and distribute anaerobically into test tubes in 10.0mL volumes using 80% N_2 + 10% CO_2 + 10% H_2. Stopper the tubes anaerobically. Autoclave for 15 min at 15 psi pressure–121°C. Cool to room temperature. Aseptically add 0.1mL of sterile Na_2CO_3 solution and 0.1mL of sterile $Na_2S·9H_2O$ solution to each tube. Mix thoroughly.

Use: For the cultivation and maintenance of microorganisms that can utilize cellobiose as sole carbon source, such as *Clostridium cellulovorans*.

Anaerobic Citrate Medium

Composition per liter:
Ferric citrate 17.0g
Sodium acetate 6.8g
$NaHCO_3$ 2.5g
NH_4Cl 1.5g
$NaH_2PO_4·H_2O$ 0.6g
KCl 0.1g
Trace elements solution 10.0mL
Wolfe's vitamin solution 10.0mL
pH 7.0 ± 0.2 at 25°C

Trace Elements Solution:
Composition per liter:
Na_2WO_4 .... 25.0mg
$NiCl_2·6H_2O$ 24.0mg
Wolfe's mineral solution 1.0L

Wolfe's Mineral Solution:
Composition per liter:
$MgSO_4·7H_2O$ 3.0g
Nitrilotriacetic acid 1.5g
NaCl 1.0g
$MnSO_4·H_2O$ 0.5g
$CaCl_2$ 0.1g
$CoCl_2·6H_2O$ 0.1g
$FeSO_4·7H_2O$ 0.1g
$ZnSO_4·7H_2O$ 0.1g
$AlK(SO_4)_2·12H_2O$ 0.01g
$CuSO_4·5H_2O$ 0.01g
H_3BO_3 .... 0.01g
$Na_2MoO_4·2H_2O$ 0.01g

Preparation of Wolfe's Mineral Solution: Add nitrilotriacetic acid to approximately 500.0mL of water and adjust to pH 6.5 with KOH to dissolve the compound. Bring volume to 1.0L with remaining water and add remaining compounds one at a time.

Preparation of Trace Elements Solution: Combine components. Mix thoroughly.

Wolfe's Vitamin Solution:
Composition per liter:
Pyridoxine·HCl 10.0mg
p-Aminobenzoic acid 5.0mg
Lipoic acid 5.0mg
Nicotinic acid 5.0mg
Riboflavin 5.0mg
Thiamine·HCl 5.0mg
Calcium DL-pantothenate 5.0mg
Biotin 2.0mg
Folic acid 2.0mg
Vitamin B_{12} 0.1mg

Preparation of Wolfe's Vitamin Solution: Add components to distilled/deionized water and bring volume to 1.0L. Mix thoroughly.

Preparation of Medium: Prepare and dispense medium under 80% N_2 + 20% CO_2. Add ferric citrate to 500.0mL of boiling distilled/deionized water. Mix thoroughly. Cool to room temperature while sparging with 80% N_2 + 20% CO_2.

Add remaining components. Add distilled/deionized water and bring volume to 1.0L. Mix thoroughly. Adjust pH to 7.0 with NaOH. Continue sparging with 80% N_2 + 20% CO_2. Anaerobically distribute into tubes. Autoclave for 15 min at 15 psi pressure–121°C.

Use: For the cultivation of *Geobacter metallireducens.*

Anaerobranca gottschalkii Medium (DSMZ Medium 895)

Composition per 1020mL:

NaCl	10.0g
$(NH_4)_2SO_4$	1.0g
K_2HPO_4	0.5g
L-Cysteine	0.5g
NH_4Cl	0.4g
Yeast extract	0.25g
Tryptone	0.25g
$Na_2S_2O_3 \cdot 5H_2O$	0.1g
$MgSO_4 \cdot 7H_2O$	0.1g
$CaCl_2 \cdot 2H_2O$	0.05g
$FeSO_4 \cdot 7H_2O$	2.0mg
Resazurin	0.5mg
Na_2CO_3solution	50.0mL
Soluble starch solution	20.0mL
Trace elements solution	10.0mL
Vitamin solution, 10 fold conc.	1.0mL

pH 9.4 ± 0.2 at 25°C

Trace Elements Solution:
Composition per liter:

$MgSO_4 \cdot 7H_2O$	3.0g
Nitrilotriacetic acid	1.5g
NaCl	1.0g
$MnSO_4 \cdot 2H_2O$	0.5g
$CoSO_4 \cdot 7H_2O$	0.18g
$ZnSO_4 \cdot 7H_2O$	0.18g
$CaCl_2 \cdot 2H_2O$	0.1g
$FeSO_4 \cdot 7H_2O$	0.1g
$NiCl_2 \cdot 6H_2O$	0.025g
$KAl(SO_4)_2 \cdot 12H_2O$	0.02g
H_3BO_3	0.01g
$Na_2MoO_4 \cdot 4H_2O$	0.01g
$CuSO_4 \cdot 5H_2O$	0.01g
$Na_2SeO_3 \cdot 5H_2O$	0.3mg

Preparation of Trace Elements Solution: Add nitrilotriacetic acid to 500.0mL of distilled/deionized water. Dissolve by adjusting pH to 6.5 with KOH. Add remaining components. Add distilled/deionized water to 1.0L. Mix thoroughly.

Vitamin Solution:
Composition per 100mL:

Pyridoxine-HCl	10.0mg
Thiamine-HCl·$2H_2O$	5.0mg
Riboflavin	5.0mg
Nicotinic acid	5.0mg
D-Ca-pantothenate	5.0mg
p-Aminobenzoic acid	5.0mg
Lipoic acid	5.0mg
Biotin	2.0mg
Folic acid	2.0mg
Vitamin B_{12}	0.1mg

Preparation of Vitamin Solution: Add components to distilled/deionized water and bring volume to 100.0mL. Mix thoroughly. Sparge with 80% H_2 + 20% CO_2. Filter sterilize.

Na_2CO_3 Solution:
Composition per 100.0mL:

Na_2CO_3	5.0g

Preparation of Na_2CO_3 Solution: Add Na_2CO_3 to distilled/deionized water and bring volume to 100.0mL. Mix thoroughly. Sparge with 100% N_2. Autoclave for 15 min at 15 psi pressure–121°C.

Starch Solution:
Composition per 50.0mL:

Starch, soluble	5.0g

Preparation of Starch Solution: Add starch to distilled/deionized water and bring volume to 50.0mL. Mix thoroughly. Sparge with 100% N_2. Autoclave for 15 min at 15 psi pressure–121°C.

Preparation of Medium: Add components, except starch solution, Na_2CO_3 solution, and L-cysteine, to distilled/deionized water and bring volume to 970.0mL. Mix thoroughly. Gently heat and bring to boiling. Cool to 25°C while sparging with 100% N_2. Add 0.5g L-cysteine. Mix thoroughly. Distribute to anaerobe tubes or bottles. Autoclave for 15 min at 15 psi pressure–121°C. Cool to 25°C. Aseptically and anaerobically add, per liter of medium, 20.0mL sterile starch solution, and 50.0mL sterile Na_2CO_3 solution. Final pH is 9.3-9.5.

Use: For the cultivation of *Anaerobranca gottschalkii,* a novel thermoalkaliphilic bacterium that grows anaerobically at high pH and temperature.

Anaerobranca Medium

Composition per 1015.0mL:

Yeast extract	5.0g
$Na_2HPO_4 \cdot 2H_2O$	3.9g
Sodium fumarate	1.5g
KCl	0.5g
KH_2PO_4	0.5g
L-Cysteine·HCl·H_2O	0.125g
$Na_2S \cdot 9H_2O$	0.125g
Wolfe's vitamin solution	10.0mL
Wolfe's mineral solution	5.0mL

pH 8.5 ± 0.2 at 25°C

Wolfe's Vitamin Solution:
Composition per liter:

Pyridoxine·HCl	10.0mg
p-Aminobenzoic acid	5.0mg
Lipoic acid	5.0mg
Nicotinic acid	5.0mg
Riboflavin	5.0mg
Thiamine·HCl	5.0mg
Calcium DL-pantothenate	5.0mg
Biotin	2.0mg
Folic acid	2.0mg
Vitamin B_{12}	0.1mg

Preparation of Wolfe's Vitamin Solution: Add components to distilled/deionized water and bring volume to 1.0L. Mix thoroughly. Filter sterilize.

Wolfe's Mineral Solution:
Composition per liter:

$MgSO_4 \cdot 7H_2O$	3.0g
Nitrilotriacetic acid	1.5g
NaCl	1.0g
$MnSO_4 \cdot 2H_2O$	0.5g
$CoCl_2 \cdot 6H_2O$	0.1g
$ZnSO_4 \cdot 7H_2O$	0.1g
$CaCl_2 \cdot 2H_2O$	0.1g

$FeSO_4 \cdot 7H_2O$ 0.1g
$NiCl_2 \cdot 6H_2O$ 0.025g
$KAl(SO_4)_2 \cdot 12H_2O$ 0.02g
$CuSO_4 \cdot 5H_2O$ 0.01g
H_3BO_3 .... 0.01g
$Na_2MoO_4 \cdot 2H_2O$ 0.01g
$Na_2SeO_3 \cdot 5H_2O$ 0.3mg

Preparation of Wolfe's Mineral Solution: Add nitrilotriacetic acid to 500.0mL of distilled/deionized water. Adjust pH to 6.5 with KOH. Add remaining components. Add distilled/deionized water to 1.0L. Adjust pH to 6.8.

Preparation of Medium: Prepare and dispense medium under 100% N_2. Add components, except Wolfe's vitamin solution, to distilled/deionized water and bring volume to 990.0mL. Mix thoroughly. Adjust pH to 8.5. Sparge with 100% N_2. Autoclave for 15 min at 15 psi pressure–121°C. Aseptically and anaerobically add 10.0mL of sterile Wolfe's vitamin solution. Mix thoroughly. Aseptically and anaerobically distribute into sterile tubes or bottles.

Use: For the cultivation of *Anaerobranca horikoshii.*

Anaerocellum Medium

Composition per liter:
Cellobiose or starch 5.0g
$NaHCO_3$ 1.5g
$Na_2S \cdot 9H_2O$ 0.5g
Yeast extract 0.5g
$CaCl_2 \cdot 2H_2O$ 0.33g
KCl 0.33g
KH_2PO_4 .... 0.33g
$MgCl_2 \cdot 6H_2O$ 0.33g
NH_4Cl 0.33g
Resazurin 0.5mg
$NaHCO_3$ solution 100.0mL
Cellobiose or starch solution 50.0mL
Vitamin solution 10.0mL
$Na_2S \cdot 9H_2O$ solution 10.0mL
Trace elements solution SL-10 .... 1.0mL
pH 7.1–7.3 at 25°C

$NaHCO_3$ Solution:
Composition per 100.0mL:
$NaHCO_3$ 5.0g

Preparation of $NaHCO_3$ Solution: Add $NaHCO_3$ to distilled/deionized water and bring volume to 100.0mL. Mix thoroughly. Filter sterilize. Gas under 80% N_2 + 20% CO_2.

Cellobiose or Starch Solution:
Composition per 50.0mL:
Cellobiose or starch 5.0g

Preparation of Cellobiose or Starch Solution: Add cellobiose or starch to distilled/deionized water and bring volume to 50.0mL. Mix thoroughly. Filter sterilize. Gas under 100% N_2.

$Na_2S \cdot 9H_2O$ Solution:
Composition per 10.0mL:
$Na_2S \cdot 9H_2O$ 0.5g

Preparation of $Na_2S \cdot 9H_2O$ Solution: Add $Na_2S \cdot 9H_2O$ to distilled/deionized water and bring volume to 10.0mL. Mix thoroughly. Gas under 100% N_2. Autoclave for 15 min at 15 psi pressure–121°C.

Vitamin Solution:
Composition per liter:
Pyridoxine·HCl 10.0mg
Calcium DL-pantothenate 5.0mg
Lipoic acid 5.0mg
Nicotinic acid 5.0mg
p-Aminobenzoic acid 5.0mg
Riboflavin 5.0mg
Thiamine·HCl 5.0mg
Biotin 2.0mg
Folic acid 2.0mg
Vitamin B_{12} 0.1mg

Preparation of Vitamin Solution: Add components to distilled/deionized water and bring volume to 1.0L. Mix thoroughly. Gas under 100% N_2. Autoclave for 15 min at 15 psi pressure–121°C.

Trace Elements Solution SL-10:
Composition per liter:
$FeCl_2 \cdot 4H_2O$ 1.5g
$CoCl_2 \cdot 6H_2O$ 190.0mg
$MnCl_2 \cdot 4H_2O$ 100.0mg
$ZnCl_2$ 70.0mg
$Na_2MoO_4 \cdot 2H_2O$ 36.0mg
$NiCl_2 \cdot 6H_2O$ 24.0mg
H_3BO_3 .... 6.0mg
$CuCl_2 \cdot 2H_2O$ 2.0mg
HCl (25% solution) 10.0mL

Preparation of Trace Elements Solution SL-10: Add $FeCl_2 \cdot 4H_2O$ to 10.0mL of HCl solution. Mix thoroughly. Add distilled/deionized water and bring volume to 1.0L. Add remaining components. Mix thoroughly.

Preparation of Medium: Add components, except $NaHCO_3$ solution, cellobiose or starch solution, and $Na_2S \cdot 9H_2O$ solution, to distilled/deionized water and bring volume to 830.0mL. Mix thoroughly. Gently heat and bring to boiling. Continue boiling for 3–4 min. Allow to cool to room temperature under 80% N_2 + 20% CO_2. Distribute into bottles under 80% N_2 + 20% CO_2. Autoclave for 15 min at 15 psi pressure–121°C. Aseptically and anaerobically add sterile $NaHCO_3$ solution, sterile cellobiose or starch solution, and sterile $Na_2S \cdot 9H_2O$ solution. Mix thoroughly.

Use: For the cultivation and maintenance of *Anaerocellum thermophilum* and *Dictyoglomus turgidus.*

Anaerocellum Medium

Composition per liter:
$NaHCO_3$ 1.5g
$Na_2S \cdot 9H_2O$ 0.5g
$CaCl_2 \cdot 2H_2O$ 0.33g
KCl 0.33g
KH_2PO_4 .... 0.33g
$MgCl_2 \cdot 6H_2O$ 0.33g
NH_4Cl 0.33g
Yeast extract 0.2g
Resazurin 0.5mg
$NaHCO_3$ solution 100.0mL
Cellobiose or starch solution 50.0mL
Vitamin solution 10.0mL
$Na_2S \cdot 9H_2O$ solution 10.0mL
Trace elements solution SL-10 .... 1.0mL
pH 7.1–7.3 at 25°C

$NaHCO_3$ Solution:
Composition per 100.0mL:
$NaHCO_3$ 5.0g

Preparation of $NaHCO_3$ Solution: Add $NaHCO_3$ to distilled/deionized water and bring volume to 100.0mL. Mix thoroughly. Filter sterilize. Gas under 80% N_2 + 20% CO_2.

Cellobiose or Starch Solution:
Composition per 50.0mL:
Cellobiose or starch 5.0g

Preparation of Cellobiose or Starch Solution: Add cellobiose or starch to distilled/deionized water and bring volume to 50.0mL. Mix thoroughly. Filter sterilize. Gas under 100% N_2.

$Na_2S{\cdot}9H_2O$ Solution:
Composition per 10.0mL:
$Na_2S{\cdot}9H_2O$ 0.5g

Preparation of $Na_2S{\cdot}9H_2O$ Solution: Add $Na_2S{\cdot}9H_2O$ to distilled/deionized water and bring volume to 10.0mL. Mix thoroughly. Gas under 100% N_2. Autoclave for 15 min at 15 psi pressure–121°C.

Vitamin Solution:
Composition per liter:
Pyridoxine·HCl 10.0mg
Calcium DL-pantothenate 5.0mg
Lipoic acid 5.0mg
Nicotinic acid 5.0mg
p-Aminobenzoic acid 5.0mg
Riboflavin 5.0mg
Thiamine·HCl 5.0mg
Biotin 2.0mg
Folic acid 2.0mg
Vitamin B_{12} 0.1mg

Preparation of Vitamin Solution: Add components to distilled/deionized water and bring volume to 1.0L. Mix thoroughly. Gas under 100% N_2. Autoclave for 15 min at 15 psi pressure–121°C.

Trace Elements Solution SL-10:
Composition per liter:
$FeCl_2{\cdot}4H_2O$ 1.5g
$CoCl_2{\cdot}6H_2O$ 190.0mg
$MnCl_2{\cdot}4H_2O$ 100.0mg
$ZnCl_2$ 70.0mg
$Na_2MoO_4{\cdot}2H_2O$ 36.0mg
$NiCl_2{\cdot}6H_2O$ 24.0mg
H_3BO_3 .... 6.0mg
$CuCl_2{\cdot}2H_2O$ 2.0mg
HCl (25% solution) 10.0mL

Preparation of Trace Elements Solution SL-10: Add $FeCl_2{\cdot}4H_2O$ to 10.0mL of HCl solution. Mix thoroughly. Add distilled/deionized water and bring volume to 1.0L. Add remaining components. Mix thoroughly.

Preparation of Medium: Add components, except $NaHCO_3$ solution, cellobiose or starch solution, and $Na_2S{\cdot}9H_2O$ solution, to distilled/deionized water and bring volume to 830.0mL. Mix thoroughly. Gently heat and bring to boiling. Continue boiling for 3–4 min. Allow to cool to room temperature under 80% N_2 + 20% CO_2. Distribute into bottles under 80% N_2 + 20% CO_2. Autoclave for 15 min at 15 psi pressure–121°C. Aseptically and anaerobically add sterile $NaHCO_3$ solution, sterile cellobiose or starch solution, and sterile $Na_2S{\cdot}9H_2O$ solution. Mix thoroughly.

Use: For the cultivation and maintenance of *Anaerocellum thermophilum* and *Dictyoglomus turgidus.*

Anaerovibrio burkinabensis Medium

Composition per 1001.0mL:
Solution A 870.0mL
Solution C 100.0mL
Solution D 10.0mL
Solution E (Vitamin solution) 10.0mL
Solution F 10.0mL
Solution G 10.0mL
Solution B (Trace elements solution SL-10) 1.0mL
pH 6.8–7.2 at 25°C

Solution A:
Composition per 870.0mL:
Na_2SO_4 .... 3.0g
NaCl 1.0g
KCl 0.5g
$MgCl_2{\cdot}6H_2O$ 0.4g
NH_4Cl 0.3g
KH_2PO_4 .... 0.2g
$CaCl_2{\cdot}2H_2O$ 0.15g
Resazurin 1.0mg

Preparation of Solution A: Add components to distilled/deionized water and bring volume to 870.0mL. Mix thoroughly. Gently heat and bring to boiling. Continue boiling for 3-4 min. Allow to cool to room temperature while gassing under 80% N_2 + 20% CO_2. Continue gassing until pH reaches below 6.0. Seal the flask under 80% N_2 + 20% CO_2. Autoclave for 15 min at 15 psi pressure–121°C.

Solution B (Trace Elements Solution SL-10):
Composition per liter:
$FeCl_2{\cdot}4H_2O$ 1.5g
$CoCl_2{\cdot}6H_2O$ 190.0mg
$MnCl_2{\cdot}4H_2O$ 100.0mg
$ZnCl_2$ 70.0mg
$Na_2MoO_4{\cdot}2H_2O$ 36.0mg
$NiCl_2{\cdot}6H_2O$ 24.0mg
H_3BO_3 .... 6.0mg
$CuCl_2{\cdot}2H_2O$ 2.0mg
HCl (25% solution) 10.0mL

Preparation of Solution B (Trace Elements Solution SL-10): Add $FeCl_2{\cdot}4H_2O$ to 10.0mL of HCl solution. Mix thoroughly. Add distilled/deionized water and bring volume to 1.0L. Add remaining components. Mix thoroughly. Gas under 100% N_2. Autoclave for 15 min at 15 psi pressure–121°C.

Solution C:
Composition per 100.0mL:
$NaHCO_3$ 5.0g

Preparation of Solution C: Add $NaHCO_3$ to distilled/deionized water and bring volume to 100.0mL. Mix thoroughly. Filter sterilize. Gas under 80% N_2 + 20% CO_2.

Solution D:
Composition per 10.0mL:
Sodium lactate 2.5g

Preparation of Solution D: Add sodium lactate· to distilled/deionized water and bring volume to 10.0mL. Mix thoroughly. Gas under 100% N_2. Autoclave for 15 min at 15 psi pressure–121°C.

Solution E (Vitamin Solution):
Composition per liter:
Pyridoxine·HCl 10.0mg
Calcium DL-pantothenate 5.0mg
Lipoic acid 5.0mg
Nicotinic acid 5.0mg
p-Aminobenzoic acid 5.0mg
Riboflavin 5.0mg
Thiamine·HCl 5.0mg
Biotin 2.0mg
Folic acid 2.0mg
Vitamin B_{12} 0.1mg

Preparation of Solution E (Vitamin Solution): Add components to distilled/deionized water and bring volume to 1.0L. Mix thoroughly. Gas under 100% N_2. Autoclave for 15 min at 15 psi pressure–121°C.

Solution F:
Composition per 10.0mL:
$Na_2S \cdot 9H_2O$ 0.4g

Preparation of Solution F: Add $Na_2S \cdot 9H_2O$ to distilled/deionized water and bring volume to 10.0mL. Mix thoroughly. Gas under 100% N_2. Autoclave for 15 min at 15 psi pressure–121°C.

Solution G:
Composition per 10.0mL:
Yeast extract 1.0g

Preparation of Solution F: Add yeast extract to distilled/deionized water and bring volume to 10.0mL. Mix thoroughly. Gas under 100% N_2. Autoclave for 15 min at 15 psi pressure–121°C.

Preparation of Medium: Aseptically and anaerobically combine solution A with solution B, solution C, solution D, solution E, solution F, and solution G, in that order. Mix thoroughly. Anaerobically distribute into sterile tubes or flasks under 80% N_2 + 20% CO_2.

Use: For the cultivation and maintenance of *Anaerovibrio burkinabensis,* a lactate-fermenting bacterium isolated from rice field soils.

Ancylobacter-Spirosoma Agar

Composition per liter:
Agar 20.0g
Glucose 1.0g
Peptone 1.0g
Yeast extract 1.0g
pH 6.8 ± 0.2 at 25°C

Preparation of Medium: Add components to distilled/deionized water and bring volume to 1.0L. Mix thoroughly. Adjust pH to 6.8. Gently heat and bring to boiling. Distribute into tubes or flasks. Autoclave for 15 min at 15 psi pressure–121°C. Pour into sterile Petri dishes or leave in tubes.

Use: For the cultivation of *Ancylobacter aquaticus, Ancylobacter* species, *Aquaspirillum metamorphum, Aquaspirillum serpens, Flectobacillus major, Methylobacterium mesophilicum, Runella slithyformis, Shewanella putrefaciens,* and *Spirosoma linguale.*

Ancylobacter-Spirosoma Medium (DSMZ Medium 7)

Composition per liter:
Agar 15.0g
Glucose 1.0g
Peptone 1.0g
Yeast extract 1.0g
pH 7.1 ± 0.2 at 25°C

Preparation of Medium: Add components to distilled/deionized water and bring volume to 1.0L. Mix thoroughly. Gently heat and bring to boiling. Distribute into tubes or flasks. Autoclave for 15 min at 15 psi pressure–121°C. Pour into sterile Petri dishes or leave in tubes.

Use: For the cultivation and maintenance of *Spirosoma linguale.*

Anderson's Marine Agar

Composition per liter:
Agar 15.0g
Peptone 2.5g
Yeast extract 2.5g
$FePO_4$ 0.1g
Filtered, aged seawater 750.0mL
pH 7.4–7.6 at 25°C

Preparation of Medium: Add components to distilled/deionized water and bring volume to 1.0L. Mix thoroughly. Gently heat and bring to boiling. Adjust pH to 7.4–7.6. Distribute into tubes or flasks. Autoclave for 15 min at 15 psi pressure–121°C. Pour into sterile Petri dishes or leave in tubes.

Use: For the cultivation and maintenance of *Vibrio* species.

Anderson's Marine Broth

Composition per liter:
Peptone 2.5g
Yeast extract 2.5g
$FePO_4$ 0.1g
Filtered, aged seawater 750.0mL
pH 7.4–7.6 at 25°C

Preparation of Medium: Add components to distilled/deionized water and bring volume to 1.0L. Mix thoroughly. Adjust pH to 7.4–7.6. Distribute into tubes or flasks. Autoclave for 15 min at 15 psi pressure–121°C.

Use: For the cultivation of *Vibrio* species.

Anderson's Marine Medium

Composition per liter:
Peptone 2.5g
Yeast extract 2.5g
$FePO_4$ 0.1g
Filtered, aged seawater 750.0mL
pH 7.4 ± 0.2 at 25°C

Preparation of Medium: Add components, except seawater, to distilled/deionized water and bring volume to 250.0mL. Mix thoroughly. Autoclave for 15 min at 15 psi pressure–121°C. Aseptically add 750.0mL of sterile aged seawater. Mix thoroughly. Bring pH to 7.4. Aseptically distribute into sterile tubes or flasks.

Use: For the cultivation of *Flavobacterium* species, *Micrococcus* species, *Planococcus* species, *Pseudomonas fluorescens*, and *Vibrio* species.

Andrade's Broth

Composition per liter:
Pancreatic digest of gelatin 10.0g
NaCl 5.0g
Beef extract 3.0g
Andrade's indicator 10.0mL
Carbohydrate solution 50.0mL
pH 7.4 ± 0.2 at 25°C

Source: This medium is available as a prepared medium from BD Diagnostic Systems, in tubes containing adonitol, arabinose, cellobiose, glucose, dulcitol, fructose, galactose, inositol, lactose, maltose, mannitol, raffinose, rhamnose, salicin, sorbitol, sucrose, trehalose, or xylose.

Andrade's Indicator
Composition per 100.0mL:
NaOH (1*N* solution) 16.0mL
Acid Fuchsin 0.1g

Preparation of Andrade's Indicator: Add Acid Fuchsin to NaOH solution and bring volume to 100.0mL with distilled/deionized water.

Carbohydrate Solution:
Composition per 100.0mL:

Carbohydrate	10.0g

Preparation of Carbohydrate Solution: Add carbohydrate to distilled/deionized water and bring volume to 100.0mL. Adonitol, arabinose, cellobiose, glucose, dulcitol, fructose, galactose, inositol, lactose, maltose, mannitol, raffinose, rhamnose, salicin, sorbitol, sucrose, trehalose, xylose, or other carbohydrates may be used. Mix thoroughly. Filter sterilize.

Preparation of Medium: Add components, except carbohydrate solution, to distilled/deionized water and bring volume to 1.0L. Mix thoroughly. Gently heat and bring to boiling. Distribute in 10.0mL volumes into test tubes containing inverted Durham tubes. Autoclave for 15 min at 15 psi pressure–121°C. Cool to 25°C. Add 0.5mL of sterile carbohydrate solution to each tube.

Caution: Acid Fuchsin is a potential carcinogen and care must be taken to avoid inhalation of the powdered dye and contact with the skin.

Use: For the determination of carbohydrate fermentation reactions of microorganisms, particularly members of the Enterobacteriaceae. A Durham tube is used to collect gas produced during the fermentation reaction. Acid production is indicated by a pink color.

AO Agar

Composition per liter:

Agar	11.0g
Sodium acetate	0.5g
Pancreatic digest of casein	0.5g
Yeast extract	0.5g
Beef extract	0.2g

pH 7.2 ± 0.2 at 25°C

Preparation of Medium: Add components to distilled/deionized water and bring volume to 1.0L. Mix thoroughly. Gently heat and bring to boiling. Distribute into tubes or flasks. Autoclave for 15 min at 15 psi pressure–121°C. Pour into sterile Petri dishes or leave in tubes.

Use: For the isolation and cultivation of *Cytophaga* species, *Herpetosiphon* species, *Saprospira* species, and *Flexithrix* species.

Aolpha Medium

Composition per 1041.0mL:

NaCl	100.0g
Agar	15.0g
$MgSO_4 \cdot 7H_2O$	9.5g
$MgCl_2 \cdot 6H_2O$	5.0g
KCl	5.0g
Peptone	5.0g
Yeast extract	1.0g
$CaCl_2 \cdot 2H_2O$	0.2g
$(NH_4)_2SO_4$	0.1g
KNO_3	0.1g
Metals solution	20.0mL
Phosphate solution	20.0mL
Vitamin solution	1.0mL

pH 7.0 ± 0.2 at 25°C

Metals Solution:
Composition per liter:

$MgSO_4 \cdot 7H_2O$	29.7g
Nitrilotriacetic acid	10.0g
$CaCl_2 \cdot 2H_2O$	3.3g
$FeSO_4 \cdot 7H_2O$	99.0mg
$Na_2MoO_4 \cdot 2H_2O$	12.7mg
"Metals 44"	50.0mL

Preparation of Metals Solution: Solubilize nitrilotriacetic acid with KOH. Dissolve remaining ingredients. Adjust pH to 7.2 with KOH or H_2SO_4. Autoclave for 15 min at 15 psi pressure–121°C. Add aseptically to sterile basal medium.

"Metals 44":
Composition per 100.0mL:

$ZnSO_4 \cdot 7H_2O$	1.1g
$FeSO_4 \cdot 7H_2O$	0.5g
EDTA	0.25g
$MnSO_4 \cdot 7H_2O$	0.154g
$CuSO_4 \cdot 5H_2O$	0.04g
$Co(NO_3)_2 \cdot 6H_2O$	0.025g
$Na_2B_4O_7 \cdot 10H_2O$	0.018g

Preparation of "Metals 44": Add components to distilled/deionized water and bring volume to 100.0mL. Mix thoroughly. Autoclave for 15 min at 15 psi pressure–121°C. Add aseptically to sterile basal medium.

Phosphate Solution:
Composition per liter:

K_2HPO_4	2.5g
KH_2PO_4	2.5g

Preparation of Phosphate Solution: Add components to distilled/deionized water and bring volume to 1.0L. Mix thoroughly. Autoclave for 15 min at 15 psi pressure–121°C. Add aseptically to sterile basal medium.

Vitamin Solution:
Composition per liter:

Pyridoxine·HCl	10.0mg
Calcium pantothenate	5.0mg
Nicotinamide	5.0mg
Riboflavin	5.0mg
Thiamine·HCl	5.0mg
Biotin	2.0mg
Folic acid	2.0mg
Cyanocobalamin	0.1mg

Preparation of Vitamin Solution: Add components to distilled/deionized water and bring volume to 1.0L. Mix thoroughly. Filter sterilize and add aseptically to sterile basal medium.

Preparation of Medium: Add components—except "Metals 44", phosphate solution, and vitamin solution—to distilled/deionized water and bring volume to 1.0L. Mix thoroughly. Gently heat and bring to boiling. Adjust pH of basal medium to 7.0. Autoclave for 15 min at 15 psi pressure–121°C. Cool to 50°C and aseptically add the "Metals 44", phosphate, and vitamin solutions.

Use: For the cultivation and maintenance of *Halomonas meridiana* and other *Halomonas* species.

Aplanobacterium Medium

Composition per liter:

Agar	20.0g
Glucose	10.0g
Peptone	5.0g
Yeast extract	5.0g

pH 7.2 ± 0.2 at 25°C

Preparation of Medium: Add components to distilled/deionized water and bring volume to 1.0L. Mix thoroughly. Gently heat and bring to boiling. Autoclave for 15 min at 15 psi pressure–121°C. Pour into sterile Petri dishes or leave in tubes.

Use: For the cultivation and maintenance of *Xanthomonas* species.

Apple Juice Yeast Extract Medium (AJYE Medium)

Composition per 1200.0mL:

Agar	30.0g
Yeast extract	10.0g
Apple juice	1.0L

pH 4.8 ± 0.2 at 25°C

Preparation of Medium: Add yeast extract to 1.0L of apple juice. Mix thoroughly. Adjust pH to 4.8. Autoclave for 10 min at 9 psi pressure–114°C. Cool to 45°–50°C. In a separate flask, add agar to 200.0mL of distilled/deionized water and bring volume to 1.0L. Mix thoroughly. Gently heat and bring to boiling. Autoclave for 15 min at 15 psi pressure–121°C. Cool to 45°–50°C. Aseptically combine the sterile apple juice solution with the sterile agar solution. Mix thoroughly. Pour into sterile Petri dishes.

Use: For the cultivation of *Zymomonas* species.

Aquabacter spiritensis Medium (LMG Medium 225)

Sodium succinate	2.0g
Yeast extract	1.0g
KH_2PO_4	1.0g
Peptone	0.4g
NH_4Cl	0.2g
NaCl	0.2g
$MgSO_4 \cdot 7H_2O$	0.2g
$CaCl_2 \cdot 2H_2O$	10.0mg
Ferric citrate	5.0mg
Vitamin solution	20.0mL
SL-6 Trace elements solution	1.0mL

pH 7.0 ± 0.2 at 25°C

SL-6 Trace Elements Solution:

Composition per liter:

H_3BO_3	0.3g
$CoCl_2 \cdot 6H_2O$	0.2g
$ZnSO_4 \cdot 7H_2O$	0.1g
$MnCl_2 \cdot 4H_2O$	0.03g
$Na_2MoO_4 \cdot H_2O$	0.03g
$NiCl_2 \cdot 6H_2O$	0.02g
$CuCl_2 \cdot 2H_2O$	0.01g

Preparation of SL-6 Trace Elements Solution: Add components to distilled/deionized water and bring volume to 1.0L. Mix thoroughly. Adjust pH to 3.4.

Vitamin Solution:

Composition per liter:

Calcium DL-pantothenate	5.0mg
Riboflavin	5.0mg
Thiamine·HCl	5.0mg
Biotin	2.0mg
Folic acid	2.0mg
Vitamin B_{12}	0.1mg

Preparation of Vitamin Solution: Add components to distilled/deionized water and bring volume to 1.0L. Mix thoroughly. Filter sterilize.

Preparation of Medium: Add components, except vitamin solution, to 980.0mL distilled/deionized water. Mix thoroughly. Autoclave for 15 min at 15 psi pressure–121°C. Cool to 25°C. Aseptically add 20.0mL sterile vitamin solution. Mix thoroughly. Aseptically distribute to sterile tubes or flasks.

Use: For the cultivation of *Aquabacter spiritensis*.

Aquaspirillum Autotrophic Agar

Composition per liter:

Noble agar	15.0g
$Na_2HPO_4 \cdot 12H_2O$	9.0g
KH_2PO_4	1.5g
NH_4Cl	1.0g
$MgSO_4 \cdot 7H_2O$	0.2g
$CaCl_2 \cdot 2H_2O$	0.01g
Ferric ammonium citrate	5.0mg
$NaHCO_3$ solution	10.0mL
Trace elements solution	3.0mL

pH 7.1 ± 0.2 at 25°C

$NaHCO_3$ Solution:

Composition per 10.0mL:

$NaHCO_3$	0.5g

Preparation of $NaHCO_3$ Solution: Add the $NaHCO_3$ to distilled/deionized water and bring volume to 10.0mL. Mix thoroughly. Filter sterilize.

Trace Elements Solution:

Composition per liter:

H_3BO_3	30.0mg
$CoCl_2 \cdot 6H_2O$	20.0mg
$ZnSO_4 \cdot 7H_2O$	10.0mg
$MnCl_2 \cdot 4H_2O$	3.0mg
$Na_2MoO_4 \cdot 2H_2O$	3.0mg
$NiCl_2 \cdot 6H_2O$	2.0mg
$CuCl_2 \cdot 2H_2O$	1.0mg

Preparation of Trace Elements Solution: Add components to distilled/deionized water and bring volume to 1.0L. Mix thoroughly.

Preparation of Medium: Add components, except $NaHCO_3$ solution, to double-distilled water and bring volume to 990.0mL. Mix thoroughly. Gently heat and bring to boiling. Autoclave for 15 min at 15 psi pressure–121°C. Cool to 45°–50°C. Aseptically add sterile $NaHCO_3$ solution. Mix thoroughly. Pour into sterile Petri dishes or distribute into sterile tubes. For autotrophic growth, incubate under 85% H_2 + 10% CO_2 + 5% O_2.

Use: For the autotrophic cultivation and maintenance of *Aquaspirillum autotrophicum*.

Aquaspirillum Heterotrophic Agar

Composition per liter:

Noble agar	15.0g
$Na_2HPO_4 \cdot 12H_2O$	9.0g
KH_2PO_4	1.5g
NH_4Cl	1.0g
Sodium succinate	1.0g
$MgSO_4 \cdot 7H_2O$	0.2g
$CaCl_2 \cdot 2H_2O$	0.01g
Ferric ammonium citrate	5.0mg
Trace elements solution	3.0mL

pH 7.1 ± 0.2 at 25°C

Trace Elements Solution:

Composition per liter:

H_3BO_3	30.0mg
$CoCl_2 \cdot 6H_2O$	20.0mg
$ZnSO_4 \cdot 7H_2O$	10.0mg
$MnCl_2 \cdot 4H_2O$	3.0mg
$Na_2MoO_4 \cdot 2H_2O$	3.0mg

$NiCl_2 \cdot 6H_2O$ 2.0mg
$CuCl_2 \cdot 2H_2O$ 1.0mg

Preparation of Trace Elements Solution: Add components to distilled/deionized water and bring volume to 1.0L. Mix thoroughly.

Preparation of Medium: Add components to double-distilled water and bring volume to 1.0L. Mix thoroughly. Gently heat and bring to boiling. Autoclave for 15 min at 15 psi pressure–121°C. Pour into sterile Petri dishes or distribute into sterile tubes.

Use: For the heterotrophic cultivation and maintenance of *Aquaspirillum autotrophicum*.

Aquaspirillum Medium (DSMZ Medium 888)

Composition per 1026mL:

Casamino acids 1.5g
$(NH_4)_2SO_4$ 1.0g
$MgSO_4 \cdot 7H_2O$ 1.0g
Agar 0.5g
$CaCl_2 \cdot 6H_2O$ 30.0mg
$Na_2H_2PO_4$ 10.0mg
Sodium succinate solution 10.0mL
Thiosulfate solution 10.0mL
Standard vitamin solution 5.0mL
Trace elements solution SL-10 .... 1.0mL
pH 7.5 ± 0.2 at 25°C

Standard Vitamin Solution:
Composition per 100mL:

Thiamine-HCl·$2H_2O$ 50.0mg
Nicotinic acid 50.0mg
Pyridoxine-HCl 50.0mg
Ca-pantothenate 50.0mg
Riboflavin 10.0mg
Vitamin B_{12} 1.0mg
Folic acid 0.2mg
Biotin 0.1mg

Preparation of Standard Vitamin Solution: Add components to distilled/deionized water and bring volume to 100.0mL. Mix thoroughly. Filter sterilize.

Trace Elements Solution SL-10:
Composition per liter:

$FeCl_2 \cdot 4H_2O$ 1.5g
H_3BO_3 .... 300.0mg
$CoCl_2 \cdot 6H_2O$ 190.0mg
$MnCl_2 \cdot 4H_2O$ 100.0mg
$ZnCl_2$ 70.0mg
$Na_2MoO_4 \cdot 2H_2O$ 36.0mg
$NiCl_2 \cdot 6H_2O$ 24.0mg
$CuCl_2 \cdot 2H_2O$ 2.0mg
HCl (25% solution) 7.7mL

Preparation of Trace Elements Solution SL-10: Add $FeCl_2 \cdot 4H_2O$ to 10.0mL of HCl solution. Mix thoroughly. Add distilled/deionized water and bring volume to 1.0L. Add remaining components. Mix thoroughly. Sparge with 100% N_2. Autoclave for 15 min at 15 psi pressure–121°C. Cool to room temperature.

Thiosulfate Solution:
Composition per 10mL:

$Na_2S_2O_3 \cdot 5H_2O$ 1.0g

Preparation of Thiosulfate Solution: Add $Na_2S_2O_3 \cdot 5H_2O$ to distilled/deionized water and bring volume to 10.0mL. Mix thoroughly. Autoclave for 15 min at 15 psi pressure–121°C. Cool to room temperature.

Sodium Succinate Solution:
Composition per 10mL:

Sodium succinate 1.0g

Preparation of Sodium Succinate Solution: Add sodium succinate to distilled/deionized water and bring volume to 10.0mL. Mix thoroughly. Autoclave for 15 min at 15 psi pressure–121°C. Cool to room temperature.

Preparation of Medium: Add components, except sodium succinate solution, thiosulfate solution, standard vitamin solution, and trace elements solution SL-10, to distilled/deionized water and bring volume to 1.0L. Mix thoroughly. Autoclave for 15 min at 15 psi pressure–121°C. Cool to room temperature. Aseptically add 10.0mL sterile sodium succinate solution, 1.0mL sterile thiosulfate solution, 5.0mL sterile standard vitamin solution, and 1.0mL sterile trace elements solution SL-10. Mix thoroughly. Adjust pH to 7.5. Aseptically distribute into sterile tubes or flasks.

Use: For the cultivation of *Aquaspirillum* spp.

Archaeoglobus Medium

Composition per liter:

NaCl 18.0g
$NaHCO_3$ 5.0g
$MgCl_2 \cdot 6H_2O$ 4.0g
$MgSO_4 \cdot 7H_2O$ 3.45g
Sodium L-lactate 1.5g
Yeast extract 0.5g
KCl 0.34g
NH_4Cl 0.25g
$CaCl_2 \cdot 2H_2O$ 0.14g
K_2HPO_4 .... 0.14g
$Fe(NH_4)_2(SO_4)_2 \cdot 7H_2O$ 2.0mg
Resazurin 1.0mg
$Na_2S \cdot 9H_2O$ solution 25.0mL
Trace elements solution 10.0mL
pH 6.9 ± 0.2 at 25°C

Trace Elements Solution:
Composition per liter:

$MgSO_4 \cdot 7H_2O$ 3.0g
Nitrilotriacetic acid 1.5g
NaCl 1.0g
$MnSO_4 \cdot 2H_2O$ 0.5g
$CoSO_4 \cdot 7H_2O$ 0.18g
$ZnSO_4 \cdot 7H_2O$ 0.18g
$FeSO_4 \cdot 7H_2O$ 0.1g
$CaCl_2 \cdot 2H_2O$ 0.1g
$NiCl_2 \cdot 6H_2O$ 0.025g
$KAl(SO_4)_2 \cdot 12H_2O$ 0.02g
$CuSO_4 \cdot 5H_2O$ 0.01g
H_3BO_3 .... 0.01g
$Na_2MoO_4 \cdot 2H_2O$ 0.01g
$Na_2SeO_3 \cdot 5H_2O$ 0.3mg

$Na_2S \cdot 9H_2O$ Solution:
Composition per 50.0mL:

$Na_2S \cdot 9H_2O$ 1.0g

Preparation of $Na_2S \cdot 9H_2O$ Solution: Prepare and dispense solution anaerobically under with 80% N_2 + 20% CO_2. Add $Na_2S \cdot 9H_2O$ to distilled/deionized water and bring volume to 50.0mL. Mix thoroughly. Adjust pH to 7.0. Autoclave for 15 min at 15 psi pressure–121°C.

Preparation of Trace Elements Solution: Add nitrilotriacetic acid to approximately 500.0mL of distilled/deionized water. Dissolve by adding KOH and adjust pH to 6.5. Add remaining components. Bring volume to 1.0L with additional distilled/deionized water. Adjust pH to 7.0 with KOH.

Preparation of Medium: Add components, except $NaHCO_3$ and $Na_2S·9H_2O$ solution, to distilled/deionized water and bring volume to 1.0L. Mix well and heat to boiling for a few min. Cool rapidly to room temperature while gassing with 80% N_2 + 20% CO_2. Add $NaHCO_3$ and adjust pH to 6.9. Distribute anaerobically under 80% N_2 + 20% CO_2 and pressurize sealed containers up to 2 bar pressure. Autoclave for 15 min at 15 psi pressure–121°C. Prior to inoculation of cultures, add 0.25mL of sterile $Na_2S·9H_2O$ solution to each tube containing 9.75mL of sterile basal medium.

Use: For the cultivation and maintenance of *Archaeoglobus fulgidus*.

Archaeoglobus profundus Medium

Composition per liter:

NaCl	18.0g
$MgCl_2·6H_2O$	4.0g
$MgSO_4·7H_2O$	3.45g
Na_2SO_4	2.7g
Sodium acetate	1.0g
$NaHCO_3$	1.0g
Yeast extract	0.5g
KCl	0.34g
NH_4Cl	0.25g
$CaCl_2·2H_2O$	0.14g
K_2HPO_4	0.14g
$Fe(NH_4)_2(SO_4)_2·7H_2O$	2.0mg
Resazurin	1.0mg
$Na_2S·9H_2O$ solution	25.0mL
Trace elements solution	10.0mL

pH 6.5 ± 0.2 at 25°C

Trace Elements Solution:

Composition per liter:

$MgSO_4·7H_2O$	3.0g
Nitrilotriacetic acid	1.5g
NaCl	1.0g
$MnSO_4·2H_2O$	0.5g
$CoSO_4·7H_2O$	0.18g
$ZnSO_4·7H_2O$	0.18g
$FeSO_4·7H_2O$	0.1g
$CaCl_2·2H_2O$	0.1g
$NiCl_2·6H_2O$	0.025g
$KAl(SO_4)_2·12H_2O$	0.02g
$CuSO_4·5H_2O$	0.01g
H_3BO_3	0.01g
$Na_2MoO_4·2H_2O$	0.01g
$Na_2SeO_3·5H_2O$	0.3mg

$Na_2S·9H_2O$ Solution:

Composition per 50.0mL:

$Na_2S·9H_2O$	1.0g

Preparation of $Na_2S·9H_2O$ Solution: Prepare and dispense solution anaerobically under 80% N_2 + 20% CO_2. Add $Na_2S·9H_2O$ to distilled/deionized water and bring volume to 50.0mL. Mix thoroughly. Adjust pH to 7.0. Autoclave for 15 min at 15 psi pressure–121°C.

Preparation of Trace Elements Solution: Add nitrilotriacetic acid to approximately 500.0mL of distilled/deionized water. Dissolve by adding KOH and adjust pH to 6.5. Add remaining components. Bring volume to 1.0L with additional distilled/deionized water. Adjust pH to 7.0 with KOH.

Preparation of Medium: Add components, except $NaHCO_3$ and $Na_2S·9H_2O$ solution, to distilled/deionized water and bring volume to 1.0L. Mix well and heat to boiling for a few min. Cool rapidly to room temperature while gassing with 80% N_2 + 20% CO_2. Add $NaHCO_3$ and adjust pH to 6.9. Distribute anaerobically under 80% N_2 + 20% CO_2 and pressurize sealed containers up to 2 bar pressure. Autoclave for 15 min at 15 psi pressure–121°C. Prior to inoculation of cultures, add 0.25mL of sterile $Na_2S·9H_2O$ solution to each tube containing 9.75mL of sterile basal medium.

Use: For the cultivation and maintenance of *Archaeoglobus profundus*.

Archaeoglobus veneficus Medium (DSMZ Medium 796)

Composition per liter:

NaCl	18.0g
$MgCl_2·6H_2O$	7.15g
$NaHCO_3$	5.0g
KCl	0.33g
NH_4Cl	0.25g
$K_2HPO_4·3H_2O$	0.18g
$CaCl_2·2H_2O$	0.14g
$Fe(NH_4)_2(SO_4)_2·6H_2O$	2.0mg
Resazurin	0.5mg
Na_2SO_3solution	20.0mL
$Na_2S·9H_2O$ solution	10.0mL
Na-acetate solution	4.0mL
Trace elements solution	1.0mL

pH 1.0 ± 0.2 at 25°C

$Na_2S·9H_2O$ Solution:

Composition per 20.0mL:

$Na_2S·9H_2O$	0.5g

Preparation of $Na_2S·9H_2O$ Solution: Add $Na_2S·9H_2O$ to distilled/deionized water and bring volume to 20.0mL. Sparge with N_2. Autoclave for 15 min at 15 psi pressure–121°C. Cool to 25°C. Store anaerobically.

Na_2SO_3 Solution:

Composition per 10.0mL:

Na_2SO_3	0.5g

Preparation of Na_2SO_3 Solution: Add Na_2SO_3 to distilled/deionized water and bring volume to 10.0mL. Mix thoroughly. Sparge with 100% N_2. Filter sterilize.

Na-acetate Solution:

Composition per 10.0mL:

Na-acetate	2.5g

Preparation of Na-acetate Solution: Add Na-acetate to distilled/deionized water and bring volume to 10.0mL. Mix thoroughly. Sparge with 100% N_2. Filter sterilize.

Trace Element Solution:

Composition per liter:

NaCl	5.0g
$MnCl_2·4H_2O$	2.9g
$(NH_4)_2Ni(SO_4)_2$	1.0g
$FeSO_4·7H_2O$	0.5g
$CoCl_2·6H_2O$	0.5g
$CaCl_2·2H_2O$	0.5g
$ZnSO_4·7H_2$	0.5g
$CuSO_4·5H_2O$	0.05g
H_3BO_3	0.05g
$KAl(SO_4)_2·12H_2O$	0.05g
$Na_2MO_4·4H_2O$	0.05g

$Na_2WO_4 \cdot 2H_2O$ 0.05g
$Na_2SeO_3 \cdot 5H_2O$ 0.05g

Preparation of Trace Element Solution: Add components to distilled/deionized water and bring volume to 1.0L. Mix thoroughly. Sparge with 100% N_2.

Preparation of Medium: Prepare and dispense medium under 80% N_2 + 20% CO_2 gas mixture. Add components, except Na-acetate solution, $Na_2S \cdot 9H_2O$ solution, and $NaHCO_3$, to 986.0mL distilled/deionized water. Mix thoroughly. Gently heat and bring to boiling. Boil for 5 min. Cool to 25°C while sparging with 80% N_2 + 20% CO_2. Add the solid $NaHCO_3$. Equilibrate with the 80% N_2 + 20% CO_2 gas. Adjust pH to 7.0 with HCl. Add 10.0mL $Na_2S \cdot 9H_2O$ solution. Distribute into serum bottles under 80% N_2 + 20% CO_2 gas; 25.0mL medium in 100mL serum bottles. Adjust pH to 1.0. Autoclave for 15 min at 15 psi pressure–121°C. Cool to 25°C. Aseptically and anaerobically inject 0.1mL sterile Na-acetate solution and 0.5mL Na_2SO_3 solution per 25.0mL medium. Mix thoroughly. After inoculation pressurize to 1 bar atmosphere with 80% H_2 + 20% CO_2 gas.

Use: For the cultivation of *Archaeoglobus veneficus.*

Archangium violaceum Medium

Composition per liter:

Monosodium glutamate	1.0g
L-Leucine	0.5g
L-Tyrosine	0.5g
L-Isoleucine	0.3g
L-Proline	0.25g
$MgSO_4 \cdot 7H_2O$	0.2g
L-Lysine	0.15g
L-Arginine	0.1g
L-Asparagine	0.1g
L-Serine	0.1g
L-Threonine	0.1g
L-Valine	0.1g
L-Alanine	0.05g
L-Glycine	0.05g
L-Histidine	0.05g
L-Methionine	0.05g
$Ca_3(PO_4)_2$	0.02g
KCl	0.02g
Tris(hydroxymethyl)aminomethane buffer (0.02m solution, pH 7.5)	1.0L

pH 7.5 ± 0.2 at 25°C

Preparation of Medium: Add solid components to 1.0L of Tris buffer. Mix thoroughly. Filter sterilize. Aseptically distribute into tubes or flasks.

Use: For the cultivation of *Archangium violaceum.*

Artificial Deep Lake Medium

Composition per liter:

NaCl	180.0g
$MgCl_2 \cdot 6H_2O$	75.0g
Noble agar	15.0g
Sodium succinate	10.0g
$MgSO_4 \cdot 7H_2O$	7.4g
KCl	7.4g
$CaCl_2 \cdot 2H_2O$	1.0g
Yeast extract	1.0g
Vitamin solution	10.0mL

pH 7.4 ± 0.2 at 25°C

Vitamin Solution:

Composition per liter:

Biotin	30.0mg
Cyanocobalamin	20.0mg
Thiamine·HCl	10.0mg

Preparation of Vitamin Solution: Add components to distilled/deionized water and bring volume to 1.0L. Mix thoroughly. Filter sterilize and add aseptically to sterile basal medium.

Preparation of Medium: Add components, except vitamin solution, to distilled/deionized water and bring volume to 990.0mL. Mix thoroughly. Gently heat and bring to boiling. Adjust medium to pH 7.4. Autoclave for 15 min at 15 psi pressure–121°C. Cool to 50°C. Aseptically add 10.0mL of vitamin solution. Pour into sterile Petri dishes or leave in tubes.

Use: For the cultivation and maintenance of *Halobacterium lacusprofundi.*

Artificial Organic Lake Peptone Medium

Composition per 1001.0mL:

NaCl	30.0g
$MgSO_4 \cdot 7H_2O$	9.5g
KCl	5.0g
Peptone	5.0g
Yeast extract	1.0g
$CaCl_2 \cdot 2H_2O$	0.2g
KNO_3	0.1g
$(NH_4)_2SO_4$	0.1g
Modified Hutner's basal salts	20.0mL
Phosphate supplement	20.0mL
Artificial organic lake vitamin solution	1.0mL

pH 7.3 ± 0.2 at 25°C

Modified Hutner's Basal Salts:

Composition per liter:

$MgSO_4 \cdot 7H_2O$	29.7g
Nitrilotriacetic acid	10.0g
$CaCl_2 \cdot 2H_2O$	3.34g
$FeSO_4 \cdot 7H_2O$	99.0mg
Ammonium molybdate	9.25mg
"Metals 44"	50.0mL

Preparation of Modified Hutner's Basal Salts: Dissolve the nitrilotracetic acid first and neutralize the solution with KOH. Add other components and adjust the pH to 7.2 with KOH or H_2SO_4. There may be a slight precipitate. Store at 5°C.

"Metals 44":

Composition per liter:

$ZnSO_4 \cdot 7H_2O$	1.1g
$FeSO_4 \cdot 7H_2O$	0.5g
$CuSO_4 \cdot 5H_2O$	0.04g
EDTA	0.25g
$MnSO_4 \cdot 7H_2O$	0.154g
$Co(NO_3)_2 \cdot 6H_2O$	0.025g
$Na_2B_4O_7 \cdot 10H_2O$	0.018g

Preparation of "Metals 44": Add components to distilled/deionized water and bring volume to 100.0mL. Mix thoroughly. Autoclave for 15 min at 15 psi pressure–121°C. Add aseptically to sterile modified Hutner's basal salts solution.

Phosphate Supplement:

Composition per liter:

K_2HPO_4	2.5g
KH_2PO_4	2.5g

Preparation of Phosphate Supplement: Add components to distilled/deionized water and bring volume to

20.0mL. Mix thoroughly. Autoclave for 15 min at 15 psi pressure–121°C.

Artificial Organic Lake Vitamin Solution:
Composition per liter:

Pyridoxine·HCl	10.0mg
Calcium DL-pantothenate	5.0mg
Nicotinamide	5.0mg
Riboflavin	5.0mg
Thiamine·HCl	5.0mg
Biotin	2.0mg
Folic acid	2.0mg
Cyanocobalamin	0.1mg

Preparation of Artificial Organic Lake Vitamin Solution: Add components to distilled/deionized water and bring volume to 1.0L. Mix thoroughly. Filter sterilize. Store at 5°C.

Preparation of Medium: Add components, except modified Hutner's basal salts solution, phosphate supplement solution, and vitamin solution, to distilled/deionized water and bring volume to 959.0mL. Mix thoroughly. Bring pH to 7.3. Autoclave for 15 min at 15 psi pressure–121°C. Cool to 50°C. Aseptically add 20.0mL of sterile modified Hutner's basal salts solution, 20.0mL of sterile phosphate supplement solution, and 1.0mL of sterile artificial organic lake vitamin solution. Mix thoroughly. Aseptically distribute into sterile tubes or flasks.

Use: For the cultivation of *Halomonas meridiana.*

Ashbey's Nitrogen-Free Agar

Composition per liter:

Agar	15.0g
Mannitol	15.0g
$CaCl_2 \cdot 2H_2O$	0.2g
K_2HPO_4	0.2g
$MgSO_4 \cdot 7H_2O$	0.2g
MoO_3 (10% solution)	0.1mL
$FeCl_3$ (10% solution	0.05mL

pH 7.2 ± 0.2 at 25°C

Preparation of Medium: Add components to distilled/deionized water and bring volume to 1.0L. Mix thoroughly. Gently heat and bring to boiling. Distribute into tubes or flasks. Autoclave for 15 min at 15 psi pressure–121°C. Pour into sterile Petri dishes or leave in tubes.

Use: For the isolation and cultivation of bacteria, such as *Azotobacter* species and cyanobacteria, that can utilize atmospheric N_2 as sole nitrogen source.

ASN-III Agar

Composition per liter:

NaCl	25.0g
$MgSO_4 \cdot 7H_2O$	3.5g
$MgCl_2 \cdot 6H_2O$	2.0g
$NaNO_3$	0.75g
$K_2HPO_4 \cdot 3H_2O$	0.75g
$CaCl_2 \cdot 2H_2O$	0.5g
KCl	0.5g
Na_2CO_3	0.02g
Citric acid	3.0mg
Ferric ammonium citrate	3.0mg
Magnesium EDTA	0.5mg
Vitamin B_{12}	10.0μg
Agar solution	100.0mL
A-5 trace metals	1.0mL

pH 7.3 ± 0.2 at 25°C

Agar Solution:
Composition per 100.0mL:

Noble agar	10.0g

Preparation of Agar Solution: Add agar to glass distilled water and bring volume to 100.0mL. Mix thoroughly. Gently heat and bring to boiling. Autoclave for 15 min at 15 psi pressure–121°C. Cool to 45°–50°C.

A-5 Trace Metals:
Composition per liter:

H_3BO_3	2.86g
$MnCl_2 \cdot 4H_2O$	1.81g
$ZnSO_4 \cdot 7H_2O$	0.222g
$CuSO_4 \cdot 5H_2O$	0.079g
$Co(NO_3)_2 \cdot 6H_2O$	0.049g
$Na_2MoO_4 \cdot 2H_2O$	0.039g

Preparation of A-5 Trace Metals: Add components to distilled/deionized water and bring volume to 1.0L. Mix thoroughly.

Preparation of Medium: Add components, except agar solution, to glass distilled water and bring volume to 900.0mL. Mix well and heat gently until dissolved. Filter sterilize. Warm to 45°–50°C. Aseptically add agar solution. Mix thoroughly. Pour into sterile Petri dishes or distribute into sterile tubes.

Use: For the cultivation of *Xenococcus* species. For the isolation of cyanobacteria from marine habitats.

Asparagine Broth

Composition per liter:

DL-Asparagine	30.0g
K_2HPO_4	1.0g
$MgSO_4 \cdot 7H_2O$	0.5g

pH 6.9–7.2 at 25°C

Preparation of Medium: Add components to distilled/deionized water and bring volume to 1.0L. Mix well until dissolved. Adjust pH to between 6.9 and 7.2. Distribute into tubes or flasks. Autoclave for 15 min at 15 psi pressure–121°C.

Use: For a presumptive test medium in the differentiation of nonfermentative Gram-negative bacteria, especially *Pseudomonas aeruginosa.* For use in the multiple-tube technique in the microbiological analysis of recreational waters.

Asticcacaulis Medium

Composition per liter:

Agar	15.0g
Pancreatic digest of casein	0.5g
Yeast extract	0.5g
Sodium acetate	0.2g

pH 6.8 ± 0.2 at 25°C

Preparation of Medium: Add components to distilled/deionized water and bring volume to 1.0L. Mix thoroughly. Gently heat and bring to boiling. Distribute into tubes or flasks. Autoclave for 15 min at 15 psi pressure–121°C. Pour into sterile Petri dishes or leave in tubes.

Use: For the isolation and cultivation of *Asticcacaulis* species.

ASTM Nutrient Salts Agar (American Society for Testing and Materials Nutrient Salts Agar)

Composition per liter:

Agar	15.0g

KH_2PO_4	0.7g
K_2HPO_4	0.7g
$MgSO_4 \cdot 7H_2O$	0.7g
NH_4NO_3	1.0g
NaCl	5.0mg
$FeSO_4 \cdot 7H_2O$	2.0mg
$ZnSO_4$	2.0mg
$MnSO_4 \cdot H_2O$	1.0mg

pH 6.5 ± 0.2 at 25°C

Preparation of Medium: Add components to tap water and bring volume to 1.0L. Mix thoroughly. Gently heat and bring to boiling. Distribute into tubes or flasks. Autoclave for 15 min at 15 psi pressure–121°C. Pour into sterile Petri dishes or leave in tubes.

Use: For determination of the susceptibility of plastics to fungal degradation.

ASW Medium (Artificial Seawater Medium)

Composition per liter:

NaCl	27.0g
Agar	15.0g
$MgSO_4 \cdot 7H_2O$	6.6g
$MgCl_2 \cdot 6H_2O$	5.6g
$CaCl_2 \cdot 2H_2O$	1.5g
KNO_3	1.0g
KH_2PO_4	0.07g
$NaHCO_3$	0.04g
Tris-HCl buffer (1.0*M*, pH 7.6)	20.0mL
Chelated iron solution	1.0mL
Trace metal solution	1.0mL

Trace Metal Solution:
Composition per 100.0mL:

H_3BO_3	60.0mg
$MnCl_2 \cdot 4H_2O$	40.0mg
$(NH_4)_6Mo_7O_{24} \cdot 4H_2O$	37.0mg
$CuCl_2 \cdot 2H_2O$	4.0mg
$ZnCl_2$	4.0mg
$CoCl_2 \cdot 6H_2O$	1.5mg

Preparation of Trace Metal Solution: Add components to distilled/deionized water and bring volume to 100.0mL. Mix thoroughly.

Chelated Iron Solution:
Composition per 100.0mL:

$FeCl_3 \cdot 4H_2O$	240.0mg
EDTA	14.6g

Preparation of Chelated Iron Solution: Add EDTA to distilled/deionized water and bring volume to 100.0mL. Mix thoroughly. Adjust pH to 7.6. Add $FeCl_3 \cdot 4H_2O$. Mix thoroughly.

Preparation of Medium: Add components to distilled/deionized water and bring volume to 1.0L. Mix thoroughly. Gently heat and bring to boiling. Distribute into tubes or flasks. Autoclave for 15 min at 15 psi pressure–121°C. Pour into sterile Petri dishes or leave in tubes.

Use: For the cultivation of *Porphyridium purpureum.*

Atlas Oil Agar

Composition per liter:

Bushnell-Haas agar	990.0mL
Oil	10.0mL

pH 7.0 ± 0.2 at 25°C

Bushnell-Haas Agar:
Composition per 990.0mL:

Agar	15.0g
KH_2PO_4	1.0g
K_2HPO_4	1.0g
NH_4NO_3	1.0g
$MgSO_4 \cdot 7H_2O$	0.2g
$FeCl_3$	0.05g
$CaCl_2 \cdot 2H_2O$	0.02g

Preparation of Bushnell-Haas Agar: Add components to distilled/deionized water and bring volume to 990.0mL. Mix thoroughly. Gently heat and bring to boiling. Autoclave for 15 min at 15 psi pressure–121°C. Cool to 60°C.

Preparation of Medium: Filter sterilize oil. Aseptically add 10.0mL of sterile oil to 990.0mL of cooled, sterile Bushnell-Haas agar. Put mixture into a sterile blender container. Blend on low speed to minimize the incorporation of air into the medium. Pour into sterile Petri dishes.

Use: For the cultivation and enumeration of hydrocarbon-utilizing bacteria by direct plating of water and sediment samples.

AT5N Medium

Composition per liter:

$CaCO_3$	10.0g
$(NH_4)_2SO_4$	1.5g
K_2HPO_4	0.5g
$MgSO_4$	50.0mg
$KHCO_3$	30.0mg
$CaCl_2 \cdot 2H_2O$	20.0mg

Preparation of Medium: Add components to tap water and bring volume to 1.0L. Mix thoroughly. Gently heat and bring to boiling. Distribute into tubes or flasks. Autoclave for 15 min at 15 psi pressure–121°C.

Use: For the cultivation of bacteria that oxidize ammonia, especially those from wastewater.

Aureomycin® Rose Bengal Glucose Peptone Agar

Composition per liter:

Agar	20.0g
Glucose	10.0g
Peptone	5.0g
KH_2PO_4	1.0g
$MgSO_4 \cdot 7H_2O$	0.5g
Rose Bengal	0.035g
Aureomycin solution	200.0mL

pH 5.4 ± 0.2 at 25°C

Aureomycin Solution:
Composition per 200.0mL:

Aureomycin·HCl	0.07g

Preparation of Aureomycin Solution: Add aureomycin·HCl to distilled/deionized water and bring volume to 200.0mL. Mix thoroughly. Filter sterilize.

Preparation of Medium: Add components, except aureomycin solution, to distilled/deionized water and bring volume to 800.0mL. Mix thoroughly. Gently heat and bring to boiling. Autoclave for 15 min at 15 psi pressure–121°C. Cool to 45°–50°C. Aseptically add 200.0mL of sterile aureomycin solution. Mix thoroughly. Pour into sterile Petri dishes or distribute into sterile tubes.

Use: For the cultivation and enumeration of fungi isolated from sewage and polluted waters.

Autotrophic *Nitrobacter* Medium (DSMZ Medium 756c)

Composition per liter:

$NaNO_2$ 2.0g
Stock solution 100.0mL
Trace elements solution 1.0mL

pH 7.5 ± 0.2 at 25°C

Stock Solution:

Composition per liter:

NaCl 5.0g
KH_2PO_4 .... 1.5g
$MgSO_4 \cdot 7H_2O$ 0.5g
$CaCO_3$ 0.07g

Preparation of Stock Solution: Add components to distilled/deionized water and bring volume to 1.0L. Mix thoroughly.

Trace Elements Solution:

Composition per liter:

$FeSO_4 \cdot 7H_2O$ 97.3mg
H_3BO_3 .... 49.4mg
$ZnSO_4 \cdot 7H_2O$ 43.1mg
$(NH_4)_6Mo_7O_{24} \cdot 4H_2O$ 37.1mg
$MnSO_4 \cdot 2H_2O$ 33.8mg
$CuSO_4 \cdot 5H_2O$ 25.0mg

Preparation of Trace Elements Solution: Add components to distilled/deionized water and bring volume to 1.0L. Mix thoroughly.

Preparation of Medium: Add components to distilled/deionized water and bring volume to 1.0L. Mix thoroughly. Adjust pH to 8.6. Distribute into tubes or flasks. Autoclave for 15 min at 15 psi pressure–121°C. Allow to stand for 2-3 days so that pH adjusts itself to 7.4-7.6.

Use: For the cultivation of *Nitrobacter winogradskyi*.

Autotrophic *Nitrobacter* Medium (LMG Medium 247)

Composition per liter:

$NaNO_2$ 2.0g
Stock solution 100.0mL
Trace elements solution 1.0mL

pH 7.5 ± 0.2 at 25°C

Stock Solution:

Composition per liter:

NaCl 5.0g
KH_2PO_4 .... 1.5g
$MgSO_4 \cdot 7H_2O$ 0.5g
$CaCO_3$ 0.07g

Preparation of Stock Solution: Add components to distilled/deionized water and bring volume to 1.0L. Mix thoroughly.

Trace Elements Solution:

Composition per liter:

$(NH_4)Mo7O_2$ 437.10mg
$FeSO_4 \cdot 7H_2O$ 97.30mg
$ZnSO_4 \cdot 7H_2O$ 43.10mg
H_3BO_3 .... 39.40mg
$MnSO_4 \cdot H_2O$ 33.80mg
$CuSO_2 \cdot 5H_2O$ 25.00mg

Preparation of Trace Elements Solution: Add components to distilled/deionized water and bring volume to 1.0L. Mix thoroughly.

Preparation of Medium: Add components to distilled/deionized water and bring volume to 1.0L. Mix thoroughly. Adjust pH to 8.6 with NaOH. Distribute into tubes or flasks. Autoclave for 15 min at 15 psi pressure–121°C. Allow the medium to stand for 2-3 days so that the pH can adjust itself to pH 7.4-7.6.

Use: For the cultivation of autotrophic *Nitrobacter* spp.

Azide Blood Agar

Composition per liter:

Agar 15.0g
Pancreatic digest of casein 5.0g
Peptic digest of animal tissue 5.0g
NaCl 5.0g
Beef extract 3.0g
NaN_3 .... 0.2g
Sheep blood, defibrinated 50.0mL

pH 7.2 ± 0.2 at 25°C

Source: This medium is available as a premixed powder from BD Diagnostic Systems, and Oxoid Unipath.

Caution: Sodium azide is toxic. Azides also react with metals and disposal must be highly diluted.

Preparation of Medium: Add components, except sheep blood, to distilled/deionized water and bring volume to 950.0mL. Mix thoroughly. Gently heat and bring to boiling. Autoclave for 15 min at 15 psi pressure–121°C. Cool to 45–50°C. Aseptically add 50.0mL of sterile defibrinated sheep blood. Pour into sterile Petri dishes or distribute into sterile tubes. Allow tubes to cool in a slanted position.

Use: For the isolation and differentiation of streptococci and staphylococci from specimens containing mixed flora and from nonclinical specimens such as water and sewage.

Azide Blood Agar with Crystal Violet (Packer's Agar)

Composition per liter:

Agar 15.0g
Pancreatic digest of casein 5.0g
Peptic digest of animal tissue 5.0g
NaCl 5.0g
Beef extract 3.0g
NaN_3 .... 0.9g
Crystal Violet 2.0mg
Sheep blood, defibrinated 50.0mL

pH 7.2 ± 0.2 at 25°C

Caution: Sodium azide is toxic. Azides also react with metals and disposal must be highly diluted.

Preparation of Medium: Add components, except sheep blood, to distilled/deionized water and bring volume to 950.0mL. Mix thoroughly. Gently heat and bring to boiling. Autoclave for 15 min at 15 psi pressure–121°C. Cool to 45°–50°C. Aseptically add 50.0mL of sterile defibrinated sheep blood. Pour into sterile Petri dishes or distribute into sterile tubes. Allow tubes to cool in a slanted position.

Use: For the isolation and enumeration of fecal streptococci from nonclinical specimens such as water. Also used for the isolation of *Streptococcus pneumoniae* and *Erysipelothrix rhusiopathiae*.

Azide Broth (Azide Glucose Broth) (Azide Dextrose Broth)

Composition per liter:

Pancreatic digest of casein 15.0g
Glucose 7.5g
NaCl 7.5g

Beef extract....................4.5g
NaN_3....................0.2g

pH 7.2 ± 0.2 at 25°C

Source: This medium is available as a premixed powder from BD Diagnostic Systems.

Caution: Sodium azide is toxic. Azides also react with metals and disposal must be highly diluted.

Preparation of Medium: Add components to distilled/deionized water and bring volume to 1.0L. Mix thoroughly. Gently heat and bring to boiling. Distribute into tubes or flasks. Autoclave for 15 min at 15 psi pressure–121°C. Prepare double-strength broth for samples larger than 1.0mL.

Use: For the detection and enrichment of fecal streptococci in water and sewage. Also used in the multiple-tube technique as a presumptive test for the presence of fecal streptococci.

Azide Broth, Rothe (Azide Glucose Broth, Rothe) (Azide Dextrose Broth, Rothe)

Composition per liter:

Peptone....................20.0g
Glucose....................5.0g
NaCl....................5.0g
K_2HPO_4....................2.7g
KH_2PO_4....................2.7g
NaN_3....................0.2g

pH 6.8 ± 0.2 at 25°C

Source: This medium is available as a premixed powder from Oxoid Unipath.

Caution: Sodium azide is toxic. Azides also react with metals and disposal must be highly diluted.

Preparation of Medium: Add components to distilled/deionized water and bring volume to 1.0L. Mix thoroughly. Gently heat and bring to boiling. Distribute into tubes or flasks. Autoclave for 15 min at 15 psi pressure–121°C. Prepare double-strength broth for samples larger than 1.0mL.

Use: For the detection of enterococci in water and sewage.

Azide Citrate Broth

Composition per liter:

Pancreatic digest of casein....................20.0g
Sodium citrate....................10.0g
Yeast extract....................5.0g
Glucose....................5.0g
NaCl....................5.0g
K_2HPO_4....................4.0g
KH_2PO_4....................1.5g
NaN_3....................0.25g

pH 7.0 ± 0.2 at 25°C

Caution: Sodium azide is toxic. Azides also react with metals and disposal must be highly diluted.

Preparation of Medium: Add components to distilled/deionized water and bring volume to 1.0L. Mix thoroughly. Gently heat and bring to boiling. Distribute into tubes or flasks. Autoclave for 15 min at 15 psi pressure–118°C. Prepare double-strength broth for samples larger than 1.0mL.

Use: For the detection and enrichment of fecal streptococci in water and sewage.

Azoarcus Medium (LMG Medium 202)

Composition per liter:

Solution A....................750.0mL
Phosphate buffer solution....................250.0mL

pH 6.8 ± 0.2 at 25°C

Solution A:

Composition per 750.0mL:

Malic acid....................5.0g
KOH....................4.5g
$MgSO_4 \cdot 7H_2O$....................0.2g
NaCl....................0.1g
$CaCl_2$....................20.0mg
$MnSO_4 \cdot H_2O$....................10.0mg
$Na_2MoO_4 \cdot 2H_2O$....................2.0mg
Ferric EDTA solution....................10.0mL

Ferric EDTA Solution:

Composition per liter:

Ferric EDTA....................0.066g

Preparation of Ferric EDTA Solution: Add ferric EDTA to distilled/deionized water and bring volume to 10.0mL. Mix thoroughly.

Preparation of Solution A: Add 5.0g malic acid to 500.0mL distilled/deionized water. Adjust pH to 7.0 with KOH (approximate amount of 4.5g). Add other components. Mix thoroughly. Bring volume to 750.0mL. Adjust pH to 6.8. Autoclave for 15 min at 15 psi pressure–121°C. Cool to 25°C.

Phosphate Buffer Solution:

Composition per liter:

$Na_2HPO_4 \cdot 2H_2O$....................5.8g
KH_2PO_4....................4.5g

Preparation of Phosphate Buffer Solution: Add components to distilled/deionized water and bring volume to 1.0L. Mix thoroughly. Adjust pH to 6.8. Autoclave for 15 min at 15 psi pressure–121°C. Cool to 25°C.

Preparation of Medium: Aseptically combine 750.0mL sterile solution A with 250.0mL sterile phosphate buffer solution. Aseptically distribute to sterile tubes or flasks.

Use: For the cultivation of *Azoarcus indigens*.

Azoarcus VM Medium (LMG Medium 252)

Composition per liter:

Agar....................15.0g
Beef extract....................3.0g
DL-malic acid....................2.5g
KOH....................2.5g
KH_2PO_4....................1.5g
NaCl....................1.1g
K_2HPO_4....................1.0g
Yeast extract....................1.0g
$MgSO_4 \cdot 7H_2O$....................0.2g
$CaCl_2$....................0.2g
Fe EDTA....................66.0mg
$MnSO_4 \cdot H_2O$....................10.0mg
$Na_2MoO_4 \cdot 2H_2O$....................2.0mg
Biotin....................0.1mg
NH_4Cl....................0.5mg

pH 6.8 ± 0.2 at 25°C

Preparation of Medium: Add components to distilled/deionized water and bring volume to 1.0L. Mix thoroughly. Adjust pH to 6.8. Gently heat and bring to boiling. Distribute into tubes or flasks. Autoclave for 15 min at 15 psi pressure–121°C. Pour into sterile Petri dishes or leave in tubes.

Use: For the cultivation and maintenance of *Azospira oryzae* and *Azonexus fungiphilus.*

Azorhizobium caulinodans Agar (LMG 119)

Composition per liter:

Agar	15.0g
Beef extract	5.0g
Peptone	5.0g
Sucrose	5.0g
Yeast extract	1.0g
$MgSO_4$	0.24g

Preparation of Medium: Add components to distilled/deionized water and bring volume to 1.0L. Mix thoroughly. Gently heat and bring to boiling. Distribute into tubes or flasks. Autoclave for 15 min at 15 psi pressure–121°C. Pour into sterile Petri dishes or leave in tubes.

Use: For the cultivation and maintenance of *Azorhizobium caulinodans.*

Azorhizophilus paspali Agar

Composition per liter:

Agar	20.0g
Sucrose	20.0g
$Na_2MoO_4 \cdot 2H_2O$	0.5g
$MgSO_4 \cdot 7H_2O$	0.2g
KH_2PO_4	0.15g
K_2HPO_4	0.05g
$FeCl_3$	0.01g

pH 6.9 ± 0.2 at 25°C

Preparation of Medium: Add components to distilled/deionized water and bring volume to 1.0L. Mix thoroughly. Gently heat and bring to boiling. Distribute into tubes or flasks. Autoclave for 15 min at 15 psi pressure–121°C. Pour into sterile Petri dishes or leave in tubes.

Use: For the cultivation and maintenance of *Azorhizophilus paspali.*

Azospirillum amazonense Medium (LGI Medium)

Composition per liter:

Sucrose	5.0g
Agar	1.75g
KH_2PO_4	0.6g
K_2HPO_4	0.2g
$MgSO_4 \cdot 7H_2O$	0.2g
$CaCl_2 \cdot 2H_2O$	0.02g
$FeCl_3$	0.01g
$Na_2MoO_4 \cdot 2H_2O$	2.0mg
Bromothymol Blue solution	5.0mL

pH 6.0 ± 0.2 at 25°C

Bromothymol Blue Solution:
Composition per 100.0mL:

Bromothymol Blue	0.5g

Preparation of Bromothymol Blue Solution: Add Bromothymol Blue to 100.0mL of 0.2*N* KOH. Mix thoroughly.

Preparation of Medium: Add components to distilled/deionized water and bring volume to 1.0L. Mix thoroughly. Gently heat and bring to boiling. Distribute into tubes or flasks. Autoclave for 15 min at 15 psi pressure–121°C. Pour into sterile Petri dishes or leave in tubes.

Use: For the cultivation and maintenance of *Azospirillum amazonense.*

Azospirillum lipoferum Agar Medium

Composition per liter:

Glucose	20.0g
Agar	15.0g
K_2HPO_4	0.8g
$MgSO_4 \cdot 7H_2O$	0.5g
KH_2PO_4	0.2g
$FeCl_3 \cdot 6H_2O$	0.1g
Yeast extract	0.1g
$CaCl_2 \cdot 2H_2O$	0.02g
$Na_2MoO_4 \cdot 2H_2O$	0.02g

pH 6.9 ± 0.2 at 25°C

Preparation of Medium: Add components to distilled/deionized water and bring volume to 1.0L. Mix thoroughly. Gently heat and bring to boiling. Distribute into tubes or flasks. Autoclave for 15 min at 15 psi pressure–121°C. Pour into sterile Petri dishes or leave in tubes.

Use: For the cultivation of *Azospirillum lipoferum.*

Azospirillum lipoferum Medium

Composition per liter:

Calcium malate	10.0g
K_2HPO_4	1.0g
$MgSO_4 \cdot 7H_2O$	0.5g
$CaCl_2 \cdot 2H_2O$	0.02g

pH 6.5 ± 0.2 at 25°C

Preparation of Medium: Add components to distilled/deionized water and bring volume to 1.0L. Mix thoroughly. Distribute into tubes or flasks. Autoclave for 15 min at 15 psi pressure–121°C.

Use: For the isolation and cultivation of *Azospirillum lipoferum.*

Azospirillum Medium

Composition per liter:

Sodium malate	5.0g
Agar	1.75g
KH_2PO_4	0.4g
$MgSO_4 \cdot 7H_2O$	0.2g
K_2HPO_4	0.1g
NaCl	0.1g
$CaCl_2 \cdot 2H_2O$	0.02g
$FeCl_3$	0.01g
$Na_2MoO_4 \cdot 2H_2O$	2.0mg
Bromothymol Blue solution	5.0mL

pH 6.8 ± 0.2 at 25°C

Bromothymol Blue Solution:
Composition per 10.0mL:

Bromothymol Blue	0.5g
Ethanol	10.0mL

Preparation of Bromothymol Blue Solution: Add Bromothymol Blue to 10.0mL of ethanol. Mix thoroughly.

Preparation of Medium: Add components to distilled/deionized water and bring volume to 1.0L. Mix thoroughly. Distribute into tubes or flasks. Autoclave for 15 min at 15 psi pressure–121°C.

Use: For the cultivation of *Azospirillum* species isolated from roots.

Azotobacter Agar

Composition per liter:

Agar	15.0g

Sucrose	10.0g
$MgSO_4 \cdot 7H_2O$	0.2g
KH_2PO_4	0.15g
K_2HPO_4	0.05g
$CaCl_2$	0.02g
Na_2MoO_4	0.002g
$FeCl_3$	1.0μg

Preparation of Medium: Add components to distilled/deionized water and bring volume to 1.0L. Mix thoroughly. Gently heat and bring to boiling. Distribute into tubes or flasks. Autoclave for 15 min at 15 psi pressure–121°C. Pour into sterile Petri dishes or leave in tubes.

Use: For the cultivation and maintenance of *Azorhizophilus paspali.*

Azotobacter Agar, Modified I

Composition per liter:

Agar	15.0g
Sucrose	10.0g
Glucose	10.0g
$MgSO_4 \cdot 7H_2O$	0.2g
KH_2PO_4	0.15g
$CaSO_4 \cdot 2H_2O$	0.1g
K_2HPO_4	0.05g
$CaCl_2$	0.02g
Na_2MoO_4	2.0mg
$FeCl_3$	1.0mg
$Na_2MoO_4 \cdot 2H_2O$	1.0mg

pH 7.2 ± 0.2 at 25°C

Preparation of Medium: Add components to distilled/deionized water and bring volume to 1.0L. Mix thoroughly. Gently heat and bring to boiling. Adjust pH to 7.2. Distribute into tubes or flasks. Autoclave for 15 min at 15 psi pressure–121°C. Pour into sterile Petri dishes or leave in tubes.

Use: For the cultivation and maintenance of *Azotobacter* species.

Azotobacter Agar, Modified II

Composition per liter:

Sucrose	20.0g
Agar	15.0g
KH_2PO_4	0.15g
$MgSO_4 \cdot 7H_2O$	0.2g
K_2HPO_4	0.05g
$CaCl_2$	0.02g
Na_2MoO_4	2.0mg
$FeCl_3$	1.0mg
$Na_2MoO_4 \cdot 2H_2O$	1.0mg

pH 6.2 ± 0.2 at 25°C

Preparation of Medium: Add components to distilled/deionized water and bring volume to 1.0L. Mix thoroughly. Gently heat and bring to boiling. Adjust pH to 6.2. Distribute into tubes or flasks. Autoclave for 15 min at 15 psi pressure–121°C. Pour into sterile Petri dishes or leave in tubes.

Use: For the cultivation and maintenance of *Azotobacter* species and *Beijerinckia derxii.*

Azotobacter Basal Agar

Composition per liter:

Agar	15.0g
K_2HPO_4	1.0g
$MgSO_4 \cdot 7H_2O$	0.2g
NaCl	0.2g
$FeSO_4 \cdot 7H_2O$	5.0mg
Soil extract	100.0mL

pH 7.2 ± 0.2 at 25°C

Soil Extract:

Composition per 200.0mL:

African Violet soil	0.5g
Na_2CO_3	0.5g

Preparation of Soil Extract: Add components to tap water and bring volume to 200.0mL. Autoclave for 60 min at 15 psi pressure–121°C. Filter through Whatman filter paper.

Preparation of Medium: Add components, including filtered soil extract, to tap water and bring volume to 1.0L. Mix thoroughly. Gently heat and bring to boiling. Distribute into tubes or flasks. Autoclave for 15 min at 15 psi pressure–121°C. Pour into sterile Petri dishes or leave in tubes.

Use: For the cultivation of a variety of bacteria, including *Azomonas* species, *Azotobacter* species, and others when a carbon source is added.

Azotobacter Broth

Composition per liter:

Glucose	10.0g
$CaCO_3$	5.0g
K_2HPO_4	0.9g
$CaCl_2 \cdot 2H_2O$	0.1g
$MgSO_4 \cdot 7H_2O$	0.1g
KH_2PO_4	0.1g
$FeSO_4 \cdot 7H_2O$	10.0mg
$Na_2MoO_4 \cdot 2H_2O$	5.0mg

pH 7.3 ± 0.2 at 25°C

Preparation of Medium: Add components to distilled/deionized water and bring volume to 1.0L. Mix thoroughly. Gently heat and bring to boiling. Distribute into tubes or flasks. Autoclave for 15 min at 15 psi pressure–121°C. Pour into sterile Petri dishes or leave in tubes.

Use: For the cultivation and maintenance of *Azotobacter beijerinckii, Azotobacter chroococcum, Azotobacter vinelandii,* and *Derxia gummosa.*

Azotobacter Broth, Modified I

Composition per liter:

Sucrose	10.0g
Glucose	10.0g
$MgSO_4 \cdot 7H_2O$	0.2g
KH_2PO_4	0.15g
$CaSO_4 \cdot 2H_2O$	0.1g
K_2HPO_4	0.05g
$CaCl_2$	0.02g
Na_2MoO_4	2.0mg
$FeCl_3$	1.0mg
$Na_2MoO_4 \cdot 2H_2O$	1.0mg

pH 7.2 ± 0.2 at 25°C

Preparation of Medium: Add components to distilled/deionized water and bring volume to 1.0L. Mix thoroughly. Gently heat and bring to boiling. Adjust pH to 7.2. Distribute into tubes or flasks. Autoclave for 15 min at 15 psi pressure–121°C.

Use: For the cultivation of *Azotobacter* species.

Azotobacter Broth, Modified II

Composition per liter:

Sucrose	20.0g
KH_2PO_4	0.15g

$MgSO_4 \cdot 7H_2O$ 0.2g
K_2HPO_4 .... 0.05g
$CaCl_2$ 0.02g
Na_2MoO_4 .... 2.0mg
$FeCl_3$ 1.0mg
$Na_2MoO_4 \cdot 2H_2O$ 1.0mg
pH 6.2 ± 0.2 at 25°C

Preparation of Medium: Add components to distilled/deionized water and bring volume to 1.0L. Mix thoroughly. Gently heat and bring to boiling. Adjust pH to 6.2. Distribute into tubes or flasks. Autoclave for 15 min at 15 psi pressure–121°C.

Use: For the cultivation of *Azotobacter* species and *Beijerinckia derxii.*

Azotobacter chroococcum Agar

Composition per liter:
Agar 20.0g
$CaCO_3$ 20.0g
Glucose 20.0g
K_2HPO_4 .... 0.8g
$MgSO_4 \cdot 7H_2O$ 0.5g
KH_2PO_4 .... 0.2g
$FeCl_3 \cdot 6H_2O$ 0.1g
$Na_2MoO_4 \cdot 2H_2O$ 0.05g
pH 7.4–7.6 at 25°C

Preparation of Medium: Add components to distilled/deionized water and bring volume to 1.0L. Mix thoroughly. Gently heat and bring to boiling. Distribute into tubes or flasks. Autoclave for 15 min at 15 psi pressure–121°C. Pour into sterile Petri dishes or leave in tubes.

Use: For the cultivation and maintenance of *Azotobacter chroococcum.*

Azotobacter Medium

Composition per liter:
Agar 15.0g
$CaCO_3$ 5.0g
K_2HPO_4 .... 0.9g
$CaCl_2 \cdot 2H_2O$ 0.1g
KH_2PO_4 .... 0.1g
$MgSO_4 \cdot 7H_2O$ 0.1g
$FeSO_4 \cdot 7H_2O$ 0.01g
$Na_2MoO_4 \cdot 2H_2O$ 5.0mg
Glucose solution 25.0mL
Mannitol solution 25.0mL
pH 7.3 ± 0.2 at 25°C

Glucose Solution:
Composition per 25.0mL:
D-Glucose 5.0g

Preparation of Glucose Solution: Add glucose to distilled/deionized water and bring volume to 25.0mL. Mix thoroughly. Filter sterilize. Warm to 50°–55°C.

Mannitol Solution:
Composition per 25.0mL:
Mannitol 5.0g

Preparation of Mannitol Solution: Add mannitol to distilled/deionized water and bring volume to 25.0mL. Mix thoroughly. Filter sterilize. Warm to 50°–55°C.

Preparation of Medium: Add components, except glucose solution and mannitol solution, to distilled/deionized water and bring volume to 950.0mL. Mix thoroughly. Autoclave for 15 min at 15 psi pressure–121°C. Cool to 50°–55°C. Aseptically add 25.0mL of sterile glucose solution and 25.0mL of sterile mannitol solution. Mix thoroughly. Pour into sterile Petri dishes or distribute into sterile tubes.

Use: For the cultivation and maintenance of *Azotobacter* species.

Azotobacter Medium (ATCC Medium 14)

Composition per liter:
Sucrose 20.0g
Agar 15.0g
K_2HPO_4 .... 0.8g
Yeast extract 0.5g
KH_2PO_4 .... 0.2g
$MgSO_4 \cdot 7H_2O$ 0.2g
$CaSO_4 \cdot 2H_2O$ 0.1g
$FeCl_3$ 1.0mg
$Na_2MoO_4 \cdot 2H_2O$ 1.0mg
pH 7.2 ± 0.2 at 25°C

Preparation of Medium: Add components to distilled/deionized water and bring volume to 1.0L. Mix thoroughly. Gently heat and bring to boiling. Distribute into tubes or flasks. Autoclave for 15 min at 15 psi pressure–121°C. Pour into sterile Petri dishes or leave in tubes.

Use: For the cultivation of a variety of bacteria, including *Azomonas* species, *Azotobacter* species, *Beijerinckia derxii, Pseudomonas azotocolligans,* and *Rhodococcus erythropolis.*

Azotobacter Medium (ATCC Medium 240)

Composition per liter:
Agar 15.0g
$MgSO_4 \cdot 7H_2O$ 0.2g
KH_2PO_4 .... 0.15g
K_2HPO_4 .... 0.05g
$CaCl_2$ 0.02g
$Na_2MoO_4 \cdot 2H_2O$ 2.0mg
$FeCl_3$ 1.0mg
pH 7.2 ± 0.2 at 25°C

Preparation of Medium: Add components to distilled/deionized water and bring volume to 1.0L. Mix thoroughly. Gently heat and bring to boiling. Distribute into tubes or flasks. Autoclave for 15 min at 15 psi pressure–121°C. Pour agar medium into sterile Petri dishes or leave in tubes.

Use: For the cultivation and maintenance of a variety of bacteria, including *Azotobacter* species.

Azotobacter Medium (ATCC Medium 1771)

Composition per liter:
Agar 15.0g
Glucose 10.0g
KH_2PO_4 .... 0.22g
$CaSO_4 \cdot 2H_2O$ 0.1g
$MgSO_4 \cdot 7H_2O$ 0.098g
NaCl 0.058g
K_2HPO_4 .... 0.058g
$FeSO_4 \cdot 7H_2O$ 5.0mg
$Na_2MoO_4 \cdot 2H_2O$ 0.2mg
pH 7.2 ± 0.2 at 25°C

Preparation of Medium: Add components to distilled/deionized water and bring volume to 1.0L. Mix thoroughly. Gently heat and bring to boiling. Distribute into tubes or flasks. Autoclave for 15 min at 15 psi pressure–121°C. Pour into sterile Petri dishes or leave in tubes.

Use: For the cultivation and maintenance of a variety of bacteria, including *Azotobacter* species.

Azotobacter paspali Medium

Composition per liter:

Agar	20.0g
Sucrose	20.0g
$CaCO_3$	1.0g
$MgSO_4 \cdot 7H_2O$	0.2g
KH_2PO_4	0.15g
K_2HPO_4	0.05g
$CaCl_2$	0.02g
$Na_2MoO_4 \cdot 2H_20$	2.0mg
Bromothymol Blue solution	10.0mL
$FeCl_3$ (10% solution)	0.1mL

pH 7.0 ± 0.2 at 25°C

Bromothymol Blue Solution:

Composition per 10.0mL:

Bromothymol Blue	0.5g
Ethanol	10.0mL

Preparation of Bromothymol Blue Solution: Add Bromothymol Blue to 10.0mL of ethanol. Mix thoroughly.

Preparation of Medium: Add components to distilled/deionized water and bring volume to 1.0L. Mix thoroughly. Gently heat and bring to boiling. Distribute into tubes or flasks. Autoclave for 15 min at 15 psi pressure–121°C. Pour into sterile Petri dishes or leave in tubes.

Use: For the cultivation and maintenance of *Azotobacter paspali.*

Azotobacter Supplement (ATCC Medium 11)

Composition per liter:

Agar	15.0g
K_2HPO_4	1.0g
$MgSO_4 \cdot 7H_2O$	0.2g
NaCl	0.2g
$FeSO_4 \cdot 7H_2O$	5.0mg
Soil extract	100.0mL
Glucose solution	100.0mL

pH 7.6 ± 0.2 at 25°C

Soil Extract:

Composition per 200.0mL:

African Violet soil	0.5g
Na_2CO_3	0.5g

Preparation of Soil Extract: Add components to tap water and bring volume to 200.0mL. Autoclave for 60 min at 15 psi pressure–121°C. Filter through Whatman filter paper.

Glucose Solution:

Composition per 100.0mL:

Glucose	20.0g

Preparation of Glucose Solution: Add glucose to distilled/deionized water and bring volume to 100.0mL. Mix thoroughly. Filter sterilize.

Preparation of Medium: Add components, except glucose solution, to tap water and bring volume to 900.0mL. Mix thoroughly. Adjust pH to 7.6. Autoclave for 15 min at 15 psi pressure–121°C. Cool to 50°–55°C. Aseptically add 100.0mL of sterile glucose solution. Mix thoroughly. Pour into sterile Petri dishes or leave in tubes.

Use: For the cultivation of *Azomonas agilis* and *Azotobacter chroococcum.*

Azotobacter Supplement (ATCC Medium 12)

Composition per liter:

Agar	15.0g
K_2HPO_4	1.0g
$MgSO_4 \cdot 7H_2O$	0.2g
NaCl	0.2g
$FeSO_4 \cdot 7H_2O$	5.0mg
Soil extract	100.0mL
Mannitol solution	100.0mL

pH 7.6 ± 0.2 at 25°C

Soil Extract:

Composition per 200.0mL:

African Violet soil	0.5g
Na_2CO_3	0.5g

Preparation of Soil Extract: Add components to distilled/deionized water and bring volume to 200.0mL. Autoclave for 60 min at 15 psi pressure–121°C. Filter through Whatman filter paper.

Mannitol Solution:

Composition per 100.0mL:

Mannitol	20.0g

Preparation of Mannitol Solution: Add mannitol to distilled/deionized water and bring volume to 100.0mL. Mix thoroughly. Filter sterilize.

Preparation of Medium: Add components, except mannitol solution, to tap water and bring volume to 900.0mL. Mix thoroughly. Adjust pH to7.6. Autoclave for 15 min at 15 psi pressure–121°C. Cool to 50°–55°C. Aseptically add 100.0mL of sterile mannitol solution. Mix thoroughly. Pour into sterile Petri dishes or leave in tubes.

Use: For the cultivation of *Azotobacter* species and *Azomonas* species.

Azotobacter Supplement (ATCC Medium 13)

Composition per liter:

Agar	15.0g
K_2HPO_4	1.0g
$MgSO_4 \cdot 7H_2O$	0.2g
NaCl	0.2g
$FeSO_4 \cdot 7H_2O$	5.0mg
Soil extract	100.0mL
Glucose solution	100.0mL

pH 6.0 ± 0.2 at 25°C

Soil Extract:

Composition per 200.0mL:

African Violet soil	0.5g
Na_2CO_3	0.5g

Preparation of Soil Extract: Add components to tap water and bring volume to 200.0mL. Autoclave for 60 min at 15 psi pressure–121°C. Filter through Whatman filter paper.

Glucose Solution:

Composition per 100.0mL:

Glucose	20.0g

Preparation of Glucose Solution: Add glucose to distilled/deionized water and bring volume to 100.0mL. Mix thoroughly. Filter sterilize.

Preparation of Medium: Add components, except glucose solution, to tap water and bring volume to 900.0mL. Mix thoroughly. Adjust pH to 6.0. Autoclave for 15 min at 15 psi pressure–121°C. Cool to 50°–55°C. Aseptically add sterile glucose solution. Mix thoroughly. Pour into sterile Petri dishes or leave in tubes.

Use: For the cultivation of *Beijerinckia* species.

Azotobacter Supplement (ATCC Medium 15)

Composition per liter:

Agar 15.0g
K_2HPO_4 .... 1.0g
$MgSO_4 \cdot 7H_2O$ 0.2g
NaCl 0.2g
$FeSO_4 \cdot 7H_2O$ 5.0mg
Soil extract 100.0mL
Mannitol solution 100.0mL

pH 6.0 ± 0.2 at 25°C

Soil Extract:

Composition per 200.0mL:

African Violet soil 0.5g
Na_2CO_3 .... 0.5g

Preparation of Soil Extract: Add components to distilled/deionized water and bring volume to 200.0mL. Autoclave for 60 min at 15 psi pressure–121°C. Filter through Whatman filter paper.

Mannitol Solution:

Composition per 100.0mL:

Mannitol 20.0g

Preparation of Mannitol Solution: Add mannitol to distilled/deionized water and bring volume to 100.0mL. Mix thoroughly. Filter sterilize.

Preparation of Medium: Add components, except mannitol solution, to tap water and bring volume to 900.0mL. Mix thoroughly. Adjust pH to 6.0. Autoclave for 15 min at 15 psi pressure–121°C. Cool to 50°–55°C. Aseptically add 100.0mL of sterile mannitol solution. Mix thoroughly. Pour into sterile Petri dishes or leave in tubes.

Use: For the cultivation of *Azomonas macrocytogenes.*

Azotobacter vinelandii Medium

Composition per liter:

Sodium benzoate 1.0g
K_2HPO_4 .... 0.5g
Mannitol 0.5g

Preparation of Medium: Add components to distilled/deionized water and bring volume to 1.0L. Mix thoroughly. Distribute into tubes or flasks. Autoclave for 15 min at 15 psi pressure–121°C.

Use: For the cultivation of *Azotobacter vinelandii* from water samples.

BA Medium with Cellobiose

Composition per liter:

$NaHCO_3$ 2.6g
NH_4Cl 1.0g
Yeast extract 0.75g
$K_2HPO_4 \cdot 3H_2O$ 0.4g
$MgCl_2 \cdot 6H_2O$ 0.1g
NaCl 0.1g
$CaCl_2 \cdot 2H_2O$ 0.05g
Resazurin 0.5mg
Cellobiose solution 50.0mL
$Na_2S \cdot 9H_2O$ solution 10.0mL
Wolfe's mineral solution 10.0mL
Wolfe's vitamin solution 10.0mL

pH 6.9–7.0 at 25°C

Cellobiose Solution:

Composition per 50.0mL:

Cellobiose 4.0g

Preparation of Cellobiose Solution: Add cellobiose to distilled/deionized water and bring volume to 50.0mL. Mix thoroughly. Filter sterilize.

$Na_2S \cdot 9H_2O$ Solution:

Composition per 10.0mL:

$Na_2S \cdot 9H_2O$ 0.25g

Preparation of $Na_2S \cdot 9H_2O$ Solution: Add $Na_2S \cdot 9H_2O$ to distilled/deionized water and bring volume to 10.0mL. Mix thoroughly. Sparge with 100% N_2. Autoclave for 15 min at 15 psi pressure–121°C. Before use, neutralize to pH 7.0 with sterile HCl.

Wolfe's Mineral Solution:

Composition per liter:

$MgSO_4 \cdot 7H_2O$ 3.0g
Nitrilotriacetic acid 1.5g
NaCl 1.0g
$MnSO_4 \cdot 2H_2O$ 0.5g
$CoCl_2 \cdot 6H_2O$ 0.1g
$ZnSO_4 \cdot 7H_2O$ 0.1g
$CaCl_2 \cdot 2H_2O$ 0.1g
$FeSO_4 \cdot 7H_2O$ 0.1g
$NiCl_2 \cdot 6H_2O$ 0.025g
$KAl(SO_4)_2 \cdot 12H_2O$ 0.02g
$CuSO_4 \cdot 5H_2O$ 0.01g
H_3BO_3 .... 0.01g
$Na_2MoO_4 \cdot 2H_2O$ 0.01g
$Na_2SeO_3 \cdot 5H_2O$ 0.3mg

Preparation of Wolfe's Mineral Solution: Add nitrilotriacetic acid to 500.0mL of distilled/deionized water. Adjust pH to 6.5 with KOH. Add remaining components. Add distilled/deionized water to 1.0L. Adjust pH to 6.8.

Wolfe's Vitamin Solution:

Composition per liter:

Pyridoxine·HCl 10.0mg
p-Aminobenzoic acid 5.0mg
Lipoic acid 5.0mg
Nicotinic acid 5.0mg
Riboflavin 5.0mg
Thiamine·HCl 5.0mg
Calcium DL-pantothenate 5.0mg
Biotin 2.0mg
Folic acid 2.0mg
Vitamin B_{12} 0.1mg

Preparation of Wolfe's Vitamin Solution: Add components to distilled/deionized water and bring volume to 1.0L. Mix thoroughly. Filter sterilize.

Preparation of Medium: Prepare and dispense medium under 80% N_2 + 20% CO_2 gas mixture. Add components, except cellobiose solution, $Na_2S \cdot 9H_2O$ solution, Wolfe's mineral solution, and Wolfe's vitamin solution, to distilled/deionized water and bring volume to 920.0mL. Mix thoroughly. Sparge with 80% N_2 + 20% CO_2 gas mixture. Autoclave for 15 min at 15 psi pressure–121°C. Aseptically and anaerobically add 50.0mL of sterile cellobiose solution, 10.0mL of sterile Wolfe's mineral solution, 10.0mL of sterile Wolfe's vitamin solution, and 10.0mL of sterile $Na_2S \cdot 9H_2O$ solution. Mix thoroughly. Aseptically and anaerobically distribute into sterile tubes or bottles.

Use: For the cultivation of *Caldicellulosiruptor lactoaceticus.*

Baar's Medium for Sulfate Reducers

Composition per liter:

Sodium lactate.......... 3.5g
$MgSO_4 \cdot 7H_2O$.......... 2.0g
K_2HPO_4.......... 1.0g
$CaSO_4$.......... 1.0g
NH_4Cl.......... 0.5g
Ferrous ammonium sulfate solution.......... 10.0mL
Yeast extract solution.......... 10.0mL

pH 7.5 ± 0.2 at 25°C

Ferrous Ammonium Sulfate Solution:
Composition per 10.0mL:

$Fe(NH_4)_2(SO_4)_2$.......... 0.5g

Preparation of Ferrous Ammonium Sulfate Solution: Add $Fe(NH_4)_2(SO_4)_2$ to distilled/deionized water and bring volume to 10.0mL. Mix thoroughly. Autoclave for 15 min at 15 psi pressure–121°C.

Yeast Extract Solution:
Composition per 10.0mL:

Yeast extract.......... 1.0g

Preparation of Yeast Extract Solution: Add yeast extract to distilled/deionized water and bring volume to 10.0mL. Mix thoroughly. Autoclave for 15 min at 15 psi pressure–121°C.

Preparation of Medium: Add components, except ferrous ammonium sulfate solution and yeast extract solution, to tap water and bring volume to 980.0mL. Mix thoroughly. Gently heat and bring to boiling. Autoclave for 15 min at 15 psi pressure–121°C. Cool to 45°–50°C. Aseptically add 10.0mL of sterile ferrous ammonium sulfate solution and sterile yeast extract solution. Mix thoroughly. Aseptically distribute into tubes or flasks.

Use: For the cultivation and maintenance of *Desulfotomaculum nigrificans.*

Baar's Medium for Sulfate Reducers, Modified

Composition per 1020.0mL:

Component I.......... 400.0mL
Component III.......... 400.0mL
Component II.......... 200.0mL
Ferrous ammonium sulfate solution.......... 20.0mL

pH 7.5 ± 0.2 at 25°C

Component I:
Composition per 400.0mL:

Sodium citrate.......... 5.0g
$MgSO_4$.......... 2.0g
$CaSO_4$.......... 1.0g
NH_4Cl.......... 1.0g

Preparation of Component I: Add components to distilled/deionized water and bring volume to 400.0mL. Mix thoroughly. Adjust pH to 7.5. Autoclave for 15 min at 15 psi pressure–121°C.

Component II:
Composition per 200.0mL:

K_2HPO_4.......... 0.5g

Preparation of Component II: Add K_2HPO_4 to distilled/deionized water and bring volume to 200.0mL. Mix thoroughly. Adjust pH to 7.5. Autoclave for 15 min at 15 psi pressure–121°C.

Component III:
Composition per 400.0mL:

Sodium lactate.......... 3.5g
Yeast extract.......... 1.0g

Preparation of Component III: Add components to distilled/deionized water and bring volume to 400.0mL. Mix thoroughly. Adjust pH to 7.5. Autoclave for 15 min at 15 psi pressure–121°C.

Ferrous Ammonium Sulfate Solution:
Composition per 20.0mL:

$Fe(NH_4)_2(SO_4)_2$.......... 1.0g

Preparation of Ferrous Ammonium Sulfate Solution: Add $Fe(NH_4)_2(SO_4)_2$ to distilled/deionized water and bring volume to 20.0mL. Mix thoroughly. Filter sterilize.

Preparation of Medium: Aseptically combine component I, component II, and component III. Mix thoroughly. Distribute 5.0mL volumes into tubes under 97% N_2 + 3% H_2. Add medium to tubes while still warm to exclude as much O_2 as possible. Aseptically add 0.1mL of sterile ferrous ammonium sulfate solution to 5.0mL of medium immediately prior to inoculation.

Use: For the cultivation and maintenance of *Desulfovibrio, Desulfobulbus, Desulfotomaculum,* and *Thermodesulfobacterium* species.

Baar's Medium for Sulfate Reducers, Modified with 2.5% NaCl

Composition per 1020.0mL:

Component I.......... 400.0mL
Component III.......... 400.0mL
Component II.......... 200.0mL
Ferrous ammonium sulfate solution.......... 20.0mL

pH 7.5 ± 0.2 at 25°C

Component I:
Composition per 400.0mL:

NaCl.......... 25.0g
Sodium citrate.......... 5.0g
$MgSO_4$.......... 2.0g
$CaSO_4$.......... 1.0g
NH_4Cl.......... 1.0g

Preparation of Component I: Add components to distilled/deionized water and bring volume to 400.0mL. Mix thoroughly. Adjust pH to 7.5. Autoclave for 15 min at 15 psi pressure–121°C.

Component II:
Composition per 200.0mL:

K_2HPO_4.......... 0.5g

Preparation of Component II: Add K_2HPO_4 to distilled/deionized water and bring volume to 200.0mL. Mix thoroughly. Adjust pH to 7.5. Autoclave for 15 min at 15 psi pressure–121°C.

Component III:
Composition per 400.0mL:

Sodium lactate.......... 3.5g
Yeast extract.......... 1.0g

Preparation of Component III: Add components to distilled/deionized water and bring volume to 400.0mL. Mix thoroughly. Adjust pH to 7.5. Autoclave for 15 min at 15 psi pressure–121°C.

Ferrous Ammonium Sulfate Solution:
Composition per 20.0mL:

$Fe(NH_4)_2(SO_4)_2$.......... 1.0g

Preparation of Ferrous Ammonium Sulfate Solution: Add $Fe(NH_4)_2(SO_4)_2$ to distilled/deionized water and bring volume to 20.0mL. Mix thoroughly. Filter sterilize.

Preparation of Medium: Aseptically combine component I, component II, and component III. Mix thoroughly. Distribute 5.0mL volumes into tubes under 97% N_2 + 3% H_2. Add medium to tubes while still warm to exclude as much O_2 as possible. Aseptically add 0.1mL of sterile ferrous ammonium sulfate solution to 5.0mL of medium immediately prior to inoculation.

Use: For the cultivation of *Desulfovibrio africanus* and other *Desulfovibrio* species that prefer 2.5% NaCl.

Bacillus acidocaldarius Agar

Composition per liter:

Solution A	500.0mL
Solution B	500.0mL

pH 3.0–4.0 at 25°C

Solution A:

Composition per 500.0mL:

KH_2PO_4	3.0g
Yeast extract	1.0g
Glucose	1.0g
$MgSO_4 \cdot 7H_2O$	0.5g
$CaCl_2 \cdot 2H_2O$	0.25g
$(NH_4)_2SO_4$	0.2g

Preparation of Solution A: Add components to distilled/deionized water and bring volume to 500.0mL. Adjust pH to 3.0–4.0. Mix thoroughly. Autoclave for 10 min at 15 psi pressure–121°C. Cool to 50°–55°C.

Solution B:

Composition per 500.0mL:

Agar	30.0g

Preparation of Solution B: Add agar to distilled/deionized water and bring volume to 500.0mL. Mix thoroughly. Gently heat and bring to boiling. Autoclave for 15 min at 15 psi pressure–121°C. Cool to 50°–55°C.

Preparation of Medium: Aseptically mix 500.0mL of solution A and 500.0mL of solution B. Mix thoroughly. Aseptically adjust pH to 3.0–4.0. Pour into sterile Petri dishes or distribute into sterile tubes.

Use: For the cultivation and maintenance of *Bacillus (Alicyclobacillus) acidocaldarius.*

Bacillus acidoterrestris Agar

Composition per 1001.0mL:

Solution A	500.0mL
Solution C	500.0mL
Solution B (Trace elements solution SL-6)	1.0mL

pH 4.0 ± 0.2 at 25°C

Solution A:

Composition per 500.0mL:

Glucose	5.0g
KH_2PO_4	3.0g
Yeast extract	2.0g
$MgSO_4 \cdot 7H_2O$	0.5g
$CaCl_2 \cdot 2H_2O$	0.25g
$(NH_4)_2SO_4$	0.2g

Preparation of Solution A: Add components to distilled/deionized water and bring volume to 500.0mL. Mix thoroughly. Adjust pH to 4.0. Autoclave for 15 min at 15 psi pressure–121°C.

Solution C:

Composition per 500.0mL:

Agar	15.0g

Preparation of Solution C: Add agar to distilled/deionized water and bring volume to 500.0mL. Gently heat and bring to boiling. Autoclave for 15 min at 15 psi pressure–121°C. Cool to 50°–55°C.

Solution B (Trace Elements Solution SL-6):

Composition per liter:

$MnCl_2 \cdot 4H_2O$	0.5g
H_3BO_3	0.3g
$CoCl_2 \cdot 6H_2O$	0.2g
$ZnSO_4 \cdot 7H_2O$	0.1g
$Na_2MoO_4 \cdot 2H_2O$	0.03g
$NiCl_2 \cdot 6H_2O$	0.02g
$CuCl_2 \cdot 2H_2O$	0.01g

Preparation of Solution B (Trace Elements Solution SL-6): Add components to distilled/deionized water and bring volume to 1.0L. Mix thoroughly. Autoclave for 15 min at 15 psi pressure–121°C.

Preparation of Medium: Aseptically combine 500.0mL of sterile solution A, 500.0mL of sterile solution C, and 1.0mL of sterile solution B. Mix thoroughly. Pour into sterile Petri dishes or distribute into sterile tubes.

Use: For the cultivation and maintenance of *Bacillus acidoterrestris, Alicyclobacillus acidoterrestris,* and *Alicyclobacillus cycloheptanicus.*

Bacillus Agar

Composition per liter:

Agar	20.0g
$(NH_4)_2SO_4$	1.3g
Glucose	1.0g
Yeast extract	1.0g
KH_2PO_4	0.37g
$MgSO_4 \cdot 7H_2O$	0.25g
$CaCl_2 \cdot 2H_2O$	0.07g
$FeCl_3$	0.02g

pH 4.0 ± 0.2 at 25°C

Preparation of Medium: Add components to distilled/deionized water and bring volume to 500.0mL. Mix thoroughly. Gently heat and bring to boiling. Adjust pH to 3.5. Prepare a separate agar solution by adding 20.0g/500.0mL of distilled/deionized water. Autoclave solutions separately for 15 min at 15 psi pressure–121°C. Cool to 50°–55°C. Aseptically combine both solutions. This procedure avoids acid hydrolysis of the agar. Pour into sterile Petri dishes or leave in tubes.

Use: For the cultivation of acidophilic *Bacillus* species such as *Bacillus acidocaldarius*.

Bacillus Agar, Modified

Composition per liter:

Agar	20.0g
Glucose	1.0g
Yeast extract	1.0g
KH_2PO_4	0.6g
$MgSO_4 \cdot 7H_2O$	0.5g
$CaCl_2 \cdot 2H_2O$	0.25g
$(NH_4)_2SO_4$	0.2g

pH 3.0–4.0 at 25°C

Preparation of Medium: Add components, except agar and glucose, to distilled/deionized water and bring volume to 500.0mL. Mix thoroughly. Gently heat and bring to boiling. Adjust pH to 3.5. Prepare a separate agar and glucose solution by adding 20.0g of agar and 1.0g of glucose to 500.0mL of distilled/deionized water. Autoclave solutions separately for 15 min at 15 psi pressure–121°C. Cool to 50°–55°C. Aseptically combine both solutions. This procedure avoids acid hydrolysis of the agar. Pour into sterile Petri dishes or leave in tubes.

Use: For the cultivation of acidophilic *Bacillus* species such as *Bacillus acidocaldarius*.

Bacillus benzoevorans Agar

Composition per liter:

Agar	15.0g
Yeast extract	6.0g
Peptone	5.0g
NaCl	5.0g
$Na_2HPO_4 \cdot 12H_2O$	3.6g
Sodium benzoate	2.0g
Beef extract	1.0g
KH_2PO_4	0.98g
NH_4Cl	0.5g
$MgSO_4 \cdot 7H_2O$	0.03g
Trace elements solution	0.2mL

pH 7.0–7.2 at 25°C

Trace Elements Solution:

Composition per 100.0mL:

$FeSO_4 \cdot 7H_2O$	0.1g
$MnCl_2 \cdot 4H_2O$	0.1g
$ZnSO_4 \cdot 7H_2O$	0.1g

Preparation of Trace Elements Solution: Add components to distilled/deionized water and bring volume to 100.0mL. Mix thoroughly.

Preparation of Medium: Add components to distilled/deionized water and bring volume to 1.0L. Mix thoroughly. Gently heat and bring to boiling. Distribute into tubes or flasks. Autoclave for 15 min at 15 psi pressure–121°C. Pour into sterile Petri dishes or leave in tubes.

Use: For the cultivation and maintenance of *Bacillus benzoevorans.*

Bacillus Broth

Composition per liter:

$(NH_4)_2SO_4$	1.3g
Glucose	1.0g
Yeast extract	1.0g
KH_2PO_4	0.37g
$MgSO_4 \cdot 7H_2O$	0.25g
$CaCl_2 \cdot 2H_2O$	0.07g
$FeCl_3$	0.02g

pH 4.0 ± 0.2 at 25°C

Preparation of Medium: Add components to distilled/deionized water and bring volume to 1.0L. Mix thoroughly. Gently heat and bring to boiling. Adjust pH to 4.0 with 10*N* H_2SO_4. Distribute into tubes or flasks. Autoclave for 15 min at 15 psi pressure–121°C.

Use: For the cultivation of acidophilic *Bacillus* species such as *Bacillus acidocaldarius*.

Bacillus cycloheptanicus Agar

Composition per 1001.0mL:

Solution A	500.0mL
Solution C	500.0mL
Solution B (Trace Elements solution SL-6)	1.0mL

pH 4.0 ± 0.2 at 25°C

Solution A:

Composition per liter:

Yeast extract	5.0g
Glucose	5.0g
KH_2PO_4	3.0g
$MgSO_4 \cdot 7H_2O$	0.5g
$CaCl_2 \cdot 2H_2O$	0.25g
$(NH_4)_2SO_4$	0.2g

Preparation of Solution A: Add components to distilled/deionized water and bring volume to 1.0L. Mix thoroughly. Adjust pH to 4.0. Autoclave for 15 min at 15 psi pressure–121°C.

Solution C:

Composition per 500.0mL:

Agar	15.0g

Preparation of Solution C: Add agar to distilled/deionized water and bring volume to 500.0mL. Gently heat and bring to boiling. Autoclave for 15 min at 15 psi pressure–121°C. Cool to 50°–55°C.

Solution B (Trace Elements Solution SL-6):

Composition per liter:

$MnCl_2 \cdot 4H_2O$	0.5g
H_3BO_3	0.3g
$CoCl_2 \cdot 6H_2O$	0.2g
$ZnSO_4 \cdot 7H_2O$	0.1g
$Na_2MoO_4 \cdot 2H_2O$	0.03g
$NiCl_2 \cdot 6H_2O$	0.02g
$CuCl_2 \cdot 2H_2O$	0.01g

Preparation of Solution B (Trace Elements Solution SL-6): Add components to distilled/deionized water and bring volume to 1.0L. Mix thoroughly. Autoclave for 15 min at 15 psi pressure–121°C.

Preparation of Medium: Aseptically combine 500.0mL of sterile solution A, 500.0mL of sterile solution C, and 1.0mL of sterile solution B. Mix thoroughly. Pour into sterile Petri dishes or distribute into sterile tubes.

Use: For the cultivation and maintenance of *Bacillus cycloheptanicus, Alicyclobacillus acidoterrestris,* and *Alicyclobacillus cycloheptanicus.*

Bacillus halodenitrificans Agar (LMG Medium 142)

Composition per liter:

NaCl	100.0g
Agar	15.0g
Sodium acetate·$3H_2O$	10.0g
Na_2HPO_4	3.8g
KH_2PO_4	1.3g
$(NH_4)_2SO_4$	1.0g
$Mg(NO_3)_2 \cdot 6H_2O$	1.0g
Yeast extract	1.0g
Magnesium nitrate solution	100.0mL

pH 7.2 ± 0.2 at 25°C

Magnesium Nitrate Solution:

Composition per 100.0mL:

$Mg(NO_3)_2 \cdot 6H_2O$	1.0g

Preparation of Magnesium Nitrate Solution: Add $Mg(NO_3)_2 \cdot 6H_2O$ to distilled/deionized water and bring volume to 100.0mL. Mix thoroughly. Filter sterilize.

Preparation of Medium: Add components, except magnesium nitrate solution, to distilled/deionized water and bring volume to 900.0mL. Mix thoroughly. Adjust pH to 7.2 with KOH. Autoclave for 15 min at 15 psi pressure–121°C.

Cool to 45°-50°C. Aseptically add 100.0mL sterile magnesium nitrate solution. Mix thoroughly. Aseptically pour into sterile Petri dishes or distribute into sterile tubes.

Use: For the cultivation of *Bacillus halodenitrificans.*

Bacillus Medium (ATCC Medium 21)

Composition per liter:

Glycerol	20.0g
L-Glutamic acid	4.0g
Citric acid	2.0g
K_2HPO_4	0.5g
Ferric ammonium citrate	0.5g
$MgSO_4$	0.5g

pH 7.4 ± 0.2 at 25°C

Preparation of Medium: Add components to tap water and bring volume to 1.0L. Mix thoroughly. Gently heat and bring to boiling. Distribute into tubes or flasks. Autoclave for 15 min at 15 psi pressure–121°C.

Use: For the cultivation of *Bacillus licheniformis.*

Bacillus Medium (ATCC Medium 455)

Composition per liter:

Soluble starch	30.0g
Agar	20.0g
Polypeptone™	5.0g
Yeast extract	5.0g

Preparation of Medium: Add components to distilled/deionized water and bring volume to 1.0L. Mix thoroughly. Gently heat and bring to boiling. Distribute into tubes or flasks. Autoclave for 15 min at 15 psi pressure–121°C. Swirl medium to resuspend starch. Pour into sterile Petri dishes or leave in tubes.

Use: For the cultivation and maintenance of *Bacillus subtilis.* Also used to detect amylase-producing microorganisms.

Bacillus pasteurii Agar

Composition per liter:

Urea	20.0g
Agar	15.0g
Peptone	5.0g
Meat extract	3.0g

pH 7.0 ± 0.2 at 25°C

Preparation of Medium: Add components to distilled/deionized water and bring volume to 1.0L. Gently heat and bring to boiling. Adjust pH to 7.0. Autoclave for 15 min at 15 psi pressure–121°C. Pour into sterile Petri dishes or distribute into sterile tubes.

Use: For the cultivation and maintenance of *Bacillus pasteurii* and *Sporosarcina ureae.*

Bacillus polymyxa Agar

Composition per liter:

Agar	20.0g
Starch, soluble	10.0g
Peptone	5.0g
Yeast extract	5.0g

pH 7.2 ± 0.2 at 25°C

Preparation of Medium: Add components to distilled/deionized water and bring volume to 1.0L. Mix thoroughly. Gently heat and bring to boiling. Distribute into tubes or flasks. Autoclave for 15 min at 15 psi pressure–121°C. Pour into sterile Petri dishes or leave in tubes.

Use: For the cultivation and maintenance of *Bacillus macerans, Bacillus polymyxa,* and *Bacillus thermoglucosidasius.*

Bacillus popilliae Medium

Composition per liter:

Yeast extract	10.0g
Acid hydrolysate of casein	7.95g
K_2HPO_4	3.0g
Beef extract	1.36g
Trehalose	1.0g
Starch	0.68g

pH 7.3 ± 0.1 at 25°C

Preparation of Medium: Add components to distilled/deionized water and bring to 1.0L. Mix thoroughly. Gently heat until dissolved. Do not overheat. Filter sterilize. Aseptically distribute into sterile tubes or flasks.

Use: For the cultivation of *Bacillus popilliae.*

Bacillus Pullulan Salts

Composition per liter:

Pullulan	2.5g
NaCl	1.0g
NH_4Cl	1.0g
KH_2PO_4	0.5g
$MgSO_4·7H_2O$	0.5g
Yeast extract	0.1g
$CaCl_2·2H_2O$	0.05g
Trace mineral solution	10.0mL
Vitamin solution	10.0mL

pH 6.0 ± 0.2 at 25°C

Trace Mineral Solution:

Composition per liter:

$CoCl_2·6H_2O$	0.2g
$FeSO_4·7H_2O$	0.13g
$ZnCl_2·2H_2O$	0.1g
$MnCl_2·4H_2O$	0.1g
$CaCl_2·2H_2O$	20.0mg
Na_2SeO_3	20.0mg
$Na_2WO_4·2H_2O$	20.0mg
$NaMoO_4·2H_2O$	1.0mg
H_3BO_3	0.5mg
$CuSO_4·5H_2O$	0.4mg
KI	0.1mg

Preparation of Trace Mineral Solution: Add components to distilled/deionized water and bring volume to 1.0L. Mix thoroughly.

Vitamin Solution:

Composition per liter:

Pyridoxine·HCl	10.0mg
Thiamine·HCl	5.0mg
Riboflavin	5.0mg
Nicotinic acid	5.0mg
Calcium pantothenate	5.0mg
p-Aminobenzoic acid	5.0mg
Thioctic acid	5.0mg
Biotin	2.0mg
Folic acid	2.0mg
Cyanocobalamin	0.1mg

Preparation of Vitamin Solution: Add components to distilled/deionized water and bring volume to 1.0L. Mix thoroughly. Filter sterilize.

Preparation of Medium: Add components, except vitamin solution, to distilled/deionized water and bring volume to 990.0mL. Mix thoroughly. Gently heat and bring to boiling. Adjust pH to 6.0. Autoclave for 15 min at 15 psi pressure–121°C. Cool to 25°C. Aseptically add sterile vitamin solution. Mix thoroughly. Aseptically distribute into sterile tubes or flasks.

Use: For the cultivation and maintenance of *Bacillus* species that can degrade pullulan.

Bacillus schlegelii Agar (LMG Medium 85)

Composition per liter:

Agar	30.0g
$Na_2HPO_4 \cdot 12\ H_2O$	9.0g
KH_2PO_4	1.5g
Sodium pyruvate	1.5g
NH_4Cl	1.0g
$MgSO_4 \cdot 7H_2O$	0.2g
$MnSO_4 \cdot H_2O$	10.0mg
$CaCl_2 \cdot 2H_2O$	10.0mg
Ferric ammonium citrate	5.0mg
Trace elements solution	3.0mL

pH 7.1 ± 0.2 at 25°C

Trace Elements Solution:

Composition per liter:

H_3BO_3	0.3g
$CoCl_2 \cdot 6H_2O$	0.2g
$ZnSO_4 \cdot 7H_2O$	0.1g
$Na_2MoO_4 \cdot 2H_2O$	30.0mg
$MnCl_2 \cdot 4H_2O$	30.0mg
$NiCl_2 \cdot 6H_2O$	20.0mg
$CuCl_2 \cdot 2H_2O$	10.0mg

Preparation of Trace Elements Solution: Add components to distilled/deionized water and bring volume to 1.0L. Mix thoroughly.

Preparation of Medium: Add components to distilled/deionized water and bring volume to 1.0L. Mix thoroughly. Distribute into tubes or flasks. Gently heat and bring to boiling. Distribute into tubes or flasks. Autoclave for 15 min at 15 psi pressure–121°C. Pour into sterile Petri dishes or leave in tubes.

Use: For the cultivation of *Bacillus schlegelii.*

Bacillus schlegelii Chemolithotrophic Growth Medium (DSMZ Medium 261)

Composition per liter:

$Na_2HPO_4 \cdot 2H_2O$	4.5g
KH_2PO_4	1.5g
NH_4Cl	1.0g
$MgSO_4 \cdot 7H_2O$	0.2g
$MnSO_4 \cdot 2H_2O$	0.01g
$CaCl_2 \cdot 2H_2O$	0.01g
Ferric ammonium citrate	5.0mg
Trace elements solution SL-6	3.0mL

pH 7.1 ± 0.2 at 25°C

Trace Elements Solution SL-6:

Composition per liter:

$MnCl_2 \cdot 4H_2O$	0.5g
H_3BO_3	0.3g
$CoCl_2 \cdot 6H_2O$	0.2g
$ZnSO_4 \cdot 7H_2O$	0.1g
$Na_2MoO_4 \cdot 2H_2O$	0.03g
$NiCl_2 \cdot 6H_2O$	0.02g
$CuCl_2 \cdot 2H_2O$	0.01g

Preparation of Trace Elements Solution SL-6: Add components to distilled/deionized water and bring volume to 1.0L. Mix thoroughly.

Preparation of Medium: Add components to distilled/deionized water and bring volume to 1.0L. Mix thoroughly. Adjust pH to 7.1. Distribute into tubes or flasks. Autoclave for 15 min at 15 psi pressure–121°C.

Use: For the chemolithotrophic cultivation of *Bacillus schlegelii.* Incubation is at 65°C under an atmosphere of 5% O_2, 10% CO_2, 45% H_2.

Bacillus schlegelii Heterotrophic Growth Medium (DSMZ Medium 260)

Composition per liter:

$Na_2HPO_4 \cdot 2H_2O$	4.5g
Na-Pyruvate	1.5g
KH_2PO_4	1.5g
NH_4Cl	1.0g
$MgSO_4 \cdot 7H_2O$	0.2g
$MnSO_4 \cdot 2H_2O$	0.01g
$CaCl_2 \cdot 2H_2O$	0.01g
Ferric ammonium citrate	5.0mg
Trace elements solution SL-6	3.0mL

pH 7.1 ± 0.2 at 25°C

Trace Elements Solution SL-6:

Composition per liter:

$MnCl_2 \cdot 4H_2O$	0.5g
H_3BO_3	0.3g
$CoCl_2 \cdot 6H_2O$	0.2g
$ZnSO_4 \cdot 7H_2O$	0.1g
$Na_2MoO_4 \cdot 2H_2O$	0.03g
$NiCl_2 \cdot 6H_2O$	0.02g
$CuCl_2 \cdot 2H_2O$	0.01g

Preparation of Trace Elements Solution SL-6: Add components to distilled/deionized water and bring volume to 1.0L. Mix thoroughly.

Preparation of Medium: Add components to distilled/deionized water and bring volume to 1.0L. Mix thoroughly. Adjust pH to 7.1. Distribute into tubes or flasks. Autoclave for 15 min at 15 psi pressure–121°C.

Use: For the heterotrophic cultivation of *Bacillus schlegelii.* Incubation is at 65°C.

Bacillus selenitireducens Medium (DSMZ Medium 968)

Composition per liter:

NaCl	90.0g
Na_2CO_3	10.6g
$NaHCO_3$	4.2g
Na-lactate	1.70g
$NaNO_3$	1.25g
Yeast extract	0.2g
K_2HPO_4	0.15g
$(NH_4)SO_4$	0.1g
KH_2PO_4	0.08g
$MgSO_4$	25.0mg
Resazurin	0.5mg
Cysteine solution	10.0mL
$Na_2S \cdot 9H_2O$ solution	10.0mL
Trace elements solution SL-10	1.0mL
Selenite-tungstate solution	1.0mL

pH 9.8 ± 0.2 at 25°C

Cysteine Solution:
Composition per 10.0mL:
L-Cysteine·HCl·H_2O 0.25g

Preparation of Cysteine Solution: Add L-cysteine·HCl·H_2O to distilled/deionized water and bring volume to 10.0mL. Mix thoroughly. Sparge with 100% N_2. Autoclave for 15 min at 15 psi pressure–121°C.

$Na_2S·9H_2O$ Solution:
Composition per 10.0mL:
$Na_2S·9H_2O$ 0.25g

Preparation of $Na_2S·9H_2O$ Solution: Add $Na_2S·9H_2O$ to distilled/deionized water and bring volume to 10.0mL. Sparge with N_2. Autoclave for 15 min at 15 psi pressure–121°C. Cool to 25°C.

Trace Elements Solution SL-10:
Composition per liter:
$FeCl_2·4H_2O$ 1.5g
$CoCl_2·6H_2O$ 190.0mg
$MnCl_2·4H_2O$ 100.0mg
$ZnCl_2$ 70.0mg
$Na_2MoO_4·2H_2O$ 36.0mg
$NiCl_2·6H_2O$ 24.0mg
H_3BO_3 .... 6.0mg
$CuCl_2·2H_2O$ 2.0mg
HCl (25% solution) 10.0mL

Preparation of Trace Elements Solution SL-10: Add $FeCl_2·4H_2O$ to 10.0mL of HCl solution. Mix thoroughly. Add distilled/deionized water and bring volume to 1.0L. Add remaining components. Mix thoroughly. Sparge with 80% N_2 + 20% CO_2. Autoclave for 15 min at 15 psi pressure–121°C.

Selenite-Tungstate Solution:
Composition per liter:
NaOH 0.5g
$Na_2WO_4·2H_2O$ 4.0mg
$Na_2SeO_3·5H_2O$ 3.0mg

Preparation of Selenite-Tungstate Solution: Add components to distilled/deionized water and bring volume to 1.0L. Mix thoroughly. Sparge with 100% N_2. Filter sterilize.

Preparation of Medium: Add components, except $NaHCO_3$, Na_2CO_3, cysteine solution, and $Na_2S·9H_2O$ solution, to distilled/deionized water and bring volume to 980.0mL. Mix thoroughly. Gently heat and bring to boiling. Boil for 3 min. Cool to room temperature while sparging with 80% N_2 gas. Add solid $NaHCO_3$ and Na_2CO_3. Adjust pH to 9.8. Distribute to anaerobe tubes or bottles. Autoclave for 15 min at 15 psi pressure–121°C. Cool to room temperature. Aseptically and anaerobically add per liter 10.0mL sterile cysteine solution and 10.0mL sterile $Na_2S·9H_2O$ solution. Mix thoroughly.

Use: For the cultivation of *Bacillus selenitireducens* and *Bacillus arseniciselenatis.*

Bacillus stearothermophilus Broth

Composition per liter:
Pancreatic digest of casein 10.0g
Yeast extract 5.0g
K_2HPO_4 .... 2.0g
pH 7.2 ± 0.2 at 25°C

Preparation of Medium: Add components to distilled/deionized water and bring volume to 1.0L. Mix thoroughly. Distribute into tubes or flasks. Autoclave for 15 min at 15 psi pressure–121°C.

Use: For the cultivation of *Bacillus stearothermophilus.*

Bacillus stearothermophilus Defined Broth

Composition per 100.0mL:
Mineral salts solution 10.0mL
Potassium phosphate buffer 5.0mL
L-Glutamate·HCl (1% solution) 4.0mL
L-Leucine (1% solution) 1.64mL
L-Lysine·HCl (1% solution) 1.4mL
L-Serine (1% solution) 1.4mL
L-Aspartate (1% solution) 1.3mL
L-Valine (1% solution) 1.26mL
Biotin (0.01% solution) 1.0mL
Glucose (20% solution) 1.0mL
L-Isoleucine (1% solution) 1.0mL
L-Proline (1% solution) 1.0mL
Nicotinic acid (0.01% solution) 1.0mL
Thiamine·HCl (0.01% solution) 1.0mL
L-Phenylalanine (1% solution) 0.86mL
L-Alanine (1% solution) 0.84mL
L-Threonine (1% solution) 0.84mL
L-Arginine·HCl (1% solution) 0.64mL
L-Tyrosine (1% solution) 0.56mL
L-Methionine (1% solution) 0.52mL
Glycine (1% solution) 0.5mL
L-Asparagine·H_2O (1% solution) 0.5mL
L-Cystine (1% solution) 0.5mL
L-Glutamine (1% solution) 0.5mL
L-Histidine·HCl·H_2O (1% solution) 0.42mL
L-Tryptophan (1% solution) 0.3mL
$CaCl_2$ (5% solution) 0.01mL
$FeCl_3·6H_2O$ (0.05% solution) 0.01mL
$MnCl_2$ (10mm solution) 0.01mL
$ZnSO_4·7H_2O$ (5% solution) 0.01mL
pH 7.3 ± 0.2 at 25°C

Mineral Salts Solution:
Composition per liter:
NaCl 10.0g
NH_4Cl 10.0g
$MgSO_4$ 4.0g

Preparation of Mineral Salts Solution: Add components to distilled/deionized water and bring volume to 1.0L. Mix thoroughly.

Potassium Phosphate Buffer:
Composition per 500.0mL:
K_2HPO_4 .... 125.0g
KH_2PO_4 .... 30.0g

Preparation of Potassium Phosphate Buffer: Add components to distilled/deionized water and bring volume to 500.0mL. Mix thoroughly.

Preparation of Medium: Add components to distilled/deionized water and bring volume to 100.0mL. Mix thoroughly. Filter sterilize.

Use: For the cultivation of *Bacillus stearothermophilus* in a chemically defined medium.

Bacillus thermoalcalophilus Medium

Composition per liter:
Yeast extract 10.0g
Sodium acetate 3.0g
KCl 1.8g
Na_2SO_4 .... 0.4g

K_2HPO_4 .... 0.3g
KH_2PO_4 .... 0.3g
$MgSO_4$ 0.2g
$FeSO_4$ 0.01g
pH 8.2 ± 0.2 at 25°C

Preparation of Medium: Add components to distilled/deionized water and bring volume to 1.0L. Mix thoroughly. Adjust pH to 8.2. Distribute into tubes or flasks. Autoclave for 15 min at 15 psi pressure–121°C.

Use: For the cultivation of *Bacillus thermoalcalophilus.*

Bacillus thermoantarcticus Medium

Composition per liter:
Yeast extract 6.0g
NaCl 3.0g
Soil extract 500.0mL
pH 5.6–5.8 at 25°C

Soil Extract:
Composition per liter:
Garden soil, air dried 400.0g

Preparation of Soil Extract: Pass 400.0g of air-dried garden soil through a coarse sieve. Add soil to 960.0mL of tap water. Mix thoroughly. Autoclave for 60 min at 15 psi pressure–121°C. Cool to room temperature. Allow residue to settle. Decant supernatant solution. Filter through Whatman filter paper. Distribute into bottles in 200.0mL volumes. Autoclave for 15 min at 15 psi pressure–121°C. Store at room temperature until clear.

Preparation of Medium: Add components to distilled/deionized water and bring volume to 1.0L. Mix thoroughly. Adjust pH to 5.6–5.8. Distribute into tubes or flasks. Autoclave for 15 min at 15 psi pressure–121°C.

Use: For the cultivation of *Bacillus thermoantarcticus.*

Bacillus thermoglucosidasius Agar

Composition per liter:
Agar 30.0g
Soluble starch 10.0g
Peptone 5.0g
Beef extract 3.0g
KH_2PO_4 .... 3.0g
Yeast extract 3.0g
pH 7.0 ± 0.2 at 25°C

Preparation of Medium: Add components to distilled/deionized water and bring volume to 1.0L. Gently heat and bring to boiling. Adjust pH to 7.0. Autoclave for 15 min at 15 psi pressure–121°C. Pour into sterile Petri dishes or distribute into sterile tubes.

Use: For the cultivation and maintenance of *Bacillus thermoglucosidasius.*

Bacillus thermoleovorans Medium

Composition per liter:
n-Heptadecane 1.0g
$(NH_4)_2HPO_4$ 1.0g
Yeast extract 1.0g
KCl 0.2g
$MgSO_4 \cdot 7H_2O$ 0.2g

Preparation of Medium: Add components to distilled/deionized water and bring volume to 1.0L. Mix thoroughly. Distribute into tubes or flasks. Autoclave for 15 min at 15 psi pressure–121°C.

Use: For the cultivation of *Bacillus thermoleovorans.*

Bacillus thuringiensis Medium

Composition per liter:
Glucose 3.0g
$(NH_4)_2SO_4$ 2.0g
Yeast extract 2.0g
$K_2HPO_4 \cdot 3H_2O$ 0.5g
$MgSO_4 \cdot 7H_2O$ 0.2g
$CaCl_2 \cdot 2H_2O$ 0.08g
$MnSO_4 \cdot 4H_2O$ 0.05g
pH 7.3 ± 0.2 at 25°C

Preparation of Medium: Add components to distilled/deionized water and bring volume to 1.0L. Mix thoroughly. Adjust pH to 7.3. Distribute into tubes or flasks. Autoclave for 15 min at 15 psi pressure–121°C.

Use: For the cultivation of *Bacillus thuringiensis.*

Bacillus tusciae Medium

Composition per liter:
$Na_2HPO_4 \cdot 2H_2O$ 2.9g
KH_2PO_4 .... 2.3g
NH_4Cl 1.0g
$MgSO_4 \cdot 7H_2O$ 0.5g
$NaHCO_3$ 0.5g
$Fe(NH_4)$ citrate 0.05g
$CaCl_2 \cdot 2H_2O$ 0.01g
$MnSO_4 \cdot H_2O$ 0.01g
Ferric ammonium citrate solution 20.0mL
Trace elements solution SL-6 .... 5.0mL
pH 4.0 ± 0.2 at 25°C

Ferric Ammonium Citrate Solution:
Composition per 20.0mL:
Ferric ammonium citrate 0.05g

Preparation of Ferric Ammonium Citrate Solution: Add ferric ammonium citrate to distilled/deionized water and bring volume to 20.0mL. Mix thoroughly. Autoclave for 15 min at 15 psi pressure–121°C.

Trace Elements Solution SL-6:
Composition per liter:
$MnCl_2 \cdot 4H_2O$ 0.5g
H_3BO_3 .... 0.3g
$CoCl_2 \cdot 6H_2O$ 0.2g
$ZnSO_4 \cdot 7H_2O$ 0.1g
$Na_2MoO_4 \cdot 2H_2O$ 0.03g
$NiCl_2 \cdot 6H_2O$ 0.02g
$CuCl_2 \cdot 2H_2O$ 0.01g

Preparation of Trace Elements Solution SL-6: Add components to distilled/deionized water and bring volume to 1.0L. Mix thoroughly. Autoclave for 15 min at 15 psi pressure–121°C.

Preparation of Medium: Add components, except ferric ammonium citrate solution and trace elements solution SL-6, to distilled/deionized water and bring volume to 975.0mL. Mix thoroughly. Gently heat and bring to boiling. Adjust pH to 4.0. Autoclave for 15 min at 15 psi pressure–121°C. Cool to 50°–55°C. Aseptically add 20.0mL of sterile ferric ammonium citrate solution and 5.0mL of sterile trace elements solution SL-6. Mix thoroughly. Aseptically distribute into sterile tubes or flasks. For chemolithotropic growth, incubate the culture under 2% O_2 + 10% CO_2 + 60% H_2 + 28% N_2.

Use: For the chemolithotrophic growth of *Bacillus tusciae.*

Bacillus tusciae Medium

Composition per liter:

Agar	15.0g
$Na_2HPO_4 \cdot 2H_2O$	2.9g
KH_2PO_4	2.3g
NH_4Cl	1.0g
$MgSO_4 \cdot 7H_2O$	0.5g
$NaHCO_3$	0.5g
$Fe(NH_4)$ citrate	0.05g
$CaCl_2 \cdot 2H_2O$	0.01g
$MnSO_4 \cdot H_2O$	0.01g
Ferric ammonium citrate solution	20.0mL
Carbon source	10.0mL
Trace elements solution SL-6	5.0mL

pH 4.0 ± 0.2 at 25°C

Ferric Ammonium Citrate Solution:
Composition per 20.0mL:

Ferric ammonium citrate	0.05g

Preparation of Ferric Ammonium Citrate Solution: Add ferric ammonium citrate to distilled/deionized water and bring volume to 20.0mL. Mix thoroughly. Autoclave for 15 min at 15 psi pressure–121°C.

Carbon Source:
Composition per 10.0mL:

Carbohydrate	2.0g
or organic acid	1.0g

Preparation of Carbon Source: Add carbohydrate or organic acid to distilled/deionized water and bring volume to 10.0mL. Mix thoroughly. Filter sterilize.

Trace Elements Solution SL-6:
Composition per liter:

$MnCl_2 \cdot 4H_2O$	0.5g
H_3BO_3	0.3g
$CoCl_2 \cdot 6H_2O$	0.2g
$ZnSO_4 \cdot 7H_2O$	0.1g
$Na_2MoO_4 \cdot 2H_2O$	0.03g
$NiCl_2 \cdot 6H_2O$	0.02g
$CuCl_2 \cdot 2H_2O$	0.01g

Preparation of Trace Elements Solution SL-6: Add components to distilled/deionized water and bring volume to 1.0L. Mix thoroughly. Autoclave for 15 min at 15 psi pressure–121°C.

Preparation of Medium: Add components, except ferric ammonium citrate solution, trace elements solution SL-6, and carbon source, to distilled/deionized water and bring volume to 965.0mL. Mix thoroughly. Gently heat and bring to boiling. Adjust pH to 4.0. Autoclave for 15 min at 15 psi pressure–121°C. Cool to 50°–55°C. Aseptically add 20.0mL of sterile ferric ammonium citrate solution, 10.0mL of sterile carbon source, and 5.0mL of sterile trace elements solution SL-6. Mix thoroughly. Aseptically distribute into sterile tubes or flasks.

Use: For the heterotrophic growth of *Bacillus tusciae.*

Bacterial Cell Agar (BCA)

Composition per liter:

Tryptose	17.36g
Agar	15.0g
NaCl	8.68g
Beef extract	5,2g
Yeast extract	1.7g

pH 7.3 ± 0.2 at 25°C

Preparation of Medium: Add components, except agar, to distilled/deionized water and bring volume to 1.0L. Mix thoroughly. Autoclave for 15 min at 15 psi pressure–121°C. Cool to 30°C. Inoculate with a culture of *Aeromonas hydrophila*. Incubate with shaking at 30°C for 72 hr. Centrifuge culture in 40.0mL volumes at 10,000 × g for 10 min. Wash the cells four times in sterile 0.85% saline. Resuspend the cell pellet in 25.0mL of distilled/deionized water. Autoclave for 15 min at 15 psi pressure–121°C. Cool to 45°–50°C. In a separate flask, add 15.0g of agar to 1.0L of distilled/deionized water. Mix thoroughly. Gently heat and bring to boiling. Autoclave for 15 min at 15 psi pressure–121°C. Cool to 45°–50°C. Aseptically combine 25.0mL of washed cells and 250.0mL of cooled, sterile agar solution. Mix thoroughly. Pour into sterile Petri dishes.

Use: For the cultivation of freshwater *Myxobacterium* species.

Bacteroides cellulosolvens Medium

Composition per liter:

Cellobiose or cellulose	5.0g
$NaHCO_3$	2.0g
NH_4Cl	0.68g
K_2HPO_4	0.3g
L-Cysteine·HCl·H_2O	0.25g
$Na_2S \cdot 9H_2O$	0.25g
KH_2PO_4	0.18g
$(NH_4)_2SO_4$	0.15g
$MgSO_4 \cdot 7H_2O$	0.12g
$CaCl_2 \cdot 2H_2O$	0.06g
$FeSO_4 \cdot 7H_2O$	0.02g
Resazurin	1.0mg
Trace elements solution	10.0mL
Vitamin solution	10.0mL

pH 7.0 ± 0.2 at 25°C

Trace Elements Solution:
Composition per liter:

$MgSO_4 \cdot 7H_2O$	3.0g
Nitrilotriacetic acid	1.5 g
$CaCl_2 \cdot 2H_2O$	1.0g
NaCl	1.0g
$MnSO_4 \cdot 2H_2O$	0.5 g
$CoSO_4 \cdot 7H_2O$	0.18 g
$ZnSO_4 \cdot 7H_2O$	0.18 g
$FeSO_4 \cdot 7H_2O$	0.1g
$NiCl_2 \cdot 6H_2O$	0.025 g
$KAl(SO_4)_2 \cdot 12H_2O$	0.02g
$CuSO_4 \cdot 5H_2O$	0.01g
H_3BO_3	0.01g
$Na_2MoO_4 \cdot 2H_2O$	0.01g
$Na_2SeO_3 \cdot 5H_2O$	0.3 mg

Preparation of Trace Elements Solution: Add nitrilotriacetic acid to approximately 500.0mL distilled/deionized water. Dissolve by adding KOH and adjust pH to 6.5. Add remaining components. Bring volume to 1.0L with additional distilled/deionized water. Adjust pH to 7.0 with KOH.

Vitamin Solution:
Composition per liter:

Pyridoxine·HCl	10.0mg
Calcium DL-pantothenate	5.0mg
Lipoic acid	5.0mg
Nicotinic acid	5.0mg
p-Aminobenzoic acid	5.0mg
Riboflavin	5.0mg
Thiamine·HCl	5.0mg

Biotin 2.0mg
Folic acid 2.0mg
Vitamin B_{12} 0.1mg

Preparation of Vitamin Solution: Add components to distilled/deionized water and bring volume to 1.0L. Mix thoroughly.

Cellobiose Solution:
Composition per 50.0mL:
D-Cellobiose (or cellulose) 5.0g

Preparation of Cellobiose Solution: Add cellobiose (or cellulose) to distilled/deionized water and bring volume to 50.0mL. Mix thoroughly. Filter sterilize.

Preparation of Medium: Add components, except cellobiose solution, to distilled/deionized water and bring volume to 950.0mL. Mix thoroughly. Autoclave for 15 min at 15 psi pressure–121°C. Aseptically add 50.0mL of sterile cellobiose (or cellulose) solution. Mix thoroughly. Aseptically distribute into sterile tubes or flasks.

Use: For the cultivation and maintenance of *Bacteroides cellulosolvens*.

Baird-Parker Agar

Composition per liter:
Agar 17.0g
Glycine 12.0g
Sodium pyruvate 10.0g
Pancreatic digest of casein 10.0g
Beef extract 5.0g
LiCl 5.0g
Yeast extract 1.0g
pH 7.0 ± 0.2 at 25°C

Source: This medium is available as a premixed powder from Oxoid Unipath and BD Diagnostic Systems.

Preparation of Medium: Add components to distilled/deionized water and bring volume to 1.0L. Mix thoroughly. Gently heat and bring to boiling. Autoclave for 15 min at 15 psi pressure–121°C. Cool to 45°–50°C. Pour into sterile Petri dishes.

Use: Used as a base for the preparation of egg-tellurite-glycine-pyruvate agar for the selective isolation and enumeration of coagulase-positive staphylococci from soil, air, and other materials.

Baird-Parker Agar, Supplemented

Composition per liter:
Agar 17.0g
Glycine 12.0g
Sodium pyruvate 10.0g
Pancreatic digest of casein 10.0g
Beef extract 5.0g
LiCl 5.0g
Yeast extract 1.0g
RPF supplement 100.0mL
pH 7.0 ± 0.2 at 25°C

RPF Supplement:
Composition per 100.0mL:
Bovine fibrinogen 3.75g
Trypsin inhibitor 25.0mg
K_2TeO_3 .... 25.0mg
Rabbit plasma 25.0mL

Caution: Potassium tellurite is toxic.

Preparation of RPF Supplement: Add components to distilled/deionized water and bring volume to 100.0mL. Mix thoroughly. Filter sterilize.

Preparation of Medium: Add components, except RPF supplement, to distilled/deionized water and bring volume to 900.0mL. Mix thoroughly. Gently heat and bring to boiling. Autoclave for 15 min at 15 psi pressure–121°C. Cool to 45°–50°C. Aseptically add 100.0mL of filter-sterilized RPF supplement. Mix thoroughly but gently. Pour into sterile Petri dishes.

Use: For the selective isolation and enumeration of coagulase-positive staphylococci from soil, air, and other materials. For the differentiation and identification of staphylococci on the basis of their ability to coagulate plasma. Colonies surrounded by an opaque zone of coagulated plasma are diagnostic for *Staphylococcus aureus*.

BAM Agar (ATCC Medium 1655)

Composition per liter:
Agar 30.0g
Glucose 5.0g
KH_2PO_4 .... 3.0g
Yeast extract 1.0g
$MgSO_4 \cdot 7H_2O$ 0.5g
$CaCl_2 \cdot 2H_2O$ 0.25g
$(NH_4)_2SO_4$ 0.2g
Trace elements 1.0mL
pH 4.0 ± 0.2 at 25°C

Trace Elements:
Composition per liter:
$CaCl_2 \cdot 2H_2O$ 0.66g
$Na_2MoO_4 \cdot 2H_2O$ 0.3g
$ZnSO_4 \cdot 7H_2O$ 0.18g
$CoCl_2 \cdot 6H_2O$ 0.18g
$CuSO_4 \cdot 5H_2O$ 0.16g
$MnSO_4 \cdot 4H_2O$ 0.15g
H_3BO_3 .... 0.1g

Preparation of Trace Elements: Add components to 1.0L of distilled/deionized water. Mix thoroughly.

Preparation of Medium: Add components, except agar, to distilled/deionized water and bring volume to 800.0mL. Mix thoroughly. Gently heat and bring to boiling. Adjust medium to pH 4.0 with H_2SO_4. Add agar to 200.0mL of distilled/deionized water. Autoclave agar separately to avoid acid hydrolysis. Autoclave for 15 min at 15 psi pressure–121°C. Mix the two solutions together. Pour into sterile Petri dishes or distribute into sterile tubes.

Use: For the cultivation and maintenance of *Bacillus acidoterrestris*.

BAM SM Agar (ATCC Medium 1656)

Composition per liter:
Agar 20.0g
Yeast extract 6.0g
Glucose 5.0g
KH_2PO_4 .... 3.0g
$MgSO_4 \cdot 7H_2O$ 0.5g
$CaCl_2 \cdot 2H_2O$ 0.25g
$(NH_4)_2SO_4$ 0.2g
Trace elements 1.0mL
pH 4.0 ± 0.2 at 25°C

Trace Elements:
Composition per liter:
$CaCl_2 \cdot 2H_2O$ 0.66g
$Na_2MoO_4 \cdot 2H_2O$ 0.3g
$ZnSO_4 \cdot 7H_2O$ 0.18g
$CoCl_2 \cdot 6H_2O$ 0.18g
$CuSO_4 \cdot 5H_2O$ 0.16g
$MnSO_4 \cdot 4H_2O$ 0.15g
H_3BO_3 .... 0.1g

Preparation of Trace Elements: Add components to 1.0L of distilled/deionized water. Mix thoroughly.

Preparation of Medium: Add components, except agar, to distilled/deionized water and bring volume to 800.0mL. Mix thoroughly. Gently heat and bring to boiling. Adjust medium to pH 4.0 with H_2SO_4. Add agar to 200.0mL of distilled/deionized water. Autoclave agar separately to avoid acid hydrolysis. Autoclave for 15 min at 15 psi pressure–121°C. Mix the two solutions together. Pour into sterile Petri dishes or distribute into sterile tubes.

Use: For the cultivation and maintenance of *Bacillus cycloheptanicus.*

BAM SM Agar, Modified

Composition per liter:
Agar 30.0g
Glucose 5.0g
KH_2PO_4 .... 3.0g
Yeast extract 1.0g
$MgSO_4 \cdot 7H_2O$ 0.5g
$CaCl_2 \cdot 2H_2O$ 0.25g
$(NH_4)_2SO_4$ 0.2g
Trace elements 1.0mL
pH 4.0 ± 0.2 at 25°C

Trace Elements:
Composition per liter:
$CaCl_2 \cdot 2H_2O$ 0.66g
$Na_2MoO_4 \cdot 2H_2O$ 0.30g
$ZnSO_4 \cdot 7H_2O$ 0.18g
$CoCl_2 \cdot 6H_2O$ 0.18g
$CuSO_4 \cdot 5H_2O$ 0.16g
$MnSO_4 \cdot 4H_2O$ 0.15g
H_3BO_3 .... 0.10g

Preparation of Trace Elements: Add components to 1.0L of distilled/deionized water. Mix thoroughly.

Preparation of Medium: Add components, except agar, to distilled/deionized water and bring volume to 800.0mL. Mix thoroughly. Gently heat and bring to boiling. Adjust medium to pH 4.0 with H_2SO_4. Add agar to 200.0mL of distilled/deionized water. Autoclave agar separately to avoid acid hydrolysis. Autoclave for 15 min at 15 psi pressure–121°C. Mix the two solutions together. Pour into sterile Petri dishes or distribute into sterile tubes.

Use: For the cultivation and maintenance of *Bacillus cycloheptanicus.*

Basal Mineral Medium

Composition per liter:
NH_4Cl 0.8g
K_2HPO_4 .... 0.7g
$MgSO_4 \cdot 7H_2O$ 0.01g
Disodium EDTA 9.2mg
$FeSO_4 \cdot 7H_2O$ 7.0mg
$CaSO4 \cdot 2H_2O$ 2.0mg
H_3BO_3 .... 0.1mg
$ZnSO_4 \cdot 7H_2O$ 0.1mg
$MnSO_4 \cdot 4H_2O$ 0.02mg
$Co(NO_3)_2$ 0.01mg
$NaMoO_4 \cdot 2H_2O$ 0.01mg
$CuSO_4 \cdot 5H_2O$ 0.5μg

Preparation of Medium: Add components to distilled/deionized water and bring volume to 1.0L. Mix thoroughly. Filter sterilize.

Use: For the cultivation of *Beggiatoa* species.

Basal Thermophile Medium

Composition per liter:
Solution 1 .... 850.0mL
Solution 2 .... 100.0mL
Solution 3 .... 50.0mL

Solution 1:
Composition per 850.0mL:
Pancreatic digest of casein 10.0g
K_2HPO_4 .... 1.5g
NH_4Cl 0.9g
KH_2PO_4 .... 0.75g
$MgCl_2 \cdot 6H_2O$ 0.2g
Trace elements solution 9.0mL
Vitamin solution 5.0mL
Resazurin (0.2% solution) 1.0mL
$FeSO_4 \cdot 7H_2O$ (10% solution) 0.03mL

Preparation of Solution 1: Add components to distilled/deionized water and bring volume to 850.0mL. Mix thoroughly. Autoclave for 45 min at 15 psi pressure–121°C. Cool to 45°–50°C.

Solution 2:
Composition per 100.0mL:
Yeast extract 3.0g

Preparation of Solution 2: Add yeast extract to distilled/deionized water and bring volume to 100.0mL. Mix thoroughly. Autoclave for 45 min at 15 psi pressure–121°C. Cool to 45°–50°C.

Solution 3:
Composition per 50.0mL:
Glucose 5.0g

Preparation of Solution 3: Add glucose to distilled/deionized water and bring volume to 50.0mL. Mix thoroughly. Autoclave for 45 min at 15 psi pressure–121°C. Cool to 45°–50°C.

Trace Elements Solution:
Composition per liter:
Nitrilotriacetic acid 12.5g
NaCl 1.0g
$FeCl_3 \cdot 4H_2O$ 0.2g
$MnCl_2 \cdot 4H_2O$ 0.1g
$CaCl_2 \cdot 2H_2O$ 0.1g
$ZnCl_2$ 0.1g
$CuCl_2$ 0.02g
Na_2SeO_3 .... 0.02g
$CoCl_2 \cdot 6H_2O$ 0.017g
H_3BO_3 .... 0.01g
$Na_2MoO_4 \cdot 2H_2O$ 0.01g

Preparation of Trace Elements Solution: Add nitrilotriacetic acid to 100.0mL of distilled/deionized water. Adjust pH to 6.5 with KOH. Add remaining components and bring volume to 1.0L. Mix thoroughly.

Wolfe's Vitamin Solution:
Composition per liter:
Pyridoxine·HCl 10.0mg

Thiamine·HCl 5.0mg
Riboflavin 5.0mg
Nicotinic acid 5.0mg
Calcium pantothenate 5.0mg
p-Aminobenzoic acid 5.0mg
Thioctic acid 5.0mg
Biotin 2.0mg
Folic acid 2.0mg
Cyanocobalamin 0.1mg

Preparation of Wolfe's Vitamin Solution: Add components to distilled/deionized water and bring volume to 1.0L. Mix thoroughly.

$Na_2S{\cdot}9H_2O$ Solution:
Composition per 100.0mL:
$Na_2S{\cdot}9H_2O$ 10.0g

Preparation of $Na_2S{\cdot}9H_2O$ Solution: Add $Na_2S{\cdot}9H_2O$ to distilled/deionized water and bring volume to 100.0mL. Mix thoroughly. Autoclave for 15 min at 15 psi pressure–121°C.

Preparation of Medium: Aseptically combine solution 1, solution 2, and solution 3 under 100% N_2. Distribute into tubes in 10.0mL volumes under 100% N_2. Immediately prior to inoculation, aseptically add 0.1mL of sterile $Na_2S{\cdot}9H_2O$ solution to each tube.

Use: For the cultivation and maintenance of *Clostridium* species, *Fervidobacterium nodosum*, and *Thermoanaerobium brockii.*

BC Medium
(Medium for *Acetivibrio cellulolyticus)*

Composition per liter:
Cellulose powder 3.0g
$NaHCO_3$ 2.0g
Mineral solution 1 .... 75.0mL
Mineral solution 2 .... 75.0mL
Cysteine-sulfide reducing solution 12.8mL
$FeSO_4{\cdot}7H_2O$ solution 10.0mL
Vitamin mixture 10.0mL
Wolfe's mineral solution 10.0mL
Resazurin (0.1% solution) 1.0mL
pH 7.6 ± 0.2 at 25°C

Caution: This medium contains sodium sulfide and may produce toxic H_2S gas. Prepare in a chemical fume hood.

Mineral Solution 1:
Composition per liter:
K_2HPO_4 .... 3.9g

Preparation of Mineral Solution 1: Add K_2HPO_4 to distilled/deionized water and bring volume to 1.0L. Mix thoroughly.

Mineral Solution 2:
Composition per liter:
NH_4Cl 12.0g
Na_2SO_4 .... 2.5g
KH_2PO_4 .... 2.4g
$MgSO_4{\cdot}7H_2O$ 1.2g
$CaCl_2{\cdot}2H_2O$ 0.8g

Preparation of Mineral Solution 2: Add components to distilled/deionized water and bring volume to 1.0L. Mix thoroughly.

$FeSO_4{\cdot}7H_2O$ Solution:
Composition per 100.0mL:
$FeSO_4{\cdot}7H_2O$ 0.2g

Preparation of $FeSO_4{\cdot}7H_2O$ Solution: Dissolve $FeSO_4{\cdot}7H_2O$ in 100.0mL of distilled/deionized water. Add three drops of concentrated HCl. Mix thoroughly.

Vitamin Mixture:
Composition per liter:
Pyridoxine·HCl 10.0mg
Thiamine·HCl 5.0mg
Cyanocobalamin 5.0mg
Lipoic acid (thioctic acid) 5.0mg
Biotin 2.0mg
p-Aminobenzoic acid 0.5mg

Preparation of Vitamin Mixture: Add components to distilled/deionized water and bring volume to 1.0L. Store below −20°C.

Wolfe's Mineral Solution:
Composition per liter
$MgSO_4{\cdot}7H_2O$ 3.0g
Nitriloacetic acid 1.5g
$MnSO_4{\cdot}H_2O$ 0.5g
NaCl 1.0g
$FeSO_4{\cdot}7H_2O$ 0.1g
$CoCl_2{\cdot}6H_2O$ 0.1g
$CaCl_2$ 0.1g
$ZnSO_4{\cdot}7H_2O$ 0.1g
$CuSO_4{\cdot}5H_2O$ 0.01g
$AlK(SO_4)_2{\cdot}12H_2O$ 0.01g
H_3BO_3 .... 0.01g
$Na_2MoO_4{\cdot}2H_2O$ 0.01g

Preparation of Wolfe's Mineral Solution: Add nitrilotriacetic acid to 500.0mL of distilled/deionized water and adjust to pH 6.5 with KOH to dissolve. Bring volume to 1.0L with distilled/deionized water. Add remaining components one at a time.

Cysteine-Sulfide Reducing Solution:
Composition per 200.0mL:
L-Cysteine·HCl·H_2O 2.5g
$Na_2S{\cdot}9H_2O$ 2.5g

Preparation of Cysteine-Sulfide Reducing Solution: Add L-cysteine·HCl·H_2O to 50.0mL of distilled/deionized water. Quickly adjust pH to 10 with fresh 3*N* NaOH and flush under 100% N_2. Add $Na_2S{\cdot}9H_2O$. Bring volume to 200.0mL with distilled/deionized water. Boil under 100% N_2. Transfer anaerobically to tubes or flasks and stopper. Autoclave for 15 min at 15 psi pressure–121°C.

Preparation of Medium: Add cellulose and $NaHCO_3$ to distilled/deionized water and bring volume to 800.0mL. Add all other components except cysteine-sulfide reducing solution. Heat and boil under 90% N_2 + 10% CO_2. Cool and continue flushing under 90% N_2 + 10% CO_2. The pH should be 7.6 at room temperature; do not adjust. Add 8.0mL of cysteine-sulfide reducing solution. Add 4.8mL more of cysteine-sulfide reducing solution. Distribute anaerobically into tubes in 7.0mL volumes and cap.

Use: For the cultivation and maintenance of *Acetivibrio cellulolyticus, Acetivibrio cellulosolvens, Bacteroides cellulosolvens,* and other cellulose-degrading microorganisms.

BCP Azide Broth
(Bromcresol Purple Azide Broth)

Composition per liter:
Casein peptone 10.0g
Yeast extract 10.0g

D-Glucose 5.0g
NaCl 5.0g
K_2HPO_4 .... 2.7g
KH_2PO_4 .... 2.7g
NaN_3 .... 0.5g
Bromcresol Purple 0.032g
pH 6.9 ± 0.2 at 25°C

Caution: Sodium azide is toxic. Azides also react with metals and disposal must be highly diluted.

Preparation of Medium: Add components distilled/deionized water to 1.0L. Mix thoroughly. Gently heat to boiling. Distribute into tubes or flasks. Autoclave for 15 min at 15 psi pressure–121°C.

Use: For use in the confirmation test for the presence of fecal streptococci in water and wastewater.

BCYE Agar (BCYE Alpha Base) (Buffered Charcoal Yeast Extract Agar)

Composition per liter:
Agar 15.0g
Yeast extract 10.0g
ACES buffer (2-[(2-Amino-2-oxoethyl)-amino]-ethane sulfonic acid) 10.0g
Charcoal, activated 2.0g
α-Ketoglutarate 1.0g
L-Cysteine·HCl·H_2O 0.4g
$Fe_4(P_2O_7)_3 \cdot 9H_2O$ 0.25g
pH 6.9 ± 0.2 at 25°C

Source: This medium is available as a premixed powder from BD Diagnostic Systems.

Preparation of Medium: Add components, except cysteine, to distilled/deionized water and bring volume to 1.0L. Mix thoroughly. Adjust medium to pH 6.9 with 1*N* KOH. Heat gently and bring to boil for 1 min. Autoclave for 15 min at 15 psi pressure–121°C. Cool to 50°–55°C. Add 4.0mL of a 10% solution of L-cysteine·HCl·H_2O that has been filter sterilized. Mix thoroughly. Pour into sterile Petri dishes with constant agitation to keep charcoal in suspension.

Use: For the isolation, cultivation, and maintenance of *Legionella pneumophila* and other *Legionella* species from environmental water samples.

BCYE Differential Agar (Buffered Charcoal Yeast Extract Differential Agar)

Composition per liter:
Agar 15.0g
Yeast extract 10.0g
ACES buffer (2-[(2-Amino-2-oxoethyl)-amino]-ethane sulfonic acid) 10.0g
Charcoal, activated 2.0g
α-Ketoglutarate 1.0g
L-Cysteine·HCl·H_2O 0.4g
$Fe_4(P_2O_7)_3 \cdot 9H_2O$ 0.25g
Bromcresol Purple 0.01g
Bromothymol Blue 0.01g
pH 6.9 ± 0.2 at 25°C

Source: This medium is available as a premixed powder from BD Diagnostic Systems.

Preparation of Medium: Add components, except L-cysteine·HCl·H_2O, to distilled/deionized water and bring volume to 1.0L. Mix thoroughly. Adjust medium to pH 6.9 with 1*N* KOH. Heat gently and bring to boil for 1 min. Autoclave for 15 min at 15 psi pressure–121°C. Cool to 50°–55°C. Add 4.0mL of a 10% solution of L-cysteine·HCl·H_2O that has been filter sterilized. Mix thoroughly. Pour into sterile Petri dishes with constant agitation to keep charcoal in suspension.

Use: For the isolation, cultivation, and maintenance of *Legionella pneumophila* and other *Legionella* species from environmental water samples. For the presumptive differential identification of *Legionella* species based on colony color and morphology. *Legionella pneumophila* appears as light blue/green colonies. *Legionella micdadei* appears as blue/gray or dark blue colonies.

BCYE Selective Agar with CCVC (Buffered Charcoal Yeast Extract Selective Agar with Cephalothin, Colistin, Vancomycin, and Cycloheximide)

Composition per 1014.0mL:
Agar 15.0g
Yeast extract 10.0g
ACES buffer (2-[(2-Amino-2-oxoethyl)-amino]-ethane sulfonic acid) 10.0g
Charcoal, activated 2.0g
α-Ketoglutarate 1.0g
$Fe_4(P_2O_7)_3 \cdot 9H_2O$ 0.25g
Antibiotic solution 10.0mL
Cysteine·HCl·H_2O solution 4.0mL
pH 6.9 ± 0.2 at 25°C

Source: This medium is available as a premixed powder from BD Diagnostic Systems.

L-Cysteine·HCl·H_2O Solution:

Composition per 10.0mL:
L-Cysteine·HCl·H_2O 1.0g

Preparation of L-Cysteine·HCl·H_2O Solution: Add L-cysteine·HCl·H_2O to distilled/deionized water and bring volume to 10.0mL. Mix thoroughly. Filter sterilize.

Antibiotic Solution:

Composition per 10.0mL:
Cycloheximide 80.0mg
Colistin 16.0mg
Cephalothin 4.0mg
Vancomycin 0.5mg

Preparation of Antibiotic Solution: Add components to distilled/deionized water and bring volume to 10.0mL. Mix thoroughly. Filter sterilize.

Preparation of Medium: Add components, except L-cysteine and antibiotic solutions, to distilled/deionized water and bring volume to 1.0L. Mix thoroughly. Adjust medium to pH 6.9 with 1*N* KOH. Heat gently and bring to boil for 1 min. Autoclave for 15 min at 15 psi pressure–121°C. Cool to 50°–55°C. Add 4.0mL of L-cysteine·HCl·H_2O solution and 10.0mL of sterile antibiotic solution. Mix thoroughly. Pour into sterile Petri dishes with constant agitation to keep charcoal in suspension.

Use: For the isolation, cultivation, and maintenance of *Legionella pneumophila* and other *Legionella* species from environmental samples. For the selective recovery of *Legionella pneumophila* while reducing contaminating microorganisms from environmental water samples.

BCYE Selective Agar with GPVA (Buffered Charcoal Yeast Extract Selective Agar with Glycine, Polymyxin B, Vancomycin, and Anisomycin)

Composition per 1014.0mL:

Agar	15.0g
Yeast extract	10.0g
ACES buffer (2-[(2-Amino-2-oxoethyl)-amino]-ethane sulfonic acid)	10.0g
Charcoal, activated	2.0g
α-Ketoglutarate	1.0g
$Fe_4(P_2O_7)_3 \cdot 9H_2O$	0.25g
Antibiotic solution	10.0mL
L-Cysteine·HCl·H_2O solution	4.0mL

pH 6.9 ± 0.2 at 25°C

L-Cysteine·HCl·H_2O Solution:
Composition per 10.0mL:

L-Cysteine·HCl·H_2O	1.0g

Preparation of L-Cysteine·HCl·H_2O Solution: Add L-cysteine·HCl·H_2O to distilled/deionized water and bring volume to 10.0mL. Mix thoroughly. Filter sterilize.

Antibiotic Solution:
Composition per 10.0mL:

Glycine	3.0g
Anisomycin	0.08g
Vancomycin	5.0mg
Polymyxin B	100,000U

Preparation of Antibiotic Solution: Add components to distilled/deionized water and bring volume to 10.0mL. Mix thoroughly. Filter sterilize.

Preparation of Medium: Add components, except L-cysteine·HCl·H_2O solution and antibiotic solution, to distilled/deionized water and bring volume to 1.0L. Mix thoroughly. Adjust medium to pH 6.9 with 1*N* KOH. Heat gently and bring to boil for 1 min. Autoclave for 15 min at 15 psi pressure–121°C. Cool to 50°–55°C. Add 4.0mL of L-cysteine·HCl·H_2O solution and 10.0mL of sterile antibiotic solution. Mix thoroughly. Pour into sterile Petri dishes with constant agitation to keep charcoal in suspension.

Use: For the isolation, cultivation, and maintenance of *Legionella pneumophila* and other *Legionella* species from environmental samples. For the selective recovery of *Legionella pneumophila* while reducing contaminating microorganisms from potable water samples.

BCYE Selective Agar with GVPC (Buffered Charcoal Yeast Extract Selective Agar with Glycine, Vancomycin, Polymyxin B, and Cycloheximide)

Composition per 1014.0mL:

Agar	15.0g
Yeast extract	10.0g
ACES buffer (2-[(2-Amino-2-oxoethyl)-amino]-ethane sulfonic acid)	10.0g
Charcoal, activated	2.0g
α-Ketoglutarate	1.0g
$Fe_4(P_2O_7)_3 \cdot 9H_2O$	0.25g
Antibiotic solution	10.0mL
L-Cysteine·HCl·H_2O solution	4.0mL

pH 6.9 ± 0.2 at 25°C

Source: This medium is available as a premixed powder from Oxoid Unipath.

L-Cysteine·HCl·H_2O Solution:
Composition per 10.0mL:

L-Cysteine·HCl·H_2O	1.0g

Preparation of L-Cysteine·HCl·H_2O Solution: Add L-cysteine·HCl·H_2O to distilled/deionized water and bring volume to 10.0mL. Mix thoroughly. Filter sterilize.

Antibiotic Solution:
Composition per 10.0mL:

Glycine	3.0g
Cycloheximide	0.08g
Vancomycin	1.0mg
Polymyxin B	79,200U

Preparation of Antibiotic Solution: Add components to distilled/deionized water and bring volume to 10.0mL. Mix thoroughly. Filter sterilize.

Preparation of Medium: Add components, except L-cysteine·HCl·H_2O solution and antibiotic solution, to distilled/deionized water and bring volume to 1.0L. Mix thoroughly. Adjust medium to pH 6.9 with 1*N* KOH. Heat gently and bring to boil for 1 min. Autoclave for 15 min at 15 psi pressure–121°C. Cool to 50°–55°C. Add 4.0mL of L-cysteine·HCl·H_2O solution and 10.0mL of sterile antibiotic solution. Mix thoroughly. Pour into sterile Petri dishes with constant agitation to keep charcoal in suspension.

Use: For the isolation, cultivation, and maintenance of *Legionella pneumophila* from environmental samples. For the selective recovery of *Legionella pneumophila* while reducing contaminating microorganisms from potable water samples.

BCYE Selective Agar with PAC (Buffered Charcoal Yeast Extract Selective Agar with Polymyxin B, Anisomycin, and Cefamandole)

Composition per 1014.0mL:

Agar	15.0g
Yeast extract	10.0g
ACES buffer (2-[(2-Amino-2-oxoethyl)-amino]-ethane sulfonic acid)	10.0g
Charcoal, activated	2.0g
α-Ketoglutarate	1.0g
$Fe_4(P_2O_7)_3 \cdot 9H_2O$	0.25g
Antibiotic solution	10.0mL
L-Cysteine·HCl·H_2O solution	4.0mL

pH 6.9 ± 0.2 at 25°C

Source: This medium is available as a premixed powder from BD Diagnostic Systems.

L-Cysteine·HCl·H_2O Solution:
Composition per 10.0mL:

L-Cysteine·HCl·H_2O	1.0g

Preparation of L-Cysteine·HCl·H_2O Solution: Add L-cysteine·HCl·H_2O to distilled/deionized water and bring volume to 10.0mL. Mix thoroughly. Filter sterilize.

Antibiotic Solution:
Composition per 10.0mL:

Polymyxin B	80,000 units
Anisomycin	80.0mg
Cefamandole	2.0mg

Preparation of Antibiotic Solution: Add components to distilled/deionized water and bring volume to 10.0mL. Mix thoroughly. Filter sterilize.

Preparation of Medium: Add components, except L-cysteine·HCl·H_2O solution and antibiotic solution, to distilled/deionized water and bring volume to 1.0L. Mix thoroughly. Adjust medium to pH 6.9 with 1*N* KOH. Heat gently and bring to boil for 1 min. Autoclave for 15 min at 15 psi pressure–121°C. Cool to 50°–55°C. Add 4.0mL of L-cysteine·HCl·H_2O solution and 10.0mL of sterile antibiotic solution. Mix thoroughly. Pour into sterile Petri dishes with constant agitation to keep charcoal in suspension.

Use: For the isolation, cultivation, and maintenance of *Legionella pneumophila* and other *Legionella* species from environmental samples. For the selective recovery of *Legionella pneumophila* while reducing contaminating microorganisms from potable water samples.

BCYE Selective Agar with PAV (Buffered Charcoal Yeast Extract Selective Agar with Polymyxin B, Anisomycin, and Vancomycin) (Wadowsky–Yee Medium)

Composition per 1014.0mL:

Agar 15.0g
Yeast extract 10.0g
ACES buffer (2-[(2-Amino-2-oxoethyl)-amino]-ethane sulfonic acid) 10.0g
Charcoal, activated 2.0g
α-Ketoglutarate 1.0g
$Fe_4(P_2O_7)_3·9H_2O$ 0.25g
Antibiotic solution 10.0mL
L-Cysteine·HCl·H_2O solution 4.0mL
pH 6.9 ± 0.2 at 25°C

Source: This medium is available as a premixed powder from BD Diagnostic Systems.

L-Cysteine·HCl·H_2O Solution:
Composition per 10.0mL:

L-Cysteine·HCl·H_2O 1.0g

Preparation of L-Cysteine·HCl·H_2O Solution: Add L-cysteine·HCl·H_2O to distilled/deionized water and bring volume to 10.0mL. Mix thoroughly. Filter sterilize.

Antibiotic Solution:
Composition per 10.0mL:

Polymyxin B .... 40,000 units
Anisomycin 80.0mg
Vancomycin 0.5mg

Preparation of Antibiotic Solution: Add components to distilled/deionized water and bring volume to 10.0mL. Mix thoroughly. Filter sterilize.

Preparation of Medium: Add components, except L-cysteine and antibiotic solution, to distilled/deionized water and bring volume to 1.0L. Mix thoroughly. Adjust medium to pH 6.9 with 1*N* KOH. Heat gently and bring to boil for 1 min. Autoclave for 15 min at 15 psi pressure–121°C. Cool to 50°–55°C. Add 4.0mL of L-cysteine·HCl·H_2O solution and 10.0mL of sterile antibiotic solution. Mix thoroughly. Pour into sterile Petri dishes with constant agitation to keep charcoal in suspension.

Use: For the isolation, cultivation, and maintenance of *Legionella pneumophila* and other *Legionella* species from environmental samples. For the selective recovery of *Legionella pneumophila* while reducing contaminating microorganisms from potable water samples.

BCYEα with Alb (Buffered Charcoal Yeast Extract Agar with Albumin)

Composition per liter:

Agar 15.0g
Yeast extract 10.0g
ACES buffer (2-[(2-Amino-2-oxoethyl)-amino]-ethane sulfonic acid) 10.0g
Charcoal, activated 2.0g
α-Ketoglutarate 1.0g
Bovine serum albumin solution 10.0mL
L-Cysteine·HCl·H_2O solution 10.0mL
$Fe_4(P_2O_7)_3·9H_2O$ solution 10.0mL
pH 6.9 ± 0.2 at 25°C

Bovine Serum Albumin Solution:
Composition per 10.0mL:

Bovine serum albumin 0.1g

Preparation of Bovine Serum Albumin Solution: Add bovine serum albumin to distilled/deionized water and bring volume to 10.0mL. Mix thoroughly. Filter sterilize.

L-Cysteine·HCl·H_2O Solution:
Composition per 10.0mL:

L-Cysteine·HCl·H_2O 0.4g

Preparation of L-Cysteine·HCl·H_2O Solution: Add L-cysteine·HCl·H_2O to distilled/deionized water and bring volume to 10.0mL. Mix thoroughly. Filter sterilize.

$Fe_4(P_2O_7)_3·9H_2O$ Solution:
Composition per 10.0mL:

$Fe_4(P_2O_7)_3·9H_2O$ 0.25g

Preparation of $Fe_4(P_2O_7)_3·9H_2O$ Solution: Add $Fe_4(P_2O_7)_3·9H_2O$ to distilled/deionized water and bring volume to 10.0mL. Mix thoroughly. Filter sterilize.

Preparation of Medium: Add components—except L-cysteine·HCl·H_2O solution, $Fe_4(P_2O_7)_3·9H_2O$ solution, and bovine serum albumin solution—to distilled/deionized water and bring volume to 970.0mL. Mix thoroughly. Adjust medium to pH 6.9 with 1*N* KOH. Heat gently and bring to boil for 1 min. Autoclave for 15 min at 15 psi pressure–121°C. Cool to 50°–55°C. Aseptically add the L-cysteine·HCl·H_2O solution, $Fe_4(P_2O_7)_3·9H_2O$ solution, and 10.0mL of sterile bovine serum albumin solution. Mix thoroughly. Pour into sterile Petri dishes with constant agitation to keep charcoal in suspension.

Use: For the isolation, cultivation, and maintenance of *Legionella pneumophila* and other *Legionella* species from environmental samples.

BCYEα without L-Cysteine (Buffered Charcoal Yeast Extract Agar without L-Cysteine)

Composition per liter:

Agar 15.0g
Yeast extract 10.0g
ACES buffer (2-[(2-Amino-2-oxoethyl)-amino]-ethane sulfonic acid) 10.0g
Charcoal, activated 2.0g
α-Ketoglutarate 1.0g
$Fe_4(P_2O_7)_3·9H_2O$ solution 10.0mL
pH 6.9 ± 0.2 at 25°C

$Fe_4(P_2O_7)_3·9H_2O$ Solution:
Composition per 10.0mL:

$Fe_4(P_2O_7)_3·9H_2O$ 0.25g

Preparation of $Fe_4(P_2O_7)_3·9H_2O$ Solution: Add $Fe_4(P_2O_7)_3·9H_2O$ to distilled/deionized water and bring volume to 10.0mL. Mix thoroughly. Filter sterilize.

Preparation of Medium: Add components, except $Fe_4(P_2O_7)_3·9H_2O$ solution, to distilled/deionized water and bring volume to 990.0mL. Mix thoroughly. Adjust medium to pH 6.9 with 1N KOH. Heat gently and bring to boil for 1 min. Autoclave for 15 min at 15 psi pressure–121°C. Cool to 50°–55°C. Aseptically add 10.0mL of sterile $Fe_4(P_2O_7)_3 \cdot 9H_2O$ solution. Mix thoroughly. Pour into sterile Petri dishes with constant agitation to keep charcoal in suspension.

Use: For the isolation, cultivation, and maintenance of *Legionella pneumophila* and other *Legionella* species from environmental samples.

Bdellovibrio Medium

Composition per Petri dish:

Base layer agar	10.0mL
Semisolid agar	10.0mL
Host medium	1.0mL

Host Medium:

Composition per liter:

Yeast extract	3.0g
Peptone	0.6g

Preparation of Host Medium: Add components to distilled/deionized water and bring volume to 1.0L. Mix thoroughly. Adjust pH to 7.2 Distribute into tubes in 10.0mL volumes. Autoclave for 15 min at 15 psi pressure–121°C.

Base Layer Agar:

Composition per liter:

Agar	19.0g
Yeast extract	3.0g
Peptone	0.6g

Preparation of Base Layer Agar: Add components to distilled/deionized water and bring volume to 1.0L. Mix thoroughly. Gently heat and bring to boiling. Adjust pH to 7.2 Distribute into tubes in 10.0mL volumes. Autoclave for 15 min at 15 psi pressure–121°C.

Semisolid Agar:

Composition per liter:

Agar	6.0g
Yeast extract	3.0g
Peptone	0.6g

Preparation of Semisolid Agar: Add components to distilled/deionized water and bring volume to 1.0L. Mix thoroughly. Gently heat and bring to boiling. Adjust pH to 7.2 Distribute into tubes in 10.0mL volumes. Autoclave for 15 min at 15 psi pressure–121°C.

Preparation of Medium: Inoculate appropriate bacterial host into 10.0mL of host medium. Hosts include *Erwinia amylovora, Escherichia coli, Serratia marcescens*, or *Pseudomonas putida*. Incubate host culture for 24–48 hr at 30°C. Melt the base layer agar and semisolid agar. Pour the base layer agar into a sterile Petri dish. Allow base layer agar to solidify. Cool the semisolid agar to 40°–45°C. Add 1.0mL of the previously grown host culture. Mix thoroughly. Pour over the solidified base layer agar.

Use: For the cultivation of *Bdellovibrio bacteriovorus* and *Bdellovibrio starrii.*

Beef Extract Agar

Composition per liter:

Agar	15.0g
Peptone	5.0g
Beef extract	3.0g

pH 7.4 ± 0.2 at 25°C

Preparation of Medium: Add components to distilled/deionized water and bring volume to 1.0L. Mix thoroughly. Heat gently and bring to boiling. Distribute into tubes or flasks. Autoclave for 15 min at 15 psi pressure–121°C. Pour into Petri dishes or leave in tubes.

Use: For the cultivation and maintenance of a wide variety of microorganisms. Recommended for the culture of microorganisms from water.

Beef Extract Broth

Composition per liter:

Peptone	5.0g
Beef extract	3.0g

pH 7.4 ± 0.2 at 25°C

Preparation of Medium: Add components to distilled/deionized water and bring volume to 1.0L. Mix thoroughly. Heat gently and bring to boiling. Distribute into tubes or flasks. Autoclave for 15 min at 15 psi pressure–121°C.

Use: For the cultivation and maintenance of a wide variety of microorganisms. Recommended for the culture of microorganisms from water.

Beggiatoa and *Thiothrix* Medium

Composition per liter:

$CaSO_4 \cdot 2H_2O$ (saturated solution)	20.0mL
NH_4Cl (4% solution)	5.0mL
Trace elements	5.0mL
K_2HPO_4 (1% solution)	1.0mL
$MgSO_4 \cdot 7H_2O$ (1% solution)	1.0mL

Trace Elements:

Composition per liter:

EDTA solution	20.0mL
$Co(NO_3)_2$ (0.01% solution)	10.0mL
$CuSO_4 \cdot 5H_2O$ (0.00005% solution)	10.0mL
H_3BO_3 (0.1% solution)	10.0mL
$MnSO_4 \cdot 4H_2O$ (0.02% solution)	10.0mL
$Na_2MoO_4 \cdot 2H_2O$ (0.01% solution)	10.0mL
$ZnSO_4 \cdot 7H_2O$ (0.1% solution)	10.0mL

Preparation of Trace Elements: Add components to distilled/deionized water and bring volume to 1.0L. Mix thoroughly.

EDTA Solution:

Composition per 100.0mL:

$FeSO_4$	7.0g
EDTA	2.0g
HCl, concentrated	1.0mL

Preparation of EDTA Solution: Add EDTA and $FeSO_4$ to concentrated HCl. Mix thoroughly. Carefully add to distilled/deionized water and bring volume to 100.0mL.

Preparation of Medium: Add components to distilled/deionized water and bring volume to 1.0L. Mix thoroughly. Distribute into tubes or flasks. Autoclave for 15 min at 15 psi pressure–121°C.

Use: For the cultivation of *Beggiatoa* species and myxotrophic *Thiothrix* species.

Beggiatoa Medium (ATCC Medium 138)

Composition per liter:

Yeast extract	2.0g
Agar	2.0g
Sodium acetate	0.5g
$CaCl_2$	0.1g
Catalase	10,000U

pH 7.2 ± 0.2 at 25°C

Preparation of Medium: Add components, except catalase, to tap water and bring volume to 1.0L. Mix thoroughly. Autoclave for 15 min at 15 psi pressure–121°C. Cool to 45°–50°C. Aseptically add 10,000 units of sterile catalase.

Use: For the cultivation and maintenance of *Beggiatoa alba* and *Vitreoscilla* species.

Beggiatoa Medium (ATCC Medium 1193)

Composition per liter:

Sodium sulfide	0.5g
Sodium acetate	0.01g
Yeast extract	0.01g
Nutrient broth	0.01g

pH 7.5 ± 0.2 at 25°C

Preparation of Medium: Add components to distilled/deionized water and bring volume to 1.0L. Mix thoroughly. Autoclave for 15 min at 15 psi pressure–121°C. Distribute into tubes or flasks.

Use: For the cultivation of *Beggiatoa alba.*

Beijerinckia Agar

Composition per liter:

Agar	15.0g
K_2HPO_4	0.8g
KH_2PO_4	0.2g
$MgSO_4 \cdot 7H_2O$	0.1g
$FeSO_4 \cdot 7H_2O$	20.0mg
$Na_2MoO_4 \cdot 2H_2O$	5.0mg
$ZnSO_4 \cdot 6H_2O$	5.0mg
$CuSO_4 \cdot 6H_2O$	4.0mg
$MnSO_4 \cdot 6H_2O$	2.0mg
Glucose solution	50.0mL

pH 6.5 ± 0.2 at 25°C

Glucose Solution:
Composition per 50.0mL:

D-Glucose	10.0g

Preparation of Glucose Solution: Add glucose to distilled/deionized water and bring volume to 50.0mL. Mix thoroughly. Autoclave for 15 min at 15 psi pressure–121°C.

Preparation of Medium: Add components, except glucose solution, to distilled/deionized water and bring volume to 950.0mL. Mix thoroughly. Gently heat and bring to boiling. Autoclave for 15 min at 15 psi pressure–121°C. Aseptically add 10.0mL of sterile glucose solution. Mix thoroughly. Pour into sterile Petri dishes or distribute into sterile tubes

Use: For the cultivation and maintenance of *Beijerinckia derxii, Beijerinckia fluminensis, Beijerinckia indica, Beijerinckia mobilis, Beijerinckia* species, and *Clostridium barkeri.*

Beijerinckia Medium

Composition per liter:

Glucose	20.0g
KH_2PO_4	1.0g
$MgSO_4 \cdot 7H_2O$	0.5g

pH 5.0 ± 0.2 at 25°C

Preparation of Medium: Add components to distilled/deionized water and bring volume to 1.0L. Mix thoroughly. Distribute into tubes or flasks. Autoclave for 15 min at 15 psi pressure–121°C.

Use: For the cultivation of *Beijerinckia* species.

Beijerinckia Medium, Modified

Composition per liter:

Agar	15.0g
Glucose	10.0g
K_2HPO_4	0.8g
KH_2PO_4	0.2g
$MgSO_4 \cdot 7H_2O$	0.1g
$FeSO_4 \cdot 7H_2O$	20.0mg
$MnSO_4 \cdot H_2O$	1.3mg
$ZnSO_4 \cdot 7H_2O$	5.0mg
$CuSO_4 \cdot 5H_2O$	4.0mg
$Na_2MoO_4 \cdot 2H_2O$	5.0mg

pH 6.5 ± 0.2 at 25°C

Preparation of Medium: Add components to distilled/deionized water and bring volume to 1.0L. Mix thoroughly. Gently heat and bring to boiling. Distribute into tubes or flasks. Autoclave for 15 min at 15 psi pressure–121°C. Pour into sterile Petri dishes or leave in tubes.

Use: For the isolation and cultivation of *Beijerinckia derxii, Beijerinckia fluminensis, Beijerinckia indica,* and *Beijerinckia mobilis.*

Beijerinck's *Thiobacillus* Medium

Composition per liter:

Noble agar	20.0g
Na_2HPO_4	0.2g
$MgCl_2$	0.1g
NH_4Cl	0.1g
$Na_2S_2O_3$ solution	100.0mL
$NaHCO_3$ solution	10.0mL

pH 7.0–7.2 at 25°C

$Na_2S_2O_3$ Solution:
Composition per 100.0mL:

$Na_2S_2O_3$	5.0g

Preparation of $Na_2S_2O_3$ Solution: Add $Na_2S_2O_3$ to distilled/deionized water and bring volume to 100.0mL. Mix thoroughly. Filter sterilize.

$NaHCO_3$ Solution:
Composition per 10.0mL:

$NaHCO_3$	1.0g

Preparation of $NaHCO_3$ Solution: Add $NaHCO_3$ to distilled/deionized water and bring volume to 10.0mL. Mix thoroughly. Filter sterilize.

Preparation of Medium: Add components, except $Na_2S_2O_3$ solution and $NaHCO_3$ solution, to distilled/deionized water and bring volume to 890.0mL. Mix thoroughly. Autoclave for 15 min at 15 psi pressure–121°C. Aseptically add 100.0mL of sterile $Na_2S_2O_3$ solution and 10.0mL of sterile $NaHCO_3$ solution. Mix thoroughly. Pour into sterile Petri dishes or leave in tubes.

Use: For the cultivation and maintenance of *Thiobacillus thermophilica.*

Benzoate Medium

Composition per liter:

Noble agar	20.0g
NaCl	5.0g
$(NH_4)_2HPO_4$	3.0g
Sodium benzoate	3.0g
KH_2PO_4	1.2g
Yeast extract	0.5g
$MgSO_4 \cdot 7H_2O$	0.2g
Benzoate solution	25.0mL

Benzoate Solution:
Composition per 25.0mL:
Sodium benzoate 3.0g

Preparation of Benzoate Solution: Add sodium benzoate to distilled/deionized water and bring volume to 25.0mL. Mix thoroughly. Filter sterilize.

Preparation of Medium: Add components except benzoate solution to distilled/deionized water and bring volume to 975.0mL. Mix thoroughly. Heat gently to boiling. Autoclave for 15 min at 15 psi pressure–121°C. Cool to 45°–50°C. Aseptically add 25.0mL sterile benzoate solution. Mix thoroughly and pour into sterile Petri dishes or leave in tubes.

Use: For the cultivation of *Pseudomonas putida* and other microorganisms which can utilize benzoate as a carbon source.

Benzoate Medium II

Composition per 1.5L:
Noble agar 30.0g
$(NH_4)_2HPO_4$ 3.0g
NaCl 1.67g
KH_2PO_4 .... 1.2g
Yeast extract 0.5g
$MgSO_4 \cdot 7H_2O$ 0.2g
$FeSO_4 \cdot 7H_2O$ 0.1g
Benzoate solution 25.0mL

Benzoate Solution:
Composition per 25.0mL:
Sodium benzoate 1.0g

Preparation of Benzoate Solution: Add sodium benzoate to distilled/deionized water and bring volume to 25.0mL. Mix thoroughly. Filter sterilize.

Preparation of Medium: Add components, except agar and sodium benzoate, to distilled/deionized water and bring volume to 600.0mL. Mix thoroughly. Autoclave for 15 min at 15 psi pressure–121°C. Cool to 45°–50°C. In a separate flask, add agar to distilled/deionized water and bring volume to 375.0mL. Mix thoroughly. Gently heat and bring to boiling. Autoclave for 15 min at 15 psi pressure–121°C. Cool to 45°–50°C. Aseptically combine the two autoclave-sterilized solutions. Mix thoroughly. Aseptically add the sterile benzoate solution. Mix thoroughly. Pour into sterile Petri dishes or leave in tubes.

Use: For the cultivation of *Pseudomonas putida* and other microorganisms that can utilize benzoate as a carbon source.

Benzoate Minimal Salts Medium

Composition per liter:
K_2HPO_4 .... 10.0g
$NaNH_4HPO_4 \cdot 4H_2O$ 3.5g
$MgSO_4 \cdot 7H_2O$ 0.2g
Citric acid, anhydrous 0.2g
Benzoate solution 25.0mL
pH 7.0 ± 0.2 at 25°C

Benzoate Solution:
Composition per 25.0mL:
Sodium benzoate 2.5g

Preparation of Benzoate Solution: Add sodium benzoate to distilled/deionized water and bring volume to 25.0mL. Mix thoroughly. Filter sterilize.

Preparation of Medium: Add components to distilled/deionized water and bring volume to 950.0mL. Mix thoroughly. Adjust pH to 7.0. Autoclave for 15 min at 15 psi pressure–121°C. Cool to 45°C. Aseptically add 25.0mL of sterile benzoate solution. Mix thoroughly. Aseptically distribute into sterile tubes or flasks.

Use: For the cultivation of microorganisms that can utilize benzoate as a carbon source.

Benzoate Nitrate Salts Medium (BNS)

Composition per liter:
Solution A 700.0mL
Solution B 300.0mL
pH 8.2 ± 0.2 at 25°C

Solution A:
Composition per 700.0mL:
KNO_3 .... 2.0g
Sodium benzoate 1.0g
NH_4Cl 0.3g
Phosphate buffer solution 200.0mL

Phosphate Buffer Solution:
Composition per 200.0mL:
K_2HPO_4 .... 5.12g
KH_2PO_4 .... 1.5g

Preparation of Phosphate Buffer: Add components to distilled/deionized water and bring volume to 200.0mL. Mix thoroughly. Adjust pH to 9.0 with KOH.

Preparation of Solution A: Add components to distilled/deionized water and bring volume to 700.0mL. Mix thoroughly. Autoclave for 15 min at 15 psi pressure–121°C. Cool to room temperature.

Solution B:
Composition per 300.0mL:
$MgSO_4 \cdot 7H_2O$ 0.2g
$CaCl_2$ 10.0mg
Trace metals solution 1.0mL

Preparation of Solution B: Add components to distilled/deionized water and bring volume to 300.0mL. Mix thoroughly. Autoclave for 15 min at 15 psi pressure–121°C. Cool to room temperature.

Trace Metals Solution:
Composition per 300.0mL:
$MnSO_4 \cdot H_2O$ 50.0mg
$ZnSO_4 \cdot 7H_2O$ 50.0mg
$Co(NO_3)_2 \cdot 6H_2O$ 10.0mg
$CuSO_4$ 10.0mg
$Na_2B_4O_7 \cdot 10H_2O$ 10.0mg
$Na_2MoO_4 \cdot 2H_2O$ 0.2mg
Ferric EDTA solution 10.0mL

Preparation of Trace Metals Solution: Add components to distilled/deionized water and bring volume to 1.0L. Mix thoroughly.

Ferric EDTA Solution:
Composition per 550.0mL:
EDTA 17.9g
$FeSO_4 \cdot 7H_2O$ 13.7g
KOH 3.23g

Preparation of Ferric EDTA Solution: Add EDTA and KOH to distilled/deionized water and bring volume to 186.0mL. Mix thoroughly. In a separate flask, add $FeSO_4 \cdot 7H_2O$ to distilled/deionized water and bring volume to 364.0mL. Mix thoroughly. Combine the two solutions. Sparge with air overnight to oxidize the Fe^{2+} to Fe^{3+}. Store in the dark.

Preparation of Medium: Aseptically combine 700.0mL of sterile solution A with 300.0mL of sterile solution B. Adjust pH to 8.2. Aseptically distribute into sterile screw-capped tubes. Fill tubes completely.

Use: For the cultivation of *Alcaligenes xylosoxydans*.

BG 11 Agar (Medium BG 11 for Cyanobacteria)

Composition per liter:

Agar	10.0g
$NaNO_3$	1.5g
$MgSO_4 \cdot 7H_2O$	0.075g
K_2HPO_4	0.04g
$CaCl_2 \cdot 2H_2O$	0.036g
Na_2CO_3	0.02g
Citric acid	6.0mg
Ferric ammonium citrate	6.0mg
Disodium EDTA	1.0mg
Trace metal mix A5	1.0mL

pH 7.1 ± 0.2 at 25°C

Trace Metal Mix A5:

Composition per liter:

H_3BO_3	2.86g
$MnCl_2 \cdot 4H_2O$	1.81g
$Na_2MoO_4 \cdot 2H_2O$	0.39g
$ZnSO_4 \cdot 7H_2O$	0.222g
$CuSO_4 \cdot 5H_2O$	0.079g
$Co(NO_3)_2 \cdot 6H_2O$	0.049g

Preparation of Trace Metal Mix A5: Add components to distilled/deionized water and bring volume to 1.0L. Mix thoroughly.

Preparation of Medium: Add components to distilled/deionized water and bring volume to 1.0L. Mix thoroughly. Heat gently to boiling. Distribute into tubes or flasks. Autoclave for 15 min at 15 psi pressure–121°C. For solid medium, pour into sterile Petri dishes or leave in tubes.

Use: For the cultivation and maintenance of a wide variety of cyanobacteria.

BG 11 Marine Agar (Medium BG 11 for Marine Cyanobacteria)

Composition per liter:

Agar	10.0g
NaCl	10.0g
$NaNO_3$	1.5g
$MgSO_4 \cdot 7H_2O$	0.075g
K_2HPO_4	0.04g
$CaCl_2 \cdot 2H_2O$	0.036g
Na_2CO_3	0.02g
Citric acid	6.0mg
Ferric ammonium citrate	6.0mg
EDTA disodium salt	1.0mg
Vitamin B_{12} solution	100.0mL
Trace metal mix A5	1.0mL

pH 7.1 ± 0.2 at 25°C

Trace Metal Mix A5:

Composition per liter:

H_3BO_3	2.86g
$MnCl_2 \cdot 4H_2O$	1.81g
$Na_2MoO_4 \cdot 2H_2O$	0.39g
$ZnSO_4 \cdot 7H_2O$	0.222g
$CuSO_4 \cdot 5H_2O$	0.079g
$Co(NO_3)_2 \cdot 6H_2O$	0.049g

Preparation of Trace Metal Mix A5: Add components to distilled/deionized water and bring volume to 1.0L. Mix thoroughly.

Vitamin B_{12} Solution:

Composition per 100.0mL:

Vitamin B_{12}	1.0μg

Preparation of Vitamin B_{12} Solution: Add Vitamin B_{12} to distilled/deionized water and bring volume to 100.0mL. Mix thoroughly. Filter sterilize.

Preparation of Medium: Add components, except Vitamin B_{12} solution, to distilled/deionized water and bring volume to 900.0mL. Mix thoroughly. Heat gently to boiling. Autoclave for 15 min at 15 psi pressure–121°C. Aseptically add 100.0mL of sterile Vitamin B_{12} solution. Mix thoroughly. Pour into sterile Petri dishes or leave in tubes.

Use: For the cultivation and maintenance of *Synechococcus* species. For the isolation of cyanobacteria from freshwater habitats.

BG 11 Marine Broth (Medium BG 11 for Marine Cyanobacteria)

Composition per liter:

NaCl	10.0g
$NaNO_3$	1.5g
$MgSO_4 \cdot 7H_2O$	0.075g
K_2HPO_4	0.04g
$CaCl_2 \cdot 2H_2O$	0.036g
Na_2CO_3	0.02g
Citric acid	6.0mg
Ferric ammonium citrate	6.0mg
EDTA disodium salt	1.0mg
Vitamin B_{12} solution	100.0mL
Trace metal mix A5	1.0mL

pH 7.1 ± 0.2 at 25°C

Trace Metal Mix A5:

Composition per liter:

H_3BO_3	2.86g
$MnCl_2 \cdot 4H_2O$	1.81g
$Na_2MoO_4 \cdot 2H_2O$	0.39g
$ZnSO_4 \cdot 7H_2O$	0.222g
$CuSO_4 \cdot 5H_2O$	0.079g
$Co(NO_3)_2 \cdot 6H_2O$	0.049g

Preparation of Trace Metal Mix A5: Add components to distilled/deionized water and bring volume to 1.0L. Mix thoroughly.

Vitamin B_{12} Solution:

Composition per 100.0mL:

Vitamin B_{12}	1.0μg

Preparation of Vitamin B_{12} Solution: Add Vitamin B_{12} to distilled/deionized water and bring volume to 100.0mL. Mix thoroughly. Filter sterilize.

Preparation of Medium: Add components, except Vitamin B_{12} solution, to distilled/deionized water and bring volume to 900.0mL. Mix thoroughly. Heat gently to boiling. Autoclave for 15 min at 15 psi pressure–121°C. Aseptically add 100.0mL of sterile Vitamin B_{12} solution. Mix thoroughly. Distribute into sterile tubes or flasks.

Use: For the cultivation and maintenance of *Synechococcus* species. For the isolation of cyanobacteria from freshwater habitats.

BG 11 Uracil Agar

Composition per liter:

Agar	10.0g

Uracil	2.8g
$NaNO_3$	1.5g
$MgSO_4 \cdot 7H_2O$	0.075g
K_2HPO_4	0.04g
$CaCl_2 \cdot 2H_2O$	0.036g
Na_2CO_3	0.02g
Citric acid	6.0mg
Ferric ammonium citrate	6.0mg
EDTA disodium salt	1.0mg
Trace metal mix A5	1.0mL

pH 7.1 ± 0.2 at 25°C

Trace Metal Mix A5:
Composition per liter:

H_3BO_3	2.86g
$MnCl_2 \cdot 4H_2O$	1.81g
$Na_2MoO_4 \cdot 2H_2O$	0.39g
$ZnSO_4 \cdot 7H_2O$	0.222g
$CuSO_4 \cdot 5H_2O$	0.079g
$Co(NO_3)_2 \cdot 6H_2O$	0.049g

Preparation of Trace Metal Mix A5: Add components to distilled/deionized water and bring volume to 1.0L. Mix thoroughly.

Preparation of Medium: Add components to distilled/deionized water and bring volume to 1.0L. Mix thoroughly. Heat gently to boiling. Distribute into tubes or flasks. Autoclave for 15 min at 15 psi pressure–121°C. Pour into sterile Petri dishes.

Use: For the cultivation and maintenance of *Anabaena variabilis.*

Bile Salts Gelatin Agar

Composition per 100.0mL:

Gelatin	3.0g
Agar	1.5g
Pancreatic digest of casein	1.0g
NaCl	1.0g
Sodium taurocholate	0.5g
Na_2CO_3	0.1g
Water	100.0mL

pH 8.5 ± 0.2 at 25°C

Preparation of Medium: Add components to distilled/deionized water and bring volume to 1.0L. Mix thoroughly. Gently heat and bring to boiling. Distribute into tubes or flasks. Autoclave for 15 min at 15 psi pressure–121°C. Pour into sterile Petri dishes or leave in tubes.

Use: For the cultivation of *Vibrio cholerae.*

Biphenyl Agar (DSMZ Medium 457d)

Composition per liter:

Agar	15.0g
Na_2HPO_4	2.44g
KH_2PO_4	1.52g
$(NH_4)_2SO_4$	0.5g
Biphenyl	0.25g
$MgSO_4 \cdot 7H_2O$	0.2g
$CaCl_2 \cdot 2H_2O$	0.05g
Trace elements solution SL-4	10.0mL

pH 6.9 ± 0.2 at 25°C

Trace Elements Solution SL-4:
Composition per liter:

EDTA	0.5g
$FeSO_4 \cdot 7H_2O$	0.2g
Trace elements solution SL-6	100.0mL

Trace Elements Solution SL-6:
Composition per liter:

H_3BO_3	0.3g
$CoCl_2 \cdot 6H_2O$	0.2g
$ZnSO_4 \cdot 7H_2O$	0.1g
$MnCl_2 \cdot 4H_2O$	0.03g
$Na_2MoO_4 \cdot H_2O$	0.03g
$NiCl_2 \cdot 6H_2O$	0.02g
$CuCl_2 \cdot 2H_2O$	0.01g

Preparation of Trace Elements Solution SL-6: Add components to distilled/deionized water and bring volume to 1.0L. Mix thoroughly. Adjust pH to 3.4.

Preparation of Trace Elements Solution SL-4: Add components to distilled/deionized water and bring volume to 1.0L. Mix thoroughly.

Biphenyl Solution:
Composition per liter:

Biphenyl	10.0g

Preparation of Biphenyl Solution: Add biphenyl to 1.0L ethanol. Mix thoroughly. Filter sterilize using a cellulose filter membrane.

Preparation of Medium: Add components, except biphenyl solution, to 1.0L distilled/deionized water. Adjust pH to 6.9. Heat and gently bring to boiling. Autoclave for 15 min at 15 psi pressure–121°C. Cool to 50°C. Add an aliquot of the biphenyl solution to the lid of a sterile Petri dish so that the final concentration will be approximately 0.25g/L biphenyl, and let the ethanol evaporate so that the crystals of biphenyl coat the lid of the Petri dish. Aseptically add sterile agar medium to the crystal-layered Petri dish.

Use: For the cultivation of biphenyl-utilizing bacteria.

Bismuth Sulfite Broth (m-Bismuth Sulfite Broth)

Composition per liter:

$Bi_2(SO_3)_3$	16.0g
Pancreatic digest of casein	10.0g
Peptic digest of animal tissue	10.0g
Beef extract	10.0g
Glucose	10.0g
Na_2HPO_4	8.0g
$FeSO_4 \cdot 7H_2O$	0.6g

pH 7.7 ± 0.2 at 25°C

Preparation of Medium: Add components to distilled/deionized water and bring volume to 1.0L. Mix thoroughly and heat with frequent agitation until boiling. Boil for 1 min. Do not autoclave. Cool to 45°–50°C. Mix to disperse the precipitate and aseptically distribute into sterile tubes or flasks. Use 2.0–2.2mL of medium for each membrane filter.

Use: For the selective isolation of *Salmonella typhi* and other enteric bacilli and for the detection of *Salmonella* by the membrane filter method.

Blastobacter denitrificans Agar (LMG Medium 157)

Composition per liter:

Agar	15.0g
Tryptone	2.0g
Lab Lemco beef extract	0.5g
Yeast extract	0.5g
Sodium acetate	0.2g
Glucose solution	10.0mL

pH 7.3 ± 0.2 at 25°C

Glucose Solution:
Composition per 10.0mL:
Glucose 2.5g

Preparation of Glucose Solution: Add glucose to 10.0mL of distilled/deionized water. Mix thoroughly. Filter sterilize.

Preparation of Medium: Add compents, except glucose solution, to distilled/deionized water and bring volume to 990.0mL. Mix thoroughly. Autoclave for 15 min at 15 psi pressure–121°C. Cool to 45°-50°C. Aseptically add 10.0mL glucose solution. Mix thoroughly. Aseptically pour into sterile Petri dishes or distribute into sterile tubes.

Use: For the cultivation of *Blastobacter denitrificans.*

Blood Agar Base with 2.5% NaCl

Composition per liter:
Beef heart, infusion from 500.0g
NaCl 30.0g
Agar 15.0g
Tryptose 10.0g
pH 6.8 ± 0.2 at 25°C

Preparation of Medium: Add components to distilled/deionized water and bring volume to 1.0L. Mix thoroughly. Heat with frequent agitation and boil for 1 min to completely dissolve. Autoclave for 15 min at 15 psi pressure–121°C. Cool the basal medium to 45°–50°C. Aseptically add sterile, defibrinated blood to a final concentration of 5%. Mix thoroughly and pour into sterile Petri dishes.

Use: For the cultivation of *Paracoccus halodenitrificans.*

Blood Glucose Cystine Agar

Composition per 100.0mL:
Nutrient agar 85.0mL
Glucose cystine solution 10.0mL
Human blood, fresh 5.0mL
pH 6.8 ± 0.2 at 25°C

Nutrient Agar:
Composition per liter:
Agar 15.0g
Pancreatic digest of gelatin 5.0g
Beef extract 3.0g

Source: Nutrient agar is available as a premixed powder from BD Diagnostic Systems.

Preparation of Nutrient Agar: Add components to distilled/deionized water and bring volume to 1.0L. Mix thoroughly. Gently heat while stirring and bring to boiling. Distribute into tubes or flasks. Autoclave for 15 min at 15 psi pressure–121°C. Cool to 45°–50°C.

Glucose Cystine Solution:
Composition per 50.0mL:
Glucose 12.5g
L-Cystine·HCl 0.5g

Preparation of Glucose Cystine Solution: Add components to distilled/deionized water and bring volume to 50.0mL. Mix thoroughly. Filter sterilize.

Preparation of Medium: To 85.0mL of cooled, sterile agar solution, aseptically add 10.0mL of sterile glucose cystine solution and 5.0mL of human blood. Mix thoroughly. Pour into sterile Petri dishes or distribute into sterile tubes.

Use: For the cultivation of *Francisella tularensis.*

Blue-Green Agar

Composition per liter:
Agar 10.0g
$NaNO_3$ 1.5g
$MgSO_4 \cdot 7H_2O$ 0.075g
K_2HPO_4 .... 0.04g
$CaCl_2 \cdot 2H_2O$ 0.036g
Na_2CO_3 .... 0.02g
Citric acid 6.0mg
Ferric ammonium citrate 6.0mg
EDTA disodium salt 1.0mg
Vitamin B_{12} solution 50.0mL
Trace metal mix A5 .... 1.0mL
pH 7.1 ± 0.2 at 25°C

Trace Metal Mix A5:
Composition per liter:
H_3BO_3 .... 2.86g
$MnCl_2 \cdot 4H_2O$ 1.81g
$Na_2MoO_4 \cdot 2H_2O$ 0.39g
$ZnSO_4 \cdot 7H_2O$ 0.222g
$CuSO_4 \cdot 5H_2O$ 0.079g
$Co(NO_3)_2 \cdot 6H_2O$ 0.049g

Preparation of Trace Metal Mix A5: Add components to distilled/deionized water and bring volume to 1.0L. Mix thoroughly.

Vitamin B_{12} Solution:
Composition per 50.0mL:
Vitamin B_{12} 0.01g

Preparation of Vitamin B_{12} Solution: Add Vitamin B_{12} to distilled/deionized water and bring volume to 50.0mL. Mix thoroughly. Filter sterilize.

Preparation of Medium: Add components, except Vitamin B_{12} solution, to glass-distilled water and bring volume to 950.0mL. Mix thoroughly. Heat gently and bring to boiling. Autoclave for 15 min at 15 psi pressure–121°C. Cool the basal medium to 45°–50°C. Add Vitamin B_{12} solution. Mix thoroughly. Pour into sterile Petri dishes or distribute into sterile tubes.

Use: For the cultivation and maintenance of *Synechococcus* species.

Blue-Green Nitrogen-Fixing Agar

Composition per liter:
Noble agar 10.0g
$MgSO_4 \cdot 7H_2O$ 0.075g
K_2HPO_4 .... 0.04g
$CaCl_2 \cdot 2H_2O$ 0.036g
Na_2CO_3 .... 0.02g
Citric acid 6.0mg
Ferric ammonium citrate 6.0mg
EDTA disodium salt 1.0mg
Trace metal mix A5 .... 1.0mL
pH 7.1 ± 0.2 at 25°C

Trace Metal Mix A5:
Composition per liter:
H_3BO_3 .... 2.86g
$MnCl_2 \cdot 4H_2O$ 1.81g
$Na_2MoO_4 \cdot 2H_2O$ 0.39g
$ZnSO_4 \cdot 7H_2O$ 0.222g
$CuSO_4 \cdot 5H_2O$ 0.079g
$Co(NO_3)_2 \cdot 6H_2O$ 0.049g

Preparation of Trace Metal Mix A5: Add components to distilled/deionized water and bring volume to 1.0L. Mix thoroughly.

Preparation of Medium: Add components to glass distilled water and bring volume to 1.0L. Mix thoroughly. Heat gently and bring to boiling. Autoclave for 15 min at 15 psi pressure–121°C. Check pH after autoclaving and readjust if necessary. Pour into sterile Petri dishes or distribute into sterile tubes.

Use: For the cultivation and maintenance of *Calothrix, Fischerella*, and *Nostoc* species.

BMPA-α Medium (Edelstein BMPA-α Medium)

Composition per liter:

Agar	13.0g
Yeast extract	10.0g
ACES buffer (2-[(2-Amino-2-oxoethyl)-amino]-ethane sulfonic acid)	2.0g
Charcoal, activated	2.0g
α-Ketoglutarate	0.2g
$Fe_4(P_2O_7)_3 \cdot 9H_2O$	0.05g
Antibiotic inhibitor	10.0mL
L-Cysteine·HCl·H_2O solution	10.0mL

pH 6.9 ± 0.2 at 25°C

Source: This medium is available as premixed vials from Oxoid Unipath.

Antibiotic Inhibitor:
Composition per 10.0mL:

Anisomycin	0.08g
Cefamandole	4.0mg
Polymyxin B	80,000U

Preparation of Antibiotic Inhibitor: Add components to distilled/deionized water and bring volume to 10.0mL. Mix thoroughly. Filter sterilize.

L-Cysteine·HCl·H_2O Solution:
Composition per 10.0mL:

L-Cysteine·HCl·H_2O	0.08g

Preparation of L-Cysteine·HCl·H_2O Solution: Add L-cysteine·HCl·H_2O to distilled/deionized water and bring volume to 10.0mL. Mix thoroughly. Filter sterilize.

Preparation of Medium: Add components, except cysteine and antibiotic inhibitor, to distilled/deionized water and bring volume to 980.0mL. Mix thoroughly. Adjust medium to pH 6.9 with 1*N* KOH. Heat gently and bring to boiling for 1 min. Autoclave for 15 min at 15 psi pressure–121°C. Cool to 50°–55°C. Add 10.0mL of the sterile L-cysteine·HCl·H_2O solution and 10.0mL of the sterile antibiotic solution. Mix thoroughly. Pour into sterile Petri dishes with constant agitation to keep charcoal in suspension.

Use: For the selective isolation and cultivation of *Legionella pneumophila* and other *Legionella* species.

BMS Agar

Composition per liter:

Agar	20.0g
L-Malic acid	2.5g
Sucrose	2.5g
KOH	2.0g
Potato extract solution	950.0mL
Bromothymol Blue	1.0mL
Vitamin solution	1.0mL

pH 7.0 ± 0.2 at 25°C

Potato Extract Solution:
Composition per liter:

Potatoes, washed, peeled, and sliced	200.0g

Preparation of Potato Extract Solution: Wash, peel, and slice several large potatoes. Place the poltato slices in a gauze bag. Place the bag with the potatoes in 1.0L of distilled/deionized water. Boil for 30 min. Filter through cotton.

Bromothymol Blue Solution:
Composition per 100.0mL:

Bromothymol Blue	0.5g
Ethanol, 95%	100.0mL

Preparation of Bromothymol Blue Solution: Add Bromothymol Blue to 100.0mL of 95% ethanol. Mix thorougly.

Vitamin Solution:
Composition per 100.0mL:

Biotin	10.0g
Pyridoxine	20.0mg

Preparation of Vitamin Solution: Add components to distilled/deionized water and bring volume to 100.0mL. Mix thoroughly. Filter sterilize.

Preparation of Medium: Dissolve 2.5g of L-malic acid in 50.0mL of distilled/deionized water. Add 1.0mL of Bromothymol Blue solution. Adjust pH to 7.0 by adding KOH so that the solution is green. Add 950.0mL of potato extract solution. Mix thoroughly. Add 20.0g of agar and 2.5g of sucrose. Mix thoroughly. Gently heat and bring to boiling. Autoclave for 15 min at 15 psi pressure–121°C. Cool to 50°–55°C. Aseptically add 1.0mL of sterile vitamin solution. Mix thoroughly. Pour into sterile Petri dishes or distribute into sterile tubes.

Use: For the cultivation and maintenance of *Azospirillum lipoferum.*

Bosea thiooxidans Medium (DSMZ Medium 763)

Composition per liter:

$Na_2S_2O_3 \cdot 5H_2O$	5.0g
Na-succinate	5.0g
Na_2HPO_4	4.0g
KH_2PO_4	1.5g
$MgCl_2$	0.1g
Na-glutamate	0.5g
Yeast extract	0.1g

pH 7.5-8.5 at 25°C

Preparation of Medium: Add components to distilled/deionized water and bring volume to 1.0L. Mix thoroughly. Distribute into tubes or flasks. Autoclave for 15 min at 15 psi pressure–121°C.

Use: For the cultivation of *Bosea thiooxidans.*

Bovine Albumin Tween 80 Medium, Ellinghausen and McCullough, Modified (Albumin Fatty Acid Broth, *Leptospira* Medium)

Composition per liter:

Basal medium	900.0mL
Albumin fatty acid supplement	100.0mL

Basal Medium:
Composition per liter:

Na_2HPO_4, anhydrous	1.0g
NaCl	1.0g
KH_2PO_4, anhydrous	0.3g
NH_4Cl (25% solution)	1.0mL
Glycerol (10% solution)	1.0mL

Sodium pyruvate (10% solution) 1.0mL
Thiamine·HCl (0.5% solution)..................................1.0mL
pH 7.4 ± 0.2 at 25°C

Preparation of Basal Medium: Add components to distilled/deionized water and bring volume to 1.0L. Mix thoroughly. Adjust pH to 7.4. Gently heat and bring to boiling. Autoclave for 15 min at 15 psi pressure–121°C. Cool to 25°C.

Albumin Fatty Acid Supplement:
Composition per 200.0mL:
Bovine albumin fraction V 20.0g
Polysorbate (Tween) 80 (10% solution) 25.0mL
$FeSO_4 \cdot 7H_2O$ (0.5% solution)..................................20.0mL
$CaCl_2 \cdot 2H_2O$ (1.5% solution)....................................2.0mL
$MgCl_2 \cdot 2H_2O$ (1.5% solution)....................................2.0mL
Vitamin B_{12} (0.2% solution) 2.0mL
$ZnSO_4 \cdot 7H_2O$ (0.4% solution) 2.0mL
$CuSO_4 \cdot 5H_2O$ (0.3% solution) 0.2mL

Preparation of Albumin Fatty Acid Supplement: Add bovine albumin to 100.0mL of distilled/deionized water. Mix thoroughly. Add remaining components while stirring. Adjust pH to 7.4. Bring volume to 200.0mL with distilled/deionized water. Filter sterilize. Store at –20°C.

Preparation of Medium: Aseptically combine 100.0mL of sterile albumin fatty acid supplement and 900.0mL of sterile basal medium. Mix thoroughly. Aseptically distribute into sterile tubes or flasks.

Use: For the cultivation of *Leptospira* species.

Bovine Albumin Tween 80 Semisolid Medium, Ellinghausen and McCullough, Modified (Albumin Fatty Acid Semisolid Medium, Modified)

Composition per liter:
Basal medium 900.0mL
Albumin fatty acid supplement.............................100.0mL

Basal Medium:
Composition per liter:
Agar 2.2g
Na_2HPO_4, anhydrous 1.0g
NaCl 1.0g
KH_2PO_4, anhydrous 0.3g
NH_4Cl (25% solution)..1.0mL
Glycerol (10% solution)...1.0mL
Sodium pyruvate (10% solution) 1.0mL
Thiamine·HCl (0.5% solution)..................................1.0mL
pH 7.4 ± 0.2 at 25°C

Preparation of Basal Medium: Add components to distilled/deionized water and bring volume to 1.0L. Mix thoroughly. Adjust pH to 7.4. Gently heat and bring to boiling. Autoclave for 15 min at 15 psi pressure–121°C. Cool to 25°C.

Albumin Fatty Acid Supplement:
Composition per 200.0mL:
Bovine albumin fraction V 20.0g
Polysorbate (Tween) 80 (10% solution) 25.0mL
$FeSO_4 \cdot 7H_2O$ (0.5% solution)..................................20.0mL
$CaCl_2 \cdot 2H_2O$ (1.5% solution)....................................2.0mL
$MgCl_2 \cdot 2H_2O$ (1.5% solution)....................................2.0mL
Vitamin B_{12} (0.2% solution) 2.0mL
$ZnSO_4 \cdot 7H_2O$ (0.4% solution) 2.0mL
$CuSO_4 \cdot 5H_2O$ (0.3% solution) 0.2mL

Preparation of Albumin Fatty Acid Supplement: Add bovine albumin to 100.0mL of distilled/deionized water. Mix thoroughly. Add remaining components while stirring. Adjust pH to 7.4. Bring volume to 200.0mL with distilled/deionized water. Filter sterilize. Store at –20°C.

Preparation of Medium: Aseptically combine 100.0mL of sterile albumin fatty acid supplement and 900.0mL of sterile basal medium. Mix thoroughly. Aseptically distribute into sterile tubes or flasks.

Use: For the cultivation of *Leptospira* species.

Bovine Serum Albumin Tween 80 Agar (BSA Tween 80 Agar)

Composition per liter:
Basal medium 900.0mL
Albumin supplement 100.0mL

Basal Medium:
Composition per liter:
Agar 11.0g
Na_2HPO_4 ....1.0g
NaCl 1.0g
KH_2PO_4 ....0.3g
Glycerol (10% solution) 1.0mL
NH_4Cl (25% solution) 1.0mL
Sodium pyruvate (10% solution).............................1.0mL
Thiamine (0.5% solution 1.0mL

Preparation of Basal Medium: Add components to distilled/deionized water and bring volume to 1.0L. Mix thoroughly. Adjust pH to 7.4. Autoclave for 15 min at 15 psi pressure–121°C. Cool to 25°C.

Albumin Supplement:
Composition per 100.0mL:
Bovine albumin 10.0g
Tween 80 (10% solution) 12.5mL
$FeSO_4$ (0.5% solution).. 10.0mL
$MgCl_2$-$CaCl_2$ solution 1.0mL
Cyanocobalamin (0.02% solution).......................... 1.0mL
$ZnSO_4$ (0.4% solution) 1.0mL

Preparation of Albumin Supplement: Add components to distilled/deionized water and bring volume to 100.0mL. Mix thoroughly. Adjust pH to 7.4. Filter sterilize.

$MgCl_2$-$CaCl_2$ Solution:
Composition per 100.0mL:
$CaCl_2 \cdot 2H_2O$ 1.5g
$MgCl_2 \cdot 6H_2O$ 1.5g

Preparation of $MgCl_2$-$CaCl_2$ Solution: Add components to distilled/deionized water and bring volume to 100.0mL. Mix thoroughly.

Preparation of Medium: To 900.0mL of cooled, sterile basal medium, aseptically add 100.0mL of sterile albumin supplement. Mix thoroughly. Aseptically distribute into sterile tubes or flasks.

Use: For the cultivation and maintenance of *Leptospira* species.

Brackish Acetate

Composition per liter:
Sodium acetate 1.0g
KNO_3 ....1.0g
$NaH_2PO_4 \cdot 2H_2O$ 0.05g
Artificial seawater 250.0mL

Modified Hutner's basal salts 20.0mL
Vitamin solution..10.0mL
pH 7.2 ± 0.2 at 25°C

Artificial Seawater:

Composition per liter:

NaCl 23.5g
$MgCl_2$ 5.0g
Na_2SO_4 ....3.9g
$CaCl_2$ 1.1g
KCl 0.66g
$NaHCO_3$ 0.19g
KBr 0.1g
H_3BO_3 ....0.026g
$SrCl_2$ 0.024g
NaF 3.0mg

Preparation of Artificial Seawater: Add components to distilled/deionized water and bring volume to 100.0mL. Mix thoroughly.

Modified Hutner's Basal Salts:

Composition per liter:

$MgSO_4.7H_2O$ 29.7g
Nitrilotriacetic acid 10.0g
$CaCl_2 \cdot 2H_2O$ 3.34g
$FeSO_4 \cdot 7H_2O$ 0.1g
$(NH_4)_2MoO_4$ 9.25mg
"Metals 44" 50.0mL

Preparation of Modified Hutner's Basal Salts: Dissolve the nitrilotriacetic acid first and neutralize the solution with KOH. Add other components and adjust the pH to 7.2 with KOH or H_2SO_4. There may be a slight precipitate. Store at 5°C.

"Metals 44":

Composition per 100.0mL:

$ZnSO_4 \cdot 7H_2O$ 1.1g
$FeSO_4 \cdot 7H_2O$ 0.5g
EDTA 0.25g
$MnSO_4 \cdot 7H_2O$ 0.154g
$CuSO_4 \cdot 5H_2O$ 0.04g
$Co(NO_3)_2 \cdot 6H_2O$ 0.025g
$Na_2B_4O_7 \cdot 10H_2O$ 0.018g

Preparation of "Metals 44": Add components to distilled/deionized water and bring volume to 100.0mL. Mix thoroughly. Autoclave for 15 min at 15 psi pressure–121°C. Add aseptically to sterile basal medium.

Vitamin Solution:

Composition per liter:

Thiamine·HCl 5.0mg
D-Calcium pantothenate 5.0mg
Riboflavin 5.0mg
Biotin 2.0mg
Folic acid 2.0mg
Vitamin B_{12} 0.1mg

Preparation of Vitamin Solution: Add components to distilled/deionized water and bring volume to 1.0L. Mix thoroughly. Filter sterilize and add aseptically to sterile basal medium.

Preparation of Medium: Add a few drops of H_2SO_4 to the distilled water to retard precipitation of the metal salts. Add components, except for "Metals 44" and vitamin solutions, to 250.0mL of artificial seawater and 720.0mL of distilled/deionized water. Adjust pH to 7.2. Distribute into tubes or flasks. Sterilize by autoclaving for 15 min at 15 lbs pressure–121°C. Aseptically add "Metals 44" and vitamin solutions. Mix thoroughly. Aseptically distribute into sterile tubes or flasks.

Use: For the cultivation of *Filomicrobium fusiforme*.

Brackish *Prosthecomicrobium* Medium

Composition per liter:

Agar 15.0g
Peptone 0.25g
Yeast extract 0.25g
Glucose 0.25g
Artificial seawater 250.0mL
Modified Hutner's basal salts 20.0mL
Vitamins 10.0mL
pH 7.2 ± 0.2 at 25°C

Artificial Seawater:

Composition per liter:

NaCl 23.477g
$MgCl_2$ 4.981g
Na_2SO_4 ....3.917g
$CaCl_2$ 1.102g
KCl 0.664g
$NaHCO_3$ 0.192g
KBr 0.096g
H_3BO_3 ....0.026g
$SrCl_2$ 0.024g
NaF 3.0mg

Preparation of Artificial Seawater: Add components to distilled/deionized water and bring volume to 100.0mL. Mix thoroughly.

Modified Hutner's Basal Salts:

Composition per liter:

$MgSO_4.7H_2O$ 29.7g
Nitrilotriacetic acid 10.0g
$CaCl_2 \cdot 2H_2O$ 3.34g
$FeSO_4 \cdot 7H_2O$ 0.1g
$(NH_4)_2MoO_4$ 9.25mg
"Metals 44" 50.0mL

Preparation of Modified Hutner's Basal Salts: Dissolve the nitrilotriacetic acid first and neutralize the solution with KOH. Add other ingredients and readjust the pH with KOH and/or H_2SO_4 to 7.2. There may be a slight precipitate. Store at 5°C.

"Metals 44":

Composition per 100.0mL:

$ZnSO_4 \cdot 7H_2O$ 1.1g
$FeSO_4 \cdot 7H_2O$ 0.5g
EDTA 0.25g
$MnSO_4 \cdot 7H_2O$ 0.154g
$CuSO_4 \cdot 5H_2O$ 0.04g
$Co(NO_3)_2 \cdot 6H_2O$ 0.025g
$Na_2B_4O_7 \cdot 10H_2O$ 0.018g

Preparation of "Metals 44": Add components to distilled/deionized water and bring volume to 100.0mL. Mix thoroughly. Autoclave for 15 min at 15 psi pressure–121°C. Add aseptically to sterile basal medium.

Vitamin Solution:

Composition per liter:

Thiamine·HCl 5.0mg
D-Calcium pantothenate 5.0mg
Riboflavin 5.0mg
Biotin 2.0mg
Folic acid 2.0mg
Vitamin B_{12} 0.1mg

Preparation of Vitamin Solution: Add components to distilled/deionized water and bring volume to 1.0L. Mix thoroughly. Filter sterilize and add aseptically to sterile basal medium.

Preparation of Medium: Add a few drops of H_2SO_4 to the distilled water to retard precipitation of the metal salts. Add components, except "Metals 44" and vitamin solutions, to 250.0mL of artificial seawater and 720.0mL of distilled/deionized water. Adjust pH to 7.2. Distribute into tubes or flasks. Sterilize by autoclaving for 15 min at 15 lbs pressure–121°C. Aseptically add "Metals 44" and vitamin solutions. Mix thoroughly. Aseptically distribute into sterile tubes or flasks.

Use: For the cultivation of *Prosthecomicrobium litoralum.*

Brain Heart Infusion with 3% Sodium Chloride

Composition per liter:

NaCl	30.0g
Pancreatic digest of gelatin	14.5g
Brain heart, solids from infusion	6.0g
Peptic digest of animal tissue	6.0g
Glucose	3.0g
Na_2HPO_4	2.5g

pH 7.4 ± 0.2 at 25°C

Preparation of Medium: Add components to distilled/deionized water and bring volume to 1.0L. Mix thoroughly. Distribute into tubes or flasks. Autoclave for 15 min at 15 psi pressure–121°C.

Use: For the cultivation of *Vibrio parahaemolyticus.*

Brain Heart Infusion Soil Extract Medium

Composition per liter:

Yeast extract	20.0g
Pancreatic digest of casein	16.0g
Brain heart, solids from infusion	8.0g
Peptic digest of animal tissue	5.0g
NaCl	5.0g
Na_2HPO_4	2.5g
Glucose	2.0g
Soil extract	250.0mL
Vitamin B_{12} solution	1.0mL

pH 7.2 ± 0.2 at 25°C

Soil Extract:
Composition per 400.0mL:

African Violet soil	1.0g
Na_2CO_3	1.0g

Preparation of Soil Extract: Autoclave for 60 min at 15 psi pressure–121°C. Filter through paper before using in medium.

Vitamin B_{12} Solution:
Composition per 1.0mL:

Vitamin B_{12}	2.0μg

Preparation of Vitamin B_{12} Solution: Add Vitamin B_{12} to distilled/deionized water and bring volume to 1.0mL. Mix thoroughly. Filter sterilize.

Preparation of Medium: Add components, except glucose, yeast extract, and Vitamin B_{12} solution, to tap water and bring volume to 799.0mL. Mix thoroughly. Autoclave for 15 min at 15 psi pressure–121°C. Add yeast extract and glucose to 200.0mL of tap water. Filter sterilize and add aseptically to cooled, sterile basal medium. Aseptically add 1.0mL of Vitamin B_{12} solution. Mix thoroughly. Aseptically distribute into sterile tubes or flasks.

Use: For the cultivation of a wide variety of microorganisms, including bacteria, yeasts, and molds, especially fastidious species from soil. For the isolation of *Histoplasma capsulatum* and other pathogenic fungi, including *Coccidioides immitis,* from soils.

Brevibacterium Medium (ATCC Medium 677)

Composition per liter:

Agar	30.0g
KH_2PO_4	2.0g
Na_2HPO_4	2.0g
$(NH_4)_2SO_4$	2.0g
Yeast extract	2.0g
Tween™ 60	2.0g
$MgSO_4·7H_2O$	0.2g
$FeSO_4.7H_2O$	0.1g
$MnSO_4$	0.01g
n-Hexadecane	50.0mL

pH 7.0 ± 0.2 at 25°C

Preparation of Medium: Add components to distilled/deionized water and bring volume to 1.0L. Mix thoroughly. Blend for 30 min in a blender to disperse the *n*-hexadecane. Distribute into tubes or flasks. Autoclave for 20 min at 15 psi pressure–121°C.

Use: For the cultivation and maintenance of *Brevibacterium alkanophilum* and other microorganisms that can utilize hexadecane as a carbon source.

Brevibacterium Medium (ATCC Medium 681)

Composition per liter:

Glucose	10.0g
Peptone	5.0g
Yeast extract	5.0g

pH 5.0–6.0 at 25°C

Preparation of Medium: Add components to distilled/deionized water and bring volume to 1.0L. Mix thoroughly. Distribute into tubes or flasks. Autoclave for 15 min at 15 psi pressure–121°C.

Use: For the cultivation and maintenance of *Brevibacterium* spp. and *Enterobacter cloacae.*

Brevundimonas Agar (LMG Medium 221)

Composition per liter:

Agar	15.0g
Yeast extract	10.0g
Peptone	2.0g
$MgSO_4·7H_2O$	0.2g
$CaCl_2$	0.1g
Riboflavin solution	5.0mL

Riboflavin Solution:
Composition per 10.0mL:

Riboflavin	2.0mg

Preparation of Riboflavin Solution: Add riboflavin to 10.0mL of distilled/deionized water. Mix thoroughly. Filter sterilize.

Preparation of Medium: Add components, except riboflavin, to distilled/deionized water and bring volume to 995.0mL. Mix thoroughly. Gently heat and bring to boiling. Autoclave for 15 min at 15 psi pressure–121°C. Cool to

45°–50°C. Aseptically add 5.0mL sterile riboflavin. Mix thoroughly. Pour into sterile Petri dishes or distribute into sterile tubes.

Use: For the cultivation of *Brevundimonas* spp. and *Caulobacter henricii.*

Brilliant Green Bile Agar

Composition per liter:

Noble agar	10.15g
Pancreatic digest of gelatin	8.25g
Lactose	1.9g
Na_2SO_3	0.205g
Basic Fuchsin	0.078g
Erioglaucine	0.065g
$FeCl_3$	0.0295g
KH_2PO_4	0.015g
Oxgall, dehydrated	2.95mg
Brilliant Green	0.03mg

pH 6.9 ± 0.2 at 25°C

Source: This medium is available as a premixed powder from BD Diagnostic Systems.

Caution: Basic Fuchsin is a potential carcinogen and care must be taken to avoid inhalation of the powdered dye and contamination of the skin.

Preparation of Medium: Add components to distilled/deionized water and bring volume to 1.0L. For plating 10.0mL samples, prepare the medium double strength. Mix thoroughly. Gently heat and bring to boiling. Distribute into tubes or flasks. Autoclave for 15 min at 15 psi pressure–121°C. Pour into sterile Petri dishes. Care should be taken to avoid exposure of the prepared medium to light.

Use: For the detection and enumeration of coliform bacteria in materials of sanitary importance such as water and sewage. *Escherichia coli* appears as dark red colonies with a pink halo. *Enterobacter* species appear as pink colonies.

Brilliant Green Bile Broth (Brilliant Green Lactose Bile Broth)

Composition per liter:

Oxgall, dehydrated	20.0g
Lactose	10.0g
Pancreatic digest of gelatin	10.0g
Brilliant Green	0.013g

pH 7.2 ± 0.2 at 25°C

Source: This medium is available as a premixed powder from BD Diagnostic Systems and Oxoid Unipath.

Preparation of Medium: Add components to distilled/deionized water and bring volume to1.0L. Mix thoroughly. Distribute into tubes containing inverted Durham tubes, in 10.0mL amounts for testing 1.0mL or less of sample. Autoclave for 12 min (not longer than 15 min) at 15 psi pressure–121°C. After sterilization, cool the broth rapidly. Medium is sensitive to light.

Use: For the detection of coliform microorganisms in water and wastewater as well as in other materials of sanitary importance. Turbidity in the broth and gas in the Durham tube are positive indications of *Escherichia coli.*

Brilliant Green Bile Broth with MUG

Composition per liter:

Oxgall, dehydrated	20.0g
Lactose	10.0g
Pancreatic digest of gelatin	10.0g
MUG (4-Methyl umbelliferyl-β-D-glucuronide)	0.05g
Brilliant Green	0.013g

pH 7.2 ± 0.2 at 25°C

Source: This medium is available as a premixed powder from BD Diagnostic Systems.

Preparation of Medium: Add components to distilled/deionized water and bring volume to1.0L. Mix thoroughly. Distribute into tubes containing inverted Durham tubes, in 10.0mL amounts for testing 1.0mL or less of sample. Autoclave for 12 min (not longer than 15 min) at 15 psi pressure–121°C. After sterilization, cool the broth rapidly.

Use: For the detection of coliform microorganisms in water and wastewater, as well as in other materials of sanitary importance. The presence of *Eschrichia coli* and other coliforms is determined by the presence of fluorescence in the tube.

Brilliant Green Broth (m-Brilliant Green Broth)

Composition per liter:

Proteose peptone No. 3	20.0g
Lactose	20.0g
Sucrose	20.0g
NaCl	10.0g
Yeast extract	6.0g
Phenol Red	0.16g
Brilliant Green	0.025g

pH 6.9 ± 0.2 at 25°C

Source: This medium is available as a premixed powder from BD Diagnostic Systems.

Preparation of Medium: Add components to distilled/deionized water and bring volume to 1.0L. Mix thoroughly. Gently heat with frequent mixing. Boil for 1 min. Cool to 25°C. Add 2.0mL to each sterile absorbent filter used.

Use: For the selective isolation and differentiation of *Salmonella* from polluted water by the membrane filter method.

Bromcresol Purple Broth

Composition per liter:

Peptone	10.0g
NaCl	5.0g
Beef extract	3.0g
Bromcresol Purple	0.04g
Carbohydrate solution	10.0mL

pH 7.0 ± 0.2 at 25°C

Carbohydrate Solution:

Composition per 10.0mL:

Carbohydrate	5.0g

Preparation of Carbohydrate Solution: Add carbohydrate to distilled/deionized water and bring volume to 10.0mL. Mix thoroughly. Filter sterilize.

Preparation of Medium: Add components to distilled/deionized water and bring volume to 1.0L. Mix thoroughly. Gently heat and bring to boiling. Distribute into test tubes that contain an inverted Durham tube. Autoclave for 10 min at 15 psi pressure–121°C.

Use: For the differentiation of a variety of microorganisms based on their fermentation of specific carbohydrates. Bacteria that ferment the specific carbohydrate turn the medium yellow. When bacteria produce gas, the gas is trapped in the Durham tube.

Bromcresol Purple Dextrose Broth (BCP Broth)

Composition per liter:

Glucose	10.0g
Peptone	5.0g
Beef extract	3.0g
Bromcresol Purple solution	2.0mL

pH 7.0 ± 0.2 at 25°C

Bromcresol Purple Solution:
Composition per 10.0mL:

Bromcresol Purple	0.16g
Ethanol (95% solution)	10.0mL

Preparation of Bromcresol Purple Solution: Add Bromcresol Purple to 10.0mL of ethanol. Mix thoroughly.

Preparation of Medium: Add components to distilled/deionized water and bring volume to 1.0L. Mix thoroughly. Distribute into tubes in 12–15mL volumes. Autoclave for 15 min at 15 psi pressure–121°C.

Use: For the cultivation and differentiation of bacteria based on their ability to ferment glucose. Bacteria that ferment glucose turn the medium yellow.

Bromothymol Blue Agar

Composition per liter:

Agar	11.0g
Peptone	10.0g
NaCl	5.0g
Yeast extract	5.0g
Lactose (33% solution)	27.0mL
Bromothymol Blue (1% solution)	10.0mL
Sodium thiosulfate (50% solution)	2.0mL
Glucose (33% solution)	1.2mL
Maranil solution (5% solution)	1.0mL

pH 7.7–7.8 at 25°C

Preparation of Medium: Add agar, peptone, NaCl, and yeast extract to distilled/deionized water and bring volume to 1.0L. Mix thoroughly. Adjust pH to 8.0. Autoclave for 20 min at 15 psi pressure–121°C. Cool to 45°–50°C. Filter sterilize separately the lactose solution, Bromothymol Blue solution, sodium thiosulfate solution, glucose solution, and maranil solution. To the cooled, sterile agar solution aseptically add 27.0mL of sterile lactose solution, 10.0mL of sterile Bromothymol Blue solution, 2.0mL of sterile sodium thiosulfate solution, 1.2mL of sterile glucose solution and1.0mL of sterile maranil solution. Mix thoroughly. Adjust pH to 7.7–7.8. Pour into sterile Petri dishes or distribute into sterile tubes.

Use: For the selective isolation and cultivation of members of the *Enterobacteriaceae*.

BSR Medium

Composition per liter:

Beef heart, solids from infusion	500.0g
Sorbitol	70.0g
Sucrose	10.0g
Tryptose	10.0g
NaCl	5.0g
Fructose	1.0g
Glucose	1.0g
Phenol Red	0.02g
Horse serum	100.0mL

pH 7.6 ± 0.2 at 25°C

Preparation of Medium: Add components, except horse serum, to distilled/deionized water and bring volume to 900.0mL. Mix thoroughly. Autoclave for 15 min at 15 psi pressure–121°C. Cool to 45°–50°C. Aseptically add 100.0mL of horse serum. Mix thoroughly. Aseptically distribute into sterile tubes or flasks.

Use: For the cultivation of *Spiroplasma citri*.

BSTSY Agar

Composition per liter:

Pancreatic digest of casein	17.0g
Agar	15.0g
NaCl	5.0g
Yeast extract	4.0g
Papaic digest of soybean meal	3.0g
K_2HPO_4	2.5g
Glucose	2.5g
Bovine serum	100.0mL

pH 7.3 ± 0.2 at 25°C

Preparation of Medium: Add components, except bovine serum, to distilled/deionized water and bring volume to 900.0mL. Mix thoroughly. Gently heat and bring to boiling. Autoclave for 15 min at 15 psi pressure–121°C. Cool to 45°–50°C. Aseptically add sterile bovine serum. Mix thoroughly. Pour into sterile Petri dishes or distribute into sterile tubes.

Use: For the isolation and cultivation of *Simonsiella* species and *Alysiella* species.

Buffered Charcoal Yeast Extract Differential Agar (DIFF/BCYE)

Composition per 1014.0mL:

Agar	17.0g
ACES (2-[(2-Amino-2-oxoethyl)-amino]-ethane sulfonic acid) buffer	10.0g
Yeast extract	10.0g
Charcoal, activated	1.5g
$Fe_4(P_2O_7)_3 \cdot 9H_2O$	0.25g
Bromcresol Purple	0.01g
Bromothymol Blue	0.01g
Antibiotic solution	10.0mL
L-Cysteine·HCl·H_2O solution	4.0mL

pH 6.9 ± 0.2 at 25°C

Antibiotic Solution:
Composition per 10.0mL:

Vancomycin	1.0mg
Polymyxin B	50,000U

Preparation of Antibiotic Solution: Add components to distilled/deionized water and bring volume to 10.0mL. Mix thoroughly. Filter sterilize.

L-Cysteine·HCl·H_2O Solution:
Composition per 10.0mL:

L-Cysteine·HCl·H_2O	1.0g

Preparation of L-Cysteine·HCl·H_2O Solution: Add L-cysteine·HCl·H_2O to distilled/deionized water and bring volume to 10.0mL. Mix thoroughly. Filter sterilize.

Preparation of Medium: Add components, except L-cysteine·HCl·H_2O solution and antibiotic solution, to distilled/deionized water and bring volume to 1.0L. Mix thoroughly. Adjust medium to pH 6.9 with 1*N* KOH. Heat gently and bring to boil for 1 min. Autoclave for 15 min at 15 psi pressure–121°C. Cool to 50°–55°C. Add 4.0mL of sterile L-cysteine·HCl·H_2O solution and 10.0mL of sterile antibiotic solution. Mix thoroughly. Pour into sterile Petri dishes with constant agitation to keep charcoal in suspension.

Use: For the isolation, cultivation, and maintenance of *Legionella pneumophila* and other *Legionella* species from environmental samples. For the selective recovery of *Legionella pneumophila* while reducing contaminating microorganisms from environmental water samples.

Buffered Marine Yeast Medium

Composition per liter:

NaCl	24.0g
Agar	20.0g
Yeast extract	5.0g
1*M* Phosphate buffer, pH 6.8	20.0mL
Hutner's mineral base	20.0mL
KOH (1*N*)	7.0mL

pH 6.8 ± 0.2 at 25°C

1*M* Phosphate Buffer, pH 6.8:
Composition per liter:

$K_2H_2PO_4$	85.4g
$NaH_2PO_4 \cdot H_2O$	70.4g

Preparation of 1*M* Phosphate Buffer, pH 6.8: Add components to distilled/deionized water and bring volume to 1.0L. Mix thoroughly. Adjust pH to 6.8.

Hutner's Mineral Base:
Composition per liter:

$MgSO_4.7H_2O$	29.7g
Nitrilotriacetic acid	10.0g
$CaCl_2 \cdot 2H_2O$	3.34g
$FeSO_4 \cdot 7H_2O$	0.01g
$(NH_4)_2MoO_4$	9.25mg
"Metals 44"	50.0mL

Preparation of Hutner's Mineral Base: Initially add a few drops of H_2SO_4 to the distilled water to retard precipitation. Dissolve the nitrilotriacetic acid first and neutralize the solution with KOH. Add other ingredients and adjust the pH to 7.2 with KOH and/or H_2SO_4. There may be a slight precipitate. Store at 5°C.

"Metals 44":
Composition per 100.0mL:

$ZnSO_4 \cdot 7H_2O$	1.1g
$FeSO_4 \cdot 7H_2O$	0.5g
EDTA	0.25g
$MnSO_4 \cdot 7H_2O$	0.154g
$CuSO_4 \cdot 5H_2O$	0.04g
$Co(NO_3)_2 \cdot 6H_2O$	0.025g
$Na_2B_4O_7 \cdot 10H_2O$	0.018g

Preparation of "Metals 44": Add components to distilled/deionized water and bring volume to 100.0mL. Mix thoroughly. Autoclave for 15 min at 15 psi pressure–121°C. Add aseptically to sterile basal medium.

Preparation of Medium: Add components to distilled/deionized water and bring volume to 1.0L. Mix thoroughly. Distribute into tubes or flasks. Autoclave for 15 min at 15 psi pressure–121°C.

Use: For the cultivation and maintenance of *Pseudomonas* species.

Burke's Modified Nitrogen-Free Medium

Composition per liter:

$MgSO_4 \cdot 7H_2O$	0.2g
Na_2HPO_4	0.19g
$NaHCO_3$	0.05g
$CaSO_4 \cdot 2H_2O$	0.02g
KH_2PO_4	0.011g
$SrCl_2 \cdot 6H_2O$	0.01g
NaCl	0.01g
Adenine	0.01g
$FeSO_4.7H_2O$	6.0mg
Na_2MoO_3	0.5mg

pH 7.8 ± 0.2 at 25°C

Preparation of Medium: Add components to distilled/deionized water and bring volume to 1.0L. Mix thoroughly. Distribute into tubes or flasks. Autoclave for 15 min at 15 psi pressure–121°C.

Use: For the cultivation of *Azotobacter vinelandii.*

Burke's Modified Nitrogen-Free Medium with Benzoate

Composition per liter:

Sodium benzoate	0.72g
$MgSO_4 \cdot 7H_2O$	0.2g
Na_2HPO_4	0.189g
$NaHCO_3$	0.05g
$CaSO_4 \cdot 2H_2O$	0.02g
KH_2PO_4	0.011g
$SrCl_2 \cdot 6H_2O$	0.01g
NaCl	0.01g
Adenine	0.01g
$FeSO_4.7H_2O$	6.0mg
Na_2MoO_3	0.5mg

pH 7.8 ± 0.2 at 25°C

Preparation of Medium: Add components to distilled/deionized water and bring volume to 1.0L. Mix thoroughly. Distribute into tubes or flasks. Autoclave for 15 min at 15 psi pressure–121°C.

Use: For the cultivation of *Pseudomonas* species and other microorganisms which can utilize benzoate as sole carbon source.

Bushnell-Haas Agar

Composition per liter:

Agar	15.0g
KH_2PO_4	1.0g
K_2HPO_4	1.0g
NH_4NO_3	1.0g
$MgSO_4 \cdot 7H_2O$	0.2g
$FeCl_3$	0.05g
$CaCl_2 \cdot 2H_2O$	0.02g

pH 7.0 ± 0.2 at 25°C

Preparation of Medium: Add components to distilled/deionized water and bring volume to 1.0L. Mix thoroughly. Gently heat and bring to boiling. Distribute into tubes or flasks. Autoclave for 15 min at 15 psi pressure–121°C. Pour into sterile Petri dishes or leave in tubes. For use in cultivating hydrocarbon-utilizing bacteria, layer 0.1-1.0% hydrocarbon on agar surface or aseptically add sterile hydrocarbon to cooled agar prior to pouring plates.

Use: For examining fuels for microbial contamination and for studying hydrocarbon utilization by microorganisms. Also for the cultivation of *Nocardia* species.

Butanediol Medium

Composition per liter:

$NaH_2PO_4 \cdot H_2O$	2.1g
1,4-Butanediol	1.0g
NaCl	1.0g
NH_4Cl	1.0g
$CaCl_2 \cdot 2H_2O$	0.5g
$MgSO_4 \cdot 7H_2O$	0.5g

K_2HPO_4 .... 0.3g
Yeast extract 0.2g
Modified Wolfe's mineral solution 10.0mL
pH 7.0 ± 0.2 at 25°C

Modified Wolfe's Mineral Solution:
Composition per liter:
$MgSO_4 \cdot 7H_2O$ 3.0g
Nitrilotriacetic acid 1.5g
NaCl 1.0g
$MnSO_4 \cdot H_2O$ 0.5g
$CaCl_2$ 0.1g
$CoCl_2 \cdot 6H_2O$ 0.1g
$FeSO_4 \cdot 7H_2O$ 0.1g
$ZnSO_4 \cdot 7H_2O$ 0.1g
$AlK(SO_4)_2 \cdot 12H_2O$ 0.01g
$CuSO_4 \cdot 5H_2O$ 0.01g
H_3BO_3 .... 0.01g
$Na_2MoO_4 \cdot 2H_2O$ 0.01g
Na_2SeO_3 .... 0.01g
$NaWO_4 \cdot 2H_2O$ 0.01g
$NiCl_2 \cdot 6H_2O$ 0.01g

Preparation of Modified Wolfe's Mineral Solution: Add nitrilotriacetic acid to 500.0mL of distilled/deionized water. Adjust pH to 6.5 with KOH. Add remaining components one at a time. Add distilled/deionized water to 1.0L. Adjust pH to 6.8.

Preparation of Medium: Add components to distilled/deionized water and bring volume to 1.0L. Mix thoroughly. Adjust pH to 7.0. Distribute into tubes or flasks. Autoclave for 15 min at 15 psi pressure–121°C.

Use: For the cultivation of *Pseudomonas putida.*

Butyrivibrio species Medium

Composition per 1001.0mL:
Na_2CO_3 .... 4.0g
Pancreatic digest of casein 2.0g
Yeast extract 2.0g
K_2HPO_4 .... 0.3g
Hemin 1.0mg
Resazurin 1.0mg
Rumen fluid, clarified 150.0mL
Minerals solution 75.0mL
Carbohydrate solution 20.0mL
L-Cysteine·HCl·H_2O solution 10.0mL
$Na_2S \cdot 9H_2O$ solution 10.0mL
Volatile fatty acid mixture 3.1mL
pH 6.7 ± 0.2 at 25°C

Minerals Solution:
Composition per liter:
NaCl 12.0g
KH_2PO_4 .... 6.0g
$(NH_4)_2SO_4$ 6.0g
$MgSO_4 \cdot 7H_2O$ 2.5g
$CaCl_2 \cdot 2H_2O$ 1.6g

Preparation of Minerals Solution: Add components to distilled/deionized water and bring volume to 1.0L. Mix thoroughly.

L-Cysteine·HCl·H_2O Solution:
Composition per 10.0mL:
L-Cysteine·HCl·H_2O 0.25g

Preparation of L-Cysteine·HCl·H_2O Solution: Add L-cysteine·HCl·H_2O to distilled/deionized water and bring volume to 10.0mL. Mix thoroughly. Sparge with 100% CO_2. Autoclave for 15 min at 15 psi pressure–121°C.

$Na_2S \cdot 9H_2O$ Solution:
Composition per 10.0mL:
$Na_2S \cdot 9H_2O$ 0.25g

Preparation of $Na_2S \cdot 9H_2O$ Solution: Add $Na_2S \cdot 9H_2O$ to distilled/deionized water and bring volume to 10.0mL. Mix thoroughly. Sparge with 100% CO_2. Autoclave for 15 min at 15 psi pressure–121°C.

Carbohydrate Solution:
Composition per 20.0mL:
Glucose 1.0g
Cellobiose 1.0g
Glycerol 1.0g
Maltose 1.0g
Starch, soluble 1.0g

Preparation of Carbohydrate Solution: Add components to distilled/deionized water and bring volume to 20.0mL. Mix thoroughly. Sparge under 100% CO_2. Autoclave for 15 min at 15 psi pressure–121°C.

Volatile Fatty Acid Mixture:
Composition per 7.75mL:
Acetic acid 4.25mL
Propionic acid 1.50mL
Butyric acid 1.0mL
DL-2-Methyl butyric acid 0.25mL
iso-Butyric acid 0.25mL
iso-Valeric acid 0.25mL
n-Valeric acid 0.25mL

Preparation of Volatile Fatty Acid Mixture: Combine components. Mix thoroughly.

Preparation of Medium: Prepare and dispense medium under 100% CO_2. Add components, except carbohydrate solution, Na_2CO_3, L-cysteine·HCl·H_2O solution, and $Na_2S \cdot 9H_2O$ solution, to distilled/deionized water and bring volume to 960.0mL Mix thoroughly. Gently heat and bring to boiling. Continue boiling for 5 min. Cool to room temperature while sparging with 100% CO_2. Add Na_2CO_3. Continue sparging with 100% CO_2 until pH reaches 6.8. Distribute into rubber-stoppered tubes under 100% CO_2. Autoclave for 15 min at 15 psi pressure–121°C. Autoclave for 15 min at 15 psi pressure–121°C. Aseptically and anaerobically add 20.0mL of sterile carbohydrate solution, 10.0mL of sterile L-cysteine·HCl·H_2O solution, and 10.0mL of sterile $Na_2S \cdot 9H_2O$ solution or, using a syringe, inject the appropriate amount of sterile carbohydrate solution, sterile $Na_2S \cdot 9H_2O$ solution, and sterile L-cysteine·HCl·H_2O solution into individual tubes containing medium.

Use: For the cultivation of *Butyrivibrio* species.

BY Agar Medium (ATCC Medium 2038)

Composition per liter:
Agar 15.0g
Yeast extract 5.0g
Pancreatic digest of casein 5.0g
Beef extract 5.0g
NaCl 2.5g
K_2HPO_4 .... 0.1g
$MgSO_4 \cdot 7H_2O$ 0.05g
pH 7.2 ± 0.2 at 25°C

Preparation of Medium: Add components to distilled/deionized water and bring volume to 1.0L. Mix thoroughly. Gently heat and bring to boiling. Distribute into tubes or

flasks. Autoclave for 15 min at 15 psi pressure–121°C. Pour into sterile Petri dishes or leave in tubes.

Use: For the cultivation and maintenance of *Paracoccus thiocyanatus.*

BYEB
(Buffered Yeast Extract Broth)

Composition per liter:

ACES buffer (2-[(2-Amino-2-oxoethyl)-amino]-ethane sulfonic acid)	10.0g
Yeast extract	10.0g
α-Ketoglutarate	1.0g
L-Cysteine·HCl·H_2O	0.4g
$Fe_4(P_2O_7)_3·9H_2O$	0.25g

pH 6.9 ± 0.2 at 25°C

Preparation of Medium: Add components to distilled/deionized water and bring volume to 1.0L. Mix thoroughly. Adjust pH to 6.9. Filter sterilize. Aseptically distribute into sterile tubes or flasks.

Use: For the cultivation of *Legionella pneumophila.*

C/10 Agar

Composition per liter:

Agar	15.0g
Pancreatic digest of casein	3.0g
$CaCl_2·2H_2O$	1.36g

pH 7.2 ± 0.2 at 25°C

Preparation of Medium: Add components to distilled/deionized water and bring volume to 1.0L. Mix thoroughly. Gently heat and bring to boiling. Adjust pH to 7.2. Distribute into tubes or flasks. Autoclave for 15 min at 15 psi pressure–121°C. Pour into sterile Petri dishes or leave in tubes.

Use: For the cultivation of *Cytophaga flevensis, Flexibacter filiformis, Myxococcus fulvus,* and *Myxococcus xanthus.*

C 3G *Spiroplasma* Medium

Composition per liter:

Sucrose	100.0g
Phenol Red	10.0mg
PPLO broth without Crystal Violet	500.0mL
Horse serum	150.0mL
Fresh yeast extract solution	50.0mL
CMRL-1066 medium	5.0mL

pH 7.5 ± 0.2 at 25°C

Source: PPLO broth without Crystal Violet is available as a premixed powder from BD Diagnostic Systems.

PPLO Broth without Crystal Violet:
Composition per 500.0mL:

Beef heart, infusion from	11.52g
Peptone	2.32g
NaCl	1.15g

Preparation of PPLO Broth without Crystal Violet: Add components to distilled/deionized water and bring volume to 500.0mL. Mix thoroughly.

Fresh Yeast Extract Solution:
Composition per 100.0mL:

Baker's yeast, live, pressed, starch-free	25.0g

Preparation of Fresh Yeast Extract Solution: Add the live Baker's yeast to 100.0mL of distilled/deionized water. Autoclave for 90 min at 15 psi pressure–121°C. Allow to stand. Remove supernatant solution. Adjust pH to 6.6–6.8. Filter sterilize.

CMRL-1066 Medium:
Composition per liter:

NaCl	6.8g
$NaHCO_3$	2.2g
D-Glucose	1.0g
KCl	0.4g
L-Cysteine·HCl·H_2O	0.26g
$CaCl_2$, anhydrous	0.2g
$MgSO_4·7H_2O$	0.2g
$NaH_2PO_4·H_2O$	0.14g
L-Glutamine	0.1g
Sodium acetate·$3H_2O$	0.083g
L-Glutamic acid	0.075g
L-Arginine·HCl	0.07g
L-Lysine·HCl	0.07g
L-Leucine	0.06g
Glycine	0.05g
Ascorbic acid	0.05g
L-Proline	0.04g
L-Tyrosine	0.04g
L-Aspartic acid	0.03g
L-Threonine	0.03g
L-Alanine	0.025g
L-Phenylalanine	0.025g
L-Serine	0.025g
L-Valine	0.025g
L-Cystine	0.02g
L-Histidine·HCl·H_2O	0.02g
L-Isoleucine	0.02g
Phenol Red	0.02g
L-Methionine	0.015g
Deoxyadenosine	0.01g
Deoxycytidine	0.01g
Deoxyguanosine	0.01g
Glutathione, reduced	0.01g
Thymidine	0.01g
Hydroxy-L-proline	0.01g
L-Tryptophan	0.01g
Nicotinamide adenine dinucleotide	7.0mg
Tween 80	5.0mg
Sodium glucoronate·H_2O	4.2mg
Coenzyme A	2.5mg
Cocarboxylase	1.0mg
Flavin adenine dinucleotide	1.0mg
Nicotinamide adenine dinucleotide phosphate	1.0mg
Uridine triphosphate	1.0mg
Choline chloride	0.5mg
Cholesterol	0.2mg
5-Methyldeoxycytidine	0.1mg
Inositol	0.05mg
p-Aminobenzoic acid	0.05mg
Niacin	0.025mg
Niacinamide	0.025mg
Pyridoxine	0.025mg
Pyridoxal·HCl	0.025mg
Biotin	0.01mg
D-Calcium pantothenate	0.01mg
Folic acid	0.01mg
Riboflavin	0.01mg
Thiamine·HCl	0.01mg

Source: CMRL-1066 medium is available as a premixed powder from BD Diagnostics.

Preparation of CMRL-1066 Medium: Add components to distilled/deionized water and bring volume to 1.0L. Mix thoroughly. Adjust pH to 7.2. Filter sterilize.

Preparation of Medium: Add components—except horse serum, fresh yeast extract, and CMRL medium—to distilled/deionized water and bring volume to 795.0mL. Adjust pH to 7.5. Autoclave for 15 min at 15 psi pressure–121°C. Aseptically add 150.0mL of sterile horse serum, 50.0mL of sterile fresh yeast extract solution, and 5.0mL of sterile CMRL medium. Distribute into sterile tubes or flasks.

Use: For the cultivation and maintenance of *Spiroplasma* species.

C 3N *Spiroplasma* Medium

Composition per 100.0mL:

Sucrose	12.0g
Phenol Red	10.0mg
PPLO broth without Crystal Violet	50.0mL
Horse serum	20.0mL
Fresh yeast extract solution	5.0mL
CMRL-1066 medium	0.5mL

pH 7.5 ± 0.2 at 25°C

PPLO Broth without Crystal Violet:
Composition per 500.0mL:

Beef heart, infusion from	11.52g
Peptone	2.32g
NaCl	1.15g

Source: PPLO broth without Crystal Violet is available as a premixed powder from BD Diagnostic Systems.

Preparation of PPLO Broth without Crystal Violet: Add components to distilled/deionized water and bring volume to 500.0mL. Mix thoroughly.

Fresh Yeast Extract Solution:
Composition per 100.0mL:

Baker's yeast, live, pressed, starch-free	25.0g

Preparation of Fresh Yeast Extract Solution: Add the live Baker's yeast to 100.0mL of distilled/deionized water. Autoclave for 90 min at 15 psi pressure–121°C. Allow to stand. Remove supernatant solution. Adjust pH to 6.6–6.8. Filter sterilize.

CMRL-1066 Medium:
Composition per liter:

NaCl	6.8g
$NaHCO_3$	2.2g
D-Glucose	1.0g
KCl	0.4g
L-Cysteine·HCl·H_2O	0.26g
$CaCl_2$, anhydrous	0.2g
$MgSO_4 \cdot 7H_2O$	0.2g
$NaH_2PO_4 \cdot H_2O$	0.14g
L-Glutamine	0.1g
Sodium acetate·$3H_2O$	0.083g
L-Glutamic acid	0.075g
L-Arginine·HCl	0.07g
L-Lysine·HCl	0.07g
L-Leucine	0.06g
Glycine	0.05g
Ascorbic acid	0.05g
L-Proline	0.04g
L-Tyrosine	0.04g
L-Aspartic acid	0.03g
L-Threonine	0.03g
L-Alanine	0.025g
L-Phenylalanine	0.025g
L-Serine	0.025g
L-Valine	0.025g
L-Cystine	0.02g
L-Histidine·HCl·H_2O	0.02g
L-Isoleucine	0.02g
Phenol Red	0.02g
L-Methionine	0.015g
Deoxyadenosine	0.01g
Deoxycytidine	0.01g
Deoxyguanosine	0.01g
Glutathione, reduced	0.01g
Thymidine	0.01g
Hydroxy-L-proline	0.01g
L-Tryptophan	0.01g
Nicotinamide adenine dinucleotide	7.0mg
Tween 80	5.0mg
Sodium glucoronate·H_2O	4.2mg
Coenzyme A	2.5mg
Cocarboxylase	1.0mg
Flavin adenine dinucleotide	1.0mg
Nicotinamide adenine dinucleotide phosphate	1.0mg
Uridine triphosphate	1.0mg
Choline chloride	0.5mg
Cholesterol	0.2mg
5-Methyldeoxycytidine	0.1mg
Inositol	0.05mg
p-Aminobenzoic acid	0.05mg
Niacin	0.025mg
Niacinamide	0.025mg
Pyridoxine	0.025mg
Pyridoxal·HCl	0.025mg
Biotin	0.01mg
D-Calcium pantothenate	0.01mg
Folic acid	0.01mg
Riboflavin	0.01mg
Thiamine·HCl	0.01mg

Source: CMRL-1066 medium is available as a premixed powder from BD Diagnostics.

Preparation of CMRL-1066 Medium: Add components to distilled/deionized water and bring volume to 1.0L. Mix thoroughly. Adjust pH to 7.2. Filter sterilize.

Preparation of Medium: Add components—except horse serum, fresh yeast extract, and CMRL medium—to distilled/deionized water and bring volume to 75.0mL. Adjust pH to 7.5. Autoclave for 15 min at 15 psi pressure–121°C. Aseptically add 20.0mL of sterile horse serum, 5.0mL of sterile yeast extract, and 0.5mL of sterile CMRL medium. Distribute into sterile tubes or flasks.

Use: For the cultivation and maintenance of *Spiroplasma kunkelii.*

CAL Agar
(Cellobiose Arginine Lysine Agar)
(Yersinia Isolation Agar)

Composition per liter:

Agar	20.0g
L-Arginine·HCl	6.5g
L-Lysine·HCl	6.5g
NaCl	5.0g
Cellobiose	3.5g
Yeast extract	3.0g
Sodium deoxycholate	1.5g
Neutral Red	0.03g

Preparation of Medium: Add components to distilled/deionized water and bring volume to 1.0L. Mix thoroughly. Heat to boiling. Do not autoclave. Pour into sterile Petri dishes.

Use: For the isolation and characterization of *Yersinia enterocolitica* and enumeration of *Yersinia enterocolitica* from water and other liquid specimens.

Caldicellulosiruptor Medium

Composition per liter:

Pancreatic digest of casein	2.0g
K_2HPO_4	1.5g
Cellobiose	1.0g
Yeast extract	1.0g
NaCl	0.9g
NH_4Cl	0.9g
KH_2PO_4	0.75g
L-Cysteine·HCl	0.75g
$MgCl_2 \cdot 6H_2O$	0.4g
$FeCl_3 \cdot 6H_2O$	2.5mg
Resazurin	0.5mg
Trace elements solution SL-10	1.0mL

pH 7.2 ± 0.2 at 25°C

Trace Elements Solution SL-10:

Composition per liter:

$FeCl_2 \cdot 4H_2O$	1.5g
$CoCl_2 \cdot 6H_2O$	190.0mg
$MnCl_2 \cdot 4H_2O$	100.0mg
$ZnCl_2$	70.0mg
$Na_2MoO_4 \cdot 2H_2O$	36.0mg
$NiCl_2 \cdot 6H_2O$	24.0mg
H_3BO_3	6.0mg
$CuCl_2 \cdot 2H_2O$	2.0mg
HCl (25% solution)	10.0mL

Preparation of Trace Elements Solution SL-10: Add $FeCl_2 \cdot 4H_2O$ to 10.0mL of HCl solution. Mix thoroughly. Add distilled/deionized water and bring volume to 1.0L. Add remaining components. Mix thoroughly. Sparge with 100% N_2. Autoclave for 15 min at 15 psi pressure–121°C.

Preparation of Medium: Prepare and dispense medium under 100% N_2. Add components to distilled/deionized water and bring volume to 1.0L. Mix thoroughly. Sparge with 100% N_2. Anaerobically distribute into tubes or flasks. Autoclave for 15 min at 15 psi pressure–121°C.

Use: For the cultivation of *Caldicellulosiruptor saccharolyticus.*

Caldicellulosiruptor Medium

Composition per 1001.0mL:

Pancreatic digest of casein	2.0g
K_2HPO_4	1.5g
Cellulose	1.0g
Cellobiose	1.0g
Yeast extract	1.0g
NaCl	0.9g
NH_4Cl	0.9g
KH_2PO_4	0.75g
$MgCl_2 \cdot 6H_2O$	0.4g
$FeCl_3 \cdot 6H_2O$	2.5mg
L-Cysteine·HCl	0.75g
Resazurin	0.5mg

pH 7.2 ± 0.2 at 25°C

Trace Elements Solution SL-10:

Composition per liter:

$FeCl_2 \cdot 4H_2O$	1.5g
$CoCl_2 \cdot 6H_2O$	190.0mg
$MnCl_2 \cdot 4H_2O$	100.0mg
$ZnCl_2$	70.0mg
$Na_2MoO_4 \cdot 2H_2O$	36.0mg
$NiCl_2 \cdot 6H_2O$	24.0mg
H_3BO_3	6.0mg
$CuCl_2 \cdot 2H_2O$	2.0mg
HCl (25% solution)	10.0mL

Preparation of Trace Elements Solution SL-10: Add $FeCl_2 \cdot 4H_2O$ to 10.0mL of HCl solution. Mix thoroughly. Add distilled/deionized water and bring volume to 1.0L. Add remaining components. Mix thoroughly. Sparge with 100% N_2. Autoclave for 15 min at 15 psi pressure–121°C.

Preparation of Medium: Prepare and dispense medium under 100% N_2. Add components to distilled/deionized water and bring volume to 1.0L. Mix thoroughly. Sparge with 100% N_2. Anaerobically distribute into tubes or flasks. Autoclave for 15 min at 15 psi pressure–121°C.

Use: For the cultivation of *Caldicellulosiruptor saccharolyticus.*

Camphor Minimal Medium

Composition per liter:

Agar	20.0g
K_2HPO_4	4.4g
NH_4Cl	2.1g
KH_2PO_4	1.7g
100× salt solution	10.0mL

100× Salt Solution:

Composition per liter:

$MgSO_4$	19.5g
$FeSO_4 \cdot 7H_2O$	5.0g
$MnSO_4 \cdot H_2O$	5.0g
Ascorbic acid	1.0g
$CaCl_2 \cdot 2H_2O$	0.3g

Preparation of 100× Salt Solution: Add components to distilled/deionized water and bring volume to 1.0L. Mix thoroughly.

Preparation of Medium: Add components to distilled/deionized water and bring volume to 1.0L. Gently heat and bring to boiling. Autoclave for 15 min at 15 psi pressure–121°C. Pour into sterile Petri dishes. Allow to cool to room temperature. Invert Petri dishes. Spread 0.2mL of 2m D-(+) camphor solution in methylene chloride (CH_2Cl_2) on the inside cover of each plate.

Use: For the cultivation and maintenance of *Pseudomonas putida.*

Carbohydrate Fermentation Broth

Composition per liter:

Peptone	10.0g
NaCl	5.0g
Meat extract	3.0g
Carbohydrate solution	50.0mL
Andrade's indicator	10.0mL

pH 7.1 ± 0.2 at 25°C

Andrade's Indicator:

Composition per 100.0mL:

Acid Fuchsin	0.1 g
NaOH (1*N* solution)	16.0mL

Preparation of Andrade's Indicator: Add components to distilled/deionized water and bring volume to 100.0mL. Mix thoroughly.

Carbohydrate Solution:

Composition per 100.0mL:

Carbohydrate	10.0g

Preparation of Carbohydrate Solution: Add carbohydrate to distilled/deionized water and bring volume to

100.0mL. Adonitol, arabinose, cellobiose, glucose, dulcitol, fructose, galactose, inositol, lactose, maltose, mannitol, raffinose, rhamnose, salicin, sorbitol, sucrose, trehalose, xylose, or other carbohydrates may be used. Mix thoroughly. Filter sterilize.

Caution: Acid Fuchsin is a potential carcinogen and care must be taken to avoid inhalation of the powdered dye and contact with the skin.

Preparation of Medium: Add components, except carbohydrate solution, to distilled/deionized water and bring volume to 1.0L. Mix thoroughly. Gently heat and bring to boiling. Distribute in 10.0mL volumes into test tubes containing inverted Durham tubes. Autoclave for 15 min at 15 psi pressure–121°C. Cool to 25°C. Add 0.5mL of sterile carbohydrate solution to each tube.

Use: For the determination of carbohydrate fermentation reactions of microorganisms, particularly members of the Enterobacteriaceae. A Durham tube is used to collect gas produced during the fermentation reaction. Acid production is indicated by a pink reaction.

Carbon Monoxide Oxidizers Agar, Modified

Composition per 1001.0mL:

Agar	12.0g
$Na_2HPO_4 \cdot 12H_2O$	4.5g
Sodium acetate	3.0g
NH_4Cl	1.5g
KH_2PO_4	0.75g
$MgSO_4 \cdot 7H_2O$	0.2g
$CaCl_2 \cdot 2H_2O$	30.0mg
Ferric ammonium citrate	18.0mg
Trace elements solution	1.0mL
$NaHCO_3$ solution	10.0mL
Thiamine·HCl solution	10.0mL

Trace Elements Solution:
Composition per liter:

$Na_2MoO_4 \cdot 2H_2O$	0.9g
H_3BO_3	0.3g
$CoCl_2 \cdot 6H_2O$	0.2g
$ZnSO_4 \cdot 7H_2O$	0.1g
$MnCl_2 \cdot 4H_2O$	30.0mg
Na_2SeO_3	20.0mg
$NiCl_2 \cdot 6H_2O$	20.0mg
$CuCl_2 \cdot 2H_2O$	10.0mg

Preparation of Trace Elements Solution: Add components to distilled/deinonized water and bring volume to 1.0L. Mix thoroughly.

$NaHCO_3$ Solution:
Composition per 10.0mL:

$NaHCO_3$	1.0g

Preparation of $NaHCO_3$ Solution: Add $NaHCO_3$ to distilled/deionized water and bring volume to 10.0mL. Filter sterilize.

Thiamine·HCl Solution:
Composition per 10.0mL:

Thiamine·HCl	20.0μg

Preparation of Thiamine·HCl Solution: Add thiamine·HCl to distilled/deionized water and bring volume to 10.0mL. Filter sterilize.

Preparation of Medium: Add components, except $NaHCO_3$ solution and thiamine·HCl solution, to distilled/deionized water and bring volume to 980.0mL. Mix thoroughly. Gently heat and bring to boiling. Autoclave for 15 min at 15 psi pressure–121°C. Cool to 50°–55°C. Aseptically add 10.0mL of sterile $NaHCO_3$ solution and 10.0mL of sterile thiamine·HCl solution. Mix thoroughly. Pour into sterile Petri dishes or distribute into sterile tubes.

Use: For the cultivation of *Carbophilus carboxidus* and *Zavarzinia compransoris.*

Carbon Utilization Test

Composition per liter:

Ionagar	10.0g
NH_4Cl	1.0g
$MgSO_4 \cdot 7H_2O$	0.5g
Ferric ammonium citrate	0.05g
$CaCl_2$	0.5mg
Sodium potassium phosphate buffer (0.33*M* solution, pH 6.8)	1.0L
Carbon source	10.0mL

pH 6.8 ± 0.2 at 25°C

Carbon Source:
Composition per 10.0mL:

Carbon source	1.0g

Preparation of Carbon Source: Add carbon source to distilled/deionized water and bring volume to 10.0mL. Mix thoroughly. Filter sterilize.

Preparation of Medium: Add components, except carbon source, to distilled/deionized water and bring volume to 990.0mL. Mix thoroughly. Gently heat and bring to boiling. Autoclave for 15 min at 15 psi pressure–121°C. Cool to 45°–50°C. Aseptically add sterile carbon source. Mix thoroughly. Pour into sterile Petri dishes or distribute into sterile tubes.

Use: For the cultivation and differentiation of *Pseudomonas* species based on their ability to utilize a specific carbon source.

Carbonate-Buffered Medium CMB4 with Glucose

Composition per 1002.0mL:

NaCl	4.0g
$NaHCO_3$	2.5g
Glucose	1.0g
$MgCl_2 \cdot 6H_2O$	0.8g
KCl	0.5g
NH_4Cl	0.3g
KH_2PO_4	0.2g
Resazurin	1.0mg
Modified Wolfe's mineral solution	10.0mL
Wolfe's vitamin solution	10.0mL
Sulfide-calcium solution	2.0mL

pH 6.9 ± 0.2 at 25°C

Modified Wolfe's Mineral Solution:
Composition per liter:

$MgSO_4 \cdot 7H_2O$	3.0g
Nitrilotriacetic acid	1.5g
NaCl	1.0g
$MnSO_4 \cdot H_2O$	0.5g
$CaCl_2$	0.1g
$CoCl_2 \cdot 6H_2O$	0.1g
$FeSO_4 \cdot 7H_2O$	0.1g
$ZnSO_4 \cdot 7H_2O$	0.1g
$AlK(SO_4)_2 \cdot 12H_2O$	0.01g
$CuSO_4 \cdot 5H_2O$	0.01g
H_3BO_3	0.01g
$Na_2MoO_4 \cdot 2H_2O$	0.01g

Na_2SeO_3	0.01g
$NaWO_4·2H_2O$	0.01g
$NiCl_2·6H_2O$	0.01g

Preparation of Modified Wolfe's Mineral Solution: Add nitrilotriacetic acid to 500.0mL of distilled/deionized water. Adjust pH to 6.5 with KOH. Add remaining components one at a time. Add distilled/deionized water to 1.0L. Adjust pH to 6.8.

Wolfe's Vitamin Solution:
Composition per liter:

Pyridoxine·HCl	10.0mg
p-Aminobenzoic acid	5.0mg
Lipoic acid	5.0mg
Nicotinic acid	5.0mg
Riboflavin	5.0mg
Thiamine·HCl	5.0mg
Calcium DL-pantothenate	5.0mg
Biotin	2.0mg
Folic acid	2.0mg
Vitamin B_{12}	0.1mg

Preparation of Wolfe's Vitamin Solution: Add components to distilled/deionized water and bring volume to 1.0L. Mix thoroughly.

Sulfide-Calcium Solution:
Composition per liter:

$Na_2S·9H_2O$	36.0g
$CaCl_2·2H_2O$	15.0g

Preparation of Sulfide-Calcium Solution: Add components to distilled/deionized water and bring volume to 1.0L. Mix thoroughly. Adjust pH to 7.2. Sparge with 100% N_2. Anaerobically distribute into tubes. Autoclave at 121°C for 15 min.

Preparation of Medium: Prepare and dispense medium under 80% N_2 + 20% CO_2. Add components, except $NaHCO_3$ and sulfide-calcium solution, to distilled/deionized water and bring volume to 1.0L. Mix thoroughly. Gently heat and bring to boiling. Continue boiling for 3 min. Cool to room temperature while sparging with 80% N_2 + 20% CO_2. Add $NaHCO_3$. Mix thoroughly. Anaerobically distribute 10.0mL volumes into anaerobic tubes. Autoclave for 15 min at 15 psi pressure–121°C. Aseptically and anaerobically add 0.02mL of sterile sulfide-calcium solution to each tube. Mix thoroughly. Adjust pH to 6.9.

Use: For the cultivation of *Spirochaeta thermophila.*

Carboxydobacterium Medium

Composition per liter:

$Na_2HPO_4·12H_2O$	9.0g
KH_2PO_4	1.5g
NH_4Cl	1.5g
$MgSO_4·7H_2O$	0.2g
$CaCl_2·2H_2O$	20.0mg
Ferric ammonium citrate	1.2mg
TS2 trace elements solution	1.0mL

TS2 Trace Elements Solution:
Composition per liter:

$Na_2MoO_4·2H_2O$	0.9g
H_3BO_3	0.3g
$CoCl_2·6H_2O$	0.2g
$ZnSO_4·7H_2O$	0.1g
$MnCl_2·4H_2O$	30.0mg
Na_2SeO_3	20.0mg
$NiCl_2·6H_2O$	20.0mg
$CuCl_2·2H_2O$	10.0mg

Preparation of TS2 Trace Elements Solution: Add components to distilled/deionized water and bring volume to 1.0L. Mix thoroughly.

Caution: Carbon monoxide (CO) is a toxic gas.

Preparation of Medium: Add components to distilled/deionized water and bring volume to 1.0L. Mix thoroughly. Distribute into tubes or flasks. Autoclave for 15 min at 15 psi pressure–121°C. After inoculation, incubate in an atmosphere of 50% CO + 50% air.

Use: For the autotrophic cultivation of *Oligotropha carboxidovorans.*

Carboxydobacterium Medium

Composition per liter:

$Na_2HPO_4·12H_2O$	9.0g
Sodium acetate	3.0g
KH_2PO_4	1.5g
NH_4Cl	1.5g
$MgSO_4·7H_2O$	0.2g
$CaCl_2·2H_2O$	20.0mg
Ferric ammonium citrate	1.2mg
TS2 trace elements solution	1.0mL

TS2 Trace Elements Solution:
Composition per liter:

$Na_2MoO_4·2H_2O$	0.9g
H_3BO_3	0.3g
$CoCl_2·6H_2O$	0.2g
$ZnSO_4·7H_2O$	0.1g
$MnCl_2·4H_2O$	30.0mg
Na_2SeO_3	20.0mg
$NiCl_2·6H_2O$	20.0mg
$CuCl_2·2H_2O$	10.0mg

Preparation of TS2 Trace Elements Solution: Add components to distilled/deionized water and bring volume to 1.0L. Mix thoroughly.

Preparation of Medium: Add components to distilled/deionized water and bring volume to 1.0L. Mix thoroughly. Distribute into tubes or flasks. Autoclave for 15 min at 15 psi pressure–121°C. After inoculation, incubate in air.

Use: For the organotrophic cultivation of *Oligotropha carboxidovorans.*

Carboxydothermus Medium

Composition per 1030.0mL:

$MgCl_2·6H_2O$	0.52g
KCl	0.33g
KH_2PO_4	0.33g
NH_4Cl	0.33g
$CaCl_2·2H_2O$	0.29g
Resazurin	0.5mg
Trace elements solution SL-4	10.0mL
Vitamin solution	10.0mL
Yeast extract solution	10.0mL
$Na_2S·9H_2O$ solution	10.0mL

pH 6.9 ± 0.2 at 25°C

Trace Elements Solution SL-4:
Composition per liter:

EDTA	0.5g
$FeSO_4·7H_2O$	0.2g
Trace elements solution SL-6	100.0mL

Preparation of Trace Elements Solution SL-4: Add components to distilled/deionized water and bring volume to 1.0L. Mix thoroughly.

Trace Elements Solution SL-6:
Composition per liter:
$MnCl_2 \cdot 4H_2O$ 0.5g
H_3BO_3 .... 0.3g
$CoCl_2 \cdot 6H_2O$ 0.2g
$ZnSO_4 \cdot 7H_2O$ 0.1g
$Na_2MoO_4 \cdot 2H_2O$ 0.03g
$NiCl_2 \cdot 6H_2O$ 0.02g
$CuCl_2 \cdot 2H_2O$ 0.01g

Preparation of Trace Elements Solution SL-6: Add components to distilled/deionized water and bring volume to 1.0L. Mix thoroughly.

Vitamin Solution:
Composition per liter:
Pyridoxine·HCl 10.0mg
Calcium DL-pantothenate 5.0mg
Lipoic acid 5.0mg
Nicotinic acid 5.0mg
p-Aminobenzoic acid 5.0mg
Riboflavin 5.0mg
Thiamine·HCl 5.0mg
Biotin 2.0mg
Folic acid 2.0mg
Vitamin B_{12} 0.1mg

Preparation of Vitamin Solution: Add components to distilled/deionized water and bring volume to 1.0L. Mix thoroughly. Filter sterilize. Sparge with 100% N_2.

Yeast Extract Solution:
Composition per 10.0mL:
Yeast extract 10.0mg

Preparation of Yeast Extract Solution: Add yeast extract to distilled/deionized water and bring volume to 10.0mL. Mix thoroughly. Autoclave under 100% N_2 for 15 min at 15 psi pressure–121°C.

$Na_2S \cdot 9H_2O$ Solution:
Composition per 10.0mL:
$Na_2S \cdot 9H_2O$ 0.7g

Preparation of $Na_2S \cdot 9H_2O$ Solution: Prepare and dispense solution anaerobically under 100% N_2. Add $Na_2S \cdot 9H_2O$ to distilled/deionized water and bring volume to 10.0mL. Mix thoroughly. Adjust pH to 7.0. Autoclave for 15 min at 15 psi–121°C.

Preparation of Medium: Add components, except vitamin solution, yeast extract solution, and $Na_2S \cdot 9H_2O$ solution, to distilled/deionized water and bring volume to 1.0L. Mix thoroughly. Sparge under 100% N_2 for 10 min. Autoclave under 100% N_2 for 15 min at 15 psi pressure–121°C. Aseptically and anaerobically combine with 10.0mL of sterile vitamin solution, 10.0mL of sterile yeast extract solution, and 10.0mL of sterile $Na_2S \cdot 9H_2O$ solution. Mix thoroughly. Pressurize inoculation flask with CO (carbon monoxide) gas at 2 bar pressure.

Caution: Carbon monoxide is a toxic gas.

Use: For the cultivation and maintenance of *Carboxydothermus hydrogenoformans.*

Carcboxydobrachium pacificum Medium (DSMZ Medium 902)

Composition per 1050mL:
NaCl 20.0g
$MgSO_4 \cdot 7H_2O$ 3.9g
KCl 0.7g
$CaCl_2 \cdot 2H_2O$ 0.4g
NH_4Cl 0.3g
Na_2HPO_4 .... 0.15g
Yeast extract 0.05g
Na_2SiO_3 .... 0.03g
Resazurin 0.5mg
Vitamin solution 10.0mL
$NaHCO_3$ solution 10.0mL
$Na_2S \cdot 9H_2O$ solution 10.0mL
L-Cysteine solution 10.0mL
Pyruvate solution 10.0mL
Trace elements solution SL-10 .... 1.0mL
pH 7.0 ± 0.2 at 25°C

Pyruvate Solution:
Composition per 10.0mL:
Na-pyruvate 2.5g

Preparation of Pyruvate Solution: Add pyruvate to distilled/deionized water and bring volume to 10.0mL. Mix thoroughly. Sparge with 100% N_2. Autoclave for 15 min at 15 psi pressure–121°C.

L-Cysteine Solution:
Composition per 10.0mL:
L-Cysteine·HCl·H_2O 0.45g

Preparation of L-Cysteine Solution: Add L-cysteine·HCl·H_2O to distilled/deionized water and bring volume to 10.0mL. Mix thoroughly. Sparge with 100% N_2. Autoclave for 15 min at 15 psi pressure–121°C.

$Na_2S \cdot 9H_2O$ Solution:
Composition per 10.0mL:
$Na_2S \cdot 9H_2O$ 0.45g

Preparation of $Na_2S \cdot 9H_2O$ Solution: Add $Na_2S \cdot 9H_2O$ to distilled/deionized water and bring volume to 10.0mL. Mix thoroughly. Sparge with 100% N_2. Autoclave for 15 min at 15 psi pressure–121°C.

$NaHCO_3$ Solution:
Composition per 10.0mL:
$NaHCO_3$ 0.5g

Preparation of $NaHCO_3$ Solution: Add $NaHCO_3$ to distilled/deionized water and bring volume to 10.0mL. Mix thoroughly. Autoclave for 15 min at 15 psi pressure–121°C. Cool to 25°C. Must be prepared freshly.

Trace Elements Solution SL-10:
Composition per liter:
$FeCl_2 \cdot 4H_2O$ 1.5g
$CoCl_2 \cdot 6H_2O$ 190.0mg
$MnCl_2 \cdot 4H_2O$ 100.0mg
$ZnCl_2$ 70.0mg
$Na_2MoO_4 \cdot 2H_2O$ 36.0mg
$NiCl_2 \cdot 6H_2O$ 24.0mg
H_3BO_3 .... 6.0mg
$CuCl_2 \cdot 2H_2O$ 2.0mg
HCl (25% solution) 10.0mL

Preparation of Trace Elements Solution SL-10: Add $FeCl_2 \cdot 4H_2O$ to 10.0mL of HCl solution. Mix thoroughly. Add distilled/deionized water and bring volume to 1.0L. Add remaining components. Mix thoroughly. Sparge with 100% N_2. Autoclave for 15 min at 15 psi pressure–121°C.

Vitamin Solution:
Composition per liter:
Pyridoxine-HCl 10.0mg
Thiamine-HCl·$2H_2O$ 5.0mg
Riboflavin 5.0mg
Nicotinic acid 5.0mg
D-Ca-pantothenate 5.0mg

p-Aminobenzoic acid	5.0mg
Lipoic acid	5.0mg
Biotin	2.0mg
Folic acid	2.0mg
Vitamin B_{12}	0.1mg

Preparation of Vitamin Solution: Add components to distilled/deionized water and bring volume to 1.0L. Mix thoroughly. Sparge with 80% H_2 + 20% CO_2. Filter sterilize.

Preparation of Medium: Prepare and dispense medium under 100% N_2. Add components, except vitamin solution, $NaHCO_3$ solution, pyruvate solution, L-Cysteine-HCl·H_2O solution, and $Na_2S·9H_2O$ solution, to distilled/deionized water and bring volume to 1.0L. Mix thoroughly. Adjust pH to 7.0. Sparge with 100% N_2 of at least 30 min. Distribute into anaerobe tubes or bottles. Autoclave for 15 min at 15 psi pressure–121°C. Aseptically and anaerobically add per liter, 10.0mL vitamin solution, 10.0mL $NaHCO_3$ solution, 10.0mL pyruvate solution, 10.0mL L-Cysteine-HCl·H_2O solution, and 10.0mL $Na_2S·9H_2O$. Mix thoroughly. The final pH should be 7.0.

Use: For the cultivation of *Carboxydibrachium pacificum (Carboxydobrachium pacificum).*

Caryophanon latum Medium

Composition per liter:

Papaic digest of soybean meal	2.0g
Pancreatic digest of casein	2.0g
Yeast extract	2.0g
K_2HPO_4	1.0g
Sodium acetate	1.0g
$MgSO_4·7H_2O$	0.27g
Sodium glutamate	0.1g
Thiamine·HCl	0.2mg
Biotin	0.05mg
Tris/HCl-buffer 10m*M*, pH 7.8	1000.0mL

pH 7.2 ± 0.2 at 25°C

Preparation of Medium: Combine components. Mix thoroughly. Distribute into tubes or flasks. Autoclave for 15 min at 15 psi pressure–121°C.

Use: For the cultivation and maintenance of *Caryophanon latum* and *Vitreoscilla stercoraria.*

CAS Medium

Composition per liter:

Pancreatic digest of casein	10.0g
$MgSO_4·7H_2O$	1.0g
K_2HPO_4	0.25g

pH 6.8 ± 0.2 at 25°C

Preparation of Medium: Add components to distilled/deionized water and bring volume to 1.0L. Mix thoroughly. Distribute into tubes or flasks. Autoclave for 15 min at 15 psi pressure–121°C.

Use: For the cultivation of myxobacteria.

Casamino Acids Glucose Medium (CAGV Medium)

Composition per liter:

Agar	20.0g
Glucose	1.0g
Vitamin-free casamino acids	1.0g
Solution A (Mineral Salts)	20.0mL
Vitamin solution	10.0mL

pH 7.2 ± 0.2 at 25°C

Solution A (Mineral Salts):
Composition per liter:

$MgSO_4·7H_2O$	29.7g
$NaMoO_4·2H_2O$	12.67g
Nitrilotriacetic acid	10.0g
$CaCl_2·2H_20$	3.34g
$FeSO_4·7H_2O$	0.1g
Solution B (Metallic salts)	50.0mL

Preparation of Solution A (Mineral Salts): Add nitrilotriacetic acid to 500.0mL of distilled/deionized water. Dissolve by adjusting pH to 6.5 with KOH. Add remaining components. Readjust pH to 7.2 with H_2SO_4 or KOH. Add distilled/deionized water to 1.0L.

Solution B (Metallic Salts):
Composition per 100.0mL:

$ZnSO_4·7H_2O$	1.1g
$FeSO_4·7H_20$	0.5g
Ethylenediaminetetraacetic acid	0.3g
$MnSO_4·H_2O$	0.3g
$CuSO_4·5H_2O$	0.04g
$CoCL_2·6H_20$	0.02g
$Na_2B_4O_7·10H_20$	0.02g

Preparation of Solution B (Metallic Salts): Add a few drops of H_2SO_4 to distilled/deionized water to inhibit precipitate formation. Add components to acidified distilled/deionized water and bring volume to 100.0mL. Mix thoroughly.

Vitamin Solution:
Composition per liter:

Pyridoxine·HCL	0.01g
Calcium pantothenate	5.0mg
Nicotinamide	5.0mg
Riboflavin	5.0mg
Thiamine·HCl	5.0mg
Biotin	2.0mg
Folic acid	2.0mg
Vitamin B_{12}	0.1mg

Preparation of Vitamin Solution: Add components to distilled/deionized water and bring volume to 1.0L. Mix thoroughly. Filter sterilize.

Preparation of Medium: Add components to distilled/deionized water and bring volume to 1.0L. Mix thoroughly. Gently heat and bring to boiling. Distribute into tubes or flasks. Autoclave for 15 min at 15 psi pressure–121°C. Pour into sterile Petri dishes or leave in tubes.

Use: For the cultivation and maintenance of *Microcyclus aquaticus.*

Casein Medium

Composition per liter:

NaCl	250.0g
Agar	20.0g
$MgCl_2·6H_2O$	20.0g
Casein hydrolysate	5.0g
Yeast extract	5.0g
KCl	2.0g
$CaCl_2·2H_2O$	0.2g

pH 7.4 ± 0.2 at 25°C

Preparation of Medium: Add components to distilled/deionized water and bring volume to 950.0mL. Mix thoroughly. Gently heat to boiling. Adjust pH to 7.4. Bring volume to 1.0L with distilled/deionized water. Distribute into tubes or flasks. Autoclave for 15 min at 15 psi pressure–121°C. Pour into sterile Petri dishes or leave in tubes.

Use: For the cultivation and maintenance of *Halobacterium* species and other halophilic bacteria.

Casitone Agar

Composition per liter:

Pancreatic digest of casein	20.0g
Agar	15.0g
$MgSO_4 \cdot 7H_2O$	1.0g
Potassium phosphate buffer (0.01*M* solution, pH 7.2)	1.0L

pH 7.2 ± 0.2 at 25°C

Preparation of Medium: Combine components. Mix thoroughly. Gently heat to boiling. Distribute into tubes or flasks. Adjust pH to 7.2. Autoclave for 15 min at 15 psi pressure–121°C. Pour into sterile Petri dishes or leave in tubes.

Use: For the cultivation and maintenance of *Myxococcus* species.

Casitone Glycerol Yeast Autolysate Broth (CGY Autolysate Broth)

Composition per liter:

Pancreatic digest of casein	5.0g
Yeast autolysate	1.0g
Glycerol	10.0mL

Preparation of Medium: Add components to distilled/deionized water and bring volume to 1.0L. Mix thoroughly. Distribute into tubes or flasks. Autoclave for 15 min at 15 psi pressure–121°C.

Use: For the isolation, cultivation, and enumeration, of iron and sulfur bacteria from the *Sphaerotilus* group.

Castenholz D Medium (Medium D)

Composition per liter:

$NaNO_3$	0.7g
Na_2HPO_4	0.11g
KNO_3	0.1g
$MgSO_4 \cdot 7H_2O$	0.1g
Nitrilotriacetic acid	0.1g
$CaSO_4 \cdot 2H_2O$	0.06g
NaCl	8.0mg
$FeCl_3$ solution	1.0mL
Micronutrient solution	0.5mL

pH 7.5 ± 0.2 at 25°C

$FeCl_3$ Solution:

Composition per liter:

$FeCl_3 \cdot 6H_2O$	2.28g

Preparation of $FeCl_3$ Solution: Add $FeCl_3 \cdot 6H_2O$ to distilled/deionized water and bring volume to 1.0L. Mix thoroughly.

Micronutrient Solution:

Composition per liter:

$MnSO_4 \cdot H_2O$	2.28g
H_3BO_3	0.5g
$ZnSO_4 \cdot 7H_2O$	0.5g
$CoCl_2 \cdot 6H_2O$	0.025g
$CuSO_4 \cdot 5H_2O$	0.025g
$Na_2MoO_4 \cdot 2H_2O$	0.025g
H_2SO_4	0.5mL

Preparation of Micronutrient Solution: Add components to distilled/deionized water and bring volume to 1.0L. Mix thoroughly.

Preparation of Medium: Add nitrilotriacetic acid to 500.0mL of distilled/deionized water. Dissolve by adjusting pH to 6.5 with KOH. Add remaining components. Mix thoroughly. Readjust pH to 7.5. Bring volume to 1.0L with distilled/deionized water. Mix thoroughly. Distribute into tubes or flasks. Autoclave for 15 min at 15 psi pressure–121°C.

Use: For the isolation of cyanobacteria, including thermophilic species. For the cultivation of *Chloroflexus* species and *Fischerella* species.

Castenholz D Medium, Modified (Medium D, Modified)

Composition per liter:

NaCl	160.0g
$NaNO_3$	0.69g
Na_2HPO_4	0.111g
KNO_3	0.103g
$MgSO_4 \cdot 7H_2O$	0.1g
Nitrilotriacetic acid	0.1g
$CaSO_4 \cdot 2H_2O$	0.06g
$FeCl_3$	0.3mg
Trace metal solution, Castenholz	1.0mL

pH 7.5 ± 0.2 at 25°C

Trace Metal Solution, Castenholz:

Composition per liter:

$MnSO_4 \cdot H_2O$	2.28g
H_3BO_3	0.5g
$ZnSO_4 \cdot 7H_2O$	0.5g
$Co(NO_3)_2 \cdot 6H_2O$	0.025g
$CuSO_4 \cdot 5H_2O$	0.025g
$Na_2MoO_4 \cdot 2H_2O$	0.025g
H_2SO_4	0.5mL

Preparation of Trace Metal Solution, Castenholz: Add components to distilled/deionized water and bring volume to 1.0L. Mix thoroughly.

Preparation of Medium: Add nitrilotriacetic acid to 500.0mL of distilled/deionized water. Dissolve by adjusting pH to 6.5 with KOH. Add remaining components. Mix thoroughly. Readjust pH to 7.5. Bring volume to 1.0L with distilled/deionized water. Mix thoroughly. Distribute into screw-capped tubes or flasks. Autoclave for 15 min at 15 psi pressure–121°C.

Use: For the isolation of halophilic cyanobacteria.

Castenholz DG Medium (Medium DG)

Composition per liter:

Glycyl-glycine buffer	0.8g
$NaNO_3$	0.7g
Na_2HPO_4	0.11g
KNO_3	0.1g
$MgSO_4 \cdot 7H_2O$	0.1g
Nitrilotriacetic acid	0.1g
$CaSO_4 \cdot 2H_2O$	0.06g
NaCl	8.0mg
$FeCl_3$ solution	1.0mL
Micronutrient solution	0.5mL

pH 7.5 ± 0.2 at 25°C

$FeCl_3$ Solution:

Composition per liter:

$FeCl_3 \cdot 6H_2O$	2.28g

Preparation of $FeCl_3$ Solution: Add $FeCl_3 \cdot 6H_2O$ to distilled/deionized water and bring volume to 1.0L. Mix thoroughly.

Micronutrient Solution:
Composition per liter:

$MnSO_4 \cdot H_2O$	2.28g
H_3BO_3	0.5g
$ZnSO_4 \cdot 7H_2O$	0.5g
$CoCl_2 \cdot 6H_2O$	0.025g
$CuSO_4 \cdot 5H_2O$	0.025g
$Na_2MoO_4 \cdot 2H_2O$	0.025g
H_2SO_4	0.5mL

Preparation of Micronutrient Solution: Add components to distilled/deionized water and bring volume to 1.0L. Mix thoroughly.

Preparation of Medium: Add nitrilotriacetic acid to 500.0mL of distilled/deionized water. Dissolve by adjusting pH to 6.5 with KOH. Add remaining components. Mix thoroughly. Readjust pH to 8.1. Bring volume to 1.0L with distilled/deionized water. Mix thoroughly. Distribute into tubes or flasks. Autoclave for 15 min at 15 psi pressure–121°C.

Use: For the isolation of cyanobacteria, including thermophilic species.

Castenholz DGN Medium (Medium DGN)

Composition per liter:

Glycyl-glycine buffer	0.8g
$NaNO_3$	0.7g
NH_4Cl	0.2g
Na_2HPO_4	0.11g
KNO_3	0.1g
$MgSO_4 \cdot 7H_2O$	0.1g
Nitrilotriacetic acid	0.1g
$CaSO_4 \cdot 2H_2O$	0.06g
NaCl	8.0mg
$FeCl_3$ solution	1.0mL
Micronutrient solution	0.5mL

pH 7.5 ± 0.2 at 25°C

$FeCl_3$ Solution:
Composition per liter:

$FeCl_3 \cdot 6H_2O$	2.28g

Preparation of $FeCl_3$ Solution: Add $FeCl_3 \cdot 6H_2O$ to distilled/deionized water and bring volume to 1.0L. Mix thoroughly.

Micronutrient Solution:
Composition per liter:

$MnSO_4 \cdot H_2O$	2.28g
H_3BO_3	0.5g
$ZnSO_4 \cdot 7H_2O$	0.5g
$CoCl_2 \cdot 6H_2O$	0.025g
$CuSO_4 \cdot 5H_2O$	0.025g
$Na_2MoO_4 \cdot 2H_2O$	0.025g
H_2SO_4	0.5mL

Preparation of Micronutrient Solution: Add components to distilled/deionized water and bring volume to 1.0L. Mix thoroughly.

Preparation of Medium: Add nitrilotriacetic acid to 500.0mL of distilled/deionized water. Dissolve by adjusting pH to 6.5 with KOH. Add remaining components. Mix thoroughly. Readjust pH to 8.2. Bring volume to 1.0L with distilled/deionized water. Mix thoroughly. Distribute into tubes or flasks. Autoclave for 15 min at 15 psi pressure–121°C.

Use: For the isolation of cyanobacteria, including thermophilic species.

Castenholz Medium

Composition per liter:

Tryptone	1.0g
Yeast extract	1.0g
$NaNO_3$	689.0mg
$Na_2HPO_4 \cdot 2H_2O$	140.0mg
KNO_3	103.0mg
$MgSO_4 \cdot 7H_2O$	100.0mg
Nitrilotriacetic acid	100.0mg
$CaSO_4 \cdot 2H_2O$	60.0mg
NaCl	8.0mg
$MnSO_4 \cdot H_2O$	2.2mg
H_3BO_3	0.5mg
$ZnSO_4 \cdot 7H_2O$	0.5mg
$FeCl_3 \cdot 6H_2O$	0.47mg
$CoCl_2 \cdot 6H_2O$	46.0μg
$CuSO_4 \cdot 5H_2O$	25.04μg
$Na_2MoO_4 \cdot 2H_2O$	25.0μg

pH 8.2 ± 0.2 at 25°C

Preparation of Medium: Combine components. Mix thoroughly. Adjust pH to 8.2 with NaOH. Distribute into tubes or flasks. Autoclave for 15 min at 15 psi pressure–121°C.

Use: For the cultivation of *Thermus aquaticus*.

Castenholz ND Medium (Medium ND)

Composition per liter:

Na_2HPO_4	0.11g
$MgSO_4 \cdot 7H_2O$	0.1g
Nitrilotriacetic acid	0.1g
$CaSO_4 \cdot 2H_2O$	0.06g
NaCl	8.0mg
$FeCl_3$ solution	1.0mL
Micronutrient solution	0.5mL

pH 7.5 ± 0.2 at 25°C

$FeCl_3$ Solution:
Composition per liter:

$FeCl_3 \cdot 6H_2O$	2.28g

Preparation of $FeCl_3$ Solution: Add $FeCl_3 \cdot 6H_2O$ to distilled/deionized water and bring volume to 1.0L. Mix thoroughly.

Micronutrient Solution:
Composition per liter:

$MnSO_4 \cdot H_2O$	2.28g
H_3BO_3	0.5g
$ZnSO_4 \cdot 7H_2O$	0.5g
$CoCl_2 \cdot 6H_2O$	0.025g
$CuSO_4 \cdot 5H_2O$	0.025g
$Na_2MoO_4 \cdot 2H_2O$	0.025g
H_2SO_4	0.5mL

Preparation of Micronutrient Solution: Add components to distilled/deionized water and bring volume to 1.0L. Mix thoroughly.

Preparation of Medium: Add nitrilotriacetic acid to 500.0mL of distilled/deionized water. Dissolve by adjusting pH to 6.5 with KOH. Add remaining components. Mix thoroughly. Readjust pH to 8.2. Bring volume to 1.0L with distilled/deionized water. Mix thoroughly. Distribute into tubes or flasks. Autoclave for 15 min at 15 psi pressure–121°C.

Use: For the isolation of cyanobacteria, including thermophilic species, that require reduced nitrogen concentrations.

Castenholz TYE Medium (Castenholz Trypticase Yeast Extract Medium)

Composition per liter:

Castenholz salts, 2× 500.0mL
1% TYE 100.0mL

pH 7.6 ± 0.2 at 25°C

Castenholz Salts, 2×:
Composition per liter:

Agar 30.0g
$NaNO_3$ 1.4g
Na_2HPO_4 .... 0.22g
KNO_3 .... 0.21g
Nitrilotriacetic acid 0.2g
$MgSO_4 \cdot 7H_2O$ 0.2g
$CaSO_4 \cdot 2H_2O$ 0.12g
NaCl 0.016g
$FeCl_3$ (0.03% solution) 2.0mL
Nitsch's trace elements 2.0mL

Preparation of Castenholz Salts, 2×: Add components to distilled/deionized water and bring volume to 1.0L. Mix thoroughly. Gently heat and bring to boiling. Adjust pH to 8.2. Autoclave for 15 min at 15 psi pressure–121°C.

Nitsch's Trace Elements:
Composition per liter:

$MnSO_4$ 2.2g
H_3BO_3 .... 0.5g
$ZnSO_4$ 0.5g
$CoCl_2 \cdot 6H_2O$ 0.046g
Na_2MoO_4 .... 0.025g
$CuSO_4$ 0.016g
H_2SO_4 .... 0.5mL

Preparation of Nitsch's Trace Elements: Add components to distilled/deionized water and bring volume to 1.0L. Mix thoroughly.

1% TYE:
Composition per liter:

Pancreatic digest of casein 10.0g
Yeast extract 10.0g

Preparation of 1% TYE: Add components to distilled/deionized water and bring volume to 1.0L. Mix thoroughly. Autoclave for 15 min at 15 psi pressure–121°C.

Preparation of Medium: Aseptically combine 500.0mL of sterile Castenholz salts, 2×, 100.0mL of sterile 1% TYE, and 400.0mL of sterile distilled/deionized water. Adjust pH to 7.6.

Use: For the cultivation and maintenance of *Thermonema lapsum* and *Thermus* species.

Castenholz TYE Medium with 2% Trypticase Yeast Extract

Composition per liter:

Castenholz salts, 2× 500.0mL
2% TYE 100.0mL

pH 7.6 ± 0.2 at 25°C

Castenholz Salts, 2×:
Composition per liter:

Agar 30.0g
$NaNO_3$ 1.4g
Na_2HPO_4 .... 0.22g
KNO_3 .... 0.21g
$MgSO_4 \cdot 7H_2O$ 0.2g
Nitrilotriacetic acid 0.2g
$CaSO_4 \cdot 2H_2O$ 0.12g
NaCl 0.016g
$FeCl_3$ solution (0.03% solution) 2.0mL
Nitsch's trace elements 2.0mL

Preparation of Castenholz Salts, 2×: Add components to distilled/deionized water and bring volume to 1.0L. Mix thoroughly. Gently heat and bring to boiling. Adjust pH to 8.2. Autoclave for 15 min at 15 psi pressure–121°C.

Nitsch's Trace Elements:
Composition per liter:

$MnSO_4$ 2.2g
H_3BO_3 .... 0.5g
$ZnSO_4$ 0.5g
$CoCl_2 \cdot 6H_2O$ 0.046g
Na_2MoO_4 .... 0.025g
$CuSO_4$ 0.016g
H_2SO_4 .... 0.5mL

Preparation of Nitsch's Trace Elements: Add components to distilled/deionized water and bring volume to 1.0L. Mix thoroughly.

2% TYE:
Composition per liter:

Pancreatic digest of casein 20.0g
Yeast extract 20.0g

Preparation of 2% TYE: Add components to distilled/deionized water and bring volume to 1.0L. Mix thoroughly. Autoclave for 15 min at 15 psi pressure–121°C.

Preparation of Medium: Aseptically combine 500.0mL of sterile Castenholz salts, 2×, 100.0mL of sterile 2% TYE, and 400.0mL of sterile distilled/deionized water. Adjust pH to 7.6.

Use: For the cultivation and maintenance of *Thermus* species.

Caulobacter Medium

Composition per liter:

Agar 10.0g
Peptone 2.0g
Yeast extract 1.0g
$MgSO_4 \cdot 7H_2O$ 0.2g
Riboflavin 1.0mg

pH 7.0 ± 0.2 at 25°C

Preparation of Medium: Add components to tap water and bring volume to 1.0L. Mix thoroughly. Gently heat and bring to boiling. Distribute into tubes or flasks. Autoclave for 15 min at 15 psi pressure–121°C. Pour into sterile Petri dishes or leave in tubes.

Use: For the cultivation of *Caulobacter* species from fresh water.

Caulobacter Medium

Composition per liter:

Agar 10.0g
Peptone 0.5g
Seawater, filtered 1.0L

pH 7.0 ± 0.2 at 25°C

Preparation of Medium: Combine components. Mix thoroughly. Gently heat and bring to boiling. Distribute into tubes or flasks. Autoclave for 15 min at 15 psi pressure–121°C. Pour into sterile Petri dishes or leave in tubes.

Use: For the cultivation of *Caulobacter* species from marine isolates.

Caulobacter Medium

Composition per liter:

Glucose 1.0g
Peptone 1.0g
Yeast extract 1.0g
Salt solution 100.0mL

Salt Solution:

Composition per 100.0mL:

EDTA 0.1g
KNO_3 .... 0.1g
K_2HPO_4 .... 0.066g
$MgSO_4$ 0.033g
$FeSO_4 \cdot 7H_2O$ 9.3mg
$NaBO_3 \cdot 4H_2O$ 2.63mg
$MgCl_2 \cdot 4H_2O$ 1.81mg
$CaCl_2$ 1.2mg
$(NH_4)_6Mo_7O_{24} \cdot 7H_2O$ 1.0mg
$ZnSO_4 \cdot 7H_2O$ 0.22mg
$CuSO_4 \cdot 5H_2O$ 0.079mg
$Co(NO_3)_2 \cdot H_2O$ 0.02mg

Preparation of Salt Solution: Add components to distilled/deionized water and bring volume to 100.0mL. Mix thoroughly.

Preparation of Medium: Add components to distilled/deionized water and bring volume to 1.0L. Mix thoroughly. Distribute into tubes or flasks. Autoclave for 15 min at 15 psi pressure–121°C.

Use: For the enrichment of *Stella* species from polluted waters.

Caulobacter Medium

Composition per liter:

Agar 10.0g
Peptone 2.0g
Yeast extract 1.0g
$MgSO_4 \cdot 7H_2O$ 0.2g

Preparation of Medium: Add components to tap water and bring volume to 1.0L. Mix thoroughly. Gently heat to boiling. Distribute into tubes or flasks. Autoclave for 15 min at 15 psi pressure–121°C. Pour into sterile Petri dishes or leave in tubes.

Use: For the cultivation and maintenance of *Asticcacaulis excentricus, Caulobacter* species, *Labrys monachus, Pedomicrobium* species, *Pirellula staleyi, Pseudomonas carboxydohydrogena,* and *Stella* species.

Caulobacter Medium II

Composition per liter:

Peptone 10.0g
Yeast extract 3.0g
Seawater 1.0L

pH 7.2–7.4 at 25°C

Preparation of Medium: Add components to filtered aged seawater and bring volume to 1.0L. Mix thoroughly. Adjust pH to 7.2–7.4. Distribute into tubes or flasks. Autoclave for 15 min at 15 psi pressure–121°C.

Use: For the cultivation of *Caulobacter halobacteroides* and *Caulobacter maris.*

Caulobacter Medium with Riboflavin

Composition per liter:

Peptone 10.0g
Yeast extract 3.0g
Riboflavin 1.0mg
Seawater 1.0L

pH 7.2–7.4 at 25°C

Preparation of Medium: Add components to filtered aged seawater and bring volume to 1.0L. Mix thoroughly. Adjust pH to 7.3. Distribute into tubes or flasks. Autoclave for 15 min at 15 psi pressure–121°C.

Use: For the cultivation of *Caulobacter vibrioides.*

CCVC Medium (Cephalothin Cycloheximide Vancomycin Colistin Medium

Composition per liter:

BCYE-alpha base 990.0mL
Antibiotic supplement 10.0mL

pH 6.9 ± 0.2 at 25°C

Source: This medium is available as a premixed powder from BD Diagnostic Systems.

BCYE-Alpha Base:

Composition per liter:

Agar 15.0g
Yeast extract 10.0g
ACES buffer (2-[(2-Amino-2-oxoethyl)-amino]-ethane sulfonic acid) 10.0g
Charcoal, activated 2.0g
α-Ketoglutarate 1.0g
$Fe_4(P_2O_7)_3 \cdot 9H_2O$ 0.25g
L-Cysteine·HCl·H_2O solution 10.0mL

L-Cysteine·HCl·H_2O Solution:

Composition per 10.0mL:

L-Cysteine·HCl·H_2O 0.4g

Preparation of L-Cysteine·HCl·H_2O Solution: Add L-cysteine·HCl·H_2O to distilled/deionized water and bring volume to 10.0mL. Mix thoroughly. Filter sterilize.

Preparation of BCYE-Alpha Base: Add components, except L-cysteine·HCl·H_2O solution, to distilled/deionized water and bring volume to 990.0mL. Mix thoroughly. Adjust medium to pH 6.9 with 1*N* KOH. Heat gently and bring to boiling for 1 min. Autoclave for 15 min at 15 psi pressure–121°C. Cool to 50°–55°C. Add 4.0mL of L-cysteine·HCl·H_2O solution. Mix thoroughly.

Antibiotic Supplement Solution:

Composition per 10.0mL:

Cycloheximide 80.0mg
Colistin 16.0mg
Cephalothin 4.0mg
Vancomycin 0.5mg

Preparation of Antibiotic Supplement Solution: Add components to 10.0mL of distilled/deionized water. Filter sterilize.

Preparation of Medium: To cooled BCYE-alpha base, add 10.0mL sterile antibiotic supplement. Mix thoroughly. Adjust pH to 6.9 with sterile 1*N* KOH. Pour into sterile Petri dishes with constant agitation to keep charcoal in suspension.

Use: For the selective isolation and cultivation of *Legionella* species from environmental samples.

Cellulolytic Agar with Sea Salts

Composition per liter:

Agar 20.0g
NH_4Cl 2.0g
K_2HPO_4 .... 1.65g

Yeast extract	1.2g
L-Cysteine·HCl·H_2O	0.5g
Cellulose suspension	200.0mL
Filtered seawater	200.0mL
Mineral solution	150.0mL
Resazurin (0.1% solution)	1.0mL

pH 7.2 ± 0.2 at 25°C

Cellulose Suspension:
Composition per 200.0mL:

Cellulose powder, Whatman CF11	8.0g

Preparation of Cellulose Suspension: Add cellulose powder to 200.0mL of distilled/deionized water and mix thoroughly.

Mineral Solution:
Composition per liter:

NaCl	6.0g
$(NH_4)_2SO_4$	6.0g
$CaCl_2$	0.6g
$MgSO_4$	0.6g

Preparation of Mineral Solution: Add components to distilled/deionized water and bring volume to 1.0L. Mix thoroughly.

Preparation of Medium: Prepare and dispense medium anaerobically under 100% N_2. Add components to distilled/deionized water and bring volume to 1.0L. Mix thoroughly. Adjust pH to 7.2 with 5M NaOH. Distribute into tubes or flasks. Autoclave for 15 min at 15 psi pressure–121°C.

Use: For the cultivation and maintenance of *Clostridium papyrosolvens* and other marine bacteria that can utilize cellulose as a carbon source.

Cellulolytic Agar for Thermophiles

Composition per liter:

Agar	30.0g
K_2HPO_4	1.65g
NH_4SO_4	1.6g
Yeast extract	1.0g
NaCl	0.96g
L-Cysteine·HCl·H_2O	0.5g
$CaCl_2$	0.096g
$MgSO_4$	0.096g
Cellulose suspension	200.0mL
Resazurin (0.1% solution)	1.0mL

pH 7.2 ± 0.2 at 25°C

Cellulose Suspension:
Composition per 200.0mL:

Cellulose powder, Whatman CF11	8.0g

Preparation of Cellulose Suspension: Add cellulose powder to 200.0mL of distilled/deionized water and mix thoroughly.

Preparation of Medium: Prepare and dispense medium anaerobically in 100% N_2. Add components to distilled/deionized water and bring volume to 1.0L. Mix thoroughly. Adjust pH to 7.2 with 5m NaOH. Distribute into tubes or flasks. Autoclave for 15 min at 15 psi pressure–121°C.

Use: For cultivation of *Clostridium stercorarium* and other bacteria that can utilize cellulose as a carbon source.

Cellulolytic Broth with Sea Salts

Composition per liter:

K_2HPO_4	1.65g
NH_4Cl	1.0g
Yeast extract	0.6g
L-Cysteine·HCl·H_2O	0.5g
Filtered seawater	200.0mL
Mineral solution	150.0mL
Resazurin (0.1% solution)	1.0mL

pH 7.2 ± 0.2 at 25°C

Mineral Solution:
Composition per liter:

NaCl	6.0g
$(NH_4)_2SO_4$	6.0g
$CaCl_2$	0.6g
$MgSO_4$	0.6g

Preparation of Mineral Solution: Add components to distilled/deionized water and bring volume to 1.0L. Mix thoroughly.

Preparation of Medium: Prepare and dispense medium anaerobically in 100% N_2 atmosphere. Add components to distilled/deionized water and bring volume to 1.0L. Adjust pH to 7.2 with 5m NaOH. Distribute into tubes or flasks that contain cellulose as a strip (4.5cm × 1.0cm) of Whatman No. 1 filter paper. Autoclave for 15 min at 15 psi pressure–121°C.

Use: For the cultivation and maintenance of *Clostridium papyrosolvens* and other marine bacteria that can utilize cellulose as a carbon source.

Cellulolytic Medium

Composition per liter:

$NaHCO_3$	2.06g
Cellulose	2.0g
NH_4Cl	0.68g
K_2HPO_4	0.296g
KH_2PO_4	0.18g
$(NH_4)_2SO_4$	0.15g
$MgSO_4 \cdot 7H_2O$	0.12g
$CaCl_2 \cdot 2H_2O$	61.0mg
$FeSO_4 \cdot 7H_2O$	21.0mg
Nitrilotriacetic acid	15.0mg
NaCl	10.0mg
MnSO4·H_2O	5.0mg
$CoCl_2 \cdot H_2O$	1.0mg
Resazurin	1.0mg
$ZnSO_4 \cdot 7H_2O$	1.0mg
$CuSO_4 \cdot 5H_2O$	0.1mg
H_3BO_3	0.1mg
$KAl(SO_4)_2 \cdot 12H_2O$	0.1mg
$Na_2MoO_4 \cdot 2H_2O$	0.1mg
Wolfe's vitamin solution	10.0mL
L-Cysteine·HCl·H_2O solution	5.0mL

pH 7.5 ± 0.2 at 25°C

Wolfe's Vitamin Solution:
Composition per liter:

Pyridoxine·HCl	10.0mg
Thiamine·HCl	5.0mg
Riboflavin	5.0mg
Nicotinic acid	5.0mg
Calcium DL-pantothenate	5.0mg
p-Aminobenzoic acid	5.0mg
Thioctic acid	5.0mg
Biotin	2.0mg
Folic acid	2.0mg
Cyanocobalamin	100.0μg

Preparation of Wolfe's Vitamin Solution: Add components to distilled/deionized water and bring volume to 1.0L. Mix thoroughly. Filter sterilize.

L-Cysteine·HCl·H_2O Solution:
Composition per 20.0mL:
L-Cysteine·HCl·H_2O 1.0g

Preparation of L-Cysteine·HCl·H_2O Solution: Add L-cysteine·HCl·H_2O to distilled/deionized water and bring volume to 20.0mL. Mix thoroughly. Filter sterilize.

Preparation of Medium: Add components, except Wolfe's vitamin solution and L-cysteine·HCl·H_2O solution, to distilled/deionized water and bring volume to 985.0mL. Mix thoroughly. Autoclave for 15 min at 15 psi pressure–121°C. Aseptically add 10.0mL of sterile Wolfe's vitamin solution and 5.0mL of sterile L-cysteine·HCl·H_2O solution. Mix thoroughly. Aseptically distribute into sterile tubes or flasks.

Use: For the cultivation of cellulolytic bacteria.

Cellulolytic Medium with Rumen Fluid and Soluble Starch

Composition per liter:
Basal medium 975.0mL
Alkaline solution 25.0mL
pH 6.8 ± 0.2 at 25°C

Basal Medium:
Composition per 975.0mL:
Agar 15.0g
$NaHCO_3$ 6.37g
Pancreatic digest of casein 5.0g
Cellobiose 5.0g
Soluble starch 5.0g
NaCl 0.9g
$(NH_4)_2SO_4$ 0.9g
K_2HPO_4 .... 0.45g
KH_2PO_4 .... 0.45g
$MgSO_4·7H_2O$ 0.18g
$CaCl_2$ 0.09g
Resazurin 1.0mg

Preparation of Basal Medium: Add components to distilled/deionized water and bring volume to 975.0mL. Mix thoroughly. Gently heat and bring to boiling under a gas phase of 98% CO_2 + 2% H_2. Cool slightly.

Alkaline Solution:
Composition per 25.0mL:
L-Cysteine·HCl·H_2O 0.25g
$Na_2S·9H_2O$ 0.25g

Preparation of Alkaline Solution: Add components to 25.0mL of distilled/deionized water. Mix thoroughly. Prepare freshly.

Preparation of Medium: Prepare 975.0mL of basal medium. Heat to boiling and cool to 25°C. Add 25.0mL of freshly prepared alkaline solution. Distribute into tubes using anaerobic techniques under a gas phase of 98% CO_2 + 2% H_2. Autoclave for 15 min at 15 psi pressure–121°C. Adjust pH to 6.8.

Use: For the cultivation of *Selenomonas ruminantium* and *Succinimonas amylolytica*.

Cellulomonas fermentans Medium

Composition per liter:
Yeast extract 5.0g
K_2HPO_4 .... 2.21g
KH_2PO_4 .... 1.5g
$(NH_4)_2SO_4$ 1.3g
$MgCl_2·6H_2O$ 0.1g
$CaCl_2·2H_2O$ 0.02g
$FeSO_4·7H_2O$ 1.25mg
Resazurin 1.0mg
Cellobiose solution 50.0mL
$NaHCO_3$ solution 10.0mL
L-Cysteine·HCl solution 10.0mL
pH 7.4 ± 0.2 at 25°C

Cellobiose Solution:
Composition per 50.0mL:
D-Cellobiose (or cellulose) 5.0g

Preparation of Cellobiose Solution: Add cellobiose (or cellulose) to distilled/deionized water and bring volume to 50.0mL. Mix thoroughly. Autoclave under 100% N_2 for 15 min at 15 psi pressure–121°C.

$NaHCO_3$ Solution:
Composition per 10.0mL:
$NaHCO_3$ 0.8g

Preparation of $NaHCO_3$ Solution: Add $NaHCO_3$ to distilled/deionized water and bring volume to 10.0mL. Mix thoroughly. Autoclave under 100% N_2 for 15 min at 15 psi pressure–121°C.

L-Cysteine·HCl Solution:
Composition per 10.0mL:
L-Cysteine·HCl 0.5g

Preparation of L-Cysteine·HCl Solution: Add L-cysteine·HCl to distilled/deionized water and bring volume to 10.0mL. Mix thoroughly. Autoclave under 100% N_2 for 15 min at 15 psi pressure–121°C.

Preparation of Medium: Add components, except cellobiose solution, $NaHCO_3$ solution, and L-cysteine·HCl solution, to distilled/deionized water and bring volume to 930.0mL. Mix thoroughly. Sparge under 100% N_2 for 10 min. Autoclave under 100% N_2 for 15 min at 15 psi pressure–121°C. Aseptically and anaerobically add 50.0mL of sterile cellobiose solution, 10.0mL of sterile $NaHCO_3$ solution, and 10.0mL of sterile L-cysteine·HCl solution. Mix thoroughly.

Use: For the cultivation and maintenance of *Cellulomonas fermentans*.

Cellulomonas PTYG Medium (*Cellulomonas* Peptone Tryptone Yeast Extract Glucose Medium)

Composition per liter:
Agar 15.0g
Glucose 5.0g
Peptone 5.0g
Pancreatic digest of casein 5.0g
Yeast extract 5.0g

Preparation of Medium: Add components to distilled/deionized water and bring volume to 1.0L. Mix thoroughly. Gently heat and bring to boiling. Distribute into tubes or flasks. Autoclave for 15 min at 15 psi pressure–121°C. Pour into sterile Petri dishes or leave in tubes.

Use: For the cultivation and maintenance of *Cellulomonas* species.

Cellulose Anaerobe Medium (LMG Medium 94)

Composition per liter:
$NaHCO_3$ 2.1g
Cellulose 2.0g
NH_4Cl 0.68g
KH_2PO_4 .... 0.18g

$(NH_4)_2SO_4$ 0.15g
$MgSO_4 \cdot 7H_2O$ 0.12g
K_2HPO_4 .... 296.0mg
$CaCl_2 \cdot 2H_2O$ 61.0mg
$FeSO_4 \cdot 7H_2O$ 21.0mg
Nitrilotriacetic acid 15.0mg
NaCl 10.0mg
$MnSO_4 \cdot H_2O$ 5.0mg
$CoCl_2 \cdot 6H_2O$ 1.0mg
$ZnSO_4 \cdot 7H_2O$ 1.0mg
Resazurin 1.0mg
$CuSO_4 \cdot 5H_2O$ 0.1mg
$KAl(SO_4)_2 \cdot 12H_2O$ 0.1mg
H_3BO_3 .... 0.1mg
$Na_2MoO_4 \cdot 2H_2O$ 0.1mg
Vitamin solution 5.0mL
Cysteine hydrochloride solution 5.0mL
pH 7.1 ± 0.2 at 25°C

Vitamin Solution:
Composition per liter:
Pyridoxine·HCl 10.0mg
Calcium DL-pantothenate 5.0mg
Lipoic acid 5.0mg
Nicotinic acid 5.0mg
p-Aminobenzoic acid 5.0mg
Riboflavin 5.0mg
Thiamine·HCl 5.0mg
Biotin 2.0mg
Folic acid 2.0mg
Vitamin B_{12} 0.1mg

Preparation of Vitamin Solution: Add components to distilled/deionized water and bring volume to 1.0L. Mix thoroughly. Filter sterilize.

L-Cysteine Solution:
Composition per 10.0mL:
L-Cysteine·HCl·H_2O 0.5g

Preparation of L-Cysteine Solution: Add L-cysteine·HCl·H_2O to distilled/deionized water and bring volume to 10.0mL. Mix thoroughly. Sparge with 100% N_2. Autoclave for 15 min at 15 psi pressure–121°C.

Preparation of Medium: Add nitrilotriacetic acid to 500.0mL of distilled/deionized water. Adjust pH to 6.5 with KOH. Add remaining components except vitamin solution and cysteine hydrochloride solution. Add 485.0mL distilled/deionized water. Adjust pH to 7.1. Autoclave for 15 min at 15 psi pressure–121°C. Cool to 50°C. Aseptically add 10.0mL of sterile vitamin solution and 5.0mL sterile cysteine hydrochloride solution. Mix thoroughly. Aeptically distribute to tubes or flasks.

Use: For the cultivation of cellulose-utilizing anerobic bacteria.

Cellulose Broth

Composition per liter:
Cellulose, powdered 1.0g
K_2HPO_4 .... 1.0g
$(NH_4)_2SO_4$ 1.0g
$MgSO_4 \cdot 7H_2O$ 0.2g
$CaCl_2 \cdot 2H_2O$ 0.1g
$FeCl_3$ 0.02g
pH 7.0–7.5 at 25°C

Preparation of Medium: Add cellulose to 100.0mL of distilled/deionized water. Mix thoroughly. In a separate flask, add remaining components to distilled/deionized water and bring volume to 900.0mL. Mix thoroughly. Autoclave both solutions separately for 15 min at 15 psi pressure–121°C. Cool to 45°–50°C. Aseptically combine the two sterile solutions. Mix thoroughly. Aseptically distribute into sterile tubes or flasks.

Use: For the isolation and cultivation of *Cytophaga* species, *Herpetosiphon* species, *Saprospira* species, and *Flexithrix* species.

Cellulose Overlay Agar

Composition per plate:
Stan 5 agar 15.0mL
Cellulose overlay agar 5.0mL

Stan 5 Agar:
Composition per liter:
Solution B 650.0mL
Solution A 350.0mL

Solution A:
Composition per 350.0mL:
$CaCl_2 \cdot 2H_2O$ 1.0g
$(NH_4)_2SO_4$ 1.0g
$MgSO_4 \cdot 7H_2O$ 1.0g
Trace elements solution 1.0mL

Preparation of Solution A: Add components to distilled/deionized water and bring volume to 350.0mL. Mix thoroughly. Gently heat and bring to boiling. Autoclave for 15 min at 15 psi pressure–121°C. Cool to 45°–50°C.

Trace Elements Solution:
Composition per liter:
EDTA 8.0g
$MnCl_2 \cdot 4H_2O$ 0.1g
$CoCl_2$ 0.02g
KBr 0.02g
$ZnCl_2$ 0.02g
$CuSO_4$ 0.01g
H_3BO_3 .... 0.01g
$NaMoO_4 \cdot 2H_2O$ 0.01g
$BaCl_2$ 5.0mg
LiCl 5.0mg
$SnCl_2 \cdot 2H_2O$ 5.0mg

Preparation of Trace Elements Solution: Add components to distilled/deionized water and bring volume to 1.0L. Mix thoroughly.

Solution B:
Composition per 650.0mL:
Agar 10.0g
K_2HPO_4 .... 1.0g

Preparation of Solution B: Add components to distilled/deionized water and bring volume to 650.0mL. Mix thoroughly. Gently heat and bring to boiling. Autoclave for 15 min at 15 psi pressure–121°C. Cool to 45°–50°C.

Preparation of Stan 5 Agar: Aseptically combine 350.0mL of cooled, sterile solution A and 650.0mL of cooled, sterile solution B. Mix thoroughly.

Cellulose Overlay Agar:
Composition per liter:
Solution A 350.0mL
Solution B 650.0mL

Solution A:
Composition per 350.0mL:
$CaCl_2 \cdot 2H_2O$ 1.0g
$(NH_4)_2SO_4$ 1.0g
$MgSO_4 \cdot 7H_2O$ 1.0g
Trace elements solution 1.0mL

Preparation of Solution A: Add components to distilled/deionized water and bring volume to 350.0mL. Mix thoroughly. Gently heat and bring to boiling. Autoclave for 15 min at 15 psi pressure–121°C. Cool to 45°–50°C.

Trace Elements Solution:
Composition per liter:

EDTA	8.0g
$MnCl_2 \cdot 4H_2O$	0.1g
$CoCl_2$	0.02g
KBr	0.02g
$ZnCl_2$	0.02g
$CuSO_4$	0.01g
H_3BO_3	0.01g
$NaMoO_4 \cdot 2H_2O$	0.01g
$BaCl_2$	5.0mg
LiCl	5.0mg
$SnCl_2 \cdot 2H_2O$	5.0mg

Preparation of Trace Elements Solution: Add components to distilled/deionized water and bring volume to 1.0L. Mix thoroughly.

Solution B:
Composition per 650.0mL:

Agar	10.0g
K_2HPO_4	1.0g

Preparation of Solution B: Add components to distilled/deionized water and bring volume to 650.0mL. Mix thoroughly. Gently heat and bring to boiling. Autoclave for 15 min at 15 psi pressure–121°C. Cool to 45°–50°C.

Preparation of Cellulose Overlay Agar: Aseptically combine 350.0mL of cooled, sterile solution A and 650.0mL of cooled, sterile solution B. Mix thoroughly.

Preparation of Medium: Pour cooled, sterile Stan 5 agar into sterile Petri dishes in 15.0mL volumes. Allow agar to solidify. Overlay each plate with 5.0mL of cellulose overlay agar.

Use: For the cultivation of myxobacteria.

Cellvibrio Medium

Composition per liter:

Agar	15.0g
$NaNO_3$	5.0g
K_2HPO_4	1.0g
$MgSO_4 \cdot 7H_2O$	0.5g
KCl	0.5g
$FeSO_4 \cdot 7H_2O$	10.0mg

pH 7.2 ± 0.2 at 25°C

Preparation of Medium: Add components to distilled/deionized water and bring volume to 1.0L. Mix thoroughly. Gently heat and bring to boiling. Distribute into tubes or flasks. Autoclave for 15 min at 15 psi pressure–121°C. Pour into sterile Petri dishes or leave in tubes.

Use: For the cultivation and maintenance of *Cellvibrio mixtus, Cytophaga aurantiaca, Cytophaga hutchinsonii,* and *Sporocytophaga myxococcoides.*

CENS Medium (DSMZ Medium 748)

Composition per liter:

Soytone	4.0g
NH_4Cl	1.0g
K_2HPO_4	0.9g
KH_2PO_4	0.6g
$Na_2S_2O_3 \cdot 5H_2O$	0.5g
$MgSO_4 \cdot 7H_2O$	0.2g
$CaCl_2 \cdot 2H_2O$	75.0mg
EDTA	5.0mg
Vitamin B_{12}	20.0μg
Biotin	15.0μg
Na-pyruvate solution	10.0mL
Chelated iron solution	2.0mL
True Blue Trace elements	1.0mL

pH 7.0 ± 0.2 at 25°C

Na-pyruvate Solution:
Composition per 10.0mL:

Na-pyruvate	2.2g

Preparation of Na-pyruvate Solution: Add Na-pyruvate to distilled/deionized water and bring volume to 10.0mL. Mix thoroughly. Sparge with 100% N_2. Filter sterilize.

True Blue Trace Elements:
Composition per 250mL:

EDTA	2.5g
$MnCl_2 \cdot 4H_2O$	0.2g
H_3BO_3	0.1g
$Na_2MoO_4 \cdot 4H_2O$	0.1g
$ZnCl_2$	50.0mg
$NiCl_2 \cdot 6H_2O$	50.0mg
$CoCl_2 \cdot 2H_2O$	20.0mg
$CuCl_2 \cdot 2H_2O$	10.0mg
Na_2SeO_3	5.0mg
$NaVO_3 \cdot H_2O$	5.0mg

Preparation of True Blue Trace Elements: Add components one at a time to approximately 200.0mL distilled/deionized water. Bring final volume to 250.0mL with additional distilled/deionized water.

Chelated Iron Solution:
Composition per 900mL:

EDTA	2.0g
$FeCl_2 \cdot 4H_2O$	1.0g
HCl	3.0mL

Preparation of Chelated Iron Solution: Add components to distilled/deionized water and bring volume to 900.0mL. Mix thoroughly.

Preparation of Medium: Add components, except Na-pyruvate solution, to distilled/deionized water and bring volume to 990.0L. Mix thoroughly. Autoclave for 15 min at 15 psi pressure–121°C. Cool to 25°C. Aseptically add 10.0mL sterile Na-pyruvate solution. Aseptically distribute into sterile tubes or flasks.

Use: For the cultivation of *Rhodocista centenaria=Rhodospirillum centenum.*

Centenum Medium

Composition per liter of tap water:

Agar	20.0g
Yeast extract	10.0g
Sodium pyruvate	2.2g
K_2HPO_4	1.0g
$MgSO_4$	0.5g
Vitamin B_{12}	0.02mg

pH 7.0–7.2 at 25°C

Preparation of Medium: Add components to distilled/deionized water and bring volume to 1.0L. Mix thoroughly. Gently heat and bring to boiling. Distribute into tubes or flasks. Autoclave for 15 min at 15 psi pressure–121°C. Pour into sterile Petri dishes or leave in tubes.

Use: For the cultivation and maintenance of *Rhodospirillum* species.

Cetrimide Agar, Non-USP

Composition per liter:

Beef heart, solids from infusion	500.0g
Agar	15.0g
Tryptose	10.0g
NaCl	5.0g
Cetrimide	0.9g

pH 7.2 ± 0.2 at 25°C

Preparation of Medium: Add components to distilled/deionized water and bring volume to 1.0L. Mix thoroughly. Gently heat and bring to boiling. Distribute into tubes or flasks. Autoclave for 15 min at 13 psi pressure–118°C. Pour into sterile Petri dishes or leave in tubes.

Use: For the selective isolation, cultivation, and identification of *Pseudomonas aeruginosa* and other Gram-negative, nonfermentative bacteria.

Cetrimide Agar, USP (Pseudosel® Agar)

Composition per liter:

Pancreatic digest of gelatin	20.0g
Agar	13.6g
K_2SO_4	10.0g
$MgCl_2$	1.4g
Cetrimide	0.3g
Glycerol	10.0mL

pH 7.2 ± 0.2 at 25°C

Source: This medium is available as a premixed powder from BD Diagnostic Systems.

Preparation of Medium: Add components to distilled/deionized water and bring volume to 1.0L. Mix thoroughly. Gently heat and bring to boiling. Distribute into tubes or flasks. Autoclave for 15 min at 13 psi pressure–118°C. Pour into sterile Petri dishes or leave in tubes.

Use: For the selective isolation, cultivation, and identification of *Pseudomonas aeruginosa* and other Gram-negative, nonfermentative bacteria.

CHCA Salts Medium (Cyclohexane Carboxylic Acid Salts Medium)

Composition per liter:

K_2HPO_4	3.5g
KH_2PO_4	1.5g
Cyclohexane carboxylic acid	1.0g
NH_4NO_3	1.0g
$MgSO_4 \cdot 7H_2O$	0.5g
$FeSO_4 \cdot 7H_2O$	0.1g
Yeast extract	0.1g
$CaCl_2 \cdot 2H_2O$	0.01g
$Na_2MoO_2 \cdot 2H_2O$	0.01g
$ZnSO_4 \cdot 7H_2O$	0.01g

pH 7.0 ± 0.2 at 25°C

Preparation of Medium: Add components to distilled/deionized water and bring volume to 1.0L. Mix thoroughly. Adjust pH to 7.0. Gently heat and bring to boiling. Distribute into tubes or flasks. Autoclave for 15 min at 15 psi pressure–121°C. Pour into sterile Petri dishes or leave in tubes.

Use: For the cultivation and maintenance of bacteria that can utilize cyclohexane carboxylic acid as a carbon source. For the cultivation and maintenance of *Arthrobacter globiformis*.

Chitin Agar

Composition per liter:

Agar	15.0g
Chitin, precipitated	3.0g
$(NH_4)_2SO_4$	2.0g
Na_2HPO_4	1.1g
KH_2PO_4	0.7g
$MgSO_4 \cdot 7H_2O$	0.2g
$FeSO_4$	1.0mg
$MnSO_4$	1.0mg

Chitin, Precipitated:

Composition per 2.5L:

Chitin	40.0g
HCl, concentrated	400.0mL

Preparation of Chitin, Precipitated: Add chitin to 400.0mL of cold concentrated HCl. Add this solution to 2.0L of distilled/deionized water at 5°C. Filter the solution through Whatman #1 filter paper. Dialyze the precipitated chitin against tap water for 12 hr. Adjust the pH to 7.0 with KOH.

Preparation of Medium: Add components to distilled/deionized water and bring volume to 1.0L. Mix thoroughly. Gently heat and bring to boiling. Distribute into tubes or flasks. Autoclave for 15 min at 15 psi pressure–121°C. Pour into sterile Petri dishes or leave in tubes.

Use: For the isolation and cultivation of *Cytophaga* species, *Herpetosiphon* species, *Saprospira* species, and *Flexithrix* species.

Chitin Agar

Composition per liter:

Agar	20.0g
Chitin	4.0g
K_2HPO_4	0.7g
$MgSO_4 \cdot 7H_2O$	0.5g
KH_2PO_4	0.3g
$FeSO_4 \cdot 7H_2O$	0.01g
$MnCl_2 \cdot 4H_2O$	0.001g
$ZnSO_4 \cdot 7H_2O$	0.001g

pH 8.0 ± 0.2 at 25°C

Preparation of Medium: Add components to distilled/deionized water and bring volume to 1.0L. Mix thoroughly. Gently heat and bring to boiling. Distribute into tubes or flasks. Autoclave for 15 min at 15 psi pressure–121°C. Pour into sterile Petri dishes or leave in tubes.

Use: For the selective isolation and cultivation of streptomycetes.

Chlamydomonas Enriched Medium

Composition per liter:

Agar	15.0g
Sodium acetate, anhydrous	2.0g
Tryptose	2.0g
Yeast extract	2.0g

Preparation of Medium: Add components to distilled/deionized water and bring volume to 1.0L. Mix thoroughly. Gently heat and bring to boiling. Distribute into tubes or flasks. Autoclave for 20 min at 15 psi pressure–121°C. Pour into sterile Petri dishes or leave in tubes.

Use: For the cultivation of *Chlamydomonas reinhardtii*.

Chlorinated Fatty Acid Medium

Composition per 1001.0mL:

$(NH_4)_2SO_4$	0.5g

$MgSO_4 \cdot 7H_2O$ 0.1g
$Ca(NO_3)_2$ 0.075g
2-Chloropropionate 0.54g
Phosphate buffer (20m*M* solution, pH 7.2) 1.0L
Trace elements solution 1.0mL
pH 7.2 ± 0.2 at 25°C

Trace Elements Solution:
Composition per liter:
$FeSO_4 \cdot 7H_2O$ 1.0g
$MnSO_4 \cdot H_2O$ 1.0g
$Co(NO_3)_2 \cdot 6H_2O$ 0.25g
$CuCl_2 \cdot 2H_2O$ 0.25g
$Na_2MoO_4 \cdot 2H_2O$ 0.25g
$ZnCl_2$ 0.25g
H_3BO_3 .... 0.1g
NH_4VO_3 .... 0.1g
$NiSO_4 \cdot 6H_2O$ 0.1g

Preparation of Trace Elements Solution: Add components to distilled/deionized water and bring volume to 1.0L. Mix thoroughly.

Preparation of Medium: Combine components. Mix thoroughly. Distribute into tubes or flasks. Autoclave for 15 min at 15 psi pressure–121°C.

Use: For the cultivation of *Alcaligenes* species.

Chlorobiaceae Medium 1

Composition per 4990.0mL:
Solution 1 .... 4.0L
O_2-free water 860.0mL
$NaHCO_3$ solution 100.0mL
$Na_2S \cdot 9H_2O$ solution 20.0mL
Trace elements solution 5.0mL
Vitamin B_{12} solution 5.0mL
pH 6.8 ± 0.2 at 25°C

Solution 1:
Composition per 4.0L:
$MgSO_4 \cdot 7H_2O$ 2.5g
KCl 1.7g
KH_2PO_4 .... 1.7g
NH_4Cl 1.7g
$CaCl_2 \cdot 2H_2O$ 1.25g

Preparation of Solution 1: Add components to distilled/deionized water and bring volume to 4.0L. Mix thoroughly. Autoclave for 45 min at 15 psi pressure–121°C. Cool to 25°C under 100% N_2. Saturate with CO_2 by stirring under 100% CO_2 for 30 min.

O_2-Free Water:
Composition per 860.0mL:
H_2O 860.0mL

Preparation of O_2-Free Water: Autoclave H_2O for 15 min at 15 psi pressure–121°C. Cool to 25°C under 100% N_2.

$NaHCO_3$ Solution:
Composition per 100.0mL:
$NaHCO_3$ 7.5g

Preparation of $NaHCO_3$ Solution: Add the $NaHCO_3$ to distilled/deionized water and bring volume to 100.0mL. Mix thoroughly. Gas with 100% CO_2 for 20 min. Filter sterilize with positive CO_2 pressure.

$Na_2S \cdot 9H_2O$ Solution:
Composition per 100.0mL:
$Na_2S \cdot 9H_2O$ 10.0g

Preparation of $Na_2S \cdot 9H_2O$ Solution: Add $Na_2S \cdot 9H_2O$ to distilled/deionized water. Mix thoroughly. Gas with 100% N_2 for 15 min in a screw-capped bottle. Tightly close cap. Autoclave for 15 min at 15 psi pressure–121°C. Cool to 25°C.

Trace Elements Solution:
Composition per liter:
$FeCl_2 \cdot 4H_2O$ 1.5g
$CoCl_2 \cdot 6H_2O$ 0.19g
$MnCl_2 \cdot 4H_2O$ 0.1g
$ZnCl_2$ 0.07g
H_3BO_3 .... 0.06g
$NaMoO_4 \cdot 2H_2O$ 0.04g
$CuCl_2 \cdot 2H_2O$ 0.02g
$NiCl_2 \cdot 6H_20$ 0.02g
HCl (25% solution) 6.5mL

Preparation of Trace Elements Solution: Add components to distilled/deionized water and bring volume to 1.0L. Mix thoroughly. Autoclave for 15 min at 15 psi pressure–121°C. Cool to 25°C.

Vitamin B_{12} Solution:
Composition per 100.0mL:
Vitamin B_{12} 2.0mg

Preparation of Vitamin B_{12} Solution: Add Vitamin B_{12} to distilled/deionized water and bring volume to 100.0mL. Mix thoroughly. Filter sterilize.

Preparation of Medium: To 4.0L of sterile, CO_2-saturated solution 1, aseptically add the remaining components. Mix thoroughly. Adjust pH to 6.8. Aseptically distribute into sterile 100.0mL bottles using positive pressure of 95% N_2 + 5% CO_2. Completely fill bottles with medium except for a pea-sized air bubble.

Use: For the isolation and cultivation of members of the Chlorobiaceae.

Chlorobiaceae Medium 2

Composition per 1051.0mL:
Solution 1 .... 950.0mL
$Na_2S \cdot 9H_2O$ solution 60.0mL
$NaHCO_3$ solution 40.0mL
Vitamin B_{12} solution 1.0mL
pH 6.8 ± 0.2 at 25°C

Solution 1:
Composition per 950.0mL:
KH_2PO_4 .... 1.0g
NH_4Cl 0.5g
$MgSO_4 \cdot 7H_2O$ 0.4g
$CaCl_2 \cdot 2H_2O$ 0.05g
Trace elements solution SL-8 .... 1.0mL

Preparation of Solution 1: Add components to distilled/deionized water and bring volume to 950.0mL. Mix thoroughly. Autoclave for 15 min at 15 psi pressure–121°C. Cool to 45°–50°C.

Trace Elements Solution SL-8:
Composition per liter:
Disodium EDTA 5.2g
$FeCl_2 \cdot 4H_2O$ 1.5g
$CoCl_2 \cdot 6H_2O$ 0.19g
$MnCl_2 \cdot 4H_2O$ 0.1g
$ZnCl_2$ 0.07g
H_3BO_3 .... 0.06g
$NaMoO_4 \cdot 2H_2O$ 0.04g
$CuCl_2 \cdot 2H_2O$ 0.02g
$NiCl_2 \cdot 6H_20$ 0.02g

Preparation of Trace Elements Solution SL-8: Add components to distilled/deionized water and bring volume to 1.0L. Mix thoroughly.

$Na_2S·9H_2O$ Solution:
Composition per 100.0mL:

$Na_2S·9H_2O$	5.0g

Preparation of $Na_2S·9H_2O$ Solution: Add $Na_2S·9H_2O$ to distilled/deionized water and bring volume to 100.0mL. Autoclave for 15 min at 15 psi pressure–121°C. Cool to 45°–50°C.

$NaHCO_3$ Solution:
Composition per 100.0mL:

$NaHCO_3$	5.0g

Preparation of $NaHCO_3$ Solution: Add $NaHCO_3$ to distilled/deionized water and bring volume to 100.0mL. Mix thoroughly. Filter sterilize.

Vitamin B_{12} Solution:
Composition per 100.0mL:

Vitamin B_{12}	2.0mg

Preparation of Vitamin B_{12} Solution: Add Vitamin B_{12} to distilled/deionized water and bring volume to 100.0mL. Mix thoroughly. Filter sterilize.

Preparation of Medium: To 950.0mL of cooled, sterile solution 1, aseptically add 60.0mL of sterile $Na_2S·9H_2O$ solution, 40.0mL of sterile $NaHCO_3$ solution, and 1.0mL of sterile Vitamin B_{12} solution. Mix thoroughly. Adjust pH to 6.8 with sterile H_2SO_4 or Na_2CO_3. Aseptically distribute into sterile 50.0mL or 100.0mL bottles with metal screw-caps and rubber seals. Completely fill bottles with medium except for a pea-sized air bubble.

Use: For the isolation and cultivation of freshwater and soil members of the Chlorobiaceae.

Chlorobium thiosulfatophilum Medium

Composition per 1050.0mL:

KH_2PO_4	1.0g
NH_4Cl	1.0g
$MgCl_2·6H_2O$	0.5g
Solution A	20.0mL
Solution B	20.0mL
Solution C	10.0mL
Trace elements solution	1.0mL

pH 7.0 ± 0.2 at 25°C

Solution A:
Composition per 100.0mL:

$NaHCO_3$	10.0g

Preparation of Solution A: Add $NaHCO_3$ to distilled/deionized water and bring volume to 100.0mL. Mix thoroughly. Autoclave for 15 min at 15 psi pressure–121°C.

Solution B:
Composition per 100.0mL:

$Na_2S·9H_2O$	10.0g

Preparation of Solution B: Add $Na_2S·9H_2O$ to distilled/deionized water and bring volume to 100.0mL. Mix thoroughly. Autoclave for 15 min at 15 psi pressure–121°C.

Solution C:
Composition per 100.0mL:

$Na_2S_2O_3·9H_2O$	10.0g

Preparation of Solution C: Add $Na_2S_2O_3·9H_2O$ to distilled/deionized water and bring volume to 100.0mL. Mix thoroughly. Autoclave for 15 min at 15 psi pressure–121°C.

Trace Elements Solution:
Composition per liter:

$FeCl_3·6H_2O$	2.7g
H_3BO_3	0.1g
$ZnSO_4·7H_2O$	0.1g
$Co(NO_3)_2·6H_2O$	50.0mg
$CuSO_4·5H_2O$	5.0mg
$MnCl_2·6H_2O$	5.0mg

Preparation of Trace Elements Solution: Add components to distilled/deionized water and bring volume to 1.0L. Mix thoroughly.

Preparation of Medium: Add components, except solution A, solution B, and solution C, to distilled/deionized water and bring volume to 1.0L Mix thoroughly. Bring pH to 7.0–7.2 with H_3PO_4. Distribute into tubes or flasks. Autoclave for 15 min at 15 psi pressure–121°C. Aseptically add 0.2mL of sterile solution A, 0.2mL of sterile solution B, and 0.1mL of sterile solution C for each 10.0mL of medium. Mix thoroughly. Use immediately.

Use: For the cultivation and maintenance of *Chlorobium limnicola*.

Chlorobutane Medium

Composition per 1002.0mL:

NH_4NO_3	4.0g
KH_2PO_4	1.5g
$Na_2HPO_4·12H_2O$	1.5g
$CaSO_4·2H_2O$	10.0mg
$MgSO_4·7H_2O$	10.0mg
$FeSO_4·7H_2O$	5.0mg
Yeast extract	5.0mg
1-Chlorobutane	2.0mL

pH 7.0 ± 0.2 at 25°C

Preparation of 1-Chlorobutane: Filter sterilize.

Preparation of Medium: Add components, except 1-chlorobutane, to distilled/deionized water and bring volume to 1.0L. Mix thoroughly. Adjust pH to 7.0. Autoclave for 15 min at 15 psi pressure–121°C. Aseptically add 2.0mL of sterile 1-chlorobutane. Mix thoroughly. Aseptically distribute into sterile tubes or flasks.

Use: For the cultivation of *Corynebacterium* species.

Chloroflexus Agar

Composition per liter:

Agar	15.0g
Glycyl-glycine	0.5g
Yeast extract	0.5g
Na_2S	0.5g
NH_4Cl	0.2g
$MgSO_4·7H_2O$	0.1g
Nitrilotriacetic acid	0.1g
$NaNO_3$	0.689g
Na_2HPO_4	0.111g
KNO_3	0.103g
$CaSO_4·2H_2O$	0.06g
NaCl	8.0mg
$FeCl_3$ solution	1.0mL
Micronutrient solution	1.0mL

pH 8.2–8.4 at 25°C

$FeCl_3$ Solution:
Composition per liter:

$FeCl_3$	0.29g

Preparation of $FeCl_3$ Solution: Add $FeCl_3$ to distilled/deionized water and bring volume to 1.0L. Mix thoroughly.

Micronutrient Solution:
Composition per liter:

$MnSO_4 \cdot 7H_2O$ 2.28g
H_3BO_3 .... 0.5g
$ZnSO_4 \cdot 7H_2O$ 0.5g
$CoCl_2 \cdot 6H_2O$ 0.045g
$CuSO_4 \cdot 2H_2O$ 0.025g
$Na_2MoO_4 \cdot 2H_2O$ 0.025g
H_2SO_4, concentrated 0.5mL

Preparation of Micronutrient Solution: Add components to distilled/deionized water and bring volume to 1.0L. Mix thoroughly.

Preparation of Medium: Add components, except Na_2S, to distilled/deionized water and bring volume to 1.0L. Mix thoroughly. Adjust pH to 8.2–8.4. Add Na_2S. Readjust pH to 8.2–8.4. Gently heat and bring to boiling. Distribute into tubes or flasks. Autoclave for 15 min at 15 psi pressure–121°C. Pour into sterile Petri dishes or leave in tubes.

Use: For the cultivation of *Chloroflexus aurantiacus*.

Chloroflexus aggregans Medium (DSMZ Medium 87a)

Composition per 1061.0mL:

Yeast extract 1.0g
Glycyl-glycine 1.0g
$NaNO_3$ 0.5g
$Na_2HPO_4 \cdot 7H_2O$ 0.1g
$MgSO_4 \cdot 7H_2O$ 0.1g
KNO_3 .... 0.1g
NaCl 0.1g
$CaCl_2 \cdot 2H_2O$ 0.05g
Neutralized sulfide solution 11.0mL
Ferric citrate solution 5.0mL
Trace elements solution SL-6 .... 1.0mL
pH 8.2 ± 0.2 at 25°C

Ferric Citrate Solution:
Composition per 100.0mL:

Ferric citrate 0.1mg

Preparation of Ferric Citrate Solution: Add ferric citrate to distilled/deionized water and bring volume to 100.0mL. Mix thoroughly. Sparge under 100% N_2 gas for 3 min.

Trace Elements Solution SL-6:
Composition per liter:

$MnCl_2 \cdot 4H_2O$ 0.5g
H_3BO_3 .... 0.3g
$CoCl_2 \cdot 6H_2O$ 0.2g
$ZnSO_4 \cdot 7H_2O$ 0.1g
$Na_2MoO_4 \cdot 2H_2O$ 0.03g
$NiCl_2 \cdot 6H_2O$ 0.02g
$CuCl_2 \cdot 2H_2O$ 0.01g

Preparation of Trace Elements Solution SL-6: Add components to distilled/deionized water and bring volume to 1.0L. Mix thoroughly.

Neutralized Sulfide Solution:
Composition per 100.0mL:

$Na_2S \cdot 9H_2O$ 1.5g

Preparation of Neutralized Sulfide Solution: Add $Na_2S \cdot 9H_2O$ to distilled/deionized water in a 250mL screw-capped bottle fitted with a butyl rubber septum and bring volume to 100.0mL. Add a magnetic stir bar. Mix thoroughly. Sparge under 100% N_2 gas for 3 min. Autoclave for 15 min at 15 psi pressure–121°C. Cool to room temperature. Adjust pH to about 7.3 with sterile 2*M* H_2SO_4. Do not open the bottle to add H_2SO_4; use a sterile syringe. Stir the solution continuously to avoid precipitation of elemental sulfur. The final solution should be clear and yellow in color.

Preparation of Medium: Add components, except neutralized sulfide solution, to distilled/deionized water and bring volume to 1050.0mL Mix thoroughly. Adjust pH to 8.2. Gently heat and bring to boiling. Continue boiling for 3–4 min under 100% N_2. Distribute 90.0mL of medium into 100mL screw-capped bottles with rubber septa under 100% N_2. Autoclave for 15 min at 15 psi pressure–121°C. Cool to room temperature. Using a sterile syringe, inject 1.0mL of neutralized sulfide solution into each bottle. Incubate the culture at 50°C at a light intensity of 300–500 lux. For heavy cell suspension supplement periodically with sterile yeast extract solution to yield a final concentration of 0.1%.

Use: For the growth and maintenance of *Chloroflexus aggregans* and *Roseiflexus castenholzii*.

Chloroflexus Medium, Modified

Composition per 1001.0 mL:

Glycyl-glycine 1.0g
Yeast extract 1.0g
$NaNO_3$ 0.5g
KNO_3 .... 0.1g
$MgSO_4 \cdot 7H_2O$ 0.1g
$Na_2HPO_4 \cdot 2H_2O$ 0.1g
NaCl 0.1g
$CaCl_2 \cdot 2H_2O$ 0.05g
Neutralized sulfide solution 11.0mL
Ferric citrate solution 1.0mL
Trace elements solution SL-6 .... 1.0mL
pH 8.2 ± 0.2 at 25°C

Neutralized Sulfide Solution:
Composition per 100.0mL:

$Na_2S \cdot 9H_2O$ 1.5g

Preparation of Neutralized Sulfide Solution: Add $Na_2S \cdot 9H_2O$ to distilled/deionized water in a 250mL screw-capped bottle fitted with a butyl rubber septum and bring volume to 100.0mL. Add a magnetic stir bar. Mix thoroughly. Sparge under 100% N_2 gas for 3 min. Autoclave for 15 min at 15 psi pressure–121°C. Cool to room temperature. Adjust pH to about 7.3 with sterile 2*M* H_2SO_4. Do not open the bottle to add H_2SO_4; use a sterile syringe. Stir the solution continuously to avoid precipitation of elemental sulfur. The final solution should be clear and yellow in color.

Ferric Citrate Solution:
Composition per 100.0mL:

Ferric citrate 0.1g

Preparation of Ferric Citrate Solution: Add ferric citrate to distilled/deionized water and bring volume to 100.0mL. Mix thoroughly.

Trace Elements Solution SL-6:
Composition per liter:

$MnCl_2 \cdot 4H_2O$ 0.5g
H_3BO_3 .... 0.3g
$CoCl_2 \cdot 6H_2O$ 0.2g
$ZnSO_4 \cdot 7H_2O$ 0.1g
$Na_2MoO_4 \cdot 2H_2O$ 0.03g
$NiCl_2 \cdot 6H_2O$ 0.02g
$CuCl_2 \cdot 2H_2O$ 0.01g

Preparation of Trace Elements Solution SL-6: Add components to distilled/deionized water and bring volume to 1.0L. Mix thoroughly.

Preparation of Medium: Add components, except neutralized sulfide solution, to distilled/deionized water and bring volume to 990.0mL Mix thoroughly. Gently heat and bring to boiling. Continue boiling for 3–4 min under 100% N_2. Distribute 90.0mL of medium into 100mL screw-capped bottles under 100% N_2. Autoclave for 15 min at 15 psi pressure–121°C. Cool to room temperature. Using a sterile syringe, inject 1.0mL of neutralized sulfide solution into each bottle.

Use: For the growth and maintenance of *Chloroflexus aurantiacus.*

Chlorohydroxybenzoic Acid Medium

Composition per liter:

$K_2HPO_4 \cdot 3H_2O$	4.25g
NH_4Cl	2.0g
$NaH_2PO_4 \cdot H_2O$	1.0g
5-Chloro-2-hydroxybenzoic acid	0.5g
$MgSO_4 \cdot 7H_2O$	0.2g
Nitrilotriacetic acid	0.1g
$FeSO_4 \cdot 7H_2O$	0.012g
$MnSO_4 \cdot H_2O$	3.0mg
$ZnSO_4 \cdot 7H_2O$	3.0mg
$CoSO_4$	1.0mg

pH 7.0-7.4 at 25°C

Preparation of Medium: Add 5-chloro-2-hydroxybenzoic acid to 800.0mL of distilled/deionized water. Adjust pH to 7.0 with NaOH. Add remaining components and bring volume to 1.0L. Distribute into tubes or flasks. Autoclave for 15 min at 15 psi pressure–121°C.

Use: For the cultivation of bacteria that can utilize 5-chloro-hydroxybenzoic acid. For the cultivation of ATCC strain 35944.

Cholera Medium TCBS

Composition per liter:

Sucrose	20.0g
Agar	14.0g
Peptone	10.0g
NaCl	10.0g
Sodium citrate	10.0g
$Na_2S_2O_3 \cdot 5H_2O$	10.0g
Ox bile	8.0g
Yeast extract	5.0g
Ferric citrate	1.0g
Bromothymol Blue	0.04g
Thymol Blue	0.04g

pH 8.6 ± 0.2 at 25°C

Source: This medium is available as a premixed powder from Oxoid Unipath.

Preparation of Medium: Add components to distilled/deionized water and bring volume to 1.0L. Mix thoroughly. Gently heat and bring to boiling. Do not autoclave. Pour into sterile Petri dishes. Dry agar plates before using.

Use: For the growth of *Vibrio cholerae, Vibrio parahaemolyticus,* and other *Vibrio* species.

Chopped Meat Carbohydrate Medium with Rumen Fluid

Composition per 1390.0mL:

Peptone	30.0g
K_2HPO_4	5.0g
Yeast extract	5.0g
Glucose	4.0g
Cellobiose	1.0g
Maltose	1.0g
Starch	1.0g
L-Cysteine·HCl·H_2O	0.5g
Chopped meat extract filtrate	1.0L
Chopped meat extract solids	200.0mL
Rumen fluid	150.0mL
Resazurin (0.025% solution)	4.0mL

pH 7.0 ± 0.2 at 25°C

Chopped Meat Extract:
Composition per liter:

Beef or horse meat	500.0g
NaOH (1*N* solution)	25.0mL

Preparation of Chopped Meat Extract: Use lean beef or horse meat. Remove fat and connective tissue. Grind. Add meat and NaOH to distilled/deionized water and bring volume to 1.0L. Gently heat and bring to boiling while stirring. Cool to 25°C. Remove fat from surface. Filter. Reserve ground meat particles and filtrate. Add distilled/deionized water to filtrate and bring volume to 1.0L.

Preparation of Medium: To 1.0L of chopped meat extract filtrate, add the remaining components, except L-cysteine·HCl·H_2O and chopped meat solids. Mix thoroughly. Gently heat to boiling. Cool to room temperature. Add the L-cysteine·HCl·H_2O. Adjust pH to 7.0. Distribute 1 part chopped meat solids (by volume) and 5 parts of liquid (by volume) into tubes under O_2-free 97% N_2 + 3% H_2. Cap with rubber stoppers and place tubes in a press. Autoclave for 15 min at 15 psi pressure–121°C with fast exhaust.

Use: For the cultivation of anaerobic bacteria, including *Fusobacterium prausnitzii, Eubacterium* species, and *Prevotella ruminicola.*

Chopped Meat Carbohydrate Medium with Rumen Fluid (ATCC Medium 1016)

Composition per 1390.0mL:

Peptone	30.0g
K_2HPO_4	5.0g
Yeast extract	5.0g
Cellobiose	1.0g
Maltose	1.0g
Starch	1.0g
L-Cysteine·HCl·H_2O	0.5g
Chopped meat extract filtrate	1.0L
Chopped meat extract solids	200.0mL
Rumen fluid	150.0mL
Resazurin (0.025% solution)	4.0mL

pH 7.0 ± 0.2 at 25°C

Chopped Meat Extract:
Composition per liter:

Beef or horse meat	500.0g
NaOH (1*N* solution)	25.0mL

Preparation of Chopped Meat Extract: Use lean beef or horse meat. Remove fat and connective tissue. Grind. Add meat and NaOH to distilled/deionized water and bring volume to 1.0L. Gently heat and bring to boiling while stirring. Cool to 25°C. Remove fat from surface. Filter. Reserve ground meat particles and filtrate. Add distilled/deionized water to filtrate and bring volume to 1.0L.

Preparation of Medium: To 1.0L of chopped meat extract filtrate, add the remaining components, except L-cysteine·HCl·H_2O and chopped meat solids. Mix thoroughly. Gently heat to boiling. Cool to room temperature. Add the L-cysteine·HCl·H_2O. Adjust pH to 7.0. Distribute 1 part chopped meat solids (by volume) and 5 parts of liquid (by volume) into tubes under O_2-free 97% N_2 + 3% H_2. Cap with rubber stoppers and place tubes in a press. Autoclave for 15 min at 15 psi pressure–121°C with fast exhaust.

Use: For the cultivation of anaerobic bacteria, including *Butyrivibrio crossotus, Eubacterium* species, and *Ruminococcus* species.

CHROMagar *E. coli*

Composition per liter:
Proprietary.

Source: CHROMagar *E. coli* is available from CHROMagar Microbiology.

Preparation of Medium: Add components to distilled/deionized water and bring volume to 1.0L. Mix thoroughly. Gently heat in a boiling water bath or steam bath. Shake periodically during heating to dissolve components. Heat long enough with shaking every 5 min to ensure complete dissolution. Do not overheat. Adding tellurite can increase specificity. Cool to 45-50°C. Pour into sterile Petri dishes.

Use: For the differentiation and presumptive identification of *Escherichia coli* which forms blue colonies.

CHROMagar ECC

Composition per liter:
Proprietary.

Source: CHROMagar EEC is available from CHROMagar Microbiology.

Preparation of Medium: Add components to distilled/deionized water and bring volume to 1.0L. Mix thoroughly. Gently heat in a boiling water bath or steam bath. Shake periodically during heating to dissolve components. Heat long enough with shaking every 5 min to ensure complete dissolution. Do not overheat. Adding tellurite can increase specificity. Cool to 45-50°C. Pour into sterile Petri dishes.

Use: For the differentiation and presumptive identification of *Escherichia coli* and other coliform bacteria which form red colonies.

CHROMagar O157

Composition per liter:
Proprietary.

Source: CHROMagar O157 is available from CHROMagar Microbiology.

Preparation of Medium: Add components to distilled/deionized water and bring volume to 1.0L. Mix thoroughly. Gently heat in a boiling water bath or steam bath. Shake periodically during heating to dissolve components. Heat long enough with shaking every 5 min to ensure complete dissolution. Do not overheat. Adding tellurite can increase specificity. Cool to 45-50°C. Pour into sterile Petri dishes.

Use: For the differentiation and presumptive identification of *Escherichia coli* O157.

CHROMagar *Vibrio*

Composition per liter:
Proprietary.

Preparation of Medium: Add components to distilled/deionized water and bring volume to 1.0L. Mix thoroughly. Gently heat in a boiling water bath or steam bath. Shake periodically during heating to dissolve components. Heat long enough with shaking every 5 min to ensure complete dissolution. Do not overheat. Adding tellurite can increase specificity. Cool to 45-50°C. Pour into sterile Petri dishes.

Source: CHROMagar *Vibrio* is available from CHROMagar Microbiology.

Use: For the differentiation and presumptive identification of *Vibrio parahaemolyiticus* which form mauve colonies; *Vibrio cholerae* form turquoise blue colonies and *Vibrio alginolyitcus* colonies are colorless.

Chromatiaceae Medium 1

Composition per 4990.0mL:

Solution 1	4.0L
O_2-free water	860.0mL
$NaHCO_3$ solution	100.0mL
$Na_2S{\cdot}9H_2O$ solution	20.0mL
Trace elements solution	5.0mL
Vitamin B_{12} solution	5.0mL

pH 7.3 ± 0.2 at 25°C

Solution 1:
Composition per 4.0L:

$MgSO_4{\cdot}7H_2O$	2.5g
KCl	1.7g
KH_2PO_4	1.7g
NH_4Cl	1.7g
$CaCl_2{\cdot}2H_2O$	1.25g

Preparation of Solution 1: Add components to distilled/deionized water and bring volume to 4.0L. Mix thoroughly. Autoclave for 45 min at 15 psi pressure–121°C. Cool to 25°C under 100% N_2. Saturate with CO_2 by stirring under 100% CO_2 for 30 min.

O_2-Free Water:
Composition per 860.0mL:

H_2O	860.0mL

Preparation of O_2-Free Water: Autoclave H_2O for 15 min at 15 psi pressure–121°C. Cool to 25°C under 100% N_2.

$NaHCO_3$ Solution:
Composition per 100.0mL:

$NaHCO_3$	7.5g

Preparation of $NaHCO_3$ Solution: Add the $NaHCO_3$ to distilled/deionized water and bring volume to 100.0mL. Mix thoroughly. Gas with 100% CO_2 for 20 min. Filter sterilize with positive CO_2 pressure.

$Na_2S{\cdot}9H_2O$ Solution:
Composition per 100.0mL:

$Na_2S{\cdot}9H_2O$	10.0g

Preparation of $Na_2S{\cdot}9H_2O$ Solution: Add $Na_2S{\cdot}9H_2O$ to distilled/deionized water. Mix thoroughly. Gas with 100% N_2 for 15 min in a screw-capped bottle. Tightly close cap. Autoclave for 15 min at 15 psi pressure–121°C. Cool to 25°C.

Trace Elements Solution:
Composition per liter:

$FeCl_2{\cdot}4H_2O$	1.5g
$CoCl_2{\cdot}6H_2O$	0.19g
$MnCl_2{\cdot}4H_2O$	0.1g
$ZnCl_2$	0.07g
H_3BO_3	0.06g
$NaMoO_4{\cdot}2H_2O$	0.04g
$CuCl_2{\cdot}2H_2O$	0.02g
$NiCl_2{\cdot}6H_20$	0.02g
HCl (25% solution)	6.5mL

Preparation of Trace Elements Solution: Add components to distilled/deionized water and bring volume to

1.0L. Mix thoroughly. Autoclave for 15 min at 15 psi pressure–121°C. Cool to 25°C.

Vitamin B_{12} Solution:
Composition per 100.0mL:

Vitamin B_{12}	2.0mg

Preparation of Vitamin B_{12} Solution: Add components to distilled/deionized water and bring volume to 100.0mL. Mix thoroughly. Filter sterilize.

Preparation of Medium: To 4.0L of sterile, CO_2-saturated solution 1, aseptically add the remaining components. Mix thoroughly. Adjust pH to 7.3. Aseptically distribute into sterile 100.0mL bottles using positive pressure of 95% N_2 + 5% CO_2. Completely fill bottles with medium except for a pea-sized air bubble.

Use: For the isolation and cultivation of members of the Chlorobiaceae.

Chromatiaceae Medium 2

Composition per 1051.0mL:

Solution 1	950.0mL
$Na_2S \cdot 9H_2O$ solution	60.0mL
$NaHCO_3$ solution	40.0mL
Vitamin B_{12} solution	1.0mL

pH 7.3 ± 0.2 at 25°C

Solution 1:
Composition per 950.0mL:

KH_2PO_4	1.0g
NH_4Cl	0.5g
$MgSO_4 \cdot 7H_2O$	0.4g
$CaCl_2 \cdot 2H_2O$	0.05g
Trace elements solution SL-8	1.0mL

Preparation of Solution 1: Add components to distilled/deionized water and bring volume to 950.0mL. Mix thoroughly. Autoclave for 15 min at 15 psi pressure–121°C. Cool to 45°–50°C.

Trace Elements Solution SL-8:
Composition per liter:

Disodium EDTA	5.2g
$FeCl_2 \cdot 4H_2O$	1.5g
$CoCl_2 \cdot 6H_2O$	0.19g
$MnCl_2 \cdot 4H_2O$	0.1g
$ZnCl_2$	0.07g
H_3BO_3	0.06g
$NaMoO_4 \cdot 2H_2O$	0.04g
$CuCl_2 \cdot 2H_2O$	0.02g
$NiCl_2 \cdot 6H_20$	0.02g

Preparation of Trace Elements Solution SL-8: Add components to distilled/deionized water and bring volume to 1.0L. Mix thoroughly.

$Na_2S \cdot 9H_2O$ Solution:
Composition per 100.0mL:

$Na_2S \cdot 9H_2O$	5.0g

Preparation of $Na_2S \cdot 9H_2O$ Solution: Add $Na_2S \cdot 9H_2O$ to distilled/deionized water and bring volume to 100.0mL. Autoclave for 15 min at 15 psi pressure–121°C. Cool to 45°–50°C.

$NaHCO_3$ Solution:
Composition per 100.0mL:

$NaHCO_3$	5.0g

Preparation of $NaHCO_3$ Solution: Add $NaHCO_3$ to distilled/deionized water and bring volume to 100.0mL. Mix thoroughly. Filter sterilize.

Vitamin B_{12} Solution:
Composition per 100.0mL:

Vitamin B_{12}	2.0mg

Preparation of Vitamin B_{12} Solution: Add Vitamin B_{12} to distilled/deionized water and bring volume to 100.0mL. Mix thoroughly. Filter sterilize.

Preparation of Medium: To 950.0mL of cooled, sterile solution 1, aseptically add 60.0mL of sterile $Na_2S \cdot 9H_2O$ solution, 40.0mL of sterile $NaHCO_3$ solution, and 1.0mL of sterile Vitamin B_{12} solution. Mix thoroughly. Adjust pH to 7.3 with sterile H_2SO_4 or Na_2CO_3. Aseptically distribute into sterile 50.0mL or 100.0mL bottles with metal screw-caps and rubber seals. Completely fill bottles with medium except for a pea-sized air bubble.

Use: For the isolation and cultivation of freshwater and soil members of the Chromatiaceae.

Chromatium Medium (ATCC Medium 1449)

Composition per liter:

KH_2PO_4	0.5g
NH_4Cl	0.4g
$MgSO_4 \cdot 7H_2O$	0.2g
$CaCl_2 \cdot 2H_2O$	0.05g
Disodium EDTA	0.01g
Trace elements	1.0mL
$NaHCO_3$ solution	50.0mL
$Na_2S \cdot 9H_2O$ solution	50.0mL
Sodium pyruvate solution	50.0mL

pH 7.0 ± 0.2 at 25°C

Trace Elements:
Composition per liter:

Disodium EDTA	5.2g
$FeCl_2 \cdot 4H_2O$	1.5g
$CoCl_2 \cdot 6H_2O$	0.19g
$Na_2MoO_4 \cdot 2H_2O$	0.188g
$MnCl_2 \cdot 4H_2O$	0.1g
$ZnCl_2$	0.07g
$VOSO_4 \cdot 2H_2O$	0.03g
$NiCl_2 \cdot 6H_2O$	0.025g
H_3BO_3	6.0mg
$CuCl_2 \cdot 2H_2O$	2.0mg
Na_2SeO_3	2.0mg

Preparation of Trace Elements: Add components to distilled/deionized water and bring volume to 1.0L. Mix thoroughly.

$NaHCO_3$ Solution:
Composition per 50.0mL:

$NaHCO_3$	2.0g

Preparation of $NaHCO_3$ Solution: Add $NaHCO_3$ to distilled/deionized water and bring volume to 50.0mL. Filter sterilize. Use freshly prepared solution.

$Na_2S \cdot 9H_2O$ Solution:
Composition per 50.0mL:

$Na_2S \cdot 9H_2O$	1.0g

Preparation of $Na_2S \cdot 9H_2O$ Solution: Add $Na_2S \cdot 9H_2O$ to distilled/deionized water and bring volume to 50.0mL. Autoclave for 15 min at 15 psi pressure–121°C. Use freshly prepared solution.

Sodium Pyruvate Solution:
Composition per 50.0mL:

Sodium pyruvate	0.5g

Preparation of Sodium Pyruvate Solution: Add $NaHCO_3$ to distilled/deionized water and bring volume to 50.0mL. Filter sterilize. Use freshly prepared solution. Sodium acetate may be substituted for the sodium pyruvate.

Preparation of Medium: Add components, except $NaHCO_3$ solution, $Na_2S·9H_2O$ solution, and sodium pyruvate solution, to distilled deionized water and bring volume to 850.0mL. Autoclave for 15 min at 15 psi pressure–121°C. Cool to room temperature. Add the sterile $NaHCO_3$ solution, the sterile $Na_2S·9H_2O$ solution, and the sterile sodium pyruvate solution, in that order. Adjust the pH to 7.0. Distribute into screw-capped tubes or flasks. Fill to capacity.

Use: For the cultivation and maintenance of *Chromatium* species.

Chromatium Medium with NaCl

Composition per 1051.0mL:

Solution 1	950.0mL
$Na_2S·9H_2O$ solution	60.0mL
$NaHCO_3$ solution	40.0mL
Vitamin B_{12} solution	1.0mL

pH 3-4 at 25°C

Solution 1:

Composition per 950.0mL:

NaCl	30.0g
KH_2PO_4	1.0g
NH_4Cl	0.5g
$MgSO_4·7H_2O$	0.4g
$CaCl_2·2H_2O$	0.05g
Trace elements solution SL-8	1.0mL

Preparation of Solution 1: Add components to distilled/deionized water and bring volume to 950.0mL. Mix thoroughly. Autoclave for 15 min at 15 psi pressure–121°C. Cool to 45°–50°C.

Trace Elements Solution SL-8:

Composition per liter:

Disodium EDTA	5.2g
$FeCl_2·4H_2O$	1.5g
$CoCl_2·6H_2O$	0.19g
$MnCl_2·4H_2O$	0.10g
$ZnCl_2$	0.07g
H_3BO_3	0.06g
$NaMoO_4·2H_2O$	0.04g
$CuCl_2·2H_2O$	0.02g
$NiCl_2·6H_20$	0.02g

Preparation of Trace Elements Solution SL-8: Add components to distilled/deionized water and bring volume to 1.0L. Mix thoroughly.

$Na_2S·9H_2O$ Solution:

Composition per 100.0mL:

$Na_2S·9H_2O$	5.0g

Preparation of $Na_2S·9H_2O$ Solution: Add $Na_2S·9H_2O$ to distilled/deionized water and bring volume to 100.0mL. Autoclave for 15 min at 15 psi pressure–121°C. Cool to 45°–50°C.

$NaHCO_3$ Solution:

Composition per 100.0mL:

$NaHCO_3$	5.0g

Preparation of $NaHCO_3$ Solution: Add $NaHCO_3$ to distilled/deionized water and bring volume to 100.0mL. Mix thoroughly. Filter sterilize.

Vitamin B_{12} Solution:

Composition per 100.0mL:

Vitamin B_{12}	2.0mg

Preparation of Vitamin B_{12} Solution: Add Vitamin B_{12} to distilled/deionized water and bring volume to 100.0mL. Mix thoroughly. Filter sterilize.

Preparation of Medium: To 950.0mL of cooled, sterile solution 1, aseptically add 60.0mL of sterile $Na_2S·9H_2O$ solution, 40.0mL of sterile $NaHCO_3$ solution, and 1.0mL of sterile Vitamin B_{12} solution. Mix thoroughly. Adjust pH to 7.3 with sterile H_2SO_4 or Na_2CO_3. Aseptically distribute into sterile 50.0mL or 100.0mL bottles with metal screw-caps and rubber seals. Completely fill bottles with medium except for a pea-sized air bubble.

Use: For the cultivation of *Ectothiorhodospira marismortui* and *Ectothiorhodospira mobilis.*

Chromatium salexigens Medium

Composition per 4990.0mL:

Solution A	4000.0mL
Solution B	860.0mL
Solution C (Vitamin B_{12} solution)	5.0mL
Solution D (Trace elements solution SL-12B)	5.0mL
Solution E	100.0mL
Solution F	20.0mL

pH 7.3 ± 0.2 at 25°C

Solution A:

Composition per 4000.0mL:

NaCl	100.0g
$MgCl_2·6H_2O$	3.0g
$MgSO_4$	2.5g
KH_2PO_4	1.7g
NH_4Cl	1.7g
KCl	1.7g
$CaCl_2·2H_2O$	1.25g
Sodium acetate	0.5g
$Na_2S_2O_3$	0.5g

Preparation of Solution A: Add components to distilled/deionized water and bring volume to 4.0L. Mix thoroughly. Adjust pH to 6.0. Dispense into a 5-liter flask with four openings at the top (two openings are in a central silicon rubber stopper and two openings are gas-tight screw caps). Add a teflon-coated magnetic stir bar to the flask. Autoclave for 45 min at 15 psi pressure–121°C. Cool to room temperature under 100% N_2 at 0.05–0.1 atm pressure (use a manometer to measure low pressure).

Solution B:

Composition per 860.0mL:

Distilled/deionized water	860.0mL

Preparation of Solution B: Add 860.0mL of distilled/deionized water to a cotton-stoppered flask. Autoclave for 20 min at 15 psi–121°C. Cool to room temperature under 100% N_2 in an anaerobic jar.

Solution C (Vitamin B_{12} Solution):

Composition per 5.0mL:

Vitamin B_{12}	1.0mg

Preparation of Solution C (Vitamin B_{12} Solution): Add Vitamin B_{12} to distilled/deionized water and bring volume to 5.0mL. Mix thoroughly. Filter sterilize.

Solution D (Trace Elements Solution SL-12B):
Composition per liter:
Disodium ethylendiamine-tetraacetate
(Disodium EDTA) 3.0g
$FeSO_4 \cdot 7H_2O$ 1.1g
H_3BO_3 .... 0.3g
$CoCl_2 \cdot 6H_2O$ 0.19g
$MnCl_2 \cdot 2H_2O$ 50.0mg
$ZnCl_2$ 42.0mg
$NiCl_2 \cdot 6H_2O$ 24.0mg
$Na_2MoO_4 \cdot 2H_2O$ 18.0mg
$CuCl_2 \cdot 2H_2O$ 2.0mg

Preparation of Solution D (Trace Elements Solution SL-12B): Add components to distilled/deionized water and bring volume to 1.0L. Mix thoroughly. Autoclave for 15 min at 15 psi–121°C.

Solution E:
Composition per 100.0mL:
$NaHCO_3$ 7.5g

Preparation of Solution E: Add $NaHCO_3$ to distilled/deionized water and bring volume to 100.0mL. Mix thoroughly. Sparge with 100% CO_2 until saturated. Filter sterilize under 100% CO_2 into a sterile, gas-tight 100.0mL screw-capped bottle.

Solution F:
Composition per 100.0mL:
$Na_2S \cdot 9H_2O$ 10.0g

Preparation of Solution F: Add $Na_2S \cdot 9H_2O$ to distilled/deionized water and bring volume to 100.0mL. Mix thoroughly. Dispense into a screw-capped bottle. Sparge with 100% N_2 for 3–4 min. Autoclave for 15 min at 15 psi pressure–121°C.

Preparation of Medium: Saturate cooled solution A under 100% CO_2 at 0.05–0.1 atm pressure for 30 min with magnetic stirring. Add 860.0mL of solution B, 5.0mL of solution C, 5.0mL of solution D, 100.0mL of solution E, and 20.0mL of solution F through one of the screw-capped openings under 95% N_2 and 5% CO_2 with magnetic stirring. Adjust pH to 7.3 with sterile 2*M* HCl or sterile 2*M* Na_2CO_3 solution. Aseptically and anaerobically distribute the medium through the medium outlet tube into sterile 100.0mL bottles under 95% N_2 and 5% CO_2 at 0.05–0.1 atm pressure. Leave a small gas bubble in each bottle to accommodate pressure changes. After 24 hr, the iron in the medium will precipitate out of solution as black flocs.

Use: For the growth and maintenance of *Chromatium salexigens.*

Chromatium tepidum Medium

Composition per 1001.0mL:
NH_4Cl 400.0mg
NaCl 400.0mg
$MgSO_4 \cdot 7H_2O$ 200.0mg
$CaCl_2 \cdot 2H_2O$ 50.0mg
Disodium ethylendiamine-tetraacetate
(Disodium EDTA) 10.0mg
Ammonium acetate
(or sodium pyruvate) 0.5mg
KH_2PO_4 .... 0.5mg
$NaHCO_3$ solution 20.0mL
$Na_2S \cdot 9H_2O$ solution 20.0mL
Trace elements solution 1.0mL
pH 7.0 ± 0.2 at 25°C

$NaHCO_3$ Solution:
Composition per 20.0mL:
$NaHCO_3$ 2.0g

Preparation of $NaHCO_3$ Solution: Add $NaHCO_3$ to distilled/deionized water and bring volume to 20.0mL. Mix thoroughly. Autoclave for 15 min at 15 psi pressure–121°C.

$Na_2S \cdot 9H_2O$ Solution:
Composition per 20.0mL:
$Na_2S \cdot 9H_2O$ 1.0g

Preparation of $Na_2S \cdot 9H_2O$ Solution: Add $Na_2S \cdot 9H_2O$ to distilled/deionized water and bring volume to 20.0mL. Mix thoroughly. Autoclave for 15 min at 15 psi pressure–121°C.

Trace Elements Solutions:
Composition per liter:
Disodium ethylendiamine-tetraacetate
(Disodium EDTA) 5.2g
$FeCl_2 \cdot 4H_2O$ 1.5g
$CoCl_2 \cdot 6H_2O$ 190.0mg
$Na_2MoO_4 \cdot 2H_2O$ 188.0mg
$MnCl_2 \cdot 4H_2O$ 100.0mg
$ZnCl_2$ 70.0mg
$VOSO_4 \cdot 2H_2O$ 30.0mg
$NiCl_2 \cdot 6H_2O$ 25.0mg
H_3BO_3 .... 6.0mg
$CuCl_2 \cdot 2H_2O$ 2.0mg
$Na_2WO_4 \cdot 2H_2O$ 2.0mg

Preparation of Trace Elements Solutions: Add components to distilled/deionized water and bring volume to 1.0L. Mix thoroughly.

Preparation of Medium: Add components, except $NaHCO_3$ solution and $Na_2S \cdot 9H_2O$ solution, to distilled/deionized water and bring volume to 960.0mL. Mix thoroughly. Autoclave for 15 min at 15 psi pressure–121°C. Aseptically add 20.0mL of sterile $NaHCO_3$ solution and 20.0mL of sterile $Na_2S \cdot 9H_2O$ solution. Mix thoroughly. Adjust pH to 7.0. Distribute into sterile screw-capped tubes or bottles so that medium completely fills container.

Use: For the growth and maintenance of *Chromatium tepidum.*

Chromatium/Thiocapsa Medium

Composition per 1060.0mL:
KH_2PO_4 .... 1.0g
NH_4Cl 1.0g
$MgCl_2 \cdot 6H_2O$ 0.5g
Solution A 20.0mL
Solution B 20.0mL
Solution C 10.0mL
Solution D 10.0mL
Trace elements solution 1.0mL
pH 7.0 ± 0.2 at 25°C

Solution A:
Composition per 100.0mL:
$NaHCO_3$ 10.0g

Preparation of Solution A: Add $NaHCO_3$ to distilled/deionized water and bring volume to 100.0mL. Mix thoroughly. Autoclave for 15 min at 15 psi pressure–121°C.

Solution B:
Composition per 100.0mL:
$Na_2S \cdot 9H_2O$ 10.0g

Preparation of Solution B: Add $Na_2S{\cdot}9H_2O$ to distilled/deionized water and bring volume to 100.0mL. Mix thoroughly. Autoclave for 15 min at 15 psi pressure–121°C.

Solution C:
Composition per 100.0mL:
$Na_2S_2O_3{\cdot}9H_2O$ 10.0g

Preparation of Solution C: Add $Na_2S_2O_3{\cdot}9H_2O$ to distilled/deionized water and bring volume to 100.0mL. Mix thoroughly. Autoclave for 15 min at 15 psi pressure–121°C.

Solution D:
Composition per 100.0mL:
Sodium malate 10.0g

Preparation of Solution D: Add sodium malate to distilled/deionized water and bring volume to 100.0mL. Mix thoroughly. Autoclave for 15 min at 15 psi pressure–121°C.

Trace Elements Solution:
Composition per liter:
$FeCl_3{\cdot}6H_2O$ 2.7g
H_3BO_3 .... 0.1g
$ZnSO_4{\cdot}7H_2O$ 0.1g
$Co(NO_3)_2{\cdot}6H_2O$ 50.0mg
$CuSO_4{\cdot}5H_2O$ 5.0mg
$MnCl_2{\cdot}6H_2O$ 5.0mg

Preparation of Trace Elements Solution: Add components to distilled/deionized water and bring volume to 1.0L. Mix thoroughly.

Preparation of Medium: Add components, except solution A, solution B, solution C, and solution D, to distilled/deionized water and bring volume to 1.0L Mix thoroughly. Bring pH to 7.0–7.2 with H_3PO_4. Distribute into tubes or flasks. Autoclave for 15 min at 15 psi pressure–121°C. Aseptically add 0.2mL of sterile solution A, 0.2mL of sterile solution B, 0.1mL of sterile solution C, and 0.1mL of sterile solution D for each 10.0mL of medium. Mix thoroughly. Use immediately.

Use: For the cultivation of *Chlorobium limnicola* and *Chromatium* species.

Chromobacterium Medium

Composition per liter:
NaCl 30.0g
$MgCl_2$ 10.8g
$MgSO_4$ 5.4g
Peptone 5.0g
$CaCl_2$ 1.0g
KCl 0.7g
pH 7.0 ± 0.2 at 25°C

Preparation of Medium: Add components to distilled/deionized water and bring volume to 1.0L. Mix thoroughly. Distribute into tubes or flasks. Autoclave for 15 min at 15 psi pressure–121°C.

Use: For the cultivation and maintenance of *Chromobacterium* species and *Alteromonas luteoviolacea.*

Chu's No. 10 Medium

Composition per liter:
Agar 15.0g
$Ca(NO_3)_2{\cdot}4H_2O$ 0.232g
$Na_2SiO_3{\cdot}5H_2O$ 0.044g
$MgSO_4{\cdot}7H_2O$ 0.025g
Na_2CO_3 .... 0.02g
K_2HPO_4 .... 0.01g
Citric acid 3.5mg
Ferric citrate 3.5mg

Preparation of Medium: Add components to distilled/deionized water and bring volume to 1.0L. Mix thoroughly. Gently heat to boiling. Distribute into tubes or flasks. Autoclave for 15 min at 15 psi pressure–121°C. Pour into sterile Petri dishes or leave in tubes.

Use: For the cultivation and maintenance of *Anabaena* species and *Plectomena boryanum.*

Chu's No. 10 Medium, Modified

Composition per liter:
Agar 15.0g
$Ca(NO_3)_2{\cdot}4H_2O$ 0.232g
$Na_2SiO_3{\cdot}5H_2O$ 0.044g
$MgSO_4{\cdot}7H_2O$ 0.025g
Na_2CO_3 .... 0.02g
K_2HPO_4 .... 0.01g
Citric acid 3.5mg
Ferric citrate 3.5mg
Metal solution 1.0mL

Metal Solution:
Composition per liter:
H_3BO_3 .... 2.4g
$MnCl_2{\cdot}4H_2O$ 1.4g
$ZnCl_2$ 0.4g
$CoCl_2{\cdot}6H_2O$ 0.02g
$CuCl_2{\cdot}2H_2O$ 0.1mg

Preparation of Metal Solution: Add components to distilled/deionized water and bring volume to 1.0L. Mix thoroughly.

Preparation of Medium: Add components to distilled/deionized water and bring volume to 1.0L. Mix thoroughly. Gently heat to boiling. Distribute into tubes or flasks. Autoclave for 15 min at 15 psi pressure–121°C. Pour into sterile Petri dishes or leave in tubes.

Use: For the cultivation and maintenance of *Anabaena* species and *Plectomena boryanum.*

Chu's No. 11 Medium, Modified

Composition per liter:
$NaNO_3$ 1.5g
$MgSO_4{\cdot}7H_2O$ 0.08g
$Na_2SiO_3{\cdot}9H_2O$ 0.06g
$CaCl_2{\cdot}2H_2O$ 0.04g
$K_2HPO_4{\cdot}3H_2O$ 0.04g
Na_2CO_3 .... 0.02g
Citric acid 6.0mg
Ferric ammonium citrate 6.0mg
EDTA 1.0mg
Seawater 999.0mL
Trace metal solution A5 with cobalt 1.0mL
pH 7.5 ± 0.2 at 25°C

Trace Metal Solution A5 with Cobalt:
Composition per liter:
H_3BO_3 .... 2.86g
$MnCl_2{\cdot}4H_2O$ 1.81g
$Na_2MoO_4{\cdot}2H_2O$ 0.39g
$ZnSO_4{\cdot}7H_2O$ 0.222g
$CuSO_4{\cdot}H_2O$ 0.079g
$Co(NO_3)_2{\cdot}6H_2O$ 0.049g

Preparation of Trace Metal Solution A5 with Cobalt: Add components to distilled/deionized water and bring volume to 1.0L. Mix thoroughly.

Preparation of Medium: Add components to seawater and bring volume to 1.0L. Mix thoroughly. Gently heat and bring to boiling. Distribute into tubes or flasks. Autoclave for 15 min at 15 psi pressure–121°C.

Use: For the isolation and cultivation of cyanobacteria from marine habitats.

CIN Agar (*Yersinia* Selective Agar) (Cefsulodin Irgasan® Novobiocin Agar)

Composition per liter:

Mannitol	20.0g
Agar	12.0g
Pancreatic digest of gelatin	10.0g
Beef extract	5.0g
Peptic digest of animal tissue	5.0g
Sodium pyruvate	2.0g
Yeast extract	2.0g
NaCl	1.0g
Sodium deoxycholate	0.5g
Neutral Red	0.03g
Cefsulodin	0.015g
Irgasan (triclosan)	4.0mg
Novobiocin	2.5mg
Crystal Violet	1.0mg

pH 7.4 ± 0.2 at 25°C

Source: This medium is available as a premixed powder from BD Diagnostic Systems.

Preparation of Medium: Add components, except cefsulodin and novobiocin, to distilled/deionized water and bring volume to 1.0L. Heat, mixing continuously, until boiling. Do not autoclave. Cool to 45°–50°C. Aseptically add cefsulodin and novobiocin. Mix thoroughly. Pour into sterile Petri dishes or distribute into sterile tubes.

Use: For the selective isolation and differentiation of *Yersinia enterocolitica* from a variety of nonclinical specimens based on mannitol fermentation. *Yersinia enterocolitica* appears as "bull's eye" colonies with deep red centers surrounded by a transparent periphery.

Citrate Medium, Koser's Modified

Composition per liter:

NaCl	5.0g
Citric acid	2.0g
$(NH_4)H_2PO_4$	1.0g
K_2HPO_4	1.0g
$MgSO_4 \cdot 7H_2O$	0.2g

pH 6.8 ± 0.2 at 25°C

Preparation of Medium: Add components to distilled/deionized water and bring volume to 1.0L. Mix thoroughly. Adjust pH to 6.8. Distribute into tubes in 5.0mL volumes. Autoclave for 15 min at 15 psi pressure–121°C.

Use: For the cultivation and differentiation of bacteria based on their ability to utilize citrate as a carbon source.

Citrate Phosphate Buffered Glucose Medium

Composition per liter:

Solution A	750.0mL
Solution C	200.0mL
Solution B	50.0mL

pH 3.5 ± 0.2 at 25°C

Solution A:

Composition per 750.0mL:

$(NH_4)_2SO_4$	3.0g
Citric acid, anhydrous	1.92g
Na_2HPO_4	1.23g
$MgSO_4 \cdot 7H_2O$	0.5g
KCl	0.1g
$Ca(NO_3)_2 \cdot 4H_2O$	0.02g
$FeSO_4 \cdot 7H_2O$	0.01g

Preparation of Solution A: Add components to distilled/deionized water and bring volume to 750.0mL. Mix thoroughly.

Solution B:

Composition per 50.0mL:

Glucose	10.0g
Yeast extract	1.0g

Preparation of Solution B: Add components to distilled/deionized water and bring volume to 50.0mL. Mix thoroughly.

Solution C:

Composition per 200.0mL:

Agarose (electrophoresis grade)	6.0g

Preparation of Solution C: Add agarose to distilled/deionized water and bring volume to 200.0mL. Mix thoroughly.

Preparation of Medium: Prepare solutions A, B, and C. Autoclave solutions separately for 15 min at 15 psi pressure–121°C. Cool to 50°–55°C. Combine solutions A, B, and C. Mix thoroughly. Immediately distribute into sterile tubes or flasks.

Use: For the cultivation and maintenance of *Acidiphilium organovorum*.

CK Agar

Composition per liter:

Agar	15.0g
Glucose	5.0g
KNO_3	2.0g
$CaCl_2$	1.5g
$MgSO_4 \cdot 7H_2O$	1.5g
K_2HPO_4	0.25g
Ferric citrate	0.02g

Preparation of Medium: Add components to distilled/deionized water and bring volume to 1.0L. Mix thoroughly. Gently heat and bring to boiling. Distribute into tubes or flasks. Autoclave for 15 min at 15 psi pressure–121°C. Pour into sterile Petri dishes or leave in tubes.

Use: For the cultivation of myxobacteria.

CK1 Medium

Composition per liter:

$MgSO_4 \cdot 7H_2O$	3.0g
KNO_3	2.0g
$CaCl_2$	1.4g
Ferric citrate	0.02g
Glucose solution	100.0mL
K_2HPO_4 solution	10.0mL

Glucose Solution:

Composition per 100.0mL:

D-Glucose	10.0g

Preparation of Glucose Solution: Add D-glucose to distilled/deionized water and bring volume to 100.0mL. Mix thoroughly. Autoclave for 15 min at 15 psi pressure–121°C. Cool to 25°C.

K_2HPO_4 Solution:
Composition per 10.0mL:
K_2HPO_4 .... 2.5mg

Preparation of K_2HPO_4 Solution: Add K_2HPO_4 to distilled/deionized water and bring volume to 10.0mL. Mix thoroughly. Autoclave for 15 min at 15 psi pressure–121°C. Cool to 25°C.

Preparation of Medium: Add components, except glucose solution and K_2HPO_4 solution, to distilled/deionized water and bring volume to 890.0mL. Mix thoroughly. Autoclave for 15 min at 15 psi pressure–121°C. Cool to 25°C. Aseptically add sterile glucose solution and K_2HPO_4 solution. Mix thoroughly. Aseptically distribute into sterile tubes or flasks.

Use: For the cultivation of myxobacteria.

Clostridium beijerinckii Agar

Composition per liter:
Agar 20.0g
K_2HPO_4 .... 5.0g
Proteose peptone No. 3 .... 5.0g
Sodium thioglycolate 5.0g
Yeast extract 5.0g
Polygalacturonic acid solution 50.0mL
pH 7.5 ± 0.2 at 25°C

Polygalacturonic Acid Solution:
Composition per liter:
Polygalacturonic acid (or pectin) 4.6g

Preparation of Polygalacturonic Acid Solution: Dissolve polygalacturonic acid or pectin in small amounts of distilled/deionized water neutralized with 10% NaOH to pH 7.2. Mix intensively. Bring volume to 50.0mL with distilled/deionized water.

Preparation of Medium: Add components to distilled/deionized water and bring volume to 1.0L. Mix thoroughly. Adjust pH to 7.5. Gently heat and bring to boiling. Distribute into tubes or flasks. Autoclave for 15 min at 15 psi pressure–121°C. Pour into sterile Petri dishes or leave in tubes.

Use: For the cultivation and maintenance of *Clostridium beijerinckii.*

Clostridium bryantii Medium

Composition per liter:
NaCl 21.0g
$MgCl_2 \cdot 6H_2O$ 3.1g
Na_2SO_4 .... 3.0g
KCl 0.5g
NH_4Cl 0.3g
KH_2PO_4 .... 0.2g
$CaCl_2 \cdot 2H_2O$ 0.15g
Resazurin 1.0mg
$NaHCO_3$ solution 20.0mL
$Na_2S \cdot 9H_2O$ solution 20.0mL
Sodium caproate solution 20.0mL
Vitamin solution 20.0mL
Trace elements solution SL-10 .... 1.0mL
pH 7.2 ± 0.2 at 25°C

$NaHCO_3$ Solution:
Composition per 20.0mL:
$NaHCO_3$ 5.0g

Preparation of $NaHCO_3$ Solution: Add $NaHCO_3$ to distilled/deionized water and bring volume to 20.0mL. Mix thoroughly. Sparge with 80% N_2 + 20% CO_2. Autoclave for 15 min at 15 psi pressure–121°C.

$Na_2S \cdot 9H_2O$ Solution:
Composition per 20.0mL:
$Na_2S \cdot 9H_2O$ 0.4g

Preparation of $Na_2S \cdot 9H_2O$ Solution: Add $Na_2S \cdot 9H_2O$ to distilled/deionized water and bring volume to 20.0mL. Mix thoroughly. Sparge with 80% N_2 + 20% CO_2. Autoclave for 15 min at 15 psi pressure–121°C.

Sodium Caproate Solution:
Composition per 20.0mL:
Sodium caproate 1.4g

Preparation of Sodium Caproate Solution: Add sodium caproate to distilled/deionized water and bring volume to 20.0mL. Mix thoroughly. Sparge with 80% N_2 + 20% CO_2. Autoclave for 15 min at 15 psi pressure–121°C.

Vitamin Solution:
Composition per 20.0mL:
Thiamine·HCl 100.0μg
p-Aminobenzoic acid 40.0μg
D(+)-Biotin 10.0μg

Preparation of Vitamin Solution: Add components to distilled/deionized water and bring volume to 20.0mL. Mix thoroughly. Filter sterilize. Sparge with 80% N_2 + 20% CO_2.

Trace Elements Solution SL-10:
Composition per liter:
$FeCl_2 \cdot 4H_2O$ 1.5g
$CoCl_2 \cdot 6H_2O$ 190.0mg
$MnCl_2 \cdot 4H_2O$ 100.0mg
$ZnCl_2$ 70.0mg
$Na_2MoO_4 \cdot 2H_2O$ 36.0mg
$NiCl_2 \cdot 6H_2O$ 24.0mg
H_3BO_3 .... 6.0mg
$CuCl_2 \cdot 2H_2O$ 2.0mg
HCl (25% solution) 10.0mL

Preparation of Trace Elements Solution SL-10: Add $FeCl_2 \cdot 4H_2O$ to 10.0mL of HCl solution. Mix thoroughly. Add distilled/deionized water and bring volume to 1.0L. Add remaining components. Mix thoroughly. Sparge with 80% N_2 + 20% CO_2. Autoclave for 15 min at 15 psi pressure–121°C.

Preparation of Medium: Add components, except $NaHCO_3$ solution, $Na_2S \cdot 9H_2O$ solution, sodium caproate solution, and trace elements solution SL-10, to distilled/deionized water and bring volume to 919.0mL. Mix thoroughly. Sparge with 80% N_2 + 20% CO_2. Autoclave for 15 min at 15 psi pressure–121°C. Aseptically and anaerobically add 20.0mL of sterile $NaHCO_3$ solution, 20.0mL of sterile $Na_2S \cdot 9H_2O$ solution, 20.0mL of sterile sodium caproate solution, and 1.0mL of sterile trace elements solution SL-10. Mix thoroughly. Aseptically and anaerobically distribute into sterile screw-capped bottles.

Use: For the cultivation and maintenance of *Syntrophospora bryantii.*

Clostridium cellobioparum Agar

Composition per 1025.0mL:
Ground meat, fat free 500.0g
Pancreatic digest of casein 30.0g
Agar 15.0g
K_2HPO_4 .... 5.0g
Yeast extract 5.0g
Glucose 4.0g
Cellobiose 1.0g
Maltose 1.0g

Soluble starch 1.0g
L-Cysteine·HCl·H_2O 0.5g
Resazurin 1.0mg
NaOH (1*N* solution) 25.0mL
pH 7.0 ± 0.2 at 25°C

Preparation of Medium: Remove fat and connective tissue from lean beef or horse meat. Grind meat finely. Add ground meat to 25.0mL of 1*N* NaOH. Add 1.0L of distilled/deionized water. Gently heat and bring to boiling. Continue boiling for 15 min while stirring. Cool to room temperature. Skim fat off surface. Filter suspension and retain the filtrate and the meat particles. Bring volume of filtrate to 1.0L with distilled/deionized water. Add remaining components, except L-cysteine·HCl·H_2O. Mix thoroughly. Gently heat and bring to boiling. Cool to 50°–55°C. Add L-cysteine·HCl·H_2O. Adjust pH to 7.0. Distribute 7.0mL of agar into tubes containing meat particles (use 1 part meat particles to 4–5 parts fluid). Autoclave for 15 min at 15 psi pressure–121°C.

Use: For the cultivation of *Clostridium cellobioparum.*

Clostridium cellobioparum Medium

Composition per 1010.0mL:
NaCl 1.0g
K_2HPO_4 .... 0.5g
KH_2PO_4 .... 0.5g
$(NH_4)_2SO_4$ 0.5g
$CaCl_2·2H_2O$ 0.1g
$MgSO_4·7H_2O$ 0.1g
Resazurin 1.0mg
Rumen fluid, clarified 300.0mL
Cellobiose solution 50.0mL
$NaHCO_3$ solution 30.0mL
$Na_2S·9H_2O$ solution 20.0mL
L-Cysteine·HCl solution 10.0mL
pH 6.8 ± 0.2 at 25°C

Cellobiose Solution:
Composition per 50.0mL:
D-Cellobiose 5.0g

Preparation of Cellobiose Solution: Add cellobiose to distilled/deionized water and bring volume to 50.0mL. Mix thoroughly. Sparge under 100% N_2 gas for 3 min. Filter sterilize. Store under N_2 gas.

$NaHCO_3$ Solution:
Composition per 30.0mL:
$NaHCO_3$ 10.0g

Preparation of $NaHCO_3$ Solution: Add $NaHCO_3$ to distilled/deionized water and bring volume to 30.0mL. Mix thoroughly. Sparge with 100% N_2. Autoclave for 15 min at 15 psi pressure–121°C.

$Na_2S·9H_2O$ Solution:
Composition per 20.0mL:
$Na_2S·9H_2O$ 0.25g

Preparation of $Na_2S·9H_2O$ Solution: Add $Na_2S·9H_2O$ to distilled/deionized water and bring volume to 20.0mL. Mix thoroughly. Sparge with 100% N_2. Autoclave for 15 min at 15 psi pressure–121°C.

L-Cysteine·HCl Solution:
Composition per 10.0mL:
L-Cysteine·HCl 0.25g

Preparation of L-Cysteine·HCl Solution: Add L-cysteine·HCl to distilled/deionized water and bring volume to 10.0mL. Mix thoroughly. Autoclave under 100% N_2 for 15 min at 15 psi pressure–121°C.

Preparation of Medium: Add components, except cellobiose solution, $NaHCO_3$ solution, $Na_2S·9H_2O$ solution, and L-cysteine·HCl solution, to distilled/deionized water and bring volume to 900.0mL. Mix thoroughly. Sparge with 100% CO_2. Autoclave for 15 min at 15 psi pressure–121°C. Aseptically and anaerobically add 50.0mL of sterile cellobiose solution, 30.0mL of sterile $NaHCO_3$ solution, 20.0mL of sterile $Na_2S·9H_2O$ solution, and 10.0mL of sterile L-cysteine·HCl solution. Mix thoroughly. Aseptically and anaerobically distribute into sterile screw-capped bottles.

Use: For the cultivation and maintenance of *Clostridium cellobioparum* and *Clostridium polysaccharolyticum.*

Clostridium Cellulolytic Medium

Composition per liter:
Agar 20.0g
Cellulose 7.5g
$K_2HPO_4·3H_2O$ 2.9g
Yeast extract 2.0g
KH_2PO_4 .... 1.5g
$(NH_4)_2SO_4$ 1.3g
$FeSO_4$ 1.25g
L-Cysteine·HCl·H_2O 1.0g
$MgCl_2·6H_2O$ 1.0g
$CaCl_2·2H_2O$ 0.15g
Resazurin 2.0mg
pH 7.5 ± 0.2 at 25°C

Preparation of Medium: Add components, except L-cysteine·HCl·H_2O, to distilled/deionized water and bring volume to 1.0L. Mix thoroughly. Heat to boiling. Adjust pH to 7.5. Prereduce under 100% N_2. Add L-cysteine·HCl·H_2O. Distribute into tubes under 100% N_2. Cap tubes with rubber stoppers. Autoclave for 15 min at 15 psi pressure–121°C.

Use: For the cultivation and maintenance of *Clostridium cellulolyticum* and other bacteria that can degrade cellulose.

Clostridium Cellulose Medium

Composition per liter:
Agar 20.0g
Filter paper (or 5.0g Avicel) 10.0g
$CaCO_3$ 5.0g
Polypeptone™ 5.0g
$Na_2CO_3·10H_2O$ 4.0g
K_2HPO_4 .... 2.2g
Yeast extract 2.0g
KH_2PO_4 .... 1.5g
$(NH_4)_2SO_4$ 1.3g
$MgCl_2·6H_2O$ 1.0g
L-Cysteine·HCl·H_2O 0.5g
$CaCl_2$ 0.15g
$FeSO_4·7H_2O$ 6.0mg
pH 7.0 ± 0.2 at 25°C

Preparation of Medium: Add components to distilled/deionized water and bring volume to 1.0L. Mix thoroughly. Heat to boiling. Autoclave for 15 min at 15 psi pressure–121°C. Pour into sterile Petri dishes or distribute into sterile tubes.

Use: For the cultivation and maintenance of *Clostridium cellulolyticum* and other bacteria that can degrade cellulose.

Clostridium cellulovorans Medium

Composition per liter:
$K_2HPO_4·3H_2O$ 1.0g

NH_4Cl 1.0g
KCl 0.5g
$MgSO_4 \cdot 7H_2O$ 0.5g
Pancreatic digest of casein 0.5g
Yeast extract 0.5g
L-Cysteine·HCl·H_2O 0.15g
Resazurin 1.0mg
Cellulose, MN 300
or cellobiose solution 50.0mL
Na_2CO_3 solution 30.0mL
Rumen fluid, clarified 20.0mL
$Na_2S \cdot 9H_2O$ 20.0mL
Trace elements solution SL-10 .... 1.0mL
pH 7.0 ± 0.2 at 25°C

Cellobiose Solution:
Composition per 50.0mL:
Cellulose, MN 300 or D-cellobiose 5.0g

Preparation of Cellobiose Solution: Add cellulose or cellobiose to distilled/deionized water and bring volume to 50.0mL. Mix thoroughly. Sparge under 100% N_2 gas for 3 min. Filter sterilize. Store under N_2 gas.

Na_2CO_3 Solution:
Composition per 30.0mL:
Na_2CO_3 .... 1.0g

Preparation of Na_2CO_3 Solution: Add Na_2CO_3 to distilled/deionized water and bring volume to 30.0mL. Mix thoroughly. Sparge with 100% N_2. Autoclave for 15 min at 15 psi pressure–121°C.

$Na_2S \cdot 9H_2O$ Solution:
Composition per 20.0mL:
$Na_2S \cdot 9H_2O$ 0.15g

Preparation of $Na_2S \cdot 9H_2O$ Solution: Add $Na_2S \cdot 9H_2O$ to distilled/deionized water and bring volume to 20.0mL. Mix thoroughly. Sparge with 100% N_2. Autoclave for 15 min at 15 psi pressure–121°C.

Trace Elements Solution SL-10:
Composition per liter:
$FeCl_2 \cdot 4H_2O$ 1.5g
$CoCl_2 \cdot 6H_2O$ 190.0mg
$MnCl_2 \cdot 4H_2O$ 100.0mg
$ZnCl_2$ 70.0mg
$Na_2MoO_4 \cdot 2H_2O$ 36.0mg
$NiCl_2 \cdot 6H_2O$ 24.0mg
H_3BO_3 .... 6.0mg
$CuCl_2 \cdot 2H_2O$ 2.0mg
HCl (25% solution) 10.0mL

Preparation of Trace Elements Solution SL-10: Add $FeCl_2 \cdot 4H_2O$ to 10.0mL of HCl solution. Mix thoroughly. Add distilled/deionized water and bring volume to 1.0L. Add remaining components. Mix thoroughly.

Preparation of Medium: Add components, except cellobiose solution, Na_2CO_3 solution, and $Na_2S \cdot 9H_2O$ solution, to distilled/deionized water and bring volume to 900.0mL. Mix thoroughly. Sparge with 100% N_2. Autoclave for 15 min at 15 psi pressure–121°C. Aseptically and anaerobically add 50.0mL of sterile cellobiose solution, 30.0mL of sterile Na_2CO_3 solution, and 20.0mL of sterile $Na_2S \cdot 9H_2O$ solution. Mix thoroughly. Aseptically and anaerobically distribute into sterile screw-capped bottles.

Use: For the cultivation and maintenance of *Clostridium cellulovorans.*

Clostridium formicoaceticum Agar

Composition per liter:
Agar 15.0g
K_2HPO_4 .... 10.0g
Yeast extract 5.0g
Sodium thioglycolate 0.75g
Pyridoxine·HCl 1.0mg
Resazurin 1.0mg
Fructose solution 50.0mL
$NaHCO_3$ solution 30.0mL
Trace elements solution SL-4 .... 10.0mL
pH 8.0 ± 0.2 at 25°C

Trace Elements Solution SL-4:
Composition per liter:
EDTA 0.5g
$FeSO_4 \cdot 7H_2O$ 0.2g
Trace elements solution SL-6 .... 100.0mL

Trace Elements Solutions SL-6:
Composition per liter:
$MnCl_2 \cdot 4H_2O$ 0.5g
H_3BO_3 .... 0.3g
$CoCl_2 \cdot 6H_2O$ 0.2g
$ZnSO_4 \cdot 7H_2O$ 0.1g
$Na_2MoO_4 \cdot 2H_2O$ 0.03g
$NiCl_2 \cdot 6H_2O$ 0.02g
$CuCl_2 \cdot 2H_2O$ 0.01g

Preparation of Trace Elements Solution SL-6: Add components to distilled/deionized water and bring volume to 1.0L. Mix thoroughly.

Preparation of Trace Elements Solution SL-4: Add components to distilled/deionized water and bring volume to 1.0L. Mix thoroughly.

Fructose Solution:
Composition per 50.0mL:
Fructose 5.0g

Preparation of Fructose Solution: Add fructose to distilled/deionized water and bring volume to 50.0mL. Mix thoroughly. Sparge under 100% N_2 gas for 3 min. Filter sterilize. Store under N_2 gas.

$NaHCO_3$ Solution:
Composition per 30.0mL:
$NaHCO_3$ 10.0g

Preparation of $NaHCO_3$ Solution: Add $NaHCO_3$ to distilled/deionized water and bring volume to 30.0mL. Mix thoroughly. Sparge with 100% CO_2. Autoclave for 15 min at 15 psi pressure–121°C.

Preparation of Medium: Add components, except fructose solution and $NaHCO_3$ solution, to distilled/deionized water and bring volume to 920.0mL. Mix thoroughly. Gently heat and bring to boiling. Cool to 50°C while sparging with 100% CO_2. Autoclave for 15 min at 15 psi pressure–121°C. Aseptically and anaerobically add 50.0mL of sterile fructose solution and 30.0mL of sterile $NaHCO_3$ solution. Mix thoroughly. Adjust pH to 8.0. Aseptically and anaerobically pour into sterile Petri dishes or distribute into sterile screw-capped bottles under 100% CO_2.

Use: For the cultivation and maintenance of *Clostridium formicoaceticum.*

Clostridium halophilum Medium

Composition per liter:
Solution A 900.0mL
Solution B 80.0mL

Solution C 10.0mL
Solution D 10.0mL

pH 8.3 ± 0.2 at 25°C

Solution A:

Composition per 900.0mL:

NaCl 60.0g
Betaine 5.86g
$MgSO_4 \cdot 7H_2O$ 5.0g
L-Alanine 2.2g
NH_4Cl 1.0g
Yeast extract 1.0g
$CaCl_2 \cdot 2H_2O$ 25.0mg
Resazurin 1.0mg
$Na_2SeO_3 \cdot 5H_2O$ 15.0μg
Wolfe's vitamin solution 10.0mL
Trace elements solution SL-10 .... 1.0mL

Wolfe's Vitamin Solution:

Composition per liter:

Pyridoxine·HCl 10.0mg
p-Aminobenzoic acid 5.0mg
Lipoic acid 5.0mg
Nicotinic acid 5.0mg
Riboflavin 5.0mg
Thiamine·HCl 5.0mg
Calcium DL-pantothenate 5.0mg
Biotin 2.0mg
Folic acid 2.0mg
Vitamin B_{12} 0.1mg

Preparation of Wolfe's Vitamin Solution: Add components to distilled/deionized water and bring volume to 1.0L. Mix thoroughly.

Trace Elements Solution SL-10:

Composition per liter:

$FeCl_2 \cdot 4H_2O$ 1.5g
$CoCl_2 \cdot 6H_2O$ 190.0mg
$MnCl_2 \cdot 4H_2O$ 100.0mg
$ZnCl_2$ 70.0mg
$Na_2MoO_4 \cdot 2H_2O$ 36.0mg
$NiCl_2 \cdot 6H_2O$ 24.0mg
H_3BO_3 .... 6.0mg
$CuCl_2 \cdot 2H_2O$ 2.0mg
HCl (25% solution) 10.0mL

Preparation of Trace Elements Solution SL-10: Add $FeCl_2 \cdot 4H_2O$ to 10.0mL of HCl solution. Mix thoroughly. Add distilled/deionized water and bring volume to 1.0L. Add remaining components. Mix thoroughly.

Preparation of Solution A: Prepare and dispense medium under 80% N_2 + 20% CO_2. Add components to distilled/deionized water and bring volume to 900.0mL. Mix thoroughly. Gently heat and bring to boiling. Continue boiling for 3 min. Cool to room temperature while sparging with 80% N_2 + 20% CO_2.

Solution B:

Composition per 80.0mL:

$NaHCO_3$ 5.0g

Preparation of Solution B: Add $NaHCO_3$ to distilled/deionized water and bring volume to 80.0mL. Mix thoroughly. Sparge with 100% N_2 for 20 min.

Solution C:

Composition per 10.0mL:

K_2HPO_4 .... 0.358g
KH_2PO_4 .... 0.223g

Preparation of Solution C: Add components to distilled/deionized water and bring volume to 10.0mL. Mix thoroughly. Sparge with 100% N_2. Autoclave for 15 min at 15 psi pressure–121°C.

Solution D:

Composition per 10.0mL:

$Na_2S \cdot 9H_2O$ 0.3g

Preparation of Solution D: Add components to distilled/deionized water and bring volume to 10.0mL. Mix thoroughly. Sparge with 100% N_2. Autoclave for 15 min at 15 psi pressure–121°C.

Preparation of Medium: Prepare and dispense medium under 80% N_2 + 20% CO_2. Anaerobically combine 900.0mL of cooled solution A with 80.0mL of sparged solution B. Mix thoroughly. Adjust pH to 8.3. Anaerobically distribute 9.8mL volumes into anaerobe tubes. Autoclave for 15 min at 15 psi pressure–121°C. Aseptically and anaerobically add 0.1mL of sterile solution C and 0.1mL of sterile solution D to each tube. Mix thoroughly.

Use: For the cultivation of *Clostridium halophilum*.

Clostridium hydroxybenzoicum Medium

Composition per 1055.0mL:

NaCl 10.0g
Yeast extract 10.0g
L-Arginine·HCl 2.1g
L-Lysine·HCl 1.8g
Glycine 0.75g
NH_4Cl 0.5g
L-Cysteine·HCl 0.4g
$Na_2S \cdot 9H_2O$ 0.4g
$MgCl_2 \cdot 7H_2O$ 0.1g
$CaCl_2 \cdot 2H_2O$ 0.025g
Resazurin 1.0mg
$Na_2WO_4 \cdot 2H_2O$ 0.05mg
Sodium/potassium phosphate buffer
(0.02*M* solution, pH 7.0) 1.0L
Wolfe's vitamin solution 50.0mL
Wolfe's mineral solution 5.0mL

pH 7.0 ± 0.2 at 25°C

Wolfe's Vitamin Solution:

Composition per liter:

Pyridoxine·HCl 10.0mg
p-Aminobenzoic acid 5.0mg
Lipoic acid 5.0mg
Nicotinic acid 5.0mg
Riboflavin 5.0mg
Thiamine·HCl 5.0mg
Calcium DL-pantothenate 5.0mg
Biotin 2.0mg
Folic acid 2.0mg
Vitamin B_{12} 0.1mg

Preparation of Wolfe's Vitamin Solution: Add components to distilled/deionized water and bring volume to 1.0L. Mix thoroughly.

Wolfe's Mineral Solution:

Composition per liter:

$MgSO_4 \cdot 7H_2O$ 3.0g
Nitrilotriacetic acid 1.5g
NaCl 1.0g
$MnSO_4 \cdot 2H_2O$ 0.5g
$CoCl_2 \cdot 6H_2O$ 0.1g
$ZnSO_4 \cdot 7H_2O$ 0.1g
$CaCl_2 \cdot 2H_2O$ 0.1g
$FeSO_4 \cdot 7H_2O$ 0.1g

$NiCl_2·6H_2O$	0.025g
$KAl(SO_4)_2·12H_2O$	0.02g
$CuSO_4·5H_2O$	0.01g
H_3BO_3	0.01g
$Na_2MoO_4·2H_2O$	0.01g
$Na_2SeO_3·5H_2O$	0.3mg

Preparation of Wolfe's Mineral Solution: Add nitrilotriacetic acid to 500.0mL of distilled/deionized water. Adjust pH to 6.5 with KOH. Add remaining components. Add distilled/deionized water to 1.0L. Adjust pH to 6.8.

Preparation of Medium: Prepare and dispense medium under 100% N_2. Combine components. Mix thoroughly. Sparge with 100% N_2. Anaerobically distribute into tubes or flasks. Autoclave for 15 min at 15 psi pressure–121°C.

Use: For the cultivation of *Clostridium hydroxybenzoicum.*

Clostridium kluyveri Agar

Composition per 100.0mL:

Potassium acetate	1.0g
Sodium thioglycolate	50.0mg
K_2HPO_4	31.0mg
$NH_4·Cl$	25.0mg
KH_2PO_4	23.0mg
$MgSO_4·7H_2O$	20.0mg
$FeSO_4·7H_2O$	2.0mg
$MnSO_4·H_2O$	2.0mg
$CaCl_2·2H_2O$	1.0mg
$Na_2MoO_4·2H_2O$	0.2mg
Agar	50.0μg
Resazurin	50.0μg
p-Aminobenzoic acid	20.0μg
Biotin	10.0μg
$CaCO_3$	variable
Ethanol	2.0mL

pH 7.0 ± 0.2 at 25°C

Preparation of Medium: Add components, except sodium thioglycolate, ethanol, and $CaCO_3$, to distilled/deionized water and bring volume to 100.0mL. Mix thoroughly. Gently heat and bring to boiling. Continue boiling for 5 min. Cool rapidly to 50°C. Add sodium thioglycolate and ethanol. Distribute into tubes containing a small amount of $CaCO_3$. Autoclave for 15 min at 15 psi pressure–121°C. Store anaerobically.

Use: For the cultivation and maintenance of *Clostridium kluyveri.*

Clostridium lentocellum Agar

Composition per 1201.0mL:

Agar	30.0g
K_2HPO_4	1.65g
NH_4SO_4	1.6g
Yeast extract	1.0g
NaCl	0.96g
L-Cysteine·HCl	0.5g
$CaCl_2$	96.0mg
$MgSO_4$	96.0mg
Cellulose suspension	200.0mL
Resazurin (0.1% solution)	1.0mL

pH 7.2 ± 0.2 at 25°C

Cellulose Suspension:

Composition per 100.0mL:

Whatman CF cellulose powder	4.0g

Preparation of Cellulose Suspension: Add components to distilled/deionized water and bring volume to 100.0mL. Mix thoroughly.

Preparation of Medium: Add components to distilled/deionized water and bring volume to 1.0L. Mix thoroughly. Adjust pH to 7.2. Gently heat and bring to boiling. Distribute into tubes or flasks. Autoclave for 15 min at 15 psi pressure–121°C. Pour into sterile Petri dishes or leave in tubes.

Use: For the cultivation of *Clostridium lentocellum* and other *Clostridium* species.

Clostridium litorale Medium

Composition per liter:

Solution A	900.0mL
Solution B	80.0mL
Solution C	10.0mL
Solution D	10.0mL

pH 8.3 ± 0.2 at 25°C

Solution A:

Composition per 900.0mL:

NaCl	10.0g
Betaine	5.86g
L-Alanine	2.2g
NH_4Cl	1.0g
Yeast extract	1.0g
$MgSO_4·7H_2O$	0.5g
$CaCl_2·2H_2O$	25.0mg
Resazurin	1.0mg
$Na_2SeO_3·5H_2O$	15.0μg
Wolfe's vitamin solution	10.0mL
Trace elements solution SL-10	1.0mL

Wolfe's Vitamin Solution:

Composition per liter:

Pyridoxine·HCl	10.0mg
p-Aminobenzoic acid	5.0mg
Lipoic acid	5.0mg
Nicotinic acid	5.0mg
Riboflavin	5.0mg
Thiamine·HCl	5.0mg
Calcium DL-pantothenate	5.0mg
Biotin	2.0mg
Folic acid	2.0mg
Vitamin B_{12}	0.1mg

Preparation of Wolfe's Vitamin Solution: Add components to distilled/deionized water and bring volume to 1.0L. Mix thoroughly.

Trace Elements Solution SL-10:

Composition per liter:

$FeCl_2·4H_2O$	1.5g
$CoCl_2·6H_2O$	190.0mg
$MnCl_2·4H_2O$	100.0mg
$ZnCl_2$	70.0mg
$Na_2MoO_4·2H_2O$	36.0mg
$NiCl_2·6H_2O$	24.0mg
H_3BO_3	6.0mg
$CuCl_2·2H_2O$	2.0mg
HCl (25% solution)	10.0mL

Preparation of Trace Elements Solution SL-10: Add $FeCl_2·4H_2O$ to 10.0mL of HCl solution. Mix thoroughly. Add distilled/deionized water and bring volume to 1.0L. Add remaining components. Mix thoroughly.

Preparation of Solution A: Prepare and dispense medium under 80% N_2 + 20% CO_2. Add components to distilled/deionized water and bring volume to 900.0mL. Mix thoroughly. Gently heat and bring to boiling. Continue boiling for 3 min. Cool to room temperature while sparging with 80% N_2 + 20% CO_2.

Solution B:
Composition per 80.0mL:
$NaHCO_3$ 5.0g

Preparation of Solution B: Add $NaHCO_3$ to distilled/deionized water and bring volume to 80.0mL. Mix thoroughly. Sparge with 100% N_2 for 20 min.

Solution C:
Composition per 10.0mL:
K_2HPO_4 .... 0.358g
KH_2PO_4 .... 0.223g

Preparation of Solution C: Add components to distilled/deionized water and bring volume to 10.0mL. Mix thoroughly. Sparge with 100% N_2. Autoclave for 15 min at 15 psi pressure–121°C.

Solution D:
Composition per 10.0mL:
$Na_2S \cdot 9H_2O$ 0.3g

Preparation of Solution D: Add components to distilled/deionized water and bring volume to 10.0mL. Mix thoroughly. Sparge with 100% N_2. Autoclave for 15 min at 15 psi pressure–121°C.

Preparation of Medium: Prepare and dispense medium under 80% N_2 + 20% CO_2. Anaerobically combine 900.0mL of cooled solution A with 80.0mL of sparged solution B. Mix thoroughly. Adjust pH to 8.3. Anaerobically distribute 9.8mL volumes into anaerobe tubes. Autoclave for 15 min at 15 psi pressure–121°C. Aseptically and anaerobically add 0.1mL of sterile solution C and 0.1mL of sterile solution D to each tube. Mix thoroughly.

Use: For the cultivation of *Clostridium litorale.*

Clostridium ljungdahlii Medium (DSMZ Medium 879)

Composition per liter:
NH_4Cl 1.0g
Yeast extract 1.0g
NaCl 0.8g
$MgSO_4 \cdot 7H_2O$ 0.2g
KCl 0.1g
KH_2PO_4 .... 0.1g
$CaCl_2 \cdot 2H_2O$ 0.02g
$Na_2WO_4 \cdot 2H_2O$ 0.20mg
Fructose solution 50.0mL
Trace elements solution 10.0mL
Vitamin solution 10.0mL
$NaHCO_3$ solution 10.0mL
Cysteine solution 10.0mL
$Na_2S \cdot 9H_2O$ solution 10.0mL
pH 5.9 ± 0.2 at 25°C

$NaHCO_3$ Solution:
Composition per 10.0mL:
$NaHCO_3$ 1.0g

Preparation of $NaHCO_3$ Solution: Add $NaHCO_3$ to distilled/deionized water and bring volume to 10.0mL. Mix thoroughly. Sparge with 80% N_2 + 20% CO_2. Filter sterilize.

Fructose Solution:
Composition per 50.0mL:
Fructose 5.0g

Preparation of Fructose Solution: Add fructose to distilled/deionized water and bring volume to 50.0mL. Mix thoroughly. Sparge with 100% N_2. Filter sterilize.

$Na_2S \cdot 9H_2O$ Solution:
Composition per 10.0mL:
$Na_2S \cdot 9H_2O$ 0.3g

Preparation of $Na_2S \cdot 9H_2O$ Solution: Add $Na_2S \cdot 9H_2O$ to distilled/deionized water and bring volume to 10.0mL. Sparge with N_2. Autoclave for 15 min at 15 psi pressure–121°C. Cool to 25°C. Store anaerobically.

L-Cysteine Solution:
Composition per 10.0mL:
L-Cysteine·HCl·H_2O 0.3g

Preparation of L-Cysteine Solution: Add L-cysteine·HCl·H_2O to distilled/deionized water and bring volume to 10.0mL. Mix thoroughly. Sparge with 100% N_2. Autoclave for 15 min at 15 psi pressure–121°C.

Trace Elements Solution:
Composition per liter:
$MgSO_4 \cdot 7H_2O$ 3.0g
Nitrilotriacetic acid 1.5g
NaCl 1.0g
$MnSO_4 \cdot 2H_2O$ 0.5g
$CoSO_4 \cdot 7H_2O$ 0.18g
$ZnSO_4 \cdot 7H_2O$ 0.18g
$CaCl_2 \cdot 2H_2O$ 0.1g
$FeSO_4 \cdot 7H_2O$ 0.1g
$NiCl_2 \cdot 6H_2O$ 0.025g
$KAl(SO_4)_2 \cdot 12H_2O$ 0.02g
H_3BO_3 .... 0.01g
$Na_2MoO_4 \cdot 4H_2O$ 0.01g
$CuSO_4 \cdot 5H_2O$ 0.01g
$Na_2SeO_3 \cdot 5H_2O$ 0.3mg

Preparation of Trace Elements Solution: Add nitrilotriacetic acid to 500.0mL of distilled/deionized water. Dissolve by adjusting pH to 6.5 with KOH. Add remaining components. Add distilled/deionized water to 1.0L. Mix thoroughly.

Vitamin Solution:
Composition per liter:
Pyridoxine-HCl 10.0mg
Thiamine-HCl·$2H_2O$ 5.0mg
Riboflavin 5.0mg
Nicotinic acid 5.0mg
D-Ca-pantothenate 5.0mg
p-Aminobenzoic acid 5.0mg
Lipoic acid 5.0mg
Biotin 2.0mg
Folic acid 2.0mg
Vitamin B_{12} 0.1mg

Preparation of Vitamin Solution: Add components to distilled/deionized water and bring volume to 1.0L. Mix thoroughly. Sparge with 80% H_2 + 20% CO_2. Filter sterilize.

Preparation of Medium: Prepare and dispense medium under 80% N_2 + 20% CO_2 gas atmosphere. Add components, except $NaHCO_3$ solution, fructose solution, cysteine solution, $Na_2S \cdot 9H_2O$ solution, vitamin solution, and trace elements solution SL-10, to distilled/deionized water and bring volume to 900.0mL. Mix thoroughly. Gently heat and bring to boiling. Boil for 10 min. Cool to room temperature while sparging with 80% N_2 + 20% CO_2. Autoclave for 15 min at 15 psi pressure–121°C. Aseptically and anaerobically add 50.0mL fructose solution, 10.0mL $NaHCO_3$ solution, 10.0mL cysteine solution, 10.0mL $Na_2S \cdot 9H_2O$ solution, 10.0mL vitamin solution, and 10.0mL trace elements solution SL-10. Mix thoroughly. Final pH is 5.9. Aseptically and anaerobically distribute into sterile tubes or bottles.

Use: For the cultivation of *Clostridium ljungdahlii.*

Clostridium M1 Medium

Composition per liter:

NaCl	10.0g
Betaine·H_2O	6.0g
$NaHCO_3$	5.0g
L-Alanine	2.2g
NH_4Cl	1.0g
Yeast extract	1.0g
$MgSO_4 \cdot 7H_2O$	0.5g
$CaCl_2 \cdot 2H_2O$	25.0mg
Resazurin	1.0mg
$Na_2SeO_3 \cdot 5H_2O$	15.0μg
Phosphate solution	100.0mL
Vitamin solution	10.0mL
$Na_2S \cdot 9H_2O$ solution	10.0mL
Trace elements solution SL-10	1.0mL

pH 7.3 ± 0.2 at 25°C

Phosphate Solution:

Composition per 100.0mL:

K_2HPO_4	0.358g
KH_2PO_4	0.223g

Preparation of Phosphate Solution: Add components to distilled/deionized water and bring volume to 100.0mL. Mix thoroughly. Sparge with 100% N_2. Autoclave for 15 min at 15 psi pressure–121°C.

$Na_2S \cdot 9H_2O$ Solution:

Composition per 10.0mL:

$Na_2S \cdot 9H_2O$	0.3g

Preparation of $Na_2S \cdot 9H_2O$ Solution: Add $Na_2S \cdot 9H_2O$ to distilled/deionized water and bring volume to 10.0mL. Mix thoroughly. Sparge with 100% N_2. Autoclave for 15 min at 15 psi pressure–121°C.

Vitamin Solution:

Composition per liter:

Pyridoxine·HCl	10.0mg
Calcium DL-pantothenate	5.0mg
Lipoic acid	5.0mg
Nicotinic acid	5.0mg
p-Aminobenzoic acid	5.0mg
Riboflavin	5.0mg
Thiamine·HCl	5.0mg
Biotin	2.0mg
Folic acid	2.0mg
Vitamin B_{12}	0.1mg

Preparation of Vitamin Solution: Add components to distilled/deionized water and bring volume to 1.0L. Mix thoroughly. Filter sterilize. Sparge with 100% N_2.

Trace Elements Solution SL-10:

Composition per liter:

$FeCl_2 \cdot 4H_2O$	1.5g
$CoCl_2 \cdot 6H_2O$	190.0mg
$MnCl_2 \cdot 4H_2O$	100.0mg
$ZnCl_2$	70.0mg
$Na_2MoO_4 \cdot 2H_2O$	36.0mg
$NiCl_2 \cdot 6H_2O$	24.0mg
H_3BO_3	6.0mg
$CuCl_2 \cdot 2H_2O$	2.0mg
HCl (25% solution)	10.0mL

Preparation of Trace Elements Solution SL-10: Add $FeCl_2 \cdot 4H_2O$ to 10.0mL of HCl solution. Mix thoroughly. Add distilled/deionized water and bring volume to 1.0L. Add remaining components. Mix thoroughly.

Preparation of Medium: Add components, except phosphate solution and $Na_2S \cdot 9H_2O$ solution, to distilled/deionized water and bring volume to 890.0mL. Mix thoroughly. Sparge with 80% N_2 + 20% CO_2. Autoclave for 15 min at 15 psi pressure–121°C. Aseptically and anaerobically add 100.0mL of sterile phosphate solution and 10.0mL of sterile $Na_2S \cdot 9H_2O$ solution. Mix thoroughly. Aseptically and anaerobically distribute into sterile screw-capped bottles under 80% N_2 + 20% CO_2.

Use: For the cultivation and maintenance of *Clostridium halophilum* and *Clostridium litorale.*

Clostridium papyrosolvens Medium

Composition per liter:

K_2HPO_4	1.65g
NH_4Cl	1.0g
Yeast extract	0.6g
L-Cysteine·HCl	0.5g
Resazurin	1.0mg
Seawater, filtered	200.0mL
Mineral salt solution	150.0mL
Cellobiose solution	50.0mL

pH 7.2 ± 0.2 at 25°C

Mineral Salt Solution:

Composition per liter:

$(NH_4)_2SO_4$	6.0g
NaCl	6.0g
$MgSO_4 \cdot 7H_2O$	1.2g
$CaCl_2 \cdot 2H_2O$	0.8g

Preparation of Mineral Salt Solution: Add components to distilled/deionized water and bring volume to 1.0L. Mix thoroughly.

Cellobiose Solution:

Composition per 50.0mL:

D-Cellobiose	5.0g

Preparation of Cellobiose Solution: Add cellobiose to distilled/deionized water and bring volume to 50.0mL. Mix thoroughly. Sparge under 100% N_2 gas for 3 min. Filter sterilize.

Preparation of Medium: Add components, except cellobiose solution, to distilled/deionized water and bring volume to 950.0mL. Mix thoroughly. Adjust pH to 7.2 with 5*N* NaOH. Sparge with 100% N_2. Autoclave for 15 min at 15 psi pressure–121°C. Aseptically and anaerobically add 50.0mL of sterile cellobiose solution. Mix thoroughly. Aseptically and anaerobically distribute into sterile screw-capped bottles under 100% N_2.

Use: For the cultivation and maintenance of *Clostridium papyrosolvens.*

Clostridium pfennigii Medium

Composition per 1001.0mL:

Solution A	890.0mL
Solution B	100.0mL
Solution C	10.0mL
Solution D	1.0mL

pH 7.0–7.2 at 25°C

Solution A:

Composition per 890.0mL:

Sodium vanillate	2.0g
Yeast extract	2.0g
Resazurin	1.0mg

Rumen fluid, clarified 267.0mL
Mineral solution 50.0mL
Vitamin solution 5.0mL
Trace elements solution SL-10 .... 1.0mL

Preparation of Solution A: Add components to distilled/deionized water and bring volume to 890.0mL. Mix thoroughly. Adjust pH to 6.9. Sparge with 80% N_2 + 20% CO_2 for 20 min. Distribute 8.9mL into anaerobic tubes under 80% N_2 + 20% CO_2. Autoclave under 80% N_2 + 20% CO_2 for 15 min at 15 psi pressure–121°C.

Mineral Solution:
Composition per liter:
KH_2PO_4 .... 10.0g
NaCl 8.0g
NH_4Cl 8.0g
$MgCl_2 \cdot 6H_2O$ 6.6g
$CaCl_2 \cdot 2H_2O$ 1.0g

Preparation of Mineral Solution: Add components to distilled/deionized water and bring volume to 1.0L. Mix thoroughly.

Vitamin Solution:
Composition per liter:
Pyridoxine·HCl 6.2mg
Nicotinic acid 2.5mg
p-Aminobenzoic acid 1.25mg
Thiamine·HCl 1.25mg
Pantothenic acid 0.62mg
Biotin 0.25mg

Preparation of Vitamin Solution: Add components to distilled/deionized water and bring volume to 1.0L. Adjust pH to 7.0. Mix thoroughly.

Trace Elements Solution SL-10:
Composition per liter:
$FeCl_2 \cdot 4H_2O$ 1.5g
$CoCl_2 \cdot 6H_2O$ 190.0mg
$MnCl_2 \cdot 4H_2O$ 100.0mg
$ZnCl_2$ 70.0mg
$Na_2MoO_4 \cdot 2H_2O$ 36.0mg
$NiCl_2 \cdot 6H_2O$ 24.0mg
H_3BO_3 .... 6.0mg
$CuCl_2 \cdot 2H_2O$ 2.0mg
HCl (25% solution) 10.0mL

Preparation of Trace Elements Solution SL-10: Add $FeCl_2 \cdot 4H_2O$ to 10.0mL of HCl solution. Mix thoroughly. Add distilled/deionized water and bring volume to 1.0L. Add remaining components. Mix thoroughly.

Solution B:
Composition per 100.0mL:
$NaHCO_3$ 5.0g

Preparation of Solution B: Add $NaHCO_3$ to distilled/deionized water and bring volume to 100.0mL. Mix thoroughly. Filter sterilize. Sparge with 80% N_2 + 20% CO_2 for 20 min.

Solution C:
Composition per 10.0mL:
L-Cysteine 0.24g

Preparation of Solution C: Add L-cysteine to distilled/deionized water and bring volume to 10.0mL. Mix thoroughly. Autoclave under 80% N_2 + 20% CO_2 for 15 min at 15 psi pressure–121°C.

Solution D:
Composition per 1.0mL:
$Na_2S \cdot 9H_2O$ 78.0mg

Preparation of Solution D: Add $Na_2S \cdot 9H_2O$ to distilled/deionized water and bring volume to 1.0mL. Mix thoroughly. Autoclave under 80% N_2 + 20% CO_2 for 15 min at 15 psi pressure–121°C.

Preparation of Medium: To each tube containing 8.9mL of sterile solution A, add (using a syringe) 1.0mL of sterile solution B, 0.1mL of sterile solution C, and 0.01mL of sterile solution D.

Use: For the cultivation and maintenance of *Clostridium pfennigii.*

Clostridium propionicum Medium

Composition per 1007.5mL:
Yeast extract 4.0g
L-Alanine 3.0g
Peptone 3.0g
L-Cysteine·HCl 0.3g
$MgSO_4 \cdot 7H_2O$ 0.1g
$FeSO_4 \cdot 7H_2O$ 0.018g
Resazurin 1.0mg
Potassium phosphate buffer solution, 1*M*, pH 7.1 .... 5.0mL
$CaSO_4$, saturated solution 2.5mL

Preparation of Medium: Add components to distilled/deionized water and bring volume to 1.0L. Mix thoroughly. Bring pH to 7.1. Sparge with 100% N_2 for 20 min. Distribute into tubes or bottles under 100% N_2. Autoclave under 100% N_2 for 15 min at 15 psi pressure–121°C.

Use: For the cultivation and maintenance of *Clostridium propionicum.*

Clostridium sticklandii Medium

Composition per liter:
Yeast extract 5.0g
L-Arginine·HCl 2.0g
L-Lysine·HCl 2.0g
NH_4Cl 2.0g
Sodium formate 2.0g
K_2HPO_4 .... 1.75g
$MgSO_4 \cdot 7H_2O$ 0.2g
$CaCl_2 \cdot 2H_2O$ 10.0mg
$FeSO_4 \cdot 7H_2O$ 10.0mg
$Na_2S \cdot H_2O$ solution 10.0mL
pH 7.0 ± 0.2 at 25°C

$Na_2S \cdot 9H_2O$ Solution:
Composition per 10.0mL:
$Na_2S \cdot 9H_2O$ 0.3g

Preparation of $Na_2S \cdot 9H_2O$ Solution: Add $Na_2S \cdot 9H_2O$ to distilled/deionized water and bring volume to 10.0mL. Mix thoroughly. Autoclave for 15 min at 15 psi pressure–121°C.

Preparation of Medium: Add components, except $Na_2S \cdot H_2O$ solution, to distilled/deionized water and bring volume to 1.0L. Mix thoroughly. Adjust pH to 7.0. Distribute into tubes or flasks. Autoclave for 15 min at 15 psi pressure–121°C. Aseptically add 0.1mL of sterile $Na_2S \cdot H_2O$ solution to each 10.0mL of medium.

Use: For the cultivation of *Clostridium sticklandii.*

Clostridium termitidis Medium

Composition per liter:

NaCl 1.0g
KCl 0.5g
Yeast extract 0.5g
$MgCl_2 \cdot 6H_2O$ 0.4g
NH_4Cl 0.3g
KH_2PO_4 .... 0.2g
$CaCl_2 \cdot 2H_2O$ 0.15g
Resazurin 1.0mg
Trace elements solution SL-10 .... 1.0mL
Cellobiose solution 50.0mL
$NaHCO_3$ solution 20.0mL
$Na_2S \cdot 9H_2O$ solution 10.0mL

pH 7.0 ± 0.2 at 25°C

Trace Elements Solution SL-10:

Composition per liter:

$FeCl_2 \cdot 4H_2O$ 1.5g
$CoCl_2 \cdot 6H_2O$ 190.0mg
$MnCl_2 \cdot 4H_2O$ 100.0mg
$ZnCl_2$ 70.0mg
$Na_2MoO_4 \cdot 2H_2O$ 36.0mg
$NiCl_2 \cdot 6H_2O$ 24.0mg
H_3BO_3 .... 6.0mg
$CuCl_2 \cdot 2H_2O$ 2.0mg
HCl (25% solution) 10.0mL

Preparation of Trace Elements Solution SL-10: Add $FeCl_2 \cdot 4H_2O$ to 10.0mL of HCl solution. Mix thoroughly. Add distilled/deionized water and bring volume to 1.0L. Add remaining components. Mix thoroughly.

Cellobiose Solution:

Composition per 50.0mL:

D-Cellobiose 5.0g

Preparation of Cellobiose Solution: Add cellobiose to distilled/deionized water and bring volume to 50.0mL. Mix thoroughly. Sparge under 100% N_2 gas for 3 min. Filter sterilize.

$NaHCO_3$ Solution:

Composition per 20.0mL:

$NaHCO_3$ 4.5g

Preparation of $NaHCO_3$ Solution: Add $NaHCO_3$ to distilled/deionized water and bring volume to 20.0mL. Mix thoroughly. Sparge with 100% N_2. Autoclave for 15 min at 15 psi pressure–121°C.

$Na_2S \cdot 9H_2O$ Solution:

Composition per 10.0mL:

$Na_2S \cdot 9H_2O$ 0.3g

Preparation of $Na_2S \cdot 9H_2O$ Solution: Add $Na_2S \cdot 9H_2O$ to distilled/deionized water and bring volume to 10.0mL. Mix thoroughly. Sparge with 100% N_2. Autoclave for 15 min at 15 psi pressure–121°C.

Preparation of Medium: Add components, except cellobiose solution, $NaHCO_3$ solution, and $Na_2S \cdot 9H_2O$ solution, and bring volume to 920.0mL. Mix thoroughly. Sparge with 80% N_2 + 20% CO_2 until pH reaches below 6.0. Autoclave for 15 min at 15 psi pressure–121°C. Aseptically and anaerobically add 50.0mL of sterile cellobiose solution, 20.0mL of sterile $NaHCO_3$ solution, and 10.0mL of sterile $Na_2S \cdot 9H_2O$ solution. Mix thoroughly. Aseptically and anaerobically distribute into sterile screw-capped bottles under 80% N_2 + 20% CO_2.

Use: For the cultivation and maintenance of *Clostridium termitidis*.

Clostridium thermoaceticum Medium (TYE-CO) (DSMZ Medium 316)

Composition per liter:

Trypticase 10.0g
Yeast extract 3.0g
$Na_2HPO_4 \cdot 12H_2O$ 2.8g
NH_4Cl 1.0g
KH_2PO_4 .... 0.3g
$MgCl_2 \cdot 6H_2O$ 0.2g
$FeSO_4 \cdot 7H_2O$ 1.0mg
Resazurin 1.0mg
Trace elements solution 10.0mL
$Na_2S \cdot 9H_2O$ solution 10.0mL
Vitamin solution 5.0mL

pH 7.0 ± 0.2 at 25°C

Vitamin Solution:

Composition per liter:

Pyridoxine-HCl 10.0mg
Thiamine-HCl·$2H_2O$ 5.0mg
Riboflavin 5.0mg
Nicotinic acid 5.0mg
D-Ca-pantothenate 5.0mg
p-Aminobenzoic acid 5.0mg
Lipoic acid 5.0mg
Biotin 2.0mg
Folic acid 2.0mg
Vitamin B_{12} 0.1mg

Preparation of Vitamin Solution: Add components to distilled/deionized water and bring volume to 1.0L. Mix thoroughly. Sparge with 80% H_2 + 20% CO_2. Filter sterilize.

Trace Elements Solution:

Composition per liter:

$MgSO_4 \cdot 7H_2O$ 3.0g
Nitrilotriacetic acid 1.5g
NaCl 1.0g
$MnSO_4 \cdot 2H_2O$ 0.5g
$CoSO_4 \cdot 7H_2O$ 0.18g
$ZnSO_4 \cdot 7H_2O$ 0.18g
$CaCl_2 \cdot 2H_2O$ 0.1g
$FeSO_4 \cdot 7H_2O$ 0.1g
$NiCl_2 \cdot 6H_2O$ 0.025g
$KAl(SO_4)_2 \cdot 12H_2O$ 0.02g
H_3BO_3 .... 0.01g
$Na_2MoO_4 \cdot 4H_2O$ 0.01g
$CuSO_4 \cdot 5H_2O$ 0.01g
$Na_2SeO_3 \cdot 5H_2O$ 0.3mg

Preparation of Trace Elements Solution: Add nitrilotriacetic acid to 500.0mL of distilled/deionized water. Dissolve by adjusting pH to 6.5 with KOH. Add remaining components. Add distilled/deionized water to 1.0L. Mix thoroughly.

$Na_2S \cdot 9H_2O$ Solution:

Composition per 10mL:

$Na_2S \cdot 9H_2O$ 0.6g

Preparation of $Na_2S \cdot 9H_2O$ Solution: Add $Na_2S \cdot 9H_2O$ to distilled/deionized water and bring volume to 10.0mL. Mix thoroughly. Autoclave under 100% N_2 for 15 min at 15 psi pressure–121°C. Cool to room temperature.

Preparation of Medium: Add components, except vitamin solution, and $Na_2S \cdot 9H_2O$ solution, to distilled/deionized water and bring volume to 985.0mL. Mix thoroughly. Sparge with 100% CO_2. Autoclave for 15 min at 15 psi pressure–121°C. Cool to 25°C while sparging with 100%

CO_2. Aseptically and anaerobically add 10.0mL vitamin solution, and 10.0mL of sterile $Na_2S{\cdot}9H_2O$ solution. Mix thoroughly. Aseptically and anaerobically distribute into sterile tubes or flasks.

Use: For the cultivation of *Moorella thermoaceticum=Clostridium thermaceticum.*

Clostridium thermocellum Medium (LMG Medium 42)

Composition per liter:

Agar	30.0g
Cellulose	10.0g
Sodium-beta-glycerophophate	6.0g
K_2HPO_4	5.5g
Yeast extract	4.5g
$MgCl_2{\cdot}6H_2O$	2.6g
KH_2PO_4	1.43g
$(NH_4)_2SO_4$	1.3g
$CaCl_2{\cdot}2H_2O$	0.13g
Glutathione	0.25g
$FeSO_4{\cdot}7H_2O$	1.1mg
Resazurin	1.0mg

pH 7.1 ± 0.2 at 25°C

Preparation of Medium: Add components to distilled/deionized water and bring volume to 1.0L under 95% N_2+ 5% CO_2 gas atmosphere. Mix thoroughly and sparge with 95% N_2+ 5% CO_2 gas. Gently heat and bring to boiling. Distribute into tubes or flasks. Autoclave for 15 min at 15 psi pressure–121°C. Pour into sterile Petri dishes or leave in tubes.

Use: For the cultivation and maintenance of *Clostridium thermocellum.*

Clostridium thermohydrosulfuricum Medium

Composition per liter:

Sucrose	10.0g
Pancreatic digest of casein	10.0g
Yeast extract	2.0g
$FeSO_4{\cdot}7H_2O$	0.2g
Na_2SO_3	0.2g
$Na_2S_2O_3{\cdot}5H_2O$	0.08g
Resazurin	1.0mg

pH 6.8 ± 0.2 at 25°C

Preparation of Medium: Add components to distilled/deionized water and bring volume to 1.0L. Mix thoroughly. Sparge with 100% N_2 for 15 min. Autoclave for 30 min at 15 psi pressure–121°C.

Use: For the cultivation and maintenance of *Clostridium thermohydrosulfuricum, Clostridium thermosaccharolyticum, Thermoanaerobacter ethanolicus,* and *Thermoanaerobacter thermohydrosulfuricus.*

Clostridium thermosuccinogenes Medium

Composition per 1011.0mL:

Inulin	5.0g
NaCl	1.2g
$MgCl_2{\cdot}6H_2O$	0.4g
KCl	0.3g
NH_4Cl	0.27g
KH_2PO_4	0.21g
$CaCl_2{\cdot}2H_2O$	0.15g
Na_2SO_4	0.1g
Resazurin	1.0mg
Na_2HPO_4 solution	20.0mL
Vitamin solution	10.0mL
Yeast extract solution	10.0mL
Casamino acids solution	10.0mL
$NaHCO_3$ solution	10.0mL
$Na_2S{\cdot}9H_2O$ solution	10.0mL
Trace elements solution SL-10	1.0mL

pH 7.0 ± 0.2 at 25°C

Na_2HPO_4 Solution:

Composition per 20.0mL:

Na_2HPO_4	2.66g

Preparation of Na_2HPO_4 Solution: Add Na_2HPO_4 to distilled/deionized water and bring volume to 20.0mL. Mix thoroughly. Sparge with 100% N_2. Autoclave for 15 min at 15 psi pressure–121°C.

Vitamin Solution:

Composition per liter:

Pyridoxine·HCl	10.0mg
Calcium DL-pantothenate	5.0mg
Lipoic acid	5.0mg
Nicotinic acid	5.0mg
p-Aminobenzoic acid	5.0mg
Riboflavin	5.0mg
Thiamine·HCl	5.0mg
Biotin	2.0mg
Folic acid	2.0mg
Vitamin B_{12}	0.1mg

Preparation of Vitamin Solution: Add components to distilled/deionized water and bring volume to 1.0L. Adjust pH to 7.0. Mix thoroughly. Sparge with 80% N_2 + 20% CO_2. Autoclave for 15 min at 15 psi pressure–121°C.

Yeast Extract Solution:

Composition per 10.0mL:

Yeast extract	0.03 g

Preparation of Yeast Extract Solution: Add yeast extract to distilled/deionized water and bring volume to 10.0mL. Mix thoroughly. Autoclave under 100% N_2 for 15 min at 15 psi pressure–121°C.

Casamino Acids Solution:

Composition per 10.0mL:

Casamino acids	0.03 g

Preparation of Casamino Acids Solution: Add casamino acids to distilled/deionized water and bring volume to 10.0mL. Mix thoroughly. Autoclave under 100% N_2 for 15 min at 15 psi pressure–121°C.

$NaHCO_3$ Solution:

Composition per 10.0mL:

$NaHCO_3$	1.0mg

Preparation of $NaHCO_3$ Solution: Add $NaHCO_3$ to distilled/deionized water and bring volume to 10.0mL. Mix thoroughly. Sparge with 80% N_2 + 20% CO_2. Autoclave for 15 min at 15 psi pressure–121°C.

$Na_2S{\cdot}9H_2O$ Solution:

Composition per 10.0mL:

$Na_2S{\cdot}9H_2O$	0.15mg

Preparation of $Na_2S{\cdot}9H_2O$ Solution: Add $Na_2S{\cdot}9H_2O$ to distilled/deionized water and bring volume to 10.0mL. Mix thoroughly. Sparge with 100% N_2. Autoclave for 15 min at 15 psi pressure–121°C.

Trace Elements Solution SL-10:

Composition per liter:

$FeCl_2{\cdot}4H_2O$	1.5g
$CoCl_2{\cdot}6H_2O$	190.0mg

$MnCl_2·4H_2O$ 100.0mg
$ZnCl_2$ 70.0mg
$Na_2MoO_4·2H_2O$ 36.0mg
$NiCl_2·6H_2O$ 24.0mg
H_3BO_3 .... 6.0mg
$CuCl_2·2H_2O$ 2.0mg
HCl (25% solution) 10.0mL

Preparation of Trace Elements Solution SL-10: Add $FeCl_2·4H_2O$ to 10.0mL of HCl solution. Mix thoroughly. Add distilled/deionized water and bring volume to 1.0L. Add remaining components. Mix thoroughly. Sparge with 100% N_2. Autoclave for 15 min at 15 psi pressure–121°C.

Preparation of Medium: Add components, except Na_2HPO_4 solution, vitamin solution, yeast extract solution, casamino acids solution, $NaHCO_3$ solution, $Na_2S·9H_2O$ solution, and trace elements solution SL-10, to distilled/deionized water and bring volume to 930.0mL. Mix thoroughly. Sparge with 80% N_2 + 100% CO_2. Autoclave for 15 min at 15 psi pressure–121°C. Aseptically and anaerobically add 20.0mL of sterile Na_2HPO_4 solution, 10.0mL of sterile vitamin solution, 10.0mL of sterile yeast extract solution, 10.0mL of sterile casamino acids solution, 10.0mL of sterile $NaHCO_3$ solution, 10.0mL of sterile $Na_2S·9H_2O$ solution, and 1.0mL of sterile trace elements solution SL-10. Aseptically and anaerobically distribute into tubes or bottles.

Use: For the cultivation and maintenance of *Clostridium thermosuccinogenes.*

Clostridium thermosulfurogenes Medium

Composition per 1015.0mL:

$Na_2HPO_4·12H_2O$ 5.3g
NH_4Cl 1.0g
Yeast extract 1.0g
KH_2PO_4 .... 0.3g
$MgCl_2·6H_2O$ 0.2g
$FeSO_4·7H_2O$ 1.5mg
Resazurin 1.0mg
Glucose solution 50.0mL
Trace elements solution 10.0mL
$Na_2S·9H_2O$ solution 10.0mL
Vitamin solution 5.0mL

pH 6.0 ± 0.2 at 25°C

Glucose Solution:

Composition per 50.0mL:

D-Glucose 5.0g

Preparation of Glucose Solution: Add glucose to distilled/deionized water and bring volume to 50.0mL. Mix thoroughly. Sparge with 100% N_2. Autoclave for 15 min at 15 psi pressure–121°C.

Trace Elements Solution:

Composition per liter:

Nitrilotriacetic acid 12.5g
NaCl 1.0g
$FeCl_3·4H_2O$ 0.2g
$MnCl_2·4H_2O$ 0.1g
$CaCl_2·2H_2O$ 0.1g
$ZnCl_2$ 0.02g
$CuCl_2$ 0.02g
Na_2SeO_3 .... 0.02g
$CoCl_2·6H_2O$ 0.017g
H_3BO_3 .... 0.01g
$Na_2MoO_4·2H_2O$ 0.01g

Preparation of Trace Elements Solution: Add nitrilotriacetic acid to 500.0mL of distilled/deionized water. Adjusting pH to 6.5 with KOH. Add remaining components. Add distilled/deionized water to 1.0L.

Vitamin Solution:

Composition per liter:

Pyridoxine·HCl 10.0mg
Calcium DL-pantothenate 5.0mg
Lipoic acid 5.0mg
Nicotinic acid 5.0mg
p-Aminobenzoic acid 5.0mg
Riboflavin 5.0mg
Thiamine·HCl 5.0mg
Biotin 2.0mg
Folic acid 2.0mg
Vitamin B_{12} 0.1mg

Preparation of Vitamin Solution: Add components to distilled/deionized water and bring volume to 1.0L. Adjust pH to 7.0. Mix thoroughly. Sparge with 100% N_2.

$Na_2S·9H_2O$ Solution:

Composition per 10.0mL:

$Na_2S·9H_2O$ 0.5g

Preparation of $Na_2S·9H_2O$ Solution: Add $Na_2S·9H_2O$ to distilled/deionized water and bring volume to 10.0mL. Mix thoroughly. Sparge with 100% N_2. Autoclave for 15 min at 15 psi pressure–121°C. Adjust pH to 7.0 with 1*N* HCl before use.

Preparation of Medium: Add components, except glucose solution and $Na_2S·9H_2O$ solution, and bring volume to 940.0mL. Mix thoroughly. Sparge with 80% N_2 + 100% CO_2. Autoclave for 15 min at 15 psi pressure–121°C. Aseptically and anaerobically add 50.0mL of sterile glucose solution and 10.0mL of sterile $Na_2S·9H_2O$ solution. Mix thoroughly. Aseptically and anaerobically distribute into tubes or bottles.

Use: For the cultivation and maintenance of *Thermoanaerobacterium thermosulfurogenes.*

CM4 Medium

Composition per liter:

Cellobiose 6.0g
Yeast extract 5.0g
K_2HPO_4 .... 2.9g
KH_2PO_4 .... 1.5g
$(NH_4)_2SO_4$ 1.3g
NaCl 1.0g
$MgCl_2$ 0.75g
Sodium thioglycolate 0.5g
$CaCl_2$ 0.0132g
Resazurin (1.0% solution) 0.2mL
$FeSO_4$ (1.25% solution) 0.1mL

Preparation of Medium: Add components to distilled/deionized water and bring volume to 1.0L. Mix thoroughly. Gently heat until boiling. Boil until color changes from red to colorless, indicating a reduced state. Cool. Distribute into tubes or flasks under 97% N_2 + 3% H_2. Cap with rubber stoppers. Autoclave for 15 min at 15 psi pressure–121°C. Pour into sterile Petri dishes or leave in tubes.

Use: For the cultivation and maintenance of *Clostridium* species and other bacteria that can utilize cellobiose as a carbon source.

Colby and Zatman Agar

Composition per liter:

Agar, noble 20.0g
K_2HPO_4 .... 1.2g
KH_2PO_4 .... 0.62g
$(NH_4)_2SO_4$ 0.5g
$MgSO_4 \cdot 7H_2O$ 0.2g
NaCl 0.1g
$CaCl_2 \cdot 6H_2O$ 0.05g
$ZnSO_4 \cdot 7H_2O$ 70.0μg
H_3BO_3 .... 10.0μg
$MnSO_4 \cdot 5H_2O$ 10.0μg
$Na_2MoO_4 \cdot 2H_2O$ 10.0μg
$CoCl_2 \cdot 6H_2O$ 5.0μg
$CuSO_4 \cdot 5H_2O$ 5.0μg
$FeCl_3 \cdot 6H_2O$ 1.0mg
Trimethylamine solution 10.0mL

pH 7.0 ± 0.2 at 25°C

Trimethylamine Solution

Composition per 10.0mL:

Trimethylamine 1.0g

Preparation of Trimethylamine Solution: Add trimethylamine to distilled/deionized water and bring volume to 10.0mL. Mix thoroughly. Filter sterilize.

Preparation of Medium: Add components, except trimethylamine solution, to distilled/deionized water and bring volume to 990.0mL. Mix thoroughly. Adjust pH to 7.0. Gently heat and bring to boiling. Autoclave for 15 min at 15 psi pressure–121°C. Cool to 50°C. Aseptically add 10.0mL of sterile trimethylamine solution. Mix thoroughly. Pour into sterile Petri dishes or distribute into sterile tubes.

Use: For the cultivation of *Aminobacter aminovorans, Bacillus* species, *Hyphomicrobium aestuarii, Hyphomicrobium facilis, Hyphomicrobium* species, *Hyphomicrobium variabile, Hyphomicrobium zavarzinii, Methylobacterium extorquens, Methylobacterium* species, and *Methylophilus methylotrophus.*

Colloidal Chitin Agar

Composition per liter:

Agar 20.0g
Chitin, colloidal 4.0g
K_2HPO_4 .... 0.7g
$MgSO_4 \cdot 5H_2O$ 0.5g
KH_2PO_4 .... 0.3g
$FeSO_4 \cdot 7H_2O$ 0.01g
$MnCl_2$ 1.0mg
$ZnSO_4$ 1.0mg

pH 7.0 ± 0.2 at 25°C

Preparation of Medium: Add components to distilled/deionized water and bring volume to 1.0L. Mix thoroughly. Gently heat and bring to boiling. Distribute into tubes or flasks. Autoclave for 15 min at 15 psi pressure–121°C. Pour into sterile Petri dishes or leave in tubes.

Use: For the isolation and cultivation of *Micromonospora* species from water, soil, or sediment. For the germination of spores of *Micromonospora* species.

Colwella psychroerythrus Medium

Composition per liter:

NaCl 29.0g
$MgCl_2 \cdot 6H_2O$ 8.0g
Pancreatic digest of casein 8.0g
KH_2PO_4 .... 5.4g
$CaCl_2 \cdot 6H_2O$ 33.0mg
$FeCl_2 \cdot 4H_2O$ 2.0mg

pH 7.0 ± 0.2 at 25°C

Preparation of Medium: Add components to distilled/deionized water and bring volume to 1.0L. Mix thoroughly. Distribute into tubes or flasks. Autoclave for 15 min at 15 psi pressure–121°C.

Use: For the cultivation of *Aminobacter aminovorans, Bacillus* species, *Hyphomicrobium aestuarii, Hyphomicrobium facilis, Hyphomicrobium* species, *Hyphomicrobium variabile, Hyphomicrobium zavarzinii, Methylobacterium extorquens, Methylobacterium* species, and *Methylophilus methylotrophus.*

Cook's *Cytophaga* Agar

Composition per liter:

Agar 10.0g
Pancreatic digest of casein 2.0g

pH 7.3 ± 0.2 at 25°C

Preparation of Medium: Add components to distilled/deionized water and bring volume to 1.0L. Mix thoroughly. Gently heat and bring to boiling. Distribute into tubes or flasks. Autoclave for 15 min at 15 psi pressure–121°C. Pour into sterile Petri dishes or leave in tubes.

Use: For the cultivation of *Lysobacter antibioticus, Lysobacter brunescens, Lysobacter enzymogenes, Lysobacter gummosus*, and other *Lysobacter* species.

Coprinus Medium

Composition per 1026mL:

Agar 20.0g
Glucose 20.0g
Asparagine 2.0g
Pancreatic digest of casein 0.75g
Yeast extract 0.75g
Malt extract 0.60g
Salt solution 25.0mL
Thiamine solution 1.0mL

pH 6.8 ± 0.2 at 25°C

Salt Solution:

Composition per 500mL:

Na_2HPO_4 .... 45.0g
KH_2PO_4 .... 20.0g
Ammonium tartrate 10.0g
$Na_2SO_4 \cdot 10H_2O$ 5.6g

Preparation of Salt Solution: Add components to distilled/deionized water and bring volume to 500mL. Mix thoroughly. Filter sterilize.

Thiamine Solution:

Composition per 100mL:

Thiamine 10.0mg

Preparation of Thiamine Solution: Add thiamine to distilled/deionized water and bring volume to 100.0mL. Mix thoroughly. Filter sterilize.

Preparation of Medium: Add components, except salt solution and thiamine solution, to distilled/deionized water and bring volume to 1.0L. Mix thoroughly. Gently heat and bring to boiling. Autoclave for 15 min at 15 psi pressure–121°C. Cool to 45°–50°C. Aseptically add 25.0mL of sterile salt solution and 1.0mL of sterile thiamine solution. Mix thoroughly. Pour into sterile Petri dishes or distribute into sterile tubes.

Use: For the cultivation and maintenance of *Coprinus cinereus, Dendrophoma obscurans,* and *Trichophyton violaceum.*

Coprothermobacter proteolyticus Medium

Composition per 1168.1mL:

Yeast extract	2.0g
Trypticase	2.0g
NaOH solution	1.0L
Gelatin solution	113.0mL
Na_2S solution	22.6mL
Solution A	10.0mL
Mineral salts solution	10.0mL
Solution B	2.0mL
Resazurin solution	0.5mL

NaOH Solution:

Composition per liter:

NaOH	4.0g

Preparation of NaOH Solution: Add NaOH to distilled/deionized water and bring volume to 1.0L. Mix thoroughly.

Gelatin Solution:

Composition per 100.0mL:

Gelatin	3.0g

Preparation of Gelatin Solution: Gently heat 100.0mL of distilled/deionized water to 80°C. Sparge with 100% N_2 for 15 min. Add the gelatin. Mix thoroughly. Sparge with 100% N_2 for 10 min. Autoclave for 15 min at 15 psi pressure–121°C.

Na_2S Solution:

Composition per 100.0mL:

Na_2S	2.5g
Distilled water	100 ml

Preparation of Na_2S Solution: Gently heat 100.0mL of distilled/deionized water to 100°C. Boil for 5 min. Sparge with 100% N_2 for 15 min. Add the Na_2S. Mix thoroughly. Sparge with 100% N_2 for 10 min. Autoclave for 15 min at 15 psi pressure–121°C.

Solution A:

Composition per liter:

NH_4Cl	100.0g
$MgCl_2 \cdot H_2O$	100.0g
$CaCl_2 \cdot 2H_2O$	40.0g

Preparation of Solution A: Add components to distilled/deionized water and bring volume to 1.0L. Mix thoroughly. Adjust pH to 4 with HCl.

Mineral Salts Solution:

Composition per liter:

$EDTA \cdot 2H_2O$	0.5g
$CoCl_2 \cdot H_2O$	0.15g
$MnCl_2 \cdot 4H_2O$	0.1g
$FeSO_4 \cdot 7H_2O$	0.1g
$ZnCl_2$	0.1g
$AlCl_3 \cdot H_2O$	40mg
$Na_2WO_4 \cdot 2H_2O$	30mg
$CuCl_2 \cdot 2H_2O$	20mg
$NiSO_4 \cdot H_2O$	20mg
H_2SeO_3	10mg
H_3BO_4	10mg
$NaMoO_4 \cdot 2H_2O$	10mg

Preparation of Mineral Salts Solution: Add components to distilled/deionized water and bring volume to 1.0L. Mix thoroughly. Adjust pH to 3 with HCl.

Solution B:

Composition per liter:

$K_2HPO_4 \cdot 3H_2O$	200.0g

Preparation of Solution B: Add $K_2HPO_4 \cdot 3H_2O$ to distilled/deionized water and bring volume to 1.0L. Mix thoroughly.

Resazurin Solution:

Composition per 100.0mL:

Resazurin	0.2g

Preparation of Resazurin Solution: Add resazurin to distilled/deionized water and bring volume to 100.0mL. Mix thoroughly.

Preparation of Medium: Sparge 1.0L of NaOH solution with 100% CO_2 for 30 min. Add 2.0g of yeast extract and 2.0g of Trypticase. Mix thoroughly. Add 10.0mL of solution A, 2.0mL of solution B, 0.5mL of resazurin solution, and 10.0mL of mineral salts solution with pipets which have been flushed for a few times with 100% N_2. Mix thoroughly. Anaerobically distribute 9.0mL volumes into anaerobic tubes fitted with butyl rubber stoppers. Autoclave for 15 min at 15 psi pressure–121°C. One hour prior to inoculation, add 1.0mL of sterile gelatin solution and 0.2mL of sterile Na_2S solution to each 9.0mL of medium.

Use: For the cultivation of *Coprothermobacter proteolyticus.*

Cornmeal Agar (CMA)

Composition per liter:

Agar	20.0g
Cornmeal polenta	15.0g

pH 7.0 ± 0.2 at 25°C

Preparation of Medium: Add cornmeal polenta to distilled/deionized water and bring volume to 1.0L. Mix thoroughly. Gently heat and bring to boiling. Continue boiling for 30 min. Filter through Whatman #1 filter paper. Add agar to filtrate. Gently heat and bring to boiling. Distribute into tubes or flasks. Autoclave for 10 min at 15 psi pressure–121°C. Pour into sterile Petri dishes or leave in tubes.

Use: For the cultivation and maintenance of many filamentous fungi.

Cornmeal Agar with Soil Extract

Composition per liter:

Cornmeal	50.0g
Agar	7.5g
Soil extract	50.0mL

Soil Extract:

Composition per 200.0mL:

African Violet soil	77.0g
Na_2CO_3	0.2g

Preparation of Soil Extract: Add components to 200.0mL of distilled/deionized water. Mix thoroughly. Autoclave for 60 min at 15 psi pressure–121°C. Filter through paper and reserve filtrate.

Preparation of Medium: Add cornmeal to distilled/deionized water and bring volume to 800.0mL. Leave overnight in refrigerator. Heat to 60°C for 1 hr. Add 50.0mL of soil extract. Bring volume to 1.0L with distilled/deionized water. Add agar. Gently heat and bring to boiling. Autoclave for 15 min at 15 psi pressure–121°C. Pour into sterile Petri dishes or distribute into sterile tubes.

Use: For the cultivation and maintenance of *Helicodendron tubulosum, Microsporum distortum, Mortierella humilis, Mortierella hygrophila, Mortierella minutissima,* and *Nigrospora sphaerica.*

Cornmeal *Phytophthora* Isolation Medium No. 1

Composition per liter:

Agar	15.0g
Cornmeal, solids from infusion	2.0g
Vancomycin	0.2g
Pentachloronitrobenzene (PCNB)	0.1g
Pimaricin	0.01g

pH 5.6–6.0 at 25°C

Preparation of Medium: Add components, except pimaricin and vancomycin, to distilled/deionized water and bring volume to 1.0L. Mix thoroughly. Gently heat until boiling. Autoclave for 15 min at 15 psi pressure–121°C. Aseptically add pimaricin and vancomycin. Mix thoroughly. Pour into sterile Petri dishes.

Use: For the cultivation of *Phytophthora* species.

Cornmeal *Phytophthora* Isolation Medium No. 2

Composition per liter:

Agar	15.0g
Cornmeal, solids from infusion	2.0g
Vancomycin	0.3g
Pentachloronitrobenzene (PCNB)	0.025g
Pimaricin	5.0mg

pH 5.6–6.0 at 25°C

Preparation of Medium: Add components, except pimaricin and vancomycin, to distilled/deionized water and bring volume to 1.0L. Mix thoroughly. Gently heat until boiling. Autoclave for 15 min at 15 psi pressure–121°C. Aseptically add pimaricin and vancomycin. Mix thoroughly. Pour into sterile Petri dishes.

Use: For the cultivation of *Phytophthora* species.

Cornmeal Yeast Extract Seawater Agar (ATCC Medium 2422)

Composition per liter:

Instant Ocean	17.5g
Agar	15.0g
Yeast extract	1.0g
Cornmeal infusion	400.0mL

pH 7.2–7.5 at 25°C

Cornmeal Infusion:

Composition per liter:

Yellow cornmeal	50.0g

Preparation of Cornmeal Infusion: Add cornmeal to distilled/deionized water and bring volume to 1.0L. Gently heat and bring to boiling. Simmer for 10 minutes. Filter through cheesecloth. Return volume to 1.0 liter.

Preparation of Medium: Add instant ocean, agan, and yeast extract to 400.0mL cornmeal infusion and bring volume to 1.0L with distilled/deionized water. Gently heat and bring to boiling. Autoclave for 15 min at 15 psi pressure–121°C. Pour into sterile Petri dishes or distribute into sterile tubes.

Use: For the cultivation and maintenance of fungi.

Corynebacterium Agar

Composition per liter:

Agar	15.0g
Beef extract	10.0g
Peptone	10.0g
NaCl	5.0g

pH 7.2 ± 0.2 at 25°C

Preparation of Medium: Add components to distilled/deionized water and bring volume to 1.0L. Mix thoroughly. Adjust pH to 7.2. Gently heat and bring to boiling. Distribute into tubes or flasks. Autoclave for 15 min at 15 psi pressure–121°C. Pour into sterile Petri dishes or leave in tubes.

Use: For the cultivation and maintenance of *Brevibacterium helvolum, Brevibacterium linens, Brochothrix thermosphacta, Cellulomonas cellasea, Corynebacterium ammoniagenes, Corynebacterium callunae, Corynebacterium glutamicum,* other *Corynebacterium* species, *Curtobacterium flaccumfaciens, Deinococcus radiodurans, Microbacterium laevaniformans, Mycobacterium vaccae, Rhodococcus equi, Rhodococcus fascians, Sporolactobacillus inulinus,* and *Streptococcus mutans.*

Corynebacterium Medium with Salt (DSMZ Medium 229)

Composition per liter:

NaCl	65.0g
Agar	15.0g
Casein peptone, tryptic digest	10.0g
Yeast extract	5.0g
Glucose	5.0g

pH 7.3 ± 0.2 at 25°C

Preparation of Medium: Add components to distilled/deionized water and bring volume to 1.0L. Mix thoroughly. Gently heat and bring to boiling. Distribute into tubes or flasks. Autoclave for 15 min at 15 psi pressure–121°C. Pour into sterile Petri dishes or leave in tubes.

Use: For the cultivation and maintenance of *Nesterenkonia halobia=Micrococcus halobius.*

Cow Manure Agar

Composition per liter:

Cow manure	50.0g
Agar	15.0g

Preparation of Medium: Add cow manure to tap water and bring volume to 1.0L. Gently heat and bring to boiling. Boil for 1 hr. Filter through cheese cloth. Filter through Whatman filter paper. Bring volume to 1.0L with tap water. Add agar. Mix thoroughly. Gently heat and bring to boiling. Distribute into tubes or flasks. Autoclave for 15 min at 15 psi pressure–121°C. Pour into sterile Petri dishes or leave in tubes.

Use: For the cultivation of *Streptomyces* species.

CP Medium for *Coprothermobacter proteolyticus*

Composition per 1010.0mL:

$NaHCO_3$	8.4g
Pancreatic digest of casein	2.0g
Yeast extract	2.0g
$MgCl_2·6H_2O$	1.0g
NH_4Cl	1.0g
$CaCl_2·2H_2O$	0.4g
$K_2HPO_4·3H_2O$	0.4g
Resazurin	0.5mg
Gelatin solution	100.0mL
$Na_2S·9H_2O$ solution	10.0mL
Wolfe's mineral solution	10.0mL

pH 7.0 ± 0.2 at 25°C

Gelatin Solution:
Composition per 100.0mL:
Gelatin 3.0g

Preparation of Gelatin Solution: Add gelatin to distilled/deionized water and bring volume to 100.0mL. Mix thoroughly. Sparge with 100% N_2. Autoclave for 15 min at 15 psi pressure–121°C.

$Na_2S\cdot 9H_2O$ Solution:
Composition per 10.0mL:
$Na_2S\cdot 9H_2O$ 0.5g

Preparation of $Na_2S\cdot 9H_2O$ Solution: Add $Na_2S\cdot 9H_2O$ to distilled/deionized water and bring volume to 10.0mL. Mix thoroughly. Sparge with 100% N_2. Autoclave for 15 min at 15 psi pressure–121°C. Before use, neutralize to pH 7.0 with sterile HCl.

Wolfe's Mineral Solution:
Composition per liter:
$MgSO_4\cdot 7H_2O$ 3.0g
Nitrilotriacetic acid 1.5g
NaCl 1.0g
$MnSO_4\cdot 2H_2O$ 0.5g
$CoCl_2\cdot 6H_2O$ 0.1g
$ZnSO_4\cdot 7H_2O$ 0.1g
$CaCl_2\cdot 2H_2O$ 0.1g
$FeSO_4\cdot 7H_2O$ 0.1g
$NiCl_2\cdot 6H_2O$ 0.025g
$KAl(SO_4)_2\cdot 12H_2O$ 0.02g
$CuSO_4\cdot 5H_2O$ 0.01g
H_3BO_3 .... 0.01g
$Na_2MoO_4\cdot 2H_2O$ 0.01g
$Na_2SeO_3\cdot 5H_2O$ 0.3mg

Preparation of Wolfe's Mineral Solution: Add nitrilotriacetic acid to 500.0mL of distilled/deionized water. Adjust pH to 6.5 with KOH. Add remaining components. Add distilled/deionized water to 1.0L. Adjust pH to 6.8.

Preparation of Medium: Prepare medium anaerobically under 100% CO_2. Add components, except gelatin solution, $Na_2S\cdot 9H_2O$ solution, and Wolfe's mineral solution, to distilled/deionized water and bring volume to 880.0mL. Mix thoroughly. Gently heat and bring to boiling. Cool to room temperature while sparging with 100% CO_2. Sparge with 100% CO_2 for 20 min. Adjust pH to 7.0. Autoclave for 15 min at 15 psi pressure–121°C. Aseptically and anaerobically add 100.0mL of sterile gelatin solution, 10.0mL of sterile $Na_2S\cdot 9H_2O$ solution, and 10.0mL of sterile Wolfe's mineral solution to each tube. Mix thoroughly.

Use: For the cultivation of *Coprothermobacter proteolyticus*.

CT Agar

Composition per liter:
Agar 20.0g
Pancreatic digest of casein 20.0g
$MgSO_4\cdot 7H_2O$ 2.0g
Potassium phosphate
buffer (0.02*M* solution, pH 7.6) 500.0mL
pH 7.6 ± 0.2 at 25°C

Preparation of Medium: Add agar, pancreatic digest of casein, and $MgSO_4\cdot 7H_2O$ to distilled/deionized water and bring volume to 500.0mL. Mix thoroughly. Gently heat and bring to boiling. Autoclave agar–pancreatic digest of casein-$MgSO_4\cdot 7H_2O$ solution and potassium phosphate buffer solution separately for 15 min at 15 psi pressure–121°C. Cool to 25°C. Aseptically combine the two solutions. Aseptically add sterile components. Mix thoroughly. Pour into sterile Petri dishes or distribute into sterile tubes.

Use: For the cultivation of myxobacteria.

CT Medium

Composition per liter:
Agar 15.0g
Pancreatic digest of casein 10.0g
Yeast extract 3.5g
$MgSO_4$ 0.96g

Preparation of Medium: Add components to distilled/deionized water and bring volume to 1.0L. Mix thoroughly. Gently heat until boiling. Distribute into tubes or flasks. Autoclave for 15 min at 15 psi pressure–121°C. Pour into sterile Petri dishes or leave in tubes.

Use: For the cultivation and maintenance of *Stigmatella aurantiaca*.

CTT Medium

Composition per liter:
Agar 15.0g
Pancreatic digest of casein 10.0g
Tris(hydroxymethyl)aminomethane buffer 1.21g
Potassium phosphate
buffer (1 m*M*, pH 7.6) 1.0L
Magnesium sulfate solution 10.0mL
pH 7.6 ± 0.2 at 25°C

Magnesium Sulfate Solution:
Composition per 10.0mL:
$MgSO_4\cdot 7H_2O$ 2.0g

Preparation of Magnesium Sulfate Solution: Add $MgSO_4\cdot 7H_2O$ to 10.0mL of distilled/deionized water. Mix thoroughly.

Preparation of Medium: Combine components. Mix thoroughly. Adjust pH to 7.6. Gently heat and bring to boiling. Distribute into tubes or flasks. Autoclave for 15 min at 15 psi pressure–121°C. Pour into sterile Petri dishes or leave in tubes.

Use: For the cultivation of myxobacteria.

CY Agar

Composition per liter:
Agar 15.0g
Pancreatic digest of casein 3.0g
$CaCl_2\cdot 2H_2O$ 1.0g
Yeast extract 1.0g
Cyanocobalamin 0.5mg
pH 7.2 ± 0.2 at 25°C

Preparation of Medium: Add components to distilled/deionized water and bring volume to 1.0L. Mix thoroughly. Gently heat and bring to boiling. Distribute into tubes or flasks. Autoclave for 15 min at 15 psi pressure–121°C. Pour into sterile Petri dishes or leave in tubes.

Use: For the cultivation of myxobacteria.

CYC Medium

Composition per liter:
Sucrose 30.0g
Casamino acids, vitamin free 6.0g
$NaNO_3$ 3.0g
Yeast extract 2.0g
K_2HPO_4 .... 1.0g
$MgSO_4\cdot 7H_2O$ 0.5g
KCl 0.5g

$FeSO_4 \cdot 7H_2O$ 0.01g
Antibiotic solution 10.0mL

pH 7.2 ± 0.2 at 25°C

Antibiotic Solution:
Composition per 10.0mL:
Cycloheximide 0.05g
Novobiocin 0.025g

Preparation of Antibiotic Solution: Add components to distilled/deionized water and bring volume to 10.0mL. Mix thoroughly. Filter sterilize.

Preparation of Medium: Add components, except antibiotic solution, to distilled/deionized water and bring volume to 990.0mL. Mix thoroughly. Gently heat and bring to boiling. Autoclave for 15 min at 15 psi pressure–121°C. Cool to 45°–50°C. Aseptically add sterile antibiotic solution. Mix thoroughly. Aseptically distribute into sterile tubes.

Use: For the isolation and cultivation of *Thermoactinomyces* species.

CYC Medium, Cross and Attwell Modification (DSMZ Medium 550)

Composition per liter:
Sucrose 30.0g
Agar 15.0g
Casamino acids 6.1g
$NaNO_3$ 3.0g
Yeast extract 2.0g
K_2HPO_4 .... 1.0g
$MgSO_4 \cdot 7H_2O$ 0.5g
KCl 0.5g
Tryptophan 0.02g
$FeSO_4 \cdot 7H_2O$ 0.01g

pH 7.2 ± 0.2 at 25°C

Preparation of Medium: Add components to distilled/deionized water and bring volume to 1.0L. Mix thoroughly. Distribute into tubes or flasks. Autoclave for 15 min at 15 psi pressure–121°C. Pour into sterile Petri dishes or leave in tubes.

Use: For the cultivation and maintenance of *Thermomonospora curvata, Thermobifida fusca, Thermoactinomyces vulgaris, Saccharopolyspora thermophila,* and *Thermobifida alba.*

Cyclohexanecarboxylic Acid Agar

Composition per liter:
Agar, noble 15.0g
Cyclohexanecarboxylic acid 5.0g
$(NH_4)_2SO_4$ 1.0g
KH_2PO_4 .... 1.0g
K_2HPO_4 .... 1.0g
$MgSO_4 \cdot 7H_2O$ 0.2g
Yeast extract 0.1g
$FeSO_4 \cdot 7H_2O$ 10.0mg
$CaCl_2 \cdot 2H_2O$ 2.0mg
$MnSO_4 \cdot 4H_2O$ 2.0mg
$ZnSO_4 \cdot 7H_2O$ 2.0mg

pH7.2 ± 0.2 at 25°C

Preparation of Medium: Add components to distilled/deionized water and bring volume to 1.0L. Mix thoroughly. Adjust pH to 7.2. Gently heat and bring to boiling. Distribute into tubes or flasks. Autoclave for 15 min at 15 psi pressure–121°C. Pour into sterile Petri dishes or leave in tubes.

Use: For the cultivation and maintenance of *Corynebacterium cyclohexanicum* and *Saccharomyces cerevisiae.*

Cyclohexanecarboxylic Acid Medium

Composition per liter:
K_2HPO_4 .... 3.5g
Cyclohexanecarboxylic acid 2.0g
KH_2PO_4 .... 1.5g
NH_4NO_3 .... 1.0g
$MgSO_4 \cdot 7H_2O$ 0.5g
Yeast extract 0.1g
$CaCl_2 \cdot 2H_2O$ 0.01g
$FeCl_3 \cdot 6H_2O$ 0.01g
$NaMoO_4 \cdot 7H_2O$ 0.01g
$ZnSO_4 \cdot 7H_2O$ 0.01g

pH 7.0 ± 0.2 at 25°C

Preparation of Medium: Add components to distilled/deionized water and bring volume to 1.0L. Mix thoroughly. Distribute into tubes or flasks. Autoclave for 15 min at 15 psi pressure–121°C.

Use: For the cultivation and maintenance of *Alcaligenes faecalis* and other bacteria that can utilize cyclohexanecarboxylic acid as a carbon source.

Cyclohexanone Medium

Composition per liter:
NH_4NO_3 .... 3.0g
K_2HPO_4 .... 0.25g
$MgSO_4 \cdot 7H_2O$ 0.2g
$CaCl_2 \cdot 2H_2O$ 0.01g
$FeCl_3 \cdot 6H_2O$ 1.0mg
Cyclohexanone 1.0mL

Preparation of Medium: Add components, except cyclohexanone, to distilled/deionized water and bring volume to 999.0mL. Mix thoroughly. Distribute into tubes or flasks. Autoclave for 20 min at 15 psi pressure–121°C. Filter sterilize cyclohexanone. Aseptically add 1.0mL of cyclohexanone. Mix thoroughly.

Use: For the cultivation and maintenance of *Nocardia* species and other bacteria that can utilize cyclohexanone as a carbon source.

CYE-ACES Agar (DSMZ Medium 585)

Composition per liter:
Solution A 490.0mL
Solution B 490.0mL
Solution C 10.0mL
Solution D 10.0mL

pH 6.9 ± 0.1 at 25°C

Solution A:
Composition per 490mL:
Yeast extract 10.0g
ACES (*N*-2-acetamido-2-aminoethane-sulfonic acid) 10.0g
Activated charcoal 2.0g

Preparation of Solution A: Add components to distilled/deionized water and bring volume to 490.0mL. Mix thoroughly. Gently heat and bring to boiling. Autoclave for 15 min at 15 psi pressure–121°C. Cool to 50°C.

Solution B:
Composition per 490mL:
Agar 15.0g

Preparation of Solution B: Add agar to distilled/deionized water and bring volume to 490.0mL. Mix thoroughly. Gently heat and bring to boiling. Autoclave for 15 min at 15 psi pressure–121°C. Cool to 50°C.

Solution C:

Composition per 10mL:

L-Cysteine·HCl·H_2O 0.4g

Preparation of Solution C: Add L-cysteine·HCl·H_2O to distilled/deionized water and bring volume to 10.0mL. Mix thoroughly. Filter sterilize.

Solution D:

Composition per 10mL:

$Fe_4(PO_4)_2$ 0.25g

Preparation of Solution D: Add $Fe_4(PO_4)_2$ to distilled/deionized water and bring volume to 10.0mL. Heat to 50-55°C to dissolve $Fe_4(PO_4)_2$. Mix thoroughly. Filter sterilize. Store in the dark. Do not use if chemical loses its green color and becomes brown or yellow.

Preparation of Medium: Add 10.0mL solution C and then 10.0mL solution D to 490.0mL solution A. Adjust the pH 6.9 ± 0.05 at 50°C by adding 4.0 to 4.5 ml of sterile 1.0 N KOH. The pH of the medium is critical. Finally, add 490.0mL solution B. Mix thoroughly. Swirl medium in flask during dispensing to petri dishes or tubes to keep charcoal suspended. Pour into Petri dishes or aseptically distribute into sterile tubes.

Use: For the cultivation of *Afipia clevelandensis, Afipia broomeae, Afipia felis, Legionella pneumophila, Legionella longbeachae,* and *Xylella fastidiosa.*

CYE Agar (Charcoal Yeast Extract Agar)

Composition per liter:

Agar 17.0g
Yeast extract 10.0g
Charcoal, activated, acid-washed 2.0g
L-Cysteine·HCl·H_2O solution 10.0mL
$Fe_4(P_2O_7)_3$ solution 10.0mL
pH 6.9 ± .05 at 50°C

L-Cysteine·HCl·H_2O Solution:

Composition per 10.0mL:

L-Cysteine·HCl·H_2O 0.4g

Preparation of L-Cysteine·HCl·H_2O solution: Add L-cysteine·HCl·H_2O to distilled/deionized water and bring volume to 10.0mL. Mix thoroughly. Filter sterilize.

$Fe_4(P_2O_7)_3$ Solution:

Composition per liter:

$Fe_4(P_2O_7)_3$ 0.25g

Preparation of $Fe_4(P_2O_7)_3$ Solution: Add soluble $Fe_4(P_2O_7)_3$ to distilled/deionized water and bring volume to 10.0mL. Mix thoroughly. Filter sterilize. The soluble $Fe_4(P_2O_7)_3$ must be kept dry and in the dark. Do not use if brown or yellow. Prepare solutions freshly. Do not heat over 60°C to dissolve. The mixture dissolves readily in a 50°C water bath.

Preparation of Medium: Add components, except L-cysteine·HCl·H_2O solution and $Fe_4(P_2O_7)_3$ solution, to distilled/deionized water and bring volume to 980.0mL. Mix thoroughly. Gently heat to boiling. Autoclave for 15 min at 15 psi pressure–121°C. Cool to 50°C. Add 10.0mL of sterile L-cysteine·HCl·H_2O solution and 10.0mL of sterile $Fe_4(P_2O_7)_3$ solution. Adjust pH to 6.9 at 50°C by adding 4.0–4.5mL of 1.0*N* KOH. This is a critical step. Mix thoroughly. Pour in 20.0mL volumes into sterile Petri dishes. Swirl medium while pouring to keep charcoal in suspension.

Use: For the cultivation and maintenance of *Legionella* species and *Tatlockia micdadei.*

CYE Agar, Buffered (Charcoal Yeast Extract Agar, Buffered)

Composition per liter:

Agar 17.0g
ACES buffer (*N*-2-acetamido-2-aminoethane sulfonic acid) 10.0g
Yeast extract 10.0g
Charcoal, activated, acid-washed 2.0g
L-Cysteine·HCl·H_2O solution 10.0mL
$Fe_4(P_2O_7)_3$ solution 10.0mL
pH 6.9 ± .05 at 50°C

L-Cysteine·HCl·H_2O Solution:

Composition per 10.0mL:

L-Cysteine·HCl·H_2O 0.4g

Preparation of L-Cysteine·HCl·H_2O Solution: Add L-cysteine·HCl·H_2O to distilled/deionized water and bring volume to 10.0mL. Mix thoroughly. Filter sterilize.

$Fe_4(P_2O_7)_3$ Solution:

Composition per liter:

$Fe_4(P_2O_7)_3$ 0.25g

Preparation of $Fe_4(P_2O_7)_3$ Solution: Add soluble $Fe_4(P_2O_7)_3$ to distilled/deionized water and bring volume to 10.0mL. Mix thoroughly. Filter sterilize. The soluble $Fe_4(P_2O_7)_3$ must be kept dry and in the dark. Do not use if brown or yellow. Prepare solutions freshly. Do not heat over 60°C to dissolve. The mixture dissolves readily in a 50°C water bath.

Preparation of Medium: Add components, except L-cysteine·HCl·H_2O solution and $Fe_4(P_2O_7)_3$ solution, to distilled/deionized water and bring volume to 980.0mL. Mix thoroughly. Gently heat to boiling. Autoclave for 15 min at 15 psi pressure–121°C. Cool to 50°C. Add 10.0mL of sterile L-cysteine·HCl·H_2O solution and 10.0mL of sterile $Fe_4(P_2O_7)_3$ solution. Adjust pH to 6.9 at 50°C by adding 4.0–4.5mL of 1.0 *N* KOH. This is a critical step. Mix thoroughly. Pour in 20.0mL volumes into sterile Petri dishes. Swirl medium while pouring to keep charcoal in suspension.

Use: For the cultivation and maintenance of *Legionella* species and *Xylella fastidiosa.*

CYG Agar

Composition per liter:

Agar 15.0g
Pancreatic digest of casein 3.0g
$CaCl_2·2H_2O$ 1.0g
Yeast extract 1.0g
Cyanocobalamin 0.5mg
Glucose solution 100.0mL
pH 7.2 ± 0.2 at 25°C

Glucose Solution:

Composition per 100.0mL:

D-Glucose 5.0g

Preparation of Glucose Solution: Add D-glucose to distilled/deionized water and bring volume to 100.0mL. Mix thoroughly. Autoclave for 15 min at 15 psi pressure–121°C. Cool to 25°C.

Preparation of Medium: Add components, except glucose solution, to distilled/deionized water and bring volume to 900.0mL. Mix thoroughly. Gently heat and bring to boiling. Autoclave for 15 min at 15 psi pressure–121°C. Cool to 45°–50°C. Aseptically add sterile glucose solution. Mix thoroughly. Pour into sterile Petri dishes or distribute into sterile tubes.

Use: For the isolation and cultivation of *Cytophaga* species, *Herpetosiphon* species, *Saprospira* species, and *Flexithrix* species.

Cytophaga Agarase Agar (ATCC Medium 793)

Composition per liter:

Agar	15.0g
KH_2PO_4	1.0g
$MgSO_4 \cdot 7H_2O$	0.5g
NH_4Cl	0.5g
$CaCl_2 \cdot H_2O$	0.02g
Vishniac and Santer trace element mixture	0.2mL

pH 7.2 ± 0.2 at 25°C

Vishniac and Santer Trace Element Mixture:
Composition per liter:

Ethylenediamine tetraacetic acid (EDTA)	50.0g
$ZnSO_4 \cdot 7H_2O$	22.0g
$CaCl_2$	5.54g
$MnCl_2 \cdot 4H_2O$	5.06g
$FeSO_4 \cdot 7H_2O$	4.99g
$CoCl_2 \cdot 6H_2O$	1.61g
$CuSO_4 \cdot 5H_2O$	1.57g
$(NH_4)_6Mo_7O_{24} \cdot 4H_2O$	1.1g

Preparation of Vishniac and Santer Trace Element Mixture: Add components to distilled/deionized water and bring volume to 1.0L. Adjust pH to 6.0 with KOH. Mix thoroughly.

Preparation of Medium: Add components to distilled/deionized water and bring volume to 1.0L. Adjust pH to 7.2. Mix thoroughly. Distribute into tubes or flasks. Autoclave for 15 min at 15 psi pressure–121°C. Pour into sterile Petri dishes or leave in tubes.

Use: For the cultivation and maintenance of *Cytophaga flevensis*.

Cytophaga agarovorans Agar (LMG Medium 99)

Composition per 1001.0mL:

NaCl	30.0g
Agar	15.0g
KH_2PO_4	1.0g
$MgSO_4 \cdot 7H_2O$	1.0g
NH_4Cl	1.0g
Yeast extract	1.0g
$CaCl_2 \cdot 2H_2O$	50.0mg
$FeCl_3 \cdot H_2O$	1.25mg
Glucose solution	10.0mL
$NaHCO_3$ solution	10.0mL
$Na_2S \cdot 9H_2O$ solution	1.0mL

Glucose Solution:
Composition per 10.0mL:

D-Glucose	1.0g

Preparation of Glucose Solution: Add glucose to distilled/deionized water and bring volume to 10.0mL. Mix thoroughly. Filter sterilize.

$NaHCO_3$ Solution:
Composition per 100.0mL:

$NaHCO_3$	5.0g

Preparation of $NaHCO_3$ Solution: Add $NaHCO_3$ to distilled/deionized water and bring volume to 100.0mL. Mix thoroughly. Filter sterilize.

$Na_2S \cdot 9H_2O$ Solution:
Composition per 10.0mL:

$Na_2S \cdot 9H_2O$	1.0g

Preparation of $Na_2S \cdot 9H_2O$ Solution: Add $Na_2S \cdot 9H_2O$ to distilled/deionized water and bring volume to 10.0mL. Filter sterilize.

Preparation of Medium: Add components, except glucose solution, $NaHCO_3$ solution, and $Na_2S \cdot 9H_2O$ solution, to distilled/deionized water and bring volume to 980.0mL. Mix thoroughly. Gently heat and bring to boiling. Autoclave for 15 min at 15 psi pressure–121°C. Cool to 50°–55°C. Aseptically add 10.0mL of sterile glucose solution, 10.0mL of sterile $NaHCO_3$ solution, and 1.0mL of sterile $Na_2S \cdot 9H_2O$ solution. Mix thoroughly. Pour into sterile Petri dishes or distribute into sterile tubes.

Use: For the cultivation of *Cytophaga agarovorans*.

Cytophaga fermentans Medium

Composition per liter:

NaCl	30.0g
Agar	5.0g
$NaHCO_3$	5.0g
KH_2PO_4	1.0g
NH_4Cl	1.0g
$MgCl_2 \cdot 6H_2O$	0.5g
Yeast extract	0.3g
$Na_2S \cdot 9H_2O$	0.1g
$CaCl_2$	0.04g
Ferric citrate (4m*M* solution)	5.0mL
Trace elements solution	2.0mL

pH 7.0 ± 0.2 at 25°C

Trace Elements Solution:
Composition per 100.0mL:

H_3BO_3	0.28g
$MnSO_4 \cdot 6H_2O$	0.21g
$Na_2MoO_4 \cdot 2H_2O$	0.075g
$Zn(NO_3)_2 \cdot 6H_2O$	0.025g
$CoCl_2 \cdot 6H_2O$	0.02g
$Cu(NO_3)_2 \cdot 3H_2O$	0.02g

Preparation of Trace Elements Solution: Add components to distilled/deionized water and bring volume to 100.0mL. Mix thoroughly.

Preparation of Medium: Add components to distilled/deionized water and bring volume to 1.0L. Mix thoroughly. Gently heat and bring to boiling. Distribute into tubes or flasks. Autoclave for 15 min at 15 psi pressure–121°C.

Use: For the cultivation of agar-digesting *Cytophaga fermentans*.

Cytophaga hutchinsonii Agar

Composition per liter:

Agar	15.0g
Pancreatic digest of casein	3.0g
$CaCl_2 \cdot 2H_2O$	1.36g
Yeast extract	1.0g
Cellobiose solution	50.0mL

pH 7.2 ± 0.2 at 25°C

Cellobiose Solution:
Composition per 50.0mL:

D-Cellobiose 6.0g

Preparation of Cellobiose Solution: Add cellobiose to distilled/deionized water and bring volume to 50.0mL. Mix thoroughly. Filter sterilize.

Preparation of Medium: Add components, except cellobiose solution, to distilled/deionized water and bring volume to 950.0mL. Mix thoroughly. Gently heat and bring to boiling. Autoclave for 15 min at 15 psi pressure–121°C. Cool to 50°–55°C. Aseptically add 50.0mL of sterile cellobiose solution. Mix thoroughly. Pour into sterile Petri dishes or distribute into sterile tubes.

Use: For the cultivation and maintenance of *Cytophaga aurantiaca* and *Cytophaga hutchinsonii.*

Cytophaga Marine Medium (DSMZ Medium 172)

Composition per liter:

NaCl 24.7g
Agar 15.0g
$MgSO_4 \cdot 7H_2O$ 6.3g
$MgCl_2 \cdot 6H_2O$ 4.6g
$CaCl_2 \cdot 2H_2O$ 1.2g
Yeast extract 1.0g
Tryptone 1.0g
KCl 0.7g
Sodium bicarbonate solution 10.0mL
Calcium chloride solution 10.0mL
pH 7.0 ± 0.2 at 25°C

Sodium Bicarbonate Solution:
Composition per 10.0mL:

$NaHCO_3$ 0.2g

Preparation of Sodium Bicarbonate Solution: Add $NaHCO_3$ to distilled/deionized water and bring volume to 10.0mL. Mix thoroughly. Autoclave for 15 min at 15 psi pressure–121°C.

Calcium Chloride Solution:
Composition per 10.0mL:

$CaCl_2 \cdot 2H_2O$ 1.2g

Preparation of Calcium Chloride Solution: Add $CaCl_2 \cdot 2H_2O$ to distilled/deionized water and bring volume to 10.0mL. Mix thoroughly. Autoclave for 15 min at 15 psi pressure–121°C.

Preparation of Medium: Add components, except sodium bicarbonate and calcium chloride solution to distilled/deionized water and bring volume to 900.0mL. Mix thoroughly. Gently heat and bring to boiling. Autoclave for 15 min at 15 psi pressure–121°C. Cool to 45-50°C. Aseptically add 10.0mL of sterile bicarbonate solution and 10.0mL sterile calcium chloride solution. Adjust pH to 7.2. Mix thoroughly. Pour into sterile Petri dishes or distribute into sterile tubes.

Use: For the cultivation and maintenance of *Cellulophaga lytica, Cytophaga latercula, Marinilabilia salmonicolor, Saprospira grandis, Cytophaga marinoflava, Persicobacter diffluens, Flammeovirga aprica, Flexibacter tractuosus, Microscilla sp., Marinilabilia salmonicolor, Flexibacter litoralis, Flexithrix dorotheae,* and *Cellulophaga lytica.*

Cytophaga Medium

Composition per liter:

NaCl 30.0g
Agar 15.0g
KH_2PO_4 ..... 1.0g
NH_4Cl 1.0g
Yeast extract 1.0g
$MgSO_4$ 0.5g
$CaCl_2$ 0.04g
$FeCl_3 \cdot 6H_2O$ 1.25mg
$NaHCO_3$ solution 100.0mL
Glucose solution 10.0mL
$Na_2S \cdot 9H_2O$ solution 1.0mL

Glucose Solution:
Composition per 100.0mL:

Glucose 10.0g

Preparation of Glucose Solution: Add glucose to distilled/deionized water and bring volume to 100.0mL. Mix thoroughly. Autoclave for 15 min at 15 psi pressure–121°C.

$NaHCO_3$ Solution:
Composition per 100.0mL:

$NaHCO_3$ 5.0g

Preparation of $NaHCO_3$ Solution: Add the $NaHCO_3$ to distilled/deionized water and bring volume to 100.0mL. Mix thoroughly. Filter sterilize.

$Na_2S \cdot 9H_2O$ Solution:
Composition per 100.0mL:

$Na_2S \cdot 9H_2O$ 10.0g

Preparation of $Na_2S \cdot 9H_2O$ Solution: Add $Na_2S \cdot 9H_2O$ to distilled/deionized water and bring volume to 100.0mL. Mix thoroughly. Autoclave for 15 min at 15 psi pressure–121°C.

Preparation of Medium: Add components, except $NaHCO_3$ solution, glucose solution, and $Na_2S \cdot 9H_2O$ solution, to distilled/deionized water and bring volume to 889.0mL. Autoclave for 15 min at 15 psi pressure–121°C. Cool to 50°C. Aseptically add sterile $NaHCO_3$ solution, sterile glucose solution, and sterile $Na_2S \cdot 9H_2O$ solution. Mix thoroughly. Pour into sterile Petri dishes or distribute into sterile tubes.

Use: For the cultivation of *Cytophaga agarovorans.*

Czapek Agar (ATCC Medium 312)

Composition per liter:

Sucrose 30.0g
Agar 15.0g
$NaNO_3$ 3.0g
K_2HPO_4 ..... 1.0g
KCl 0.5g
$MgSO_4 \cdot 7H_2O$ 0.5g
$FeSO_4 \cdot 7H_2O$ 0.01g
pH 7.3 ± 0.2 at 25°C

Preparation of Medium: Add components, except sucrose, to distilled/deionized water and bring volume to 900.0mL. Mix thoroughly. Distribute into tubes or flasks. In a separate flask, add sucrose to distilled/deionized water and bring volume to 100.0mL. Mix thoroughly. Autoclave both solutions separately for 15 min at 15 psi pressure–121°C. Cool to 50°C. Combine the sterile solutions. Mix thoroughly. Pour into sterile Petri dishes or distribute into sterile tubes.

Use: For the cultivation and maintenance of *Streptomyces* species. For the cultivation of Actinoplanaceae.

Czapek Dox Agar, Modified

Composition per liter:

Sucrose 30.0g

Agar 12.0g
$NaNO_3$ 2.0g
Magnesium glycerophosphate 0.5g
KCl 0.5g
K_2SO_4 .... 0.35g
$FeSO_4$ 0.01g

pH 6.8 ± 0.2 at 25°C

Source: This medium is available as a premixed powder from Oxoid Unipath.

Preparation of Medium: Add components to distilled/deionized water and bring volume to 1.0L. Mix thoroughly. Distribute into tubes or flasks. Autoclave for 15 min at 15 psi pressure–121°C. Pour into sterile Petri dishes or leave in tubes.

Use: For the cultivation and maintenance of numerous fungal species.

Czapek Dox Broth

Composition per liter:

Sucrose 30.0g
$NaNO_3$ 3.0g
K_2HPO_4 .... 1.0g
$MgSO_4 \cdot 7H_2O$ 0.5g
KCl 0.5g
$FeSO_4 \cdot 7H_2O$ 0.01g

pH 7.3 ± 0.2 at 25°C

Source: This medium is available as a premixed powder from BD Diagnostic Systems.

Preparation of Medium: Add components to distilled/deionized water and bring volume to 1.0L. Mix thoroughly. Distribute into tubes or flasks. Autoclave for 15 min at 15 psi pressure–121°C.

Use: For the cultivation and maintenance of a variety of fungal and bacterial species that can use nitrate as sole nitrogen source.

Czapek Solution Agar with Sucrose

Composition per liter:

Sucrose 200.0g
Agar 20.0g
$NaNO_3$ 3.0g
K_2HPO_4 .... 1.0g
KCl 0.5g
$MgSO_4 \cdot 7H_2O$ 0.5g
$FeSO_4 \cdot 7H_2O$ 10.0mg

Preparation of Medium: Add components to distilled/deionized water and bring volume to 1.0L. Mix thoroughly. Gently heat and bring to boiling. Distribute into tubes or flasks. Autoclave for 15 min at 15 psi pressure–121°C. Pour into sterile Petri dishes or leave in tubes.

Use: For the cultivation and maintenance of osmophilic fungi.

Decarboxylase Base, Møller

Composition per liter:

Amino acid 10.0g
Beef extract 5.0g
Peptone 5.0g
Glucose 0.5g
Bromcresol Purple 0.01g
Cresol Red 5.0mg
Pyridoxal 5.0mg
Mineral oil 200.0mL

pH 6.0 ± 0.2 at 25°C

Source: This medium is available as a premixed powder from BD Diagnostic Systems.

Preparation of Medium: Add components, except mineral oil, to distilled/deionized water and bring volume to 1.0L. For amino acid, use L-arginine, L-lysine, or L-ornithine. Mix thoroughly. Distribute into screw-capped tubes in 5.0mL volumes. Autoclave medium and mineral oil separately for 15 min at 15 psi pressure–121°C. After inoculation, overlay medium with 1.0mL of sterile mineral oil per tube.

Use: For the cultivation and differentiation of bacteria based on their ability to decarboxylate the amino acid. Bacteria that decarboxylate arginine, lysine, or ornithine turn the medium turbid purple.

Decarboxylase Medium Base, Falkow

Composition per liter:

Amino acid 5.0g
Peptone 5.0g
Yeast extract 3.0g
Glucose 1.0g
Bromcresol Purple 0.02g
Mineral oil 200.0mL

pH 6.8 ± 0.2 at 25°C

Source: This medium is available as a premixed powder from BD Diagnostic Systems.

Preparation of Medium: Add components, except mineral oil, to distilled/deionized water and bring volume to 1.0L. For amino acid, use L-arginine, L-lysine, or L-ornithine. Mix thoroughly. Distribute into screw-capped tubes in 5.0mL volumes. Autoclave medium and mineral oil separately for 15 min at 15 psi pressure–121°C. After inoculation, overlay medium with 1.0mL of sterile mineral oil per tube.

Use: For the cultivation and differentiation of bacteria based on their ability to decarboxylate the amino acid. Bacteria that decarboxylate arginine, lysine, or ornithine turn the medium turbid purple.

Decarboxylase Medium, Ornithine Modified

Composition per liter:

L-Ornithine 10.0g
Meat peptone 5.0g
Yeast extract 3.0g
Bromcresol Purple solution 5.0mL

pH 5.5 ± 0.2 at 25°C

Bromcresol Purple Solution:
Composition per 100.0mL:

Bromcresol Purple 0.2g
Ethanol 50.0mL

Preparation of Bromcresol Purple Solution: Add Bromcresol Purple to ethanol. Mix thoroughly. Bring volume to 100.0mL with distilled/deionized water. Mix thoroughly. Filter sterilize.

Preparation of Medium: Add components to distilled/deionized water and bring volume to 1.0L. Mix thoroughly. Gently heat until dissolved. Adjust pH to 5.5 with HCl or NaOH. Distribute into screw-capped tubes. Autoclave for 15 min at 15 psi pressure–121°C.

Use: For the cultivation and differentiation of bacteria based on their ability to decarboxylate ornithine. Bacteria that decarboxylate ornithine turn the medium turbid purple.

Deferribacter Medium (DSMZ Medium 935)

Composition per liter:

NaCl	25.0g
Sulfur, powdered	10.0g
Na-acetate	2.0g
KNO_3	0.5g
NH_4Cl	0.33g
KCl	0.33g
$CaCl_2·2H_2O$	0.33g
$MgCl_2·6H_2O$	0.33g
KH_2PO_4	0.33g
Yeast extract	0.15g
$NaHCO_3$ solution	10.0mL
Trace elements solution	10.0mL
Vitamin solution	10.0mL

pH 7.0 ± 0.2 at 25°C

$NaHCO_3$ Solution:
Composition per 10.0mL:

$NaHCO_3$	0.3g

Preparation of $NaHCO_3$ Solution: Add $NaHCO_3$ to distilled/deionized water and bring volume to 10.0mL. Mix thoroughly. Autoclave for 15 min at 15 psi pressure–121°C. Cool to 25°C. Must be prepared freshly.

Trace Elements Solution:
Composition per liter:

$MgSO_4·7H_2O$	3.0g
Nitrilotriacetic acid	1.5g
NaCl	1.0g
$MnSO_4·2H_2O$	0.5g
$CoSO_4·7H_2O$	0.18g
$ZnSO_4·7H_2O$	0.18g
$CaCl_2·2H_2O$	0.1g
$FeSO_4·7H_2O$	0.1g
$NiCl_2·6H_2O$	0.025g
$KAl(SO_4)_2·12H_2O$	0.02g
H_3BO_3	0.01g
$Na_2MoO_4·4H_2O$	0.01g
$CuSO_4·5H_2O$	0.01g
$Na_2SeO_3·5H_2O$	0.3mg

Preparation of Trace Elements Solution: Add nitrilotriacetic acid to 500.0mL of distilled/deionized water. Dissolve by adjusting pH to 6.5 with KOH. Add remaining components. Add distilled/deionized water to 1.0L. Mix thoroughly.

Vitamin Solution:
Composition per liter:

Pyridoxine-HCl	10.0mg
Thiamine-HCl·$2H_2O$	5.0mg
Riboflavin	5.0mg
Nicotinic acid	5.0mg
D-Ca-pantothenate	5.0mg
p-Aminobenzoic acid	5.0mg
Lipoic acid	5.0mg
Biotin	2.0mg
Folic acid	2.0mg
Vitamin B_{12}	0.1mg

Preparation of Vitamin Solution: Add components to distilled/deionized water and bring volume to 1.0L. Mix thoroughly. Sparge with 80% H_2 + 20% CO_2. Filter sterilize.

Preparation of Medium: Add components, except vitamin solution, $NaHCO_3$ solution, and sulfur to 980.0mL distilled/deionized water. Gently heat and bring to boiling. Boil for 3 min. Cool to room temperature while sparging with 100% N_2. Adjust pH to 7.0. Anaerobically under 100% N_2 distribute into tubes of bottles containing the sulfur (0.1g sulfur per 10mL medium). Autoclave for 20 min at 110°C. Aseptically and anaerobically add 10.0mL sterile vitamin solution and 10.0mL sterile $NaHCO_3$ solution. Mix thoroughly. The final pH should be 7.0.

Use: For the cultivation of *Deferribacter desulfuricans* and *Deferribacter thermophilus.*

Defined Glucose Medium EMSY-1

Composition per liter:

Na_2HPO_4	1.79g
KH_2PO_4	1.7g
Citric acid	0.5g
NH_4Cl	0.43g
$MgSO_4·7H_2O$	0.41g
$CaCl_2·2H_2O$	0.04g
NaCl	0.03g
$FeCl_3·6H_2O$	4.84mg
Glucose solution	100.0mL
Yeast extract solution	10.0mL
TK6-3 solution	1.0mL

pH 7.2 ± 0.2 at 25°C

Glucose Solution:
Composition per 100.0mL:

Glucose	10.0g

Preparation of Glucose Solution: Add glucose to distilled/deionized water and bring volume to 100.0mL. Mix thoroughly. Filter sterilize.

Yeast Extract Solution:
Composition per 10.0mL:

Yeast extract	0.4g

Preparation of Yeast Extract Solution: Add yeast extract to distilled/deionized water and bring volume to 10.0mL. Mix thoroughly. Filter sterilize.

TK6-3 Solution:
Composition per liter:

$ZnSO_4·7H_2O$	1.45g
$CuSO_4·5H_2O$	0.76g
$MnSO_4·H_2O$	0.31g
H_3BO_3	0.19g
$Na_2MoO_4·2H_2O$	0.17g
KI	0.04g
H_2SO_4 (1*N* solution)	1.0mL

Preparation of TK6-3 Solution: Add components to distilled/deionized water and bring volume to 1.0L. Mix thoroughly.

Preparation of Medium: Add components, except glucose solution and yeast extract solution, to distilled/deionized water and bring volume to 890.0mL. Mix thoroughly. Gently heat and bring to boiling. Autoclave for 15 min at 15 psi pressure–121°C. Cool rapidly to 25°C. Aseptically add 100.0mL of sterile glucose solution and 10.0mL of sterile yeast extract solution. Mix thoroughly. Aseptically distribute into sterile tubes or flasks.

Use: For the cultivation and maintenance of *Xanthomonas campestris*.

Defined Medium with Povidone Iodine

Composition per 1025.0mL:

Basal solution	1.0L
Solution B	10.0mL
Solution C	10.0mL
Solution A	5.0mL

Basal Solution:

Composition per liter:

Agar	20.0g
Na_2HPO_4	4.8g
KH_2PO_4	4.4g
NH_4Cl	1.0g
$MgSO_4 \cdot 7H_2O$	0.5g

Preparation of Basal Solution: Add components to distilled/deionized water and bring volume to 1.0L. Mix thoroughly. Gently heat and bring to boiling. Autoclave for 15 min at 15 psi pressure–121°C. Cool to 45°–50°C.

Solution A:

Composition per 100.0mL:

Ferric ammonium citrate	1.0g
$CaCl_2 \cdot 2H_2O$	0.1g

Preparation of Solution A: Add components to distilled/deionized water and bring volume to 100.0mL. Mix thoroughly. Filter sterilize.

Solution B:

Composition per 100.0mL:

D-Glucose	10.0g

Preparation of Solution B: Add glucose to distilled/deionized water and bring volume to 100.0mL. Mix thoroughly. Filter sterilize.

Solution C:

Composition per 100.0mL:

Povidone-iodine	0.1g

Preparation of Solution C: Add povidone-iodine to distilled/deionized water and bring volume to 100.0mL. Mix thoroughly. Filter sterilize.

Preparation of Medium: To 1.0L of cooled, sterile basal solution, aseptically add 5.0mL of sterile solution A, 10.0mL of sterile solution B, and 10.0mL of sterile solution C. Mix thoroughly. Pour into sterile Petri dishes or distribute into sterile tubes.

Use: For the cultivation and maintenance of *Pseudomonas aeruginosa* and *Pseudomonas (Burkholderia) cepacia.*

Defined Medium for *Rhodopseudomonas*

Composition per liter:

Malic acid	4.0g
$(NH_4)_2SO_4$	1.0g
K_2HPO_4	0.9g
KH_2PO_4	0.6g
$MgSO_4 \cdot 7H_2O$	0.2g
$CaCl_2 \cdot 2H_2O$	0.075g
EDTA	0.02g
$FeSO_4 \cdot 7H_2O$	0.012g
Thiamine	1.0mg
Biotin	0.015mg
Trace elements	1.0mL

pH 6.8 ± 0.2 at 25°C

Trace Elements:

Composition per 250.0mL:

H_3BO_3	0.7g
$MnSO_4 \cdot H_2O$	0.4g
$Na_2MoO_4 \cdot 2H_2O$	0.19g
$ZnSO_4 \cdot 7H_2O$	0.06g
$CoCl_2 \cdot 6H_2O$	0.05g
$Cu(NO_3)_2 \cdot 3H_2O$	0.01g

Preparation of Trace Elements: Add components to distilled/deionized water and bring volume to 250.0mL. Mix thoroughly.

Preparation of Medium: Add components to distilled/deionized water and bring volume to 1.0L. Mix thoroughly. Adjust pH to 6.8. Distribute into tubes or flasks. Autoclave for 15 min at 15 psi pressure–121°C.

Use: For the cultivation and maintenance of *Rhodobacter capsulatus.*

Dehalobacter restrictus Medium (DSMZ Medium 732)

Composition per 1046mL:

Solution A	900.0mL
Solution B	100.0mL
Solution G	15.0mL
Solution C	10.0mL
Solution E	10.0mL
Solution F	10.0mL
Solution D	1.0mL

pH 7.2 ± 0.2 at 25°C

Solution A:

Composition per liter:

K_2HPO_4	0.653g
Na-acetate	0.460g
$NaH_2PO_4 \cdot H_2O$	0.173g
Peptone	0.1g
Resazurin	0.5mg

Preparation of Solution A: Add components to distilled/deionized water and bring volume to 1.0 L Mix thoroughly. Gently heat and bring to boiling. Cool to room temperature under 80% H_2 + 20% CO_2 gas. Distribute 9 ml volumes into 50 mL serum bottles under 80% H_2 + 20% CO_2 gas. Pressurize closed bottles with H_2 + CO_2 gas to 0.5 bar overpressure. Autoclave for 15 min at 15 psi pressure–121°C.

Solution B:

Composition per 100mL:

$NaHCO_3$	3.730g
NH_4HCO_3	0.443g

Preparation of Solution B: Add components to distilled/deionized water and bring volume to 100.0 mL in bottles. Mix thoroughly. Flush solution with 80% N_2 + 20% CO_2 gas. Close bottles. Autoclave for 15 min at 15 psi pressure–121°C.

Solution C:

Composition per 10mL:

$MgCl_2 \cdot 6H_2O$	0.12g
$CaCl_2 \cdot 2H_2O$	0.11g

Preparation of Solution C: Add components to distilled/deionized water and bring volume to 10.0 mL in bottles. Mix thoroughly. Flush solution with 100% N_2 for 20 min. Close bottles. Autoclave for 15 min at 15 psi pressure–121°C.

Solution D:

Composition per 10mL:

Na_2-EDTA	5.0mg
$FeCl_2 \cdot 4H_2O$	5.0mg
$AlCl_3$	0.1mg
Trace elements solution SL-10	10.0mL

Trace Elements Solution SL-10:

Composition per liter:

$FeCl_2 \cdot 4H_2O$	1.5g
$CoCl_2 \cdot 6H_2O$	190.0mg
$MnCl_2 \cdot 4H_2O$	100.0mg
$ZnCl_2$	70.0mg
$Na_2MoO_4 \cdot 2H_2O$	36.0mg

$NiCl_2 \cdot 6H_2O$	24.0mg
H_3BO_3	6.0mg
$CuCl_2 \cdot 2H_2O$	2.0mg
HCl (25% solution)	10.0mL

Preparation of Trace Elements Solution SL-10: Add $FeCl_2 \cdot 4H_2O$ to 10.0mL of HCl solution. Mix thoroughly. Add distilled/deionized water and bring volume to 1.0L. Add remaining components. Mix thoroughly. Sparge with 100% N_2. Autoclave for 15 min at 15 psi pressure–121°C.

Preparation of Solution D: Add Na_2-EDTA, $FeCl_2 \cdot 4H_2O$, and $AlCl_3$ to 10.0 mL trace solution SL-10 in a Hungate bottle. Mix thoroughly. Flush solution with 100% N_2 for 20 min. Close bottles. Autoclave for 15 min at 15 psi pressure–121°C.

Solution E:
Composition per liter:

Vitamin solution	900.0mL
Seven vitamin solution	100.0mL

Vitamin Solution:
Composition per liter:

Pyridoxine-HCl	10.0mg
Thiamine-HCl·$2H_2O$	5.0mg
Riboflavin	5.0mg
Nicotinic acid	5.0mg
D-Ca-pantothenate	5.0mg
p-Aminobenzoic acid	5.0mg
Lipoic acid	5.0mg
Biotin	2.0mg
Folic acid	2.0mg
Vitamin B_{12}	0.1mg

Preparation of Vitamin Solution: Add components to distilled/deionized water and bring volume to 1.0L. Mix thoroughly.

Seven Vitamin Solution:
Composition per liter:

Pyridoxine hydrochloride	300.0mg
Thiamine-HCl·$2H_2O$	200.0mg
Nicotinic acid	200.0mg
Vitamin B_{12}	100.0mg
Calcium pantothenate	100.0mg
p-Aminobenzoic acid	80.0mg
D(+)-Biotin	20.0mg

Preparation of Seven Vitamin Solution: Add components to distilled/deionized water and bring volume to 1.0L. Sparge with 100% N_2. Mix thoroughly.

Preparation of Solution E: Combine 900.0mL vitamin solution and 100.0mL seven vitamin solution. Mix thoroughly. Sparge with 100% N_2. Filter sterilize.

Solution F:
Composition per 10.0mL:

$Na_2S \cdot 9H_2O$	0.3g

Preparation of Solution F: Add $Na_2S \cdot 9H_2O$ to distilled/deionized water and bring volume to 10.0mL. Mix thoroughly. Sparge with 100% N_2. Autoclave for 15 min at 15 psi pressure–121°C.

Solution G:

Hexadecane	45.0mL
Tetrachloroethene	5.0mL

Preparation of Solution G: Using a syringe, aseptically inject 5.0mL sterile tetrachloroethene to the 45.0mL sterile hexadecane in the 100mL serum bottle. Sparge with 100% N_2. Autoclave for 15 min at 15 psi pressure–121°C.

Hexadecane:

Hexadecane	45.0mL

Preparation of Hexadecane: Add hexadecane to a 100mL serum bottle. Sparge with 100% N_2. Autoclave for 15 min at 15 psi pressure–121°C.

Tetrachloroethene:

Tetrachloroethene	10.0mL

Preparation of Tetrachloroethene: Add tetrachloroethene to a 10mL serum bottle. Sparge with 100% N_2. Autoclave for 15 min at 15 psi pressure–121°C.

Preparation of Medium: Add 9mL sterile solution A to a 50 mL sterile bottle. Then add by injection 1.0mL sterile solution B, 0.1mL sterile solution C, 0.01mL sterile solution D, 0.1mL sterile solution E, and 0.1mL sterile solution F. Inoculate the culture into the medium. Then add by injection 0.15mL sterile solution G.

Use: For the cultivation of *Dehalobacter restrictus* and *Sulfurospirillum halorespiran*.

Deleya halophila Medium

Composition per liter:

NaCl	81.0g
$MgSO_4$	19.6g
Yeast extract	10.0g
Proteose peptone No.3	5.0g
$MnCl_2$	4.0g
KCl	2.0g
Glucose	1.0g
$CaCl_2$	0.47g
NaBr	0.026g
$NaHCO_3$ solution	10.0mL

pH 7.5 ± 0.2 at 25°C

$NaHCO_3$ Solution
Composition per 10.0mL:

$NaHCO_3$	0.06g

Preparation of $NaHCO_3$ Solution: Add $NaHCO_3$ to distilled/deionized water and bring volume to 10.0mL. Mix thoroughly. Filter sterilize.

Preparation of Medium: Add components, except $NaHCO_3$ solution, to distilled/deionized water and bring volume to 990.0mL. Mix thoroughly. Adjust pH to 7.5. Autoclave for 15 min at 15 psi pressure–121°C. Cool to 50°C. Aseptically add 10.0mL of sterile $NaHCO_3$ solution. Mix thoroughly. Aseptically distribute into sterile tubes or flasks.

Use: For the cultivation of *Deleya halophila.*

Demi-Fraser Broth

Composition per liter:

NaCl	20.0g
Tryptose	10.0g
Na_2HPO_4	9.6g
Beef extract	5.0g
Yeast extract	5.0g
LiCl	3.0g
KH_2PO_4	1.35g
Esculin	1.0g
Acriflavin·HCl	12.5mg
Nalidixic acid	10.0mg
Ferric ammonium citrate supplement	10.0mL

pH 7.2 ± 0.2 at 25°C

Source: This medium is available as a premixed powder and supplement from BD Diagnostic Systems.

Ferric Ammonium Citrate Supplement:
Composition per 10.0mL:
Ferric ammonium citrate 0.5g

Preparation of Ferric Ammonium Citrate Supplement: Add ferric ammonium citrate to distilled/deionized water and bring volume to 10.0mL. Mix thoroughly. Filter sterilize.

Preparation of Medium: Add components, except ferric ammonium citrate supplement, to distilled/deionized water and bring volume to 990.0mL. Mix thoroughly. Autoclave for 15 min at 15 psi pressure–121°C. Aseptically add 10.0mL of sterile ferric ammonium citrate supplement. Mix thoroughly. Aseptically distribute into sterile tubes or flasks.

Use: For the cultivation of *Listeria* species from environmental samples.

Denitrovibrio Medium (DSMZ Medium 881)

Composition per 1032mL:
NaCl 20.0g
$MgCl_2 \cdot 6H_2O$ 3.0g
KH_2PO_4 .... 1.0g
$NaNO_3$ 0.7g
KCl 0.5g
NH_4Cl 0.25g
$CaCl_2 \cdot 2H_2O$ 0.15g
Na_2SO_4 .... 0.02g
Resazurin 0.5mg
Na-acetate solution 10.0mL
$NaHCO_3$ solution 10.0mL
$Na_2S \cdot 9H_2O$ solution 10.0mL
Seven vitamin solution 1.0mL
Trace elements solution SL-10 .... 1.0mL
pH 6.8-7.2 at 25°C

Na-Acetate Solution:
Composition per 10.0mL:
Na-acetate 1.64g

Preparation of Na-Acetate Solution: Add Na-acetate to distilled/deionized water and bring volume to 10.0mL. Mix thoroughly. Sparge with 100% N_2. Filter sterilize.

$Na_2S \cdot 9H_2O$ Solution:
Composition per 10mL:
$Na_2S \cdot 9H_2O$ 0.5g

Preparation of $Na_2S \cdot 9H_2O$ Solution: Add $Na_2S \cdot 9H_2O$ to distilled/deionized water and bring volume to 10.0mL. Mix thoroughly. Autoclave under 100% N_2 for 15 min at 15 psi pressure–121°C. Cool to room temperature.

$NaHCO_3$ Solution:
Composition per 10.0mL:
$NaHCO_3$ 2.5g

Preparation of $NaHCO_3$ Solution: Add $NaHCO_3$ to distilled/deionized water and bring volume to 10.0mL. Mix thoroughly. Sparge with 80% N_2 + 20% CO_2. Filter sterilize.

Seven Vitamin Solution:
Composition per liter:
Pyridoxine hydrochloride 300.0mg
Thiamine-HCl·$2H_2O$ 200.0mg
Nicotinic acid 200.0mg
Vitamin B_{12} 100.0mg
Calcium pantothenate 100.0mg
p-Aminobenzoic acid 80.0mg
D(+)-Biotin 20.0mg

Preparation of Seven Vitamin Solution: Add components to distilled/deionized water and bring volume to 1.0L. Sparge with 100% N_2. Mix thoroughly. Filter sterilize.

Trace Elements Solution SL-10:
Composition per liter:
$FeCl_2 \cdot 4H_2O$ 1.5g
$CoCl_2 \cdot 6H_2O$ 190.0mg
$MnCl_2 \cdot 4H_2O$ 100.0mg
$ZnCl_2$ 70.0mg
$Na_2MoO_4 \cdot 2H_2O$ 36.0mg
$NiCl_2 \cdot 6H_2O$ 24.0mg
H_3BO_3 .... 6.0mg
$CuCl_2 \cdot 2H_2O$ 2.0mg
HCl (25% solution) 10.0mL

Preparation of Trace Elements Solution SL-10: Add $FeCl_2 \cdot 4H_2O$ to 10.0mL of HCl solution. Mix thoroughly. Add distilled/deionized water and bring volume to 1.0L. Add remaining components. Mix thoroughly. Sparge with 80% N_2 + 20% CO_2. Autoclave for 15 min at 15 psi pressure–121°C.

Preparation of Medium: Prepare and dispense medium under 100% N_2 gas atmosphere. Add components, except $NaHCO_3$ solution, Na-acetate solution, $Na_2S \cdot 9H_2O$ solution, seven vitamin solution, and trace elements solution SL-10, to distilled/deionized water and bring volume to 949.0mL. Mix thoroughly. Adjust pH to 6.8-7.2. Sparge with 80% N_2 + 20% CO_2. Autoclave for 15 min at 15 psi pressure–121°C. Aseptically and anaerobically add 10.0mL $NaHCO_3$ solution, 10.0mL Na-acetate solution, 10.0mL $Na_2S \cdot 9H_2O$ solution, 1.0mL seven vitamin solution, and 1.0mL trace elements solution SL-10. Mix thoroughly. Aseptically and anaerobically distribute into sterile tubes or bottles.

Use: For the cultivation of *Denitrovibrio acetiphilus*.

Deoxycholate Agar

Composition per liter:
Agar 16.0g
Lactose 10.0g
NaCl 5.0g
Pancreatic digest of casein 5.0g
Peptic digest of animal tissue 5.0g
K_2HPO_4 .... 2.0g
Ferric citrate 1.0g
Sodium citrate 1.0g
Sodium deoxycholate 1.0g
Neutral Red 0.033g
pH 7.3 ± 0.2 at 25°C

Source: This medium is available as a premixed powder from BD Diagnostic Systems.

Preparation of Medium: Add components to distilled/deionized water and bring volume to 1.0L. Mix thoroughly. Gently heat and bring to boiling. Do not autoclave. Cool to 45°–50°C. Pour into sterile Petri dishes.

Use: For the selective isolation, cultivation, enumeration, and differentiation of Gram-negative enteric microorganisms from a variety of clinical and nonclinical specimens. *Escherichia coli* appears as large, flat, rose-red colonies. *Enterobacter* and *Klebsiella* species appear as large, mucoid, pale colonies with a pink center. *Proteus* and *Salmonella* species appear as large, colorless to tan colonies. *Shigella* species appear as colorless to pink colonies.

Pseudomonas species appear as irregular colorless to brown colonies.

Deoxycholate Agar (Desoxycholate Agar)

Composition per liter:

Agar	15.0g
Lactose	10.0g
Peptone	10.0g
NaCl	5.0g
K_2HPO_4	2.0g
Ferric citrate	1.0g
Sodium citrate	1.0g
Sodium deoxycholate	1.0g
Neutral Red	0.03g

pH 7.3 ± 0.2 at 25°C

Source: This medium is available as a premixed powder from Oxoid Unipath and BD Diagnostic Systems.

Preparation of Medium: Add components to distilled/deionized water and bring volume to 1.0L. Mix thoroughly. Gently heat and bring to boiling. Do not autoclave. Cool to 50°C. Pour into sterile Petri dishes.

Use: For the selective isolation, cultivation, enumeration, and differentiation of Gram-negative enteric microorganisms from a variety of clinical and nonclinical specimens. *Escherichia coli* appears as large, flat, rose-red colonies. *Enterobacter* and *Klebsiella* species appear as large, mucoid, pale colonies with a pink center. *Proteus* and *Salmonella* species appear as large, colorless to tan colonies. *Shigella* species appear as colorless to pink colonies. *Pseudomonas* species appear as irregular colorless to brown colonies.

Deoxycholate Citrate Agar

Composition per liter:

Sodium citrate	50.0g
Agar	15.0g
Lactose	10.0g
Beef extract	5.0g
Peptone	5.0g
$Na_2S_2O_3 \cdot 5H_2O$	5.0g
Sodium deoxycholate	2.5g
Ferric citrate	1.0g
Neutral Red	0.025g

pH 7.3 ± 0.2 at 25°C

Source: This medium is available as a premixed powder from Oxoid Unipath.

Preparation of Medium: Add components to distilled/deionized water and bring volume to 1.0L. Mix thoroughly. Gently heat and bring to boiling. Do not autoclave. Cool to 45°–50°C. Pour into sterile Petri dishes. Dry the agar surface before use.

Use: For the selective isolation and cultivation of enteric pathogens, especially *Salmonella* and *Shigella* species.

Dermabacter Medium

Composition per liter:

Pancreatic digest of casein	10.0g
Glucose	5.0g
NaCl	5.0g
Yeast extract	5.0g

pH 7.4 ± 0.2 at 25°C

Preparation of Medium: Add components to distilled/deionized water and bring volume to 1.0L. Mix thoroughly. Distribute into tubes or flasks. Autoclave for 15 min at 15 psi pressure–121°C.

Use: For the cultivation and maintenance of *Dermabacter hominus*.

Desulfacinum hydrothermale Medium (DSMZ Medium 875)

Composition per 1004mL:

Solution A	920.0mL
Solution C ($NaHCO_3$ solution)	50.0mL
Solution F	13.0mL
Solution D (Seven vitamin solution)	10.0mL
Solution E	10.0mL
Solution B (Trace elements solution SL-10)	1.0mL

pH 7.0-7.3 at 25°C

Solution A:

Composition per 920mL:

NaCl	10.4g
$MgSO_4 \cdot 7H_2O$	2.72g
$MgCl_2 \cdot 6H_2O$	2.24g
$CaCl_2 \cdot 2H_2O$	0.56g
KCl	0.29g
NH_4Cl	0.1g
KH_2PO_4	0.08g
Resazurin	0.5mg

Preparation of Solution A: Add components to distilled/deionized water and bring volume to 920.0mL. Mix thoroughly. Sparge with 100% N_2. Autoclave for 15 min at 15 psi pressure–121°C.

Solution B (Trace Elements Solution SL-10):

Composition per liter:

$FeCl_2 \cdot 4H_2O$	1.5g
$CoCl_2 \cdot 6H_2O$	190.0mg
$MnCl_2 \cdot 4H_2O$	100.0mg
$ZnCl_2$	70.0mg
$Na_2MoO_4 \cdot 2H_2O$	36.0mg
$NiCl_2 \cdot 6H_2O$	24.0mg
H_3BO_3	6.0mg
$CuCl_2 \cdot 2H_2O$	2.0mg
HCl (25% solution)	10.0mL

Preparation of Solution B (Trace Elements Solution SL-10): Add $FeCl_2 \cdot 4H_2O$ to 10.0mL of HCl solution. Mix thoroughly. Add distilled/deionized water and bring volume to 1.0L. Add remaining components. Mix thoroughly. Sparge with 100% N_2. Autoclave for 15 min at 15 psi pressure–121°C.

Solution C ($NaHCO_3$ Solution):

Composition per 100.0mL:

$NaHCO_3$	5.0g

Preparation of Solution C ($NaHCO_3$ Solution): Add $NaHCO_3$ to distilled/deionized water and bring volume to 100.0mL. Mix thoroughly. Sparge with 80% N_2 + 20% CO_2. Autoclave for 15 min at 15 psi pressure–121°C.

Solution D (Seven Vitamin Solution):

Composition per liter:

Pyridoxine hydrochloride	300.0mg
Thiamine-HCl·$2H_2O$	200.0mg
Nicotinic acid	200.0mg
Vitamin B_{12}	100.0mg
Calcium pantothenate	100.0mg
p-Aminobenzoic acid	80.0mg
D(+)-Biotin	20.0mg

Preparation of Solution D (Seven Vitamin Solution): Add components to distilled/deionized water and bring volume to 1.0L. Sparge with 100% N_2. Mix thoroughly. Filter sterilize.

Solution E:
Composition per 10.0mL:
Na-lactate 2.5g

Preparation of Solution E: Add Na-lactate to distilled/deionized water and bring volume to 10.0mL. Mix thoroughly. Sparge with 100% N_2. Autoclave for 15 min at 15 psi pressure–121°C.

Solution F:
Composition per 100.0mL:
$Na_2S \cdot 9H_2O$ 3.0g

Preparation of Solution F: Add $Na_2S \cdot 9H_2O$ to distilled/deionized water and bring volume to 100.0mL. Mix thoroughly. Sparge with 100% N_2. Autoclave for 15 min at 15 psi pressure–121°C.

Preparation of Medium: Prepare and dispense medium under 80% N_2 + 20% CO_2 gas atmosphere. Add 50.0mL sterile solution C, 13.0mL sterile solution F, 10.0mL sterile solution D, 10.0mL sterile solution E, and 1.0mL sterile solution B, to 920.0mL sterile solution A. Mix thoroughly. The pH of the completed medium should be 7.0-7.3. Aseptically and anaerobically distribute into sterile tubes or bottles.

Use: For the cultivation of *Desulfacinum hydrothermale.*

Desulfitobacterium dehalogenans Medium

Composition per liter:
Solution A 955.0mL
Solution B 25.0mL
Solution C 20.0mL

Solution A:
Composition per 955.0mL:
Na_2HPO_4 .... 2.2g
Yeast extract 2.0g
3-Chloro-4-hydroxy-phenylacetic acid 1.5g
L-Cysteine·HCl·H_2O 0.7g
NH_4Cl 0.5g
KH_2PO_4 .... 0.44g
$MgCl_2 \cdot 6H_2O$ 0.2g
$CaCl_2$ 25.0mg
Wolfe's mineral solution 10.0mL

Wolfe's Mineral Solution:
Composition per liter:
$MgSO_4 \cdot 7H_2O$ 3.0g
Nitrilotriacetic acid 1.5g
NaCl 1.0g
$MnSO_4 \cdot H_2O$ 0.5g
$FeSO_4 \cdot 7H_2O$ 0.1g
$CoCl_2 \cdot 6H_2O$ 0.1g
$CaCl_2$ 0.1g
$ZnSO_4 \cdot 7H_2O$ 0.1g
$CuSO_4 \cdot 5H_2O$ 0.01g
$AlK(SO_4)_2 \cdot 12H_2O$ 0.01g
H_3BO_3 .... 0.01g
$Na_2MoO_4 \cdot 2H_2O$ 0.01g

Preparation of Wolfe's Mineral Solution: Add nitrilotriacetic acid to 500.0mL of distilled/deionized water. Adjust pH to 6.5 with KOH. Add remaining components one at a time. Add distilled/deionized water to 1.0L.

Preparation of Solution A: Add components, except L-cysteine·HCl·H_2O, to distilled/deionized water and bring volume to 955.0mL. Mix thoroughly. Gently heat and bring to boiling. Cool to room temperature while sparging with 90% N_2 + 10% CO_2. Adjust pH to 7.3. Add L-cysteine·HCl·H_2O. Autoclave for 15 min at 15 psi pressure–121°C.

Solution B:
Composition per 25.0mL:
Sodium pyruvate 2.2g

Preparation of Solution B: Add sodium pyruvate to distilled/deionized water and bring volume to 25.0mL. Mix thoroughly. Filter sterilize. Sparge with 100% N_2.

Solution C:
Composition per 20.0mL:
$NaHCO_3$ 1.0g

Preparation of Solution C: Add $NaHCO_3$ to distilled/deionized water and bring volume to 20.0mL. Mix thoroughly. Filter sterilize. Sparge with 100% CO_2.

Preparation of Medium: Aseptically and anaerobically combine 955.0mL of sterile solution A with 25.0mL of sterile solution B and 20.0mL of sterile solution C. Mix thoroughly. Aseptically and anaerobically distribute into sterile tubes or flasks.

Use: For the cultivation of *Desulfitobacterium dehalogenans.*

Desulfitobacterium hafniense Medium

Composition per 1005.0mL:
$NaHCO_3$ 2.6g
NH_4Cl 1.0g
Yeast extract 1.0g
$K_2HPO_4 \cdot 3H_2O$ 0.4g
$MgCl_2 \cdot 6H_2O$ 0.1g
NaCl 0.1g
$CaCl_2 \cdot 2H_2O$ 0.05g
Resazurin 0.5mg
$Na_2S \cdot 9H_2O$ solution 10.0mL
Sodium pyruvate solution 10.0mL
Wolfe's vitamin solution 10.0mL
$Na_2S_2O_3$ solution 5.0mL
Selenite-tungstate solution 1.0mL
Trace elements solution SL-10 with EDTA 1.0mL
pH 7.5 ± 0.2 at 25°C

$Na_2S \cdot 9H_2O$ Solution:
Composition per 10.0mL:
$Na_2S \cdot 9H_2O$ 0.3g

Preparation of $Na_2S \cdot 9H_2O$ Solution: Add $Na_2S \cdot 9H_2O$ to distilled/deionized water and bring volume to 10.0mL. Mix thoroughly. Sparge with 100% N_2. Autoclave for 15 min at 15 psi pressure–121°C.

Sodium Pyruvate Solution:
Composition per 10.0mL:
Sodium pyruvate 2.5g

Preparation of Sodium Pyruvate Solution: Add sodium pyruvate to distilled/deionized water and bring volume to 10.0mL. Mix thoroughly. Sparge with 100% N_2. Autoclave for 15 min at 15 psi pressure–121°C.

Wolfe's Vitamin Solution:
Composition per liter:
Pyridoxine·HCl 10.0mg
p-Aminobenzoic acid 5.0mg
Lipoic acid 5.0mg
Nicotinic acid 5.0mg
Riboflavin 5.0mg
Thiamine·HCl 5.0mg

Calcium DL-pantothenate 5.0mg
Biotin 2.0mg
Folic acid 2.0mg
Vitamin B_{12} 0.1mg

Preparation of Wolfe's Vitamin Solution: Add components to distilled/deionized water and bring volume to 1.0L. Mix thoroughly. Filter sterilize. Sparge with 100% N_2.

$Na_2S_2O_3$ Solution:
Composition per 10.0mL:
$Na_2S_2O_3 \cdot 5H_2O$ 2.5g

Preparation of $Na_2S_2O_3$ Solution: Add $Na_2S_2O_3 \cdot 5H_2O$ to distilled/deionized water and bring volume to 10.0mL. Mix thoroughly. Sparge with 100% N_2. Autoclave for 15 min at 15 psi pressure–121°C.

Selenite-Tungstate Solution:
Composition per liter:
NaOH 0.5g
$Na_2WO_4 \cdot 2H_2O$ 4.0mg
$Na_2SeO_3 \cdot 5H_2O$ 3.0mg

Preparation of Selenite-Tungstate Solution: Add components to distilled/deionized water and bring volume to 1.0L. Mix thoroughly.

Trace Elements Solution SL-10 with EDTA:
Composition per liter:
$FeCl_2 \cdot 4H_2O$ 1.5g
Disodium EDTA 0.5g
$CoCl_2 \cdot 6H_2O$ 190.0mg
$MnCl_2 \cdot 4H_2O$ 100.0mg
$ZnCl_2$ 70.0mg
$Na_2MoO_4 \cdot 2H_2O$ 36.0mg
$NiCl_2 \cdot 6H_2O$ 24.0mg
H_3BO_3 .... 6.0mg
$CuCl_2 \cdot 2H_2O$ 2.0mg
HCl (25% solution) 10.0mL

Preparation of Trace Elements Solution SL-10 with EDTA: Add $FeCl_2 \cdot 4H_2O$ to 10.0mL of HCl solution. Mix thoroughly. Add distilled/deionized water and bring volume to 1.0L. Add remaining components. Mix thoroughly.

Preparation of Medium: Prepare and dispense medium under 80% N_2 + 20% CO_2. Add components, except $NaHCO_3$, $Na_2S \cdot 9H_2O$ solution, sodium pyruvate solution, vitamin solution, and $Na_2S_2O_3 \cdot 5H_2O$ solution, to distilled/deionized water and bring volume to 970.0mL. Mix thoroughly. Gently heat and bring to boiling. Continue boiling for 3 min. Cool to room temperature while sparging with 80% N_2 + 20% CO_2. Add $NaHCO_3$. Mix thoroughly. Adjust pH to 7.0. Anaerobically distribute 9.7mL volumes into anaerobic tubes. Autoclave for 15 min at 15 psi pressure–121°C. Aseptically and anaerobically add 0.1mL of sterile $Na_2S \cdot 9H_2O$ solution, 0.1mL of sterile sodium pyruvate solution, 0.1mL of sterile vitamin solution, and 0.05mL of sterile $Na_2S_2O_3 \cdot 5H_2O$ solution to each tube. Mix thoroughly.

Use: For the cultivation of *Desulfitobacterium hafniense.*

Desulfitobacterium Medium (DSMZ Medium 663)

Composition per liter:
KH_2PO_4 .... 5.44g
Yeast extract 1.0g
NH_4Cl 0.5g
$MgCl_2 \cdot 2H_2O$ 0.18g
$CaCl_2 \cdot 2H_2O$ 0.032g
Resazurin 0.5mg
Vitamin solution 20.0mL
Na-pyruvate solution 10.0mL
Na-thiosulfate solution 10.0mL
Cysteine solution 10.0mL
$Na_2S \cdot 9H_2O$ solution 10.0mL
Trace elements solution 5.0mL
pH 7.5 ± 0.2 at 25°C

Na-pyruvate Solution:
Composition per 10.0mL:
Na-pyruvate 2.5g

Preparation of Na-pyruvate Solution: Add Na-pyruvate to distilled/deionized water and bring volume to 10.0mL. Mix thoroughly. Sparge with 100% N_2. Filter sterilize.

Na-thiosulfate Solution:
Composition per 10.0mL:
$Na_2S_2O_3 \cdot 5H_2O$ 2.5g

Preparation of Na-thiosulfate Solution: Add 2.5g $Na_2S_2O_3 \cdot 5H_2O$ to distilled/deionized water and bring volume to 10.0mL. Mix thoroughly. Sparge with 100% N_2. Filter sterilize.

$Na_2S \cdot 9H_2O$ Solution:
Composition per 10mL:
$Na_2S \cdot 9H_2O$ 0.4g

Preparation of $Na_2S \cdot 9H_2O$ Solution: Add $Na_2S \cdot 9H_2O$ to distilled/deionized water and bring volume to 10.0mL. Mix thoroughly. Autoclave under 100% N_2 for 15 min at 15 psi pressure–121°C. Cool to room temperature.

Cysteine Solution:
Composition per 10.0mL:
L-Cysteine·HCl·H_2O 0.4g

Preparation of Cysteine Solution: Add L-cysteine·HCl·H_2O to distilled/deionized water and bring volume to 10.0mL. Mix thoroughly. Sparge with 100% N_2. Autoclave for 15 min at 15 psi pressure–121°C.

Trace Elements Solution:
Composition per liter:
Nitrilotriacetic acid 12.8g
$FeCl_3 \cdot 6H_2O$ 1.35g
NaCl 1.0g
$CoCl_2 \cdot 4H_2O$ 0.24g
$NiCl_2 \cdot 6H_2O$ 0.12g
$MnCl_2 \cdot 4H_2O$ 0.1g
$CaCl_2 \cdot 2H_2O$ 0.1g
$ZnCl_2$ 0.1g
$Na_2SeO_3 \cdot 5H_2O$ 0.026g
$CuCl_2 \cdot 2H_2O$ 0.025g
$Na_2MoO_4 \cdot 4H_2O$ 0.024g
H_3BO_3 .... 0.01g

Preparation of Trace Elements Solution: Add nitrilotriacetic acid to 500.0mL of distilled/deionized water. Dissolve by adjusting pH to 6.5 with KOH. Add remaining components. Add distilled/deionized water to 1.0L. Mix thoroughly. Adjust pH to 6.8.

Vitamin Solution:
Composition per liter:
Pyridoxine-HCl 10.0mg
Thiamine-HCl·$2H_2O$ 5.0mg
Riboflavin 5.0mg
Nicotinic acid 5.0mg
D-Ca-pantothenate 5.0mg

p-Aminobenzoic acid 5.0mg
Lipoic acid 5.0mg
Biotin 2.0mg
Folic acid 2.0mg
Vitamin B_{12} 0.1mg

Preparation of Vitamin Solution: Add components to distilled/deionized water and bring volume to 1.0L. Mix thoroughly. Sparge with 80% H_2 + 20% CO_2. Filter sterilize.

Preparation of Medium: Prepare and dispense medium under an oxygen-free 100% N_2. Add components, except vitamin solution, cysteine solution, Na-pyruvate solution, Na-thiosulfate solution, and $Na_2S{\cdot}9H_2O$ solution, to distilled/deionized water and bring volume to 940.0mL. Mix thoroughly. Sparge with 100% N_2. Autoclave for 15 min at 15 psi pressure–121°C. Cool to 25°C. Aseptically and anaerobically add 20.0mL sterile vitamin solution, 10.0mL of sterile cysteine solution 10.0mL sterile, Na-pyruvate solution, 10.0mL sterile Na-thiosulfate solution, and 10.0mL of sterile $Na_2S{\cdot}9H_2O$ solution. Mix thoroughly. Adjust pH to 7.5. Aseptically and anaerobically distribute into sterile tubes or flasks.

Use: For the cultivation of *Desulfitobacterium dehalogenans.*

Desulfitobacterium PCE Medium (DSMZ Medium 717)

Composition per liter:
$(NH_4)H_2PO_4$ 2.88g
$MgSO_4{\cdot}7H_2O$ 0.1g
Yeast extract 0.1g
$Ca(NO_3)_2{\cdot}4H_2O$ 0.05g
Resazurin 0.1mg
$NaHCO_3$ solution 50.0mL
KOH solution 20.0mL
Na-lactate 20.0mL
Na-fumarate 20.0mL
Vitamin solution 10.0mL
$Na_2S{\cdot}9H_2O$ solution 3.3mL
Seven vitamin solution 1.0mL
Trace elements solution 1.0mL
Selenite-tungstate solution 1.0mL
pH 7.1 ± 0.2 at 25°C

Selenite-Tungstate Solution
Composition per liter:
NaOH 0.5g
$Na_2WO_4{\cdot}2H_2O$ 4.0mg
$Na_2SeO_3{\cdot}5H_2O$ 3.0mg

Preparation of Selenite-Tungstate Solution: Add components to distilled/deionized water and bring volume to 1.0L. Mix thoroughly. Sparge with 100% N_2. Filter sterilize.

$Na_2S{\cdot}9H_2O$ Solution:
Composition per 10mL:
$Na_2S{\cdot}9H_2O$ 0.5g

Preparation of $Na_2S{\cdot}9H_2O$ Solution: Add $Na_2S{\cdot}9H_2O$ to distilled/deionized water and bring volume to 10.0mL. Mix thoroughly. Autoclave under 100% N_2 for 15 min at 15 psi pressure–121°C. Cool to room temperature.

$NaHCO_3$ Solution:
Composition per 100.0mL:
$NaHCO_3$ 5.0g

Preparation of $NaHCO_3$ Solution: Add $NaHCO_3$ to distilled/deionized water and bring volume to 100.0mL. Mix thoroughly. Sparge with 80% N_2 + 20% CO_2. Filter sterilize.

KOH Solution:
Composition per 100.0mL:
KOH 10.0g

Preparation of KOH Solution: Add KOH to distilled/deionized water and bring volume to 100.0mL. Mix thoroughly. Sparge with 100% N_2. Filter sterilize.

Trace Elements Solution SL-10:
Composition per liter:
$FeCl_2{\cdot}4H_2O$ 1.5g
$CoCl_2{\cdot}6H_2O$ 190.0mg
$MnCl_2{\cdot}4H_2O$ 100.0mg
$ZnCl_2$ 70.0mg
$Na_2MoO_4{\cdot}2H_2O$ 36.0mg
$NiCl_2{\cdot}6H_2O$ 24.0mg
H_3BO_3 .... 6.0mg
$CuCl_2{\cdot}2H_2O$ 2.0mg
HCl (25% solution) 10.0mL

Preparation of Trace Elements Solution SL-10: Add $FeCl_2{\cdot}4H_2O$ to 10.0mL of HCl solution. Mix thoroughly. Add distilled/deionized water and bring volume to 1.0L. Add remaining components. Mix thoroughly. Sparge with 80% N_2 + 20% CO_2. Autoclave for 15 min at 15 psi pressure–121°C.

Vitamin Solution:
Composition per liter:
Pyridoxine-HCl 10.0mg
Thiamine-HCl·$2H_2O$ 5.0mg
Riboflavin 5.0mg
Nicotinic acid 5.0mg
D-Ca-pantothenate 5.0mg
p-Aminobenzoic acid 5.0mg
Lipoic acid 5.0mg
Biotin 2.0mg
Folic acid 2.0mg
Vitamin B_{12} 0.1mg

Preparation of Vitamin Solution: Add components to distilled/deionized water and bring volume to 1.0L. Mix thoroughly. Sparge with 80% H_2 + 20% CO_2. Filter sterilize.

Seven Vitamin Solution:
Composition per liter:
Pyridoxine hydrochloride 300.0mg
Thiamine-HCl·$2H_2O$ 200.0mg
Nicotinic acid 200.0mg
Vitamin B_{12} 100.0mg
Calcium pantothenate 100.0mg
p-Aminobenzoic acid 80.0mg
D(+)-Biotin 20.0mg

Preparation of Seven Vitamin Solution: Add components to distilled/deionized water and bring volume to 1.0L. Sparge with 100% N_2. Mix thoroughly. Filter sterilize.

Na-lactate Solution:
Composition per 100.0mL:
Na-lactate 25.0g

Preparation of Na-lactate Solution: Add Na-lactate to distilled/deionized water and bring volume to 100.0mL. Mix thoroughly. Sparge with 100% N_2. Autoclave for 15 min at 15 psi pressure–121°C. Cool to room temperature.

Na-fumarate Solution:
Composition per 100.0mL:
Na-fumarate 16.0g

Preparation of Na-fumarate Solution: Add Na-fumarate to distilled/deionized water and bring volume to 100.0mL. Mix thoroughly. Sparge with 100% N_2. Autoclave for 15 min at 15 psi pressure–121°C. Cool to room temperature.

Preparation of Medium: Prepare and dispense medium under 80% N_2 + 20% CO_2 gas atmosphere. Add components, except $NaHCO_3$ solution, $Na_2S·9H_2O$ solution, KOH solution, Na-lactate solution, Na-fumarate solution, vitamin solution, seven vitamin solution, selenite-tungstate solution, and trace elements solution SL-10, to distilled/deionized water and bring volume to 873.7mL. Mix thoroughly. Adjust pH to 7.0-7.2. Sparge with 80% N_2 + 20% CO_2. Autoclave for 15 min at 15 psi pressure–121°C. Aseptically and anaerobically add 50.0mL $NaHCO_3$ solution, 3.3mL $Na_2S·9H_2O$ solution, 20.0mL KOH solution, 20.0mL Na-lactate solution, 20.0mL Na-fumarate solution, 10.0mL vitamin solution, 1.0mL seven vitamin solution, 1.0mL selenite-tungstate solution, and 1.0mL trace elements solution SL-10. Mix thoroughly. Aseptically and anaerobically distribute into sterile tubes or bottles.

Use: For the cultivation of *Desulfitobacterium* sp. and *Desulfitobacterium hafniense.*

Desulfobacca Medium (DSMZ Medium 728)

Composition per liter:
$NaHCO_3$ 4.0g
Na_2SO_4 .... 3.0g
Na-acetate 1.64g
$Na_2HPO_4·2H_2O$ 0.53g
KH_2PO_4 .... 0.41g
NH_4Cl 0.3g
NaCl 0.3g
$CaCl_2·2H_2O$ 0.11g
$MgCl_2·6H_2O$ 0.1g
Resazurin 0.5mg
$Na_2S·9H_2O$ solution 10.0mL
$CaCl_2$ solution 10.0mL
Trace elements solution SL-10 .... 1.0mL
Selenite-tungstate solution 1.0mL
Seven vitamin solution 0.2mL
pH 7.1 ± 0.2 at 25°C

Seven Vitamin Solution:
Composition per liter:
Pyridoxine hydrochloride 300.0mg
Thiamine-HCl·$2H_2O$ 200.0mg
Nicotinic acid 200.0mg
Vitamin B_{12} 100.0mg
Calcium pantothenate 100.0mg
p-Aminobenzoic acid 80.0mg
D(+)-Biotin 20.0mg

Preparation of Seven Vitamin Solution: Add components to distilled/deionized water and bring volume to 1.0L. Sparge with 100% N_2. Mix thoroughly. Filter sterilize.

Trace Elements Solution SL-10:
Composition per liter:
$FeCl_2·4H_2O$ 1.5g
$CoCl_2·6H_2O$ 190.0mg
$MnCl_2·4H_2O$ 100.0mg
$ZnCl_2$ 70.0mg
$Na_2MoO_4·2H_2O$ 36.0mg
$NiCl_2·6H_2O$ 24.0mg
H_3BO_3 .... 6.0mg
$CuCl_2·2H_2O$ 2.0mg
HCl (25% solution) 10.0mL

Preparation of Trace Elements Solution SL-10: Add $FeCl_2·4H_2O$ to 10.0mL of HCl solution. Mix thoroughly. Add distilled/deionized water and bring volume to 1.0L. Add remaining components. Mix thoroughly. Sparge with 80% N_2 + 20% CO_2. Autoclave for 15 min at 15 psi pressure–121°C.

$Na_2S·9H_2O$ Solution:
Composition per 10mL:
$Na_2S·9H_2O$ 0.5g

Preparation of $Na_2S·9H_2O$ Solution: Add $Na_2S·9H_2O$ to distilled/deionized water and bring volume to 10.0mL. Mix thoroughly. Autoclave under 100% N_2 for 15 min at 15 psi pressure–121°C. Cool to room temperature.

Selenite-Tungstate Solution
Composition per liter:
NaOH 0.5g
$Na_2WO_4·2H_2O$ 4.0mg
$Na_2SeO_3·5H_2O$ 3.0mg

Preparation of Selenite-Tungstate Solution: Add components to distilled/deionized water and bring volume to 1.0L. Mix thoroughly. Sparge with 100% N_2. Filter sterilize.

$CaCl_2$ Solution:
Composition per 10.0mL:
$CaCl_2·2H_2O$ 0.11g

Preparation of $CaCl_2$ Solution: Add $CaCl_2·2H_2O$ to distilled/deionized water and bring volume to 10.0mL. Mix thoroughly. Sparge with 80% N_2 + 20% CO_2. Filter sterilize.

Preparation of Medium: Prepare and dispense medium under 80% N_2 + 20% CO_2 gas atmosphere. Add components, except $Na_2S·9H_2O$ solution, $CaCl_2$ solution, seven vitamin solution, selenite-tungstate solution, and trace elements solution SL-10, to distilled/deionized water and bring volume to 977.8mL. Mix thoroughly. Adjust pH to 7.0-7.2. Sparge with 80% N_2 + 20% CO_2. Autoclave for 15 min at 15 psi pressure–121°C. Aseptically and anaerobically add 10.0mL $Na_2S·9H_2O$ solution, 10.0mL $CaCl_2$ solution, 0.2mL seven vitamin solution, 1.0mL selenite-tungstate solution, and 1.0mL trace elements solution SL-10. Mix thoroughly. Aseptically and anaerobically distribute into sterile tubes or bottles.

Use: For the cultivation of *Desulfobacca acetoxidans.*

Desulfobacter Medium

Composition per 1001.0mL:
Solution A 870.0mL
Solution C 100.0mL
Solution D 10.0mL
Solution E (Vitamin solution) 10.0mL
Solution F 10.0mL
Solution B (Trace elements solution SL-10) 1.0mL
pH 7.1–7.4 at 25°C

Solution A:
Composition per 870.0mL:
NaCl 21.0g
$MgCl_2·6H_2O$ 3.1g
Na_2SO_4 .... 3.0g

KCl 0.5g
NH_4Cl 0.3g
KH_2PO_4 .... 0.2g
$CaCl_2 \cdot 2H_2O$ 0.15g
Resazurin 1.0mg

Preparation of Solution A: Add components to distilled/deionized water and bring volume to 870.0mL. Mix thoroughly. Gently heat and bring to boiling. Continue boiling for 3-4 min. Allow to cool to room temperature while gassing under 80% N_2 + 20% CO_2. Continue gassing until pH reaches below 6.0. Seal the flask under 80% N_2 + 20% CO_2. Autoclave for 15 min at 15 psi pressure–121°C.

Solution B (Trace Elements Solution SL-10):
Composition per liter:
$FeCl_2 \cdot 4H_2O$ 1.5g
$CoCl_2 \cdot 6H_2O$ 190.0mg
$MnCl_2 \cdot 4H_2O$ 100.0mg
$ZnCl_2$ 70.0mg
$Na_2MoO_4 \cdot 2H_2O$ 36.0mg
$NiCl_2 \cdot 6H_2O$ 24.0mg
H_3BO_3 .... 6.0mg
$CuCl_2 \cdot 2H_2O$ 2.0mg
HCl (25% solution) 10.0mL

Preparation of Solution B (Trace Elements Solution SL-10): Add $FeCl_2 \cdot 4H_2O$ to 10.0mL of HCl solution. Mix thoroughly. Add distilled/deionized water and bring volume to 1.0L. Add remaining components. Mix thoroughly. Gas under 100% N_2. Autoclave for 15 min at 15 psi pressure–121°C.

Solution C:
Composition per 100.0mL:
$NaHCO_3$ 5.0g

Preparation of Solution C: Add $NaHCO_3$ to distilled/deionized water and bring volume to 100.0mL. Mix thoroughly. Filter sterilize. Gas under 80% N_2 + 20% CO_2.

Solution D:
Composition per 10.0mL:
Sodium acetate·$3H_2O$ 2.5g

Preparation of Solution D: Add sodium acetate to distilled/deionized water and bring volume to 10.0mL. Mix thoroughly. Gas under 100% N_2. Autoclave for 15 min at 15 psi pressure–121°C.

Solution E (Vitamin Solution):
Composition per liter:
Pyridoxine·HCl 10.0mg
Calcium DL-pantothenate 5.0mg
Lipoic acid 5.0mg
Nicotinic acid 5.0mg
p-Aminobenzoic acid 5.0mg
Riboflavin 5.0mg
Thiamine·HCl 5.0mg
Biotin 2.0mg
Folic acid 2.0mg
Vitamin B_{12} 0.1mg

Preparation of Solution E (Vitamin Solution): Add components to distilled/deionized water and bring volume to 1.0L. Mix thoroughly. Gas under 100% N_2. Autoclave for 15 min at 15 psi pressure–121°C.

Solution F:
Composition per 10.0mL:
$Na_2S \cdot 9H_2O$ 0.4g

Preparation of Solution F: Add $Na_2S \cdot 9H_2O$ to distilled/deionized water and bring volume to 10.0mL. Mix thoroughly. Gas under 100% N_2. Autoclave for 15 min at 15 psi pressure–121°C.

Preparation of Medium: Aseptically and anaerobically combine solution A with solution B, solution C, solution D, solution E, and solution F, in that order. Mix thoroughly. Anaerobically distribute into sterile tubes or flasks under 80% N_2 + 20% CO_2.

Use: For the cultivation and maintenance of *Desulfobacter* species and *Malonomonas rubra.*

Desulfobacter postgatei Medium (DSMZ Medium 193)

Composition per 1001.0mL:
Solution A 870.0mL
Solution C 100.0mL
Solution D 10.0mL
Solution E (Vitamin solution) 10.0mL
Solution F 10.0mL
Solution B (Trace elements solution SL-10) 1.0mL
pH 7.1-7.4 at 25°C

Solution A:
Composition per 870.0mL:
NaCl 7.0g
Na_2SO_4 .... 3.0g
$MgCl_2 \cdot 6H_2O$ 1.3g
KCl 0.5g
NH_4Cl 0.3g
KH_2PO_4 .... 0.2g
$CaCl_2 \cdot 2H_2O$ 0.15g
Resazurin 1.0mg

Preparation of Solution A: Add components to distilled/deionized water and bring volume to 870.0mL Mix thoroughly.

Solution B (Trace Elements Solution SL-10):
Composition per liter:
$FeCl_2 \cdot 4H_2O$ 1.5g
$CoCl_2 \cdot 6H_2O$ 190.0mg
$MnCl_2 \cdot 4H_2O$ 100.0mg
$ZnCl_2$ 70.0mg
$Na_2MoO_4 \cdot 2H_2O$ 36.0mg
$NiCl_2 \cdot 6H_2O$ 24.0mg
H_3BO_3 .... 6.0mg
$CuCl_2 \cdot 2H_2O$ 2.0mg
HCl (25% solution) 10.0mL

Preparation of Solution B (Trace Elements Solution SL-10): Add $FeCl_2 \cdot 4H_2O$ to 10.0mL of HCl solution. Mix thoroughly. Add distilled/deionized water and bring volume to 1.0L. Add remaining components. Mix thoroughly. Sparge with 100% N_2. Autoclave for 15 min at 15 psi pressure–121°C.

Solution C:
Composition per 100.0mL:
$NaHCO_3$ 5.0g

Preparation of Solution C: Add $NaHCO_3$ to distilled/deionized water and bring volume to 100.0mL Mix thoroughly. Filter sterilize. Flush with 80% N_2 + 20% CO_2 to remove dissolved oxygen.

Solution D:
Composition per 10.0mL:
Na-acetate·$3H_2O$ 2.5g

Preparation of Solution D: Add Na-acetate·$3H_2O$ to distilled/deionized water and bring volume to 10.0mL. Mix thoroughly. Sparge with 100% N_2. Autoclave for 15 min at 15 psi pressure–121°C.

Solution E (Vitamin Solution):
Composition per liter:

Pyridoxine-HCl	10.0mg
Thiamine-HCl·$2H_2O$	5.0mg
Riboflavin	5.0mg
Nicotinic acid	5.0mg
D-Ca-pantothenate	5.0mg
p-Aminobenzoic acid	5.0mg
Lipoic acid	5.0mg
Biotin	2.0mg
Folic acid	2.0mg
Vitamin B_{12}	0.10mg

Solution E (Vitamin Solution): Add components to distilled/deionized water and bring volume to 1.0L. Mix thoroughly. Sparge with 100% N_2. Autoclave for 15 min at 15 psi pressure–121°C.

Solution F:
Composition per 10.0mL:

$Na_2S·9H_2O$	0.4g

Preparation of Solution F: Add $Na_2S·9H_2O$ to distilled/deionized water and bring volume to 10.0mL. Mix thoroughly. Sparge with 100% N_2. Autoclave for 15 min at 15 psi pressure–121°C.

Preparation of Medium: Gently heat solution A and bring to boiling. Boil solution A for a few minutes. Cool to room temperature. Gas with 80% N_2 + 20% CO_2 gas mixture to reach a pH below 6. Autoclave for 15 min at 15 psi pressure–121°C. Cool to room temperature. Sequentially add 1.0mL solution B, 100.0mL solution C, 10.0mL solution D, 10.0mL solution E, and 10.0mL solution F. Distribute anaerobically under 80% N_2 + 20% CO_2 into appropriate vessels. Addition of 10–20mg sodium dithionite per liter from a 5% (w/v) solution, freshly prepared under N_2 and filter-sterilized, may stimulate growth.

Use: For the cultivation of *Desulfobacter postgatei, Paracoccus solventivorans,* and *Desulfotomaculum* sp.

Desulfobacter sp. Medium (DSMZ Medium 195)

Composition per 991.0mL:

Solution A	870.0mL
Solution C	100.0mL
Solution D	10.0mL
Solution E (Vitamin solution)	10.0mL
Solution B (Trace elements solution SL-10)	1.0mL

pH 7.1-7.4 at 25°C

Solution A:
Composition per 870.0mL:

NaCl	21.0g
$MgCl_2·6H_2O$	3.1g
Na_2SO_4	3.0g
KCl	0.5g
NH_4Cl	0.3g
KH_2PO_4	0.2g
$CaCl_2·2H_2O$	0.15g
Resazurin	1.0mg

Preparation of Solution A: Add components to distilled/deionized water and bring volume to 870.0mL Mix thoroughly.

Solution B (Trace Elements Solution SL-10):
Composition per liter:

$FeCl_2·4H_2O$	1.5g
$CoCl_2·6H_2O$	190.0mg
$MnCl_2·4H_2O$	100.0mg
$ZnCl_2$	70.0mg
$Na_2MoO_4·2H_2O$	36.0mg
$NiCl_2·6H_2O$	24.0mg
H_3BO_3	6.0mg
$CuCl_2·2H_2O$	2.0mg
HCl (25% solution)	10.0mL

Preparation of Solution B (Trace Elements Solution SL-10): Add $FeCl_2·4H_2O$ to 10.0mL of HCl solution. Mix thoroughly. Add distilled/deionized water and bring volume to 1.0L. Add remaining components. Mix thoroughly. Sparge with 100% N_2. Autoclave for 15 min at 15 psi pressure–121°C.

Solution C:
Composition per 100.0mL:

$NaHCO_3$	5.0g

Preparation of Solution C: Add $NaHCO_3$ to distilled/deionized water and bring volume to 100.0mL Mix thoroughly. Filter sterilize. Flush with 80% N_2 + 20% CO_2 to remove dissolved oxygen.

Solution D:
Composition per 10.0mL:

Na-acetate·$3H_2O$	2.5g

Preparation of Solution D: Add Na-acetate·$3H_2O$ to distilled/deionized water and bring volume to 10.0mL. Mix thoroughly. Sparge with 100% N_2. Autoclave for 15 min at 15 psi pressure–121°C.

Solution E (Vitamin Solution):
Composition per liter:

Pyridoxine-HCl	10.0mg
Thiamine-HCl·$2H_2O$	5.0mg
Riboflavin	5.0mg
Nicotinic acid	5.0mg
D-Ca-pantothenate	5.0mg
p-Aminobenzoic acid	5.0mg
Lipoic acid	5.0mg
Biotin	2.0mg
Folic acid	2.0mg
Vitamin B_{12}	0.10mg

Preparation of Solution E (Vitamin Solution): Add components to distilled/deionized water and bring volume to 1.0L. Mix thoroughly. Sparge with 100% N_2. Autoclave for 15 min at 15 psi pressure–121°C.

Solution F:
Composition per 10.0mL:

$Na_2S·9H_2O$	0.4g

Preparation of Solution F: Add $Na_2S·9H_2O$ to distilled/deionized water and bring volume to 10.0mL. Mix thoroughly. Sparge with 100% N_2. Autoclave for 15 min at 15 psi pressure–121°C.

Preparation of Medium: Gently heat solution A and bring to boiling. Boil solution A for a few minutes. Cool to room temperature. Gas with 80% N_2 + 20% CO_2 gas mixture to reach a pH below 6. Autoclave for 15 min at 15 psi pressure–121°C. Cool to room temperature. Sequentially add 1.0mL solution B, 100.0mL solution C, 10.0mL solution D, 10.0mL solution E, and 10.0mL solution F. Distribute anaerobically under 80% N_2 + 20% CO_2 into appropriate vessels. Addition of 10–20mg sodium dithionite per liter from a 5% (w/v) solution, freshly prepared under N_2 and filter-sterilized, may stimulate growth.

Use: For the cultivation of *Desulfobacter* species.

Desulfobacterium anilini Medium

Composition per 1011.0mL:

Solution A 870.0mL
Solution C 100.0mL
Solution D 10.0mL
Solution E (Vitamin solution) 10.0mL
Solution F 10.0mL
Solution G 10.0mL
Solution B (Trace elements solution SL-10) 1.0mL
pH 7.1–7.4 at 25°C

Solution A:

Composition per 870.0mL:

NaCl 7.0g
Na_2SO_4 .... 3.0g
$MgCl_2 \cdot 6H_2O$ 1.3g
KCl 0.5g
NH_4Cl 0.3g
KH_2PO_4 .... 0.2g
$CaCl_2 \cdot 2H_2O$ 0.15g
Resazurin 1.0mg

Preparation of Solution A: Add components to distilled/deionized water and bring volume to 870.0mL. Mix thoroughly. Gently heat and bring to boiling. Continue boiling for 3-4 min. Allow to cool to room temperature while gassing under 80% N_2 + 20% CO_2. Continue gassing until pH reaches below 6.0. Seal the flask under 80% N_2 + 20% CO_2. Autoclave for 15 min at 15 psi pressure–121°C.

Solution B (Trace Elements Solution SL-10):

Composition per liter:

$FeCl_2 \cdot 4H_2O$ 1.5g
$CoCl_2 \cdot 6H_2O$ 190.0mg
$MnCl_2 \cdot 4H_2O$ 100.0mg
$ZnCl_2$ 70.0mg
$Na_2MoO_4 \cdot 2H_2O$ 36.0mg
$NiCl_2 \cdot 6H_2O$ 24.0mg
H_3BO_3 .... 6.0mg
$CuCl_2 \cdot 2H_2O$ 2.0mg
HCl (25% solution) 10.0mL

Preparation of Solution B (Trace Elements Solution SL-10): Add $FeCl_2 \cdot 4H_2O$ to 10.0mL of HCl solution. Mix thoroughly. Add distilled/deionized water and bring volume to 1.0L. Add remaining components. Mix thoroughly. Gas under 100% N_2. Autoclave for 15 min at 15 psi pressure–121°C.

Solution C:

Composition per 100.0mL:

$NaHCO_3$ 5.0g

Preparation of Solution C: Add $NaHCO_3$ to distilled/deionized water and bring volume to 100.0mL. Mix thoroughly. Filter sterilize. Gas under 80% N_2 + 20% CO_2.

Solution D:

Composition per 10.0mL:

Sodium acetate·$3H_20$ 2.5g

Preparation of Solution D: Add sodium acetate to distilled/deionized water and bring volume to 10.0mL. Mix thoroughly. Gas under 100% N_2. Autoclave for 15 min at 15 psi pressure–121°C.

Solution E (Vitamin Solution):

Composition per liter:

Pyridoxine·HCl 10.0mg
Calcium DL-pantothenate 5.0mg
Lipoic acid 5.0mg
Nicotinic acid 5.0mg
p-Aminobenzoic acid 5.0mg
Riboflavin 5.0mg
Thiamine·HCl 5.0mg
Biotin 2.0mg
Folic acid 2.0mg
Vitamin B_{12} 0.1mg

Preparation of Solution E (Vitamin Solution): Add components to distilled/deionized water and bring volume to 1.0L. Mix thoroughly. Gas under 100% N_2. Autoclave for 15 min at 15 psi pressure–121°C.

Solution F:

Composition per 10.0mL:

$Na_2S \cdot 9H_2O$ 0.4g

Preparation of Solution F: Add $Na_2S \cdot 9H_2O$ to distilled/deionized water and bring volume to 10.0mL. Mix thoroughly. Gas under 100% N_2. Autoclave for 15 min at 15 psi pressure–121°C.

Solution G:

Composition per 10.0mL:

Phenol 94.0mg

Preparation of Solution G: Add phenol to distilled/deionized water and bring volume to 10.0mL. Mix thoroughly. Filter sterilize. Gas under 80% N_2 + 20% CO_2. Prepare solution freshly.

Preparation of Medium: Aseptically and anaerobically combine solution A with solution B, solution C, solution D, solution E, solution F, and solution G, in that order. Mix thoroughly. Anaerobically distribute into sterile tubes or flasks under 80% N_2 + 20% CO_2.

Use: For the cultivation of *Desulfobacterium anilini.*

Desulfobacterium anilini Medium

Composition per 1002.0mL:

Solution A 870.0mL
Solution C 100.0mL
Solution D 10.0mL
Solution E (Vitamin solution) 10.0mL
Solution F 10.0mL
Solution B (Trace elements solution SL-10) 1.0mL
Solution G 1.0mL
pH 7.1–7.4 at 25°C

Solution A:

Composition per 870.0mL:

NaCl 7.0g
Na_2SO_4 .... 3.0g
$MgCl_2 \cdot 6H_2O$ 1.3g
KCl 0.5g
NH_4Cl 0.3g
KH_2PO_4 .... 0.2g
$CaCl_2 \cdot 2H_2O$ 0.15g
Resazurin 1.0mg

Preparation of Solution A: Add components to distilled/deionized water and bring volume to 870.0mL. Mix thoroughly. Gently heat and bring to boiling. Continue boiling for 3-4 min. Allow to cool to room temperature while gassing under 80% N_2 + 20% CO_2. Continue gassing until pH reaches below 6.0. Seal the flask under 80% N_2 + 20% CO_2. Autoclave for 15 min at 15 psi pressure–121°C.

Solution B (Trace Elements Solution SL-10):

Composition per liter:

$FeCl_2 \cdot 4H_2O$ 1.5g
$CoCl_2 \cdot 6H_2O$ 190.0mg
$MnCl_2 \cdot 4H_2O$ 100.0mg
$ZnCl_2$ 70.0mg

$Na_2MoO_4 \cdot 2H_2O$	36.0mg
$NiCl_2 \cdot 6H_2O$	24.0mg
H_3BO_3	6.0mg
$CuCl_2 \cdot 2H_2O$	2.0mg
HCl (25% solution)	10.0mL

Preparation of Solution B (Trace Elements Solution SL-10): Add $FeCl_2 \cdot 4H_2O$ to 10.0mL of HCl solution. Mix thoroughly. Add distilled/deionized water and bring volume to 1.0L. Add remaining components. Mix thoroughly. Gas under 100% N_2. Autoclave for 15 min at 15 psi pressure–121°C.

Solution C:
Composition per 100.0mL:

$NaHCO_3$	5.0g

Preparation of Solution C: Add $NaHCO_3$ to distilled/deionized water and bring volume to 100.0mL. Mix thoroughly. Filter sterilize. Gas under 80% N_2 + 20% CO_2.

Solution D:
Composition per 10.0mL:

Phenol	0.094g

Preparation of Solution D: Add phenol to distilled/deionized water and bring volume to 10.0mL. Mix thoroughly. Gas under 100% N_2. Filter sterilize.

Solution E (Vitamin Solution):
Composition per 100.0mL

Nicotinic acid amide	35.0mg
Thiamine dichloride	30.0mg
p-Aminobenzoic acid	20.0mg
Biotin	10.0mg
Calcuim pantothenate	10.0mg
Pyridoxal·HCl	10.0mg
Vitamin B_{12}	5.0mg

Preparation of Solution E (Vitamin Solution): Add components to distilled/deionized water and bring volume to 100.0mL. Mix thoroughly. Gas under 100% N_2. Autoclave for 15 min at 15 psi pressure–121°C.

Solution F:
Composition per 10.0mL:

$Na_2S \cdot 9H_2O$	0.4g

Preparation of Solution F: Add $Na_2S \cdot 9H_2O$ to distilled/deionized water and bring volume to 10.0mL. Mix thoroughly. Gas under 100% N_2. Autoclave for 15 min at 15 psi pressure–121°C.

Solution G:
Composition per liter:

NaOH	0.5g
$Na_2WO_4 \cdot 2H_2O$	4.0mg
$Na_2SeO_3 \cdot 5H_2O$	3.0mg

Preparation of Solution G: Add components to distilled/deionized water and bring volume to 1.0L. Mix thoroughly. Gas under 100% N_2. Autoclave for 15 min at 15 psi pressure–121°C.

Preparation of Medium: Aseptically and anaerobically combine solution A with solution B, solution C, solution D, solution E, solution F, and solution G, in that order. Mix thoroughly. Anaerobically distribute into sterile tubes or flasks under 80% N_2 + 20% CO_2

Use: For the cultivation and maintenance of *Desulfobacterium anilini.*

Desulfobacterium catecholicum Medium

Composition per 1001.0mL:

Solution A	870.0mL
Solution C	100.0mL
Solution D	10.0mL
Solution E (Vitamin solution)	10.0mL
Solution F	10.0mL
Solution B (Trace elements solution SL-10)	1.0mL

pH 7.1–7.4 at 25°C

Solution A:
Composition per 870.0mL:

Na_2SO_4	3.0g
NaCl	1.0g
KCl	0.5g
$MgCl_2 \cdot 6H_2O$	0.4g
NH_4Cl	0.3g
KH_2PO_4	0.2g
$CaCl_2 \cdot 2H_2O$	0.15g
Resazurin	1.0mg

Preparation of Solution A: Add components to distilled/deionized water and bring volume to 870.0mL. Mix thoroughly. Gently heat and bring to boiling. Continue boiling for 3-4 min. Allow to cool to room temperature while gassing under 80% N_2 + 20% CO_2. Continue gassing until pH reaches below 6.0. Seal the flask under 80% N_2 + 20% CO_2. Autoclave for 15 min at 15 psi pressure–121°C.

Solution B (Trace Elements Solution SL-10):
Composition per liter:

$FeCl_2 \cdot 4H_2O$	1.5g
$CoCl_2 \cdot 6H_2O$	190.0mg
$MnCl_2 \cdot 4H_2O$	100.0mg
$ZnCl_2$	70.0mg
$Na_2MoO_4 \cdot 2H_2O$	36.0mg
$NiCl_2 \cdot 6H_2O$	24.0mg
H_3BO_3	6.0mg
$CuCl_2 \cdot 2H_2O$	2.0mg
HCl (25% solution)	10.0mL

Preparation of Solution B (Trace Elements Solution SL-10): Add $FeCl_2 \cdot 4H_2O$ to 10.0mL of HCl solution. Mix thoroughly. Add distilled/deionized water and bring volume to 1.0L. Add remaining components. Mix thoroughly. Gas under 100% N_2. Autoclave for 15 min at 15 psi pressure–121°C.

Solution C:
Composition per 100.0mL:

$NaHCO_3$	2.5g

Preparation of Solution C: Add $NaHCO_3$ to distilled/deionized water and bring volume to 100.0mL. Mix thoroughly. Filter sterilize. Gas under 80% N_2 + 20% CO_2.

Solution D:
Composition per 10.0mL:

Sodium benzoate	0.37g
Catechol	0.055g

Preparation of Solution D: Add sodium benzoate and catechol to distilled/deionized water and bring volume to 10.0mL. Mix thoroughly. Gas under 100% N_2. Filter sterilize.

Solution E (Vitamin Solution):
Composition per liter:

Pyridoxine·HCl	10.0mg
Calcium DL-pantothenate	5.0mg
Lipoic acid	5.0mg
Nicotinic acid	5.0mg
p-Aminobenzoic acid	5.0mg

Riboflavin 5.0mg
Thiamine·HCl 5.0mg
Biotin 2.0mg
Folic acid.. 2.0mg
Vitamin B_{12} 0.1mg

Preparation of Solution E (Vitamin Solution): Add components to distilled/deionized water and bring volume to 1.0L. Mix thoroughly. Gas under 100% N_2. Autoclave for 15 min at 15 psi pressure–121°C.

Solution F:
Composition per 10.0mL:
$Na_2S·9H_2O$.. 0.4g

Preparation of Solution F: Add $Na_2S·9H_2O$ to distilled/deionized water and bring volume to 10.0mL. Mix thoroughly. Gas under 100% N_2. Autoclave for 15 min at 15 psi pressure–121°C.

Preparation of Medium: Aseptically and anaerobically combine solution A with solution B, solution C, solution D, solution E, and solution F, in that order. Mix thoroughly. Anaerobically distribute into sterile tubes or flasks under 80% N_2 + 20% CO_2.

Use: For the cultivation and maintenance of *Desulfobacterium catecholicum.*

Desulfobacterium cetonicum Medium

Composition per 1011.0mL:
Solution A.. 950.0mL
Solution C 10.0mL
Solution D.. 40.0mL
Solution E 10.0mL
Solution B (Trace elements solution SL-10)............. 1.0mL
pH 7.2–7.4 at 25°C

Solution A:
Composition per 950.0mL:
NaCl.. 10.0g
Na_2SO_4.. 2.8g
KH_2PO_4.. 0.7g
$MgCl_2·6H_2O$.. 0.3g
NH_4Cl.. 0.3g
$CaCl_2·2H_2O$.. 0.05g

Preparation of Solution A: Add components to distilled/deionized water and bring volume to 950.0mL. Mix thoroughly. Sparge with 80% N_2 + 20% CO_2. Autoclave for 15 min at 15 psi pressure–121°C.

Solution B (Trace Elements Solution SL-10):
Composition per liter:
$FeCl_2·4H_2O$.. 1.5g
$CoCl_2·6H_2O$ 190.0mg
$MnCl_2·4H_2O$.. 100.0mg
$ZnCl_2$.. 70.0mg
$Na_2MoO_4·2H_2O$ 36.0mg
$NiCl_2·6H_2O$.. 24.0mg
H_3BO_3.. 6.0mg
$CuCl_2·2H_2O$ 2.0mg
HCl (25% solution).. 10.0mL

Preparation of Solution B (Trace Elements Solution SL-10): Add $FeCl_2·4H_2O$ to 10.0mL of HCl solution. Mix thoroughly. Add distilled/deionized water and bring volume to 1.0L. Add remaining components. Mix thoroughly. Sparge with 100% N_2. Autoclave for 15 min at 15 psi pressure–121°C.

Solution C:
Composition per 10.0mL:
Sodium butyrate.. 1.2g

Preparation of Solution C: Add sodium butyrate to distilled/deionized water and bring volume to 10.0mL. Mix thoroughly. Sparge with 100% N_2. Autoclave for 15 min at 15 psi pressure–121°C.

Solution D:
Composition per 40.0mL:
$NaHCO_3$.. 2.0g

Preparation of Solution D: Add $NaHCO_3$ to distilled/deionized water and bring volume to 40.0mL. Mix thoroughly. Sparge with 100% N_2. Autoclave for 15 min at 15 psi pressure–121°C.

Solution E:
Composition per 10.0mL:
$Na_2S·9H_2O$.. 0.3g

Preparation of Solution E: Add $Na_2S·9H_2O$ to distilled/deionized water and bring volume to 10.0mL. Mix thoroughly. Sparge with 100% N_2. Autoclave for 15 min at 15 psi pressure–121°C.

Preparation of Medium: Prepare and dispense medium under 80% N_2 + 20% CO_2. Aseptically and anaerobically combine 950.0mL of sterile solution A with 1.0mL of sterile solution B, 10.0mL of sterile solution C, 4.0mL of sterile solution D, and 10.0mL of sterile solution E. Mix thoroughly. Check that final pH is 7.2–7.4.

Use: For the cultivation of *Desulfobacterium cetonicum.*

Desulfobacterium indolicum Medium

Composition per 1002.4mL:
Solution A 900.0mL
Solution C.. 50.0mL
Solution D 30.0mL
Solution E (Wolfe's vitamin solution).................... 10.0mL
Solution G 10.0mL
Solution B (Trace elements solution SL-10)............ 1.0mL
Solution F 1.0mL
Solution H 0.4mL
pH 7.6 ± 0.2 at 25°C

Solution A:
Composition per 900.0mL:
NaCl 21.0g
$MgCl_2·6H_2O$.. 3.0g
Na_2SO_4 .... 3.0g
KCl.. 0.5g
NH_4Cl.. 0.3g
KH_2PO_4 .... 0.2g
$CaCl_2·2H_2O$ 0.15g
Resazurin.. 1.0mg

Preparation of Solution A: Prepare and dispense solution anaerobically under 80% N_2 + 20% CO_2. Add components to distilled/deionized water and bring volume to 900.0mL. Mix thoroughly. Gently heat and bring to boiling. Continue boiling until resazurin turns colorless, indicating reduction. Cap with rubber stoppers. Autoclave for 15 min at 15 psi pressure–121°C. Cool to 45°–50°C.

Solution B (Trace Elements Solution SL-10):
Composition per liter:
$FeCl_2·4H_2O$.. 1.5g
$CoCl_2·6H_2O$ 0.19g
$MnCl_2·4H_2O$ 0.1g
$ZnCl_2$ 0.07g

$Na_2MoO_4 \cdot 2H_2O$ 0.036g
$NiCl_2 \cdot 6H_2O$ 0.024g
H_3BO_3 .... 6.0mg
$CuCl_2 \cdot 2H_2O$ 2.0mg
HCl (25% solution) 10.0mL

Preparation of Solution B (Trace Elements Solution SL-10): Add the $FeCl_2 \cdot 4H_2O$ to 10.0mL of HCl solution. Mix thoroughly. Bring volume to approximately 900.0mL with distilled/deionized water. Mix thoroughly. Adjust pH to 6.0 with NaOH. Bring volume to 1.0L with distilled/deionized water. Filter sterilize. Aseptically gas under 100% N_2 for 20 min.

Solution C:
Composition per 50.0mL:
$NaHCO_3$ 2.5g

Preparation of Solution C: Add $NaHCO_3$ to distilled/deionized water and bring volume to 50.0mL. Mix thoroughly. Filter sterilize. Aseptically gas under 80% N_2 + 20% CO_2 for 20 min.

Solution D:
Composition per 107.7mL:
Indole 0.3g
NaCl (30% solution) 7.0mL
$MgCl_2 \cdot 6H_2O$ (40% solution) 0.7mL

Preparation of Solution D: Prepare and dispense all solutions anaerobically under 100% N_2. Add indole to distilled/deionized water and bring volume to 100.0mL. Mix thoroughly. Gently heat while stirring until dissolved. Prepare the NaCl solution and the $MgCl_2 \cdot 6H_2O$ solution separately. Autoclave the three solutions separately for 15 min at 15 psi pressure–121°C. Cool to 45°–50°C. To 100.0mL of sterile indole solution, aseptically and anaerobically add 7.0mL of sterile NaCl solution and 0.7mL of sterile $MgCl_2 \cdot 6H_2O$ solution. Mix thoroughly.

Solution E (Wolfe's Vitamin Solution):
Composition per liter:
Pyridoxine·HCl 0.01g
Thiamine·HCl 5.0mg
Riboflavin 5.0mg
Nicotinic acid 5.0mg
Calcium pantothenate 5.0mg
p-Aminobenzoic acid 5.0mg
Thioctic acid 5.0mg
Biotin 2.0mg
Folic acid 2.0mg
Cyanocobalamin 0.1mg

Preparation of Solution E (Wolfe's Vitamin Solution): Add components to distilled/deionized water and bring volume to 1.0L. Mix thoroughly. Filter sterilize. Aseptically gas under 100% N_2 for 20 min.

Solution F:
Composition per liter:
$Na_2SeO_3 \cdot 5H_2O$ 3.0mg
NaOH (0.01*M* solution) 1.0L

Preparation of Solution F: Add $Na_2SeO_3 \cdot 5H_2O$ to 1.0L of NaOH solution. Mix thoroughly. Filter sterilize. Aseptically gas under 100% N_2 for 20 min.

Solution G:
Composition per 10.0mL:
$Na_2S \cdot 9H_2O$ 0.4g

Preparation of Solution G: Add $Na_2S \cdot 9H_2O$ to distilled/deionized water and bring volume to 10.0mL. Gas under 100% N_2 for 20 min. Cap with a rubber stopper. Autoclave for 15 min at 15 psi pressure–121°C. Cool to 25°C.

Solution H:
Composition per 10.0mL:
$Na_2S_2O_4$ 0.5g

Preparation of Solution H: Add $Na_2S_2O_4$ to distilled/deionized water and bring volume to 10.0mL. Mix thoroughly. Filter sterilize. Aseptically gas under 100% N_2 for 20 min. Prepare solution freshly.

Preparation of Medium: To 900.0mL of cooled, sterile solution A, aseptically and anaerobically add in the following order: 1.0mL of sterile solution B, 50.0mL of sterile solution C, 10.0mL of sterile solution E, 1.0mL of sterile solution F, and 10.0mL of sterile solution G. Mix thoroughly. Immediately prior to inoculation, aseptically and anaerobically add 30.0mL of sterile solution D and 0.4mL of sterile solution H. Mix thoroughly. Aseptically and anaerobically distribute into sterile tubes or flasks.

Use: For the cultivation and maintenance of *Desulfobacterium indolicum*.

Desulfobacterium Medium

Composition per 1002.4mL:
Solution A 930.0mL
Solution C 50.0mL
Solution D (Wolfe's vitamin solution) 10.0mL
Solution F 10.0mL
Solution B (Trace elements solution SL-10) 1.0mL
Solution E 1.0mL
Solution G 0.4mL
pH 7.0 ± 0.2 at 25°C

Solution A:
Composition per 930.0mL:
NaCl 21.0g
$MgCl_2 \cdot 6H_2O$ 3.0g
Na_2SO_4 .... 3.0g
KCl 0.5g
NH_4Cl 0.3g
KH_2PO_4 .... 0.2g
$CaCl_2 \cdot 2H_2O$ 0.15g
Resazurin 1.0mg

Preparation of Solution A: Prepare and dispense solution anaerobically under 80% N_2 + 20% CO_2. Add components to distilled/deionized water and bring volume to 930.0mL. Mix thoroughly. Gently heat and bring to boiling. Continue boiling until resazurin turns colorless, indicating reduction. Cap with rubber stoppers. Autoclave for 15 min at 15 psi pressure–121°C. Cool to 45°–50°C.

Solution B (Trace Elements Solution SL-10):
Composition per liter:
$FeCl_2 \cdot 4H_2O$ 1.5g
$CoCl_2 \cdot 6H_2O$ 0.19g
$MnCl_2 \cdot 4H_2O$ 0.1g
$ZnCl_2$ 0.07g
$Na_2MoO_4 \cdot 2H_2O$ 0.036g
$NiCl_2 \cdot 6H_2O$ 0.024g
H_3BO_3 .... 6.0mg
$CuCl_2 \cdot 2H_2O$ 2.0mg
HCl (25% solution) 10.0mL

Preparation of Solution B (Trace Elements Solution SL-10): Add the $FeCl_2 \cdot 4H_2O$ to 10.0mL of HCl solution. Mix thoroughly. Bring volume to approximately 900.0mL with distilled/deionized water. Mix thoroughly. Adjust pH to 6.0 with NaOH. Bring volume to 1.0L with distilled/deionized water. Filter sterilize. Aseptically gas under 100% N_2 for 20 min.

Solution C:
Composition per 50.0mL:
$NaHCO_3$ 2.5g

Preparation of Solution C: Add $NaHCO_3$ to distilled/deionized water and bring volume to 50.0mL. Mix thoroughly. Filter sterilize. Aseptically gas under 80% N_2 + 20% CO_2 for 20 min.

Solution D (Wolfe's Vitamin Solution):
Composition per liter:
Pyridoxine·HCl 0.01g
Thiamine·HCl 5.0mg
Riboflavin 5.0mg
Nicotinic acid 5.0mg
Calcium pantothenate 5.0mg
p-Aminobenzoic acid 5.0mg
Thioctic acid 5.0mg
Biotin 2.0mg
Folic acid 2.0mg
Cyanocobalamin 0.1mg

Preparation of Solution D (Wolfe's Vitamin Solution): Add components to distilled/deionized water and bring volume to 1.0L. Mix thoroughly. Filter sterilize. Aseptically gas under 100% N_2 for 20 min.

Solution E:
Composition per liter:
$Na_2SeO_3 \cdot 5H_2O$ 3.0mg
NaOH (0.01*M* solution) 1.0L

Preparation of Solution E: Add $Na_2SeO_3 \cdot 5H_2O$ to 1.0L of NaOH solution. Mix thoroughly. Filter sterilize. Aseptically gas under 100% N_2 for 20 min.

Solution F:
Composition per 10.0mL:
$Na_2S \cdot 9H_2O$ 0.4g

Preparation of Solution F: Add $Na_2S \cdot 9H_2O$ to distilled/deionized water and bring volume to 10.0mL. Gas under 100% N_2 for 20 min. Cap with a rubber stopper. Autoclave for 15 min at 15 psi pressure–121°C. Cool to 25°C.

Solution G:
Composition per 10.0mL:
$Na_2S_2O_4$ 0.5g

Preparation of Solution G: Add $Na_2S_2O_4$ to distilled/deionized water and bring volume to 10.0mL. Mix thoroughly. Filter sterilize. Aseptically gas under 100% N_2 for 20 min. Prepare solution freshly.

Preparation of Medium: To 900.0mL of cooled, sterile solution A, aseptically and anaerobically add in the following order: 1.0mL of sterile solution B, 50.0mL of sterile solution C, 10.0mL of sterile solution D, 1.0mL of sterile solution E, and 10.0mL of sterile solution F. Mix thoroughly. Immediately prior to inoculation, aseptically and anaerobically add 0.4mL of sterile solution G. Mix thoroughly. Aseptically and anaerobically distribute into sterile tubes or flasks.

Use: For the cultivation and maintenance of *Desulfobacterium autotrophicum*.

Desulfobacterium Medium with Lactate

Composition per 1002.4mL:
Solution A 930.0mL
Solution C 50.0mL
Solution D (Wolfe's vitamin solution) 10.0mL
Solution F 10.0mL
Solution B (Trace elements solution SL-10) 1.0mL
Solution E 1.0mL
Solution G 0.4mL
pH 7.0 ± 0.2 at 25°C

Solution A:
Composition per 930.0mL:
NaCl 21.0g
$MgCl_2 \cdot 6H_2O$ 3.0g
Na_2SO_4 .... 3.0g
Lactic acid, sodium salt 1.1g
KCl 0.5g
NH_4Cl 0.3g
KH_2PO_4 .... 0.2g
$CaCl_2 \cdot 2H_2O$ 0.15g
Resazurin 1.0mg

Preparation of Solution A: Prepare and dispense solution anaerobically under 80% N_2 + 20% CO_2. Add components to distilled/deionized water and bring volume to 930.0mL. Mix thoroughly. Gently heat and bring to boiling. Continue boiling until resazurin turns colorless, indicating reduction. Cap with rubber stoppers. Autoclave for 15 min at 15 psi pressure–121°C. Cool to 45°–50°C.

Solution B (Trace Elements Solution SL-10):
Composition per liter:
$FeCl_2 \cdot 4H_2O$ 1.5g
$CoCl_2 \cdot 6H_2O$ 0.19g
$MnCl_2 \cdot 4H_2O$ 0.10g
$ZnCl_2$ 0.070g
$Na_2MoO_4 \cdot 2H_2O$ 0.036g
$NiCl_2 \cdot 6H_2O$ 0.024g
H_3BO_3 .... 6.0mg
$CuCl_2 \cdot 2H_2O$ 2.0mg
HCl (25% solution) 10.0mL

Preparation of Solution B (Trace Elements Solution SL-10): Add the $FeCl_2 \cdot 4H_2O$ to 10.0mL of HCl solution. Mix thoroughly. Bring volume to approximately 900.0mL with distilled/deionized water. Mix thoroughly. Adjust pH to 6.0 with NaOH. Bring volume to 1.0L with distilled/deionized water. Filter sterilize. Aseptically gas under 100% N_2 for 20 min.

Solution C:
Composition per 50.0mL:
$NaHCO_3$ 2.5g

Preparation of Solution C: Add $NaHCO_3$ to distilled/deionized water and bring volume to 50.0mL. Mix thoroughly. Filter sterilize. Aseptically gas under 80% N_2 + 20% CO_2 for 20 min.

Solution D (Wolfe's Vitamin Solution):
Composition per liter:
Pyridoxine·HCl 0.01g
Thiamine·HCl 5.0mg
Riboflavin 5.0mg
Nicotinic acid 5.0mg
Calcium pantothenate 5.0mg
p-Aminobenzoic acid 5.0mg
Thioctic acid 5.0mg
Biotin 2.0mg
Folic acid 2.0mg
Cyanocobalamin 0.1mg

Preparation of Solution D (Wolfe's Vitamin Solution): Add components to distilled/deionized water and bring volume to 1.0L. Mix thoroughly. Filter sterilize. Aseptically gas under 100% N_2 for 20 min.

Solution E:
Composition per liter:

$Na_2SeO_3 \cdot 5H_2O$	3.0mg
NaOH (0.01*M* solution)	1.0L

Preparation of Solution E: Add $Na_2SeO_3 \cdot 5H_2O$ to 1.0L of NaOH solution. Mix thoroughly. Filter sterilize. Aseptically gas under 100% N_2 for 20 min.

Solution F:
Composition per 10.0mL:

$Na_2S \cdot 9H_2O$	0.4g

Preparation of Solution F: Add $Na_2S \cdot 9H_2O$ to distilled/deionized water and bring volume to 10.0mL. Gas under 100% N_2 for 20 min. Cap with a rubber stopper. Autoclave for 15 min at 15 psi pressure–121°C. Cool to 25°C.

Solution G:
Composition per 10.0mL:

$Na_2S_2O_4$	0.5g

Preparation of Solution G: Add $Na_2S_2O_4$ to distilled/deionized water and bring volume to 10.0mL. Mix thoroughly. Filter sterilize. Aseptically gas under 100% N_2 for 20 min. Prepare solution freshly.

Preparation of Medium: To 900.0mL of cooled, sterile solution A, aseptically and anaerobically add, in the following order, 1.0mL of sterile solution B, 50.0mL of sterile solution C, 10.0mL of sterile solution D, 1.0mL of sterile solution E, and 10.0mL of sterile solution F. Mix thoroughly. Immediately prior to inoculation, aseptically and anaerobically add 0.4mL of sterile solution G. Mix thoroughly. Aseptically and anaerobically distribute into sterile tubes or flasks.

Use: For the cultivation of *Desulfobacterium autotrophicum.*

Desulfobacterium Medium, Modified

Composition per 1002.4mL:

Solution A	920.0mL
Solution C	50.0mL
Solution D	10.0mL
Solution E (Wolfe's vitamin solution)	10.0mL
Solution G	10.0mL
Solution B (Trace elements solution SL-10)	1.0mL
Solution F	1.0mL
Solution H	0.4mL

pH 7.0 ± 0.2 at 25°C

Solution A:
Composition per 920.0mL:

NaCl	21.0g
$MgCl_2 \cdot 6H_2O$	3.0g
Na_2SO_4	3.0g
KCl	0.5g
NH_4Cl	0.3g
KH_2PO_4	0.2g
$CaCl_2 \cdot 2H_2O$	0.15g
Resazurin	1.0mg

Preparation of Solution A: Prepare and dispense solution anaerobically under 80% N_2 + 20% CO_2. Add components to distilled/deionized water and bring volume to 920.0mL. Mix thoroughly. Gently heat and bring to boiling. Continue boiling until resazurin turns colorless, indicating reduction. Cap with rubber stoppers. Autoclave for 15 min at 15 psi pressure–121°C. Cool to 45°–50°C.

Solution B (Trace Elements Solution SL-10):
Composition per liter:

$FeCl_2 \cdot 4H_2O$	1.5g
$CoCl_2 \cdot 6H_2O$	0.19g
$MnCl_2 \cdot 4H_2O$	0.1g
$ZnCl_2$	0.07g
$Na_2MoO_4 \cdot 2H_2O$	0.036g
$NiCl_2 \cdot 6H_2O$	0.024g
H_3BO_3	6.0mg
$CuCl_2 \cdot 2H_2O$	2.0mg
HCl (25% solution)	10.0mL

Preparation of Solution B (Trace Elements Solution SL-10): Add the $FeCl_2 \cdot 4H_2O$ to 10.0mL of HCl solution. Mix thoroughly. Bring volume to approximately 900.0mL with distilled/deionized water. Mix thoroughly. Adjust pH to 6.0 with NaOH. Bring volume to 1.0L with distilled/deionized water. Filter sterilize. Aseptically gas under 100% N_2 for 20 min.

Solution C:
Composition per 50.0mL:

$NaHCO_3$	2.5g

Preparation of Solution C: Add $NaHCO_3$ to distilled/deionized water and bring volume to 50.0mL. Mix thoroughly. Filter sterilize. Aseptically gas under 80% N_2 + 20% CO_2 for 20 min.

Solution D:
Composition per 10.0mL:

Sodium acetate·$3H_2O$	2.5g

Preparation of Solution D: Prepare and dispense solution anaerobically under 80% N_2 + 20% CO_2. Add sodium acetate to distilled/deionized water and bring volume to 10.0mL. Mix thoroughly. Cap with rubber stopper. Autoclave for 15 min at 15 psi pressure–121°C. Cool to 45°–50°C.

Solution E (Wolfe's Vitamin Solution):
Composition per liter:

Pyridoxine·HCl	0.01g
Thiamine·HCl	5.0mg
Riboflavin	5.0mg
Nicotinic acid	5.0mg
Calcium pantothenate	5.0mg
p-Aminobenzoic acid	5.0mg
Thioctic acid	5.0mg
Biotin	2.0mg
Folic acid	2.0mg
Cyanocobalamin	0.1mg

Preparation of Solution E (Wolfe's Vitamin Solution): Add components to distilled/deionized water and bring volume to 1.0L. Mix thoroughly. Filter sterilize. Aseptically gas under 100% N_2 for 20 min.

Solution F:
Composition per liter:

$Na_2SeO_3 \cdot 5H_2O$	3.0mg
NaOH (0.01*M* solution)	1.0L

Preparation of Solution F: Add $Na_2SeO_3 \cdot 5H_2O$ to 1.0L of NaOH solution. Mix thoroughly. Filter sterilize. Aseptically gas under 100% N_2 for 20 min.

Solution G:
Composition per 10.0mL:

$Na_2S \cdot 9H_2O$	0.4g

Preparation of Solution G: Add $Na_2S·9H_2O$ to distilled/deionized water and bring volume to 10.0mL. Gas under 100% N_2 for 20 min. Cap with a rubber stopper. Autoclave for 15 min at 15 psi pressure–121°C. Cool to 25°C.

Solution H:

Composition per 10.0mL:

$Na_2S_2O_4$ 0.5g

Preparation of Solution H: Add $Na_2S_2O_4$ to distilled/deionized water and bring volume to 10.0mL. Mix thoroughly. Filter sterilize. Aseptically gas under 100% N_2 for 20 min. Prepare solution freshly.

Preparation of Medium: To 920.0mL of cooled, sterile solution A, aseptically and anaerobically add in the following order: 1.0mL of sterile solution B, 50.0mL of sterile solution C, 10.0mL of sterile solution D, 10.0mL of sterile solution E, 1.0mL of sterile solution F, and 10.0mL of sterile solution G. Mix thoroughly. Immediately prior to inoculation, aseptically and anaerobically add 0.4mL of sterile solution H. Mix thoroughly. Aseptically and anaerobically distribute into sterile tubes or flasks.

Use: For the cultivation and maintenance of *Desulfobacter curvatus* and *Desulfobacter latus*.

Desulfobacterium oleovorans Medium

Composition per 1154.0mL:

Solution A 1.0L
Solution H 67.0mL
Solution D 50.0mL
Solution I 13.0mL
Solution E 10.0mL
Solution G 10.0mL
Solution C (Selenite-tungstate solution) 2.0mL
Solution B (Trace elements solution SL-10) 1.0mL
Solution F 1.0mL

pH 7.2 ± 0.2 at 25°C

Solution A:

Composition per liter:

Na_2SO_4 .... 4.0g
NaCl 1.0g
$MgCl_2·6H_2O$ 0.4g
NH_4Cl 0.25g
KH_2PO_4 .... 0.2g
$CaCl_2·2H_2O$ 0.1g

Preparation of Solution A: Add components to distilled/deionized water and bring volume to 1.0L. Mix thoroughly. Gently heat and bring to boiling. Continue boiling for 3-4 min. Allow to cool to room temperature while gassing under 80% N_2 + 20% CO_2. Continue gassing until pH reaches below 6.0. Seal the flask under 80% N_2 + 20% CO_2. Autoclave for 15 min at 15 psi pressure–121°C.

Solution B (Trace Elements Solution SL-10):

Composition per liter:

$FeCl_2·4H_2O$ 1.5g
$CoCl_2·6H_2O$ 190.0mg
$MnCl_2·4H_2O$ 100.0mg
$ZnCl_2$ 70.0mg
$Na_2MoO_4·2H_2O$ 36.0mg
$NiCl_2·6H_2O$ 24.0mg
H_3BO_3 .... 6.0mg
$CuCl_2·2H_2O$ 2.0mg
HCl (25% solution) 10.0mL

Preparation of Solution B (Trace Elements Solution SL-10): Add $FeCl_2·4H_2O$ to 10.0mL of HCl solution. Mix thoroughly. Add distilled/deionized water and bring volume to 1.0L. Add remaining components. Mix thoroughly. Gas under 100% N_2. Autoclave for 15 min at 15 psi pressure–121°C.

Solution C (Selenite-Tungstate Solution):

Composition per liter:

NaOH 0.5g
$Na_2WO_4·2H_2O$ 4.0mg
$Na_2SeO_3·5H_2O$ 3.0mg

Preparation of Solution C (Selenite-Tungstate Solution): Add components to distilled/deionized water and bring volume to 1.0L. Mix thoroughly. Gas under 100% N_2. Autoclave for 15 min at 15 psi pressure–121°C.

Solution D:

Composition per 50.0mL:

$NaHCO_3$ 2.5g

Preparation of Solution D: Add $NaHCO_3$ to distilled/deionized water and bring volume to 50.0mL. Mix thoroughly. Gas under 80% N_2 + 20% CO_2. Autoclave for 15 min at 15 psi pressure–121°C.

Solution E:

Composition per liter:

Pyridoxine·HCl 10.0mg
Calcium DL-pantothenate 5.0mg
Lipoic acid 5.0mg
Nicotinic acid 5.0mg
p-Aminobenzoic acid 5.0mg
Riboflavin 5.0mg
Thiamine·HCl 5.0mg
Biotin 2.0mg
Folic acid 2.0mg
Vitamin B_{12} 0.1mg

Preparation of Solution E: Add components to distilled/deionized water and bring volume to 1.0L. Mix thoroughly. Gas under 100% N_2. Filter sterilize.

Solution F:

Composition per 10.0mL:

Vitamin B_{12} 0.5mg

Preparation of Solution F: Add Vitamin B_{12} to distilled/deionized water and bring volume to 10.0mL. Mix thoroughly. Gas under 100% N_2. Filter sterilize.

Solution G:

Composition per 80.0mL:

Stearic acid 2.85g
NaOH (4.0*M* solution) 2.5mL

Preparation of Solution G: Add components to distilled/deionized water and bring volume to 80.0mL. Mix thoroughly. Gas under 100% N_2. In a closed bottle, heat in a boiling water bath. Shake until stearic acid dissolves. Autoclave for 15 min at 15 psi pressure–121°C. On storage, solution will solidify and should be remelted before use.

Solution H:

Composition per liter:

NaCl 286.4g
$MgCl_2·6H_2O$ 44.7g
$CaCl_2·2H_2O$ 2.2g

Preparation of Solution H: Add components to distilled/deionized water and bring volume to 80.0mL. Mix thoroughly. Gas under 100% N_2. Autoclave for 15 min at 15 psi pressure–121°C.

Solution I:
Composition per 20.0mL:
$Na_2S{\cdot}9H_2O$ 0.6g

Preparation of Solution I: Add $Na_2S{\cdot}9H_2O$ to distilled/deionized water and bring volume to 10.0mL. Mix thoroughly. Gas under 100% N_2. Autoclave for 15 min at 15 psi pressure–121°C.

Preparation of Medium: To 1.0L of sterile solution A, add in order: 1.0mL of sterile solution B, 2.0mL of sterile solution C, 50.0mL of sterile solution D, 10.0mL of sterile solution E, 1.0mL of sterile solution F, 10.0mL of sterile solution G, 67.0mL of sterile solution H, and 13.0mL of sterile solution I. Mix thoroughly. Final pH of the medium should be 7.2. Prior to inoculation, add 10.0–20.0mg of sodium dithionate to 1.0L of medium.

Use: For the cultivation and maintenance of *Desulfobacterium oleovorans.*

Desulfobacterium phenolicum Medium

Composition per 1002.4mL:
Solution A 930.0mL
Solution C 50.0mL
Solution E (Wolfe's vitamin solution) 10.0mL
Solution G 10.0mL
Solution D 4.0mL
Solution B (Trace elements solution SL-10) 1.0mL
Solution F 1.0mL
Solution H 0.4mL
pH 7.0 ± 0.2 at 25°C

Solution A:
Composition per 920.0mL:
NaCl 21.0g
$MgCl_2{\cdot}6H_2O$ 3.0g
Na_2SO_4 .... 3.0g
KCl 0.5g
NH_4Cl 0.3g
KH_2PO_4 .... 0.2g
$CaCl_2{\cdot}2H_2O$ 0.15g
Resazurin 1.0mg

Preparation of Solution A: Prepare and dispense solution anaerobically under 80% N_2 + 20% CO_2. Add components to distilled/deionized water and bring volume to 920.0mL. Mix thoroughly. Gently heat and bring to boiling. Continue boiling until resazurin turns colorless, indicating reduction. Cap with rubber stoppers. Autoclave for 15 min at 15 psi pressure–121°C. Cool to 45°–50°C.

Solution B (Trace Elements Solution SL-10):
Composition per liter:
$FeCl_2{\cdot}4H_2O$ 1.5g
$CoCl_2{\cdot}6H_2O$ 0.19g
$MnCl_2{\cdot}4H_2O$ 0.1g
$ZnCl_2$ 0.07g
$Na_2MoO_4{\cdot}2H_2O$ 0.036g
$NiCl_2{\cdot}6H_2O$ 0.024g
H_3BO_3 .... 6.0mg
$CuCl_2{\cdot}2H_2O$ 2.0mg
HCl (25% solution) 10.0mL

Preparation of Solution B (Trace Elements Solution SL-10): Add $FeCl_2{\cdot}4H_2O$ to 10.0mL of HCl solution. Mix thoroughly. Bring volume to 900.0mL with distilled/deionized water. Mix thoroughly. Adjust pH to 6.0 with NaOH. Bring volume to 1.0L with distilled/deionized water. Filter sterilize. Aseptically gas under 100% N_2 for 20 min.

Solution C:
Composition per 50.0mL:
$NaHCO_3$ 2.5g

Preparation of Solution C: Add $NaHCO_3$ to distilled/deionized water and bring volume to 50.0mL. Mix thoroughly. Filter sterilize. Aseptically gas under 80% N_2 + 20% CO_2 for 20 min.

Solution D:
Composition per 10.0mL:
Sodium benzoate 1.0g
Phenol 0.1g

Preparation of Solution D: Add components to distilled/deionized water and bring volume to 10.0mL. Mix thoroughly. Filter sterilize. Aseptically gas under 100% N_2 for 20 min.

Solution E (Wolfe's Vitamin Solution):
Composition per liter:
Pyridoxine·HCl 0.01g
Thiamine·HCl 5.0mg
Riboflavin 5.0mg
Nicotinic acid 5.0mg
Calcium pantothenate 5.0mg
p-Aminobenzoic acid 5.0mg
Thioctic acid 5.0mg
Biotin 2.0mg
Folic acid 2.0mg
Cyanocobalamin 0.1mg

Preparation of Solution E (Wolfe's Vitamin Solution): Add components to distilled/deionized water and bring volume to 1.0L. Mix thoroughly. Filter sterilize. Aseptically gas under 100% N_2 for 20 min.

Solution F:
Composition per liter:
$Na_2SeO_3{\cdot}5H_2O$ 3.0mg
NaOH (0.01*M* solution) 1.0L

Preparation of Solution F: Add $Na_2SeO_3{\cdot}5H_2O$ to 1.0L of NaOH solution. Mix thoroughly. Filter sterilize. Aseptically gas under 100% N_2 for 20 min.

Solution G:
Composition per 10.0mL:
$Na_2S{\cdot}9H_2O$ 0.4g

Preparation of Solution G: Add $Na_2S{\cdot}9H_2O$ to distilled/deionized water and bring volume to 10.0mL. Gas under 100% N_2 for 20 min. Cap with a rubber stopper. Autoclave for 15 min at 15 psi pressure–121°C. Cool to 25°C.

Solution H:
Composition per 10.0mL:
$Na_2S_2O_4$ 0.5g

Preparation of Solution H: Add $Na_2S_2O_4$ to distilled/deionized water and bring volume to 10.0mL. Mix thoroughly. Filter sterilize. Aseptically gas under 100% N_2 for 20 min. Prepare solution freshly.

Preparation of Medium: To 920.0mL of cooled, sterile solution A, aseptically and anaerobically add in the following order: 1.0mL of sterile solution B, 50.0mL of sterile solution C, 10.0mL of sterile solution D, 10.0mL of sterile solution E, 1.0mL of sterile solution F, and 10.0mL of sterile solution G. Mix thoroughly. Immediately prior to inoculation aseptically and anaerobically add 0.4mL of sterile solution H. Mix thoroughly. Aseptically and anaerobically distribute into sterile tubes or flasks.

Use: For the cultivation and maintenance of *Desulfobacterium phenolicum.*

Desulfobacula toluolica Medium (DSMZ Medium 383b)

Composition per 1013.5mL:

Solution A	930.0mL
Solution C	50.0mL
Solution D	10.0mL
Solution E	10.0mL
Solution G	10.0mL
Solution B	1.0mL
Solution F	1.0mL
Selenite-tungstate solution	1.0mL
Vitamin B_{12} solution	0.5mL

pH 7.0 at 25°C

Solution A:

Composition per 930.0mL:

NaCl	21.0g
Na_2SO_4	3.0g
$MgCl_2·6H_2O$	3.0g
KCl	0.5g
NH_4Cl	0.3g
KH_2PO_4	0.2g
$CaCl_2·2H_2O$	0.15g
Resazurin	1.0mg

Preparation of Solution A: Add components to distilled/deionized water and bring volume to 930.0mL. Mix thoroughly. Sparge with 80% N_2 + 20% CO_2 gas until saturated. Autoclave for 15 min at 15 psi pressure–121°C. Cool to 25°C.

Solution B:

Composition per liter:

Na-EDTA	5.2g
$FeCl_2·4H_2O$	1.5g
H_3BO_3	300.0mg
$CoCl_2·6H_2O$	190.0mg
$MnCl_2·4H_2O$	100.0mg
$ZnCl_2$	70.0mg
$Na_2MoO_4·2H_2O$	36.0mg
$NiCl_2·6H_2O$	24.0mg
$CuCl_2·2H_2O$	2.0mg
HCl (25% solution)	7.7mL

Preparation of Solution B: Add $FeCl_2·4H_2O$ to 7.7mL of HCl solution. Mix thoroughly. Add distilled/deionized water and bring volume to 1.0L. Add remaining components. Mix thoroughly. Adjust pH to 6.0. Sparge with 100% N_2. Autoclave for 15 min at 15 psi pressure–121°C.

Solution C:

Composition per 100.0mL:

$NaHCO_3$	5.0g

Preparation of Solution C: Add $NaHCO_3$ to distilled/deionized water and bring volume to 100.0mL. Mix thoroughly. Sparge with 100% CO_2 until saturated, approximately 20 minutes. Filter sterilize under 100% CO_2 into a sterile, gas-tight 100.0mL screw-capped bottle.

Solution D:

Composition per 10.0mL:

Na-benzoate	0.4g

Preparation of Solution D: Add Na_2-benzoate to distilled/deionized water and bring volume to 10.0mL. Sparge with N_2. Filter sterilize. Store anaerobically.

Solution E:

Composition per liter:

Pyridoxine-HCl	10.0mg
Thiamine-HCl·$2H_2O$	5.0mg
Riboflavin	5.0mg
Nicotinic acid	5.0mg
D-Ca-pantothenate	5.0mg
p-Aminobenzoic acid	5.0mg
Lipoic acid	5.0mg
Biotin	2.0mg
Folic acid	2.0mg
Vitamin B_{12}	0.1mg

Preparation of Solution E: Add components to distilled/deionized water and bring volume to 1.0L. Mix thoroughly. Sparge with 100% N_2. Filter sterilize.

Solution F:

Composition per liter:

NaOH	0.5g
$Na_2SeO_3·5H_2O$	3.0mg

Preparation of Solution F: Add components to distilled/deionized water and bring volume to 1.0L. Mix thoroughly. Sparge with 100% N_2. Filter sterilize.

Solution G:

Composition per 10.0mL:

$Na_2S·9H_2O$	0.4g

Preparation of Solution G: Add $Na_2S·9H_2O$ to distilled/deionized water and bring volume to 10.0mL. Sparge with N_2. Autoclave for 15 min at 15 psi pressure–121°C. Cool to 25°C. Store anaerobically.

Vitamin B_{12} Solution:

Composition per 100.0mL:

Vitamin B_{12}	10.0mg

Vitamin B_{12} Solution: Add Vitamin B_{12} to distilled/deionized water and bring volume to 100.0mL. Mix thoroughly. Sparge under 100% N_2 gas for 3 min. Filter sterilize.

Selenite-Tungstate Solution:

Composition per liter:

NaOH	0.5g
$Na_2WO_4·2H_2O$	4.0mg
$Na_2SeO_3·5H_2O$	3.0mg

Preparation of Selenite-Tungstate Solution: Add components to distilled/deionized water and bring volume to 1.0L. Mix thoroughly. Sparge with 100% N_2. Filter sterilize.

Preparation of Medium: Add solution B, solution C, solution D, solution E, Vitamin $B_{12,}$ solution F, selenite-tungstate solution, and solution G to solution A in that order under N_2 gas. Adjust the pH to 7.0.

Use: For the cultivation of *Desulfobacula toluolica.*

Desulfobulbus Medium

Composition per 1001.0mL:

Solution A	870.0mL
Solution C	100.0mL
Solution D	10.0mL
Solution E (Vitamin solution)	10.0mL
Solution F	10.0mL
Solution B (Trace elements solution SL-10)	1.0mL

pH 7.1–7.4 at 25°C

Solution A:
Composition per 870.0mL:

Na_2SO_4	3.0g
NaCl	1.0g
KCl	0.5g
$MgCl_2 \cdot 6H_2O$	0.4g
NH_4Cl	0.3g
KH_2PO_4	0.2g
$CaCl_2 \cdot 2H_2O$	0.15g
Resazurin	1.0mg

Preparation of Solution A: Add components to distilled/deionized water and bring volume to 870.0mL. Mix thoroughly. Gently heat and bring to boiling. Continue boiling for 3-4 min. Allow to cool to room temperature while gassing under 80% N_2 + 20% CO_2. Continue gassing until pH reaches below 6.0. Seal the flask under 80% N_2 + 20% CO_2. Autoclave for 15 min at 15 psi pressure–121°C.

Solution B (Trace Elements Solution SL-10):
Composition per liter:

$FeCl_2 \cdot 4H_2O$	1.5g
$CoCl_2 \cdot 6H_2O$	190.0mg
$MnCl_2 \cdot 4H_2O$	100.0mg
$ZnCl_2$	70.0mg
$Na_2MoO_4 \cdot 2H_2O$	36.0mg
$NiCl_2 \cdot 6H_2O$	24.0mg
H_3BO_3	6.0mg
$CuCl_2 \cdot 2H_2O$	2.0mg
HCl (25% solution)	10.0mL

Preparation of Solution B (Trace Elements Solution SL-10): Add $FeCl_2 \cdot 4H_2O$ to 10.0mL of HCl solution. Mix thoroughly. Add distilled/deionized water and bring volume to 1.0L. Add remaining components. Mix thoroughly. Gas under 100% N_2. Autoclave for 15 min at 15 psi pressure–121°C.

Solution C:
Composition per 100.0mL:

$NaHCO_3$	5.0g

Preparation of Solution C: Add $NaHCO_3$ to distilled/deionized water and bring volume to 100.0mL. Mix thoroughly. Filter sterilize. Gas under 80% N_2 + 20% CO_2.

Solution D:
Composition per 10.0mL:

Sodium propionate	2.5g

Preparation of Solution D: Add sodium propionate to distilled/deionized water and bring volume to 10.0mL. Mix thoroughly. Gas under 100% N_2. Autoclave for 15 min at 15 psi pressure–121°C.

Solution E (Vitamin Solution):
Composition per liter:

Pyridoxine·HCl	10.0mg
Calcium DL-pantothenate	5.0mg
Lipoic acid	5.0mg
Nicotinic acid	5.0mg
p-Aminobenzoic acid	5.0mg
Riboflavin	5.0mg
Thiamine·HCl	5.0mg
Biotin	2.0mg
Folic acid	2.0mg
Vitamin B_{12}	0.1mg

Preparation of Solution E (Vitamin Solution): Add components to distilled/deionized water and bring volume to 1.0L. Mix thoroughly. Gas under 100% N_2. Autoclave for 15 min at 15 psi pressure–121°C.

Solution F:
Composition per 10.0mL:

$Na_2S \cdot 9H_2O$	0.4g

Preparation of Solution F: Add $Na_2S \cdot 9H_2O$ to distilled/deionized water and bring volume to 10.0mL. Mix thoroughly. Gas under 100% N_2. Autoclave for 15 min at 15 psi pressure–121°C.

Preparation of Medium: Aseptically and anaerobically combine 870.0mL of sterile solution A with 1.0mL of sterile solution B, 100.0mL of sterile solution C, 10.0mL of sterile solution D, 10.0mL of sterile solution E, and 10.0mL of sterile solution F, in that order. Mix thoroughly. Anaerobically distribute into sterile tubes or flasks under 100% N_2.

Use: For the cultivation and maintenance of *Desulfobulbus* species.

Desulfobulbus Medium

Composition per 1001.0mL:

Solution A	870.0mL
Solution C	100.0mL
Solution D	10.0mL
Solution E (Vitamin solution)	10.0mL
Solution F	10.0mL
Solution B (Trace elements solution SL-10)	1.0mL

pH 7.1–7.4 at 25°C

Solution A:
Composition per 870.0mL:

NaCl	7.0g
Na_2SO_4	3.0g
$MgCl_2 \cdot 6H_2O$	1.3g
KCl	0.5g
NH_4Cl	0.3g
KH_2PO_4	0.2g
$CaCl_2 \cdot 2H_2O$	0.15g
Resazurin	1.0mg

Preparation of Solution A: Add components to distilled/deionized water and bring volume to 870.0mL. Mix thoroughly. Gently heat and bring to boiling. Continue boiling for 3-4 min. Allow to cool to room temperature while gassing under 80% N_2 + 20% CO_2. Continue gassing until pH reaches below 6.0. Seal the flask under 80% N_2 + 20% CO_2. Autoclave for 15 min at 15 psi pressure–121°C.

Solution B (Trace Elements Solution SL-10):
Composition per liter:

$FeCl_2 \cdot 4H_2O$	1.5g
$CoCl_2 \cdot 6H_2O$	190.0mg
$MnCl_2 \cdot 4H_2O$	100.0mg
$ZnCl_2$	70.0mg
$Na_2MoO_4 \cdot 2H_2O$	36.0mg
$NiCl_2 \cdot 6H_2O$	24.0mg
H_3BO_3	6.0mg
$CuCl_2 \cdot 2H_2O$	2.0mg
HCl (25% solution)	10.0mL

Preparation of Solution B (Trace Elements Solution SL-10): Add $FeCl_2 \cdot 4H_2O$ to 10.0mL of HCl solution. Mix thoroughly. Add distilled/deionized water and bring volume to 1.0L. Add remaining components. Mix thoroughly. Gas under 100% N_2. Autoclave for 15 min at 15 psi pressure–121°C.

Solution C:
Composition per 100.0mL:

$NaHCO_3$	5.0g

Preparation of Solution C: Add $NaHCO_3$ to distilled/deionized water and bring volume to 100.0mL. Mix thoroughly. Filter sterilize. Gas under 80% N_2 + 20% CO_2.

Solution D:
Composition per 10.0mL:

Sodium propionate 1.5g

Preparation of Solution D: Add sodium propionate to distilled/deionized water and bring volume to 10.0mL. Mix thoroughly. Gas under 100% N_2. Autoclave for 15 min at 15 psi pressure–121°C.

Solution E (Vitamin Solution):
Composition per liter:

Pyridoxine·HCl 10.0mg
Calcium DL-pantothenate 5.0mg
Lipoic acid 5.0mg
Nicotinic acid 5.0mg
p-Aminobenzoic acid 5.0mg
Riboflavin 5.0mg
Thiamine·HCl 5.0mg
Biotin 2.0mg
Folic acid 2.0mg
Vitamin B_{12} 0.1mg

Preparation of Solution E (Vitamin Solution): Add components to distilled/deionized water and bring volume to 1.0L. Mix thoroughly. Gas under 100% N_2. Autoclave for 15 min at 15 psi pressure–121°C.

Solution F:
Composition per 10.0mL:

$Na_2S \cdot 9H_2O$ 0.4g

Preparation of Solution F: Add $Na_2S \cdot 9H_2O$ to distilled/deionized water and bring volume to 10.0mL. Mix thoroughly. Gas under 100% N_2. Autoclave for 15 min at 15 psi pressure–121°C.

Preparation of Medium: Aseptically and anaerobically combine solution A with solution B, solution C, solution D, solution E, and solution F, in that order. Mix thoroughly. Anaerobically distribute into sterile tubes or flasks under 80% N_2 + 20% CO_2.

Use: For the cultivation and maintenance of *Desulfobulbus* species and *Streptomyces* species.

Desulfobulbus sp. Medium (DSMZ Medium 196)

Composition per 1001.0mL:

Solution A 870.0mL
Solution C 100.0mL
Solution D 10.0mL
Solution E (Vitamin solution) 10.0mL
Solution F 10.0mL
Solution B (Trace elements solution SL-10) 1.0mL
pH 7.1-7.4 at 25°C

Solution A:
Composition per 870.0mL:

NaCl 21.0g
$MgCl_2 \cdot 6H_2O$ 3.1g
Na_2SO_4 .... 3.0g
KH_2PO_4 .... 0.2g
NH_4Cl 0.3g
KCl 0.5g
$CaCl_2 \cdot 2H_2O$ 0.15g
Resazurin 1.0mg

Preparation of Solution A: Add components to distilled/deionized water and bring volume to 870.0mL Mix thoroughly.

Solution B (Trace Elements Solution SL-10):
Composition per liter:

$FeCl_2 \cdot 4H_2O$ 1.5g
$CoCl_2 \cdot 6H_2O$ 190.0mg
$MnCl_2 \cdot 4H_2O$ 100.0mg
$ZnCl_2$ 70.0mg
$Na_2MoO_4 \cdot 2H_2O$ 36.0mg
$NiCl_2 \cdot 6H_2O$ 24.0mg
H_3BO_3 .... 6.0mg
$CuCl_2 \cdot 2H_2O$ 2.0mg
HCl (25% solution) 10.0mL

Preparation of Solution B (Trace Elements Solution SL-10): Add $FeCl_2 \cdot 4H_2O$ to 10.0mL of HCl solution. Mix thoroughly. Add distilled/deionized water and bring volume to 1.0L. Add remaining components. Mix thoroughly. Sparge with 100% N_2. Autoclave for 15 min at 15 psi pressure–121°C.

Solution C:
Composition per 100.0mL:

$NaHCO_3$ 5.0g

Preparation of Solution C: Add $NaHCO_3$ to distilled/deionized water and bring volume to 100.0mL. Mix thoroughly. Filter sterilize. Flush with 80% N_2 + 20% CO_2 to remove dissolved oxygen.

Solution D:
Composition per 10.0mL:

Na-propionate 1.5g

Preparation of Solution D: Add Na-propionate to distilled/deionized water and bring volume to 10.0mL. Mix thoroughly. Sparge with 100% N_2. Autoclave for 15 min at 15 psi pressure–121°C.

Solution E (Vitamin Solution):
Composition per liter:

Pyridoxine-HCl 10.0mg
Thiamine-HCl·$2H_2O$ 5.0mg
Riboflavin 5.0mg
Nicotinic acid 5.0mg
D-Ca-pantothenate 5.0mg
p-Aminobenzoic acid 5.0mg
Lipoic acid 5.0mg
Biotin 2.0mg
Folic acid 2.0mg
Vitamin B_{12} 0.10mg

Preparation of Solution E (Vitamin Solution): Add components to distilled/deionized water and bring volume to 1.0L. Mix thoroughly. Sparge with 100% N_2. Autoclave for 15 min at 15 psi pressure–121°C.

Solution F:
Composition per 10.0mL:

$Na_2S \cdot 9H_2O$ 0.4g

Preparation of Solution F: Add $Na_2S \cdot 9H_2O$ to distilled/deionized water and bring volume to 10.0mL. Mix thoroughly. Sparge with 100% N_2. Autoclave for 15 min at 15 psi pressure–121°C.

Preparation of Medium: Gently heat solution A and bring to boiling. Boil solution A for a few minutes. Cool to room temperature. Gas with 80% N_2 + 20% CO_2 gas mixture to reach a pH below 6. Autoclave for 15 min at 15 psi pressure–121°C. Cool to room temperature. Sequentially add 1.0mL solution B, 100.0mL solution C, 10.0mL solution D, 10.0mL solution E, and 10.0mL solution F. Distribute anaerobically under 80% N_2 + 20% CO_2 into appropriate vessels. Addition of 10-20mg sodium dithionite per liter from a 5% (w/v) solution, freshly prepared under N_2 and filter-sterilized, may stimulate growth.

Use: For the cultivation of *Desulfosarcina variabilis.*

Desulfocapsa sulfoexigens Medium (DSMZ Medium 195b)

Composition per 1001.0mL:

Solution A	890.0mL
Solution C	100.0mL
Solution D	10.0mL
Solution B (Trace elements solution SL-10)	1.0mL

pH 7.2 ± 0.2 at 25°C

Solution A:

Composition per 890.0mL:

NaCl	21.0g
$MgCl_2·6H_2O$	3.1g
Na_2SO_4	3.0g
$FeCl_3·6H_2O$	2.7g
KCl	0.5g
NH_4Cl	0.3g
KH_2PO_4	0.2g
$CaCl_2·2H_2O$	0.15g
Resazurin	1.0mg

Preparation of Solution A: Dissolve 2.7g $FeCl_3·6H_2O$ in 890 ml distilled/deionized water. Adjust pH to 7 with 1*N* NaOH. Add remaining components. Mix thoroughly. Sparge with 80% N_2 + 20% CO_2 gas mixture.

Solution B (Trace Elements Solution SL-10):

Composition per liter:

$FeCl_2·4H_2O$	1.5g
$CoCl_2·6H_2O$	190.0mg
$MnCl_2·4H_2O$	100.0mg
$ZnCl_2$	70.0mg
$Na_2MoO_4·2H_2O$	36.0mg
$NiCl_2·6H_2O$	24.0mg
H_3BO_3	6.0mg
$CuCl_2·2H_2O$	2.0mg
HCl (25% solution)	10.0mL

Preparation of Solution B (Trace Elements Solution SL-10): Add $FeCl_2·4H_2O$ to 10.0mL of HCl solution. Mix thoroughly. Add distilled/deionized water and bring volume to 1.0L. Add remaining components. Mix thoroughly. Sparge with 100% N_2. Autoclave for 15 min at 15 psi pressure–121°C.

Solution C:

Composition per 100.0mL:

$NaHCO_3$	5.0g

Preparation of Solution C: Add $NaHCO_3$ to distilled/deionized water and bring volume to 100.0mL Mix thoroughly. Filter sterilize. Flush with 80% N_2 + 20% CO_2 to remove dissolved oxygen.

Solution D:

Composition per 10.0mL:

Na-thiosulfate	5.0g

Preparation of Solution D: Add Na-thiosulfate to distilled/deionized water and bring volume to 10.0mL. Mix thoroughly. Filter sterilize. Flush with 80% N_2 + 20% CO_2 to remove dissolved oxygen.

Preparation of Medium: Gently heat solution A and bring to boiling. Boil solution A for a few minutes. Cool to room temperature. Gas with 80% N_2 + 20% CO_2 gas mixture to reach a pH below 6. Autoclave for 15 min at 15 psi pressure–121°C. Cool to room temperature. Sequentially add 1.0mL solution B, 100.0mL solution C, and 10.0mL solution D. Adjust pH to 7.2 with sodium bicarbonate or sodium carbonate. Distribute anaerobically under 80% N_2 + 20% CO_2 into appropriate vessels. Alternately distribute solution A to tubes prior to autoclaving. Dispense 8.9 mL amounts under 80% N_2 + 20% CO_2 into Hungate tubes. Seal and autoclave or 15 min at 15 psi pressure–121°C. Before use aseptically add appropriate amounts of remaining solutions to each tube from sterile anaerobic solutions.

Use: For the cultivation of *Desulfocapsa sulfoexigens.*

Desulfococcus amylolyticus Medium

Composition per 1011.0mL:

Sulfur, powdered	10.0g
Starch	5.0g
$NaHCO_3$	0.8g
$MgCl_2·6H_2O$	0.7g
$Na_2S·9H_2O$	0.5g
NaCl	0.5g
$CaCl_2·2H_2O$	0.44g
KCl	0.33g
KH_2PO_4	0.33g
NH_4Cl	0.33g
Yeast extract	0.2g
Resazurin	1.0mg
Vitamin solution	10.0mL
Trace elements solution SL-10	1.0mL

pH 6.2–6.4 at 25°C

Vitamin Solution:

Composition per liter:

Pyridoxine-HCl	10.0mg
Calcium DL-pantothenate	5.0mg
Lipoic acid	5.0mg
Nicotinic acid	5.0mg
p-Aminobenzoic acid	5.0mg
Riboflavin	5.0mg
Thiamine-HCl	5.0mg
Biotin	2.0mg
Folic acid	2.0mg
Vitamin B_{12}	0.1mg

Preparation of Vitamin Solution: Add components to distilled/deionized water and bring volume to 1.0L. Mix thoroughly. Gas under 100% N_2. Autoclave for 15 min at 15 psi pressure–121°C.

Trace Elements Solution SL-10:

Composition per liter:

$FeCl_2·4H_2O$	1.5g
$CoCl_2·6H_2O$	190.0mg
$MnCl_2·4H_2O$	100.0mg
$ZnCl_2$	70.0mg
$Na_2MoO_4·2H_2O$	36.0mg
$NiCl_2·6H_2O$	24.0mg
H_3BO_3	6.0mg
$CuCl_2·2H_2O$	2.0mg
HCl (25% solution)	10.0mL

Preparation of Trace Elements Solution SL-10: Add $FeCl_2$ to 10.0mL of HCl. Mix thoroughly. Add distilled/deionized water and bring volume to 1.0L. Add remaining components. Mix thoroughly. Gas under 100% N_2. Autoclave for 15 min at 15 psi pressure–121°C.

Preparation of Medium: Add components to distilled/deionized water and bring volume to 1.0L. Mix thoroughly.

Gas under 80% N_2 + 20% CO_2. Autoclave for 15 min at 15 psi pressure–121°C.

Use: For the cultivation and maintenance of *Desulfurococcus amylolyticus*.

Desulfococcus Medium

Composition per 1001.0mL:

Solution A	870.0mL
Solution C	100.0mL
Solution D	10.0mL
Solution E (Vitamin Solution)	10.0mL
Solution F	10.0mL
Solution B (Trace Elements Solution SL-10)	1.0mL

pH 7.1–7.4 at 25°C

Solution A:

Composition per 870.0mL:

NaCl	7.0g
Na_2SO_4	3.0g
$MgCl_2 \cdot 6H_2O$	1.3g
KCl	0.5g
NH_4Cl	0.3g
KH_2PO_4	0.2g
$CaCl_2 \cdot 2H_2O$	0.15g
Resazurin	1.0mg
$Na_2SeO_3 \cdot 5H_2O$	3.0μg

Preparation of Solution A: Add components to distilled/deionized water and bring volume to 870.0mL. Mix thoroughly. Gently heat and bring to boiling. Continue boiling for 3-4 min. Allow to cool to room temperature while gassing under 80% N_2 + 20% CO_2. Continue gassing until pH reaches below 6.0. Seal the flask under 80% N_2 + 20% CO_2. Autoclave for 15 min at 15 psi pressure–121°C.

Solution B (Trace Elements Solution SL-10):

Composition per liter:

$FeCl_2 \cdot 4H_2O$	1.5g
$CoCl_2 \cdot 6H_2O$	190.0mg
$MnCl_2 \cdot 4H_2O$	100.0mg
$ZnCl_2$	70.0mg
$Na_2MoO_4 \cdot 2H_2O$	36.0mg
$NiCl_2 \cdot 6H_2O$	24.0mg
H_3BO_3	6.0mg
$CuCl_2 \cdot 2H_2O$	2.0mg
HCl (25% solution)	10.0mL

Preparation of Solution B (Trace Elements Solution SL-10): Add $FeCl_2 \cdot 4H_2O$ to 10.0mL of HCl solution. Mix thoroughly. Add distilled/deionized water and bring volume to 1.0L. Add remaining components. Mix thoroughly. Gas under 100% N_2. Autoclave for 15 min at 15 psi pressure–121°C.

Solution C:

Composition per 100.0mL:

$NaHCO_3$	5.0g

Preparation of Solution C: Add $NaHCO_3$ to distilled/deionized water and bring volume to 100.0mL. Mix thoroughly. Filter sterilize. Gas under 80% N_2 + 20% CO_2.

Solution D:

Composition per 10.0mL:

Sodium benzoate	0.6g

Preparation of Solution D: Add sodium benzoate to distilled/deionized water and bring volume to 10.0mL. Mix thoroughly. Gas under 100% N_2. Autoclave for 15 min at 15 psi pressure–121°C.

Solution E (Vitamin Solution):

Composition per liter:

Pyridoxine·HCl	10.0mg
Calcium DL-pantothenate	5.0mg
Lipoic acid	5.0mg
Nicotinic acid	5.0mg
p-Aminobenzoic acid	5.0mg
Riboflavin	5.0mg
Thiamine·HCl	5.0mg
Biotin	2.0mg
Folic acid	2.0mg
Vitamin B_{12}	0.1mg

Preparation of Solution E (Vitamin Solution): Add components to distilled/deionized water and bring volume to 1.0L. Mix thoroughly. Gas under 100% N_2. Autoclave for 15 min at 15 psi pressure–121°C.

Solution F:

Composition per 10.0mL:

$Na_2S \cdot 9H_2O$	0.4g

Preparation of Solution F: Add $Na_2S \cdot 9H_2O$ to distilled/deionized water and bring volume to 10.0mL. Mix thoroughly. Gas under 100% N_2. Autoclave for 15 min at 15 psi pressure–121°C.

Preparation of Medium: Aseptically and anaerobically combine solution A with solution B, solution C, solution D, solution E, and solution F, in that order. Mix thoroughly. Anaerobically distribute into sterile tubes or flasks under 80% N_2 + 20% CO_2.

Use: For the cultivation and maintenance of *Desulfococcus multivorans*.

Desulfococcus multivorans Medium

Composition per liter:

NaCl	10.0g
$MgSO_4 \cdot 7H_2O$	2.0g
$CaSO_4$	1.0g
NH_4Cl	1.0g
Yeast extract	1.0g
K_2HPO_4	0.5g
Sodium lactate (70% solution)	3.5mL

Preparation of Medium: Add components, except $FeSO_4 \cdot 7H_2O$, ascorbic acid, and thioglycollic acid, to distilled/deionized water and bring volume to 1.0L. Mix thoroughly. Sparge with 80% N_2 + 20% CO_2 for 10–15 min. Add $FeSO_4 \cdot 7H_2O$, ascorbic acid, and thioglycollic acid. Mix thoroughly. Continue to sparge with 80% N_2 + 20% CO_2 and adjust pH to 7.4. Anaerobically distribute into tubes or flasks. Autoclave for 10 min at 10 psi pressure–115°C.

Use: For the cultivation of *Desulfococcus multivorans*.

Desulfococcus niacini Medium

Composition per 1001.0mL:

Solution A	870.0mL
Solution C	100.0mL
Solution D	10.0mL
Solution E (Vitamin solution)	10.0mL
Solution F	10.0mL
Solution B (Trace elements solution SL-10)	1.0mL

pH 7.4 ± 0.2 at 25°C

Solution A:

Composition per 870.0mL:

NaCl	7.0g
Na_2SO_4	3.0g

$MgCl_2 \cdot 6H_2O$ 1.3g
KCl 0.5g
NH_4Cl 0.3g
KH_2PO_4 .... 0.2g
$CaCl_2 \cdot 2H_2O$ 0.15g
Resazurin 1.0mg
$Na_2SeO_3 \cdot 5H_2O$ 3.0μg

Preparation of Solution A: Add components to distilled/deionized water and bring volume to 870.0mL. Mix thoroughly. Gently heat and bring to boiling. Continue boiling for 3-4 min. Allow to cool to room temperature while gassing under 80% N_2 + 20% CO_2. Continue gassing until pH reaches below 6.0. Seal the flask under 80% N_2 + 20% CO_2. Autoclave for 15 min at 15 psi pressure–121°C.

Solution B (Trace Elements Solution SL-10):
Composition per liter:
$FeCl_2 \cdot 4H_2O$ 1.5g
$CoCl_2 \cdot 6H_2O$ 190.0mg
$MnCl_2 \cdot 4H_2O$ 100.0mg
$ZnCl_2$ 70.0mg
$Na_2MoO_4 \cdot 2H_2O$ 36.0mg
$NiCl_2 \cdot 6H_2O$ 24.0mg
H_3BO_3 .... 6.0mg
$CuCl_2 \cdot 2H_2O$ 2.0mg
HCl (25% solution) 10.0mL

Preparation of Solution B (Trace Elements Solution SL-10): Add $FeCl_2 \cdot 4H_2O$ to 10.0mL of HCl solution. Mix thoroughly. Add distilled/deionized water and bring volume to 1.0L. Add remaining components. Mix thoroughly. Gas under 100% N_2. Autoclave for 15 min at 15 psi pressure–121°C.

Solution C:
Composition per 100.0mL:
$NaHCO_3$ 5.0g

Preparation of Solution C: Add $NaHCO_3$ to distilled/deionized water and bring volume to 100.0mL. Mix thoroughly. Filter sterilize. Gas under 80% N_2 + 20% CO_2.

Solution D:
Composition per 10.0mL:
Sodium nicotinate 0.82g

Preparation of Solution D: Add sodium nicotinate to distilled/deionized water and bring volume to 10.0mL. Mix thoroughly. Gas under 100% N_2. Autoclave for 15 min at 15 psi pressure–121°C.

Solution E (Vitamin Solution):
Composition per liter:
Pyridoxine·HCl 10.0mg
Calcium DL-pantothenate 5.0mg
Lipoic acid 5.0mg
Nicotinic acid 5.0mg
p-Aminobenzoic acid 5.0mg
Riboflavin 5.0mg
Thiamine·HCl 5.0mg
Biotin 2.0mg
Folic acid 2.0mg
Vitamin B_{12} 0.1mg

Preparation of Solution E (Vitamin Solution): Add components to distilled/deionized water and bring volume to 1.0L. Mix thoroughly. Gas under 100% N_2. Autoclave for 15 min at 15 psi pressure–121°C.

Solution F:
Composition per 10.0mL:
$Na_2S \cdot 9H_2O$ 0.4g

Preparation of Solution F: Add $Na_2S \cdot 9H_2O$ to distilled/deionized water and bring volume to 10.0mL. Mix thoroughly. Gas under 100% N_2. Autoclave for 15 min at 15 psi pressure–121°C.

Preparation of Medium: Aseptically and anaerobically combine solution A with solution B, solution C, solution D, solution E, and solution F, in that order. Mix thoroughly. Adjust pH to 7.4. Anaerobically distribute into sterile tubes or flasks under 80% N_2 + 20% CO_2.

Use: For the cultivation and maintenance of *Desulfococcus niacini.*

Desulfohalobium Medium

Composition per 1010.0mL:
NaCl 100.0g
$MgCl_2 \cdot 6H_2O$ 20.0g
KCl 4.0g
Na_2SO_4 .... 3.0g
$CaCl_2 \cdot 2H_2O$ 2.7g
NH_4Cl 1.0g
Sodium acetate 1.0g
Trypticase 1.0g
Yeast extract 1.0g
K_2HPO_4 .... 0.3g
KH_2PO_4 .... 0.3g
Sodium (L)-lactate 2.5g
Resazurin 1.0mg
$Na_2SeO_3 \cdot 5H_2O$ 3.0μg
$Na_2S \cdot 9H_2O$ solution 10.0mL
Trace elements solution SL-10 .... 1.0mL
pH 7.0 ± 0.2 at 25°C

Trace Elements Solution SL-10:
Composition per liter:
$FeCl_2 \cdot 4H_2O$ 1.5g
$CoCl_2 \cdot 6H_2O$ 190.0mg
$MnCl_2 \cdot 4H_2O$ 100.0mg
$ZnCl_2$ 70.0mg
$Na_2MoO_4 \cdot 2H_2O$ 36.0mg
$NiCl_2 \cdot 6H_2O$ 24.0mg
H_3BO_3 .... 6.0mg
$CuCl_2 \cdot 2H_2O$ 2.0mg
HCl (25% solution) 10.0mL

Preparation of Trace Elements Solution SL-10: Add $FeCl_2 \cdot 4H_2O$ to 10.0mL of HCl solution. Mix thoroughly. Add distilled/deionized water and bring volume to 1.0L. Add remaining components. Mix thoroughly. Gas under 100% N_2. Autoclave for 15 min at 15 psi pressure–121°C.

$Na_2S \cdot 9H_2O$ Solution:
Composition per 10.0mL:
$Na_2S \cdot 9H_2O$ 1.0mg

Preparation of $Na_2S \cdot 9H_2O$ Solution: Add $Na_2S \cdot 9H_2O$ to distilled/deionized water and bring volume to 10.0mL. Mix thoroughly. Gas under 100% N_2. Autoclave for 15 min at 15 psi pressure–121°C.

Preparation of Medium: Add components, except $Na_2S \cdot 9H_2O$ solution, to distilled/deionized water and bring volume to 1.0L. Mix thoroughly. Gas under 100% N_2. Autoclave for 15 min at 15 psi pressure–121°C. Aseptically and anaerobically add 10.0mL of sterile $Na_2S \cdot 9H_2O$ solution. Mix thoroughly. Aseptically and anaerobically distribute into sterile tubes or flasks.

Use: For the cultivation and maintenance of *Methanohalophilus oregonense.*

Desulfomicrobium WHB Medium

Composition per 1003.0mL:

Solution A 870.0mL
Solution C 100.0mL
Solution D 10.0mL
Solution E (Vitamin solution) 10.0mL
Solution F 10.0mL
Solution B (Trace elements solution SL-10) 1.0mL
Solution G 1.0mL
Solution H 1.0mL

pH 7.1–7.4 at 25°C

Solution A:

Composition per 870.0mL:

NaCl 21.0g
$MgCl_2 \cdot 6H_2O$ 3.1g
Na_2SO_4 ....3.0g
KCl 0.5g
NH_4Cl 0.3g
KH_2PO_4 ....0.2g
$CaCl_2 \cdot 2H_2O$ 0.15g
Resazurin 1.0mg

Preparation of Solution A: Add components to distilled/deionized water and bring volume to 870.0mL. Mix thoroughly. Gently heat and bring to boiling. Continue boiling for 3-4 min. Allow to cool to room temperature while gassing under 80% N_2 + 20% CO_2. Continue gassing until pH reaches below 6.0. Seal the flask under 80% N_2 + 20% CO_2. Autoclave for 15 min at 15 psi pressure–121°C.

Solution B (Trace Elements Solution SL-10):

Composition per liter:

$FeCl_2 \cdot 4H_2O$ 1.5g
$CoCl_2 \cdot 6H_2O$ 190.0mg
$MnCl_2 \cdot 4H_2O$ 100.0mg
$ZnCl_2$ 70.0mg
$Na_2MoO_4 \cdot 2H_2O$ 36.0mg
$NiCl_2 \cdot 6H_2O$ 24.0mg
H_3BO_3 ....6.0mg
$CuCl_2 \cdot 2H_2O$ 2.0mg
HCl (25% solution) 10.0mL

Preparation of Solution B (Trace Elements Solution SL-10): Add $FeCl_2 \cdot 4H_2O$ to 10.0mL of HCl solution. Mix thoroughly. Add distilled/deionized water and bring volume to 1.0L. Add remaining components. Mix thoroughly. Sparge with 100% N_2. Autoclave for 15 min at 15 psi pressure–121°C.

Solution C:

Composition per 100.0mL:

$NaHCO_3$ 5.0g

Preparation of Solution C: Add $NaHCO_3$ to distilled/deionized water and bring volume to 100.0mL. Mix thoroughly. Filter sterilize. Sparge with 80% N_2 + 20% CO_2.

Solution D:

Composition per 10.0mL:

Sodium lactate 4.0g

Preparation of Solution D: Add sodium lactate to distilled/deionized water and bring volume to 10.0mL. Mix thoroughly. Sparge with 100% N_2. Autoclave for 15 min at 15 psi pressure–121°C.

Solution E (Vitamin Solution):

Composition per liter:

Pyridoxine·HCl 10.0mg
Calcium DL-pantothenate 5.0mg
Lipoic acid 5.0mg
Nicotinic acid 5.0mg
p-Aminobenzoic acid 5.0mg
Riboflavin 5.0mg
Thiamine·HCl 5.0mg
Biotin 2.0mg
Folic acid 2.0mg
Vitamin B_{12} 0.1mg

Preparation of Solution E (Vitamin Solution): Add components to distilled/deionized water and bring volume to 1.0L. Mix thoroughly. Filter sterilize. Sparge with 100% N_2.

Solution F:

Composition per 10.0mL:

$Na_2S \cdot 9H_2O$ 0.4g

Preparation of Solution F: Add $Na_2S \cdot 9H_2O$ to distilled/deionized water and bring volume to 10.0mL. Mix thoroughly. Sparge with 100% N_2. Autoclave for 15 min at 15 psi pressure–121°C.

Solution G (Selenite-Tungstate Solution):

Composition per liter:

NaOH 0.5g
$Na_2WO_4 \cdot 2H_2O$ 4.0mg
$Na_2SeO_3 \cdot 5H_2O$ 3.0mg

Preparation of Solution G (Selenite-Tungstate Solution): Add components to distilled/deionized water and bring volume to 1.0L. Mix thoroughly. Sparge with 100% N_2. Autoclave for 15 min at 15 psi pressure–121°C.

Solution H (Seven Vitamin):

Composition per liter:

Pyridoxine·HCl 0.3g
Thiamine·HCl 0.2g
Nicotinic acid 0.2g
Calcium DL-pantothenate 0.1g
Vitamin B_{12} 0.1g
p-Aminobenzoic acid 80.0mg
Biotin 20.0mg

Preparation of Solution H (Seven Vitamin): Add components to distilled/deionized water and bring volume to 1.0L. Mix thoroughly. Filter sterilize. Sparge with 100% N_2.

Preparation of Medium: Aseptically and anaerobically combine solution A with solution B, solution C, solution D, solution E, solution F, solution G, and solution H, in that order. Mix thoroughly. Anaerobically distribute into sterile tubes or flasks under 80% N_2 + 20% CO_2.

Use: For the cultivation of *Desulfomicrobium* species.

Desulfomonile Medium

Composition per 1002.0mL:

Solution A 870.0mL
Solution C 100.0mL
Solution D 10.0mL
Solution E (Vitamin solution) 10.0mL
Solution F 10.0mL
Solution B (Trace elements solution SL-10) 1.0mL
Solution G 1.0mL

pH 6.8–7.0 at 25°C

Solution A:

Composition per 870.0mL:

Na_2SO_4 ....3.0g
NaCl 1.0g
KCl 0.5g
$MgCl_2 \cdot 6H_2O$ 0.4g
NH_4Cl 0.3g

KH_2PO_4 .... 0.2g
$CaCl_2 \cdot 2H_2O$ 0.15g
Resazurin 1.0mg

Preparation of Solution A: Add components to distilled/deionized water and bring volume to 870.0mL. Mix thoroughly. Gently heat and bring to boiling. Continue boiling for 3-4 min. Allow to cool to room temperature while gassing under 80% N_2 + 20% CO_2. Continue gassing until pH reaches below 6.0. Seal the flask under 80% N_2 + 20% CO_2. Autoclave for 15 min at 15 psi pressure–121°C.

Solution B (Trace Elements Solution SL-10):
Composition per liter:
$FeCl_2 \cdot 4H_2O$ 1.5g
$CoCl_2 \cdot 6H_2O$ 190.0mg
$MnCl_2 \cdot 4H_2O$ 100.0mg
$ZnCl_2$ 70.0mg
$Na_2MoO_4 \cdot 2H_2O$ 36.0mg
$NiCl_2 \cdot 6H_2O$ 24.0mg
H_3BO_3 .... 6.0mg
$CuCl_2 \cdot 2H_2O$ 2.0mg
HCl (25% solution) 10.0mL

Preparation of Solution B (Trace Elements Solution SL-10): Add $FeCl_2 \cdot 4H_2O$ to 10.0mL of HCl solution. Mix thoroughly. Add distilled/deionized water and bring volume to 1.0L. Add remaining components. Mix thoroughly. Gas under 100% N_2. Autoclave for 15 min at 15 psi pressure–121°C.

Solution C:
Composition per 100.0mL:
$NaHCO_3$ 2.5g

Preparation of Solution C: Add $NaHCO_3$ to distilled/deionized water and bring volume to 100.0mL. Mix thoroughly. Filter sterilize. Gas under 80% N_2 + 20% CO_2.

Solution D:
Composition per 10.0mL:
Sodium pyruvate 4.0g

Preparation of Solution D: Add sodium propionate to distilled/deionized water and bring volume to 10.0mL. Mix thoroughly. Gas under 100% N_2. Autoclave for 15 min at 15 psi pressure–121°C.

Solution E (Vitamin Solution):
Composition per liter:
Nicotinamide 50.0mg
1,4-Naphthoquinone 20.0mg
Pyridoxine·HCl 5.0mg
Calcium DL-pantothenate 5.0mg
Thioctic acid 5.0mg
p-Aminobenzoic acid 5.0mg
Riboflavin 5.0mg
Thiamine·HCl 5.0mg
Biotin 5.0mg
Folic acid 5.0mg
Vitamin B_{12} 5.0mg
Hemin 5.0mg

Preparation of Solution E (Vitamin Solution): Add 1,4-naphthoquinone and hemin to 10.0mL of 0.1*N* NaOH. Mix thoroughly. Add remaining components and bring volume to 1.0L with distilled/deionized water. Mix thoroughly. Gas under 100% N_2. Autoclave for 15 min at 15 psi pressure–121°C.

Solution F:
Composition per 10.0mL:
$Na_2S \cdot 9H_2O$ 0.4g

Preparation of Solution F: Add $Na_2S \cdot 9H_2O$ to distilled/deionized water and bring volume to 10.0mL. Mix thoroughly. Gas under 100% N_2. Autoclave for 15 min at 15 psi pressure–121°C.

Solution G:
Composition per 10.0mL:
NaOH 0.5g
$Na_2WO_4 \cdot 2H_2O$ 4.0mg
$Na_2SeO_3 \cdot 5H_2O$ 3.0mg

Preparation of Solution G: Add components to distilled/deionized water and bring volume to 10.0mL. Mix thoroughly. Gas under 100% N_2. Autoclave for 15 min at 15 psi pressure–121°C.

Preparation of Medium: Aseptically and anaerobically combine 870.0mL of sterile solution A with 1.0mL of sterile solution B, 100.0mL of sterile solution C, 10.0mL of sterile solution D, 10.0mL of sterile solution E, and 10.0mL of sterile solution F, in that order. Mix thoroughly. Anaerobically distribute into sterile tubes or flasks under 100% N_2. Add 50.0mg/L of sodium dithionite prior to inoculation.

Use: For the cultivation and maintenance of *Desulfomonile tiedjei.*

Desulfomonile tiedjei Medium

Composition per liter:
$NaHCO_3$ 3.0g
PIPES (Piperazine-*N,N′*-bis-2-ethanesulfonic acid) buffer 1.5g
Na_2SO_4 .... 1.42g
Yeast extract 1.0g
Mineral solution 20.0mL
Trace metal solution 10.0mL
$Na_2S_2O_4$ solution 10.0mL
Vitamin solution 10.0mL
Sodium pyruvate solution 10.0mL
Resazurin (0.1% solution) 1.0mL
pH 7.3 ± 0.2 at 25°C

Mineral Solution:
Composition per liter:
NH_4Cl 50.0g
NaCl 40.0g
$MgCl_2 \cdot 6H_2O$ 8.3g
KCl 5.0g
KH_2PO_4 .... 5.0g
$CaCl_2 \cdot 2H_2O$ 1.0g

Preparation of Mineral Solution: Add components to distilled/deionized water and bring volume to 1.0L. Mix thoroughly.

Trace Metal Solution:
Composition per liter:
Nitrilotriacetic acid 2.0g
$MnSO_4 \cdot H_2O$ 1.0g
$Fe(NH_4)_2(SO_4)_2 \cdot 6H_2O$ 0.8g
$CoCl_2 \cdot 6H_2O$ 0.2g
$ZnSO_4 \cdot 7H_2O$ 0.2g
$CuCl_2 \cdot 2H_2O$ 0.02g
$Na_2MoO_4 \cdot H_2O$ 0.02g
Na_2SeO_4 .... 0.02g
Na_2WO_4 .... 0.02g
$NiCl_2 \cdot 6H_2O$ 0.02g

Preparation of Trace Metal Solution: Add nitrilotriacetic acid to 500.0mL of distilled/deionized water. Dissolve by adjusting pH to 6.5 with KOH. Add remaining components. Add distilled/deionized water to 1.0L.

Vitamin Solution:
Composition per liter:
Nicotinamide....0.05g
1,4-Naphthoquinone....0.02g
p-Aminobenzoic acid....5.0mg
Biotin....5.0mg
Calcium pantothenate....5.0mg
Cyanocobalamin....5.0mg
Folic acid....5.0mg
Hemin....5.0mg
Pyridoxine·HCl....5.0mg
Riboflavin....5.0mg
Thioctic acid....5.0mg
NaOH (0.1*N* solution)....5.0mL

Preparation of Vitamin Solution: Add thioctic acid, 1,4-naphthoquinone, and hemin to 5.0mL of 0.1*N* NaOH solution. Mix thoroughly. Bring volume to 1.0L with distilled/deionized water. Add remaining components. Mix thoroughly.

$Na_2S_2O_4$ Solution:
Composition per 10.0mL:
$Na_2S_2O_4$....0.087g

Preparation of $Na_2S_2O_4$ Solution: Add $Na_2S_2O_4$ to distilled/deionized water and bring volume to 10.0mL. Filter sterilize. Prepare freshly.

Sodium Pyruvate Solution:
Composition per 10.0mL:
Sodium pyruvate....4.4g

Preparation of Sodium Pyruvate Solution: Add sodium pyruvate to distilled/deionized water and bring volume to 10.0mL. Filter sterilize.

Preparation of Medium: Add PIPES buffer, Na_2SO_4, yeast extract, mineral solution, and trace metal solution to distilled/deionized water and bring volume to 970.0mL. Mix thoroughly. Adjust pH to 7.3 with HCl. Add $NaHCO_3$ and resazurin. Gently heat and bring to boiling under 80% N_2 + 20% CO_2. Replace headspace with 2 atm pressure of the same gas phase. Autoclave for 15 min at 15 psi pressure–121°C. Cool to 25°C. Anaerobically and aseptically add sterile vitamin solution, sodium pyruvate solution, and $Na_2S_2O_4$ solution. Mix thoroughly.

Use: For the cultivation and maintenance of *Desulfomonile tiedjei*.

Desulfomusa hansenii Medium (DSMZ Medium 916)

Composition per 992.0mL:
Solution A....870.0mL
Solution D....50.0mL
Solution F....50.0mL
Solution E (Vitamin solution)....10.0mL
Solution G....10.0mL
Solution B (Trace elements solution SL-10)....1.0mL
Solution C (Selenite-tungstate solution)....1.0mL
pH 7.2 ± 0.2 at 25°C

Solution A:
Composition per 870mL:
NaCl....20.0g
Na_2SO_4....1.42g
KCl....0.67g
NH_4Cl....0.1g
KH_2PO_4....0.01g
$MgSO_4·7H_2O$....0.02g
Resazurin....0.5mg

Preparation of Solution A: Add components to 870.0mL with distilled/deionized water. Mix thoroughly. Gently heat and bring to boiling. Boil for 3 min. Cool to room temperature while sparging with 80% N_2 + 20% CO_2. Autoclave for 15 min at 15 psi pressure–121°C. Cool to room temperature.

Solution B (Trace Elements Solution SL-10):
Composition per liter:
$FeCl_2·4H_2O$....1.5g
$CoCl_2·6H_2O$....190.0mg
$MnCl_2·4H_2O$....100.0mg
$ZnCl_2$....70.0mg
$Na_2MoO_4·2H_2O$....36.0mg
$NiCl_2·6H_2O$....24.0mg
H_3BO_3....6.0mg
$CuCl_2·2H_2O$....2.0mg
HCl (25% solution)....10.0mL

Preparation of Solution B (Trace Elements Solution SL-10): Add $FeCl_2·4H_2O$ to 10.0mL of HCl solution. Mix thoroughly. Add distilled/deionized water and bring volume to 1.0L. Add remaining components. Mix thoroughly. Sparge with 100% N_2. Autoclave for 15 min at 15 psi pressure–121°C.

Solution C (Selenite-Tungstate Solution):
Composition per liter:
NaOH....0.5g
$Na_2WO_4·2H_2O$....4.0mg
$Na_2SeO_3·5H_2O$....3.0mg

Preparation of Solution C (Selenite-Tungstate Solution): Add components to distilled/deionized water and bring volume to 1.0L. Mix thoroughly. Sparge with 100% N_2. Autoclave for 15 min at 15 psi pressure–121°C.

Solution D:
Composition per 50.0mL:
$NaHCO_3$....5.0g

Preparation of Solution D: Add $NaHCO_3$ to distilled/deionized water and bring volume to 50.0mL. Mix thoroughly. Sparge with 100% N_2. Autoclave for 15 min at 15 psi pressure–121°C.

Solution E (Vitamin Solution):
Composition per liter:
Pyridoxine-HCl....10.0mg
Thiamine-HCl·$2H_2O$....5.0mg
Riboflavin....5.0mg
Nicotinic acid....5.0mg
D-Ca-pantothenate....5.0mg
p-Aminobenzoic acid....5.0mg
Lipoic acid....5.0mg
Biotin....2.0mg
Folic acid....2.0mg
Vitamin B_{12}....0.1mg

Preparation of Solution E (Vitamin Solution): Add components to distilled/deionized water and bring volume to 1.0L. Mix thoroughly. Sparge with 100% N_2. Filter sterilize.

Solution F:
Composition per 50mL:
$MgCl_2·6H_2O$....10.6g
$CaCl_2·2H_2O$....1.52g

Preparation of Solution F: Add components to distilled/deionized water and bring volume to 50.0mL. Mix thoroughly. Sparge with 100% N_2. Autoclave for 15 min at 15 psi pressure–121°C.

Solution G:
Composition per 10.0mL:
$FeCl_2 \cdot 4H_2O$ 0.52g

Preparation of Solution G: Add $FeCl_2 \cdot 4H_2O$ to distilled/deionized water and bring volume to 10.0mL. Mix thoroughly. Adjust pH to 2.0. Sparge with 100% N_2. Autoclave for 15 min at 15 psi pressure–121°C.

Solution H:
Composition per 10.0mL:
Na-propionate 0.96g

Preparation of Solution H: Add Na-propionate to distilled/deionized water and bring volume to 10.0mL. Mix thoroughly. Sparge with 100% N_2. Autoclave for 15 min at 15 psi pressure–121°C.

Preparation of Medium: Aseptically and anaerobially under 80% N_2 + 20% CO_2 sequentially add to 870.0mL sterile solution A, 1.0mL solution B, 1.0mL solution C, 50.0mL solution D, 10.0mL solution E, 50.0mL solution F, and 10.0mL solution G. Aseptically and anaerobically distribute under 80% N_2 + 20% CO_2 into appropriate vessels. The final pH should be 7.2. Addition of 10-20mg sodium dithionite per liter from a 5% (w/v) solution, freshly prepared under N_2 and filter-sterilized, may stimulate growth.

Use: For the cultivation of *Desulfomusa hansenii.*

Desulfonatronovibrio Medium (DSMZ Medium 742)

Composition per liter:
$NaHCO_3$ 15.0g
Na_2CO_3 .... 10.0g
NaCl 10.0g
Na_2SO_4 .... 3.0g
NH_4Cl 1.0g
Na_2HPO_4 .... 0.2g
KCl 0.2g
Resazurin 0.5mg
Yeast extract solution 10.0mL
Na-formate solution 10.0mL
Trace elements solution 10.0mL
Vitamin solution 10.0mL
$Na_2S \cdot 9H_2O$ solution 10.0mL
pH 9.6 ± 0.2 at 25°C

Na-Formate Solution:
Composition per 10.0mL:
Na-formate 5.0g

Preparation of Na-Formate Solution: Add Na-formate to distilled/deionized water and bring volume to 10.0mL. Mix thoroughly. Sparge with 100% N_2. Autoclave for 15 min at 15 psi pressure–121°C.

Yeast Extract Solution:
Composition per 10.0mL:
Yeast extract 1.5g

Preparation of Yeast Extract Solution: Add yeast extract to distilled/deionized water and bring volume to 10.0mL. Mix thoroughly. Sparge with 100% N_2. Autoclave under 100% N_2 for 15 min at 15 psi pressure–121°C. Cool to room temperature.

$Na_2S \cdot 9H_2O$ Solution:
Composition per 10.0mL:
$Na_2S \cdot 9H_2O$ 1.0g

Preparation of $Na_2S \cdot 9H_2O$ Solution: Add $Na_2S \cdot 9H_2O$ to distilled/deionized water and bring volume to 10.0mL. Sparge with N_2. Autoclave for 15 min at 15 psi pressure–121°C. Cool to 25°C.

Trace Elements Solution:
Composition per liter:
$MgSO_4 \cdot 7H_2O$ 3.0g
Nitrilotriacetic acid 1.5g
NaCl 1.0g
$MnSO_4 \cdot 2H_2O$ 0.5g
$CoSO_4 \cdot 7H_2O$ 0.18g
$ZnSO_4 \cdot 7H_2O$ 0.18g
$CaCl_2 \cdot 2H_2O$ 0.1g
$FeSO_4 \cdot 7H_2O$ 0.1g
$NiCl_2 \cdot 6H_2O$ 0.025g
$KAl(SO_4)_2 \cdot 12H_2O$ 0.02g
H_3BO_3 .... 0.01g
$Na_2MoO_4 \cdot 4H_2O$ 0.01g
$CuSO_4 \cdot 5H_2O$ 0.01g
$Na_2SeO_3 \cdot 5H_2O$ 0.3mg

Preparation of Trace Elements Solution: Add nitrilotriacetic acid to 500.0mL of distilled/deionized water. Dissolve by adjusting pH to 6.5 with KOH. Add remaining components. Add distilled/deionized water to 1.0L. Mix thoroughly. Filter sterilize.

Vitamin Solution:
Composition per liter:
Pyridoxine-HCl 10.0mg
Thiamine-HCl·$2H_2O$ 5.0mg
Riboflavin 5.0mg
Nicotinic acid 5.0mg
D-Ca-pantothenate 5.0mg
p-Aminobenzoic acid 5.0mg
Lipoic acid 5.0mg
Biotin 2.0mg
Folic acid 2.0mg
Vitamin B_{12} 0.1mg

Preparation of Vitamin Solution: Add components to distilled/deionized water and bring volume to 1.0L. Mix thoroughly. Sparge with 100% N_2. Filter sterilize.

Preparation of Medium: Prepare and dispense medium under an oxygen-free 100% N_2. Add components, except vitamin solution, yeast extract solution, Na-formate solution, trace elements solution, and $Na_2S \cdot 9H_2O$ solution, to distilled/deionized water and bring volume to 940.0mL. Mix thoroughly. Sparge with 100% N_2. Autoclave for 15 min at 15 psi pressure–121°C. Cool to 25°C. Aseptically and anaerobically add 10.0mL sterile vitamin solution, 10.0mL of sterile yeast extract solution, 10.0mL sterile Na-formate solution, 10.0mL sterile trace elements solution, and 10.0mL of sterile $Na_2S \cdot 9H_2O$ solution. Mix thoroughly. Adjust pH to 9.6. Aseptically and anaerobically distribute into sterile tubes or flasks.

Use: For the cultivation of *Desulfonatronovibrio hydrogenovorans.*

Desulfonatronum Medium (DSMZ Medium 813)

Composition per 1010mL:
NaCl 10.0g
Na_2SO_4 .... 5.0g
Na_2CO_3 .... 3.5g
NH_4Cl 1.0g
Yeast extract 1.0g
$Na_2S \cdot 9H_2O$ 0.5g
KH_2PO_4 .... 0.2g
KCl 0.2g

$MgCl_2 \cdot 6H_2O$ 0.1g
Resazurin 0.5mg
Na-formate solution 10.0mL
Vitamin solution 10.0mL
Trace elements solution SL-10 .... 1.0mL
pH 8.9 ± 0.2 at 25°C

Na-Formate Solution:
Composition per 10.0mL:
Na-formate 4.0g

Preparation of Na-Formate Solution: Add Na-formate to distilled/deionized water and bring volume to 10.0mL. Mix thoroughly. Sparge with 100% N_2. Autoclave for 15 min at 15 psi pressure–121°C.

Vitamin Solution:
Composition per liter:
Pyridoxine-HCl 10.0mg
Thiamine-HCl·$2H_2O$ 5.0mg
Riboflavin 5.0mg
Nicotinic acid 5.0mg
D-Ca-pantothenate 5.0mg
p-Aminobenzoic acid 5.0mg
Lipoic acid 5.0mg
Biotin 2.0mg
Folic acid 2.0mg
Vitamin B_{12} 0.1mg

Preparation of Vitamin Solution: Add components to distilled/deionized water and bring volume to 1.0L. Mix thoroughly. Sparge with 80% H_2 + 20% CO_2. Filter sterilize.

Trace Elements Solution SL-10:
Composition per liter:
$FeCl_2 \cdot 4H_2O$ 1.5g
$CoCl_2 \cdot 6H_2O$ 190.0mg
$MnCl_2 \cdot 4H_2O$ 100.0mg
$ZnCl_2$ 70.0mg
$Na_2MoO_4 \cdot 2H_2O$ 36.0mg
$NiCl_2 \cdot 6H_2O$ 24.0mg
H_3BO_3 .... 6.0mg
$CuCl_2 \cdot 2H_2O$ 2.0mg
HCl (25% solution) 10.0mL

Preparation of Trace Elements Solution SL-10: Add $FeCl_2 \cdot 4H_2O$ to 10.0mL of HCl solution. Mix thoroughly. Add distilled/deionized water and bring volume to 1.0L. Add remaining components. Mix thoroughly. Sparge with 100% N_2. Autoclave for 15 min at 15 psi pressure–121°C.

Preparation of Medium: Prepare and dispense medium under an oxygen-free 100% N_2. Add components, except Na-formate solution and $Na_2S \cdot 9H_2O$, to distilled/deionized water and bring volume to 1.0L. Mix thoroughly. Sparge with 100% N_2 for 30 min. Add the $Na_2S \cdot 9H_2O$. Mix thoroughly. Adjust pH to 8.8-9.0. Dispense into Hungate tubes under 100% N_2. Autoclave for 15 min at 15 psi pressure–121°C. Cool to 25°C. Aseptically and anaerobically add Na-formate solution, 0.1mL per 10.0mL medium.

Use: For the cultivation of *Desulfonatronum lacustre.*

Desulfonema ishimotoi Medium (DSMZ Medium 739)

Composition per 957.6mL:
Solution A 850.0mL
Agar solution 50.0mL
Solution C 50.0mL
Solution H 20.0mL
Solution D 10.0mL
Solution F (Vitamin solution) 10.0mL
Solution I 10.0mL
Solution G (Artificial sediment) 6.6mL
Solution E 1.0mL
Solution B (Trace elemens solution SL-10) 1.0mL
pH 6.9 ± 0.2 at 25°C

Solution A:
Composition per 800.0mL:
NaCl 21.0g
$MgCl_2 \cdot 6H_2O$ 5.5g
Na_2SO_4 .... 3.0g
$CaCl_2 \cdot 2H_2O$ 1.35g
KCl 0.5g
NH_4Cl 0.3g
KH_2PO_4 .... 0.2g
Resazurin 0.5mg
$Na_2SeO_3 \cdot 5H_2O$ 3μg

Preparation of Solution A: Add components to distilled/deionized water and bring volume to 800.0mL. Mix thoroughly.

Solution B (Trace Elements Solution SL-10):
Composition per liter:
$FeCl_2 \cdot 4H_2O$ 1.5g
$CoCl_2 \cdot 6H_2O$ 190.0mg
$MnCl_2 \cdot 4H_2O$ 100.0mg
$ZnCl_2$ 70.0mg
$Na_2MoO_4 \cdot 2H_2O$ 36.0mg
$NiCl_2 \cdot 6H_2O$ 24.0mg
H_3BO_3 .... 6.0mg
$CuCl_2 \cdot 2H_2O$ 2.0mg
HCl (25% solution) 10.0mL

Preparation of Solution B (Trace Elements Solution SL-10): Add $FeCl_2 \cdot 4H_2O$ to 10.0mL of HCl solution. Mix thoroughly. Add distilled/deionized water and bring volume to 1.0L. Add remaining components. Mix thoroughly. Sparge with 100% N_2. Autoclave for 15 min at 15 psi pressure–121°C.

Solution C:
Composition per 100.0mL:
$NaHCO_3$ 5.0g

Preparation of Solution C: Add $NaHCO_3$ to distilled/deionized water and bring volume to 100.0mL Mix thoroughly. Filter sterilize. Flush with 80% N_2 + 20% CO_2 to remove dissolved oxygen.

Solution D:
Composition per 10.0mL:
Na-acetate·$3H_2O$ 2.5g
Isobutyric acid 0.18g
Na_2-succinate 0.1g

Preparation of Solution D: Add components to distilled/deionized water and bring volume to 10.0mL. Mix thoroughly. Sparge with 100% N_2. Filter sterilize.

Solution E:
Composition per 10.0mL:
Na_2-succinate 1.0g

Preparation of Solution E: Add Na_2-succinate to distilled/deionized water and bring volume to 10.0mL. Mix thoroughly. Sparge with 100% N_2. Autoclave for 15 min at 15 psi pressure–121°C.

Solution F (Vitamin Solution):
Composition per liter:
Pyridoxine-HCl 10.0mg
Vitamin B_{12} 5.1mg

Thiamine-HCl·$2H_2O$....................5.0mg
Riboflavin....................5.0mg
Nicotinic acid....................5.0mg
D-Ca-pantothenate....................5.0mg
p-Aminobenzoic acid....................5.0mg
Lipoic acid....................5.0mg
Biotin....................2.0mg
Folic acid....................2.0mg

Preparation of Solution F (Vitamin Solution): Add components to distilled/deionized water and bring volume to 1.0L. Mix thoroughly. Filter sterilize. Sparge with 100% N_2.

Solution G (Artificial Sediment):
Composition per 6.6mL:
$AlCl_3$·$6H_2O$, 4.9% (w/v)....................5.0mL
Na_2CO_3, 10.6% (w/v)....................1.6mL

Preparation of Solution G (Artificial Sediment): Combine components. Mix thoroughly. Sparge with 100% N_2. Autoclave for 15 min at 15 psi pressure–121°C.

Solution H:
Composition per 20.0mL:
Rumen fluid, clarified....................20.0mL

Preparation of Solution H: Sparge with 100% N_2. Autoclave for 15 min at 15 psi pressure–121°C.

Solution I:
Composition per 10.0mL:
Na_2S·$9H_2O$....................0.4g

Preparation of Solution I: Add Na_2S·$9H_2O$ to distilled/deionized water and bring volume to 10.0mL. Mix thoroughly. Sparge with 100% N_2. Autoclave for 15 min at 15 psi pressure–121°C.

Preparation of Medium: Gently heat solution A and bring to boiling. Boil solution A for a few minutes. Cool to room temperature. Gas with 80% N_2 + 20% CO_2 gas mixture to reach a pH below 6. Autoclave for 15 min at 15 psi pressure–121°C. Add 50.0mL hot agar solution. Mix thoroughly. Cool to 50°C. Sequentially add 1.0mL solution B, 100.0mL solution C, 10.0mL solution D, 1.0mL solution E, 10.0mL solution F, 6.6mL solution G, 20.0mL solution H, and 10.0mL solution I. Distribute anaerobically under 80% N_2 + 20% CO_2 into appropriate vessels. The pH should be 6.9. Addition of 10-20mg sodium dithionite per liter from a 5% (w/v) solution, freshly prepared under N_2 and filter-sterilized, may stimulate growth.

Use: For the cultivation of *Desulfonema ishimotonii.*

Desulfonema limicola Medium (DSMZ Medium 201)

Composition per 1007.6mL:
Solution A....................850.0mL
Solution C....................100.0mL
Solution H....................20.0mL
Solution D....................10.0mL
Solution F (Vitamin solution)....................10.0mL
Solution I....................10.0mL
Solution G (Artificial sediment)....................6.6mL
Solution E....................1.0mL
Solution B (Trace Elements Solution SL-10)....................1.0mL
pH 7.6 ± 0.2 at 25°C

Solution A:
Composition per 870.0mL:
NaCl....................13.0g
Na_2SO_4....................3.0g
$MgCl_2$·$6H_2O$....................2.2g
KCl....................0.5g
KH_2PO_4....................0.2g
NH_4Cl....................0.3g
$CaCl_2$·$2H_2O$....................0.15g
Resazurin....................0.5mg

Preparation of Solution A: Add components to distilled/deionized water and bring volume to 870.0mL. Mix thoroughly.

Solution B (Trace Elements Solution SL-10):
Composition per liter:
$FeCl_2$·$4H_2O$....................1.5g
$CoCl_2$·$6H_2O$....................190.0mg
$MnCl_2$·$4H_2O$....................100.0mg
$ZnCl_2$....................70.0mg
Na_2MoO_4·$2H_2O$....................36.0mg
$NiCl_2$·$6H_2O$....................24.0mg
H_3BO_3....................6.0mg
$CuCl_2$·$2H_2O$....................2.0mg
HCl (25% solution)....................10.0mL

Preparation of Solution B (Trace Elements Solution SL-10): Add $FeCl_2$·$4H_2O$ to 10.0mL of HCl solution. Mix thoroughly. Add distilled/deionized water and bring volume to 1.0L. Add remaining components. Mix thoroughly. Sparge with 100% N_2. Autoclave for 15 min at 15 psi pressure–121°C.

Solution C:
Composition per 100.0mL:
$NaHCO_3$....................5.0g

Preparation of Solution C: Add $NaHCO_3$ to distilled/deionized water and bring volume to 100.0mL. Mix thoroughly. Filter sterilize. Flush with 80% N_2 + 20% CO_2 to remove dissolved oxygen.

Solution D:
Composition per 10.0mL:
Na-acetate·$3H_2O$....................2.5g

Preparation of Solution D: Add Na-acetate·$3H_2O$ to distilled/deionized water and bring volume to 10.0mL. Mix thoroughly. Sparge with 100% N_2. Autoclave for 15 min at 15 psi pressure–121°C.

Solution E:
Composition per 10.0mL:
Na_2-succinate....................1.0g

Preparation of Solution E: Add Na_2-succinate to distilled/deionized water and bring volume to 10.0mL. Mix thoroughly. Sparge with 100% N_2. Autoclave for 15 min at 15 psi pressure–121°C.

Solution F (Vitamin Solution):
Composition per liter:
Pyridoxine-HCl....................10.0mg
Thiamine-HCl·$2H_2O$....................5.0mg
Riboflavin....................5.0mg
Nicotinic acid....................5.0mg
D-Ca-pantothenate....................5.0mg
p-Aminobenzoic acid....................5.0mg
Lipoic acid....................5.0mg
Biotin....................2.0mg
Folic acid....................2.0mg
Vitamin B_{12}....................0.10mg

Preparation of Solution F (Vitamin Solution): Add components to distilled/deionized water and bring volume to 1.0L. Mix thoroughly. Filter sterilize. Sparge with 100% N_2.

Solution G (Artificial Sediment):

Composition per 6.6mL:

$AlCl_3·6H_2O$, 4.9% (w/v) 5.0mL
Na_2CO_3, 10.6% (w/v) 1.6mL

Preparation of Solution G (Artificial Sediment): Combine components. Mix thoroughly. Sparge with 100% N_2. Autoclave for 15 min at 15 psi pressure–121°C.

Solution H:

Composition per 20.0mL:

Rumen fluid, clarified 20.0mL

Preparation of Solution H: Sparge with 100% N_2. Autoclave for 15 min at 15 psi pressure–121°C.

Solution I:

Composition per 10.0mL:

$Na_2S·9H_2O$ 0.4g

Preparation of Solution I: Add $Na_2S·9H_2O$ to distilled/deionized water and bring volume to 10.0mL. Mix thoroughly. Sparge with 100% N_2. Autoclave for 15 min at 15 psi pressure–121°C.

Preparation of Medium: Gently heat solution A and bring to boiling. Boil solution A for a few minutes. Cool to room temperature. Gas with 80% N_2 + 20% CO_2 gas mixture to reach a pH below 6. Autoclave for 15 min at 15 psi pressure–121°C. Cool to room temperature. Sequentially add 1.0mL solution B, 100.0mL solution C, 10.0mL solution D, 1.0mL solution E, 10.0mL solution F, 6.6mL solution G, 20.0mL solution H, and 10.0mL solution I. Distribute anaerobically under 80% N_2 + 20% CO_2 into appropriate vessels. The pH should be 7.6. Addition of 10-20mg sodium dithionite per liter from a 5% (w/v) solution, freshly prepared under N_2 and filter-sterilized, may stimulate growth.

Use: For the cultivation of *Desulfonema limicola.*

Desulfonema magnum Medium (DSMZ Medium 202)

Composition per 957.6mL:

Solution A 850.0mL
Solution C 50.0mL
Solution H 20.0mL
Solution D 10.0mL
Solution F (Vitamin solution) 10.0mL
Solution I 10.0mL
Solution G (Artificial sediment) 6.6mL
Solution E 1.0mL
Solution B (Trace elements solution SL-10) 1.0mL
pH 6.9 ± 0.2 at 25°C

Solution A:

Composition per 870.0mL:

NaCl 21.0g
$MgCl_2·6H_2O$ 5.5g
Na_2SO_4 .... 3.0g
$CaCl_2·2H_2O$ 1.35g
KCl 0.5g
KH_2PO_4 .... 0.2g
NH_4Cl 0.3g
Resazurin 0.5mg
$Na_2SeO_3·5H_2O$ 3μg

Preparation of Solution A: Add components to distilled/deionized water and bring volume to 870.0mL. Mix thoroughly.

Solution B (Trace Elements Solution SL-10):

Composition per liter:

$FeCl_2·4H_2O$ 1.5g
$CoCl_2·6H_2O$ 190.0mg
$MnCl_2·4H_2O$ 100.0mg
$ZnCl_2$ 70.0mg
$Na_2MoO_4·2H_2O$ 36.0mg
$NiCl_2·6H_2O$ 24.0mg
H_3BO_3 .... 6.0mg
$CuCl_2·2H_2O$ 2.0mg
HCl (25% solution) 10.0mL

Preparation of Solution B (Trace Elements Solution SL-10): Add $FeCl_2·4H_2O$ to 10.0mL of HCl solution. Mix thoroughly. Add distilled/deionized water and bring volume to 1.0L. Add remaining components. Mix thoroughly. Sparge with 100% N_2. Autoclave for 15 min at 15 psi pressure–121°C.

Solution C:

Composition per 100.0mL:

$NaHCO_3$ 5.0g

Preparation of Solution C: Add $NaHCO_3$ to distilled/deionized water and bring volume to 100.0mL Mix thoroughly. Filter sterilize. Flush with 80% N_2 + 20% CO_2 to remove dissolved oxygen.

Solution D:

Composition per 10.0mL:

Na-acetate 0.6g

Preparation of Solution D: Add Na-acetate·$3H_2O$ to distilled/deionized water and bring volume to 10.0mL. Mix thoroughly. Sparge with 100% N_2. Autoclave for 15 min at 15 psi pressure–121°C.

Solution E:

Composition per 10.0mL:

Na_2-succinate 1.0g

Preparation of Solution D: Add Na_2-succinate to distilled/deionized water and bring volume to 10.0mL. Mix thoroughly. Sparge with 100% N_2. Autoclave for 15 min at 15 psi pressure–121°C.

Solution F (Vitamin Solution):

Composition per liter:

Pyridoxine-HCl 10.0mg
Vitamin B_{12} 5.1mg
Thiamine-HCl·$2H_2O$ 5.0mg
Riboflavin 5.0mg
Nicotinic acid 5.0mg
D-Ca-pantothenate 5.0mg
p-Aminobenzoic acid 5.0mg
Lipoic acid 5.0mg
Biotin 2.0mg
Folic acid 2.0mg

Preparation of Solution F (Vitamin Solution): Add components to distilled/deionized water and bring volume to 1.0L. Mix thoroughly. Filter sterilize. Sparge with 100% N_2.

Solution G (Artificial Sediment):

Composition per 6.6mL:

$AlCl_3·6H_2O$, 4.9% (w/v) 5.0mL
Na_2CO_3, 10.6% (w/v) 1.6mL

Preparation of Solution G (Artificial Sediment): Combine components. Mix thoroughly. Sparge with 100% N_2. Autoclave for 15 min at 15 psi pressure–121°C.

Solution H:
Composition per 20.0mL:
Rumen fluid, clarified 20.0mL

Preparation of Solution H: Sparge with 100% N_2. Autoclave for 15 min at 15 psi pressure–121°C.

Solution I:
Composition per 10.0mL:
$Na_2S·9H_2O$ 0.4g

Preparation of Solution I: Add $Na_2S·9H_2O$ to distilled/deionized water and bring volume to 10.0mL. Mix thoroughly. Sparge with 100% N_2. Autoclave for 15 min at 15 psi pressure–121°C.

Preparation of Medium: Gently heat solution A and bring to boiling. Boil solution A for a few minutes. Cool to room temperature. Gas with 80% N_2 + 20% CO_2 gas mixture to reach a pH below 6. Autoclave for 15 min at 15 psi pressure–121°C. Cool to room temperature. Sequentially add 1.0mL solution B, 100.0mL solution C, 10.0mL solution D, 1.0mL solution E, 10.0mL solution F, 6.6mL solution G, 20.0mL solution H, and 10.0mL solution I. Distribute anaerobically under 80% N_2 + 20% CO_2 into appropriate vessels. The pH should be 6.9. Addition of 10-20mg sodium dithionite per liter from a 5% (w/v) solution, freshly prepared under N_2 and filter-sterilized, may stimulate growth.

Use: For the cultivation of *Desulfonema magnum.*

Desulforhabdus amnigenus Medium (DSMZ Medium 708)

Composition per liter:
Na_2SO_4 .... 2.8g
$Na_2HPO_4·2H_2O$ 0.53g
KH_2PO_4 .... 0.41g
NH_4Cl 0.3g
NaCl 0.3g
$CaCl_2·2H_2O$ 0.11g
$MgCl_2·6H_2O$ 0.1g
Resazurin 0.5mg
$Na_2S·9H_2O$ solution 10.0mL
Vitamin solution 10.0mL
$NaHCO_3$ solution 10.0mL
Na-propionate solution 10.0mL
Selenite-tungstate solution 1.0mL
Seven vitamin solution 1.0mL
Trace elements solution SL-10 .... 1.0mL
pH 7.2-7.6 at 25°C

Na-propionate Solution:
Composition per 10.0mL:
Na-propionate 2.0g

Preparation of Na-propionate Solution: Add Na-propionate to distilled/deionized water and bring volume to 10.0mL. Mix thoroughly. Sparge with 100% N_2. Autoclave for 15 min at 15 psi pressure–121°C.

Selenite-Tungstate Solution:
Composition per liter:
NaOH 0.5g
$Na_2WO_4·2H_2O$ 4.0mg
$Na_2SeO_3·5H_2O$ 3.0mg

Preparation of Selenite-Tungstate Solution: Add components to distilled/deionized water and bring volume to 1.0L. Mix thoroughly. Sparge with 100% N_2. Filter sterilize.

$Na_2S·9H_2O$ Solution:
Composition per 10mL:
$Na_2S·9H_2O$ 0.5g

Preparation of $Na_2S·9H_2O$ Solution: Add $Na_2S·9H_2O$ to distilled/deionized water and bring volume to 10.0mL. Mix thoroughly. Autoclave under 100% N_2 for 15 min at 15 psi pressure–121°C. Cool to room temperature.

$NaHCO_3$ Solution:
Composition per 10.0mL:
$NaHCO_3$ 4.0g

Preparation of $NaHCO_3$ Solution: Add $NaHCO_3$ to distilled/deionized water and bring volume to 10.0mL. Mix thoroughly. Sparge with 80% N_2 + 20% CO_2. Filter sterilize.

Trace Elements Solution SL-10:
Composition per liter:
$FeCl_2·4H_2O$ 1.5g
$CoCl_2·6H_2O$ 190.0mg
$MnCl_2·4H_2O$ 100.0mg
$ZnCl_2$ 70.0mg
$Na_2MoO_4·2H_2O$ 36.0mg
$NiCl_2·6H_2O$ 24.0mg
H_3BO_3 .... 6.0mg
$CuCl_2·2H_2O$ 2.0mg
HCl (25% solution) 10.0mL

Preparation of Trace Elements Solution SL-10: Add $FeCl_2·4H_2O$ to 10.0mL of HCl solution. Mix thoroughly. Add distilled/deionized water and bring volume to 1.0L. Add remaining components. Mix thoroughly. Sparge with 80% N_2 + 20% CO_2. Autoclave for 15 min at 15 psi pressure–121°C.

Vitamin Solution:
Composition per liter:
Pyridoxine-HCl 10.0mg
Thiamine-HCl·$2H_2O$ 5.0mg
Riboflavin 5.0mg
Nicotinic acid 5.0mg
D-Ca-pantothenate 5.0mg
p-Aminobenzoic acid 5.0mg
Lipoic acid 5.0mg
Biotin 2.0mg
Folic acid 2.0mg
Vitamin B_{12} 0.1mg

Preparation of Vitamin Solution: Add components to distilled/deionized water and bring volume to 1.0L. Mix thoroughly. Sparge with 80% H_2 + 20% CO_2. Filter sterilize.

Seven Vitamin Solution:
Composition per liter:
Pyridoxine hydrochloride 300.0mg
Thiamine-HCl·$2H_2O$ 200.0mg
Nicotinic acid 200.0mg
Vitamin B_{12} 100.0mg
Calcium pantothenate 100.0mg
p-Aminobenzoic acid 80.0mg
D(+)-Biotin 20.0mg

Preparation of Seven Vitamin Solution: Add components to distilled/deionized water and bring volume to 1.0L. Sparge with 100% N_2. Mix thoroughly. Filter sterilize.

Preparation of Medium: Prepare and dispense medium under 80% N_2 + 20% CO_2 gas atmosphere. Add components, except $NaHCO_3$ solution, $Na_2S·9H_2O$ solution, Na-propionate solution, vitamin solution, seven vitamin solution, selenite-tungstate solution, and trace elements solution SL-10, to distilled/deionized water and bring volume to 947.0mL. Mix thoroughly. Adjust pH to 7.2-7.6. Sparge with 80% N_2 + 20% CO_2. Autoclave for 15 min at 15 psi pressure–121°C. Aseptically and anaerobically add 10.0mL $NaHCO_3$ solution, 10.0mL $Na_2S·9H_2O$ solution, 10.0mL Na-propionate solution, 10.0mL vitamin solution, 1.0mL seven vitamin solution, 1.0mL selenite-tungstate solution, and 1.0mL trace elements solution SL-10. Mix thoroughly. Aseptically and anaerobically distribute into sterile tubes or bottles.

Use: For the cultivation of *Desulforhabdus amnigena (Desulforhabdus amnigenus).*

Desulfosarcina Medium

Composition per 1001.0mL:

Solution A	870.0mL
Solution C	100.0mL
Solution D	10.0mL
Solution E (Vitamin solution)	10.0mL
Solution F	10.0mL
Solution B (Trace elements solution SL-10)	1.0mL

pH 7.1–7.4 at 25°C

Solution A:
Composition per 870.0mL:

NaCl	7.0g
Na_2SO_4	3.0g
$MgCl_2·6H_2O$	1.3g
KCl	0.5g
NH_4Cl	0.3g
KH_2PO_4	0.2g
$CaCl_2·2H_2O$	0.15g
Resazurin	1.0mg
$Na_2SeO_3·5H_2O$	3.0μg

Preparation of Solution A: Add components to distilled/deionized water and bring volume to 870.0mL. Mix thoroughly. Gently heat and bring to boiling. Continue boiling for 3-4 min. Allow to cool to room temperature while gassing under 80% N_2 + 20% CO_2. Continue gassing until pH reaches below 6.0. Seal the flask under 80% N_2 + 20% CO_2. Autoclave for 15 min at 15 psi pressure–121°C.

Solution B (Trace Elements Solution SL-10):
Composition per liter:

$FeCl_2·4H_2O$	1.5g
$CoCl_2·6H_2O$	190.0mg
$MnCl_2·4H_2O$	100.0mg
$ZnCl_2$	70.0mg
$Na_2MoO_4·2H_2O$	36.0mg
$NiCl_2·6H_2O$	24.0mg
H_3BO_3	6.0mg
$CuCl_2·2H_2O$	2.0mg
HCl (25% solution)	10.0mL

Preparation of Solution B (Trace Elements Solution SL-10): Add $FeCl_2·4H_2O$ to 10.0mL of HCl solution. Mix thoroughly. Add distilled/deionized water and bring volume to 1.0L. Add remaining components. Mix thoroughly. Gas under 100% N_2. Autoclave for 15 min at 15 psi pressure–121°C.

Solution C:
Composition per 100.0mL:

$NaHCO_3$	5.0g

Preparation of Solution C: Add $NaHCO_3$ to distilled/deionized water and bring volume to 100.0mL. Mix thoroughly. Filter sterilize. Gas under 80% N_2 + 20% CO_2.

Solution D:
Composition per 10.0mL:

Sodium benzoate	0.6g

Preparation of Solution D: Add sodium benzoate to distilled/deionized water and bring volume to 10.0mL. Mix thoroughly. Gas under 100% N_2. Autoclave for 15 min at 15 psi pressure–121°C.

Solution E (Vitamin Solution):
Composition per liter:

Pyridoxine·HCl	10.0mg
Calcium DL-pantothenate	5.0mg
Lipoic acid	5.0mg
Nicotinic acid	5.0mg
p-Aminobenzoic acid	5.0mg
Riboflavin	5.0mg
Thiamine·HCl	5.0mg
Biotin	2.0mg
Folic acid	2.0mg
Vitamin B_{12}	0.1mg

Preparation of Solution E (Vitamin Solution): Add components to distilled/deionized water and bring volume to 1.0L. Mix thoroughly. Gas under 100% N_2. Autoclave for 15 min at 15 psi pressure–121°C.

Solution F:
Composition per 10.0mL:

$Na_2S·9H_2O$	0.4g

Preparation of Solution F: Add $Na_2S·9H_2O$ to distilled/deionized water and bring volume to 10.0mL. Mix thoroughly. Gas under 100% N_2. Autoclave for 15 min at 15 psi pressure–121°C.

Preparation of Medium: Aseptically and anaerobically combine solution A with solution B, solution C, solution D, solution E, and solution F, in that order. Mix thoroughly. Anaerobically distribute into sterile tubes or flasks under 80% N_2 + 20% CO_2.

Use: For the cultivation and maintenance of *Desulfosarcina variabilis.*

Desulfosarcina variabilis Medium

Composition per 1009.0mL:

Solution A	850.0mL
Solution C	100.0mL
Solution G	20.0mL
Solution D	10.0mL
Solution E (Wolfe's vitamin solution)	10.0mL
Solution H	10.0mL
Solution F	6.6mL
Solution B (Trace elements solution SL-10)	1.0mL
Solution I	0.4mL

pH 7.6 ± 0.2 at 25°C

Solution A:
Composition per 920.0mL:

NaCl	13.5g
Na_2SO_4	3.0g
$MgCl_2·6H_2O$	2.2g
KCl	0.5g
NH_4Cl	0.3g
KH_2PO_4	0.2g
$CaCl_2·2H_2O$	0.15g
$Na_2SeO_3·5H_2O$	3.0μg
Resazurin	0.5mg

Preparation of Solution A: Prepare and dispense solution anaerobically under 80% N_2 + 20% CO_2. Add components to distilled/deionized water and bring volume to

920.0mL. Mix thoroughly. Gently heat and bring to boiling. Continue boiling until resazurin turns colorless, indicating reduction, and a pH of 6.0 is reached. Cap with rubber stoppers. Autoclave for 15 min at 15 psi pressure–121°C. Cool to 25°C.

Solution B (Trace Elements Solution SL-10):
Composition per liter:

$FeCl_2 \cdot 4H_2O$	1.5g
$CoCl_2 \cdot 6H_2O$	0.19g
$MnCl_2 \cdot 4H_2O$	0.10g
$ZnCl_2$	0.070g
$Na_2MoO_4 \cdot 2H_2O$	0.036g
$NiCl_2 \cdot 6H_2O$	0.024g
H_3BO_3	6.0mg
$CuCl_2 \cdot 2H_2O$	2.0mg
HCl (25% solution)	10.0mL

Preparation of Solution B (Trace Elements Solution SL-10): Add the $FeCl_2 \cdot 4H_2O$ to 10.0mL of HCl solution. Mix thoroughly. Bring volume to approximately 900.0mL with distilled/deionized water. Mix thoroughly. Adjust pH to 6.0 with NaOH. Bring volume to 1.0L with distilled/deionized water. Filter sterilize. Aseptically gas under 100% N_2 for 20 min.

Solution C:
Composition per 100.0mL:

$NaHCO_3$	5.0g

Preparation of Solution C: Add $NaHCO_3$ to distilled/deionized water and bring volume to 100.0mL. Mix thoroughly. Filter sterilize. Aseptically gas under 80% N_2 + 20% CO_2 for 20 min.

Solution D:
Composition per 10.0mL:

Sodium benzoate	0.6g

Preparation of Solution D: Prepare and dispense solution anaerobically under 80% N_2 + 20% CO_2. Add sodium benzoate to distilled/deionized water and bring volume to 10.0mL. Mix thoroughly. Cap with a rubber stopper. Autoclave for 15 min at 15 psi pressure–121°C. Cool to 25°C.

Solution E (Wolfe's Vitamin Solution):
Composition per liter:

Pyridoxine·HCl	0.01g
Thiamine·HCl	5.0mg
Riboflavin	5.0mg
Nicotinic acid	5.0mg
Calcium pantothenate	5.0mg
p-Aminobenzoic acid	5.0mg
Thioctic acid	5.0mg
Biotin	2.0mg
Folic acid	2.0mg
Cyanocobalamin	0.1mg

Preparation of Solution E (Wolfe's Vitamin Solution): Add components to distilled/deionized water and bring volume to 1.0L. Mix thoroughly. Filter sterilize. Aseptically gas under 100% N_2 for 20 min.

Solution F:
Composition per 6.6mL:

$AlCl_3 \cdot 6H_2O$ (4.9% solution)	5.0mL
Na_2CO_3 (10.6% solution)	1.6mL

Preparation of Solution F: Combine both solutions. Mix thoroughly. Gas with 100% N_2. Cap with a rubber stopper. Autoclave for 15 min at 15 psi pressure–121°C. Cool to 25°C.

Solution G:
Composition per 10.0mL:

Rumen fluid, clarified	20.0mL

Preparation of Solution G: Gas rumen fluid under 100% N_2 for 20 min. Cap with a rubber stopper. Autoclave for 15 min at 15 psi pressure–121°C. Cool to 25°C.

Solution H:
Composition per 10.0mL:

$Na_2S \cdot 9H_2O$	0.4g

Preparation of Solution H: Add $Na_2S \cdot 9H_2O$ to distilled/deionized water and bring volume to 10.0mL. Gas under 100% N_2 for 20 min. Cap with a rubber stopper. Autoclave for 15 min at 15 psi pressure–121°C. Cool to 25°C.

Solution I:
Composition per 10.0mL:

$Na_2S_2O_4$	0.5g

Preparation of Solution I: Add $Na_2S_2O_4$ to distilled/deionized water and bring volume to 10.0mL. Mix thoroughly. Filter sterilize. Aseptically gas under 100% N_2 for 20 min. Prepare solution freshly.

Preparation of Medium: To 850.0mL of cooled, sterile solution A, aseptically and anaerobically add in the following order: 1.0mL of sterile solution B, 1000.0mL of sterile solution C, 10.0mL of sterile solution D, 10.0mL of sterile solution E, 6.6mL of sterile solution F, 20.0mL of sterile solution G, and 10.0mL of sterile solution H. Mix thoroughly. Immediately prior to inoculation, aseptically and anaerobically add 0.4mL of sterile solution I. Mix thoroughly. Aseptically and anaerobically distribute into sterile tubes or flasks.

Use: For the cultivation and maintenance of *Desulfosarcina variabilis.*

Desulfotomaculum acetoxidans Medium

Composition per 1011.0mL:

Solution A	1.0L
Solution B	10.0mL
Vitamin solution	1.0mL

pH 7.0 ± 0.2 at 25°C

Solution A:
Composition per liter:

$NaHCO_3$	4.5g
Na_2SO_4	2.84g
Sodium acetate	1.4g
Sodium butyrate	1.4g
NaCl	1.17g
Yeast extract	1.0g
$MgCl_2 \cdot 6H_2O$	0.4g
KCl	0.3g
NH_4Cl	0.27g
KH_2PO_4	0.2g
$CaCl_2 \cdot 2H_2O$	0.15g
Resazurin	0.5mg
Trace elements solution	1.0mL

Preparation of Solution A: Add components, except $NaHCO_3$ and vitamin solution, and bring volume to 1.0L. Mix thoroughly. Gently heat and bring to boiling. Continue boiling for 3-4 min. Allow to cool to room temperature while gassing under O_2-free 80% N_2 + 20% CO_2. Add $NaHCO_3$ and continue gassing with O_2-free 80% N_2 + 20% CO_2 until pH reaches 6.9–7.1. Seal the flask under 80% N_2 + 20% CO_2. Autoclave for 15 min at 15 psi pressure–121°C.

Trace Elements Solution:
Composition per liter:

$FeCl_2 \cdot 4H_2O$	1.5g
$CoCl_2 \cdot 6H_2O$	120.0mg
$MnCl_2 \cdot 4H_2O$	100.0mg
$ZnCl_2$	68.0mg
H_3BO_3	62.0mg
$Na_2MoO_4 \cdot 2H_2O$	24.0mg
$NiCl_2 \cdot 6H_2O$	24.0mg
$CuCl_2 \cdot 2H_2O$	17.0mg
HCl (25% solution)	10.0mL

Preparation of Trace Elements Solution: Add $FeCl_2 \cdot 4H_2O$ to 10.0mL of HCl solution. Mix thoroughly. Add distilled/deionized water and bring volume to 1.0L. Add remaining components. Mix thoroughly. Gas under 100% N_2.

Vitamin Solution:
Composition per 100.0mL:

Thiamine·HCl	10.0mg
p-Aminobenzoic acid	4.0mg
D(+)-Biotin	1.0mg

Preparation of Vitamin Solution: Add components to distilled/deionized water and bring volume to 100.0mL. Mix thoroughly. Filter sterilize. Gas under 100% N_2.

Solution B:
Composition per 10.0mL:

$Na_2S \cdot 9H_2O$	0.36g

Preparation of Solution B: Add $Na_2S \cdot 9H_2O$ to distilled/deionized water and bring volume to 10.0mL. Mix thoroughly. Gas under 100% N_2. Autoclave for 15 min at 15 psi pressure–121°C.

Preparation of Medium: To 1.0L of sterile solution A, add 10.0mL of sterile solution B and 1.0mL of sterile vitamin solution. Mix thoroughly.

Use: For the cultivation and maintenance of *Desulfotomaculum acetoxidans.*

Desulfotomaculum acetoxidans Medium

Composition per 1020.0mL:

Solution A	1.0L
$Na_2S \cdot 9H_2O$ solution	10.0mL
Wolfe's vitamin solution	10.0mL

Solution A:
Composition per liter:

$NaHCO_3$	4.5g
Na_2SO_4	2.84g
Sodium acetate	1.4g
Sodium butyrate	1.4g
NaCl	1.17g
Yeast extract	1.0g
$MgCl_2 \cdot 6H_2O$	0.4g
KCl	0.3g
NH_4Cl	0.27g
KH_2PO_4	0.2g
$CaCl_2 \cdot 2H_2O$	0.15g
Resazurin	0.5mg
Trace elements solution SL-7	1.0mL

Trace Elements Solution SL-7:
Composition per 1010.0mL:

$FeCl_2 \cdot 4H_2O$	1.5g
$CoCl_2 \cdot 6H_2O$	190.0mg
$MnCl_2 \cdot 4H_2O$	100.0mg
$ZnCl_2$	70.0mg
H_3BO_3	62.0mg
$Na_2MoO_4 \cdot 2H_2O$	36.0mg
$NiCl_2 \cdot 6H_2O$	24.0mg
$CuCl_2 \cdot 2H_2O$	17.0mg
Hydrochloric acid, 25%	10.0mL

Preparation of Trace Elements Solution SL-7: Add $FeCl_2 \cdot 4H_2O$ to 10.0mL of HCl solution. Mix thoroughly. Add distilled/deionized water and bring volume to 1.0L. Add remaining components. Mix thoroughly.

Preparation of Solution A: Prepare and dispense medium under 80% N_2 + 20% CO_2. Add components, except $NaHCO_3$ and bring volume to 1.0L. Mix thoroughly. Gently heat and bring to boiling. Cool to room temperature while sparging with 80% N_2 + 20% CO_2. Add $NaHCO_3$. Mix thoroughly. Continue sparging with 80% N_2 + 20% CO_2 until pH stabilizes at 6.9-7.1. Anaerobically distribute into tubes in 10.0mL amounts under an atmosphere of 80% N_2 + 20% CO_2. Autoclave for 15 min at 15 psi pressure–121°C. Cool to room temperature.

$Na_2S \cdot 9H_2O$ Solution:
Composition per 10.0mL:

$Na_2S \cdot 9H_2O$	0.36g

Preparation of $Na_2S \cdot 9H_2O$ Solution: Add $Na_2S \cdot 9H_2O$ to distilled/deionized water and bring volume to 10.0mL. Mix thoroughly. Sparge with 100% N_2. Autoclave for 15 min at 15 psi pressure–121°C. Before use, neutralize to pH 7.0 with sterile HCl.

Wolfe's Vitamin Solution:
Composition per liter:

Pyridoxine·HCl	10.0mg
p-Aminobenzoic acid	5.0mg
Lipoic acid	5.0mg
Nicotinic acid	5.0mg
Riboflavin	5.0mg
Thiamine·HCl	5.0mg
Calcium DL-pantothenate	5.0mg
Biotin	2.0mg
Folic acid	2.0mg
Vitamin B_{12}	0.1mg

Preparation of Wolfe's Vitamin Solution: Add components to distilled/deionized water and bring volume to 1.0L. Mix thoroughly. Filter sterilize. Sparge with 100% N_2.

Preparation of Medium: Aseptically and anaerobically add 0.1mL of sterile $Na_2S \cdot 9H_2O$ solution and 0.1mL of sterile Wolfe's vitamin solution to each tube containing 10.0mL of solution A. Prepare immediately prior to use.

Use: For the cultivation of *Desulfotomaculum acetoxidans.*

Desulfotomaculum alkaliphilum Medium (DSMZ Medium 866)

Composition per liter:

$NaHCO_3$	8.0g
Na_2SO_4	5.0g
Na-formate	5.0g
NaCl	5.0g
Yeast extract	1.5g
NH_4Cl	1.0g
Na_2CO_3	0.5g
$Na_2S \cdot 9H_2O$	0.5g
KH_2PO_4	0.2g
KCl	0.2g
$MgCl_2 \cdot 6H_2O$	0.1g
Vitamin solution	10.0mL
Trace elements solution SL-10	1.0mL

pH 8.7-9.0 at 25°C

Vitamin Solution:
Composition per liter:

Pyridoxine-HCl	10.0mg
Thiamine-HCl·$2H_2O$	5.0mg
Riboflavin	5.0mg
Nicotinic acid	5.0mg
D-Ca-pantothenate	5.0mg
p-Aminobenzoic acid	5.0mg
Lipoic acid	5.0mg
Biotin	2.0mg
Folic acid	2.0mg
Vitamin B_{12}	0.1mg

Preparation of Vitamin Solution: Add components to distilled/deionized water and bring volume to 1.0L. Mix thoroughly. Sparge with 80% H_2 + 20% CO_2. Filter sterilize.

Trace Elements Solution SL-10:
Composition per liter:

$FeCl_2·4H_2O$	1.5g
$CoCl_2·6H_2O$	190.0mg
$MnCl_2·4H_2O$	100.0mg
$ZnCl_2$	70.0mg
$Na_2MoO_4·2H_2O$	36.0mg
$NiCl_2·6H_2O$	24.0mg
H_3BO_3	6.0mg
$CuCl_2·2H_2O$	2.0mg
HCl (25% solution)	10.0mL

Preparation of Trace Elements Solution SL-10: Add $FeCl_2·4H_2O$ to 10.0mL of HCl solution. Mix thoroughly. Add distilled/deionized water and bring volume to 1.0L. Add remaining components. Mix thoroughly. Sparge with 80% N_2 + 20% CO_2. Autoclave for 15 min at 15 psi pressure–121°C.

Preparation of Medium: Add components, except $NaHCO_3$ and $Na_2S·9H_2O$, to distilled/deionized water and bring volume to 1.0L. Mix thoroughly. Gently heat and bring to boiling. Boil for 5 min. Cool to room temperature while sparging with 100% N_2. Add $NaHCO_3$ and $Na_2S·9H_2O$. Mix thoroughly. Distribute into anaerobe tubes or bottles. Autoclave for 15 min at 15 psi pressure–121°C.

Use: For the cultivation of *Desulfotomaculum alkaliphilum.*

Desulfotomaculum geothermicum Medium

Composition per 1001.0mL:

Solution A	870.0mL
Solution C	100.0mL
Solution D	10.0mL
Solution E (Vitamin solution)	10.0mL
Solution F	10.0mL
Solution B (Trace elements solution SL-10)	1.0mL

pH 7.1–7.4 at 25°C

Solution A:
Composition per 870.0mL:

NaCl	21.0g
$MgCl_2·6H_2O$	3.1g
Na_2SO_4	3.0g
KCl	0.5g
NH_4Cl	0.3g
KH_2PO_4	0.2g
$CaCl_2·2H_2O$	0.15g
Resazurin	1.0mg

Preparation of Solution A: Add components to distilled/deionized water and bring volume to 870.0mL. Mix thoroughly. Gently heat and bring to boiling. Continue boiling for 3-4 min. Allow to cool to room temperature while gassing under 80% N_2 + 20% CO_2. Continue gassing until pH reaches below 6.0. Seal the flask under 80% N_2 + 20% CO_2. Autoclave for 15 min at 15 psi pressure–121°C.

Solution B (Trace Elements Solution SL-10):
Composition per liter:

$FeCl_2·4H_2O$	1.5g
$CoCl_2·6H_2O$	190.0mg
$MnCl_2·4H_2O$	100.0mg
$ZnCl_2$	70.0mg
$Na_2MoO_4·2H_2O$	36.0mg
$NiCl_2·6H_2O$	24.0mg
H_3BO_3	6.0mg
$CuCl_2·2H_2O$	2.0mg
HCl (25% solution)	10.0mL

Preparation of Solution B (Trace Elements Solution SL-10): Add $FeCl_2·4H_2O$ to 10.0mL of HCl solution. Mix thoroughly. Add distilled/deionized water and bring volume to 1.0L. Add remaining components. Mix thoroughly. Gas under 100% N_2. Autoclave for 15 min at 15 psi pressure–121°C.

Solution C:
Composition per 100.0mL:

$NaHCO_3$	5.0g

Preparation of Solution C: Add $NaHCO_3$ to distilled/deionized water and bring volume to 100.0mL. Mix thoroughly. Filter sterilize. Gas under 80% N_2 + 20% CO_2.

Solution D:
Composition per 10.0mL:

Sodium lactate	2.5g

Preparation of Solution D: Add Na lactate to distilled/deionized water and bring volume to 10.0mL. Mix thoroughly. Gas under 100% N_2. Autoclave for 15 min at 15 psi pressure–121°C.

Solution E (Vitamin Solution):
Composition per liter:

Pyridoxine·HCl	10.0mg
Calcium DL-pantothenate	5.0mg
Lipoic acid	5.0mg
Nicotinic acid	5.0mg
p-Aminobenzoic acid	5.0mg
Riboflavin	5.0mg
Thiamine·HCl	5.0mg
Biotin	2.0mg
Folic acid	2.0mg
Vitamin B_{12}	0.1mg

Preparation of Solution E (Vitamin Solution): Add components to distilled/deionized water and bring volume to 1.0L. Mix thoroughly. Gas under 100% N_2. Autoclave for 15 min at 15 psi pressure–121°C.

Solution F:
Composition per 10.0mL:

$Na_2S·9H_2O$	0.05g

Preparation of Solution F: Add $Na_2S·9H_2O$ to distilled/deionized water and bring volume to 10.0mL. Mix thoroughly. Gas under 100% N_2. Autoclave for 15 min at 15 psi pressure–121°C.

Preparation of Medium: Aseptically and anaerobically combine solution A with solution B, solution C, solution D, solution E, and solution F, in that order. Mix thoroughly. Anaerobically distribute into sterile tubes or flasks under 80% N_2 + 20% CO_2.

Use: For the cultivation and maintenance of *Desulfotomaculum geothermicum.*

Desulfotomaculum Groll Medium (DSMZ Medium 124a)

Composition per liter:

$NaHCO_3$	4.5g
Na_2SO_4	2.84g
Na-acetate	1.4g
Na-butyrate	1.4g
NaCl	1.17g
Yeast extract	1.0g
$MgCl_2 \cdot 6H_2O$	0.4g
KCl	0.3g
NH_4Cl	0.27g
KH_2PO_4	0.2g
$CaCl_2 \cdot 2H_2O$	0.15g
Resazurin	0.5mg
$Na_2S \cdot 9H_2O$ solution	10.0mL
Substrate solution	10.0mL
Selenite solution	10.0mL
Vitamin solution	1.0mL
Trace elements solution	1.0mL

pH 7.0 ± 0.2 at 25°C

$Na_2S \cdot 9H_2O$ Solution:

Composition per 100.0mL:

$Na_2S \cdot 9H_2O$	3.6g

Preparation of $Na_2S \cdot 9H_2O$ Solution: Add $Na_2S \cdot 9H_2O$ to distilled/deionized water and bring volume to 100.0mL. Sparge with N_2. Autoclave for 15 min at 15 psi pressure–121°C. Cool to 25°C. Store anaerobically.

Trace Elements Solution:

Composition per liter:

$FeCl_2 \cdot 4H_2O$	1.5g
$CoCl_2 \cdot 6H_2O$	120.0mg
$MnCl_2 \cdot 4H_2O$	100.0mg
$ZnCl_2$	68.0mg
$Na_2MoO_4 \cdot 2H_2O$	24.0mg
$NiCl_2 \cdot 6H_2O$	24.0mg
H_3BO_3	62.0mg
$CuCl_2 \cdot 2H_2O$	17.0mg
HCl (0.5*M*)	1.0L

Preparation of Trace Elements Solution: Add $FeCl_2 \cdot 4H_2O$ to 1.0L of 0.5*M* HCl. Mix thoroughly. Add remaining components. Mix thoroughly. Sparge with 100% N_2. Autoclave for 15 min at 15 psi pressure–121°C.

Vitamin Solution:

Composition 100.0mL:

Thiamine-HCl·$2H_2O$	10.0mg
p-Aminobenzoic acid	4.0mg
D(+)-Biotin	1.0mg

Preparation of Vitamin Solution: Add components to distilled/deionized water and bring volume to 100.0mL. Mix thoroughly. Sparge with 80% N_2 + 20% CO_2. Filter sterilize.

Selenite Solution:

Composition per 10.0mL:

Sodium selenite	3.0μg

Preparation of Selenite Solution: Add sodium selenite to distilled/deionized water and bring volume to 10.0mL. Mix thoroughly. Sparge with 80% N_2 + 20% CO_2. Autoclave for 15 min at 15 psi pressure–121°C.

Substrate Solution:

Composition per 10.0mL:

Sodium benzoate	0.6g

Preparation of Substrate Solution: Add sodium benzoate to distilled/deionized water and bring volume to 10.0mL. Mix thoroughly. Sparge with 80% N_2 + 20% CO_2. Autoclave for 15 min at 15 psi pressure–121°C.

Preparation of Medium: Prepare and dispense medium under an oxygen-free 80% N_2 + 20% CO_2 gas mixture. Add components, except substrate solution, selenite solution, vitamin solution, and $Na_2S \cdot 9H_2O$ solution, to distilled/deionized water and bring volume to 1.0L. Mix thoroughly. Sparge with 80% N_2 + 20% CO_2. Heat gently and bring to boiling. Cool to room temperature. Add $NaHCO_3$. Continue sparging with 80% N_2 + 20% CO_2 until an equilibrium pH of 6.9-7.1 is reached. Autoclave for 15 min at 15 psi pressure–121°C. Cool to 25°C while sparging with 80% N_2 + 20% CO_2. Aseptically and anaerobically add 10.0mL of sterile substrate solution, 10.0mL of sterile selenite solution, 1.0mL of sterile vitamin solution, and 10.0mL of sterile $Na_2S \cdot 9H_2O$ solution. Mix thoroughly. Aseptically and anaerobically distribute into sterile tubes or flasks. Alternately the medium can be distributed to tubes under anaerobic conditionsand autoclaved in tubes prior to addition of substrate solution, selenite solution, vitamin solution, and $Na_2S \cdot 9H_2O$ solution. Appropriate amounts of these solutions can then be added to each tube to yield the desired concentrations. Additions are performed aseptically and anaerobically under an oxygen-free 80% N_2 + 20% CO_2 gas mixture.

Use: For the cultivation of *Desulfotomaculum gibsoniae.*

Desulfotomaculum halophilum Medium (DSMZ Medium 815)

Composition per liter:

Iron, powder	150.0g
NaCl	40.0g
$MgCl_2 \cdot 6H_2O$	8.0g
$CaCl_2 \cdot 2H_2O$	6.0g
Na-lactate	3.6g
Na_2SO_4	3.0g
MOPS	3.0g
KCl	2.0g
Yeast extract	1.0g
NH_4Cl	0.3g
KH_2PO_4	0.2g
$SrCl_2 \cdot 6H_2O$	0.1g
Resazurin	0.5mg
$NaHCO_3$ solution	50.0mL
Trace elements solution SL-12	1.0mL

pH 7.2 ± 0.2 at 25°C

Trace Elements Solution SL-12:

Composition per liter:

Na_2-EDTA	5.2g
$FeCl_2 \cdot 4H_2O$	1.5g
$CoCl_2 \cdot 6H_2O$	190.0mg
$MnCl_2 \cdot 4H_2O$	100.0mg
$ZnCl_2$	70.0mg
$Na_2MoO_4 \cdot 2H_2O$	36.0mg
$NiCl_2 \cdot 6H_2O$	24.0mg
H_3BO_3	6.0mg
$CuCl_2 \cdot 2H_2O$	2.0mg
HCl (25% solution)	10.0mL

Preparation of Trace Elements Solution SL-12: Add $FeCl_2 \cdot 4H_2O$ to 10.0mL of HCl solution. Mix thoroughly. Add distilled/deionized water and bring volume to 1.0L. Add remaining components. Mix thoroughly. Adjust pH to 6.0. Sparge with 80% N_2 + 20% CO_2. Autoclave for 15 min at 15 psi pressure–121°C.

$NaHCO_3$ Solution:
Composition per 100.0mL:

$NaHCO_3$	10.0g

Preparation of $NaHCO_3$ Solution: Add $NaHCO_3$ to distilled/deionized water and bring volume to 100.0mL. Mix thoroughly. Sparge with 80% N_2 + 20% CO_2. Filter sterilize.

Preparation of Medium: Prepare and dispense medium under 100% N_2. Add components, except iron and $NaHCO_3$ solution, to distilled/deionized water and bring volume to 950.0L. Mix thoroughly. Sparge with 100% N_2. Adjust pH to 6.0. Dispense under 100% N_2 into tubes or bottles containing 1.5g iron per 10.0mL medium. Autoclave for 30 min at 105°C. Cool to 25°C. Aseptically and anaerobically add sterile $NaHCO_3$ solution, 0.5mL per 10.0mL medium. Final pH is 7.2.

Use: For the cultivation of *Desulfotomaculum halophilum* and *Desulfocella halophila.*

Desulfotomaculum sapomandens Medium

Composition per 1019.7mL:

Solution A	966.0mL
Solution B	20.0mL
Solution C	10.0mL
Solution D	10.0mL
Solution F	10.0mL
Solution H	1.7mL
Solution G	1.0mL
Solution E	1.0mL

pH 7.2–7.5 at 25°C

Solution A:
Composition per 966.0mL:

Na_2SO_4	3.0g
NaCl	1.0g
$MgCl_2 \cdot 6H_2O$	0.4g
NH_4Cl	0.3g
KH_2PO_4	0.2g
$CaCl_2 \cdot 2H_2O$	0.15g
Resazurin	1.0mg
Trace elements solution SL-10	1.0mL

Trace Elements Solution SL-10:
Composition per liter:

$FeCl_2 \cdot 4H_2O$	1.5g
$CoCl_2 \cdot 6H_2O$	190.0mg
$MnCl_2 \cdot 4H_2O$	100.0mg
$ZnCl_2$	70.0mg
$Na_2MoO_4 \cdot 2H_2O$	36.0mg
$NiCl_2 \cdot 6H_2O$	24.0mg
H_3BO_3	6.0mg
$CuCl_2 \cdot 2H_2O$	2.0mg
HCl (25% solution)	10.0mL

Preparation of Trace Elements Solution SL-10: Add $FeCl_2 \cdot 4H_2O$ to 10.0mL of HCl solution. Mix thoroughly. Add distilled/deionized water and bring volume to 1.0L. Add remaining components. Mix thoroughly. Gas under 100% N_2. Autoclave for 15 min at 15 psi pressure–121°C.

Solution B:
Composition per 20.0mL:

$NaHCO_3$	1.0g

Preparation of Solution B: Add $NaHCO_3$ to distilled/deionized water and bring volume to 20.0mL. Mix thoroughly. Gas under 100% N_2. Autoclave for 15 min at 15 psi pressure–121°C.

Solution C:
Composition per 10.0mL:

Ethanol	1.0mL

Preparation of Solution C: Add ethanol to distilled/deionized water and bring volume to 10.0mL. Mix thoroughly. Filter sterilize.

Solution D:
Composition per 10.0mL:

Sodium benzoate	0.7g

Preparation of Solution D: Add sodium benzoate to distilled/deionized water and bring volume to 10.0mL. Mix thoroughly. Filter sterilize.

Solution E:
Composition per 1.0mL:

Rumen fluid, clarified	1.0mL

Preparation of Solution E: Gas under 100% N_2. Autoclave for 15 min at 15 psi pressure–121°C.

Solution F:
Composition per liter:

Pyridoxine·HCl	10.0mg
Calcium DL-pantothenate	5.0mg
Lipoic acid	5.0mg
Nicotinic acid	5.0mg
p-Aminobenzoic acid	5.0mg
Riboflavin	5.0mg
Thiamine·HCl	5.0mg
Biotin	2.0mg
Folic acid	2.0mg
Vitamin B_{12}	0.1mg

Preparation of Solution F: Add components to distilled/deionized water and bring volume to 1.0L. Mix thoroughly. Gas under 100% N_2. Filter sterilize.

Solution G:
Composition per 1.0mL:

Sodium dithionite	0.025g

Preparation of Solution G: Add sodium dithionite to distilled/deionized water and bring volume to 1.0mL. Mix thoroughly. Gas under 100% N_2. Filter sterilize.

Solution H:
Composition per 10.0mL:

$Na_2S \cdot 9H_2O$	0.3g

Preparation of Solution H: Add $Na_2S \cdot 9H_2O$ to distilled/deionized water and bring volume to 10.0mL. Mix thoroughly. Gas under 100% N_2. Autoclave for 15 min at 15 psi pressure–121°C.

Preparation of Medium: Aseptically and anaerobically combine in the following order: 966.0mL of sterile solution A with 20.0.mL of sterile solution B, 10.0mL of sterile solution C, 10.0mL of sterile solution D, 1.0mL of sterile solution E, 10.0mL of sterile solution F, 1.0mL of sterile solution G, and 1.7mL of sterile solution H. Mix thoroughly.

Use: For the cultivation and maintenance of *Desulfotomaculum sapomandens.*

Desulfotomaculum sp. Medium I (DSMZ Medium 63a)

Composition per liter:

Solution A	980.0mL
Solution B	10.0mL
Solution C	10.0mL

pH 6.5-7.0 at 25°C

Solution A:

Composition per 980.0mL:

Na-pyruvate 5.0g
Na-acetate 2.0g
$MgSO_4 \cdot 7H_2O$ 2.0g
Yeast extract 1.0g
NH_4Cl 1.0g
Na_2SO_4 .... 1.0g
K_2HPO_4 .... 0.5g
$CaCl_2 \cdot 2H_2O$ 0.1g
Resazurin 1.0mg

Preparation of Solution A: Add components to distilled/deionized water and bring volume to 980.0mL. Mix thoroughly.

Solution B:

Composition per 10.0mL:

$FeSO_4 \cdot 7H_2O$ 0.5g

Preparation of Solution B: Add $FeSO_4 \cdot 7H_2O$ to distilled/deionized water and bring volume to 10.0mL. Mix thoroughly.

Solution C:

Composition per 10.0mL:

Na-thioglycolate 0.1g
Ascorbic acid 0.1g

Preparation of Solution C: Add components to distilled/deionized water and bring volume to 10.0mL. Mix thoroughly.

Preparation of Medium: Bring solution A to a boil for a few minutes. Cool to room temperature while gassing with oxygen-free N_2 gas. Add solutions B and C. Adjust pH to 6.5-7.0. Immediately distribute under N_2 into anaerobic tubes. During distribution continuously swirl the medium to keep the grey precipitate suspended. Autoclave for 15 min at 15 psi pressure–121°C.

Use: For the cultivation of *Desulfotomaculum* spp.

Desulfotomaculum species Medium II

Composition per 1001.0mL:

Solution A 870.0mL
Solution C 100.0mL
Solution D 10.0mL
Solution E (Seven vitamin solution) 10.0mL
Solution F 10.0mL
Solution B (Trace elements solution SL-10) 1.0mL
pH 7.1–7.4 at 25°C

Solution A:

Composition per 870.0mL:

NaCl 7.0g
$MgCl_2 \cdot 6H_2O$ 1.3g
Na_2SO_4 .... 0.7g
KCl 0.5g
NH_4Cl 0.3g
KH_2PO_4 .... 0.2g
$CaCl_2 \cdot 2H_2O$ 0.15g
Resazurin 1.0mg

Preparation of Solution A: Add components to distilled/deionized water and bring volume to 870.0mL. Mix thoroughly. Gently heat and bring to boiling. Continue boiling for 3-4 min. Allow to cool to room temperature while gassing under 80% N_2 + 20% CO_2. Continue gassing until pH reaches below 6.0. Seal the flask under 80% N_2 + 20% CO_2. Autoclave for 15 min at 15 psi pressure–121°C.

Solution B (Trace Elements Solution SL-10):

Composition per liter:

$FeCl_2 \cdot 4H_2O$ 1.5g
$CoCl_2 \cdot 6H_2O$ 190.0mg
$MnCl_2 \cdot 4H_2O$ 100.0mg
$ZnCl_2$ 70.0mg
$Na_2MoO_4 \cdot 2H_2O$ 36.0mg
$NiCl_2 \cdot 6H_2O$ 24.0mg
H_3BO_3 .... 6.0mg
$CuCl_2 \cdot 2H_2O$ 2.0mg
HCl (25% solution) 10.0mL

Preparation of Solution B (Trace Elements Solution SL-10): Add $FeCl_2 \cdot 4H_2O$ to 10.0mL of HCl solution. Mix thoroughly. Add distilled/deionized water and bring volume to 1.0L. Add remaining components. Mix thoroughly. Sparge with 100% N_2. Autoclave for 15 min at 15 psi pressure–121°C.

Solution C:

Composition per 100.0mL:

$NaHCO_3$ 5.0g

Preparation of Solution C: Add $NaHCO_3$ to distilled/deionized water and bring volume to 100.0mL. Mix thoroughly. Filter sterilize. Sparge with 80% N_2 + 20% CO_2.

Solution D:

Composition per 10.0mL:

3,4,5-Trimethoxybenzoate 0.42g

Preparation of Solution D: Add 3,4,5-trimethoxybenzoic acid to distilled/deionized water and bring volume to 10.0mL. Mix thoroughly. Sparge with 100% N_2. Autoclave for 15 min at 15 psi pressure–121°C.

Solution E (Seven Vitamin Solution):

Composition per liter:

Pyridoxine·HCl 0.3g
Thiamine·HCl 0.2g
Nicotinic acid 0.2g
Calcium DL-pantothenate 0.1g
Vitamin B_{12} 0.1g
p-Aminobenzoic acid 80.0mg
Biotin 20.0mg

Preparation of Solution E (Seven Vitamin Solution): Add components to distilled/deionized water and bring volume to 1.0L. Mix thoroughly. Sparge with 100% N_2. Autoclave for 15 min at 15 psi pressure–121°C.

Solution F:

Composition per 10.0mL:

$Na_2S \cdot 9H_2O$ 0.4g

Preparation of Solution F: Add $Na_2S \cdot 9H_2O$ to distilled/deionized water and bring volume to 10.0mL. Mix thoroughly. Sparge with 100% N_2. Autoclave for 15 min at 15 psi pressure–121°C.

Preparation of Medium: Aseptically and anaerobically combine solution A with solution B, solution C, solution D, solution E, and solution F, in that order. Mix thoroughly. Anaerobically distribute into sterile tubes or flasks under 80% N_2 + 20% CO_2.

Use: For the cultivation of *Desulfotomaculum* species.

Desulfotomaculum thermosapovorans Medium

Composition per 1015.0mL:

Solution A 900.0mL
Solution B 100.0mL
Solution C 10.0mL

Solution D 5.0mL
Solution E 0.5mL
pH 7.0–7.2 at 25°C

Solution A:
Composition per 900.0mL:

NaCl	8.0g
Na_2SO_4	3.0g
Sodium butyrate	2.2g
NH_4Cl	1.0g
KCl	0.5g
Sodium acetate	0.5g
$MgCl_2 \cdot 6H_2O$	0.4g
Yeast extract	0.4g
KH_2PO_4	0.2g
$CaCl_2 \cdot 2H_2O$	0.15g
$Na_2SeO_3 \cdot 5H_2O$	0.003mg
Trace elements solution SL-10	1.5mL

Trace Elements Solution SL-10:
Composition per liter:

$FeCl_2 \cdot 4H_2O$	1.5g
$CoCl_2 \cdot 6H_2O$	190.0mg
$MnCl_2 \cdot 4H_2O$	100.0mg
$ZnCl_2$	70.0mg
$Na_2MoO_4 \cdot 2H_2O$	36.0mg
$NiCl_2 \cdot 6H_2O$	24.0mg
H_3BO_3	6.0mg
$CuCl_2 \cdot 2H_2O$	2.0mg
HCl (25% solution)	10.0mL

Preparation of Trace Elements Solution SL-10: Add $FeCl_2 \cdot 4H_2O$ to 10.0mL of HCl solution. Mix thoroughly. Add distilled/deionized water and bring volume to 1.0L. Add remaining components. Mix thoroughly. Sparge with 100% N_2. Autoclave for 15 min at 15 psi pressure–121°C.

Preparation of Solution A: Add components to distilled/deionized water and bring volume to 900.0mL. Mix thoroughly. Sparge with 80% N_2 + 20% CO_2.

Solution B:
Composition per 100.0mL:

$NaHCO_3$	5.0g

Preparation of Solution B: Add $NaHCO_3$ to distilled/deionized water and bring volume to 10.0mL. Mix thoroughly. Sparge with 100% N_2. Autoclave for 15 min at 15 psi pressure–121°C.

Solution C:
Composition per liter:

Pyridoxine·HCl	10.0mg
p-Aminobenzoic acid	5.0mg
Lipoic acid	5.0mg
Nicotinic acid	5.0mg
Riboflavin	5.0mg
Thiamine·HCl	5.0mg
Calcium DL-pantothenate	5.0mg
Biotin	2.0mg
Folic acid	2.0mg
Vitamin B_{12}	0.1mg

Preparation of Solution C: Add components to distilled/deionized water and bring volume to 1.0L. Mix thoroughly. Filter sterilize.

Solution D:
Composition per 5.0mL:

$Na_2S \cdot 9H_2O$	0.2g

Preparation of Solution D: Add $Na_2S \cdot 9H_2O$ to distilled/deionized water and bring volume to 5.0mL. Mix thoroughly. Sparge with 100% N_2. Autoclave for 15 min at 15 psi pressure–121°C. Before use, neutralize to pH 7.0 with sterile HCl.

Solution E:
Composition per 1.0mL:

$Na_2S_2O_4$ (sodium dithionite)	20.0mg

Preparation of Solution E: Add $Na_2S_2O_4$ to distilled/deionized water and bring volume to 1.0mL. Mix thoroughly. Sparge with 100% N_2. Autoclave for 15 min at 15 psi pressure–121°C.

Preparation of Medium: Prepare medium anaerobically under 80% N_2 + 20% CO_2. Aseptically and anaerobically combine 900.0mL of sterile solution A, 100.0mL of sterile solution B, 10.0mL of sterile solution C, 5.0mL of sterile solution D, and 0.5 mL of sterile solution E. Mix thoroughly. Adjust pH to 7.0–7.2.

Use: For the cultivation of *Desulfotomaculum thermosapovorans.*

Desulfovibrio aespoeensis Medium (DSMZ Medium 721)

Composition per 1004.0mL:

Solution A	870.0mL
Solution C	100.0mL
Solution D	10.0mL
Solution E (Vitamin solution)	10.0mL
Solution F	10.0mL
Selenite-tungstate solution	2.0mL
Solution B (Trace elements solution SL-10)	1.0mL
Seven vitamin solution	1.0mL

pH 7.3-7.5 at 25°C

Solution A:
Composition per 870.0mL:

NaCl	7.0g
Na_2SO_4	3.0g
$MgCl_2 \cdot 6H_2O$	1.3g
KCl	0.5g
NH_4Cl	0.3g
KH_2PO_4	0.2g
$CaCl_2 \cdot 2H_2O$	0.15g
Resazurin	1.0mg

Preparation of Solution A: Add components to distilled/deionized water and bring volume to 870.0mL Mix thoroughly.

Solution B (Trace Elements Solution SL-10):
Composition per liter:

$FeCl_2 \cdot 4H_2O$	1.5g
$CoCl_2 \cdot 6H_2O$	190.0mg
$MnCl_2 \cdot 4H_2O$	100.0mg
$ZnCl_2$	70.0mg
$Na_2MoO_4 \cdot 2H_2O$	36.0mg
$NiCl_2 \cdot 6H_2O$	24.0mg
H_3BO_3	6.0mg
$CuCl_2 \cdot 2H_2O$	2.0mg
HCl (25% solution)	10.0mL

Preparation of Solution B (Trace Elements Solution SL-10): Add $FeCl_2 \cdot 4H_2O$ to 10.0mL of HCl solution. Mix thoroughly. Add distilled/deionized water and bring volume to 1.0L. Add remaining components. Mix thoroughly. Sparge with 100% N_2. Autoclave for 15 min at 15 psi pressure–121°C.

Solution C:
Composition per 100.0mL:

$NaHCO_3$	5.0g

Preparation of Solution C: Add $NaHCO_3$ to distilled/deionized water and bring volume to 100.0mL Mix thoroughly. Filter sterilize. Flush with 80% N_2 + 20% CO_2 to remove dissolved oxygen.

Solution D:

Composition per 10.0mL:

Na-lactate 2.5g

Preparation of Solution D: Add Na-lactate to distilled/deionized water and bring volume to 10.0mL. Mix thoroughly. Sparge with 100% N_2. Autoclave for 15 min at 15 psi pressure–121°C.

Solution E (Vitamin Solution):

Composition per liter:

Pyridoxine-HCl 10.0mg
Thiamine-HCl·$2H_2O$ 5.0mg
Riboflavin 5.0mg
Nicotinic acid 5.0mg
D-Ca-pantothenate 5.0mg
p-Aminobenzoic acid 5.0mg
Lipoic acid 5.0mg
Biotin 2.0mg
Folic acid 2.0mg
Vitamin B_{12} 0.10mg

Solution E Vitamin Solution: Add components to distilled/deionized water and bring volume to 1.0L. Mix thoroughly. Sparge with 100% N_2. Autoclave for 15 min at 15 psi pressure–121°C.

Solution F:

Composition per 10.0mL:

$Na_2S·9H_2O$ 0.4g

Preparation of Solution F: Add $Na_2S·9H_2O$ to distilled/deionized water and bring volume to 10.0mL. Mix thoroughly. Sparge with 100% N_2. Autoclave for 15 min at 15 psi pressure–121°C.

Seven Vitamin Solution:

Composition per liter:

Pyridoxine hydrochloride 300.0mg
Thiamine-HCl·$2H_2O$ 200.0mg
Nicotinic acid 200.0mg
Vitamin B_{12} 100.0mg
Calcium pantothenate 100.0mg
p-Aminobenzoic acid 80.0mg
D(+)-Biotin 20.0mg

Preparation of Seven Vitamin Solution: Add components to distilled/deionized water and bring volume to 1.0L. Sparge with 100% N_2. Mix thoroughly. Filter sterilize.

Selenite-Tungstate Solution:

Composition per liter:

NaOH 0.5g
$Na_2WO_4·2H_2O$ 4.0mg
$Na_2SeO_3·5H_2O$ 3.0mg

Preparation of Selenite-Tungstate Solution: Add components to distilled/deionized water and bring volume to 1.0L. Mix thoroughly. Sparge with 100% N_2. Filter sterilize.

Preparation of Medium: Gently heat solution A and bring to boiling. Boil solution A for a few minutes. Cool to room temperature. Gas with 80% N_2 + 20% CO_2 gas mixture to reach a pH below 6. Autoclave for 15 min at 15 psi pressure–121°C. Cool to room temperature. Sequentially add 1.0mL solution B, 100.0mL solution C, 10.0mL solution D, 10.0mL solution E, 10.0mL solution F, 2.0mL selenite-tungstate solution, and 1.0mL seven vitamin soluiton. Distribute anaerobically under 80% N_2 + 20% CO_2 into appropriate vessels.

Use: For the cultivation of *Desulfovibrio aespoeensis.*

Desulfovibrio alcoholovorans Medium

Composition per liter:

Solution A 980.0mL
Solution B 10.0mL
Solution C 10.0mL
pH 7.8 ± 0.2 at 25°C

Solution A:

Composition per 980.0mL:

$MgSO_4·7H_2O$ 2.0g
1,2-Propanediol 1.5g
Na_2SO_4 .... 1.0g
NH_4Cl 1.0g
Yeast extract 1.0g
K_2HPO_4 .... 0.5g
$CaCl_2·2H_2O$ 0.1g
$Na_2SeO_3·5H_2O$ 3.0mg
Resazurin 1.0mg

Preparation of Solution A: Add components to distilled/deionized water and bring volume to 980.0mL. Mix thoroughly. Gently heat and bring to boiling. Continue boiling for 3-4 min. Allow to cool to room temperature while gassing under 100% N_2.

Solution B:

Composition per 10.0mL:

$FeSO_4·7H_2O$ 0.5g

Preparation of Solution B: Add $FeSO_4·7H_2O$ to distilled/deionized water and bring volume to 10.0mL. Mix thoroughly. Gas under 100% N_2. Autoclave for 15 min at 15 psi pressure–121°C.

Solution C:

Composition per 10.0mL:

Ascorbic acid 0.1g
Sodium thioglycolate 0.1g

Preparation of Solution C: Add components to distilled/deionized water and bring volume to 10.0mL. Mix thoroughly. Gas under 100% N_2. Autoclave for 15 min at 15 psi pressure–121°C.

Preparation of Medium: To 980.0mL of cooled solution A, anaerobically add 10.0mL of solution B and 10.0mL of solution C. Mix thoroughly. Adjust pH to 7.8 with NaOH. Distribute into tubes or flasks. During distribution, swirl the medium to keep the precipitate in suspension. Autoclave for 15 min at 15 psi pressure–121°C.

Use: For the cultivation and maintenance of *Desulfovibrio alcoholovorans.*

Desulfovibrio asponium Medium

Composition per 1004.0mL:

Solution A 870.0mL
Solution C 100.0mL
Solution D 10.0mL
Solution E (Vitamin solution) 10.0mL
Solution F 10.0mL
Solution G (Selenite-tungstate solution) 2.0mL
Solution B (Trace elements solution SL-10) 1.0mL
Solution H (Seven vitamin solution) 1.0mL
pH 7.3–7.5 at 25°C

Solution A:

Composition per 870.0mL:

NaCl 7.0g

Na_2SO_4 .... 3.0g
$MgCl_2 \cdot 6H_2O$ 1.3g
KCl 0.5g
NH_4Cl 0.3g
KH_2PO_4 .... 0.2g
$CaCl_2 \cdot 2H_2O$ 0.15g
Resazurin 1.0mg

Preparation of Solution A: Add components to distilled/deionized water and bring volume to 870.0mL. Mix thoroughly. Gently heat and bring to boiling. Continue boiling for 3-4 min. Allow to cool to room temperature while gassing under 80% N_2 + 20% CO_2. Continue gassing until pH reaches below 6.0. Seal the flask under 80% N_2 + 20% CO_2. Autoclave for 15 min at 15 psi pressure–121°C.

Solution B (Trace Elements Solution SL-10):
Composition per liter:
$FeCl_2 \cdot 4H_2O$ 1.5g
$CoCl_2 \cdot 6H_2O$ 190.0mg
$MnCl_2 \cdot 4H_2O$ 100.0mg
$ZnCl_2$ 70.0mg
$Na_2MoO_4 \cdot 2H_2O$ 36.0mg
$NiCl_2 \cdot 6H_2O$ 24.0mg
H_3BO_3 .... 6.0mg
$CuCl_2 \cdot 2H_2O$ 2.0mg
HCl (25% solution) 10.0mL

Preparation of Solution B (Trace Elements Solution SL-10): Add $FeCl_2 \cdot 4H_2O$ to 10.0mL of HCl solution. Mix thoroughly. Add distilled/deionized water and bring volume to 1.0L. Add remaining components. Mix thoroughly. Sparge with 100% N_2. Autoclave for 15 min at 15 psi pressure–121°C.

Solution C:
Composition per 100.0mL:
$NaHCO_3$ 5.0g

Preparation of Solution C: Add $NaHCO_3$ to distilled/deionized water and bring volume to 100.0mL. Mix thoroughly. Filter sterilize. Sparge with 80% N_2 + 20% CO_2.

Solution D:
Composition per 10.0mL:
Sodium lactate 2.5g

Preparation of Solution D: Add sodium lactate to distilled/deionized water and bring volume to 10.0mL. Mix thoroughly. Sparge with 100% N_2. Autoclave for 15 min at 15 psi pressure–121°C.

Solution E (Vitamin Solution):
Composition per liter:
Pyridoxine·HCl 10.0mg
Calcium DL-pantothenate 5.0mg
Lipoic acid 5.0mg
Nicotinic acid 5.0mg
p-Aminobenzoic acid 5.0mg
Riboflavin 5.0mg
Thiamine·HCl 5.0mg
Biotin 2.0mg
Folic acid 2.0mg
Vitamin B_{12} 0.1mg

Preparation of Solution E (Vitamin Solution): Add components to distilled/deionized water and bring volume to 1.0L. Mix thoroughly. Sparge with 100% N_2. Autoclave for 15 min at 15 psi pressure–121°C.

Solution F:
Composition per 10.0mL:
$Na_2S \cdot 9H_2O$ 0.4g

Preparation of Solution F: Add $Na_2S \cdot 9H_2O$ to distilled/deionized water and bring volume to 10.0mL. Mix thoroughly. Sparge with 100% N_2. Autoclave for 15 min at 15 psi pressure–121°C.

Solution G (Selenite-Tungstate Solution):
Composition per liter:
NaOH 0.5g
$Na_2WO_4 \cdot 2H_2O$ 4.0mg
$Na_2SeO_3 \cdot 5H_2O$ 3.0mg

Preparation of Solution G (Selenite-Tungstate Solution): Add components to distilled/deionized water and bring volume to 1.0L. Mix thoroughly. Sparge with 100% N_2. Autoclave for 15 min at 15 psi pressure–121°C.

Solution H (Seven Vitamin Solution):
Composition per liter:
Pyridoxine·HCl 0.3g
Thiamine·HCl 0.2g
Nicotinic acid 0.2g
Calcium DL-pantothenate 0.1g
Vitamin B_{12} 0.1g
p-Aminobenzoic acid 80.0mg
Biotin 20.0mg

Preparation of Solution H (Seven Vitamin Solution): Add components to distilled/deionized water and bring volume to 1.0L. Mix thoroughly. Filter sterilize. Sparge with 100% N_2.

Preparation of Medium: Aseptically and anaerobically combine solution A with solution B, solution C, solution D, solution E, and solution F, in that order. Mix thoroughly. Anaerobically distribute into sterile tubes or flasks under 80% N_2 + 20% CO_2.

Use: For the cultivation of *Desulfovibrio asponium.*

Desulfovibrio baarsii Medium

Composition per 1009.0mL:
Solution A 850.0mL
Solution C 100.0mL
Solution G 20.0mL
Solution D 10.0mL
Solution E (Wolfe's vitamin solution) 10.0mL
Solution H 10.0mL
Solution F 6.6mL
Solution B (Trace elements solution SL-10) 1.0mL
Solution I 0.4mL
pH 7.6 ± 0.2 at 25°C

Solution A:
Composition per 920.0mL:
Na_2SO_4 .... 3.0g
NaCl 1.0g
KCl 0.5g
$MgCl_2 \cdot 6H_2O$ 0.4g
NH_4Cl 0.3g
KH_2PO_4 .... 0.2g
$CaCl_2 \cdot 2H_2O$ 0.15g
Resazurin 0.5mg

Preparation of Solution A: Prepare and dispense solution anaerobically under 80% N_2 + 20% CO_2. Add components to distilled/deionized water and bring volume to 920.0mL. Mix thoroughly. Gently heat and bring to boiling. Continue boiling until resazurin turns colorless, indicating reduction, and a pH of 6.0 is reached. Cap with rubber stoppers. Autoclave for 15 min at 15 psi pressure–121°C. Cool to 25°C.

Solution B (Trace Elements Solution SL-10):
Composition per liter:

$FeCl_2·4H_2O$	1.5g
$CoCl_2·6H_2O$	0.19g
$MnCl_2·4H_2O$	0.10g
$ZnCl_2$	0.070g
$Na_2MoO_4·2H_2O$	0.036g
$NiCl_2·6H_2O$	0.024g
H_3BO_3	6.0mg
$CuCl_2·2H_2O$	2.0mg
HCl (25% solution)	10.0mL

Preparation of Solution B (Trace Elements Solution SL-10): Add the $FeCl_2·4H_2O$ to 10.0mL of HCl solution. Mix thoroughly. Bring volume to approximately 900.0mL with distilled/deionized water. Mix thoroughly. Adjust pH to 6.0 with NaOH. Bring volume to 1.0L with distilled/deionized water. Filter sterilize. Aseptically gas under 100% N_2 for 20 min.

Solution C:
Composition per 100.0mL:

$NaHCO_3$	5.0g

Preparation of Solution C: Add $NaHCO_3$ to distilled/deionized water and bring volume to 100.0mL. Mix thoroughly. Filter sterilize. Aseptically gas under 80% N_2 + 20% CO_2 for 20 min.

Solution D:
Composition per 10.0mL:

Sodium butyrate	0.7g
Sodium caproate	0.3g
Sodium octanoate	0.15g

Preparation of Solution D: Prepare and dispense solution anaerobically under 80% N_2 + 20% CO_2. Add components to distilled/deionized water and bring volume to 10.0mL. Mix thoroughly. Cap with a rubber stopper. Autoclave for 15 min at 15 psi pressure–121°C. Cool to 25°C.

Solution E (Wolfe's Vitamin Solution):
Composition per liter:

Pyridoxine·HCl	0.01g
Thiamine·HCl	5.0mg
Riboflavin	5.0mg
Nicotinic acid	5.0mg
Calcium pantothenate	5.0mg
p-Aminobenzoic acid	5.0mg
Thioctic acid	5.0mg
Biotin	2.0mg
Folic acid	2.0mg
Cyanocobalamin	0.1mg

Preparation of Solution E (Wolfe's Vitamin Solution): Add components to distilled/deionized water and bring volume to 1.0L. Mix thoroughly. Filter sterilize. Aseptically gas under 100% N_2 for 20 min.

Solution F:
Composition per 6.6mL:

$AlCl_3·6H_2O$ (4.9% solution)	5.0mL
Na_2CO_3 (10.6% solution)	1.6mL

Preparation of Solution F: Combine both solutions. Mix thoroughly. Gas with 100% N_2. Cap with a rubber stopper. Autoclave for 15 min at 15 psi pressure–121°C. Cool to 25°C.

Solution G:
Composition per 10.0mL:

Rumen fluid, clarified	20.0mL

Preparation of Solution G: Gas rumen fluid under 100% N_2 for 20 min. Cap with a rubber stopper. Autoclave for 15 min at 15 psi pressure–121°C. Cool to 25°C.

Solution H:
Composition per 10.0mL:

$Na_2S·9H_2O$	0.4g

Preparation of Solution H: Add $Na_2S·9H_2O$ to distilled/deionized water and bring volume to 10.0mL. Gas under 100% N_2 for 20 min. Cap with a rubber stopper. Autoclave for 15 min at 15 psi pressure–121°C. Cool to 25°C.

Solution I:
Composition per 10.0mL:

$Na_2S_2O_4$	0.5g

Preparation of Solution I: Add $Na_2S_2O_4$ to distilled/deionized water and bring volume to 10.0mL. Mix thoroughly. Filter sterilize. Aseptically gas under 100% N_2 for 20 min. Prepare solution freshly.

Preparation of Medium: To 850.0mL of cooled, sterile solution A, aseptically and anaerobically add in the following order: 1.0mL of sterile solution B, 1000.0mL of sterile solution C, 10.0mL of sterile solution D, 10.0mL of sterile solution E, 6.6mL of sterile solution F, 20.0mL of sterile solution G, and 10.0mL of sterile solution H. Mix thoroughly. Immediately prior to inoculation, aseptically and anaerobically add 0.4mL of sterile solution I. Mix thoroughly. Aseptically and anaerobically distribute into sterile tubes or flasks.

Use: For the cultivation and maintenance of *Desulfovibrio baarsii.*

Desulfovibrio Brackish Medium (DSMZ Medium 410)

Composition per liter:

Solution A	980.0mL
Solution B	10.0mL
Solution C	10.0mL

Solution A:
Composition per 980.0mL:

NaCl	10.0g
$MgSO_4·7H_2O$	2.0g
DL-Na-lactate	2.0g
Yeast extract	1.0g
NH_4Cl	1.0g
Na_2SO_4	1.0g
K_2HPO_4	0.5g
$CaCl_2·2H_2O$	0.1g
Resazurin	1.0mg

pH 7.8 ± 0.2 at 25°C

Preparation of Solution A: Add components to distilled/deionized water and bring volume to 980.0mL. Mix thoroughly.

Solution B:
Composition per 10.0mL:

$FeSO_4·7H_2O$	0.5g

Preparation of Solution B: Add $FeSO_4·7H_2O$ to distilled/deionized water and bring volume to 10.0mL. Mix thoroughly.

Solution C:
Composition per 10.0mL:

Na-thioglycolate	0.1g
Ascorbic acid	0.1g

Preparation of Solution C: Add components to distilled/deionized water and bring volume to 10.0mL. Mix thoroughly.

Preparation of Medium: Bring solution A to a boil for a few minutes. Cool to room temperature while gassing with oxygen-free N_2 gas. Add solutions B and C. Adjust pH to 7.8 with NaOH. Immediately distribute under N_2 into anaerobic tubes. During distribution continuously swirl the medium to keep the grey precipitate suspended. Autoclave for 15 min at 15 psi pressure–121°C.

Use: For the cultivation of *Desulfovibrio giganteus (Desulfobacter giganteus).*

Desulfovibrio carbinolicus Medium

Composition per 1001.0mL:

Solution A	870.0mL
Solution C	100.0mL
Solution D	10.0mL
Solution E (Vitamin solution)	10.0mL
Solution F	10.0mL
Solution B (Trace elements solution SL-10)	1.0mL

pH 7.1–7.4 at 25°C

Solution A:
Composition per 870.0mL:

Na_2SO_4	3.0g
NaCl	1.0g
KCl	0.5g
$MgCl_2 \cdot 6H_2O$	0.4g
NH_4Cl	0.3g
KH_2PO_4	0.2g
$CaCl_2 \cdot 2H_2O$	0.15g
Yeast extract	0.1g
Casamino acids	0.1g
Resazurin	1.0mg

Preparation of Solution A: Add components to distilled/deionized water and bring volume to 870.0mL. Mix thoroughly. Gently heat and bring to boiling. Continue boiling for 3-4 min. Allow to cool to room temperature while gassing under 80% N_2 + 20% CO_2. Continue gassing until pH reaches below 6.0. Seal the flask under 80% N_2 + 20% CO_2. Autoclave for 15 min at 15 psi pressure–121°C.

Solution B (Trace Elements Solution SL-10):
Composition per liter:

$FeCl_2 \cdot 4H_2O$	1.5g
$CoCl_2 \cdot 6H_2O$	190.0mg
$MnCl_2 \cdot 4H_2O$	100.0mg
$ZnCl_2$	70.0mg
$Na_2MoO_4 \cdot 2H_2O$	36.0mg
$NiCl_2 \cdot 6H_2O$	24.0mg
H_3BO_3	6.0mg
$CuCl_2 \cdot 2H_2O$	2.0mg
HCl (25% solution)	10.0mL

Preparation of Solution B (Trace Elements Solution SL-10): Add $FeCl_2 \cdot 4H_2O$ to 10.0mL of HCl solution. Mix thoroughly. Add distilled/deionized water and bring volume to 1.0L. Add remaining components. Mix thoroughly. Gas under 100% N_2. Autoclave for 15 min at 15 psi pressure–121°C.

Solution C:
Composition per 100.0mL:

$NaHCO_3$	5.0g

Preparation of Solution C: Add $NaHCO_3$ to distilled/deionized water and bring volume to 100.0mL. Mix thoroughly. Filter sterilize. Gas under 80% N_2 + 20% CO_2.

Solution D:
Composition per 10.0mL:

Sodium propionate	0.7g

Preparation of Solution D: Add sodium propionate to distilled/deionized water and bring volume to 10.0mL. Mix thoroughly. Gas under 100% N_2. Autoclave for 15 min at 15 psi pressure–121°C.

Solution E (Vitamin Solution):
Composition per liter:

Pyridoxine·HCl	10.0mg
Calcium DL-pantothenate	5.0mg
Lipoic acid	5.0mg
Nicotinic acid	5.0mg
p-Aminobenzoic acid	5.0mg
Riboflavin	5.0mg
Thiamine·HCl	5.0mg
Biotin	2.0mg
Folic acid	2.0mg
Vitamin B_{12}	0.1mg

Preparation of Solution E (Vitamin Solution): Add components to distilled/deionized water and bring volume to 1.0L. Mix thoroughly. Gas under 100% N_2. Autoclave for 15 min at 15 psi pressure–121°C.

Solution F:
Composition per 10.0mL:

$Na_2S \cdot 9H_2O$	0.4g

Preparation of Solution F: Add $Na_2S \cdot 9H_2O$ to distilled/deionized water and bring volume to 10.0mL. Mix thoroughly. Gas under 100% N_2. Autoclave for 15 min at 15 psi pressure–121°C.

Preparation of Medium: Aseptically and anaerobically combine 870.0mL of sterile solution A with 1.0mL of sterile solution B, 100.0mL of sterile solution C, 10.0mL of sterile solution D, 10.0mL of sterile solution E, and 10.0mL of sterile solution F, in that order. Mix thoroughly. Anaerobically distribute into sterile tubes or flasks under 100% N_2.

Use: For the cultivation and maintenance of *Desulfovibrio carbinolicus.*

Desulfovibrio Choline Medium (DSMZ Medium 272)

Composition per liter:

Solution A	980.0mL
Solution B	10.0mL
Solution C	10.0mL

Solution A:
Composition per 980.0mL:

Choline hydrochloride	5.0g
$MgSO_4 \cdot 7H_2O$	2.0g
Yeast extract	1.0g
NH_4Cl	1.0g
Na_2SO_4	1.0g
K_2HPO_4	0.5g
$CaCl_2 \cdot 2H_2O$	0.1g
Resazurin	1.0mg

pH 7.8 ± 0.2 at 25°C

Preparation of Solution A: Add components to distilled/deionized water and bring volume to 980.0mL. Mix thoroughly.

Solution B:
Composition per 10.0mL:

$FeSO_4 \cdot 7H_2O$	0.5g

Preparation of Solution B: Add $FeSO_4 \cdot 7H_2O$ to distilled/deionized water and bring volume to 10.0mL. Mix thoroughly.

Solution C:

Composition per 10.0mL:

Na-thioglycolate 0.1g
Ascorbic acid 0.1g

Preparation of Solution C: Add components to distilled/deionized water and bring volume to 10.0mL. Mix thoroughly.

Preparation of Medium: Bring solution A to a boil for a few minutes. Cool to room temperature while gassing with oxygen-free N_2 gas. Add solutions B and C. Adjust pH to 7.8 with NaOH. Immediately distribute under N_2 into anaerobic tubes. During distribution continuously swirl the medium to keep the grey precipitate suspended. Autoclave for 15 min at 15 psi pressure–121°C.

Use: For the cultivation of *Desulfovibrio* spp.

Desulfovibrio gabonensis Medium

Composition per 1002.0mL:

NaCl 50.0g
$MgCl_2 \cdot 6H_2O$ 3.3g
Na_2SO_4 .... 3.0g
$MgSO_4 \cdot 7H_2O$ 1.6g
KCl 0.3g
NH_4Cl 0.3g
KH_2PO_4 .... 0.2g
$CaCl_2 \cdot 2H_2O$ 0.1g
Yeast extract 0.1g
Resazurin 0.5mg
Sodium lactate solution 10.0mL
$NaHCO_3$ solution 10.0mL
$Na_2S \cdot 9H_2O$ solution 10.0mL
Trace elements solution SL-10 with EDTA 1.0mL
Seven vitamin solution 1.0mL
pH 7.0–7.2 at 25°C

Sodium Lactate Solution:

Composition per 10.0mL:

Sodium lactate 2.5g

Preparation of Sodium Lactate Solution: Add sodium lactate to distilled/deionized water and bring volume to 10.0mL. Mix thoroughly. Sparge with 100% N_2. Autoclave for 15 min at 15 psi pressure–121°C.

$NaHCO_3$ Solution:

Composition per 10.0mL:

$NaHCO_3$ 2.5g

Preparation of $NaHCO_3$ Solution: Add $NaHCO_3$ to distilled/deionized water and bring volume to 10.0mL. Mix thoroughly. Sparge with 80% N_2 + 20% CO_2. Autoclave for 15 min at 15 psi pressure–121°C.

$Na_2S \cdot 9H_2O$ Solution:

Composition per 10.0mL:

$Na_2S \cdot 9H_2O$ 0.2g

Preparation of $Na_2S \cdot 9H_2O$ Solution: Add $Na_2S \cdot 9H_2O$ to distilled/deionized water and bring volume to 10.0mL. Mix thoroughly. Sparge with 100% N_2. Autoclave for 15 min at 15 psi pressure–121°C.

Trace Elements Solution SL-10 with EDTA:

Composition per liter:

Disodium EDTA 3.0g
$FeCl_2 \cdot 4H_2O$ 1.5g
$CoCl_2 \cdot 6H_2O$ 190.0mg
$MnCl_2 \cdot 4H_2O$ 100.0mg
$ZnCl_2$ 70.0mg
$Na_2MoO_4 \cdot 2H_2O$ 36.0mg
$NiCl_2 \cdot 6H_2O$ 24.0mg
H_3BO_3 .... 6.0mg
$CuCl_2 \cdot 2H_2O$ 2.0mg

Preparation of Trace Elements Solution SL-10 with EDTA: Add components to distilled/deionized water and bring volume to 1.0L. Mix thoroughly. Adjust pH to 6.0.

Seven Vitamin Solution:

Composition per liter:

Pyridoxine·HCl 0.3g
Thiamine·HCl 0.2g
Nicotinic acid 0.2g
Calcium DL-pantothenate 0.1g
Vitamin B_{12} 0.1g
p-Aminobenzoic acid 80.0mg
Biotin 20.0mg

Preparation of Seven Vitamin Solution: Add components to distilled/deionized water and bring volume to 1.0L. Mix thoroughly. Filter sterilize. Sparge with 100% N_2.

Preparation of Medium: Prepare and dispense medium under 80% N_2% + 20% CO_2. Add components, except sodium lactate solution, $NaHCO_3$ solution, $Na_2S \cdot 9H_2O$ solution, trace elements solution SL-10 with EDTA, and seven vitamin solution, to distilled/deionized water and bring volume to 970.0mL. Mix thoroughly. Gently heat and bring to boiling. Continue boiling for 3 min. Cool to room temperature while sparging with 80% N_2 + 20% CO_2. Anaerobically distribute 9.7mL volumes into anaerobic tubes. Autoclave for 15 min at 15 psi pressure–121°C. Aseptically add 0.1mL of sterile sodium lactate solution, 0.1mL of sterile $NaHCO_3$ solution, 0.1mL of sterile $Na_2S \cdot 9H_2O$ solution, 0.01mL of sterile trace elements solution SL-10 with EDTA, and 0.01mL of sterile seven vitamin solution to each tube. Mix thoroughly.

Use: For the cultivation of *Desulfovibrio gabonensis.*

Desulfovibrio giganteus Medium

Composition per 1001.0mL:

Solution A 870.0mL
Solution C 100.0mL
Solution D 10.0mL
Solution E (Vitamin solution) 10.0mL
Solution F 10.0mL
Solution B (Trace elements solution SL-10) 1.0mL
pH 7.5 ± 0.2 at 25°C

Solution A:

Composition per 870.0mL:

NaCl 20.0g
Na_2SO_4 .... 3.0g
KCl 0.5g
$MgCl_2 \cdot 6H_2O$ 0.4g
NH_4Cl 0.3g
KH_2PO_4 .... 0.2g
$CaCl_2 \cdot 2H_2O$ 0.15g
Resazurin 1.0mg

Preparation of Solution A: Add components to distilled/deionized water and bring volume to 870.0mL. Mix thoroughly. Gently heat and bring to boiling. Continue boiling for 3-4 min. Allow to cool to room temperature while gassing under 80% N_2 + 20% CO_2. Continue gassing until pH reaches below 6.0. Seal the flask under 80% N_2 + 20% CO_2. Autoclave for 15 min at 15 psi pressure–121°C.

Solution B (Trace Elements Solution SL-10):
Composition per liter:

$FeCl_2·4H_2O$	1.5g
$CoCl_2·6H_2O$	190.0mg
$MnCl_2·4H_2O$	100.0mg
$ZnCl_2$	70.0mg
$Na_2MoO_4·2H_2O$	36.0mg
$NiCl_2·6H_2O$	24.0mg
H_3BO_3	6.0mg
$CuCl_2·2H_2O$	2.0mg
HCl (25% solution)	10.0mL

Preparation of Solution B (Trace Elements Solution SL-10): Add $FeCl_2·4H_2O$ to 10.0mL of HCl solution. Mix thoroughly. Add distilled/deionized water and bring volume to 1.0L. Add remaining components. Mix thoroughly. Gas under 100% N_2. Autoclave for 15 min at 15 psi pressure–121°C.

Solution C:
Composition per 100.0mL:

$NaHCO_3$	5.0g

Preparation of Solution C: Add $NaHCO_3$ to distilled/deionized water and bring volume to 100.0mL. Mix thoroughly. Filter sterilize. Gas under 80% N_2 + 20% CO_2.

Solution D:
Composition per 10.0mL:

Sodium lactate	1.5g

Preparation of Solution D: Add sodium acetate to distilled/deionized water and bring volume to 10.0mL. Mix thoroughly. Gas under 100% N_2. Autoclave for 15 min at 15 psi pressure–121°C.

Solution E (Vitamin Solution):
Composition per liter:

Pyridoxine·HCl	10.0mg
Calcium DL-pantothenate	5.0mg
Lipoic acid	5.0mg
Nicotinic acid	5.0mg
p-Aminobenzoic acid	5.0mg
Riboflavin	5.0mg
Thiamine·HCl	5.0mg
Biotin	2.0mg
Folic acid	2.0mg
Vitamin B_{12}	0.1mg

Preparation of Solution E (Vitamin Solution): Add components to distilled/deionized water and bring volume to 1.0L. Mix thoroughly. Gas under 100% N_2. Autoclave for 15 min at 15 psi pressure–121°C.

Solution F:
Composition per 10.0mL:

$Na_2S·9H_2O$	0.4g

Preparation of Solution F: Add $Na_2S·9H_2O$ to distilled/deionized water and bring volume to 10.0mL. Mix thoroughly. Gas under 100% N_2. Autoclave for 15 min at 15 psi pressure–121°C.

Preparation of Medium: Aseptically and anaerobically combine solution A with solution B, solution C, solution D, solution E, and solution F, in that order. Mix thoroughly. Anaerobically distribute into sterile tubes or flasks under 80% N_2 + 20% CO_2.

Use: For the cultivation and maintenance of *Desulfovibrio giganteus.*

Desulfovibrio gigas Medium

Composition per 1001.0mL:

Solution A	950.0mL
Solution B	40.0mL
Solution C	6.0mL
Solution D (Vitamin solution)	5.0mL

pH 7.2 ± 0.2 at 25°C

Solution A:
Composition per 950.0mL:

Na_2SO_4	2.0g
Sodium (L)-lactate	2.0g
KH_2PO_4	1.0g
NH_4Cl	0.5g
$MgSO_4·7H_2O$	0.4g
$CaCl_2·2H_2O$	0.1g
H_2SO_4 (1*M* solution)	1.0mL
Trace elements solution SL-6	1.0mL

Trace Elements Solution SL-6:
Composition per liter:

$MnCl_2·4H_2O$	0.5g
H_3BO_3	0.3g
$CoCl_2·6H_2O$	0.2g
$ZnSO_4·7H_2O$	0.1g
$Na_2MoO_4·2H_2O$	0.03g
$NiCl_2·6H_2O$	0.02g
$CuCl_2·2H_2O$	0.01g

Preparation of Trace Elements Solution SL-6: Add components to distilled/deionized water and bring volume to 1.0L. Mix thoroughly. Gas under 100% N_2. Autoclave for 15 min at 15 psi pressure–121°C.

Preparation of Solution A: Add components to distilled/deionized water and bring volume to 950.0mL. Mix thoroughly. Gently heat and bring to boiling. Continue boiling for 3-4 min. Allow to cool to room temperature while gassing under 80% N_2 + 20% CO_2. Seal the flask under 80% N_2 + 20% CO_2. Autoclave for 15 min at 15 psi pressure–121°C.

Solution B:
Composition per 40.0mL:

$NaHCO_3$	2.0g

Preparation of Solution B: Add $NaHCO_3$ to distilled/deionized water and bring volume to 40.0mL. Mix thoroughly. Filter sterilize. Gas under 80% N_2 + 20% CO_2.

Solution C:
Composition per 10.0mL:

$Na_2S·9H_2O$	0.5g

Preparation of Solution C: Add $Na_2S·9H_2O$ to distilled/deionized water and bring volume to 10.0mL. Mix thoroughly. Gas under 100% N_2. Autoclave for 15 min at 15 psi pressure–121°C.

Solution D (Vitamin Solution):
Composition per liter:

Pyridoxine·HCl	62.5g
Nicotinic acid	25.0mg
p-Aminobenzoic acid	12.5mg
Thiamine·HCl	12.5mg
Calcium DL-pantothenate	6.5mg
Biotin	2.5mg

Preparation of Solution D (Vitamin Solution): Add components to distilled/deionized water and bring volume to 1.0L. Mix thoroughly. Gas under 100% N_2. Autoclave for 15 min at 15 psi pressure–121°C.

Preparation of Medium: Aseptically and anaerobically combine 950.0mL of sterile solution A with 40.0mL of sterile solution B, 6.0mL of sterile solution C, and 5.0mL of sterile solution D. Adjust pH to 7.2. Mix thoroughly. Anaerobically distribute into sterile tubes or flasks under 80% N_2 + 20% CO_2

Use: For the cultivation and maintenance of *Desulfovibrio gigas*.

Desulfovibrio halophilus Medium

Composition per 1154.0mL:

Solution A	1.0L
Solution H	67.0mL
Solution D	50.0mL
Solution I	13.0mL
Solution E	10.0mL
Solution G	10.0mL
Solution C (Selenite-tungstate solution)	2.0mL
Solution B (Trace elements solution SL-10)	1.0mL
Solution F	1.0mL

pH 6.8 ± 0.2 at 25°C

Solution A:

Composition per liter:

Na_2SO_4	4.0g
NH_4Cl	0.25g
KH_2PO_4	0.2g
$CaCl_2 \cdot 2H_2O$	0.1g

Preparation of Solution A: Add components to distilled/deionized water and bring volume to 1.0L. Mix thoroughly. Gently heat and bring to boiling. Continue boiling for 3-4 min. Allow to cool to room temperature while gassing under 80% N_2 + 20% CO_2. Continue gassing until pH reaches below 6.0. Seal the flask under 80% N_2 + 20% CO_2. Autoclave for 15 min at 15 psi pressure–121°C.

Solution B (Trace Elements Solution SL-10):

Composition per liter:

$FeCl_2 \cdot 4H_2O$	1.5g
$CoCl_2 \cdot 6H_2O$	190.0mg
$MnCl_2 \cdot 4H_2O$	100.0mg
$ZnCl_2$	70.0mg
$Na_2MoO_4 \cdot 2H_2O$	36.0mg
$NiCl_2 \cdot 6H_2O$	24.0mg
H_3BO_3	6.0mg
$CuCl_2 \cdot 2H_2O$	2.0mg
HCl (25% solution)	10.0mL

Preparation of Solution B (Trace Elements Solution SL-10): Add $FeCl_2 \cdot 4H_2O$ to 10.0mL of HCl solution. Mix thoroughly. Add distilled/deionized water and bring volume to 1.0L. Add remaining components. Mix thoroughly. Gas under 100% N_2. Autoclave for 15 min at 15 psi pressure–121°C.

Solution C (Selenite-Tungstate Solution):

Composition per liter:

NaOH	0.5g
$Na_2WO_4 \cdot 2H_2O$	4.0mg
$Na_2SeO_3 \cdot 5H_2O$	3.0mg

Preparation of Solution C (Selenite-Tungstate Solution): Add components to distilled/deionized water and bring volume to 1.0L. Mix thoroughly. Gas under 100% N_2. Autoclave for 15 min at 15 psi pressure–121°C.

Solution D:

Composition per 50.0mL:

$NaHCO_3$	2.5g

Preparation of Solution D: Add $NaHCO_3$ to distilled/deionized water and bring volume to 50.0mL. Mix thoroughly. Gas under 80% N_2 + 20% CO_2. Autoclave for 15 min at 15 psi pressure–121°C.

Solution E:

Composition per liter:

Pyridoxine·HCl	10.0mg
Calcium DL-pantothenate	5.0mg
Lipoic acid	5.0mg
Nicotinic acid	5.0mg
p-Aminobenzoic acid	5.0mg
Riboflavin	5.0mg
Thiamine·HCl	5.0mg
Biotin	2.0mg
Folic acid	2.0mg
Vitamin B_{12}	0.1mg

Preparation of Solution E: Add components to distilled/deionized water and bring volume to 1.0L. Mix thoroughly. Gas under 100% N_2. Filter sterilize.

Solution F:

Composition per 10.0mL:

Vitamin B_{12}	0.5mg

Preparation of Solution F: Add Vitamin B_{12} to distilled/deionized water and bring volume to 10.0mL. Mix thoroughly. Gas under 100% N_2. Filter sterilize.

Solution G:

Composition per 80.0mL:

Sodium-(L)-lactate	2.25g

Preparation of Solution G: Add sodium-(L)-lactate to distilled/deionized water and bring volume to 80.0mL. Mix thoroughly. Gas under 100% N_2. In a closed bottle, heat in a boiling water bath. Shake until stearic acid dissolves. Autoclave for 15 min at 15 psi pressure–121°C. On storage, solution will solidify and should be remelted before use.

Solution H:

Composition per liter:

NaCl	70.4g
$MgCl_2 \cdot 6H_2O$	3.0g
$CaCl_2 \cdot 2H_2O$	2.2g

Preparation of Solution H: Add components to distilled/deionized water and bring volume to 80.0mL. Mix thoroughly. Gas under 100% N_2. Autoclave for 15 min at 15 psi pressure–121°C.

Solution I:

Composition per 20.0mL:

$Na_2S \cdot 9H_2O$	0.15g

Preparation of Solution I: Add $Na_2S \cdot 9H_2O$ to distilled/deionized water and bring volume to 10.0mL. Mix thoroughly. Gas under 100% N_2. Autoclave for 15 min at 15 psi pressure–121°C.

Preparation of Medium: To 1.0L of sterile solution A, add in the following order: 1.0mL of sterile solution B, 2.0mL of sterile solution C, 50.0mL of sterile solution D, 10.0mL of sterile solution E, 1.0mL of sterile solution F, 10.0mL of sterile solution G, 67.0mL of sterile solution H, and 13.0mL of sterile solution I. Mix thoroughly. Final pH of medium should be 7.2. Prior to inoculation, add 10.0-20.0mg of sodium dithionate to 1.0L of medium.

Use: For the cultivation and maintenance of *Desulfovibrio halophilus*.

Desulfovibrio halophilus Medium

Composition per liter:

NaCl	70.0g
$MgCl_2·6H_2O$	3.0g
Na_2SO_4	3.0g
$NaHCO_3$	2.5g
KCl	0.3g
NH_4Cl	0.3g
KH_2PO_4	0.2g
$Na_2S·9H_2O$	0.2g
$CaCl_2·2H_2O$	0.15g
Wolfe's vitamin solution	10.0mL
Sodium lactate	3.7mL
Trace elements solution SL-10	1.0mL

pH 6.9–7.1 at 25°C

Trace Elements Solution SL-10:
Composition per liter:

$FeCl_2·4H_2O$	1.5g
$CoCl_2·6H_2O$	190.0mg
$MnCl_2·4H_2O$	100.0mg
$ZnCl_2$	70.0mg
$Na_2MoO_4·2H_2O$	36.0mg
$NiCl_2·6H_2O$	24.0mg
H_3BO_3	6.0mg
$CuCl_2·2H_2O$	2.0mg
HCl (25% solution)	10.0mL

Preparation of Trace Elements Solution SL-10: Add $FeCl_2·4H_2O$ to 10.0mL of HCl solution. Mix thoroughly. Add distilled/deionized water and bring volume to 1.0L. Add remaining components. Mix thoroughly.

Wolfe's Vitamin Solution:
Composition per liter:

Pyridoxine·HCl	10.0mg
p-Aminobenzoic acid	5.0mg
Lipoic acid	5.0mg
Nicotinic acid	5.0mg
Riboflavin	5.0mg
Thiamine·HCl	5.0mg
Calcium DL-pantothenate	5.0mg
Biotin	2.0mg
Folic acid	2.0mg
Vitamin B_{12}	0.1mg

Preparation of Wolfe's Vitamin Solution: Add components to distilled/deionized water and bring volume to 1.0L. Mix thoroughly.

Preparation of Medium: Prepare and dispense medium under 90% N_2 + 10% CO_2. Add components, except $NaHCO_3$ and $Na_2S·9H_2O$, to distilled/deionized water and bring volume to 1.0L. Mix thoroughly. Gently heat and bring to boiling. Continue boiling for 3 min. Cool to room temperature while sparging with 90% N_2 + 10% CO_2. Add $NaHCO_3$ and $Na_2S·9H_2O$. Mix thoroughly. Anaerobically distribute into tubes. Autoclave for 15 min at 15 psi pressure–121°C.

Use: For the cultivation of *Desulfovibrio halophilus.*

Desulfovibrio inopinatus Medium (DSMZ Medium 799)

Composition per 1008.0mL:

Solution A	870.0mL
Solution C	100.0mL
Solution D	10.0mL
Solution E (Vitamin solution)	10.0mL
Solution F	10.0mL
Yeast extract solution	5.0mL
Solution B (Trace elements solution SL-10)	1.0mL
Seven vitamin solution	1.0mL
Selenite-tungstate solution	1.0mL

pH 7.1-7.4 at 25°C

Solution A:
Composition per 870.0mL:

NaCl	7.0g
Na_2SO_4	3.0g
$MgCl_2·6H_2O$	1.3g
KCl	0.5g
NH_4Cl	0.3g
KH_2PO_4	0.2g
$CaCl_2·2H_2O$	0.15g
Resazurin	1.0mg

Preparation of Solution A: Add components to distilled/deionized water and bring volume to 870.0mL Mix thoroughly.

Solution B (Trace Elements Solution SL-10):
Composition per liter:

$FeCl_2·4H_2O$	1.5g
$CoCl_2·6H_2O$	190.0mg
$MnCl_2·4H_2O$	100.0mg
$ZnCl_2$	70.0mg
$Na_2MoO_4·2H_2O$	36.0mg
$NiCl_2·6H_2O$	24.0mg
H_3BO_3	6.0mg
$CuCl_2·2H_2O$	2.0mg
HCl (25% solution)	10.0mL

Preparation of Solution B (Trace Elements Solution SL-10): Add $FeCl_2·4H_2O$ to 10.0mL of HCl solution. Mix thoroughly. Add distilled/deionized water and bring volume to 1.0L. Add remaining components. Mix thoroughly. Sparge with 100% N_2. Autoclave for 15 min at 15 psi pressure–121°C.

Solution C:
Composition per 100.0mL:

$NaHCO_3$	5.0g

Preparation of Solution C: Add $NaHCO_3$ to distilled/deionized water and bring volume to 100.0mL Mix thoroughly. Filter sterilize. Flush with 80% N_2 + 20% CO_2 to remove dissolved oxygen.

Solution D:
Composition per 10.0mL:

Na-pyruvate	2.5g

Preparation of Solution D: Add Na-pyruvate to distilled/deionized water and bring volume to 10.0mL. Mix thoroughly. Sparge with 100% N_2. Autoclave for 15 min at 15 psi pressure–121°C.

Solution E (Vitamin Solution):
Composition per liter:

Pyridoxine-HCl	10.0mg
Thiamine-HCl·$2H_2O$	5.0mg
Riboflavin	5.0mg
Nicotinic acid	5.0mg
D-Ca-pantothenate	5.0mg
p-Aminobenzoic acid	5.0mg
Lipoic acid	5.0mg
Biotin	2.0mg
Folic acid	2.0mg
Vitamin B_{12}	0.10mg

Preparation of Solution E (Vitamin Solution): Add components to distilled/deionized water and bring volume to 1.0L. Mix thoroughly. Sparge with 100% N_2. Autoclave for 15 min at 15 psi pressure–121°C.

Solution F:

Composition per 10.0mL:

$Na_2S·9H_2O$ 0.4g

Preparation of Solution F: Add $Na_2S·9H_2O$ to distilled/deionized water and bring volume to 10.0mL. Mix thoroughly. Sparge with 100% N_2. Autoclave for 15 min at 15 psi pressure–121°C.

Seven Vitamin Solution:

Composition per liter:

Pyridoxine hydrochloride 300.0mg
Thiamine-HCl·$2H_2O$ 200.0mg
Nicotinic acid 200.0mg
Vitamin B_{12} 100.0mg
Calcium pantothenate 100.0mg
p-Aminobenzoic acid 80.0mg
D(+)-Biotin 20.0mg

Preparation of Seven Vitamin Solution: Add components to distilled/deionized water and bring volume to 1.0L. Sparge with 100% N_2. Mix thoroughly. Filter sterilize.

Selenite-Tungstate Solution:

Composition per liter:

NaOH 0.5g
$Na_2WO_4·2H_2O$ 4.0mg
$Na_2SeO_3·5H_2O$ 3.0mg

Preparation of Selenite-Tungstate Solution: Add components to distilled/deionized water and bring volume to 1.0L. Mix thoroughly. Sparge with 100% N_2. Filter sterilize.

Yeast Extract Solution:

Composition per 10.0mL:

Yeast extract 1.0g

Preparation of Yeast Extract Solution: Add yeast extract to distilled/deionized water and bring volume to 10.0mL. Mix thoroughly. Sparge with 100% N_2. Autoclave under 100% N_2 for 15 min at 15 psi pressure–121°C. Cool to room temperature.

Preparation of Medium: Gently heat solution A and bring to boiling. Boil solution A for a few minutes. Cool to room temperature. Gas with 80% N_2 + 20% CO_2 gas mixture to reach a pH below 6. Autoclave for 15 min at 15 psi pressure–121°C. Cool to room temperature. Sequentially add 1.0mL solution B, 100.0mL solution C, 10.0mL solution D, 10.0mL solution E, 10.0mL solution F, 5.0mL yeast extract solution, 1.0mL selenite-tungstate solution, and 1.0ml seven vitamin solution. Distribute aseptically and anaerobically under 80% N_2 + 20% CO_2 into sterile tubes or bottles.

Use: For the cultivation of *Desulfovibrio inopinatus.*

Desulfovibrio magneticus Medium (DSMZ Medium 896)

Composition per liter:

Na-fumarate 0.58g
Na-pyruvate 0.44g
KH_2PO_4 .... 0.2g
NH_4Cl 0.06g
Cysteine-HCl·H_2O 0.05g
Vitamin solution 8.0mL
Trace elements solution 4.0mL
Fe(III)quinate solution 2.0mL

pH 7.0 ± 0.2 at 25°C

Trace Elements Solution:

Composition per liter:

$MgSO_4·7H_2O$ 3.0g
Nitrilotriacetic acid 1.5g
NaCl 1.0g
$MnSO_4·2H_2O$ 0.5g
$CoSO_4·7H_2O$ 0.18g
$ZnSO_4·7H_2O$ 0.18g
$CaCl_2·2H_2O$ 0.1g
$FeSO_4·7H_2O$ 0.1g
$NiCl_2·6H_2O$ 0.025g
$KAl(SO_4)_2·12H_2O$ 0.02g
H_3BO_3 .... 0.01g
$Na_2MoO_4·4H_2O$ 0.01g
$CuSO_4·5H_2O$ 0.01g
$Na_2SeO_3·5H_2O$ 0.3mg

Preparation of Trace Elements Solution: Add nitrilotriacetic acid to 500.0mL of distilled/deionized water. Dissolve by adjusting pH to 6.5 with KOH. Add remaining components. Add distilled/deionized water to 1.0L. Mix thoroughly.

Vitamin Solution:

Composition per liter:

Pyridoxine-HCl 10.0mg
Thiamine-HCl·$2H_2O$ 5.0mg
Riboflavin 5.0mg
Nicotinic acid 5.0mg
D-Ca-pantothenate 5.0mg
p-Aminobenzoic acid 5.0mg
Lipoic acid 5.0mg
Biotin 2.0mg
Folic acid 2.0mg
Vitamin B_{12} 0.1mg

Preparation of Vitamin Solution: Add components to distilled/deionized water and bring volume to 1.0L. Mix thoroughly. Sparge with 80% H_2 + 20% CO_2. Filter sterilize.

Ferric Quinate Solution:

Composition per 100.0mL:

$FeCl_3·6H_2O$ 0.45g
Quinic acid 0.19g

Preparation of Ferric Quinate Solution: Add components to distilled/deionized water and bring volume to 100.0mL. Sparge with N_2. Autoclave for 15 min at 15 psi pressure–121°C. Cool to 25°C.

Preparation of Medium: Add components, except vitamin solution and ferric quinate solution to distilled/deionized water and bring volume to 990.0mLL. Purge medium with N_2 gas for 10 min. Mix thoroughly. Autoclave for 15 min at 15 psi pressure–121°C. Cool to 25°C. Aseptically and anerobically add 8.0mL vitamin solution and 2.0mL ferric quinate solution. Mix thoroughly. Adjust pH to 7.0. Purge medium with N_2 gas for 10 min. Under the same atmosphere, aseptically distribute medium to sterile tubes or bottles.

Use: For the cultivation of *Desulfovibrio magneticus.*

Desulfovibrio Marine Medium (DSMZ Medium 163)

Composition per liter:

Solution A 980.0mL
Solution B 10.0mL
Solution C 10.0mL

pH 7.8 ± 0.2 at 25°C

Solution A:

Composition per 980.0mL:

NaCl	25.0g
DL-Na-lactate	2.0g
$MgSO_4 \cdot 7H_2O$	2.0g
Yeast extract	1.0g
NH_4Cl	1.0g
Na_2SO_4	1.0g
K_2HPO_4	0.5g
$CaCl_2 \cdot 2H_2O$	0.1g
Resazurin	1.0mg

Preparation of Solution A: Add components to 980.0mL distilled/deionized water. Mix thoroughly.

Solution B:

Composition per 10.0mL:

$FeSO_4 \cdot 7H_2O$	0.5g

Preparation of Solution B: Add components to 10.0mL distilled/deionized water. Mix thoroughly.

Solution C:

Composition per 10.0mL:

Na-thioglycolate	0.1g
Ascorbic acid	0.1g

Preparation of Solution C: Add components to 10.0mL distilled/deionized water. Mix thoroughly.

Preparation of Medium: Bring solution A to a boil for a few minutes. Cool to room temperature while gassing with oxygen-free N_2 gas. Add solutions B and C. Mix thoroughly. Adjust pH to 7.8 with NaOH. Distribute under N_2 into anaerobic tubes. During distribution continuously swirl the medium to keep the grey precipitate suspended. Autoclave for 15 min at 15 psi pressure–121°C.

Use: For the cultivation of *Desulfovibrio vulgaris, Desulfovibrio desulfuricans, Desulfovibrio senezii,* and *Desulfovibrio vietnamensis.*

Desulfovibrio Medium

Composition per 1056.5mL:

$(NH_4)_2SO_4$	5.3g
Sodium acetate	2.0g
NaCl	1.0g
KH_2PO_4	0.5g
$MgSO_4 \cdot 7H_2O$	0.2g
$CaCl_2 \cdot 2H_2O$	0.1g
Na_2CO_3 solution	50.0mL
Solution 1	10.0mL
Solution 2	1.0mL

pH 7.2 ± 0.2 at 25°C

Solution 1:

Composition per liter:

Nitrilotriacetic acid	12.8g
$FeCl_2 \cdot 4H_2O$	0.3g
$CoCl_2 \cdot 6H_2O$	0.17g
$MnCl_2 \cdot 4H_2O$	0.1g
$ZnCl_2$	0.1g
$CuCl_2$	0.02g
H_3BO_3	0.01g
$Na_2MoO_4 \cdot 2H_2O$	0.01g

Preparation of Solution 1: Add nitrilotriacetic acid to 500.0mL of distilled/deionized water. Dissolve by adjusting pH to 6.5 with NaOH. Add remaining components. Readjust pH to 7.2 with H_2SO_4 or NaOH. Add distilled/deionized water to 1.0L.

Solution 2:

Composition per 100.0mL:

Resazurin	0.2g

Preparation of Solution 2: Add resazurin to distilled/deionized water and bring volume to 100.0mL. Mix thoroughly.

Na_2CO_3 Solution:

Composition per 100.0mL:

Na_2CO_3	8.0g

Preparation of Na_2CO_3 Solution Solution: Add Na_2CO_3 to distilled/deionized water and bring volume to 100.0mL. Mix thoroughly. Filter sterilize. Gas with 100% N_2 for 20 min.

HCl Solution:

Composition per 100.0mL:

HCl	25.0mL

Preparation of HCl Solution: Add HCl to distilled/deionized water and bring volume to 100.0mL. Mix thoroughly. Autoclave for 15 min at 15 psi pressure–121°C. Cool to 25°C. Gas with 100% N_2 for 20 min.

$Na_2S_2O_4$ Solution:

Composition per 100.0mL:

$Na_2S_2O_4$	8.7g

Preparation of $Na_2S_2O_4$ Solution: Add $Na_2S_2O_4$ to distilled/deionized water and bring volume to 100.0mL. Mix thoroughly. Autoclave for 15 min at 15 psi pressure–121°C. Cool to 25°C. Gas with 100% N_2 for 20 min.

Preparation of Medium: Add components—except Na_2CO_3 solution, HCl solution, and $Na_2S_2O_4$ solution—to distilled/deionized water and bring volume to 1.0L. Mix thoroughly. Gently heat and bring to boiling. Autoclave for 15 min at 15 psi pressure–121°C. Cool to 45°–50°C. Anaerobically and aseptically add 50.0mL of sterile Na_2CO_3 solution, 5.5mL of sterile HCl solution, and 1.0mL of sterile $Na_2S_2O_4$ solution. Mix thoroughly. Anaerobically and aseptically distribute into sterile tubes or flasks.

Use: For the isolation, cultivation, and enrichment of *Desulfovibrio* species.

Desulfovibrio Medium

Composition per liter of tap water:

Agar	15.0g
Glucose	5.0g
Peptone	5.0g
Beef extract	3.0g
$MgSO_4$	1.5g
Na_2SO_4	1.5g
Yeast extract	0.2g
$Fe(NH_4)_2(SO_4)_2$	0.1g

pH 7.0 ± 0.2 at 25°C

Preparation of Medium: Sterilize by autoclaving for 15 min at 15 lbs pressure (121°C).

Use: For the cultivation and maintenance of *Desulfomaculum nigrificans, Desulfovibrio desulfuricans,* and *Desulfovibrio gigas.*

Desulfovibrio Medium

Composition per liter:

Solution A	980.0mL
Solution B	10.0mL
Solution C	10.0mL

pH 7.8 ± 0.2 at 25°C

Solution A:

Composition per 980.0mL:

DL-Sodium lactate 2.0g
$MgSO_4 \cdot 7H_2O$ 2.0g
Na_2SO_4 .... 1.0g
NH_4Cl 1.0g
Yeast extract 1.0g
K_2HPO_4 .... 0.5g
$CaCl_2 \cdot 2H_2O$ 0.1g
Resazurin 1.0mg

Preparation of Solution A: Add components to distilled/deionized water and bring volume to 980.0mL. Mix thoroughly. Adjust pH to 7.4.

Solution B:

Composition per 10.0mL:

$FeSO_4 \cdot 7H_2O$ 0.5g

Preparation of Solution B: Add $FeSO_4 \cdot 7H_2O$ to distilled/deionized water and bring volume to 10.0mL. Mix thoroughly.

Solution C:

Composition per 10.0mL:

Ascorbic acid 0.1g
Sodium thioglycolate 0.1g

Preparation of Solution C: Add components to distilled/deionized water and bring volume to 10.0mL. Mix thoroughly.

Preparation of Medium: Combine 980.0mL of solution A, 10.0mL of solution B, and 10.0mL of solution C. Mix thoroughly. Adjust pH to 7.8. Distribute into tubes or flasks. Autoclave for 15 min at 15 psi pressure–121°C.

Use: For the cultivation and maintenance of *Desulfovibrio desulfuricans*, *Desulfovibrio giganteus*, and *Desulfovibrio vulgaris*.

Desulfovibrio Medium with Lactate

Composition per liter:

Agar 15.0g
Lactate 10.0g
Glucose 5.0g
Peptone 5.0g
Beef extract 3.0g
$MgSO_4$ 1.5g
Na_2SO_4 .... 1.5g
Yeast extract 0.2g
$Fe(NH_4)_2(SO_4)_2$ 0.1g

pH 7.0 ± 0.2 at 25°C

Preparation of Medium: Add components to tap water and bring volume to 1.0L. Mix thoroughly. Gently heat and bring to boiling. Distribute into tubes or flasks. Autoclave for 15 min at 15 psi pressure–121°C. Pour into sterile Petri dishes or leave in tubes.

Use: For the cultivation and maintenance of *Desulfovibrio desulfuricans.*

Desulfovibrio Medium with NaCl

Composition per liter:

NaCl 30.0g
Agar 15.0g
Glucose 5.0g
Peptone 5.0g
Beef extract 3.0g
$MgSO_4$ 1.5g
Na_2SO_4 .... 1.5g
Yeast extract 0.2g
$Fe(NH_4)_2(SO_4)_2$ 0.1g

pH 7.0 ± 0.2 at 25°C

Preparation of Medium: Add components to tap water and bring volume to 1.0L. Mix thoroughly. Gently heat and bring to boiling. Distribute into tubes or flasks. Autoclave for 15 min at 15 psi pressure–121°C. Pour into sterile Petri dishes or leave in tubes.

Use: For the cultivation and maintenance of *Desulfovibrio desulfuricans* and *Desulfovibrio salexigens.*

Desulfovibrio MG-1 Medium (DSMZ Medium 615)

Composition per liter:

Na_2SO_4 .... 4.5g
Glycerol 2.0g
NH_4Cl 1.0g
Yeast extract 1.0g
Na_3-citrate·$2H_2O$ 0.6g
KH_2PO_4 .... 0.5g
Na-thioglycolate 0.1g
$MgSO_4 \cdot 7H_2O$ 0.06g
$CaCl_2 \cdot 2H_2O$ 0.04g
$FeSO_4 \cdot 7H_2O$ 4.0mg
Resazurin 0.5mg

pH 6.9 ± 0.2 at 25°C

Preparation of Medium: Prepare and dispense medium under 100% N_2 gas atmosphere. Add components to distilled/deionized water and bring volume to 1.0L. Mix thoroughly. Adjust pH to 7.5. Distribute into tubes or flasks. Autoclave for 15 min at 15 psi pressure–121°C.

Use: For the cultivation of *Desulfovibrio* sp.

Desulfovibrio sapovorans Medium

Composition per 1009.0mL:

Solution A 850.0mL
Solution C 100.0mL
Solution G 20.0mL
Solution D 10.0mL
Solution E (Wolfe's vitamin solution) 10.0mL
Solution H 10.0mL
Solution F 6.6mL
Solution B (Trace elements solution SL-10) 1.0mL
Solution I 0.4mL

pH 7.7 ± 0.2 at 25°C

Solution A:

Composition per 920.0mL:

Na_2SO_4 .... 3.0g
NaCl 1.0g
KCl 0.5g
$MgCl_2 \cdot 6H_2O$ 0.4g
NH_4Cl 0.3g
KH_2PO_4 .... 0.2g
$CaCl_2 \cdot 2H_2O$ 0.15g
Resazurin 0.5mg

Preparation of Solution A: Prepare and dispense solution anaerobically under 90% N_2 + 10% CO_2. Add components to distilled/deionized water and bring volume to 920.0mL. Mix thoroughly. Gently heat and bring to boiling. Continue boiling until resazurin turns colorless, indicating reduction, and a pH of 6.0 is reached. Cap with rubber stoppers. Autoclave for 15 min at 15 psi pressure–121°C. Cool to 25°C.

Solution B (Trace Elements Solution SL-10):
Composition per liter:

$FeCl_2 \cdot 4H_2O$	1.5g
$CoCl_2 \cdot 6H_2O$	0.19g
$MnCl_2 \cdot 4H_2O$	0.10g
$ZnCl_2$	0.070g
$Na_2MoO_4 \cdot 2H_2O$	0.036g
$NiCl_2 \cdot 6H_2O$	0.024g
H_3BO_3	6.0mg
$CuCl_2 \cdot 2H_2O$	2.0mg
HCl (25% solution)	10.0mL

Preparation of Solution B (Trace Elements Solution SL-10): Add the $FeCl_2 \cdot 4H_2O$ to 10.0mL of HCl solution. Mix thoroughly. Bring volume to approximately 900.0mL with distilled/deionized water. Mix thoroughly. Adjust pH to 6.0 with NaOH. Bring volume to 1.0L with distilled/deionized water. Filter sterilize. Aseptically gas under 100% N_2 for 20 min.

Solution C:
Composition per 100.0mL:

$NaHCO_3$	5.0g

Preparation of Solution C: Add $NaHCO_3$ to distilled/deionized water and bring volume to 100.0mL. Mix thoroughly. Filter sterilize. Aseptically gas under 90% N_2 + 10% CO_2 for 20 min.

Solution D:
Composition per 10.0mL:

Sodium butyrate	0.7g
Sodium caproate	0.3g
Sodium octanoate	0.15g

Preparation of Solution D: Prepare and dispense solution anaerobically under 90% N_2 + 10% CO_2. Add components to distilled/deionized water and bring volume to 10.0mL. Mix thoroughly. Cap with a rubber stopper. Autoclave for 15 min at 15 psi pressure–121°C. Cool to 25°C.

Solution E (Wolfe's Vitamin Solution):
Composition per liter:

Pyridoxine·HCl	0.01g
Thiamine·HCl	5.0mg
Riboflavin	5.0mg
Nicotinic acid	5.0mg
Calcium pantothenate	5.0mg
p-Aminobenzoic acid	5.0mg
Thioctic acid	5.0mg
Biotin	2.0mg
Folic acid	2.0mg
Cyanocobalamin	0.1mg

Preparation of Solution E (Wolfe's Vitamin Solution): Add components to distilled/deionized water and bring volume to 1.0L. Mix thoroughly. Filter sterilize. Aseptically gas under 100% N_2 for 20 min.

Solution F:
Composition per 6.6mL:

$AlCl_3 \cdot 6H_2O$ (4.9% solution)	5.0mL
Na_2CO_3 (10.6% solution)	1.6mL

Preparation of Solution F: Combine both solutions. Mix thoroughly. Gas with 100% N_2. Cap with a rubber stopper. Autoclave for 15 min at 15 psi pressure–121°C. Cool to 25°C.

Solution G:
Composition per 10.0mL:

Rumen fluid, clarified	20.0mL

Preparation of Solution G: Gas rumen fluid under 100% N_2 for 20 min. Cap with a rubber stopper. Autoclave for 15 min at 15 psi pressure–121°C. Cool to 25°C.

Solution H:
Composition per 10.0mL:

$Na_2S \cdot 9H_2O$	0.4g

Preparation of Solution H: Add $Na_2S \cdot 9H_2O$ to distilled/deionized water and bring volume to 10.0mL. Gas under 100% N_2 for 20 min. Cap with a rubber stopper. Autoclave for 15 min at 15 psi pressure–121°C. Cool to 25°C.

Solution I:
Composition per 10.0mL:

$Na_2S_2O_4$	0.5g

Preparation of Solution I: Add $Na_2S_2O_4$ to distilled/deionized water and bring volume to 10.0mL. Mix thoroughly. Filter sterilize. Aseptically gas under 100% N_2 for 20 min. Prepare solution freshly.

Preparation of Medium: Prepare and dispense medium under 90% N_2 + 10% CO_2. To 850.0mL of cooled, sterile solution A, aseptically and anaerobically add in the following order: 1.0mL of sterile solution B, 1000.0mL of sterile solution C, 10.0mL of sterile solution D, 10.0mL of sterile solution E, 6.6mL of sterile solution F, 20.0mL of sterile solution G, and 10.0mL of sterile solution H. Mix thoroughly. Immediately prior to inoculation, aseptically and anaerobically add 0.4mL of sterile solution I. Mix thoroughly. Aseptically and anaerobically distribute into sterile tubes or flasks.

Use: For the cultivation and maintenance of *Desulfovibrio sapovorans*.

Desulfovibrio sax Medium (DSMZ Medium 383a)

Composition per 1022.6mL:

Solution A	930.0mL
Solution C	50.0mL
Solution E	20.0mL
Solution D	10.0mL
Solution G	10.0mL
Solution B	1.0mL
Solution F	1.0mL
Vitamin B_{12} solution	0.5mL
Yeast extract solution	0.1mL

pH 7.3 at 25°C

Solution A:
Composition per 930.0mL:

NaCl	21.0g
Na_2SO_4	3.0g
$MgCl_2 \cdot 6H_2O$	3.0g
KCl	0.5g
NH_4Cl	0.3g
KH_2PO_4	0.2g
$CaCl_2 \cdot 2H_2O$	0.15g
Resazurin	1.0mg

Preparation of Solution A: Add components to distilled/deionized water and bring volume to 930.0mL. Mix thoroughly. Sparge with 80% N_2 + 20% CO_2 gas until saturated. Autoclave for 15 min at 15 psi pressure–121°C. Cool to 25°C.

Solution B:
Composition per liter:

$FeCl_2 \cdot 4H_2O$	1.5g
H_3BO_3	300.0mg

$CoCl_2·6H_2O$ 190.0mg
$MnCl_2·4H_2O$ 100.0mg
$ZnCl_2$ 70.0mg
$Na_2MoO_4·2H_2O$ 36.0mg
$NiCl_2·6H_2O$ 24.0mg
$CuCl_2·2H_2O$ 2.0mg
HCl (25% solution) 7.7mL

Preparation of Solution B: Add $FeCl_2·4H_2O$ to 10.0mL of HCl solution. Mix thoroughly. Add distilled/deionized water and bring volume to 1.0L. Add remaining components. Mix thoroughly. Sparge with 100% N_2. Autoclave for 15 min at 15 psi pressure–121°C.

Solution C:
Composition per 100.0mL:
$NaHCO_3$ 5.0g

Preparation of Solution C: Add $NaHCO_3$ to distilled/deionized water and bring volume to 100.0mL. Mix thoroughly. Sparge with 100% CO_2 until saturated, approximately 20 minutes. Filter sterilize under 100% CO_2 into a sterile, gas-tight 100.0mL screw-capped bottle.

Solution D:
Composition per 10.0mL:
Na-benzoate 0.5g

Preparation of Solution D: Add Na-benzoate to distilled/deionized water and bring volume to 10.0mL. Sparge with N_2. Filter sterilize. Store anaerobically.

Solution E:
Composition per liter:
Pyridoxine-HCl 10.0mg
Thiamine-HCl·$2H_2O$ 5.0mg
Riboflavin 5.0mg
Nicotinic acid 5.0mg
D-Ca-pantothenate 5.0mg
p-Aminobenzoic acid 5.0mg
Lipoic acid 5.0mg
Biotin 2.0mg
Folic acid 2.0mg
Vitamin B_{12} 0.1mg

Preparation of Solution E: Add components to distilled/deionized water and bring volume to 1.0L. Mix thoroughly. Sparge with 100% N_2. Filter sterilize.

Solution F:
Composition per liter:
NaOH 0.5g
$Na_2SeO_3·5H_2O$ 3.0mg

Preparation of Solution F: Add components to distilled/deionized water and bring volume to 1.0L. Mix thoroughly. Sparge with 100% N_2. Filter sterilize.

Solution G:
Composition per 10.0mL:
$Na_2S·9H_2O$ 0.4g

Preparation of Solution G: Add $Na_2S·9H_2O$ to distilled/deionized water and bring volume to 10.0mL. Sparge with N_2. Autoclave for 15 min at 15 psi pressure–121°C. Cool to 25°C. Store anaerobically.

Vitamin B_{12} Solution:
Composition per 100.0mL:
Vitamin B_{12} 10.0mg

Vitamin B_{12} Solution: Add Vitamin B_{12} to distilled/deionized water and bring volume to 100.0mL. Mix thoroughly. Sparge under 100% N_2 gas for 3 min. Filter sterilize.

Yeast Extract Solution:
Composition per 10.0mL:
Yeast extract 1.0g

Preparation of Yeast Extract Solution: Add yeast extract to distilled/deionized water and bring volume to 10.0mL. Mix thoroughly. Sparge with 100% N_2. Autoclave under 100% N_2 for 15 min at 15 psi pressure–121°C. Cool to room temperature.

Preparation of Medium: Add solution B, solution C, solution D, solution E, Vitamin B_{12} solution, yeast extract solution, solution F, and solution G to solution A in that order under N_2 gas. Adjust the pH to 7.3.

Use: For the cultivation of *Desulfotignum balticum (Desulfoarculus* sp.*)*.

Desulfovibrio SHV Medium

Composition per 1003.0mL:
Solution A 870.0mL
Solution C 100.0mL
Solution D 10.0mL
Solution E (Vitamin solution) 10.0mL
Solution F 10.0mL
Solution B (Trace elements solution SL-10) 1.0mL
Solution G (Selenite-tungstate solution) 1.0mL
Solution H (Seven vitamin solution) 1.0mL
pH 7.1–7.4 at 25°C

Solution A:
Composition per 870.0mL:
NaCl 7.0g
Na_2SO_4 .... 3.0g
$MgCl_2·6H_2O$ 1.3g
KCl 0.5g
NH_4Cl 0.3g
KH_2PO_4 .... 0.2g
$CaCl_2·2H_2O$ 0.15g
Resazurin 1.0mg

Preparation of Solution A: Add components to distilled/deionized water and bring volume to 870.0mL. Mix thoroughly. Gently heat and bring to boiling. Continue boiling for 3-4 min. Allow to cool to room temperature while gassing under 80% N_2 + 20% CO_2. Continue gassing until pH reaches below 6.0. Seal the flask under 80% N_2 + 20% CO_2. Autoclave for 15 min at 15 psi pressure–121°C.

Solution B (Trace Elements Solution SL-10):
Composition per liter:
$FeCl_2·4H_2O$ 1.5g
$CoCl_2·6H_2O$ 190.0mg
$MnCl_2·4H_2O$ 100.0mg
$ZnCl_2$ 70.0mg
$Na_2MoO_4·2H_2O$ 36.0mg
$NiCl_2·6H_2O$ 24.0mg
H_3BO_3 .... 6.0mg
$CuCl_2·2H_2O$ 2.0mg
HCl (25% solution) 10.0mL

Preparation of Solution B (Trace Elements Solution SL-10): Add $FeCl_2·4H_2O$ to 10.0mL of HCl solution. Mix thoroughly. Add distilled/deionized water and bring volume to 1.0L. Add remaining components. Mix thoroughly. Sparge with 100% N_2. Autoclave for 15 min at 15 psi pressure–121°C.

Solution C:
Composition per 100.0mL:
$NaHCO_3$ 5.0g

Preparation of Solution C: Add $NaHCO_3$ to distilled/deionized water and bring volume to 100.0mL. Mix thoroughly. Filter sterilize. Sparge with 80% N_2 + 20% CO_2.

Solution D:

Composition per 10.0mL:

Sodium lactate	4.0g

Preparation of Solution D: Add sodium lactate to distilled/deionized water and bring volume to 10.0mL. Mix thoroughly. Sparge with 100% N_2. Autoclave for 15 min at 15 psi pressure–121°C.

Solution E (Vitamin Solution):

Composition per liter:

Pyridoxine·HCl	10.0mg
Calcium DL-pantothenate	5.0mg
Lipoic acid	5.0mg
Nicotinic acid	5.0mg
p-Aminobenzoic acid	5.0mg
Riboflavin	5.0mg
Thiamine·HCl	5.0mg
Biotin	2.0mg
Folic acid	2.0mg
Vitamin B_{12}	0.1mg

Preparation of Solution E (Vitamin Solution): Add components to distilled/deionized water and bring volume to 1.0L. Mix thoroughly. Sparge with 100% N_2. Autoclave for 15 min at 15 psi pressure–121°C.

Solution F:

Composition per 10.0mL:

$Na_2S{\cdot}9H_2O$	0.4g

Preparation of Solution F: Add $Na_2S{\cdot}9H_2O$ to distilled/deionized water and bring volume to 10.0mL. Mix thoroughly. Sparge with 100% N_2. Autoclave for 15 min at 15 psi pressure–121°C.

Solution G (Selenite-Tungstate Solution):

Composition per liter:

NaOH	0.5g
$Na_2WO_4{\cdot}2H_2O$	4.0mg
$Na_2SeO_3{\cdot}5H_2O$	3.0mg

Preparation of Solution G (Selenite-Tungstate Solution): Add components to distilled/deionized water and bring volume to 1.0L. Mix thoroughly. Sparge with 100% N_2. Autoclave for 15 min at 15 psi pressure–121°C.

Solution H (Seven Vitamin Solution) :

Composition per liter:

Pyridoxine·HCl	0.3g
Thiamine·HCl	0.2g
Nicotinic acid	0.2g
Calcium DL-pantothenate	0.1g
Vitamin B_{12}	0.1g
p-Aminobenzoic acid	80.0mg
Biotin	20.0mg

Preparation of Solution H (Seven Vitamin Solution): Add components to distilled/deionized water and bring volume to 1.0L. Mix thoroughly.

Preparation of Medium: Aseptically and anaerobically combine solution A with solution B, solution C, solution D, solution E, solution F, solution G, and solution H, in that order. Mix thoroughly. Anaerobically distribute into sterile tubes or flasks under 80% N_2 + 20% CO_2.

Use: For the cultivation of *Desulfovibrio* species.

Desulfovibrio sp. Medium (DSMZ Medium 200)

Composition per 1001.0mL:

Solution A	870.0mL
Solution C	100.0mL
Solution D	10.0mL
Solution E (Vitamin solution)	10.0mL
Solution F	10.0mL
Solution B (Trace elements solution SL-10)	1.0mL

pH 6.8-7.0 at 25°C

Solution A:

Composition per 870.0mL:

NaCl	20.0g
$MgCl_2{\cdot}6H_2O$	3.1g
Na_2SO_4	3.0g
KH_2PO_4	0.2g
NH_4Cl	0.3g
KCl	0.5g
$CaCl_2{\cdot}2H_2O$	0.15g
Resazurin	1.0mg

Preparation of Solution A: Add components to distilled/deionized water and bring volume to 870.0mL. Mix thoroughly.

Solution B (Trace Elements Solution SL-10):

Composition per liter:

$FeCl_2{\cdot}4H_2O$	1.5g
$CoCl_2{\cdot}6H_2O$	190.0mg
$MnCl_2{\cdot}4H_2O$	100.0mg
$ZnCl_2$	70.0mg
$Na_2MoO_4{\cdot}2H_2O$	36.0mg
$NiCl_2{\cdot}6H_2O$	24.0mg
H_3BO_3	6.0mg
$CuCl_2{\cdot}2H_2O$	2.0mg
HCl (25% solution)	10.0mL

Preparation of Solution B (Trace Elements Solution SL-10): Add $FeCl_2{\cdot}4H_2O$ to 10.0mL of HCl solution. Mix thoroughly. Add distilled/deionized water and bring volume to 1.0L. Add remaining components. Mix thoroughly. Sparge with 100% N_2. Autoclave for 15 min at 15 psi pressure–121°C.

Solution C:

Composition per 100.0mL:

$NaHCO_3$	5.0g

Preparation of Solution C: Add $NaHCO_3$ to distilled/deionized water and bring volume to 100.0mL Mix thoroughly. Filter sterilize. Flush with 80% N_2 + 20% CO_2 to remove dissolved oxygen.

Solution D:

Composition per 10.0mL:

Na-butyrate	0.7g
Na-caproate	0.3g
Na-octanoate	0.15g

Preparation of Solution D: Add components to distilled/deionized water and bring volume to 10.0mL. Mix thoroughly. Sparge with 100% N_2. Autoclave for 15 min at 15 psi pressure–121°C.

Solution E (Vitamin Solution):

Composition per liter:

Pyridoxine-HCl	10.0mg
Thiamine-HCl·$2H_2O$	5.0mg
Riboflavin	5.0mg
Nicotinic acid	5.0mg
D-Ca-pantothenate	5.0mg
p-Aminobenzoic acid	5.0mg

Lipoic acid 5.0mg
Biotin 2.0mg
Folic acid 2.0mg
Vitamin B_{12} 0.10mg

Preparation of Solution E (Vitamin Solution): Add components to distilled/deionized water and bring volume to 1.0L. Mix thoroughly. Sparge with 100% N_2. Autoclave for 15 min at 15 psi pressure–121°C.

Solution F:

Composition per 10.0mL:

$Na_2S·9H_2O$ 0.4g

Preparation of Solution F: Add $Na_2S·9H_2O$ to distilled/deionized water and bring volume to 10.0mL. Mix thoroughly. Sparge with 100% N_2. Autoclave for 15 min at 15 psi pressure–121°C.

Preparation of Medium: Gently heat solution A and bring to boiling. Boil solution A for a few minutes. Cool to room temperature. Gas with 80% N_2 + 20% CO_2 gas mixture to reach a pH below 6. Autoclave for 15 min at 15 psi pressure–121°C. Cool to room temperature. Sequentially add 1.0mL solution B, 100.0mL solution C, 10.0mL solution D, 10.0mL solution E, and 10.0mL solution F. Distribute anaerobically under 80% N_2 + 20% CO_2 into appropriate vessels.

Use: For the cultivation of *Desulfovibrio* sp.

Desulfovibrio sulfodismutans Medium (DSMZ Medium 386)

Composition per 1002mL:

Solution A 920.0mL
Solution C 50.0mL
Solution D 10.0mL
Solution F 10.0mL
Solution G 10.0mL
Solution B 1.0mL
Solution E 1.0mL

pH 7.1-7.4 at 25°C

Solution A:

Composition per 920mL:

NaCl 1.0g
KCl 0.5g
$MgCl_2·6H_2O$ 0.4g
KH_2PO_4 .... 0.2g
NH_4Cl 0.3g
$CaCl_2·2H_2O$ 0.15g

Preparation of Solution A: Add components to distilled/deionized water and bring volume to 920.0mL. Mix thoroughly. Sparge with 80% N_2 + 20% CO_2 gas until saturated. Autoclave for 15 min at 15 psi pressure–121°C. Cool to 25°C.

Solution B:

Composition per liter:

$FeCl_2·4H_2O$ 1.5g
H_3BO_3 .... 300.0mg
$CoCl_2·6H_2O$ 190.0mg
$MnCl_2·4H_2O$ 100.0mg
$ZnCl_2$ 70.0mg
$Na_2MoO_4·2H_2O$ 36.0mg
$NiCl_2·6H_2O$ 24.0mg
$CuCl_2·2H_2O$ 2.0mg
HCl (25% solution) 7.7mL

Preparation of Solution B: Add $FeCl_2·4H_2O$ to 10.0mL of HCl solution. Mix thoroughly. Add distilled/deionized water and bring volume to 1.0L. Add remaining components. Mix thoroughly. Sparge with 100% N_2. Autoclave for 15 min at 15 psi pressure–121°C.

Solution C:

Composition per 100.0mL:

$NaHCO_3$ 5.0g

Preparation of Solution C: Add $NaHCO_3$ to distilled/deionized water and bring volume to 100.0mL. Mix thoroughly. Sparge with 80% N_2 + 20% CO_2 gas until saturated, approximately 20 minutes. Filter sterilize under 100% CO_2 into a sterile, gas-tight 100.0mL screw-capped bottle.

Solution D:

Composition per 10.0mL:

Na-acetate·$3H_2O$ 0.3g

Preparation of Solution D: Add Na_2-acetate to distilled/deionized water and bring volume to 10.0mL. Sparge with N_2. Filter sterilize. Store anaerobically.

Solution E:

Composition per 1mL:

Ca-D-pantothenate 50.0μg
D(+)-Biotin 10.0μg

Preparation of Solution E: Add components to distilled/deionized water and bring volume to 1.0mL. Mix thoroughly. Sparge with 100% N_2. Filter sterilize.

Solution F:

Composition per 10.0mL:

$Na_2S·9H_2O$ 0.4g

Preparation of Solution F: Add $Na_2S·9H_2O$ to distilled/deionized water and bring volume to 10.0mL. Sparge with N_2. Autoclave for 15 min at 15 psi pressure–121°C. Cool to 25°C. Store anaerobically.

Solution G:

Composition per liter:

NaOH 0.5g
$Na_2SeO_3·5H_2O$ 3.0mg

Preparation of Solution G: Add components to distilled/deionized water and bring volume to 1.0L. Mix thoroughly. Sparge with 100% N_2. Filter sterilize.

Preparation of Medium: Add solution B, solution C, solution D, solution E, solution F, and solution G to solution A in that order under 80% N_2 + 20% CO_2 gas. Adjust the pH to 7.1-7.4. When growth has started feed culture again with same amount of solution G. After a further two days repeat feeding once more.

Use: For the cultivation of *Desulfovibrio sulfodismutans.*

Desulfovibrio zosterae Medium (DSMZ Medium 383c)

Composition per 1023.5mL:

Solution A 930.0mL
Solution C 50.0mL
Solution E 20.0mL
Solution D 10.0mL
Solution G 10.0mL
Solution B 1.0mL
Solution F 1.0mL
Selenite-tungstate solution 1.0mL
Vitamin B_{12} solution 0.5mL

pH 7.3 at 25°C

Solution A:

Composition per 930.0mL:

NaCl 21.0g
Na_2SO_4 .... 3.0g

$MgCl_2 \cdot 6H_2O$ 3.0g
KCl 0.5g
NH_4Cl 0.3g
KH_2PO_4 .... 0.2g
$CaCl_2 \cdot 2H_2O$ 0.15g
Resazurin 1.0mg

Preparation of Solution A: Add components to distilled/deionized water and bring volume to 930.0mL. Mix thoroughly. Sparge with 80% N_2 + 20% CO_2 gas until saturated. Autoclave for 15 min at 15 psi pressure–121°C. Cool to 25°C.

Solution B:
Composition per liter:
$FeCl_2 \cdot 4H_2O$ 1.5g
H_3BO_3 .... 300.0mg
$CoCl_2 \cdot 6H_2O$ 190.0mg
$MnCl_2 \cdot 4H_2O$ 100.0mg
$ZnCl_2$ 70.0mg
$Na_2MoO_4 \cdot 2H_2O$ 36.0mg
$NiCl_2 \cdot 6H_2O$ 24.0mg
$CuCl_2 \cdot 2H_2O$ 2.0mg
HCl (25% solution) 7.7mL

Preparation of Solution B: Add $FeCl_2 \cdot 4H_2O$ to 10.0mL of HCl solution. Mix thoroughly. Add distilled/deionized water and bring volume to 1.0L. Add remaining components. Mix thoroughly. Sparge with 100% N_2. Autoclave for 15 min at 15 psi pressure–121°C.

Solution C:
Composition per 100.0mL:
$NaHCO_3$ 5.0g

Preparation of Solution C: Add $NaHCO_3$ to distilled/deionized water and bring volume to 100.0mL. Mix thoroughly. Sparge with 100% CO_2 until saturated, approximately 20 min. Filter sterilize under 100% CO_2 into a sterile, gas-tight 100.0mL screw-capped bottle.

Solution D:
Composition per 10.0mL:
Na-lactate 1.25g

Preparation of Solution D: Add Na_2-lactate to distilled/deionized water and bring volume to 10.0mL. Sparge with N_2. Filter sterilize. Store anaerobically.

Solution E:
Composition per liter:
Pyridoxine hydrochloride 300.0mg
Thiamine-HCl·$2H_2O$ 200.0mg
Nicotinic acid 200.0mg
Vitamin B_{12} 100.0mg
Calcium pantothenate 100.0mg
p-Aminobenzoic acid 80.0mg
D(+)-Biotin 20.0mg

Preparation of Solution E: Add components to distilled/deionized water and bring volume to 1.0L. Mix thoroughly. Sparge with 100% N_2. Filter sterilize.

Solution F:
Composition per liter:
NaOH 0.5g
$Na_2SeO_3 \cdot 5H_2O$ 3.0mg

Preparation of Solution F: Add components to distilled/deionized water and bring volume to 1.0L. Mix thoroughly. Sparge with 100% N_2. Filter sterilize.

Solution G:
Composition per 10.0mL:
$Na_2S \cdot 9H_2O$ 0.4g

Preparation of Solution G: Add $Na_2S \cdot 9H_2O$ to distilled/deionized water and bring volume to 10.0mL. Sparge with N_2. Autoclave for 15 min at 15 psi pressure–121°C. Cool to 25°C. Store anaerobically.

Vitamin B_{12} Solution:
Composition per 100.0mL:
Vitamin B_{12} 10.0mg

Vitamin B_{12} Solution: Add Vitamin B_{12} to distilled/deionized water and bring volume to 100.0mL. Mix thoroughly. Sparge under 100% N_2 gas for 3 min. Filter sterilize.

Selenite-Tungstate Solution
Composition per liter:
NaOH 0.5g
$Na_2WO_4 \cdot 2H_2O$ 4.0mg
$Na_2SeO_3 \cdot 5H_2O$ 3.0mg

Preparation of Selenite-Tungstate Solution: Add components to distilled/deionized water and bring volume to 1.0L. Mix thoroughly. Sparge with 100% N_2. Filter sterilize.

Preparation of Medium: Add solution B, solution C, solution D, solution E, Vitamin B_{12} solution, selenite-tunstate solution, solution F, and solution G to solution A in that order under N_2 gas. Adjust the pH to 7.3.

Use: For the cultivation of *Desulfovibrio zosterae (Desulfovibrio* sp.*)*.

Desulfovigra adipica Medium (DSMZ Medium 868)

Composition per 2L:
Solution A 940.0mL
Solution E 50.0mL
Solution K 50.0mL
Solution G 20.0mL
Solution F 10.0mL
Solution M (Vitamin solution) 10.0mL
Solution H 7.0mL
Solution B 1.0mL
Solution C 1.0mL
Solution D 1.0mL
Solution I 1.0mL
Solution J (Trace elements solution SL-10) 1.0mL
Solution L (Selenite-tungstate solution) 1.0mL
Solution N .... variable
pH 7.1 ± 0.2 at 25°C

Solution A:
Composition per 940.0mL:
NaCl 1.0g
KCl 0.5g
$MgCl_2 \cdot 6H_2O$ 0.4g
NH_4Cl 0.25g
KH_2PO_4 .... 0.2g
$CaCl_2 \cdot 2H_2O$ 0.15g
Resazurin 0.5mg

Preparation of Solution A: Prepare under 80% N_2 + 20% CO_2 gas atmosphere. Add componentsto distilled/deionized water and bring volume to 940.0mL. Mix thoroughly. Adjust pH to 7.2. Sparge with 80% N_2 + 20% CO_2. Autoclave for 15 min at 15 psi pressure–121°C. Cool to 25°C.

Solution B:
Composition per liter:

$FeCl_2 \cdot 4H_2O$	1.5g
$CoCl_2 \cdot 6H_2O$	190.0mg
$MnCl_2 \cdot 4H_2O$	100.0mg
$ZnCl_2$	70.0mg
$Na_2MoO_4 \cdot 2H_2O$	36.0mg
$NiCl_2 \cdot 6H_2O$	24.0mg
H_3BO_3	6.0mg
$CuCl_2 \cdot 2H_2O$	2.0mg
HCl (25% solution)	10.0mL

Preparation of Solution B: Add $FeCl_2 \cdot 4H_2O$ to 10.0mL of HCl solution. Mix thoroughly. Add distilled/deionized water and bring volume to 1.0L. Add remaining components. Mix thoroughly. Sparge with 80% N_2 + 20% CO_2. Autoclave for 15 min at 15 psi pressure–121°C. Cool to 25°C.

Solution C:
Composition per liter:

Pyridoxine hydrochloride	300.0mg
Thiamine-HCl·$2H_2O$	200.0mg
Nicotinic acid	200.0mg
Vitamin B_{12}	100.0mg
Calcium pantothenate	100.0mg
p-Aminobenzoic acid	80.0mg
D(+)-Biotin	20.0mg

Preparation of Solution C: Add components to distilled/deionized water and bring volume to 1.0L. Sparge with 100% N_2. Mix thoroughly. Filter sterilize.

Solution D:
Composition per liter:

NaOH	0.5g
$Na_2WO_4 \cdot 2H_2O$	4.0mg
$Na_2SeO_3 \cdot 5H_2O$	3.0mg

Preparation of Solution D: Add components to distilled/deionized water and bring volume to 1.0L. Mix thoroughly. Sparge with 100% N_2. Filter sterilize.

Solution E:
Composition per 100.0mL:

$NaHCO_3$	5.0g

Preparation of Solution E: Add $NaHCO_3$ to distilled/deionized water and bring volume to 100.0mL. Mix thoroughly. Sparge with 100% N_2 gas mixture. Autoclave for 15 min at 15 psi pressure–121°C. Cool to 25°C.

Solution F:
Composition per 10mL:

Yeast extract	1.0g

Preparation of Solution F: Add yeast extract to distilled/deionized water and bring volume to 10.0mL. Mix thoroughly. Sparge with 100% N_2 gas mixture. Autoclave for 15 min at 15 psi pressure–121°C. Cool to 25°C.

Solution G:
Composition per 20mL:

$Na_2S \cdot 9H_2O$	0.625g

Preparation of Solution G: Add $Na_2S \cdot 9H_2O$ to distilled/deionized water and bring volume to 20.0mL. Mix thoroughly. Autoclave under 100% N_2 for 15 min at 15 psi pressure–121°C. Cool to 25°C.

Solution H:
Composition per 10mL:

Na_2SO_4	1.0g

Preparation of Solution H: Add Na_2SO_4 to distilled/deionized water and bring volume to 10.0mL. Mix thoroughly. Sparge with 100% N_2 gas mixture. Autoclave for 15 min at 15 psi pressure–121°C. Cool to 25°C.

Solution I:
Composition per 10mL:

Propanol	1.0g

Preparation of Solution I: Add propanol to distilled/deionized water and bring volume to 10.0mL. Mix thoroughly. Sparge with 100% N_2 gas mixture. Filter sterilize.

Solution J (Trace Elements Solution SL-10):
Composition per liter:

$FeCl_2 \cdot 4H_2O$	1.5g
$CoCl_2 \cdot 6H_2O$	190.0mg
$MnCl_2 \cdot 4H_2O$	100.0mg
$ZnCl_2$	70.0mg
$Na_2MoO_4 \cdot 2H_2O$	36.0mg
$NiCl_2 \cdot 6H_2O$	24.0mg
H_3BO_3	6.0mg
$CuCl_2 \cdot 2H_2O$	2.0mg
HCl (25% solution)	10.0mL

Preparation of Solution J (Trace Elements Solution SL-10): Add $FeCl_2 \cdot 4H_2O$ to 10.0mL of HCl solution. Mix thoroughly. Add distilled/deionized water and bring volume to 1.0L. Add remaining components. Mix thoroughly. Sparge with 80% N_2 + 20% CO_2. Autoclave for 15 min at 15 psi pressure–121°C.

Solution K:
Composition per 100.0mL:

$NaHCO_3$	5.0g

Preparation of Solution K: Add $NaHCO_3$ to distilled/deionized water and bring volume to 100.0mL. Mix thoroughly. Sparge with 100% N_2 gas mixture. Autoclave for 15 min at 15 psi pressure–121°C. Cool to 25°C.

Solution L (Selenite-Tungstate Solution):
Composition per liter:

NaOH	0.5g
$Na_2WO_4 \cdot 2H_2O$	4.0mg
$Na_2SeO_3 \cdot 5H_2O$	3.0mg

Preparation of Solution L (Selenite-Tungstate Solution): Add components to distilled/deionized water and bring volume to 1.0L. Mix thoroughly. Sparge with 100% N_2. Filter sterilize.

Solution M (Vitamin Solution):
Composition per liter:

Pyridoxine-HCl	10.0mg
Thiamine-HCl·$2H_2O$	5.0mg
Riboflavin	5.0mg
Nicotinic acid	5.0mg
D-Ca-pantothenate	5.0mg
p-Aminobenzoic acid	5.0mg
Lipoic acid	5.0mg
Biotin	2.0mg
Folic acid	2.0mg
Vitamin B_{12}	0.1mg

Preparation of Solution M (Vitamin Solution): Add components to distilled/deionized water and bring volume to 1.0L. Mix thoroughly. Sparge with 80% H_2 + 20% CO_2. Filter sterilize.

Solution N:
Composition per 100mL:

Na_2CO_3	5.0g

Preparation of Solution N: Add Na_2CO_3 to distilled/deionized water and bring volume to 100.0mL. Mix thoroughly. Sparge with 100% N_2 gas mixture. Filter sterilize.

Preparation of Medium: Prepare and dispense medium under 80% N_2 + 20% CO_2 gas atmosphere. Sequentially add 1.0mL solution B, 1.0mL solution C, 1.0mL solution D, 50.0mL solution E, 10.0mL solution F, 20.0mL solution G, 7.0mL solution H, 1.0mL solution I, 1.0mL solution J, 50.0mL solution K, 1.0mL solution L, and 10.0mL solution M, to 940.0mL solution A. Mix thoroughly. Adjust pH to 7.1 with solution N. Distribute anaerobically under 80% N_2 + 20% CO_2 into appropriate vessels.

Use: For the cultivation of *Desulfovigra adipica (Desulfobacterium* sp.*)*.

Desulfurella Medium

Composition per liter:

Sulfur, powdered	10.0g
Sodium acetate	5.0g
$CaCl_2 \cdot 2H_2O$	0.33g
KCl	0.33g
KH_2PO_4	0.33g
$MgCl_2 \cdot 6H_2O$	0.33g
NH_4Cl	0.33g
Yeast extract	0.1g
Resazurin	1.0 mg
$NaHCO_3$ solution	40.0mL
$Na_2S \cdot 9H_2O$ solution	10.0mL
Wolfe's vitamin solution	10.0mL
Trace elements solution SL-10	1.0 mL

pH 6.8–7.0 at 25°C

$NaHCO_3$ Solution:
Composition per 40.0mL:

$NaHCO_3$	2.0g

Preparation of $NaHCO_3$ Solution: Add $NaHCO_3$ to distilled/deionized water and bring volume to 40.0mL. Mix thoroughly. Gas under 100% N_2. Autoclave for 15 min at 15 psi pressure–121°C.

$Na_2S \cdot 9H_2O$ Solution:
Composition per 10.0mL:

$Na_2S \cdot 9H_2O$	0.5g

Preparation of $Na_2S \cdot 9H_2O$ Solution: Add $Na_2S \cdot 9H_2O$ to distilled/deionized water and bring volume to 10.0mL. Mix thoroughly. Sparge with 100% N_2. Autoclave for 15 min at 15 psi pressure–121°C.

Wolfe's Vitamin Solution:
Composition per liter:

Pyridoxine·HCl	10.0mg
Calcium D-(+)-pantothenate	5.0mg
Nicotinic acid	5.0mg
p-Aminobenzoic acid	5.0mg
Riboflavin	5.0mg
Thiamine·HCl	5.0mg
Thioctic acid	5.0mg
Biotin	2.0mg
Folic acid	2.0mg
Cyanocobalamine	100.0μg

Preparation of Wolfe's Vitamin Solution: Add components to distilled/deionized water and bring volume to 1.0L. Mix thoroughly. Filter sterilize.

Trace Elements Solution SL-10:
Composition per liter:

$FeCl_2 \cdot 4H_2O$	1.5g
$CoCl_2 \cdot 6H_2O$	190.0mg
$MnCl_2 \cdot 4H_2O$	100.0mg
$ZnCl_2$	70.0mg
$Na_2MoO_4 \cdot 2H_2O$	36.0mg
$NiCl_2 \cdot 6H_2O$	24.0mg
H_3BO_3	6.0mg
$CuCl_2 \cdot 2H_2O$	2.0mg
HCl (25% solution)	10.0mL

Preparation of Trace Elements Solution SL-10: Add $FeCl_2 \cdot 4H_2O$ to 10.0mL of HCl solution. Mix thoroughly. Add distilled/deionized water and bring volume to 1.0L. Add remaining components. Mix thoroughly. Sparge with 100% N_2. Autoclave for 15 min at 15 psi pressure–121°C.

Preparation of Medium: Prepare and dispense medium under 80% N_2 + 20% CO_2. Add components, except $NaHCO_3$, Wolfe's vitamin solution, and $Na_2S \cdot 9H_2O$ solution, to distilled/deionized water and bring volume to 940.0mL. Mix thoroughly. Adjust pH to 5.9. Do not autoclave. Sterilize medium by heating to 100°C for 1 hr on three consecutive days. Prior to inoculation, aseptically and anaerobically add 40.0mL of sterile $NaHCO_3$ solution, 10.0mL of sterile Wolfe's vitamin solution, and 10.0mL of sterile $Na_2S \cdot 9H_2O$ solution. Mix thoroughly. Final pH should be 6.8–7.0.

Use: For the cultivation of *Desulfurella* species, especially *Desulfurella acetivorans*.

Desulfurella II Medium (DSMZ Medium 480c)

Composition per liter:

MOPS [3-(*N*-morpholino) propane sulfonic acid]	3.0g
Sulfur, powder	1.0g
NH_4Cl	0.33g
$CaCl_2 \cdot 2H_2O$	0.33g
$MgCl_2 \cdot 6H_2O$	0.33g
KCl	0.33g
KH_2PO_4	0.33g
Yeast extract	0.1g
Resazurin	1.0mg
$NaHCO_3$ solution	40.0mL
$Na_2S \cdot 9H_2O$ solution	10.0mL
Vitamin solution	10.0mL
Substrate solution	10.0mL
Trace elements solution SL-10	1.0mL

pH 6.9 ± 0.2 at 25°C

$NaHCO_3$ Solution:
Composition per 10.0mL:

$NaHCO_3$	2.0g

Preparation of $NaHCO_3$ Solution: Add $NaHCO_3$ to distilled/deionized water and bring volume to 10.0mL. Mix thoroughly. Autoclave for 15 min at 15 psi pressure–121°C. Cool to 25°C. Must be prepared freshly.

Trace Elements Solution SL-10:
Composition per liter:

$FeCl_2 \cdot 4H_2O$	1.5g
$CoCl_2 \cdot 6H_2O$	190.0mg
$MnCl_2 \cdot 4H_2O$	100.0mg
$ZnCl_2$	70.0mg
$Na_2MoO_4 \cdot 2H_2O$	36.0mg
$NiCl_2 \cdot 6H_2O$	24.0mg
H_3BO_3	6.0mg
$CuCl_2 \cdot 2H_2O$	2.0mg
HCl (25% solution)	10.0mL

Preparation of Trace Elements Solution SL-10: Add $FeCl_2 \cdot 4H_2O$ to 10.0mL of HCl solution. Mix thoroughly.

Add distilled/deionized water and bring volume to 1.0L. Add remaining components. Mix thoroughly. Sparge with 100% N_2. Autoclave for 15 min at 15 psi pressure–121°C.

Vitamin Solution:

Composition per liter:

Pyridoxine-HCl	10.0mg
Thiamine-HCl·$2H_2O$	5.0mg
Riboflavin	5.0mg
Nicotinic acid	5.0mg
D-Ca-pantothenate	5.0mg
p-Aminobenzoic acid	5.0mg
Lipoic acid	5.0mg
Biotin	2.0mg
Folic acid	2.0mg
Vitamin B_{12}	0.1mg

Preparation of Vitamin Solution: Add components to distilled/deionized water and bring volume to 1.0L. Mix thoroughly. Sparge with 80% H_2 + 20% CO_2. Filter sterilize.

$Na_2S·9H_2O$ Solution:

Composition per 10.0mL:

$Na_2S·9H_2O$	0.5g

Preparation of $Na_2S·9H_2O$ Solution: Add $Na_2S·9H_2O$ to distilled/deionized water and bring volume to 10.0mL. Sparge with N_2. Autoclave for 15 min at 15 psi pressure–121°C. Cool to 25°C. Store anaerobically.

Substrate Solution:

Composition per 10.0mL:

Na-lactate	2.5g

Preparation of Substrate Solution: Add Na-lactate to distilled/deionized water and bring volume to 10.0mL. Sparge with N_2. Autoclave for 15 min at 15 psi pressure–121°C. Cool to 25°C. Store anaerobically.

Preparation of Medium: Prepare and dispense medium under an oxygen-free 80% N_2 + 20% CO_2 gas mixture. Add components, except sulfur, substrate solution, vitamin solution, $NaHCO_3$ solution, and $Na_2S·9H_2O$ solution to 930.0mL distilled/deionized water. Mix thoroughly. Sparge for 30 min with 80% N_2 + 20% CO_2. Adjust pH to 5.9 with concentrated NaOH. Distribute under 80% N_2 + 20% CO_2 into anaerobic tubes or bottles containing sulfur powder (100 mg S per 10.0mL medium). Autoclave 20 min at 110°C. Sparge with 80% N_2 + 20% CO_2. Aseptically and anaerobically add 10.0mL sterile $Na_2S·9H_2O$ solution, 10.0mL sterile vitamin solution, 10.0mL substrate solution, and 40.0mL sterile $NaHCO_3$ solution per liter medium.

Use: For the cultivation of *Thermoproteus uzoniensis* and *Desulfurella kamchatkensis*.

Desulfurella multipotens Medium (DSMZ Medium 480a)

Composition per liter:

Sulfur, powder	10.0g
Na-butyrate	5.0g
NH_4Cl	0.33g
$CaCl_2·2H_2O$	0.33g
$MgCl_2·6H_2O$	0.33g
KCl	0.33g
KH_2PO_4	0.33g
Yeast extract	0.1g
Resazurin	1.0mg
$NaHCO_3$ solution	40.0mL
$Na_2S·9H_2O$ solution	10.0mL
Vitamin solution	10.0mL
Trace elements solution SL-10	1.0mL

pH 6.9 ± 0.2 at 25°C

$NaHCO_3$ Solution:

Composition per 10.0mL:

$NaHCO_3$	2.0g

Preparation of $NaHCO_3$ Solution: Add $NaHCO_3$ to distilled/deionized water and bring volume to 10.0mL. Mix thoroughly. Autoclave for 15 min at 15 psi pressure–121°C. Cool to 25°C. Must be prepared freshly.

Trace Elements Solution SL-10:

Composition per liter:

$FeCl_2·4H_2O$	1.5g
$CoCl_2·6H_2O$	190.0mg
$MnCl_2·4H_2O$	100.0mg
$ZnCl_2$	70.0mg
$Na_2MoO_4·2H_2O$	36.0mg
$NiCl_2·6H_2O$	24.0mg
H_3BO_3	6.0mg
$CuCl_2·2H_2O$	2.0mg
HCl (25% solution)	10.0mL

Preparation of Trace Elements Solution SL-10: Add $FeCl_2·4H_2O$ to 10.0mL of HCl solution. Mix thoroughly. Add distilled/deionized water and bring volume to 1.0L. Add remaining components. Mix thoroughly. Sparge with 100% N_2. Autoclave for 15 min at 15 psi pressure–121°C.

Vitamin Solution:

Composition per liter:

Pyridoxine-HCl	10.0mg
Thiamine-HCl·$2H_2O$	5.0mg
Riboflavin	5.0mg
Nicotinic acid	5.0mg
D-Ca-pantothenate	5.0mg
p-Aminobenzoic acid	5.0mg
Lipoic acid	5.0mg
Biotin	2.0mg
Folic acid	2.0mg
Vitamin B_{12}	0.1mg

Preparation of Vitamin Solution: Add components to distilled/deionized water and bring volume to 1.0L. Mix thoroughly. Sparge with 80% H_2 + 20% CO_2. Filter sterilize.

$Na_2S·9H_2O$ Solution:

Composition per 10.0mL:

$Na_2S·9H_2O$	0.5g

Preparation of $Na_2S·9H_2O$ Solution: Add $Na_2S·9H_2O$ to distilled/deionized water and bring volume to 10.0mL. Sparge with N_2. Autoclave for 15 min at 15 psi pressure–121°C. Cool to 25°C. Store anaerobically.

Preparation of Medium: Prepare and dispense medium under an oxygen-free 80% N_2 + 20% CO_2 gas mixture. Add components, except vitamin solution, $NaHCO_3$ solution, and $Na_2S·9H_2O$ solution to 940.0mL distilled/deionized water. Mix thoroughly. Sparge with 80% N_2 + 20% CO_2. Adjust pH to 5.9 with concentrated NaOH. Sterilize medium by heating for 1 h at 90-100°C on three subsequent days. Sparge with 80% N_2 + 20% CO_2. Before use, aseptically and anaerobically add 10.0mL sterile vitamin solution, 10.0mL sterile $Na_2S·9H_2O$ solution, and 40.0mL sterile $NaHCO_3$ solution. Mix thoroughly. Aseptically and anaerobically distribute into sterile tubes or flasks.

Use: For the cultivation of *Desulfurella multipotens*.

Desulfurococcus Medium

Composition per 1300.0mL:

Solution C 500.0mL
Solution B 450.0mL
Solution A 300.0mL
Solution D 50.0mL

Solution A:

Composition per 300.0mL:

$(NH_4)_2SO_4$ 1.3g
KH_2PO_4 .... 0.28g
$MgSO_4 \cdot 7H_2O$ 0.25g
$CaCl_2 \cdot 2H_2O$ 0.07g
$FeSO_4 \cdot 7H_2O$ 0.028g
$Na_2B_4O_7 \cdot 10H_2O$ 4.5mg
$MnCl_2 \cdot 4H_2O$ 1.8mg
$ZnSO_4 \cdot 7H_2O$ 0.22mg
$CuCl_2 \cdot 2H_2O$ 0.05mg
$Na_2MoO_4 \cdot 2H_2O$ 0.03mg
$VOSO_4 \cdot 2H_2O$ 0.03mg
$CoSO_4 \cdot 7H_2O$ 0.01mg

Preparation of Solution A: Add components to distilled/deionized water and bring volume to 300.0mL. Mix thoroughly. Gently heat and bring to boiling. Autoclave for 15 min at 15 psi pressure–121°C. Cool to 25°C. Gas under 100% N_2 for 20 min.

Solution B:

Composition per 450.0mL:

Sulfur 5.0g

Preparation of Solution B: Add sulfur to distilled/deionized water and bring volume to 450.0mL. Autoclave for 30 min at 0 psi pressure–100°C on three consecutive days. Gas under 100% N_2 for 20 min.

Solution C:

Composition per 500.0mL:

Pancreatic digest of casein 2.0g
Yeast extract 2.0g
Resazurin 1.0mg

Preparation of Solution C: Add components to distilled/deionized water and bring volume to 500.0mL. Mix thoroughly. Gently heat and bring to boiling. Autoclave for 15 min at 15 psi pressure–121°C. Cool to 25°C. Gas under 100% N_2 for 20 min.

Solution D:

$Na_2S \cdot 9H_2O$ 0.5g

Preparation of Solution D: Add components to distilled/deionized water and bring volume to 50.0mL. Mix thoroughly. Autoclave for 15 min at 15 psi pressure–121°C. Cool to 25°C. Gas under 100% N_2 for 20 min.

Preparation of Medium: Aseptically combine solutions A–D under nitrogen gas. Seal containers with butyl rubber stoppers.

Use: For the cultivation and maintenance of *Desulfurococcus mobilis* and *Desulfurococcus mucosus.*

Desulfuromonas acetexigenes Medium (DSMZ Medium 647)

Composition per liter:

Sulfur, powdered 5.0g
KH_2PO_4 .... 1.0g
NH_4Cl 0.5g
$MgSO_4 \cdot 7H_2O$ 0.4g
$CaCl_2 \cdot 2H_2O$ 0.1g
$NaHCO_3$ solution 33.0mL
Na-pyruvate solution 10.0mL
$Na_2S \cdot 9H_2O$ solution 10.0mL
Seven vitamin solution 1.0mL
Trace elements solution SL-10 .... 1.0mL

pH 7.2 ± 0.2 at 25°C

$Na_2S \cdot 9H_2O$ Solution:

Composition per 10mL:

$Na_2S \cdot 9H_2O$ 0.5g

Preparation of $Na_2S \cdot 9H_2O$ Solution: Add $Na_2S \cdot 9H_2O$ to distilled/deionized water and bring volume to 10.0mL. Mix thoroughly. Autoclave under 100% N_2 for 15 min at 15 psi pressure–121°C. Cool to room temperature.

$NaHCO_3$ Solution:

Composition per 100.0mL:

$NaHCO_3$ 5.0g

Preparation of $NaHCO_3$ Solution: Add $NaHCO_3$ to distilled/deionized water and bring volume to 100.0mL. Mix thoroughly. Sparge with 80% N_2 + 20% CO_2. Filter sterilize.

Na-pyruvate Solution:

Composition per 10.0mL:

Na-pyruvate 0.6g

Preparation of Na-pyruvate Solution: Add Na-pyruvate to distilled/deionized water and bring volume to 10.0mL. Mix thoroughly. Sparge with 100% N_2. Filter sterilize.

Seven Vitamin Solution:

Composition per liter:

Pyridoxine hydrochloride 300.0mg
Thiamine-HCl·$2H_2O$ 200.0mg
Nicotinic acid 200.0mg
Vitamin B_{12} 100.0mg
Calcium pantothenate 100.0mg
p-Aminobenzoic acid 80.0mg
D(+)-Biotin 20.0mg

Preparation of Seven Vitamin Solution: Add components to distilled/deionized water and bring volume to 1.0L. Sparge with 100% N_2. Mix thoroughly. Filter sterilize.

Trace Elements Solution SL-10:

Composition per liter:

$FeCl_2 \cdot 4H_2O$ 1.5g
$CoCl_2 \cdot 6H_2O$ 190.0mg
$MnCl_2 \cdot 4H_2O$ 100.0mg
$ZnCl_2$ 70.0mg
$Na_2MoO_4 \cdot 2H_2O$ 36.0mg
$NiCl_2 \cdot 6H_2O$ 24.0mg
H_3BO_3 .... 6.0mg
$CuCl_2 \cdot 2H_2O$ 2.0mg
HCl (25% solution) 10.0mL

Preparation of Trace Elements Solution SL-10: Add $FeCl_2 \cdot 4H_2O$ to 10.0mL of HCl solution. Mix thoroughly. Add distilled/deionized water and bring volume to 1.0L. Add remaining components. Mix thoroughly. Sparge with 80% N_2 + 20% CO_2. Autoclave for 15 min at 15 psi pressure–121°C.

Preparation of Medium: Sulfur is sterilized by steaming for 3 hr on each of 3 successive days. Prepare and dispense medium under 80% N_2 + 20% CO_2 gas atmosphere. Add components, except sulfur, $NaHCO_3$ solution, Na-pyruvate solution, $Na_2S \cdot 9H_2O$ solution, seven vitamin solution, and trace elements solution SL-10, to distilled/deionized water and bring volume to 945.0mL. Mix thoroughly. Adjust pH to 7.2. Sparge with 80% N_2 + 20% CO_2. Autoclave for 15 min at 15 psi pressure–121°C. Aseptically and anaerobically add 5.0g sterile sulfur, 33.0mL $NaHCO_3$ solution, 10.0mL Na-pyruvate solution, 10.0mL $Na_2S·9H_2O$ solution, 1.0mL seven vitamin solution, and 1.0mL trace elements solution SL-10. Mix thoroughly. Aseptically and anaerobically distribute into sterile tubes or bottles.

Use: For the cultivation of *Desulfuromonas thiophila* and *Desulfuromonas acetexigens.*

Desulfuromonas acetoxidans Medium

Composition per 1001.0mL:

Fumaric acid 1.5g
KH_2PO_4 .... 1.0g
NH_4Cl 0.5g
Sodium acetate 0.5g
$MgSO_4·7H_2O$ 0.4g
$CaCl_2·2H_2O$ 0.1g
Resazurin 0.5mg
$NaHCO_3$ solution 40.0mL
Trace elements solution SL-4 .... 10.0mL
$Na_2S·9H_2O$ solution 6.0mL
Vitamin solution 5.0mL

pH 7.5 ± 0.2 at 25°C

$NaHCO_3$ Solution:

Composition per 40.0mL:

$NaHCO_3$ 2.0g

Preparation of $NaHCO_3$ Solution: Add $NaHCO_3$ to distilled/deionized water and bring volume to 40.0mL. Mix thoroughly. Gas under 80% N_2 + 20% CO_2. Autoclave for 15 min at 15 psi pressure–121°C.

Trace Elements Solution SL-4:

Composition per liter:

EDTA 0.5g
$FeSO_4·7H_2O$ 0.2g
Trace elements solution SL-6 .... 100.0mL

Trace Elements Solution SL-6:

Composition per liter:

$MnCl_2·4H_2O$ 0.5g
H_3BO_3 .... 0.3g
$CoCl_2·6H_2O$ 0.2g
$ZnSO_4·7H_2O$ 0.1g
$Na_2MoO_4·2H_2O$ 0.03g
$NiCl_2·6H_2O$ 0.02g
$CuCl_2·2H_2O$ 0.01g

Preparation of Trace Elements Solution SL-6: Add components to distilled/deionized water and bring volume to 1.0L. Mix thoroughly. Gas under 100% N_2. Autoclave for 15 min at 15 psi pressure–121°C.

Preparation of Trace Elements Solution SL-4: Add components to distilled/deionized water and bring volume to 100.0mL. Mix thoroughly.

$Na_2S·9H_2O$ Solution:

Composition per 6.0mL:

$Na_2S·9H_2O$ 0.3g

Preparation of $Na_2S·9H_2O$ Solution: Add $Na_2S·9H_2O$ to distilled/deionized water and bring volume to 10.0mL. Mix thoroughly. Gas under 100% N_2. Autoclave for 15 min at 15 psi pressure–121°C.

Vitamin Solution:

Composition per liter:

Pyridoxine·HCl 62.5g
Nicotinic acid 25.0mg
p-Aminobenzoic acid 12.5mg
Thiamine·HCl 12.5mg
Calcium DL-pantothenate 6.5mg
Biotin 2.5mg

Preparation of Vitamin Solution: Add components to distilled/deionized water and bring volume to 1.0L. Mix thoroughly. Adjust pH to 7.5. Gas under 100% N_2. Filter sterilize.

Preparation of Medium: Add components, except $NaHCO_3$ solution, $Na_2S·9H_2O$ solution, and vitamin solution, to distilled/deionized water and bring volume to 1.0L. Mix thoroughly. Adjust pH to 5.0. Sparge with 100% N_2. Autoclave for 15 min at 15 psi pressure–121°C. Aseptically and anaerobically add 40.0mL of sterile $NaHCO_3$ solution, 6.0mL of sterile $Na_2S·9H_2O$ solution, and 5.0mL of sterile vitamin solution. Mix thoroughly. Aseptically and anaerobically distribute into sterile screw-capped bottles or tubes. Fill completely, leaving only a small gas bubble.

Use: For the cultivation and maintenance of *Desulfuromonas acetoxidans.*

Desulfuromonas Medium

Composition per 1051.0mL:

Elemental sulfur slurry 10.0g
Solution 1 .... 1.0L
Solution 3 .... 40.0mL
Solution 4 .... 6.0mL
Solution 5 .... 5.0mL
Solution 2 .... 1.0mL

pH 7.2 ± 0.2 at 25°C

Solution 1:

Composition per liter:

NaCl 20.0g
$MgCl_2·6H_2O$ 3.0g
KH_2PO_4 .... 1.0g
NH_4Cl 0.3g
$CaCl_2·2H_2O$ 0.1g
HCl (2*N* solution) 4.0mL

Preparation of Solution 1: Add components to distilled/deionized water and bring volume to 1.0L. Mix thoroughly. Autoclave for 15 min at 15 psi pressure–121°C. Cool to 25°C.

Solution 2:

Composition per liter:

Disodium EDTA 5.2g
$CoCl_2·6H_2O$ 1.9g
$FeCl_2·4H_2O$ 1.5g
$MnCl_2·4H_2O$ 1.0g
$ZnCl_2$ 0.7g
H_3BO_3 .... 0.62g
$Na_2MoO_4·2H_2O$ 0.36g
$NiCl_2·6H_2O$ 0.24g
$CuCl_2·2H_2O$ 0.17g

pH 6.5 ± 0.2 at 25°C

Preparation of Solution 2: Add components to distilled/deionized water and bring volume to 1.0L. Mix thoroughly. Adjust pH to 6.5. Autoclave for 15 min at 15 psi pressure–121°C. Cool to 25°C.

Solution 3:

Composition per 100.0mL:

$NaHCO_3$ 10.0g

Preparation of Solution 3: Add $NaHCO_3$ to distilled/deionized water and bring volume to 100.0mL. Mix thoroughly. Autoclave for 15 min at 15 psi pressure–121°C. Cool to 25°C.

Solution 4:

Composition per 100.0mL:

$Na_2S \cdot 9H_2O$ 5.0g

Preparation of Solution 4: Add $Na_2S \cdot 9H_2O$ to distilled/deionized water and bring volume to 100.0mL. Mix thoroughly. Autoclave for 15 min at 15 psi pressure–121°C. Cool to 25°C.

Solution 5:

Composition per 200.0mL:

Pyridoxamine·HCl 0.01g
Nicotinic acid 4.0mg
p-Aminobenzoic acid 2.0mg
Thiamine 2.0mg
Cyanocobalamin 1.0mg
Pantothenic acid 1.0mg
Biotin 0.5mg

Preparation of Solution 5: Add components to distilled/deionized water and bring volume to 200.0mL. Mix thoroughly. Filter sterilize.

Elemental Sulfur Slurry:

Composition per 10.0g:

Sulfur flowers 10.0g

Preparation of Elemental Sulfur Slurry: Add highly purified sulfur flowers to a mortar and grind to a fine powder. Add sufficient distilled/deionized water to produce a slurry. Distribute into 100.0mL screw-capped bottles in 20.0mL volumes. Autoclave for 30 min at 10 psi pressure–115°C. Decant supernatant solution. Reserve sulfur slurry.

Preparation of Medium: To 1.0L of cooled, sterile solution 1, aseptically add 1.0mL of sterile solution 2, 40.0mL of sterile solution 3, 6.0mL of sterile solution 4, and 5.0mL of sterile solution 5. Mix thoroughly. Adjust pH to 7.2. Aseptically distribute into sterile 50.0mL screw-capped bottles. Fill bottles completely with medium except for a pea-sized air bubble. Aseptically add a pea-sized piece of sulfur slurry to each 50.0mL of medium.

Use: For the isolation and cultivation of marine *Desulfuromonas* species.

Desulfuromonas Medium

Composition per 1031.0mL:

Elemental sulfur slurry 10.0g
Solution 1 .... 1.0L
Solution 3 .... 20.0mL
Solution 4 .... 6.0mL
Solution 5 .... 5.0mL
Solution 2 .... 1.0mL

pH 7.2 ± 0.2 at 25°C

Solution 1:

Composition per liter:

KH_2PO_4 .... 1.0g
$MgCl_2 \cdot 6H_2O$ 0.4g
NH_4Cl 0.3g
$CaCl_2 \cdot 2H_2O$ 0.1g
HCl (2*N* solution) 4.0mL

Preparation of Solution 1: Add components to distilled/deionized water and bring volume to 1.0L. Mix thoroughly. Autoclave for 15 min at 15 psi pressure–121°C. Cool to 25°C.

Solution 2:

Composition per liter:

Disodium EDTA 5.2g
$CoCl_2 \cdot 6H_2O$ 1.9g
$FeCl_2 \cdot 4H_2O$ 1.5g
$MnCl_2 \cdot 4H_2O$ 1.0g
$ZnCl_2$ 0.7g
H_3BO_3 .... 0.62g
$Na_2MoO_4 \cdot 2H_2O$ 0.36g
$NiCl_2 \cdot 6H_2O$ 0.24g
$CuCl_2 \cdot 2H_2O$ 0.17g

Preparation of Solution 2: Add components to distilled/deionized water and bring volume to 1.0L. Mix thoroughly. Adjust pH to 6.5. Autoclave for 15 min at 15 psi pressure–121°C. Cool to 25°C.

Solution 3:

Composition per 100.0mL:

$NaHCO_3$ 10.0g

Preparation of Solution 3: Add $NaHCO_3$ to distilled/deionized water and bring volume to 100.0mL. Mix thoroughly. Autoclave for 15 min at 15 psi pressure–121°C. Cool to 25°C.

Solution 4:

Composition per 100.0mL:

$Na_2S \cdot 9H_2O$ 5.0g

Preparation of Solution 4: Add $Na_2S \cdot 9H_2O$ to distilled/deionized water and bring volume to 100.0mL. Mix thoroughly. Autoclave for 15 min at 15 psi pressure–121°C. Cool to 25°C.

Solution 5:

Composition per 200.0mL:

Pyridoxamine·HCl 0.01g
Nicotinic acid 4.0mg
p-Aminobenzoic acid 2.0mg
Thiamine 2.0mg
Cyanocobalamin 1.0mg
Pantothenic acid 1.0mg
Biotin 0.5mg

Preparation of Solution 5: Add components to distilled/deionized water and bring volume to 200.0mL. Mix thoroughly. Filter sterilize.

Elemental Sulfur Slurry:

Composition per 10.0g:

Sulfur flowers 10.0g

Preparation of Elemental Sulfur Slurry: Add highly purified sulfur flowers to a mortar and grind to a fine powder. Add sufficient distilled/deionized water to produce a slurry. Distribute into 100.0mL screw-capped bottles in 20.0mL volumes. Autoclave for 30 min at 10 psi pressure–115°C. Decant supernatant solution. Reserve sulfur slurry.

Preparation of Medium: To 1.0L of cooled, sterile solution 1, aseptically add 1.0mL of sterile solution 2, 40.0mL of sterile solution 3, 6.0mL of sterile solution 4, and 5.0mL of sterile solution 5. Mix thoroughly. Adjust pH to 7.2. Aseptically distribute into sterile 50.0mL screw-capped bottles. Fill bottles completely with medium except for a pea-sized air bubble. Aseptically add a pea-sized piece of sulfur slurry to each 50.0mL of medium.

Use: For the isolation and cultivation of freshwater *Desulfuromonas* species.

Desulfuromonas succinoxidans Medium

Composition per liter:

NaCl ... 20.0g
$MgCl_2 \cdot 6H_2O$ ... 2.0g
KH_2PO_4 ... 1.0g
$MgSO_4 \cdot 7H_2O$ ... 1.0g
NH_4Cl ... 0.3g
$CaCl_2 \cdot 2H_2O$ ... 0.1g
Resazurin ... 0.5mg
$NaHCO_3$ solution ... 50.0mL
Disodium fumarate solution ... 10.0mL
Sodium acetate solution ... 10.0mL
$Na_2S \cdot 9H_2O$ solution ... 10.0mL
Trace elements solution SL-10 ... 1.0mL
Seven vitamin solution ... 1.0mL

pH 7.2 ± 0.2 at 25°C

$NaHCO_3$ Solution:

Composition per 50.0mL:

$NaHCO_3$ ... 2.5g

Preparation of $NaHCO_3$ Solution: Add $NaHCO_3$ to distilled/deionized water and bring volume to 50.0mL. Mix thoroughly. Sparge with 80% N_2 + 20% CO_2. Autoclave for 15 min at 15 psi pressure–121°C.

Disodium Fumarate Solution:

Composition per 10.0mL:

Disodium fumarate ... 1.6g

Preparation of Disodium Fumarate Solution: Add disodium fumarate to distilled/deionized water and bring volume to 10.0mL. Mix thoroughly. Sparge with 100% N_2. Autoclave for 15 min at 15 psi pressure–121°C.

Sodium Acetate Solution:

Composition per 10.0mL:

Sodium acetate ... 0.8g

Preparation of Sodium Acetate Solution: Add sodium acetate to distilled/deionized water and bring volume to 10.0mL. Mix thoroughly. Sparge with 100% N_2. Autoclave for 15 min at 15 psi pressure–121°C.

$Na_2S \cdot 9H_2O$ Solution:

Composition per 10.0mL:

$Na_2S \cdot 9H_2O$ ... 0.5g

Preparation of $Na_2S \cdot 9H_2O$ Solution: Add $Na_2S \cdot 9H_2O$ to distilled/deionized water and bring volume to 10.0mL. Mix thoroughly. Sparge with 100% N_2. Autoclave for 15 min at 15 psi pressure–121°C.

Trace Elements Solution SL-10:

Composition per liter:

$FeCl_2 \cdot 4H_2O$ ... 1.5g
$CoCl_2 \cdot 6H_2O$ ... 190.0mg
$MnCl_2 \cdot 4H_2O$ ... 100.0mg
$ZnCl_2$ ... 70.0mg
$Na_2MoO_4 \cdot 2H_2O$ ... 36.0mg
$NiCl_2 \cdot 6H_2O$ ... 24.0mg
H_3BO_3 ... 6.0mg
$CuCl_2 \cdot 2H_2O$ ... 2.0mg
HCl (25% solution) ... 10.0mL

Preparation of Trace Elements Solution SL-10: Add $FeCl_2 \cdot 4H_2O$ to 10.0mL of HCl solution. Mix thoroughly. Add distilled/deionized water and bring volume to 1.0L. Add remaining components. Mix thoroughly.

Seven Vitamin Solution:

Composition per liter:

Pyridoxine·HCl ... 0.3g
Thiamine·HCl ... 0.2g
Nicotinic acid ... 0.2g
Calcium DL-pantothenate ... 0.1g
Vitamin B_{12} ... 0.1g
p-Aminobenzoic acid ... 80.0mg
Biotin ... 20.0mg

Preparation of Seven Vitamin Solution: Add components to distilled/deionized water and bring volume to 1.0L. Mix thoroughly. Filter sterilize. Sparge with 100% N_2.

Preparation of Medium: Prepare and dispense medium under 80% N_2 + 20% CO_2. Add components, except $NaHCO_3$ solution, disodium fumarate solution, sodium acetate solution, $Na_2S \cdot 9H_2O$ solution, and seven vitamin solution, to distilled/deionized water and bring volume to 920.0mL. Mix thoroughly. Sparge with 80% N_2 + 20% CO_2. Autoclave for 15 min at 15 psi pressure–121°C. Aseptically and anaerobically add 50.0mL of sterile $NaHCO_3$ solution, 10.0mL of sterile disodium fumarate solution, 10.0mL of sterile sodium acetate solution, 10.0mL of sterile $Na_2S \cdot 9H_2O$ solution, and 1.0mL of sterile seven vitamin solution. Mix thoroughly.

Use: For the cultivation of *Desulfuromonas succinoxidans.*

Desultobacterium Medium (DSMZ Medium 383)

Composition per 1012mL:

Solution A ... 930.0mL
Solution C ... 50.0mL
Solution D ... 10.0mL
Solution E ... 10.0mL
Solution G ... 10.0mL
Solution B ... 1.0mL
Solution F ... 1.0mL

pH 7.0 at 25°C

Solution A:

Composition per 930.0mL:

NaCl ... 21.0g
Na_2SO_4 ... 3.0g
$MgCl_2 \cdot 6H_2O$ ... 3.0g
KCl ... 0.5g
NH_4Cl ... 0.3g
KH_2PO_4 ... 0.2g
$CaCl_2 \cdot 2H_2O$ ... 0.15g
Resazurin ... 1.0mg

Preparation of Solution A: Add components to distilled/deionized water and bring volume to 930.0mL. Mix thoroughly. Sparge with 80% N_2 + 20% CO_2 gas until saturated. Autoclave for 15 min at 15 psi pressure–121°C. Cool to 25°C.

Solution B:

Composition per liter:

$FeCl_2 \cdot 4H_2O$ ... 1.5g
$CoCl_2 \cdot 6H_2O$ ... 190.0mg
$MnCl_2 \cdot 4H_2O$ ... 100.0mg
$ZnCl_2$ ... 70.0mg
$Na_2MoO_4 \cdot 2H_2O$ ... 36.0mg
$NiCl_2 \cdot 6H_2O$ ... 24.0mg
H_3BO_3 ... 300.0mg
$CuCl_2 \cdot 2H_2O$ ... 2.0mg
HCl (25% solution) ... 7.7mL

Preparation of Solution B: Add $FeCl_2 \cdot 4H_2O$ to 10.0mL of HCl solution. Mix thoroughly. Add distilled/deionized water and bring volume to 1.0L. Add remaining components. Mix thoroughly. Sparge with 100% N_2. Autoclave for 15 min at 15 psi pressure–121°C.

Solution C:
Composition per 100.0mL:
$NaHCO_3$ 5.0g

Preparation of Solution C: Add $NaHCO_3$ to distilled/deionized water and bring volume to 100.0mL. Mix thoroughly. Sparge with 100% CO_2 until saturated, approximately 20 minutes. Filter sterilize under 100% CO_2 into a sterile, gas-tight 100.0mL screw-capped bottle.

Solution D:
Composition per 10.0mL:
Na-benzoate 0.5g

Preparation of Solution D: Add Na_2-benzoate to distilled/deionized water and bring volume to 10.0mL. Sparge with N_2. Filter sterilize. Store anaerobically.

Solution E:
Composition per liter:
Pyridoxine-HCl 10.0mg
Thiamine-HCl·$2H_2O$ 5.0mg
Riboflavin 5.0mg
Nicotinic acid 5.0mg
D-Ca-pantothenate 5.0mg
p-Aminobenzoic acid 5.0mg
Lipoic acid 5.0mg
Biotin 2.0mg
Folic acid 2.0mg
Vitamin B_{12} 0.1mg

Preparation of Solution E: Add components to distilled/deionized water and bring volume to 1.0L. Mix thoroughly. Sparge with 100% N_2. Filter sterilize.

Solution F:
Composition per liter:
NaOH 0.5g
$Na_2SeO_3·5H_2O$ 3.0mg

Preparation of Solution F: Add components to distilled/deionized water and bring volume to 1.0L. Mix thoroughly. Sparge with 100% N_2. Filter sterilize.

Solution G:
Composition per 10.0mL:
$Na_2S·9H_2O$ 0.4g

Preparation of Solution G: Add $Na_2S·9H_2O$ to distilled/deionized water and bring volume to 10.0mL. Sparge with N_2. Autoclave for 15 min at 15 psi pressure–121°C. Cool to 25°C. Store anaerobically.

Preparation of Medium: Add solutions B, C, D, E, F, and G to solution A in that order under N_2 gas. Adjust the pH to 7.0.

Use: For the cultivation of *Desulfotobacterium* spp.

Dethiosulfovibrio II Medium (DSMZ Medium 906)

Composition per liter:
NaCl 20.0g
Yeast extract 5.0g
$MgCl_2·6H_2O$ 3.0g
Na_3-citrate·$5H_2O$ 3.0g
Peptone 2.0g
KH_2PO_4 .... 1.0g
Resazurin 0.5mg
Calcium chloride solution 10.0mL
$Na_2S·9H_2O$ solution 10.0mL
$Na_2S_2O_3$solution 10.0mL
Trace elements solution SL-10 .... 1.0mL
pH 6.7-6.8 at 25°C

$Na_2S_2O_3$ Solution:
Composition per 10mL:
$Na_2S_2O_3·5H_2O$ 2.5g

Preparation of $Na_2S_2O_3$ Solution: Add $Na_2S_2O_3·5H_2O$ to distilled/deionized water and bring volume to 10.0mL. Mix thoroughly. Autoclave under 100% N_2 for 15 min at 15 psi pressure–121°C. Cool to room temperature.

$Na_2S·9H_2O$ Solution:
Composition per 10.0mL:
$Na_2S·9H_2O$ 0.5g

Preparation of $Na_2S·9H_2O$ Solution: Add $Na_2S·9H_2O$ to distilled/deionized water and bring volume to 10.0mL. Mix thoroughly. Sparge with 100% N_2. Autoclave for 15 min at 15 psi pressure–121°C.

Trace Elements Solution SL-10:
Composition per liter:
$FeCl_2·4H_2O$ 1.5g
$CoCl_2·6H_2O$ 190.0mg
$MnCl_2·4H_2O$ 100.0mg
$ZnCl_2$ 70.0mg
$Na_2MoO_4·2H_2O$ 36.0mg
$NiCl_2·6H_2O$ 24.0mg
H_3BO_3 .... 6.0mg
$CuCl_2·2H_2O$ 2.0mg
HCl (25% solution) 10.0mL

Preparation of Trace Elements Solution SL-10: Add $FeCl_2·4H_2O$ to 10.0mL of HCl solution. Mix thoroughly. Add distilled/deionized water and bring volume to 1.0L. Add remaining components. Mix thoroughly. Sparge with 80% N_2 + 20% CO_2. Autoclave for 15 min at 15 psi pressure–121°C.

Calcium Chloride Solution:
Composition per 10.0mL:
$CaCl_2·2H_2O$ 0.2g

Preparation of Calcium Chloride Solution: Add $CaCl_2·2H_2O$ to distilled/deionized water and bring volume to 10.0mL. Mix thoroughly. Sparge with 100% N_2. Autoclave for 15 min at 15 psi pressure–121°C.

Preparation of Medium: Add components, except calcium chloride solution, trace elements solution SL-10, $Na_2S·9H_2O$ solution, and $Na_2S_2O_3$ solution, to distilled/deionized water and bring volume to 969.0mL. Mix thoroughly. Sparge with 100% N_2 for 30 min. Distribute under 100% N_2 into anaerobe tubes or bottles. Autoclave for 15 min at 15 psi pressure–121°C. Aseptically and anaerobically per 1.0L of medium add 10.0mL calcium chloride solution, 1.0mL trace elements solution SL-10, 10.0mL $Na_2S·9H_2O$ solution, and 10.0mL $Na_2S_2O_3$ solution. Mix thoroughly. The final pH should be 6.7-6.8.

Use: For the cultivation of *Dethiosulfovibrio* spp.

Dethiosulfovibrio peptidovorans Medium (DSMZ Medium 786)

Composition per 1085mL:
NaCl 30.0g
Trypticase 5.0g
$MgCl_2·6H_2O$ 3.0g
Yeast extract 1.0g
NH_4Cl 1.0g
L-Cysteine 0.5g
Na-acetate 0.5g
K_2HPO_4 .... 0.3g
KH_2PO_4 .... 0.3g
$CaCl_2·2H_2O$ 0.1g

KCl 0.1g
Resazurin 0.5mg
$NaHCO_3$ solution 50.0mL
Na-thiosulfate solution 20.0mL
$Na_2S·9H_2O$ solution 15.0mL
Trace elements solution 10.0mL
pH 7.3 ± 0.2 at 25°C

Trace Elements Solution:
Composition per liter:
$MgSO_4·7H_2O$ 3.0g
Nitrilotriacetic acid 1.5g
NaCl 1.0g
$MnSO_4·2H_2O$ 0.5g
$CoSO_4·7H_2O$ 0.18g
$ZnSO_4·7H_2O$ 0.18g
$CaCl_2·2H_2O$ 0.1g
$FeSO_4·7H_2O$ 0.1g
$NiCl_2·6H_2O$ 0.025g
$KAl(SO_4)_2·12H_2O$ 0.02g
H_3BO_3 .... 0.01g
$Na_2MoO_4·4H_2O$ 0.01g
$CuSO_4·5H_2O$ 0.01g
$Na_2SeO_3·5H_2O$ 0.3mg

Preparation of Trace Elements Solution: Add nitrilotriacetic acid to 500.0mL of distilled/deionized water. Dissolve by adjusting pH to 6.5 with KOH. Add remaining components. Add distilled/deionized water to 1.0L. Mix thoroughly.

$Na_2S·9H_2O$ Solution:
Composition per 20mL:
$Na_2S·9H_2O$ 0.6g

Preparation of $Na_2S·9H_2O$ Solution: Add $Na_2S·9H_2O$ to distilled/deionized water and bring volume to 20.0mL. Mix thoroughly. Autoclave under 100% N_2 for 15 min at 15 psi pressure–121°C. Cool to room temperature.

$NaHCO_3$ Solution:
Composition per 100.0mL:
$NaHCO_3$ 10.0g

Preparation of $NaHCO_3$ Solution: Add $NaHCO_3$ to distilled/deionized water and bring volume to 100.0mL. Mix thoroughly. Sparge with 80% N_2 + 20% CO_2. Filter sterilize.

Na-thiosulfate Solution:
Composition per 20.0mL:
$Na_2S_2O_3·5H_2O$ 5.0g

Preparation of Na-thiosulfate Solution: Add $Na_2S_2O_3·5H_2O$ to distilled/deionized water and bring volume to 20.0mL. Mix thoroughly. Sparge with 100% N_2. Filter sterilize.

Preparation of Medium: Prepare and dispense medium under 80% N_2 + 20% CO_2. Add components, except Na-thiosulfate solution, $NaHCO_3$ solution, and $Na_2S·9H_2O$ solution, to distilled/deionized water and bring volume to 1.0L. Mix thoroughly. Sparge with 80% N_2 + 20% CO_2. Autoclave for 15 min at 15 psi pressure–121°C. Cool to 25°C. Aseptically and anaerobically add 20.0mL Na-thiosulfate solution, 50.0mL $NaHCO_3$ solution, and 15.0mL $Na_2S·9H_2O$ solution. Mix thoroughly. Adjust pH to 7.3. Aseptically and anaerobically distribute into sterile tubes or flasks.

Use: For the cultivation of *Dethiosulfovibrio peptidovorans.*

Dextran Agar

Composition per liter:
Minimal mineral base solution 700.0mL
Agar solution 200.0mL
Dextran-deoxyglucose solution 100.0mL
pH 4.0 ± 0.2 at 25°C

Minimal Mineral Base Solution:
Composition per 700.0mL:
$(NH_4)_2SO_4$ 5.0g
KH_2PO_4 .... 1.5g
$CaCl_2$ 0.1g
$MnSO_4$ 0.1g
NaCl 0.1g
Yeast extract 0.1g

Preparation of Minimal Mineral Base Solution: Add components to distilled/deionized water and bring volume to 700.0mL. Mix thoroughly. Gently heat and bring to boiling. Before autoclaving pH is 4.0. Autoclave for 15 min at 15 psi pressure–121°C. Cool to 45°–50°C.

Agar Solution:
Composition per 200.0mL:
Agar 10.0g

Preparation of Agar Solution: Add agar to distilled/deionized water and bring volume to 200.0mL. Mix thoroughly. Gently heat and bring to boiling. Autoclave for 15 min at 15 psi pressure–121°C. Cool to 45°–50°C.

Dextran-Deoxyglucose Solution:
Composition per 100.0mL:
Dextran 10.0g
2-Deoxy-D-glucose 0.5g

Preparation of Dextran-Deoxyglucose Solution: Add components to distilled/deionized water and bring volume to 100.0mL. Mix thoroughly. Filter sterilize.

Preparation of Medium: Aseptically combine 700.0mL of sterile minimal mineral base solution, 200.0mL of sterile agar solution, and 100.0mL of sterile dextran-deoxyglucose solution. Mix thoroughly. Pour into sterile Petri dishes or distribute into sterile tubes.

Use: For the cultivation of *Lipomyces starkeyi.*

Dextrose Broth

Composition per liter:
Pancreatic digest of casein 10.0g
Glucose 5.0g
NaCl 5.0g
pH 7.3 ± 0.2 at 25°C

Source: This medium is available as a premixed powder from BD Diagnostic Systems.

Preparation of Medium: Add components to distilled/deionized water and bring volume to 1.0L. Mix thoroughly. Distribute into tubes or flasks. Autoclave for 15 min at 15 psi pressure–121°C.

Use: For the cultivation and differentiation of microorganisms based on their ability to ferment glucose. If desired, a Durham tube may be added to the test tubes to determine gas production.

Dextrose Soil Agar (DSA)

Composition per liter:
Soil 150.0g
Agar 20.0g
Glucose 5.0g

Preparation of Medium: Add soil to distilled/deionized water and bring volume to 1.0L. Mix thoroughly. Autoclave for 60 min at 15 psi pressure–121°C. Filter through Whatman #1 filter paper. Bring volume to filtrate to 1.0L with distilled/deionized water. Mix thoroughly. Add agar and glucose. Gently heat and bring to boiling. Distribute into tubes or flasks. Autoclave for 15 min at 15 psi pressure–121°C. Pour into sterile Petri dishes or leave in tubes.

Use: For the cultivation of *Chaetomium globosum.*

Diazotrophic Medium (RBA)

Composition per 1008.0mL:

Solution A	903.0mL
Solution B	50.0mL
Solution C	50.0mL
Solution D	5.0mL

pH 7.3 ± 0.2 at 25°C

Solution A:

Composition per 903.0mL:

Agar	15.0g
K_2HPO_4	0.9g
$CaCl_2 \cdot 2H_2O$	0.1g
KH_2PO_4	0.1g
$MgSO_4 \cdot 7H_2O$	0.1g
NaCl	0.1g
$FeSO_4 \cdot 7H_2O$	0.01g
$MnSO_4 \cdot H_2O$	5.0mg
$NaVO_3 \cdot 2H_2O$	5.0mg
$Na_2MoO_4 \cdot 2H_2O$	0.5mg
Trace elements solution SL-6	3.0mL

Trace Elements Solution SL-6:

Composition per liter:

$MnCl_2 \cdot 4H_2O$	0.5g
H_3BO_3	0.3g
$CoCl_2 \cdot 6H_2O$	0.2g
$ZnSO_4 \cdot 7H_2O$	0.1g
$Na_2MoO_4 \cdot 2H_2O$	0.03g
$NiCl_2 \cdot 6H_2O$	0.02g
$CuCl_2 \cdot 2H_2O$	0.01g

Preparation of Trace Elements Solution SL-6: Add components to distilled/deionized water and bring volume to 1.0L. Mix thoroughly. Adjust pH to 7.3. Autoclave for 15 min at 15 psi pressure–121°C.

Preparation of Solution A: Add components to distilled/deionized water and bring volume to 903.0mL. Mix thoroughly. Gently heat and bring to boiling. Adjust pH to 7.3. Autoclave for 15 min at 15 psi pressure–121°C. Cool to 50°–55°C. Pour into sterile Petri dishes or distribute into sterile tubes.

Solution B:

Composition per 50.0mL:

DL-Malate	2.0g
Disodium succinate	1.0g
Yeast extract	0.05g

Preparation of Solution B: Add components to distilled/deionized water and bring volume to 50.0mL. Mix thoroughly. Adjust pH to 7.3. Autoclave for 15 min at 15 psi pressure–121°C.

Solution C:

Composition per 50.0mL:

D-Mannitol	2.0g
D-Glucose	2.0g
Sodium pyruvate	1.0g

Preparation of Solution C: Add components to distilled/deionized water and bring volume to 50.0mL. Mix thoroughly. Adjust pH to 7.3. Filter sterilize.

Solution D:

Composition per liter:

Pyridoxine·HCl	62.5g
Nicotinic acid	25.0mg
p-Aminobenzoic acid	12.5mg
Thiamine·HCl	12.5mg
Calcium DL-pantothenate	6.5mg
Biotin	2.5mg

Preparation of Solution D: Add components to distilled/deionized water and bring volume to 1.0L. Mix thoroughly. Adjust pH to 7.5. Gas under 100% N_2. Filter sterilize.

Preparation of Medium: To 903.0mL of sterile solution A, aseptically add 50.0mL of sterile solution B, 50.0mL of sterile solution C, and 5.0mL of sterile solution D. Mix thoroughly. Final pH should be 7.3. Pour into sterile Petri dishes or distribute into sterile tubes.

Use: For the cultivation and maintenance of *Arthrobacter* species, *Azomonas* species, *Azorhizophilus paspali,* and *Azotobacter* species.

Dibenzothiophene Mineral Medium

Composition per liter:

Beef extract	10.0g
Na_2HPO_4	3.0g
KH_2PO_4	2.0g
NH_4Cl	2.0g
Dibenzothiophene	0.5g
$MgCl_2 \cdot 6H_2O$	0.2g
$FeCl_3 \cdot 6H_2O$	0.028g

Preparation of Medium: Add components to distilled/deionized water and bring volume to 1.0L. Mix thoroughly. Distribute into tubes or flasks. Autoclave for 15 min at 15 psi pressure–121°C.

Use: For the cultivation of bacteria that can metabolize dibenzothiophene.

Dichloroacetic Acid Medium No. 1

Composition per liter:

Yeast extract	10.0g
Glucose	5.0g
$(NH_4)_2PO_4$	1.5g
K_2HPO_4	1.0g
2,4-Dichloroacetic acid	0.75g
$MgSO_4 \cdot 7H_2O$	0.2g
$Fe_2(SO_4)_3 \cdot 5H_2O$	0.01g
$ZnSO_4 \cdot 7H_2O$	2.0mg

pH 7.0 ± 0.2 at 25°C

Preparation of Medium: Add components to distilled/deionized water and bring volume to 1.0L. Mix thoroughly. Distribute into tubes or flasks. Autoclave for 15 min at 15 psi pressure–121°C.

Use: For the cultivation of *Rhodococcus* species.

Dichloroacetic Acid Medium No. 2

Composition per liter:

Yeast extract	10.0g
Glucose	5.0g
$(NH_4)_2PO_4$	1.5g
K_2HPO_4	1.0g
$MgSO_4 \cdot 7H_2O$	0.2g

$Fe_2(SO_4)_3 \cdot 5H_2O$ 0.01g
$ZnSO_4 \cdot 7H_2O$ 0.002g
2,4-Dichloroacetic acid 10.0mg
pH 7.0 ± 0.2 at 25°C

Preparation of Medium: Add components to distilled/deionized water and bring volume to 1.0L. Mix thoroughly. Distribute into tubes or flasks. Autoclave for 15 min at 15 psi pressure–121°C.

Use: For the cultivation of *Pseudomonas* species.

Dichloromethane Medium for *Hyphomicrobium*

Composition per liter:
$K_2HPO_4 \cdot 3H_2O$ 4.1g
KH_2PO_4 .... 1.4g
$MgSO_4 \cdot 7H_2O$ 0.2g
$(NH_4)_2SO_4$ 0.2g
Dichloromethane (methylene chloride) 1.0mL
Trace elements solution 1.0mL
pH 7.2 ± 0.2 at 25°C

Trace Elements Solution:
Composition per liter:
$Ca(NO_3)_2$ 25.0g
$FeSO_4 \cdot 7H_2O$ 1.0g
H_3BO_3 .... 1.0g
$MnSO_4 \cdot H_2O$ 1.0g
$Co(NO_3)_2 \cdot 6H_2O$ 0.25g
$CuCl_2 \cdot 2H_2O$ 0.25g
$(NH_4)_6Mo_7O_{24} \cdot 4H_2O$ 0.25g
$ZnCl_2$ 0.25g
NH_4VO_3 .... 0.1g

Preparation of Medium: Filter sterilize dichloromethane. Add components, except dichloromethane, to distilled/deionized water and bring volume to 999.0mL. Mix thoroughly. Gently heat and bring to boiling. Adjust pH to 7.2. Autoclave for 15 min at 15 psi pressure–121°C. Cool to 45°–50°C. Aseptically add sterile dichloromethane. Mix thoroughly. Aseptically distribute into sterile tubes or flasks.

Use: For the cultivation and maintenance of *Hyphomicrobium* species

Dichotomicrobium thermohalophilum Agar

Composition per liter:
Agar 18.0g
Disodium DL-malate 1.0g
Yeast extract 1.0g
Artificial seawater, 3× 960.0mL
Hutner's basal salts solution 20.0mL
$NaHCO_3$ solution 20.0mL
pH 7.0–7.2 at 25°C

Artificial Seawater, 3×:
Composition per liter:
NaCl 70.43g
$MgCl_2 \cdot 6H_2O$ 31.86g
Na_2SO_4 .... 11.75g
$CaCl_2 \cdot 2H_2O$ 4.35g
$NaHCO_3$ 2.88g
KCl 1.99g
KBr 0.29g
H_3BO_3 .... 0.08g

Preparation of Artificial Seawater, 3×: Add components to distilled/deionized water and bring volume to 1.0L. Mix thoroughly.

Hutner's Basal Salts Solution:
Composition per liter:
$MgSO_4 \cdot 7H_2O$ 29.7g
Nitrilotriacetic acid 10.0g
$CaCl_2 \cdot 2H_2O$ 3.335g
$FeSO_4 \cdot 7H_2O$ 99.0mg
$(NH_4)_6MoO_7O_{24} \cdot 4H_2O$ 9.25mg
"Metals 44" 50.0mL

"Metals 44":
Composition per 100.0mL:
$ZnSO_4 \cdot 7H_2O$ 1.095g
$FeSO_4 \cdot 7H_2O$ 0.5g
Sodium EDTA 0.25g
$MnSO_4 \cdot H2O$ 0.154g
$CuSO_4 \cdot 5H_2O$ 39.2mg
$Co(NO_3)_2 \cdot 6H_2O$ 24.8mg
$Na_2B_4O_7 \cdot 10H_2O$ 17.7mg

Preparation of "Metals 44": Add sodium EDTA to distilled/deionized water and bring volume to 90.0mL. Mix thoroughly. Add a few drops of concentrated H_2SO_4 to retard precipitation of heavy metal ions. Add remaining components. Mix thoroughly. Bring volume to 100.0mL with distilled/deionized water.

Preparation of Hutner's Basal Salts Solution: Add nitrilotriacetic acid to 500.0mL of distilled/deionized water. Adjust pH to 6.5 with KOH. Add remaining components. Add distilled/deionized water to 1.0L. Adjust pH to 6.8.

$NaHCO_3$ Solution:
Composition per 20.0mL:
$NaHCO_3$ 3.0g

Preparation of $NaHCO_3$ Solution: Add $NaHCO_3$ to distilled/deionized water and bring volume to 20.0mL. Mix thoroughly. Filter sterilize.

Preparation of Medium: Add components, except $NaHCO_3$ solution, to distilled/deionized water and bring volume to 980.0mL. Mix thoroughly. Gently heat and bring to boiling. Autoclave for 15 min at 15 psi pressure–121°C. Cool to 50°–55°C. Aseptically add 20.0mL of sterile $NaHCO_3$ solution. Mix thoroughly. Pour into sterile Petri dishes or distribute into sterile tubes.

Use: For the cultivation and maintenance of *Dichotomicrobium thermohalophilum.*

Dictyoglomus Medium

Composition per liter:
Soluble starch 5.0g
$Na_2HPO_4 \cdot 12H_2O$ 4.2g
Polypeptone 2.0g
Yeast extract 2.0g
KH_2PO_4 .... 1.5g
L-Cysteine·HCl·H_2O 1.0g
Na_2CO_3 .... 1.0g
NH_4Cl 0.5g
$MgCl_2 \cdot 6H_2O$ 0.38g
$CaCl_2$ 0.05g
$Fe(NH_4)_2(SO_4)_2 \cdot 6H_2O$ 0.039g
Resazurin 2.0mg
Trace metals 10.0mL
Wolfe's vitamin solution 10.0mL
pH 7.2 ± 0.2 at 25°C

Trace Metals:
Composition per liter:
$CoCl_2 \cdot 6H_2O$ 0.29g
$ZnSO_4 \cdot 7H_2O$ 0.28g

$Na_2MoO_4 \cdot 2H_2O$ 0.24g
$MnCl_2 \cdot 4H_2O$ 0.2g
Na_2SeO_3 .... 0.017g

Preparation of Trace Metals: Add components to distilled/deionized water and bring volume to 1.0L. Adjust pH to 6.0 with KOH. Mix thoroughly.

Wolfe's Vitamin Solution:
Composition per liter:
Pyridoxine·HCl 0.01g
Thiamine·HCl 5.0mg
Riboflavin 5.0mg
Nicotinic acid 5.0mg
Calcium pantothenate 5.0mg
p-Aminobenzoic acid 5.0mg
Thioctic acid 5.0mg
Biotin 2.0mg
Folic acid 2.0mg
Cyanocobalamin 0.1mg

Preparation of Wolfe's Vitamin Solution: Add components to distilled/deionized water and bring volume to 1.0L. Mix thoroughly. Filter sterilize.

Preparation of Medium: Prepare and dispense medium under 100% N_2. Add components, except Wolfe's vitamin solution, to distilled/deionized water and bring volume to 990.0mL. Mix thoroughly. Gently heat and bring to boiling. Continue boiling until resazurin turns colorless. Autoclave for 15 min at 15 psi pressure–121°C. Cool to 25°C under 100% N_2. Aseptically add sterile Wolfe's vitamin solution. Mix thoroughly. Adjust pH to 7.2 if necessary. Aseptically and anaerobically distribute into sterile tubes or flasks.

Use: For the cultivation and maintenance of *Dictyoglomus thermophilum*.

Dictyostelium Medium

Composition per liter:
Glucose 15.4g
Agar 15.0g
Peptone 14.3g
Yeast extract 7.15g
$Na_2HPO_4 \cdot 12H_2O$ 1.28g
KH_2PO_4 .... 0.49g
pH 6.7 ± 0.2 at 25°C

Preparation of Medium: Add components to distilled/deionized water and bring volume to 1.0L. Mix thoroughly. Gently heat and bring to boiling. Adjust pH to 6.7. Distribute into tubes or flasks. Autoclave for 15 min at 15 psi pressure–121°C. Pour into sterile Petri dishes or leave in tubes.

Use: For the cultivation of *Dictyostelium discoideum* and *Fusarium acuminatum*.

Dilute Potato Medium

Composition per 1090.0mL:
Glucose 1.0g
Na_2HPO_4 .... 0.12g
$Ca(NO_3)_2 \cdot 4H_2O)$ 0.05g
Peptone 0.05g
Potato decoction 100.0mL
pH 6.8 ± 0.2 at 25°C

Potato Decoction:
Composition per liter:
Potato 20.0g

Preparation of Potato Decoction: Peel and dice potato. Add to 1.0L of distilled/deionized water. Gently heat and bring to boiling. Continue boiling for 30 min. Filter through Whatman #1 filter paper. Bring volume of filtrate to 1.0L with distilled/deionized water.

Preparation of Medium: Add components to distilled/deionized water and bring volume to 1090.0mL. Mix thoroughly. Adjust pH to 6.8. Distribute into tubes or flasks. Autoclave for 15 min at 15 psi pressure–121°C.

Use: For the cultivation and maintenance of *Rhizobacter daucus*.

Dinoflagellate Medium

Composition per 1020.0mL:
Seawater solution 1.0L
Basal solution 20.0mL
pH 7.8 ± 0.2 at 25°C

Seawater Solution:
Composition per 1100mL:
Seawater 1010.0mL

Preparation of Seawater Solution: Add seawater to distilled/deionized water and bring volume to 1100.0mL. Mix thoroughly. Adjust pH to 7.8. Autoclave for 15 min at 15 psi pressure–121°C. Cool to 25°C.

Basal Solution:
Composition per 100.0mL:
Buffer salts solution 25.0mL
Fe solution 25.0mL
Vitamin solution 25.0mL
Metal solution 25.0mL

Preparation of Basal Solution: Adjust final pH to 7.8.

Buffer Salts Solution:
Composition per 25.0mL:
Tris-HCl 500.0mg
$NaNO_3$ 350.0mg
Sodium glycerophosphate·$6H_2O$ 50.0mg

Preparation of Buffer Salts Solution: Add components to distilled/deionized water and bring volume to 25.0mL. Adjust ph to 7.8. Mix thoroughly.

Fe Solution:
Composition per 500.0mL:
$Fe(NH_4)_2(SO_4)_2 \cdot 6H_2O$ 351.0mg
EDTA 330.0mg

Preparation of Fe Solution: Add components to distilled/deionized water and bring volume to 500.0mL. Mix thoroughly.

Vitamin Solution:
Composition per 25.0mL:
Vitamin B_{12} 10.0μg
Biotin 5.0μg
Thiamine 0.5mg

Preparation of Vitamin Solution: Add components to distilled/deionized water and bring volume to 25.0mL. Mix thoroughly.

Metal Solution:
Composition per 25.0mL:
H_3BO_3 .... 114.0mg
EDTA 100.0mg
$MnSO_4 \cdot 4H_2O$ 16.4mg
$FeCl_3 \cdot 6H_2O$ 4.9mg
$ZnSO_4 \cdot 7H_2O$ 2.2mg
$CoSO_4 \cdot 7H_2O$ 0.48mg

Preparation of Metal Solution: Add components, in the order listed, to distilled/deionized water and bring volume to 25.0mL. Mix thoroughly. Adjust pH to 7.5.

Preparation of Basal Solution: Combine 25.0mL buffer salts solution, 25.0mL Fe solution, 25.0mL vitamin solution, and 25.0mL metal solution. Adjust pH to 7.8. Mix thoroughly. Autoclave for 15 min at 15 psi pressure–121°C. Cool to 25°C.

Preparation of Medium: Aseptically combine 20.0mL of sterile basal solution with 1.0L of sterile seawater solution. Mix thoroughly. Aseptically distribute into sterile, screw-capped tubes or flasks.

Use: For the cultivation of *Amphidinium carteri.*

DM Medium

Composition per liter:

Starch, soluble	5.0g
$MgSO_4 \cdot 7H_2O$	0.5g
K_2HPO_4	0.25g

Preparation of Medium: Add components to distilled/deionized water and bring volume to 1.0L. Mix thoroughly. Distribute into tubes or flasks. Autoclave for 15 min at 15 psi pressure–121°C.

Use: For the cultivation of myxobacteria.

DNase Medium

Composition per liter:

Agar	15.0g
Pancreatic digest of casein	10.0g
Peptic digest of animal tissue	10.0g
L-Arabinose	10.0g
NaCl	5.0g
Deoxyribonucleic acid	2.0g
Methyl Green	0.09g
Phenol Red	0.05g
Antibiotic solution	10.0mL

pH 7.3 ± 0.2 at 25°C

Antibiotic Solution:

Composition per 10.0mL:

Cephalothin	0.01g
Ampicillin	5.0mg
Colistimethate	5.0mg
Amphotericin B	2.5mg

Preparation of Antibiotic Solution: Add components to distilled/deionized water and bring volume to 10.0mL. Mix thoroughly. Filter sterilize.

Preparation of Medium: Add components, except antibiotic solution, to distilled/deionized water and bring volume to 990.0mL. Mix thoroughly. Gently heat and bring to boiling. Autoclave for 15 min at 15 psi pressure–121°C. Cool to 45°–50°C. Aseptically add sterile components. Mix thoroughly. Pour into sterile Petri dishes or distribute into sterile tubes.

Use: For the isolation and cultivation of *Serratia marcescens.*

DPM Medium (DSMZ Medium 737)

Composition per liter:

Na-propionate	1.20g
KH_2PO_4	1.0g
$MgSO_4 \cdot 7H_2O$	0.1g
$CaCl_2 \cdot 2H_2O$	0.01g
Chelated iron solution	1.8mL
Trace elements solution	1.0mL

pH 6.8 ± 0.2 at 25°C

Chelated Iron Solution:

Composition per liter:

Na_2-EDTA	7.56g
$FeSO_4 \cdot 5H_2O$	5.54g

Preparation of Chelated Iron Solution: Add components to distilled/deionized water and bring volume to 1.0L. Mix thoroughly.

Trace Elements Solution:

Composition per liter:

H_3BO_3	2.860
$MnCl_2 \cdot 4H_2O$	1.81g
$ZnSO_4 \cdot 7H_2O$	0.22g
$CuSO_4 \cdot 5H_2O$	0.08g
$Na_2MoO_4 \cdot 4H_2O$	0.025g
$CoCl_2$	0.025g

Preparation of Trace Elements Solution: Add components to distilled/deionized water and bring volume to 1.0L. Mix thoroughly.

Preparation of Medium: Add components to distilled/deionized water and bring volume to 1.0L. Mix thoroughly. Distribute into tubes or flasks. Autoclave for 15 min at 15 psi pressure–121°C.

Use: For the cultivation of *Frankia* sp.

DSIC Medium, Modified (DSMZ Medium 747)

Composition per 994.0mL:

Solution A	910.0mL
Solution B	70.0mL
Solution C	14.0mL

pH 7.0-7.1 at 25°C

Solution A:

Composition per 960.0mL:

NaCl	125.0g
K_2SO_4	2.5g
Na-acetate	2.0g
Yeast extract	0.75g
KH_2PO_4	0.6g
NH_4Cl	0.5g
$Na_2S_2O_3 \cdot 5H_2O$	0.1g
MOPS buffer	10.0mL
Vitamin B_{12} solution	1.0mL
Trace elements solution SL-10	1.0mL

Vitamin B_{12} Solution:

Composition per 100.0mL:

Vitamin B_{12}	10.0mg

Vitamin B_{12} Solution: Add Vitamin B_{12} to distilled/deionized water and bring volume to 100.0mL. Mix thoroughly. Sparge under 100% N_2 gas for 3 min.

Trace Elements Solution SL-10:

Composition per liter:

$FeCl_2 \cdot 4H_2O$	1.5g
$CoCl_2 \cdot 6H_2O$	190.0mg
$MnCl_2 \cdot 4H_2O$	100.0mg
$ZnCl_2$	70.0mg
$Na_2MoO_4 \cdot 2H_2O$	36.0mg
$NiCl_2 \cdot 6H_2O$	24.0mg
H_3BO_3	6.0mg
$CuCl_2 \cdot 2H_2O$	2.0mg
HCl (25% solution)	10.0mL

Preparation of Trace Elements Solution SL-10: Add $FeCl_2 \cdot 4H_2O$ to 10.0mL of HCl solution. Mix thoroughly. Add distilled/deionized water and bring volume to 1.0L.

Add remaining components. Mix thoroughly. Sparge with 80% N_2 + 20% CO_2. Autoclave for 15 min at 15 psi pressure–121°C.

MOPS Buffer:
Composition per 10mL:
MOPS [3-(*N*-morpholino) propane sulfonic acid] 2.1g
Na-acetate 0.3g
EDTA 0.1g

Preparation of MOPS Buffer: Add components to distilled/deionized water and bring volume to 10.0mL. Mix thoroughly. Sparge with 100% N_2. Adjust to pH 7.2. Filter sterilize.

Preparation of Solution A: Add components to distilled/deionized water and bring volume to 960.0mL. Mix thoroughly. Sparge with 100% N_2. Gently heat and bring to boiling while continuing to sparge with 100% N_2. Distribute about 13mL aliquots in 15mL Hungate tubes. Autoclave for 15 min at 15 psi pressure–121°C. Cool to 25°C.

Solution B:
Composition per 70.0mL:
$MgCl_2 \cdot 6H_2O$ 10.0g
$CaCl_2 \cdot 2H_2O$ 0.2g

Preparation of Solution B: Add components to distilled/deionized water and bring volume to 70.0mL. Mix thoroughly. Adjust pH to 7.0. Sparge with 100% N_2. Gently heat and bring to boiling while continuing to sparge with 100% N_2. Distribute into a screw-capped bottle. Autoclave for 15 min at 15 psi pressure–121°C. Cool to 25°C.

Solution C:
Composition per 14.0mL:
$NaHCO_3$ 1.0g

Preparation of Solution C: Add $NaHCO_3$ to distilled/deionized water and bring volume to 14.0mL. Mix thoroughly. Sparge with 80% N_2 + 20% CO_2. Filter sterilize.

Preparation of Medium: Inject 1.0 ml of solution B and 0.2 ml of solution C in each tube of solution A.

Use: For the cultivation of *Rhodovibrio sodomensis=Rhodospirillum sodomense.*

Dubos Agar with Filter Paper

Composition per liter:
Agar 15.0g
K_2HPO_4 .... 1.0g
KCl 0.5g
$MgSO_4 \cdot 7H_2O$ 0.5g
$NaNO_3$ 0.5g
$FeSO_4 \cdot 7H_2O$ 0.01g
pH 7.2 ± 0.2 at 25°C

Preparation of Medium: Add components to distilled/deionized water and bring volume to 1.0L. Mix thoroughly. Gently heat and bring to boiling. Adjust pH to 7.2. Autoclave for 15 min at 15 psi pressure–121°C. Pour into sterile Petri dishes. Lay sterile strips of Whatman #1 filter paper on the surface of the agar.

Use: For the cultivation and maintenance of *Cytophaga hutchinsonii.*

Dubos Mineral Medium

Composition per liter:
K_2HPO_4 .... 1.0g
KCl 0.5g
$MgSO_4 \cdot 7H_2O$ 0.5g
$NaNO_3$ 0.5g
$FeSO_4 \cdot 7H_2O$ 0.01g
pH 7.2 ± 0.2 at 25°C

Preparation of Medium: Add components to distilled/deionized water and bring volume to 1.0L. Mix thoroughly. Distribute into tubes or flasks. Autoclave for 15 min at 15 psi pressure–121°C.

Use: For the isolation and cultivation of *Cytophaga* species, *Herpetosiphon* species, *Saprospira* species, and *Flexithrix* species.

Dubos Salts Agar

Composition per liter:
Agar 15.0g
K_2HPO_4 .... 1.0g
KCl 0.5g
$MgSO_4 \cdot 7H_2O$ 0.5g
$NaNO_3$ 0.5g
$FeSO_4 \cdot 7H_2O$ 0.01g
Filter paper strips, sterile .... variable
pH 7.2 ± 0.2 at 25°C

Preparation of Medium: Add components, except filter paper strips, to distilled/deionized water and bring volume to 1.0L. Mix thoroughly. Gently heat and bring to boiling. Autoclave for 15 min at 15 psi pressure–121°C. Pour into sterile Petri dishes. Aseptically place sterile filter paper strips onto the surface of the solidified medium.

Use: For the cultivation and maintenance of *Alteromonas* species, *Cellvibrio mixtus, Cellvibrio* species, *Cytophaga aurantiaca, Cytophaga hutchinsonii, Pseudomonas* species, and *Sporocytophaga myxococcoides.*

Dubos Salts Agar with 1% NaCl

Composition per liter:
Agar 15.0g
NaCl 10.0g
K_2HPO_4 .... 1.0g
KCl 0.5g
$MgSO_4 \cdot 7H_2O$ 0.5g
$NaNO_3$ 0.5g
$FeSO_4 \cdot 7H_2O$ 10.0mg
Filter paper strips .... 1 strip per tube
pH 7.2 ± 0.2 at 25°C

Preparation of Medium: Add components to distilled/deionized water and bring volume to 1.0L. Mix thoroughly. Gently heat and bring to boiling. Distribute into tubes. Autoclave for 15 min at 15 psi pressure–121°C. Allow tubes to cool in a slanted position. Aseptically add a strip of sterile filter paper to the surface of each slant.

Use: For the cultivation of *Cytophaga* species.

Dubos Salts Agar with Yeast Extract

Composition per liter:
Agar 15.0g
K_2HPO_4 .... 1.0g
Yeast extract 0.5g
KCl 0.5g
$MgSO_4 \cdot 7H_2O$ 0.5g
$NaNO_3$ 0.5g
$FeSO_4 \cdot 7H_2O$ 0.01g
Filter paper strips, sterile .... variable
pH 7.2 ± 0.2 at 25°C

Preparation of Medium: Add components, except filter paper strips, to distilled/deionized water and bring volume to 1.0L. Mix thoroughly. Gently heat and bring to boiling. Autoclave for 15 min at 15 psi pressure–121°C. Pour into sterile Petri dishes. Aseptically place sterile filter paper strips onto the surface of the solidified medium.

Use: For the cultivation and maintenance of *Cellulomonas* species and *Cellvibrio* species.

Dubos Salts Broth

Composition per liter:

K_2HPO_4 .... 1.0g
KCl 0.5g
$MgSO_4 \cdot 7H_2O$ 0.5g
$NaNO_3$ 0.5g
$FeSO_4 \cdot 7H_2O$ 0.01g
Filter paper strips, sterile .... variable
pH 7.2 ± 0.2 at 25°C

Preparation of Medium: Add components, except filter paper strips, to distilled/deionized water and bring volume to 1.0L. Mix thoroughly. Distribute into tubes containing a filter paper strip (filter paper strip should protrude above the surface of the broth). Autoclave for 15 min at 15 psi pressure–121°C.

Use: For the cultivation of *Cytophaga aurantiaca* and *Pseudomonas* species.

Dubos Salts Broth with Yeast Extract

Composition per liter:

K_2HPO_4 .... 1.0g
Yeast extract 0.5g
KCl 0.5g
$MgSO_4 \cdot 7H_2O$ 0.5g
$NaNO_3$ 0.5g
$FeSO_4 \cdot 7H_2O$ 0.01g
pH 7.2 ± 0.2 at 25°C

Preparation of Medium: Add components to distilled/deionized water and bring volume to 1.0L. Mix thoroughly. Distribute into tubes or flasks. Autoclave for 15 min at 15 psi pressure–121°C.

Use: For the cultivation of *Cytophaga aurantiaca.*

Dubos Salts Broth with Yeast Extract

Composition per liter:

K_2HPO_4 .... 1.0g
Yeast extract 0.5g
KCl 0.5g
$MgSO_4 \cdot 7H_2O$ 0.5g
$NaNO_3$ 0.5g
$FeSO_4 \cdot 7H_2O$ 0.01g
Filter paper strips, sterile .... variable
pH 7.2 ± 0.2 at 25°C

Preparation of Medium: Add components, except filter paper strips, to distilled/deionized water and bring volume to 1.0L. Mix thoroughly. Distribute into tubes containing a filter paper strip (filter paper strip should protrude above the surface of the broth). Autoclave for 15 min at 15 psi pressure–121°C.

Use: For the cultivation of *Cellulomonas* species and *Cellvibrio* species.

Dung Extract Agar
(DSMZ Medium 781)

Composition per liter:

Agar 15.0g
Malt extract 5.0g
$(NO_3)_2 \cdot 4H_2O$ 0.72g
$MgSO_4 \cdot 7H_2O$ 0.5g
K_2HPO_4 .... 0.25g
Peptone 0.1g
Dung extract 100.0mL
pH 6.9 ± 0.2 at 25°C

Dung Extract:

Composition per 150.0mL:

Horse dung .... variable

Preparation of Dung Extract: Add an average sized piece of horse dung to150.0mL water. Gently heat and bring to boiling. Boil for 2h in a water bath. Filter. Use immediately.

Preparation of Medium: Add components to distilled/deionized water and bring volume to 1.0L. Mix thoroughly. Gently heat and bring to boiling. Distribute into tubes or flasks. Autoclave for 15 min at 15 psi pressure–121°C. Pour into sterile Petri dishes or leave in tubes.

Use: For the cultivation and maintenance of *Panaeolus cyanescens.*

E Agar
(m-E Agar)

Composition per liter:

Yeast extract 30.0g
Agar 15.0g
NaCl 15.0g
Pancreatic digest of gelatin 10.0g
Esculin 1.0g
Nalidixic acid 0.25g
NaN_3 .... 0.15g
Cycloheximide 0.05g
TTC solution 15.0mL
pH 7.1 ± 0.2 at 25°C

TTC Solution:

Composition per 15.0mL:

2,3,5-Triphenyltetrazolium chloride 0.15g

Preparation of TTC Solution: Add triphenyltetrazolium chloride to distilled/deionized water and bring volume to 15.0mL. Mix thoroughly. Filter sterilize.

Preparation of Medium: Add components, except TTC solution, to distilled/deionized water and bring volume to 1.0L. Mix thoroughly. Gently heat and bring to boiling. Autoclave for 15 min at 15 psi pressure–121°C. Cool to 45°–50°C. Aseptically add sterile TTC solution. Mix thoroughly. Pour into sterile Petri dishes or distribute into sterile tubes.

Caution: Sodium azide is toxic. Azides also react with metals and disposal must be highly diluted.

Use: For the isolation, cultivation, and enumeration of enterococci in water by the membrane filter method. It is used in conjunction with esculin iron agar.

E Medium for Anaerobes
with 0.1% Cellobiose

Composition per 100.0mL:

Cellobiose 0.1g
Glucose 0.05g
L-Cysteine·HCl·H_2O 0.05g
Maltose 0.05g
$(NH_4)_2SO_4$ 0.05g
Peptone 0.05g
Soluble starch 0.05g
Yeast extract 0.05g
Salts solution 50.0mL

Rumen fluid 30.0mL
Resazurin solution 0.4mL

pH 7.0 ± 0.2 at 25°C

Salts Solution:
Composition per liter:

$NaHCO_3$ 10.0g
NaCl 2.0g
K_2HPO_4 ....1.0g
KH_2PO_4 ....1.0g
$CaCl_2$, anhydrous 0.2g
$MgSO_4$ 0.2g

Preparation of Salts Solution: Add $CaCl_2$ and $MgSO_4$ to approximately 300.0mL of distilled/deionized water. Mix thoroughly. Bring volume to 800.0mL with distilled/deionized water. Add remaining components. Mix thoroughly. Bring volume to 1.0L with distilled/deionized water. Mix thoroughly. Store at 4°C.

Rumen Fluid:
Composition per 100.0mL:

Rumen fluid 100.0mL

Preparation of Rumen Fluid: Obtain the rumen contents from a cow that has been fed an alfalfa-hay ration. Filter rumen contents through two layers of cheesecloth. Store under 100% CO_2 in the refrigerator. The particulate material will settle out. Use only the supernatant liquid.

Resazurin Solution:
Composition per 44.0mL:

Resazurin 0.011g

Preparation of Resazurin Solution: Add resazurin to distilled/deionized water and bring volume to 44.0mL. Mix thoroughly.

Preparation of Medium: Add components, except L-cysteine·HCl·H_2O, to distilled/deionized water and bring volume to 100.0mL. Mix thoroughly. Gently heat and bring to boiling. Continue boiling until resazurin turns colorless, indicating reduction. Cool in an ice-water bath under 100% CO_2. Add the L-cysteine·HCl·H_2O. Adjust pH to 7.0 with 8*N* NaOH or 5*N* HCl. Anaerobically distribute into tubes under O_2-free 100% N_2. Cap tubes with butyl rubber stoppers. Place tubes in a press. Autoclave for 12 min at 15 psi pressure–121°C with fast exhaust.

Use: For the cultivation and maintenance of *Eubacterium cellulosolvens* and *Fibrobacter inyrdyinslid.*

E Medium for Anaerobes with Filtered Rumen Fluid and 0.1% Cellobiose

Composition per 100.0mL:

Cellobiose 0.1g
Glucose 0.05g
L-Cysteine·HCl·H_2O 0.05g
Maltose 0.05g
$(NH_4)_2SO_4$ 0.05g
Peptone 0.05g
Soluble starch 0.05g
Yeast extract 0.05g
Salts solution 50.0mL
Rumen fluid, filtered 30.0mL
Resazurin solution 0.4mL

pH 7.0 ± 0.2 at 25°C

Salts Solution:
Composition per liter:

$NaHCO_3$ 10.0g
NaCl 2.0g
K_2HPO_4 ....1.0g
KH_2PO_4 ....1.0g
$CaCl_2$ (anhydrous) 0.2g
$MgSO_4$ 0.2g

Preparation of Salts Solution: Add $CaCl_2$ and $MgSO_4$ to approximately 300.0mL of distilled/deionized water. Mix thoroughly. Bring volume to 800.0mL with distilled/deionized water. Add remaining components. Mix thoroughly. Bring volume to 1.0L with distilled/deionized water. Mix thoroughly. Store at 4°C.

Rumen Fluid:
Composition per 100.0mL:

Rumen fluid 100.0mL

Preparation of Rumen Fluid: Obtain the rumen contents from a cow that has been fed an alfalfa-hay ration. Filter rumen contents through two layers of cheesecloth. Store under 100% CO_2 in the refrigerator. The particulate material will settle out. Use only the supernatant liquid. Filter through a 0.20μm filter.

Resazurin Solution:
Composition per 44.0mL:

Resazurin 0.011g

Preparation of Resazurin Solution: Add resazurin to distilled/deionized water and bring volume to 44.0mL. Mix thoroughly.

Preparation of Medium: Add components, except L-cysteine·HCl·H_2O, to distilled/deionized water and bring volume to 100.0mL. Mix thoroughly. Gently heat and bring to boiling. Continue boiling until resazurin turns colorless, indicating reduction. Cool in an ice-water bath under 100% CO_2. Add the L-cysteine·HCl·H_2O. Adjust pH to 7.0 with 8*N* NaOH or 5*N* HCl. Anaerobically distribute into tubes under O_2-free 100% N_2. Cap tubes with butyl rubber stoppers. Place tubes in a press. Autoclave for 12 min at 15 psi pressure–121°C with fast exhaust.

Use: For the cultivation and maintenance of the *Fibrobacter* species.

E Medium for Anaerobes, Modified

Composition per 103.6mL:

L-Cysteine·HCl·H_2O 0.05g
$(NH_4)_2SO_4$ 0.05g
Peptone 0.05g
Yeast extract 0.05g
Salts solution 50.0mL
Rumen fluid 30.0mL
Potassium phosphate buffer (1*M*, pH 6.5) 2.8mL
Hemin solution 1.0mL
Glucose-maltose solution 1.4mL
Starch solution 1.4mL
Resazurin (0.025% solution) 0.4mL
Vitamin K_3 solution 0.2mL

pH 6.5 ± 0.2 at 25°C

Salts Solution:
Composition per liter:

$NaHCO_3$ 10.0g
NaCl 2.0g
K_2HPO_4 ....1.0g
KH_2PO_4 ....1.0g
$CaCl_2$, anhydrous 0.2g
$MgSO_4$ 0.2g

Preparation of Salts Solution: Add $CaCl_2$ and $MgSO_4$ to approximately 300.0mL of distilled/deionized water. Mix thoroughly. Bring volume to 800.0mL with distilled/deionized water. Add remaining components. Mix thoroughly. Bring volume to 1.0L with distilled/deionized water. Mix thoroughly. Store at 4°C.

Rumen Fluid:
Composition per 100.0mL:
Rumen fluid 100.0mL

Preparation of Rumen Fluid: Obtain the rumen contents from a cow that has been fed an alfalfa-hay ration. Filter rumen contents through two layers of cheesecloth. Store under 100% CO_2 in the refrigerator. The particulate material will settle out. Use only the supernatant liquid.

Hemin Solution:
Composition per 100.0mL:
Hemin 0.05g
NaOH (1*N* solution) 1.0mL

Preparation of Hemin Solution: Add hemin to 1.0mL of 1*N* NaOH solution. Mix thoroughly. Bring volume to 100.0mL with distilled/deionized water. Autoclave for 15 min at 15 psi pressure–121°C. Cool to 45°–50°C.

Glucose-Maltose Solution:
Composition per 10.0mL:
Glucose 0.5g
Maltose 0.5g

Preparation of Glucose-Maltose Solution: Add components to distilled/deionized water and bring volume to 10.0mL. Mix thoroughly. Filter sterilize.

Starch Solution:
Composition per 10.0mL:
Starch, soluble 0.5g

Preparation of Starch Solution: Add starch to distilled/deionized water and bring volume to 10.0mL. Mix thoroughly. Autoclave for 15 min at 15 psi pressure–121°C.

Resazurin Solution:
Composition per 44.0mL:
Resazurin 0.011g

Preparation of Resazurin Solution: Add resazurin to distilled/deionized water and bring volume to 44.0mL. Mix thoroughly.

Vitamin K_3 Solution:
Composition per 25.0mL:
Vitamin K_3 (menadione) 0.0125g
Ethanol, absolute 25.0mL

Preparation of Vitamin K_1 Solution: Add Vitamin K_3 to 99.0mL of ethanol. Mix thoroughly.

Preparation of Medium: Filter sterilize potassium phosphate buffer. Add components—except L-cysteine·HCl·H_2O, Vitamin K_3 solution, potassium phosphate buffer, glucose-maltose solution, and starch solution—to distilled/deionized water and bring volume to 98.0mL. Mix thoroughly. Gently heat and bring to boiling. Continue boiling until resazurin turns colorless, indicating reduction. Cool in an ice-water bath under O_2-free 97% N_2 + 3% H_2. Add the L-cysteine·HCl·H_2O and Vitamin K_3 solution. Mix thoroughly. Adjust pH to 6.5 with 8*N* NaOH or 5*N* HCl. Anaerobically distribute into tubes under O_2-free 97% N_2 + 3% H_2 in 7.0mL volumes. Cap tubes with butyl rubber stoppers. Place tubes in a press. Autoclave for 12 min at 15 psi pressure–121°C with fast exhaust. Immediately prior to inoculation, aseptically add 0.2mL of filter-sterilized potassium phosphate buffer, 0.1mL of sterile glucose-maltose solution, and 0.1mL of sterile starch solution to each tube. Mix thoroughly.

Use: For the cultivation and maintenance of *Bacteroides* species, *Butyrivibrio fibrisolvens, Clostridium methylpentosum, Eubacterium ruminantium, Lachnospira multipara, Micromonospora ruminantium, Prevotella ruminicola, Propionibacterium acidipropionici, Selenomonas* species, *Succinivibrio dextrinosolvens,* and *Treponema* species.

E Medium for Anaerobes with 0.3% Phloroglucinol

Composition per 110.4mL:
$(NH_4)_2SO_4$ 0.5g
L-Cysteine·HCl·H_2O 0.05g
Soluble starch 0.05g
Salts solution 50.0mL
Rumen fluid 30.0mL
Phloroglucinol solution 30.0mL
Resazurin solution 0.4mL
pH 6.6 ± 0.2 at 25°C

Salts Solution:
Composition per liter:
$NaHCO_3$ 10.0g
NaCl 2.0g
K_2HPO_4 .... 1.0g
KH_2PO_4 .... 1.0g
$CaCl_2$, anhydrous 0.2g
$MgSO_4$ 0.2g

Preparation of Salts Solution: Add $CaCl_2$ and $MgSO_4$ to approximately 300.0mL of distilled/deionized water. Mix thoroughly. Bring volume to 800.0mL with distilled/deionized water. Add remaining components. Mix thoroughly. Bring volume to 1.0L with distilled/deionized water. Mix thoroughly. Store at 4°C.

Rumen Fluid:
Composition per 100.0mL:
Rumen fluid 100.0mL

Preparation of Rumen Fluid: Obtain the rumen contents from a cow that has been fed an alfalfa-hay ration. Filter rumen contents through two layers of cheesecloth. Store under 100% CO_2 in the refrigerator. The particulate material will settle out. Use only the supernatant liquid.

Phloroglucinol Solution:
Composition per 100.0mL:
Phloroglucinol 1.0g

Preparation of Phloroglucinol Solution: Add phloroglucinol to distilled/deionized water and bring volume to 100.0mL. Mix thoroughly. Filter sterilize. Keep away from light.

Resazurin Solution:
Composition per 44.0mL:
Resazurin 0.011g

Preparation of Resazurin Solution: Add resazurin to distilled/deionized water and bring volume to 44.0mL. Mix thoroughly.

Preparation of Medium: Add components, except L-cysteine·HCl·H_2O, to distilled/deionized water and bring volume to 100.0mL. Mix thoroughly. Gently heat and bring to boiling. Continue boiling until resazurin turns colorless, indicating reduction. Cool in an ice-water bath under 100% CO_2. Add the L-cysteine·HCl·H_2O. Adjust pH to 6.6 with 8*N* NaOH or 5*N* HCl. Anaerobically distribute into tubes under O_2-free 100% N_2. Cap tubes with butyl rubber stoppers. Place tubes in a press. Autoclave for 12 min at 15 psi pressure–121°C with fast exhaust.

Use: For the cultivation and maintenance of *Coprococcus* species.

E Medium for Anaerobes with 0.2% Rutin

Composition per 110.4mL:

$(NH_4)_2SO_4$	0.5g
L-Cysteine·HCl·H_2O	0.05g
Soluble starch	0.05g
Salts solution	50.0mL
Rumen fluid	30.0mL
Rutin solution	30.0mL
Resazurin solution	0.4mL

pH 6.6 ± 0.2 at 25°C

Salts Solution:
Composition per liter:

$NaHCO_3$	10.0g
NaCl	2.0g
K_2HPO_4	1.0g
KH_2PO_4	1.0g
$CaCl_2$, anhydrous	0.2g
$MgSO_4$	0.2g

Preparation of Salts Solution: Add $CaCl_2$ and $MgSO_4$ to approximately 300.0mL of distilled/deionized water. Mix thoroughly. Bring volume to 800.0mL with distilled/deionized water. Add remaining components. Mix thoroughly. Bring volume to 1.0L with distilled/deionized water. Mix thoroughly. Store at 4°C.

Rumen Fluid:
Composition per 100.0mL:

Rumen fluid	100.0mL

Preparation of Rumen Fluid: Obtain the rumen contents from a cow that has been fed an alfalfa-hay ration. Filter rumen contents through two layers of cheesecloth. Store under 100% CO_2 at 4°C. The particulate material will settle out. Use the liquid.

Rutin Solution:
Composition per 100.0mL:

Rutin	0.2g

Preparation of Rutin Solution: Add rutin to distilled/deionized water and bring volume to 100.0mL. Mix thoroughly. Filter sterilize.

Resazurin Solution:
Composition per 44.0mL:

Resazurin	0.011g

Preparation of Resazurin Solution: Add resazurin to distilled/deionized water and bring volume to 44.0mL. Mix thoroughly.

Preparation of Medium: Add components, except L-cysteine·HCl·H_2O, to distilled/deionized water and bring volume to 100.0mL. Mix thoroughly. Gently heat and bring to boiling. Continue boiling until resazurin turns colorless, indicating reduction. Cool in an ice-water bath under 100% CO_2. Add the L-cysteine·HCl·H_2O. Adjust pH to 6.6 with 8*N* NaOH or 5*N* HCl. Anaerobically distribute into tubes under O_2-free 100% N_2. Cap tubes with butyl rubber stoppers. Place tubes in a press. Autoclave for 12 min at 15 psi pressure–121°C with fast exhaust.

Use: For the cultivation of *Butyrivibrio* species.

EC Broth (*Escherichia coli* Broth) (EC Medium)

Composition per liter:

Pancreatic digest of casein	20.0g
Lactose	5.0g
NaCl	5.0g
K_2HPO_4	4.0g
Bile salts mixture	1.5g
KH_2PO_4	1.5g

pH 6.9 ± 0.2 at 25°C

Source: This medium is available as a premixed powder from BD Diagnostic Systems.

Preparation of Medium: Add components to distilled/deionized water and bring volume to 1.0L. Mix thoroughly. Distribute into test tubes that contain an inverted Durham tube. Autoclave for 12 min at 15 psi pressure–121°C. Cool broth as quickly as possible.

Use: For the cultivation and differentiation of coliform bacteria at 37°C and of *Escherichia coli* at 45.5°C.

EC Broth with MUG

Composition per liter:

Pancreatic digest of casein	20.0g
Lactose	5.0g
NaCl	5.0g
K_2HPO_4	4.0g
Bile salts mixture	1.5g
KH_2PO_4	1.5g
4-Methylumbeliferyl-β-D-glucuronide (MUG)	0.05g

pH 6.9 ± 0.2 at 25°C

Source: This medium is available as a premixed powder from BD Diagnostic Systems.

Preparation of Medium: Add components to distilled/deionized water and bring volume to 1.0L. Mix thoroughly. Distribute into test tubes that contain an inverted Durham tube in 10.0mL volumes. Autoclave for 15 min at 15 psi pressure–121°C.

Use: For the detection of *Escherichia coli* in water samples by a fluorogenic procedure.

EC Medium, Modified with Novobiocin

Composition per liter:

Tryptone	20.0g
NaCl	5.0g
Lactose	5.0g
K_2HPO_4	4.0g
KH_2PO_4	1.5g
Bile salts	1.12g
Novobiocin supplement	10.0mL

pH 6.9 ± 0.2 at 25°C

Source: This medium is available as a premixed powder and supplement from BD Diagnostic Systems.

Novobiocin Supplement:
Composition per 10.0mL:
Sodium novobiocin 20.0mg

Preparation of Novobiocin Supplement: Add sodium novobiocin to distilled/deionized water and bring volume to 10.0mL. Mix thoroughly. Filter sterilize.

Preparation of Medium: Add components, except novobiocin supplement, to distilled/deionized water and bring volume to 990.0mL. Mix thoroughly. Autoclave for 15 min at 15 psi pressure–121°C. Aseptically add 10.0mL of sterile novobiocin supplement. Mix thoroughly. Aseptically distribute into sterile tubes or flasks.

Use: For the cultivation of *Escherichia coli* O157:H7.

Echinamoeba Agar (ATCC Medium 2339)

Composition per liter:
Agar 18.0g
$FeCl_2·6H_2O$ 0.552g
$MgSO_4·7H_2O$ 0.5g
$CaSO_4·2H_2O$ 0.49g
$NaHCO_3$ 0.37g
$CaCl_2·2H_2O$ 0.24g
Yeast extract 0.2g
KCl 8.6mg
KNO_3 .... 0.16mg
Mineral solution 10.0mL
Glycerol 1.0mL
pH 7.0 ± 0.2 at 25°C

Mineral Solution:
Composition per liter:
$ZnSO_4·7H_2O$ 3.5g
$Na_2MoO_4·2H_2O$ 0.4g
$MnCl_2·4H_2O$ 0.25g
$(NH_4)_2Ni(SO_4)_2·6H_2O$ 0.2g
LiCl 0.3g
$SnCl_2·2H_2O$ 0.1

Preparation of Mineral Solution: Add components to distilled/deionized water and bring volume to 1.0L. Mix thoroughly.

Preparation of Medium: Add components to distilled/deionized water and bring volume to 1.0L. Mix thoroughly. Adjust pH to 7.0 with NaOH. Distribute into tubes or flasks. Autoclave for 15 min at 15 psi pressure–121°C.

Use: For the cultivation and maintenance of *Echinamoeba* spp., e.g., *Echinamoeba thermarum* n. sp., an extremely thermophilic amoeba thriving in hot springs.

Echinamoeba Agar

Composition per liter:
Agar 18.0g
$FeCl_2·6H_2O$ 0.552g
$MgSO_4·7H_2O$ 0.5g
$CaSO_4·2H_2O$ 0.49g
$NaHCO_3$ 0.37g
$CaCl_2·2H_2O$ 0.24g
Yeast extract 0.2g
KCl 8.6mg
KNO_3 .... 0.16mg
Mineral solution 10.0mL
Glycerol 1.0mL
pH 7.0 ± 0.2 at 25°C

Mineral Solution:
Composition per liter:
$ZnSO_4·7H_2O$ 3.5g
$Na_2MoO_4·2H_2O$ 0.4g
$MnCl_2·4H_2O$ 0.25g
$(NH_4)_2Ni(SO_4)_2·6H_2O$ 0.2g
LiCl 0.3g
$SnCl_2·2H_2O$ 0.1g

Preparation of Mineral Solution: Add components to distilled/deionized water and bring volume to 1.0L. Mix thoroughly.

Preparation of Medium: Add components to distilled/deionized water and bring volume to 1.0L. Mix thoroughly. Adjust pH to 7.0 with NaOH. Gently heat and bring to boiling. Distribute into tubes or flasks. Autoclave for 15 min at 15 psi pressure–121°C. Pour into sterile Petri dishes or leave in tubes.

Use: For the cultivation and maintenance of *Echinamoeba* spp., e.g., *Echinamoeba thermarum* n. sp., an extremely thermophilic amoeba thriving in hot springs.

Echinamoeba Broth

Composition per liter:
$FeCl_2·6H_2O$ 0.552g
$MgSO_4·7H_2O$ 0.5g
$CaSO_4·2H_2O$ 0.49g
$NaHCO_3$ 0.37g
$CaCl_2·2H_2O$ 0.24g
Yeast extract 0.2g
KCl 8.6mg
KNO_3 .... 0.16mg
Mineral solution 10.0mL
Glycerol 1.0mL
pH 7.0 ± 0.2 at 25°C

Mineral Solution:
Composition per liter:
$ZnSO_4·7H_2O$ 3.5g
$Na_2MoO_4·2H_2O$ 0.4g
$MnCl_2·4H_2O$ 0.25g
$(NH_4)_2Ni(SO_4)_2·6H_2O$ 0.2g
LiCl 0.3g
$SnCl_2·2H_2O$ 0.1g

Preparation of Mineral Solution: Add components to distilled/deionized water and bring volume to 1.0L. Mix thoroughly.

Preparation of Medium: Add components to distilled/deionized water and bring volume to 1.0L. Mix thoroughly. Adjust pH to 7.0 with NaOH. Distribute into tubes or flasks. Autoclave for 15 min at 15 psi pressure–121°C.

Use: For the cultivation and maintenance of *Echinamoeba* spp., such as *Echinamoeba thermarum*, an extremely thermophilic amoeba thriving in hot springs.

ECM Agar

Composition per liter:
Agar 15.0g
NaCl 6.0g
Escherichia coli cells, washed 1.0g
$MgSO_4·7H_2O$ 0.5g

Preparation of Medium: Add components to distilled/deionized water and bring volume to 1.0L. Mix thoroughly. Gently heat and bring to boiling. Distribute into tubes or

flasks. Autoclave for 15 min at 15 psi pressure–121°C. Pour into sterile Petri dishes or leave in tubes.

Use: For the cultivation of myxobacteria.

Ectothiorhodospira abdelmalekii Medium

Composition per 1010.0mL:

NaCl	120.0g
Na_2SO_4	15.0g
$NaHCO_3$	10.0g
Na_2CO_3	5.0g
Sodium acetate	1.0g
KH_2PO_4	0.8g
NH_4Cl	0.8g
$MgCl_2 \cdot 6H_2O$	0.1g
$Na_2S \cdot 9H_2O$ solution	10.0mL
Trace elements solution SLA	1.0mL
Vitamin solution	1.0mL

pH 8.5 ± 0.2 at 25°C

$Na_2S \cdot 9H_2O$ Solution:

Composition per 10.0mL:

$Na_2S \cdot 9H_2O$	0.5g

Preparation of $Na_2S \cdot 9H_2O$ Solution: Add $Na_2S \cdot 9H_2O$ to distilled/deionized water and bring volume to 10.0mL. Mix thoroughly. Gas under 100% N_2. Autoclave for 15 min at 15 psi pressure–121°C.

Trace Elements Solution SLA:

Composition per liter:

$CuCl_2 \cdot 2H_2O$	10.0g
$FeCl_2 \cdot 4H_2O$	1.8g
H_3BO_3	0.5g
$CoCl_2 \cdot 6H_2O$	0.25g
$ZnCl_2$	0.1g
$MnCl_2 \cdot 4H_2O$	70.0mg
$Na_2MoO_4 \cdot 2H_2O$	30.0mg
$Na_2SeO_3 \cdot 5H_2O$	10.0mg
$NiCl_2 \cdot 6H_2O$	10.0mg

Preparation of Trace Elements Solution SLA: Add components to distilled/deionized water and bring volume to 1.0L. Mix thoroughly. Adjust pH to 2.0–3.0.

Vitamin Solution VA:

Composition per 100.0mL:

Nicotinic acid amide	35.0mg
Thiamine dichloride	30.0mg
p-Aminobenzoic acid	20.0mg
Biotin	10.0mg
Calcuim DL-pantothenate	10.0mg
Pyridoxal·HCl	10.0mg
Vitamin B_{12}	5.0mg

Preparation of Vitamin Solution VA: Add components to distilled/deionized water and bring volume to 100.0mL. Mix thoroughly. Filter sterilize.

Preparation of Medium: Add components, except $Na_2S \cdot 9H_2O$ solution, to distilled/deionized water and bring volume to 1.0L. Mix thoroughly. Adjust pH to 8.5. Filter sterilize. Aseptically add 10.0mL of sterile $Na_2S \cdot 9H_2O$ solution. Mix thoroughly. Distribute into sterile tubes or flasks.

Use: For the cultivation and maintenance of *Ectothiorhodospira abdelmalekii.*

Ectothiorhodospira halochloris Medium

Composition per liter:

NaCl	180.0g
Na_2SO_4	20.0g
$NaHCO_3$	14.0g
Na_2CO_3	6.0g
$Na_2S \cdot 9H_2O$	1.0g
Sodium succinate	1.0g
NH_4Cl	0.8g
KH_2PO_4	0.5g
Yeast extract	0.5g
$MgCl_2 \cdot 6H_2O$	0.1g
$CaCl_2 \cdot 2H_2O$	0.05g
Vitamin solution VA	1.0mL
Trace elements solution SLA	1.0mL

pH 8.5–8.7 at 25°C

Vitamin Solution VA:

Composition per liter:

Nicotinamide	0.04g
Thiamine dichloride	0.03g
p-Aminobenzoic acid	0.02g
Biotin	0.01g
Calcium pantothenate	0.01g
Pyridoxal chloride	0.01g
Vitamin B_{12}	5.0mg

Preparation of Vitamin Solution VA: Add components to distilled/deionized water and bring volume to 1.0L. Mix thoroughly.

Trace Elements Solution SLA:

Composition per liter:

$FeCl_2 \cdot 4H_2O$	1.8g
H_3BO_3	0.5g
$CoCl_2 \cdot 6H_2O$	0.25g
$ZnCl_2$	0.1g
$MnCl_2 \cdot 4H_2O$	0.07g
$NaMoO_4 \cdot 2H_2O$	0.03g
$CuCl_2 \cdot 2H_2O$	0.01g
Na_2SeO_3	0.01g
$NiCl_2 \cdot 6H_20$	0.01g

Preparation of Trace Elements Solution SLA: Add components to distilled/deionized water and bring volume to 1.0L. Mix thoroughly. Adjust pH to 3.0 with 2*N* HCl.

Preparation of Medium: Add components, except trace elements solution SLA, to distilled/deionized water and bring volume to 999.0mL. Mix thoroughly. Filter sterilize. Aseptically add 1.0mL of sterile trace elements solution SLA. Mix thoroughly. Aseptically distribute into flasks or bottles. Completely fill bottles with medium except for a pea-sized air bubble.

Use: For the enrichment and isolation of *Ectothiorhodospira halochloris.*

Ectothiorhodospira halophila Medium

Composition per liter:

NaCl	200.0g
NH_4Cl	0.4g
$(NH_4)_2SO_4$	0.1g
Na_2CO_3 solution	100.0mL
Tris buffer (1*M* solution, pH 7.5)	30.0mL
Solution C	5.0mL
Potassium phosphate buffer (1*M* solution, pH 7.5)	3.0mL
Additional solution	2.5mL

pH 7.4–8.0 at 25°C

Na_2CO_3 Solution:
Composition per 100.0mL:
Na_2CO_3 .... 10.0g

Preparation of Na_2CO_3 Solution: Add Na_2CO_3 to distilled/deionized water and bring volume to 100.0mL. Mix thoroughly. Autoclave for 15 min at 15 psi pressure–121°C. Cool to 25°C.

Solution C:
Composition per liter:
$MgCl_2 \cdot 6H_2O$ 24.0g
$CaCl_2 \cdot 2H_2O$ 3.3g
$FeCl_3 \cdot 4H_2O$ 1.1g
$(NH_4)_6Mo_7O_{24} \cdot 4H_2O$ 0.1g
Nitrilotriacetic acid 10.0mg
Trace elements solution 50.0mL

Preparation of Solution C: Add nitrilotriacetic acid to 500.0mL of distilled/deionized water. Dissolve by adjusting pH to 6.5 with KOH. Add remaining components. Readjust pH to 7.2 with H_2SO_4 or KOH. Add distilled/deionized water to 1.0L.

Trace Elements Solution:
Composition per 100.0mL:
$ZnCl_2$ 0.52g
EDTA 0.25g
$MnCl_2 \cdot 4H_2O$ 0.08g
$FeCl_3 \cdot 4H_2O$ 0.03g
$Co(NO_3)_2 \cdot 6H_2O$ 0.02g
$CuCl_2 \cdot 2H_2O$ 0.02g
H_3BO_3 .... 0.01g

Preparation of Trace Elements Solution: Add components to distilled/deionized water and bring volume to 1.0L. Mix thoroughly. Adjust pH to 3.0 with 2*N* HCl.

Additional Solution:
Composition per 50.0mL:
$NaS_2O_3 \cdot 6H_2O$ 6.0g
Sodium succinate 5.0g
Sodium ascorbate 1.0g

Preparation of Additional Solution: Add components to distilled/deionized water and bring volume to 50.0mL. Mix thoroughly. Filter sterilize.

Preparation of Medium: Add components, except Na_2CO_3 solution and additional solution, to distilled/deionized water and bring volume to 900.0mL. Autoclave for 15 min at 15 psi pressure–121°C. Cool to 45°–50°C. Aseptically adjust pH to 7.4–7.8 with filter-sterilized HCl. Aseptically distribute into 50.0mL screw-capped bottles. Fill each bottle almost to the top, leaving a space of 2.8mL in the neck. Aseptically add 2.5mL of sterile additional solution to each bottle. Mix thoroughly.

Use: For the isolation and cultivation of *Ectothiorhodospira halophila.*

Ectothiorhodospira Medium

Composition per liter:
NaCl 180.0g
Na_2SO_4 .... 20.0g
$NaHCO_3$ 14.0g
Na_2CO_3 .... 6.0g
Sodium succinate 1.0g
NH_4Cl 0.8g
KH_2PO_4 .... 0.5g
$MgCl_2 \cdot 6H_2O$ 0.1g
$CaCl_2 \cdot 2H_2O$ 0.05g
Feeding solution 10.0mL
Trace elements solution SLA 1.0mL
Vitamin solution VA 1.0mL
pH 8.5–8.7 at 25°C

Feeding Solution:
Composition per 100.0mL:
NaCl 10.0g
$NaHCO_3$ 10.0g
$Na_2S \cdot 9H_2O$ 5.0g

Preparation of Feeding Solution: Add components to distilled/deionized water and bring volume to 100.0mL. Mix thoroughly. Filter sterilize.

Trace Elements Solution SLA:
Composition per liter:
$FeCl_2 \cdot 4H_2O$ 1.8g
H_3BO_3 .... 0.5g
$CoCl_2 \cdot 6H_2O$ 0.25g
$ZnCl_2$ 0.1g
$MnCl_2 \cdot 4H_2O$ 0.07g
$NaMoO_4 \cdot 2H_2O$ 0.03g
$CuCl_2 \cdot 2H_2O$ 0.01g
Na_2SeO_3 .... 0.01g
$NiCl_2 \cdot 6H_2O$ 0.01g

Preparation of Trace Elements Solution SLA: Add components to distilled/deionized water and bring volume to 1.0L. Mix thoroughly. Adjust pH to 3.0 with 2*N* HCl.

Vitamin Solution VA:
Composition per liter:
Nicotinamide 0.04g
Thiamine dichloride 0.03g
p-Aminobenzoic acid 0.02g
Biotin 0.01g
Calcium pantothenate 0.01g
Pyridoxal chloride 0.01g
Vitamin B_{12} 5.0mg

Preparation of Vitamin Solution VA: Add components to distilled/deionized water and bring volume to 1.0L. Mix thoroughly.

Preparation of Medium: Add components, except trace elements solution SLA and feeding solution, to distilled/deionized water and bring volume to 999.0mL. Mix thoroughly. Filter sterilize. Aseptically add 1.0mL of sterile trace elements solution SLA. Mix thoroughly. Aseptically distribute into flasks or bottles. Completely fill bottles with medium except for a pea-sized air bubble. Prior to inoculation, aseptically remove a sufficient amount of medium to permit the addition of feeding medium. Add 1.0mL of feeding solution per each 100.0mL of medium.

Use: For the isolation and cultivation of *Ectothiorhodospira halochloris* and *Ectothiorhodospira halophila.*

Ectothiorhodospira Medium

Composition per liter:
NaCl 130.0g
Na_2SO_4 .... 10.0g
Sodium acetate 2.0g
KH_2PO_4 .... 0.8g
Sodium carbonate buffer, (1*M*, pH 9.0) 200.0mL
$MgCl_2 \cdot 6H_2O$ solution 10.0mL
$Na_2S \cdot 9H_2O$ solution 6.0mL

$CaCl_2 \cdot 2H_2O$ solution	5.0mL
NH_4Cl solution	4.0mL
SLA trace elements	1.0mL
VA vitamin solution	1.0mL

pH 9.0 ± 0.2 at 25°C

$MgCl_2 \cdot 6H_2O$ Solution:
Composition per 10.0mL:

$MgCl_2 \cdot 6H_2O$	0.1g

Preparation of $MgCl_2 \cdot 6H_2O$ Solution: Add 0.1g of $MgCl_2 \cdot 6H_2O$ to distilled/deionized water and bring volume to 10.0mL. Mix thoroughly. Filter sterilize.

$Na_2S \cdot 9H_2O$ Solution:
Composition per 10.0mL:

$Na_2S \cdot 9H_2O$	0.5g

Preparation of $Na_2S \cdot 9H_2O$ Solution: Add $Na_2S \cdot 9H_2O$ to distilled/deionized water and bring volume to 10.0mL. Mix thoroughly. Filter sterilize. Use freshly prepared solution.

$CaCl_2 \cdot 2H_2O$ Solution:
Composition per 10.0mL:

$CaCl_2 \cdot 2H_2O$	0.1g

Preparation of $CaCl_2 \cdot 2H_2O$ Solution: Add $CaCl_2 \cdot 2H_2O$ to distilled/deionized water and bring volume to 10.0mL. Mix thoroughly. Filter sterilize.

NH_4Cl Solution:
Composition per 10.0mL:

NH_4Cl	2.0g

Preparation of NH_4Cl Solution: Add NH_4Cl to distilled/deionized water and bring volume to 10.0mL. Mix thoroughly. Filter sterilize.

SLA Trace Elements:
Composition per liter:

$FeCl_2 \cdot 4H_2O$	1.8g
H_3BO_3	0.5g
$CoCl_2 \cdot 6H_2O$	0.25g
$ZnCl_2$	0.1g
$MnCl_2 \cdot 4H_2O$	0.07g
$Na_2MoO_4 \cdot 2H_2O$	0.03g
$CuCl_2 \cdot 2H_2O$	0.01g
$Na_2SeO_3 \cdot 5H_2O$	0.01g
$NiCl_2 \cdot 6H_2O$	0.01g

Preparation of SLA Trace Elements: Add components to distilled/deionized water and bring volume to 1.0L. Mix thoroughly. Adjust pH to 2–3.

VA Vitamin Solution:
Composition per 500.0mL:

Nicotinamide	0.175g
Thiamine·HCl	0.15g
p-Aminobenzoic acid	0.1g
Biotin	0.05g
Calcium pantothenate	0.05g
Pyridoxine·2HCl	0.05g
Cyanocobalamin	0.025g

Preparation of VA Vitamin Solution: Add components to distilled/deionized water and bring volume to 500.0mL. Mix thoroughly.

Preparation of Medium: Add components—except $MgCl_2 \cdot 6H_2O$ solution, $Na_2S \cdot 9H_2O$ solution, $CaCl_2 \cdot 2H_2O$ solution, and NH_4Cl solution—to distilled/deionized water and bring volume to 975.0mL. Mix thoroughly. Gently heat and bring to boiling. Autoclave for 15 min at 15 psi pressure–121°C. Cool to 25°C. Aseptically add 10.0mL of sterile $MgCl_2 \cdot 6H_2O$ solution, 6.0mL of sterile $Na_2S \cdot 9H_2O$ solution, 5.0mL of sterile $CaCl_2 \cdot 2H_2O$ solution, and 4.0mL of sterile NH_4Cl solution. Mix thoroughly. Aseptically distribute into culture bottles. Incubate for 2 days before inoculation.

Use: For the cultivation and maintenance of *Ectothiorhodospira abdelmalekii* and *Ectothiorhodospira halochloris.*

EIA Substrate

Composition per liter:

Agar	15.0g
Esculin	1.0g
Ferric citrate	0.5g

pH 7.1 ± 0.1 at 25°C

Source: This medium is available as a premixed powder from BD Diagnostic Systems.

Preparation of Medium: Add components to distilled/deionized water and bring volume to 1.0L. Mix thoroughly. Gently heat and bring to boiling. Adjust pH to 7.1. Distribute into tubes or flasks. Autoclave for 15 min at 15 psi pressure–121°C. Pour into sterile Petri dishes.

Use: For the cultivation and enumeration of marine enterococci by the membrane filter method.

Eijkman Lactose Medium

Composition per liter:

Pancreatic digest of casein	15.0g
K_2HPO_4	10.0g
KH_2PO_4	4.0g
Lactose	3.0g
NaCl	2.5g

pH 6.8 ± 0.1 at 25°C

Preparation of Medium: Add components to distilled/deionized water and bring volume to 1.0L. Mix thoroughly. Distribute into test tubes that contain an inverted Durham tube. Autoclave for 15 min at 15 psi pressure–121°C.

Use: For the cultivation and differentiation of *Escherichia coli* from other coliform organisms based on their ability to ferment lactose and produce gas.

Ekho Lake Strains Medium (DSMZ Medium 621a)

Composition per liter:

Agar	15.0g
Peptone	0.25g
Yeast extract	0.25g
Artificial seawater	250.0mL
Hutner's basal salts	20.0mL
Glucose solution	10.0mL
Vitamin solution	5.0mL

pH 7.3 ± 0.2 at 25°C

Hutner's Basal Salts Solution:
Composition per liter:

$MgSO_4 \cdot 7H_2O$	29.7g
Nitrilotriacetic acid	10.0g
$CaCl_2 \cdot 2H_2O$	3.335g
$FeSO_4 \cdot 7H_2O$	99.0mg
$(NH_4)_6MoO_7O_{24} \cdot 4H_2O$	9.25mg
"Metals 44"	50.0mL

"Metals 44":

Composition per 100.0mL:

$ZnSO_4 \cdot 7H_2O$	1.095g
$FeSO_4 \cdot 7H_2O$	0.5g
Sodium EDTA	0.25g
$MnSO_4 \cdot H2O$	0.154g
$CuSO_4 \cdot 5H_2O$	39.2mg
$Co(NO_3)_2 \cdot 6H_2O$	24.8mg
$Na_2B_4O_7 \cdot 10H_2O$	17.7mg

Preparation of "Metals 44": Add sodium EDTA to distilled/deionized water and bring volume to 90.0mL. Mix thoroughly. Add a few drops of concentrated H_2SO_4 to retard precipitation of heavy metal ions. Add remaining components. Mix thoroughly. Bring volume to 100.0mL with distilled/deionized water.

Preparation of Hutner's Basal Salts Solution: Add nitrilotriacetic acid to 500.0mL of distilled/deionized water. Adjust pH to 6.5 with KOH. Add remaining components. Add distilled/deionized water to 1.0L. Adjust pH to 6.8.

Artificial Seawater:

Composition per liter:

NaCl	23.477g
$MgCl_2 \cdot 6H_2O$	4.981g
Na_2SO_4	3.917g
$CaCl_2$	1.12g
KCl	664.0mg
$NaHCO_3$	192.0mg
H_3BO_3	26.0mg
$SrCl_2$	24.0mg
KBr	6.0mg
NaF	3.0mg

Preparation of Artificial Seawater: Add components to distilled/deionized water and bring volume to 1.0L. Mix thoroughly. Filter sterilize.

Vitamin Solution:

Composition per liter:

Pyridoxine-HCl	20.0mg
Riboflavin	10.0mg
Nicotinamide	10.0mg
Thiamine-HCl·$2H_2O$	10.0mg
Ca-pantothenate	10.0mg
p-Aminobenzoic acid	10.0mg
Biotin	4.0mg
Folic acid	4.0mg
Vitamin B_{12}	0.2mg

Preparation of Vitamin Solution: Add components to distilled/deionized water and bring volume to 1.0L. Mix thoroughly. Filter sterilize. Store in the dark at 5°C.

Glucose Solution:

Composition per 10mL:

Glucose	0.25g

Preparation of Glucose Solution: Add glucose to distilled/deionized water and bring volume to 10.0mL. Mix thoroughly. Filter sterilize.

Preparation of Medium: Add components, except artificial seawater, glucose solution, and vitamin solution to distilled/deionized water and bring volume to 735.0mL. Mix thoroughly. Adjust pH to 7.5. Autoclave for 20 min at 15 psi pressure–121°C. Cool to 60°C. Warm artificial seawater to 55°C. Aseptically add 250.0mL warm artificial seawater. Mix thoroughly. Adjust pH to 7.3. Aseptically add 10.0mL glucose solution and 5.0 mL vitamin solution. Mix thoroughly. Aseptically distribute into sterile tubes or flasks.

Use: For the cultivation of *Nocardioides aquaticus, Antarctobacter heliothermus, Sulfitobacter brevis, Roseovarius tolerans, Staleya guttiformis, Roseovarius tolerans, Friedmanniella lacustris,* and *Nesterenkonia lacusekhoensis.*

EMB Agar (Eosin Methylene Blue Agar)

Composition per liter:

Agar	13.5g
Pancreatic digest of casein	10.0g
Lactose	5.0g
Sucrose	5.0g
K_2HPO_4	2.0g
Eosin Y	0.4g
Methylene Blue	0.065g

pH 7.2 ± 0.2 at 25°C

Source: This medium is available as a premixed powder from BD Diagnostic Systems.

Preparation of Medium: Add components to distilled/deionized water and bring volume to 1.0L. Mix thoroughly. Gently heat and bring to boiling. Distribute into tubes or flasks. Autoclave for 15 min at 15 psi pressure–121°C. Pour into sterile Petri dishes.

Use: For the isolation, cultivation, and differentiation of Gram-negative enteric bacteria based on lactose fermentation. Bacteria that ferment lactose, especially the coliform bacterium *Escherichia coli*, appear as colonies with a green metallic sheen or blue-black to brown color. Bacteria that do not ferment lactose appear as colorless or transparent, light purple colonies.

EMB Agar, Modified (Eosin Methylene Blue Agar, Modified)

Composition per liter:

Agar	15.0g
Lactose	10.0g
Pancreatic digest of gelatin	10.0g
K_2HPO_4	2.0g
Eosin Y	0.4g
Methylene Blue	0.065g

pH 6.8 ± 0.2 at 25°C

Source: This medium is available as a premixed powder from Oxoid Unipath.

Preparation of Medium: Add components to distilled/deionized water and bring volume to 1.0L. Mix thoroughly. Gently heat and bring to boiling. Distribute into tubes or flasks. Autoclave for 15 min at 15 psi pressure–121°C. Cool to 60°C. Shake medium to oxidize methylene blue. Pour into sterile Petri dishes. Swirl flask while pouring plates to distribute precipitate.

Use: For the isolation, cultivation, and differentiation of Gram-negative enteric bacteria based on lactose fermentation. Bacteria that ferment lactose, especially the coliform bacterium *Escherichia coli*, appear as colonies with a green metallic sheen or blue-black to brown color. Bacteria that do not ferment lactose appear as colorless or transparent, light purple colonies.

Emerson Agar
(ATCC Medium 199)

Composition per liter:

Agar	20.0g
Glucose	10.0g
Beef extract	4.0g
Pancreatic digest of gelatin	4.0g
NaCl	2.5g
Yeast extract	1.0g

pH 7.0 ± 0.2 at 25°C

Source: This medium is available as a premixed powder from BD Diagnostic Systems.

Preparation of Medium: Add components to distilled/deionized water and bring volume to 1.0L. Mix thoroughly. Gently heat and bring to boiling. Distribute into tubes or flasks. Autoclave for 15 min at 13 psi pressure–118°C. Pour into sterile Petri dishes or leave in tubes.

Use: For the isolation, cultivation, and maintenance of members of the Actinomycetaceae, Streptomycetaceae, and molds. For the cultivation and maintenance of *Arthrobacter* species, *Microbispora rosea, Micromonospora coerulea, Mycobacterium* species, *Nocardia asteroides, Nocardiopsis dassonvillei, Pseudonocardia thermophila, Staphylococcus epidermidis, Streptomyces flaveus, Streptomyces olivaceus, Streptomyces thermoviolaceus, Streptomyces thermovulgaris,* and *Streptomyces vendargensis.*

Endo Agar

Composition per liter:

Agar	15.0g
Lactose	10.0g
Peptic digest of animal tissue	10.0g
K_2HPO_4	3.5g
Na_2SO_3	2.5g
Basic Fuchsin	0.5g

pH 7.4 ± 0.2 at 25°C

Source: This medium is available as a premixed powder from BD Diagnostic Systems.

Caution: Basic Fuchsin is a potential carcinogen and care must be taken to avoid inhalation of the powdered dye and contact with the skin.

Preparation of Medium: Add components to distilled/deionized water and bring volume to 1.0L. Mix thoroughly. Gently heat and bring to boiling. Autoclave for 15 min at 15 psi pressure–121°C. Cool to 45°–50°C. Pour into sterile Petri dishes. Swirl flask while pouring plates to keep precipitate in suspension. Protect from the light.

Use: For the selective isolation, cultivation, and differentiation of coliform and other enteric microorganisms based on their ability to ferment lactose. Lactose-fermenting bacteria appear as dark red colonies with a gold metallic sheen. Lactose-nonfermenting bacteria appear as colorless or translucent colonies.

Endo Agar

Composition per liter:

Agar	10.0g
Lactose	10.0g
Peptic digest of animal tissue	10.0g
K_2HPO_4	3.5g
Na_2SO_3	2.5g
Basic Fuchsin solution	4.0mL

pH 7.5 ± 0.2 at 25°C

Source: This medium is available as a premixed powder from Oxoid Unipath.

Basic Fuchsin Solution:

Composition per 10.0mL:

Basic Fuchsin	1.0g
Ethanol (95% solution)	10.0mL

Preparation of Basic Fuchsin Solution: Add Basic Fuchsin to 10.0mL of ethanol. Mix thoroughly.

Caution: Basic Fuchsin is a potential carcinogen and care must be taken to avoid inhalation of the powdered dye and contact with the skin.

Preparation of Medium: Add components to distilled/deionized water and bring volume to 1.0L. Mix thoroughly. Gently heat and bring to boiling. Autoclave for 15 min at 15 psi pressure–121°C. Cool to 45°–50°C. Pour into sterile Petri dishes. Swirl flask while pouring plates to keep precipitate in suspension. Protect from the light.

Use: For the selective isolation, cultivation, and differentiation of coliform and other enteric microorganisms based on their ability to ferment lactose. Lactose-fermenting bacteria appear as dark red colonies with a gold metallic sheen. Lactose-nonfermenting bacteria appear as colorless or translucent colonies.

Endo Agar, LES
(Endo Agar, Laurance Experimental Station)
(m-Endo Agar, LES)
(m-LES, Endo Agar)

Composition per liter:

Agar	14.0g
Lactose	9.4g
Peptones (pancreatic digest of casein 65% and yeast extract 35%)	7.5g
NaCl	3.7g
Pancreatic digest of casein	3.7g
Peptic digest of animal tissue	3.7g
K_2HPO_4	3.3g
Na_2SO_3	1.6g
Yeast extract	1.2g
KH_2PO_4	1.0g
Basic Fuchsin	0.8g
Sodium lauryl sulfate	0.05g
Ethanol	20.0mL

pH 7.2 ± 0.2 at 25°C

Source: This medium is available as a premixed powder from BD Diagnostic Systems.

Caution: Basic Fuchsin is a potential carcinogen and care must be taken to avoid inhalation of the powdered dye and contact with the skin.

Preparation of Medium: Add ethanol to approximately 900.0mL of distilled/deionized water. Add remaining components. Bring volume to 1.0L with distilled/deionized water. Mix thoroughly. Gently heat and bring to boiling. Autoclave for 15 min at 15 psi pressure–121°C. Pour into sterile 60mm Petri dishes in 4.0mL volumes. Protect from the light.

Use: For the cultivation and enumeration of coliform bacteria by the membrane filter method.

Endo Broth
(m-Endo Broth)

Composition per liter:

Lactose	12.5g
Peptone	10.0g
NaCl	5.0g
Pancreatic digest of casein	5.0g
Peptic digest of animal tissue	5.0g
K_2HPO_4	4.375g
Na_2SO_3	2.1g
Yeast extract	1.5g
KH_2PO_4	1.375g
Basic Fuchsin	1.05g
Sodium deoxycholate	0.1g
Ethanol (95% solution)	20.0mL

pH 7.2 ± 0.1 at 25°C

Source: This medium is available as a premixed powder from BD Diagnostic Systems.

Caution: Basic Fuchsin is a potential carcinogen and care must be taken to avoid inhalation of the powdered dye and contact with the skin.

Preparation of Medium: Add ethanol to approximately 900.0mL of distilled/deionized water. Add remaining components. Bring volume to 1.0L with distilled/deionized water. Mix thoroughly. Gently heat and bring to boiling. Rapidly cool broth below 45°C. Do not autoclave. Use 1.8–2.0mL for each filter pad. Protect from the light. Prepare broth freshly.

Use: For the cultivation and enumeration of coliform bacteria from water by the membrane filter method.

Enriched *Cytophaga* Agar

Composition per liter:

Agar	15.0g
Pancreatic digest of casein	2.0g
Beef extract	0.5g
Yeast extract	0.5g
Sodium acetate	0.2g

pH 7.2 ± 0.2 at 25°C

Preparation of Medium: Add components to distilled/deionized water and bring volume to 1.0L. Mix thoroughly. Gently heat and bring to boiling. Distribute into tubes or flasks. Autoclave for 15 min at 15 psi pressure–121°C. Pour into sterile Petri dishes or leave in tubes.

Use: For the cultivation of *Cytophaga arvensicola, Cytophaga johnsonae, Cytophaga psychrophila, Cytophaga* species, *Cytophaga succinicans, Flavobacterium aquatile, Flavobacterium branchiophila, Flexibacter aurantiacus, Flexibacter canadensis, Flexibacter* species, *Pseudomonas echinoides, Psychrobacter immobilis, Xanthobacter autotrophicus,* and *Zoogloea ramigera.*

Enrichment Broth
for *Aeromonas hydrophila*

Composition per liter:

NaCl	5.0g
Maltose	3.5g
Yeast extract	3.0g
Bile salts No. 3	1.0g
L-Cysteine·HCl·H_2O	0.3g
Bromothymol Blue	0.03g
Novobiocin	5.0mg

pH 7.0 ± 0.2 at 25°C

Preparation of Medium: Add components to distilled/deionized water and bring volume to 1.0L. Mix thoroughly. Distribute into tubes or flasks. Autoclave for 15 min at 15 psi pressure–121°C.

Use: For the cultivation and enrichment of *Aeromonas hydrophila.*

Entamoeba Medium
(Endamoeba Medium)

Composition per liter:

Liver infusion	272.0g
Rice powder	14.2g
Agar	11.0g
Proteose peptone	5.5g
Sodium glycerophosphate	3.0g
NaCl	2.7g
Horse serum	50.0mL

pH 7.0 ± 0.2 at 25°C

Source: This medium is available as a premixed powder from BD Diagnostic Systems.

Rice Powder:
Composition per 15.0g:

Rice powder	15.0g

Preparation of Rice Powder: Sterilize rice powder at 160°C for 60 min. Do not overheat or rice powder will scorch.

Preparation of Medium: Add components, except horse serum and rice powder, to distilled/deionized water and bring volume to 994.0mL. Mix thoroughly. Gently heat and bring to boiling. Distribute into tubes in 7.0mL volumes. Autoclave for 15 min at 15 psi pressure–121°C. Allow tubes to cool in a slanted position. Aseptically add enough sterile horse serum to each tube to cover about half the slant. Aseptically add 0.1g of sterile rice powder to each tube.

Use: For the cultivation of *Entamoeba histolytica.*

Enteric Fermentation Base
(Fermentation Base for *Campylobacter*)

Composition per liter:

Peptic digest of animal tissue	10.0g
NaCl	5.0g
Beef extract	3.0g
Carbohydrate solution	100.0mL
Andrade's indicator	10.0mL

pH 7.2 ± 0.1 at 25°C

Source: This medium is available as a premixed powder from BD Diagnostic Systems.

Carbohydrate Solution:
Composition per 100.0mL:

Carbohydrate	10.0g

Preparation of Carbohydrate Solution: Add carbohydrate to distilled/deionized water and bring volume to 100.0mL. Mix thoroughly. Filter sterilize. Glucose, lactose, mannitol, sucrose, adonitol, arabinose, cellobiose, dulcitol, glycerol, inositol, salicin, xylose, or other carbohydrates may be used. For the preparation of expensive carbohydrate solutions (adonitol, arabinose, cellobiose, dulcitol, glycerol, inositol, salicin, or xylose), 5.0g of carbohydrate per 100.0mL of distilled/deionized water may be used.

Andrade's Indicator:
Composition per 100.0mL:

NaOH (1*N* solution)	16.0mL
Acid Fuchsin	0.1g

Caution: Acid Fuchsin is a potential carcinogen and care must be taken to avoid inhalation of the powdered dye and contact with the skin.

Preparation of Andrade's Indicator: Add components to distilled/deionized water and bring volume to 100.0mL. Mix thoroughly.

Preparation of Medium: Add components, except carbohydrate solution, to distilled/deionized water and bring volume to 900.0mL. Mix thoroughly. Gently heat and bring to boiling. Distribute into tubes that contain an inverted Durham tube in 9.0mL volumes. Autoclave for 15 min at 15 psi pressure–121°C. Cool to 25°C. Aseptically add 1.0mL of sterile carbohydrate solution per tube. Mix thoroughly.

Use: For the cultivation and differentiation of a variety of bacteria based on their ability to ferment different carbohydrates. Bacteria that produce acid from carbohydrate fermentation turn the medium dark pink to red. Bacteria that produce gas have a bubble trapped in the Durham tube.

Enterobacter Medium

Composition per 800.0mL:

Casein hydrolysate	2.0g
K_2HPO_4	1.4g
K_2SO_4	1.0g
Yeast extract	1.0g
KH_2PO_4	0.6g
$MgSO_4$	0.5g
Glycerol	20.0mL

Preparation of Medium: Add components to distilled/deionized water and bring volume to 800.0mL. Mix thoroughly. Distribute into tubes or flasks. Autoclave for 15 min at 15 psi pressure–121°C.

Use: For the cultivation and maintenance of *Enterobacter* species and *Klebsiella pneumoniae.*

Enterococci Confirmatory Agar

Composition per liter:

Agar	15.0g
Glucose	5.0g
Pancreatic digest of casein	5.0g
Yeast extract	5.0g
NaN_3	0.4g
Methylene Blue	10.0mg
Enterococci confirmatory broth	variable

pH 8.0 ± 0.2 at 25°C

Source: This medium is available as a premixed powder from BD Diagnostic Systems.

Caution: Sodium azide is toxic. Azides also react with metals and disposal must be highly diluted.

Preparation of Medium: Add components to distilled/deionized water and bring volume to 1.0L. Mix thoroughly. Gently heat and bring to boiling. Distribute into tubes. Autoclave for 15 min at 15 psi pressure–121°C. Allow tubes to cool in a slanted position. Add sufficient amount of Enterococci confirmatory broth (see below) to cover half the slant.

Use: For the identification of enterococci from water by the confirmatory test.

Enterococci Presumptive Broth

Composition per liter:

Glucose	5.0g
Pancreatic digest of casein	5.0g
Yeast extract	5.0g
NaN_3	0.4g
Bromothymol Blue	32.0mg

pH 8.4 ± 0.2 at 25°C

Source: This medium is available as a premixed powder from BD Diagnostic Systems.

Caution: Sodium azide is toxic. Azides also react with metals and disposal must be highly diluted.

Preparation of Medium: Add components to distilled/deionized water and bring volume to 1.0L. Mix thoroughly. Distribute into tubes or flasks. Autoclave for 15 min at 15 psi pressure–121°C.

Use: For the isolation and identification of enterococci from water by the presumptive test. Bacteria that produce acid and turn the medium yellow and turbid after incubation at 45°C are presumptive enterococci.

Enterococcus Agar (m-*Enterococcus* Agar) (Azide Agar)

Composition per liter:

Pancreatic digest of casein	15.0g
Agar	10.0g
Papaic digest of soybean meal	5.0g
Yeast extract	5.0g
KH_2PO_4	4.0g
Glucose	2.0g
NaN_3	0.4g
Triphenyltetrazolium chloride	0.1g

pH 7.2 ± 0.2 at 25°C

Source: This medium is available as a premixed powder from BD Diagnostic Systems.

Caution: Sodium azide is toxic. Azides also react with metals and disposal must be highly diluted.

Preparation of Medium: Add components to distilled/deionized water and bring volume to 1.0L. Mix thoroughly. Gently heat and bring to boiling. Cool to 45°–50°C. Do not autoclave. Pour into sterile Petri dishes.

Use: For the isolation, cultivation, and enumeration of entercocci in water and sewage by the membrane filter method. For the direct plating of specimens for the detection and enumeration of fecal streptococci.

Erwinia amylovora Selective Medium

Composition per liter:

Agar	20.0g
Mannitol	10.0g
L-Asparagine	3.0g
Sodium taurocholate	2.5g
K_2HPO_4	2.0g
Nicotinic acid	0.5g
$MgSO_4 \cdot 7H_2O$	0.2g
Nitrilotriacetic acid	10.0mL
Actidione (cycloheximide) solution	10.0mL
Bromothymol Blue	9.0mL

Neutral Red 2.5mL
$TlNO_3$ solution 1.75mL
Tergitol 7 .... 0.1mL
pH 7.2–7.3 at 25°C

Cycloheximide Solution:
Composition per 10.0mL:
Cycloheximide 0.05g

Preparation of Cycloheximide Solution: Add cycloheximide to distilled/deionized water and bring volume to 10.0mL. Mix thoroughly. Filter sterilize.

$TlNO_3$ Solution:
Composition per 10.0mL:
$TlNO_3$ 0.1g

Preparation of $TlNO_3$ Solution: Add $TlNO_3$ to distilled/deionized water and bring volume to 10.0mL. Mix thoroughly. Filter sterilize.

Preparation of Medium: Add components, except $TlNO_3$ solution and cycloheximide solution, to distilled/deionized water and bring volume to 988.25mL. Mix thoroughly. Adjust pH to 7.2–7.3. Gently heat and bring to boiling. Autoclave for 15 min at 15 psi pressure–121°C. Cool to 45°–50°C. Aseptically add 1.75mL of sterile $TlNO_3$ solution and 10.0mL of sterile cycloheximide solution. Mix thoroughly. Pour into sterile Petri dishes or distribute into sterile tubes.

Use: For the isolation and cultivation of *Erwinia amylovora.*

Erwinia Fermentation Medium

Composition per liter:
Lactose 45.0g
K_2HPO_4 .... 3.6g
KH_2PO_4 .... 1.8g
Yeast extract 1.8g
$(NH_4)_2SO_4$ 1.46g
$MgSO_4 \cdot 7H_2O$ 0.6g
$CaCl_2 \cdot 2H_2O$ 0.04g
$FeSO_4 \cdot 7H_2O$ 1.9mg
$CoCl_2$ 1.0mg
$CuSO_4 \cdot 5H_2O$ 1.0mg
$MnSO_4 \cdot H_2O$ 1.0mg
$Na_2MoO_4 \cdot 2H_2O$ 1.0mg
$ZnSO_4 \cdot 7H_2O$ 1.0mg
pH 7.0 ± 0.2 at 25°C

Preparation of Medium: Add components to distilled/deionized water and bring volume to 1.0L. Mix thoroughly. Adjust pH to 7.0 with NaOH (approximately 0.18g). Distribute into tubes or flasks. Autoclave for 15 min at 15 psi pressure–121°C.

Use: For the cultivation of *Rahnella aquatilis.*

Erwinia Medium D3

Composition per liter:
Agar 15.0g
Arabinose 10.0g
Sucrose 10.0g
LiCl 7.0g
Casein hydrolysate 5.0g
NaCl 5.0g
$MgSO_4 \cdot 7H_2O$ 0.3g
Acid Fuchsin 0.1g
Bromothymol Blue 0.06g
Sodium dodecyl sulfate 0.05g
pH 7.0 ± 0.2 at 25°C

Caution: Acid Fuchsin is a potential carcinogen and care must be taken to avoid inhalation of the powdered dye and contact with the skin.

Preparation of Medium: Add components to distilled/deionized water and bring volume to 1.0L. Mix thoroughly. Gently heat and bring to boiling. Adjust pH to 8.2. Autoclave for 15 min at 15 psi pressure–121°C. Cool to 45°–50°C. The pH after autoclaving should be 7.0. Pour into sterile Petri dishes or distribute into sterile tubes.

Use: For the isolation and cultivation of *Erwinia* species.

Erwinia Selective Medium

Composition per liter:
Agar 15.0g
$(NH_4)_2SO_4$ 5.0g
K_2HPO_4 .... 2.0g
Eosin Y 0.4g
Methylene Blue 0.065g
Glycerol 10.0mL
Antibiotic solution 10.0mL

Antibiotic Solution:
Composition per 10.0mL:
Cycloheximide 0.25g
Novobiocin 0.04g
Neomycin sulfate 0.04g

Preparation of Antibiotic Solution: Add components to distilled/deionized water and bring volume to 10.0mL. Mix thoroughly. Filter sterilize.

Preparation of Medium: Add components, except antibiotic solution, to distilled/deionized water and bring volume to 990.0mL. Mix thoroughly. Gently heat and bring to boiling. Autoclave for 15 min at 15 psi pressure–121°C. Cool to 45°–50°C. Aseptically add sterile antibiotic solution. Mix thoroughly. Pour into sterile Petri dishes or distribute into sterile tubes.

Use: For the selective isolation and cultivation of *Erwinia* species.

Erwinia tracheiphila Agar

Composition per liter:
Agar 15.0g
Glucose 10.0g
Peptone 10.0g
Beef extract 5.0g
Yeast extract 1.0g
pH 7.4 ± 0.2 at 25°C

Preparation of Medium: Add components to distilled/deionized water and bring volume to 1.0L. Mix thoroughly. Gently heat and bring to boiling. Distribute into tubes or flasks. Autoclave for 15 min at 15 psi pressure–121°C. Pour into sterile Petri dishes or leave in tubes.

Use: For the cultivation of *Erwinia tracheiphila.*

Erythrobacter longus Medium

Composition per liter:
Peptone 2.0g
Proteose peptone No. 3 .... 1.0g
Papaic digest of soybean meal 1.0g
Yeast extract 1.0g

Artificial seawater 700.0mL
Ferric citrate solution 2.0mL
pH 7.5 ± 0.2 at 25°C

Ferric Citrate Solution:
Composition per 10.0mL:
Ferric citrate 0.5g

Preparation of Ferric Citrate Solution: Add ferric citrate to distilled/deionized water and bring volume to 10.0mL. Mix thoroughly.

Preparation of Medium: Add components to distilled/deionized water and bring volume to 1.0L. Mix thoroughly. Distribute into tubes or flasks. Autoclave for 15 min at 15 psi pressure–121°C.

Use: For the cultivation of *Erythrobacter longus.*

Erythromicrobium roseococcus Medium (DSMZ Medium 767)

Composition per 1001mL:
Na-acetate 1.0g
Yeast extract 1.0g
Peptone 1.0g
$MgSO_4 \cdot 7H_2O$ 0.5g
NH_4Cl 0.3g
K_2HPO_4 .... 0.3g
KCl 0.3g
$CaCl_2 \cdot 2H_2O$ 0.05g
Trace elements SL-6 .... 1.0mL
Vitamin B_{12} solution 1.0mL
pH 7.5-7.8 at 25°C

Vitamin B_{12} Solution:
Composition per 100.0mL:
Vitamin B_{12} 2.0mg

Vitamin B_{12} Solution: Add Vitamin B_{12} to distilled/deionized water and bring volume to 100.0mL. Mix thoroughly. Filter sterilize.

Trace Elements Solution SL-6:
Composition per liter:
$MnCl_2 \cdot 4H_2O$ 0.5g
H_3BO_3 .... 0.3g
$CoCl_2 \cdot 6H_2O$ 0.2g
$ZnSO_4 \cdot 7H_2O$ 0.1g
$Na_2MoO_4 \cdot 2H_2O$ 0.03g
$NiCl_2 \cdot 6H_2O$ 0.02g
$CuCl_2 \cdot 2H_2O$ 0.01g

Preparation of Trace Elements Solution SL-6: Add components to distilled/deionized water and bring volume to 1.0L. Mix thoroughly. Autoclave for 15 min at 15 psi pressure–121°C.

Preparation of Medium: Add components, except vitamin solution, to distilled/deionized water and bring volume to 1.0L. Mix thoroughly. Adjust pH to 7.5-7.8. Autoclave for 15 min at 15 psi pressure–121°C. Cool to room temperature. Aseptically add 10mL vitamin solution. Mix thoroughly. Aseptically distribute into sterile tubes or bottles.

Use: For the cultivation of *Erythrobacter litoralis, Erythromicrobium ramosum,* and *Roseococcus thiosulfatophilus.*

Esculin Agar

Composition per liter:
Agar 15.0g
Pancreatic digest of casein 13.0g
NaCl 5.0g
Yeast extract 5.0g
Heart muscle, solids from infusion 2.0g
Esculin 1.0g
Ferric citrate 0.5g
pH 7.3 ± 0.2 at 25°C

Preparation of Medium: Add components to distilled/deionized water and bring volume to 1.0L. Mix thoroughly. Gently heat and bring to boiling. Distribute into screw-capped tubes in 3.0mL volumes. Autoclave for 15 min at 15 psi pressure–121°C. Allow tubes to cool in a slanted position.

Use: For the cultivation and differentiation of bacteria based on their ability to hydrolyze esculin and produce H_2S. Bacteria that hydrolyze esculin appear as colonies surrounded by a reddish-brown to dark brown zone. Bacteria that produce H_2S appear as black colonies.

Esculin Iron Agar

Composition per liter:
Agar 15.0g
Esculin 1.0g
Ferric ammonium citrate 0.5g
pH 7.1 ± 0.2 at 25°C

Source: This medium is available as a premixed powder from BD Diagnostic Systems.

Preparation of Medium: Add components to distilled/deionized water and bring volume to 1.0L. Mix thoroughly. Gently heat and bring to boiling. Distribute into tubes or flasks. Autoclave for 15 min at 15 psi pressure–121°C. Pour into sterile Petri dishes.

Use: For the cultivation and identification of enterococci based on their ability to hydrolyze esculin. Used in conjunction with E agar and the membrane filter method.

Ethyl Violet Azide Broth (EVA Broth)

Composition per liter:
Pancreatic digest of casein 13.5g
Yeast extract 6.5g
Glucose 5.0g
NaCl 5.0g
K_2HPO_4 .... 2.7g
KH_2PO_4 .... 2.7g
NaN_3 .... 0.4g
Ethyl Violet 0.83mg
pH 7.0 ± 0.2 at 25°C

Source: This medium is available as a premixed powder from BD Diagnostic Systems.

Caution: Sodium azide is toxic. Azides also react with metals and disposal must be highly diluted.

Preparation of Medium: Add components to distilled/deionized water and bring volume to 1.0L. Mix thoroughly. Gently heat and bring to boiling. Distribute into tubes in 10.0mL volumes. Autoclave for 15 min at 15 psi pressure–121°C.

Use: For the isolation, cultivation, and enumeration of enterococci from water and other specimens. Fecal enterococci turn the medium turbid with a purple sediment on the bottom of the tube.

Ethyl Violet Azide Broth (EVA Broth)

Composition per liter:

Tryptose	20.0g
Glucose	5.0g
NaCl	5.0g
K_2HPO_4	2.7g
KH_2PO_4	2.7g
NaN_3	0.4g
Ethyl Violet	0.83mg

pH 7.0 ± 0.2 at 25°C

Source: This medium is available as a premixed powder from BD Diagnostic Systems.

Caution: Sodium azide is toxic. Azides also react with metals and disposal must be highly diluted.

Preparation of Medium: Add components to distilled/deionized water and bring volume to 1.0L. Mix thoroughly. Gently heat and bring to boiling. Distribute into tubes in 10.0mL volumes. Autoclave for 15 min at 15 psi pressure–121°C.

Use: For the isolation, cultivation, and enumeration of enterococci from water and other specimens. Fecal enterococci turn the medium turbid with a purple sediment on the bottom of the tube.

Ethyl Violet Azide Broth (EVA Broth)

Composition per liter:

Tryptose	20.0g
Glucose	5.0g
NaCl	5.0g
K_2HPO_4	2.7g
KH_2PO_4	2.7g
NaN_3	0.3g
Ethyl Violet	0.5mg

pH 6.8 ± 0.2 at 25°C

Source: This medium is available as a premixed powder from Oxoid Unipath.

Preparation of Medium: Add components to distilled/deionized water and bring volume to 1.0L. Mix thoroughly. Gently heat and bring to boiling. Distribute into tubes in 10.0mL volumes. Autoclave for 15 min at 15 psi pressure–121°C.

Caution: Sodium azide is toxic. Azides also react with metals and disposal must be highly diluted.

Use: For the isolation, cultivation, and enumeration of enterococci from water and other specimens. Fecal enterococci turn the medium turbid with a purple sediment on the bottom of the tube.

Eubacterium Medium

Composition per liter:

Pancreatic digest of casein	20.0g
Agar	15.0g
Meat extract	15.0g
Glucose	5.0g
$Na_2HPO_4 \cdot 12H_2O$	4.0g
L-Cysteine·HCl	0.5g

pH 7.4 ± 0.2 at 25°C

Preparation of Medium: Add components to distilled/deionized water and bring volume to 1.0L. Mix thoroughly. Gently heat and bring to boiling. Distribute into tubes or flasks. Autoclave for 15 min at 15 psi pressure–121°C. Pour into sterile Petri dishes or leave in tubes. Use freshly prepared medium.

Use: For the cultivation of *Eubacterium* species.

Eubacterium oxidoreducens Medium

Composition per 1001.0mL:

Solution A	890.0mL
Solution B	100.0mL
Solution C	10.0mL
Solution D	1.0mL

pH 7.0–7.2 at 25°C

Solution A:

Composition per 890.0mL:

Crotonic acid	5.0g
Yeast extract	2.0g
Resazurin	1.0mg
Mineral solution	50.0mL
Vitamin solution	5.0mL
Trace elements solution SL-10	1.0mL

Preparation of Solution A: Add components to distilled/deionized water and bring volume to 890.0mL. Mix thoroughly. Adjust pH to 6.9. Sparge with 80% N_2 + 20% CO_2 for 20 min. Distribute 8.9mL into anaerobic tubes under 80% N_2 + 20% CO_2. Autoclave under 80% N_2 + 20% CO_2 for 15 min at 15 psi pressure–121°C.

Mineral Solution:

Composition per liter:

KH_2PO_4	10.0g
$MgCl_2 \cdot 6H_2O$	6.6g
NaCl	8.0g
NH_4Cl	8.0g
$CaCl_2 \cdot 2H_2O$	1.0g

Preparation of Mineral Solution: Add components to distilled/deionized water and bring volume to 1.0L. Mix thoroughly.

Vitamin Solution:

Composition per liter:

Pyridoxine·HCl	6.2mg
Nicotinic acid	2.5mg
p-Aminobenzoic acid	1.25mg
Thiamine·HCl	1.25mg
Pantothenic acid	0.62mg
Biotin	0.25mg

Preparation of Vitamin Solution: Add components to distilled/deionized water and bring volume to 1.0L. Adjust pH to 7.0. Mix thoroughly.

Trace Elements Solution SL-10:

Composition per liter:

$FeCl_2 \cdot 4H_2O$	1.5g
$CoCl_2 \cdot 6H_2O$	190.0mg
$MnCl_2 \cdot 4H_2O$	100.0mg
$ZnCl_2$	70.0mg
$Na_2MoO_4 \cdot 2H_2O$	36.0mg
$NiCl_2 \cdot 6H_2O$	24.0mg
H_3BO_3	6.0mg
$CuCl_2 \cdot 2H_2O$	2.0mg
HCl (25% solution)	10.0mL

Preparation of Trace Elements Solution SL-10: Add $FeCl_2 \cdot 4H_2O$ to 10.0mL of HCl solution. Mix thoroughly. Add distilled/deionized water and bring volume to 1.0L. Add remaining components. Mix thoroughly.

Solution B:
Composition per 100.0mL:

$NaHCO_3$	5.0g

Preparation of Solution B: Add $NaHCO_3$ to distilled/deionized water and bring volume to 100.0mL. Mix thoroughly. Filter sterilize. Sparge with 80% N_2 + 20% CO_2 for 20 min.

Solution C:
Composition per 10.0mL:

L-Cysteine	0.24g

Preparation of Solution C: Add L-cysteine to distilled/deionized water and bring volume to 10.0mL. Mix thoroughly. Autoclave under 80% N_2 + 20% CO_2 for 15 min at 15 psi pressure–121°C.

Solution D:
Composition per 1.0mL:

$Na_2S·9H_2O$	78.0mg

Preparation of Solution D: Add $Na_2S·9H_2O$ to distilled/deionized water and bring volume to 1.0mL. Mix thoroughly. Autoclave under 80% N_2 + 20% CO_2 for 15 min at 15 psi pressure–121°C.

Preparation of Medium: To each tube containing 8.9mL of sterile solution A, add (using a syringe) 1.0mL of sterile solution B, 0.1mL of sterile solution C, and 0.01mL of sterile solution D.

Use: For the cultivation and maintenance of *Eubacterium oxidoreducans*.

Exiguobacterium Medium

Composition per liter:

Beef extract	10.0g
Peptone	10.0g
NaCl	5.0g
Glucose	5.0g
Yeast extract	3.0g

pH 8.0 ± 0.2 at 25°C

Preparation of Medium: Add components to distilled/deionized water and bring volume to 1.0L. Mix thoroughly. Adjust pH to 8.0. Distribute into tubes or flasks. Autoclave for 15 min at 15 psi pressure–121°C.

Use: For the cultivation and maintenance of *Exiguobacterium aurantiacum*.

Extracted Hay Medium

Composition:

Hay or grass	50.0g

Preparation of Medium: Add hay or grass to 1.0L of distilled/deionized water. Gently heat and bring to boiling. Continue boiling for 30 min. Rinse with cold water twice. Add 1.0L of distilled/deionized water, boil 30 min and rinse. Repeat this process at least 5 times. Dry the extracted hay or grass. Add 10–30 blades of extracted hay or grass to a large test tube. Autoclave for 15 min at 15 psi pressure–121°C.

Use: For the isolation and cultivation of *Beggiatoa* species and myxotrophic *Thiothrix* species.

EYGA Agar

Composition per liter:

Agar	12.0g
K_2HPO_4	1.1g
Glucose	1.0g
Yeast extract	1.0g
KH_2PO_4	0.86g
$(NH_4)_2SO_4$	0.5g
$MgSO_4·7H_2O$	0.2g
NaCl	0.1g
$CaCl_2$	0.025g
Vitamin B_{12}	2.0μg
EDTA/trace elements mix	3.0mL

pH 6.8 ± 0.2 at 25°C

EDTA/Trace Elements Mix:
Composition per 600.0mL:

EDTA	5.0g
$ZnSO_4·7H_2O$	2.2g
$MnSO_4·4H_2O$	0.57g
$FeSO_4·7H_2O$	0.50g
$CoCl_2·6H_2O$	0.161g
$CuSO_4·5H_2O$	0.157g
$Na_2MoO_4·2H_2O$	0.151g

Preparation of EDTA/Trace Elements Mix: Add components to distilled/deionized water and bring volume to 600.0mL. Mix thoroughly.

Preparation of Medium: Add components to distilled/deionized water and bring volume to 1.0L. Mix thoroughly. Adjust pH to 6.8. Distribute into tubes or flasks. Autoclave for 15 min at 15 psi pressure–121°C.

Use: For the cultivation and maintenance of *Arthrobacter* species.

Fay and Barry Medium

Composition per liter:

Amino acid	10.0g
Peptone	5.0g
Yeast extract	3.0g
Bromcresol Purple solution	5.0mL

pH 5.5 ± 0.2 at 25°C

Bromcresol Purple Solution:
Composition per 100.0mL:

Bromcresol Purple	0.2g
Ethanol	50.0mL

Preparation of Bromcresol Purple Solution: Add Bromcresol Purple to 50.0mL of absolute ethanol. Add distilled/deionized water and bring volume to 100.0mL. Mix thoroughly.

Preparation of Medium: Add components to distilled/deionized water and bring volume to 1.0L. The amino acid may be L-arginine, L-ornithine, or L-lysine, depending on which amino acid decarboxylase activity is being measured. Mix thoroughly. Distribute into tubes or flasks. Autoclave for 15 min at 15 psi pressure–121°C.

Use: For the determination of decarboxylase activities of *Aeromonas* species.

FB Medium (DSMZ Medium 980)

Composition per liter:

NH_4Cl	0.54g
$MgCl_2·6H_2O$	0.2g
$CaCl_2·2H_2O$	0.15g
KH_2PO_4	0.14g

Resazurin 0.5mg
$NaHCO_3$ solution 10.0mL
Yeast extract solution 10.0mL
Na-crotonate solution 10.0mL
Cysteine solution 10.0mL
$Na_2S \cdot 9H_2O$ solution 10.0mL
Vitamin solution 10.0mL
Trace elements solution SL-9 .... 1.0mL
Selenite-tungstate solution 1.0mL
pH 7.0 ± 0.2 at 25°C

Na-Crotonate Solution:
Composition per 10.0mL:
Na-crotonate 0.86g

Preparation of Na-Crotonate Solution: Add Na-crotonate to distilled/deionized water and bring volume to 10.0mL. Mix thoroughly. Sparge with 100% N_2. Autoclave for 15 min at 15 psi pressure–121°C.

Cysteine Solution:
Composition per 10.0mL:
L-Cysteine·HCl·H_2O 0.25g

Preparation of Cysteine Solution: Add L-cysteine·HCl·H_2O to distilled/deionized water and bring volume to 10.0mL. Mix thoroughly. Sparge with 100% N_2. Autoclave for 15 min at 15 psi pressure–121°C.

Trace Elements Solution SL-9:
Composition per liter:
$MgSO_4 \cdot 7H_2O$ 3.0g
Nitrilotriacetic acid 1.5g
NaCl 1.0g
$MnSO_4 \cdot 2H_2O$ 0.5g
$CoSO_4 \cdot 7H_2O$ 0.18g
$ZnSO_4 \cdot 7H_2O$ 0.18g
$CaCl_2 \cdot 2H_2O$ 0.1g
$FeSO_4 \cdot 7H_2O$ 0.1g
$NiCl_2 \cdot 6H_2O$ 0.025g
$KAl(SO_4)_2 \cdot 12H_2O$ 0.02g
H_3BO_3 .... 0.01g
$Na_2MoO_4 \cdot 4H_2O$ 0.01g
$CuSO_4 \cdot 5H_2O$ 0.01g
$Na_2SeO_3 \cdot 5H_2O$ 0.3mg

Preparation of Trace Elements Solution SL-9: Add nitrilotriacetic acid to 500.0mL of distilled/deionized water. Dissolve by adjusting pH to 6.5 with KOH. Add remaining components. Add distilled/deionized water to 1.0L. Mix thoroughly.

Vitamin Solution:
Composition per liter:
Pyridoxine-HCl 10.0mg
Thiamine-HCl·$2H_2O$ 5.0mg
Riboflavin 5.0mg
Nicotinic acid 5.0mg
D-Ca-pantothenate 5.0mg
p-Aminobenzoic acid 5.0mg
Lipoic acid 5.0mg
Biotin 2.0mg
Folic acid 2.0mg
Vitamin B_{12} 0.1mg

Preparation of Vitamin Solution: Add components to distilled/deionized water and bring volume to 1.0L. Mix thoroughly. Sparge with 80% H_2 + 20% CO_2. Filter sterilize.

$Na_2S \cdot 9H_2O$ Solution:
Composition per 10.0mL:
$Na_2S \cdot 9H_2O$ 0.25g

Preparation of $Na_2S \cdot 9H_2O$ Solution: Add $Na_2S \cdot 9H_2O$ to distilled/deionized water and bring volume to 10.0mL. Sparge with N_2. Autoclave for 15 min at 15 psi pressure–121°C. Cool to 25°C. Store anaerobically.

$NaHCO_3$ Solution:
Composition per 10.0mL:
$NaHCO_3$ 2.5g

Preparation of $NaHCO_3$ Solution: Add $NaHCO_3$ to distilled/deionized water and bring volume to 10.0mL. Mix thoroughly. Sparge with 80% N_2 + 20% CO_2. Filter sterilize.

Yeast Extract Solution:
Composition per 10.0mL:
Yeast extract 0.2g

Preparation of Yeast Extract Solution: Add yeast extract to distilled/deionized water and bring volume to 10.0mL. Mix thoroughly. Sparge with 100% N_2. Filter sterilize.

Preparation of Medium: Add components, except Na-crotonate solution, yeast extract solution, cysteine solution, vitamin solution, $NaHCO_3$ solution, and $Na_2S \cdot 9H_2O$ solution, to distilled/deionized water and bring volume to 940.0mL. Mix thoroughly. Gently heat and bring to boiling. Boil for 3 min. Cool to 25°C while sparging with 80% N_2 + 20% CO_2. Distribute to anaerobe tubes or bottles under 80% N_2 + 20% CO_2. Autoclave for 15 min at 15 psi pressure–121°C. Aseptically and anaerobically add per liter of medium, 10.0mL sterile cysteine solution, 10.0mL sterile Na-crotonate solution, 10.0mL sterile vitamin solution, 10.0mL sterile yeast extract solution, 10.0mL sterile $NaHCO_3$ solution, and 10.0mL sterile $Na_2S \cdot 9H_2O$ solution. Mix thoroughly. The final pH should be 7.0.

Use: For the cultivation of *Sporotomaculum syntrophicum*.

FC Agar
(Fecal Coliform Agar)
(m-FC Agar)
(m-Fecal Coliform Agar)

Composition per liter:
Agar 15.0g
Lactose 12.5g
NaCl 5.0g
Proteose peptone No. 3 .... 5.0g
Yeast extract 3.0g
Bile salts 1.5g
Aniline Blue 0.1g
Rosolic acid solution 10.0mL
pH 7.4 ± 0.2 at 25°C

Source: This medium is available as a premixed powder from BD Diagnostic Systems.

Rosolic Acid Solution:
Composition per 100.0mL:
Rosolic acid 1.0g

Preparation of Rosolic Acid Solution: Add rosolic acid to 0.2*N* NaOH and bring volume to 100.0L. Mix thoroughly.

Preparation of Medium: Add 10.0mL rosolic acid solution to 950.0mL distilled/deionized water. Mix thoroughly. Add other components and bring volume to 1.0L with distilled/deionized water. Mix thoroughly. Gently heat and bring to boiling with frequent mixing. Do not autoclave. Pour into sterile Petri dishes or leave in tubes.

Use: For the cultivation of fecal coliform bacteria from waters and the enumeration of coliform bacteria using the membrane filtration method.

FC Agar (Fecal Coliform Agar) (m-FC Agar) (m-Fecal Coliform Agar)

Composition per liter:

Agar	15.0g
Lactose	12.5g
Tryptose	10.0g
NaCl	5.0g
Proteose peptone No. 3	5.0g
Yeast extract	3.0g
Bile salts	1.5g
Aniline Blue	0.1g
Rosolic acid solution	10.0mL

pH 7.4 ± 0.2 at 25°C

Rosolic Acid Solution:
Composition per 100.0mL:

Rosolic acid	1.0g

Preparation of Rosolic Acid Solution: Add rosolic acid to 0.2*N* NaOH and bring volume to 100.0L. Mix thoroughly.

Preparation of Medium: Add 10.0mL rosolic acid solution to 950.0mL of distilled/deionized water. Mix thoroughly. Add other components and bring volume to 1.0L with distilled/deionized water. Mix thoroughly. Gently heat and bring to boiling with frequent mixing. Do not autoclave. Pour into sterile Petri dishes or leave in tubes.

Use: For the cultivation of fecal coliform bacteria from waters and the enumeration of coliform bacteria using the membrane filtration method.

FC Broth (Fecal Coliform Broth) (m-FC Broth) (m-Fecal Coliform Broth)

Composition per liter:

Lactose	12.5g
Tryptose	10.0g
NaCl	5.0g
Proteose peptone No. 3	5.0g
Yeast extract	3.0g
Bile salts	1.5g
Aniline Blue	0.1g
Rosolic acid solution	10.0mL

pH 7.4 ± 0.2 at 25°C

Rosolic Acid Solution:
Composition per 100.0mL:

Rosolic acid	1.0g

Preparation of Rosolic Acid Solution: Add rosolic acid to 0.2*N* NaOH and bring volume to 100.0L. Mix thoroughly.

Preparation of Medium: Add 10.0mL of rosolic acid solution to 950.0mL of distilled/deionized water. Mix thoroughly. Add other components and bring volume to 1.0L with distilled/deionized water. Mix thoroughly. Gently heat and bring to boiling with frequent mixing. Do not autoclave. Pour into sterile Petri dishes or leave in tubes.

Use: For the cultivation of fecal coliform bacteria from waters and the enumeration of coliform bacteria using the membrane filtration method.

FC Broth (Fecal Coliform Broth) (m-FC Broth) (m-Fecal Coliform Broth)

Composition per liter:

Lactose	12.5g
NaCl	5.0g
Proteose peptone No. 3	5.0g
Yeast extract	3.0g
Bile salts	1.5g
Aniline Blue	0.1g
Rosolic acid solution	10.0mL

pH 7.4 ± 0.2 at 25°C

Source: This medium is available as a premixed powder from BD Diagnostic Systems.

Rosolic Acid Solution:
Composition per 100.0mL:

Rosolic acid	1.0g

Preparation of Rosolic Acid Solution: Add rosolic acid to 0.2*N* NaOH and bring volume to 100.0L. Mix thoroughly.

Preparation of Medium: Add 10.0mL of rosolic acid solution to 950.0mL of distilled/deionized water. Mix thoroughly. Add other components and bring volume to 1.0L with distilled/deionized water. Mix thoroughly. Gently heat and bring to boiling with frequent mixing. Do not autoclave. Pour into sterile Petri dishes or leave in tubes.

Use: For the cultivation of fecal coliform bacteria from waters and the enumeration of coliform bacteria using the membrane filtration method.

Fe(III) Lactate Nutrient Agar

Composition per liter:

Agar	15.0g
Peptone	5.0g
NaCl	5.0g
Yeast extract	2.0g
Beef extract	1.0g
Fe(III)-lactate solution	25.0mL

pH 7.2 ± 0.2 at 25°C

Fe(III)-Lactate Solution:
Composition per 30.0mL:

$FeCl_3 \cdot 6H_2O$ solution	20.0mL
Sodium lactate solution	10.0mL

Preparation of Fe(III)-Lactate Solution: Aseptically combine the component solutions. Mix thoroughly.

$FeCl_3 \cdot 6H_2O$ Solution:
Composition per 100.0mL:

$FeCl_3 \cdot 6H_2O$	5.0g

Preparation of $FeCl_3 \cdot 6H_2O$ Solution: Add $FeCl_3 \cdot 6H_2O$ to distilled/deionized water and bring volume to 100.0mL. Mix thoroughly. Filter sterilize.

Sodium Lactate Solution:
Composition per 100.0mL:

Sodium lactate	5.0g

Preparation of Sodium Lactate Solution: Add sodium lactate to distilled/deionized water and bring volume to 100.0mL. Mix thoroughly. Filter sterilize.

Preparation of Medium: Add components, except Fe(III)-lactate solution, to distilled/deionized water and bring volume to 975.0L. Mix thoroughly. Gently heat and bring to boiling. Autoclave for 15 min at 15 psi pressure–

121°C. Cool to 50°–55°C. Aseptically add 25.0mL of filter-sterilized Fe(III)-lactate solution. Mix thoroughly. Pour into sterile Petri dishes or distribute into sterile tubes.

Use: For the cultivation of *Shewanella putrefaciens*.

Fecal Coliform Agar, Modified (m-Fecal Coliform Agar, Modified) (FCIC)

Composition per liter:

Agar	15.0g
Inositol	10.0g
Tryptose	10.0g
Proteose peptone No. 3	5.0g
NaCl	5.0g
Yeast extract	3.0g
Bile salts No. 3	1.5g
Aniline Blue	0.1g

pH 7.4 ± 0.2 at 25°C

Preparation of Medium: Add components and bring volume to 1.0L. Mix thoroughly. Gently heat and bring to boiling. Do not autoclave. Cool to 50°C. Adjust pH to 7.4. Pour into sterile Petri dishes in 20.0mL volumes. Allow surface of plates to dry before using.

Use: For the isolation, cultivation and enumeration of *Klebsiella* species using the membrane filter method.

Fecal Coliform Agar, Modified

Composition per liter:

Agar	15.0g
Lactose	12.5g
Tryptose	10.0g
Proteose peptone No. 3	5.0g
NaCl	5.0g
Yeast extract	3.0g
Bile salts No. 3	1.5g
Aniline Blue	0.1g

pH 7.4 ± 0.2 at 25°C

Preparation of Medium: Add components and bring volume to 1.0L. Mix thoroughly. Gently heat and bring to boiling. Do not autoclave. Cool to 50°C. Adjust pH to 7.4. Pour into sterile Petri dishes in 20.0mL volumes. Allow surface of plates to dry before using.

Use: For the isolation, cultivation, and identification of stressed fecal coliform microorganisms based on their ability to ferment lactose. Lactose-fermenting bacteria turn the medium blue.

Feodorov Medium

Composition per liter:

Mannitol or glucose	20.0g
Marine salts mixture	18.0g
$CaCO_3$	0.5g
K_2HPO_4	0.3g
$MgSO_4$	0.3g
$CaHPO_4$	0.2g
K_2SO_4	0.2g
$FeCl_3$	0.1g
Trace elements solution	1.0mL

Trace Elements Solution:

Composition per 100.0mL:

H_3BO_3	0.5g
$(NH_4)_6Mo_7O_{24}\cdot 4H_2O$	0.5g
KI	0.05g
NaBr	0.05g
$Al_2(SO_4)_3\cdot 18H_2O$	0.03g
$ZnSO_4$	0.02g

Preparation of Trace Elements Solution: Add components to distilled/deionized water and bring volume to 100.0mL. Mix thoroughly.

Preparation of Medium: Add components to distilled/deionized water and bring volume to 1.0L. Mix thoroughly. Distribute into tubes or flasks. Autoclave for 15 min at 15 psi pressure–121°C.

Use: For the cultivation and maintenance of *Azotobacter vinelandii*.

Ferric Citrate Medium

Composition per liter:

Ferric citrate	13.7g
Sodium lactate (60% solution)	5.6g
$NaHCO_3$	2.5g
NH_4Cl	1.5g
NaH_2PO_4	0.6g
KCl	0.1g
Wolfe's mineral solution	10.0mL
Wolfe's vitamin solution	10.0mL

pH 7.0 ± 0.2 at 25°C

Wolfe's Vitamin Solution:

Composition per liter:

Pyridoxine·HCl	10.0mg
Calcium D-(+)-pantothenate	5.0mg
Nicotinic acid	5.0mg
p-Aminobenzoic acid	5.0mg
Riboflavin	5.0mg
Thiamine·HCl	5.0mg
Thioctic acid	5.0mg
Biotin	2.0mg
Folic acid	2.0mg
Cyanocobalamin	0.1mg

Preparation of Wolfe's Vitamin Solution: Add components to distilled/deionized water and bring volume to 1.0L. Mix thoroughly. Filter sterilize.

Wolfe's Mineral Solution:

Composition per liter:

$MgSO_4\cdot 7H_2O$	3.0g
Nitrilotriacetic acid	1.5g
NaCl	1.0g
$MnSO_4\cdot H_2O$	0.5g
$CaCl_2$	0.1g
$CoCl_2\cdot 6H_2O$	0.1g
$FeSO_4\cdot 7H_2O$	0.1g
$ZnSO_4\cdot 7H_2O$	0.1g
$AlK(SO_4)_2\cdot 12H_2O$	0.01g
$CuSO_4\cdot 5H_2O$	0.01g
H_3BO_3	0.01g
$Na_2MoO_4\cdot 2H_2O$	0.01g

Preparation of Wolfe's Mineral Solution: Add nitrilotriacetic acid to 500.0mL of distilled/deionized water. Adjust pH to 6.5 with KOH. Add remaining components sequentially. Add distilled/deionized water to 1.0L. Mix thoroughly.

Preparation of Medium: Add ferric citrate to distilled/deionized water and bring volume to 1.0L. Gently heat and bring to boiling. Continue boiling until ferric citrate is dissolved. Cool to room temperature. Adjust to pH 6.6 with 10*N* NaOH. Add remaining components. Mix thoroughly. Sparge with 80% N_2 + 20% CO_2. Anaerobically distribute into tubes or flasks. Autoclave for 15 min at 15 psi pressure–121°C. Final pH should be 7.0.

Use: For the cultivation of *Aeromonas encheleia* and *Shewanella alga.*

Ferroglobus placidus Medium (DSMZ Medium 730)

Composition per 1020mL:

NaCl	18.0g
$NaHCO_3$	10.0g
$MgCl_2·6H_2O$	4.3g
KNO_3	1.0g
KCl	0.34g
NH_4Cl	0.24g
$CaCl_2·2H_2O$	0.14g
$K_2HPO_4·3H_2O$	0.14g
Resazurin	0.5mg
Trace elements solution	10.0mL
Vitamin solution	10.0mL
$Na_2S·9H_2O$ solution	10.0mL
Na-pyruvate solution	10.0mL

pH 7.0 ± 0.2 at 25°C

Na-pyruvate Solution:

Composition per 10.0mL:

Na-pyruvate	1.0g

Preparation of Na-pyruvate Solution: Add Na-pyruvate to distilled/deionized water and bring volume to 10.0mL. Mix thoroughly. Sparge with 100% N_2. Filter sterilize.

Trace Elements Solution:

Composition per liter:

$MgSO_4·7H_2O$	3.0g
Nitrilotriacetic acid	1.5g
NaCl	1.0g
$MnSO_4·2H_2O$	0.5g
$CoSO_4·7H_2O$	0.18g
$ZnSO_4·7H_2O$	0.18g
$CaCl_2·2H_2O$	0.1g
$FeSO_4·7H_2O$	0.1g
$NiCl_2·6H_2O$	0.025g
$KAl(SO_4)_2·12H_2O$	0.02g
H_3BO_3	0.01g
$Na_2MoO_4·4H_2O$	0.01g
$CuSO_4·5H_2O$	0.01g
$Na_2SeO_3·5H_2O$	0.3mg

Preparation of Trace Elements Solution: Add nitrilotriacetic acid to 500.0mL of distilled/deionized water. Dissolve by adjusting pH to 6.5 with KOH. Add remaining components. Add distilled/deionized water to 1.0L. Mix thoroughly.

Vitamin Solution:

Composition per liter:

Pyridoxine-HCl	10.0mg
Thiamine-HCl·$2H_2O$	5.0mg
Riboflavin	5.0mg
Nicotinic acid	5.0mg
D-Ca-pantothenate	5.0mg
p-Aminobenzoic acid	5.0mg
Lipoic acid	5.0mg
Biotin	2.0mg
Folic acid	2.0mg
Vitamin B_{12}	0.1mg

Preparation of Vitamin Solution: Add components to distilled/deionized water and bring volume to 1.0L. Mix thoroughly. Sparge with 80% H_2 + 20% CO_2. Filter sterilize.

$Na_2S·9H_2O$ Solution:

Composition per 10.0mL:

$Na_2S·9H_2O$	0.5g

Preparation of $Na_2S·9H_2O$ Solution: Add $Na_2S·9H_2O$ to distilled/deionized water and bring volume to 10.0mL. Sparge with N_2. Autoclave for 15 min at 15 psi pressure–121°C. Cool to 25°C. Store anaerobically.

Preparation of Medium: Add components, except $Na_2S·9H_2O$ solution, vitamin solution, and Na-pyruvate solution, to distilled/deionized water and bring volume to 990.0mL. Mix thoroughly. Flush medium with N_2 for 20 min. Adjust medium pH to 7.0 with 4*N* H_2SO_4. Add 10.0mL of $Na_2S·9H_2O$ solution. Mix thoroughly. Readjust the medium pH to 7.0 with H_2SO_4, while flushing the gas phase only with N_2. Dispense 10 mL volumes into 100 mL serum bottles with rubber stoppers under N_2. Replace gas phase by 80% H_2 + 20% CO_2 gas mixture and finally pressurize the bottles to 2 bar gas overpressure. Autoclave for 15 min at 15 psi pressure–121°C. Aseptically inject via syringe 0.1mL vitamin mix and 0.1mL Na-pyruvate solution into each tube containing 10.0mL of the autoclaved medium.

Use: For the cultivation of *Ferroglobus placidus.*

Ferroplasma acidiphilum Medium (DSMZ Medium 874)

Composition per 1001.6mL:

Solution A	950.0mL
Solution B	10.0mL
Solution C	1.6mL

pH 1.7 ± 0.2 at 25°C

Solution A:

Composition per 950mL:

$MgSO_4·7H_2O$	0.4g
$(NH_4)_2SO_4$	0.2g
KCl	0.1g
K_2HPO_4	0.1g

Preparation of Solution A: Add components to distilled/deionized water and bring volume to 950.0mL. Mix thoroughly.

Solution B:

Composition per 50mL:

$FeSO_4·7H_2O$	25.0g
H_2SO_4, 1*N*	40.0mL

Preparation of Solution B: Add components to distilled/deionized water and bring volume to 50.0mL. Mix thoroughly.

Solution C:

Composition per 10mL:

Yeast extract	1.0g

Preparation of Solution C: Add yeast extract to distilled/deionized water and bring volume to 10.0mL. Mix thoroughly. Autoclave for 15 min at 15 psi pressure–121°C. Cool to 25°C.

Preparation of Medium: Add 50.0mL solution B to 950.0mL solution A. Mix thoroughly. Adjust pH to 1.6-1.8 with H_2SO_4. Filter sterilize. Aseptically add 1.6 mL of sterile solution C. Pour into sterile Petri dishes or leave in tubes. Mix thoroughly. Aseptically distribute to sterile tubes or flasks.

Use: For the cultivation of *Ferroplasma acidiphilum (Ferriplasma acidophilum).*

Ferrous Sulfide Agar

Composition per 1200.0mL:

Agar layer 1.0L
Liquid overlay 200.0mL

Agar Layer:
Composition per liter:

Agar 30.0g
FeS washed precipitate supension 500.0mL

Preparation of Agar Layer: Add agar to distilled/deionized water and bring volume to 500.0mL. Mix thoroughly. Gently heat and bring to boiling. Autoclave for 15 min at 15 psi pressure–121°C. Cool to 45°–50°C. Heat FeS washed precipitate suspension to 45°–50°C. Mix thoroughly. Aseptically add 500.0mL of sterile FeS washed precipitate supension to 500.0mL of sterile agar at 45°–50°C. Mix thoroughly.

FeS Washed Precipitate Suspension:
Composition per 500.0mL:

$Fe(NH_4)_2(SO_4)_2 \cdot 6H_2O$ 78.4g
$Na_2S \cdot 9H_2O$ 15.6g

Preparation of FeS Washed Precipitate Suspension: Add $Na_2S \cdot 9H_2O$ and $Fe(NH_4)_2(SO_4)_2$ to 500.0mL boiling distilled/deionized water. Let precipitate settle from the hot solution in a completely filled and stoppered bottle. Wash precipitate four times by decanting supernatant and replacing each time with 500.0mL of boiling distilled/deionized water. Store FeS washed precipitate suspension in a completely filled 500.0mL glass-stoppered bottle.

Liquid Overlay:
Composition per liter:

$(NH_4)_2Cl$ 1.0g
K_2HPO_4 .... 0.5g
$MgSO_4 \cdot 7H_2O$ 0.2g
$CaCl_2$ 0.1g

Preparation of Liquid Overlay: Add components to distilled/deionized water and bring volume to 1.0L. Mix thoroughly. Autoclave for 15 min at 15 psi pressure–121°C. Cool to 25°C. Aseptically bubble 100% CO_2 for 15 sec.

Preparation of Medium: Aseptically distribute agar layer into sterile tubes in 10.0mL volumes. Allow tubes to cool in a slanted poistion. Aseptically add 2.0mL of sterile liquid overlay to each tube.

Use: For the enumeration, enrichment, and isolation of iron and sulfur bacteria, including *Gallionella ferruginea.*

Fervidobacterium islandicum Medium

Composition per liter:

$(NH_4)_2SO_4$ 1.3g
Yeast extract 1.0g
KH_2PO_4 .... 0.28g
$MgSO_4 \cdot 7H_2O$ 0.25g
$CaCl_2 \cdot 2H_2O$ 0.07g
$FeCl_3 \cdot 6H_2O$ 0.02g
$Na_2B_4 \cdot 10H_2O$ 4.5mg
$MnCl_2 \cdot 4H_2O$ 1.8mg
Resazurin 1.0mg
$ZnSO_4 \cdot 7H_2O$ 0.22mg
$CuCl_2 \cdot 2H_2O$ 0.05mg
$Na_2MoO_4 \cdot 2H_2O$ 0.03mg
$VOSO_4 \cdot 2H_2O$ 0.03mg
$CoSO_4$ 0.01mg
Glucose solution 20.0mL
$Na_2S \cdot 9H_2O$ solution 10.0mL

pH 7.0 ± 0.2 at 25°C

Glucose Solution:
Composition per 20.0mL:

Glucose 2.0g

Preparation of Glucose Solution: Add glucose to distilled/deionized water and bring volume to 20.0mL. Mix thoroughly. Sparge with 100% N_2 for 3–4 min. Autoclave for 15 min at 15 psi pressure–121°C.

$Na_2S \cdot 9H_2O$ Solution:
Composition per 10.0mL:

$Na_2S \cdot 9H_2O$ 0.5g

Preparation of $Na_2S \cdot 9H_2O$ Solution: Add $Na_2S \cdot 9H_2O$ to distilled/deionized water and bring volume to 10.0mL. Mix thoroughly. Gas under 100% N_2. Autoclave for 15 min at 15 psi pressure–121°C.

Preparation of Medium: Add components, except glucose solution and $Na_2S \cdot 9H_2O$ solution, to distilled/deionized water and bring volume to 970.0mL. Mix thoroughly. Adjust pH to 7.0. Sparge with 100% N_2. Autoclave for 15 min at 15 psi pressure–121°C. Aseptically and anaerobically add 20.0mL of sterile glucose solution and 10.0mL of sterile $Na_2S \cdot 9H_2O$ solution. Mix thoroughly.

Use: For the cultivation of *Fervidobacterium islandicum.*

Fervidobacterium Medium

Composition per liter:

Pancreatic digest of casein 10.0g
Glucose 5.0g
Yeast extract 3.0g
K_2HPO_4 .... 1.5g
NH_4Cl 0.9g
KH_2PO_4 .... 0.75g
$MgCl_2 \cdot 6H_2O$ 0.2g
$Na_2S \cdot 9H_2O$ solution 10.0mL
Trace elements solution 9.0mL
Wolfe's vitamin solution 5.0mL
Resazurin (0.2% solution) 1.0mL
$FeSO_4 \cdot 7H_2O$ (10% solution) 0.03mL

pH 7.0 ± 0.1 at 25°C

Trace Elements Solution:
Composition per liter:

Nitrilotriacetic acid 12.5g
NaCl 1.0g
$FeCl_3 \cdot 4H_2O$ 0.2g
$CaCl_2 \cdot 2H_2O$ 0.1g
$MnCl_2 \cdot 4H_2O$ 0.1g
$ZnCl_2$ 0.1g
$CuCl_2$ 0.02g
Na_2SeO_3 .... 0.02g
$CoCl_2 \cdot 6H_2O$ 0.017g
H_3BO_3 .... 0.01g
$Na_2MoO_4 \cdot 2H_2O$ 0.01g

Preparation of Trace Elements Solution: Add nitrilotriacetic acid to 500.0mL of distilled/deionized water. Dissolve by adjusting pH to 6.5 with KOH. Add remaining components. Add distilled/deionized water to 1.0L. Filter sterilize. Maintain under an atmosphere of 100% N_2.

Wolfe's Vitamin Solution:
Composition per liter:

Pyridoxine·HCl 0.01g
Thiamine·HCl 5.0mg
Riboflavin 5.0mg

Nicotinic acid	5.0mg
Calcium pantothenate	5.0mg
p-Aminobenzoic acid	5.0mg
Thioctic acid	5.0mg
Biotin	2.0mg
Folic acid	2.0mg
Cyanocobalamin	0.1mg

Preparation of Wolfe's Vitamin Solution: Add components to distilled/deionized water and bring volume to 1.0L. Mix thoroughly. Filter sterilize. Maintain under an atmosphere of 100% N_2.

$Na_2S·9H_2O$ Solution:
Composition per 10.0mL:

$Na_2S·9H_2O$	0.5g

Preparation of $Na_2S·9H_2O$ Solution: Add $Na_2S·9H_2O$ to distilled/deionized water and bring volume to 10.0mL. Mix thoroughly. Filter sterilize. Maintain under an atmosphere of 100% N_2.

Preparation of Medium: Add components, except sodium sulfide solution, trace elements solution, and Wolfe's vitamin solution, to distilled/deionized water and bring volume to 976.0mL. Mix thoroughly. Autoclave for 15 min at 15 psi pressure–121°C. Cool under an atmosphere of 100% N_2. Aseptically add 9.0mL of trace elements solution and 5.0mL of Wolfe's vitamin solution under an atmosphere of 100% N_2. Mix thoroughly. Aseptically distribute into sterile tubes or flasks under an atmosphere of 100% N_2. Add $Na_2S·9H_2O$ solution just prior to use to a concentration 0.1%.

Use: For the cultivation and maintenance of *Clostridium* species, *Fervidobacterium nodosum, Fervidobacterium islandicum,* and *Thermoanaerobium brockii.*

F-G Agar
(Feeley-Gorman Agar)

Composition per liter:

Casein, acid hydrolyzed	17.5g
Agar	17.0g
Beef extract	3.0g
Starch	1.5g
L-Cysteine solution	10.0mL
$Fe_4(P_2O_7)_3$ solution	10.0mL

pH 6.9 ± 0.05 at 25°C

L-Cysteine Solution:
Composition per 10.0mL:

L-Cysteine·HCl·H_2O	0.4g

Preparation of L-Cysteine Solution: Add L-cysteine·HCl·H_2O to distilled/deionized water and bring volume to 10.0mL. Mix thoroughly. Filter sterilize.

$Fe_4(P_2O_7)_3$ Solution:
Composition per 10.0mL:

$Fe_4(P_2O_7)_3$	0.25g

Preparation of $Fe_4(P_2O_7)_3$ Solution: Add $Fe_4(P_2O_7)_3$ to distilled/deionized water and bring volume to 10.0mL. Mix thoroughly. Filter sterilize.

Preparation of Medium: Add components, except L-cysteine solution and $Fe_4(P_2O_7)_3$ solution, to distilled/deionized water and bring volume to 980.0mL. Mix thoroughly. Gently heat and bring to boiling. Autoclave for 15 min at 15 psi pressure–121°C. Cool to 45°–50°C. Aseptically add 10.0mL of L-cysteine solution. Mix thoroughly. Aseptically add 10.0mL of $Fe_4(P_2O_7)_3$ solution. Mix thoroughly. Adjust pH to 6.9. Pour into sterile Petri dishes or distribute into sterile tubes.

Use: For the isolation and cultivation of *Legionella pneumophila.*

F-G Agar with Selenium
(Feeley-Gorman Agar with Selenium)

Composition per liter:

Casein, acid hydrolyzed	17.5g
Agar	17.0g
Beef extract	3.0g
Starch	1.5g
L-Cysteine solution	10.0mL
$Fe_4(P_2O_7)_3$ solution	10.0mL
$Na_2SeO_3·5H_2O$ solution	10.0mL

pH 6.9 ± 0.05 at 25°C

L-Cysteine Solution:
Composition per 10.0mL:

L-Cysteine·HCl·H_2O	0.4g

Preparation of L-Cysteine Solution: Add L-cysteine·HCl·H_2O to distilled/deionized water and bring volume to 10.0mL. Mix thoroughly. Filter sterilize.

$Fe_4(P_2O_7)_3$ Solution:
Composition per 10.0mL:

$Fe_4(P_2O_7)_3$	0.25g

Preparation of $Fe_4(P_2O_7)_3$ Solution: Add $Fe_4(P_2O_7)_3$ to distilled/deionized water and bring volume to 10.0mL. Mix thoroughly. Filter sterilize.

$Na_2SeO_3·5H_2O$ Solution:
Composition per 10.0mL:

$Na_2SeO_3·5H_2O$	0.01g

Preparation of $Na_2SeO_3·5H_2O$ Solution: Add 0.1g of $Na_2SeO_3·5H_2O$ to distilled/deionized water and bring volume to 10.0mL. Mix thoroughly. Filter sterilize.

Preparation of Medium: Add components—except L-cysteine solution, $Fe_4(P_2O_7)_3$ solution, and $Na_2SeO_3·5H_2O$ solution—to distilled/deionized water and bring volume to 970.0mL. Mix thoroughly. Gently heat and bring to boiling. Autoclave for 15 min at 15 psi pressure–121°C. Cool to 45°–50°C. Aseptically add 10.0mL of sterile L-cysteine solution. Mix thoroughly. Aseptically add 10.0mL of sterile $Fe_4(P_2O_7)_3$ solution and 10.0mL of sterile $Na_2SeO_3·5H_2O$ solution. Mix thoroughly. Adjust pH to 6.9. Pour into sterile Petri dishes or distribute into sterile tubes.

Use: For the isolation and cultivation of *Legionella pneumophila.*

FGTC Agar

Composition per liter:

Pancreatic digest of casein	15.0g
Agar	15.0g
Papaic digest of soybean meal	5.0g
NaCl	5.0g
KH_2PO_4	5.0g
Amylose Azure	3.0g
Galactose	1.0g
Thallous acetate	0.5g
MUG (4-Methylumbelliferyl- α-D-galactoside	0.1g
$NaHCO_3$ solution	20.0mL
Gentamicin solution	2.5mL
Tween 80	0.75mL

pH 7.3 ± 0.2 at 25°C

Gentamicin Solution:
Composition per 10.0mL:
Gentamicin 0.01g

Preparation of Gentamicin Solution: Add gentamicin to distilled/deionized water and bring volume to 10.0mL. Mix thoroughly.

$NaHCO_3$ Solution:
Composition per 20.0mL:
$NaHCO_3$ 2.0g

Preparation of $NaHCO_3$ Solution: Add the $NaHCO_3$ to distilled/deionized water and bring volume to 20.0mL. Mix thoroughly. Filter sterilize. Use freshly prepared solution.

Preparation of Medium: Add components, except $NaHCO_3$ solution, to distilled/deionized water and bring volume to 980.0mL. Mix thoroughly. Gently heat and bring to boiling. Autoclave for 15 min at 15 psi pressure–121°C. Cool to 50°C. Aseptically add sterile $NaHCO_3$ solution. Mix thoroughly. Pour into sterile Petri dishes.

Use: For the cultivation, differentiation, and enumeration of *Enterococcus* species based on starch hydrolysis and production of fluorescence. Bacteria that hydrolyze starch, such as *Streptococcus bovis*, appear as colonies surrounded by a clear zone. Bacteria that produce fluorescence, such as *Streptococcus bovis* and *Enterococcus faecium*, appear as colonies surrounded by a zone of bluish fluorescence when viewed under a long-wave UV lamp. Other bacteria, such as *Enterococcus faecalis and Enterococcus avium*, do not hydrolyze starch or produce fluorescence.

Fish Peptone Broth

Composition per liter:
Maltose 5.0g
NaCl 5.0g
Peptone 5.0g
Pancreatic digest of casein 5.0g
Yeast extract 5.0g
Trout tissue extract solution 50.0mL
pH 7.0 ± 0.2 at 25°C

Trout Tissue Extract Solution:
Composition per liter:
Fish (brook trout) 500.0g
Pepsin 1.0g
HCl, concentrated 15.0mL

Preparation of Trout Tissue Extract: Add 1.0L of distilled/deionized water to brook trout and blend for 20–30 min. Add 1.0g of pepsin and 15.0mL of concentrated HCl to digest the trout proteins. Incubate for 12 hr at 45°C. Adjust pH to 7.0. Allow solids to settle. Filter sterilize. Do not autoclave. Store at 5°C.

Preparation of Medium: Add components, except trout tissue extract, to distilled/deionized water and bring volume to 950.0L. Mix thoroughly. Gently heat and bring to boiling. Autoclave for 15 min at 10 psi pressure–118°C. Cool to 45°–50°C. Aseptically add 50.0mL of sterile trout tissue extract. Mix thoroughly. Aseptically distribute into sterile tubes or flasks.

Use: For the cultivation of *Aeromonas salmonicida*.

Flavobacterium aquatile Medium (DSMZ Medium 102)

Composition per liter:
Agar 15.0g
Na-caseinate 2.0g
Proteose peptone 1.0g
Yeast extract 0.5g
K_2HPO_4 .... 0.5g
pH 7.4 ± 0.2 at 25°C

Preparation of Medium: Add components to distilled/deionized water and bring volume to 1.0L. Mix thoroughly. Adjust pH to 7.4. Gently heat and bring to boiling. Distribute into tubes or flasks. Autoclave for 15 min at 15 psi pressure–121°C.

Use: For the cultivation and maintenance of *Flavobacterium aquatile.*

Flavobacterium M1 Agar

Composition per liter:
Agar 15.0g
Proteose peptone 5.0g
NaCl 3.0g
Beef extract 2.0g
Yeast extract 1.0g
pH 7.0–7.2 at 25°C

Preparation of Medium: Add components to distilled/deionized water and bring volume to 1.0L. Mix thoroughly. Gently heat and bring to boiling. Distribute into tubes or flasks. Autoclave for 15 min at 15 psi pressure–121°C. Pour into sterile Petri dishes or leave in tubes.

Use: For the cultivation and maintenance of *Flavobacterium indolthelicum.*

Flavobacterium Medium

Composition per liter:
Na_2SO_4 .... 1.0g
Pancreatic digest of casein 1.0g
Yeast extract 1.0g
pH 6.0 ± 0.2 at 25°C

Preparation of Medium: Add components to distilled/deionized water and bring volume to 1.0L. Mix thoroughly. Adjust pH to 6.0 with H_2SO_4. Distribute into tubes or flasks. Autoclave for 15 min at 15 psi pressure–121°C.

Use: For the cultivation and maintenance of *Flavobacterium acidurans.*

Flavobacterium Medium (ATCC Medium 65)

Composition per liter:
Agar 12.0g
Sodium caseinate 2.0g
Peptone 1.0g
K_2HPO_4 .... 0.5g
Yeast extract 0.5g
pH 7.4 ± 0.2 at 25°C

Preparation of Medium: Add components to distilled/deionized water and bring volume to 1.0L. Mix thoroughly. Gently heat and bring to boiling. Distribute into tubes or flasks. Autoclave for 15 min at 15 psi pressure–121°C. Pour into sterile Petri dishes or leave in tubes.

Use: For the cultivation and maintenance of *Flavobacterium aquatile.*

Flavobacterium Medium (ATCC Medium 1687)

Composition per liter:
Sodium glutamate 4.0g
K_2HPO_4 .... 0.65g
$NaNO_3$ 0.5g

KH_2PO_4 ..... 0.19g
$MgSO_4 \cdot 7H_2O$ 0.1g
$FeSO_4$ solution 2.0mL
pH 7.4 ± 0.2 at 25°C

$FeSO_4$ Solution:
Composition per 10.0mL:
$FeSO_4 \cdot 7H_2O$ 0.03g

Preparation of $FeSO_4$ Solution: Add $FeSO_4$ to distilled/deionized water and bring volume to 10.0mL. Mix thoroughly. Filter sterilize.

Preparation of Medium: Add components, except $FeSO_4$ solution, to distilled/deionized water and bring volume to 998.0mL. Mix thoroughly. Autoclave for 15 min at 15 psi pressure–121°C. Cool to 25°C. Aseptically add 2.0mL of sterile $FeSO_4$ solution. Mix thoroughly. Adjust pH to 7.4. Aseptically distribute into sterile tubes or flasks.

Use: For the cultivation of *Flavobacterium* species.

Flavobacterium Medium M1

Composition per liter:
Agar 12.0g
Proteose peptone 5.0g
NaCl 3.0g
Beef extract 2.0g
Yeast extract 0.2g
pH 7.2–7.4 at 25°C

Preparation of Medium: Add components to distilled/deionized water and bring volume to 1.0L. Mix thoroughly. Gently heat and bring to boiling. Distribute into tubes or flasks. Autoclave for 15 min at 15 psi pressure–121°C. Pour into sterile Petri dishes or leave in tubes.

Use: For the isolation and cultivation of *Flavobacterium* species.

Flavobacterium Medium with Thiamine

Composition per liter:
Agar 12.0g
Sodium caseinate 2.0g
Peptone 1.0g
K_2HPO_4 ..... 0.5g
Yeast extract 0.5g
Thiamine·HCl 10.0mg
pH 7.4 ± 0.2 at 25°C

Preparation of Medium: Add components to distilled/deionized water and bring volume to 1.0L. Mix thoroughly. Gently heat and bring to boiling. Distribute into tubes or flasks. Autoclave for 15 min at 15 psi pressure–121°C. Pour into sterile Petri dishes or leave in tubes.

Use: For the cultivation and maintenance of *Flavobacterium aquatile* and *Flavobacterium lutescens*.

Flavobacterium resinovorum Agar (LMG Medium 216)

Composition per liter:
Agar 15.0g
Lab-Lemco beef extract 10.0g
Peptone 10.0g
NaCl 5.0g
pH 7.3 ± 0.2 at 25°C

Preparation of Medium: Add components to tap water and bring volume to 1.0L. Mix thoroughly. Gently heat and bring to boiling. Distribute into tubes or flasks. Autoclave for 15 min at 15 psi pressure–121°C. Pour into sterile Petri dishes or leave in tubes.

Use: For the cultivation and maintenance of *Flavobacterium resinovorum.*

Fletcher Medium

Composition per liter:
Agar 1.5g
NaCl 0.5g
Peptone 0.3g
Beef extract 0.2g
Rabbit serum 50.0mL
pH 7.9 ± 0.1 at 25°C

Source: This medium is available as a premixed powder from BD Diagnostic Systems.

Preparation of Medium: Add components, except rabbit serum, to distilled/deionized water and bring volume to 950.0mL. Mix thoroughly. Gently heat and bring to boiling. Autoclave for 15 min at 15 psi pressure–121°C. Cool to 50°–55°C. Aseptically add 50.0mL of sterile rabbit serum. Mix thoroughly. Aseptically distribute into sterile tubes or flasks.

Use: For the isolation, cultivation, and maintenance of cultures of *Leptospira* species.

Fletcher Medium with Fluorouracil (Fluorouracil *Leptospira* Medium)

Composition per liter:
Agar 1.5g
NaCl 0.5g
Peptone 0.3g
Beef extract 0.2g
Rabbit serum 50.0mL
Fluorouracil solution 20.0mL
pH 7.9 ± 0.1 at 25°C

Fluorouracil Solution:
Composition per 100.0mL:
Fluorouracil 10.0g

Preparation of Fluorouracil Solution: Add fluorouracil to 50.0mL of distilled/deionized water. Add 1.0mL of 2*N* NaOH and bring volume to 100.0mL. Gently heat to 56°C for 2 hr. Adjust pH to 7.4–7.6 with NaOH. Mix thoroughly. Filter sterilize.

Preparation of Medium: Add components, except rabbit serum and fluorouracil solution, to distilled/deionized water and bring volume to 930.0mL. Mix thoroughly. Gently heat and bring to boiling. Autoclave for 15 min at 15 psi pressure–121°C. Cool to 50°–55°C. Aseptically add 80.0mL of sterile rabbit serum. Mix thoroughly. Aseptically distribute into sterile tubes or flasks. Immediately prior to use, add 0.1mL of fluorouracil solution per 5.0mL of medium.

Use: For the isolation, cultivation, and maintenance of cultures of *Leptospira* species.

Flexibacter Agar (LMG Medium 60)

Composition per liter:
Agar 15.0g
Casamino acids 1.0g
Tris buffer 1.0g
$MgSO_4 \cdot 7H_2O$ 0.1g
KNO_3 ..... 0.1g
$CaCl_2 \cdot 2H_2O$ 0.1g
Sodium-β-glycerophosphate 0.1g
Thiamine 1.0mg
Vitamin B_{12} 1.0μg

Glucose solution 10.0mL
Vitamin solution.. 10.0mL
Trace elements solution 1.0mL
pH 7.5 ± 0.2 at 25°C

Glucose Solution:
Composition per 100.0mL:
Glucose 10.0g

Preparation of Glucose Solution: Add glucose to distilled/deionized water and bring volume to 100.0mL. Mix thoroughly. Filter sterilize.

Trace Elements Solution:
Composition per liter:
H_3BO_3 .... 2.85g
$FeSO_4·7H_2O$ 2.49g
$MnCl_2·4H_2O$ 1.8g
Sodium tartrate.. 1.77g
$CaCl_2·2H_2O$ 40.4mg
$CuCl_2·2H_2O$ 26.9mg
$ZnCl_2$ 20.8mg
$Na_2MoO_4·2H_2O$ 25.2mg

Preparation of Trace Elements Solution: Add components to distilled/deionized water and bring volume to 1.0L. Mix thoroughly. Filter sterilize.

Vitamin Solution:
Composition per 10.0mL:
Thiamine 10.0mg
Vitamin B_{12} 10.0μg

Preparation of Vitamin Solution: Add components to distilled/deionized water and bring volume to 10.0mL. Mix thoroughly. Filter sterilize.

Preparation of Medium: Add components, except vitamin solution, glucose solution, and trace elements solution, to 979.0mL distilled/deionized water. Mix thoroughly. Gently heat and bring to boiling. Autoclave for 15 min at 15 psi pressure–121°C. Cool to 45°–50°C. Aseptically add 10.0mL of sterile vitamin solution, 10.0mL of sterile glucose solution, and 1.0mL of sterile trace elements solution. Mix thoroughly. Pour into sterile Petri dishes or distribute into sterile tubes.

Use: For the cultivation of *Flexibacter* spp.

Flexibacter Agar

Composition per liter:
Agar 15.0g
Sodium glutamate 5.0g
$MgSO_4·7H_2O$ 1.0g
Yeast extract.. 1.0g
Glucose solution 20.0mL
pH 7.2 ± 0.2 at 25°C

Glucose Solution:
Composition per 20.0mL:
Glucose 2.0g

Preparation of Glucose Solution: Add glucose to distilled/deionized water and bring volume to 20.0mL. Mix thoroughly. Sparge with 100% N_2 for 3–4 min. Autoclave for 15 min at 15 psi pressure–121°C.

Preparation of Medium: Add components, except glucose solution, to distilled/deionized water and bring volume to 980.0mL. Mix thoroughly. Gently heat and bring to boiling. Autoclave for 15 min at 15 psi pressure–121°C. Aseptically add 20.0mL of sterile glucose solution. Mix thoroughly. Pour into sterile Petri dishes or distribute into sterile tubes.

Use: For the cultivation of *Flexibacter elegans.*

Flexibacter canadensis Agar

Composition per 1001.0mL:
Agar 10.0g
Casamino acids 1.0g
Tris... 1.0g
$CaCl_2·2H_2O$ 0.1g
KNO_3 .... 0.1g
$MgSO_4·7H_2O$ 0.1g
Sodium glycerophosphate 0.1g
Thiamine HCl 1.0mg
Vitamin B_{12} 1.0μg
Glucose solution 10.0mL
Trace elements solution SL-10 .... 1.0mL
pH 7.5 ± 0.2 at 25°C

Glucose Solution:
Composition per 10.0mL:
Glucose 1.0g

Preparation of Glucose Solution: Add glucose to distilled/deionized water and bring volume to 10.0mL. Mix thoroughly. Sparge with 100% N_2 for 3–4 min. Autoclave for 15 min at 15 psi pressure–121°C.

Trace Elements Solution SL-10:
Composition per liter:
$FeCl_2·4H_2O$ 1.5g
$CoCl_2·6H_2O$ 190.0mg
$MnCl_2·4H_2O$ 100.0mg
$ZnCl_2$ 70.0mg
$Na_2MoO_4·2H_2O$ 36.0mg
$NiCl_2·6H_2O$ 24.0mg
H_3BO_3 ....6.0mg
$CuCl_2·2H_2O$ 2.0mg
HCl (25% solution) 10.0mL

Preparation of Trace Elements Solution SL-10: Add $FeCl_2·4H_2O$ to 10.0mL of HCl solution. Mix thoroughly. Add distilled/deionized water and bring volume to 1.0L. Add remaining components. Mix thoroughly.

Preparation of Medium: Add components, except glucose solution, to distilled/deionized water and bring volume to 990.0mL. Mix thoroughly. Gently heat and bring to boiling. Autoclave for 15 min at 15 psi pressure–121°C. Aseptically add 10.0mL of sterile glucose solution. Mix thoroughly. Pour into sterile Petri dishes or distribute into sterile tubes.

Use: For the cultivation of *Flexibacter canadensis.*

Flexibacter Medium

Composition per liter:
Agar 15.0g
Tris(hydroxymethyl)amino
methane buffer 1.0g
Casamino acids 1.0g
$MgSO_4·7H_2O$ 0.1g
KNO_3 .... 0.1g
$CaCl_2·2H_2O$ 0.1g
Sodium β-glycerophosphate 0.1g
Thiamine.. 1.0mg
Cobalamin 1.0μg
Glucose solution 10.0mL
Trace elements solution HO-LE 1.0mL
pH 7.5 ± 0.2 at 25°C

Glucose Solution:
Composition per 10.0mL:
D-Glucose 1.0g

Preparation of Glucose Solution: Add glucose to distilled/deionized water and bring volume to 10.0mL. Mix thoroughly. Filter sterilize.

Trace Elements Solution HO-LE:
Composition per liter:

H_3BO_3	2.85g
$MnCl_2 \cdot 4H_2O$	1.8g
Sodium tartrate	1.77g
$FeSO_4 \cdot 7H_2O$	1.36g
$CoCl_2 \cdot 6H_2O$	0.04g
$CuCl_2 . 2H_2O$	0.027g
$Na_2MoO_4 \cdot 2H_2O$	0.025g
$ZnCl_2$	0.020g

Preparation of Trace Elements Solution HO-LE: Add components to distilled/deionized water and bring volume to 1.0L. Mix thoroughly. Filter sterilize.

Preparation of Medium: Add components, except glucose solution, to distilled/deionized water and bring volume to 990.0mL. Mix thoroughly. Gently heat and bring to boiling. Autoclave for 15 min at 15 psi pressure–121°C. Aseptically add 10.0mL of sterile glucose solution. Mix thoroughly. Pour into sterile Petri dishes or distribute into sterile tubes.

Use: For the cultivation and maintenance of *Flexibacter* species.

Flexibacter Medium (ATCC Medium 1559)

Composition per liter:

Solution B	700.0mL
Solution A	300.0mL

Solution A:
Composition per 300.0mL:

Pancreatic digest of casein	0.5g
Yeast extract	0.5g
Beef extract	0.2g
Sodium acetate	0.2g

Preparation of Solution A: Add components to distilled/deionized water and bring volume to 1.0L. Mix thoroughly. Autoclave for 15 min at 15 psi pressure–121°C. Cool to 45°–50°C.

Solution B:

Aged seawater	700.0 mL

Preparation of Solution B: Allow seawater to sit for 7 days. Autoclave for 15 min at 15 psi pressure–121°C. Cool to 45°–50°C.

Preparation of Medium: Aseptically add 300.0mL of sterile solution A to 700.0mL of sterile solution B at 45°–50°C. Mix thoroughly. Aseptically distribute into sterile tubes or flasks.

Use: For the cultivation of *Flexibacter maritimus.*

Flexibacter polymorphus Medium (LMG 108)

Composition per liter:

Agar	15.0g
Monosodium glutamate	5.0g
Tryptone	1.0g
Casamino acids, vitamin-free	1.0g
Sodium glycerophosphate	0.1g
Vitamin B_{12}	1.0μg
Trace elements solution HO-LE	1.0mL
Seawater, filtered, aged	1.0L

Trace Elements Solution HO-LE:
Composition per liter:

H_3BO_3	2.85g
$MnCl_2 \cdot 4H_2O$	1.8g
Sodium tartrate	1.77g
$FeSO_4 \cdot 7H_2O$	1.36g
$CoCl_2 \cdot 6H_2O$	0.04g
$CuCl_2 . 2H_2O$	0.027g
$Na_2MoO_4 \cdot 2H_2O$	0.025g
$ZnCl_2$	0.02g

Preparation of Trace Elements Solution HO-LE: Add components to distilled/deionized water and bring volume to 1.0L. Mix thoroughly.

Preparation of Medium: Combine components. Mix thoroughly. Distribute into tubes or flasks. Autoclave for 15 min at 15 psi pressure–121°C.

Use: For the cultivation and maintenance of *Flexibacter polymorphus.*

Flexibacterium Medium

Composition per 1060.0mL:

Yeast extract	1.0g
$Ca(NO_3)_2 \cdot 4H_2O$	0.1g
K_2HPO_4	0.02g
Seawater, filtered	1.0L
Glucose solution	50.0mL
Trace elements	10.0mL

pH 7.0 ± 0.2 at 25°C

Glucose Solution:
Composition per 50.0mL:

Glucose	1.0g

Preparation of Glucose Solution: Add glucose to distilled/deionized water and bring volume to 50.0mL. Mix thoroughly. Autoclave for 15 min at 15 psi pressure–121°C. Cool to 25°C.

Trace Elements:
Composition per liter:

$FeSO_4 \cdot 7H_2O$	0.5mg
$ZnSO_4 \cdot 7H_2O$	0.3mg
H_3BO_3	0.1mg
$CoCl_2 \cdot 6H_2O$	0.1mg
$CuSO_4 \cdot 5H_2O$	0.1mg
$MnSO_4 \cdot 4H_2O$	0.1mg
$Na_2MoO_4 \cdot 2H_2O$	0.1mg

Preparation of Trace Elements: Add components to distilled/deionized water and bring volume to 1.0L. Mix thoroughly.

Preparation of Medium: Combine components, except glucose solution. Mix thoroughly. Adjust pH to 7.2. Gently heat and bring to boiling. Autoclave for 15 min at 15 psi pressure–121°C. Cool to 45°–50°C. Aseptically add sterile glucose solution. Mix thoroughly. Aseptically distribute into sterile tubes or bottles.

Use: For the cultivation of *Flexibacter litoralis* and *Flexibacter marinus.*

Flexibacterium Medium

Composition per 1050.0mL:

Tris(hydroxymethyl)aminomethane buffer	1.0g
Yeast extract	1.0g
$CaCl_2 \cdot 2H_2O$	0.1g
KCl	0.1g
$MgSO_4 \cdot 7H_2O$	0.1g
Sodium glycerophosphate	0.1g

$NaNO_3$	0.1g
Cobalamin	1.0μg
Glucose solution	50.0mL
Trace elements	10.0mL

pH 7.5 ± 0.2 at 25°C

Glucose Solution:

Composition per 50.0mL:

Glucose	1.0g

Preparation of Glucose Solution: Add glucose to distilled/deionized water and bring volume to 50.0mL. Mix thoroughly. Autoclave for 15 min at 15 psi pressure–121°C. Cool to 25°C.

Trace Elements:

Composition per liter:

$FeSO_4 \cdot 7H_2O$	0.5mg
$ZnSO_4 \cdot 7H_2O$	0.3mg
H_3BO_3	0.1mg
$CoCl_2 \cdot 6H_2O$	0.1mg
$CuSO_4 \cdot 5H_2O$	0.1mg
$MnSO_4 \cdot 4H_2O$	0.1mg
$Na_2MoO_4 \cdot 2H_2O$	0.1mg

Preparation of Trace Elements: Add components to distilled/deionized water and bring volume to 1.0L. Mix thoroughly.

Preparation of Medium: Add components, except glucose solution, to distilled/deionized water and bring volume to 1.0L. Mix thoroughly. Adjust pH to 7.5. Autoclave for 15 min at 15 psi pressure–121°C. Cool to 45°–50°C. Aseptically add sterile glucose solution. Mix thoroughly. Aseptically distribute into sterile tubes or flasks.

Use: For the cultivation of *Saprospira thermalis*, *Flexibacter elegans*, and *Flexibacter rubrum*.

Flexithrix Marine Agar (LMG Medium 61)

Composition per liter:

Agar	15.0g
Tryptone	5.0g
Yeast extract	5.0g
Tris buffer	1.0g
KNO_3	0.5g
Sodium-beta-glycerophosphate	0.1g
Trace elements solution	1.0mL

pH 7.4 ± 0.2 at 25°C

Trace Elements Solution:

Composition per liter:

H_3BO_3	2.85g
$FeSO_4 \cdot 7H_2O$	2.49g
$MnCl_2 \cdot 4H_2O$	1.8g
Sodium tartrate	1.77g
$CaCl_2 \cdot 2H_2O$	40.4mg
$CuCl_2 \cdot 2H_2O$	26.9mg
$ZnCl_2$	20.8mg
$Na_2MoO_4 \cdot 2H_2O$	25.2mg

Preparation of Trace Elements Solution: Add components to distilled/deionized water and bring volume to 1.0L. Mix thoroughly.

Preparation of Medium: Add components to filtered aged seawater and bring volume to 1.0L. Mix thoroughly. Gently heat and bring to boiling. Distribute into tubes or flasks. Autoclave for 15 min at 15 psi pressure–121°C. Pour into sterile Petri dishes or leave in tubes.

Use: For the cultivation of *Flexithrix dorotheae*.

FlGlyM Medium (DSMZ Medium 298b)

Composition per liter:

NaCl	1.0g
KCl	0.5g
$MgCl_2 \cdot 6H_2O$	0.4g
NH_4Cl	0.25g
KH_2PO_4	0.2g
$CaCl_2 \cdot 2H_2O$	0.15g
Resazurin	1.0mg
$NaHCO_3$ solution	50.0mL
$Na_2S \cdot 9H_2O$ solution	10.0mL
Na-glycolate solution	10.0mL
L-Cysteine solution	10.0mL
Na-acetate solution	7.0mL
Trace elements solution SL-10	1.0mL
Vitamin solution	0.5mL

pH 7.2 ± 0.2 at 25°C

$Na_2S \cdot 9H_2O$ Solution:

Composition per 10mL:

$Na_2S \cdot 9H_2O$	0.36g

Preparation of $Na_2S \cdot 9H_2O$ Solution: Add $Na_2S \cdot 9H_2O$ to distilled/deionized water and bring volume to 10.0mL. Mix thoroughly. Autoclave under 100% N_2 for 15 min at 15 psi pressure–121°C. Cool to room temperature.

$NaHCO_3$ Solution:

Composition per 100.0mL:

$NaHCO_3$	10.0g

Preparation of $NaHCO_3$ Solution: Add $NaHCO_3$ to distilled/deionized water and bring volume to 100.0mL. Mix thoroughly. Sparge with 80% N_2 + 20% CO_2. Filter sterilize.

Na-Acetate Solution:

Composition per 10.0mL:

Na-acetate	0.25g

Preparation of Na-Acetate Solution: Add Na-acetate to distilled/deionized water and bring volume to 10.0mL. Mix thoroughly. Sparge with 100% N_2. Filter sterilize.

Na-Glycolate Solution:

Composition per 10.0mL:

Na-glycolate	1.0g

Preparation of Na-Glycolate Solution: Add Na-glycolate to distilled/deionized water and bring volume to 10.0mL. Mix thoroughly. Sparge with 100% N_2. Filter sterilize.

L-Cysteine Solution:

Composition per 10.0mL:

$L\text{-Cysteine} \cdot HCl \cdot H_2O$	0.3g

Preparation of L-Cysteine Solution: Add L-cysteine·$HCl \cdot H_2O$ to distilled/deionized water and bring volume to 10.0mL. Mix thoroughly. Sparge with 100% N_2. Autoclave for 15 min at 15 psi pressure–121°C.

Trace Elements Solution SL-10:

Composition per liter:

$FeCl_2 \cdot 4H_2O$	1.5g
H_3BO_3	300.0mg
$CoCl_2 \cdot 6H_2O$	190.0mg
$MnCl_2 \cdot 4H_2O$	100.0mg
$ZnCl_2$	70.0mg
$Na_2MoO_4 \cdot 2H_2O$	36.0mg
$NiCl_2 \cdot 6H_2O$	24.0mg

$CuCl_2 \cdot 2H_2O$ 2.0mg
HCl (25% solution) 10.0mL

Preparation of Trace Elements Solution SL-10: Add $FeCl_2 \cdot 4H_2O$ to 10.0mL of HCl solution. Mix thoroughly. Add distilled/deionized water and bring volume to 1.0L. Add remaining components. Mix thoroughly. Sparge with 80% N_2 + 20% CO_2. Autoclave for 15 min at 15 psi pressure–121°C.

Vitamin Solution:
Composition per liter:
Pyridoxine hydrochloride 300.0mg
Thiamine-HCl·$2H_2O$ 200.0mg
Nicotinic acid 200.0mg
Vitamin B_{12} 100.0mg
Calcium pantothenate 100.0mg
p-Aminobenzoic acid 80.0mg
Folic Acid 30.0mg
D(+)-Biotin 20.0mg
α-Lipoic acid 10.0mg

Preparation of Vitamin Solution: Add components to distilled/deionized water and bring volume to 1.0L. Mix thoroughly. Filter sterilize.

Preparation of Medium: Prepare and dispense medium under 80% N_2 + 20% CO_2 gas atmosphere. Add components, except $NaHCO_3$ solution, $Na_2S \cdot 9H_2O$ solution, L-cysteine solution, Na-acetate solution, Na-glycolate solution, vitamin solution, and trace elements solution SL-10, to distilled/deionized water and bring volume to 979.0mL. Mix thoroughly. Adjust pH to 7.2. Sparge with 80% N_2 + 20% CO_2. Autoclave for 15 min at 15 psi pressure–121°C. Aseptically and anaerobically add 10.0mL $NaHCO_3$ solution, 10.0mL $Na_2S \cdot 9H_2O$ solution, 10.0mL L-cysteine solution, 7.0mL Na-acetate solution, 10.0mL Na-glycolate solution, 1.0mL trace elements solution SL-10, and 0.5mL vitamin solution. Mix thoroughly. Aseptically and anaerobically distribute into sterile tubes or bottles. After inoculation, flush and repressurize the gas head space of culture bottles with sterile 80% N_2 + 20% CO_2 to 1 bar overpressure.

Use: For the cultivation of *Syntrophobotulus* spp.

Flo Agar

Composition per liter:
Agar 14.0g
Pancreatic digest of casein 10.0g
Peptic digest of animal tissue 10.0g
K_2HPO_4 .... 1.5g
$MgSO_4 \cdot 7H_2O$ 1.5g
pH 7.2 ± 0.2 at 25°C

Source: This medium is available as a premixed powder from BD Diagnostic Systems.

Preparation of Medium: Add components to distilled/deionized water and bring volume to 1.0L. Mix thoroughly. Gently heat and bring to boiling. Distribute into tubes or flasks. Autoclave for 15 min at 15 psi pressure–121°C. Pour into sterile Petri dishes or leave in tubes.

Use: For cultivation of fluorescent *Pseudomonas* species.

Fluorescent Pectolytic Agar (FPA Medium)

Composition per liter:
Proteose peptone No. 3 .... 20.0g
Agar 15.0g
Pectin 5.0g
K_2HPO_4 .... 1.5g
$MgSO_4 \cdot 7H_2O$ 0.73g
Antibiotic solution 10.0mL

Antibiotic Solution:
Composition per 10.0mL:
Cycloheximide 0.075g
Novobiocin 0.045g
Penicillin G 75,000U
Ethanol 1.0mL

Preparation of Antibiotic Solution: Add components to 1.0mL of ethanol. Mix thoroughly. Let stand for 30 min. Bring volume to 10.0mL with distilled/deionized water. Mix thoroughly. Filter sterilize.

Preparation of Medium: Add components, except antibiotic solution, to distilled/deionized water and bring volume to 990.0mL. Mix thoroughly. Gently heat and bring to boiling. Autoclave for 15 min at 15 psi pressure–121°C. Cool to 45°–50°C. Aseptically add sterile antibiotic solution. Mix thoroughly. Pour into sterile Petri dishes.

Use: For the cultivation of fluorescent *Pseudomonas* species that are pectinolytic.

FN Medium (Fluorescence Denitrification Medium)

Composition per liter:
Agar 15.0g
Proteose peptone No. 3 .... 10.0g
KNO_3 .... 2.0g
K_2HPO_4 .... 1.5g
$MgSO_4 \cdot 7H_2O$ 1.5g
$NaNO_2$ 0.5g
pH 7.2 ±0.2 at 25°C

Preparation of Medium: Add components to distilled/deionized water and bring volume to 1.0L. Mix thoroughly. Gently heat and bring to boiling. Distribute into tubes or flasks. Autoclave for 15 min at 15 psi pressure–121°C. Pour into sterile Petri dishes or leave in tubes.

Use: For the differentiation of pseudomonads from other nonfermentative bacilli. Denitrification from nitrate or nitrite is indicated by the formation of gas bubbles in the solid medium. *Pseudomonas aeruginosa, Pseudomonas mendocina,* and *Pseudomonas denitrificans* are positive for denitrification. Fluorescein production is indicated by fluorescence under UV light. *Pseudomonas aeruginosa* is positive for fluorescein production; *Pseudomonas denitrificans* does not produce fluorescein.

FPA Medium

Composition per liter:
Proteose peptone No. 3 .... 20.0g
Agar 15.0g
Pectin, citrus 5.0g
K_2HPO_4 .... 1.5g
$MgSO_4 \cdot 7H_2O$ 1.5g
Antibiotic solution 10.0mL
pH 7.0 ± 0.2 at 25°C

Antibiotic Solution:
Composition per 10.0mL:
Cycloheximide 0.075g
Novobiocin 0.045g
Penicillin G 75,000U

Preparation of Antibiotic Solution: Add components to distilled/deionized water and bring volume to 10.0mL. Mix thoroughly. Filter sterilize.

Preparation of Medium: Add components, except agar and antibiotic solution, to distilled/deionized water and bring volume to 990.0mL. Mix thoroughly. Adjust pH to 7.0. Add agar. Mix thoroughly. Gently heat and bring to boiling. Autoclave for 15 min at 15 psi pressure–121°C. Cool to 45°–50°C. Aseptically add sterile antibiotic solution. Mix thoroughly. Pour into sterile Petri dishes or distribute into sterile tubes.

Use: For the isolation and cultivation of *Pseudomonas* species that cause soft rot.

Frankia Agar

Composition per liter:

Starch, soluble	20.0g
Agar	15.0g
Glucose	10.0g
N-Z-Amine	5.0g
Yeast extract	5.0g
$CaCO_3$	1.0g

pH 7.0 ± 0.2 at 25°C

Preparation of Medium: Add components to distilled/deionized water and bring volume to 1.0L. Mix thoroughly. Gently heat and bring to boiling. Adjust pH to 7.0 with NaOH. Distribute into tubes or flasks. Autoclave for 15 min at 15 psi pressure–121°C. Pour into sterile Petri dishes or leave in tubes.

Use: For the cultivation and maintenance of *Frankia alni.*

Frankia Isolation Medium

Composition per liter:

Sucrose	40.0g
$Ca(NO_3)_2 \cdot 4H_2O$	0.242g
KNO_3	0.085g
KCl	0.061g
$MgSO_4 \cdot 7H_2O$	0.042g
KH_2PO_4	0.02g
$MnSO_4 \cdot H_2O$	4.5mg
$FeCl_3 \cdot 6H_2O$	2.5mg
H_3BO_3	1.5mg
$ZnSO_4 \cdot 7H_2O$	1.5mg
Nicotinic acid	0.5mg
Pyridoxine·HCl	0.5mg
$Na_2MoO_4 \cdot 2H_2O$	0.25mg
Thiamine·HCl	0.1mg
$CuSO_4 \cdot 5H_2O$	0.04mg
Mannitol solution	10.0mL
Supplement solution	10.0mL

pH 5.5 ± 0.2 at 25°C

Mannitol Solution:

Composition per 100.0mL:

Mannitol	11.84g

Preparation of Mannitol Solution: Add mannitol to distilled/deionized water and bring volume to 100.0mL. Mix thoroughly. Filter sterilize.

Supplement Solution:

Composition per 10.0mL:

L-Glutamic acid	0.185g
L-Arginine	0.174g
L-Glutamine	0.146g
L-Aspartic acid	0.133g
L-Asparagine	0.132g
Glycine	0.075g
Urea	0.06g
Naphthaloneacetic acid	2.0mg
Zeatin	1.0μg

Preparation of Supplement Solution: Add components to distilled/deionized water and bring volume to 10.0mL. Mix thoroughly. Filter sterilize.

Preparation of Medium: Add components, except mannitol solution and supplement solution, to distilled/deionized water and bring volume to 980.0mL. Mix thoroughly. Adjust pH to 5.5. Gently heat and bring to boiling. Autoclave for 15 min at 15 psi pressure–121°C. Cool to 45°–50°C. Aseptically add 10.0mL of sterile mannitol solution and 10.0mL of sterile supplement solution. Mix thoroughly. Aseptically distribute into sterile tubes or flasks.

Use: For the isolation and cultivation of *Frankia* species from root nodules.

Fraser Broth

Composition per liter:

NaCl	20.0g
Na_2HPO_4	12.0g
Beef Extract	5.0g
Proteose peptone	5.0g
Pancreatic digest of casein	5.0g
Yeast extract	5.0g
LiCl	3.0g
KH_2PO_4	1.35g
Esculin	1.0g
Fraser supplement solution	10.0mL

pH 7.2 ± 0.2 at 25°C

Source: This medium is available as a premixed powder from BD Diagnostic Systems and Oxoid Unipath.

Fraser Supplement Solution:

Composition per 10.0mL:

Ferric ammonium citrate	0.5g
Acriflavine·HCl	0.25g
Nalidixic acid	0.1g
Ethanol	5.0mL

Preparation of Fraser Supplement Solution Solution: Add components to distilled/deionized water and bring volume to 10.0mL. Mix thoroughly. Filter sterilize.

Preparation of Medium: Add components, except Fraser supplement solution, to distilled/deionized water and bring volume to 990.0mL. Mix thoroughly. Gently heat and bring to boiling. Autoclave for 15 min at 15 psi pressure–121°C. Cool to 45°–50°C. Aseptically add sterile Fraser supplement solution. Mix thoroughly. Aseptically distribute into sterile tubes or flasks.

Use: For the isolation of *Listeria* species from environmental species.

Fraser Secondary Enrichment Broth

Composition per liter:

NaCl	20.0g
Na_2HPO_4	12.0g
Beef extract	5.0g
Proteose peptone	5.0g
Pancreatic digest of casein	5.0g
Yeast extract	5.0g
LiCl	3.0g
KH_2PO_4	1.35g
Esculin	1.0g
Acriflavin solution	10.0mL
Ferric ammonium citrate solution	10.0mL
Nalidixic acid solution	1.0mL

Ferric Ammonium Citrate Solution:
Composition per 10.0mL:
Ferric ammonium citrate 0.5g

Preparation of Ferric Ammonium Citrate Solution: Add ferric ammonium citrate to distilled/deionized water and bring volume to 10.0mL. Mix thoroughly. Filter sterilize.

Acriflavin Solution:
Composition per 10.0mL:
Acriflavin 0.025g

Preparation of Acriflavin Solution: Add acriflavin to distilled/deionized water and bring volume to 10.0mL. Mix thoroughly. Filter sterilize.

Nalidixic Acid Solution:
Composition per 10.0mL:
Nalidixic acid 0.04g
NaOH (0.1*N* solution) 10.0mL

Preparation of Nalidixic Acid Solution: Add nalidixic acid to 10.0mL of NaOH solution. Mix thoroughly. Filter sterilize.

Preparation of Medium: Add components, except acriflavin solution and ferric ammonium citrate solution, to distilled/deionized water and bring volume to 980.0mL. Mix thoroughly. Gently heat and bring to boiling. Distribute into tubes in 10.0mL volumes. Autoclave for 12 min at 15 psi pressure–121°C. Cool rapidly to 25°C. Immediately prior to inoculation, aseptically add 0.1mL of sterile acriflavin solution and 0.1mL of ferric ammonium citrate solution to each tube. Mix thoroughly.

Use: For the isolation, cultivation, and enrichment of *Listeria monocytogenes* from environmental specimens based on esculin hydrolysis. Bacteria that hydrolyze esculin appear as black colonies.

Freshwater *Ameba* Medium

Composition per liter:
Agar 10.0g
Malt extract 0.1g
Yeast extract 0.1g

Preparation of Medium: Add components to distilled/deionized water and bring volume to 1.0L. Mix thoroughly. Gently heat and bring to boiling. Distribute into tubes or flasks. Autoclave for 15 min at 15 psi pressure–121°C. Pour into sterile Petri dishes or leave in tubes.

Use: For the cultivation of *Acanthamoeba astronyxis, Acanthamoeba castellanii, Acanthamoeba griffini, Acanthamoeba pearcei, Acanthamoeba polyphaga, Acanthamoeba rhysodes, Acanthamoeba stevensoni, Acanthamoeba tubiashi, Capsellina* species, *Cochliopodium actinophora, Cochliopodium bilimbosum, Hartmannella limax, Hartmannella vermiformis, Naegleria fowleri, Naegleria gruberi, Naegleria lovaniensis, Naegleria minor, Naegleria thorntoni, Paraflabellula reniformis, Protacanthamoeba caledonica, Rosculus* species, *Saccamoeba limax, Tetramitus rostratus, Vahlkampfia inornata*, and *Vannella miroides*.

FSM Selective Medium

Composition per liter:
Agar 18.0g
Peptone 10.0g
Glucose 4.0g
$(NH_4)_2SO_4$ 1.32g
K_2HPO_4 .... 1.18g
Casamino acids 1.0g
Yeast extract 1.0g
KH_2PO_4 .... 0.44g
$MgSO_4 \cdot 7H_2O$ 0.2g
$FeC_6H_5O_7 \cdot 5H_2O$ 3.0mg
Citric acid 1.9mg
$ZnSO_4 \cdot 7H_2O$ 1.6mg
$MnSO_4 \cdot H_2O$ 1.5mg
2,3,5-Triphenyltetrazolium·HCl solution 10.0mL
Benomyl solution 10.0mL
Polymyxin B solution 10.0mL
Chloroneb solution 10.0mL
Dichloran solution 10.0mL
Bacitracin solution 10.0mL
Cycloheximide solution 10.0mL
Pentachloronitrobenzene solution 10.0mL
Pimaricin solution 10.0mL
Tyrothricin solution 10.0mL
Vancomycin solution 10.0mL
Chloromycetin solution 10.0mL
Penicillin G solution 10.0mL

2,3,5-Triphenyltetrazolium·HCl Solution:
Composition per 10.0mL:
2,3,5-Triphenyltetrazolium·HCl 0.5mg

Preparation of 2,3,5-Triphenyltetrazolium·HCl Solution: Add 2,3,5-triphenyltetrazolium·HCl to distilled/deionized water and bring volume to 10.0mL. Mix thoroughly. Autoclave for 7 min at 15 psi pressure–121°C.

Benomyl Solution:
Composition per 10.0mL:
Benomyl 0.5mg

Preparation of Benomyl Solution: Add benomyl to distilled/deionized water and bring volume to 10.0mL. Mix thoroughly. Filter sterilize.

Polymyxin B Solution:
Composition per 10.0mL:
Polymyxin B 0.1mg

Preparation of Polymyxin B Solution: Add polymyxin B to distilled/deionized water and bring volume to 10.0mL. Mix thoroughly. Filter sterilize.

Chloroneb Solution:
Composition per 10.0mL:
Chloroneb 0.1mg

Preparation of Chloroneb Solution: Add chloroneb to distilled/deionized water and bring volume to 10.0mL. Mix thoroughly. Filter sterilize.

Dichloran Solution:
Composition per 10.0mL:
Dichloran 0.1mg

Preparation of Dichloran Solution: Add dichloran to distilled/deionized water and bring volume to 10.0mL. Mix thoroughly. Filter sterilize.

Bacitracin Solution:
Composition per 10.0mL:
Bacitracin 0.05mg

Preparation of Bacitracin Solution: Add bacitracin to distilled/deionized water and bring volume to 10.0mL. Mix thoroughly. Filter sterilize.

Cycloheximide Solution:
Composition per 10.0mL:
Cycloheximide 0.05mg

Preparation of Cycloheximide Solution: Add cycloheximide to distilled/deionized water and bring volume to 10.0mL. Mix thoroughly. Filter sterilize.

Pentachloronitrobenzene Solution:
Composition per 10.0mL:
Pentachloronitrobenzene 0.03mg

Preparation of Pentachloronitrobenzene Solution: Add pentachloronitrobenzene to distilled/deionized water and bring volume to 10.0mL. Mix thoroughly. Filter sterilize.

Pimaricin Solution:
Composition per 10.0mL:
Pimaricin 0.02mg

Preparation of Pimaricin Solution: Add pimaricin to distilled/deionized water and bring volume to 10.0mL. Mix thoroughly. Filter sterilize.

Tyrothricin Solution:
Composition per 10.0mL:
Tyrothricin 0.02mg

Preparation of Tyrothricin Solution: Add tyrothricin to distilled/deionized water and bring volume to 10.0mL. Mix thoroughly. Filter sterilize.

Vancomycin Solution:
Composition per 10.0mL:
Vancomycin 0.01mg

Preparation of Vancomycin Solution: Add vancomycin to distilled/deionized water and bring volume to 10.0mL. Mix thoroughly. Filter sterilize.

Chloromycetin Solution:
Composition per 10.0mL:
Chloromycetin 5.0μg

Preparation of Chloromycetin Solution: Add chloromycetin to distilled/deionized water and bring volume to 10.0mL. Mix thoroughly. Filter sterilize.

Penicillin G Solution:
Composition per 10.0mL:
Penicillin G 1.0μg

Preparation of Penicillin G Solution: Add penicillin G to distilled/deionized water and bring volume to 10.0mL. Mix thoroughly. Filter sterilize.

Preparation of Medium: Add components—except 2,3,5-triphenyltetrazolium chloride solution, benomyl solution, polymyxin B solution, chloroneb solution, dichloran solution, bacitracin solution, cycloheximide solution, pentachloronitrobenzene solution, pimaricin solution, tyrothricin solution, vancomycin solution, chloromycetin solution, and penicillin G solution—to distilled/deionized water and bring volume to 870.0mL. Mix thoroughly. Gently heat and bring to boiling. Autoclave for 15 min at 15 psi pressure–121°C. Cool to 45°–50°C. Aseptically add 10.0mL each of sterile 2,3,5-triphenyltetrazolium chloride solution, benomyl solution, polymyxin B solution, chloroneb solution, dichloran solution, bacitracin solution, cycloheximide solution, pentachloronitrobenzene solution, pimaricin solution, tyrothricin solution, vancomycin solution, chloromycetin solution, and penicillin G solution. Mix thoroughly. Pour into sterile Petri dishes. Dry plates for 24 hr at 30°C.

Use: For the isolation and cultivation of *Pseudomonas solanacearum* from soil.

FTX Broth

Composition per 1001.0mL:
Sodium glutamate 10.0g
Glucose 2.0g
Tris 2.0g
Sodium glycerophosphate 0.1g
Artificial seawater 1.0L
Trace elements solution 1.0mL
pH 8.0 ± 0.2 at 25°C

Artificial Seawater:
Composition per liter:
NaCl 24.7g
$MgSO_4 \cdot 7H_2O$ 6.3g
$MgCl_2 \cdot 6H_2O$ 4.6g
$CaCl_2$ 1.0g
KCl 0.7g
$NaHCO_3$ 0.2g

Preparation of Artificial Seawater: Add components to distilled/deionized water and bring volume to 1.0L. Mix thoroughly.

Trace Elements Solution:
Composition per liter:
Disodium EDTA 8.0g
$MnCl_2 \cdot 4H_2O$ 0.1g
$CoCl_2 \cdot 6H_2O$ 0.02g
KBr 0.02g
KI 0.02g
$ZnCl_2$ 0.02g
$CuSO_4$ 0.01g
H_3BO_3 .... 0.01g
$Na_2MoO_4 \cdot 2H_2O$ 0.01g
LiCl 5.0mg
$SnCl_2 \cdot 2H_2O$ 5.0mg

Preparation of Trace Elements Solution: Add components to distilled/deionized water and bring volume to 1.0L. Mix thoroughly.

Preparation of Medium: Add components to 1.0L of artificial seawater. Mix thoroughly. Gently heat and bring to boiling. Distribute into tubes or flasks. Autoclave for 15 min at 15 psi pressure–121°C.

Use: For the isolation and cultivation of *Cytophaga* species, *Herpetosiphon* species, *Saprospira* species, and *Flexithrix* species.

FUF Medium (DSMZ Medium 318a)

Composition per liter:
$KHCO_3$ 2.0g
NH_4Cl 1.0g
NaCl 0.6g
KH_2PO_4 .... 0.3g
$MgCl_2 \cdot 6H_2O$ 0.1g
$CaCl_2 \cdot 2H_2O$ 0.08g
Resazurin 1.0mg
HEPES solution 50.0mL
Furoic acid solution 50.0mL
Trace elements solution 10.0mL
Vitamin solution 10.0mL
Cysteine solution 10.0mL
$Na_2S \cdot 9H_2O$ solution 10.0mL
pH 6.8 ± 0.2 at 25°C

Vitamin Solution:
Composition per liter:
Pyridoxine-HCl 10.0mg
Thiamine-HCl·$2H_2O$ 5.0mg

Riboflavin 5.0mg
Nicotinic acid 5.0mg
D-Ca-pantothenate 5.0mg
p-Aminobenzoic acid 5.0mg
Lipoic acid 5.0mg
Biotin 2.0mg
Folic acid 2.0mg
Vitamin B_{12} 0.1mg

Preparation of Vitamin Solution: Add components to distilled/deionized water and bring volume to 1.0L. Mix thoroughly. Sparge with 80% H_2 + 20% CO_2. Filter sterilize.

Cysteine Solution:
Composition per 10.0mL:
L-Cysteine·HCl·H_2O 0.3g

Preparation of Cysteine Solution: Add L-cysteine·HCl·H_2O to distilled/deionized water and bring volume to 10.0mL. Mix thoroughly. Sparge with 100% N_2. Autoclave for 15 min at 15 psi pressure–121°C.

$Na_2S·9H_2O$ Solution:
Composition per 10mL:
$Na_2S·9H_2O$ 0.3g

Preparation of $Na_2S·9H_2O$ Solution: Add $Na_2S·9H_2O$ to distilled/deionized water and bring volume to 10.0mL. Mix thoroughly. Autoclave under 100% N_2 for 15 min at 15 psi pressure–121°C. Cool to room temperature.

Trace Elements Solution:
Composition per liter:
Nitrilotriacetic acid 12.8g
$FeCl_3·6H_2O$ 1.35g
NaCl 1.0g
$NiCl_2·6H_2O$ 0.12g
$MgCl_2·4H_2O$ 0.1g
$CaCl_2·2H_2O$ 0.1g
$ZnCl_2$ 0.1g
$Na_2SeO_3·5H_2O$ 0.026g
$CuCl_2·2H_2O$ 0.025g
$CoCl_2·6H_2O$ 0.024g
$Na_2MoO_4·4H_2O$ 0.024g
H_3BO_3 .... 0.01g

Preparation of Trace Elements Solution: Add nitrilotriacetic acid to 200.0mL of distilled/deionized water. Dissolve by adjusting pH to 6.5 with KOH. Add remaining components. Add distilled/deionized water to 1.0L. Mix thoroughly.

Furoic Acid Solution:
Composition per 100.0mL:
2-Furoic acid 4.4g

Preparation of Furoic Acid Solution: Add furoic acid to distilled/deionized water and bring volume to 100.0mL. Mix thoroughly. Adjust pH to 7.0 with NaOH. Sparge with 100% N_2. Filter sterilize.

HEPES Solution:
Composition per 100.0mL:
HEPES 6.2g

Preparation of HEPES Solution: Add HEPES to distilled/deionized water and bring volume to 100.0mL. Mix thoroughly. Adjust pH to 7.0. Sparge with 100% N_2. Filter sterilize.

Preparation of Medium: Prepare and dispense medium under 80% N_2 + 20% CO_2 gas atmosphere. Add components, except cysteine solution, furoic acid solution, $Na_2S·9H_2O$ solution, and vitamin solution, to distilled/deionized water and bring volume to 920.0mL. Mix thoroughly. Adjust pH to 7.0. Sparge with 80% N_2 + 20% CO_2. Autoclave for 15 min at 15 psi pressure–121°C. Aseptically and anaerobically add 10.0mL cysteine solution, 50.0mL furoic acid solution, 10.0mL $Na_2S·9H_2O$ solution, and 10.0mL vitamin solution. Mix thoroughly. Aseptically and anaerobically distribute into sterile tubes or bottles. After inoculation, flush and repressurize the gas head space of culture bottles with sterile 80% N_2 + 20% CO_2 to 1 bar overpressure.

Use: For the cultivation of *Methanosarcina thermophila (Methanosarcina* sp.*)* and *Methanosarcina mazei=Methanococcus mazei (Methanosarcina frisia).*

Fumarate Medium (DSMZ Medium 195a)

Composition per 991.0mL:
Solution A 870.0mL
Solution C 100.0mL
Solution D 10.0mL
Solution E (Vitamin solution) 10.0mL
Solution B (Trace elements solution Sl-10) 1.0mL
pH 7.4 ± 0.2 at 25°C

Solution A:
NaCl 21.0g
$MgCl_2·6H_2O$ 3.1g
Na_2SO_4 .... 3.0g
Na_2-fumarate 2.5g
KCl 0.5g
NH_4Cl 0.3g
KH_2PO_4 .... 0.2g
$CaCl_2·2H_2O$ 0.15g
Resazurin 1.0mg

Preparation of Solution A: Add components to distilled/deionized water and bring volume to 870.0mL. Mix thoroughly.

Solution B (Trace Elements Solution SL-10):
Composition per liter:
$FeCl_2·4H_2O$ 1.5g
$CoCl_2·6H_2O$ 190.0mg
$MnCl_2·4H_2O$ 100.0mg
$ZnCl_2$ 70.0mg
$Na_2MoO_4·2H_2O$ 36.0mg
$NiCl_2·6H_2O$ 24.0mg
H_3BO_3 .... 6.0mg
$CuCl_2·2H_2O$ 2.0mg
HCl (25% solution) 10.0mL

Preparation of Solution B (Trace Elements Solution SL-10): Add $FeCl_2·4H_2O$ to 10.0mL of HCl solution. Mix thoroughly. Add distilled/deionized water and bring volume to 1.0L. Add remaining components. Mix thoroughly. Sparge with 100% N_2. Autoclave for 15 min at 15 psi pressure–121°C.

Solution C:
Composition per 100.0mL:
$NaHCO_3$ 5.0g

Preparation of Solution C: Add $NaHCO_3$ to distilled/deionized water and bring volume to 100.0mL Mix thoroughly. Filter sterilize. Flush with 80% N_2 + 20% CO_2 to remove dissolved oxygen.

Solution D:
Composition per 100.0mL:

Resorcinol	1.1g

Preparation of Solution D: Add resorcinol to distilled/deionized water and bring volume to 100.0mL. Mix thoroughly. Sparge with 100% N_2. Filter sterilize. Use freshly prepared.

Solution E (Vitamin Solution):
Composition per liter:

Pyridoxine-HCl	10.0mg
Thiamine-HCl·$2H_2O$	5.0mg
Riboflavin	5.0mg
Nicotinic acid	5.0mg
D-Ca-pantothenate	5.0mg
p-Aminobenzoic acid	5.0mg
Lipoic acid	5.0mg
Biotin	2.0mg
Folic acid	2.0mg
Vitamin B_{12}	0.10mg

Preparation of Solution E (Vitamin Solution): Add components to distilled/deionized water and bring volume to 1.0L. Mix thoroughly. Sparge with 100% N_2. Autoclave for 15 min at 15 psi pressure–121°C.

Solution F:
Composition per 10.0mL:

$Na_2S·9H_2O$	0.4g

Preparation of Solution F: Add $Na_2S·9H_2O$ to distilled/deionized water and bring volume to 10.0mL. Mix thoroughly. Sparge with 100% N_2. Autoclave for 15 min at 15 psi pressure–121°C.

Preparation of Medium: Gently heat solution A and bring to boiling. Boil solution A for a few minutes. Cool to room temperature. Gas with 80% N_2 + 20% CO_2 gas mixture to reach a pH below 6. Autoclave for 15 min at 15 psi pressure–121°C. Cool to room temperature. Sequentially add 1.0mL solution B, 100.0mL solution C, 10.0mL solution D, 10.0mL solution E, and 10.0mL solution F. Adjust pH to 7.4 with sodium bicarbonate or sodium carbonate. Distribute anaerobically under 80% N_2 + 20% CO_2 into appropriate vessels. Addition of 10–20mg sodium dithionite per liter from a 5% (w/v) solution, freshly prepared under N_2 and filter-sterilized, may stimulate growth. During growth the culture can be fed with the resorcinol solution.

Use: For the cultivation of *Desulfuromusa kysingii, Desulfuromusa bakii,* and *Desulfuromusa succinoxidans.*

Fundibacter jadensis Medium (DSMZ Medium 821)

Composition per liter:

Peptone	2.5g
Meat extract	1.5g
Na-Acetate	1.0g
Artificial seawater, concentrated	190.0mL

pH 7.2 ± 0.2 at 25°C

Artificial Seawater, Concentrated:

NaCl	99.4g
$MgSO_4·7H_2O$	23.76g
$MgCl_2·6H_2O$	18.12g
$CaCl_2·2H_2O$	5.2g
KCl	2.56g

Preparation of Artificial Seawater, Concentrated: Add components to distilled/deionized water and bring volume to 1.0L. Mix thoroughly. Adjust pH to 7.1-7.3. Autoclave for 15 min at 15 psi pressure–121°C. Cool to room temperature.

Preparation of Medium: Add components, except concentrated artificial seawater, to distilled/deionized water and bring volume to 810.0mL. Mix thoroughly. Adjust pH to 7.1-7.3. Autoclave for 15 min at 15 psi pressure–121°C. Cool to 60°C. Add 190.0mL concentrated artificial sea water. Mix thoroughly. Aseptically distribute into tubes or flasks.

Use: For the cultivation of *Alcanivorax jadensis.*

Furoate Agar

Composition per liter:

Agar	20.0g
2-Furoic acid	2.0g
K_2HPO_4	1.0g
NH_4Cl	1.0g
$MgSO_4·7H_2O$	0.1g

pH 7.0 ± 0.2 at 25°C

Preparation of Medium: Add components to distilled/deionized water and bring volume to 1.0L. Mix thoroughly. Gently heat and bring to boiling. Distribute into tubes or flasks. Autoclave for 15 min at 15 psi pressure–121°C. Pour into sterile Petri dishes or leave in tubes.

Use: For the cultivation and maintenance of *Bacillus megaterium* and *Pseudomonas* species.

Fusibacter paucivorans Medium (DSMZ Medium 853)

Composition per 1068mL:

NaCl	30.0g
Na-Thiosulfate·$5H_2O$	3.16g
$MgCl_2·6H_2O$	3.0g
Yeast extract	1.0g
Trypticase	1.0g
NH_4Cl	1.0g
KCl	1.0g
Na-acetate·$3H_2O$	0.5g
Cysteine-HCl·H_2O	0.5g
K_2HPO_4	0.3g
KH_2PO_4	0.3g
$CaCl_2·2H_2O$	0.1g
Resazurin	0.5mg
$NaHCO_3$ solution	40.0mL
Glucose solution	18.0mL
$Na_2S·9H_2O$ solution	10.0mL
Trace elements solution	10.0mL

pH 7.3 ± 0.2 at 25°C

Trace Elements Solution:
Composition per liter:

$MgSO_4·7H_2O$	3.0g
Nitrilotriacetic acid	1.5g
NaCl	1.0g
$MnSO_4·2H_2O$	0.5g
$CoSO_4·7H_2O$	0.18g
$ZnSO_4·7H_2O$	0.18g
$CaCl_2·2H_2O$	0.1g
$FeSO_4·7H_2O$	0.1g
$NiCl_2·6H_2O$	0.025g
$KAl(SO_4)_2·12H_2O$	0.02g
H_3BO_3	0.01g
$Na_2MoO_4·4H_2O$	0.01g
$CuSO_4·5H_2O$	0.01g
$Na_2SeO_3·5H_2O$	0.3mg

Preparation of Trace Elements Solution: Add nitrilotriacetic acid to 500.0mL of distilled/deionized water. Dissolve by adjusting pH to 7.0 with KOH. Add remaining components. Add distilled/deionized water to 1.0L. Mix thoroughly.

$NaHCO_3$ Solution:
Composition per 100.0mL:

$NaHCO_3$	10.0g

Preparation of $NaHCO_3$ Solution: Add $NaHCO_3$ to distilled/deionized water and bring volume to 100.0mL. Mix thoroughly. Autoclave for 15 min at 15 psi pressure–121°C. Cool to 25°C. Must be prepared freshly.

$Na_2S{\cdot}9H_2O$ Solution:
Composition per 10mL:

$Na_2S{\cdot}9H_2O$	0.3g

Preparation of $Na_2S{\cdot}9H_2O$ Solution: Add $Na_2S{\cdot}9H_2O$ to distilled/deionized water and bring volume to 10.0mL. Mix thoroughly. Autoclave under 100% N_2 for 15 min at 15 psi pressure–121°C. Cool to room temperature.

Glucose Solution:
Composition per 50mL:

Glucose	10.0g

Preparation of Glucose Solution: Add glucose to distilled/deionized water and bring volume to 50.0mL. Mix thoroughly. Filter sterilize.

Preparation of Medium: Prepare and dispense medium under 80% N_2 + 20% CO_2 gas mixture. Add components, except glucose solution, $NaHCO_3$ solution, and $Na_2S{\cdot}9H_2O$ solution, to distilled/deionized water and bring volume to 1.0L. Mix thoroughly. Adjust pH to 7.0. Sparge with 80% N_2 + 20% CO_2. Autoclave for 15 min at 15 psi pressure–121°C. Cool to 25°C. Aseptically and anaerobically add 18.0mL sterile glucose solution, 10.0mL sterile $Na_2S{\cdot}9H_2O$ solution, and 40.0mL sterile $NaHCO_3$ solution. Mix thoroughly. Adjust pH to 7.3. Aseptically and anaerobically distribute into tubes or bottles.

Use: For the cultivation of *Fusibacter paucivorans.*

Fusobacterium Medium

Composition per liter:

Agar	15.0g
Pancreatic digest of casein	15.0g
Glucose	5.0g
NaCl	5.0g
Yeast extract	5.0g
L-Cysteine	0.75g
Crystal Violet	0.01g
Bovine serum	50.0mL
Streptomycin solution	10.0mL

pH 7.2 ± 0.2 at 25°C

Streptomycin Solution:
Composition per 10.0mL:

Streptomycin	0.01g

Preparation of Streptomycin Solution: Add streptomycin to distilled/deionized water and bring volume to 10.0mL. Mix thoroughly. Filter sterilize.

Preparation of Medium: Add components, except bovine serum and streptomycin solution, to distilled/deionized water and bring volume to 940.0mL. Mix thoroughly. Gently heat and bring to boiling. Autoclave for 15 min at 15 psi pressure–121°C. Cool to 45°–50°C. Aseptically add 50.0mL of sterile bovine serum and 10.0mL of sterile streptomycin solution. Mix thoroughly. Pour into sterile Petri dishes or distribute into sterile tubes.

Use: For the cultivation of *Fusobacterium* species.

Fusobacterium necrophorum Medium

Composition per 500.0mL:

Pancreatic digest of casein	16.0g
Agar	8.3g
Biosate	4.0g
$MgSO_4$	2.5g
Na_2HPO_4	2.5g
Thiotone	2.0g
Glucose	0.5g
Egg yolk emulsion, 50%	45.0mL
Crystal Violet solution	25.0mL
Phenylethyl alcohol solution	1.35mL

pH 7.3 ± 0.2 at 25°C

Egg Yolk Emulsion, 50%:
Composition per 100.0mL:

Chicken egg yolks	11
Whole chicken egg	1
NaCl (0.9% solution)	50.0mL

Preparation of Egg Yolk Emulsion, 50%: Soak eggs with 1:100 dilution of saturated mercuric chloride solution for 1 min. Crack 11 eggs, separating yolks from whites. Mix egg yolks with 1 chicken egg. Beat to form emulsion. Combine 50.0mL of egg yolk emulsion and 50.0mL of 0.9% NaCl solution. Mix thoroughly.

Crystal Violet Solution:
Composition per 25.0mL:

Crystal Violet	0.0115g

Preparation of Crystal Violet Solution: Aseptically add Crystal Violet to sterile distilled/deionized water and bring volume to 25.0mL. Mix thoroughly.

Phenylethyl Alcohol Solution:
Composition per 100.0mL:

Phenylethyl alcohol	0.27g

Preparation of Phenylethyl Alcohol Solution: Aseptically add phenylethyl alcohol to sterile distilled/deionized water and bring volume to 100.0mL. Mix thoroughly.

Preparation of Medium: Add components—except egg yolk emulsion, 50%, Crystal Violet solution, and phenylethyl alcohol solution—to distilled/deionized water and bring volume to 428.65mL. Mix thoroughly. Gently heat and bring to boiling. Autoclave for 15 min at 15 psi pressure–121°C. Cool to 45°–50°C. Aseptically add 45.0mL of sterile egg yolk emulsion, 50%, 25.0mL of Crystal Violet solution, and 1.35mL of sterile phenylethyl alcohol solution. Mix thoroughly. Pour into sterile Petri dishes or distribute into sterile tubes.

Use: For the isolation and cultivation of *Fusobacterium necrophorum*.

FWM Medium

Composition per 1013.0mL:

Solution A	940.0mL
Solution E ($NaHCO_3$ solution)	50.0mL
Solution F (Substrate solution)	10.0mL
Solution G ($Na_2S{\cdot}9H_2O$ solution)	10.0mL
Solution B (Trace elements solution Sl-10)	1.0mL
Solution C (Seven vitamin solution)	1.0mL
Solution D (Selenite-tungstate solution)	1.0mL

pH 7.2–7.4 at 25°C

Solution A:

Composition per 940.0mL:

NaCl	1.0g
KCl	0.5g
$MgCl_2 \cdot 6H_2O$	0.4g
NH_4Cl	0.25g
KH_2PO_4	0.2g
$CaCl_2 \cdot 2H_2O$	0.15g
Resazurin	0.5mg

Preparation of Solution A: Prepare and dispense under 80% N_2 + 20% CO_2. Add components to distilled/deionized water and bring volume to 940.0mL. Mix thoroughly. Autoclave for 15 min at 15 psi pressure–121°C.

Solution B (Trace Elements Solution SL-10):

Composition per liter:

$FeCl_2 \cdot 4H_2O$	1.5g
$CoCl_2 \cdot 6H_2O$	190.0mg
$MnCl_2 \cdot 4H_2O$	100.0mg
$ZnCl_2$	70.0mg
$Na_2MoO_4 \cdot 2H_2O$	36.0mg
$NiCl_2 \cdot 6H_2O$	24.0mg
H_3BO_3	6.0mg
$CuCl_2 \cdot 2H_2O$	2.0mg
HCl (25% solution)	10.0mL

Preparation of Solution B (Trace Elements Solution SL-10): Prepare and dispense under 100% N_2. Add $FeCl_2 \cdot 4H_2O$ to 10.0mL of HCl solution. Mix thoroughly. Add distilled/deionized water and bring volume to 1.0L. Add remaining components. Mix thoroughly. Autoclave for 15 min at 15 psi pressure–121°C.

Solution C (Seven Vitamin Solution):

Composition per liter:

Pyridoxine·HCl	300.0mg
Nicotinic acid	200.0mg
Thiamine·HCl	200.0mg
Calcium DL-pantothenate	100.0mg
Cyanocobalamine	100.0mg
p-Aminobenzoic acid	80.0mg
D(+) Biotin	20.0mg

Preparation of Solution C (Seven Vitamin Solution): Add components to distilled/deionized water and bring volume to 1.0L. Mix thoroughly. Filter sterilize. Sparge with 100% N_2.

Solution D (Selenite-Tungstate Solution):

Composition per liter:

NaOH	0.5g
$Na_2WO_4 \cdot 2H_2O$	4.0mg
$Na_2SeO_3 \cdot 5H_2O$	3.0mg

Preparation of Solution D (Selenite-Tungstate Solution): Add components to distilled/deionized water and bring volume to 1.0L. Mix thoroughly. Filter sterilize. Sparge with 100% N_2.

Solution E ($NaHCO_3$ Solution):

Composition per 50.0mL:

$NaHCO_3$	2.5g

Preparation of Solution E ($NaHCO_3$ Solution): Add $NaHCO_3$ to distilled/deionized water and bring volume to 50.0mL. Mix thoroughly. Gas under 100% N_2. Autoclave for 15 min at 15 psi pressure–121°C.

Solution F (Substrate Solution):

Composition per 10.0mL:

Sodium-DL-3-hydroxybutyrate	1.5g

Preparation of Solution F (Substrate Solution): Add sodium-DL-3-hydroxybutyrate to distilled/deionized water and bring volume to 50.0mL. Mix thoroughly. Gas under 100% N_2. Autoclave for 15 min at 15 psi pressure–121°C.

Solution G ($Na_2S \cdot 9H_2O$ Solution):

Composition per 10.0mL:

$Na_2S \cdot 9H_2O$	0.3g

Preparation of Solution G ($Na_2S \cdot 9H_2O$ Solution): Add $Na_2S \cdot 9H_2O$ to distilled/deionized water and bring volume to 10.0mL. Mix thoroughly. Gas under 100% N_2. Autoclave for 15 min at 15 psi pressure–121°C.

Preparation of Medium: To 940.0mL of sterile solution A, aseptically and anaerobically add 1.0mL of sterile solution B, 1.0mL of sterile solution C, 1.0mL of sterile solution D, 50.0mL of sterile solution E, 10.0mL of sterile solution F, and 10.0mL of sterile solution G. Mix thoroughly. Aseptically and anaerobically distribute into sterile tubes or flasks.

Use: For the cultivation and maintenance of *Clostridium homopropionicum* and *Desulfococcus biacutus.*

FWM Medium

Composition per 1013.0mL:

Solution A	940.0mL
Solution E ($NaHCO_3$ solution)	50.0mL
Solution F (Substrate solution)	10.0mL
Solution G ($Na_2S \cdot 9H_2O$ solution)	10.0mL
Solution B (Trace elements solution Sl-10)	1.0mL
Solution C (Seven vitamin solution)	1.0mL
Solution D (Selenite-tungstate solution)	1.0mL

pH 7.2–7.4 at 25°C

Solution A:

Composition per 940.0mL:

NaCl	1.0g
KCl	0.5g
$MgCl_2 \cdot 6H_2O$	0.4g
NH_4Cl	0.25g
KH_2PO_4	0.2g
$CaCl_2 \cdot 2H_2O$	0.15g
Resazurin	0.5mg

Preparation of Solution A: Prepare and dispense under 80% N_2 + 20% CO_2. Add components to distilled/deionized water and bring volume to 940.0mL. Mix thoroughly. Autoclave for 15 min at 15 psi pressure–121°C.

Solution B (Trace Elements Solution SL-10):

Composition per liter:

$FeCl_2 \cdot 4H_2O$	1.5g
$CoCl_2 \cdot 6H_2O$	190.0mg
$MnCl_2 \cdot 4H_2O$	100.0mg
$ZnCl_2$	70.0mg
$Na_2MoO_4 \cdot 2H_2O$	36.0mg
$NiCl_2 \cdot 6H_2O$	24.0mg
H_3BO_3	6.0mg
$CuCl_2 \cdot 2H_2O$	2.0mg
HCl (25% solution)	10.0mL

Preparation of Solution B (Trace Elements Solution SL-10): Prepare and dispense under 100% N_2. Add $FeCl_2 \cdot 4H_2O$ to 10.0mL of HCl solution. Mix thoroughly. Add distilled/deionized water and bring volume to 1.0L. Add remaining components. Mix thoroughly. Autoclave for 15 min at 15 psi pressure–121°C.

Solution C (Seven Vitamin Solution):
Composition per liter:
Pyridoxine·HCl 300.0mg
Nicotinic acid 200.0mg
Thiamine·HCl 200.0mg
Calcium DL-pantothenate 100.0mg
Cyanocobalamine 100.0mg
p-Aminobenzoic acid 80.0mg
D(+)-Biotin 20.0mg

Preparation of Solution C (Seven Vitamin Solution): Add components to distilled/deionized water and bring volume to 1.0L. Mix thoroughly. Filter sterilize. Sparge with 100% N_2.

Solution D (Selenite-Tungstate Solution):
Composition per liter:
NaOH 0.5g
$Na_2WO_4 \cdot 2H_2O$ 4.0mg
$Na_2SeO_3 \cdot 5H_2O$ 3.0mg

Preparation of Solution D (Selenite-Tungstate Solution): Add components to distilled/deionized water and bring volume to 1.0L. Mix thoroughly. Filter sterilize. Sparge with 100% N_2.

Solution E ($NaHCO_3$ Solution):
Composition per 50.0mL:
$NaHCO_3$ 2.5g

Preparation of Solution E ($NaHCO_3$ Solution): Add $NaHCO_3$ to distilled/deionized water and bring volume to 50.0mL. Mix thoroughly. Gas under 100% N_2. Autoclave for 15 min at 15 psi pressure–121°C.

Solution F (Substrate Solution):
Composition per 10.0mL:
Xylan or xylose 2.0g

Preparation of Solution F (Substrate Solution): Add 2.0g of xylan or 2.0g of xylose to distilled/deionized water and bring volume to 10.0mL. Mix thoroughly. Gas under 100% N_2. Autoclave for 15 min at 15 psi pressure–121°C.

Solution G ($Na_2S \cdot 9H_2O$ Solution):
Composition per 10.0mL:
$Na_2S \cdot 9H_2O$ 0.3g

Preparation of Solution G ($Na_2S \cdot 9H_2O$ Solution): Add $Na_2S \cdot 9H_2O$ to distilled/deionized water and bring volume to 10.0mL. Mix thoroughly. Gas under 100% N_2. Autoclave for 15 min at 15 psi pressure–121°C.

Preparation of Medium: To 940.0mL of sterile solution A, aseptically and anaerobically add 1.0mL of sterile solution B, 1.0mL of sterile solution C, 1.0mL of sterile solution D, 50.0mL of sterile solution E, 10.0mL of sterile solution F, and 10.0mL of sterile solution G. Mix thoroughly. Aseptically and anaerobically distribute into sterile tubes or flasks.

Use: For the cultivation and maintenance of *Desulfococcus biacutus.*

FWN Medium, Modified with Xylan

Composition per 1020.0mL:
Solution A 950.0mL
Solution C ($NaHCO_3$ solution) 50.0mL
Solution D ($Na_2S \cdot 9H_2O$ solution) 10.0mL
Solution B (Wolfe's vitamin solution) 10.0mL

Solution A:
Composition per 950.0mL:
Xylan 2.0g
NaCl 1.0g
KCl 0.5g
$MgCl_2 \cdot 6H_2O$ 0.4g
NH_4Cl 0.25g
KH_2PO_4 .... 0.2g
$CaCl_2 \cdot 2H_2O$ 0.15g
Resazurin 0.5mg
Modified Wolfe's mineral solution 10.0mL

Modified Wolfe's Mineral Solution:
Composition per liter:
$MgSO_4 \cdot 7H_2O$ 3.0g
Nitrilotriacetic acid 1.5g
NaCl 1.0g
$MnSO_4 \cdot H_2O$ 0.5g
$CaCl_2$ 0.1g
$CoCl_2 \cdot 6H_2O$ 0.1g
$FeSO_4 \cdot 7H_2O$ 0.1g
$ZnSO_4 \cdot 7H_2O$ 0.1g
$AlK(SO_4)_2 \cdot 12H_2O$ 0.01g
$CuSO_4 \cdot 5H_2O$ 0.01g
H_3BO_3 .... 0.01g
$Na_2MoO_4 \cdot 2H_2O$ 0.01g
Na_2SeO_3 .... 0.01g
$NaWO_4 \cdot 2H_2O$ 0.01g
$NiCl_2 \cdot 6H_2O$ 0.01g

Preparation of Modified Wolfe's Mineral Solution: Add nitrilotriacetic acid to 500.0mL of distilled/deionized water. Adjust pH to 6.5 with KOH. Add remaining components one at a time. Add distilled/deionized water to 1.0L. Adjust pH to 6.8.

Preparation of Solution A: Prepare and dispense under 80% N_2 + 20% CO_2. Add components to distilled/deionized water and bring volume to 950.0mL. Mix thoroughly. Autoclave for 15 min at 15 psi pressure–121°C.

Solution B (Wolfe's Vitamin Solution):
Composition per liter:
Pyridoxine·HCl 10.0mg
p-Aminobenzoic acid 5.0mg
Lipoic acid 5.0mg
Nicotinic acid 5.0mg
Riboflavin 5.0mg
Thiamine·HCl 5.0mg
Calcium DL-pantothenate 5.0mg
Biotin 2.0mg
Folic acid 2.0mg
Vitamin B_{12} 0.1mg

Preparation of Solution B (Wolfe's Vitamin Solution): Add components to distilled/deionized water and bring volume to 1.0L. Mix thoroughly. Sparge with 100% N_2. Filter sterilize.

Solution C ($NaHCO_3$ Solution):
Composition per 50.0mL:
$NaHCO_3$ 2.5g

Preparation of Solution C ($NaHCO_3$ Solution): Add $NaHCO_3$ to distilled/deionized water and bring volume to 50.0mL. Mix thoroughly. Sparge with 100% N_2. Autoclave for 15 min at 15 psi pressure–121°C.

Solution D ($Na_2S \cdot 9H_2O$ Solution):
Composition per 10.0mL:
$Na_2S \cdot 9H_2O$ 0.3g

Preparation of Solution D ($Na_2S \cdot 9H_2O$ Solution): Add $Na_2S \cdot 9H_2O$ to distilled/deionized water and bring volume to 10.0mL. Mix thoroughly. Sparge with 100% N_2. Autoclave for 15 min at 15 psi pressure–121°C. Before use, neutralize to pH 7.0 with sterile HCl.

Preparation of Medium: To 950.0mL of sterile solution A, aseptically and anaerobically add 10.0mL of sterile solution B, 50.0mL of sterile solution C, and 10.0mL of sterile solution D. Mix thoroughly. Aseptically and anaerobically distribute into sterile tubes or flasks under 100% N_2.

Use: For the cultivation of *Cytophaga xylanolytica.*

FX A Broth

Composition per liter:

Pancreatic digest of casein	10.0g
Yeast extract	2.0g
$MgSO_4 \cdot 7H_2O$	1.0g

pH 7.0 ± 0.2 at 25°C

Preparation of Medium: Add components to distilled/deionized water and bring volume to 1.0L. Mix thoroughly. Distribute into tubes or flasks. Autoclave for 15 min at 15 psi pressure–121°C.

Use: For the isolation and cultivation of *Cytophaga* species, *Herpetosiphon* species, *Saprospira* species, and *Flexithrix* species.

GC Agar with Streptomycin and Chloramphenicol

Composition per 1030.0mL:

GC agar base, 2×	500.0mL
Hemoglobin solution	500.0mL
Supplement solution	10.0mL
Streptomcyin solution	10.0mL
Chloramphenicol solution	10.0mL

pH 7.2 ± 0.2 at 25°C

GC Agar Base, 2×:
Composition per 500.0mL:

Agar	10.0g
Pancreatic digest of casein	7.5g
Peptic digest of animal tissue	7.5g
NaCl	5.0g
K_2HPO_4	4.0g
Cornstarch	1.0g
KH_2PO_4	1.0g

Source: GC agar base is available as a premixed powder from BD Diagnostic Systems. This base may be replaced by GC medium base available from BD Diagnostic Systems.

Preparation of GC Agar Base, 2×: Add components to distilled/deionized water and bring volume to 500.0mL. Mix thoroughly. Gently heat until boiling. Autoclave for 15 min at 15 psi pressure–121°C. Cool to 45°–50°C.

Hemoglobin Solution:
Composition per 500.0mL:

Bovine hemoglobin	10.0g

Preparation of Hemoglobin Solution: Add bovine hemoglobin to distilled/deionized water and bring volume to 500.0mL. Mix thoroughly. Autoclave for 15 min at 15 psi pressure–121°C. Cool to 45°–50°C.

Supplement Solution:
Composition per liter:

Glucose	100.0g
L-Cysteine·HCl	25.9g
L-Glutamine	10.0g
L-Cystine	1.1g
Adenine	1.0g
Nicotinamide adenine dinucleotide	0.25g
Vitamin B_{12}	0.1g
Thiamine pyrophosphate	0.1g
Guanine·HCl	0.03g
$Fe(NO_3)_3 \cdot 6H_2O$	0.02g
p-Aminobenzoic acid	0.013g
Thiamine·HCl	3.0mg

Source: The supplement solution (IsoVitaleX enrichment) is available from BD Diagnostic Systems. This enrichment may be replaced by supplement VX from BD Diagnostic Systems.

Preparation of Supplement Solution: Add components to distilled/deionized water and bring volume to 1.0L. Mix thoroughly. Filter sterilize.

Streptomycin Solution:
Composition per 10.0mL:

Streptomycin	0.25g

Preparation of Streptomycin Solution: Add streptomycin to distilled/deionized water and bring volume to 10.0mL. Mix thoroughly. Filter sterilize.

Chloramphenicol Solution:
Composition per 10.0mL:

Chloramphenicol	25.0mg

Preparation of Chloramphenicol Solution: Add chloramphenicol to distilled/deionized water and bring volume to 10.0mL. Mix thoroughly. Filter sterilize.

Preparation of Medium: To 500.0mL of sterile GC agar base, aseptically add 500.0mL of sterile hemoglobin solution at 45°–50°C. Mix thoroughly. Aseptically add 10.0mL of sterile supplement solution, 10.0mL of sterile ampicillin solution, and 10.0mL of sterile tetracycline solution. Mix thoroughly. Pour into sterile Petri dishes or distribute into sterile tubes.

Use: For the cultivation of *Azorhizophilus paspali.*

Gelatin Agar

Composition per liter:

Agar	15.0g
Gelatin	15.0g
Peptone	4.0g
Yeast extract	1.0g

pH 7.2 ± 0.2 at 25°C

Preparation of Medium: Add components to distilled/deionized water and bring volume to 1.0L. Mix thoroughly. Gently heat and bring to boiling. Distribute into tubes or flasks. Autoclave for 15 min at 15 psi pressure–121°C. Pour into sterile Petri dishes or leave in tubes.

Use: For the cultivation of a variety of heterotrophic bacteria based upon their utilization of gelatin.

General Salts Medium for Estuarine Methanogens

Composition per 410.8mL:

Agar	8.0g
NaCl	3.6g
$NaHCO_3$	2.0g
Complete salts solution	200.0mL
Cysteine-sulfide reducing agent	16.0mL
Wolfe's mineral solution	4.0mL
Vitamin solution	4.0mL
Yeast extract–Trypticase solution	4.0mL
Sodium acetate (25% solution)	2.0mL

$Fe(NH_4)_2SO_4$ (0.2% solution) 0.4mL
Resazurin (0.1% solution) 0.4mL

pH 7.0 ± 0.2 at 25°C

Complete Salts Solution:
Composition per liter:

$MgSO_4 \cdot 7H_2O$	6.9g
$MgCl_2 \cdot 6H_2O$	5.5g
KCl	0.67g
NH_4Cl	0.5g
$CaCl_2 \cdot 2H_2O$	0.28g
K_2HPO_4	0.28g

Preparation of Complete Salts Solution: Add components to distilled/deionized water and bring volume to 1.0L. Mix thoroughly.

Cysteine-Sulfide Reducing Agent:
Composition per 400.0mL:

L-Cysteine·HCl·H_2O	5.0g
Na_2S (12.5% solution)	40.0mL
NaOH (1*N* solution)	30.0mL

Preparation of Cysteine-Sulfide Reducing Agent: Add distilled/deionized water to a 500.0mL round-bottomed flask. Add freshly prepared NaOH solution. Gently heat and bring to boiling under 100% N_2. Remove gassing probe. Add L-cysteine·HCl·H_2O. Add freshly prepared Na_2S solution. Renew gassing for several minutes. Cap with rubber stoppers. Distribute into 8.0mL/18mm Hungate tubes.

Yeast Extract-Trypticase Solution:
Composition per 100.0mL:

Yeast extract	20.0g
Pancreatic digest of casein	20.0g

Preparation of Yeast Extract-Trypticase Solution: Add components to distilled/deionized water and bring volume to 100.0mL. Mix thoroughly.

Wolfe's Mineral Solution:
Composition per liter:

$MgSO_4 \cdot 7H_2O$	3.0g
Nitriloacetic acid	1.5g
NaCl	1.0g
$MnSO_4 \cdot H_2O$	0.5g
$FeSO_4 \cdot 7H_2O$	0.1g
$CoCl_2 \cdot 6H_2O$	0.1g
$CaCl_2$	0.1g
$ZnSO_4 \cdot 7H_2O$	0.1g
$CuSO_4 \cdot 5H_2O$	0.01g
$AlK(SO_4)_2 \cdot 12H_2O$	0.01g
H_3BO_3	0.01g
$Na_2MoO_4 \cdot 2H_2O$	0.01g

Preparation of Wolfe's Mineral Solution: Add nitrilotriacetic acid to 500.0mL of distilled/deionized water. Dissolve by adjusting pH to 6.5 with KOH. Add remaining components. Add distilled/deionized water to 1.0L.

Preparation of Medium: Add components, except cysteine-sulfide reducing agent, to distilled/deionized water and bring volume to 410.8mL. Mix thoroughly. Adjust pH to 7.0. Gently heat and bring to boiling under 80% N_2 + 20% CO_2. Add cysteine-sulfide reducing agent. Continue boiling until resazurin turns colorless, indicating reduction. Distribute anaerobically into culture tubes with aluminum crimp seals. Autoclave for 15 min at 15 psi pressure–121°C.

Use: For the cultivation and maintenance of *Methanococcus deltae, Methanococcus vannielii, Methanococcus voltae, Methanogenium cariaci, Methanogenium marisnigri,* and *Methanogenium olentangyi*.

Geo Medium

Composition per liter:

Agar	15.0g
$CaCO_3$	1.0g
Glucose	1.0g
Starch, soluble	1.0g
Yeast extract	1.0g

pH 7.2 ± 0.2 at 25°C

Preparation of Medium: Add components to distilled/deionized water and bring volume to 1.0L. Mix thoroughly. Gently heat and bring to boiling. Adjust pH to 7.2. Distribute into tubes or flasks. Autoclave for 15 min at 15 psi pressure–121°C. Pour into sterile Petri dishes or leave in tubes.

Use: For the cultivation of *Geodermatophilus obscurus*.

Geobacter Medium (DSMZ Medium 579)

Composition per liter:

Fe(III) citrate	13.7g
$NaHCO_3$	2.5g
Na-acetate	2.5g
NH_4Cl	1.5g
NaH_2PO_4	0.6g
KCl	0.1g
$Na_2WO_4 \cdot 2H_2O$	0.25mg
Vitamin solution	10.0mL
Trace elements solution	10.0mL

pH 6.7-7.0 at 25°C

Trace Elements Solution:
Composition per liter:

$MgSO_4 \cdot 7H_2O$	3.0g
Nitrilotriacetic acid	1.5g
NaCl	1.0g
$MnSO_4 \cdot 2H_2O$	0.5g
$CoSO_4 \cdot 7H_2O$	0.18g
$ZnSO_4 \cdot 7H_2O$	0.18g
$CaCl_2 \cdot 2H_2O$	0.1g
$FeSO_4 \cdot 7H_2O$	0.1g
$NiCl_2 \cdot 6H_2O$	0.025g
$KAl(SO_4)_2 \cdot 12H_2O$	0.02g
H_3BO_3	0.01g
$Na_2MoO_4 \cdot 4H_2O$	0.01g
$CuSO_4 \cdot 5H_2O$	0.01g
$Na_2SeO_3 \cdot 5H_2O$	0.3mg

Preparation of Trace Elements Solution: Add nitrilotriacetic acid to 500.0mL of distilled/deionized water. Dissolve by adjusting pH to 6.5 with KOH. Add remaining components. Add distilled/deionized water to 1.0L. Mix thoroughly.

Vitamin Solution:
Composition per liter:

Pyridoxine-HCl	10.0mg
Thiamine-HCl·$2H_2O$	5.0mg
Riboflavin	5.0mg
Nicotinic acid	5.0mg
D-Ca-pantothenate	5.0mg
p-Aminobenzoic acid	5.0mg
Lipoic acid	5.0mg
Biotin	2.0mg
Folic acid	2.0mg
Vitamin B_{12}	0.1mg

Preparation of Vitamin Solution: Add components to distilled/deionized water and bring volume to 1.0L. Mix thoroughly. Sparge with 80% H_2 + 20% CO_2. Filter sterilize.

Preparation of Medium: Prepare and dispense medium under an oxygen-free 80% N_2 + 20% CO_2 gas mixture. Dissolve ferric citrate in 900.0mL distilled/deionized water by heating and adjust to pH 6.0. Add other components, except vitamin solution, to distilled/deionized water and bring volume to 990.0mL. Mix thoroughly. Sparge with 80% N_2 + 20% CO_2. Autoclave for 15 min at 15 psi pressure–121°C. Cool to 25°C. Aseptically and anaerobically add 10.0mL vitamin solution. Mix thoroughly. Aseptically and anaerobically distribute into sterile tubes or flasks.

Use: For the cultivation of *Geobacter metallireducens* and *Geobacter grbiciae.*

Geobacter Medium (DSMZ Medium 826)

Composition per 1050mL:

$NaHCO_3$	2.5g
NH_4Cl	1.5g
Na-acetate	0.82g
Na_2HPO_4	0.6g
KCl	0.1g
Resazurin	0.5mg
Na-fumarate solution	50.0mL
Trace elements solution	10.0mL
Vitamin solution	10.0mL
Selenite-tungstate solution	1.0mL

pH 6.8 ± 0.2 at 25°C

Na-Fumarate Solution:

Composition per 100.0mL:

Na-fumarate	16.0g

Preparation of Na-Fumarate Solution: Add Na-fumarate to distilled/deionized water and bring volume to 100.0mL. Mix thoroughly. Sparge with 100% N_2. Autoclave for 15 min at 15 psi pressure–121°C.

Selenite-Tungstate Solution

Composition per liter:

NaOH	0.5g
$Na_2WO_4 \cdot 2H_2O$	4.0mg
$Na_2SeO_3 \cdot 5H_2O$	3.0mg

Preparation of Selenite-Tungstate Solution: Add components to distilled/deionized water and bring volume to 1.0L. Mix thoroughly. Sparge with 100% N_2. Filter sterilize.

Trace Elements Solution:

Composition per liter:

$MgSO_4 \cdot 7H_2O$	3.0g
Nitrilotriacetic acid	1.5g
NaCl	1.0g
$MnSO_4 \cdot 2H_2O$	0.5g
$CoSO_4 \cdot 7H_2O$	0.18g
$ZnSO_4 \cdot 7H_2O$	0.18g
$CaCl_2 \cdot 2H_2O$	0.1g
$FeSO_4 \cdot 7H_2O$	0.1g
$NiCl_2 \cdot 6H_2O$	0.025g
$KAl(SO_4)_2 \cdot 12H_2O$	0.02g
H_3BO_3	0.01g
$Na_2MoO_4 \cdot 4H_2O$	0.01g
$CuSO_4 \cdot 5H_2O$	0.01g
$Na_2SeO_3 \cdot 5H_2O$	0.3mg

Preparation of Trace Elements Solution: Add nitrilotriacetic acid to 500.0mL of distilled/deionized water. Dissolve by adjusting pH to 6.5 with KOH. Add remaining components. Add distilled/deionized water to 1.0L. Mix thoroughly.

Vitamin Solution:

Composition per liter:

Pyridoxine-HCl	10.0mg
Thiamine-HCl·$2H_2O$	5.0mg
Riboflavin	5.0mg
Nicotinic acid	5.0mg
D-Ca-pantothenate	5.0mg
p-Aminobenzoic acid	5.0mg
Lipoic acid	5.0mg
Biotin	2.0mg
Folic acid	2.0mg
Vitamin B_{12}	0.1mg

Preparation of Vitamin Solution: Add components to distilled/deionized water and bring volume to 1.0L. Mix thoroughly. Sparge with 80% H_2 + 20% CO_2. Filter sterilize.

Preparation of Medium: Prepare and dispense medium under 80% N_2 + 20% CO_2 gas mixture. Add components, except $NaHCO_3$ and fumarate solution, to distilled/deionized water and bring volume to 1.0L. Mix thoroughly. Gently heat and bring to boiling. Boil for 5 min. Cool to room temperature while sparging with 80% N_2 + 20% CO_2. Add the $NaHCO_3$. Equilibarte with the 80% N_2 + 20% CO_2 gas mixture to reach a pH of 6.8. Distribute into anaerobe tubes under 80% N_2 + 20% CO_2. Autoclave for 15 min at 15 psi pressure–121°C. Cool to room temperature. Aseptically and anaerobically add 0.5mL Na-fumarate solution per 10.0mL medium. The pH should be 6.8.

Use: For the cultivation of *Geobacter sulfurreducens, Geobacter bremensis (Geobacter* sp.*), Geobacter chapellei, Geobacter hydrogenophilus,* and *Geothrix fermentans.*

Geodermatophilus obscurus Medium

Composition per liter:

Agar	20.0g
Malt extract, purified solids	15.0g
Starch, soluble	10.0g
Sucrose	10.0g
Yeast extract	5.0g
$CaCO_3$	2.0g

Preparation of Medium: Add components to distilled/deionized water and bring volume to 1.0L. Mix thoroughly. Gently heat and bring to boiling. Distribute into tubes or flasks. Autoclave for 15 min at 15 psi pressure–121°C. Pour into sterile Petri dishes or leave in tubes.

Use: For the isolation and cultivation of *Geodermatophilus obscurus.*

George's Medium, Modified

Composition per liter:

Agar	15.0g
Peptone	1.0g
KNO_3	0.2g
K_2HPO_4	0.02g
$MgSO_4 \cdot 7H_2O$	0.02g
Ferric citrate	0.035mg

Preparation of Medium: Add components to tap water and bring volume to 1.0L. Mix thoroughly. Gently heat and bring to boiling. Distribute into tubes. Autoclave for 15 min at 15 psi pressure–121°C. Pour into sterile Petri dishes or leave in tubes.

Use: For the cultivation of a variety of algae.

GFY Agar

Composition per liter:

Agar	15.0g
Glucose	5.0g
Fructose	5.0g
Yeast extract	5.0g
$CaCO_3$	3.0g
Vitamin B_{12}	2.0mg

Preparation of Medium: Add components to distilled/deionized water and bring volume to 1.0L. Mix thoroughly. Gently heat and bring to boiling. Distribute into tubes or flasks. Autoclave for 15 min at 15 psi pressure–121°C. Pour into sterile Petri dishes or leave in tubes.

Use: For the cultivation and maintenance of *Flexibacter* species.

Gisa Agar

Composition per liter:

Agar	18.0g
$(NH_4)_2SO_4$	5.0g
Calcium glucoisosaccharinate	2.0g
KH_2PO_4	1.0g
$MgSO_4·7H_2O$	0.5g
NaCl	0.1g
$CaCl_2·2H_2O$	0.1g
Inositol	2.0mg
KI	1.0mg
H_3BO_3	0.5mg
$ZnSO_4·7H_2O$	0.4mg
$MnSO_4·4H_2O$	0.4mg
Thiamine·HCl	0.4mg
Pyroxidine·HCl	0.4mg
Niacin	0.4mg
Calcium pantothenate	0.4mg
p-Aminobenzoic acid	0.2mg
Riboflavin	0.2mg
$FeCl_3$	0.2mg
$Na_2MoO_4·4H_2O$	0.2mg
$CuSO_4·5H_2O$	0.04mg
Folic acid	2.0μg
Biotin	2.0μg

pH 5.5 ± 0.2 at 25°C

Preparation of Medium: Add components to distilled/deionized water and bring volume to 1.0L. Mix thoroughly. Gently heat and bring to boiling. Distribute into tubes or flasks. Autoclave for 15 min at 15 psi pressure–121°C. Pour into sterile Petri dishes or leave in tubes.

Use: For the cultivation and maintenance of *Ancylobacter aquaticus, Pseudomonas* species, *Xanthobacter autotrophicus,* and *Xanthobacter* species.

Gliding Medium

Composition per liter:

Pancreatic digest of casein	0.5g
Yeast extract	0.5g

pH 7.0 ± 0.2 at 25°C

Preparation of Medium: Add components to distilled/deionized water and bring volume to 1.0L. Mix thoroughly. Distribute into tubes or flasks. Autoclave for 15 min at 15 psi pressure–121°C.

Use: For the cultivation and maintenance of the gliding bacteria *Cytophaga* species and *Flexibacter columnaris*.

Gluconacetobacter johannae Gluconacetobacter azotocaptans Medium (DSMZ Medium 920)

Composition per liter:

K_2HPO_4	4.81g
MES buffer	4.4g
Yeast extract	2.7g
Glucose	2.7g
Mannitol	1.8g
KH_2PO_4	0.65g

pH 6.7 ± 0.2 at 25°C

Preparation of Medium: Add components to distilled/deionized water and bring volume to 1.0L. Mix thoroughly. Distribute into tubes or flasks. Autoclave for 15 min at 15 psi pressure–121°C.

Use: For the cultivation of *Gluconacetobacter azotocaptans* and *Gluconacetobacter johannae.*

Gluconate Peptone Broth

Composition per liter:

Potassium gluconate	40.0g
Casein peptone	1.5g
K_2HPO_4	1.0g
Yeast extract	1.0g

pH 7.0 ± 0.2 at 25°C

Preparation of Medium: Add components to distilled/deionized water and bring volume to 1.0L. Mix thoroughly. Distribute into tubes or flasks. Autoclave for 15 min at 15 psi pressure–121°C.

Use: For the cultivation and differentiation of Gram-negative bacteria based on their ability to oxidize gluconate to 2-ketogluconate. For the differentiation of fluorescent *Pseudomonas* species. After inoculation with bacteria and 48 hr of growth in this medium, Benedict's reagent is added. Bacteria that produce the reducing sugar 2-ketogluconate turn the reagent yellow-orange to orange-red.

Glucose Agar, 9K

Composition per liter:

$(NH_4)_2SO_4$	3.0g
KH_2PO_4	0.5g
$MgSO_4·7H_2O$	0.5g
KCl	0.1g
$Ca(NO_3)_2$	0.0125g
$FeSO_4·7H_2O$	0.01mg
Agar solution	500.0mL
Glucose solution	100.0mL

pH 4.5 ± 0.2 at 25°C

Agar Solution:

Composition per 500.0mL:

Agar	15.0g

Preparation of Agar Solution: Add agar to distilled/deionized water and bring volume to 500.0mL. Mix thoroughly. Autoclave for 15 min at 15 psi pressure–121°C. Cool to 55°C.

Glucose Solution:

Composition per 100.0mL:

Glucose	10.0g

Preparation of Glucose Solution: Add glucose to distilled/deionized water and bring volume to 100.0mL. Mix thoroughly. Filter sterilize.

Preparation of Medium: Add components, except agar solution and glucose solution, to distilled/deionized water and bring volume to 400.0mL. Mix thoroughly. Adjust pH to 4.5 with H_2SO_4. Autoclave for 15 min at 15 psi pressure–121°C. Cool to 55°C. Aseptically add 500.0mL of sterile agar solution and 100.0mL of sterile glucose solution. Mix thoroughly. Pour into sterile Petri dishes or distribute into sterile tubes.

Use: For the cultivation of *Thiobacillus acidophilus.*

Glucose Agar with 25%Glucose

Composition per liter:

Glucose	250.0g
Agar	25.0g
Polypeptone	5.0g
Malt extract	3.0g
Yeast extract	3.0g

Preparation of Medium: Add components to distilled/deionized water and bring volume to 1.0L. Mix thoroughly. Gently heat and bring to boiling. Distribute into tubes or flasks. Autoclave for 15 min at 15 psi pressure–121°C. Pour into sterile Petri dishes or leave in tubes.

Use: For the cultivation and maintenance of osmophilic yeasts and bacteria.

Glucose Broth

Composition per 800.0mL:

Agar	10.0g
Beef extract	10.0g
Peptone	10.0g
NaCl	5.0g
Glucose solution	200.0mL

pH 7.0 ± 0.2 at 25°C

Glucose Solution:

Composition per 200.0mL:

Glucose	20.0g

Preparation of Glucose Solution: Add glucose to distilled/deionized water and bring volume to 200.0mL. Mix thoroughly. Filter sterilize.

Preparation of Medium: Add components, except glucose solution, to distilled/deionized water and bring volume to 800.0mL. Mix thoroughly. Gently heat and bring to boiling. Adjust pH to 7.0. Autoclave for 15 min at 15 psi pressure–121°C. Cool to 45°C. Aseptically add 200.0mL of sterile glucose solution. Mix thoroughly. Aseptically distribute into sterile tubes or flasks.

Use: For the cultivation of *Pseudomonas* species.

Glucose Broth, 9K

Composition per liter:

Glucose	10.0g
$(NH_4)_2SO_4$	3.0g
KH_2PO_4	0.5g
$MgSO_4·7H_2O$	0.5g
KCl	0.1g
$Ca(NO_3)_2$	0.0125g
$FeSO_4·7H_2O$	0.01mg
Glucose solution	100.0mL

pH 3.5 ± 0.2 at 25°C

Glucose Solution:

Composition per 100.0mL:

Glucose	10.0g

Preparation of Glucose Solution: Add glucose to distilled/deionized water and bring volume to 100.0mL. Mix thoroughly. Filter sterilize.

Preparation of Medium: Add components, except glucose solution, to distilled/deionized water and bring volume to 900.0mL. Mix thoroughly. Adjust pH to 3.5 with H_2SO_4. Autoclave for 15 min at 15 psi pressure–121°C. Cool to 25°C. Aseptically add 100.0mL of sterile glucose solution. Mix thoroughly. Aseptically distribute into sterile tubes or flasks.

Use: For the cultivation of *Thiobacillus acidophilus.*

Glucose Nitrogen-Free Salt Agar

Composition per liter:

Agar	15.0g
$CaCO_3$	1.0g
K_2HPO_4	1.0g
$MgSO_4·7H_2O$	0.2g
NaCl	0.2g
$FeSO_4·7H_2O$	0.1g
$Na_2MoO_4·2H_2O$	5.0mg
Glucose solution	100.0mL

pH 7.0 ± 0.2 at 25°C

Glucose Solution:

Composition per 100.0mL:

Glucose	10.0g

Preparation of Glucose Solution: Add glucose to distilled/deionized water and bring volume to 100.0mL. Mix thoroughly. Filter sterilize.

Preparation of Medium: Add components, except glucose solution, to distilled/deionized water and bring volume to 900.0mL. Mix thoroughly. Gently heat and bring to boiling. Autoclave for 15 min at 15 psi pressure–121°C. Cool to 45°–50°C. Aseptically add 100.0mL of sterile glucose solution. Mix thoroughly. Pour into sterile Petri dishes or distribute into sterile tubes.

Use: For the cultivation of *Azotobacter* species.

Glucose Nitrogen-Free Salt Solution

Composition per liter:

$CaCO_3$	1.0g
K_2HPO_4	1.0g
$MgSO_4·7H_2O$	0.2g
NaCl	0.2g
$FeSO_4·7H_2O$	0.1g
$Na_2MoO_4·2H_2O$	5.0mg
Glucose solution	100.0mL

pH 7.0 ± 0.2 at 25°C

Glucose Solution:

Composition per 100.0mL:

Glucose	10.0g

Preparation of Glucose Solution: Add glucose to distilled/deionized water and bring volume to 100.0mL. Mix thoroughly. Filter sterilize.

Preparation of Medium: Add components, except glucose solution, to distilled/deionized water and bring volume to 900.0mL. Mix thoroughly. Gently heat and bring to boiling. Autoclave for 15 min at 15 psi pressure–121°C. Cool to 45°–50°C. Aseptically add 100.0mL of sterile glucose solution. Mix thoroughly. Aseptically distribute into sterile tubes or flasks.

Use: For the cultivation of *Azotobacter* species.

Glucose Peptone Agar

Composition per liter:

Peptone....20.0g
Agar....15.0g
Glucose....10.0g
NaCl....5.0g

pH 7.2 ± 0.2 at 25°C

Preparation of Medium: Add components to distilled/deionized water and bring volume to 1.0L. Mix thoroughly. Gently heat and bring to boiling. Distribute into tubes or flasks. Autoclave for 15 min at 15 psi pressure–121°C. Pour into sterile Petri dishes or leave in tubes.

Use: For the cultivation of *Agrobacterium* species.

Glucose Yeast Chalk Agar

Composition per liter:

Chalk....40.0g
Agar....15.0g
Glucose....5.0g
Yeast extract....5.0g

Preparation of Medium: Add components to distilled/deionized water and bring volume to 1.0L. Mix thoroughly. Gently heat and bring to boiling. Distribute into tubes or flasks. Autoclave for 15 min at 15 psi pressure–121°C. Pour into sterile Petri dishes or leave in tubes.

Use: For the cultivation and maintenance of *Xanthomonas* species.

Glucose Yeast Extract Agar

Composition per liter:

$CaCO_3$....20.0g
Glucose....20.0g
Agar....17.0g
Yeast extract....10.0g

Preparation of Medium: Add components to distilled/deionized water and bring volume to 1.0L. Mix thoroughly. Gently heat and bring to boiling. Distribute into tubes or flasks. Autoclave for 15 min at 15 psi pressure–121°C. Pour into sterile Petri dishes or leave in tubes.

Use: For the cultivation of *Agrobacterium* species, *Clostridium* species, *Erwinia* species, *Pseudomonas* species, and *Xanthomonas campestris*.

Glucose Yeast Extract Iron Agar (LMG Medium 153)

Composition per liter:

Agar....15.0g
Glucose....10.0g
Yeast extract....5.0g
Iron solution....20.0mL

pH 7.1 ± 0.2 at 25°C

Iron Solution:

Composition per 100.0mL:

$FeCl_3$....0.03g

Preparation of Iron Solution: Add $FeCl_3$ to distilled/deionized water and bring volume to 100.0mL. Mix thoroughly. Filter sterilize.

Preparation of Medium: Add components, except iron solution, to distilled/deionized water and bring volume to 980.0mL. Mix thoroughly. Autoclave for 15 min at 15 psi pressure–121°C. Cool to 45°-50°C. Aseptically add 20.0mL sterile iron solution. Mix thoroughly. Aseptically pour into sterile Petri dishes or distribute into sterile tubes.

Use: For the cultivation of *Pseudomonas denitrificans*.

Glucose Yeast Extract Medium (ATCC Medium 985)

Composition per liter:

Agar....15.0g
Yeast extract....3.0g
Glucose....1.0g

pH 7.0 ± 0.2 at 25°C

Preparation of Medium: Add components to distilled/deionized water and bring volume to 1.0L. Mix thoroughly. Gently heat and bring to boiling. Distribute into tubes or flasks. Autoclave for 15 min at 15 psi pressure–121°C. Pour into sterile Petri dishes or leave in tubes.

Use: For the cultivation and maintenance of *Acinetobacter tartarogenes, Agrobacterium viscosum,* and *Pseudomonas* species.

Glucose Yeast Medium with Calcium Carbonate

Composition per liter:

Agar....15.0g
$CaCO_3$....7.5g
Peptone....5.0g
Yeast extract....5.0g
Glucose....3.0g

pH 7.0 ± 0.2 at 25°C

Preparation of Medium: Add components to distilled/deionized water and bring volume to 1.0L. Mix thoroughly. Gently heat and bring to boiling. Distribute into tubes or flasks. Adjust pH to 6.3. Autoclave for 15 min at 15 psi pressure–121°C. Pour into sterile Petri dishes or leave in tubes.

Use: For the cultivation and maintenance of *Erwinia herbicola* and *Bacillus* species.

Glucose Yeast Peptone Medium

Composition per liter:

Glucose....10.0g
Peptone....5.0g
Yeast extract....5.0g

pH 5.0 ± 1.0 at 25°C

Preparation of Medium: Add components to distilled/deionized water and bring volume to 1.0L. Mix thoroughly. Adjust pH to 5.0. Distribute into tubes or flasks. Autoclave for 15 min at 15 psi pressure–121°C.

Use: For the cultivation of *Enterobacter cloacae*.

Glutamate Medium

Composition per liter:

Solution A....500.0mL
Solution B....250.0mL
Solution C....250.0mL

Solution A:

Composition per 500.0mL:

Mannitol....10.0g
K_2HPO_4....0.22g

Preparation of Solution A: Add components to distilled/deionized water and bring volume to 500.0mL. Mix thoroughly. Autoclave for 15 min at 15 psi pressure–121°C. Cool to 25°C.

Solution B:

Composition per 250.0mL:

$MgSO_4·7H_2O$....0.1g
$CaCl_2·6H_2O$....0.08g
$FeCl_3·6H_2O$....0.05g

Preparation of Solution B: Add components to distilled/deionized water and bring volume to 250.0mL. Mix thoroughly. Autoclave for 15 min at 15 psi pressure–121°C. Cool to 25°C.

Solution C:

Composition per 250.0mL:

Sodium glutamate	1.1g
Calcium pantothenate	0.5mg
Thiamine·HCl	0.1mg
Biotin	0.5μg

Preparation of Solution C: Add components to distilled/deionized water and bring volume to 250.0mL. Mix thoroughly. Filter sterilize.

Preparation of Medium: Aseptically combine 500.0mL of cooled, sterile solution A, 250.0mL of cooled, sterile solution B, and 250.0mL of sterile solution C. Mix thoroughly. Aseptically distribute into sterile tubes or flasks.

Use: For the isolation of *Rhizobium* species.

Glutamate Medium (ATCC Medium 820)

Composition per liter:

Agar	15.0g
Sodium glutamate	5.0g
KH_2PO_4	1.0g
$MgSO_4 \cdot 7H_2O$	0.2g
KCl	0.1g
Glucose solution	100.0mL

pH 6.5 ± 0.2 at 25°C

Glucose Solution:

Composition per 100.0mL:

Glucose	10.0g

Preparation of Glucose Solution: Add glucose to distilled/deionized water and bring volume to 100.0mL. Mix thoroughly. Filter sterilize.

Preparation of Medium: Add components, except glucose solution, to distilled/deionized water and bring volume to 900.0mL. Mix thoroughly. Gently heat and bring to boiling. Autoclave for 15 min at 15 psi pressure–121°C. Cool to 45°–50°C. Aseptically add 100.0mL of sterile glucose solution. Mix thoroughly. Pour into sterile Petri dishes or distribute into sterile tubes.

Use: For the cultivation and maintenance of *Pseudomonas* species.

Glycerol Agar

Composition per 1070.0mL:

Agar	15.0g
Peptone	5.0g
Beef extract	3.0g
Soil extract	1.0L
Glycerol	70.0mL

pH 7.0 ± 0.2 at 25°C

Soil Extract:

Composition per liter:

Soil, airdried	1.0Kg

Preparation of Soil Extract: Sift soil through a #9 mesh screen. Add to 2.4L of tap water. Mix thoroughly. Autoclave for 60 min at 15 psi pressure–121°C. Cool to 25°C. Filter through Whatman #1 filter paper. Bring volume to 1.0L with tap water.

Preparation of Medium: Combine components. Mix thoroughly. Gently heat and bring to boiling. Autoclave for 15 min at 15 psi pressure–121°C. Pour into sterile Petri dishes or distribute into sterile tubes.

Use: For the selective isolation and cultivation of *Nocardia* species and *Rhodococcus* species.

Glycerol Agar

Composition per liter:

Beef heart, solids from infusion	250.0g
Glycerol	60.0g
Agar	15.0g
Pancreatic digest of gelatin	5.0g
Tryptose	5.0g
Beef extract	3.0g
NaCl	2.5g

pH 7.3 ± 0.2 at 25°C

Preparation of Medium: Add components to distilled/deionized water and bring volume to 1.0L. Mix thoroughly. Gently heat and bring to boiling. Distribute into tubes or flasks. Autoclave for 15 min at 15 psi pressure–121°C. Pour into sterile Petri dishes or leave in tubes.

Use: For the cultivation and maintenance of *Bacillus subtilis, Enterococcus faecalis, Erwinia chrysanthemi, Gordona rubropertinctus, Mycobacterium* species, *Nocardia brevicatena, Rhodococcus equi,* and *Rhodococcus rhodochrous.*

Glycerol Arginine Agar

Composition per liter:

Agar	15.0g
Glycerol	12.5g
Arginine	1.0g
K_2HPO_4	1.0g
NaCl	1.0g
$MgSO_4 \cdot 7H_2O$	0.5g
$Fe_2(SO_4)_3 \cdot 6H_2O$	0.01g
$CuSO_4 \cdot 5H_2O$	1.0mg
$MnSO_4 \cdot H_2O$	1.0mg
$ZnSO_4 \cdot 7H_2O$	1.0mg

Preparation of Medium: Add components to distilled/deionized water and bring volume to 1.0L. Mix thoroughly. Gently heat and bring to boiling. Distribute into tubes or flasks. Autoclave for 15 min at 15 psi pressure–121°C. Pour into sterile Petri dishes or leave in tubes.

Use: For the selective isolation and cultivation of streptomycetes.

Glycerol Chalk Agar

Composition per liter:

NaCl	30.0g
Agar	15.0g
Glycerol	10.0g
$CaCO_3$	5.0g
Peptone	5.0g
Yeast extract	3.0g

Preparation of Medium: Add components to distilled/deionized water and bring volume to 1.0L. Mix thoroughly. Gently heat and bring to boiling. Autoclave for 15 min at 15 psi pressure–121°C. Pour into sterile Petri dishes. Swirl flask while dispensing medium to keep $CaCO_3$ in suspension.

Use: For the cultivation of *Photobacterium* species and *Lucibacterium* species.

GÖ1 Medium

Composition per liter:

Sucrose 17.1g
Sodium acetate 10.0g
NaCl 2.25g
Pancreatic digest of casein 2.0g
Yeast extract 2.0g
$NaHCO_3$ 0.85g
$MgSO_4 \cdot 7H_2O$ 0.5g
NH_4Cl 0.5g
K_2HPO_4 .... 0.348g
$CaCl_2 \cdot 2H_2O$ 0.25g
KH_2PO_4 .... 0.227g
$FeSO_4 \cdot 7H_2O$ 2.0mg
Resazurin 1.0mg
$NaHCO_3$ solution 40.0mL
Vitamin solution 10.0mL
L-Cysteine·HCl·H_2O solution 10.0mL
$Na_2S \cdot 9H_2O$ solution 10.0mL
Trace elements solution SL-10 .... 1.0mL
pH 6.5–6.8 at 25°C

$NaHCO_3$ Solution:
Composition per 50.0mL:
$NaHCO_3$ 5.0g

Preparation of $NaHCO_3$ Solution: Add $NaHCO_3$ to distilled/deionized water and bring volume to 50.0mL. Mix thoroughly. Sparge with 80% N_2 + 20% CO_2. Autoclave for 15 min at 15 psi pressure–121°C.

Vitamin Solution:
Composition per liter:
Pyridoxine·HCl 10.0mg
DL-Calcium pantothenate 5.0mg
Lipoic acid 5.0mg
Nicotinic acid 5.0mg
p-Aminobenzoic acid 5.0mg
Riboflavin 5.0mg
Thiamine·HCl 5.0mg
Biotin 2.0mg
Folic acid 2.0mg
Vitamin B_{12} 0.1mg

Preparation of Vitamin Solution: Add components to distilled/deionized water and bring volume to 1.0L. Mix thoroughly. Filter sterilize. Sparge with 80% N_2 + 20% CO_2.

L-Cysteine·HCl·H_2O Solution:
Composition per 10.0mL:
L-Cysteine·HCl·H_2O 0.3g

Preparation of L-Cysteine·HCl·H_2O Solution: Add L-cysteine·HCl·H_2O to distilled/deionized water and bring volume to 10.0mL. Mix thoroughly. Autoclave under 100% N_2 for 15 min at 15 psi pressure–121°C.

$Na_2S \cdot 9H_2O$ Solution:
Composition per 10.0mL:
$Na_2S \cdot 9H_2O$ 0.3g

Preparation of $Na_2S \cdot 9H_2O$ Solution: Add $Na_2S \cdot 9H_2O$ to distilled/deionized water and bring volume to 10.0mL. Mix thoroughly. Sparge with 100% N_2. Autoclave for 15 min at 15 psi pressure–121°C.

Trace Elements Solution SL-10:
Composition per liter:
$FeCl_2 \cdot 4H_2O$ 1.5g
$CoCl_2 \cdot 6H_2O$ 190.0mg
$MnCl_2 \cdot 4H_2O$ 100.0mg
$ZnCl_2$ 70.0mg
$Na_2MoO_4 \cdot 2H_2O$ 36.0mg
$NiCl_2 \cdot 6H_2O$ 24.0mg
H_3BO_3 .... 6.0mg
$CuCl_2 \cdot 2H_2O$ 2.0mg
HCl (25% solution) 10.0mL

Preparation of Trace Elements Solution SL-10: Add $FeCl_2 \cdot 4H_2O$ to 10.0mL of HCl solution. Mix thoroughly. Add distilled/deionized water and bring volume to 1.0L. Add remaining components. Mix thoroughly. Sparge with 100% N_2.

Preparation of Medium: Add components, except $NaHCO_3$ solution, vitamin solution, L-cysteine·HCl·H_2O solution, and $Na_2S \cdot 9H_2O$ solution, to distilled/deionized water and bring volume to 920.0mL. Mix thoroughly. Sparge under 80% N_2 + 20% CO_2 for 3–4 min. Autoclave for 15 min at 15 psi pressure–121°C. Aseptically and anaerobically add 40.0mL of sterile $NaHCO_3$ solution, 20.0mL of sterile vitamin solution, 10.0mL of sterile L-cysteine·HCl·H_2O solution, and 10.0mL of sterile $Na_2S \cdot 9H_2O$ solution. Mix thoroughly. Aseptically and anaerobically distribute into sterile screw-capped bottles under 80% N_2 + 20% CO_2.

Use: For the cultivation of *Methanosarcina mazei.*

Gorbenko Medium

Composition per liter:
Agar 16.2g
Peptone 0.9g
NaCl 0.9g
Yeast extract 0.36g
Beef extract 0.18g
Lake water 1000.0mL
pH 7.4 ± 0.2 at 25°C

Preparation of Medium: Combine components. Mix thoroughly. Gently heat and bring to boiling. Distribute into tubes or flasks. Autoclave for 15 min at 15 psi pressure–121°C. Pour into sterile Petri dishes or leave in tubes.

Use: For the cultivation of *Gordona rubropertinctus, Gordona terrae,* and heterotrophic marine bacteria.

Gorham's Medium for Algae

Composition per liter:
$NaNO_3$ 0.496g
$MgSO_4 \cdot 7H_2O$ 0.075g
$Na_2SiO_3 \cdot 9H_2O$ 0.058g
K_2HPO_4 .... 0.039g
$CaCl_2 \cdot 2H_2O$ 0.036g
Na_2CO_3 .... 0.02g
Citric acid 6.0mg
EDTA 6.0mg
Ferric citrate 6.0mg
pH 7.5 ± 0.5 at 25°C

Preparation of Medium: Add components to distilled/deionized water and bring volume to 1.0L. Mix thoroughly. Distribute into tubes or flasks. Autoclave for 15 min at 15 psi pressure–121°C.

Use: For the cultivation and maintenance of *Anabaena flos-aquae, Selenastrum capricornutum* and *Microcystis aeruginosa.*

GPVA Medium

Composition per liter:
Agar 15.0g
Yeast extract 10.0g

ACES buffer (2-[(2-Amino-2-oxoethyl)-amino]-ethane sulfonic acid) 10.0g
Glycine 3.0g
Charcoal, activated 2.0g
α-Ketoglutarate 1.0g
$Fe_4(P_2O_7)_3 \cdot 9H_2O$ 0.25g
Antibiotic inhibitor solution 10.0mL
pH 6.9 ± 0.2 at 25°C

Antibiotic Inhibitor Solution:
Composition per 10.0mL:
Anisomycin 0.08g
Vancomycin 5.0mg
Polymyxin B 100,000U

Preparation of Antibiotic Inhibitor Solution: Add components to distilled/deionized water and bring volume to 10.0mL. Mix thoroughly. Filter sterilize.

Preparation of Medium: Add components, except antibiotic inhibitor solution, to distilled/deionized water and bring volume to 990.0mL. Mix thoroughly. Gently heat and bring to boiling. Autoclave for 15 min at 15 psi pressure–121°C. Cool to 45°–50°C. Adjust pH to 6.9. Aseptically add 10.0mL of sterile antibiotic inhibitor solution. Mix thoroughly. Pour into sterile Petri dishes or distribute into sterile tubes.

Use: For the isolation and cultivation of *Legionella* species from environmental waters.

GPY Salts Medium (Glucose Peptone Yeast Extract Salts Medium)

Composition per liter:
Glucose 1.0g
Peptone 0.5g
Yeast extract 0.1g
Modified Hutner's mineral base 20.0mL

Hutner's Mineral Base:
Composition per liter:
$MgSO_4 \cdot 7H_2O$ 29.7g
Nitrilotriacetic acid 10.0g
$CaCl_2 \cdot 2H_2O$ 3.34g
$FeSO_4 \cdot 7H_2O$ 99.0mg
$(NH_4)_2MoO_4$ 9.25mg
"Metals 44" 50.0mL

Preparation of Hutner's Mineral Base: Add nitrilotriacetic acid to 500.0mL of distilled/deionized water. Dissolve by adjusting pH to 6.5 with KOH. Add remaining components. Readjust pH to 7.2 with H_2SO_4 or KOH. Add distilled/deionized water to 1.0L. Store at 5°C.

"Metals 44":
Composition per 100.0mL:
$ZnSO_4 \cdot 7H_2O$ 1.1g
$FeSO_4 \cdot 7H_2O$ 0.5g
EDTA 0.25g
$MnSO_4 \cdot 7H_2O$ 0.154g
$CuSO_4 \cdot 5H_2O$ 0.04g
$Co(NO_3)_2 \cdot 6H_2O$ 0.025g
$Na_2B_4O_7 \cdot 10H_2O$ 0.018g

Preparation of "Metals 44": Add a few drops of H_2SO_4 to distilled/deionized water to inhibit precipitate formation. Add components to acidified distilled/deionized water and bring volume to 100.0mL. Mix thoroughly.

Preparation of Medium: Add components to distilled/deionized water and bring volume to 1.0L. Mix thoroughly. Distribute into tubes or flasks. Autoclave for 15 min at 15 psi pressure–121°C.

Use: For the cultivation and maintenance of *Prosthecobacter fusiformis*.

GYM + Seawater (DSMZ Medium 871)

Composition per liter:
Sea salts 32.0g
Agar 15.0g
Malt extract 10.0g
Glucose 4.0g
Yeast extract 4.0g
$CaCO_3$ 2.0g
pH 7.0-7.4 at 25°C

Preparation of Medium: Add sea salts to distilled/deionized water and bring volume to 1.0L. Mix thoroughly. Add remaining components. Mix thoroughly. Distribute into tubes or flasks. Autoclave for 15 min at 15 psi pressure–121°C.

Use: For the cultivation of *Nocardiopsis aegyptia (Nocardiopsis* sp.*)* and *Nocardiopsis halotolerans (Nocardiopsis* sp.*)*.

GYM Starch Agar (DSMZ Medium 214)

Composition per liter:
Starch 20.0g
Agar 12.0g
Malt extract 10.0g
Glucose 4.0g
Yeast extract 4.0g
$CaCO_3$ 2.0g
pH 7.2 ± 0.2 at 25°C

Preparation of Medium: Add components, except agar, to distilled/deionized water and bring volume to 1.0L. Mix thoroughly. Adjust pH to 7.2. Add agar. Gently heat and bring to boiling. Distribute into tubes or flasks. Autoclave for 15 min at 15 psi pressure–121°C. Pour into sterile Petri dishes or leave in tubes.

Use: For the cultivation and maintenance of *Amycolatopsis orientalis* subsp. *Orientalis, Nocardia* spp., *Mycobacterium* spp., *Pseudonocardia* spp., *Saccharothrix* spp., *Kineosporia aurantiaca, Kitasatospora setae, Oerskovia turbata (Cellulomonas turbata), Cellulosimicrobium cellulans,* and *Thermoactinomyces dichotomicus.*

GYM *Streptomyces* Agar (DSMZ Medium 65)

Composition per liter:
Agar 12.0g
Malt extract 10.0g
Glucose 4.0g
Yeast extract 4.0g
$CaCO_3$ 2.0g
pH 7.2 ± 0.2 at 25°C

Preparation of Medium: Add components, except agar, to distilled/deionized water and bring volume to 1.0L. Mix thoroughly. Adjust pH to 7.2. Add agar. Gently heat and bring to boiling. Distribute into tubes or flasks. Autoclave for 15 min at 15 psi pressure–121°C. Pour into sterile Petri dishes or leave in tubes.

Use: For the cultivation and maintenance of *Streptomyces* spp.

Hagedorn and Holt Selective Medium

Composition per liter:

NaCl	20.0g
Agar	15.0g
Yeast extract	2.0g
Pancreatic digest of casein	1.7g
Agar	1.5g
NaCl	0.5g
Papaic digest of soybean meal	0.3g
K_2HPO_4	0.25g
Glucose	0.25g
Cycloheximide	0.1g
Methyl Red	0.15mg

pH 7.3 ± 0.2 at 25°C

Preparation of Medium: Add components to distilled/deionized water and bring volume to 1.0L. Mix thoroughly. Gently heat and bring to boiling. Distribute into tubes or flasks. Autoclave for 15 min at 15 psi pressure–121°C. Pour into sterile Petri dishes or leave in tubes.

Use: For the selective isolation of *Arthrobacter* spp. in soil.

Haloalkaliphilic Agar

Composition per liter:

Solution A	900.0mL
Solution B	100.0mL

pH 8.5–9.5 at 25°C

Solution A:

Composition per 900.0mL:

NaCl	200.0g
Agar	25.0g
Yeast extract	10.0g
Casamino acids	7.5g
Sodium citrate	3.0g
KCl	2.0g
$MgSO_4 \cdot 7H_2O$	1.0g
$FeSO_4 \cdot 7H_2O$	0.05g
$MnSO_4 \cdot 4H_2O$	0.25mg

Preparation of Solution A: Add components, except NaCl, to distilled/deionized water and bring volume to 900.0mL. Mix thoroughly. Gently heat and bring to boiling. Add 200.0g of NaCl. Mix thoroughly. Autoclave for 15 min at 15 psi pressure–121°C. Cool to 50°–55°C.

Solution B:

Composition per 100.0mL:

Na_2CO_3	5.0g

Preparation of Solution B: Dissolve 5.0g of Na_2CO_3 in distilled/deionized water and bring volume to 100.0mL. Mix thoroughly. Autoclave for 15 min at 15 psi pressure–121°C. Cool to 50°–55°C.

Preparation of Medium: Aseptically mix 900.0mL of solution A and 100.0mL of solution B. Mix thoroughly. Aseptically adjust pH to 8.5–9.5. Pour into sterile Petri dishes or distribute into sterile tubes.

Use: For the cultivation and maintenance of *Natronobacterium gregoryi, Natronobacterium magadii, Natronobacterium pharaonis,* and *Natronococcus occultus.*

Haloanaerobacter chitinovorans Medium

Composition per 1001.0mL:

NaCl	10.0g
$MgSO_4 \cdot 7H_2O$	9.6g
$MgCl_2 \cdot 6H_2O$	7.0g
Chitin	5.0g
KCl	3.8g
Na_2CO_3	1.0g
NH_4Cl	1.0g
Yeast extract	1.0g
$CaCl_2 \cdot 2H_2O$	0.5g
$K_2HPO_4 \cdot 3H_2O$	0.4g
Resazurin	0.001g
Na_2CO_3	20.0mL
$Na_2S \cdot 9H_2O$ solution	10.0mL
L-Cysteine·HCl·H_2O solution	10.0mL
Trace elements solution SL-6	1.0mL

pH 7.2 ± 0.2 at 25°C

Na_2CO_3 Solution:

Composition per 20.0mL:

Na_2CO_3	3.0g

Preparation of Na_2CO_3 Solution: Add Na_2CO_3 to distilled/deionized water and bring volume to 20.0mL. Mix thoroughly. Filter sterilize. Sparge with 100% N_2.

$Na_2S \cdot 9H_2O$ Solution:

Composition per 10.0mL:

$Na_2S \cdot 9H_2O$	0.5g

Preparation of $Na_2S \cdot 9H_2O$ Solution: Add $Na_2S \cdot 9H_2O$ to distilled/deionized water and bring volume to 10.0mL. Mix thoroughly. Sparge with 100% N_2. Autoclave for 15 min at 15 psi pressure–121°C. Before use, neutralize to pH 7.0 with sterile HCl.

L-Cysteine·HCl·H_2O Solution:

Composition per 10.0mL:

L-Cysteine·HCl·H_2O	0.5g

Preparation of L-Cysteine·HCl·H_2O Solution: Add L-cysteine·HCl·H_2O to distilled/deionized water and bring volume to 10.0mL. Mix thoroughly. Sparge with 100% N_2. Autoclave for 15 min at 15 psi pressure–121°C.

Trace Elements Solution SL-6:

Composition per liter:

$MnCl_2 \cdot 4H_2O$	0.5g
H_3BO_3	0.3g
$CoCl_2 \cdot 6H_2O$	0.2g
$ZnSO_4 \cdot 7H_2O$	0.1g
$Na_2MoO_4 \cdot 2H_2O$	0.03g
$NiCl_2 \cdot 6H_2O$	0.02g
$CuCl_2 \cdot 2H_2O$	0.01g

Preparation of Trace Elements Solution SL-6: Add components to distilled/deionized water and bring volume to 1.0L. Mix thoroughly.

Preparation of Medium: Prepare and dispense medium under 100% N_2. Add components, except Na_2CO_3 solution, substrate solution, $Na_2S \cdot 9H_2O$ solution, and L-cysteine·HCl·H_2O solution, to distilled/deionized water and bring volume to 940.0mL. Mix thoroughly. Adjust pH to 7.2. Gently heat and bring to boiling. Cool to room temperature while sparging with 100% N_2. Autoclave for 15 min at 15 psi pressure–121°C. Aseptically and anaerobically add 20.0mL of sterile $NaHCO_3$ solution, 20.0mL of sterile substrate solution, 10.0mL of sterile $Na_2S \cdot 9H_2O$ solution, and 10.0mL of sterile L-cysteine·HCl·H_2O solution. Mix thoroughly. Aseptically distribute into sterile tubes or flasks.

Use: For the cultivation of *Haloanaerobacter chitinovorans.*

Haloanaerobium alcaliphilum Medium (DSMZ Medium 807)

Composition per liter:

NaCl	100.0g

$MgSO_4 \cdot 7H_2O$ 17.0g
Trypticase™ 10.0g
$NaHCO_3$ 4.1g
Cysteine-HCl·H_2O 0.5g
Resazurin 1.0mg
Solution A 50.0mL
Solution B 50.0mL
Yeast extract solution 50.0mL
Glucose solution 20.0mL
Trace elements solution 10.0mL
$Na_2S \cdot 9H_2O$ solution 10.0mL
pH 7.0 ± 0.1 at 25°C

$Na_2S \cdot 9H_2O$ Solution:
Composition per 10mL:
$Na_2S \cdot 9H_2O$ 0.25g

Preparation of $Na_2S \cdot 9H_2O$ Solution: Add $Na_2S \cdot 9H_2O$ to distilled/deionized water and bring volume to 10.0mL. Mix thoroughly. Autoclave under 100% N_2 for 15 min at 15 psi pressure–121°C. Cool to room temperature.

Glucose Solution:
Composition per 20mL:
Glucose 5.0g

Preparation of Glucose Solution: Add glucose to distilled/deionized water and bring volume to 20.0mL. Mix thoroughly. Filter sterilize.

Yeast Extract Solution:
Composition per 50mL:
Yeast extract 10.g

Preparation of Yeast Extract Solution: Add yeast extract to distilled/deionized water and bring volume to 50.0mL. Mix thoroughly. Autoclave under 100% N_2 for 15 min at 15 psi pressure–121°C. Cool to room temperature.

Solution A:
Composition per liter:
K_2HPO_4 .... 6.0g

Preparation of Solution A: Add K_2HPO_4 to distilled/deionized water and bring volume to 1.0L. Mix thoroughly.

Solution B:
Composition per liter:
NaCl 12.0g
KH_2PO_4 .... 6.0g
$(NH_4)SO_4$ 6.0g
$MgSO_4 \cdot 7H_2O$ 2.6g
NH_4Cl 2.5g
$CaCl_2 \cdot 2H_2O$ 0.28g
K_2HPO_4 .... 0.28g

Preparation of Solution B: Add components to distilled/deionized water and bring volume to 1.0L. Mix thoroughly.

Trace Elements Solution:
Composition per liter:
$MgSO_4 \cdot 7H_2O$ 3.0g
Nitrilotriacetic acid 1.5g
NaCl 1.0g
$MnSO_4 \cdot 2H_2O$ 0.5g
$CoSO_4 \cdot 7H_2O$ 0.18g
$ZnSO_4 \cdot 7H_2O$ 0.18g
$CaCl_2 \cdot 2H_2O$ 0.1g
$FeSO_4 \cdot 7H_2O$ 0.1g
$NiCl_2 \cdot 6H_2O$ 0.025g
$KAl(SO_4)_2 \cdot 12H_2O$ 0.02g
H_3BO_3 .... 0.01g
$Na_2MoO_4 \cdot 4H_2O$ 0.01g
$CuSO_4 \cdot 5H_2O$ 0.01g
$Na_2SeO_3 \cdot 5H_2O$ 0.3mg

Preparation of Trace Elements Solution: Add nitrilotriacetic acid to 500.0mL of distilled/deionized water. Dissolve by adjusting pH to 6.5 with KOH. Add remaining components. Add distilled/deionized water to 1.0L. Mix thoroughly.

Preparation of Medium: Prepare and dispense medium under 80% N_2 + 20% CO_2 gas atmosphere. Add components, except $NaHCO_3$, $Na_2S \cdot 9H_2O$ solution, cysteine-HCl·H_2O, yeast extract solution, and glucose solution, to distilled/deionized water and bring volume to 920.0mL. Mix thoroughly. Gently heat and bring to boiling. Boil for 5 min. Cool to room temperature under 80% N_2 + 20% CO_2 gas atmosphere. Add 4.1g $NaHCO_3$ and 0.5g cysteine-HCl·H_2O. Adjust pH to 7.0. Autoclave for 15 min at 15 psi pressure–121°C. Aseptically and anaerobically add 20.0mL sterile glucose solution, 50.0mL sterile yeast extract solution, and 10.0mL sterile $Na_2S \cdot 9H_2O$. Mix thoroughly. Aseptically and anaerobically distribute into sterile tubes or bottles.

Use: For the cultivation of *Halanaerobium alcaliphilum.*

Haloanaerobium congolense Medium (DSMZ Medium 933)

Composition per 1080mL:
NaCl 100.0g
$MgCl_2 \cdot 6H_2O$ 10.0g
Trypticase 1.0g
NH_4Cl 1.0g
KCl 1.0g
Na-acetate 0.5g
Cysteine 0.5g
K_2HPO_4 .... 0.3g
KH_2PO_4 .... 0.3g
$CaCl_2 \cdot 2H_2O$ 0.1g
Resazurin 0.01g
Glucose solution 20.0mL
Thiosulfate solution 20.0mL
$Na_2S \cdot 9H_2O$ solution 20.0mL
$NaHCO_3$ solution 20.0mL
Trace elements solution 1.0mL
pH 7.0 ± 0.2 at 25°C

Trace Elements Solution:
Composition per liter:
$MgSO_4 \cdot 7H_2O$ 3.0g
Nitrilotriacetic acid 1.5g
NaCl 1.0g
$MnSO_4 \cdot 2H_2O$ 0.5g
$CoSO_4 \cdot 7H_2O$ 0.18g
$ZnSO_4 \cdot 7H_2O$ 0.18g
$CaCl_2 \cdot 2H_2O$ 0.1g
$FeSO_4 \cdot 7H_2O$ 0.1g
$NiCl_2 \cdot 6H_2O$ 0.025g
$KAl(SO_4)_2 \cdot 12H_2O$ 0.02g
H_3BO_3 .... 0.01g
$Na_2MoO_4 \cdot 4H_2O$ 0.01g
$CuSO_4 \cdot 5H_2O$ 0.01g
$Na_2SeO_3 \cdot 5H_2O$ 0.3mg

Preparation of Trace Elements Solution: Add nitrilotriacetic acid to 500.0mL of distilled/deionized water. Dissolve by adjusting pH to 6.5 with KOH. Add remaining components. Add distilled/deionized water to 1.0L. Mix thoroughly.

$Na_2S \cdot 9H_2O$ Solution:
Composition per 10.0mL:
$Na_2S \cdot 9H_2O$ 0.2g

Preparation of $Na_2S \cdot 9H_2O$ Solution: Add $Na_2S \cdot 9H_2O$ to distilled/deionized water and bring volume to 10.0mL. Sparge with 100% N_2. Autoclave for 15 min at 15 psi pressure–121°C. Cool to 25°C.

$NaHCO_3$ Solution:
Composition per 50.0mL:
$NaHCO_3$ 5.0g

Preparation of $NaHCO_3$ Solution: Add $NaHCO_3$ to distilled/deionized water and bring volume to 50.0mL. Mix thoroughly. Sparge with 80% N_2 + 20% CO_2. Autoclave for 15 min at 15 psi pressure–121°C. Cool to 25°C.

Thiosulfate Solution:
Composition per 20mL:
$Na_2S_2O_3 \cdot 5H_2O$ 5.0g

Preparation of Thiosulfate Solution: Add $Na_2S_2O_3 \cdot 5H_2O$ to distilled/deionized water and bring volume to 20.0mL. Mix thoroughly. Sparge with 100% N_2. Autoclave for 15 min at 15 psi pressure–121°C. Cool to room temperature.

Glucose Solution:
Composition per 20.0mL:
Glucose 3.5g

Preparation of Glucose Solution: Add glucose to distilled/deionized water and bring volume to 20.0mL. Mix thoroughly. Sparge with 100% N_2. Filter sterilize.

Preparation of Medium: Prepare and dispense medium under 80% N_2 + 20% CO_2 gas atmosphere. Add components, except $NaHCO_3$ solution, glucose solution, $Na_2S \cdot 9H_2O$ solution, and thiosulfate solution, to distilled/deionized water and bring volume to 1.0L. Mix thoroughly. Gently heat and bring to boiling. Boil for 5 min. Cool to room temperature while sparging with 80% N_2 + 20% CO_2. Adjust pH to 7.0. Distribute into anaerobe tubes or bottles. Autoclave for 15 min at 15 psi pressure–121°C. Cool to room temperature. Aseptically and anaerobically add per 10.0mL medium, 0.2mL $NaHCO_3$ solution, 0.2mL glucose solution, 0.2mL $Na_2S \cdot 9H_2O$ solution, and 0.2mL thiosulfate solution. Mix thoroughly. The final pH should be 7.0.

Use: For the cultivation of *Thermococcus waiotapuensis.*

Haloanaerobium lacusroseus Medium (DSMZ Medium 764)

Composition per liter:
NaCl 200.0g
KCl 4.0g
$MgCl_2 \cdot 6H_2O$ 2.0g
Yeast extract 1.0g
NH_4Cl 1.0g
Na-acetate 1.0g
Trypticase™ 0.5g
K_2HPO_4 .... 0.3g
KH_2PO_4 .... 0.3g
$CaCl_2 \cdot 2H_2O$ 0.2g
Resazurin 0.001g
Trace elements SL-6 .... 1.0mL
Glucose solution 100.0mL
$NaHCO_3$ solution 50.0mL
$Na_2S \cdot 9H_2O$ solution 10.0mL
Dithionite solution 5.0mL
pH 7.0 ± 0.2 at 25°C

Glucose Solution:
Composition per 100mL:
Glucose 17.4g

Preparation of Glucose Solution: Add glucose to distilled/deionized water and bring volume to 100.0mL. Mix thoroughly. Filter sterilize.

Trace Elements Solution SL-6:
Composition per liter:
$MnCl_2 \cdot 4H_2O$ 0.5g
H_3BO_3 .... 0.3g
$CoCl_2 \cdot 6H_2O$ 0.2g
$ZnSO_4 \cdot 7H_2O$ 0.1g
$Na_2MoO_4 \cdot 2H_2O$ 0.03g
$NiCl_2 \cdot 6H_2O$ 0.02g
$CuCl_2 \cdot 2H_2O$ 0.01g

Preparation of Trace Elements Solution SL-6: Add components to distilled/deionized water and bring volume to 1.0L. Mix thoroughly. Autoclave for 15 min at 15 psi pressure–121°C.

$Na_2S \cdot 9H_2O$ Solution:
Composition per 10.0mL:
$Na_2S \cdot 9H_2O$ 0.2g

Preparation of $Na_2S \cdot 9H_2O$ Solution: Add $Na_2S \cdot 9H_2O$ to distilled/deionized water and bring volume to 10.0mL. Mix thoroughly. Sparge with 100% N_2. Autoclave for 15 min at 15 psi pressure–121°C. Neutralize to pH 7.0 with sterile HCl.

$NaHCO_3$ Solution:
Composition per 100.0mL:
$NaHCO_3$ 10.0g

Preparation of $NaHCO_3$ Solution: Add $NaHCO_3$ to distilled/deionized water and bring volume to 100.0mL. Mix thoroughly. Sparge with 80% N_2 + 20% CO_2. Filter sterilize.

Dithionite Solution:
Composition per 10mL:
Na-dithionite 2.0mg

Preparation of Dithionite Solution: Add Na-dithionite to distilled/deionized water and bring volume to 10.0mL. Mix thoroughly. Sparge with 100% N_2. Filter sterilize.

Preparation of Medium: Prepare and dispense medium under 80% N_2 + 20% CO_2 gas atmosphere. Add components, except glucose solution, $NaHCO_3$ solution, dithionite solution, and $Na_2S \cdot 9H_2O$ solution, to distilled/deionized water and bring volume to 835.0mL. Mix thoroughly. Adjust pH to 7.0. Gently heat and bring to boiling. Cool while sparging with 80% N_2 + 20% CO_2. Distribute into Hungate tubes under 80% N_2 + 20% CO_2. Autoclave for 15 min at 15 psi pressure–121°C. Cool to room temperature. Aseptically and anaerobically inject glucose solution (0.25mL/10/mL medium), dithionite solution (0.05mL/10/mL medium), $NaHCO_3$ solution (0.5mL/10/mL medium), and $Na_2S \cdot 9H_2O$ solution (0.1mL/10/mL medium). Aseptically and anaerobically distribute into sterile tubes or bottles.

Use: For the cultivation of *Haloanaerobium lacusrosei (Haloanaerobium lacusroseus).*

Haloanaerobium praevalens Medium

Composition per liter:
NaCl 130.0g
Agar 20.0g
Yeast extract 2.0g
Pancreatic digest of casein 2.0g

NH_4Cl 0.5g
$MgSO_4·7H_2O$ 0.5g
K_2HPO_4 .... 0.35g
$CaCl_2·2H_2O$ 0.25g
KH_2PO_4 .... 0.23g
$FeSO_4·7H_2O$ 2.0mg
$NaHCO_3$ solution 20.0mL
L-Cysteine-sulfide reducing agent 20.0mL
Wolfe's vitamin solution 10.0mL
Methanol 10.0mL
Resazurin (0.025% solution) 4.0mL
Trace elements SL-6 .... 3.0mL
pH 6.8 ± 0.2 at 25°C

$NaHCO_3$ Solution:
Composition per 20.0mL:
$NaHCO_3$ 850.0mg

Preparation of $NaHCO_3$ Solution: Add $NaHCO_3$ to distilled/deionized water and bring volume to 20.0mL. Mix thoroughly. Filter sterilize. Gas with 100% CO_2 for 20 min.

L-Cysteine-Sulfide Reducing Agent:
Composition per 20.0mL:
L-Cysteine·HCl·H_2O 0.3g
$Na_2S·9H_2O$ 0.3g

Preparation of L-Cysteine-Sulfide Reducing Agent: Add L-cysteine·HCl·H_2O to 10.0mL of distilled/deionized water. Mix thoroughly. In a separate tube, add $Na_2S·9H_2O$ to 10.0mL of distilled/deionized water. Mix thoroughly. Gas both solutions with 100% N_2 and cap tubes. Autoclave both solutions for 15 min at 15 psi pressure–121°C using fast exhaust. Cool to 50°C. Aseptically combine the two solutions under 100% N_2.

Wolfe's Vitamin Solution:
Composition per liter:
Pyridoxine·HCl 10.0mg
Thiamine·HCl 5.0mg
Riboflavin 5.0mg
Nicotinic acid 5.0mg
Calcium pantothenate 5.0mg
p-Aminobenzoic acid 5.0mg
Thioctic acid 5.0mg
Biotin 2.0mg
Folic acid 2.0mg
Cyanocobalamin 100.0μg

Preparation of Wolfe's Vitamin Solution: Add components to distilled/deionized water and bring volume to 1.0L. Mix thoroughly. Filter sterilize.

Trace Elements Solution SL-6:
Composition per liter:
H_3BO_3 .... 0.3g
$CoCl_2·6H_2O$ 0.2g
$ZnSO_4·7H_2O$ 0.1g
$MnCl_2·4H_2O$ 0.03g
$Na_2MoO_4·H_2O$ 0.03g
$NiCl_2·6H_2O$ 0.02g
$CuCl_2.2H_2O$ 0.01g

Preparation of Trace Elements Solution SL-6: Add components to distilled/deionized water and bring volume to 1.0L. Mix thoroughly. Adjust pH to 3.4.

Preparation of Medium: Add components, except $NaHCO_3$ solution, L-cysteine-sulfide reducing agent, Wolfe's vitamin solution, and methanol, to distilled/deionized water and bring volume to 940.0mL. Mix thoroughly. Autoclave for 15 min at 15 psi pressure–121°C. Cool under 80% N_2 + 20% CO_2. Aseptically and anaerobically add the sterile $NaHCO_3$ solution, the sterile L-cysteine-sulfide reducing agent, the sterile Wolfe's vitamin solution, and filter-sterilized methanol. Mix thoroughly. Adjust pH to 6.8. Aseptically and anaerobically distribute into sterile tubes or flasks.

Use: For the cultivation and maintenance of *Haloanaerobium praevalens*.

Haloanaerobium salsugo Medium

Composition per liter:
NaCl 90.0g
Purified agar (if necessary) 20.0g
Casamino acids 5.0g
Yeast extract 5.0g
Dipotassium PIPES (piperazine-*N,N'*-bis[2-ethanesulfonic acid]) buffer 1.5g
Resazurin 1.0mg
Glucose solution 50.0mL
L-Cysteine-sulfide reducing solution 20.0mL
Mineral solution 20.0mL
Wolfe's vitamin solution 10.0mL
Modified Wolfe's mineral solution 5.0mL
pH 6.0–7.0 at 25°C

Preparation of Medium: Prepare and dispense medium under 100% N_2. Add components, except glucose solution and L-cysteine-sulfide reducing solution, to distilled/deionized water and bring volume to 950.0mL Mix thoroughly. Gently heat and bring to boiling. Continue boiling for 3 min. Cool to room temperature while sparging with 100% N_2. Adjust pH to 6.0–7.0. Add 20.0mL of L-cysteine-sulfide reducing solution. Mix thoroughly. Anaerobically distribute 9.5mL volumes into anaerobic tubes. Autoclave for 15 min at 15 psi pressure–121°C. Aseptically and anaerobically add 0.5mL of sterile glucose solution to each tube. Mix thoroughly.

Glucose Solution:
Composition per 50.0mL:
D-Glucose 10.0g

Preparation of Glucose Solution: Add glucose to distilled/deionized water and bring volume to 50.0mL. Mix thoroughly. Sparge with 100% N_2. Autoclave for 15 min at 15 psi pressure–121°C.

Mineral Solution:
Composition per liter:
NH_4Cl 50.0g
NaCl 40.0g
$MgSO_4·7H_2O$ 10.0g
KCl 5.0g
KH_2PO_4 .... 5.0g
$CaCl_2·2H_2O$ 2.0g

Wolfe's Vitamin Solution:
Composition per liter:
Pyridoxine·HCl 10.0mg
p-Aminobenzoic acid 5.0mg
Lipoic acid 5.0mg
Nicotinic acid 5.0mg
Riboflavin 5.0mg
Thiamine·HCl 5.0mg
Calcium DL-pantothenate 5.0mg
Biotin 2.0mg
Folic acid 2.0mg
Vitamin B_{12} 0.1mg

Preparation of Wolfe's Vitamin Solution: Add components to distilled/deionized water and bring volume to 1.0L. Mix thoroughly.

Modified Wolfe's Mineral Solution:
Composition per liter:

$MgSO_4 \cdot 7H_2O$	3.0g
Nitrilotriacetic acid	1.5g
NaCl	1.0g
$MnSO_4 \cdot H_2O$	0.5g
$CaCl_2$	0.1g
$CoCl_2 \cdot 6H_2O$	0.1g
$FeSO_4 \cdot 7H_2O$	0.1g
$ZnSO_4 \cdot 7H_2O$	0.1g
$AlK(SO_4)_2 \cdot 12H_2O$	0.01g
$CuSO_4 \cdot 5H_2O$	0.01g
H_3BO_3	0.01g
$Na_2MoO_4 \cdot 2H_2O$	0.01g
Na_2SeO_3	0.01g
$NaWO_4 \cdot 2H_2O$	0.01g
$NiCl_2 \cdot 6H_2O$	0.01g

Preparation of Modified Wolfe's Mineral Solution: Add nitrilotriacetic acid to 500.0mL of distilled/deionized water. Adjust pH to 6.5 with KOH. Add remaining components one at a time. Add distilled/deionized water to 1.0L. Adjust pH to 6.8.

L-Cysteine-Sulfide Reducing Solution:
Composition per 200.0mL:

L-Cysteine·HCl·H_2O	5.0g
$Na_2S \cdot 9H_2O$	5.0g
NaOH	1.25g

Preparation of L-Cysteine-Sulfide Reducing Solution: Add NaOH to distilled/deionized water and bring volume to 200.0mL. Mix thoroughly. Gently heat and bring to boiling. Cool to room temperature while sparging with 100% N_2. Add L-cysteine·HCl·H_2O and $Na_2S \cdot 9H_2O$. Mix thoroughly. Anaerobically distribute into tubes. Autoclave for 15 min at 15 psi pressure–121°C.

Use: For the cultivation of *Haloanaerobium salsugo.*

Haloarcula japonica Medium

Composition per liter:

NaCl	200.0g
$MgSO_4 \cdot 7H_2O$	20.0g
Yeast extract	10.0g
Casamino acids	7.5g
Trisodium citrate·$2H_2O$	3.0g
KCl	2.0g
$FeSO_4 \cdot 7H_2O$	50.0mg
$MnCl_2 \cdot 4H_2O$	0.36mg

Preparation of Medium: Add components to distilled/deionized water and bring volume to 1.0L. Mix thoroughly. Distribute into tubes or flasks. Autoclave for 15 min at 15 psi pressure–121°C.

Use: For the cultivation of *Haloarcula japonica.*

Haloarcula marismortui Medium

Composition per liter:

NaCl	208.0g
$MgSO_4 \cdot 7H_2O$	46.6g
Yeast extract	10.0g
$CaCl_2$	0.5g
$MnCl_2$	0.125g

Preparation of Medium: Add components to distilled/deionized water and bring volume to 1.0L. Mix thoroughly. Distribute into tubes or flasks. Autoclave for 15 min at 15 psi pressure–121°C.

Use: For the cultivation of *Haloarcula marismortui.*

Haloarcula vallismortis Synthetic Medium

Composition per 1029.0mL:

Basal salts solution	1.0L
Glucose solution	20.0mL
NH_4Cl solution	5.0mL
$FeSO_4 \cdot 6H_2O$ solution	2.0mL
K_2HPO_4 solution	2.0mL

pH 7.5 ± 0.2 at 25°C

Basal Salts Solution:
Composition per liter:

NaCl	200.0g
$MgSO_4 \cdot 7H_2O$	36.0g
Tris[hydroxymethyl]aminomethane	6.0g
KCl	4.0g
$CaCl_2 \cdot 2H_2O$	1.0g

Preparation of Basal Salts Solution: Add components to distilled/deionized water and bring volume to 1.0L. Mix thoroughly. Adjust pH to 7.5 with HCl. Autoclave for 15 min at 15 psi pressure–121°C.

$FeSO_4 \cdot 6H_2O$ Solution:
Composition per 100.0mL:

$FeSO_4 \cdot 6H_2O$	0.2g
HCl (1.0m*M* solution)	100.0mL

Preparation of $FeSO_4 \cdot 6H_2O$ Solution: Combine components. Mix thoroughly. Filter sterilize.

K_2HPO_4 Solution:
Composition per 100.0mL:

K_2HPO_4	5.0g

Preparation of K_2HPO_4 Solution: Combine components. Mix thoroughly. Filter sterilize.

NH_4Cl Solution:
Composition per 100.0mL:

NH_4Cl	20.0g

Preparation of NH_4Cl Solution: Combine components. Mix thoroughly. Filter sterilize.

Glucose Solution:
Composition per 100.0mL:

D-Glucose	25.0g

Preparation of Glucose Solution: Add glucose to distilled/deionized water and bring volume to 100.0mL. Mix thoroughly. Filter sterilize.

Preparation of Glucose Solution: Aseptically combine 1.0L of sterile basal salts solution with 20.0mL of sterile glucose solution, 5.0mL of sterile NH_4Cl solution, 2.0mL of sterile K_2HPO_4 solution, and 2.0mL of sterile $FeSO_4 \cdot 6H_2O$ solution. Mix thoroughly. Aseptically distribute into sterile tubes or flasks.

Use: For the cultivation of *Haloarcula vallismortis.*

Halobacillus Medium (DSMZ Medium 755)

Composition per liter:

NaCl	100.0g
$MgSO_4 \cdot 7H_2O$	5.0g
Peptone, casein digest	5.0g
Yeast extract	3.0g

pH 7.5 ± 0.2 at 25°C

Preparation of Medium: Add components to distilled/deionized water and bring volume to 1.0L. Mix thoroughly. Adjust pH to 7.5. Distribute into tubes or flasks. Autoclave for 15 min at 15 psi pressure–121°C.

Use: For the cultivation of *Halobacillus trueperi* and *Halobacillus litoralis.*

Halobacteria Medium

Composition per liter:

NaCl	220.0g
Agar	10.0g
$MgSO_4 \cdot 7H_2O$	10.0g
Casein hydrolysate	5.0g
KCl	5.0g
Disodium citrate	3.0g
KNO_3	1.0g
Yeast extract	1.0g
$CaCl_2 \cdot 6H_2O$	0.2g

pH 7.2–7.4 at 25°C

Preparation of Medium: Add components to distilled/deionized water and bring volume to 1.0L. Mix thoroughly. Gently heat until dissolved. Adjust pH to 7.2–7.4. Distribute into tubes or flasks. Autoclave for 15 min at 15 psi pressure–121°C.

Use: For the cultivation and enumeration of halobacteria.

Halobacteriaceae Medium 1

Composition per liter:

Salt, crude solar	250.0g
$MgSO_4 \cdot 7H_2O$	20.0g
KCl	5.0g
Pancreatic digest of casein	5.0g
Yeast extract	5.0g
$CaCl_2 \cdot 6H_2O$	0.2g

pH 7.0 ± 0.2 at 25°C

Preparation of Medium: Add components to distilled/deionized water and bring volume to 1.0L. Mix thoroughly. Gently heat until dissolved. Adjust pH to 7.0. Distribute into tubes or flasks. Autoclave for 15 min at 15 psi pressure–121°C.

Use: For the axenic cultivation of members of the Halobacteriaceae.

Halobacteriaceae Medium 2

Composition per liter:

NaCl	250.0g
$MgSO_4 \cdot 7H_2O$	20.0g
Yeast extract	10.0g
Casamino acids	7.5g
Trisodium citrate	3.0g
KCl	2.0g
$FeCl_2$	2.3mg

pH 7.5–7.8 at 25°C

Preparation of Medium: Add components to distilled/deionized water and bring volume to 1.0L. Mix thoroughly. Gently heat until dissolved. Adjust pH to 7.5–7.8. Distribute into tubes or flasks. Autoclave for 15 min at 15 psi pressure–121°C.

Use: For the axenic cultivation of halobacteria and halococci.

Halobacteriaceae Medium 3

Composition per liter:

NaCl	240.0g
L-Glutamine	15.0g
KCl	5.0g
K_2SO_4	5.0g
$MgCl_2 \cdot 6H_2O$	5.0g
$MgSO_4$, anhydrous	5.0g
NH_4Cl	5.0g
Pancreatic digest of casein	5.0g
Yeast extract	5.0g
K_2HPO_4	0.5g
L-Arginine	0.5g
L-Isoleucine	0.25g
L-Leucine	0.25g
L-Lysine	0.25g
L-Proline	0.25g
L-Valine	0.25g
Cytidylic acid	0.2g
$CaCl_2 \cdot 2H_2O$	0.1g
L-Methionine	0.1g
L-Tyrosine	0.1g
L-Phenylalanine	0.05g
$FeCl_2 \cdot 6H_2O$	5.0mg

pH 6.8 ± 0.2 at 25°C

Preparation of Medium: Add components to distilled/deionized water and bring volume to 1.0L. Mix thoroughly. Gently heat until dissolved. Adjust pH to 6.8. Distribute into tubes or flasks. Autoclave for 15 min at 15 psi pressure–121°C.

Use: For the cultivation of some halobacteria and halococci.

Halobacteriaceae Medium 4

Composition per liter:

NaCl	250.0g
$MgSO_4 \cdot 7H_2O$	20.0g
NH_4Cl	5.0g
L-Glutamic acid	1.3g
DL-Valine	1.0g
Glycerol	1.0g
L-Lysine	0.85g
L-Leucine	0.8g
DL-Serine	0.61g
DL-Threonine	0.5g
DL-Isoleucine	0.44g
DL-Alanine	0.43g
L-Arginine	0.4g
DL-Methionine	0.37g
DL-Phenylalanine	0.26g
L-Tyrosine	0.2g
Adenylic acid	0.1g
KNO_3	0.1g
Uridylic acid	0.1g
Glycine	0.06g
KH_2PO_4	0.05g
K_2HPO_4	0.05g
L-Cysteine	0.05g
L-Proline	0.05g
Sodium citrate	0.05g
$FeCl_2$	2.3mg
$CaCl_2 \cdot 2H_2O$	0.7mg
$ZnSO_4 \cdot 7H_2O$	0.44mg
$MnSO_4 \cdot H_2O$	0.3mg
$CuSO_4 \cdot 5H_2O$	0.05mg

pH 6.2 ± 0.2 at 25°C

Preparation of Medium: Add components to distilled/deionized water and bring volume to 1.0L. Mix thoroughly. Gently heat until dissolved. Adjust pH to 6.2. Distribute into tubes or flasks. Autoclave for 15 min at 15 psi pressure–121°C.

Use: For the cultivation of members of the Halobacteriaceae.

Halobacterium Agar

Composition per liter:

NaCl	250.0g
Agar	20.0g
$MgSO_4 \cdot 7H_2O$	20.0g
Yeast extract	10.0g
Casamino acids	7.5g
Trisodium citrate	3.0g
KCl	2.0g
$FeSO_4 \cdot 7H_2O$	0.05g
$MnSO_4 \cdot H_2O$	0.2mg

pH 7.4 ± 0.2 at 25°C

Preparation of Medium: Add components, except agar, to distilled/deionized water and bring volume to 1.0L. Mix thoroughly. Add agar. Gently heat and bring to boiling. Distribute into tubes or flasks. Autoclave for 15 min at 15 psi pressure–121°C. Pour into sterile Petri dishes or leave in tubes.

Use: For the cultivation and maintenance of *Haloarcula* species, *Halobacterium* species, *Halococcus morrhuae, Haloferax mediterranei,* and *Haloferax volcanii.*

Halobacterium denitrificans Medium

Composition per liter:

NaCl	176.0g
Agar	20.0g
$MgCl_2 \cdot 6H_2O$	20.0g
HEPES (*N*-2-Hydroxyethylpiperazine-*N′*-2-ethanesulfonic acid) buffer	11.9g
Yeast extract	5.0g
Hy-Case SF (Humko-Sheffield)	2.0g
KCl	2.0g
$CaCl_2 \cdot 2H_2O$	0.1g

pH 6.7 ± 0.2 at 25°C

Preparation of Medium: Add components to distilled/deionized water and bring volume to 1.0L. Mix thoroughly. Gently heat and bring to boiling. Adjust pH to 6.7. Distribute into tubes or flasks. Autoclave for 15 min at 15 psi pressure–121°C. Pour into sterile Petri dishes or leave in tubes.

Use: For the aerobic cultivation and maintenance of *Haloferax (Halobacterium) denitrificans.*

Halobacterium denitrificans Medium

Composition per liter:

NaCl	176.0g
Agar	20.0g
$MgCl_2 \cdot 6H_2O$	20.0g
HEPES (*N*-2-Hydroxyethylpiperazine-*N′*-2-ethanesulfonic acid) buffer	11.9g
KNO_3	5.0g
Yeast extract	5.0g
Hy-Case SF (Humko-Sheffield)	2.0g
KCl	2.0g
$CaCl_2 \cdot 2H_2O$	0.1g

pH 6.7 ± 0.2 at 25°C

Preparation of Medium: Add components to distilled/deionized water and bring volume to 1.0L. Mix thoroughly. Gently heat and bring to boiling. Adjust pH to 6.7. Distribute into tubes or flasks. Autoclave for 15 min at 15 psi pressure–121°C. Pour into sterile Petri dishes or leave in tubes.

Use: For the anerobic cultivation and maintenance of *Haloferax (Halobacterium) denitrificans.*

Halobacterium halobium Defined Medium

Composition per liter:

NaCl	250.0g
$MgSO_4 \cdot 7H_2O$	20.0g
L-Glutamic acid	1.3g
KCl	1.0g
Glycerol	1.0g
KCl	1.0g
L-Valine	1.0g
L-Lysine	0.85g
L-Leucine	0.8g
$CaCl_2 \cdot 7H_2O$	0.71g
L-Serine	0.6g
Sodium citrate·$2H_2O$	0.5g
L-Proline	0.5g
L-Threonine	0.5g
L-Alanine	0.43g
L-Arginine	0.4g
L-Methionine	0.37g
L-Phenylalanine	0.26g
L-Tyrosine	0.2g
KH_2PO_4	0.15g
K_2HPO_4	0.15g
KNO_3	0.1g
KNO_3	0.1g
Glycine	0.06g
L-Cysteine·HCl·H_2O	0.05g
L-Isoleucine	0.044g
$FeCl_2 \cdot 5H_2O$	2.3mg
$ZnSO_4 \cdot 7H_2O$	0.44mg
$CuSO_4 \cdot 5H_2O$	0.005mg

Preparation of Medium: Add components to distilled/deionized water and bring volume to 1.0L. Mix thoroughly. Distribute into tubes or flasks. Autoclave for 15 min at 15 psi pressure–121°C.

Use: For the cultivation of *Halobacterium halobium.*

Halobacterium halobium/ Halobacterium salinarium Medium

Composition per liter:

NaCl	250.0g
$MgSO_4 \cdot 7H_2O$	20.0g
Pancreatic digest of casein	5.0g
Yeast extract	3.0g
Trisodium citrate·$2H_2O$	3.0g
KCl	2.0g
Trace metals solution	0.1mL

Trace Metals Solution:

Composition per liter:

$ZnSO_4 \cdot 7H_2O$	1.32g
$MnSO_4 \cdot H_2O$	0.34g
$Fe(NH_4)_2SO_4 \cdot 7H_2O$	0.78g
$CuSO_4 \cdot 5H_2O$	0.14g

Preparation of Trace Metals Solution: Add components to distilled/deionized water and bring volume to 1.0L. Mix thoroughly.

Preparation of Medium: Add components to distilled/deionized water and bring volume to 1.0L. Mix thoroughly. Distribute into tubes or flasks. Autoclave for 15 min at 15 psi pressure–121°C.

Use: For the cultivation of *Halobacterium halobium* and *Halobacterium salinarium.*

Halobacterium/Halococcus Medium

Composition per liter:

Solution A....500.0mL
Salt solution....500.0mL

Solution A:

Composition per 500.0mL:

Skim milk....50.0g

Preparation of Solution A: Add 50.0g of skim milk to distilled/deionized water and bring volume to 500.0mL. Mix thoroughly.

Salt Solution:

Composition per 500.0mL:

$MgSO_4 \cdot 7H_2O$....10.0g
KNO_3....2.0g

Preparation of Salt Solution: Add components to distilled/deionized water and bring volume to 500.0mL. Mix thoroughly.

Preparation of Medium: Add components to distilled/deionized water and bring volume to 1.0L. Mix thoroughly. Distribute into tubes or flasks. Autoclave for 15 min at 15 psi pressure–121°C.

Use: For the cultivation of *Haloarcula vallismortis, Halobacterium cutirubrum, Halobacterium halobium, Halobacterium saccharovorum, Halobacterium salinarium,* and *Halococcus morrhuae.*

Halobacterium Medium

Composition per liter:

Solution 1....500.0mL
Solution 2....500.0mL

pH 7.0 ± 0.2 at 25°C

Solution 1:

Composition per 500.0mL:

Yeast extract....10.0g
Pancreatic digest of casein....2.5g

Preparation of Solution 1: Add components to distilled/deionized water and bring volume to 500.0mL. Mix thoroughly. Gently heat and bring to boiling. Adjust pH to 7.0. Autoclave for 15 min at 15 psi pressure–121°C. Cool to 45°–50°C.

Solution 2:

Composition per 500.0mL:

NaCl....250.0g
$MgSO_4 \cdot 7H_2O$....10.0g
KCl....5.0g
$CaCl_2 \cdot 6H_2O$....0.2g

Preparation of Solution 2: Add components to distilled/deionized water and bring volume to 500.0mL. Mix thoroughly. Gently heat and bring to boiling. Autoclave for 15 min at 15 psi pressure–121°C. Cool to 45°–50°C.

Preparation of Medium: Aseptically combine sterile solution 1 and sterile solution 2. Mix thoroughly. Aseptically distribute into sterile tubes or flasks.

Use: For the cultivation of *Halobacterium salinarium.*

Halobacterium Medium (ATCC Medium 974)

Composition per 100.0mL:

Solution 1....75.0mL
Solution 2....25.0mL

pH 6.8 ± 0.2 at 25°C

Solution 1:

Composition per 75.0mL:

NaCl....12.5g
$MgCl_2 \cdot 6H_2O$....5.0g
K_2SO_4....0.5g
$CaCl_2 \cdot 6H_2O$....0.02g

Preparation of Solution 1: Add components to distilled/deionized water and bring volume to 75.0mL. Mix thoroughly. Adjust pH to 6.8. Autoclave for 15 min at 15 psi pressure–121°C. Cool to 45°–50°C.

Solution 2:

Composition per 25.0mL:

Agar....2.0g
Pancreatic digest of casein....0.5g
Yeast extract....0.5g

Preparation of Solution 2: Add components to distilled/deionized water and bring volume to 25.0mL. Mix thoroughly. Adjust pH to 6.8. Autoclave for 15 min at 15 psi pressure–121°C. Cool to 45°–50°C.

Preparation of Medium: Aseptically combine sterile solution 1 and sterile solution 2. Mix thoroughly. Pour into sterile Petri dishes or distribute into sterile tubes.

Use: For the cultivation and maintenance of *Haloferax volcanii.*

Halobacterium Medium (ATCC Medium 1176)

Composition per liter:

NaCl....156.0g
$MgSO_4 \cdot 7H_2O$....20.0g
$MgCl_2 \cdot 6H_2O$....13.0g
Yeast extract....5.0g
KCl....4.0g
$CaCl_2 \cdot 6H_2O$....1.0g
Glucose....1.0g
NaBr....0.5g
$NaHCO_3$....0.2g

pH 7.0 ± 0.2 at 25°C

Preparation of Medium: Add components to distilled/deionized water and bring volume to 1.0L. Mix thoroughly. Distribute into tubes or flasks. Autoclave for 15 min at 15 psi pressure–121°C.

Use: For the cultivation of *Haloferax mediterranei.*

Halobacterium Medium (ATCC Medium 1270)

Composition per liter:

NaCl....194.0g
$MgSO_4$....24.0g
$MgCl_2$....16.0g
KCl....5.0g
Yeast extract....5.0g
$CaCl_2$....1.0g
NaBr....0.5g
$NaHCO_3$....0.2g

pH 7.3 ± 0.2 at 25°C

Preparation of Medium: Add components to distilled/deionized water and bring volume to 1.0L. Mix thoroughly. Distribute into tubes or flasks. Autoclave for 15 min at 15 psi pressure–121°C.

Use: For the cultivation of *Haloarcula hispanica* and *Haloferax gibbonsii.*

Halobacterium pharaonis Medium

Composition per liter:

NaCl	250.0g
Agar	20.0g
Casamino acids	15.0g
Trisodium citrate·$2H_2O$	3.0g
Glutamic acid	2.5g
$MgSO_4 \cdot 7H_2O$	2.5g
KCl	2.0g

pH 8.5 ± 0.2 at 25°C

Preparation of Medium: Add components to distilled/deionized water and bring volume to 1.0L. Mix thoroughly. Gently heat and bring to boiling. Adjust pH to 6.0. Autoclave for 15 min at 15 psi pressure–121°C. Cool to 50°C. Readjust pH to 8.5. Pour into sterile Petri dishes or distribute into sterile tubes.

Use: For the cultivation and maintenance of *Natronobacterium (Halobacterium) pharaonis.*

Halobacterium saccharovorum Medium

Composition per liter:

NaCl	250.0g
MgCl·$6H_2O$	20.0g
Glucose	10.0g
Casimino acids	5.0g
Yeast extract	2.5g
KCl	2.0g
$CaCl_2 \cdot 7H_2O$	0.2g

pH 7.35 ± 0.2 at 25°C

Preparation of Medium: Add components to distilled/deionized water and bring volume to 1.0L. Mix thoroughly. Distribute into tubes or flasks. Autoclave for 15 min at 15 psi pressure–121°C.

Use: For the cultivation of *Halobacterium saccharovorum.*

Halobacterium soldomense Medium

Composition per liter:

$MgCl_2 \cdot 6H_2O$	160.0g
NaCl	125.0g
K_2SO_4	5.0g
Soluble starch	2.0g
Peptone	1.0g
Yeast extract	1.0g
$CaCl_2 \cdot 2H_2O$	0.13g

pH 7.0 ± 0.2 at 25°C

Preparation of Medium: Add components to distilled/deionized water and bring volume to 1.0L. Mix thoroughly. Distribute into tubes or flasks. Autoclave for 15 min at 15 psi pressure–121°C.

Use: For the cultivation of *Halobacterium soldomense.*

Halobacterium Starch Medium

Composition per liter:

$MgCl_2 \cdot 6H_2O$	160.0g
NaCl	125.0g
K_2SO_4	5.0g
Soluble starch	2.0g
Peptone	1.0g
Yeast extract	1.0g
$CaCl_2 \cdot 2H_2O$	0.13g

pH 7.0 ± 0.2 at 25°C

Preparation of Medium: Add components to distilled/deionized water and bring volume to 1.0L. Mix thoroughly. Distribute into tubes or flasks. Autoclave for 15 min at 15 psi pressure–121°C.

Use: For the cultivation and maintenance of *Haloarcula marismortui.*

Halobacterium volcanii Medium

Composition per 100.0mL:

NaCl	25.0g
$MgSO_4 \cdot 7H_2O$	1.0g
KCl	0.5g
Glycine	0.2g
$CaCl_2 \cdot 6H_2O$	0.02g
Yeast autolysate	1.0mL

pH 7.0 ± 0.2 at 25°C

Preparation of Medium: Add components to distilled/deionized water and bring volume to 100.0mL. Mix thoroughly. Adjust pH to 7.0. Distribute into tubes or flasks. Autoclave for 15 min at 15 psi pressure–121°C.

Use: For the specific enrichment of *Halobacterium volcanii.*

Halobacteroides acetoethylicus Medium

Composition per liter:

NaCl	100.0g
Pancreatic digest of casein	10.0g
Yeast extract	3.0g
K_2HPO_4	1.5g
NH_4Cl	0.9g
KH_2PO_4	0.75g
$MgCl_2 \cdot 6H_2O$	0.2g
Glucose solution	25.0mL
$Na_2S \cdot 9H_2O$ (10% solution)	10.0mL
Trace elements solution	9.0mL
Vitamin solution	5.0mL
Resazurin (0.025% solution)	4.0mL
$FeSO_4 \cdot 7H_2O$ (10% solution)	0.03mL

pH 7.3 ± 0.2 at 25°C

Glucose Solution:

Composition per 100.0mL:

Glucose	20.0g

Preparation of Glucose Solution: Add glucose to distilled/deionized water and bring volume to 100.0mL. Mix thoroughly. Filter sterilize. Sparge with 100% N_2.

Trace Elements Solution:

Composition per liter:

Nitrilotriacetic acid	12.5g
NaCl	1.0g
$FeCl_3 \cdot 4H_2O$	0.2g
$MnCl_2 \cdot 4H_2O$	0.1g
$CaCl_2 \cdot 2H_2O$	0.1g
$ZnCl_2$	0.1g
$CuCl_2$	0.02g
Na_2SeO_3	0.02g
$CoCl_2 \cdot 6H_2O$	0.017g
H_3BO_3	0.01g
$Na_2MoO_4 \cdot 2H_2O$	0.01g

Preparation of Trace Elements Solution: Add nitrilotriacetic acid to 500.0mL of distilled/deionized water. Adjust pH to 6.5 with KOH. Add remaining components. Add distilled/deionized water to 1.0L.

Wolfe's Vitamin Solution:

Composition per liter:

Pyridoxine·HCl	10.0mg
Thiamine·HCl	5.0mg
Riboflavin	5.0mg
Nicotinic acid	5.0mg
Calcium DL-pantothenate	5.0mg

p-Aminobenzoic acid	5.0mg
Thioctic acid	5.0mg
Biotin	2.0mg
Folic acid	2.0mg
Cyanocobalamin	100.0μg

Preparation of Wolfe's Vitamin Solution: Add components to distilled/deionized water and bring volume to 1.0L. Mix thoroughly.

Preparation of Medium: Add components, except glucose solution, to distilled/deionized water and bring volume to 975.0mL. Mix thoroughly. Autoclave for 15 min at 15 psi pressure–121°C. While still hot, aseptically add 25.0mL of the sterile glucose solution under 97% N_2 + 3% H_2. Adjust pH to 7.3 if necessary. Aseptically and anaerobically distribute into tubes. Cap with rubber stoppers.

Use: For the cultivation and maintenance of *Halobacteroides acetoethylicus.*

Halobacteroides/Haloincola Medium

Composition per liter:

NaCl	100.0g
Peptone	5.0g
$CaCl_2$	0.33g
KCl	0.33g
KH_2PO_4	0.33g
$MgCl_2$	0.33g
NH_4Cl	0.33g
Resazurin	2.0mg
Glucose solution	30.0mL
$NaHCO_3$ solution	10.0mL
$Na_2S·9H_2O$ solution	10.0mL
Wolfe's vitamin solution	10.0mL
Trace elements solution SL-10	1.0mL

pH 7.5 ± 0.2 at 25°C

Glucose Solution:
Composition per 30.0mL:

D-Glucose	5.0g

Preparation of Glucose Solution: Add glucose to distilled/deionized water and bring volume to 30.0mL. Mix thoroughly. Filter sterilize. Sparge with 100% N_2.

$NaHCO_3$ Solution:
Composition per 10.0mL:

$NaHCO_3$	1.5g

Preparation of $NaHCO_3$ Solution: Add $NaHCO_3$ to distilled/deionized water and bring volume to 10.0mL. Mix thoroughly. Filter sterilize. Sparge with 100% N_2.

$Na_2S·9H_2O$ Solution:
Composition per 10.0mL:

$Na_2S·9H_2O$	0.5g

Preparation of $Na_2S·9H_2O$ Solution: Add $Na_2S·9H_2O$ to distilled/deionized water and bring volume to 10.0mL. Mix thoroughly. Sparge with 100% N_2. Autoclave for 15 min at 15 psi pressure–121°C.

Wolfe's Vitamin Solution:
Composition per liter:

Pyridoxine·HCl	10.0mg
p-Aminobenzoic acid	5.0mg
Lipoic acid	5.0mg
Nicotinic acid	5.0mg
Riboflavin	5.0mg
Thiamine·HCl	5.0mg
Calcium DL-pantothenate	5.0mg
Biotin	2.0mg
Folic acid	2.0mg
Vitamin B_{12}	0.1mg

Preparation of Wolfe's Vitamin Solution: Add components to distilled/deionized water and bring volume to 1.0L. Mix thoroughly. Filter sterilize.

Trace Elements Solution SL-10:
Composition per liter:

$FeCl_2·4H_2O$	1.5g
$CoCl_2·6H_2O$	190.0mg
$MnCl_2·4H_2O$	100.0mg
$ZnCl_2$	70.0mg
$Na_2MoO_4·2H_2O$	36.0mg
$NiCl_2·6H_2O$	24.0mg
H_3BO_3	6.0mg
$CuCl_2·2H_2O$	2.0mg
HCl (25% solution)	10.0mL

Preparation of Trace Elements Solution SL-10: Add $FeCl_2·4H_2O$ to 10.0mL of HCl solution. Mix thoroughly. Add distilled/deionized water and bring volume to 1.0L. Add remaining components. Mix thoroughly. Sparge with 100% N_2. Autoclave for 15 min at 15 psi pressure–121°C.

Preparation of Medium: Prepare and dispense medium under 80% N_2 + 20% CO_2. Add components, except glucose solution, $NaHCO_3$ solution, $Na_2S·9H_2O$ solution, and Wolfe's vitamin solution, to distilled/deionized water and bring volume to 940.0mL Mix thoroughly. Gently heat and bring to boiling. Cool to room temperature while sparging with 80% N_2 + 20% CO_2. Autoclave for 15 min at 15 psi pressure–121°C. Aseptically and anaerobically add 30.0mL of sterile glucose solution, 10.0mL of sterile $NaHCO_3$ solution, 10.0mL of sterile $Na_2S·9H_2O$ solution, and 10.0mL of sterile Wolfe's vitamin solution. Mix thoroughly. Check that final pH is 7.5.

Use: For the cultivation of *Haloanaerobium (Haloincola) saccharolyticum.*

Halobacteroides Medium

Composition per 990.0mL:

NaCl	88.0g
$MgCl_2·6H_2O$	20.0g
$CaCl_2·2H_2O$	7.4g
Yeast extract	5.0g
KCl	3.7g
L-Cysteine·HCl·H_2O	0.5g
Resazurin	1.0mg
Glucose (10% solution)	50.0mL
Sodium PIPES (Piperazine-*N,N'*-bis-2-ethane sulfonate buffer, 1*M*, pH 6.8–7.0)	40.0mL

pH 6.8–7.0 at 25°C

Preparation of Medium: Filter sterilize glucose solution and PIPES buffer solution separately. Add remaining components—except glucose solution, PIPES buffer solution, and L-cysteine·HCl·H_2O—to distilled/deionized water and bring volume to 900.0mL. Mix thoroughly. Gently heat and bring to boiling under 100% N_2. Add L-cysteine·HCl·H_2O. Mix thoroughly. Autoclave for 15 min at 15 psi pressure–121°C. Cool to 45°–50°C. Aseptically add 50.0mL of sterile glucose solution and 40.0mL of sterile PIPES buffer solution. Mix thoroughly. Aseptically and anaerobically distribute into sterile tubes or flasks.

Use: For the cultivation and maintenance of *Halobacteroides halobius* and *Sporohalobacter marismortui.*

Halobacteroides Medium

Composition per liter:

NaCl	150.0g
Sucrose	5.0
$MgSO_4 \cdot 7H_2O$	4.0g
$NaHCO_3$	2.0g
$Na_2S \cdot 9H_2O$	0.5g
Yeast extract	0.5g
$CaCl_2 \cdot 2H_2O$	0.33g
KCl	0.33g
KH_2PO_4	0.33g
$MgCl_2 \cdot 6H_2O$	0.33g
NH_4Cl	0.33g
Resazurin	2.0mg
Wolfe's mineral solution	10.0mL
Wolfe's vitamin solution	10.0mL

pH 7.0 ± 0.2 at 25°C

Wolfe's Mineral Solution:

Composition per liter:

$MgSO_4 \cdot 7H_2O$	3.0g
Nitrilotriacetic acid	1.5g
NaCl	1.0g
$MnSO_4 \cdot 2H_2O$	0.5g
$CoCl_2 \cdot 6H_2O$	0.1g
$ZnSO_4 \cdot 7H_2O$	0.1g
$CaCl_2 \cdot 2H_2O$	0.1g
$FeSO_4 \cdot 7H_2O$	0.1g
$NiCl_2 \cdot 6H_2O$	0.025g
$KAl(SO_4)_2 \cdot 12H_2O$	0.02g
$CuSO_4 \cdot 5H_2O$	0.01g
H_3BO_3	0.01g
$Na_2MoO_4 \cdot 2H_2O$	0.01g
$Na_2SeO_3 \cdot 5H_2O$	0.3mg

Preparation of Wolfe's Mineral Solution: Add nitrilotriacetic acid to 500.0mL of distilled/deionized water. Adjust pH to 6.5 with KOH. Add remaining components one at a time. Add distilled/deionized water to 1.0L. Adjust pH to 6.8.

Wolfe's Vitamin Solution:

Composition per liter:

Pyridoxine·HCl	10.0mg
p-Aminobenzoic acid	5.0mg
Lipoic acid	5.0mg
Nicotinic acid	5.0mg
Riboflavin	5.0mg
Thiamine·HCl	5.0mg
Calcium DL-pantothenate	5.0mg
Biotin	2.0mg
Folic acid	2.0mg
Vitamin B_{12}	0.1mg

Preparation of Wolfe's Vitamin Solution: Add components to distilled/deionized water and bring volume to 1.0L. Mix thoroughly.

Preparation of Medium: Prepare and dispense medium under 80% N_2 + 10% CO_2 + 10% H_2. Add components, except $NaHCO_3$ and $Na_2S \cdot 9H_2O$, to distilled/deionized water and bring volume to 1.0L. Mix thoroughly. Gently heat and bring to boiling. Continue boiling for 3 min. Cool to room temperature while sparging with 80% N_2 + 10% CO_2 + 10% H_2. Add $NaHCO_3$ and $Na_2S \cdot 9H_2O$. Mix thoroughly. Adjust pH to 7.0. Anaerobically distribute into tubes or flasks. Autoclave for 15 min at 15 psi pressure–121°C.

Use: For the cultivation of *Haloanaerobacter lacunaris.*

Halobaculum gomorrense Medium (DSMZ Medium 823)

Composition per liter:

$MgCl_2 \cdot 6H_2O$	160.0g
NaCl	125.0g
K_2SO_4	5.0g
Starch	2.0g
Yeast extract	1.0g
Casamino acids	1.0g
$CaCl_2 \cdot 2H_2O$	0.1g

pH 7.0 ± 0.2 at 25°C

Preparation of Medium: Add components to distilled/deionized water and bring volume to 1.0L. Mix thoroughly. Distribute into tubes or flasks. Autoclave for 15 min at 15 psi pressure–121°C.

Use: For the cultivation of *Halobaculum gomorrense (Haloferax gomorrae).*

Halobius Medium

Composition per liter:

NaCl	116.0g
Agar	20.0g
$MgSO_4 \cdot 7H_2O$	20.0g
Yeast extract	10.0g
Vitamin-free casamino acids	7.5g
Sodium citrate	3.0g
KCl	2.0g
$FeCl_2$	0.023g

pH 6.2 ± 0.2 at 25°C

Preparation of Medium: Add components to distilled/deionized water and bring volume to 1.0L. Mix thoroughly. Gently heat and bring to boiling. Distribute into tubes or flasks. Autoclave for 15 min at 15 psi pressure–121°C. Pour into sterile Petri dishes or leave in tubes.

Use: For the cultivation and maintenance of *Micrococcus halobius*.

Halocella cellulolytica Medium

Composition per liter:

NaCl	150.0g
Cellobiose or microcrystalline cellulose	5.0g
Yeast extract	2.0g
$MgCl_2 \cdot 6H_2O$	3.6g
$CaCl_2$	0.33g
KCl	0.33g
KH_2PO_4	0.33g
NH_4Cl	0.33g
Resazurin	2.0mg
$NaHCO_3$ solution	10.0mL
$Na_2S \cdot 9H_2O$ solution	10.0mL
Wolfe's vitamin solution	10.0mL
Trace elements solution SL-10	1.0mL

pH 7.0 ± 0.2 at 25°C

$NaHCO_3$ Solution:

Composition per 10.0mL:

$NaHCO_3$	2.5g

Preparation of $NaHCO_3$ Solution: Add $NaHCO_3$ to distilled/deionized water and bring volume to 10.0mL. Mix thoroughly. Filter sterilize. Sparge with 100% N_2.

$Na_2S \cdot 9H_2O$ Solution:

Composition per 10.0mL:

$Na_2S \cdot 9H_2O$	0.5g

Preparation of $Na_2S \cdot 9H_2O$ Solution: Add $Na_2S \cdot 9H_2O$ to distilled/deionized water and bring volume to 10.0mL.

Mix thoroughly. Sparge with 100% N_2. Autoclave for 15 min at 15 psi pressure–121°C.

Wolfe's Vitamin Solution:

Composition per liter:

Pyridoxine·HCl	10.0mg
p-Aminobenzoic acid	5.0mg
Lipoic acid	5.0mg
Nicotinic acid	5.0mg
Riboflavin	5.0mg
Thiamine·HCl	5.0mg
Calcium DL-pantothenate	5.0mg
Biotin	2.0mg
Folic acid	2.0mg
Vitamin B_{12}	0.1mg

Preparation of Wolfe's Vitamin Solution: Add components to distilled/deionized water and bring volume to 1.0L. Mix thoroughly. Filter sterilize.

Trace Elements Solution SL-10:

Composition per liter:

$FeCl_2 \cdot 4H_2O$	1.5g
$CoCl_2 \cdot 6H_2O$	190.0mg
$MnCl_2 \cdot 4H_2O$	100.0mg
$ZnCl_2$	70.0mg
$Na_2MoO_4 \cdot 2H_2O$	36.0mg
$NiCl_2 \cdot 6H_2O$	24.0mg
H_3BO_3	6.0mg
$CuCl_2 \cdot 2H_2O$	2.0mg
HCl (25% solution)	10.0mL

Preparation of Trace Elements Solution SL-10: Add $FeCl_2 \cdot 4H_2O$ to 10.0mL of HCl solution. Mix thoroughly. Add distilled/deionized water and bring volume to 1.0L. Add remaining components. Mix thoroughly. Sparge with 100% N_2. Autoclave for 15 min at 15 psi pressure–121°C.

Preparation of Medium: Prepare and dispense medium under 80% N_2 + 20% CO_2. Add components, except $NaHCO_3$ solution, $Na_2S \cdot 9H_2O$ solution, and Wolfe's vitamin solution, to distilled/deionized water and bring volume to 970.0mL. Mix thoroughly. Gently heat and bring to boiling. Cool to room temperature while sparging with 80% N_2 + 20% CO_2. Autoclave for 15 min at 15 psi pressure–121°C. Aseptically and anaerobically add 10.0mL of sterile $NaHCO_3$ solution, 10.0mL of sterile $Na_2S \cdot 9H_2O$ solution, and 10.0mL of sterile Wolfe's vitamin solution. Mix thoroughly. Check that final pH is 7.0.

Use: For the cultivation of *Halocella cellulolytica.*

Halococcus Agar

Composition per liter:

Solution A	500.0mL
Solution B	100.0mL
Solution C	400.0mL

pH 8.4 ± 0.2 at 25°C

Solution A:

Composition per 500.0mL:

Skim milk powder	50.0g

Preparation of Solution A: Add skim milk powder to distilled/deionized water and bring volume to 500.0mL. Autoclave for 15 min at 8 psi pressure–112°C. Cool to 50°-55°C.

Solution B:

Composition per 100.0mL:

NaCl	200.0g
KNO_3	2.0g
$MgSO_4 \cdot 7H_2O$	10.0mg
Ferric citrate	1.0μg

Preparation of Solution B: Add components to distilled/deionized water and bring volume to 100.0mL. Autoclave for 15 min at 15 psi pressure–121°C. Cool to 50°-55°C.

Solution C:

Composition per 400.0mL:

Agar	25.0g
Glycerol	10.0g
Neopeptone	5.0g

Preparation of Solution C: Add components to distilled/deionized water and bring volume to 400.0mL. Gently heat and bring to boiling. Autoclave for 15 min at 15 psi pressure–121°C. Cool to 50-55°C.

Preparation of Medium: Aseptically combine 100.0mL of sterile solution B with 400.0mL of sterile solution C. Mix thoroughly. Adjust pH to 8.4. Aseptically add 500.0mL of sterile solution A. Mix thoroughly. Pour into sterile Petri dishes or distribute into sterile tubes.

Use: For the cultivation and maintenance of *Halococcus morrhuae.*

Halococcus dombrowskii Medium (DSMZ Medium 954)

Composition per liter:

NaCl	200.0g
Agar	20.0g
$MgCl_2 \cdot 6H_2O$	20.0g
TRIS	12.1g
Casamino acids	5.0g
Yeast extract	5.0g
KCl	2.0g
$CaCl_2 \cdot 2H_2O$	0.2g

pH 7.4 ± 0.2 at 25°C

Preparation of Medium: Add components, except agar, to distilled/deionized water and bring volume to 1.0L. Mix thoroughly. Adjust pH to 7.4. Add agar. Gently heat and bring to boiling. Distribute into tubes or flasks. Autoclave for 15 min at 15 psi pressure–121°C. Pour into sterile Petri dishes or leave in tubes.

Use: For the cultivation and maintenance of *Halococcus dombrowskii.*

Halodurans Medium

Composition per liter:

NaCl	150.0g
Agar	20.0g
$MgSO_4 \cdot 7H_2O$	20.0g
Yeast extract	10.0g
Casamino acids	7.5g
Sodium citrate	3.0g
KCl	2.0g
$FeCl_2$	0.023g

pH 7.0 ± 0.2 at 25°C

Preparation of Medium: Add components to distilled/deionized water and bring volume to 1.0L. Mix thoroughly. Gently heat and bring to boiling. Distribute into tubes or flasks. Autoclave for 15 min at 15 psi pressure–121°C. Pour into sterile Petri dishes or leave in tubes.

Use: For the cultivation and maintenance of *Micrococcus varians.*

Haloferax mediterranei Medium

Composition per liter:

NaCl	195.0g
$MgSO_4 \cdot 7H_2O$	49.4g
$MgCl_2 \cdot 6H_2O$	34.6g
Yeast extract	5.0g
$CaCl_2 \cdot 2H_2O$	0.92g
NaBr	0.58g
KCl	0.5g
$NaHCO_3$	0.17g

pH 7.2 ± 0.2 at 25°C

Preparation of Medium: Add components to distilled/deionized water and bring volume to 1.0L. Mix thoroughly. Distribute into tubes or flasks. Autoclave for 15 min at 15 psi pressure–121°C.

Use: For the cultivation of *Haloferax mediterranei.*

Haloferax mediterranei Minimal Medium I

Composition per liter:

NaCl	156.0g
$MgSO_4 \cdot 7H_2O$	20.0g
$MgCl_2 \cdot 6H_2O$	13.0g
Glucose	10.0g
KCl	4.0g
NaH_4Cl	2.0g
$CaCl_2 \cdot 2H_2O$	1.0g
NaBr	0.58g
KH_2PO_4	0.5g
$NaHCO_3$	0.2g
$FeCl_2 \cdot 6H_2O$	0.005g

pH 7.2 ± 0.2 at 25°C

Preparation of Medium: Add components to distilled/deionized water and bring volume to 1.0L. Mix thoroughly. Distribute into tubes or flasks. Autoclave for 15 min at 15 psi pressure–121°C.

Use: For the cultivation of *Haloferax mediterranei.*

Haloferax mediterranei Minimal Medium II

Composition per liter:

NaCl	160.0g
$MgCl_2 \cdot 7H_2O$	20.0g
Sodium glutamate·H_2O	20.0g
Sucrose	10.0g
KCl	4.0g

pH 7.2 ± 0.2 at 25°C

Preparation of Medium: Add components to distilled/deionized water and bring volume to 1.0L. Mix thoroughly. Distribute into tubes or flasks. Autoclave for 15 min at 15 psi pressure–121°C.

Use: For the cultivation of *Haloferax mediterranei.*

Haloferax volcanii Low-Salt Medium

Composition per liter:

NaCl	125.0g
$MgCl_2 \cdot 6H_2O$	45.0g
$MgSO_4 \cdot 7H_2O$	10.0g
KCl	10.0g
$CaCl_2 \cdot 2H_2O$	10.0g
Pancreatic digest of casein	3.0g
Yeast extract	1.34g

pH 7.2 ± 0.2 at 25°C

Preparation of Medium: Add components to distilled/deionized water and bring volume to 1.0L. Mix thoroughly. Distribute into tubes or flasks. Autoclave for 15 min at 15 psi pressure–121°C.

Use: For the cultivation of *Haloferax volcanii.*

Haloferax volcanii Medium

Composition per liter:

NaCl	206.0g
$MgSO_4 \cdot 7H_2O$	37.0g
KCl	3.7g
Yeast tryptone solution	100.0mL
Tris[hydroxymethyl]aminomethane·HCl (1*M* solution, pH 7.2)	50.0mL
$CaCl_2 \cdot 2H_2O$ solution	5.0mL
$MnCl_2$ solution	1.7mL

pH 7.2 ± 0.2 at 25°C

Yeast Tryptone Solution:

Composition per 100.0mL:

Tryptone	5.0g
Yeast extract	3.0g

Preparation of Yeast Tryptone Solution: Add components to distilled/deionized water and bring volume to 100.0mL. Mix thoroughly.

$CaCl_2 \cdot 2H_2O$ Solution:

Composition per 100.0mL:

$CaCl_2 \cdot 2H_2O$	10.0g

Preparation of $CaCl_2 \cdot 2H_2O$ Solution: Add $CaCl_2 \cdot 2H_2O$ to distilled/deionized water and bring volume to 100.0mL. Mix thoroughly.

$MnCl_2 \cdot 6H_2O$ Solution:

Composition per liter:

$MnCl_2 \cdot 6H_2O$	75.0mg

Preparation of $MnCl_2 \cdot 6H_2O$ Solution: Add $MnCl_2 \cdot 6H_2O$ to distilled/deionized water and bring volume to 1.0L. Mix thoroughly.

Preparation of Medium: Add components to distilled/deionized water and bring volume to 1.0L. Mix thoroughly. Distribute into tubes or flasks. Autoclave for 15 min at 15 psi pressure–121°C.

Use: For the cultivation of *Haloferax volcanii.*

Haloferax volcanii Minimal Medium

Composition per liter:

NaCl	206.0g
$MgSO_4 \cdot 7H_2O$	36.9g
Glycerol	45.0mL
Sodium succinate solution	5.0mL
KCl solution	5.0mL
$CaCl_2$ solution	5.0mL
NH_4Cl solution	5.0mL
$MnCl_2$ solution	1.7mL
K_2HPO_4 solution	2.0mL
Trace elements solution	1.0mL

pH 7.2 ± 0.2 at 25°C

Sodium Succinate Solution:

Composition per 100.0mL:

Sodium succinate	10.0g

Preparation of Sodium Succinate Solution: Add sodium succinate to distilled/deionized water and bring volume to 100.0mL. Mix thoroughly.

KCl Solution:

Composition per 100.0mL:

KCl	7.45g

Preparation of KCl Solution: Add KCl to distilled/deionized water and bring volume to 100.0mL. Mix thoroughly.

$CaCl_2 \cdot 2H_2O$ Solution:
Composition per 100.0mL:
$CaCl_2 \cdot 2H_2O$ 10.0g

Preparation of $CaCl_2 \cdot 2H_2O$ Solution: Add $CaCl_2 \cdot 2H_2O$ to distilled/deionized water and bring volume to 100.0mL. Mix thoroughly. Filter sterilize.

NH_4Cl Solution:
Composition per 100.0mL:
NH_4Cl 5.35g

Preparation of NH_4Cl Solution: Add NH_4Cl to distilled/deionized water and bring volume to 100.0mL. Mix thoroughly.

$MnCl_2 \cdot 6H_2O$ Solution:
Composition per liter:
$MnCl_2 \cdot 6H_2O$ 75.0mg

Preparation of $MnCl_2 \cdot 6H_2O$ Solution: Add $MnCl_2 \cdot 6H_2O$ to distilled/deionized water and bring volume to 1.0L. Mix thoroughly.

K_2HPO_4 Solution:
Composition per 100.0mL:
K_2HPO_4 .... 8.7g

Preparation of K_2HPO_4 Solution: Add K_2HPO_4 to distilled/deionized water and bring volume to 100.0mL. Mix thoroughly. Adjust pH to 7.0.

Trace Elements Solution:
Composition per 100.0mL:
$FeSO_4$ 334.0mg
$MnCl_2$ 36.0mg
$ZnSO_4$ 44.0mg
$CuSO_4$ 5.0mg

Preparation of Trace Elements Solution: Add components to distilled/deionized water and bring volume to 100.0mL. Mix thoroughly.

Preparation of Medium: Add components, except $CaCl_2$ solution, to distilled/deionized water and bring volume to 995.0mL. Mix thoroughly. Adjust pH to 7.2. Autoclave for 15 min at 15 psi pressure–121°C. Aseptically add 5.0mL of sterile $CaCl_2$ solution. Mix thoroughly. Aseptically distribute into sterile tubes or flasks.

Use: For the cultivation of *Haloferax volcanii.*

Halomethanococcus Medium (*Methanohalophilus* Medium)

Composition per 1030.0mL:
Trimethylamine·HCl 2.5g
Na_2CO_3 .... 2.0g
$NaHCO_3$ 2.0g
Casamino acids 0.5g
L-Cysteine·HCl·H_2O 0.5g
$Na_2S \cdot 9H_2O$ solution 0.5g
NH_4Cl 0.5g
Pancreatic digest of casein 0.5g
Yeast extract 0.5g
K_2HPO_4 .... 0.2g
Ammonium-2-mercaptoethanesulfonate 1.0mg
Artificial brine 1.0L
Wolfe's vitamin solution 10.0mL
Wolfe's mineral solution 10.0mL
Volatile acids solution 10.0mL
pH 7.1 ± 0.2 at 25°C

$Na_2S \cdot 9H_2O$ Solution:
Composition per 10.0mL:
$Na_2S \cdot 9H_2O$ 2.0g

Preparation of $Na_2S \cdot 9H_2O$ Solution: Add $Na_2S \cdot 9H_2O$ to distilled/deionized water and bring volume to 10.0mL. Mix thoroughly. Filter sterilize.

Artificial Brine:
Composition per liter:
NaCl 80.7g
$MgCl_2 \cdot 6H_2O$ 35.1g
Na_2SO_4 .... 12.9g
KCl 5.7g
$CaCl_2$ 0.55g
$LiCl_2$ 0.13g
H_3BO_3 .... 0.12g

Preparation of Artificial Brine: Add components to distilled/deionized water and bring volume to 1.0L. Mix thoroughly.

Wolfe's Vitamin Solution:
Composition per liter:
Pyridoxine·HCl 0.01g
Thiamine·HCl 5.0mg
Riboflavin 5.0mg
Nicotinic acid 5.0mg
Calcium pantothenate 5.0mg
p-Aminobenzoic acid 5.0mg
Thioctic acid 5.0mg
Biotin 2.0mg
Folic acid 2.0mg
Cyanocobalamin 0.1mg

Preparation of Wolfe's Vitamin Solution: Add components to distilled/deionized water and bring volume to 1.0L. Mix thoroughly. Filter sterilize. Store at 4°C.

Wolfe's Mineral Solution:
Composition per liter:
$MgSO_4 \cdot 7H_2O$ 3.0g
Nitriloacetic acid 1.5g
NaCl 1.0g
$MnSO_4 \cdot H_2O$ 0.5g
$FeSO_4 \cdot 7H_2O$ 0.1g
$CoCl_2 \cdot 6H_2O$ 0.1g
$CaCl_2$ 0.1g
$ZnSO_4 \cdot 7H_2O$ 0.1g
$CuSO_4 \cdot 5H_2O$ 0.01g
$AlK(SO_4)_2 \cdot 12H_2O$ 0.01g
H_3BO_3 .... 0.01g
$Na_2MoO_4 \cdot 2H_2O$ 0.01g

Preparation of Wolfe's Mineral Solution: Add nitrilotriacetic acid to 500.0mL of distilled/deionized water. Dissolve by adjusting pH to 6.5 with KOH. Add remaining components. Add distilled/deionized water to 1.0L.

Volatile Acids Solution:
Composition per liter:
α-Methylbutyric acid 0.5mL
Isobutyric acid 0.5mL
Isovaleric acid 0.5mL
Valeric acid 0.5mL

Preparation of Volatile Acids Solution: Add components to distilled/deionized water and bring volume to 1.0L. Mix thoroughly.

Preparation of Medium: Combine components, except the L-cysteine·HCl·H_2O, trimethylamine·HCl, and $Na_2S·9H_2O$ solution. Mix thoroughly. Gently heat and bring to boiling. Add L-cysteine·HCl·H_2O. Mix thoroughly. Cool in an ice-water bath under 80% N_2 + 20% CO_2. Add trimethylamine·HCl. Mix thoroughly. Adjust pH to 7.1. Aseptically and anaerobically distribute into tubes in 10.0mL volumes under 80% N_2 + 20% CO_2. Autoclave for 15 min at 15 psi pressure–121°C. Immediately prior to inoculation, aseptically add 0.25mL of sterile $Na_2S·9H_2O$ solution to each tube. Mix thoroughly.

Use: For the cultivation and maintenance of *Methanohalophilus mahii.*

Halomonas desiderata Medium (DSMZ Medium 762)

Composition per liter:

Glucose 5.0g
KNO_3 .... 2.0g
KH_2PO_4 .... 1.0g
$MgCl_2·6H_2O$ 0.2g
Carbonate solution 100.0mL

pH 9.5-10.0 at 25°C

Carbonate Solution

Composition per 100mL:

Na_2CO_3 .... 5.4g
$NaHCO_3$ 4.2g

Preparation of Carbonate Solution: Add components to distilled/deionized water and bring volume to 100.0mL. Mix thoroughly. Filter sterilize.

Preparation of Medium: Add components, except carbonate solution, to distilled/deionized water and bring volume to 900.0mL. Mix thoroughly. Autoclave for 15 min at 15 psi pressure–121°C. Cool to room temperature. Aseptically add 100.0mL sterile carbonate solution. Mix thoroughly. Aseptically distribute into sterile tubes or bottles.

Use: For the cultivation of *Halomonas desiderata.*

Halomonas magadiensis Medium (DSMZ Medium 971)

Composition per liter:

Agar 20.0g
Glucose 10.0g
Peptone 5.0g
Yeast extract 5.0g
KH_2PO_4 .... 1.0g
$MgSO_4·7H_2O$ 0.2g
NaCl solution 200.0mL
Na_2CO_3 solution 100.0mL

pH 10.0 ± 0.2 at 25°C

NaCl Solution:

Composition per 200.0mL:

NaCl 40.0g

Preparation of NaCl Solution: Add NaCl to distilled/deionized water and bring volume to 200.0mL. Mix thoroughly. Autoclave for 15 min at 15 psi pressure–121°C. Cool to 60°C.

Na_2CO_3 Solution:

Composition per 100.0mL:

Na_2CO_3 .... 10.0g

Preparation of Na_2CO_3 Solution: Add Na_2CO_3 to distilled/deionized water and bring volume to 100.0mL. Mix thoroughly. Autoclave for 15 min at 15 psi pressure–121°C. Cool to 60°C.

Preparation of Medium: Add components, except NaCl solution and Na_2CO_3 solution, to distilled/deionized water and bring volume to 700.0mL. Mix thoroughly. Gently heat and bring to boiling. Autoclave for 15 min at 15 psi pressure–121°C. Cool to 60°C. Aseptically add 200.0mL NaCl solution and 100.0mL Na_2CO_3 solution. Mix thoroughly. Pour into sterile Petri dishes or distribute into sterile tubes.

Use: For the cultivation of *Halomonas magadiensis.*

Halomonas Medium

Composition per liter:

NaCl 80.0g
$MgSO_4·7H_2O$ 20.0g
Casamino acids with vitamins 7.5g
Proteose peptone No. 3 .... 5.0g
Sodium citrate 3.0g
Yeast extract 1.0g
K_2HPO_4 .... 0.5g
$Fe(NH_4)_2(SO_4)_2·6H_2O$ 0.05g

pH 7.0 ± 0.2 at 25°C

Preparation of Medium: Add components to distilled/deionized water and bring volume to 1.0L. Mix thoroughly. Adjust pH to 7.0 with KOH. Distribute into tubes or flasks. Autoclave for 15 min at 15 psi pressure–121°C.

Use: For the cultivation and maintenance of *Halomonas elongata.*

Halomonas pantelleriense Agar (DSMZ Medium 752)

Composition per liter:

NaCl 100.0g
Agar 20.0g
Yeast extract 10.0g
Na_2-citrate 3.0g
KCl 2.0g
$MgSO_4·7H_2O$ 1.0g
Na_2CO_3 solution 25.0mL
$MnCl_2$ solution 1.0mL
$FeSO_4$ solution 1.0mL

pH 9.0 ± 0.2 at 25°C

Na_2CO_3 Solution:

Composition per 25.0mL:

Na_2CO_3 .... 2.5g

Preparation of Na_2CO_3 Solution: Add Na_2CO_3 to distilled/deionized water and bring volume to 25.0mL. Mix thoroughly.

$FeSO_4$ Solution:

Composition per 100.0mL:

$FeSO_4·7H_2O$ 5.0g

Preparation of $FeSO_4$ Solution: Add $FeSO_4·7H_2O$ to distilled/deionized water and bring volume to 100.0mL. Mix thoroughly.

$MnCl_2$ Solution:

Composition per 100.0mL:

$MnCl_2·4H_2O$ 0.036g

Preparation of $MnCl_2$ Solution: Add $MnCl_2·4H_2O$ to distilled/deionized water and bring volume to 100.0mL. Mix thoroughly.

Preparation of Medium: Add components, except Na_2CO_3 soltuion to distilled/deionized water and bring volume to 990.0L. Mix thoroughly. Adjust pH to 9.0. Autoclave for 15 min at 15 psi pressure–121°C. Aseptically add

10.0mL sterile Na_2CO_3 solution. Mix thoroughly. Pour into Petri dishes or distribute into sterile tubes.

Use: For the cultivation of *Halomonas pantelleriensis (Halomonas pantelleriense).*

Halomonas subglaciescola Medium (Artificial Organic Lake Medium)

Composition per 1001.0mL:

NaCl	80.0g
$MgSO_4 \cdot 7H_2O$	9.5g
Yeast extract	1.0g
KCl	0.5g
$CaCl_2 \cdot 2H_2O$	0.2g
KNO_3	0.1g
$(NH_4)_2SO_4$	0.1g
Hutner's basal salts solution	20.0mL
Phosphate supplement	20.0mL
Vitamin solution	1.0mL

pH 8.0 ± 0.2 at 25°C

Hutner's Basal Salts Solution:

Composition per liter:

$MgSO_4 \cdot 7H_2O$	29.7g
Nitrilotriacetic acid	10.0g
$CaCl_2 \cdot 2H_2O$	3.335g
$FeSO_4 \cdot 7H_2O$	99.0mg
$(NH_4)_6MoO_7O_{24} \cdot 4H_2O$	9.25mg
"Metals 44"	50.0mL

"Metals 44":

Composition per 100.0mL:

$ZnSO_4 \cdot 7H_2O$	1.095g
$FeSO_4 \cdot 7H_2O$	0.5g
Sodium EDTA	0.25g
$MnSO_4 \cdot H2O$	0.154g
$CuSO_4 \cdot 5H_2O$	39.2mg
$Co(NO_3)_2 \cdot 6H_2O$	24.8mg
$Na_2B_4O_7 \cdot 10H_2O$	17.7mg

Preparation of "Metals 44": Add sodium EDTA to distilled/deionized water and bring volume to 90.0mL. Mix thoroughly. Add a few drops of concentrated H_2SO_4 to retard precipitation of heavy metal ions. Add remaining components. Mix thoroughly. Bring volume to 100.0mL with distilled/deionized water.

Preparation of Hutner's Basal Salts Solution: Add nitrilotriacetic acid to 500.0mL of distilled/deionized water. Adjust pH to 6.5 with KOH. Add remaining components. Add distilled/deionized water to 1.0L. Adjust pH to 6.8. Filter sterilize.

Phosphate Supplement:

Composition per 20.0mL:

K_2HPO_4	50.0mg
KH_2PO_4	50.0mg

Preparation of Phosphate Supplement: Add components to distilled/deionized water and bring volume to 20.0mL. Mix thoroughly. Filter sterilize.

Vitamin Solution:

Composition per liter:

Cyanocobalamin	10.0mg
Pyridoxine·HCl	10.0mg
Thiamine·HCl	10.0mg
Calcium DL-pantothenate	5.0mg
Nicotinamide	5.0mg
Biotin	2.0mg
Folic acid	2.0mg

Preparation of Vitamin Solution: Add components to distilled/deionized water and bring volume to 1.0L. Mix thoroughly. Filter sterilize.

Preparation of Medium: Add components, except Hutner's basal salts solution, phosphate supplement, and vitamin solution, to distilled/deionized water and bring volume to 960.0mL. Mix thoroughly. Adjust pH to 8.0. Autoclave for 15 min at 15 psi pressure–121°C. Aseptically add 20.0mL of sterile Hutner's basal salts solution, 20.0mL of sterile phosphate supplement, and 1.0mL of sterile vitamin solution. Mix thoroughly. Aseptically distribute into sterile tubes or flasks.

Use: For the cultivation of *Halomonas subglaciescola.*

Halonatronum saccharophilum Medium (DSMZ Medium 932)

Composition per liter:

Na_2CO_3	68.0g
NaCl	50.0g
$NaHCO_3$	38.0g
NH_4Cl	0.5g
KCl	0.2g
KH_2PO_4	0.2g
$MgCl_2$	0.1g
Resazurin	0.01g
Sucrose solution	50.0mL
$Na_2S \cdot 9H_2O$ solution	10.0mL
Yeast extract solution	10.0mL
Vitamin solution	10.0mL
Trace elements	1.0mL

pH 9.5-10.0 at 25°C

Sucrose Solution:

Composition per 50.0mL:

Sucrose	5.0g

Preparation of Sucrose Solution: Add sucrose to distilled/deionized water and bring volume to 50.0mL. Mix thoroughly. Sparge with 100% N_2. Autoclave for 15 min at 15 psi pressure–121°C. Cool to room temperature.

Yeast Extract Solution:

Composition per 10.0mL:

Yeast extract	0.2g

Preparation of Yeast Extract Solution: Add yeast extract to distilled/deionized water and bring volume to 10.0mL. Mix thoroughly. Sparge with 100% N_2. Autoclave under 100% N_2 for 15 min at 15 psi pressure–121°C. Cool to room temperature.

$Na_2S \cdot 9H_2O$ Solution:

Composition per 10mL:

$Na_2S \cdot 9H_2O$	0.7g

Preparation of $Na_2S \cdot 9H_2O$ Solution: Add $Na_2S \cdot 9H_2O$ to distilled/deionized water and bring volume to 10.0mL. Mix thoroughly. Autoclave under 100% N_2 for 15 min at 15 psi pressure–121°C. Cool to room temperature.

Trace Elements Solution:

Composition per liter:

$MgSO_4 \cdot 7H_2O$	3.0g
Nitrilotriacetic acid	1.5g
NaCl	1.0g
$MnSO_4 \cdot 2H_2O$	0.5g
$CoSO_4 \cdot 7H_2O$	0.18g
$ZnSO_4 \cdot 7H_2O$	0.18g
$CaCl_2 \cdot 2H_2O$	0.1g
$FeSO_4 \cdot 7H_2O$	0.1g

$NiCl_2 \cdot 6H_2O$ 0.025g
$KAl(SO_4)_2 \cdot 12H_2O$ 0.02g
H_3BO_3 .... 0.01g
$Na_2MoO_4 \cdot 4H_2O$ 0.01g
$CuSO_4 \cdot 5H_2O$ 0.01g
$Na_2SeO_3 \cdot 5H_2O$ 0.3mg

Preparation of Trace Elements Solution: Add nitrilotriacetic acid to 500.0mL of distilled/deionized water. Dissolve by adjusting pH to 6.5 with KOH. Add remaining components. Add distilled/deionized water to 1.0L. Mix thoroughly.

Vitamin Solution:
Composition per liter:
Pyridoxine-HCl 10.0mg
Thiamine-HCl·$2H_2O$ 5.0mg
Riboflavin 5.0mg
Nicotinic acid 5.0mg
D-Ca-pantothenate 5.0mg
p-Aminobenzoic acid 5.0mg
Lipoic acid 5.0mg
Biotin 2.0mg
Folic acid 2.0mg
Vitamin B_{12} 0.1mg

Preparation of Vitamin Solution: Add components to distilled/deionized water and bring volume to 1.0L. Mix thoroughly. Sparge with 80% H_2 + 20% CO_2. Filter sterilize.

Preparation of Medium: Prepare and dispense medium under 100% N_2 gas atmosphere. Add components, except $NaHCO_3$, NH_4Cl, Na_2CO_3, sucrose solution, $Na_2S \cdot 9H_2O$ solution, yeast extract solution, and vitamin solution, to distilled/deionized water and bring volume to 920.0mL. Mix thoroughly. Gently heat and bring to boiling. Boil for 5 min. Cool to room temperature while sparging with 100% N_2. Add solid $NaHCO_3$, NH_4Cl, and Na_2CO_3. Mix thoroughly. Distribute into anaerobe tubes or bottles. Aseptically and anaerobically add per liter of medium 50.0mL sucrose solution, 10.0mL yeast extract solution, 10.0mL $Na_2S \cdot 9H_2O$ solution, and 10.0mL vitamin solution. Final pH 9.5-10.0. Autoclave for 15 min at 15 psi pressure–121°C.

Use: For the cultivation of *Halonatronum saccharophilum.*

Halophile Agar

Composition per liter:
NaCl 156.0g
Agar 20.0g
$MgSO_4 \cdot 7H_2O$ 20.0g
$MgCl_2 \cdot 6H_2O$ 13.0g
Yeast extract 10.0g
KCl 4.0g
$CaCl_2 \cdot 6H_2O$ 1.0g
NaBr 0.5g
$NaHCO_3$ 0.2g
pH 7.0 ± 0.2 at 25°C

Preparation of Medium: Add components to distilled/deionized water and bring volume to 1.0L. Mix thoroughly. Adjust pH to 7.0. Distribute into tubes or flasks. Autoclave for 15 min at 15 psi pressure–121°C.

Use: For the cultivation of *Haloarcula hispanica, Haloferax gibbonsii, Marinococcus halophilus, Planococcus* species, and *Sporosarcina halophila.*

Halophile Agar I

Composition per liter:
NaCl 250.0g
$MgSO_4 \cdot 7H_2O$ 20.0g
Agar 15.0g
Yeast extract 10.0g
Casamino acids 7.5g
Trisodium citrate 3.0g
KCl 2.0g
$FeSO_4 \cdot 7H_2O$ 0.05g
$MnSO_4 \cdot 4H_2O$ 0.25mg
pH 7.4 ± 0.2 at 25°C

Preparation of Medium: Add components, except agar, to distilled/deionized water and bring volume to 1.0L. Mix thoroughly. Adjust pH to 7.4. Add agar. Gently heat and bring to boiling. Do not autoclave. Sterilize by steaming at 100°C for 30 min. Aseptically distribute into sterile tubes or flasks.

Use: For the cultivation and maintenance of *Halobacterium halobium, Halobacterium salinarium,* and *Halobacterium trapanicum.*

Halophile Medium

Composition per liter:
NaCl 30.0g
Agar 15.0g
Peptone 5.0g
Yeast extract 2.0g
Beef extract 1.0g
pH 7.4 ± 0.2 at 25°C

Preparation of Medium: Add components to distilled/deionized water and bring volume to 1.0L. Mix thoroughly. Gently heat and bring to boiling. Distribute into tubes or flasks. Autoclave for 15 min at 15 psi pressure–121°C. Pour into sterile Petri dishes or leave in tubes.

Use: For the cultivation and maintenance of *Bacillus tusciae, Chromobacterium maris-mortui, Flavobacterium tirrenicum, Flavobacterium uliginosum, Halomonas halmophila, Pseudomonas beijerinckii, Vibrio alginolyticus, Vibrio parahaemolyticus,* and *Vibrio proteolyticus.*

Halophile Medium

Composition per liter:
NaCl 100.0g
KCl 5.0g
$MgCl_2 \cdot 6H_2O$ 5.0g
$MgSO_4 \cdot 7H_2O$ 5.0g
NH_4Cl 5.0g
Peptone solution (15% solution) 30.0mL
Yeast extract solution (15% solution) 30.0mL
Ferric citrate solution (1% solution) 10.0mL
Trace elements solution 5.0mL

Trace Elements Solution:
Composition per liter:
$ZnSO_4 \cdot 7H_2O$ 0.22g
$MgCl_2 \cdot 4H_2O$ 0.18g
$CoCl_2 \cdot 6H_2O$ 0.01g
$Na_2MoO_4 \cdot H_2O$ 6.3mg
$CuSO_4 \cdot 5H_2O$ 1.0mg

Preparation of Trace Elements Solution: Add components to distilled/deionized water and bring volume to 1.0L. Mix thoroughly.

Preparation of Medium: Add components to distilled/deionized water and bring volume to 1.0L. Mix thoroughly. Gently heat and bring to boiling. Distribute into tubes or flasks. Autoclave for 15 min at 15 psi pressure–121°C.

Use: For the cultivation of *Rhodospirillum salinarum.*

Halophile Medium III

Composition per liter:

NaCl	95.3g
$MgSO_4 \cdot 7H_2O$	81.3g
Glycerol	5.0g
NH_4Cl	1.0g
K_2HPO_4 solution	100.0mL
Wolfe's vitamin solution	10.0mL
Trace elements solution SL-10	1.0mL

pH 7.5 ± 0.2 at 25°C

Wolfe's Vitamin Solution:

Composition per liter:

Pyridoxine·HCl	10.0mg
Calcium D-(+)-pantothenate	5.0mg
Nicotinic acid	5.0mg
p-Aminobenzoic acid,	5.0mg
Riboflavin	5.0mg
Thiamine·HCl	5.0mg
Thioctic acid	5.0mg
Biotin	2.0mg
Folic acid	2.0mg
Cyanocobalamin	100.0μg

Preparation of Wolfe's Vitamin Solution: Add components to distilled/deionized water and bring volume to 1.0L. Mix thoroughly. Filter sterilize.

Trace Elements Solution SL-10:

Composition per liter:

$FeCl_2 \cdot 4H_2O$	1.5g
$CoCl_2 \cdot 6H_2O$	190.0mg
$MnCl_2 \cdot 4H_2O$	100.0mg
$ZnCl_2$	70.0mg
$Na_2MoO_4 \cdot 2H_2O$	36.0mg
$NiCl_2 \cdot 6H_2O$	24.0mg
H_3BO_3	6.0mg
$CuCl_2 \cdot 2H_2O$	2.0mg
HCl (25% solution)	10.0mL

Preparation of Trace Elements Solution SL-10: Add $FeCl_2 \cdot 4H_2O$ to 10.0mL of HCl solution. Mix thoroughly. Add distilled/deionized water and bring volume to 1.0L. Add remaining components. Mix thoroughly. Sparge with 100% N_2. Autoclave for 15 min at 15 psi pressure–121°C.

K_2HPO_4 Solution:

Composition per 100.0mL:

K_2HPO_4	1.0g

Preparation of K_2HPO_4 Solution: Add K_2HPO_4 to distilled/deionized water and bring volume to 100.0mL. Mix thoroughly. Autoclave for 15 min at 15 psi pressure–121°C. Cool to 25°C

Preparation of Medium: Add components, except K_2HPO_4 solution, Wolfe's vitamin solution, and trace elments solution SL-10, to distilled/deionized water and bring volume to 889.0mL. Mix thoroughly. Autoclave for 15 min at 15 psi pressure–121°C. Cool to 25°C. Aseptically add 1000.0mL of sterile K_2HPO_4 solution, 10.0mL of sterile Wolfe's vitamin solution, and 1.0mL of sterile trace elements solution SL-10. Mix thoroughly. Aseptically adjust pH to 7.5. Aseptically distribute into sterile tubes or flasks.

Use: For the cultivation and maintenance of *Halovibrio variabilis.*

Halophilic Agar (HA)

Composition per liter:

NaCl	250.0g
$MgSO_4 \cdot 7H_2O$	25.0g
Agar	20.0g
Casamino acids	10.0g
Yeast extract	10.0g
Proteose peptone	5.0g
Trisodium citrate	3.0g
KCl	2.0g

pH 7.2 ± 0.2 at 25°C

Preparation of Medium: Combine the ingredients with distilled water and heat to boiling to dissolve completely. Autoclave at 121°C for 15 min.

Use: For the isolation and cultivation of halophilic microorganisms, such as *Pseudomonas* species and *Flavobacterium* species, from fish.

Halophilic Broth (HB)

Composition per liter:

NaCl	250.0g
$MgSO_4 \cdot 7H_2O$	25.0g
Casamino acids	10.0g
Yeast extract	10.0g
Proteose peptone	5.0g
Trisodium citrate	3.0g
KCl	2.0g

pH 7.2 ± 0.2 at 25°C

Preparation of Medium: Add components to distilled/deionized water and bring volume to 1.0L. Mix thoroughly. Gently heat and bring to boiling. Distribute into tubes or flasks. Autoclave for 15 min at 15 psi pressure–121°C.

Use: For the isolation and cultivation of halophilic microorganisms, such as *Pseudomonas* species and *Flavobacterium* species, from fish.

Halophilic *Chromatium* Medium

Composition per 1060.0mL:

NaCl	60.0g
KH_2PO_4	1.0g
NH_4Cl	1.0g
$MgCl_2 \cdot 6H_2O$	0.5g
Solution A	20.0mL
Solution B	20.0mL
Solution C	10.0mL
Solution D	10.0mL
Trace elements solution	1.0mL

pH 7.0 ± 0.2 at 25°C

Solution A:

Composition per 100.0mL:

$NaHCO_3$	10.0g

Preparation of Solution A: Add $NaHCO_3$ to distilled/deionized water and bring volume to 100.0mL. Mix thoroughly. Autoclave for 15 min at 15 psi pressure–121°C.

Solution B:

Composition per 100.0mL:

$Na_2S \cdot 9H_2O$	10.0g

Preparation of Solution B: Add $Na_2S \cdot 9H_2O$ to distilled/deionized water and bring volume to 100.0mL. Mix thoroughly. Autoclave for 15 min at 15 psi pressure–121°C.

Solution C:

Composition per 100.0mL:

$Na_2S_2O_3 \cdot 9H_2O$	10.0g

Preparation of Solution C: Add $Na_2S_2O_3 \cdot 9H_2O$ to distilled/deionized water and bring volume to 100.0mL. Mix thoroughly. Autoclave for 15 min at 15 psi pressure–121°C.

Solution D:

Composition per 100.0mL:

Sodium malate 10.0g

Preparation of Solution D: Add sodium malate to distilled/deionized water and bring volume to 100.0mL. Mix thoroughly. Autoclave for 15 min at 15 psi pressure–121°C.

Trace Elements Solution:

Composition per liter:

$FeCl_3 \cdot 6H_2O$ 2.7g
H_3BO_3 .... 0.1g
$ZnSO_4 \cdot 7H_2O$ 0.1g
$Co(NO_3)_2 \cdot 6H_2O$ 50.0mg
$CuSO_4 \cdot 5H_2O$ 5.0mg
$MnCl_2 \cdot 6H_2O$ 5.0mg

Preparation of Trace Elements Solution: Add components to distilled/deionized water and bring volume to 1.0L. Mix thoroughly.

Preparation of Medium: Add components, except solution A, solution B, solution C, and solution D, to distilled/deionized water and bring volume to 1.0L Mix thoroughly. Bring pH to 7.0–7.2 with H_3PO_4. Distribute into tubes or flasks. Autoclave for 15 min at 15 psi pressure–121°C. Aseptically add 0.2mL of sterile solution A, 0.2mL of sterile solution B, 0.1mL of sterile solution C, and 0.1mL of sterile solution D for each 10.0mL of medium. Mix thoroughly. Use immediately.

Use: For the cultivation of halophilic *Chromatium* species.

Halophilic *Clostridium* Agar

Composition per liter:

L-Cysteine·HCl·H_2O 0.5g
Solution 1 .... 1.0L
Solution 2 .... 100.0mL

pH 6.2–7.0 at 25°C

Solution 1:

Composition per liter:

NaCl 105.0g
Agar 20.0g
KCl 7.5g
$CaCO_3$ 5.0g
L-Glutamic acid 4.0g
Soluble starch 2.0g
Casamino acids 2.0g
Nutrient broth 2.0g
Yeast extract 2.0g
$FeSO_4 \cdot 7H_2O$ 2.0mg
Resazurin 1.0mg
NaOH (2.5*N* solution) 12.5mL
Wolfe's vitamin solution 10.0mL
Wolfe's mineral solution 10.0mL

Preparation of Solution 1: Add components, except $CaCO_3$, to distilled/deionized water and bring volume to 1.0L. Mix thoroughly. Gently heat and bring to boiling. When all components have dissolved, add the $CaCO_3$. Mix thoroughly.

Wolfe's Vitamin Solution:

Composition per liter:

Pyridoxine·HCl 0.01g
Thiamine·HCl 5.0mg
Riboflavin 5.0mg
Nicotinic acid 5.0mg
Calcium pantothenate 5.0mg
p-Aminobenzoic acid 5.0mg
Thioctic acid 5.0mg
Biotin 2.0mg
Folic acid 2.0mg
Cyanocobalamin 0.1mg

Preparation of Wolfe's Vitamin Solution: Add components to distilled/deionized water and bring volume to 1.0L. Mix thoroughly.

Wolfe's Mineral Solution:

Composition per liter

$MgSO_4 \cdot 7H_2O$ 3.0g
Nitriloacetic acid 1.5g
NaCl 1.0g
$MnSO_4 \cdot H_2O$ 0.5g
$FeSO_4 \cdot 7H_2O$ 0.1g
$CoCl_2 \cdot 6H_2O$ 0.1g
$CaCl_2$ 0.1g
$ZnSO_4 \cdot 7H_2O$ 0.1g
$CuSO_4 \cdot 5H_2O$ 0.01g
$AlK(SO_4)_2 \cdot 12H_2O$ 0.01g
H_3BO_3 .... 0.01g
$Na_2MoO_4 \cdot 2H_2O$ 0.01g

Preparation of Wolfe's Mineral Solution: Add nitrilotriacetic acid to 500.0mL of distilled/deionized water. Dissolve by adjusting pH to 6.5 with KOH. Add remaining components. Add distilled/deionized water to 1.0L.

Solution 2:

Composition per 100.0mL:

$MgCl_2 \cdot 6H_2O$ 20.3g
$CaCl_2 \cdot 2H_2O$ 7.35g

Preparation of Solution 2: Add components to distilled/deionized water and bring volume to 100.0mL. Mix thoroughly. Gas with 100% N_2 for 20 min. Autoclave for 15 min at 15 psi pressure–121°C. Cool to 45°–50°C.

Preparation of Medium: Gently heat 1.0L of solution 1 and bring to boiling under 100% N_2. Add the L-cysteine·HCl·H_2O. Continue boiling until resazurin turns colorless, indicating reduction. The volume of solution 1 should be about 900.0mL. Anaerobically distribute into tubes in 9.0mL volumes under 100% N_2. Cap tubes with rubber stoppers. Place tubes in a press. Autoclave for 15 min at 15 psi pressure–121°C with fast exhaust. Cool to 50°C. Aseptically add 1.0mL of sterile solution 2 to each tube. In the presence of $CaCO_3$, the pH may be higher than 7.0. Do not adjust pH.

Use: For the cultivation and maintenance of *Sporohalobacter lortetii*.

Halophilic *Halobacterium* Medium

Composition per liter:

NaCl 200.0g
$MgSO_4 \cdot 7H_2O$ 37.0g
$CaCl_2 \cdot 2H_2O$ 0.7g
KCl 0.5g
$MnCl_2 \cdot 4H_2O$ 0.05g
Yeast extract 100.0mL

pH 7.0 ± 0.2 at 25°C

Preparation of Medium: Add components to distilled/deionized water and bring volume to 1.0L. Mix thoroughly. Gently heat until dissolved. Adjust pH to 7.0. Distribute into tubes or flasks. Autoclave for 15 min at 15 psi pressure–121°C.

Use: For the cultivation of extremely halophilic *Halobacterium* species.

Halophilic Nutrient Agar (LMG Medium 220)

Composition per liter:

NaCl 60.0g
Agar 15.0g
Casein peptone, tryptic digest 10.0g
Yeast extract 5.0g
Glucose 5.0g

pH 7.3 ± 0.2 at 25°C

Preparation of Medium: Add components to tap water and bring volume to 1.0L. Mix thoroughly. Gently heat and bring to boiling. Distribute into tubes or flasks. Autoclave for 15 min at 15 psi pressure–121°C. Pour into sterile Petri dishes or leave in tubes.

Use: For the cultivation and maintenance of halophilic bacteria.

Halophilic Synthetic Medium

Composition per liter:

Glucose 0.1g
KNO_3 .... 0.05g
$FePO_4$ 0.01g
Artificial seawater 100.0mL

Artificial Seawater:
Composition per 100.0mL:

NaCl 2.4g
$MgCl_2 \cdot 6H_2O$ 1.1g
Na_2SO_4 .... 0.4g
$CaCl_2 \cdot 6H_2O$ 0.2g
KCl 0.07g
$NaHCO_3$ 0.02g
KBr 0.01g
$SrCl_2 \cdot 6H_2O$ 4.0mg
H_3BO_3 .... 3.0mg
$Na_2SiO_3 \cdot 9H_2O$ 0.5mg
NaF 0.3mg

Preparation of Artificial Seawater: Add components to distilled/deionized water and bring volume to 100.0mL. Mix thoroughly.

Preparation of Medium: Add components to distilled/deionized water and bring volume to 1.0L. Mix thoroughly. Distribute into tubes or flasks. Autoclave for 15 min at 15 psi pressure–121°C.

Use: For the cultivation of halophilic bacteria.

Halorhabdus utahensis Medium (DSMZ Medium 927)

Composition per 1003.25mL:

NaCl 270.0g
$MgSO_4 \cdot 7H_2O$ 20.0g
Tris-HCl 12.0g
KCl 5.0g
NH_4Cl 2.0g
Glucose 2.0g
Yeast extract 1.0g
NaBr 0.1g
Phosphate solution 2.5mL
Calcium chloride solution 0.5mL
Iron chloride manganese chloride solution 0.25mL

pH 7.6 ± 0.2 at 25°C

Potassium Phosphate Solution:
Composition per 10.0mL:

KH_2PO_4 .... 0.5g

Preparation of Potassium Phosphate Solution: Add KH_2PO_4 to distilled/deionized water and bring volume to 10.0mL. Mix thoroughly. Autoclave for 15 min at 15 psi pressure–121°C. Cool to room temperature.

Calcium Chloride Solution:
Composition per 10.0mL:

$CaCl_2 \cdot 2H_2O$ 1.0g

Preparation of Calcium Chloride Solution: Add $CaCl_2 \cdot 2H_2O$ to distilled/deionized water and bring volume to 10.0mL. Mix thoroughly. Autoclave for 15 min at 15 psi pressure–121°C. Cool to room temperature.

Iron Chloride Manganese Chloride Solution:
Composition per 10.0mL:

$FeCl_2 \cdot 4H_2O$ 0.2g
$MnCl_2 \cdot 4H_2O$ 0.2g

Preparation of Iron Chloride Manganese Chloride Solution: Add components to distilled/deionized water and bring volume to 10.0mL. Mix thoroughly. Autoclave for 15 min at 15 psi pressure–121°C. Cool to room temperature.

Preparation of Medium: Add components, except potassium phosphate solution, calcium chloride solution, and iron chloride manganese chloride solution, to distilled/deionized water and bring volume to 1.0L. Mix thoroughly. Adjust pH to 7.6 with 5*M* NaOH. Autoclave for 15 min at 15 psi pressure–121°C. Cool to room temperature. Aseptically add 2.5mL phosphate solution, 0.5mL calcium chloride solution, and 0.25mL iron chloride manganese chloride solution. Mix thoroughly. Aseptically distribute to sterile tubes or flasks.

Use: For the cultivation of *Halorhabdus utahensis.*

Halothermothrix orenii Medium (DSMZ Medium 761)

Composition per liter:

NaCl 100.0g
Glucose 10.0g
KCl 4.0g
$MgCl_2 \cdot 6H_2O$ 2.0g
NH_4Cl 1.0g
Na-acetate 1.0g
Trypticase™ 0.5g
K_2HPO_4 .... 0.3g
KH_2PO_4 .... 0.3g
$CaCl_2 \cdot 2H_2O$ 0.2g
Resazurin 0.001g
$NaHCO_3$ solution 50.0mL
$Na_2S \cdot 9H_2O$ solution 10.0mL
Dithionite solution 5.0mL
Trace elements SL-6 .... 1.0mL

pH 7.0 ± 0.2 at 25°C

Trace Elements Solution SL-6:
Composition per liter:

$MnCl_2 \cdot 4H_2O$ 0.5g
H_3BO_3 .... 0.3g
$CoCl_2 \cdot 6H_2O$ 0.2g
$ZnSO_4 \cdot 7H_2O$ 0.1g
$Na_2MoO_4 \cdot 2H_2O$ 0.03g
$NiCl_2 \cdot 6H_2O$ 0.02g
$CuCl_2 \cdot 2H_2O$ 0.01g

Preparation of Trace Elements Solution SL-6: Add components to distilled/deionized water and bring volume to 1.0L. Mix thoroughly. Autoclave for 15 min at 15 psi pressure–121°C.

$Na_2S·9H_2O$ Solution:
Composition per 10.0mL:
$Na_2S·9H_2O$ 0.2g

Preparation of $Na_2S·9H_2O$ Solution: Add $Na_2S·9H_2O$ to distilled/deionized water and bring volume to 10.0mL. Mix thoroughly. Sparge with 100% N_2. Autoclave for 15 min at 15 psi pressure–121°C. Neutralize to pH 7.0 with sterile HCl.

$NaHCO_3$ Solution:
Composition per 100.0mL:
$NaHCO_3$ 10.0g

Preparation of $NaHCO_3$ Solution: Add $NaHCO_3$ to distilled/deionized water and bring volume to 100.0mL. Mix thoroughly. Sparge with 80% N_2 + 20% CO_2. Filter sterilize.

Dithionite Solution:
Composition per 10mL:
Na-dithionite 2.0mg

Preparation of Dithionite Solution: Add Na-dithionite to distilled/deionized water and bring volume to 10.0mL. Mix thoroughly. Sparge with 100% N_2. Filter sterilize.

Preparation of Medium: Prepare and dispense medium under 80% N_2 + 20% CO_2 gas atmosphere. Add components, except $NaHCO_3$ solution, dithionite solution, and $Na_2S·9H_2O$ solution, to distilled/deionized water and bring volume to 935.0mL. Mix thoroughly. Adjust pH to 7.0. Gently heat and bring to boiling. Cool while sparging with 80% N_2 + 20% CO_2. Distribute into Hungate tubes under 80% N_2 + 20% CO_2. Autoclave for 15 min at 15 psi pressure–121°C. Cool to room temperature. Aseptically and anaerobically inject $NaHCO_3$ solution (0.5mL/10/mL medium), dithionite solution (0.05mL/10/mL medium), and $Na_2S·9H_2O$ solution (0.1mL/10/mL medium). Aseptically and anaerobically distribute into sterile tubes or bottles.

Use: For the cultivation of *Halothermothrix orenii.*

Halothiobacillus Medium (DSMZ Medium 864)

Composition per 1030mL:
NaCl 29.0g
$MgSO_4·7H_2O$ 1.5g
$(NH_4)_2SO_4$ 1.0g
KCl 0.7g
$CaCl_2·2H_2O$ 0.42g
Bromothymol Blue 4.0mg
Na-thiosulfate solution 20.0mL
Phosphate solution 10.0mL
Trace elements solution 1.0mL
pH 6.5-7.0 at 25°C

Trace Elements Solution:
Composition per liter:
$FeCl_2·4H_2O$ 1.5g
Na_2-EDTA 0.5g
$CoCl_2·6H_2O$ 190.0mg
$MnCl_2·4H_2O$ 100.0mg
$ZnCl_2$ 70.0mg
$Na_2MoO_4·2H_2O$ 36.0mg
$NiCl_2·6H_2O$ 24.0mg
H_3BO_3 .... 6.0mg
$CuCl_2·2H_2O$ 2.0mg
HCl (25% solution) 10.0mL

Preparation of Trace Elements Solution: Add $FeCl_2·4H_2O$ to 10.0mL of HCl solution. Mix thoroughly. Add distilled/deionized water and bring volume to 1.0L. Add remaining components. Mix thoroughly. Adjust pH to 7.0. Sparge with 80% N_2 + 20% CO_2. Autoclave for 15 min at 15 psi pressure–121°C.

Na-thiosulfate Solution:
Composition per 20.0mL:
$Na_2S_2O_3·5H_2O$ 5.0g

Preparation of Na-thiosulfate Solution: Add 5.0g of $Na_2S_2O_3·5H_2O$ to distilled/deionized water and bring volume to 20.0mL. Mix thoroughly. Sparge with 100% N_2. Filter sterilize.

Phosphate Solution:
Composition per 10.0mL:
K_2HPO_4 .... 0.5g

Preparation of Phosphate Solution: Add K_2HPO_4 to distilled/deionized water and bring volume to 10.0mL. Mix thoroughly. Sparge with 100% N_2. Filter sterilize.

Preparation of Medium: Add components, except phosphate solution and Na-thiosulfate solution, to distilled/deionized water and bring volume to 1.0L. Mix thoroughly. Adjust pH to 6.7. Autoclave for 15 min at 15 psi pressure–121°C. Cool to room temperature. Aseptically add 20.0mL sterile Na-thiosulfate solution and 10.0mL sterile phosphate solution. Mix thoroughly. Aseptically distribute into sterile tubes or flasks.

Use: For the cultivation of *Halothiobacillus kellyi.*

Halovibrio variabilis Medium

Composition per liter:
NaCl 95.0g
$MgSO_4·7H_2O$ 81.0g
Yeast extract 7.5g
Proteose peptone 2.5g
KCl 1.0g
Trace elements solution SL-4 .... 10.0mL
Vitamin solution 10.0mL
pH 7.5 ± 0.2 at 25°C

Vitamin Solution:
Composition per liter:
Pyridoxine HCl 1.0mg
Lipoic acid 0.5mg
Nicotinic acid 0.5mg
p-Aminobenzoic acid 0.5mg
Pantothenic acid 0.5mg
Riboflavin 0.5mg
Thiamine HCl 0.5mg
Biotin 0.2mg
Folic acid 0.2mg
Cyanocobalamin 0.01mg

Preparation of Vitamin Solution: Add components to distilled/deionized water and bring volume to 1.0L. Mix thoroughly. Filter sterilize. Store at 5°C.

Trace Elements Solution SL-4:
Composition per 900.0mL:
EDTA 0.5g
$FeSO_4·7H_2O$ 0.2g
Trace elements solution SL-6 .... 100.0mL

Trace Elements Solution SL-6:
Composition per liter:

H_3BO_3	0.3g
$CoCl_2 \cdot 6H_2O$	0.2g
$ZnSO_4 \cdot 7H_2O$	0.1g
$MnCl_2 \cdot 4H_2O$	0.03g
$CuCl_2 \cdot 2H_2O$	0.01g
$NiCl_2 \cdot 6H_2O$	0.02g
$Na_2MoO_4 \cdot 2H_2O$	0.03g

Preparation of Trace Elements Solution SL-6: Add components to distilled/deionized water and bring volume to 1.0L. Mix thoroughly.

Preparation of Trace Elements Solution SL-4: Add components to distilled/deionized water and bring volume to 1.0L. Mix thoroughly. Filter sterilize.

Preparation of Medium: Add components, except trace elements solution SL-4 and vitamin solution, to distilled/deionized water and bring volume to 980.0mL. Mix thoroughly. Autoclave for 15 min at 15 psi pressure–121°C. Aseptically add 10.0mL of sterile trace elements solution SL-4 and 10.0mL of sterile vitamin solution. Mix thoroughly. Aseptically distribute into sterile tubes or flasks.

Use: For the cultivation of *Halovibrio variabilis*.

Harpo's Htye

Composition per liter:

Pancreatic digest of casein	5.0g
HEPES (*N*-[2-Hydroxyethyl]piperazine-*N'*-2-ethanesulfonic acid) buffer	4.0g
Yeast extract	2.0g

pH 6.8–7.0 at 25°C

Preparation of Medium: Add components to distilled/deionized water and bring volume to 1.0L. Mix thoroughly. Adjust pH to 7.0–7.2. Distribute into tubes or flasks. Autoclave for 15 min at 15 psi pressure–121°C.

Use: For the cultivation of *Cytophaga arvensicola* and *Flexibacter columnaris*.

Harpo's HTYEM Marine Medium

Pancreatic digest of casein	5.0g
HEPES	4.0g
Yeast extract	2.0g
Artificial seawater	1.0L

pH 7.5 ± 0.2 at 25°C

Artificial Seawater:
Composition per liter:

NaCl	27.5g
$MgSO_4 \cdot 7H_2O$	6.78g
$MgCl_2 \cdot 6H_2O$	5.38g
$CaCL_2 \cdot 2H_2O$	1.4g
KCl	0.72g
$NaHCO_3$	0.2g

Preparation of Artificial Seawater: Add components to distilled/deionized water and bring volume to 1.0L. Mix thoroughly.

Preparation of Medium: Add components to artificial seawater and bring volume to 1.0L. Mix thoroughly. Adjust pH to 7.5 with NaOH. Distribute into tubes or flasks. Autoclave for 15 min at 15 psi pressure–121°C.

Use: For the cultivation of *Cytophaga fermentans, Cytophaga latercula, Cytophaga uliginosa,* and *Microscilla aggregans*.

Hay Infusion Agar

Composition per liter:

Hay, partially decomposed	50.0g
Agar	15.0g
K_2HPO_4	2.0g

pH 6.2 ± 0.3 at 25°C

Preparation of Medium: Add hay to distilled/deionized water and bring volume to 1.0L. Autoclave for 30 min at 15 psi pressure–121°C. Filter through paper and reserve filtrate. Add distilled/deionized water to filtrate and bring volume to 1.0L. Mix thoroughly. Add agar and K_2HPO_4. Mix thoroughly. Gently heat and bring to boiling. Adjust pH to 6.2. Distribute into tubes or flasks. Autoclave for 15 min at 15 psi pressure–121°C. Pour into sterile Petri dishes or leave in tubes.

Use: For the cultivation and maintenance of *Alternaria* species, *Caulochytrium protostelioides, Choanephora infundibulifera, Dipsacomyces acuminosporus, Eremascus albus, Eremascus fertilis, Eurotium chevalieri, Eurotium halophilicum, Eurotium herbariorum, Mortierella bisporalis, Protostelium irregularis,* and *Saksenaea vasiformis*.

HE Medium (Hay Extract Medium)

Composition per liter:

Agar	10.0g
Peptone	1.0g
Yeast extract	1.0g
Hay extract solution	500.0mL

pH 6.5 ± 0.2 at 25°C

Hay Extract Solution:
Composition per liter:

Hay, dried	50.0g

Preparation of Hay Extract Solution: Add dried barn hay to distilled/deionized water and bring volume to 1.0L. Mix thoroughly. Gently heat and bring to boiling. Filter through Whatman #40 filter paper.

Preparation of Medium: Add components to distilled/deionized water and bring volume to 1.0L. Mix thoroughly. Gently heat and bring to boiling. Distribute into tubes or flasks. Autoclave for 15 min at 15 psi pressure–121°C. Pour into sterile Petri dishes or leave in tubes.

Use: For the isolation and cultivation of *Spirochaeta aurantia*.

Heart Infusion Agar

Composition per liter:

Beef heart, infusion from	500.0g
Agar	15.0g
Tryptose	10.0g
NaCl	5.0g

pH 7.4 ± 0.2 at 25°C

Source: This medium is available as a premixed powder from BD Diagnostic Systems.

Preparation of Medium: Add components to distilled/deionized water and bring volume to 1.0L. Mix thoroughly. Gently heat and bring to boiling. Distribute into tubes or flasks. Autoclave for 15 min at 15 psi pressure–121°C. Pour into sterile Petri dishes or leave in tubes.

Use: For the isolation and cultivation of a wide variety of fastidious microorganisms. For the cultivation and maintenance of *Bacillus anthracis, Bacillus cereus, Bacillus mycoides, Serratia rubidaea, Staphylococcus aureus, Tsatumella ptyseos,* and *Vibrio vulnificus*.

Hektoen Enteric Agar

Composition per liter:

Agar	13.5g
Lactose	12.0g
Peptic digest of animal tissue	12.0g
Sucrose	12.0g
Bile salts	9.0g
NaCl	5.0g
$Na_2S_2O_3$	5.0g
Yeast extract	3.0g
Salicin	2.0g
Ferric ammonium citrate	1.5g
Acid Fuchsin	0.1g
Bromothymol Blue	0.064g

pH 7.6 ± 0.2 at 25°C

Source: This medium is available as a premixed powder from BD Diagnostic Systems, and Oxoid Unipath.

Caution: Acid Fuchsin is a potential carcinogen and care must be taken to avoid inhalation of the powdered dye and contact with the skin.

Preparation of Medium: Add components to distilled/deionized water and bring volume to 1.0L. Mix thoroughly. Gently heat while stirring until components are dissolved. Do not autoclave. Pour into sterile Petri dishes. Allow agar to solidify with the Petri dish covers partially off.

Use: For the isolation and differentiation of *Salmonella* and *Shigella*. Bacteria that ferment lactose or sucrose appear as yellow to orange colonies. Bacteria that produce H_2S appear as colonies with black centers.

Heliobacillus mobilis Medium

Composition per 966.0mL:

Yeast extract	10.0g
$MgSO_4$	0.1g
EDTA	2.0mg
Sodium pyruvate solution	100.0mL
Trace elements solution B	10.0mL
K_2HPO_4 solution	5.0mL
Trace elements solution A	1.0mL

pH 7.1 ± 0.2 at 25°C

Sodium Pyruvate Solution:
Composition per 100.0mL:

Sodium pyruvate	1.1g

Preparation of Sodium Pyruvate Solution: Add sodium pyruvate to distilled/deionized water and bring volume to 100.0mL. Mix thoroughly. Filter sterilize.

Trace Elements Solution B:
Composition per 100.0mL:

$CaCl_2 \cdot 2H_2O$	0.3g
Ferric ammonium citrate	0.2g

Preparation of Trace Elements Solution B: Add components to distilled/deionized water and bring volume to 100.0mL. Mix thoroughly. Filter sterilize.

K_2HPO_4 Solution:
Composition per 100.0mL:

K_2HPO_4	4.0g

Preparation of K_2HPO_4 Solution: Add K_2HPO_4 to distilled/deionized water and bring volume to 100.0mL. Mix thoroughly. Filter sterilize.

Trace Elements Solution A:
Composition per 100.0mL:

H_3BO_3	2.86g
$MnCl_2 \cdot 4H_2O$	1.81g
$Na_2MoO_4 \cdot 2H_2O$	0.39g
$ZnSO_4 \cdot 7H_2O$	0.222g
$CuSO_4 \cdot 5H_2O$	0.079g
$Co(NO_3)_2 \cdot 6H_2O$	49.4mg

Preparation of Trace Elements Solution A: Add components to distilled/deionized water and bring volume to 100.0mL. Mix thoroughly. Filter sterilize.

Preparation of Medium: Add components, except sodium pyruvate solution, trace elements solution B, K_2HPO_4 solution, and trace elements solution A, to distilled/deionized water and bring volume to 850.0mL. Mix thoroughly. Adjust pH to 7.1. Autoclave for 15 min at 15 psi pressure–121°C. Cool to room temperature. Aseptically add 100.0mL of sterile sodium pyruvate solution, 10.0mL of sterile trace elements solution B, 5.0mL of sterile K_2HPO_4 solution, and 1.0mL of sterile trace elements solution A. Mix thoroughly. Aseptically distribute into sterile tubes or flasks.

Use: For the cultivation and maintenance of *Heliobacillus mobilis.*

Heliobacterium chlorum Medium

Composition per liter:

Yeast extract	10.0g
K_2HPO_4	1.0g
$MgSO_4 \cdot 7H_2O$	1.0g
Sodium ascorbate	0.5g

pH 6.8 ± 0.2 at 25°C

Preparation of Medium: Add components, except sodium ascorbate, to distilled/deionized water and bring volume to 1.0L. Mix thoroughly. Gently heat and bring to boiling. Sparge with 100% N_2 and continue boiling for 3-4 min. Add sodium ascorbate and continue to sparge with 100% N_2. Adjust pH to 6.8. Under 100% N_2, immediately dispense 45.0mL of medium into 50.0mL screw-capped tubes fitted with rubber septa. Tighten screw caps. Autoclave for 15 min at 15 psi pressure–121°C.

Use: For the cultivation and maintenance of *Heliobacillus mobilis* and *Heliobacterium chlorum.*

Heliorestis Medium (DSMZ Medium 886)

Composition per liter:

Na-acetate	1.0g
$MgCl_2 \cdot 6H_2O$	0.6g
Yeast extract	0.5g
KH_2PO_4	0.5g
NaCl	0.5g
Resazurin	0.2g
$CaCl_2$	0.1g
Vitamin B_{12}	20.0μg
Biotin	20.0μg
Solution A	50.0mL
Trace elements SL-6	1.0mL

pH 9.0-9.5 at 25°C

Solution A:
Composition per 50mL:

Na_2CO_3	2.5g
$NaHCO_3$	2.5g
NH_4Cl	0.5g
$Na_2S \cdot 9H_2O$	0.4g

Preparation of Solution A: Add components to distilled/deionized water and bring volume to 50.0mL. Mix thoroughly. Sparge with 100% N_2. Autoclave for 15 min at 15 psi pressure–121°C. Cool to 25°C.

Trace Elements Solution SL-6:

Composition per liter:

$MnCl_2·4H_2O$	0.5g
H_3BO_3	0.3g
$CoCl_2·6H_2O$	0.2g
$ZnSO_4·7H_2O$	0.1g
$Na_2MoO_4·2H_2O$	0.03g
$NiCl_2·6H_2O$	0.02g
$CuCl_2·2H_2O$	0.01g

Preparation of Trace Elements Solution SL-6: Add components to distilled/deionized water and bring volume to 1.0L. Mix thoroughly.

Preparation of Medium: Add components, except solution A, to distilled/deionized water and bring volume to 950.0mL. Mix thoroughly. Autoclave for 15 min at 15 psi pressure–121°C. Cool to 25°C. Aseptically add 50.0mL solution A. Mix thoroughly. Adjust pH to 9.0-9.5. Aseptically distribute to sterile tubes or flasks.

Use: For the cultivation of *Heliorestis baculata.*

Herbaspirillum Agar

Composition per liter:

Agar	15.0g
KH_2PO_4	4.0g
$MgSO_4·7H_2O$	0.2g
K_2HPO_4	0.1g
NaCl	0.1g
Yeast extract	0.05g
$CaCl_2$	0.02g
$FeCl_2·6H_2O$	0.01g
$NaMoO_4·2H_2O$	2.0mg
Solution A	50.0mL

pH 7.0 ± 0.2 at 25°C

Solution A:

Composition per 50.0mL

Sodium malate	5.0g

Preparation of Solution A: Add sodium malate to distilled/deionized water and bring volume to 50.0mL. Mix thoroughly. Adjust pH to 7.0. Filter sterilize.

Preparation of Medium: Add components, except solution A, to distilled/deionized water and bring volume to 950.0mL. Mix thoroughly. Gently heat and bring to boiling. Autoclave for 15 min at 15 psi pressure–121°C. Aseptically add 50.0mL of sterile solution A. Mix thoroughly. Aseptically pour into sterile Petri dishes or distribute into sterile tubes.

Use: For the cultivation and maintenance of *Herbaspirillum seropedicae.*

Herpetosiphon giganteus Medium

Composition per liter:

Pancreatic digest of casein	3.0g
Yeast extract	1.0g
$CaCl_2·2H_2O$	0.5g

pH 7.2 ± 0.2 at 25°C

Preparation of Medium: Add components to distilled/deionized water and bring volume to 1.0L. Mix thoroughly. Adjust pH to 7.2. Distribute into tubes or flasks. Autoclave for 15 min at 15 psi pressure–121°C.

Use: For the cultivation of *Herpetosiphon giganteus.*

HESNW Medium

Composition per 1011.0mL:

Natural seawater	1.0L
Enrichment solution	10.0mL
Vitamin solution	1.0mL

Enrichment Solution:

Composition per liter:

$NaNO_3$	4.667g
$Na_2SiO_3·9H_2O$	3.000g
Sodium glycerophosphate	0.667g
EDTA·$2H_2O$	0.553g
H_3BO_3	0.380g
$Fe(NH_4)_2(SO_4)_2·6H_2O$	0.234g
$MnSO_4·4H_2O$	0.054g
$FeCl_3·6H_2O$	0.016g
$ZnSO_4·7H_2O$	7.3mg
$CoSO_4·7H_2O$	1.6mg

Preparation of Enrichment Solution: Add $Na_2SiO_3·9H_2O$ to distilled/deionized water. Mix thoroughly. Neutralize $Na_2SiO_3·9H_2O$ with 1*N* HCl. Add 500.0mL of distilled/deionized water. Mix thoroughly. Add remaining components and bring volume to 1.0L with distilled/deionized water. Mix thoroughly. Filter sterilize.

Vitamin Solution:

Composition per liter:

Thiamine	0.1g
Vitamin B_{12}	2.0mg
Biotin	1.0mg

Preparation of Vitamin Solution: Add components to distilled/deionized water and bring volume to 1.0L. Mix thoroughly. Filter sterilize.

Preparation of Medium: Allow natural seawater to age for 2 months. Filter sterilize. Aseptically add 10.0mL of sterile enrichment solution and 1.0mL of sterile vitamin solution. Mix thoroughly. Aseptically distribute into sterile tubes or flasks.

Use: For the cultivation of *Amphiprora hyalina, Chlamydomonas hedleyi, Chlamydomonas provasolii, Chlorella saccharophila, Chroomonas salina, Pavlova lutheri,* and *Trichosphaerium* species.

Heterotrophic Hyperthermophilic Archaea Medium

Composition per liter:

Pancreatic digest of casein	3.0g
Glucose	3.0g
Yeast extract	3.0g
Artificial seawater	990.0mL
$Na_2S·9H_2O$ solution	10.0mL

pH 7.2 ± 0.2 at 25°C

Artificial Seawater:

Composition per liter:

NaCl	20.0g
$MgSO_4·7H_2O$	6.0g
$MgCl_2·6H_2O$	3.0g
$(NH_4)_2SO_4$	1.0g
$NaHCO_3$	0.2g
$CaCl_2·2H_2O$	0.3g
KCl	0.5g
KH_2PO_4	0.42g
NaBr	0.05g
$SrCl_2·6H_2O$	0.02g
$Fe(NH_4)$ citrate	0.01g
Wolfe's mineral solution	5.0mL
Vitamin solution	5.0mL

Wolfe's Mineral Solution:

Composition per liter:

$MgSO_4·7H_2O$	3.0g
Nitrilotriacetic acid	1.5g
NaCl	1.0g
$MnSO_4·2H_2O$	0.5g
$CoCl_2·6H_2O$	0.1g
$ZnSO_4·7H_2O$	0.1g
$CaCl_2·2H_2O$	0.1g
$FeSO_4·7H_2O$	0.1g
$NiCl_2·6H_2O$	0.025g
$KAl(SO_4)_2·12H_2O$	0.02g
$CuSO_4·5H_2O$	0.01g
H_3BO_3	0.01g
$Na_2MoO_4·2H_2O$	0.01g
$Na_2SeO_3·5H_2O$	0.3mg

Preparation of Wolfe's Mineral Solution: Add nitrilotriacetic acid to 500.0mL of distilled/deionized water. Adjust pH to 6.5 with KOH. Add remaining components one at a time. Add distilled/deionized water to 1.0L. Adjust pH to 6.8.

Vitamin Solution:

Composition per liter:

Niacin	10.0mg
Pantothenate	10.0mg
Lipoic acid	10.0mg
p-Aminobenzoic acid	10.0mg
Thiamine (B_1)	10.0mg
Riboflavin (B_2)	10.0mg
Pyridoxine (B_6)	10.0mg
Cobalamin (B_{12})	10.0mg
Biotin	4.0mg
Folic acid	4.0mg

Preparation of Vitamin Solution: Add components to distilled/deionized water and bring volume to 1.0L. Mix thoroughly.

Preparation of Artificial Seawater: Add components to distilled/deionized water and bring volume to 1.0L. Mix thoroughly.

$Na_2S·9H_2O$ Solution:

Composition per 10.0mL:

$Na_2S·9H_2O$	0.8g

Preparation of $Na_2S·9H_2O$ Solution: Add $Na_2S·9H_2O$ to distilled/deionized water and bring volume to 10.0mL. Mix thoroughly. Sparge with 100% N_2. Autoclave for 15 min at 15 psi pressure–121°C. Before use, neutralize to pH 7.0 with sterile HCl.

Preparation of Medium: Add components, except $Na_2S·9H_2O$ solution, to artificial seawater and bring volume to 990.0mL. Mix thoroughly. Filter sterilize. Sparge with 100% N_2 for 20 min. Aseptically and anaerobically add 10.0mL of sterile $Na_2S·9H_2O$ solution. Mix thoroughly. Adjust medium pH to 7.2 by adding sterile anaerobic HCl. Aseptically and anaerobically distribute into sterile tubes or bottles.

Use: For the cultivation of hyperthermophilic archaea.

Heterotrophic Medium H3P (DSMZ Medium 428)

Composition per 1010mL:

Solution B	855.0mL
Solution A	50.0mL
Solution E	50.0mL
Solution D	30.0mL
Solution C	20.0mL
Standard vitamin solution	5.0mL

Solution A:

Composition per 50mL:

$Na_2HPO_4·2H_2O$	2.9g
KH_2PO_4	2.3g

Preparation of Solution A: Add components to distilled/deionized water and bring volume to 50.0mL. Mix thoroughly. Autoclave for 15 min at 15 psi pressure–121°C. Cool to 25°C.

Solution B:

Composition per 855mL:

NH_4Cl	1.0g
$MgSO_4·7H_2O$	0.5g
$CaCl_2·2H_2O$	0.010g
$MnCl_2·4H_2O$	0.005g
$NaVO_2·H_2O$	0.005g
Trace elements solution SL-6	5.0mL

Preparation of Solution B: Add components to distilled/deionized water and bring volume to 855.0mL. Mix thoroughly. Autoclave for 15 min at 15 psi pressure–121°C. Cool to 25°C.

Trace Elements Solution SL-6:

Composition per liter:

$MnCl_2·4H_2O$	0.5g
H_3BO_3	0.3g
$CoCl_2·6H_2O$	0.2g
$ZnSO_4·7H_2O$	0.1g
$Na_2MoO_4·2H_2O$	0.03g
$NiCl_2·6H_2O$	0.02g
$CuCl_2·2H_2O$	0.01g

Preparation of Trace Elements Solution SL-6: Add components to distilled/deionized water and bring volume to 1.0L. Mix thoroughly. Autoclave for 15 min at 15 psi pressure–121°C. Cool to room temperature.

Solution C:

Composition per 20mL:

Ferric ammonium citrate	0.050g
Distilled water	20.0mL

Preparation of Solution C: Add ferric ammonium citrate to distilled/deionized water and bring volume to 20.0mL. Mix thoroughly. Autoclave for 15 min at 15 psi pressure–121°C. Cool to 25°C.

Solution D:

Composition per 30mL:

Yeast extract	1.0g
Na-acetate	1.0g
Na_2-succinate	1.0g
DL-Malate	1.0g

Preparation of Solution D: Add components to distilled/deionized water and bring volume to 30.0mL. Mix thoroughly. Adjust to pH 7.0. Autoclave for 15 min at 15 psi pressure–121°C. Cool to 25°C.

Solution E:

Composition per 50mL:

D-Glucose	2.0g
Na-lactate	1.0g
Na-pyruvate	1.0g
D-Mannitol	1.0g

Preparation of Solution E: Add components to distilled/deionized water and bring volume to 50.0mL. Mix thoroughly. Adjust to pH 7.0. Filter sterilize.

Standard Vitamin Solution:

Composition per 100mL:

Thiamine-HCl·$2H_2O$	50.0mg
Nicotinic acid	50.0mg
Pyridoxine-HCl	50.0mg
Ca-pantothenate	50.0mg
Riboflavin	10.0mg
Vitamin B_{12}	1.0mg
Folic acid	0.2mg
Biotin	0.1mg

Preparation of Standard Vitamin Solution: Add components to distilled/deionized water and bring volume to 100.0mL. Mix thoroughly. Filter sterilize.

Preparation of Medium: Aseptically mix solutions A, B, C, D, E, and standard vitamin solution. Aseptically distribute to flasks or tubes.

Use: For the cultivation and maintenance of *Bacillus* spp., *Pseudomonas* spp., *Xanthomonas campestris* pvar. *Campestris, Xanthobacter* sp., *Aquaspirillum arcticum, Herbaspirillum seropedicae, Sphingobium chlorophenolicum=Sphingomonas chlorophenolica,* and *Azoarcus* sp. Also H3P is a heterotrophic medium for growth, purity checking, and isolation of a broad spectrum of aerobic bacteria.

Heterotrophic Medium for *Hydrogenomonas*

Composition per liter:

Agar	15.0g
Tryptose	5.0g
Cornstarch	2.0g
Sodium succinate·$6H_2O$	2.0g
Sodium glutamate	1.0g
Yeast extract	1.0g
Sodium citrate·$2H_2O$	0.5g
Sodium acetate·$3H_2O$	0.3g
KH_2PO_4	0.2g
$MgSO_4$	0.1g

pH 6.8–7.2 at 25°C

Preparation of Medium: Add components to distilled/deionized water and bring volume to 1.0L. Mix thoroughly. Gently heat and bring to boiling. Distribute into tubes or flasks. Autoclave for 15 min at 15 psi pressure–121°C. Pour into sterile Petri dishes or leave in tubes.

Use: For the heterotrophic cultivation of *Hydrogenomonas* species.

Heterotrophic Medium for Hydrogen-Oxidizing Bacteria

Composition per 1010.0mL:

Solution A	900.0mL
Solution C	100.0mL
Solution B	10.0mL

Solution A:

Composition per 900.0mL:

Noble agar	17.0g
Na_2HPO_4·$12H_2O$	9.0g
KH_2PO_4	1.5g
NH_4Cl	1.0g
$MgSO_4$·$7H_2O$	0.2g
Trace elements solution SL-6	1.0mL

Preparation of Solution A: Add components to distilled/deionized water and bring volume to 900.0mL. Mix thoroughly. Autoclave for 15 min at 15 psi pressure–121°C. Cool to 45°–50°C.

Trace Elements Solution SL-6:

Composition per liter:

H_3BO_3	0.3g
$CoCl_2$·$6H_2O$	0.2g
$ZnSO_4$·$7H_2O$	0.1g
$MnCl_2$·$4H_2O$	0.03g
Na_2MoO_4·H_2O	0.03g
$NiCl_2$·$6H_2O$	0.02g
$CuCl_2.2H_2O$	0.01g

Preparation of Trace Elements Solution SL-6: Add components to distilled/deionized water and bring volume to 1.0L. Mix thoroughly. Adjust pH to 3.4.

Solution B:

Composition per 10.0mL:

$CaCl_2$·$2H_2O$	0.01g
Ferric ammonium citrate	5.0mg

Preparation of Solution B: Add components to distilled/deionized water and bring volume to 10.0mL. Mix thoroughly. Autoclave for 15 min at 15 psi pressure–121°C. Cool to 45°–50°C.

Solution C:

Composition per 100.0mL:

Sodium 3-hydroxybutyrate	2.0g

Preparation of Solution C: Add sodium 3-hydroxybutyrate to distilled/deionized water and bring volume to 100.0mL. Mix thoroughly. Filter sterilize. Warm to 45°–50°C.

Preparation of Medium: Aseptically combine 900.0mL of sterile solution A, 10.0mL of sterile solution B, and 100.0mL of sterile solution C. Mix thoroughly. Pour into sterile Petri dishes or distribute into sterile tubes.

Use: For the heterotrophic cultivation and maintenance of *Xanthobacter agilis*.

Heterotrophic *Nitrobacter* Medium (DSMZ Medium 756)

Composition per liter:

Yeast extract	1.5g
Peptone	1.5g
Na-pyruvate	0.55g
Stock solution	100.0mL
Trace elements solution	1.0mL

pH 7.4 ± 0.2 at 25°C

Stock Solution:

Composition per liter:

NaCl	5.0g
KH_2PO_4	1.5g
$MgSO_4$·$7H_2O$	0.5g
$CaCO_3$	0.07g

Preparation of Stock Solution: Add components to distilled/deionized water and bring volume to 1.0L. Mix thoroughly.

Trace Elements Solution:

Composition per liter:

$FeSO_4$·$7H_2O$	97.3mg
H_3BO_3	49.4mg
$ZnSO_4$·$7H_2O$	43.1mg
$(NH_4)_6Mo_7O_{24}$·$4H_2O$	37.1mg
$MnSO_4$·$2H_2O$	33.8mg
$CuSO_4$·$5H_2O$	25.0mg

Preparation of Trace Elements Solution: Add components to distilled/deionized water and bring volume to 1.0L. Mix thoroughly.

Preparation of Medium: Add components to distilled/deionized water and bring volume to 1.0L. Mix thoroughly. Adjust pH to 7.4. Distribute into tubes or flasks. Autoclave for 15 min at 15 psi pressure–121°C.

Use: For the cultivation of *Nitrospira moscoviensis, Nitrobacter hamburgensis, Nitrobacter vulgaris,* and *Nitrobacter winogradskyi.*

Heterotrophic *Nitrobacter* Medium (LMG Medium 245)

Composition per liter:

Yeast extract 1.5g
Peptone 1.5g
Na-pyruvate 0.55g
Stock solution 100.0mL
Trace elements solution 1.0mL

pH 7.4 ± 0.2 at 25°C

Stock Solution:

Composition per liter:

NaCl 5.0g
KH_2PO_4 .... 1.5g
$MgSO_4 \cdot 7H_2O$ 0.5g
$CaCO_3$ 0.07g

Preparation of Stock Solution: Add components to distilled/deionized water and bring volume to 1.0L. Mix thoroughly.

Trace Elements Solution:

Composition per liter:

$(NH_4)Mo7O_2$ 437.1mg
$FeSO_4 \cdot 7H_2O$ 97.3mg
$ZnSO_4 \cdot 7H_2O$ 43.1mg
H_3BO_3 .... 39.4mg
$MnSO_4 \cdot H_2O$ 33.8mg
$CuSO_2 \cdot 5H_2O$ 25.0mg

Preparation of Trace Elements Solution: Add components to distilled/deionized water and bring volume to 1.0L. Mix thoroughly.

Preparation of Medium: Add components to distilled/deionized water and bring volume to 1.0L. Mix thoroughly. Adjust pH to 7.4 with NaOH. Distribute into tubes or flasks. Autoclave for 15 min at 15 psi pressure–121°C.

Use: For the cultivation of heterotrophic *Nitrobactger* spp.

Hexamita Medium

Composition per liter:

TYGM-9 medium 250.0mL
Sonneborn's *Paramecium* medium 750.0mL

TYGM-9 Medium:

Composition per liter:

NaCl 7.5g
K_2HPO_4 .... 2.8g
Casein digest 2.0g
Gastric mucin 2.0g
Yeast extract 1.0g
KH_2PO_4 .... 0.4g
Bovine serum, heat inactivated 30.0mL
Rice starch solution 30.0mL
Tween solution 0.5mL

pH 7.4 ± 0.2 at 25°C

Tween Solution:

Composition per 100.0mL:

Tween 80 .... 10.0mL

Preparation of Tween Solution: Add Tween 80 to absolute ethanol and bring volume to 100.0mL. Mix thoroughly. Filter sterilize.

Rice Starch Solution:

Composition per 100.0mL:

Rice starch 5.0g
Phosphate buffered saline solution 100.0mL

Phosphate Buffered Saline Solution:

Composition per liter:

NaCl 9.0g
$Na_2HPO_4 \cdot 7H_2O$ 0.795g
KH_2PO_4 .... 0.114g

Preparation of Phosphate Buffered Saline Solution: Add components to distilled/deionized water and bring volume to 1.0L. Mix thoroughly. Adjust pH to 7.4. Autoclave for 15 min at 15 psi pressure–121°C. Cool to 25°C.

Preparation of Rice Starch Solution: Heat sterilize rice starch at 150°C for 2 hr. Aseptically add 100.0mL of sterile phosphate-buffered saline solution. Mix thoroughly. Use immediately.

Preparation of TYGM-9 Medium: Add components, except rice starch solution, Tween solution, and bovine serum, to distilled/deionized water and bring volume to 939.5mL. Mix thoroughly. Autoclave for 15 min at 15 psi pressure–121°C. Cool to 25°C. Aseptically add 30.0mL of sterile bovine serum, 30.0mL of sterile rice starch solution, and 0.5mL of sterile Tween solution. Mix thoroughly. Aseptically distribute into sterile, screw-capped tubes or flasks.

Sonneborn's *Paramecium* Medium:

Composition per liter:

Solution 1 .... 1.0L
Klebsiella pneumoniae cultured on solution 2 .... variable

Solution 1:

Composition per liter:

Rye grass cerophyll 2.5g
Na_2HPO_4 .... 0.5g

Source: Cerophyll can be obtained from Ward's Natural Science Establishment, Inc. Dairy Goat Nutrition distributes Grass Media Culture, which is equivalent. Cereal Leaf Product from Sigma Chemical is similar to cerophyll.

Preparation of Solution 1: Add cerophyll to distilled/deionized water and bring volume to 1.0L. Mix thoroughly. Gently heat and bring to boil. Boil for 5 min. Filter through Whatman #1 filter paper. Add 0.5g of Na_2HPO_4. Bring volume to 1.0L with distilled/deionized water. Mix thoroughly. Distribute 10.0mL volumes into tubes. Autoclave for 15 min at 15 psi pressure–121°C.

Solution 2:

Composition per liter:

Agar 20.0g
Yeast extract 4.0g
Glucose 0.16g

Preparation of Solution 2: Add components to distilled/deionized water and bring volume to 1.0L. Mix thoroughly. Gently heat and bring to boiling. Distribute 5.0mL into tubes. Autoclave for 15 min at 15 psi pressure–121°C. Allow tubes to cool in a slanted position.

Preparation of Sonneborn's *Paramecium* Medium: Inoculate the surface of agar slants of solution 2 with a culture of *Klebsiella pneumoniae*. Incubate at 37°C for

24–48 hr. Scrape cells from the surface of the agar slants and add to 10.0mL of solution 1. Incubate at 30°C for 24 hr.

Preparation of Medium: Aseptically combine 3.0mL of sterile TYGM-9 medium with 9.0mL of Sonneborn's *Paramecium* medium in 16 × 125mm screw-capped test tubes.

Use: For the cultivation of *Hexamita inflata, Hexamita pusilla, Mastigamoeba invertens,* and *Trepomonas agilis.*

Hippea Medium (DSMZ Medium 854)

Composition per 1010mL:

NaCl	25.0g
Sulfur, powdered	10.0g
Na-acetate	5.0g
MOPS [3-(*N*-morpholino) propane sulfonic acid]	3.0g
$Na_2S·9H_2O$	0.5g
NH_4Cl	0.33g
$CaCl_2·2H_2O$	0.33g
$MgCl_2·6H_2O$	0.33g
KCl	0.33g
KH_2PO_4	0.33g
Yeast extract	0.1g
Resazurin	0.5mg
Trace elements solution	10.0mL
Vitamin solution	10.0mL

pH 6.1 ± 0.2 at 25°C

Trace Elements Solution:

Composition per liter:

$MgSO_4·7H_2O$	3.0g
Nitrilotriacetic acid	1.5g
NaCl	1.0g
$MnSO_4·2H_2O$	0.5g
$CoSO_4·7H_2O$	0.18g
$ZnSO_4·7H_2O$	0.18g
$CaCl_2·2H_2O$	0.1g
$FeSO_4·7H_2O$	0.1g
$NiCl_2·6H_2O$	0.025g
$KAl(SO_4)_2·12H_2O$	0.02g
H_3BO_3	0.01g
$Na_2MoO_4·4H_2O$	0.01g
$CuSO_4·5H_2O$	0.01g
$Na_2SeO_3·5H_2O$	0.3mg

Preparation of Trace Elements Solution: Add nitrilotriacetic acid to 500.0mL of distilled/deionized water. Dissolve by adjusting pH to 6.5 with KOH. Add remaining components. Add distilled/deionized water to 1.0L. Mix thoroughly.

Vitamin Solution:

Composition per liter:

Pyridoxine-HCl	10.0mg
Thiamine-HCl·$2H_2O$	5.0mg
Riboflavin	5.0mg
Nicotinic acid	5.0mg
D-Ca-pantothenate	5.0mg
p-Aminobenzoic acid	5.0mg
Lipoic acid	5.0mg
Biotin	2.0mg
Folic acid	2.0mg
Vitamin B_{12}	0.1mg

Preparation of Vitamin Solution: Add components to distilled/deionized water and bring volume to 1.0L. Mix thoroughly. Sparge with 80% H_2 + 20% CO_2. Filter sterilize.

Preparation of Medium: Add components, except vitamin solution, sulfur, and $Na_2S·9H_2O$, to distilled/deionized water and bring volume to 1.0L. Mix thoroughly. Adjust pH to 6.0. Sparge the medium with 80% N_2 + 20% CO_2 gas mixture for 30 min. Add $Na_2S·9H_2O$. Mix thoroughly. Readjust the pH to 6.0–6.2. Dispense medium under 80% N_2 + 20% CO_2 gas mixture into anaerobe tubes or bottles containing 100.0mg sulfur powder per 10mL medium. Autoclave 20 min at 110°C. Prior to use inject 0.1mL sterile vitamin solution per 10.0mL medium.

Use: For the cultivation of *Hippea maritima.*

Hirschia Medium

Composition per liter:

Pancreatic digest of casein	5.0g
HEPES	4.0g
Yeast extract	2.0g
Artificial seawater	250.0mL
Glucose solution	100.0mL

pH 7.4 ± 0.2 at 25°C

Glucose Solution:

Composition per 100.0mL:

Glucose	0.25g

Preparation of Glucose Solution: Add glucose to distilled/deionized water and bring volume to 100.0mL. Mix thoroughly. Filter sterilize.

Artificial Seawater:

Composition per liter:

NaCl	27.5g
$MgCl_2·6H_2O$	5.38g
$MgSO_4·7H_2O$	6.78g
KCl	0.72g
$NaHCO_3$	0.2g
$CaCL_2·2H_2O$	1.4g

Preparation of Artificial Seawater: Add components to distilled/deionized water and bring volume to 1.0L. Mix thoroughly.

Preparation of Medium: Add components, except glucose solution, to distilled/deionized water and bring volume to 1.0L. Mix thoroughly. Adjust pH to 7.4. Autoclave for 15 min at 15 psi pressure–121°C. Cool to 25°C. Aseptically add 100.0mL of sterile glucose solution. Mix thoroughly. Aseptically distribute into sterile tubes or flasks.

Use: For the cultivation of *Hirschia baltica.*

Histoplasma capsulatum Agar

Composition per liter:

Agar	12.5g
Glucose	10.0g
Citric acid	10.0g
Potato starch	2.0g
α-Ketoglutaric acid	1.0g
L-Cystine·HCl·H_2O	1.0g
Glutathione, reduced	0.5g
L-Asparagine	0.1g
L-Tryptophan	0.02g
Solution 1	250.0mL
Solution 3	40.0mL
Solution 2	10.0mL
Solution 4	10.0mL
Solution 8	10.0mL
Solution 5	1.0mL
Solution 6	0.1mL
Solution 7	0.1mL

pH 6.5 ± 0.2 at 25°C

Solution 1:
Composition per liter:

KH_2PO_4	8.0g
$(NH_4)_2SO_4$	8.0g
$MgSO_4 \cdot 7H_2O$	0.86g
$CaCl_2$, anhydrous	0.08g
$ZnSO_4 \cdot 7H_2O$	0.05g

Preparation of Solution 1: Add components to distilled/deionized water and bring volume to 500.0mL. Mix thoroughly. Bring volume to 1.0L with distilled/deionized water. Store at 5°C.

Solution 2:
Composition per liter:

$FeSO_4 \cdot 7H_2O$	5.7g
$MnCl_2 \cdot 6H_2O$	0.8g
$NaMoO_4 \cdot 2H_2O$	0.15g
HCl, concentrated	1.0mL

Preparation of Solution 2: Add 1.0mL of concentrated HCl to 100.0mL of distilled water in a 1.0L volumetric flask. Dissolve each component completely in the sequence given. Bring volume to 1.0L with distilled/deionized water. Store at 5°C. Discard if red color or red precipitate appears.

Solution 3:
Composition per 100.0mL:

Casein, acid-hydrolyzed, vitamin-free	10.0g

Preparation of Solution 3: Add casein to distilled/deionized water and bring volume to 100.0mL.

Solution 4:
Composition per liter:

Calcium pantothenate	0.2g
Inositol	0.2g
Riboflavin	0.2g
Thiamine·HCl	0.2g
Nicotinamide	0.1g
Biotin	0.01g

Preparation of Solution 4: Add components to distilled/deionized water and bring volume to 1.0L. Mix thoroughly. Store at –20°C.

Solution 5:
Composition per 100.0mL:

Hemin	0.2g
NH_4OH, concentrated	0.3mL

Preparation of Solution 5: Add hemin to approximately 30.0mL of distilled/deionized water. Add NH_4OH. Mix thoroughly until dissolved. Bring volume to 100.0mL with distilled/deionized water. Store at 5° C.

Solution 6:
Composition per 10.0mL:

DL-Thioctic acid	0.01g
Ethanol (95% solution)	10.0mL

Preparation of Solution 6: Add DL-thioctic acid to 10.0mL of ethanol. Mix thoroughly. Store at –20°C.

Solution 7:
Composition per 10.0mL:

Coenzyme A	0.01g
$Na_2S \cdot 5H_2O$ (0.05% solution)	0.2mL

Preparation of Solution 7: Prepare $Na_2S \cdot 5H_2O$ solution in freshly boiled distilled/deionized water. Add coenzyme A to 9.8mL of distilled/deionized water. Mix thoroughly. Add freshly prepared $Na_2S \cdot 5H_2O$ solution. Mix thoroughly. Store the solution at –20°C.

Solution 8:
Composition per 100.0mL:

Oleic acid	0.1g

Preparation of Solution 8: Add oleic acid to 50.0mL of distilled/deionized water. Adjust pH to 9.0 with NaOH. Gently heat until dissolved. Bring volume to 100.0mL with distilled/deionized water. Store at 5°C.

Preparation of Medium: Add components—except agar, potato starch, and solution 8—to distilled/deionized water and bring volume to 400.0mL. Mix thoroughly. Adjust pH to 6.5 with 20% KOH solution. Filter sterilize. In a separate flask, add potato starch to 50.0mL of distilled/deionized water. Add the starch solution to 450.0mL of boiling distilled/deionized water. Add 10.0mL of solution 8 and the agar. Mix thoroughly. Autoclave for 15 min at 15 psi pressure–121°C. Cool to 70°C. Aseptically combine the two sterile solutions. Pour into sterile Petri dishes or distribute into sterile tubes.

Use: For the cultivation and maintenance of *Histoplasma capsulatum* in the yeast phase. For the cultivation of *Histoplasma duboisii, Blastomyces dermatitidis,* and *Sprotrichum schenckii.*

HP 6 Agar

Composition per liter:

Agar	15.0g
Sodium glutaminate	10.0g
$MgSO_4 \cdot 7H_2O$	1.0g
Yeast extract	1.0g
Cyanocobalamin	0.5mg
Glucose solution	100.0mL

pH 7.2 ± 0.2 at 25°C

Glucose Solution:
Composition per 100.0mL:

D-Glucose	5.0g

Preparation of Glucose Solution: Add D-glucose to distilled/deionized water and bring volume to 100.0mL. Mix thoroughly. Autoclave for 15 min at 15 psi pressure–121°C. Cool to 25°C.

Preparation of Medium: Add components, except glucose solution, to distilled/deionized water and bring volume to 900.0mL. Mix thoroughly. Gently heat and bring to boiling. Autoclave for 15 min at 15 psi pressure–121°C. Cool to 45°–50°C. Aseptically add sterile glucose solution. Mix thoroughly. Pour into sterile Petri dishes or distribute into sterile tubes.

Use: For the isolation and cultivation of *Cytophaga* species, *Herpetosiphon* species, *Saprospira* species, and *Flexithrix* species.

HP 74 Broth

Composition per liter:

Sodium glutaminate	10.0g
$MgSO_4 \cdot 7H_2O$	2.0g
Yeast extract	2.0g
Glucose solution	100.0mL
Phosphate buffer solution	20.0mL

pH 6.5 ± 0.2 at 25°C

Glucose Solution:
Composition per 100.0mL:

D-Glucose	10.0g

Preparation of Glucose Solution: Add D-glucose to distilled/deionized water and bring volume to 100.0mL.

Mix thoroughly. Autoclave for 15 min at 15 psi pressure–121°C. Cool to 25°C.

Phosphate Buffer Solution:
Composition per 100.0mL:

K_2HPO_4	6.81g

Preparation of Phosphate Buffer Solution: Add K_2HPO_4 to distilled/deionized water and bring volume to 100.0mL. Mix thoroughly. Adjust pH to 6.5. Autoclave for 15 min at 15 psi pressure–121°C. Cool to 25°C.

Preparation of Medium: Add components, except glucose solution and phosphate buffer solution, to distilled/deionized water and bring volume to 880.0mL. Mix thoroughly. Gently heat and bring to boiling. Autoclave for 15 min at 15 psi pressure–121°C. Cool to 45°–50°C. Aseptically add 100.0mL of sterile glucose solution and 20.0mL of sterile phosphate buffer solution. Mix thoroughly. Aseptically distribute into sterile tubes or flasks.

Use: For the isolation and cultivation of *Cytophaga* species, *Herpetosiphon* species, *Saprospira* species, and *Flexithrix* species.

HP 101 Halophile Medium

Composition per liter:

NaCl	100.0g
Agar	20.0g
Peptone	10.0g
$MgSO_4 \cdot 7H_2O$	4.3g
$NaNO_3$	2.0g
Yeast extract	1.0g

pH 7.2 ± 0.2 at 25°C

Preparation of Medium: Add components to distilled/deionized water and bring volume to 1.0L. Mix thoroughly. Gently heat and bring to boiling. Distribute into tubes or flasks. Autoclave for 15 min at 15 psi pressure–121°C. Pour into sterile Petri dishes or leave in tubes.

Use: For the cultivation and maintenance of *Pseudomonas* species.

HP Medium

Composition per liter:

Pancreatic digest of soybean meal	20.0g
Beef extract	10.0g
Yeast extract	6.0g
Ammonium citrate	5.0g
Tween 80	0.5g
$MgSO_4 \cdot 7H_2O$	0.2g
$MnSO_4 \cdot 4H_2O$	0.05g
$FeSO_4 \cdot 7H_2O$	0.04g
Glucose solution	10.0mL
Tetracycline solution	10.0mL

Glucose Solution:
Composition per 100.0mL:

Glucose	10.0g

Preparation of Glucose Solution: Add glucose to distilled/deionized water and bring volume to 100.0mL. Mix thoroughly. Filter sterilize.

Tetracycline Solution:
Composition per 100.0mL:

Tetracycline	10.0g

Preparation of Tetracycline Solution: Add tetracycline to distilled/deionized water and bring volume to 100.0mL. Mix thoroughly. Filter sterilize.

Preparation of Medium: Add components, except glucose solution and tetracycline solution, to distilled/deionized water and bring volume to 990.0mL. Mix thoroughly. Gently heat and bring to boiling. Autoclave for 15 min at 15 psi pressure–121°C. Cool to 45°–50°C. Aseptically add sterile glucose solution and tetracycline solution. Mix thoroughly. Aseptically distribute into sterile tubes or flasks.

Use: For the cultivation and enumeration of *Leuconostoc* species.

HPC Agar (Heterotrophic Plate Count Agar) (m-HPC Agar)

Composition per liter:

Gelatin	25.0g
Pancreatic digest of gelatin	20.0g
Agar	15.0g
Glycerol	10.0mL

pH 7.1 ± 0.2 at 25°C

Source: This medium is available from BD Diagnostic Systems.

Preparation of Medium: Add components, except glycerol, to distilled/deionized water and bring volume to 990.0mL. Mix thoroughly. Gently heat and bring to boiling. Add glycerol. Mix thoroughly. Autoclave for 15 min at 15 psi pressure–121°C. Cool to 45°–50°C. Pour into sterile Petri dishes.

Use: For the the cultivation and enumeration of microorganisms from potable water sources, swimming pools, and other water specimens by the membrane filter method and heterotrophic plate count technique.

Hugh-Leifson's Glucose Broth

Composition per liter:

NaCl	30.0g
Glucose	10.0g
Agar	3.0g
Peptone	2.0g
Yeast extract	0.5g
Bromcresol Purple	0.015g

pH 7.4 ± 0.2 at 25°C

Preparation of Medium: Add components to distilled/deionized water and bring volume to 1.0L. Mix thoroughly. Gently heat while stirring and bring to boiling. Adjust pH to 7.4. Distibute into tubes or flasks. Autoclave for 15 min at 15 psi pressure–121°C.

Use: For the cultivation and differentiation of bacteria based on their ability to ferment glucose. Bacteria that ferment glucose turn the medium yellow.

Hungate's Habitat-Simulating Medium

Composition per 1140.2mL:

Rumen fluid	333.0mL
Mineral solution A	167.0mL
Mineral solution B	167.0mL
$NaHCO_3$ solution	53.0mL
L-Cysteine·HCl solution	10.6mL
Substrate solution	10.6mL
Resazurin solution	1.0mL

Mineral Solution A:
Composition per liter:

NaCl ... 6.0g
KH_2PO_4 ... 3.0g
$(NH_4)_2SO_4$ ... 3.0g
$CaCl_2$ ... 0.6g
$MgSO_4$ ... 0.6g

Preparation of Solution A: Add components to distilled/deionized water and bring volume to 1.0L. Mix thoroughly.

Mineral Solution B:

K_2HPO_4 ... 3.0

Preparation of Solution B: Add K_2HPO_4 to distilled/deionized water and bring volume to 1.0L. Mix thoroughly.

Resazurin Solution:
Composition per 100.0mL:

Resazurin ... 0.1g

Preparation of Resazurin Solution: Add resazurin to distilled/deionized water and bring volume to 100.0L. Mix thoroughly.

L-Cysteine·HCl Solution:
Composition per 100.0mL:

L-Cysteine·HCl ... 3.0g

Preparation of L-Cysteine·HCl Solution: Add L-cysteine·HCl to O_2-free distilled/deionized water and bring volume to 100.0L. Mix thoroughly. Gently heat and bring to boiling. Continue boiling for 2 min. Cool to 25°C under 100% N_2. Seal tube with a stopper that is wired in place. Autoclave for 15 min at 15 psi pressure–121°C. Cool to 25°C.

$NaHCO_3$ Solution:
Composition per 10.0mL:

$NaHCO_3$ ... 1.0g

Preparation of $NaHCO_3$ Solution: Add $NaHCO_3$ to O_2-free distilled/deionized water and bring volume to 10.0mL. Mix thoroughly. Filter sterilize. Gas with 100% CO_2 for 15 min.

Substrate Solution:
Composition per 100.0mL:

Sugar ... 10.0g

Preparation of Substrate Solution: Add sugar to O_2-free distilled/deionized water. Mix thoroughly. Gas with 100% N_2 for 15 min. Autoclave for 15 min at 15 psi pressure–121°C. Cool to 45°–50°C.

Preparation of Medium: Add 167.0mL of solution A, 167.0mL of solution B, and 1.0mL of resazurin solution to distilled/deionized water and bring volume to 733.0mL. Mix thoroughly. Gently heat and bring to boiling. Continue boiling until resazurin turns colorless, indicating reduction. Bring volume back to 733.0mL (some evaporation will have occurred) with O_2-free distilled/deionized water. Cool to 45°–50°C under O_2-free 100% CO_2. Anaerobically add rumen fluid. Anaerobically distribute into tubes in 10.0mL volumes. Cap with butyl rubber stoppers. Place tubes in a press. Autoclave for 15 min at 15 psi pressure–121°C. Cool to 25°C. Immediately prior to inoculation, aseptically and anaerobically add 0.1mL of sterile L-cysteine·HCl solution, 0.5mL of sterile $NaHCO_3$ solution, and 0.1mL of substrate solution per 10.0mL of medium in each tube.

Use: For the cultivation of *Bacteroides* species from rumens.

HY Agar for *Flavobacterium*

Composition per liter:

Agar ... 8.0g
Glutamic acid ... 5.0g
K_2HPO_4 ... 0.1g
$MgSO_4 \cdot 7H_2O$ ... 0.1g
pH 7.3 ± 0.2 at 25°C

Preparation of Medium: Add components to distilled/deionized water and bring volume to 1.0L. Glutamic acid may be replaced by 1.0g of folic acid if desired. Mix thoroughly. Gently heat and bring to boiling. Distribute into tubes or flasks. Autoclave for 15 min at 15 psi pressure–121°C.

Use: For the cultivation of *Flavobacterium* species.

Hydrogen-Oxidizing Bacteria Medium

Composition per 1020.0mL:

Solution I ... 1.0L
Solution II ... 10.0mL
Solution III ... 10.0mL

Solution I:
Composition per liter:

$Na_2HPO_4 \cdot 12H_2O$ ... 9.0g
KH_2PO_4 ... 1.5g
NH_4Cl ... 1.0g
$MgSO_4 \cdot 7H_2O$ ... 0.2g
Trace elements solution ... 1.0mL

Preparation of Solution I: Add components to distilled/deionized water and bring volume to 1.0L. Mix thoroughly. Gently heat until dissolved. Autoclave for 15 min at 15 psi pressure–121°C. Cool to 25°C.

Trace Elements Solution:
Composition per liter:

H_3BO_3 ... 0.3g
$CoCl_2 \cdot 6H_2O$ ... 0.2g
$ZnSO_4 \cdot 7H_2O$ ... 0.1g
$MnCl_2 \cdot 4H_2O$ ... 0.03g
$NaMoO_4 \cdot 2H_2O$ ... 0.03g
$NiCl_2 \cdot 6H_2O$ ... 0.02g
$CuCl_2 \cdot 2H_2O$ ... 0.01g

Preparation of Trace Elements Solution: Add components to distilled/deionized water and bring volume to 1.0L. Mix thoroughly.

Solution II:
Composition per 100.0mL:

$CaCl_2 \cdot 2H_2O$ ... 0.1g
Ferric ammonium citrate ... 0.05g

Preparation of Solution II: Add components to distilled/deionized water and bring volume to 100.0mL. Mix thoroughly. Autoclave for 15 min at 15 psi pressure–121°C. Cool to 25°C.

Solution III:
Composition per 100.0mL:

$NaHCO_3$ ... 5.0g

Preparation of Solution III: Add $NaHCO_3$ to distilled/deionized water and bring volume to 100.0mL. Mix thoroughly. Filter sterilize.

Preparation of Medium: Aseptically combine 1.0L of cooled, sterile solution I, 10.0mL of cooled, sterile solution II, and 10.0mL of sterile solution III. Mix thoroughly. Aseptically distribute into sterile tubes or flasks.

Use: For the cultivation of hydrogen-oxidizing bacteria.

Hydrogenobacter acidophilus Medium (DSMZ Medium 743)

Composition per liter:

Sulfur	5.0g
$(NH_4)_2SO_4$	1.0g
K_2HPO_4	1.0g
NaCl	1.0g
$MgSO_4 \cdot 7H_2O$	0.3g
$FeSO_4 \cdot 7H_2O$	1.0mg
$CaCl_2$	1.0mg
$NiSO_4 \cdot 6H_2O$	0.06mg
Trace elements solution	0.5mL

pH 3.0 ± 0.2 at 25°C

Trace Elements Solution:

Composition per liter:

$ZnSO_4 \cdot 7H_2O$	28.0mg
MoO_3	4.0mg
H_3BO_3	4.0mg
$MnSO_4 \cdot 5H_2O$	4.0mg
$CoCl_2 \cdot 6H_2O$	4.0mg
$CuSO_4 \cdot 5H_2O$	2.0mg

Preparation of Trace Elements Solution: Add components to distilled/deionized water and bring volume to 1.0L. Mix thoroughly. Sparge with 80% H_2 + 20% CO_2.

Preparation of Medium: Autoclave sulfur for 15 min at 9 psi pressure–113°C. Add components, except sulfur to distilled/deionized water and bring volume to 1.0L. Mix thoroughly. Autoclave for 15 min at 15 psi pressure–121°C. Add 5.0g sterile sulfur. Mix thoroughly by swirling. Adjust pH to 3.0 with HCl. Aseptically distribute into sterile tubes or flasks.

Use: For the cultivation of *Hydrogenobaculum acidophilum=Hydrogenobacter acidophilus.*

Hydrogenobacter halophilus Medium (DSMZ Medium 744)

Composition per liter:

NaCl	29.3g
K_2HPO_4	2.5g
$(NH_4)_2SO_4$	2.0g
KH_2PO_4	0.5g
$MgSO_4 \cdot 7H_2O$	0.2g
$CaCl_2$	10.0mg
$FeSO_4 \cdot 7H_2O$	10.0mg
$NiSO_4 \cdot 7H_2O$	0.6mg
Trace elements solution	0.5mL

pH 6.9 ± 0.2 at 25°C

Trace Elements Solution:

Composition per liter:

$ZnSO_4 \cdot 7H_2O$	28.0mg
MoO_3	4.0mg
H_3BO_3	4.0mg
$MnSO_4 \cdot 5H_2O$	4.0mg
$CoCl_2 \cdot 6H_2O$	4.0mg
$CuSO_4 \cdot 5H_2O$	2.0mg

Preparation of Trace Elements Solution: Add components to distilled/deionized water and bring volume to 1.0L. Mix thoroughly. Sparge with 80% H_2 + 20% CO_2.

Preparation of Medium: Add components to distilled/deionized water and bring volume to 1.0L. Mix thoroughly. Autoclave for 15 min at 15 psi pressure–121°C. Adjust pH to 6.9. Aseptically distribute into sterile tubes or flasks.

Use: For the cultivation of *Hydrogenovibrio marinus.*

Hydrogenobacter thermophilus Medium

Composition per liter:

Na_2HPO_4	4.5g
KH_2PO_4	1.5g
NH_4NO_3	1.0g
NaCl	1.0g
$MgSO_4 \cdot 7H_2O$	0.2g
$CaCl_2$	10.0mg
$FeSO_4 \cdot 7H_2O$	10.0mg
$NiSO_4 \cdot 7H_2O$	0.06mg
Trace elements solution	2.0mL

Trace Elements Solution:

Composition per liter:

$ZnSO_4 \cdot 7H_2O$	7.0mg
MoO_3	1.0mg
H_3BO_3	1.0mg
$MnSO_4 \cdot H_2O$	1.0mg
$CoCl_2 \cdot 6H_2O$	1.0mg
$CuSO_4 \cdot 5H_2O$	0.5mg

Preparation of Trace Elements Solution: Add components to distilled/deionized water and bring volume to 1.0L. Mix thoroughly.

Preparation of Medium: Add components to distilled/deionized water and bring volume to 1.0L. Mix thoroughly. Distribute into tubes or flasks. Autoclave for 15 min at 15 psi pressure–121°C.

Use: For the cultivation of *Calderobacterium hydrogenophilum* and *Hydrogenobacter thermophilus.*

Hydrogenothermus hirschii Medium (DSMZ Medium 783)

Composition per liter:

$MgSO_4 \cdot 7H_2O$	7.0g
$MgCl_2 \cdot 6H_2O$	5.5g
$NaHCO_3$	2.0g
KCl	0.65g
$CaCl_2 \cdot 2H_2O$	0.5g
Sulfur, powdered	0.5g
NH_4Cl	0.15g
K_2HPO_4	0.15g
NaBr	0.1g
Trace elements solution	10.0mL
$CaCO_3$ solution	5.0mL

pH 7.0 ± 0.2 at 25°C

$CaCO_3$ Solution:

Composition per 10mL:

$CaCO_3$	1.0g

Preparation of $CaCO_3$ Solution: Add $CaCO_3$ to 10.0mL of distilled/deionized water. Mix thoroughly. Autoclave for 15 min at 15 psi pressure–121°C. Cool to room temperature.

Trace Elements Solution:

Composition per liter:

$MgSO_4 \cdot 7H_2O$	3.0g
Nitrilotriacetic acid	1.5g
NaCl	1.0g
$MnSO_4 \cdot 2H_2O$	0.5g
$CoSO_4 \cdot 7H_2O$	0.18g
$ZnSO_4 \cdot 7H_2O$	0.18g
$CaCl_2 \cdot 2H_2O$	0.1g
$FeSO_4 \cdot 7H_2O$	0.1g
$NiCl_2 \cdot 6H_2O$	0.025g
$KAl(SO_4)_2 \cdot 12H_2O$	0.02g
H_3BO_3	0.01g
$Na_2MoO_4 \cdot 4H_2O$	0.01g

$CuSO_4·5H_2O$ 0.01g
$Na_2SeO_3·5H_2O$ 0.3mg

Preparation of Trace Elements Solution: Add nitrilotriacetic acid to 500.0mL of distilled/deionized water. Dissolve by adjusting pH to 6.5 with KOH. Add remaining components. Add distilled/deionized water to 1.0L. Mix thoroughly.

Preparation of Medium: Prepare the medium aerobically. Add sulfur to 900.0mL distilled/deionized water. Dissolve sulfur using Ultra-Turrax-dispersing instrument. Add remaining components. Bring volume to 1.0L with distilled/deionized water. Adjust pH to 7.0 using H_2SO_4. Add 20.0mL medium into 100mL serum bottles. Seal with a rubber stopper. Change atmosphere to 80% H_2 + 20% CO_2 with an overpressure of 2 atmospheres. Autoclave for 20 min at 15 psi pressure–121°C. Cool to room temperature. Inject 20.0mL filter sterilized air and 0.1mL sterile $CaCO_3$ solution. Shake to mix.

Use: For the cultivation of *Hydrogenophilus hirschii (Hydrogenothermophilus hirschii).*

Hydroxybenzoate Agar (*p*-Hydroxybenzoate Agar)

Composition per liter:

Agar 20.0g
$(NH_4)_2HPO_4$ 3.0g
p-hydroxybenzoic acid 3.0g
K_2HPO_4 .... 1.2g
NaCl 0.5g
$MgSO_4·7H_2O$ 0.2g
$FeSO_4·7H_2O$ 0.1g

pH 7.0 ± 0.2 at 25°C

Preparation of Medium: Add components to distilled/deionized water and bring volume to 1.0L. Mix thoroughly. Gently heat and bring to boiling. Distribute into tubes or flasks. Autoclave for 15 min at 15 psi pressure–121°C. Pour into sterile Petri dishes or leave in tubes.

Use: For the cultivation of *p*-hydroxybenzoate-utilizing bacteria.

Hyperthermus butylicus Medium

Composition per 1010.0mL:

NaCl 17.0g
Pancreatic digest of casein 6.0g
Sulfur, powdered 6.0g
$MgSO_4·7H_2O$ 3.5g
$MgCl_2·6H_2O$ 2.75g
$NiCl_2·6H_2O$ 2.0g
Yeast extract 2.0g
$CaCl_2·2H_2O$ 0.75g
KH_2PO_4 .... 0.5g
NH_4Cl 0.5g
KCl 0.325g
NaBr 0.05g
H_3BO_3 .... 0.015g
$(NH_4)_2SO_4$ 10.0mg
$SrCl_2·6H_2O$ 7.5mg
Citric acid 5.0mg
KI 2.5mg
Resazurin 1.0mg
Trace elements solution 10.0mL
$Na_2S·9H_2O$ solution 10.0mL

pH 6.0–6.5 at 25°C

Trace Elements Solution:
Composition per liter:

$MgSO_4·7\ H_2O$ 3.0g
Nitrilotriacetic acid .... 1.5 g
$CaCl_2·2\ H_2O$ 1.0g
NaCl 1.0g
$MnSO_4·2\ H_2O$ 0.5g
$CoSO_4·7\ H_2O$ 0.18g
$ZnSO_4·7\ H_2O$ 0.18g
$FeSO_4·7\ H_2O$ 0.1g
$NiCl_2·6\ H_2O$ 0.025g
$KAl(SO_4)_2·12\ H_2O$ 0.02g
$CuSO_4·5\ H_2O$ 0.01g
H_3BO_3 .... 0.01g
$Na_2MoO_4·2\ H_2O$ 0.01g
$Na_2SeO_3·5\ H_2O$ 0.3mg

Preparation of Trace Elements Solution: Add nitrilotriacetic acid to 500.0mL of distilled/deionized water. Adjust pH to 6.5 with KOH. Add remaining components. Adjust pH to 7.0 with KOH. Add distilled/deionized water to 1.0L.

$Na_2S·9H_2O$ Solution:
Composition per 10.0mL:

$Na_2S·9H_2O$ 0.5g

Preparation of $Na_2S·9H_2O$ Solution: Add $Na_2S·9H_2O$ to distilled/deionized water and bring volume to 10.0mL. Mix thoroughly. Gas under 80% N_2 + 20% CO_2. Autoclave for 15 min at 15 psi pressure–121°C.

Preparation of Medium: Add components, except $Na_2S·9H_2O$ solution, to distilled/deionized water and bring volume to 1.0L. Mix thoroughly. Adjust pH to 6.0–6.5 with 6*N* H_2SO_4. Sparge wih 100% N_2. Sterilize by bringing to 90°C for 60 min on 3 consecutive days. Immediately prior to inoculation, add 10.0mL of sterile $Na_2S·9H_2O$ solution. Mix thoroughly.

Use: For the cultivation and maintenance of *Hyperthermus butylicus*.

Hyphomicrobium Enrichment Medium

Composition per 100.0mL:

KNO_3 .... 0.04g
$Na_2HPO_4·7H_2O$ 0.02g
$MgSO_4·7H_2O$ 0.48mg
$FeCl_3·7H_2O$ 0.02mg
$MnCl_2·4H_2O$ 0.01mg

pH 7.2 ± 0.2 at 25°C

Preparation of Medium: Add components to distilled/deionized water and bring volume to 1.0L. Mix thoroughly. Adjust pH to 7.2. Distribute into tubes or flasks. Autoclave for 15 min at 15 psi pressure–121°C.

Use: For the cultivation and enrichment of *Hyphomicrobium* species.

Hyphomicrobium Medium

Composition per liter:

Agar 15.0g
Na_2HPO_4 .... 2.13g
KH_2PO_4 .... 1.36g
$MgSO_4·7H_2O$ 0.2g
$CaCl_2·2H_2O$ 9.95mg
$FeSO_4·7H_2O$ 5.0mg
$MnSO_4·4H_2O$ 2.5mg
$Na_2MoO_4·2H_2O$ 2.5mg

Urea solution 30.0mL
Methanol 4.0mL

Urea Solution:
Composition per 100.0mL:
Urea 20.0g

Preparation of Urea Solution: Add urea to distilled/deionized water and bring volume to 100.0mL. Mix thoroughly. Filter sterilize.

Preparation of Medium: Filter sterilize methanol. Add components, except urea solution and methanol, to distilled/deionized water and bring volume to 966.0mL. Mix thoroughly. Gently heat and bring to boiling. Autoclave for 15 min at 15 psi pressure–121°C. Cool to 45°–50°C. Aseptically add sterile urea solution and sterile methanol. Mix thoroughly. Aseptically distribute into sterile tubes or bottles.

Use: For the cultivation of *Hyphomicrobium* species.

Hyphomicrobium methylovorum Medium

Composition per liter:
Agar 15.0g
$(NH_4)_2HPO_4$ 3.0g
NaCl 1.0g
$MgSO_4 \cdot 7H_2O$ 0.2g
$FeSO_4 \cdot 7H_2O$ 10.0mg
$MnSO_4 \cdot 2H_2O$ 5.0mg
Tap water 1.0L
Methanol 10.0mL
Vitamin mixture 5.0mL

Vitamin Mixture:
Composition per liter:
Inositol 200.0mg
Choline 100.0mg
Calcium DL-pantothenate 40.0mg
Niacin 40.0mg
Pyridoxine·HCl 40.0mg
Riboflavin 40.0mg
p-Aminobenzoic acid 20.0mg
Thiamine·HCl 20.0mg
Biotin 0.2mg
Folic acid 0.2mg
Cyanocobalamin 2.0μg

Preparation of Vitamin Mixture: Add components to distilled/deionized water and bring volume to 1.0L. Mix thoroughly. Filter sterilize.

Preparation of Methanol: Filter sterilize 10.0mL of methanol.

Preparation of Medium: Add components, except methanol and vitamin mixture, to distilled/deionized water and bring volume to 985.0mL. Mix thoroughly. Autoclave for 15 min at 15 psi pressure–121°C. Aseptically add 10.0mL of sterile methanol and 5.0mL of sterile vitamin mixture. Mix thoroughly. Aseptically distribute into sterile tubes or flasks.

Use: For the cultivation of *Hyphomicrobium methylovorum.*

Hyphomicrobium Strain X Agar

Composition per liter:
Agar 15.0g
Methylamine·HCl 3.4g
K_2HPO_4 .... 1.55g
$(NH_4)_2SO_4$ 1.0g
$NaH_2PO_4 \cdot H_2O$ 0.5g
$MgSO_4 \cdot 7H_2O$ 0.2g
Trace elements solution 0.2mL
pH 7.2 ± 0.2 at 25°C

Trace Elements Solution:
Composition per liter:
Disodium EDTA 50.0g
$ZnSO_4 \cdot 7H_2O$ 22.0g
$CaCl_2 \cdot 2H_2O$ 5.54g
$MnCl_2 \cdot 4H_2O$ 5.06g
$FeSO_4 \cdot 7H_2O$ 5.0g
$CoCl_2 \cdot 6H_2O$ 1.61g
$CuSO_4 \cdot 5H_2O$ 1.57g
$(NH_4)6Mo_7O_{24} \cdot 4H_2O$ 1.1g

Preparation of Trace Elements Solution: Add components to distilled/deionized water and bring volume to 1.0L. Adjust pH to 7.0 with KOH.

Preparation of Medium: Add components to distilled/deionized water and bring volume to 1.0L. Mix thoroughly. Adjust pH to 7.2. Gently heat and bring to boiling. Distribute into tubes or flasks. Autoclave for 15 min at 15 psi pressure–121°C. Pour into sterile Petri dishes or leave in tubes.

Use: For the cultivation and maintenance of *Hyphomicrobium* species.

Hyphomonas Enrichment Medium

Composition per liter:
Peptone 0.05g
Yeast extract 0.05g

Preparation of Medium: Add components to distilled/deionized water and bring volume to 1.0L. Mix thoroughly. Distribute into tubes or flasks. Autoclave for 15 min at 15 psi pressure–121°C.

Use: For the isolation and cultivation of *Hyphomonas* species.

Hyphomonas Medium

Composition per liter:
Pancreatic digest of casein 2.0g
$MgCl_2 \cdot 2H_2O$ 2.0g
Yeast extract 1.0g
pH 7.5 ± 0.2 at 25°C

Preparation of Medium: Add components to distilled/deionized water and bring volume to 1.0L. Mix thoroughly. Adjust pH to 7.5 using indicator paper. Distribute into tubes or flasks. Autoclave for 15 min at 15 psi pressure–121°C. Autoclave for 15 min at 15 psi pressure–121°C.

Use: For the cultivation and maintenence of *Hyphomonas polymorpha.*

IFO Agar

Composition per liter:
Agar 20.0g
$(NH_4)_2HPO_4$ 3.0g
NaCl 1.0g
$MgSO_4 \cdot 7H_2O$ 0.2g
$FeSO_4 \cdot 7H_2O$ 10.0mg
$MnSO_4 \cdot 4\text{-}6H_2O$ 5.0mg
Riboflavin 0.02mg
Calcium pantothenate 0.02mg
Pyridoxine·HCl 0.02mg
Nicotinic acid 0.02mg
p-Aminobenzoic acid 0.01mg
Thiamine·HCl 0.01mg

Biotin 1.0μg
Methanol 10.0mL

pH 7.0 ± 0.2 at 25°C

Preparation of Medium: Add components, except agar and methanol, to distilled/deionized water and bring volume to 490.0mL. Mix thoroughly. Autoclave for 15 min at 15 psi pressure–121°C. Cool to 45°–50°C. In a separate flask, add agar to distilled/deionized water and bring volume to 500.0mL. Mix thoroughly. Gently heat and bring to boiling. Autoclave for 15 min at 15 psi pressure–121°C. Cool to 45°–50°C. Aseptically combine the two sterile solutions. Aseptically add 10.0mL of filter-sterilized methanol. Mix thoroughly. Adjust pH to 7.0. Pour into sterile Petri dishes or distribute into sterile tubes.

Use: For the cultivation and maintenance of *Hyphomicrobium methylovorum.*

IFO Broth

Composition per liter:

$(NH_4)_2HPO_4$ 3.0g
NaCl 1.0g
$MgSO_4·7H_2O$ 0.2g
$FeSO_4·7H_2O$ 10.0mg
$MnSO_4·4\text{-}6H_2O$ 5.0mg
Riboflavin 20.0μg
Calcium pantothenate 20.0μg
Pyridoxine·HCl 20.0μg
Nicotinic acid 20.0μg
p-Aminobenzoic acid 10.0μg
Thiamine·HCl 10.0μg
Biotin 1.0μg
Methanol 10.0mL

pH 7.0 ± 0.2 at 25°C

Preparation of Medium: Add components, except methanol, to distilled/deionized water and bring volume to 990.0mL. Mix thoroughly. Autoclave for 15 min at 15 psi pressure–121°C. Aseptically add 10.0mL of filter-sterilized methanol. Mix thoroughly. Adjust pH to 7.0. Aseptically distribute into sterile tubes or flasks.

Use: For the cultivation and maintenance of *Hyphomicrobium methylovorum.*

IFO Medium 802

Composition per liter:

Polypeptone 10.0g
Yeast extract 2.0g
$MgSO_4·7H_2O$ 1.0g

pH 7.0 ± 0.2 at 25°C

Preparation of Medium: Add components to distilled/deionized water and bring volume to 1.0L. Mix thoroughly. Adjust pH to 7.0. Distribute into tubes or flasks. Autoclave for 15 min at 15 psi pressure–121°C.

Use: For the cultivation of *Sphingomonas asaccharolytica, Sphingomonas pruni, Sphingomonas mali,* and *Sphingomonas rosa.*

Ignicoccus Medium (DSMZ Medium 897)

Composition per liter:

NaCl 13.65g
Sulfur, powdered 5.0g
$MgSO_4·7H_2O$ 3.5g
$MgCl_2·6H_2O$ 2.75g
Meat extract 1.0g
KH_2PO_4 .... 0.5g
$CaCl_2·2H_2O$ 0.38g
KCl 0.33g
$(NH_4)_2SO_4$ 0.25g
NaBr 0.05g
H_3BO_3 .... 15.0mg
SrCl3·6H2O 7.50mg
KI 0.05mg
Resazurin 0.5mg
$Na_2S·9H_2O$ solution 10.0mL

pH 5.5 ± 0.2 at 25°C

$Na_2S·9H_2O$ Solution:

Composition per 10.0mL:

$Na_2S·9H_2O$ 0.2g

Preparation of $Na_2S·9H_2O$ Solution: Add 0.2g of $Na_2S·9H_2O$ to distilled/deionized water and bring volume to 10.0mL. Sparge with N_2. Autoclave for 15 min at 15 psi pressure–121°C. Cool to 25°C. Store anaerobically.

Preparation of Medium: Add components, except $Na_2S·9H_2O$ solution to distilled/deionized water and bring volume to 990.0mL. Sparge medium with N_2 gas for 30-60 min. Mix thoroughly. Add 10.0mL $Na_2S·9H_2O$ solution. Mix thoroughly. Adjust pH to 5.5 with H_2SO_4. Distribute into tubes or bottles under 80% H_2 and 20% CO_2 gas mixture. Heat the vessels containing medium in boiling water for 1 hr before inoculation. After inoculation pressurize the vessels with 80% H_2 and 20% CO_2 gas mixture to 2 bar overpressure.

Use: For the cultivation of *Ignicoccus islandicus* and *Ignicoccus pacificus.*

Imhoff's Medium, Modified

Composition per liter:

NaCl 30.0g
$NaHCO_3$ 3.0g
KH_2PO_4 .... 1.0g
NH_4Cl 1.0g
Sodium acetate 1.0g
Na_2SO_4 .... 0.7g
$MgCl_2·6H_2O$ 0.5g
Sodium ascorbate 0.5g
$CaCl_2·2H_2O$ 0.1g
Yeast extract 0.1g
Sodium sulfide solution 10.0mL
SLA trace elements solution 1.0mL
VA vitamin solution 1.0mL

pH 6.9–7.0 at 25°C

Sodium Sulfide Solution:

Composition per 100.0mL:

$Na_2S·9H_2O$ 2.0g

Preparation of Sodium Sulfide Solution: Add 2.0g of $Na_2S·9H_2O$ to distilled/deionized water and bring volume to 100.0mL. Mix thoroughly. Autoclave for 15 min at 15 psi pressure–121°C under N_2. Maintain under 100% N_2.

SLA Trace Elements Solution:

Composition per liter:

$FeCl_2·4H_2O$ 1.8g
H_3BO_3 .... 0.5g
$CoCl_2·6H_2O$ 0.25g
$ZnCl_2$ 0.1g
$MnCl_2·4H_2O$ 0.07g
$Na_2MoO_4·2H_2O$ 0.03g
$CuCl_2·2H_2O$ 0.01g
$Na_2SeO_3·5H_2O$ 0.01g
$NiCl_2·6H_2O$ 0.01g

Preparation of SLA Trace Elements Solution: Add components to distilled/deionized water and bring volume to 1.0L. Mix thoroughly. Adjust pH to pH 2–3. Filter sterilize.

VA Vitamin Solution:
Composition per 500.0mL:

Nicotinamide	0.17g
Thiamine·HCl	0.15g
p-Aminobenzoic acid	0.1g
Biotin	0.05g
Calcium pantothenate	0.05g
Pyridoxine·2HCl	0.05g
Cyanocobalamin	0.02g

Preparation of VA Vitamin Solution: Add components to distilled/deionized water and bring volume to 500.0mL. Mix thoroughly. Filter sterilize.

Preparation of Medium: Add components—except sodium sulfide solution, SLA trace elements solution, and VA vitamin solution—to distilled/deionized water and bring volume to 988.0mL. Mix thoroughly. Autoclave for 15 min at 15 psi pressure–121°C. Cool to 25°C. Aseptically add 1.0mL of sterile SLA trace elements solution and 1.0mL of sterile VA vitamin solution. Aseptically add 10.0mL of sterile sodium sulfide solution. Mix thoroughly. Aseptically distribute into sterile tubes or flasks.

Use: For the cultivation and maintenance of *Rhodobacter adriaticus* and *Rhodobacter sulfidophilus.*

Imidazole Utilization Medium

Composition per liter:

Imidazole	5.0g
KH_2PO_4	0.5g
$MgSO_4 \cdot 7H_2O$	0.5g
$CaCl_2$	3.0mg
$FeSO_4 \cdot 7H_2O$	3.0mg
Molybdenum solution	1.0mL
Trace elements solution	1.0mL

pH 6.0 ± 0.2 at 25°C

Molybdenum Solution:
Composition per 18.0mL:

$Na_2MoO_4 \cdot 2H_2O$	0.5mg

Preparation of Molybdenum Solution: Add components to distilled/deionized water and bring volume to 18.0mL. Mix thoroughly. Filter sterilize.

Trace Elements Solution:
Composition per 18.0mL:

H_3BO_3	11.0mg
$MnCl_2 \cdot 4H_2O$	7.0mg
$Al_2(SO_4)_3 \cdot 18\ H_2O$	1.94mg
$Co(NO_3)_2 \cdot 6H_2O$	1.0mg
$CuSO_4 \cdot 5H_2O$	1.0mg
$NiSO_4 \cdot 6H_2O$	1.0mg
$ZnSO_4 \cdot H_2O$	0.62mg
KBr	0.5mg
KI	0.5mg
LiCl	0.5mg
$SnCl_2 \cdot 2H_2O$	0.5mg

Preparation of Trace Elements Solution: Add components to distilled/deionized water and bring volume to 18.0mL. Mix thoroughly. Filter sterilize.

Preparation of Medium: Add components, except molybdenum solution and trace elements solution, to distilled/deionized water and bring volume to 998.0mL. Mix thoroughly. Distribute into tubes or flasks. Autoclave for 15 min at 15 psi pressure–121°C. Cool to 25°C. Aseptically add 1.0mL of molybdenum solution and 1.0mL of trace elements solution. Mix thoroughly. Adjust pH to 6.0 with phosphoric acid. Mix thoroughly. Aseptically distribute into sterile tubes or flasks.

Use: For the cultivation and maintenance of *Pseudo-monas* species.

Inorganic Salts-Maltose Medium (DSMZ Medium 754)

Composition per liter:

Yeast extract	4.0g
Peptone	2.0g
Inorganic salt solution	980.0mL
Maltose solution	20.0mL

Maltose Solution:
Composition per liter:

Maltose	5.0g

Preparation of Maltose Solution: Add maltose to 20.0mL distilled/deionized water. Mix thoroughly. Filter sterilize.

Inorganic Salt Solution:
Composition per liter:

$MgSO_4 \cdot 7H_2O$	49.37g
NaCl	43.8g
$CaCl_2 \cdot 2H_2O$	1.29g

Preparation of Inorganic Salt Solution: Add components in the order $CaCl_2 \cdot 2H_2O$, NaCl, $MgSO_4 \cdot 7H_2O$ to 900.0mL distilled/deionized water. After addition of each mix thorougly to prevent precipitation. Add distilled/deionized water bring volume to 1.0L. Mix thoroughly.

Preparation of Medium: Add components except maltose solution to 980.0mL inorganic salt solution. Mix thoroughly. Autoclave for 15 min at 15 psi pressure–121°C. Cool to 25°C. Aseptically add 20.0mL sterile maltose solution. Mix thoroughly. Aseptically distribute to sterile tubes or flasks.

Use: For the cultivation of *Spirochaeta halophila.*

Inositol Brilliant Green Bile Salts Agar (IBB Agar) (*Pleisomonas* Differential Agar)

Composition per liter:

Agar	15.0g
meso-Inositol	10.0g
Proteose peptone	10.0g
Bile salts No. 3	8.5g
Meat extract	5.0g
NaCL	5.0g
Neutral Red (2% solution)	1.25mL
Brilliant Green (0.1% solution)	0.33mL

pH 7.2 ± 0.1 at 25°C

Preparation of Medium: Add components to distilled/deionized water and bring volume to 1.0L. Mix thoroughly. Gently heat and bring to boiling. Distribute into tubes or flasks. Autoclave for 15 min at 15 psi pressure–121°C. Pour into sterile Petri dishes or leave in tubes.

Use: For the isolation of *Aeromonas* and *Plesiomonas* species.

Iron Bacteria Isolation Medium

Composition per liter:

Agar	10.0g
$(NH_4)_2SO_4$	0.5g

Glucose 0.15g
$CaCO_3$ 0.1g
K_2HPO_4 .... 0.05g
$MgSO_4 \cdot 7H_2O$ 0.05g
KCl 0.05g
$Ca(NO_3)_2$ 0.01g
Vitamin solution 10.0mL

Vitamin Solution:
Composition per 10.0mL:
Thiamine 0.4mg
Cyanocobalamin 0.01mg

Preparation of Vitamin Solution: Add components to distilled/deionized water and bring volume to 10.0mL. Mix thoroughly. Filter sterilize.

Preparation of Medium: Add components, except vitamin solution, to distilled/deionized water and bring volume to 990.0mL. Mix thoroughly. Gently heat and bring to boiling. Autoclave for 15 min at 15 psi pressure–121°C. Cool to 45°–50°C. Aseptically add 10.0mL of vitamin solution. Mix thoroughly. Pour into sterile Petri dishes or distribute into sterile tubes.

Use: For the isolation of iron bacteria.

Iron-Oxidizing Medium

Composition per liter:
$(NH_4)_2SO_4$ 3.0g
K_2HPO_4 .... 0.5g
$MgSO_4 \cdot 7H_2O$ 0.5g
KCl 0.1g
$Ca(NO_3)_2$ 0.01g
$FeSO_4 \cdot 7H_2O$ solution 300.0mL
H_2SO_4 (10*N*) 1.0mL
pH 3.0–3.6 at 25°C

$FeSO_4 \cdot 7H_2O$ Solution:
Composition per 300.0mL:
$FeSO_4 \cdot 7H_2O$ 44.22g

Preparation of $FeSO_4 \cdot 7H_2O$ Solution: Add 44.22g of $FeSO_4 \cdot 7H_2O$ to distilled/deionized water and bring volume to 300.0mL. Mix thoroughly. Autoclave for 15 min at 15 psi pressure–121°C. Cool to 25°C.

Preparation of Medium: Add components, except $FeSO_4 \cdot 7H_2O$ solution, to distilled/deionized water and bring volume to 700.0mL. Mix thoroughly. Gently heat and bring to boiling. Autoclave for 15 min at 15 psi pressure–121°C. Cool to 25°C. Aseptically add 300.0mL of sterile $FeSO_4 \cdot 7H_2O$ solution. Mix thoroughly. Aseptically distribute into sterile tubes or flasks.

Use: For the enumeration, isolation, and cultivation of iron and sulfur bacteria, such as *Thiobacillus ferrooxidans*.

Iron Sulfite Agar

Composition per liter:
Agar 12.0g
Pancreatic digest of casein 10.0g
Ferric citrate 0.5g
$Na_2S \cdot 9H_2O$ 0.5g
pH 7.1 ± 0.2 at 25°C

Source: This medium is available as a premixed powder from Oxoid Unipath.

Preparation of Medium: Add components to distilled/deionized water and bring volume to 1.0L. Mix thoroughly. Gently heat and bring to boiling. Distribute into tubes or flasks. Autoclave for 15 min at 15 psi pressure–121°C. Mix thoroughly. Pour into sterile Petri dishes or leave in tubes.

Use: For the detection of thermophilic anaerobic organisms.

ISM Agar

Composition per liter:
$MgSO_4 \cdot 7H_2O$ 49.2g
NaCl 43.5g
Agar 7.5g
Yeast exract 4.0g
Peptone 2.0g
$CaCl_2 \cdot 2H_2O$ 1.5g
Maltose solution 100.0mL

Maltose Solution:
Composition per 100.0mL:
Maltose 5.0g

Preparation of Maltose Solution: Add maltose to distilled/deionized water and bring volume to 100.0mL. Mix thoroughly. Filter sterilize.

Preparation of Medium: Add components, except maltose solution, to distilled/deionized water and bring volume to 900.0mL. Mix thoroughly. Gently heat and bring to boiling. Autoclave for 15 min at 15 psi pressure–121°C. Cool to 45°–50°C. Aseptically add sterile maltose solution. Mix thoroughly. Pour into sterile Petri dishes or distribute into sterile tubes.

Use: For the cultivation and maintenance of *Spirochaeta halophila*.

Isoleucine Hydroxamate Medium

Composition per liter:
Agar 15.0g
K_2HPO_4 .... 7.0g
Glucose 5.0g
KH_2PO_4 .... 3.0g
L-Isoleucine hydroxamate 1.0g
$(NH_4)_2SO_4$ 1.0g
pH 7.0 ± 0.2 at 25°C

Preparation of Medium: Add components to distilled/deionized water and bring volume to 1.0L. Mix thoroughly. Gently heat and bring to boiling. Distribute into tubes or flasks. Autoclave for 15 min at 15 psi pressure–121°C. Pour into sterile Petri dishes or leave in tubes.

Use: For the cultivation and maintenance of *Serratia marcescens*.

Isonema Medium

Composition per liter:
Pancreatic digest of casin 1.0g
Seawater 990.0mL
Horse serum, heat inactivated 10.0mL

Preparation of Medium: Add pancreatic digest of casein to seawater and bring volume to 990.0mL. Mix thoroughly. Autoclave for 15 min at 15 psi pressure–121°C. Cool to 25°C. Aseptically add 10.0mL of heat-inactivated horse serum. Mix thoroughly. Aseptically distribute into sterile tubes or flasks.

Use: For the cultivation of *Isonema papillatum*.

Isosphaera Agar

Composition per 1000.5mL:
Solution A 800.0mL
Solution B 100.0mL

Solution C 100.0mL
Vitamin solution 0.5mL

pH 7.6 ± 0.2 at 25°C

Solution A:

Composition per 800.0mL:

Agar, noble 15.0g
NaCl 0.25g
$(NH_4)_2SO_4$ 0.125g
KCl 0.125g
$MgSO_4·7H_2O$ 0.1g
$CaCl_2·2H_2O$ 80.0mg
KH_2PO_4 .... 75.0mg
$FeCl_3$ 73.0μg
Trace elements solution SL-7 .... 2.5mL

Trace Elements Solution SL-7:

Composition per liter:

$FeCl_2·4H_2O$ 1.5g
$CoCl_2·6H_2O$ 190.0mg
$MnCl_2·4H_2O$ 100.0mg
$ZnCl_2$ 70.0mg
H_3BO_3 .... 62.0mg
$Na_2MoO_4·2H_2O$ 36.0mg
$NiCl_2·6H_2O$ 24.0mg
$CuCl_2·2H_2O$ 17.0mg
HCl (25% solution) 10.0mL

Preparation of Trace Elements Solution SL-7: Add $FeCl_2·4H_2O$ to 10.0mL of HCl solution. Mix thoroughly. Add distilled/deionized water and bring volume to 1.0L. Add remaining components. Mix thoroughly.

Preparation of Solution A: Add agar to 400.0mL of distilled/deionized water. In another flask, add remaining components to 400.0mL of distilled/deionized water. Mix thoroughly. Adjust pH to 7.6 with NaOH. Remove any precipitate that forms by filtering through Whatman #1 filter paper. Autoclave both solutions separately for 15 min at 15 psi pressure–121°C. Aseptically combine the 2 sterile solutions. Cool to 50°–55°C.

Solution B:

Composition per 100.0mL:

$NaHCO_3$ 0.42g

Preparation of Solution B: Add $NaHCO_3$ to distilled/deionized water and bring volume to 100.0mL. Mix thoroughly. Filter sterilize. Warm to 50°–55°C.

Solution C:

Composition per 100.5mL:

Glucose 0.25g
Casamino acids 0.25g

Preparation of Solution C: Add components to distilled/deionized water and bring volume to 100.0mL. Mix thoroughly. Filter sterilize. Warm to 50°–55°C.

Vitamin Solution:

Composition per 100.0mL:

Nicotinic acid 200.0mg
Thiamine HCl 200.0mg
p-Aminobenzoic acid 20.0mg
Biotin 2.0mg
Vitamin B_{12} 25.0μg

Preparation of Vitamin Solution: Add components to distilled/deionized water and bring volume to 100.0mL. Mix thoroughly. Filter sterilize. Warm to 50°–55°C.

Preparation of Medium: Aseptically combine 800.0mL of sterile solution A with 100.0mL of sterile solution B, 100.0mL of sterile solution C, and 0.5mL of sterile vitamin solution. Mix thoroughly. Pour into sterile Petri dishes or distribute into sterile tubes.

Use: For the cultivation of *Isosphaera pallida*, a gliding, budding eubacterium from hot springs.

Isosphaera pallida Medium (DSMZ Medium 765)

Composition per 1005mL:

Solution A 900.0mL
$NaHCO_3$ solution 100.0mL
Vitamin solution 5.0mL

pH 7.9 ± 0.2 at 25°C

Solution A:

Composition per 900mL:

KCl 4.0g
$MgSO_4·7H_2O$ 2.0g
$(NH_4)_2SO_4$ 1.0g
NaCl 1.0g
KH_2PO_4 .... 0.3g
Glucose 0.25g
Casamino acids 0.25g
$CaCl_2·2H_2O$ 0.2g
$FeCl_3$ 0.292mg
Trace elements solution SL-7a 1.0mL

Trace Elements Solution SL-7a:

$CoCl_2·6H_2O$ 200.0mg
$MnCl_2·4H_2O$ 100.0mg
$ZnCl_2$ 70.0mg
H_3BO_3 .... 60.0mg
$Na_2MoO_4·4H_2O$ 40.0mg
$NiCl_2·6H_2O$ 20.0mg
$CuCl_2·2H_2O$ 20.0mg
HCl (25%) 1.0mL

Preparation of Trace Elements Solution SL-7a: Add components to distilled/deionized water and bring volume to 1.0L. Mix thoroughly.

Preparation of Solution A: Add components to distilled/deionized water and bring volume to 900.0mL. Mix thoroughly. Sparge with 95% air + 5% CO_2.

Vitamin Solution:

Composition per 100mL:

Nicotinic acid 20.0mg
Thiamine-HCl·$2H_2O$ 10.0mg
p-Aminobenzoic acid 0.2mg
Vitamin B_{12} 0.1mg

Preparation of Vitamin Solution: Add components to distilled/deionized water and bring volume to 100.0mL. Mix thoroughly. Filter sterilize.

$NaHCO_3$ Solution:

Composition per 100.0mL:

$NaHCO_3$ 4.2g

Preparation of $NaHCO_3$ Solution: Add $NaHCO_3$ to distilled/deionized water and bring volume to 100.0mL. Mix thoroughly. Sparge with 80% N_2 + 20% CO_2. Filter sterilize. Seal in serum bottle under 100% CO_2.

Preparation of Medium: Dispense under 95% air + 5% CO_2. Fill serum bottles so that the gas-to-liquid ratio is about 5:1 (v/v). Sparge with 95% air + 5% CO_2. Seal bottles. Autoclave for 15 min at 15 psi pressure–121°C. Cool to room temperature. Using a syringe, inject 10.0mL sterile $NaHCO_3$ solution/90.0mL solution A. Then inject 0.5 mL sterile vitamin solution per 100.0mL of the resulting medium.

Use: For the cultivation of *Isosphaera pallida=Isocystis pallida*, a gliding, budding eubacterium from hot springs.

JB Medium with Glucose

Composition per liter:

Yeast extract	15.0g
Pancreatic digest of casein	5.0g
K_2HPO_4	3.0g
Glucose	2.0g

pH 7.3–7.5 at 25°C

Preparation of Medium: Add components to distilled/deionized water and bring volume to 1.0L. Mix thoroughly. Gently heat and bring to boiling. Distribute into tubes or flasks. Autoclave for 15 min at 15 psi pressure–121°C. Pour into sterile Petri dishes or leave in tubes.

Use: For the cultivation and maintenance of *Bacillus popilliae*.

Johnson's Marine Medium

Composition per liter:

Peptone	5.0g
Yeast extract	1.0g
$Na_2S_2O_3$	0.3g
$FeSO_4 \cdot 7H_2O$	0.2g
Filtered, aged seawater	750.0mL

Preparation of Medium: Add components to distilled/deionized water and bring volume to 1.0L. Mix thoroughly. Distribute into tubes or flasks. Autoclave for 15 min at 15 psi pressure–121°C.

Use: For the cultivation of marine bacteria.

K101 *Flexibacter* Medium

Composition per liter:

Agar	10.0g
Casamino acids	1.0g
Glucose	1.0g
Tris(hydroxymethyl)aminomethane buffer	1.0g
$CaCl_2$	0.1g
KNO_3	0.1g
$MgSO_4 \cdot 7H_2O$	0.1g
Sodium glycerophosphate	0.1g
Thiamine·HCl	1.0mg
Cyanocobalamin	1.0μg
Trace elements solution HO-LE	1.0mL

pH 7.5 ± 0.2 at 25°C

Trace Elements Solution HO-LE:

Composition per liter:

H_3BO_3	2.85g
$MnCl_2 \cdot 4H_2O$	1.8g
Sodium tartrate	1.77g
$FeSO_4 \cdot 7H_2O$	1.36g
$CoCl_2 \cdot 6H_2O$	0.04g
$CuCl_2 \cdot 2H_2O$	0.027g
$Na_2MoO_4 \cdot 2H_2O$	0.025g
$ZnCl_2$	0.02g

Preparation of Trace Elements Solution HO-LE: Add components to distilled/deionized water and bring volume to 1.0L. Mix thoroughly. Filter sterilize.

Preparation of Medium: Add components, except trace elements solution HO-LE, to distilled/deionized water and bring volume to 999.0mL. Mix thoroughly. Gently heat and bring to boiling. Autoclave for 15 min at 15 psi pressure–121°C. Cool to 45°–50°C. Aseptically add 1.0mL of trace elements solution HO-LE. Mix thoroughly. Pour into sterile Petri dishes or distribute into sterile tubes.

Use: For the cultivation and maintenance of *Cytophaga* species, *Flexibacter* species, *Herpetosiphon geysericola*, and *Myxococcus fulvus*.

Keister's Modified TYI-S-33 Medium

Composition per liter:

Pancreatic digest of casein	20.0g
Glucose	10.0g
Yeast extract	10.0g
L-Cysteine·HCl	2.0g
NaCl	2.0g
K_2HPO_4	1.0g
Bovine bile	0.75g
KH_2PO_4	0.6g
Ascorbic acid	0.2g
Ferric ammonium citrate	22.8mg
Bovine serum, heat inactivated	100.0mL

Preparation of Medium: Add components, except bovine serum, to distilled/deionized water and bring volume to 900.0mL. Mix thoroughly. Autoclave for 15 min at 15 psi pressure–121°C. Aseptically add 1.0L of sterile, heat-inactivated bovine serum. Mix thoroughly. Aseptically distribute into sterile, screw-capped tubes or flasks.

Use: For the cultivation of *Giardia cati, Giardia intestinalis,* and *Hexamita* species.

Kerosene Mineral Salts Medium

Composition per liter:

KH_2PO_4	1.0g
K_2HPO_4	1.0g
NH_4NO_3	1.0g
$MgSO_4 \cdot 7H_2O$	0.2g
$CaCl_2$	0.02g
$FeCl_3$	0.05g
Kerosene	20.0mL

pH 6.9–7.0 at 25°C

Preparation of Medium: Add components, except kerosene, to distilled/deionized water and bring volume to 1.0L. Mix thoroughly. Adjust pH to 6.9–7.0 with dilute NaOH. Distribute into tubes in 10.0mL volumes or flasks in 100.0mL volumes. Autoclave for 15 min at 15 psi pressure–121°C. Overlay tubes with 0.2mL of kerosene per tube. Overlay flasks with 2.0mL of kerosene per flask.

Use: For the cultivation of *Pseudomonas aeruginosa.*

Ketolactonate Broth

Composition per liter:

Agar	20.0g
Lactose	10.0g
Yeast extract	10.0g

Preparation of Medium: Add components to distilled/deionized water and bring volume to 1.0L. Mix thoroughly. Gently heat and bring to boiling. Distribute into tubes or flasks. Autoclave for 15 min at 15 psi pressure–121°C. Pour into sterile Petri dishes or leave in tubes.

Use: For use in the identification of agrobacteria and other bacteria based upon production of 3-ketogluconate. After incubation, Benedicts solution is added to the plates. Yellow zones around colonies indicate positive production of 3-ketogluconate.

KF *Streptococcus* Agar

Composition per liter:

Agar	20.0g
Maltose	20.0g

Proteose peptone 10.0g
Sodium glycerophosphate 10.0g
Yeast extract 10.0g
NaCl 5.5g
Lactose 1.0g
NaN_3 .... 0.4g
Bromcresol Purple 0.015g
2,3,5-Triphenyltetrazolium
chloride solution 10.0mL
pH 7.2 ± 0.2 at 25°C

Caution: Sodium azide is toxic. Azides also react with metals and disposal must be highly diluted.

Source: This medium is available as a premixed powder from BD Diagnostic Systems and Oxoid Unipath.

2,3,5-Triphenyltetrazolium Chloride Solution:
Composition per 10.0mL:
2,3,5-Triphenyltetrazolium chloride 0.1g

Preparation of 2,3,5-Triphenyltetrazolium Chloride Solution: Add 2,3,5-triphenyltetrazolium chloride to distilled/deionized water and bring volume to 10.0mL. Mix thoroughly. Filter sterilize.

Preparation of Medium: Add components, except 2,3,5-triphenyltetrazolium chloride solution, to distilled/deionized water and bring volume to 990.0mL. Mix thoroughly. Gently heat and bring to boiling. Autoclave for 15 min at 15 psi pressure–121°C. Cool to 45°–50°C. Aseptically add 2,3,5-triphenyltetrazolium chloride solution. Mix thoroughly. Pour into sterile Petri dishes or distribute into sterile tubes.

Use: For the isolation and enumeration of enterococci.

Kidney Bean Agar

Composition per liter:
Kidney beans, dry 30.0g
Agar 15.0g

Preparation of Medium: Add dry kidney beans to distilled/deionized water and bring volume to 1.0L. Autoclave for 30min at 15 psi pressure–121°C. Filter solids through cheesecloth. Add agar to filtrate. Mix thoroughly. Bring volume to 1.0L with distilled/deionized water. Gently heat and bring to boiling. Distribute into tubes or flasks. Autoclave for 15 min at 15 psi pressure–121°C. Pour into sterile Petri dishes or leave in tubes.

Use: For the cultivation and maintenance of *Conidiobolus stromoideus, Phialophora gregata, Phytophthora cactorum, Phytophthora cryptogea, Phytophthora erythroseptica, Phytophthora fragariae, Phytophthora heveae,* and *Phytophthora syringae.*

Kievskaya Agar (Medium K)

Composition per liter:
Agar 15.0g
KNO_3 .... 1.0g
KH_2PO_4 .... 1.0g
K_2HPO_4 .... 1.0g
NaCl 1.0g
$MgSO_4$ 0.2g
$CaCl_2$ 0.02g
$FeCl_3$ 1.0mg
pH 6.8–7.0 at 25°C

Preparation of Medium: Add components to distilled/deionized water and bring volume to 1.0L. Mix thoroughly. Gently heat and bring to boiling. Distribute into tubes or flasks. Autoclave for 30 min at 3 psi pressure–105°C. Pour into sterile Petri dishes or leave in tubes. After inoculation, incubate in an atmosphere of natural gas.

Use: For the cultivation of *Rhodococcus luteus, Rhodococcus rhodochrous, Rhodococcus ruber, Rhodococcus* species, and other bacteria that can grow on natural gas.

Kievskaya Broth with *n*-Hexadecane

Composition per liter:
KNO_3 .... 1.0g
KH_2PO_4 .... 1.0g
K_2HPO_4 .... 1.0g
NaCl 1.0g
$MgSO_4$ 0.2g
$CaCl_2$ 0.02g
$FeCl_3$ 1.0mg
n-Hexadecane 10.0mL
pH 6.8–7.0 at 25°C

Preparation of Medium: Add components to distilled/deionized water and bring volume to 1.0L. Mix thoroughly. Distribute into tubes or flasks. Autoclave for 15 min at 15 psi pressure–121°C.

Use: For the cultivation of *Gordona terrae, Rhodococcus erythropolis, Rhodococcus luteus,* and *Rhodococcus maris.*

King's Medium A

Composition per liter:
Proteose peptone 20.0g
Agar 15.0g
Glycerol 10.0g
K_2SO_4 .... 10.0g
$MgCl_2 \cdot 6H_2O$ 3.5g
pH 7.2–7.4 ± 0.2 at 25°C

Preparation of Medium: Add components to distilled/deionized water and bring volume to 1.0L. Mix thoroughly. Gently heat and bring to boiling. Distribute into tubes or flasks. Autoclave for 15 min at 15 psi pressure–121°C. Pour into sterile Petri dishes or leave in tubes.

Use: For the nonselective isolation, cultivation, and pigment production of *Pseudomonas.*

King's Medium B

Composition per liter:
Agar 20.0g
Proteose peptone No. 3 .... 20.0g
K_2HPO_4, anhydrous 1.5g
$MgSO_4 \cdot 7H_2O$ 1.5g
Glycerol 15.0mL
pH 7.2 ± 0.2 at 25°C

Preparation of Medium: Add components to distilled/deionized water and bring volume to 1.0L. Mix thoroughly. Gently heat and bring to boiling. Distribute into tubes or flasks. Autoclave for 15 min at 15 psi pressure–121°C. Pour into sterile Petri dishes or leave in tubes.

Use: For the nonselective isolation, cultivation, and pigment production of *Pseudomonas* species.

Kleb Agar (m-Kleb Agar)

Composition per liter:
Agar 15.0g
Proteose peptone No. 3 .... 10.0g
NaCl 5.0g
Adonitol 5.0g

Beef extract 1.0g
Aniline Blue 0.1g
Sodium lauryl sulfate 0.1g
Phenol Red 0.025g
Ethanol (95% solution) 20.0mL
Carbenicillin solution 10.0mL

pH 7.4 ± 0.2 at 25°C

Carbenicillin Solution:
Composition per 10.0mL:
Carbenicillin 0.05g

Preparation of Carbenicillin Solution: Add carbenicillin to distilled/deionized water and bring volume to 10.0mL. Mix thoroughly. Filter sterilize.

Preparation of Medium: Add components, except ethanol and carbenicillin solution, to distilled/deionized water and bring volume to 970.0mL. Mix thoroughly. Gently heat and bring to boiling. Autoclave for 15 min at 15 psi pressure–121°C. Cool to 45°–50°C. Aseptically add 20.0mL of ethanol and 10.0mL of carbenicillin solution. Mix thoroughly. Pour into sterile Petri dishes or distribute into sterile tubes.

Use: For the enumeration of bacteria from waters.

Klebsiella Medium (m-*Klebsiella* Medium)

Composition per 1041.0mL:
Agar 15.0g
Adonitol 4.0g
2× Salt solution 500.0mL
Uric acid solution 200.0mL
Phenol Red solution 10.0mL
Sodium taurocholate solution 30.0mL
Carbenicillin solution 1.0mL

2× Salt Solution:
Composition per liter:
KCl 8.0g
K_2HPO_4 .... 3.0g
NaCl 2.0g
KH_2PO_4 .... 1.0g
$MgSO_4 \cdot 7H_2O$ 0.2g

Preparation of 2× Salt Solution: Add components to distilled/deionized water and bring volume to 1.0L. Mix thoroughly.

Uric Acid Solution:
Composition per 200.0mL:
Uric acid 0.3g

Preparation of Uric Acid Solution: Dissolve uric acid in a small volume of 1*N* NaOH. Bring volume to 200.0mL with distilled/deionized water. Adjust pH to 7.1 with 1*N* HCl. Filter sterilize.

Phenol Red Solution:
Composition per 10.0mL:
Phenol Red 0.1g

Preparation of Phenol Red Solution: Add Phenol Red to sterile distilled/deionized water and bring volume to 10.0mL. Mix thoroughly.

Sodium Taurocholate Solution:
Composition per 30.0mL:
Sodium taurocholate 0.4g

Preparation of Sodium Taurocholate Solution: Add sodium taurocholate to sterile distilled/deionized water and bring volume to 30.0mL. Mix thoroughly.

Carbenicillin Solution:
Composition per 1.0mL:
Carbenicillin 5.0mg

Preparation of Carbenicillin Solution: Add carbenicillin to distilled/deionized water and bring volume to 1.0mL. Mix thoroughly. Filter sterilize.

Preparation of Medium: Add adonitol and agar to 500.0mL of 2× salt solution. Bring volume to 800.0mL with distilled/deionized water. Mix thoroughly. Gently heat and bring to boiling. Autoclave for 15 min at 15 psi pressure–121°C. Cool to 45°–50°C. Aseptically add 200.0mL of uric acid solution, 30.0mL of sodium taurocholate solution, 10.0mL of Phenol Red solution, and 1.0mL of carbenicillin solution. Mix thoroughly. Pour into sterile Petri dishes or distribute into sterile tubes.

Use: For the enumeration of *Klebsiella* species by the membrane filter method.

Kligler Iron Agar

Composition per liter:
Peptone 20.0g
Agar 12.0g
Lactose 10.0g
NaCl 5.0g
Beef extract 3.0g
Yeast extract 3.0g
Glucose 1.0g
Ferric citrate 0.3g
$Na_2S_2O_3$ 0.3g
Phenol Red 0.05g

pH 7.4 ± 0.2 at 25°C

Source: This medium is available as a premixed powder from BD Diagnostic Systems and Oxoid Unipath.

Preparation of Medium: Add components to distilled/deionized water and bring volume to 1.0L. Mix thoroughly. Gently heat and bring to boiling. Distribute into tubes. Autoclave for 15 min at 15 psi pressure–121°C. Pour into sterile Petri dishes or leave in tubes.

Use: For the differentiation and identification of Enterobacteriaceae based upon sugar fermentation and hydrogen sulfide production. Sugar fermentation is indicated by the medium turning yellow. H_2S production results in the medium turning black.

KoKo Medium

Composition per 1020.0mL:
K_2HPO_4 .... 1.6g
$NaH_2PO_4 \cdot 2H_2O$ 1.0g
Peptone, meat 1.0g
Pancreatic digest of casein 1.0g
Yeast extract 1.0g
NH_4Cl 0.5g
$MgSO_4 \cdot 6H_2O$ 0.16g
Resazurin 0.5mg
Glucose solution 100.0mL
$NaHCO_3$ solution 10.0mL
$CaCl_2 \cdot 2H_2O$ solution 10.0mL
L-Cysteine·HCl·H_2O solution 10.0mL
$Na_2S \cdot 9H_2O$ solution 10.0mL
Trace elements solution SL-4 .... 10.0mL
Wolfe's vitamin solution 10.0mL

pH 7.0 ± 0.2 at 25°C

Glucose Solution:
Composition per 100.0mL:
D-Glucose 5.0g

Preparation of Glucose Solution: Add glucose to distilled/deionized water and bring volume to 100.0mL. Mix thoroughly. Sparge with 100% N_2. Autoclave for 15 min at 15 psi pressure–121°C.

$NaHCO_3$ Solution:

Composition per 10.0mL:

$NaHCO_3$ 1.0g

Preparation of $NaHCO_3$ Solution: Add $NaHCO_3$ to distilled/deionized water and bring volume to 10.0mL. Mix thoroughly. Sparge with 100% N_2. Autoclave for 15 min at 15 psi pressure–121°C.

$CaCl_2$ Solution:

Composition per 10.0mL:

$CaCl_2 \cdot 2H_2O$ 0.06g

Preparation of $CaCl_2$ Solution: Add $CaCl_2 \cdot 2H_2O$ to distilled/deionized water and bring volume to 10.0mL. Mix thoroughly. Sparge with 100% N_2. Autoclave for 15 min at 15 psi pressure–121°C.

L-Cysteine·HCl·H_2O Solution:

Composition per 10.0mL:

L-Cysteine·HCl·H_2O 0.3g

Preparation of L-Cysteine·HCl·H_2O Solution: Add L-cysteine·HCl·H_2O to distilled/deionized water and bring volume to 10.0mL. Mix thoroughly. Sparge with 100% N_2. Autoclave for 15 min at 15 psi pressure–121°C.

$Na_2S \cdot 9H_2O$ Solution:

Composition per 10.0mL:

$Na_2S \cdot 9H_2O$ 0.3g

Preparation of $Na_2S \cdot 9H_2O$ Solution: Add $Na_2S \cdot 9H_2O$ to distilled/deionized water and bring volume to 10.0mL. Mix thoroughly. Sparge with 100% N_2. Autoclave for 15 min at 15 psi pressure–121°C. Before use, neutralize to pH 7.0 with sterile HCl.

Trace Elements Solution SL-4:

Composition per liter:

EDTA 0.5g
$FeSO_4 \cdot 7H_2O$ 0.2g
Trace elements solution SL-6 .... 100.0mL

Trace Elements Solution SL-6:

Composition per liter:

$MnCl_2 \cdot 4H_2O$ 0.5g
H_3BO_3 .... 0.3g
$CoCl_2 \cdot 6H_2O$ 0.2g
$ZnSO_4 \cdot 7H_2O$ 0.1g
$Na_2MoO_4 \cdot 2H_2O$ 0.03g
$NiCl_2 \cdot 6H_2O$ 0.02g
$CuCl_2 \cdot 2H_2O$ 0.01g

Preparation of Trace Elements Solution SL-6: Add components to distilled/deionized water and bring volume to 1.0L. Mix thoroughly.

Preparation of Trace Elements Solution SL-4: Add components to distilled/deionized water and bring volume to 1.0L. Mix thoroughly. Filter sterilize.

Wolfe's Vitamin Solution:

Composition per liter:

Pyridoxine·HCl 10.0mg
p-Aminobenzoic acid 5.0mg
Lipoic acid 5.0mg
Nicotinic acid 5.0mg
Riboflavin 5.0mg
Thiamine·HCl 5.0mg
Calcium DL-pantothenate 5.0mg
Biotin 2.0mg
Folic acid 2.0mg
Vitamin B_{12} 0.1mg

Preparation of Wolfe's Vitamin Solution: Add components to distilled/deionized water and bring volume to 1.0L. Mix thoroughly. Sparge with 100% N_2.Filter sterilize.

Preparation of Medium: Prepare and dispense medium under 100% N_2. Add components, except glucose solution, $NaHCO_3$ solution, $CaCl_2 \cdot 2H_2O$ solution, $Na_2S \cdot 9H_2O$ solution, L-cysteine·HCl·H_2O solution, and Wolfe's vitamin solution, to distilled/deionized water and bring volume to 850.0mL. Mix thoroughly. Adjust pH to 7.0. Sparge with 100% N_2. Autoclave for 15 min at 15 psi pressure–121°C. Aseptically and anaerobically add 100.0mL of sterile glucose solution, 10.0mL of sterile $NaHCO_3$ solution, 10.0mL of sterile $CaCl_2 \cdot 2H_2O$ solution, 10.0mL of sterile L-cysteine·HCl·H_2O solution, 10.0mL of sterile $Na_2S \cdot 9H_2O$ solution, and 10.0mL of sterile Wolfe's vitamin solution. Mix thoroughly. The pH should be 7.0. A buffer solution of 1% MOPS from a 10% anaerobic solution at pH 6.9–7.0 may be added aseptically and anaerobically to enhance the buffer capacity of the medium. Aseptically and anaerobically distribute into sterile tubes or bottles.

Use: For the cultivation of *Thermoanaerobacter italicus*.

Korthof Medium

Composition per 1088.0mL:

NaCl 1.4g
$Na_2HPO_4 \cdot 2H_2O$ 0.88g
Peptone 0.8g
KH_2PO_4 .... 0.24g
$CaCl_2$ 0.04g
KCl 0.04g
$NaHCO_3$ 0.02g
Rabbit serum, inactivated 80.0mL
Rabbit hemoglobin solution 8.0mL
pH 7.2 ± 0.2 at 25°C

Rabbit Hemoglobin Solution:

Composition per 20.0mL:

Rabbit blood clot 10.0mL

Preparation of Rabbit Hemoglobin Solution: Add rabbit blood clot to 10.0mL of distilled/deionized water. Lyse the clot by freezing and thawing.

Preparation of Medium: Add components, except rabbit serum and rabbit hemoglobin solution, to distilled/deionized water and bring volume to 1.0L. Mix thoroughly. Gently heat and bring to boiling. Cool to 25°C. Filter through Whatman #1 filter paper. Distribute into flasks in 100.0mL volumes. Autoclave for 15 min at 15 psi pressure–121°C. Cool to 45°–50°C. Aseptically add 8.0mL of rabbit serum and 0.8mL of rabbit hemoglobin solution to each flask. Mix thoroughly.

Use: For the cultivation of *Leptospira* species.

Koser Citrate Medium

Composition per liter:

Sodium citrate 3.0g
$NaNH_4HPO_4 \cdot 4H_2O$ 1.5g
KH_2PO_4 .... 1.0g
$MgSO_4 \cdot 7H_2O$ 0.2g
pH 6.7 ± 0.2 at 25°C

Source: This medium is available as a premixed powder from BD Diagnostic Systems.

Preparation of Medium: Add components to distilled/deionized water and bring volume to 1.0L. Mix thoroughly. Gently heat and bring to boiling. Distribute into tubes or flasks. Autoclave for 15 min at 15 psi pressure–121°C. Pour into sterile Petri dishes or leave in tubes.

Use: For the differentiation of *Escherichia coli* and *Enterobacter aerogenes* based on citrate utilization.

KYE Agar

Composition per liter:

Agar	15.0g
$NaNO_3$	2.5g
KH_2PO_4	1.0g
Yeast extract	1.0g
$MgSO_4 \cdot 7H_2O$	0.3g
$CaCl_2 \cdot 6H_2O$	0.15g
NaCl	0.1g
$FeCl_3$	10.0mg

pH 10.0 ± 0.2 at 25°C

Preparation of Medium: Add components to distilled/deionized water and bring volume to 1.0L. Mix thoroughly. Gently heat and bring to boiling. Distribute into tubes or flasks. Autoclave for 15 min at 15 psi pressure–121°C. Pour into sterile Petri dishes or leave in tubes.

Use: For the cultivation of a variety of alkaliphilic bacteria.

L and F Basal Salts, Modified with Heptadecane

Composition per liter:

NH_4Cl	2.0g
Na_2HPO_4	0.21g
$MgSO_4 \cdot 7H_2O$	0.2g
NaH_2PO_4	0.09g
KCl	0.04g
$CaCl_2$	0.015g
$FeSO_4 \cdot 7H_2O$	1.0mg
$ZnSO_4 \cdot 7H_2O$	0.07mg
H_3BO_3	0.01mg
$MnSO_4 \cdot 5H_2O$	0.01mg
MoO_3	0.01mg
$CuSO_4 \cdot 5H_2O$	5.0μg
Heptadecane	2.0mL

pH 7.2 ± 0.2 at 25°C

Preparation of Medium: Add components, except heptadecane, to distilled/deionized water and bring volume to 1.0L. Mix thoroughly. Gently heat and bring to boiling. Distribute equally into four 250.0mL volumes. Autoclave for 15 min at 15 psi pressure–121°C. Cool to 60°C. Filter sterilize heptadecane. To one 250.0mL fraction of basal salts, aseptically add 0.5mL of sterile heptadecane. Pour mixture into a sterile blender. Homogenize slowly to mix heptadecane with basal salts and not to create excess bubbles. Rapidly distribute medium to sterile screw-capped tubes. Chill tubes quickly in an ice pack or in the refrigerator. Allow tubes to solidify in a slanted position.

Use: For the cultivation and maintenance of *Thermoleophilum album* and *Thermoleophilum minutum*

L Medium (ATCC Medium 167)

Composition per liter:

Agar	20.0g
$NaNO_3$	2.0g
Na_2HPO_4	0.21g
$MgSO_4 \cdot 7H_2O$	0.2g
NaH_2PO_4	0.09g
KCl	0.04g
$CaCl_2$	0.015g
$FeSO_4 \cdot 7H_2O$	1.0mg
Salts solution	1.0mL

Salts Solution:

Composition per 100.0mL:

$ZnSO_4 \cdot 7H_2O$	7.0mg
H_3BO_3	1.0mg
$MnSO_4 \cdot 5H_2O$	1.0mg
MoO_3	1.0mg
$CuSO_4 \cdot 5H_2O$	0.5mg

Preparation of Salts Solution: Add components to distilled/deionized water and bring volume to 100.0mL. Mix thoroughly.

Preparation of Medium: Add components to distilled/deionized water and bring volume to 1.0L. Mix thoroughly. Gently heat and bring to boiling. Distribute into tubes or flasks. Autoclave for 15 min at 15 psi pressure–121°C. Pour into sterile Petri dishes or leave in tubes.

Use: For the cultivation and maintenance of *Methylococcus capsulatus* and *Pseudomonas methanica.*

L Medium with Methanol

Composition per liter:

Agar	20.0g
$NaNO_3$	2.0g
Na_2HPO_4	0.21g
$MgSO_4 \cdot 7H_2O$	0.2g
NaH_2PO_4	0.09g
KCl	0.04g
$CaCl_2$	0.015g
$FeSO_4 \cdot 7H_2O$	1.0mg
Methanol	20.0mL
Salts solution	1.0mL

Salts Solution:

Composition per 100.0mL:

$ZnSO_4 \cdot 7H_2O$	7.0mg
H_3BO_3	1.0mg
$MnSO_4 \cdot 5H_2O$	1.0mg
MoO_3	1.0mg
$CuSO_4 \cdot 5H_2O$	0.5mg

Preparation of Salts Solution: Add components to distilled/deionized water and bring volume to 100.0mL. Mix thoroughly.

Preparation of Medium: Add components, except methanol, to distilled/deionized water and bring volume to 980.0mL. Mix thoroughly. Gently heat and bring to boiling. Autoclave for 15 min at 15 psi pressure–121°C. Cool to 45°–50°C. Filter sterilize methanol. Aseptically add 20.0mL of sterile methanol to cooled, sterile basal medium. Mix thoroughly. Pour into sterile Petri dishes or distribute into sterile tubes.

Use: For the cultivation and maintenance of *Methylobacillus glycogenes.*

Lab-Lemco Agar

Composition per liter:

Agar	15.0g
Peptone	5.0g
Lab-Lemco meat extract	3.0g

pH 7.4 ± 0.2 at 25°C

Preparation of Medium: Add components to distilled/deionized water and bring volume to 1.0L. Mix thoroughly. Gently heat and bring to boiling. Distribute into tubes or

flasks. Autoclave for 15 min at 15 psi pressure–121°C. Pour into sterile Petri dishes or leave in tubes.

Use: For the cultivation and maintenance of a variety of heterotrophic microorganisms.

Lactate Agar

Composition per liter:

Yeast extract	3.0g
K_2HPO_4	2.8g
Agar	2.0g
Peptone	2.0g
KH_2PO_4	0.52g
Sodium lactate (60% solution)	10.0mL

pH 7.2 ± 0.2 at 25°C

Preparation of Medium: Add components to distilled/deionized water and bring volume to 1.0L. Mix thoroughly. Gently heat and bring to boiling. Adjust pH to 7.2. Distribute into tubes or flasks. Autoclave for 15 min at 15 psi pressure–121°C. Pour into sterile Petri dishes or leave in tubes.

Use: For the cultivation and maintenance of *Serpens flexibilis*.

Lactose Broth

Composition per liter:

Lactose	5.0g
Pancreatic digest of gelatin	5.0g
Beef extract	3.0g

pH 6.9 ± 0.2 at 25°C

Source: This medium is available as a premixed powder from BD Diagnostic Systems and Oxoid Unipath.

Preparation of Medium: Add components to distilled/deionized water and bring volume to 1.0L. Mix thoroughly. Distribute into tubes containing an inverted Durham tube in 10.0mL volumes. Autoclave for 12 min at 15 psi pressure–121°C. Cool broth quickly to 25°C. For testing water samples with 10.0mL volumes, prepare medium double strength.

Use: For the detection of lactose-fermenting, Gram-negative coliforms, as a preenrichment broth for *Salmonella* species, and in the study of lactose fermentation of bacteria in general.

Lactose Minimal Medium

Composition per liter:

Agar	20.0g
Lactose	15.0g
K_2HPO_4	5.0g
NH_4Cl	2.0g
NaCl	1.0g
$MgSO_4$	0.1g
Yeast extract	0.1g

Preparation of Medium: Add components to distilled/deionized water and bring volume to 1.0L. Mix thoroughly. Gently heat and bring to boiling. Distribute into tubes or flasks. Autoclave for 15 min at 15 psi pressure–121°C. Pour into sterile Petri dishes or leave in tubes.

Use: For the cultivation and maintenance of *Xanthomonas campestris*.

Lactose Ricinoleate Broth

Composition per liter:

Lactose	10.0g
Peptone	5.0g
Sodium ricinoleate	1.0g

pH 7.6 ± 0.2 at 25°C

Preparation of Medium: Add components to distilled/deionized water and bring volume to 1.0L. Mix thoroughly. Distribute into tubes or flasks. Autoclave for 15 min at 15 psi pressure–121°C.

Use: For the selective cultivation of members of the Enterobacteriaceae.

Lauryl Sulfate Broth (m-Lauryl Sulfate Broth)

Composition per liter:

Peptone	39.0g
Lactose	30.0g
Yeast extract	6.0g
Sodium lauryl sulfate	1.0g
Phenol Red	0.2g

pH 7.4 ± 0.2 at 25°C

Source: This medium is available as a premixed powder from Oxoid Unipath.

Preparation of Medium: Add components to distilled/deionized water and bring volume to 1.0L. Mix thoroughly. Distribute into bottles or flasks. Autoclave for 15 min at 15 psi pressure–121°C.

Use: For the cultivation and enumeration of coliform bacteria, especially *Escherichia coli*, in water by the membrane filter method.

Lauryl Sulfate Broth (Lauryl Tryptose Broth)

Composition per liter:

Pancreatic digest of casein	20.0g
Lactose	5.0g
NaCl	5.0g
K_2HPO_4	2.75g
KH_2PO_4	2.75g
Sodium lauryl sulfate	0.1g

pH 6.8 ± 0.2 at 25°C

Source: This medium is available as a premixed powder from BD Diagnostic Systems.

Preparation of Medium: Add components to distilled/deionized water and bring volume to 1.0L. Mix thoroughly. Distribute into tubes containing an inverted Durham tube in 10.0mL volumes. Autoclave for 12 min at 15 psi pressure–121°C. Cool broth quickly to 25°C. For testing water samples with 10.0mL volumes, prepare medium double strength.

Use: For the detection of coliform bacteria in a variety of specimens. Also, for the enumeration of coliform bacteria by the multiple-tube fermentation technique.

Lauryl Sulfate Broth with MUG

Composition per liter:

Pancreatic digest of casein	20.0g
Lactose	5.0g
NaCl	5.0g
K_2HPO_4	2.75g
KH_2PO_4	2.75g
Sodium lauryl sulfate	0.1g
4-Methylumbellferyl-β-D-glucuronide (MUG)	0.05g

pH 6.8 ± 0.2 at 25°C

Source: This medium is available as a premixed powder from BD Diagnostic Systems.

Preparation of Medium: Add components to distilled/deionized water and bring volume to 1.0L. Mix thoroughly.

Distribute into tubes containing an inverted Durham tube in 10.0mL volumes. Autoclave for 12 min at 15 psi pressure–121°C. Cool broth quickly to 25°C. For testing water samples with 10.0mL volumes, prepare medium double strength.

Use: For the detection of *Escherichia coli* in water and by a fluorogenic procedure.

Lauryl Tryptose Mannitol Broth with Tryptophan

Composition per liter:

Pancreatic digest of casein	20.0g
Lactose	5.0g
NaCl	5.0g
K_2HPO_4	2.75g
KH_2PO_4	2.75g
Sodium lauryl sulfate	0.1g
L-Tryptophan	0.2g

pH 6.8 ± 0.2 at 25°C

Source: This medium is available as a premixed powder from Oxoid Unipath.

Preparation of Medium: Add components to distilled/deionized water and bring volume to 1.0L. Mix thoroughly. Distribute into tubes containing an inverted Durham tube in 10.0mL volumes. Autoclave for 10 min at 10 psi pressure–115°C. Cool broth quickly to 25°C.

Use: For the detection of *Escherichia coli* in water samples.

LAVMm2 Medium

Composition per liter:

Lactalbumin hydrolysate	10.0g
Sodium acetate	5.0g
$MgCl_2 \cdot 6H_2O$	20.3mg
Nitrilotriacetic acid	19.1mg
$CaCl_2$	11.1mg
$FeSO_4$	0.152mg
Thiamine·HCl	0.05mg
Cupric acetate	0.04mg
Biotin	0.02mg

pH 8.0–8.1 at 25°C

Preparation of Medium: Add components to distilled/deionized water and bring volume to 1.0L. Mix thoroughly. Adjust pH to 7.5 with Na_2CO_3. Distribute into tubes or flasks. Autoclave for 15 min at 15 psi pressure–121°C. The pH should be 8.0–8.1 after autoclaving.

Use: For the cultivation of *Caryophanon latum*.

LB Agar

Composition per liter:

Agar	15.0g
Pancreatic digest of casein	10.0g
NaCl	5.0g
Yeast extract	5.0g
1*N* NaOH	1.0mL

pH 7.0 ± 0.2 at 25°C

Preparation of Medium: Add components to distilled/deionized water and bring volume to 1.0L. Mix thoroughly. Gently heat and bring to boiling. Adjust pH to 7.0. Distribute into tubes or flasks. Autoclave for 25 min at 15 psi pressure–121°C. Pour into sterile Petri dishes in 35–40.0mL volumes.

Use: For the cultivation of *Escherichia coli.*

Lead Acetate Agar

Composition per liter:

Agar	15.0g
Peptone	15.0g
Proteose peptone	5.0g
Glucose	1.0g
Lead acetate	0.2g
$Na_2S_2O_3$	0.08g

pH 6.6 ± 0.2 at 25°C

Preparation of Medium: Add components to distilled/deionized water and bring volume to 1.0L. Mix thoroughly. Gently heat and bring to boiling. Distribute into tubes or flasks. Autoclave for 15 min at 15 psi pressure–121°C. Pour into sterile Petri dishes or leave in tubes. Allow tubes to cool in a slanted position.

Use: For the cultivation and differentiation of Gram-negative coliform bacteria based on H_2S production. Bacteria that produce H_2S turn the medium brown.

Legionella Selective Agar

Composition per liter:

Agar	15.0g
ACES (2-[(2-Amino-2-oxoethyl)-amino]ethane sulfonic acid) buffer	10.0g
Yeast extract	10.0g
Charcoal, activated	2.0g
α-Ketoglutarate	1.0g
L-Cysteine·HCl·H_2O solution	10.0mL
$Fe_4(P_2O_7)_3$ solution	10.0mL
Antibiotic solution	10.0mL

pH 6.85–7.0 at 25°C

Source: This medium is available as a prepared medium from BD Diagnostic Systems.

L-Cysteine·HCl·H_2O Solution:
Composition per 10.0mL:

L-Cysteine·HCl·H_2O	0.4g

Preparation of L-Cysteine·HCl·H_2O Solution: Add L-cysteine·HCl·H_2O to distilled/deionized water and bring volume to 10.0mL. Mix thoroughly. Filter sterilize.

$Fe_4(P_2O_7)_3$ Solution:
Composition per 10.0mL:

$Fe_4(P_2O_7)_3$	0.25g

Preparation of $Fe_4(P_2O_7)_3$ Solution: Add $Fe_4(P_2O_7)_3$ to distilled/deionized water and bring volume to 10.0mL. Mix thoroughly. Filter sterilize.

Antibiotic Solution:
Composition per 10.0mL:

Anisomycin	10.0mg
Colistin	3.75mg
Vancomycin	2.0mg

Preparation of Antibiotic Solution: Add components to distilled/deionized water and bring volume to 10.0mL. Mix thoroughly. Filter sterilize.

Preparation of Medium: Add components—except L-cysteine·HCl·H_2O, $Fe_4(P_2O_7)_3$, and antibiotic solutions—to distilled/deionized water and bring volume to 970.0mL. Mix thoroughly. Gently heat and bring to boiling. Autoclave for 15 min at 15 psi pressure–121°C. Cool to 45°–50°C. Aseptically add sterile L-cysteine·HCl·H_2O, $Fe_4(P_2O_7)_3$, and antibiotic solutions. Mix thoroughly. Pour into sterile Petri dishes. Swirl medium while pouring to keep charcoal in suspension.

Use: *Legionella* selective agar is used in qualitative procedures for the isolation of *Legionella* species.

Legume Extract Agar

Composition per liter:

Alfalfa roots	35.0g
Agar	20.0g
Soybean meal	10.0g
Sucrose	10.0g
$CaCO_3$	5.0g
Glucose	5.0g
K_2HPO_4	1.0g
$MgSO_4 \cdot 7H_2O$	0.2g
$CaCl_2$	0.1g
NaCl	0.1g
$FeCl_3$	1.0mg

Preparation of Medium: Wash the alfalfa roots well and cut them up. Add 10.0g of soybean meal. Add three times the volume of distilled/deionized water. Steam for 1 hr. Let stand at 25°C overnight. Bring volume to 1.0L with distilled/deionized water. Filter through paper pulp. To the filtrate, add the K_2HPO_4, $CaCl_2$, $MgSO_4 \cdot 7H_2O$, NaCl, $FeCl_3$, and agar. Autoclave for 20 min at 15 psi pressure–121°C. Cool to 45°–50°C. Add the $CaCO_3$, sucrose, and glucose. Mix thoroughly. Distribute into tubes or flasks. Autoclave for 20 min at 10 psi pressure–115°C.

Use: For the cultivation of *Rhizobium* species.

Leifson Medium

Composition per liter:

Agar	15.0g
Pancreatic digest of casein	2.0g
$MgCl_2$	1.0g
Yeast extract	1.0g

pH 8.0 ± 0.2 at 25°C

Preparation of Medium: Add components to distilled/deionized water and bring volume to 1.0L. Mix thoroughly. Gently heat and bring to boiling. Adjust pH to 8.0. Distribute into tubes or flasks. Autoclave for 15 min at 15 psi pressure–121°C. Pour into sterile Petri dishes or leave in tubes.

Use: For the direct isolation and routine culturing of *Hyphomonas* species.

Leptospira Medium

Composition per liter:

$(NH_4)_2Fe(SO_4)_2 \cdot 6H_2O$	6.0g
NaH_2PO_4	0.53g
L-Asparagine	0.5g
Glycerol	0.2g
Tween 60	0.2g
$MgSO_4 \cdot 7H_2O$	0.15g
KH_2PO_4	0.069g
Tween 80	0.05g
EDTA	0.01g
$CaCO_3$	4.0mg
Thiamine·HCl	1.0mg
Vitamin B_{12}	1.0μg

pH 7.4–7.6 at 25°C

Preparation of Medium: Add components, except thiamine·HCl, to distilled/deionized water and bring volume to 990.0mL. Mix thoroughly. Gently heat and bring to boiling. Autoclave for 15 min at 15 psi pressure–121°C. Aseptically add 1.0mg of thiamine·HCl. Aseptically distribute into sterile tubes or flasks.

Use: For the cultivation of *Leptospira* species.

Leptospira Medium, EMJH (*Leptospira* Medium, Ellinghausen-McCullough/Johnson-Harris)

Composition per liter:

Na_2HPO_4	1.0g
NaCl	1.0g
KH_2PO_4	0.3g
NH_4Cl	0.25g
Thiamine	5.0mg
Rabbit serum	100.0mL

pH 7.5 ± 0.2 at 25°C

Source: This medium is available as a premixed powder from BD Diagnostic Systems.

Preparation of Medium: Add components, except rabbit serum, to distilled/deionized water and bring volume to 900.0mL. Mix thoroughly. Gently heat and bring to boiling. Autoclave for 15 min at 15 psi pressure–121°C. Cool to 25°C. Aseptically add sterile rabbit serum. Mix thoroughly. Aseptically distribute into sterile tubes or flasks.

Use: For the cultivation and maintenance of *Leptospira* species.

Leptospira Medium, Modified

Composition per liter:

Agar	1.5g
NaCl	0.5g
Peptone	0.3g
Beef extract	0.2g
Hemin solution	2.5mL
Sterile rabbit serum	100.0mL

pH 7.3 ± 0.1 at 25°C

Hemin Solution:

Composition per 100.0mL:

Hemin	0.05g
NaOH (1*N* solution)	1.0mL

Preparation of Hemin Solution: Add hemin to 1.0mL of 1*N* NaOH solution. Mix thoroughly. Bring volume to 100.0mL with distilled/deionized water. Autoclave for 15 min at 15 psi pressure–121°C. Cool to 45°–50°C.

Preparation of Medium: Add components, except hemin solution and rabbit serum, to distilled/deionized water and bring volume to 897.5mL. Mix thoroughly. Gently heat and bring to boiling. Adjust pH to 7.4. Autoclave for 15 min at 15 psi pressure–121°C. Cool to 45°–50°C. Aseptically add 2.5mL of sterile hemin solution and 100.0mL of sterile rabbit serum. Mix thoroughly. The pH of the medium should be 7.3. Store at 4°C for 24 hr. Inactivate medium at 56°C for 60 min. Aseptically distribute into sterile tubes or flasks.

Use: For the cultivation and maintenance of *Leptospira biflexa, Leptospira borgpetersenii, Leptospira interrogans, Leptospira meyeri, Leptospira noguchii, Leptospira santarosai,* and *Leptospira weili.*

Leptospira Protein-Free Medium (*Leptospira* PF Medium)

Composition per liter:

TES (*N*-Tris[hydroxymethyl]methyl-2-aminoethane sulfonic acid) buffer	1.2g
NaCl	0.9g
Sodium pyruvate	0.2g
CT-Tween 60	12.0mL
CT-Tween 40	3.0mL
$MgCl_2$-$CaCl_2$ solution	1.0mL
Cyanocobalamin (0.02% solution)	1.0mL

Glycerol (10% solution)..1.0mL
KH_2PO_4 (1% solution)..1.0mL
$MnSO_4 \cdot H_2O$ (0.1% solution)....................................1.0mL
$ZnSO_4$ (0.4% solution)..0.1mL
pH 7.6 ± 0.2 at 25°C

CT-Tween 60:
Composition per 200.0mL:
Charcoal, Norit A...40.0g
Tween 60..20.0g

Preparation of CT-Tween 60: Add Tween 60 to 200.0mL of distilled/deionized water. Mix thoroughly. While stirring, add charcoal. Stir mixture for 18 hr at 25°C. Allow charcoal to settle out of suspension for 18 hr at 4°C. Carefully decant the Tween solution off the sediment. Centrifuge the Tween solution at 10,000 × g for 1 hr. Decant supernatant solution. Pass Tween solution through a thin-channel ultrafiltration XM 100 membrane. Store stock solution at −20°C.

CT-Tween 40:
Composition per 200.0mL:
Charcoal, Norit A...40.0g
Tween 40..20.0g

Preparation of CT-Tween 40: Add Tween 40 to 200.0mL of distilled/deionized water. Mix thoroughly. While stirring, add charcoal. Stir mixture for 18 hr at 25°C. Allow charcoal to settle out of suspension for 18 hr at 4°C. Carefully decant the Tween solution off the sediment. Centrifuge the Tween solution at 10,000 × g for 1 hr. Decant supernatant solution. Pass Tween solution through a thin-channel ultrafiltration XM 100 membrane. Store stock solution at −20°C.

$MgCl_2$–$CaCl_2$ Solution:
Composition per 100.0mL:
$CaCl_2 \cdot 2H_2O$..1.5g
$MgCl_2 \cdot 6H_2O$..1.5g

Preparation of $MgCl_2$-$CaCl_2$ Solution: Add components to distilled/deionized water and bring volume to 100.0mL. Mix thoroughly.

Preparation of Medium: Add components to distilled/deionized water and bring volume to 1.0L. Mix thoroughly. Filter sterilize. Aseptically distribute into sterile tubes or flasks.

Use: For the cultivation of *Leptospira* species.

Leptospirillum ferrooxidans Medium

Composition per liter:
$FeSO_4 \cdot 7H_2O$...30.0g
$CaCl_2 \cdot 2H_2O$..0.147g
$(NH_4)_2SO_4$...0.13g
KH_2PO_4...27.0mg
$MgCl_2 \cdot 6H_2O$...25.0mg
Trace elements solution 1.0mL
pH 2.0 ± 0.2 at 25°C

Trace Elements Solution:
Composition per liter:
$CoCl_2 \cdot 6H_2O$ 0.12g
$MnCl_2 \cdot 4H_2O$..0.1g
$Na_2MoO_4 \cdot 2H_2O$ 85.2mg
$ZnCl_2$...70.0mg
H_3BO_3 ....31.0mg

Preparation of Trace Elements Solution: Add components to distilled/deionized water and bring volume to 1.0L. Mix thoroughly.

Preparation of Medium: Add H_2SO_4 to 900.0mL of distilled/deionized water and bring pH to 3.0. Add components. Mix thoroughly. Bring volume to 1.0L with distilled/deionized water. Mix thoroughly. Adjust pH to 2.0 with H_2SO_4. Distribute into tubes or flasks. Autoclave for 15 min at 15 psi pressure–121°C.

Use: For the cultivation of *Leptospirillum ferrooxidans.*

Leptospirillum HH Medium (DSMZ Medium 882)

Composition per 1001mL:
Solution A 950.0mL
Solution B...50.0mL
Trace elements solution 1.0mL
pH 1.8 ± 0.2 at 25°C

Solution A:
$CaCl_2 \cdot 2H_2O$ 147.0mg
$(NH_4)_2SO_4$..132.0mg
$MgCl_2 \cdot 6H_2O$ 53.0mg
KH_2PO_4 ....27.0mg

Preparation of Solution A: Add components to distilled/deionized water and bring volume to 950.0mL. Mix thoroughly. Adjust pH to 1.8 with 10*N* H_2SO_4. Autoclave for 30 min at 112°C. Cool to room temperature.

Solution B:
$FeSO_4 \cdot 7H_2O$ 20.0g
H_2SO_4, 0.25*N* 50.0mL

Preparation of Solution B: Add $FeSO_4 \cdot 7H_2O$ to 50.0mL 0.25*N* H_2SO_4. Mix thoroughly. The pH should be 1.2. Autoclave for 30 min at 112°C. Cool to room temperature.

Trace Elements Solution:
$ZnCl_2$ 68.0mg
$CuCl_2 \cdot 2H_2O$..67.0mg
$CoCl_2 \cdot 6H_2O$..64.0mg
$MnCl_2 \cdot 2H_2O$ 62.0mg
H_3BO_3..31.0mg
Na_2MoO_4..10.0mg

Preparation of Trace Elements Solution: Add components to distilled/deionized water and bring volume to 1.0L. Mix thoroughly. Adjust pH to 1.8 with 10*N* H_2SO_4. Autoclave for 15 min at 15 psi pressure–121°C. Cool to room temperature.

Preparation of Medium: Aseptically mix 950.0mL of solution A and 50.0mL solution B. Mix thoroughly. Aseptically add 1.0mL trace elements solution. Mix thoroughly. Adjust pH to 1.8.

Use: For the cultivation of *Acidithiobacillus ferrooxidans=Thiobacillus ferrooxidans* and *Leptospirillum* spp.

Leptothrix ochracea Medium

Composition per liter:
Agar 10.0g
Manganous acetate 0.1g
Manganese bicarbonate solution 100.0mL

Manganese Bicarbonate Solution:
Composition per 100.0mL:
$MnCO_3$ 2.0g

Preparation of Manganese Bicarbonate Solution: Add $MnCO_3$ to distilled/deionized water and bring volume to 100.0mL. Mix thoroughly. Gas with 100% CO_2 for 20 min. Filter through Whatman #1 filter paper.

Preparation of Medium: Add components to distilled/deionized water and bring volume to 1.0L. Mix thoroughly. Gently heat and bring to boiling. Distribute into tubes or flasks. Autoclave for 15 min at 15 psi pressure–121°C. Pour into sterile Petri dishes or leave in tubes.

Use: For the cultivation of *Leptothrix ochracea.*

Leptothrix 2× PYG Medium

Composition per liter:

HEPES (*N*-2-Hydroxyethyl piperazine-*N*′-2-ethanesulfonic acid) buffer 3.57g
$MgSO_4 \cdot 7H_2O$ 0.6g
Glucose 0.5g
Peptone 0.5g
Yeast extract 0.5g
$CaCl_2 \cdot 2H_2O$ 0.07g
$MnSO_4 \cdot H_2O$ 0.017g
pH 7.3 ± 0.2 at 25°C

Preparation of Medium: Add components to distilled/deionized water and bring volume to 1.0L. Mix thoroughly. Adjust pH to 7.3. Distribute into tubes or flasks. Autoclave for 15 min at 15 psi pressure–121°C.

Use: For the cultivation and maintenance of *Leptothrix discophora.*

Leptothrix Strains Medium

Composition per liter:

Agar 12.0g
Peptone 5.0g
$MgSO_4 \cdot 7H_2O$ 0.2g
Ferric ammonium citrate 0.15g
$CaCl_2$ 0.05g
$FeCl_3 \cdot 6H_2O$ 0.01g
$MnSO_4 \cdot H_2O$ 0.01g

Preparation of Medium: Add components to distilled/deionized water and bring volume to 1.0L. Mix thoroughly. Gently heat and bring to boiling. Distribute into tubes or flasks. Autoclave for 15 min at 15 psi pressure–121°C. Pour into sterile Petri dishes or leave in tubes.

Use: For the isolation and cultivation of *Leptothrix* species.

Leuconostoc Medium

Composition per liter:

$CaCO_3$ 50.0g
Malt extract 50.0g
Agar 15.0g
NaCl 2.5g
Beef extract 1.0g
Polypeptone 1.0g

Preparation of Medium: Add components to distilled/deionized water and bring volume to 1.0L. Mix thoroughly. Gently heat and bring to boiling. Distribute into tubes or flasks. Autoclave for 10 min at 15 psi pressure–121°C. Pour into sterile Petri dishes or leave in tubes.

Use: For the cultivation and maintenance of *Leuconostoc mesenteroides.*

Leuconostoc oenos Medium

Composition per liter:

Glucose 10.0g
Peptone 10.0g
Yeast extract 5.0g
$MnSO_4 \cdot 4H_2O$ 0.1g
Tomato juice 250.0mL
L-Cysteine·HCl solution 10.0mL
pH 4.8 ± 0.2 at 25°C

L-Cysteine·HCl Solution:
Composition per 10.0mL:

L-Cysteine·HCl 0.5g

Preparation of L-Cysteine·HCl Solution: Add L-cysteine·HCl to distilled/deionized water and bring volume to 10.0mL. Mix thoroughly. Filter sterilize.

Preparation of Medium: Add components, except L-cysteine·HCl solution, to distilled/deionized water and bring volume to 990.0mL. Mix thoroughly. Gently heat and bring to boiling. Autoclave for 15 min at 15 psi pressure–121°C. Cool to 25°C. Aseptically add sterile L-cysteine·HCl solution. Mix thoroughly. Aseptically distribute into sterile tubes or flasks.

Use: For the cultivation of *Leuconostoc oenos.*

Leucothrix Medium

Composition per liter:

Pancreatic digest of casein 10.0g
Synthetic seawater 1000.0mL
pH 7.8 ± 0.2 at 25°C

Synthetic Seawater:
Composition per liter:

NaCl 24.0g
$MgCl_2 \cdot 6H_2O$ 11.0g
Na_2SO_4 .... 4.0g
$CaCl_2 \cdot 6H_2O$ 2.0g
KCl 0.7g
KBr 0.1g
$SrCl_2 \cdot 6H_2O$ 0.04g
H_3BO_3 .... 0.03g
$NaSiO_3 \cdot 9H_2O$ 5.0mg
NaF 3.0mg
NH_4NO_3 .... 2.0mg
$FePO_4 \cdot 4H_2O$ 1.0mg

Preparation of Synthetic Seawater: Add components to distilled/deionized water and bring volume to 1.0L. Mix thoroughly.

Preparation of Medium: Add 10.0g of pancreatic digest of casein to 1.0L of synthetic seawater. Mix thoroughly. Adjust pH to 7.8. Distribute into tubes or flasks. Autoclave for 15 min at 15 psi pressure–121°C.

Use: For the cultivation of *Leucothrix* species.

Leucothrix mucor Medium

Composition per liter:

NaCl 11.75g
Monosodium glutamate 10.0g
$MgCl_2 \cdot 6H_2O$ 5.35g
Na_2SO_4 .... 2.0g
Sodium lactate 2.0g
$CaCl_2 \cdot 6H_2O$ 1.12g
Tris (hydroxymethyl) amino methane 0.5g
KCl 0.35g
Na_2HPO_4 .... 0.05g
pH 7.6 ± 0.2 at 25°C

Preparation of Medium: Add components to distilled/deionized water and bring volume to 1.0L. Mix thoroughly. Distribute into tubes or flasks. Autoclave for 15 min at 15 psi pressure–121°C.

Use: For the cultivation of *Leucothrix mucor.*

Levine EMB Agar (Levine Eosin Methylene Blue Agar) (Eosin Methylene Blue Agar, Levine) (LEMB Agar)

Composition per liter:

Agar	15.0g
Lactose	10.0g
Peptone	10.0g
K_2HPO_4	2.0g
Eosin Y	0.4g
Methylene Blue	0.065mg

pH 7.1 ± 0.2 at 25°C

Source: This medium is available as a premixed powder from BD Diagnostic Systems.

Preparation of Medium: Add components to distilled/deionized water and bring volume to 1.0L. Mix thoroughly. Gently heat and bring to boiling. Distribute into tubes or flasks. Autoclave for 15 min at 15 psi pressure–121°C. Pour into sterile Petri dishes or leave in tubes.

Use: For the isolation, cultivation, and differentiation of Gram-negative enteric bacteria based on lactose fermentation. Bacteria that ferment lactose, especially the coliform bacterium *Escherichia coli*, appear as colonies with a green metallic sheen or blue-black to brown color. Bacteria that do not ferment lactose appear as colorless or transparent light purple colonies.

LHET2 Medium

Composition per liter:

Solution A	500.0mL
Solution B	500.0mL

pH 2.5–3.0 at 25°C

Solution A:

Composition per 500.0mL:

$(NH_4)_2SO_4$	2.0g
K_2HPO_4	0.51g
$MgSO_4 \cdot 7H_2O$	0.5g
KCl	0.1g
Pancreatic digest of casein	0.06g
NaCl	0.02g
Papaic digest of soybean meal	0.01g

Preparation of Solution A: Add components to distilled/deionized water and bring volume to 500.0mL. Mix thoroughly. Gently heat and bring to boiling. Adjust pH to 2.5–3.0 with 1*N* H_2SO_4. Autoclave for 15 min at 15 psi pressure–121°C. Cool to 45°–50°C.

Solution B:

Composition per 500.0mL:

Agar	12.0g
Glucose	1.0g

Preparation of Solution B: Add components to distilled/deionized water and bring volume to 500.0mL. Mix thoroughly. Gently heat and bring to boiling. Autoclave for 15 min at 15 psi pressure–121°C. Cool to 45°–50°C.

Preparation of Medium: Aseptically combine sterile solution A and sterile solution B. Mix thoroughly. Pour into sterile Petri dishes or distribute into sterile tubes.

Use: For the cultivation and maintenance of *Acidiphilium cryptum*.

LHET2 Medium with Yeast Extract or Yeast Autolysate

Composition per liter:

Solution A	500.0mL
Solution B	500.0mL

pH 2.5–3.0 at 25°C

Solution A:

Composition per 500.0mL:

$(NH_4)_2SO_4$	2.0g
K_2HPO_4	0.51g
$MgSO_4 \cdot 7H_2O$	0.5g
KCl	0.1g
Yeast extract or yeast autolysate	0.1g
Pancreatic digest of casein	0.06g
NaCl	0.02g
Papaic digest of soybean meal	0.01g

Preparation of Solution A: Add components to distilled/deionized water and bring volume to 500.0mL. Mix thoroughly. Gently heat and bring to boiling. Adjust pH to 2.5–3.0 with 1*N* H_2SO_4. Autoclave for 15 min at 15 psi pressure–121°C. Cool to 45°–50°C.

Solution B:

Composition per 500.0mL:

Agar	12.0g
Glucose	1.0g

Preparation of Solution B: Add components to distilled/deionized water and bring volume to 500.0mL. Mix thoroughly. Gently heat and bring to boiling. Autoclave for 15 min at 15 psi pressure–121°C. Cool to 45°–50°C.

Preparation of Medium: Aseptically combine sterile solution A and sterile solution B. Mix thoroughly. Pour into sterile Petri dishes or distribute into sterile tubes.

Use: For the cultivation and maintenance of *Acidiphilium angustum, Aacidiphilium facilis,* and *Acidiphilium rubrum.*

Lichen Fungi Medium

Composition per liter:

Agar	20.0g
Malt extract	20.0g
Yeast extract	2.0g

Preparation of Medium: Add components to distilled/deionized water and bring volume to 1.0L. Mix thoroughly. Gently heat until boiling. Distribute into tubes or flasks. Autoclave for 15 min at 15 psi pressure–121°C. Pour into sterile Petri dishes or leave in tubes.

Use: For the cultivation of numerous fungi from lichen symbiotic relationships.

Lima Bean Agar (ATCC Medium 322)

Composition per liter:

Lima beans, infusion from 62.5g	8.0g
Agar	15.0g

pH 5.6 ± 0.2 at 25°C

Source: This medium is available as a premixed powder from BD Diagnostic Systems.

Preparation of Medium: Add components to distilled/deionized water and bring volume to 1.0L. Mix thoroughly. Gently heat and bring to boiling. Distribute into tubes or flasks. Autoclave for 15 min at 15 psi pressure–121°C. Pour into sterile Petri dishes or leave in tubes.

Use: For the cultivation of a variety of phytopathological fungi and other fungi.

Limnobacter Medium (DSMZ Medium 919)

Composition per liter:

Yeast extract	0.5g
Proteose peptone	0.5g
Casamino acids	0.5g
Glucose	0.5g
Soluble starch	0.5g
Sodium pyruvate	0.3g
K_2HPO_4	0.3g
$MgSO_4 \cdot 7H_2O$	0.05g

pH 7.2 ± 0.2 at 25°C

Preparation of Medium: Add components to distilled/deionized water and bring volume to 1.0L. Mix thoroughly. Distribute into tubes or flasks. Autoclave for 15 min at 15 psi pressure–121°C.

Use: For the cultivation of *Limnobacter thiooxidans.*

Lipovitellin Salt Mannitol Agar

Composition per liter:

NaCl	75.0g
Egg yolk	20.0g
Agar	15.0g
D-Mannitol	10.0g
Polypeptone	10.0g
Beef extract	1.0g
Phenol Red	0.025g

Preparation of Medium: Add components to distilled/deionized water and bring volume to 1.0L. Mix thoroughly. Gently heat and bring to boiling. Distribute into tubes or flasks. Autoclave for 15 min at 15 psi pressure–121°C. Pour into sterile Petri dishes or leave in tubes.

Use: For the detection of *Staphylococcus aureus* in swimming pool water based on lipovitellin-lipase activity and mannitol fermentation. *Staphylococcus aureus* and other bacteria with lipovitellin-lipase activity attack the egg yolk and appear as colonies surrounded by an opaque zone. Bacteria that ferment mannitol appear as colonies surrounded by a yellow zone.

Lobosphaera Medium

Composition per liter:

Polypeptone	10.0g
Glycerol	7.5g
Yeast extract	2.0g
K_2HPO_4	1.0g
KH_2PO_4	1.0g
$MgSO_4 \cdot 7H_2O$	0.5g

pH 6.5 ± 0.2 at 25°C

Preparation of Medium: Add components to distilled/deionized water and bring volume to 1.0L. Mix thoroughly. Adjust pH to 6.5. Filter sterilize. Aseptically distribute into sterile tubes or flasks.

Use: For the cultivation of *Lobosphaera* species.

Low Phosphate Buffered Basal Medium, Modified

Composition per 1030.0mL:

Pectin	4.0g
NH_4Cl	1.0g
Na_2HPO_4	0.72g
KH_2PO_4	0.3g
$MgCl_2 \cdot 6H_2O$	0.2g
Reducing agent	20.0mL
Yeast extract solution	10.0mL
Trace minerals	10.0mL
Vitamins	5.0mL
Resazurin (0.2% solution)	1.0mL
$FeSO_4 \cdot 7H_2O$ (2.5% solution)	0.03mL

pH 7.3 ± 0.1 at 25°C

Reducing Agent:

Composition per 20.0mL:

$Na_2S \cdot 9H_2O$	0.5g

Preparation of Reducing Agent: Add $Na_2S \cdot 9H_2O$ to distilled/deionized water and bring volume to 20.0mL. Mix thoroughly. Gas with 100% N_2 for 20 min. Cap with a rubber stopper. Autoclave for 45 min at 15 psi pressure–121°C. Use freshly prepared solution.

Yeast Extract Solution:

Composition per 10.0mL:

Yeast extract	1.0g

Preparation of Yeast Extract Solution: Add yeast extract to distilled/deionized water and bring volume to 10.0mL. Mix thoroughly. Autoclave for 45 min at 15 psi pressure–121°C. Cool to 25°C.

Trace Minerals:

Composition per liter:

Nitrilotriacetic acid	12.8g
NaCl	1.0g
$CoCl_2 \cdot 6H_2O$	0.16g
$CaCl_2 \cdot 2H_2O$	0.1g
$FeSO_4 \cdot 7H_2O$	0.1g
$MnCl_2 \cdot 4H_2O$	0.1g
$ZnCl_2$	0.1g
$NiSO_4 \cdot 6H_2O$	0.026g
$CuCl_2$	0.02g
Na_2SeO_3	0.02g
H_3BO_3	0.01g
$Na_2MoO_4 \cdot 2H_2O$	0.01g

Preparation of Trace Minerals: Add nitrilotriacetic acid to 500.0mL of distilled/deionized water. Dissolve by adjusting pH to 6.5 with KOH. Add remaining components. Add distilled/deionized water to 1.0L.

Wolfe's Vitamin Solution:

Composition per liter:

Pyridoxine·HCl	0.01g
Thiamine·HCl	5.0mg
Riboflavin	5.0mg
Nicotinic acid	5.0mg
Calcium pantothenate	5.0mg
p-Aminobenzoic acid	5.0mg
Thioctic acid	5.0mg
Biotin	2.0mg
Folic acid	2.0mg
Cyanocobalamin	0.1mg

Preparation of Wolfe's Vitamin Solution: Add components to distilled/deionized water and bring volume to 1.0L. Mix thoroughly.

Preparation of Medium: Add components, except yeast extract solution and reducing agent, to distilled/deionized water and bring volume to 1.0L. Mix thoroughly. Gently heat and bring to boiling. Cool under 90% N_2 + 10% CO_2. Anaerobically distribute into tubes in 6.0mL volumes. Autoclave for 45 min at 15 psi pressure–121°C. Aseptically add 0.06mL of sterile yeast extract solution to each tube. Mix thoroughly. Immediately prior to inoculation, aseptically add 0.12mL of sterile reducing agent to each tube. Mix thoroughly.

Use: For the cultivation and maintenance of *Clostridium thermosulfurogenes*.

LPBM Acido-Thermophile Medium

Composition per liter:

Agar	20.0g
Cellulose	5.0g
KH_2PO_4	1.0g
NH_4Cl	1.0g
Yeast extract	1.0g
Cellobiose	0.5g
$MgSO_4 \cdot 7H_2O$	0.2g
$Na_2HPO_4 \cdot 7H_2O$	0.1g
$CaCl_2 \cdot 2H_2O$	0.02g

pH 5.2 ± 0.2 at 25°C

Preparation of Medium: Add components, except cellulose and cellobiose, to distilled/deionized water and bring volume to 1.0L. Mix thoroughly. Adjust pH to 5.2 with H_3PO_4. Add cellulose and cellobiose. Mix thoroughly. Gently heat and bring to boiling. Distribute into tubes or flasks. Autoclave for 15 min at 15 psi pressure–121°C. Pour into sterile Petri dishes or leave in tubes.

Use: For the cultivation and maintenance of *Acidothermus cellulolyticus*.

LTH Medium for *Thiothrix*

Composition per liter:

HEPES (*N*-[2-Hydroxyethyl]piperazine-*N*′-2-ethanesulfonic acid) buffer	2.38.g
Sodium lactate	1.0g
$Na_2S_2O_3 \cdot 5H_2O$	0.5g
$(NH_4)_2SO_4$	0.5g
K_2HPO_4	0.11g
$MgSO_4 \cdot 7H_2O$	0.1g
$CaCl_2 \cdot 2H_2O$	0.05g
KH_2PO_4	85.0mg
EDTA	3.0mg
$FeCl_3 \cdot 6H_2O$	2.0mg
Wolfe's vitamin solution	10.0mL

pH 7.3 ± 0.2 at 25°C

Wolfe's Vitamin Solution:

Composition per liter:

Pyridoxine·HCl	10.0mg
p-Aminobenzoic acid	5.0mg
Lipoic acid	5.0mg
Nicotinic acid	5.0mg
Riboflavin	5.0mg
Thiamine·HCl	5.0mg
Calcium DL-pantothenate	5.0mg
Biotin	2.0mg
Folic acid	2.0mg
Vitamin B_{12}	0.1mg

Preparation of Wolfe's Vitamin Solution: Add components to distilled/deionized water and bring volume to 1.0L. Mix thoroughly. Filter sterilize.

Preparation of Medium: Add components, except Wolfe's vitamin solution, to distilled/deionized water and bring volume to 990.0mL. Mix thoroughly. Adjust pH to 7.3 with NaOH. Autoclave for 15 min at 15 psi pressure–121°C. Cool to room temperature. Aseptically add 10.0mL of sterile Wolfe's vitamin solution. Mix thoroughly. Aseptically distribute into sterile tubes or flasks.

Use: For the cultivation of *Thiothrix* species.

Luminous Medium

Composition per liter:

NaCl	30.0g
Agar	20.0g
NH_4Cl	5.0g
Pancreatic digest of casein	5.0g
Yeast extract	5.0g
K_2HPO_4	3.9g
KH_2PO_4	2.1g
$CaCO_3$	1.0g
$MgSO_4 \cdot 7H_2O$	1.0g
KCl	0.75g
Tris buffer (1*M* solution, pH 7.5)	50.0mL
Glycerol	3.0mL

pH 7.2 ± 0.2 at 25°C

Preparation of Medium: Add components to distilled/deionized water and bring volume to 1.0L. Mix thoroughly. Gently heat and bring to boiling. Distribute into tubes or flasks. Autoclave for 15 min at 15 psi pressure–121°C. Pour into sterile Petri dishes or leave in tubes.

Use: For the cultivation and maintenance of *Alteromonas hanedai, Photobacterium* species, *Shewanella hanedai,* and *Vibrio* species.

Lysine Decarboxylase Broth, Falkow

Composition per liter:

Peptone	5.0g
L-Lysine	5.0g
Yeast extract	3.0g
Glucose	1.0g
Bromcresol Purple	0.02g

pH 6.5–6.8 at 25°C

Preparation of Medium: Add components to distilled/deionized water and bring volume to 1.0L. Mix thoroughly. Gently heat and bring to boiling. Adjust pH to 6.5–6.8. Distribute into tubes in 5.0mL volumes. Autoclave for 15 min at 15 psi pressure–121°C.

Use: For the cultivation and differentiation of bacteria, especially *Salmonella,* based on their ability to decarboxylate lysine. Bacteria that decarboxylate lysine turn the medium turbid purple.

Lysine Decarboxylase Broth, Taylor Modification

Composition per liter:

L-Lysine	5.0g
Yeast extract	3.0g
Glucose	1.0g
Bromcresol Purple	0.02g

pH 6.1 ± 0.2 at 25°C

Source: This medium is available as a premixed powder from Oxoid Unipath.

Preparation of Medium: Add components to distilled/deionized water and bring volume to 1.0L. Mix thoroughly. Gently heat and bring to boiling. Adjust pH to 6.1. Distribute into tubes in 5.0mL volumes. Autoclave for 15 min at 15 psi pressure–121°C.

Use: For the cultivation and differentiation of bacteria, especially *Salmonella,* based on their ability to decarboxylate lysine. Bacteria that decarboxylate lysine turn the medium turbid purple.

Lysine Decarboxylase Broth, Taylor Modification (Lysine Decarboxylase Broth)

Composition per liter:

L-Lysine	5.0g
Peptone	5.0g
Yeast extract	3.0g
Glucose	1.0g
Bromcresol Purple	0.02g

pH 6.8 ± 0.2 at 25°C

Source: This medium is available as a premixed powder from BD Diagnostic Systems.

Preparation of Medium: Add components to distilled/deionized water and bring volume to 1.0L. Mix thoroughly. Gently heat and bring to boiling. Adjust pH to 6.1. Distribute into tubes in 5.0mL volumes. Autoclave for 15 min at 15 psi pressure–121°C.

Use: For the cultivation and differentiation of bacteria, especially *Salmonella*, based on their ability to decarboxylate lysine. Bacteria that decarboxylate lysine turn the medium turbid purple.

Lysine Decarboxylase Medium

Composition per liter:

Glucose	0.5g
KH_2PO_4	0.5g
L-Lysine·HCl	0.5g

pH 4.6 ± 0.2 at 25°C

Preparation of Medium: Add components to distilled/deionized water and bring volume to 1.0L. Mix thoroughly. Gently heat and bring to boiling. Adjust pH to 4.6. Autoclave for 15 min at 15 psi pressure–121°C. Aseptically distribute into sterile tubes in 1.0mL volumes.

Use: For the cultivation and differentiation of Gram-negative, nonfermentative bacteria based on their ability to decarboxylate lysine. Bacteria that decarboxylate lysine turn the medium turbid purple.

Lysine Iron Agar

Composition per liter:

Agar	13.5g
L-Lysine	10.0g
Pancreatic digest of gelatin	5.0g
Yeast extract	3.0g
Glucose	1.0g
Ferric ammonium citrate	0.5g
$Na_2S_2O_3·5H_2O$	0.04g
Bromcresol Purple	0.02g

pH 6.7 ± 0.2 at 25°C

Source: This medium is available as a premixed powder from BD Diagnostic Systems and Oxoid Unipath.

Preparation of Medium: Add components to distilled/deionized water and bring volume to 1.0L. Mix thoroughly. Gently heat while stirring and bring to boiling. Distribute into tubes in 10.0mL volumes. Autoclave for 12 min at 15 psi pressure–121°C. Allow tubes to cool in a slanted position.

Use: For the cultivation and differentiation of members of the Enterobacteriaceae based on their ability to decarboxylate lysine and to form H_2S. Bacteria that decarboxylate lysine turn the medium purple. Bacteria that produce H_2S appear as black colonies.

M3 Agar

Composition per 1020.0mL:

Agar	18.0g
Na_2HPO_4	0.732g
KH_2PO_4	0.466g
NaCl	0.29g
Sodium propionate	0.2g
$MgSO_4·7H_2O$	0.1g
$CaCO_3$	0.02g
KNO_3	0.01g
$FeSO_4·7H_2O$	0.2mg
$ZnSO_4·7H_2O$	0.18mg
$MnSO_4·4H_2O$	0.02mg
Cycloheximide solution	10.0mL
Thiamine·HCl solution	10.0mL

pH 7.0 ± 0.2 at 25°C

Cycloheximide Solution:

Composition per 10.0mL:

Cycloheximide	0.05g

Preparation of Cycloheximide Solution: Add cycloheximide to distilled/deionized water and bring volume to 10.0mL. Mix thoroughly. Filter sterilize.

Thiamine·HCl Solution:

Composition per 10.0mL:

Thiamine·HCl	4.0mg

Preparation of Thiamine·HCl Solution: Add thiamine·HCl to distilled/deionized water and bring volume to 10.0mL. Mix thoroughly. Filter sterilize.

Preparation of Medium: Add components, except cycloheximide solution and thiamine·HCl solution, to distilled/deionized water and bring volume to 980.0mL. Mix thoroughly. Gently heat and bring to boiling. Autoclave for 15 min at 15 psi pressure–121°C. Cool to 45°–50°C. Aseptically add 10.0mL of sterile cycloheximide solution and 10.0mL of thiamine·HCl solution. Mix thoroughly. Pour into sterile Petri dishes or distribute into sterile tubes.

Use: For the selective isolation and cultivation of *Nocardia* species and *Rhodococcus* species.

M Medium

Composition per liter:

Beef	5.0g
Neopeptone	4.0g
NaCl	1.6g
Glucose	0.5g
$CaCl_2$	0.06g
KH_2PO_4	0.06g
KCl	0.04g
Rabbit blood solution	200.0mL

Rabbit Blood Solution:

Composition per liter:

Rabbit blood	500.0mL

Preparation of Rabbit Blood Solution: Add 500.0mL of whole rabbit blood to 500.0mL of sterile distilled/deionized water. Freeze and thaw twice to lyse blood cells.

Preparation of Medium: Trim beef to remove fat. Add 5.0g of lean beef to 200.0mL of distilled/deionized water. Gently heat and bring to boiling. Boil for 2–3 min. Filter through Whatman #2 filter paper. Add other components, except rabbit blood solution. Bring volume to 800.0mL with distilled/deionized water. Mix thoroughly. Adjust pH to 7.2 with 1*N* NaOH. Autoclave for 15 min at 15 psi pressure–121°C. Cool to 50°–55°C. Aseptically add 200.0mL of lysed rabbit blood solution. Mix thoroughly. Aseptically distribute into sterile tubes or flasks.

Use: For the cultivation of *Herpetomonas megaseliae.*

M1 Medium

Composition per liter:

L-Leucine	2.0g
L-Alanine	1.0g
L-Isoleucine	1.0g
L-Phenylalanine	1.0g
L-Proline	1.0g
L-Tryptophane	1.0g
L-Asparagine	0.5g
L-Lysine	0.5g
L-Methionine	0.5g
L-Tyrosine	0.4g
L-Valine	0.2g
L-Serine	0.2g
$MgSO_4 \cdot 7H_2O$	0.2g
NaCl	0.2g
KH_2PO_4	0.14g
L-Arginine	0.1g
L-Cysteine	0.1g
L-Glycine	0.1g
L-Histidine	0.1g
L-Threonine	0.1g
$CaCl_2$	2.0mg
$FeCl_3 \cdot 6H_2O$	2.0mg
Tris(hydroxymethyl)aminomethane buffer (0.01*M* solution, pH 7.6)	1.0L

pH 7.6 ± 0.2 at 25°C

Preparation of Medium: Add solid components to 1.0L of Tris buffer. Mix thoroughly. Filter sterilize. Aseptically distribute into tubes or flasks.

Use: For the cultivation of *Myxococcus xanthus.*

M14 Medium

Composition per liter:

Yeast extract	1.0g
D-Glucose	1.0g
Tris(hydroxymethyl)aminomethane	0.753g
Artificial seawater	250.0mL
Modified Hutner's basal salts	20.0mL

pH 7.5 ± 0.2 at 25°C

Artificial Seawater:

Composition per liter:

NaCl	23.48g
$MgCl_2$	4.98g
Na_2SO_4	3.92g
$CaCl_2$	1.1g
KCl	0.66g
$NaHCO_3$	0.19g
H_3BO_3	0.026g
$SrCl_2$	0.024g
KBr	6.0mg
NaF	3.0mg

Preparation of Artificial Seawater: Add components to distilled/deionized water and bring volume to 1.0L. Mix thoroughly.

Modified Hutner's Basal Salts:

Composition per liter:

$MgSO_4 \cdot 7H_2O$	29.7g
Nitrilotriacetic acid	10.0g
$CaCl_2 \cdot 2H_2O$	3.34g
$FeSO_4 \cdot 7H_2O$	99.0mg
$(NH_4)_2MoO_4$	9.25mg
"Metals 44"	50.0mL

Preparation of Modified Hutner's Basal Salts: Add nitrilotriacetic acid to 500.0mL of distilled/deionized water. Dissolve by adjusting pH to 6.5 with KOH. Add remaining components. Add distilled/deionized water to 1.0L.

"Metals 44":

Composition per 100.0mL:

$ZnSO_4 \cdot 7H_2O$	1.1g
$FeSO_4 \cdot 7H_2O$	0.5g
EDTA	0.25g
$MnSO_4 \cdot 7H_2O$	0.154g
$CuSO_4 \cdot 5H_2O$	0.04g
$Co(NO_3)_2 \cdot 6H_2O$	0.025g
$Na_2B_4O_7 \cdot 10H_2O$	0.018g

Preparation of "Metals 44": Add components to distilled/deionized water and bring volume to 100.0mL. Mix thoroughly.

Preparation of Medium: Add components, except modified Hutner's basal salts, to distilled/deionized water and bring volume to 980.0mL. Mix thoroughly. Autoclave for 15 min at 15 psi pressure–121°C. Cool to room temperature. Aseptically add 20.0mL of sterile modified Hutner's basal salts. Mix thoroughly. Aseptically distribute into sterile tubes or flasks.

Use: For the cultivation and maintenance of *Pirellula marina.*

M17 Medium for *Filomicrobium fusiforme* (DSMZ Medium 768)

Composition per liter:

Na-acetate	1.0g
KNO_3	1.0g
Artificial seawater, concentrated	500.0mL
Hutner's salts	20.0mL
Vitamin solution	10.0mL

pH 7.2 ± 0.2 at 25°C

Hutner's Salts Solution:

Composition per liter:

$MgSO_4 \cdot 7H_2O$	29.7g
Nitrilotriacetic acid	10.0g
$CaCl_2 \cdot 2H_2O$	3.335g
$FeSO_4 \cdot 7H_2O$	99.0mg
$(NH_4)_6MoO_7O_{24} \cdot 4H_2O$	9.25mg
"Metals 44"	50.0mL

"Metals 44":

Composition per 100.0mL:

$ZnSO_4 \cdot 7H_2O$	1.095g
$FeSO_4 \cdot 7H_2O$	0.5g
Sodium EDTA	0.25g
$MnSO_4 \cdot H2O$	0.154g
$CuSO_4 \cdot 5H_2O$	39.2mg
$Co(NO_3)_2 \cdot 6H_2O$	24.8mg
$Na_2B_4O_7 \cdot 10H_2O$	17.7mg

Preparation of "Metals 44": Add sodium EDTA to distilled/deionized water and bring volume to 90.0mL. Mix thoroughly. Add a few drops of concentrated H_2SO_4 to retard precipitation of heavy metal ions. Add remaining components. Mix thoroughly. Bring volume to 100.0mL with distilled/deionized water.

Preparation of Hutner's Salts Solution: Add nitrilotriacetic acid to 500.0mL of distilled/deionized water. Adjust pH to 6.5 with KOH. Add remaining components. Add distilled/deionized water to 1.0L. Adjust pH to 6.8.

Artificial Seawater, Concentrated:
Composition per liter:

NaCl	70.43g
$MgCl_2 \cdot 6H_2O$	31.86g
Na_2SO_4	11.75g
$CaCl_2 \cdot 2H_2O$	4.35g
$NaHCO_3$	2.88g
KCl	1.99g
KBr	0.29g
H_3BO_3	0.08g

Preparation of Artificial Seawater: Add components to distilled/deionized water and bring volume to 1.0L. Mix thoroughly. Filter sterilize.

Vitamin Solution:
Composition per liter:

Riboflavin	5.0mg
Thiamine-HCl·$2H_2O$	5.0mg
Ca-pantothenate	5.0mg
Biotin	2.0mg
Folic acid	2.0mg
Vitamin B_{12}	0.1mg

Preparation of Vitamin Solution: Add components to distilled/deionized water and bring volume to 1.0L. Mix thoroughly. Filter sterilize.

Preparation of Medium: Add components, except artificial seawater and vitamin solution, to distilled/deionized water and bring volume to 490.0mL. Mix thoroughly. Adjust pH to 7.2. Autoclave for 15 min at 15 psi pressure–121°C. Cool to room temperature. Aseptically add 500.0mL artificial seawater and 10.0mL vitamin solution. Mix thoroughly. Aseptically and anaerobically distribute into sterile tubes or bottles.

Use: For the cultivation of *Filomicrobium fusiforme.*

M13 *Verrucomicrobium* Medium

Composition per liter:

Glucose	0.25g
Peptone	0.25g
Yeast extract	0.25g
Distilled water	670.0mL
Artificial seawater	250.0mL
Tris-HCl buffer, (0.1*M* solution, pH 7.5)	50.0mL
Modified Huntner's basal salts	20.0mL
Vitamin solution	10.0mL

pH 7.5 ± 0.2 at 25°C

Artificial Seawater:
Composition per liter:

NaCl	23.48g
$MgCl_2$	4.98g
Na_2SO_4	3.92g
$CaCl_2$	1.1g
KCl	0.66g
$NaHCO_3$	0.19g
H_3BO_3	0.026g
$SrCl_2$	0.024g
KBr	6.0mg
NaF	3.0mg

Preparation of Artificial Seawater: Add components to distilled/deionized water and bring volume to 1.0L. Mix thoroughly.

Modified Hutner's Basal Salts:
Composition per liter:

$MgSO_4 \cdot 7H_2O$	29.7g
Nitrilotriacetic acid	10.0g
$CaCl_2 \cdot 2H_2O$	3.34g
$FeSO_4 \cdot 7H_2O$	99.0mg
$(NH_4)_2MoO_4$	9.25mg
"Metals 44"	50.0mL

Preparation of Modified Hutner's Basal Salts: Add nitrilotriacetic acid to 500.0mL of distilled/deionized water. Dissolve by adjusting pH to 6.5 with KOH. Add remaining components. Add distilled/deionized water to 1.0L.

"Metals 44":
Composition per 100.0mL:

$ZnSO_4 \cdot 7H_2O$	1.1g
$FeSO_4 \cdot 7H_2O$	0.5g
EDTA	0.25g
$MnSO_4 \cdot 7H_2O$	0.154g
$CuSO_4 \cdot 5H_2O$	0.04g
$Co(NO_3)_2 \cdot 6H_2O$	0.025g
$Na_2B_4O_7 \cdot 10H_2O$	0.018g

Preparation of "Metals 44": Add components to distilled/deionized water and bring volume to 100.0mL. Mix thoroughly.

Vitamin Solution:
Composition per liter:

D-Calcium pantothenate	5.0mg
Riboflavin	5.0mg
Thiamine·HCl	5.0mg
Biotin	2.0mg
Folic acid	2.0mg
Vitamin B_{12}	0.1mg

Preparation of Vitamin Solution: Add components to distilled/deionized water and bring volume to 1.0L. Mix thoroughly. Filter sterilize.

Preparation of Medium: Add components, except modified Hutner's basal salts, to distilled/deionized water and bring volume to 980.0mL. Mix thoroughly. Autoclave for 15 min at 15 psi pressure–121°C. Cool to room temperature. Aseptically add 20.0mL of sterile modified Hutner's basal salts. Mix thoroughly. Aseptically distribute into sterile tubes or flasks.

Use: For the cultivation and maintenance of *Verrucomicrobium spinosum.*

M1A Medium

Composition per 1001.0mL:

Sorbitol	23.3g
Peptone	6.0g
Sucrose	3.3g
Pancreatic digest of casein	3.3g
Beef heart infusion	2.0g
Glucose	1.3g
Yeast extract	1.0g
Fructose	0.3g
Phenol Red	20.0mg
Schneider's *Drosophila* medium	533.0mL
Fetal calf serum, heat inactivated	167.0mL
Fresh yeast extract solution	33.0mL
Penicillin solution	8.0mL

Schneider's *Drosophila* Medium:
Composition per liter:

$MgSO_4 \cdot 7H_2O$	3.7g

NaCl 2.1g
Yeast extract 2.0g
Trehalose 2.0g
D-Glucose 2.0g
L-Glutamine 1.8g
L-Lysine·HCl 1.7g
L-Proline 1.7g
KCl 1.6g
$Na_2HPO_4 \cdot 7H_2O$ 1.3g
L-Glutamic acid 0.8g
L-Methionine 0.8g
$CaCl_2$, anhydrous 0.6g
KH_2PO_4 .... 0.5g
β-Alanine 0.5g
L-Tyrosine 0.5g
L-Arginine 0.4g
L-Aspartic acid 0.4g
L-Histidine 0.4g
L-Threonine 0.4g
$NaHCO_3$ 0.4g
Glycine 0.3g
L-Serine 0.3g
L-Valine 0.3g
L-Isoleucine 0.2g
L-Leucine 0.2g
L-Phenylalanine 0.2g
α-Ketoglutaric acid 0.2g
Fumaric acid 0.1g
Malic acid 0.1g
Succinic acid 0.1g
L-Cystine 0.1g
L-Tryptophan 0.1g
L-Cysteine 0.06g

Preparation of Schneider's *Drosophila* Medium: Add components to distilled/deionized water and bring volume to 1.0L. Mix thoroughly. Filter sterilize.

Penicillin Solution:
Composition per 10.0mL:
Penicillin 2,500,000U

Preparation of Penicillin Solution: Add penicillin to distilled/deionized water and bring volume to 10.0mL. Filter sterilize.

Fresh Yeast Extract Solution:
Composition per 100.0mL:
Baker's yeast, live, pressed, starch-free 25.0g

Preparation of Fresh Yeast Extract Solution: Add the live Baker's yeast to 100.0mL of distilled/deionized water. Autoclave for 90 min at 15 psi pressure–121°C. Allow to stand. Remove supernatant solution. Adjust pH to 6.6–6.8. Filter sterilize.

Preparation of Medium: Add components—except Schneider's *Drosophila* medium, fetal calf serum, fresh yeast extract solution, and penicillin solution— to distilled/deionized water and bring volume to 260.0mL. Mix thoroughly. Gently heat and bring to boiling. Autoclave for 15 min at 15 psi pressure–121°C. Cool to 45°–50°C. Aseptically add 533.0mL of sterile Schneider's *Drosophila* medium, 167.0mL of sterile fetal calf serum, 33.0mL of sterile fresh yeast extract solution, and 8.0mL of sterile penicillin solution. Mix thoroughly. Pour into sterile Petri dishes or distribute into sterile tubes.

Use: For the isolation and cultivation of *Spiroplasma* species that cause corn stunt.

mAB1 Medium

Composition per 1003.0mL:
NaCl 20.0g
$MgCl_2 \cdot 6H_2O$ 3.0g
Na_2SO_4 .... 3.0g
KCl 0.5g
NH_4Cl 0.25g
Yeast extract 0.2g
KH_2PO_4 .... 0.2g
Sodium benzoate 0.15g
$CaCl_2 \cdot 2H_2O$ 0.15g
Resazurin 1.0mg
Wolfe's vitamin solution 20.0mL
$NaHCO_3$ solution 10.0mL
$Na_2S \cdot 9H_2O$ solution 10.0mL
Na_2SeO_3/Na_2WO_4 solution 1.0mL
Sodium dithionite solution 1.0mL
Trace elements solution SL-10 .... 1.0mL
pH 7.2 ± 0.2 at 25°C

Wolfe's Vitamin Solution:
Composition per liter:
Pyridoxine·HCl 10.0mg
p-Aminobenzoic acid 5.0mg
Lipoic acid 5.0mg
Nicotinic acid 5.0mg
Riboflavin 5.0mg
Thiamine·HCl 5.0mg
Calcium DL-pantothenate 5.0mg
Biotin 2.0mg
Folic acid 2.0mg
Vitamin B_{12} 0.1mg

Preparation of Wolfe's Vitamin Solution: Add components to distilled/deionized water and bring volume to 1.0L. Mix thoroughly. Filter sterilize.

$NaHCO_3$ Solution:
Composition per 10.0mL:
$NaHCO_3$ 2.5g

Preparation of $NaHCO_3$ Solution: Add $NaHCO_3$ to distilled/deionized water and bring volume to 10.0mL. Mix thoroughly. Filter sterilize. Sparge with 80% N_2 + 20% CO_2.

$Na_2S \cdot 9H_2O$ Solution:
Composition per 10.0mL:
$Na_2S \cdot 9H_2O$ 0.3g

Preparation of $Na_2S \cdot 9H_2O$ Solution: Add $Na_2S \cdot 9H_2O$ to distilled/deionized water and bring volume to 10.0mL. Mix thoroughly. Sparge with 100% N_2. Autoclave for 15 min at 15 psi pressure–121°C.

Na_2SeO_3/Na_2WO_4 Solution:
Composition per liter:
NaOH 0.5g
$Na_2WO_4 \cdot 2H_2O$ 4.0mg
$Na_2SeO_3 \cdot 5H_2O$ 3.0mg

Preparation of Na_2SeO_3/Na_2WO_4 Solution: Add components to distilled/deionized water and bring volume to 1.0L. Mix thoroughly. Sparge with 100% N_2. Autoclave for 15 min at 15 psi pressure–121°C.

Sodium Dithionite Solution:
Composition per 10.0mL:
Sodium dithioninium 0.2g

Preparation of Sodium Dithionite Solution: Add sodium dithioninium to distilled/deionized water and bring volume to 10.0mL. Mix thoroughly. Sparge with 100% N_2. Autoclave for 15 min at 15 psi pressure–121°C.

Trace Elements Solution SL-10:
Composition per liter:

$FeCl_2·4H_2O$	1.5g
$CoCl_2·6H_2O$	190.0mg
$MnCl_2·4H_2O$	100.0mg
$ZnCl_2$	70.0mg
$Na_2MoO_4·2H_2O$	36.0mg
$NiCl_2·6H_2O$	24.0mg
H_3BO_3	6.0mg
$CuCl_2·2H_2O$	2.0mg
HCl (25% solution)	10.0mL

Preparation of Trace Elements Solution SL-10: Prepare and dispense under 80% N_2 + 20% CO_2. Add $FeCl_2·4H_2O$ to 10.0mL of HCl solution. Mix thoroughly. Add distilled/deionized water and bring volume to 1.0L. Add remaining components. Mix thoroughly. Sparge with 80% N_2 + 20% CO_2. Autoclave for 15 min at 15 psi pressure–121°C.

Preparation of Medium: Prepare medium and dispense under 80% N_2 + 20% CO_2. Add components, except Wolfe's vitamin solution, $NaHCO_3$ solution, sodium dithionite solution, $Na_2S·9H_2O$ solution, Na_2SeO_3/Na_2WO_4 solution, and trace elements solution SL-10, to distilled/deionized water and bring volume to 960.0mL. Mix thoroughly. Adjust pH to 7.2. Sparge with 80% N_2 + 20% CO_2. Autoclave for 15 min at 15 psi pressure–121°C. Aseptically and anaerobically add 20.0mL of Wolfe's vitamin solution, 10.0mL of sterile $NaHCO_3$ solution, 10.0mL of sterile $Na_2S·9H_2O$ solution, 1.0mL Na_2SeO_3/Na_2WO_4 solution, 1.0mL of sterile sodium dithionite solution, and 1.0mL of sterile trace elements solution SL-10. Mix thoroughly. Aseptically and anaerobically distribute into sterile tubes or flasks.

Use: For the cultivation of *Desulfotomaculum* species.

MacConkey Agar

Composition per liter:

Pancreatic digest of gelatin	17.0g
Agar	13.5g
Lactose	10.0g
NaCl	5.0g
Bile salts	1.5g
Pancreatic digest of casein	1.5g
Peptic digest of animal tissue	1.5g
Neutral Red	0.03g
Crystal Violet	1.0mg

pH 7.1 ± 0.2 at 25°C

Source: This medium is available as a premixed powder from BD Diagnostic Systems.

Preparation of Medium: Add components to distilled/deionized water and bring volume to 1.0L. Mix thoroughly. Gently heat while stirring until boiling. Autoclave for 15 min at 15 psi pressure–121°C. Pour into sterile Petri dishes or distribute into sterile tubes.

Use: For the selective isolation, cultivation, and differentiation of coliforms and enteric pathogens based on the ability to ferment lactose. Lactose-fermenting organisms appear as red to pink colonies. Lactose-nonfermenting organisms appear as colorless or transparent colonies.

MacConkey Agar

Composition per liter:

Peptone	20.0g
Agar	12.0g
Lactose	10.0g
Bile salts	5.0g
NaCl	5.0g
Neutral Red	0.075g

pH 7.4 ± 0.2 at 25°C

Source: This medium is available as a premixed powder from Oxoid Unipath.

Preparation of Medium: Add components to distilled/deionized water and bring volume to 1.0L. Mix thoroughly. Gently heat while stirring until boiling. Autoclave for 15 min at 15 psi pressure–121°C. Pour into sterile Petri dishes or distribute into sterile tubes.

Use: For the selective isolation, cultivation, and differentiation of coliforms and enteric pathogens based on the ability to ferment lactose. Lactose-fermenting organisms appear as red-to-pink colonies. Lactose-nonfermenting organisms appear as colorless or transparent colonies.

MacConkey Agar, CS

Composition per liter:

Peptone	17.0g
Agar	13.5g
Lactose	10.0g
NaCl	5.0g
Proteose peptone	3.0g
Bile salts	1.5g
Neutral Red	0.03g
Crystal Violet	1.0mg

pH 7.1 ± 0.2 at 25°C

Source: This medium is available as a prepared medium from BD Diagnostic Systems.

Preparation of Medium: Add components to distilled/deionized water and bring volume to 1.0L. Mix thoroughly. Gently heat while stirring until boiling. Autoclave for 15 min at 15 psi pressure–121°C. Pour into sterile Petri dishes or distribute into sterile tubes.

Use: For the cultivation and differentiation of lactose-fermenting and nonfermenting Gram-negative bacteria while also controlling the swarming of *Proteus* species, if present. Lactose-fermenting organisms appear as red-to-pink colonies. Lactose-nonfermenting organisms appear as colorless or transparent colonies.

MacConkey Broth

Composition per liter:

Pancreatic digest of gelatin	20.0g
Lactose	10.0g
Oxgall	5.0g
Bromcresol Purple	0.02g

pH 7.3 ± 0.2 at 25°C

Source: This medium is available as a premixed powder from BD Diagnostic Systems.

Preparation of Medium: Add components to distilled/deionized water and bring volume to 1.0L. If testing 10.0mL samples, prepare medium double strength. Mix thoroughly. Gently heat while stirring until boiling. Distribute into test tubes containing inverted Durham tubes. Autoclave for 15 min at 15 psi pressure–121°C.

Use: For the selective isolation and cultivation of coliforms in water.

MacConkey Broth

Composition per liter:

Peptone	20.0g
Lactose	10.0g

Bile salts 5.0g
NaCl 5.0g
Neutral Red 0.075g

pH 7.4 ± 0.2 at 25°C

Source: This medium is available as a premixed powder from Oxoid Unipath.

Preparation of Medium: Add components to distilled/deionized water and bring volume to 1.0L. If testing 10.0mL samples, prepare medium double strength. Mix thoroughly. Gently heat while stirring until boiling. Distribute into test tubes containing inverted Durham tubes. Autoclave for 15 min at 15 psi pressure–121°C.

Use: For the selective isolation and cultivation of coliforms in water.

MacConkey Broth, Purple

Composition per liter:

Peptone 20.0g
Lactose 10.0g
Bile salts 5.0g
NaCl 5.0g
Bromcresol Purple 0.01g

pH 7.4 ± 0.2 at 25°C

Source: This medium is available as a premixed powder or tablets from Oxoid Unipath.

Preparation of Medium: Add components to distilled/deionized water and bring volume to 1.0L. If testing 10.0mL samples, prepare medium double strength. Mix thoroughly. Gently heat while stirring until boiling. Distribute into test tubes containing inverted Durham tubes. Autoclave for 15 min at 15 psi pressure–121°C.

Use: For the selective isolation and cultivation of coliforms in water.

Magnetic *Spirillum* Growth Medium, Revised (MSGM, Revised)

Composition per liter:

Agar 1.3g
KH_2PO_4 .... 0.68g
Tartaric acid 0.37g
Succinic acid 0.37g
$NaNO_3$ 0.12g
Sodium acetate 0.05g
Ascorbic acid 0.035g
Wolfe's vitamin solution 10.0mL
Wolfe's mineral solution 5.0mL
Ferric quinate solution 2.0mL
Resazurin (0.1% solution) 0.45mL

pH 6.75 ± 0.2 at 25°C

Wolfe's Vitamin Solution:

Composition per liter:

Pyridoxine·HCl 10.0mg
Thiamine·HCl 5.0mg
Riboflavin 5.0mg
Nicotinic acid 5.0mg
Calcium pantothenate 5.0mg
p-Aminobenzoic acid 5.0mg
Thioctic acid 5.0mg
Biotin 2.0mg
Folic acid 2.0mg
Cyanocobalamin 100.0μg

Preparation of Wolfe's Vitamin Solution: Add components to distilled/deionized water and bring volume to 1.0L. Mix thoroughly.

Wolfe's Mineral Solution:

Composition per liter:

$MgSO_4 \cdot 7H_2O$ 3.0g
Nitriloacetic acid 1.5g
NaCl 1.0g
$MnSO_4 \cdot H_2O$ 0.5g
$FeSO_4 \cdot 7H_2O$ 0.1g
$CoCl_2 \cdot 6H_2O$ 0.1g
$CaCl_2$ 0.1g
$ZnSO_4 \cdot 7H_2O$ 0.1g
$CuSO_4 \cdot 5H_2O$ 0.01g
$AlK(SO_4)_2 \cdot 12H_2O$ 0.01g
H_3BO_3 .... 0.01g
$Na_2MoO_4 \cdot 2H_2O$ 0.01g

Preparation of Wolfe's Mineral Solution: Add nitrilotriacetic acid to 500.0mL of distilled/deionized water. Dissolve by adjusting pH to 6.5 with KOH. Add remaining components. Add distilled/deionized water to 1.0L.

Ferric Quinate Solution:

Composition per 100.0mL:

$FeCl_3$ 0.27g
Quinic acid 0.19g

Preparation of Ferric Quinate Solution: Add components to distilled/deionized water and bring volume to 1.0L. Mix thoroughly. Autoclave for 15 min at 15 psi pressure–121°C.

Preparation of Medium: To 1.0L of distilled/deionized water add components in the following order: Wolfe's vitamin solution, Wolfe's mineral solution, ferric quinate solution, resazurin, KH_2PO_4, $NaNO_3$, ascorbic acid, tartaric acid, succinic acid, sodium acetate, and agar. Mix thoroughly after each addition. Adjust pH to 6.75 with NaOH. Autoclave for 15 min at 15 psi pressure–121°C. Aseptically distribute into sterile screw-capped tubes. Fill tubes to capacity with medium. Use a heavy inoculum in each tube and do not introduce a headspace of air. Screw down caps tightly.

Use: For the cultivation and maintenance of *Aquaspirillum magnetotacticum*.

Magnetospirillum Medium (DSMZ Medium 380)

Composition per liter:

KH_2PO_4 .... 0.68g
L(+)-Tartaric acid 0.37g
Succinic acid 0.37g
$NaNO_3$ 0.12g
Na-thioglycolate 0.05g
Na-acetate 0.05g
Resazurin 0.5mg
Vitamin solution 10.0mL
Trace elements solution 5.0mL
Ferric quinate solution 2.0mL

pH 6.8 ± 0.2 at 25°C

Trace Elements Solution:

Composition per liter:

$MgSO_4 \cdot 7H_2O$ 3.0g
Nitrilotriacetic acid 1.5g
NaCl 1.0g
$MnSO_4 \cdot 2H_2O$ 0.5g
$CoSO_4 \cdot 7H_2O$ 0.18g
$ZnSO_4 \cdot 7H_2O$ 0.18g

$CaCl_2·2H_2O$ 0.1g
$FeSO_4·7H_2O$ 0.1g
$NiCl_2·6H_2O$ 0.025g
$KAl(SO_4)_2·12H_2O$ 0.02g
H_3BO_3 ...... 0.01g
$Na_2MoO_4·4H_2O$ 0.01g
$CuSO_4·5H_2O$ 0.01g
$Na_2SeO_3·5H_2O$ 0.3mg

Preparation of Trace Elements Solution: Add nitrilotriacetic acid to 500.0mL of distilled/deionized water. Dissolve by adjusting pH to 6.5 with KOH. Add remaining components. Add distilled/deionized water to 1.0L. Mix thoroughly.

Vitamin Solution:
Composition per liter:
Pyridoxine-HCl 10.0mg
Thiamine-HCl·$2H_2O$ 5.0mg
Riboflavin 5.0mg
Nicotinic acid 5.0mg
D-Ca-pantothenate 5.0mg
p-Aminobenzoic acid 5.0mg
Lipoic acid 5.0mg
Biotin 2.0mg
Folic acid 2.0mg
Vitamin B_{12} 0.1mg

Preparation of Vitamin Solution: Add components to distilled/deionized water and bring volume to 1.0L. Mix thoroughly. Sparge with 80% H_2 + 20% CO_2. Filter sterilize.

Ferric Quinate Solution:
Composition per 100.0mL:
$FeCl_3·6H_2O$ 0.45g
Quinic acid 0.19g

Preparation of Ferric Quinate Solution: Add components to distilled/deionized water and bring volume to 100.0mL. Sparge with N_2. Autoclave for 15 min at 15 psi pressure–121°C. Cool to 25°C.

Preparation of Medium: Add components, except vitamin solution and ferric quinate solution, to distilled/deionized water and bring volume to 988.0mL. Purge medium with N_2 gas for 10 min. Mix thoroughly. Autoclave for 15 min at 15 psi pressure–121°C. Cool to 25°C. Aseptically and anerobically add 10.0mL vitamin solution and 2.0mL ferric quinate solution. Mix thoroughly. Purge medium with N_2 gas for 10 min. Under the same atmosphere, anaerobically fill tubes to 1/3 of their volume and seal. Autoclave at 121°C for 15 min. Before inoculation, add sterile air (with hypodermic syringe through the rubber closure) to 1% O_2 concentration in the gas phase.

Use: For the cultivation of *Magnetospirillum magnetotacticum=Aquaspirillum magnetotacticum,* and *Magnetospirillum gryphiswaldense.*

Magnetospirillum 2 Medium

Composition per liter:
Sodium acetate 1.0g
K_2HPO_4 ...... 0.5g
Sodium thioglycolate 0.5g
NH_4Cl 0.1g
Yeast extract 0.1g
Ferric citrate 20.0μg
pH 6.8 ± 0.2 at 25°C

Preparation of Medium: Add components to distilled/deionized water and bring volume to 1.0L. Mix thoroughly. Distribute into screw-capped tubes in 5.0mL volumes. Autoclave for 15 min at 15 psi pressure–121°C. Allow medium to stand upright at room temperature for 2 to 3 days before inoculation. Do not shake.

Use: For the cultivation and maintenance of *Magnetospirillum gryphiswaldense.*

Magnetospirillum Semi-Solid Medium (DSMZ Medium 380)

Composition per liter:
Agar 1.3g
KH_2PO_4 ...... 0.68g
L(+)-Tartaric acid 0.37g
Succinic acid 0.37g
$NaNO_3$ 0.12g
Na-thioglycolate 0.05g
Na-acetate 0.05g
Resazurin 0.5mg
Vitamin solution 10.0mL
Trace elements solution 5.0mL
Ferric quinate solution 2.0mL
pH 6.8 ± 0.2 at 25°C

Trace Elements Solution:
Composition per liter:
$MgSO_4·7H_2O$ 3.0g
Nitrilotriacetic acid 1.5g
NaCl 1.0g
$MnSO_4·2H_2O$ 0.5g
$CoSO_4·7H_2O$ 0.18g
$ZnSO_4·7H_2O$ 0.18g
$CaCl_2·2H_2O$ 0.1g
$FeSO_4·7H_2O$ 0.1g
$NiCl_2·6H_2O$ 0.025g
$KAl(SO_4)_2·12H_2O$ 0.02g
H_3BO_3 ...... 0.01g
$Na_2MoO_4·4H_2O$ 0.01g
$CuSO_4·5H_2O$ 0.01g
$Na_2SeO_3·5H_2O$ 0.3mg

Preparation of Trace Elements Solution: Add nitrilotriacetic acid to 500.0mL of distilled/deionized water. Dissolve by adjusting pH to 6.5 with KOH. Add remaining components. Add distilled/deionized water to 1.0L. Mix thoroughly.

Vitamin Solution:
Composition per liter:
Pyridoxine-HCl 10.0mg
Thiamine-HCl·$2H_2O$ 5.0mg
Riboflavin 5.0mg
Nicotinic acid 5.0mg
D-Ca-pantothenate 5.0mg
p-Aminobenzoic acid 5.0mg
Lipoic acid 5.0mg
Biotin 2.0mg
Folic acid 2.0mg
Vitamin B_{12} 0.1mg

Preparation of Vitamin Solution: Add components to distilled/deionized water and bring volume to 1.0L. Mix thoroughly. Sparge with 80% H_2 + 20% CO_2. Filter sterilize.

Ferric Quinate Solution:
Composition per 100.0mL:
$FeCl_3·6H_2O$ 0.45g
Quinic acid 0.19g

Preparation of Ferric Quinate Solution: Add components to distilled/deionized water and bring volume to

100.0mL. Sparge with N_2. Autoclave for 15 min at 15 psi pressure–121°C. Cool to 45°C.

Preparation of Medium: Add components, except vitamin solution and ferric quinate solution, to distilled/deionized water and bring volume to 988.0mL. Purge medium with N_2 gas for 10 min. Mix thoroughly. Autoclave for 15 min at 15 psi pressure–121°C. Cool to 45°C. Aseptically and anerobically add 10.0mL vitamin solution and 2.0mL ferric quinate solution. Mix thoroughly. Purge medium with N_2 gas for 10 min. Dispense 12mL of medium per 16 x 150mm anaerobe screw cap tube under N_2 gas. Prior to inoculation, remove caps briefly under air, tighten the caps again and wait several hours to establish oxygen gradients. The medium should be slightly pink in color. Strongly reduced conditions will not support growth of the organism. During growth O_2 will be consumed, resazurin decolorized, and the pH increased. Feed oxygen (by adding air) and succinic acid from sterile 0.05*M* solution (to maintain pH below 7.0). If higher densities of magnetic cell are wanted, ferric quinate also can be fed. For transfer use cell material which has been concentrated at the glass wall of the culture vessel by means of a magnetic rod attached outside.

Use: For the cultivation of *Magnetospirillum magnetotacticum=Aquaspirillum magnetotacticum,* and *Magnetospirillum gryphiswaldense.*

Malachite Green Broth

Composition per liter:

Peptone	15.0g
Beef extract	9.0g
Malachite Green	0.01mg

pH 7.3 ± 0.2 at 25°C

Preparation of Medium: Add components to distilled/deionized water and bring volume to 1.0L. Mix thoroughly. Distribute into tubes or flasks. Autoclave for 15 min at 15 psi pressure–121°C.

Use: For the cultivation of *Pseudomonas aeruginosa.*

Malonate Broth

Composition per liter:

Sodium malonate	3.0g
NaCl	2.0g
$(NH_4)_2SO_4$	2.0g
K_2HPO_4	0.6g
KH_2PO_4	0.4g
Bromothymol Blue	0.025g

pH 6.7 ± 0.2 at 25°C

Source: This medium is available as a premixed powder from BD Diagnostic Systems.

Preparation of Medium: Add components to distilled/deionized water and bring volume to 1.0L. Mix thoroughly. Distribute into tubes or flasks. Autoclave for 15 min at 15 psi pressure–121°C. Avoid introduction of carbon and nitrogen from other sources.

Use: For the cultivation and differentiation of coliforms and other enteric organisms, particularly *Enterobacter* and *Escherichia,* based on their ability to utilize malonate as the sole carbon source and ammonium sulfate as the sole nitrogen source. Malonate-utilizing organisms turn the medium blue.

Malonate Broth, Ewing Modified

Composition per liter:

Sodium malonate	3.0g
NaCl	2.0g
$(NH_4)_2SO_4$	2.0g
Yeast extract	1.0g
Glucose	0.25g
K_2HPO_4	0.6g
KH_2PO_4	0.4g
Bromothymol Blue	0.025g

pH 6.7 ± 0.2 at 25°C

Source: This medium is available as a premixed powder from BD Diagnostic Systems.

Preparation of Medium: Add components to distilled/deionized water and bring volume to 1.0L. Mix thoroughly. Distribute into tubes or flasks. Autoclave for 15 min at 15 psi pressure–121°C.

Use: For the cultivation and differentiation of coliforms and other enteric organisms, particularly *Enterobacter* and *Escherichia,* based on their ability to utilize malonate as a carbon source and ammonium sulfate as a nitrogen source. The small amount of yeast extract and glucose encourages the growth of some organisms that may be distressed or fail to respond. Malonate-utilizing organisms turn the medium blue.

Malt Agar

Composition per liter:

Malt extract	30.0g
Agar	15.0g

pH 5.5 ± 0.2 at 25°C

Source: This medium is available as a premixed powder from BD Diagnostic Systems.

Preparation of Medium: Add components to distilled/deionized water and bring volume to 1.0L. Mix thoroughly. Gently heat while stirring until boiling. Distribute into tubes or flasks. Autoclave for 15 min at 15 psi pressure–118°C. Do not overheat or agar will not harden. Pour into sterile Petri dishes or distribute into sterile tubes.

Use: For the cultivation of yeasts and molds.

Manganese Acetate Agar

Composition per liter:

Agar, highly purified	10.0g
Manganous acetate	0.1g

pH 7.0 ± 0.2 at 25°C

Preparation of Medium: Add manganous acetate to distilled/deionized water and bring volume to 1.0L. Mix thoroughly. Adjust pH to 7.0. Add agar. Steam the medium to dissolve agar. Distribute into screw-capped tubes or bottles. Autoclave for 15 min at 15 psi pressure–121°C. Allow tubes or bottles to cool in a slanted position.

Use: For the cultivation of manganese-oxidizing bacteria.

Manganese Agar No. 1 (Mn Agar No. 1)

Composition per liter:

Agar	10.0g
$MnCO_3$	2.0g
Beef extract	1.0g
$Fe(NH_4)_2(SO_4)_2$	0.15g
Sodium citrate	0.15g
Yeast extract	0.075g
Cyanocobalamin solution	10.0mL

Cyanocobalamin Solution:
Composition per 10.0mL:

Cyanocobalamin	0.005mg

Preparation of Cyanocobalamin Solution: Add cyanocobalamin to distilled/deionized water and bring volume to 10.0mL. Mix thoroughly. Filter sterilize.

Preparation of Medium: Add components, except cyanocobalamin, to distilled/deionized water and bring volume to 990.0mL. Mix thoroughly. Autoclave for 15 min at 15 psi pressure–121°C. Cool to 45°–50°C. Aseptically add 10.0mL of the sterile cyanocobalamin solution. Mix thoroughly. Pour into sterile Petri dishes or distribute into sterile tubes.

Use: For the isolation and cultivation of iron and sulfur bacteria. Also used to differentiate *Leptothrix (Sphaerotilus) discophorus* from *Sphaerotilus natans*.

Manganese Agar No. 2 (Mn Agar No. 2)

Composition per liter:

Agar	15.0g
$MnSO_4 \cdot H_2O$	10.0mg

Preparation of Medium: Add components to distilled/deionized water and bring volume to 1.0L. Mix thoroughly. Gently heat and bring to boiling. Distribute into tubes or flasks. Autoclave for 15 min at 15 psi pressure–121°C. Pour into sterile Petri dishes or leave in tubes. Use freshly prepared solution.

Use: For the enumeration, enrichment, and isolation of iron and sulfur bacteria. For the isolation and cultivation of *Leptothrix* species from water.

Manganese Medium for *Pseudomonas* species

Composition per liter:

Noble agar	10.0g
$MnCO_3$	1.0g
$Fe(NH_4)_2(SO_4)_2 \cdot 6H_2O$	0.15g
Sodium citrate	0.15g
Yeast extract	0.075g
$Na_4P_2O_7 \cdot 10H_2O$	0.05g

pH 6.8 ± 0.2 at 25°C

Preparation of Medium: Add components to distilled/deionized water and bring volume to 1.0L. Mix thoroughly. Gently heat and bring to boiling. Distribute into tubes or flasks. Autoclave for 15 min at 15 psi pressure–121°C. Pour into sterile Petri dishes or leave in tubes.

Use: For the cultivation and maintenance of *Pseudomonas putida* and other *Pseudomonas* species.

Mannitol Yeast Extract Medium (LMG 135)

Composition per liter:

Agar	20.0g
Mannitol	10.0g
Yeast extract	1.0g
KH_2PO_4	0.5g
NaCl	0.1g
$CaCl_2 \cdot 2H_2O$ solution	1.0mL
$FeCl_3 \cdot 6H_2O$ solution	1.0mL

$CaCl_2 \cdot 2H_2O$ Solution:

Composition per 10.0mL:

$CaCl_2 \cdot 2H_2O$	5.28mg

Preparation of $CaCl_2 \cdot 2H_2O$ Solution: Add 5.28g of $CaCl_2 \cdot 2H_2O$ to distilled/deionized water and bring volume to 10.0mL. Mix thoroughly.

$FeCl_3 \cdot 6H_2O$ Solution:

Composition per 10.0mL:

$FeCl_3 \cdot 6H_2O$	0.66mg

Preparation of $FeCl_3 \cdot 6H_2O$ Solution: Add $FeCl_3 \cdot 6H_2O$ to distilled/deionized water and bring volume to 10.0mL. Mix thoroughly.

Preparation of Medium: Add components to distilled/deionized water and bring volume to 1.0L. Mix thoroughly. Gently heat and bring to boiling. Distribute into tubes or flasks. Autoclave for 15 min at 15 psi pressure–121°C. Pour into sterile Petri dishes or leave in tubes.

Use: For the cultivation and maintenance of *Rhizobium* species and *Bradyrhizobium* species.

Marine Agar (DSMZ Medium 123)

Composition per liter:

Agar	15.0g
Tryptone	10.0g
Peptone	5.0g
Yeast extract	1.0g
Synthetic seawater	1.0L

pH 7.8 ± 0.2 at 25°C

Synthetic Seawater:

Composition per liter:

NaCl	24.0g
$MgCl_2 \cdot 6H_2O$	11.0g
Na_2SO_4	4.0g
$CaCl_2 \cdot 6H_2O$	2.0g
KCl	0.7g
KBr	0.1g
$SrCl_2 \cdot 6H_2O$	0.04g
H_3BO_3	0.03g
$NaSiO_3 \cdot 9H_2O$	5.0mg
NaF	3.0mg
NH_4NO_3	2.0mg
$Fe_3PO_4 \cdot 4H_2O$	1.0mg

Preparation of Synthetic Seawater: Add components to distilled water and bring volume to 1.0L. Mix thoroughly.

Preparation of Medium: Add agar, tryptone, peptone, and yeast extract to synthetic seawater and bring volume to 1.0L. Mix thoroughly. Adjust pH to 7.8. Gently heat and bring to boiling. Distribute into tubes or flasks. Autoclave for 15 min at 15 psi pressure–121°C. Pour into sterile Petri dishes or leave in tubes.

Use: For the cultivation and maintenance of *Halobacillus halophilus, Halomonas spp., Vibrio harveyi, Cobetia marina,* and *Ruegeria atlantica.*

Marine Agar 2216 (DSMZ Medium 604)

Composition per liter:

NaCl	19.45g
Agar	15.0g
$MgCl_2$	8.8g
Peptone	5.0g
Na_2SO_3	3.24g
$CaCl_2$	1.8g
Yeast extract	1.0g
KCl	0.55g
$NaHCO_3$	0.16g
Ferric citrate	0.1g
KBr	0.08g
$SrCl_2$	0.03g
H_3BO_3	0.02g

Na_2HPO_4 .... 8.0mg
Na_2SiO_3 .... 4.0mg
NaF 2.4mg
NH_4NO_3 .... 1.6mg
pH 7.6 ± 0.2 at 25°C

Source: This medium is available as a premixed powder from BD Diagnostic Systems.

Preparation of Medium: Add components to distilled/deionized water and bring volume to 1.0L. Mix thoroughly. Gently heat while stirring and bring to boiling. Distribute into tubes or flasks. Autoclave for 15 min at 15 psi pressure–121°C. Pour into sterile Petri dishes or leave in tubes.

Use: For the cultivation and maintenance of *Hyphomonas* spp., *Oceanospirillum* spp., *Hyphomicrobium indicum, Psychroflexus gondwanensis=Flavobacterium gondwanense, Salegentibacter salegens=Flavobacterium salegens, Psychromonas antarctica, Sulfitobacter mediterraneus, Thalassomonas viridans, Vibrio* spp., *Marinospirillum minutulum=Oceanospirillum minutulum, Terasakiella pusilla=Oceanospirillum pusillum, Pseudoalteromonas atlantica=Alteromonas atlantica, Pseudomonas atlantica, Roseobacter* spp., *Erythrobacter longus, Pseudospirillum japonicum=Oceanospirillum japonicum, Marinobacter hydrocarbonoclasticus (Pseudomonas nautica), Psychrobacter spp.,* and *Moritella japonica.* For the isolation, cultivation, and maintenance of a wide variety of heterotrophic marine bacteria.

Marine Agar with Biphenyl

Composition per liter:
NaCl 19.45g
Agar 15.0g
$MgCl_2$ 8.8g
Peptone 5.0g
Na_2SO_3 .... 3.24g
$CaCl_2$ 1.8g
Yeast extract 1.0g
KCl 0.55g
$NaHCO_3$ 0.16g
Ferric citrate 0.1g
KBr 0.08g
$SrCl_2$ 0.03g
H_3BO_3 .... 0.02g
Na_2HPO_4 .... 8.0mg
Na_2SiO_3 .... 4.0mg
NaF 2.4mg
NH_4NO_3 .... 1.6mg
Biphenyl 1.0mg
pH 7.6 ± 0.2 at 25°C

Preparation of Medium: Add components, except biphenyl, to distilled/deionized water and bring volume to 1.0L. Mix thoroughly. Gently heat and bring to boiling. Distribute into tubes or flasks. Autoclave for 15 min at 15 psi pressure–121°C. Pour into sterile Petri dishes or leave in tubes. After agar solidifies, aseptically add a few crystals of biphenyl to each plate.

Use: For the cultivation and maintenance of biphenyl-utilizing marine bacteria, such as *Cycloclasticus pugetii,*

Marine Agar with κ- and λ-Carrageenan

Composition per 1070.0mL:
Solution A 1.0L
Solution B 60.0mL
Solution C 10.0mL
pH 7.2 ± 0.2 at 25°C

Solution A:
Composition per liter:
NaCl 25.0g
Agar 15.0g
$MgSO_4 \cdot 7H_2O$ 5.0g
Casamino acids 2.5g
$NaNO_3$ 2.0g
κ-Carrageenan 1.25g
λ-Carrageenan 1.25g
$CaCl_2 \cdot 2H_2O$ 0.2g
KCl 0.1g

Preparation of Solution A: Add components to distilled/deionized water and bring volume to 1.0L. Mix thoroughly. Gently heat and bring to boiling. Autoclave for 15 min at 15 psi pressure–121°C.

Solution B:
Composition per 100.0mL:
$Na_2HPO_4 \cdot 2H_2O$ 3.56g

Preparation of Solution B: Add component to distilled/deionized water and bring volume to 100.0mL. Mix thoroughly. Autoclave for 15 min at 15 psi pressure–121°C.

Solution C:
Composition per 100.0mL:
$FeSO_4 \cdot 7H_2O$ 0.3g

Preparation of Solution C: Add component to distilled/deionized water and bring volume to 100.0mL. Mix thoroughly. Autoclave for 15 min at 15 psi pressure–121°C.

Preparation of Medium: Aseptically add 60.0mL of sterile solution B and 10.0mL of sterile solution C to 1.0L of sterile solution A. Mix thoroughly. Pour into sterile Petri dishes or distribute into sterile tubes.

Use: For the cultivation and maintenance of *Pseudomonas carrageenovora.*

Marine Agar with Naphthalene

Composition per liter:
NaCl 19.45g
Agar 15.0g
$MgCl_2$ 8.8g
Peptone 5.0g
Na_2SO_3 .... 3.24g
$CaCl_2$ 1.8g
Yeast extract 1.0g
KCl 0.55g
$NaHCO_3$ 0.16g
Ferric citrate 0.1g
KBr 0.08g
$SrCl_2$ 0.03g
H_3BO_3 .... 0.02g
Na_2HPO_4 .... 8.0mg
Na_2SiO_3 .... 4.0mg
NaF 2.4mg
NH_4NO_3 .... 1.6mg
Naphthalene 1mg
pH 7.6 ± 0.2 at 25°C

Preparation of Medium: Add components, except naphthalene, to distilled/deionized water and bring volume to 1.0L. Mix thoroughly. Gently heat and bring to boiling. Distribute into tubes or flasks. Autoclave for 15 min at 15 psi pressure–121°C. Pour into sterile Petri dishes or leave in tubes. After agar solidifies, aseptically add a few crystals of naphthalene to each plate.

Use: For the cultivation and maintenance of naphthalene-utilizing marine bacteria.

Marine Agar with Sulfur (ATCC Medium 1922)

Composition per liter:

NaCl	19.45g
Sulfur	10.0g
$MgCl_2$	8.8g
Peptone	5.0g
Na_2SO_3	3.24g
$CaCl_2$	1.8g
Yeast extract	1.0g
KCl	0.55g
$NaHCO_3$	0.16g
Ferric citrate	0.1g
KBr	0.08g
$SrCl_2$	0.03g
H_3BO_3	0.02g
Na_2HPO_4	8.0mg
Na_2SiO_3	4.0mg
NaF	2.4mg
NH_4NO_3	1.6mg

pH 7.6 ± 0.2 at 25°C

Preparation of Sulfur: Autoclave sulfur for 15 min at 0 psi pressure–100°C on 3 successive days.

Preparation of Medium: Prepare anaerobically under a gas phase of 80% N_2 + 10% CO_2 + 10% H_2. Add components, except sulfur, to distilled/deionized water and bring volume to 1.0L. Mix thoroughly. Gently heat while stirring and bring to boiling. Autoclave for 15 min at 15 psi pressure–121°C. Cool to 50–55°C. Aseptically add 10.0g of sterile sulfur. Mix thoroughly. Aseptically and anaerobically, under a gas phase of 80% N_2 + 10% CO_2 + 10% H_2, distribute into sterile tubes.

Use: For the cultivation and maintenance of *Thermococcus litoralis.*

Marine Broth 2216 (LMG Medium 164)

Composition per liter:

NaCl	19.45g
$MgCl_2$	8.8g
Peptone	5.0g
Na_2SO_3	3.24g
$CaCl_2$	1.8g
Yeast extract	1.0g
KCl	0.55g
$NaHCO_3$	0.16g
Ferric citrate	0.1g
KBr	0.08g
$SrCl_2$	0.03g
H_3BO_3	0.02g
Na_2HPO_4	8.0mg
Na_2SiO_3	4.0mg
NaF	2.4mg
NH_4NO_3	1.6mg

pH 7.6 ± 0.2 at 25°C

Source: This medium is available as a premixed powder from BD Diagnostic Systems.

Preparation of Medium: Add components to distilled/deionized water and bring volume to 1.0L. Mix thoroughly. Gently heat while stirring and bring to boiling. Distribute into tubes or flasks. Autoclave for 15 min at 15 psi pressure–121°C.

Use: For the cultivation of *Vibrio liquefaciens* and for the isolation, cultivation, and maintenance of a wide variety of heterotrophic marine bacteria.

Marine Broth with Biphenyl

Composition per liter:

NaCl	19.45g
$MgCl_2$	8.8g
Peptone	5.0g
Na_2SO_3	3.24g
$CaCl_2$	1.8g
Yeast extract	1.0g
KCl	0.55g
$NaHCO_3$	0.16g
Ferric citrate	0.1g
KBr	0.08g
$SrCl_2$	0.03g
H_3BO_3	0.02g
Na_2HPO_4	8.0mg
Na_2SiO_3	4.0mg
NaF	2.4mg
NH_4NO_3	1.6mg
Biphenyl	1.0mg

pH 7.6 ± 0.2 at 25°C

Preparation of Medium: Add components, except biphenyl, to distilled/deionized water and bring volume to 1.0L. Mix thoroughly. Distribute into tubes or flasks. Autoclave for 15 min at 15 psi pressure–121°C. Aseptically add a few crystals of biphenyl to each tube or flask.

Use: For the cultivation of biphenyl-utilizing marine bacteria.

Marine Broth with κ- and λ-Carrageenan

Composition per 1070.0mL:

Solution A	1.0L
Solution B	60.0mL
Solution C	10.0mL

pH 7.2 ± 0.2 at 25°C

Solution A:

Composition per liter:

NaCl	25.0g
$MgSO_4 \cdot 7H_2O$	5.0g
Casamino acids	2.5g
$NaNO_3$	2.0g
κ-Carrageenan	1.25g
λ-Carrageenan	1.25g
$CaCl_2 \cdot 2H_2O$	0.2g
KCl	0.1g

Preparation of Solution A: Add components to distilled/deionized water and bring volume to 1.0L. Mix thoroughly. Gently heat and bring to boiling. Autoclave for 15 min at 15 psi pressure–121°C.

Solution B:

Composition per 100.0mL:

$Na_2HPO_4 \cdot 2H_2O$	3.56g

Preparation of Solution B: Add component to distilled/deionized water and bring volume to 100.0mL. Mix thoroughly. Autoclave for 15 min at 15 psi pressure–121°C.

Solution C:

Composition per 100.0mL:

$FeSO_4 \cdot 7H_2O$	0.3g

Preparation of Solution C: Add component to distilled/deionized water and bring volume to 100.0mL. Mix thoroughly. Autoclave for 15 min at 15 psi pressure–121°C.

Preparation of Medium: Aseptically add 60.0mL of sterile solution B and 10.0mL of sterile solution C to 1.0L of sterile solution A. Mix thoroughly. Distribute into sterile tubes or flasks.

Use: For the cultivation and maintenance of *Pseudomonas carrageenovora.*

Marine Broth with Naphthalene

Composition per liter:

NaCl	19.45g
$MgCl_2$	8.8g
Peptone	5.0g
Na_2SO_3	3.24g
$CaCl_2$	1.8g
Yeast extract	1.0g
KCl	0.55g
$NaHCO_3$	0.16g
Ferric citrate	0.1g
KBr	0.08g
$SrCl_2$	0.03g
H_3BO_3	0.02g
Na_2HPO_4	8.0mg
Na_2SiO_3	4.0mg
NaF	2.4mg
NH_4NO_3	1.6mg
Naphthalene	1mg

pH 7.6 ± 0.2 at 25°C

Preparation of Medium: Add components, except biphenyl, to distilled/deionized water and bring volume to 1.0L. Mix thoroughly. Distribute into tubes or flasks. Autoclave for 15 min at 15 psi pressure–121°C. Aseptically add a few crystals of naphthalene to each tube or flask.

Use: For the cultivation of naphthalene-utilizing marine bacteria.

Marine Broth with Sulfur

Composition per liter:

NaCl	19.45g
Sulfur	10.0g
$MgCl_2$	8.8g
Peptone	5.0g
Na_2SO_3	3.24g
$CaCl_2$	1.8g
Yeast extract	1.0g
KCl	0.55g
$NaHCO_3$	0.16g
Ferric citrate	0.1g
KBr	0.08g
$SrCl_2$	0.03g
H_3BO_3	0.02g
Na_2HPO_4	8.0mg
Na_2SiO_3	4.0mg
NaF	2.4mg
NH_4NO_3	1.6mg

pH 7.6 ± 0.2 at 25°C

Preparation of Sulfur: Autoclave for 15 min at 0 psi pressure–100°C on 3 successive days.

Preparation of Medium: Prepare anaerobically under a gas phase of 80% N_2 + 10% CO_2 + 10% H_2. Add components, except sulfur, to distilled/deionized water and bring volume to 1.0L. Mix thoroughly. Gently heat while stirring and bring to boiling. Autoclave for 15 min at 15 psi pressure–121°C. Cool to 50°C. Aseptically add 10.0g of sulfur. Mix thoroughly. Aseptically and anaerobically, under a gas phase of 80% N_2 + 10% CO_2 + 10% H_2, distribute into sterile tubes.

Use: For the cultivation of *Thermococcus litoralis.*

Marine *Caulobacter* Medium

Composition per liter:

Proteose peptone	10.0g
Yeast extract	3.0g
Artificial seawater	1.0L

pH 7.2–7.4 at 25°C

Artificial Seawater:
Composition per liter:

Commercially available marine aquarium salts mixture	variable

Preparation of Artificial Seawater: Add commercially available marine aquarium salts mixture to distilled/deionized water and bring volume to 1.0L. Mix thoroughly.

Preparation of Medium: Combine components. Mix thoroughly. Distribute into tubes or flasks. Autoclave for 15 min at 15 psi pressure–121°C.

Use: For the cultivation of *Caulobacter halobacteroides* and *Caulobacter maris.*

Marine Chlorobiaceae Medium 2

Composition per 1051.0mL:

Solution 1	950.0mL
$Na_2S·9H_2O$ solution	60.0mL
$NaHCO_3$ solution	40.0mL
Vitamin B_{12} solution	1.0mL

pH 6.8 ± 0.2 at 25°C

Solution 1:
Composition per 950.0mL:

NaCl	20.0g
$MgSO_4·7H_2O$	3.0g
KH_2PO_4	1.0g
NH_4Cl	0.5g
$CaCl_2·2H_2O$	0.05g
Trace elements solution SL-8	1.0mL

Preparation of Solution 1: Add components to distilled/deionized water and bring volume to 950.0mL. Mix thoroughly. Autoclave for 15 min at 15 psi pressure–121°C. Cool to 45°–50°C.

Trace Elements Solution SL-8:
Composition per liter:

Disodium EDTA	5.2g
$FeCl_2·4H_2O$	1.5g
$CoCl_2·6H_2O$	0.19g
$MnCl_2·4H_2O$	0.1g
$ZnCl_2$	0.07g
H_3BO_3	0.06g
$NaMoO_4·2H_2O$	0.04g
$CuCl_2·2H_2O$	0.02g
$NiCl_2·6H_20$	0.02g

Preparation of Trace Elements Solution SL-8: Add components to distilled/deionized water and bring volume to 1.0L. Mix thoroughly.

$Na_2S·9H_2O$ Solution:
Composition per 100.0mL:

$Na_2S·9H_2O$	5.0g

Preparation of $Na_2S·9H_2O$ Solution: Add $Na_2S·9H_2O$ to distilled/deionized water and bring volume to 100.0mL. Autoclave for 15 min at 15 psi pressure–121°C. Cool to 45°–50°C.

$NaHCO_3$ Solution:
Composition per 100.0mL:

$NaHCO_3$	5.0g

Preparation of $NaHCO_3$ Solution: Add $NaHCO_3$ to distilled/deionized water and bring volume to 100.0mL. Mix thoroughly. Filter sterilize.

Vitamin B_{12} Solution:
Composition per 100.0mL:

Vitamin B_{12}	2.0mg

Preparation of Vitamin B_{12} Solution: Add Vitamin B_{12} to distilled/deionized water and bring volume to 100.0mL. Mix thoroughly. Filter sterilize.

Preparation of Medium: To 950.0mL of cooled, sterile solution 1, aseptically add 60.0mL of sterile $Na_2S \cdot 9H_2O$ solution, 40.0mL of sterile $NaHCO_3$ solution, and 1.0mL of sterile Vitamin B_{12} solution. Mix thoroughly. Adjust pH to 6.8 with sterile H_2SO_4 or Na_2CO_3. Aseptically distribute into sterile 50.0mL or 100.0mL bottles with metal screw-caps and rubber seals. Completely fill bottles with medium except for a pea-sized air bubble.

Use: For the isolation and cultivation of marine members of the Chlorobiaceae.

Marine Chromatiaceae Medium 2

Composition per 1051.0mL:

Solution 1	950.0mL
$Na_2S \cdot 9H_2O$ solution	60.0mL
$NaHCO_3$ solution	40.0mL
Vitamin B_{12} solution	1.0mL

pH 7.3 ± 0.2 at 25°C

Solution 1:
Composition per 950.0mL:

NaCl	20.0g
$MgSO_4 \cdot 7H_2O$	3.0g
KH_2PO_4	1.0g
NH_4Cl	0.5g
$CaCl_2 \cdot 2H_2O$	0.05g
Trace elements solution SL-8	1.0mL

Preparation of Solution 1: Add components to distilled/deionized water and bring volume to 950.0mL. Mix thoroughly. Autoclave for 15 min at 15 psi pressure–121°C. Cool to 45°–50°C.

Trace Elements Solution SL-8:
Composition per liter:

Disodium EDTA	5.2g
$FeCl_2 \cdot 4H_2O$	1.5g
$CoCl_2 \cdot 6H_2O$	0.19g
$MnCl_2 \cdot 4H_2O$	0.1g
$ZnCl_2$	0.07g
H_3BO_3	0.06g
$NaMoO_4 \cdot 2H_2O$	0.04g
$CuCl_2 \cdot 2H_2O$	0.02g
$NiCl_2 \cdot 6H_20$	0.02g

Preparation of Trace Elements Solution SL-8: Add components to distilled/deionized water and bring volume to 1.0L. Mix thoroughly.

$Na_2S \cdot 9H_2O$ Solution:
Composition per 100.0mL:

$Na_2S \cdot 9H_2O$	5.0g

Preparation of $Na_2S \cdot 9H_2O$ Solution: Add $Na_2S \cdot 9H_2O$ to distilled/deionized water and bring volume to 100.0mL. Autoclave for 15 min at 15 psi pressure–121°C. Cool to 45°–50°C.

$NaHCO_3$ Solution:
Composition per 100.0mL:

$NaHCO_3$	5.0g

Preparation of $NaHCO_3$ Solution: Add $NaHCO_3$ to distilled/deionized water and bring volume to 100.0mL. Mix thoroughly. Filter sterilize.

Vitamin B_{12} Solution:
Composition per 100.0mL:

Vitamin B_{12}	2.0mg

Preparation of Vitamin B_{12} Solution: Add Vitamin B_{12} to distilled/deionized water and bring volume to 100.0mL. Mix thoroughly. Filter sterilize.

Preparation of Medium: To 950.0mL of cooled, sterile solution 1, aseptically add 60.0mL of sterile $Na_2S \cdot 9H_2O$ solution, 40.0mL of sterile $NaHCO_3$ solution, and 1.0mL of sterile Vitamin B_{12} solution. Mix thoroughly. Adjust pH to 7.3 with sterile H_2SO_4 or Na_2CO_3. Aseptically distribute into sterile 50.0mL or 100.0mL bottles with metal screw-caps and rubber seals. Completely fill bottles with medium except for a pea-sized air bubble.

Use: For the isolation and cultivation of marine members of the Chromatiaceae.

Marine *Cytophaga* Agar

Composition per liter:

Agar	15.0g
Nutrient broth	8.0g
Yeast extract	5.0g
Salt solution	1.0L

Salt Solution:
Composition per liter:

NaCl	12.86g
$MgCl_2$	2.48g
KCl	0.75g
$CaCl_2$	0.56g
$Fe(SO_4)_2(NH_4)_2$	0.048g

Preparation of Salt Solution: Add components to distilled/deionized water and bring volume to 1.0L. Mix thoroughly.

Preparation of Medium: Add solid components to 1.0L of salt solution. Mix thoroughly. Gently heat while stirring and bring to boiling. Distribute into tubes or flasks. Autoclave for 15 min at 15 psi pressure–121°C. Pour into sterile Petri dishes or leave in tubes.

Use: For the cultivation and maintenance of *Cytophaga* species.

Marine *Cytophaga* Medium A

Composition per liter:

Agar	15.0g
Pancreatic digest of casein	2.0g
Beef extract	0.5g
Yeast extract	0.5g
Sodium acetate	0.2g
Seawater	700.0mL

pH 7.2 ± 0.2 at 25°C

Preparation of Medium: Add components to distilled/deionized water and bring volume to 1.0L. Mix thoroughly. Gently heat and bring to boiling. Distribute into tubes or flasks. Autoclave for 15 min at 15 psi pressure–121°C. Pour into sterile Petri dishes or leave in tubes.

Use: For the cultivation of *Flexibacter maritimus*.

Marine *Cytophaga* Medium B

Composition per liter:

Agar	15.0g
Pancreatic digest of casein	2.0g

Beef extract 0.5g
Yeast extract 0.5g
Sodium acetate 0.2g
Seawater 500.0mL
pH 7.2 ± 0.2 at 25°C

Preparation of Medium: Add components to distilled/deionized water and bring volume to 1.0L. Mix thoroughly. Gently heat and bring to boiling. Distribute into tubes or flasks. Autoclave for 15 min at 15 psi pressure–121°C. Pour into sterile Petri dishes or leave in tubes.

Use: For the cultivation of *Vibrio ordalii.*

Marine *Cytophaga* Medium C

Composition per liter:

Agar 15.0g
Pancreatic digest of casein 2.0g
Beef extract 0.5g
Yeast extract 0.5g
Sodium acetate 0.2g
pH 7.2 ± 0.2 at 25°C

Preparation of Medium: Add components to seawater and bring volume to 1.0L. Mix thoroughly. Gently heat and bring to boiling. Distribute into tubes or flasks. Autoclave for 15 min at 15 psi pressure–121°C. Pour into sterile Petri dishes or leave in tubes.

Use: For the cultivation of *Cytophaga agarovorans, Cytophaga fermentans,* and *Cytophaga salmonicolor.*

Marine *Desulfovibrio* Medium

Composition per liter:

Solution A 980.0mL
Solution B 10.0mL
Solution C 10.0mL
pH 7.8 ± 0.2 at 25°C

Solution A:

Composition per 980.0mL:

NaCl 25.0g
DL-Sodium lactate 2.0g
$MgSO_4 \cdot 7H_2O$ 2.0g
Na_2SO_4 .... 1.0g
NH_4Cl 1.0g
Yeast extract 1.0g
K_2HPO_4 .... 0.5g
$CaCl_2 \cdot 2H_2O$ 0.1g
Resazurin 1.0mg

Preparation of Solution A: Add components to distilled/deionized water and bring volume to 980.0mL. Mix thoroughly. Gently heat and bring to boiling. Continue boiling for 3–4 min. Allow to cool to room temperature while gassing under 100% N_2.

Solution B:

Composition per 10.0mL:

$FeSO_4 \cdot 7H_2O$ 0.5g

Preparation of Solution B: Add $FeSO_4 \cdot 7H_2O$ to distilled/deionized water and bring volume to 10.0mL. Mix thoroughly.

Solution C:

Composition per 10.0mL:

Ascorbic acid 0.1g
Sodium thioglycolate 0.1g

Preparation of Solution C: Add components to distilled/deionized water and bring volume to 10.0mL. Mix thoroughly.

Preparation of Medium: To 980.0mL of cooled solution A, anaerobically add 10.0mL of solution B and 10.0mL of solution C. Mix thoroughly. Adjust pH to 7.8 with NaOH. Distribute into tubes or flasks. During distribution, swirl the medium to keep the precipitate in suspension. Autoclave for 15 min at 15 psi pressure–121°C.

Use: For the cultivation and maintenance of *Desulfovibrio desulfuricans, Desulfovibrio salexigens,* and *Desulfovibrio vulgaris.*

Marine Flagellate Medium

Composition per 15.0mL:

Rice grains 2.0g
Seawater 15.0mL

Preparation of Medium: Autoclave rice grains for 15 min at 15 psi pressure–121°C. Add 2.0g of sterile rice grains to 15.0mL of filter-sterilized seawater. Aseptically distribute into T-25 tissue culture flasks.

Use: For the cultivation of *Acanthoecopsis unguiculata, Amastigomonas* species, *Bicosoeca vacillans, Bodo designis, Bodo variabilis, Caecitellus parvulus, Choanoeca perplexa, Codosiga gracilis, Diaphanoeca grandis, Entosiphon* species, *Goniomonas* species, *Procryptobia* species, *Pseudobodo tremulans, Rhynchomonas nasuta, Salpingoeca urceolata, Stephanoeca diplocostata,* and *Stephanopogon apogon.*

Marine Flagellate Medium with B-Vitamins

Composition per liter:

Seawater 990.0mL
Vitamin solution 10.0mL

Vitamin Solution:

Composition per 100.0mL:

Thiamine·HCl 0.15g
Calcium D-(+)-pantothenate 0.05g
Nicotinamide 0.05g
Pyridoxal·HCl 0.05g
Riboflavin 0.05g
Folic acid 0.025g
Pyridoxamine·HCl 0.025g
Biotin 12.5mg

Preparation of Vitamin Solution: Add components to distilled/deionized water and bring volume to 100.0mL. Mix thoroughly. Filter sterilize.

Preparation of Medium: Allow natural seawater to age for 2 months. Filter sterilize. Aseptically add 100.0mL of sterile vitamin solution. Mix thoroughly. Aseptically distribute into sterile tubes or flasks.

Use: For the cultivation of *Oikomonas* species.

Marine Glucose Trypticase Yeast Extract Agar (MGTY Agar)

Composition per liter:

Agar 8.0g
Glucose 2.0g
Pancreatic digest of casein 1.0g
Yeast extract 1.0g
L-Cysteine·HCl·H_2O 0.5g
Seawater 750.0mL
Tris-HCl buffer (5.0 m*M*, pH 7.5) 50.0mL
Resazurin (0.1% solution) 1.0mL
pH 7.5 ± 0.2 at 25°C

Preparation of Medium: Add components to distilled/deionized water and bring volume to 1.0L. Mix thoroughly. Gently heat while stirring and bring to boiling. Distribute into tubes or flasks under 97% N_2 + 3% H_2. Cap with rubber stoppers and place tubes in a press. Autoclave for 15 min at 15 psi pressure–121°C with fast exhaust.

Use: For the cultivation and maintenance of *Spirochaeta isovalerica.*

Marine *Methanogenium* Alcohol Medium

Composition per 1003.0mL:

NaCl	21.0g
$MgCl_2 \cdot 6H_2O$	3.0g
NaCl	1.0g
KCl	0.5g
$MgCl_2 \cdot 6H_2O$	0.5g
NH_4Cl	0.4g
Sodium acetate·$3H_2O$	0.4g
KH_2PO_4	0.2g
$CaCl_2 \cdot 2H_2O$	0.1g
$NaHCO_3$ solution	60.0mL
2-Propanol	5.0mL
$Na_2S \cdot 9H_2O$ solution	3.0mL
Cyanocobalamin solution	1.0mL
Selenite-molybdate-tungstate solution	1.0mL
Thiamine solution	1.0mL
Trace elements solution	1.0mL
Vitamin solution	1.0mL

Trace Elements Solution:
Composition per 100.0mL:

$FeSO_4 \cdot 7H_2O$	1400.0mg
$ZnSO_4 \cdot 7H_2O$	145.0mg
$CoCl_2 \cdot 6H_2O$	120.0mg
$MnCl_2 \cdot 4H_2O$	100.0mg
$NiCl_2 \cdot 6H_2O$	50.0mg
H_3BO_3	6.0mg
$CuSO_4 \cdot 5H_2O$	3.0mg
HCl (25%,w/v)	8.0mL

Preparation of Trace Elements Solution: Add components to distilled/deionized water and bring volume to 1.0L. Mix thoroughly. Sparge with 100% N_2. Autoclave for 15 min at 15 psi pressure–121°C.

Selenite-Molybdate-Tungstate Solution:
Composition per liter:

NaOH	0.2g
$Na_2MoO_4 \cdot 2H_2O$	40.0mg
$Na_2WO_4 \cdot 2H_2O$	33.0mg
$Na_2SeO_3 \cdot 2H_2O$	5.0mg

Preparation of Selenite-Molybdate-Tungstate Solution: Add components to distilled/deionized water and bring volume to 1.0L. Mix thoroughly. Sparge with 100% N_2. Autoclave for 15 min at 15 psi pressure–121°C.

$NaHCO_3$ Solution:
Composition per liter:

$NaHCO_3$	84.0g

Preparation of $NaHCO_3$ Solution: Add $NaHCO_3$ to distilled/deionized water and bring volume to 10.0mL. Mix thoroughly. Sparge with 80% N_2 + 20% CO_2. Autoclave for 15 min at 15 psi pressure–121°C.

$Na_2S \cdot 9H_2O$ Solution:
Composition per 100.0mL:

$Na_2S \cdot 9H_2O$	2.5g
NaOH	1 pellet

Preparation of $Na_2S \cdot 9H_2O$ Solution: Bring 100.0mL of distilled/deionized water to boiling. Cool to room temperature while sparging with 100%N_2. Dissolve 1 pellet of NaOH in the anaerobic water. Weigh out a little more than 2.5g of $Na_2S \cdot 9H_2O$. Briefly rinse the crystals in distilled/deionized water. Dry the crystals by blotting on paper towels or filter paper. Add 2.5g of washed $Na_2S \cdot 9H_2O$ crystals to 100.0mL of anaerobic NaOH solution. Distribute into serum bottles fitted with butyl rubber stoppers and aluminum seals. Do not grease stoppers. Pressurize to 60kPa with 100% N_2. Autoclave for 15 min at 15 psi pressure–121°C. Store at room temperature in an anaerobic chamber.

Preparation of 2-Propanol: Filter sterilize 10.0mL of 2-propanol. Sparge with 100% N_2.

Vitamin Solution:
Composition per liter:

Sodium 2-mercaptoethanesulfonate	0.25g
Pyridoxine·HCl	0.15g
Calcium pantothenate	0.1g
Nicotinic acid	0.1g
p-Aminobenzoic acid	40.0mg
Biotin	10.0mg
Potassium phosphate buffer (25m*M* solution, pH 7.0)	1.0L

Preparation of Vitamin Solution: Combine components. Mix thoroughly. Filter sterilize. Sparge with 100% N_2.

Thiamine Solution:
Composition per liter:

Thiamine·HCl	0.1g
Sodium phosphate buffer (0.1*M* solution, pH 3.6)	1.0L

Preparation of Thiamine Solution: Combine components. Mix thoroughly. Filter sterilize. Sparge with 100% N_2.

Cyanocobalamin Solution:
Composition per liter:

Cyanocobalamin	50.0mg

Preparation of Cyanocobalamin Solution: Add cyanocobalamin to distilled/deionized water and bring volume to 1.0L. Mix thoroughly. Filter sterilize. Sparge with 100% N_2.

Preparation of Medium: Prepare and dispense medium under 80% N_2 + 20% CO_2. Add components, except $NaHCO_3$ solution, 2-propanol, $Na_2S \cdot 9H_2O$ solution, cyanocobalamin solution, selenite-molybdate-tungstate solution, thiamine solution, trace elements solution, and vitamin solution, to distilled/deionized water and bring volume to 930.0mL. Mix thoroughly. Sparge with 80% N_2 + 20% CO_2. Autoclave for 15 min at 15 psi pressure–121°C. Aseptically and anaerobically add 60.0mL of sterile $NaHCO_3$ solution, 5.0mL of sterile 2-propanol, 3.0mL of sterile $Na_2S \cdot 9H_2O$ solution, 1.0mL of sterile cyanocobalamin solution, 1.0mL of sterile selenite-molybdate-tungstate solution, 1.0mL of sterile thiamine solution, 1.0mL of sterile trace elements solution, and 1.0mL of sterile vitamin solution. Mix thoroughly. Aseptically and anaerobically distribute into sterile tubes or bottles.

Use: For the cultivation of marine *Methanogenium* species.

Marine Methanol Medium

Composition per liter:

NaCl	20.0g
$(NH_4)_2SO_4$	2.0g

K_2HPO_4	2.0g
KH_2PO_4	1.0g
$MgSO_4 \cdot 7H_2O$	0.3g
Methanol	10.0mL
Vitamin B_{12} solution	10.0mL
Trace metals solution	1.0mL

pH 7.0 ± 0.2 at 25°C

Vitamin B_{12} Solution:
Composition per 100.0mL:

Vitamin B_{12}	10.0μg

Preparation of Vitamin B_{12} Solution: Add the Vitamin B_{12} to distilled/deionized water and bring volume to 100.0mL. Adjust pH to 5. Autoclave for 15 min at 15 psi pressure–121°C.

Trace Metals Solution:
Composition per liter:

$ZnSO_4 \cdot 7H_2O$	1.4g
$MnSO_4 \cdot H_2O$	0.84g
$FeSO_4 \cdot 7H_2O$	0.28g
$CuSO_4 \cdot 5H_2O$	0.25g
$Na_2MoO_4 \cdot 2H_2O$	0.24g
$CoCl_2 \cdot 6H_2O$	0.24g
$CaCl_2 \cdot 2H_2O$	0.15g

Preparation of Trace Metals Solution: Add components to distilled/deionized water and bring volume to 1.0L. Mix thoroughly.

Preparation of Medium: Add components, except Vitamin B_{12} solution and methanol, to distilled/deionized water and bring volume to 980.0mL. Adjust pH to 7.0 with NaOH. Autoclave for 15 min at 15 psi pressure–121°C. Filter sterilize methanol. Aseptically add sterile Vitamin B_{12} solution and filter-sterilized methanol. Distribute into sterile tubes or flasks.

Use: For the cultivation and maintenance of *Methylophaga thalassica.*

Marine Methylotroph Agar

Composition per 1003.0mL:

Agarose	12.0g
Bis (2-hydroxyethyl) aminotris (hydroxymethyl) methane	2.0g
KH_2PO_4	0.14g
Ferric ammonium citrate	0.06g
Methanol	2.0mL
Vitamin B_{12} solution	1.0mL

pH 7.4 ± 0.2 at 25°C

Vitamin B_{12} Solution:
Composition per 100.0mL:

Vitamin B_{12}	0.1mg

Preparation of Vitamin Solution: Add Vitamin B_{12} to distilled/deionized water and bring volume to 100.0mL. Mix thoroughly. Filter sterilize. Store at 5°C.

Preparation of Medium: Add components, except methanol and Vitamin B_{12} solution, to distilled/deionized water and bring volume to 1.0L. Mix thoroughly. Adjust pH to 7.4. Gently heat and bring to boiling. Autoclave for 15 min at 15 psi pressure–121°C. Cool to 50°C. Aseptically add 2.0mL of filter-sterilized methanol and 1.0mL of sterile Vitamin B_{12} solution. Mix thoroughly. Pour into sterile Petri dishes or distribute into sterile tubes.

Use: For the cultivation of *Alteromonas* species, *Methylophaga marina, Methylophaga thalassica,* and *Methylophilus* species.

Marine Peptone Succinate Salts Medium (PSS Medium)

Composition per liter:

Peptone	10.0g
Succinic acid	1.0g
$(NH_4)_2SO_4$	1.0g
$MgSO_4 \cdot 7H_2O$	1.0g
$FeCl_3 \cdot 6H_2O$	2.0mg
$MnSO_4 \cdot H_2O$	2.0mg
Synthetic seawater	1.0L

pH 6.8 ± 0.2 at 25°C

Synthetic Seawater:
Composition per liter:

NaCl	27.5g
$MgCl_2$	5.0g
$MgSO_4 \cdot 7H_2O$	2.0g
KCl	1.0g
$CaCl_2$	0.5g
$FeSO_4$	1.0mg

Preparation of Synthetic Seawater: Add components to distilled/deionized water and bring volume to 1.0L. Mix thoroughly.

Preparation of Medium: Add components to 1.0L of synthetic seawater. Mix thoroughly. Gently heat while stirring and bring to boiling. Adjust pH to 6.8 with KOH. Distribute into tubes or flasks. Autoclave for 15 min at 15 psi pressure–121°C.

Use: For the cultivation and maintenance of *Oceanospirillum beijerinckii* and *Oceanospirillum multiglobuliferum.*

Marine Peptone Yeast Medium with Magnesium Sulfate

Composition per liter:

NaCl	20.0g
Peptone	10.0g
$MgSO_4 \cdot 7H_2O$	2.0g
$(NH_4)_2SO_4$	2.0g
Yeast extract	1.0g

pH 7.0 ± 0.2 at 25°C

Preparation of Medium: Add components to distilled/deionized water and bring volume to 1.0L. Mix thoroughly. Distribute into tubes or flasks. Autoclave for 15 min at 15 psi pressure–121°C.

Use: For the cultivation and maintenance of *Oceanospirillum pusillum.*

Marine *Pseudomonas* Medium

Composition per liter:

Agar	15.0g
Nutrient broth	8.0g
Yeast extract	5.0g
Salt solution	1.0L

Salt Solution:
Composition per liter:

NaCl	12.86g
$MgCl_2$	2.48g
KCl	0.75g
$CaCl_2$	0.56g
$Fe(SO_4)_2(NH_4)_2$	0.048g

Preparation of Salt Solution: Add components to distilled/deionized water and bring volume to 1.0L. Mix thoroughly.

Preparation of Medium: Add components to 1.0L of salt solution. Mix thoroughly. Gently heat and bring to boiling. Distribute into tubes or flasks. Autoclave for 15 min at 15 psi pressure–121°C. Pour into sterile Petri dishes or leave in tubes.

Use: For the cultivation and maintenance of *Alteromonas haloplanktis*.

Marine *Rhodococcus* Medium

Composition per liter:

Yeast extract 10.0g
Malt extract 4.0g
Glucose 4.0g
Seawater 750.0mL

Preparation of Medium: Add components to distilled/deionized water and bring volume to 1.0L. Mix thoroughly. Gently heat while stirring and bring to boiling. Distribute into tubes or flasks. Autoclave for 15 min at 15 psi pressure–121°C.

Use: For the cultivation and maintenance of *Rhodococcus marinonascens*.

Marine *Rhodopseudomonas* Medium

Composition per liter:

NaCl 30.4g
Yeast extract 1.0g
Disodium succinate 1.0g
KH_2PO_4 .... 0.5g
$MgSO_4 \cdot 7H_2O$ 0.4g
NH_4Cl 0.4g
$CaCl_2 \cdot 2H_2O$ 0.05g
Ferric citrate (0.1% solution) 5.0mL
Trace elements SL-6 .... 1.0mL
Ethanol 0.5mL

pH 6.8 ± 0.2 at 25°C

Trace Elements Solution SL-6:

Composition per liter:

H_3BO_3 .... 0.3g
$CoCl_2 \cdot 6H_2O$ 0.2g
$ZnSO_4 \cdot 7H_2O$ 0.1g
$MnCl_2 \cdot 4H_2O$ 0.03g
$Na_2MoO_4 \cdot H_2O$ 0.03g
$NiCl_2 \cdot 6H_2O$ 0.02g
$CuCl_2 \cdot 2H_2O$ 0.01g

Preparation of Trace Elements Solution SL-6: Add components to distilled/deionized water and bring volume to 1.0L. Mix thoroughly. Adjust pH to 3.4.

Preparation of Medium: Add components to distilled/deionized water and bring volume to 1.0L. Mix thoroughly. Gently heat while stirring and bring to boiling. Distribute into tubes or flasks. Autoclave for 15 min at 15 psi pressure–121°C.

Use: For the cultivation of *Rhodopseudomonas marina*.

Marine Salts Medium

Composition per liter:

NaCl 81.0g
Yeast extract 10.0g
$MgSO_4$ 9.6g
$MgCl_2$ 7.0g
Proteose peptone No.3 .... 5.0g
KCl 2.0g
Glucose 1.0g
$CaCl_2$ 0.36g
$NaHCO_3$ 0.06g
NaBr 0.026g

pH 7.0 ± 0.2 at 25°C

Preparation of Medium: Add components to distilled/deionized water and bring volume to 1.0L. Mix thoroughly. Adjust pH to 7.0. Distribute into tubes or flasks. Autoclave for 15 min at 15 psi pressure–121°C.

Use: For the cultivation of marine bacteria.

Marine Spirochete Medium

Composition per liter:

Cellobiose 2.0g
Peptone 2.0g
Yeast extract 1.0g
Sodium thioglycolate 1.0g
Seawater, charcoal filtered 800.0mL

pH 7.5 ± 0.2 at 25°C

Preparation of Medium: Add components, except sodium thioglycolate, to glass-distilled water and bring volume to 1.0L. Mix thoroughly. Bubble 100% N_2 into medium for 1.5 min. Add sodium thioglycolate. Adjust pH to 7.5 with 10*N* KOH. Distribute into tubes or flasks. Autoclave for 15 min at 15 psi pressure–121°C.

Use: For the cultivation and maintenance of *Spirochaeta bajacaliforniensis*.

Marine *Thermococcus* Medium (DSMZ Medium 760)

Composition per liter:

NaCl 19.45g
$MgCl_2$ 8.8g
Sulfur 5.0g
Peptone 5.0g
Na_2SO_3 .... 3.24g
$CaCl_2$ 1.8g
Yeast extract 1.0g
KCl 0.55g
$NaHCO_3$ 0.16g
Ferric citrate 0.1g
KBr 0.08g
$SrCl_2$ 0.03g
H_3BO_3 .... 0.02g
Na_2HPO_4 .... 8.0mg
Na_2SiO_3 .... 4.0mg
NaF 2.4mg
NH_4NO_3 .... 1.6mg
$Na_2S \cdot 9H_2O$ solution 0.5mL

pH 6.0 ± 0.2 at 25°C

$Na_2S \cdot 9H_2O$ Solution:

Composition per 10.0mL:

$Na_2S \cdot 9H_2O$ 1.0g

Preparation of $Na_2S \cdot 9H_2O$ Solution: Add $Na_2S \cdot 9H_2O$ to distilled/deionized water and bring volume to 10.0mL. Mix thoroughly. Sparge with 100% N_2. Autoclave for 15 min at 15 psi pressure–121°C. Neutralize to pH 7.0 with sterile HCl.

Preparation of Medium: Add components, except sulfur and $Na_2S \cdot 9H_2O$ solution, to distilled/deionized water and bring volume to 1.0L. Mix thoroughly. Filter through normal filter paper. An iron sediment will collect in the filter. Gently heat while stirring and bring to boiling. Boil for 5 min. Cool under an anaerobic gas mixture of N_2. Adjust pH to 6.0. Distribute the medium into Hungate tubes or serum bottles containing finely divided sulfur (0.5% w/v). Seal the tubes or bottles under the same anaerobic gas used when cooling the medium. Sterilize the medium at 100°C for 3 hr on 3 consecutive days. Reduce the medium by adding 10% neutralized $Na_2S·9H_2O$ solution to a final concentration of 0.05%. The medium should not give a heavy black precipitate. If it does the iron sediment was not adequately removed by filtering in the intial stages and the medium should be made again, making sure that the iron is removed by filtering.

Use: For the cultivation, and maintenance of *Thermococcus aegaeus*.

Marinithermus hydrothermalis Medium (DSMZ Medium 973)

Composition per liter:

NaCl	30.0g
$MgCl_2·6H_2O$	4.18g
$MgSO_4·7H_2O$	3.4g
Yeast extract	1.0g
Tryptone	1.0g
KCl	0.33g
NH_4Cl	0.25g
K_2HPO_4	0.14g
$CaCl_2·2H_2O$	0.14g
$Fe(NH_4)_2(SO_4)_2·6H_2O$	10.0mg
$NiCl_2·6H_2O$	0.5mg
$Na2Se_3·5H_2O$	0.5mg
Trace elements solution	10.0mL

pH 7.0 ± 0.2 at 25°C

Trace Elements Solution:
Composition per liter:

$MgSO_4·7H_2O$	3.0g
Nitrilotriacetic acid	1.5g
NaCl	1.0g
$MnSO_4·2H_2O$	0.5g
$CoSO_4·7H_2O$	0.18g
$ZnSO_4·7H_2O$	0.18g
$CaCl_2·2H_2O$	0.1g
$FeSO_4·7H_2O$	0.1g
$NiCl_2·6H_2O$	0.025g
$KAl(SO_4)_2·12H_2O$	0.02g
H_3BO_3	0.01g
$Na_2MoO_4·4H_2O$	0.01g
$CuSO_4·5H_2O$	0.01g
$Na_2SeO_3·5H_2O$	0.3mg

Preparation of Trace Elements Solution: Add nitrilotriacetic acid to 500.0mL of distilled/deionized water. Dissolve by adjusting pH to 6.5 with KOH. Add remaining components. Add distilled/deionized water to 1.0L. Mix thoroughly.

Preparation of Medium: Add components to distilled/deionized water and bring volume to 1.0L. Mix thoroughly. Distribute into tubes or flasks. Autoclave for 15 min at 15 psi pressure–121°C.

Use: For the cultivation of *Marinithermus hydrothermalis*.

Marinitoga Medium (DSMZ Medium 904)

Composition per 1045mL:

Sea salts	30.0g
PIPES	6.0g
Yeast extract	1.0g
Tryptone	1.0g
Resazurin	0.5mg
Glucose solution	25.0mL
$Na_2S·9H_2O$ solution	10.0mL
L-Cysteine solution	10.0mL

pH 7.0 ± 0.2 at 25°C

Glucose Solution:
Composition per 25.0mL:

Sucrose	2.5g

Preparation of Glucose Solution: Add sucrose to distilled/deionized water and bring volume to 25.0mL. Mix thoroughly. Sparge with 100% N_2. Autoclave for 15 min at 15 psi pressure–121°C.

L-Cysteine Solution:
Composition per 10.0mL:

L-Cysteine·HCl·H_2O	0.5g

Preparation of L-Cysteine Solution: Add L-cysteine·HCl·H_2O to distilled/deionized water and bring volume to 10.0mL. Mix thoroughly. Sparge with 100% N_2. Autoclave for 15 min at 15 psi pressure–121°C.

$Na_2S·9H_2O$ Solution:
Composition per 10.0mL:

$Na_2S·9H_2O$	0.5g

Preparation of $Na_2S·9H_2O$ Solution: Add $Na_2S·9H_2O$ to distilled/deionized water and bring volume to 10.0mL. Mix thoroughly. Sparge with 100% N_2. Autoclave for 15 min at 15 psi pressure–121°C.

Preparation of Medium: Prepare and dispense medium under parge with 100% N_2. Add components, except glucose solution, L-Cysteine-HCl·H_2O solution, and $Na_2S·9H_2O$ solution, to distilled/deionized water and bring the volume to 1.0L. Mix thoroughly. Adjust pH to 7.0. Distribute into anaerobe tubes or bottles. Autoclave for 15 min at 15 psi pressure–121°C. Aseptically and anaerobically add per liter, 50.0mL glucose solution, 10.0mL L-Cysteine-HCl·H_2O solution, and 10.0mL $Na_2S·9H_2O$. Mix thoroughly. The final pH should be 7.0.

Use: For the cultivation of *Marinitoga camini* and *Caloranaerobacter azorensis*.

Marinitoga piezophila Medium (DSMZ Medium 945)

Composition per liter:

NaCl	30.0g
Yeast extract	5.0g
Trypticase	5.0g
MES	1.95g
NH_4Cl	1.0g
Na-acetate	0.83g
K_2HPO_4	0.3g
KH_2PO_4	0.3g
$MgCl_2·6H_2O$	0.2g
$CaCl_2·2H_2O$	0.1g
KCl	0.1g
Resazurin	0.5mg
Maltose solution	100.0mL
$Na_2S·9H_2O$ solution	10.0mL
Cysteine solution	10.0mL

pH 6.0 ± 0.2 at 25°C

Maltose Solution:
Composition per 100.0mL:

Maltose	4.96g

Preparation of Maltose Solution: Add maltose to distilled/deionized water and bring volume to 100.0mL. Mix thoroughly. Sparge with 100% N_2. Filter sterilize.

Cysteine Solution:
Composition per 10.0mL:

L-Cysteine·HCl·H_2O	0.3g

Preparation of L-Cysteine Solution: Add L-cysteine·HCl·H_2O to distilled/deionized water and bring volume to 10.0mL. Mix thoroughly. Sparge with 100% N_2. Autoclave for 15 min at 15 psi pressure–121°C. Cool to 25°C.

$Na_2S·9H_2O$ Solution:
Composition per 10.0mL:

$Na_2S·9H_2O$	0.3g

Preparation of $Na_2S·9H_2O$ Solution: Add $Na_2S·9H_2O$ to distilled/deionized water and bring volume to 10.0mL. Sparge with N_2. Autoclave for 15 min at 15 psi pressure–121°C. Cool to 25°C. Store anaerobically.

Preparation of Medium: Add components, except maltose solution, vitamin solution, $NaHCO_3$ solution, and $Na_2S·9H_2O$ solution to 880.0mL distilled/deionized water. Mix thoroughly. Sparge for 30 min with 100% N_2. Adjust pH to 6.0 with concentrated NaOH. Distribute under 100% N_2 into anaerobic tubes or bottles. Autoclave for 15 min at 15 psi pressure–121°C. Cool to 25°C. Aseptically and anaerobically add 100.0mL sterile maltose solution, 10.0mL sterile $Na_2S·9H_2O$ solution and 10.0mL sterile cysteine solution per liter medium. Mix thoroughly.

Use: For the cultivation of *Marinitoga piezophila.*

Marinobacter Medium (DSMZ Medium 941)

Composition per liter:

NaCl	6.0g
NH_4Cl	1.0g
Na-acetate	1.0g
$MgSO_4·7H_2O$	0.2g
KCl	0.1g
KH_2PO_4	0.1g
Peptone	0.1g
$CaCl_2·2H_2O$	0.04g
Trace elements solution SL-7	1.0mL
Vitamin solution, concentrated	1.0mL

pH 7.2 ± 0.2 at 25°C

Trace Elements Solution SL-7:
Composition per liter:

$FeCl_2·7H_2O$	1.5g
$CoCl_2·6H_2O$	190.0mg
$MnCl_2·4H_2O$	100.0mg
$ZnCl_2$	70.0mg
$Na_2MoO_4·2H_2O$	36.0mg
$NiCl_2·6H_2O$	24.0mg
H_3BO_3	62.0mg
$CuCl_2·2H_2O$	17.0mg
HCl (25% solution)	6.5mL

Preparation of Trace Elements Solution SL-7: Add $FeCl_2·7H_2O$ to 10.0mL of HCl solution. Mix thoroughly. Add distilled/deionized water and bring volume to 1.0L. Add remaining components. Mix thoroughly. Sparge with 80% N_2 + 20% CO_2. Autoclave for 15 min at 15 psi pressure–121°C.

Vitamin Solution, Concentrated:
Composition per 100mL:

Pyridoxine-HCl	10.0mg
Thiamine-HCl·$2H_2O$	5.0mg
Riboflavin	5.0mg
Nicotinic acid	5.0mg
D-Ca-pantothenate	5.0mg
p-Aminobenzoic acid	5.0mg
Lipoic acid	5.0mg
Biotin	2.0mg
Folic acid	2.0mg
Vitamin B_{12}	0.1mg

Preparation of Vitamin Solution, Concentrated: Add components to distilled/deionized water and bring volume to 100.0mL. Mix thoroughly. Filter sterilize.

Preparation of Medium: Add components to distilled/deionized water and bring volume to 1.0L. Mix thoroughly. Distribute into tubes or flasks. Autoclave for 15 min at 15 psi pressure–121°C.

Use: For the cultivation of *Marionobacter* sp.

Marinococcus albus Agar (LMG Medium 212)

Composition per liter:

NaCl	81.0g
Agar	15.0g
Yeast	10.0g
$MgSO_4·7H_2O$	9.6g
$MgCl_2·6H_2O$	7.0g
Proteose peptone	5.0g
KCl	2.0g
Glucose	1.0g
$CaCl_2$	0.36g
NaB	226.0mg
$NaHCO_3$	60.0mg

pH 7.2 ± 0.2 at 25°C

Preparation of Medium: Add components to distilled/deionized water and bring volume to 1.0L. Mix thoroughly. Gently heat and bring to boiling. Distribute into tubes or flasks. Autoclave for 15 min at 15 psi pressure–121°C. Pour into sterile Petri dishes or leave in tubes.

Use: For the cultivation and maintenance of *Marinococcus albus.*

Marinomonas vaga Medium (DSMZ Medium 617)

Composition per liter:

NaCl	30.0g
Agar	15.0g
Beef extract	10.0g
Peptone	10.0g

pH 7.1 ± 0.2 at 25°C

Preparation of Medium: Add components to tap water and bring volume to 1.0L. Mix thoroughly. Gently heat and bring to boiling. Distribute into tubes or flasks. Autoclave for 15 min at 15 psi pressure–121°C. Pour into sterile Petri dishes or leave in tubes.

Use: For the cultivation and maintenance of *Marinomonas communis=Alteromonas communis* and *Marinomonas vaga=Alteromonas vaga.*

MB Medium (DSMZ Medium 924)

Composition per liter:

NaCl	10.0g
$NaHCO_3$	4.0g
Yeast extract	2.0g
Trypticase	2.0g
NH_4Cl	1.0g
$MgCl_2·6H_2O$	1.0g
KCl	0.5g

$CaCl_2·2H_2O$ 0.4g
K_2HPO_4 .... 0.4g
Resazurin 0.5mg
Sodium formate solution 50.0mL
$Na_2S·9H_2O$ solution 10.0mL
Cysteine-HCl·H_2O solution 10.0mL
Vitamin solution 10.0mL
Trace elements solution 10.0mL
Selenite-tungstate solution 1.0mL
pH 7.2 ± 0.2 at 25°C

Cysteine Solution:
Composition per 10.0mL:
L-Cysteine·HCl·H_2O 0.25g

Preparation of Cysteine Solution: Add L-cysteine·HCl·H_2O to distilled/deionized water and bring volume to 10.0mL. Mix thoroughly. Sparge with 100% N_2. Autoclave for 15 min at 15 psi pressure–121°C.

Sodium Formate Solution:
Composition per 50.0mL:
Na-formate 6.8g

Preparation of Sodium Formate Solution: Add sodium formate to distilled/deionized water and bring volume to 50.0mL. Mix thoroughly. Sparge with 100% N_2. Filter sterilize.

Selenite-Tungstate Solution:
Composition per liter:
NaOH 0.5g
$Na_2WO_4·2H_2O$ 4.0mg
$Na_2SeO_3·5H_2O$ 3.0mg

Preparation of Selenite-Tungstate Solution: Add components to distilled/deionized water and bring volume to 1.0L. Mix thoroughly. Sparge with 100% N_2. Filter sterilize.

$Na_2S·9H_2O$ Solution:
Composition per 10.0mL:
$Na_2S·9H_2O$ 0.25g

Preparation of $Na_2S·9H_2O$ Solution: Add $Na_2S·9H_2O$ to distilled/deionized water and bring volume to 10.0mL. Sparge with N_2. Autoclave for 15 min at 15 psi pressure–121°C. Cool to 25°C. Store anaerobically.

Trace Elements Solution:
Composition per liter:
$MgSO_4·7H_2O$ 3.0g
Nitrilotriacetic acid 1.5g
NaCl 1.0g
$MnSO_4·2H_2O$ 0.5g
$CoSO_4·7H_2O$ 0.18g
$ZnSO_4·7H_2O$ 0.18g
$CaCl_2·2H_2O$ 0.1g
$FeSO_4·7H_2O$ 0.1g
$NiCl_2·6H_2O$ 0.025g
$KAl(SO_4)_2·12H_2O$ 0.02g
H_3BO_3 .... 0.01g
$Na_2MoO_4·4H_2O$ 0.01g
$CuSO_4·5H_2O$ 0.01g
$Na_2SeO_3·5H_2O$ 0.3mg

Preparation of Trace Elements Solution: Add nitrilotriacetic acid to 500.0mL of distilled/deionized water. Dissolve by adjusting pH to 6.5 with KOH. Add remaining components. Add distilled/deionized water to 1.0L. Mix thoroughly.

Vitamin Solution:
Composition per liter:
Pyridoxine-HCl 10.0mg
Thiamine-HCl·$2H_2O$ 5.0mg
Riboflavin 5.0mg
Nicotinic acid 5.0mg
D-Ca-pantothenate 5.0mg
p-Aminobenzoic acid 5.0mg
Lipoic acid 5.0mg
Biotin 2.0mg
Folic acid 2.0mg
Vitamin B_{12} 0.1mg

Preparation of Vitamin Solution: Add components to distilled/deionized water and bring volume to 1.0L. Mix thoroughly. Sparge with 80% H_2 + 20% CO_2. Filter sterilize.

Preparation of Medium: Prepare and dispense medium under an oxygen-free 80% N_2 + 20% CO_2 gas mixture. Add components, except sodium formate solution, $NaHCO_3$, cysteine solution, and $Na_2S·9H_2O$ solution, to distilled/deionized water and bring volume to 920.0mL. Mix thoroughly. Gently heat and bring to boiling. Boil for 3 min. Cool to 25°C while sparging with 80% N_2 + 20% CO_2. Add solid $NaHCO_3$. Mix thoroughly. Adjust pH to 6.8-7.0. Distribute into tubes or bottles. Autoclave for 15 min at 15 psi pressure–121°C. Aseptically and anaerobically add per liter 50.0mL sterile sodium formate solution, 10.0mL of sterile cysteine solution, and 10.0mL of sterile $Na_2S·9H_2O$ solution. Mix thoroughly. The final pH should be 7.2.

Use: For the cultivation of *Methanocalculus taiwanensis, Methanococcus voltae (Methanococcus voltaei),* and *Methanofollis aquaemaris.*

MD 1 Medium

Composition per liter:
Pancreatic digest of casein 3.0g
$MgSO_4·7H_2O$ 2.0g
$CaCl_2$ 0.5g
Trace elements solution 1.0mL
Vitamin B_{12} solution 1.0mL

Trace Elements Solution:
Composition per liter:
EDTA 8.0g
$MnCl_2·4H_2O$ 0.1g
$CoCl_2$ 0.02g
KBr 0.02g
$ZnCl_2$ 0.02g
$CuSO_4$ 0.01g
H_3BO_3 .... 0.01g
$NaMoO_4·2H_2O$ 0.01g
$BaCl_2$ 5.0mg
LiCl 5.0mg
$SnCl_2·2H_2O$ 5.0mg

Preparation of Trace Elements Solution: Add components to distilled/deionized water and bring volume to 1.0L. Mix thoroughly.

Vitamin B_{12} Solution:
Composition per 10.0mL:
Vitamin B_{12} 5.0mg

Preparation of Vitamin B_{12} Solution: Add Vitamin B_{12} to distilled/deionized water and bring volume to 10.0mL. Mix thoroughly.

Preparation of Medium: Add components to distilled/deionized water and bring volume to 1.0L. Mix thoroughly.

Distribute into tubes or flasks. Autoclave for 15 min at 15 psi pressure–121°C.

Use: For the cultivation of myxobacteria.

Me15% MH Medium (DSMZ Medium 582)

Composition per liter:

NaCl	121.5g
$MgSO_4$	14.4g
$MgCl_2$	10.5g
Yeast extract	10.0g
Proteose peptone no. 3	5.0g
KCl	3.0g
Glucose	1.0g
$CaCl_2$	0.54g
NaBr	0.039g
$NaHCO_3$ solution	10.0mL

pH 7.5 ± 0.2 at 25°C

$NaHCO_3$ Solution:
Composition per 100.0mL:

$NaHCO_3$	0.9g

Preparation of $NaHCO_3$ Solution: Add $NaHCO_3$ to distilled/deionized water and bring volume to 100.0mL. Mix thoroughly. Sparge with 80% N_2 + 20% CO_2. Autoclave for 15 min at 15 psi pressure–121°C. Cool to 25°C.

Preparation of Medium: Add components, except $NaHCO_3$ solution, to distilled/deionized water and bring volume to 990.0mL. Mix thoroughly. Autoclave for 15 min at 15 psi pressure–121°C. Cool to 25°C. Aseptically add 10.0mL $NaHCO_3$ solution. Mix thoroughly. Aseptically distribute into sterile tubes or flasks.

Use: For the cultivation of *Bacillus halophilus.*

MED IIa

Composition per liter:

Tris buffer stock solution	10.0mL
$CaCl_2$ (5.0% solution)	10.0mL
$MgSO_4 \cdot 7H_2O$ (3.33% solution)	1.0mL

pH 7.2 ± 0.2 at 25°C

Tris Buffer Stock Solution:
Composition per 500.0mL:

Tris(hydroxymethyl)aminomethane·HCl	35.01g
Tris(hydroxymethyl)aminomethane	3.35g

Preparation of Tris Buffer Stock Solution: Add components to distilled/deionized water and bring volume to 500.0mL. Mix thoroughly. Adjust pH to 7.2.

Preparation of Medium: Add components to distilled/deionized water and bring volume to 1.0L. Mix thoroughly. Distribute into tubes or flasks. Autoclave for 20 min at 15 psi pressure–121°C.

Use: For the cultivation and maintenance of *Vampirovibrio chlorellavorus.*

Medium A

Composition per liter:

D-Glucose	20.0g
Agar	20.0g
Yeast extract	10.0g
Biotin	1.0mg
Calcium pantothenate	1.0mg

pH 7.3 ± 0.2 at 25°C

Preparation of Medium: Add components, except biotin and calcium pantothenate, to distilled/deionized water and bring volume to 990.0mL. Mix thoroughly. Gently heat and bring to boiling. Autoclave for 15 min at 15 psi pressure–121°C. Cool to 45°–50°C. Add biotin and calcium pantothenate to distilled/deionized water and bring volume to 10.0mL. Mix thoroughly. Filter sterilize. Aseptically add the sterile biotin and calcium pantothenate solution to the cooled sterile basal medium. Mix thoroughly. Pour into sterile Petri dishes or distribute into sterile tubes.

Use: For the cultivation and maintenance of *Zymomonas mobilis.*

Medium 2A

Composition per liter:

Arginine	10.0g
NaCl	5.0g
Agar	4.0g
Peptone	1.0g
$K_2HPO_4 \cdot 3H_2O$	0.3g
Phenol Red	0.01g

pH 7.2–7.4 at 25°C

Preparation of Medium: Add components to distilled/deionized water and bring volume to 1.0L. Mix thoroughly. Gently heat and bring to boiling. Distribute into tubes. Autoclave for 15 min at 15 psi pressure–121°C.

Use: For the cultivation and differentiation of *Pseudomonas* species based on their production of arginine dihydrolase activity.

Medium AS4

Composition per liter:

Sucrose	80.0g
PPLO broth without Crystal Violet	500.0mL
Horse serum	200.0mL
Phenol Red (0.5% solution)	5.0mL

pH 7.2 ± 0.2 at 25°C

PPLO Broth without Crystal Violet:
Composition per 500.0mL:

Beef heart, solids from infusion	11.53g
Peptone	2.33g
NaCl	1.15g

Source: PPLO broth without Crystal Violet is available as a premixed powder from BD Diagnostic Systems.

Preparation of PPLO Broth without Crystal Violet: Add components to distilled/deionized water and bring volume to 500.0mL.Mix thoroughly. Beef heart for infusion may be substituted; 100.0g of beef heart for infusion is equivalent to 500.0g of fresh heart tissue.

Preparation of Medium: Add components, except horse serum, to distilled/deionized water and bring volume to 800.0mL. Mix thoroughly. Adjust pH to 7.2. Autoclave for 10 min at 15 psi pressure–121°C. Cool to 45°–50°C. Aseptically add 200.0mL of noninactivated, sterile horse serum. Mix thoroughly. Aseptically distribute into sterile tubes or flasks.

Use: For the cultivation and maintenance of *Spiroplasma melliferum.*

Medium for Ammonia Oxidizers

Composition per liter:

$MgSO_4 \cdot 7H_2O$	0.2g
$(NH_4)_2SO_4$	0.13g
K_2HPO_4	0.09g
$CaCl_2 \cdot 2H_2O$	0.02g
Chelated iron	1.0mg
$MnCl_2 \cdot 4H_2O$	0.2mg

$Na_2MoO_4 \cdot 2H_2O$ 0.1mg
$ZnSO_4 \cdot 7H_2O$ 0.1mg
$CuSO_4 \cdot 5H_2O$ 0.02mg
$CoCl_2 \cdot 6H_2O$ 2.0μg

Preparation of Medium: Add components to distilled/deionized water and bring volume to 1.0L. Mix thoroughly. Distribute into tubes or flasks. Autoclave for 15 min at 15 psi pressure–121°C.

Use: For the isolation, cultivation, and enrichment of ammonia-oxidizing bacteria from soil.

Medium for Ammonia Oxidizers, Brackish

Composition per liter:
$CaCO_3$ 5.0g
NH_4Cl 0.5g
K_2HPO_4 .... 0.05g
Seawater 400.0mL

Preparation of Medium: Add components to distilled/deionized water and bring volume to 1.0L. Mix thoroughly. Distribute into tubes or flasks. Autoclave for 15 min at 15 psi pressure–121°C.

Use: For the isolation, cultivation, and enrichment of ammonia-oxidizing bacteria from brackish specimens.

Medium for Ammonia Oxidizers, Marine

Composition per liter:
$(NH_4)_2SO_4$ 1.32g
$MgSO_4 \cdot 7H_2O$ 0.2g
Chelated iron 0.13g
K_2HPO_4 .... 0.11g
$CaCl_2 \cdot 2H_2O$ 0.02g
$ZnSO_4 \cdot 7H_2O$ 0.1mg
$CuSO_4 \cdot 5H_2O$ 0.02mg
$CoCl_2 \cdot 6H_2O$ 2.0μg
$MnCl_2 \cdot 4H_2O$ 2.0μg
$Na_2MoO_4 \cdot 2H_2O$ 1.0μg
Seawater 1.0L

Preparation of Medium: Combine components. Mix thoroughly. Distribute into tubes or flasks. Autoclave for 15 min at 15 psi pressure–121°C.

Use: For the isolation, cultivation, and enrichment of marine ammonia-oxidizing bacteria.

Medium for Ammonia-Oxidizing Bacteria

Composition per liter:
$(NH_4)_2SO_4$ 235.0mg
KH_2PO_4 .... 200.0mg
$CaCl_2 \cdot 2H_2O$ 40.0mg
$MgSO_4 \cdot 7H_2O$ 40.0mg
Iron-EDTA-Phenol Red solution 1.0mL
Na_2CO_3 solution .... variable

Na_2CO_3 Solution:
Composition per 100.0mL:
Na_2CO_3 .... 5.0g

Preparation of Na_2CO_3 Solution: Add Na_2CO_3 to distilled/deionized water and bring volume to 100.0mL. Mix thoroughly. Autoclave for 15 min at 15 psi pressure–121°C.

Iron-EDTA-Phenol Red Solution:
Composition per 100.0mL:
$FeSO_4 \cdot 7H_2O$ 50.0mg
Sodium EDTA 50.0mg
Phenol Red 50.0mg

Preparation of Iron-EDTA-Phenol Red Solution: Add components to distilled/deionized water and bring volume to 100.0mL. Mix thoroughly.

Preparation of Medium: Add components, except Na_2CO_3 solution, to distilled/deionized water and bring volume to 1.0L. Mix thoroughly. Distribute into tubes or flasks. Autoclave for 15 min at 15 psi pressure–121°C. Add enough sterile Na_2CO_3 solution to turn the medium pale pink. During incubation and growth of bacteria, add additional sterile Na_2CO_3 solution to restore the pale pink color. Growth is complete when no further color change is observed.

Use: For the cultivation of *Nitrosolobus multiformis* and *Nitrosomonas europaea.*

Medium B for Sulfate Reducers (Postgate's Medium B for Sulfate Reducers)

Composition per liter:
Sodium lactate 3.5g
$MgSO_4 \cdot 7H_2O$ 2.0g
NH_4Cl 1.0g
$CaSO_4$ 1.0g
Yeast extract 1.0g
KH_2PO_4 .... 0.5g
$FeSO_4 \cdot 7H_2O$ 0.5g
Ascorbic acid 0.1g
Thioglycollic acid 0.1g
pH 7.0–7.5 at 25°C

Preparation of Medium: Add components, except ascorbic acid and thioglycollic acid, to tap water and bring volume to 1.0L. For marine bacteria, NaCl may be added or seawater used in place of tap water. Mix thoroughly. Adjust pH to 7.0–7.5. Thioglycolate and ascorbate should be added immediately prior to sterilization. Distribute into tubes or flasks. Autoclave for 15 min at 15 psi pressure–121°C.

Use: For the isolation, cultivation, and maintenance of *Desulfovibrio* species and *Desulfotomaculum* species. This medium turns black as a result of H_2S production due to bacterial growth.

Medium for *Bacillus schlegelii*

Composition per liter:
Agar 15.0g
$Na_2HPO_4 \cdot 2H_2O$ 2.9g
KH_2PO_4 .... 2.3g
NH_4Cl 1.0g
$MgSO_4 \cdot 7H_2O$ 0.5g
$NaHCO_3$ 0.5g
$CaCl_2 \cdot 2H_2O$ 0.01g
$MnSO_4 \cdot H_2O$ 10.0mg
Ferric ammonium citrate solution 20.0mL
Trace elements solution SL-6 .... 5.0mL
pH 6.8 ± 0.2 at 25°C

Ferric Ammonium Citrate Solution:
Composition per 20.0mL:
Ferric ammonium citrate 0.05g

Preparation of Ferric Ammonium Citrate Solution: Add ferric ammonium citrate to distilled/deionized water and bring volume to 20.0mL. Mix thoroughly. Autoclave for 15 min at 15 psi pressure–121°C.

Trace Elements Solution SL-6:
Composition per liter:
$MnCl_2 \cdot 4H_2O$ 0.5g
H_3BO_3 .... 0.3g

$CoCl_2 \cdot 6H_2O$ 0.2g
$ZnSO_4 \cdot 7H_2O$ 0.1g
$Na_2MoO_4 \cdot 2H_2O$ 0.03g
$NiCl_2 \cdot 6H_2O$ 0.02g
$CuCl_2 \cdot 2H_2O$ 0.01g

Preparation of Trace Elements Solution SL-6: Add components to distilled/deionized water and bring volume to 1.0L. Mix thoroughly.

Preparation of Medium: Add components, except ferric ammonium citrate solution, to distilled/deionized water and bring volume to 980.0mL. Mix thoroughly. Gently heat and bring to boiling. Autoclave for 15 min at 15 psi pressure–121°C. Aseptically add 20.0mL of sterile ferric ammonium citrate solution. Mix thoroughly. Pour into sterile Petri dishes or distribute into sterile tubes.

Use: For the chemolithotrophic growth of *Bacillus schlegelii.*

Medium for *Bacillus stearothermophilus*

Composition per 1001.0mL:
NH_4Cl 1.0g
K_2HPO_4 .... 0.5g
Yeast extract 0.2g
Casamino acids 0.1g
$MgSO_4 \cdot 7H_2O$ 0.02g
Phenol solution 100.0mL
Trace elements solution SL-4 .... 1.0mL
pH 7.4 ± 0.2 at 25°C

Phenol Solution:
Composition per 100.0mL:
Phenol 0.47g

Preparation of Phenol Solution: Add phenol to distilled/deionized water and bring volume to 100.0mL. Mix thoroughly. Filter sterilize.

Trace Elements Solution SL-4:
Composition per liter:
EDTA 0.5g
$FeSO_4 \cdot 7H_2O$ 0.2g
Trace elements solution SL-6 .... 100.0mL

Trace Elements Solution SL-6:
Composition per liter:
$MnCl_2 \cdot 4H_2O$ 0.5g
H_3BO_3 .... 0.3g
$CoCl_2 \cdot 6H_2O$ 0.2g
$ZnSO_4 \cdot 7H_2O$ 0.1g
$Na_2MoO_4 \cdot 2H_2O$ 0.03g
$NiCl_2 \cdot 6H_2O$ 0.02g
$CuCl_2 \cdot 2H_2O$ 0.01g

Preparation of Trace Elements Solution SL-6: Add components to distilled/deionized water and bring volume to 1.0L. Mix thoroughly. Autoclave for 15 min at 15 psi pressure–121°C.

Preparation of Trace Elements Solution SL-4: Add components to distilled/deionized water and bring volume to 1.0L. Mix thoroughly. Filter sterilize.

Preparation of Medium: Add components, except phenol solution and trace elements solution SL-4, to distilled/deionized water and bring volume to 900.0mL. Mix thoroughly. Autoclave for 15 min at 15 psi pressure–121°C. Aseptically add 100.0mL of sterile phenol solution and 1.0mL of sterile trace elements solution SL-4. Mix thoroughly. Aseptically distribute into sterile tubes or flasks.

Use: For the cultivation of *Bacillus stearothermophilus.*

Medium with Biphenyl (DSMZ Medium 457d)

Composition per liter:
Na_2HPO_4 .... 2.44g
KH_2PO_4 .... 1.52g
$(NH_4)_2SO_4$ 0.5g
Biphenyl 0.25g
$MgSO_4 \cdot 7H_2O$ 0.2g
$CaCl_2 \cdot 2H_2O$ 0.05g
Trace elements solution SL-4 .... 10.0mL
pH 6.9 ± 0.2 at 25°C

Trace Elements Solution SL-4:
Composition per liter:
EDTA 0.5g
$FeSO_4 \cdot 7H_2O$ 0.2g
Trace elements solution SL-6 .... 100.0mL

Trace Elements Solution SL-6:
Composition per liter:
H_3BO_3 .... 0.3g
$CoCl_2 \cdot 6H_2O$ 0.2g
$ZnSO_4 \cdot 7H_2O$ 0.1g
$MnCl_2 \cdot 4H_2O$ 0.03g
$Na_2MoO_4 \cdot H_2O$ 0.03g
$NiCl_2 \cdot 6H_2O$ 0.02g
$CuCl_2 \cdot 2H_2O$ 0.01g

Preparation of Trace Elements Solution SL-6: Add components to distilled/deionized water and bring volume to 1.0L. Mix thoroughly. Adjust pH to 3.4.

Preparation of Trace Elements Solution SL-4: Add components to distilled/deionized water and bring volume to 1.0L. Mix thoroughly.

Biphenyl Solution:
Composition per liter:
Biphenyl 10.0g

Preparation of Biphenyl Solution: Add biphenyl to 1.0L ethanol. Mix thoroughly. Filter sterilize using a cellulose filter membrane.

Preparation of Medium: Add components, except biphenyl solution, to 1.0L distilled/deionized water. Adjust pH to 6.9. Autoclave for 15 min at 15 psi pressure–121°C. Cool to room temperature. Add an aliquot of the biphenyl solution to a sterile flask so that the final concentration will be 0.25g/L biphenyl, and let the ethanol evaporate. Aseptically add sterile medium to the crystal-layered flask.

Use: For the cultivation of biphenyl utilizing bacteria.

Medium C for Sulfate Reducers (Postgate's Medium C for Sulfate Reducers)

Composition per liter:
Sodium lactate 6.0g
Na_2SO_4 .... 4.5g
NH_4Cl 1.0g
Yeast extract 1.0g
KH_2PO_4 .... 0.5g
Sodium citrate·$2H_2O$ 0.3g
$CaCl_2 \cdot 6H_2O$ 0.06g
$MgSO_4 \cdot 7H_2O$ 0.06g
$FeSO_4 \cdot 7H_2O$ 0.004g
pH 7.5 ± 0.2 at 25°C

Preparation of Medium: Add components to distilled/deionized water and bring volume to 1.0L. For marine bacteria, NaCl may be added or seawater used in place of distilled/deionized water. Mix thoroughly. Adjust pH to 7.5.

Distribute into tubes or flasks. Autoclave for 15 min at 15 psi pressure–121°C.

Use: For detection, culturing and storage of *Desulfovibrio* species and many *Desulfotomaculum* species. This medium should be used when a clear culture medium is desired such as for chemostat culture. This medium may be cloudy after sterilization but usually clears on cooling. It turns black as a result of H_2S production due to bacterial growth.

Medium for Carbon Monoxide Oxidizers

Composition per liter:

Agar	12.0g
$Na_2HPO_4 \cdot 12H_2O$	4.5g
NH_4Cl	1.5g
KH_2PO_4	0.75g
$MgSO_4 \cdot 7H_2O$	0.2g
$CaCl_2 \cdot 2H_2O$	0.03g
Ferric ammonium citrate	0.018g
Trace elements solution SL-6	1.0mL

pH 7.0 ± 0.2 at 25°C

Trace Elements Solution SL-6:

Composition per liter:

$MnCl_2 \cdot 4H_2O$	0.5g
H_3BO_3	0.3g
$CoCl_2 \cdot 6H_2O$	0.2g
$ZnSO_4 \cdot 7H_2O$	0.1g
$Na_2MoO_4 \cdot 2H_2O$	0.03g
$NiCl_2 \cdot 6H_2O$	0.02g
$CuCl_2 \cdot 2H_2O$	0.01g

Preparation of Trace Elements Solution SL-6: Add components to distilled/deionized water and bring volume to 1.0L. Mix thoroughly. Autoclave for 15 min at 15 psi pressure–121°C.

Preparation of Medium: Add components to distilled/deionized water and bring volume to 1.0L. Mix thoroughly. Gently heat and bring to boiling. Autoclave for 15 min at 15 psi pressure–121°C. Pour into sterile Petri dishes or distribute into sterile tubes. After inoculation, incubate in an atmosphere of 80% CO +10% O_2 + 10% N_2.

Use: For the chemoautotrophic cultivation and maintenance of *Alcaligenes* species, *Pseudomonas carboxydohydrogena,* and other *Pseudomonas* species.

Medium for Chlorate Respirers (DSMZ Medium 908)

Composition per liter:

Solution A	1.0L
Solution B	10.0mL
Solution C	10.0mL
Vitamin solution	5.0mL
Trace elements solution SL-10	1.0mL

pH 7.2 ± 0.2 at 25°C

Solution A:

Composition per liter:

$NaHCO_3$	2.5g
Na-Acetate	1.36g
$NaClO_3$	1.0g
NaH_2PO_4	0.6g
NH_4Cl	0.25g
KCl	0.1g

Preparation of Solution A: Add components to distilled/deionized water and bring volume to 1.0L. Mix thoroughly. Sparge with 80% N_2 + 20% CO_2. Autoclave for 15 min at 15 psi pressure–121°C. Cool to room temperature.

Solution B:

Composition per 10mL:

$MgSO_4$	30.0mg
$CaCl_2 \cdot 2H_2O$	10.0mg

Preparation of Solution B: Add components to distilled/deionized water and bring volume to 10.0mL. Mix thoroughly. Sparge with 80% N_2 + 20% CO_2. Autoclave for 15 min at 15 psi pressure–121°C. Cool to room temperature.

Solution C:

Na_2MoO_4	25.0mg
$Na_2WO_4 \cdot 2H_2O$	25.0mg

Preparation of Solution B: Add components to distilled/deionized water and bring volume to 10.0mL. Mix thoroughly. Sparge with 80% N_2 + 20% CO_2. Autoclave for 15 min at 15 psi pressure–121°C. Cool to room temperature.

Vitamin Solution:

Composition per liter:

Vitamin B_{12}	50.0mg
Pantothenic acid	50.0mg
Riboflavin	50.0mg
Alpha-lipoic acid	50.0mg
p-aminobenzoic acid	50.0mg
Thiamine-HCl·$2H_2O$	50.0mg
Nicotinic acid	25.0mg
Nicotine amide	25.0mg
Biotin	20.0mg
Folic acid	20.0mg
Pyridoxamine-HCl	10.0mg

Preparation of Vitamin Solution: Add components to distilled/deionized water and bring volume to 1.0L. Mix thoroughly. Filter sterilize.

Trace Elements Solution SL-10:

Composition per liter:

$FeCl_2 \cdot 4H_2O$	1.5g
H_3BO_3	300.0mg
$CoCl_2 \cdot 6H_2O$	190.0mg
$MnCl_2 \cdot 4H_2O$	100.0mg
$ZnCl_2$	70.0mg
$Na_2MoO_4 \cdot 2H_2O$	36.0mg
$NiCl_2 \cdot 6H_2O$	24.0mg
$CuCl_2 \cdot 2H_2O$	2.0mg
HCl (25% solution)	7.7mL

Preparation of Trace Elements Solution SL-10: Add $FeCl_2 \cdot 4H_2O$ to 10.0mL of HCl solution. Mix thoroughly. Add distilled/deionized water and bring volume to 1.0L. Add remaining components. Mix thoroughly. Sparge with 100% N_2. Autoclave for 15 min at 15 psi pressure–121°C. Cool to room temperature.

Preparation of Medium: Aseptically and anaerobically combine 1000.0mL solution A, 10.0mL solution B, and 10.0mL solution C. Aseptically and anaerobically add 5.0mL vitamin solution and 1.0mL trace elements solution SL-10. Mix thoroughly. The pH should be 7.2.

Use: For the cultivation of *Dechloromonas agitata* and *Azospira oryzae.*

Medium with Chloroacrylic Acid (DSMZ Medium 457c)

Composition per liter:

Na_2HPO_4	2.44g
KH_2PO_4	1.52g
$(NH_4)_2SO_4$	0.5g

$MgSO_4 \cdot 7H_2O$ 0.2g
$CaCl_2 \cdot 2H_2O$ 0.05g
Chloroacrylic acid solution 20.0mL
Trace elements solution SL-4 .... 10.0mL
pH 6.9 ± 0.2 at 25°C

Trace Elements Solution SL-4:
Composition per liter:
EDTA 0.5g
$FeSO_4 \cdot 7H_2O$ 0.2g
Trace elements solution SL-6 .... 100.0mL

Trace Elements Solution SL-6:
Composition per liter:
H_3BO_3 .... 0.3g
$CoCl_2 \cdot 6H_2O$ 0.2g
$ZnSO_4 \cdot 7H_2O$ 0.1g
$MnCl_2 \cdot 4H_2O$ 0.03g
$Na_2MoO_4 \cdot H_2O$ 0.03g
$NiCl_2 \cdot 6H_2O$ 0.02g
$CuCl_2 \cdot 2H_2O$ 0.01g

Preparation of Trace Elements Solution SL-6: Add components to distilled/deionized water and bring volume to 1.0L. Mix thoroughly. Adjust pH to 3.4.

Preparation of Trace Elements Solution SL-4: Add components to distilled/deionized water and bring volume to 1.0L. Mix thoroughly.

Chloroacrylic Acid Solution:
Composition per liter:
3-Chloroacrylic acid 4.0g

Preparation of Chloroacrylic Acid Solution: Add 3-Chloroacrylic acid to distilled/deionized water and bring volume to 1.0L. Mix thoroughly. Adjust pH to 7.0. Filter sterilize.

Preparation of Medium: Add components, except chloroacrylic acid solution, to 1.0L distilled/deionized water. Adjust pH to 6.9. Autoclave for 15 min at 15 psi pressure–121°C. Cool to room temperature. Aseptically add 20.0mL sterile chloroacrylic acid solution. Mix thoroughly. Aseptically distribute to sterile tubes or flasks.

Use: For the cultivation of chloroacrylic acid-utilizing *Burkholderia* sp. *(Burkholderia cepacia), Rhodococcus erythropolis (Arthrobacter picolinophilus,* and *Nocardia* spp.

Medium for *Chlorobium ferrooxidans* (DSMZ Medium 29a)

Composition per 5.0L:
Solution A 4.0L
Solution B 860.0mL
Solution E 100.0mL
Solution F 30.0mL
Solution C 5.0mL
Solution D 5.0mL
pH 6.8 at 25°C

Solution A:
Composition per 4.0L:
$MgSO_4$ 2.5g
KH_2PO_4 .... 1.7g
NH_4Cl 1.7g
KCl 1.7g
$CaCl_2 \cdot 2H_2O$ 1.25g
Na-acetate 0.82g

Preparation of Solution A: Add components to 4.0L distilled water. Mix thoroughly. Autoclave for 45 min at 15 psi pressure–121°C in 5-liter special bottle or flask with four openings at the top, together with a teflon-coated magnetic bar. In this 5-liter bottle, two openings are for tubes in the central, silicon rubber stopper; one is a short, gas-inlet tube with a sterile cotton filter; and the other is an outlet tube for medium, which reaches the bottom of the vessel at one end and has, at the other end, a silicon rubber tube with a pinch cock and a bell for aseptic dispensing of the medium into bottles. The other two openings have gas-tight screw caps; one of these openings is for the addition of sterile solutions and the other serves as a gas outlet. After autoclaving, cool solution A to room temperature under a N_2 atmosphere with a positive pressure of 0.05–0.1 atm (a manometer for low pressure will be required). Saturate the cold medium with CO_2 by magnetic stirring for 30 min under a CO_2 atmosphere of 0.05–0.1 atm.

Solution B:
Distilled water 860.0mL

Preparation of Solution B: Autoclave distilled water for 15 min at 15 psi pressure–121°C in a cotton-stoppered Erlenmeyer flask. Cool to room temperature under an atmosphere of N_2 in an anaerobic jar.

Solution C:
Composition per 100.0mL:
Vitamin B_{12} 2.0mg

Preparation of Solution C: Add Vitamin B_{12} to distilled/deionized water and bring volume to 100.0mL. Mix thoroughly. Sparge under 100% N_2 gas for 3 min. Filter sterilize Store under N_2 gas.

Solution D:
Composition per liter:
$FeCl_2 \cdot 4H_2O$ 1.5g
H_3BO_3 .... 300.0mg
$CoCl_2 \cdot 6H_2O$ 190.0mg
$MnCl_2 \cdot 4H_2O$ 100.0mg
$ZnCl_2$ 70.0mg
$Na_2MoO_4 \cdot 2H_2O$ 36.0mg
$NiCl_2 \cdot 6H_2O$ 24.0mg
$CuCl_2 \cdot 2H_2O$ 2.0mg
HCl (25% solution) 7.7mL

Preparation of Solution D: Add $FeCl_2 \cdot 4H_2O$ to 10.0mL of HCl solution. Mix thoroughly. Add distilled/deionized water and bring volume to 1.0L. Add remaining components. Mix thoroughly. Sparge with 100% N_2. Autoclave for 15 min at 15 psi pressure–121°C.

Solution E:
Composition per 100.0mL:
$NaHCO_3$ 4.2g

Preparation of Solution E: Add $NaHCO_3$ to distilled/deionized water and bring volume to 100.0mL. Mix thoroughly. Sparge with 100% CO_2 until saturated. Filter sterilize under 100% CO_2 into a sterile, gas-tight 100.0mL screw-capped bottle.

Solution F:
Composition per 100.0mL:
$FeSO_4$ 25.0g

Preparation of Solution F: Add $FeSO_4$ to distilled/deionized water and bring volume to 100.0mL. Mix thoroughly. Sparge with 100% CO_2 until saturated. Filter sterilize under 100% CO_2 into a sterile, gas-tight 100.0mL screw-capped bottle.

Preparation of Medium: Add solutions B, C, D, and E to solution A through one of the screw-cap openings against a stream of either N_2 gas or better, a mixture of 95% N_2 and

5% CO_2 while the medium is magnetically stirred. Adjust the pH of the medium with sterile HCl or Na_2CO_3 solution (2 *M* solutions) to pH 6.8. Distribute the medium aseptically through the medium outlet tube into sterile, 100mL bottles (with metal caps and autoclavable rubber seals) using the positive gas pressure (0.05 - 0.1 atm) of the N_2/CO_2 gas mixture: Leave a small air bubble in each bottle to meet possible pressure changes. The tightly sealed, screw-cap bottles can be stored for several weeks or months in the dark.

Use: For the cultivation of *Chlorobium ferrooxidans*.

Medium D4

Composition per liter:

Agar	15.0g
Sucrose	10.0g
NH_4Cl	5.0g
Na_2HPO_4, anhydrous	2.3g
Pancreatic digest of casein	1.0g
Sodium dodecyl sulfate	0.6g
Glycerol	10.0mL

Preparation of Medium: Add components to distilled/deionized water and bring volume to 1.0L. Mix thoroughly. Gently heat and bring to boiling. Distribute into tubes or flasks. Autoclave for 15 min at 15 psi pressure–121°C. Pour into sterile Petri dishes or leave in tubes.

Use: For the selective isolation and cultivation of *Pseudomonas syringae*.

Medium D for Sulfate Reducers (Postgate's Medium D for Sulfate Reducers)

Composition per liter:

Sodium pyruvate	3.5g
$MgCl_2·6H_2O$	1.6g
NH_4Cl	1.0g
Yeast extract	1.0g
KH_2PO_4	0.5g
$CaCl_2·2H_2O$	0.1g
$FeSO_4·7H_2O$	0.004g

pH 7.5 ± 0.2 at 25°C

Preparation of Medium: Add components to distilled/deionized water and bring volume to 1.0L. Malate or fumarate may also be used as a carbon source. For marine bacteria, NaCl may be added or seawater used in place of distilled/deionized water. Mix thoroughly. Adjust pH to 7.5. Filter sterilize. Aseptically distribute into sterile tubes or flasks.

Use: For the cultivation of *Desulfovibrio* species and *Desulfotomaculum* species that can grow in the absence of sulfate.

Medium D for Sulfate Reducers (Postgate's Medium D for Sulfate Reducers)

Composition per liter:

$MgCl_2·6H_2O$	1.6g
Choline chloride	1.0g
NH_4Cl	1.0g
Yeast extract	1.0g
KH_2PO_4	0.5g
$CaCl_2·2H_2O$	0.1g
$FeSO_4·7H_2O$	0.004g

pH 7.5 ± 0.2 at 25°C

Preparation of Medium: Add components to distilled/deionized water and bring volume to 1.0L. Malate or fumarate may also be used as a carbon source. For marine bacteria, NaCl may be added or seawater used in place of distilled/deionized water. Mix thoroughly. Adjust pH to 7.5. Filter sterilize. Aseptically distribute into sterile tubes or flasks.

Use: For the cultivation of *Desulfovibrio* species and *Desulfotomaculum* species that can grow in the absence of sulfate.

Medium D for *Thermus*

Composition per liter:

Pancreatic digest of casein	1.0g
Yeast extract	1.0g
$NaNO_3$	0.7g
KNO_3	0.1g
$MgSO_4·7H_2O$	0.1g
Na_2HPO_4	0.1g
Nitrilotriacetic acid	0.1g
$CaSO_4·2H_2O$	0.06g
NaCl	8.0mg
$MnSO_4·H_2O$	2.2mg
$ZnSO_4·7H_2O$	0.5mg
H_3BO_3	0.5mg
$FeCl_3$	0.28mg
$Na_2MoO_4·2H_2O$	0.03mg
$CuSO_4$	0.02mg

pH 8.2 ± 0.2 at 25°C

Preparation of Medium: Add nitrilotriacetic acid to 500.0mL of distilled/deionized water. Dissolve by adjusting pH to 6.5 with KOH. Add remaining components. Readjust pH to 8.2 with H_2SO_4 or KOH. Add distilled/deionized water to 1.0L. Distribute into tubes or flasks. Autoclave for 15 min at 15 psi pressure–121°C.

Use: For the cultivation of *Thermus* species.

Medium D for *Thermus*, Modified

Composition per liter:

Agar	25.0g
Pancreatic digest of casein	1.0g
Yeast extract	1.0g
Salt solution	100.0mL

pH 8.2 ± 0.2 at 25°C

Salt Solution:

Composition per liter:

$NaNO_3$	6.89g
$Na_2HPO_4·12H_2O$	2.8g
KNO_3	1.03g
Nitrilotriacetic acid	1.0g
$MgSO_4·7H_2O$	1.0g
$CaSO_4·2H_2O$	0.6g
NaCl	0.08g
$FeCl_3·6H_2O$ solution	10.0mL
Trace elements solution	10.0mL

$FeCl_3·6H_2O$ Solution:

Composition per 100.0mL:

$FeCl_3·6H_2O$	47.0mg

Preparation of $FeCl_3·6H_2O$ Solution: Add 47.0mg of $FeCl_3·6H_2O$ to distilled/deionized water and bring volume to 100.0mL. Mix thoroughly.

Trace Elements Solution:

Composition per liter:

$MnSO_4·4H_2O$	1.7g
$ZnSO_4·7H_2O$	0.5g
H_3BO_3	0.5g
$CoCl_2·6H_2O$	46.0mg
$CuSO_4·5H_2O$	25.0mg

$Na_2MoO_4 \cdot 2H_2O$ 25.0mg
H_2SO_4 .... 0.5mL

Preparation of Trace Elements Solution: Add components to distilled/deionized water and bring volume to 1.0L. Mix thoroughly.

Preparation of Salt Solution: Add nitrilotriacetic acid to 500.0mL of distilled/deionized water. Dissolve by adjusting pH to 6.5 with KOH. Add remaining components. Readjust pH to 8.2 with H_2SO_4 or KOH. Add distilled/deionized water to 1.0L.

Preparation of Medium: Add components to distilled/deionized water and bring volume to 1.0L. Mix thoroughly. Gently heat and bring to boiling. Adjust pH to 8.2. Distribute into tubes or flasks. Autoclave for 15 min at 15 psi pressure–121°C. Pour into sterile Petri dishes or leave in tubes.

Use: For the cultivation of *Thermus aquaticus*.

Medium for *Ectothiorhodospira*

Composition per 1001.0mL:
Basal medium 800.0mL
Solution C 200.0mL
Vitamin solution B 1.0mL

Basal Medium:
Composition per 800.0mL:
NaCl 180.0g
Na_2SO_4 .... 20.0g
Na_2CO_3 .... 6.0g
$Na_2S \cdot 9H_2O$ 1.0g
Sodium succinate 1.0g
NH_4Cl 0.8g
KH_2PO_4 .... 0.5g
Yeast extract 0.5g
$MgCl_2 \cdot 6H_2O$ 0.1g
$CaCl_2 \cdot 7H_2O$ 0.05g
Trace elements solution A 1.0mL
pH 8.5 ± 0.2 at 25°C

Trace Elements Solution A:
Composition per liter:
$FeCl_2 \cdot 4H_2O$ 1.8g
H_3BO_3 .... 500.0mg
$CoCl_2 \cdot 6H_2O$ 250.0mg
$ZnCl_2$ 100.0mg
$MnCl_2 \cdot 4H_2O$ 70.0mg
$Na_2MoO_4 \cdot 2H_2O$ 30.0mg
$CuCl_2 \cdot 2H_2O$ 10.0mg
$NiCl_2 \cdot 6H_2O$ 10.0mg
$Na_2SeO_3 \cdot 5H_2O$ 10.0mg

Preparation of Trace Elements Solution A: Add components to distilled/deionized water and bring volume to 990.0mL. Mix thoroughly. Adjust pH to 3 with 1*N* HCl. Bring volume to 1.0L with distilled/deionized water.

Preparation of Basal Solution: Add components to distilled/deionized water and bring volume to 800.0mL. Mix thoroughly. Adjust pH to 8.5. Distribute into screw-capped bottles. Autoclave for 15 min at 14 psi pressure–120°C.

Vitamin Solution B:
Composition per 100.0mL:
Nicotinamide 35.0mg
Thiamine dichloride 30.0mg
p-Aminobenzoic acid 20.0mg
Biotin 10.0mg
Calcium DL-pantothenate 10.0mg
Pyridoxal·HCl 10.0mg
Vitamin B_{12} 5.0mg

Preparation of Vitamin Solution B: Add components to distilled/deionized water and bring volume to 100.0mL. Mix thoroughly. Filter sterilize.

Solution C:
Composition per 200.0mL:
$NaHCO_3$ 14.0g

Preparation of Solution C: Add $NaHCO_3$ to distilled/deionized water and bring volume to 200.0mL. Mix thoroughly. Filter sterilize.

Preparation of Medium: To 800.0mL of sterile basal solution, aseptically add 200.0mL of sterile solution C and 1.0mL of sterile vitamin solution B. Mix thoroughly.

Use: For the cultivation and maintenance of *Ectothiorhodospira halochloris*.

Medium for *Erythrobacter longus* (DSMZ Medium 695)

Composition per liter:
Peptone 2.0g
Soytone 1.0g
Yeast extract 1.0g
Proteose peptone no.3 .... 1.0g
Ferric citrate solution 2.0mL
Artificial seawater 700.0mL
pH 7.5 ± 0.2 at 25°C

Artificial Seawater:
Composition per liter:
NaCl 23.477g
$MgCl_2 \cdot 6H_2O$ 4.981g
Na_2SO_4 .... 3.917g
$CaCl_2$ 1.12g
KCl 664.0mg
$NaHCO_3$ 192.0mg
H_3BO_3 .... 26.0mg
$SrCl_2$ 24.0mg
KBr 6.0mg
NaF 3.0mg

Preparation of Artificial Seawater: Add components to distilled/deionized water and bring volume to 1.0L. Mix thoroughly. Filter sterilize.

Ferric Citrate Solution:
Composition per 10.0mL:
Ferric citrate 0.5g

Preparation of Ferric Citrate Solution: Add ferric citrate to distilled/deionized water and bring volume to 10.0mL. Mix thoroughly.

Preparation of Medium: Add components, except artificial seawater, to distilled/deionized water and bring volume to 300.0mL. Mix thoroughly. Adjust pH to 7.5. Aseptically add 700.0mL artificial seawater. Mix thoroughly. Aseptically and anaerobically distribute into sterile tubes or bottles.

Use: For the cultivation of *Erythrobacter longus*.

Medium E for Sulfate Reducers (Postgate's Medium E for Sulfate Reducers)

Composition per liter:
Agar 15.0g
Sodium lactate 3.5g
$MgCl_2 \cdot 6H_2O$ 2.0g
NH_4Cl 1.0g

Na_2SO_4 .... 1.0g
$CaCl_2 \cdot 2H_2O$ 1.0g
Yeast extract 1.0g
KH_2PO_4 .... 0.5g
Ascorbic acid 0.1g
Thioglycollic acid 0.1g
$FeSO_4 \cdot 7H_2O$ 0.004g
pH 7.6 ± 0.2 at 25°C

Preparation of Medium: Add components, except ascorbic acid and thioglycollic acid, to tap water and bring volume to 1.0L. For marine bacteria, NaCl may be added or seawater used in place of tap water. Mix thoroughly. Gently heat and bring to boiling. Adjust pH to 7.6. Thioglycolate and ascorbate should be added immediately prior to sterilization. Distribute into screw-capped tubes or flasks. Autoclave for 15 min at 15 psi pressure–121°C.

Use: For the cultivation and enumeration of *Desulfovibrio* species and *Desulfotomaculum* species as black colonies in deep agar cultures. Also used for the isolation of pure cultures of *Desulfovibrio* species and *Desulfotomaculum* species.

Medium F

Composition per liter:
$MgSO_4 \cdot 7H_2O$ 0.5g
$(NH_4)_2SO_4$ 0.15g
KCl 0.05g
KH_2PO_4 .... 0.05g
$Ca(NO_3)_2$ 0.01g
$FeSO_4 \cdot 7H_2O$ solution 10.0mL
pH 3.5 ± 0.2 at 25°C

$FeSO_4 \cdot 7H_2O$ Solution:

Composition per 10.0mL:
$FeSO_4 \cdot 7H_2O$ 1.0g

Preparation of $FeSO_4 \cdot 7H_2O$ Solution: Add the $FeSO_4 \cdot 7H_2O$ to distilled/deionized water and bring volume to 10.0mL. Mix thoroughly. Filter sterilize.

Preparation of Medium: Add components, except $FeSO_4 \cdot 7H_2O$ solution, to tap water and bring volume to 990.0mL. Mix thoroughly. Gently heat until dissolved. Adjust pH to 3.5. Autoclave for 15 min at 15 psi pressure–121°C. Cool to 45°–50°C. Aseptically add 10.0mL of sterile $FeSO_4 \cdot 7H_2O$ solution. Mix thoroughly. Aseptically distribute into sterile tubes or flasks.

Use: For the cultivation of *Thiobacillus* species.

Medium F for Sulfate Reducers (Postgate's Medium F for Sulfate Reducers)

Composition per liter:
Agar 12.0g
Pancreatic digest of casein 10.0g
Sodium lactate 3.5g
Ferrous citrate 0.5g
Na_2SO_3 .... 0.5g
$MgSO_4 \cdot 7H_2O$ 0.2g
Ascorbic acid 0.1g
Sodium thioglycolate 0.1g
pH 7.1 ± 0.2 at 25°C

Preparation of Medium: Add components, except ascorbic acid and thioglycollic acid, to tap water and bring volume to 1.0L. For marine bacteria, NaCl may be added or seawater used in place of tap water. Mix thoroughly. Gently heat and bring to boiling. Adjust pH to 7.1. Thioglycolate and ascorbate should be added immediately prior to sterilization. Distribute into screw-capped tubes or flasks. Autoclave for 15 min at 15 psi pressure–121°C.

Use: For isolation and cultivation of *Desulfotomaculum nigrificans, Desulfovibrio* species, and other *Desulfotomaculum* species especially in food. These bacteria form black colonies in deep agar cultures.

Medium with Fluoranthene (DSMZ Medium 457b)

Composition per liter:
Na_2HPO_4 .... 2.44g
KH_2PO_4 .... 1.52g
$(NH_4)_2SO_4$ 0.5g
$MgSO_4 \cdot 7H_2O$ 0.2g
Tween 80 .... 0.2g
Fluoaranthene 0.1g
$CaCl_2 \cdot 2H_2O$ 0.05g
Trace elements solution SL-4 .... 10.0mL
pH 6.9 ± 0.2 at 25°C

Trace Elements Solution SL-4:

Composition per liter:
EDTA 0.5g
$FeSO_4 \cdot 7H_2O$ 0.2g
Trace elements solution SL-6 .... 100.0mL

Trace Elements Solution SL-6:

Composition per liter:
H_3BO_3 .... 0.3g
$CoCl_2 \cdot 6H_2O$ 0.2g
$ZnSO_4 \cdot 7H_2O$ 0.1g
$MnCl_2 \cdot 4H_2O$ 0.03g
$Na_2MoO_4 \cdot H_2O$ 0.03g
$NiCl_2 \cdot 6H_2O$ 0.02g
$CuCl_2 \cdot 2H_2O$ 0.01g

Preparation of Trace Elements Solution SL-6: Add components to distilled/deionized water and bring volume to 1.0L. Mix thoroughly. Adjust pH to 3.4.

Preparation of Trace Elements Solution SL-4: Add components to distilled/deionized water and bring volume to 1.0L. Mix thoroughly.

Fluoaranthene Solution:

Composition per liter:
Fluoranthene 2.0g

Preparation of Fluoaranthene Solution: Add fluoranthene to 1.0L acetone. Mix thoroughly. Filter sterilize using a cellulose filter membrane.

Preparation of Medium: Add components, except fluoranthene solution, to 1.0L distilled/deionized water. Adjust pH to 6.9. Autoclave for 15 min at 15 psi pressure–121°C. Cool to room temperature. Add an aliquot of the fluoranthene solution to a sterile flask so that the final concentration will be 0.1g/L fluoranthene, and let the acetone evaporate. Aseptically add sterile medium to the crystal-layered flask.

Use: For the cultivation of fluoranthene-utilizing *Pseudomonas frederiksbergensis Sphingomonas* sp. *(Pseudomonas paucimobilis),* and other bacteria.

Medium for Freshwater Flexibacteria

Composition per 1002.0mL:
Casamino acids 1.0g
$MgSO_4 \cdot 7H_2O$ 1.0g
Tris (hydroxymethyl) amino methane 1.0g
$CaCl_2 \cdot 2H_2O$ 0.1g
KNO_3 .... 0.1g

Sodium glycerophosphate....0.1g
Thiamine....1.0mg
Cobalamine....1.0μg
Glucose solution....1.0mL
Trace elements solution....1.0mL

pH 7.5 ± 0.2 at 25°C

Glucose Solution:

Composition per 100.0mL:

Glucose....10.0g

Preparation of Glucose Solution: Add glucose to distilled/deionized water and bring volume to 100.0mL. Mix thoroughly. Filter sterilize.

Trace Elements Solution:

Composition per liter:

$ZnCl_2$....20.8g
H_3BO_3....2.85g
$MnCl_2 \cdot 4H_2O$....1.8g
Sodium tartrate....1.77g
$FeSO_4$....1.36g
$CoCl_2 \cdot 6H_2O$....40.4mg
$CuCl_2 \cdot 2H_2O$....26.9mg
$Na_2MoO_4 \cdot 2H_2O$....25.2mg

Preparation of Trace Elements Solution: Add components to distilled/deionized water and bring volume to 1.0L. Mix thoroughly. Autoclave for 15 min at 15 psi pressure–121°C.

Preparation of Medium: Add components, except glucose solution and trace elements solution, to distilled/deionized water and bring volume to 1.0L. Mix thoroughly. Autoclave for 15 min at 15 psi pressure–121°C. Aseptically add 1.0mL of sterile glucose solution and 1.0mL of sterile trace elements solution. Mix thoroughly. Aseptically distribute into sterile tubes or flasks.

Use: For the cultivation of *Cytophaga psychrophila, Flectobacillus major, Flexibacter aurantiacus, Flexibacter aurantiacus, Flexibacter elegans, Flexibacter flexilis, Flexibacter roseolus, Flexibacter ruber, Flexibacter sancti,* and *Herpetosiphon geysericola.*

Medium G for Sulfate Reducers (Postgate's Medium G for Sulfate Reducers)

Composition per 1015.2mL:

Solution 1....970.0mL
Solution 4....30.0mL
Solution 8A, 8B, 8C, 8D or 8E....10.0mL
Solution 5....3.0mL
Solution 2....1.0mL
Solution 3....1.0mL
Solution 6....0.1mL
Solution 7....0.1mL

pH 7.2 ± 0.2 at 25°C

Solution 1:

Composition per 970.0mL:

Na_2SO_4....3.0g
NaCl....1.2g
$MgCl_2 \cdot 6H_2O$....0.4g
KCl....0.3g
NH_4Cl....0.3g
KH_2PO_4....0.2g
$CaCl_2 \cdot 2H_2O$....0.15g

Preparation of Solution 1: Add components to distilled/deionized water and bring volume to 970.0mL. Mix thoroughly. Adjust pH to 7.2 with 2*N* HCl. Autoclave for 15 min at 15 psi pressure–121°C. Cool to 25°C.

Solution 2:

Composition per 10.0mL:

NaOH....5.0mg
Na_2SeO_3....0.03mg

Preparation of Solution 2: Add NaOH and Na_2SeO_3 to distilled/deionized water and bring volume to 10.0mL. Mix thoroughly. Autoclave for 15 min at 15 psi pressure–121°C. Cool to 25°C.

Solution 3:

Composition per liter:

$FeCl_2 \cdot 4H_2O$....1.5g
$CoCl_2 \cdot 6H_2O$....0.12g
$MnCl_2 \cdot 4H_2O$....0.1g
$ZnCl_2$....0.07g
H_3BO_3....0.06g
$NiCl_2 \cdot 6H_2O$....0.025g
$NaMoO_4 \cdot 2H_2O$....0.025g
$CuCl_2 \cdot 2H_2O$....0.015g

Preparation of Solution 3: Add components to distilled/deionized water and bring volume to 1.0L. Mix thoroughly. Autoclave for 15 min at 15 psi pressure–121°C. Cool to 25°C.

Solution 4:

Composition per 30.0mL:

$NaHCO_3$....2.55g

Preparation of Solution 4: Add $NaHCO_3$ to distilled/deionized water and bring volume to 30.0mL. Mix thoroughly. Gas with 100% CO_2 for 10–15 min. Filter sterilize.

Solution 5:

Composition per 3.0mL:

$Na_2S \cdot 9H_2O$....0.36g

Preparation of Solution 5: Add $Na_2S \cdot 9H_2O$ to distilled/deionized water and bring volume to 3.0mL. Mix thoroughly. Gas with 100% N_2 for 5–10 min. Cap tube with a rubber stopper. Autoclave for 15 min at 15 psi pressure–121°C. Cool to 25°C.

Solution 6:

Composition per 100.0mL:

Thiamine·HCl....0.01g
Cyanocobalamin....5.0mg
p-Aminobenzoic acid....5.0mg
Biotin....1.0mg

Preparation of Solution 6: Add components to distilled/deionized water and bring volume to 100.0mL. Mix thoroughly. Filter sterilize.

Solution 7:

Composition per 100.0mL:

Succinic acid....0.6g
Isobutyric acid....0.5g
Valeric acid....0.5g
2-Methylbutyric acid....0.5g
3-Methylbutyric acid....0.5g
Caproic acid....0.2g

Preparation of Solution 7: Add components to distilled/deionized water and bring volume to 100.0mL. Mix thoroughly. Adjust pH to 9.0 with NaOH. Autoclave for 15 min at 15 psi pressure–121°C. Cool to 25°C.

Solution 8A:

Composition per 100.0mL:

Sodium acetate·$3H_2O$....20.0g

Solution 8B:
Composition per 100.0mL:
Propionic acid 7.0g

Solution 8C:
Composition per 100.0mL:
n-Butyric acid 8.0g

Solution 8D:
Composition per 100.0mL:
Benzoic acid 5.0g

Solution 8E:
Composition per 100.0mL:
n-Palmitic acid 5.0g

Preparation of Solutions 8A–E: Add the appropriate amount of component to distilled/deionized water and bring volume to 100.0mL. Mix thoroughly. Adjust pH to 9.0 with NaOH. Autoclave for 15 min at 15 psi pressure–121°C. Cool to 25°C.

Preparation of Medium: To 970.0mL of cooled, sterile solution 1, aseptically add 1.0mL of sterile solution 2, 1.0mL of sterile solution 3, 30.0mL of sterile solution 4, 3.0mL of sterile solution 5, 0.1mL of sterile solution 6, 0.1mL of sterile solution 7, and 10.0mL of sterile solution 8A, 8B, 8C, 8D, or 8E. Mix thoroughly. Aseptically distribute into sterile tubes or flasks.

Use: For the isolation and cultivation of *Desulfovibrio baarsii, Desulfovibrio sapovorans, Desulfobacter* species, *Desulfonema* species, *Desulfobulbus* species, and *Desulfotomaculum acetoxidans.*

Medium for Halophilic Bacilli

Composition per liter:
NaCl 100.0g
Casamino acids 10.0g
Yeast extract 10.0g
pH 7.0 ± 0.2 at 25°C

Preparation of Medium: Add components to distilled/deionized water and bring volume to 1.0L. Mix thoroughly. Distribute into tubes or flasks. Autoclave for 15 min at 15 psi pressure–121°C.

Use: For the cultivation of halophilic *Bacillus* species.

Medium for Hydrocarbon-Degrading Bacteria

Composition per 1020.0mL:
NH_4Cl 0.5g
$MgSO_4·7H_2O$ 0.5g
NaCl 0.4g
Hydrocarbon 20.0mL
KH_2PO_4 solution 0.5mL
$Na_2HPO_4·H_2O$ solution 0.5mL

KH_2PO_4 Solution:
Composition per 100.0mL:
KH_2PO_4 .... 10.0g

Preparation of KH_2PO_4 Solution: Add KH_2PO_4 to distilled/deionized water and bring volume to 100.0mL. Mix thoroughly. Autoclave for 15 min at 15 psi pressure–121°C. Cool to 25°C.

$Na_2HPO_4·H_2O$ Solution:
Composition per 100.0mL:
$Na_2HPO_4·H_2O$ 10.0g

Preparation of $Na_2HPO_4·H_2O$ Solution: Add 10.0g of $Na_2HPO_4·H_2O$ to distilled/deionized water and bring volume to 100.0mL. Mix thoroughly. Autoclave for 15 min at 15 psi pressure–121°C. Cool to 25°C.

Preparation of Medium: Add components—except hydrocarbon, KH_2PO_4 solution, and $Na_2HPO_4·H_2O$ solution, to distilled/deionized water and bring volume to 999.0mL. Mix thoroughly. Gently heat and bring to boiling. Autoclave for 15 min at 15 psi pressure–121°C. Cool to 45°–50°C. Aseptically add 0.5mL of sterile KH_2PO_4 solution and 0.5mL of the sterile $Na_2HPO_4·H_2O$ solution. Mix thoroughly. Aseptically distribute into sterile tubes in 10.0mL volumes. Add 0.2mL of sterile hydrocarbon to each tube.

Use: For the cultivation and enumeration of hydrocarbon-degrading bacteria in fresh water.

Medium for Hydrocarbon-Degrading Bacteria (Naphthalene Mineral Salts Medium)

Composition per liter:
K_2HPO_4 .... 1.0g
$(NH_4)_2SO_4$ 1.0g
$MgSO_4·7H_2O$ 0.3g
$CaCl_2$ 0.1g
$FeSO_4·7H_2O$ 0.02g
Naphthalene 2.0mL
pH 7.0 ± 0.2 at 25°C

Preparation of Medium: Add components, except naphthalene, to distilled/deionized water and bring volume to 998.0mL. Mix thoroughly. Gently heat and bring to boiling. Autoclave for 15 min at 15 psi pressure–121°C. Cool to 45°–50°C. Aseptically add 2.0mL of sterile naphthalene to 20.0mL of sterile basal salts. Ultrasonically homogenize the solution. Add the naphthalene–basal salts homogenate back to the remainder of the sterile basal salts medium. Mix thoroughly. Aseptically distribute into sterile tubes or flasks.

Use: For the cultivation and enrichment of hydrocarbon-degrading bacteria.

Medium M71

Composition per liter:
Agar 20.0g
Peptone 10.0g
Glucose 5.0g
H_3BO_3 .... 1.0g
Pancreatic digest of casein 1.0g
Cycloheximide 0.05g
2,3,5-Triphenyltetrazolium·HCl solution 10.0mL

2,3,5-Triphenyltetrazolium·HCl Solution:
Composition per 10.0mL:
2,3,5-Triphenyltetrazolium·HCl 0.05g

Preparation of 2,3,5-Triphenyltetrazolium·HCl Solution: Add 2,3,5-triphenyltetrazolium·HCl to distilled/deionized water and bring volume to 10.0mL. Mix thoroughly. Autoclave for 15 min at 15 psi pressure–121°C.

Preparation of Medium: Add components, except 2,3,5-triphenyltetrazolium·HCl solution, to distilled/deionized water and bring volume to 990.0mL. Mix thoroughly. Gently heat and bring to boiling. Autoclave for 15 min at 15 psi pressure–121°C. Cool to 45°–50°C. Aseptically add 10.0mL of sterile 2,3,5-triphenyltetrazolium·HCl solution. Mix thoroughly. Pour into sterile Petri dishes.

Use: For the selective isolation and cultivation of *Pseudomonas syringae.*

Medium 523M

Composition per liter:
Agar 15.0g
Sucrose 10.0g

Casamino acids 2.0g
K_2HPO_4 .... 2.0g
Yeast extract 2.0g
$MgSO_4 \cdot 7H_2O$ 0.3g

Preparation of Medium: Add components to distilled/deionized water and bring volume to 1.0L. Mix thoroughly. Gently heat and bring to boiling. Distribute into tubes or flasks. Autoclave for 15 min at 15 psi pressure–121°C. Pour into sterile Petri dishes or leave in tubes.

Use: For the cultivation of *Clavibacter toxicus.*

Medium for Marine Flexibacteria

Composition per 1001.0mL:

Pancreatic digest of casein 5.0g
Yeast extract 5.0g
Tris (hydroxymethyl) amino methane 1.0g
KNO_3 .... 0.5g
Sodium glycerophosphate 0.1g
Trace elements solution 1.0mL

pH 7.0 ± 0.2 at 25°C

Trace Elements Solution:
Composition per liter:

$ZnCl_2$ 20.8g
H_3BO_3 .... 2.85g
$MnCl_2 \cdot 4H_2O$ 1.8g
Sodium tartrate 1.77g
$FeSO_4$ 1.36g
$CoCl_2 \cdot 6H_2O$ 40.4mg
$CuCl_2 \cdot 2H_2O$ 26.9mg
$Na_2MoO_4 \cdot 2H_2O$ 25.2mg

Preparation of Trace Elements Solution: Add components to distilled/deionized water and bring volume to 1.0L. Mix thoroughly. Autoclave for 15 min at 15 psi pressure–121°C.

Preparation of Medium: Add components, except trace elements solution, to distilled/deionized water and bring volume to 1.0L. Mix thoroughly. Autoclave for 15 min at 15 psi pressure–121°C. Aseptically add 1.0mL of sterile trace elements solution. Mix thoroughly. Aseptically distribute into sterile tubes or flasks.

Use: For the cultivation of *Cytophaga aprica, Cytophaga diffluens, Cytophaga johnsonae, Cytophaga lytica, Cytophaga* species, *Flexibacter aggregans, Flexibacter aurantiacus, Flexibacter litoralis, Flexibacter tractuosus, Flexithrix dorotheae, Herpetosiphon cohaerens, Herpetosiphon nigricans, Herpetosiphon persicus, Microscilla arenaria, Microscilla furvescens, Microscilla marina, Microscilla sericea,* and *Saprospira grandis.*

Medium for Marine Methylotrophs (DSMZ Medium 750)

Composition per liter:

NaCl 25.0g
Agar 20.0g
Peptone 10.0g
Beef extract 7.0g
K_2HPO_4 .... 1.0g
$(NH_4)_2SO_4$ 1.0g
Methanol 10.0mL

pH 7.0 ± 0.2 at 25°C

Preparation of Medium: Add components, except methanol, to distilled/deionized water and bring volume to 990.0mL. Mix thoroughly. Gently heat and bring to boiling. Autoclave for 15 min at 15 psi pressure–121°C. Cool to 50°C. Aseptically add 10.0mL filter sterilized methanol. Mix thoroughly. Pour into sterile Petri dishes or distribute into sterile tubes.

Use: For the cultivation and maintenance of *Methylophaga marina* and *Methylophaga thalassica.*

Medium N for Sulfate Reducers (Postgate's Medium N for Sulfate Reducers)

Composition per liter:

$(NH_4)_2SO_4$ 7.0g
Sodium lactate 6.0g
NH_4Cl 1.0g
Yeast extract 1.0g
KH_2PO_4 .... 0.5g
Sodium citrate·$2H_2O$ 0.3g
$FeSO_4 \cdot 7H_2O$ 0.1g
$CaCl_2 \cdot 6H_2O$ 0.06g
$MgSO_4 \cdot 7H_2O$ 0.06g

pH 7.5 ± 0.2 at 25°C

Preparation of Medium: Add components to distilled/deionized water and bring volume to 1.0L. For marine bacteria, NaCl may be added or seawater used in place of distilled/deionized water. Mix thoroughly. Adjust pH to 7.5. Distribute into tubes or flasks. Autoclave for 15 min at 15 psi pressure–121°C.

Use: For the detection, culturing, and storage of *Desulfovibrio* species and many *Desulfotomaculum* species. This medium should be used when a clear culture medium is desired such as for chemostat culture. This medium may be cloudy after sterilization but usually clears on cooling. It turns black as a result of H_2S production due to bacterial growth.

Medium for Nitrite Oxidizers

Composition per liter:

$KHCO_3$ 1.5g
KH_2PO_4 .... 0.5g
K_2HPO_4 .... 0.5g
KNO_2 .... 0.3g
$MgSO_4 \cdot 7H_2O$ 0.2g
NaCl 0.2g
$CaCl_2 \cdot 2H_2O$ 0.01g
$FeSO_4 \cdot 7H_2O$ 0.01g

Preparation of Medium: Add components to distilled/deionized water and bring volume to 1.0L. Mix thoroughly. Distribute into tubes or flasks. Autoclave for 15 min at 15 psi pressure–121°C.

Use: For the isolation, cultivation, and enrichment of nitrate-oxidizing bacteria.

Medium for Nitrite Oxidizers, Marine

Composition per liter:

$MgSO_4 \cdot 7H_2O$ 0.1g
$NaNO_2$ 0.07g
$CaCl_2 \cdot 2H_2O$ 6.0mg
K_2HPO_4 .... 1.74mg
Chelated iron 1.0mg
$MnCl_2 \cdot 4H_2O$ 66.0μg
$Na_2MoO_4 \cdot 2H_2O$ 30.0μg
$ZnSO_4 \cdot 7H_2O$ 30.0μg
$CuSO_4 \cdot 5H_2O$ 6.0μg
$CoCl_2 \cdot 6H_2O$ 0.6μg
Seawater 700.0mL

Preparation of Medium: Add components to distilled/deionized water and bring volume to 1.0L. Mix thoroughly.

Distribute into tubes or flasks. Autoclave for 15 min at 15 psi pressure–121°C.

Use: For the isolation, cultivation, and enrichment of marine nitrate-oxidizing bacteria.

Medium for Osmophilic Fungi (M 40 Y)

Composition per liter:

Sucrose	400.0g
Agar	20.0g
Malt extract	20.0g
Yeast extract	5.0g

Preparation of Medium: Add components to distilled/deionized water and bring volume to 1.0L. Mix thoroughly. Gently heat and bring to boiling. Distribute into tubes or flasks. Autoclave for 15 min at 15 psi pressure–121°C. Pour into sterile Petri dishes or leave in tubes.

Use: For the cultivation of osmophilic fungi.

Medium with Phenanthrene (DSMZ Medium 457b)

Composition per liter:

Na_2HPO_4	2.44g
KH_2PO_4	1.52g
$(NH_4)_2SO_4$	0.5g
Phenanthrene	0.1g
$MgSO_4 \cdot 7H_2O$	0.2g
Tween 80	0.2g
$CaCl_2 \cdot 2H_2O$	0.05g
Trace elements solution SL-4	10.0mL

pH 6.9 ± 0.2 at 25°C

Trace Elements Solution SL-4:

Composition per liter:

EDTA	0.5g
$FeSO_4 \cdot 7H_2O$	0.2g
Trace elements solution SL-6	100.0mL

Preparation of Trace Elements Solution SL-4: Add components to distilled/deionized water and bring volume to 1.0L. Mix thoroughly.

Trace Elements Solution SL-6:

Composition per liter:

H_3BO_3	0.3g
$CoCl_2 \cdot 6H_2O$	0.2g
$ZnSO_4 \cdot 7H_2O$	0.1g
$MnCl_2 \cdot 4H_2O$	0.03g
$Na_2MoO_4 \cdot H_2O$	0.03g
$NiCl_2 \cdot 6H_2O$	0.02g
$CuCl_2 \cdot 2H_2O$	0.01g

Preparation of Trace Elements Solution SL-6: Add components to distilled/deionized water and bring volume to 1.0L. Mix thoroughly. Adjust pH to 3.4.

Phenanthrene Solution:

Composition per liter:

Phenanthrene	2.0g

Preparation of Phenanthrene Solution: Add phenanthrene to 1.0L acetone. Mix thoroughly. Filter sterilize using a cellulose filter membrane.

Preparation of Medium: Add components, except phenanthrene solution, to 1.0L distilled/deionized water. Adjust pH to 6.9. Autoclave for 15 min at 15 psi pressure–121°C. Cool to room temperature. Add an aliquot of the phenanthrene solution to a sterile flask so that the final concentration will be 0.1g/L phenanthrene, and let the acetone evaporate. Aseptically add sterile medium to the crystal-layered flask.

Use: For the cultivation of phenanthrene-utilizing *Sphingomonas* sp. *(Pseudomonas paucimobilis), Pseudomonas frederiksbergensis,* and other bacteria.

Medium for *Prosthecomicrobium* and *Ancalomicrobium*

Composition per liter:

Agar	15.0g
Peptone	0.1g
Hutner's modified salts solution	20.0mL
Vitamin solution	10.0mL

Hutner's Mineral Base:

Composition per liter:

$MgSO_4 \cdot 7H_2O$	29.7g
Nitrilotriacetic acid	10.0g
$CaCl_2 \cdot 2H_2O$	3.34g
$FeSO_4 \cdot 7H_2O$	0.1g
$(NH_4)_2MoO_4$	9.25mg
"Metals 44"	50.0mL

Preparation of Hutner's Mineral Base: Add nitrilotriacetic acid to 500.0mL of distilled/deionized water. Dissolve by adjusting pH to 6.5 with KOH. Add remaining components. Add distilled/deionized water to 1.0L.

"Metals 44":

Composition per 100.0mL:

$ZnSO_4 \cdot 7H_2O$	1.1g
$FeSO_4 \cdot 7H_2O$	0.5g
EDTA	0.25g
$MnSO_4 \cdot 7H_2O$	0.154g
$CuSO_4 \cdot 5H_2O$	0.04g
$Co(NO_3)_2 \cdot 6H_2O$	0.025g
$Na_2B_4O_7 \cdot 10H_2O$	0.018g

Preparation of "Metals 44": Add components to distilled/deionized water and bring volume to 100.0mL. Mix thoroughly.

Vitamin Solution:

Composition per liter:

Pyridoxine·HCl	0.01g
Calcium pantothenate	5.0mg
Nicotinamide	5.0mg
Riboflavin	5.0mg
Thiamine HCl	5.0mg
Biotin	2.0mg
Folic acid	2.0mg
Vitamin B_{12}	0.1mg

Preparation of Vitamin Solution: Add components to distilled/deionized water and bring volume to 1.0L. Mix thoroughly. Filter sterilize.

Preparation of Medium: Add components, except vitamin solution, to distilled/deionized water and bring volume to 990.0mL. Mix thoroughly. Gently heat and bring to boiling. Autoclave for 15 min at 15 psi pressure–121°C. Cool to 45°–50°C. Aseptically add sterile vitamin solution. Mix thoroughly. Pour into sterile Petri dishes or distribute into sterile tubes.

Use: For the isolation of *Prosthecomicrobium* species and *Ancalomicrobium* species.

Medium R

Composition per liter:

$Na_2S_2O_3 \cdot 5H_2O$	5.0g
KNO_3	2.0g

$MgCl_2 \cdot 6H_2O$ ... 0.5g
NH_4Cl ... 0.5g
KH_2PO_4 solution ... 10.0mL
$NaHCO_3$ solution ... 10.0mL
$FeSO_4 \cdot 7H_2O$ solution ... 10.0mL
pH 7.0 ± 0.2 at 25°C

KH_2PO_4 Solution:
Composition per 10.0mL:
KH_2PO_4 ... 2.0g

Preparation of KH_2PO_4 Solution: Add KH_2PO_4 to distilled/deionized water and bring volume to 10.0mL. Mix thoroughly. Filter sterilize.

$NaHCO_3$ Solution:
Composition per 10.0mL:
$NaHCO_3$ ... 1.0g

Preparation of $NaHCO_3$ Solution: Add the $NaHCO_3$ to distilled/deionized water and bring volume to 10.0mL. Mix thoroughly. Filter sterilize.

$FeSO_4 \cdot 7H_2O$ Solution:
Composition per 10.0mL:
$FeSO_4 \cdot 7H_2O$ ... 10.0mg

Preparation of $FeSO_4 \cdot 7H_2O$ Solution: Add 10.0mg of $FeSO_4 \cdot 7H_2O$ to distilled/deionized water and bring volume to 10.0mL. Mix thoroughly. Filter sterilize.

Preparation of Medium: Add components, except KH_2PO_4 solution, $NaHCO_3$ solution, and $FeSO_4 \cdot 7H_2O$ solution, to tap water and bring volume to 970.0mL. Mix thoroughly. Gently heat until dissolved. Adjust pH to 7.0. Autoclave for 15 min at 15 psi pressure–121°C. Cool to 45°–50°C. Aseptically add 10.0mL of sterile KH_2PO_4 solution, 10.0mL of $NaHCO_3$ solution, and 10.0mL of $FeSO_4 \cdot 7H_2O$ solution. Mix thoroughly. Aseptically distribute into sterile tubes or flasks.

Use: For the cultivation of *Thiobacillus denitrificans.*

Medium S

Composition per liter:
$Na_2S_2O_3 \cdot 5H_2O$ ... 5.0g
$(NH_4)_2SO_4$ ... 4.0g
KH_2PO_4 ... 4.0g
$MgSO_4$ ... 0.5g
$CaCl_2$ ... 0.25g
$FeSO_4$ ... 0.01g

Preparation of Medium: Add components to distilled/deionized water and bring volume to 1.0L. Mix thoroughly. Distribute into tubes or flasks. Autoclave for 15 min at 15 psi pressure–121°C.

Use: For the cultivation of *Thiobacillus* species.

Medium SP 4

Composition per liter:
Pancreatic digest of casein ... 11.0g
Peptone ... 5.3g
Glucose ... 5.0g
NaCl ... 0.875g
Beef extract ... 0.525g
Yeast extract ... 0.525g
Beef heart, solids from infusion ... 0.35g
Fetal bovine serum, heat inactivated ... 170.0mL
Yeast extract solution ... 100.0mL
CMRL 1066, 10× solution ... 50.0mL
Fresh yeast extract solution ... 35.0mL
Phenol Red solution ... 20.0mL
Penicillin solution ... 10.0mL
pH 7.6 ± 0.2 at 25°C

Yeast Extract Solution:
Composition per 100.0mL:
Yeast extract ... 2.0g

Preparation of Yeast Extract Solution: Add yeast extract to distilled/deionized water and bring volume to 100.0mL. Mix thoroughly. Autoclave for 15 min at 15 psi pressure–121°C.

CMRL 1066, 10X Solution:
Composition per liter:
NaCl ... 6.8g
$NaHCO_3$ ... 2.2g
D-Glucose ... 1.0g
KCl ... 0.4g
L-Cysteine·HCl·H_2O ... 0.26g
$CaCl_2$, anhydrous ... 0.2g
$MgSO_4 \cdot 7H_2O$ ... 0.2g
$NaH_2PO_4 \cdot H_2O$ ... 0.14g
L-Glutamine ... 0.1g
Sodium acetate·$3H_2O$ ... 0.083g
L-Glutamic acid ... 0.075g
L-Arginine·HCl ... 0.07g
L-Lysine·HCl ... 0.07g
L-Leucine ... 0.06g
Glycine ... 0.05g
Ascorbic acid ... 0.05g
L-Proline ... 0.04g
L-Tyrosine ... 0.04g
L-Aspartic acid ... 0.03g
L-Threonine ... 0.03g
L-Alanine ... 0.025g
L-Phenylalanine ... 0.025g
L-Serine ... 0.025g
L-Valine ... 0.025g
L-Cystine ... 0.02g
L-Histidine·HCl·H_2O ... 0.02g
L-Isoleucine ... 0.02g
Phenol Red ... 0.02g
L-Methionine ... 0.015g
Deoxyadenosine ... 0.01g
Deoxycytidine ... 0.01g
Deoxyguanosine ... 0.01g
Glutathione, reduced ... 0.01g
Thymidine ... 0.01g
Hydroxy-L-proline ... 0.01g
L-Tryptophan ... 0.01g
Nicotinamide adenine dinucleotide ... 7.0mg
Tween™ 80 ... 5.0mg
Sodium glucoronate·H_2O ... 4.2mg
Coenzyme A ... 2.5mg
Cocarboxylase ... 1.0mg
Flavin adenine dinucleotide ... 1.0mg
Nicotinamide adenine
dinucleotide phosphate ... 1.0mg
Uridine triphosphate ... 1.0mg
Choline chloride ... 0.5mg
Cholesterol ... 0.2mg
5-Methyldeoxycytidine ... 0.1mg
Inositol ... 0.05mg
p-Aminobenzoic acid ... 0.05mg
Niacin ... 0.025mg
Niacinamide ... 0.025mg
Pyridoxine ... 0.025mg
Pyridoxal·HCl ... 0.025mg

Biotin 0.01mg
D-Calcium pantothenate 0.01mg
Folic acid 0.01mg
Riboflavin 0.01mg
Thiamine·HCl 0.01mg

Source: CMRL 1066, 10× medium is available as a premixed powder from BD Diagnostics.

Preparation of CMRL 1066, 10X Solution: Add components to distilled/deionized water and bring volume to 1.0L. Mix thoroughly. Adjust pH to 7.2. Filter sterilize.

Fresh Yeast Extract Solution:
Composition per 100.0mL:
Baker's yeast, live, pressed, starch-free 25.0g

Preparation of Fresh Yeast Extract Solution: Add the live Baker's yeast to 100.0mL of distilled/deionized water. Autoclave for 90 min at 15 psi pressure–121°C. Allow to stand. Remove supernatant solution. Adjust pH to 6.6–6.8. Filter sterilize.

Phenol Red Solution:
Composition per 100.0mL:
Phenol Red 0.01g

Preparation of Phenol Red Solution: Add Phenol Red to distilled/deionized water and bring volume to 10.0mL. Mix thoroughly. Filter sterilize.

Penicillin Solution:
Composition per 10.0mL:
Penicillin 1,000,000U

Preparation of Penicillin Solution: Add penicillin to distilled/deionized water and bring volume to 10.0mL. Filter sterilize.

Preparation of Medium: Add components—except fetal bovine serum, yeast extract solution, CMRL 1066, 10× solution, fresh yeast extract solution, Phenol Red solution, and penicillin solution—to distilled/deionized water and bring volume to 615.0mL. Mix thoroughly. Gently heat and bring to boiling. Autoclave for 15 min at 15 psi pressure–121°C. Cool to 45°–50°C. Aseptically add 170.0mL of sterile fetal bovine serum, 100.0mL of sterile yeast extract solution, 50.0mL of sterile CMRL 1066, 10× solution, 35.0mL of sterile fresh yeast extract solution, 20.0mL of sterile Phenol Red solution, and 10.0mL of sterile penicillin solution. Mix thoroughly. Aseptically distribute into sterile tubes or flasks.

Use: For the isolation and cultivation of *Spiroplasma* species from ticks.

Medium for Sulfate Reducers (Postgate's Medium for Sulfate Reducers) (ATCC Medium 1283)

Composition per liter:
Part A 869.0mL
Part C 100.0mL
Part D 10.0mL
Part E 10.0mL
Part F 10.0mL
Part B 1.0mL

pH 7.7 ± 0.2 at 25°C

Part A:
Composition per 869.0mL:
Na_2SO_4 .... 3.0g
NaCl 1.0g
KCl 0.5g
$MgCl_2·6H_2O$ 0.4g
NH_4Cl 0.3g
KH_2PO_4 .... 0.2g
$CaCl_2·2H_2O$ 0.15g

Preparation of Part A: Add components to distilled/deionized water and bring volume to 869.0mL. Mix thoroughly. Prepare and autoclave part A under 90% N_2 + 10% CO_2. Autoclave for 15 min at 15 psi pressure–121°C. Cool to room temperature.

Part B:
Composition per liter:
$FeCl_2·4H_2O$ 1.5g
$CoCl_2·6H_2O$ 0.19g
$MnCl_2·4H_2O$ 0.1g
$ZnCl_2$ 0.07g
H_3BO_3 .... 0.06g
$Na_2MoO_4·2H_2O$ 0.04g
$NiCl_2·6H_2O$ 0.02g
$CuCl_2·2H_2O$ 0.02g
HCl, 25% 10.0mL

Preparation of Part B: Add the $FeCl_2·4H_2O$ to the HCl. Add distilled/deionized water and bring volume to 1.0L. Add remaining components. Mix thoroughly. Autoclave under 100% N_2 for 15 min at 15 psi pressure–121°C. Cool to room temperature.

Part C:
Composition per 100.0mL:
$NaHCO_3$ 5.0g

Preparation of Part C: Add the $NaHCO_3$ to distilled/deionized water and bring volume to 100.0mL. Mix thoroughly. Filter sterilize. Gas with 90% N_2 + 10% CO_2 to remove residual O_2.

Part D:
Composition per 10.0mL:
Sodium butyrate 0.7g
Sodium caproate 0.3g
Sodium octanoate 0.15g

Preparation of Part D: Add components to distilled/deionized water and bring volume to 10.0mL. Mix thoroughly. Autoclave under 100% N_2 for 15 min at 15 psi pressure–121°C. Cool to room temperature.

Part E:
Composition per 10.0mL:
Yeast extract 1.0g
Thiamine·HCl 100.0μg
p-Aminobenzoic acid 40.0μg
D(+)-Biotin 10.0μg

Preparation of Part E: Add components to distilled/deionized water and bring volume to 10.0mL. Mix thoroughly. Autoclave under 100% N_2 for 15 min at 15 psi pressure–121°C. Cool to room temperature.

Part F:
Composition per 10.0mL:
$Na_2S·9H_2O$ 0.4g

Preparation of Part F: Add $Na_2S·9H_2O$ to distilled/deionized water and bring volume to 10.0mL. Mix thoroughly. Autoclave under 100% N_2 for 15 min at 15 psi pressure–121°C. Cool to room temperature.

Preparation of Medium: To 869.0mL of sterile cooled Part A, aseptically add the remaining sterile solutions in the following order: part B, part C, part D, part E, and part F. Mix thoroughly. Adjust pH to 7.7. Anaerobically distribute under 80% N_2 + 20% CO_2 into sterile tubes or flasks.

Use: For the cultivation and maintenance of *Desulfovibrio baarsii* and *Desulfovibrio sapovorans*.

Medium for Thermophilic Actinomycetes

Composition per liter:

Agar	20.0g
Soluble starch	10.0g
Maize extract	5.0g
NaCl	5.0g
Peptone	5.0g
$CaCl_2$	0.5g

pH 6.5 ± 0.2 at 25°C

Preparation of Medium: Add components to distilled/deionized water and bring volume to 1.0L. Mix thoroughly. Gently heat and bring to boiling. Autoclave for 15 min at 15 psi pressure–121°C. Pour into sterile Petri dishes or distribute into sterile tubes.

Use: For the cultivation of thermophilic actinomycetes.

Megasphaera Medium

Composition per liter:

Yeast extract	4.0g
K_2HPO_4	3.2g
KH_2PO_4	1.6g
Agar	1.0g
NH_4Cl	0.5g
Sodium thioglycolate	0.45g
$CaCl_2$	0.2g
$MgCl_2$	0.2g
Sodium lactate (60% solution)	16.0mL

pH 7.0 ± 0.2 at 25°C

Preparation of Medium: Add components to distilled/deionized water and bring volume to 1.0L. Mix thoroughly. Gently heat and bring to boiling. Distribute into tubes or flasks. Autoclave for 15 min at 15 psi pressure–121°C.

Use: For the cultivation and maintenance of *Megasphaera elsdenii*.

Melissococcus pluton Medium

Composition per liter:

Glucose	10.0g
Neopeptone	5.0g
Peptone	2.5g
Yeast extract	2.5g
Soluble starch	2.0g
Pancreatic digest of casein	2.0g
L-Cysteine·HCl·H_2O	0.25g
Phosphate buffer (1*M*, pH 6.7)	50.0mL

pH 7.2 ± 0.2 at 25°C

Preparation of Medium: Add components to distilled/deionized water and bring volume to 1.0L. Mix thoroughly. Adjust pH to 7.2. Gently heat and bring to boiling. Distribute into tubes or flasks that have been flushed with 90% N_2 + 10% CO_2. Cap with butyl rubber stoppers. Place tubes in a press. Autoclave for 15 min at 15 psi pressure–121°C.

Use: For the cultivation and maintenance of *Melissococcus pluton*.

Membrane Lauryl Sulfate Broth

Composition per liter:

Peptone	39.0g
Lactose	30.0g
Yeast extract	6.0g
Sodium lauryl sulfate	1.0g
Phenol Red	0.2g

pH 7.4 ± 0.2 at 25°C

Preparation of Medium: Add components to distilled/deionized water and bring volume to 1.0L. Mix thoroughly. Distribute into tubes or flasks. Autoclave for 15 min at 15 psi pressure–121°C.

Use: For the enumeration of coliform organisms and *Escherichia coli* in water.

Metal Acetate Agar

Composition per liter:

Agar	15.0g
Sodium acetate	2.0g
Beijerinck's solution	50.0mL
Phosphate buffer solution	50.0mL
Trace elements solution	1.0mL

pH 6.8 ± 0.2 at 25°C

Beijerinck's Solution:

Composition per liter:

NH_4Cl	10.0g
$MgSO_4·7H_2O$	0.4g
$CaCl_2·2H_2O$	0.2g

Preparation of Beijerinck's Solution: Add the $CaCl_2·2H_2O$ to distilled/deionized water and bring volume to 500.0mL. Mix thoroughly. Add the remaining components to distilled/deionized water and bring volume to 500.0mL in a separate flask. Combine the two solutions.

Phosphate Buffer Solution:

Composition per liter:

K_2HPO_4	28.8g
KH_2PO_4	14.4g

Preparation of Phosphate Buffer Solution: Add components to distilled/deionized water and bring volume to 1.0L. Mix thoroughly. Adjust pH to 6.8. Autoclave for 15 min at 15 psi pressure–121°C.

Trace Elements Solution:

Composition per liter:

EDTA	50.0g
H_3BO_3 solution	200.0mL
$ZnSO_4·7H_2O$ solution	100.0mL
$CoCl_2·6H_2O$ solution	50.0mL
$CuSO_4·5H_2O$ solution	50.0mL
$FeSO_4·7H_2O$ solution	50.0mL
$MnCl_2·4H_2O$ solution	50.0mL
$(NH_4)_6Mo_7O_{24}·4H_2O$ solution	50.0mL

H_3BO_3 Solution:

Composition per 200.0mL:

H_3BO_3	11.4g

Preparation of H_3BO_3 Solution: Add H_3BO_3 to distilled/deionized water and bring volume to 200.0mL. Mix thoroughly.

$ZnSO_4·7H_2O$ Solution:

Composition per 100.0mL:

$ZnSO_4·7H_2O$	22.0g

Preparation of $ZnSO_4·7H_2O$ Solution: Add $ZnSO_4·7H_2O$ to distilled/deionized water and bring volume to 100.0mL. Mix thoroughly.

$MnCl_2·4H_2O$ Solution:

Composition per 50.0mL:

$MnCl_2·4H_2O$	5.06g

Preparation of $MnCl_2·4H_2O$ Solution: Add 4.99g of $MnCl_2·4H_2O$ to distilled/deionized water and bring volume to 50.0mL. Mix thoroughly.

$FeSO_4·7H_2O$ Solution:
Composition per 50.0mL:
$FeSO_4·7H_2O$ 4.99g

Preparation of $FeSO_4·7H_2O$ Solution: Add 4.99g of $FeSO_4·7H_2O$ to distilled/deionized water and bring volume to 50.0mL. Mix thoroughly.

$CoCl_2·6H_2O$ Solution:
Composition per 50.0mL:
$CoCl_2·6H_2O$ 1.61g

Preparation of $CoCl_2·6H_2O$ Solution: Add 1.61g of $CoCl_2·6H_2O$ to distilled/deionized water and bring volume to 50.0mL. Mix thoroughly.

$CuSO_4·5H_2O$ Solution:
Composition per 50.0mL:
$CuSO_4·5H_2O$ 1.57g

Preparation of $CuSO_4·5H_2O$ Solution: Add 1.57g of $CuSO_4·5H_2O$ to distilled/deionized water and bring volume to 50.0mL. Mix thoroughly.

$(NH_4)_6Mo_7O_{24}·4H_2O$ Solution:
Composition per 50.0mL:
$(NH_4)_6Mo_7O_{24}·4H_2O$ 1.1g

Preparation of $(NH_4)_6Mo_7O_{24}·4H_2O$ Solution: Add $(NH_4)_6Mo_7O_{24}·4H_2O$ to distilled/deionized water and bring volume to 50.0mL. Mix thoroughly.

Preparation of Trace Elements Solution: Add EDTA to distilled/deionized water and bring volume to 250.0mL. Mix thoroughly. Gently heat and bring to boiling. Continue boiling until dissolved. Add 200.0mL of H_3BO_3 solution, 100.0mL of $ZnSO_4·7H_2O$ solution, 50.0mL of $MnCl_2·4H_2O$ solution, 50.0mL of $FeSO_4·7H_2O$ solution, 50.0mL of $CoCl_2·6H_2O$ solution, 50.0mL of $CuSO_4·5H_2O$ solution, and 50.0mL of $(NH_4)_6Mo_7O_{24}·4H_2O$ solution. Gently heat and bring to boiling. Cool to 70°C. Adjust pH to 6.8 with hot (70°C) 20% KOH solution. Add distilled/deionized water and bring volume to 1.0L. Allow solution to stand in a 2.0L cotton-stoppered flask at room temperature until the solution turns purple (approximately 2 weeks). Filter using two layers of Whatman #1 filter paper. Filter until clear.

Preparation of Medium: Add components, except phosphate buffer solution, to distilled/deionized water and bring volume to 950.0mL. Mix thoroughly. Gently heat and bring to boiling. Autoclave for 15 min at 15 psi pressure–121°C. Cool to 50°–55°C. Aseptically add 50.0mL of sterile phosphate solution. Mix thoroughly. Pour into sterile Petri dishes or distribute into sterile tubes.

Use: For the cultivation and maintenance of *Chlamydomonas reinhardtii.*

Metallogenium Cultivation Broth

Composition per liter:
Starch, hydrolyzed 20.0g
$MnCO_3$ 0.5g

$MnCO_3$:
Composition per 100.0mL:
$MnCl_2$ 20.0g
$NaHCO_3$ (25% solution) 25.0mL

Preparation of $MnCO_3$: Add $MnCl_2$ to distilled/deionized water and bring volume to 100.0mL. Mix thoroughly. Add $NaHCO_3$ solution. Filter through Whatman #1 filter paper. Save the $MnCO_3$ precipitate. Wash and store under distilled/deionized water.

Preparation of Medium: Hydrolyze starch with HCl. Add components to distilled/deionized water and bring volume to 1.0L. Mix thoroughly. Distribute into tubes or flasks. Autoclave for 15 min at 15 psi pressure–121°C.

Use: For the cultivation of *Metallogenium* species.

Metallogenium Isolation Agar

Composition per liter:
Agar 15.0g
Manganese acetate 0.1g

Preparation of Medium: Add components to distilled/deionized water and bring volume to 1.0L. Mix thoroughly. Gently heat and bring to boiling. Distribute into tubes or flasks. Autoclave for 15 min at 15 psi pressure–121°C. Pour into sterile Petri dishes or leave in tubes.

Use: For the isolation and cultivation of *Metallogenium* species.

Metallogenium Medium

Composition per liter:
$MnCO_3$ 2.0g
Starch, hydrolyzed 1.0g
DNA 0.01g
Catalase 5.0mg
Mycoplasma broth base 100.0mL
Yeast extract, ultrafiltrate 100.0mL
Horse serum 10.0mL

***Mycoplasma* Broth Base:**
Composition per liter:
Pancreatic digest of casein 7.0g
NaCl 5.0g
Beef extract 3.0g
Yeast extract 3.0g
Beef heart, solids from infusion 2.0g

Preparation of *Mycoplasma* Broth Base: Add components to distilled/deionized water and bring volume to 1.0L. Mix thoroughly. Autoclave for 15 min at 15 psi pressure–121°C. Cool to 25°C.

$MnCO_3$:
Composition per 100.0mL:
$MnCl_2$ 20.0g
$NaHCO_3$ (25% solution) 25.0mL

Preparation of $MnCO_3$: Add $MnCl_2$ to distilled/deionized water and bring volume to 100.0mL. Mix thoroughly. Add $NaHCO_3$ solution. Filter through Whatman #1 filter paper. Save the $MnCO_3$ precipitate. Wash and store under distilled/deionized water.

Preparation of Medium: Add $MnCO_3$, hydrolyzed starch, and DNA to 25.0mL of distilled/deionized water. Mix thoroughly. Autoclave for 15 min at 15 psi pressure–121°C. Cool to 45°–50°C. Aseptically add 100.0mL of sterile *Mycoplasma* broth base, 100.0mL of ultrafiltrate of yeast extract, 10.0mL of horse serum, and 5.0mg of catalase. Mix thoroughly. Aseptically distribute into sterile tubes or flasks.

Use: For the cultivation of *Metallogenium* species.

Metallosphaera Medium

Composition per liter:
$(NH_4)_2SO_4$ 1.3g
Yeast extract 1.0g

KH_2PO_4 .... 0.28g
$MgSO_4·7H_2O$ 0.25g
$CaCl_2·2H_2O$ 0.07g
$FeCl_3·6H_2O$ 0.02g
$Na_2B_4·10H_2O$ 4.5mg
$MnCl_2·4H_2O$ 1.8mg
$ZnSO_4·7H_2O$ 0.22mg
$CuCl_2·2H_2O$ 0.05mg
$Na_2MoO_4·2H_2O$ 0.03mg
$VOSO_4·2H_2O$ 0.03mg
$CoSO_4$ 0.01mg

Preparation of Medium: Add components to distilled/deionized water and bring volume to 1.0L. Mix thoroughly. Adjust pH to 2.0 using 10*N* H_2SO_4. Distribute into tubes or flasks. Autoclave for 15 min at 15 psi pressure–121°C.

Use: For the cultivation of *Metallosphaera sedula.*

Methanobacillus Medium

Composition per liter:
KH_2PO_4 .... 9.0g
K_2HPO_4 .... 6.0g
NH_4Cl 5.0g
$MgCl_2$ 1.0g
$CaCl_2$ 0.01g
$FeSO_4·7H_2O$ 0.01g
Ethanol 10.0mL
pH 7.4 ± 0.2 at 25°C

Preparation of Medium: Filter sterilize ethanol. Add components, except ethanol, to tap water and bring volume to 990.0mL. Mix thoroughly. Gently heat until dissolved. Autoclave for 20 min at 10psi pressure–115°C. Cool to 45°–50°C. Aseptically add sterile ethanol. Mix thoroughly. Aseptically distribute into sterile tubes or flasks.

Use: For the selective isolation and cultivation of *Methanobacillus* species from mixed cultures.

Methanobacteria Medium

Composition per liter:
Mineral solution 2 .... 50.0mL
Sodium carbonate solution 50.0mL
Mineral solution 1 .... 25.0mL
L-Cysteine-sulfide reducing agent 20.0mL
Wolfe's mineral solution 10.0mL
Vitamin solution 10.0mL
Resazurin (0.025% solution) 4.0mL
pH 7.2 ± 0.2 at 25°C

Mineral Solution 1:
Composition per liter:
K_2HPO_4 .... 6.0g

Preparation of Medium: Add K_2HPO_4 to distilled/deionized water and bring volume to 1.0L. Mix thoroughly.

Mineral Solution 2:
Composition per liter:
NaCl 12.0g
KH_2PO_4 .... 6.0g
$(NH_4)_2SO_4$ 6.0g
$MgSO_4·7H_2O$ 2.4g
$CaCl_2·2H_2O$ 1.6g

Preparation of Mineral Solution 2: Add components to distilled/deionized water and bring volume to 1.0L. Mix thoroughly.

Sodium Carbonate Solution:
Composition per 100.0mL:
Na_2CO_3 .... 8.0g

Preparation of Sodium Carbonate Solution: Add Na_2CO_3 to distilled/deionized water and bring volume to 100.0mL. Mix thoroughly.

L-Cysteine-Sulfide Reducing Agent:
Composition per 20.0mL:
L-Cysteine·HCl·H_2O 0.3g
$Na_2S·9H_2O$ 0.3g

Preparation of L-Cysteine-Sulfide Reducing Agent: Add L-cysteine·HCl·H_2O to 10.0mL of distilled/deionized water. Mix thoroughly. In a separate tube, add $Na_2S·9H_2O$ to 10.0mL of distilled/deionized water. Mix thoroughly. Gas both solutions with 100% N_2 and cap tubes. Autoclave both solutions for 15 min at 15 psi pressure–121°C using fast exhaust. Cool to 50°C. Aseptically combine the two solutions under 100% N_2.

Wolfe's Mineral Solution:
Composition per liter
$MgSO_4·7H_2O$ 3.0g
Nitriloacetic acid 1.5g
NaCl 1.0g
$MnSO_4·H_2O$ 0.5g
$FeSO_4·7H_2O$ 0.1g
$CoCl_2·6H_2O$ 0.1g
$CaCl_2$ 0.1g
$ZnSO_4·7H_2O$ 0.1g
$CuSO_4·5H_2O$ 0.01g
$AlK(SO_4)_2·12H_2O$ 0.01g
H_3BO_3 .... 0.01g
$Na_2MoO_4·2H_2O$ 0.01g

Preparation of Wolfe's Mineral Solution: Add nitrilotriacetic acid to 500.0mL of distilled/deionized water. Dissolve by adjusting pH to 6.5 with KOH. Add remaining components. Add distilled/deionized water and bring volume to 1.0L.

Wolfe's Vitamin Solution:
Composition per liter:
Pyridoxine·HCl 10.0mg
Thiamine·HCl 5.0mg
Riboflavin 5.0mg
Nicotinic acid 5.0mg
Calcium pantothenate 5.0mg
p-Aminobenzoic acid 5.0mg
Thioctic acid 5.0mg
Biotin 2.0mg
Folic acid 2.0mg
Cyanocobalamin 100.0μg

Preparation of Wolfe's Vitamin Solution: Add components to distilled/deionized water and bring volume to 1.0L. Mix thoroughly. Filter sterilize.

Preparation of Medium: Add components, except vitamin solution and L-cysteine-sulfide reducing agent, to distilled/deionized water and bring volume to 970.0mL. Mix thoroughly. Autoclave for 15 min at 15 psi pressure–121°C. Cool under 80% N_2 + 20% CO_2. Aseptically add the sterile vitamin solution and then the sterile L-cysteine-sulfide reducing agent. Adjust the pH to 7.2. Distribute aseptically and anaerobically into sterile tubes.

Use: For the cultivation and maintenance of *Acetogenium kivui, Methanobacterium formicicum, Methanobacterium thermoautotrophicum,* and *Methanobrevibacter arboriphilicus.*

Methanobacteria Medium with Glucose and Yeast Extract

Composition per liter:

Glucose	5.0g
Yeast extract	2.0g
Mineral solution 2	50.0mL
Sodium carbonate solution	50.0mL
Mineral solution 1	25.0mL
L-Cysteine-sulfide reducing agent	20.0mL
Wolfe's mineral solution	10.0mL
Vitamin solution	10.0mL
Resazurin (0.025% solution)	4.0mL

pH 7.2 ± 0.2 at 25°C

Mineral Solution 1:
Composition per liter:

K_2HPO_4	6.0g

Preparation of Mineral Solution 1: Add K_2HPO_4 to distilled/deionized water and bring volume to 1.0L. Mix thoroughly.

Mineral Solution 2:
Composition per liter:

NaCl	12.0g
KH_2PO_4	6.0g
$(NH_4)_2SO_4$	6.0g
$MgSO_4 \cdot 7H_2O$	2.4g
$CaCl_2 \cdot 2H_2O$	1.6g

Preparation of Mineral Solution 2: Add components to distilled/deionized water and bring volume to 1.0L. Mix thoroughly.

Sodium Carbonate Solution:
Composition per 100.0mL:

Na_2CO_3	8.0g

Preparation of Sodium Carbonate Solution: Add Na_2CO_3 to distilled/deionized water and bring volume to 100.0mL. Mix thoroughly.

L-Cysteine-Sulfide Reducing Agent:
Composition per 20.0mL:

L-Cysteine·HCl·H_2O	0.3g
$Na_2S \cdot 9H_2O$	0.3g

Preparation of L-Cysteine-Sulfide Reducing Agent: Add L-cysteine·HCl·H_2O to 10.0mL of distilled/deionized water. Mix thoroughly. In a separate tube, add $Na_2S \cdot 9H_2O$ to 10.0mL of distilled/deionized water. Mix thoroughly. Gas both solutions with 100% N_2 and cap tubes. Autoclave both solutions for 15 min at 15 psi pressure–121°C using fast exhaust. Cool to 50°C. Aseptically combine the two solutions under 100% N_2.

Wolfe's Mineral Solution:
Composition per liter

$MgSO_4 \cdot 7H_2O$	3.0g
Nitriloacetic acid	1.5g
NaCl	1.0g
$MnSO_4 \cdot H_2O$	0.5g
$FeSO_4 \cdot 7H_2O$	0.1g
$CoCl_2 \cdot 6H_2O$	0.1g
$CaCl_2$	0.1g
$ZnSO_4 \cdot 7H_2O$	0.1g
$CuSO_4 \cdot 5H_2O$	0.01g
$AlK(SO_4)_2 \cdot 12H_2O$	0.01g
H_3BO_3	0.01g
$Na_2MoO_4 \cdot 2H_2O$	0.01g

Preparation of Wolfe's Mineral Solution: Add nitrilotriacetic acid to 500.0mL of distilled/deionized water. Dissolve by adjusting pH to 6.5 with KOH. Add remaining components. Add distilled/deionized water to 1.0L.

Wolfe's Vitamin Solution:
Composition per liter:

Pyridoxine·HCl	10.0mg
Thiamine·HCl	5.0mg
Riboflavin	5.0mg
Nicotinic acid	5.0mg
Calcium pantothenate	5.0mg
p-Aminobenzoic acid	5.0mg
Thioctic acid	5.0mg
Biotin	2.0mg
Folic acid	2.0mg
Cyanocobalamin	100.0μg

Preparation of Wolfe's Vitamin Solution: Add components to distilled/deionized water and bring volume to 1.0L. Mix thoroughly. Filter sterilize.

Preparation of Medium: Add components, except vitamin solution and L-cysteine-sulfide reducing agent, to distilled/deionized water and bring volume to 970.0mL. Mix thoroughly. Autoclave for 15 min at 15 psi pressure–121°C. Cool under 80% N_2 + 20% CO_2. Aseptically add the sterile vitamin solution and then the sterile L-cysteine-sulfide reducing agent. Adjust the pH to 7.2. Distribute aseptically and anaerobically into sterile tubes.

Use: For the cultivation and maintenance of *Clostridium saccharolyticum, Clostridium thermoaceticum,* and *Clostridium thermohydrosulfuricum.*

Methanobacteria Medium with Xylose, Yeast Extract, and Tryptone

Composition per liter:

Pancreatic digest of casein	10.0g
Xylose	5.0g
Yeast extract	3.0g
Mineral solution 2	50.0mL
Sodium carbonate solution	50.0mL
Mineral solution 1	25.0mL
L-Cysteine-sulfide reducing agent	20.0mL
Wolfe's mineral solution	10.0mL
Vitamin solution	10.0mL
Resazurin (0.025% solution)	4.0mL

pH 7.2 ± 0.2 at 25°C

Mineral Solution 1:
Composition per liter:

K_2HPO_4	6.0g

Preparation of Medium: Add K_2HPO_4 to distilled/deionized water and bring volume to 1.0L. Mix thoroughly.

Mineral Solution 2:
Composition per liter:

NaCl	12.0g
KH_2PO_4	6.0g
$(NH_4)_2SO_4$	6.0g
$MgSO_4 \cdot 7H_2O$	2.4g
$CaCl_2 \cdot 2H_2O$	1.6g

Preparation of Mineral Solution 2: Add components to distilled/deionized water and bring volume to 1.0L. Mix thoroughly.

Sodium Carbonate Solution:
Composition per 100.0mL:

Na_2CO_3	8.0g

Preparation of Sodium Carbonate Solution: Add Na_2CO_3 to distilled/deionized water and bring volume to 100.0mL. Mix thoroughly.

L-Cysteine-Sulfide Reducing Agent:
Composition per 20.0mL:

L-Cysteine·HCl·H_2O 0.3g
$Na_2S·9H_2O$ 0.3g

Preparation of L-Cysteine-Sulfide Reducing Agent: Add L-cysteine·HCl·H_2O to 10.0mL of distilled/deionized water. Mix thoroughly. In a separate tube, add $Na_2S·9H_2O$ to 10.0mL of distilled/deionized water. Mix thoroughly. Gas both solutions with 100% N_2 and cap tubes. Autoclave both solutions for 15 min at 15 psi pressure–121°C using fast exhaust. Cool to 50°C. Aseptically combine the two solutions under 100% N_2.

Wolfe's Mineral Solution:
Composition per liter:

$MgSO_4·7H_2O$ 3.0g
Nitriloacetic acid 1.5g
NaCl 1.0g
$MnSO_4·H_2O$ 0.5g
$FeSO_4·7H_2O$ 0.1g
$CoCl_2·6H_2O$ 0.1g
$CaCl_2$ 0.1g
$ZnSO_4·7H_2O$ 0.1g
$CuSO_4·5H_2O$ 0.01g
$AlK(SO_4)_2·12H_2O$ 0.01g
H_3BO_3 .... 0.01g
$Na_2MoO_4·2H_2O$ 0.01g

Preparation of Wolfe's Mineral Solution: Add nitrilotriacetic acid to 500.0mL of distilled/deionized water. Dissolve by adjusting pH to 6.5 with KOH. Add remaining components. Add distilled/deionized water to 1.0L.

Wolfe's Vitamin Solution:
Composition per liter:

Pyridoxine·HCl 10.0mg
Thiamine·HCl 5.0mg
Riboflavin 5.0mg
Nicotinic acid 5.0mg
Calcium pantothenate 5.0mg
p-Aminobenzoic acid 5.0mg
Thioctic acid 5.0mg
Biotin 2.0mg
Folic acid 2.0mg
Cyanocobalamin 100.0μg

Preparation of Wolfe's Vitamin Solution: Add components to distilled/deionized water and bring volume to 1.0L. Mix thoroughly. Filter sterilize.

Preparation of Medium: Add components, except vitamin solution and L-cysteine-sulfide reducing agent, to distilled/deionized water and bring volume to 970.0mL. Mix thoroughly. Autoclave for 15 min at 15 psi pressure–121°C. Cool under 80% N_2 + 20% CO_2. Aseptically add the sterile vitamin solution and then the sterile L-cysteine-sulfide reducing agent. Adjust the pH to 7.2. Distribute aseptically and anaerobically into sterile tubes.

Use: For the cultivation and maintenance of *Thermobacteroides acetoethylicus.*

Methanobacteria Medium with Yeast Extract, Sodium Acetate, and Methanol

Composition per liter:

Glucose 5.0g
Sodium acetate 4.1g
Yeast extract 2.0g
Mineral solution 2 .... 50.0mL
Sodium carbonate solution 50.0mL
Mineral solution 1 .... 25.0mL
L-cysteine-sulfide reducing agent 20.0mL
Wolfe's mineral solution 10.0mL
Vitamin solution 10.0mL
Methanol 4.0mL
Resazurin (0.025% solution) 4.0mL

pH 7.2 ± 0.2 at 25°C

Mineral Solution 1:
Composition per liter:

K_2HPO_4 .... 6.0g

Preparation of Mineral Solution 1: Add K_2HPO_4 to distilled/deionized water and bring volume to 1.0L. Mix thoroughly.

Mineral Solution 2:
Composition per liter:

NaCl 12.0g
KH_2PO_4 .... 6.0g
$(NH_4)_2SO_4$ 6.0g
$MgSO_4·7H_2O$ 2.4g
$CaCl_2·2H_2O$ 1.6g

Preparation of Mineral Solution 2: Add components to distilled/deionized water and bring volume to 1.0L. Mix thoroughly.

Sodium Carbonate Solution:
Composition per 100.0mL:

Na_2CO_3 .... 8.0g

Preparation of Sodium Carbonate Solution: Add Na_2CO_3 to distilled/deionized water and bring volume to 100.0mL. Mix thoroughly.

L-Cysteine-Sulfide Reducing Agent:
Composition per 20.0mL:

L-Cysteine·HCl·H_2O 300.0mg
$Na_2S·9H_2O$ 300.0mg

Preparation of L-Cysteine-Sulfide Reducing Agent: Add L-cysteine·HCl·H_2O to 10.0mL of distilled/deionized water. Mix thoroughly. In a separate tube, add $Na_2S·9H_2O$ to 10.0mL of distilled/deionized water. Mix thoroughly. Gas both solutions with 100% N_2 and cap tubes. Autoclave both solutions for 15 min at 15 psi pressure–121°C using fast exhaust. Cool to 50°C. Aseptically combine the two solutions under 100% N_2.

Wolfe's Mineral Solution:
Composition per liter:

$MgSO_4·7H_2O$ 3.0g
Nitriloacetic acid 1.5g
NaCl 1.0g
$MnSO_4·H_2O$ 0.5g
$FeSO_4·7H_2O$ 0.1g
$CoCl_2·6H_2O$ 0.1g
$CaCl_2$ 0.1g
$ZnSO_4·7H_2O$ 0.1g
$CuSO_4·5H_2O$ 0.01g
$AlK(SO_4)_2·12H_2O$ 0.01g
H_3BO_3 .... 0.01g
$Na_2MoO_4·2H_2O$ 0.01g

Preparation of Wolfe's Mineral Solution: Add nitrilotriacetic acid to 500.0mL of distilled/deionized water. Dissolve by adjusting pH to 6.5 with KOH. Add remaining components. Add distilled/deionized water to 1.0L.

Wolfe's Vitamin Solution:
Composition per liter:

Pyridoxine·HCl 10.0mg
Thiamine·HCl 5.0mg
Riboflavin 5.0mg
Nicotinic acid 5.0mg
Calcium pantothenate 5.0mg
p-Aminobenzoic acid 5.0mg
Thioctic acid 5.0mg
Biotin 2.0mg
Folic acid 2.0mg
Cyanocobalamin 100.0μg

Preparation of Wolfe's Vitamin Solution: Add components to distilled/deionized water and bring volume to 1.0L. Mix thoroughly. Filter sterilize.

Preparation of Medium: Add components, except vitamin solution, L-cysteine-sulfide reducing agent, and methanol, to distilled/deionized water and bring volume to 970.0mL. Mix thoroughly. Autoclave for 15 min at 15 psi pressure–121°C. Cool under 80% N_2 + 20% CO_2. Filter sterilize methanol. Aseptically add 4.0mL of sterile methanol to cooled, sterile basal medium. Aseptically add the sterile vitamin solution and then the sterile L-cysteine-sulfide reducing agent. Adjust the pH to 7.2. Distribute aseptically and anaerobically into sterile tubes.

Use: For the cultivation and maintenance of *Butyribacterium methylotrophicum*.

Methanobacterium alcaliphilum Medium

Composition per liter:

$NaHCO_3$ 10.0g
Yeast extract 2.0g
Peptone 2.0g
NH_4Cl 1.0g
L-Cysteine·HCl·H_2O 0.5g
K_2HPO_4 .... 0.4g
$MgCl_2 \cdot 6H_2O$ 0.1g
$CaCl_2$ 0.02g
Resazurin 1.0mg
Salt solution 5.0mL

pH 8.4 ± 0.2 at 25°C

Salt Solution:
Composition per 100.0mL:

Sodium EDTA·$2H_2O$ 0.1g
$CoCl_2 \cdot 6H_2O$ 0.03g
$MnCl_2 \cdot 4H_2O$ 0.02g
$ZnCl_2$ 0.02g
$AlCl_3 \cdot 6H_2O$ 8.0mg
$CuCl_2 \cdot 2H_2O$ 4.0mg
$NiSO_4 \cdot 6H_2O$ 4.0mg
Na_2SeO_3 .... 2.7mg
$FeSO_4 \cdot 7H_2O$ 2.0mg
H_3BO_3 .... 2.0mg
$NaMoO_4 \cdot 2H_2O$ 2.0mg

Preparation of Salt Solution: Add components to distilled/deionized water and bring volume to 100.0mL. Mix thoroughly.

Preparation of Medium: Add components, except $NaHCO_3$, yeast extract, peptone, and L-cysteine·HCl·H_2O, to distilled/deionized water and bring volume to 990.0mL. Gently heat and bring to boiling under O_2-free 100% N_2. Continue boiling until rezasurin becomes pale, indicating partial reduction. Add the yeast extract, peptone, and L-cysteine·HCl·H_2O and continue boiling under O_2-free 100% N_2 until rezasurin becomes colorless, indicating complete reduction. Cool to room temperature under O_2-free 100% N_2. Add $NaHCO_3$ to 10.0mL of distilled/deionized water. Mix thoroughly. Gas with O_2-free 100% N_2 in a sealed tube. Add reduced $NaHCO_3$ solution to cooled reduced medium. Distribute anaerobically into tubes. Cap with butyl rubber stoppers and secure with closures. Autoclave for 15 min at 15 psi pressure–121°C with fast exhaust.

Use: For the cultivation and maintenance of *Methanobacterium alcaliphilum*.

Methanobacterium Enrichment Medium

Composition per liter:

$CaCO_3$ 100.0g
K_2HPO_4 .... 5.0g
$(NH_4)_2SO_4$ 0.3g
$MgSO_47H_2O$ 0.1g
$FeSO_4 \cdot 7H_2O$ 0.02g
Na_2CO_3 solution 10.0mL
$Na_2S \cdot 9H_2O$ solution 10.0mL
Ethanol 10.0mL
Yeast autolysate 5.0mL

pH 7.2 ± 0.2 at 25°C

Na_2CO_3 Solution:
Composition per 10.0mL:

$NaHCO_3$ 0.5g

Preparation of Na_2CO_3 Solution: Add Na_2CO_3 to distilled/deionized water and bring volume to 10.0mL. Mix thoroughly. Filter sterilize.

$Na_2S \cdot 9H_2O$ Solution:
Composition per 10.0mL:

$Na_2S \cdot 9H_2O$ 0.1g

Preparation of $Na_2S \cdot 9H_2O$ Solution: Add $Na_2S \cdot 9H_2O$ to distilled/deionized water and bring volume to 10.0mL. Mix thoroughly. Filter sterilize.

Preparation of Medium: Filter sterilize ethanol. Add components, except Na_2CO_3 solution, $Na_2S \cdot 9H_2O$ solution, and ethanol, to distilled/deionized water and bring volume to 970.0mL. Mix thoroughly. Gently heat and bring to boiling. Autoclave for 15 min at 15 psi pressure–121°C. Cool to 45°–50°C. Aseptically add sterile ethanol, Na_2CO_3 solution, and $Na_2S \cdot 9H_2O$ solution. Mix thoroughly. Aseptically distribute into sterile tubes or flasks.

Use: For the cultivation and enrichment of *Methanobacterium* species.

Methanobacterium espanolae Medium

Composition per 1020.0mL:

Sodium acetate·$3H_2O$ 2.5g
Na_2CO_3 .... 0.5g
$(NH_4)_2SO_4$ 0.45g
K_2HPO_4 .... 0.29g
KH_2PO_4 .... 0.18g
$MgSO_4 \cdot 7H_2O$ 0.12g
$CaCl_2 \cdot 2H_2O$ 0.06g
NaCl 0.05g
L-Cysteine·HCl 0.25g
$Na_2S \cdot 9H_2O$ 0.25g
Resazurin 1.0mg
Trace elements solution 10.0mL
Vitamin solution 10.0mL

pH 5.5 ± 0.2 at 25°C

Trace Elements Solution:
Composition per liter:

$MgSO_4 \cdot 7H_2O$	3.0g
Nitrilotriacetic acid	1.5g
NaCl	1.0g
$MnSO_4 \cdot 2H_2O$	0.5g
$CoSO_4 \cdot 7H_2O$	0.18g
$ZnSO_4 \cdot 7H_2O$	0.18g
$CaCl_2 \cdot 2H_2O$	0.1g
$FeSO_4 \cdot 7H_2O$	0.1g
$NiCl_2 \cdot 6H_2O$	0.025g
$KAl(SO_4)_2 \cdot 12H_2O$	0.02g
$CuSO_4 \cdot 5H_2O$	0.01g
H_3BO_3	0.01g
$Na_2MoO_4 \cdot 2H_2O$	0.01g
$Na_2SeO_3 \cdot 5H_2O$	0.3mg

Preparation of Trace Elements Solution: Add nitrilotriacetic acid to 500.0mL of distilled/deionized water. Adjust pH to 6.5 with KOH. Add remaining components. Add distilled/deionized water to 1.0L.

Vitamin Solution:
Composition per liter:

Pyridoxine·HCl	10.0mg
Calcium DL-pantothenate	5.0mg
Lipoic acid	5.0mg
Nicotinic acid	5.0mg
p-Aminobenzoic acid	5.0mg
Riboflavin	5.0mg
Thiamine·HCl	5.0mg
Biotin	2.0mg
Folic acid	2.0mg
Vitamin B_{12}	0.1mg

Preparation of Vitamin Solution: Add components to distilled/deionized water and bring volume to 1.0L. Adjust pH to 7.0. Mix thoroughly. Sparge with 80% N_2 + 20% CO_2.

Preparation of Medium: Prepare and dispense the medium anaerobically with 80% N_2 and 20% CO_2. Add components to distilled/deionized water and bring volume to 1.0L. Mix thoroughly. Adjust pH to 5.5–6.0. Anaerobically distribute into tubes or flasks. Autoclave for 15 min at 15 psi pressure–121°C. Check the pH of the medium after autoclaving.

Use: For the cultivation of *Methanobacterium espanolae.*

Methanobacterium formicicum Medium

Composition per liter:

$NaCH_3CO_2$	2.0g
$(NH_4)_2SO_4$	1.48g
NaCl	0.45g
L-Cysteine·HCl·H_2O	0.27g
$CaCl_2 \cdot 2H_2O$	0.06g
$Fe(NH_4)(SO_4)_2$	0.06g
$MgSO_4$	45.0mg
Na_2MoO_4	24.0mg
K_2HPO_4	21.0mg
KH_2PO_4	21.0mg
Resazurin	1.0mg
Na_2SeO_3	0.2mg
$Na_2S \cdot 9H_2O$ solution	1.0mL

pH 6.8 ± 0.2 at 25°C

$Na_2S \cdot 9H_2O$ Solution:
Composition per 100.0mL:

$Na_2S \cdot 9H_2O$	25.0g

Preparation of $Na_2S \cdot 9H_2O$ Solution: Bring 100.0mL of distilled/deionized water to boiling. Cool to room temperature while sparging with 100%N_2. Add $Na_2S \cdot 9H_2O$ to the 100.0mL of anaerobic water. Mix thoroughly. Distribute into serum bottles fitted with butyl rubber stoppers and aluminum seals. Do not grease stoppers. Autoclave for 20 min at 15 psi pressure–121°C.

Preparation of Medium: Add components, except $Na_2S \cdot 9H_2O$ solution, to distilled/deionized water and bring volume to 1.0L. Mix thoroughly. Adjust pH to 6.8 with concentrated HCl. Distribute into tubes or flasks. Autoclave for 15 min at 15 psi pressure–121°C. Cool medium while sparging with 100% N_2. Prior to inoculation, aseptically and anaerobically add 1.0mL of sterile $Na_2S \cdot 9H_2O$ solution per liter of medium. Repeat the addition of 1.0mL of sterile $Na_2S \cdot 9H_2O$ solution per liter of medium every 48 hr during growth.

Use: For the cultivation of *Methanobacterium formicicum.*

Methanobacterium Medium

Composition per 1010.0mL:

$NaHCO_3$	4.0g
Sodium formate	2.0g
Sodium acetate	1.0g
Yeast extract	1.0g
L-Cysteine·HCl	0.5g
KH_2PO_4	0.5g
$Na_2S \cdot 9H_2O$	0.5g
$MgSO_4 \cdot 7H_2O$	0.4g
NaCl	0.4g
NH_4Cl	0.4g
$CaCl_2 \cdot 2H_2O$	0.05g
$FeSO_4 \cdot 7H_2O$	2.0mg
Resazurin	1.0mg
Sludge fluid	50.0mL
Fatty acid mixture	20.0mL
Trace elements solution SL-10	1.0mL

pH 6.7 ± 0.2 at 25°C

Sludge Fluid:
Composition per 100.0mL:

Preparation of Sludge Fluid: To 100.0mL of sludge from an anaerobic digester, add 0.4g of yeast extract. Sparge with 100% N_2 for a few minutes. Incubate at 37°C for 24 hr. Centrifuge the sludge at 13,000 × g for 15 min. Decant the clear supernatant solution. Sparge with 100% N_2 for a few minutes. Store in screw-capped bottles at room temperature in the dark.

Fatty Acid Mixture:
Composition per 20.0mL:

α-Methylbutyric acid	0.5g
Isobutyric acid	0.5g
Isovaleric acid	0.5g
Valeric acid	0.5g

Preparation of Fatty Acid Mixture: Add components to distilled/deionized water and bring volume to 20.0mL. Mix thoroughly. Adjust pH to 7.5 with concentrated NaOH.

Trace Elements Solution SL-10:
Composition per liter:

$FeCl_2 \cdot 4H_2O$	1.5g
$CoCl_2 \cdot 6H_2O$	190.0mg
$MnCl_2 \cdot 4H_2O$	100.0mg
$ZnCl_2$	70.0mg
$Na_2MoO_4 \cdot 2H_2O$	36.0mg
$NiCl_2 \cdot 6H_2O$	24.0mg
H_3BO_3	6.0mg
$CuCl_2 \cdot 2H_2O$	2.0mg
HCl (25% solution)	10.0mL

Preparation of Trace Elements Solution SL-10: Add $FeCl_2 \cdot 4H_2O$ to 10.0mL of HCl solution. Mix thoroughly. Add distilled/deionized water and bring volume to 1.0L. Add remaining components. Mix thoroughly. Autoclave for 15 min at 15 psi pressure–121°C.

Preparation of Medium: Prepare and dispense medium anaerobically under 80% H_2 + 20% CO_2. Add components to distilled/deionized water and bring volume to 1010.0mL. Mix thoroughly. Sparge with 80% H_2 + 20% CO_2. Autoclave for 15 min at 15 psi pressure–121°C.

Use: For the cultivation and maintenance of *Methanobacterium* species.

Methanobacterium II Medium (DSMZ Medium 825)

Composition per 1050mL:

Yeast extract	1.0g
NH_4Cl	1.0g
NaCl	0.6g
Cysteine-HCl·H_2O	0.5g
Sodium acetate	0.5g
K_2HPO_4	0.3g
KH_2PO_4	0.3g
$MgCl_2 \cdot 6H_2O$	0.2g
$CaCl_2 \cdot 2H_2O$	0.1g
KCl	0.1g
Resazurin	0.5mg
$NaHCO_3$ solution	40.0mL
Trace elements solution	10.0mL
Vitamin solution	10.0mL
$Na_2S \cdot 9H_2O$ solution	10.0mL

pH 7.0 ± 0.2 at 25°C

Trace Elements Solution:
Composition per liter:

$MgSO_4 \cdot 7H_2O$	3.0g
Nitrilotriacetic acid	1.5g
NaCl	1.0g
$MnSO_4 \cdot 2H_2O$	0.5g
$CoSO_4 \cdot 7H_2O$	0.18g
$ZnSO_4 \cdot 7H_2O$	0.18g
$CaCl_2 \cdot 2H_2O$	0.1g
$FeSO_4 \cdot 7H_2O$	0.1g
$NiCl_2 \cdot 6H_2O$	0.025g
$KAl(SO_4)_2 \cdot 12H_2O$	0.02g
H_3BO_3	0.01g
$Na_2MoO_4 \cdot 4H_2O$	0.01g
$CuSO_4 \cdot 5H_2O$	0.01g
$Na_2SeO_3 \cdot 5H_2O$	0.3mg

Preparation of Trace Elements Solution: Add nitrilotriacetic acid to 500.0mL of distilled/deionized water. Dissolve by adjusting pH to 6.5 with KOH. Add remaining components. Add distilled/deionized water to 1.0L. Mix thoroughly.

Vitamin Solution:
Composition per liter:

Pyridoxine-HCl	10.0mg
Thiamine-HCl·$2H_2O$	5.0mg
Riboflavin	5.0mg
Nicotinic acid	5.0mg
D-Ca-pantothenate	5.0mg
p-Aminobenzoic acid	5.0mg
Lipoic acid	5.0mg
Biotin	2.0mg
Folic acid	2.0mg
Vitamin B_{12}	0.1mg

Preparation of Vitamin Solution: Add components to distilled/deionized water and bring volume to 1.0L. Mix thoroughly. Sparge with 80% H_2 + 20% CO_2. Filter sterilize.

$Na_2S \cdot 9H_2O$ Solution:
Composition per 10.0mL:

$Na_2S \cdot 9H_2O$	0.3g

Preparation of $Na_2S \cdot 9H_2O$ Solution: Add $Na_2S \cdot 9H_2O$ to distilled/deionized water and bring volume to 10.0mL. Sparge with N_2. Autoclave for 15 min at 15 psi pressure–121°C. Cool to 25°C. Store anaerobically.

$NaHCO_3$ Solution:
Composition per 100.0mL:

$NaHCO_3$	10.0g

Preparation of $NaHCO_3$ Solution: Add $NaHCO_3$ to distilled/deionized water and bring volume to 100.0mL. Mix thoroughly. Autoclave for 15 min at 15 psi pressure–121°C. Cool to 25°C. Must be prepared freshly.

Preparation of Medium: Prepare and dispense medium under 80% H_2 + 20% CO_2 gas mixture. Add components, except $NaHCO_3$ solution and $Na_2S \cdot 9H_2O$ solution, to 1.0L distilled/deionized water. Mix thoroughly. Sparge with 80% H_2 + 20% CO_2. Dispense 5.0mL aliquots into Hungate tubes under 80% H_2 + 20% CO_2 gas mixture. Autoclave for 15 min at 15 psi pressure–121°C. Prior to use inject, for each 5.0mL medium, 0.2mL sterile $NaHCO_3$ solution, and 0.05mL sterile $Na_2S \cdot 9H_2O$ solution. The pH should be 7.0. After inoculation, pressurize culture vessels to 2 bar 80% H_2 + 20% CO_2 overpressure.

Use: For the cultivation of *Methanobacterium formicicum, Methanosarcina barkeri, Methanosarcina mazei=Methanococcus mazei (Methanosarcina frisia), Methanobacterium bryantii, and Methanobacterium oryzae (Methanobacterium* sp.*).*

Methanobacterium ruminantium Medium

Composition per liter:

$NaHCO_3$	6.0g
NaCl	2.0g
L-Cysteine·HCl·H_2O	1.0g
$K_2HPO_4 \cdot 3H_2O$	1.0g
KH_2PO_4	1.0g
NH_4Cl	1.0g
$CaCl_2 \cdot 2H_2O$	0.1g
$MgSO_4 \cdot 7H_2O$	0.1g
Resazurin	1.0mg
Rumen fluid	300.0mL
$Na_2S \cdot 9H_2O$ solution	10.0mL

6.8 ± 0.2 at 25°C

$Na_2S \cdot 9H_2O$ Solution:
Composition per 10.0mL:

$Na_2S \cdot 9H_2O$	0.25g

Preparation of $Na_2S \cdot 9H_2O$ Solution: Add $Na_2S \cdot 9H_2O$ to distilled/deionized water and bring volume to 10.0mL. Mix thoroughly. Autoclave for 15 min at 15 psi pressure–121°C. Cool to 25°C.

Preparation of Medium: Prepare and distribute medium anaerobically under 80% H_2 + 20% CO_2. Add components, except rumen fluid and $Na_2S \cdot 9H_2O$ solution, to distilled/deionized water and bring volume to 690.0mL. Mix thoroughly. Gently heat and bring to boiling. Continue boiling until resazurin turns colorless, indicating reduction. Autoclave for 15 min at 15 psi pressure–121°C. Cool to

25°C. Aseptically add 10.0mL of sterile $Na_2S·9H_2O$ solution and 300.0mL of sterile rumen fluid. Mix thoroughly. Aseptically and anaerobically distribute into sterile tubes or flasks.

Use: For the cultivation of *Methanobacterium ruminantium.*

Methanobacterium subterraneum Medium (DSMZ Medium 814)

Composition per liter:

Na-formate	3.2g
Na-acetate	1.0g
Tryptone	1.0g
K_2HPO_4	0.5g
NaCl	0.45g
NH_4Cl	0.4g
$MgCl_2·6H_2O$	0.03g
$FeSO_4·7H_2O$	0.003g
Resazurin	0.5mg
Trace elements solution	10.0mL
Cysteine solution	10.0mL
$Na_2S·9H_2O$ solution	10.0mL
$NaHCO_3$ solution	10.0mL
Seven vitamin solution	1.0mL

pH 8.3 ± 0.2 at 25°C

$NaHCO_3$ Solution:

Composition per 10.0mL:

$NaHCO_3$	2.0g

Preparation of $NaHCO_3$ Solution: Add $NaHCO_3$ to distilled/deionized water and bring volume to 10.0mL. Mix thoroughly. Sparge with 100% N_2. Autoclave for 15 min at 15 psi pressure–121°C. Cool to 25°C. Must be prepared freshly.

Cysteine Solution:

Composition per 10.0mL:

L-Cysteine-HCl·H_2O	0.25g

Preparation of Cysteine Solution: Add L-cysteine·HCl·H_2O to distilled/deionized water and bring volume to 10.0mL. Mix thoroughly. Sparge with 100% N_2. Autoclave for 15 min at 15 psi pressure–121°C.

$Na_2S·9H_2O$ Solution:

Composition per 10.0mL:

$Na_2S·9H_2O$	0.25g

Preparation of $Na_2S·9H_2O$ Solution: Add $Na_2S·9H_2O$ to distilled/deionized water and bring volume to 10.0mL. Sparge with N_2. Autoclave for 15 min at 15 psi pressure–121°C. Cool to 25°C. Store anaerobically.

Trace Elements Solution:

Composition per liter:

$MgSO_4·7H_2O$	3.0g
Nitrilotriacetic acid	1.5g
NaCl	1.0g
$MnSO_4·2H_2O$	0.5g
$CoSO_4·7H_2O$	0.18g
$ZnSO_4·7H_2O$	0.18g
$CaCl_2·2H_2O$	0.1g
$FeSO_4·7H_2O$	0.1g
$NiCl_2·6H_2O$	0.025g
$KAl(SO_4)_2·12H_2O$	0.02g
H_3BO_3	0.01g
$Na_2MoO_4·4H_2O$	0.01g
$CuSO_4·5H_2O$	0.01g
$Na_2SeO_3·5H_2O$	0.3mg

Preparation of Trace Elements Solution: Add nitrilotriacetic acid to 500.0mL of distilled/deionized water. Dissolve by adjusting pH to 6.5 with KOH. Add remaining components. Add distilled/deionized water to 1.0L. Mix thoroughly.

Seven Vitamin Solution:

Composition per liter:

Pyridoxine hydrochloride	300.0mg
Thiamine-HCl·$2H_2O$	200.0mg
Nicotinic acid	200.0mg
Vitamin B_{12}	100.0mg
Calcium pantothenate	100.0mg
p-Aminobenzoic acid	80.0mg
D(+)-Biotin	20.0mg

Preparation of Seven Vitamin Solution: Add components to distilled/deionized water and bring volume to 1.0L. Sparge with 100% N_2. Mix thoroughly. Filter sterilize.

Preparation of Medium: Prepare and dispense medium under 100% N_2. Add components, except $Na_2S·9H_2O$ solution, seven vitamin solution, cysteine solution, and $NaHCO_3$ solution, to distilled/deionized water and bring volume to 969.0mL. Mix thoroughly. Flush medium with N_2 for 5 min. Adjust medium pH to 8.3. Autoclave for 15 min at 15 psi pressure–121°C. Aseptically and anaerobically add 1.0mL seven vitamin mix, 10.0mL cysteine solution, 10.0mL $Na_2S·9H_2O$, and 10.0mL $NaHCO_3$ solution. Mix thoroughly. Adjust pH to 8.3. Aseptically and anaerobically distribute into sterile tubes or bottles.

Use: For the cultivation of *Methanobacterium subterraneum.*

Methanobacterium thermoautotrophicum Marburg Medium

Composition per liter:

L-Cysteine·HCl·H_2O	12.5g
$Na_2S·9H_2O$	12.5g
KH_2PO_4	6.8g
NH_4Cl	2.1g
Na_2CO_3 (1*M* solution)	32.0mL
Trace elements solution	10.0mL
Resazurin (0.2% solution)	0.3mL

Trace Elements Solution:

Composition per liter:

$MgCl_2·6H_2O$	4.0g
Nitrilotriacetic acid	3.0g
$FeCl_2·4H_2O$	1.0g
$NiCl_2·6H_2O$	0.12g
$CoCl_2·6H_2O$	0.02g
$NaMoO_4·2H_2O$	0.02g

Preparation of Trace Elements Solution: Add nitrilotriacetic acid to 500.0mL of distilled/deionized water. Adjust pH to 6.5 with NaOH. Add remaining components. Mix thoroughly. Add distilled/deionized water to 1.0L. Adjust pH to 7.0.

Preparation of Medium: Prepare and dispense medium under 80% H_2 + 20% CO_2. Add components, except L-cysteine·HCl·H_2O and $Na_2S·9H_2O$, to distilled/deionized water and bring volume to 1.0L. Mix thoroughly. Gently heat and bring to boiling. Continue boiling for 3 min. Cool to room temperature while sparging with 80% H_2 + 20% CO_2. Add L-cysteine·HCl·H_2O and $Na_2S·9H_2O$. Mix thoroughly. Anaerobically distribute into serum bottles fitted with butyl rubber stoppers and aluminum seals. Do not grease stoppers. Autoclave for 15 min at 15 psi pressure–121°C.

Use: For the cultivation of *Methanobacterium thermoautotrophicum.*

Methanobacterium thermoautotrophicum Medium (DSMZ Medium 131)

Composition per liter:

Na_2CO_3	4.0g
$(NH_4)_2SO_4$	1.5g
NaCl	0.6g
KH_2PO_4	0.3g
K_2HPO_4	0.15g
$MgSO_4 \cdot 7H_2O$	0.12g
$CaCl_2 \cdot 2H_2O$	0.08g
$FeSO_4 \cdot 7H_2O$	4.0mg
Resazurin	1.0mg
Vitamin solution	10.0mL
Trace elements solution	10.0mL
L-Cysteine solution	10.0mL
$Na_2S \cdot 9H_2O$ solution	10.0mL

pH 7.2 ± 0.2 at 25°C

Vitamin Solution:

Composition per liter:

Pyridoxine-HCl	10.0mg
Thiamine-HCl·$2H_2O$	5.0mg
Riboflavine	5.0mg
Nicotinic acid	5.0mg
Ca-pantothenate	5.0mg
Biotin	2.0mg
Folic acid	2.0mg
p-Aminobenzoic acid	1.0mg
Vitamin B_{12}	0.01mg

Preparation of Vitamin Solution: Add components to distilled/deionized water and bring volume to 1.0L. Mix thoroughly. Sparge with 80% H_2 + 20% CO_2. Filter sterilize.

Trace Elements Solution:

$MgSO_4 \cdot 7H_2O$	6.2g
NaCl	1.0g
Na_2-EDTA	0.64g
$MnSO_4 \cdot 4H_2O$	0.55g
$ZnSO_4 \cdot 7H_2O$	0.18g
$CoCl_2 \cdot 6H_2O$	0.17g
$CaCl_2 \cdot 2H_2O$	0.13g
$FeSO_4 \cdot 7H_2O$	0.1g
$CuSO_4$	0.05g
$NiCl_2 \cdot 6H_2O$	0.025g
$KAl(SO_4)_2 \cdot 12H_2O$	0.018g
$Na_2MoO_4 \cdot 4H_2O$	0.011g
H_3BO_3	0.01g

Preparation of Trace Elements Solution: Add Na_2-EDTA to 500.0mL distilled/deionized water. Mix thoroughly. Add other components and bring volume to 1.0L with distilled/deionized water. Mix thoroughly. Sparge with 80% H_2 + 20% CO_2. Autoclave for 15 min at 15 psi pressure–121°C.

$Na_2S \cdot 9H_2O$ Solution:

Composition per 10.0mL:

$Na_2S \cdot 9H_2O$	1.5g

Preparation of $Na_2S \cdot 9H_2O$ Solution: Add $Na_2S \cdot 9H_2O$ to distilled/deionized water and bring volume to 10.0mL. Sparge with N_2. Autoclave for 15 min at 15 psi pressure–121°C. Cool to 25°C. Store anaerobically.

L-Cysteine Solution:

Composition per 10.0mL:

L-Cysteine·HCl·H_2O	1.5g

Preparation of L-Cysteine Solution: Add L-cysteine·HCl·H_2O to distilled/deionized water and bring volume to 10.0mL. Mix thoroughly. Sparge with 100% N_2. Autoclave for 15 min at 15 psi pressure–121°C.

Preparation of Medium: Prepare and dispense medium under an oxygen-free 80% H_2 + 20% CO_2 gas mixture. Add components, except vitamin solution, L-cysteine solution, and $Na_2S \cdot 9H_2O$ solution, to distilled/deionized water and bring volume to 1.0L. Mix thoroughly. Sparge with 80% H_2 + 20% CO_2. Autoclave for 15 min at 15 psi pressure–121°C. Cool to 25°C while sparging with 80% H_2 + 20% CO_2. Aseptically and anaerobically add 10.0mL vitamin solution, 10.0mL of sterile L-cysteine solution, and 10.0mL of sterile $Na_2S \cdot 9H_2O$ solution. Mix thoroughly. Aseptically and anaerobically distribute into sterile tubes or flasks. Alternately the medium can be distributed to tubes under anaerobic conditions and autoclaved in tubes prior to addition of substrate solution, vitamin solution, and $Na_2S \cdot 9H_2O$ solution. Appropriate amounts of these solutions can then be added to each tube by syringes to yield the desired concentrations.

Use: For the cultivation of *Ectothiorhodospira shaposhnikovii* and *Methanothermobacter thermautotrophicus.*

Methanobacterium thermoautotrophicum Medium

Composition per liter:

Na_2CO_3	4.0g
$(NH_4)_2SO_4$	1.5g
NaCl	0.6g
KH_2PO_4	0.3g
K_2HPO_4	0.15g
$MgSO_4 \cdot 7H_2O$	0.12g
$CaCl_2 \cdot 2H_2O$	0.08g
$FeSO_4 \cdot 7H_2O$	4.0mg
Resazurin	1.0mg
Trace elements solution	10.0mL
Vitamin solution	10.0mL
L-Cysteine·HCl solution	10.0mL
$Na_2S \cdot 9H_2O$ solution	10.0mL

pH 7.2 ± 0.2 at 25°C

Trace Elements Solution:

Composition per liter:

NaCl	1.0g
Disodium EDTA	0.64g
$MnSO_4 \cdot 4H_2O$	0.55g
$MgSO_4 \cdot 7H_2O$	0.2g
$ZnSO_4 \cdot 7H_2O$	0.18g
$CoCl_2 \cdot 6H_2O$	0.17g
$CuSO_4$	0.15g
$CaCl_2 \cdot 2H_2O$	0.13g
$FeSO_4 \cdot 7H_2O$	0.1g
$NiCl_2 \cdot H_2O$	0.025g
$KAl(SO_4)_2 \cdot 12H_2O$	0.018g
$Na_2MoO_4 \cdot 2H_2O$	0.011g
H_3BO_3	0.01g

Preparation of Trace Elements Solution: Add disodium EDTA to 500.0mL of distilled/deionized water. Mix thoroughly to dissolve. Add remaining components. Bring volume to 1.0L with distilled/deionized water.

Vitamin Solution:

Composition per liter:

Pyridoxine·HCl	10.0mg
Calcium DL-pantothenate	5.0mg
Nicotinic acid	5.0mg
Robiflavin	5.0mg
Thiamine·HCl	5.0mg

Biotin 2.0mg
Folic acid 2.0mg
p-Aminobenzoic acid 1.0mg
Cyanocobalamin 0.01mg

Preparation of Vitamin Solution: Add components to distilled/deionized water and bring volume to 1.0L. Adjust pH to 7.0. Mix thoroughly.

L-Cysteine·HCl Solution:
Composition per 10.0mL:
L-Cysteine·HCl 1.5g

Preparation of L-Cysteine·HCl Solution: Add L-cysteine·HCl to distilled/deionized water and bring volume to 10.0mL. Mix thoroughly. Autoclave under 80% N_2 + 20% CO_2 for 15 min at 15 psi pressure–121°C.

$Na_2S·9H_2O$ Solution:
Composition per 10.0mL:
$Na_2S·9H_2O$ 1.5g

Preparation of $Na_2S·9H_2O$ Solution: Add $Na_2S·9H_2O$ to distilled/deionized water and bring volume to 10.0mL. Mix thoroughly. Sparge with 80% N_2 + 20% CO_2. Autoclave for 15 min at 15 psi pressure–121°C.

Preparation of Medium: Prepare and dispense medium under 80% H_2 + 20% CO_2. Add components, except L-cysteine·HCl solution and $Na_2S·9H_2O$ solution, to distilled/deionized water and bring volume to 980.0mL. Mix thoroughly. Anaerobically distribute into tubes or flasks fitted with butyl rubber stoppers. Autoclave for 15 min at 15 psi pressure–121°C. Anaerobically add 10.0mL of sterile L-cysteine·HCl solution and 10.0mL of sterile $Na_2S·9H_2O$ solution to each liter of medium or, using a syringe, inject the appropriate amount of sterile L-cysteine·HCl solution and sterile $Na_2S·9H_2O$ solution into individual tubes containing medium.

Use: For the cultivation and maintenance of *Methanobacterium thermoautotrophicus* and *Methanoculleus oldenburgensis*.

Methanobacterium thermoautotrophicum Medium, Taylor and Pirt

Composition per liter:
Na_2CO_3 .... 4.0g
$(NH_4)_2SO_4$ 3.0g
NaCl 1.2g
KH_2PO_4 .... 0.6g
L-Cysteine·HCl·H_2O 0.5g
K_2HPO_4 .... 0.3g
Nitrilotriacetic acid 0.03g
$CoCl_2$ 0.02g
$CaCl_2$ 0.01g
$FeSO_4$ 0.01g
$MgSO_4$ 0.01g
$MnSO_4$ 0.01g
$ZnSO_4$ 2.0mg
Resazurin 1.0mg
$AlK(SO_4)_2$ 0.2mg
$CuSO_4$ 0.2mg
H_3BO_3 .... 0.2mg
Na_2MoO_4 .... 0.2mg
$Na_2S·9H_2O$ solution 10.0mL
pH 7.2 ± 0.2 at 25°C

$Na_2S·9H_2O$ Solution:
Composition per 10.0mL:
$Na_2S·9H_2O$ 0.5g

Preparation of $Na_2S·9H_2O$ Solution: Add $Na_2S·9H_2O$ to distilled/deionized water and bring volume to 10.0mL. Mix thoroughly. Autoclave for 15 min at 15 psi pressure–121°C. Cool to 25°C.

Preparation of Medium: Prepare and distribute medium anaerobically under 80% H_2 + 20% CO_2. Add components, except $Na_2S·9H_2O$ solution, to distilled/deionized water and bring volume to 990.0mL. Mix thoroughly. Gently heat and bring to boiling. Continue boiling until resazurin turns colorless, indicating reduction. Autoclave for 15 min at 15 psi pressure–121°C. Cool to 25°C. Aseptically add 10.0mL of sterile $Na_2S·9H_2O$ solution. Mix thoroughly. Aseptically and anaerobically distribute into sterile tubes or flasks.

Use: For the cultivation of *Methanobacterium thermoautotrophicum*.

Methanobacterium thermoautotrophicum MS Medium

Composition per liter:
$NaHCO_3$ 8.4g
Pancreatic digest of casein 2.0g
Yeast extract 2.0g
$MgCl_2·6H_2O$ 1.0g
NH_4Cl 1.0g
K_2HPO_4 .... 0.84g
$CaCl_2·2H_2O$ 0.4g
Reazurin 0.001g
Trace minerals solution 1.0mL
pH 7.1 ± 0.2 at 25°C

Trace Minerals Solution:
Composition per liter:
Disodium EDTA·$2H_2O$ 0.5g
$ZnSO_4·2H_2O$ 0.21g
$CoCl_2·6H_2O$ 0.15g
$AlK(SO_4)_2$ 0.14g
$FeSO_4·7H_2O$ 0.1g
$MnSO_4·H_2O$ 0.1g
$Na_2WO_4·2H_2O$ 0.03g
$CuCl_2·2H_2O$ 0.02g
$NiCl_2·6H_2O$ 0.02g
H_3BO_3 .... 0.01g
$Na_2MoO_2·2H_2O$ 0.01g
Na_2SeO_4 .... 0.01g

Preparation of Trace Minerals Solution: Add components, except $MnSO_4·H_2O$, to distilled/deionized water and bring volume to 1.0L. Mix thoroughly. Adjust pH to 7.1. Add $MnSO_4·H_2O$. Mix thoroughly.

Preparation of Medium: Prepare and dispense medium under 80% H_2 + 20% CO_2. Add components to distilled/deionized water and bring volume to 1.0L. Mix thoroughly. Gently heat and bring to boiling. Continue boiling for 3 min. Cool to room temperature while sparging with 80% H_2 + 20% CO_2. Adjust pH to 7.1 with 6*N* HCl. Anaerobically distribute into serum bottles fitted with butyl rubber stoppers and aluminum seals. Do not grease stoppers. Autoclave for 15 min at 15 psi pressure–121°C. Store at room temperature in an anaerobic chamber.

Use: For the cultivation of *Methanobacterium thermoautotrophicum*.

Methanobacterium wolfei Medium

Composition per liter:
Na_2CO_3 .... 4.0g
Sodium acetate·$3H_2O$ 1.0g

L-Cysteine·HCl	0.5g
NaCl	0.5g
$Na_2S \cdot 9H_2O$	0.5g
K_2HPO_4	0.26g
KH_2PO_4	0.26g
$(NH_4)_2SO_4$	0.26g
$MgSO_4 \cdot 7H_2O$	0.1g
$CaCl_2 \cdot 2H_2O$	0.07g
$NaWO_4$	2.6mg
$FeSO_4$	2.0mg
Resazurin	1.0mg
Vitamin solution	10.0mL
Trace elements solution	10.0mL

pH 7.2 ± 0.2 at 25°C

Trace Elements Solution:
Composition per liter:

Nitriloacetic acid	9.5g
$MgCl_2$	4.06g
$FeCl_2$	0.99g
$NiCl_2 \cdot 6H_2O$	0.12g
Na_2MoO_4	0.024g
$CoCl_2 \cdot 6H_2O$	0.023g

Preparation of Trace Elements Solution: Add nitrilotriacetic acid to 500.0mL of distilled/deionized water. Adjust pH to 6.5 with KOH. Add remaining components one at a time. Add distilled/deionized water to 1.0L. Mix thoroughly. Sparge with 100% N_2. Adjust pH to 7.0 with 10*N* NaOH.

Vitamin Solution:
Composition per liter:

Calcium DL-pantothenate	5.0g
Vitamin B_{12}	0.1g
Pyridoxine·HCl	10.0mg
p-Aminobenzoic acid	5.0mg
Lipoic acid	5.0mg
Nicotinic acid	5.0mg
Riboflavin	5.0mg
Thiamine·HCl	5.0mg
Biotin	2.0mg
Folic acid	2.0mg

Preparation of Vitamin Solution: Add components to distilled/deionized water and bring volume to 1.0L. Mix thoroughly. Sparge with 100% N_2.

Preparation of Medium: Prepare and dispense medium under 80% H_2 + 20% CO_2. Add components to distilled/deionized water and bring volume to 1.0L. Mix thoroughly. Adjust pH to 7.2. Sparge with 80% H_2 + 20% CO_2. Autoclave for 15 min at 15 psi pressure–121°C. Aseptically and anaerobically distribute into sterile tubes or bottles.

Use: For the cultivation of *Methanobacterium wolfei*.

Methanobrevibacter curvatus Medium (DSMZ Medium 734)

Composition per 1010mL:

NaCl	1.0g
KCl	0.5g
Casamino acids	0.5g
Yeast extract	0.5g
$MgCl_2 \cdot 6H_2O$	0.4g
NH_4Cl	0.3g
KH_2PO_4	0.2g
Na_2SO_4	0.15g
$CaCl_2 \cdot 2H_2O$	0.1g
Resazurin	0.5mg
$NaHCO_3$ solution	40.0mL
Dithionite solution	10.0mL
Trace elements solution SL-10	1.0mL
Selenite-tungstate solution	1.0mL
Seven vitamin solution	1.0mL

pH 7.4 ± 0.2 at 25°C

$NaHCO_3$ Solution:
Composition per 40.0mL:

$NaHCO_3$	5.8g

Preparation of $NaHCO_3$ Solution: Add $NaHCO_3$ to distilled/deionized water and bring volume to 40.0mL. Mix thoroughly. Autoclave for 15 min at 15 psi pressure–121°C. Cool to 25°C. Must be prepared freshly.

Selenite-Tungstate Solution:
Composition per liter:

NaOH	0.5g
$Na_2WO_4 \cdot 2H_2O$	4.0mg
$Na_2SeO_3 \cdot 5H_2O$	3.0mg

Preparation of Selenite-Tungstate Solution: Add components to distilled/deionized water and bring volume to 1.0L. Mix thoroughly. Sparge with 100% N_2. Filter sterilize.

Dithionite Solution:
Composition per 10mL:

Na-dithionite	2.0mg

Preparation of Dithionite Solution: Add Na-dithionite to distilled/deionized water and bring volume to 10.0mL. Mix thoroughly. Sparge with 100% N_2. Filter sterilize.

Trace Elements Solution SL-10:
Composition per liter:

$FeCl_2 \cdot 4H_2O$	1.5g
$CoCl_2 \cdot 6H_2O$	190.0mg
$MnCl_2 \cdot 4H_2O$	100.0mg
$ZnCl_2$	70.0mg
$Na_2MoO_4 \cdot 2H_2O$	36.0mg
$NiCl_2 \cdot 6H_2O$	24.0mg
H_3BO_3	6.0mg
$CuCl_2 \cdot 2H_2O$	2.0mg
HCl (25% solution)	10.0mL

Preparation of Trace Elements Solution SL-10: Add $FeCl_2 \cdot 4H_2O$ to 10.0mL of HCl solution. Mix thoroughly. Add distilled/deionized water and bring volume to 1.0L. Add remaining components. Mix thoroughly. Sparge with 80% N_2 + 20% CO_2. Autoclave for 15 min at 15 psi pressure–121°C.

Seven Vitamin Solution:
Composition per liter:

Pyridoxine hydrochloride	300.0mg
Thiamine-HCl·$2H_2O$	200.0mg
Nicotinic acid	200.0mg
Vitamin B_{12}	100.0mg
Calcium pantothenate	100.0mg
p-Aminobenzoic acid	80.0mg
D(+)-Biotin	20.0mg

Preparation of Seven Vitamin Solution: Add components to distilled/deionized water and bring volume to 1.0L. Sparge with 100% N_2. Mix thoroughly. Filter sterilize.

Preparation of Medium: Prepare and dispense medium under 80% H_2 + 20% CO_2 gas atmosphere. Add components, except $NaHCO_3$ solution, seven vitamin solution, selenite-tungstate solution, and trace elements solution SL-10, to distilled/deionized water and bring volume to 947.0mL. Mix thoroughly. Adjust pH to 7.6. Sparge with 80% H_2 + 20% CO_2. Autoclave for 15 min at 15 psi pressure–121°C. Aseptically and anaerobically add 40.0mL $NaHCO_3$ solution, 1.0mL selenite-tungstate solution, 1.0mL seven vitamin solution, and 1.0mL trace elements solution SL-10. Mix thoroughly. Aseptically and anaerobically distribute into sterile tubes or bottles. Adjust pH to 7.6. Prior to inoculation add dithionite solution (0.1mL per 10mL medium) as reductant.

Use: For the cultivation of *Methanobrevibacter curvatus.*

Methanocalculus halotolerans Medium (DSMZ Medium 905)

Composition per 1010mL:

NaCl	50.0g
NH_4Cl	1.0g
Na-acetate	0.5g
Yeast extract	0.5g
Trypticase	0.5g
K_2HPO_4	0.3g
KH_2PO_4	0.3g
KCl	0.17g
Resazurin	0.5mg
Magnesium chloride solution	30.0mL
$NaHCO_3$ solution	20.0mL
Trace elements solution	10.0mL
Calcium chloride solution	10.0mL
$Na_2S{\cdot}9H_2O$ solution	10.0mL
L-Cysteine solution	10.0mL

pH 7.2-7.6 at 25°C

Magnesium Chloride Solution:
Composition per 30.0mL:

$MgCl_2{\cdot}6H_2O$	3.2g

Preparation of Magnesium Chloride Solution: Add $MgCl_2{\cdot}6H_2O$ to distilled/deionized water and bring volume to 30.0mL. Mix thoroughly. Sparge with 100% N_2. Autoclave for 15 min at 15 psi pressure–121°C.

Calcium Chloride Solution:
Composition per 10.0mL:

$CaCl_2{\cdot}2H_2O$	0.6g

Preparation of Calcium Chloride Solution: Add $CaCl_2{\cdot}2H_2O$ to distilled/deionized water and bring volume to 10.0mL. Mix thoroughly. Sparge with 100% N_2. Autoclave for 15 min at 15 psi pressure–121°C.

L-Cysteine Solution:
Composition per 10.0mL:

L-Cysteine·HCl·H_2O	0.5g

Preparation of L-Cysteine Solution: Add L-cysteine·HCl·H_2O to distilled/deionized water and bring volume to 10.0mL. Mix thoroughly. Sparge with 100% N_2. Autoclave for 15 min at 15 psi pressure–121°C.

$Na_2S{\cdot}9H_2O$ Solution:
Composition per 10.0mL:

$Na_2S{\cdot}9H_2O$	0.3g

Preparation of $Na_2S{\cdot}9H_2O$ Solution: Add $Na_2S{\cdot}9H_2O$ to distilled/deionized water and bring volume to 10.0mL. Mix thoroughly. Sparge with 100% N_2. Autoclave for 15 min at 15 psi pressure–121°C.

$NaHCO_3$ Solution:
Composition per 20.0mL:

$NaHCO_3$	2.0g

Preparation of $NaHCO_3$ Solution: Add $NaHCO_3$ to distilled/deionized water and bring volume to 20.0mL. Mix thoroughly. Sparge with 80% N_2 + 20% CO_2. Autoclave for 15 min at 15 psi pressure–121°C. Cool to 25°C. Must be prepared freshly.

Trace Elements Solution:
Composition per liter:

$MgSO_4{\cdot}7H_2O$	3.0g
Nitrilotriacetic acid	1.5g
NaCl	1.0g
$MnSO_4{\cdot}2H_2O$	0.5g
$CoSO_4{\cdot}7H_2O$	0.18g
$ZnSO_4{\cdot}7H_2O$	0.18g
$CaCl_2{\cdot}2H_2O$	0.1g
$FeSO_4{\cdot}7H_2O$	0.1g
$NiCl_2{\cdot}6H_2O$	0.025g
$KAl(SO_4)_2{\cdot}12H_2O$	0.02g
H_3BO_3	0.01g
$Na_2MoO_4{\cdot}4H_2O$	0.01g
$CuSO_4{\cdot}5H_2O$	0.01g
$Na_2SeO_3{\cdot}5H_2O$	0.3mg

Preparation of Trace Elements Solution: Add nitrilotriacetic acid to 500.0mL of distilled/deionized water. Dissolve by adjusting pH to 6.5 with KOH. Add remaining components. Add distilled/deionized water to 1.0L. Mix thoroughly.

Preparation of Medium: Prepare and dispense medium under sparge with 80% N_2 + 20% CO_2. Add components, except $NaHCO_3$ solution, magnesium chloride solution, calcium chloride solution, L-cysteine-HCl·H_2O solution, and $Na_2S{\cdot}9H_2O$ solution, to distilled/deionized water and bring volume to 950.0mL. Mix thoroughly. Gently heat and bring to boiling. Boil for 3 min. Cool to room temperature while sparging with 80% N_2 + 20% CO_2. Add 10.0mL L-cysteine-HCl·H_2O solution and 20.0mL $NaHCO_3$ solution. Mix thoroughly. Adjust pH to 7.5. Distribute into anaerobe tubes or bottles. Autoclave for 15 min at 15 psi pressure–121°C. Aseptically and anaerobically add per liter, 30.0mL magnesium chloride solution, 10.0mL calcium chloride solution, and 10.0mL $Na_2S{\cdot}9H_2O$. Mix thoroughly. After inoculation pressurize vessels with 80% H_2 + 20% CO_2 gas mixture to 1 bar overpressure and add sulfide from a sterile, anaerobic stock solution. The final pH of the medium should be 7.2–7.6.

Use: For the cultivation of *Methanocalculus halotolerans.*

Methanocalculus pumilus Medium (DSMZ Medium 892)

Composition per 1080mL:

NaCl	10.0g
Yeast extract	2.0g
Trypticase	2.0g
NH_4Cl	0.9g
K_2HPO_4	0.4g
$MgCl_2{\cdot}6H_2O$	0.36g
Resazurin	0.5mg
Na_2CO_3 solution	50.0mL
$Na_2S{\cdot}9H_2O$ solution	15.0mL
Cysteine-HCl·H_2O	15.0mL
Vitamin solution	10.0mL
Trace elements solution	10.0mL

Cysteine Solution:
Composition per 15.0mL:

L-Cysteine·HCl·H_2O	0.5g

Preparation of Cysteine Solution: Add L-cysteine·HCl·H_2O to distilled/deionized water and bring volume to 15.0mL. Mix thoroughly. Sparge with 100% N_2. Autoclave for 15 min at 15 psi pressure–121°C.

Na_2CO_3 Solution:
Composition per 50.0mL:

Na_2CO_3	5.0g

Preparation of Na_2CO_3 Solution: Add Na_2CO_3 to distilled/deionized water and bring volume to 10.0mL. Mix thoroughly. Sparge with 80% N_2 + 20% CO_2. Filter sterilize.

$Na_2S{\cdot}9H_2O$ Solution:
Composition per 15mL:

$Na_2S{\cdot}9H_2O$	0.5g

Preparation of $Na_2S{\cdot}9H_2O$ Solution: Add $Na_2S{\cdot}9H_2O$ to distilled/deionized water and bring volume to 15.0mL. Mix thoroughly. Autoclave under 100% N_2 for 15 min at 15 psi pressure–121°C. Cool to room temperature.

Trace Elements Solution:
Composition per liter:

$MgSO_4{\cdot}7H_2O$	3.0g
Nitrilotriacetic acid	1.5g
NaCl	1.0g
$MnSO_4{\cdot}2H_2O$	0.5g
$CoSO_4{\cdot}7H_2O$	0.18g
$ZnSO_4{\cdot}7H_2O$	0.18g
$CaCl_2{\cdot}2H_2O$	0.1g
$FeSO_4{\cdot}7H_2O$	0.1g
$NiCl_2{\cdot}6H_2O$	0.025g
$KAl(SO_4)_2{\cdot}12H_2O$	0.02g
H_3BO_3	0.01g
$Na_2MoO_4{\cdot}4H_2O$	0.01g
$CuSO_4{\cdot}5H_2O$	0.01g
$Na_2SeO_3{\cdot}5H_2O$	0.3mg

Preparation of Trace Elements Solution: Add nitrilotriacetic acid to 500.0mL of distilled/deionized water. Dissolve by adjusting pH to 6.5 with KOH. Add remaining components. Add distilled/deionized water to 1.0L. Mix thoroughly.

Vitamin Solution:
Composition per liter:

Pyridoxine-HCl	10.0mg
Thiamine-HCl·$2H_2O$	5.0mg
Riboflavin	5.0mg
Nicotinic acid	5.0mg
D-Ca-pantothenate	5.0mg
p-Aminobenzoic acid	5.0mg
Lipoic acid	5.0mg
Biotin	2.0mg
Folic acid	2.0mg
Vitamin B_{12}	0.1mg

Preparation of Vitamin Solution: Add components to distilled/deionized water and bring volume to 1.0L. Mix thoroughly. Sparge with 80% H_2 + 20% CO_2. Filter sterilize.

Preparation of Medium: Add components, except Na_2CO_3 solution, $Na_2S{\cdot}9H_2O$ solution, and cysteine solution, to distilled/deionized water and bring volume to 1.0L. Mix thoroughly. Sparge with 100% N_2 for 30 min. Autoclave for 15 min at 15 psi pressure–121°C. Aseptically distribute 5.0mL aliquots into Hungate tubes under 100% N_2. Aseptically and anaerobically add per 5.0mL medium 0.25mL Na_2CO_3 solution, 0.075mL $Na_2S{\cdot}9H_2O$ solution, and 0.075mL L-cysteine solution. Mix thoroughly. Replace N_2 atmosphere with atmosphere of 80% H_2 + 20% CO_2. Repeat atmosphere replacement several times with overpressurization. The initial pH of 9.0 will decrease over a 30 min period to 7.3-7.5. After inoculation use atmosphere of 80% H_2 + 20% CO_2 to 1.5 bar overpressure.

Use: For the cultivation of *Methanocalculus pumilus.*

Methanococcoides Medium

Composition per liter:

NaCl	18.0g
$NaHCO_3$	5.0g
$MgCl_2{\cdot}6H_2O$	4.0g
$MgSO_4{\cdot}7H_2O$	3.45g
Trimethylamine·HCl	3.0g
Trypticase	2.0g
Yeast extract	2.0g
Sodium acetate	1.0g
L-Cysteine·HCl	0.5g
$Na_2S{\cdot}9H_2O$	0.5g
KCl	0.335g
NH_4Cl	0.25g
$CaCl_2{\cdot}2H_2O$	0.14g
K_2HPO_4	0.14g
$Fe(NH_4)_2(SO_4)_2{\cdot}7H_2O$	2.0mg
Resazurin	1.0mg
Trace elements solution	10.0mL
Vitamin solution	10.0mL

pH 7.0 ± 0.2 at 25°C

Trace Elements Solution:
Composition per liter:

$MgSO_4{\cdot}7H_2O$	3.0g
Nitrilotriacetic acid	1.5g
NaCl	1.0g
$MnSO_4{\cdot}2H_2O$	0.5g
$CoSO_4{\cdot}7H_2O$	0.18g
$ZnSO_4{\cdot}7H_2O$	0.18g
$CaCl_2{\cdot}2H_2O$	0.1g
$FeSO_4{\cdot}7H_2O$	0.1g
$NiCl_2{\cdot}6H_2O$	0.025g
$KAl(SO_4)_2{\cdot}12H_2O$	0.02g
$CuSO_4{\cdot}5H_2O$	0.01g
H_3BO_3	0.01g
$Na_2MoO_4{\cdot}2H_2O$	0.01g
$Na_2SeO_3{\cdot}5H_2O$	0.3mg

Preparation of Trace Elements Solution: Add nitrilotriacetic acid to 500.0mL of distilled/deionized water. Adjust pH to 6.5 with KOH. Add remaining components. Adjust pH to 7.0. Add distilled/deionized water to 1.0L.

Vitamin Solution:
Composition per liter:

Pyridoxine·HCl	10.0mg
Calcium DL-pantothenate	5.0mg
Lipoic acid	5.0mg
Nicotinic acid	5.0mg
p-Aminobenzoic acid	5.0mg
Riboflavin	5.0mg
Thiamine·HCl	5.0mg
Biotin	2.0mg
Folic acid	2.0mg
Vitamin B_{12}	0.1mg

Preparation of Vitamin Solution: Add components to distilled/deionized water and bring volume to 1.0L. Adjust pH to 7.0. Mix thoroughly.

Preparation of Medium: Prepare the medium anaerobically under 80% H_2 + 20% CO_2. Add components to distilled/deionized water and bring volume to 1.0L. Mix thoroughly. Sparge with 80% H_2 + 20% CO_2. Autoclave for 15 min at 15 psi pressure–121°C.

Use: For the cultivation and maintenance of *Methanococoides methylutens.*

Methanococcus deltae Medium

Composition per liter:

NaCl	35.0g
$NaHCO_3$	5.0g
$MgCl_2·6H_2O$	4.0g
NH_4Cl	2.7g
Sodium acetate	2.5g
L-Cysteine·HCl	0.3g
K_2HPO_4	0.3g
KH_2PO_4	0.3g
$Na_2S·9H_2O$	0.3g
$MgSO_4·7H_2O$	0.13g
Resazurin	1.0mg
$(NH_4)_2SO_4$	0.3mg
Trace elements solution	10.0mL
Vitamin solution	10.0mL
L-Cysteine·HCl solution	10.0mL
$Na_2S·9H_2O$ solution	10.0mL

pH 6.9 ± 0.2 at 25°C

Trace Elements Solution:
Composition per liter:

$MgSO_4·7H_2O$	3.0g
Nitrilotriacetic acid	1.5g
NaCl	1.0g
$MnSO_4·2H_2O$	0.5g
$CoSO_4·7H_2O$	0.18g
$ZnSO_4·7H_2O$	0.18g
$CaCl_2·2H_2O$	0.1g
$FeSO_4·7H_2O$	0.1g
$NiCl_2·6H_2O$	0.025g
$KAl(SO_4)_2·12H_2O$	0.02g
$CuSO_4·5H_2O$	0.01g
H_3BO_3	0.01g
$Na_2MoO_4·2H_2O$	0.01g
$Na_2SeO_3·5H_2O$	0.3mg

Preparation of Trace Elements Solution: Add nitrilotriacetic acid to 500.0mL of distilled/deionized water. Adjust pH to 6.5 with KOH. Add remaining components. Adjust pH to 7.0. Add distilled/deionized water to 1.0L.

Vitamin Solution:
Composition per liter:

Pyridoxine·HCl	10.0mg
Calcium DL-pantothenate	5.0mg
Lipoic acid	5.0mg
Nicotinic acid	5.0mg
p-Aminobenzoic acid	5.0mg
Riboflavin	5.0mg
Thiamine·HCl	5.0mg
Biotin	2.0mg
Folic acid	2.0mg
Vitamin B_{12}	0.1mg

Preparation of Vitamin Solution: Add components to distilled/deionized water and bring volume to 1.0L. Adjust pH to 7.0. Mix thoroughly.

L-Cysteine·HCl Solution:
Composition per 10.0mL:

L-Cysteine·HCl	0.3g

Preparation of L-Cysteine·HCl Solution: Add L-cysteine·HCl to distilled/deionized water and bring volume to 10.0mL. Mix thoroughly. Autoclave under 100% N_2 for 15 min at 15 psi pressure–121°C.

$Na_2S·9H_2O$ Solution:
Composition per 10.0mL:

$Na_2S·9H_2O$	0.3g

Preparation of $Na_2S·9H_2O$ Solution: Add $Na_2S·9H_2O$ to distilled/deionized water and bring volume to 10.0mL. Mix thoroughly. Sparge with 100% N_2. Autoclave for 15 min at 15 psi pressure–121°C.

Preparation of Medium: Prepare and dispense medium under 80% H_2 + 20% CO_2. Add components, except L-cysteine·HCl solution and $Na_2S·9H_2O$ solution, to distilled/deionized water and bring volume to 980.0mL. Mix thoroughly. Anaerobically distribute into tubes or flasks fitted with butyl rubber stoppers. Autoclave for 15 min at 15 psi pressure–121°C. Anaerobically add 10.0mL of sterile L-cysteine·HCl solution and 10.0mL of sterile $Na_2S·9H_2O$ solution to each liter of medium or, using a syringe, inject the appropriate amount of sterile L-cysteine·HCl solution and sterile $Na_2S·9H_2O$ solution into individual tubes containing medium.

Use: For the cultivation and maintenance of *Methanococcus deltae.*

Methanococcus jannaschii Medium

Composition per liter:

NaCl	30.0g
$MgSO_4·7H_2O$	3.40g
$MgCl_2·2H_2O$	2.7g
$NaHCO_3$	1.0g
$Na_2S·9H_2O$	0.5g
KCl	0.33g
NH_4Cl	0.25g
$CaCl_2·2H_2O$	0.14g
K_2HPO_4	0.14g
$Fe(NH_4)_2(SO_4)_2·6H_2O$	0.01g
Resazurin	1.0mg
$Na_2SeO_3·5H_2O$	0.5mg
$NiCl_2·6H_2O$	0.5mg
Trace elements solution	10.0mL
Vitamin solution	10.0mL
L-Cysteine·HCl solution	10.0mL
$Na_2S·9H_2O$ solution	10.0mL

pH 6.0 ± 0.2 at 25°C

Trace Elements Solution:
Composition per liter:

$MgSO_4·7H_2O$	3.0g
Nitrilotriacetic acid	1.5g
NaCl	1.0g
$MnSO_4·2H_2O$	0.5g
$CoSO_4·7H_2O$	0.18g
$ZnSO_4·7H_2O$	0.18g
$CaCl_2·2H_2O$	0.1g
$FeSO_4·7H_2O$	0.1g
$NiCl_2·6H_2O$	0.025g
$KAl(SO_4)_2·12H_2O$	0.02g
$CuSO_4·5H_2O$	0.01g
H_3BO_3	0.01g
$Na_2MoO_4·2H_2O$	0.01g
$Na_2SeO_3·5H_2O$	0.3mg

Preparation of Trace Elements Solution: Add nitrilotriacetic acid to 500.0mL of distilled/deionized water. Adjust pH to 6.5 with KOH. Add remaining components. Adjust pH to 7.0. Add distilled/deionized water to 1.0L.

Vitamin Solution:
Composition per liter:

Pyridoxine·HCl	10.0mg

Calcium DL-pantothenate 5.0mg
Lipoic acid 5.0mg
Nicotinic acid 5.0mg
p-Aminobenzoic acid 5.0mg
Riboflavin 5.0mg
Thiamine·HCl 5.0mg
Biotin 2.0mg
Folic acid 2.0mg
Vitamin B_{12} 0.1mg

Preparation of Vitamin Solution: Add components to distilled/deionized water and bring volume to 1.0L. Adjust pH to 7.0. Mix thoroughly.

L-Cysteine·HCl Solution:
Composition per 10.0mL:
L-Cysteine·HCl 0.5g

Preparation of L-Cysteine·HCl Solution: Add L-cysteine·HCl to distilled/deionized water and bring volume to 10.0mL. Mix thoroughly. Autoclave under 100% N_2 for 15 min at 15 psi pressure–121°C.

$Na_2S{\cdot}9H_2O$ Solution:
Composition per 10.0mL:
$Na_2S{\cdot}9H_2O$ 0.3g

Preparation of $Na_2S{\cdot}9H_2O$ Solution: Add $Na_2S{\cdot}9H_2O$ to distilled/deionized water and bring volume to 10.0mL. Mix thoroughly. Sparge with 100% N_2. Autoclave for 15 min at 15 psi pressure–121°C.

Preparation of Medium: Prepare and dispense medium under 80% H_2 + 20% CO_2. Add components, except L-cysteine·HCl solution and $Na_2S{\cdot}9H_2O$ solution, to distilled/deionized water and bring volume to 980.0mL. Mix thoroughly. Anaerobically distribute into tubes or flasks. Autoclave for 15 min at 15 psi pressure–121°C. Anaerobically add 10.0mL of sterile L-cysteine·HCl solution and 10.0mL of sterile $Na_2S{\cdot}9H_2O$ solution to each liter of medium or, using a syringe, inject the appropriate amount of sterile L-cysteine·HCl solution and sterile $Na_2S{\cdot}9H_2O$ solution into individual tubes containing medium.

Use: For the cultivation and maintenance of *Methanococcus* species.

Methanococcus McC Medium

Composition per 1100.0mL:
$NaHCO_3$ 5.0g
Yeast extract 2.0g
L-Cysteine·HCl·H_2O 0.5g
General salts solution 500.0mL
NaCl solution 75.0mL
$Na_2S{\cdot}9H_2O$ solution 20.0mL
K_2HPO_4 solution 10.0mL
Trace minerals solution 10.0mL
Sodium acetate solution 10.0mL
Iron stock solution 5.0mL
Resazurin solution 1.0mL

General Salts Solution:
Composition per liter:
$MgSO_4{\cdot}7H_2O$ 6.9g
$MgCl_2{\cdot}6H_2O$ 5.5g
NH_4Cl 1.0g
KCl 0.67g
$CaCl_2{\cdot}2H_2O$ 0.28g

Preparation of General Salts Solution: Add components to distilled/deionized water and bring volume to 1.0L. Mix thoroughly.

NaCl Solution:
Composition per 100.0mL:
NaCl 29.3g

Preparation of NaCl Solution: Add NaCl to distilled/deionized water and bring volume to 100.0mL. Mix thoroughly.

$Na_2S{\cdot}9H_2O$ Solution:
Composition per 100.0mL:
NaOH .... 1 pellet
$Na_2S{\cdot}9H_2O$ 2.5g

Preparation of $Na_2S{\cdot}9H_2O$ Solution: Bring 100.0mL of distilled/deionized water to boiling. Cool to room temperature while sparging with 100%N_2. Dissolve 1 pellet of NaOH in the anaerobic water. Weigh out a little more than 2.5g of $Na_2S{\cdot}9H_2O$. Briefly rinse the crystals in distilled/deionized water. Dry the crystals by blotting on paper towels or filter paper. Add 2.5g of washed $Na_2S{\cdot}9H_2O$ crystals to 100.0mL of anaerobic NaOH solution. Distribute into serum bottles fitted with butyl rubber stoppers and aluminum seals. Do not grease stoppers. Pressurize to 60kPa with 100% N_2. Autoclave for 15 min at 15 psi pressure–121°C. Store at room temperature in an anaerobic chamber.

K_2HPO_4 Solution:
Composition per 100.0mL:
K_2HPO_4 .... 1.4g

Preparation of K_2HPO_4 Solution: Add K_2HPO_4 to distilled/deionized water and bring volume to 100.0mL. Mix thoroughly.

Trace Minerals Solution:
Composition per liter:
Nitrilotriacetic acid 1.5g
$Na_2WO_4{\cdot}2H_2O$ 1.0g
$Fe(NH_4)_2(SO_4)_2{\cdot}6H_2O$ 0.2g
Na_2SeO_3 .... 0.2g
$Na_2MoO_4{\cdot}2H_2O$ 0.1g
$Mn_4{\cdot}2H_2O$ 0.1g
$Zn_4{\cdot}7H_2O$ 0.1g
$NiCl_2{\cdot}7H_2O$ 0.025g
$CuSO_4{\cdot}5H_2O$ 0.01g

Preparation of Trace Minerals Solution: Add nitrilotriacetic acid to 500.0mL of distilled/deionized water. Adjust pH to 6.5 with KOH. Add remaining components. Add distilled/deionized water to 1.0L. Adjust pH to 7.0.

Sodium Acetate Solution:
Composition per 100.0mL:
Sodium acetate·$3H_2O$ 13.6g

Preparation of Sodium Acetate Solution: Add sodium acetate·$3H_2O$ to distilled/deionized water and bring volume to 100.0mL. Mix thoroughly.

Iron Stock Solution:
Composition per 100.0mL:
$Fe(NH_4)_2(SO_4)_2{\cdot}6H_2O$ 0.2g

Preparation of Iron Stock Solution: Add $Fe(NH_4)_2(SO_4)_2{\cdot}6H_2O$ to 5.0mL of distilled H_2O containing 2 drops of concentrated HCl. Mix thoroughly. When the $Fe(NH_4)_2(SO_4)_2{\cdot}6H_2O$ has dissolved, bring the volume to 100.0mL with distilled/deionized water.

Resazurin Solution:
Composition per 10.0mL:
Resazurin 10.0mg

Preparation of Resazurin Solution: Add resazurin to distilled/deionized water and bring volume to 10.0mL. Mix thoroughly.

Preparation of Medium: Prepare and dispense medium under 80% H_2 + 20% CO_2. Add components, except $NaHCO_3$ and $Na_2S \cdot 9H_2O$ solution, to distilled/deionized water and bring volume to 1080.0mL. Mix thoroughly. Gently heat and bring to boiling. Continue boiling for 3 min. Cool to room temperature while sparging with 80% H_2 + 20% CO_2. Add $NaHCO_3$. Mix thoroughly. Anaerobically distribute 9.8mL volumes into anaerobic tubes. Autoclave for 15 min at 15 psi pressure–121°C. Aseptically and anaerobically add 0.2mL of sterile $Na_2S \cdot 9H_2O$ solution to each tube. Mix thoroughly.

Use: For the cultivation of *Methanococcus* species.

Methanococcus McN Medium

Composition per 1100.0mL:

$NaHCO_3$	5.0g
L-Cysteine·HCl·H_2O	0.5g
General salts solution	500.0mL
NaCl solution	75.0mL
$Na_2S \cdot 9H_2O$ solution	20.0mL
K_2HPO_4 solution	10.0mL
Trace minerals solution	10.0mL
Iron stock solution	5.0mL
Resazurin solution	1.0mL

General Salts Solution:
Composition per liter:

$MgSO_4 \cdot 7H_2O$	6.9g
$MgCl_2 \cdot 6H_2O$	5.5g
NH_4Cl	1.0g
KCl	0.67g
$CaCl_2 \cdot 2H_2O$	0.28g

Preparation of General Salts Solution: Add components to distilled/deionized water and bring volume to 1.0L. Mix thoroughly.

NaCl Solution:
Composition per 100.0mL:

NaCl	29.3g

Preparation of NaCl Solution: Add NaCl to distilled/deionized water and bring volume to 100.0mL. Mix thoroughly.

$Na_2S \cdot 9H_2O$ Solution:
Composition per 100.0mL:

$Na_2S \cdot 9H_2O$	2.5g
NaOH	1 pellet

Preparation of $Na_2S \cdot 9H_2O$ Solution: Bring 100.0mL of distilled/deionized water to boiling. Cool to room temperature while sparging with 100%N_2. Dissolve 1 pellet of NaOH in the anaerobic water. Weigh out a little more than 2.5g of $Na_2S \cdot 9H_2O$. Briefly rinse the crystals in distilled/deionized water. Dry the crystals by blotting on paper towels or filter paper. Add 2.5g of washed $Na_2S \cdot 9H_2O$ crystals to 100.0mL of anaerobic NaOH solution. Distribute into serum bottles fitted with butyl rubber stoppers and aluminum seals. Do not grease stoppers. Pressurize to 60kPa with 100% N_2. Autoclave for 15 min at 15 psi pressure–121°C. Store at room temperature in an anaerobic chamber.

K_2HPO_4 Solution:
Composition per 100.0mL:

K_2HPO_4	1.4g

Preparation of K_2HPO_4 Solution: Add K_2HPO_4 to distilled/deionized water and bring volume to 100.0mL. Mix thoroughly.

Trace Minerals Solution:
Composition per liter:

Nitrilotriacetic acid	1.5g
$Na_2WO_4 \cdot 2H_2O$	1.0g
$Fe(NH_4)_2(SO_4)_2 \cdot 6H_2O$	0.2g
Na_2SeO_3	0.2g
$Na_2MoO_4 \cdot 2H_2O$	0.1g
$Mn_4 \cdot 2H_2O$	0.1g
$Zn_4 \cdot 7H_2O$	0.1g
$NiCl_2 \cdot 7H_2O$	0.025g
$CuSO_4 \cdot 5H_2O$	0.01g

Preparation of Trace Minerals Solution: Add nitrilotriacetic acid to 500.0mL of distilled/deionized water. Adjust pH to 6.5 with KOH. Add remaining components. Add distilled/deionized water to 1.0L. Adjust pH to 7.0.

Iron Stock Solution:
Composition per 100.0mL:

$Fe(NH_4)_2(SO_4)_2 \cdot 6H_2O$	0.2g

Preparation of Iron Stock Solution: Add $Fe(NH_4)_2(SO_4)_2 \cdot 6H_2O$ to 5.0mL of distilled H_2O containing 2 drops of concentrated HCl. Mix thoroughly. When the $Fe(NH_4)_2(SO_4)_2 \cdot 6H_2O$ has dissolved, bring the volume to 100.0mL with distilled/deionized water.

Resazurin Solution:
Composition per 10.0mL:

Resazurin	10.0mg

Preparation of Resazurin Solution: Add resazurin to distilled/deionized water and bring volume to 10.0mL. Mix thoroughly.

Preparation of Medium: Prepare and dispense medium under 80% H_2 + 20% CO_2. Add components, except $NaHCO_3$ and $Na_2S \cdot 9H_2O$ solution, to distilled/deionized water and bring volume to 1080.0mL. Mix thoroughly. Gently heat and bring to boiling. Continue boiling for 3 min. Cool to room temperature while sparging with 80% H_2 + 20% CO_2. Add $NaHCO_3$. Mix thoroughly. Anaerobically distribute 9.8mL volumes into anaerobic tubes. Autoclave for 15 min at 15 psi pressure–121°C. Aseptically and anaerobically add 0.2mL of sterile $Na_2S \cdot 9H_2O$ solution to each tube. Mix thoroughly.

Use: For the cultivation of *Methanococcus* species.

Methanococcus McNail Medium

Composition per 1100.0mL:

$NaHCO_3$	5.0g
L-Leucine	1.0g
L-Isoleucine	0.5g
L-Cysteine·HCl·H_2O	0.5g
General salts solution	500.0mL
NaCl solution	75.0mL
$Na_2S \cdot 9H_2O$ solution	20.0mL
K_2HPO_4 solution	10.0mL
Trace minerals solution	10.0mL
Sodium acetate solution	10.0mL
Pantoyllactone solution	10.0mL
Iron stock solution	5.0mL
Resazurin solution	1.0mL

General Salts Solution:
Composition per liter:

$MgSO_4 \cdot 7H_2O$	6.9g
$MgCl_2 \cdot 6H_2O$	5.5g

NH_4Cl 1.0g
KCl 0.67g
$CaCl_2 \cdot 2H_2O$ 0.28g

Preparation of General Salts Solution: Add components to distilled/deionized water and bring volume to 1.0L. Mix thoroughly.

NaCl Solution:
Composition per 100.0mL:
NaCl 29.3g

Preparation of NaCl Solution: Add NaCl to distilled/deionized water and bring volume to 100.0mL. Mix thoroughly.

$Na_2S \cdot 9H_2O$ Solution:
Composition per 100.0mL:
NaOH .... 1 pellet
$Na_2S \cdot 9H_2O$ 2.5g

Preparation of $Na_2S \cdot 9H_2O$ Solution: Bring 100.0mL of distilled/deionized water to boiling. Cool to room temperature while sparging with 100%N_2. Dissolve 1 pellet of NaOH in the anaerobic water. Weigh out a little more than 2.5g of $Na_2S \cdot 9H_2O$. Briefly rinse the crystals in distilled/deionized water. Dry the crystals by blotting on paper towels or filter paper. Add 2.5g of washed $Na_2S \cdot 9H_2O$ crystals to 100.0mL of anaerobic NaOH solution. Distribute into serum bottles fitted with butyl rubber stoppers and aluminum seals. Do not grease stoppers. Pressurize to 60kPa with 100% N_2. Autoclave for 15 min at 15 psi pressure–121°C. Store at room temperature in an anaerobic chamber.

K_2HPO_4 Solution:
Composition per 100.0mL:
K_2HPO_4 .... 1.4g

Preparation of K_2HPO_4 Solution: Add K_2HPO_4 to distilled/deionized water and bring volume to 100.0mL. Mix thoroughly.

Trace Minerals Solution:
Composition per liter:
Nitrilotriacetic acid 1.5g
$Na_2WO_4 \cdot 2H_2O$ 1.0g
$Fe(NH_4)_2(SO_4)_2 \cdot 6H_2O$ 0.2g
Na_2SeO_3 .... 0.2g
$Na_2MoO_4 \cdot 2H_2O$ 0.1g
$Mn_4 \cdot 2H_2O$ 0.1g
$Zn_4 \cdot 7H_2O$ 0.1g
$NiCl_2 \cdot 7H_2O$ 0.025g
$CuSO_4 \cdot 5H_2O$ 0.01g

Preparation of Trace Minerals Solution: Add nitrilotriacetic acid to 500.0mL of distilled/deionized water. Adjust pH to 6.5 with KOH. Add remaining components. Add distilled/deionized water to 1.0L. Adjust pH to 7.0.

Sodium Acetate Solution:
Composition per 100.0mL:
Sodium acetate·$3H_2O$ 13.6g

Preparation of Sodium Acetate Solution: Add sodium acetate·$3H_2O$ to distilled/deionized water and bring volume to 100.0mL. Mix thoroughly.

Pantoyllactone Solution:
Composition per 100.0mL:
Pantoyllactone 0.013g

Preparation of Pantoyllactone Solution: Add pantoyllactone to distilled/deionized water and bring volume to 100.0mL. Mix thoroughly.

Iron Stock Solution:
Composition per 100.0mL:
$Fe(NH_4)_2(SO_4)_2 \cdot 6H_2O$ 0.2g

Preparation of Iron Stock Solution: Add $Fe(NH_4)_2(SO_4)_2 \cdot 6H_2O$ to 5.0mL of distilled H_2O containing 2 drops of concentrated HCl. Mix thoroughly. When the $Fe(NH_4)_2(SO_4)_2 \cdot 6H_2O$ has dissolved, bring the volume to 100.0mL with distilled/deionized water.

Resazurin Solution:
Composition per 10.0mL:
Resazurin 10.0mg

Preparation of Resazurin Solution: Add resazurin to distilled/deionized water and bring volume to 10.0mL. Mix thoroughly.

Preparation of Medium: Prepare and dispense medium under 80% H_2 + 20% CO_2. Add components, except $NaHCO_3$ and $Na_2S \cdot 9H_2O$ solution, to distilled/deionized water and bring volume to 1080.0mL. Mix thoroughly. Gently heat and bring to boiling. Continue boiling for 3 min. Cool to room temperature while sparging with 80% H_2 + 20% CO_2. Add $NaHCO_3$. Mix thoroughly. Anaerobically distribute 9.8mL volumes into anaerobic tubes. Autoclave for 15 min at 15 psi pressure–121°C. Aseptically and anaerobically add 0.2mL of sterile $Na_2S \cdot 9H_2O$ solution to each tube. Mix thoroughly.

Use: For the cultivation of *Methanococcus* species.

Methanococcus Medium

Composition per liter:
NaCl 18.0g
$Mg_2SO_4 \cdot 7H_2O$ 3.45g
$MgCl_2 \cdot 2H_2O$ 2.75g
Pancreatic digest of casein 2.0g
Yeast extract 2.0g
Sodium acetate 1.0g
L-Cysteine·HCl 0.5g
$Na_2S \cdot 9H_2O$ 0.5g
NH_4HCO_3 .... 0.5g
KCl 0.335g
NH_4Cl 0.225g
$CaCl_2 \cdot 2H_2O$ 0.14g
KH_2PO_4 .... 0.14g
Calcium DL-pantothenate 5.0mg
Na_2SeO_3 .... 2.0mg
$FeSO_4 \cdot 2H_2O$ 1.0mg
Resazurin 1.0mg
Trace minerals stock solution 10mL

pH 6.5 ± 0.2 at 25°C

Trace Minerals Stock Solution:
Composition per liter:
Nitriloacetic acid 1.5g
$MnSO_4 \cdot H_2O$ 0.5g
$CoCl_2$ 0.1g
$ZnSO_4$ 0.1g
$AlK(SO_4)_2 \cdot 12H_2O$ 0.01g
$CuSO_4 \cdot 5H_2O$ 0.01g
H_3BO_3 .... 0.01g
$Na_2MoO_4 \cdot 2H_2O$ 0.01g
$NiCl_2$ 0.01g

Preparation of Trace Minerals Stock Solution: Add nitrilotriacetic acid to 500.0mL of distilled/deionized water. Adjust pH to 6.5 with KOH. Add remaining components one at a time. Add distilled/deionized water to 1.0L. Adjust pH to 6.8.

Preparation of Medium: Prepare medium anaerobically under 80% N_2 + 20% CO_2. Add components, except Na_2CO_3, L-cysteine·HCl, and $Na_2S·9H_2O$, to distilled/deionized water and bring volume to 1.0L. Mix thoroughly. Gently heat and bring to boiling. Cool while sparging with 80% N_2 + 20% CO_2. Add Na_2CO_3, L-cysteine·HCl, and $Na_2S·9H_2O$. Dispense into tubes, bottles, or flasks under an atmosphere of 80% H_2 + 20% CO_2. Seal with butyl rubber stoppers secured with aluminum crimp seals. Autoclave for 15 min at 15 psi pressure–121°C. Cool to room temperature. Pressurize the head space to 69kPA with 80% H_2 + 20% CO_2.

Use: For the cultivation of *Methanococcus* species.

Methanococcus vannielii Medium

Composition per 1020.0mL:

Solution A	500.0mL
Inorganic salts solution	500.0mL
$Na_2S·9H_2O$ solution	10.0mL
Na_2CO_3 solution	10.0mL

Solution A:

Composition per 500.0mL:

Sodium formate	10.0g
Phenol Red	3.0mg
Methylene Blue	2.0mg

Preparation of Solution A: Add components to distilled/deionized water and bring volume to 500.0mL. Mix thoroughly. Autoclave for 15 min at 15 psi pressure–121°C. Cool to 25°C.

Inorganic Salts Solution:

Composition per 500.0mL:

$K_2HPO_4·3H_2O$	1.45g
NH_4Cl	1.0g
KH_2PO_4	0.75g
$MgCl_2·6H_2O$	0.2g
Nitrilotriacetic acid	0.04g
$CaCl_2·2H_2O$	0.02g
$FeCl_2·4H_2O$	3.6mg
$CoCl_2·6H_2O$	1.5mg
$MnCl_2$·4H2O	0.9mg
$ZnCl_2$	0.9mg
H_3BO_2	0.17mg
$Na_2MoO_4·2H_2O$	0..09mg

Preparation of Inorganic Salts Solution: Add nitrilotriacetic acid to 250.0mL of distilled/deionized water. Dissolve by adjusting pH to 6.5 with KOH. Add remaining components. Readjust pH to 7.2 with H_2SO_4 or KOH. Add distilled/deionized water to 500.0mL. Filter sterilize.

$Na_2S·9H_2O$ Solution:

Composition per 10.0mL:

$Na_2S·9H_2O$	0.3g

Preparation of $Na_2S·9H_2O$ Solution: Add $Na_2S·9H_2O$ to distilled/deionized water and bring volume to 10.0mL. Mix thoroughly. Autoclave for 15 min at 15 psi pressure–121°C. Cool to 25°C.

Na_2CO_3 Solution:

Composition per 10.0mL:

Na_2CO_3	2.5g

Preparation of Na_2CO_3 Solution: Add Na_2CO_3 to distilled/deionized water and bring volume to 10.0mL. Mix thoroughly. Autoclave for 15 min at 15 psi pressure–121°C. Cool to 25°C.

Preparation of Medium: Prepare and distribute medium anaerobically under 80% N_2 + 20% CO_2. Aseptically and anaerobically combine 500.0mL of sterile inorganic salts solution, 500.0mL of sterile solution A, 10.0mL of sterile $Na_2S·9H_2O$ solution, and 10.0mL of sterile Na_2CO_3 solution. Mix thoroughly. Aseptically and anaerobically distribute into sterile tubes or flasks.

Use: For the isolation and cultivation of *Methanococcus vannielii* from marine mud.

Methanococcus vannielii Medium

Composition per liter:

Sodium formate	15.0g
K_2HPO_4	3.48g
$CoCl_2·6H_2O$	2.38g
NH_4Cl	1.0g
L-Cysteine·HCl·H_2O	0.3g
$MgSO_4·7H_2O$	0.2g
$CaCl_2·2H_2O$	0.01g
$FeSO_4·7H_2O$	0.01g
$MnSO_4·H_2O$	7.5mg
$Na_2MoO_4·2H_2O$	7.5mg
Na_2SeO_3	1.7mg
$Na_2S·9H_2O$ solution	10.0mL

$Na_2S·9H_2O$ Solution:

Composition per 10.0mL:

$Na_2S·9H_2O$	0.15g

Preparation of $Na_2S·9H_2O$ Solution: Add $Na_2S·9H_2O$ to distilled/deionized water and bring volume to 10.0mL. Mix thoroughly. Autoclave for 15 min at 15 psi pressure–121°C. Cool to 25°C.

Preparation of Medium: Prepare and distribute medium anaerobically under 100% N_2. Add components, except $Na_2S·9H_2O$ solution, to distilled/deionized water and bring volume to 990.0mL. Mix thoroughly. Gently heat and bring to boiling. Continue boiling until resazurin turns colorless, indicating reduction. Autoclave for 15 min at 15 psi pressure–121°C. Cool to 25°C. Aseptically add 10.0mL of sterile $Na_2S·9H_2O$ solution. Mix thoroughly. Aseptically and anaerobically distribute into sterile tubes or flasks.

Use: For the cultivation of *Methanococcus vannielii.*

Methanococcus voltae BD Medium

Composition 1003.0mL:

L-Leucine	50.0g
NaCl	18.0g
Sodium panthothenate	5.0g
$MgSO_4·7H_2O$	3.48g
$MgCl_2·6H_2O$	2.75g
$CH_3COONa·3H_2O$	1.0g
KCl	0.34g
NH_4Cl	0.26g
$CaCl_2·2H_2O$	0.14g
K_2PO_4	0.14g
L-Isoleucine	0.1g
L-Cysteine/Na_2S solution	17.5mL
Trace minerals solution	10.0mL
Vitamin solution	10.0mL
Na_2CO_3 solution	6.0mL
$FeSO_4·7H_2O$ solution	4.5mL

Trace Minerals Solution:

Composition per liter:

Nitrilotriacetic acid	1.5g
$MnSO_4·7H_2O$	0.5g
$NiCl_2·6H_2O$	0.12g

$CoCl_2 \cdot 6H_2O$	0.1g
$FeSO_4 \cdot 7H_2O$	0.1g
Resazurin	0.1g
$ZnSO_4 \cdot 7H_2O$	0.1g
$AlK(SO_4)_2 \cdot 5H_2O$	0.01g
$CuSO_4 \cdot 5H_2O$	0.01g
H_3BO_3	0.01g

Preparation of Trace Minerals Solution: Add nitrilotriacetic acid to 500.0mL of distilled/deionized water. Adjust pH to 6.5 with KOH. Add remaining components one at a time. Add distilled/deionized water to 1.0L. Adjust pH to 6.8.

Vitamin Solution:
Composition per liter:

Pyridoxine·HCl	10.0mg
p-Aminobenzoic acid	5.0mg
Lipoic acid	5.0mg
Nicotinic acid	5.0mg
Riboflavin	5.0mg
Thiamine·HCl	5.0mg
Biotin	2.0mg
Folic acid	2.0mg
Vitamin B_{12}	0.5mg

Preparation of Vitamin Solution: Add components to distilled/deionized water and bring volume to 1.0L. Mix thoroughly. Sparge with 100% N_2. Filter sterilize.

L-Cysteine/Na_2S Solution:
Composition per liter:

L-Cysteine·HCl	1.25g
$Na_2S \cdot 9H_2O$	1.25g

Preparation of L-Cysteine/Na_2S Solution: Add L-Cysteine·HCl and $Na_2S \cdot 9H_2O$ to distilled/deionized water and bring volume to 10.0mL. Mix thoroughly. Sparge with 100% N_2. Autoclave for 15 min at 15 psi pressure–121°C. Before use, neutralize to pH 7.0 with sterile HCl.

Na_2CO_3 Solution:
Composition per 100.0mL:

Na_2CO_3	1.0g

Preparation of Na_2CO_3 Solution: Add Na_2CO_3 to distilled/deionized water and bring volume to 100.0mL. Mix thoroughly. Sparge with 80% N_2 + 20% CO_2. Autoclave for 15 min at 15 psi pressure–121°C.

$FeSO_4 \cdot 7H_2O$ Solution:
Composition per 100.0mL:

$FeSO_4 \cdot 7H_2O$	0.1g

Preparation of $FeSO_4 \cdot 7H_2O$ Solution: Add $FeSO_4 \cdot 7H_2O$ to distilled/deionized water and bring volume to 100.0mL. Mix thoroughly. Filter sterilize. Sparge with 100% N_2.

Preparation of Medium: Prepare and dispense medium under 80% H_2 + 20% CO_2. Add components, except L-cysteine/Na_2S solution, Na_2CO_3 solution, and $FeSO_4 \cdot 7H_2O$ solution, to distilled/deionized water and bring volume to 975.0mL. Mix thoroughly. Gently heat and bring to boiling while sparging with 80% H_2 + 20% CO_2. Cool to room temperature while sparging with 80% H_2 + 20% CO_2. Add 17.5mL of L-cysteine/$Na_2S \cdot 9H_2O$ solution and 6.0mL of Na_2CO_3 solution. Anaerobically dispense into tubes or bottles in 10.0mL aliquots under an atmosphere of 80% H_2 + 20% CO_2. Seal with butyl rubber stoppers secured with aluminum crimp seals. Autoclave for 15 min at 15 psi pressure–121°C. Immediately prior to inoculation, aseptically and anaerobically add 0.05mL of sterile $FeSO_4 \cdot 7H_2O$ solution to each tube.

Use: For the cultivation of *Methanococcus voltae.*

Methanocorpusculum Medium (DSMZ Medium 279)

Composition per liter:

Na-acetate	4.0g
$NaHCO_3$	4.0g
Na-formate	2.0g
Yeast extract	1.0g
KH_2PO_4	0.5g
$MgSO_4 \cdot 7H_2O$	0.4g
NaCl	0.4g
NH_4Cl	0.4g
$CaCl_2 \cdot 2H_2O$	0.05g
$FeSO_4 \cdot 7H_2O$	0.002g
Resazurin	0.001g
$NiCl_2 \cdot 6H_2O$	24.0mg
Sludge fluid	50.0mL
Fatty acid mixture	20.0mL
L-Cysteine solution	10.0mL
$Na_2S \cdot 9H_2O$	10.0mL
Trace elements solution SL-10	1.0mL

pH 6.7-7.0

Sludge Fluid:
Composition per 500.0mL:

Yeast extract	2.0g
Sludge	500.0mL

Preparation of Sludge Fluid: Add yeast extract to sludge from an anaerobic digester. Gas with nitrogen gas for a few minutes. Incubate at 37°C for 24 hours. Centrifuge the sludge at 13,000 x g. Autoclave for 15 min at 15 psi pressure–121°C. The resulting, clear supernatant in screw-capped vessels under nitrogen gas. The sludge fluid can be stored at room temperature in the dark.

Fatty Acid Mixture:
Composition per 20.0mL:

Valeric acid	0.5g
Isovaleric acid	0.5g
α-Methylbutyric acid	0.5g
Isobutyric acid	0.5g
Distilled water	20.0mL

Preparation of Fatty Acid Mixture: Add components to 20.0mL distilled/deionized water. Mix thoroughly.

Trace Elements Solution SL-10:
Composition per liter:

$FeCl_2 \cdot 4H_2O$	1.5g
$CoCl_2 \cdot 6H_2O$	190.0mg
$MnCl_2 \cdot 4H_2O$	100.0mg
$ZnCl_2$	70.0mg
$Na_2MoO_4 \cdot 2H_2O$	36.0mg
$NiCl_2 \cdot 6H_2O$	24.0mg
H_3BO_3	6.0mg
$CuCl_2 \cdot 2H_2O$	2.0mg
HCl (25% solution)	10.0mL

Preparation of Trace Elements Solution SL-10: Add $FeCl_2 \cdot 4H_2O$ to 10.0mL of HCl solution. Mix thoroughly. Add distilled/deionized water and bring volume to 1.0L. Add remaining components. Mix thoroughly. Sparge with 100% N_2. Autoclave for 15 min at 15 psi pressure–121°C.

$Na_2S \cdot 9H_2O$ Solution:
Composition per 10.0mL:

$Na_2S \cdot 9H_2O$	0.5g

Preparation of $Na_2S{\cdot}9H_2O$ Solution: Add $Na_2S{\cdot}9H_2O$ to distilled/deionized water and bring volume to 10.0mL. Mix thoroughly. Sparge with 100% N_2. Autoclave for 15 min at 15 psi pressure–121°C. Before use, neutralize to pH 7.0 with sterile HCl.

L-Cysteine Solution:
Composition per 10.0mL:

L-Cysteine·HCl·H_2O	0.5g

Preparation of L-Cysteine Solution: Add L-cysteine·HCl·H_2O to distilled/deionized water and bring volume to 10.0mL. Mix thoroughly. Sparge with 100% N_2. Autoclave for 15 min at 15 psi pressure–121°C.

Preparation of Medium: Prepare and dispense medium under 80% H_2 + 20% CO_2 gas atmosphere. Add components, except L-cysteine solution, $Na_2S{\cdot}9H_2O$ solution, and trace elements solution SL-10, to distilled/deionized water and bring volume to 920.0mL. Mix thoroughly. Adjust pH to 6.8. Sparge with 80% H_2 + 20% CO_2. Autoclave for 15 min at 15 psi pressure–121°C. Aseptically and anaerobically add 10.0mL L-cysteine solution, 10.0mL $Na_2S{\cdot}9H_2O$ solution, and 1.0mL trace elements solution SL-10. Mix thoroughly. Aseptically and anaerobically distribute into sterile tubes or bottles. After inoculation, flush and repressurize the gas head space of culture bottles with sterile 80% H_2 + 20% CO_2 to 1 bar overpressure. Alternately, the medium without L-Cysteine solution, $Na_2S{\cdot}9H_2O$ solution, and trace elements solution SL-10, can be distributed to tubes anaerobically prior to autoclaving. After autoclaving in tubes the appropriate volumes of the individual solutions can be injected through the stoppers so that the final concentrations of the medium are achieved.

Use: For the cultivation of *Methanococcoides* spp. and *Methanolobus bombayensis.*

Methanoculleus olentangyi Medium

Composition per liter:

$NaHCO_3$	5.0g
NH_4Cl	2.7g
Sodium acetate	2.5g
NaCl	0.61g
K_2HPO_4	0.3g
KH_2PO_4	0.3g
$MgSO_4{\cdot}7H_2O$	0.13g
Resazurin	1.0mg
$(NH_4)_2SO_4$	0.3mg
Trace elements solution	10.0mL
Vitamin solution	10.0mL
L-Cysteine·HCl solution	10.0mL
$Na_2S{\cdot}9H_2O$ solution	10.0mL

pH 6.9 ± 0.2 at 25°C

Trace Elements Solution:
Composition per liter:

$MgSO_4{\cdot}7H_2O$	3.0g
Nitrilotriacetic acid	1.5g
NaCl	1.0g
$MnSO_4{\cdot}2H_2O$	0.5g
$CoSO_4{\cdot}7H_2O$	0.18g
$ZnSO_4{\cdot}7H_2O$	0.18g
$CaCl_2{\cdot}2H_2O$	0.1g
$FeSO_4{\cdot}7H_2O$	0.1g
$KAl(SO_4)_2{\cdot}12H_2O$	0.02g
$CuSO_4{\cdot}5H_2O$	0.01g
H_3BO_3	0.01g
$Na_2MoO_4{\cdot}2H_2O$	0.01g
$NiCl_2{\cdot}6H_2O$	0.025g
$Na_2SeO_3{\cdot}5H_2O$	0.3mg

Preparation of Trace Elements Solution: Add nitrilotriacetic acid to 500.0mL of distilled/deionized water. Adjust pH to 6.5 with KOH. Add remaining components. Add distilled/deionized water to 1.0L.

Vitamin Solution:
Composition per liter:

Pyridoxine·HCl	10.0mg
Calcium DL-pantothenate	5.0mg
Lipoic acid	5.0mg
Nicotinic acid	5.0mg
p-Aminobenzoic acid	5.0mg
Riboflavin	5.0mg
Thiamine·HCl	5.0mg
Biotin	2.0mg
Folic acid	2.0mg
Vitamin B_{12}	0.1mg

Preparation of Vitamin Solution: Add components to distilled/deionized water and bring volume to 1.0L. Mix thoroughly.

L-Cysteine·HCl Solution:
Composition per 10.0mL:

L-Cysteine·HCl	0.3g

Preparation of L-Cysteine·HCl Solution: Add L-cysteine·HCl to distilled/deionized water and bring volume to 10.0mL. Mix thoroughly. Autoclave under 100% N_2 for 15 min at 15 psi pressure–121°C.

$Na_2S{\cdot}9H_2O$ Solution:
Composition per 10.0mL:

$Na_2S{\cdot}9H_2O$	0.3g

Preparation of $Na_2S{\cdot}9H_2O$ Solution: Add $Na_2S{\cdot}9H_2O$ to distilled/deionized water and bring volume to 10.0mL. Mix thoroughly. Sparge with 100% N_2. Autoclave for 15 min at 15 psi pressure–121°C.

Preparation of Medium: Prepare and dispense medium under 80% H_2 + 20% CO_2. Add components, except L-cysteine·HCl solution and $Na_2S{\cdot}9H_2O$ solution, to distilled/deionized water and bring volume to 980.0mL. Mix thoroughly. Anaerobically distribute into tubes or flasks. Autoclave for 15 min at 15 psi pressure–121°C. Anaerobically add 10.0mL of sterile L-cysteine·HCl solution and 10.0mL of sterile $Na_2S{\cdot}9H_2O$ solution to each liter of medium or, using a syringe, inject the appropriate amount of sterile L-cysteine·HCl solution and sterile $Na_2S{\cdot}9H_2O$ solution into individual tubes containing medium.

Use: For the cultivation and maintenance of *Methanoculleus olentangyi.*

Methanogen Enrichment Medium, Barker

Composition per liter:

$CaCO_3$	20.0g
NH_4Cl	1.0g
$K_2HPO_4{\cdot}3H_2O$	0.4g
$MgCl_2{\cdot}6H_2O$	0.1g
Methanol	20.0mL

pH 7.0 ± 0.2 at 25°C

Preparation of Medium: Add components, except methanol and $CaCO_3$, to distilled/deionized water and bring volume to 1.0L. Mix thoroughly. Gently heat and bring to boiling. Autoclave for 15 min at 15 psi pressure–121°C. Cool to 25°C. Aseptically add filter-sterilized methanol solution. Mix thoroughly. Add 1.0g of $CaCO_3$ to each of 50.0mL screw-capped bottles. Autoclave for 15 min at 15 psi pressure–121°C. Cool to 25°C. Fill each bottle to capacity with enrichment medium.

Use: For the cultivation of methanogenic bacteria.

Methanogen Medium

Composition per 106.0mL:

$CaCO_3$	10.0g
Calcium acetate	2.0g
NH_4Cl	0.1g
$K_2HPO_4·3H_2O$	0.04g
$MgCl_2·6H_2O$	0.01g
$Na_2S·9H_2O$ solution	3.0mL
Na_2CO_3 solution	3.0mL

$Na_2S·9H_2O$ Solution:
Composition per 10.0mL:

$Na_2S·9H_2O$	0.1g

Preparation of $Na_2S·9H_2O$ Solution: Add $Na_2S·9H_2O$ to distilled/deionized water and bring volume to 10.0mL. Mix thoroughly. Autoclave for 15 min at 15 psi pressure–121°C. Cool to 25°C.

Na_2CO_3 Solution:
Composition per 10.0mL:

Na_2CO_3	0.5g

Preparation of Na_2CO_3 Solution: Add Na_2CO_3 to distilled/deionized water and bring volume to 10.0mL. Mix thoroughly. Autoclave for 15 min at 15 psi pressure–121°C. Cool to 25°C.

Preparation of Medium: Prepare and distribute medium anaerobically under 100% N_2. Add components, except $Na_2S·9H_2O$ solution and Na_2CO_3 solution, to distilled/deionized water and bring volume to 100.0mL. Mix thoroughly. Autoclave for 15 min at 15 psi pressure–121°C. Cool to 25°C. Aseptically add 3.0mL of sterile $Na_2S·9H_2O$ solution and 3.0mL of sterile Na_2CO_3 solution. Mix thoroughly. Aseptically and anaerobically distribute into sterile tubes or flasks.

Use: For the cultivation and enrichment of acetate-utilizing methanogenic bacteria.

Methanogen Medium, Zeikus

Composition per 1010.0mL:

Inorganic salts solution	500.0mL
Vitamin solution	500.0mL
$Na_2S·9H_2O$ solution	10.0mL

pH 7.0 ± 0.2 at 25°C

Inorganic Salts Solution:
Composition per 500.0mL:

$K_2HPO_4·3H_2O$	1.45g
NH_4Cl	1.0g
KH_2PO_4	0.75g
$MgCl_2·6H_2O$	0.2g
Nitrilotriacetic acid	0.04g
$CaCl_2·2H_2O$	0.02g
$FeCl_2·4H_2O$	3.6mg
$CoCl_2·6H_2O$	1.5mg
$MnCl_2$·4H2O	0.9mg
$ZnCl_2$	0.9mg
H_3BO_2	0.17mg
$Na_2MoO_4·2H_2O$	.09mg

Preparation of Inorganic Salts Solution: Add nitrilotriacetic acid to 250.0mL of distilled/deionized water. Dissolve by adjusting pH to 6.5 with KOH. Add remaining components. Readjust pH to 7.2 with H_2SO_4 or KOH. Add distilled/deionized water to 500.0mL. Filter sterilize.

Vitamin Solution:
Composition per 500.0mL:

Pyridoxine·HCl	1.0mg
p-Aminobenzoic acid	0.5mg
Ca-D-pantothenate	0.5mg
Nicotinic acid	0.5mg
Riboflavin	0.5mg
Thiamine·HCl	0.5mg
Thioctic acid	0.5mg
Biotin	0.2mg
Folic acid	0.2mg
Vitamin B_{12}	0.01mg

Preparation of Vitamin Solution: Add components to distilled/deionized water and bring volume to 500.0mL. Mix thoroughly. Filter sterilize.

$Na_2S·9H_2O$ Solution:
Composition per 10.0mL:

$Na_2S·9H_2O$	0.3g

Preparation of $Na_2S·9H_2O$ Solution: Add $Na_2S·9H_2O$ to distilled/deionized water and bring volume to 10.0mL. Mix thoroughly. Autoclave for 15 min at 15 psi pressure–121°C. Cool to 25°C.

Preparation of Medium: Prepare and distribute medium anaerobically under 95% N_2 + 5% CO_2. Aseptically and anaerobically combine 500.0mL of sterile inorganic salts solution, 500.0mL of sterile vitamin solution, and 10.0mL of sterile $Na_2S·9H_2O$ solution. Mix thoroughly. Aseptically and anaerobically distribute into sterile tubes or flasks.

Use: For the cultivation of methanogenic bacteria.

Methanogenium aggregans Medium (DSMZ Medium 321)

Composition per liter:

Na-formate	5.0g
Na_2CO_3	1.5g
Na-acetate	1.0g
Trypticase	1.0g
Yeast extract	1.0g
NH_4Cl	1.0g
$K_2HPO_4·3H_2O$	0.4g
$MgCl_2·6H_2O$	0.4g
Resazurin	1.0mg
Mineral solution	50.0mL
Cysteine-solution	10.0mL
$Na_2S·9H_2O$ solution	10.0mL
Trace elements solution	10.0mL
Sludge fluid	5.0mL

pH 6.8 ± 0.2 at 25°C

Mineral Solution:
Composition per liter:

NaCl	12.0g
KH_2PO_4	6.0g
$(NH_4)_2SO_4$	6.0g
$MgSO_4·7H_2O$	2.6g
$CaCl_2·2H_2O$	0.16g

Preparation of Mineral Solution: Add components to distilled/deionized water and bring volume to 1.0L. Mix thoroughly.

Trace Elements Solution:
Composition per liter:

$MgSO_4·7H_2O$	3.0g
Nitrilotriacetic acid	1.5g
NaCl	1.0g
$MnSO_4·2H_2O$	0.5g

$CoSO_4 \cdot 7H_2O$ 0.18g
$ZnSO_4 \cdot 7H_2O$ 0.18g
$CaCl_2 \cdot 2H_2O$ 0.1g
$FeSO_4 \cdot 7H_2O$ 0.1g
$NiCl_2 \cdot 6H_2O$ 0.025g
$KAl(SO_4)_2 \cdot 12H_2O$ 0.02g
H_3BO_3 .... 0.01g
$Na_2MoO_4 \cdot 4H_2O$ 0.01g
$CuSO_4 \cdot 5H_2O$ 0.01g
$Na_2SeO_3 \cdot 5H_2O$ 0.3mg

Preparation of Trace Elements Solution: Add nitrilotriacetic acid to 500.0mL of distilled/deionized water. Dissolve by adjusting pH to 6.5 with KOH. Add remaining components. Add distilled/deionized water to 1.0L. Mix thoroughly.

Cysteine Solution:
Composition per 10.0mL:
L-Cysteine·HCl·H_2O 0.2g

Preparation of Cysteine Solution: Add L-cysteine·HCl·H_2O to distilled/deionized water and bring volume to 10.0mL. Mix thoroughly. Sparge with 100% N_2. Autoclave for 15 min at 15 psi pressure–121°C.

$Na_2S \cdot 9H_2O$ Solution:
Composition per 10mL:
$Na_2S \cdot 9H_2O$ 0.2g

Preparation of $Na_2S \cdot 9H_2O$ Solution: Add $Na_2S \cdot 9H_2O$ to distilled/deionized water and bring volume to 10.0mL. Mix thoroughly. Autoclave under 100% N_2 for 15 min at 15 psi pressure–121°C. Cool to room temperature.

Sludge Fluid:
Composition per 500.0mL:
Yeast extract 2.0g
Sludge 500.0mL

Preparation of Sludge Fluid: Add yeast extract to sludge from an anaerobic digester. Gas with nitrogen gas for a few minutes. Incubate at 37°C for 24 hours. Centrifuge the sludge at 13,000 x g. Autoclave for 15 min at 15 psi pressure–121°C. The resulting, clear supernatant in screw-capped vessels under nitrogen gas. The sludge fluid can be stored at room temperature in the dark.

Preparation of Medium: Prepare and dispense medium under an oxygen-free 80% H_2 + 20% CO_2 gas mixture. Add components, except sludge fluid, cysteine solution, and $Na_2S \cdot 9H_2O$ solution, to distilled/deionized water and bring volume to 975.0mL. Mix thoroughly. Sparge with 80% H_2 + 20% CO_2. Autoclave for 15 min at 15 psi pressure–121°C. Cool to 25°C while sparging with 80% H_2 + 20% CO_2. Aseptically and anaerobically add 5.0mL sterile sludge fluid, 10.0mL of sterile cysteine solution, and 10.0mL of sterile $Na_2S \cdot 9H_2O$ solution. Mix thoroughly. Aseptically and anaerobically distribute into sterile tubes or flasks.

Use: For the cultivation of *Methanocorpusculum aggregans=Methanogenium aggregans.*

Methanogenium Alcohol Medium

Composition per 1003.0mL:
NaCl 1.0g
KCl 0.5g
$MgCl_2 \cdot 6H_2O$ 0.5g
NH_4Cl 0.4g
Sodium acetate·$3H_2O$ 0.4g
KH_2PO_4 .... 0.2g
$CaCl_2 \cdot 2H_2O$ 0.1g
$NaHCO_3$ solution 60.0mL
2-Propanol 5.0mL
$Na_2S \cdot 9H_2O$ solution 3.0mL
Cyanocobalamin solution 1.0mL
Selenite-molybdate-tungstate solution 1.0mL
Thiamine solution 1.0mL
Trace elements solution 1.0mL
Vitamin solution 1.0mL

Trace Elements Solution:
Composition per 100.0mL:
$FeSO_4 \cdot 7H_2O$ 1400.0mg
$ZnSO_4 \cdot 7H_2O$ 145.0mg
$CoCl_2 \cdot 6H_2O$ 120.0mg
$MnCl_2 \cdot 4H_2O$ 100.0mg
$NiCl_2 \cdot 6H_2O$ 50.0mg
H_3BO_3 .... 6.0mg
$CuSO_4 \cdot 5H_2O$ 3.0mg
HCl (25%,w/v) 8.0mL

Preparation of Trace Elements Solution: Add components to distilled/deionized water and bring volume to 1.0L. Mix thoroughly. Sparge with 100% N_2. Autoclave for 15 min at 15 psi pressure–121°C.

Selenite-Molybdate-Tungstate Solution:
Composition per liter:
NaOH 0.2g
$Na_2MoO_4 \cdot 2H_2O$ 40.0mg
$Na_2WO_4 \cdot 2H_2O$ 33.0mg
$Na_2SeO_3 \cdot 2H_2O$ 5.0mg

Preparation of Selenite-Molybdate-Tungstate Solution: Add components to distilled/deionized water and bring volume to 1.0L. Mix thoroughly. Sparge with 100% N_2. Autoclave for 15 min at 15 psi pressure–121°C.

$NaHCO_3$ Solution:
Composition per liter:
$NaHCO_3$ 84.0g

Preparation of $NaHCO_3$ Solution: Add $NaHCO_3$ to distilled/deionized water and bring volume to 10.0mL. Mix thoroughly. Sparge with 80% N_2 + 20% CO_2. Autoclave for 15 min at 15 psi pressure–121°C.

$Na_2S \cdot 9H_2O$ Solution:
Composition per 100.0mL:
$Na_2S \cdot 9H_2O$ 2.5g
NaOH .... 1 pellet

Preparation of $Na_2S \cdot 9H_2O$ Solution: Bring 100.0mL of distilled/deionized water to boiling. Cool to room temperature while sparging with 100%N_2. Dissolve 1 pellet of NaOH in the anaerobic water. Weigh out a little more than 2.5g of $Na_2S \cdot 9H_2O$. Briefly rinse the crystals in distilled/deionized water. Dry the crystals by blotting on paper towels or filter paper. Add 2.5g of washed $Na_2S \cdot 9H_2O$ crystals to 100.0mL of anaerobic NaOH solution. Distribute into serum bottles fitted with butyl rubber stoppers and aluminum seals. Do not grease stoppers. Pressurize to 60kPa with 100% N_2. Autoclave for 15 min at 15 psi pressure–121°C. Store at room temperature in an anaerobic chamber.

Preparation of 2-Propanol: Filter sterilize 10.0mL of 2-propanol. Sparge with 100% N_2.

Vitamin Solution:
Composition per liter:
Sodium 2-mercaptoethanesulfonate 0.25g
Pyridoxine·HCl 0.15g
Calcium pantothenate 0.1g
Nicotinic acid 0.1g

p-Aminobenzoic acid 40.0mg
Biotin 10.0mg
Potassium phosphate buffer (25m*M* solution, pH 7.0) . 1.0L

Preparation of Vitamin Solution: Combine components. Mix thoroughly. Filter sterilize. Sparge with 100% N_2.

Thiamine Solution:
Composition per liter:
Thiamine·HCl 0.1g
Sodium phosphate buffer (0.1*M* solution, pH 3.6) 1.0L

Preparation of Thiamine Solution: Combine components. Mix thoroughly. Filter sterilize. Sparge with 100% N_2.

Cyanocobalamin Solution:
Composition per liter:
Cyanocobalamin 50.0mg

Preparation of Cyanocobalamin Solution: Add cyanocobalamin to distilled/deionized water and bring volume to 1.0L. Mix thoroughly. Filter sterilize. Sparge with 100% N_2.

Preparation of Medium: Prepare and dispense medium under 80% N_2 + 20% CO_2. Add components, except $NaHCO_3$ solution, 2-propanol, $Na_2S·9H_2O$ solution, cyanocobalamin solution, selenite-molybdate-tungstate solution, thiamine solution, trace elements solution, and vitamin solution, to distilled/deionized water and bring volume to 930.0mL. Mix thoroughly. Sparge with 80% N_2 + 20% CO_2. Autoclave for 15 min at 15 psi pressure–121°C. Aseptically and anaerobically add 60.0mL of sterile $NaHCO_3$ solution, 5.0mL of sterile 2-propanol, 3.0mL of sterile $Na_2S·9H_2O$ solution, 1.0mL of sterile cyanocobalamin solution, 1.0mL of sterile selenite-molybdate-tungstate solution, 1.0mL of sterile thiamine solution, 1.0mL of sterile trace elements solution, and 1.0mL of sterile vitamin solution. Mix thoroughly. Aseptically and anaerobically distribute into sterile tubes or bottles.

Use: For the cultivation of *Methanogenium* species.

Methanogenium bourgense Medium (DSMZ Medium 322)

Composition per liter:
Na-formate 5.0g
Na_2CO_3 .... 1.5g
Na-acetate 1.0g
Trypticase peptone 1.0g
Yeast extract 1.0g
NH_4Cl 1.0g
$K_2HPO_4·3H_2O$ 0.4g
$MgCl_2·6H_2O$ 0.1g
Resazurin 1.0mg
Cysteine-solution 10.0mL
$Na_2S·9H_2O$ solution 10.0mL
pH 6.0-7.0 at 25°C

Cysteine Solution:
Composition per 10.0mL:
L-Cysteine·HCl·H_2O 0.5g

Preparation of Cysteine Solution: Add L-cysteine·HCl·H_2O to distilled/deionized water and bring volume to 10.0mL. Mix thoroughly. Sparge with 100% N_2. Autoclave for 15 min at 15 psi pressure–121°C.

$Na_2S·9H_2O$ Solution:
Composition per 10mL:
$Na_2S·9H_2O$ 0.2g

Preparation of $Na_2S·9H_2O$ Solution: Add $Na_2S·9H_2O$ to distilled/deionized water and bring volume to 10.0mL. Mix thoroughly. Autoclave under 100% N_2 for 15 min at 15 psi pressure–121°C. Cool to room temperature.

Preparation of Medium: Prepare and dispense medium under an oxygen-free 80% H_2 + 20% CO_2 gas mixture. Add components, except cysteine solution and $Na_2S·9H_2O$ solution, to distilled/deionized water and bring volume to 980.0mL. Mix thoroughly. Sparge with 80% H_2 + 20% CO_2. Autoclave for 15 min at 15 psi pressure–121°C. Cool to 25°C while sparging with 80% H_2 + 20% CO_2. Aseptically and anaerobically add 10.0mL of sterile cysteine solution and 10.0mL of sterile $Na_2S·9H_2O$ solution. Mix thoroughly. Aseptically and anaerobically distribute into sterile tubes or flasks.

Use: For the cultivation of *Methanoculleus bourgensis=Methanogenium bourgense.*

Methanogenium CV Medium

Composition per liter:
NaCl 18.0g
Isopropanol 7.5g
$NaHCO_3$ 5.0g
$MgCl_2·6H_2O$ 4.0g
$MgSO_4·7H_2O$ 3.45g
Trypticase™ 2.0g
Yeast extract 2.0g
Sodium acetate 1.0g
KCl 0.335g
NH_4Cl 0.25g
$CaCl_2·2H_2O$ 0.14g
K_2HPO_4 .... 0.14g
L-Cysteine·HCl 0.5g
$Na_2S·9H_2O$ 0.4g
$Fe(NH_4)_2(SO_4)_2·7H_2O$ 2.0mg
Resazurin 1.0mg
$Na_2WoO_4·2H_2O$ 0.03mg
Trace elements solution 10.0mL
Vitamin solution 10.0mL
pH 6.8–7.0 at 25°C

Trace Elements Solution:
Composition per liter:
$MgSO_4·7H_2O$ 3.0g
Nitrilotriacetic acid 1.5g
NaCl 1.0g
$MnSO_4·2H_2O$ 0.5g
$CoSO_4·7H_2O$ 0.18g
$ZnSO_4·7H_2O$ 0.18g
$CaCl_2·2H_2O$ 0.1g
$FeSO_4·7H_2O$ 0.1g
$NiCl_2·6H_2O$ 0.025g
$KAl(SO_4)_2·12H_2O$ 0.02g
$CuSO_4·5H_2O$ 0.01g
H_3BO_3 .... 0.01g
$Na_2MoO_4·2H_2O$ 0.01g
$Na_2SeO_3·5H_2O$ 0.3mg

Preparation of Trace Elements Solution: Add nitrilotriacetic acid to 500.0mL of distilled/deionized water. Adjust pH to 6.5 with KOH. Add remaining components. Adjust pH to 7.0. Add distilled/deionized water to 1.0L.

Vitamin Solution:
Composition per liter:

Pyridoxine·HCl	10.0mg
Calcium DL-pantothenate	5.0mg
Nicotinic acid	5.0mg
p-Aminobenzoic acid	5.0mg
Lipoic acid	5.0mg
Riboflavin	5.0mg
Thiamine·HCl	5.0mg
Biotin	2.0mg
Folic acid	2.0mg
Vitamin B_{12}	0.1mg

Preparation of Vitamin Solution: Add components to distilled/deionized water and bring volume to 1.0L. Mix thoroughly. Filter sterilize. Sparge with 80% H_2 + 20% CO_2.

Preparation of Medium: Prepare and dispense medium under 80% H_2 + 20% CO_2. Add components, except vitamin solution, to distilled/deionized water and bring volume to 990.0mL. Mix thoroughly. Adjust pH to 6.8–7.0. Anaerobically distribute into tubes or flasks fitted with butyl rubber stoppers. Autoclave for 15 min at 15 psi pressure–121°C. Anaerobically add 10.0mL of sterile vitamin solution to each liter of medium or, using a syringe, inject the appropriate amount of sterile vitamin solution into individual tubes containing medium.

Use: For the cultivation and maintenance of *Methanogenium organophilum.*

Methanogenium Medium

Composition per liter:

$NaHCO_3$	4.0g
Sodium acetate	4.0g
Sodium formate	2.0g
Yeast extract	1.0g
L-Cysteine·HCl	0.5g
KH_2PO_4	0.5g
$Na_2S \cdot 9H_2O$	0.5g
$MgSO_4 \cdot 7H_2O$	0.4g
NaCl	0.4g
NH_4Cl	0.4g
$CaCl_2 \cdot 2H_2O$	0.05g
$NiCl_2 \cdot 6H_2O$	24.0mg
$FeSO_4 \cdot 7H_2O$	2.0mg
Resazurin	1.0mg
Sludge fluid	50.0mL
Fatty acid mixture	20.0mL
Trace elements solution SL-10	1.0mL

pH 6.7 ± 0.2 at 25°C

Sludge Fluid:
Composition per 100.0mL:

Sludge	100.0mL
Yeast extract	0.4g

Preparation of Sludge Fluid: To 100.0mL of sludge from an anaerobic digester, add 0.4g of yeast extract. Sparge with 100% N_2 for a few minutes. Incubate at 37°C for 24 hr. Centrifuge the sludge at 13,000 × g for 15 min. Decant the clear supernatant solution. Sparge with 100% N_2 for a few minutes. Store in screw-capped bottles at room temperature in the dark.

Fatty Acid Mixture:
Composition per 20.0mL:

α-Methylbutyric acid	0.5g
Isobutyric acid	0.5g
Isovaleric acid	0.5g
Valeric acid	0.5g

Preparation of Fatty Acid Mixture: Add components to distilled/deionized water and bring volume to 20.0mL. Mix thoroughly. Adjust pH to 7.5 with concentrated NaOH.

Trace Elements Solution SL-10:
Composition per liter:

$FeCl_2 \cdot 4H_2O$	1.5g
$CoCl_2 \cdot 6H_2O$	190.0mg
$MnCl_2 \cdot 4H_2O$	100.0mg
$ZnCl_2$	70.0mg
$Na_2MoO_4 \cdot 2H_2O$	36.0mg
$NiCl_2 \cdot 6H_2O$	24.0mg
H_3BO_3	6.0mg
$CuCl_2 \cdot 2H_2O$	2.0mg
HCl (25% solution)	10.0mL

Preparation of Trace Elements Solution SL-10: Add $FeCl_2 \cdot 4H_2O$ to 10.0mL of HCl solution. Mix thoroughly. Add distilled/deionized water and bring volume to 1.0L. Add remaining components. Mix thoroughly. Sparge with 100% N_2. Autoclave for 15 min at 15 psi pressure–121°C.

Preparation of Medium: Prepare and dispense medium anaerobically under 80% H_2 + 20% CO_2. Add components to distilled/deionized water and bring volume to 1.0L. Mix thoroughly. Sparge with 80% H_2 + 20% CO_2. Autoclave for 15 min at 15 psi pressure–121°C.

Use: For the cultivation and maintenance of *Methanobacterium* species, *Methanococcus* species, *Methanocorpusculum* species, *Methanoculleus* species, *Methanogenium* species, *Methanoplanus* species, and *Thermotoga* species.

Methanogenium olentangyi Medium (DSMZ Medium 287)

Composition per liter:

$NaHCO_3$	5.0g
NH_4Cl	2.7g
Na-acetate	2.5g
NaCl	0.61g
K_2HPO_4	0.3g
KH_2PO_4	0.3g
$(NH_4)_2SO_4$	0.3g
$CaCl_2 \cdot H_2O$	0.14g
$MgSO_4 \cdot 7H_2O$	0.13g
Resazurin	1.0mg
Trace elements solution	10.0mL
Vitamin solution	10.0mL
Cysteine-HCl·H_2O solution	10.0mL
$Na_2S \cdot 9H_2O$ solution	10.0mL

pH 6.9 ± 0.2 at 25°C

Trace Elements Solution:
Composition per liter:

$MgSO_4 \cdot 7H_2O$	3.0g
Nitrilotriacetic acid	1.5g
NaCl	1.0g
$MnSO_4 \cdot 2H_2O$	0.5g
$CoSO_4 \cdot 7H_2O$	0.18g
$ZnSO_4 \cdot 7H_2O$	0.18g
$FeSO_4 \cdot 7H_2O$	0.1g
$CaCl_2 \cdot 2H_2O$	0.1g
$NiCl_2 \cdot 6H_2O$	0.025g
$KAl(SO_4)_2 \cdot 12H_2O$	0.02g
$CuSO_4 \cdot 5H_2O$	0.01g
H_3BO_3	0.01g
$Na_2MoO_4 \cdot 4H_2O$	0.01g
$Na_2SeO_3 \cdot 5H_2O$	0.3mg

Preparation of Trace Elements Solution: Add nitrilotriacetic acid to 500.0mL of distilled/deionized water. Adjust pH to 6.5 with KOH. Add remaining components. Add distilled/deionized water to 1.0L. Adjust pH to 7.0 with KOH.

Vitamin Solution:
Composition per liter:

Pyridoxine-HCl	10.0mg
Thiamine-HCl·$2H_2O$	5.0mg
Riboflavin	5.0mg
Nicotinic acid	5.0mg
D-Ca-pantothenate	5.0mg
p-Aminobenzoic acid	5.0mg
Lipoic acid	5.0mg
Biotin	2.0mg
Folic acid	2.0mg
Vitamin B_{12}	0.1mg

Preparation of Vitamin Solution: Add components to distilled/deionized water and bring volume to 1.0L. Mix thoroughly. Filter sterilize.

Cysteine-HCl·H_2O Solution:
Composition per 10mL:

Cysteine-HCl·H_2O	0.3g

Preparation of L-Cysteine Solution: Add L-cysteine·HCl·H_2O to distilled/deionized water and bring volume to 10.0mL. Mix thoroughly. Sparge with 100% N_2. Autoclave for 15 min at 15 psi pressure–121°C. Cool to room temperature.

$Na_2S·9H_2O$ Solution:
Composition per 10mL:

$Na_2S·9H_2O$	0.3g

Preparation of $Na_2S·9H_2O$ Solution: Add $Na_2S·9H_2O$ to distilled/deionized water and bring volume to 10.0mL. Mix thoroughly. Autoclave under 100% N_2 for 15 min at 15 psi pressure–121°C. Cool to room temperature.

Preparation of Medium: Prepare and dispense medium under 80% H_2 + 20% CO_2 gas atmosphere. Add components, except cysteine-HCl·H_2O solution, $Na_2S·9H_2O$ solution, and vitamin solution, to distilled/deionized water and bring volume to 970.0mL. Mix thoroughly. Adjust pH to 6.9. Sparge with 80% H_2 + 20% CO_2. Autoclave for 15 min at 15 psi pressure–121°C. Aseptically and anaerobically add 10.0mL cysteine-HCl·H_2O solution, 10.0mL $Na_2S·9H_2O$ solution, and 10.0mL vitamin solution. Mix thoroughly. Aseptically and anaerobically distribute into sterile tubes or bottles. After inoculation, flush and repressurize the gas head space of culture bottles with sterile 80% H_2 + 20% CO_2 to 1 bar overpressure.

Use: For the cultivation of *Methanoculleus olentangyi=Methanogenium olentangyi.*

Methanogenium thermophilicum Medium

Composition per liter:

NaCl	11.7g
Pancreatic digest of casein	6.0g
Sodium formate	5.0g
$MgCl_2·6H_2O$	4.0g
$NaHCO_3$	4.0g
Yeast extract	2.0g
L-Cysteine·HCl	0.5g
$Na_2S·9H_2O$	0.25g
K_2HPO_4	0.14g
KH_2PO_4	0.14g
$CaCl_2·2H_2O$	0.075g
Resazurin	1.0mg
Vitamin solution	10.0mL

Preparation of Medium: Add components, except L-cysteine·HCl and $Na_2S·9H_2O$, to distilled/deionized water and bring volume to 1.0L. Mix thoroughly. Bubble the solution with a stream of oxygen-free 80% N_2 + 20% CO_2 for 20 minutes. Adjust the pH to 7.2 with 2*M* KOH. Add the L-cysteine·HCl and $Na_2S·9H_2O$. Mix thoroughly. Tube the medium under oxygen-free 80% N_2 + 20% CO_2 in either serum tubes or bottles. Autoclave for 15 min at 15 psi pressure–121°C. After sterilization, the medium may form a precipitate. Allow it to cool and mix thoroughly to bring the precipitate back into solution.

Use: For the cultivation and maintenance of *Methanogenium thermophilicum.*

Methanohalobium Medium

Composition per liter:

NaCl	250.0g
$MgSO_4·7H_2O$	4.0g
$CaCl_2·2H_2O$	0.33g
KCl	0.33g
KH_2PO_4	0.33g
$MgCl_2·6H_2O$	0.33g
NH_4Cl	0.33g
Resazurin	1.0mg
$NaHCO_3$ solution	100.0mL
Trimethylamine·HCl solution	20.0mL
$Na_2S·9H_2O$ solution	10.0mL
Yeast extract solution	10.0mL
Vitamin solution	10.0mL
Trace elements solution SL-10	10.0mL
Na_2CO_3 solution	variable

pH 7.4 ± 0.2 at 25°C

$NaHCO_3$ Solution:
Composition per 10.0mL:

$NaHCO_3$	5.0g

Preparation of $NaHCO_3$ Solution: Add $NaHCO_3$ to distilled/deionized water and bring volume to 10.0mL. Mix thoroughly. Sparge with 80% N_2 + 20% CO_2. Autoclave for 15 min at 15 psi pressure–121°C.

Trimethylamine·HCl Solution:
Composition per 20.0mL:

Trimethylamine.HCl	0.2g

Preparation of Trimethylamine·HCl Solution: Add trimethylamine·HCl to distilled/deionized water and bring volume to 20.0mL. Mix thoroughly. Sparge with 100% N_2. Autoclave for 15 min at 15 psi pressure–121°C.

$Na_2S·9H_2O$ Solution:
Composition per 10.0mL:

$Na_2S·9H_2O$	0.5g

Preparation of $Na_2S·9H_2O$ Solution: Add $Na_2S·9H_2O$ to distilled/deionized water and bring volume to 10.0mL. Mix thoroughly. Sparge with 100% N_2. Autoclave for 15 min at 15 psi pressure–121°C.

Yeast Extract Solution:
Composition per 5.0mL:

Yeast extract	50.0mg

Preparation of Yeast Extract Solution: Add yeast extract to distilled/deionized water and bring volume to 5.0mL. Mix thoroughly. Sparge with 100% N_2. Autoclave for 15 min at 15 psi pressure–121°C.

Vitamin Solution:
Composition per liter:

Pyridoxine·HCl 10.0mg
Calcium DL-pantothenate 5.0mg
Nicotinic acid 5.0mg
p-Aminobenzoic acid 5.0mg
Lipoic acid 5.0mg
Riboflavin 5.0mg
Thiamine·HCl 5.0mg
Biotin 2.0mg
Folic acid 2.0mg
Vitamin B_{12} 0.1mg

Preparation of Vitamin Solution: Add components to distilled/deionized water and bring volume to 1.0L. Mix thoroughly. Filter sterilize. Sparge with 80% H_2 + 20% CO_2.

Trace Elements Solution SL-10:
Composition per liter:

$FeCl_2 \cdot 4H_2O$ 1.5g
$CoCl_2 \cdot 6H_2O$ 190.0mg
$MnCl_2 \cdot 4H_2O$ 100.0mg
$ZnCl_2$ 70.0mg
$Na_2MoO_4 \cdot 2H_2O$ 36.0mg
$NiCl_2 \cdot 6H_2O$ 24.0mg
H_3BO_3 .... 6.0mg
$CuCl_2 \cdot 2H_2O$ 2.0mg
HCl (25% solution) 10.0mL

Preparation of Trace Elements Solution SL-10: Add $FeCl_2 \cdot 4H_2O$ to 10.0mL of HCl solution. Mix thoroughly. Add distilled/deionized water and bring volume to 1.0L. Add remaining components. Mix thoroughly. Sparge with 100% N_2. Autoclave for 15 min at 15 psi pressure–121°C.

Na_2CO_3 Solution:
Composition per 10.0mL:

Na_2CO_3 .... 0.5g

Preparation of Na_2CO_3 Solution: Add Na_2CO_3 to distilled/deionized water and bring volume to 10.0mL. Mix thoroughly. Sparge with 100% N_2. Autoclave for 15 min at 15 psi pressure–121°C.

Preparation of Medium: Prepare and dispense medium under 80% H_2 + 20% CO_2. Add components, except $NaHCO_3$ solution, trimethylamine·HCl solution, $Na_2S \cdot 9H_2O$ solution, yeast extract solution, vitamin solution, trace elements solution SL-10, and Na_2CO_3 solution, to distilled/deionized water and bring volume to 840.0mL. Mix thoroughly. Autoclave for 15 min at 15 psi pressure–121°C. Anaerobically add 100.0mL of sterile $NaHCO_3$ solution, 20.0mL of sterile trimethylamine·HCl solution, 10.0mL of sterile $Na_2S \cdot 9H_2O$ solution, 10.0mL of sterile yeast extract solution, 10.0mL of sterile vitamin solution, and 10.0mL of sterile trace elements solution SL-10. Mix thoroughly. Add a sufficient quantity of sterile Na_2CO_3 solution to bring the pH to 7.4.

Use: For the cultivation and maintenance of *Methanohalobium evestigatus*.

Methanohalophilus euhalobius Medium

Composition per 1010.0mL:

NaCl 60.0g
$MgCl_2 \cdot 6H_2O$ 2.4g
Yeast extract 2.0g
KCl 1.0g
Disodium EDTA 0.5g
NH_4Cl 0.5g
KH_2PO_4 .... 0.4g
Resazurin 0.5mg
Trimethylamine solution 50.0mL
$NaHCO_3$ solution 40.0mL
$CaCl_2$ solution 20.0mL
$Na_2S \cdot 9H_2O$ solution 15.0mL
L-Cysteine·HCl solution 15.0mL
Wolfe's mineral solution 10.0mL
Wolfe's vitamin solution 5.0mL

pH 6.8–7.0 at 25°C

Trimethylamine Solution:
Composition per 50.0mL:

Trimethylamine·HCl 10.0g

Preparation of Trimethylamine Solution: Add trimethylamine·HCl to distilled/deionized water and bring volume to 50.0mL. Mix thoroughly. Sparge with 100% N_2. Autoclave for 15 min at 15 psi pressure–121°C.

$NaHCO_3$ Solution:
Composition per 40.0mL:

$NaHCO_3$ 4.0g

Preparation of $NaHCO_3$ Solution: Add $NaHCO_3$ to distilled/deionized water and bring volume to 40.0mL. Mix thoroughly. Sparge with 80% N_2 + 20% CO_2. Autoclave for 15 min at 15 psi pressure–121°C.

$CaCl_2$ Solution:
Composition per 20.0mL:

$CaCl_2 \cdot 2H_2O$ 2.0g

Preparation of $CaCl_2$ Solution: Add $CaCl_2 \cdot 2H_2O$ to distilled/deionized water and bring volume to 20.0mL. Mix thoroughly. Sparge with 80% N_2 + 20% CO_2 gas mixture. Autoclave for 15 min at 15 psi pressure–121°C.

$Na_2S \cdot 9H_2O$ Solution:
Composition per 20.0mL:

$Na_2S \cdot 9H_2O$ 0.6g

Preparation of $Na_2S \cdot 9H_2O$ Solution: Add $Na_2S \cdot 9H_2O$ to distilled/deionized water and bring volume to 20.0mL. Mix thoroughly. Sparge with 100% N_2. Autoclave for 15 min at 15 psi pressure–121°C. Before use, neutralize to pH 7.0 with sterile HCl.

L-Cysteine·HCl Solution:
Composition per 20.0mL:

L-Cysteine·HCl·H_2O 0.6g

Preparation of L-Cysteine·HCl Solution: Add L-cysteine·HCl·H_2O to distilled/deionized water and bring volume to 20.0mL. Mix thoroughly. Sparge with 100% N_2. Autoclave for 15 min at 15 psi pressure–121°C.

Wolfe's Mineral Solution:
Composition per liter:

$MgSO_4 \cdot 7H_2O$ 3.0g
Nitrilotriacetic acid 1.5g
NaCl 1.0g
$MnSO_4 \cdot 2H_2O$ 0.5g
$CoCl_2 \cdot 6H_2O$ 0.1g
$ZnSO_4 \cdot 7H_2O$ 0.1g
$CaCl_2 \cdot 2H_2O$ 0.1g
$FeSO_4 \cdot 7H_2O$ 0.1g
$NiCl_2 \cdot 6H_2O$ 0.025g
$KAl(SO_4)_2 \cdot 12H_2O$ 0.02g
$CuSO_4 \cdot 5H_2O$ 0.01g
H_3BO_3 .... 0.01g
$Na_2MoO_4 \cdot 2H_2O$ 0.01g
$Na_2SeO_3 \cdot 5H_2O$ 0.3mg

Preparation of Wolfe's Mineral Solution: Add nitrilotriacetic acid to 500.0mL of distilled/deionized water. Adjust pH to 6.5 with KOH. Add remaining components. Add distilled/deionized water to 1.0L. Adjust pH to 6.8.

Wolfe's Vitamin Solution:
Composition per liter:

Pyridoxine·HCl	10.0mg
p-Aminobenzoic acid	5.0mg
Lipoic acid	5.0mg
Nicotinic acid	5.0mg
Riboflavin	5.0mg
Thiamine·HCl	5.0mg
Calcium DL-pantothenate	5.0mg
Biotin	2.0mg
Folic acid	2.0mg
Vitamin B_{12}	0.1mg

Preparation of Wolfe's Vitamin Solution: Add components to distilled/deionized water and bring volume to 1.0L. Mix thoroughly. Filter sterilize.

Preparation of Medium: Prepare and dispense medium under 80% N_2 + 20% CO_2 gas mixture. Add components, except trimethylamine solution, $NaHCO_3$ solution, $CaCl_2$ solution solution, L-Cysteine·HCl solution, and $Na_2S·9H_2O$ solution, to distilled/deionized water and bring volume to 860.0mL. Mix thoroughly. Adjust pH to 6.8–7.0. Sparge with 80% N_2 + 20% CO_2 gas mixture. Autoclave for 15 min at 15 psi pressure–121°C. Aseptically and anaerobically add 50.0mL of sterile trimethylamine solution, 40.0mL of sterile $NaHCO_3$ solution, 20.0mL of sterile $CaCl_2$ solution, 15.0mL of sterile L-cysteine·HCl solution, and 15.0mL of sterile $Na_2S·9H_2O$ solution. Mix thoroughly. Aseptically and anaerobically distribute into sterile tubes or bottles.

Use: For the cultivation of *Methanohalophilus euhalobius.*

Methanohalophilus halophilus Medium

Composition per 1011.0mL:

NaCl	70.0g
KCl	0.33g
KH_2PO_4	0.33g
NH_4Cl	0.33g
Yeast extract	0.05g
K_2SO_4	0.01g
Resazurin	1.0mg
Methylamine·HCl solution	20.0mL
$MgCl_2/CaCl_2$ solution	10.0mL
$NaHCO_3$ solution	10.0mL
Vitamin solution	10.0mL
L-Cysteine·HCl solution	10.0mL
$Na_2S·9H_2O$ solution	10.0mL
Trace elements solution SL-10	1.0mL

pH 7.2 ± 0.2 at 25°C

$MgCl_2/CaCl_2$ Solution:
Composition per 10.0mL:

$MgCl_2·6H_2O$	0.33g
$CaCl_2·2H_2O$	0.33g

Preparation of $MgCl_2/CaCl_2$ Solution: Add components to distilled/deionized water and bring volume to 10.0mL. Mix thoroughly. Sparge with 100% N_2. Autoclave for 15 min at 15 psi pressure–121°C.

$NaHCO_3$ Solution:
Composition per 10.0mL:

$NaHCO_3$	5.0g

Preparation of $NaHCO_3$ Solution: Add $NaHCO_3$ to distilled/deionized water and bring volume to 10.0mL. Mix thoroughly. Sparge with 80% N_2 + 20% CO_2. Autoclave for 15 min at 15 psi pressure–121°C.

Vitamin Solution:
Composition per liter:

Pyridoxine·HCl	10.0mg
Calcium DL-pantothenate	5.0mg
Nicotinic acid	5.0mg
p-Aminobenzoic acid	5.0mg
Riboflavin	5.0mg
Thiamine·HCl	5.0mg
Biotin	2.0mg
Folic acid	2.0mg
Vitamin B_{12}	0.1mg

Preparation of Vitamin Solution: Add components to distilled/deionized water and bring volume to 1.0L. Mix thoroughly. Filter sterilize. Sparge with 80% H_2 + 20% CO_2.

Methylamine·HCl Solution:
Composition per 20.0mL:

Methylamine·HCl	5.0g

Preparation of Methylamine·HCl Solution: Add methylamine·HCl to distilled/deionized water and bring volume to 20.0mL. Mix thoroughly. Sparge with 100% N_2. Autoclave for 15 min at 15 psi pressure–121°C.

L-Cysteine·HCl Solution:
Composition per 10.0mL:

L-Cysteine·HCl	0.2g

Preparation of L-Cysteine·HCl Solution: Add L-cysteine·HCl to distilled/deionized water and bring volume to 10.0mL. Mix thoroughly. Autoclave under 100% N_2 for 15 min at 15 psi pressure–121°C.

$Na_2S·9H_2O$ Solution:
Composition per 10.0mL:

$Na_2S·9H_2O$	0.2g

Preparation of $Na_2S·9H_2O$ Solution: Add $Na_2S·9H_2O$ to distilled/deionized water and bring volume to 10.0mL. Mix thoroughly. Sparge with 100% N_2. Autoclave for 15 min at 15 psi pressure–121°C.

Trace Elements Solution SL-10:
Composition per liter:

$FeCl_2·4H_2O$	1.5g
$CoCl_2·6H_2O$	190.0mg
$MnCl_2·4H_2O$	100.0mg
$ZnCl_2$	70.0mg
$Na_2MoO_4·2H_2O$	36.0mg
$NiCl_2·6H_2O$	24.0mg
H_3BO_3	6.0mg
$CuCl_2·2H_2O$	2.0mg
HCl (25% solution)	10.0mL

Preparation of Trace Elements Solution SL-10: Add $FeCl_2·4H_2O$ to 10.0mL of HCl solution. Mix thoroughly. Add distilled/deionized water and bring volume to 1.0L. Add remaining components. Mix thoroughly. Sparge with 100% N_2. Autoclave for 15 min at 15 psi pressure–121°C.

Preparation of Medium: Prepare and dispense medium under 80% H_2 + 20% CO_2. Add components, except $MgCl_2/CaCl_2$ solution, $NaHCO_3$ solution, vitamin solution, methylamine·HCl solution, L-cysteine·HCl solution, and $Na_2S·9H_2O$ solution, to distilled/deionized water and bring volume to 840.0mL. Mix thoroughly. Adjust pH to 6.9–7.0. Sparge with 80% H_2 + 20% CO_2. Autoclave for 15 min at 15 psi pressure–121°C. Anaerobically add 10.0mL of sterile $MgCl_2/CaCl_2$ solution, 10.0mL of sterile $NaHCO_3$ solution,

10.0mL of sterile vitamin solution, 20.0mL of sterile methylamine·HCl solution, 10.0mL of sterile L-cysteine·HCl solution, and 10.0mL of sterile $Na_2S·9H_2O$ solution. Mix thoroughly.

Use: For the cultivation and maintenance of *Methanohalophilus halophilus*.

Methanohalophilus mahii Medium

Composition per 1010.0mL:

NaCl	87.0g
$MgCl_2·6H_2O$	6.0g
$NaHCO_3$	4.0g
Trypticase	2.0g
Yeast extract	2.0g
KCl	1.5g
NH_4Cl	1.0g
$CaCl_2·2H_2O$	0.4g
$K_2HPO_4·3H_2O$	0.4g
Coenzyme M (mercaptoethane sulfonic acid)	0.2g
Resazurin	1.0mg
Trace elements solution	10.0mL
$Na_2S·9H_2O$ solution	10.0mL
Methanol	4.0mL

pH 7.0 ± 0.2 at 25°C

Trace Elements Solution:
Composition per liter:

$MgSO_4·7H_2O$	3.0g
Nitrilotriacetic acid	1.5g
NaCl	1.0g
$MnSO_4·2H_2O$	0.5g
$FeSO_4·7H_2O$	0.1g
$CoSO_4·7H_2O$	0.18g
$ZnSO_4·7H_2O$	0.18g
$CaCl_2·2H_2O$	0.1g
$KAl(SO_4)_2·12H_2O$	0.02g
$CuSO_4·5H_2O$	0.01g
H_3BO_3	0.01g
$Na_2MoO_4·2H_2O$	0.01g
$NiCl_2·6H_2O$	0.025g
$Na_2SeO_3·5H_2O$	0.3mg

Preparation of Trace Elements Solution: Add nitrilotriacetic acid to 500.0mL of distilled/deionized water. Adjust pH to 6.5 with KOH. Add remaining components. Adjust pH to 7.0. Add distilled/deionized water to 1.0L.

$Na_2S·9H_2O$ Solution:
Composition per 10.0mL:

$Na_2S·9H_2O$	0.25g

Preparation of $Na_2S·9H_2O$ Solution: Add $Na_2S·9H_2O$ to distilled/deionized water and bring volume to 10.0mL. Mix thoroughly. Sparge with 100% N_2. Autoclave for 15 min at 15 psi pressure–121°C.

Preparation of Medium: Prepare and dispense medium under 80% N_2 + 20% CO_2. Add components, except $Na_2S·9H_2O$ solution, to distilled/deionized water and bring volume to 1.0L. Mix thoroughly. Sparge with 80% N_2 + 20% CO_2. Autoclave for 15 min at 15 psi pressure–121°C. Aseptically and anaerobically add 10.0mL of sterile $Na_2S·9H_2O$ solution. Mix thoroughly.

Use: For the cultivation of *Methanohalophilus mahii.*

Methanohalophilus oregonense Medium

Composition per 1010.0mL:

NaCl	29.0g
Trimethylamine·HCl	2.0g
Trypticase	2.0g
Yeast extract	2.0g
$MgCl_2·6H_2O$	1.7g
KCl	1.5g
NH_4Cl	1.0g
Resazurin	1.0g
Coenzyme M (mercaptoethane sulfonic acid)	0.5g
$K_2HPO_4·3H_2O$	0.4g
$Na_2S·9H_2O$ solution	10.0mL
Trace elements solution	10.0mL
Na_2CO_3 solution	variable

pH 8.5 ± 0.2 at 25°C

Trace Elements Solution:
Composition per liter:

$MgSO_4·7H_2O$	3.0g
Nitrilotriacetic acid	1.5g
NaCl	1.0g
$MnSO_4·2H_2O$	0.5g
$CoSO_4·7H_2O$	0.18g
$ZnSO_4·7H_2O$	0.18g
$CaCl_2·2H_2O$	0.1g
$FeSO_4·7H_2O$	0.1g
$KAl(SO_4)_2·12H_2O$	0.02g
$CuSO_4·5H_2O$	0.01g
H_3BO_3	0.01g
$Na_2MoO_4·2H_2O$	0.01g
$NiCl_2·6H_2O$	0.025g
$Na_2SeO_3·5H_2O$	0.3mg

Preparation of Trace Elements Solution: Add nitrilotriacetic acid to 500.0mL of distilled/deionized water. Adjust pH to 6.5 with KOH. Add remaining components. Adjust pH to 7.0. Add distilled/deionized water to 1.0L.

$Na_2S·9H_2O$ Solution:
Composition per 10.0mL:

$Na_2S·9H_2O$	0.3g

Preparation of $Na_2S·9H_2O$ Solution: Add $Na_2S·9H_2O$ to distilled/deionized water and bring volume to 10.0mL. Mix thoroughly. Sparge with 100% N_2. Autoclave for 15 min at 15 psi pressure–121°C.

Na_2CO_3 Solution:
Composition per 10.0mL:

Na_2CO_3	1.0g

Preparation of Na_2CO_3 Solution: Add Na_2CO_3 to distilled/deionized water and bring volume to 10.0mL. Mix thoroughly. Sparge with 100% N_2. Autoclave for 15 min at 15 psi pressure–121°C.

Preparation of Medium: Prepare and dispense medium under 80% N_2 + 20% CO_2. Add components, except $Na_2S·9H_2O$ solution, to distilled/deionized water and bring volume to 1.0L. Mix thoroughly. Sparge with 80% N_2 + 20% CO_2. Autoclave for 15 min at 15 psi pressure–121°C. Aseptically and anaerobically add 10.0mL of sterile $Na_2S·9H_2O$ solution. Mix thoroughly. Adjust pH to 8.5 with a sufficient quantity of sterile Na_2CO_3 solution (approximately 0.1mL per 10.0mL of medium).

Use: For the cultivation and maintenance of *Methanohalophilus oregonense*.

Methanohalophilus zhilinae Medium

Composition per liter:

NaCl	40.0g
$MgCl_2·6H_2O$	3.5g
$MgSO_4·7H_2O$	3.0g

Trypticase	2.0g
Yeast extract	2.0g
KCl	1.0g
NH_4Cl	1.0g
L-Cysteine·HCl	0.5g
K_2HPO_4	0.4g
Resazurin	1.0mg
$Na_2SeO_3 \cdot 5H_2O$	0.1mg
Trimethylamine·HCl solution	20.0mL
$NaHCO_3$ solution	10.0mL
Na_2CO_3 solution	10.0mL
$Na_2S \cdot 9H_2O$ solution	10.0mL
Trace elements solution	5.0mL

pH 9.2 ± 0.2 at 25°C

Trimethylamine·HCl Solution:
Composition per 20.0mL:

Trimethylamine·HCl	2.0g

Preparation of Trimethylamine·HCl Solution: Add trimethylamine·HCl to distilled/deionized water and bring volume to 20.0mL. Mix thoroughly. Sparge with 100% N_2. Autoclave for 15 min at 15 psi pressure–121°C.

$NaHCO_3$ Solution:
Composition per 10.0mL:

$NaHCO_3$	0.5g

Preparation of $NaHCO_3$ Solution: Add $NaHCO_3$ to distilled/deionized water and bring volume to 10.0mL. Mix thoroughly. Sparge with 100% N_2. Autoclave for 15 min at 15 psi pressure–121°C.

Na_2CO_3 Solution:
Composition per 10.0mL:

Na_2CO_3	2.0g

Preparation of Na_2CO_3 Solution: Add Na_2CO_3 to distilled/deionized water and bring volume to 10.0mL. Mix thoroughly. Sparge with 100% N_2. Autoclave for 15 min at 15 psi pressure–121°C.

$Na_2S \cdot 9H_2O$ Solution:
Composition per 10.0mL:

$Na_2S \cdot 9H_2O$	0.25g

Preparation of $Na_2S \cdot 9H_2O$ Solution: Add $Na_2S \cdot 9H_2O$ to distilled/deionized water and bring volume to 10.0mL. Mix thoroughly. Sparge with 100% N_2. Autoclave for 15 min at 15 psi pressure–121°C.

Trace Elements Solution:
Composition per liter:

$MgSO_4 \cdot 7H_2O$	3.0g
Nitrilotriacetic acid	1.5g
NaCl	1.0g
$MnSO_4 \cdot 2H_2O$	0.5g
$CoSO_4 \cdot 7H_2O$	0.18g
$ZnSO_4 \cdot 7H_2O$	0.18g
$CaCl_2 \cdot 2H_2O$	0.1g
$FeSO_4 \cdot 7H_2O$	0.1g
$KAl(SO_4)_2 \cdot 12H_2O$	0.02g
$CuSO_4 \cdot 5H_2O$	0.01g
H_3BO_3	0.01g
$Na_2MoO_4 \cdot 2H_2O$	0.01g
$NiCl_2 \cdot 6H_2O$	0.025g
$Na_2SeO_3 \cdot 5H_2O$	0.3mg

Preparation of Trace Elements Solution: Add nitrilotriacetic acid to 500.0mL of distilled/deionized water. Adjust pH to 6.5 with KOH. Add remaining components. Adjust pH to 7.0. Add distilled/deionized water to 1.0L.

Preparation of Medium: Prepare and dispense medium under 100% N_2. Add components, except trimethylamine·HCl solution, $NaHCO_3$ solution, Na_2CO_3 solution, L-cysteine·HCl, and $Na_2S \cdot 9H_2O$ solution, to distilled/deionized water and bring volume to 950.0mL. Mix thoroughly. Gently heat and bring to boiling. Continue boiling for 5 min. Allow to cool to room temperature while sparging with 100% N_2. Add L-cysteine·HCl. Distribute medium into tubes and seal. Autoclave for 15 min at 15 psi pressure–121°C. Anaerobically add 20.0mL of sterile trimethylamine·HCl solution, 10.0mL of sterile $NaHCO_3$ solution, 10.0mL of sterile Na_2CO_3 solution, and 10.0mL of sterile $Na_2S \cdot 9H_2O$ solution. Mix thoroughly. Check pH.

Use: For the cultivation and maintenance of *Methanohalophilus zhilinae*.

Methanol Agar (LMG Medium 72)

Composition per liter:

Agar	15.0g
K_2HPO_4	1.2g
KH_2PO_4	0.62g
$(NH_4)_2SO_4$	0.5g
$MgSO_4 \cdot 7H_2O$	0.2g
NaCl	0.1g
$CaCl_2 \cdot 2H_2O$	34.0mg
$FeCl_3 \cdot 6H_2O$	1.0mg
Methanol	10.0mL
Trace elements solution	1.0mL

pH 7.0 ± 0.2 at 25°C

Trace Elements Solution:
Composition per liter:

$ZnSO_4 \cdot 7H_2O$	70.0mg
$Na_2MoO_4 \cdot 2H_2O$	10.0mg
H_3BO_3	10.0mg
$MnSO_4 \cdot H_2O$	7.0mg
$CuSO_4 \cdot 5H_2O$	5.0mg
$CoCl_2 \cdot 6H_2O$	5.0mg

Preparation of Trace Elements Solution: Add components to distilled/deionized water and bring volume to 1.0L. Mix thoroughly.

Preparation of Medium: Add components, except methanol, to 990.0mL distilled/deionized water. Mix thoroughly. Gently heat and bring to boiling. Autoclave for 15 min at 15 psi pressure–121°C. Cool to 45°C. Aseptically add 10.0mL sterile methanol. Mix thoroughly. Pour into sterile Petri dishes or distribute into sterile tubes.

Use: For the cultivation of *Methylobacterium* spp., *Methylobacillus glycogens*, and *Methylophilus methylotrophus*.

Methanol Ammonium Salts Medium

Composition per liter:

$MgSO_4 \cdot 7H_2O$	1.0g
NH_4Cl	0.5g
Na_2HPO_4	0.33g
KH_2PO_4	0.26g
$CaCl_2$	0.2g
Ferrous EDTA	5.0mg
$Na_2MoO_4 \cdot 2H_2O$	2.0mg
$FeSO_4 \cdot 7H_2O$	500.0μg
$ZnSO_4 \cdot 7H_2O$	400.0μg
EDTA	250.0μg
$CoCl_2 \cdot 6H_2O$	50.0μg
$MnCl_2 \cdot 4H_2O$	20.0μg
H_3BO_4	15.0μg

$NiCl_2·6H_2O$ 10.0μg
Methanol 5.0mL

pH 6.8 ± 0.2 at 25°C

Preparation of Medium: Add Na_2HPO_4 and KH_2PO_4 to distilled/deionized water and bring volume to 100.0mL. Mix thoroughly. In a separate container, add remaining components, except methanol, to distilled/deionized water and bring volume to 895.0mL. Mix thoroughly. Autoclave both solutions for 15 min at 15 psi pressure–121°C. Cool to room temperature. Filter sterilize methanol. Aseptically add the sterile phosphate solution and the sterile methanol to the cooled, sterile basal medium. Mix thoroughly. Aseptically distribute into sterile tubes or flasks.

Use: For the maintenance and cultivation of *Methylomonas methylotrophus.*

Methanol Medium (ATCC Medium 436)

Composition per liter:

Agar 15.0g
K_2HPO_4 .... 7.0g
$(NH_4)_2SO_4$ 3.0g
KH_2PO_4 .... 2.0g
$MgSO_4·7H_2O$ 0.5g
Yeast extract 0.2g
$FeSO_4·7H_2O$ 0.01g
$MnSO_4·H_2O$ 8.0mg
Biotin 0.2μg
Thiamine·HCl 0.2μg
Methanol 10.0mL

pH 7.0 ± 0.2 at 25°C

Preparation of Medium: Add components, except methanol, to distilled/deionized water and bring volume to 990.0mL. Mix thoroughly. Gently heat and bring to boiling. Autoclave for 15 min at 15 psi pressure–121°C. Cool to 50°–55°C. Filter sterilize methanol. Aseptically add the sterile methanol to the cooled, sterile basal medium. Mix thoroughly. Aseptically distribute into sterile tubes or flasks.

Use: For the cultivation and maintenance of *Ancylobacter* species, *Methanomonas methylovora*, and *Methylobacterium* species.

Methanol Medium for *Achromobacter*

Composition per liter:

NH_4Cl 5.0g
KH_2PO_4 .... 2.0g
NaCl 0.5g
$MgSO_4$ 0.2g
Yeast extract 0.2g
$FeSO_4$ 2.0mg
$MnCl_2$ 2.0mg
Methanol 20.0mL

pH 7.0 ± 0.2 at 25°C

Preparation of Medium: Add components, except methanol, to distilled/deionized water and bring volume to 980.0mL. Mix thoroughly. Autoclave for 15 min at 15 psi pressure–121°C. Cool to 50°–55°C. Filter sterilize methanol. Aseptically add the sterile methanol to the cooled, sterile basal medium. Mix thoroughly. Aseptically distribute into sterile tubes or flasks.

Use: For the cultivation and maintenance of *Achromobacter methanolophila, Methylobacterium rhodesianum, Pseudomonas insueta,* and *Pseudomonas polysaccharogenes.*

Methanol Medium with 1% Peptone

Composition per liter:

Agar 15.0g
Peptone 10.0g
K_2HPO_4 .... 7.0g
$(NH_4)_2SO_4$ 3.0g
KH_2PO_4 .... 2.0g
$MgSO_4·7H_2O$ 0.5g
Yeast extract 0.2g
$FeSO_4·7H_2O$ 0.01g
$MnSO_4·H_2O$ 8.0mg
Biotin 0.2μg
Thiamine·HCl 0.2μg
Methanol 10.0mL

pH 7.0 ± 0.2 at 25°C

Preparation of Medium: Add components, except methanol, to distilled/deionized water and bring volume to 990.0mL. Mix thoroughly. Autoclave for 15 min at 15 psi pressure–121°C. Cool to 50°–55°C. Filter sterilize methanol. Aseptically add the sterile methanol to the cooled, sterile basal medium. Mix thoroughly. Aseptically distribute into sterile tubes or flasks.

Use: For the cultivation and maintenance of *Methylobacterium* species.

Methanol Mineral Salts Medium

Composition per liter:

Agar 20.0g
$(NH_4)_2SO_4$ 2.0g
NH_4Cl 2.0g
$(NH_4)_2HPO_4$ 2.0g
Yeast extract 2.0g
KH_2PO_4 .... 1.0g
K_2HPO_4 .... 1.0g
$MgSO_4·7H_2O$ 0.5g
$Fe_2SO_4·7H_2O$ 0.01g
$CaCl_2·2H_2O$ 0.01g
Methanol 10.0mL

pH 7.0 ± 0.2 at 25°C

Preparation of Medium: Add components, except methanol, to distilled/deionized water and bring volume to 990.0mL. Mix thoroughly. Gently heat and bring to boiling. Autoclave for 15 min at 15 psi pressure–121°C. Cool to 50°–55°C. Filter sterilize methanol. Aseptically add the sterile methanol to the cooled, sterile basal medium. Mix thoroughly. Aseptically distribute into sterile Petri dishes or sterile tubes.

Use: For the cultivation and maintenance of *Pseudomonas viscogena.*

Methanol Salts Medium

Composition per liter:

Agar, noble 20.0g
K_2HPO_4 .... 1.2g
KH_2PO_4 .... 0.62g
$(NH_4)_2SO_4$ 0.5g
$MgSO_4·7H_2O$ 0.2g
NaCl 0.1g
$CaCl_2·6H_2O$ 0.05g
$ZnSO_4·7H_2O$ 70.0μg
H_3BO_3 .... 10.0μg
$MnSO_4·5H_2O$ 10.0μg
$Na_2MoO_4·2H_2O$ 10.0μg
$CoCl_2·6H_2O$ 5.0μg
$CuSO_4·5H_2O$ 5.0μg

$FeCl_3 \cdot 6H_2O$ 1.0mg
Methanol 1.0mL

pH 7.0 ± 0.2 at 25°C

Preparation of Medium: Add components, except methanol, to distilled/deionized water and bring volume to 1.0L. Mix thoroughly. Gently heat and bring to boiling. Autoclave for 15 min at 15 psi pressure–121°C. Cool to 50°C. Aseptically add 1.0mL of filter-sterilized methanol. Mix thoroughly. Pour into sterile Petri dishes or distribute into sterile tubes.

Use: For the cultivation of *Hyphomicrobium* species, *Methylobacillus glycogenes, Methylobacterium extorquens, Methylobacterium fujisawaense, Methylobacterium mesophilicum, Methylobacterium organophilum, Methylobacterium radiotolerans, Methylobacterium rhodesianum, Methylobacterium rhodinum, Methylobacterium* species, *Methylobacterium zatmanii, Methylomonas* species, *Methylophilus methylotrophus, Methylovorus glucosotrophus, Paracoccus* species, *Protaminobacter thiaminophaga*, and *Pseudomonas insueta.*

Methanol Urea Mineral Salts Medium

Composition per liter:

Na_2HPO4 .... 2.13g
KH_2PO_4 .... 1.36g
$(NH_4)_2SO_4$ 0.5g
$MgSO_4 \cdot 7H_2O$ 0.2g
$CaCl_2 \cdot 2H_2O$ 0.01g
$FeSO_4 \cdot 7H_2O$ 5.0mg
$MnSO_4 \cdot 5H_2O$ 2.5mg
$NaMoO_4 \cdot 2H_2O$ 2.5mg
Urea solution 30.0mL
Methanol 5.0mL

Urea Solution:

Composition per 50.0mL:

Urea 10.0g

Preparation of Urea Solution: Add urea to distilled/deionized water and bring volume to 50.0mL. Mix thoroughly. Filter sterilize.

Preparation of Methanol: Filter sterilize 5.0mL of methanol using a teflon filter.

Preparation of Medium: Add components, except urea solution and methanol, to distilled/deionized water and bring volume to 965.0mL. Mix thoroughly. Autoclave for 15 min at 15 psi pressure–121°C. Aseptically add 30.0mL of sterile urea solution and 5.0mL of sterile methanol. Mix thoroughly. Aseptically distribute into sterile tubes or flasks.

Use: For the cultivation of *Hyphomicrobium vulgare.*

Methanol-Utilizing Bacteria Medium B

Composition per liter:

Na_2HPO_4 .... 3.0g
$(NH_4)_2SO_4$ 3.0g
KH_2PO_4 .... 1.4g
$MgSO_4 \cdot 7H_2O$ 0.2g
$CaCl_2 \cdot 2H_2O$ 30.0mg
Ferric citrate 30.0mg
$MnCl_2 \cdot 4H_2O$ 5.0mg
$ZnSO_4 \cdot 7H_2O$ 5.0mg
$CuSO_4 \cdot 5H_2O$ 0.5mg
Thiamine·HCl 0.4mg
Methanol 10.0mL

pH 7.1 ± 0.2 at 25°C

Preparation of Medium: Add components to distilled/deionized water and bring volume to 1.0L. Mix thoroughly. Adjust pH to 7.1. Distribute into tubes or flasks. Autoclave for 15 min at 15 psi pressure–121°C.

Use: For the cultivation of methanol-utilizing bacteria.

Methanol-Utilizing Bacteria Medium D

Composition per liter:

Na_2HPO_4 .... 3.0g
$(NH_4)_2SO_4$ 3.0g
KH_2PO_4 .... 1.4g
Yeast extract 0.2g
$MgSO_4 \cdot 7H_2O$ 0.2g
$CaCl_2 \cdot 2H_2O$ 30.0mg
Ferric citrate 30.0mg
$MnCl_2 \cdot 4H_2O$ 5.0mg
$ZnSO_4 \cdot 7H_2O$ 5.0mg
$CuSO_4 \cdot 5H_2O$ 0.5mg
Methanol 10.0mL

pH 9.0 ± 0.2 at 25°C

Preparation of Medium: Add components to distilled/deionized water and bring volume to 1.0L. Mix thoroughly. Autoclave for 15 min at 15 psi pressure–121°C. Aseptically adjust pH to 9.0 with filter-sterilized 10% $NaCO_3$ solution. Mix thoroughly. Aseptically distribute into sterile tubes or flasks.

Use: For the cultivation of *Paracoccus alcaliphilus.*

Methanol-Utilizing Bacteria Medium E

Composition per 1010.0mL:

NaCl 30.0g
Na_2HPO_4 .... 3.0g
$(NH_4)_2SO_4$ 3.0g
KH_2PO_4 .... 1.4g
$MgSO_4 \cdot 7H_2O$ 0.2g
$CaCl_2 \cdot 2H_2O$ 30.0mg
Ferric citrate 30.0mg
$MnCl_2 \cdot 4H_2O$ 5.0mg
$ZnSO_4 \cdot 7H_2O$ 5.0mg
$CuSO_4 \cdot 5H_2O$ 0.5mg
Thiamine·HCl 0.4mg
Methanol 10.0mL
Vitamin B_{12} 10.0μg

pH 9.0 ± 0.2 at 25°C

Preparation of Medium: Add components to distilled/deionized water and bring volume to 1.0L. Mix thoroughly. Autoclave for 15 min at 15 psi pressure–121°C. Aseptically adjust pH to 9.0 with filter-sterilized 10% $NaCO_3$ solution. Mix thoroughly. Aseptically distribute into sterile tubes or flasks.

Use: For the cultivation of *Methylophaga marina* and *Methylophaga thalassica.*

Methanolobus Medium

Composition per liter:

NaCl 18.0g
$NaHCO_3$ 5.0g
$MgSO_4 \cdot 7H_2O$ 3.45g
$MgCl_2 \cdot 6H_2O$ 2.75g
L-Cysteine·HCl·H_2O 0.5g
$Na_2S \cdot 9H_2O$ 0.5g
KCl 0.335g
NH_4Cl 0.25g
$CaCl_2 \cdot 2H_2O$ 0.14g
K_2HPO_4 .... 0.14g
$Fe(NH_4)_2(SO_4)_2 \cdot 6H_2O$ 2.0mg

Resazurin 1.0mg
Wolfe's mineral solution 10.0mL
Wolfe's vitamin solution 10.0mL
Methanol 5.0mL
pH 6.5 ± 0.2 at 25°C

Wolfe's Mineral Solution:
Composition per liter:
$MgSO_4 \cdot 7H_2O$ 3.0g
Nitriloacetic acid 1.5g
NaCl 1.0g
$MnSO_4 \cdot H_2O$ 0.5g
$FeSO_4 \cdot 7H_2O$ 0.1g
$CoCl_2 \cdot 6H_2O$ 0.1g
$CaCl_2$ 0.1g
$ZnSO_4 \cdot 7H_2O$ 0.1g
$CuSO_4 \cdot 5H_2O$ 0.01g
$AlK(SO_4)_2 \cdot 12H_2O$ 0.01g
H_3BO_3 .... 0.01g
$Na_2MoO_4 \cdot 2H_2O$ 0.01g

Preparation of Wolfe's Mineral Solution: Add nitrilotriacetic acid to 500.0mL of distilled/deionized water. Dissolve by adjusting pH to 6.5 with KOH. Add remaining components. Add distilled/deionized water to 1.0L.

Wolfe's Vitamin Solution:
Composition per liter:
Pyridoxine·HCl 10.0mg
Thiamine·HCl 5.0mg
Riboflavin 5.0mg
Nicotinic acid 5.0mg
Calcium pantothenate 5.0mg
p-Aminobenzoic acid 5.0mg
Thioctic acid 5.0mg
Biotin 2.0mg
Folic acid 2.0mg
Cyanocobalamin 100.0μg

Preparation of Wolfe's Vitamin Solution: Add components to distilled/deionized water and bring volume to 1.0L. Mix thoroughly. Filter sterilize.

Preparation of Medium: Prepare and dispense medium under 80% N_2 + 20% CO_2. Add components, except methanol and Wolfe's vitamin solution, to distilled/deionized water and bring volume to 985.0mL. Mix thoroughly. Autoclave for 15 min at 15 psi pressure–121°C. Cool under 80% N_2 + 20% CO_2. Aseptically add sterile Wolfe's vitamin solution and sterile methanol. Adjust pH to 6.5. Aseptically and anaerobically distribute into sterile tubes or flasks.

Use: For the cultivation and maintenance of *Methanolobus siciliae* and *Methanolobus tindarius*.

Methanolobus 2 Medium

Composition per liter:
NaCl 18.0g
$NaHCO_3$ 5.0g
$MgCl_2 \cdot 6H_2O$ 4.0g
$MgSO_4 \cdot 7H_2O$ 3.45g
Trypticase 2.0g
Yeast extract 2.0g
Sodium acetate 1.0g
L-Cysteine·HCl 0.5g
$Na_2S \cdot 9H_2O$ 0.4g
KCl 0.335g
NH_4Cl 0.25g
$CaCl_2 \cdot 2H_2O$ 0.14g
K_2HPO_4 .... 0.14g
$Fe(NH_4)_2(SO_4)_2 \cdot 7H_2O$ 2.0mg
Resazurin 1.0mg
$Na_2WoO_4 \cdot 2H_2O$ 0.03mg
Trace elements solution 10.0mL
Vitamin solution 10.0mL
Methanol solution 5.0mL
pH 6.8–7.0 at 25°C

Trace Elements Solution:
Composition per liter:
$MgSO_4 \cdot 7H_2O$ 3.0g
Nitrilotriacetic acid 1.5g
NaCl 1.0g
$MnSO_4 \cdot 2H_2O$ 0.5g
$CoSO_4 \cdot 7H_2O$ 0.18g
$ZnSO_4 \cdot 7H_2O$ 0.18g
$CaCl_2 \cdot 2H_2O$ 0.1g
$FeSO_4 \cdot 7H_2O$ 0.1g
$NiCl_2 \cdot 6H_2O$ 0.025g
$KAl(SO_4)_2 \cdot 12H_2O$ 0.02g
$CuSO_4 \cdot 5H_2O$ 0.01g
H_3BO_3 .... 0.01g
$Na_2MoO_4 \cdot 2H_2O$ 0.01g
$Na_2SeO_3 \cdot 5H_2O$ 0.3mg

Preparation of Trace Elements Solution: Add nitrilotriacetic acid to 500.0mL of distilled/deionized water. Adjust pH to 6.5 with KOH. Add remaining components. Adjust pH to 7.0. Add distilled/deionized water to 1.0L.

Vitamin Solution:
Composition per liter:
Pyridoxine·HCl 10.0mg
Calcium DL-pantothenate 5.0mg
Nicotinic acid 5.0mg
p-Aminobenzoic acid 5.0mg
Lipoic acid 5.0mg
Riboflavin 5.0mg
Thiamine·HCl 5.0mg
Biotin 2.0mg
Folic acid 2.0mg
Vitamin B_{12} 0.1mg

Preparation of Vitamin Solution: Add components to distilled/deionized water and bring volume to 1.0L. Mix thoroughly. Filter sterilize. Sparge with 80% H_2 + 20% CO_2.

Methanol Solution:
Composition per 5.0mL:
Methanol 5.0mL

Preparation of Methanol Solution: Sparge 5.0mL of methanol with 100% N_2. Autoclave for 15 min at 15 psi pressure–121°C.

Preparation of Medium: Prepare and dispense medium under 80% N_2 + 20% CO_2. Add components, except vitamin solution, to distilled/deionized water and bring volume to 990.0mL. Mix thoroughly. Adjust pH to 6.9–7.0. Anaerobically distribute into tubes or flasks fitted with butyl rubber stoppers. Autoclave for 15 min at 15 psi pressure–121°C. Anaerobically add 10.0mL of sterile vitamin solution and 10.0mL of sterile methanol to each liter of medium or, using a syringe, inject the appropriate amount of sterile vitamin solution and sterile methanol into individual tubes containing medium.

Use: For the cultivation and maintenance of *Methanolobus siciliae*, *Methanolobus vulcani*, and *Methanosarcina* species.

Methanolobus taylori Medium (DSMZ Medium 490a)

Composition per liter:

NaCl	29.0g
Yeast extract	2.0g
Trypticase peptone	2.0g
$MgCl_2·6H_2O$	1.7g
KCl	1.5g
NH_4Cl	1.0g
Resazurin	1.0g
Mercaptoethanesulfonic acid (coenzyme M)	0.5g
$K_2HPO_4·3H_2O$	0.4g
Trace elements solution	10.0mL
Trimethylamine·HCl	10.0mL
$Na_2S·9H_2O$ solution	10.0mL
$NaHCO_3$ solution	10.0mL

pH 8.5 ± 0.2 at 25°C

Trace Elements Solution:
Composition per liter:

$MgSO_4·7H_2O$	3.0g
Nitrilotriacetic acid	1.5g
NaCl	1.0g
$MnSO_4·2H_2O$	0.5g
$CoSO_4·7H_2O$	0.18g
$ZnSO_4·7H_2O$	0.18g
$CaCl_2·2H_2O$	0.1g
$FeSO_4·7H_2O$	0.1g
$NiCl_2·6H_2O$	0.025g
$KAl(SO_4)_2·12H_2O$	0.02g
H_3BO_3	0.01g
$Na_2MoO_4·4H_2O$	0.01g
$CuSO_4·5H_2O$	0.01g
$Na_2SeO_3·5H_2O$	0.3mg

Preparation of Trace Elements Solution: Add nitrilotriacetic acid to 500.0mL of distilled/deionized water. Dissolve by adjusting pH to 6.5 with KOH. Add remaining components. Add distilled/deionized water to 1.0L. Mix thoroughly.

$Na_2S·9H_2O$ Solution:
Composition per 10.0mL:

$Na_2S·9H_2O$	0.3g

Preparation of $Na_2S·9H_2O$ Solution: Add $Na_2S·9H_2O$ to distilled/deionized water and bring volume to 10.0mL. Sparge with N_2. Autoclave for 15 min at 15 psi pressure–121°C. Cool to 25°C. Store anaerobically.

Trimethylamine Solution:
Composition per 10.0mL:

Trimethylamine·HCl	2.0g

Preparation of Trimethylamine Solution: Add trimethylamine·HCl to distilled/deionized water and bring volume to 10.0mL. Mix thoroughly. Autoclave for 15 min at 15 psi pressure–121°C. Cool to 25°C.

$NaHCO_3$ Solution:
Composition per 10.0mL:

$NaHCO_3$	2.0g

Preparation of $NaHCO_3$ Solution: Add $NaHCO_3$ to distilled/deionized water and bring volume to 10.0mL. Mix thoroughly. Autoclave for 15 min at 15 psi pressure–121°C. Cool to 25°C. Should be freshly prepared.

Preparation of Medium: Prepare and dispense medium under an oxygen-free 100% N_2 gas mixture. Add components, except trimethylamine solution, $NaHCO_3$ solution, and $Na_2S·9H_2O$ solution, to 970.0mL distilled/deionized water. Mix thoroughly. Sparge for 20 min with 100% N_2. Adjust pH to 8.0. Autoclave for 15 min at 15 psi pressure–121°C. Cool to 25°C. Aseptically and anaerobically add 10.0mL sterile trimethylamine solution, and 10.0mL $Na_2S·9H_2O$ solution. Adjust pH to 8.5 using sterile $NaHCO_3$ solution, approximately 10.0mL per liter medium. Aseptically and anaerbobically distribute to sterile tubes or flasks under 100% N_2.

Use: For the cultivation of *Methanolobus taylorii.*

Methanomicrobium Medium

Composition per liter:

$NaHCO_3$	2.0g
Yeast extract	1.0g
Pancreatic digest of casein	1.0g
NaCl	0.6g
L-Cysteine·HCl·H_2O	0.5g
$Na_2S·9H_2O$	0.5g
K_2HPO_4	0.3g
KH_2PO_4	0.3g
$(NH_4)_2SO_4$	0.3g
$MgSO_4·7H_2O$	0.13g
$CaCl_2·2H_2O$	8.0mg
$FeSO_4·7H_2O$	2.0mg
Rumen fluid, clarified	300.0mL
Fatty acid mixture	20.0mL
Wolfe's mineral solution	10.0mL
Wolfe's vitamin solution	10.0mL
Resazurin (0.1% solution)	1.0mL

pH 6.5 ± 0.2 at 25°C

Fatty Acid Mixture:
Composition per liter:

Valeric acid	0.7mL
Isovaleric acid	0.7mL
α-Methylbutyric acid	0.5mL
Isobutyric acid	0.5mL

Preparation of Fatty Acid Mixture: Add components to distilled/deionized water and bring volume to 1.0L. Mix thoroughly.

Wolfe's Mineral Solution:
Composition per liter:

$MgSO_4·7H_2O$	3.0g
Nitriloacetic acid	1.5g
NaCl	1.0g
$MnSO_4·H_2O$	0.5g
$FeSO_4·7H_2O$	0.1g
$CoCl_2·6H_2O$	0.1g
$CaCl_2$	0.1g
$ZnSO_4·7H_2O$	0.1g
$CuSO_4·5H_2O$	0.01g
$AlK(SO_4)_2·12H_2O$	0.01g
H_3BO_3	0.01g
$Na_2MoO_4·2H_2O$	0.01g

Preparation of Wolfe's Mineral Solution: Add nitrilotriacetic acid to 500.0mL of distilled/deionized water. Dissolve by adjusting pH to 6.5 with KOH. Add remaining components. Add distilled/deionized water to 1.0L.

Wolfe's Vitamin Solution:
Composition per liter:

Pyridoxine·HCl	10.0mg
Thiamine·HCl	5.0mg
Riboflavin	5.0mg
Nicotinic acid	5.0mg
Calcium pantothenate	5.0mg
p-Aminobenzoic acid	5.0mg
Thioctic acid	5.0mg

Biotin 2.0mg
Folic acid 2.0mg
Cyanocobalamin 100.0μg

Preparation of Wolfe's Vitamin Solution: Add components to distilled/deionized water and bring volume to 1.0L. Mix thoroughly.

Preparation of Medium: Prepare and dispense medium under 80% N_2 + 20% CO_2. Add components to distilled/deionized water and bring volume to 1.0L. Mix thoroughly. Adjust pH to 6.5. Distribute into tubes or flasks under 80% N_2 + 20% CO_2. Cap with rubber stoppers. Autoclave for 15 min at 15 psi pressure–121°C.

Use: For the cultivation and maintenance of *Methanomicrobium mobile.*

Methanomicrobium mobile Medium (DSMZ Medium 161)

Composition per liter:
$NaHCO_3$ 2.0g
Yeast extract 1.0g
Trypticase 1.0g
NaCl 0.6g
Cysteine-HCl·H_2O 0.5g
$Na_2S·9H_2O$ solution 0.5g
K_2HPO_4 .... 0.3g
KH_2PO_4 .... 0.3g
$(NH_4)_2SO_4$ 0.3g
$MgSO_4·7H_2O$ 0.13g
$CaCl_2·2H_2O$ 0.008g
$FeSO_4·7H_2O$ 0.002g
Resazurin 0.001g
Rumen fluid, clarified 300.0mL
Fatty acid mixture 20.0mL
Trace elements solution 10.0mL
Vitamin solution 10.0mL
pH 6.5 ± 0.2 at 25°C

Trace Elements Solution:
Composition per liter:
$MgSO_4·7H_2O$ 3.0g
Nitrilotriacetic acid 1.5g
NaCl 1.0g
$MnSO_4·2H_2O$ 0.5g
$CoSO_4·7H_2O$ 0.18g
$ZnSO_4·7H_2O$ 0.18g
$CaCl_2·2H_2O$ 0.1g
$FeSO_4·7H_2O$ 0.1g
$NiCl_2·6H_2O$ 0.025g
$KAl(SO_4)_2·12H_2O$ 0.02g
H_3BO_3 .... 0.01g
$Na_2MoO_4·4H_2O$ 0.01g
$CuSO_4·5H_2O$ 0.01g
$Na_2SeO_3·5H_2O$ 0.3mg

Preparation of Trace Elements Solution: Add nitrilotriacetic acid to 500.0mL of distilled/deionized water. Dissolve by adjusting pH to 6.5 with KOH. Add remaining components. Add distilled/deionized water to 1.0L. Mix thoroughly.

Vitamin Solution:
Composition per liter:
Pyridoxine-HCl 10.0mg
Thiamine-HCl·$2H_2O$ 5.0mg
Riboflavin 5.0mg
Nicotinic acid 5.0mg
D-Ca-pantothenate 5.0mg
p-Aminobenzoic acid 5.0mg
Lipoic acid 5.0mg
Biotin 2.0mg
Folic acid 2.0mg
Vitamin B_{12} 0.1mg

Preparation of Vitamin Solution: Add components to distilled/deionized water and bring volume to 1.0L. Mix thoroughly. Sparge with 80% H_2 + 20% CO_2. Filter sterilize.

Fatty Acid Mixture:
Composition per 20.0mL:
Valeric acid 0.5g
Isovaleric acid 0.5g
α-Methylbutyric acid 0.5g
Isobutyric acid 0.5g
Distilled water 20.0mL

Preparation of Fatty Acid Mixture: Add components to 20.0mL distilled/deionized water. Mix thoroughly.

Preparation of Medium: Prepare and dispense medium under an oxygen-free 80% H_2 + 20% CO_2 gas mixture. Add components, except vitamin solution, to 990.0mL distilled/deionized water. Mix thoroughly. Sparge with 80% H_2 + 20% CO_2. Adjust pH to 6.5 with concentrated NaOH. Autoclave for 15 min at 15 psi pressure–121°C. Cool to 25°C while sparging with 80% H_2 + 20% CO_2. Aseptically and anaerobically add 10.0mL sterile vitamin solution. Mix thoroughly. Aseptically and anaerobically distribute into sterile tubes or flasks. Alternately the medium can be distributed to tubes under anaerobic conditions and autoclaved in tubes prior to addition of vitamin solution. Appropriate amounts of the vitamin solution can then be added to each tube by syringes. After inoculation, pressurize culture vessels to 2 bar 80% H_2 + 20% CO_2 overpressure.

Use: For the cultivation of *Methanomicrobium mobile.*

Methanomicrobium paynteri Medium

Composition per liter:
3-(*N*-morpholino) propane
sulfonic acid (MOPS buffer) 20.93g
NaCl 6.31g
$NaHCO_3$ 5.0g
Sodium acetate·$3H_2O$ 4.14g
$MgSO_4·7H_2O$ 3.40g
$MgCl_2·2H_2O$ 2.75g
NH_4Cl 1.5g
KCl 0.34g
$CaCl_2·2H_2O$ 0.14g
K_2HPO_4 .... 0.14g
$Fe(NH_4)_2(SO_4)_2·6H_2O$ 2.0mg
Resazurin 1.0mg
Trace elements solution 10.0mL
Vitamin solution 10.0mL
L-Cysteine·HCl solution 10.0mL
$Na_2S·9H_2O$ solution 10.0mL
pH 7.0 ± 0.2 at 25°C

Trace Elements Solution:
Composition per liter:
$MgSO_4·7H_2O$ 3.0g
NaCl 1.0g
Nitrilotriacetic acid 1.5g
$MnSO_4·2H_2O$ 0.5g
$CoSO_4·7H_2O$ 0.18g
$ZnSO_4·7H_2O$ 0.18g
$CaCl_2·2H_2O$ 0.1g
$FeSO_4·7H_2O$ 0.1g
$KAl(SO_4)_2·12H_2O$ 0.02g

$CuSO_4 \cdot 5H_2O$ 0.01g
H_3BO_3 .... 0.01g
$Na_2MoO_4 \cdot 2H_2O$ 0.01g
$NiCl_2 \cdot 6H_2O$ 0.025g
$Na_2SeO_3 \cdot 5H_2O$ 0.3mg

Preparation of Trace Elements Solution: Add nitrilotriacetic acid to 500.0mL of distilled/deionized water. Adjust pH to 6.5 with KOH. Add remaining components. Adjust pH to 7.0. Add distilled/deionized water to 1.0L.

Vitamin Solution:
Composition per liter:

Pyridoxine·HCl 10.0mg
Calcium DL-pantothenate 5.0mg
Lipoic acid 5.0mg
Nicotinic acid 5.0mg
p-Aminobenzoic acid 5.0mg
Riboflavin 5.0mg
Thiamine·HCl 5.0mg
Biotin 2.0mg
Folic acid 2.0mg
Vitamin B_{12} 0.1mg

Preparation of Vitamin Solution: Add components to distilled/deionized water and bring volume to 1.0L. Mix thoroughly. Filter sterilize. Sparge with 80% H_2 + 20% CO_2.

L-Cysteine·HCl Solution:
Composition per 10.0mL:

L-Cysteine·HCl 0.5g

Preparation of L-Cysteine·HCl Solution: Add L-cysteine·HCl to distilled/deionized water and bring volume to 10.0mL. Mix thoroughly. Autoclave under 100% N_2 for 15 min at 15 psi pressure–121°C.

$Na_2S \cdot 9H_2O$ Solution:
Composition per 10.0mL:

$Na_2S \cdot 9H_2O$ 0.5g

Preparation of $Na_2S \cdot 9H_2O$ Solution: Add $Na_2S \cdot 9H_2O$ to distilled/deionized water and bring volume to 10.0mL. Mix thoroughly. Sparge with 100% N_2. Autoclave for 15 min at 15 psi pressure–121°C.

Preparation of Medium: Prepare and dispense medium under 80% H_2 + 20% CO_2. Add components, except MOPS buffer, L-cysteine·HCl solution, and $Na_2S \cdot 9H_2O$ solution, to distilled/deionized water and bring volume to 890.0mL. Mix thoroughly. Sparge with O_2-free 80% H_2 + 20% CO_2. In a separate flask, add MOPS buffer to distilled/deionized water and bring volume to 90.0mL. Adjust pH to 7.0 with 2*M* KOH. Add the 90.0mL of MOPS solution to the 890.0mL of medium and continue sparging with O_2-free 80% H_2 + 20% CO_2. Anaerobically distribute into tubes or flasks fitted with butyl rubber stoppers. Autoclave for 15 min at 15 psi pressure–121°C. Anaerobically add 10.0mL of sterile L-cysteine·HCl solution and 10.0mL of sterile $Na_2S \cdot 9H_2O$ solution to each liter of medium or, using a syringe, inject the appropriate amount of sterile L-cysteine·HCl solution and sterile $Na_2S \cdot 9H_2O$ solution into individual tubes containing medium.

Use: For the cultivation and maintenance of *Methanomicrobium paynteri*.

Methanomicrococcus Medium (DSMZ Medium 120b)

Composition 1112.0mL:

NaCl 2.25g
Yeast extract 2.0g
Casitone 2.0g
NH_4Cl 0.5g
$MgSO_4 \cdot 7H_2O$ 0.5g
K_2HPO_4 .... 0.348g
$CaCl_2 \cdot 2H_2O$ 0.25g
KH_2PO_4 .... 0.227g
$FeSO_4 \cdot 7H_2O$ 0.002g
Resazurin 0.001g
$NaHCO_3$ solution 40.0mL
Methanol solution 20.0mL
L-Cysteine-HCl·H_2O solution 15.0mL
$Na_2S \cdot 9H_2O$ solution 15.0mL
Na-acetate solution 10.0mL
Vitamin solution 10.0mL
Coenzyme M solution 1.0mL
Trace elements solution SL-10 .... 1.0mL

pH 7.2 ± 0.2 at 25°C

Vitamin Solution:
Composition per liter:

Pyridoxamine-HCl 10.0mg
Pantothenic acid 5.0mg
Riboflavin 5.0mg
Alpha-lipoic acid 5.0mg
p-Aminobenzoic acid 5.0mg
Thiamine-HCl·$2H_2O$ 5.0mg
Nicotinic acid 5.0mg
Biotin 2.0mg
Folic acid 2.0mg
Vitamin B_{12} 0.1mg

Preparation of Vitamin Solution: Add components to distilled/deionized water and bring volume to 1.0L. Mix thoroughly. Sparge with 80% N_2 + 20% CO_2. Filter sterilize.

Trace Elements Solution SL-10:
Composition per liter:

$FeCl_2 \cdot 4H_2O$ 1.5g
$CoCl_2 \cdot 6H_2O$ 190.0mg
$MnCl_2 \cdot 4H_2O$ 100.0mg
$ZnCl_2$ 70.0mg
$Na_2MoO_4 \cdot 2H_2O$ 36.0mg
$NiCl_2 \cdot 6H_2O$ 24.0mg
H_3BO_3 .... 6.0mg
$CuCl_2 \cdot 2H_2O$ 2.0mg
HCl (25% solution) 10.0mL

Preparation of Trace Elements Solution SL-10: Add $FeCl_2 \cdot 4H_2O$ to 10.0mL of HCl solution. Mix thoroughly. Add distilled/deionized water and bring volume to 1.0L. Add remaining components. Mix thoroughly. Sparge with 80% N_2 + 20% CO_2. Autoclave for 15 min at 15 psi pressure–121°C.

$NaHCO_3$ Solution:
Composition per 100.0mL:

$NaHCO_3$ 5.0g

Preparation of $NaHCO_3$ Solution: Add $NaHCO_3$ to distilled/deionized water and bring volume to 100.0mL. Mix thoroughly. Sparge with 80% N_2 + 20% CO_2. Autoclave for 15 min at 15 psi pressure–121°C.

$Na_2S \cdot 9H_2O$ Solution:
Composition per 100.0mL:

$Na_2S \cdot 9H_2O$ 3.0g

Preparation of $Na_2S \cdot 9H_2O$ Solution: Add $Na_2S \cdot 9H_2O$ to distilled/deionized water and bring volume to 100.0mL. Mix thoroughly. Sparge with 100% N_2. Autoclave for 15 min at 15 psi pressure–121°C. Before use, neutralize to pH 7.0 with sterile HCl.

Methanol Solution:
Composition per 100.0mL:
Methanol 50.0mL

Preparation of Methanol Solution: Add methanol to distilled/deionized water and bring volume to 100.0mL. Mix thoroughly. Sparge with 100% N_2. Autoclave for 15 min at 15 psi pressure–121°C.

Na-Acetate Solution:
Composition per 100.0mL:
Na-acetate 25.0g

Preparation of Na-Acetate Solution: Add Na-acetate to distilled/deionized water and bring volume to 100.0mL. Mix thoroughly. Sparge with 100% N_2. Autoclave for 15 min at 15 psi pressure–121°C.

Coenzyme M Solution:
Composition per 10.0mL:
Coenzyme M 0.1g

Preparation of Coenzyme M Solution: Add coenzyme M to distilled/deionized water and bring volume to 10.0mL. Mix thoroughly. Sparge with 100% N_2. Autoclave for 15 min at 15 psi pressure–121°C.

L-Cysteine Solution:
Composition per 100.0mL:
L-Cysteine·HCl·H_2O 3.0g

Preparation of L-Cysteine Solution: Add L-cysteine·HCl·H_2O to distilled/deionized water and bring volume to 100.0mL. Mix thoroughly. Sparge with 100% N_2. Autoclave for 15 min at 15 psi pressure–121°C.

Preparation of Medium: Prepare and dispense medium under 80% N_2 + 20% CO_2 gas atmosphere. Add components, except vitamin solution, $NaHCO_3$ solution, methanol solution, L-cysteine-HCl·H_2O solution, Na_2S·$9H_2O$ solution, Na-acetate solution, coenzyme M solution, and trace elements solution SL-10, to distilled/deionized water and bring volume to 1.0L. Mix thoroughly. Adjust pH to 7.2. Sparge with 80% N_2 + 20% CO_2. Autoclave for 15 min at 15 psi pressure–121°C. Aseptically and anaerobically add 10.0mL vitamin solution, 40.0mL $NaHCO_3$ solution, 20.0mL methanol solution, 15.0mL L-Cysteine-HCl·H_2O solution, 15.0mL Na_2S·$9H_2O$ solution, 10.0mL Na-acetate solution, 1.0mL coenzyme M solution, and 1.0mL trace elements solution SL-10. Mix thoroughly. Aseptically and anaerobically distribute into sterile tubes or bottles. After inoculation, flush and repressurize the gas head space of culture bottles with sterile 80% H_2 + 20% CO_2 to 1 bar overpressure.

Use: For the cultivation of *Methanomicrococcus blatticola.*

Methanomonas Autotrophic Medium

Composition per liter:
$NaNO_3$ 2.0g
Na_2HPO_4 .... 0.21g
$MgSO_4$·$7H_2O$ 0.2g
NaH_2PO_4 .... 0.09g
KCl 0.04g
$CaCl_2$ 0.015g
$FeSO_4$·$7H_2O$ 1.0mg
$ZnSO_4$·$7H_2O$ 0.3mg
$CuSO_4$·$5H_2O$ 0.2mg
H_3BO_3 .... 0.06mg
$MnSO_4$·H_2O 0.03mg
MoO_3 .... 0.015mg

Preparation of Medium: Add components to distilled/deionized water and bring volume to 1.0L. Mix thoroughly. Gently heat until dissolved. Distribute into tubes or flasks. Autoclave for 15 min at 15 psi pressure–121°C.

Use: For the autotrophic cultivation of *Methanomonas* species.

Methanopyrus Medium

Composition per liter:
NaCl 11.80g
$MgCl_2$·$6H_2O$ 4.50g
$MgSO_4$·$7H_2O$ 1.75g
Na_2SO_4 .... 0.81g
$CaCl_2$·$2H_2O$ 0.78g
KCl 0.30g
KH_2PO_4 .... 0.09g
K_2HPO_4·$3H_2O$ 0.07g
Na_2WO_4·$2H_2O$ 2.0mg
$(NH_4)_2Fe(SO_4)_2$·$6H_2O$ 2.0mg
$(NH_4)_2Ni(SO_4)_2$ 2.0mg
Resazurin 0.2mg
Marine trace elements 10.0mL
Trace elements solution 10.0mL
Vitamin solution 10.0mL
$NaHCO_3$ solution 10.0mL
Na_2S·$9H_2O$ solution 10.0mL
pH 6.5 ± 0.2 at 25°C

Marine Trace Elements:
Composition per liter:
NaBr 4.0g
$SrCl_2$·$6H_2O$ 1.8g
H_3BO_3 .... 1.3g
Sodium silicate 100.0mg
NaF 60.0mg
KNO_3 .... 40.0mg
KI 1.25mg
Na_2HPO_4·$3H_2O$ 0.25mg

Preparation of Marine Trace Elements: Add components to distilled/deionized water and bring volume to 1.0L. Mix thoroughly.

Trace Elements Solution:
Composition per liter:
$MgSO_4$·$7H_2O$ 3.0g
Nitrilotriacetic acid 1.5g
NaCl 1.0g
$MnSO_4$·$2H_2O$ 0.5g
$CoSO_4$·$7H_2O$ 0.18g
$ZnSO_4$·$7H_2O$ 0.18g
$CaCl_2$·$2H_2O$ 0.1g
$FeSO_4$·$7H_2O$ 0.1g
$NiCl_2$·$6H_2O$ 0.025g
$KAl(SO_4)_2$·$12H_2O$ 0.02g
$CuSO_4$·$5H_2O$ 0.01g
H_3BO_3 .... 0.01g
Na_2MoO_4·$2H_2O$ 0.01g
Na_2SeO_3·$5H_2O$ 0.3mg

Preparation of Trace Elements Solution: Add nitrilotriacetic acid to 500.0mL of distilled/deionized water. Adjust pH to 6.5 with KOH. Add remaining components. Adjust pH to 7.0. Add distilled/deionized water to 1.0L.

Vitamin Solution:
Composition per liter:
Pyridoxine·HCl 10.0mg

Calcium DL-pantothenate 5.0mg
Lipoic acid 5.0mg
Nicotinic acid 5.0mg
p-Aminobenzoic acid 5.0mg
Riboflavin 5.0mg
Thiamine·HCl 5.0mg
Biotin 2.0mg
Folic acid 2.0mg
Vitamin B_{12} 0.1mg

Preparation of Vitamin Solution: Add components to distilled/deionized water and bring volume to 1.0L. Mix thoroughly. Filter sterilize. Sparge with 80% H_2 + 20% CO_2.

$NaHCO_3$ Solution:
Composition per 10.0mL:
$NaHCO_3$ 0.5g

Preparation of $NaHCO_3$ Solution: Add $NaHCO_3$ to distilled/deionized water and bring volume to 10.0mL. Mix thoroughly. Sparge with 80% N_2 + 20% CO_2. Autoclave for 15 min at 15 psi pressure–121°C.

$Na_2S·9H_2O$ Solution:
Composition per 10.0mL:
$Na_2S·9H_2O$ 0.5g

Preparation of $Na_2S·9H_2O$ Solution: Add $Na_2S·9H_2O$ to distilled/deionized water and bring volume to 10.0mL. Mix thoroughly. Sparge with 100% N_2. Autoclave for 15 min at 15 psi pressure–121°C.

Preparation of Medium: Prepare and dispense medium under 80% H_2 + 20% CO_2. Add components, except vitamin solution, $NaHCO_3$ solution, and $Na_2S·9H_2O$ solution, to distilled/deionized water and bring volume to 970.0mL. Mix thoroughly. Anaerobically distribute into tubes or flasks fitted with butyl rubber stoppers. Autoclave for 15 min at 15 psi pressure–121°C. Anaerobically add 10.0mL of sterile vitamin solution, 10.0mL of sterile $NaHCO_3$ solution, and 10.0mL of sterile $Na_2S·9H_2O$ solution to each liter of medium or, using a syringe, inject the appropriate amount of sterile vitamin solution, sterile $NaHCO_3$ solution, and sterile $Na_2S·9H_2O$ solution into individual tubes containing medium. Check that final pH of medium is 6.5.

Use: For the cultivation and maintenance of *Methanopyrus kandleri*.

Methanosarcina acetivorans Medium

Composition per liter:
NaCl 23.4g
Agar 10.0g
$MgSO_4$ 6.3g
Na_2CO_3 .... 5.0g
Trimethylamine·HCl 3.0g
Yeast extract 1.0g
NH_4Cl 1.0g
KCl 0.8g
Na_2HPO_4 .... 0.6g
L-Cysteine·HCl·H_2O 0.25g
$Na_2S·9H_2O$ 0.25g
$CaCl_2·2H_2O$ 0.14g
Resazurin 1.0mg
Wolfe's mineral solution 10.0mL
pH 7.2 ± 0.2 at 25°C

Wolfe's Mineral Solution:
Composition per liter:
$MgSO_4·7H_2O$ 3.0g
Nitriloacetic acid 1.5g
NaCl 1.0g
$MnSO_4·H_2O$ 0.5g
$FeSO_4·7H_2O$ 0.1g
$CoCl_2·6H_2O$ 0.1g
$CaCl_2$ 0.1g
$ZnSO_4·7H_2O$ 0.1g
$CuSO_4·5H_2O$ 0.01g
$AlK(SO_4)_2·12H_2O$ 0.01g
H_3BO_3 .... 0.01g
$Na_2MoO_4·2H_2O$ 0.01g

Preparation of Wolfe's Mineral Solution: Add nitrilotriacetic acid to 500.0mL of distilled/deionized water. Dissolve by adjusting pH to 6.5 with KOH. Add remaining components. Add distilled/deionized water to 1.0L.

Preparation of Medium: Add components, except $Na_2S·9H_2O$, to glass-distilled water and bring volume to 990.0mL. Mix thoroughly. Methanol or methylamine·HCl may be substituted for the trimethylamine·HCl at a concentration of 50 m*M*. Heat gently and bring to boiling. Adjust pH to 7.2 with 6*N* HCl. Autoclave for 5 min at 10 psi pressure–115°C. Cool to 50°C under 80% N_2 + 20% CO_2. If a large precipitate is present, add a small amount of HCl and mix thoroughly. Add $Na_2S·9H_2O$. Mix thoroughly. Distribute into tubes under 80% N_2 + 20% CO_2. Cap with butyl rubber stoppers. Autoclave for 15 min at 15 psi pressure–121°C. A precipitate will form but resolubilizes as the medium cools. Invert tubes as they are cooling to facilitate resolubilization. Allow tubes to cool in a slanted position.

Use: For the cultivation and maintenance of *Methanococcoides methylutens* and *Methanosarcina acetivorans*.

Methanosarcina barkeri Medium

Composition per liter:
$NaHCO_3$ 2.5g
NaCl 0.46g
Yeast extract 0.24g
KH_2PO_4 .... 0.23g
K_2HPO_4 .... 0.23g
$(NH_4)_2SO_4$ 0.23g
$MgCl_2·6H_2O$ 0.09g
$CaCl_2·2H_2O$ 0.06g
$NiCl_2·6H_2O$ 2.0mg
Methanol 10.0mL
Trace elements solution 10.0mL
Vitamin solution 10.0mL
L-Cysteine·HCl·H_2O solution 10.0mL
$Na_2S·9H_2O$ solution 10.0mL
Resazurin (0.01% solution) 1.0mL

Preparation of Methanol: Filter sterilize 10.0mL of methanol.

Trace Elements Solution:
Composition per liter:
$MgSO_4·5H_2O$ 3.0g
Nitrilotriacetic acid 1.5g
NaCl 1.0g
$MnSO_4·2H_2O$ 0.5g
$NaS_4·SeO_3·5H_2O$ 0.3g
$NiCl_2·6H_2O$ 0.25g
$CoSO_4·7H_2O$ 0.18g
$ZnSO_4·7H_2O$ 0.18g
$CaCl_2·7H_2O$ 0.1g
$FeSO_4·7H_2O$ 0.1g
$KAl(SO_4)_2·12H_2O$ 0.02g
$CuSO_4·5H_2O$ 0.01g

H_3BO_3 .... 0.01g
$Na_2MoO_4·2H_2O$ 0.01g

Preparation of Trace Elements Solution: Add nitrilotriacetic acid to 500.0mL of distilled/deionized water. Adjust pH to 6.5 with KOH. Add remaining components. Mix thoroughly. Add distilled/deionized water to 1.0L. Adjust pH to 6.8.

Vitamin Solution:
Composition per liter:
Calcium DL-pantothenate 5.0g
Vitamin B_{12} 0.1g
Pyridoxine·HCl 10.0mg
p-Aminobenzoic acid 5.0mg
Lipoic acid 5.0mg
Nicotinic acid 5.0mg
Riboflavin 5.0mg
Thiamine·HCl 5.0mg
Biotin 2.0mg
Folic acid 2.0mg

Preparation of Vitamin Solution: Add components to distilled/deionized water and bring volume to 1.0L. Mix thoroughly.

L-Cysteine·HCl·H_2O Solution:
Composition per 100.0mL:
L-Cysteine·HCl·H_2O 2.5g

Preparation of L-Cysteine·HCl·H_2O Solution: Bring 100.0mL of distilled/deionized water to boiling. Cool to room temperature while sparging with 100% N_2. Add L-cysteine·HCl·H_2O to the 100.0mL of anaerobic water. Distribute into serum bottles fitted with butyl rubber stoppers and aluminum seals. Do not grease stoppers. Autoclave for 20 min at 15 psi pressure–121°C.

$Na_2S·9H_2O$ Solution:
Composition per 100.0mL:
NaOH .... 1 pellet
$Na_2S·9H_2O$ 2.5g

Preparation of $Na_2S·9H_2O$ Solution: Bring 100.0mL of distilled/deionized water to boiling. Cool to room temperature while sparging with 100% N_2. Dissolve 1 pellet of NaOH in the anaerobic water. Weigh out a little more than 2.5g of $Na_2S·9H_2O$. Briefly rinse the crystals in distilled/deionized water. Dry the crystals by blotting on paper towels or filter paper. Weigh out 2.5g of washed $Na_2S·9H_2O$ crystals. Add to the 100.0mL of anaerobic NaOH solution. Distribute into serum bottles fitted with butyl rubber stoppers and aluminum seals. Do not grease stoppers. Pressurize to 60kPa with 100% N_2. Autoclave for 15 min at 15 psi pressure–121°C. Store at room temperature in an anaerobic chamber.

Preparation of Medium: Prepare and dispense medium under 80% N_2 + 20% CO_2. Add components, except methanol, L-cysteine·HCl·H_2O solution, and $Na_2S·9H_2O$ solution, to distilled/deionized water and bring volume to 970.0mL. Mix thoroughly. Gently heat and bring to boiling. Continue boiling for 3 min. Cool to room temperature while sparging with 80% H_2 + 20% CO_2. Anaerobically distribute 9.7mL volumes into anaerobic tubes. Autoclave for 20 min at 15 psi pressure–121°C. Aseptically and anaerobically add 0.1mL of sterile methanol, 0.1mL of sterile L-cysteine·HCl·H_2O solution, and 0.1mL of sterile $Na_2S·9H_2O$ solution to each tube. Mix thoroughly.

Use: For the cultivation of *Methanosarcina barkeri*.

Methanosarcina BCYT Medium (DSMZ Medium 318)

Composition per liter:
$KHCO_3$ 2.0g
NH_4Cl 1.0g
NaCl 0.6g
Yeast extract 0.5g
Trypticase 0.5g
KH_2PO_4 .... 0.3g
$MgCl_2·6H_2O$ 0.1g
$CaCl_2·2H_2O$ 0.08g
Resazurin 1.0mg
Cysteine-solution 10.0mL
$Na_2S·9H_2O$ solution 10.0mL
Trace elements solution 10.0mL
Vitamin solution 10.0mL
Methanol 5.0mL
pH 6.8 ± 0.2 at 25°C

Vitamin Solution:
Composition per liter:
Pyridoxine-HCl 10.0mg
Thiamine-HCl·$2H_2O$ 5.0mg
Riboflavin 5.0mg
Nicotinic acid 5.0mg
D-Ca-pantothenate 5.0mg
p-Aminobenzoic acid 5.0mg
Lipoic acid 5.0mg
Biotin 2.0mg
Folic acid 2.0mg
Vitamin B_{12} 0.1mg

Preparation of Vitamin Solution: Add components to distilled/deionized water and bring volume to 1.0L. Mix thoroughly. Sparge with 80% H_2 + 20% CO_2. Filter sterilize.

Trace Elements Solution:
Composition per liter:
$MgSO_4·7H_2O$ 3.0g
Nitrilotriacetic acid 1.5g
NaCl 1.0g
$MnSO_4·2H_2O$ 0.5g
$CoSO_4·7H_2O$ 0.18g
$ZnSO_4·7H_2O$ 0.18g
$CaCl_2·2H_2O$ 0.1g
$FeSO_4·7H_2O$ 0.1g
$NiCl_2·6H_2O$ 0.025g
$KAl(SO_4)_2·12H_2O$ 0.02g
H_3BO_3 .... 0.01g
$Na_2MoO_4·4H_2O$ 0.01g
$CuSO_4·5H_2O$ 0.01g
$Na_2SeO_3·5H_2O$ 0.3mg

Preparation of Trace Elements Solution: Add nitrilotriacetic acid to 500.0mL of distilled/deionized water. Dissolve by adjusting pH to 6.5 with KOH. Add remaining components. Add distilled/deionized water to 1.0L. Mix thoroughly.

$Na_2S·9H_2O$ Solution:
Composition per 10mL:
$Na_2S·9H_2O$ 0.3g

Preparation of $Na_2S·9H_2O$ Solution: Add $Na_2S·9H_2O$ to distilled/deionized water and bring volume to 10.0mL. Mix thoroughly. Autoclave under 100% N_2 for 15 min at 15 psi pressure–121°C. Cool to room temperature.

Cysteine Solution:
Composition per 10.0mL:

L-Cysteine·HCl·H_2O	0.3g

Preparation of Cysteine Solution: Add L-cysteine·HCl·H_2O to distilled/deionized water and bring volume to 10.0mL. Mix thoroughly. Sparge with 100% N_2. Autoclave for 15 min at 15 psi pressure–121°C.

Preparation of Medium: Add components, except methanol, vitamin solution, cysteine solution, and $Na_2S·9H_2O$ solution, to distilled/deionized water and bring volume to 985.0mL. Mix thoroughly. Sparge with 100% CO_2. Autoclave for 15 min at 15 psi pressure–121°C. Cool to 25°C while sparging with 100% CO_2. Aseptically and anaerobically add 5.0mL filter sterilized methanol, 10.0mL vitamin solution, 10.0mL cysteine solution, and 10.0mL of sterile $Na_2S·9H_2O$ solution. Mix thoroughly. Aseptically and anaerobically distribute into sterile tubes or flasks.

Use: For the cultivation of *Methanosarcina* spp.

Methanosarcina DPB Medium

Composition per 1001.0mL:

Sodium acetate·$3H_2O$	4.1g
NH_4Cl	1.4g
K_2HPO_4	1.3g
KH_2PO_4	1.3g
$MgSO_4$	0.5g
NaCl	0.5g
L-Cysteine·HCl·H_2O	0.27g
$CaCl_2·2H_2O$	0.06g
$Fe(NH_4)_2(SO_4)_2$	0.01g
Antifoam C	10.0mL
Trace elements solution	10.0mL
Vitamin solution	10.0mL
$Na_2S·9H_2O$ solution	1.0mL

pH 6.8 ± 0.2 at 25°C

Source: Antifoam C is available from Sigma Chemical Co.

Trace Elements Solution:
Composition per liter:

$MgSO_4·5H_2O$	3.0g
Nitrilotriacetic acid	1.5g
NaCl	1.0g
$MnSO_4·2H_2O$	0.5g
$NaS_4·SeO_3·5H_2O$	0.3g
$NiCl_2·6H_2O$	0.25g
$CoSO_4·7H_2O$	0.18g
$ZnSO_4·7H_2O$	0.18g
$CaCl_2·7H_2O$	0.1g
$FeSO_4·7H_2O$	0.1g
$KAl(SO_4)_2·12H_2O$	0.02g
$CuSO_4·5H_2O$	0.01g
H_3BO_3	0.01g
$Na_2MoO_4·2H_2O$	0.01g

Preparation of Trace Elements Solution: Add nitrilotriacetic acid to 500.0mL of distilled/deionized water. Adjust pH to 6.5 with KOH. Add remaining components. Mix thoroughly. Add distilled/deionized water to 1.0L. Adjust pH to 6.8.

Vitamin Solution:
Composition per liter:

Calcium DL-pantothenate	5.0g
Vitamin B_{12}	0.1g
Pyridoxine·HCl	10.0mg
p-Aminobenzoic acid	5.0mg
Lipoic acid	5.0mg
Nicotinic acid	5.0mg
Riboflavin	5.0mg
Thiamine·HCl	5.0mg
Biotin	2.0mg
Folic acid	2.0mg

Preparation of Vitamin Solution: Add components to distilled/deionized water and bring volume to 1.0L. Mix thoroughly.

$Na_2S·9H_2O$ Solution:
Composition per 100.0mL:

$Na_2S·9H_2O$	25.0g

Preparation of $Na_2S·9H_2O$ Solution: Bring 100.0mL of distilled/deionized water to boiling. Cool to room temperature while sparging with 100%N_2. Add $Na_2S·9H_2O$ to the 100.0mL of anaerobic water. Mix thoroughly. Distribute into serum bottles fitted with butyl rubber stoppers and aluminum seals. Do not grease stoppers. Autoclave for 20 min at 15 psi pressure–121°C.

Preparation of Medium: Add components, except $Na_2S·9H_2O$ solution, to distilled/deionized water and bring volume to 1.0L. Mix thoroughly. Adjust pH to 6.8 with concentrated HCl. Distribute into tubes or flasks. Autoclave for 15 min at 15 psi pressure–121°C. Cool medium while sparging with 100% N_2. Prior to inoculation, aseptically and anaerobically add 1.0mL of sterile $Na_2S·9H_2O$ solution per liter of medium. Repeat the addition of 1.0mL of sterile $Na_2S·9H_2O$ solution per liter of medium every 48 hr during growth.

Use: For the cultivation of *Methanosarcina* species.

Methanosarcina frisia Medium

Composition per liter:

NaCl	18.0g
$NaHCO_3$	5.0g
$MgCl_2·6H_2O$	4.0g
$MgSO_4·7H_2O$	3.45g
Trypticase	2.0g
Yeast extract	2.0g
Sodium acetate	1.0g
KCl	0.335g
NH_4Cl	0.25g
$CaCl_2·2H_2O$	0.14g
K_2HPO_4	0.14g
L-Cysteine·HCl	0.5g
$Na_2S·9H_2O$	0.5g
$Fe(NH_4)_2(SO_4)_2·7H_2O$	2.0mg
Resazurin	1.0mg
Trace elements solution	10.0mL
Vitamin solution	10.0mL
Methanol solution	2.0mL

pH 6.8–7.0 at 25°C

Trace Elements Solution:
Composition per liter:

$MgSO_4·7H_2O$	3.0g
Nitrilotriacetic acid	1.5g
NaCl	1.0g
$MnSO_4·2H_2O$	0.5g
$CoSO_4·7H_2O$	0.18g
$ZnSO_4·7H_2O$	0.18g
$CaCl_2·2H_2O$	0.1g
$FeSO_4·7H_2O$	0.1g
$KAl(SO_4)_2·12H_2O$	0.02g
$CuSO_4·5H_2O$	0.01g
H_3BO_3	0.01g
$Na_2MoO_4·2H_2O$	0.01g

$NiCl_2 \cdot 6H_2O$ 0.025g
$Na_2SeO_3 \cdot 5H_2O$ 0.3mg

Preparation of Trace Elements Solution: Add nitrilotriacetic acid to 500.0mL of distilled/deionized water. Adjust pH to 6.5 with KOH. Add remaining components. Adjust pH to 7.0. Add distilled/deionized water to 1.0L.

Vitamin Solution:
Composition per liter:

Pyridoxine·HCl	10.0mg
Calcium DL-pantothenate	5.0mg
Lipoic acid	5.0mg
Nicotinic acid	5.0mg
p-Aminobenzoic acid	5.0mg
Riboflavin	5.0mg
Thiamine·HCl	5.0mg
Biotin	2.0mg
Folic acid	2.0mg
Vitamin B_{12}	0.1mg

Preparation of Vitamin Solution: Add components to distilled/deionized water and bring volume to 1.0L. Mix thoroughly. Sparge with 80% H_2 + 20% CO_2.

Methanol Solution:
Composition per 10.0mL:
Methanol 10.0mL

Preparation of Methanol Solution: Sparge 10.0mL of methanol with 100% N_2. Autoclave for 15 min at 15 psi pressure–121°C.

Preparation of Medium: Prepare and dispense medium under 80% N_2 + 20% CO_2. Add components, except methanol solution, to distilled/deionized water and bring volume to 1.0L. Mix thoroughly. Anaerobically distribute into tubes or flasks fitted with butyl rubber stoppers. Autoclave for 15 min at 15 psi pressure–121°C. Anaerobically add 10.0mL of sterile methanol solution to each liter of medium or, using a syringe, inject the appropriate amount of sterile methanol solution into individual tubes containing medium.

Use: For the cultivation and maintenance of *Methanosarcina mazei.*

Methanosarcina mazei Alpha Basal Medium

Composition per liter:

$NaHCO_3$	4.4g
Pancreatic digest of casein	2.0g
Yeast extract	2.0g
NH_4Cl	1.0g
$Na_2S \cdot 6H_2O$	0.5g
K_2HPO_4	0.4g
Sodium acetate·$3H_2O$	0.27g
$MgCl_2 \cdot 6H_2O$	0.08g
$CaCl_2 \cdot 2H_2O$	0.04g
$CoCl_2 \cdot 6H_2O$	1.5mg
$FeSO_4 \cdot 7H_2O$	1.0mg
$MnCl_2 \cdot 4H_2O$	1.0mg
Resazurin	1.0mg
H_3BO_4	0.1mg
$NaMoO_4 \cdot 2H_2O$	0.1mg
$ZnCl_2$	0.1mg

pH 7.2 ± 0.2 at 25°C

Preparation of Medium: Prepare and dispense medium under 70% N_2 + 30% CO_2. Add components to distilled/deionized water and bring volume to 1.0L. Mix thoroughly. Adjust pH to 7.2. Sparge with 70% N_2 + 30% CO_2. Autoclave for 15 min at 15 psi pressure–121°C. Aseptically and anaerobically distribute into sterile tubes or bottles.

Use: For the cultivation of *Methanosarcina mazei.*

Methanosarcina mazei Medium (DSMZ Medium 120)

Composition 1112.0mL:

NaCl	2.25g
Yeast extract	2.0g
Casitone	2.0g
$NaHCO_3$	0.85g
NH_4Cl	0.5g
$MgSO_4 \cdot 7H_2O$	0.5g
K_2HPO_4	0.348g
$CaCl_2 \cdot 2H_2O$	0.25g
KH_2PO_4	0.227g
$FeSO_4 \cdot 7H_2O$	0.002g
Resazurin	0.001g
Methanol solution	20.0mL
L-Cysteine-HCl·H_2O solution	15.0mL
$Na_2S \cdot 9H_2O$ solution	15.0mL
Na-acetate solution	10.0mL
Vitamin solution	10.0mL
$NaHCO_3$ solution	10.0mL
Trace elements solution SL-10	1.0mL

pH 6.5-6.8 at 25°C

Vitamin Solution:
Composition per liter:

Pyridoxamine-HCl	10.0mg
Pantothenic acid	5.0mg
Riboflavin	5.0mg
Alpha-lipoic acid	5.0mg
p-Aminobenzoic acid	5.0mg
Thiamine-HCl·$2H_2O$	5.0mg
Nicotinic acid	5.0mg
Biotin	2.0mg
Folic acid	2.0mg
Vitamin B_{12}	0.1mg

Preparation of Vitamin Solution: Add components to distilled/deionized water and bring volume to 1.0L. Mix thoroughly. Sparge with 80% N_2 + 20% CO_2. Filter sterilize.

Trace Elements Solution SL-10:
Composition per liter:

$FeCl_2 \cdot 4H_2O$	1.5g
$CoCl_2 \cdot 6H_2O$	190.0mg
$MnCl_2 \cdot 4H_2O$	100.0mg
$ZnCl_2$	70.0mg
$Na_2MoO_4 \cdot 2H_2O$	36.0mg
$NiCl_2 \cdot 6H_2O$	24.0mg
H_3BO_3	6.0mg
$CuCl_2 \cdot 2H_2O$	2.0mg
HCl (25% solution)	10.0mL

Preparation of Trace Elements Solution SL-10: Add $FeCl_2 \cdot 4H_2O$ to 10.0mL of HCl solution. Mix thoroughly. Add distilled/deionized water and bring volume to 1.0L. Add remaining components. Mix thoroughly. Sparge with 80% N_2 + 20% CO_2. Autoclave for 15 min at 15 psi pressure–121°C.

$NaHCO_3$ Solution:
Composition per 100.0mL:
$NaHCO_3$ 5.0g

Preparation of $NaHCO_3$ Solution: Add $NaHCO_3$ to distilled/deionized water and bring volume to 100.0mL. Mix thoroughly. Sparge with 80% N_2 + 20% CO_2. Autoclave for 15 min at 15 psi pressure–121°C.

$Na_2S{\cdot}9H_2O$ Solution:
Composition per 100.0mL:
$Na_2S{\cdot}9H_2O$ 3.0g

Preparation of $Na_2S{\cdot}9H_2O$ Solution: Add $Na_2S{\cdot}9H_2O$ to distilled/deionized water and bring volume to 100.0mL. Mix thoroughly. Sparge with 100% N_2. Autoclave for 15 min at 15 psi pressure–121°C. Before use, neutralize to pH 7.0 with sterile HCl.

Methanol Solution:
Composition per 100.0mL:
Methanol 50.0mL

Preparation of Methanol Solution: Add methanol to distilled/deionized water and bring volume to 100.0mL. Mix thoroughly. Sparge with 100% N_2. Autoclave for 15 min at 15 psi pressure–121°C.

Na-Acetate Solution:
Composition per 100.0mL:
Na-acetate 25.0g

Preparation of Na-Acetate Solution: Add Na-acetate to distilled/deionized water and bring volume to 100.0mL. Mix thoroughly. Sparge with 100% N_2. Autoclave for 15 min at 15 psi pressure–121°C.

L-Cysteine Solution:
Composition per 100.0mL:
L-Cysteine·HCl·H_2O 3.0g

Preparation of L-Cysteine Solution: Add L-cysteine·HCl·H_2O to distilled/deionized water and bring volume to 100.0mL. Mix thoroughly. Sparge with 100% N_2. Autoclave for 15 min at 15 psi pressure–121°C.

Preparation of Medium: Prepare and dispense medium under 80% N_2 + 20% CO_2 gas atmosphere. Add components, except vitamin solution, $NaHCO_3$ solution, methanol solution, L-cysteine-HCl·H_2O solution, $Na_2S{\cdot}9H_2O$ solution, Na-acetate solution, and trace elements solution SL-10, to distilled/deionized water and bring volume to 1.0L. Mix thoroughly. Adjust pH to 6.5. Sparge with 80% N_2 + 20% CO_2. Autoclave for 15 min at 15 psi pressure–121°C. Aseptically and anaerobically add 10.0mL vitamin solution, 10.0mL $NaHCO_3$ solution, 20.0mL methanol solution, 15.0mL L-cysteine-HCl·H_2O solution, 15.0mL $Na_2S{\cdot}9H_2O$ solution, 10.0mL Na-acetate solution, and 1.0mL trace elements solution SL-10. Mix thoroughly. Aseptically and anaerobically distribute into sterile tubes or bottles. After inoculation, flush and repressurize the gas head space of culture bottles with sterile 80% H_2 + 20% CO_2 to 1 bar overpressure.

Use: For the cultivation of *Mathanosarcina mazei.*

Methanosarcina Medium (BCYT)

Composition per liter:
$KHCO_3$ 2.0g
NH_4Cl 1.0g
NaCl 0.6g
Pancreatic digest of casein 0.5g
Yeast extract 0.5g
KH_2PO_4 .... 0.3g
$MgCl_2{\cdot}6H_2O$ 0.1g
$CaCl_2{\cdot}2H_2O$ 0.08g
Resazurin 1.0mg
Trace elements solution 10.0mL
Vitamin solution 10.0mL
L-Cysteine·HCl·H_2O solution 10.0mL
$Na_2S{\cdot}9H_2O$ solution 10.0mL
Methanol 5.0mL
pH 6.8 ± 0.2 at 25°C

Trace Elements Solution:
Composition per liter:
Nitrilotriacetic acid 12.8g
$FeCl_3{\cdot}6H_2O$ 1.35g
NaCl 1.0g
$NiCl_2{\cdot}6H_2O$ 0.12g
$MnCl_2{\cdot}4H_2O$ 0.1g
$CaCl_2{\cdot}2H_2O$ 0.1g
$ZnCl_2$ 0.1g
$Na_2SeO_3{\cdot}5H_2O$ 0.026g
$CuCl_2{\cdot}2H_2O$ 0.025g
$CoCl_2{\cdot}6H_2O$ 0.024g
$Na_2MoO_4{\cdot}2H_2O$ 0.024g
H_3BO_3 .... 0.01g

Preparation of Trace Elements Solution: Add nitrilotriacetic acid to approximately 500.0mL of distilled/deionized water. Dissolve by adding KOH and adjust pH to 6.5. Add remaining components. Bring volume to 1.0L with additional distilled/deionized water. Adjust pH to 7.0 with KOH.

Vitamin Solution:
Composition per liter:
Pyridoxine·HCl 10.0mg
Calcium DL-pantothenate 5.0mg
Lipoic acid 5.0mg
Nicotinic acid 5.0mg
p-Aminobenzoic acid 5.0mg
Riboflavin 5.0mg
Thiamine·HCl 5.0mg
Biotin 2.0mg
Folic acid 2.0mg
Vitamin B_{12} 0.1mg

Preparation of Vitamin Solution: Add components to distilled/deionized water and bring volume to 1.0L. Mix thoroughly. Filter sterilize. Sparge with 80% N_2 + 20% CO_2.

L-Cysteine·HCl·H_2O Solution:
Composition per 10.0mL:
L-Cysteine·HCl·H_2O 0.25g

Preparation of L-Cysteine·HCl·H_2O Solution: Add L-cysteine·HCl·H_2O to distilled/deionized water and bring volume to 10.0mL. Mix thoroughly. Autoclave under 80% N_2 + 20% CO_2 for 15 min at 15 psi pressure–121°C.

$Na_2S{\cdot}9H_2O$ Solution:
Composition per 10.0mL:
$Na_2S{\cdot}9H_2O$ 0.3g

Preparation of $Na_2S{\cdot}9H_2O$ Solution: Add $Na_2S{\cdot}9H_2O$ to distilled/deionized water and bring volume to 10.0mL. Mix thoroughly. Sparge with 80% N_2 + 20% CO_2. Autoclave for 15 min at 15 psi pressure–121°C.

Preparation of Medium: Add components, except vitamin solution, L-cysteine·HCl·H_2O solution, and $Na_2S{\cdot}9H_2O$ solution, to distilled/deionized water and bring volume to 960.0mL. Mix thoroughly. Sparge under 80% N_2 + 20% CO_2 for 3–4 min. Autoclave for 15 min at 15 psi pressure–121°C. Aseptically and anaerobically add 20.0mL of sterile vitamin solution, 10.0mL of sterile L-cysteine·HCl·H_2O solution, and 10.0mL of sterile $Na_2S{\cdot}9H_2O$ solution. Mix thoroughly. Aseptically and anaerobically distribute into sterile screw-capped bottles under 80% N_2 + 20% CO_2.

Use: For the cultivation and maintenance of *Methanosarcina mazei* and *Methanosarcina thermophila*.

Methanosarcina Medium

Composition per liter:

NaCl	2.25g
Pancreatic digest of casein	2.0g
Yeast extract	2.0g
$NaHCO_3$	0.85g
$MgSO_4 \cdot 7H_2O$	0.5g
NH_4Cl	0.5g
K_2HPO_4	0.348g
$CaCl_2 \cdot 2H_2O$	0.25g
KH_2PO_4	0.227g
$FeSO_4 \cdot 7H_2O$	2.0mg
Resazurin	1.0mg
Methanol solution	10.0mL
Vitamin solution	10.0mL
L-Cysteine·HCl·H_2O solution	10.0mL
$Na_2S \cdot 9H_2O$ solution	10.0mL
Trace elements solution SL-10	1.0mL

pH 6.5–6.8 at 25°C

Methanol Solution:

Composition per 10.0mL:

Methanol	5.0mL

Preparation of Methanol Solution: Add methanol to distilled/deionized water and bring volume to 10.0mL. Sparge with 100% N_2. Autoclave for 15 min at 15 psi pressure–121°C.

Vitamin Solution:

Composition per liter:

Pyridoxine·HCl	10.0mg
Calcium DL-pantothenate	5.0mg
Lipoic acid	5.0mg
Nicotinic acid	5.0mg
p-Aminobenzoic acid	5.0mg
Riboflavin	5.0mg
Thiamine·HCl	5.0mg
Biotin	2.0mg
Folic acid	2.0mg
Vitamin B_{12}	0.1mg

Preparation of Vitamin Solution: Add components to distilled/deionized water and bring volume to 1.0L. Mix thoroughly. Filter sterilize. Sparge with 80% N_2 + 20% CO_2.

L-Cysteine·HCl·H_2O Solution:

Composition per 10.0mL:

L-Cysteine·HCl·H_2O	0.3g

Preparation of L-Cysteine·HCl·H_2O Solution: Add L-cysteine·HCl·H_2O to distilled/deionized water and bring volume to 10.0mL. Mix thoroughly. Autoclave under 100% N_2 for 15 min at 15 psi pressure–121°C.

$Na_2S \cdot 9H_2O$ Solution:

Composition per 10.0mL:

$Na_2S \cdot 9H_2O$	0.3g

Preparation of $Na_2S \cdot 9H_2O$ Solution: Add $Na_2S \cdot 9H_2O$ to distilled/deionized water and bring volume to 10.0mL. Mix thoroughly. Sparge with 100% N_2. Autoclave for 15 min at 15 psi pressure–121°C.

Trace Elements Solution SL-10:

Composition per liter:

$FeCl_2 \cdot 4H_2O$	1.5g
$CoCl_2 \cdot 6H_2O$	190.0mg
$MnCl_2 \cdot 4H_2O$	100.0mg
$ZnCl_2$	70.0mg
$Na_2MoO_4 \cdot 2H_2O$	36.0mg
$NiCl_2 \cdot 6H_2O$	24.0mg
H_3BO_3	6.0mg
$CuCl_2 \cdot 2H_2O$	2.0mg
HCl (25% solution)	10.0mL

Preparation of Trace Elements Solution SL-10: Add $FeCl_2 \cdot 4H_2O$ to 10.0mL of HCl solution. Mix thoroughly. Add distilled/deionized water and bring volume to 1.0L. Add remaining components. Mix thoroughly. Sparge with 100% N_2.

Preparation of Medium: Add components, except methanol solution, vitamin solution, L-cysteine·HCl·H_2O solution, and $Na_2S \cdot 9H_2O$ solution, to distilled/deionized water and bring volume to 960.0mL. Mix thoroughly. Sparge under 80% N_2 + 20% CO_2 for 3–4 min. Autoclave for 15 min at 15 psi pressure–121°C. Aseptically and anaerobically add 10.0mL of sterile methanol, 10.0mL of sterile vitamin solution, 10.0mL of sterile L-cysteine·HCl·H_2O solution, and 10.0mL of sterile $Na_2S \cdot 9H_2O$ solution. Mix thoroughly. Aseptically and anaerobically distribute into sterile screw-capped bottles under 80% N_2 + 20% CO_2.

Use: For the cultivation and maintenance of *Methanosarcina* species.

Methanosarcina MP Medium

Composition per 1015.0mL:

NH_4Cl	1.0g
Yeast extract	1.0g
L-Cysteine·HCl	0.5g
K_2HPO_4	0.4g
$MgCl_2 \cdot 6H_2O$	0.2g
Resazurin	1.0mg
Mineral solution	50.0mL
Trace elements solution	10.0mL
Na_2CO_3 solution	10.0mL
$Na_2S \cdot 9H_2O$ solution	10.0mL
Methanol	5.0mL

pH 6.8 ± 0.2 at 25°C

Mineral Solution:

Composition per liter:

NaCl	12.0g
KH_2PO_4	6.0g
$(NH_4)_2SO_4$	6.0g
$MgSO_4 \cdot 7H_2O$	2.6g
$CaCl_2 \cdot 2H_2O$	0.16g

Preparation of Mineral Solution: Add components to distilled/deionized water and bring volume to 1.0L. Mix thoroughly.

Trace Elements Solution:

Composition per liter:

$MgSO_4 \cdot 7H_2O$	3.0g
Nitrilotriacetic acid	1.5g
NaCl	1.0g
$MnSO_4 \cdot 2H_2O$	0.5g
$CoSO_4 \cdot 7H_2O$	0.18g
$ZnSO_4 \cdot 7H_2O$	0.18g
$CaCl_2 \cdot 2H_2O$	0.1g
$FeSO_4 \cdot 7H_2O$	0.1g
$NiCl_2 \cdot 6H_2O$	0.025g
$KAl(SO_4)_2 \cdot 12H_2O$	0.02g
$CuSO_4 \cdot 5H_2O$	0.01g
H_3BO_3	0.01g
$Na_2MoO_4 \cdot 2H_2O$	0.01g
$Na_2SeO_3 \cdot 5H_2O$	0.3mg

Preparation of Trace Elements Solution: Add nitrilotriacetic acid to 500.0mL of distilled/deionized water. Adjust pH to 6.5 with KOH. Add remaining components. Adjust pH to 7.0. Add distilled/deionized water to 1.0L.

Na_2CO_3 Solution:
Composition per 10.0mL:
Na_2CO_3 .... 1.0g

Preparation of Na_2CO_3 Solution: Add Na_2CO_3 to distilled/deionized water and bring volume to 10.0mL. Mix thoroughly. Sparge with 100% N_2. Autoclave for 15 min at 15 psi pressure–121°C.

$Na_2S{\cdot}9H_2O$ Solution:
Composition per 10.0mL:
$Na_2S{\cdot}9H_2O$ 0.25g

Preparation of $Na_2S{\cdot}9H_2O$ Solution: Add $Na_2S{\cdot}9H_2O$ to distilled/deionized water and bring volume to 10.0mL. Mix thoroughly. Sparge with 100% N_2. Autoclave for 15 min at 15 psi pressure–121°C.

Preparation of Medium: Prepare and dispense medium under 80% N_2 + 20% CO_2. Add components, except L-cysteine·HCl, Na_2CO_3 solution, and $Na_2S{\cdot}9H_2O$ solution, to distilled/deionized water and bring volume to 980.0mL. Mix thoroughly. Gently heat and bring to boiling. Continue boiling for 5 min. Cool to room temperature while sparging with 80% N_2 + 20% CO_2. Add L-cysteine·HCl. Mix thoroughly. Adjust pH to 6.8–7.0. Autoclave for 15 min at 15 psi pressure–121°C. Aseptically and anaerobically add 10.0mL of sterile Na_2CO_3 solution and 10.0mL of sterile $Na_2S{\cdot}9H_2O$ solution. Mix thoroughly.

Use: For the cultivation and maintenance of *Methanosarcina* species.

Methanosarcina Nitrogen-Fixing Medium

Composition per 1005.0mL:
NaH_2PO_4 .... 1.38g
K_2HPO_4 .... 0.34g
$MgCl_2{\cdot}2H_2O$ 0.1g
Yeast extract 0.1g
$CaCl_2{\cdot}2H_2O$ 0.05g
Resazurin 0.001g
Trace elements solution 10.0mL
Methanol 2.0mL
$Na_2S{\cdot}9H_2O$ 2.0mL
$NaHCO_3$ 1.0mL
pH 7.0 ± 0.2 at 25°C

Trace Elements Solution:
Composition per liter:
Nitrilotriacetic acid 4.5g
$FeCl_2{\cdot}7H_2O$ 0.4g
H_3BO_3 .... 0.19g
$CoCl_2{\cdot}6H_2O$ 0.17g
$MnCl_2{\cdot}4H_2O$ 0.1g
$ZnCl_2$ 0.1g
$NiCl_2{\cdot}6H_2O$ 0.02g
$Na_2MoO_4{\cdot}6H_2O$ 0.01g

Preparation of Trace Elements Solution: Add components to distilled/deionized water and bring volume to 1.0L. Mix thoroughly. Sparge with 100% N_2. Adjust pH to 7.0 with 10*N* NaOH. Autoclave for 15 min at 15 psi pressure–121°C.

Preparation of Methanol: Sparge 2-propanol with 100% N_2. Filter sterilize.

$Na_2S{\cdot}9H_2O$ Solution:
Composition per 10.0mL:
$Na_2S{\cdot}9H_2O$ 2.0g
NaOH 0.1g

Preparation of $Na_2S{\cdot}9H_2O$ Solution: Add $Na_2S{\cdot}9H_2O$ and NaOH to distilled/deionized water and bring volume to 10.0mL. Mix thoroughly. Sparge with 100% N_2. Autoclave for 15 min at 15 psi pressure–121°C. Before use, neutralize to pH 7.0 with sterile HCl.

$NaHCO_3$ Solution:
Composition per 10.0mL:
$NaHCO_3$ 1.0g

Preparation of $NaHCO_3$ Solution: Add $NaHCO_3$ to distilled/deionized water and bring volume to 10.0mL. Mix thoroughly. Sparge with 80% N_2 + 20% CO_2. Autoclave for 15 min at 15 psi pressure–121°C.

Preparation of Medium: Prepare and dispense medium under 100% N_2. Add components, except methanol, $Na_2S{\cdot}9H_2O$ solution, and $NaHCO_3$ solution, to distilled/deionized water and bring volume to 1.0L. Mix thoroughly. Adjust pH to 7.5. Sparge with 100% N_2. Autoclave for 15 min at 15 psi pressure–121°C. Aseptically and anaerobically add 2.0mL of sterile methanol, 2.0mL of sterile $Na_2S{\cdot}9H_2O$ solution, and 2.0mL of sterile $NaHCO_3$ solution. Mix thoroughly. Adjust pH to 7.0 by adding sterile anaerobic 10*N* NaOH. Aseptically and anaerobically distribute into sterile tubes or bottles.

Use: For the cultivation of nitrogen-fixing *Methanosarcina* species.

Methanosarcina semesiae Medium (DSMZ Medium 865)

Composition per 1214mL:
Solution A 950.0mL
$NaHCO_3$ solution 50.0mL
Solution E 30.0mL
$Na_2S{\cdot}9H_2O$ solution 15.0mL
Solution D 10.0mL
Solution F 10.0mL
Solution G 10.0mL
Trace elements solution 10.0mL
Vitamin solution 10.0mL
Methanol solution 5.0mL
Solution B 1.0mL
Solution C 1.0mL
Seven vitamin solution 1.0mL
Dithionite solution 1.0mL
pH 7.2 ± 0.2 at 25°C

Solution A:
Composition per 950.0mL:
KH_2PO_4 .... 1.4g
NH_4Cl 0.5g
$MgCl_2{\cdot}6H_2O$ 0.2g
$CaCl_2{\cdot}2H_2O$ 0.15g
Resazurin 1.0mg

Preparation of Solution A: Add components to distilled/deionized water and bring volume to 950.0mL. Mix thoroughly. Sparge with 80% N_2 + 20% CO_2. Filter sterilize.

Solution B:
Composition per liter:
$FeCl_2{\cdot}4H_2O$ 1.5g
$CoCl_2{\cdot}6H_2O$ 190.0mg
$MnCl_2{\cdot}4H_2O$ 100.0mg

$ZnCl_2$ 70.0mg
$Na_2MoO_4 \cdot 2H_2O$ 36.0mg
$NiCl_2 \cdot 6H_2O$ 24.0mg
H_3BO_3 .... 6.0mg
$CuCl_2 \cdot 2H_2O$ 2.0mg
HCl (25% solution) 10.0mL

Preparation of Solution B: Add $FeCl_2 \cdot 4H_2O$ to 10.0mL of HCl solution. Mix thoroughly. Add distilled/deionized water and bring volume to 1.0L. Add remaining components. Mix thoroughly. Sparge with 80% N_2 + 20% CO_2. Autoclave for 15 min at 15 psi pressure–121°C.

Solution C:
Composition per liter:
NaOH 0.5g
$Na_2WO_4 \cdot 2H_2O$ 4.0mg
$Na_2SeO_3 \cdot 5H_2O$ 3.0mg

Preparation of Solution C: Add components to distilled/deionized water and bring volume to 1.0L. Mix thoroughly. Sparge with 100% N_2. Filter sterilize.

Solution D:
Composition per liter:
Pyridoxine-HCl 10.0mg
Thiamine-HCl·$2H_2O$ 5.0mg
Riboflavin 5.0mg
Nicotinic acid 5.0mg
D-Ca-pantothenate 5.0mg
p-Aminobenzoic acid 5.0mg
Lipoic acid 5.0mg
Biotin 2.0mg
Folic acid 2.0mg
Vitamin B_{12} 0.1mg

Preparation of Solution D: Add components to distilled/deionized water and bring volume to 1.0L. Mix thoroughly. Sparge with 80% H_2 + 20% CO_2. Filter sterilize.

Solution E:
Composition per 100.0mL:
$NaHCO_3$ 5.0g

Preparation of Solution E: Add $NaHCO_3$ to distilled/deionized water and bring volume to 150.0mL. Mix thoroughly. Sparge with 80% N_2 + 20% CO_2. Filter sterilize.

Solution F:
Composition per 10.0mL:
Na_2-maleate 1.6g

Preparation of Solution F: Add Na_2-maleate to distilled/deionized water and bring volume to 150.0mL. Mix thoroughly. Sparge with 80% N_2 + 20% CO_2. Filter sterilize.

Solution G:
Composition per 10mL:
$Na_2S \cdot 9H_2O$ 0.25g

Preparation of Solution G: Add $Na_2S \cdot 9H_2O$ to distilled/deionized water and bring volume to 10.0mL. Mix thoroughly. Autoclave under 100% N_2 for 15 min at 15 psi pressure–121°C. Cool to room temperature.

$NaHCO_3$ Solution:
Composition per 100.0mL:
$NaHCO_3$ 5.0g

Preparation of $NaHCO_3$ Solution: Add $NaHCO_3$ to distilled/deionized water and bring volume to 100.0mL. Mix thoroughly. Sparge with 100% N_2. Autoclave for 15 min at 15 psi pressure–121°C. Cool to 25°C.

Trace Elements Solution:
Composition per liter:
$MgSO_4 \cdot 7H_2O$ 3.0g
Nitrilotriacetic acid 1.5g
NaCl 1.0g
$MnSO_4 \cdot 2H_2O$ 0.5g
$CoSO_4 \cdot 7H_2O$ 0.18g
$ZnSO_4 \cdot 7H_2O$ 0.18g
$CaCl_2 \cdot 2H_2O$ 0.1g
$FeSO_4 \cdot 7H_2O$ 0.1g
$NiCl_2 \cdot 6H_2O$ 0.025g
$KAl(SO_4)_2 \cdot 12H_2O$ 0.02g
H_3BO_3 .... 0.01g
$Na_2MoO_4 \cdot 4H_2O$ 0.01g
$CuSO_4 \cdot 5H_2O$ 0.01g
$Na_2SeO_3 \cdot 5H_2O$ 0.3mg

Preparation of Trace Elements Solution: Add nitrilotriacetic acid to 500.0mL of distilled/deionized water. Dissolve by adjusting pH to 6.5 with KOH. Add remaining components. Add distilled/deionized water to 1.0L. Mix thoroughly.

Vitamin Solution:
Composition per liter:
Pyridoxine-HCl 10.0mg
Thiamine-HCl·$2H_2O$ 5.0mg
Riboflavin 5.0mg
Nicotinic acid 5.0mg
D-Ca-pantothenate 5.0mg
p-Aminobenzoic acid 5.0mg
Lipoic acid 5.0mg
Biotin 2.0mg
Folic acid 2.0mg
Vitamin B_{12} 0.1mg

Preparation of Vitamin Solution: Add components to distilled/deionized water and bring volume to 1.0L. Mix thoroughly. Sparge with 80% H_2 + 20% CO_2. Filter sterilize.

Seven Vitamin Solution:
Composition per liter:
Pyridoxine hydrochloride 300.0mg
Thiamine-HCl·$2H_2O$ 200.0mg
Nicotinic acid 200.0mg
Vitamin B_{12} 100.0mg
Calcium pantothenate 100.0mg
p-Aminobenzoic acid 80.0mg
D(+)-Biotin 20.0mg

Preparation of Seven Vitamin Solution: Add components to distilled/deionized water and bring volume to 1.0L. Sparge with 100% N_2. Mix thoroughly. Filter sterilize.

Methanol Solution:
Composition per 100.0mL:
Methanol 50.0mL

Preparation of Methanol Solution: Add methanol to distilled/deionized water and bring volume to 100.0mL. Mix thoroughly. Sparge with 100% N_2. Autoclave for 15 min at 15 psi pressure–121°C.

$Na_2S \cdot 9H_2O$ Solution:
Composition per 20mL:
$Na_2S \cdot 9H_2O$ 0.6g

Preparation of $Na_2S \cdot 9H_2O$ Solution: Add $Na_2S \cdot 9H_2O$ to distilled/deionized water and bring volume to 20.0mL. Mix thoroughly. Autoclave under 100% N_2 for 15 min at 15 psi pressure–121°C. Cool to room temperature.

Dithionite Solution
Composition per 10mL:
Na-dithionite 0.25g

Preparation of Dithionite Solution: Add Na-dithionite to distilled/deionized water and bring volume to 10.0mL. Mix thoroughly. Sparge with 100% N_2. Filter sterilize.

Preparation of Medium: Prepare and dispense medium under 80% N_2 + 20% CO_2. Combine component solutions A-G. Do not add $NaHCO_3$ solution, $Na_2S{\cdot}9H_2O$ solution, trace elements solution, vitamin solution, methanol solution, seven vitamin solution, and dithionite solution. Mix thoroughly. Flush medium with 80% N_2 + 20% CO_2 for 5 min. Distribute into serum bottles. Autoclave for 15 min at 15 psi pressure–121°C. Cool to 25°C. For every 10.0mL medium aseptically and anaerobically add from sterile anaerobic solution: 0.5mL $NaHCO_3$ solution, 0.1mL trace elements solution, 0.1mL vitamin solution, 0.01mL seven vitamin solution, 0.01mL dithionite solution, 0.15mL $Na_2S{\cdot}9H_2O$ solution, and 0.05mL methanol solution. Final pH should be 7.0.

Use: For the cultivation of *Methanosarcina semesiae.*

Methanosarcina Thermophilic Medium (DSMZ Medium 164)

Composition 1051.0mL:
NaCl 2.25g
Yeast extract 2.0g
Casitone 2.0g
NH_4Cl 0.5g
$MgSO_4{\cdot}7H_2O$ 0.5g
K_2HPO_4 .... 0.348g
$CaCl_2{\cdot}2H_2O$ 0.25g
KH_2PO_4 .... 0.227g
$FeSO_4{\cdot}7H_2O$ 0.002g
Resazurin 0.001g
Rumen fluid, clarified 50.0mL
Methanol solution 10.0mL
Vitamin solution 10.0mL
$NaHCO_3$ 10.0mL
L-Cysteine solution 10.0mL
$Na_2S{\cdot}9H_2O$ solution 10.0mL
Trace elements solution SL-10 .... 1.0mL
pH 6.5-6.8 at 25°C

Vitamin Solution:
Composition per liter:
Pyridoxine-HCl 10.0mg
Thiamine-HCl·$2H_2O$ 5.0mg
Riboflavin 5.0mg
Nicotinic acid 5.0mg
D-Ca-pantothenate 5.0mg
p-Aminobenzoic acid 5.0mg
Lipoic acid 5.0mg
Biotin 2.0mg
Folic acid 2.0mg
Vitamin B_{12} 0.1mg

Preparation of Vitamin Solution: Add components to distilled/deionized water and bring volume to 1.0L. Mix thoroughly. Sparge with 80% N_2 + 20% CO_2. Filter sterilize.

Methanol Solution:
Composition per 100.0mL:
Methanol 50.0mL

Preparation of Methanol Solution: Add methanol to distilled/deionized water and bring volume to 100.0mL. Mix thoroughly. Sparge with 100% N_2. Autoclave for 15 min at 15 psi pressure–121°C.

$Na_2S{\cdot}9H_2O$ Solution:
Composition per 100.0mL:
$Na_2S{\cdot}9H_2O$ 3.0g

Preparation of $Na_2S{\cdot}9H_2O$ Solution: Add $Na_2S{\cdot}9H_2O$ to distilled/deionized water and bring volume to 100.0mL. Mix thoroughly. Sparge with 100% N_2. Autoclave for 15 min at 15 psi pressure–121°C. Before use, neutralize to pH 7.0 with sterile HCl.

$NaHCO_3$ Solution:
Composition per 100.0mL:
$NaHCO_3$ 8.5g

Preparation of $NaHCO_3$ Solution: Add $NaHCO_3$ to distilled/deionized water and bring volume to 100.0mL. Mix thoroughly. Autoclave for 15 min at 15 psi pressure–121°C. Cool to 25°C. Must be prepared freshly.

Trace Elements Solution SL-10:
Composition per liter:
$FeCl_2{\cdot}4H_2O$ 1.5g
H_3BO_3 .... 300.0mg
$CoCl_2{\cdot}6H_2O$ 190.0mg
$MnCl_2{\cdot}4H_2O$ 100.0mg
$ZnCl_2$ 70.0mg
$Na_2MoO_4{\cdot}2H_2O$ 36.0mg
$NiCl_2{\cdot}6H_2O$ 24.0mg
$CuCl_2{\cdot}2H_2O$ 2.0mg
HCl (25% solution) 10.0mL

Preparation of Trace Elements Solution SL-10: Add $FeCl_2{\cdot}4H_2O$ to 10.0mL of HCl solution. Mix thoroughly. Add distilled/deionized water and bring volume to 1.0L. Add remaining components. Mix thoroughly. Sparge with 100% N_2. Autoclave for 15 min at 15 psi pressure–121°C.

L-Cysteine Solution:
Composition per 10.0mL:
L-Cysteine·HCl·H_2O 0.3g

Preparation of L-Cysteine Solution: Add L-cysteine·HCl·H_2O to distilled/deionized water and bring volume to 10.0mL. Mix thoroughly. Sparge with 100% N_2. Autoclave for 15 min at 15 psi pressure–121°C.

Preparation of Medium: Prepare and dispense medium under 80% N_2 + 20% CO_2 gas atmosphere. Add components, except vitamin solution, $NaHCO_3$ solution, methanol solution, L-cysteine-HCl·H_2O solution, $Na_2S{\cdot}9H_2O$ solution, and trace elements solution SL-10, to distilled/deionized water and bring volume to 1.0L. Mix thoroughly. Adjust pH to 6.5. Sparge with 80% N_2 + 20% CO_2. Autoclave for 15 min at 15 psi pressure–121°C. Aseptically and anaerobically add 10.0mL vitamin solution, 10.0mL $NaHCO_3$ solution, 10.0mL methanol solution, 10.0mL L-cysteine-HCl·H_2O solution, 10.0mL $Na_2S{\cdot}9H_2O$ solution, and 1.0mL trace elements solution SL-10. Mix thoroughly. Aseptically and anaerobically distribute into sterile tubes or bottles. After inoculation, flush and repressurize the gas head space of culture bottles with sterile 80% H_2 + 20% CO_2 to 1 bar overpressure. Alternately, the medium without vitamin solution, $NaHCO_3$ solution, methanol solution, L-Cysteine-HCl·H_2O solution, $Na_2S{\cdot}9H_2O$ solution, and trace elements solution SL-10, can be distributed to tubes anaerobically prior to autoclaving. After autoclaving in tubes the appropriate volumes of the individual solutions can be injected through the stoppers so that the final concentrations of the medium are achieved.

Use: For the cultivation of *Methanosarcina thermophila.*

Methanosphaera Medium I

Composition per liter:

KH_2PO_4	2.8g
Trypticase	2.0g
Yeast extract	2.0g
NH_4Cl	1.0g
K_2HPO_4	0.6g
NaCl	0.6g
Sodium acetate	0.5g
Sodium formate	0.5g
$(NH_4)_2SO_4$	0.3g
$MgSO_4 \cdot 7H_2O$	0.15g
$CaCl_2 \cdot 2H_2O$	0.076g
$FeSO_4 \cdot 7H_2O$	3.0mg
Na_2SeO_4	1.9mg
Resazurin	1.0mg
$NiCl_2 \cdot 6H_2O$	0.7mg
Rumen fluid, clarified	100.0mL
Trace elements solution	10.0mL
L-Cysteine·HCl·H_2O solution	10.0mL
$Na_2S \cdot 9H_2O$ solution	10.0mL
Methanol solution	5.0mL
Vitamin solution	1.0mL

pH 6.5 ± 0.2 at 25°C

Trace Elements Solution:
Composition per liter:

$MgSO_4 \cdot 7H_2O$	3.0g
Nitrilotriacetic acid	1.5g
NaCl	1.0g
$MnSO_4 \cdot 2H_2O$	0.5g
$CoSO_4 \cdot 7H_2O$	0.18g
$ZnSO_4 \cdot 7H_2O$	0.18g
$CaCl_2 \cdot 2H_2O$	0.1g
$FeSO_4 \cdot 7H_2O$	0.1g
$NiCl_2 \cdot 6H_2O$	0.025g
$KAl(SO_4)_2 \cdot 12H_2O$	0.02g
$CuSO_4 \cdot 5H_2O$	0.01g
H_3BO_3	0.01g
$Na_2MoO_4 \cdot 2H_2O$	0.01g
$Na_2SeO_3 \cdot 5H_2O$	0.3mg

Preparation of Trace Elements Solution: Add nitrilotriacetic acid to 500.0mL of distilled/deionized water. Adjust pH to 6.5 with KOH. Add remaining components. Adjust pH to 7.0. Add distilled/deionized water to 1.0L.

L-Cysteine·HCl·H_2O Solution:
Composition per 10.0mL:

L-Cysteine·HCl·H_2O	0.875g

Preparation of L-Cysteine·HCl·H_2O Solution: Add L-cysteine·HCl·H_2O to distilled/deionized water and bring volume to 10.0mL. Mix thoroughly. Autoclave under 80% N_2 + 20% CO_2 for 15 min at 15 psi pressure–121°C.

$Na_2S \cdot 9H_2O$ Solution:
Composition per 10.0mL:

$Na_2S \cdot 9H_2O$	0.375g

Preparation of $Na_2S \cdot 9H_2O$ Solution: Add $Na_2S \cdot 9H_2O$ to distilled/deionized water and bring volume to 10.0mL. Mix thoroughly. Sparge with 80% N_2 + 20% CO_2. Autoclave for 15 min at 15 psi pressure–121°C.

Methanol Solution:
Composition per 10.0mL:

Methanol	10.0mL

Preparation of Methanol Solution: Sparge 10.0mL of methanol with 100% N_2. Autoclave for 15 min at 15 psi pressure–121°C.

Vitamin Solution:
Composition per 10.0mL:

Calcium-D-pantothenate	20.0mg
Nicotinamide	20.0mg
Pyridoxine·HCl	20.0mg
Riboflavin	20.0mg
Thiamine·HCl	20.0mg
Biotin	10.0mg
p-Aminobenzoic acid	1.0mg
Folic acid	0.5mg
Vitamin B_{12}	0.2mg

Preparation of Vitamin Solution: Add components to distilled/deionized water and bring volume to 1.0L. Mix thoroughly. Filter sterilize. Sparge with 80% N_2 + 20% CO_2.

Preparation of Medium: Add components, except methanol solution, L-cysteine·HCl·H_2O solution, and $Na_2S \cdot 9H_2O$ solution, to distilled/deionized water and bring volume to 975.0mL. Mix thoroughly. Gently heat and bring to boiling. Continue boiling for 5 min. Cool to room temperature while sparging with 100% N_2. Autoclave for 15 min at 15 psi pressure–121°C. Aseptically and anaerobically add 10.0mL of sterile L-cysteine·HCl·H_2O solution, 10.0mL of sterile $Na_2S \cdot 9H_2O$ solution, and 5.0mL of sterile methanol solution. Mix thoroughly. Aseptically and anaerobically distribute into sterile screw-capped bottles under 100% N_2.

Use: For the cultivation and maintenance of *Methanosphaera stadtmanae.*

Methanosphaera Medium II

Composition per 1005.5mL:

$NaHCO_3$	3.0g
Sodium acetate·$3H_2O$	3.0g
Trypticase	1.0g
Yeast extract	1.0g
$Fe(NH_4)_2(SO_4)_2 \cdot 7H_2O$	2.0mg
Resazurin	1.0mg
$NiCl_2 \cdot 6H_2O$	0.2mg
$Na_2SeO_3 \cdot 5H_2O$(1m*M*)	0.1mg
Mineral solution 1	40.0mL
Mineral solution 2	40.0mL
Trace elements solution	10.0mL
Vitamin solution	10.0mL
L-Cysteine·HCl·H_2O solution	10.0mL
$Na_2S \cdot 9H_2O$ solution	10.0mL
Methanol	5.0mL

pH 7.0 ± 0.2 at 25°C

Mineral Solution 1:
Composition per liter:

K_2HPO_4	6.0g

Preparation of Mineral Solution 1: Add K_2HPO_4 to distilled/deionized water and bring volume to 1.0L. Mix thoroughly.

Mineral Solution 2:
Composition per liter:

NaCl	12.0g
KH_2PO_4	6.0g
$(NH_4)_2SO_4$	6.0g
$MgSO_4 \cdot 7H_2O$	2.6g
$CaCl_2 \cdot 2H_2O$	0.16g

Preparation of Mineral Solution 2: Add components to distilled/deionized water and bring volume to 1.0L. Mix thoroughly.

Trace Elements Solution:
Composition per liter:

$MgSO_4 \cdot 7H_2O$	3.0g
Nitrilotriacetic acid	1.5g
NaCl	1.0g
$MnSO_4 \cdot 2H_2O$	0.5g
$CoSO_4 \cdot 7H_2O$	0.18g
$ZnSO_4 \cdot 7H_2O$	0.18g
$CaCl_2 \cdot 2H_2O$	0.1g
$FeSO_4 \cdot 7H_2O$	0.1g
$NiCl_2 \cdot 6H_2O$	0.025g
$KAl(SO_4)_2 \cdot 12H_2O$	0.02g
$CuSO_4 \cdot 5H_2O$	0.01g
H_3BO_3	0.01g
$Na_2MoO_4 \cdot 2H_2O$	0.01g
$Na_2SeO_3 \cdot 5H_2O$	0.3mg

Preparation of Trace Elements Solution: Add nitrilotriacetic acid to 500.0mL of distilled/deionized water. Adjust pH to 6.5 with KOH. Add remaining components. Adjust pH to 7.0. Add distilled/deionized water to 1.0L.

Vitamin Solution:
Composition per liter:

Pyridoxine·HCl	10.0mg
Calcium DL-pantothenate	5.0mg
Lipoic acid	5.0mg
Nicotinic acid	5.0mg
p-Aminobenzoic acid	5.0mg
Riboflavin	5.0mg
Thiamine·HCl	5.0mg
Biotin	2.0mg
Folic acid	2.0mg
Vitamin B_{12}	0.1mg

Preparation of Vitamin Solution: Add components to distilled/deionized water and bring volume to 1.0L. Mix thoroughly. Sparge with 80% N_2 + 20% CO_2.

L-Cysteine·HCl·H_2O Solution:
Composition per 10.0mL:

L-Cysteine·HCl·H_2O	0.3g

Preparation of L-Cysteine·HCl·H_2O Solution: Add L-cysteine·HCl·H_2O to distilled/deionized water and bring volume to 10.0mL. Mix thoroughly. Autoclave under 80% N_2 + 20% CO_2 for 15 min at 15 psi pressure–121°C.

$Na_2S \cdot 9H_2O$ Solution:
Composition per 10.0mL:

$Na_2S \cdot 9H_2O$	0.3g

Preparation of $Na_2S \cdot 9H_2O$ Solution: Add $Na_2S \cdot 9H_2O$ to distilled/deionized water and bring volume to 10.0mL. Mix thoroughly. Sparge with 80% N_2 + 20% CO_2. Autoclave for 15 min at 15 psi pressure–121°C.

Methanol Solution:
Composition per 10.0mL:

Methanol	10.0mL

Preparation of Methanol Solution: Sparge 10.0mL of methanol with 100% N_2. Autoclave for 15 min at 15 psi pressure–121°C.

Preparation of Medium: Add components, except L-cysteine·HCl·H_2O solution, $Na_2S \cdot 9H_2O$ solution, and methanol solution, to distilled/deionized water and bring volume to 975.0mL. Mix thoroughly. Gently heat and bring to boiling. Continue boiling for 5 min. Cool to room temperature while sparging with 100% N_2. Autoclave for 15 min at 15 psi pressure–121°C. Aseptically and anaerobically add 10.0mL of sterile L-cysteine·HCl·H_2O solution, 10.0mL of sterile $Na_2S \cdot 9H_2O$ solution, and 5.0mL of sterile methanol solution. Mix thoroughly. Aseptically and anaerobically distribute into sterile screw-capped bottles under 100% N_2.

Use: For the cultivation and maintenance of *Methanospaera cuniculi.*

Methanospirillum hungatei JMA Medium

Composition per liter:

Sodium acetate	2.5g
NH_4Cl	1.9g
Mineral solution 1	75.0mL
Mineral solution 2	75.0mL
Wolfe's mineral solution	10.0mL
Vitamin solution	10.0mL
L-Cysteine·HCl·H_2O solution	10.0mL
$Na_2S \cdot 9H_2O$ solution	10.0mL
$FeSO_4 \cdot 6H_2O$ solution	6.0mL
Na_2CO_3 solution	4.8mL
Resazurin	1.0mL
$NiCl_2 \cdot 6H_2O$ solution	0.1mL

$FeSO_4 \cdot 6H_2O$ Solution:
Composition per 100.0mL:

$FeSO_4 \cdot 6H_2O$	0.2g

Preparation of $FeSO_4 \cdot 6H_2O$ Solution: Add 0.2g of $FeSO_4 \cdot 6H_2O$ to distilled/deionized water and bring volume to 100.0mL. Mix thoroughly.

$NiCl_2 \cdot 6H_2O$ Solution:
Composition per liter:

$NiCl_2 \cdot 6H_2O$	1.2g

Preparation of $NiCl_2 \cdot 6H_2O$ Solution: Add 1.2g of $NiCl_2 \cdot 6H_2O$ to distilled/deionized water and bring volume to 1.0L. Mix thoroughly.

Mineral Solution 1:
Composition per liter:

K_2HPO_4	39.0g

Preparation of Mineral Solution 1: Add K_2HPO_4 to distilled/deionized water and bring volume to 1.0L. Mix thoroughly.

Mineral Solution 2:
Composition per liter:

KH_2PO_4	24.0g
$(NH_4)_2SO_4$	6.0g
$MgSO_4 \cdot 7H_2O$	1.2g
$CaCl_2 \cdot 2H_2O$	0.79g
NaCl	0.59g

Preparation of Mineral Solution 2: Add components to distilled/deionized water and bring volume to 1.0L. Mix thoroughly.

Wolfe's Mineral Solution:
Composition per liter:

$MgSO_4 \cdot 7H_2O$	3.0g
Nitrilotriacetic acid	1.5g
NaCl	1.0g
$MnSO_4 \cdot 2H_2O$	0.5g
$CoCl_2 \cdot 6H_2O$	0.1g
$ZnSO_4 \cdot 7H_2O$	0.1g
$CaCl_2 \cdot 2H_2O$	0.1g
$FeSO_4 \cdot 7H_2O$	0.1g
$NiCl_2 \cdot 6H_2O$	0.025g
$KAl(SO_4)_2 \cdot 12H_2O$	0.02g

$CuSO_4·5H_2O$ 0.01g
H_3BO_3 .... 0.01g
$Na_2MoO_4·2H_2O$ 0.01g
$Na_2SeO_3·5H_2O$ 0.3mg

Preparation of Wolfe's Mineral Solution: Add nitrilotriacetic acid to 500.0mL of distilled/deionized water. Adjust pH to 6.5 with KOH. Add remaining components one at a time. Add distilled/deionized water to 1.0L. Adjust pH to 6.8.

Vitamin Solution:
Composition per liter:
Calcium DL-pantothenate 5.0g
Vitamin B_{12} 0.1g
Pyridoxine·HCl 10.0mg
p-Aminobenzoic acid 5.0mg
Lipoic acid 5.0mg
Nicotinic acid 5.0mg
Riboflavin 5.0mg
Thiamine·HCl 5.0mg
Biotin 2.0mg
Folic acid 2.0mg
Vitamin B_{12} 0.4mg

Preparation of Vitamin Solution: Add components to distilled/deionized water and bring volume to 1.0L. Mix thoroughly. Sparge with 100% N_2. Filter sterilize.

Na_2CO_3 Solution:
Composition per liter:
Na_2CO_3 .... 0.84g

Preparation of Na_2CO_3 Solution: Add Na_2CO_3 to distilled/deionized water and bring volume to 10.0mL. Mix thoroughly. Sparge with 80% N_2 + 20% CO_2. Autoclave for 15 min at 15 psi pressure–121°C.

$Na_2S·9H_2O$ Solution:
Composition per 10.0mL:
$Na_2S·9H_2O$ 0.2g

Preparation of $Na_2S·9H_2O$ Solution: Add $Na_2S·9H_2O$ to distilled/deionized water and bring volume to 10.0mL. Mix thoroughly. Sparge with 100% N_2. Autoclave for 15 min at 15 psi pressure–121°C. Before use, neutralize to pH 7.0 with sterile HCl.

L-Cysteine·HCl·H_2O Solution:
Composition per 10.0mL:
L-Cysteine·HCl·H_2O 0.22g

Preparation of L-Cysteine·HCl·H_2O Solution: Add L-cysteine·HCl·H_2O to distilled/deionized water and bring volume to 10.0mL. Mix thoroughly. Sparge with 100% N_2. Autoclave for 15 min at 15 psi pressure–121°C.

Preparation of Medium: Prepare and dispense medium under 80% N_2 + 20% CO_2. Add components, except Na_2CO_3 solution, L-cysteine·HCl solution, and $Na_2S·9H_2O$ solution, to distilled/deionized water and bring volume to 975.2mL. Mix thoroughly. Gently heat and bring to boiling. Cool while sparging with 80% N_2 + 20% CO_2. Autoclave for 15 min at 15 psi pressure–121°C. Aseptically and anaerobically add 4.8mL of sterile Na_2CO_3 solution, 10.0mL of sterile L-cysteine·HCl solution, and 10.0mL of sterile $Na_2S·9H_2O$ solution. Mix thoroughly. Aseptically and anaerobically distribute into sterile tubes or bottles.

Use: For the cultivation of *Methanospirillum hungatei.*

Methanospirillum hungatei Medium

Composition per 100.0mL:
Na_2CO_3 .... 0.4g
Sodium formate 0.2g
Pancreatic digest of casein 0.2g
Yeast extract 0.2g
NaCl 0.05g
L-Cysteine·HCl·H_2O 0.03g
K_2HPO_4 .... 0.02g
KH_2PO_4 .... 0.02g
$(NH_4)_2SO_4$ 0.02g
$MgSO_4·7H_2O$ 9.0mg
$CaCl_2·2H_2O$ 6.0mg
Resazurin 0.1mg
$Na_2S·9H_2O$ solution 10.0mL
Vitamin solution 1.0mL
Trace metal solution 1.0mL
pH 7.0 ± 0.2 at 25°C

$Na_2S·9H_2O$ Solution:
Composition per 10.0.mL:
$Na_2S·9H_2O$ 0.03g

Preparation of $Na_2S·9H_2O$ Solution: Add $Na_2S·9H_2O$ to distilled/deionized water and bring volume to 10.0mL. Mix thoroughly. Autoclave for 15 min at 15 psi pressure–121°C. Cool to 25°C.

Vitamin Solution:
Composition per 1000.0mL:
Pyridoxine·HCl 1.0mg
p-Aminobenzoic acid 0.5mg
Calcium-D-pantothenate 0.5mg
Nicotinic acid 0.5mg
Riboflavin 0.5mg
Thiamine·HCl 0.5mg
Thioctic acid 0.5mg
Biotin 0.2mg
Folic acid 0.2mg
Vitamin B_{12} 0.01mg

Preparation of Vitamin Solution: Add components to distilled/deionized water and bring volume to 1.0L. Mix thoroughly. Filter sterilize.

Trace Metal Solution:
Composition per liter:
$K_2HPO_4·3H_2O$ 9.0g
K_2HPO_4 .... 6.0g
NH_4Cl 5.0g
$MgCl_2·6H_2O$ 1.0g
$CaCl_2·2H_2O$ 0.01g

Preparation of Trace Metal Solution: Add components to distilled/deionized water and bring volume to 1.0L. Mix thoroughly.

Preparation of Medium: Prepare and distribute medium anaerobically under 80% H_2 + 20% CO_2. Add components, except $Na_2S·9H_2O$ solution, to distilled/deionized water and bring volume to 90.0mL. Mix thoroughly. Gently heat and bring to boiling. Continue boiling until resazurin turns colorless, indicating reduction. Autoclave for 15 min at 15 psi pressure–121°C. Cool to 25°C. Aseptically add 10.0mL of sterile $Na_2S·9H_2O$ solution. Mix thoroughly. Aseptically and anaerobically distribute into sterile tubes or flasks.

Use: For the cultivation of *Methanospirillum hungatei.*

Methanospirillum hungatei SAM Medium

Composition per liter:
Na_2CO_3 .... 2.63
Sodium acetate·$3H_2O$ 2.5g
$(NH_4)_2SO_4$ 0.45g

K_2HPO_4	0.3g
KH_2PO_4	0.18g
$MgSO_4 \cdot 9H_2O$	0.12g
$CaCl_2 \cdot 2H_2O$	0.06g
NaCl	0.05g
Na_2CO_3 solution	20.0mL
Trace minerals solution	10.0mL
Vitamin solution	10.0mL
L-Cysteine/Na_2S solution	10.0mL
$FeSO_4 \cdot 7H_2O$ solution	5.0mL
Resazurin solution	0.2mL

pH 7.1 ± 0.2 at 25°C

Na_2CO_3 Solution:
Composition per liter:

Na_2CO_3	100.0g

Preparation of Na_2CO_3 Solution: Bring 1.0L of distilled/deionized water to boiling. Cool to room temperature while sparging with 80% H_2 + 20% CO_2. Distribute into serum bottles fitted with butyl rubber stoppers and aluminum seals. Do not grease stoppers. Autoclave for 15 min at 15 psi pressure–121°C. Store at room temperature in an anaerobic chamber.

Trace Minerals Solution:
Composition per liter:

Nitrolotriacetic acid	1.5g
$MnSO_4 \cdot 2H_2O$	0.5g
$Na_2MoO_4 \cdot 2H_2O$	0.24g
$Na_2WO_4 \cdot 2H_2O$	0.165g
Na_2SeO_3	0.15g
$CoCl_2 \cdot 6H_2O$	0.1g
$NiCl_2 \cdot 6H_2O$	0.1g
$ZnSO_4 \cdot 7H_2O$	0.1g
$AIK(SO_4)_2 \cdot 12H_2O$	0.01g
$CuSO_4 \cdot 5H_2O$	0.01g
H_3BO_3	0.01g

Preparation of Trace Minerals Solution: Add nitrilotriacetic acid to 500.0mL of distilled/deionized water. Adjust pH to 6.5 with KOH. Add remaining components. Mix thoroughly. Add distilled/deionized water to 1.0L. Adjust pH to 7.0.

Vitamin Solution:
Composition per liter:

Vitamin B_{12}	0.1g
Pyridoxine·HCl	10.0mg
p-Aminobenzoic acid	5.0mg
Lipoic acid	5.0mg
Nicotinic acid	5.0mg
Riboflavin	5.0mg
Thiamine·HCl	5.0mg
Biotin	2.0mg
Folic acid	2.0mg
Vitamin B_{12}	0.4mg

Preparation of Vitamin Solution: Add components to distilled/deionized water and bring volume to 1.0L. Mix thoroughly.

L-Cysteine/Na_2S Solution:
Composition per liter:

L-Cysteine·HCl	1.25g
$Na_2S \cdot 7H_2O$	1.25g

Preparation of L-Cysteine/Na_2S Solution: Gently heat and bring 75.0mL of distilled/deionized water to boiling. Add L-cysteine·HCl. Adjust pH to 10.0 with 3*N* NaOH. Add $Na_2S \cdot 9H_2O$. Mix thoroughly. Gently heat and bring to boiling. Cool to room temperature while sparging with 100% N_2. Bring volume to 100.0mL with distilled/deionized water. Gently heat and bring to boiling. Cool to 60°C while sparging with 100% N_2. Anaerobically distribute into serum bottles. Autoclave for 15 min at 15 psi pressure–121°C.

$FeSO_4 \cdot 6H_2O$ Solution:
Composition per 100.0mL:

$FeSO_4 \cdot 6H_2O$	0.2g

Preparation of $FeSO_4 \cdot 6H_2O$ Solution: Add 0.2g of $FeSO_4 \cdot 6H_2O$ to distilled/deionized water and bring volume to 100.0mL. Mix thoroughly.

Resazurin Solution:
Composition per 10.0mL:

Resazurin	10.0mg

Preparation of Resazurin Solution: Add resazurin to distilled/deionized water and bring volume to 10.0mL. Mix thoroughly.

Preparation of Medium: Prepare and dispense medium under 80% H_2 + 20% CO_2. Add components, except Na_2CO_3, Na_2CO_3 solution and L-cysteine/Na_2S solution, to distilled/deionized water and bring volume to 1.0L. Mix thoroughly. Adjust pH to 7.0. Gently heat and bring to boiling. Continue boiling for 3 min. Cool to 60°C while sparging with 80% H_2 + 20% CO_2. Add Na_2CO_3 and L-cysteine/Na_2S solution. Mix thoroughly. Anaerobically distribute 10.0mL volumes into anaerobic tubes. Autoclave for 20 min at 15 psi pressure–121°C. Aseptically and anaerobically add 0.2mL of sterile Na_2CO_3 solution to each tube. Mix thoroughly. Adjust pH to 7.1.

Use: For the cultivation of *Methanospirillum hungatei.*

Methanospirillum SK Medium

Composition per liter:

NaCl	18.0g
Isopropanol	7.5g
$NaHCO_3$	5.0g
$MgCl_2 \cdot 6H_2O$	4.0g
$MgSO_4 \cdot 7H_2O$	3.45g
Trypticase	2.0g
Yeast extract	2.0g
Sodium acetate	1.0g
L-Cysteine·HCl	0.5g
KCl	0.335g
NH_4Cl	0.25g
$CaCl_2 \cdot 2H_2O$	0.14g
K_2HPO_4	0.14g
$Fe(NH_4)_2(SO_4)_2 \cdot 7H_2O$	2.0mg
Resazurin	1.0mg
Trace elements solution	10.0mL
Vitamin solution	10.0mL
$Na_2S \cdot 9H_2O$ solution	10.0mL

pH 7.0–7.4 at 25°C

Trace Elements Solution:
Composition per liter:

$MgSO_4 \cdot 7H_2O$	3.0g
Nitrilotriacetic acid	1.5g
NaCl	1.0g
$MnSO_4 \cdot 2H_2O$	0.5g
$CoSO_4 \cdot 7H_2O$	0.18g
$ZnSO_4 \cdot 7H_2O$	0.18g
$CaCl_2 \cdot 2H_2O$	0.1g
$FeSO_4 \cdot 7H_2O$	0.1g
$NiCl_2 \cdot 6H_2O$	0.025g
$KAl(SO_4)_2 \cdot 12H_2O$	0.02g
$CuSO_4 \cdot 5H_2O$	0.01g

H_3BO_3 .... 0.01g
$Na_2MoO_4·2H_2O$ 0.01g
$Na_2SeO_3·5H_2O$ 0.3mg

Preparation of Trace Elements Solution: Add nitrilotriacetic acid to approximately 500.0mL of distilled/deionized water. Dissolve by adding KOH and adjust pH to 6.5. Add remaining components. Bring volume to 1.0L with additional distilled/deionized water. Adjust pH to 7.0 with KOH.

Vitamin Solution:
Composition per liter:
Pyridoxine·HCl 10.0mg
Calcium DL-pantothenate 5.0mg
Lipoic acid 5.0mg
Nicotinic acid 5.0mg
p-Aminobenzoic acid 5.0mg
Riboflavin 5.0mg
Thiamine·HCl 5.0mg
Biotin 2.0mg
Folic acid 2.0mg
Vitamin B_{12} 0.1mg

Preparation of Vitamin Solution: Add components to distilled/deionized water and bring volume to 1.0L. Mix thoroughly. Sparge with 80% N_2 + 20% CO_2.

$Na_2S·9H_2O$ Solution:
Composition per 10.0mL:
$Na_2S·9H_2O$ 0.4g

Preparation of $Na_2S·9H_2O$ Solution: Add $Na_2S·9H_2O$ to distilled/deionized water and bring volume to 10.0mL. Mix thoroughly. Sparge with 80% N_2 + 20% CO_2. Autoclave for 15 min at 15 psi pressure–121°C.

Preparation of Medium: Add components, except $Na_2S·9H_2O$ solution, to distilled/deionized water and bring volume to 990.0mL. Mix thoroughly. Sparge with 80% N_2 + 20% CO_2. Autoclave for 15 min at 15 psi pressure–121°C. Aseptically and anaerobically add 10.0mL of sterile $Na_2S·9H_2O$ solution. Mix thoroughly.

Use: For the cultivation and maintenance of *Methanospirillum hungatei.*

Methanothermus fervidus Medium

Composition per liter:
Na_2SO_4 .... 3.4g
$NaHCO_3$ 2.0g
Trypticase 2.0g
Yeast extract 2.0g
L-Cysteine·HCl 0.5g
$Na_2S·9H_2O$ 0.5g
$FeSO_4·7H_2O$ 2.0mg
$Ni(NH_4)_2(SO_4)_2$ 2.0mg
Resazurin 1.0mg
Mineral solution 1 .... 37.5mL
Mineral solution 2 .... 37.5mL
Trace elements solution 10.0mL
Vitamin solution 10.0mL
pH 6.5 ± 0.2 at 25°C

Mineral Solution 1:
Composition per liter:
K_2HPO_4 .... 6.0g

Preparation of Mineral Solution 1: Add K_2HPO_4 to distilled/deionized water and bring volume to 1.0L. Mix thoroughly.

Mineral Solution 2:
Composition per liter:
NaCl 12.0g
K_2HPO_4 .... 6.0g
$(NH_4)_2SO_4$ 6.0g
$MgSO_4·7H_2O$ 2.4g
$CaCl_2·2H_2O$ 1.6g

Preparation of Mineral Solution 2: Add components to distilled/deionized water and bring volume to 1.0L. Mix thoroughly.

Trace Elements Solution:
Composition per liter:
$MgSO_4·7H_2O$ 3.0g
Nitrilotriacetic acid 1.5g
NaCl 1.0g
$MnSO_4·2H_2O$ 0.5g
$CoSO_4·7H_2O$ 0.18g
$ZnSO_4·7H_2O$ 0.18g
$CaCl_2·2H_2O$ 0.1g
$FeSO_4·7H_2O$ 0.1g
$NiCl_2·6H_2O$ 0.025g
$KAl(SO_4)_2·12H_2O$ 0.02g
$CuSO_4·5H_2O$ 0.01g
H_3BO_3 .... 0.01g
$Na_2MoO_4·2H_2O$ 0.01g
$Na_2SeO_3·5H_2O$ 0.3mg

Preparation of Trace Elements Solution: Add nitrilotriacetic acid to approximately 500.0mL of distilled/deionized water. Dissolve by adding KOH and adjust pH to 6.5. Add remaining components. Bring volume to 1.0L with additional distilled/deionized water. Adjust pH to 7.0 with KOH.

Vitamin Solution:
Composition per liter:
Pyridoxine·HCl 10.0mg
Calcium DL-pantothenate 5.0mg
Lipoic acid 5.0mg
Nicotinic acid 5.0mg
p-Aminobenzoic acid 5.0mg
Riboflavin 5.0mg
Thiamine·HCl 5.0mg
Biotin 2.0mg
Folic acid 2.0mg
Vitamin B_{12} 0.1mg

Preparation of Vitamin Solution: Add components to distilled/deionized water and bring volume to 1.0L. Mix thoroughly. Sparge with 80% N_2 + 20% CO_2.

Preparation of Medium: Add components, except $NaHCO_3$ and $Na_2S·9H_2O$, to distilled/deionized water and bring volume to 1.0L. Adjust pH to 6.5 with 10*N* H_2SO_4. Add $NaHCO_3$ and sparge with 80% N_2 + 20% CO_2 for 15 min. Add $Na_2S·9H_2O$. Mix thoroughly. Anaerobically distribute 20.0mL of medium into 100mL alkali-rich soda lime glass bottles. Pressurize to 2 bar with 80% H_2 + 20% CO_2. Autoclave for 15 min at 15 psi pressure–121°C.

Use: For the cultivation and maintenance of *Methanothermus fervidus.*

Methanothrix Medium

Composition per liter:
Sodium acetate 6.8g
$KHCO_3$ 4.0g
NH_4Cl 1.0g
NaCl 0.6g
KH_2PO_4 .... 0.3g

$MgCl_2·6H_2O$ 0.1g
$CaCl_2·2H_2O$ 0.08g
Resazurin 1.0mg
Trace elements solution 10.0mL
Vitamin solution 10.0mL
L-Cysteine·HCl·H_2O solution 10.0mL
$Na_2S·9H_2O$ solution 10.0mL
pH 7.0 ± 0.2 at 25°C

Trace Elements Solution:
Composition per liter:
Nitrilotriacetic acid 12.8g
$FeCl_3·6H_2O$ 1.35g
NaCl 1.0g
$NiCl_2·6H_2O$ 0.12g
$CaCl_2·2H_2O$ 0.10g
$MnCl_2·4H_2O$ 0.10g
$ZnCl_2$ 0.10g
$Na_2SeO_3·5H_2O$ 0.026g
$CuCl_2·2H_2O$ 0.025g
$CoCl_2·6H_2O$ 0.024g
$Na_2MoO_4·2H_2O$ 0.024g
H_3BO_3 ..... 0.01g

Preparation of Trace Elements Solution: Add nitrilotriacetic acid to approximately 500.0mL of distilled/deionized water. Dissolve by adding KOH and adjust pH to 6.5. Add remaining components. Bring volume to 1.0L with additional distilled/deionized water. Adjust pH to 7.0 with KOH.

Vitamin Solution:
Composition per liter:
Pyridoxine·HCl 10.0mg
Calcium DL-pantothenate 5.0mg
Lipoic acid 5.0mg
Nicotinic acid 5.0mg
p-Aminobenzoic acid 5.0mg
Riboflavin 5.0mg
Thiamine·HCl 5.0mg
Biotin 2.0mg
Folic acid 2.0mg
Vitamin B_{12} 0.1mg

Preparation of Vitamin Solution: Add components to distilled/deionized water and bring volume to 1.0L. Mix thoroughly. Filter sterilize. Sparge with 80% N_2 + 20% CO_2.

L-Cysteine·HCl·H_2O Solution:
Composition per 10.0mL:
L-Cysteine·HCl·H_2O 0.3g

Preparation of L-Cysteine·HCl·H_2O Solution: Add L-cysteine·HCl·H_2O to distilled/deionized water and bring volume to 10.0mL. Mix thoroughly. Autoclave under 80% N_2 + 20% CO_2 for 15 min at 15 psi pressure–121°C.

$Na_2S·9H_2O$ Solution:
Composition per 10.0mL:
$Na_2S·9H_2O$ 0.3g

Preparation of $Na_2S·9H_2O$ Solution: Add $Na_2S·9H_2O$ to distilled/deionized water and bring volume to 10.0mL. Mix thoroughly. Sparge with 80% N_2 + 20% CO_2. Autoclave for 15 min at 15 psi pressure–121°C.

Preparation of Medium: Add components, except vitamin solution, L-cysteine·HCl·H_2O solution, and $Na_2S·9H_2O$ solution, to distilled/deionized water and bring volume to 970.0mL. Mix thoroughly. Autoclave for 15 min at 15 psi pressure–121°C. Aseptically and anaerobically add 10.0mL of sterile vitamin solution, 10.0mL of sterile L-cysteine·HCl·H_2O solution, and 10.0mL of sterile $Na_2S·9H_2O$ solution. Mix thoroughly. Check that final pH is 7.0. Use 20% inoculum.

Use: For the cultivation and maintenance of *Methanosaeta concilii.*

Methylamine Salts Medium

Composition per liter:
Agar 15.0g
Methylamine·HCl 6.75g
K_2HPO_4 ..... 2.12g
KH_2PO_4 ..... 1.0g
Solution A 5.0mL
Solution B 1.0mL
pH 7.0 ± 0.2 at 25°C

Solution A:
Composition per 100.0mL:
$MgSO_4·7H_2O$ 2.0g
$CaCl_2·2H_2O$ 0.2g
$FeSO_4·7H_2O$ 0.2g

Preparation of Solution A: Add components to distilled/deionized water and bring volume to 100.0mL. Mix thoroughly.

Solution B:
Composition per 100.0mL:
$MnSO_4·7H_2O$ 0.05g
$Na_2MoO_4·2H_2O$ 0.05g

Preparation of Solution B: Add components to distilled/deionized water and bring volume to 100.0mL. Mix thoroughly.

Preparation of Medium: Add components to distilled/deionized water and bring volume to 1.0L. Mix thoroughly. Gently heat and bring to boiling. Distribute into tubes or flasks. Autoclave for 15 min at 15 psi pressure–121°C. Pour into sterile Petri dishes or leave in tubes.

Use: For the cultivation and maintenance of *Methylobacterium extorquens* and *Pseudomonas* species.

Methylobacterium Agar

Composition per liter:
Agar 12.0g
KNO_3 ..... 1.0g
Na_2HPO_4 ..... 0.23g
$MgSO_4·7H_2O$ 0.2g
NaH_2PO_4 ..... 0.07g
$CaCl_2·2H_2O$ 0.02g
$FeSO_4·7H_2O$ 1.0mg
$ZnSO_4·7H_2O$ 70.0μg
H_3BO_3 ..... 10.0μg
$MnSO_4·5H_2O$ 10.0μg
MoO_3 ..... 10.0μg
$CuSO_4·5H_2O$ 5.0μg
Methanol, filter sterilized 5.0mL
pH 6.8 ± 0.2 at 25°C

Preparation of Medium: Add components, except methanol, to distilled/deionized water and bring volume to 995.0mL. Mix thoroughly. Gently heat and bring to boiling. Autoclave for 15 min at 15 psi pressure–121°C. Cool to 50°–55°C. Aseptically add 5.0mL of filter-sterilized methanol. Mix thoroughly. Pour into sterile Petri dishes or distribute into sterile tubes.

Use: For the cultivation and maintenance of *Methylobacterium organophilum.*

Methylobacterium Medium (DSMZ Medium 125)

Composition per liter:

Agar	12.0g
KNO_3	1.0g
Na_2HPO_4	0.23g
$MgSO_4 \cdot 7H_2O$	0.2g
NaH_2PO_4	0.07g
$CaCl_2 \cdot 2H_2O$	0.02g
$FeSO_4 \cdot 7H_2O$	1.0mg
$ZnSO_4 \cdot 7H_2O$	70.0μg
MoO_3	10.0μg
H_3BO_3	10.0μg
$MnSO_4 \cdot 5H_2O$	10.0μg
$CuSO_4 \cdot 5H_2O$	5.0μg
Methanol	15.0mL

pH 4.0-5.4 at 25°C

Preparation of Medium: Add components to distilled/deionized water and bring volume to 1.0L. Mix thoroughly. Adjust pH to 6.8. Gently heat and bring to boiling. Distribute into tubes or flasks. Autoclave for 15 min at 15 psi pressure–121°C. Pour into sterile Petri dishes or leave in tubes.

Use: For the cultivation and maintenance of *Acidomonas methanolica* and *Methanomonas methylovora.*

Methylobacterium Medium

Composition per liter:

Agar	15.0g
K_2HPO_4	1.2g
KH_2PO_4	0.62g
$(NH_4)_2SO_4$	0.5g
$MgSO_4 \cdot 7H_2O$	0.2g
NaCl	0.1g
$CaCl_2 \cdot 2H_2O$	34.0mg
$FeCl_3 \cdot H_2O$	1.0mg
Trace elements solution	1.0mL
Methanol, filter sterilized	10.0mL

pH 7.0 ± 0.2 at 25°C

Trace Elements Solution:

Composition per liter:

$ZnSO_4 \cdot 7H_2O$	70.0mg
H_3BO_3	10.0mg
$Na_2MoO_4 \cdot 2H_2O$	10.0mg
$MnSO_4 \cdot H_2O$	7.0mg
$CoCl_2 \cdot H_2O$	5.0mg
$CuSO_4 \cdot 5H_2O$	5.0mg

Preparation of Medium: Add components, except methanol, to distilled/deionized water and bring volume to 990.0mL. Mix thoroughly. Gently heat and bring to boiling. Autoclave for 15 min at 15 psi pressure–121°C. Cool to 50°–55°C. Aseptically add 10.0mL of sterile methanol. Mix thoroughly. Pour into sterile Petri dishes or distribute into sterile tubes.

Use: For the cultivation and maintenance of *Methylobacterium* species.

Methylobacterium thiocyanatum Medium (DSMZ Medium 805)

Composition per liter:

$Na_2HPO_4 \cdot 2H_2O$	7.9g
Glucose	4.5g
K_2HPO_4	1.5g
KSCN	0.25g
$MgSO_4 \cdot 7H_2O$	0.1g
Iron sulfate solution	1.0mL

pH 7.1 ± 0.1 at 25°C

Iron Sulfate Solution:

Composition per 10.0mL:

$FeSO_4 \cdot 7H_2O$	0.2g

Preparation of Iron Sulfate Solution: Add 0.2g of $FeSO_4 \cdot 7H_2O$ to distilled/deionized water and bring volume to 10.0mL. Mix thoroughly.

Preparation of Medium: Add components to distilled/deionized water and bring to 1.0L. Mix thoroughly. Distribute into tubes or flasks. Autoclave for 10 min at 10 psi pressure–115°C.

Use: For the cultivation of *Methylobacterium thiocyanatum.*

Methylocapsa acidophila Medium (DSMZ Medium 922)

Composition per liter:

KH_2PO_4	100.0mg
$MgSO_4 \cdot 7H_2O$	50.0mg
$CaCl_2 \cdot 2H_2O$	10.0mg
Trace elements solution	1.0mL

pH 4.5-5.8 at 25°C

Trace Elements Solution:

EDTA	5.0g
$FeSO_4 \cdot 7H_2O$	2.0g
$CoCl_2 \cdot 6H_2O$	0.2g
$CuCl_2 \cdot 5H_2O$	0.1g
$ZnSO_4 \cdot 7H_2O$	0.1g
Na_2MoO_4	0.03g
$NiCl_2 \cdot 6H_2O$	0.02g

Preparation of Trace Elements Solution: Add components to distilled/deionized water and bring volume to 1.0L. Mix thoroughly.

Preparation of Medium: Add components to distilled/deionized water and bring volume to 1.0L. Mix thoroughly. Distribute into tubes or flasks. Autoclave for 15 min at 15 psi pressure–121°C. The final pH should be 4.5-5.8. The medium is fairly weakly buffered so the pH should be checked before and after autoclaving. Incubate under atmosphere of 10-30% methane.

Use: For the cultivation of *Methylocapsa acidiphila.*

Methylocapsa acidophila Medium (DSMZ Medium 922)

Composition per liter:

KNO_3	100.0mg
KH_2PO_4	100.0mg
$MgSO_4 \cdot 7H_2O$	50.0mg
$CaCl_2 \cdot 2H_2O$	10.0mg
Trace elements	1.0mL

pH 4.5-5.8 at 25°C

Trace Elements:

EDTA	5.0g
$FeSO_4 \cdot 7H_2O$	2.0g
$CoCl_2 \cdot 6H_2O$	0.2g
$CuCl_2 \cdot 5H_2O$	0.1g
$ZnSO_4 \cdot 7H_2O$	0.1g
Na_2MoO_4	0.03g
$NiCl_2 \cdot 6H_2O$	0.02g

Preparation of Trace Elements: Add components to distilled/deionized water and bring volume to 1.0L. Mix thoroughly.

Preparation of Medium: Add components to distilled/deionized water and bring volume to 1.0L. Mix thoroughly. Distribute into tubes or flasks. Autoclave for 15 min at 15 psi pressure–121°C. The final pH should be 4.5-5.8. The medium is fairly weakly buffered so the pH should be checked before and after autoclaving. Incubate under atmosphere of 10-30% methane.

Use: For the cultivation and enhanced growth of *Methylocapsa acidophila.*

Methylococcus Medium

Composition per liter:

Agar 8.0g
$NaNO_3$ (20% solution) 10.0mL
L-F salts solution 10.0mL
Sodium-potassium phosphate buffer 6.5mL

pH 7.1 ± 0.2 at 25°C

Sodium-Potassium Phosphate Buffer:
Composition per liter:

KH_2PO_4 .... 136.0g
NaOH 28.8g

Preparation of Sodium-Potassium Phosphate Buffer: Add components to distilled/deionized water and bring volume to 1.0L. Mix thoroughly. Adjust pH to 7.1.

L-F Salts Solution:
Composition per liter:

$MgSO_4 \cdot 7H_2O$ (10% solution) 200.0mL
$CaCl_2 \cdot 2H_2O$ (10% solution) 20.0mL
$FeSO_4$ (10% solution) 10.0mL
$ZnSO_4 \cdot 7H_2O$ (1% solution) 4.9mL
H_3BO_3 (1% solution) 0.6mL
$MnSO_4 \cdot H_2O$ (1% solution) 0.27mL
$CuSO_4 \cdot 5H_2O$ (1% solution) 0.2mL

Preparation of L-F Salts Solution: Filter sterilize $FeSO_4$ solution immediately prior to use. Add all components to distilled/deionized water and bring volume to 1.0L. Mix thoroughly.

Preparation of Medium: Add components to distilled/deionized water and bring volume to 1.0L. Mix thoroughly. Adjust pH to 7.1. Autoclave for 15 min at 15 psi pressure–121°C. Pour into sterile Petri dishes or leave in tubes.

Use: For the cultivation and maintenance of *Methylococcus* species.

Methylophaga Agar

Composition per 103.0mL:

Agar solution 50.0mL
Mineral base, 2× 50.0mL
Solution T 2.0mL
Vitamin B_{12} solution 1.0mL
Methanol 0.3mL

pH 7.3 ± 0.2 at 25°C

Agar Solution:
Composition per 500.0mL:

Agar 15.0g

Preparation of Agar Solution: Add agar to distilled/deionized water and bring volume to 500.0mL. Mix thoroughly. Autoclave for 15 min at 15 psi pressure–121°C. Cool to 50°C.

Mineral Base, 2×:
Composition per 500.0mL:

NaCl 24.0g
$MgCl_2 \cdot 6H_2O$ 3.0g
$MgSO_4 \cdot 7H_2O$ 2.0g
$CaCl_2 \cdot 2H_2O$ 1.0g
KCl 0.5g
Bis-Tris buffer (bis[2-Hydroxyethyl]imino-tris[hydroxymethyl]-methane) 0.5g
Wolfe's mineral solution 10.0mL

Preparation of Mineral Base, 2×: Add components to distilled/deionized water and bring volume to 500.0mL. Mix thoroughly. Adjust pH to 7.3. Autoclave for 15 min at 15 psi pressure–121°C. Cool to 50°C.

Wolfe's Mineral Solution:
Composition per liter:

$MgSO_4 \cdot 7H_2O$ 3.0g
Nitriloacetic acid 1.5g
NaCl 1.0g
$MnSO_4 \cdot H_2O$ 0.5g
$FeSO_4 \cdot 7H_2O$ 0.1g
$CoCl_2 \cdot 6H_2O$ 0.1g
$CaCl_2$ 0.1g
$ZnSO_4 \cdot 7H_2O$ 0.1g
$CuSO_4 \cdot 5H_2O$ 0.01g
$AlK(SO_4)_2 \cdot 12H_2O$ 0.01g
H_3BO_3 .... 0.01g
$Na_2MoO_4 \cdot 2H_2O$ 0.01g

Preparation of Wolfe's Mineral Solution: Add nitrilotriacetic acid to 500.0mL of distilled/deionized water. Dissolve by adjusting pH to 6.5 with KOH. Add remaining components. Add distilled/deionized water to 1.0L.

Solution T:
Composition per 100.0mL:

NH_4Cl 10.0g
Bis-Tris buffer (bis[2-Hydroxyethyl]imino-tris[hydroxymethyl]-methane) 10.0g
KH_2PO_4 .... 0.7g
Ferric ammonium citrate 0.3g

Preparation of Solution T: Add components to distilled/deionized water and bring volume to 100.0mL. Mix thoroughly. Adjust pH to 7.3. Autoclave for 15 min at 15 psi pressure–121°C.

Vitamin B_{12} Solution:
Composition per 10.0mL:

Vitamin B_{12} 1.0μg

Preparation of Vitamin B_{12} Solution: Add Vitamin B_{12} to 10.0mL of distilled/deionized water. Mix thoroughly. Filter sterilize.

Preparation of Medium: Aseptically mix 50.0mL of the sterile agar solution with 50.0mL of the sterile mineral base, 2×. Aseptically combine the sterile solution T and sterile Vitamin B_{12} solution with the sterile mineral base. Filter sterilize methanol and add to basal medium. Pour into sterile Petri dishes or distribute into sterile tubes.

Use: For the cultivation and maintenance of *Methylophaga marina.*

Methylophaga alcalica Agar (DSMZ Medium 976)

Composition per liter:

NaCl 30.0g
Agar 20.0g
KH_2PO_4 .... 1.0g
KNO_3 .... 1.0g
$MgSO_4 \cdot 7H_2O$ 0.22g
Na_2CO_3 solution 50.0mL
Methanol solution 50.0mL
Trace elements solution 1.0mL

pH 9.5 ± 0.2 at 25°C

Methanol Solution:
Composition per 50.0mL:
Methanol 10.0mL

Preparation of Methanol Solution: Add methanol to distilled/deionized water and bring volume to 50.0mL. Mix thoroughly. Autoclave for 15 min at 15 psi pressure–121°C.

Na_2CO_3 Solution:
Composition per 50.0mL:
Na_2CO_3 .... 5.0g

Preparation of Na_2CO_3 Solution: Add Na_2CO_3 to distilled/deionized water and bring volume to 50.0mL. Mix thoroughly. Autoclave for 15 min at 15 psi pressure–121°C.

Trace Elements Solution:
Composition per liter:
Ferric citrate 30.0mg
$CaCl_2 \cdot 2H_2O$ 30.0mg
$MgCl_2 \cdot 4H_2O$ 5.0mg
$ZnSO_4 \cdot 7H_2O$ 5.0mg
$CuSO_4 \cdot 5H_2O$ 0.5mg

Preparation of Trace Elements Solution: Add components to distilled/deionized water and bring volume to 1.0L. Mix thoroughly.

Preparation of Medium: Add components, except methanol solution and Na_2CO_3 solution, to distilled/deionized water and bring volume to 900.0mL. Mix thoroughly. Gently heat and bring to boiling. Autoclave for 15 min at 15 psi pressure–121°C. Cool to 55°C. Aseptically add 50.0mL warm sterile Na_2CO_3 solution and 50.0mL warm sterile methanol solution. Mix thoroughly. Pour into Petri dishes or aseptically distribute into sterile tubes.

Use: For the cultivation and maintenance of *Methylophaga alcalica.*

Methylophaga Medium

Composition per liter:
NaCl 25.0g
Agar 20.0g
Peptone 10.0g
Beef extract 7.0g
K_2HPO_4 .... 1.0g
$(NH_4)_2SO_4$ 1.0g
Methanol, filter sterilized 10.0mL
pH 7.0 ± 0.2 at 25°C

Preparation of Medium: Add components, except methanol, to distilled/deionized water and bring volume to 990.0mL. Mix thoroughly. Gently heat and bring to boiling. Autoclave for 15 min at 15 psi pressure–121°C. Cool to 50°–55°C. Aseptically add 10.0mL of sterile methanol. Mix thoroughly. Pour into sterile Petri dishes or distribute into sterile tubes.

Use: For the cultivation and maintenance of *Methylophaga marina* and *Methylophaga thalassica.*

Methylophaga sulfidovorans Medium (DSMZ Medium 951)

Composition per liter:
NaCl 15.0g
Na_2CO_3 .... 2.0g
$MgSO_4 \cdot 7H_2O$ 1.0g
$(NH_4)_2SO_4$ 0.5g
$CaCl_2 \cdot 6H_2O$ 0.33g
KCl 0.2g
KH_2PO_4 .... 0.02g
DMS (Dimethylsulphide) 62.0mg
$FeSO_4 \cdot 7H_2O$ 1.0mg
Trace elements solution SL10 .... 1.0mL
Vitamin solution 1.0mL
pH 7.5 ± 0.3 at 25°C

Trace Elements Solution SL-10:
Composition per liter:
$FeCl_2 \cdot 4H_2O$ 1.5g
$CoCl_2 \cdot 6H_2O$ 190.0mg
$MnCl_2 \cdot 4H_2O$ 100.0mg
$ZnCl_2$ 70.0mg
$Na_2MoO_4 \cdot 2H_2O$ 36.0mg
$NiCl_2 \cdot 6H_2O$ 24.0mg
H_3BO_3 .... 6.0mg
$CuCl_2 \cdot 2H_2O$ 2.0mg
HCl (25% solution) 10.0mL

Preparation of Trace Elements Solution SL-10: Add $FeCl_2 \cdot 4H_2O$ to 10.0mL of HCl solution. Mix thoroughly. Add distilled/deionized water and bring volume to 1.0L. Add remaining components. Mix thoroughly. Sparge with 80% N_2 + 20% CO_2.

Vitamin Solution:
Composition per liter:
Pyridoxine-HCl 500.0mg
Nicotinic acid 200.0mg
Thiamine 100.0mg
p-Aminobenzoic acid 100.0mg
Pantothenate 50.0mg
Biotin 20.0mg
Riboflavine 10.0mg
Vitamin B_{12} 10.0mg

Preparation of Vitamin Solution: Add components to distilled/deionized water and bring volume to 1.0L. Mix thoroughly. Sparge with 80% H_2 + 20% CO_2.

Preparation of Medium: Add components to distilled/deionized water and bring volume to 1.0L. Mix thoroughly. Adjust pH to 7.5. Distribute into tubes or bottles. Autoclave for 15 min at 15 psi pressure–121°C.

Use: For the cultivation of *Methylophaga sulfidovorans.*

Methylophaga thalassica Agar (LMG Medium 73)

Composition per liter:
NaCl 25.0g
Agar 20.0g
Peptone 10.0g
Lab Lemco beef extract 7.0g
K_2HPO_4 .... 1.0g
$(NH_4)_2SO_4$ 1.0g
pH 7.0 ± 0.2 at 25°C

Preparation of Medium: Add components, except methanol, to 990.0mL distilled/deionized water. Mix thoroughly. Gently heat and bring to boiling. Autoclave for 15 min at 15 psi pressure–121°C. Cool to 45°C. Aseptically add 10.0mL sterile methanol. Mix thoroughly. Pour into sterile Petri dishes or distribute into sterile tubes.

Use: For the cultivation of *Methylophaga thalassica.*

Methylosarcina quisquillarum Methylosarcina fibrata Medium (DSMZ Medium 921)

Composition per 1012.1mL:
Solution 1 .... 100.0mL
Phosphate buffer 10.0mL
Solution 3 .... 1.0mL

Trace elements 1.0mL
Solution 2 .... 0.1mL

pH 7.0 ± 0.2 at 25°C

Solution 1 (10x NMS Salts):
Composition per liter:

KNO_3 .... 10.0g
$MgSO_4·6H_2O$ 10.0g
$CaCl_2·2H_2O$ 2.0g

Preparation of Solution 1 (10x NMS Salts): Add components to 700.0mL distilled/deionized water. Mix thoroughly. Bring volume to 1.0L with distilled/deionized water. Mix thoroughly.

Solution 2 Fe EDTA:
Composition per liter:

Fe EDTA 3.8g

Preparation of Solution 2 Fe EDTA: Add Fe EDTA to distilled/deionized water and bring volume to 1.0L. Mix thoroughly.

Solution 3 Sodium Molybdate:
Composition per liter:

$Na_2MoO_4·4H_2O$ 0.26g

Preparation of Solution 3 Sodium Molybdate: Add $Na_2MoO_4·4H_2O$ to distilled/deionized water and bring volume to 1.0L. Mix thoroughly.

Trace Elements:
Composition per 100mL:

$CuSO_4·5H_2O$ 100.0mg
$FeSO_4·7H_2O$ 50.0mg
$ZnSO_4·7H_2O$ 40.0mg
EDTA disodium salt 25.0mg
$CoCl_2·6H_2O$ 5.0mg
$MnCl_2·4H_2O$ 2.0mg
H_3BO_3 .... 1.5mg
$NiCl_2·6H_2O$ 1.0mg

Preparation of Trace Elements: Add components to distilled/deionized water and bring volume to 100.0mL. Mix thoroughly.

Phosphate Buffer:
Composition per liter:

$Na_2HPO_4·12H_2O$ 71.6g
KH_2PO_4 .... 26.0g

Preparation of Phosphate Buffer: Add components to 800.0mL distilled/deionized water. Mix thoroughly. Adjust pH to 6.8. Bring volume to 1.0L. Mix thoroughly. Autoclave for 15 min at 15 psi pressure–121°C. Cool to 55°C.

Preparation of Medium: Add 100.0mL solution 1 to distilled/deionized water and bring volume to 1.0L. Mix thoroughly. Add 1.0mL of solution 3, 1.0mL of the trace elements, and 0.1mL of solution 2. Autoclave for 15 min at 15 psi pressure–121°C. Cool to 55°C. Aseptically add 10.0mL phosphate buffer. Mix thoroughly. Aseptically distribute to sterile tubes or bottles.

Use: For the cultivation of *Methylosarcina fibrata* and *Methylosarcina quisquillarum.*

Methylotrophic *Arthrobacter Hyphomicrobium* Medium (DSMZ Medium 939)

Composition per liter:

$Na_2HPO_4·2H_2O$ 7.9g
Dimethylsulfone 1.9g
KH_2PO_4 .... 1.5g
NH_4Cl 0.8g
$MgSO_4·7H_2O$ 0.1g
Trace elements solution 10.0mL

pH 7.2-7.5 at 25°C

Trace Elements Solution:
Composition per liter:

EDTA 50.0g
NaOH 9.0g
$CaCl_2·2H_2O$ 7.34g
$FeSO_4·7H_2O$ 5.0g
$MnCl_2·4H_2O$ 2.5g
$ZnSO_4·7H_2O$ 1.0g
$CoCl_2·6H_2O$ 0.5g
$NH_4(MoO_4)$ 0.5g
$CuSO_4·5H_2O$ 0.2g

Preparation of Trace Elements Solution: Add Na_2-EDTA to 400.0mL distilled/deionized water. Mix thoroughly. Add 9.0g NaOH. Mix thoroughly. Individually dissolve each of the other components in 40.0mL distilled/deionized water. Add each of the other dissolved components to the EDTA solution. and bring volume to 1.0L with distilled/deionized water. Mix thoroughly. Adjust the pH to 6.0 with 1*M* NaOH.

Preparation of Medium: Add components to distilled/deionized water and bring volume to 1.0L. Mix thoroughly. Distribute into tubes or flasks. Autoclave for 15 min at 15 psi pressure–121°C.

Use: For the cultivation of *Hyphomicrobium sulfonivorans.*

Methylpyridine Medium

Composition per 1002.0mL:

K_2HPO_4 .... 0.61g
KH_2PO_4 .... 0.39g
KCl 0.25g
Yeast extract 0.1g
Wolfe's mineral solution 10.0mL
2-Methylpyridine 1.0mL

Wolfe's Mineral Solution:
Composition per liter:

$MgSO_4·7H_2O$ 3.0g
Nitrilotriacetic acid 1.5g
NaCl 1.0g
$MnSO_4·2H_2O$ 0.5g
$CoCl_2·6H_2O$ 0.1g
$ZnSO_4·7H_2O$ 0.1g
$CaCl_2·2H_2O$ 0.1g
$FeSO_4·7H_2O$ 0.1g
$NiCl_2·6H_2O$ 0.025g
$KAl(SO_4)_2·12H_2O$ 0.02g
$CuSO_4·5H_2O$ 0.01g
H_3BO_3 .... 0.01g
$Na_2MoO_4·2H_2O$ 0.01g
$Na_2SeO_3·5H_2O$ 0.3mg

Preparation of Wolfe's Mineral Solution: Add nitrilotriacetic acid to 500.0mL of distilled/deionized water. Adjust pH to 6.5 with KOH. Add remaining components one at a time. Add distilled/deionized water to 1.0L. Adjust pH to 6.8.

Preparation of Medium: Add components, except 2-methylpyridine, to distilled/deionized water and bring volume to 1.0L. Mix thoroughly. Autoclave for 15 min at 15 psi pressure–121°C. Cool to room temperature. In a fume hood, aseptically add 1.0mL of 2-methylpyridine. Mix thoroughly. Aseptically distribute into sterile tubes or flasks. Use polyurethane foam closures to eliminate odors caused by volatilization of 2-methylpyridine.

Use: For the cultivation of *Arthrobacter* species.

MG Medium

Composition per liter:

NaCl	100.0g
$MgSO_4 \cdot 7H_2O$	3.45g
$MgCl_2 \cdot 6H_2O$	2.75g
Sodium acetate	1.0g
KCl	0.335g
NH_4Cl	0.25g
$CaCl_2 \cdot 2H_2O$	0.14g
$K_2HPO_4 \cdot 3H_2O$	0.14g
Resazurin	1.0mg
$NaHCO_3$ solution	80.0mL
Trimethylamine·HCl solution	20.0mL
Na_2CO_3 solution	10.0mL
Trace elements solution	10.0mL
Vitamin solution	10.0mL
L-Cysteine·HCl solution	10.0mL
$Na_2S \cdot 9H_2O$ solution	10.0mL

pH 6.9 ± 0.2 at 25°C

$NaHCO_3$ Solution:
Composition per 100.0mL:

$NaHCO_3$	5.0g

Preparation of $NaHCO_3$ Solution: Add $NaHCO_3$ to distilled/deionized water and bring volume to 100.0mL. Mix thoroughly. Autoclave for 15 min at 15 psi pressure–121°C.

Trimethylamine·HCl Solution:
Composition per 20.0mL:

Trimethylamine·HCl	5.0g

Preparation of Trimethylamine·HCl Solution: Add trimethylamine·HCl to distilled/deionized water and bring volume to 20.0mL. Mix thoroughly. Sparge with 100% N_2. Autoclave for 15 min at 15 psi pressure–121°C.

Na_2CO_3 Solution:
Composition per 10.0mL:

Na_2CO_3	0.5g

Preparation of Na_2CO_3 Solution: Add Na_2CO_3 to distilled/deionized water and bring volume to 10.0mL. Mix thoroughly. Sparge with 100% N_2. Autoclave for 15 min at 15 psi pressure–121°C.

Trace Elements Solution:
Composition per liter:

$MgSO_4 \cdot 7H_2O$	3.0g
Nitrilotriacetic acid	1.5g
NaCl	1.0g
$MnSO_4 \cdot 2H_2O$	0.5g
$CoSO_4 \cdot 7H_2O$	0.18g
$ZnSO_4 \cdot 7H_2O$	0.18g
$CaCl_2 \cdot 2H_2O$	0.1g
$FeSO_4 \cdot 7H_2O$	0.1g
$NiCl_2 \cdot 6H_2O$	0.025g
$KAl(SO_4)_2 \cdot 12H_2O$	0.02g
$CuSO_4 \cdot 5H_2O$	0.01g
H_3BO_3	0.01g
$Na_2MoO_4 \cdot 2H_2O$	0.01g
$Na_2SeO_3 \cdot 5H_2O$	0.3mg

Preparation of Trace Elements Solution: Add nitrilotriacetic acid to approximately 500.0mL distilled/deionized water. Dissolve by adding KOH and adjust pH to 6.5. Add remaining components. Bring volume to 1.0L with additional distilled/deionized water. Adjust pH to 7.0 with KOH.

Vitamin Solution:
Composition per liter:

Pyridoxine·HCl	10.0mg
Calcium DL-pantothenate	5.0mg
Lipoic acid	5.0mg
Nicotinic acid	5.0mg
p-Aminobenzoic acid	5.0mg
Riboflavin	5.0mg
Thiamine·HCl	5.0mg
Biotin	2.0mg
Folic acid	2.0mg
Vitamin B_{12}	0.1mg

Preparation of Vitamin Solution: Add components to distilled/deionized water and bring volume to 1.0L. Mix thoroughly. Filter sterilize. Sparge with 100% N_2.

L-Cysteine·HCl Solution:
Composition per 10.0mL:

L-Cysteine·HCl	0.5g

Preparation of L-Cysteine·HCl Solution: Add L-cysteine·HCl to distilled/deionized water and bring volume to 10.0mL. Mix thoroughly. Autoclave under 100% N_2 for 15 min at 15 psi pressure–121°C.

$Na_2S \cdot 9H_2O$ Solution:
Composition per 10.0mL:

$Na_2S \cdot 9H_2O$	0.5g

Preparation of $Na_2S \cdot 9H_2O$ Solution: Add $Na_2S \cdot 9H_2O$ to distilled/deionized water and bring volume to 10.0mL. Mix thoroughly. Sparge with 100% N_2. Autoclave for 15 min at 15 psi pressure–121°C.

Preparation of Medium: Add components, except $NaHCO_3$ solution, trimethylamine·HCl solution, Na_2CO_3 solution, vitamin solution, L-cysteine·HCl solution, and $Na_2S \cdot 9H_2O$ solution, to distilled/deionized water and bring volume to 860.0mL. Mix thoroughly. Sparge with 100% N_2 for 20 min. Then sparge with 80% N_2 + 20% CO_2 for 10 min. Anaerobically distribute into tubes or bottles. Autoclave for 15 min at 15 psi pressure–121°C. Aseptically and anaerobically add 80.0mL of sterile $NaHCO_3$ solution, 20.0mL of sterile trimethylamine·HCl solution, 10.0mL of sterile Na_2CO_3 solution, 10.0mL of sterile vitamin solution, 10.0mL of sterile L-cysteine·HCl solution, and 10.0mL of sterile $Na_2S \cdot 9H_2O$ solution. Mix thoroughly.

Use: For the cultivation and maintenance of *Methanohalobium* species, *Methanohalophilus halophilus*, and *Methanohalophilus* species.

MG Medium

Composition per liter:

NaCl	150.0g
$MgSO_4 \cdot 7H_2O$	3.45g
$MgCl_2 \cdot 6H_2O$	2.75g
Sodium acetate	1.0g
KCl	0.335g
NH_4Cl	0.25g
$CaCl_2 \cdot 2H_2O$	0.14g
$K_2HPO_4 \cdot 3H_2O$	0.14g
Resazurin	1.0mg
$NaHCO_3$ solution	80.0mL
Trimethylamine·HCl solution	20.0mL
Na_2CO_3 solution	10.0mL
Trace elements solution	10.0mL
Vitamin solution	10.0mL

L-Cysteine·HCl solution 10.0mL
$Na_2S \cdot 9H_2O$ solution 10.0mL

pH 6.9 ± 0.2 at 25°C

$NaHCO_3$ Solution:

Composition per 100.0mL:

$NaHCO_3$ 5.0g

Preparation of $NaHCO_3$ Solution: Add $NaHCO_3$ to distilled/deionized water and bring volume to 100.0mL. Mix thoroughly. Autoclave for 15 min at 15 psi pressure–121°C.

Trimethylamine·HCl Solution:

Composition per 20.0mL:

Trimethylamine·HCl 5.0g

Preparation of Trimethylamine·HCl Solution: Add trimethylamine·HCl to distilled/deionized water and bring volume to 20.0mL. Mix thoroughly. Sparge with 100% N_2. Autoclave for 15 min at 15 psi pressure–121°C.

Na_2CO_3 Solution:

Composition per 10.0mL:

Na_2CO_3 .... 0.5g

Preparation of Na_2CO_3 Solution: Add Na_2CO_3 to distilled/deionized water and bring volume to 10.0mL. Mix thoroughly. Sparge with 100% N_2. Autoclave for 15 min at 15 psi pressure–121°C.

Trace Elements Solution:

Composition per liter:

$MgSO_4 \cdot 7H_2O$ 3.0g
Nitrilotriacetic acid 1.5g
NaCl 1.0g
$MnSO_4 \cdot 2H_2O$ 0.5g
$CoSO_4 \cdot 7H_2O$ 0.18g
$ZnSO_4 \cdot 7H_2O$ 0.18g
$CaCl_2 \cdot 2H_2O$ 0.1g
$FeSO_4 \cdot 7H_2O$ 0.1g
$NiCl_2 \cdot 6H_2O$ 0.025g
$KAl(SO_4)_2 \cdot 12H_2O$ 0.02g
$CuSO_4 \cdot 5H_2O$ 0.01g
H_3BO_3 .... 0.01g
$Na_2MoO_4 \cdot 2H_2O$ 0.01g
$Na_2SeO_3 \cdot 5H_2O$ 0.3mg

Preparation of Trace Elements Solution: Add nitrilotriacetic acid to approximately 500.0mL of distilled/deionized water. Dissolve by adding KOH and adjust pH to 6.5. Add remaining components. Bring volume to 1.0L with additional distilled/deionized water. Adjust pH to 7.0 with KOH.

Vitamin Solution:

Composition per liter:

Pyridoxine·HCl 10.0mg
Calcium DL-pantothenate 5.0mg
Lipoic acid 5.0mg
Nicotinic acid 5.0mg
p-Aminobenzoic acid 5.0mg
Riboflavin 5.0mg
Thiamine·HCl 5.0mg
Biotin 2.0mg
Folic acid 2.0mg
Vitamin B_{12} 0.1mg

Preparation of Vitamin Solution: Add components to distilled/deionized water and bring volume to 1.0L. Mix thoroughly. Filter sterilize. Sparge with 100% N_2.

L-Cysteine·HCl Solution:

Composition per 10.0mL:

L-Cysteine·HCl 0.5g

Preparation of L-Cysteine·HCl Solution: Add L-cysteine·HCl to distilled/deionized water and bring volume to 10.0mL. Mix thoroughly. Autoclave under 100% N_2 for 15 min at 15 psi pressure–121°C.

$Na_2S \cdot 9H_2O$ Solution:

Composition per 10.0mL:

$Na_2S \cdot 9H_2O$ 0.5g

Preparation of $Na_2S \cdot 9H_2O$ Solution: Add $Na_2S \cdot 9H_2O$ to distilled/deionized water and bring volume to 10.0mL. Mix thoroughly. Sparge with 100% N_2. Autoclave for 15 min at 15 psi pressure–121°C.

Preparation of Medium: Add components, except $NaHCO_3$ solution, trimethylamine·HCl solution, Na_2CO_3 solution, vitamin solution, L-cysteine·HCl solution, and $Na_2S \cdot 9H_2O$ solution, to distilled/deionized water and bring volume to 860.0mL. Mix thoroughly. Sparge with 100% N_2 for 20 min. Then sparge with 80% N_2 + 20% CO_2 for 10 min. Anaerobically distribute into tubes or bottles. Autoclave for 15 min at 15 psi pressure–121°C. Aseptically and anaerobically add 80.0mL of sterile $NaHCO_3$ solution, 20.0mL of sterile trimethylamine·HCl solution, 10.0mL of sterile Na_2CO_3 solution, 10.0mL of sterile vitamin solution, 10.0mL of sterile L-cysteine·HCl solution, and 10.0mL of sterile $Na_2S \cdot 9H_2O$ solution. Mix thoroughly.

Use: For the cultivation and maintenance of *Methanohalophilus* species.

MH IH Agar

Composition per liter:

Solution A 490.0mL
Solution B 490.0mL
Supplement solution 20.0mL

pH 6.9 ± 0.2 at 25°C

Solution A:

Composition per 490.0mL:

Beef infusion 300.0g
Acid hydrolysate of casein 17.5g
Agar 17.0g
Starch 1.5g

Preparation of Solution A: Add components to distilled/deionized water and bring volume to 490.0mL. Mix thoroughly. Gently heat and bring to boiling. Autoclave for 15 min at 15 psi pressure–121°C. Cool to 45°–50°C.

Solution B:

Composition per 490.0mL:

Hemoglobin 10.0g

Preparation of Solution B: Add hemoglobin to distilled/deionized water and bring volume to 490.0mL. Mix thoroughly. Gently heat and bring to boiling. Autoclave for 15 min at 15 psi pressure–121°C. Cool to 45°–50°C.

Supplement Solution:

Composition per liter:

Glucose 100.0g
L-Cysteine·HCl 25.9g
L-Glutamine 10.0g
L-Cystine 1.1g
Adenine 1.0g
Nicotinamide adenine dinucleotide 0.25g
Vitamin B_{12} 0.1g
Thiamine pyrophosphate 0.1g

Guanine·HCl 0.03g
$Fe(NO_3)_3 \cdot 6H_2O$ 0.02g
p-Aminobenzoic acid 0.013g
Thiamine·HCl 3.0mg

Source: The supplement solution IsoVitaleX enrichment is available from BD Diagnostic Systems. This enrichment may be replaced by supplement VX from BD Diagnostic Systems.

Preparation of Supplement Solution: Add components to distilled/deionized water and bring volume to 1.0L. Mix thoroughly. Filter sterilize.

Preparation of Medium: Aseptically combine cooled, sterile solution A and cooled, sterile solution B. Mix thoroughly. Adjust pH to 6.9 with sterile 1*N* HCl or sterile 1*N* KOH. Aseptically add 20.0mL of sterile supplement solution. Pour into sterile Petri dishes or distribute into sterile tubes.

Use: For the cultivation and differentiation of *Legionella* species.

MH Medium

Composition per liter:
NaCl 60.7g
Agar 20.0g
$MgCl_2 \cdot 6H_2O$ 15.0g
Yeast extract 10.0g
$MgSO_4 \cdot 7H_2O$ 7.4g
Proteose peptone No. 3 .... 5.0g
KCl 1.5g
Glucose 1.0g
$CaCl_2$ 0.27g
$NaHCO_3$ 0.45g
NaBr 0.19g

Preparation of Medium: Add components to distilled/deionized water and bring volume to 1.0L. Mix thoroughly. Gently heat and bring to boiling. Distribute into tubes or flasks. Autoclave for 15 min at 15 psi pressure–121°C. Pour into sterile Petri dishes or leave in tubes.

Use: For the cultivation and maintenance of *Deleya salina* and *Volcaniella eurihalina*.

MH Medium

Composition per liter:
NaCl 60.7g
$MgCl_2 \cdot 6H_2O$ 15.0g
Yeast extract 10.0g
$MgSO_4 \cdot 7H_2O$ 7.4g
Proteose peptone No. 3 .... 5.0g
KCl 1.5g
Glucose 1.0g
$CaCl_2$ 0.27g
$NaHCO_3$ 0.045g
NaBr 0.019g
pH 7.2 ± 0.2 at 25°C

Preparation of Medium: Add components to distilled/deionized water and bring volume to 1.0L. Mix thoroughly. Adjust pH to 7.2. Distribute into tubes or flasks. Autoclave for 15 min at 15 psi pressure–121°C.

Use: For the cultivation of *Halomonas eurihalina*.

MH Medium, 2%

Composition per liter:
KH_2PO_4 .... 2.0g
$(NH_4)_2SO_4$ 2.0g
NaCl 0.5g
$MgSO_4 \cdot 7H_2O$ 0.025g
$FeSO_4 \cdot 7H_2O$ 2.0mg
Glucose solution 20.0mL
Methanol, filter sterilized 20.0mL
pH 7.0–7.5 at 25°C

Glucose Solution:
Composition per 20.0mL:
Glucose 1.5g

Preparation of Glucose Solution: Add glucose to distilled/deionized water and bring volume to 20.0mL. Mix thoroughly. Filter sterilize.

Preparation of Medium: Add components, except glucose solution and methanol, to distilled/deionized water and bring volume to 960.0mL. Mix thoroughly. Autoclave for 15 min at 15 psi pressure–121°C. Aseptically add 20.0mL of sterile glucose solution and 20.0mL of filter-sterilized methanol. Mix thoroughly. Aseptically distribute into sterile tubes or flasks.

Use: For the cultivation of *Methylobacillus fructoseoxidans* and *Methylophilus glucoseoxidans*.

MH Medium, 10% (LMG Medium 270)

Composition per liter:
NaCl 81.0g
Agar 15.0g
Yeast extract 10.0g
$MgSO_4 \cdot 7H_2O$ 9.6g
$MgCl_2 \cdot 6H_2O$ 7.0g
Proteose peptone no.3 .... 5.0g
KCl 2.0g
Glucose 1.0g
$CaCl_2$ 0.54g
NaBr 26.0mg
$NaHCO_3$ solution 10.0mL
pH 7.5 ± 0.2 at 25°C

$NaHCO_3$ Solution:
Composition per 10.0mL:
$NaHCO_3$ 0.06g

Preparation of $NaHCO_3$ Solution: Add $NaHCO_3$ to 10.0mL of distilled/deionized water. Mix thoroughly. Filter sterilize.

Preparation of Medium: Add components except $NaHCO_3$ solution, to distilled/deionized water and bring volume to 990.0mL. Mix thoroughly. Autoclave for 15 min at 15 psi pressure–121°C. Cool to 45°C. Aseptically add 10.0mL sterile $NaHCO_3$ solution. Pour into sterile Petri dishes or aseptically distribute into sterile tubes.

Use: For cultivation and maintenance of *Chromohalobacter israelensis* and *Chromohalobacter canadensis*.

MH Medium, 10%

Composition per liter:
NaCl 81.0g
Yeast extract 10.0g
$MgSO_4 \cdot 7H_2O$ 9.6g
$MgCl_2 \cdot 6H_2O$ 7.0g
Proteose peptone No. 3 .... 5.0g
KCl 2.0g
Glucose 1.0g
$CaCl_2$ 0.36g
NaBr 0.026g
$NaHCO_3$ solution 10.0mL
pH 7.5 ± 0.2 at 25°C

$NaHCO_3$ Solution:
Composition per 10.0mL:

$NaHCO_3$	0.06g

Preparation of $NaHCO_3$ Solution: Add $NaHCO_3$ to distilled/deionized water and bring volume to 10.0mL. Mix thoroughly. Filter sterilize.

Preparation of Medium: Add components, except $NaHCO_3$ solution, to distilled/deionized water and bring volume to 990.0mL. Mix thoroughly. Autoclave for 15 min at 15 psi pressure–121°C. Aseptically add 10.0mL of sterile $NaHCO_3$ solution. Mix thoroughly. Aseptically distribute into sterile tubes or flasks.

Use: For the cultivation of *Zygomonas mobilis, Salinicoccus hispanicus, Salinicoccus roseus,* and *Pseudomonas beijerinckii.*

MH Medium, 15% (LMG Medium 258)

Composition per liter:

NaCl	121.5g
$MgSO_4·7H_2O$	14.4g
$MgCl_2$	10.5g
Yeast extract	10.0g
Proteose peptone no.3	5.0g
KCl	3.0g
Glucose	1.0g
$CaCl_2$	0.54g
NaBr	39.0mg
$NaHCO_3$ solution	10.0mL

pH 7.5 ± 0.2 at 25°C

$NaHCO_3$ Solution:
Composition per 10.0mL:

$NaHCO_3$	0.09g

Preparation of $NaHCO_3$ Solution: Add $NaHCO_3$ to 10.0mL of distilled/deionized water. Mix thoroughly. Filter sterilize.

Preparation of Medium: Add components, except $NaHCO_3$ solution, to distilled/deionized water and bring volume to 990.0mL. Mix thoroughly. Autoclave for 15 min at 15 psi pressure–121°C. Aseptically add 10.0mL sterile $NaHCO_3$ solution. Aseptically distribute into sterile tubes or flasks.

Use: For cultivation and maintenance of *Bacillus halophilus.*

Micrococcus/Sarcina Medium

Composition per liter:

Agar	16.0g
Pancreatic digest of casein	5.0g
Sodium succinate·$6H_2O$	2.0g
Starch	2.0g
Yeast autolysate	1.0g
Sodium citrate·$2H_2O$	0.5g
Sodium acetate·$3H_2O$	0.3g
K_2HPO_4	0.2g

pH 7.0 ± 0.2 at 25°C

Preparation of Medium: Add components to distilled/deionized water and bring volume to 1.0L. Mix thoroughly. Gently heat and bring to boiling. Distribute into tubes or flasks. Autoclave for 15 min at 15 psi pressure–121°C. Pour into sterile Petri dishes or leave in tubes.

Use: For the cultivation and maintenance of *Micrococcus luteus* and *Sarcina* species.

Microcyclus eburneus Medium

Composition per liter:

K_2HPO_4	7.0g
$(NH_4)SO_4$	3.0g
KH_2PO_4	2.0g
$MgSO_4·7H_2O$	0.5g
Yeast extract	0.2g
Thiamine·HCl	0.2mg
Biotin	0.02mg
$FeSO_4·7H_2O$	2.0μg
$MnSO_4·4H_2O$	2.0μg

Preparation of Medium: Add components to distilled/deionized water and bring volume to 1.0L. Mix thoroughly. Distribute into tubes or flasks. Autoclave for 15 min at 15 psi pressure–121°C.

Use: For the cultivation of *Microcyclus eburneus.*

Microcyclus major Medium

Composition per liter:

Glucose	1.0g
Peptone	1.0g
KNO_3	0.1g
K_2HPO_4	0.07g
$MgSO_4·7H_2O$	0.03g
Trace elements solution	1.0mL

Trace Elements Solution:
Composition per liter:

Disodium EDTA	10.0g
$FeSO_4·7H_20$	9.3g
$NaBO_3·4H_2O$	2.6g
$MnCl_2·4H_2O$	1.8g
$CaCl_2$	1.2g
$(NH_4)_6Mo_7O_{24}·4H_2O$	1.0g
$ZnSO_4·7H_2O$	0.2g
$CuSO_4·5H_2O$	0.08g
$Co(NO_3)_2·H_2O$	0.02g

Preparation of Trace Elements Solution: Add components to distilled/deionized water and bring volume to 1.0L. Mix thoroughly.

Preparation of Medium: Add components to distilled/deionized water and bring volume to 1.0L. Mix thoroughly. Distribute into tubes or flasks. Autoclave for 15 min at 15 psi pressure–121°C.

Use: For the cultivation of *Microcyclus major.*

Microcyclus marinus Medium

Composition per liter:

NaCl	23.5g
$MgCl_2$	5.0g
Na_2SO_4	4.0g
$CaCl_2·2H_2O$	1.5g
KCl	0.7g
$NaHCO_3$	0.2g

Preparation of Medium: Add components to distilled/deionized water and bring volume to 1.0L. Mix thoroughly. Distribute into tubes or flasks. Autoclave for 15 min at 15 psi pressure–121°C.

Use: For the cultivation of *Microcyclus marinus.*

Microcyclus Medium

Composition per liter:

Agar	15.0g
Glucose	5.0g

Peptone...5.0g
Yeast extract...5.0g
pH 6.8 ± 0.2 at 25°C

Preparation of Medium: Add components to distilled/deionized water and bring volume to 1.0L. Mix thoroughly. Gently heat and bring to boiling. Distribute into tubes or flasks. Autoclave for 15 min at 15 psi pressure–121°C. Pour into sterile Petri dishes or leave in tubes.

Use: For the cultivation and maintenance of *Flectobacillus major* and *Microcyclus* species.

Microcyclus/Spirosoma Medium

Composition per liter:
Agar..15.0g
Glucose...1.0g
Peptone...1.0g
Yeast extract...1.0g
pH 6.8–7.0 at 25°C

Preparation of Medium: Add components to distilled/deionized water and bring volume to 1.0L. Mix thoroughly. Gently heat and bring to boiling. Distribute into tubes or flasks. Autoclave for 15 min at 15 psi pressure–121°C. Pour into sterile Petri dishes or leave in tubes.

Use: For the cultivation and maintenance of *Spirosoma linguale* and *Microcyclus* species.

Microvirgula Medium (DSMZ Medium 957)

Composition per liter:
Na-succinate...1.5g
KNO_3...1.5g
$(NH_4)_2SO_4$...1.0g
KH_2PO_4..0.82g
K_2HPO_4..0.7g
$MgSO_4 \cdot 7H_2O$...0.5g
Yeast extract...0.25g
pH 7.0 ± 0.2 at 25°C

Preparation of Medium: Add components to distilled/deionized water and bring volume to 1.0L. Mix thoroughly. Sparge with 100% N_2. Distribute into tubes or bottles. Autoclave for 15 min at 15 psi pressure–121°C. The final pH should be 7.0.

Use: For the cultivation of *Microvirgula aerodenitrificans*.

MIL Medium (Motility Indole Lysine Medium)

Composition per liter:
Peptone...10.0g
Pancreatic digest of casein.........................10.0g
L-Lysine·HCl..10.0g
Yeast extract...3.0g
Agar..2.0g
Dextrose...1.0g
Ferric ammonium citrate..............................0.5g
Bromcresol Purple.......................................0.02g
pH 6.6 ± 0.2 at 25°C

Source: This medium is available as a premixed powder and prepared medium from BD Diagnostic Systems.

Preparation of Medium: Add components to distilled/deionized water and bring volume to 1.0L. Mix thoroughly. Gently heat and bring to boiling. Distribute into tubes in 5.0mL volumes. Autoclave for 15 min at 15 psi pressure–121°C.

Use: For the cultivation and differentiation of members of the Enterobacteriaceae on the basis of motility, lysine decarboxylase activity, lysine deaminase activity, and indole production.

Milk Agar

Composition per liter:
Mixture A..500.0mL
Mixture B..500.0mL

Mixture A:

Composition per 500.0mL:
Instant nonfat milk.......................................100.0g

Preparation of Mixture A: Add instant nonfat milk to distilled/deionized water and bring volume to 500.0mL. Mix thoroughly. Autoclave for 15 min at 15 psi pressure–121°C. Cool rapidly to 55°C.

Mixture B:

Composition per 500.0mL:
Agar..15.0g
Nutrient broth..12.5g
NaCl..2.5g

Preparation of Mixture B: Add components to distilled/deionized water and bring volume to 500.0mL. Mix thoroughly. Gently heat and bring to boiling. Autoclave for 15 min at 15 psi pressure–121°C. Cool rapidly to 55°C.

Preparation of Medium: Aseptically combine cooled, sterile mixture A with cooled, sterile mixture B. Mix thoroughly. Pour into sterile Petri dishes in 20.0mL volumes.

Use: For the cultivation and estimation of the numbers of *Pseudomonas aeruginosa* in water by the membrane filter method.

Mineral Agar

Composition per liter:
NH_4Cl...0.5g
$Na_2HPO_4 \cdot 7H_2O$....................................670.0mg
KH_2PO_4..340.0mg
$MgSO_4 \cdot 7H_2O$...112.0mg
$CaCl_2$..14.0mg
$ZnSO_4 \cdot 7H_2O$..5.0mg
$Na_2MoO_4 \cdot 2H_2O$.....................................2.5mg
$FeCl_3$..0.13mg
1,4-Dichlorobenzene....................................variable
pH 7.0 ± 0.2 at 25°C

Preparation of Medium: Add components, except 1,4-dichlorobenzene, to distilled/deionized water and bring volume to 1.0L. Mix thoroughly. Gently heat and bring to boiling. Distribute into tubes or flasks. Autoclave for 15 min at 15 psi pressure–121°C. Pour into sterile Petri dishes or leave in tubes. After inoculation, place Petri dishes or tubes into a desiccator. Add a few crystals of 1,4-dichlorobenzene to the desiccator.

Use: For the cultivation of dichlorobenzene-degrading *Pseudomonas* species.

Mineral Base E for Autotrophic Growth

Composition per liter:
Noble agar...15.0g
K_2HPO_4..1.2g
KH_2PO_4..0.624g
$(NH_4)_2SO_4$...0.5g
NaCl..0.1g
$CaCl_2 \cdot 6H_2O$ solution...............................10.0mL
$MgSO_4 \cdot 7H_2O$ solution.............................10.0mL

Mineral solution 1.0mL
p-Aminobenzoic acid solution 1.0mL

$CaCl_2 \cdot 6H_2O$ Solution:
Composition per liter:
$CaCl_2 \cdot 6H_2O$ 5.0g

Preparation of $CaCl_2 \cdot 6H_2O$ Solution: Add 5.0g of $CaCl_2 \cdot 6H_2O$ to distilled/deionized water and bring volume to 1.0L. Mix thoroughly. Autoclave for 15 min at 15 psi pressure–121°C.

$MgSO_4 \cdot 7H_2O$ Solution:
Composition per liter:
$MgSO_4 \cdot 7H_2O$ 20.0g

Preparation of $MgSO_4 \cdot 7H_2O$ Solution: Add 20.0g of $MgSO_4 \cdot 7H_2O$ to distilled/deionized water and bring volume to 1.0L. Mix thoroughly. Autoclave for 15 min at 15 psi pressure–121°C.

***p*-Aminobenzoic Acid Solution:**
Composition per 10.0.mL:
p-Aminobenzoic acid 100.0mg

Preparation of *p*-Aminobenzoic Acid Solution: Add *p*-aminobenzoic acid to distilled/deionized water and bring volume to 10.0mL. Mix thoroughly. Autoclave for 15 min at 15 psi pressure–121°C.

Mineral Solution:
Composition per 1000.0mL:
Disodium EDTA 1.58g
ZnSO4·7H2O 0.7g
$MnSO_4 \cdot 4H_2O$ 0.18g
$FeSO_4 \cdot 7H_2O$ 0.16g
$CoCl_2 \cdot 6H_2O$ 0.052g
$Na_2MoO_4 \cdot 2H_2O$ 0.047g
$CuSO_4 \cdot 5H_2O$ 0.047g

Preparation of Medium: Add components, except $CaCl_2 \cdot 6H_2O$ solution, $MgSO_4 \cdot 7H_2O$ solution, and *p*-aminobenzoic acid solution, to distilled/deionized water and bring volume to 979.0mL. Mix thoroughly. Autoclave for 15 min at 15 psi pressure–121°C. Cool to 50°C. Aseptically add in the following order: 10.0mL of sterile $CaCl_2 \cdot 6H_2O$ solution, 10.0mL of sterile $MgSO_4 \cdot 7H_2O$ solution, and 1.0mL of sterile *p*-aminobenzoic acid solution. Mix thoroughly. Aseptically distribute into sterile tubes or flasks. Incubate inoculated tubes in 50% CO_2.

Use: For the autotrophic cultivation and maintenance of *Pseudomonas thermocarboxydovorans*.

Mineral Base E for Heterotrophic Growth

Composition per liter:
Noble agar 15.0g
K_2HPO_4 .... 1.2g
KH_2PO_4 .... 0.624g
$(NH_4)_2SO_4$ 0.5g
NaCl 0.1g
$CaCl_2 \cdot 6H_2O$ solution 10.0mL
$MgSO_4 \cdot 7H_2O$ solution 10.0mL
Sodium pyruvate solution 10.0mL
Mineral solution 1.0mL
p-Aminobenzoic acid solution 1.0mL

$CaCl_2 \cdot 6H_2O$ Solution:
Composition per liter:
$CaCl_2 \cdot 6H_2O$ 5.0g

Preparation of $CaCl_2 \cdot 6H_2O$ Solution: Add 5.0g of $CaCl_2 \cdot 6H_2O$ to distilled/deionized water and bring volume to 1.0L. Mix thoroughly. Autoclave for 15 min at 15 psi pressure–121°C.

$MgSO_4 \cdot 7H_2O$ Solution:
Composition per liter:
$MgSO_4 \cdot 7H_2O$ 20.0g

Preparation of $MgSO_4 \cdot 7H_2O$ Solution: Add 20.0g of $MgSO_4 \cdot 7H_2O$ to distilled/deionized water and bring volume to 1.0L. Mix thoroughly. Autoclave for 15 min at 15 psi pressure–121°C.

Sodium Pyruvate Solution:
Composition per 10.0mL:
Sodium pyruvate 2.0g

Preparation of Sodium Pyruvate Solution: Add sodium pyruvate to distilled/deionized water and bring volume to 10.0mL. Mix thoroughly. Filter sterilize.

***p*-Aminobenzoic Acid Solution:**
Composition per 10.0mL:
p-Aminobenzoic acid 100.0mg

Preparation of *p*-Aminobenzoic Acid Solution: Add *p*-aminobenzoic acid to distilled/deionized water and bring volume to 10.0mL. Mix thoroughly. Autoclave for 15 min at 15 psi pressure–121°C.

Mineral Solution:
Composition per 100.0mL:
Disodium EDTA 1.58g
ZnSO4·7H2O 0.7g
$MnSO_4 \cdot 4H_2O$ 0.18g
$FeSO_4 \cdot 7H_2O$ 0.16g
$CoCl_2 \cdot 6H_2O$ 0.052g
$Na_2MoO_4 \cdot 2H_2O$ 0.047g
$CuSO_4 \cdot 5H_2O$ 0.047g

Preparation of Medium: Add components, except $CaCl_2 \cdot 6H_2O$ solution, $MgSO_4 \cdot 7H_2O$ solution, sodium pyruvate solution, and *p*-aminobenzoic acid solution to distilled/deionized water and bring volume to 969.0mL. Mix thoroughly. Autoclave for 15 min at 15 psi pressure–121°C. Cool to 45°–50°C. Aseptically add in the following order: 10.0mL of the sterile $CaCl_2 \cdot 6H_2O$ solution, 10.0mL of the sterile $MgSO_4 \cdot 7H_2O$ solution, 10.0mL of sterile sodium pyruvate solution, and 1.0mL of sterile *p*-aminobenzoic acid solution. Mix thoroughly. Aseptically distribute into sterile tubes or flasks.

Use: For the heterotrophic cultivation and maintenance of *Pseudomonas thermocarboxydovorans*.

Mineral Lactate Medium

Composition per liter:
$K_2HPO_4 \cdot 3H_2O$ 1.13g
NaCl 1.0g
NH_4Cl 1.0g
KH_2PO_4 .... 0.88g
$MgSO_4 \cdot 7H_2O$ 0.5g
Sodium lactate 0.5g
$CaCl_2 \cdot 2H_2O$ 5.0mg
Trace elements solution 1.2mL
pH 7.0 ± 0.2 at 25°C

Trace Elements Solution:
Composition per liter:
Disodium EDTA 50.0g
$ZnSO_4 \cdot 7H_2O$ 22.0g
$CaCl_2 \cdot 2H_2O$ 5.54g

$MnCl_2 \cdot 4H_2O$ 5.06g
$FeSO_4 \cdot 7H_2O$ 5.0g
$CoCl_2 \cdot 6H_2O$ 1.61g
$CuSO_4 \cdot 5H_2O$ 1.57g
$(NH_4)6Mo_7O_{24} \cdot 4H_2O$ 1.1g

Preparation of Trace Elements Solution: Add components to distilled/deionized water and bring volume to 1.0L. Mix thoroughly. Adjust pH to 7.0 with KOH.

Preparation of Medium: Add components to distilled/deionized water and bring volume to 1.0L. Mix thoroughly. Adjust pH to 7.0. Distribute into tubes or flasks. Autoclave for 15 min at 15 psi pressure–121°C.

Use: For the cultivation and maintenance of *Pseudomonas* species and *Spirillum* species.

Mineral Medium

Composition per liter:
Yeast extract 2.0g
Mineral base 5X 200.0mL
Trace elements solution SL-6 .... 1.0mL
Thiamine·HCl 3.0μg
Biotin 0.2μg
pH 6.8 ± 0.2 at 25°C

Mineral Base 5X:
Composition per liter:
NaCl 5.0g
NH_4Cl 2.0g
KH_2PO_4 .... 1.35g
$MgSO_4 \cdot 7H_2O$ 1.0g
K_2HPO_4 .... 0.87g
$CaCl_2$ 0.05g
$FeCl_3 \cdot 6H_2O$ 1.25mg

Preparation of Mineral Base 5X: Add components to distilled/deionized water and bring volume to 1.0L. Mix thoroughly.

Trace Elements Solution SL-6:
Composition per liter:
H_3BO_3 .... 0.3g
$CoCl_2 \cdot 6H_2O$ 0.2g
$ZnSO_4 \cdot 7H_2O$ 0.1g
$MnCl_2 \cdot 4H_2O$ 0.03g
$Na_2MoO_4 \cdot H_2O$ 0.03g
$NiCl_2 \cdot 6H_2O$ 0.02g
$CuCl_2 \cdot 2H_2O$ 0.01g

Preparation of Trace Elements Solution SL-6: Add components to distilled/deionized water and bring volume to 1.0L. Mix thoroughly. Adjust pH to 3.4.

Preparation of Medium: Add components to distilled/deionized water and bring volume to 1.0L. Mix thoroughly. Adjust pH to 6.8. Distribute into tubes or flasks. Autoclave for 15 min at 15 psi pressure–121°C.

Use: For the cultivation of *Arthrobacter* species.

Mineral Medium

Composition per liter:
NH_4Cl 0.5g
Yeast extract 0.2g
1,4-Dichlorobenzene 0.1g
$Na_2HPO_4 \cdot 7H_2O$ 670.0mg
KH_2PO_4 .... 340.0mg
$MgSO_4 \cdot 7H_2O$ 112.0mg
$CaCl_2$ 14.0mg
$ZnSO_4 \cdot 7H_2O$ 5.0mg
$Na_2MoO_4 \cdot 2H_2O$ 2.5mg
$FeCl_3$ 0.13mg
pH 7.0 ± 0.2 at 25°C

Preparation of Medium: Add components to distilled/deionized water and bring volume to 1.0L. Mix thoroughly. Distribute into tubes or flasks. Autoclave for 15 min at 15 psi pressure–121°C.

Use: For the cultivation of dichlorobenzene-degrading *Pseudomonas* species.

Mineral Medium with 3-Aminophenol (DSMZ Medium 465f)

Composition per liter:
$Na_2HPO_4 \cdot 2H_2O$ 3.5g
KH_2PO_4 .... 1.0g
$(NH_4)_2SO_4$ 0.5g
$MgCl_2 \cdot 6H_2O$ 0.1g
$Ca(NO_3)_2 \cdot 4H_2O$ 0.05g
Aminophenol solution 10.0mL
Trace elements solution SL-4 .... 1.0mL
pH 7.25 ± 0.2 at 25°C

Trace Elements Solution SL-4:
Composition per liter:
EDTA 0.5g
$FeSO_4 \cdot 7H_2O$ 0.2g
Trace elements solution SL-6 .... 100.0mL

Trace Elements Solution SL-6:
Composition per liter:
H_3BO_3 .... 0.3g
$CoCl_2 \cdot 6H_2O$ 0.2g
$ZnSO_4 \cdot 7H_2O$ 0.1g
$MnCl_2 \cdot 4H_2O$ 0.03g
$Na_2MoO_4 \cdot H_2O$ 0.03g
$NiCl_2 \cdot 6H_2O$ 0.02g
$CuCl_2 \cdot 2H_2O$ 0.01g

Preparation of Trace Elements Solution SL-6: Add components to distilled/deionized water and bring volume to 1.0L. Mix thoroughly. Adjust pH to 3.4.

Preparation of Trace Elements Solution SL-4: Add components to distilled/deionized water and bring volume to 1.0L. Mix thoroughly.

Aminophenol Solution:
Composition per 100.0mL:
3-Aminophenol 1.0g

Preparation of Aminophenol Solution: Add 100.0mL boiling water to 1.0g aminophenol crystals. Stir the solution to mix thoroughly. Cool to room temperature. Sterilize by filtration.

Preparation of Medium: Add components, except aminophenol solution, to 990.0mL distilled/deionized water. Adjust pH to 7.25. Autoclave for 15 min at 15 psi pressure–121°C. Aseptically add 10.0mL aminophenol solution. Mix thorougly. Aseptically distribute to sterile tubes or flasks.

Use: For the cultivation of *Arthrobacter* sp. and other aminophenol utilizing bacteria.

Mineral Medium with Antipyrin

Composition per liter:
Antipyrin 1.0g
$Na_2HPO_4 \cdot 12H_2O$ 0.7g
$(NH_4)_2HPO_4$ 0.7g
KH_2PO_4 .... 0.3g
$(NH_4)H_2PO_4$ 0.3g
$MgSO_4 \cdot 7H_2O$ 0.25g

$(NH_4)_2SO_4$ 0.1g
$CaCl_2·6H_2O$ 0.05g
H_3BO_3 .... 0.5mg
$MnSO_4·4H_2O$ 0.4mg
$ZnSO_4·7H_2O$ 0.4mg
$FeCl_3·6H_2O$ 0.2mg
$(NH_4)_6Mo_7O_{24}·4H_2O$ 0.2mg
KI 0.1mg
$CuSO_4·5H_2O$ 0.04mg
Vitamin solution 20.0mL
pH 6.8–7.0 at 25°C

Vitamin Solution:
Composition per 20.0mL:
Biotin 0.1mg
Vitamin B_{12} 0.03mg

Preparation of Vitamin Solution: Add biotin and Vitamin B_{12} to 20.0mL of distilled/deionized water. Mix thoroughly. Filter sterilize.

Preparation of Medium: Add components, except vitamin solution, to distilled/deionized water and bring volume to 980.0mL. Mix thoroughly. Adjust pH to 6.8–7.0 with 1*N* NaOH. Autoclave for 20 min at 15 psi pressure–121°C. Cool to 45°–50°C. Aseptically add the sterile vitamin solution. Mix thoroughly. Distribute into sterile tubes or flasks.

Use: For the cultivation and maintenance of *Phenylobacterium immobile.*

Mineral Medium with Atrazine (DSMZ Medium 465i)

Composition per liter:
$Na_2HPO_4·2H_2O$ 3.5g
Na-citrate 1.0g
KH_2PO_4 .... 1.0g
$MgCl_2·6H_2O$ 0.1g
$CaCl_2$ 0.05g
Atrazine solution 10.0mL
Trace elements solution SL-4 .... 1.0mL
pH 7.25 ± 0.2 at 25°C

Trace Elements Solution SL-4:
Composition per liter:
EDTA 0.5g
$FeSO_4·7H_2O$ 0.2g
Trace elements solution SL-6 .... 100.0mL

Preparation of Trace Elements Solution SL-4: Add components to distilled/deionized water and bring volume to 1.0L. Mix thoroughly.

Trace Elements Solution SL-6:
Composition per liter:
H_3BO_3 .... 0.3g
$CoCl_2·6H_2O$ 0.2g
$ZnSO_4·7H_2O$ 0.1g
$MnCl_2·4H_2O$ 0.03g
$Na_2MoO_4·H_2O$ 0.03g
$NiCl_2·6H_2O$ 0.02g
$CuCl_2·2H_2O$ 0.01g

Preparation of Trace Elements Solution SL-6: Add components to distilled/deionized water and bring volume to 1.0L. Mix thoroughly. Adjust pH to 3.4.

Atrazine Solution:
Composition per 10.0mL:
Atrazine 100mg

Preparation of Atrazine Solution: Add (2-chloro-4(ethylamino)-6-(isopropylamino)-1,3,5-triazine) to 10.0mL methanol. Mix thoroughly. Shortly sonicate to reduce particle size.

Preparation of Medium: Add components, except atrazine solution, to 990.0mLL distilled/deionized water. Adjust pH to 7.25. Autoclave for 15 min at 15 psi pressure–121°C. Cool to room temperature. Aseptically add 10.0mL atrazine solution. Mix thoroughly. Aseptically distribute to sterile tubes or flasks.

Use: For the cultivation of *Pseudomonas* sp. and other atrazine-utilizing bacteria.

Mineral Medium with Benzylcyanide (DSMZ Medium 465d)

Composition per1liter:
$Na_2HPO_4·2H_2O$ 3.5g
KH_2PO_4 .... 1.0g
$MgCl_2·6H_2O$ 0.1g
$Ca(NO_3)_2·4H_2O$ 0.05g
Benzylcyanide solution 10.0mL
Glucose solution 10.0mL
Trace elements solution SL-4 .... 1.0mL
pH 7.25 ± 0.2 at 25°C

Trace Elements Solution SL-4:
Composition per liter:
EDTA 0.5g
$FeSO_4·7H_2O$ 0.2g
Trace elements solution SL-6 .... 100.0mL

Trace Elements Solution SL-6:
Composition per liter:
H_3BO_3 .... 0.3g
$CoCl_2·6H_2O$ 0.2g
$ZnSO_4·7H_2O$ 0.1g
$MnCl_2·4H_2O$ 0.03g
$Na_2MoO_4·H_2O$ 0.03g
$NiCl_2·6H_2O$ 0.02g
$CuCl_2·2H_2O$ 0.01g

Preparation of Trace Elements Solution SL-6: Add components to distilled/deionized water and bring volume to 1.0L. Mix thoroughly. Adjust pH to 3.4.

Preparation of Trace Elements Solution SL-4: Add components to distilled/deionized water and bring volume to 1.0L. Mix thoroughly.

Glucose Solution:
Composition per 10.0mL:
Glucose 1.8g

Preparation of Glucose Solution: Add glucose to 10.0mL distilled/deionized water. Mix thoroughly. Filter sterilize.

Benzylcyanide Solution:
Composition per 100mL:
Benzylcyanide 0.12g

Preparation of Benzylcyanide Solution: Add benzylcyanide to 10.0mL distilled/deionized water. Mix thoroughly. Do not sterilize.

Preparation of Medium: Add components, except benzylcyanide solution and glucose solution, to 980.0mL distilled/deionized water. Adjust pH to 7.25. Autoclave for 15 min at 15 psi pressure–121°C. Cool to room temperature Aseptically add 10.0mL glucose solution and 10.0mL benzylcyanide solution to the medium. Mix thoroughly. Aseptically distribute the medium to sterile tubes or flasks.

Use: For the cultivation of *Pseudomonas* sp., *Rhodococcus erythropolis,* and other benzylcyanide-utilizing bacteria.

Mineral Medium, Brunner

Composition per liter:

Na_2HPO_4	2.44g
KH_2PO_4	1.52g
$(NH_4)_2SO_4$	0.5g
$MgSO_4 \cdot 7H_2O$	0.2g
$CaCl_2 \cdot 2H_2O$	0.05g
Trace elements solution SL-4	10.0mL

pH 6.9 ± 0.2 at 25°C

Trace Elements Solution SL-4:
Composition per liter:

EDTA	0.5g
$FeSO_4 \cdot 7H_2O$	0.2g
Trace elements solution SL-6	100.0mL

Trace Elements Solution SL-6:
Composition per liter:

$MnCl_2 \cdot 4H_2O$	0.5g
H_3BO_3	0.3g
$CoCl_2 \cdot 6H_2O$	0.2g
$ZnSO_4 \cdot 7H_2O$	0.1g
$Na_2MoO_4 \cdot 2H_2O$	0.03g
$NiCl_2 \cdot 6H_2O$	0.02g
$CuCl_2 \cdot 2H_2O$	0.01g

Preparation of Trace Elements Solution SL-6: Add components to distilled/deionized water and bring volume to 1.0L. Mix thoroughly.

Preparation of Trace Elements Solution SL-4: Add components to distilled/deionized water and bring volume to 1.0L. Mix thoroughly.

Preparation of Medium: Add components to distilled/deionized water and bring volume to 1.0L. Mix thoroughly. Distribute into tubes or flasks. Autoclave for 15 min at 15 psi pressure–121°C.

Use: For the cultivation and maintenance of *Alcaligenes* species, *Bacillus benzoevorans*, *Bacillus gordonae*, *Comamonas acidovorans*, *Hyphomicrobium* species, *Moraxella* species, *Nocardia* species, *Pseudomonas* species, *Rhodococcus* species, *Sphingomonas* species, and *Xanthobacter* species.

Mineral Medium with Camphor

Composition per liter:

$Na_2HPO_4 \cdot 12H_2O$	9.0g
Ferric ammonium citrate	5.0g
$MnSO_4 \cdot H_2O$	3.0g
NH_4Cl	2.0g
KH_2PO_4	1.5g
$MgSO_4 \cdot 7H_2O$	0.2g
$ZnSO_4 \cdot 7H_2O$	0.2g
Titriplex I	10.0mg
$CoSO_4$	10.0μg
Camphor crumbs	variable

pH 7.0 ± 0.2 at 25°C

Preparation of Medium: Add components, except camphor crumbs, to distilled/deionized water and bring volume to 1.0L. Mix thoroughly. Distribute into tubes or flasks. Autoclave for 15 min at 15 psi pressure–121°C. Inoculate tubes or flasks and place in a dessicator jar in which crumbs of camphor will be evaporated.

Use: For the cultivation of bacteria that can utilize camphor as sole carbon source.

Mineral Medium with Chloridazon

Composition per liter:

Chloridazon	1.0g
$Na_2HPO_4 \cdot 12H_2O$	0.7g
$(NH_4)_2HPO_4$	0.7g
KH_2PO_4	0.3g
$(NH_4)H_2PO_4$	0.3g
$MgSO_4 \cdot 7H_2O$	0.25g
$(NH_4)_2SO_4$	0.1g
$CaCl_2 \cdot 6H_2O$	0.05g
H_3BO_3	0.5mg
$MnSO_4 \cdot 4H_2O$	0.4mg
$ZnSO_4 \cdot 7H_2O$	0.4mg
$FeCl_3 \cdot 6H_2O$	0.2mg
$(NH_4)_6Mo_7O_{24} \cdot 4H_2O$	0.2mg
KI	0.1mg
$CuSO_4 \cdot 5H_2O$	0.04mg
Vitamin solution	20.0mL

pH 6.8–7.0 at 25°C

Vitamin Solution:
Composition per 20.0mL:

Biotin	0.1mg
Vitamin B_{12}	0.03mg

Preparation of Vitamin Solution: Add biotin and Vitamin B_{12} to 20.0mL of distilled/deionized water. Mix thoroughly. Filter sterilize.

Preparation of Medium: Add components, except vitamin solution, to distilled/deionized water and bring volume to 980.0mL. Mix thoroughly. Adjust pH to 6.8–7.0 with 1*N* NaOH. Autoclave for 20 min at 15 psi pressure–121°C. Cool to 45°–50°C. Aseptically add the sterile vitamin solution. Mix thoroughly. Distribute into sterile tubes or flasks.

Use: For the cultivation and maintenance of *Phenylobacterium immobile*.

Mineral Medium for Chemolithotrophic Growth

Composition per 985.0mL:

Agar	15.0g
$Na_2HPO_4 \cdot 2H_2O$	2.9g
KH_2PO_4	2.3g
NH_4Cl	1.0g
$MgSO_4 \cdot 7H_2O$	0.5g
$NaHCO_3$	0.5g
$CaCl_2 \cdot 2H_2O$	0.01g
Ferric ammonium citrate solution	20.0mL
Trace elements solution SL-6	5.0mL

pH 6.8 ± 0.2 at 25°C

Ferric Ammonium Citrate Solution:
Composition per 20.0mL:

Ferric ammonium citrate	0.05g

Preparation of Ferric Ammonium Citrate Solution: Add ferric ammonium citrate to distilled/deionized water and bring volume to 20.0mL. Mix thoroughly. Autoclave for 15 min at 15 psi pressure–121°C.

Trace Elements Solution SL-6:
Composition per liter:

$MnCl_2 \cdot 4H_2O$	0.5g
H_3BO_3	0.3g
$CoCl_2 \cdot 6H_2O$	0.2g
$ZnSO_4 \cdot 7H_2O$	0.1g
$Na_2MoO_4 \cdot 2H_2O$	0.03g
$NiCl_2 \cdot 6H_2O$	0.02g
$CuCl_2 \cdot 2H_2O$	0.01g

Preparation of Trace Elements Solution SL-6: Add components to distilled/deionized water and bring volume to 1.0L. Mix thoroughly.

Preparation of Medium: Add components, except ferric ammonium citrate solution, to distilled/deionized water and bring volume to 980.0mL. Mix thoroughly. Gently heat and bring to boiling. Autoclave for 15 min at 15 psi pressure–121°C. Cool to 50°–55°C. Aseptically add 20.0mL of sterile ferric ammonium citrate solution. Mix thoroughly. Pour into sterile Petri dishes or distribute into sterile tubes.

Use: For the chemolithotrophic growth and cultivation of a wide variety of bacteria.

Mineral Medium with 2-Chlorobenzoate (DSMZ Medium 457a)

Composition per liter:

Na_2HPO_4	2.44g
KH_2PO_4	1.52g
$(NH_4)_2SO_4$	0.5g
$MgSO_4 \cdot 7H_2O$	0.2g
Tween 80	0.2g
$CaCl_2 \cdot 2H_2O$	0.05g
Trace elements solution SL-4	10.0mL
Chlorobenzoate solution	10.0mL

pH 7.4 ± 0.2 at 25°C

Trace Elements Solution SL-4:
Composition per liter:

EDTA	0.5g
$FeSO_4 \cdot 7H_2O$	0.2g
Trace elements solution SL-6	100.0mL

Trace Elements Solution SL-6:
Composition per liter:

H_3BO_3	0.3g
$CoCl_2 \cdot 6H_2O$	0.2g
$ZnSO_4 \cdot 7H_2O$	0.1g
$MnCl_2 \cdot 4H_2O$	0.03g
$Na_2MoO_4 \cdot H_2O$	0.03g
$NiCl_2 \cdot 6H_2O$	0.02g
$CuCl_2 \cdot 2H_2O$	0.01g

Preparation of Trace Elements Solution SL-6: Add components to distilled/deionized water and bring volume to 1.0L. Mix thoroughly. Adjust pH to 3.4.

Preparation of Trace Elements Solution SL-4: Add components to distilled/deionized water and bring volume to 1.0L. Mix thoroughly.

Chlorobenzoate Solution:
Composition per liter:

2-Chlorobenzoic acid	78.3g

Preparation of Chlorobenzoate Solution: Add 2-chlorobenzoic acid to distilled/deionized water and bring volume to 1.0L. Mix thoroughly. Slowly add concentrated NaOH to adjust pH to 7.4. Filter sterilize.

Preparation of Medium: Add components, except chlorobenzoate solution, to 990.0mL distilled/deionized water. Adjust pH to 7.4. Autoclave for 15 min at 15 psi pressure–121°C. Cool to room temperature. Aseptically add 10.0mL chlorobenzoate solution. Mix thoroughly. Aseptically distribute to sterile tubes or flasks.

Use: For the cultivation of chlorobenzoate-utilizing bacteria.

Mineral Medium with Crude Oil

Composition per 100.0mL:

K_2HPO_4	0.45g
$(NH_4)_2SO_4$	0.1g
$MgSO_4 \cdot 7H_2O$	0.02g
NaCl	0.01g
$CaCl_2$	0.01g
$FeCl_3$	0.002g
Crude oil	5.0mL

pH 7.2 ± 0.3 at 25°C

Preparation of Medium: Add components, except crude oil, to distilled/deionized water and bring volume to 1.0L. Mix thoroughly. Autoclave for 15 min at 15 psi pressure–121°C. Cool to 50°C. Aseptically add 5.0mL of filter-sterilized crude oil. Mix thoroughly. Aseptically distribute into sterile tubes or flasks.

Use: For the cultivation of *Acinetobacter baumannii*.

Mineral Medium with Cyanuric Acid as Nitrogen Source (DSMZ Medium 465g)

Composition per liter:

$Na_2HPO_4 \cdot 2H_2O$	3.5g
KH_2PO_4	1.0g
$MgCl_2 \cdot 6H_2O$	0.1g
$Ca(NO_3)_2 \cdot 4H_2O$	0.05g
Cyanuric acid solution	10.0mL
Vitamin solution	10.0mL
Glycerol solution	10.0mL
Trace elements solution SL-4	1.0mL

pH 7.25 ± 0.2 at 25°C

Trace Elements Solution SL-4:
Composition per liter:

EDTA	0.5g
$FeSO_4 \cdot 7H_2O$	0.2g
Trace elements solution SL-6	100.0mL

Trace Elements Solution SL-6:
Composition per liter:

H_3BO_3	0.3g
$CoCl_2 \cdot 6H_2O$	0.2g
$ZnSO_4 \cdot 7H_2O$	0.1g
$MnCl_2 \cdot 4H_2O$	0.03g
$Na_2MoO_4 \cdot H_2O$	0.03g
$NiCl_2 \cdot 6H_2O$	0.02g
$CuCl_2 \cdot 2H_2O$	0.01g

Preparation of Trace Elements Solution SL-6: Add components to distilled/deionized water and bring volume to 1.0L. Mix thoroughly. Adjust pH to 3.4.

Preparation of Trace Elements Solution SL-4: Add components to distilled/deionized water and bring volume to 1.0L. Mix thoroughly.

Vitamin Solution:
Composition per liter:

Vitamin B_{12}	50.0mg
Pantothenic acid	50.0mg
Riboflavin	50.0mg
Alpha-lipoic acid	50.0mg
p-Aminobenzoic acid	50.0mg
Thiamine-HCl·$2H_2O$	50.0mg
Nicotinic acid	25.0mg
Nicotine amide	25.0mg
Biotin	20.0mg
Folic acid	20.0mg
Pyridoxamine-HCl	10.0mg

Preparation of Vitamin Solution: Add components to distilled/deionized water and bring volume to 1.0L. Mix thoroughly. Filter sterilize.

Cyanuric Acid Solution:
Composition per 10.0mL:
Cyanuric acid 645mg

Preparation of Cyanuric Acid Solution: Add cyanuric acid to 10.0mL distilled/deionized water. Mix thoroughly. Adjust pH to 7.0. Filter sterilize.

Glycerol Solution:
Composition per 10.0mL:
Glycerol 5.5g

Preparation of Glycerol Solution: Add glycerol acid to distilled/deionized water. Mix thoroughly. Filter sterilize.

Preparation of Medium: Add components, except cyanuric acid solution, glycerol solution, and vitamin solution to 1.0L distilled/deionized water. Adjust pH to 7.25. Autoclave for 15 min at 15 psi pressure–121°C. Aseptically add 10.0mL cyanuric acid solution, 10.0mL glycerol solution, and 10.0mL vitamin solution. Mix thoroughly. Aseptically distribute to sterile tubes or flasks.

Use: For the cultivation of *Gordonia rubripertincta=Rhodococcus rubropertinctus* and other cyanuric acid-utilizing bacteria.

Mineral Medium with Dichlorobenzoate

Composition per liter:
Na_2HPO_4 .... 2.78g
KH_2PO_4 .... 2.78g
$(NH_4)_2SO_4$ 1.0g
Hutner's mineral base 20.0mL
2,4-Dichlorobenzoate solution 10.0mL
pH 6.8 ± 0.2 at 25°C

Hutner's Mineral Base:
Composition per liter:
$MgSO_4 \cdot 7H_2O$ 29.7g
Nitrilotriacetic acid 10.0g
$CaCl_2 \cdot 2H_2O$ 3.34g
$FeSO_4 \cdot 7H_2O$ 99.0mg
$(NH_4)_2MoO_4$ 9.25mg
"Metals 44" 50.0mL

Preparation of Hutner's Mineral Base: Add nitrilotriacetic acid to 500.0mL of distilled/deionized water. Dissolve by adjusting pH to 6.5 with KOH. Add remaining components. Readjust pH to 7.2 with H_2SO_4 or KOH. Add distilled/deionized water to 1.0L.

"Metals 44":
Composition per 100.0mL:
$ZnSO_4 \cdot 7H_2O$ 1.1g
$FeSO_4 \cdot 7H_2O$ 0.5g
EDTA 0.25g
$MnSO_4 \cdot 7H_2O$ 0.154g
$CuSO_4 \cdot 5H_2O$ 0.04g
$Co(NO_3)_2 \cdot 6H_2O$ 0.025g
$Na_2B_4O_7 \cdot 10H_2O$ 0.018g

Preparation of "Metals 44": Add a few drops of H_2SO_4 to distilled/deionized water to inhibit precipitate formation. Add components to acidified distilled/deionized water and bring volume to 100.0mL. Mix thoroughly.

2,4-Dichlorobenzoate Solution:
Composition per 10.0mL:
2,4-Dichlorobenzoate 5.0mg

Preparation of 2,4-Dichlorobenzoate Solution: Add 2,4-dichlorobenzoate to 10.0mL of distilled/deionized water. Mix thoroughly. Filter sterilize.

Preparation of Medium: Add components, except 2,4-dichlorobenzoate solution, to distilled/deionized water and bring volume to 990.0mL. Mix thoroughly. Adjust pH to 6.8 with 1*N* KOH. Autoclave for 15 min at 15 psi pressure–121°C. Cool to 45°–50°C. Aseptically add the sterile 2,4-dichlorobenzoate solution. Mix thoroughly. Distribute into sterile tubes or flasks.

Use: For the cultivation and maintenance of *Actinomyces viscosus.*

Mineral Medium with Dichloromethane (DSMZ Medium 465c)

Composition per liter:
$Na_2HPO_4 \cdot 2H_2O$ 3.5g
KH_2PO_4 .... 1.0g
$(NH_4)_2SO_4$ 0.5g
$MgCl_2 \cdot 6H_2O$ 0.1g
$Ca(NO_3)_2 \cdot 4H_2O$ 0.05g
Bromothymol Blue 50.0mg
Methanol 10.0mL
Trace elements solution SL-4 .... 1.0mL
Dichloromethane .... variable
pH 7.25 ± 0.2 at 25°C

Trace Elements Solution SL-4:
Composition per liter:
EDTA 0.5g
$FeSO_4 \cdot 7H_2O$ 0.2g
Trace elements solution SL-6 .... 100.0mL

Trace Elements Solution SL-6:
Composition per liter:
H_3BO_3 .... 0.3g
$CoCl_2 \cdot 6H_2O$ 0.2g
$ZnSO_4 \cdot 7H_2O$ 0.1g
$MnCl_2 \cdot 4H_2O$ 0.03g
$Na_2MoO_4 \cdot H_2O$ 0.03g
$NiCl_2 \cdot 6H_2O$ 0.02g
$CuCl_2 \cdot 2H_2O$ 0.01g

Preparation of Trace Elements Solution SL-6: Add components to distilled/deionized water and bring volume to 1.0L. Mix thoroughly. Adjust pH to 3.4.

Preparation of Trace Elements Solution SL-4: Add components to distilled/deionized water and bring volume to 1.0L. Mix thoroughly.

Preparation of Medium: Add components, except methanol and dichloromethane, to 1.0L distilled/deionized water. Adjust pH to 7.25. Autoclave for 15 min at 15 psi pressure–121°C. Cool to room temperature. Inoculate the medium and supply methanol via incubation atmosphere (up to 10.0mL methanol per liter medium). Subsequently feed small amounts of dichloromethane (toxic to bacteria at 2 mmol per liter or less) via incubation atmosphere. Readjust pH of the culture with sterile 1*M* NaOH as necessary.

Use: For the cultivation of *Methylophilus leisingeri*, *Rhizobium radiobacter,* and other dichloromethane-utilizing bacteria.

Mineral Medium with 2,4-Dichlorophenoxyacetic Acid

Composition per liter:
$Na_2HPO_4 \cdot 12H_2O$ 9.0g
Ferric ammonium citrate 5.0g
$MnSO_4 \cdot H_2O$ 3.0g
NH_4Cl 2.0g
KH_2PO_4 .... 1.5g
$MgSO_4 \cdot 7H_2O$ 0.2g

$ZnSO_4 \cdot 7H_2O$ 0.2g
Titriplex I 10.0mg
$CoSO_4$ 10.0μg
2,4-Dichlorophenoxyacetic acid solution 10.0mL
pH 7.0 ± 0.2 at 25°C

2,4-Dichlorophenoxyacetic Acid Solution:
Composition per 10.0mL:
2,4-Dichlorophenoxyacetic acid 0.5g

Preparation of 2,4-Dichlorophenoxyacetic Acid Solution: Add 2,4-dichlorophenoxyacetic acid to distilled/deionized water and bring volume to 10.0mL. Mix thoroughly. Filter sterilize.

Preparation of Medium: Add components, except 2,4-dichlorophenoxyacetic acid solution, to distilled/deionized water and bring volume to 990.0mL. Mix thoroughly. Autoclave for 15 min at 15 psi pressure–121°C. Aseptically add 10.0mL of sterile 2,4-dichlorophenoxyacetic acid solution. Mix thoroughly. Aseptically distribute into sterile tubes or flasks.

Use: For the cultivation of bacteria that can utilize 2,4-dichlorophenoxyacetic acid as sole carbon source.

Mineral Medium with 2,4-Dichlorotoluene

Composition per liter:
NH_4Cl 0.5g
Yeast extract 0.1g
2,4-Dichlorotoluene 0.1g
$Na_2HPO_4 \cdot 7H_2O$ 670.0mg
KH_2PO_4 .... 340.0mg
$MgSO_4 \cdot 7H_2O$ 112.0mg
$CaCl_2$ 14.0mg
$ZnSO_4 \cdot 7H_2O$ 5.0mg
$Na_2MoO_4 \cdot 2H_2O$ 2.5mg
$FeCl_3$ 0.13mg
pH 7.0 ± 0.2 at 25°C

Preparation of Medium: Add components to distilled/deionized water and bring volume to 1.0L. Mix thoroughly. Distribute into tubes or flasks. Autoclave for 15 min at 15 psi pressure–121°C.

Use: For the cultivation of 2,4-dichlorotoluene-degrading *Pseudomonas* species.

Mineral Medium H-3

Composition per liter:
Agar (if needed) 20.0g
$Na_2HPO_4 \cdot 2H_2O$ 2.9g
KH_2PO_4 .... 2.3g
NH_4Cl 1.0g
$MgSO_4 \cdot 7H_2O$ 0.5g
$NaHCO_3$ 0.5g
$CaCl_2 \cdot 2H_2O$ 0.01g
Ferric ammonium citrate solution 20.0mL
Trace elements solution 5.0mL

Ferric Ammonium Citrate Solution:
Composition per 20.0mL:
Ferric ammonium citrate 0.05g

Preparation of Ferric Ammonium Citrate Solution: Add ferric ammonium citrate to distilled/deionized water and bring volume to 20.0mL. Mix thoroughly. Autoclave for 15 min at 15 psi pressure–121°C.

Trace Elements Solution:
Composition per liter:
H_3BO_3 .... 0.3g
$CoCl_3 \cdot 6H_2O$ 0.2g
$ZnSO_4 \cdot 7H_2O$ 0.1g
$MnCl_2 \cdot 4H_2O$ 0.03g
$Na_2MoO_4 \cdot 2H_2O$ 0.03g
$CuCl_2 \cdot 2H_2O$ 0.01g
$NiCl_2 \cdot 6H_2O$ 0.002g

Preparation of Trace Elements Solution: Add components to distilled/deionized water and bring volume to 1.0L. Mix thoroughly. For chemolithotrophic growth, incubate the culture under an atmosphere of 2% (v/v) O_2, 10% CO_2, 60% H_2, and 28% N_2.

Preparation of Medium: Add components, except ferric ammonium citrate solution, to distilled/deionized water and bring volume to 980.0mL. Mix thoroughly. Gently heat and bring to boiling. Autoclave for 15 min at 15 psi pressure–121°C. Cool to 50°–55°C. Aseptically add 20.0mL of sterile ferric ammonium citrate solution. Mix thoroughly. Pour into sterile Petri dishes or distribute into sterile tubes. Incubate cultures in 60% H_2+ 28% N_2 + 10% CO_2 + 2% O_2.

Use: For the chemolithotrophic growth of *Alcaligenes eutrophus.*

Mineral Medium for Hydrogen Bacteria

Composition per liter:
Agar 15.0g
$Na_2HPO_4 \cdot 2H_2O$ 2.9g
KH_2PO_4 .... 2.3g
NH_4Cl 1.0g
$MgSO_4 \cdot 7H_2O$ 0.5g
$NaHCO_3$ 0.5g
$CaCl_2 \cdot 2H_2O$ 0.01g
Ferric ammonium citrate solution 20.0mL

Ferric Ammonium Citrate Solution:
Composition per 20.0mL:
Ferric ammonium citrate 0.05g

Preparation of Ferric Ammonium Citrate Solution: Add ferric ammonium citrate to 20.0mL of distilled/deionized water. Filter sterilize.

Preparation of Medium: Add components, except ferric ammonium citrate solution, to distilled/deionized water and bring volume to 980.0mL. Mix thoroughly. Gently heat and bring to boiling. Autoclave for 15 min at 15 psi pressure–121°C. Cool to 50°C. Aseptically add the sterile ferric ammonium citrate solution. Mix thoroughly. Pour into sterile Petri dishes or distribute into sterile tubes. Incubate inoculated medium at 30°C under 60% H_2 + 25% N_2 + 10% CO_2 + 5% O_2.

Use: For the cultivation and maintenance of *Alcaligenes eutrophus, Hydrogenophaga flava,* and *Hydrogenophaga pseudoflava.*

Mineral Medium with 2-Hydroxybiphenyl (DSMZ Medium 465a)

Composition per1010mL:
$Na_2HPO_4 \cdot 2H_2O$ 3.5g
KH_2PO_4 .... 1.0g
$(NH_4)_2SO_4$ 0.5g
$MgCl_2 \cdot 6H_2O$ 0.1g
$Ca(NO_3)_2 \cdot 4H_2O$ 0.05g
Hydroxybiphenyl solution 10.0mL
Trace elements solution SL-4 .... 1.0mL
pH 7.25 ± 0.2 at 25°C

Trace Elements Solution SL-4:
Composition per liter:
EDTA 0.5g
$FeSO_4 \cdot 7H_2O$ 0.2g
Trace elements solution SL-6 .... 100.0mL

Trace Elements Solution SL-6:
Composition per liter:
H_3BO_3 .... 0.3g
$CoCl_2 \cdot 6H_2O$ 0.2g
$ZnSO_4 \cdot 7H_2O$ 0.1g
$MnCl_2 \cdot 4H_2O$ 0.03g
$Na_2MoO_4 \cdot H_2O$ 0.03g
$NiCl_2 \cdot 6H_2O$ 0.02g
$CuCl_2 \cdot 2H_2O$ 0.01g

Preparation of Trace Elements Solution SL-6: Add components to distilled/deionized water and bring volume to 1.0L. Mix thoroughly. Adjust pH to 3.4.

Preparation of Trace Elements Solution SL-4: Add components to distilled/deionized water and bring volume to 1.0L. Mix thoroughly.

Hydroxybiphenyl Solution:
Composition per 100.0mL:
2-Hydroxybiphenyl 5.0g

Preparation of Hydroxybiphenyl Solution: Add 2-hydroxybiphenyl to 100.0mL ethanol. Mix thoroughly. Filter sterilize.

Preparation of Medium: Add components, except hydroxybiphenyl solution, to 1.0L distilled/deionized water. Adjust pH to 7.25. Autoclave for 15 min at 15 psi pressure–121°C. Aseptically add hydroxybiphenyl solution to a sterile culture vessel so that the final concentration of 2-hydroxybiphenyl will be 0.5g/L medium. Let the ethanol evaporate under sterile conditions. Aseptically add the liquid medium to the culture vessel to achieve the appropriate concentration of biphenyl.

Use: For the cultivation of *Pseudomonas* sp. and other hydroxybiphenyl-utilizing bacteria.

Mineral Medium, Nagel and Andreesen

Composition per liter:
$MgSO_4 \cdot 7H_2O$ 0.5g
NH_4Cl 0.3g
NaCl 0.05g
$CaCl_2$ 0.01g
$MnSO_4$ 0.01g
Phosphate solution 50.0mL
Vitamin solution 5.0mL
Trace elements solution SL-10 .... 1.0mL
pH 7.5 ± 0.2 at 25°C

Phosphate Solution:
Composition per 50.0mL:
$Na_2HPO_4 \cdot 2H_2O$ 1.45g
KH_2PO_4 .... 0.25g

Preparation of Phosphate Solution: Add components to distilled/deionized water and bring volume to 50.0mL. Mix thoroughly. Adjust pH to 7.5. Autoclave for 15 min at 15 psi pressure–121°C.

Vitamin Solution:
Composition per liter:
Folic acid 20.0g
α-Lipoic acid 50.0mg
p-Aminobenzoic acid 50.0mg
Pantothenic acid 50.0mg
Riboflavin 50.0mg
Thamine·HCl 50.0mg
Vitamin B_{12} 50.0mg
Nicotine amide 25.0mg
Biotin 20.0mg
Nicotinic acid 20.0mg
Pyridoxamine·HCl 10.0mg

Preparation of Vitamin Solution: Add components to distilled/deionized water and bring volume to 1.0L Mix thoroughly. Stir for 2 hr. Filter sterilize.

Trace Elements Solution SL-10:
Composition per liter:
$FeCl_2 \cdot 4H_2O$ 1.5g
$CoCl_2 \cdot 6H_2O$ 190.0mg
$MnCl_2 \cdot 4H_2O$ 100.0mg
$ZnCl_2$ 70.0mg
$Na_2MoO_4 \cdot 2H_2O$ 36.0mg
$NiCl_2 \cdot 6H_2O$ 24.0mg
H_3BO_3 .... 6.0mg
$CuCl_2 \cdot 2H_2O$ 2.0mg
HCl (25% solution) 10.0mL

Preparation of Trace Elements Solution SL-10: Add $FeCl_2 \cdot 4H_2O$ to 10.0mL of HCl solution. Mix thoroughly. Add distilled/deionized water and bring volume to 1.0L. Add remaining components. Mix thoroughly. Autoclave for 15 min at 15 psi pressure–121°C.

Preparation of Medium: Add components, except phosphate solution, vitamin solution, and trace elements solution SL-10, to distilled/deionized water and bring volume to 944.0mL. Mix thoroughly. Autoclave for 15 min at 15 psi pressure–121°C. Aseptically add 50.0mL of sterile phosphate solution 5.0mL of sterile vitamin solution and 1.0mL of sterile trace elements solution SL-10. Mix thoroughly. Aseptically distribute into sterile tubes or flasks.

Use: For the cultivation of *Arthrobacter* species, *Mycobacterium* species, *Paracoccus denitrificans*, *Pseudomonas putida*, and *Rhodococcus* species.

Mineral Medium with Naphthalene

Composition per liter:
$Na_2HPO_4 \cdot 12H_2O$ 9.0g
Ferric ammonium citrate 5.0g
$MnSO_4 \cdot H_2O$ 3.0g
NH_4Cl 2.0g
KH_2PO_4 .... 1.5g
$MgSO_4 \cdot 7H_2O$ 0.2g
$ZnSO_4 \cdot 7H_2O$ 0.2g
Titriplex I 10.0mg
$CoSO_4$ 10.0μg
Naphthalene crumbs .... variable
pH 7.0 ± 0.2 at 25°C

Preparation of Medium: Add components, except naphthalene crumbs, to distilled/deionized water and bring volume to 1.0L. Mix thoroughly. Distribute into tubes or flasks. Autoclave for 15 min at 15 psi pressure–121°C. Inoculate tubes or flasks and place in a dessicator jar in which crumbs of naphthalene will be evaporated.

Use: For the cultivation of bacteria that can utilize naphthalene as sole carbon source.

Mineral Medium with *o*-Nitrophenol (DSMZ Medium 461c)

Composition per liter:
$MgSO_4 \cdot 7H_2O$ 0.5g
NH_4Cl 0.3g
NaCl 0.05g

$CaCl_2$ 0.01g
$MnSO_4$ 0.01g
o-Nitrophenol solution 200.0mL
Phosphate solution 100.0mL
Trace elements solution SL-10 .... 1.0mL

pH 7.2 ± 0.2 at 25°C

Trace Elements Solution SL-10:
Composition per liter:
$FeCl_2 \cdot 4H_2O$ 1.5g
H_3BO_3 .... 300.0mg
$CoCl_2 \cdot 6H_2O$ 190.0mg
$MnCl_2 \cdot 4H_2O$ 100.0mg
$ZnCl_2$ 70.0mg
$Na_2MoO_4 \cdot 2H_2O$ 36.0mg
$NiCl_2 \cdot 6H_2O$ 24.0mg
$CuCl_2 \cdot 2H_2O$ 2.0mg
HCl (25% solution) 7.7mL

Preparation of Trace Elements Solution SL-10: Add $FeCl_2 \cdot 4H_2O$ to 10.0mL of HCl solution. Mix thoroughly. Add distilled/deionized water and bring volume to 1.0L. Add remaining components. Mix thoroughly. Sparge with 100% N_2. Autoclave for 15 min at 15 psi pressure–121°C. Cool to room temperature.

Phosphate Solution:
Composition 100mL:
$Na_2HPO_4 \cdot 2H_2O$ 1.45g
KH_2PO_4 .... 0.25g

Preparation of Phosphate Solution: Add components to distilled/deionized water and bring volume to 100.0mL. Mix thoroughly. Autoclave for 15 min at 15 psi pressure–121°C. Cool to room temperature.

***o*-Nitrophenol Solution:**
Composition liter:
o-Nitrophenol 0.5g

Preparation of *o*-Nitrophenol Solution: Dissolve *o*-nitrophenol in phosphate buffer (50 mM, pH 7.5) and bring volume to 1.0L. Sterilize by filtration.

Preparation of Medium: Add components, except phosphate solution and *o*-nitrophenol solution, to distilled/deionized water and bring volume to 700.0mL. Mix thoroughly. Autoclave for 15 min at 15 psi pressure–121°C. Cool to room temperature. Aseptically add 100.0mL phosphate solution and 200.0mL *o*-nitrophenol solution. Mix thoroughly. Aseptically distribute into sterile tubes or flasks.

Use: For the cultivation of *o*-nitrophenol-utilizing bacteria.

Mineral Medium with PCP (DSMZ Medium 465b)

Composition per liter:
$Na_2HPO_4 \cdot 2H_2O$ 3.5g
KH_2PO_4 .... 1.0g
$(NH_4)_2SO_4$ 0.5g
$MgCl_2 \cdot 6H_2O$ 0.1g
$Ca(NO_3)_2 \cdot 4H_2O$ 0.05g
Pentachlorophenol solution 100.0mL
Trace elements solution SL-4 .... 1.0mL

pH 7.25 ± 0.2 at 25°C

Trace Elements Solution SL-4:
Composition per liter:
EDTA 0.5g
$FeSO_4 \cdot 7H_2O$ 0.2g
Trace elements solution SL-6 .... 100.0mL

Trace Elements Solution SL-6:
Composition per liter:
H_3BO_3 .... 0.3g
$CoCl_2 \cdot 6H_2O$ 0.2g
$ZnSO_4 \cdot 7H_2O$ 0.1g
$MnCl_2 \cdot 4H_2O$ 0.03g
$Na_2MoO_4 \cdot H_2O$ 0.03g
$NiCl_2 \cdot 6H_2O$ 0.02g
$CuCl_2 \cdot 2H_2O$ 0.01g

Preparation of Trace Elements Solution SL-6: Add components to distilled/deionized water and bring volume to 1.0L. Mix thoroughly. Adjust pH to 3.4.

Preparation of Trace Elements Solution SL-4: Add components to distilled/deionized water and bring volume to 1.0L. Mix thoroughly.

Pentachlorophenol Solution:
Composition per liter:
Pentachlorophenol 1.0g

Preparation of Pentachlorophenol Solution: Add pentachlorophenol to 1.0L 0.1*M* NaOH. Mix thoroughly. Filter sterilize.

Preparation of Medium: Add components, except pentachlorophenol solution, to 900.0mL distilled/deionized water. Adjust pH to 7.25. Autoclave for 15 min at 15 psi pressure–121°C. Aseptically add 100.0mL sterile pentachlorophenol solution. Mix thoroughly. Aseptically distribute to sterile tubes or flasks.

Use: For the cultivation of *Sphingobium chlorophenolicum=Sphingomonas chlorophenolica* and other pentachlorophenol-utilizing bacteria.

Mineral Medium, pH 7.25

Composition per liter:
$Na_2HPO_4 \cdot 2H_2O$ 3.5g
KH_2PO_4 .... 1.0g
$(NH_4)_2SO_4$ 0.5g
$MgCl_2 \cdot 6H_2O$ 0.1g
$Ca(NO_3)_2 \cdot 4H_2O$ 0.05g
Trace elements solution SL-4 .... 1.0mL

pH 7.25 ± 0.2 at 25°C

Trace Elements Solution SL-4:
Composition per liter:
EDTA 0.5g
$FeSO_4 \cdot 7H_2O$ 0.2g
Trace elements solution SL-6 .... 100.0mL

Trace Elements Solution SL-6:
Composition per liter:
$MnCl_2 \cdot 4H_2O$ 0.5g
H_3BO_3 .... 0.3g
$CoCl_2 \cdot 6H_2O$ 0.2g
$ZnSO_4 \cdot 7H_2O$ 0.1g
$Na_2MoO_4 \cdot 2H_2O$ 0.03g
$NiCl_2 \cdot 6H_2O$ 0.02g
$CuCl_2 \cdot 2H_2O$ 0.01g

Preparation of Trace Elements Solution SL-6: Add components to distilled/deionized water and bring volume to 1.0L. Mix thoroughly.

Preparation of Trace Elements Solution SL-4: Add components to distilled/deionized water and bring volume to 1.0L. Mix thoroughly.

Preparation of Medium: Add components to distilled/deionized water and bring volume to 1.0L. Mix thoroughly.

Distribute into tubes or flasks. Autoclave for 15 min at 15 psi pressure–121°C.

Use: For the cultivation of *Azotobacter* species, *Pseudomonas* species, and *Sphingomonas* species.

Mineral Medium with Phenol

Composition per liter:

Phenol	1.0g
K_2HPO_4	1.0g
NH_4NO_3	1.0g
$(NH_4)_2SO_4$	0.5g
$MgSO_4$	0.5g
KH_2PO_4	0.5g
NaCl	0.5g
$CaCl_2$	0.02g
$FeSO_4$	0.02g
Wolfe's mineral solution	10.0mL

Wolfe's Mineral Solution:

Composition per liter

$MgSO_4·7H_2O$	3.0g
Nitriloacetic acid	1.5g
NaCl	1.0g
$MnSO_4·H_2O$	0.5g
$FeSO_4·7H_2O$	0.1g
$CoCl_2·6H_2O$	0.1g
$CaCl_2$	0.1g
$ZnSO_4·7H_2O$	0.1g
$CuSO_4·5H_2O$	0.01g
$AlK(SO_4)_2·12H_2O$	0.01g
H_3BO_3	0.01g
$Na_2MoO_4·2H_2O$	0.01g

Preparation of Wolfe's Mineral Solution: Add nitrilotriacetic acid to 500.0mL of distilled/deionized water. Dissolve by adjusting pH to 6.5 with KOH. Add remaining components. Add distilled/deionized water to 1.0L.

Preparation of Medium: Add components, except phenol, to distilled/deionized water and bring volume to 1.0L. Mix thoroughly. Autoclave for 15 min at 15 psi pressure–121°C. Aseptically add the phenol. Mix thoroughly. Distribute into sterile tubes or flasks.

Use: For the cultivation and maintenance of *Pseudomonas putida.*

Mineral Medium with Phenylacetate

Composition per liter:

$Na_2HPO_4·12H_2O$	9.0g
Ferric ammonium citrate	5.0g
$MnSO_4·H_2O$	3.0g
NH_4Cl	2.0g
KH_2PO_4	1.5g
$MgSO_4·7H_2O$	0.2g
$ZnSO_4·7H_2O$	0.2g
Titriplex I	10.0mg
$CoSO_4$	10.0μg
Phenylacetic acid solution	10.0mL

pH 7.0 ± 0.2 at 25°C

Phenylacetic Acid Solution:

Composition per 10.0mL:

Phenylacetic acid	0.5g

Preparation of Phenylacetic Acid Solution: Add phenylacetic acid to distilled/deionized water and bring volume to 10.0mL. Mix thoroughly. Filter sterilize.

Preparation of Medium: Add components, except phenylacetic acid solution, to distilled/deionized water and bring volume to 990.0mL. Mix thoroughly. Autoclave for 15 min at 15 psi pressure–121°C. Aseptically add 10.0mL of sterile phenylacetic acid solution. Mix thoroughly. Aseptically distribute into sterile tubes or flasks.

Use: For the cultivation of bacteria that can utilize phenylacetic acid as sole carbon source.

Mineral Medium with Pyrrolic Acid (DSMZ Medium 461b)

Composition per liter:

$MgSO_4·7H_2O$	0.5g
NH_4Cl	0.3g
NaCl	0.05g
$CaCl_2$	0.01g
$MnSO_4$	0.01g
Phosphate solution	100.0mL
Pyrrolic acid solution	10.0mL
Trace elements solution SL-10	1.0mL

pH 7.2 ± 0.2 at 25°C

Trace Elements Solution SL-10:

Composition per liter:

$FeCl_2·4H_2O$	1.5g
H_3BO_3	300.0mg
$CoCl_2·6H_2O$	190.0mg
$MnCl_2·4H_2O$	100.0mg
$ZnCl_2$	70.0mg
$Na_2MoO_4·2H_2O$	36.0mg
$NiCl_2·6H_2O$	24.0mg
$CuCl_2·2H_2O$	2.0mg
HCl (25% solution)	7.7mL

Preparation of Trace Elements Solution SL-10: Add $FeCl_2·4H_2O$ to 10.0mL of HCl solution. Mix thoroughly. Add distilled/deionized water and bring volume to 1.0L. Add remaining components. Mix thoroughly. Sparge with 100% N_2. Autoclave for 15 min at 15 psi pressure–121°C. Cool to room temperature.

Phosphate Solution:

Composition 100mL:

$Na_2HPO_4·2H_2O$	1.45g
KH_2PO_4	0.25g

Preparation of Phosphate Solution: Add components to distilled/deionized water and bring volume to 100.0mL. Mix thoroughly. Autoclave for 15 min at 15 psi pressure–121°C. Cool to room temperature.

Pyrrolic Acid Solution:

Composition liter:

Pyrrolic acid	10.0g

Preparation of Pyrrolic Acid Solution: Dissolve pyrrolic acid in phosphate buffer (50 mM, pH 7.5) and bring volume to 1.0L. Sterilize by filtration.

Preparation of Medium: Add components, except phosphate solution and pyrrolic acid solution, to distilled/deionized water and bring volume to 890.0mL. Mix thoroughly. Autoclave for 15 min at 15 psi pressure–121°C. Cool to room temperature. Aseptically add 100.0mL phosphate solution and 10.0mL pyrrolic acid solution. Mix thoroughly. Aseptically distribute into sterile tubes or flasks

Use: For the cultivation of pyrrolic acid-utilizing bacteria.

Mineral Medium with Quinoline (DSMZ Medium 461a)

Composition per liter:

$MgSO_4·7H_2O$	0.5g
NH_4Cl	0.3g

NaCl 0.05g
$CaCl_2$ 0.01g
$MnSO_4$ 0.01g
Phosphate solution 100.0mL
Quinoline emulsion 100.0mL
Vitamin solution 5.0mL
Trace elements solution. SL-10 .... 1.0mL
pH 7.2 ± 0.2 at 25°C

Trace Elements Solution SL-10:
Composition per liter:
$FeCl_2·4H_2O$ 1.5g
H_3BO_3 .... 300.0mg
$CoCl_2·6H_2O$ 190.0mg
$MnCl_2·4H_2O$ 100.0mg
$ZnCl_2$ 70.0mg
$Na_2MoO_4·2H_2O$ 36.0mg
$NiCl_2·6H_2O$ 24.0mg
$CuCl_2·2H_2O$ 2.0mg
HCl (25% solution) 7.7mL

Preparation of Trace Elements Solution SL-10: Add $FeCl_2·4H_2O$ to 10.0mL of HCl solution. Mix thoroughly. Add distilled/deionized water and bring volume to 1.0L. Add remaining components. Mix thoroughly. Sparge with 100% N_2. Autoclave for 15 min at 15 psi pressure–121°C. Cool to room temperature.

Vitamin Solution:
Composition per liter:
Vitamin B_{12} 50.0mg
Pantothenic acid 50.0mg
Riboflavin 50.0mg
Alpha-lipoic acid 50.0mg
p-Aminobenzoic acid 50.0mg
Thiamine-HCl·$2H_2O$ 50.0mg
Nicotinic acid 25.0mg
Nicotine amide 25.0mg
Biotin 20.0mg
Folic acid 20.0mg
Pyridoxamine-HCl 10.0mg

Preparation of Vitamin Solution: Add components to distilled/deionized water and bring volume to 1.0L. Mix thoroughly. Filter sterilize.

Phosphate Solution:
Composition 100mL:
$Na_2HPO_4·2H_2O$ 1.45g
KH_2PO_4 .... 0.25g

Preparation of Phosphate Solution: Add components to distilled/deionized water and bring volume to 100.0mL. Mix thoroughly. Autoclave for 15 min at 15 psi pressure–121°C. Cool to room temperature.

Quinoline Emulsion:
Composition liter:
Quinoline 3.0g

Preparation of Quinoline Emulsion: Prepare an emulsion of quinoline in 50 mM phosphate buffer by stirring or ultrasonication. Add quinoline acid to phosphate buffer (50 mM, pH 7.5) and bring volume to 1.0L. Stir vigorously or sonicate to form emulsion. Autoclave or 15 min at 15 psi pressure–121°C in a gas tight vessel. Cool to room temperature.

Preparation of Medium: Add components, except phosphate solution, vitamin solution, and quinoline emulsion to distilled/deionized water and bring volume to 795.0mL. Mix thoroughly. Autoclave for 15 min at 15 psi pressure–121°C. Cool to room temperature. Aseptically add 100.0mL phosphate solution, 5.0mL vitamin solution, and 100.0mL quinoline emulsion. Mix thoroughly. Aseptically distribute into sterile tubes or flasks

Use: For the cultivation of *Rhodococcus rhodochrous (Rhodococcus roseus)* and other quinoline-utilizing bacteria.

Mineral Medium with Salicylate

Composition per liter:
$Na_2HPO_4·12H_2O$ 9.0g
Ferric ammonium citrate 5.0g
$MnSO_4·H_2O$ 3.0g
NH_4Cl 2.0g
KH_2PO_4 .... 1.5g
$MgSO_4·7H_2O$ 0.2g
$ZnSO_4·7H_2O$ 0.2g
Titriplex I 10.0mg
$CoSO_4$ 10.0μg
Salicylic acid solution 10.0mL
pH 7.0 ± 0.2 at 25°C

Salicylic Acid Solution:
Composition per 10.0mL:
Salicylic acid 0.5g

Preparation of Salicylic Acid Solution: Add salicylic acid to distilled/deionized water and bring volume to 10.0mL. Mix thoroughly. Filter sterilize.

Preparation of Medium: Add components, except salicylic acid solution, to distilled/deionized water and bring volume to 990.0mL. Mix thoroughly. Autoclave for 15 min at 15 psi pressure–121°C. Aseptically add 10.0mL of sterile salicylic acid solution. Mix thoroughly. Aseptically distribute into sterile tubes or flasks.

Use: For the cultivation of bacteria that can utilize salicylic acid as sole carbon source.

Mineral Medium with Sulfobenzoic Acid (DSMZ Medium 465e)

Composition per liter:
$Na_2HPO_4·2H_2O$ 3.5g
KH_2PO_4 .... 1.0g
$(NH_4)_2SO_4$ 0.5g
$MgCl_2·6H_2O$ 0.1g
$Ca(NO_3)_2·4H_2O$ 0.05g
2-Sulfobenzoic acid 50.0mL
Vitamin solution 2.5mL
Trace elements solution SL-4 .... 1.0mL
pH 7.25 ± 0.2 at 25°C

Trace Elements Solution SL-4:
Composition per liter:
EDTA 0.5g
$FeSO_4·7H_2O$ 0.2g
Trace elements solution SL-6 .... 100.0mL

Trace Elements Solution SL-6:
Composition per liter:
H_3BO_3 .... 0.3g
$CoCl_2·6H_2O$ 0.2g
$ZnSO_4·7H_2O$ 0.1g
$MnCl_2·4H_2O$ 0.03g
$Na_2MoO_4·H_2O$ 0.03g
$NiCl_2·6H_2O$ 0.02g
$CuCl_2·2H_2O$ 0.01g

Preparation of Trace Elements Solution SL-6: Add components to distilled/deionized water and bring volume to 1.0L. Mix thoroughly. Adjust pH to 3.4.

Preparation of Trace Elements Solution SL-4: Add components to distilled/deionized water and bring volume to 1.0L. Mix thoroughly.

Vitamin Solution:
Composition per 100mL:

Pyridoxamine	5.0mg
Vitamin B_{12}	2.0mg
Nicotinic acid	2.0mg
Thiamine-HCl·$2H_2O$	1.0mg
p-Aminobenzoate	1.0mg
Ca-pantothenate	0.5mg
Biotin	0.2mg

Preparation of Vitamin Solution: Add components to distilled/deionized water and bring volume to 100.0mL. Mix thoroughly. Filter sterilize.

2-Sulfobenzoic Acid Solution:
Composition per 100.0mL:

2-Sulfobenzoic acid	2.0g

Preparation of 2-Sulfobenzoic Acid Solution: Add 2-sulfobenzoic acid to 100.0mL distilled/deionized water. Mix thoroughly. Filter sterilize.

Preparation of Medium: Add components, except vitamin solution and 2-sulfobenzoic acid solution, to 947.5mL distilled/deionized water. Adjust pH to 7.25. Autoclave for 15 min at 15 psi pressure–121°C. Cool to room temperature. Aseptically add 2-sulfobenzoic acid solution and vitamin solution. Mix thoroughly. Aseptically distribute to sterile tubes or flasks.

Use: For the cultivation of sulfobenzoic acid-utilizing bacteria.

Mineral Medium with Toluene

Composition per liter:

$Na_2HPO_4 \cdot 12H_2O$	9.0g
Ferric ammonium citrate	5.0g
$MnSO_4 \cdot H_2O$	3.0g
NH_4Cl	2.0g
KH_2PO_4	1.5g
$MgSO_4 \cdot 7H_2O$	0.2g
$ZnSO_4 \cdot 7H_2O$	0.2g
Titriplex I	10.0mg
$CoSO_4$	10.0μg
Toluene	1 drop

pH 7.0 ± 0.2 at 25°C

Preparation of Medium: Add components, except toluene, to distilled/deionized water and bring volume to 1.0L. Mix thoroughly. Distribute into tubes or flasks. Autoclave for 15 min at 15 psi pressure–121°C. Inoculate tubes or flasks and place in a dessicator jar in which 1 drop of toluene will be evaporated.

Use: For the cultivation of bacteria that can utilize toluene as sole carbon source.

Mineral Medium with Vitamins

Composition per 1002.5mL:

Mineral medium	1000.0mL
Schlegel's vitamin solution	2.5mL

Mineral Medium:
Composition per 1010.0mL:

Na_2HPO_4	2.44g
KH_2PO_4	1.52g
$(NH_4)_2SO_4$	0.5g
$MgSO_4 \cdot 7H_2O$	0.2g
$CaCl_2 \cdot 2H_2O$	0.05g
Trace elements solution SL-4	10.0mL

pH 6.9 ± 0.2 at 25°C

Trace Elements Solution SL-4:
Composition per liter:

EDTA	0.5g
$FeSO_4 \cdot 7H_2O$	0.2g
Trace elements solution SL-6	100.0mL

Trace Elements Solution SL-6:
Composition per liter:

$MnCl_2 \cdot 4H_2O$	0.5g
H_3BO_3	0.3g
$CoCl_2 \cdot 6H_2O$	0.2g
$ZnSO_4 \cdot 7H_2O$	0.1g
$Na_2MoO_4 \cdot 2H_2O$	0.03g
$NiCl_2 \cdot 6H_2O$	0.02g
$CuCl_2 \cdot 2H_2O$	0.01g

Preparation of Trace Elements Solution SL-6: Add components to distilled/deionized water and bring volume to 1.0L. Mix thoroughly.

Preparation of Trace Elements Solution SL-4: Add components to distilled/deionized water and bring volume to 1.0L. Mix thoroughly.

Schlegel's Vitamin Solution:
Composition per 100.0mL:

Nicotinic acid	2.0g
Pyridoxamine	5.0mg
Cyanocobalamin	2.0mg
p-Aminobenzoate	1.0mg
Thiamine	1.0mg
Calcium DL-pantothenate	0.5mg
Biotin	0.2mg

Preparation of Schlegel's Vitamin Solution: Add components to distilled/deionized water and bring volume to 100.0mL. Filter sterilize.

Preparation of Medium: Add components, except Schlegel's vitamin solution, to distilled/deionized water and bring volume to 1.0L. Mix thoroughly. Autoclave for 15 min at 15 psi pressure–121°C. Aseptically add 2.5mL of sterile Schlegel's vitamin solution. Mix thoroughly. Aseptically distribute into sterile tubes or flasks.

Use: For the cultivation of *Pseudomonas chlororaphis* and *Rhodococcus* species.

Mineral Salts Agar

Composition per liter:

Agar	15.0g
$NaNO_3$	2.0g
K_2HPO_4	1.2g
$MgSO_4$	0.5g
KCl	0.5g
KH_2PO_4	0.14g
Yeast extract	0.02g
$Fe_2(SO_4)_3 \cdot H_2O$	0.01g

pH 7.2 ± 0.2 at 25°C

Preparation of Medium: Add components to distilled/deionized water and bring volume to 1.0L. Mix thoroughly. Adjust pH to 7.2. Gently heat and bring to boiling. Distribute into tubes. Autoclave for 15 min at 15 psi pressure–121°C. Allow tubes to cool in a slanted position. Add a strip of sterile filter paper onto cooled slant. Inoculate organisms on filter paper.

Use: For the cultivation and maintenance of *Cytophaga aurantiaca* and *Sporocytophaga myxococcoides*.

Mineral Salts Enrichment Medium

Composition per liter:

KH_2PO_4	1.36g
$(NH_4)_2SO_4$	0.5g
$MgSO_4 \cdot 7H_2O$	0.2g
$CaCl_2 \cdot 2H_2O$	0.01g
$FeSO_4 \cdot 7H_2O$	5.0mg
$MnSO_4 \cdot 7H_2O$	2.5mg
$Na_2MoO_4 \cdot 2H_2O$	2.5mg
Na_2HPO_4	2.13mg

pH 7.2 ± 0.2 at 25°C

Preparation of Medium: Add components to distilled/deionized water and bring volume to 1.0L. Mix thoroughly. Distribute into tubes or flasks. Autoclave for 15 min at 15 psi pressure–121°C.

Use: For the enrichment and cultivation of *Caulobacter* species.

Mineral Salts Medium

Composition per liter:

Na_2HPO_4	4.0g
KH_2PO_4	1.5g
NH_4Cl	1.0g
$MgSO_4 \cdot 7H_2O$	0.2g
Ferric ammonium citrate	5.0mg
Modified Hoagland trace elements solution	1.0mL

pH 7.0 ± 0.2 at 25°C

Modified Hoagland Trace Elements Solution:
Composition per 3.6L:

H_3BO_3	11.0g
$MnCl_2 \cdot 4H_2O$	7.0g
$AlCl_3$	1.0g
$CoCl_2$	1.0g
$CuCl_2$	1.0g
KI	1.0g
$NiCl_2$	1.0g
$ZnCl_2$	1.0g
$BaCl_2$	0.5g
KBr	0.5g
LiCl	0.5g
Na_2MoO_4	0.5g
$SeCl_4$	0.5g
$SnCl_2 \cdot 2H_2O$	0.5g
$NaVO_3 \cdot H_2O$	0.1g

Preparation of Modified Hoagland Trace Elements Solution: Prepare each component as a separate solution. Dissolve each salt in approximately 100.0mL of distilled/deionized water. Adjust the pH of each solution to below 7.0. Combine all the salt solutions and bring the volume to 3.6L with distilled/deionized water. Adjust the pH to 3–4. A yellow precipitate may form after mixing. After a few days, it will turn into a fine white precipitate. Mix the solution thoroughly before using.

Preparation of Medium: Add components to distilled/deionized water and bring volume to 1.0L. Mix thoroughly. Distribute into tubes or flasks. Autoclave for 15 min at 15 psi pressure–121°C.

Use: For the cultivation and maintenance of *Rhodococcus rhodochrous*.

Mineral Salts Medium with Butane

Composition per liter of tap water:

$(NH_4)_2HPO_4$	8.0g
$Na_2HPO_4 \cdot 12H_2O$	2.5g
KH_2PO_4	2.0g
$MgSO_4 \cdot 7H_2O$	0.5g
Yeast extract	100.0mg
$CaCl_2 \cdot 2H_2O$	60.0mg
$FeSO_4 \cdot 7H_2O$	30.0mg
$MnCl_2 \cdot 4H_2O$	60.0μg
$CuSO_4 \cdot 5H_2O$	15.0μg

pH 7.1 ± 0.2 at 25°C

Preparation of Medium: Add components to distilled/deionized water and bring volume to 1.0L. Mix thoroughly. Distribute into tubes or flasks. Autoclave for 15 min at 15 psi pressure–121°C. Incubate inoculated medium in 88% air + 7% *n*-butane + 5% CO_2.

Use: For the cultivation and maintenance of *Pseudomonas butanovora*.

Mineral Salts Medium with Methanol

Composition per liter:

$NaNH_4HPO_4 \cdot 4H_2O$	1.74g
$NaH_2PO_4 \cdot H_2O$	0.54g
$MgSO_4 \cdot 7H_2O$	0.2g
KCl	0.04g
$FeSO_4 \cdot 7H_2O$	5.0mg
Methanol	5.0mL
Trace mineral solution	1.0mL

pH 7.2 ± 0.2 at 25°C

Trace Mineral Solution:
Composition per liter:

H_3BO_3	2.86g
$MnCl_2 \cdot 4H_2O$	1.81g
$ZnSO_4 \cdot 7H_2O$	0.22g
$CuSO_4 \cdot 5H_2O$	0.08g
$CoCl_2 \cdot 6H_2O$	0.06g
$Na_2MoO_4 \cdot 2H_2O$	25.0mg

Preparation of Trace Mineral Solution: Add components to distilled/deionized water and bring volume to 1.0L. Mix thoroughly.

Preparation of Medium: Add components, except methanol, to distilled/deionized water and bring volume to 1.0L. Mix thoroughly. Distribute into tubes or flasks. Autoclave for 15 min at 15 psi pressure–121°C. Cool to 50°C. Filter sterilize methanol. Aseptically add sterile methanol to cooled, sterile basal medium.

Use: For the cultivation and maintenance of *Rhodococcus rhodochrous*.

Mineral Salts Medium with Methanol and Yeast Extract

Composition per liter:

$NaNH_4HPO_4 \cdot 4H_2O$	1.74g
$NaH_2PO_4 \cdot H_2O$	0.54g
$MgSO_4 \cdot 7H_2O$	0.2g
Yeast extract	0.2g
KCl	0.04g
$FeSO_4 \cdot 7H_2O$	5.0mg
Methanol	5.0mL
Trace mineral solution	1.0mL

pH 7.2 ± 0.2 at 25°C

Trace Mineral Solution:
Composition per liter:

H_3BO_3	2.86g

$MnCl_2 \cdot 4H_2O$ 1.81g
$ZnSO_4 \cdot 7H_2O$ 0.22g
$CuSO_4 \cdot 5H_2O$ 0.08g
$CoCl_2 \cdot 6H_2O$ 0.06g
$Na_2MoO_4 \cdot 2H_2O$ 25.0mg

Preparation of Trace Mineral Solution: Add components to distilled/deionized water and bring volume to 1.0L. Mix thoroughly.

Preparation of Medium: Add components, except methanol, to distilled/deionized water and bring volume to 1.0L. Mix thoroughly. Autoclave for 15 min at 15 psi pressure–121°C. Cool to 50°C. Filter sterilize methanol. Aseptically add sterile methanol to cooled, sterile basal medium. Aseptically distribute into sterile tubes or flasks.

Use: For the cultivation and maintenance of *Pseudomonas* species.

Mineral Salts Medium for Thermophiles

Composition per liter:
$NaNO_3$ 0.25g
NH_4Cl 0.25g
Na_2HPO_4 .... 210.0mg
$MgSO_4 \cdot 7H_2O$ 200.0mg
NaH_2PO_4 .... 90.0mg
KCl 40.0mg
$CaCl_2$ 15.0mg
$FeSO_4$ 1.0mg
Trace minerals solution 10.0mL
n-Heptadecane 1.0mL

Trace Minerals Solution:
Composition per liter:
$ZnSO_4 \cdot 7H_2O$ 7.0mg
H_3BO_4 .... 1.0mg
MoO_3 .... 1.0mg
$CuSO_4 \cdot 5H_2O$ 500.0μg
$CoSO_4 \cdot 7H_2O$ 18.0μg
$MnSO_4 \cdot 5H_2O$ 7.0μg

Preparation of Trace Minerals Solution: Add components to distilled/deionized water and bring volume to 1.0L. Mix thoroughly.

Preparation of Medium: Add components to distilled/deionized water and bring volume to 1.0L. Mix thoroughly. Distribute into tubes or flasks. Autoclave for 15 min at 15 psi pressure–121°C.

Use: For the cultivation of *Bacillus thermoleovorans.*

Mineral Salts Peptonized Milk Agar (SPMA)

Composition per liter:
Agar 15.0g
Milk, peptonized 1.0g
Mineral solution 100.0mL

Mineral Solution:
Composition per 100.0mL:
$MgSO_4 \cdot 7H_2O$ 0.5g
$CaCl_2$ 0.25g
K_2HPO_4 .... 0.25g
$(NH_4)_2SO_4$ 0.1g
$FeCl_3 \cdot 6H_2O$ 0.01g
$MnCl_2$ 0.1mg

Preparation of Mineral Solution: Add components to distilled/deionized water and bring volume to 100.0mL. Mix thoroughly. Filter sterilize.

Preparation of Medium: Add components, except mineral solution, to distilled/deionized water and bring volume to 900.0mL. Mix thoroughly. Gently heat and bring to boiling. Autoclave for 15 min at 15 psi pressure–121°C. Cool to 45°–50°C. Aseptically add 100.0mL of sterile mineral solution. Mix thoroughly. Pour into sterile Petri dishes or distribute into sterile tubes.

Use: For the cultivation of freshwater *Myxobacterium* species.

Mineral Salts for Thermophiles (L Salts for Thermophiles)

Composition per liter:
$NaNO_3$ 0.25g
NH_4Cl 0.25g
Na_2HPO_4 .... 0.21g
$MgSO_4 \cdot 7H_2O$ 0.2g
NaH_2PO_4 .... 0.09g
KCl 0.04g
$CaCl_2$ 0.02g
$FeSO_4$ 1.0mg
Trace mineral solution 10.0mL
n-Heptadecane 1.0mL

Trace Mineral Solution:
Composition per liter:
$ZnSO_4 \cdot 7H_2O$ 7.0mg
H_3BO_4 .... 1.0mg
MoO_3 .... 1.0mg
$CuSO_4 \cdot 5H_2O$ 500.0μg
$CoSO_4 \cdot 7H_2O$ 18.0μg
$MnSO_4 \cdot 5H_2O$ 7.0μg

Preparation of Trace Mineral Solution: Add components to distilled/deionized water and bring volume to 1.0L. Mix thoroughly.

Preparation of Medium: Add components, except *n*-heptadecane, to distilled/deionized water and bring volume to 1.0L. Mix thoroughly. Autoclave for 15 min at 15 psi pressure–121°C. Aseptically add the *n*-heptadecane. Mix thoroughly. Distribute into sterile tubes or flasks.

Use: For the cultivation and maintenance of *Bacillus thermoleovorans.*

Minerals Modified Medium

Composition per liter:
Lactose 20.0g
Sodium glutamate 12.7g
NH_4Cl 5.0g
K_2HPO_4 .... 1.8g
Sodium formate 0.5g
$MgSO_4 \cdot 7H_2O$ 0.2g
L-Aspartic acid 0.048g
L-Cystine 0.04g
L-Arginine 0.04g
Ferric ammonium citrate 0.02g
$CaCl_2 \cdot 2H_2O$ 0.02g
Bromcresol Purple 0.02g
Thiamine 2.0mg
Nicotinic acid 2.0mg
Pantothenic acid 2.0mg
pH 6.7 ± 0.2 at 25°C

Source: This medium is available as a premixed powder from Oxoid Unipath.

Preparation of Medium: Add NH_4Cl to distilled/deionized water and bring volume to 800.0mL. Add remaining components and bring volume to 1.0L. Mix thoroughly. Adjust pH to 6.7. Distribute into tubes or flasks. Autoclave for 10 min at 10 psi pressure–116°C. Check pH after autoclaving. This medium is double strength.

Use: For the enumeration of coliform bacteria in water.

Minimal Medium for Denitrifying Bacteria

Composition per liter:

Solution A	980.0mL
Solution B	10.0mL
Solution C	10.0mL

Solution A:

Composition per 980.0mL:

KNO_3	5.0g
Carbon source	4.0g
$(NH_4)_2SO_4$	1.0g
$K_2HPO_4 \cdot 3H_2O$	0.87g
KH_2PO_4	0.54g

Preparation of Solution A: Add components to distilled/deionized water and bring volume to 1.0L. Mix thoroughly. Autoclave for 15 min at 15 psi pressure–121°C. Cool to 25°C.

Solution B:

Composition per 100.0mL:

$MgSO_4 \cdot 7H_2O$	2.0g

Preparation of Solution B: Add $MgSO_4 \cdot 7H_2O$ to distilled/deionized water and bring volume to 100.0mL. Mix thoroughly. Autoclave for 15 min at 15 psi pressure–121°C. Cool to 25°C.

Solution C:

Composition per 100.0mL:

$CaCl_2 \cdot 2H_2O$	0.2g
$FeSO_4 \cdot 7H_2O$	0.1g
$MnSO_4 \cdot H_2O$	0.05g
$CuSO_4 \cdot 5H_2O$	0.01g
$Na_2MoO_4 \cdot 2H_2O$	0.01g
HCl (0.1*N* solution)	100.0mL

Preparation of Solution C: Combine components. Mix thoroughly. Autoclave for 15 min at 15 psi pressure–121°C. Cool to 25°C.

Preparation of Medium: Aseptically combine 980.0mL of cooled sterile solution A, 10.0mL of cooled sterile solution B, and 10.0mL of cooled sterile solution C. Mix thoroughly. Aseptically distribute into sterile tubes or flasks.

Use: For the isolation and cultivation of denitrifying bacteria.

Mixotrophic *Nitrobacter* Medium (DSMZ Medium 756a)

Composition per liter:

$NaNO_2$	2.0g
Yeast extract	1.5g
Peptone	1.5g
Na-pyruvate	0.55g
Stock solution	100.0mL
Trace elements solution	1.0mL

pH 7.4 ± 0.2 at 25°C

Stock Solution:

Composition per liter:

NaCl	5.0g
KH_2PO_4	1.5g
$MgSO_4 \cdot 7H_2O$	0.5g
$CaCO_3$	0.07g

Preparation of Stock Solution: Add components to distilled/deionized water and bring volume to 1.0L. Mix thoroughly.

Trace Elements Solution:

Composition per liter:

$FeSO_4 \cdot 7H_2O$	97.3mg
H_3BO_3	49.4mg
$ZnSO_4 \cdot 7H_2O$	43.1mg
$(NH_4)_6Mo_7O_{24} \cdot 4H_2O$	37.1mg
$MnSO_4 \cdot 2H_2O$	33.8mg
$CuSO_4 \cdot 5H_2O$	25.0mg

Preparation of Trace Elements Solution: Add components to distilled/deionized water and bring volume to 1.0L. Mix thoroughly.

Preparation of Medium: Add components to distilled/deionized water and bring volume to 1.0L. Mix thoroughly. Adjust pH to 7.4. Distribute into tubes or flasks. Autoclave for 15 min at 15 psi pressure–121°C.

Use: For the cultivation of *Nitrobacter vulgaris.*

Mixotrophic *Nitrobacter* Medium (LMG Medium 246)

Composition per liter:

$NaNO_2$	2.0g
Yeast extract	1.50g
Peptone	1.50g
Na-pyruvate	0.55g
Stock solution	100.0mL
Trace elements solution	1.0mL

pH 7.4 ± 0.2 at 25°C

Stock Solution:

Composition per liter:

NaCl	5.0g
KH_2PO_4	1.5g
$MgSO_4 \cdot 7H_2O$	0.5g
$CaCO_3$	0.07g

Preparation of Stock Solution: Add components to distilled/deionized water and bring volume to 1.0L. Mix thoroughly.

Trace Elements Solution:

Composition per liter:

$(NH_4)Mo7O_2$	437.10mg
$FeSO_4 \cdot 7H_2O$	97.30mg
$ZnSO_4 \cdot 7H_2O$	43.10mg
H_3BO_3	39.40mg
$MnSO_4 \cdot H_2O$	33.80mg
$CuSO_2 \cdot 5H_2O$	25.00mg

Preparation of Trace Elements Solution: Add components to distilled/deionized water and bring volume to 1.0L. Mix thoroughly.

Preparation of Medium: Add components to distilled/deionized water and bring volume to 1.0L. Mix thoroughly. Adjust pH to 7.4 with NaOH. Distribute into tubes or flasks. Autoclave for 15 min at 15 psi pressure–121°C.

Use: For the cultivation of mixotrophic *Nitrobacter* spp.

Mixotrophic *Nitrobacter* Medium, 10% (DSMZ Medium 756b)

Composition per liter:

$NaNO_2$	2.0g
Yeast extract	0.15g
Peptone	0.15g
Na-pyruvate	0.055g

Stock solution..100.0mL
Trace elements solution 1.0mL
pH 8.6 ± 0.2 at 25°C

Stock Solution:
Composition per liter:
NaCl 5.0g
KH_2PO_4 ....1.5g
$MgSO_4·7H_2O$ 0.5g
$CaCO_3$ 0.07g

Preparation of Stock Solution: Add components to distilled/deionized water and bring volume to 1.0L. Mix thoroughly.

Trace Elements Solution:
Composition per liter:
$FeSO_4·7H_2O$ 97.3mg
H_3BO_3 ....49.4mg
$ZnSO_4·7H_2O$ 43.1mg
$(NH_4)_6Mo_7O_{24}·4H_2O$ 37.1mg
$MnSO_4·2H_2O$ 33.8mg
$CuSO_4·5H_2O$ 25.0mg

Preparation of Trace Elements Solution: Add components to distilled/deionized water and bring volume to 1.0L. Mix thoroughly.

Preparation of Medium: Add components to distilled/deionized water and bring volume to 1.0L. Mix thoroughly. Adjust pH to 8.6. Distribute into tubes or flasks. Autoclave for 15 min at 15 psi pressure–121°C.

Use: For the cultivation of *Nitrobacter hamburgensis* and *Nitrobacter vulgaris*.

ML Medium (Minimal Lactate Medium)

Composition per liter:
Sodium lactate...5.0g
$MgSO_4·7H_2O$ 2.0g
NH_4Cl 1.0g
Na_2SO_4 ....1.0g
Yeast extract..1.0g
K_2HPO_4 ....0.5g
L-Cysteine 0.5g
$CaCl_2·6H_2O$ 0.1g
Resazurin 1.0mg
$NaHCO_3$ solution 25.0mL
$FeSO_4·7H_2O$ solution 25.0mL
pH 6.8 ± 0.2 at 25°C

$NaHCO_3$ Solution:
Composition per 25.0mL:
$NaHCO_3$ 4.0g

Preparation of $NaHCO_3$ Solution: Add $NaHCO_3$ to distilled/deionized water and bring volume to 25.0mL. Mix thoroughly. Filter sterilize. Gas with O_2-free 97% N_2 + 3% H_2. Cap with a rubber stopper.

$FeSO_4·7H_2O$ Solution:
Composition per 25.0mL:
$FeSO_4·7H_2O$ 4.0mg

Preparation of $FeSO_4·7H_2O$ Solution: Add $FeSO_4·7H_2O$ to distilled/deionized water and bring volume to 25.0mL. Mix thoroughly. Filter sterilize. Gas with O_2-free 97% N_2 + 3% H_2. Cap with a rubber stopper.

Preparation of Medium: Add components, except $NaHCO_3$ solution and $FeSO_4·7H_2O$ solution, to distilled/deionized water and bring volume to 1.0L. Gently heat and bring to boiling under O_2-free 97% N_2 + 3% H_2. Adjust pH to 6.8. Continue boiling until rezasurin becomes colorless, indicating reduction. Distribute anaerobically under O_2-free 97% N_2 + 3% H_2 into tubes in 10.0mL volumes. Cap with rubber stoppers. Place tubes in a press. Autoclave for 15 min at 15 psi pressure–121°C. Cool to room temperature. Prior to inoculation, add 0.25mL of sterile $NaHCO_3$ solution and 0.25mL of sterile $FeSO_4·7H_2O$ solution to each test tube containing 10.0mL of sterile basal medium.

Use: For the cultivation and maintenance of *Desulfovibrio* species.

MMA Salts Medium

Composition per liter:
Agar, noble 20.0g
K_2HPO_4 ....1.2g
KH_2PO_4 ....0.62g
$(NH_4)_2SO_4$ 0.5g
$MgSO_4·7H_2O$ 0.2g
NaCl 0.1g
$CaCl_2·6H_2O$ 0.05g
$ZnSO_4·7H_2O$ 70.0μg
H_3BO_3 ....10.0μg
$MnSO_4·5H_2O$ 10.0μg
$Na_2MoO_4·2H_2O$ 10.0μg
$CoCl_2·6H_2O$ 5.0μg
$CuSO_4·5H_2O$ 5.0μg
$FeCl_3·6H_2O$ 1.0mg
Monomethylamine solution.................................10.0mL
pH 7.0 ± 0.2 at 25°C

Monomethylamine Solution:
Composition per 10.0mL:
Monomethylamine 1.0g

Preparation of Monomethylamine Solution: Add monomethylamine to distilled/deionized water and bring volume to 10.0mL. Mix thoroughly. Filter sterilize.

Preparation of Medium: Add components, except monomethylamine solution, to distilled/deionized water and bring volume to 990.0mL. Mix thoroughly. Gently heat and bring to boiling. Autoclave for 15 min at 15 psi pressure–121°C. Cool to 50°C. Aseptically add 10.0mL of sterile methylamine solution. Mix thoroughly. Pour into sterile Petri dishes or distribute into sterile tubes.

Use: For the cultivation of *Hyphomicrobium* species and *Methylophilus methylotrophus*.

MMB Medium

Composition per liter:
Glucose 1.0g
$(NH_4)_2SO_4$ 0.25g
Peptone 0.15g
Yeast extract 0.15g
Glucose solution 20.0mL
Hutner's basal salts solution...............................20.0mL
Vitamin solution 10.0mL
pH 7.2 ± 0.2 at 25°C

Glucose Solution:
Composition per 20.0mL:
Glucose 1.5g

Preparation of Glucose Solution: Add glucose to distilled/deionized water and bring volume to 20.0mL. Mix thoroughly. Filter sterilize.

Hutner's Basal Salts Solution:
Composition per liter:
$MgSO_4·7H_2O$ 29.7g
Nitrilotriacetic acid 10.0g

$CaCl_2 \cdot 2H_2O$ 3.335g
$FeSO_4 \cdot 7H_2O$ 99.0mg
$(NH_4)_6MoO_7O_{24} \cdot 4H_2O$ 9.25mg
"Metals 44" 50.0mL

"Metals 44":
Composition per 100.0mL:
$ZnSO_4 \cdot 7H_2O$ 1.095g
$FeSO_4 \cdot 7H_2O$ 0.5g
Sodium EDTA 0.25g
$MnSO_4 \cdot H2O$ 0.154g
$CuSO_4 \cdot 5H_2O$ 39.2mg
$Co(NO_3)_2 \cdot 6H_2O$ 24.8mg
$Na_2B_4O_7 \cdot 10H_2O$ 17.7mg

Preparation of "Metals 44": Add sodium EDTA to distilled/deionized water and bring volume to 90.0mL. Mix thoroughly. Add a few drops of concentrated H_2SO_4 to retard precipitation of heavy metal ions. Add remaining components. Mix thoroughly. Bring volume to 100.0mL with distilled/deionized water.

Preparation of Hutner's Basal Salts Solution: Add nitrilotriacetic acid to 500.0mL of distilled/deionized water. Adjust pH to 6.5 with KOH. Add remaining components. Add distilled/deionized water to 1.0L. Adjust pH to 6.8.

Vitamin Solution:
Composition per 100.0mL:
Vitamin B_{12} 0.01mg
Calcium DL-pantothenate 0.5mg
Nicotinamide 0.5mg
Pyridoxine·HCl 0.5mg
Riboflavin 0.5mg
Thiamine·HCl 0.5mg
Biotin 0.2mg
Folic acid 0.2mg

Preparation of Vitamin Solution: Add components to distilled/deionized water and bring volume to 100.0mL. Mix thoroughly. Filter sterilize.

Preparation of Medium: Add components, except glucose solution and vitamin solution, to distilled/deionized water and bring volume to 970.0mL. Mix thoroughly. Autoclave for 15 min at 15 psi pressure–121°C. Aseptically add 20.0mL of sterile glucose solution and 20.0mL of sterile vitamin solution. Mix thoroughly. Aseptically distribute into sterile tubes or flasks.

Use: For the cultivation of *Angulomicrobium tetraedrale, Labrys monachus, Prosthecomicrobium polyspheroidum,* and *Aquabacter spiritensis.*

MMS Medium for *Thermotoga neapolitana*

Composition per liter:
NaCl 6.93g
Starch 5.0g
$MgSO_4 \cdot 7H_2O$ 1.75g
$MgCl_2 \cdot 6H_2O$ 1.38g
KH_2PO_4 .... 0.5g
$Na_2S \cdot 9H_2O$ 0.5g
$CaCl_2$ 0.38g
KCl 0.16g
NaBr 25.0mg
H_3BO_3 .... 7.5mg
$SrCl_2 \cdot 6H_2O$ 3.8mg
$(NH_4)_2Ni(SO_4)_2$ 2.0mg
Resazurin 1.0mg
KI 0.025mg
Wolfe's mineral solution 15.0mL
pH 6.5 ± 0.2 at 25°C

Wolfe's Mineral Solution:
Composition per liter:
$MgSO_4 \cdot 7H_2O$ 3.0g
Nitriloacetic acid 1.5g
NaCl 1.0g
$MnSO_4 \cdot H_2O$ 0.5g
$FeSO_4 \cdot 7H_2O$ 0.1g
$CoCl_2 \cdot 6H_2O$ 0.1g
$CaCl_2$ 0.1g
$ZnSO_4 \cdot 7H_2O$ 0.1g
$CuSO_4 \cdot 5H_2O$ 0.01g
$AlK(SO_4)_2 \cdot 12H_2O$ 0.01g
H_3BO_3 .... 0.01g
$Na_2MoO_4 \cdot 2H_2O$ 0.01g

Preparation of Wolfe's Mineral Solution: Add nitrilotriacetic acid to 500.0mL of distilled/deionized water. Dissolve by adjusting pH to 6.5 with KOH. Add remaining components. Add distilled/deionized water to 1.0L.

Preparation of Medium: Prepare and dispense medium under 80% N_2 and 20% CO_2. Add components to distilled/deionized water and bring volume to 1.0L. Mix thoroughly. Adjust pH to 6.5 with H_2SO_4. Distribute into tubes or flasks. Autoclave for 15 min at 15 psi pressure–121°C.

Use: For the cultivation and maintenance of *Thermotoga neapolitana.*

MN Marine Medium

Composition per liter:
Noble agar 10.0g
$NaNO_3$ 0.75g
$MgSO_4 \cdot 7H_2O$ 0.04g
$CaCl_2 \cdot 2H_2O$ 0.02g
$K_2HPO_4 \cdot 3H_2O$ 0.02g
Na_2CO_3 .... 0.02g
Citric acid 3.0mg
Ferric ammonium citrate 3.0mg
Disodium potassium EDTA 0.5mg
Trace metals A-5 .... 1.0mL
pH 8.5 ± 0.2 at 25°C

A-5 Trace Metal Mix:
Composition per liter:
H_3BO_3 .... 2.86g
$MnCl_2 \cdot 4H_2O$ 1.81g
$ZnSO_4 \cdot 7H_2O$ 0.222g
$CuSO_4 \cdot 5H_2O$ 0.079g
$Na_2MoO_4 \cdot 2H_2O$ 0.039g
$Co(NO_3)_2 \cdot 6H_2O$ 0.049g

Preparation of A-5 Trace Metal Mix: Add components to distilled/deionized water and bring volume to 1.0L. Mix thoroughly.

Preparation of Medium: Add components to 750.0mL of seawater and bring volume to 1L with glass-distilled water. Mix thoroughly. Gently heat and bring to boiling. Autoclave for 15 min at 15 psi pressure–121°C. After autoclaving, adjust pH to 8.5 with KOH.

Use: For the cultivation and maintenance of marine cyanobacteria.

MN Marine Medium with Vitamin B_{12}

Composition per liter:
Noble agar 10.0g

$NaNO_3$	0.75g
$MgSO_4 \cdot 7H_2O$	0.04g
$CaCl_2 \cdot 2H_2O$	0.02g
$K_2HPO_4 \cdot 3H_2O$	0.02g
Na_2CO_3	0.02g
Citric acid	3.0mg
Ferric ammonium citrate	3.0mg
Disodium potassium EDTA	0.5mg
Vitamin B_{12}	20.0μg
Trace metals A-5	1.0mL

pH 8.5 ± 0.2 at 25°C

A-5 Trace Metal Mix:
Composition per liter:

H_3BO_3	2.86g
$MnCl_2 \cdot 4H_2O$	1.81g
$ZnSO_4 \cdot 7H_2O$	0.222g
$CuSO_4 \cdot 5H_2O$	0.079g
$Na_2MoO_4 \cdot 2H_2O$	0.039g
$Co(NO_3)_2 \cdot 6H_2O$	0.049g

Preparation of A-5 Trace Metal Mix: Add components to distilled/deionized water and bring volume to 1.0L. Mix thoroughly.

Preparation of Medium: Add components to 750.0mL of seawater and bring volume to 1.0L with glass-distilled water. Mix thoroughly. Gently heat and bring to boiling. Autoclave for 15 min at 15 psi pressure–121°C. After autoclaving, adjust pH to 8.5 with KOH.

Use: For the cultivation and maintenance of *Dermocarpa* species, *Dermocarpella* species, *Myxosarcina* species, *Phormidium* species, *Pleurocapsa* species, *Synechococcus* species, *Synechocystis* species, and *Xenococcus* species.

Moderate Halophilic Medium (HM)

Composition per liter:

NaCl	81.0g
Agar	20.0g
Yeast extract	10.0g
$MgSO_4 \cdot 7H_2O$	9.6g
$MgCl_2 \cdot 6H_2O$	7.0g
Proteose peptone No. 3	5.0g
KCl	2.0g
Glucose	1.0g
$CaCl_2 \cdot 2H_2O$	0.36g
$NaHCO_3$	0.06g
NaBr	0.026g

pH 7.2 ± 0.2 at 25°C

Preparation of Medium: Add components to distilled/deionized water and bring volume to 1.0L. Mix thoroughly. Gently heat and bring to boiling. Distribute into tubes or flasks. Autoclave for 15 min at 15 psi pressure–121°C. Pour into sterile Petri dishes or leave in tubes.

Use: For the cultivation and maintenance of *Bacillus halophilus, Halococcus saccharolyticus, Marinococcus albus, Marinococcus halophilus,* and *Marinococcus hispanicus.*

Modified PYNFH Medium

Composition per liter:

Peptone	10.0g
Yeast extract	10.0g
Yeast nucleic acid	1.0g
Folic acid	15.0mg
Hemin	1.0mg
Fetal bovine serum, heat inactivated	100.0mL
Buffer solution	20.0mL

pH 6.5 ± 0.2 at 25°C

Buffer Solution:
Composition per liter:

Na_2HPO_4	25.0g
KH_2PO_4	18.1g

Preparation of Buffer Solution: Add components to distilled/deionized water and bring volume to 1.0L. Mix thoroughly. Adjust pH to 6.5. Autoclave for 15 min at 15 psi pressure–121°C. Cool to 25°C.

Preparation of Medium: Add components, except buffer solution and fetal bovine serum, to distilled/deionized water and bring volume to 880.0mL. Mix thoroughly. Autoclave for 15 min at 15 psi pressure–121°C. Cool to 25°C. Aseptically add 20.0mL of sterile buffer solution and 100.0mL of sterile, heat-inactivated fetal bovine serum. Mix thoroughly. Aseptically distribute into sterile tubes or flasks.

Use: For the cultivation of *Dexiostoma campyla, Hartmannella vermiformis, Naegleria australiensis, Naegleria fowleri, Naegleria gruberi, Naegleria jadini, Phytomonas davidi, Tetrahymena* species, *Vahlkampfia avara,* and *Willaertia magna.*

Modified Semisolid Rappaport-Vassiliadis Medium (MSRV Medium)

Composition per liter:

$MgCl_2$, anhydrous	10.93g
NaCl	7.34g
Casein hydrolysate	4.59g
Tryptose	4.59g
Agar	2.7g
KH_2PO_4	1.47g
Malachite Green oxalate	0.037g
Novobiocin	10.0mL

pH 5.2 ± 0.2 at 25°C

Source: This medium is available as a premixed powder from Oxoid Unipath.

Novobiocin Solution:
Composition per 10.0mL:

Novobiocin	0.02g

Preparation of Novobiocin Solution: Add novobiocin to 10.0mL of distilled/deionized water. Mix thoroughly. Filter sterilize.

Preparation of Medium: Add components, except novobiocin solution, to distilled/deionized water and bring volume to 990.0mL. Mix thoroughly. Gently heat to boiling. Do not autoclave. Cool to 45°–50°C. Aseptically add 10.0mL of sterile novobiocin solution. Mix thoroughly. Pour into sterile Petri dishes. Air-dry plates for at least 1 hr.

Use: For the isolation and cultivation of motile *Salmonella* species from environmental samples.

Modified Shieh Agar (LMG Medium 215)

Composition per liter:

Agar	15.0g
Peptone	5.0g
Yeast extract	1.0g
$MgSO_4 \cdot 7H_2O$	0.3g
K_2HPO_4	0.1g
KH_2PO_4	50.0mg
$NaHCO_3$	50.0mg
Na-acetate	10.0mg
$BaCl_2 \cdot H_2O$	10.0mg

$CaCl_2 \cdot 2H_2O$ .... 6.7 mg
$FeSO_4 \cdot 7H_2O$ 1.0mg
pH 7.3 ± 0.2 at 25°C

Preparation of Medium: Add components to distilled/deionized water and bring volume to 1.0L. Mix thoroughly. Gently heat and bring to boiling. Distribute into tubes or flasks. Autoclave for 15 min at 15 psi pressure–121°C. Pour into sterile Petri dishes or leave in tubes.

Use: For the cultivation and maintenance of *Flavobacterium* spp., *Flexibacter* spp., *Chitinophaga pinensis,* and *Flectobacillus major.*

Modified *Thermus* Medium (DSMZ Medium 630)

Composition per liter:
Agar 28.0g
Yeast extract 2.5g
Tryptone 2.5g
$MgCl_2 \cdot 6H_2O$ 200.0mg
Nitrilotriacetic acid 100.0mg
$CaSO_4 \cdot 2H_2O$ 40.0mg
Phosphate solution 100.0mL
Ferric citrate solution 0.5mL
Trace elements solution 0.5mL
pH 7.2 ± 0.2 at 25°C

Phosphate Solution:
Composition liter:
$Na_2HPO_4 \cdot 12H_2O$ 43.0g
KH_2PO_4 .... 5.44g

Preparation of Phosphate Solution: Add components to distilled/deionized water and bring volume to 1.0L. Mix thoroughly. Adjust pH to 7.2. Autoclave for 15 min at 15 psi pressure–121°C. Cool to 50°C.

Ferric Citrate Solution:
Composition liter:
Ferric citrate 2.5g

Preparation of Ferric Citrate Solution: Add ferric citrate to distilled/deionized water and bring volume to 1.0L. Mix thoroughly.

Trace Elements Solution:
Composition per liter:
Nitrilotriacetic acid 12.8g
$FeCl_2 \cdot 4H_2O$ 1.0g
$MnCl_2 \cdot 4H_2O$ 0.5g
$CoCl_2 \cdot 4H_2O$ 0.3g
$Na_2MoO_4 \cdot 4H_2O$ 50.0mg
$CuCl_2 \cdot 2H_2O$ 50.0mg
$NiCl_2 \cdot 6H_2O$ 20.0mg
H_3BO_3 .... 20.0mg

Preparation of Trace Elements Solution: Add nitrilotriacetic acid to 500.0mL of distilled/deionized water. Dissolve by adjusting pH to 7.0 with KOH. Add remaining components. Add distilled/deionized water to 1.0L. Mix thoroughly. Adjust pH to 6.8.

Preparation of Medium: Add components, except phosphate solution, to distilled/deionized water and bring volume to 900.0mL. Mix thoroughly. Autoclave for 15 min at 15 psi pressure–121°C. Cool to 50°C. Aseptically add 100.0mL phosphate solution. Mix thoroughly. Adjust pH to 7.2. Pour into Petri dishes or aseptically distribute into sterile tubes.

Use: For the cultivation and maintenance of *Thermus* sp., *Rhodothermus marinus,* and *Albidovulum inexpectatum.*

Modified VWM Medium (DSMZ Medium 503a)

Composition per 1013.4mL:
Solution A 940.0mL
Solution E 50.0mL
Solution F 10.0mL
Solution G 10.0mL
Solution B 1.0mL
Solution C 1.0mL
Solution D 1.0mL
Solution H 0.4mL
pH 7.3 ± 0.2 at 25°C

Solution A:
Composition per 940.0mL:
NaCl 1.0g
KCl 0.5g
$MgCl_2 \cdot 6H_2O$ 0.4g
NH_4Cl 0.25g
KH_2PO_4 .... 0.2g
$CaCl_2 \cdot 2H_2O$ 0.15g
Resazurin 0.5mg

Preparation of Solution A: Prepare under 80% N_2 + 20% CO_2 gas atmosphere. Add components to distilled/deionized water and bring volume to 940.0mL. Mix thoroughly. Adjust pH to 7.2. Sparge with 80% N_2 + 20% CO_2. Autoclave for 15 min at 15 psi pressure–121°C. Cool to 25°C.

Solution B:
Composition per liter:
$FeCl_2 \cdot 4H_2O$ 1.5g
$CoCl_2 \cdot 6H_2O$ 190.0mg
$MnCl_2 \cdot 4H_2O$ 100.0mg
$ZnCl_2$ 70.0mg
$Na_2MoO_4 \cdot 2H_2O$ 36.0mg
$NiCl_2 \cdot 6H_2O$ 24.0mg
H_3BO_3 .... 6.0mg
$CuCl_2 \cdot 2H_2O$ 2.0mg
HCl (25% solution) 10.0mL

Preparation of Solution B: Add $FeCl_2 \cdot 4H_2O$ to 10.0mL of HCl solution. Mix thoroughly. Add distilled/deionized water and bring volume to 1.0L. Add remaining components. Mix thoroughly. Sparge with 80% N_2 + 20% CO_2. Autoclave for 15 min at 15 psi pressure–121°C. Cool to 25°C.

Solution C:
Composition per liter:
Pyridoxine hydrochloride 300.0mg
Thiamine-HCl·$2H_2O$ 200.0mg
Nicotinic acid 200.0mg
Vitamin B_{12} 100.0mg
Calcium pantothenate 100.0mg
p-Aminobenzoic acid 80.0mg
D(+)-Biotin 20.0mg

Preparation of Solution C: Add components to distilled/deionized water and bring volume to 1.0L. Sparge with 100% N_2. Mix thoroughly. Filter sterilize.

Solution D:
Composition per liter:
NaOH 0.5g
$Na_2WO_4 \cdot 2H_2O$ 4.0mg
$Na_2SeO_3 \cdot 5H_2O$ 3.0mg

Preparation of Solution D: Add components to distilled/deionized water and bring volume to 1.0L. Mix thoroughly. Sparge with 100% N_2. Filter sterilize.

Solution E:

Composition per 100.0mL:

$NaHCO_3$ 5.0g

Preparation of Solution E: Add $NaHCO_3$ to distilled/deionized water and bring volume to 100.0mL. Mix thoroughly. Sparge with 100% N_2 gas mixture. Autoclave for 15 min at 15 psi pressure–121°C. Cool to 25°C.

Solution F:

Composition per 10mL:

Taurine 2.5g

Preparation of Solution F: Add taurine to distilled/deionized water and bring volume to 100.0mL. Mix thoroughly. Sparge with 100% N_2 gas mixture. Autoclave for 15 min at 15 psi pressure–121°C. Cool to 25°C.

Solution G:

Composition per 10mL:

$Na_2S \cdot 9H_2O$ 0.125g

Preparation of Solution G: Add $Na_2S \cdot 9H_2O$ to distilled/deionized water and bring volume to 10.0mL. Mix thoroughly. Autoclave under 100% N_2 for 15 min at 15 psi pressure–121°C. Cool to 25°C.

Solution H:

Composition per 10mL:

Na-dithionite 0.5g

Preparation of Solution H: Add Na-dithionite to distilled/deionized water and bring volume to 10.0mL. Mix thoroughly. Autoclave under 100% N_2 for 15 min at 15 psi pressure–121°C. Cool to 25°C.

Preparation of Medium: Prepare and dispense medium under 80% N_2 + 20% CO_2 gas atmosphere. Sequentially add 1.0mL solution B, 1.0mL solution C, 1.0mL solution D, 50.0mL solution E, 10.0mL solution F, 10.0mL solution G, and 0.4mL solution H, to 940.0mL solution A. Distribute anaerobically under 80% N_2 + 20% CO_2 into appropriate vessels. The pH should be 7.2-7.4.

Use: For the cultivation of *Desulfonispora thiosulfatigenes.*

Møller Decarboxylase Broth

Composition per liter:

Amino acid 10.0g
Peptic digest of animal tissue 5.0g
Beef extract 5.0g
Glucose 0.5g
Bromcresol Purple 0.01g
Cresol Red 5.0mg
Pyridoxal 5.0mg

pH 6.0 ± 0.2 at 25°C

Source: This medium is available as a premixed powder from BD Diagnostic Systems.

Preparation of Medium: Add components to distilled/deionized water and bring volume to 1.0L. Use L-lysine, L-arginine, or L-ornithine. Mix thoroughly. Gently heat until dissolved. Distribute into screw-capped tubes in 5.0mL volumes. Autoclave for 15 min at 15 psi pressure–121°C. A slight precipitate may form in the ornithine broth.

Use: For the differentiation of Gram-negative enteric bacteria based on the production of arginine dihydrolase, lysine decarboxylase, or ornithine decarboxylase.

Møller KCN Broth Base

Composition per liter:

Na_2HPO_4 .... 5.64g
NaCl 5.0g
Pancreatic digest of casein 1.5g
Peptic digest of animal tissue 1.5g
KH_2PO_4 .... 0.225g
KCN solution 0.15mL

pH 7.6 ± 0.2 at 25°C

Source: This medium is available as a premixed powder from BD Diagnostic Systems.

KCN Solution:

Composition per 100.0mL:

KCN 0.5g

Preparation of KCN Solution: Add KCN to 100.0mL of cold distilled/deionized water. Mix thoroughly and cap. Do not mouth pipette.

Caution: Cyanide is toxic.

Preparation of Medium: Add components, except KCN solution, to distilled/deionized water and bring volume to 1.0L. Mix thoroughly. Autoclave for 15 min at 15 psi pressure–121°C. Cool to room temperature. Prior to use, add 0.15mL of KCN solution. Mix thoroughly. Aseptically distribute into sterile tubes.

Use: For the differentiation of Gram-negative enteric bacteria on the basis of their ability to grow in the presence of cyanide.

Moraxella Medium (LMG Medium 204)

Composition per liter:

Special peptone 23.0g
Agar 15.0g
Glucose 5.0g
NaCl 5.0g
Soluble starch 1.0g
Cysteine hydrochloride 0.3g
Sheep blood, sterile defrinated 50.0mL

pH 7.1 ± 0.2 at 25°C

Source: Special peptone is available from Oxoid Unipath.

Preparation of Medium: Add components, except sheep blood, to 950.0mL distilled/deionized water and bring volume to 1.0L. Mix thoroughly. Gently heat and bring to boiling. Autoclave for 15 min at 15 psi pressure–121°C. Cool to 45°–50°C. Aseptically add 50.0mL sterile sheep blood. Mix thoroughly. Pour into sterile Petri dishes or distribute into sterile tubes.

Use: For the cultivation and maintenance of *Moraxella osloensis, Moraxella atlantae,* and *Cellulomonas hominis.*

MP Agar

Composition per liter:

Agar 15.0g
Sodium acetate 0.1g
Basal medium 1.0L
Sodium sulfide solution 3.0mL

pH 7.0–7.5 at 25°C

Basal Medium:

Composition per liter:

$CaSO_4 \cdot 2H_2O$ (saturated solution) 20.0mL
NH_4Cl (4% solution) 5.0mL
Trace elements solution 5.0mL

$MgSO_4·7H_2O$ (1% solution) 1.0mL
K_2HPO_4 (1% solution) 1.0mL

Preparation of Basal Medium: Add components to distilled/deionized water and bring volume to 1.0L. Mix thoroughly.

Trace Elements Solution:
Composition per liter:

Ferrous EDTA solution 20.0mL
$ZnSO_4·7H_2O$ (0.1% solution) 10.0mL
$MnSO_4·4H_2O$ (0.02% solution) 10.0mL
$CuSO_4·5H_2O$ (0.00005% solution) 10.0mL
H_3BO_3 (0.1% solution) 10.0mL
$Co(NO_3)_2$ or
$CoCl_2·6H_2O$ (0.01% solution) 10.0mL
$Na_2MoO_4·2H_2O$ (0.01% solution) 10.0mL

Preparation of Trace Elements Solution: Add components to distilled/deionized water and bring volume to 1.0L. Mix thoroughly.

Ferrous EDTA Solution:
Composition per 100.0mL:

$FeSO_4·7H_2O$ 7.0g
EDTA 2.0g
HCl, concentrated 1.0mL

Preparation of Ferrous EDTA Solution: Add components to distilled/deionized water and bring volume to 100.0mL. Mix thoroughly.

Sodium Sulfide Solution:
Composition per 10.0mL:

$Na_2S·9H_20$ 1.0g

Preparation of Sodium Sulfide Solution: Add $Na_2S·9H_20$ to distilled/deionized water and bring volume to 10.0mL. Mix thoroughly. Autoclave for 15 min at 15 psi pressure–121°C. Prepare freshly.

Preparation of Medium: Add sodium acetate and agar to 1.0L of basal medium. Mix thoroughly. Adjust pH to 7.0–7.5. Gently heat and bring to boiling. Autoclave for 15 min at 15 psi pressure–121°C. Cool to 45°–50°C. Aseptically add 3.0mL of sterile sodium sulfide solution immediately prior to dispensing. Mix thoroughly. Pour into sterile Petri dishes or distribute into sterile screw-capped tubes.

Use: For the isolation and cultivation of *Beggiatoa* species and myxotrophic strains of *Thiothrix* species from water and environmental sources.

MP 5 Medium
(Mineral Pectin 5 Medium)

Composition per liter:

Agar solution 500.0mL
Basal medium 250.0mL
Mineral solution 250.0mL

pH 5.0–6.0 at 25°C

Agar Solution:
Composition per 500.0mL:

Agar 15.0g

Preparation of Agar Solution: Add agar to distilled/deionized water and bring volume to 500.0mL. Mix thoroughly. Gently heat and bring to boiling. Autoclave for 15 min at 15 psi pressure–121°C. Cool to 45°–50°C.

Basal Medium:
Composition per 250.0mL:

Na_2HPO_4 .... 6.0g
Pectin, citrus or apple 5.0g
KH_2PO_4 .... 4.0g
NH_4SO_4 .... 2.0g
Yeast extract 1.0g

Preparation of Basal Medium: Add components to distilled/deionized water and bring volume to 250.0mL. Mix thoroughly. Gently heat and bring to boiling.

Mineral Solution:
Composition per 250.0mL:

$FeSO_4$ (0.1% solution) 1.0mL
$MgSO_4·7H_2O$ (20% solution) 1.0mL
$CaCl_2·2H_2O$ (0.1% solution) 1.0mL
H_3BO_3 (0.001% solution) 1.0mL
$MnSO_4·H_2O$ (0.001% solution) 1.0mL
$ZnSO_4·7H_2O$ (0.007% solution 1.0mL
$CuSO_4·5H_2O$ (0.005% solution) 1.0mL
MoO_3 (0.001% solution) 1.0mL

Preparation of Mineral Solution: Add components to distilled/deionized water and bring volume to 250.0mL. Mix thoroughly.

Preparation of Medium: Combine 250.0mL of basal medium and 250.0mL of mineral solution. Mix thoroughly. Adjust pH to 5.0–6.0 with 1*N* HCl. Autoclave the basal medium-mineral solution and agar solution separately for 15 min at 15 psi pressure–121°C. Cool to 45°–50°C. Aseptically combine the two sterile solutions. Mix thoroughly. Pour immediately into sterile Petri dishes to prevent hydrolysis of the agar.

Use: For the cultivation of microorganisms that produce polygalactanase.

MP 7 Medium
(Mineral Pectin 7 Medium)

Composition per liter:

Basal medium 500.0mL
Mineral solution 500.0mL

pH 7.2 ± 0.2 at 25°C

Basal Medium:
Composition per 500.0mL:

Agar 15.0g
Na_2HPO_4 .... 6.0g
Pectin, citrus or apple 5.0g
KH_2PO_4 .... 4.0g
NH_4SO_4 .... 2.0g
Yeast extract 1.0g

Preparation of Basal Medium: Add components to distilled/deionized water and bring volume to 500.0mL. Mix thoroughly. Gently heat and bring to boiling.

Mineral Solution:
Composition per 500.0mL:

$FeSO_4$ (0.1% solution) 1.0mL
$MgSO_4·7H_2O$ (20% solution) 1.0mL
$CaCl_2·2H_2O$ (0.1% solution) 1.0mL
H_3BO_3 (0.001% solution) 1.0mL
$MnSO_4·H_2O$ (0.001% solution) 1.0mL
$ZnSO_4·7H_2O$ (0.007% solution 1.0mL
$CuSO_4·5H_2O$ (0.005% solution 1.0mL
MoO_3 (0.001% solution) 1.0mL

Preparation of Mineral Solution: Add components to distilled/deionized water and bring volume to 500.0mL. Mix thoroughly.

Preparation of Medium: Combine 500.0mL of basal medium and 500.0mL of mineral solution. Mix thoroughly.

Adjust pH to 7.2. Autoclave for 15 min at 15 psi pressure–121°C. Cool to 50°C. Pour into sterile Petri dishes.

Use: For the cultivation of microorganisms that produce pectate lyase.

MPH Agar (Milk Protein Hydrolysate Agar)

Composition per liter:

Agar	15.0g
Casein hydrolysate	9.0g
Glucose	1.0g

pH 7.0 ± 0.2 at 25°C

Source: This medium is available as a premixed powder from BD Diagnostic Systems.

Preparation of Medium: Add components to distilled/deionized water and bring volume to 1.0L. Mix thoroughly. Gently heat while stirring and bring to boiling. Autoclave for 15 min at 15 psi pressure–121°C. Aseptically distribute into sterile tubes. Cool to 43°–45°C before using.

Use: For use in the enumeration of bacteria in water.

MPOB Medium

Composition per 1012.0mL:

Disodium fumarate	3.2g
$Na_2HPO_4 \cdot 2H_2O$	0.53g
KH_2PO_4	0.41g
NaCl	0.3g
NH_4Cl	0.3g
Yeast extract	0.2g
$CaCl_2 \cdot 2H_2O$	0.11g
$MgCl_2 \cdot 6H_2O$	0.10g
Resazurin	0.5mg
$NaHCO_3$ solution	20.0mL
$Na_2S \cdot 9H_2O$ solution	10.0mL
Wolfe's vitamin solution	10.0mL
Selenite-tungstate solution	1.0mL
Trace elements solution SL-10	1.0mL

pH 7.0–7.2 at 25°C

$NaHCO_3$ Solution:

Composition per 20.0mL:

$NaHCO_3$	4.0g

Preparation of $NaHCO_3$ Solution: Add $NaHCO_3$ to distilled/deionized water and bring volume to 20.0mL. Mix thoroughly. Sparge with 80% N_2 + 20% CO_2. Autoclave for 15 min at 15 psi pressure–121°C.

$Na_2S \cdot 9H_2O$ Solution:

Composition per 10.0mL:

$Na_2S \cdot 9H_2O$	0.5g

Preparation of $Na_2S \cdot 9H_2O$ Solution: Add $Na_2S \cdot 9H_2O$ to distilled/deionized water and bring volume to 10.0mL. Mix thoroughly. Sparge with 100% N_2. Autoclave for 15 min at 15 psi pressure–121°C.

Wolfe's Vitamin Solution:

Composition per liter:

Pyridoxine·HCl	10.0mg
p-Aminobenzoic acid	5.0mg
Lipoic acid	5.0mg
Nicotinic acid	5.0mg
Riboflavin	5.0mg
Thiamine·HCl	5.0mg
Calcium DL-pantothenate	5.0mg
Biotin	2.0mg
Folic acid	2.0mg
Vitamin B_{12}	0.1mg

Preparation of Wolfe's Vitamin Solution: Add components to distilled/deionized water and bring volume to 1.0L. Mix thoroughly. Filter sterilize.

Selenite-Tungstate Solution:

Composition per liter:

NaOH	0.5g
$Na_2WO_4 \cdot 2H_2O$	4.0mg
$Na_2SeO_3 \cdot 5H_2O$	3.0mg

Preparation of Selenite-Tungstate Solution: Add components to distilled/deionized water and bring volume to 1.0L. Mix thoroughly. Sparge with 100% N_2. Autoclave for 15 min at 15 psi pressure–121°C.

Trace Elements Solution SL-10:

Composition per liter:

$FeCl_2 \cdot 4H_2O$	1.5g
$CoCl_2 \cdot 6H_2O$	190.0mg
$MnCl_2 \cdot 4H_2O$	100.0mg
$ZnCl_2$	70.0mg
$Na_2MoO_4 \cdot 2H_2O$	36.0mg
$NiCl_2 \cdot 6H_2O$	24.0mg
H_3BO_3	6.0mg
$CuCl_2 \cdot 2H_2O$	2.0mg
HCl (25% solution)	10.0mL

Preparation of Trace Elements Solution SL-10: Add $FeCl_2 \cdot 4H_2O$ to 10.0mL of HCl solution. Mix thoroughly. Add distilled/deionized water and bring volume to 1.0L. Add remaining components. Mix thoroughly. Sparge with 100% N_2. Autoclave for 15 min at 15 psi pressure–121°C.

Preparation of Medium: Prepare medium under 80% N_2 + 20% CO_2. Add components, except $NaHCO_3$ solution and $Na_2S \cdot 9H_2O$ solution, to distilled/deionized water and bring volume to 970.0mL. Mix thoroughly. Sparge with 80% N_2 + 20% CO_2. Autoclave for 15 min at 15 psi pressure–121°C. Aseptically and anaerobically add 20.0mL of sterile $NaHCO_3$ solution and 10.0mL of sterile $Na_2S \cdot 9H_2O$ solution. Mix thoroughly. Aseptically and anaerobically distribute into sterile tubes or flasks. After inoculation, bring culture bottles to 0.7 bar 80% N_2 + 20% CO_2 overpressure.

Use: For the cultivation of *Syntrophobacter* species.

MPSS Broth

Composition per liter:

Peptone	5.0g
$MgSO_4 \cdot 7H_2O$	1.0g
$(NH_4)_2SO_4$	1.0g
Succinic acid	1.0g
$FeCl_3 \cdot 6H_2O$ (0.2% solution)	1.0mL
$MnSO_4 \cdot H_2O$ (0.2% solution)	1.0mL

pH 6.8 ± 0.2 at 25°C

Preparation of Medium: Add components to distilled/deionized water and bring volume to 1.0L. Mix thoroughly. Distribute into tubes or flasks. Autoclave for 15 min at 15 psi pressure–121°C.

Use: For the cultivation of *Spirillum volutans*.

MPY Agar (Maltose Peptone Yeast Extract Medium) (ATCC Medium 518)

Composition per liter:

Agar	10.0g
Maltose	2.0g
Peptone	2.0g

Yeast extract 1.0g
Potassium phosphate buffer (1*M*, pH 7.5) 10.0mL

pH 7.5 ± 0.2 at 25°C

Preparation of Medium: Add components, except potassium phosphate buffer, to distilled/deionized water and bring volume to 990.0mL. Mix thoroughly. Gently heat and bring to boiling. Autoclave for 15 min at 15 psi pressure–121°C. Cool to 45°–50°C. Filter sterilize potassium phosphate bufffer. Aseptically add sterile potassium phosphate buffer to sterile, cooled basal medium. Distribute into sterile tubes or flasks.

Use: For the cultivation and maintenance of *Spirochaeta aurantia*.

MRVP Broth (Methyl Red Voges-Proskauer Broth)

Composition per liter:

Glucose 5.0g
KH_2PO_4 .... 5.0g
Pancreatic digest of casein 3.5g
Peptic digest of animal tissue 3.5g

pH 6.9 ± 0.2 at 25°C

Source: Available as a premixed powder from BD Diagnostic Systems and as a prepared medium from BD Diagnostic Systems.

Preparation of Medium: Add components to distilled/deionized water and bring volume to 1.0L. Mix thoroughly. Distribute into tubes or flasks. Autoclave for 15 min at 15 psi pressure–121°C.

Use: For the differentiation of bacteria based on acid production (Methyl Red test) and acetoin production (Voges-Proskauer reaction).

MRVP Medium (Methyl Red Voges-Proskauer Medium)

Composition per liter:

Glucose 5.0g
Peptone 5.0g
Phosphate buffer 5.0g

pH 7.5 ± 0.2 at 25°C

Source: This medium is available as a premixed powder from Oxoid Unipath.

Preparation of Medium: Add components to distilled/deionized water and bring volume to 1.0L. Mix thoroughly. Distribute into tubes or flasks. Autoclave for 15 min at 15 psi pressure–121°C.

Use: For the differentiation of bacteria based on acid production (Methyl Red test) and acetoin production (Voges-Proskauer reaction).

MS Agar

Composition per liter:

Agar 15.0g
Peptone 1.0g
Yeast extract 1.0g
Glucose 1.0g

pH 6.8–7.2 at 25°C

Preparation of Medium: Add components to distilled/deionized water and bring volume to 1.0L. Mix thoroughly. Gently heat and bring to boiling. Distribute into tubes or flasks. Autoclave for 15 min at 15 psi pressure–121°C. Pour into sterile Petri dishes or leave in tubes.

Use: For the cultivation of *Runella slithyformis*.

MS 1 Agar

Composition per liter:

Agar 15.0g
Seawater 1.0L

Preparation of Medium: Add agar to 1.0L of natural seawater. Mix thoroughly. Gently heat and bring to boiling. Distribute into tubes or flasks. Autoclave for 15 min at 15 psi pressure–121°C. Pour into sterile Petri dishes or leave in tubes.

Use: For the isolation and cultivation of *Cytophaga* species, *Herpetosiphon* species, *Saprospira* species, and *Flexithrix* species.

MS 3 Agar

Composition per liter:

Agar 15.0g
$(NH_4)_2SO_4$ 1.0g
Seawater 1.0L

Preparation of Medium: Add agar to 500.0mL of natural seawater. Mix thoroughly. Gently heat and bring to boiling. In a separate flask, add $(NH_4)_2SO_4$ to 500.0mL of natural seawater. Mix thoroughly. Autoclave both solutions separately for 15 min at 15 psi pressure–121°C. Aseptically combine the two sterile solutions. Pour into sterile Petri dishes or distribute into sterile tubes.

Use: For the isolation and cultivation of *Cytophaga* species, *Herpetosiphon* species, *Saprospira* species, and *Flexithrix* species.

MS 4 Agar

Composition per liter:

Agar 15.0g
Glucose 2.0g
$(NH_4)_2SO_4$ 1.0g
Seawater 1.0L

Preparation of Medium: Add agar to 500.0mL of natural seawater. Mix thoroughly. Gently heat and bring to boiling. Add $(NH_4)_2SO_4$ to 250.0mL of natural seawater. Mix thoroughly. Add glucose to 250.0mL of natural seawater. Mix thoroughly. Autoclave the three solutions separately for 15 min at 15 psi pressure–121°C. Aseptically combine the three sterile solutions. Pour into sterile Petri dishes or distribute into sterile tubes.

Use: For the isolation and cultivation of *Cytophaga* species, *Herpetosiphon* species, *Saprospira* species, and *Flexithrix* species.

MS Medium for *Acidiphilium cryptum*

Composition per liter:

$(NH_4)_2SO_4$ 2.0g
$MgSO_4 \cdot 7H_2O$ 0.25g
K_2HPO_4 .... 0.1g
KCl 0.1g
Glucose solution 20.0mL
Yeast extract solution 10.0mL

pH 3.0 ± 0.2 at 25°C

Glucose Solution:

Composition per 100.0mL:

D-Glucose 10.0g

Preparation of Glucose Solution: Add glucose to distilled/deionized water and bring volume to 100.0mL. Mix thoroughly. Filter sterilize.

Yeast Extract Solution:
Composition per 10.0mL:

Yeast extract	1.0g

Preparation of Yeast Extract Solution: Add yeast extract to distilled/deionized water and bring volume to 10.0mL. Mix thoroughly. Autoclave for 15 min at 15 psi pressure–121°C.

Preparation of Medium: Add components, except glucose solution and yeast extract solution, to distilled/deionized water and bring volume to 970.0mL. Mix thoroughly. Adjust pH to 3.0 with 2*N* H_2SO_4. Autoclave for 15 min at 15 psi pressure–121°C. Aseptically add 20.0mL of sterile glucose solution and 10.0mL of sterile yeast extract solution. Mix thoroughly. Aseptically distribute into sterile tubes or flasks.

Use: For the cultivation of *Acidiphilium cryptum*.

MS Medium for *Leptospirillum* species

Composition per liter:

$(NH_4)_2SO_4$	2.0g
$MgSO_4 \cdot 7H_2O$	0.25g
K_2HPO_4	0.1g
KCl	0.1g
$FeSO_4$ solution	50.0mL

pH 1.6 ± 0.2 at 25°C

Fe_2SO_4 Solution:
Composition per 10.0mL:

$FeSO_4 \cdot 7H_2O$	4.0g

Preparation of Fe_2SO_4 Solution: Add $FeSO_4 \cdot 7H_2O$ to distilled/deionized water and bring volume to 10.0mL. Mix thoroughly. Sparge with 100% N_2. Filter sterilize.

Preparation of Medium: Add components, except $FeSO_4$ solution, to distilled/deionized water and bring volume to 950.0mL. Mix thoroughly. Adjust pH to 1.6 with 4*N* H_2SO_4. Autoclave for 15 min at 15 psi pressure–121°C. Aseptically add 50.0mL of sterile $FeSO_4$ solution. Mix thoroughly. Aseptically distribute into sterile tubes or flasks.

Use: For the cultivation of *Leptospirillum* species.

MS Medium for Methanogens

Composition per 340.0mL:

Agar	8.0g
$NaHCO_3$	2.4g
L-Cysteine-sulfide reducing agent	16.0mL
Mineral solution 1	15.0mL
Mineral solution 2	15.0mL
Sodium formate (20% solution)	6.0mL
Yeast extract-soybean casein solution	4.0mL
Sodium acetate (25% solution)	4.0mL
Wolfe's vitamin solution	4.0mL
Wolfe's mineral solution	4.0mL
$FeSO_4 \cdot 7H_2O$ (0.2% solution)	0.4mL
Resazurin (0.1% solution)	0.4mL

pH 7.0 ± 0.2 at 25°C

L-Cysteine-Sulfide Reducing Agent:
Composition per 400.0mL:

L-Cysteine·HCl·H_2O	5.0g
Na_2S (12.5% solution)	40.0mL
NaOH (1*N* solution)	30.0mL

Preparation of L-Cysteine-Sulfide Reducing Agent: Add distilled/deionized water to a 500.0mL round-bottomed flask. Add freshly prepared NaOH solution. Gently heat and bring to boiling under 100% N_2. Remove gassing probe. Add L-cysteine·HCl·H_2O. Add freshly prepared Na_2S solution. Renew gassing for several minutes. Cap with rubber stoppers. Distribute into 8.0mL/18.0mm Hungate tubes.

Mineral Solution 1:
Composition per liter:

K_2HPO_4	6.0g

Preparation of Mineral Solution 1: Add K_2HPO_4 to distilled/deionized water and bring volume to 1.0L. Mix thoroughly.

Mineral Solution 2:
Composition per liter:

NaCl	12.0g
KH_2PO_4	6.0g
$(NH_4)_2SO_4$	6.0g
$MgSO_4 \cdot 7H_2O$	2.6g
$CaCl_2 \cdot 2H_2O$	0.16g

Preparation of Mineral Solution 2: Add components to distilled/deionized water and bring volume to 1.0L. Mix thoroughly.

Yeast Extract-Soybean Casein Solution:
Composition per 100.0mL:

Yeast extract	20.0g
Pancreatic digest of casein	20.0g

Preparation of Yeast Extract-Soybean Casein Solution: Add components to distilled/deionized water and bring volume to 100.0mL. Mix thoroughly.

Wolfe's Mineral Solution:
Composition per liter:

$MgSO_4 \cdot 7H_2O$	3.0g
Nitriloacetic acid	1.5g
NaCl	1.0g
$MnSO_4 \cdot H_2O$	0.5g
$FeSO_4 \cdot 7H_2O$	0.1g
$CoCl_2 \cdot 6H_2O$	0.1g
$CaCl_2$	0.1g
$ZnSO_4 \cdot 7H_2O$	0.1g
$CuSO_4 \cdot 5H_2O$	0.01g
$AlK(SO_4)_2 \cdot 12H_2O$	0.01g
H_3BO_3	0.01g
$Na_2MoO_4 \cdot 2H_2O$	0.01g

Preparation of Wolfe's Mineral Solution: Add nitrilotriacetic acid to 500.0mL of distilled/deionized water. Dissolve by adjusting pH to 6.5 with KOH. Add remaining components. Add distilled/deionized water to 1.0L.

Wolfe's Vitamin Solution:
Composition per liter:

Pyridoxine·HCl	10.0mg
Thiamine·HCl	5.0mg
Riboflavin	5.0mg
Nicotinic acid	5.0mg
Calcium pantothenate	5.0mg
p-Aminobenzoic acid	5.0mg
Thioctic acid	5.0mg
Biotin	2.0mg
Folic acid	2.0mg
Cyanocobalamin	100.0μg

Preparation of Wolfe's Vitamin Solution: Add components to distilled/deionized water and bring volume to 1.0L. Mix thoroughly.

Preparation of Medium: Add components to distilled/deionized water and bring volume to 408.0mL. Gently heat and bring to boiling under 80% N_2 + 20% CO_2. Continue boiling until rezasurin turns colorless, indicating reduction.

Adjust pH to 7.0. Anaerobically distribute into tubes under 80% N_2 + 20% CO_2. Cap with rubber stoppers and aluminum crimp closures. Autoclave for 15 min at 15 psi pressure–121°C.

Use: For the cultivation and maintenance of *Methanobacterium thermoautotrophicum, Methanobacterium wolfei, Methanobrevibacter smithii, Methanogenium bourgense,* and *Methanogenium* species.

MS Medium for *Thiobacillus caldus*

Composition per liter:

Sulfur, powdered	5.0g
$(NH_4)_2SO_4$	2.0g
$MgSO_4 \cdot 7H_2O$	0.25g
K_2HPO_4	0.1g
KCl	0.1g
Yeast extract solution	10.0mL

pH 3.5 ± 0.2 at 25°C

Preparation of Sulfur: Sterilize powdered elemental sulfur by steaming for 3 hr at 0 psi pressure–100°C on 3 successive days.

Yeast Extract Solution:

Composition per 10.0mL:

Yeast extract	0.2g

Preparation of Yeast Extract Solution: Add yeast extract to distilled/deionized water and bring volume to 10.0mL. Mix thoroughly. Autoclave for 15 min at 15 psi pressure–121°C.

Preparation of Medium: Add components, except elemental sulfur and yeast extract solution, to distilled/deionized water and bring volume to 990.0mL. Mix thoroughly. Adjust pH to 3.5 with 1*N* sulfuric acid. Autoclave for 15 min at 15 psi pressure–121°C. Aseptically add 5.0g of sterile elemental sulfur and 10.0mL of sterile yeast extract solution. Mix thoroughly. Aseptically distribute into sterile tubes or flasks.

Use: For the cultivation of *Thiobacillus caldus.*

MS Medium for *Thiobacillus ferrooxidans*

Composition per liter:

$(NH_4)_2SO_4$	2.0g
$MgSO_4 \cdot 7H_2O$	0.25g
K_2HPO_4	0.1g
KCl	0.1g
$FeSO_4$ solution	50.0mL

pH 2.2 ± 0.2 at 25°C

Fe_2SO_4 Solution:

Composition per 10.0mL:

$FeSO_4 \cdot 7H_2O$	4.0g

Preparation of Fe_2SO_4 Solution: Add $FeSO_4 \cdot 7H_2O$ to distilled/deionized water and bring volume to 10.0mL. Mix thoroughly. Sparge with 100% N_2. Filter sterilize.

Preparation of Medium: Add components, except $FeSO_4$ solution, to distilled/deionized water and bring volume to 950.0mL. Mix thoroughly. Adjust pH to 2.2 with 4*N* H_2SO_4. Autoclave for 15 min at 15 psi pressure–121°C. Aseptically add 50.0mL of sterile $FeSO_4$ solution. Mix thoroughly. Aseptically distribute into sterile tubes or flasks.

Use: For the cultivation of *Thiobacillus ferrooxidans.*

MS Medium for *Thiobacillus thiooxidans*

Composition per liter:

Sulfur, powdered	5.0g
$(NH_4)_2SO_4$	2.0g
$MgSO_4 \cdot 7H_2O$	0.25g
K_2HPO_4	0.1g
KCl	0.1g

pH 3.5 ± 0.2 at 25°C

Preparation of Sulfur: Sterilize powdered elemental sulfur by steaming for 3 hr at 0 psi pressure–100°C on 3 successive days.

Preparation of Medium: Add components, except elemental sulfur, to distilled/deionized water and bring volume to 1.0L. Mix thoroughly. Adjust pH to 3.5 with 1*N* sulfuric acid. Autoclave for 15 min at 15 psi pressure–121°C. Aseptically add 5.0g of sterile elemental sulfur. Mix thoroughly. Aseptically distribute into sterile tubes or flasks.

Use: For the cultivation of *Thiobacillus thiooxidans.*

MSA-Fe Medium

Composition per 1010.0mL:

NaCl	5.8g
Pancreatic digest of casein	2.0g
Yeast extract	2.0g
$MgCl_2 \cdot 6H_2O$	1.0g
NH_4Cl	1.0g
Mercaptoethanesulfonic acid	0.5g
$K_2HPO_4 \cdot 3H_2O$	0.4g
Resazurin	0.5mg
Trace elements solution	10.0mL

pH 7.5 ± 0.2 at 25°C

Trace Elements Solution:

Composition per 100.0mL:

Disodium EDTA·$2H_2O$	50.0mg
$CoCl_2 \cdot 6H_2O$	15.0mg
$FeSO_4 \cdot 7H_2O$	10.0mg
$MnCl_2 \cdot 4H_2O$	10.0mg
$ZnCl_2$	10.0mg
$AlCl_3 \cdot 6H_2O$	4.0mg
$Na_2WO_4 \cdot 2H_2O$	3.0mg
$CuCl_2 \cdot 2H_2O$	2.0mg
$NiSO_4 \cdot 6H_2O$	2.0mg
H_2SeO_3	1.0mg
H_3BO_3	1.0mg
$Na_2MoO_4 \cdot 2H_2O$	1.0mg

Preparation of Trace Elements Solution: Add components to distilled/deionized water and bring volume to 100.0mL. Mix thoroughly.

Preparation of Medium: Prepare and dispense medium under 100% N_2. Add components to distilled/deionized water and bring volume to 1.0L. Mix thoroughly. Adjust pH to 7.5 with 2*N* NaOH. Sparge with 100% N_2 for 30 min. Anaerobically distribute into tubes. Autoclave for 15 min at 15 psi pressure–121°C.

Use: For the cultivation of *Bacillus infermus.*

MSV AcS Agar

Composition per liter:

$Na_2S \cdot 9H_2O$	0.187g
Sodium acetate	0.15g
MSV agar	1.0L

pH 7.2–7.5 at 25°C

MSV Agar:

Composition per liter:

Agar	12.0g
$(NH_4)_2SO_4$	0.5g
K_2HPO_4	0.11g
KH_2PO_4	0.085g

$MgSO_4 \cdot 7H_2O$ 0.05g
$CaCl_2 \cdot 2H_2O$ 0.05g
EDTA 3.0mg
$FeCl_3 \cdot H_2O$ 2.0mg
Vitamin mix 1.0mL

Preparation of MSV Agar: Add components to distilled/deionized water and bring volume to 1.0L. Mix thoroughly.

Vitamin Mix:
Composition per 100.0mL:
Calcium pantothenate 0.01g
Niacin 0.01g
Pyridoxine 0.01g
p-Aminobenzoic acid 0.01g
Cocarboxylase 0.01g
Inositol 0.01g
Thiamine 0.01g
Riboflavin 0.01g
Biotin 0.5mg
Cyanocobalamin 0.5mg
Folic acid 0.5mg

Preparation of Vitamin Mix: Add components to distilled/deionized water and bring volume to 100.0mL. Mix thoroughly.

Preparation of Medium: To 1.0L of MSV agar add sodium acetate and $Na_2S \cdot 9H_2O$. Adjust pH to 7.2–7.5. Gently heat to boiling. Distribute into tubes or flasks. Autoclave for 15 min at 15 psi pressure–121°C. Pour into sterile Petri dishes or leave in tubes.

Use: For the isolation, cultivation, and enrichment of heterotrophic strains of *Thiothrix* species from water and environmental sources.

MSV Agar

Composition per liter:
Agar 12.0g
$(NH_4)_2SO_4$ 0.5g
K_2HPO_4 .... 0.11g
KH_2PO_4 .... 0.085g
$MgSO_4 \cdot 7H_2O$ 0.05g
$CaCl_2 \cdot 2H_2O$ 0.05g
EDTA 3.0mg
$FeCl_3 \cdot H_2O$ 2.0mg
Vitamin mix 1.0mL
pH 7.2–7.5 at 25°C

Vitamin Mix:
Composition per 100.0mL:
Calcium pantothenate 0.01g
Niacin 0.01g
Pyridoxine 0.01g
p-Aminobenzoic acid 0.01g
Cocarboxylase 0.01g
Inositol 0.01g
Thiamine 0.01g
Riboflavin 0.01g
Biotin 0.5mg
Cyanocobalamin 0.5mg
Folic acid 0.5mg

Preparation of Vitamin Mix: Add components to distilled/deionized water and bring volume to 100.0mL. Mix thoroughly.

Preparation of Medium: Add components to distilled/deionized water and bring volume to 1.0L. Mix thoroughly. Adjust pH to 7.2–7.5. Gently heat to boiling. Distribute into tubes or flasks. Autoclave for 15 min at 15 psi pressure–121°C. Pour into sterile Petri dishes or leave in tubes.

Use: For the isolation, cultivation, and enrichment of heterotrophic strains of *Thiothrix* species from water and environmental sources.

MSV Broth

Composition per liter:
$(NH_4)_2SO_4$ 0.5g
K_2HPO_4 .... 0.11g
KH_2PO_4 .... 0.085g
$MgSO_4 \cdot 7H_2O$ 0.05g
$CaCl_2 \cdot 2H_2O$ 0.05g
EDTA 3.0mg
$FeCl_3 \cdot H_2O$ 2.0mg
Vitamin mix 1.0mL
pH 7.2–7.5 at 25°C

Vitamin Mix:
Composition per 100.0mL:
Calcium pantothenate 0.01g
Niacin 0.01g
Pyridoxine 0.01g
p-Aminobenzoic acid 0.01g
Cocarboxylase 0.01g
Inositol 0.01g
Thiamine 0.01g
Riboflavin 0.01g
Biotin 0.5mg
Cyanocobalamin 0.5mg
Folic acid 0.5mg

Preparation of Vitamin Mix: Add components to distilled/deionized water and bring volume to 100.0mL. Mix thoroughly.

Preparation of Medium: Add components to distilled/deionized water and bring volume to 1.0L. Mix thoroughly. Adjust pH to 7.2–7.5. Distribute into tubes or flasks. Autoclave for 15 min at 15 psi pressure–121°C.

Use: For the isolation, cultivation, and enrichment of heterotrophic strains of *Thiothrix* species from water and environmental sources.

MSV GS Agar

Composition per liter:
$Na_2S \cdot 9H_2O$ 0.187g
Glucose 0.15g
MSV agar 1.0L
pH 7.2–7.5 at 25°C

MSV Agar:
Composition per liter:
Agar 12.0g
$(NH_4)_2SO_4$ 0.5g
K_2HPO_4 .... 0.11g
KH_2PO_4 .... 0.085g
$MgSO_4 \cdot 7H_2O$ 0.05g
$CaCl_2 \cdot 2H_2O$ 0.05g
EDTA 3.0mg
$FeCl_3 \cdot H_2O$ 2.0mg
Vitamin mix 1.0mL

Preparation of MSV Agar: Add components to distilled/deionized water and bring volume to 1.0L. Mix thoroughly.

Vitamin Mix:
Composition per 100.0mL:
Calcium pantothenate 0.01g
Niacin 0.01g

Pyridoxine 0.01g
p-Aminobenzoic acid 0.01g
Cocarboxylase 0.01g
Inositol 0.01g
Thiamine 0.01g
Riboflavin 0.01g
Biotin 0.5mg
Cyanocobalamin 0.5mg
Folic acid 0.5mg

Preparation of Vitamin Mix: Add components to distilled/deionized water and bring volume to 100.0mL. Mix thoroughly.

Preparation of Medium: To 1.0L of MSV agar add glucose and $Na_2S·9H_2O$. Adjust pH to 7.2–7.5. Gently heat to boiling. Distribute into tubes or flasks. Autoclave for 15 min at 15 psi pressure–121°C. Pour into sterile Petri dishes or leave in tubes.

Use: For the isolation, cultivation, and enrichment of heterotrophic strains of *Thiothrix* species from water and environmental sources.

MSV I Agar

Composition per liter:

Glucose 0.15g
MSV agar 1.0L

pH 7.2–7.5 at 25°C

MSV Agar:

Composition per liter:

Agar 12.0g
$(NH_4)_2SO_4$ 0.5g
K_2HPO_4 .... 0.11g
KH_2PO_4 .... 0.085g
$MgSO_4·7H_2O$ 0.05g
$CaCl_2·2H_20$ 0.05g
EDTA 3.0mg
$FeCl_3·H_2O$ 2.0mg
Vitamin mix 1.0mL

Preparation of MSV Agar: Add components to distilled/deionized water and bring volume to 1.0L. Mix thoroughly.

Vitamin Mix:

Composition per 100.0mL:

Calcium pantothenate 0.01g
Niacin 0.01g
Pyridoxine 0.01g
p-Aminobenzoic acid 0.01g
Cocarboxylase 0.01g
Inositol 0.01g
Thiamine 0.01g
Riboflavin 0.01g
Biotin 0.5mg
Cyanocobalamin 0.5mg
Folic acid 0.5mg

Preparation of Vitamin Mix: Add components to distilled/deionized water and bring volume to 100.0mL. Mix thoroughly.

Preparation of Medium: To 1.0L of MSV agar, add glucose. Adjust pH to 7.2–7.5. Gently heat to boiling. Distribute into tubes or flasks. Autoclave for 15 min at 15 psi pressure–121°C. Pour into sterile Petri dishes or leave in tubes.

Use: For the isolation, cultivation, and enrichment of heterotrophic strains of *Thiothrix* species from water and environmental sources.

MSV LT Agar

Composition per liter:

Sodium lactate 0.5g
$Na_2S_2O_3$ 0.5g
MSV agar 1.0L

pH 7.2–7.5 at 25°C

MSV Agar:

Composition per liter:

Agar 12.0g
$(NH_4)_2SO_4$ 0.5g
K_2HPO_4 .... 0.11g
KH_2PO_4 .... 0.085g
$MgSO_4·7H_2O$ 0.05g
$CaCl_2·2H_20$ 0.05g
EDTA 3.0mg
$FeCl_3·H_2O$ 2.0mg
Vitamin mix 1.0mL

Preparation of MSV Agar: Add components to distilled/deionized water and bring volume to 1.0L. Mix thoroughly.

Vitamin Mix:

Composition per 100.0mL:

Calcium pantothenate 0.01g
Niacin 0.01g
Pyridoxine 0.01g
p-Aminobenzoic acid 0.01g
Cocarboxylase 0.01g
Inositol 0.01g
Thiamine 0.01g
Riboflavin 0.01g
Biotin 0.5mg
Cyanocobalamin 0.5mg
Folic acid 0.5mg

Preparation of Vitamin Mix: Add components to distilled/deionized water and bring volume to 100.0mL. Mix thoroughly.

Preparation of Medium: To 1.0L of MSV agar, add sodium lactate and $Na_2S_2O_3$. Adjust pH to 7.2–7.5. Gently heat to boiling. Distribute into tubes or flasks. Autoclave for 15 min at 15 psi pressure–121°C. Pour into sterile Petri dishes or leave in tubes.

Use: For the isolation, cultivation, and enrichment of heterotrophic strains of *Thiothrix* species from water and environmental sources.

MSV S Agar

Composition per liter:

$Na_2S·9H_2O$ 0.187g
MSV agar 1.0L

pH 7.2–7.5 at 25°C

MSV Agar:

Composition per liter:

Agar 12.0g
$(NH_4)_2SO_4$ 0.5g
K_2HPO_4 .... 0.11g
KH_2PO_4 .... 0.085g
$MgSO_4·7H_2O$ 0.05g
$CaCl_2·2H_20$ 0.05g
EDTA 3.0mg
$FeCl_3·H_2O$ 2.0mg
Vitamin mix 1.0mL

Preparation of MSV Agar: Add components to distilled/deionized water and bring volume to 1.0L. Mix thoroughly.

Vitamin Mix:
Composition per 100.0mL:

Calcium pantothenate	0.01g
Niacin	0.01g
Pyridoxine	0.01g
p-Aminobenzoic acid	0.01g
Cocarboxylase	0.01g
Inositol	0.01g
Thiamine	0.01g
Riboflavin	0.01g
Biotin	0.5mg
Cyanocobalamin	0.5mg
Folic acid	0.5mg

Preparation of Vitamin Mix: Add components to distilled/deionized water and bring volume to 100.0mL. Mix thoroughly.

Preparation of Medium: To 1.0L of MSV agar, add $Na_2S{\cdot}9H_2O$. Adjust pH to 7.2–7.5. Gently heat to boiling. Distribute into tubes or flasks. Autoclave for 15 min at 15 psi pressure–121°C. Pour into sterile Petri dishes or leave in tubes.

Use: For the isolation, cultivation, and enrichment of heterotrophic strains of *Thiothrix* species from water and environmental sources.

MSV SS Agar

Composition per liter:

$Na_2S{\cdot}9H_2O$	0.187g
Sucrose	0.15g
MSV agar	1.0L

pH 7.2–7.5 at 25°C

MSV Agar:
Composition per liter:

Agar	12.0g
$(NH_4)_2SO_4$	0.5g
K_2HPO_4	0.11g
KH_2PO_4	0.085g
$MgSO_4{\cdot}7H_2O$	0.05g
$CaCl_2{\cdot}2H_20$	0.05g
EDTA	3.0mg
$FeCl_3{\cdot}H_2O$	2.0mg
Vitamin mix	1.0mL

Preparation of MSV Agar: Add components to distilled/deionized water and bring volume to 1.0L. Mix thoroughly.

Vitamin Mix:
Composition per 100.0mL:

Calcium pantothenate	0.01g
Niacin	0.01g
Pyridoxine	0.01g
p-Aminobenzoic acid	0.01g
Cocarboxylase	0.01g
Inositol	0.01g
Thiamine	0.01g
Riboflavin	0.01g
Biotin	0.5mg
Cyanocobalamin	0.5mg
Folic acid	0.5mg

Preparation of Vitamin Mix: Add components to distilled/deionized water and bring volume to 100.0mL. Mix thoroughly.

Preparation of Medium: To 1.0L of MSV agar, add $Na_2S{\cdot}9H_2O$ and sucrose. Adjust pH to 7.2–7.5. Gently heat to boiling. Distribute into tubes or flasks. Autoclave for 15 min at 15 psi pressure–121°C. Pour into sterile Petri dishes or leave in tubes.

Use: For the isolation, cultivation, and enrichment of heterotrophic strains of *Thiothrix* species from water and environmental sources.

MSV SUC Agar

Composition per liter:

Sodium succinate	0.15g
MSV agar	1.0L

pH 7.2–7.5 at 25°C

MSV Agar:
Composition per liter:

Agar	12.0g
$(NH_4)_2SO_4$	0.5g
K_2HPO_4	0.11g
KH_2PO_4	0.085g
$MgSO_4{\cdot}7H_2O$	0.05g
$CaCl_2{\cdot}2H_20$	0.05g
EDTA	3.0mg
$FeCl_3{\cdot}H_2O$	2.0mg
Vitamin mix	1.0mL

Preparation of MSV Agar: Add components to distilled/deionized water and bring volume to 1.0L. Mix thoroughly.

Vitamin Mix:
Composition per 100.0mL:

Calcium pantothenate	0.01g
Niacin	0.01g
Pyridoxine	0.01g
p-Aminobenzoic acid	0.01g
Cocarboxylase	0.01g
Inositol	0.01g
Thiamine	0.01g
Riboflavin	0.01g
Biotin	0.5mg
Cyanocobalamin	0.5mg
Folic acid	0.5mg

Preparation of Vitamin Mix: Add components to distilled/deionized water and bring volume to 100.0mL. Mix thoroughly.

Preparation of Medium: To 1.0L of MSV agar, add sodium succinate. Adjust pH to 7.2–7.5. Gently heat to boiling. Distribute into tubes or flasks. Autoclave for 15 min at 15 psi pressure–121°C. Pour into sterile Petri dishes or leave in tubes.

Use: For the isolation, cultivation, and enrichment of heterotrophic strains of *Thiothrix* species from water and environmental sources.

MSVP Agar

Composition per 984.0mL:

Agar, noble	15.0g
HEPES (*N*-[2-Hydroxyethyl]piperazine-*N'*-2-ethanesulfonic acid) buffer	2.383g
$(NH_4)_2SO_4$	0.24g
$CaCl_2{\cdot}2H_2O$	0.06g
$MgSO_4{\cdot}7H_2O$	0.06g
Na_2HPO_4	0.03g
KH_2PO_4	0.02g
$FeSO_4$ (10m*M* solution)	1.0mL
Sodium pyruvate solution	5.0mL
Vitamin solution	1.0mL

pH 7.2 ± 0.2 at 25°C

Sodium Pyruvate Solution:
Composition per 50.0mL:
Sodium pyruvate 10.0g

Preparation of Sodium Pyruvate Solution: Add sodium pyruvate to distilled/deionized water and bring volume to 50.0mL. Mix thoroughly. Filter sterilize.

Vitamin Solution:
Composition per liter:
Pyridoxine·HCl 100.0mg
p-Aminobenzoic acid 50.0mg
D-(+)-Calcium pantothenate 50.0mg
Nicotinic acid 50.0mg
Riboflavin 50.0mg
Thiamine·HCl 50.0mg
Biotin 20.0mg
Folic acid 20.0mg
Vitamin B_{12} 1.0mg

Preparation of Vitamin Solution: Add components to distilled/deionized water and bring volume to 1.0L. Mix thoroughly. Filter sterilize.

Preparation of Medium: Add components, except sodium pyruvate solution and vitamin solution, to distilled/deionized water and bring volume to 994.0mL. Mix thoroughly. Gently heat and bring to boiling. Adjust pH to 7.2. Autoclave for 15 min at 15 psi pressure–121°C. Cool to 50°–55°C. Aseptically add 5.0mL of sterile sodium pyruvate solution and 1.0mL of sterile vitamin solution. Mix thoroughly. Pour into sterile Petri dishes or distribute into sterile tubes.

Use: For the cultivation of *Leptothrix discophora.*

MTP4 Medium

Composition per 1001.0mL:
Solution A 870.0mL
Solution C 100.0mL
Solution D (Vitamin solution) 10.0mL
Solution E 10.0mL
Solution B (Trace elements solution SL-10) 1.0mL
Methanol 1.0mL
Methanethiol gas 1–2.0mL
pH 7.1–7.4 at 25°C

Solution A:
Composition per 870.0mL:
NaCl 21.0g
$MgCl_2 \cdot 6H_2O$ 3.1g
KCl 0.5g
NH_4Cl 0.3g
KH_2PO_4 .... 0.2g
$CaCl_2 \cdot 2H_2O$ 0.15g
Resazurin 1.0mg

Preparation of Solution A: Add components to distilled/deionized water and bring volume to 870.0mL. Mix thoroughly. Gently heat and bring to boiling. Continue boiling for 3-4 min. Allow to cool to room temperature while gassing under 80% N_2 + 20% CO_2. Continue gassing until pH reaches below 6.0. Seal the flask under 80% N_2 + 20% CO_2. Autoclave for 15 min at 15 psi pressure–121°C.

Solution B (Trace Elements Solution SL-10):
Composition per liter:
$FeCl_2 \cdot 4H_2O$ 1.5g
$CoCl_2 \cdot 6H_2O$ 190.0mg
$MnCl_2 \cdot 4H_2O$ 100.0mg
$ZnCl_2$ 70.0mg
$Na_2MoO_4 \cdot 2H_2O$ 36.0mg
$NiCl_2 \cdot 6H_2O$ 24.0mg
H_3BO_3 .... 6.0mg
$CuCl_2 \cdot 2H_2O$ 2.0mg
HCl (25% solution) 10.0mL

Preparation of Solution B (Trace Elements Solution SL-10): Add $FeCl_2 \cdot 4H_2O$ to 10.0mL of HCl solution. Mix thoroughly. Add distilled/deionized water and bring volume to 1.0L. Add remaining components. Mix thoroughly. Gas under 100% N_2. Autoclave for 15 min at 15 psi pressure–121°C.

Solution C:
Composition per 100.0mL:
$NaHCO_3$ 5.0g

Preparation of Solution C: Add $NaHCO_3$ to distilled/deionized water and bring volume to 100.0mL. Mix thoroughly. Filter sterilize. Gas under 80% N_2 + 20% CO_2.

Solution D (Vitamin Solution):
Composition per liter:
Pyridoxine·HCl 10.0mg
Calcium DL-pantothenate 5.0mg
Lipoic acid 5.0mg
Nicotinic acid 5.0mg
p-Aminobenzoic acid 5.0mg
Riboflavin 5.0mg
Thiamine·HCl 5.0mg
Biotin 2.0mg
Folic acid 2.0mg
Vitamin B_{12} 0.1mg

Preparation of Solution D (Vitamin Solution): Add components to distilled/deionized water and bring volume to 1.0L. Mix thoroughly. Gas under 100% N_2. Autoclave for 15 min at 15 psi pressure–121°C.

Solution E:
Composition per 10.0mL:
$Na_2S \cdot 9H_2O$ 0.4g

Preparation of Solution E: Add $Na_2S \cdot 9H_2O$ to distilled/deionized water and bring volume to 10.0mL. Mix thoroughly. Gas under 100% N_2. Autoclave for 15 min at 15 psi pressure–121°C.

Preparation of Medium: Aseptically and anaerobically combine solution A with solution B, solution C, solution D, and solution E, in that order. Mix thoroughly. Anaerobically distribute into sterile tubes or flasks under 80% N_2 + 20% CO_2. Prior to inoculation, aseptically and anaerobically add 1.0mL of filter-sterilized methanol and 1.0-2.0mL of methanethiol gas to each liter of medium. Addition of 10-20mg of sodium dithionite per liter (e.g., from a 5% solution, freshly prepared under N_2 and filter-sterilized) may stimulate growth at the beginning.

Use: For the cultivation and maintenance of *Methanosarcina* species.

MV Medium

Composition per 1001.0mL:
Na_2SO_4 .... 2.0g
$MgSO_4 \cdot 7H_2O$ 1.0g
$Na_2S_2O_35H_2O$ 1.0g
NH_4Cl 1.0g
Yeast extract 1.0g
KH_2PO_4 .... 0.5g
$CaCl_2 \cdot 2H_2O$ 0.1g
Resazurin 0.5mg
Wolfe's vitamin solution 10.0mL
Sodium malate solution 10.0mL

Sodium pyruvate solution....................10.0mL
$NaHCO_3$ solution....................10.0mL
$Na_2S{\cdot}9H_2O$ solution....................10.0mL
Trace elements solution SL-10....................1.0mL
pH 7.0–7.2 at 25°C

Sodium Malate Solution:
Composition per 100.0mL:
Sodium malate....................1.0g

Preparation of Sodium Malate Solution: Add sodium malate to distilled/deionized water and bring volume to 10.0mL. Mix thoroughly. Sparge with 100% N_2. Autoclave for 15 min at 15 psi pressure–121°C.

Sodium Pyruvate Solution:
Composition per 100.0mL:
Sodium pyruvate....................1.0g

Preparation of Sodium Pyruvate Solution: Add sodium pyruvate to distilled/deionized water and bring volume to 10.0mL. Mix thoroughly. Sparge with 100% N_2. Autoclave for 15 min at 15 psi pressure–121°C.

$NaHCO_3$ Solution:
Composition per 10.0mL:
$NaHCO_3$....................2.0g

Preparation of $NaHCO_3$ Solution: Add $NaHCO_3$ to distilled/deionized water and bring volume to 10.0mL. Mix thoroughly. Sparge with 80% N_2 + 20% CO_2. Autoclave for 15 min at 15 psi pressure–121°C.

$Na_2S{\cdot}9H_2O$ Solution:
Composition per 10.0mL:
$Na_2S{\cdot}9H_2O$....................0.75μg

Preparation of $Na_2S{\cdot}9H_2O$ Solution: Add $Na_2S{\cdot}9H_2O$ to distilled/deionized water and bring volume to 10.0mL. Mix thoroughly. Sparge with 100% N_2. Autoclave for 15 min at 15 psi pressure–121°C. Before use, neutralize to pH 7.0 with sterile HCl.

Wolfe's Vitamin Solution:
Composition per liter:
Pyridoxine·HCl....................10.0mg
p-Aminobenzoic acid....................5.0mg
Lipoic acid....................5.0mg
Nicotinic acid....................5.0mg
Riboflavin....................5.0mg
Thiamine·HCl....................5.0mg
Calcium DL-pantothenate....................5.0mg
Biotin....................2.0mg
Folic acid....................2.0mg
Vitamin B_{12}....................0.1mg

Preparation of Wolfe's Vitamin Solution: Add components to distilled/deionized water and bring volume to 1.0L. Mix thoroughly. Sparge with 100% N_2. Filter sterilize.

Trace Elements Solution SL-10:
Composition per liter:
$FeCl_2{\cdot}4H_2O$....................1.5g
$CoCl_2{\cdot}6H_2O$....................190.0mg
$MnCl_2{\cdot}4H_2O$....................100.0mg
$ZnCl_2$....................70.0mg
$Na_2MoO_4{\cdot}2H_2O$....................36.0mg
$NiCl_2{\cdot}6H_2O$....................24.0mg
H_3BO_3....................6.0mg
$CuCl_2{\cdot}2H_2O$....................2.0mg
HCl (25% solution)....................10.0mL

Preparation of Trace Elements Solution SL-10: Add $FeCl_2{\cdot}4H_2O$ to 10.0mL of HCl solution. Mix thoroughly. Add distilled/deionized water and bring volume to 1.0L. Add remaining components. Mix thoroughly. Sparge with 100% N_2. Autoclave for 15 min at 15 psi pressure–121°C.

Preparation of Medium: Prepare and dispense medium under 100% N_2. Add components, except Wolfe's vitamin solution, sodium malate solution, sodium pyruvate solution, $NaHCO_3$ solution, $Na_2S{\cdot}9H_2O$ solution, and trace elements solution SL-10, to distilled/deionized water and bring volume to 950.0mL. Mix thoroughly. Adjust pH to 7.0–7.2. Gently heat and bring to boiling. Cool while sparging with 100% N_2. Autoclave for 15 min at 15 psi pressure–121°C. Aseptically and anaerobically add 10.0mL of sterile Wolfe's vitamin solution, 10.0mL of sterile sodium malate solution, 10.0mL of sterile sodium pyruvate solution, 10.0mL of sterile $NaHCO_3$ solution, 10.0mL of sterile $Na_2S{\cdot}9H_2O$ solution, and 1.0mL of sterile trace elements solution SL-10. Mix thoroughly. Aseptically and anaerobically distribute into sterile tubes or bottles.

Use: For the cultivation of *Desulfotomaculum orientis.*

MWY Medium (Wadowsky-Yee Medium, Modified)

Composition per liter:
Agar....................13.0g
Yeast extract....................10.0g
Glycine....................3.0g
ACES buffer (2-[(2-Amino-2-oxoethyl)-amino]-ethane sulfonic acid)....................2.0g
Charcoal, activated....................2.0g
α-Ketoglutarate....................0.2g
$Fe_4(P_2O_7)_3{\cdot}9H_2O$....................0.05g
Bromcresol Purple....................0.01g
Bromcresol Blue....................0.01g
Antibiotic inhibitor....................10.0mL
L-Cysteine·HCl·H_2O solution....................10.0mL
pH 6.9 ± 0.2 at 25°C

Antibiotic Inhibitor:
Composition per 10.0mL:
Anisomycin....................0.16g
Cefamandole....................4.0mg
Vancomycin....................1.0mg
Polymyxin B....................130,000U

Preparation of Antibiotic Inhibitor: Add components to distilled/deionized water and bring volume to 10.0mL. Mix thoroughly. Filter sterilize.

L-Cysteine·HCl·H_2O Solution:
Composition per 10.0mL:
L-Cysteine·HCl·H_2O....................0.08g

Preparation of L-Cysteine·HCl·H_2O Solution: Add L-cysteine·HCl·H_2O to distilled/deionized water and bring volume to 10.0mL. Mix thoroughly. Filter sterilize.

Preparation of Medium: Add components, except L-cysteine and antibiotic inhibitor, to distilled/deionized water and bring volume to 980.0mL. Mix thoroughly. Adjust medium to pH 6.9 with 1*N* KOH. Heat gently and bring to boiling for 1 min. Autoclave for 15 min at 15 psi pressure–121°C. Cool to 50°–55°C. Add 10.0mL of the sterile L-cysteine·HCl·H_2O solution and 10.0mL of the sterile antibiotic solution. Mix thoroughly. Pour into sterile Petri dishes with constant agitation to keep charcoal in suspension.

Use: For the selective isolation and cultivation of *Legionella pneumophila* and other *Legionella* species.

MY Agar

Composition per liter:

Agar 15.0g
Sodium acetate 0.1g
Yeast extract 0.1g
Pancreatic digest of gelatin 0.06g
Beef extract 0.04g
Basal medium 1.0L
Sodium sulfide solution 3.0mL
pH 7.0–7.5 at 25°C

Basal Medium:

Composition per liter:

$CaSO_4 \cdot 2H_2O$ (saturated solution) 20.0mL
NH_4Cl (4% solution) 5.0mL
Trace elements solution 5.0mL
K_2HPO_4 (1% solution) 1.0mL
$MgSO_4 \cdot 7H_2O$ (1% solution) 1.0mL

Preparation of Basal Medium: Add components to distilled/deionized water and bring volume to 1.0L. Mix thoroughly.

Trace Elements Solution:

Composition per liter:

Ferrous EDTA solution 20.0mL
$ZnSO_4 \cdot 7H_2O$ (0.1% solution) 10.0mL
$MnSO_4 \cdot 4H_2O$ (0.02% solution) 10.0mL
$CuSO_4 \cdot 5H_2O$ (0.00005% solution) 10.0mL
H_3BO_3 (0.1% solution) 10.0mL
$Co(NO_3)_2$ or
$CoCl_2 \cdot 6H_2O$ (0.01% solution) 10.0mL
$Na_2MoO_4 \cdot 2H_2O$ (0.01% solution) 10.0mL

Preparation of Trace Elements Solution: Add components to distilled/deionized water and bring volume to 1.0L. Mix thoroughly.

Ferrous EDTA Solution:

Composition per 100.0mL:

$FeSO_4 \cdot 7H_2O$ 7.0g
EDTA 2.0g
HCl, concentrated 1.0mL

Preparation of Ferrous EDTA Solution: Add components to distilled/deionized water and bring volume to 100.0mL. Mix thoroughly.

Sodium Sulfide Solution:

Composition per 10.0mL:

$Na_2S \cdot 9H_20$ 1.0g

Preparation of Sodium Sulfide Solution: Add $Na_2S \cdot 9H_20$ to distilled/deionized water and bring volume to 10.0mL. Mix thoroughly. Autoclave for 15 min at 15 psi pressure–121°C. Prepare freshly.

Preparation of Medium: Add sodium acetate, beef extract, pancreatic digest of gelatin, yeast extract, and agar to 1.0L of basal medium. Mix thoroughly. Adjust pH to 7.0–7.5. Gently heat and bring to boiling. Autoclave for 15 min at 15 psi pressure–121°C. Cool to 45°–50°C. Aseptically add 3.0mL of sterile sodium sulfide solution immediately prior to dispensing. Mix thoroughly. Pour into sterile Petri dishes or distribute into sterile screw-capped tubes.

Use: For the isolation and cultivation of *Beggiatoa* species and myxotrophic strains of *Thiothrix* species from water and environmental sources.

MY20 Agar

Composition per liter:

Glucose 200.0g
Agar 20.0g
Peptone 5.0g
Yeast extract 3.0g
Malt extract 3.0g

Preparation of Medium: Add components to distilled/deionized water and bring volume to 1.0L. Mix thoroughly. Gently heat and bring to boiling. Distribute into tubes or flasks. Autoclave for 15 min at 15 psi pressure–121°C. Pour into sterile Petri dishes or leave in tubes.

Use: For the cultivation of a variety of osmophilic heterotrophic bacteria and fungi.

MY40 Agar

Composition per liter:

Sucrose 400.0g
Agar 20.0g
Malt extract 20.0g
Yeast extract 5.0g

Preparation of Medium: Add components to distilled/deionized water and bring volume to 1.0L. Mix thoroughly. Gently heat and bring to boiling. Distribute into tubes or flasks. Autoclave for 15 min at 15 psi pressure–121°C. Pour into sterile Petri dishes or leave in tubes.

Use: For the cultivation of a variety of osmophilic heterotrophic bacteria and fungi.

MY60 Agar

Composition per liter:

Sucrose 600.0g
Agar 20.0g
Malt extract 20.0g
Yeast extract 5.0g

Preparation of Medium: Add components to distilled/deionized water and bring volume to 1.0L. Mix thoroughly. Gently heat and bring to boiling. Distribute into tubes or flasks. Autoclave for 15 min at 15 psi pressure–121°C. Pour into sterile Petri dishes or leave in tubes.

Use: For the cultivation of a variety of osmophilic heterotrophic bacteria and fungi.

Mycobacterium Medium

Composition per liter:

Noble agar 15.0g
$(NH_4)_2SO_4$ 1.0g
Na_2HPO_4 ..... 0.5g
KH_2PO_4 ..... 0.5g
$MgSO_4$ 0.2g
$FeSO_4 \cdot 7H_2O$ 5.0mg
$MnSO_4$ 2.0mg
Liquid paraffin 5.0mL

Preparation of Medium: Add components, except agar, to distilled/deionized water and bring volume to 1.0L. Homogenize in a blender. Add agar. Gently heat and bring to boiling. Distribute into tubes or flasks. Autoclave for 15 min at 15 psi pressure–121°C. Pour into sterile Petri dishes or leave in tubes.

Use: For the cultivation and maintenance of *Mycobacterium paraffinicum*.

Mycobacterium Yeast Extract Medium

Composition per liter:

Agar 15.0g
Pancreatic digest of casein 5.0g
Yeast extract 2.5g
Glucose 1.0g

Preparation of Medium: Add components to distilled/deionized water and bring volume to 1.0L. Mix thoroughly. Gently heat and bring to boiling. Distribute into tubes or flasks. Autoclave for 15 min at 15 psi pressure–121°C. Pour into sterile Petri dishes or leave in tubes.

Use: For the cultivation and maintenance of *Mycobacterium* species and *Rhodococcus* species.

Mycorrhiza Medium

Composition per liter:

Agar	15.0g
Glucose	4.0g
Ammonium tartrate	1.0g
Malt extract	1.0g
KH_2PO_4	0.2g
$MgSO_4·7H_2O$	0.1g
$CaCl_2·2H_2O$	26.0mg
NaCl	20.0mg
Inositol	10.0mg
$ZnSO_4·7H_2O$	0.88mg
$MnSO_4·4H_2O$	0.81mg
$FeCl_3·6H_2O$	0.8mg
Nicotinamide	100.0μg
p-Aminobenzoic acid	100.0μg
Pantothenic acid	100.0μg
Pyridoxine	100.0μg
Thiamine	100.0μg
Biotin	25.0μg
Riboflavin	25.0μg

Preparation of Medium: Add components to distilled/deionized water and bring volume to 1.0L. Mix thoroughly. Gently heat until boiling. Distribute into tubes or flasks. Autoclave for 10 min at 15 psi pressure–121°C. Pour into sterile Petri dishes or leave in tubes.

Use: For the cultivation and maintenance of *Thelephora terrestris*.

Myxobacteria Medium

Composition per liter:

Agar	15.0g
Skim milk powder	5.0g
Yeast extract	0.5g

Preparation of Medium: Add components to distilled/deionized water and bring volume to 1.0L. Mix thoroughly. Do not adjust pH. Gently heat and bring to boiling. Distribute into tubes or flasks. Autoclave for 15 min at 15 psi pressure–121°C.

Use: For the cultivation and maintenance of *Archangium primigenium, Chondrococcus macrosporus*, and *Myxococcus coralloides*.

Myxococcus flavescens Medium

Composition per liter:

Agar	15.0g
Soluble starch	5.0g
Casitone	2.5g
Galactose	1.0g
Raffinose	1.0g
Sucrose	1.0g
Yeast extract	1.0g
$MgSO_4·7H_2O$	0.5g
K_2HPO_4	0.25g

pH 6.0 ± 0.2 at 25°C

Preparation of Medium: Add components to distilled/deionized water and bring volume to 1.0L. Mix thoroughly. Adjust pH to 6.0. Gently heat and bring to boiling. Distribute into tubes or flasks. Autoclave for 15 min at 15 psi pressure–121°C. Pour into sterile Petri dishes or leave in tubes.

Use: For the cultivation and maintenance of *Myxococcus flavescens*.

Myxococcus Medium

Composition per liter:

Agar	12.0g
Pancreatic digest of casein	1.0g
Meat extract	1.0g
Glucose solution	50.0mL

pH 7.2 ± 0.2 at 25°C

Glucose Solution:

Composition per 50.0mL:

Glucose	1.0g

Preparation of Glucose Solution: Add glucose to distilled/deionized water and bring volume to 50.0mL. Mix thoroughly. Autoclave for 15 min at 15 psi pressure–121°C. Cool to 25°C.

Preparation of Medium: Add components, except glucose solution, to distilled/deionized water and bring volume to 950.0mL. Mix thoroughly. Adjust pH to 7.2. Gently heat and bring to boiling. Autoclave for 15 min at 15 psi pressure–121°C. Cool to 45°–50°C. Aseptically add sterile glucose solution. Mix thoroughly. Pour into sterile Petri dishes or distribute into sterile tubes or bottles. Allow tubes or bottles to cool in a slanted position.

Use: For the cultivation of *Myxococcus* species.

Myxococcus xanthus Medium

Agar	20.0g
Pancreatic digest of casein	10.0g
$MgSO_4·7H_2O$	0.5g
K_2HPO_4	0.148g
KH_2PO_4	0.017g

pH 7.6 ± 0.2 at 25°C

Preparation of Medium: Add components to distilled/deionized water and bring volume to 1.0L. Mix thoroughly. Gently heat and bring to boiling. Distribute into tubes or flasks. Autoclave for 15 min at 15 psi pressure–121°C. Pour into sterile Petri dishes or leave in tubes.

Use: For the cultivation and maintenance of *Myxococcus xanthus*.

Nannocystis Agar

Composition per liter:

Agar	15.0g
$CaCl_2·2H_2O$	1.0g

pH 7.2 ± 0.2 at 25°C

Preparation of Medium: Add components to distilled/deionized water and bring volume to 1.0L. Mix thoroughly. Gently heat and bring to boiling. Autoclave for 15 min at 15 psi pressure–121°C. Pour into sterile Petri dishes. After the agar has solidified, overlay the surface with 0.5mL of a suspension of dead (autoclaved) *Escherichia coli* cells.

Use: For the cultivation and maintenance of *Nannocystis* species.

Naphthalene Medium

Composition per liter:

NH_4NO_3	2.5g
$Na_2HPO_4·2H_2O$	1.0g
Naphthalene	0.64g

$MgSO_4 \cdot 7H_2O$ 0.5g
$Fe(SO_4)_3 \cdot 5H_2O$ 0.01g
$Co(NO_3)_2 \cdot 6H_2O$ 5.0mg
$CaCl_2 \cdot 2H_2O$ 1.0mg
KH_2PO_4 .... 0.5mg
$MnSO_4 \cdot 2H_2O$ 0.1mg
$(NH_4)_6Mo_7O_{24} \cdot 4H_2O$ 0.1mg

Preparation of Medium: Add components to distilled/deionized water and bring volume to 1.0L. Mix thoroughly. Distribute into tubes or flasks. Autoclave for 15 min at 15 psi pressure–121°C.

Use: For the cultivation of *Pseudomonas alcaligenes*.

Naphthalene Sulfonic Acid Medium

Composition per 1004.0mL:
$Na_2HPO_4 \cdot 2H_2O$ 3.5g
KH_2PO_4 .... 1.0g
NH_4Cl 0.31g
$MgCl_2 \cdot 6H_2O$ 0.1g
$Ca(NO_3)_2 \cdot 4H_2O$ 0.05g
Solution A 100.0mL
Solution B 3.0mL
Trace elements solution SL-4 .... 1.0mL
pH 7.0 ± 0.2 at 25°C

Solution A:
Composition per liter:
Glucose 3.0g
Glycerol 3.0g
Sodium succinate 3.0g

Preparation of Solution A: Add components to distilled/deionized water and bring volume to 1.0L. Mix thoroughly. Filter sterilize.

Solution B:
Composition per liter:
Naphthalene sulfonic acid 2.3g

Preparation of Solution B: Add naphthalene sulfonic acid to distilled/deionized water and bring volume to 1.0L. Mix thoroughly. Filter sterilize.

Trace Elements Solution SL-4:
Composition per liter:
EDTA 0.5g
$FeSO_4 \cdot 7H_2O$ 0.2g
Trace elements solution SL-6 .... 100.0mL

Preparation of Trace Elements Solution SL-4: Add components to distilled/deionized water and bring volume to 1.0L. Mix thoroughly. Filter sterilize.

Trace Elements Solution SL-6:
Composition per liter:
$MnCl_2 \cdot 4H_2O$ 0.5g
H_3BO_3 .... 0.3g
$CoCl_2 \cdot 6H_2O$ 0.2g
$ZnSO_4 \cdot 7H_2O$ 0.1g
$Na_2MoO_4 \cdot 2H_2O$ 0.03g
$NiCl_2 \cdot 6H_2O$ 0.02g
$CuCl_2 \cdot 2H_2O$ 0.01g

Preparation of Trace Elements Solution SL-6: Add components to distilled/deionized water and bring volume to 1.0L. Mix thoroughly.

Preparation of Medium: Add components, except solution A, solution B, and trace elements solution SL-4, to distilled/deionized water and bring volume to 900.0mL. Mix thoroughly. Autoclave for 15 min at 15 psi pressure–121°C. Aseptically add 200.0mL of sterile solution A, 3.0mL of sterile solution B, and 1.0mL of sterile trace elements solution SL-4. Mix thoroughly. Aseptically distribute into sterile tubes or flasks.

Use: For the cultivation of *Pseudomonas putida*.

Natroniella Medium (DSMZ Medium 784)

Composition per liter:
Na_2CO_3 .... 68.3g
NaCl 15.7g
NH_4Cl 1.0g
KCl 0.2g
KH_2PO_4 .... 0.2g
Yeast extract 0.2g
$MgCl_2 \cdot 6H_2O$ 0.1g
Resazurin 0.5mg
$NaHCO_3$ solution 100.0mL
Vitamin solution 10.0mL
$Na_2S \cdot 9H_2O$ solution 10.0mL
Ethanol solution 10.0mL
Trace elements solution SL-4 .... 1.0mL
pH 9.7-10.0 at 25°C

$Na_2S \cdot 9H_2O$ Solution:
Composition per 10mL:
$Na_2S \cdot 9H_2O$ 1.0g

Preparation of $Na_2S \cdot 9H_2O$ Solution: Add $Na_2S \cdot 9H_2O$ to distilled/deionized water and bring volume to 10.0mL. Mix thoroughly. Autoclave under 100% N_2 for 15 min at 15 psi pressure–121°C. Cool to room temperature.

$NaHCO_3$ Solution:
Composition per 100.0mL:
$NaHCO_3$ 38.3g

Preparation of $NaHCO_3$ Solution: Add $NaHCO_3$ to distilled/deionized water and bring volume to 100.0mL. Mix thoroughly. Sparge with 80% N_2 + 20% CO_2. Filter sterilize.

Ethanol Solution:
Composition per 10.0mL:
Ethanol 5.0mL

Preparation of Ethanol Solution: Add ethanol to distilled/deionized water and bring volume to 10.0mL. Mix thoroughly. Sparge with 100% N_2. Filter sterilize.

Vitamin Solution:
Composition per liter:
Pyridoxine-HCl 10.0mg
Thiamine-HCl·$2H_2O$ 5.0mg
Riboflavin 5.0mg
Nicotinic acid 5.0mg
D-Ca-pantothenate 5.0mg
p-Aminobenzoic acid 5.0mg
Lipoic acid 5.0mg
Biotin 2.0mg
Folic acid 2.0mg
Vitamin B_{12} 0.1mg

Preparation of Vitamin Solution: Add components to distilled/deionized water and bring volume to 1.0L. Mix thoroughly. Sparge with 80% H_2 + 20% CO_2. Filter sterilize.

Trace Elements Solution SL-4:
Composition per liter:
EDTA 0.5g
$FeSO_4 \cdot 7H_2O$ 0.2g
Trace elements solution SL-6 .... 100.0mL

Trace Elements Solution SL-6:
Composition per liter:
H_3BO_3 .... 0.3g
$CoCl_2 \cdot 6H_2O$ 0.2g
$ZnSO_4 \cdot 7H_2O$ 0.1g
$MnCl_2 \cdot 4H_2O$ 0.03g
$Na_2MoO_4 \cdot H_2O$ 0.03g
$NiCl_2 \cdot 6H_2O$ 0.02g
$CuCl_2 \cdot 2H_2O$ 0.01g

Preparation of Trace Elements Solution SL-6: Add components to distilled/deionized water and bring volume to 1.0L. Mix thoroughly. Adjust pH to 3.4.

Preparation of Trace Elements Solution SL-4: Add components to distilled/deionized water and bring volume to 1.0L. Mix thoroughly.

Preparation of Medium: Prepare and dispense medium under 100% N_2. Add components, except vitamin solution, $NaHCO_3$ solution, ethanol, and $Na_2S \cdot 9H_2O$ solution, to distilled/deionized water and bring volume to 870.0mL. Mix thoroughly. Gently heat and bring to boiling. Cool to room temperature while sparging with 100% N_2. Autoclave for 15 min at 15 psi pressure–121°C. Cool to 25°C. Aseptically and anaerobically add 10.0mL sterile vitamin solution, 100.0mL of sterile $NaHCO_3$ solution, 10.0mL sterile ethanol solution, and 10.0mL of sterile $Na_2S \cdot 9H_2O$ solution. Mix thoroughly. Adjust pH to 9.7-10.0. Aseptically and anaerobically distribute into sterile tubes or flasks.

Use: For the cultivation of *Natroniella acetigena (Acetohalobium* sp.*)*.

Natronincola histidinovorans Medium (DSMZ Medium 930)

Composition per liter:
NaCl 80.0g
Na_2CO_3 .... 6.83g
$NaHCO_3$ 3.83g
NH_4Cl 1.0g
KCl 0.2g
KH_2PO_4 .... 0.2g
$MgCl_2 \cdot 6H_2O$ 0.1g
Resazurin 0.01g
Histidine solution 50.0mL
$Na_2S \cdot 9H_2O$ solution 10.0mL
Yeast extract solution 10.0mL
Vitamin solution 2.0mL
Trace elements solution 1.0mL
pH 8.9 ± 0.2 at 25°C

Histidine Solution:
Composition per 50.0mL:
Histidine 5.0g

Preparation of Histidine Solution: Add histidine to distilled/deionized water and bring volume to 50.0mL. Mix thoroughly. Sparge with 100% N_2. Autoclave for 15 min at 15 psi pressure–121°C. Cool to room temperature.

Yeast Extract Solution:
Composition per 10.0mL:
Yeast extract 0.2g

Preparation of Yeast Extract Solution: Add yeast extract to distilled/deionized water and bring volume to 10.0mL. Mix thoroughly. Sparge with 100% N_2. Autoclave under 100% N_2 for 15 min at 15 psi pressure–121°C. Cool to room temperature.

$Na_2S \cdot 9H_2O$ Solution:
Composition per 10mL:
$Na_2S \cdot 9H_2O$ 0.2g

Preparation of $Na_2S \cdot 9H_2O$ Solution: Add $Na_2S \cdot 9H_2O$ to distilled/deionized water and bring volume to 10.0mL. Mix thoroughly. Autoclave under 100% N_2 for 15 min at 15 psi pressure–121°C. Cool to room temperature.

Trace Elements Solution:
Composition per liter:
$MgSO_4 \cdot 7H_2O$ 3.0g
Nitrilotriacetic acid 1.5g
NaCl 1.0g
$MnSO_4 \cdot 2H_2O$ 0.5g
$CoSO_4 \cdot 7H_2O$ 0.18g
$ZnSO_4 \cdot 7H_2O$ 0.18g
$CaCl_2 \cdot 2H_2O$ 0.1g
$FeSO_4 \cdot 7H_2O$ 0.1g
$NiCl_2 \cdot 6H_2O$ 0.025g
$KAl(SO_4)_2 \cdot 12H_2O$ 0.02g
H_3BO_3 .... 0.01g
$Na_2MoO_4 \cdot 4H_2O$ 0.01g
$CuSO_4 \cdot 5H_2O$ 0.01g
$Na_2SeO_3 \cdot 5H_2O$ 0.3mg

Preparation of Trace Elements Solution: Add nitrilotriacetic acid to 500.0mL of distilled/deionized water. Dissolve by adjusting pH to 6.5 with KOH. Add remaining components. Add distilled/deionized water to 1.0L. Mix thoroughly.

Vitamin Solution:
Composition per liter:
Pyridoxine-HCl 10.0mg
Thiamine-HCl·$2H_2O$ 5.0mg
Riboflavin 5.0mg
Nicotinic acid 5.0mg
D-Ca-pantothenate 5.0mg
p-Aminobenzoic acid 5.0mg
Lipoic acid 5.0mg
Biotin 2.0mg
Folic acid 2.0mg
Vitamin B_{12} 0.1mg

Preparation of Vitamin Solution: Add components to distilled/deionized water and bring volume to 1.0L. Mix thoroughly. Sparge with 80% H_2 + 20% CO_2. Filter sterilize.

Preparation of Medium: Prepare and dispense medium under 100% N_2 gas atmosphere. Add components, except $NaHCO_3$, NH_4Cl, Na_2CO_3, histidine solution, $Na_2S \cdot 9H_2O$ solution, yeast extract solution, and vitamin solution, to distilled/deionized water and bring volume to 928.0mL. Mix thoroughly. Gently heat and bring to boiling. Boil for 5 min. Cool to room temperature while sparging with 100% N_2. Add solid $NaHCO_3$, NH_4Cl, and Na_2CO_3. Mix thoroughly. Distribute into anaerobe tubes or bottles. Autoclave for 15 min at 15 psi pressure–121°C. Aseptically and anaerobically add per liter of medium 50.0mL histidine solution, 10.0mL yeast extract solution, 10.0mL $Na_2S \cdot 9H_2O$ solution, and 2.0mL vitamin solution. The final pH should be 8.9.

Use: For the cultivation of *Natronincola histidinovorans.*

Natronobacteria Medium

Composition per liter:
NaCl 200.0g
Agar 20.0g

Yeast extract	5.0g
Casamino acids	5.0g
KH_2PO_4	1.0g
KCl	1.0g
NH_4Cl	1.0g
Sodium glutamate	1.0g
$MgSO_4 \cdot 7H_2O$	0.24g
$CaSO_4 \cdot 2H_2O$	0.17g
Na_2CO_3 solution	100.0mL
Trace elements solution SL-6	1.0mL

pH 9.0 ± 0.2 at 25°C

Na_2CO_3 Solution:
Composition per 100.0mL:

Na_2CO_3	5.0g

Preparation of Na_2CO_3 Solution: Add Na_2CO_3 to distilled/deionized water and bring volume to 100.0mL. Mix thoroughly. Autoclave for 15 min at 15 psi pressure–121°C. Cool to 50°C.

Trace Elements Solution SL-6:
Composition per liter:

H_3BO_3	0.3g
$CoCl_2 \cdot 6H_2O$	0.2g
$ZnSO_4 \cdot 7H_2O$	0.1g
$MnCl_2 \cdot 4H_2O$	0.03g
$Na_2MoO_4 \cdot H_2O$	0.03g
$NiCl_2 \cdot 6H_2O$	0.02g
$CuCl_2 \cdot 2H_2O$	0.01g

Preparation of Trace Elements Solution SL-6: Add components to distilled/deionized water and bring volume to 1.0L. Mix thoroughly. Adjust pH to 3.4.

Preparation of Medium: Add components, except Na_2CO_3 solution, to distilled/deionized water and bring volume to 900.0mL. Mix thoroughly. Gently heat and bring to boiling. Autoclave for 15 min at 15 psi pressure–121°C. Cool to 45°–50°C. Aseptically add sterile Na_2CO_3 solution. Mix thoroughly. Adjust pH to 9.0, if necessary. Pour into sterile Petri dishes or distribute into sterile tubes.

Use: For the cultivation and maintenance of *Natronobacterium gregoryi, Natronobacterium magadii, Natronobacterium pharaonis,* and *Natronococcus occultus.*

Natronobacterium pharaonis Medium

Composition per liter:

NaCl	250.0g
Casamino acids	15.0g
Sodium citrate	3.0g
Glutamic acid	2.5g
$MgSO_4 \cdot 7H_2O$	2.0g
KCl	2.0g

pH 8.5 ± 0.2 at 25°C

Preparation of Medium: Add components to distilled/deionized water and bring volume to 1.0L. Mix thoroughly. Adjust pH to 8.5. Distribute into tubes or flasks. Autoclave for 15 min at 15 psi pressure–121°C.

Use: For the cultivation of *Natronobacterium pharaonis.*

Nautilia Medium (DSMZ Medium 946)

Composition per liter:

Sulfur	10.0g
NH_4Cl	0.33g
KH_2PO_4	0.33g
Resazurin	0.5mg
Synthetic seawater, concentrated	500.0mL
$NaHCO_3$ solution	50.0mL
$Na_2S \cdot 9H_2O$ solution	20.0mL
Sodium formate solution	15.0mL
Trace elements solution	10.0mL
Vitamin solution	10.0mL
Selenite-tungstate solution	1.0mL

pH 6.8 ± 0.2 at 25°C

Synthetic Seawater, Concentrated:

NaCl	55.4g
$MgSO_4 \cdot 7H_2O$	14.0g
$MgCl_2 \cdot 6H_2O$	11.0g
$CaCl_2 \cdot 2H_2O$	1.5g
KCl	1.3g
NaBr	0.2g
H_3BO_3	0.06g
$SrCl_2 \cdot 6H_2O$	0.03g
Na_3-citrate	20.0mg
KI	0.1mg

Trace Elements Solution:
Composition per liter:

$MgSO_4 \cdot 7H_2O$	3.0g
Nitrilotriacetic acid	1.5g
NaCl	1.0g
$MnSO_4 \cdot 2H_2O$	0.5g
$CoSO_4 \cdot 7H_2O$	0.18g
$ZnSO_4 \cdot 7H_2O$	0.18g
$CaCl_2 \cdot 2H_2O$	0.1g
$FeSO_4 \cdot 7H_2O$	0.1g
$NiCl_2 \cdot 6H_2O$	0.025g
$KAl(SO_4)_2 \cdot 12H_2O$	0.02g
H_3BO_3	0.01g
$Na_2MoO_4 \cdot 4H_2O$	0.01g
$CuSO_4 \cdot 5H_2O$	0.01g
$Na_2SeO_3 \cdot 5H_2O$	0.3mg

Preparation of Trace Elements Solution: Add nitrilotriacetic acid to 500.0mL of distilled/deionized water. Dissolve by adjusting pH to 6.5 with KOH. Add remaining components. Add distilled/deionized water to 1.0L. Mix thoroughly.

Vitamin Solution:
Composition per liter:

Pyridoxine-HCl	10.0mg
Thiamine-HCl·$2H_2O$	5.0mg
Riboflavin	5.0mg
Nicotinic acid	5.0mg
D-Ca-pantothenate	5.0mg
p-Aminobenzoic acid	5.0mg
Lipoic acid	5.0mg
Biotin	2.0mg
Folic acid	2.0mg
Vitamin B_{12}	0.1mg

Preparation of Vitamin Solution: Add components to distilled/deionized water and bring volume to 1.0L. Mix thoroughly. Sparge with 80% H_2 + 20% CO_2. Filter sterilize.

Sodium Formate Solution:
Composition per 50.0mL:

Na-formate	10.0g

Preparation of Sodium Formate Solution: Add sodium formate to distilled/deionized water and bring volume to 50.0mL. Mix thoroughly. Sparge with 100% N_2. Filter sterilize.

$Na_2S·9H_2O$ Solution:
Composition per 20mL:
$Na_2S·9H_2O$ 0.6g

Preparation of $Na_2S·9H_2O$ Solution: Add $Na_2S·9H_2O$ to distilled/deionized water and bring volume to 20.0mL. Mix thoroughly. Autoclave under 100% N_2 for 15 min at 15 psi pressure–121°C. Cool to room temperature.

$NaHCO_3$ Solution:
Composition per 100.0mL:
$NaHCO_3$ 5.0g

Preparation of $NaHCO_3$ Solution: Add $NaHCO_3$ to distilled/deionized water and bring volume to 100.0mL. Mix thoroughly. Sparge with 80% N_2 + 20% CO_2. Filter sterilize.

Selenite-Tungstate Solution
Composition per liter:
NaOH 0.5g
$Na_2WO_4·2H_2O$ 4.0mg
$Na_2SeO_3·5H_2O$ 3.0mg

Preparation of Selenite-Tungstate Solution: Add components to distilled/deionized water and bring volume to 1.0L. Mix thoroughly. Sparge with 100% N_2. Filter sterilize.

Preparation of Sulfur: Sterilize sulfur by steaming for 3 hr on each of 3 successive days.

Preparation of Medium: Prepare and dispense medium under 80% N_2 + 20% CO_2 gas atmosphere. Add components, except sulfur, $NaHCO_3$ solution, sodium formate solution, $Na_2S·9H_2O$ solution, vitamin solution, selenite-tungstate solution, and trace elements solution, to distilled/deionized water and bring volume to 894.0mL. Mix thoroughly. Adjust pH to 6.8. Sparge with 80% N_2 + 20% CO_2. Autoclave for 15 min at 15 psi pressure–121°C. Aseptically and anaerobically add 10.0g steam sterilized sulfur, 50.0mL $NaHCO_3$ solution, 15.0mL sodium formate solution, 20.0mL $Na_2S·9H_2O$ solution, 10.0mL vitamin solution, 1.0mL selenite-tungstate solution, and 10.0mL trace elements solution. Mix thoroughly. Adjust pH to 6.8. Aseptically and anaerobically distribute into sterile tubes or bottles.

Use: For the cultivation of *Caldithrix abyssi* and *Nautilia lithotrophica.*

NBA Medium

Composition per liter:
Pancreatic digest of gelatin 5.0g
Casamino acids 5.0g
Beef extract 3.0g
Yeast extract 1.0g
pH 6.8 ± 0.2 at 25°C

Preparation of Medium: Add components to distilled/deionized water and bring volume to 1.0L. Mix thoroughly. Distribute into tubes or flasks. Autoclave for 15 min at 15 psi pressure–121°C.

Use: For the cultivation of *Bdellovibrio* species.

NBY Medium (Nutrient Broth Yeast Extract Medium)

Composition per liter:
Nutrient broth, dehydrated 8.0g
Yeast extract 2.0g
K_2HPO_4 ..... 2.0g
KH_2PO_4 ..... 0.5g
Glucose solution 50.0mL
$MgSO_4·7H_2O$ (1*M* solution) 1.0mL

Glucose Solution:
Composition per 50.0mL:
D-Glucose 5.0g

Preparation of Glucose Solution: Add glucose to distilled/deionized water and bring volume to 50.0mL. Mix thoroughly. Filter sterilize.

Preparation of Medium: Add components, except glucose solution, to distilled/deionized water and bring volume to 950.0mL. Mix thoroughly. Gently heat and bring to boiling. Autoclave for 15 min at 15 psi pressure–121°C. Cool to 45°–50°C. Aseptically add sterile glucose solution. Mix thoroughly. Pour into sterile Petri dishes or distribute into sterile tubes.

Use: For the cultivation and maintenance of *Curtobacterium flaccumfaciens* and *Pseudomonas syringae.*

Nevskia Medium (DSMZ Medium 828)

Composition per liter:
Na-lactate 0.56g
$MgSO_4·7H_2O$ 0.05g
$CaCl_2·2H_2O$ 0.014g
Potassium phosphate buffer, 1*M*, pH 7 ..... 27.5mL
Trace elements solution SL-9 ..... 0.5mL
Seven vitamin solution 0.5mL
pH 7.1 ± 0.2 at 25°C

Trace Elements Solution SL-9:
Composition per liter:
Nitrilotriacetic acid 12.8g
$FeCl_2·4H_2O$ 3.5g
$CoCl_2·6H_2O$ 190.0mg
$MnCl_2·4H_2O$ 100.0mg
$ZnCl_2$ 70.0mg
$Na_2MoO_4·2H_2O$ 36.0mg
$NiCl_2·6H_2O$ 24.0mg
H_3BO_3 ..... 6.0mg
$CuCl_2·2H_2O$ 2.0mg
HCl (25% solution) 10.0mL

Preparation of Trace Elements Solution SL-9: Add $FeCl_2·4H_2O$ to 10.0mL of HCl solution. Mix thoroughly. Add distilled/deionized water and bring volume to 1.0L. Add remaining components. Mix thoroughly. Adjust pH to 6.0. Sparge with 80% N_2 + 20% CO_2. Autoclave for 15 min at 15 psi pressure–121°C.

Seven Vitamin Solution:
Composition per liter:
Pyridoxine hydrochloride 300.0mg
Thiamine-HCl·$2H_2O$ 200.0mg
Nicotinic acid 200.0mg
Vitamin B_{12} 100.0mg
Calcium pantothenate 100.0mg
p-Aminobenzoic acid 80.0mg
D(+)-Biotin 20.0mg

Preparation of Seven Vitamin Solution: Add components to distilled/deionized water and bring volume to 1.0L. Sparge with 100% N_2. Mix thoroughly. Filter sterilize.

Preparation of Medium: Add components, except seven vitamin solution, to distilled/deionized water and bring volume to 999.5mL. Mix thoroughly. Adjust pH to 7.1. Autoclave for 15 min at 15 psi pressure–121°C. Cool to room temperature. Aseptically add 0.5mL seven vitamin solution.

Mix thoroughly. Aseptically distribute into sterile tubes or flasks.

Use: For the cultivation of *Nevskia ramosa.*

Nine K Medium (9K Medium)

Composition per liter:

$FeSO_4 \cdot 7H_2O$	50.0g
$(NH_4)_2SO_4$	3.0g
$Ca(NO_3)_2$	1.0g
K_2HPO_4	0.5g
$MgSO_4 \cdot 7H_2O$	0.5g
KCl	0.1g
H_2SO_4, 10*N*	1.0mL

pH 3.0 ± 0.2 at 25°C

Preparation of Medium: Add components to distilled/deionized water and bring volume to 1.0L. Mix thoroughly. Adjust pH to 3.0. Distribute into tubes or flasks. Autoclave for 15 min at 15 psi pressure–121°C.

Use: For the cultivation of *Thiobacillus ferrooxidans.*

Nitrate Agar

Composition per liter:

Agar	12.0g
Peptone	5.0g
Beef extract	3.0g
KNO_3	1.0g

pH 6.8 ± 0.2 at 25°C

Preparation of Medium: Add components to distilled/deionized water and bring volume to 1.0L. Mix thoroughly. Gently heat and bring to boiling. Distribute into tubes. Autoclave for 15 min at 15 psi pressure–121°C. Allow tubes to cool in a slanted position.

Use: For the differentiation of aerobic and facultative Gram-negative microorganisms based on their ability to reduce nitrate. Test for nitrates with sulfanilic acid and α-naphthylamine reagents. Bacteria that reduce nitrate to nitrite turn the reagents red or pink.

Nitrate Liquid Medium

Composition per liter:

Solution A	500.0mL
Solution B	250.0mL
Solution C	250.0mL

Solution A:
Composition per 500.0mL:

Mannitol	10.0g
KNO_3	0.6g
$Na_2HPO_4 \cdot 12H_2O$	0.45g
Na_2SO_4	0.03g

Preparation of Solution A: Add components to distilled/deionized water and bring volume to 500.0mL. Mix thoroughly. Autoclave for 15 min at 15 psi pressure–121°C. Cool to 25°C.

Solution B:
Composition per 250.0mL:

$MgSO_4 \cdot 7H_2O$	0.12g
$CaCl_2 \cdot 6H_2O$	0.1g
$FeCl_3 \cdot 6H_2O$	0.01g

Preparation of Solution B: Add components to distilled/deionized water and bring volume to 250.0mL. Mix thoroughly. Autoclave for 15 min at 15 psi pressure–121°C. Cool to 25°C.

Solution C:
Composition per 250.0mL:

Calcium pantothenate	0.5mg
Thiamine·HCl	0.1mg
Biotin	0.5μg

Preparation of Solution C: Add components to distilled/deionized water and bring volume to 250.0mL. Mix thoroughly. Filter sterilize.

Preparation of Medium: Aseptically combine 500.0mL of cooled, sterile solution A, 250.0mL of cooled, sterile solution B, and 250.0mL of sterile solution C. Mix thoroughly. Aseptically distribute into sterile tubes or flasks.

Use: For the isolation and cultivation of *Rhizobium* species.

Nitrate Methanol Medium

Composition per liter:

$NaNO_3$	5.0g
K_2HPO_4	2.0g
NaCl	1.0g
$MgSO_4 \cdot 7H_2O$	0.02g
$Na_2MoO_4 \cdot H_2O$	1.0mg
Riboflavin	0.2mg
Calcium pantothenate	0.2mg
Pyridoxine·HCl	0.2mg
Nicotinic acid	0.2mg
Thiamine·HCl	0.1mg
p-Aminobenzoic acid	0.1mg
Biotin	0.01mg
Methanol	10.0mL

pH 7.0 ± 0.2 at 25°C

Preparation of Medium: Add components, except methanol, to distilled/deionized water and bring volume to 990.0mL. Mix thoroughly. Autoclave for 15 min at 15 psi pressure–121°C. Cool to 45°–50°C. Filter sterilize methanol. Aseptically add sterile methanol to cooled sterile medium. Mix thoroughly. Aseptically distribute into sterile tubes or flasks.

Use: For the cultivation and maintenance of *Methylobacterium rhodinum.*

Nitrate Mineral Salts Medium (NMS Medium)

Composition per liter:

Noble agar	12.5g
$MgSO_4 \cdot 7H_2O$	1.0g
KNO_3	1.0g
$Na_2HPO_4 \cdot 12H_2O$	0.717g
KH_2PO_4	0.272g
$CaCl_2 \cdot 6H_2O$	0.2g
Ferric ammonium EDTA	4.0mg
Trace elements solution	0.5mL

pH 6.8 ± 0.2 at 25°C

Trace Elements Solution:
Composition per liter:

Disodium EDTA	0.5g
$FeSO_4 \cdot 7H_2O$	0.2g
H_3BO_3	0.03g
$CoCl_2 \cdot 6H_2O$	0.02g
$ZnSO_4 \cdot 7H_2O$	0.01g
$MnCl_2 \cdot 4H_2O$	3.0mg
$Na_2MoO_4 \cdot 2H_2O$	3.0mg
$NiCl_2 \cdot 6H_2O$	2.0mg
$CaCl_2 \cdot 2H_2O$	1.0mg

Preparation of Trace Elements Solution: Add components to distilled/deionized water and bring volume to 1.0L. Mix thoroughly.

Preparation of Medium: Add components to distilled/deionized water and bring volume to 1.0L. Mix thoroughly. Gently heat and bring to boiling. Adjust pH to 6.8. Distribute into tubes or flasks. Autoclave for 15 min at 15 psi pressure–121°C. Pour into sterile Petri dishes or leave in tubes.

Use: For the cultivation and maintenance of *Methylobacterium* species, *Methylococcus capsulatus*, *Methylomonas agile*, and *Methylomonas methanica*.

Nitrate Mineral Salts Medium with Methanol (NMS Medium with Methanol)

Composition per liter:

Noble agar	12.5g
$MgSO_4 \cdot 7H_2O$	1.0g
KNO_3	1.0g
$Na_2HPO_4 \cdot 12H_2O$	0.717g
KH_2PO_4	0.272g
$CaCl_2 \cdot 6H_2O$	0.2g
Ferric ammonium EDTA	4.0mg
Trace elements solution	0.5mL
Methanol	1.0mL

pH 6.8 ± 0.2 at 25°C

Trace Elements Solution:

Composition per liter:

Disodium EDTA	0.5g
$FeSO_4 \cdot 7H_2O$	0.2g
H_3BO_3	0.03g
$CoCl_2 \cdot 6H_2O$	0.02g
$ZnSO_4 \cdot 7H_2O$	0.01g
$MnCl_2 \cdot 4H_2O$	3.0mg
$Na_2MoO_4 \cdot 2H_2O$	3.0mg
$NiCl_2 \cdot 6H_2O$	2.0mg
$CaCl_2 \cdot 2H_2O$	1.0mg

Preparation of Trace Elements Solution: Add components to distilled/deionized water and bring volume to 1.0L. Mix thoroughly.

Preparation of Medium: Add components, except methanol, to distilled/deionized water and bring volume to 999.0mL. Mix thoroughly. Gently heat and bring to boiling. Adjust pH to 6.8. Distribute into tubes or flasks. Autoclave for 15 min at 15 psi pressure–121°C. Cool to 45°–50°C. Filter sterilize methanol. Aseptically add sterile methanol to cooled sterile medium. Mix thoroughly. Pour into sterile Petri dishes or leave in tubes.

Use: For the cultivation and maintenance of *Methylobacterium fujisawaense*, *Methylobacterium* species, and *Methylomonas clara*.

Nitrate Reduction Broth

Composition per liter:

Pancreatic digest of gelatin	5.0g
Beef extract	3.0g
KNO_3	1.0g

pH 6.9 ± 0.2 at 25°C

Preparation of Medium: Add components to distilled/deionized water and bring volume to 1.0L. Mix thoroughly. Distribute into test tubes that contain an inverted Durham tube. Autoclave for 15 min at 15 psi pressure–121°C.

Use: For the differentiation of members of the Pseudomonadaceae based on their ability to reduce nitrate to nitrite or form N_2 gas. Test for nitrates with sulfanilic acid and α-naphthylamine reagents. Bacteria that reduce nitrate to nitrite turn the reagents red or pink.

Nitrate Reduction Broth for *Pseudomonas* and Related Genera

Composition per liter:

Peptone	5.0g
NaCl	5.0g
Yeast extract	2.0g
Beef extract	1.0g
$NaNO_3$	0.1g

pH 7.4 ± 0.2 at 25°C

Preparation of Medium: Add components to distilled/deionized water and bring volume to 1.0L. Mix thoroughly. Distribute into test tubes that contain an inverted Durham tube. Autoclave for 15 min at 15 psi pressure–121°C.

Use: For the differentiation of members of the Pseudomonadaceae based on their ability to reduce nitrate to nitrite or form N_2 gas. Test for nitrates with sulfanilic acid and α-naphthylamine reagents. Bacteria that reduce nitrate to nitrite turn the reagents red or pink.

Nitrobacter agilis Medium

Composition per liter:

$CaCO_3$	10.0g
NaCl	0.3g
Na_2CO_3	0.25g
KNO_2	0.17g
K_2HPO_4	0.14g
$MgSO_4 \cdot 7H_2O$	0.14g
$FeSO_4 \cdot 7H_2O$	0.03g
$MnSO_4 \cdot 4H_2O$	0.01g
Biotin solution	10.0mL

Biotin Solution:

Composition per 10.0mL:

Biotin	0.15g

Preparation of Biotin Solution: Add biotin to distilled/deionized water and bring volume to 10.0mL. Mix thoroughly. Filter sterilize.

Preparation of Medium: Add Na_2CO_3 to distilled/deionized water and bring volume to 200.0mL. Mix thoroughly. In a separate flask, add the remaining components, except the biotin solution, to distilled/deionized water and bring volume to 790.0mL. Autoclave the Na_2CO_3 solution and salts solution separately for 15 min at 15 psi pressure–121°C. Cool to 25°C. Aseptically combine the sterile Na_2CO_3 solution, sterile salts solution, and sterile biotin solution. Mix thoroughly. Aseptically distribute into sterile tubes or flasks.

Use: For the cultivation of *Nitrobacter agilis*.

Nitrobacter Medium 203

Composition per liter:

Solution C	1.0mL
Solution A	0.5mL
Solution B	0.5mL
Solution D	0.5mL
Solution E	0.5mL
Solution F	0.2mL

Solution A:

Composition per 100.0mL:

$CaCl_2$	2.0g

Preparation of Solution A: Add $CaCl_2$ to distilled/deionized water and bring volume to 100.0mL. Mix thoroughly.

Solution B:

Composition per 100.0mL:

$MgSO_4 \cdot 7H_2O$ 20.0g

Preparation of Solution B: Add $MgSO_4 \cdot 7H_2O$ to distilled/deionized water and bring volume to 100.0mL. Mix thoroughly.

Solution C:

Composition per 100.0mL:

Chelated iron (Sequestrene) 0.1g

Preparation of Solution C: Add chelated iron to distilled/deionized water and bring volume to 100.0mL. Mix thoroughly.

Solution D:

Composition per liter:

$MnCl_2 \cdot 4H_2O$ 0.2g
$Na_2MoO_4 \cdot 2H_2O$ 0.1g
$ZnSO_4 \cdot 7H_2O$ 0.1g
$CuSO_4 \cdot 5H_2O$ 0.02g
$CoCl_2 \cdot 6H_2O$ 2.0mg

Preparation of Solution D: Add components to distilled/deionized water and bring volume to 1.0L. Mix thoroughly.

Solution E:

Composition per 100.0mL:

$NaNO_2$ 41.4g

Preparation of Solution E: Add $NaNO_2$ to distilled/deionized water and bring volume to 100.0mL. Mix thoroughly.

Solution F:

Composition per 100.0mL:

K_2HPO_4 .... 1.74g

Preparation of Solution F: Add K_2HPO_4 to distilled/deionized water and bring volume to 100.0mL. Mix thoroughly.

Preparation of Medium: Add the appropriate volumes of solutions A–F to distilled/deionized water and bring volume to 1.0L. Mix thoroughly. Distribute into tubes or flasks. Autoclave for 15 min at 15 psi pressure–121°C.

Use: For the cultivation and maintenance of *Nitrobacter* species and *Nitrobacter winogradskyi*.

Nitrobacter Medium 204

Composition per liter:

Seawater 700.0mL
Solution C 1.0mL
Solution A 0.5mL
Solution B 0.5mL
Solution D 0.5mL
Solution E 0.5mL
Solution F 0.2mL

Solution A:

Composition per 100.0mL:

$CaCl_2$ 2.0g

Preparation of Solution A: Add $CaCl_2$ to distilled/deionized water and bring volume to 100.0mL. Mix thoroughly.

Solution B:

Composition per 100.0mL:

$MgSO_4 \cdot 7H_2O$ 20.0g

Preparation of Solution B: Add $MgSO_4 \cdot 7H_2O$ to distilled/deionized water and bring volume to 100.0mL. Mix thoroughly.

Solution C:

Composition per 100.0mL:

Chelated iron (Sequestrene) 0.1g

Preparation of Solution C: Add chelated iron to distilled/deionized water and bring volume to 100.0mL. Mix thoroughly.

Solution D:

Composition per liter:

$MnCl_2 \cdot 4H_2O$ 0.2g
$Na_2MoO_4 \cdot 2H_2O$ 0.1g
$ZnSO_4 \cdot 7H_2O$ 0.1g
$CuSO_4 \cdot 5H_2O$ 0.02g
$CoCl_2 \cdot 6H_2O$ 2.0mg

Preparation of Solution D: Add components to distilled/deionized water and bring volume to 1.0L. Mix thoroughly.

Solution E:

Composition per 100.0mL:

$NaNO_2$ 41.4g

Preparation of Solution E: Add $NaNO_2$ to distilled/deionized water and bring volume to 100.0mL. Mix thoroughly.

Solution F:

Composition per 100.0mL:

K_2HPO_4 .... 1.74g

Preparation of Solution F: Add K_2HPO_4 to distilled/deionized water and bring volume to 100.0mL. Mix thoroughly.

Preparation of Medium: Add the appropriate volumes of solutions A–F and seawater to distilled/deionized water and bring volume to 1.0L. Mix thoroughly. Distribute into tubes or flasks. Autoclave for 15 min at 15 psi pressure–121°C.

Use: For the cultivation and maintenance of *Nitrococcus mobilis*.

Nitrobacter Medium B

Composition per liter:

$NaNO_2$ 1.0g
K_2HPO_4 .... 0.5g
$MgSO_4$ 0.5g
NaCl 0.3g
$Fe_2(SO_4)_3$ 5.0mg
$MnSO_4$ 2.0mg
Marble chips .... as needed

pH 7.5 ± 0.2 at 25°C

Preparation of Medium: Add components, except marble chips, to distilled/deionized water and bring volume to 1.0L. Mix thoroughly. Autoclave for 15 min at 15 psi pressure–121°C. Cool to 25°C. Wash marble chips in distilled/deionized water. Put a few chips into test tubes. Autoclave for 60 min at 15 psi pressure–121°C. Cool to 25°C. Aseptically distribute cooled sterile medium into test tubes to cover marble chips.

Use: For the cultivation of *Nitrobacter* species.

Nitrococcus Medium

Composition per 1004.0mL:

$NaNO_2$ solution	1.0mL
K_2HPO_4 solution	1.0mL
$NaHCO_3$ solution	1.0mL
Chelated metals solution	1.0mL

pH 7.5 ± 0.1 at 25°C

$NaNO_2$ Solution:

Composition per 100.0mL:

$NaNO_2$ solution	10.0g

Preparation of $NaNO_2$ Solution: Add $NaNO_2$ to distilled/deionized water and bring volume to 100.0mL. Mix thoroughly. Filter sterilize.

K_2HPO_4 Solution:

Composition per 100.0mL:

K_2HPO_4 solution	2.5g

Preparation of K_2HPO_4 Solution: Add K_2HPO_4 to distilled/deionized water and bring volume to 100.0mL. Mix thoroughly. Filter sterilize.

$NaHCO_3$ Solution:

Composition per 100.0mL:

$NaHCO_3$ solution	5.0g

Preparation of $NaHCO_3$ Solution: Add $NaHCO_3$ to distilled/deionized water and bring volume to 100.0mL. Mix thoroughly. Filter sterilize.

Chelated Metals Solution:

Composition per liter:

EDTA	6.0g
$FeCl_3 \cdot 6H_2O$	1.0g
$MnSO_4 \cdot H_2O$	0.6g
$ZnSO_4 \cdot 7H_2O$	0.3g
$Na_2MoO_4 \cdot 2H_2O$	0.15g
$CoCl_2 \cdot 6H_2O$	4.0mg
$CuSO_4 \cdot 5H_2O$	4.0mg

Preparation of Chelated Metals Solution: Add components to distilled/deionized water and bring volume to 1.0L. Mix thoroughly. Filter sterilize.

Preparation of Medium: Adjust pH of 1.0L of seawater to pH 7.5 with NaOH. Add 1mL of chelated metals solution to the seawater. Mix thoroughly. Autoclave for 15 min at 15 psi pressure–121°C. Cool to 50°C. Aseptically add 1.0mL each of sterile $NaNO_2$, K_2HPO_4, and $NaHCO_3$ solutions. Mix thoroughly. Aseptically distribute into sterile tubes or flasks.

Use: For the cultivation of *Nitrococcus* species.

Nitrogen-Fixing Hydrocarbon Oxidizers Medium

Composition per liter:

Na_2HPO_4	0.3g
KH_2PO_4	0.2g
$MgSO_4 \cdot 7H_2O$	0.1g
$FeSO_4 \cdot 7H_2O$	5.0mg
$Na_2MoO_4 \cdot 2H_2O$	2.0mg

Preparation of Medium: Add components to distilled/deionized water and bring volume to 1.0L. Mix thoroughly. Distribute into tubes or flasks. Autoclave for 15 min at 15 psi pressure–121°C.

Use: For the cultivation and enrichment of nitrogen-fixing hydrocarbon-oxidizing bacteria.

Nitrogen-Fixing Marine Medium

Composition per liter:

Noble agar	10.0g
$MgSO_4 \cdot 7H_2O$	0.04g
$CaCl_2 \cdot 2H_2O$	0.02g
$K_2HPO_4 \cdot 3H_2O$	0.02g
Na_2CO_3	0.02g
Citric acid	3.0mg
Ferric ammonium citrate	3.0mg
Disodium potassium EDTA	0.5mg
Seawater	750.0mL
Trace metals A-5	1.0mL

pH 8.5 ± 0.2 at 25°C

Trace Metals A-5 Mix:

Composition per liter:

H_3BO_3	2.86g
$MnCl_2 \cdot 4H_2O$	1.81g
$ZnSO_4 \cdot 7H_2O$	0.222g
$CuSO_4 \cdot 5H_2O$	0.079g
$Co(NO_3)_2 \cdot 6H_2O$	0.05g
$Na_2MoO_4 \cdot 2H_2O$	0.039g

Preparation of Trace Metals A-5: Add components to distilled/deionized water and bring volume to 1.0L. Mix thoroughly.

Preparation of Medium: Add components to glass-distilled water and bring volume to 1.0L. Mix thoroughly. Gently heat and bring to boiling. Autoclave for 15 min at 15 psi pressure–121°C. Adjust pH to 8.5 with KOH. Pour into sterile Petri dishes or distribute into sterile tubes.

Use: For the cultivation and maintenance of *Anabaena* species.

Nitrogen-Free Agar

Composition per liter:

Agar	15.0g
$CaCO_3$	1.0g
K_2HPO_4	1.0g
$MgSO_4 \cdot 7H_2O$	0.2g
NaCl	0.2g
$FeSO_4 \cdot 7H_2O$	0.1g
$Na_2MoO_4 \cdot 2H_2O$	5.0mg
Glucose solution	50.0mL

pH 7.0 ± 0.2 at 25°C

Glucose Solution:

Composition per 50.0mL:

Glucose	10.0g

Preparation of Glucose Solution: Add glucose to distilled/deionized water and bring volume to 50.0mL. Mix thoroughly. Filter sterilize.

Preparation of Medium: Add components, except glucose solution and agar, to distilled/deionized water and bring volume to 950.0mL. Mix thoroughly. Add agar. Gently heat and bring to boiling. Autoclave for 15 min at 15 psi pressure–121°C. Cool to 50°–55°C. Aseptically add 50.0mL of sterile glucose solution. Mix thoroughly. Pour into sterile Petri dishes or distribute into sterile tubes.

Use: For the cultivation of *Azomonas agilis, Azomonas insignis, Azomonas macrocytogenes, Azorhizophilus paspali, Azotobacter beijerinckii, Azotobacter chroococcum, Azotobacter vinelandii, Beijerinckia acida, Beijerinckia fluminensis, Beijerinckia indica, Beijerinckia mobilis,* and *Derxia gummosa.*

Nitrogen-Free Agar (Norris Agar)

Composition per liter:

Agar	15.0g
$CaCO_3$	1.0g
K_2HPO_4	1.0g
$MgSO_4 \cdot 7H_2O$	0.2g
NaCl	0.2g
$FeSO_4 \cdot 7H_2O$	0.1g
$Na_2MoO_4 \cdot 2H_2O$	5.0mg
Glucose solution	50.0mL

Glucose Solution:

Composition per 100.0mL:

Glucose	20.0g

Preparation of Glucose Solution: Add 20.0g of glucose to distilled/deionized water and bring volume to 100.0mL. Mix thoroughly. Filter sterilize. Warm to 50°C.

Preparation of Medium: Add components, except glucose solution, to distilled/deionized water and bring volume to 950.0mL. Mix thoroughly. Adjust pH to 7.2. Gently heat and bring to boiling. Autoclave for 15 min at 15 psi pressure–121°C. Cool to 50°–55°C. Aseptically add 50.0mL of sterile glucose solution. Mix thoroughly. Pour into sterile Petri dishes or distribute into sterile tubes.

Use: For the cultivation and maintenance of *Azomonas agilis, Azotobacter chroococcum,* and *Azotobacter vinelandii.*

Nitrosococcus Medium

Composition per liter:

$(NH_4)_2SO_4$	1.32g
$MgSO_4 \cdot 7H_2O$	0.38g
$CaCl_2 \cdot 2H_2O$	0.02g
K_2HPO_4	8.7mg
Chelated iron	1.0mg
$MnCl_2 \cdot 4H_2O$	0.2mg
$Na_2MoO_4 \cdot 2H_2O$	0.1mg
$ZnSO_4 \cdot 7H_2O$	0.1mg
$CoCl_2 \cdot 6H_2O$	2.0μg
Phenol Red (0.04% solution)	3.25mL

pH 7.5–7.8 at 25°C

Preparation of Medium: Add components to filtered seawater and bring volume to 1.0L. Mix thoroughly. Adjust pH to 7.5–7.8 with 1*N* HCl. Distribute into tubes. Autoclave for 15 min at 15 psi pressure–121°C.

Use: For the cultivation of *Nitrosococcus oceanus.*

Nitrosococcus oceanus Medium

Composition per 1001.0mL:

Phenol red	5.0g
$NH_4 \cdot Cl$	0.635g
$MgSO_4 \cdot 7H_2O$	0.357g
K_2HPO_4	43.0mg
$CaCl_2 \cdot H_2O$	20.0mg
Chelated metals solution	1.0mL

pH 7.5 ± 0.2 at 25°C

Chelated Metals Solution:

Composition per liter:

EDTA	6.0g
$FeCl_3 \cdot 6H_2O$	1.0g
$MnSO_4 \cdot H_2O$	0.6g
$ZnSO_4 \cdot 7H_2O$	0.3g
$Na_2MoO_4 \cdot 2H_2O$	0.15g
$CoCl_2 \cdot 6H_2O$	4.0mg
$CuSO_4 \cdot 5H_2O$	4.0mg

Preparation of Chelated Metals Solution: Add components to distilled/deionized water and bring volume to 1.0L. Mix thoroughly. Filter sterilize.

Preparation of Medium: Add components, except chelated metals solution, to filtered seawater and bring volume to 1.0L. Mix thoroughly. Adjust pH to 7.5 with sterile 0.1*M* K_2CO_3. Autoclave for 15 min at 15 psi pressure–121°C. Aseptically add 1.0mL of sterile chelated metals solution. Mix thoroughly. Aseptically distribute into sterile tubes or flasks.

Use: For the cultivation of *Nitrosococcus oceanus.*

Nitrosolobus Medium (ATCC Medium 438)

Composition per liter:

$(NH_4)_2SO_4$	1.65g
$MgSO_4 \cdot 7H_2O$	0.2g
K_2HPO_4	0.087g
$CaCl_2 \cdot 2H_2O$	0.02g
Phenol Red	5.0mg
Disodium EDTA	1.0mg
$MnCl_2 \cdot 4H_2O$	0.2mg
$Na_2MoO_4 \cdot 2H_2O$	0.1mg
$ZnSO_4 \cdot 7H_2O$	0.1mg
$CuSO_4 \cdot 5H_2O$	0.02mg
$CoCl_2 \cdot 6H_2O$	2.0μg

pH 7.5 ± 0.2 at 25°C

Preparation of Medium: Add components to distilled/deionized water and bring volume to 1.0L. Mix thoroughly. Adjust pH to 7.5 with 0.1*M* K_2CO_3. Distribute into tubes or flasks. Autoclave for 15 min at 15 psi pressure–121°C.

Use: For the cultivation and maintenance of *Nitrosolobus multiformis.*

Nitrosolobus Medium (ATCC Medium 929)

Composition per liter:

$(NH_4)_2SO_4$	1.32g
$MgSO_4 \cdot 7H_2O$	0.38g
K_2HPO_4	0.087g
$CaCl_2 \cdot 2H_2O$	0.02g
Chelated iron	1.0mg
$MnCl_2 \cdot 4H_2O$	0.2mg
$Na_2MoO_4 \cdot 2H_2O$	0.1mg
$ZnSO_4 \cdot 7H_2O$	0.1mg
$CoCl_2 \cdot 6H_2O$	2.0μg
Phenol Red (0.5% solution)	0.25mL

pH 7.5 ± 0.2 at 25°C

Preparation of Medium: Add components to distilled/deionized water and bring volume to 1.0L. Mix thoroughly. Adjust pH to 7.5 with 0.1*M* K_2CO_3. Distribute into tubes or flasks. Autoclave for 15 min at 15 psi pressure–121°C.

Use: For the cultivation and maintenance of *Nitrosolobus multiformis.*

Nitrosomonas europaea Medium

Composition per liter:

$(NH_4)_2SO_4$	1.7g
$MgSO_4 \cdot 7H_2O$	0.2g
$CaCl_2 \cdot 2H_2O$	0.02g
K_2HPO_4	0.015g
Ferric EDTA	1.0mg
Trace elements solution	1.0mL

pH 7.5 ± 0.2 at 25°C

Trace Elements Solution:
Composition per 100.0mL:

$MnCl_2 \cdot 4H_2O$	0.02g
$Na_2MoO_4 \cdot 2H_2O$	0.01g
$ZnSO_4 \cdot 7H_2O$	0.01g
$CuSO_4 \cdot 5H_2O$	2.0mg
$CoCl_2 \cdot 6H_2O$	0.2mg

Preparation of Trace Elements Solution: Add components to distilled/deionized water and bring volume to 1.0L. Mix thoroughly.

Preparation of Medium: Add components to distilled/deionized water and bring volume to 1.0L. Mix thoroughly. Adjust pH to 7.5 with K_2CO_3. Distribute into tubes or flasks. Autoclave for 15 min at 15 psi pressure–121°C. After inoculation, maintain pH at 7.5–7.8 with sterile 50% K_2CO_3 solution.

Use: For the cultivation and maintenance of *Nitrosomonas europaea*.

Nitrosomonas Medium

Composition per liter:

$(NH_4)_2SO_4$	3.0g
K_2HPO_4	0.5g
$MgSO_4 \cdot 7H_2O$	0.05g
$CaCl_2 \cdot 2H_2O$	4.0mg
Cresol Red (0.0005% solution)	25.0mL
Ferric EDTA solution	0.1mL

pH 8.2–8.4 at 25°C

Ferric EDTA Solution:
Composition per 100.0mL:

$FeSO_4 \cdot 7H_2O$	0.5g
Disodium EDTA	0.14g
H_2SO_4, concentrated	0.05mL

Preparation of Ferric EDTA Solution: Add components to distilled/deionized water and bring volume to 100.0mL. Mix thoroughly.

Preparation of Medium: Add $CaCl_2 \cdot 2H_2O$ and $MgSO_4 \cdot 7H_2O$ to distilled/deionized water and bring volume to 500.0mL. Mix thoroughly. In a separate flask, add remaining components to distilled/deionized water and bring volume to 500.0mL. Mix thoroughly. Autoclave both solutions separately for 15 min at 15 psi pressure–121°C. Cool to 25°C. Aseptically combine the two sterile solutions. Mix thoroughly. Aseptically distribute into sterile tubes or flasks. After inoculation, maintain pH at 8.2–8.4 with sterile 50% K_2CO_3 solution.

Use: For the cultivation and maintenance of *Nitrosomonas europaea*.

Nitrospira moscoviensis Medium (DSMZ Medium 756d)

Composition per liter:

$NaNO_2$	0.5g
Stock solution	100.0mL
Trace elements solution	1.0mL

pH 8.6 ± 0.2 at 25°C

Stock Solution:
Composition per liter:

NaCl	5.0g
KH_2PO_4	1.5g
$MgSO_4 \cdot 7H_2O$	0.5g
$CaCO_3$	0.07g

Preparation of Stock Solution: Add components to distilled/deionized water and bring volume to 1.0L. Mix thoroughly.

Trace Elements Solution:
Composition per liter:

$FeSO_4 \cdot 7H_2O$	97.3mg
H_3BO_3	49.4mg
$ZnSO_4 \cdot 7H_2O$	43.1mg
$(NH_4)_6Mo_7O_{24} \cdot 4H_2O$	37.1mg
$MnSO_4 \cdot 2H_2O$	33.8mg
$CuSO_4 \cdot 5H_2O$	25.0mg

Preparation of Trace Elements Solution: Add components to distilled/deionized water and bring volume to 1.0L. Mix thoroughly.

Preparation of Medium: Add components to distilled/deionized water and bring volume to 1.0L. Mix thoroughly. Adjust pH to 8.6. Distribute into tubes or flasks. Autoclave for 15 min at 15 psi pressure–121°C. Allow to stand for 2-3 days so that pH adjusts itself to 7.4-7.6.

Use: For the cultivation of *Nitrospira moscoviensis*.

NSMP, Modified

Composition per liter:

Casamino acids	5.0g
Glucose	2.0g
KH_2PO_4	0.86g
Sodium citrate	0.6g
K_2HPO_4	0.55g
$MgCl_2 \cdot 6H_2O$	0.43g
$CaCl_2$	0.1g
$MnCl_2 \cdot 4H_2O$	0.016g
$ZnCl_2$	7.0mg
$FeCl_3$	3.0mg

pH 6.5 ± 0.2 at 25°C

Preparation of Medium: Add components to distilled/deionized water and bring volume to 1.0L. Mix thoroughly. Distribute into tubes or flasks. Autoclave for 15 min at 15 psi pressure–121°C.

Use: For the cultivation and maintenance of *Bacillus thuringiensis*.

Nutrient Agar (LMG Medium 160)

Composition per liter:

Agar	3.0g
Lab-Lemco beef extract	1.0g
Peptone	1.0g
NaCl	0.5g

pH 7.3 ± 0.2 at 25°C

Preparation of Medium: Add components to distilled/deionized water and bring volume to 1.0L. Mix thoroughly. Distribute into tubes or flasks. Autoclave for 15 min at 15 psi pressure–121°C.

Use: For the cultivation of heterotrophic bacteria.

Nutrient Agar

Composition per liter:

Agar	15.0g
Peptone	5.0g
NaCl	5.0g
Yeast extract	2.0g
Beef extract	1.0g

pH 7.4 ± 0.2 at 25°C

Source: This medium is available as a premixed powder from Oxoid Unipath.

Preparation of Medium: Add components to distilled/deionized water and bring volume to 1.0L. Mix thoroughly. Gently heat and bring to boiling. Distribute into tubes or flasks. Autoclave for 15 min at 15 psi pressure–121°C. Pour into sterile Petri dishes or leave in tubes.

Use: For the cultivation and maintenance of a wide variety of microorganisms.

Nutrient Agar (ATCC Medium 3)

Composition per liter:

Agar 15.0g
Pancreatic digest of gelatin 5.0g
Beef extract 3.0g
pH 6.8 ± 0.2 at 25°C

Source: This medium is available as a premixed powder from BD Diagnostic Systems.

Preparation of Medium: Add components to distilled/deionized water and bring volume to 1.0L. Mix thoroughly. Gently heat while stirring and bring to boiling. Distribute into tubes or flasks. Autoclave for 15 min at 15 psi pressure–121°C. Pour into sterile Petri dishes or leave in tubes.

Use: For the cultivation of a wide variety of bacteria and for the enumeration of organisms in water, sewage, feces, and other materials.

Nutrient Agar pH 6.8 (LMG Medium 2)

Composition per liter:

Agar 15.0g
Peptone 5.0g
NaCl 5.0g
$Na_2HPO_4 \cdot 12\ H_2O$ 2.39g
Yeast extract 2.0g
Lab-Lemco beef extract 1.0g
KH_2PO_4 .... 0.45g
pH 6.8 ± 0.2 at 25°C

Preparation of Medium: Add components to distilled/deionized water and bring volume to 1.0L. Mix thoroughly. Adjust pH to 6.8. Distribute into tubes or flasks. Autoclave for 15 min at 15 psi pressure–121°C.

Use: For the cultivation and maintenance of *Acidovorax facilis, Pseudomonas geniculata, Arthrobacter siderocapsulatus, Delftia acidovorans, Pseudomonas* spp., and various other heterotrophic bacteria.

Nutrient Agar pH 9.0 (ATCC Medium 2029)

Composition per liter:

Agar 15.0g
Pancreatic digest of gelatin 5.0g
Beef extract 3.0g
pH 9.0 ± 0.1 at 25°C

Source: This medium is available as a premixed powder from BD Diagnostic Systems.

Preparation of Medium: Add components to distilled/deionized water and bring volume to 1.0L. Mix thoroughly. Adjust pH to 9.0. Gently heat while stirring and bring to boiling. Distribute into tubes or flasks. Autoclave for 15 min at 15 psi pressure–121°C. Pour into sterile Petri dishes or leave in tubes.

Use: For the cultivation of alkilophilic bacteria.

Nutrient Agar pH 10.0 (ATCC Medium 2030)

Composition per liter:

Agar 15.0g
Pancreatic digest of gelatin 5.0g
Beef extract 3.0g
pH 10.0 ± 0.1 at 25°C

Source: This medium is available as a premixed powder from BD Diagnostic Systems.

Preparation of Medium: Add components to distilled/deionized water and bring volume to 1.0L. Mix thoroughly. Adjust pH to 10.0. Gently heat while stirring and bring to boiling. Distribute into tubes or flasks. Autoclave for 15 min at 15 psi pressure–121°C. Pour into sterile Petri dishes or leave in tubes.

Use: For the cultivation of alkilophilic bacteria.

Nutrient Agar with 1% Methanol (ATCC Medium 620)

Composition per liter:

Agar 15.0g
Pancreatic digest of gelatin 5.0g
Beef extract 3.0g
Methanol 10.0mL
pH 6.8 ± 0.2 at 25°C

Source: Nutrient agar is available as a premixed powder from BD Diagnostic Systems.

Preparation of Medium: Filter sterilize methanol. Add components, except methanol, to distilled/deionized water and bring volume to 990.0mL. Mix thoroughly. Gently heat and bring to boiling. Autoclave for 15 min at 15 psi pressure–121°C. Cool to 45°–50°C. Aseptically add sterile methanol. Mix thoroughly. Pour into sterile Petri dishes or distribute into sterile tubes.

Use: For the cultivation and maintenance of *Bacillus* species, *Methylomonas clara*, and *Pseudomonas methanolica.*

Nutrient Agar with 2% Methanol (ATCC Medium 628)

Composition per liter:

Agar 15.0g
Pancreatic digest of gelatin 5.0g
Beef extract 3.0g
Methanol 20.0mL
pH 6.8 ± 0.2 at 25°C

Source: Nutrient agar is available as a premixed powder from BD Diagnostic Systems.

Preparation of Medium: Filter sterilize methanol. Add components, except methanol, to distilled/deionized water and bring volume to 980.0mL. Mix thoroughly. Gently heat and bring to boiling. Autoclave for 15 min at 15 psi pressure–121°C. Cool to 45°–50°C. Aseptically add sterile methanol. Mix thoroughly. Pour into sterile Petri dishes or distribute into sterile tubes.

Use: For the cultivation and maintenance of *Pseudomonas* species.

Nutrient Agar with 0.5% NaCl

Composition per liter:

Agar 15.0g
NaCl 5.0g

Pancreatic digest of gelatin 5.0g
Beef extract 3.0g
pH 6.8 ± 0.2 at 25°C

Source: Nutrient agar is available as a premixed powder from BD Diagnostic Systems.

Preparation of Medium: Add components to distilled/deionized water and bring volume to 1.0L. Mix thoroughly. Gently heat and bring to boiling. Distribute into tubes or flasks. Autoclave for 15 min at 15 psi pressure–121°C. Pour into sterile Petri dishes or leave in tubes.

Use: For the cultivation and maintenance of *Agrobacterium tumefaciens*, *Escherichia coli*, *Pseudomonas aeruginosa*, *Salmonella choleraesuis*, *Shigella dysenteriae*, *Shigella flexneri*, *Vibrio* species, and *Yersinia* species.

Nutrient Agar with 0.5% NaCl, pH 9.5–10.0 (LMG Medium 253)

Composition per liter:

Agar 15.0g
NaCl 10.0g
Peptone 5.0g
Yeast extract 2.0g
Lab-Lemco beef extract 1.0g
pH 9.5-10.0

Preparation of Medium: Add components to distilled/deionized water and bring volume to 1.0L. Mix thoroughly. Gently heat and bring to boiling. Distribute into tubes or flasks. Autoclave for 15 min at 15 psi pressure–121°C. Cool to 45°-50°C. Adjust pH to 9.5-10.0 with sterile Na_2CO_3 solution. Pour into sterile Petri dishes or leave in tubes.

Use: For the cultivation and maintenance of *Bacillus* spp.

Nutrient Agar with 1.5% NaCl

Composition per liter:

NaCl 15.0g
Agar 15.0g
Pancreatic digest of gelatin 5.0g
Beef extract 3.0g
pH 6.8 ± 0.2 at 25°C

Source: Nutrient agar is available as a premixed powder from BD Diagnostic Systems.

Preparation of Medium: Add components to distilled/deionized water and bring volume to 1.0L. Mix thoroughly. Gently heat and bring to boiling. Distribute into tubes or flasks. Autoclave for 15 min at 15 psi pressure–121°C. Pour into sterile Petri dishes or leave in tubes.

Use: For the cultivation and maintenance of *Photobacterium leiognathi*, *Pseudomonas fluorescens*, and *Vibrio natriegens*.

Nutrient Agar with 2% NaCl

Composition per liter:

NaCl 20.0g
Agar 15.0g
Pancreatic digest of gelatin 5.0g
Beef extract 3.0g
pH 6.8 ± 0.2 at 25°C

Preparation of Medium: Add components to distilled/deionized water and bring volume to 1.0L. Mix thoroughly. Gently heat and bring to boiling. Distribute into tubes or flasks. Autoclave for 15 min at 15 psi pressure–121°C. Pour into sterile Petri dishes or leave in tubes.

Use: For the cultivation and maintenance of *Vibrio alginolyticus*, *Vibrio bivalvii*, *Vibrio mediterranei*, *Vibrio natriegens*, *Vibrio ordali*, *Vibrio orientalis*, *Vibrio parahaemolyticus*, *Vibrio pelagigius*, and *Vibrio vulnificus*.

Nutrient Agar with 3% NaCl

Composition per liter:

NaCl 30.0g
Agar 15.0g
Pancreatic digest of gelatin 5.0g
Beef extract 3.0g
pH 6.8 ± 0.2 at 25°C

Source: Nutrient agar is available as a premixed powder from BD Diagnostic Systems.

Preparation of Medium: Add components to distilled/deionized water and bring volume to 1.0L. Mix thoroughly. Gently heat and bring to boiling. Distribute into tubes or flasks. Autoclave for 15 min at 15 psi pressure–121°C. Pour into sterile Petri dishes or leave in tubes.

Use: For the cultivation and maintenance of *Bacillus* species, *Alteromonas nigrifaciens*, *Halococcus* species, *Planococcus citreus*, *Pseudomonas beijerinckii*, *Staphylococcus* species, *Streptococcus pyogenes*, and *Vibrio* species.

Nutrient Agar with 5% NaCl, pH 9.5–10.0 (LMG Medium 254)

Composition per liter:

NaCl 55.0g
Agar 15.0g
Peptone 5.0g
Yeast extract 2.0g
Lab-Lemco beef extract 1.0g
pH 9.5-10.0

Preparation of Medium: Add components to distilled/deionized water and bring volume to 1.0L. Mix thoroughly. Gently heat and bring to boiling. Distribute into tubes or flasks. Autoclave for 15 min at 15 psi pressure–121°C. Cool to 45°-50°C. Adjust pH to 9.5-10.0 with sterile Na_2CO_3 solution. Pour into sterile Petri dishes or leave in tubes.

Use: For the cultivation and maintenance of *Bacillus haloalkaliphilus*.

Nutrient Agar with 10% NaCl

Composition per liter:

NaCl 100.0g
Agar 15.0g
Pancreatic digest of gelatin 5.0g
Beef extract 3.0g
pH 6.8 ± 0.2 at 25°C

Source: Nutrient agar is available as a premixed powder from BD Diagnostic Systems.

Preparation of Medium: Add components to distilled/deionized water and bring volume to 1.0L. Mix thoroughly. Gently heat and bring to boiling. Distribute into tubes or flasks. Autoclave for 15 min at 15 psi pressure–121°C. Pour into sterile Petri dishes or leave in tubes.

Use: For the cultivation and maintenance of *Paracoccus halodenitrificans* and *Micrococcus* species.

Nutrient Agar with 10% NaCl and Maltose

Composition per liter:

NaCl 100.0g
Agar 15.0g
Maltose 10.0g

Pancreatic digest of gelatin 5.0g
Beef extract 3.0g

pH 6.8 ± 0.2 at 25°C

Source: Nutrient agar is available as a premixed powder from BD Diagnostic Systems.

Preparation of Medium: Add components to distilled/deionized water and bring volume to 1.0L. Mix thoroughly. Gently heat and bring to boiling. Distribute into tubes or flasks. Autoclave for 15 min at 15 psi pressure–121°C. Pour into sterile Petri dishes or leave in tubes.

Use: For the cultivation and maintenance of *Paracoccus halodenitrificans* and *Micrococcus* species.

Nutrient Agar with Phytone

Composition per liter:

Agar 15.0g
Phytone 10.0g
Pancreatic digest of gelatin 5.0g
Beef extract 3.0g

pH 6.8 ± 0.2 at 25°C

Preparation of Medium: Add components to distilled/deionized water and bring volume to 1.0L. Mix thoroughly. Gently heat while stirring and bring to boiling. Distribute into tubes or flasks. Autoclave for 15 min at 15 psi pressure–121°C. Pour into sterile Petri dishes or leave in tubes.

Use: For the cultivation of a wide variety of bacteria.

Nutrient Agar with Soil Extract

Composition per liter:

Agar 15.0g
Pancreatic digest of gelatin 5.0g
Beef extract 3.0g
Soil extract 250.0mL

pH 6.8 ± 0.2 at 25°C

Source: Nutrient agar is available as a premixed powder from BD Diagnostic Systems.

Soil Extract:

Composition per 300.0mL:

African Violet soil 115.5g
Na_2CO_3 .... 0.3g

Preparation of Soil Extract: Add components to tap water and bring volume to 300.0mL. Autoclave for 60 min at 15 psi pressure–121°C. Filter through Whatman filter paper.

Preparation of Medium: Add components to distilled/deionized water and bring volume to 1.0L. Mix thoroughly. Gently heat and bring to boiling. Distribute into tubes or flasks. Autoclave for 15 min at 15 psi pressure–121°C. Pour into sterile Petri dishes.

Use: For the cultivation and maintenance of *Auerobacterium* species, *Bacillus* species, and *Saccharomonospora viridis.*

Nutrient Agar with 2% Sucrose (ATCC Medium 1297)

Composition per liter:

Sucrose .... 20.0 g
Agar 15.0g
Pancreatic digest of gelatin 5.0g
Beef extract 3.0g

pH 6.8 ± 0.2 at 25°C

Preparation of Medium: Add components to distilled/deionized water and bring volume to 1.0L. Mix thoroughly. Gently heat and bring to boiling. Distribute into tubes or flasks. Autoclave for 15 min at 15 psi pressure–121°C. Pour into sterile Petri dishes or leave in tubes.

Use: For the cultivation of osmophilic bacteria

Nutrient Broth

Composition per liter:

Peptone 5.0g
NaCl 5.0g
Yeast extract 2.0g
Beef extract 1.0g

pH 7.4 ± 0.2 at 25°C

Source: This medium is available as a premixed powder from Oxoid Unipath.

Preparation of Medium: Add components to distilled/deionized water and bring volume to 1.0L. Mix thoroughly. Distribute into tubes or flasks. Autoclave for 15 min at 15 psi pressure–121°C.

Use: For the cultivation of a wide variety of nonfastidious microorganisms.

Nutrient Broth

Composition per liter:

Pancreatic digest of gelatin 5.0g
Beef extract 3.0g

pH 6.9 ± 0.2 at 25°C

Source: This medium is available as a premixed powder from BD Diagnostic Systems.

Preparation of Medium: Add components to distilled/deionized water and bring volume to 1.0L. Mix thoroughly. Distribute into tubes or flasks. Autoclave for 15 min at 15 psi pressure–121°C.

Use: For the cultivation of a wide variety of nonfastidious microorganisms.

Nutrient Broth with 6% NaCl

Composition per liter:

NaCl 60.0g
Pancreatic digest of gelatin 5.0g
Beef extract 3.0g

pH 6.8 ± 0.2 at 25°C

Preparation of Medium: Add components to distilled/deionized water and bring volume to 1.0L. Mix thoroughly. Distribute into tubes or flasks. Autoclave for 15 min at 15 psi pressure–121°C.

Use: For the cultivation of organisms in water, sewage, feces, and other materials. For the cultivation and maintenance of *Paracoccus halodenitrificans.*

Nutrient Broth with Rifampicin

Composition per liter:

Peptone 5.0g
NaCl 5.0g
Yeast extract 2.0g
Beef extract 1.0g
Rifampicin solution 10.0mL

pH 7.4 ± 0.2 at 25°C

Rifampicin Solution:

Composition per 10.0mL:

Rifampicin 10.0mg

Preparation of Rifampicin Solution: Add rifampicin to distilled/deionized water and bring volume to 10.0mL. Mix thoroughly. Filter sterilize.

Preparation of Medium: Add components, except rifampicin solution, to distilled/deionized water and bring volume to 990.0mL. Mix thoroughly. Adjust pH to 7.4. Autoclave for 15 min at 15 psi pressure–121°C. Cool to room temperature. Aseptically add 10.0mL of sterile rifampicin solution. Mix thoroughly. Aseptically distribute into sterile tubes or flasks.

Use: For the cultivation of *Agrobacterium tumefaciens.*

Nutrient Gelatin

Composition per liter:

Gelatin	120.0g
Pancreatic digest of gelatin	5.0g
Beef extract	3.0g

pH 6.8 ± 0.2 at 25°C

Source: This medium is available as a premixed powder from BD Diagnostic Systems and Oxoid Unipath.

Preparation of Medium: Add components to distilled/deionized water and bring volume to 1.0L. Mix thoroughly. Gently heat while stirring to 50°C. Distribute into tubes. Autoclave for 15 min at 15 psi pressure–121°C.

Use: For the cultivation and differentiation of bacteria based on their ability to liquefy gelatin.

Nutrient Soil Extract Agar (LMG Medium 30)

Composition per liter:

Agar	15.0g
Yeast extract	2.0g
Peptone	5.0g
NaCl	5.0g
Lab-Lemco beef extract	1.0g
Soil extract	1.0L

pH 7.4 ± 0.2 at 25°C

Soil Extract:

Composition per liter:

Soil	1.0kg
$CaCO_3$	2.0g

Preparation of Soil Extract: Add 1.0kg soil to 1.0L tap water. Autoclave for 30 min at 15 psi pressure–121°C. Add 2.0g $CaCO_3$. Filter through Whatman filter paper. Bring volume to 1.0L with tap water. Mix thoroughly.

Preparation of Medium: Add components to 1.0L soil extract. Mix thoroughly. Gently heat and bring to boiling. Distribute into tubes or flasks. Autoclave for 15 min at 15 psi pressure–121°C. Pour into sterile Petri dishes or leave in tubes.

Use: For the cultivation and maintenance of soil bacteria.

Nutrient Yeast Glucose Medium

Composition per liter:

Glucose	10.0g
Pancreatic digest of gelatin	5.0g
Yeast extract	5.0g
Beef extract	3.0g

pH 6.8 ± 0.2 at 25°C

Preparation of Medium: Add components to distilled/deionized water and bring volume to 1.0L. Mix thoroughly. Distribute into tubes or flasks. Autoclave for 15 min at 15 psi pressure–121°C.

Use: For the cultivation and maintenance of *Erwinia amylovora.*

NWRI Agar (HPC Agar)

Composition per liter:

Agar	15.0g
Peptone	3.0g
Soluble casein	0.5g
K_2HPO_4	0.2g
$MgSO_4$	0.05g
$FeCl_3$	1.0mg

pH 7.2 ± 0.2 at 25°C.

Preparation of Medium: Add components to distilled/deionized water and bring volume to 1.0L. Mix thoroughly. Gently heat and bring to boiling. Adjust pH to 7.2. Distribute into tubes or flasks. Autoclave for 15 min at 15 psi pressure–121°C. Pour into sterile Petri dishes.

Use: For estimation of the number of live heterotrophic bacteria in water using the heterotrophic plate count technique.

Oat Flake Medium (DSMZ Medium 189)

Composition per liter:

Oat flakes	30.0g
Agar	15.0g

pH 7.2 ± 0.2 at 25°C

Preparation of Medium: Add rolled oats to 500.0mL distilled/deionized water. Gently heat and bring to boiling. Boil for 10 min. Filter. Add agar and bring volume to 1.0L with distilled/deionized water. Mix thoroughly. Gently heat and bring to boiling. Mix by shaking. Distribute into tubes or flasks. Autoclave for 20 min at 15 psi pressure–121°C. Mix by shaking. Pour into sterile Petri dishes or leave in tubes.

Use: For the cultivation and maintenance of actinomycetes and fungi.

Oceanithermus profundus Medium (DSMZ Medium 975)

Composition per liter:

NaCl	30.0g
HEPES	2.38g
NH_4Cl	0.33g
KCl	0.33g
Potassium nitrate solution	10.0mL
Calcium chloride solution	10.0mL
Magnesium chloride solution	10.0mL
Yeast extract solution	10.0mL
Sucrose solution	10.0mL
Tryptone solution	10.0mL
Vitamin solution	1.0mL
Trace elements solution	1.0mL

pH 7.0-7.5 at 25°C

Potassium Nitrate Solution:

Composition per 10.0mL:

KNO_3	0.33g

Preparation of Potassium Nitrate Solution: Add KNO_3 to distilled/deionized water and bring volume to 10.0mL. Mix thoroughly. Autoclave for 15 min at 15 psi pressure–121°C. Cool to room temperature.

Magnesium Chloride Solution:

Composition per 10.0mL:

$MgCl_2 \cdot 6H_2O$	0.33g

Preparation of Magnesium Chloride Solution: Add $MgCl_2 \cdot 6H_2O$ to distilled/deionized water and bring volume to 10.0mL. Mix thoroughly. Autoclave for 15 min at 15 psi pressure–121°C. Cool to room temperature.

Calcium Chloride Solution:
Composition per 10.0mL:
$CaCl_2 \cdot 2H_2O$ 0.33g

Preparation of Calcium Chloride Solution: Add $CaCl_2 \cdot 2H_2O$ to distilled/deionized water and bring volume to 10.0mL. Mix thoroughly. Autoclave for 15 min at 15 psi pressure–121°C. Cool to room temperature.

Yeast Extract Solution:
Composition per 10.0mL:
Yeast extract 0.2g

Preparation of Yeast Extract Solution: Add yeast extract to distilled/deionized water and bring volume to 10.0mL. Mix thoroughly. Autoclave for 15 min at 15 psi pressure–121°C. Cool to room temperature.

Sucrose Solution:
Composition per 10.0mL:
Sucrose 2.0g

Preparation of Sucrose Solution: Add sucrose to distilled/deionized water and bring volume to 10.0mL. Mix thoroughly. Autoclave for 15 min at 15 psi pressure–121°C. Cool to room temperature.

Tryptone Solution:
Composition per 10.0mL:
Tryptone 1.0g

Preparation of Tryptone Solution: Add tryptone to distilled/deionized water and bring volume to 10.0mL. Mix thoroughly. Autoclave for 15 min at 15 psi pressure–121°C. Cool to room temperature.

Trace Elements Solution:
Composition per liter:
$MgSO_4 \cdot 7H_2O$ 3.0g
Nitrilotriacetic acid 1.5g
NaCl 1.0g
$MnSO_4 \cdot 2H_2O$ 0.5g
$CoSO_4 \cdot 7H_2O$ 0.18g
$ZnSO_4 \cdot 7H_2O$ 0.18g
$CaCl_2 \cdot 2H_2O$ 0.1g
$FeSO_4 \cdot 7H_2O$ 0.1g
$NiCl_2 \cdot 6H_2O$ 0.025g
$KAl(SO_4)_2 \cdot 12H_2O$ 0.02g
H_3BO_3 .... 0.01g
$Na_2MoO_4 \cdot 4H_2O$ 0.01g
$CuSO_4 \cdot 5H_2O$ 0.01g
$Na_2SeO_3 \cdot 5H_2O$ 0.3mg

Preparation of Trace Elements Solution: Add nitrilotriacetic acid to 500.0mL of distilled/deionized water. Dissolve by adjusting pH to 6.5 with KOH. Add remaining components. Add distilled/deionized water to 1.0L. Mix thoroughly. Sparge with 100% N_2. Autoclave for 15 min at 15 psi pressure–121°C. Cool to room temperature.

Vitamin Solution:
Composition per liter:
Pyridoxine-HCl 10.0mg
Thiamine-HCl·$2H_2O$ 5.0mg
Riboflavin 5.0mg
Nicotinic acid 5.0mg
D-Ca-pantothenate 5.0mg
p-Aminobenzoic acid 5.0mg
Lipoic acid 5.0mg
Biotin 2.0mg
Folic acid 2.0mg
Vitamin B_{12} 0.1mg

Preparation of Vitamin Solution: Add components to distilled/deionized water and bring volume to 1.0L. Mix thoroughly. Sparge with 80% H_2 + 20% CO_2. Filter sterilize.

Preparation of Medium: Prepare and dispense medium under 100% N_2 gas atmosphere. Add components, except yeast extract solution, sucrose solution, calcium chloride solution, magnesium chloride solution, potassium nitrate solution, tryptone solution, vitamin solution, and trace elements solution, to distilled/deionized water and bring volume to 938.0mL. Mix thoroughly. distribute into anaerobe tubes or bottles. Autoclave for 15 min at 15 psi pressure–121°C. Aseptically and anaerobically add per liter, 10.0mL sucrose solution, 10.0mL yeast extract solution, 10.0mL potassium nitrate solution, 10.0mL tryptone solution, 10.0mL magnesium chloride solution, 10.0ml calcium chloride solution, 1.0mL vitamin solution, and 1.0mL trace elements solution. Mix thoroughly. Adjust pH to 7.0-7.5

Use: For the cultivation of *Oceanithermus profundus.*

Oil Agar Medium

Composition per liter:
Agar, purified 20.0g
NaCl 10.0g
Oil powder 10.0g
NH_4NO_3 .... 1.0g
$MgSO_4$ 0.5g
Amphotericin B solution 10.0mL
K_2HPO_4 solution 7.0mL
KH_2PO_4 solution 3.0mL
$FeCl_3$ 0.1mL

Oil Powder:
Composition per 10.0g:
Hydrocarbon 10.0g
Silica gel 10.0g
Diethyl ether 30.0mL

Preparation of Oil Powder: Add 10.0g of hydrocarbon to 30.0mL of diethyl ether. Mix thoroughly. Add 10.0g of silica gel. Allow ether to evaporate.

Amphotericin B Solution:
Composition per 10.0mL:
Amphotericin B 0.01g

Preparation of Amphotericin B Solution: Add amphotericin B to distilled/deionized water and bring volume to 10.0mL. Mix thoroughly. Filter sterilize.

K_2HPO_4 Solution:
Composition per 100.0mL:
K_2HPO_4 .... 10.0g

Preparation of K_2HPO_4 Solution: Add K_2HPO_4 to distilled/deionized water and bring volume to 100.0mL. Mix thoroughly. Autoclave for 15 min at 15 psi pressure–121°C. Cool to 25°C.

KH_2PO_4 Solution:
Composition per 100.0mL:
KH_2PO_4 .... 10.0g

Preparation of KH_2PO_4 Solution: Add KH_2PO_4 to distilled/deionized water and bring volume to 100.0mL. Mix thoroughly. Autoclave for 15 min at 15 psi pressure–121°C. Cool to 25°C.

Preparation of Medium: Add components—except amphotericin B solution, K_2HPO_4 solution, and KH_2PO_4 solution, to distilled/deionized water and bring volume to 980.0mL. Mix thoroughly. Gently heat and bring to boiling. Autoclave for 15 min at 15 psi pressure–121°C. Cool to 45°–50°C. Aseptically add 10.0mL of sterile amphotericin B solution, 7.0mL of sterile K_2HPO_4 solution, and 3.0mL of sterile KH_2PO_4 solution. Mix thoroughly. Pour into sterile Petri dishes or distribute into sterile tubes.

Use: For the cultivation and enumeration of hydrocarbon-utilizing bacteria by direct plating of estuarine water and sediment samples.

OSrt Broth (ATCC Medium 2340)

Composition per liter:

Yeast extract	2.0 g
Glycerol	10.0 mL

Preparation of Medium: Add components to distilled/deionized water and bring volume to 1.0L. Mix thoroughly. Distribute into flasks or tubes. Autoclave for 15 min at 15 psi pressure–121°C.

Use: For the cultivation and maintenance of *Geodermatophilus obscurus* subspecies *utahensis*.

Oxidation-Fermentation Medium (OF Medium)

Composition per liter:

NaCl	5.0g
Agar	2.5g
Pancreatic digest of casein	2.0g
K_2HPO_4	0.3g
Bromothymol Blue	0.03g
Carbohydrate solution	100.0mL

pH 6.8 ± 0.1 at 25°C

Source: This medium is available as a premixed powder from BD Diagnostic Systems.

Carbohydrate Solution:
Composition per 100.0mL:

Carbohydrate	10.0g

Preparation of Carbohydrate Solution: Add carbohydrate to distilled/deionized water and bring volume to 100.0mL. Mix thoroughly. Filter sterilize.

Preparation of Medium: Add components, except carbohydrate solution, to distilled/deionized water and bring volume to 900.0mL. Mix thoroughly. Gently heat and bring to boiling. Autoclave for 15 min at 15 psi pressure–121°C. Cool to 45°–50°C. Aseptically add 100.0mL of sterile carbohydrate solution. Mix thoroughly. Pour into sterile Petri dishes or distribute into sterile tubes.

Use: For differentiating Gram-negative bacteria based upon determining the oxidative and fermentative metabolism of carbohydrates.

Oxidation-Fermentation Medium, Hugh-Leifson's (Hugh-Leifson's Oxidation Fermentation Medium)

Composition per liter:

NaCl	5.0g
Agar	3.0g
Peptone	2.0g
K_2HPO_4	0.3g
Carbohydrate solution	100.0mL
Bromothymol Blue solution (0.2%)	15.0mL

pH 7.1 ± 0.2 at 25°C

Carbohydrate Solution:
Composition per 100.0mL:

Carbohydrate	10.0g

Preparation of Carbohydrate Solution: Add carbohydrate to distilled/deionized water and bring volume to 100.0mL. Mix thoroughly. Filter sterilize.

Preparation of Medium: Add components, except carbohydrate solution, to distilled/deionized water and bring volume to 900.0mL. Mix thoroughly. Gently heat and bring to boiling. Autoclave for 15 min at 15 psi pressure–121°C. Cool to 45°–50°C. Aseptically add 100.0mL of sterile carbohydrate solution. Mix thoroughly. Pour into sterile Petri dishes or distribute into sterile tubes.

Use: For differentiating Gram-negative bacteria, such as *Vibrio* species, based upon determining the oxidative and fermentative metabolism of carbohydrates. Bacteria that ferment the carbohydrate turn the medium yellow.

Oxidation-Fermentation Medium, King's (King's OF Medium)

Composition per liter:

Base	900.0mL
Carbohydrate solution	100.0mL

Base:
Composition per liter:

Agar	3.0g
Pancreatic digest of casein	2.0g
Carbohydrate solution	100.0mL
Phenol Red (1.5% solution)	2.0mL

pH to 7.3 ± 0.2

Carbohydrate Solution:
Composition per 100.0mL:

Carbohydrate	10.0g

Preparation of Carbohydrate Solution: Add carbohydrate to distilled/deionized water and bring volume to 100.0mL. Mix thoroughly. Filter sterilize.

Preparation of Medium: Add components, except carbohydrate solution, to distilled/deionized water and bring volume to 900.0mL. Mix thoroughly. Gently heat and bring to boiling. Autoclave for 15 min at 15 psi pressure–121°C. Cool to 45°–50°C. Aseptically add 100.0mL of sterile carbohydrate solution. Mix thoroughly. Pour into sterile Petri dishes or distribute into sterile tubes.

Use: For differentiating bacteria based upon determining the oxidative and fermentative metabolism of carbohydrates. Bacteria that ferment the carbohydrate turn the medium yellow.

OZR Medium

Composition per liter:

Agar	15.0g
Pancreatic digest of casein	1.0g
Yeast extract	1.0g

Preparation of Medium: Add components to seawater and bring volume to 1.0L. Mix thoroughly. Gently heat and bring to boiling. Distribute into tubes or flasks. Autoclave for 15 min at 15 psi pressure–121°C. Pour into sterile Petri dishes or leave in tubes.

Use: For the cultivation and maintenance of *Leucothrix mucor.*

PA-C Agar (mPA-C Agar)

Composition per liter:

Agar	12.0g
L-Lysine·HCl	5.0g
NaCl	5.0g
$Na_2S_2O_3$	5.0g
Yeast extract	2.0g
$MgSO_4 \cdot 7H_2O$	1.5g
Lactose	1.25g
Sucrose	1.25g
Xylose	1.25g
Ferric ammonium citrate	0.8g
Phenol Red	0.08g
Nalidixic acid	0.037g
Kanamycin	8.0mg

pH 7.2 ± 0.1 at 25°C

Source: This medium is available as a premixed powder from BD Diagnostic Systems.

Preparation of Medium: Add components to distilled/deionized water and bring volume to 1.0L. Mix thoroughly. Gently heat and bring to boiling. Distribute into tubes or flasks. Autoclave for 15 min at 15 psi pressure–121°C. Pour into sterile Petri dishes or leave in tubes.

Use: For the selective recovery and enumeration of *Pseudomonas aeruginosa* from water samples.

Paracoccus alcaliphilus Medium (DSMZ Medium 772)

Composition per liter:

$(NH_4)_2SO_4$	3.0g
Na_2HPO_4	3.0g
Yeast extract	2.0g
KH_2PO_4	1.4g
$MgSO_4 \cdot 7H_2O$	0.2g
Fe-citrate	30.0mg
$CaCl_2 \cdot 2H_2O$	30.0mg
$MnCl_2 \cdot 4H_2O$	5.0mg
$ZnSO_4 \cdot 7H_2O$	5.0mg
$CuSO_4 \cdot 2H_2O$	0.5mg
Methanol	10.0mL
$NaHCO_3$ solution	variable

pH 9.0 ± 0.2 at 25°C

$NaHCO_3$ Solution:

Composition per 50.0mL:

$NaHCO_3$	5.0g

Preparation of $NaHCO_3$ Solution: Add $NaHCO_3$ to distilled/deionized water and bring volume to 50.0mL. Mix thoroughly. Sparge with 80% N_2 + 20% CO_2. Filter sterilize.

Preparation of Medium: Add components, except $NaHCO_3$ solution and methanol, to distilled/deionized water and bring volume to 990.0mL. Mix thoroughly. Autoclave for 15 min at 15 psi pressure–121°C. Filter sterilize 10.0mL of methanol. Aseptically add the 10.0mL filter sterilized methanol. Mix thoroughly. Adjust pH to 9.0 using the sterile $NaHCO_3$ solution.

Use: For the cultivation of *Paracoccus alcaliphilus.*

Paracoccus aminophilus Paracoccus aminovorans Medium (DSMZ Medium 774)

Composition per liter:

Agar	20.0g
Peptone	5.0g
Yeast extract	5.0g
Glucose	5.0g

pH 7.0 ± 0.2 at 25°C

Preparation of Medium: Add components to distilled/deionized water and bring volume to 1.0L. Mix thoroughly. Gently heat and bring to boiling. Distribute into tubes or flasks. Autoclave for 15 min at 15 psi pressure–121°C. Pour into sterile Petri dishes or leave in tubes.

Use: For the cultivation and maintenance of *Paracoccus aminovorans* and *Paracoccus aminophilus.*

Paracoccus halodenitrificans Agar

Composition per liter:

NaCl	60.0g
Agar	15.0g
Peptone	5.0g
NaCl	5.0g
Yeast extract	4.0g
Beef extract	1.0g

pH 7.2 ± 0.2 at 25°C

Preparation of Medium: Add components to distilled/deionized water and bring volume to 1.0L. Mix thoroughly. Gently heat and bring to boiling. Distribute into tubes or flasks. Autoclave for 15 min at 15 psi pressure–121°C. Pour into sterile Petri dishes or leave in tubes.

Use: For the cultivation and maintenance of *Deleya aquamarina, Deleya halophila, Deleya venusta, Halomonas halmophila, Marinococcus communis, Micrococcus halobius, Micrococcus varians,* and *Paracoccus halodenitrificans.*

Paracoccus kocurii Medium (DSMZ Medium 773)

Composition per liter:

$NaH_2PO_4 \cdot 2H_2O$	1.71g
K_2HPO_4	1.41g
$(NH_4)_2SO_4$	1.0g
Hutner's salts solution	20.0mL
Tetramethyl ammonium chloride solution	10.0mL
Thiamine hydrochloride solution	10.0mL

pH 6.9 ± 0.2 at 25°C

Tetramethyl Ammonium Chloride Solution:

Composition per 10.0mL:

Tetramethyl ammonium chloride	1.0g

Preparation of Tetramethyl Ammonium Chloride Solution: Add tetramethyl ammonium chloride to distilled/deionized water and bring volume to 10.0mL. Mix thoroughly. Filter sterilize.

Thiamine Hydrochloride Solution:

Composition per 10.0mL:

Thiamine-HCl·$2H_2O$	0.5mg

Preparation of Thiamine Hydrochloride Solution: Add thiamine-HCl·$2H_2O$ to distilled/deionized water and bring volume to 10.0mL. Mix thoroughly. Filter sterilize.

Hutner's Salts Solution:

Composition per liter:

$MgSO_4 \cdot 7H_2O$	29.7g
Nitrilotriacetic acid	10.0g

$CaCl_2 \cdot 2H_2O$ 3.335g
$FeSO_4 \cdot 7H_2O$ 99.0mg
$(NH_4)_6MoO_7O_{24} \cdot 4H_2O$ 9.25mg
"Metals 44" 50.0mL

"Metals 44":

Composition per 100.0mL:

$ZnSO_4 \cdot 7H_2O$ 1.095g
$FeSO_4 \cdot 7H_2O$ 0.5g
Sodium EDTA 0.25g
$MnSO_4 \cdot H2O$ 0.154g
$CuSO_4 \cdot 5H_2O$ 39.2mg
$Co(NO_3)_2 \cdot 6H_2O$ 24.8mg
$Na_2B_4O_7 \cdot 10H_2O$ 17.7mg

Preparation of "Metals 44": Add sodium EDTA to distilled/deionized water and bring volume to 90.0mL. Mix thoroughly. Add a few drops of concentrated H_2SO_4 to retard precipitation of heavy metal ions. Add remaining components. Mix thoroughly. Bring volume to 100.0mL with distilled/deionized water.

Preparation of Hutner's Salts Solution: Add nitrilotriacetic acid to 500.0mL of distilled/deionized water. Adjust pH to 6.5 with KOH. Add remaining components. Add distilled/deionized water to 1.0L. Adjust pH to 6.8.

Preparation of Medium: Add components, except tetramethyl ammonium chloride solution and thiamine hydrochloride solution, to distilled/deionized water and bring volume to 980.0L. Mix thoroughly. Autoclave for 15 min at 15 psi pressure–121°C. Cool to room temperature. Aseptically add 10.0mL sterile tetramethyl ammonium chloride solution and 10.0mL sterile thiamine hydrochloride solution. Mix thoroughly. Adjust pH to 6.9.

Use: For the cultivation of *Paracoccus kocurii.*

Paraffin Agar

Composition per liter:

Agar 15.0g
K_2HPO_4 .... 6.0g
NH_4NO_3 .... 4.0g
KH_2PO_4 .... 2.0g
Paraffin, liquid 1.0g
$ZnSO_4 \cdot 7H_2O$ 0.049g
$MnCl_2 \cdot 4H_2O$ 0.046g
$FeSO_4 \cdot 7H_2O$ 5.4mg
$CuSO_4 \cdot 5H_2O$ 2.5mg
$Na_2B_4O_7 \cdot 10H_2O$ 0.94mg
$(NH_4)_6Mo_7O_{24} \cdot 4H_2O$ 0.2mg

Preparation of Medium: Add components to distilled/deionized water and bring volume to 1.0L. Mix thoroughly. Gently heat and bring to boiling. Distribute into tubes or flasks. Autoclave for 15 min at 15 psi pressure–121°C. Pour into sterile Petri dishes or leave in tubes.

Use: For the selective isolation and cultivation of streptomycetes.

Paramecium Medium

Composition per liter:

Solution C 500.0mL
Solution A 10.0mL
Solution B 1.0mL

Solution A:

Composition per liter:

Thiamine·HCl 1.5g
Calcium pantothenate 1.0g
Nicotinamide 0.5g
Pyridoxal·HCl 0.5g
Riboflavin 0.5g
Folic acid 0.5g
α-Lipoic acid 0.1g
Biotin 0.1mg

Preparation of Solution A: Add components to distilled/deionized water and bring volume to 1.0L. Mix thoroughly. Distribute while stirring into screw-capped tubes in 10.0mL volumes. Store at −20°C. Thaw as needed.

Solution B:

Composition per 100.0mL:

TEM-4T (Hachmeister, Pittsburgh) 10.0g
Stigmasterol 0.5g
Ethanol, absolute 100.0mL

Preparation of Solution B: Add TEM-4T and stigmasterol to 100.0mL of hot ethanol. Mix thoroughly. Store at 4°C.

Solution C:

Composition per 500.0mL:

Proteose peptone 10.0g
Pancreatic digest of casein 5.0g
Ribonucleic acid 1.0g
$MgSO_4 \cdot 7H_2O$ 0.5g

Preparation of Solution C: Add components to distilled/deionized water and bring volume to 500.0mL. Mix thoroughly.

Preparation of Medium: Combine 500.0mL of solution C, 10.0mL of solution A, and 1.0mL of solution B. Mix thoroughly. Bring volume to 1.0L with distilled/deionized water. Adjust pH to 7.0–7.2 with 0.1*N* NaOH. Autoclave for 20 min at 15 psi pressure–121°C. Cool to 45°–50°C. Aseptically distribute into sterile tubes or flasks.

Use: For the cultivation of *Paramecium* species to be used as host cells by bacterial symbionts.

Payne, Seghal, and Gibbons Medium

Composition per liter:

NaCl 250.0g
$MgSO_4 \cdot 7H_2O$ 20.0g
Yeast extract 10.0g
Casamino acids 7.5g
Trisodium citrate 3.0g
KCl 2.0g
$FeCl_2 \cdot 4H_2O$ 36.0mg
$MnCl_2 \cdot 4H_2O$ 0.36mg

pH 7.4 ± 0.2 at 25°C

Preparation of Medium: Add components to distilled/deionized water and bring volume to 1.0L. Mix thoroughly. Adjust ph to 7.4. Distribute into tubes or flasks. Autoclave for 15 min at 15 psi pressure–121°C.

Use: For the cultivation of *Haloarcula californiae, Haloarcula hispanica, Haloarcula japonica, Haloarcula marismortui, Haloarcula sinaiiensis, Haloarcula vallismortis, Halobacterium cutirubrum, Halobacterium distributum, Halobacterium lacusprofundi, Halobacterium saccharovorum, Haloferax gibbonsii, Haloferax mediterranei, Halobacterium salinarium, Halobacterium simoncinii, Halobacterium* species, *Halobacterium trapanicum, Halobacterium volcanii, Halococcus morrhuae, Halococcus saccharolyticus, Halococcus* species, *Halococcus turkmenicus, Haloferax denitrificans, Halomonas elongata,* and *Salinicoccus roseus.*

PB90-1 Medium
(DSMZ Medium 298g)

Composition per liter:

NaCl	1.0g
KCl	0.5g
$MgCl_2 \cdot 6H_2O$	0.4g
NH_4Cl	0.25g
KH_2PO_4	0.2g
$CaCl_2 \cdot 2H_2O$	0.15g
Resazurin	1.0mg
$NaHCO_3$ solution	10.0mL
Butanediol solution	10.0mL
$Na_2S \cdot 9H_2O$ solution	10.0mL
Vitamin solution	10.0mL
Seven vitamin solution	10.0mL
Glucose solution	10.0mL
Trace elements solution SL-10	1.0mL

pH 7.2 ± 0.2 at 25°C

$Na_2S \cdot 9H_2O$ Solution:

Composition per 10mL:

$Na_2S \cdot 9H_2O$	0.36g

Preparation of $Na_2S \cdot 9H_2O$ Solution: Add $Na_2S \cdot 9H_2O$ to distilled/deionized water and bring volume to 10.0mL. Mix thoroughly. Autoclave under 100% N_2 for 15 min at 15 psi pressure–121°C. Cool to room temperature.

$NaHCO_3$ Solution:

Composition per 10.0mL:

$NaHCO_3$	2.5g

Preparation of $NaHCO_3$ Solution: Add $NaHCO_3$ to distilled/deionized water and bring volume to 10.0mL. Mix thoroughly. Sparge with 80% N_2 + 20% CO_2. Filter sterilize.

Butanediol Solution:

Composition per 10.0mL:

2,3 butanediol	0.9g

Preparation of Butanediol Solution: Add butanediol to distilled/deionized water and bring volume to 10.0mL. Mix thoroughly. Sparge with 100% N_2. Filter sterilize.

Trace Elements Solution SL-10:

Composition per liter:

$FeCl_2 \cdot 4H_2O$	1.5g
$CoCl_2 \cdot 6H_2O$	190.0mg
$MnCl_2 \cdot 4H_2O$	100.0mg
$ZnCl_2$	70.0mg
$Na_2MoO_4 \cdot 2H_2O$	36.0mg
$NiCl_2 \cdot 6H_2O$	24.0mg
H_3BO_3	6.0mg
$CuCl_2 \cdot 2H_2O$	2.0mg
HCl (25% solution)	10.0mL

Preparation of Trace Elements Solution SL-10: Add $FeCl_2 \cdot 4H_2O$ to 10.0mL of HCl solution. Mix thoroughly. Add distilled/deionized water and bring volume to 1.0L. Add remaining components. Mix thoroughly. Sparge with 80% N_2 + 20% CO_2. Autoclave for 15 min at 15 psi pressure–121°C.

Vitamin Solution:

Composition per liter:

Pyridoxine-HCl	10.0mg
Thiamine-HCl·$2H_2O$	5.0mg
Riboflavin	5.0mg
Nicotinic acid	5.0mg
D-Ca-pantothenate	5.0mg
p-Aminobenzoic acid	5.0mg
Lipoic acid	5.0mg
Biotin	2.0mg
Folic acid	2.0mg
Vitamin B_{12}	0.1mg

Preparation of Vitamin Solution: Add components to distilled/deionized water and bring volume to 1.0L. Mix thoroughly. Sparge with 80% H_2 + 20% CO_2. Filter sterilize.

Seven Vitamin Solution:

Composition per liter:

Pyridoxine hydrochloride	300.0mg
Thiamine-HCl·$2H_2O$	200.0mg
Nicotinic acid	200.0mg
Vitamin B_{12}	100.0mg
Calcium pantothenate	100.0mg
p-Aminobenzoic acid	80.0mg
D(+)-Biotin	20.0mg

Preparation of Seven Vitamin Solution: Add components to distilled/deionized water and bring volume to 1.0L. Sparge with 100% N_2. Mix thoroughly. Filter sterilize.

Glucose Solution:

Composition per 10.0mL:

Glucose	0.7g

Preparation of Glucose Solution: Add glucose to distilled/deionized water and bring volume to 10.0mL. Mix thoroughly. Sparge with 100% N_2. Filter sterilize.

Preparation of Medium: Prepare and dispense medium under 80% N_2 + 20% CO_2 gas atmosphere. Add components, except $NaHCO_3$ solution, butanediol solution, $Na_2S \cdot 9H_2O$ solution, vitamin solution, seven vitamin solution, glucose solution, and trace elements solution SL-10, to distilled/deionized water and bring volume to 939.0mL. Mix thoroughly. Adjust pH to 7.2. Sparge with 80% N_2 + 20% CO_2. Autoclave for 15 min at 15 psi pressure–121°C. Aseptically and anaerobically add 10.0mL $NaHCO_3$ solution, 10.0mL butanediol solution, 10.0mL $Na_2S \cdot 9H_2O$ solution, and 1.0mL 10.0mL vitamin solution, 10.0mL seven vitamin solution, 10.0mL glucose solution, trace elements solution SL-10. Mix thoroughly. Aseptically and anaerobically distribute into sterile tubes or bottles. After inoculation, flush and repressurize the gas head space of culture bottles with sterile 80% N_2 + 20% CO_2 to 1 bar overpressure.

Use: For the cultivation of *Opitutus terrae.*

Pectobacterium carotovorum Medium
(LMG Medium 172)

Composition per liter:

Glucose	5.0g
Peptone	5.0g
Yeast extract	3.0g

pH 7.2 ± 0.2 at 25°C

Preparation of Medium: Add components to distilled/deionized water and bring volume to 1.0L. Mix thoroughly. Distribute into tubes or flasks. Autoclave for 15 min at 15 psi pressure–121°C.

Use: For the cultivation of *Pectobacterium carotovorum* subsp. *Odoriferum.*

Pelobacter acetylenicus Medium

Composition per liter:

NaCl	20.0g
$MgCl_2 \cdot 6H_2O$	3.0g

KCl ... 0.5g
NH_4Cl ... 0.25g
KH_2PO_4 ... 0.2g
$CaCl_2 \cdot 2H_2O$ ... 0.15g
Resazurin ... 1.0mg
$NaHCO_3$ solution ... 10.0mL
$Na_2S \cdot 9H_2O$ solution ... 10.0mL
Acetoin solution ... 10.0mL
Trace elements solution SL-10 ... 1.0mL
pH 7.2 ± 0.2 at 25°C

$NaHCO_3$ Solution:
Composition per 10.0mL:
$NaHCO_3$ ... 2.5g

Preparation of $NaHCO_3$ Solution: Add $NaHCO_3$ to distilled/deionized water and bring volume to 10.0mL. Mix thoroughly. Sparge with 80% N_2 + 20% CO_2. Autoclave for 15 min at 15 psi pressure–121°C.

$Na_2S \cdot 9H_2O$ Solution:
Composition per 10.0mL:
$Na_2S \cdot 9H_2O$ ... 0.36g

Preparation of $Na_2S \cdot 9H_2O$ Solution: Add $Na_2S \cdot 9H_2O$ to distilled/deionized water and bring volume to 10.0mL. Mix thoroughly. Sparge with 100% N_2. Autoclave for 15 min at 15 psi pressure–121°C.

Acetoin Solution:
Composition per 10.0mL:
Acetoin ... 1.0g

Preparation of Acetoin Solution: Add acetoin to distilled/deionized water and bring volume to 10.0mL. Mix thoroughly. Sparge with 100% N_2. Autoclave for 15 min at 15 psi pressure–121°C.

Trace Elements Solution SL-10:
Composition per liter:
$FeCl_2 \cdot 4H_2O$ ... 1.5g
$CoCl_2 \cdot 6H_2O$ ... 0.19g
$MnCl_2 \cdot 4H_2O$ ... 100.0mg
$ZnCl_2$ ... 70.0mg
$Na_2MoO_4 \cdot 2H_2O$ ... 36.0mg
$NiCl_2 \cdot 6H_2O$ ... 24.0mg
H_3BO_3 ... 6.0mg
$CuCl_2 \cdot 2H_2O$ ... 2.0mg
HCl (25% solution) ... 10.0mL

Preparation of Trace Elements Solution SL-10: Add $FeCl_2 \cdot 4H_2O$ to 10.0mL of HCl solution. Mix thoroughly. Add distilled/deionized water and bring volume to 1.0L. Add remaining components. Mix thoroughly.

Preparation of Medium: Prepare and dispense medium under 80% N_2 + 20% CO_2. Add components, except $NaHCO_3$ solution, $Na_2S \cdot 9H_2O$ solution, and acetoin solution, to distilled/deionized water and bring volume to 970.0mL. Mix thoroughly. Sparge with 80% N_2 + 20% CO_2. Autoclave for 15 min at 15 psi pressure–121°C. Aseptically and anaerobically add 10.0mL of sterile $NaHCO_3$ solution, 10.0mL of sterile $Na_2S \cdot 9H_2O$ solution, and 10.0mL of sterile acetoin solution or, using a syringe, inject the appropriate amount of sterile $NaHCO_3$ solution, sterile $Na_2S \cdot 9H_2O$ solution, and sterile acetoin solution into individual tubes containing medium.

Use: For the cultivation and maintenance of *Pelobacter acetylenicus*.

Pelobacter acidigallici Medium

Composition per 1001.0mL:
NaCl ... 20.0g
$MgCl_2 \cdot 6H_2O$ ... 3.0g
KCl ... 0.5g
NH_4Cl ... 0.25g
KH_2PO_4 ... 0.2g
$CaCl_2 \cdot 2H_2O$ ... 0.15g
Resazurin ... 1.0mg
$NaHCO_3$ solution ... 10.0mL
$Na_2S \cdot 9H_2O$ solution ... 10.0mL
Gallic acid solution ... 10.0mL
Trace elements solution SL-10 ... 1.0mL
pH 7.2 ± 0.2 at 25°C

$NaHCO_3$ Solution:
Composition per 10.0mL:
$NaHCO_3$ ... 2.5g

Preparation of $NaHCO_3$ Solution: Add $NaHCO_3$ to distilled/deionized water and bring volume to 10.0mL. Mix thoroughly. Filter sterilize. Sparge with 80% N_2 + 20% CO_2.

$Na_2S \cdot 9H_2O$ Solution:
Composition per 10.0mL:
$Na_2S \cdot 9H_2O$ ... 0.36g

Preparation of $Na_2S \cdot 9H_2O$ Solution: Add $Na_2S \cdot 9H_2O$ to distilled/deionized water and bring volume to 10.0mL. Mix thoroughly. Sparge with 100% N_2. Autoclave for 15 min at 15 psi pressure–121°C.

Gallic Acid Solution:
Composition per 10.0mL:
Gallic acid ... 1.0g

Preparation of Gallic Acid Solution: Add gallic acid to distilled/deionized water and bring volume to 10.0mL. Mix thoroughly. Filter sterilize. Sparge with 80% N_2 + 20% CO_2.

Trace Elements Solution SL-10:
Composition per liter:
$FeCl_2 \cdot 4H_2O$ ... 1.5g
$CoCl_2 \cdot 6H_2O$ ... 0.19g
$MnCl_2 \cdot 4H_2O$ ... 100.0mg
$ZnCl_2$ ... 70.0mg
$Na_2MoO_4 \cdot 2H_2O$ ... 36.0mg
$NiCl_2 \cdot 6H_2O$ ... 24.0mg
H_3BO_3 ... 6.0mg
$CuCl_2 \cdot 2H_2O$ ... 2.0mg
HCl (25% solution) ... 10.0mL

Preparation of Trace Elements Solution SL-10: Add $FeCl_2 \cdot 4H_2O$ to 10.0mL of HCl solution. Mix thoroughly. Add distilled/deionized water and bring volume to 1.0L. Add remaining components. Mix thoroughly.

Preparation of Medium: Prepare and dispense medium under 80% H_2 + 20% CO_2. Add components, except $NaHCO_3$ solution, $Na_2S \cdot 9H_2O$ solution, gallic acid solution, and trace elements solution SL-10, to distilled/deionized water and bring volume to 970.0mL. Mix thoroughly. Sparge with 80% N_2 + 20% CO_2. Autoclave for 15 min at 15 psi pressure–121°C. Aseptically and anaerobically add 10.0mL of sterile $NaHCO_3$ solution, 10.0mL of sterile $Na_2S \cdot 9H_2O$ solution, 10.0mL of sterile gallic acid solution, and 1.0mL of sterile trace elements solution SL-10 or, using a syringe, inject the appropriate amount of sterile $NaHCO_3$ solution, sterile $Na_2S \cdot 9H_2O$ solution, sterile gallic acid solution, and sterile trace elements solution SL-10 into individual tubes containing medium.

Use: For the cultivation and maintenance of *Pelobacter acidigallici*.

Pelobacter carbinolicus Medium

Composition per 1001.0mL:

NaCl	20.0g
$MgCl_2·6H_2O$	3.0g
KCl	0.5g
NH_4Cl	0.25g
KH_2PO_4	0.2g
$CaCl_2·2H_2O$	0.15g
Resazurin	1.0mg
$NaHCO_3$ solution	10.0mL
$Na_2S·9H_2O$ solution	10.0mL
2,3-Butanediol solution	10.0mL
Trace elements solution SL-10	1.0mL

pH 7.2 ± 0.2 at 25°C

$NaHCO_3$ Solution:

Composition per 10.0mL:

$NaHCO_3$	2.5g

Preparation of $NaHCO_3$ Solution: Add $NaHCO_3$ to distilled/deionized water and bring volume to 10.0mL. Mix thoroughly. Filter sterilize. Sparge with 80% N_2 + 20% CO_2.

$Na_2S·9H_2O$ Solution:

Composition per 10.0mL:

$Na_2S·9H_2O$	0.36g

Preparation of $Na_2S·9H_2O$ Solution: Add $Na_2S·9H_2O$ to distilled/deionized water and bring volume to 10.0mL. Mix thoroughly. Sparge with 100% N_2. Autoclave for 15 min at 15 psi pressure–121°C.

2,3-Butanediol Solution:

Composition per 10.0mL:

2,3-Butanediol	0.68g

Preparation of 2,3-Butanediol Solution: Add 2,3-butanediol to distilled/deionized water and bring volume to 10.0mL. Mix thoroughly. Filter sterilize. Sparge with 80% N_2 + 20% CO_2.

Trace Elements Solution SL-10:

Composition per liter:

$FeCl_2·4H_2O$	1.5g
$CoCl_2·6H_2O$	0.19g
$MnCl_2·4H_2O$	100.0mg
$ZnCl_2$	70.0mg
$Na_2MoO_4·2H_2O$	36.0mg
$NiCl_2·6H_2O$	24.0mg
H_3BO_3	6.0mg
$CuCl_2·2H_2O$	2.0mg
HCl (25% solution)	10.0mL

Preparation of Trace Elements Solution SL-10: Add $FeCl_2·4H_2O$ to 10.0mL of HCl solution. Mix thoroughly. Add distilled/deionized water and bring volume to 1.0L. Add remaining components. Mix thoroughly.

Preparation of Medium: Prepare and dispense medium under 80% H_2 + 20% CO_2. Add components, except $NaHCO_3$ solution, $Na_2S·9H_2O$ solution, 2,3-butanediol solution, and trace elements solution SL-10, to distilled/deionized water and bring volume to 970.0mL. Mix thoroughly. Sparge with 80% N_2 + 20% CO_2. Autoclave for 15 min at 15 psi pressure–121°C. Aseptically and anaerobically add 10.0mL of sterile $NaHCO_3$ solution, 10.0mL of sterile $Na_2S·9H_2O$ solution, 10.0mL of sterile 2,3-butanediol solution, and 1.0mL of sterile trace elements solution SL-10 or, using a syringe, inject the appropriate amount of sterile $NaHCO_3$ solution, sterile $Na_2S·9H_2O$ solution, sterile 2,3-butanediol solution, and sterile trace elements solution SL-10 into individual tubes containing medium.

Use: For the cultivation and maintenance of *Pelobacter carbinolicus*.

Pelobacter Medium

Composition per 1025.0mL:

$KHCO_3$	4.5g
NH_4Cl	1.0g
NaCl	0.6g
Trypticase	0.5g
Yeast extract	0.5g
KH_2PO_4	0.3g
$MgCl_2·6H_2O$	0.1g
$CaCl_2·2H_2O$	0.08g
Resazurin	1.0mg
Trace elements solution	10.0mL
Vitamin solution	10.0mL
Sodium gallate solution	10.0mL
L-Cysteine·HCl·H_2O solution	10.0mL
$Na_2S·9H_2O$ solution	10.0mL

pH 7.3 ± 0.2 at 25°C

Trace Elements Solution:

Composition per liter:

Nitrilotriacetic acid	12.8g
$FeCl_3·6H_2O$	1.35g
NaCl	1.0g
$NiCl_2·6H_2O$	0.12g
$CaCl_2·2H_2O$	0.10g
$MnCl_2·4H_2O$	0.10g
$ZnCl_2$	0.10g
$Na_2SeO_3·5H_2O$	0.026g
$CuCl_2·2H_2O$	0.025g
$CoCl_2·6H_2O$	0.024g
$Na_2MoO_4·2H_2O$	0.024g
H_3BO_3	0.01g

Preparation of Trace Elements Solution: Add nitrilotriacetic acid to 500.0mL of distilled/deionized water. Adjust pH to 6.5 with KOH. Add remaining components. Adjust pH to 7.0. Add distilled/deionized water to 1.0L.

Vitamin Solution:

Composition per liter:

Biotin	2.0mg
Folic acid	2.0mg
Pyridoxine·HCl	10.0mg
Thiamine·HCl	5.0mg
Riboflavin	5.0mg
Nicotinic acid	5.0mg
Calcium DL-pantothenate	5.0mg
Vitamin B_{12}	0.1mg
p-Aminobenzoic acid	5.0mg
Lipoic acid	5.0mg

Preparation of Vitamin Solution: Add components to distilled/deionized water and bring volume to 1.0L. Mix thoroughly. Filter sterilize. Sparge with 80% N_2 + 20% CO_2.

Sodium Gallate Solution:

Composition per 10.0mL:

Gallic acid	1.88g
NaOH (1*M* solution)	variable

Preparation of Sodium Gallate Solution: Add gallic acid to distilled/deionized water and bring volume to 8.0mL. Mix thoroughly. Add sufficient NaOH solution to bring pH to 7.3. Filter sterilize. Sparge with 80% N_2 + 20% CO_2.

L-Cysteine·HCl·H_2O Solution:
Composition per 10.0mL:
L-Cysteine·HCl·H_2O 0.3g

Preparation of L-Cysteine·HCl·H_2O Solution: Add L-cysteine·HCl·H_2O to distilled/deionized water and bring volume to 10.0mL. Mix thoroughly. Sparge with 100% N_2. Autoclave for 15 min at 15 psi pressure–121°C.

$Na_2S·9H_2O$ Solution:
Composition per 10.0mL:
$Na_2S·9H_2O$ 0.3g

Preparation of $Na_2S·9H_2O$ Solution: Add $Na_2S·9H_2O$ to distilled/deionized water and bring volume to 10.0mL. Mix thoroughly. Sparge with 100% N_2. Autoclave for 15 min at 15 psi pressure–121°C.

Preparation of Medium: Prepare and dispense medium under 80% N_2 + 20% CO_2. Add components, except $NaHCO_3$ solution, $Na_2S·9H_2O$ solution, sodium gallate solution, and vitamin solution, to distilled/deionized water and bring volume to 960.0mL. Mix thoroughly. Sparge with 80% N_2 + 20% CO_2. Autoclave for 15 min at 15 psi pressure–121°C. Aseptically and anaerobically add 10.0mL of sterile $NaHCO_3$ solution, 10.0mL of sterile $Na_2S·9H_2O$ solution, 10.0mL of sterile sodium gallate solution, and 10.0mL of sterile vitamin solution or, using a syringe, inject the appropriate amount of sterile $NaHCO_3$ solution, sterile $Na_2S·9H_2O$ solution, sterile sodium gallate solution, and sterile vitamin solution into individual tubes containing medium.

Use: For the cultivation and maintenance of *Pelobacter acidigallici*.

Pelobacter Medium with Gallic Acid

Composition per liter:
Solution A 950.0mL
Solution B 25.0mL
Solution C 25.0mL
pH 7.2 ± 0.2 at 25°C

Solution A:
Composition per liter:
NaCl 20.0g
$MgCl_2·6H_2O$ 3.65g
$NaHCO_3$ 2.5g
KCl 0.5g
NH_4Cl 0.25g
KH_2PO_4 .... 0.2g
$CaCl_2·2H_2O$ 0.15g
Resazurin 0.5mg
Modified Wolfe's mineral solution 10.0mL
Wolfe's vitamin solution 10.0mL

Preparation of Solution A: Prepare and dispense solution under 80% N_2 + 20% CO_2. Add components, except $NaHCO_3$, to distilled/deionized water and bring volume to 950.0mL. Mix thoroughly. Gently heat and bring to boiling. Continue boiling for 3 min. Cool to room temperature while sparging with 80% N_2 + 20% CO_2. Add $NaHCO_3$. Mix thoroughly. Anaerobically distribute 9.5mL volumes into anaerobic tubes. Autoclave for 15 min at 15 psi pressure–121°C.

Modified Wolfe's Mineral Solution:
Composition per liter:
$MgSO_4·7H_2O$ 3.0g
Nitrilotriacetic acid 1.5g
NaCl 1.0g
$MnSO_4·H_2O$ 0.5g
$CaCl_2$ 0.1g
$CoCl_2·6H_2O$ 0.1g
$FeSO_4·7H_2O$ 0.1g
$ZnSO_4·7H_2O$ 0.1g
$AlK(SO_4)_2·12H_2O$ 0.01g
$CuSO_4·5H_2O$ 0.01g
H_3BO_3 .... 0.01g
$Na_2MoO_4·2H_2O$ 0.01g
Na_2SeO_3 .... 0.01g
$NaWO_4·2H_2O$ 0.01g
$NiCl_2·6H_2O$ 0.01g

Preparation of Modified Wolfe's Mineral Solution: Add nitrilotriacetic acid to 500.0mL of distilled/deionized water. Adjust pH to 6.5 with KOH. Add remaining components one at a time. Add distilled/deionized water to 1.0L. Adjust pH to 6.8.

Wolfe's Vitamin Solution:
Composition per liter:
Pyridoxine·HCl 10.0mg
p-Aminobenzoic acid 5.0mg
Lipoic acid 5.0mg
Nicotinic acid 5.0mg
Riboflavin 5.0mg
Thiamine·HCl 5.0mg
Calcium DL-pantothenate 5.0mg
Biotin 2.0mg
Folic acid 2.0mg
Vitamin B_{12} 0.1mg

Preparation of Wolfe's Vitamin Solution: Add components to distilled/deionized water and bring volume to 1.0L. Mix thoroughly.

Solution B:
Composition per 25.0mL:
Gallic acid 0.85g

Preparation of Solution B: Prepare solution B immediately prior to use. Add gallic acid to distilled/deionized water and bring volume to 25.0mL. Mix thoroughly. Rapidly adjust pH to 6.5. Filter sterilize. Sparge with 100% N_2.

Solution C:
Composition per 25.0mL:
$Na_2S·9H_2O$ 0.4g

Preparation of Solution C: Add $Na_2S·9H_2O$ to distilled/deionized water and bring volume to 25.0mL. Mix thoroughly. Sparge with 100% N_2. Autoclave for 15 min at 15 psi pressure–121°C.

Preparation of Medium: Aseptically and anaerobically add 0.25mL of sterile solution B and 0.25mL of sterile solution C to each tube containing 9.5mL of sterile solution A. Mix thoroughly. Adjust pH to 7.2.

Use: For the cultivation of *Pelobacter acidigallici* and *Pelobacter massiliensis*.

Pelobacter propionicus Medium (DSMZ Medium 298)

Composition per liter:
NaCl 1.0g
KCl 0.5g
$MgCl_2·6H_2O$ 0.4g
NH_4Cl 0.25g
KH_2PO_4 .... 0.2g
$CaCl_2·2H_2O$ 0.15g
Resazurin 1.0mg
$NaHCO_3$ solution 10.0mL

Butanediol solution 10.0mL
$Na_2S \cdot 9H_2O$ solution 10.0mL
Trace elements solution SL-10 ....1.0mL

pH 7.2 ± 0.2 at 25°C

$Na_2S \cdot 9H_2O$ Solution:

Composition per 10mL:

$Na_2S \cdot 9H_2O$ 0.36g

Preparation of $Na_2S \cdot 9H_2O$ Solution: Add $Na_2S \cdot 9H_2O$ to distilled/deionized water and bring volume to 10.0mL. Mix thoroughly. Autoclave under 100% N_2 for 15 min at 15 psi pressure–121°C. Cool to room temperature.

$NaHCO_3$ Solution:

Composition per 10.0mL:

$NaHCO_3$ 2.5g

Preparation of $NaHCO_3$ Solution: Add $NaHCO_3$ to distilled/deionized water and bring volume to 10.0mL. Mix thoroughly. Sparge with 80% N_2 + 20% CO_2. Filter sterilize.

Butanediol Solution:

Composition per 10.0mL:

2,3 butanediol 0.9g

Preparation of Butanediol Solution: Add butanediol to distilled/deionized water and bring volume to 10.0mL. Mix thoroughly. Sparge with 100% N_2. Filter sterilize.

Trace Elements Solution SL-10:

Composition per liter:

$FeCl_2 \cdot 4H_2O$ 1.5g
$CoCl_2 \cdot 6H_2O$ 190.0mg
$MnCl_2 \cdot 4H_2O$ 100.0mg
$ZnCl_2$ 70.0mg
$Na_2MoO_4 \cdot 2H_2O$ 36.0mg
$NiCl_2 \cdot 6H_2O$ 24.0mg
H_3BO_3 ....6.0mg
$CuCl_2 \cdot 2H_2O$ 2.0mg
HCl (25% solution) 10.0mL

Preparation of Trace Elements Solution SL-10: Add $FeCl_2 \cdot 4H_2O$ to 10.0mL of HCl solution. Mix thoroughly. Add distilled/deionized water and bring volume to 1.0L. Add remaining components. Mix thoroughly. Sparge with 80% N_2 + 20% CO_2. Autoclave for 15 min at 15 psi pressure–121°C.

Preparation of Medium: Prepare and dispense medium under 80% N_2 + 20% CO_2 gas atmosphere. Add components, except $NaHCO_3$ solution, butanediol solution, $Na_2S \cdot 9H_2O$ solution, and trace elements solution SL-10, to distilled/deionized water and bring volume to 969.0mL. Mix thoroughly. Adjust pH to 7.2. Sparge with 80% N_2 + 20% CO_2. Autoclave for 15 min at 15 psi pressure–121°C. Aseptically and anaerobically add 10.0mL $NaHCO_3$ solution, 10.0mL butanediol solution, 10.0mL $Na_2S \cdot 9H_2O$ solution, and 1.0mL trace elements solution SL-10. Mix thoroughly. Aseptically and anaerobically distribute into sterile tubes or bottles. After inoculation, flush and repressurize the gas head space of culture bottles with sterile 80% N_2 + 20% CO_2 to 1 bar overpressure.

Use: For the cultivation of *Pelobacter propionicus.*

Pelobacter venetianus Marine Medium (DSMZ Medium 296)

Composition per liter:

NaCl 20.0g
$MgCl_2 \cdot 6H_2O$ 3.0g
KCl 0.5g
NH_4Cl 0.25g
KH_2PO_4 ....0.2g
$CaCl_2 \cdot 2H_2O$ 0.15g
Resazurin 1.0mg
$NaHCO_3$ solution 10.0mL
Polyethylene glycol solution 10.0mL
$Na_2S \cdot 9H_2O$ solution 10.0mL
Trace elements solution SL-10 ....1.0mL

pH 7.2 ± 0.2 at 25°C

$Na_2S \cdot 9H_2O$ Solution:

Composition per 10mL:

$Na_2S \cdot 9H_2O$ 0.36g

Preparation of $Na_2S \cdot 9H_2O$ Solution: Add $Na_2S \cdot 9H_2O$ to distilled/deionized water and bring volume to 10.0mL. Mix thoroughly. Autoclave under 100% N_2 for 15 min at 15 psi pressure–121°C. Cool to room temperature.

$NaHCO_3$ Solution:

Composition per 10.0mL:

$NaHCO_3$ 2.5g

Preparation of $NaHCO_3$ Solution: Add $NaHCO_3$ to distilled/deionized water and bring volume to 10.0mL. Mix thoroughly. Sparge with 80% N_2 + 20% CO_2. Filter sterilize.

Polyethylene Glycol Solution:

Composition per 10.0mL:

Polyethylene glycol
(molecular weight 106 - 20000) 1.0g

Preparation of Polyethylene Glycol Solution: Add polyethylene glycol to distilled/deionized water and bring volume to 10.0mL. Mix thoroughly. Sparge with 100% N_2. Filter sterilize.

Trace Elements Solution SL-10:

Composition per liter:

$FeCl_2 \cdot 4H_2O$ 1.5g
$CoCl_2 \cdot 6H_2O$ 190.0mg
$MnCl_2 \cdot 4H_2O$ 100.0mg
$ZnCl_2$ 70.0mg
$Na_2MoO_4 \cdot 2H_2O$ 36.0mg
$NiCl_2 \cdot 6H_2O$ 24.0mg
H_3BO_3 ....6.0mg
$CuCl_2 \cdot 2H_2O$ 2.0mg
HCl (25% solution) 10.0mL

Preparation of Trace Elements Solution SL-10: Add $FeCl_2 \cdot 4H_2O$ to 10.0mL of HCl solution. Mix thoroughly. Add distilled/deionized water and bring volume to 1.0L. Add remaining components. Mix thoroughly. Sparge with 80% N_2 + 20% CO_2. Filter sterilize.

Preparation of Medium: Prepare and dispense medium under 80% N_2 + 20% CO_2 gas atmosphere. Add components, except $NaHCO_3$ solution, polyethylene glycol solution, $Na_2S \cdot 9H_2O$ solution, and trace elements solution SL-10, to distilled/deionized water and bring volume to 969.0mL. Mix thoroughly. Adjust pH to 7.2. Sparge with 80% N_2 + 20% CO_2. Autoclave for 15 min at 15 psi pressure–121°C. Aseptically and anaerobically add 10.0mL $NaHCO_3$ solution, 10.0mL polyethylene glycol solution, 10.0mL $Na_2S \cdot 9H_2O$ solution, and 1.0mL trace elements solution SL-10. Mix thoroughly. Aseptically and anaerobically distribute into sterile tubes or bottles. After inoculation, flush and repressurize the gas head space of culture bottles with sterile 80% N_2 + 20% CO_2 to 1 bar overpressure.

Use: For the cultivation of *Pelobacter venetianus.*

Pelotomaculum Medium (DSMZ Medium 960)

Composition per liter:

$NaHCO_3$	2.5g
NH_4Cl	0.54g
$MgCl_2 \cdot 6H_2O$	0.2g
$CaCl_2 \cdot 2H_2O$	0.15g
KH_2PO_4	0.14g
Yeast extract	0.1g
Resazurin	0.5mg
Na-pyruvate solution	20.0mL
Cysteine solution	10.0mL
$Na_2S \cdot 9H_2O$ solution	10.0mL
Vitamin solution	5.0mL
Trace elements solution	1.0mL
Selenite-tungstate solution	1.0mL

pH 7.0 ± 0.2 at 25°C

Na-Pyruvate Solution:

Composition per 20mL:

Na-pyruvate	2.2g

Preparation of Na-Pyruvate Solution: Add Na-pyruvate to distilled/deionized water and bring volume to 20.0mL. Mix thoroughly. Autoclave under 100% N_2 for 15 min at 15 psi pressure–121°C. Cool to room temperature.

$Na_2S \cdot 9H_2O$ Solution:

Composition per 10mL:

$Na_2S \cdot 9H_2O$	0.3g

Preparation of $Na_2S \cdot 9H_2O$ Solution: Add $Na_2S \cdot 9H_2O$ to distilled/deionized water and bring volume to 10.0mL. Mix thoroughly. Sparge with 100% N_2. Autoclave for 15 min at 15 psi pressure–121°C. Cool to room temperature.

Cysteine Solution:

Composition per 10.0mL:

L-Cysteine·HCl·H_2O	0.3g

Preparation of Cysteine Solution: Add L-cysteine·HCl·H_2O to distilled/deionized water and bring volume to 10.0mL. Mix thoroughly. Sparge with 100% N_2. Autoclave for 15 min at 15 psi pressure–121°C. Cool to room temperature.

Vitamin Solution:

Composition per liter:

Pyridoxine-HCl	10.0mg
Thiamine-HCl·$2H_2O$	5.0mg
Riboflavin	5.0mg
Nicotinic acid	5.0mg
D-Ca-pantothenate	5.0mg
p-Aminobenzoic acid	5.0mg
Lipoic acid	5.0mg
Biotin	2.0mg
Folic acid	2.0mg
Vitamin B_{12}	0.1mg

Preparation of Vitamin Solution: Add components to distilled/deionized water and bring volume to 1.0L. Mix thoroughly. Sparge with 80% H_2 + 20% CO_2. Filter sterilize.

Trace Elements Solution:

Composition per liter:

$MgSO_4 \cdot 7H_2O$	3.0g
Nitrilotriacetic acid	1.5g
NaCl	1.0g
$MnSO_4 \cdot 2H_2O$	0.5g
$CoSO_4 \cdot 7H_2O$	0.18g
$ZnSO_4 \cdot 7H_2O$	0.18g
$CaCl_2 \cdot 2H_2O$	0.1g
$FeSO_4 \cdot 7H_2O$	0.1g
$NiCl_2 \cdot 6H_2O$	0.025g
$KAl(SO_4)_2 \cdot 12H_2O$	0.02g
H_3BO_3	0.01g
$Na_2MoO_4 \cdot 4H_2O$	0.01g
$CuSO_4 \cdot 5H_2O$	0.01g
$Na_2SeO_3 \cdot 5H_2O$	0.3mg

Preparation of Trace Elements Solution: Add nitrilotriacetic acid to 500.0mL of distilled/deionized water. Dissolve by adjusting pH to 6.5 with KOH. Add remaining components. Add distilled/deionized water to 1.0L. Mix thoroughly.

Selenite-Tungstate Solution

Composition per liter:

NaOH	0.5g
$Na_2WO_4 \cdot 2H_2O$	4.0mg
$Na_2SeO_3 \cdot 5H_2O$	3.0mg

Preparation of Selenite-Tungstate Solution: Add components to distilled/deionized water and bring volume to 1.0L. Mix thoroughly. Sparge with 100% N_2. Filter sterilize.

Preparation of Medium: Add components, except Na-pyruvate solution, cysteine solution, and $Na_2S \cdot 9H_2O$ solution, to distilled/deionized water and bring volume to 960.0mL. Mix thoroughly. Sparge with 80% N_2 + 20% CO_2. Equilibrate with this gas mixture to reach pH 7.0. Distribute into anaerobe tubes or bottles. Autoclave for 15 min at 15 psi pressure–121°C. Aseptically and anaerobically add per liter of medium 20.0mL sterile Na-pyruvate solution, 10.0mL sterile cysteine solution, and 10.0mL of sterile $Na_2S \cdot 9H_2O$ solution. Mix thoroughly.

Use: For the cultivation of *Pelotomaculum thermopropionicum.*

Pentachlorophenol Medium

Composition per liter:

NH_4NO_3	2.5g
$Na_2HPO_4 \cdot 2H_2O$	1.0g
$MgSO_4 \cdot 7H_2O$	0.5g
$Fe(SO_4)_3 \cdot 5H_2O$	0.01g
$Co(NO_3)_2 \cdot 6H_2O$	0.005g
$CaCl_2 \cdot 2H_2O$	1.0mg
Pentachlorophenol	1.0mg
KH_2PO_4	0.5mg
$MnSO_4 \cdot 2H_2O$	0.1mg
$(NH_4)_6Mo_7O_{24} \cdot 4H_2O$	0.1mg

Preparation of Medium: Add components to distilled/deionized water and bring volume to 1.0L. Mix thoroughly. Distribute into tubes or flasks. Autoclave for 15 min at 15 psi pressure–121°C.

Use: For the cultivation of *Flavobacterium* species.

Pentachlorophenol Medium

Composition per 1007.0mL:

Sodium glutamate	4.0g
K_2HPO_4	0.65g
$NaNO_3$	0.5g
KH_2PO_4	0.19g
$MgSO_4 \cdot 7H_2O$	0.1g
Pentachlorophenol solution	5.0mL
$FeSO_4$ solution	2.0mL

pH 7.3 ± 0.1 at 25°C

Pentachlorophenol Solution:
Composition per 100.0mL:
Pentachlorophenol 1.0g
NaOH (0.5*N* solution) 100.0mL

Preparation of Pentachlorophenol Solution: Add pentachlorophenol to 100.0mL of NaOH solution. Mix thoroughly. Filter sterilize.

$FeSO_4$ Solution:
Composition per 100.0mL:
$FeSO_4$ solution 2.5g

Preparation of $FeSO_4$ Solution: Add $FeSO_4$ to distilled/deionized water and bring volume to 100.0mL. Mix thoroughly. Filter sterilize.

Preparation of Medium: Add components, except pentachlorophenol solution and $FeSO_4$ solution, to distilled/deionized water and bring volume to 1.0L. Mix thoroughly. Adjust pH to 7.3–7.4. Autoclave for 15 min at 15 psi pressure–121°C. Aseptically add 2.0mL of sterile $FeSO_4$ solution. Mix thoroughly. Aseptically distribute into sterile flasks. Inoculate flasks and place on a shaker at 200 rpm at 25°–30°C. Monitor growth with a spectrophotometer at 560 nm. When absorbance at 560 nm (A_{560}) is 0.5, add 5.0mL of sterile pentachlorophenol solution per liter of medium.

Use: For the cultivation of *Pseudomonas mendocina*.

Peptone Succinate Agar

Composition per liter:
Agar 15.0g
Peptone... 5.0g
Succinic acid 1.68g
$MgSO_4 \cdot 7H_2O$ 1.0g
$(NH_4)_2SO_4$ 1.0g
$FeCl_3 \cdot 6H_2O$ 2.0mg
$MnSO_4 \cdot H_2O$ 2.0mg
pH 7.0 ± 0.2 at 25°C

Preparation of Medium: Add components to distilled/deionized water and bring volume to 1.0L. Mix thoroughly. Gently heat and bring to boiling. Distribute into tubes or flasks. Autoclave for 15 min at 15 psi pressure–121°C. Pour into sterile Petri dishes or leave in tubes.

Use: For the cultivation and maintenance of *Aquaspirillum bengal, Aquaspirillum dispar,* and *Spirillum volutans*.

Peptone Succinate Agar

Composition per liter:
Peptone... 5.0g
Succinic acid 1.68g
Agar 1.5g
$MgSO_4 \cdot 7H_2O$ 1.0g
NH_4SO_4 .... 1.0g
$FeCl_3 \cdot 6H_2O$ 2.0mg
$MnSO_4 \cdot H_2O$ 2.0mg
pH 7.0 ± 0.2 at 25°C

Preparation of Medium: Add components to distilled/deionized water and bring volume to 1.0L. Mix thoroughly. Adjust pH to 7.0. Distribute into tubes or flasks. Autoclave for 15 min at 15 psi pressure–121°C.

Use: For the cultivation of *Aquaspirillum serpens*.

Peptone Succinate Agar in Seawater

Composition per liter:
Peptone... 5.0g
Succinic acid 1.68g
Agar 1.5g
$MgSO_4 \cdot 7H_2O$ 1.0g
$(NH_4)_2SO_4$ 1.0g
$FeCl_3 \cdot 6H_2O$ 2.0mg
$MnSO_4 \cdot H_2O$ 2.0mg
Seawater 1.0L
pH 7.0 ± 0.2 at 25°C

Preparation of Medium: Combine components. Mix thoroughly. Gently heat and bring to boiling. Distribute into tubes or flasks. Autoclave for 15 min at 15 psi pressure–121°C.

Use: For the cultivation and maintenance of *Oceanospirillum maris*.

Peptone Succinate Salts Broth (PSS Broth)

Composition per 100.0mL:
Peptone... 1.0g
$MgSO_4 \cdot 7H_2O$ 0.1g
$(NH_4)_2SO_4$ 0.1g
Succinic acid 0.1g
$FeCl_3 \cdot 6H_2O$ 0.2mg
$MnSO_4 \cdot H_2O$ 0.2mg
pH 6.8 ± 0.2 at 25°C

Preparation of Medium: Add components to distilled/deionized water and bring volume to 1.0L. Mix thoroughly. Adjust pH to 6.8 with KOH. Distribute into tubes or flasks. Autoclave for 15 min at 15 psi pressure–121°C.

Use: For the cultivation of *Spirillum* species.

Peptone Succinate Salts Medium (PSS Medium)

Composition per liter:
Peptone... 10.0g
$MgSO_4 \cdot 7H_2O$ 1.0g
$(NH_4)_2SO_4$ 1.0g
Succinic acid 1.0g
$FeCl_3 \cdot 6H_2O$ 2.0mg
$MnSO_4 \cdot H_2O$ 2.0mg
pH 6.8 ± 0.2 at 25°C

Preparation of Medium: Add solid components to synthetic seawater and bring volume to 1.0L. Mix thoroughly. Adjust pH to 6.8 with 2*N* KOH. Distribute into tubes or flasks. Autoclave for 15 min at 15 psi pressure–121°C.

Use: For the cultivation and maintenance of *Aquaspirillum anulus*.

Peptone Succinate Salts in Seawater

Composition per liter:
Peptone... 10.0g
$MgSO_4 \cdot 7H_2O$ 1.0g
$(NH_4)_2SO_4$ 1.0g
Succinic acid 1.0g
$FeCl_3 \cdot 6H_2O$ 2.0mg
$MnSO_4 \cdot H_2O$ 2.0mg
Synthetic seawater 1.0L
pH 6.8 ± 0.2 at 25°C

Synthetic Seawater:
Composition per liter:
NaCl 27.5g
$MgCl_2$ 5.0g
$MgSO_4$ 2.0g
KCl 1.0g
$CaCl_2$ 0.5g
$FeSO_4$ 1.0μg

Preparation of Synthetic Seawater: Add components to distilled/deionized water and bring volume to 1.0L. Mix thoroughly.

Preparation of Medium: Add solid components to synthetic seawater and bring volume to 1.0L. Mix thoroughly. Adjust pH to 6.8 with 2*N* KOH. Distribute into tubes or flasks. Autoclave for 15 min at 15 psi pressure–121°C.

Use: For the cultivation and maintenance of *Oceanospirillum maris*.

Peptone Sucrose Broth

Composition per liter:

Sucrose	20.0g
Peptone	10.0g

Preparation of Medium: Add components to distilled/deionized water and bring volume to 1.0L. Mix thoroughly. Distribute into tubes or flasks. Autoclave for 15 min at 15 psi pressure–121°C.

Use: For the cultivation and maintenance of *Xanthomonas campestris*.

Peptone Yeast Extract Agar (ATCC Medium 526)

Composition per liter:

Agar	15.0g
Peptone	10.0g
Yeast extract	3.0g

Preparation of Medium: Add components to distilled/deionized water and bring volume to 1.0L. Mix thoroughly. Gently heat and bring to boiling. Distribute into tubes or flasks. Autoclave for 15 min at 15 psi pressure–121°C. Pour into sterile Petri dishes or leave in tubes.

Use: For the cultivation and maintenance of *Bdellovibrio bacteriovorus* and *Bdellovibrio stolpii*.

Peptone Yeast Extract Medium (ATCC Medium 1366)

Composition per liter:

Peptone	10.0g
NaCl	5.0g
Yeast extract	5.0g

pH 7.2 ± 0.2 at 25°C

Preparation of Medium: Add components to distilled/deionized water and bring volume to 1.0L. Mix thoroughly. Gently heat and bring to boiling. Adjust pH to 7.2. Distribute into tubes or flasks. Autoclave for 15 min at 15 psi pressure–121°C.

Use: For the cultivation and maintenance of *Xenorhabdus nematophilus*.

Peptone Yeast Extract Medium (PY Medium) (ATCC Medium 1524)

Composition per 950.0mL:

Yeast extract	10.0g
Peptone	5.0g
Pancreatic digest of casein	5.0g
L-Cysteine·HCl·H_2O	0.5g
Salt solution	40.0mL
Hemin solution	10.0mL
Resazurin (0.025% solution)	4.0mL
Vitamin K_1 solution	0.2mL

pH 7.0 ± 0.2 at 25°C

Salt Solution:

Composition per liter:

$NaHCO_3$	10.0g
NaCl	2.0g
K_2HPO_4	1.0g
KH_2PO_4	1.0g
$CaCl_2$, anhydrous	0.2g
$MgSO_4$	0.2g

Preparation of Salt Solution: Add $CaCl_2$ and $MgSO_4$ to 300.0mL of distilled/deionized water. Mix thoroughly until dissolved. Bring volume to 800.0mL with distilled/deionized water. Add remaining components while stirring. Bring volume to 1.0L. Mix thoroughly. Store at 4°C.

Hemin Solution:

Composition per 100.0mL:

Hemin	0.05g
NaOH (1*N* solution)	1.0mL

Preparation of Hemin Solution: Add hemin to NaOH solution and bring volume to 100.0mL with distilled/deionized water. Mix thoroughly. Autoclave for 15 min at 15 psi pressure–121°C. Cool to 45°–50°C.

Vitamin K_1 Solution:

Composition per 30.0mL:

Vitamin K_1	0.15g
Ethanol (95% solution)	30.0mL

Preparation of Vitamin K_1 Solution: Add Vitamin K_1 to ethanol. Mix thoroughly. Filter sterilize.

Preparation of Medium: Add components, except Vitamin K_1 solution, hemin solution, and L-cysteine·HCl·H_2O, to distilled/deionized water and bring volume to 939.8mL. Gently heat and bring to boiling under 80% N_2 + 10% H_2 + 10% CO_2. Continue boiling until resazurin turns colorless, indicating reduction. Cool to 45°–50°C. Add Vitamin K_1 solution, hemin solution, and L-cysteine·HCl·H_2O. Adjust pH to 7.0. Distribute into tubes under 80% N_2 + 10% H_2 + 10% CO_2. Cap with rubber stoppers. Place tubes in a press. Autoclave for 15 min at 15 psi pressure–121°C with fast exhaust.

Use: For the cultivation and maintenance of *Megasphaera cerevisiae* and *Clostridium* species.

Peptone Yeast Medium with Magnesium Sulfate (DSMZ Medium 790)

Composition per liter:

Peptone	10.0g
Yeast extract, dehydrated	1.0g
$MgSO_4 \cdot 7H_2O$	2.0g
$(NH_4)_2SO_4$	2.0g

pH 7.0 ± 0.2 at 25°C

Preparation of Medium: Add components to distilled/deionized water and bring volume to 1.0L. Mix thoroughly. Distribute into tubes or flasks. Autoclave for 15 min at 15 psi pressure–121°C.

Use: For the cultivation of *Aquaspirillum psychrophilum* and *Aquaspirillum peregrinum* subsp. *integrum*.

Peptone Yeast Medium with $MgSO_4$

Composition per liter:

Peptone	10.0g
$MgSO_4 \cdot 7H_2O$	2.0g
$(NH_4)_2SO_4$	2.0g
Yeast extract	1.0g

pH 7.0 ± 0.2 at 25°C

Preparation of Medium: Add components to distilled/deionized water and bring volume to 1.0L. Mix thoroughly. Distribute into tubes or flasks. Autoclave for 15 min at 15 psi pressure–121°C.

Use: For the cultivation and maintenance of *Aquaspirillum itersonii, Aquaspirillum peregrinum,* and *Aquaspirillum psychrophilum.*

Peptonized Milk Agar (PMA Medium)

Composition per liter:

Agar 15.0g
Milk, peptonized 1.0g

Preparation of Medium: Add components to distilled/deionized water and bring volume to 1.0L. Mix thoroughly. Gently heat and bring to boiling. Distribute into tubes or flasks. Autoclave for 15 min at 15 psi pressure–121°C. Pour into sterile Petri dishes or leave in tubes.

Use: For the cultivation of freshwater *Myxobacterium* species.

Petrotoga Medium (ATCC 1881)

Composition per liter:

NaCl 20.0g
Sodium PIPES (piperazine-*N,N'*-bis[2-ethanesulfonic acid]) buffer 5.24g
Pancreatic digest of casein 5.0g
Yeast extract 2.0g
Resazurin 1.0g
Soluble starch 1.0g
L-Cysteine·HCl·H_2O 0.5g

pH 7.2 ± 0.2 at 25°C

Preparation of Medium: Add components, except L-cysteine·HCl·H_2O, to distilled/deionized water and bring volume to 1.0L. Mix thoroughly. Adjust pH to 7.4. Gently heat and bring to boiling. Continue boiling for 3 min. Cool to room temperature while sparging with 100% N_2. Add L-cysteine·HCl·H_2O. Mix thoroughly. Anaerobically distribute into tubes. Autoclave for 15 min at 15 psi pressure–121°C.

Use: For the cultivation of *Petrotoga miotherma.*

Pfennig's Medium I, Modified for Marine Purple Sulfur Bacteria (DSMZ Medium 28)

Composition per 5.0L:

Solution A 4.0L
Solution B 860.0mL
Solution E 100.0mL
Solution F 20.0mL
Solution C 5.0mL
Solution D 5.0mL

pH 7.3 at 25°C

Solution A:

Composition per 4.0L:

NaCl 100.0g
$MgSO_4$ 15.0g
KH_2PO_4 .... 1.7g
NH_4Cl 1.7g
KCl 1.7g
$CaCl_2 \cdot 2H_2O$ 1.25g

Preparation of Solution A: Add components to 4.0L distilled water. Mix thoroughly. Autoclave for 45 min at 15 psi pressure–121°C in 5-liter special bottle or flask with 4 openings at the top, together with a teflon-coated magnetic bar. In this 5-liter bottle, 2 openings for tubes are in the central, silicon rubber stopper–a short, gas-inlet tube with a sterile cotton filter; and an outlet tube for medium, which reaches the bottom of the vessel at one end and has, at the other end, a silicon rubber tube with a pinch cock and a bell for aseptic dispensing of the medium into bottles. The other two openings have gas-tight screw caps—one of these openings is for the addition of sterile solutions, and the other serves as a gas outlet. After autoclaving, cool solution A to room temperature under a N_2 atmosphere with a positive pressure of 0.05–0.1 atm (a manometer for low pressure will be required). Saturate the cold medium with CO_2 by magnetic stirring for 30 min under a CO_2 atmosphere of 0.05–0.1 atm.

Solution B:

Distilled water 860.0mL

Preparation of Solution B: Autoclave distilled water for 15 min at 15 psi pressure–121°C in a cotton-stoppered Erlenmeyer flask. Cool to room temperature under an atmosphere of N_2 in an anaerobic jar.

Solution C:

Composition per 100.0mL:

Vitamin B_{12} 2.0mg

Preparation of Solution C: Add Vitamin B_{12} to distilled/deionized water and bring volume to 100.0mL. Mix thoroughly. Sparge under 100% N_2 gas for 3 min. Filter sterilize. Store under N_2 gas.

Solution D:

Composition per liter:

Disodium ethylendiamine-tetraacetate (Disodium EDTA) 3.0g
$FeSO_4 \cdot 7H_2O$ 1.1g
H_3BO_3 .... 0.3g
$CoCl_2 \cdot 6H_2O$ 0.19g
$MnCl_2 \cdot 2H_2O$ 50.0mg
$ZnCl_2$ 42.0mg
$NiCl_2 \cdot 6H_2O$ 24.0mg
$Na_2MoO_4 \cdot 2H_2O$ 18.0mg
$CuCl_2 \cdot 2H_2O$ 2.0mg

Preparation of Solution D: Add components to distilled/deionized water and bring volume to 1.0L. Mix thoroughly. Autoclave for 15 min at 15 psi pressure–121°C.

Solution E:

Composition per 100.0mL:

$NaHCO_3$ 7.5g

Preparation of Solution E: Add $NaHCO_3$ to distilled/deionized water and bring volume to 100.0mL. Mix thoroughly. Sparge with 100% CO_2 until saturated. Filter sterilize under 100% CO_2 into a sterile, gas-tight 100.0mL screw-capped bottle.

Solution F:

Composition per 100.0mL:

$Na_2S \cdot 9H_2O$ 10.0g

Preparation of Solution F: Add $Na_2S \cdot 9H_2O$ to distilled/deionized water in a 250.0mL screw-capped bottle fitted with a butyl rubber septum and bring volume to 100.0mL. Mix thoroughly. Sparge under 100% N_2 gas for 3 min. Autoclave for 15 min at 15 psi pressure–121°C. Cool to room temperature.

Neutralized Sulfide Solution:

Composition per 100.0mL:

$Na_2S \cdot 9H_2O$ 1.5g

Preparation of Neutralized Sulfide Solution: Add $Na_2S \cdot 9H_2O$ to distilled/deionized water in a 250.0mL screw-capped bottle fitted with a butyl rubber septum, and bring volume to 100.0mL. Add a magnetic stir bar. Mix thoroughly. Sparge under 100% N_2 gas for 3 min. Autoclave for 15 min at 15 psi pressure–121°C. Cool to room temperature. Adjust pH to about 7.3 with sterile 2*M* H_2SO_4. Do not open the bottle to add H_2SO_4; use a sterile syringe. Stir the solution continuously to avoid precipitation of elemental sulfur. The final solution should be clear and yellow in color.

Preparation of Medium: Add solutions B, C, D, E, and F to solution A through one of the screw-cap openings against a stream of either N_2 gas or better, a mixture of 95% N_2 and 5% CO_2 while the medium is magnetically stirred. Adjust the pH of the medium with sterile HCl or Na_2CO_3 solution (2 *M* solutions) to pH 7.3. Distribute the medium aseptically through the medium outlet tube into sterile, 100mL bottles (with metal caps and autoclavable rubber seals) using the positive gas pressure (0.05 - 0.1 atm) of the N_2/CO_2 gas mixture. Leave a small air bubble in each bottle to meet possible pressure changes. The tightly sealed, screw-capped bottles can be stored for several weeks or months in the dark. During the first 24 hr, the iron of the medium precipitates in the form of black flocs. No other sediment should arise in the otherwise clear medium. Incubate in the light using a tungsten lamp. Feed periodically with neutralized solution of sodium sulfide to replenish sulfide with other supplement solutions.

Use: For the cultivation of marine purple sulfur bacteria.

Pfennig's Medium I, Modified for Purple Sulfur Bacteria (DSMZ Medium 28)

Composition per 5.0L:

Solution A	4.0L
Solution B	860.0mL
Solution E	100.0mL
Solution F	20.0mL
Solution C	5.0mL
Solution D	5.0mL

pH 7.3 at 25°C

Solution A:

Composition per 4.0L:

$MgSO_4$	2.5g
KH_2PO_4	1.7g
NH_4Cl	1.7g
KCl	1.7g
$CaCl_2 \cdot 2H_2O$	1.25g

Preparation of Solution A: Add components to 4.0L distilled water. Mix thoroughly. Autoclave for 45 min at 15 psi pressure–121°C in 5-liter special bottle or flask with 4 openings at the top, together with a teflon-coated magnetic bar. In this 5-liter bottle, 2 openings for tubes are in the central, silicon rubber stopper—a short, gas-inlet tube with a sterile cotton filter; and an outlet tube for medium, which reaches the bottom of the vessel at one end and has, at the other end, a silicon rubber tube with a pinch cock and a bell for aseptic dispensing of the medium into bottles. The other two openings have gas-tight screw caps—one of these openings is for the addition of sterile solutions, and the other serves as a gas outlet. After autoclaving, cool solution A to room temperature under a N_2 atmosphere with a positive pressure of 0.05–0.1 atm (a manometer for low pressure will be required). Saturate the cold medium with CO_2 by magnetic stirring for 30 min under a CO_2 atmosphere of 0.05–0.1 atm.

Solution B:

Distilled water	860.0mL

Preparation of Solution B: Autoclave distilled water for 15 min at 15 psi pressure–121°C in a cotton-stoppered Erlenmeyer flask. Cool to room temperature under an atmosphere of N_2 in an anaerobic jar.

Solution C:

Composition per 100.0mL:

Vitamin B_{12}	2.0mg

Preparation of Solution C: Add Vitamin B_{12} to distilled/deionized water and bring volume to 100.0mL. Mix thoroughly. Sparge under 100% N_2 gas for 3 min. Filter sterilize. Store under N_2 gas.

Solution D:

Composition per liter:

Disodium ethylendiamine-tetraacetate (Disodium EDTA)	3.0g
$FeSO_4 \cdot 7H_2O$	1.1g
H_3BO_3	0.3g
$CoCl_2 \cdot 6H_2O$	0.19g
$MnCl_2 \cdot 2H_2O$	50.0mg
$ZnCl_2$	42.0mg
$NiCl_2 \cdot 6H_2O$	24.0mg
$Na_2MoO_4 \cdot 2H_2O$	18.0mg
$CuCl_2 \cdot 2H_2O$	2.0mg

Preparation of Solution D: Add components to distilled/deionized water and bring volume to 1.0L. Mix thoroughly. Autoclave for 15 min at 15 psi pressure–121°C.

Solution E:

Composition per 100.0mL:

$NaHCO_3$	7.5g

Preparation of Solution E: Add $NaHCO_3$ to distilled/deionized water and bring volume to 100.0mL. Mix thoroughly. Sparge with 100% CO_2 until saturated. Filter sterilize under 100% CO_2 into a sterile, gas-tight 100.0mL screw-capped bottle.

Solution F:

Composition per 100.0mL:

$Na_2S \cdot 9H_2O$	10.0g

Preparation of Solution F: Add $Na_2S \cdot 9H_2O$ to distilled/deionized water in a 250.0mL screw-capped bottle fitted with a butyl rubber septum and bring volume to 100.0mL. Mix thoroughly. Sparge under 100% N_2 gas for 3 min. Autoclave for 15 min at 15 psi pressure–121°C. Cool to room temperature.

Neutralized Sulfide Solution:

Composition per 100.0mL:

$Na_2S \cdot 9H_2O$	1.5g

Preparation of Neutralized Sulfide Solution: Add $Na_2S \cdot 9H_2O$ to distilled/deionized water in a 250.0mL screw-capped bottle fitted with a butyl rubber septum and bring volume to 100.0mL. Add a magnetic stir bar. Mix thoroughly. Sparge under 100% N_2 gas for 3 min. Autoclave for 15 min at 15 psi pressure–121°C. Cool to room temperature. Adjust pH to about 7.3 with sterile 2*M* H_2SO_4. Do not open the bottle to add H_2SO_4; use a sterile syringe. Stir the solution continuously to avoid precipitation of elemental sulfur. The final solution should be clear and yellow in color.

Preparation of Medium: Add solutions B, C, D, E, and F to solution A through one of the screw-cap openings against a stream of either N_2 gas or better, a mixture of 95% N_2, and 5% CO_2 while the medium is magnetically stirred. Adjust the pH of the medium with sterile HCl or Na_2CO_3 solution (2 *M* solutions) to pH 7.3. Distribute the medium aseptically through the medium outlet tube into sterile, 100mL bottles (with metal caps and autoclavable rubber seals) using the positive gas pressure (0.05–0.1 atm) of the N_2/CO_2 gas mixture. Leave a small air bubble in each bottle to meet possible pressure changes. The tightly sealed, screw-capped bottles can be stored for several weeks or months in the dark. During the first 24 hr, the iron of the medium precipitates in the form of black flocs. No other sediment should arise in the otherwise clear medium. Incubate in the light using a tungsten lamp. Feed periodically with neutralized solution of sodium sulfide to replenish sulfide with other supplement solutions.

Use: For the cultivation of purple sulfur bacteria.

Pfennig's Medium 1 with 1% Salt

Composition per 4990.0mL:

Solution A	4000.0mL
Solution B	860.0mL
Solution E	100.0mL
Solution F	20.0mL
Solution C (Vitamin B_{12} solution)	5.0mL
Solution D (Trace elements solution SL-12B	5.0mL

pH 7.3 ± 0.2 at 25°C

Solution A:

Composition per 4000.0mL:

NaCl	50.0g
$CaCl_2 \cdot 2H_2O$	1.25g
KH_2PO_4	1.7g
NH_4Cl	1.7g
KCl	1.7g
$MgSO_4$	2.5g

Preparation of Solution A: Add components to distilled/deionized water and bring volume to 4.0L. Mix thoroughly. Adjust pH to 6.0. Dispense into a 5-liter flask with 4 openings at the top (two openings are in a central silicon rubber stopper and 2 openings are gas-tight screw caps). Add a teflon-coated magnetic stir bar to the flask. Autoclave for 45 min at 15 psi pressure–121°C. Cool to room temperature under 100% N_2 at 0.05–0.1 atm pressure (use a manometer to measure low pressure).

Solution B:

Composition per 860.0mL:

Distilled/deionized water	860.0mL

Preparation of Solution B: Add 860.0mL of distilled/deionized water to a cotton-stoppered flask. Autoclave for 20 min at 15 psi pressure–121°C. Cool to room temperature under 100% N_2 in an anaerobic jar.

Solution C (Vitamin B_{12} Solution):

Composition per 5.0mL:

Vitamin B_{12}	1.0mg

Preparation of Solution C (Vitamin B_{12} Solution): Add Vitamin B_{12} to distilled/deionized water and bring volume to 5.0mL. Mix thoroughly. Filter sterilize.

Solution D (Trace Elements Solution SL-12B):

Composition per liter:

Disodium ethylendiamine-tetraacetate (Na_2-EDTA)	3.0g
$FeSO_4 \cdot 7H_2O$	1.1g
H_3BO_3	0.3g
$CoCl_2 \cdot 6H_2O$	0.19g
$MnCl_2 \cdot 2H_2O$	50.0mg
$ZnCl_2$	42.0mg
$NiCl_2 \cdot 6H_2O$	24.0mg
$Na_2MoO_4 \cdot 2H_2O$	18.0mg
$CuCl_2 \cdot 2H_2O$	2.0mg

Preparation of Solution D (Trace Elements Solution SL-12B): Add components to distilled/deionized water and bring volume to 1.0L. Mix thoroughly. Autoclave for 15 min at 15 psi pressure–121°C.

Solution E:

Composition per 100.0mL:

$NaHCO_3$	7.5g

Preparation of Solution E: Add $NaHCO_3$ to distilled/deionized water and bring volume to 100.0mL. Mix thoroughly. Sparge with 100% CO_2 until saturated. Filter sterilize under 100% CO_2 into a sterile, gas-tight 100.0mL screw-capped bottle.

Solution F:

Composition per 100.0mL:

$Na_2S \cdot 9H_2O$	10.0g

Preparation of Solution F: Add $Na_2S \cdot 9H_2O$ to distilled/deionized water and bring volume to 100.0mL. Mix thoroughly. Dispense into a screw-capped bottle. Sparge with 100% N_2 for 3–4 min. Autoclave for 15 min at 15 psi pressure–121°C.

Preparation of Medium: Saturate cooled solution A under 100% CO_2 at 0.05–0.1 atm pressure for 30 min with magnetic stirring. Add 860.0mL of solution B, 5.0mL of solution C, 5.0mL of solution D, 100.0mL of solution E, and 20.0mL of solution F through one of the screw-capped openings under 95% N_2 and 5% CO_2 with magnetic stirring. Adjust pH to 7.3 with sterile 2*M* HCl or sterile 2*M* Na_2CO_3 solution. Aseptically and anaerobically distribute the medium through the medium outlet tube into sterile 100.0mL bottles under 95% N_2 and 5% CO_2 at 0.05–0.1 atm pressure. Leave a small gas bubble in each bottle to accommodate pressure changes. After 24 hr, the iron in the medium will precipitate out of solution as black flocs.

Use: For the cultivation and maintenance of *Ectothiorhodospira mobilis*.

Pfennig's Medium 1 with 3% Salt

Composition per 4990.0mL:

Solution A	4000.0mL
Solution B	860.0mL
Solution E	100.0mL
Solution F	20.0mL
Solution C (Vitamin B_{12} solution)	5.0mL
Solution D (Trace elements solution SL-12B	5.0mL

pH 7.3 ± 0.2 at 25°C

Solution A:

Composition per 4000.0mL:

NaCl	150.0g
$CaCl_2 \cdot 2H_2O$	1.25g
KH_2PO_4	1.7g
NH_4Cl	1.7g
KCl	1.7g
$MgSO_4$	2.5g

Preparation of Solution A: Add components to distilled/deionized water and bring volume to 4.0L. Mix thoroughly. Adjust pH to 6.0. Dispense into a 5-liter flask with 4 openings at the top (2 openings are in a central silicone rubber stopper and 2 openings are gas-tight screw caps). Add a teflon-coated magnetic stir bar to the flask. Autoclave for 45 min at 15 psi pressure–121°C. Cool to room temperature under 100% N_2 at 0.05–0.1 atm pressure (use a manometer to measure low pressure).

Solution B:

Composition per 860.0mL:

Distilled/deionized water 860.0mL

Preparation of Solution B: Add 860.0mL of distilled/deionized water to a cotton-stoppered flask. Autoclave for 20 min at 15 psi pressure–121°C. Cool to room temperature under 100% N_2 in an anaerobic jar.

Solution C (Vitamin B_{12} Solution):

Composition per 5.0mL:

Vitamin B_{12} 1.0mg

Preparation of Solution C (Vitamin B_{12} Solution): Add Vitamin B_{12} to distilled/deionized water and bring volume to 5.0mL. Mix thoroughly. Filter sterilize.

Solution D (Trace Elements Solution SL-12B):

Composition per liter:

Disodium ethylendiamine-tetraacetate
(Na_2-EDTA) 3.0g
$FeSO_4 \cdot 7H_2O$ 1.1g
H_3BO_3 .... 0.3g
$CoCl_2 \cdot 6H_2O$ 0.19g
$MnCl_2 \cdot 2H_2O$ 50.0mg
$ZnCl_2$ 42.0mg
$NiCl_2 \cdot 6H_2O$ 24.0mg
$Na_2MoO_4 \cdot 2H_2O$ 18.0mg
$CuCl_2 \cdot 2H_2O$ 2.0mg

Preparation of Solution D (Trace Elements Solution SL-12B): Add components to distilled/deionized water and bring volume to 1.0L. Mix thoroughly. Autoclave for 15 min at 15 psi pressure–121°C.

Solution E:

Composition per 100.0mL:

$NaHCO_3$ 7.5g

Preparation of Solution E: Add $NaHCO_3$ to distilled/deionized water and bring volume to 100.0mL. Mix thoroughly. Sparge with 100% CO_2 until saturated. Filter sterilize under 100% CO_2 into a sterile, gas-tight 100.0mL screw-capped bottle.

Solution F:

Composition per 100.0mL:

$Na_2S \cdot 9H_2O$ 10.0g

Preparation of Solution F: Add $Na_2S \cdot 9H_2O$ to distilled/deionized water and bring volume to 100.0mL. Mix thoroughly. Dispense into a screw-capped bottle. Sparge with 100% N_2 for 3–4 min. Autoclave for 15 min at 15 psi pressure–121°C.

Preparation of Medium: Saturate cooled solution A under 100% CO_2 at 0.05–0.1 atm pressure for 30 min with magnetic stirring. Add 860.0mL of solution B, 5.0mL of solution C, 5.0mL of solution D, 100.0mL of solution E, and 20.0mL of solution F through one of the screw-capped openings under 95% N_2 and 5% CO_2 with magnetic stirring. Adjust pH to 7.3 with sterile 2*M* HCl or sterile 2*M* Na_2CO_3 solution. Aseptically and anaerobically distribute the medium through the medium outlet tube into sterile 100.0mL bottles under 95% N_2 and 5% CO_2 at 0.05–0.1 atm pressure. Leave a small gas bubble in each bottle to accommodate pressure changes. After 24 hr, the iron in the medium will precipitate out of solution as black flocs.

Use: For the cultivation and maintenance of *Chromatium gracile, Thiocystis gelatinosa,* and *Thiocystis violacea.*

Pfennig's Medium 1 with Yeast Extract

Composition per 4990.0mL:

Solution A 4000.0mL
Solution B 860.0mL
Solution E 100.0mL
Solution F 20.0mL
Solution C (Vitamin B_{12} solution) 5.0mL
Solution D (Trace elements solution SL-12B 5.0mL

pH 7.3 ± 0.2 at 25°C

Solution A:

Composition per 4000.0mL:

$CaCl_2 \cdot 2H_2O$ 1.25g
KH_2PO_4 .... 1.7g
NH_4Cl 1.7g
KCl 1.7g
$MgSO_4$ 2.5g
Yeast extract 2.5g

Preparation of Solution A: Add components to distilled/deionized water and bring volume to 4.0L. Mix thoroughly. Adjust pH to 6.0. Dispense into a 5-liter flask with 4 openings at the top (2 openings are in a central silicon rubber stopper and 2 openings are gas-tight screw caps). Add a teflon-coated magnetic stir bar to the flask. Autoclave for 45 min at 15 psi pressure–121°C. Cool to room temperature under 100% N_2 at 0.05–0.1 atm pressure (use a manometer to measure low pressure).

Solution B:

Composition per 860.0mL:

Distilled/deionized water 860.0mL

Preparation of Solution B: Add 860.0mL of distilled/deionized water to a cotton-stoppered flask. Autoclave for 20 min at 15 psi pressure–121°C. Cool to room temperature under 100% N_2 in an anaerobic jar.

Solution C (Vitamin B_{12} Solution):

Composition per 5.0mL:

Vitamin B_{12} 1.0mg

Preparation of Solution C (Vitamin B_{12} Solution): Add Vitamin B_{12} to distilled/deionized water and bring volume to 5.0mL. Mix thoroughly. Filter sterilize.

Solution D (Trace Elements Solution SL-12B):

Composition per liter:

Disodium ethylendiamine-tetraacetate
(Na_2-EDTA) 3.0g
$FeSO_4 \cdot 7H_2O$ 1.1g
H_3BO_3 .... 0.3g
$CoCl_2 \cdot 6H_2O$ 0.19g
$MnCl_2 \cdot 2H_2O$ 50.0mg
$ZnCl_2$ 42.0mg
$NiCl_2 \cdot 6H_2O$ 24.0mg
$Na_2MoO_4 \cdot 2H_2O$ 18.0mg
$CuCl_2 \cdot 2H_2O$ 2.0mg

Preparation of Solution D (Trace Elements Solution SL-12B): Add components to distilled/deionized water and bring volume to 1.0L. Mix thoroughly. Autoclave for 15 min at 15 psi pressure–121°C.

Solution E:

Composition per 100.0mL:

$NaHCO_3$ 7.5g

Preparation of Solution E: Add $NaHCO_3$ to distilled/deionized water and bring volume to 100.0mL. Mix thoroughly. Sparge with 100% CO_2 until saturated. Filter sterilize under 100% CO_2 into a sterile, gas-tight 100.0mL screw-capped bottle.

Solution F:

Composition per 100.0mL:

$Na_2S \cdot 9H_2O$ 10.0g

Preparation of Solution F: Add $Na_2S \cdot 9H_2O$ to distilled/deionized water and bring volume to 100.0mL. Mix thoroughly. Dispense into a screw-capped bottle. Sparge with 100% N_2 for 3–4 min. Autoclave for 15 min at 15 psi pressure–121°C.

Preparation of Medium: Saturate cooled solution A under 100% CO_2 at 0.05–0.1 atm pressure for 30 min with magnetic stirring. Add 860.0mL of solution B, 5.0mL of solution C, 5.0mL of solution D, 100.0mL of solution E, and 20.0mL of solution F through one of the screw-capped openings under 95% N_2 and 5% CO_2 with magnetic stirring. Adjust pH to 7.3 with sterile 2*M* HCl or sterile 2*M* Na_2CO_3 solution. Aseptically and anaerobically distribute the medium through the medium outlet tube into sterile 100.0mL bottles under 95% N_2 and 5% CO_2 at 0.05–0.1 atm pressure. Leave a small gas bubble in each bottle to accommodate pressure changes. After 24 hr, the iron in the medium will precipitate out of solution as black flocs.

Use: For the cultivation and maintenance of *Amoebobacter pendens*.

Pfennig's Medium II Modified for Green Sulfur Bacteria (DSMZ Medium 29)

Composition per 5.0L:

Solution A 4.0L
Solution B 860.0mL
Solution E 100.0mL
Solution F 30.0mL
Solution C 5.0mL
Solution D 5.0mL

pH 6.8 at 25°C

Solution A:

Composition per 4.0L:

$MgSO_4$ 2.5g
KH_2PO_4 .... 1.7g
NH_4Cl 1.7g
KCl 1.7g
$CaCl_2 \cdot 2H_2O$ 1.25g

Preparation of Solution A: Add components to 4.0L distilled water. Mix thoroughly. Autoclave for 45 min at 15 psi pressure–121°C in 5-liter special bottle or flask with 4 openings at the top, together with a teflon-coated magnetic bar. In this 5-liter bottle, 2 openings for tubes are in the central, silicone rubber stopper, a short, gas-inlet tube with a sterile cotton filter—and an outlet tube for medium, which reaches the bottom of the vessel at one end and has, at the other end, a silicon rubber tube with a pinch cock and a bell for aseptic dispensing of the medium into bottles. The other two openings have gas-tight screw caps—one of these openings is for the addition of sterile solutions and the other serves as a gas outlet. After autoclaving, cool solution A to room temperature under a N_2 atmosphere with a positive pressure of 0.05–0.1 atm (a manometer for low pressure will be required). Saturate the cold medium with CO_2 by magnetic stirring for 30 min under a CO_2 atmosphere of 0.05–0.1 atm.

Solution B:

Distilled water 860.0mL

Preparation of Solution B: Autoclave distilled water for 15 min at 15 psi pressure–121°C in a cotton-stoppered Erlenmeyer flask. Cool to room temperature under an atmosphere of N_2 in an anaerobic jar.

Solution C:

Composition per 100.0mL:

Vitamin B_{12} 2.0mg

Preparation of Solution C: Add Vitamin B_{12} to distilled/deionized water and bring volume to 100.0mL. Mix thoroughly. Sparge under 100% N_2 gas for 3 min. Filter sterilize. Store under N_2 gas.

Solution D:

Composition per liter:

$FeCl_2 \cdot 4H_2O$ 1.5g
$CoCl_2 \cdot 6H_2O$ 190.0mg
$MnCl_2 \cdot 4H_2O$ 100.0mg
$ZnCl_2$ 70.0mg
$Na_2MoO_4 \cdot 2H_2O$ 36.0mg
$NiCl_2 \cdot 6H_2O$ 24.0mg
H_3BO_3 .... 300.0mg
$CuCl_2 \cdot 2H_2O$ 2.0mg
HCl (25% solution) 7.7mL

Preparation of Solution D: Add $FeCl_2 \cdot 4H_2O$ to 10.0mL of HCl solution. Mix thoroughly. Add distilled/deionized water and bring volume to 1.0L. Add remaining components. Mix thoroughly. Sparge with 100% N_2. Autoclave for 15 min at 15 psi pressure–121°C.

Solution E:

Composition per 100.0mL:

$NaHCO_3$ 7.5g

Preparation of Solution E: Add $NaHCO_3$ to distilled/deionized water and bring volume to 100.0mL. Mix thoroughly. Sparge with 100% CO_2 until saturated. Filter sterilize under 100% CO_2 into a sterile, gas-tight 100.0mL screw-capped bottle.

Solution F:

Composition per 100.0mL:

$Na_2S \cdot 9H_2O$ 10.0g

Preparation of Solution F: Add $Na_2S \cdot 9H_2O$ to distilled/deionized water in a 250.0mL screw-capped bottle fitted with a butyl rubber septum and bring volume to 100.0mL. Mix thoroughly. Sparge under 100% N_2 gas for 3 min. Autoclave for 15 min at 15 psi pressure–121°C. Cool to room temperature.

Neutralized Sulfide Solution:

Composition per 100.0mL:

$Na_2S \cdot 9H_2O$ 1.5g

Preparation of Neutralized Sulfide Solution: Add $Na_2S \cdot 9H_2O$ to distilled/deionized water in a 250.0mL screw-capped bottle fitted with a butyl rubber septum and bring volume to 100.0mL. Add a magnetic stir bar. Mix thoroughly. Sparge under 100% N_2 gas for 3 min. Autoclave for 15 min at 15 psi pressure–121°C. Cool to room temperature. Adjust pH to about 6.8 with sterile 2*M* H_2SO_4. Do not open the bottle to add H_2SO_4; use a sterile syringe. Stir the solution continuously to avoid precipitation of elemental sulfur. The final solution should be clear and yellow in color.

Preparation of Medium: Add solutions B, C, D, E, and F to solution A through one of the screw-cap openings against a stream of either N_2 gas or better, a mixture of 95% N_2 and 5% CO_2 while the medium is magnetically stirred. Adjust the pH of the medium with sterile HCl or Na_2CO_3

solution (2 *M* solutions) to pH 6.8. Distribute the medium aseptically through the medium outlet tube into sterile, 100mL bottles (with metal caps and autoclavable rubber seals) using the positive gas pressure (0.05–0.1 atm) of the N_2/CO_2 gas mixture. Leave a small air bubble in each bottle to meet possible pressure changes. The tightly sealed, screw-capped bottles can be stored for several weeks or months in the dark. During the first 24 hr, the iron of the medium precipitates in the form of black flocs. No other sediment should arise in the otherwise clear medium. Incubate in the light using a tungsten lamp. Feed periodically with neutralized solution of sodium sulfideto replenish sulfide with other supplement solutions.

Use: For the cultivation of green sulfur bacteria.

Pfennig's Medium II with Salt

Composition per 5000.0mL:

Solution A	4000.0mL
Solution B	860.0mL
Solution E	100.0mL
Solution F	30.0mL
Solution C (Vitamin B_{12} solution)	5.0mL
Solution D (Trace elements solution SL-10B)	5.0mL

pH 6.8 ± 0.2 at 25°C

Solution A:

Composition per 4000.0mL:

NaCl	50.0g
$CaCl_2 \cdot 2H_2O$	1.25g
KH_2PO_4	1.7g
NH_4Cl	1.7g
KCl	1.7g
$MgSO_4$	2.5g

Preparation of Solution A: Add components to distilled/deionized water and bring volume to 4.0L. Mix thoroughly. Adjust pH to 6.0. Dispense into a 5-liter flask with 4 openings at the top (2 openings are in a central silicon rubber stopper and 2 openings are gas-tight screw caps). Add a teflon-coated magnetic stir bar to the flask. Autoclave for 45 min at 15 psi pressure–121°C. Cool to room temperature under 100% N_2 at 0.05–0.1 atm pressure (use a manometer to measure low pressure).

Solution B:

Composition per 860.0mL:

Distilled/deionized water	860.0mL

Preparation of Solution B: Add 860.0mL of distilled/deionized water to a cotton-stoppered flask. Autoclave for 20 min at 15 psi pressure–121°C. Cool to room temperature under 100% N_2 in an anaerobic jar.

Solution C (Vitamin B_{12} Solution):

Composition per 5.0mL:

Vitamin B_{12}	1.0mg

Preparation of Solution C (Vitamin B_{12} Solution): Add Vitamin B_{12} to distilled/deionized water and bring volume to 5.0mL. Mix thoroughly. Filter sterilize.

Solution D (Trace Elements Solution SL-10B):

Composition per liter:

$FeCl_2 \cdot 4H_2O$	1.5g
$CoCl_2 \cdot 6H_2O$	0.19g
$MnCl_2 \cdot 4H_2O$	100.0mg
$ZnCl_2$	70.0mg
$Na_2MoO_4 \cdot 2H_2O$	36.0mg
$NiCl_2 \cdot 6H_2O$	24.0mg
H_3BO_3	6.0mg
$CuCl_2 \cdot 2H_2O$	2.0mg
HCl (25% solution)	10.0mL

Preparation of Solution D (Trace Elements Solution SL-10): Add $FeCl_2 \cdot 4H_2O$ to 10.0mL of HCl solution. Mix thoroughly. Add distilled/deionized water and bring volume to 1.0L. Add remaining components. Mix thoroughly.

Solution E:

Composition per 100.0mL:

$NaHCO_3$	7.5g

Preparation of Solution E: Add $NaHCO_3$ to distilled/deionized water and bring volume to 100.0mL. Mix thoroughly. Sparge with 100% CO_2 until saturated. Filter sterilize under 100% CO_2 into a sterile, gas-tight 100.0mL screw-capped bottle.

Solution F:

Composition per 100.0mL:

$Na_2S \cdot 9H_2O$	10.0g

Preparation of Solution F: Add $Na_2S \cdot 9H_2O$ to distilled/deionized water and bring volume to 100.0mL. Mix thoroughly. Dispense into a screw-capped bottle. Sparge with 100% $N_2$for 3–4 min. Autoclave for 15 min at 15 psi pressure–121°C.

Preparation of Medium: Saturate cooled solution A under 100% CO_2 at 0.05–0.1 atm pressure for 30 min with magnetic stirring. Add 860.0mL of solution B, 5.0mL of solution C, 5.0mL of solution D, 100.0mL of solution E, and 20.0mL of solution F through one of the screw-capped openings under 95% N_2 and 5% CO_2 with magnetic stirring. Adjust pH to 6.8 with sterile 2*M* HCl or sterile 2*M* Na_2CO_3 solution. Aseptically and anaerobically distribute the medium through the medium outlet tube into sterile 100.0mL bottles under 95% N_2 and 5% CO_2 at 0.05–0.1 atm pressure. Leave a small gas bubble in each bottle to accommodate pressure changes. After 24 hr, the iron in the medium will precipitate out of solution as black flocs.

Use: For the cultivation and maintenance of *Chlorobium phaeobacteroides, Chlorobium vibrioforme, Pelodictyon luteolum, Pelodictyon phaeum,* and *Prosthecochloris aestuarii.*

PFS Medium (Peptone Fumarate Sulfate Medium)

Composition per liter:

Peptone	10.0g
Fumaric acid	2.0g
$(NH_4)_2SO_4$	1.0g
$MgSO_4 \cdot 7H_2O$	0.5g
$FeCl_3 \cdot 6H_2O$	0.2mg
$MnSO_4 \cdot H_2O$	0.2mg

pH 7.0 ± 0.2 at 25°C

Preparation of Medium: Add components to distilled/deionized water and bring volume to 1.0L. Mix thoroughly. Adjust pH to 7.0 with KOH. Distribute into tubes or flasks. Autoclave for 15 min at 15 psi pressure–121°C.

Use: For the cultivation and maintenance of *Aquaspirillum fasciculus.*

Phenol Red Agar

Composition per liter:

Agar	15.0g
Pancreatic digest of casein	10.0g
NaCl	5.0g

Phenol Red 0.018g
Carbohydrate solution 20.0mL

pH 7.4 ± 0.2 at 25°C

Source: This medium is available as a premixed powder from BD Diagnostic Systems.

Carbohydrate Solution:

Composition per 20.0mL:

Carbohydrate 5.0–10.0g

Preparation of Carbohydrate Solution: Add carbohydrate to distilled/deionized water and bring volume to 20.0mL. Mix thoroughly. Filter sterilize.

Preparation of Medium: Add components, except carbohydrate solution, to distilled/deionized water and bring volume to 980.0mL. Mix thoroughly. Adjust pH to 7.4 if necessary. Autoclave for 15 min at 15 psi pressure–121°C. Cool to 45°–50°C. Aseptically add 20.0mL of sterile carbohydrate solution. Pour into sterile Petri dishes or distribute into sterile tubes. Allow tubes to cool in a slanted position.

Use: For the determination of fermentation reactions. Bacteria that can ferment the added carbohydrate turn the medium yellow.

Phenol Red Glucose Broth

Composition per liter:

Pancreatic digest of casein 10.0g
Glucose 5.0g
NaCl 5.0g
Phenol Red 0.018g

pH 7.3 ± 0.2 at 25°C

Source: This medium is available as a premixed powder from BD Diagnostic Systems.

Preparation of Medium: Add components to distilled/deionized water and bring volume to 1.0L. Mix thoroughly. Adjust pH to 7.3 if necessary. Distribute into tubes containing an inverted Durham tube. Fill each tube with 10.0mL of medium. Autoclave for 15 min at 13 psi pressure–118°C.

Use: For determination of the ability of a microorganism to ferment glucose. Fermentation is determined by the production of acid (when broth turns yellow) and formation of gas (when bubble trapped in Durham tube).

Phenol Red Lactose Agar

Composition per liter:

Agar 15.0g
Lactose 10.0g
Proteose peptone No. 3 .... 10.0g
NaCl 5.0g
Beef extract 1.0g
Phenol Red 25.0mg

pH 7.4 ± 0.2 at 25°C

Source: This medium is available as a premixed powder from BD Diagnostic Systems.

Preparation of Medium: Add components to distilled/deionized water and bring volume to 1.0L. Mix thoroughly. Gently heat and bring to boiling. Distribute into tubes or flasks. Autoclave for 15 min at 13 psi pressure–118°C. Pour into sterile Petri dishes or leave in tubes. Allow tubes to cool in a slanted position.

Use: For determination of the ability of a microorganism to ferment lactose. Fermentation is determined by the production of acid (when medium turns yellow).

Phenol Red Mannitol Agar

Composition per liter:

Agar 15.0g
Mannitol 10.0g
Proteose peptone No. 3 .... 10.0g
NaCl 5.0g
Beef extract 1.0g
Phenol Red 25.0mg

pH 7.4 ± 0.2 at 25°C

Source: This medium is available as a premixed powder from BD Diagnostic Systems.

Preparation of Medium: Add components to distilled/deionized water and bring volume to 1.0L. Mix thoroughly. Gently heat and bring to boiling. Distribute into tubes or flasks. Autoclave for 15 min at 13 psi pressure–118°C. Pour into sterile Petri dishes or leave in tubes. Allow tubes to cool in a slanted position.

Use: For determination of the ability of a microorganism to ferment mannitol. Fermentation is determined by the production of acid (when medium turns yellow).

Phenol Red Sucrose Broth

Composition per liter:

Pancreatic digest of casein 10.0g
NaCl 5.0g
Sucrose 5.0g
Phenol Red 0.018g

pH 7.3 ± 0.2 at 25°C

Source: This medium is available as a premixed powder from BD Diagnostic Systems.

Preparation of Medium: Add components to distilled/deionized water and bring volume to 1.0L. Mix thoroughly. Adjust pH to 7.3 if necessary. Distribute into tubes containing an inverted Durham tube. Fill each tube with 10.0mL of medium. Autoclave for 15 min at 13 psi pressure–118°C.

Use: For determination of the ability of a microorganism to ferment sucrose. Fermentation is determined by the production of acid (when broth turns yellow) and formation of gas (when bubble trapped in Durham tube).

Phosphate Mineral Salts Medium with Octane

Composition per liter:

$(NH_4)_2HPO_4$ 10.0g
K_2HPO_4 .... 5.0g
Na_2SO_4 .... 0.5g
Octane 10.0mL

Preparation of Medium: Add components, except octane, to tap water and bring volume to 990.0mL. Mix thoroughly. Autoclave for 15 min at 15 psi pressure–121°C. Prior to inoculation, filter sterilize octane. Aseptically add sterile octane to sterile medium. Aseptically distribute into sterile tubes or flasks.

Use: For the cultivation and maintenance of *Pseudomonas oleovorans*.

Photobacterium Agar (DSMZ Medium 32)

Composition per liter:

Seawater aquarium salt 33.0g
Agar 20.0g
Tris 6.0g
Yeast extract 5.0g
Tryptone 5.0g

NH_4Cl 5.0g
Glycerol 3.0g
$CaCO_3$ 1.0g
pH 7.2-7.5 at 25°C

Preparation of Medium: Add components to distilled/deionized water and bring volume to 1.0L. Mix thoroughly. Adjust pH to 7.2 - 7.5. Gently heat and bring to boiling. Distribute into tubes or flasks. Autoclave for 15 min at 15 psi pressure–121°C. Pour into sterile Petri dishes or leave in tubes.

Use: For the cultivation and maintenance of *Vibrio fischeri.*

Photobacterium Broth

Composition per liter:
NaCl 30.0g
Sodium glycerol phosphate 23.5g
Pancreatic digest of casein 5.0g
KH_2PO_4 .... 3.0g
Yeast extract 2.5g
$CaCO_3$ 1.0g
NH_4Cl 0.3g
$MgSO_4 \cdot 7H_2O$ 0.3g
$FeCl_3$ 0.01g
pH 7.0 ± 0.2 at 25°C

Source: This medium is available as a premixed powder from BD Diagnostic Systems.

Preparation of Medium: Add components to distilled/deionized water and bring volume to 1.0L. Mix thoroughly. Distribute into tubes or flasks to form a shallow layer of medium. Autoclave for 15 min at 15 psi pressure–121°C.

Use: For the cultivation and demonstration of luminescence by photobacteria. For the cultivation and maintenance of *Alteromonas hanedai, Photobacterium phosphoreum, Shewanella hanedai, Vibrio fischeri, Vibrio harveyi,* and other *Vibrio* species.

Photobacterium MPY Medium

Composition per liter:
NaCl 28.2g
$MgSO_4 \cdot 7H_2O$ 6.9g
$MgCl_2 \cdot 6H_2O$ 5.5g
Peptone 5.0g
Yeast extract 3.0g
$CaCl_2 \cdot 2H_2O$ 1.5g
KCl 0.7g
pH 7.4 ± 0.2 at 25°C

Preparation of Medium: Add components to distilled/deionized water and bring volume to 1.0L. Mix thoroughly. Distribute into tubes or flasks. Autoclave for 15 min at 15 psi pressure–121°C.

Use: For the cultivation and maintenance of *Photobacterium leiognathi.*

Phthalic Acid Medium

Composition per liter:
Solution 1 .... 400.0mL
Solution 2 .... 400.0mL
Potassium hydrogen phthalate solution 200.0mL
pH 6.8 ± 0.2 at 25°C

Solution 1:
Composition per 400.0mL:
KH_2PO_4 .... 9.1g
$(NH_4)_2SO_4$ 1.2g

Preparation of Solution 1: Add components to distilled/deionized water and bring volume to 400.0mL. Mix thoroughly. Adjust pH to 6.8 with KOH. Autoclave for 15 min at 15 psi pressure–121°C. Cool to 25°C.

Solution 2:
Composition per 400.0mL:
$MgSO_4 \cdot 7H_2O$ 0.4g
$FeSO_4 \cdot 7H_2O$ 0.01g

Preparation of Solution 2: Add components to distilled/deionized water and bring volume to 400.0mL. Mix thoroughly. Adjust pH to 6.8 with KOH. Autoclave for 15 min at 15 psi pressure–121°C. Cool to 25°C.

Potassium Hydrogen Phthalate Solution:
Composition per 200.0mL:
Potassium hydrogen phthalate 1.0g

Preparation of Medium: Add component to distilled/deionized water and bring volume to 200.0mL. Mix thoroughly. Adjust pH to 6.8 with KOH. Autoclave for 15 min at 15 psi pressure–121°C. Cool to 25°C.

Preparation of Medium: Aseptically combine the three sterile solutions. Mix thoroughly. Aseptically distribute into sterile tubes or flasks.

Use: For the cultivation and maintenance of *Pseudomonas cepacia.*

Pilobolus Agar

Composition per liter:
Agar 15.0g
Sodium acetate 10.0g
Yeast extract 2.0g
K_2HPO_4 .... 1.0g
$(NH_4)_2SO_4$ 0.66g
$MgSO_4 \cdot 7H_2O$ 0.5g
Thiamine 10.0mg
Hemin 10.0mg
NaOH (0.1*N* solution) 37.5mL
pH 8.0 ± 0.2 at 25°C

Preparation of Medium: Add hemin to NaOH solution and mix thoroughly to dissolve. Add remaining components and bring volume to 1.0L with distilled/deionized water. Mix thoroughly. Gently heat and bring to boiling. Distribute into tubes or flasks. Autoclave for 15 min at 15 psi pressure–121°C. Pour into sterile Petri dishes or leave in tubes.

Use: For the cultivation and maintenance of *Pilobolus* species.

Pilobolus Medium (DSMZ Medium 192)

Composition per liter:
Agar 15.0g
Na-acetate 10.0g
Yeast extract 2.0g
K_2HPO_4 .... 1.0g
$(NH_4)_2SO_4$ 0.66g
$MgSO_4 \cdot 7H_2O$ 0.5g
Thiamine-HCl·$2H_2O$ 10.0mg
Haemin solution .... 37.5 mL
pH 8.0 ± 0.2 at 25°C

Haemin Solution:
Composition per 10.0mL:
Haemin 10.0mg

Preparation of Haemin Solution: Add haemin to 37.5 mL 0.1*N* NaOH. Mix thoroughly.

Preparation of Medium: Add components to distilled/deionized water and bring volume to 1.0L. Mix thoroughly. Adjust pH to 8.0. Gently heat and bring to boiling. Distribute into tubes or flasks. Autoclave for 15 min at 15 psi pressure–121°C. Pour into sterile Petri dishes or leave in tubes.

Use: For the cultivation and maintenance of *Pilobolus sphaerosporus.*

Pirellula marina Medium M14

Composition per liter:

Glucose	1.0g
Yeast extract	1.0g
Artificial seawater	680.0mL
Tris·HCl buffer, (0.1*M* solution, pH 7.5)	50.0mL
Hutner's basal salts	20.0mL
Vitamin solution	10.0mL

pH 7.5 ± 0.2 at 25°C

Artificial Seawater:

Composition per liter:

NaCl	23.47g
$MgCl_2 \cdot 6H_2O$	10.64g
Na_2SO_4	3.92g
$CaCl_2$	1.1g
KCl	664.0mg
$NaHCO_3$	192.0mg
KBr	96.0mg
H_3BO_3	26.0mg
$SrCl_2$	24.0mg
NaF	3.0mg

Preparation of Artificial Seawater: Add components to distilled/deionized water and bring volume to 1.0L. Mix thoroughly.

Vitamin Solution:

Composition per liter:

Nicotinamide	9.0mg
Calcium DL-pantothenate	5.0mg
Riboflavin	5.0mg
Thiamine·HCl	5.0mg
Biotin	2.0mg
Folic acid	2.0mg
Cyanocobalamin	0.1mg

Preparation of Vitamin Solution: Add components to distilled/deionized water and bring volume to 1.0L. Mix thoroughly. Filter sterilize.

Preparation of Medium: Add components, except vitamin solution, to distilled/deionized water and bring volume to 990.0mL. Mix thoroughly. Autoclave for 15 min at 15 psi pressure–121°C. Aseptically add 10.0mL of sterile vitamin solution. Mix thoroughly. Aseptically distribute into sterile tubes or flasks.

Use: For the cultivation of *Pirellula marina.*

Plaice Medium

Composition per liter:

Fresh plaice, minced	200.0g
NaCl	120.0g
Peptone	20.0g

pH 7.3 ± 0.2 at 25°C

Preparation of Medium: Soak plaice in 4.0L of water and allow to stand for 2 hr. Boil for 1 hr and filter. Add peptone and NaCl. Adjust pH. Boil for a few minutes and filter. Mix thoroughly. Distribute into tubes or flasks. Autoclave for 15 min at 15 psi pressure–121°C.

Use: For the cultivation of bioluminescent bacteria.

Planctomyces Medium

Composition per liter:

Glucose	1.0g
$(NH_4)_2SO_4$	0.25g
Peptone	0.15g
Yeast extract	0.15g
Seawater, aged filtered	1.0L

Preparation of Medium: Add components to distilled/deionized water and bring volume to 1.0L. Mix thoroughly. Distribute into tubes or flasks. Autoclave for 15 min at 15 psi pressure–121°C.

Use: For the cultivation of *Planctomyces* species.

Plant *Mycoplasma* Agar

Composition per 291.2mL:

Schneider's *Drosophila* medium	160.0mL
Solution 1	70.0mL
Fetal calf serum	50.0mL
Fresh yeast extract solution	10.0mL
Phenol Red (0.5% solution)	1.2mL

pH 7.4 ± 0.2 at 25°C

Schneider's *Drosophila* Medium:

Composition per liter:

$MgSO_4 \cdot 7H_2O$	3.7g
NaCl	2.1g
Yeast extract	2.0g
Trehalose	2.0g
D-Glucose	2.0g
L-Glutamine	1.8g
L-Lysine·HCl	1.7g
L-Proline	1.7g
KCl	1.6g
$Na_2HPO_4 \cdot 7H_2O$	1.3g
L-Glutamic acid	0.8g
L-Methionine	0.8g
$CaCl_2$, anhydrous	0.6g
KH_2PO_4	0.5g
β-Alanine	0.5g
L-Tyrosine	0.5g
L-Arginine	0.4g
L-Aspartic acid	0.4g
L-Histidine	0.4g
L-Threonine	0.4g
$NaHCO_3$	0.4g
Glycine	0.3g
L-Serine	0.3g
L-Valine	0.3g
L-Isoleucine	0.2g
L-Leucine	0.2g
L-Phenylalanine	0.2g
α-Ketoglutaric acid	0.2g
Fumaric acid	0.1g
Malic acid	0.1g
Succinic acid	0.1g
L-Cystine	0.1g
L-Tryptophan	0.1g
L-Cysteine	0.06g

Preparation of Schneider's *Drosophila* Medium: Add components to distilled/deionized water and bring volume to 1.0L. Mix thoroughly. Filter sterilize.

Solution 1:

Composition per 70.0mL:

Sorbitol	7.0g
Noble agar	5.0g

Beef heart, solids from infusion 5.0g
Peptone 1.8g
Pancreatic digest of casein 1.0g
Sucrose 1.0g
NaCl 0.5g
D-Fructose 0.1g
D-Glucose 0.1g

Preparation of Solution 1: Add components to distilled/deionized water and bring volume to 70.0mL. Mix thoroughly. Adjust pH to 7.8 with 1*N* NaOH. Autoclave for 15 min at 15 psi pressure–121°C. Cool to 50°C.

Fresh Yeast Extract Solution:
Composition per 100.0mL:
Baker's yeast, live, pressed, starch-free 25.0g

Preparation of Fresh Yeast Extract Solution: Add the live Baker's yeast to 100.0mL of distilled/deionized water. Autoclave for 90 min at 15 psi pressure–121°C. Allow to stand. Remove supernatant solution. Adjust pH to 6.6–6.8.

Preparation of Medium: Bring fetal calf serum and Phenol Red solution to 56°C. Rapidly bring Schneider's *Drosophila* medium to 37°C. Rapidly combine the components. Mix thoroughly. Pour into sterile Petri dishes or distribute into sterile tubes or flasks.

Use: For the cultivation and maintenance of *Spiroplasma floricola, Spiroplasma kunkelii, Spiroplasma melliferum,* and *Spiroplasma* species.

Plates with Fluoranthene (DSMZ Medium 462a)

Composition per 1052mL:
Agar 15.0g
Na_2HPO_4 .... 2.44g
KH_2PO_4 .... 1.52g
$(NH_4)_2SO_4$ 0.5g
$MgSO_4 \cdot 7H_2O$ 0.2g
$CaCl_2 \cdot 2H_2O$ 0.05g
Fluoranthene solution 50.0mL
Trace elements solution SL-4 .... 10.0mL
Vitamin solution 2.5mL
pH 7.1 ± 0.2 at 25°C

Trace Elements Solution SL-4:
Composition per liter:
EDTA 0.5g
$FeSO_4 \cdot 7H_2O$ 0.2g
Trace elements solution SL-6 .... 100.0mL

Trace Elements Solution SL-6:
Composition per liter:
H_3BO_3 .... 0.3g
$CoCl_2 \cdot 6H_2O$ 0.2g
$ZnSO_4 \cdot 7H_2O$ 0.1g
$MnCl_2 \cdot 4H_2O$ 0.03g
$Na_2MoO_4 \cdot H_2O$ 0.03g
$NiCl_2 \cdot 6H_2O$ 0.02g
$CuCl_2 \cdot 2H_2O$ 0.01g

Preparation of Trace Elements Solution SL-6: Add components to distilled/deionized water and bring volume to 1.0L. Mix thoroughly. Adjust pH to 3.4.

Preparation of Trace Elements Solution SL-4: Add components to distilled/deionized water and bring volume to 1.0L. Mix thoroughly.

Vitamin Solution:
Composition per liter:
Pyridoxamine 5.0mg
Vitamin B_{12} 2.0mg
Nicotinic acid 2.0mg
p-Aminobenzoate 1.0mg
Thiamine–$HCl \cdot 2H_2O$ 1.0mg
Ca-pantothenate 0.5mg
Biotin 0.2mg

Preparation of Vitamin Solution: Add components to distilled/deionized water and bring volume to 1.0L. Mix thoroughly. Filter sterilize.

Fluoranthene Solution:
Composition per 100.0mL:
Fluoranthene 10.0g

Preparation of Fluoranthene Solution: Add fluoranthene to 100.0mL acetone. Mix thoroughly. Filter sterilize.

Preparation of Medium: Add components, except vitamin solution and fluoranthene solution, to 1.0L distilled/deionized water. Adjust pH to 7.1. Gently heat and bring to boil. Autoclave for 15 min at 15 psi pressure–121°C. Pour into sterile Petri dishes (20.0mL per Petri dish). Allow plates to dry for 48 hr. Aseptically spread 1.0mL of fluoranthene solution onto the surface of the dried plates. Allow the acetone to evaporate under sterile condition.

Use: For the cultivation of *Mycobacterium gilvum, Sphingomonas* sp. *(Sphingomonas paucimobilis),* unclassified bacterium *(Gordona* sp.*), Mycobacterium* sp., *Sphingomonas* sp., and other biphenyl utilizing bacteria.

PMY Medium

Composition per liter:
Agar 15.0g
Glucose 10.0g
NaCl 5.0g
Polypeptone 5.0g
Beef extract 2.0g
Yeast extract 1.0g
$MgSO_4 \cdot 7H_2O$ 0.5g
pH 7.0 ± 0.2 at 25°C

Preparation of Medium: Add components to distilled/deionized water and bring volume to 1.0L. Mix thoroughly. Gently heat and bring to boiling. Distribute into tubes or flasks. Autoclave for 15 min at 15 psi pressure–121°C. Pour into sterile Petri dishes or leave in tubes.

Use: For the cultivation and maintenance of *Xanthomonas campestris.*

Postgate's Medium

Composition per liter:
$MgSO_4 \cdot 7H_2O$ 2.0g
$CaSO_4$ 1.0g
NH_4Cl 1.0g
Yeast extract 1.0g
K_2HPO_4 .... 0.5g
Sodium lactate (70% solution) 3.5mL

Preparation of Medium: Add components, except $FeSO_4 \cdot 7H_2O$, ascorbic acid, and thioglycollic acid, to distilled/deionized water and bring volume to 1.0L. Mix thoroughly. Sparge with 80% N_2 + 20% CO_2 for 10–15 min. Add $FeSO_4 \cdot 7H_2O$, ascorbic acid, and thioglycollic acid. Mix thoroughly. Continue to sparge with 80% N_2 + 20% CO_2 and adjust pH to 7.4. Anaerobically distribute into tubes or flasks. Autoclave for 10 min at 10 psi pressure–115°C.

Use: For the cultivation of *Desulfobulbus proprionicus, Desulfotomaculum nigrificans, Desulfotomaculum orientis,*

Desulfotomaculum ruminis, Desulfovibrio africanus, Desulfovibrio desulfuricans, Desulfovibrio gigas, Desulfovibrio multispirans, Desulfovibrio species, and *Desulfovibrio vulgaris.*

Potato P-YE *Thermus* Medium

Composition per liter:

Agar	20.0g
Peptone	5.0g
Yeast extract	0.2g
Potatoes, infusion from	200.0mL

pH 7.8 ± 0.2 at 25°C

Potatoes, Infusion From:
Composition per 500.0mL:

Potatoes	300.0g

Preparation of Medium: Add components to distilled/deionized water and bring volume to 1.0L. Mix thoroughly. Gently heat and bring to boiling. Distribute into tubes or flasks. Autoclave for 15 min at 15 psi pressure–121°C. Pour into sterile Petri dishes or leave in tubes.

Use: For the cultivation and maintenance of *Thermus ruber.*

PP Medium

Composition per liter:

Proteose peptone	10.0g
Pancreatic digest of peptone	10.0g
Ribonucleic acid from *Torula* yeast	1.0g
Asolectin	0.2g
Artificial seawater	167.0mL
Vitamin solution	2.0mL

Artificial Seawater:
Composition per 167.0mL:

Aqua-Marin sea salts	6.95g

Source: Aqua-Marin sea salts are available from Aquatrol, Inc., Anaheim, CA.

Preparation of Artificial Seawater: Add Aqua-Marin sea salts to distilled/deionized water and bring volume to 167.0mL. Mix thoroughly. Filter sterilize.

Vitamin Solution:
Composition per 100.0mL:

Thiamine·HCl	150.0mg
Calcium D-(+)-pantothenate	100.0mg
Folic acid	50.0mg
Nicotinamide	50.0mg
Pyridoxal·HCl	50.0mg
Riboflavin	50.0mg
DL-6 Thioctic acid	1.0mg
Biotin solution	10.0mL

Preparation of Vitamin Solution: Add components to distilled/deionized water and bring volume to 100.0mL. Mix thoroughly. Filter sterilize. For long-term storage, preserve under nitrogen at –20°C.

Biotin Solution:
Composition per 10.0mL:

Biotin	0.01mg

Preparation of Biotin Solution: Add biotin to 10.0mL of absolute ethanol. Mix thoroughly.

Preparation of Medium: Add ascolectin to 500.0mL of distilled/deionized water. Gently heat to 80°C. Mix thoroughly. Add other components, except artificial seawater and vitamin solution, to distilled/deionized water and bring volume to 831.0mL. Mix thoroughly. Adjust pH to 7.2. Autoclave for 15 min at 15 psi pressure–121°C. Aseptically add 167.0mL of sterile artificial seawater and 2.0mL of sterile vitamin solution. Mix thoroughly. Aseptically distribute into sterile tubes or flasks.

Use: For the cultivation of *Potomacus pottsi.*

PPB, Modified Caldwell and Bryant

Composition per liter:

Pancreatic digest of casein	2.0g
Yeast extract	2.0g
Cellobiose	1.0g
Glucose	1.0g
Maltose	1.0g
Starch	1.0g
Resazurin	1.0mg
Rumen fluid, clarified	150.0mL
Mineral solution I	100.0mL
Mineral solution II	100.0mL
Na_2CO_3 solution	50.0mL
Hemin solution	10.0mL
L-Cysteine·HCl solution	10.0mL
Volatile fatty acid mixture	3.1mL

pH 6.8 ± 0.2 at 25°C

Mineral Solution I:
Composition per 100.0mL:

K_2HPO_4	0.2g

Preparation of Mineral Solution I: Add K_2HPO_4 to distilled/deionized water and bring volume to 100.0mL. Mix thoroughly.

Mineral Solution II:
Composition per 100.0mL:

NaCl	0.4g
$(NH_4)_2SO_4$	0.4g
KH_2PO_4	0.3g
$MgSO_4·7H_2O$	0.09g
$CaCl_2$	0.05g

Preparation of Mineral Solution II: Add components to distilled/deionized water and bring volume to 100.0mL. Mix thoroughly.

Na_2CO_3 Solution:
Composition per 100.0mL:

Na_2CO_3	8.0g

Preparation of Na_2CO_3 Solution: Add Na_2CO_3 to distilled/deionized water and bring volume to 100.0mL. Mix thoroughly. Sparge with 100% CO_2. Autoclave for 15 min at 15 psi pressure–121°C.

Hemin Solution:
Composition per 100.0mL:

Hemin	1.0g
NaOH (1*N* solution)	10.0mL

Preparation of Hemin Solution: Add components to 100.0mL of distilled/deionized water. Mix thoroughly.

L-Cysteine·HCl Solution:
Composition per 10.0mL:

L-Cysteine·HCl	0.25g

Preparation of L-Cysteine·HCl Solution: Add L-cysteine·HCl to distilled/deionized water and bring volume to 10.0mL. Mix thoroughly. Autoclave under 100% N_2 for 15 min at 15 psi pressure–121°C.

Volatile Fatty Acid Mixture:
Composition per 31.0mL:

Acetic acid	17.0mL
Propionic acid	6.0mL

Butyric acid 4.0mL
DL–α-Methylbutyric acid 1.0mL
Isobutyric acid 1.0mL
Isovaleric acid 1.0mL
n-Valeric acid 1.0mL

Preparation of Volatile Fatty Acid Mixture: Combine components. Mix thoroughly. Store under 100% N_2.

Preparation of Medium: Prepare and dispense medium under 100% CO_2. Add components, except L-cysteine·HCl solution and Na_2CO_3 solution, to distilled/deionized water and bring volume to 930.0mL. Mix thoroughly. Sparge with 100% CO_2. Adjust pH to 6.8 with 1*N* NaOH. Distribute anaerobically 9.3mL volumes into Hungate tubes. Autoclave for 15 min at 15 psi pressure–121°C. Aseptically and anaerobically add 0.2mL of sterile L-cysteine·HCl solution and 0.5mL of sterile Na_2CO_3 solution. Check that final pH is 6.8. (Note: if not properly gassed with 100% CO_2, the medium pH can be as high as 9.5.)

Use: For the cultivation of *Anaerovibrio lipolytica, Bacteroides* species, *Butyrivibrio crossotus, Butyrivibrio fibrisolvens, Eubacterium cellulosolvans, Eubacterium ruminantium, Fibrobacter succinogenes, Lachnospira multiparus, Megasphaera elsdenii, Rhodococcus torques, Ruminobacter amylophilus, Ruminococcus albus, Ruminococcus bromii, Ruminococcus flavifaciens, Selenomonas ruminantium, Succinomonus amylolytica, Succinovibrio dextrinisolvens,* and *Veillonella parvula.*

PPES II Medium

Composition per liter:

Agar 15.0g
Polypeptone 2.0g
Yeast extract 1.0g
Papaic digest of soybean meal 1.0g
Proteose peptone No. 3 .... 1.0g
Ferric phosphate, soluble 0.1g
Marine mud extract 100.0mL

pH 7.6 ± 0.2 at 25°C

Preparation of Medium: Add components to distilled/deionized water and bring volume to 1.0L. Mix thoroughly. Gently heat and bring to boiling. Distribute into tubes or flasks. Autoclave for 15 min at 15 psi pressure–121°C. Pour into sterile Petri dishes or leave in tubes.

Use: For the cultivation of *Erythrobacter longus, Haloferax mediterranei,* and *Roseobacter denitrificans.*

PPGA Medium

Composition per liter:

Agar 18.0g
Glucose 5.0g
Peptone 5.0g
NaCl 3.0g
Na_2HPO_4 .... 1.2g
KH_2PO_4 .... 0.5g
Potato decoction 1.0L

pH 7.0 ± 0.2 at 25°C

Potato Decoction:

Composition per liter:

Potatoes 200.0g

Preparation of Potato Decoction: Peel and dice potatoes. Add 1.0L of distilled/deionized water. Gently heat and bring to boiling. Continue boiling for 20 min. Filter through two layers of cheesecloth. Bring volume of filtrate to 1.0L with distilled/deionized water.

Preparation of Medium: Combine components. Gently heat and bring to boiling. Adjust pH to 7.0. Distribute into tubes or flasks. Autoclave for 15 min at 15 psi pressure–121°C. Pour into sterile Petri dishes or leave in tubes.

Use: For the cultivation of *Burkholderia glumae* and *Burkholderia plantarii.*

Presence-Absence Broth (P-A Broth)

Composition per liter:

Pancreatic digest of casein 10.0g
Lactose 7.5g
Pancreatic digest of gelatin 5.0g
Beef extract 3.0g
NaCl 2.5g
K_2HPO_4 .... 1.375g
KH_2PO_4 .... 1.375g
Sodium lauryl sulfate 0.05g
Bromcresol Purple 8.5mg

pH 6.8 ± 0.2 at 25°C

Source: This medium is available as a premixed powder from BD Diagnostic Systems.

Preparation of Medium: Add components to distilled/deionized water and bring volume to 333.0mL. Mix thoroughly. Distribute into screw-capped 250.0mL milk dilution bottles in 50.0mL volumes. Autoclave for 15 min at 15 psi pressure–121°C.

Use: For the detection of coliform bacteria in water from treatment plants or distribution systems using the presence-absence coliform test.

Propionigenium maris Medium

Composition per 1014.0mL:

Solution A 940.0mL
Solution E ($NaHCO_3$ solution) 50.0mL
Solution F (Substrate solution) 10.0mL
Solution G ($Na_2S·9H_2O$ solution) 10.0mL
Solution B (Trace elements solution SL-10) 2.0mL
Solution C (Seven vitamin solution) 1.0mL
Solution D (Selenite-tungstate solution) 1.0mL

pH 7.2–7.4 at 25°C

Solution A:

Composition per 940.0mL:

NaCl 1.0g
Yeast extract 0.5g
KCl 0.5g
$MgCl_2·6H_2O$ 0.4g
NH_4Cl 0.25g
KH_2PO_4 .... 0.2g
$CaCl_2·2H_2O$ 0.15g
Resazurin 0.5mg

Preparation of Solution A: Prepare and dispense under 80% N_2 + 20% CO_2. Add components to distilled/deionized water and bring volume to 940.0mL. Mix thoroughly. Autoclave for 15 min at 15 psi pressure–121°C.

Solution B (Trace Elements Solution SL-10):

Composition per liter:

$FeCl_2·4H_2O$ 1.5g
$CoCl_2·6H_2O$ 190.0mg
$MnCl_2·4H_2O$ 100.0mg
$ZnCl_2$ 70.0mg
$Na_2MoO_4·2H_2O$ 36.0mg
$NiCl_2·6H_2O$ 24.0mg
H_3BO_3 .... 6.0mg

$CuCl_2 \cdot 2H_2O$ 2.0mg
HCl (25% solution) 10.0mL

Preparation of Solution B (Trace Elements Solution SL-10): Prepare and dispense under 100% N_2. Add $FeCl_2 \cdot 4H_2O$ to 10.0mL of HCl solution. Mix thoroughly. Add distilled/deionized water and bring volume to 1.0L. Add remaining components. Mix thoroughly. Autoclave for 15 min at 15 psi pressure–121°C.

Solution C (Seven Vitamin Solution):
Composition per liter:
Pyridoxine·HCl 300.0mg
Nicotinic acid 200.0mg
Thiamine·HCl 200.0mg
Calcium pantothenate 100.0mg
Cyanocobalamine 100.0mg
p-Aminobenzoic acid 80.0mg
D(+)-Biotin 20.0mg

Preparation of Solution C (Seven Vitamin Solution): Add components to distilled/deionized water and bring volume to 1.0L. Mix thoroughly. Filter sterilize. Sparge with 100% N_2.

Solution D (Selenite-Tungstate Solution):
Composition per liter:
NaOH 0.5g
$Na_2WO_4 \cdot 2H_2O$ 4.0mg
$Na_2SeO_3 \cdot 5H_2O$ 3.0mg

Preparation of Solution D (Selenite-Tungstate Solution): Add components to distilled/deionized water and bring volume to 1.0L. Mix thoroughly. Filter sterilize. Sparge with 100% N_2.

$Na_2S \cdot 9H_2O$ Solution:
Composition per 10.0mL:
$Na_2S \cdot 9H_2O$ 0.6g

Preparation of $Na_2S \cdot 9H_2O$ Solution: Add $Na_2S \cdot 9H_2O$ to distilled/deionized water and bring volume to 10.0mL. Mix thoroughly. Sparge with 100% N_2. Autoclave for 15 min at 15 psi pressure–121°C.

Solution E ($NaHCO_3$ Solution):
Composition per 50.0mL:
$NaHCO_3$ 2.5g

Preparation of Solution E ($NaHCO_3$ Solution): Add $NaHCO_3$ to distilled/deionized water and bring volume to 50.0mL. Mix thoroughly. Sparge with 100% N_2. Autoclave for 15 min at 15 psi pressure–121°C.

Solution F (Substrate Solution):
Composition per 10.0mL:
Disodium succinate 2.5g

Preparation of Solution F (Substrate Solution): Add disodium succinate to distilled/deionized water and bring volume to 10.0mL. Mix thoroughly. Sparge with 100% N_2. Autoclave for 15 min at 15 psi pressure–121°C.

Solution G ($Na_2S \cdot 9H_2O$ Solution):
Composition per 10.0mL:
$Na_2S \cdot 9H_2O$ 0.3g

Preparation of Solution G ($Na_2S \cdot 9H_2O$ Solution): Add $Na_2S \cdot 9H_2O$ to distilled/deionized water and bring volume to 10.0mL. Mix thoroughly. Sparge with 100% N_2. Autoclave for 15 min at 15 psi pressure–121°C.

Preparation of Medium: To 940.0mL of sterile solution A, aseptically and anaerobically add 1.0mL of sterile solution B, 1.0mL of sterile solution C, 1.0mL of sterile solution D, 50.0mL of sterile solution E, 10.0mL of sterile solution F, and 10.0mL of sterile solution G. Mix thoroughly. Aseptically and anaerobically distribute into sterile tubes or flasks.

Use: For the cultivation of *Propionigenium maris.*

Propionivibrio/Acetivibrio/Formivibrio Medium

Composition per 1012.0mL:
Solution A 950.0mL
Solution E 30.0mL
Solution D (Vitamin solution) 10.0mL
Solution F 10.0mL
Solution G 10.0mL
Solution B (Trace elements solution SL-10) 1.0mL
Solution C (Selenite-tungstate solution) 1.0mL
pH 6.7 ± 0.2 at 25°C

Solution A:
Composition per 950.0mL:
KH_2PO_4 .... 1.4g
NH_4Cl 0.5g
$MgCl_2 \cdot 6H_2O$ 0.2g
$CaCl_2 \cdot 2H_2O$ 0.15g
Resazurin 1.0mg

Preparation of Solution A: Add components to distilled/deionized water and bring volume to 950.0mL. Mix thoroughly. Sparge with 80% N_2 + 20% CO_2. Autoclave for 15 min at 15 psi pressure–121°C.

Solution B (Trace Elements Solution SL-10):
Composition per liter:
$FeCl_2 \cdot 4H_2O$ 1.5g
$CoCl_2 \cdot 6H_2O$ 190.0mg
$MnCl_2 \cdot 4H_2O$ 100.0mg
$ZnCl_2$ 70.0mg
$Na_2MoO_4 \cdot 2H_2O$ 36.0mg
$NiCl_2 \cdot 6H_2O$ 24.0mg
H_3BO_3 .... 6.0mg
$CuCl_2 \cdot 2H_2O$ 2.0mg
HCl (25% solution) 10.0mL

Preparation of Solution B (Trace Elements Solution SL-10): Add FeCl2·4H2O to 10.0mL of HCl solution. Mix thoroughly. Add distilled/deionized water and bring volume to 1.0L. Add remaining components. Mix thoroughly. Sparge with 80% N_2 + 20% CO_2. Autoclave for 15 min at 15 psi pressure–121°C.

Solution C (Selenite-Tungstate Solution):
Composition per liter:
NaOH 0.5g
$Na_2WO_4 \cdot 2H_2O$ 4.0mg
$Na_2SeO_3 \cdot 5H_2O$ 3.0mg

Preparation of Solution C (Selenite-Tungstate Solution): Add components to distilled/deionized water and bring volume to 1.0L. Mix thoroughly. Sparge with 80% N_2 + 20% CO_2. Autoclave for 15 min at 15 psi pressure–121°C.

Solution D (Vitamin Solution):
Composition per liter:
Pyridoxine·HCl 10.0mg
Calcium DL-pantothenate 5.0mg
Lipoic acid 5.0mg
Nicotinic acid 5.0mg
p-Aminobenzoic acid 5.0mg
Riboflavin 5.0mg
Thiamine·HCl 5.0mg
Biotin 2.0mg

Folic acid....2.0mg
Vitamin B_{12}....0.1mg

Preparation of Solution D (Vitamin Solution): Add components to distilled/deionized water and bring volume to 1.0L. Mix thoroughly. Sparge with 80% N_2 + 20% CO_2. Filter sterilize.

Solution E:

Composition per 30.0mL:

$NaHCO_3$....1.5g

Preparation of Solution E: Add $NaHCO_3$ to distilled/deionized water and bring volume to 30.0mL. Mix thoroughly. Filter sterilize. Sparge with 80% N_2 + 20% CO_2.

Solution F:

Composition per 10.0mL:

Disodium maleate....1.6g

Preparation of Solution F: Add disodium maleate to distilled/deionized water and bring volume to 10.0mL. Mix thoroughly. Sparge with 80% N_2 + 20% CO_2. Autoclave for 15 min at 15 psi pressure–121°C.

Solution G:

Composition per 10.0mL:

$Na_2S \cdot 9H_2O$....0.25g

Preparation of Solution G: Add $Na_2S \cdot 9H_2O$ to distilled/deionized water and bring volume to 10.0mL. Mix thoroughly. Gas under 100% N_2. Autoclave for 15 min at 15 psi pressure–121°C.

Preparation of Medium: Aseptically and anaerobically combine 950.0mL of sterile solution A with 1.0mL of sterile solution B, 1.0mL of sterile solution C, 10.0mL of sterile solution D, 30.0mL of sterile solution E, 10.0mL of sterile solution F, and 10.0mL of sterile solution G, in that order. Mix thoroughly. Final pH should be 6.7–6.8. Aseptically and anaerobically distribute into sterile tubes or flasks under 80% N_2 + 20% CO_2.

Use: For the cultivation and maintenance of *Propionivibrio dicarboxylicus*.

Propionivibrio/Acetivibrio/Formivibrio Medium

Composition per 1012.0mL:

Solution A....950.0mL
Solution E....30.0mL
Solution D (Vitamin solution)....10.0mL
Solution F....10.0mL
Solution G....10.0mL
Solution B (Trace elements solution SL-10)....1.0mL
Solution C (Selenite-tungstate solution)....1.0mL
pH 7.7–7.9 at 25°C

Solution A:

Composition per 950.0mL:

KH_2PO_4....1.4g
NH_4Cl....0.5g
$MgCl_2 \cdot 6H_2O$....0.2g
$CaCl_2 \cdot 2H_2O$....0.15g
Yeast extract....50.0mg
Resazurin....1.0mg

Preparation of Solution A: Add components to distilled/deionized water and bring volume to 950.0mL. Mix thoroughly. Sparge with 100% N_2. Autoclave for 15 min at 15 psi pressure–121°C.

Solution B (Trace Elements Solution SL-10):

Composition per liter:

$FeCl_2 \cdot 4H_2O$....1.5g
$CoCl_2 \cdot 6H_2O$....190.0mg
$MnCl_2 \cdot 4H_2O$....100.0mg
$ZnCl_2$....70.0mg
$Na_2MoO_4 \cdot 2H_2O$....36.0mg
$NiCl_2 \cdot 6H_2O$....24.0mg
H_3BO_3....6.0mg
$CuCl_2 \cdot 2H_2O$....2.0mg
HCl (25% solution)....10.0mL

Preparation of Solution B (Trace Elements Solution SL-10): Add $FeCl_2 \cdot 4H_2O$ to 10.0mL of HCl solution. Mix thoroughly. Add distilled/deionized water and bring volume to 1.0L. Add remaining components. Mix thoroughly. Sparge with 100% N_2. Autoclave for 15 min at 15 psi pressure–121°C.

Solution C (Selenite-Tungstate Solution):

Composition per liter:

NaOH....0.5g
$Na_2WO_4 \cdot 2H_2O$....4.0mg
$Na_2SeO_3 \cdot 5H_2O$....3.0mg

Preparation of Solution C (Selenite-Tungstate Solution): Add components to distilled/deionized water and bring volume to 1.0L. Mix thoroughly. Sparge with 100% N_2. Autoclave for 15 min at 15 psi pressure–121°C.

Solution D (Vitamin Solution):

Composition per liter:

Pyridoxine·HCl....10.0mg
Calcium DL-pantothenate....5.0mg
Lipoic acid....5.0mg
Nicotinic acid....5.0mg
p-Aminobenzoic acid....5.0mg
Riboflavin....5.0mg
Thiamine·HCl....5.0mg
Biotin....2.0mg
Folic acid....2.0mg
Vitamin B_{12}....0.1mg

Preparation of Solution D (Vitamin Solution): Add components to distilled/deionized water and bring volume to 1.0L. Mix thoroughly. Sparge with 100% N_2. Filter sterilize.

Solution E:

Composition per 30.0mL:

$NaHCO_3$....1.5g

Preparation of Solution E: Add $NaHCO_3$ to distilled/deionized water and bring volume to 30.0mL. Mix thoroughly. Filter sterilize. Sparge with 100% N_2.

Solution F:

Composition per 10.0mL:

Cinnamic acid....1.6g
NaOH (1*N* solution)....variable

Preparation of Solution F: Add cinnamic acid to distilled/deionized water and bring volume to 8.0mL. Mix thoroughly. Add sufficient quantity of 1*N* NaOH solution to bring pH to 7.8. Sparge with 100% N_2. Autoclave for 15 min at 15 psi pressure–121°C.

Solution G:

Composition per 10.0mL:

$Na_2S \cdot 9H_2O$....0.25g

Preparation of Solution G: Add $Na_2S \cdot 9H_2O$ to distilled/deionized water and bring volume to 10.0mL. Mix thoroughly. Sparge with 100% N_2. Autoclave for 15 min at 15 psi pressure–121°C.

Preparation of Medium: Aseptically and anaerobically under 100% N_2 combine 950.0mL of sterile solution A with 1.0mL of sterile solution B, 1.0mL of sterile solution C, 10.0mL of sterile solution D, 30.0mL of sterile solution E, 10.0mL of sterile solution F, and 10.0mL of sterile solution G, in that order. Mix thoroughly. Final pH should be 7.7–7.9. If necessary, add about 15.0mL of sterile anaerobic 5% Na_2CO_3 solution to 1.0L of medium to adjust pH. Aseptically and anaerobically distribute into sterile tubes or flasks under 100% N_2.

Use: For the cultivation and maintenance of *Acetivibrio multivorans.*

Prototheca Isolation Agar (PIM)

Composition per liter:

Agar	20.0g
Glucose	10.0g
Potassium hydrogen phthalate	10.0g
NaOH	0.9g
NH_4Cl	0.3g
5-Fluorocytosine	0.25g
KH_2PO_4	0.2g
$MgSO_4$	0.1g
Thiamine·HCl	0.001g

pH 5.1 ± 0.1 at 25°C

Preparation of Medium: Add components to distilled/deionized water and bring volume to 1.0L. Mix thoroughly. Adjust pH to 5.1. Gently heat and bring to boiling. Distribute into tubes or flasks. Autoclave for 15 min at 15 psi pressure–121°C. Pour into sterile Petri dishes or leave in tubes.

Use: For the cultivation of *Prototheca moriformis* and *Prototheca ulmea*.

Provasoli Medium

Composition per liter:

NaCl	11.75g
$MgCl_2·6H_2O$	5.35g
Na_2SO_4	2.0g
$CaCl_2·2H_2O$	0.75g
Tris(hydroxymethyl)aminomethane	0.5g
KCl	0.35g
Na_2HPO_4	0.05g

pH 7.6 ± 0.2 at 25°C

Preparation of Medium: Add components to distilled/deionized water and bring volume to 1.0L. Mix thoroughly. Distribute into tubes or flasks. Autoclave for 15 min at 15 psi pressure–121°C.

Use: For the isolation and cultivation of *Leucothrix* species from marine habitats.

Pseudoamycolata halophobica Medium

Composition per liter:

Glucose	5.0g
Peptone	5.0g
Yeast extract	3.0g
K_2HPO_4	0.2g

pH 6.8 ± 0.2 at 25°C

Preparation of Medium: Add components to distilled/deionized water and bring volume to 1.0L. Mix thoroughly. Adjust pH to 6.8. Distribute into tubes or flasks. Autoclave for 15 min at 15 psi pressure–121°C.

Use: For the cultivation of *Pseudoamycolata halophobica*.

Pseudobutyrivibrio Medium

Composition per liter:

Disodium succinate	5.0g
Yeast extract	5.0g
NaCl	0.45g
$(NH_4)_2SO_4$	0.45g
K_2HPO_4	0.225g
KH_2PO_4	0.225g
$MgSO_4·7H_2O$	0.09g
$CaCl_2·2H_2O$	0.06g
Indigocarmine	5.0mg
Rumen fluid, clarified	400.0mL
Glucose solution	20.0mL
$NaHCO_3$ solution	10.0mL
L-Cysteine·HCl·H_2O solution	10.0mL
$Na_2S·9H_2O$ solution	10.0mL

pH 6.6–6.8at 25°C

Glucose Solution:

Composition per 20.0mL:

D-Glucose	5.0g

Preparation of Glucose Solution: Add glucose to distilled/deionized water and bring volume to 20.0mL. Mix thoroughly. Sparge with 100% N_2. Autoclave for 15 min at 15 psi pressure–121°C.

$NaHCO_3$ Solution:

Composition per 10.0mL:

$NaHCO_3$	6.4g

Preparation of $NaHCO_3$ Solution: Add $NaHCO_3$ to distilled/deionized water and bring volume to 10.0mL. Mix thoroughly. Sparge with 100% N_2. Autoclave for 15 min at 15 psi pressure–121°C.

L-Cysteine·HCl·H_2O Solution:

Composition per 10.0mL:

L-Cysteine·HCl·H_2O	0.3g

Preparation of L-Cysteine·HCl·H_2O Solution: Add L-cysteine·HCl·H_2O to distilled/deionized water and bring volume to 10.0mL. Mix thoroughly. Sparge with 100% N_2. Autoclave for 15 min at 15 psi pressure–121°C.

$Na_2S·9H_2O$ Solution:

Composition per 10.0mL:

$Na_2S·9H_2O$	0.3g

Preparation of $Na_2S·9H_2O$ Solution: Add $Na_2S·9H_2O$ to distilled/deionized water and bring volume to 10.0mL. Mix thoroughly. Sparge with 100% N_2. Autoclave for 15 min at 15 psi pressure–121°C. Before use, neutralize to pH 7.0 with sterile HCl.

Preparation of Medium: Add components, except glucose solution, L-cysteine·HCl·H_2O solution, $Na_2S·9H_2O$ solution, and $NaHCO_3$ solution, to distilled/deionized water and bring volume to 960.0mL. Mix thoroughly. Gently heat and bring to boiling. Cool to room temperature while sparging with 100% CO_2. Autoclave for 15 min at 15 psi pressure–121°C. Aseptically and anaerobically add 10.0mL of sterile glucose solution, 10.0mL of sterile L-cysteine·HCl solution, 10.0mL of sterile $Na_2S·9H_2O$ solution, and 10.0mL of sterile $NaHCO_3$ solution to each tube. Mix thoroughly.

Use: For the cultivation of *Pseudobutyrivibrio ruminis*.

Pseudomonas aeruginosa Agar (*PA* Agar) (m–PA Agar) (m-*Pseudomonas aeruginosa* Agar)

Composition per liter:

Agar	15.0g
$Na_2S_2O_3$	6.8g
L-Lysine·HCl	5.0g
NaCl	5.0g
Xylose	2.5g
Yeast extract	2.0g
Lactose	1.25g
Sucrose	1.25g
Ferric ammonium citrate	0.8g
Sulfapyridine	0.176g
Cycloheximide	0.15g
Phenol Red	0.08g
Nalidixic acid	0.037g
Kanamycin	8.5mg

pH 7.1 ± 0.2 at 25°C

Preparation of Medium: Add components, except sulfapyridine, cycloheximide, nalidixic acid, and kanamycin, to distilled/deionized water and bring volume to 1.0L. Mix thoroughly. Adjust pH to 6.5. Autoclave for 15 min at 15 psi pressure–121°C. Cool to 55°–60°C. Readjust pH to 7.1. Aseptically add the sulfapyridine, cycloheximide, nalidixic acid, and kanamycin. Mix thoroughly. Pour into 50mm × 12mm Petri dishes in 3.0mL volumes.

Use: For the cultivation and estimation of numbers of *Pseudomonas aeruginosa* in water by the membrane filter method.

Pseudomonas Agar F

Composition per liter:

Proteose peptone No. 3	20.0g
Agar	15.0g
Glycerol	10.0g
Pancreatic digest of casein	10.0g
K_2HPO_4	1.5g
$MgSO_4·7H_2O$	0.73g

pH 7.0 ± 0.2 at 25°C

Preparation of Medium: Add components to distilled/deionized water and bring volume to 1.0L. Mix thoroughly. Gently heat and bring to boiling. Distribute into tubes or flasks. Autoclave for 15 min at 15 psi pressure–121°C. Pour into sterile Petri dishes or leave in tubes.

Use: For the cultivation and observation of fluorescein production in *Pseudomonas* species.

Pseudomonas Agar P

Composition per liter:

Proteose peptone No. 3	20.0g
Agar	15.0g
Glycerol	10.0g
K_2HPO_4	10.0g
$MgCl_2·6H_2O$	1.4g

pH 7.0 ± 0.2 at 25°C

Source: This medium is available as a premixed powder from BD Diagnostic Systems.

Preparation of Medium: Add components to distilled/deionized water and bring volume to 1.0L. Mix thoroughly. Gently heat and bring to boiling. Distribute into tubes or flasks. Autoclave for 15 min at 15 psi pressure–121°C. Pour into sterile Petri dishes or leave in tubes.

Use: For the isolation, cultivation, and differentiation of *Pseudomonas aeruginosa* on the basis of pigment production.

Pseudomonas Basal Mineral Medium

Composition per liter:

K_2HPO_4	12.5g
KH_2PO_4	3.8g
$(NH_4)_2SO_4$	1.0g
$MgSO_4·7H_2O$	0.1g
Carbon source (0.8*M* solution)	100.0mL
Trace elements solution	5.0mL

pH 7.2 ± 0.2 at 25°C

Trace ElementsSolution:

Composition per liter:

H_3BO_3	0.232g
$ZnSO_4·7H_2O$	0.174g
$FeSO_4(NH_4)_2SO_4·6H_2O$	0.116g
$CoSO_4·7H_2O$	0.096g
$(NH_4)_6Mo_7O_{24}·4H_2O$	0.022g
$CuSO_4·5H_2O$	8.0mg
$MnSO_4·4H_2O$	8.0mg

Preparation of Trace Elements Solution: Add components to distilled/deionized water and bring volume to 1.0L. Mix thoroughly.

Carbon Source:

Composition per 100.0mL:

Glucose	14.4g

Preparation of Carbon Source: Add glucose to distilled/deionized water and bring volume to 100.0mL. Mix thoroughly. Filter sterilize. Other carbon sources may replace glucose. Prepare 0.8*M* carbon source solution.

Preparation of Medium: Add components, except carbon source, to distilled/deionized water and bring volume to 900.0mL. Mix thoroughly. Gently heat and bring to boiling. Autoclave for 15 min at 15 psi pressure–121°C. Cool to 45°–50°C. Aseptically add 100.0mL of sterile carbon source. Mix thoroughly. Aseptically distribute into sterile tubes or flasks.

Use: For the cultivation and differentiation of *Pseudomonas* species based on their ability to grow on different carbon sources.

Pseudomonas bathycetes Medium

Composition per liter:

NaCl	24.0g
Proteose peptone	10.0g
$MgSO_4·7H_2O$	7.0g
$MgCl_2$	5.3g
Yeast extract	3.0g
KCl	0.7g

pH 7.2–7.4 at 25°C

Preparation of Medium: Add components to distilled/deionized water and bring volume to 1.0L. Mix thoroughly. Distribute into tubes or flasks. Autoclave for 15 min at 15 psi pressure–121°C.

Use: For the cultivation and maintenance of *Alteromonas haloplanktis, Alteromonas nigrifaciens, Pseudomonas bathycetes,* and *Pseudomonas elongata.*

Pseudomonas CFC Agar

Composition per liter:

Pancreatic digest of gelatin	16.0g
Agar	11.0g
Pancreatic digest of casein	10.0g

K_2SO_4 .... 10.0g
$MgCl_2 \cdot 6H_2O$ 1.4g
CFC selective supplement 10.0mL
Glycerol 10.0mL
pH 7.1 ± 0.2 at 25°C

Source: This medium is available as a premixed powder from Oxoid Unipath.

CFC Selective Supplement:
Composition per 10.0mL:
Cephaloridine 0.05g
Fucidin 0.01g
Cetrimide 0.01g

Preparation of CFC Selective Supplement: Add components to distilled/deionized water and bring volume to 10.0mL. Mix thoroughly. Filter sterilize.

Preparation of Medium: Add components, except CFC selective supplement, to distilled/deionized water and bring volume to 990.0mL. Mix thoroughly. Gently heat and bring to boiling. Autoclave for 15 min at 15 psi pressure–121°C. Cool to 45°–50°C. Aseptically add sterile CFC selective supplement. Mix thoroughly. Pour into sterile Petri dishes or distribute into sterile tubes.

Use: For the selective isolation and cultivation of *Pseudomonas* species.

Pseudomonas chlorotidismutans Medium (DSMZ Medium 944)

Composition per 1001mL:
Solution A 900.0mL
Solution B 50.0mL
Solution C 50.0mL
Vitamin solution 1.0mL
pH 9.0 ± 0.2 at 25°C

Solution A:
Composition per 900mL:
Na-acetate·$3H_2O$ 2.72g
$NaClO_3$ 1.06g
KH_2PO_4 .... 0.41g
Na_2HPO_4 .... 0.53g
Resazurin 0.5mg
Selenite-tungstate solution 4.0mL
Trace elements solution SL-10 .... 1.0mL

Trace Elements Solution SL-10:
Composition per liter:
$FeCl_2 \cdot 4H_2O$ 1.5g
$CoCl_2 \cdot 6H_2O$ 190.0mg
$MnCl_2 \cdot 4H_2O$ 100.0mg
$ZnCl_2$ 70.0mg
$Na_2MoO_4 \cdot 2H_2O$ 36.0mg
$NiCl_2 \cdot 6H_2O$ 24.0mg
H_3BO_3 .... 6.0mg
$CuCl_2 \cdot 2H_2O$ 2.0mg
HCl (25% solution) 10.0mL

Preparation of Trace Elements Solution SL-10: Add $FeCl_2 \cdot 4H_2O$ to 10.0mL of HCl solution. Mix thoroughly. Add distilled/deionized water and bring volume to 1.0L. Add remaining components. Mix thoroughly. Sparge with 80% N_2 + 20% CO_2.

Selenite-Tungstate Solution
Composition per liter:
NaOH 0.5g
$Na_2WO_4 \cdot 2H_2O$ 4.0mg
$Na_2SeO_3 \cdot 5H_2O$ 3.0mg

Preparation of Selenite-Tungstate Solution: Add components to distilled/deionized water and bring volume to 1.0L. Mix thoroughly. Sparge with 100% N_2.

Preparation of Solution A: Add components to distilled/deionized water and bring volume to 900.0mL. Mix thoroughly. Sparge with 100% N_2. Autoclave for 15 min at 15 psi pressure–121°C. Cool to room temperature.

Solution B:
Composition per 50mL:
$CaCl_2$ 0.11g
$MgCl_2$ 0.10g

Preparation of Solution B: Add components to distilled/deionized water and bring volume to 50.0mL. Mix thoroughly. Sparge with 100% N_2. Autoclave for 15 min at 15 psi pressure–121°C. Cool to room temperature.

Solution C:
$NaHCO_3$ 3.73g
$Na_2S \cdot 9H_2O$ 0.5g
NH_4HCO_3 .... 0.44g

Preparation of Solution C: Add components to distilled/deionized water and bring volume to 50.0mL. Mix thoroughly. Sparge with 100% N_2. Autoclave for 15 min at 15 psi pressure–121°C. Cool to room temperature.

Vitamin Solution:
Composition per liter:
Vitamin B_{12} 50.0mg
Pantothenic acid 50.0mg
Riboflavin 50.0mg
Alpha-lipoic acid 50.0mg
p-Aminobenzoic acid 50.0mg
Thiamine-HCl·$2H_2O$ 50.0mg
Nicotinic acid 25.0mg
Nicotine amide 25.0mg
Biotin 20.0mg
Folic acid 20.0mg
Pyridoxamine-HCl 10.0mg

Preparation of Vitamin Solution: Add components to distilled/deionized water and bring volume to 1.0L. Mix thoroughly. Filter sterilize.

Preparation of Medium: Prepare and dispense medium under 100% N_2 gas. Asetpically and anaerobically combine 900.0mL sterile solution A, 50.0mL sterile solution B, 50.0mL sterile solution C, and 1.0mL sterile vitamin solution. Mix thoroughly. The pH should be 9.0. Aseptically and anaerobically distribute into sterile tubes or flasks.

Use: For the cultivation of *Pseudomonas chlorotidismutans (Pseudomonas stutzeri).*

Pseudomonas CN Agar

Composition per liter:
Pancreatic digest of gelatin 16.0g
Agar 11.0g
Pancreatic digest of casein 10.0g
K_2SO_4 .... 10.0g
$MgCl_2 \cdot 6H_2O$ 1.4g
CN selective supplement 10.0mL
Glycerol 10.0mL

CN Selective Supplement:
Cetrimide 0.1g
Sodium nalidixate 7.5mg
pH 7.1 ± 0.2 at 25°C

Source: This medium is available as a premixed powder from Oxoid Unipath.

Preparation of Medium: Add components, except CN selective supplement, to distilled/deionized water and bring volume to 990.0mL. Mix thoroughly. Gently heat and bring to boiling. Autoclave for 15 min at 15 psi pressure–121°C. Cool to 45°–50°C. Aseptically add sterile CN selective supplement. Mix thoroughly. Pour into sterile Petri dishes or distribute into sterile tubes.

Use: For the selective isolation and cultivation of *Pseudomonas* species.

Pseudomonas denitrificans Medium (LMG 153)

Composition per liter:

Agar 15.0g
Glucose 10.0g
Yeast extract 5.0g
$FeCl_3$ solution 20.0mL

$FeCl_3$ Solution:
Composition per 100.0mL:

$FeCl_3$ 0.03g

Preparation of $FeCl_3$ Solution: Add $FeCl_3$ to distilled/deionized water and bring volume to 100.0mL. Mix thoroughly. Filter sterilize.

Preparation of Medium: Add components, except $FeCl_3$ solution, to distilled/deionized water and bring volume to 980.0mL. Mix thoroughly. Gently heat and bring to boiling. Autoclave for 15 min at 15 psi pressure–121°C. Cool to 45°–50°C. Aseptically add 20.0mL of sterile $FeCl_3$ solution. Mix thoroughly. Pour into sterile Petri dishes or distribute into sterile tubes.

Use: For the cultivation and maintenance of *Pseudomonas* species.

Pseudomonas Denitrification Medium

Composition per liter:

Glycerol 10.0g
KNO_3 .... 10.0g
Yeast extract 3.0g
$(NH_4)_2SO_4$ 1.5g
Agar 1.0g
$K_2HPO_4 \cdot 3H_2O$ 0.8g
$MgSO_4 \cdot 7H_2O$ 0.5g
KH_2PO_4 .... 0.2g
$CaCl_2$ 0.1g

pH 7.2 ± 0.2 at 25°C

Preparation of Medium: Add components to distilled/deionized water and bring volume to 1.0L. Mix thoroughly. Distribute into tubes in 10.0mL volumes. Autoclave for 15 min at 15 psi pressure–121°C.

Use: For the cultivation and differentiation of *Pseudomonas* species based on their ability to produce pyocin and other fluorescent pigments during denitrification.

Pseudomonas halophila Medium

Composition per liter:

Solution A 890.0mL
Solution B 100.0mL
Vitamin solution 10.0mL

pH 7.0 ± 0.2 at 25°C

Solution A:
Composition per 890.0mL:

NaCl 46.8g
$MgSO_4 \cdot 7H_2O$ 39.4g
Glycerol 5.0g
NH_4Cl 1.0g
Trace elements solution SL-10 .... 1.0mL

Trace Elements Solution SL-10:
Composition per liter:

$FeCl_2 \cdot 4H_2O$ 1.5g
$CoCl_2 \cdot 6H_2O$ 0.19g
$MnCl_2 \cdot 4H_2O$ 100.0mg
$ZnCl_2$ 70.0mg
$Na_2MoO_4 \cdot 2H_2O$ 36.0mg
$NiCl_2 \cdot 6H_2O$ 24.0mg
H_3BO_3 .... 6.0mg
$CuCl_2 \cdot 2H_2O$ 2.0mg
HCl (25% solution) 10.0mL

Preparation of Trace Elements Solution SL-10: Add $FeCl_2 \cdot 4H_2O$ to 10.0mL of HCl solution. Mix thoroughly. Add distilled/deionized water and bring volume to 1.0L. Add remaining components. Mix thoroughly.

Preparation of Solution A: Add components to distilled/deionized water and bring volume to 890.0mL. Mix thoroughly. Autoclave for 15 min at 15 psi pressure–121°C. Cool to room temperature.

Solution B:
Composition per 100.0mL:

KH_2PO_4 .... 1.0g

Preparation of Solution B: Add KH_2PO_4 to distilled/deionized water and bring volume to 100.0mL. Mix thoroughly. Autoclave for 15 min at 15 psi pressure–121°C. Cool to room temperature.

Vitamin Solution:
Composition per liter:

Pyridoxine·HCl 10.0mg
Calcium pantothenate 5.0mg
Nicotinic acid 5.0mg
Robiflavin 5.0mg
Thiamine·HCl 5.0mg
Biotin 2.0mg
Folic acid 2.0mg
p-Aminobenzoic acid 1.0mg
Cyanocobalamin 0.01mg

Preparation of Vitamin Solution: Add components to distilled/deionized water and bring volume to 1.0L. Mix thoroughly. Filter sterilize.

Preparation of Medium: Aseptically combine 890.0mL of cooled sterile solution A, 100.0mL of cooled sterile solution B, and 10.0mL of sterile vitamin solution. Mix thoroughly. Adjust pH to 7.0 with sterile 6*N* NaOH or HCl solutions.

Use: For the cultivation and maintenance of *Pseudomonas halophila*.

Pseudomonas Isolation Agar

Composition per liter:

Peptone 20.0g
Agar 13.6g
K_2SO_4 .... 10.0g
$MgCl_2 \cdot 6H_2O$ 1.4g
Irgasan (triclosan) 0.025g
Glycerol 20.0mL

pH 7.0 ± 0.2 at 25°C

Source: This medium is available as a premixed powder from BD Diagnostic Systems.

Preparation of Medium: Add components to distilled/deionized water and bring volume to 1.0L. Mix thoroughly. Gently heat and bring to boiling. Distribute into tubes or

flasks. Autoclave for 15 min at 15 psi pressure–121°C. Pour into sterile Petri dishes or leave in tubes.

Use: For the isolation and cultivation of *Pseudomonas* species.

Pseudomonas Medium (ATCC Medium 59)

Composition per liter:

K_2HPO_4	1.15g
NH_4NO_3	1.0g
Yeast extract	1.0g
KH_2PO_4	0.625g
$MgSO_4 \cdot 7H_2O$	0.02g
Pyrrolidine	4.0mL

pH 7.0 ± 0.2 at 25°C

Preparation of Medium: Add pyrrolidine to approximately 500.0mL of distilled/deionized water. Mix thoroughly. Adjust pH to 7.0. Add remaining components. Bring volume to 1.0L with distilled/deionized water. Distribute into tubes or flasks. Autoclave for 15 min at 15 psi pressure–121°C.

Use: For the cultivation and maintenance of *Pseudomonas fluorescens.*

Pseudomonas Medium (ATCC Medium 609)

Composition per liter:

Agar	15.0g
K_2HPO_4	8.71g
Nitrilotriacetic acid	1.91g
Na_2SO_4	0.57g
$MgSO_4$	0.25g
$FeSO_4$	0.5mg
$Ca(NO_3)_2$	0.5mg

pH 6.5 ± 0.2 at 25°C

Preparation of Medium: Add nitrilotriacetic acid to approximately 500.0mL of distilled/deionized water. Mix thoroughly. Adjust pH to 6.5. Add remaining components. Bring volume to 1.0L with distilled/deionized water. Gently heat and bring to boiling. Distribute into tubes or flasks. Autoclave for 15 min at 15 psi pressure–121°C. Pour into sterile Petri dishes or leave in tubes.

Use: For the cultivation and maintenance of *Pseudomonas* species.

Pseudomonas Medium A

Composition per liter:

Peptone	20.0g
Agar	15.0g
Glycerol	10.0g
K_2SO_4	10.0g
$MgCl_2$	1.4g

pH 7.2 ± 0.2 at 25°C

Preparation of Medium: Add components to distilled/deionized water and bring volume to 1.0L. Mix thoroughly. Gently heat and bring to boiling. Distribute into tubes or flasks. Autoclave for 10 min at 10 psi pressure–115°C. Pour into sterile Petri dishes or leave in tubes.

Use: For the cultivation and production of pyocyanin by *Pseudomonas* species.

Pseudomonas Medium B

Composition per liter:

Peptone	20.0g
Agar	15.0g
Glycerol	10.0g
$MgSO_4 \cdot 7H_2O$	1.5g
K_2HPO_4 solution	100.0mL

pH 7.2 ± 0.2 at 25°C

K_2HPO_4 Solution:
Composition per 100.0mL:

K_2HPO_4	1.5g

Preparation of K_2HPO_4 Solution: Add K_2HPO_4 to distilled/deionized water and bring volume to 100.0mL. Mix thoroughly. Autoclave for 15 min at 15 psi pressure–121°C. Cool to 45°–50°C.

Preparation of Medium: Add components, except K_2HPO_4 solution, to distilled/deionized water and bring volume to 900.0mL. Mix thoroughly. Gently heat and bring to boiling. Autoclave for 15 min at 15 psi pressure–121°C. Cool to 45°–50°C. Aseptically add 100.0mL of sterile K_2HPO_4 solution. Mix thoroughly. Pour into sterile Petri dishes or distribute into sterile tubes.

Use: For the cultivation and observation of fluorescin production by *Pseudomonas* species.

Pseudomonas Medium No. 2

Composition per liter:

Agar	15.0g
$Na_2HPO_4 \cdot 12H_2O$	6.0g
Succinic acid	5.0g
KH_2PO_4	2.4g
NH_4Cl	1.0g
$MgSO_4 \cdot 7H_2O$	0.5g
$CaCl_2 \cdot 6H_2O$	0.01g
$FeCl_3 \cdot 6H_2O$	0.01g

pH 6.8 ± 0.2 at 25°C

Preparation of Medium: Add components to distilled/deionized water and bring volume to 1.0L. Mix thoroughly. Adjust pH to 6.8. Gently heat and bring to boiling. Distribute into tubes or flasks. Autoclave for 15 min at 15 psi pressure–121°C. Pour into sterile Petri dishes or leave in tubes.

Use: For the cultivation of *Pseudomonas* species and *Psychrobacter immobilis.*

Pseudomonas pickettii Medium

Composition per liter:

Agar	15.0g
Peptone	5.0g
NaCl	5.0g
$Na_2HPO_4 \cdot 12H_2O$	2.39g
Yeast extract	2.0g
Beef extract	1.0g
K_2HPO_4	0.45g

pH 6.8 ± 0.2 at 25°C

Preparation of Medium: Add components to distilled/deionized water and bring volume to 1.0L. Mix thoroughly. Gently heat and bring to boiling. Distribute into tubes or flasks. Autoclave for 15 min at 15 psi pressure–121°C. Pour into sterile Petri dishes or leave in tubes.

Use: For the cultivation of *Burkholderia pickettii.*

Pseudomonas saccharophila Medium

Composition per 1015.0mL:

Agar	20.0g
Na_2HPO_4	4.8g
KH_2PO_4	4.4g
NH_4Cl	1.0g
$MgSO_4 \cdot 7H_2O$	0.5g

Solution A 5.0mL
Solution B 10.0mL

Solution A:

Composition per 100.0mL:

Ferric ammonium citrate 1.0g
$CaCl_2$ 0.1g

Preparation of Solution A: Add components to distilled/deionized water and bring volume to 100.0mL. Mix thoroughly. Filter sterilize.

Solution B:

Composition per 100.0mL:

Sucrose 10.0g

Preparation of Solution B: Add sucrose to distilled/deionized water and bring volume to 100.0mL. Mix thoroughly. Filter sterilize.

Preparation of Medium: Add components, except solution A and solution B, to distilled/deionized water and bring volume to 1.0L. Mix thoroughly. Gently heat and bring to boiling. Autoclave for 15 min at 15 psi pressure–121°C. Cool to 45°–50°C. Aseptically add sterile solution A and sterile solution B. Mix thoroughly. Pour into sterile Petri dishes or distribute into sterile tubes.

Use: For the cultivation and maintenance of *Pseudomonas saccharophila* and other *Pseudomonas* species.

Pseudomonas solanacearum Medium

Composition per liter:

Agar 17.0g
Peptone 10.0g
Glucose 5.0g
Pancreatic digest of casein 1.0g

Preparation of Medium: Add components to distilled/deionized water and bring volume to 1.0L. Mix thoroughly. Gently heat and bring to boiling. Distribute into tubes or flasks. Autoclave for 15 min at 15 psi pressure–121°C. Pour into sterile Petri dishes or leave in tubes.

Use: For the cultivation and maintenance of *Pseudomonas solanacearum*.

Pseudomonas syngii Medium

Composition per liter:

Agar 15.0g
Acid casein hydrolysate 7.5g
Sucrose 2.0g
$MgSO_4 \cdot 7H_2O$ 250.0mg
K_2HPO_4 .... 500.0mg
Ammonium ferricitrate solution 20.0mL

Ammonium Ferricitrate Solution:

Composition per 20.0mL:

Ammonium ferricitrate 0.25g

Preparation of Ammonium Ferricitrate Solution: Add ammonium ferricitrate to distilled/deionized water and bring volume to 20.0mL. Mix thoroughly. Filter sterilize.

Preparation of Medium: Add components, except ammonium ferricitrate solution, to distilled/deionized water and bring volume to 980.0mL. Mix thoroughly. Gently heat and bring to boiling. Autoclave for 15 min at 15 psi pressure–121°C. Cool to 50°–55°C. Aseptically add 20.0mL of sterile ammonium ferricitrate solution. Mix thoroughly. Pour into sterile Petri dishes or distribute into sterile tubes.

Use: For the cultivation and maintenance of *Pseudomonas syzygii*.

Pseudomonas syringae Selective Medium

Composition per liter:

Agar 15.0g
L-Proline 5.0g
$MgSO_4 \cdot 7H_2O$ 0.2g
K_2HPO_4 .... 0.08g
KH_2PO_4 .... 0.02g
$MnSO_4 \cdot 4H_2O$ solution 10.0mL

pH 6.8 ± 0.2 at 25°C

$MnSO_4 \cdot 4H_2O$ Solution:

Composition per 10.0mL:

$MnSO_4 \cdot 4H_2O$ 2.1g

Preparation of $MnSO_4 \cdot 4H_2O$ Solution: Add 2.1g of $MnSO_4 \cdot 4H_2O$ to distilled/deionized water and bring volume to 10.0mL. Mix thoroughly. Autoclave for 15 min at 15 psi pressure–121°C.

Preparation of Medium: Add components, except $MnSO_4 \cdot 4H_2O$ solution, to distilled/deionized water and bring volume to 990.0mL. Mix thoroughly. Gently heat and bring to boiling. Adjust pH to 6.8. Autoclave for 10 min at 10 psi pressure–115°C. Cool to 45°–50°C. Aseptically add sterile $MnSO_4 \cdot 4H_2O$ solution. Mix thoroughly. Pour into sterile Petri dishes.

Use: For the selective isolation and cultivation of *Pseudomonas syringae*.

PT Agar

Composition per liter:

Agar 15.0g
Pancreatic digest of casein 4.0g
Yeast extract 4.0g
$MgSO_4 \cdot 7H_2O$ 2.0g
$CaCl_2 \cdot 2H_2O$ 1.0g

pH 7.2 ± 0.2 at 25°C

Preparation of Medium: Add components to distilled/deionized water and bring volume to 1.0L. Mix thoroughly. Gently heat and bring to boiling. Distribute into tubes or flasks. Autoclave for 15 min at 15 psi pressure–121°C. Pour into sterile Petri dishes or leave in tubes.

Use: For the cultivation of myxobacteria.

PTYG Medium (LMG Medium 238)

Composition per liter:

Agar 15.0g
Peptone 5.0g
Tryptone 5.0g
Yeast extract 5.0g
Glucose 5.0g

pH 7.2 ± 0.2 at 25°C

Preparation of Medium: Add components to distilled/deionized water and bring volume to 1.0L. Mix thoroughly. Gently heat and bring to boiling. Distribute into tubes or flasks. Autoclave for 15 min at 15 psi pressure–121°C. Pour into sterile Petri dishes or leave in tubes.

Use: For the cultivation and maintenance of *Sphingobium herbicidovorans* and *Sphingomonas pruni*.

PTYG Medium (DSMZ Medium 914)

Composition per liter:

Glucose 10.0g
Peptone 5.0g
Tryptone 5.0g

Yeast extract 5.0g
$MgSO_4 \cdot 7H_2O$ 0.6g
$CaCl_2$ 0.06g
pH 7.0 ± 0.2 at 25°C

Preparation of Medium: Add components to distilled/deionized water and bring volume to 1.0L. Mix thoroughly. Distribute into tubes or flasks. Autoclave for 15 min at 15 psi pressure–121°C.

Use: For the cultivation of *Kineococcus radiotolerans.*

Purple Agar

Composition per liter:
Agar 15.0g
Proteose peptone No. 3 .... 10.0g
NaCl 5.0g
Beef extract 1.0g
Bromcresol Purple 0.02g
Carbohydrate solution 20.0mL
pH 6.8 ± 0.2 at 25°C

Source: This medium is available as a premixed powder from BD Diagnostic Systems.

Carbohydrate Solution:
Composition per 20.0mL:
Carbohydrate 10.0g

Preparation of Carbohydrate Solution: Add carbohydrate to distilled/deionized water and bring volume to 20.0mL. For expensive carbohydrates, 5.0g may be used instead of 10.0g. Mix thoroughly. Filter sterilize.

Preparation of Medium: Add components, except carbohydrate solution, to distilled/deionized water and bring volume to 980.0mL. Mix thoroughly. Gently heat and bring to boiling. Distribute into tubes in 9.8mL volumes. Autoclave for 15 min at 15 psi pressure–121°C. Cool to 45°–50°C. Aseptically add 0.2mL of sterile carbohydrate solution to each tube. Mix thoroughly. Allow tubes to cool in a slanted position.

Use: For the preparation of carbohydrate media used in fermentation studies for the identification of bacteria, especially members of the Enterobacteriaceae. Bacteria that can ferment the carbohydrate turn the medium yellow.

Purple Broth
(Purple Carbohydrate Broth)

Composition per liter:
Proteose peptone No. 3 .... 10.0g
NaCl 5.0g
Beef extract 1.0g
Bromcresol Purple 0.015g
Carbohydrate solution 20.0mL
pH 6.8 ± 0.2 at 25°C

Source: This medium is available as a premixed powder from BD Diagnostic Systems.

Carbohydrate Solution:
Composition per 20.0mL:
Carbohydrate 10.0g

Preparation of Carbohydrate Solution: Add carbohydrate to distilled/deionized water and bring volume to 20.0mL. For expensive carbohydrates, 5.0g may be used instead of 10.0g. Mix thoroughly. Filter sterilize.

Preparation of Medium: Add components, except carbohydrate solution, to distilled/deionized water and bring volume to 980.0mL. Mix thoroughly. Gently heat and bring to boiling. Distribute into tubes in 9.8mL volumes. Autoclave for 15 min at 15 psi pressure–121°C. Cool to 25°C. Aseptically add 0.2mL of sterile carbohydrate solution to each tube. Mix thoroughly.

Use: For the preparation of carbohydrate media used in fermentation studies for the identification of bacteria, especially members of the Enterobacteriaceae. Bacteria that can ferment the carbohydrate turn the medium yellow.

Purple Lactose Agar

Composition per liter:
Agar 10.0g
Lactose 10.0g
Peptone 5.0g
Beef extract 3.0g
Bromcresol Purple 0.025g
pH 6.8 ± 0.1 at 25°C

Source: This medium is available as a premixed powder from BD Diagnostic Systems.

Preparation of Medium: Add components to distilled/deionized water and bring volume to 1.0L. Mix thoroughly. Gently heat and bring to boiling. Distribute into tubes or flasks. Autoclave for 15 min at 15 psi pressure–121°C. Pour into sterile Petri dishes or leave in tubes. Allow tubes to cool in a slanted position.

Use: For the detection and differentiation of members of the Enterobacteriaceae. Bacteria that can ferment lactose turn the medium yellow.

PW Medium
(LMG 182)

Composition per 1100.0mL:
Agar 12.0g
Papaic digest of soybean meal 4.0g
KH_2PO_4 .... 1.2g
K_2HPO_4 .... 1.0g
Trypticase peptone 1.0g
$MgSO_4 \cdot 7H_2O$ 0.4g
Glutamine solution 50.0mL
Bovine serum albumin solution 30.0mL
Phenol Red (0.2% solution) 10.0mL
Solution A 10.0mL

Glutamine Solution:
Composition per 50.0mL:
L-Glutamine 4.0g

Preparation of Glutamine Solution: Add glutamine to distilled/deionized water and bring volume to 50.0ml. Mix thoroughly. Filter sterilize.

Bovine Serum Albumin Solution:
Composition per 50.0mL:
Bovine serum albumin, fraction V 10.0g

Preparation of Bovine Serum Albumin Solution: Add bovine serum albumin to distilled/deionized water and bring volume to 50.0mL. Mix thoroughly. Filter sterilize.

Solution A:
Composition per 101.0mL:
NaOH (0.05*N* solution) 100.0mL
Hemin chloride 0.1g

Preparation of Medium: Add components, except glutamine solution and bovine serum albumin solution, to distilled/deionized water and bring volume to 920.0mL. Mix thoroughly. Gently heat and bring to boiling. Autoclave for 15 min at 15 psi pressure–121°C. Cool to 50°–55°C. Aseptically add 50.0mL of sterile glutamine solution and 10.0mL of sterile bovine serum albumin solution. Mix thoroughly. Pour into sterile Petri dishes or distribute into sterile tubes.

Use: For the cultivation and maintenance of *Xylella fastidiosa.*

PY CMC Medium (Peptone Yeast Extract Carboxymethyl Cellulose Medium)

Composition per liter:

Agar	15.0g
Carboxymethyl cellulose	10.0g
NaCl	5.0g
Polypeptone	5.0g
Yeast extract	5.0g
$MgSO_4 \cdot 7H_2O$	2.0g
KH_2PO_4	1.0g
Na_2CO_3 solution	100.0mL

pH 9.5 ± 0.2 at 25°C

Na_2CO_3 Solution:

Composition per 100.0mL:

Na_2CO_3	10.0g

Preparation of Na_2CO_3 Solution: Add Na_2CO_3 to distilled/deionized water and bring volume to 100.0mL. Mix thoroughly. Autoclave for 15 min at 15 psi pressure–121°C. Cool to 45°–50°C.

Preparation of Medium: Add components, except Na_2CO_3 solution, to distilled/deionized water and bring volume to 900.0mL. Mix thoroughly. Gently heat and bring to boiling. Autoclave for 15 min at 15 psi pressure–121°C. Cool to 45°–50°C. Aseptically add sterile Na_2CO_3 solution. Mix thoroughly. Adjust pH to 9.5 if necessary. Pour into sterile Petri dishes or distribute into sterile tubes.

Use: For the cultivation and maintenance of alkalophilic *Bacillus* species.

PYb Agar

Composition per liter:

Agar	20.0g
Proteose peptone	1.0g
Yeast extract	1.0g
KH_2PO_4 solution	32.0mL
Na_2HPO_4 solution	8.0mL
$CaCl_2$ solution	4.0mL
$MgSO_4 \cdot 7H_2O$ solution	2.5mL

pH 6.5 ± 0.5 at 25°C

$CaCl_2$ Solution:

Composition per 100.0mL:

$CaCl_2$	0.75g

Preparation of $CaCl_2$ Solution: Add $CaCl_2$ to distilled/deionized water and bring volume to 100.0mL. Mix thoroughly. Adjust pH to 6.5. Autoclave for 25 min at 15 psi pressure–121°C. Cool to 50°–55°C.

$MgSO_4 \cdot 7H_2O$ Solution:

Composition per 100.0mL:

$MgSO_4 \cdot 7H_2O$	9.8g

Preparation of $MgSO_4 \cdot 7H_2O$ Solution: Add 9.8g of $MgSO_4 \cdot 7H_2O$ to distilled/deionized water and bring volume to 100.0mL. Mix thoroughly. Adjust pH to 6.5. Autoclave for 25 min at 15 psi pressure–121°C. Cool to 50°–55°C.

Na_2HPO_4 Solution:

Composition per 100.0mL:

Na_2HPO_4	6.7g

Preparation of Na_2HPO_4 Solution: Add Na_2HPO_4 to distilled/deionized water and bring volume to 100.0mL. Mix thoroughly. Adjust pH to 6.5. Autoclave for 25 min at 15 psi pressure–121°C. Cool to 50°–55°C.

KH_2PO_4 Solution:

Composition per 100.0mL:

KH_2PO_4	3.4g

Preparation of KH_2PO_4 Solution: Add KH_2PO_4 to distilled/deionized water and bring volume to 100.0mL. Mix thoroughly. Adjust pH to 6.5. Autoclave for 25 min at 15 psi pressure–121°C. Cool to 50°–55°C.

Preparation of Medium: Add components, except KH_2PO_4 solution, Na_2HPO_4 solution, $CaCl_2$ solution, and $MgSO_4 \cdot 7H_2O$ solution, to distilled/deionized water and bring volume to 953.5mL. Mix thoroughly. Gently heat and bring to boiling. Autoclave for 15 min at 15 psi pressure–121°C. Cool to 50°–55°C. Aseptically add 32.0mL of sterile KH_2PO_4 solution, 8.0mL of sterile Na_2HPO_4 solution, 4.0mL of sterile $CaCl_2$ solution, and 2.5mL of sterile $MgSO_4 \cdot 7H_2O$ solution. Mix thoroughly. Pour into sterile Petri dishes or distribute into sterile tubes.

Use: For the cultivation of *Acanthamoeba hatchetti, Acanthamoeba jacobsi, Acanthamoeba polyphaga, Echinamoeba exundans, Naegleria gruberi, Paratetramitus jugosus, Rhizamoeba* species, *Tetramitus rostratus*, and *Vahlkampfia lobospinosa.*

PYCS Medium

Composition per liter:

KH_2PO_4	1.0g
$(NH_4)_2SO_4$	1.0g
$MgCl_2 \cdot 6H_2O$	0.2g
NaCl	0.2g
$CaCl_2 \cdot 2H_2O$	45.0mg
Fructose solution	50.0mL
$NaHCO_3$ solution	10.0mL
Wolfe's mineral solution	10.0mL

pH 7.2 ± 0.2 at 25°C

Fructose Solution:

Composition per 50.0mL:

D-Fructose	5.0g

Preparation of Fructose Solution: Add D-fructose to distilled/deionized water and bring volume to 50.0mL. Mix thoroughly. Filter sterilize.

$NaHCO_3$ Solution:

Composition per 10.0mL:

$NaHCO_3$	1.68g

Preparation of $NaHCO_3$ Solution: Add $NaHCO_3$ to distilled/deionized water and bring volume to 10.0mL. Mix thoroughly. Filter sterilize.

Wolfe's Mineral Solution:

Composition per liter:

$MgSO_4 \cdot 7H_2O$	3.0g
Nitrilotriacetic acid	1.5g
NaCl	1.0g
$MnSO_4 \cdot H_2O$	0.5g
$CaCl_2$	0.1g
$CoCl_2 \cdot 6H_2O$	0.1g
$FeSO_4 \cdot 7H_2O$	0.1g
$ZnSO_4 \cdot 7H_2O$	0.1g
$AlK(SO_4)_2 \cdot 12H_2O$	0.01g
$CuSO_4 \cdot 5H_2O$	0.01g
H_3BO_3	0.01g
$Na_2MoO_4 \cdot 2H_2O$	0.01g

Preparation of Wolfe's Mineral Solution: Add nitrilotriacetic acid to approximately 500.0mL of water and adjust to pH 6.5 with KOH to dissolve the compound. Bring volume to 1.0L with remaining water and add remaining components one at a time.

Preparation of Medium: Add components, except fructose solution and $NaHCO_3$ solution, to distilled/deionized water and bring volume to 940.0mL. Mix thoroughly. Autoclave for 15 min at 15 psi pressure–121°C. Aseptically add 50.0mL of sterile fructose solution and 10.0mL of sterile $NaHCO_3$ solution. Mix thoroughly. Adjust pH to 7.2. Aseptically distribute into sterile tubes or flasks. Fill containers to capacity.

Use: For the cultivation of *Rhodoferax fermentans* and *Thiocapsa halophila*.

PYE Medium

Composition per liter:

Yeast extract	4.0g
Sodium pyruvate	2.2g
K_2HPO_4	1.0g
$(NH_4)_2SO_4$	1.0g
$MgSO_4 \cdot 7H_2O$	0.2g
$Na_2S_2O_3 \cdot 5H_2O$	0.2g
$CaCl_2 \cdot 2H_2O$	0.02g
Trace elements solution SL-6	1.0mL

Trace Elements Solution SL-6:
Composition per liter:

$MnCl_2 \cdot 4H_2O$	0.5g
H_3BO_3	0.3g
$CoCl_2 \cdot 6H_2O$	0.2g
$ZnSO_4 \cdot 7H_2O$	0.1g
$Na_2MoO_4 \cdot 2H_2O$	0.03g
$NiCl_2 \cdot 6H_2O$	0.02g
$CuCl_2 \cdot 2H_2O$	0.01g

Preparation of Trace Elements Solution SL-6: Add components to distilled/deionized water and bring volume to 1.0L. Mix thoroughly.

Preparation of Medium: Add components to distilled/deionized water and bring volume to 1.0L. Mix thoroughly. Distribute into tubes or flasks. Autoclave for 15 min at 15 psi pressure–121°C.

Use: For the cultivation of *Heliobacterium modestocaldum.*

PYEA Agar

Composition per liter:

Agar	15.0g
Peptone	10.0g
Yeast extract	10.0g
NaCl	5.0g

pH 7.2 ± 0.2 at 25°C

Preparation of Medium: Add components to distilled/deionized water and bring volume to 1.0L. Mix thoroughly. Gently heat and bring to boiling. Distribute into tubes or flasks. Autoclave for 15 min at 15 psi pressure–121°C. Pour into sterile Petri dishes or leave in tubes.

Use: For the cultivation and maintenance of *Blastobacter natatorius* and *Deinobacter grandis*.

PYG Agar

Composition per liter:

Agar	20.0g
Proteose peptone	20.0g
Yeast extract	1.0g
Glucose solution	50.0mL
Sodium citrate solution	34.0mL
Ferric ammonium sulfate	10.0mL
KH_2PO_4 solution	10.0mL
$MgSO_4 \cdot 7H_2O$ solution	10.0mL
Na_2HPO_4 solution	10.0mL
$CaCl_2$ solution	8.0mL

pH 6.5 ± 0.5 at 25°C

Glucose Solution:
Composition per 100.0mL:

Glucose	36.0g

Preparation of Glucose Solution: Add glucose to distilled/deionized water and bring volume to 100.0mL. Mix thoroughly. Filter sterilize. Warm to 55°C.

$MgSO_4 \cdot 7H_2O$ Solution:
Composition per 100.0mL:

$MgSO_4 \cdot 7H_2O$	9.8g

Preparation of $MgSO_4 \cdot 7H_2O$ Solution: Add 9.8g of $MgSO_4 \cdot 7H_2O$ to distilled/deionized water and bring volume to 100.0mL. Mix thoroughly. Adjust pH to 6.5. Autoclave for 25 min at 15 psi pressure–121°C. Cool to 50°–55°C.

Ferric Ammonium Sulfate Solution:
Composition per 100.0mL:

$Fe(NH_4)_2(SO_4)_2 \cdot 6H_2O$	0.135g

Preparation of Ferric Ammonium Sulfate Solution: Add $Fe(NH_4)_2(SO_4)_2 \cdot 6H_2O$ to distilled/deionized water and bring volume to 100.0mL. Mix thoroughly. Adjust pH to 6.5. Autoclave for 25 min at 15 psi pressure–121°C. Cool to 50°–55°C.

Na_2HPO_4 Solution:
Composition per 100.0mL:

Na_2HPO_4	6.7g

Preparation of Na_2HPO_4 Solution: Add Na_2HPO_4 to distilled/deionized water and bring volume to 100.0mL. Mix thoroughly. Adjust pH to 6.5. Autoclave for 25 min at 15 psi pressure–121°C. Cool to 50°–55°C.

Sodium Citrate Solution:
Composition per 100.0mL:

Sodium citrate·$2H_2O$	2.9g

Preparation of Sodium Citrate Solution: Add sodium citrate·$2H_2O$ to distilled/deionized water and bring volume to 100.0mL. Mix thoroughly. Adjust pH to 6.5. Autoclave for 25 min at 15 psi pressure–121°C. Cool to 50°–55°C.

KH_2PO_4 Solution:
Composition per 100.0mL:

KH_2PO_4	3.4g

Preparation of KH_2PO_4 Solution: Add KH_2PO_4 to distilled/deionized water and bring volume to 100.0mL. Mix thoroughly. Adjust pH to 6.5. Autoclave for 25 min at 15 psi pressure–121°C. Cool to 50°–55°C.

$CaCl_2$ Solution:
Composition per 100.0mL:

$CaCl_2$	0.75g

Preparation of $CaCl_2$ Solution: Add $CaCl_2$ to distilled/deionized water and bring volume to 100.0mL. Mix thoroughly. Adjust pH to 6.5. Autoclave for 25 min at 15 psi pressure–121°C. Cool to 50°–55°C.

Preparation of Medium: Add components, except glucose solution, sodium citrate solution, ferric ammonium sulfate solution, KH_2PO_4 solution, Na_2HPO_4 solution,

$CaCl_2$ solution, and $MgSO_4 \cdot 7H_2O$ solution, to distilled/deionized water and bring volume to 868.0mL. Mix thoroughly. Gently heat and bring to boiling. Autoclave for 15 min at 15 psi pressure–121°C. Cool to 50°–55°C. Aseptically add 50.0mL of sterile glucose solution, 34.0mL of sterile sodium citrate solution, 10.0mL of sterile ferric ammonium sulfate solution, 10.0mL of sterile KH_2PO_4 solution, 10.0mL of sterile Na_2HPO_4 solution, 8.0mL of sterile $CaCl_2$ solution, and 10.0mL of sterile $MgSO_4 \cdot 7H_2O$ solution. Mix thoroughly. Pour into sterile Petri dishes or distribute into sterile tubes.

Use: For the cultivation of *Acanthamoeba* spp.

PYG Medium for *Spirillum* (Peptone Yeast Extract Glucose Medium for *Spirillum*)

Composition per liter:

Agar 15.0g
Peptone 10.0g
Yeast extract 5.0g
Glucose 3.0g

pH 7.2 ± 0.2 at 25°C

Glucose Solution:

Composition per 10.0mL:

D-Glucose 3.0g

Preparation of Glucose Solution: Add glucose to distilled/deionized water and bring volume to 10.0mL. Mix thoroughly. Filter sterilize.

Preparation of Medium: Add components, except glucose solution, to distilled/deionized water and bring volume to 990.0mL. Mix thoroughly. Gently heat and bring to boiling. Autoclave for 15 min at 15 psi pressure–121°C. Cool to 45°–50°C. Aseptically add sterile glucose solution. Mix thoroughly. Pour into sterile Petri dishes or distribute into sterile tubes.

Use: For the cultivation and maintenance of *Spirillum pleomorphum.*

PYGS Agar (Peptone Yeast Glucose Seawater Agar) (ATCC Medium 1973)

Composition per liter:

Agar 15.0g
Glucose 3.0g
Peptone 1.25g
Yeast extract 1.25g
Seawater 25.0mL

pH 7.3 ± 0.2 at 25°C

Preparation of Medium: Add components to cold distilled/deionized water and bring volume to 1.0L. Mix thoroughly. Gently heat and bring to boiling. Distribute into tubes or flasks. Autoclave for 15 min at 15 psi pressure–121°C. Pour into sterile Petri dishes or leave in tubes.

Use: For the cultivation of a variety of marine bacteria.

PYGV Agar

Composition per liter:

Agar 15.0g
Peptone 0.25g
Yeast extract 0.25g
Hutner's basal salts solution 20.0mL
Glucose solution 10.0mL
Vitamin solution 2× 5.0mL

pH 7.5 ± 0.2 at 25°C

Hutner's Basal Salts Solution:

Composition per liter:

$MgSO_4 \cdot 7H_2O$ 29.7g
Nitrilotriacetic acid 10.0g
$CaCl_2 \cdot 2H_2O$ 3.335g
$FeSO_4 \cdot 7H_2O$ 99.0mg
$(NH_4)_6MoO_7O_{24} \cdot 4H_2O$ 9.25mg
"Metals 44" 50.0mL

"Metals 44":

Composition per 100.0mL:

$ZnSO_4 \cdot 7H_2O$ 1.095g
$FeSO_4 \cdot 7H_2O$ 0.5g
Sodium EDTA 0.25g
$MnSO_4 \cdot H2O$ 0.154g
$CuSO_4 \cdot 5H_2O$ 39.2mg
$Co(NO_3)_2 \cdot 6H_2O$ 24.8mg
$Na_2B_4O_7 \cdot 10H_2O$ 17.7mg

Preparation of "Metals 44": Add sodium EDTA to distilled/deionized water and bring volume to 90.0mL. Mix thoroughly. Add a few drops of concentrated H_2SO_4 to retard precipitation of heavy metal ions. Add remaining components. Mix thoroughly. Bring volume to 100.0mL with distilled/deionized water.

Preparation of Hutner's Basal Salts Solution: Add nitrilotriacetic acid to 500.0mL of distilled/deionized water. Adjust pH to 6.5 with KOH. Add remaining components. Add distilled/deionized water to 1.0L. Adjust pH to 6.8.

Glucose Solution:

Composition per 10.0mL:

D-Glucose 0.25g

Preparation of Glucose Solution: Add glucose to distilled/deionized water and bring volume to 10.0mL. Mix thoroughly. Filter sterilize.

Vitamin Solution 2×:

Composition per liter:

Pyridoxine·HCl 20.0mg
p-Aminobenzoic acid 10.0mg
Calcium DL-pantothenate 10.0mg
Nicotinamide 10.0mg
Riboflavin 10.0mg
Thiamine·HCl 10.0mg
Biotin 4.0mg
Folic acid 4.0mg
Vitamin B_{12} 0.2mg

Preparation of Vitamin Solution 2×: Add components to distilled/deionized water and bring volume to 1.0L. Mix thoroughly. Filter sterilize. Store in the dark at 5°C.

Preparation of Medium: Add components, except glucose solution and vitamin solution 2×, to distilled/deionized water and bring volume to 985.0mL. Mix thoroughly. Gently heat and bring to boiling. Adjust pH to 7.5 with 6*N* KOH (approximately 6 drops). Autoclave for 20 min at 15 psi pressure–121°C. Cool to 50°–55°C. Aseptically add 10.0mL of sterile glucose solution and 5.0mL of sterile vitamin solution 2×. Mix thoroughly. Pour into sterile Petri dishes or distribute into sterile tubes.

Use: For the cultivation of *Blastobacter denitrificans, Planctomyces limnophilus,* and *Gemmobacter aquatilis.*

PYGV Marine Medium (Peptone Yeast Extract Glucose Vitamin Marine Medium)

Composition per liter:

Agar 15.0g

Peptone 0.25g
Yeast extract 0.25g
Mineral salt solution 20.0mL
Glucose solution 10.0mL
Vitamin solution 5.0mL

pH 7.5 ± 0.2 at 25°C

Mineral Salt Solution:

Composition per liter:

$MgSO_4.7H_2O$ 29.7g
Nitrilotriacetic acid 10.0g
$CaCl_2·2H_2O$ 3.34g
$FeSO_4·7H_2O$ 0.099g
$Na_2MoO_4·2H_2O$ 0.013g
"Metals 44" 50.0mL

Preparation of Mineral Salt Solution: Add nitrilotriacetic acid to 500.0mL of distilled/deionized water. Dissolve by adjusting pH to 6.5 with KOH. Add remaining components. Add distilled/deionized water to 1.0L. Readjust pH to 7.2.

"Metals 44":

Composition per 100.0mL:

$ZnSO_4·7H_2O$ 1.1g
$FeSO_4·7H_2O$ 0.5g
EDTA 0.25g
$MnSO_4·7H_2O$ 0.154g
$CuSO_4·5H_2O$ 0.04g
$Co(NO_3)_2·6H_2O$ 0.025g
$Na_2B_4O_7·10H_2O$ 0.018g

Preparation of "Metals 44": Add a few drops of H_2SO_4 to distilled/deionized water to inhibit precipitate formation. Add components to acidified distilled/deionized water and bring volume to 100.0mL. Mix thoroughly.

Glucose Solution:

Composition per 100.0mL:

D-Glucose 2.5g

Preparation of Glucose Solution: Add glucose to distilled/deionized water and bring volume to 100.0mL. Mix thoroughly. Filter sterilize.

Vitamin Solution:

Composition per liter:

Pyridoxine·HCl 0.02g
p-Aminobenzoic acid 0.01g
Calcium D-pantothenate 0.01g
Nicotinamide 0.01g
Riboflavin 0.01g
Thiamine·HCl 0.01g
Biotin 4.0mg
Folic acid 4.0mg
Cyanocobalamin 0.2mg

Preparation of Vitamin Solution: Add components to distilled/deionized water and bring volume to 1.0L. Mix thoroughly. Filter sterilize.

Preparation of Medium: Add components, except glucose solution and vitamin solution, to seawater and bring volume to 985.0mL. Mix thoroughly. Gently heat and bring to boiling. Autoclave for 15 min at 15 psi pressure–121°C. Cool to 45°–50°C. Aseptically add 10.0mL of sterile glucose solution and 5.0mL of sterile vitamin solution. Mix thoroughly. Adjust pH to 7.5 with sterile KOH if necessary. Pour into sterile Petri dishes or distribute into sterile tubes.

Use: For the cultivation and maintenance of *Planctomyces brasiliensis*.

PYGV Medium (Peptone Yeast Extract Glucose Vitamin Medium)

Composition per liter:

Agar 15.0g
Peptone 0.25g
Yeast extract 0.25g
Mineral salt solution 20.0mL
Glucose solution 10.0mL
Vitamin solution 5.0mL

pH 7.5 ± 0.2 at 25°C

Mineral Salt Solution:

Composition per liter:

$MgSO_4.7H_2O$ 29.7g
Nitrilotriacetic acid 10.0g
$CaCl_2·2H_2O$ 3.34g
$FeSO_4·7H_2O$ 99.0mg
$Na_2MoO_4·2H_2O$ 12.67mg
"Metals 44" 50.0mL

Preparation of Mineral Salt Solution: Add nitrilotriacetic acid to 500.0mL of distilled/deionized water. Dissolve by adjusting pH to 6.5 with KOH. Add remaining components. Add distilled/deionized water to 1.0L. Readjust pH to 7.2.

"Metals 44":

Composition per 100.0mL:

$ZnSO_4·7H_2O$ 1.1g
$FeSO_4·7H_2O$ 0.5g
EDTA 0.25g
$MnSO_4·7H_2O$ 0.154g
$CuSO_4·5H_2O$ 0.04g
$Co(NO_3)_2·6H_2O$ 0.025g
$Na_2B_4O_7·10H_2O$ 0.018g

Preparation of "Metals 44": Add a few drops of H_2SO_4 to distilled/deionized water to inhibit precipitate formation. Add components to acidified distilled/deionized water and bring volume to 100.0mL. Mix thoroughly.

Glucose Solution:

Composition per 100.0mL:

D-Glucose 2.5g

Preparation of Glucose Solution: Add glucose to distilled/deionized water and bring volume to 100.0mL. Mix thoroughly. Filter sterilize.

Vitamin Solution:

Composition per liter:

Pyridoxine·HCl 0.02g
p-Aminobenzoic acid 0.01g
Calcium D-pantothenate 0.01g
Nicotinamide 0.01g
Riboflavin 0.01g
Thiamine·HCl 0.01g
Biotin 4.0mg
Folic acid 4.0mg
Cyanocobalamin 0.2mg

Preparation of Vitamin Solution: Add components to distilled/deionized water and bring volume to 1.0L. Mix thoroughly. Filter sterilize.

Preparation of Medium: Add components, except glucose solution and vitamin solution, to distilled/deionized water and bring volume to 985.0mL. Mix thoroughly. Gently heat and bring to boiling. Autoclave for 15 min at 15 psi pressure–121°C. Cool to 45°–50°C. Aseptically add 10.0mL of sterile glucose solution and 5.0mL of sterile vitamin solution. Mix thoroughly. Adjust pH to 7.5 with sterile KOH if necessary. Pour into sterile Petri dishes or distribute into sterile tubes.

Use: For the cultivation and maintenance of *Blastobacter aggregatus, Blastobacter capsulatus, Blastobacter denitrificans,* and *Planctomyces limnophilus.*

PYGV Medium

Composition per liter:

Agar	15.0g
Peptone	0.25g
Yeast extract	0.25g
Mineral solution	20.0mL
Glucose solution	10.0mL
Vitamin solution	5.0mL

pH 7.5 ± 0.2 at 25°C

Mineral Solution:

Composition per liter:

$MgSO_4·7H_20$	29.7g
$NaMoO_4·2H_2O$	12.67g
Nitrilotriacetic acid	10.0g
$CaCl_2·2H_2O$	3.34g
$FeSO_4·7H_2O$	0.1g
"Metals 44" solution	50.0mL

Preparation of Mineral Solution: Add nitrilotriacetic acid to 500.0mL of distilled/deionized water. Dissolve by adjusting pH to 6.5 with KOH. Add remaining components. Readjust pH to 7.2 with H_2SO_4 or KOH. Add distilled/deionized water to 1.0L. Store at 5°C.

"Metals 44":

Composition per 100.0mL:

$ZnSO_4·7H_2O$	1.1g
$FeSO_4·7H_2O$	0.5g
EDTA	0.25g
$MnSO_4·7H_2O$	0.154g
$CuSO_4·5H_2O$	0.04g
$Co(NO_3)_2·6H_2O$	0.025g
$Na_2B_4O_7·10H_2O$	0.018g

Preparation of "Metals 44": Add components to distilled/deionized water and bring volume to 100.0mL. Mix thoroughly.

Glucose Solution:

Composition per 100.0mL:

D-Glucose	2.5g

Preparation of Glucose Solution: Add D-glucose to distilled/deionized water and bring volume to 100.0mL. Mix thoroughly. Filter sterilize.

Vitamin Solution:

Composition per liter:

Pyridoxin·HCl	0.02g
p-Aminobenzoic acid	0.01g
Ca-panthothenate	0.01g
Nicotinamide	0.01g
Riboflavin	0.01g
Thiamine·HCl	0.01g
Biotin	4.0mg
Folic acid	4.0mg
Vitamin B_{12}	0.2mg

Preparation of Vitamin Solution: Add components to distilled/deionized water and bring volume to 1.0L.

Preparation of Medium: Add components, except glucose solution and vitamin solution, to distilled/deionized water and bring volume to 985.0mL. Mix thoroughly. Gently heat and bring to boiling. Autoclave for 20 min at 15 psi pressure–121°C. Cool to 60°C. Aseptically add 10.0mL of sterile glucose solution and 5.0mL of sterile vitamin solution. Mix thoroughly. Pour into sterile Petri dishes or distribute into sterile tubes.

Use: For the enrichment of *Stella* species from polluted waters.

Pyridine Medium

Composition per 1001.0mL:

K_2HPO_4	0.61g
KH_2PO_4	0.39g
KCl	0.25g
Yeast extract	0.15g
$MgSO_4·7H_2O$	0.13g
Pyridine	1.0mL
Trace elements solution	1.0mL

Trace Elements Solution:

Composition per liter:

$FeSO_4·7H_2O$	40.0mg
$MnSO_4·4H_2O$	40.0mg
$ZnSO_4·7H_2O$	20.0mg
$CuSO_4·5H_2O$	5.0mg
$Na_2MoO_4·2H_2O$	5.0mg
$CoCl_2·6H_2O$	4.0mg
$CaCl_2·2H_2O$	0.4mg
NaCl	1.0g

Preparation of Trace Elements Solution: Add components to distilled/deionized water and bring volume to 1.0L. Mix thoroughly.

Preparation of Medium: Add components, except pyridine, to distilled/deionized water and bring volume to 1.0L. Mix thoroughly. Autoclave for 15 min at 15 psi pressure–121°C. Cool to room temperature. In a fume hood, aseptically add 1.0mL of pyridine. Mix thoroughly. Aseptically distribute into sterile tubes or flasks. Use polyurethane foam closures to eliminate odors caused by volatilization of pyridine.

Use: For the cultivation of *Micrococcus luteus.*

Pyrobaculum Medium

Composition per liter:

$Na_2S_2O_3·5H_2O$	2.0g
$(NH_4)_2SO_4$	1.3g
Peptone	0.5g
$Na_2S·9H_2O$	0.5g
KH_2PO_4	0.28g
$MgSO_4·7H_2O$	0.25g
Yeast extract	0.2g
$CaCl_2·2H_2O$	0.07g
$FeCl_3·6H_2O$	0.02g
Resazurin	1.0mg
$MnCl_2·4H_2O$	1.8mg
$Na_2B_4·10H_2O$	4.5mg
$ZnSO_4·7H_2O$	0.22mg
$CuCl_2·2H_2O$	0.05mg
$Na_2MoO_4·2H_2O$	0.03mg
$VOSO_4·2H_2O$	0.03mg
$CoSO_4$	0.01mg

Preparation of Medium: Add components, except peptone, yeast extract, and $Na_2S·9H_2O$, to distilled/deionized water and bring volume to 1.0L. Mix thoroughly. Bring pH to 6.0 using 8*N* NaOH. Sparge with 100% N_2 for 30 min. Add peptone, yeast extract, and $Na_2S·9H_2O$. Bring pH back to 6.0 using 10*N* H_2SO_4. Anaerobically distribute into sterile tubes or flasks under 100% N_2. Do not autoclave medium.

If not used immediately, heat the medium to 90°C for 60 min on each of 3 consecutive days.

Use: For the cultivation and maintenance of *Pyrobaculum islandicum*.

Pyrococcus endeavori Medium ES4

Composition per 3.0L:

Solution A	1.0L
Solution B	1.0L
Solution C	1.0L

Solution A:

Composition per liter:

NaCl	47.8g
Na_2SO_4	8.0g
KCl	1.4g
$NaHCO_3$	0.4g
KBr	0.2g
H_3BO_3	0.06g

Preparation of Solution A: Add components to distilled/deionized water and bring volume to 1.0L. Mix thoroughly. Autoclave for 15 min at 15 psi pressure–121°C.

Solution B:

Composition per liter:

$MgCl_2 \cdot 6H_2O$	21.6g
$CaCl_2 \cdot 2H_2O$	3.0g
$SrCl_2 \cdot 6H_2O$	0.05g

Preparation of Solution B: Add components to distilled/deionized water and bring volume to 1.0L. Mix thoroughly. Autoclave for 15 min at 15 psi pressure–121°C.

Solution C:

Composition per liter:

Sodium acetate	50.0g
NH_4Cl	12.5g
K_2HPO_4	7.0g

Preparation of Solution C: Add components to distilled/deionized water and bring volume to 1.0L. Mix thoroughly. Autoclave for 15 min at 15 psi pressure–121°C.

Preparation of Medium: Aseptically combine 1.0L of sterile solution A with 1.0L of sterile solution B and 1.0L of sterile solution C. Mix thoroughly. Aseptically distribute into sterile tubes or flasks.

Use: For the cultivation of *Pyrococcus endeavori*.

Pyrococcus furiosus Medium

Composition per liter:

NaCl	13.8g
Pancreatic digest of casein	5.0g
Yeast extract	5.0g
Maltose	5.0g
$MgSO_4$	3.5g
$MgCl_2$	2.75g
KH_2PO_4	0.5g
$CaCl_2$	0.75g
KCl	0.325g
NaBr	50.0mg
KI	50.0mg
H_3BO_3	15.0mg
$SrCl_2$	7.5mg
Citric acid	5.0mg
Resazurin	2.5mg
Mineral solution	10.0mL

pH 6.8 ± 0.2 at 25°C

Mineral Solution:

Composition per liter:

Nitriloacetic acid	1.0g
$MnSO_4$	0.5g
$FeCl_3 \cdot 6H_2O$	1.1g
$Na_2WO_4 \cdot 2H_2O$	0.3g
EDTA	0.292g
$NiCl_2 \cdot 6H_2O$	0.2g
$CoSO_4 \cdot 7H_2O$	0.1g
$ZnSO_4 \cdot 7H_2O$	0.1g
$CuSO_4 \cdot 5H_2O$	0.01g
$Na_2MoO_4 \cdot 2H_2O$	0.01g

Preparation of Mineral Solution: Add nitrilotriacetic acid to 500.0mL of distilled/deionized water. Adjust pH to 6.5 with KOH. Add remaining components. Mix thoroughly. Add distilled/deionized water to 1.0L. Adjust pH to 6.8.

Preparation of Medium: Add components to distilled/deionized water and bring volume to 1.0L. Mix thoroughly. Adjust pH to 6.8. Distribute into tubes or flasks. Autoclave for 15 min at 15 psi pressure–121°C.

Use: For the cultivation of high cell concentrations of *Pyrococcus furiosus*.

Pyrococcus Medium

Composition per liter:

Sulfur	30.0g
NaCl	13.85g
Peptone	5.0g
$MgSO_4 \cdot 7H_2O$	3.5g
$MgCl_2 \cdot 6H_2O$	2.75g
Yeast extract	1.0g
$CaCl_2$	0.75g
KH_2PO_4	0.5g
KCl	0.325g
NaBr	0.05g
H_3BO_3	15.0mg
$SrCl_2 \cdot 6H_2O$	7.5mg
Citric acid	5.0mg
$(NH_4)_2Ni(SO_4)_2$	2.0mg
Resazurin	1.0mg
KI	0.05mg
Trace minerals solution	10.0mL
$Na_2S \cdot 9H_2O$ solution	10.0mL

Trace Minerals Solution:

Compostion per liter:

$MgSO_4 \cdot 7H_2O$	3.0g
Nitrilotracetic acid	1.5g
NaCl	1.0g
$MnSO_4 \cdot xH_2O$	0.5g
$CaCl_2 \cdot 2H_2O$	0.1g
$CoSO_4$ (or $CoCl_2$)	0.1g
$FeSO_4 \cdot 7H_2O$	0.1g
$ZnSO_4$	0.1g
$AlK(SO_4)_2$	0.01g
$CuSO_4 \cdot 5H_2O$	0.01g
H_3BO_3	0.01g
$Na_2MoSO_42H_2O$	0.01g

Preparation of Trace Minerals Solution: Add nitrilotriacetic acid to 500.0mL of distilled/deionized water. Adjust pH to 6.5 with KOH. Add remaining components. Add distilled/deionized water to 1.0L. Adjust pH to 7.0.

$Na_2S \cdot 9H_2O$ Solution:

Composition per 10.0mL:

$Na_2S \cdot 9H_2O$	0.3g

Preparation of $Na_2S·9H_2O$ Solution: Add $Na_2S·9H_2O$ to distilled/deionized water and bring volume to 10.0mL. Mix thoroughly. Sparge with 100% N_2. Autoclave for 15 min at 15 psi pressure–121°C.

Preparation of Medium: Add components, except $Na_2S·9H_2O$ solution, to distilled/deionized water and bring volume to 1.0L. Mix thoroughly. Adjust pH to 6.5 with H_2SO_4. Do not autoclave. Sterilize by steaming at 100°C for 30 min on 3 consecutive days. Before inoculation, add 10.0mL of sterile $Na_2S·9H_2O$ solution. Mix thoroughly.

Use: For the cultivation and maintenance of *Pyrococcus furiosus* and *Pyrococcus woesei*.

Pyrococcus/Staphylothermus Medium

Composition per 1010.0mL:

Sulfur, powdered	30.0g
NaCl	13.85g
Peptone	5.0g
$MgSO_4·7H_2O$	3.5g
$MgCl_2·6H_2O$	2.75g
$NiCl_2·6H_2O$	2.0g
Yeast extract	1.0g
$CaCl_2·2H_2O$	0.75g
KH_2PO_4	0.5g
KCl	0.325g
NaBr	0.05g
H_3BO_3	0.015g
$(NH_4)_2SO_4$	10.0mg
$SrCl_2·6H_2O$	7.5mg
Citric acid	5.0mg
Resazurin	1.0mg
KI	0.05mg
Trace elements solution	10.0mL
$Na_2S·9H_2O$ solution	10.0mL

pH 6.5 ± 0.2 at 25°C

Trace Elements Solution:
Composition per liter:

$MgSO_4·7H_2O$	3.0g
Nitrilotriacetic acid	1.5g
NaCl	1.0g
$MnSO_4·2H_2O$	0.5g
$CoSO_4·7H_2O$	0.18g
$ZnSO_4·7H_2O$	0.18g
$CaCl_2·2H_2O$	0.1g
$FeSO_4·7H_2O$	0.1g
$NiCl_2·6H_2O$	0.025g
$KAl(SO_4)_2·12H_2O$	0.02g
$CuSO_4·5H_2O$	0.01g
H_3BO_3	0.01g
$Na_2MoO_4·2H_2O$	0.01g
$Na_2SeO_3·5H_2O$	0.3mg

Preparation of Trace Elements Solution: Add nitrilotriacetic acid to 500.0mL of distilled/deionized water. Adjust pH to 6.5 with KOH. Add remaining components. Adjust pH to 7.0. Add distilled/deionized water to 1.0L.

$Na_2S·9H_2O$ Solution:
Composition per 10.0mL:

$Na_2S·9H_2O$	0.5g

Preparation of $Na_2S·9H_2O$ Solution: Add $Na_2S·9H_2O$ to distilled/deionized water and bring volume to 10.0mL. Mix thoroughly. Sparge with 100% N_2. Autoclave for 15 min at 15 psi pressure–121°C.

Preparation of Medium: Prepare and dispense medium under 100% N_2. Add components, except $Na_2S·9H_2O$ solution, to distilled/deionized water and bring volume to 1.0L. Mix thoroughly. Gently heat and bring to boiling. Continue boiling for 5 min. Cool to room temperature while sparging with 100% N_2. Bring pH to 6.5 using 10*N* H_2SO_4. Anaerobically distribute into tubes or flasks. Autoclave for 15 min at 15 psi pressure–121°C. Immediately prior to inoculation, add 0.1mL of sterile $Na_2S·9H_2O$ solution to each 10.0mL of medium. Check that final pH is 6.5.

Use: For the cultivation and maintenance of *Pyrococcus furiosus, Pyrococcus woesei,* and *Staphylothermus marinus.*

Pyrodictium abyssi Medium

Composition per liter:

Sulfur, powdered	30.0g
NaCl	13.85g
$MgSO_4·7H_2O$	3.5g
$MgCl_2·6H_2O$	2.75g
$CaCl_2·2H_2O$	0.75g
$Na_2S·9H_2O$	0.5g
KH_2PO_4	0.5g
Yeast extract	0.5g
KCl	0.325g
NaBr	0.05g
H_3BO_3	0.015g
$(NH_4)_2SO_4$	10.0mg
$SrCl_2·6H_2O$	7.5mg
$NiCl_2·6H_2O$	2.0mg
Resazurin	1.0mg
$Na_2WO_4·2H_2O$	0.1mg
KI	0.05mg
Trace elements solution	10.0mL

pH 5.5–6.0 at 25°C

Trace Elements Solution:
Composition per liter:

$MgSO_4·7H_2O$	3.0g
Nitrilotriacetic acid	1.5g
NaCl	1.0g
$MnSO_4·2H_2O$	0.5g
$CoSO_4·7H_2O$	0.18g
$ZnSO_4·7H_2O$	0.18g
$CaCl_2·2H_2O$	0.1g
$FeSO_4·7H_2O$	0.1g
$NiCl_2·6H_2O$	0.025g
$KAl(SO_4)_2·12H_2O$	0.02g
$CuSO_4·5H_2O$	0.01g
H_3BO_3	0.01g
$Na_2MoO_4·2H_2O$	0.01g
$Na_2SeO_3·5H_2O$	0.3mg

Preparation of Trace Elements Solution: Add nitrilotriacetic acid to 500.0mL of distilled/deionized water. Adjust pH to 6.5 with KOH. Add remaining components. Adjust pH to 7.0. Add distilled/deionized water to 1.0L.

Preparation of Medium: Prepare and dispense medium under 80% H_2 + 20% CO_2. Add components, except $Na_2S·9H_2O$, to distilled/deionized water and bring volume to 1.0L. Mix thoroughly. Gently heat and bring to boiling. Continue boiling for 5 min. Cool to room temperature while sparging with 80% H_2 + 20% CO_2. Add $Na_2S·9H_2O$. Mix thoroughly. Bring pH to 5.5 using 10*N* H_2SO_4. Anaerobically distribute into tubes or flasks making sure to evenly distribute sulfur. Do not autoclave. For immediate use, heat the medium in a boiling water bath for 60 min prior to inoculation. For storage of medium, heat medium in a boiling water bath for 60 min on 3 consecutive days. Store at room temperature.

Use: For the cultivation and maintenance of *Pyrodictium abyssi.*

Pyrodictium Medium

Composition per liter:

Sulfur, powdered	30.0g
NaCl	13.85g
$MgSO_4 \cdot 7H_2O$	3.5g
$MgCl_2 \cdot 6H_2O$	2.75g
Yeast extract	2.0g
$CaCl_2 \cdot 2H_2O$	0.75g
$Na_2S \cdot 9H_2O$	0.5g
KH_2PO_4	0.5g
KCl	0.325g
NaBr	0.05g
H_3BO_3	0.015g
$(NH_4)_2SO_4$	10.0mg
$SrCl_2 \cdot 6H_2O$	7.5mg
Citric acid	5.0mg
$NiCl_2 \cdot 6H_2O$	2.0mg
Resazurin	1.0mg
KI	0.05mg
Trace elements solution	10.0mL

pH 5.5 ± 0.2 at 25°C

Trace Elements Solution:

Composition per liter:

$MgSO_4 \cdot 7H_2O$	3.0g
Nitrilotriacetic acid	1.5g
NaCl	1.0g
$MnSO_4 \cdot 2H_2O$	0.5g
$CoSO_4 \cdot 7H_2O$	0.18g
$ZnSO_4 \cdot 7H_2O$	0.18g
$CaCl_2 \cdot 2H_2O$	0.1g
$FeSO_4 \cdot 7H_2O$	0.1g
$NiCl_2 \cdot 6H_2O$	0.025g
$KAl(SO_4)_2 \cdot 12H_2O$	0.02g
$CuSO_4 \cdot 5H_2O$	0.01g
H_3BO_3	0.01g
$Na_2MoO_4 \cdot 2H_2O$	0.01g
$Na_2SeO_3 \cdot 5H_2O$	0.3mg

Preparation of Trace Elements Solution: Add nitrilotriacetic acid to 500.0mL of distilled/deionized water. Adjust pH to 6.5 with KOH. Add remaining components. Adjust pH to 7.0. Add distilled/deionized water to 1.0L.

Preparation of Medium: Prepare and dispense medium under 80% H_2 + 20% CO_2. Add components, except $Na_2S \cdot 9H_2O$, to distilled/deionized water and bring volume to 1.0L. Mix thoroughly. Gently heat and bring to boiling. Continue boiling for 5 min. Cool to room temperature while sparging with 80% H_2 + 20% CO_2. Add $Na_2S \cdot 9H_2O$. Mix thoroughly. Bring pH to 5.5 using 10*N* H_2SO_4. Anaerobically distribute into tubes or flasks, making sure to evenly distribute sulfur. Do not autoclave. For immediate use, heat the medium in a boiling water bath for 60 min prior to inoculation. For storage of medium, heat medium in a boiling water bath for 60 min on 3 consecutive days. Store at room temperature.

Use: For the cultivation and maintenance of *Pyrodictium brockii, Pyrodictium occultum*, and a consortium consisting of *Lactobacillus brevis, Streptococcus lactis*, and *Saccharomyces cerevisiae.*

Pyrolobus fumarii Medium (DSMZ Medium 792)

Composition per liter:

NaCl	13.850g
$MgSO_4 \cdot 7H_2O$	3.5g
$MgCl_2 \cdot 6H_2O$	2.75g
KNO_3	1.0g
KH_2PO_4	0.5g
$CaCl_2 \cdot 2H_2O$	0.375g
KCl	0.325g
NaBr	0.05g
H_3BO_3	0.015g
$SrCl_2 \cdot 6H_2O$	7.5mg
Resazurin	1.0mg
Trace elements solution	10.0mL
$Na_2S \cdot 9H_2O$ solution	10.0mL
KI solution	0.05mL

pH 5.5 ± 0.2 at 25°C

Trace Elements Solution:

$MgSO_4 \cdot 7H_2O$	3.0g
NaCl	1.0g
$MnSO_4 \cdot 2H_2O$	0.5g
$ZnSO_4 \cdot 7H_2O$	0.18g
$CoSO_4 \cdot 7H_2O$	0.18g
$FeSO_4 \cdot 7H_2O$	0.1g
$CaCl_2 \cdot 2H_2O$	0.1g
$NiCl_2 \cdot 6H_2O$	0.025g
$KAl(SO_4)_2 \cdot 12H_2O$	0.02g
$CuSO_4 \cdot 5H_2O$	0.01g
H_3BO_3	0.01g
$Na_2MoO_4 \cdot 4H_2O$	0.01g
$Na_2WO_4 \cdot 2H_2O$	0.01g
$Na_2SeO_3 \cdot 5H_2O$	0.30mg

Preparation of Trace Elements Solution: Add components to distilled/deionized water and bring volume to 1.0L. Mix thoroughly. Adjust pH to 1.0 with H_2SO_4.

$Na_2S \cdot 9H_2O$ Solution:

Composition per 10mL:

$Na_2S \cdot 9H_2O$	0.3g

Preparation of $Na_2S \cdot 9H_2O$ Solution: Add $Na_2S \cdot 9H_2O$ to distilled/deionized water and bring volume to 10.0mL. Mix thoroughly. Autoclave under 100% N_2 for 15 min at 15 psi pressure–121°C. Cool to room temperature.

KI Solution:

Composition per 10mL:

KI	5.0mg

Preparation of KI Solution: Add KI to distilled/deionized water and bring volume to 10.0mL. Mix thoroughly. Autoclave under 100% N_2 for 15 min at 15 psi pressure–121°C. Cool to room temperature.

Preparation of Medium: Add components, except $Na_2S \cdot 9H_2O$ solution, to distilled/deionized water and bring volume to 990.0mL. Mix thoroughly. Gently heat and bring to boiling. Cool to room temperature while sparging with 80% H_2 + 20% CO_2. Distribute into serum bottles under 80% H_2 + 20% CO_2, e.g., 20mL into 120mL serum bottles. Autoclave for 15 min at 15 psi pressure–121°C. Cool to 25°C. Aseptically inject $Na_2S \cdot 9H_2O$ solution, 0.2mL per 20mL medium. Mix thoroughly. Adjust pH to 5.5. After inoculation pressurize vials to 2 bar overpressure with 0% H_2 + 20% CO_2 gas mixture.

Use: For the cultivation of *Pyrolobus fumarii.*

Pyrrolidone Agar

Composition per liter:

Noble agar	21.0g
K_2HPO_4	5.65g
KH_2PO_4	2.95g
$MgSO_4 \cdot 7H_2O$	1.0g
Pyrrolidone carboxylic acid solution	30.0mL

NaOH solution 30.0mL
Trace metals 6.3mL

Pyrrolidone Carboxylic Acid Solution:

Composition per 300.0mL:

Pyrrolidone carboxylic acid 50.0g

Preparation of Pyrrolidone Carboxylic Acid Solution: Add pyrrolidone carboxylic acid to distilled/deionized water and bring volume to 300.0mL. Mix thoroughly. Filter sterilize.

NaOH Solution:

Composition per 100.0mL:

NaOH 5.0g

Preparation of Resazurin Solution: Add NaOH to distilled/deionized water and bring volume to 100.0mL. Mix thoroughly. Filter sterilize.

Trace Metals:

Composition per 100.0mL:

$FeSO_4 \cdot 7H_2O$	0.18g
$MnCl_2 \cdot 2H_2O$	0.13g
$CuSO_4 \cdot 5H_2O$	0.1g
$ZnSO_4 \cdot 7H_2O$	0.02g

Preparation of Trace Metals: Add a few drops of H_2SO_4 to distilled/deionized water to inhibit precipitate formation. Add components to acidified distilled/deionized water and bring volume to 100.0mL. Mix thoroughly.

Preparation of Medium: Add components, except pyrrolidone carboxylic acid solution and NaOH solution, to distilled/deionized water and bring volume to 940.0mL. Mix thoroughly. Gently heat and bring to boiling. Autoclave for 15 min at 15 psi pressure–121°C. Cool to 45°–50°C. Aseptically add 30.0mL of sterile pyrrolidone carboxylic acid solution and 30.0mL of sterile NaOH solution. Mix thoroughly. Pour into sterile Petri dishes or distribute into sterile tubes.

Use: For the cultivation and maintenance of *Pseudomonas fluorescens.*

Quinoline Medium

Composition per 1000.2mL:

K_2HPO_4	0.61g
KH_2PO_4	0.39g
KCl	0.25g
Yeast extract	0.1g
Wolfe's mineral solution	10.0mL
Quinoline	0.2mL

Wolfe's Mineral Solution:

Composition per liter:

$MgSO_4 \cdot 7H_2O$	3.0g
Nitrilotriacetic acid	1.5g
NaCl	1.0g
$MnSO_4 \cdot 2H_2O$	0.5g
$CoCl_2 \cdot 6H_2O$	0.1g
$ZnSO_4 \cdot 7H_2O$	0.1g
$CaCl_2 \cdot 2H_2O$	0.1g
$FeSO_4 \cdot 7H_2O$	0.1g
$NiCl_2 \cdot 6H_2O$	0.025g
$KAl(SO_4)_2 \cdot 12H_2O$	0.02g
$CuSO_4 \cdot 5H_2O$	0.01g
H_3BO_3	0.01g
$Na_2MoO_4 \cdot 2H_2O$	0.01g
$Na_2SeO_3 \cdot 5H_2O$	0.3mg

Preparation of Wolfe's Mineral Solution: Add nitrilotriacetic acid to 500.0mL of distilled/deionized water. Adjust pH to 6.5 with KOH. Add remaining components one at a time. Add distilled/deionized water to 1.0L. Adjust pH to 6.8.

Preparation of Medium: Add components, except quinoline, to distilled/deionized water and bring volume to 1.0L. Mix thoroughly. Autoclave for 15 min at 15 psi pressure–121°C. Cool to room temperature. In a fume hood, aseptically add 0.2mL of quinoline. Mix thoroughly. Aseptically distribute into sterile tubes or flasks. Use polyurethane foam closures to eliminate odors caused by volatilization of quinoline.

Use: For the cultivation of *Rhodococcus* species.

Quinolinic Acid Medium

Composition per liter:

Quinolinic acid	1.5g
K_2HPO_4	1.1g
NH_4NO_3	1.0g
KH_2PO_4	0.5g
$MgSO_4 \cdot 7H_2O$	0.25g

Preparation: Add quinolinic acid to distilled/deionized water and bring volume to 900.0mL. Mix thoroughly. Bring pH to 7.0 with NaOH. Add other components. Bring volume to 1.0L. Mix thoroughly. Distribute into tubes or flasks. Autoclave for 15 min at 15 psi pressure–121°C.

Use: For the cultivation of microorganisms that can utilize quinolinic acid as sole carbon source.

R Agar with 3% NaCl

Composition per liter:

NaCl	30.0g
Agar	20.0g
Peptone	10.0g
Casamino acids	5.0g
Malt extract	5.0g
Yeast extract	5.0g
Beef extract	2.0g
Glycerol	2.0g
$MgSO_4 \cdot 7H_2O$	1.0g
Tween 80	50.0mg

pH 7.2 ± 0.2 at 25°C

Preparation of Medium: Add components to distilled/deionized water and bring volume to 1.0L. Mix thoroughly. Gently heat and bring to boiling. Distribute into tubes or flasks. Autoclave for 15 min at 15 psi pressure–121°C. Pour into sterile Petri dishes or leave in tubes.

Use: For the cultivation and maintenance of *Rhodococcus marinonascens.*

R Agar with 5% NaCl

Composition per liter:

NaCl	50.0g
Agar	20.0g
Peptone	10.0g
Casamino acids	5.0g
Malt extract	5.0g
Yeast extract	5.0g
Beef extract	2.0g
Glycerol	2.0g
$MgSO_4 \cdot 7H_2O$	1.0g
Tween 80	50.0mg

pH 7.2 ± 0.2 at 25°C

Preparation of Medium: Add components to distilled/deionized water and bring volume to 1.0L. Mix thoroughly. Gently heat and bring to boiling. Distribute into tubes or

flasks. Autoclave for 15 min at 15 psi pressure–121°C. Pour into sterile Petri dishes or leave in tubes.

Use: For the cultivation and maintenance of *Marinococcus albus, Marinococcus halophilus,* and other *Marinococcus* species.

R8 Medium (DSMZ Medium 912)

Composition per liter:

$NaHCO_3$	2.52g
MOPS	2.1g
NaCl	1.0g
Glucose	0.9g
$MgCl_2 \cdot 6H_2O$	0.5g
$Na_2S \cdot 9H_2O$	0.6g
Cysteine-HCl	0.4g
NH_4Cl	0.3g
KCl	0.3g
K_2HPO_4	0.25g
KH_2PO_4	0.2g
Yeast extract	0.19g
Peptone	0.19g
Na-Pantothenate	0.1g
$CaCl_2 \cdot 2H_2O$	0.015g
Resazurin	0.5mg
Rumen fluid	50.0mL
Na-Pantothenate solution	10.0mL
Trace elements solution SL-10	1.0mL

pH 7.2 ± 0.2 at 25°C

Trace Elements Solution SL-10:
Composition per liter:

$FeCl_2 \cdot 4H_2O$	1.5g
$CoCl_2 \cdot 6H_2O$	190.0mg
$MnCl_2 \cdot 4H_2O$	100.0mg
$ZnCl_2$	70.0mg
$Na_2MoO_4 \cdot 2H_2O$	36.0mg
$NiCl_2 \cdot 6H_2O$	24.0mg
H_3BO_3	6.0mg
$CuCl_2 \cdot 2H_2O$	2.0mg
HCl (25% solution)	10.0mL

Preparation of Trace Elements Solution SL-10: Add $FeCl_2 \cdot 4H_2O$ to 10.0mL of HCl solution. Mix thoroughly. Add distilled/deionized water and bring volume to 1.0L. Add remaining components. Mix thoroughly. Sparge with 100% N_2.

Rumen Fluid:
Composition per 50.0mL:

Rumen fluid, clarified	50.0mL

Preparation of Rumen Fluid: Sparge clarified rumen fluid with 100% N_2. Autoclave for 15 min at 15 psi pressure–121°C.

Na-Pantothenate Solution:
Composition per 10.0mL:

Na-Pantothenate	0.1g

Preparation of Na-Pantothenate Solution: Add Na-pantothenate to distilled/deionized water and bring volume to 10.0mL. Mix thoroughly. Sparge with 100% N_2. Filter sterilize.

Preparation of Medium: Prepare and dispense medium under 80% N_2 + 20% CO_2 gas atmosphere. Add components, except $NaHCO_3$, pantothenate solution, $Na_2S \cdot 9H_2O$, cysteine-HCl, and rumen fluid, to distilled/deionized water and bring volume to 940.0mL. Mix thoroughly. Gently heat and bring to boiling. Boil for 3 min. Cool to room temperature while sparging with 80% N_2 + 20% CO_2. Add solid bicarbonate, sodium sulfide and cysteine-HCl. Adjust pH to 7.2. Distribute under 80% N_2 + 20% CO_2 atmosphere into anaerobe tubes or bottles. Autoclave for 15 min at 15 psi pressure–121°C. Aseptically and anaerobically add per liter of medium, 50.0mL rumen fluid and 10.0mL Na-pantothenate solution. Mix thoroughly. The final pH of the medium should be 7.2.

Use: For the cultivation of *Spirochaeta* spp.

R2A Agar

Composition per liter:

Agar	15.0g
Yeast extract	0.5g
Acid hydrolysate of casein	0.5g
Glucose	0.5g
Soluble starch	0.5g
K_2HPO_4	0.3g
Sodium pyruvate	0.3g
Pancreatic digest of casein	0.25g
Peptic digest of animal tissue	0.25g
$MgSO_4$, anhydrous	0.024g

pH 7.2 ± 0.2 at 25°C

Source: This medium is available as a premixed powder from BD Diagnostic Systems.

Preparation of Medium: Add components to distilled/deionized water and bring volume to 1.0L. Mix thoroughly. Gently heat with mixing and bring to boiling. Distribute into tubes or flasks. Autoclave for 15 min at 15 psi pressure–121°C. Do not overheat. Pour into sterile Petri dishes or leave in tubes.

Use: For use in standard methods for pour plate, spread plate, and membrane filter analysis to enumerate heterotrophic bacteria from potable waters.

R2A Agar, Modified

Composition per liter:

Agar	15.0g
NH_4Cl	0.8g
KNO_3	0.505g
Casamino acids	0.5g
Glucose	0.5g
Peptone	0.5g
Sodium pyruvate	0.5g
Starch, soluble	0.5g
Yeast extract	0.5g
K_2HPO_4	0.4g
KH_2PO_4	0.25g
$MgCl_2 \cdot 6H_2O$	20.0mg
$CaCl_2 \cdot 2H_2O$	15.0mg
$FeSO_4 \cdot 7H_2O$	7.0mg
$MnCl_2 \cdot 4H_2O$	5.0mg
Na_2SO_4	5.0mg
$CoCl_2 \cdot 6H_2O$	0.5mg
H_3BO_3	0.5mg
$NiSO_4 \cdot 6H_2O$	0.5mg
$ZnCl_2$	0.5mg
$CuCl_2 \cdot 2H_2O$	0.3mg
$Na_2MoO_4 \cdot 2H_2O$	10.0μg

pH 7.0 ± 0.2 at 25°C

Preparation of Medium: Add components to distilled/deionized water and bring volume to 1.0L. Mix thoroughly. Adjust pH to 7.0. Gently heat and bring to boiling. Distribute into tubes or flasks. Autoclave for 15 min at 15 psi pressure–121°C. Pour into sterile Petri dishes or leave in tubes.

Use: For the cultivation of *Azoarcus tolulyticus.*

R3A Agar

Composition per liter:

Agar	15.0g
Yeast extract	1.0g
Acid hydrolysate of casein	1.0g
Glucose	1.0g
Soluble starch	1.0g
K_2HPO_4	0.6g
Sodium pyruvate	0.6g
Pancreatic digest of casein	0.5g
Peptic digest of animal tissue	0.5g
$MgSO_4$, anhydrous	0.048g

pH 7.2 ± 0.2 at 25°C

Source: This medium is available as a premixed powder from BD Diagnostic Systems.

Preparation of Medium: Add components to distilled/deionized water and bring volume to 1.0L. Mix thoroughly. Gently heat with mixing and bring to boiling. Distribute into tubes or flasks. Autoclave for 15 min at 15 psi pressure–121°C. Do not overheat. Pour into sterile Petri dishes or leave in tubes.

Use: For the cultivation and maintenance of heterotrophic bacteria from potable waters.

Rabbit Dung Agar

Composition per liter:

Rabbit dung	20.0g
Agar	15.0g

pH 7.2 ± 0.2 at 25°C

Preparation of Medium: Add rabbit dung to 1.0L of distilled/deionized water. Gently heat and bring to boiling. Continue boiling for 20 min. Filter through Whatman #1 filter paper. Bring volume of filtrate to 1.0L with distilled/deionized water. Add agar. Adjust pH to 7.2. Distribute into tubes or flasks. Autoclave for 15 min at 15 psi pressure–121°C. Pour into sterile Petri dishes or leave in tubes.

Use: For the cultivation of myxobacteria.

R8AH Medium (DSMZ Medium 651)

Composition per liter:

Malic acid	2.5g
$(NH_4)_2SO_4$	1.25g
Yeast extract	1.0g
K_2HPO_4	0.9g
KH_2PO_4	0.6g
$MgSO_4 \cdot 7H_2O$	0.2g
$CaCl_2 \cdot 2H_2O$	0.07g
EDTA	0.02g
Ferric citrate	0.01g
Vitamin solution	7.5mL
Trace elements solution	1.0mL

pH 6.9 ± 0.2 at 25°C

Trace Elements Solution:

Composition per 100.0mL:

Ferric citrate	0.3g
EDTA	0.05g
$CaCl_2 \cdot 2H_2O$	0.02g
$MnSO_4 \cdot H_2O$	2.0mg
$(NH_4)_6Mo_7O_{24} \cdot 4H_2O$	2.0mg
H_3BO_3	1.0mg
$CuSO_4 \cdot 5H_2O$	1.0mg
$ZnSO_4$	1.0mg

Preparation of Trace Elements Solution: Add components to distilled/deionized water and bring volume to 100.0mL. Mix thoroughly.

Vitamin Solution:

Composition per liter:

Thiamine·HCl	0.4g
Nicotinic acid	0.2g
Nicotinamide	0.2g
Biotin	8.0mg

Preparation of Vitamin Solution: Add components to distilled/deionized water and bring volume to 1.0L. Mix thoroughly.

Preparation of Medium: Add malic acid to approximately 500.0mL of distilled/deionized water. Adjust pH to 6.9 with NaOH. Add remaining components. Bring volume to 1.0L with distilled/deionized water. Mix thoroughly. Adjust pH to 6.9. Distribute into tubes or flasks. Autoclave for 15 min at 15 psi pressure–121°C.

Use: For the cultivation and maintenance of *Rhodopseudomonas palustris, Rhodobacter sphaeroides, Rhodocyclus tenuis, Rhodopseudomonas rutila, Rhodospirillum photometricum,* and *Rhodospirillum rubrum.*

Raper *Achyla* Medium No. 1

Composition per liter:

Agar	20.0g
Lentil (hot water extract)	10.0g
Starch, soluble	3.0g
Peptone	1.0g
$CaCl_2$	1.0μg
$FeCl_3$	1.0μg
KH_2PO_4	1.0μg
$MgSO_4$	1.0μg
$ZnSO_4$	1.0μg

Preparation of Medium: Add components to distilled/deionized water and bring volume to 1.0L. Mix thoroughly. Gently heat and bring to boiling. Distribute into tubes or flasks. Autoclave for 15 min at 15 psi pressure–121°C. Pour into sterile Petri dishes or leave in tubes.

Use: For the cultivation of *Achyla* species.

Raper *Achyla* Medium No. 2

Composition per liter:

Agar	20.0g
Starch, soluble	3.0g
Inositol	1.0g
Peptone	1.0g

Preparation of Medium: Add components to distilled/deionized water and bring volume to 1.0L. Mix thoroughly. Gently heat and bring to boiling. Distribute into tubes or flasks. Autoclave for 15 min at 15 psi pressure–121°C. Pour into sterile Petri dishes or leave in tubes.

Use: For the cultivation of *Achyla* species.

Rappaport-Vassiliadis Enrichment Broth (RV Enrichment Broth)

Composition per 1110.0mL:

NaCl	8.0g
Papaic digest of soybean meal	5.0g
KH_2PO_4	1.6g
Magnesium chloride solution	100.0mL
Malachite Green solution	10.0mL

pH 5.2 ± 0.2 at 25°C

Source: This medium is available as a premixed powder from Oxoid Unipath.

Magnesium Chloride Solution:
Composition per 100.0mL:

$MgCl_2 \cdot 6H_2O$	40.0g

Preparation of Magnesium Chloride Solution: Add $MgCl_2 \cdot 6H_2O$ to distilled/deionized water and bring volume to 100.0mL. Mix thoroughly. Autoclave for 15 min at 15 psi pressure–121°C. Cool to 45°–50°C.

Malachite Green Solution:
Composition per 10.0mL:

Malachite Green oxalate	0.04g

Preparation of Malachite Green Solution: Add Malachite Green to distilled/deionized water and bring volume to 10.0mL. Mix thoroughly. Autoclave for 15 min at 15 psi pressure–121°C. Cool to 45°–50°C.

Preparation of Medium: Add components to distilled/deionized water and bring volume to 1.0L. Mix thoroughly. Distribute into tubes in 10.0mL volumes. Autoclave for 15 min at 10 psi pressure–115°C.

Use: For the isolation and cultivation of *Salmonella* species from environmental specimens.

Rappaport-Vassiliadis R10 Broth

Composition per liter:

$MgCl_2$, anhydrous	13.4g
NaCl	7.2g
Papaic digest of soybean meal	4.54g
KH_2PO_4	1.45g
Malachite Green oxalate	0.036g

pH 5.1 ± 0.2 at 25°C

Preparation of Medium: Add components to distilled/deionized water and bring volume to 1.0L. Mix thoroughly. Distribute into screw-capped tubes in 10.0mL volumes. Autoclave for 15 min at 10 psi pressure–116°C.

Use: For the isolation and cultivation of *Salmonella* species from environmental specimens.

Rappaport-Vassiliadis Soya Peptone Broth (RVS Broth)

Composition per liter:

$MgCl_2$, anhydrous	13.58g
NaCl	7.2g
Papaic digest of soybean meal	4.5g
KH_2PO_4	1.26g
K_2HPO_4	0.18g
Malachite Green	0.036g

pH 5.2 ± 0.2 at 25°C

Source: This medium is available as a premixed powder from Oxoid Unipath.

Preparation of Medium: Add components to distilled/deionized water and bring volume to 1.0L. Mix thoroughly. Distribute into screw-capped tubes in 10.0mL volumes. Autoclave for 15 min at 10 psi pressure–115°C.

Use: For the isolation and cultivation of *Salmonella* species from environmental specimens.

Raymond's Medium

Composition per liter:

Na_2HPO_4	3.0g
NaCl	3.0g
NH_4NO_3	2.0g
KH_2PO_4	2.0g
$MgSO_4$	0.2g
Na_2CO_3	0.1g
$MnSO_4$	0.02g
$CaCl_2$	0.01g
$FeSO_4$	0.01g
n-Hexadecane	10.0mL

pH 6.8–7.0 at 25°C

Preparation of Medium: Add components to distilled/deionized water and bring volume to 1.0L. Mix thoroughly. Distribute into tubes or flasks. Autoclave for 20 min at 15 psi pressure–121°C.

Use: For the cultivation of *Rhodococcus erythropolis, Rhodococcus luteus, Rhodococcus maris*, and other bacteria that can utilize *n*-hexadecane as a carbon source.

RF Medium

Composition per liter:

Yeast extract	0.05g
Peptone	0.05g
$(NH_4)_2SO_4$	0.05g
L-Cysteine·HCl·H_2O	0.05g
Salt solution	50.0mL
Rumen fluid, clarified	30.0mL
Resazurin (1% solution)	0.1mL

pH 7.4 ± 0.2 at 25°C

Salt Solution:
Composition per liter:

$NaHCO_3$	10.0g
NaCl	2.0g
K_2HPO_4	1.0g
KH_2PO_4	1.0g
$CaCl_2$, anhydrous	0.2g
$MgSO_4$	0.2g

Preparation of Salts Solution: Add $CaCl_2$ and $MgSO_4$ to 300.0mL of distilled/deionized water. Mix thoroughly until dissolved. Bring volume to 800.0mL with distilled/deionized water. Add remaining components while stirring. Bring volume to 1.0L. Mix thoroughly. Store at 4°C.

Preparation of Medium: Add components to distilled/deionized water and bring volume to 1.0L. Mix thoroughly. Adjust pH to 6.2–6.3 with 4*N* HCl. Gently heat and bring to boiling under 100% N_2. Anaerobically distribute into tubes in 7.0mL volumes. Cap with rubber stoppers. Place tubes in a press. Autoclave for 20 min at 15 psi pressure–121°C with fast exhaust. The pH of the medium should be 7.4 after autoclaving.

Use: For the cultivation and maintenance of *Treponema bryantii*.

RGCA Medium

Composition per 300.3mL:

Rumen fluid	120.0mL
Solution IV	65.0mL
Mineral solution I	45.0mL
Mineral solution II	45.0mL
Na_2CO_3 solution	20.0mL
L-Cysteine·HCl·H_2O solution	5.0mL
Solution III	0.3mL

pH 6.6 ± 0.2 at 25°C

Mineral Solution I:
Composition per 100.0mL:

K_2HPO_4	0.3g

Preparation of Mineral Solution I: Add K_2HPO_4 to distilled/deionized water and bring volume to 100.0mL. Mix thoroughly.

Mineral Solution II:
Composition per 100.0mL:

$(NH_4)_2SO_4$	0.6g
NaCl	0.6g
K_2HPO_4	0.3g
$MgSO_4$	0.06g
$CaCl_2$	0.06g

Preparation of Mineral Solution II: Add K_2HPO_4 to distilled/deionized water and bring volume to 100.0mL. Mix thoroughly.

Solution III:
Composition per 10.0mL:

Resazurin	0.01g

Preparation of Solution III: Add resazurin to 10.0mL of distilled/deionized water. Mix thoroughly.

Solution IV:
Composition per 65.0mL:

Agar	4.5g
Glucose	0.6g
Cellobiose	0.6g

Preparation of Solution IV: Add components to distilled/deionized water and bring volume to 65.0mL. Mix thoroughly.

L-Cysteine·HCl·H_2O Solution:
Composition per 100.0mL:

L-Cysteine·HCl·H_2O	3.0g

Preparation of L-Cysteine·HCl·H_2O Solution: Add L-cysteine·HCl·H_2O to distilled/deionized water and bring volume to 100.0mL. Mix thoroughly. Filter sterilize.

Na_2CO_3 Solution:
Composition per 100.0mL:

Na_2CO_3	6.0g

Preparation of Na_2CO_3 Solution: Add Na_2CO_3 to distilled/deionized water and bring volume to 100.0mL. Mix thoroughly. Filter sterilize.

Rumen Fluid:
Composition per 120.0mL:

Rumen fluid	120.0mL

Preparation of Rumen Fluid: Filter rumen contents, obtained from a cow on an alfalfa-hay concentrate ration, through two layers of cheesecloth to remove larger particles. Store under CO_2 in quart milk bottles in the refrigerator. Much of the particulate matter settles out. Use the supernatant fluid.

Preparation of Medium: Combine 45.0mL of mineral solution I, 45.0mL of mineral solution II, 0.3mL of solution III, and 65.0mL of solution IV in a 500.0mL flask. Gently heat and bring to boiling. Add 120.0mL of rumen fluid. Gently heat and bring to boiling under 100% CO_2. Cap with a rubber stopper and wire the stopper secure. Autoclave for 20 min at 15 psi pressure–121°C. Cool to 45°–50°C. Remove stopper and gas with 100% CO_2 to eliminate O_2. Aseptically add 5.0mL of sterile L-cysteine·HCl·H_2O solution and 20.0mL of sterile Na_2CO_3 solution. Mix thoroughly. Aseptically and anaerobically distribute into tubes under 100% CO_2 in 6.0mL volumes. Cap with rubber stoppers.

Use: For the cultivation and maintenance of *Ruminococcus albus, Ruminococcus flavifaciens,* and *Succinimonas amylolytica.*

Rhamnose Salts Medium

Composition per liter:

Rhamnose	10.0g
Yeast extract	3.0g
K_2HPO_4	2.9g
KH_2PO_4	2.1g
$NH_4 \cdot Cl$	2.0g
$MgSO_4 \cdot 7H_2O$	0.4g
NaCl	30.0mg
$CaCl_2$	3.0mg
$FeSO_4 \cdot 7H_2O$	1.0mg

pH 7.0 ± 0.2 at 25°C

Preparation of Medium: Add components to distilled/deionized water and bring volume to 1.0L. Mix thoroughly. Distribute into tubes or flasks. Autoclave for 15 min at 15 psi pressure–121°C.

Use: For the cultivation of *Rhodococcus chlorophenolicus.*

Rhizobium Agar (LMG 201)

Composition per liter:

Agar	20.0g
Mannitol	10.0g
Yeast extract	1.0g
Sodium glutamate	0.5g
KH_2PO_4	0.5g
$MgSO_4 \cdot 7H_2O$	0.1g
$CaCl_2 \cdot 2H_2O$	40.0mg
$FeCl_3$	4.0mg

pH 6.8 ± 0.2 at 25°C

Preparation of Medium: Add components to distilled/deionized water and bring volume to 1.0L. Mix thoroughly. Adjust pH to 6.8. Gently heat and bring to boiling. Distribute into tubes or flasks. Autoclave for 15 min at 15 psi pressure–121°C. Pour into sterile Petri dishes or leave in tubes.

Use: For the cultivation and maintenance of *Rhizobium fredii, Rhizobium galegae, Rhizobium leguminosarum, Rhizobium loti, Rhizobium meliloti,* and *Rhizobium tropici.*

Rhizobium BIII Defined Agar

Composition per liter:

Agar	13.0g
Mannitol	10.0g
Sodium glutamate	1.1g
K_2HPO_4	0.23g
$MgSO_4 \cdot 7H_2O$	0.1g
Trace elements stock	1.0mL
Vitamin stock	1.0mL

pH 7.0 ± 0.2 at 25°C

Trace Elements Stock:
Composition per liter:

Nitrilotriacetic acid	7.0g
$CaCl_2 \cdot 2H_2O$	6.62g
H_3BO_3	0.145g
$FeSO_4 \cdot 7H_2O$	0.125g
Na_2MoO_4	0.125g
$ZnSO_4 \cdot 7H_2O$	0.108g
$CoSO_4 \cdot 7H_2O$	0.07g
$CuSO_4 \cdot 5H_2O$	5.0mg
$MnCl_2 \cdot 4H_2O$	4.3mg

Preparation of Trace Elements Stock: Add each of the components to 500.0mL of distilled/deionized water in the following order: $CaCl_2 \cdot 2H_2O$, H_3BO_3, $FeSO_4 \cdot 7H_2O$, $CoSO_4 \cdot 7H_2O$, $CuSO_4 \cdot 5H_2O$, $MnCl_2 \cdot 4H_2O$, $ZnSO_4 \cdot 7H_2O$, and Na_2MoO_4. Adjust pH to 5.0. Add nitrilotriacetic acid. Bring volume to 1.0L with distilled/deionized water.

Vitamin Stock:
Composition per liter:

Inositol	0.12g
p-Aminobenzoic acid	0.02g
Biotin	0.02g
Calcium pantothenate	0.02g
Nicotinic acid	0.02g
Pyridoxine·HCl	0.02g
Riboflavin	0.02g
Thiamine·HCl	0.02g
Sodium phosphate buffer (50.0mM solution, pH 7.0)	1.0L

Preparation of Vitamin Stock: Combine components. Mix thoroughly. Filter sterilize. Store at 4°C in the dark.

Preparation of Medium: Add components, except vitamin stock, to distilled/deionized water and bring volume to 999.0mL. Mix thoroughly. Gently heat and bring to boiling. Autoclave for 15 min at 15 psi pressure–121°C. Cool to 45°–50°C. Aseptically add 1.0mL of sterile vitamin stock. Mix thoroughly. Pour into sterile Petri dishes or distribute into sterile tubes.

Use: For the isolation and cultivation of *Rhizobium* species from root nodules.

Rhizobium japonicum Agar

Composition per liter:

Agar	15.0g
Mannitol	10.0g
Yeast extract	1.0g
Soil extract	200.0mL

Soil Extract:
Composition per liter:

African Violet soil	77.0g
Na_2CO_3	0.2g

Preparation of Soil Extract: Add components to 1.0L of tap water. Autoclave for 15 min at 15 psi pressure–121°C. Filter through Whatman filter paper. Bring volume to 1.0L with tap water.

Preparation of Medium: Add components to distilled/deionized water and bring volume to 1.0L. Mix thoroughly. Gently heat and bring to boiling. Distribute into tubes or flasks. Autoclave for 15 min at 15 psi pressure–121°C. Pour into sterile Petri dishes or leave in tubes.

Use: For the cultivation and maintenance of *Bradyrhizobium japonicum.*

Rhizobium Medium 1

Composition per liter:

Agar	15.0g
Yeast extract	10.0g
K_2HPO_4	0.5g
$MgSO_4 \cdot 7H_2O$	0.2g
NaCl	0.2g
$FeCl_3 \cdot 6H_2O$	0.002g

pH 7.2 ± 0.2 at 25°C

Preparation of Medium: Add components, except agar, to distilled/deionized water and bring volume to 1.0L. Mix thoroughly. Adjust pH to 7.2. Add agar. Gently heat and bring to boiling. Distribute into tubes or flasks. Autoclave for 15 min at 15 psi pressure–121°C. Pour into sterile Petri dishes or leave in tubes.

Use: For the cultivation of members of the Rhizobiaceae.

Rhizobium Medium 2

Composition per liter:

Agar	15.0g
Glycerol	4.6g
$CaSO_4$	1.3g
K_2HPO_4	1.0g
L-Arabinose	1.0g
Yeast extract	1,0g
KNO_3	0.7g
$MgSO_4 \cdot 7H_2O$	0.36g
$FeCl_3 \cdot 6H_2O$	4.0mg

pH 7.2 ± 0.2 at 25°C

Preparation of Medium: Add components, except agar, to distilled/deionized water and bring volume to 1.0L. Mix thoroughly. Adjust pH to 7.2. Add agar. Gently heat and bring to boiling. Distribute into tubes or flasks. Autoclave for 15 min at 15 psi pressure–121°C. Pour into sterile Petri dishes or leave in tubes.

Use: For the cultivation of members of the Rhizobiaceae.

Rhizobium X Medium

Composition per liter:

Agar	15.0g
Mannitol	10.0g
Yeast extract	1.0g
Soil extract	200.0mL

pH 7.2 ± 0.2 at 25°C

Soil Extract:
Composition per 200.0mL:

African Violet soil	77.0g
Na_2CO_3	0.2g

Preparation of Soil Extract: Add components to tap water and bring volume to 200.0mL. Autoclave for 60 min at 15 psi pressure–121°C. Filter through Whatman #1 filter paper.

Preparation of Medium: Add components to distilled/deionized water and bring volume to 1.0L. Mix thoroughly. Gently heat and bring to boiling. Distribute into tubes or flasks. Autoclave for 15 min at 15 psi pressure–121°C. Pour into sterile Petri dishes or leave in tubes.

Use: For the cultivation and maintenance of *Bradyrhizobium japonicum, Rhizobium* species, and *Sinorhizobium xinjiangensis.*

Rhizobium X Medium with Thiram

Composition per liter:

Agar	15.0g
Mannitol	10.0g
Yeast extract	1.0g
Soil extract	200.0mL
Thiram solution	10.0mL

pH 7.2 ± 0.2 at 25°C

Thiram Solution:
Composition per 10.0mL:

Thiram	1.0mg
Ethanol, absolute	10.0mL

Preparation of Thiram Solution: Add thiram to 10.0mL of absolute ethanol. Mix thoroughly. Filter sterilize.

Soil Extract:

Composition per 200.0mL:

African Violet soil 77.0g
Na_2CO_3 .... 0.2g

Preparation of Soil Extract: Add components to tap water and bring volume to 200.0mL.

Preparation of Medium: Add components, except thiram solution, to distilled/deionized water and bring volume to 990.0mL. Mix thoroughly. Gently heat and bring to boiling. Autoclave for 15 min at 15 psi pressure–121°C. Cool to 50°C. Aseptically add 10.0mL of sterile thiram solution. Pour into sterile Petri dishes or distribute into sterile tubes.

Use: For the cultivation and maintenance of *Bradyrhizobium japonicum, Rhizobium* species, and *Sinorhizobium xinjiangensis.*

Rhizoctonia Isolation Medium

Composition per liter:

Agar 20.0g
K_2HPO_4 .... 1.0g
KCl 0.5g
$MgSO_4 \cdot 7H_2O$ 0.5g
$NaNO_2$ 0.2g
$FeSO_4 \cdot 7H_2O$ 0.01g
Dexon® solution 10.0mL
Antibiotic solution 10.0mL
Gallic acid solution 10.0mL

Antibiotic Solution:

Composition per 10.0mL:

Chloramphenicol 0.05g
Streptomycin 0.05g

Preparation of Antibiotic Solution: Add components to distilled/deionized water and bring volume to 10.0mL. Mix thoroughly. Filter sterilize.

Dexon® Solution:

Composition per 10.0mL:

Dexon (Chemagro®) wettable powder 0.09g

Preparation of Dexon Solution: Add Dexon® to distilled/deionized water and bring volume to 10.0mL. Mix thoroughly. Filter sterilize.

Gallic Acid Solution:

Composition per 10.0mL:

Gallic acid 0.4g

Preparation of Gallic Acid Solution: Add gallic acid to distilled/deionized water and bring volume to 10.0mL. Mix thoroughly. Filter sterilize.

Preparation of Medium: Add components, except Dexon solution, antibiotic solution, and gallic acid solution, to distilled/deionized water and bring volume to 970.0mL. Mix thoroughly. Gently heat and bring to boiling. Autoclave for 15 min at 15 psi pressure–121°C. Cool to 45°–50°C. Aseptically add sterile Dexon solution, sterile antibiotic solution, and sterile gallic acid solution. Mix thoroughly. Pour into sterile Petri dishes or distribute into sterile tubes.

Use: For the isolation and cultivation of *Rhizoctonia* species.

Rhizomonas Medium

Composition per liter:

Noble agar 11.0g
Pancreatic digest of casein 5.0g
Glucose 2.5g
K_2HPO_4 .... 1.0g
$MgSO_4 \cdot 7H_2O$ 0.5g
KNO_3 .... 0.5g
$Ca(NO_3)_2 \cdot 4H_2O$ 0.06g

pH 7.2 ± 0.2 at 25°C

Preparation of Medium: Add components to distilled/deionized water and bring volume to 1.0L. Mix thoroughly. Gently heat and bring to boiling. Adjust pH to 7.2. Distribute into tubes or flasks. Autoclave for 15 min at 15 psi pressure–121°C. Pour into sterile Petri dishes or leave in tubes.

Use: For the cultivation and maintenance of *Rhizomonas suberifaciens.*

Rhizomonas suberifaciens Medium

Composition per liter:

Pancreatic digest of casein 5.0g
$K_2HPO_4 \cdot 3H_2O$ 1.3g
Agar,, noble 1.1g
KNO_3 .... 0.5g
$MgSO_4 \cdot 7H_2O$ 0.5g
$Ca(NO_3)_2 \cdot 4H_2O$ 60.0mg

pH 7.2 ± 0.2 at 25°C

Preparation of Medium: Add components to distilled/deionized water and bring volume to 1.0L. Mix thoroughly. Gently heat and bring to boiling. Adjust pH to 7.2. Distribute into tubes or flasks. Autoclave for 15 min at 15 psi pressure–121°C. Pour into sterile Petri dishes or leave in tubes.

Use: For the cultivation of *Rhizomonas suberifaciens.*

Rhodobacter adriaticus Medium

Composition per 1001.0mL:

NaCl 25.0g
$NaHCO_3$ 3.0g
K_2HPO_4 .... 1.0g
NH_4Cl 1.0g
$MgCl_2 \cdot 6H_2O$ 0.5g
Sodium ascorbate 0.5g
$CaCl_2 \cdot 2H_2O$ 0.1g
Trace elements solution SLA 1.0mL
Vitamin solution 1.0mL

pH 7.0 ± 0.2 at 25°C

Trace Elements Solution SLA:

Composition per liter:

$CuCl_2 \cdot 2H_2O$ 10.0g
$FeCl_2 \cdot 4H_2O$ 1.8g
H_3BO_3 .... 0.5g
$CoCl_2 \cdot 6H_2O$ 0.25g
$ZnCl_2$ 0.1g
$MnCl_2 \cdot 4H_2O$ 70.0mg
$Na_2MoO_4 \cdot 2H_2O$ 30.0mg
$Na_2SeO_3 \cdot 5H_2O$ 10.0mg
$NiCl_2 \cdot 6H_2O$ 10.0mg

Preparation of Trace Elements Solution SLA: Add components to distilled/deionized water and bring volume to 1.0L. Mix thoroughly. Bring pH to 2.0–3.0.

Vitamin Solution:

Composition per liter:

Nicotinamide 35.0mg
Thiamine·HCl 30.0mg
p-Aminobenzoic acid 20.0mg
Pyridoxal·HCl 10.0mg
Calcium DL-pantothenate 10.0mg
Biotin 10.0mg
Vitamin B_{12} 5.0mg

Preparation of Vitamin Solution: Add components to distilled/deionized water and bring volume to 1.0L. Mix thoroughly. Filter sterilize.

Preparation of Medium: Add components, except vitamin solution, to distilled/deionized water and bring volume to 1.0L. Mix thoroughly. Autoclave for 15 min at 15 psi pressure–121°C. Aseptically add 1.0mL of sterile vitamin solution. Mix thoroughly. Aseptically distribute into sterile tubes or flasks.

Use: For the cultivation and maintenance of *Rhodobacter adraiticus*.

Rhodobacter Medium (LMG Medium 80)

Composition per liter:

Yeast extract	1.0g
Di-sodium succinate	1.0g
KH_2PO_4	0.5g
$MgSO_4 \cdot 7H2O$	0.4g
NaCl	0.4g
NH_4Cl	0.4g
$CaCl_2 \cdot 2H_2O$	50.0mg
Ferric citrate solution	5.0mL
Trace elements solution	1.0mL
Ethanol	0.5mL

pH 5.8 ± 0.2 at 25°C

Ferric Citrate Solution :
Composition per 100.0mL:

Ferric citrate	0.1g

Preparation of Ferric Citrate Solution: Add ferric citrate to distilled/deionized water and bring volume to 100.0mL. Mix thoroughly.

Trace Elements Solution:
Composition per liter:

H_3BO_3	0.3g
$CoCl_2 \cdot 6H_2O$	0.2g
$ZnSO_4 \cdot 7H_2O$	0.1g
$Na_2MoO_4 \cdot 2H_2O$	30.0mg
$MnCl_2 \cdot 4H_2O$	30.0mg
$NiCl_2 \cdot 6H_2O$	20.0mg
$CuCl_2 \cdot 2H_2O$	10.0mg

Preparation of Trace Elements Solution: Add components to distilled/deionized water and bring volume to 1.0L. Mix thoroughly.

Preparation of Medium: Add components to distilled/deionized water and bring volume to 1.0L. Mix thoroughly. Distribute 40mL medium into 50mL screw-capped bottles. Flush each bottle for 1 to 2 min with nitrogen gas and then close immediately with a rubber septum and screw-cap. Autoclave for 15 min at 15 psi pressure–121°C. Sterile syringes are used to inoculate and remove the samples. Incubate in light using a tungsten lamp.

Use: For the cultivation of *Rhodobacter capsulatus, Rhodobacter sphaeroides,* and *Rhodospirillum rubrum.*

Rhodobacter veldkampii Medium (DSMZ Medium 867)

Composition per 2780mL:

Solution 1	1540.0mL
Solution 3	1000.0mL
Solution 4	120.0mL
Solution 5	120.0mL

pH 4.0 ± 0.1 at 25°C

Solution 1:
Composition per 2500mL:

$CaCl_2$	2.0g

Preparation of Solution 1: Add $CaCl_2$ to distilled/deionized water and bring volume to 2.5L. Mix thoroughly.

Solution 3:
Composition per liter:

$NaHCO_3$	4.5g
Solution 2	100.0mL

Preparation of Solution 3: Add $NaHCO_3$ to distilled/deionized water and bring volume to 900.0mL. Mix thoroughly. Sparge with gaseous CO_2 through for at least 30 min. Add 100.0mL solution 2. Immediately filter sterilize using CO_2 pressure to push the liquid through (no suction).

Solution 2:
Composition per 100mL:

Sodium ascorbate	2.4g
KH_2PO_4	1.0g
KCl	1.0g
NH_4Cl	0.8g
$MgCl_2 \cdot 6H_2O$	0.8g
Heavy metal solution	50.0mL
Vitamin solution	15.0mL
Vitamin B_{12} solution	3.0mL

Preparation of Solution 2: Add components to distilled/deionized water and bring volume to 100.0mL. Mix thoroughly.

Heavy Metal Solution:
Composition per liter:

EDTA	1.50g
$FeSO_4 \cdot 7H_2O$	0.2g
$ZnSO_4 \cdot 7H_2O$	0.1g
$MnCl_2 \cdot 7H_2O$	0.02g
Modified Hoagland trace elements solution	6.0mL

Preparation of Heavy Metal Solution: Add components to distilled/deionized water and bring volume to 1.0L. Mix thoroughly.

Modified Hoagland Trace Elements Solution:
Composition per 3.6L:

H_3BO_3	11.0g
$MnCl_2 \cdot 4H_2O$	7.0g
$ZnCl_2$	1.0g
$CuCl_2$	1.0g
$NiCl_2$	1.0g
$CoCl_2$	1.0g
$AlCl_3$	1.0g
KI	1.0g
KBr	0.5g
LiCl	0.5g
$SnCl_2 \cdot 2H_2O$	0.5g
$BaCl_2$	0.5g
Na_2MoO_4	0.5g
Na_2SeO_3	0.5g
$NaVO_3 \cdot H_2O$	0.1g

Preparation of Modified Hoagland Trace Elements Solution: Add components sequentially to distilled/deionized water and bring final volume to 3.6L. Mix thoroughly after adding each component until dissolved. Adjust pH to just below 7.0. Adjust the final pH to 3-4. The flaky yellow precipitate which is formed after mixing transforms after standing for one or a few days into a very fine white precipitate. Mix thoroughly before use.

Vitamin B_{12} Solution:
Composition per 3mL:
Vitamin B_{12} (cyanocobalamine) 2.0mg

Preparation of Vitamin B_{12} Solution: Add Vitamin B_{12} to distilled/deionized water and bring volume to 3.0mL. Mix thoroughly.

Vitamin Solution:
Composition per 100mL:
Pyridoxamine·2HCl 5.0mg
Nicotinic acid 2.0mg
Thiamine 1.0mg
Pantothenic acid 0.5mg
Biotin 0.2mg
p-Aminobenzoic acid 0.1mg

Preparation of Vitamin Solution: Add components to distilled/deionized water and bring volume to 100.0mL. Mix thoroughly. Sparge with 100% CO_2 for 30 min.

Solution 4:
Composition per 200mL:
$Na_2S \cdot 9H_2O$ 3.0g

Preparation of Solution 4: Add $Na_2S \cdot 9H_2O$ to distilled/deionized water in a flask with a magnetic stirrer and bring volume to 200.0mL. Mix thoroughly. Autoclave under 100% N_2 for 15 min at 15 psi pressure–121°C. Cool to room temperature. Partially neutralize the sterilized solution by adding, on a magnetic stirrer, drop by drop, 2.0mL sterile 2*M* H_2SO_4.

Solution 5:
Composition per 200mL:
Na-acetate 6.0g

Preparation of Solution 5: Add Na-acetate to distilled/deionized water and bring volume to 200.0mL. Mix thoroughly. Sparge with 100% N_2 for 5 min. Autoclave under 100% N_2 for 15 min at 15 psi pressure–121°C. Cool to room temperature.

Preparation of Medium: Distribute solution 1 in 77.0mL amounts into 20 127mL screw-capped Boston round bottles. Autoclave for 15 min at 15 psi pressure–121°C. Cool to room temperature. Aseptically add 50.0mL sterile solution 3 to each of the 20 127mL bottles containing 77.0mL of sterile solution 1 so that the solution completely fills the bottle. Mix thoroughly. Remove 6.0mL of the medium from the completely filled bottles. Add 6.0mL of neutralized solution 4 so that the bottles are again completely filled. Mix thoroughly. Remove 6.0mL of the medium from the completely filled bottles. Add 6.0mL of solution 5 so that the bottles are again completely filled. Mix thoroughly. Adjust pH to 4.0. Allow the bottles to stand overnight to develop a hazy, white precipitate before inoculating. Mix the solution thoroughly before use. To inoculate, remove 6.0mL of completed medium and replace it with an equal volume of inoculum. Grow cultures under tungsten light.

Use: For the cultivation of *Rhodobacter veldkampii.*

Rhodobium Medium (DSMZ Medium 745)

Composition per liter:
NaCl 50.2g
Na-DL-malate 3.6g
Yeast extract 1.0g
KH_2PO_4 ..... 1.0g
$(NH_4)_2SO_4$ 1.0g
$MgCl_2 \cdot 6H_2O$ 0.2g
$CaCl_2 \cdot 2H_2O$ 0.05g
$Na_2S \cdot 9H_2O$ solution 10.0mL
Trace elements solution SL-8 ..... 1.0mL
pH 6.8 ± 0.2 at 25°C

Trace Elements Solution SL-8:
Composition per liter:
Na_2-EDTA 5.2g
$FeCl_2 \cdot 4H_2O$ 1.5g
$CoCl_2 \cdot 6H_2O$ 190.0mg
$MnCl_2 \cdot 4H_2O$ 100.0mg
$ZnCl_2$ 70.0mg
H_3BO_3 ..... 62.0mg
$Na_2MoSO_4 \cdot 2H_2O$ 36.0mg
$NiCl_2 \cdot 6H_2O$ 24.0mg
$CuCl_2 \cdot 2H_2O$ 17.0mg

Preparation of Trace Elements Solution SL-8: Add components to distilled/deionized water and bring volume to 1.0L. Mix thoroughly. Sparge with 100% N_2.

$Na_2S \cdot 9H_2O$ Solution:
Composition per 10mL:
$Na_2S \cdot 9H_2O$ 0.5g

Preparation of $Na_2S \cdot 9H_2O$ Solution: Add $Na_2S \cdot 9H_2O$ to distilled/deionized water and bring volume to 10.0mL. Mix thoroughly. Adjust pH to 6.8. Autoclave under 100% N_2 for 15 min at 15 psi pressure–121°C. Cool to room temperature.

Preparation of Medium: Add components, except $Na_2S \cdot 9H_2O$ solution, to distilled/deionized water and bring volume to 1.0L. Mix thoroughly. Adjust to pH 6.8. Gently heat and bring to boiling. Cool to room temperature while sparging with 100% N_2. Autoclave for 15 min at 15 psi pressure–121°C. Cool to room temperature. Add 10.0mL sterile $Na_2S \cdot 9H_2O$ solution. Mix thoroughly. Aseptically and anaerobically under 100% N_2 distribute into sterile screw-cap tubes or flasks.

Use: For the cultivation of *Rhodobium orientis (Rhodovulum orientis).*

Rhodocyclus Medium (LMG Medium 82)

Composition per liter:
Yeast extract 1.0g
Ammonium acetate 0.5g
KH_2PO_4 ..... 0.5g
$MgSO_4 \cdot 7H2O$ 0.4g
NaCl 0.4g
NH_4Cl 0.4g
$CaCl_2 \cdot 2H_2O$ 50.0mg
Ferric citrate solution 5.0mL
Trace elements solution 1.0mL
Vitamin solution 1.0mL
Ethanol 0.5mL
pH 5.8 ± 0.2 at 25°C

Ferric Citrate Solution:
Composition per 100.0mL:
Ferric citrate 0.1g

Preparation of Ferric Citrate Solution: Add ferric citrate to distilled/deionized water and bring volume to 100.0mL. Mix thoroughly.

Trace Elements Solution:
Composition per liter:
H_3BO_3 ..... 0.3g
$CoCl_2 \cdot 6H_2O$ 0.2g
$ZnSO_4 \cdot 7H_2O$ 0.1g

$Na_2MoO_4 \cdot 2H_2O$ 30.0mg
$MnCl_2 \cdot 4H_2O$ 30.0mg
$NiCl_2 \cdot 6H_2O$ 20.0mg
$CuCl_2 \cdot 2H_2O$ 10.0mg

Preparation of Trace Elements Solution: Add components to distilled/deionized water and bring volume to 1.0L. Mix thoroughly.

Vitamin Solution:
Composition per 100.0mL:
Vitamin B_{12} 2.0mg

Preparation of Vitamin Solution: Add Vitamin B_{12} to 100.0mL of distilled/deionized water. Mix thoroughly.

Preparation of Medium: Add components to distilled/deionized water and bring volume to 1.0L. Mix thoroughly. Distribute 40.0mL medium into 50mL screw-capped bottles. Flush each bottle for 1 to 2 min with nitrogen gas and then close immediately with rubber septa and screw-caps. Autoclave for 15 min at 15 psi pressure–121°C. Sterile syringes are used to inoculate and remove the samples. Incubate in light using a tungsten lamp.

Use: For the cultivation of *Rhodocyclus purpureus.*

Rhodocyclus purpureus Medium (DSMZ Medium 44)

Composition per 1056.9 mL:
Na_2-succinate 1.0g
(NH_4)-acetate 0.5g
KH_2PO_4 .... 0.5g
$MgSO_4 \cdot 7H_2O$ 0.4g
NaCl 0.4g
NH_4Cl 0.4g
Yeast extract 0.3g
$CaCl_2 \cdot 2H_2O$ 0.05g
Ferric citrate solution 5.0mL
Trace elements solution SL-6 .... 1.0mL
Ethanol 0.5mL
Vitamin B_{12} solution 0.4mL
Sulfide solution .... variable
pH 6.8 ± 0.2 at 25°C

Trace Elements Solution SL-6:
Composition per liter:
$MnCl_2 \cdot 4H_2O$ 0.5g
H_3BO_3 .... 0.3g
$CoCl_2 \cdot 6H_2O$ 0.2g
$ZnSO_4 \cdot 7H_2O$ 0.1g
$Na_2MoO_4 \cdot 2H_2O$ 0.03g
$NiCl_2 \cdot 6H_2O$ 0.02g
$CuCl_2 \cdot 2H_2O$ 0.01g

Preparation of Trace Elements Solution SL-6: Add components to distilled/deionized water and bring volume to 1.0L. Mix thoroughly.

Vitamin B_{12} Solution:
Composition per 100.0mL:
Vitamin B_{12} 10.0mg

Preparation of Vitamin B_{12} Solution: Add Vitamin B_{12} to distilled/deionized water and bring volume to 100.0mL. Mix thoroughly. Sparge under 100% N_2 gas for 3 min.

Ferric Citrate Solution:
Composition per 10.0mL:
Ferric citrate 10.0mg

Preparation of Ferric Citrate Solution: Add ferric citrate to distilled/deionized water and bring volume to 10.0mL. Mix thoroughly. Sparge under 100% N_2 gas for 3 min.

Neutralized Sulfide Solution:
Composition per 100.0mL:
$Na_2S \cdot 9H_2O$ 1.5g

Preparation of Neutralized Sulfide Solution: Add $Na_2S \cdot 9H_2O$ to distilled/deionized water in a 250.0mL screw-capped bottle fitted with a butyl rubber septum and bring volume to 100.0mL. Add a magnetic stir bar. Mix thoroughly. Sparge under 100% N_2 gas for 3 min. Autoclave for 15 min at 15 psi pressure–121°C. Cool to room temperature. Adjust pH to about 7.0 with sterile 2*M* H_2SO_4. Do not open the bottle to add H_2SO_4; use a sterile syringe. Stir the solution continuously to avoid precipitation of elemental sulfur. The final solution should be clear and yellow in color.

Preparation of Medium: Add components, except neutralized sulfide solution, to 1050.0mL distilled/deionized water. Mix thoroughly. Gently heat and bring to boiling. Boil for 3–4 min under a stream of 100% N_2. Distribute 45.0mL of the prepared medium into 50.0mL screw-capped tubes that have been flushed with 100% N_2. Autoclave for 15 min at 15 psi pressure–121°C. Cool to room temperature. Before inoculation, aseptically and anaerobically add 0.25–0.50mL of neutralized sulfide solution. Sterile syringes are used to inoculate and remove samples. Incubate in the light using a tungsten lamp.

Use: For the cultivation and maintenance of brown and other oxygen sensitive Rhodospirillaceae.

Rhodomicrobium Medium (LMG Medium 79

Composition per liter:
Di-sodium succinate 1.0g
KH_2PO_4 .... 0.5g
$MgSO_4 \cdot$7H2O 0.4g
NaCl 0.4g
NH_4Cl 0.4g
Yeast extract 0.2g
$CaCl_2 \cdot 2H_2O$ 50.0mg
Ferric citrate solution 5.0mL
Trace elements solution 1.0mL
pH 5.7 ± 0.2 at 25°C

Ferric Citrate Solution :
Composition per 100.0mL:
Ferric citrate 0.1g

Preparation of Ferric Citrate Solution: Add ferric citrate to distilled/deionized water and bring volume to 100.0mL. Mix thoroughly.

Trace Elements Solution:
Composition per liter:
H_3BO_3 .... 0.3g
$CoCl_2 \cdot 6H_2O$ 0.2g
$ZnSO_4 \cdot 7H_2O$ 0.1g
$Na_2MoO_4 \cdot 2H_2O$ 30.0mg
$MnCl_2 \cdot 4H_2O$ 30.0mg
$NiCl_2 \cdot 6H_2O$ 20.0mg
$CuCl_2 \cdot 2H_2O$ 10.0mg

Preparation of Trace Elements Solution: Add components to distilled/deionized water and bring volume to 1.0L. Mix thoroughly.

Preparation of Medium: Add components to distilled/deionized water and bring volume to 1.0L. Mix thoroughly. Distribute 40.0mL medium into 50mL screw-capped bottles. Flush each bottle for 1 to 2 min with nitrogen gas and then close immediately with rubber septa and screw-caps. Autoclave for 15 min at 15 psi pressure–121°C. Sterile syringes are used to inoculate and remove the samples. Incubate in light using a tungsten lamp.

Use: For the cultivation of *Rhodomicrobium vannielii* and *Rhodoblastus acidophilus.*

Rhodopila globiformis Medium

Composition per liter:

Mannitol	1.5g
Sodium gluconate	0.56g
KH_2PO_4	0.4g
NaCl	0.4g
$MgCl_2 \cdot 6H_2O$	0.4g
NH_4Cl	0.4g
$Na_2S_2O_3 \cdot 5H_2O$	0.2g
$CaCl_2 \cdot 2H_2O$	0.05g
Ferric citrate	5.0mg
VA vitamins	1.0mL
Trace elements solution SL-6	1.0mL

pH 4.9 ± 0.2 at 25°C

VA Vitamins:
Composition per 500.0mL:

Nicotinamide	0.175g
Thiamine·HCl	0.15g
p-Aminobenzoic acid	0.1g
Biotin	0.05g
Pyridoxine·2HCl	0.05g
Calcium pantothenate	0.05g
Cyanocobalamin	0.025g

Preparation of VA Vitamins: Add components to distilled/deionized water and bring volume to 500.0mL. Mix thoroughly.

Trace Elements Solution SL-6:
Composition per liter:

H_3BO_3	0.3g
$CoCl_2 \cdot 6H_2O$	0.2g
$ZnSO_4 \cdot 7H_2O$	0.1g
$MnCl_2 \cdot 4H_2O$	0.03g
$Na_2MoO_4 \cdot H_2O$	0.03g
$NiCl_2 \cdot 6H_2O$	0.02g
$CuCl_2 \cdot 2H_2O$	0.01g

Preparation of Trace Elements Solution SL-6: Add components to distilled/deionized water and bring volume to 1.0L. Mix thoroughly. Adjust pH to 3.4.

Preparation of Medium: Add components to distilled/deionized water and bring volume to 1.0L. Mix thoroughly. Adjust pH to 4.9. Distribute into tubes or flasks. Autoclave for 15 min at 15 psi pressure–121°C.

Use: For the cultivation and maintenance of *Rhodopila globiformis.*

Rhodopseudomonas blastica Medium

Composition per liter:

Sodium pyruvate	1.5g
Sodium hydrogen malate	1.5g
Yeast extract	1.0g
NH_4Cl	0.5g
$MgSO_4 \cdot 7H_2O$	0.4g
NaCl	0.4g
$CaCl_2 \cdot 2H_2O$	0.05g
Sodium phosphate buffer (0.1*M*, pH 6.8)	50.0mL

pH 6.8 ± 0.2 at 25°C

Preparation of Medium: Add components, except sodium pyruvate solution, sodium hydrogen malate solution, and sodium phosphate buffer, to distilled/deionized water and bring volume to 950.0mL. Mix thoroughly. Gently heat and bring to boiling. Adjust pH to 6.8 with KOH. Autoclave for 15 min at 15 psi pressure–121°C. Cool to 45°–50°C. Filter sterilize the sodium pyruvate solution, sodium hydrogen malate solution, and sodium phosphate buffer solution. Aseptically add 1.5g of sterile sodium pyruvate solution, 1.5g of sodium hydrogen malate solution, and 50.0mL of sodium phosphate buffer solution to cooled basal medium. Mix thoroughly. Pour into sterile Petri dishes or distribute into sterile tubes.

Use: For the cultivation and maintenance of *Rhodopseudomonas blastica* and other *Rhodopseudomonas* species.

Rhodopseudomonas globiformis Medium

Composition per 1002.0mL:

Mannitol	1.5g
KH_2PO_4	0.5g
Sodium gluconate	0.5g
$MgSO_4 \cdot 7H_2O$	0.4g
NaCl	0.4g
NH_4Cl	0.4g
Yeast extract	0.25g
$CaCl_2 \cdot 2H_2O$	0.05g
Ferric citrate solution	5.0mL
$Na_2S_2O_3$ solution	2.0mL
Biotin solution	1.0mL
p-Aminobenzoic acid solution	1.0mL
Trace elements solution SL-6	1.0mL

pH 4.9 ± 0.2 at 25°C

Ferric Citrate Solution :
Composition per 10.0mL:

Ferric citrate	0.01g

Preparation of Ferric Citrate Solution: Add ferric citrate to distilled/deionized water and bring volume to 10.0mL. Mix thoroughly. Autoclave for 15 min at 15 psi pressure–121°C.

$Na_2S_2O_3$ Solution :
Composition per 10.0mL:

$Na_2S_2O_3$	1.0g

Preparation of $Na_2S_2O_3$ Solution: Add $Na_2S_2O_3$ to distilled/deionized water and bring volume to 10.0mL. Mix thoroughly. Autoclave for 15 min at 15 psi pressure–121°C.

Biotin Solution:
Composition per 10.0mL:

Biotin	0.2mg

Preparation of Biotin Solution: Add biotin to distilled/deionized water and bring volume to 10.0mL. Mix thoroughly. Autoclave for 15 min at 15 psi pressure–121°C.

***p*-Aminobenzoic Acid Solution:**
Composition per 10.0mL:

p-Aminobenzoic acid	1.0mg

Preparation of *p*-Aminobenzoic Acid Solution: Add *p*-aminobenzoic acid to distilled/deionized water and bring volume to 10.0mL. Mix thoroughly. Autoclave for 15 min at 15 psi pressure–121°C.

Trace Elements Solution SL-6:
Composition per liter:

$MnCl_2 \cdot 4H_2O$	0.5g
H_3BO_3	0.3g

$CoCl_2·6H_2O$ 0.2g
$ZnSO_4·7H_2O$ 0.1g
$Na_2MoO_4·2H_2O$ 0.03g
$NiCl_2·6H_2O$ 0.02g
$CuCl_2·2H_2O$ 0.01g

Preparation of Trace Elements Solution SL-6: Add components to distilled/deionized water and bring volume to 1.0L. Mix thoroughly.

Preparation of Medium: Add components, except $Na_2S_2O_3$ solution, ferric citrate solution, biotin solution, and *p*-aminobenzoic acid solution, to distilled deionized water and bring volume to 993.0mL. Mix thoroughly. Adjust pH to 4.9. Add 5.0mL of ferric citrate solution, 1.0mL of biotin solution, and 1.0mL of *p*-aminobenzoic acid solution. Mix thoroughly. Distribute into screw-capped tubes or bottles. Autoclave for 15 min at 15 psi pressure–121°C. Allow to cool to room temperature. Aseptically add 0.2mL of sterile $Na_2S_2O_3$ solution to each 100.0mL of medium. Mix thoroughly.

Use: For the cultivation and maintenance of *Rhodopila globiformis*.

Rhodopseudomonas julia Medium

Composition per liter:

$NaHCO_3$ 3.0g
NaCl 1.0g
KH_2PO_4 .... 1.0g
NH_4Cl 1.0g
Sodium acetate 1.0g
Na_2SO_4 .... 0.7g
$MgCl_2·6H_2O$ 0.5g
Sodium ascorbate 0.5g
$CaCl_2·2H_2O$ 0.1g
Yeast extract 0.1g
$Na_2S·9H_2O$ solution 10.0mL
SLA trace elements solution 1.0mL
VA vitamin solution 1.0mL

pH 6.9–7.0 at 25°C

$Na_2S·9H_2O$ Solution:

Composition per 10.0mL:

$Na_2S·9H_2O$ 0.156g

Preparation of $Na_2S·9H_2O$ Solution: Add $Na_2S·9H_2O$ to distilled/deionized water and bring volume to 10.0mL. Mix thoroughly. Autoclave for 15 min at 15 psi pressure–121°C. Before use, neutralize to pH 7.0 with sterile HCl.

SLA Trace Elements Solution:

Composition per liter:

$FeCl_2·4H_2O$ 1.8g
H_3BO_3 .... 0.5g
$CoCl_2·6H_2O$ 0.25g
$ZnCl_2$ 0.1g
$MnCl_2·4H_2O$ 70.0mg
$Na_2MoO_4·2H_2O$ 30.0mg
$CuCl_2·2H_2O$ 10.0mg
$Na_2SeO_3·5H_2O$ 10.0mg
$NiCl_2·6H_2O$ 10.0mg

Preparation of SLA Trace Elements Solution: Add components to distilled/deionized water and bring volume to 1.0L. Mix thoroughly. Adjust pH to 2.0–3.0.

VA Vitamin Solution:

Composition per 500.0mL:

Nicotinamide 0.175g
Thiamine·HCl 0.15g
p-Aminobenzoic acid 0.1g
Biotin 50.0mg
Calcium D-(+)-pantothenate 50.0mg
Pyridoxine·2HCl 50.0mg
Cyanocobalamin 25.0mg

Preparation of VA Vitamin Solution: Add components to distilled/deionized water and bring volume to 1.0L. Mix thoroughly. Filter sterilize.

Preparation of Medium: Add components, except $Na_2S·9H_2O$ solution, to distilled/deionized water and bring volume to 990.0mL. Mix thoroughly. Adjust pH to 6.9–7.0. Autoclave for 15 min at 15 psi pressure–121°C. Aseptically add 10.0mL of sterile $Na_2S·9H_2O$ solution. Mix thoroughly. Aseptically distribute into sterile tubes or flasks.

Use: For the cultivation of *Rhodopseudomonas julia.*

Rhodopseudomonas Medium (ATCC Medium 543)

Composition per liter:

Sodium succinate 2.5g
$(NH_4)_2SO_4$ 1.25g
K_2HPO_4 .... 0.9g
KH_2PO_4 .... 0.6g
Yeast extract 0.5g
$MgSO_4·7H_2O$ 0.2g
$CaCl_2$ 0.07g
Ferric citrate 3.0mg
Ethylenediamine tetraacetate (EDTA) 2.0mg

pH 7.0 ± 0.2 at 25°C

Preparation of Medium: Add components to distilled/deionized water and bring volume to 1.0L. Mix thoroughly. Distribute into tubes or flasks. Autoclave for 15 min at 15 psi pressure–121°C.

Use: For the cultivation and maintenance of *Rhodopseudomonas* species.

Rhodopseudomonas rutila Medium

Composition per liter:

Sodium glutamate 2.0g
Sodium L-malate 2.0g
Yeast extract 2.0g
KH_2PO_4 .... 1.0g
$NaHCO_3$ 0.5g
$MgSO_4·7H_2O$ 0.2g
$CaCl_2·2H_2O$ 0.1g
$MnSO_4·H_2O$ 2.0mg
Biotin 1.0mg
Nicotinic acid 1.0mg
Thiamine·HCl 1.0mg
$CoCl_2·6H_2O$ 0.5mg
$FeSO_4·7H_2O$ 0.5mg

Preparation of Medium: Add components to distilled/deionized water and bring volume to 1.0L. Mix thoroughly. Distribute into tubes or flasks. Autoclave for 15 min at 15 psi pressure–121°C.

Use: For the cultivation of *Rhodopseudomonas palustris* and *Rhodopseudomonas rutila*.

Rhodopseudomonas sulfoviridis Medium

Composition per 1050.0 mL:

Ammonium acetate 1.5g
Sodium malate 1.0g
KH_2PO_4 .... 0.5g
$Na_2S_2O_3$ 0.5g

$MgSO_4.7H_2O$ 0.4g
NaCl 0.4g
NH_4Cl 0.4g
Yeast extract 0.3g
Disodium succinate 0.25g
$CaCl_2 \cdot 2H_2O$ 0.05g
Ferric citrate solution 5.0mL
Trace elements solution SL-6 ..... 1.0mL
Ethanol 0.5mL
Vitamin B_{12} solution 0.4mL
Neutralized sulfide solution ..... variable
pH 6.8 ± 0.2 at 25°C

Ferric Citrate Solution:
Composition per 10.0mL:
Ferric citrate 10.0mg

Preparation of Ferric Citrate Solution: Add ferric citrate to distilled/deionized water and bring volume to 10.0mL. Mix thoroughly. Sparge under 100% N_2 for 3 min. Autoclave for 15 min at 15 psi pressure–121°C. Store under N_2.

Trace Elements Solution SL-6:
Composition per liter:
$MnCl_2 \cdot 4H_2O$ 0.5g
H_3BO_3 ..... 0.3g
$CoCl_2 \cdot 6H_2O$ 0.2g
$ZnSO_4 \cdot 7H_2O$ 0.1g
$Na_2MoO_4 \cdot 2H_2O$ 0.03g
$NiCl_2 \cdot 6H_2O$ 0.02g
$CuCl_2 \cdot 2H_2O$ 0.01g

Preparation of Trace Elements Solution SL-6: Add components to distilled/deionized water and bring volume to 1.0L. Mix thoroughly.

Vitamin B_{12} Solution:
Composition per 100.0mL:
Vitamin B_{12} 10.0mg

Preparation of Vitamin B_{12} Solution: Add Vitamin B_{12} to distilled/deionized water and bring volume to 100.0mL. Mix thoroughly. Sparge under 100% N_2 for 3 min. Autoclave for 15 min at 15 psi pressure–121°C. Store under N_2.

Neutralized Sulfide Solution:
Composition per 100.0mL:
$Na_2S \cdot 9H_2O$ 1.5g

Preparation of Neutralized Sulfide Solution: Add $Na_2S \cdot 9H_2O$ to distilled/deionized water in a 250mL screw-capped bottle fitted with a butyl rubber septum and bring volume to 100.0mL. Add a magnetic stir bar. Mix thoroughly. Sparge under 100% N_2 for 3 min. Autoclave for 15 min at 15 psi pressure–121°C. Cool to room temperature. Adjust pH to about 7.3 with sterile 2*M* H_2SO_4. Do not open the bottle to add H_2SO_4; use a sterile syringe. Stir the solution continuously to avoid precipitation of elemental sulfur. The final solution should be clear and yellow in color.

Preparation of Medium: Add components, except neutralized sulfide solution, to distilled/deionized water and bring volume to 1050.0mL. Mix thoroughly. Gently heat and bring to boiling. Boil for 3–4 min under a stream of 100% N_2. Distribute 45.0mL of the prepared medium into 50.0mL screw-capped tubes that have been flushed with 100% N_2. Autoclave for 15 min at 15 psi pressure–121°C. Cool to room temperature. Before inoculation, aseptically and anaerobically add 0.25–0.50mL of neutralized sulfide solution to each tube.

Use: For the cultivation and maintenance of *Rhodopseudomonas sulfoviridis*.

Rhodospirillaceae Enrichment Medium

Composition per liter:
Dicarboxylic acid substrate 1.0g
KH_2PO_4 ..... 0.5g
NaCl 0.4g
NH_4Cl 0.4g
$MgSO_4 \cdot 7H_2O$ 0.2g
Yeast extract 0.2g
$CaCl_2 \cdot 2H_2O$ 0.05g
Ferric citrate solution 5.0mL
Trace elements solution SL-7 ..... 1.0mL
Vitamin B_{12} solution 1.0mL
pH 6.8 ± 0.2 at 25°C

Ferric Citrate Solution:
Composition per 100.0mL:
Ferric citrate 0.1g

Preparation of Ferric Citrate Solution: Add ferric citrate to distilled/deionized water and bring volume to 100.0mL. Mix thoroughly.

Trace Elements Solution SL-7:
Composition per liter:
$CoCl_2 \cdot 6H_2O$ 0.2g
$MnCl_2 \cdot 4H_2O$ 0.1g
$ZnCl_2$ 0.07g
H_3BO_3 ..... 0.06g
$NaMoO_4 \cdot 2H_2O$ 0.04g
$CuCl_2 \cdot 2H_2O$ 0.02g
$NiCl_2 \cdot 6H_2O$ 0.02g
HCl (25% solution) 1.0mL

Preparation of Trace Elements Solution SL-7: Add components to distilled/deionized water and bring volume to 1.0L. Mix thoroughly.

Vitamin B_{12} Solution:
Composition per 100.0mL:
Vitamin B_{12} 1.0mg

Preparation of Vitamin B_{12} Solution: Add Vitamin B_{12} to distilled/deionized water and bring volume to 100.0mL. Mix thoroughly.

Preparation of Medium: Add components to distilled/deionized water and bring volume to 1.0L. Succinic acid or glutaric acid may be used for the dicarboxylic acid substrate. Mix thoroughly. Adjust pH to 6.8. Distribute into tubes or flasks. Autoclave for 15 min at 15 psi pressure–121°C.

Use: For the enrichment and isolation of members of the Rhodospirillaceae.

Rhodospirillaceae Medium

Composition per liter:
Succinic acid 1.0g
KH_2PO_4 ..... 0.5g
NaCl 0.4g
NH_4Cl 0.4g
$MgSO_4 \cdot 7H_2O$ 0.2g
Yeast extract 0.2g
$CaCl_2 \cdot 2H_2O$ 0.05g
Ferric citrate solution 5.0mL
Vitamin B_{12} solution 1.0mL
Trace elements solution SL7 ..... 1.0mL
pH 6.8 ± 0.2 at 25°C

Ferric Citrate Solution:
Composition per 100.0mL:
Ferric citrate 0.1g

Preparation of Ferric Citrate Solution: Add ferric citrate to distilled/deionized water and bring volume to 100.0mL. Mix thoroughly. Filter sterilize.

Vitamin B_{12} Solution:
Composition per 100.0mL:
Vitamin B_{12} 1.0mg

Preparation of Vitamin B_{12} Solution: Add Vitamin B_{12} to distilled/deionized water and bring volume to 100.0mL. Mix thoroughly. Filter sterilize

Trace Elements Solution SL7:
Composition per liter:
$MnCl_2 \cdot 4H_2O$ 100.0mg
ZnCl 70.0mg
H_3Bo_3 .... 60.0mg
$NaMoO_4 \cdot 2H_2O$ 40.0mg
$CoCl_2 \cdot 2H_2O$ 20.0mg
$CuCl_2 \cdot 2H_2O$ 20.0mg
$NiCl_2 \cdot 6H_2O$ 20.0mg
HCl (25%) 1.0mL

Preparation of Trace Elements Solution SL7: Add $MnCl_2 \cdot 4H_2O$ to 1.0mL of HCl solution. Mix thoroughly. Add distilled/deionized water and bring volume to 1.0L. Add remaining components. Mix thoroughly. Sparge with 100% N_2. Autoclave for 15 min at 15 psi pressure–121°C.

Preparation of Medium: Add components, except Fe-citrate solution, Vitamin B_{12} solution, and trace elements solution SL7, to distilled/deionized water and bring volume to 993.0mL. Mix thoroughly. Adjust pH to 6.8. Autoclave for 30 min at 15 psi pressure–121°C. Aseptically add 5.0mL of sterile ferric citrate solution, 1.0mL of sterile Vitamin B_{12} solution, and 1.0mL of sterile trace elements solution SL7. Mix thoroughly. Aseptically distribute into sterile screw-capped tubes under anaerobic condition. Tighten screw caps.

Use: For the cultivation of *Rhodospirillum* species and other members of the family Rhodospirillaceae.

Rhodospirillaceae Medium, Modified

Composition per 1050.0 mL:
Ammonium acetate 1.5g
KH_2PO_4 .... 0.5g
$MgSO_4.7H_2O$ 0.4g
NaCl 0.4g
NH_4Cl 0.4g
Yeast extract 0.3g
Disodium succinate 0.25g
$CaCl_2 \cdot 2H_2O$ 0.05g
Ferric citrate solution 5.0mL
Trace elements solution SL-6 .... 1.0mL
Ethanol 0.5mL
Vitamin B_{12} solution 0.4mL
Neutralized sulfide solution .... variable
pH 6.8 ± 0.2 at 25°C

Ferric Citrate Solution:
Composition per 10.0mL:
Ferric citrate 10.0mg

Preparation of Ferric Citrate Solution: Add ferric citrate to distilled/deionized water and bring volume to 10.0mL. Mix thoroughly. Sparge under 100% N_2 for 3 min. Autoclave for 15 min at 15 psi pressure–121°C. Store under N_2 gas.

Trace Elements Solution SL-6:
Composition per liter:
$MnCl_2 \cdot 4H_2O$ 0.5g
H_3BO_3 .... 0.3g
$CoCl_2 \cdot 6H_2O$ 0.2g
$ZnSO_4 \cdot 7H_2O$ 0.1g
$Na_2MoO_4 \cdot 2H_2O$ 0.03g
$NiCl_2 \cdot 6H_2O$ 0.02g
$CuCl_2 \cdot 2H_2O$ 0.01g

Preparation of Trace Elements Solution SL-6: Add components to distilled/deionized water and bring volume to 1.0L. Mix thoroughly.

Vitamin B_{12} Solution:
Composition per 100.0mL:
Vitamin B_{12} 10.0mg

Preparation of Vitamin B_{12} Solution: Add Vitamin B_{12} to distilled/deionized water and bring volume to 100.0mL. Mix thoroughly. Sparge under 100% N_2 gas for 3 min. Autoclave for 15 min at 15 psi pressure–121°C. Store under N_2 gas.

Neutralized Sulfide Solution:
Composition per 100.0mL:
$Na_2S \cdot 9H_2O$ 1.5g

Preparation of Neutralized Sulfide Solution: Add $Na_2S \cdot 9H_2O$ to distilled/deionized water in a 250mL screw-capped bottle fitted with a butyl rubber septum and bring volume to 100.0mL. Add a magnetic stir bar. Mix thoroughly. Sparge under 100% N_2 gas for 3 min. Autoclave for 15 min at 15 psi pressure–121°C. Cool to room temperature. Adjust pH to about 7.3 with sterile 2*M* H_2SO_4. Do not open the bottle to add H_2SO_4; use a sterile syringe. Stir the solution continuously to avoid precipitation of elemental sulfur. The final solution should be clear and yellow in color.

Preparation of Medium: Add components, except neutralized sulfide solution, to distilled/deionized water and bring volume to 1050.0mL. Mix thoroughly. Gently heat and bring to boiling. Boil for 3–4 min under a stream of 100% N_2. Distribute 45.0mL of the prepared medium into 50.0mL screw-capped tubes that have been flushed with 100% N_2. Autoclave for 15 min at 15 psi pressure–121°C. Cool to room temperature. Before inoculation, aseptically and anaerobically add 0.25–0.50mL of neutralized sulfide solution to each tube.

Use: For the cultivation and maintenance of *Ectothiorhodospira marismortui, Rhodobacter capsulatus, Rhodobacter sphaeroides, Rhodobacter sulfidophilus, Rhodocyclus tenuis, Rhodopseudomonas blastica, Rhodopseudomonas marina, Rhodopseudomonas palustris, Rhodopseudomonas rosea, Rhodopseudomonas viridis, Rhodospirillum fulvum, Rhodospirillum molischianum, Rhodospirillum photometricum, Rhodospirillum rubrum, Rhodospirillum salexigens,* and *Rubrivivax gelatinosus.*

Rhodospirillum Medium (ATCC Medium 1308)

Composition per liter:
Yeast extract 1.0g
Disodium succinate 1.0g
KH_2PO_4 .... 0.5g
Sodium ascorbate 0.5g
$MgSO_4 \cdot 7H_2O$ 0.4g
NaCl 0.4g
NH_4Cl 0.4g
$CaCl_2 \cdot 2H_2O$ 0.05g
Ferric citrate (0.1% solution) 5.0mL

Trace elements solution SL-6 .... 1.0mL
Ethanol 0.5mL

pH 6.0 ± 0.2 at 25°C

Trace Elements Solution SL-6:
Composition per liter:

H_3BO_3 .... 0.3g
$CoCl_2 \cdot 6H_2O$ 0.2g
$ZnSO_4 \cdot 7H_2O$ 0.1g
$MnCl_2 \cdot 4H_2O$ 0.03g
$Na_2MoO_4 \cdot H_2O$ 0.03g
$NiCl_2 \cdot 6H_2O$ 0.02g
$CuCl_2 \cdot 2H_2O$ 0.01g

Preparation of Trace Elements Solution SL-6: Add components to distilled/deionized water and bring volume to 1.0L. Mix thoroughly. Adjust pH to 3.4.

Preparation of Medium: Add components to distilled/deionized water and bring volume to 1.0L. Mix thoroughly. Adjust pH to 6.0. Distribute into tubes or flasks. Autoclave for 15 min at 15 psi pressure–121°C.

Use: For the cultivation and maintenance of *Rhodospirillum* species.

Rhodospirillum Medium, Modified I

Composition per liter:

Disodium succinate 1.0g
KH_2PO_4 .... 0.5g
$MgSO_4 \cdot 7H_2O$ 0.4g
NaCl 0.4g
NH_4Cl 0.4g
Yeast extract 0.2g
$CaCl_2 \cdot 2H_2O$ 0.05g
Ferric citrate (0.1% solution) 5.0mL
Trace elements solution SL-6 .... 1.0mL

pH 5.7 ± 0.2 at 25°C

Trace Elements Solution SL-6:
Composition per liter:

H_3BO_3 .... 0.3g
$CoCl_2 \cdot 6H_2O$ 0.2g
$ZnSO_4 \cdot 7H_2O$ 0.10g
$MnCl_2 \cdot 4H_2O$ 0.03g
$Na_2MoO_4 \cdot H_2O$ 0.03g
$NiCl_2 \cdot 6H_2O$ 0.02g
$CuCl_2 \cdot 2H_2O$ 0.01g

Preparation of Trace Elements Solution SL-6: Add components to distilled/deionized water and bring volume to 1.0L. Mix thoroughly. Adjust pH to 3.4.

Preparation of Medium: Add components to distilled/deionized water and bring volume to 1.0L. Mix thoroughly. Adjust pH to 5.7. Distribute 40.0mL volumes into tubes or bottles. Sparge with 100% N_2 for 1–2 min. Autoclave for 15 min at 15 psi pressure–121°C.

Use: For the cultivation and maintenance of *Rhodomicrobium vannielii, Rhodopseudomonas acidophila,* and *Rhodobacter capsulatus.*

Rhodospirillum Medium, Modified II

Composition per liter:

Yeast extract 1.0g
Disodium succinate 1.0g
KH_2PO_4 .... 0.5g
$MgSO_4 \cdot 7H_2O$ 0.4g
NaCl 0.4g
NH_4Cl 0.4g
$CaCl_2 \cdot 2H_2O$ 0.05g
Ferric citrate (0.1% solution) 5.0mL
Trace elements solution SL-6 .... 1.0mL
Ethanol 0.5mL

pH 6.8 ± 0.2 at 25°C

Trace Elements Solution SL-6:
Composition per liter:

H_3BO_3 .... 0.3g
$CoCl_2 \cdot 6H_2O$ 0.2g
$ZnSO_4 \cdot 7H_2O$ 0.10g
$MnCl_2 \cdot 4H_2O$ 0.03g
$Na_2MoO_4 \cdot H_2O$ 0.03g
$NiCl_2 \cdot 6H_2O$ 0.02g
$CuCl_2 \cdot 2H_2O$ 0.01g

Preparation of Trace Elements Solution SL-6: Add components to distilled/deionized water and bring volume to 1.0L. Mix thoroughly. Adjust pH to 3.4.

Preparation of Medium: Add components to distilled/deionized water and bring volume to 1.0L. Mix thoroughly. Adjust pH to 6.8. Distribute 40.0mL volumes into tubes or bottles. Sparge with 100% N_2 for 1–2 min. Autoclave for 15 min at 15 psi pressure–121°C.

Use: For the cultivation and maintenance of *Rhodobacter sphaeroides, Rhodocyclus tenuis, Rhodopseudomonas blastica, Rhodopseudomonas palustris, Rhodopseudomonas viridis, Rhodospirillum fulvum, Rhodospirillum molischianum, Rhodospirillum photometricum, Rhodospirillum rubrum,* and *Rubrivivax gelatinosus.*

Rhodospirillum Medium, Modified III

Composition per liter:

Yeast extract 1.0g
Ammonium acetate 0.5g
KH_2PO_4 .... 0.5g
$MgSO_4 \cdot 7H_2O$ 0.4g
NaCl 0.4g
NH_4Cl 0.4g
$CaCl_2 \cdot 2H_2O$ 0.05g
Vitamin B_{12} 20.0mg
Ferric citrate (0.1% solution) 5.0mL
Trace elements solution SL-6 .... 1.0mL
Ethanol 0.5mL

pH 6.8 ± 0.2 at 25°C

Trace Elements Solution SL-6:
Composition per liter:

H_3BO_3 .... 0.3g
$CoCl_2 \cdot 6H_2O$ 0.2g
$ZnSO_4 \cdot 7H_2O$ 0.1g
$MnCl_2 \cdot 4H_2O$ 0.03g
$Na_2MoO_4 \cdot H_2O$ 0.03g
$NiCl_2 \cdot 6H_2O$ 0.02g
$CuCl_2 \cdot 2H_2O$ 0.01g

Preparation of Trace Elements Solution SL-6: Add components to distilled/deionized water and bring volume to 1.0L. Mix thoroughly. Adjust pH to 3.4.

Preparation of Medium: Add components to distilled/deionized water and bring volume to 1.0L. Mix thoroughly. Adjust pH to 6.8. Distribute 40.0mL volumes into tubes or bottles. Sparge with 100% N_2 for 1–2 min. Autoclave for 15 min at 15 psi pressure–121°C.

Use: For the cultivation and maintenance of *Rhodocyclus purpureus.*

Rhodospirillum Medium, Modified IV

Composition per liter:

NaCl 25.0g
Yeast extract 1.0g

Disodium succinate	1.0g
KH_2PO_4	0.5g
$MgSO_4 \cdot 7H_2O$	0.4g
NaCl	0.4g
NH_4Cl	0.4g
$CaCl_2 \cdot 2H_2O$	0.05g
Ferric citrate (0.1% solution)	5.0mL
Trace elements solution SL-6	1.0mL
Ethanol	0.5mL

pH 6.8 ± 0.2 at 25°C

Trace Elements Solution SL-6:
Composition per liter:

H_3BO_3	0.3g
$CoCl_2 \cdot 6H_2O$	0.2g
$ZnSO_4 \cdot 7H_2O$	0.1g
$MnCl_2 \cdot 4H_2O$	0.03g
$Na_2MoO_4 \cdot H_2O$	0.03g
$NiCl_2 \cdot 6H_2O$	0.02g
$CuCl_2 \cdot 2H_2O$	0.01g

Preparation of Trace Elements Solution SL-6: Add components to distilled/deionized water and bring volume to 1.0L. Mix thoroughly. Adjust pH to 3.4.

Preparation of Medium: Add components to distilled/deionized water and bring volume to 1.0L. Mix thoroughly. Adjust pH to 6.8. Distribute 40.0mL volumes into tubes or bottles. Sparge with 100% N_2 for 1–2 min. Autoclave for 15 min at 15 psi pressure–121°C.

Use: For the cultivation and maintenance of *Rhodobacter sulfidophilus.*

Rhodospirillum salinarum Medium

Composition per liter:

NaCl	100.0g
KCl	5.0g
$MgCl_2 \cdot 6H_2O$	5.0g
$MgSO_4 \cdot 7H_2O$	5.0g
$NH_4 \cdot Cl$	5.0g
Peptone solution	30.0mL
Yeast extract solution	30.0mL
Ferric citrate solution	10.0mL
Trace elements solution	5.0mL

Peptone Solution:
Composition per 100.0mL:

Peptone	15.0g

Preparation of Peptone Solution: Add peptone to distilled/deionized water and bring volume to 100.0mL. Mix thoroughly. Filter sterilize.

Yeast Extract Solution:
Composition per 10.0mL:

Yeast extract	15.0g

Preparation of Yeast Extract Solution: Add yeast extract to distilled/deionized water and bring volume to 100.0mL. Mix thoroughly. Filter sterilize.

Ferric Citrate Solution:
Composition per 10.0mL:

Ferric citrate	0.1g

Preparation of Ferric Citrate Solution: Add ferric citrate to distilled/deionized water and bring volume to 10.0mL. Mix thoroughly. Filter sterilize.

Trace Elements Solution:
Composition per liter:

$CoCl_2 \cdot 6H_2O$	10.0g
$ZnSO_4 \cdot 7H_2O$	220.0mg
$MgCl_2 \cdot 4H_2O$	180.0mg
$Na_2MoO_4 \cdot 2H_2O$	6.3mg
$CuSO_4 \cdot 5H_2O$	1.0mg

Preparation of Medium: Add components, except peptone solution, yeast extract solution, ferric citrate solution, and trace elements solution to distilled/deionized water and bring volume to 925.0mL. Mix thoroughly. Autoclave for 15 min at 15 psi pressure–121°C. Aseptically add 30.0mL of sterile peptone solution, 30.0mL of sterile yeast extract solution, 10.0mL of sterile ferric citrate solution, and 5.0mL of sterile trace elements solution. Mix thoroughly. Aseptically distribute into sterile tubes or flasks.

Use: For the cultivation of *Rhodospirillum salinarum.*

Rhodovulum iodosum *Rhodovulum robiginosum* Medium (DSMZ Medium 929)

Composition per liter:

NaCl	26.4g
$MgSO_4 \cdot 7H_2O$	6.8g
$MgCl_2 \cdot 6H_2O$	5.7g
$CaCl_2 \cdot 2H_2O$	1.5g
KCl	0.66g
KBr	0.09g
$NaHCO_3$ solution	30.0mL
Phosphate solution	10.0mL
Sodium acetate solution	10.0mL
Iron sulfate solution	10.0mL
NH_4Cl sollution	1.0mL
Thiosulfate solution	1.0mL
Selenite-tungstate solution	1.0mL
Trace elements solution	1.0mL
Vitamin solution	1.0mL
Vitamin B_{12} solution	1.0mL
Vitamin B_1 solution	1.0mL

pH 6.8 ± 0.2 at 25°C

Sodium Acetate Solution:
Composition per 100.0mL:

Na-acetate	4.1g

Preparation of Sodium Acetate Solution: Add sodium acetate to distilled/deionized water and bring volume to 100.0mL. Mix thoroughly. Sparge with 100% N_2. Filter sterilize.

Trace Elements Solution:
Composition per liter:

$CoCl_2 \cdot 6H_2O$	190.0mg
$ZnSO_4 \cdot 7H_2O$	144.0mg
$MnCl_2 \cdot 4H_2O$	100.0mg
$Na_2MoO_4 \cdot 4H_2O$	36.0mg
H_3BO_3	30.0mg
$NiCl_2 \cdot 6H_2O$	24.0mg
Na_2EDTA	5.2mg
$FeSO_4 \cdot 7H_2O$	2.1mg
$CuCl_2 \cdot 2H_2O$	2.0mg

Preparation of Trace Elements Solution: Add components to distilled/deionized water and bring volume to 1.0L. Mix thoroughly. Adjust pH to 6.0. Sparge with 100% N_2. Autoclave for 15 min at 15 psi pressure–121°C. Cool to room temperature.

Vitamin Solution:
Composition per 100.0mL:

Pyridoxin	15.0mg
Nicotinate	10.0mg
Pantothenate	5.0mg

Para-aminobenzoic acid 4.0mg
Biotin 1.0mg

Preparation of Vitamin Solution: Add components to distilled/deionized water and bring volume to 100.0mL. Mix thoroughly. Sparge with 100% N_2. Filter sterilize.

Vitamin B_{12} Solution:
Composition per 100.0mL:
Cyanocobalamine 5.0mg

Preparation of Vitamin B_{12} Solution: Add cyanocobalamine to distilled/deionized water and bring volume to 100.0mL. Mix thoroughly. Sparge with 100% N_2. Filter sterilize.

Vitamin B_1 Solution:
Composition per 100.0mL:
Thiamine 10.0mg

Preparation of Vitamin B_1 Solution: Add thiamine to distilled/deionized water and bring volume to 100.0mL. Mix thoroughly. Sparge with 100% N_2. Filter sterilize.

NH_4Cl Solution:
Composition per 10.0mL:
NH_4Cl 2.5g

Preparation of NH_4Cl Solution: Add NH_4Cl to distilled/deionized water and bring volume to 10.0mL. Mix thoroughly. Sparge with 100% N_2. Autoclave for 15 min at 15 psi pressure–121°C. Cool to room temperature.

Thiosulfate Solution:
Composition per 10mL:
$Na_2S_2O_3 \cdot 5H_2O$ 1.24g

Preparation of Thiosulfate Solution: Add 1.24g of $Na_2S_2O_3 \cdot 5H_2O$ to distilled/deionized water and bring volume to 10.0mL. Mix thoroughly. Sparge with 100% N_2. Autoclave for 15 min at 15 psi pressure–121°C. Cool to room temperature.

$NaHCO_3$ Solution:
Composition per 100.0mL:
$NaHCO_3$ 8.4g

Preparation of $NaHCO_3$ Solution: Add $NaHCO_3$ to distilled/deionized water and bring volume to 100.0mL. Mix thoroughly. Sparge with 80% N_2 + 20% CO_2. Autoclave for 15 min at 15 psi pressure–121°C under an atmosphere of CO_2. Cool to room temperature.

Selenite-Tungstate Solution
Composition per liter:
NaOH 0.5g
$Na_2WO_4 \cdot 2H_2O$ 4.0mg
$Na_2SeO_3 \cdot 5H_2O$ 3.0mg

Preparation of Selenite-Tungstate Solution: Add components to distilled/deionized water and bring volume to 1.0L. Mix thoroughly. Sparge with 100% N_2. Autoclave for 15 min at 15 psi pressure–121°C. Cool to room temperature.

Phosphate Solution:
Composition per 10mL:
KH_2PO_4 .... 0.4g

Preparation of Phosphate Solution: Add KH_2PO_4 to distilled/deionized water and bring volume to 10.0mL. Mix thoroughly. Sparge with 100% N_2. Autoclave for 15 min at 15 psi pressure–121°C. Cool to room temperature.

Iron Sulfate Solution:
Composition per 100mL:
$FeSO_4$ 1.52g

Preparation of Iron Sulfate Solution: Add $FeSO_4$ to distilled/deionized water and bring volume to 100.0mL. Mix thoroughly. Sparge with 100% N_2. Autoclave for 15 min at 15 psi pressure–121°C. Cool to room temperature.

Preparation of Medium: Prepare and dispense medium under 90% N_2 + 10% CO_2 gas mixture. Add components, except $NaHCO_3$ solution, phosphate solution, sodium acetate solution, iron sulfate solution, NH_4Cl solution, thiosulfate solution, selenite-tungstate solution, trace elements solution, vitamin solution, Vitamin B_{12} solution, and Vitamin B_1 solution, to distilled/deionized water and bring volume to 933.0mL. Mix thoroughly. Sparge with 90% N_2 + 10% CO_2 gas mixture. Autoclave for 15 min at 15 psi pressure–121°C. Aseptically and anaerobically add 30.0mL $NaHCO_3$ solution, 10.0mL phosphate solution, 10.0mL sodium acetate solution, 10.0mL iron sulfate solution, 1.0mL NH_4Cl solution, 1.0mL thiosulfate solution, 1.0mL selenite-tungstate solution, 1.0mL trace elements solution, 1.0mL vitamin solution, 1.0mL Vitamin B_{12} solution, and 1.0mL Vitamin B_1 solution. When the iron is added a white precipitate may form. Adjust pH to 6.8. Aseptically and anaerobically distribute into tubes or bottles.

Use: For the cultivation of *Rhodovulum iodosum* and *Rhodovulum robiginosum.*

Rhodovulum strictum Medium (DSMZ Medium 746)

Composition per liter:
Solution 1 .... 500.0mL
Solution 2 .... 500.0mL
pH 7.8 ± 0.2 at 25°C

Solution 1:
Composition 500mL:
NaCl 8.2g
Na-DL-malate 3.6g
Yeast extract 1.0g
$(NH_4)_2SO_4$ 1.0g
$Na2_S2O_3 \cdot 5H_2O$ 0.5g
$MgCl_2 \cdot 6H_2O$ 0.2g
$CaCl_2 \cdot 2H_2O$ 0.05g
Trace elements solution SL-8 .... 1.0mL

Trace Elements Solution SL-8:
Composition per liter:
Na_2-EDTA 5.2g
$FeCl_2 \cdot 4H_2O$ 1.5g
$CoCl_2 \cdot 6H_2O$ 190.0mg
$MnCl_2 \cdot 4H_2O$ 100.0mg
$ZnCl_2$ 70.0mg
H_3BO_3 .... 62.0mg
$Na_2MoSO_4 \cdot 2H_2O$ 36.0mg
$NiCl_2 \cdot 6H_2O$ 24.0mg
$CuCl_2 \cdot 2H_2O$ 17.0mg

Preparation of Trace Elements Solution SL-8: Add components to distilled/deionized water and bring volume to 1.0L. Mix thoroughly. Sparge with 100% N_2.

Preparation of Solution 1: Prepare and dispense medium under an oxygen-free 100% N_2. Add components to distilled/deionized water and bring volume to 500.0L. Mix thoroughly. Sparge with 100% N_2. Autoclave for 15 min at 15 psi pressure–121°C. Cool to 25°C.

Solution 2:
Composition per 500mL:

K_2HPO_4	1.35g
KH_2PO_4	0.35g

Preparation of Solution 2: Add components to distilled/deionized water and bring volume to 100.0mL. Mix thoroughly. Adjust pH to 7.8. Autoclave for 15 min at 15 psi pressure–121°C. Cool to room temperature.

Preparation of Medium: Aseptically and anaerobically combine 500.0mL solution 1 and 500.0mL solution 2 under N_2. Mix thoroughly. Adjust pH to 7.8. Aseptically and anaerobically distribute into sterile screw cap tubes or flasks.

Use: For the cultivation of *Rhodovulum strictum.*

Rhodovulum sulfidophilum Medium (LMG Medium 84)

Composition per liter:

NaCl	25.0g
Yeast extract	1.0g
Di-sodium succinate	1.0g
KH_2PO_4	0.5g
$MgSO_4 \cdot 7H2O$	0.4g
NaCl	0.4g
NH_4Cl	0.4g
$CaCl_2 \cdot 2H_2O$	50.0mg
Ferric citrate solution	5.0mL
Trace elements solution	1.0mL
Ethanol	0.5mL

pH 5.8 ± 0.2 at 25°C

Ferric Citrate Solution:
Composition per 100.0mL:

Ferric citrate	0.1g

Preparation of Ferric Citrate Solution: Add ferric citrate to distilled/deionized water and bring volume to 100.0mL. Mix thoroughly.

Trace Elements Solution:
Composition per liter:

H_3BO_3	0.3g
$CoCl_2 \cdot 6H_2O$	0.2g
$ZnSO_4 \cdot 7H_2O$	0.1g
$Na_2MoO_4 \cdot 2H_2O$	30.0mg
$MnCl_2 \cdot 4H_2O$	30.0mg
$NiCl_2 \cdot 6H_2O$	20.0mg
$CuCl_2 \cdot 2H_2O$	10.0mg

Preparation of Trace Elements Solution: Add components to distilled/deionized water and bring volume to 1.0L. Mix thoroughly.

Preparation of Medium: Add components to distilled/deionized water and bring volume to 1.0L. Mix thoroughly. Distribute 40.0mL medium into 50 mL screw-capped bottles. Flush each bottle for 1 to 2 minutes with nitrogen gas and then close immediately with a rubber septum and screw-cap. Autoclave for 15 min at 15 psi pressure–121°C. Sterile syringes are used to inoculate and remove the samples. Incubate in light using a tungsten lamp.

Use: For the cultivation of *Rhodovulum sulfidophilum.*

Rila Marine Medium

Composition per liter:

Agar	15.0g
Peptone	0.5g
Yeast extract	0.5g
Pancreatic digest of casein	0.5g
Marine salts mixture	800.0mL

pH 7.6-8.0 at 25°C

Preparation of Medium: Add components to distilled/deionized water and bring volume to 1.0L. Mix thoroughly. Gently heat and bring to boiling. Autoclave for 15 min at 15 psi pressure–121°C. Adjust pH to 7.6–8.0. Pour into sterile Petri dishes or distribute into sterile tubes.

Use: For the cultivation and maintenance of *Alteromonas denitrificans.*

Rimler-Shotts Medium (RS Medium)

Composition per liter:

Agar	13.5g
$Na_2S_2O_3 \cdot 5H_2O$	6.8g
L-Ornithine·HCl	6.5g
NaCl	5.0g
L-Lysine·HCl	5.0g
Maltose	3.5g
Yeast extract	3.0g
Sodium deoxycholate	1.0g
Ferric ammonium citrate	0.8g
L-Cysteine·HCl	0.3g
Bromothymol Blue	0.03g
Novobiocin solution	10.0mL

pH 7.0 ± 0.2 at 25°C

Novobiocin Solution:
Composition per 10.0mL:

Novobiocin	5.0mg

Preparation of Novobiocin Solution: Add novobiocin to distilled/deionized water and bring volume to 10.0mL. Mix thoroughly. Filter sterilize.

Preparation of Medium: Add components, except novobiocin solution, to distilled/deionized water and bring volume to 990.0mL. Mix thoroughly. Gently heat and bring to boiling. Autoclave for 15 min at 15 psi pressure–121°C. Cool to 45°–50°C. Aseptically add sterile components. Mix thoroughly. Pour into sterile Petri dishes or distribute into sterile tubes.

Use: For the selective isolation, cultivation, and presumptive identification of *Aeromonas hydrophila* and other Gram-negative bacteria based on their ability to decarboxylate lysine and ornithine, ferment maltose, and produce H_2S. Maltose-fermenting bacteria appear as yellow colonies. Bacteria that produce lysine or ornithine decarboxylase turn the medium greenish-yellow to yellow. Bacteria that produce H_2S appear as colonies with black centers.

Rippey-Cabelli Agar (RC Agar)

Composition per liter:

Agar	15.0g
Meat peptone	5.0g
Trehalose	5.0g
NaCl	3.0g
KCl	2.0g
Yeast extract	2.0g
Bromothymol Blue	0.44g
$MgSO_4 \cdot 7H_2O$	0.2g
$FeCl_3 \cdot 6H_2O$	0.1g
Sodium deoxycholate	0.1g
Ampicillin solution	10.0mL
Ethanol	10.0mL

pH 8.0 ± 0.2 at 25°C

Ampicillin Solution:
Composition per 10.0mL:
Ampicillin 0.02g

Preparation of Ampicillin Solution: Add ampicillin to distilled/deionized water and bring volume to 10.0mL. Mix thoroughly. Filter sterilize.

Preparation of Medium: Add components, except sodium deoxycholate, ampicillin solution, and ethanol, to distilled/deionized water and bring volume to 990.0mL. Mix thoroughly. Gently heat and bring to boiling. Autoclave for 15 min at 15 psi pressure–121°C. Cool to 45°–50°C. Aseptically add sodium deoxycholate, 10.0mL of sterile ampicillin solution, and 10.0mL of ethanol. Mix thoroughly. Pour into sterile Petri dishes or distribute into sterile tubes.

Use: For the isolation, cultivation, and differentiation of *Aeromonas* species and *Plesiomonas* species from water samples using the membrane filter method. This medium differentiates bacteria on the basis of trehalose fermentation. Bacteria that ferment trehalose turn the medium yellow.

RM Medium

Composition per liter:
Glucose 20.0g
Agar 15.0g
Yeast extract 10.0g
KH_2PO_4 .... 2.0g
Solution 1 .... 250.0mL
Solution 2 .... 250.0mL
Solution 3 .... 250.0mL
Solution 4 .... 250.0mL
pH 6.0 ± 0.2 at 25°C

Solution 1:
Composition per 250.0mL:
Glucose 20.0g

Preparation of Solution 1: Add glucose to distilled/deionized water and bring volume to 250.0mL. Mix thoroughly. Autoclave for 15 min at 15 psi pressure–121°C. Cool to 45°–50°C.

Solution 2:
Composition per 250.0mL:
Agar 15.0g

Preparation of Solution 2: Add agar to distilled/deionized water and bring volume to 250.0mL. Mix thoroughly. Autoclave for 15 min at 15 psi pressure–121°C. Cool to 45°–50°C.

Solution 3:
Composition per 250.0mL:
Yeast extract 10.0g

Preparation of Solution 3: Add yeast extract to distilled/deionized water and bring volume to 250.0mL. Mix thoroughly. Autoclave for 15 min at 15 psi pressure–121°C. Cool to 45°–50°C.

Solution 4:
Composition per 250.0mL:
KH_2PO_4 .... 2.0g

Preparation of Solution 4: Add KH_2PO_4 to distilled/deionized water and bring volume to 250.0mL. Mix thoroughly. Autoclave for 15 min at 15 psi pressure–121°C. Cool to 45°–50°C.

Preparation of Medium: Aseptically combine the four sterile solutions. Mix thoroughly. Adjust pH to 6.0. Pour into sterile Petri dishes or distribute into sterile tubes.

Use: For the cultivation and maintenance of *Zymomonas mobilis*.

Rose Bengal Chloramphenicol Agar

Composition per liter:
Agar 15.0g
Glucose 10.0g
Papaic digest of soybean meal 5.0g
KH_2PO_4 .... 1.0g
$MgSO_4 \cdot 7H_2O$ 0.5g
Rose Bengal 0.05g
Chloramphenicol solution 10.0mL
pH 7.0 ± 0.2 at 25°C

Source: This medium is available as a premixed powder from BD Diagnostic Systems and Oxoid Unipath.

Chloramphenicol Solution:
Composition per 10.0mL:
Chloramphenicol 0.1g

Preparation of Chloramphenicol Solution: Add chloramphenicol to distilled/deionized water and bring volume to 10.0mL. Mix thoroughly. Filter sterilize.

Preparation of Medium: Add components, except chloramphenicol solution, to distilled/deionized water and bring volume to 990.0mL. Mix thoroughly. Gently heat and bring to boiling. Autoclave for 15 min at 15 psi pressure–121°C. Cool to 45°C. Aseptically add sterile chloramphenicol solution. Mix thoroughly. Pour into sterile Petri dishes or distribute into sterile tubes.

Use: For the selective isolation, cultivation, and enumeration of yeasts and molds from environmental specimens.

Roseinatronobacter Agar (DSMZ Medium 928)

Composition per liter:
K_2HPO_4 .... 25.0g
Na_2CO_3 .... 11.0g
$NaHCO_3$ 4.0g
NaCl 2.5g
Sodium acetate 0.8g
Yeast extract 0.5g
Peptone 0.5g
KNO_3 .... 0.25g
Agar solution 500.0mL
pH 10.0 ± 0.2 at 25°C

Agar Solution:
Composition per 500.0mL:
Agar 20.0g

Preparation of Agar Solution: Add agar to distilled/deionized water and bring volume to 500.0mL. Mix thoroughly. Gently heat and bring to boiling. Autoclave for 15 min at 15 psi pressure–121°C. Cool to 55°C.

Preparation of Medium: Add components, except agar solution, to distilled/deionized water and bring volume to 500.0mL. Mix thoroughly. Autoclave for 15 min at 15 psi pressure–121°C. Cool to 55°C. Add 500.0mL sterile warm agar solution. Mix thoroughly. Pour into Petri dishes or distribute to sterile tubes.

Use: For the cultivation of *Roseinatronobacter thiooxidans*.

Rumen Bacteria Medium

Composition per 1001.0mL:

Na_2CO_3	4.0g
Trypticase	2.0g
Yeast extract	0.5g
K_2HPO_4	0.3g
Hemin	1.0mg
Resazurin	1.0mg
Minerals solution	38.0mL
Carbohydrate solution	20.0mL
L-Cysteine·HCl·H_2O solution	10.0mL
$Na_2S·9H_2O$ solution	10.0mL
Volatile fatty acid mixture	3.1mL

pH 6.7 ± 0.2 at 25°C

Minerals Solution:
Composition per liter:

NaCl	12.0g
KH_2PO_4	6.0g
$(NH_4)_2SO_4$	6.0g
$MgSO_4·7H_2O$	2.5g
$CaCl_2·2H_2O$	1.6g

Preparation of Minerals Solution: Add components to distilled/deionized water and bring volume to 1.0L. Mix thoroughly.

L-Cysteine·HCl·H_2O Solution:
Composition per 10.0mL:

L-Cysteine·HCl·H_2O	0.25g

Preparation of L-Cysteine·HCl·H_2O Solution: Add L-cysteine·HCl·H_2O to distilled/deionized water and bring volume to 10.0mL. Mix thoroughly. Sparge with 100% CO_2. Autoclave for 15 min at 15 psi pressure–121°C.

$Na_2S·9H_2O$ Solution:
Composition per 10.0mL:

$Na_2S·9H_2O$	0.25g

Preparation of $Na_2S·9H_2O$ Solution: Add $Na_2S·9H_2O$ to distilled/deionized water and bring volume to 10.0mL. Mix thoroughly. Sparge with 100% CO_2. Autoclave for 15 min at 15 psi pressure–121°C.

Carbohydrate Solution:
Composition per 20.0mL:

Glucose	0.5g
Cellobiose	0.5g
Glycerol	0.5g
Maltose	0.5g
Starch, soluble	0.5g

Preparation of Carbohydrate Solution: Add components to distilled/deionized water and bring volume to 20.0mL. Mix thoroughly. Sparge under 100% CO_2. Autoclave for 15 min at 15 psi pressure–121°C.

Volatile Fatty Acid Mixture:
Composition per 7.75mL:

Acetic acid	4.25mL
Propionic acid	1.50mL
Butyric acid	1.0mL
DL-2-Methyl butyric acid	0.25mL
iso-Butyric acid	0.25mL
iso-Valeric acid	0.25mL
n-Valeric acid	0.25mL

Preparation of Volatile Fatty Acid Mixture: Combine components. Mix thoroughly.

Preparation of Medium: Prepare and dispense medium under 100% CO_2. Add components, except carbohydrate solution, Na_2CO_3, L-cysteine·HCl·H_2O solution, and $Na_2S·9H_2O$ solution, to distilled/deionized water and bring volume to 960.0mL Mix thoroughly. Gently heat and bring to boiling. Continue boiling for 5 min. Cool to room temperature while sparging with 100% CO_2. Add Na_2CO_3. Continue sparging with 100% CO_2 until pH reaches 6.8. Distribute into rubber-stoppered tubes under 100% CO_2. Autoclave for 15 min at 15 psi pressure–121°C. Aseptically and anaerobically add 20.0mL of sterile carbohydrate solution, 10.0mL of sterile L-cysteine·HCl·H_2O solution, and 10.0mL of sterile $Na_2S·9H_2O$ solution or, using a syringe, inject the appropriate amount of sterile carbohydrate solution, sterile $Na_2S·9H_2O$ solution, and sterile L-cysteine·HCl·H_2O solution into individual tubes containing medium.

Use: For the cultivation and maintenance of *Anaerovibrio glycerini, Anaerovibrio lipolytica, Butyrivibrio fibrisolvens, Lachnospira multiparus, Succinimonas amylolytica,* and *Succinivibrio dextrinosolvens.*

Rumen Fluid Cellobiose Agar (RFC Agar)

Composition per 10.0mL:

Rumen fluid cellobiose base medium	8.9mL
$NaHCO_3$-rifampin solution	1.0mL
Cellobiose solution	0.1mL

Rumen Fluid Cellobiose Base Medium:
Composition per 89.0mL:

Noble agar	0.7g
Cysteine·HCl·H_2O	0.1g
Clarified rumen fluid	30.0mL
Salts solution A	20.0mL
Salts solution B	20.0mL
Resazurin (0.1% solution)	0.1mL

pH 6.7–7.0 at 25°C

Preparation of Rumen Fluid Cellobiose Base Medium: Add components to distilled/deionized water and bring volume to 89.0mL. Mix thoroughly. Gently heat and bring to boiling. Continue boiling until resazurin turns colorless, indicating reduction. Anaerobically distribute into tubes in 8.9mL volumes under 100% CO_2. Cap tubes with rubber stoppers. Autoclave for 15 min at 15 psi pressure–121°C. Cool to 25°C.

Salts Solution A:
Composition per liter:

$CaCl_2$	0.45g
$MgSO_4$	0.45g

Preparation of Salts Solution A: Add components to distilled/deionized water and bring volume to 1.0L. Mix thoroughly.

Salts Solution B:
Composition per liter:

NaCl	4.5g
$(NH_4)_2SO_4$	4.5g
KH_2PO_4	2.25g
K_2HPO_4	2.25g

Preparation of Salts Solution B: Add components to distilled/deionized water and bring volume to 1.0L. Mix thoroughly.

$NaHCO_3$-Rifampin Solution:
Composition per 10.0mL:

$NaHCO_3$	0.5g
Rifampin	0.1mg

Preparation of $NaHCO_3$-Rifampin Solution: Add components to distilled/deionized water and bring volume to 10.0mL. Mix thoroughly. Filter sterilize.

Cellobiose Solution:
Composition per 10.0mL:

Cellobiose	1.0g

Preparation of Cellobiose Solution: Add cellobiose to distilled/deionized water and bring volume to 10.0mL. Mix thoroughly. Filter sterilize.

Preparation of Medium: To each tube containing 8.9mL of sterile rumen fluid cellobiose base medium, aseptically add 1.0mL of sterile $NaHCO_3$-rifampin solution and 0.1mL of sterile cellobiose solution. Mix thoroughly.

Use: For the selective isolation of rumen treponemes.

Ruminobacter amylophilus Medium

Composition per liter:

Pancreatic digest of casein	10.0g
$NaHCO_3$	6.0g
Starch, soluble	5.0g
NaCl	0.9g
$(NH_4)_2SO_4$	0.9g
L-Cysteine·HCl	0.5g
K_2HPO_4	0.45g
KH_2PO_4	0.45g
$MgSO_4 \cdot 7H_2O$	0.18g
$CaCl_2 \cdot 2H_2O$	0.12g
Resazurin	1.0mg

pH 7.0 ± 0.2 at 25°C

Preparation of Medium: Prepare and dispense under 100% CO_2. Add components to distilled/deionized water and bring volume to 1.0L. Mix thoroughly. Sparge with 100% CO_2. Anaerobically distribute into tubes or flasks. Autoclave for 15 min at 15 psi pressure–121°C.

Use: For the cultivation and maintenance of *Ruminobacter amylophilus.*

Ruminococcus albus Medium

Composition per 1001.0mL:

Pancreatic digest of casein	5.0g
Na_2CO_3	4.0g
Glucose	3.0g
Cellobiose	2.0g
Yeast extract	2.0g
L-Cysteine·HCl	0.5g
Resazurin	1.0mg
Mineral solution 1	40.0mL
Mineral solution 2	40.0mL
Fatty acid mixture	1.0mL

Mineral Solution 1:
Composition per 100.0mL:

K_2HPO_4	0.6g

Preparation of Mineral Solution 1: Add K_2HPO_4 to distilled/deionized water and bring volume to 100.0mL. Mix thoroughly.

Mineral Solution 2:
Composition per 100.0mL:

$(NH_4)_2SO_4$	2.0g
NaCl	2.0g
KH_2PO_4	0.6g
$MgSO_4 \cdot 7H_2O$	0.25g
$CaCl_2 \cdot 7H_2O$	0.16g

Preparation of Mineral Solution 2: Add components to distilled/deionized water and bring volume to 100.0mL. Mix thoroughly.

Fatty Acid Mixture:
Composition per 100.0mL:

Isobutyric acid	10.0mL
Isovaleric acid	10.0mL
2-Methylbutyric acid	10.0mL

Preparation of Fatty Acid Mixture: Add components to distilled/deionized water and bring volume to 100.0mL. Sparge with 100% CO_2.

Preparation of Medium: Add components, except Na_2CO_3, L-cysteine·HCl, and fatty acid mixture, to distilled/deionized water and bring volume to 1.0L. Mix thoroughly. Gently heat and bring to boiling. Continue boiling for 5 min. Cool to room temperature while sparging with 100% CO_2. Add Na_2CO_3, L-cysteine·HCl, and fatty acid mixture. Adjust pH to 7.0. Anaerobically distribute into tubes or flasks under 100% N_2. Autoclave for 15 min at 15 psi pressure–121°C.

Use: For the cultivation and maintenance of *Ruminococcus albus.*

Ruminococcus pasteurii Medium

Composition per liter:

$NaHCO_3$	2.5g
Sodium tartrate	2.0g
NaCl	1.0g
KCl	0.5g
$MgCl_2 \cdot 6H_2O$	0.4g
$Na_2S \cdot 9H_2O$	0.36g
NH_4Cl	0.25g
KH_2PO_4	0.2g
$CaCl_2 \cdot 2H_2O$	0.15g
Resazurin	1.0mg
Trace elements solution SL-7	1.0mL

pH 7.2 ± 0.2 at 25°C

Trace Elements Solution SL-7:
Composition per liter:

$FeCl_2 \cdot 4H_2O$	1.5g
$CoCl_2 \cdot 6H_2O$	0.19g
$MnCl_2 \cdot 4H_2O$	0.1g
$ZnCl_2$	0.07g
H_3BO_3	0.062g
$Na_2MoO_4 \cdot 2H_2O$	0.036g
$NiCl_2 \cdot 6H_2O$	0.024g
$CuCl_2 \cdot 2H_2O$	0.017g
HCl (25% solution)	10.0mL

Preparation of Trace Elements Solution SL-7: Add the $FeCl_2 \cdot 4H_2O$ to the HCl. Add distilled/deionized water and bring volume to 1.0L. Add remaining components. Mix thoroughly. Autoclave for 15 min at 15 psi pressure–121°C under 100% N_2. Cool to room temperature.

$NaHCO_3$ Solution:
Composition per 10.0mL:

$NaHCO_3$	2.5g

Preparation of $NaHCO_3$ Solution: Add the $NaHCO_3$ to distilled/deionized water and bring volume to 10.0mL. Mix thoroughly. Filter sterilize.

$Na_2S \cdot 9H_2O$ Solution:
Composition per 10.0mL:

$Na_2S \cdot 9H_2O$	0.36g

Preparation of $Na_2S \cdot 9H_2O$ Solution: Add $Na_2S \cdot 9H_2O$ to distilled/deionized water and bring volume to 10.0mL. Mix thoroughly. Autoclave for 15 min at 15 psi pressure–121°C under 100% N_2.

Preparation of Medium: Add components, except $NaHCO_3$ solution, $Na_2S \cdot 9H_2O$ solution, and trace elements solution SL-7, to distilled/deionized water and bring volume to 999.0mL. Mix thoroughly. Adjust pH to 7.2. Gently heat and bring to boiling under 80% N_2 + 20% CO_2. Distribute into tubes in 9.8mL volumes under 80% N_2 + 20% CO_2. Cool to 25°C. Aseptically add 0.1mL of sterile $NaHCO_3$ solution and 0.01mL of sterile trace elements solution SL-7 to each tube. Mix thoroughly. Immediately prior to inoculation, aseptically add 0.1mL of sterile $Na_2S \cdot 9H_2O$ solution to each tube.

Use: For the cultivation and maintenance of *Ruminococcus pasteurii*.

Russell Double-Sugar Agar

Composition per liter:

Agar	15.0g
Proteose peptone No. 3	12.0g
Lactose	10.0g
NaCl	5.0g
Beef extract	1.0g
Glucose	1.0g
Phenol Red	0.025g

pH 7.5 ± 0.2 at 25°C

Preparation of Medium: Add components to distilled/deionized water and bring volume to 1.0L. Mix thoroughly. Gently heat and bring to boiling. Distribute into tubes. Autoclave for 15 min at 15 psi pressure–121°C. Allow tubes to cool in a slanted position.

Use: For the identification of Gram-negative enteric bacilli based on their fermentation of glucose and lactose. Bacteria that ferment both glucose and lactose produce a yellow slant and yellow butt. Bacteria that ferment glucose but do not ferment lactose produce a red slant and a yellow butt. Bacteria that ferment neither glucose nor lactose produce an unchanged pink-orange color.

S6 Medium for Thiobacilli

Composition per liter:

Agar	15.0g
$Na_2S_2O_3$	10.0g
KH_2PO_4	11.8g
Na_2HPO_4	1.2g
$MgSO_4 \cdot 7H_2O$	0.1g
$(NH_4)_2SO_4$	0.1g
$CaCl_2$	0.03g
$FeCl_3$	0.02g
$MnSO_4$	0.02g

Preparation of Medium: Add components to distilled/deionized water and bring volume to 1.0L. Mix thoroughly. Gently heat and bring to boiling. Distribute into tubes or flasks. Autoclave for 15 min at 15 psi pressure–121°C. Pour into sterile Petri dishes or leave in tubes.

Use: For the cultivation and maintenance of *Thiobacillus denitrificans* and *Thiobacillus thioparus*.

S8 Medium for Thiobacilli

Composition per liter:

Agar	15.0g
KH_2PO_4	11.8g
$Na_2S_2O_3$	10.0g
KNO_3	5.0g
Na_2HPO_4	1.2g
$NaHCO_3$	0.5g
$MgSO_4 \cdot 7H_2O$	0.1g
$(NH_4)_2SO_4$	0.1g
$CaCl_2$	0.03g
$FeCl_3$	0.02g
$MnSO_4$	0.02g

Preparation of Medium: Add components to distilled/deionized water and bring volume to 1.0L. Mix thoroughly. Gently heat and bring to boiling. Distribute into tubes or flasks. Autoclave for 15 min at 15 psi pressure–121°C. Pour into sterile Petri dishes or leave in tubes.

Use: For the cultivation and maintenance of *Thiobacillus neapolitanus*.

Sabouraud Agar

Composition per liter:

Neopeptone	30.0g
Agar	20.0g

Preparation of Medium: Add components to tap water and bring volume to 1.0L. Mix thoroughly. Gently heat and bring to boiling. Distribute into tubes or flasks. Autoclave for 15 min at 15 psi pressure–121°C. Pour into sterile Petri dishes or leave in tubes.

Use: For the cultivation of yeasts and molds.

Saccharococcus Agar

Composition per liter:

Agar	20.0g
Beef extract	5.0g
Sucrose	5.0g
Pancreatic digest of casein	3.0g
Glucose	1.0g

pH 6.8 ± 0.2 at 25°C

Preparation of Medium: Add components to distilled/deionized water and bring volume to 1.0L. Mix thoroughly. Gently heat and bring to boiling. Distribute into tubes or flasks. Autoclave for 15 min at 15 psi pressure–121°C. Pour into sterile Petri dishes or leave in tubes.

Use: For the cultivation and maintenance *Saccharococcus thermophilus*.

Saccharolytic Clostridia Medium

Composition per liter:

Sodium thioglycolate	1.0g
K_2HPO_4	0.8g
KH_2PO_4	0.2g
$MgSO_4 \cdot 7H_2O$	0.2g
NaCl	0.2g
$Na_2MoO_4 \cdot 2H_2O$	0.025g
Yeast extract	0.01g
$FeSO_4 \cdot 7H_2O$	0.01g
$MnSO_4 \cdot 4H_2O$	0.01g
CaCl2	0.01g
Carbohydrate solution	100.0mL
Soil extract	10.0mL
Trace elements solution	1.0mL

pH 7.2 ± 0.2 at 25°C

Carbohydrate Solution:

Composition per 100.0mL:

Glucose or sucrose	10.0g

Preparation of Carbohydrate Solution: Add glucose or sucrose to distilled/deionized water and bring volume to 100.0mL. Mix thoroughly. Filter sterilize.

Soil Extract:

Composition per 200.0mL:

Garden soil, neutral	100.0g

Preparation of Soil Extract: Add garden soil to 100.0mL of tap water. Gently heat and bring to 130°C for 60 min. Cool to 45°C. Filter through Whatman #1 filter paper. Autoclave for 15 min at 15 psi pressure–121°C. Cool to 45°–50°C.

Trace Elements Solution:

Composition per liter:

$Na_2B_4O_7 \cdot 10H_2O$	0.05g
$CoNO_3 \cdot 6H_2O$	0.05g
$CdSO_4 \cdot 2H_2O$	0.05g
$CuSO_4 \cdot 5H_2O$	0.05g
$ZnSO_4 \cdot 7H_2O$	0.05g
$MnSO_4 \cdot H_2O$	0.05g

Preparation of Trace Elements Solution: Add components to distilled/deionized water and bring volume to 1.0L. Mix thoroughly.

Preparation of Medium: Add components, except sodium thioglycolate and carbohydrate solution, to distilled/deionized water and bring volume to 900.0mL. Mix thoroughly. Gently heat and bring to boiling. Add sodium thioglycolate. Mix thoroughly. Distribute 9.5mL into test tubes that contain inverted Durham tubes. Autoclave for 15 min at 15 psi pressure–121°C. Cool to 45°–50°C. Aseptically add 0.5mL of sterile carbohydrate solution to each tube. Mix thoroughly.

Use: For the isolation of N_2-fixing, saccharolytic *Clostridium* species.

Salinibacter ruber Agar (DSMZ Medium 936)

Composition per liter:

NaCl	195.0g
$MgSO_4 \cdot 7H_2O$	49.5g
$MgCl_2 \cdot 6H_2O$	34.6g
Agar	20.0g
KCl	5.0g
$CaCl_2 \cdot 2H_2O$	1.25g
Yeast extract	1.0g
NaBr	0.625g
$NaHCO_3$	0.25g

pH 7.2 ± 0.2 at 25°C

Preparation of Medium: Add components to distilled/deionized water and bring volume to 1.0L. Mix thoroughly. Gently heat and bring to boiling. Distribute into tubes or flasks. Autoclave for 15 min at 15 psi pressure–121°C. Pour into sterile Petri dishes or leave in tubes.

Use: For the cultivation and maintenance of *Salinibacter ruber.*

Salinivibrio costicola Subspecies *vallismortis* Medium (DSMZ Medium 597)

Composition per liter:

NaCl	25.0g
$MgSO_4 \cdot 7H_2O$	9.6g
$MgCl_2 \cdot 6H_2O$	7.0g
Glucose	5.0g
KCl	3.8g
Yeast extract	1.0g
$CaCl_2 \cdot 2H_2O$	0.5g
$K_2HPO_4 \cdot 3H_2O$	0.4g
$NaHCO_3$ solution	100.0mL

pH 7.0 ± 0.2 at 25°C

$NaHCO_3$ Solution:

Composition per 100.0mL:

$NaHCO_3$	3.0g

Preparation of $NaHCO_3$ Solution: Add $NaHCO_3$ to distilled/deionized water and bring volume to 100.0mL. Mix thoroughly. Filter sterilize.

Preparation of Medium: Add components, except $NaHCO_3$ solution, to distilled/deionized water and bring volume to 900.0mL. Mix thoroughly. Autoclave for 15 min at 15 psi pressure–121°C. Cool to 25°C. Aseptically add 100.0mL $NaHCO_3$ solution. Mix thoroughly. Aseptically distribute into sterile tubes or flasks.

Use: For the cultivation of *Salinivibrio costicola* subsp. *vallismortis.*

Salmonella Rapid Test Elective Medium

Composition per liter:

Tryptone	10.0g
Na_2HPO_4	9.0g
Sodium chloride	5.0g
Casein	5.0g
KH_2PO_4	1.5g
Malachite Green	0.0025g

pH 6.5 ± 0.2 at 25°C

Preparation of Medium: Add components to distilled/deionized water and bring volume to 1.0L. Mix thoroughly. Autoclave for 15 min at 15 psi pressure–121°C.

Use: For the Oxoid *Salmonella* Rapid Test which is for the presumptive detection of motile *Salmonella* in environmental samples.

Salmonella Rapid Test Elective Medium, 2×

Composition per liter:

Tryptone	20.0g
Na_2HPO_4	18.0g
Sodium chloride	10.0g
KH_2PO_4	3.0g
Casein	10.0g
Malachite Green	0.005g

pH 6.5 ± 0.2 at 25°C

Preparation of Medium: Add components to distilled/deionized water and bring volume to 1.0L. Mix thoroughly. Autoclave for 15 min at 15 psi pressure–121°C.

Use: Use as described in the Oxoid *Salmonella* Rapid Test which is for the presumptive detection of motile *Salmonella* in environmental samples.

Salmonella Shigella Agar (SS Agar)

Composition per liter:

Agar	13.5g
Lactose	10.0g
Bile salts	8.5g
$Na_2S_2O_3$	8.5g
Sodium citrate	8.5g
Beef extract	5.0g
Pancreatic digest of casein	2.5g
Peptic digest of animal tissue	2.5g

Ferric citrate 1.0g
Neutral Red 0.025g
Brilliant Green 0.33mg

pH 7.0 ± 0.2 at 25°C

Source: This medium is available as a premixed powder from BD Diagnostic Systems and Oxoid Unipath.

Preparation of Medium: Add components to distilled/deionized water and bring volume to 1.0L. Mix thoroughly. Gently heat while stirring and bring to boiling. Do not autoclave. Cool to 45°–50°C. Pour into sterile Petri dishes in 20.0mL volumes. Allow the surface of the plates to dry before inoculation.

Use: For the selective isolation and differentiation of pathogenic enteric bacilli, especially those belonging to the genus *Salmonella.* This medium is not recommended for the primary isolation of *Shigella* species. Lactose-fermenting bacteria such as *Escherichia coli* or *Klebsiella pneumoniae* appear as small pink or red colonies. Lactose-nonfermenting bacteria—such as *Salmonella* species, *Proteus* species, and *Shigella* species—appear as colorless colonies. Production of H_2S by *Salmonella* species turns the center of the colonies black.

Salt Agar

Composition per liter:
NaCl 58.4g
Agar 15.0g
Proteose peptone 5.0g
Pancreatic digest of casein 5.0g

pH 6.9 ± 0.2 at 25°C

Preparation of Medium: Add components to distilled/deionized water and bring volume to 1.0L. Mix thoroughly. Gently heat and bring to boiling. Distribute into tubes or flasks. Autoclave for 15 min at 15 psi pressure–121°C. Pour into sterile Petri dishes or leave in tubes.

Use: For the cultivation and maintenance of *Marinococcus halophilus*.

Salt Colistin Broth

Composition per liter:
NaCl 20.0g
Peptone 10.0g
Yeast extract 3.0g
Colistin solution 10.0mL

pH 7.4 ± 0.2 at 25°C

Colistin Solution:
Composition per 10.0mL:
Colistin methane sulfonate 500,000U

Preparation of Colistin Solution: Add colistin methane sulfonate to distilled/deionized water and bring volume to 10.0mL. Mix thoroughly. Filter sterilize.

Preparation of Medium: Add components, except colistin solution, to distilled/deionized water and bring volume to 990.0mL. Mix thoroughly. Gently heat until dissolved. Autoclave for 15 min at 15 psi pressure–121°C. Cool to 25°C. Aseptically add sterile colistin solution. Mix thoroughly. Aseptically distribute into sterile tubes or flasks.

Use: For the cultivation of halophilic *Vibrio* species.

Salt Tolerance Medium

Composition per liter:
Beef heart, solids from infusion 500.0g
NaCl 65.0g
Tryptose 10.0g

pH 7.4 ± 0.2 at 25°C

Preparation of Medium: Add components to distilled/deionized water and bring volume to 1.0L. Mix thoroughly. Distribute into tubes or flasks. Autoclave for 15 min at 15 psi pressure–121°C.

Use: For testing the salt tolerance of a variety of microorganisms.

Salt Tolerance Medium, Gilardi

Composition per liter:
NaCl 65.0g
Pancreatic digest of casein 15.0g
Agar 15.0g
Papaic digest of soybean meal 5.0g

pH 7.3 ± 0.2 at 25°C

Preparation of Medium: Add components to distilled/deionized water and bring volume to 1.0L. Mix thoroughly. Gently heat and bring to boiling. Distribute into tubes or flasks. Autoclave for 15 min at 15 psi pressure–121°C. Do not overheat. Pour into sterile Petri dishes or leave in tubes.

Use: For the cultivation and maintenance of salt-tolerant, nonfermenting Gram-negative bacteria. For the differentiation of nonfermenting Gram-negative bacteria based on salt tolerance.

SAP 1 Agar

Composition per liter:
Agar 15.0g
Pancreatic digest of casein 5.0g
Yeast extract 5.0g
Artificial seawater 1.0L

pH 7.2 ± 0.2 at 25°C

Artificial Seawater:
Composition per liter:
NaCl 24.7g
$MgSO_4 \cdot 7H_2O$ 6.3g
$MgCl_2 \cdot 6H_2O$ 4.6g
$CaCl_2$ 1.0g
KCl 0.7g
$NaHCO_3$ 0.2g

Preparation of Artificial Seawater: Add components to distilled/deionized water and bring volume to 1.0L. Mix thoroughly.

Preparation of Medium: Add solid components to 1.0L of artificial seawater. Mix thoroughly. Gently heat and bring to boiling. Distribute into tubes or flasks. Autoclave for 15 min at 15 psi pressure–121°C. Pour into sterile Petri dishes or leave in tubes.

Use: For the isolation and cultivation of *Cytophaga* species, *Herpetosiphon* species, *Saprospira* species, and *Flexithrix* species.

SAP 2 Agar

Composition per liter:
Agar 15.0g
Pancreatic digest of casein 1.0g
Yeast extract 1.0g
Artificial seawater 1.0L

pH 7.2 ± 0.2 at 25°C

Artificial Seawater:
Composition per liter:
NaCl 24.7g

$MgSO_4·7H_2O$ 6.3g
$MgCl_2·6H_2O$ 4.6g
$CaCl_2$ 1.0g
KCl 0.7g
$NaHCO_3$ 0.2g

Preparation of Artificial Seawater: Add components to distilled/deionized water and bring volume to 1.0L. Mix thoroughly.

Preparation of Medium: Add solid components to 1.0L of artificial seawater. Mix thoroughly. Gently heat and bring to boiling. Distribute into tubes or flasks. Autoclave for 15 min at 15 psi pressure–121°C. Pour into sterile Petri dishes or leave in tubes.

Use: For the isolation and cultivation of *Cytophaga* species, *Herpetosiphon* species, *Saprospira* species, and *Flexithrix* species.

Saprospira grandis Medium

Composition per 1010.0mL:
Pancreatic digest of casein 5.0g
Yeast extract 5.0g
$Ca(NO_3)_2·4H_2O$ 0.1g
K_2HPO_4 .... 0.02g
Seawater, filtered 1.0L
Trace elements 10.0mL
pH 7.0 ± 0.2 at 25°C

Trace Elements:
Composition per liter:
$FeSO_4·7H_2O$ 0.5mg
$ZnSO_4·7H_2O$ 0.3mg
H_3BO_3 .... 0.1mg
$CoCl_2·6H_2O$ 0.1mg
$CuSO_4·5H_2O$ 0.1mg
$MnSO_4·4H_2O$ 0.1mg
$Na_2MoO_4·2H_2O$ 0.1mg

Preparation of Trace Elements: Add components to distilled/deionized water and bring volume to 1.0L. Mix thoroughly.

Preparation of Medium: Combine components. Mix thoroughly. Adjust pH to 7.0. Filter sterilize.

Use: For the cultivation of *Saprospira grandis*.

SB/SW Medium

Composition per liter:
NaCl 1.0g
KCl 0.5g
$MgCl_2·6H_2O$ 0.4g
NH_4Cl 0.25g
KH_2PO_4 .... 0.2g
$CaCl_2·2H_2O$ 0.15g
Resazurin 1.0mg
Sodium crotonate solution 50.0mL
$NaHCO_3$ solution 20.0mL
$Na_2S·9H_2O$ solution 10.0mL
Seven vitamin solution 10.0mL
Sodium dithionite solution 10.0mL
Trace elements solution SL-10 .... 1.0mL
pH 7.2 ± 0.2 at 25°C

Trace Elements Solution SL-10:
Composition per liter:
$FeCl_2·4H_2O$ 1.5g
$CoCl_2·6H_2O$ 190.0mg
$MnCl_2·4H_2O$ 100.0mg
$ZnCl_2$ 70.0mg
$Na_2MoO_4·2H_2O$ 36.0mg
$NiCl_2·6H_2O$ 24.0mg
H_3BO_3 .... 6.0mg
$CuCl_2·2H_2O$ 2.0mg
HCl (25% solution) 10.0mL

Preparation of Trace Elements Solution SL-10: Prepare and dispense under 80% N_2 + 20% CO_2. Add $FeCl_2·4H_2O$ to 10.0mL of HCl solution. Mix thoroughly. Add distilled/deionized water and bring volume to 1.0L. Add remaining components. Mix thoroughly. Sparge with 80% N_2 + 20% CO_2. Autoclave for 15 min at 15 psi pressure–121°C.

Sodium Crotonate Solution:
Composition per 50.0mL:
Sodium crotonate 1.1g

Preparation of Sodium Crotonate Solution: Add sodium crotonate to distilled/deionized water and bring volume to 50.0mL. Mix thoroughly. Filter sterilize. Sparge with 80% N_2 + 20% CO_2.

$NaHCO_3$ Solution:
Composition per 20.0mL:
$NaHCO_3$ 2.5g

Preparation of $NaHCO_3$ Solution: Add $NaHCO_3$ to distilled/deionized water and bring volume to 20.0mL. Mix thoroughly. Filter sterilize. Sparge with 80% N_2 + 20% CO_2.

$Na_2S·9H_2O$ Solution:
Composition per 10.0mL:
$Na_2S·9H_2O$ 0.36g

Preparation of $Na_2S·9H_2O$ Solution: Add $Na_2S·9H_2O$ to distilled/deionized water and bring volume to 10.0mL. Mix thoroughly. Sparge with 100% N_2. Autoclave for 15 min at 15 psi pressure–121°C.

Seven Vitamin Solution:
Composition per liter:
Pyridoxine·HCl 0.3g
Thiamine·HCl 0.2g
Nicotinic acid 0.2g
Calcium DL-pantothenate 0.1g
Vitamin B_{12} 0.1g
p-Aminobenzoic acid 80.0mg
Biotin 20.0mg

Preparation of Seven Vitamin Solution: Add components to distilled/deionized water and bring volume to 1.0L. Mix thoroughly.

Sodium Dithionite Solution:
Composition per 10.0mL:
Sodium dithioninium 0.2g

Preparation of Sodium Dithionite Solution: Add sodium dithioninium to distilled/deionized water and bring volume to 10.0mL. Mix thoroughly. Sparge with 100% N_2. Autoclave for 15 min at 15 psi pressure–121°C.

Preparation of Medium: Prepare medium and dispense under 80% N_2 + 20% CO_2. Add components, except sodium crotonate solution, seven vitamin solution, sodium dithionite solution, $NaHCO_3$ solution, and $Na_2S·9H_2O$ solution, to distilled/deionized water and bring volume to 910.0mL. Mix thoroughly. Sparge with 80% N_2 + 20% CO_2. Autoclave for 15 min at 15 psi pressure–121°C. Aseptically and anaerobically add 50.0mL of sterile sodium crotonate solution, 20.0mL of sterile $NaHCO_3$ solution, 10.0mL of seven vitamin solution, and 10.0mL of sterile $Na_2S·9H_2O$ solution. Mix thoroughly. Aseptically and anaerobically distribute into sterile tubes or flasks. After inoculation, add 0.1mL of sodium dithionite solution per 10.0mL of medium.

Use: For the cultivation of *Syntrophobacter buswellii.*

SB/SW Medium

Composition per liter:

NaCl 1.0g
KCl 0.5g
$MgCl_2 \cdot 6H_2O$ 0.4g
NH_4Cl 0.25g
KH_2PO_4 .... 0.2g
$CaCl_2 \cdot 2H_2O$ 0.15g
Resazurin 1.0mg
Sodium crotonate solution 50.0mL
$NaHCO_3$ solution 20.0mL
$Na_2S \cdot 9H_2O$ solution 10.0mL
Seven vitamin solution 10.0mL
Sodium dithionite solution 10.0mL
Trace elements solution SL-10 .... 1.0mL

pH 7.2 ± 0.2 at 25°C

Trace Elements Solution SL-10:
Composition per liter:

$FeCl_2 \cdot 4H_2O$ 1.5g
$CoCl_2 \cdot 6H_2O$ 190.0mg
$MnCl_2 \cdot 4H_2O$ 100.0mg
$ZnCl_2$ 70.0mg
$Na_2MoO_4 \cdot 2H_2O$ 36.0mg
$NiCl_2 \cdot 6H_2O$ 24.0mg
H_3BO_3 .... 6.0mg
$CuCl_2 \cdot 2H_2O$ 2.0mg
HCl (25% solution) 10.0mL

Preparation of Trace Elements Solution SL-10: Prepare and dispense under 80% N_2 + 20% CO_2. Add $FeCl_2 \cdot 4H_2O$ to 10.0mL of HCl solution. Mix thoroughly. Add distilled/deionized water and bring volume to 1.0L. Add remaining components. Mix thoroughly. Sparge with 80% N_2 + 20% CO_2. Autoclave for 15 min at 15 psi pressure–121°C.

Sodium Pyruvate Solution:
Composition per 50.0mL:

Sodium pyruvate 1.25g

Preparation of Sodium Pyruvate Solution: Add sodium pyruvate to distilled/deionized water and bring volume to 50.0mL. Mix thoroughly. Filter sterilize. Sparge with 80% N_2 + 20% CO_2.

$NaHCO_3$ Solution:
Composition per 20.0mL:

$NaHCO_3$ 2.5g

Preparation of $NaHCO_3$ Solution: Add $NaHCO_3$ to distilled/deionized water and bring volume to 20.0mL. Mix thoroughly. Filter sterilize. Sparge with 80% N_2 + 20% CO_2.

$Na_2S \cdot 9H_2O$ Solution:
Composition per 10.0mL:

$Na_2S \cdot 9H_2O$ 0.36g

Preparation of $Na_2S \cdot 9H_2O$ Solution: Add $Na_2S \cdot 9H_2O$ to distilled/deionized water and bring volume to 10.0mL. Mix thoroughly. Sparge with 100% N_2. Autoclave for 15 min at 15 psi pressure–121°C.

Seven Vitamin Solution:
Composition per liter:

Pyridoxine·HCl 0.3g
Thiamine·HCl 0.2g
Nicotinic acid 0.2g
Calcium DL-pantothenate 0.1g
Vitamin B_{12} 0.1g
p-Aminobenzoic acid 80.0mg
Biotin 20.0mg

Preparation of Seven Vitamin Solution: Add components to distilled/deionized water and bring volume to 1.0L. Mix thoroughly.

Sodium Dithionite Solution:
Composition per 10.0mL:

Sodium dithioninium 0.2g

Preparation of Sodium Dithionite Solution: Add sodium dithioninium to distilled/deionized water and bring volume to 10.0mL. Mix thoroughly. Sparge with 100% N_2. Autoclave for 15 min at 15 psi pressure–121°C.

Preparation of Medium: Prepare medium and dispense under 80% N_2 + 20% CO_2. Add components, except sodium pyruvate solution, seven vitamin solution, sodium dithionite solution, $NaHCO_3$ solution, and $Na_2S \cdot 9H_2O$ solution, to distilled/deionized water and bring volume to 910.0mL. Mix thoroughly. Sparge with 80% N_2 + 20% CO_2. Autoclave for 15 min at 15 psi pressure–121°C. Aseptically and anaerobically add 50.0mL of sterile sodium pyruvate solution, 20.0mL of sterile $NaHCO_3$ solution, 10.0mL of seven vitamin solution, and 10.0mL of sterile $Na_2S \cdot 9H_2O$ solution. Mix thoroughly. Aseptically and anaerobically distribute into sterile tubes or flasks. After inoculation, add 0.1mL of sodium dithionite solution per 10.0mL of medium.

Use: For the cultivation of *Syntrophobacter wolinii.*

SC Agar

Composition per liter:

Agar 15.0g
Papaic digest of soybean meal 8.0g
Cornmeal (solids from infusion) 2.0g
K_2HPO_4 .... 1.0g
KH_2PO_4 .... 1.0g
$MgSO_4 \cdot 7H_2O$ 0.2g
Hemin solution 15.0mL
Bovine serum albumin, fraction V 10.0mL
L-Cysteine·H_2O solution 10.0mL
Glucose solution 1.0mL

pH 6.6 ± 0.2 at 25°C

Hemin Solution:
Composition per 100.0mL:

Hemin 0.1g
NaOH (0.05*N* solution) 100.0mL

Preparation of Hemin Solution: Add hemin to NaOH solution. Mix thoroughly.

Bovine Serum Albumin, Fraction V Solution:
Composition per 10.0mL:

Bovine serum albumin, fraction V 2.0g

Preparation of Bovine Serum Albumin, Fraction V Solution: Add bovine serum albumin to distilled/deionized water and bring volume to 10.0mL. Mix thoroughly. Filter sterilize.

L-Cysteine·H_2O Solution:
Composition per 10.0mL:

L-Cysteine·H_2O 1.0g

Preparation of L-Cysteine·H_2O Solution: Add L-cysteine·H_2O to distilled/deionized water and bring volume to 10.0mL. Mix thoroughly. Filter sterilize.

Glucose Solution:
Composition per 10.0mL:

D-Glucose	5.0g

Preparation of Glucose Solution: Add glucose to distilled/deionized water and bring volume to 10.0mL. Mix thoroughly. Filter sterilize.

Preparation of Medium: Add components, except bovine serum albumin, L-cysteine·H_2O solution, and glucose solution, to distilled/deionized water and bring volume to 979.0mL. Mix thoroughly. Adjust pH to 6.6 with NaOH. Gently heat and bring to boiling. Autoclave for 15 min at 15 psi pressure–121°C. Cool to 45°–50°C. Aseptically add 10.0mL of sterile bovine serum albumin, 10.0mL of sterile L-cysteine·H_2O solution, and 1.0mL of sterile glucose solution. Mix thoroughly. Pour into sterile Petri dishes or distribute into sterile tubes.

Use: For the cultivation and maintenance of *Clavibacter xyli*.

SC Medium

Composition per 1021.0mL:

Agar	15.0g
Papaic digest of soybean meal	8.0g
Cornmeal, solids from infusion	2.0g
Tween 80	1.0g
K_2HPO_4	1.0g
KH_2PO_4	1.0g
$MgSO_4 \cdot 7H_2O$	0.2g
Hemin chloride solution	15.0mL
Bovine serum albumin solution	10.0mL
L-Cysteine solution	10.0mL
Glucose solution	1.0mL

pH 6.6 at 25°C

Hemin Chloride Solution:
Composition per 100.0mL:

Hemin chloride	0.1g
NaOH (0.05*N* solution)	100.0mL

Preparation of Hemin Chloride Solution: Add hemin chloride to 100.0mL of NaOH solution. Mix thoroughly.

Bovine Serum Albumin Solution:
Composition per 10.0mL:

Bovine serum albumin	2.0g

Preparation of Bovine Serum Albumin Solution: Add bovine serum albumin to distilled/deionized water and bring volume to 10.0mL. Mix thoroughly. Filter sterilize.

L-Cysteine Solution:
Composition per 10.0mL:

L-Cysteine, free base	1.0g

Preparation of L-Cysteine Solution: Add L-cysteine to distilled/deionized water and bring volume to 10.0mL. Mix thoroughly. Filter sterilize.

Glucose Solution:
Composition per 10.0mL:

Glucose	5.0g

Preparation of Glucose Solution: Add glucose to distilled/deionized water and bring volume to 10.0mL. Mix thoroughly. Autoclave for 15 min at 15 psi pressure–121°C. Cool to 25°C.

Preparation of Medium: Add components, except bovine serum albumin solution, L-cysteine solution, and glucose solution, to distilled/deionized water and bring volume to 1.0L. Mix thoroughly. Gently heat and bring to boiling. Autoclave for 15 min at 15 psi pressure–121°C. Cool to 45°–50°C. Aseptically add 10.0mL of sterile bovine serum albumin solution, 10.0mL of sterile L-cysteine solution, and 1.0mL of sterile glucose solution. Mix thoroughly. Pour into sterile Petri dishes or distribute into sterile tubes.

Use: For the isolation and cultivation of coryneform bacteria that cause ratoon stunting disease of sugarcane.

SCY Medium (Maintenance SCY Medium)

Composition per liter:

Solution A	1.0L
Solution B	200.0mL

Solution A:
Composition per liter:

Agar	10.0g
Pancreatic digest of casein	0.92g
NaCl	0.05g
Papaic digest of soybean meal	0.03g
K_2HPO_4	0.025g

pH 7.0 ± 0.2 at 25°C

Preparation of Solution A: Add components to distilled/deionized water and bring volume to 1.0L. Mix thoroughly. Gently heat and bring to boiling. Distribute into tubes in 10.0mL volumes. Autoclave for 15 min at 15 psi pressure–121°C. Allow tubes to cool in a slanted position.

Solution B:
Composition per 200.0mL:

Sucrose	2.0g
Yeast extract	0.5g
Thiamine	0.8mg
Vitamin B_{12}	0.02mg

pH 8.5 ± 0.2 at 25°C

Preparation of Solution B: Add components to slightly alkaline tap water, pH 8.5, and bring volume to 200.0mL. Mix thoroughly. Filter sterilize.

Preparation of Medium: Inoculate bacteria onto prepared slants of solution A. After inoculation of tubes, aseptically add 2.0mL of sterile solution B on top of each slant.

Use: For the cultivation and maintenance of iron bacteria. For the cultivation and maintenance of *Haliscomenobacter hydrossis*.

Seawater Agar (SWA)

Composition per liter:

Agar	15.0g
Peptone	5.0g
Yeast extract	5.0g
Beef extract	3.0g
Seawater, synthetic	1.0L

pH 7.5 ± 0.2 at 25°C

Seawater, Synthetic:
Composition per liter:

NaCl	27.0g
$MgSO_4 \cdot 7H_2O$	7.0g
Tris(hydroxymethyl)aminomethane buffer	2.0g
KCl	0.6g
$CaCl_2$	0.3g

Preparation of Seawater, Synthetic: Add components to distilled/deionized water and bring volume to 1.0L. Mix thoroughly.

Preparation of Medium: Combine components. Mix thoroughly. Gently heat and bring to boiling. Distribute into tubes or flasks. Autoclave for 15 min at 15 psi pressure–121°C. Pour into sterile Petri dishes or leave in tubes.

Use: For the isolation and cultivation of halophilic microorganisms, such as *Pseudomonas* species and *Vibrio* species, from fish.

Seawater Agar

Composition per 1.0L:

Agar	20.0g
Beef extract	10.0g
Peptone	10.0g
Seawater	750.0mL

pH 7.2 ± 0.2 at 25°C

Preparation of Medium: Add components to tap water and bring volume to 1.0L. Mix thoroughly. Gently heat and bring to boiling. Distribute into tubes or flasks. Autoclave for 15 min at 15 psi pressure–121°C. Pour into sterile Petri dishes or leave in tubes.

Use: For the selective isolation and cultivation of *Planococcus* species.

Seawater Agar (SWA)

Composition per liter:

Agar	15.0g
Peptone	5.0g
Yeast extract	5.0g
Beef extract	3.0g
Seawater, synthetic	1.0L

pH 7.5 ± 0.2 at 25°C

Seawater, Synthetic:
Composition per liter:

NaCl	24.0g
$MgSO_4 \cdot 7H_2O$	7.0g
$MgCl_2 \cdot 6H_2O$	5.3g
KCl	0.7g
$CaCl_2$	0.1g

Preparation of Seawater, Synthetic: Add components to distilled/deionized water and bring volume to 1.0L. Mix thoroughly. Adjust pH to 7.5.

Preparation of Medium: Combine components. Mix thoroughly. Gently heat and bring to boiling. Distribute into tubes or flasks. Autoclave for 15 min at 15 psi pressure–121°C. Pour into sterile Petri dishes or leave in tubes.

Use: For the isolation and cultivation of halophilic *Vibrio* species from fish.

Seawater Agar

Composition per liter:

Agar	20.0g
Beef extract	10.0g
Peptone	10.0g
Seawater	750.0mL

Artificial Seawater:
Composition per liter:

NaCl	28.13g
$MgSO_4 \cdot 7H_2O$	3.5g
$MgCl_2$	2.55g
$CaCl_2$	1.2g
KCl	0.77g
$NaHCO_3$	0.11g

pH 7.3 ± 0.2 at 25°C

Preparation of Artificial Seawater: Natural seawater is stored in the dark for at least 3 weeks to "age." If natural seawater is not available, use artificial seawater. To prepare artificial seawater, add components to distilled/deionized water and bring volume to 1.0L. Mix thoroughly.

Preparation of Medium: Add beef extract and peptone to distilled/deionized water and bring volume to 250.0mL. Mix thoroughly. Adjust pH to 7.8. Gently heat and bring to boiling. Boil for 10.0 min. Add 750.0mL of natural or artificial seawater. Mix thoroughly. Adjust pH to 7.3. Add 20.0g of agar. Mix thoroughly. Gently heat and bring to boiling. Distribute into tubes or flasks. Autoclave for 15 min at 15 psi pressure–121°C. Pour into sterile Petri dishes or leave in tubes.

Use: For the cultivation and maintenance of *Alteromonas rubra, Brevibacterium stationis, Chromohalobacter marismortui, Flectobacillus marinus, Marinococcus albus, Marinococcus halophilus, Pasteurella piscicida, Photobacterium phosphoreum, Planococcus citreus, Planococcus kocurii, Vibrio adaptatus, Vibrio campbellii, Vibrio costicola, Vibrio harveyi, Vibrio mediterranei, Vibrio natriegens,* and other *Vibrio* species.

Seawater Agar with Fetal Calf Serum

Composition per liter:

Agar	20.0g
Beef extract	10.0g
Peptone	10.0g
Seawater	750.0mL
Fetal calf serum	100.0mL

pH 7.2 ± 0.2 at 25°C

Preparation of Medium: Add components, except fetal calf serum, to tap water and bring volume to 900.0mL. Mix thoroughly. Gently heat and bring to boiling. Autoclave for 15 min at 15 psi pressure–121°C. Cool to 50°–55°C. Aseptically add 100.0mL of sterile fetal calf serum warmed to 50°–55°C. Mix thoroughly. Pour into sterile Petri dishes or distribute into sterile tubes.

Use: For the cultivation of *Aeromonas* species and *Vibrio salmonicida.*

Seawater Agar with Horse Blood

Composition per liter:

Agar	20.0g
Beef extract	10.0g
Peptone	10.0g
Seawater	750.0mL
Horse blood	100.0mL

pH 7.2 ± 0.2 at 25°C

Preparation of Medium: Add components, except horse blood, to tap water and bring volume to 900.0mL. Mix thoroughly. Gently heat and bring to boiling. Autoclave for 15 min at 15 psi pressure–121°C. Cool to 50°–55°C. Aseptically add 100.0mL of sterile horse blood warmed to 50°–55°C. Mix thoroughly. Pour into sterile Petri dishes or distribute into sterile tubes.

Use: For the cultivation of *Aeromonas* species.

Seawater Agar Medium

Composition per liter:

Agar	15.0g
Beef extract	10.0g
Peptone	10.0g
Seawater, aged	750.0mL

pH 7.2–7.3 at 25°C

Preparation of Medium: Add components to distilled/deionized water and bring volume to 1.0L. Mix thoroughly.

Gently heat and bring to boiling. Distribute into tubes or flasks. Autoclave for 15 min at 15 psi pressure–121°C. Pour into sterile Petri dishes or leave in tubes.

Use: For the isolation and cultivation of marine *Flavobacterium* species.

Seawater Agar Modified (DSMZ Medium 917)

Composition per liter:

NaCl	17.7g
Agar	15.0g
$MgSO_4 \cdot 7H_2O$	4.46g
$MgCl_2 \cdot 6H_2O$	3.4g
Peptone	2.5g
Hexadecane	2.0g
Yeast extract	1.5g
KCl	0.48g
Calcium chloride solution	10.0mL

pH 7.2 ± 0.2 at 25°C

Calcium Chloride Solution:

Composition per 10.0mL:

$CaCl_2 \cdot 2H_2O$	0.98g

Preparation of Calcium Chloride Solution: Add $CaCl_2 \cdot 2H_2O$ to distilled/deionized water and bring volume to 10.0mL. Mix thoroughly. Autoclave for 15 min at 15 psi pressure–121°C.

Preparation of Medium: Add components, except agar and calcium chloride solution, to distilled/deionized water and bring volume to 990.0mL. Adjust pH to 7.2. Mix thoroughly. Add 15.0g agar. Gently heat and bring to boiling. Autoclave for 15 min at 15 psi pressure–121°C. Cool to 50°C. Aseptically add 10.0mL sterile calcium chloride solution. Mix thoroughly. Pour into sterile Petri dishes or distribute to sterile tubes.

Use: For the cultivation and maintenance of *Muricauda ruestringensis.*

Seawater Agar with 1% Serum

Composition per liter:

Agar	12.0g
Seawater	990.0mL
Horse serum	10.0mL

Preparation of Medium: Add agar to 990.0mL of seawater. Mix thoroughly. Gently heat and bring to boiling. Autoclave for 15 min at 15 psi pressure–121°C. Cool to 45°–50°C. Aseptically add 10.0mL of sterile horse serum. Mix thoroughly. Pour into sterile Petri dishes or distribute into sterile tubes.

Use: For the cultivation and maintenance of *Basipetospora halophila, Halosphaeria retorquens, Thraustochytrium striatum,* and *Lagenidium callinectes.*

Seawater Basal Medium

Composition per liter:

$NH_4 \cdot Cl$	10.0g
Lactate	2.0g
$K_2HPO_4 \cdot 3H_2O$	75.0mg
$FeSO_4 \cdot 7H_2O$	29.0mg
Tris (hydroxymethyl) amino methane	50.0m*M*
Artificial seawater	500.0mL

Artificial Seawater:

Composition per liter:

$MgSO_4 \cdot 7H_2O$	24.65g
NaCl	23.37g
$CaCl_2 \cdot 2H_2O$	2.94g
KCl	1.49g

pH 7.3 ± 0.2 at 25°C

Preparation of Medium: Add components to distilled/deionized water and bring volume to 1.0L. Mix thoroughly. Distribute into tubes or flasks. Autoclave for 15 min at 15 psi pressure–121°C.

Use: For the cultivation of marine bacteria.

Seawater Complete Medium

Composition per liter:

Pancreatic digest of casein	5.0g
Yeast extract	3.0g
Seawater	750.0mL
Glycerol	3.0mL

Preparation of Medium: Add components to distilled/deionized water and bring volume to 1.0L. Mix thoroughly. Distribute into tubes or flasks. Autoclave for 15 min at 15 psi pressure–121°C.

Use: For the cultivation and maintenance of *Vibrio fischeri.*

Seawater Lemco Agar

Composition per liter:

Agar	15.0g
Beef extract	10.0g
Peptone	10.0g
Seawater, filtered aged	750.0mL

pH 7.3 ± 0.2 at 25°C

Preparation of Medium: Add components, except agar, to distilled/deionized water and bring volume to 1.0L. Mix thoroughly. Adjust pH to 7.8. Gently heat and bring to boiling. Boil for 3–5 min. Filter through Whatman filter paper. Adjust pH to 7.3. Add agar. Mix thoroughly. Gently heat and bring to boiling. Distribute into tubes or flasks. Autoclave for 15 min at 15 psi pressure–121°C. Pour into sterile Petri dishes or leave in tubes.

Use: For the cultivation of *Halococcus nondenitrificans.*

Seawater Medium

Composition per liter:

Agar	15.0g
Peptone	5.0g
Beef extract	2.0g
KNO_3	0.5g
Seawater, aged	1.0L

pH 7.8 ± 0.2 at 25°C

Preparation of Medium: Combine components. Mix thoroughly. Gently heat and bring to boiling. Distribute into tubes or flasks. Autoclave for 15 min at 15 psi pressure–121°C. Pour into sterile Petri dishes or leave in tubes.

Use: For the cultivation of halophilic bacteria.

Seawater 802 Medium

Composition per liter:

Solution A	500.0mL
Solution B	500.0mL

Solution A:

Composition per 500.0mL:

NaCl	27.5g
$MgCl_2 \cdot 6H_2O$	5.38g
$MgSO_4 \cdot 7H_2O$	6.78g
KCl	0.72g
$NaHCO_3$	0.2g
$CaCL_2 \cdot 2H_2O$	1.4g

Preparation of Solution A: Add components to distilled/deionized water and bring volume to 500.0mL. Mix thoroughly. Filter sterilize.

Solution B:
Composition per liter:
Rye grass cerophyll 5.0g

Preparation of Solution B: Add cerophyll to distilled/deionized water and bring volume to 1.0L. Mix thoroughly. Gently heat and bring to a boil. Boil for 5 min. Filter through Whatman #1 filter paper. Add 0.5g of Na_2HPO_4. Bring volume to 1.0L with distilled/deionized water. Mix thoroughly. Distribute 10.0mL volumes into tubes. Autoclave for 15 min at 15 psi pressure–121°C. Cool to 25°C.

Source: Cerophyll can be obtained from Ward's Natural Science Establishment, Inc. Dairy Goat Nutrition distributes Grass Media Culture, which is equivalent. Cereal Leaf Product from Sigma Chemical is similar to cerophyll.

Preparation of Medium: Aseptically add 500.0mL of sterile solution A and 500.0mL of sterile solution B. Mix thoroughly. Aseptically distribute into sterile tubes or flasks.

Use: For the cultivation of *Amastigomonas bermudensis, Ancyromonas sigmoides, Bodo curvifilus, Bodo saliens, Cafeteria roenbergensis, Cafeteria minuta, Ciliophrys infusionum, Cruzella marina, Glauconema bermudense, Helkesimastix faecicola, Jakoba libera, Massisteria marina, Monosiga brevicollis, Percolomonas cosmopolitus, Pteridomonas danica,* and *Trimyema shoalsia.*

Use: For the cultivation of *Euplotes harpa.*

Seawater *Nitrosomonas* Medium

Composition per 1003.3mL:
HEPES (*N*-[2-Hydroxyethyl]piperazine-*N'*-2-ethanesulfonic acid) buffer 4.76g
$(NH_4)_2SO_4$ 1.5g
$MgSO_4 \cdot 7H_2O$ 0.2g
$CaCl_2 \cdot 2H_2O$ 20.0mg
K_2HPO_4 .... 15.0mg
Artificial seawater 1.0L
K_2CO_3 (5% solution) 2.0mL
Trace elements solution 1.0mL
Phenol Red (0.04% solution) 0.3mL
pH 7.8 ± 0.2 at 25°C

Trace Elements Solution:
Composition per liter:
EDTA 2.06g
$FeSO_4 \cdot 7H_2O$ 1.54g
$MnCl_2 \cdot 4H_2O$ 0.2g
$Na_2MoO_4 \cdot 2H_2O$ 0.1g
$ZnSO_4 \cdot 7H_2O$ 0.1g
$CuSO_4 \cdot 5H_2O$ 20.0mg
$CoCl_2 \cdot 6H_2O$ 2.0mg
HCl, concentrated 83.0mL

Preparation of Trace Elements: Add HCl and EDTA to distilled/deionized water and and bring volume to 900.0mL. Mix thoroughly. Add remaining components. Mix thoroughly.

Artificial Seawater:
Composition per liter:
NaCl 27.5g
$MgSO_4 \cdot 7H_2O$ 6.78g
$MgCl_2 \cdot 6H_2O$ 5.38g
$CaCl_2 \cdot 2H_2O$ 1.4g
KCl 0.72g
$NaHCO_3$ 0.2g

Preparation of Artificial Seawater: Add components to distilled/deionized water and bring volume to 1.0L. Mix thoroughly.

Preparation of Medium: Combine components. Mix thoroughly. Adjust pH to 7.8. Distribute into tubes or flasks. Autoclave for 15 min at 15 psi pressure–121°C.

Use: For the cultivation of *Nitrosomonas cryotolerans.*

Seawater Nutrient Agar (SNA) (ATCC Medium 2205)

Composition per liter:
Nutrient agar, 2× 500.0mL
Artificial seawater, 2× 500.0mL
pH 7.3 ± 0.2 at 25°C

Nutrient Agar, 2×:
Composition per 500.0mL:
Agar 15.0g
Peptone 5.0g
NaCl 5.0g
Yeast extract 2.0g
Beef extract 1.0g

Preparation of Nutrient Agar, 2×: Add components to distilled/deionized water and bring volume to 500.0mL. Mix thoroughly. Gently heat and bring to boiling. Autoclave for 15 min at 15 psi pressure–121°C. Cool to 45°–50°C.

Seawater, Synthetic 2×:
Composition per 500.0mL:
NaCl 24.0g
$MgSO_4 \cdot 7H_2O$ 7.0g
$MgCl_2 \cdot 6H_2O$ 5.3g
KCl 0.7g
$CaCl_2$ 0.1g

Preparation of Seawater, Synthetic: Add components to distilled/deionized water and bring volume to500.0mL. Mix thoroughly. Adjust pH to 7.5. Filter sterilize.

Preparation of Medium: Warm synthetic seawater to 50°C. Aseptically combine sterile nutrient agar and sterile synthetic seawater. Mix thoroughly. Pour into sterile Petri dishes or distribute into sterile tubes.

Use: For the isolation and cultivation of marine bacteria.

Seawater with Serum

Composition per liter:
Agar 10.0g
Seawater 1.0L
Bovine serum, sterile 100.0mL

Preparation of Medium: Add agar to 1.0L of seawater. Mix thoroughly. Gently heat and bring to boiling. Autoclave for 15 min at 15 psi pressure–121°C. Cool to 45–50°C. Aseptically add 100.0mL of sterile liquid beef serum. Mix thoroughly. Pour into sterile Petri dishes or distribute into sterile tubes.

Use: For the cultivation and maintenance of *Basipetospora halophila, Halosphaeria retorquens, Lagenidium callinectes,* and *Thraustochytrium striatum.*

Seawater *Spirillum* Medium

Composition per liter:
Calcium lactate 10.0g

Peptone..5.0g
Beef extract..3.0g
Yeast extract..3.0g
Seawater..750.0mL

pH 7.0 ± 0.2 at 25°C

Preparation of Medium: Add components to distilled/deionized water and bring volume to 1.0L. Mix thoroughly. Adjust pH to 7.0. Distribute into tubes or flasks. Autoclave for 20 min at 10 psi pressure–115°C. A precipitate will form during autoclaving.

Use: For the cultivation of marine *Spirillum* species.

Seawater Yeast Extract Agar

Composition per liter:

Marine salts mix..37.9g
Agar..15.0g
Proteose peptone..10.0g
Yeast extract..3.0g

pH 7.2–7.4 at 25°C

Preparation of Medium: Add components to distilled/deionized water and bring volume to 1.0L. Mix thoroughly. Gently heat and bring to boiling. Distribute into tubes or flasks. Autoclave for 15 min at 15 psi pressure–121°C. Pour into sterile Petri dishes or leave in tubes.

Use: For the cultivation and maintenance of *Alteromonas* species, *Caulobacter halobacteroides*, *Caulobacter maris*, *Cytophaga marinoflava*, and *Cytophaga salmonicolor.*

Seawater Yeast Extract Broth, Modified

Composition per liter:

NaCl..23.4g
$MgSO_4 \cdot 7H_2O$..6.9g
Peptone..1.0g
Yeast extract..1.0g
KCl..0.75g

Preparation of Medium: Add components to distilled/deionized water and bring volume to 1.0L. Mix thoroughly. Distribute into tubes or flasks. Autoclave for 15 min at 15 psi pressure–121°C.

Use: For the cultivation and maintenance of *Proteus* species and *Vibrio* species.

Seawater Yeast Extract Peptone Medium

Composition per liter:

Agar..15.0g
Peptone..5.0g
Yeast extract..3.0g
Seawater, aged and filtered..750.0mL

pH 7.3 ± 0.2 at 25°C

Preparation of Medium: Add components, except agar, to distilled/deionized water and bring volume to 1.0L. Mix thoroughly. Adjust pH to 7.8. Gently heat and bring to boiling. Continue boiling for 3–5 min. Filter through Whatman filter paper. Adjust pH to 7.3. Add agar. Gently heat and bring to boiling. Distribute into tubes or flasks. Autoclave for 15 min at 15 psi pressure–121°C. Pour into sterile Petri dishes or leave in tubes.

Use: For the cultivation of *Planococcus kocurii.*

Seawater Yeast Peptone Agar (DSMZ Medium 243)

Composition per liter:

Agar..12.0g
Peptone..5.0g
Yeast extract..3.0g
Filtered, seawater..750.0mL

pH 7.3 ± 0.2 at 25°C

Artificial Seawater:

NaCl..28.13g
$MgCl_2 \cdot 6H_2O$..4.8g
$MgSO_4 \cdot 7H_2O$..3.5g
$CaCl_2 \cdot 2H_2O$..1.6g
KCl..0.77g
$NaHCO_3$..0.11g

Preparation of Artificial Seawater: Add components to distilled/deionized water and bring volume to 1.0L. Mix thoroughly. Filter sterilize.

Preparation of Medium: Add peptone and yeast extract to 250.0mL distilled/deionized water. Mix thoroughly. Adjust pH to 7.8. Boil for 5 min. Filter and readjust the pH to 7.3. Add agar. Gently heat and bring to boiling. Distribute into tubes or flasks. Autoclave for 15 min at 15 psi pressure–121°C. Cool to 45-50°C. Add 750.0mL filter sterilized seawater that has been heated to 50°C. (Note: Natural seawater is stored in the dark for at least 3 weeks to age. If natural seawater is not available use artificial seawater.) Mix thoroughly. Pour into sterile Petri dishes or leave in tubes.

Use: For the cultivation and maintenance of *Zobellia uliginosa=Cellulophaga uliginosa=Cytophaga uliginosa, Marinobacterium jannaschii=Oceanospirillum jannaschii, Vibrio harveyi=Lucibacterium harveyi, Halomonas* sp., and *Pseudoalteromonas espejiana=Alteromonas espejiana.*

Seawater Yeast Peptone Medium (DSMZ Medium 949)

Composition per liter:

Peptone..5.0g
Yeast extract..3.0g
Seawater, filtered and aged..750.0mL

pH 7.3 ± 0.2 at 25°C

Preparation of Medium: Add components to distilled/deionized water and bring volume to 1.0L. Mix thoroughly. Adjust pH to 7.3. Distribute into tubes or flasks. Autoclave for 15 min at 15 psi pressure–121°C.

Use: For the cultivation of *Shewanella algae.*

Selenomonas acidaminophila Medium

Composition per liter:

Disodium β-glycerophosphate..19.0g
Beef extract..5.0g
Lactose..5.0g
Papaic digest of soybean meal..5.0g
Sodium glutamate..3.4g
Pancreatic digest of casein..2.5g
Peptic digest of animal tissue..2.5g
Yeast extract..2.5g
Ascorbic acid..0.5g
$MgSO_4 \cdot 7H_2O$..0.25g

pH 7.15 ± 0.05 at 25°C

Preparation of Medium: Add components to distilled/deionized water and bring volume to 1.0L. Mix thoroughly. Distribute into tubes or flasks. Autoclave for 15 min at 15 psi pressure–121°C.

Use: For the cultivation and maintenance of *Selenomonas acidaminophila.*

Selenomonas ruminantium Medium

Composition per liter:

Pancreatic digest of casein 5.0g
Na_2CO_3 .... 4.0g
Sodium acetate 4.0g
Yeast extract 2.0g
Glucose 1.0g
KH_2PO_4 .... 1.0g
L-Cysteine·HCl 0.5g
Resazurin 1.0mg
n-Valeric acid 0.1mL
pH 7.0 ± 0.2 at 25°C

Preparation of Medium: Prepare and dispense medium under 100% CO_2. Add components to distilled/deionized water and bring volume to 1.0L. Mix thoroughly. Sparge with 100% CO_2. Adjust pH to 7.0. Anaerobically distribute into tubes or flasks. Autoclave for 20 min at 15 psi pressure–121°C.

Use: For the cultivation and maintenance of *Selenomonas ruminantium* and *Selenomonas* species.

Selenomonas Selective Medium (SS Medium)

Composition per 100.0mL:

Pancreatic digest of casein 0.5g
Mannitol 0.2g
$FeSO_4 \cdot 7H_2O$ 0.1g
Sodium acetate 0.1g
Yeast extract 0.1g
L-Cysteine·HCl 0.08g
Mineral solution S 4.0mL
Sodium carbonate (8% solution) 2.5mL
n-Valeric acid 0.05mL
pH 5.9–6.1 at 25°C

Mineral Solution S:
Composition per liter:

KH_2PO_4 .... 12.0g
NaCl 12.0g
$(NH_4)_2SO_4$ 6.0g
$MgSO_4 \cdot 7H_2O$ 2.5g
$CaCl_2 \cdot 2H_2O$ 1.6g

Preparation of Mineral Solution S: Add components to distilled/deionized water and bring volume to 1.0L. Mix thoroughly.

Preparation of Medium: Add components to distilled/deionized water and bring volume to 100.0mL. Mix thoroughly. Filter sterilize. Aseptically distribute into sterile tubes or flasks.

Use: For the isolation and cultivation of *Selenomonas* species.

Seven-Hour Fecal Coliform Agar (Seven-Hour FC Agar) (m-Seven-Hour Fecal Coliform Agar)

Composition per liter:

Agar 15.0g
Lactose 10.0g
NaCl 7.5g
D-Mannitol 5.0g
Proteose peptone No. 3 .... 5.0g
Yeast extract 3.0g
Bromcresol Purple 0.35g
Phenol Red 0.3g
Sodium lauryl sulfate 0.2g
Sodium deoxycholate 0.1g
pH 7.3 ± 0.1 at 25°C

Preparation of Medium: Add components to distilled/deionized water and bring volume to 1.0L. Mix thoroughly. Gently heat and bring to boiling. Continue boiling for 5 min. Cool to 55°–60°C. Adjust pH to 7.3 with 0.1*N* NaOH. Cool to 45°–50°C. Pour into sterile Petri dishes with tight-fitting lids in 5.0mL volumes. Store at 2°–10°C.

Use: For the rapid estimation of the bacteriological quality of water using the membrane filter method.

SF1 Medium

Composition per liter:

NaCl 120.0g
$MgCl_2 \cdot 6H_2O$ 7.0g
$MgSO_4 \cdot 7H_2O$ 6.0g
KCl 3.8g
Pancreatic digest of casein 2.0g
Yeast extract 2.0g
NH_4Cl 1.0g
$CaCl_2 \cdot 2H_2O$ 0.5g
L-Cysteine·HCl 0.5g
$K_2HPO_4 \cdot 3H_2O$ 0.4g
Resazurin 1.0mg
$Na_2SeO_3 \cdot 5H_2O$ 75.0μg
Na_2CO_3 solution 20.0mL
Trimethylamine·HCl solution 20.0mL
$Na_2S \cdot 9H_2O$ solution 10.0mL
Trace elements solution SL-10 .... 1.0mL
NaOH (10*M* solution) 0.6mL
pH 7.3 ± 0.2 at 25°C

Na_2CO_3 Solution:
Composition per 20.0mL:

Na_2CO_3 .... 10.0mg

Preparation of Na_2CO_3 Solution: Add Na_2CO_3 to distilled/deionized water and bring volume to 20.0mL. Mix thoroughly. Sparge under 100% N_2. Autoclave for 15 min at 15 psi pressure–121°C. Store under N_2.

Trimethylamine·HCl Solution:
Composition per 20.0mL:

Trimethylamine·HCl 10.0mg

Preparation of Trimethylamine·HCl Solution: Add trimethylamine·HCl to distilled/deionized water and bring volume to 20.0mL. Mix thoroughly. Sparge under 100% N_2. Autoclave for 15 min at 15 psi pressure–121°C. Store under N_2.

$Na_2S \cdot 9H_2O$ Solution:
Composition per 10.0mL:

$Na_2S \cdot 9H_2O$ 10.0mg

Preparation of $Na_2S \cdot 9H_2O$ Solution: Add $Na_2S \cdot 9H_2O$ to distilled/deionized water and bring volume to 10.0mL. Mix thoroughly. Sparge under 100% N_2. Autoclave for 15 min at 15 psi pressure–121°C. Store under N_2.

Trace Elements Solution SL-10:
Composition per liter:

$FeCl_2 \cdot 4H_2O$ 1.5g
$CoCl_2 \cdot 6H_2O$ 190.0mg
$MnCl_2 \cdot 4H_2O$ 100.0mg
$ZnCl_2$ 70.0mg
$Na_2MoO_4 \cdot 2H_2O$ 36.0mg
$NiCl_2 \cdot 6H_2O$ 24.0mg
H_3BO_3 .... 6.0mg

$CuCl_2 \cdot 2H_2O$ 2.0mg
HCl (25% solution) 10.0mL

Preparation of Trace Elements Solution SL-10: Add $FeCl_2 \cdot 4H_2O$ to 10.0mL of HCl solution. Mix thoroughly. Add distilled/deionized water and bring volume to 1.0L. Add remaining components. Mix thoroughly. Sparge with 80% N_2 + 20% CO_2. Autoclave for 15 min at 15 psi pressure–121°C.

Preparation of Medium: Prepare and dispense medium under 80% N_2 + 20% CO_2. Add components, except L-cysteine·HCl, NaOH, Na_2CO_3 solution, trimethylamine·HCl solution, and $Na_2S \cdot 9H_2O$ solution, to distilled/deionized water and bring volume to 950.0mL. Mix thoroughly. Gently heat and bring to boiling. Continue boiling for 5 min. Cool to room temperature while sparging with 80% N_2 + 20% CO_2. Add L-cysteine·HCl and NaOH while contiuning to sparge with 80% N_2 + 20% CO_2. Adjust pH to 6.7. Anaerobically distribute into tubes or bottles. Autoclave for 15 min at 15 psi pressure–121°C. Aseptically and anaerobically add 20.0mL of sterile trimethylamine·HCl solution, 20.0mL of sterile Na_2CO_3 solution, and 10.0mL of sterile $Na_2S \cdot 9H_2O$ solution per 950.0mL of medium. Check that final pH is 6.7.

Use: For the cultivation and maintenance of *Methanohalophilus* species.

SG Agar

Composition per liter:

Agar 15.0g
Pancreatic digest of casein 15.0g
$CaCl_2 \cdot 2H_2O$ 2.0g
$MgSO_4 \cdot 7H_2O$ 1.0g
pH 7.0 ± 0.2 at 25°C

Preparation of Medium: Add components to distilled/deionized water and bring volume to 1.0L. Mix thoroughly. Gently heat and bring to boiling. Distribute into tubes or flasks. Autoclave for 15 min at 15 psi pressure–121°C. Pour into sterile Petri dishes or leave in tubes.

Use: For the cultivation of myxobacteria.

SIM Medium

Composition per liter:

Peptone 30.0g
Agar 3.0g
Beef extract 3.0g
Peptonized iron 0.2g
$Na_2S_2O_3 \cdot 5H_2O$ 0.025g
pH 7.3 ± 0.2 at 25°C

Source: This medium is available as a premixed powder from BD Diagnostic Systems.

Preparation of Medium: Add components to distilled/deionized water and bring volume to 1.0L. Mix thoroughly. Gently heat and bring to boiling. Distribute into tubes in 15.0mL volumes. Autoclave for 15 min at 15 psi pressure–121°C. Allow tubes to cool in an upright position.

Use: For the differentiation of members of the Enterobacteriaceae based on H_2S production, indole production, and motility.

SIM Medium

Composition per liter:

Pancreatic digest of casein 20.0g
Peptic digest of animal tissue 6.1g
Agar 3.5g
$Fe(NH_4)_2(SO_4)_2 \cdot 6H_2O$ 0.2g
$Na_2S_2O_3 \cdot 5H_2O$ 0.2g
pH 7.3 ± 0.2 at 25°C

Source: This medium is available as a premixed powder from BD Diagnostic Systems and Oxoid Unipath.

Preparation of Medium: Add components to distilled/deionized water and bring volume to 1.0L. Mix thoroughly. Gently heat and bring to boiling. Distribute into tubes in 15.0mL volumes. Autoclave for 15 min at 15 psi pressure–121°C. Allow tubes to cool in an upright position.

Use: For the differentiation of members of the Enterobacteriaceae based on H_2S production, indole production, and motility.

Simmons' Citrate Agar (Citrate Agar)

Composition per liter:

Agar 15.0g
NaCl 5.0g
Sodium citrate 2.0g
K_2HPO_4 .... 1.0g
$(NH_4)H_2PO_4$ 1.0g
$MgSO_4 \cdot 7H_2O$ 0.2g
Bromothymol Blue 0.08g
pH 6.9 ± 0.2 at 25°C

Source: This medium is available as a premixed powder from BD Diagnostic Systems and Oxoid Unipath.

Preparation of Medium: Add components to distilled/deionized water and bring volume to 1.0L. Mix thoroughly. Gently heat while stirring and bring to boiling. Distribute into tubes or flasks. Autoclave for 15 min at 15 psi pressure–121°C. Pour into sterile Petri dishes or leave in tubes.

Use: For the differentiation of Gram-negative bacteria on the basis of citrate utilization. Bacteria that can utilize citrate as sole carbon source turn the medium blue.

Simonsiella Agar (LMG Medium 31)

Composition per liter:

Tryptone 17.0g
Agar 15.0g
NaCl 5.0g
Yeast extract 4.0g
Soya peptone 3.0g
K_2HPO_4 .... 2.5g
Bovine serum 100.0mL
pH 7.2 ± 0.2 at 25°C

Preparation of Medium: Add components, except bovine serum, to distilled/deionized water and bring volume to 900.0mL. Mix thoroughly. Gently heat and bring to boiling. Autoclave for 15 min at 15 psi pressure–121°C. Cool to 45°–50°C. Aseptically add 100.0mL sterile bovine serum. Mix thoroughly. Pour into sterile Petri dishes or distribute into sterile tubes.

Use: For the cultivation of *Simonsiella* spp.

Simonsiella Agar

Composition per liter:

Pancreatic digest of casein 17.0g
Agar 15.0g
NaCl 5.0g
Yeast extract 4.0g
Papaic digest of soybean meal 3.0g

K_2HPO_4 .... 2.5g
Horse serum 100.0mL

Preparation of Medium: Add components, except horse serum, to distilled/deionized water and bring volume to 900.0mL. Mix thoroughly. Gently heat and bring to boiling. Autoclave for 15 min at 15 psi pressure–121°C. Cool to 50°–55°C. Aseptically add 100.0mL of sterile horse serum warmed to 50°–55°C. Mix thoroughly. Pour into sterile Petri dishes or distribute into sterile tubes.

Use: For the cultivation and maintenance of *Simonsiella muelleri* and *Simonsiella steedae*.

Singh's Medium, Modified

Composition per liter:

NaCl 8.75g
Lactalbumin hydrolysate 8.13g
Yeast extract 6.25g
D-Glucose 5.0g
$CaCl_2 \cdot 2H_2O$ 0.25g
KCl 0.25g
$NaH_2PO_4 \cdot H_2O$ 0.25g
$NaHCO_3$ 0.15g
$MgCl_2 \cdot 6H_2O$ 0.13g
Phenol Red 0.01g
Fetal bovine serum
(heat inactivated at 56°C, 30 min) 200.0mL
pH 7.0 ± 0.2 at 25°C

Preparation of Medium: Add components to distilled/deionized water and bring volume to 1.0L. Mix thoroughly. Adjust pH to 7.0 with NaOH if necessary. Filter sterilize. Aseptically distribute into sterile tubes or flasks.

Use: For the cultivation and maintenance of *Spiroplasma* species.

SJ Agar

Composition per liter:

Agar 15.0g
K_2HPO_4 .... 1.0g
KCl 0.5g
$MgSO_4 \cdot 7H_2O$ 0.5g
$NaNO_3$ 0.5g
$FeSO_4 \cdot 7H_2O$ 0.01g
Glucose solution 100.0mL
pH 7.2 ± 0.2 at 25°C

Glucose Solution:
Composition per 100.0mL:

D-Glucose 1.0g

Preparation of Glucose Solution: Add D-glucose to distilled/deionized water and bring volume to 100.0mL. Mix thoroughly. Autoclave for 15 min at 15 psi pressure–121°C. Cool to 25°C.

Preparation of Medium: Add components, except glucose solution, to distilled/deionized water and bring volume to 900.0mL. Mix thoroughly. Gently heat and bring to boiling. Autoclave for 15 min at 15 psi pressure–121°C. Cool to 45°–50°C. Aseptically add sterile glucose solution. Mix thoroughly. Pour into sterile Petri dishes or distribute into sterile tubes.

Use: For the isolation and cultivation of *Cytophaga* species, *Herpetosiphon* species, *Saprospira* species, and *Flexithrix* species.

Skim Milk Acetate Medium

Composition per liter:

Agar 15.0g
Skim milk powder 5.0g
Yeast extract 0.5g
Sodium acetate 0.2g

Preparation of Medium: Add components to distilled/deionized water and bring volume to 1.0L. Mix thoroughly. Gently heat and bring to boiling. Distribute into tubes or flasks. Autoclave for 15 min at 15 psi pressure–121°C. Pour into sterile Petri dishes or leave in tubes.

Use: For the cultivation and maintenance of *Cytophaga johnsonae*.

Skim Milk Agar (Milk Agar) (ATCC Medium 377)

Composition per liter:

Agar 15.0g
Skim milk 8.0g

Preparation of Medium: Add components to distilled/deionized water and bring volume to 1.0L. Mix thoroughly. Gently heat and bring to boiling. Distribute into tubes or flasks. Autoclave for 15 min at 15 psi pressure–121°C. Pour into sterile Petri dishes or leave in tubes.

Use: For the cultivation and maintenance of *Herpetosiphon aurantiacus*.

SL Medium (DSMZ Medium 959)

Composition per liter:

NaCl 15.0g
PIPES 3.4g
$Na\text{-}acetate \cdot 3H_2O$ 2.72g
Yeast extract 2.0g
Trypticase 2.0g
NH_4Cl 1.0g
$MgCl_2 \cdot 6H_2O$ 0.5g
K_2HPO_4 .... 0.35g
KH_2PO_4 .... 0.35g
KCl 0.2g
$CaCl_2 \cdot 2H_2O$ 0.1g
Resazurin 0.5mg
Maltose solution 50.0mL
$Na_2S \cdot 9H_2O$ solution 10.0mL
pH 47.5 ± 0.2 at 25°C

Maltose Solution:
Composition per 100.0mL:

Maltose 3.5g

Preparation of Maltose Solution: Add maltose to distilled/deionized water and bring volume to 100.0mL. Mix thoroughly. Sparge with N_2. Autoclave for 15 min at 15 psi pressure–121°C. Cool to 25°C. Store anaerobically.

$Na_2S \cdot 9H_2O$ Solution:
Composition per 50.0mL:

$Na_2S \cdot 9H_2O$ 5.0g

Preparation of $Na_2S \cdot 9H_2O$ Solution: Add $Na_2S \cdot 9H_2O$ to distilled/deionized water and bring volume to 50.0mL. Sparge with N_2. Autoclave for 15 min at 15 psi pressure–121°C. Cool to 25°C. Store anaerobically.

Preparation of Medium: Add components, except maltose solution and $Na_2S \cdot 9H_2$ solution, to distilled/deionized water and bring volume to 940.0mL. Mix thoroughly. Gently heat and bring to boiling. Sparge with N_2. Adjust pH to 7.0. Autoclave for 15 min at 15 psi pressure–121°C. Cool to 25°C. Aseptically add 10.0mL sterile $Na_2S \cdot 9H_2O$ solution and 50.0mL sterile maltose solution. Mix thoroughly.

Aseptically and anaerobically distribute into sterile tubes or flasks under N_2. The final pH should be 7.5.

Use: For the cultivation of *Petrotoga olearia* and *Petrotoga sibirica.*

Slanetz and Bartley Medium

Composition per liter:

Tryptose	20.0g
Agar	10.0g
Yeast extract	5.0g
Na_2HPO_4 $2H_2O$	4.0g
Glucose	2.0g
NaN_3	0.4g
Tetrazolium chloride	0.1g

pH 7.2 ± 0.2 at 25°C

Source: This medium is available as a premixed powder from Oxoid Unipath.

Preparation of Medium: Add components to distilled/deionized water and bring volume to 1.0L. Mix thoroughly. Gently heat and bring to boiling. Distribute into tubes or flasks. Autoclave for 15 min at 15 psi pressure–121°C. Pour into sterile Petri dishes.

Use: For the detection and enumeration of enterococci by the membrane filter method.

Sludge Medium for Methanobacteria

Composition per liter:

$NaHCO_3$	4.0g
Sodium formate	2.0g
Sodium acetate	1.0g
Yeast extract	1.0g
L-Cysteine·HCl·H_2O	0.5g
KH_2PO_4	0.5g
$Na_2S·9H_2O$	0.5g
$MgSO_4·7H_2O$	0.4g
NaCl	0.4g
NH_4Cl	0.4g
$CaCl_2·2H_2O$	0.05g
$FeSO_4·7H_2O$	2.0mg
Resazurin	1.0mg
Sludge fluid	50.0mL
Fatty acid mixture	20.0mL
Trace elements SL-6	1.0mL

pH 6.7-7.0 at 25°C

Sludge Fluid:

Composition per 100.0mL:

Yeast extract	0.4g
Sludge	100.0mL

Preparation of Sludge Fluid: Add yeast extract to a concentration of 0.4% to sludge taken from an anaerobic digester. Gas with 100% N_2 for 5 min. Incubate at 37°C for 24 hr. Centrifuge at 13,000 × g. Remove the supernatant fluid. Gas with 100% N_2 for 5 min. Autoclave for 15 min at 15 psi pressure–121°C. Store at 25°C protected from light.

Fatty Acid Mixture:

Composition per 20.0mL:

α-Methylbutyric acid	0.5g
Isobutyric acid	0.5g
Isovaleric acid	0.5g
Valeric acid	0.5g

Preparation of Fatty Acid Mixture: Add components to distilled/deionized water and bring volume to 20.0mL. Mix thoroughly. Adjust pH to 7.5 with concentrated NaOH.

Trace Elements Solution SL-6:

Composition per liter:

H_3BO_3	0.3g
$CoCl_2·6H_2O$	0.2g
$ZnSO_4·7H_2O$	0.1g
$MnCl_2·4H_2O$	0.03g
$Na_2MoO_4·H_2O$	0.03g
$NiCl_2·6H_2O$	0.02g
$CuCl_2·2H_2O$	0.01g

Preparation of Trace Elements Solution SL-6: Add components to distilled/deionized water and bring volume to 1.0L. Mix thoroughly. Adjust pH to 3.4.

Preparation of Medium: Prepare and dispense medium under 80% N_2 + 20% CO_2. Add components to distilled/deionized water and bring volume to 1.0L. Mix thoroughly. Adjust pH to 6.7–7.0. Distribute anaerobically into tubes or bottles with aluminum seals. Autoclave for 15 min at 15 psi pressure–121°C with fast exhaust.

Use: For the cultivation of *Methanobacterium uliginosum* and *Methanobrevibacter ruminantium.*

Sludge Medium for Methanobacteria, pH 7.9

Composition per liter:

$NaHCO_3$	4.0g
Sodium formate	2.0g
Sodium acetate	1.0g
Yeast extract	1.0g
L-Cysteine·HCl·H_2O	0.5g
KH_2PO_4	0.5g
$Na_2S·9H_2O$	0.5g
$MgSO_4·7H_2O$	0.4g
NaCl	0.4g
NH_4Cl	0.4g
$CaCl_2·2H_2O$	0.05g
$FeSO_4·7H_2O$	2.0mg
Resazurin	1.0mg
Sludge fluid	50.0mL
Fatty acid mixture	20.0mL
Trace elements SL-6	1.0mL

pH 7.9 ± 0.2 at 25°C

Sludge Fluid:

Composition per 100.0mL:

Yeast extract	0.4g
Sludge	100.0mL

Preparation of Sludge Fluid: Add yeast extract to a concentration of 0.4% to sludge taken from an anaerobic digester. Gas with 100% N_2 for 5 min. Incubate at 37°C for 24 hr. Centrifuge at 13,000 × g. Remove the supernatant fluid. Gas with 100% N_2 for 5 min. Autoclave for 15 min at 15 psi pressure–121°C. Store at 25°C protected from light.

Fatty Acid Mixture:

Composition per 20.0mL:

α-Methylbutyric acid	0.5g
Isobutyric acid	0.5g
Isovaleric acid	0.5g
Valeric acid	0.5g

Preparation of Fatty Acid Mixture: Add components to distilled/deionized water and bring volume to 20.0mL. Mix thoroughly. Adjust pH to 7.5 with concentrated NaOH.

Trace Elements Solution SL-6:

Composition per liter:

H_3BO_3	0.3g
$CoCl_2·6H_2O$	0.2g

$ZnSO_4·7H_2O$ 0.1g
$MnCl_2·4H_2O$ 0.03g
$Na_2MoO_4·H_2O$ 0.03g
$NiCl_2·6H_2O$ 0.02g
$CuCl_2.2H_2O$ 0.01g

Preparation of Trace Elements Solution SL-6: Add components to distilled/deionized water and bring volume to 1.0L. Mix thoroughly. Adjust pH to 3.4.

Preparation of Medium: Prepare and dispense medium under 80% N_2 + 20% CO_2. Add components to distilled/deionized water and bring volume to 1.0L. Adjust pH to 7.9. Distribute anaerobically into tubes or bottles with aluminum seals. Autoclave for 15 min at 15 psi pressure–121°C with fast exhaust.

Use: For the cultivation of *Methanobacterium alcaliphilum* and *Methanobacterium thermoalcalip.*

SM Basal Salts Medium

Composition per liter:
Na_2HPO_4 .... 4.5g
KH_2PO_4 .... 1.5g
NH_4Cl 0.3g
$MgSO_4·7H_2O$ 0.1g
Trace metals solution 5.0mL
pH 6.0 ± 0.2 at 25°C

Trace Metals Solution:
Composition per liter:
Ethylenediamine tetraacetate 50.0g
$ZnSO_4·7H_2O$ 22.0g
$CaCl_2$ 5.54g
$MnCl_2·4H_2O$ 5.06g
$FeSO_4·7H_2O$ 4.99g
$CoCl_2·H_2O$ 1.61g
$CuSO_4·5H_2O$ 1.57g
$(NH_4)_6Mo_7O_{24}·4H_2O$ 1.1g

Preparation of Trace Elements Solution: Add components to distilled/deionized water and bring volume to 1.0L. Mix thoroughly. Adjust pH to 6.0.

Preparation of Medium: Add components to distilled/deionized water and bring volume to 1.0L. Mix thoroughly. Adjust pH to 6.0. Distribute into tubes or flasks. Autoclave for 15 min at 15 psi pressure–121°C.

Use: For the cultivation of *Thiobacillus delicatus.*

SM Medium

Composition per liter:
Na_2HPO_4 .... 4.5g
KH_2PO_4 .... 1.5g
NH_4Cl 0.3g
$MgSO_4·7H_2O$ 0.1g
$Na_2S_2O_3$ solution 100.0mL
Trace metals solution 5.0mL
pH 7.5 ± 0.2 at 25°C

$Na_2S_2O_3$ Solution
$Na_2S_2O_3$ solution 10.0g

Preparation of $Na_2S_2O_3$ Solution: Add 10.0g of $Na_2S_2O_3$ to distilled/deionized water and bring volume to 100.0mL. Mix thoroughly. Filter sterilize.

Trace Metals Solution:
Composition per liter:
Ethylenediaminetetraacetic acid 50.0g
$ZnSO_4·7H_2O$ 22.0g
$CaCl_2$ 5.54g
$MnCl_2·4H_2O$ 5.06g
$FeSO_4·7H_2O$ 4.99g
$CoCl_2·6H_2O$ 1.61g
$CuSO_4·5H_2O$ 1.57g
$(NH_4)_6Mo_7O_{24}·H_2O$ 1.10g

Preparation of Trace Metals Solution: Add components to distilled/deionized water and bring volume to 1.0L. Mix thoroughly. Adjust pH to 6.0 using KOH. Filter sterilize.

Preparation of Medium: Add components, except $Na_2S_2O_3$ solution and trace metals solution, to distilled/deionized water and bring volume to 895.0mL. Mix thoroughly. Autoclave for 15 min at 15 psi pressure–121°C. Aseptically add 100.0mL of sterile $Na_2S_2O_3$ solution and 5.0mL of sterile trace metals solution. Mix thoroughly. Aseptically distribute into sterile tubes or flasks.

Use: For the cultivation of *Thiobacillus thioparus.*

SMC Medium

Composition per liter:
Sorbitol 70.0g
Pancreatic digest of casein 17.0g
NaCl 5.0g
Beef extract 3.0g
Yeast extract 3.0g
Beef heart, solids from infusion 2.0g
Horse serum 200.0mL
Yeast extract solution 100.0mL
Phenol Red solution 20.0mL
Sucrose solution 20.0mL
L-Arginine·HCl solution 10.0mL
Fructose solution 2.0mL
Glucose solution 2.0mL
pH 7.5 ± 0.2 at 25°C

Yeast Extract Solution:
Composition per 100.0mL:
Yeast extract 25.0g

Preparation of Yeast Extract Solution: Add yeast extract to distilled/deionized water and bring volume to 100.0mL. Mix thoroughly. Autoclave for 15 min at 15 psi pressure–121°C. Cool to 45°–50°C.

Phenol Red Solution:
Composition per 100.0mL:
Phenol Red 0.01g

Preparation of Phenol Red Solution: Add Phenol Red to distilled/deionized water and bring volume to 100.0mL. Mix thoroughly. Autoclave for 15 min at 15 psi pressure–121°C. Cool to 45°–50°C.

Sucrose Solution:
Composition per 20.0mL:
Sucrose 10.0g

Preparation of Sucrose Solution: Add sucrose to distilled/deionized water and bring volume to 20.0mL. Mix thoroughly. Autoclave for 15 min at 15 psi pressure–121°C. Cool to 45°–50°C.

L-Arginine·HCl Solution:
Composition per 10.0mL:
L-Arginine·HCl 4.2g

Preparation of L-Arginine·HCl Solution: Add L-arginine·HCl to distilled/deionized water and bring volume to 10.0mL. Mix thoroughly. Autoclave for 15 min at 15 psi pressure–121°C. Cool to 45°–50°C.

Fructose Solution:
Composition per 10.0mL:
Fructose 5.0g

Preparation of Fructose Solution: Add fructose to distilled/deionized water and bring volume to 10.0mL. Mix thoroughly. Autoclave for 15 min at 15 psi pressure–121°C. Cool to 45°–50°C.

Glucose Solution:
Composition per 10.0mL:
Glucose 5.0g

Preparation of Glucose Solution: Add glucose to distilled/deionized water and bring volume to 10.0mL. Mix thoroughly. Autoclave for 15 min at 15 psi pressure–121°C. Cool to 45°–50°C.

Preparation of Medium: Add components, except horse serum, yeast extract solution, Phenol Red solution, sucrose solution, L-arginine·HCl solution, fructose solution, and glucose solution, to distilled/deionized water and bring volume to 646.0mL. Mix thoroughly. Gently heat and bring to boiling. Autoclave for 15 min at 15 psi pressure–121°C. Cool to 45°–50°C. Aseptically add 200.0mL of sterile horse serum, 100.0mL of sterile yeast extract solution, 20.0mL of sterile Phenol Red solution, 20.0mL of sterile sucrose solution, 10.0mL of sterile L-arginine·HCl solution, 2.0mL of sterile fructose solution, and 2.0mL of sterile glucose solution. Mix thoroughly. Aseptically distribute into sterile tubes or flasks.

Use: For the cultivation and maintenance of *Spiroplasma citri.*

SMC, Modified

Composition per liter:
Sorbitol 70.0g
Pancreatic digest of casein 17.0g
NaCl 5.0g
Beef extract 3.0g
Yeast extract 3.0g
Beef heart, solids from infusion 2.0g
Solution 1 .... 100.0mL
Solution 3 .... 20.0mL
Solution 2 .... 10.0mL
NaOH (1*N* solution) 6.0mL
pH 7.7–7.8 at 25°C

Solution 1:
Composition per 100.0mL:
Sucrose 10.0g
Yeast extract 2.0g
Fructose 1.0g
Glucose 1.0g
Phenol Red 0.02g

Preparation of Solution 1: Add components to distilled/deionized water and bring volume to 100.0mL. Mix thoroughly. Filter sterilize.

Solution 2:
Composition per 10.0mL:
Bovine serum albumin, fraction V 0.1g

Preparation of Solution 2: Add bovine serum albumin to distilled/deionized water and bring volume to 10.0mL. Mix thoroughly. Filter sterilize.

Solution 3:
Composition per 20.0mL:
Horse serum 20.0mL

Preparation of Solution 3: Inactivate horse serum at 56°C for 30 min. Filter sterilize.

Preparation of Medium: Add components, except solution 1, solution 2, and solution 3, to distilled/deionized water and bring volume to 870.0mL. Autoclave for 15 min at 15 psi pressure–121°C. Cool to 45°–50°C. Aseptically add 100.0mL of sterile solution 1, 20.0mL of sterile solution 3, and 10.0mL of sterile solution 2. Mix thoroughly. Adjust pH to 7.7–7.8. Aseptically distribute into sterile tubes or flasks.

Use: For the cultivation and maintenance of *Spiroplasma citri.*

SME Agar (ATCC Medium 2345)

Composition per 1010.0mL:
NaCl 27.7g
Agar 18.0g
$MgSO_4 \cdot 7H_2O$ 7.0g
$MgCl_2 \cdot 6H_2O$ 5.5g
$CaCl_2 \cdot 2H_2O$ 1.5g
KCl 0.65g
NaBr 0.1g
H_3BO_3 .... 30.0mg
$SrCl \cdot 6H_2O$ 15.0mg
KI 0.05mg
Tris·HCl buffer (1.0*M* solution, pH 7.0) 25.0mL

Preparation of Medium: Add components to distilled/deionized water and bring volume to 1.0L. Mix thoroughly. Gently heat and bring to boiling. Distribute into tubes or flasks. Autoclave for 15 min at 15 psi pressure–121°C. Pour into sterile Petri dishes or leave in tubes.

Use: For the cultivation and maintenance of *Aquifex pyrophilus.*

SME Medium, Modified

Composition per 1010.0mL:
NaCl 30.0g
$MgSO_4 \cdot 7H_2O$ 7.0g
$NaHCO_3$ 2.0g
KCl 0.65g
$CaCl_2 \cdot 2H_2O$ 0.5g
K_2HPO_4 .... 0.15g
NH_4Cl 0.15g
NaBr 0.1g
$MgCl_2 \cdot 6H_2O$ 5.5mg
Trace elements solution 10.0mL
pH 6.5 ± 0.2 at 25°C

Trace Elements Solution:
Composition per liter:
$MgSO_4 \cdot 7H_2O$ 3.0g
$(NH_4)_2Ni(SO_4)_2$ 2.0g
Nitrilotriacetic acid 1.5g
NaCl 1.0g
$MnSO_4 \cdot 2H_2O$ 0.5g
$CaCl_2 \cdot 2H_2O$ 0.1g
$FeSO_4 \cdot 7H_2O$ 0.1g
$NiCl_2 \cdot 6H_2O$ 0.025g
$KAl(SO_4)_2 \cdot 12H_2O$ 0.02g
$CoSO_4 \cdot 7H_2O$ 0.18g
$ZnSO_4 \cdot 7H_2O$ 0.18g
$CuSO_4 \cdot 5H_2O$ 0.01g
H_3BO_3 .... 0.01g
$Na_2MoO_4 \cdot 2H_2O$ 0.01g
Na_2WO_4 .... 10.0mg

Na_2SeO_4 .... 10.0mg
$Na_2SeO_3 \cdot 5H_2O$ 0.3mg

Preparation of Trace Elements Solution: Add nitrilotriacetic acid to approximately 500.0mL of distilled/deionized water. Dissolve by adding KOH and adjust pH to 6.5. Add remaining components. Bring volume to 1.0L with additional distilled/deionized water. Adjust pH to 7.0 with KOH.

Preparation of Medium: Add components to distilled/deionized water and bring volume to 1.0L. Mix thoroughly. Sparge with 100% N_2 for 20 min. Adjust pH to 6.5–6.8 with H_2SO_4. Distribute 10.0mL volumes into 120.0mL serum bottles while gassing under 100% CO_2. Stopper each serum bottle tightly. Exchange the gas phase in each serum bottle with 79.75% H_2 + 19.75% CO_2 + 0.5% O_2 and bring pressure to 300KPa. Autoclave for 15 min at 15 psi pressure–121°C.

Use: For the cultivation and maintenance of *Aquifex pyrophilus.*

Soil Extract

Composition per 200.0mL:
African Violet soil 77.0g
Na_2CO_3 .... 0.2g

Preparation of Medium: Add components to 200.0mL of distilled/deionized water. Mix thoroughly. Autoclave for 60 min at 15 psi pressure–121°C. Filter through paper and reserve filtrate.

Use: Used as a growth factor additive for the cultivaiton of soil bacteria and fungi.

Soil Extract Agar

Composition per liter:
Agar 20.0g
Glucose 1.0g
K_2HPO_4 .... 1.0g
Peptone 1.0g
Yeast extract 1.0g
Soil extract 400.0mL
Cycloheximide solution 10.0mL
pH 6.6 ± 0.2 at 25°C

Soil Extract:
Composition per liter:
Garden soil, neutral 1.0Kg

Preparation of Soil Extract: Add garden soil to 1.0L of tap water. Autoclave for 20 min at 15 psi pressure–121°C. Filter through Whatman filter paper. Bring volume to 1.0L with tap water.

Cycloheximide Solution:
Composition per 10.0mL:
Cycloheximide 0.04g

Preparation of Cycloheximide Solution: Add cycloheximide to distilled/deionized water and bring volume to 10.0mL. Mix thoroughly. Filter sterilize.

Preparation of Medium: Add components, except cycloheximide solution, to distilled/deionized water and bring volume to 990.0mL. Mix thoroughly. Gently heat and bring to boiling. Autoclave for 15 min at 15 psi pressure–121°C. Cool to 45°–50°C. Aseptically add sterile cycloheximide solution. Mix thoroughly. Pour into sterile Petri dishes or distribute into sterile tubes.

Use: For the isolation and cultivation of *Arthrobacter* species.

Soil Extract Agar

Composition per liter:
Soil 500.0g
Agar 15.0g
Glucose 2.0g
Yeast extract 1.0g
KH_2PO_4 .... 0.5g

Preparation of Medium: Add 500.0g of garden soil to 1.0L of tap water. Autoclave for 3 hr at 15 psi pressure–121°C. Filter through Whatman #2 filter paper. Add remaining components to filtrate. Bring volume to 1.0L with tap water. Gently heat and bring to boiling. Distribute into tubes in 7.0mL volumes. Autoclave for 15 min at 15 psi pressure–121°C. Allow tubes to cool in a slanted position.

Use: For the cultivation and identification of *Histoplasma capsulatum, Blastomyces dermatitidis*, and *Bacillus* species based on the formation of typical conidia.

Soil Extract Agar

Composition per liter:
Agar 15.0g
Soil extract 1.0L
pH 6.8 ± 0.2 at 25°C

Soil Extract:
Composition per liter:
Soil 400.0g

Preparation of Soil Extract: Air dry garden soil with a high content of organic matter and pass through a sieve. Weigh out 400.0g and add to 960.0mL of tap water. Autoclave for 60 min at 15 psi pressure–121°C. Cool to room temperature and allow soil to settle out. Decant supernatant solution. Filter through paper. Bring volume to 1.0L with distilled/deionized water.

Preparation of Medium: Add agar to 1.0L soil extract. Gently heat and bring to boiling. Autoclave for 15 min at 15 psi pressure–121°C. Pour into sterile Petri dishes or distribute into sterile tubes.

Use: For the cultivation of *Aureobacterium flavescens* and *Bacillus* species.

Soil Extract Glucose Yeast Extract Agar

Composition per liter:
Agar 15.0g
Glucose 2.0g
Yeast extract 1.0g
Soil extract 250.0mL
pH 6.8 ± 0.2 at 25°C

Soil Extract:
Composition per liter:
Garden soil 500.0g

Preparation of Soil Extract: Add 500.0g of garden soil to 1.0L of tap water. Autoclave for 1 hr at 15 psi pressure–121°C. Filter through Whatman #2 filter paper.

Preparation of Medium: Add components to distilled/deionized water and bring volume to 1.0L. Mix thoroughly. Gently heat and bring to boiling. Distribute into tubes or flasks. Autoclave for 15 min at 15 psi pressure–121°C. Pour into sterile Petri dishes or leave in tubes.

Use: For the cultivation and maintenance of *Streptomyces rectus.*

Soil Extract Glycerol Medium

Composition per liter:

Glycerol 20.0g
Peptone 5.0g
Beef extract 3.0g
Soil extract 150.0mL

pH 7.0 ± 0.2 at 25°C

Soil Extract:

Composition per liter:

Soil 400.0g

Preparation of Soil Extract: Allow garden soil to air dry. Add 400.0g of the air dried garden soil to 960.0mL of tap water. Autoclave for 60 min at 15 psi pressure–121°C. Allow to cool to 25°C. Let stand until settling ceases. Decant the liquid through Whatman filter paper. Autoclave for 30 min at 15 psi pressure–121°C. Cool to room temperature. Let stand until settling ceases.

Preparation of Medium: Add components to tap water and bring volume to 1.0L. Mix thoroughly. Adjust pH to 7.0. Distribute into tubes or flasks. Autoclave for 15 min at 15 psi pressure–121°C.

Use: For the cultivation and maintenance of *Mycobacterium terrae.*

Soil Extract Medium

Composition per liter:

Agar 15.0g
Pancreatic digest of gelatin 5.0g
Beef extract 3.0g
Soil extract 250.0mL

pH 6.8 ± 0.2 at 25°C

Soil Extract:

Composition per liter:

Garden soil 500.0g

Preparation of Soil Extract: Add 500.0g of garden soil to 1.0L of tap water. Autoclave for 1 hr at 15 psi pressure–121°C. Filter through Whatman #2 filter paper.

Preparation of Medium: Add components to distilled/deionized water and bring volume to 1.0L. Mix thoroughly. Gently heat and bring to boiling. Distribute into tubes or flasks. Autoclave for 15 min at 15 psi pressure–121°C. Pour into sterile Petri dishes or leave in tubes.

Use: For the cultivation and maintenance of *Streptomyces rectus.*

Soil Extract Medium

Composition per liter:

Soil 400.0g
Agar 15.0g

pH 6.8 ± 0.2 at 25°C

Preparation of Medium: Sieve air-dried garden soil with a high content of organic matter. To 400.0g of soil add 1.0L of distilled/deionized water. Autoclave for 60 min at 15 psi pressure–121°C. Cool to room temperature. Allow solids to sediment for a few hours. Carefully decant the supernatant solution. Centrifuge the supernatant solution at 10,000 × g for 15 min. Decant supernatant solution. Add agar (15.0g per liter of supernatant). Gently heat and bring to boiling. Distribute into tubes or flasks. Autoclave for 15 min at 15 psi pressure–121°C. Pour into sterile Petri dishes or leave in tubes.

Use: For the cultivation of *Arthrobacter* species, *Aureobacterium flavescens, Aureobacterium terregens,* and *Bacillus thiaminolyticus.*

Soil Extract Peptone Beef Extract Medium

Composition per liter:

Agar 15.0g
Peptone 5.0g
Beef extract 3.0g
Soil extract 1.0L

pH 7.0 ± 0.2 at 25°C

Soil Extract:

Composition per liter:

Garden soil 400.0g

Preparation of Soil Extract: Add garden soil to 1.0L of tap water. Autoclave for 1 hr at 15 psi pressure–121°C. Filter through cheesecloth and Whatman #2 filter paper. Autoclave filtrate again for 20 min at 15 psi pressure–121°C. Filter through Whatman #2 filter paper.

Preparation of Medium: Add agar, peptone, and beef extract to 1.0L of soil extract. Mix thoroughly. Gently heat and bring to boiling. Distribute into tubes or flasks. Autoclave for 15 min at 15 psi pressure–121°C. Pour into sterile Petri dishes or leave in tubes.

Use: For the cultivation and maintenance of *Oerskovia turbata* and *Oerskovia xanthineolytica.*

Soil Extract Potato Extract Medium

Composition per 510.0mL:

Malt extract 10.0g
Yeast extract 4.0g
Potato extract 250.0mL
Soil extract 250.0mL

pH 7.0 ± 0.2 at 25°C

Soil Extract:

Composition per liter:

Garden soil 400.0g

Preparation of Soil Extract: Add garden soil to 1.0L of tap water. Autoclave for 45 min at 15 psi pressure–121°C. Filter through cheesecloth.

Potato Extract:

Composition per liter:

Potatoes 400.0g

Preparation of Potato Extract: Peel and dice potatoes. Add 500.0mL of distilled/deionized water. Gently heat and bring to boiling. Continue boiling for 15 min. Filter through cheesecloth. Bring volume to 1.0L with distilled/deionized water.

Preparation of Medium: Combine components. Mix thoroughly. Distribute into tubes or flasks. Autoclave for 15 min at 15 psi pressure–121°C.

Use: For the cultivation and maintenance of *Saccharopolyspora rectivirgula.*

Soil Extract Salts Medium

Composition per liter:

Soil extract stock 625.0mL
Solution 1 .... 125.0mL
Solution 2 .... 125.0mL
Solution 3 .... 125.0mL

Soil Extract Stock:

Composition per liter:

Soil, air dried and sieved 333.3g

Preparation of Soil Extract Stock: Add soil to distilled/deionized water and bring volume to 1.0L. Mix thoroughly. Adjust pH to 8.0 with 1*N* NaOH or HCl. Autoclave for 30 min at 15 psi pressure–121°C. Allow soil to settle out. Carefully pour off supernatant. Filter through two layers of cheesecloth.

Solution 1:
Composition per liter:
K_2HPO_4 .... 1.0g

Preparation of Solution 1: Add K_2HPO_4 to distilled/deionized water and bring volume to 1.0L. Mix thoroughly.

Solution 2:
Composition per liter:
$MgSO_4 \cdot 7H_2O$ 1.0g

Preparation of Solution 2: Add $MgSO_4 \cdot 7H_2O$ to distilled/deionized water and bring volume to 1.0L. Mix thoroughly.

Solution 3:
Composition per liter:
KNO_3 .... 10.0g

Preparation of Solution 3: Add KNO_3 to distilled/deionized water and bring volume to 1.0L. Mix thoroughly.

Preparation of Medium: Combine components. Mix thoroughly. Filter sterilize. Aseptically distribute into sterile tubes or flasks.

Use: For the cultivation of *Chlamydomonas applanata, Polytoma uvella, Polytoma mirum, Polytoma ellipticum, Polytoma difficile,* and *Polytoma anomale.*

Soil Seawater Medium for Algae

Composition per liter:
HESNW medium solution 750.0mL
Soil extracts salts medium solution 250.0mL

HESNW Medium Solution:
Composition per 1011.0mL:
Natural seawater 1.0L
Enrichment solution 10.0mL
Vitamin solution 1.0mL

Preparation of HESNW Medium Solution: Allow natural seawater to age for 2 months. Filter sterilize. Aseptically add 10.0mL of sterile vitamin solution and 1.0mL enrichment solution. Mix thoroughly.

Enrichment Solution:
Composition per liter:
$NaNO_3$ 4.667g
$Na_2SiO_3 \cdot 9H_2O$ 3.000g
Sodium glycerophosphate 0.667g
$EDTA \cdot 2H_2O$ 0.553g
H_3BO_3 .... 0.380g
$Fe(NH_4)_2(SO_4)_2 \cdot 6H_2O$ 0.234g
$MnSO_4 \cdot 4H_2O$ 0.054g
$FeCl_3 \cdot 6H_2O$ 0.016g
$ZnSO_4 \cdot 7H_2O$ 7.3mg
$CoSO_4 \cdot 7H_2O$ 1.6mg

Preparation of Enrichment Solution: Add 3.000g of $Na_2SiO_3 \cdot 9H_2O$ to distilled/deionized water. Mix thoroughly. Neutralize $Na_2SiO_3 \cdot 9H_2O$ with 1*N* HCl. Add 500.0mL of distilled/deionized water. Mix thoroughly. Add remaining components and bring volume to 1.0L with distilled/deionized water. Mix thoroughly. Filter sterilize.

Vitamin Solution:
Composition per liter:
Thiamine 0.1g
Vitamin B_{12} 2.0mg
Biotin 1.0mg

Preparation of Vitamin Solution: Add components to distilled/deionized water and bring volume to 1.0L. Mix thoroughly. Filter sterilize.

Soil Extracts Salts Medium Solution:
Composition per liter:
Soil extract stock 625.0mL
Solution 1 .... 125.0mL
Solution 2 .... 125.0mL
Solution 3 .... 125.0mL

Soil Extract Stock:
Composition per liter:
Soil, air dried and sieved 333.3g

Preparation of Soil Extract Stock: Air dry soil. Sieve through fine-mesh screen. Add soil to distilled/deionized water and bring volume to 1.0L. Adjust pH to 8.0 with 1*N* NaOH or 1*N* HCl. Autoclave for 30 min at 15 psi pressure–121°C. Allow soil to settle. Decant liquid. Filter through cheesecloth.

Solution 1:
Composition per liter:
K_2HPO_4 .... 1.0g

Preparation of Solution 1: Add K_2HPO_4 to distilled/deionized water and bring volume to 1.0L. Mix thoroughly.

Solution 2:
Composition per liter:
$MgSO_4 \cdot 7H_2O$ 1.0g

Preparation of Solution 2: Add $MgSO_4 \cdot 7H_2O$ to distilled/deionized water and bring volume to 1.0L. Mix thoroughly.

Solution 3:
Composition per liter:
KNO_3 .... 10.0g

Preparation of Solution 3: Add KNO_3 to distilled/deionized water and bring volume to 1.0L. Mix thoroughly.

Preparation of Soil Extracts Salts Medium Solution: Combine 625.0mL of soil extract stock, 125.0mL of solution 1, 125.0mL of solution 2, and 125.0mL of solution 3. Filter sterilize.

Preparation of Medium: Aseptically combine 750.0mL of sterile HESNW medium solution and 250.0mL of sterile soil extracts salts medium solution. Aseptically distribute into sterile tubes or flasks.

Use: For the cultivation of *Amphora roettgeri.*

Sorogena Medium

Composition per 2000.0mL:
HI Agar 1.0L
HI/LV Broth 1.0L

HI Agar:
Composition per liter:
Agar 15.0g
Hay infusion broth 1.0L
pH 6.5 ± 0.2 at 25°C

Hay Infusion Broth:
Composition per liter:
Hay 2.5g

Preparation of Hay Infusion Broth: Add hay to distilled/deionized water and bring volume to 1.0L. Mix thoroughly. Gently heat and bring to boiling. Boil for 30 min. Filter through Whatman #1 filter paper. Bring volume to 1.0L with distilled/deionized water.

Preparation of HI Agar: Add hay to hay infusion broth and bring volume to 1.0L. Mix thoroughly. Adjust pH to 7.0 with 5% lactic acid or 1*N* NaOH. Gently heat and bring to boiling. Autoclave for 20 min at 15 psi pressure–121°C. Pour into sterile Petri dishes.

HI/LY Broth:
Composition per liter:

Lactose	0.2g
Yeast extract	0.1g
Hay infusion broth	1.0L

pH 6.0 ± 0.2 at 25°C

Hay Infusion Broth:
Composition per liter:

Hay	2.5g

Preparation of Hay Infusion Broth: Add lactose and yeast extract to distilled/deionized water and bring volume to 1.0L. Mix thoroughly. Gently heat and bring to boiling. Boil for 30 min. Filter through Whatman #1 filter paper. Bring volume to 1.0L with distilled/deionized water.

Preparation of HI/LY Broth: Add components to hay infusion broth and bring volume to 1.0L. Mix thoroughly. Adjust pH to 6.0 with 5% lactic acid. Autoclave for 15 min at 15 psi pressure–121°C. Cool to 25°C.

Preparation of Medium: Aseptically add 15.0mL of HI/LY broth as an overlay over the surface of the HI agar plates.

Use: For the cultivation of *Sorogena stoianovitchae*.

SOT Medium

Composition per liter:

$NaHCO_3$	16.8g
$NaNO_3$	2.5g
K_2SO_4	1.0g
NaCl	1.0g
K_2HPO_4	0.5g
$MgSO_4 \cdot 7H_2O$	0.2g
Disodium EDTA·$2H_2O$	0.08g
$CaCl_2 \cdot 2H_2O$	0.04g
$FeSO_4 \cdot 7H_2O$	0.01g
Trace metals mix A5	1.0mL
Trace metals mix B6, modified	1.0mL

pH 9.0 ± 0.2 at 25°C

Trace Metals Mix A5:
Composition per liter:

H_3BO_3	2.86g
$MnCl_2 \cdot 4H_2O$	1.81g
$Na_2MoO_4 \cdot 2H_2O$	0.39g
$ZnSO_4 \cdot 7H_2O$	0.222g
$CuSO_4 \cdot 5H_2O$	0.079g
$Co(NO_3)_2 \cdot 6H_2O$	0.049g

Preparation of Trace Metals Mix A5: Add components to distilled/deionized water and bring volume to 1.0L. Mix thoroughly.

Trace Metals Mix B6, Modified:
Composition per liter:

NH_4NO_3	0.23g
$K_2Cr_2(SO_4)_4 \cdot 24H_2O$	0.096g
$NiSO_4 \cdot 7H_2O$	0.048g
$Ti_2(SO_4)_3$	0.04g
$Na_2WO_4 \cdot 2H_2O$	0.018g

Preparation of Trace Metals Mix B6, Modified: Add components to distilled/deionized water and bring volume to 1.0L. Mix thoroughly.

Preparation of Medium: Add components to distilled/deionized water and bring volume to 1.0L. Mix thoroughly. Adjust pH to 9.0. Distribute into tubes or flasks. Autoclave for 15 min at 15 psi pressure–121°C.

Use: For the cultivation and maintenance of *Spirulina maxima* and *Spirulina platensis*.

Soybean Agar

Composition per liter:

White soybeans	100.0g
Agar	15.0g

Preparation of Medium: Add soybeans to 1.0L of distilled/deionized water. Soak overnight. Autoclave for 60 min at 15 psi pressure–121°C. Filter through cheesecloth. Measure volume of filtrate. Add agar to a concentration of 1.5%. Gently heat and bring to boiling. Distribute into tubes or flasks. Autoclave for 15 min at 15 psi pressure–121°C. Pour into sterile Petri dishes or leave in tubes.

Use: For the cultivation and maintenance of *Bacillus subtilis* and *Pseudomonas syringae*.

SP Agar

Composition per liter:

Agar	15.0g
Pancreatic digest of casein	2.5g
Galactose	1.0g
Raffinose	1.0g
Sucrose	1.0g
$MgSO_4 \cdot 7H_2O$	0.5g
K_2HPO_4	0.25g
Vitamin solution	2.5mL

Vitamin Solution:
Composition per liter:

Inositol	1.0g
Calcium pantothenate	0.2g
Choline hydrochloride	0.2g
Thiamine	0.1g
Nicotinamide	0.75g
Pyridoxin	0.75g
Riboflavin	0.75g
p-Aminobenzoic acid	5.0mg
Folic acid	1.0mg
Biotin	0.05mg
Vitamin B_{12}	0.05mg
Ethanol	1.0L

Preparation of Vitamin Solution: Add solid components to 1.0L of ethanol. Mix thoroughly.

Preparation of Medium: Add components to distilled/deionized water and bring volume to 1.0L. Mix thoroughly. Gently heat and bring to boiling. Distribute into tubes or flasks. Autoclave for 15 min at 15 psi pressure–121°C. Pour into sterile Petri dishes or leave in tubes.

Use: For the cultivation of myxobacteria.

SP 2 Agar

Composition per liter:

Agar	15.0g
Pancreatic digest of casein	3.0g
Yeast extract	1.0g
Sodium acetate	0.02g
Artificial seawater	1.0L

pH 7.2 ± 0.2 at 25°C

Artificial Seawater:
Composition per liter:

NaCl	24.7g

$MgSO_4 \cdot 7H_2O$ 6.3g
$MgCl_2 \cdot 6H_2O$ 4.6g
$CaCl_2$ 1.0g
KCl 0.7g
$NaHCO_3$ 0.2g

Preparation of Artificial Seawater: Add components to distilled/deionized water and bring volume to 1.0L. Mix thoroughly.

Preparation of Medium: Add solid components to 1.0L of artificial seawater. Mix thoroughly. Gently heat and bring to boiling. Distribute into tubes or flasks. Autoclave for 15 min at 15 psi pressure–121°C. Pour into sterile Petri dishes or leave in tubes.

Use: For the isolation and cultivation of *Cytophaga* species, *Herpetosiphon* species, *Saprospira* species, and *Flexithrix* species.

SP 6 Agar

Composition per liter:
Agar 15.0g
Pancreatic digest of casein 3.0g
Yeast extract 1.0g
Artificial seawater 1.0L
pH 7.2 ± 0.2 at 25°C

Artificial Seawater:
Composition per liter:
NaCl 24.7g
$MgSO_4 \cdot 7H_2O$ 6.3g
$MgCl_2 \cdot 6H_2O$ 4.6g
$CaCl_2$ 1.0g
KCl 0.7g
$NaHCO_3$ 0.2g

Preparation of Artificial Seawater: Add components to distilled/deionized water and bring volume to 1.0L. Mix thoroughly.

Preparation of Medium: Add solid components to 1.0L of artificial seawater. Mix thoroughly. Gently heat and bring to boiling. Distribute into tubes or flasks. Autoclave for 15 min at 15 psi pressure–121°C. Pour into sterile Petri dishes or leave in tubes.

Use: For the isolation and cultivation of *Cytophaga* species, *Herpetosiphon* species, *Saprospira* species, and *Flexithrix* species.

SP5 Broth

Composition per liter:
Pancreatic digest of casein 9.0g
Yeast extract 1.0g
Artificial seawater 1.0L
pH 7.2 ± 0.2 at 25°C

Artificial Seawater:
Composition per liter:
NaCl 24.7g
$MgSO_4 \cdot 7H_2O$ 6.3g
$MgCl_2 \cdot 6H_2O$ 4.6g
$CaCl_2$ 1.0g
KCl 0.7g
$NaHCO_3$ 0.2g

Preparation of Artificial Seawater: Add components to distilled/deionized water and bring volume to 1.0L. Mix thoroughly.

Preparation of Medium: Add solid components to 1.0L of artificial seawater. Mix thoroughly. Gently heat and bring to boiling. Distribute into tubes or flasks. Autoclave for 15 min at 15 psi pressure–121°C.

Use: For the isolation and cultivation of *Cytophaga* species, *Herpetosiphon* species, *Saprospira* species, and *Flexithrix* species.

SP Medium

Composition per liter:
Agar 15.0g
Soluble starch 5.0g
Pancreatic digest of casein 2.5g
Galactose 1.0g
Raffinose 1.0g
Sucrose 1.0g
$MgSO_4 \cdot 7H_2O$ 0.5g
K_2HPO_4 .... 0.25g

Preparation of Medium: Add components to distilled/deionized water and bring volume to 1.0L. Mix thoroughly. Gently heat and bring to boiling. Distribute into tubes or flasks. Autoclave for 15 min at 15 psi pressure–121°C. Pour into sterile Petri dishes or leave in tubes.

Use: For the cultivation and maintenance of *Archanigium gephyra*, *Cystobacter fuscus*, *Melittangium lichenicola*, *Myxococcus* species, *Polyangium brachy- sporum*, *Stigmatella aurantiaca*, and *Stigmatella erecta*.

SP4 Medium

Composition per liter:
Base solution 615.0mL
Fetal calf serum
(inactivated at 56°C, 1 hr) 170.0mL
Yeast extract (2% solution) 100.0mL
CMRL 1066, 10X with glutamine 50.0mL
Fresh yeast extract solution 35.0mL
Phenol Red (0.1% solution) 20.0mL
Penicillin solution 10.0mL
pH 7.0–7.4 ± 0.2 at 25°C

Base Solution:
Composition per 615.0mL:
Pancreatic digest of casein 11.2g
Noble agar 8.0g
Pancreatic digest of gelatin 5.3g
Glucose 5.0g
NaCl 0.875g
Beef extract 0.525g
Yeast extract 0.525g
Beef heart, solids from infusion 0.35g

Preparation of Base Solution: Add components to distilled/deionized water and bring volume to 615.0mL. Mix thoroughly. Adjust pH to 7.5. Gently heat and bring to boiling. Autoclave for 15 min at 15 psi pressure–121°C. Cool to 45°–50°C.

CMRL 1066, 10X with Glutamine:
Composition per liter:
NaCl 6.8g
$NaHCO_3$ 2.2g
D-Glucose 1.0g
KCl 0.4g
L-Cysteine·HCl·H_2O 0.26g
$CaCl_2$, anhydrous 0.2g
$MgSO_4 \cdot 7H_2O$ 0.2g
$NaH_2PO_4 \cdot H_2O$ 0.14g
L-Glutamine 0.1g
Sodium acetate·$3H_2O$ 0.083g
L-Glutamic acid 0.075g

L-Arginine·HCl 0.07g
L-Lysine·HCl 0.07g
L-Leucine 0.06g
Glycine 0.05g
Ascorbic acid 0.05g
L-Proline 0.04g
L-Tyrosine 0.04g
L-Aspartic acid 0.03g
L-Threonine 0.03g
L-Alanine 0.025g
L-Phenylalanine 0.025g
L-Serine 0.025g
L-Valine 0.025g
L-Cystine 0.02g
L-Histidine·HCl·H_2O 0.02g
L-Isoleucine 0.02g
Phenol Red 0.02g
L-Methionine 0.015g
Deoxyadenosine 0.01g
Deoxycytidine 0.01g
Deoxyguanosine 0.01g
Glutathione, reduced 0.01g
Thymidine 0.01g
Hydroxy-L-proline 0.01g
L-Tryptophan 0.01g
Nicotinamide adenine dinucleotide 7.0mg
Tween 80 .... 5.0mg
Sodium glucoronate·H_2O 4.2mg
Coenzyme A 2.5mg
Cocarboxylase 1.0mg
Flavin adenine dinucleotide 1.0mg
Nicotinamide adenine dinucleotide phosphate 1.0mg
Uridine triphosphate 1.0mg
Choline chloride 0.5mg
Cholesterol 0.2mg
5-Methyldeoxycytidine 0.1mg
Inositol 0.05mg
p-Aminobenzoic acid 0.05mg
Niacin 0.025mg
Niacinamide 0.025mg
Pyridoxine 0.025mg
Pyridoxal·HCl 0.025mg
Biotin 0.01mg
D-Calcium pantothenate 0.01mg
Folic acid 0.01mg
Riboflavin 0.01mg
Thiamine·HCl 0.01mg

Preparation of CMRL 1066, 10X with Glutamine: Add components to distilled/deionized water and bring volume to 1.0L. Mix thoroughly. Adjust pH to 7.2. Filter sterilize.

Fresh Yeast Extract Solution:
Composition per 100.0mL:
Baker's yeast, live, pressed, starch-free 25.0g

Preparation of Fresh Yeast Extract Solution: Add the live Baker's yeast to 100.0mL of distilled/deionized water. Autoclave for 90 min at 15 psi pressure–121°C. Allow to stand. Remove supernatant solution. Adjust pH to 6.6–6.8.

Penicillin Solution:
Composition per 10.0mL:
Penicillin G 1,000,000U

Preparation of Penicillin Solution: Add penicillin to distilled/deionized water and bring volume to 10.0mL. Mix thoroughly. Filter sterilize.

Preparation of Medium: To 615.0mL of cooled sterile base solution, aseptically add 170.0mL of sterile inactivated fetal calf serum, 100.0mL of sterile yeast extract, 50.0mL of sterile CMRL 1066, 10X with glutamine, 35.0mL of sterile fresh yeast extract solution, 20.0mL of Phenol Red solution, and 10.0mL of sterile penicillin solution. Mix thoroughly. Aseptically distribute into sterile tubes. Allow tubes to cool in a slanted position.

Use: For the cultivation of tick-derived *Mycoplasma* (*Spiroplasma*). Used for the enhanced recovery of *Mycoplasma pneumoniae, Mycoplasma alvi*, and *Mycoplasma hyopneumoniae.*

Sphaerotilus Agar (DSMZ Medium 51)

Composition per liter:
Agar 15.0g
Beef extract, Lab Lemco 5.0g
pH 7.1 ± 0.2 at 25°C

Preparation of Medium: Add components to distilled/deionized water and bring volume to 1.0L. Mix thoroughly. Adjust pH to 7.0. Gently heat and bring to boiling. Distribute into tubes. Autoclave for 15 min at 15 psi pressure–121°C. Cool in a sloping position to form slants. Cover solid slants with 2ml sterile tap water. Inoculate into the covering tap water and incubate at 20–25°C.

Use: For the cultivation and maintenance of *Sphaerotilus natans.*

Sphaerotilus CGYA Medium

Composition per liter:
Glycerol 10.0g
Pancreatic digest of casein 5.0g
Yeast extract 1.0g

Preparation of Medium: Add components to distilled/deionized water and bring volume to 1.0L. Mix thoroughly. Distribute into tubes or flasks. Autoclave for 15 min at 15 psi pressure–121°C.

Use: For the cultivation and maintenance of *Sphaerotilus natans* and *Sphaerotilus* species.

Sphaerotilus Defined Medium

Composition per liter:
Agar 15.0g
Glycerol 5.0g
Glutamic acid 0.9g
$FeSO_4 \cdot 7H_2O$ 0.5g
$MgSO_4 \cdot 7H_2O$ 0.1g
$CaCl_2 \cdot 2H_2O$ 0.03g
$ZnSO_4 \cdot 7H_2O$ 0.03g
Phosphate solution 100.0mL
pH 7.0 ± 0.2 at 25°C

Phosphate Solution:
Composition per 500.0mL:
K_2HPO_4 .... 5.7g
KH_2PO_4 .... 2.3g

Preparation of Phosphate Solution: Add components to distilled/deionized water and bring volume to 500.0mL. Mix thoroughly. Gently heat until dissolved. Autoclave for 15 min at 15 psi pressure–121°C.

Preparation of Medium: Add components, except phosphate solution, to distilled/deionized water and bring volume to 900.0mL. Mix thoroughly. Gently heat and bring to boiling. Autoclave for 10 min at 15 psi pressure–121°C. Cool to 45°–50°C. Aseptically add 100.0mL of sterile phosphate solution. Mix thoroughly. Pour into sterile Petri dishes or distribute into sterile tubes.

Use: For the cultivation of *Sphaerotilus* species.

Sphaerotilus discophorus Medium

Composition per liter:

Agar	12.0g
Peptone	5.0g
$MgSO_4 \cdot 7H_2O$	0.2g
$CaCl_2$	0.05g
$MnSO_4 \cdot H_2O$	0.05g
Ferric solution	100.0mL

pH 7.0 ± 0.2 at 25°C

Ferric Solution:

Composition per 100.0mL:

Ferric ammonium citrate	0.5g
$FeCl_3 \cdot 6H_2O$	0.01g

Preparation of Ferric Solution: Add components to distilled/deionized water and bring volume to 100.0mL. Mix thoroughly. Filter sterilize.

Preparation of Medium: Add components, except ferric solution, to tap water and bring volume to 900.0mL. Mix thoroughly. Gently heat and bring to boiling. Autoclave for 15 min at 15 psi pressure–121°C. Cool to 45°–50°C. Aseptically add sterile ferric solution. Mix thoroughly. Pour into sterile Petri dishes or distribute into sterile tubes.

Use: For the cultivation of *Sphaerotilus discophorus*.

Sphaerotilus Isolation Medium

Composition per liter:

Agar	15.0g
Glycerol	10.0g
Pancreatic digest of casein	5.0g
Yeast extract	1.0g

pH 7.0 ± 0.2 at 25°C

Preparation of Medium: Add components to distilled/deionized water and bring volume to 1.0L. Mix thoroughly. Gently heat and bring to boiling. Distribute into tubes or flasks. Autoclave for 15 min at 15 psi pressure–121°C. Pour into sterile Petri dishes or leave in tubes.

Use: For the isolation and cultivation of *Sphaerotilus* species.

Sphaerotilus/Leptothrix Agar

Composition per liter:

Agar	20.0g
Peptone	1.5g
Yeast extract	1.0g
Ferric ammonium citrate	0.5g
$MgSO_4 \cdot 7H_2O$	0.2g
$CaCl_2$	0.05g
$MnSO_4 \cdot H_2O$	0.05g
$FeCl_3 \cdot 6H_2O$	0.01g

pH 7.1 ± 0.2 at 25°C

Preparation of Medium: Add components to distilled/deionized water and bring volume to 1.0L. Mix thoroughly. Adjust pH to 7.1. Gently heat and bring to boiling. Distribute into tubes or flasks. Autoclave for 15 min at 15 psi pressure–121°C. Pour into sterile Petri dishes or leave in tubes.

Use: For the cultivation and maintenance of *Leptothrix cholodnii, Leptothrix* species, and *Sphaerotilus natans.*

Sphaerotilus and *Leptothrix* Enrichment Medium

Composition per liter:

Glucose	1.0g
Peptone	1.0g
$MgSO_4 \cdot 7H_2O$	0.2g
$FeCl_3 \cdot 6H_2O$	0.1g
$CaCl_2 \cdot 2H_2O$	0.05g

pH 7.0 ± 0.2 at 25°C

Preparation of Medium: Add components to distilled/deionized water and bring volume to 1.0L. Mix thoroughly. Distribute into tubes or flasks. Autoclave for 15 min at 15 psi pressure–121°C.

Use: For the enrichment and cultivation of *Sphaerotilus* species and *Leptothrix* species.

Sphaerotilus Leptothrix Medium (DSMZ Medium 803)

Composition per liter:

Agar	20.0g
Peptone	1.5g
Yeast extract	1.0g
Ferric ammonium citrate	0.5g
$MgSO_4 \cdot 7H_2O$	0.2g
$CaCl_2$	0.05g
$MnSO_4 \cdot 2H_2O$	0.05g
$FeCl_3 \cdot 6H_2O$	0.01g

pH 7.1 ± 0.2 at 25°C

Preparation of Medium: Add components to tap water and bring volume to 1.0L. Mix thoroughly. Gently heat and bring to boiling. Distribute into tubes or flasks. Autoclave for 15 min at 15 psi pressure–121°C. Pour into sterile Petri dishes or leave in tubes.

Use: For the cultivation and maintenance of *Leptothrix mobilis.*

Sphaerotilus Medium

Composition per liter:

Agar	15.0g
Lab-Lemco powder	5.0g

pH 7.0 ± 0.2 at 25°C

Source: Lab-Lemco powder is available from Oxoid Unipath.

Preparation of Medium: Add components to distilled/deionized water and bring volume to 1.0L. Mix thoroughly. Gently heat and bring to boiling. Distribute into tubes or flasks. Autoclave for 15 min at 15 psi pressure–121°C. Pour into sterile Petri dishes or leave in tubes.

Use: For the cultivation of *Sphaerotilus natans.*

Sphaerotilus natans Enrichment Medium

Composition per liter:

Sodium lactate	0.1g
$Na_2HPO_4 \cdot 7H_2O$	0.034g
$CaCl_2$	0.027g
$MgSO_4 \cdot 7H_2O$	0.023g
K_2HPO_4	0.022g
KH_2PO_4	8.5mg
NH_4Cl	1.7mg
$FeCl_3 \cdot 6H_2O$	0.25mg

pH 7.1–7.2 at 25°C

Preparation of Medium: Add components to distilled/deionized water and bring volume to 1.0L. Mix thoroughly. Distribute into tubes or flasks. Autoclave for 15 min at 15 psi pressure–121°C.

Use: For the enrichment and cultivation of *Sphaerotilus natans*.

Sphaerotilus natans Isolation Agar

Composition per liter:

Agar 15.0g
Casein hydrolysate 1.5g

Preparation of Medium: Add components to tap water and bring volume to 1.0L. Mix thoroughly. Gently heat and bring to boiling. Distribute into tubes or flasks. Autoclave for 15 min at 15 psi pressure–121°C. Pour into sterile Petri dishes or leave in tubes.

Use: For the isolation and cultivation of *Sphaerotilus natans*.

Spirillum gracile Medium

Composition per liter:

Agar 15.0g
Peptone 5.0g
Yeast extract 0.5g
K_2HPO_4 .... 0.1g
Tween 80 .... 0.02g

pH 7.2 ± 0.2 at 25°C

Preparation of Medium: Add components to tap water and bring volume to 1.0L. Mix thoroughly. Gently heat and bring to boiling. Distribute into tubes or flasks. Autoclave for 15 min at 15 psi pressure–121°C. Pour into sterile Petri dishes or leave in tubes.

Use: For the cultivation and maintenance of *Aquaspirillum gracile*.

Spirillum lipoferum Medium

Composition per liter:

Sodium malate 5.0g
Agar 3.5g
KH_2PO_4 .... 0.4g
$MgSO_4·7H_2O$ 0.2g
K_2HPO_4 .... 0.1g
NaCl 0.1g
$CaCl_2$ 0.02g
$FeCl_3$ 0.01g
$NaMoO_4·2H_2O$ 2.0mg
Bromothymol Blue solution 5.0mL

pH 6.8 ± 0.2 at 25°C

Bromothymol Blue Solution:

Composition per 10.0mL:

Bromothymol Blue 0.5g
Ethanol 10.0mL

Preparation of Bromothymol Blue Solution: Add Bromothymol Blue to 10.0mL of ethanol. Mix thoroughly.

Preparation of Medium: Add components to distilled/deionized water and bring volume to 1.0L. Mix thoroughly. Gently heat and bring to boiling. Distribute into tubes or flasks. Autoclave for 15 min at 15 psi pressure–121°C.

Use: For the isolation and cultivation of *Spirillum leptoferum*.

Spirillum Medium

Composition per liter:

Calcium lactate 10.0g
Peptone 5.0g
Beef extract 3.0g
Yeast extract 3.0g

pH 7.0 ± 0.2 at 25°C

Preparation of Medium: Add components to distilled/deionized water and bring volume to 1.0L. Mix thoroughly. Adjust pH to 7.0. Distribute into tubes or flasks. Autoclave for 20 min at 11 psi pressure–116°C. A precipitate will form during autoclaving.

Use: For the cultivation of *Spirillum* species.

Spirillum Medium

Composition per liter:

Peptone 10.0g
$MgSO_4·7H_2O$ 1.0g
$(NH_4)_2SO_4$ 1.0g
Succinic acid 1.0g
$FeCl_3·6H_2O$ 2.0mg
$MnSO_4·H_2O$ 2.0mg

pH 6.8 ± 0.2 at 25°C

Preparation of Medium: Add components to distilled/deionized water and bring volume to 1.0L. Mix thoroughly. Distribute into tubes or flasks. Autoclave for 15 min at 15 psi pressure–121°C.

Use: For the cultivation of *Aquaspirillum autotrophicum, Aquaspirillum dispar, Aquaspirillum peregrinum,* and *Aquaspirillum serpens.*

Spirillum Nitrogen-Fixing Medium

Composition per liter:

Sodium malate 5.0g
KH_2PO_4 .... 0.4g
$MgSO_4·7H_2O$ 0.2g
K_2HPO_4 .... 0.1g
NaCl 0.1g
Yeast extract 0.05g
$CaCl_2$ 0.02g
$FeCl_3$ 0.01g
$NaMoO_4·2H_2O$ 2.0mg

pH 7.2-7.4 ± 0.2 at 25°C

Preparation of Medium: Add components to distilled/deionized water and bring volume to 1.0L. Mix thoroughly. Distribute into tubes or flasks. Autoclave for 15 min at 15 psi pressure–121°C.

Use: For the cultivation and maintenance of *Azospirillum brasilense, Azospirillum lipoferum,* and *Herbaspirillum seropedicae.*

Spirillum volutans Defined Medium

Composition per liter:

BES (*N,N*-bis[2-hydroxyethyl]-2-aminoethane sulfonic acid) buffer 1.07g
$MgSO_4·7H_2O$ 1.0g
$(NH_4)_2SO_4$ 1.0g
Succinic acid 1.0g
L-Histidine 0.2g
L-Isoleucine 0.2g
L-Methionine 0.2g
L-Threonine 0.2g
NaCl 0.085g
L-Cystine 0.025g
K_2HPO_4 .... 0.02g
$FeCl_3·6H_2O$ 3.0mg
DL-Norepinephrine 2.0mg

$MnSO_4 \cdot H_2O$	2.0mg
$CaCO_3$	1.0mg
$ZnSO_4 \cdot 7H_2O$	0.72mg
$Na_2MoO_4 \cdot 2H_2O$	0.245mg
$CoSO_4 \cdot 7H_2O$	0.14mg
$CuSO_4 \cdot 5H_2O$	0.13mg
H_2BO_3	0.031mg

pH 6.8 ± 0.2 at 25°C

Preparation of Medium: Add components to distilled/deionized water and bring volume to 1.0L. Mix thoroughly. Adjust pH to 6.8. Distribute into tubes or flasks. Autoclave for 15 min at 15 psi pressure–121°C.

Use: For the cultivation of *Spirillum volutans.*

Spirochete Enrichment Medium

Composition per liter:

Agar	10.0g
Beef extract	1.0g
Peptone	1.0g
Yeast extract	1.0g
Seawater	500.0mL

Preparation of Medium: Add components to distilled/deionized water and bring volume to 1.0L. Mix thoroughly. Gently heat and bring to boiling. Distribute into tubes or flasks. Autoclave for 15 min at 15 psi pressure–121°C. Pour into sterile Petri dishes or leave in tubes.

Use: For the isolation of spirochetes from muds. A well is cut into the agar plate and filled with mud samples. Spirochetes migrate out of the mud into the agar surrounding the well.

Spirochete Medium (ATCC Medium 1712)

Composition per liter:

Tris(hydroxymethyl)aminomethane buffer	7.52g
Pancreatic digest of casein	1.0g
Yeast extract	1.0g
L-Cysteine·HCl·$2H_2O$	0.5g
Resazurin	1.0mg
Seawater	750.0mL
Glucose solution	20.0mL

pH 7.2 ± 0.2 at 25°C

Glucose Solution:

Composition per 20.0mL:

Glucose	2.0g

Preparation of Glucose Solution: Add glucose to distilled/deionized water and bring volume to 20.0mL. Mix thoroughly. Filter sterilize.

Preparation of Medium: Prepare and dispense medium under 100% N_2. Add components, except glucose solution, to distilled/deionized water and bring volume to 980.0mL. Mix thoroughly. Adjust pH to 7.5. Autoclave for 15 min at 15 psi pressure–121°C. Cool to 50°C. Aseptically add sterile glucose solution. Mix thoroughly. Aseptically distribute into sterile tubes or flasks.

Use: For the cultivation and maintenance of *Spirochaeta litoralis.*

Spirochete Thermophile Medium (DSMZ Medium 509)

Composition per 1012mL:

Solution A	920.0mL
Solution D	50.0mL
Solution E	20.0mL
Solution F	10.0mL
Solution G	10.0mL
Solution B	1.0mL
Solution C	1.0mL

pH 7.0 ± 0.2 at 25°C

Solution A:

Composition per 920.0mL:

NaCl	4.0g
$MgCl_2 \cdot 6H_2O$	0.8g
KCl	0.5g
NH_4Cl	0.3g
KH_2PO_4	0.2g
$CaCl_2 \cdot 2H_2O$	0.03g
Resazurin	1.0mg

Preparation of Solution A: Add components to 920.0mL distilled/deionized water. Mix thoroughly. Bring to boiling for a few minutes. Cool to room temperature while gassing with 80% N_2 + 20% CO_2 gas. Adjust pH to 6.0. Immediately distribute under N_2 into anaerobic tubes. Autoclave for 15 min at 15 psi pressure–121°C. Cool to 25°C.

Solution B:

Composition per liter:

$FeCl_2 \cdot 4H_2O$	1.5g
$CoCl_2 \cdot 6H_2O$	190.0mg
$MnCl_2 \cdot 4H_2O$	100.0mg
$ZnCl_2$	70.0mg
$Na_2MoO_4 \cdot 2H_2O$	36.0mg
$NiCl_2 \cdot 6H_2O$	24.0mg
H_3BO_3	6.0mg
$CuCl_2 \cdot 2H_2O$	2.0mg
HCl (25% solution)	10.0mL

Preparation of Solution B: Add $FeCl_2 \cdot 4H_2O$ to 10.0mL of HCl solution. Mix thoroughly. Add distilled/deionized water and bring volume to 1.0L. Add remaining components. Mix thoroughly. Sparge with 80% N_2 + 20% CO_2. Autoclave for 15 min at 15 psi pressure–121°C. Cool to 25°C.

Solution C:

Composition per liter:

NaOH	0.5g
$Na_2WO_4 \cdot 2H_2O$	4.0mg
$Na_2SeO_3 \cdot 5H_2O$	3.0mg

Preparation of Solution C: Add components to distilled/deionized water and bring volume to 1.0L. Mix thoroughly. Sparge with 100% N_2. Filter sterilize.

Solution D:

Composition per 100.0mL:

$NaHCO_3$	5.0g

Preparation of Solution D: Add $NaHCO_3$ to distilled/deionized water and bring volume to 100.0mL. Mix thoroughly. Sparge with 100% N_2 gas mixture. Autoclave for 15 min at 15 psi pressure–121°C. Cool to to 25°C.

Solution E:

Composition per liter:

Pyridoxine-HCl	10.0mg
Thiamine-HCl·$2H_2O$	5.0mg
Riboflavin	5.0mg
Nicotinic acid	5.0mg
D-Ca-pantothenate	5.0mg
p-Aminobenzoic acid	5.0mg
Lipoic acid	5.0mg
Biotin	2.0mg
Folic acid	2.0mg
Vitamin B_{12}	0.1mg

Preparation of Solution E: Add components to distilled/deionized water and bring volume to 1.0L. Mix thoroughly. Sparge with 80% N_2 + 20% CO_2. Filter sterilize.

Solution F:

Composition per 10mL:

Starch 1.0g

Preparation of Solution F: Add starch to distilled/deionized water and bring volume to 10.0mL. Mix thoroughly. Sparge with 100% N_2 gas mixture. Autoclave for 15 min at 15 psi pressure–121°C. Cool to 25°C.

Solution G:

Composition per 10mL:

$Na_2S{\cdot}9H_2O$ 0.3g

Preparation of Solution G: Add $Na_2S{\cdot}9H_2O$ to distilled/deionized water and bring volume to 10.0mL. Mix thoroughly. Autoclave under 100% N_2 for 15 min at 15 psi pressure–121°C. Cool to 25°C.

Preparation of Medium: Solution A is distributed into anaerobic tubes with rubber stoppers prior to autoclaving. Using aseptic and anaerobic conditions and syringes appropriate volumes of sterile solutions B-G are injected into each tube to yield the specified concentrations.

Use: For the cultivation of *Spirochaeta thermophila.*

Spiroplasma Agar MID

Composition per 291.2mL:

Schneider's *Drosophila* medium 160.0mL
Solution 1 .... 80.0mL
Fetal calf serum 50.0mL
Phenol Red (0.5% solution) 1.2mL

pH 7.4 ± 0.2 at 25°C

Schneider's *Drosophila* Medium:

Composition per liter:

$MgSO_4{\cdot}7H_2O$ 3.7g
NaCl 2.1g
Yeast extract 2.0g
Trehalose 2.0g
D-Glucose 2.0g
L-Glutamine 1.8g
L-Lysine·HCl 1.7g
L-Proline 1.7g
KCl 1.6g
$Na_2HPO_4{\cdot}7H_2O$ 1.3g
L-Glutamic acid 0.8g
L-Methionine 0.8g
$CaCl_2$, anhydrous 0.6g
KH_2PO_4 .... 0.5g
β-Alanine 0.5g
L-Tyrosine 0.5g
L-Arginine 0.4g
L-Aspartic acid 0.4g
L-Histidine 0.4g
L-Threonine 0.4g
$NaHCO_3$ 0.4g
Glycine 0.3g
L-Serine 0.3g
L-Valine 0.3g
L-Isoleucine 0.2g
L-Leucine 0.2g
L-Phenylalanine 0.2g
α-Ketoglutaric acid 0.2g
Fumaric acid 0.1g
Malic acid 0.1g
Succinic acid 0.1g
L-Cystine 0.1g
L-Tryptophan 0.1g
L-Cysteine 0.06g

Preparation of Schneider's *Drosophila* Medium: Add components to 1.0L of distilled/deionized water. Mix thoroughly. Filter sterilize.

Solution 1:

Composition per 80.0mL:

Sorbitol 7.0g
Noble agar 5.0g
Beef heart, solids from infusion 5.0g
Peptone 1.8g
Sucrose 1.0g
Pancreatic digest of casein 1.0g
NaCl 0.5g
D-Fructose 0.1g
D-Glucose 0.1g

Preparation of Solution 1: Add components to distilled/deionized water and bring volume to 80.0mL. Mix thoroughly. Adjust pH to 7.8 with 1*N* NaOH. Autoclave for 15 min at 15 psi pressure–121°C. Cool to 50°C.

Preparation of Medium: Bring fetal calf serum and Phenol Red solution to 56°C. Rapidly bring Schneider's *Drosophila* medium to 37°C. Rapidly combine the components. Mix thoroughly. Pour into sterile Petri dishes or distribute into sterile tubes or flasks.

Use: For the cultivation and maintenance of *Spiroplasma kunkelii* and *Spiroplasma* species.

Spiroplasma Medium

Composition per liter:

Sucrose 80.0g
Beef heart, solids from infusion 34.7g
Peptone 6.9g
NaCl 3.5g
Horse serum, heat inactivated 100.0mL

pH 7.2 ± 0.2 at 25°C

Preparation of Medium: Add components, except horse serum, to distilled/deionized water and bring volume to 900.0mL. Mix thoroughly. Autoclave for 15 min at 15 psi pressure–121°C. Cool to 25°C. Aseptically add horse serum. Mix thoroughly. Aseptically distribute into sterile tubes or flasks.

Use: For the cultivation and maintenance of *Spiroplasma* species.

Spiroplasma Medium

Composition per liter:

Sorbitol 70.0g
Pancreatic digest of casein 7.0g
Yeast extract 5.0g
NaCl 5.0g
Beef extract 3.0g
Yeast extract 3.0g
Beef heart, solids from infusion 2.0g
Fructose 1.0g
Glucose 1.0g
Phenol Red 20.0mg
Horse serum 100.0mL

pH 7.8 ± 0.2 at 25°C

Preparation of Medium: Add components to distilled/deionized water and bring volume to 900.0mL. Mix thoroughly. Gently heat and bring to boiling. Autoclave for 15 min at 15 psi pressure–121°C. Cool to 50°–55°C. Aseptically add 100.0mL of sterile horse serum. Mix thoroughly.

Use: For the cultivation of *Spiroplasma citri.*

Spiroplasma Medium with 25 mg/L of Phenol Red

Composition per liter:

Sucrose	80.0g
Beef heart, solids from infusion	34.7g
Peptone	6.9g
NaCl	3.5g
Phenol Red	25.0mg
Horse serum, heat inactivated	100.0mL

pH 7.2 ± 0.2 at 25°C

Preparation of Medium: Add components, except horse serum, to distilled/deionized water and bring volume to 900.0mL. Mix thoroughly. Gently heat and bring to boiling. Autoclave for 15 min at 15 psi pressure–121°C. Cool to 45°–50°C. Aseptically add heat-inactivated horse serum. Mix thoroughly. Aseptically distribute into sterile tubes or flasks.

Use: For the cultivation and maintenance of *Spiroplasma floricola.*

Sporobacter Medium (DSMZ Medium 711)

Composition per liter:

NH_4Cl	1.0g
NaCl	0.6g
Na-acetate·$3H_2O$	0.5g
Cysteine-HCl·H_2O	0.5g
K_2HPO_4	0.3g
KH_2PO_4	0.3g
Yeast extract	0.2g
$MgCl_2·6H_2O$	0.2g
$CaCl_2·2H_2O$	0.1g
KCl	0.1g
Resazurin	0.5mg
$NaHCO_3$ solution	40.0mL
Trimethoxycinnamate solution	10.0mL
$Na_2S·9H_2O$ solution	10.0mL
Trace elements solution SL-10	1.5mL

pH 7.1 ± 0.2 at 25°C

Trace Elements Solution SL-10:
Composition per liter:

$FeCl_2·4H_2O$	1.5g
$CoCl_2·6H_2O$	190.0mg
$MnCl_2·4H_2O$	100.0mg
$ZnCl_2$	70.0mg
$Na_2MoO_4·2H_2O$	36.0mg
$NiCl_2·6H_2O$	24.0mg
H_3BO_3	6.0mg
$CuCl_2·2H_2O$	2.0mg
HCl (25% solution)	10.0mL

Preparation of Trace Elements Solution SL-10: Add $FeCl_2·4H_2O$ to 10.0mL of HCl solution. Mix thoroughly. Add distilled/deionized water and bring volume to 1.0L. Add remaining components. Mix thoroughly. Sparge with 80% N_2 + 20% CO_2. Autoclave for 15 min at 15 psi pressure–121°C.

$Na_2S·9H_2O$ Solution:
Composition per 10mL:

$Na_2S·9H_2O$	0.3g

Preparation of $Na_2S·9H_2O$ Solution: Add $Na_2S·9H_2O$ to distilled/deionized water and bring volume to 10.0mL. Mix thoroughly. Autoclave under 100% N_2 for 15 min at 15 psi pressure–121°C. Cool to room temperature.

$NaHCO_3$ Solution:
Composition per 100.0mL:

$NaHCO_3$	10.0g

Preparation of $NaHCO_3$ Solution: Add $NaHCO_3$ to distilled/deionized water and bring volume to 100.0mL. Mix thoroughly. Sparge with 80% N_2 + 20% CO_2. Filter sterilize.

Trimethoxycinnamate Solution:
Composition per 10.0mL:

Trans-3,4,5-trimethoxycinnamate	1.2g

Preparation of Trimethoxycinnamate Solution: Add trans-3,4,5-trimethoxycinnamate to distilled/deionized water and bring volume to 10.0mL. Mix thoroughly. Neutralize with NaOH. Sparge with 100% N_2. Autoclave for 15 min at 15 psi pressure–121°C. Cool to room temperature.

Preparation of Medium: Prepare and dispense medium under 80% N_2 + 20% CO_2 gas atmosphere. Add components, except $NaHCO_3$ solution, $Na_2S·9H_2O$ solution, trimethoxycinnamate solution, and trace elements solution SL-10, to distilled/deionized water and bring volume to 938.5mL. Mix thoroughly. Adjust pH to 7.1. Sparge with 80% N_2 + 20% CO_2. Autoclave for 15 min at 15 psi pressure–121°C. Aseptically and anaerobically add 40.0mL $NaHCO_3$ solution, 10.0mL $Na_2S·9H_2O$ solution, 10.0mL trimethoxycinnamate solution, and 1.0mL trace elements solution SL-10. Mix thoroughly. Aseptically and anaerobically distribute into sterile tubes or bottles.

Use: For the cultivation of *Sporobacter termitidis.*

Sporocytophaga Medium

Composition per liter:

$NaNO_3$	2.0g
K_2HPO_4	1.2g
$MgSO_4·7H_2O$	1.0g
KCl	0.5g
KH_2PO_4	0.14g
Yeast extract	0.02g
$FeSO_4·7H_2O$	6.0mg
Filter paper strips	variable

pH 7.2 ± 0.2 at 25°C

Preparation of Medium: Add components, except filter paper strips, to distilled/deionized water and bring volume to 1.0L. Mix thoroughly. Distribute into tubes in 5.0mL volumes. Autoclave for 15 min at 15 psi pressure–121°C. Aseptically add a sterile filter paper strip to each tube so that 1.0-2.0cm of the strip extends above the medium.

Use: For the cultivation and maintenance of *Sporocytophaga myxococcoides.*

Sporohalobacter lortetii Agar

Composition per liter:

NaCl	105.0g
L-Glutamic acid	4.0g
Agar	20.0g
$CaCO_2$	5.0g
Soluble starch	2.0g
Casamino acids	2.0g
Nutrient broth	2.0g
Yeast extract	2.0g
KCl	0.75g
L-Cysteine	0.5g
$FeSO_4·7H_2O$	0.002g
Resazurin	1.0mg
$MgCl_2·6H_2O$ solution	40.0mL
$CaCl_2·2H_2O$ solution	10.0mL

Trace elements solution 10.0mL
Vitamin solution 10.0mL

pH 6.5 ± 0.2 at 25°C

$MgCl_2 \cdot 6H_2O$ Solution:

Composition per 40.0mL:

$MgCl_2 \cdot 6H_2O$ 0.01g

Preparation of $MgCl_2 \cdot 6H_2O$ Solution: Add 0.01g of $MgCl_2 \cdot 6H_2O$ to distilled/deionized water and bring volume to 40.0mL. Mix thoroughly. Sparge with 100% N_2. Autoclave for 15 min at 15 psi pressure–121°C.

$CaCl_2 \cdot 2H_2O$ Solution:

Composition per 40.0mL:

$CaCl_2 \cdot 2H_2O$ 0.01g

Preparation of $CaCl_2 \cdot 2H_2O$ Solution: Add 0.01g of $CaCl_2 \cdot 2H_2O$ to distilled/deionized water and bring volume to 40.0mL. Mix thoroughly. Sparge with 100% N_2. Autoclave for 15 min at 15 psi pressure–121°C.

Trace Elements Solution:

Composition per liter:

$MgSO_4 \cdot 7H_2O$ 3.0g
Nitrilotriacetic acid 1.5g
NaCl 1.0g
$MnSO_4 \cdot 2H_2O$ 0.5g
$CoSO_4 \cdot 7H_2O$ 0.18g
$ZnSO_4 \cdot 7H_2O$ 0.18g
$CaCl_2 \cdot 2H_2O$ 0.1g
$FeSO_4 \cdot 7H_2O$ 0.1g
$NiCl_2 \cdot 6H_2O$ 0.025g
$KAl(SO_4)_2 \cdot 12H_2O$ 0.02g
$CuSO_4 \cdot 5H_2O$ 0.01g
H_3BO_3 .... 0.01g
$Na_2MoO_4 \cdot 2H_2O$ 0.01g
$Na_2SeO_3 \cdot 5H_2O$ 0.3mg

Preparation of Trace Elements Solution: Add nitrilotriacetic acid to 500.0mL of distilled/deionized water. Adjust pH to 6.5 with KOH. Add remaining components. Add distilled/deionized water to 1.0L.

Vitamin Solution:

Composition per liter:

Pyridoxine·HCl 10.0mg
Calcium DL-pantothenate 5.0mg
Lipoic acid 5.0mg
Nicotinic acid 5.0mg
p-Aminobenzoic acid 5.0mg
Riboflavin 5.0mg
Thiamine·HCl 5.0mg
Biotin 2.0mg
Folic acid 2.0mg
Vitamin B_{12} 0.1mg

Preparation of Vitamin Solution: Add components to distilled/deionized water and bring volume to 1.0L. Mix thoroughly.

Preparation of Medium: Prepare and dispense medium under 100% N_2. Add components, except $MgCl_2 \cdot 6H_2O$ solution and $CaCl_2 \cdot 2H_2O$ solution, to distilled/deionized water and bring volume to 950.0mL. Mix thoroughly. Gently heat and bring to boiling. Sparge with 100% N_2. Autoclave for 15 min at 15 psi pressure–121°C. Aseptically add 40.0mL of sterile $MgCl_2 \cdot 6H_2O$ solution and 10.0mL of sterile $CaCl_2 \cdot 2H_2O$ solution. Mix thoroughly. Aseptically and anaerobically pour into sterile Petri dishes or distribute into sterile tubes.

Use: For the cultivation and maintenance of *Sporohalobacter lortetii.*

Sporomusa Medium

Composition per 1010.0mL:

Betaine·H_2O 6.7g
$NaHCO_3$ 4.0g
NaCl 2.25g
Pancreatic digest of casein 2.0g
Yeast extract 2.0g
$MgSO_4 \cdot 7H_2O$ 0.5g
NH_4Cl 0.5g
K_2HPO_4 .... 0.348g
$CaCl_2 \cdot 2H_2O$ 0.25g
KH_2PO_4 .... 0.227g
$FeSO_4 \cdot 7H_2O$ 2.0mg
Resazurin 1.0mg
$NaHSeO_3$ 26.3μg
Vitamin solution 10.0mL
Reducing agent solution 10.0mL
Trace elements solution SL-10 .... 1.0mL

pH 7.0 ± 0.2 at 25°C

Vitamin Solution:

Composition per liter:

Pyridoxine·HCl 10.0mg
Calcium DL-pantothenate 5.0mg
Lipoic acid 5.0mg
Nicotinic acid 5.0mg
p-Aminobenzoic acid 5.0mg
Riboflavin 5.0mg
Thiamine·HCl 5.0mg
Biotin 2.0mg
Folic acid 2.0mg
Vitamin B_{12} 0.1mg

Preparation of Vitamin Solution: Add components to distilled/deionized water and bring volume to 1.0L. Mix thoroughly.

Reducing Agent Solution:

Composition per 10.0mL:

L-Cysteine·HCl·H_2O 0.3g
$Na_2S \cdot 9H_2O$ 0.3g

Preparation of Reducing Agent Solution: Add 10.0mL of distilled/deionized water to a flask. Gently heat and bring to boiling. Continue to boil for 1.0 min while sparging with 100% N_2. Cool to room temperature. Add L-cysteine·HCl·H_2O. Mix thoroughly. Adjust pH to 9 with 5*N* NaOH. Add $Na_2S \cdot 9H_2O$. Mix thoroughly. Autoclave for 10 min at 15 psi pressure–121°C.

Trace Elements Solution SL-10:

Composition per liter:

$FeCl_2 \cdot 4H_2O$ 1.5g
$CoCl_2 \cdot 6H_2O$ 190.0mg
$MnCl_2 \cdot 4H_2O$ 100.0mg
$ZnCl_2$ 70.0mg
$Na_2MoO_4 \cdot 2H_2O$ 36.0mg
$NiCl_2 \cdot 6H_2O$ 24.0mg
H_3BO_3 .... 6.0mg
$CuCl_2 \cdot 2H_2O$ 2.0mg
HCl (25% solution) 10.0mL

Preparation of Trace Elements Solution SL-10: Add $FeCl_2 \cdot 4H_2O$ to 10.0mL of HCl solution. Mix thoroughly. Add distilled/deionized water and bring volume to 1.0L. Add remaining components. Mix thoroughly.

Preparation of Medium: Prepare and dispense medium under under 80% N_2 + 20% CO_2. Add components, except reducing agent solution, to distilled/deionized water and bring volume to 1.0L. Mix thoroughly. Sparge with 80% N_2 + 20% CO_2. Anaerobically distribute into tubes or flasks. Autoclave for 15 min at 15 psi pressure–121°C. Using a syringe, aseptically and anaerobically add sterile reducing agent to each tube (10.0mL per liter of medium).

Use: For the cultivation and maintenance of *Sporomusa* species.

Sporomusa Medium, Modified

Composition per liter:

$NaHCO_3$	4.0g
NaCl	2.25g
Pancreatic digest of casein	2.0g
Yeast extract	2.0g
$MgSO_4 \cdot 7H_2O$	0.5g
NH_4Cl	0.5g
K_2HPO_4	0.348g
$CaCl_2 \cdot 2H_2O$	0.25g
KH_2PO_4	0.227g
$FeSO_4 \cdot 7H_2O$	2.0mg
Resazurin	1.0mg
$NaHSeO_3$	26.3μg
Fructose solution	50.0mL
Reducing agent solution	10.0mL
Vitamin solution	10.0mL
Trace elements solution SL-10	1.0mL

pH 7.0 ± 0.2 at 25°C

Fructose Solution:
Composition per 50.0mL:

Fructose	10.0g

Preparation of Fructose Solution: Add fructose to distilled/deionized water and bring volume to 50.0mL. Mix thoroughly. Sparge under 100% N_2 gas for 3 min. Filter sterilize. Store under N_2 gas.

Vitamin Solution:
Composition per liter:

Pyridoxine·HCl	10.0mg
Calcium DL-pantothenate	5.0mg
Lipoic acid	5.0mg
Nicotinic acid	5.0mg
p-Aminobenzoic acid	5.0mg
Riboflavin	5.0mg
Thiamine·HCl	5.0mg
Biotin	2.0mg
Folic acid	2.0mg
Vitamin B_{12}	0.1mg

Preparation of Vitamin Solution: Add components to distilled/deionized water and bring volume to 1.0L. Mix thoroughly.

Reducing Agent Solution:
Composition per 10.0mL:

L-Cysteine·HCl·H_2O	0.3g
$Na_2S \cdot 9H_2O$	0.3g

Preparation of Reducing Agent Solution: Add 10.0mL of distilled/deionized water to a flask. Boil under N_2 gas for 1 min. Cool to room temperature. Add L-cysteine·HCl·H_2O and dissolve. Adjust to pH 9 with 5*N* NaOH. Add washed $Na_2S \cdot 9H_2O$ and dissolve. Mix thoroughly. Autoclave for 10 min at 15 psi pressure–121°C.

Trace Elements Solution SL-10:
Composition per liter:

$FeCl_2 \cdot 4H_2O$	1.5g
$CoCl_2 \cdot 6H_2O$	190.0mg
$MnCl_2 \cdot 4H_2O$	100.0mg
$ZnCl_2$	70.0mg
$Na_2MoO_4 \cdot 2H_2O$	36.0mg
$NiCl_2 \cdot 6H_2O$	24.0mg
H_3BO_3	6.0mg
$CuCl_2 \cdot 2H_2O$	2.0mg
HCl (25% solution)	10.0mL

Preparation of Trace Elements Solution SL-10: Add $FeCl_2 \cdot 4H_2O$ to 10.0mL of HCl solution. Mix thoroughly. Add distilled/deionized water and bring volume to 1.0L. Add remaining components. Mix thoroughly.

Preparation of Medium: Add components, except $NaHCO_3$ and reducing agent solution, to distilled/deionized water and bring volume to 940.0mL. Gently heat and bring to boiling. Continue boiling for 3 min. Cool to room temperature under 80% N_2 + 20% CO_2. Add solid $NaHCO_3$ and bring pH to 7.0 by gassing. Distribute anaerobically under 80% N_2 + 20% CO_2 into tubes or flasks. Autoclave for 15 min at 15 psi pressure–121°C. Prior to inoculation of cultures, aseptically and anaerobically add 0.1mL of sterile reducing agent solution and 0.5mL of sterile fructose solution to each tube containing 9.4mL of sterile basal medium.

Use: For the cultivation and maintenance of *Sporomusa acidovorans*.

Sporomusa silvacetica Medium (DSMZ Medium 777)

Composition per liter:

NaCl	2.25g
Yeast extract	1.0g
Casitone	1.0g
NH_4Cl	0.5g
$MgSO_4 \cdot 7H_2O$	0.5g
K_2HPO_4	0.348g
KH_2PO_4	0.227g
$CaCl_2 \cdot 2H_2O$	0.25g
$FeSO_4 \cdot 7H_2O$	0.002g
Resazurin	1.0mg
$NaHSeO_3$	15.1μg
Cysteine solution	10.0mL
$NaHCO_3$ solution	10.0mL
$Na_2S \cdot 9H_2O$ solution	10.0mL
Fructose solution	10.0mL
Vitamin solution	10.0mL
Trace elements solution SL-10	1.0mL

pH 6.6 ± 0.2 at 25°C

Cysteine Solution:
Composition per 10.0mL:

L-Cysteine-HCl·H_2O	0.3g

Preparation of Cysteine Solution: Add L-cysteine·HCl·H_2O to distilled/deionized water and bring volume to 10.0mL. Mix thoroughly. Sparge with 100% N_2. Autoclave for 15 min at 15 psi pressure–121°C.

$Na_2S \cdot 9H_2O$ Solution:
Composition per 10mL:

$Na_2S \cdot 9H_2O$	0.3g

Preparation of $Na_2S \cdot 9H_2O$ Solution: Add $Na_2S \cdot 9H_2O$ to distilled/deionized water and bring volume to 10.0mL. Mix thoroughly. Autoclave under 100% N_2 for 15 min at 15 psi pressure–121°C. Cool to room temperature.

$NaHCO_3$ Solution:

Composition per 10.0mL:

$NaHCO_3$ 1.5g

Preparation of $NaHCO_3$ Solution: Add $NaHCO_3$ to distilled/deionized water and bring volume to 10.0mL. Mix thoroughly. Sparge with 80% N_2 + 20% CO_2. Filter sterilize.

Fructose Solution:

Composition per 10.0mL:

Fructose 5.0g

Preparation of Fructose Solution: Add fructose to distilled/deionized water and bring volume to 10.0mL. Mix thoroughly. Sparge with 100% N_2. Autoclave for 15 min at 15 psi pressure–121°C. Cool to room temperature.

Trace Elements Solution SL-10:

Composition per liter:

$FeCl_2 \cdot 4H_2O$	1.5g
$CoCl_2 \cdot 6H_2O$	190.0mg
$MnCl_2 \cdot 4H_2O$	100.0mg
$ZnCl_2$	70.0mg
$Na_2MoO_4 \cdot 2H_2O$	36.0mg
$NiCl_2 \cdot 6H_2O$	24.0mg
H_3BO_3	6.0mg
$CuCl_2 \cdot 2H_2O$	2.0mg
HCl (25% solution)	10.0mL

Preparation of Trace Elements Solution SL-10: Add $FeCl_2 \cdot 4H_2O$ to 10.0mL of HCl solution. Mix thoroughly. Add distilled/deionized water and bring volume to 1.0L. Add remaining components. Mix thoroughly. Sparge with 80% N_2 + 20% CO_2. Autoclave for 15 min at 15 psi pressure–121°C.

Vitamin Solution:

Composition per liter:

Pyridoxine-HCl	10.0mg
Thiamine-HCl·2H_2O	5.0mg
Riboflavin	5.0mg
Nicotinic acid	5.0mg
D-Ca-pantothenate	5.0mg
p-Aminobenzoic acid	5.0mg
Lipoic acid	5.0mg
Biotin	2.0mg
Folic acid	2.0mg
Vitamin B_{12}	0.1mg

Preparation of Vitamin Solution: Add components to distilled/deionized water and bring volume to 1.0L. Mix thoroughly. Sparge with 80% H_2 + 20% CO_2. Filter sterilize.

Preparation of Medium: Prepare and dispense medium under 80% N_2 + 20% CO_2 gas atmosphere. Add components, except $NaHCO_3$ solution, fructose solution, cysteine solution, vitamin solution, and $Na_2S \cdot 9H_2O$ solution, to distilled/deionized water and bring volume to 950.0mL. Mix thoroughly. Adjust pH to 6.5-6.7. Autoclave for 15 min at 15 psi pressure–121°C. Cool to room temperature. Aseptically and anaerobically add 10.0mL $NaHCO_3$ solution, 10.0mL fructose solution, 10.0mL cysteine solution, 10.0mL vitamin solution, and 10.0mL $Na_2S \cdot 9H_2O$ solution. Mix thoroughly. Aseptically and anaerobically distribute into sterile tubes or bottles.

Use: For the cultivation of *Sporomusa silvacetica, Thermicanus aegyptius, Moorella thermoacetica=Clostridium thermaceticum,* and *Moorella mulderi.*

Sporosarcina halophila Agar

Composition per liter:

NaCl	30.0g
Agar	20.0g
$MgCl_2 \cdot 6H_2O$	5.0g
Peptone	5.0g
NaCl	5.0g
Yeast extract	2.0g
Beef extract	1.0g

pH 7.2 ± 0.2 at 25°C

Preparation of Medium: Add components to distilled/deionized water and bring volume to 1.0L. Mix thoroughly. Gently heat and bring to boiling. Distribute into tubes or flasks. Autoclave for 15 min at 15 psi pressure–121°C. Pour into sterile Petri dishes or leave in tubes.

Use: For the cultivation and maintenance of *Halomonas elongata, Halomonas halmophila, Listonella anguillara, Salinicoccus roseus, Sporosarcina halophila, Vibrio fluvialis, Vibrio furnissii, Vibrio hollisae, Vibrio ordalii,* and *Vibrio vulnificus.*

Sporosarcina ureae Medium

Composition per liter:

L-Asparagine·H_2O or L-glutamine	30.0g
KCl	3.4g
NaCl	2.92g
K_2HPO_4	0.25g
$(NH_4)_2SO_4$	0.2g
$MgSO_4 \cdot 7H_2O$	0.05g
$FeSO_4 \cdot 7H_2O$	2.5mg
$MnCl_2 \cdot 4H_2O$	0.25mg
Biotin solution	10.0mL
L-Cysteine solution	10.0mL
$(NH_4)_2SO_4$ solution	10.0mL

pH 8.7 ± 0.2 at 25°C

Biotin Solution:

Composition per 10.0mL:

D-Biotin 1.0mg

Preparation of Biotin Solution: Add biotin to distilled/deionized water and bring volume to 10.0mL. Mix thoroughly. Filter sterilize.

L-Cysteine Solution:

Composition per 10.0mL:

L-Cysteine 0.04g

Preparation of L-Cysteine Solution: Add L-cysteine to distilled/deionized water and bring volume to 10.0mL. Mix thoroughly. Filter sterilize.

$(NH_4)_2SO_4$ Solution:

Composition per 10.0mL:

$(NH_4)_2SO_4$ 0.2g

Preparation of $(NH_4)_2SO_4$ Solution: Add $(NH_4)_2SO_4$ to distilled/deionized water and bring volume to 10.0mL. Mix thoroughly. Filter sterilize.

Preparation of Medium: Add components, except biotin solution, L-cysteine solution, and $(NH_4)_2SO_4$ solution, to distilled/deionized water and bring volume to 970.0mL. Mix thoroughly. Adjust pH to 8.7 with 1*N* NaOH. Autoclave for 15 min at 15 psi pressure–121°C. Cool to 45°–50°C. Aseptically add sterile biotin solution, L-cysteine solution, and $(NH_4)_2SO_4$ solution. Mix thoroughly. Aseptically distribute into sterile tubes.

Use: For the cultivation of *Sporosarcina ureae.*

Sporosarcina ureae Medium

Composition per liter:

Agar	30.0g
Glucose	4.0g
$(NH_4)_2SO_4$	4.0g
Malt extract	3.0g
Peptone	3.0g
Yeast extract	2.0g
K_2HPO_4	1.0g
$MgSO_4$	0.8g
$CaCl_2$	0.1g
$MnSO_4 \cdot H_2O$	0.1g
$CuSO_4 \cdot 5H_2O$	0.01g
$ZnSO_4$	0.01g
$FeSO_4 \cdot 7H_2O$	1.0mg

Preparation of Medium: Add components to 1.0L of distilled/deionized water. Mix thoroughly. Gently heat and bring to boiling. Distribute into tubes or flasks. Autoclave for 15 min at 15 psi pressure–121°C. Pour into sterile Petri dishes or leave in tubes.

Use: For the cultivation and induction of sporulation of *Sporosarcina ureae.*

SRB-Psychrophile Medium (DSMZ Medium 861)

Composition per 1158mL:

NaCl	20.0g
Na_2SO_4	4.0g
$MgCl_2 \cdot 6H_2O$	3.0g
$CaCl_2 \cdot 2H_2O$	0.15g
KBr	0.09g
KCl	0.5g
Resazurin	0.5mg
NH_4Cl solution	49.5mL
KH_2PO_4 solution	49.5mL
$NaHCO_3$ solution	29.7mL
Vitamin solution	9.9mL
Substrate solution	9.9mL
$Na_2S \cdot 9H_2O$ solution	9.9mL
Dithionite solution	1.0mL
Trace elements solution SL-10	1.0mL
Selenite-tungstate solution	1.0mL

pH 7.2 ± 0.2 at 25°C

Dithionite Solution:

Composition per 10mL:

Na-dithionite	0.25g

Preparation of Dithionite Solution: Add Na-dithionite to distilled/deionized water and bring volume to 10.0mL. Mix thoroughly. Sparge with 100% N_2. Filter sterilize.

$Na_2S \cdot 9H_2O$ Solution:

Composition per 10mL:

$Na_2S \cdot 9H_2O$	0.5g

Preparation of $Na_2S \cdot 9H_2O$ Solution: Add $Na_2S \cdot 9H_2O$ to distilled/deionized water and bring volume to 10.0mL. Mix thoroughly. Autoclave under 100% N_2 for 15 min at 15 psi pressure–121°C. Cool to room temperature.

Vitamin Solution:

Composition per liter:

Pyridoxine-HCl	10.0mg
Thiamine-HCl·$2H_2O$	5.0mg
Riboflavin	5.0mg
Nicotinic acid	5.0mg
D-Ca-pantothenate	5.0mg
p-Aminobenzoic acid	5.0mg
Lipoic acid	5.0mg
Biotin	2.0mg
Folic acid	2.0mg
Vitamin B_{12}	0.1mg

Preparation of Vitamin Solution: Add components to distilled/deionized water and bring volume to 1.0L. Mix thoroughly. Sparge with 80% H_2 + 20% CO_2. Filter sterilize.

Selenite-Tungstate Solution:

Composition per liter:

NaOH	0.5g
$Na_2WO_4 \cdot 2H_2O$	4.0mg
$Na_2SeO_3 \cdot 5H_2O$	3.0mg

Preparation of Selenite-Tungstate Solution: Add components to distilled/deionized water and bring volume to 1.0L. Mix thoroughly. Sparge with 100% N_2. Filter sterilize.

Trace Elements Solution SL-10:

Composition per liter:

$FeCl_2 \cdot 4H_2O$	1.5g
$CoCl_2 \cdot 6H_2O$	190.0mg
$MnCl_2 \cdot 4H_2O$	100.0mg
$ZnCl_2$	70.0mg
$Na_2MoO_4 \cdot 2H_2O$	36.0mg
$NiCl_2 \cdot 6H_2O$	24.0mg
H_3BO_3	6.0mg
$CuCl_2 \cdot 2H_2O$	2.0mg
HCl (25% solution)	10.0mL

Preparation of Trace Elements Solution SL-10: Add $FeCl_2 \cdot 4H_2O$ to 10.0mL of HCl solution. Mix thoroughly. Add distilled/deionized water and bring volume to 1.0L. Add remaining components. Mix thoroughly. Sparge with 80% N_2 + 20% CO_2. Autoclave for 15 min at 15 psi pressure–121°C.

$NaHCO_3$ Solution:

Composition per 100.0mL:

$NaHCO_3$	10.0g

Preparation of $NaHCO_3$ Solution: Add $NaHCO_3$ to distilled/deionized water and bring volume to 100.0mL. Mix thoroughly. Sparge with 80% N_2 + 20% CO_2. Filter sterilize.

NH_4Cl Solution:

Composition per 100.0mL:

NH_4Cl	0.5g

Preparation of NH_4Cl Solution: Add NH_4Cl to distilled/deionized water and bring volume to 100.0mL. Mix thoroughly. Sparge with 100% N_2. Filter sterilize.

KH_2PO_4 Solution:

Composition per 100.0mL:

KH_2PO_4	0.4g

Preparation of KH_2PO_4 Solution: Add KH_2PO_4 to distilled/deionized water and bring volume to 100.0mL. Mix thoroughly. Sparge with 100% N_2. Filter sterilize.

Substrate Solution:

Composition per 10.0mL:

Na-acetate	1.5g

Preparation of Substrate Solution: Add Na-acetate to distilled/deionized water and bring volume to 10.0mL. Sparge with N_2. Filter sterilize.

Preparation of Medium: Prepare and dispense medium under 80% N_2 + 20% CO_2. Add components, except

$NaHCO_3$ solution, $Na_2S·9H_2O$ solution, substrate solution, selenite-tungstate solution, NH_4Cl solution, KH_2PO_4 solution, vitamin solution, dithionite solution, and trace elements solution SL-10. Distribute 30.0mL aliquots into 50mL serum bottles. Autoclave for 15 min at 15 psi pressure–121°C. Cool to room temperature. Aseptically and anaerobically for each 30.0mL medium add 0.9mL $NaHCO_3$ solution, 0.3mL $Na_2S·9H_2O$ solution, 0.3mL substrate solution, 0.03mL selenite-tungstate solution, 1.5mL NH_4Cl solution, 1.5mL KH_2PO_4 solution, 0.3mL vitamin solution, 0.03mL dithionite solution, and 0.03mL trace elements solution SL-10. Final pH is 7.2.

Use: For the cultivation of *Desulfofrigus oceanense*.

SRB-Psychrophile Medium (DSMZ Medium 861)

Composition per 1158mL:

NaCl	20.0g
Na_2SO_4	4.0g
$MgCl_2·6H_2O$	3.0g
KCl	0.5g
$CaCl_2·2H_2O$	0.15g
KBr	0.09g
Resazurin	0.5mg
NH_4Cl solution	49.5mL
KH_2PO_4 solution	49.5mL
$NaHCO_3$ solution	29.7mL
Vitamin solution	9.9mL
Substrate solution	9.9mL
$Na_2S·9H_2O$ solution	9.9mL
Dithionite solution	1.0mL
Trace elements solution SL-10	1.0mL
Selenite-tungstate solution	1.0mL

pH 7.2 ± 0.2 at 25°C

Dithionite Solution:
Composition per 10mL:

Na-dithionite	0.25g

Preparation of Dithionite Solution: Add Na-dithionite to distilled/deionized water and bring volume to 10.0mL. Mix thoroughly. Sparge with 100% N_2. Filter sterilize.

$Na_2S·9H_2O$ Solution:
Composition per 10mL:

$Na_2S·9H_2O$	0.5g

Preparation of $Na_2S·9H_2O$ Solution: Add $Na_2S·9H_2O$ to distilled/deionized water and bring volume to 10.0mL. Mix thoroughly. Autoclave under 100% N_2 for 15 min at 15 psi pressure–121°C. Cool to room temperature.

Vitamin Solution:
Composition per liter:

Pyridoxine-HCl	10.0mg
Thiamine-HCl·$2H_2O$	5.0mg
Riboflavin	5.0mg
Nicotinic acid	5.0mg
D-Ca-pantothenate	5.0mg
p-Aminobenzoic acid	5.0mg
Lipoic acid	5.0mg
Biotin	2.0mg
Folic acid	2.0mg
Vitamin B_{12}	0.1mg

Preparation of Vitamin Solution: Add components to distilled/deionized water and bring volume to 1.0L. Mix thoroughly. Sparge with 80% H_2 + 20% CO_2. Filter sterilize.

Selenite-Tungstate Solution:
Composition per liter:

NaOH	0.5g
$Na_2WO_4·2H_2O$	4.0mg
$Na_2SeO_3·5H_2O$	3.0mg

Preparation of Selenite-Tungstate Solution: Add components to distilled/deionized water and bring volume to 1.0L. Mix thoroughly. Sparge with 100% N_2. Filter sterilize.

Trace Elements Solution SL-10:
Composition per liter:

$FeCl_2·4H_2O$	1.5g
$CoCl_2·6H_2O$	190.0mg
$MnCl_2·4H_2O$	100.0mg
$ZnCl_2$	70.0mg
$Na_2MoO_4·2H_2O$	36.0mg
$NiCl_2·6H_2O$	24.0mg
H_3BO_3	6.0mg
$CuCl_2·2H_2O$	2.0mg
HCl (25% solution)	10.0mL

Preparation of Trace Elements Solution SL-10: Add $FeCl_2·4H_2O$ to 10.0mL of HCl solution. Mix thoroughly. Add distilled/deionized water and bring volume to 1.0L. Add remaining components. Mix thoroughly. Sparge with 80% N_2 + 20% CO_2. Autoclave for 15 min at 15 psi pressure–121°C.

$NaHCO_3$ Solution:
Composition per 100.0mL:

$NaHCO_3$	10.0g

Preparation of $NaHCO_3$ Solution: Add $NaHCO_3$ to distilled/deionized water and bring volume to 100.0mL. Mix thoroughly. Sparge with 80% N_2 + 20% CO_2. Filter sterilize.

NH_4Cl Solution:
Composition per 100.0mL:

NH_4Cl	0.5g

Preparation of NH_4Cl Solution: Add NH_4Cl to distilled/deionized water and bring volume to 100.0mL. Mix thoroughly. Sparge with 100% N_2. Filter sterilize.

KH_2PO_4 Solution:
Composition per 100.0mL:

KH_2PO_4	0.4g

Preparation of KH_2PO_4 Solution: Add KH_2PO_4 to distilled/deionized water and bring volume to 100.0mL. Mix thoroughly. Sparge with 100% N_2. Filter sterilize.

Substrate Solution:
Composition per 10.0mL:

Na-lactate	2.5g

Preparation of Substrate Solution: Add Na-lactate to distilled/deionized water and bring volume to 10.0mL. Sparge with N_2. Filter sterilize.

Preparation of Medium: Prepare and dispense medium under 80% N_2 + 20% CO_2. Add components, except $NaHCO_3$ solution, $Na_2S·9H_2O$ solution, substrate solution, selenite-tungstate solution, NH_4Cl solution, KH_2PO_4 solution, vitamin solution, dithionite solution, and trace elements solution SL-10. Distribute 30.0mL aliquots into 50mL serum bottles. Autoclave for 15 min at 15 psi pressure–121°C. Cool to room temperature. Aseptically and anaerobically for each 30.0mL medium add 0.9mL $NaHCO_3$ solution, 0.3mL $Na_2S·9H_2O$ solution, 0.3mL substrate solution, 0.03mL selenite-tungstate solution, 1.5mL

NH_4Cl solution, 1.5mL KH_2PO_4 solution, 0.3mL vitamin solution, 0.03mL dithionite solution, and 0.03mL trace elements solution SL-10. Final pH is 7.2.

Use: For the cultivation of *Desulfotalea arctica* and *Desulfofrigus fragile.*

SRB-Psychrophile Medium (DSMZ Medium 861)

Composition per 1158mL:

NaCl	10.0g
Na_2SO_4	4.0g
$MgCl_2 \cdot 6H_2O$	3.0g
KCl	0.5g
$CaCl_2 \cdot 2H_2O$	0.15g
KBr	0.09g
Resazurin	0.5mg
NH_4Cl solution	49.5mL
KH_2PO_4 solution	49.5mL
$NaHCO_3$ solution	29.7mL
Vitamin solution	9.9mL
Substrate solution	9.9mL
$Na_2S \cdot 9H_2O$ solution	9.9mL
Dithionite solution	1.0mL
Trace elements solution SL-10	1.0mL
Selenite-tungstate solution	1.0mL

pH 7.2 ± 0.2 at 25°C

Dithionite Solution:

Composition per 10mL:

Na-dithionite	0.25g

Preparation of Dithionite Solution: Add Na-dithionite to distilled/deionized water and bring volume to 10.0mL. Mix thoroughly. Sparge with 100% N_2. Filter sterilize.

$Na_2S \cdot 9H_2O$ Solution:

Composition per 10mL:

$Na_2S \cdot 9H_2O$	0.5g

Preparation of $Na_2S \cdot 9H_2O$ Solution: Add $Na_2S \cdot 9H_2O$ to distilled/deionized water and bring volume to 10.0mL. Mix thoroughly. Autoclave under 100% N_2 for 15 min at 15 psi pressure–121°C. Cool to room temperature.

Vitamin Solution:

Composition per liter:

Pyridoxine-HCl	10.0mg
Thiamine-HCl·$2H_2O$	5.0mg
Riboflavin	5.0mg
Nicotinic acid	5.0mg
D-Ca-pantothenate	5.0mg
p-Aminobenzoic acid	5.0mg
Lipoic acid	5.0mg
Biotin	2.0mg
Folic acid	2.0mg
Vitamin B_{12}	0.1mg

Preparation of Vitamin Solution: Add components to distilled/deionized water and bring volume to 1.0L. Mix thoroughly. Sparge with 80% H_2 + 20% CO_2. Filter sterilize.

Selenite-Tungstate Solution:

Composition per liter:

NaOH	0.5g
$Na_2WO_4 \cdot 2H_2O$	4.0mg
$Na_2SeO_3 \cdot 5H_2O$	3.0mg

Preparation of Selenite-Tungstate Solution: Add components to distilled/deionized water and bring volume to 1.0L. Mix thoroughly. Sparge with 100% N_2. Filter sterilize.

Trace Elements Solution SL-10:

Composition per liter:

$FeCl_2 \cdot 4H_2O$	1.5g
$CoCl_2 \cdot 6H_2O$	190.0mg
$MnCl_2 \cdot 4H_2O$	100.0mg
$ZnCl_2$	70.0mg
$Na_2MoO_4 \cdot 2H_2O$	36.0mg
$NiCl_2 \cdot 6H_2O$	24.0mg
H_3BO_3	6.0mg
$CuCl_2 \cdot 2H_2O$	2.0mg
HCl (25% solution)	10.0mL

Preparation of Trace Elements Solution SL-10: Add $FeCl_2 \cdot 4H_2O$ to 10.0mL of HCl solution. Mix thoroughly. Add distilled/deionized water and bring volume to 1.0L. Add remaining components. Mix thoroughly. Sparge with 80% N_2 + 20% CO_2. Autoclave for 15 min at 15 psi pressure–121°C.

$NaHCO_3$ Solution:

Composition per 100.0mL:

$NaHCO_3$	10.0g

Preparation of $NaHCO_3$ Solution: Add $NaHCO_3$ to distilled/deionized water and bring volume to 100.0mL. Mix thoroughly. Sparge with 80% N_2 + 20% CO_2. Filter sterilize.

NH_4Cl Solution:

Composition per 100.0mL:

NH_4Cl	0.5g

Preparation of NH_4Cl Solution: Add NH_4Cl to distilled/deionized water and bring volume to 100.0mL. Mix thoroughly. Sparge with 100% N_2. Filter sterilize.

KH_2PO_4 Solution:

Composition per 100.0mL:

KH_2PO_4	0.4g

Preparation of KH_2PO_4 Solution: Add KH_2PO_4 to distilled/deionized water and bring volume to 100.0mL. Mix thoroughly. Sparge with 100% N_2. Filter sterilize.

Substrate Solution:

Composition per 10.0mL:

Na-lactate	2.5g

Preparation of Substrate Solution: Add Na-lactate to distilled/deionized water and bring volume to 10.0mL. Sparge with N_2. Filter sterilize.

Preparation of Medium: Prepare and dispense medium under 80% N_2 + 20% CO_2. Add components, except $NaHCO_3$ solution, $Na_2S \cdot 9H_2O$ solution, substrate solution, selenite-tungstate solution, NH_4Cl solution, KH_2PO_4 solution, vitamin solution, dithionite solution, and trace elements solution SL-10. Distribute 30.0mL aliquots into 50mL serum bottles. Autoclave for 15 min at 15 psi pressure–121°C. Cool to room temperature. Aseptically and anaerobically for each 30.0mL medium add 0.9mL $NaHCO_3$ solution, 0.3mL $Na_2S \cdot 9H_2O$ solution, 0.3mL substrate solution, 0.03mL selenite-tungstate solution, 1.5mL NH_4Cl solution, 1.5mL KH_2PO_4 solution, 0.3mL vitamin solution, 0.03mL dithionite solution, and 0.03mL trace elements solution SL-10. Final pH is 7.2.

Use: For the cultivation of *Desulfotalea psychrophila.*

SRB-Psychrophile Medium (DSMZ Medium 861)

Composition per 1158mL:

NaCl....20.0g
Na_2SO_4....4.0g
$MgCl_2·6H_2O$....3.0g
$CaCl_2·2H_2O$....0.15g
KBr....0.09g
KCl....0.5g
Resazurin....0.5mg
NH_4Cl solution....49.5mL
KH_2PO_4 solution....49.5mL
$NaHCO_3$ solution....29.7mL
Vitamin solution....9.9mL
Substrate solution....9.9mL
$Na_2S·9H_2O$ solution....9.9mL
Dithionite solution....1.0mL
Trace elements solution SL-10....1.0mL
Selenite-tungstate solution....1.0mL

pH 7.2 ± 0.2 at 25°C

Dithionite Solution:
Composition per 10mL:

Na-dithionite....0.25g

Preparation of Dithionite Solution: Add Na-dithionite to distilled/deionized water and bring volume to 10.0mL. Mix thoroughly. Sparge with 100% N_2. Filter sterilize.

$Na_2S·9H_2O$ Solution:
Composition per 10mL:

$Na_2S·9H_2O$....0.5g

Preparation of $Na_2S·9H_2O$ Solution: Add $Na_2S·9H_2O$ to distilled/deionized water and bring volume to 10.0mL. Mix thoroughly. Autoclave under 100% N_2 for 15 min at 15 psi pressure–121°C. Cool to room temperature.

Vitamin Solution:
Composition per liter:

Pyridoxine-HCl....10.0mg
Thiamine-HCl·$2H_2O$....5.0mg
Riboflavin....5.0mg
Nicotinic acid....5.0mg
D-Ca-pantothenate....5.0mg
p-Aminobenzoic acid....5.0mg
Lipoic acid....5.0mg
Biotin....2.0mg
Folic acid....2.0mg
Vitamin B_{12}....0.1mg

Preparation of Vitamin Solution: Add components to distilled/deionized water and bring volume to 1.0L. Mix thoroughly. Sparge with 80% H_2 + 20% CO_2. Filter sterilize.

Selenite-Tungstate Solution:
Composition per liter:

NaOH....0.5g
$Na_2WO_4·2H_2O$....4.0mg
$Na_2SeO_3·5H_2O$....3.0mg

Preparation of Selenite-Tungstate Solution: Add components to distilled/deionized water and bring volume to 1.0L. Mix thoroughly. Sparge with 100% N_2. Filter sterilize.

Trace Elements Solution SL-10:
Composition per liter:

$FeCl_2·4H_2O$....1.5g
$CoCl_2·6H_2O$....190.0mg
$MnCl_2·4H_2O$....100.0mg
$ZnCl_2$....70.0mg
$Na_2MoO_4·2H_2O$....36.0mg
$NiCl_2·6H_2O$....24.0mg
H_3BO_3....6.0mg
$CuCl_2·2H_2O$....2.0mg
HCl (25% solution)....10.0mL

Preparation of Trace Elements Solution SL-10: Add $FeCl_2·4H_2O$ to 10.0mL of HCl solution. Mix thoroughly. Add distilled/deionized water and bring volume to 1.0L. Add remaining components. Mix thoroughly. Sparge with 80% N_2 + 20% CO_2. Autoclave for 15 min at 15 psi pressure–121°C.

$NaHCO_3$ Solution:
Composition per 100.0mL:

$NaHCO_3$....10.0g

Preparation of $NaHCO_3$ Solution: Add $NaHCO_3$ to distilled/deionized water and bring volume to 100.0mL. Mix thoroughly. Sparge with 80% N_2 + 20% CO_2. Filter sterilize.

NH_4Cl Solution:
Composition per 100.0mL:

NH_4Cl....0.5g

Preparation of NH_4Cl Solution: Add NH_4Cl to distilled/deionized water and bring volume to 100.0mL. Mix thoroughly. Sparge with 100% N_2. Filter sterilize.

KH_2PO_4 Solution:
Composition per 100.0mL:

KH_2PO_4....0.4g

Preparation of KH_2PO_4 Solution: Add KH_2PO_4 to distilled/deionized water and bring volume to 100.0mL. Mix thoroughly. Sparge with 100% N_2. Filter sterilize.

Substrate Solution:
Composition per 10.0mL:

Na-propionate....1.5g

Preparation of Substrate Solution: Add Na-propionate to distilled/deionized water and bring volume to 10.0mL. Sparge with N_2. Filter sterilize.

Preparation of Medium: Prepare and dispense medium under 80% N_2 + 20% CO_2. Add components, except $NaHCO_3$ solution, $Na_2S·9H_2O$ solution, substrate solution, selenite-tungstate solution, NH_4Cl solution, KH_2PO_4 solution, vitamin solution, dithionite solution, and trace elements solution SL-10. Distribute 30.0mL aliquots into 50mL serum bottles. Autoclave for 15 min at 15 psi pressure–121°C. Cool to room temperature. Aseptically and anaerobically for each 30.0mL medium add 0.9mL $NaHCO_3$ solution, 0.3mL $Na_2S·9H_2O$ solution, 0.3mL substrate solution, 0.03mL selenite-tungstate solution, 1.5mL NH_4Cl solution, 1.5mL KH_2PO_4 solution, 0.3mL vitamin solution, 0.03mL dithionite solution, and 0.03mL trace elements solution SL-10. Final pH is 7.2.

Use: For the cultivation of *Desulfofaba gelida*.

SSL Agar

Composition per liter:

Agar....2.5g
$CaCl_2·2H_2O$....1.0g
Gelatin....1.0g
KNO_3....1.0g
$MgSO_4·7H_2O$....1.0g
NaCl....1.0g
Pancreatic digest of casein....1.0g
Yeast extract....1.0g

Sodium glycerophosphate	0.1g
Cyanocobalamin	1.0μg
Trace elements solution	1.0mL

pH 7.5 ± 0.2 at 25°C

Trace Elements Solution:
Composition per liter:

Disodium EDTA	8.0g
$MnCl_2 \cdot 4H_2O$	0.1g
$CoCl_2 \cdot 6H_2O$	0.02g
KBr	0.02g
KI	0.02g
$ZnCl_2$	0.02g
$CuSO_4$	0.01g
H_3BO_3	0.01g
$Na_2MoO_4 \cdot 2H_2O$	0.01g
LiCl	5.0mg
$SnCl_2 \cdot 2H_2O$	5.0mg

Preparation of Trace Elements Solution: Add components to distilled/deionized water and bring volume to 1.0L. Mix thoroughly.

Preparation of Medium: Add components to distilled/deionized water and bring volume to 1.0L. Mix thoroughly. Gently heat and bring to boiling. Distribute into tubes or flasks. Autoclave for 15 min at 15 psi pressure–121°C. Pour into sterile Petri dishes or leave in tubes.

Use: For the isolation and cultivation of *Cytophaga* species, *Herpetosiphon* species, *Saprospira* species, and *Flexithrix* species.

StA
(Schmitthenner's Agar)

Composition per liter:

Agar	20.0g
Sucrose	2.5g
Asparagine	0.27g
KH_2PO_4	0.15g
K_2HPO_4	0.15g
$MgSO_4$	0.1g
Sitosterol	0.01g

Preparation of Medium: Add components, except agar, to distilled/deionized water and bring volume to 1.0L. Mix thoroughly. Add agar. Swirl. Gently heat and bring to boiling. Distribute into tubes or flasks. Autoclave for 15 min at 15 psi pressure–121°C. Pour into sterile Petri dishes or leave in tubes.

Use: For the cultivation and maintenance of *Pythium aphanidermatum, Pythium graminicola, Pythium myriotylum, Pythium sylvaticum,* and *Pythium ultimum.*

Staley's Maintenance Agar

Composition per liter:

Agar	15.0g
Peptone	5.0g
Yeast extract	0.5g
Hutner's basal salts solution	20.0mL
Vitamin solution	10.0mL

Hutner's Basal Salts Solution:
Composition per liter:

$MgSO_4 \cdot 7H_2O$	29.7g
Nitrilotriacetic acid	10.0g
$CaCl_2 \cdot 2H_2O$	3.335g
$FeSO_4 \cdot 7H_2O$	99.0mg
$(NH_4)_6MoO_7O_{24} \cdot 4H_2O$	9.25mg
"Metals 44"	50.0mL

"Metals 44":
Composition per 100.0mL:

$ZnSO_4 \cdot 7H_2O$	1.095g
$FeSO_4 \cdot 7H_2O$	0.5g
Sodium EDTA	0.25g
$MnSO_4 \cdot H2O$	0.154g
$CuSO_4 \cdot 5H_2O$	39.2mg
$Co(NO_3)_2 \cdot 6H_2O$	24.8mg
$Na_2B_4O_7 \cdot 10H_2O$	17.7mg

Preparation of "Metals 44": Add sodium EDTA to distilled/deionized water and bring volume to 90.0mL. Mix thoroughly. Add a few drops of concentrated H_2SO_4 to retard precipitation of heavy metal ions. Add remaining components. Mix thoroughly. Bring volume to 100.0mL with distilled/deionized water.

Preparation of Hutner's Basal Salts Solution: Add nitrilotriacetic acid to 500.0mL of distilled/deionized water. Adjust pH to 6.5 with KOH. Add remaining components. Add distilled/deionized water to 1.0L. Adjust pH to 6.8. Autoclave for 15 min at 15 psi pressure–121°C. Cool to 50°C.

Vitamin Solution:
Composition per liter:

Nicotinamide	9.0mg
Calcium DL-pantothenate	5.0mg
Riboflavin	5.0mg
Thiamine·HCl	5.0mg
Biotin	2.0mg
Folic acid	2.0mg
Cyanocobalamin	0.1mg

Preparation of Vitamin Solution: Add components to distilled/deionized water and bring volume to 1.0L. Mix thoroughly. Filter sterilize.

Preparation of Medium: Add components, except Hutner's basal salts solution and vitamin solution, to distilled/deionized water and bring volume to 970.0mL. Mix thoroughly. Gently heat and bring to boiling. Autoclave for 15 min at 15 psi pressure–121°C. Cool to 50°–55°C. Aseptically add 20.0 mL of sterile Hutner's basal salts solution and 10.0mL of sterile vitamin solution. Pour into sterile Petri dishes or distribute into sterile tubes.

Use: For the cultivation of *Gemmata obscuriglobus.*

Stan 4 Agar

Composition per liter:

Solution B	650.0mL
Solution A	350.0mL

Solution A:
Composition per 350.0mL:

$CaCl_2 \cdot 2H_2O$	1.0g
KNO_3	1.0g
$MgSO_4 \cdot 7H_2O$	1.0g
Trace elements solution	1.0mL

Preparation of Solution A: Add components to distilled/deionized water and bring volume to 350.0mL. Mix thoroughly. Gently heat and bring to boiling. Autoclave for 15 min at 15 psi pressure–121°C. Cool to 45°–50°C.

Trace Elements Solution:
Composition per liter:

EDTA	8.0g
$MnCl_2 \cdot 4H_2O$	0.1g
$CoCl_2$	0.02g
KBr	0.02g
$ZnCl_2$	0.02g
$CuSO_4$	0.01g

H_3BO_3 .... 0.01g
$NaMoO_4 \cdot 2H_2O$ 0.01g
$BaCl_2$ 5.0mg
LiCl 5.0mg
$SnCl_2 \cdot 2H_2O$ 5.0mg

Preparation of Trace Elements Solution: Add components to distilled/deionized water and bring volume to 1.0L. Mix thoroughly.

Solution B:
Composition per 650.0mL:
Agar 10.0g
K_2HPO_4 .... 1.0g

Preparation of Solution B: Add components to distilled/deionized water and bring volume to 650.0mL. Mix thoroughly. Gently heat and bring to boiling. Autoclave for 15 min at 15 psi pressure–121°C. Cool to 45°–50°C.

Preparation of Medium: Aseptically combine 350.0mL of cooled, sterile solution A and 650.0mL of cooled, sterile solution B. Mix thoroughly. Pour into sterile Petri dishes or distribute into sterile tubes.

Use: For the cultivation of myxobacteria.

Stan 5 Agar

Composition per liter:
Solution B 650.0mL
Solution A 350.0mL

Solution A:
Composition per 350.0mL:
$CaCl_2 \cdot 2H_2O$ 1.0g
$(NH_4)_2SO_4$ 1.0g
$MgSO_4 \cdot 7H_2O$ 1.0g
Trace elements solution 1.0mL

Preparation of Solution A: Add components to distilled/deionized water and bring volume to 350.0mL. Mix thoroughly. Gently heat and bring to boiling. Autoclave for 15 min at 15 psi pressure–121°C. Cool to 45°–50°C.

Trace Elements Solution:
Composition per liter:
EDTA 8.0g
$MnCl_2 \cdot 4H_2O$ 0.1g
$CoCl_2$ 0.02g
KBr 0.02g
$ZnCl_2$ 0.02g
$CuSO_4$ 0.01g
H_3BO_3 .... 0.01g
$NaMoO_4 \cdot 2H_2O$ 0.01g
$BaCl_2$ 5.0mg
LiCl 5.0mg
$SnCl_2 \cdot 2H_2O$ 5.0mg

Preparation of Trace Elements Solution: Add components to distilled/deionized water and bring volume to 1.0L. Mix thoroughly.

Solution B:
Composition per 650.0mL:
Agar 10.0g
K_2HPO_4 .... 1.0g

Preparation of Solution B: Add components to distilled/deionized water and bring volume to 650.0mL. Mix thoroughly. Gently heat and bring to boiling. Autoclave for 15 min at 15 psi pressure–121°C. Cool to 45°–50°C.

Preparation of Medium: Aseptically combine 350.0mL of cooled, sterile solution A and 650.0mL of cooled, sterile solution B. Mix thoroughly. Pour into sterile Petri dishes or distribute into sterile tubes.

Use: For the cultivation of myxobacteria.

Stan 6 Agar

Composition per liter:
Agar 10.0g
$CaCl_2 \cdot 2H_2O$ 1.0g
K_2HPO_4 .... 1.0g
$MgSO_4 \cdot 7H_2O$ 1.0g
$(NH_4)_2SO_4$ 1.0g
$FeCl_3$ 0.2g
$MnSO_4 \cdot 7H_2O$ 0.1g
Yeast extract 0.02g
Trace elements solution 1.0mL

Trace Elements Solution:
Composition per liter:
Disodium EDTA 8.0g
$MnCl_2 \cdot 4H_2O$ 0.1g
$CoCl_2 \cdot 6H_2O$ 0.02g
KBr 0.02g
KI 0.02g
$ZnCl_2$ 0.02g
$CuSO_4$ 0.01g
H_3BO_3 .... 0.01g
$Na_2MoO_4 \cdot 2H_2O$ 0.01g
LiCl 5.0mg
$SnCl_2 \cdot 2H_2O$ 5.0mg

Preparation of Trace Elements Solution: Add components to distilled/deionized water and bring volume to 1.0L. Mix thoroughly.

Preparation of Medium: Add components to distilled/deionized water and bring volume to 1.0L. Mix thoroughly. Gently heat and bring to boiling. Distribute into tubes or flasks. Autoclave for 15 min at 15 psi pressure–121°C. Pour into sterile Petri dishes or leave in tubes.

Use: For the isolation and cultivation of *Cytophaga* species, *Herpetosiphon* species, *Saprospira* species, and *Flexithrix* species.

Standard Methods Agar (Tryptone Glucose Yeast Agar) (Plate Count Agar)

Composition per liter:
Agar 15.0g
Pancretic digest of casein 5.0g
Yeast extract 2.5g
Glucose 1.0g

pH 7.0 ± 0.1 at 25°C

Source: Available as a premixed powder from BD Diagnostic Systems.

Preparation of Medium: Add components to distilled/deionized water and bring volume to 1.0L. Mix thoroughly. Gently heat and bring to boiling. Distribute into tubes or flasks. Autoclave for 15 min at 15 psi pressure–121°C. Pour into sterile Petri dishes or leave in tubes.

Use: For the cultivation and enumeration by microbial plate counts of microorganisms isolated from water and other specimens.

Stanier's Basal Medium with Pyridoxine and Yeast Extract

Composition per liter:
KH_2PO_4 .... 2.78g

Na_2HPO_4 .... 2.78g
$(NH_4)_2SO_4$ 1.0g
Yeast extract 0.2g
Hutner's mineral base 20.0mL
Pyridoxine solution 10.0mL
pH 6.8 ± 0.2 at 25°C

Pyridoxine Solution:
Composition per 10.0mL:
Pyridoxine 2.0g

Preparation of Pyridoxine Solution: Add pyridoxine to distilled/deionized water and bring volume to 10.0mL. Mix thoroughly. Filter sterilize.

Hutner's Mineral Base:
Composition per liter:
$MgSO_4 \cdot 7H_2O$ 29.7g
Nitrilotriacetic acid 10.0g
$CaCl_2 \cdot 2H_2O$ 3.34g
$FeSO_4 \cdot 7H_2O$ 99.0mg
$(NH_4)_2MoO_4$ 9.25mg
"Metals 44" 50.0mL

Preparation of Hutner's Mineral Base: Add nitrilotriacetic acid to 500.0mL of distilled/deionized water. Dissolve by adjusting pH to 6.5 with KOH. Add remaining components. Readjust pH to 7.2 with H_2SO_4 or KOH. Add distilled/deionized water to 1.0L.

"Metals 44":
Composition per 100.0mL:
$ZnSO_4 \cdot 7H_2O$ 1.1g
$FeSO_4 \cdot 7H_2O$ 0.5g
EDTA 0.25g
$MnSO_4 \cdot 7H_2O$ 0.154g
$CuSO_4 \cdot 5H_2O$ 0.04g
$Co(NO_3)_2 \cdot 6H_2O$ 0.025g
$Na_2B_4O_7 \cdot 10H_2O$ 0.018g

Preparation of "Metals 44": Add a few drops of H_2SO_4 to distilled/deionized water to inhibit precipitate formation. Add components to acidified distilled/deionized water and bring volume to 100.0mL. Mix thoroughly.

Preparation of Medium: Add components, except pyridoxine solution, to distilled/deionized water and bring volume to 990.0mL. Mix thoroughly. Adjust pH to 6.8 with 1*N* KOH. Autoclave for 15 min at 15 psi pressure–121°C. Cool to 45°–50°C. Aseptically add sterile pyridoxine solution. Mix thoroughly. Aseptically distribute into sterile tubes or flasks.

Use: For the cultivation and maintenance of *Pseudomonas* species.

Stanier's Basal Medium with Trichlorophenoxyacetate

Composition per liter:
KH_2PO_4 .... 2.78g
Na_2HPO_4 .... 2.78g
$(NH_4)_2SO_4$ 1.0g
2,4,5-Trichlorophenoxyacetate 1.0g
Hutner's mineral base 20.0mL

Hutner's Mineral Base:
Composition per liter:
$MgSO_4 \cdot 7H_2O$ 29.7g
Nitrilotriacetic acid 10.0g
$CaCl_2 \cdot 2H_2O$ 3.34g
$FeSO_4 \cdot 7H_2O$ 0.1g
$(NH_4)_2MoO_4$ 9.25mg
"Metals 44" 50.0mL

Preparation of Hutner's Mineral Base: Add nitrilotriacetic acid to 500.0mL of distilled/deionized water. Dissolve by adjusting pH to 6.5 with KOH. Add remaining components. Readjust pH to 7.2 with H_2SO_4 or KOH. Add distilled/deionized water to 1.0L. Store at 5°C.

"Metals 44":
Composition per 100.0mL:
$ZnSO_4 \cdot 7H_2O$ 1.1g
$FeSO_4 \cdot 7H_2O$ 0.5g
EDTA 0.25g
$MnSO_4 \cdot 7H_2O$ 0.154g
$CuSO_4 \cdot 5H_2O$ 0.04g
$Co(NO_3)_2 \cdot 6H_2O$ 0.025g
$Na_2B_4O_7 \cdot 10H_2O$ 0.018g

Preparation of "Metals 44": Add a few drops of H_2SO_4 to distilled/deionized water to inhibit precipitate formation. Add components to acidified distilled/deionized water and bring volume to 100.0mL. Mix thoroughly.

Preparation of Medium: Add components to distilled/deionized water and bring volume to 1.0L. Mix thoroughly. Distribute into tubes or flasks. Autoclave for 15 min at 15 psi pressure–121°C.

Use: For the cultivation and maintenance of *Pseudomonas cepacia.*

Starch Agar

Composition per liter:
Starch, soluble 20.0g
Agar 10.0g
$NaNO_3$ 2.5g
K_2HPO_4 .... 1.0g
$MgSO_4 \cdot 7H_2O$ 0.6g
$CaCl_2 \cdot 2H_2O$ 0.1g
NaCl 0.1g
$FeCl_3$ 1mg
pH 7.2 ± 0.2 at 25°C

Preparation of Medium: Add components to distilled/deionized water and bring volume to 1.0L. Mix thoroughly. Gently heat and bring to boiling. Distribute into tubes or flasks. Autoclave for 15 min at 15 psi pressure–121°C. Pour into sterile Petri dishes or leave in tubes.

Use: For the cultivation of myxobacteria.

Starch Hydrolysis Agar

Composition per liter:
Beef heart, infusion from 500.0g
Soluble starch 20.0g
Agar 15.0g
Tryptose 10.0g
NaCl 5.0g
pH 7.4 ± 0.2 at 25°C

Preparation of Medium: Add components to distilled/deionized water and bring volume to 1.0L. Mix thoroughly. Gently heat and bring to boiling. Distribute into tubes or flasks. Autoclave for 15 min at 15 psi pressure–121°C. Pour into sterile Petri dishes or leave in tubes.

Use: For the cultivation and differentiation of a variety of microorganisms based on amylase production. After incubation, starch hydrolysis is determined by the addition of Gram's or Lugol's iodine solution. Organisms that produce amylase appear as colonies surrounded by a clear zone.

Starkey's Medium C, Modified

Composition per liter:

Sodium lactate	3.5g
$MgSO_4 \cdot 7H_2O$	2.0g
Na_2SO_4	1.0g
NH_4Cl	1.0g
Yeast extract	1.0g
KH_2PO_4	0.5g
$CaCl_2 \cdot 2H_2O$	0.1g
Ferrous ammonium sulfate solution	50.0mL
L-Cysteine·HCl·H_2O solution	10.0mL

pH 7.5 ± 0.2 at 25°C

Ferrous Ammonium Sulfate Solution:
Composition per 100.0mL:

$Fe(NH_4)_2(SO_4)_2 \cdot 6H_2O$	1.0g

Preparation of Ferrous Ammonium Sulfate Solution: Add $Fe(NH_4)_2(SO_4)_2 \cdot 6H_2O$ to distilled/deionized water and bring volume to 100.0mL. Mix thoroughly. Filter sterilize.

L-Cysteine·HCl·H_2O Solution:
Composition per 10.0mL:

L-Cysteine·HCl·H_2O	0.75g

Preparation of L-Cysteine·HCl·H_2O Solution: Add L-cysteine·HCl·H_2O to distilled/deionized water and bring volume to 10.0mL. Mix thoroughly. Filter sterilize.

Preparation of Medium: Add components, except ferrous ammonium sulfate solution and L-cysteine·HCl·H_2O solution, to tap water and bring volume to 940.0mL. Mix thoroughly. Gently heat and bring to boiling. Autoclave for 15 min at 15 psi pressure–121°C. Cool to 45°–50°C. Aseptically add 50.0mL of sterile ferrous ammonium sulfate solution and 10.0mL of sterile L-cysteine·HCl·H_2O solution. Mix thoroughly. Adjust pH to 7.5 with filter-sterilized 2*N* NaOH. Pour into sterile Petri dishes or distribute into sterile tubes.

Use: For the cultivation and maintenance of *Desulfotomaculum* species and *Desulfovibrio* species.

Starkey's Medium C, Modified with Salt

Composition per liter:

NaCl	25.0g
Sodium lactate	3.5g
$MgSO_4 \cdot 7H_2O$	2.0g
Na_2SO_4	1.0g
NH_4Cl	1.0g
Yeast extract	1.0g
KH_2PO_4	0.5g
$CaCl_2 \cdot 2H_2O$	0.1g
Ferrous ammonium sulfate solution	50.0mL
L-Cysteine·HCl·H_2O solution	10.0mL

pH 7.5 ± 0.2 at 25°C

Ferrous Ammonium Sulfate Solution:
Composition per 100.0mL:

$Fe(NH_4)_2(SO_4)_2 \cdot 6H_2O$	1.0g

Preparation of Ferrous Ammonium Sulfate Solution: Add $Fe(NH_4)_2(SO_4)_2 \cdot 6H_2O$ to distilled/deionized water and bring volume to 100.0mL. Mix thoroughly. Filter sterilize.

L-Cysteine·HCl·H_2O Solution:
Composition per 10.0mL:

L-Cysteine·HCl·H_2O	0.75g

Preparation of L-Cysteine·HCl·H_2O Solution: Add L-cysteine·HCl·H_2O to distilled/deionized water and bring volume to 10.0mL. Mix thoroughly. Filter sterilize.

Preparation of Medium: Add components, except ferrous ammonium sulfate solution and L-cysteine·HCl·H_2O solution, to tap water and bring volume to 940.0mL. Mix thoroughly. Gently heat and bring to boiling. Autoclave for 15 min at 15 psi pressure–121°C. Cool to 45°–50°C. Aseptically add 50.0mL of sterile ferrous ammonium sulfate solution and 10.0mL of sterile L-cysteine·HCl·H_2O solution. Mix thoroughly. Adjust pH to 7.5 with filter-sterilized 2*N* NaOH. Pour into sterile Petri dishes or distribute into sterile tubes.

Use: For the cultivation and maintenance of halophilic *Desulfovibrio* species.

Stetteria Medium (DSMZ Medium 795)

Composition per liter:

Sulfur, powdered	10.0g
Peptone	2.0g
Yeast extract	1.0g
KH_2PO_4	0.5g
$NaHCO_3$	0.16g
$NiCl_2 \cdot 6H_2O$	3.0mg
Resazurin	0.75mg
Synthetic seawater, concentrated	500.0mL
Trace elements solution	15.0mL
$Na_2S \cdot 9H_2O$ solution	10.0mL
Selenite-tungstate solution	1.5mL

pH 7.2 ± 0.2 at 25°C

$Na_2S \cdot 9H_2O$ Solution:
Composition per 10mL:

$Na_2S \cdot 9H_2O$	0.5g

Preparation of $Na_2S \cdot 9H_2O$ Solution: Add $Na_2S \cdot 9H_2O$ to distilled/deionized water and bring volume to 10.0mL. Mix thoroughly. Autoclave under 100% N_2 for 15 min at 15 psi pressure–121°C. Cool to room temperature.

Selenite-Tungstate Solution:
Composition per liter:

NaOH	0.5g
$Na_2WO_4 \cdot 2H_2O$	4.0mg
$Na_2SeO_3 \cdot 5H_2O$	3.0mg

Preparation of Selenite-Tungstate Solution: Add components to distilled/deionized water and bring volume to 1.0L. Mix thoroughly. Sparge with 100% N_2.

Synthetic Seawater, Concentrated:

NaCl	55.4g
$MgSO_4 \cdot 7H_2O$	14.0g
$MgCl_2 \cdot 6H_2O$	11.0g
$CaCl_2 \cdot 2H_2O$	1.5g
KCl	1.3g
NaBr	0.2g
H_3BO_3	0.06g
$SrCl_2 \cdot 6H_2O$	0.03g
Na_3-citrate	20.0mg
KI	0.1mg

Preparation of Synthetic Seawater, Concentrated: Add components to distilled/deionized water and bring volume to 1.0L. Mix thoroughly. Filter sterilize.

Trace Elements Solution:
Composition per liter:

$MgSO_4 \cdot 7H_2O$	3.0g
Nitrilotriacetic acid	1.5g

NaCl 1.0g
$MnSO_4 \cdot 2H_2O$ 0.5g
$CoSO_4 \cdot 7H_2O$ 0.18g
$ZnSO_4 \cdot 7H_2O$ 0.18g
$CaCl_2 \cdot 2H_2O$ 0.1g
$FeSO_4 \cdot 7H_2O$ 0.1g
$NiCl_2 \cdot 6H_2O$ 0.025g
$KAl(SO_4)_2 \cdot 12H_2O$ 0.02g
H_3BO_3 .... 0.01g
$Na_2MoO_4 \cdot 4H_2O$ 0.01g
$CuSO_4 \cdot 5H_2O$ 0.01g
$Na_2SeO_3 \cdot 5H_2O$ 0.3mg

Preparation of Trace Elements Solution: Add nitrilotriacetic acid to 500.0mL of distilled/deionized water. Dissolve by adjusting pH to 6.5 with KOH. Add remaining components. Add distilled/deionized water to 1.0L. Mix thoroughly.

Preparation of Medium: Prepare and dispense medium under 80% N_2 + 20% CO_2 gas mixture. Add components, except synthetic seawater and $Na_2S \cdot 9H_2O$ solution, to 490.0mL distilled/deionized water. Mix thoroughly. Sparge with 80% N_2 + 20% CO_2. Autoclave for 15 min at 15 psi pressure–121°C. Cool to 25°C. Aseptically add 500.0mL filter sterilized concentrated seawater. Flush with 80% N_2 + 20% CO_2 gas mixture for 20 min. Aseptically add 10.0mL $Na_2S \cdot 9H_2O$ solution. Adjust pH to 6.0 with H_2SO_4. Mix thoroughly. Aseptically and anaerobically distribute 20mL aliquots into sterile 100mL serum bottles. Pressurize bottles to 2 bar gas overpressure with 80% N_2 + 20% CO_2. Heat at 100°C for 1.5 hr. Before use, check that the medium pH is 6.0.

Use: For the cultivation of *Stetteria hydrogenophila* and *Staphylothermus hellenicus.*

STL Broth

Composition per liter:
Casamino acids 1.0g
Glucose 1.0g
Sodium glutamate 1.0g
$CaCl_2 \cdot 2H_2O$ 0.1g
KNO_3 .... 0.1g
$MgSO_4 \cdot 7H_2O$ 0.1g
Sodium glycerophosphate 0.1g
Thiamine 1.0mg
Vitamin B_{12} 1.0μg
Trace elements solution 1.0mL
pH 7.5 ± 0.2 at 25°C

Trace Elements Solution:
Composition per liter:
Disodium EDTA 8.0g
$MnCl_2 \cdot 4H_2O$ 0.1g
$CoCl_2 \cdot 6H_2O$ 0.02g
KBr 0.02g
KI 0.02g
$ZnCl_2$ 0.02g
$CuSO_4$ 0.01g
H_3BO_3 .... 0.01g
$Na_2MoO_4 \cdot 2H_2O$ 0.01g
LiCl 5.0mg
$SnCl_2 \cdot 2H_2O$ 5.0mg

Preparation of Trace Elements Solution: Add components to distilled/deionized water and bring volume to 1.0L. Mix thoroughly.

Preparation of Medium: Add components to distilled/deionized water and bring volume to 1.0L. Mix thoroughly. Gently heat and bring to boiling. Distribute into tubes or flasks. Autoclave for 15 min at 15 psi pressure–121°C. Pour into sterile Petri dishes or leave in tubes.

Use: For the isolation and cultivation of *Cytophaga* species, *Herpetosiphon* species, *Saprospira* species, and *Flexithrix* species.

Stokes Agar

Composition per liter:
Agar 12.5g
Glucose 1.0g
Peptone 1.0g
$MgSO_4 \cdot 7H_2O$ 0.2g
$CaCl_2$ 0.05g
$FeCl_3 \cdot 6H_2O$ 0.01g

Preparation of Medium: Add components to tap water and bring volume to 1.0L. Mix thoroughly. Gently heat and bring to boiling. Distribute into tubes or flasks. Autoclave for 15 min at 15 psi pressure–121°C. Pour into sterile Petri dishes or leave in tubes.

Use: For the isolation and cultivation of *Sphaerotilus natans.*

Straw DYAA

Composition per liter:
Agar 20.0g
Glucose 10.0g
Yeast extract 1.0g
Asparagine 0.5g
$K_2HPO_4 \cdot 3H_2O$ 0.5g
$MgSO_4 \cdot 7H_2O$ 0.25g
$FeCl_3$ solution 0.5mL
Straw .... variable

$FeCl_3$ Solution:
Composition per 10.0mL:
$FeCl_3$ 1.0g

Preparation of $FeCl_3$ Solution: Add $FeCl_3$ to distilled/deionized water and bring volume to 10.0mL. Mix thoroughly.

Preparation of Medium: Add components, except straw, to distilled/deionized water and bring volume to 1.0L. Gently heat and bring to boiling. Distribute into tubes or flasks. Autoclave for 15 min at 15 psi pressure–121°C. Pour into sterile Petri dishes or leave in tubes. Autoclave straw for 15 min at 15 psi pressure–121°C. Aseptically add straw to the solidified agar.

Use: For the cultivation of *Cochliobolus sativus.*

Straw Malt Agar

Composition per liter:
Agar 15.0g
Malt extract 10.0g
Straw .... variable

Preparation of Medium: Add components, except straw, to distilled/deionized water and bring volume to 1.0L. Mix thoroughly. Gently heat and bring to boiling. Distribute into tubes or flasks. Autoclave for 15 min at 15 psi pressure–121°C. Pour into sterile Petri dishes or leave in tubes. Autoclave straw for 15 min at 15 psi pressure–121°C. Aseptically add some straw to the solidified agar.

Use: For the cultivation of *Cladosporium vignae, Cochliobolus sativus,* and *Cochliobolus victoriae.*

Stuart *Leptospira* Broth, Modified

Composition per liter:

NaCl 1.93g
Na_2HPO_4 .... 0.66g
NH_4Cl 0.34g
$MgCl_2 \cdot 6H_2O$ 0.19g
L-Asparagine 0.13g
KH_2PO_4 .... 0.08g
Glycerol 5.0mL
Rabbit serum,
inactivated at 56°C, 30 min 100.0mL
pH 7.4 ± 0.2 at 25°C

Preparation of Medium: Add each component, except rabbit serum, to distilled/deionized water in separate flasks and bring each volume to 100.0mL. Mix thoroughly. Combine the seven solutions, except the rabbit serum. Mix thoroughly. Gently heat and bring to boiling. Autoclave for 15 min at 15 psi pressure–121°C. Cool to 45°–50°C. Aseptically add sterile rabbit serum. Mix thoroughly. Aseptically distribute into sterile tubes or flasks.

Use: For the cultivation of *Leptospira* species.

Stuart Medium Base

Composition per 1100.0mL:

NaCl 1.8g
Na_2HPO_4 .... 0.67g
$MgCl_2 \cdot 6H_2O$ 0.41g
NH_4Cl 0.27g
Asparagine 0.13g
KH_2PO_4 .... 0.09g
Phenol Red 0.01g
Glycerol 5.0mL
Leptospira enrichment 100.0mL
pH 7.6 ± 0.2 at 25°C

Source: This medium is available as a premixed powder from BD Diagnostic Systems. *Leptospira* enrichment contains rabbit serum and hemoglobin and is available from BD Diagnostic Systems.

Preparation of Medium: Add components, except glycerol and *Leptospira* enrichment, to distilled/deionized water and bring volume to 995.0mL. Mix thoroughly. Add glycerol. Mix thoroughly. Autoclave for 15 min at 15 psi pressure–121°C. Cool to 45°–50°C. Aseptically add *Leptospira* enrichment. Mix thoroughly. Aseptically distribute into sterile screw-capped tubes in 10.0mL volumes.

Use: For the cultivation of *Leptospira* species.

Styrene Mineral Salts Agar

Composition per liter:

Agar 20.0g
$(NH_4)_2SO_4$ 2.0g
K_2HPO_4 .... 1.55g
$NaH_2PO \cdot 2H_2O$ 0.85g
$MgCl_2 \cdot 6H_2O$ 0.1g
EDTA 10.0mg
$FeSO_4 \cdot 7H_2O$ 5.0mg
$ZnSO_4$ 2.0mg
$CaCl_2 \cdot 2H_2O$ 1.0mg
$MnCl_2 \cdot 2H_2O$ 1.0mg
$CoCl_2 \cdot 6H_2O$ 0.4mg
$CuSO_4 \cdot 5H_2O$ 0.2mg
$Na_2MoO_4 \cdot 2H_2O$ 0.2mg

Preparation of Medium: Add components to distilled/deionized water and bring volume to 1.0L. Mix thoroughly. Gently heat and bring to boiling. Distribute into tubes or flasks. Autoclave for 15 min at 15 psi pressure–121°C. Pour into sterile Petri dishes or leave in tubes. Place plates in a desiccator. Add to the desiccator an open bottle containing 10.0mL of dibutyl phthalate and 200.0μl of styrene.

Use: For the cultivation of styrene-utilizing microorganisms.

Sucrose Teepol Tellurite Agar (STT Agar)

Composition per liter:

Agar 20.0g
Beef extract 1.0g
Peptone 1.0g
Sucrose 1.0g
NaCl 0.5g
Bromothymol Blue (0.2% solution) 2.5mL
Tellurite solution 2.5mL
Sodium lauryl sulfate
(Teepol, 0.1% solution) 0.2mL
pH 8.0 ± 0.2 at 25°C

Tellurite Solution:

Composition per 100.0mL:

K_2TeO_3 .... 0.05g

Preparation of Tellurite Solution: Add the K_2TeO_3 to distilled/deionized water and bring the volume to 100.0mL. Mix thoroughly. Filter sterilize. Use freshly prepared solution.

Caution: Potassium tellurite is toxic.

Preparation of Medium: Add components to distilled/deionized water and bring volume to 1.0L. Mix thoroughly. Gently heat and bring to boiling. Do not autoclave. Pour into sterile Petri dishes.

Use: For the selective isolation, cultivation, and differentiation of *Vibrio* species based on their ability to ferment sucrose. *Vibrio cholerae* appears as flat yellow colonies. *Vibrio parahaemolyticus* appears as elevated green-yellow mucoid colonies.

Sulfate API Broth

Composition per liter:

NaCl 10.0g
Sodium lactate 5.2g
Yeast extract 1.0g
$MgSO_4 \cdot 7H_2O$ 0.2g
Ascorbic acid 0.1g
$Fe(NH_4)_2(SO_4)_2 \cdot 6H_2O$ 0.1g
K_2HPO_4 .... 0.01g
pH 7.5± 0.2 at 25°C

Source: This medium is available as a premixed powder from BD Diagnostic Systems.

Preparation of Medium: Add the components to distilled/deionized water and bring volume to 1.0L. Mix thoroughly until dissolved. Distribute into tubes in 9.0mL volumes. Autoclave for 10 min at 15 psi pressure–121°C.

Use: For the detection, differentiation, and estimation of sulfate-reducing bacteria.

Sulfate-Reducing Bacteria Enrichment Medium

Composition per 1018.0mL:

Solution 1 .... 970.0mL
Solution 4 .... 30.0mL
Solution 6A, 6B, 6C, 6D, or 6E 10.0mL
Solution 5 .... 3.0mL

Solution 2 .... 1.0mL
Solution 3 .... 1.0mL
Solution 7 .... 1.0mL
Solution 8 .... 1.0mL
Solution 9 .... 1.0mL

pH 7.2 ± 0.2 at 25°C

Solution 1:

Composition per 970.0mL:

Na_2SO_4 .... 3.0g
NaCl 1.2g
$MgCl_2 \cdot 6H_2O$ 0.4g
KCl 0.3g
NH_4Cl 0.3g
KH_2PO_4 .... 0.2g
$CaCl_2 \cdot 2H_2O$ 0.15g

Preparation of Solution 1: Add components to distilled/deionized water and bring volume to 970.0mL. Mix thoroughly. Autoclave for 30 min at 15 psi pressure–121°C. Cool to 25°C under 90% N_2 + 10% CO_2.

Solution 2:

Composition per liter:

$FeCl_2 \cdot 4H_2O$ 1.5g
$CoCl_2 \cdot 6H_2O$ 0.12g
$MnCl_2 \cdot 4H_2O$ 0.1g
$ZnCl_2$ 0.07g
H_3BO_3 .... 0.06g
$Na_2MoO_4 \cdot 2H_2O$ 0.025g
$NiCl_2 \cdot 6H_2O$ 0.025g
$CuCl_2 \cdot 2H_2O$ 0.015g
HCl (25% solution) 6.5mL

Preparation of Solution 2: Add components to distilled/deionized water and bring volume to 1.0L. Mix thoroughly. Autoclave for 15 min at 15 psi pressure–121°C. Cool to 25°C.

Solution 3:

Composition per liter:

NaOH 0.5g
Na_2SeO_3 .... 3.0mg

Preparation of Solution 3: Add components to distilled/deionized water and bring volume to 1.0L. Mix thoroughly. Autoclave for 15 min at 15 psi pressure–121°C. Cool to 25°C.

Solution 4:

Composition per 100.0mL:

$NaHCO_3$ 8.5g

Preparation of Solution 4: Add $NaHCO_3$ to distilled/deionized water and bring volume to 100.0mL. Mix thoroughly. Saturate with 100% CO_2. Filter sterilize. Aseptically add solution to sterile, gas-tight, screw-capped bottles.

Solution 5:

Composition per 100.0mL:

$Na_2S \cdot 9H_2O$ 12.0g

Preparation of Solution 5: Add $Na_2S \cdot 9H_2O$ to distilled/deionized water and bring volume to 100.0mL. Mix thoroughly. Add solution to gas-tight, screw-capped bottles. Gas under 100% N_2 for 20 min. Close caps tightly. Autoclave for 15 min at 15 psi pressure–121°C. Cool to 25°C.

Solution 6A:

Composition per 100.0mL:

Sodium acetate·$3H_2O$ 20.0g

Preparation of Solution 6A: Add sodium acetate·$3H_2O$ to distilled/deionized water and bring volume to 100.0mL. Autoclave for 15 min at 15 psi pressure–121°C. Cool to 25°C.

Solution 6B:

Composition per 100.0mL:

n-Butyric acid 8.0g

Preparation of Solution 6B: Add *n*-butyric acid to distilled/deionized water and bring volume to 100.0mL. Adjust pH to 9.0 with NaOH. Autoclave for 15 min at 15 psi pressure–121°C. Cool to 25°C.

Solution 6C:

Composition per 100.0mL:

Propionic acid 7.0g

Preparation of Solution 6C: Add propionic acid to 100.0mL of distilled/deionized water. Adjust pH to 9.0 with NaOH. Autoclave for 15 min at 15 psi pressure–121°C. Cool to 25°C.

Solution 6D:

Composition per 100.0mL:

Benzoic acid 5.0g

Preparation of Solution 6D: Add benzoic acid to distilled/deionized water and bring volume to 100.0mL. Adjust pH to 9.0 with NaOH. Autoclave for 15 min at 15 psi pressure–121°C. Cool to 25°C.

Solution 6E:

Composition per 100.0mL:

n-Palmitic acid 5.0g
NaOH 0.78g

Preparation of Solution 6E: Add *n*-palmitic acid and NaOH to distilled/deionized water and bring volume to 100.0mL. Heat in a water bath until clear. Autoclave for 15 min at 15 psi pressure–121°C. Cool to 25°C.

Solution 7:

Composition per 100.0mL:

Thiamine 0.01g
p-Aminobenzoic acid 5.0mg
Vitamin B_{12} 5.0mg
Biotin 1.0mg

Preparation of Solution 7: Add components to distilled/deionized water and bring volume to 100.0mL. Mix thoroughly. Filter sterilize.

Solution 8:

Composition per liter:

Succinic acid 0.6g
Isobutyric acid 0.5g
2-Methylbutyric acid 0.5g
3-Methylbutyric acid 0.5g
Valeric acid 0.5g
Caproic acid 0.2g

Preparation of Solution 8: Add components to distilled/deionized water and bring volume to 100.0mL. Mix thoroughly. Adjust pH to 9.0 with NaOH. Autoclave for 15 min at 15 psi pressure–121°C. Cool to 25°C.

Solution 9:

Composition per 100.0mL:

$Na_2S_2O_4$ 3.0g

Preparation of Solution 9: Add $Na_2S_2O_4$ to 100.0mL of O_2-free distilled/deionized water. Mix thoroughly. Anaerobically filter sterilize.

Preparation of Medium: To 970.0mL of cooled, sterile solution 1, aseptically and anaerobically add 1.0mL of sterile solution 2, 1.0mL of sterile solution 3, 30.0mL of sterile solution 4, and 3.0mL of sterile solution 5. Mix thoroughly. Adjust pH to 7.2 with sterile HCl solution or sterile Na_2CO_3 solution. Aseptically and anaerobically distribute into sterile screw-capped bottles in 100.0mL volumes. Add 1.0mL of solutions 6A, 6B, 6C, 6D, or 6E to each bottle containing 100.0mL of basal medium. Add 0.1mL of solution 7, 0.1mL of solution 8, and 0.1mL of solution 9 to each bottle containing 100.0mL of basal medium. Mix thoroughly.

Use: For the isolation, cultivation, and enrichment of sulfate-reducing bacteria.

Sulfate-Reducing Bacteria Medium

Composition per 1009.0mL:

Solution A 850.0mL
Solution C 100.0mL
Solution G 20.0mL
Solution D 10.0mL
Solution E (Wolfe's vitamin solution) 10.0mL
Solution H 10.0mL
Solution F 6.6mL
Solution B
(Trace elements solution SL-10) 1.0mL
Solution I 0.4mL

pH 7.6 ± 0.2 at 25°C

Solution A:

Composition per 920.0mL:

Na_2SO_4 ....3.0g
NaCl 1.0g
KCl 0.5g
$MgCl_2 \cdot 6H_2O$ 0.4g
NH_4Cl 0.3g
KH_2PO_4 ....0.2g
$CaCl_2 \cdot 2H_2O$ 0.15g
Resazurin 0.5mg

Preparation of Solution A: Prepare and dispense solution anaerobically under 80% N_2 + 20% CO_2. Add components to distilled/deionized water and bring volume to 920.0mL. Mix thoroughly. Gently heat and bring to boiling. Continue boiling until resazurin turns colorless, indicating reduction, and a pH of 6.0 is reached. Cap with rubber stoppers. Autoclave for 15 min at 15 psi pressure–121°C. Cool to 25°C.

Solution B (Trace Elements Solution SL-10):

Composition per liter:

$FeCl_2 \cdot 4H_2O$ 1.5g
$CoCl_2 \cdot 6H_2O$ 0.19g
$MnCl_2 \cdot 4H_2O$ 0.10g
$ZnCl_2$ 0.070g
$Na_2MoO_4 \cdot 2H_2O$ 0.036g
$NiCl_2 \cdot 6H_2O$ 0.024g
H_3BO_3 ....6.0mg
$CuCl_2 \cdot 2H_2O$ 2.0mg
HCl (25% solution) 10.0mL

Preparation of Solution B (Trace Elements Solution SL-10): Add the $FeCl_2 \cdot 4H_2O$ to 10.0mL of HCl solution. Mix thoroughly. Bring volume to approximately 900.0mL with distilled/deionized water. Mix thoroughly. Adjust pH to 6.0 with NaOH. Bring volume to 1.0L with distilled/deionized water. Filter sterilize. Aseptically gas under 100% N_2 for 20 min.

Solution C:

Composition per 100.0mL:

$NaHCO_3$ 5.0g

Preparation of Solution C: Add $NaHCO_3$ to distilled/deionized water and bring volume to 100.0mL. Mix thoroughly. Filter sterilize. Aseptically gas under 80% N_2 + 20% CO_2 for 20 min.

Solution D:

Composition per 10.0mL:

Sodium propionate 1.5g

Preparation of Solution D: Prepare and dispense solution anaerobically under 80% N_2 + 20% CO_2. Add sodium propionate to distilled/deionized water and bring volume to 10.0mL. Mix thoroughly. Cap with a rubber stopper. Autoclave for 15 min at 15 psi pressure–121°C. Cool to 25°C.

Solution E (Wolfe's Vitamin Solution):

Composition per liter:

Pyridoxine·HCl 0.01g
Thiamine·HCl 5.0mg
Riboflavin 5.0mg
Nicotinic acid 5.0mg
Calcium pantothenate 5.0mg
p-Aminobenzoic acid 5.0mg
Thioctic acid 5.0mg
Biotin 2.0mg
Folic acid 2.0mg
Cyanocobalamin 0.1mg

Preparation of Solution E (Wolfe's Vitamin Solution): Add components to distilled/deionized water and bring volume to 1.0L. Mix thoroughly. Filter sterilize. Aseptically gas under 100% N_2 for 20 min.

Solution F:

Composition per 6.6mL:

$AlCl_3 \cdot 6H_2O$ (4.9% solution) 5.0mL
Na_2CO_3 (10.6% solution) 1.6mL

Preparation of Solution F: Combine both solutions. Mix thoroughly. Gas with 100% N_2. Cap with a rubber stopper. Autoclave for 15 min at 15 psi pressure–121°C. Cool to 25°C.

Solution G:

Composition per 10.0mL:

Rumen fluid, clarified 20.0mL

Preparation of Solution G: Gas rumen fluid under 100% N_2 for 20 min. Cap with a rubber stopper. Autoclave for 15 min at 15 psi pressure–121°C. Cool to 25°C.

Solution H:

Composition per 10.0mL:

$Na_2S \cdot 9H_2O$ 0.4g

Preparation of Solution H: Add $Na_2S \cdot 9H_2O$ to distilled/deionized water and bring volume to 10.0mL. Gas under 100% N_2 for 20 min. Cap with a rubber stopper. Autoclave for 15 min at 15 psi pressure–121°C. Cool to 25°C.

Solution I:

Composition per 10.0mL:

$Na_2S_2O_4$ 0.5g

Preparation of Solution I: Add $Na_2S_2O_4$ to distilled/deionized water and bring volume to 10.0mL. Mix thoroughly. Filter sterilize. Aseptically gas under 100% N_2 for 20 min. Prepare solution freshly.

Preparation of Medium: To 850.0mL of cooled, sterile solution A, aseptically and anaerobically add in the following order: 1.0mL of sterile solution B, 1000.0mL of sterile solution C, 10.0mL of sterile solution D, 10.0mL of sterile solution E, 6.6mL of sterile solution F, 20.0mL of sterile solution G, and 10.0mL of sterile solution H. Mix thoroughly. Immediately prior to inoculation, aseptically and anaerobically add 0.4mL of sterile solution I. Mix thoroughly. Aseptically and anaerobically distribute into sterile tubes or flasks.

Use: For the cultivation and maintenance of a variety of sulfate-reducing bacteria.

Sulfate-Reducing Bacteria Medium with Lactate

Composition per liter:

Solution 1 .... 980.0mL
Solution 2 .... 10.0mL
Solution 3 .... 10.0mL

pH 7.4 ± 0.2 at 25°C

Solution 1:
Composition per 980.0mL:

Sodium lactate (70% solution) 3.5g
$MgSO_4 \cdot 7H_2O$ 2.0g
NH_4Cl 1.0g
Na_2SO_4 .... 1.0g
Yeast extract 1.0g
K_2HPO_4 .... 0.5g
$CaCl_2 \cdot 2H_2O$ 0.1g

Preparation of Solution 1: Add components to distilled/deionized water and bring volume to 980.0mL. Mix thoroughly. Autoclave for 15 min at 15 psi pressure–121°C. Cool to 50°C.

Solution 2:
Composition per 10.0mL:

$FeSO_4 \cdot 7H_2O$ 0.5g

Preparation of Solution 2: Add $FeSO_4 \cdot 7H_2O$ to distilled/deionized water and bring volume to 10.0mL. Mix thoroughly. Autoclave for 15 min at 15 psi pressure–121°C. Cool to 50°C.

Solution 3:
Composition per 10.0mL:

Ascorbic acid 0.1g
Sodium thioglycolate 0.1g

Preparation of Solution 2: Add components to distilled/deionized water and bring volume to 10.0mL. Mix thoroughly. Autoclave for 15 min at 15 psi pressure–121°C. Cool to 50°C.

Preparation of Medium: Aseptically combine 980.0mL of cooled, sterile solution 1, 10.0mL of cooled, sterile solution 2, and 10.0mL of cooled, sterile solution 3. Mix thoroughly. Aseptically distribute into sterile tubes or flasks.

Use: For the enrichment and isolation of sulfate-reducing bacteria.

Sulfate-Reducing Medium

Composition per liter:

Sodium lactate 3.5g
$MgSO_4 \cdot 7H_2O$ 2.0g
Peptone 2.0g
Na_2SO_4 .... 1.5g
Beef extract 1.0g
K_2HPO_4 .... 0.5g
$CaCl_2$ 0.1g
$Fe(NH_4)_2(SO_4)_2 \cdot 6H_2O$ solution 10.0mL
Sodium ascorbate solution 10.0mL

pH 7.5 ± 0.3 at 25°C

$Fe(NH_4)_2(SO_4)_2 \cdot 6H_2O$ Solution:
Composition per 100.0mL:

$Fe(NH_4)_2(SO_4)_2 \cdot 6H_2O$ 3.92g

Preparation of $Fe(NH_4)_2(SO_4)_2 \cdot 6H_2O$ Solution: Add $Fe(NH_4)_2(SO_4)_2 \cdot 6H_2O$ to distilled/deionized water and bring volume to 100.0mL. Mix thoroughly. Filter sterilize. Use freshly prepared medium.

Sodium Ascorbate Solution:
Composition per 100.0mL:

Sodium ascorbate 0.05g

Preparation of Sodium Ascorbate Solution: Add sodium ascorbate to distilled/deionized water and bring volume to 100.0mL. Mix thoroughly. Filter sterilize. Use freshly prepared medium.

Preparation of Medium: Add components, except $Fe(NH_4)_2(SO_4)_2 \cdot 6H_2O$ solution and sodium ascorbate solution, to distilled/deionized water and bring volume to 980.0mL. Mix thoroughly. Distribute into screw-capped tubes in 10.0mL volumes. Autoclave for 15 min at 15 psi pressure–121°C. Tubes must be filled to capacity after inoculation, so prepare extra medium and sterilize in a screw-capped flask or bottle. Prior to inoculation, aseptically add 0.1mL of freshly prepared sterile $Fe(NH_4)_2(SO_4)_2 \cdot 6H_2O$ solution for each 10.0mL of medium in the tubes. Also aseptically add 0.1mL of freshly prepared sterile sodium ascorbate solution for each 10.0mL of medium in the tubes. Inoculate tubes. Fill tubes to capacity with extra sterile medium. Screw caps tight.

Use: For the isolation, cultivation, and enumeration of iron and sulfur bacteria.

Sulfite Agar

Composition per liter:

Agar 20.0g
Pancreatic digest of casein 10.0g
Na_2SO_3 .... 1.0g
Iron nails 66

pH 7.6 ± 0.2 at 25°C

Source: This medium is available as a premixed powder from BD Diagnostic Systems.

Preparation of Medium: Add components to distilled/deionized water and bring volume to 1.0L. Mix thoroughly. Gently heat and bring to boiling. Distribute into screw-capped tubes in 15.0mL volumes. Add a clean iron nail to each tube. Autoclave for 15 min at 15 psi pressure–121°C. Cool to 45°–50°C until ready to inoculate.

Use: For the detection and cultivation of thermophilic anaerobes that can produce H_2S from sulfite. Sulfite reduction appears as a blackening of the medium.

Sulfitobacter pontiacus Medium (DSMZ Medium 733)

Composition per liter:

Basal salts solution 959.0mL
Biotin solution 10.0mL
Na-acetate solution 10.0mL
Yeast extract peptone solution 10.0mL
Magnesium calcium solution 10.0mL
Trace elements solution 1.0mL

pH 7.6 ± 0.2 at 25°C

Trace Elements Solution:
Composition per liter:

$MgSO_4 \cdot 7H_2O$ 3.0g
Nitrilotriacetic acid 1.5g
NaCl 1.0g
$MnSO_4 \cdot 2H_2O$ 0.5g
$CoSO_4 \cdot 7H_2O$ 0.18g

$ZnSO_4·7H_2O$ ... 0.18g
$CaCl_2·2H_2O$ ... 0.1g
$FeSO_4·7H_2O$ ... 0.1g
$NiCl_2·6H_2O$ ... 0.025g
$KAl(SO_4)_2·12H_2O$ ... 0.02g
H_3BO_3 ... 0.01g
$Na_2MoO_4·4H_2O$ ... 0.01g
$CuSO_4·5H_2O$ ... 0.01g
$Na_2SeO_3·5H_2O$ ... 0.3mg

Preparation of Trace Elements Solution: Add nitrilotriacetic acid to 500.0mL of distilled/deionized water. Dissolve by adjusting pH to 6.5 with KOH. Add remaining components. Add distilled/deionized water to 1.0L. Mix thoroughly. Filter sterilize.

Basal Salts Solution:
Composition per liter:
HEPES ... 8.0g
K_2HPO_4 ... 1.0g
NH_4Cl ... 0.5g
NaCl ... 15.0g

Preparation of Basal Salts Solution: Add components to 1.0L of distilled/deionized water. Adjust pH to 7.5-7.8 with NaOH. Mix thoroughly. Autoclave for 15 min at 15 psi pressure–121°C. Cool to 25°C.

Biotin Solution:
Composition per 10.0mL:
Biotin ... 0.1g

Preparation of Biotin Solution: Add biotin to 10.0mL of distilled/deionized water. Mix thoroughly. Filter sterilize.

Magnesium Calcium Solution:
Composition per 10.0mL:
$MgSO_4·7H_2O$ ... 1.0g
$CaCl_2·2H_2O$ ... 0.05g

Preparation of Magnesium Calcium Solution: Add components to 10.0mL of distilled/deionized water. Mix thoroughly. Filter sterilize.

Na-Acetate Solution:
Composition per 10.0mL:
Na-acetate ... 1.6g

Preparation of Na-Acetate Solution: Add Na-acetate to 10.0mL of distilled/deionized water. Mix thoroughly. Filter sterilize.

Yeast Extract Peptone Solution:
Composition per 10.0mL:
Yeast extract ... 1.0g
Peptone ... 0.5g

Preparation of Yeast Extract Peptone Solution: Add components to 10.0mL of distilled/deionized water. Mix thoroughly. Autoclave for 15 min at 15 psi pressure–121°C. Cool to 25°C.

Preparation of Medium: Aseptically combine 959.0mL basal salts solution, 1.0mL trace elements solution, 10.0mL Na-acetate solution, 10.0mL magnesium calcium solution, 10.0mL yeast extract peptone solution, and 10.0mL biotin solution. Mix thoroughly. Aseptically distribute to sterile tubes or flasks.

Use: For the cultivation of *Sulfitobacter pontiacus.*

Sulfobacillus disulfidooxidans Medium (DSMZ Medium 812)

Composition per liter:
$(NH_4)_2SO_4$ ... 3.0g
KH_2PO_4 ... 0.5g
$MgSO_4·7H_2O$ ... 0.5g
KCl ... 0.1g
$Ca(NO_3)_2·4H_2O$ ... 0.1g
Yeast extract ... 0.1g
Glutathione solution ... 10.0mL
pH 2.25 ± 0.1 at 25°C

Glutathione Solution:
Composition per 10mL:
Glutathione ... 1.0g

Preparation of Glutathione Solution: Add glutathione to distilled/deionized water and bring volume to 10.0mL. Mix thoroughly. Filter sterilize.

Preparation of Medium: Add components, except glutathione solution, to distilled/deionized water and bring volume to 990.0mL. Mix thoroughly. Adjust pH to 2.25. Autoclave for 15 min at 15 psi pressure–121°C. Cool to room temperature. Aseptically add 10.0mL sterile glutathione solution. Mix thoroughly. Aseptically distribute into sterile tubes or bottles.

Use: For the cultivation of *Sulfobacillus disulfidooxidans.*

Sulfobacillus Medium

Composition 1020.0mL:
Solution A ... 700.0mL
Solution B ... 300.0mL
Solution C ... 20.0mL
pH 1.9–2.4 at 25°C

Solution A:
Composition per 700.0mL:
$(NH_4)_2SO_4$ ... 3.0g
KCl ... 0.1g
K_2HPO_4 ... 0.5g
$MgSO_4·7H_2O$ ... 0.5g
$Ca(NO_3)_2$ ... 0.01g

Preparation of Solution A: Add components to distilled/deionized water and bring volume to 700.0mL. Mix thoroughly. Adjust pH to 2.0–2.2 with sulfuric acid. Autoclave for 15 min at 15 psi pressure–121°C.

Solution B:
Composition per 300.0mL:
$FeSO_4·7H_2O$ ... 44.2g
H_2SO_4, 10*N* solution ... 1.0mL

Preparation of Solution B: Add components to distilled/deionized water and bring volume to 300.0mL. Mix thoroughly. Autoclave for 15 min at 15 psi pressure–121°C.

Solution C:
Composition per 20.0mL:
Yeast extract ... 0.2g

Preparation of Solution C: Add yeast extract to distilled/deionized water and bring volume to 20.0mL. Mix thoroughly. Autoclave for 15 min at 15 psi pressure–121°C.

Preparation of Medium: Aseptically combine 700.0mL of solution A, 300.0mL of solution B, and 20.0mL of solution C. Aseptically adjust pH to 1.9–2.4

Use: For the cultivation of *Sulfobacillus thermosulfidooxidans.*

Sulfolobus acidocaldarius Simplified Basal Medium

Composition per liter:
K_2SO_4 ... 6.0g
NaH_2PO_4 ... 1.0g

$MgSO_4 \cdot 7H_2O$ 0.6g
$CaCl_2 \cdot 7H_2O$ 0.2g
Trace minerals solution 0.04mL
pH 3.5 ± 0.2 at 25°C

Trace Minerals Solution:
Composition per 100.0mL:
$FeCl_3 \cdot 6H_2O$ 5.0g
$CuCl_2 \cdot 2H_2O$ 0.5g
$CoCl_2 \cdot 6H_2O$ 0.5g
$MnCl_2 \cdot 2H_2O$ 0.5g
$ZnCl_2$ 0.5g
HCl (1*N* solution) 100.0ml

Preparation of Trace Minerals Solution: Combine components. Mix thoroughly.

Preparation of Medium: Add components to distilled/deionized water and bring volume to 1.0L. Mix thoroughly. Adjust pH to 3.5 with H_2SO_4. Distribute into tubes or flasks. Autoclave for 15 min at 15 psi pressure–121°C.

Use: For the cultivation of *Sulfolobus acidocaldarius.*

Sulfolobus brierleyi Medium

Composition per liter:
Sulfur flowers 10.0g
$(NH_4)_2SO_4$ 3.0g
$K_2HPO_4 \cdot 3H_2O$ 0.5g
$MgSO_4 \cdot 7H_2O$ 0.5g
KCl 0.1g
$Ca(NO_3)$ 0.01g
Yeast extract solution 20.0mL
pH 1.5–2.5 at 25°C

Preparation of Sulfur: Autoclave sulfur at 8 psi pressure–112°C for 15 min.

Yeast Extract Solution:
Composition per 20.0mL:
Yeast extract 0.2g

Preparation of Yeast Extract Solution: Add yeast extract to distilled/deionized water and bring volume to 20.0mL. Mix thoroughly. Autoclave for 15 min at 15 psi pressure–121°C.

Preparation of Medium: Add components, except yeast extract solution and sulfur, to distilled/deionized water and bring volume to 980.0mL. Mix thoroughly. Adjust pH with 6*N* H_2SO_4 to 1.5–2.5. Autoclave for 15 min at 15 psi pressure–121°C. Aseptically add 20.0mL of sterile yeast extract solution and 10.0g of sterile sulfur. Mix thoroughly. Aseptically distribute into sterile tubes or flasks.

Use: For the cultivation of *Acidianus brierleyi.*

Sulfolobus Broth

Composition per liter:
Sucrose yeast solution 500.0mL
$CaCl_2 \cdot 2H_2O$ solution 250.0mL
Trace elements solution 250.0mL
pH 3.0–3.5 at 25°C

Sucrose Yeast Solution:
Composition per 500.0mL:
Sucrose 2.0g
Yeast extract 1.0g

Preparation of Sucrose Yeast Solution: Add components to distilled/deionized water and bring volume to 500.0mL. Mix thoroughly. Autoclave for 15 min at 15 psi pressure–121°C.

$CaCl_2 \cdot 2H_2O$ Solution:
Composition per 250.0mL:
$CaCl_2 \cdot 2H_2O$ 2.0g

Preparation of $CaCl_2 \cdot 2H_2O$ Solution: Add $CaCl_2 \cdot 2H_2O$ to distilled/deionized water and bring volume to 250.0mL. Mix thoroughly. Autoclave for 15 min at 15 psi pressure–121°C.

Trace Elements Solution:
Composition per 250.0mL:
$(NH_4)_2SO_4$ 1.3g
KH_2PO_4 .... 0.28g
$MgSO_4 \cdot 7H_2O$ 0.25g
$FeSO_4 \cdot 7H_2O$ 28.0mg
$Na_2B_4O_7 \cdot 10H_2O$ 4.5mg
$MnCl_2 \cdot 7H_2O$ 1.8mg
$ZnSO_4 \cdot 7H_2O$ 0.22mg
$CuCl_2 \cdot 2H_2O$ 0.05mg
$NaMoO_4 \cdot 2H_2O$ 0.03mg
$VOSO_4 \cdot 2H_2O$ 0.03mg
$CoSO_4 \cdot 2H_2O$ 0.01mg

Preparation of Trace Elements Solution: Add components to distilled/deionized water and bring volume to 250.0mL. Mix thoroughly. Autoclave for 15 min at 15 psi pressure–121°C.

Preparation of Medium: Aseptically combine 500.0mL of sterile sucrose yeast solution with 250.0mL of sterile $CaCl_2 \cdot 2H_2O$ solution and 250.0mL of sterile trace elements solution. Mix thoroughly. Adjust pH to 3.0–3.5 with sterile H_2SO_4. Aseptically distribute into sterile tubes or flasks.

Use: For the cultivation of *Sulfolobus* species.

Sulfolobus Medium

Composition per liter:
$(NH_4)_2SO_4$ 1.3g
Yeast extract 1.0g
KH_2PO_4 .... 0.28g
$MgSO_4 \cdot 7H_2O$ 0.25g
$CaCl_2 \cdot 2H_2O$ 0.07g
$FeCl_3 \cdot 6H_2O$ 0.02g
$Na_2B_4O_7 \cdot 10H_2O$ 4.5mg
$MnCl_2 \cdot 4H_2O$ 1.8mg
$ZnSO_4 \cdot 7H_2O$ 0.22mg
$CuCl_2 \cdot 2H_2O$ 0.05mg
$Na_2MoO_4 \cdot 2H_2O$ 0.03mg
$VOSO_4 \cdot 2H_2O$ 0.03mg
$CoSO_4$ 0.01mg
pH 2.0 ± 0.2 at 25°C

Preparation of Medium: Add components to distilled/deionized water and bring volume to 1.0L. Mix thoroughly. Adjust pH at 25°C to 2.0 with 10*N* H_2SO_4. Filter sterilize. Aseptically distribute into tubes or flasks.

Use: For the cultivation and maintenance of *Sulfolobus acidocaldarius.*

Sulfolobus Medium

Composition per liter:
Gellan sucrose yeast solution 500.0mL
$CaCl_2 \cdot 2H_2O/MgCl_2 \cdot 6H_2O$ solution 250.0mL
Trace elements solution 250.0mL
pH 3.0–3.5 at 25°C

Gellan Sucrose Yeast Solution:
Composition per 500.0mL:
Gellan gum 6.5g

Sucrose...2.0g
Yeast extract..1.0g

Source: Gellan gum is available from Kelco.

Preparation of Gellan Sucrose Yeast Solution: Add components to distilled/deionized water and bring volume to 500.0mL. Mix thoroughly. Gently heat and bring to boiling. Autoclave for 15 min at 15 psi pressure–121°C. Cool to 60°C.

$CaCl_2 \cdot 2H_2O/MgCl_2 \cdot 6H_2O$ Solution:
Composition per 250.0mL:
$CaCl_2 \cdot 2H_2O$...2.44g
$MgCl_2 \cdot 6H_2O$...2.0g

Preparation of $CaCl_2 \cdot 2H_2O/MgCl_2 \cdot 6H_2O$ Solution: Add components to distilled/deionized water and bring volume to 250.0mL. Mix thoroughly. Autoclave for 15 min at 15 psi pressure–121°C. Cool to 60°C.

Trace Elements Solution:
Composition per 250.0mL:
$(NH_4)_2SO_4$...1.3g
KH_2PO_4...0.28g
$MgSO_4 \cdot 7H_2O$...0.25g
$FeSO_4 \cdot 7H_2O$...28.0mg
$Na_2B_4O_7 \cdot 10H_2O$...4.5mg
$ZnSO_4 \cdot 7H_2O$...0.22mg
$CuCl_2 \cdot 2H_2O$...0.05mg
$NaMoO_4 \cdot 2H_2O$...0.03mg
$VOSO_4 \cdot 2H_2O$...0.03mg
$CoSO_4 \cdot 2H_2O$...0.01mg

Preparation of Trace Elements Solution: Add components to distilled/deionized water and bring volume to 250.0mL. Mix thoroughly. Autoclave for 15 min at 15 psi pressure–121°C. Cool to 60°C.

Preparation of Medium: Aseptically combine 500.0mL of sterile gellan sucrose yeast solution with 250.0mL of sterile $CaCl_2 \cdot 2H_2O/MgCl_2 \cdot 6H_2O$ solution and 250.0mL of sterile trace elements solution. Mix thoroughly. Adjust pH to 3.0–3.5 with sterile H_2SO_4. Pour into sterile Petri dishes or distribute into sterile tubes.

Use: For the cultivation of *Sulfolobus* species.

Sulfolobus Medium, Revised

Composition per liter:
$(NH_4)_2SO_4$...1.3g
Tryptone...1.0g
KH_2PO_4...0.28g
$MgSO_4 \cdot 7H_2O$...0.25g
$CaCl_2 \cdot 2H_2O$...0.07g
Yeast extract...0.05g
$FeCl_3 \cdot 6H_2O$...0.02g
$Na_2B_4O_7$...4.5mg
$MnCl_2 \cdot 4H_2O$...1.8mg
$ZnSO_4 \cdot 7H_2O$...0.22mg
$CuCl_2 \cdot H_2O$...0.05mg
$Na_2MoO_4 \cdot H_2O$...0.03mg
$VOSO_4 \cdot 2H_2O$...0.03mg
$CoSO_4$...0.01mg
pH 3.0 ± 0.2 at 25°C

Preparation of Medium: Add components to distilled/deionized water and bring volume to 1.0L. Mix thoroughly. Adjust pH at 25°C to 3.0 with 10*N* H_2SO_4. Filter sterilize. Aseptically distribute into tubes or flasks.

Use: For the cultivation and maintenance of *Sulfolobus* species.

Sulfolobus shibatae Medium

Composition per liter:
$(NH_4)_2SO_4$...1.3g
Yeast extract...1.0g
KH_2PO_4...0.28g
$MgSO_4 \cdot 7H_2O$...0.25g
$CaCl_2 \cdot 2H_2O$...0.07g
$FeCl_3 \cdot 6H_2O$...0.02g
$Na_2B_4O_7 \cdot 10H_2O$...4.5mg
$MnCl_2 \cdot 4H_2O$...1.8mg
$ZnSO_4 \cdot 7H_2O$...0.22mg
$CuCl_2 \cdot 2H_2O$...0.05mg
$Na_2MoO_4 \cdot 2H_2O$...0.03mg
$VOSO_4 \cdot 2H_2O$...0.03mg
$CoSO_4$...0.01mg
pH 3.5 ± 0.2 at 25°C

Preparation of Medium: Add components to distilled/deionized water and bring volume to 1.0L. Mix thoroughly. Adjust pH to 3.5 with 10*N* H_2SO_4. Filter sterilize. Aseptically distribute into tubes or flasks.

Use: For the cultivation of *Sulfolobus shibatae.*

Sulfolobus solfataricus Medium

Composition per liter:
KH_2PO_4...3.1g
$(NH_4)_2SO_4$...2.5g
Casamino acids...1.0g
Yeast extract...1.0g
$CaCl_2 \cdot 2H_2O$...0.25g
$MgSO_4 \cdot 7H_2O$...0.2g
$Na_2B_4O_7 \cdot 10H_2O$...4.5mg
$MnCl_2 \cdot 4H_2O$...1.8mg
$ZnSO_4 \cdot 7H_2O$...0.22mg
$CuCl_2 \cdot 2H_2O$...0.05mg
$Na_2MoO_4 \cdot 2H_2O$...0.03mg
$VOSO_4 \cdot 2H_2O$...0.03mg
$CoSO_4 \cdot 7H_2O$...0.01mg
pH 4.0–4.2 at 25°C

Preparation of Medium: Add components to distilled/deionized water and bring volume to 1.0L. Mix thoroughly. Adjust pH at 25°C to 4.0–4.2 with 10*N* H_2SO_4. Filter sterilize. Aseptically distribute into tubes or flasks.

Use: For the cultivation and maintenance of *Sulfolobus solfataricus.*

Sulfophobococcus zilligii Medium (DSMZ Medium 770)

Composition per 1035mL:
Glycine...1.5g
Na_2CO_3...230.0mg
$CaCl_2 \cdot 2H_2O$...66.0mg
Na_2-EDTA...32.0mg
$MgSO_4 \cdot 7H_2O$...31.0mg
KCl...31.0mg
$MnSO_4 \cdot 2H_2O$...2.3mg
$ZnCl_2$...2.1mg
$Na_2B_4O_7 \cdot 10H_2O$...1.8mg
Resazurin...0.5mg
Serum albumin solution...10.0mL
Dithiothreitol solution...10.0mL
Yeast extract solution...10.0mL
Iron sulfate solution...5.0mL
pH 7.6 ± 0.2 at 25°C

Dithiothreitol Solution:
Composition per 10mL:
Dithiothreitol...1.54mg

Preparation of Dithiothreitol Solution: Add dithiothreitol to distilled/deionized water and bring volume to 10.0mL. Mix thoroughly. Filter sterilize.

Iron Sulfate Solution:
Composition per 10mL:
$FeSO_4 \cdot 7H_2O$ 0.1g

Preparation of Iron Sulfate Solution: Add $FeSO_4 \cdot 7H_2O$ to distilled/deionized water and bring volume to 10.0mL. Mix thoroughly. Filter sterilize.

Serum Albumin Solution:
Composition per 10mL:
Bovine serum albumin, Fraction V 1.0g

Preparation of Serum Albumin Solution: Add bovine serum albumin, Fraction V to distilled/deionized water and bring volume to 10.0mL. Mix thoroughly. Filter sterilize.

Yeast Extract Solution:
Composition per 10.0mL:
Yeast extract 1.0g

Preparation of Yeast Extract Solution: Add yeast extract to distilled/deionized water and bring volume to 10.0mL. Mix thoroughly. Sparge with 100% N_2. Autoclave under 100% N_2 for 15 min at 15 psi pressure–121°C. Cool to room temperature.

Preparation of Medium: Prepare medium anaerobically under 100% N_2 gas. Add components, except iron sulfate solution, serum albumin solution, dithiothreitol solution and yeast extract solution, to distilled/deionized water and bring volume to 1.0L. Mix thoroughly. Gently heat and bring to boiling. Cool to 80-90°C. Adjust pH to 7.6. Cool to room temperature. Dispense 30.0mL aliquots into serum bottles. Autoclave for 15 min at 15 psi pressure–121°C. Cool to room temperature. Aseptically inject per each 30.0mL the following solutions: 0.3mL sterile yeast extract solution, 0.3mL sterile dithiothreitol solution, 0.15mL sterile iron sulfate solution, and 10.0mL sterile serum albumin solution. Final pH should be 7.6.

Use: For the cultivation of *Sulfophobococcus zilligii.*

Sulforhabdus Medium (DSMZ Medium 386a)

Composition per 1002mL:
Solution A 920.0mL
Solution C 50.0mL
Solution D 10.0mL
Solution E 10.0mL
Solution F 10.0mL
Solution B
(Trace elements solution SL-10B) 1.0mL
pH 7.2-7.5 at 25°C

Solution A:
Composition per 920mL:
NaCl 1.0g
KCl 0.5g
$MgCl_2 \cdot 6H_2O$ 0.4g
KH_2PO_4 .... 0.2g
NH_4Cl 0.3g
$CaCl_2 \cdot 2H_2O$ 0.15g

Preparation of Solution A: Add components to distilled/deionized water and bring volume to 920.0mL. Mix thoroughly. Sparge with 80% N_2 + 20% CO_2 gas until saturated. Autoclave for 15 min at 15 psi pressure–121°C. Cool to 25°C.

Solution B (Trace Elements Solution SL-10B):
Composition per liter:
$FeCl_2 \cdot 4H_2O$ 1.5g
H_3BO_3 .... 300.0mg
$CoCl_2 \cdot 6H_2O$ 190.0mg
$MnCl_2 \cdot 4H_2O$ 100.0mg
$ZnCl_2$ 70.0mg
$Na_2MoO_4 \cdot 2H_2O$ 36.0mg
$NiCl_2 \cdot 6H_2O$ 24.0mg
$CuCl_2 \cdot 2H_2O$ 2.0mg
HCl (25% solution) 7.7mL

Preparation of Solution B (Trace Elements Solution SL-10B): Add $FeCl_2 \cdot 4H_2O$ to 10.0mL of HCl solution. Mix thoroughly. Add distilled/deionized water and bring volume to 1.0L. Add remaining components. Mix thoroughly. Sparge with 100% N_2. Autoclave for 15 min at 15 psi pressure–121°C.

Solution C:
Composition per 100.0mL:
$NaHCO_3$ 5.0g

Preparation of Solution C: Add $NaHCO_3$ to distilled/deionized water and bring volume to 100.0mL. Mix thoroughly. Sparge with 80% N_2 + 20% CO_2 gas until saturated, approximately 20 min. Filter sterilize under 100% CO_2 into a sterile, gas-tight 100.0mL screw-capped bottle.

Solution D:
Composition per 10.0mL:
Na_2-acetate·$3H_2O$ 0.3g

Preparation of Solution D: Add Na_2-acetate to distilled/deionized water and bring volume to 10.0mL. Sparge with N_2. Filter sterilize. Store anaerobically.

Solution E:
Composition per 10.0mL:
$Na_2S \cdot 9H_2O$ 0.4g

Preparation of Solution E: Add $Na_2S \cdot 9H_2O$ to distilled/deionized water and bring volume to 10.0mL. Sparge with N_2. Autoclave for 15 min at 15 psi pressure–121°C. Cool to 25°C. Store anaerobically.

Solution F:
Composition per 10.0mL:
Na-thiosulfate 2.5g

Preparation of Solution F: Add Na-thiosulfate to distilled/deionized water and bring volume to 10.0mL. Mix thoroughly. Filter sterilize. Flush with 80% N_2 + 20% CO_2 to remove dissolved oxygen.

Preparation of Medium: Add solution B, solution C, solution D, solution E, and solution F to solution A in that order under 80% N_2 + 20% CO_2 gas. Adjust the pH to 7.2–7.5.

Use: For the cultivation of *Desulfocapsa thiozymogenes.*

Sulfur Medium

Composition per liter:
Sulfur, elemental 10.0g
KH_2PO_4 .... 3.0g
$MgSO_4 \cdot 7H_2O$ 0.5g
$(NH_4)_2SO_4$ 0.3g
$CaCl_2 \cdot 2H_2O$ 0.25g
$FeCl_3 \cdot 6H_2O$ 0.02g
pH 4.8± 0.2 at 25°C

Preparation of Medium: Add components, except sulfur, to distilled/deionized water and bring volume to 1.0L.

Mix thoroughly. Add 1.0g of sulfur to each of 10 250.0mL flasks. Add 100.0mL of medium to each flask. Autoclave for 30 min at 0 psi pressure–100°C on 3 consecutive days.

Use: For the isolation, cultivation, and enumeration of iron and sulfur bacteria.

Sulfurospirillum Medium

Composition per 1004.0mL:

Solution A 900.0mL
Solution C 80.0mL
Solution D 20.0mL
Solution B (Trace elements solution SL-10)....2.0mL
Solution E 2.0mL
pH 7.2 ± 0.2 at 25°C

Solution A:

Composition per 900.0mL:

KH_2PO_4....1.36g
$MgSO_4 \cdot 7H_2O$....0.37g
NH_4Cl....0.27g
$CaCl_2 \cdot 2H_2O$....0.1g

Preparation of Solution A: Add components to distilled/deionized water and bring volume to 900.0mL. Mix thoroughly. Sparge with 80% N_2 + 20% CO_2. Anaerobically distribute into tubes or flasks. Autoclave for 15 min at 15 psi pressure–121°C.

Solution B (Trace Elements Solution SL-10):

Composition per liter:

$FeCl_2 \cdot 4H_2O$....1.5g
$CoCl_2 \cdot 6H_2O$....190.0mg
$MnCl_2 \cdot 4H_2O$....100.0mg
$ZnCl_2$....70.0mg
$Na_2MoO_4 \cdot 2H_2O$....36.0mg
$NiCl_2 \cdot 6H_2O$....24.0mg
H_3BO_3....6.0mg
$CuCl_2 \cdot 2H_2O$....2.0mg
HCl (25% solution)....10.0mL

Preparation of Solution B (Trace Elements Solution SL-10): Add $FeCl_2 \cdot 4H_2O$ to 10.0mL of HCl solution. Mix thoroughly. Add distilled/deionized water and bring volume to 1.0L. Add remaining components. Mix thoroughly. Sparge with 100% N_2. Autoclave for 15 min at 15 psi pressure–121°C.

Solution C:

Composition per 80.0mL:

$NaHCO_3$....4.0g

Preparation of Solution C: Add $NaHCO_3$ to distilled/deionized water and bring volume to 80.0mL. Mix thoroughly. Sparge with 80% N_2 + 20% CO_2. Autoclave for 15 min at 15 psi pressure–121°C.

Solution D:

Composition per 20.0mL:

Sodium fumarate....4.0g

Preparation of Solution D: Add sodium fumarate to distilled/deionized water and bring volume to 20.0mL. Mix thoroughly. Sparge with 100% N_2. Autoclave for 15 min at 15 psi pressure–121°C.

Solution E:

Composition per 2.0mL:

L-Cysteine·HCl....0.063g

Preparation of Solution E: Add L-cysteine·HCl to distilled/deionized water and bring volume to 2.0mL. Mix thoroughly. Sparge with 100% N_2. Autoclave for 15 min at 15 psi pressure–121°C.

Preparation of Medium: To sterile solution A in tubes or flasks, add, using a syringe, appropriate volumes of sterile solution B, solution C, solution D, and solution E. Mix thoroughly.

Use: For the cultivation of *Sulfurospirillum deleyianum*.

Sulfurospirillum II Medium (DSMZ Medium 771)

Composition per 1080mL:

Yeast extract....1.0g
NaCl....460.0mg
K_2HPO_4....225.0mg
KH_2PO_4....225.0mg
$(NH_4)_2SO_4$....225.0mg
$MgSO_4 \cdot 7H_2O$....117.0mg
Resazurin....0.5mg
$NaHCO_3$ solution....30.0mL
$NaNO_3$ solution....10.0mL
$Na_2S \cdot 9H_2O$ solution....10.0mL
Na-lactate solution....10.0mL
Vitamin solution....10.0mL
Cysteine solution....10.0mL
Trace elements solution SL-10....1.0mL
Selenite-tungstate solution....1.0mL
pH 7.3 ± 0.2 at 25°C

Cysteine Solution:

Composition per 10.0mL:

L-Cysteine-HCl·H_2O....0.15g

Preparation of Cysteine Solution: Add L-cysteine·HCl·H_2O to distilled/deionized water and bring volume to 10.0mL. Mix thoroughly. Sparge with 100% N_2. Autoclave for 15 min at 15 psi pressure–121°C.

Selenite-Tungstate Solution

Composition per liter:

NaOH....0.5g
$Na_2WO_4 \cdot 2H_2O$....4.0mg
$Na_2SeO_3 \cdot 5H_2O$....3.0mg

Preparation of Selenite-Tungstate Solution: Add components to distilled/deionized water and bring volume to 1.0L. Mix thoroughly. Sparge with 100% N_2. Filter sterilize.

Na-lactate Solution:

Composition per 10.0mL:

Na-lactate....2.25g

Preparation of Na-lactate Solution: Add Na-lactate to distilled/deionized water and bring volume to 10.0mL. Mix thoroughly. Sparge with 100% N_2. Filter sterilize.

$NaNO_3$ Solution:

Composition per 10.0mL:

$NaNO_3$....1.7g

Preparation of $NaNO_3$ Solution: Add $NaNO_3$ to distilled/deionized water and bring volume to 10.0mL. Mix thoroughly. Sparge with 80% N_2 + 20% CO_2. Filter sterilize.

$Na_2S \cdot 9H_2O$ Solution:

Composition per 10mL:

$Na_2S \cdot 9H_2O$....0.1g

Preparation of $Na_2S \cdot 9H_2O$ Solution: Add $Na_2S \cdot 9H_2O$ to distilled/deionized water and bring volume to 10.0mL. Mix thoroughly. Autoclave under 100% N_2 for 15 min at 15 psi pressure–121°C. Cool to room temperature.

$NaHCO_3$ Solution:
Composition per 30.0mL:
$NaHCO_3$ 4.2g

Preparation of $NaHCO_3$ Solution: Add $NaHCO_3$ to distilled/deionized water and bring volume to 30.0mL. Mix thoroughly. Sparge with 80% N_2 + 20% CO_2. Filter sterilize.

Trace Elements Solution SL-10:
Composition per liter:
$FeCl_2·4H_2O$ 1.5g
$CoCl_2·6H_2O$ 190.0mg
$MnCl_2·4H_2O$ 100.0mg
$ZnCl_2$ 70.0mg
$Na_2MoO_4·2H_2O$ 36.0mg
$NiCl_2·6H_2O$ 24.0mg
H_3BO_3 .... 6.0mg
$CuCl_2·2H_2O$ 2.0mg
HCl (25% solution) 10.0mL

Preparation of Trace Elements Solution SL-10: Add $FeCl_2·4H_2O$ to 10.0mL of HCl solution. Mix thoroughly. Add distilled/deionized water and bring volume to 1.0L. Add remaining components. Mix thoroughly. Sparge with 80% N_2 + 20% CO_2.

Vitamin Solution:
Composition per liter:
Pyridoxine-HCl 10.0mg
Thiamine-HCl·$2H_2O$ 5.0mg
Riboflavin 5.0mg
Nicotinic acid 5.0mg
D-Ca-pantothenate 5.0mg
p-Aminobenzoic acid 5.0mg
Lipoic acid 5.0mg
Biotin 2.0mg
Folic acid 2.0mg
Vitamin B_{12} 0.1mg

Preparation of Vitamin Solution: Add components to distilled/deionized water and bring volume to 1.0L. Mix thoroughly. Sparge with 80% H_2 + 20% CO_2. Filter sterilize.

Preparation of Medium: Prepare and dispense medium under 80% N_2 + 20% CO_2. Add components, except $NaHCO_3$ solution, $Na_2S·9H_2O$ solution, cysteine solution, $NaNO_3$ solution, Na-lactate solution, and vitamin solution, to distilled/deionized water and bring volume to 1.0mL. Mix thoroughly. Sparge with 80% N_2 + 20% CO_2. Adjust pH to 7.3. Dispense either 10.0mL aliquots into 15mL Hungate tubes or 50.0mL aliquots into 100mL Hungate bottles. Autoclave for 15 min at 15 psi pressure–121°C. Aseptically and anaerobically inject from sterile stock solutions $NaHCO_3$, $Na_2S·9H_2O$ solution, cysteine solution, vitamin solution, sodium nitrate solution, and sodium lactate solution. Final pH should be 7.3.

Use: For the cultivation of *Sulfurospirillum arsenophilum (Geospirillum* sp.*)* and *Sulfurospirillum barnesii.*

Super MMB Medium (LMG Medium 188)

Yeast extract 1.0g
KH_2PO_4 .... 1.0g
Peptone 0.4g
Sodium succinate 0.4g
NH_4Cl 0.2g
NaCl 0.2g
$MgSO_4·7H_2O$ 0.2g
$CaCl_2·2H_2O$ 10.0mg
Ferric citrate 5.0mg
Vitamin Solution 20.0mL
SL-6 Trace Elements Solution 1.0mL
pH 7.0 ± 0.2 at 25°C

SL-6 Trace Elements Solution:
Composition per liter:
H_3BO_3 .... 0.3g
$CoCl_2·6H_2O$ 0.2g
$ZnSO_4·7H_2O$ 0.1g
$MnCl_2·4H_2O$ 0.03g
$Na_2MoO_4·H_2O$ 0.03g
$NiCl_2·6H_2O$ 0.02g
$CuCl_2·2H_2O$ 0.01g

Preparation of SL-6 Trace Elements Solution: Add components to distilled/deionized water and bring volume to 1.0L. Mix thoroughly. Adjust pH to 3.4.

Vitamin Solution:
Composition per liter:
Calcium DL-pantothenate 5.0mg
Riboflavin 5.0mg
Thiamine·HCl 5.0mg
Biotin 2.0mg
Folic acid 2.0mg
Vitamin B_{12} 0.1mg

Preparation of Vitamin Solution: Add components to distilled/deionized water and bring volume to 1.0L. Mix thoroughly. Filter sterilize.

Preparation of Medium: Add components, except vitamin solution, to 980.0mL distilled/deionized water. Mix thoroughly. Autoclave for 15 min at 15 psi pressure–121°C. Cool to 25°C. Aseptically add 20.0mL sterile vitamin solution. Mix thoroughly. Aseptically distribute to sterile tubes or flasks.

Use: For the cultivation of *Aquabacter* spp.

SWM Medium

Composition per 1014.0mL:
Solution A 940.0mL
Solution E ($NaHCO_3$ solution) 50.0mL
Solution F (Substrate solution) 10.0mL
Solution G ($Na_2S·9H_2O$ solution) 10.0mL
Solution B (Trace elements solution SL-10) 2.0mL
Solution C (Seven vitamin solution) 1.0mL
Solution D (Selenite-tungstate solution) 1.0mL
pH 7.2–7.4 at 25°C

Solution A:
Composition per 940.0mL:
NaCl 1.0g
KCl 0.5g
$MgCl_2·6H_2O$ 0.4g
NH_4Cl 0.25g
KH_2PO_4 .... 0.2g
$CaCl_2·2H_2O$ 0.15g
Resazurin 0.5mg

Preparation of Solution A: Prepare and dispense under 80% N_2 + 20% CO_2. Add components to distilled/deionized water and bring volume to 940.0mL. Mix thoroughly. Autoclave for 15 min at 15 psi pressure–121°C.

Solution B (Trace Elements Solution SL-10):
Composition per liter:
$FeCl_2·4H_2O$ 1.5g
$CoCl_2·6H_2O$ 190.0mg

$MnCl_2 \cdot 4H_2O$	100.0mg
$ZnCl_2$	70.0mg
$Na_2MoO_4 \cdot 2H_2O$	36.0mg
$NiCl_2 \cdot 6H_2O$	24.0mg
H_3BO_3	6.0mg
$CuCl_2 \cdot 2H_2O$	2.0mg
HCl (25% solution)	10.0mL

Preparation of Solution B (Trace Elements Solution SL-10): Prepare and dispense under 100% N_2. Add $FeCl_2 \cdot 4H_2O$ to 10.0mL of HCl solution. Mix thoroughly. Add distilled/deionized water and bring volume to 1.0L. Add remaining components. Mix thoroughly. Autoclave for 15 min at 15 psi pressure–121°C.

Solution C (Seven Vitamin Solution):
Composition per liter:

Pyridoxine·HCl	300.0mg
Nicotinic acid	200.0mg
Thiamine·HCl	200.0mg
Calcium pantothenate	100.0mg
Cyanocobalamine	100.0mg
p-Aminobenzoic acid	80.0mg
D(+)-Biotin	20.0mg

Preparation of Solution C (Seven Vitamin Solution): Add components to distilled/deionized water and bring volume to 1.0L. Mix thoroughly. Filter sterilize. Sparge with 100% N_2.

Solution D (Selenite-Tungstate Solution):
Composition per liter:

NaOH	0.5g
$Na_2WO_4 \cdot 2H_2O$	4.0mg
$Na_2SeO_3 \cdot 5H_2O$	3.0mg

Preparation of Solution D (Selenite-Tungstate Solution): Add components to distilled/deionized water and bring volume to 1.0L. Mix thoroughly. Filter sterilize. Sparge with 100% N_2.

$Na_2S \cdot 9H_2O$ Solution:
Composition per 10.0mL:

$Na_2S \cdot 9H_2O$	0.6g

Preparation of $Na_2S \cdot 9H_2O$ Solution: Add $Na_2S \cdot 9H_2O$ to distilled/deionized water and bring volume to 10.0mL. Mix thoroughly. Gas under 100% N_2. Autoclave for 15 min at 15 psi pressure–121°C.

Solution E ($NaHCO_3$ Solution):
Composition per 50.0mL:

$NaHCO_3$	2.5g

Preparation of Solution E ($NaHCO_3$ Solution): Add $NaHCO_3$ to distilled/deionized water and bring volume to 50.0mL. Mix thoroughly. Gas under 100% N_2. Autoclave for 15 min at 15 psi pressure–121°C.

Solution F (Substrate Solution):
Composition per 10.0mL:

Pyrogallol	0.5g

Preparation of Solution F (Substrate Solution): Add pyrogallol to distilled/deionized water and bring volume to 10.0mL. Mix thoroughly. Gas under 100% N_2. Autoclave for 15 min at 15 psi pressure–121°C.

Solution G ($Na_2S \cdot 9H_2O$ Solution):
Composition per 10.0mL:

$Na_2S \cdot 9H_2O$	0.3g

Preparation of Solution G ($Na_2S \cdot 9H_2O$ Solution): Add $Na_2S \cdot 9H_2O$ to distilled/deionized water and bring volume to 10.0mL. Mix thoroughly. Gas under 100% N_2. Autoclave for 15 min at 15 psi pressure–121°C.

Preparation of Medium: To 940.0mL of sterile solution A, aseptically and anaerobically add 1.0mL of sterile solution B, 1.0mL of sterile solution C, 1.0mL of sterile solution D, 50.0mL of sterile solution E, 10.0mL of sterile solution F, and 10.0mL of sterile solution G. Mix thoroughly. Aseptically and anaerobically distribute into sterile tubes or flasks.

Use: For the cultivation and maintenance of *Pelobacter massiliensis.*

SWMTY Marine Medium

Composition per liter:

Marine salts mix	38.0g
Agar	15.0g
Pancreatic digest of casein	2.0g
Yeast extract	2.0g
Tris(hydroxymethyl)aminomethane buffer	1.0g
KNO_3	0.5g
Sodium glycerophosphate	0.1g
Trace elements solution HO-LE	1.0mL

pH 7.0 ± 0.2 at 25°C

Trace Elements Solution HO-LE:
Composition per liter:

H_3BO_3	2.85g
$MnCl_2 \cdot 4H_2O$	1.8g
Sodium tartrate	1.77g
$FeSO_4 \cdot 7H_2O$	1.36g
$CoCl_2 \cdot 6H_2O$	0.04g
$CuCl_2 \cdot 2H_2O$	0.027g
$Na_2MoO_4 \cdot 2H_2O$	0.025g
$ZnCl_2$	0.02g

Preparation of Trace Elements Solution HO-LE: Add components to distilled/deionized water and bring volume to 1.0L. Mix thoroughly. Filter sterilize.

Preparation of Medium: Add components to distilled/deionized water and bring volume to 1.0L. Mix thoroughly. Gently heat and bring to boiling. Distribute into tubes or flasks. Autoclave for 15 min at 15 psi pressure–121°C. Pour into sterile Petri dishes or leave in tubes.

Use: For the cultivation and maintenance of a variety of heterotrophic marine bacterial species.

SYC Medium

Composition per liter:

Sucrose	10.0g
Pancreatic digest of casein	8.0g
Yeast extract	4.0g
K_2HPO_4	3.0g
$MgSO_4 \cdot 7H_2O$	0.3g

pH 7.0 ± 0.2 at 25°C

Preparation of Medium: Add components to distilled/deionized water and bring volume to 1.0L. Mix thoroughly. Distribute into tubes or flasks. Autoclave for 15 min at 15 psi pressure–121°C.

Use: For the cultivation and maintenance of *Agrobacterium tumefaciens.*

Synthetic Seawater Medium

Composition per liter:

NaCl	27.0g
$MgSO_4 \cdot 7H_2O$	7.0g
Monosodium glutamate	5.0g
Tris(hydroxymethyl)aminomethane buffer	2.0g

Glucose 1.0g
KCl 0.6g
$CaCl_2$ 0.3g
Sodium glycerophosphate 0.2g
Vitamin B_{12} 1.0μg
pH 7.5 ± 0.2 at 25°C

Preparation of Medium: Add components to distilled/deionized water and bring volume to 1.0L. Mix thoroughly. Adjust pH to 7.5. Distribute into tubes or flasks. Autoclave for 15 min at 15 psi pressure–121°C.

Use: For the cultivation and maintenance of *Leucothrix mucor.*

Syntrophobacter pfennigii Medium

Composition per liter:
NaCl 1.0g
Na_2SO_4 .... 0.7g
KCl 0.5g
$MgCl_2 \cdot 6H_2O$ 0.4g
NH_4Cl 0.25g
KH_2PO_4 .... 0.2g
$CaCl_2 \cdot 2H_2O$ 0.15g
Resazurin 1.0mg
Sodium propionate solution 50.0mL
$Na_2S \cdot 9H_2O$ solution 10.0mL
$Na_2S_2O_4$ solution 10.0mL
Trace elements solution SL-10 .... 1.0mL
Seven vitamin solution 1.0mL
$NaHCO_3$ solution .... variable
pH 7.2–7.4 at 25°C

Sodium Propionate Solution:
Composition per 50.0mL:
Sodium propionate 1.5g

Preparation of Sodium Propionate Solution: Add sodium propionate to distilled/deionized water and bring volume to 50.0mL. Mix thoroughly. Filter sterilize. Sparge with 80% N_2 + 20% CO_2.

$NaHCO_3$ Solution:
Composition per 20.0mL:
$NaHCO_3$ 2.5g

Preparation of $NaHCO_3$ Solution: Add $NaHCO_3$ to distilled/deionized water and bring volume to 20.0mL. Mix thoroughly. Filter sterilize. Sparge with 80% N_2 + 20% CO_2.

$Na_2S \cdot 9H_2O$ Solution:
Composition per 10.0mL:
$Na_2S \cdot 9H_2O$ 0.36g

Preparation of $Na_2S \cdot 9H_2O$ Solution: Add $Na_2S \cdot 9H_2O$ to distilled/deionized water and bring volume to 10.0mL. Mix thoroughly. Sparge with 100% N_2. Autoclave for 15 min at 15 psi pressure–121°C.

$Na_2S_2O_4$ Solution:
Composition per 10.0mL:
$Na_2S_2O_4 \cdot 5H_2O$ 2.0g

Preparation of $Na_2S_2O_4$ Solution: Add $Na_2S_2O_4 \cdot 5H_2O$ to distilled/deionized water and bring volume to 10.0mL. Mix thoroughly. Sparge with 100% N_2. Autoclave for 15 min at 15 psi pressure–121°C.

Trace Elements Solution SL-10:
Composition per liter:
$FeCl_2 \cdot 4H_2O$ 1.5g
$CoCl_2 \cdot 6H_2O$ 190.0mg
$MnCl_2 \cdot 4H_2O$ 100.0mg
$ZnCl_2$ 70.0mg
$Na_2MoO_4 \cdot 2H_2O$ 36.0mg
$NiCl_2 \cdot 6H_2O$ 24.0mg
H_3BO_3 .... 6.0mg
$CuCl_2 \cdot 2H_2O$ 2.0mg
HCl (25% solution) 10.0mL

Preparation of Trace Elements Solution SL-10: Prepare and dispense under 80% N_2 + 20% CO_2. Add $FeCl_2 \cdot 4H_2O$ to 10.0mL of HCl solution. Mix thoroughly. Add distilled/deionized water and bring volume to 1.0L. Add remaining components. Mix thoroughly. Sparge with 80% N_2 + 20% CO_2. Autoclave for 15 min at 15 psi pressure–121°C.

Seven Vitamin Solution:
Composition per liter:
Pyridoxine·HCl 0.3g
Thiamine·HCl 0.2g
Nicotinic acid 0.2g
Calcium DL-pantothenate 0.1g
Vitamin B_{12} 0.1g
p-Aminobenzoic acid 80.0mg
Biotin 20.0mg

Preparation of Seven Vitamin Solution: Add components to distilled/deionized water and bring volume to 1.0L. Mix thoroughly.

Preparation of Medium: Prepare medium and dispense under 80% N_2 + 20% CO_2. Add components, except sodium propionate solution, $NaHCO_3$ solution, $Na_2S \cdot 9H_2O$ solution, and $Na_2S_2O_3$ solution, to distilled/deionized water and bring volume to 930.0mL. Mix thoroughly. Sparge with 80% N_2 + 20% CO_2. Autoclave for 15 min at 15 psi pressure–121°C. Aseptically and anaerobically add 50.0mL of sterile sodium propionate solution, 10.0mL of sterile $Na_2S_2O_3$ solution, and 10.0mL of sterile $Na_2S \cdot 9H_2O$ solution. Mix thoroughly. Aseptically and anaerobically add sufficient volume of sterile $NaHCO_3$ solution to bring pH to 7.2–7.4. Aseptically and anaerobically distribute into sterile tubes or flasks.

Use: For the cultivation of *Syntrophobacter pfennigii.*

Syntrophobacter wolinii Medium

Solution A 916.0mL
Solution B 70.0mL
Solution C 10.0mL
Solution D 10.0mL
pH 7.2 ± 0.2 at 25°C

Solution A:
Composition per 916.0mL:
Na_2SO_4 .... 2.8g
Sodium propionate 1.5g
Pancreatic digest of casein 1.0g
Resazurin 1.0mg
Mineral solution 50.0mL
Rumen fluid, clarified 50.0mL
Vitamin solution 5.0mL
Trace elements solution SL-10 .... 1.0mL

Preparation of Solution A: Add components to distilled/deionized water and bring volume to 916.0mL. Adjust pH to 7.2. Gently heat and bring to boiling. Continue boiling for a few minutes. Allow to cool to room temperature under 80% N_2 + 20% CO_2. Distribute into bottles under 80% N_2 + 20% CO_2. Autoclave for 15 min at 15 psi pressure–121°C.

Mineral Solution:
Composition per liter:

Nitrilotriacetic acid	12.5g
NaCl	1.0g
$FeCl_3 \cdot 4H_2O$	0.2g
$MnCl_2 \cdot 4H_2O$	0.1g
$CaCl_2 \cdot 2H_2O$	0.1g
$ZnCl_2$	0.1g
$CuCl_2$	0.02g
Na_2SeO_3	0.02g
$CoCl_2 \cdot 6H_2O$	0.017g
H_3BO_3	0.01g
$Na_2MoO_4 \cdot 2H_2O$	0.01g

Preparation of Mineral Solution: Add nitrilotriacetic acid to 500.0mL of distilled/deionized water. Adjust pH to 6.5 with KOH. Add remaining components. Add distilled/deionized water to 1.0L. Mix thoroughly.

Vitamin Solution:
Composition per liter:

Biotin	0.25mg
Nicotinic acid	2.5mg
Thiamine·HCl	1.25mg
p-Aminobenzoic acid	1.25mg
Pantothenic acid	0.62mg
Pyridoxine·HCl	6.2mg

Preparation of Vitamin Solution: Add components to distilled/deionized water and bring volume to 1.0L. Mix thoroughly.

Trace Elements Solution SL-10:
Composition per liter:

$FeCl_2 \cdot 4H_2O$	1.5g
$CoCl_2 \cdot 6H_2O$	190.0mg
$MnCl_2 \cdot 4H_2O$	100.0mg
$ZnCl_2$	70.0mg
$Na_2MoO_4 \cdot 2H_2O$	36.0mg
$NiCl_2 \cdot 6H_2O$	24.0mg
H_3BO_3	6.0mg
$CuCl_2 \cdot 2H_2O$	2.0mg
HCl (25% solution)	10.0mL

Preparation of Trace Elements Solution SL-10: Add $FeCl_2 \cdot 4H_2O$ to 10.0mL of HCl solution. Mix thoroughly. Add distilled/deionized water and bring volume to 1.0L. Add remaining components. Mix thoroughly.

Solution B:
Composition per 70.0mL:

$NaHCO_3$	3.5g

Preparation of Solution B: Add $NaHCO_3$ to distilled/deionized water and bring volume to 70.0mL. Mix thoroughly. Filter sterilize. Sparge with 80% N_2 + 20% CO_2 for 15 min.

Solution C:
Composition per 10.0mL:

L-Cysteine·HCl	0.3g

Preparation of Solution C: Add L-cysteine·HCl to distilled/deionized water and bring volume to 10.0mL. Mix thoroughly. Sparge with 100% N_2 for 3–4 min. Autoclave under 100% N_2 for 15 min at 15 psi pressure–121°C.

Solution D:
Composition per 10.0mL:

$Na_2S \cdot 9H_2O$	0.3g

Preparation of Solution D: Add $Na_2S \cdot 9H_2O$ to distilled/deionized water and bring volume to 10.0mL. Mix thoroughly. Sparge with 100% N_2 for 3–4 min. Autoclave under 100% N_2 for 15 min at 15 psi pressure–121°C.

Preparation of Medium: To 916.0mL of sterile solution A, add 70.0mL of sterile solution B, 10.0mL of sterile solution C, and 10.0mL of sterile solution D. Mix thoroughly.

Use: For the cultivation and maintenance of *Syntrophobacter wolinii.*

Syntrophococcus sucromutans Medium

Composition per 1002.0mL:

Solution A	916.0mL
Solution C	50.0mL
Solution B	25.0mL
Solution D	10.0mL
Solution E	1.0mL

pH 7.2–7.4 at 25°C

Solution A:
Composition per 916.0mL:

Sodium formate	0.6g
Resazurin	1.0mg
Rumen fluid, clarified	300.0mL
Mineral solution	50.0mL
Vitamin solution	5.0mL
Trace elements solution SL-10	1.0mL

Preparation of Solution A: Add components to distilled/deionized water and bring volume to 916.0mL. Mix thoroughly. Adjust pH to 6.4. Autoclave for 15 min at 15 psi pressure–121°C.

Mineral Solution:
Composition per liter:

KH_2PO_4	10.0g
NaCl	8.0g
NH_4Cl	8.0g
$MgCl_2 \cdot 6H_2O$	6.6g
$CaCl_2 \cdot 2H_2O$	1.0g

Preparation of Mineral Solution: Add components to distilled/deionized water and bring volume to 1.0L. Mix thoroughly.

Vitamin Solution:
Composition per liter:

Pyridoxine·HCl	6.2mg
Nicotinic acid	2.5mg
p-Aminobenzoic acid	1.25mg
Thiamine·HCl	1.25mg
Pantothenic acid	0.62mg
Biotin	0.25mg

Preparation of Vitamin Solution: Add components to distilled/deionized water and bring volume to 1.0L. Mix thoroughly.

Trace Elements Solution SL-10:
Composition per liter:

$FeCl_2 \cdot 4H_2O$	1.5g
$CoCl_2 \cdot 6H_2O$	190.0mg
$MnCl_2 \cdot 4H_2O$	100.0mg
$ZnCl_2$	70.0mg
$Na_2MoO_4 \cdot 2H_2O$	36.0mg
$NiCl_2 \cdot 6H_2O$	24.0mg
H_3BO_3	6.0mg
$CuCl_2 \cdot 2H_2O$	2.0mg
HCl (25% solution)	10.0mL

Preparation of Trace Elements Solution SL-10: Add $FeCl_2 \cdot 4H_2O$ to 10.0mL of HCl solution. Mix thoroughly. Add distilled/deionized water and bring volume to 1.0L.

Add remaining components. Mix thoroughly. Autoclave for 15 min at 15 psi pressure–121°C.

Solution B:
Composition per 25.0mL:
Lactose 5.0g

Preparation of Solution B: Add lactose to distilled/deionized water and bring volume to 25.0mL. Mix thoroughly. Filter sterilize.

Solution C:
Composition per 50.0mL:
$NaHCO_3$ 2.5g

Preparation of Solution C: Add $NaHCO_3$ to distilled/deionized water and bring volume to 50.0mL. Mix thoroughly. Autoclave for 15 min at 15 psi pressure–121°C.

Solution D:
Composition per 10.0mL:
L-Cysteine 0.24g

Preparation of Solution D: Add L-cysteine to distilled/deionized water and bring volume to 10.0mL. Mix thoroughly. Autoclave for 15 min at 15 psi pressure–121°C.

Solution E:
Composition per 1.0mL:
$Na_2S·9H_2O$ 78.0mg

Preparation of Solution E: Add $Na_2S·9H_2O$ to distilled/deionized water and bring volume to 1.0mL. Mix thoroughly. Autoclave for 15 min at 15 psi pressure–121°C.

Preparation of Medium: Prepare and dispense medium under H_2-free 80% N_2 + 20% CO_2. Aseptically and anaerobically combine 916.0mL of sterile solution A with 25.0mL of sterile solution B, 50.0mL of sterile solution C, 10.0mL of sterile solution D, and 1.0mL of sterile solution E. Mix thoroughly. Final pH should be 6.4–6.8

Use: For the cultivation and maintenance of *Syntrophococcus sucromutans.*

Syntrophomonas bryantii Medium

Composition per 1026.0mL:
Solution A 916.0mL
Solution B 70.0mL
Solution C 10.0mL
Solution D 10.0mL
Sodium laurate solution 10.0mL
$CaCl_2·2H_2O$ solution 10.0mL
pH 7.2 ± 0.2 at 25°C

Solution A:
Composition per 916.0mL:
PIPES (Piperazine-*N,N′*-bis [2-ethanesulfonic acid]) buffer 15.12g
Na_2SO_4 .... 2.8g
Butyric acid 1.7g
Pancreatic digest of casein 1.0g
Resazurin 1.0mg
Mineral solution 50.0mL
Rumen fluid, clarified 50.0mL
Vitamin solution 5.0mL
Trace elements solution SL-10 .... 1.0mL

Mineral Solution:
Composition per liter:
Nitrilotriacetic acid 12.5g
NaCl 1.0g
$FeCl_3·4H_2O$ 0.2g
$MnCl_2·4H_2O$ 0.1g
$CaCl_2·2H_2O$ 0.1g
$ZnCl_2$ 0.1g
$CuCl_2$ 0.02g
Na_2SeO_3 .... 0.02g
$CoCl_2·6H_2O$ 0.017g
H_3BO_3 .... 0.01g
$Na_2MoO_4·2H_2O$ 0.01g

Preparation of Mineral Solution: Add nitrilotriacetic acid to 500.0mL of distilled/deionized water. Adjust pH to 6.5 with KOH. Add remaining components. Add distilled/deionized water to 1.0L. Mix thoroughly.

Vitamin Solution:
Composition per liter:
Pyridoxine·HCl 6.2mg
Nicotinic acid 2.5mg
Thiamine·HCl 1.25mg
p-Aminobenzoic acid 1.25mg
Pantothenic acid 0.62mg
Biotin 0.25mg

Preparation of Vitamin Solution: Add components to distilled/deionized water and bring volume to 1.0L. Mix thoroughly.

Trace Elements Solution SL-10:
Composition per liter:
$FeCl_2·4H_2O$ 1.5g
$CoCl_2·6H_2O$ 190.0mg
$MnCl_2·4H_2O$ 100.0mg
$ZnCl_2$ 70.0mg
$Na_2MoO_4·2H_2O$ 36.0mg
$NiCl_2·6H_2O$ 24.0mg
H_3BO_3 .... 6.0mg
$CuCl_2·2H_2O$ 2.0mg
HCl (25% solution) 10.0mL

Preparation of Trace Elements Solution SL-10: Add $FeCl_2·4H_2O$ to 10.0mL of HCl solution. Mix thoroughly. Add distilled/deionized water and bring volume to 1.0L. Add remaining components. Mix thoroughly.

Preparation of Solution A: Add components to distilled/deionized water and bring volume to 916.0mL. Adjust pH to 7.2. Gently heat and bring to boiling. Continue boiling for a few minutes. Allow to cool to room temperature under 80% N_2 + 20% CO_2. Distribute into bottles under 80% N_2 + 20% CO_2. Autoclave for 15 min at 15 psi pressure–121°C.

Solution B:
Composition per 70.0mL:
$NaHCO_3$ 3.5g

Preparation of Solution B: Add $NaHCO_3$ to distilled/deionized water and bring volume to 70.0mL. Mix thoroughly. Filter sterilize. Sparge with 80% N_2 + 20% CO_2 for 15 min.

Solution C:
Composition per 10.0mL:
L-Cysteine·HCl 0.3g

Preparation of Solution C: Add L-cysteine·HCl to distilled/deionized water and bring volume to 10.0mL. Mix thoroughly. Sparge with 100% N_2 for 3–4 min. Autoclave under 100% N_2 for 15 min at 15 psi pressure–121°C.

Solution D:
Composition per 10.0mL:
$Na_2S·9H_2O$ 0.3g

Preparation of Solution D: Add $Na_2S·9H_2O$ to distilled/deionized water and bring volume to 10.0mL. Mix thoroughly. Sparge with 100% N_2 for 3–4 min. Autoclave under 100% N_2 for 15 min at 15 psi pressure–121°C.

Sodium Laurate Solution:
Composition per 10.0mL:
Sodium laurate 2.78g

Preparation of Sodium Laurate Solution: Add sodium laurate to distilled/deionized water and bring volume to 10.0mL. Mix thoroughly. Sparge with 100% N_2. Autoclave for 15 min at 15 psi pressure–121°C.

$CaCl_2 \cdot 2H_2O$ Solution:
Composition per 40.0mL:
$CaCl_2 \cdot 2H_2O$ 1.84g

Preparation of $CaCl_2 \cdot 2H_2O$ Solution: Add $CaCl_2 \cdot 2H_2O$ to distilled/deionized water and bring volume to 40.0mL. Mix thoroughly. Sparge with 100% N_2. Autoclave for 15 min at 15 psi pressure–121°C.

Preparation of Medium: To 916.0mL of sterile solution A, add 70.0mL of sterile solution B, 10.0mL of sterile solution C, and 10.0mL of sterile solution D. Mix thoroughly. Prior to inoculation, aseptically add 10.0mL of sterile sodium laurate solution and 10.0mL of sterile $CaCl_2 \cdot 2H_2O$ solution. Mix thoroughly.

Use: For the cultivation and maintenance of *Syntrophomonas sapovorans*.

Syntrophomonas Medium

Composition per 1006.0mL:
Solution A 916.0mL
Solution B 70.0mL
Solution C 10.0mL
Solution D 10.0mL
pH 7.2 ± 0.2 at 25°C

Solution A:
Composition per 916.0mL:
Na_2SO_4 .... 2.8g
Butyric acid 1.7g
Pancreatic digest of casein 1.0g
Resazurin 1.0mg
Mineral solution 50.0mL
Rumen fluid, clarified 50.0mL
Vitamin solution 5.0mL
Trace elements solution SL-10 .... 1.0mL

Mineral Solution:
Composition per liter:
Nitrilotriacetic acid 12.5g
NaCl 1.0g
$FeCl_3 \cdot 4H_2O$ 0.2g
$MnCl_2 \cdot 4H_2O$ 0.1g
$CaCl_2 \cdot 2H_2O$ 0.1g
$ZnCl_2$ 0.1g
$CuCl_2$ 0.02g
Na_2SeO_3 .... 0.02g
$CoCl_2 \cdot 6H_2O$ 0.017g
H_3BO_3 .... 0.01g
$Na_2MoO_4 \cdot 2H_2O$ 0.01g

Preparation of Mineral Solution: Add nitrilotriacetic acid to 500.0mL of distilled/deionized water. Adjust pH to 6.5 with KOH. Add remaining components. Add distilled/deionized water to 1.0L. Mix thoroughly.

Vitamin Solution:
Composition per liter:
Pyridoxine·HCl 6.2mg
Nicotinic acid 2.5mg
Thiamine·HCl 1.25mg
p-Aminobenzoic acid 1.25mg
Pantothenic acid 0.62mg
Biotin 0.25mg

Preparation of Vitamin Solution: Add components to distilled/deionized water and bring volume to 1.0L. Mix thoroughly.

Trace Elements Solution SL-10:
Composition per liter:
$FeCl_2 \cdot 4H_2O$ 1.5g
$CoCl_2 \cdot 6H_2O$ 190.0mg
$MnCl_2 \cdot 4H_2O$ 100.0mg
$ZnCl_2$ 70.0mg
$Na_2MoO_4 \cdot 2H_2O$ 36.0mg
$NiCl_2 \cdot 6H_2O$ 24.0mg
H_3BO_3 .... 6.0mg
$CuCl_2 \cdot 2H_2O$ 2.0mg
HCl (25% solution) 10.0mL

Preparation of Trace Elements Solution SL-10: Add $FeCl_2 \cdot 4H_2O$ to 10.0mL of HCl solution. Mix thoroughly. Add distilled/deionized water and bring volume to 1.0L. Add remaining components. Mix thoroughly.

Preparation of Solution A: Add components to distilled/deionized water and bring volume to 916.0mL. Adjust pH to 7.2. Gently heat and bring to boiling. Continue boiling for a few minutes. Allow to cool to room temperature under 80% N_2 + 20% CO_2. Distribute into bottles under 80% N_2 + 20% CO_2. Autoclave for 15 min at 15 psi pressure–121°C.

Solution B:
Composition per 70.0mL:
$NaHCO_3$ 3.5g

Preparation of Solution B: Add $NaHCO_3$ to distilled/deionized water and bring volume to 70.0mL. Mix thoroughly. Filter sterilize. Sparge with 80% N_2 + 20% CO_2 for 15 min.

Solution C:
Composition per 10.0mL:
L-Cysteine·HCl 0.3g

Preparation of Solution C: Add L-cysteine·HCl to distilled/deionized water and bring volume to 10.0mL. Mix thoroughly. Sparge with 100% N_2 for 3–4 min. Autoclave under 100% N_2 for 15 min at 15 psi pressure–121°C.

Solution D:
Composition per 10.0mL:
$Na_2S \cdot 9H_2O$ 0.3g

Preparation of Solution D: Add $Na_2S \cdot 9H_2O$ to distilled/deionized water and bring volume to 10.0mL. Mix thoroughly. Sparge with 100% N_2 for 3–4 min. Autoclave under 100% N_2 for 15 min at 15 psi pressure–121°C.

Preparation of Medium: To 916.0mL of sterile solution A, add 70.0mL of sterile solution B, 10.0mL of sterile solution C, and 10.0mL of sterile solution D. Mix thoroughly.

Use: For the cultivation of *Syntrophomonas* species.

Syntrophomonas Medium, Sulfate-Free

Composition per 1006.0mL:
Solution A 916.0mL
Solution B 70.0mL
Solution C 10.0mL
Solution D 10.0mL
pH 7.2 ± 0.2 at 25°C

Solution A:
Composition per 916.0mL:

Butyric acid	1.7g
Pancreatic digest of casein	1.0g
Resazurin	1.0mg
Mineral solution	50.0mL
Rumen fluid, clarified	50.0mL
Vitamin solution	5.0mL
Trace elements solution SL-10	1.0mL

Mineral Solution:
Composition per liter:

Nitrilotriacetic acid	12.5g
NaCl	1.0g
$FeCl_3 \cdot 4H_2O$	0.2g
$MnCl_2 \cdot 4H_2O$	0.1g
$CaCl_2 \cdot 2H_2O$	0.1g
$ZnCl_2$	0.1g
$CuCl_2$	0.02g
Na_2SeO_3	0.02g
$CoCl_2 \cdot 6H_2O$	0.017g
H_3BO_3	0.01g
$Na_2MoO_4 \cdot 2H_2O$	0.01g

Preparation of Mineral Solution: Add nitrilotriacetic acid to 500.0mL of distilled/deionized water. Adjust pH to 6.5 with KOH. Add remaining components. Add distilled/deionized water to 1.0L. Mix thoroughly.

Vitamin Solution:
Composition per liter:

Pyridoxine·HCl	6.2mg
Nicotinic acid	2.5mg
Thiamine·HCl	1.25mg
p-Aminobenzoic acid	1.25mg
Pantothenic acid	0.62mg
Biotin	0.25mg

Preparation of Vitamin Solution: Add components to distilled/deionized water and bring volume to 1.0L. Mix thoroughly.

Trace Elements Solution SL-10:
Composition per liter:

$FeCl_2 \cdot 4H_2O$	1.5g
$CoCl_2 \cdot 6H_2O$	190.0mg
$MnCl_2 \cdot 4H_2O$	100.0mg
$ZnCl_2$	70.0mg
$Na_2MoO_4 \cdot 2H_2O$	36.0mg
$NiCl_2 \cdot 6H_2O$	24.0mg
H_3BO_3	6.0mg
$CuCl_2 \cdot 2H_2O$	2.0mg
HCl (25% solution)	10.0mL

Preparation of Trace Elements Solution SL-10: Add $FeCl_2 \cdot 4H_2O$ to 10.0mL of HCl solution. Mix thoroughly. Add distilled/deionized water and bring volume to 1.0L. Add remaining components. Mix thoroughly.

Preparation of Solution A: Add components to distilled/deionized water and bring volume to 916.0mL. Adjust pH to 7.2. Gently heat and bring to boiling. Continue boiling for a few minutes. Allow to cool to room temperature under 80% N_2 + 20% CO_2. Distribute into bottles under 80% N_2 + 20% CO_2. Autoclave for 15 min at 15 psi pressure–121°C.

Solution B:
Composition per 70.0mL:

$NaHCO_3$	3.5g

Preparation of Solution B: Add $NaHCO_3$ to distilled/deionized water and bring volume to 70.0mL. Mix thoroughly. Filter sterilize. Sparge with 80% N_2 + 20% CO_2 for 15 min.

Solution C:
Composition per 10.0mL:

L-Cysteine·HCl	0.3g

Preparation of Solution C: Add L-cysteine·HCl to distilled/deionized water and bring volume to 10.0mL. Mix thoroughly. Sparge with 100% N_2 for 3–4 min. Autoclave under 100% N_2 for 15 min at 15 psi pressure–121°C.

Solution D:
Composition per 10.0mL:

$Na_2S \cdot 9H_2O$	0.3g

Preparation of Solution D: Add $Na_2S \cdot 9H_2O$ to distilled/deionized water and bring volume to 10.0mL. Mix thoroughly. Sparge with 100% N_2 for 3–4 min. Autoclave under 100% N_2 for 15 min at 15 psi pressure–121°C.

Preparation of Medium: To 916.0mL of sterile solution A, add 70.0mL of sterile solution B, 10.0mL of sterile solution C, and 10.0mL of sterile solution D. Mix thoroughly.

Use: For the cultivation of *Syntrophus buswelii*.

Syntrophomonas species Medium

Composition per 1006.0mL:

Solution A	916.0mL
Solution B	70.0mL
Solution C	10.0mL
Solution D	10.0mL

pH 7.2 ± 0.2 at 25°C

Solution A:
Composition per 916.0mL:

Na_2SO_4	2.8g
Pancreatic digest of casein	1.0g
Sodium stearate	0.61g
Resazurin	1.0mg
Mineral solution	50.0mL
Rumen fluid, clarified	50.0mL
Vitamin solution	5.0mL
Trace elements solution SL-10	1.0mL

Preparation of Solution A: Add components to distilled/deionized water and bring volume to 916.0mL. Adjust pH to 7.2. Gently heat and bring to boiling. Continue boiling for a few minutes. Allow to cool to room temperature under 80% N_2 + 20% CO_2. Distribute into bottles under 80% N_2 + 20% CO_2. Autoclave for 15 min at 15 psi pressure–121°C.

Mineral Solution:
Composition per liter:

Nitrilotriacetic acid	12.5g
NaCl	1.0g
$FeCl_3 \cdot 4H_2O$	0.2g
$MnCl_2 \cdot 4H_2O$	0.1g
$CaCl_2 \cdot 2H_2O$	0.1g
$ZnCl_2$	0.1g
$CuCl_2$	0.02g
Na_2SeO_3	0.02g
$CoCl_2 \cdot 6H_2O$	0.017g
H_3BO_3	0.01g
$Na_2MoO_4 \cdot 2H_2O$	0.01g

Preparation of Mineral Solution: Add nitrilotriacetic acid to 500.0mL of distilled/deionized water. Adjust pH to 6.5 with KOH. Add remaining components. Add distilled/deionized water to 1.0L. Mix thoroughly.

Vitamin Solution:
Composition per liter:

Pyridoxine·HCl 6.2mg
Nicotinic acid 2.5mg
Thiamine·HCl 1.25mg
p-Aminobenzoic acid 1.25mg
Pantothenic acid 0.62mg
Biotin 0.25mg

Preparation of Vitamin Solution: Add components to distilled/deionized water and bring volume to 1.0L. Mix thoroughly.

Trace Elements Solution SL-10:
Composition per liter:

$FeCl_2 \cdot 4H_2O$ 1.5g
$CoCl_2 \cdot 6H_2O$ 190.0mg
$MnCl_2 \cdot 4H_2O$ 100.0mg
$ZnCl_2$ 70.0mg
$Na_2MoO_4 \cdot 2H_2O$ 36.0mg
$NiCl_2 \cdot 6H_2O$ 24.0mg
H_3BO_3 .... 6.0mg
$CuCl_2 \cdot 2H_2O$ 2.0mg
HCl (25% solution) 10.0mL

Preparation of Trace Elements Solution SL-10: Add $FeCl_2 \cdot 4H_2O$ to 10.0mL of HCl solution. Mix thoroughly. Add distilled/deionized water and bring volume to 1.0L. Add remaining components. Mix thoroughly.

Solution B:
Composition per 70.0mL:

$NaHCO_3$ 3.5g

Preparation of Solution B: Add $NaHCO_3$ to distilled/deionized water and bring volume to 70.0mL. Mix thoroughly. Filter sterilize. Sparge with 80% N_2 + 20% CO_2 for 15 min.

Solution C:
Composition per 10.0mL:

L-Cysteine·HCl 0.3g

Preparation of Solution C: Add L-cysteine·HCl to distilled/deionized water and bring volume to 10.0mL. Mix thoroughly. Sparge with 100% N_2 for 3–4 min. Autoclave under 100% N_2 for 15 min at 15 psi pressure–121°C.

Solution D:
Composition per 10.0mL:

$Na_2S \cdot 9H_2O$ 0.3g

Preparation of Solution D: Add $Na_2S \cdot 9H_2O$ to distilled/deionized water and bring volume to 10.0mL. Mix thoroughly. Sparge with 100% N_2 for 3–4 min. Autoclave under 100% N_2 for 15 min at 15 psi pressure–121°C.

Preparation of Medium: To 916.0mL of sterile solution A, add 70.0mL of sterile solution B, 10.0mL of sterile solution C, and 10.0mL of sterile solution D. Mix thoroughly.

Use: For the cultivation of *Syntrophomonas wolfei*.

Syntrophothermus Medium (DSMZ Medium 870)

Composition per liter:

$NaHCO_3$ 2.5g
NH_4Cl 0.54g
$MgCl_2 \cdot 6H_2O$ 0.2g
$CaCl_2 \cdot 2H_2O$ 0.15g
KH_2PO_4 .... 0.14g
Resazurin 0.5mg
$Na_2S \cdot 9H_2O$ solution 10.0mL
Cysteine solution 10.0mL
Vitamin solution 10.0mL
Substrate solution 10.0mL
Trace elements solution 1.0mL
Selenite-tungstate solution 1.0mL

pH 7.0 ± 0.2 at 25°C

Substrate Solution:
Composition per 10.0mL:

Na-crotonate 0.86g

Preparation of Substrate Solution: Add L-cysteine·HCl·H_2O to distilled/deionized water and bring volume to 10.0mL. Mix thoroughly. Sparge with 100% N_2. Filter sterilize.

$Na_2S \cdot 9H_2O$ Solution:
Composition per 10mL:

$Na_2S \cdot 9H_2O$ 0.3g

Preparation of $Na_2S \cdot 9H_2O$ Solution: Add $Na_2S \cdot 9H_2O$ to distilled/deionized water and bring volume to 10.0mL. Mix thoroughly. Autoclave under 100% N_2 for 15 min at 15 psi pressure–121°C. Cool to room temperature.

Cysteine Solution:
Composition per 10.0mL:

L-Cysteine·HCl·H_2O 0.3g

Preparation of Cysteine Solution: Add L-cysteine·HCl·H_2O to distilled/deionized water and bring volume to 10.0mL. Mix thoroughly. Sparge with 100% N_2. Autoclave for 15 min at 15 psi pressure–121°C.

Selenite-Tungstate Solution:
Composition per liter:

NaOH 0.5g
$Na_2WO_4 \cdot 2H_2O$ 4.0mg
$Na_2SeO_3 \cdot 5H_2O$ 3.0mg

Preparation of Selenite-Tungstate Solution: Add components to distilled/deionized water and bring volume to 1.0L. Mix thoroughly. Sparge with 100% N_2. Filter sterilize.

Trace Elements Solution:
Composition per liter:

$MgSO_4 \cdot 7H_2O$ 3.0g
Nitrilotriacetic acid 1.5g
NaCl 1.0g
$MnSO_4 \cdot 2H_2O$ 0.5g
$CoSO_4 \cdot 7H_2O$ 0.18g
$ZnSO_4 \cdot 7H_2O$ 0.18g
$CaCl_2 \cdot 2H_2O$ 0.1g
$FeSO_4 \cdot 7H_2O$ 0.1g
$NiCl_2 \cdot 6H_2O$ 0.025g
$KAl(SO_4)_2 \cdot 12H_2O$ 0.02g
H_3BO_3 .... 0.01g
$Na_2MoO_4 \cdot 4H_2O$ 0.01g
$CuSO_4 \cdot 5H_2O$ 0.01g
$Na_2SeO_3 \cdot 5H_2O$ 0.3mg

Preparation of Trace Elements Solution: Add nitrilotriacetic acid to 500.0mL of distilled/deionized water. Dissolve by adjusting pH to 6.5 with KOH. Add remaining components. Add distilled/deionized water to 1.0L. Mix thoroughly.

Vitamin Solution:
Composition per liter:

Pyridoxine-HCl 10.0mg
Thiamine-HCl·$2H_2O$ 5.0mg
Riboflavin 5.0mg
Nicotinic acid 5.0mg

D-Ca-pantothenate.......5.0mg
p-Aminobenzoic acid.......5.0mg
Lipoic acid.......5.0mg
Biotin.......2.0mg
Folic acid.......2.0mg
Vitamin B_{12}.......0.1mg

Preparation of Vitamin Solution: Add components to distilled/deionized water and bring volume to 1.0L. Mix thoroughly. Sparge with 80% H_2 + 20% CO_2. Filter sterilize.

Preparation of Medium: Prepare and dispense medium under 80% N_2 + 20% CO_2 gas atmosphere. Add components, except cysteine solution, $Na_2S{\cdot}9H_2O$ solution, and substrate solution, to distilled/deionized water and bring volume to 970.0mL. Mix thoroughly. Adjust pH to 7.0. Sparge with 80% N_2 + 20% CO_2 for 30 min. Distribute into Hungate tubes or serum bottles. Autoclave for 15 min at 15 psi pressure–121°C. For each 10.0mL medium, aseptically and anaerobically add 0.1mL cysteine solution, 0.1mL $Na_2S{\cdot}9H_2O$ solution, and 0.1mL substrate solution. Mix thoroughly.

Use: For the cultivation of *Syntrophothermus lipocalidus* DSM 12680.

Syntrophus buswellii II Medium

Composition per 1001.0mL:
Solution A.......870.0mL
Solution C.......100.0mL
Solution D.......10.0mL
Solution E (Vitamin solution).......10.0mL
Solution F.......10.0mL
Solution B (Trace elements solution SL-10).......1.0mL
pH 7.1–7.4 at 25°C

Solution A:
Composition per 870.0mL:
Na_2SO_4.......3.0g
NaCl.......1.0g
KCl.......0.5g
$MgCl_2{\cdot}6H_2O$.......0.4g
NH_4Cl.......0.3g
KH_2PO_4.......0.2g
$CaCl_2{\cdot}2H_2O$.......0.15g
Resazurin.......1.0mg

Preparation of Solution A: Add components to distilled/deionized water and bring volume to 870.0mL. Mix thoroughly. Gently heat and bring to boiling. Continue boiling for 3-4 min. Allow to cool to room temperature while gassing under 80% N_2 + 20% CO_2. Continue gassing until pH reaches below 6.0. Seal the flask under 80% N_2 + 20% CO_2. Autoclave for 15 min at 15 psi pressure–121°C.

Solution B (Trace Elements Solution SL-10):
Composition per liter:
$FeCl_2{\cdot}4H_2O$.......1.5g
$CoCl_2{\cdot}6H_2O$.......190.0mg
$MnCl_2{\cdot}4H_2O$.......100.0mg
$ZnCl_2$.......70.0mg
$Na_2MoO_4{\cdot}2H_2O$.......36.0mg
$NiCl_2{\cdot}6H_2O$.......24.0mg
H_3BO_3.......6.0mg
$CuCl_2{\cdot}2H_2O$.......2.0mg
HCl (25% solution).......10.0mL

Preparation of Solution B (Trace Elements Solution SL-10): Add $FeCl_2{\cdot}4H_2O$ to 10.0mL of HCl solution. Mix thoroughly. Add distilled/deionized water and bring volume to 1.0L. Add remaining components. Mix thoroughly. Gas under 100% N_2. Autoclave for 15 min at 15 psi pressure–121°C.

Solution C:
Composition per 100.0mL:
$NaHCO_3$.......5.0g

Preparation of Solution C: Add $NaHCO_3$ to distilled/deionized water and bring volume to 100.0mL. Mix thoroughly. Filter sterilize. Gas under 80% N_2 + 20% CO_2.

Solution D:
Composition per 10.0mL:
Sodium benzoate.......3.0g
Sodium acetate.......1.0g

Preparation of Solution D: Add sodium propionate to distilled/deionized water and bring volume to 10.0mL. Mix thoroughly. Gas under 100% N_2. Autoclave for 15 min at 15 psi pressure–121°C.

Solution E (Vitamin Solution):
Composition per liter:
Pyridoxine·HCl.......10.0mg
Calcium DL-pantothenate.......5.0mg
Lipoic acid.......5.0mg
Nicotinic acid.......5.0mg
p-Aminobenzoic acid.......5.0mg
Riboflavin.......5.0mg
Thiamine·HCl.......5.0mg
Biotin.......2.0mg
Folic acid.......2.0mg
Vitamin B_{12}.......0.1mg

Preparation of Solution E (Vitamin Solution): Add components to distilled/deionized water and bring volume to 1.0L. Mix thoroughly. Gas under 100% N_2. Autoclave for 15 min at 15 psi pressure–121°C.

Solution F:
Composition per 10.0mL:
$Na_2S{\cdot}9H_2O$.......0.4g

Preparation of Solution F: Add $Na_2S{\cdot}9H_2O$ to distilled/deionized water and bring volume to 10.0mL. Mix thoroughly. Gas under 100% N_2. Autoclave for 15 min at 15 psi pressure–121°C.

Preparation of Medium: Aseptically and anaerobically combine 870.0mL of sterile solution A with 1.0mL of sterile solution B, 100.0mL of sterile solution C, 10.0mL of sterile solution D, 10.0mL of sterile solution E, and 10.0mL of sterile solution F, in that order. Mix thoroughly. Anaerobically distribute into sterile tubes or flasks under 100% N_2.

Use: For the cultivation and maintenance of *Syntrophus buswellii.*

Syntrophus Medium

Composition per 1006.0mL:
Solution A.......916.0mL
Solution B.......70.0mL
Solution C.......10.0mL
Solution D.......10.0mL
pH 7.2 ± 0.2 at 25°C

Solution A:
Composition per 916.0mL:
Na_2SO_4.......2.8g
Sodium benzoate.......2.0g
Pancreatic digest of casein.......1.0g
Resazurin.......1.0mg
Mineral solution.......50.0mL
Rumen fluid, clarified.......50.0mL

Vitamin solution	5.0mL
Trace elements solution SL-10	1.0mL

Mineral Solution:

Composition per liter:

Nitrilotriacetic acid	12.5g
NaCl	1.0g
$FeCl_3 \cdot 4H_2O$	0.2g
$MnCl_2 \cdot 4H_2O$	0.1g
$CaCl_2 \cdot 2H_2O$	0.1g
$ZnCl_2$	0.1g
$CuCl_2$	0.02g
Na_2SeO_3	0.02g
$CoCl_2 \cdot 6H_2O$	0.017g
H_3BO_3	0.01g
$Na_2MoO_4 \cdot 2H_2O$	0.01g

Preparation of Mineral Solution: Add nitrilotriacetic acid to 500.0mL of distilled/deionized water. Adjust pH to 6.5 with KOH. Add remaining components. Add distilled/deionized water to 1.0L. Mix thoroughly.

Vitamin Solution:

Composition per liter:

Pyridoxine·HCl	6.2mg
Nicotinic acid	2.5mg
Thiamine·HCl	1.25mg
p-Aminobenzoic acid	1.25mg
Pantothenic acid	0.62mg
Biotin	0.25mg

Preparation of Vitamin Solution: Add components to distilled/deionized water and bring volume to 1.0L. Mix thoroughly.

Trace Elements Solution SL-10:

Composition per liter:

$FeCl_2 \cdot 4H_2O$	1.5g
$CoCl_2 \cdot 6H_2O$	190.0mg
$MnCl_2 \cdot 4H_2O$	100.0mg
$ZnCl_2$	70.0mg
$Na_2MoO_4 \cdot 2H_2O$	36.0mg
$NiCl_2 \cdot 6H_2O$	24.0mg
H_3BO_3	6.0mg
$CuCl_2 \cdot 2H_2O$	2.0mg
HCl (25% solution)	10.0mL

Preparation of Trace Elements Solution SL-10: Add $FeCl_2 \cdot 4H_2O$ to 10.0mL of HCl solution. Mix thoroughly. Add distilled/deionized water and bring volume to 1.0L. Add remaining components. Mix thoroughly.

Preparation of Solution A: Add components to distilled/deionized water and bring volume to 916.0mL. Adjust pH to 7.2. Gently heat and bring to boiling. Continue boiling for a few minutes. Allow to cool to room temperature under 80% N_2 + 20% CO_2. Distribute into bottles under 80% N_2 + 20% CO_2. Autoclave for 15 min at 15 psi pressure–121°C.

Solution B:

Composition per 70.0mL:

$NaHCO_3$	3.5g

Preparation of Solution B: Add $NaHCO_3$ to distilled/deionized water and bring volume to 70.0mL. Mix thoroughly. Filter sterilize. Sparge with 80% N_2 + 20% CO_2 for 15 min.

Solution C:

Composition per 10.0mL:

L-Cysteine·HCl	0.3g

Preparation of Solution C: Add L-cysteine·HCl to distilled/deionized water and bring volume to 10.0mL. Mix thoroughly. Sparge with 100% N_2 for 3–4 min. Autoclave under 100% N_2 for 15 min at 15 psi pressure–121°C.

Solution D:

Composition per 10.0mL:

$Na_2S \cdot 9H_2O$	0.3g

Preparation of Solution D: Add $Na_2S \cdot 9H_2O$ to distilled/deionized water and bring volume to 10.0mL. Mix thoroughly. Sparge with 100% N_2 for 3–4 min. Autoclave under 100% N_2 for 15 min at 15 psi pressure–121°C.

Preparation of Medium: To 916.0mL of sterile solution A, add 70.0mL of sterile solution B, 10.0mL of sterile solution C, and 10.0mL of sterile solution D. Mix thoroughly.

Use: For the cultivation and maintenance of *Syntrophus buswellii.*

Syntrophus Medium, Sulfate-Free

Composition per 1006.0mL:

Solution A	916.0mL
Solution B	70.0mL
Solution C	10.0mL
Solution D	10.0mL

pH 7.2 ± 0.2 at 25°C

Solution A:

Composition per 916.0mL:

Sodium benzoate	2.0g
Pancreatic digest of casein	1.0g
Resazurin	1.0mg
Mineral solution	50.0mL
Rumen fluid, clarified	50.0mL
Vitamin solution	5.0mL
Trace elements solution SL-10	1.0mL

Mineral Solution:

Composition per liter:

Nitrilotriacetic acid	12.5g
NaCl	1.0g
$FeCl_3 \cdot 4H_2O$	0.2g
$MnCl_2 \cdot 4H_2O$	0.1g
$CaCl_2 \cdot 2H_2O$	0.1g
$ZnCl_2$	0.1g
$CuCl_2$	0.02g
Na_2SeO_3	0.02g
$CoCl_2 \cdot 6H_2O$	0.017g
H_3BO_3	0.01g
$Na_2MoO_4 \cdot 2H_2O$	0.01g

Preparation of Mineral Solution: Add nitrilotriacetic acid to 500.0mL of distilled/deionized water. Adjust pH to 6.5 with KOH. Add remaining components. Add distilled/deionized water to 1.0L. Mix thoroughly.

Vitamin Solution:

Composition per liter:

Pyridoxine·HCl	6.2mg
Nicotinic acid	2.5mg
Thiamine·HCl	1.25mg
p-Aminobenzoic acid	1.25mg
Pantothenic acid	0.62mg
Biotin	0.25mg

Preparation of Vitamin Solution: Add components to distilled/deionized water and bring volume to 1.0L. Mix thoroughly.

Trace Elements Solution SL-10:
Composition per liter:

$FeCl_2 \cdot 4H_2O$	1.5g
$CoCl_2 \cdot 6H_2O$	190.0mg
$MnCl_2 \cdot 4H_2O$	100.0mg
$ZnCl_2$	70.0mg
$Na_2MoO_4 \cdot 2H_2O$	36.0mg
$NiCl_2 \cdot 6H_2O$	24.0mg
H_3BO_3	6.0mg
$CuCl_2 \cdot 2H_2O$	2.0mg
HCl (25% solution)	10.0mL

Preparation of Trace Elements Solution SL-10: Add $FeCl_2 \cdot 4H_2O$ to 10.0mL of HCl solution. Mix thoroughly. Add distilled/deionized water and bring volume to 1.0L. Add remaining components. Mix thoroughly.

Preparation of Solution A: Add components to distilled/deionized water and bring volume to 916.0mL. Adjust pH to 7.2. Gently heat and bring to boiling. Continue boiling for a few minutes. Allow to cool to room temperature under 80% N_2 + 20% CO_2. Distribute into bottles under 80% N_2 + 20% CO_2. Autoclave for 15 min at 15 psi pressure–121°C.

Solution B:
Composition per 70.0mL:

$NaHCO_3$	3.5g

Preparation of Solution B: Add $NaHCO_3$ to distilled/deionized water and bring volume to 70.0mL. Mix thoroughly. Filter sterilize. Sparge with 80% N_2 + 20% CO_2 for 15 min.

Solution C:
Composition per 10.0mL:

L-Cysteine·HCl	0.3g

Preparation of Solution C: Add L-cysteine·HCl to distilled/deionized water and bring volume to 10.0mL. Mix thoroughly. Sparge with 100% N_2 for 3–4 min. Autoclave under 100% N_2 for 15 min at 15 psi pressure–121°C.

Solution D:
Composition per 10.0mL:

$Na_2S \cdot 9H_2O$	0.3g

Preparation of Solution D: Add $Na_2S \cdot 9H_2O$ to distilled/deionized water and bring volume to 10.0mL. Mix thoroughly. Sparge with 100% N_2 for 3–4 min. Autoclave under 100% N_2 for 15 min at 15 psi pressure–121°C.

Preparation of Medium: To 916.0mL of sterile solution A, add 70.0mL of sterile solution B, 10.0mL of sterile solution C, and 10.0mL of sterile solution D. Mix thoroughly.

Use: For the cultivation and maintenance of *Syntrophus buswellii.*

T7 Agar Base (m-T7 Agar Base)

Composition per liter:

Lactose	20.0g
Agar	15.0g
Polyoxyethylene ether W-1	5.0g
Yeast extract	3.0g
Pancreatic digest of casein	2.5g
Peptic digest of animal tissue	2.5g
Sodium heptadecyl sulfate	0.1g
Bromothymol Blue	0.1g
Bromcresol Purple	0.1g

pH 7.4 ± 0.2 at 25°C

Source: This medium is available as a premixed powder from BD Diagnostic Systems.

Preparation of Medium: Add components to distilled/deionized water and bring volume to 1.0L. Mix thoroughly. Gently heat while stirring and bring to boiling. Distribute into tubes or flasks. Autoclave for 15 min at 15 psi pressure–121°C. Cool to 45°–50°C. The medium may be made more selective by adding 1.0mg of penicillin G per liter. Pour into sterile Petri dishes or leave in tubes.

Use: For the selective recovery and differential identification of injured coliform microorganisms from chlorinated water by the membrane filter method. For rapid estimation of the bacteriological quality of water using the membrane filter method.

T-ASW Medium

Composition per 1003.0mL:

NaCl	25.0g
$Na_2S_2O_3 \cdot 5H_2O$	2.5g
$MgSO_4 \cdot 7H_2O$	1.5g
$(NH_4)_2SO_4$	1.0g
KH_2PO_4	0.4g
$CaCl_2 \cdot 2H_2O$	0.3g
$NaHCO_3$	0.2g
Tris·HCl buffer, 0.1*M*, pH 7.5	200.0mL
Phenol Red (0.5% solution)	2.0mL
Trace elements solution	1.0mL

pH 7.5 ± 0.2 at 25°C

Trace Elements Solution:
Composition per liter:

Disodium EDTA	50.0g
$CaCl_2 \cdot 2H_2O$	5.5g
$MnCl_2 \cdot 4H_2O$	5.1g
$FeSO_4 \cdot 7H_2O$	5.0g
$ZnSO_4 \cdot 7H_2O$	2.2g
$CoCl_2 \cdot 6H_2O$	1.6g
$CuSO_4 \cdot 5H_2O$	1.6g
$(NH_4)_6Mo_7O_{24} \cdot 4H_2O$	1.1g

Preparation of Trace Elements Solution: Add components to distilled/deionized water and bring volume to 1.0L. Mix thoroughly. Adjust pH to 6.0 with KOH. Autoclave for 15 min at 15 psi pressure–121°C.

Preparation of Medium: Add components to distilled/deionized water and bring volume to 1.0L. Mix thoroughly. Adjust pH to 7.5. Filter sterilize.

Use: For the cultivation and maintenance of *Thiobacillus hydrothermalis.*

T2 Medium for *Thiobacillus*

Composition per liter:

Solution A	250.0mL
Solution B	250.0mL
Solution C	250.0mL
Solution D	250.0mL

pH 7.0 ± 0.2 at 25°C

Solution A:
Composition per 250.0mL:

$Na_2S_2O_3 \cdot 5H_2O$	5.0g
KNO_3	2.0g
NH_4Cl	1.0g

Preparation of Solution A: Add components to distilled/deionized water and bring volume to 250.0mL. Mix thoroughly. Filter sterilize.

Solution B:

Composition per 250.0mL

KH_2PO_4 .... 2.0g

Preparation of Solution B: Add KH_2PO_4 to distilled/deionized water and bring volume to 250.0mL. Mix thoroughly. Filter sterilize.

Solution C:

Composition per 250.0mL

$NaHCO_3$ 2.0g

Preparation of Solution C: Add $NaHCO_3$ to distilled/deionized water and bring volume to 250.0mL. Mix thoroughly. Filter sterilize.

Solution D:

Composition per 250.0mL

$MgSO_4 \cdot 7H_2O$ 0.8g
$FeSO_4 \cdot 7H_2O$ (2%, w/v, in 1*N* HCl) 1.0mL
Trace metal solution 1.0mL

Preparation of Solution D: Add components to distilled/deionized water and bring volume to 250.0mL. Mix thoroughly. Filter sterilize.

$FeSO_4 \cdot 7H_2O$ Solution:

Composition per 100.0mL

$FeSO_4 \cdot 7H_2O$ 2.0g
HCl (1*N* solution) 100.0mL

Preparation of $FeSO_4 \cdot 7H_2O$ Solution: Add the $FeSO_4 \cdot 7H_2O$ to the HCl solution. Mix thoroughly.

Trace Metal Solution:

Composition per liter:

EDTA 50.0g
$ZnSO_4$ 22.0g
$CaCl_2$ 5.54g
$MnCl_2$ 5.06g
$FeSO_4 \cdot 7H_2O$ 4.99g
$CoCl_2$ 1.61g
$CuSO_4$ 1.57g
$(NH_4)_2MoO_4$ 1.1g

Preparation of Trace Metal Solution: Add components to distilled/deionized water and bring volume to 1.0L. Mix thoroughly. Adjust pH to 6.0 with KOH.

Preparation of Medium: Aseptically combine the four sterile solutions: solution A, solution B, solution C, and solution D. Adjust the pH to 7.0. Aseptically distribute into sterile tubes or flasks.

Use: For the cultivation and maintenance of *Thiobacillus denitrificans* and other thiobacilli.

TCBS Agar (Thiosulfate Citrate Bile Salt Sucrose Agar)

Composition per liter:

Sucrose 20.0g
Agar 14.0g
NaCl 10.0g
Sodium citrate 10.0g
$Na_2S_2O_3$ 10.0g
Yeast extract 5.0g
Pancreatic digest of casein 5.0g
Peptic digest of animal tissue 5.0g
Oxgall 5.0g
Sodium cholate 3.0g
Ferric citrate 1.0g
Thymol Blue 0.04g
Bromothymol Blue 0.04g

pH 8.6 ± 0.2 at 25°C

Source: This medium is available as a premixed powder from BD Diagnostic Systems.

Preparation of Medium: Add components to distilled/deionized water and bring volume to 1.0L. Mix thoroughly. Gently heat while stirring and bring to boiling. Do not autoclave. Cool to 45°–50°C. Pour into sterile Petri dishes or distribute into sterile tubes.

Use: For the selective isolation of *Vibrio cholerae* and *Vibrio parahaemolyticus* from a variety of nonclinical specimens.

TCY Agar

Composition per liter:

NaCl 31.3g
Agar 15.0g
$MgCl_2 \cdot 6H_2O$ 10.8g
$CaCl_2 \cdot 2H_2O$ 1.0g
Casamino acids 1.0g
Tryptone 1.0g
KCl 0.7g
Yeast extract 0.2g

pH 7.2 ± 0.2 at 25°C

Preparation of Medium: Add components to distilled/deionized water and bring volume to 1.0L. Mix thoroughly. Gently heat and bring to boiling. Distribute into tubes or flasks. Autoclave for 15 min at 15 psi pressure–121°C. Pour into sterile Petri dishes or leave in tubes.

Use: For the cultivation and maintenance of *Flexibacter maritimus*.

TDC Medium

Composition per liter:

Agar 20.0g
$CaCO_3$ 10.0g
Glucose 5.0g
K_2HPO_4 .... 1.0g
$MgSO_4$ 1.0g

Preparation of Medium: Add components to tap water and bring volume to 1.0L. Mix thoroughly. Gently heat and bring to boiling. Distribute into tubes or flasks. Autoclave for 15 min at 15 psi pressure–121°C. Pour into sterile Petri dishes or leave in tubes.

Use: For the cultivation and maintenance of *Azotobacter beijerinckii* and other *Azotobacter* species.

TEC Agar (m-TEC Agar)

Composition per liter:

Agar 15.0g
Lactose 10.0g
NaCl 7.5g
Proteose peptone 5.0g
K_2HPO_4 .... 3.3g
Yeast extract 3.0g
KH_2PO_4 .... 1.0g
Sodium lauryl sulfate 0.2g
Sodium deoxycholate 0.1g
Bromcresol Purple 0.08g
Bromphenol Red 0.08g

pH 5.0 ± 0.2 at 25°C

Source: This medium is available as a premixed powder from BD Diagnostic Systems.

Preparation of Medium: Add components to distilled/deionized water and bring volume to 1.0L. Mix thoroughly. Gently heat and bring to boiling. Adjust pH to 5.0. Sterilization is unnecessary. Pour into sterile Petri dishes or distribute into sterile tubes or flasks. Store at 2°–8°C. Use within 1 week.

Use: For detection of *Escherichia coli* in recreational waters by the membrane filter method. This agar is used in conjunction with a urea substrate to detect urease production. After addition of the urea substrate, *Escherichia coli* appears as yellow-yellow/brown colonies when viewed under a fluorescent lamp.

Teepol Broth, Enriched (m-Teepol Broth, Enriched)

Composition per liter:

Peptone	40.0g
Lactose	30.0g
Yeast extract	6.0g
Phenol Red	0.2g
Sodium lauryl sulfate (Teepol, 0.1% solution)	4.0mL

pH 7.4 ± 0.2 at 25°C

Preparation of Medium: Add components to distilled/deionized water and bring volume to 1.0L. Mix thoroughly. Distribute into tubes or flasks. Autoclave for 15 min at 15 psi pressure–121°C.

Use: For the enumeration of coliform organisms and *Escherichia coli* in water by the membrane filter method.

Tergitol 7 Agar

Composition per liter:

Lactose	20.0g
Agar	13.0g
Peptone	10.0g
Yeast extract	6.0g
Meat extract	5.0g
Tergitol-7	0.1g
Bromothymol Blue	0.05g
TTC solution	5.0mL

pH 7.2 ± 0.2 at 25°C

Source: This medium is available as a premixed powder from Oxoid Unipath.

TTC Solution:
Composition per 100.0mL:

Triphenyltetrazolium chloride	0.05g

Preparation of TTC Solution: Add triphenyltetrazolium chloride to distilled/deionized water and bring volume to 100.0mL. Mix thoroughly. Filter sterilize.

Preparation of Medium: Add components to distilled/deionized water and bring volume to 995.0mL. Mix thoroughly. Gently heat and bring to boiling. Autoclave for 15 min at 15 psi pressure–121°C. Cool to 50°C. Aseptically add 5.0mL of sterile TTC solution. Mix thoroughly. Pour into sterile Petri dishes or distribute into sterile tubes.

Use: For the detection and enumeration of coliforms. Lactose-fermenting bacteria appear as yellow colonies. Lactose-nonfermenting bacteria appear as blue colonies.

Termitobacter Medium

Composition per 1060.0mL

NH_4Cl	1.0g
NaCl	0.6g
Sodium acetate·$3H_2O$	0.5g
L-Cysteine·HCl	0.5g
K_2HPO_4	0.3g
KH_2PO_4	0.3g
$MgCl_2·6H_2O$	0.2g
$CaCl_2·2H_2O$	0.1g
KCl	0.1g
Yeast extract	0.2g
Resazurin	0.5mg
$NaHCO_3$	40.0mL
$Na_2S·9H_2O$	10.0mL
Trans-3,4,5-trimethoxycinnamate solution	10.0mL
Trace elements solution SL-10	1.5mL

$NaHCO_3$ Solution:
Composition per 10.0mL:

$NaHCO_3$	1.0g

Preparation of $NaHCO_3$ Solution: Add $NaHCO_3$ to distilled/deionized water and bring volume to 10.0mL. Mix thoroughly. Sparge with 80% N_2 + 20% CO_2. Autoclave for 15 min at 15 psi pressure–121°C.

$Na_2S·9H_2O$ Solution:
Composition per 10.0mL:

$Na_2S·9H_2O$	0.3g

Preparation of $Na_2S·9H_2O$ Solution: Add $Na_2S·9H_2O$ to distilled/deionized water and bring volume to 10.0mL. Mix thoroughly. Sparge with 100% N_2. Autoclave for 15 min at 15 psi pressure–121°C. Before use, neutralize to pH 7.0 with sterile HCl.

Trans-3,4,5-Trimethoxycinnamate Solution:
Composition per 10.0mL:

Trans-3,4,5-trimethoxycinnamate	1.19g

Preparation of Trans-3,4,5-Trimethoxycinnamate Solution: Add trans-3,4,5-trimethoxycinnamate to distilled/deionized water and bring volume to 10.0mL. Mix thoroughly. Sparge with 100% N_2. Neutralize with NaOH. Autoclave for 15 min at 15 psi pressure–121°C.

Preparation of Medium: Prepare and dispense medium under 80% N_2 + 20% CO_2 gas mixture. Add components, except $NaHCO_3$ solution, trans-3,4,5-trimethoxycinnamate solution, and $Na_2S·9H_2O$ solution, to distilled/deionized water and bring volume to 1.0L. Mix thoroughly. Sparge with 80% N_2 + 20% CO_2 gas mixture. Autoclave for 15 min at 15 psi pressure–121°C. Aseptically and anaerobically add 40.0mL of sterile $NaHCO_3$ solution, 10.0mL of sterile trans-3,4,5-trimethoxycinnamate solution, and 10.0mL of sterile $Na_2S·9H_2O$ solution. Mix thoroughly. Aseptically and anaerobically distribute into sterile tubes or bottles.

Use: For the cultivation of *Sporobacter termitidis.*

Tetramethyl Ammonium Chloride Agar

Composition per liter:

Agar	15.0g
Tetramethyl ammonium chloride	1.0g
Thiamine·HCl	0.5g
Standard mineral base	1.0L

pH 6.5 ± 0.2 at 25°C

Standard Mineral Base:
Composition per liter:

$(NH_4)_2SO_4$	1.0g
Phosphate buffer (1*M* solution, pH 6.8)	40.0mL
Huntner's vitamin-free mineral base	20.0mL

Huntner's Vitamin-Free Mineral Base:
Composition per liter:

$MgSO_4 \cdot 7H_2O$	14.45g
Nitrilotriacetic acid	10.0g
$CaCl_2 \cdot 2H_2O$	3.335g
$FeSO_4 \cdot 7H_2O$	99.0mg
"Metals 44"	50.0mg
$(NH_4)_6Mo_7O_{24} \cdot 4H_2O$	9.25mg

Preparation of Huntner's Vitamin-Free Mineral Base: Add nitrilotriacetic acid to 500.0mL of distilled/deionized water. Adjust pH to 6.5 with KOH. Add remaining components. Add distilled/deionized water to 1.0L. Adjust pH to 7.0.

"Metals 44":
Composition per 100.0mL:

$ZnSO_4 \cdot 7H_2O$	1095.0mg
$FeSO_4 \cdot 7H_2O$	500.0mg
Ethylenediaminetetraacetic acid	250.0mg
$MnSO_4 \cdot H_2O$	154.0mg
$CuSO_4 \cdot 5H_2O$	39.2mg
$Co(NO_3)_2 \cdot 6H_2O$	24.8mg
$Na_2B_4O_7 \cdot 10H_2O$	17.7mg

Preparation of "Metals 44": Add components to distilled/deionized water and bring volume to 1.0L. Mix thoroughly.

Preparation of Medium: Add components to distilled/deionized water and bring volume to 1.0L. Mix thoroughly. Gently heat and bring to boiling. Distribute into tubes or flasks. Autoclave for 15 min at 15 psi pressure–121°C. Pour into sterile Petri dishes or leave in tubes.

Use: For the cultivation and maintenance of *Paracoccus kocurii,* a tetramethyl assimilating bacterium from sewage sludge.

Tetrathionate Broth

Composition per liter:

$Na_2S_2O_3$	40.7g
$CaCO_3$	25.0g
NaCl	4.5g
Peptone	4.5g
Yeast extract	1.8g
Beef extract	0.9g
Iodine solution	20.0mL

Iodine Solution:
Composition per 20.0mL:

Iodine	6.0g
KI	5.0g

Preparation of Iodine Solution: Add iodine and KI to distilled/deionized water and bring volume to 20.0mL. Mix thoroughly.

Preparation of Medium: Add components, except iodine solution, to distilled/deionized water and bring volume to 980.0mL. Mix thoroughly. Gently heat and bring to boiling. Do not autoclave. Cool to 40°C. Add 20.0mL of iodine solution. Mix thoroughly. Distribute into tubes in 10.0mL volumes. Use medium the same day it is prepared.

Use: For the selective isolation and enrichment of *Salmonella typhi* and other salmonellae from sewage, and other specimens.

Tetrathionate Reductase Medium

Composition per tube:

Solution I	10.0mL
Solution III	0.2mL
Solution II	0.1mL
Solution IV	0.1mL

Solution I:
Composition per liter:

$Na_2HPO_4 \cdot 12H_2O$	3.6g
KH_2PO_4	1.0g
NH_4Cl	0.5g
Peptone	0.25g
Yeast extract	0.25g
$MgSO_4 \cdot 7H_2O$	0.03g

Preparation of Solution I: Add components to distilled/deionized water and bring volume to 1.0L. Mix thoroughly. Gently heat and bring to boiling. Distribute into tubes in 10.0mL volumes. Autoclave for 15 min at 15 psi pressure–121°C. Cool to 25°C.

Solution II:
Composition per 100.0mL:

$CaCl_2 \cdot 2H_2O$	0.1g
Ferric ammonium citrate	0.05g

Preparation of Solution II: Add components to distilled/deionized water and bring volume to 100.0mL. Mix thoroughly. Gently heat and bring to boiling. Autoclave for 15 min at 15 psi pressure–121°C. Cool to 25°C.

Solution III:
Composition per 100.0mL:

Sodium succinate	15.0g

Preparation of Solution III: Add sodium succinate to distilled/deionized water and bring volume to 100.0mL. Mix thoroughly. Gently heat until dissolved. Autoclave for 15 min at 15 psi pressure–121°C. Cool to 25°C.

Solution IV:
Composition per 100.0mL:

$Na_2S_4O_6 \cdot 2H_2O$	10.0g

Preparation of Solution IV: Add $Na_2S_4O_6 \cdot 2H_2O$ to distilled/deionized water and bring volume to 100.0mL. Mix thoroughly. Sterilize by filtration. Store at 4°C.

Preparation of Medium: To each tube containing 10.0mL of sterile solution I, aseptically add 0.1mL of sterile solution II, 0.2mL of sterile solution III, and 0.1mL of sterile solution IV. Mix thoroughly. Use immediately.

Use: For the cultivation and differentiation of hydrogen-oxidizing bacteria based on their production of tetrathionate reductase.

TF Medium

Composition per liter:

NaCl	7.0g
Pancreatic digest of casein	2.0g
Yeast extract	2.0g
$MgSO_4 \cdot 7H_2O$	1.8g
K_2HPO_4	1.6g
$MgCl_2 \cdot 6H_2O$	1.4g
$Na_2HPO_4 \cdot H_2O$	1.0g
NH_4Cl	0.5g
$Na_2S \cdot 9H_2O$	0.3g
$MgSO_4 \cdot 7H_2O$	0.16g
Resazurin	0.5mg
Wolfe's mineral solution	10.0mL
Wolfe's vitamin solution	10.0mL
Glucose solution	10.0mL
$CaCl_2 \cdot 2H_2O$ solution	10.0mL
L-Cysteine·HCl·H_2O solution	10.0mL

pH 6.8–7.0 at 25°C

Wolfe's Mineral Solution:
Composition per liter:

$MgSO_4 \cdot 7H_2O$	3.0g
Nitrilotriacetic acid	1.5g
NaCl	1.0g
$MnSO_4 \cdot 2H_2O$	0.5g
$CoCl_2 \cdot 6H_2O$	0.1g
$ZnSO_4 \cdot 7H_2O$	0.1g
$CaCl_2 \cdot 2H_2O$	0.1g
$FeSO_4 \cdot 7H_2O$	0.1g
$NiCl_2 \cdot 6H_2O$	0.025g
$KAl(SO_4)_2 \cdot 12H_2O$	0.02g
$CuSO_4 \cdot 5H_2O$	0.01g
H_3BO_3	0.01g
$Na_2MoO_4 \cdot 2H_2O$	0.01g
$Na_2SeO_3 \cdot 5H_2O$	0.3mg

Preparation of Wolfe's Mineral Solution: Add nitrilotriacetic acid to 500.0mL of distilled/deionized water. Adjust pH to 6.5 with KOH. Add remaining components. Add distilled/deionized water to 1.0L. Adjust pH to 6.8.

Wolfe's Vitamin Solution:
Composition per liter:

Pyridoxine·HCl	10.0mg
p-Aminobenzoic acid	5.0mg
Lipoic acid	5.0mg
Nicotinic acid	5.0mg
Riboflavin	5.0mg
Thiamine·HCl	5.0mg
Calcium DL-pantothenate	5.0mg
Biotin	2.0mg
Folic acid	2.0mg
Vitamin B_{12}	0.1mg

Preparation of Wolfe's Vitamin Solution: Add components to distilled/deionized water and bring volume to 1.0L. Mix thoroughly. Filter sterilize.

Glucose Solution:
Composition per 10.0mL:

D-Glucose	3.0g

Preparation of Glucose Solution: Add glucose to distilled/deionized water and bring volume to 10.0mL. Mix thoroughly. Sparge with 100% N_2. Autoclave for 15 min at 15 psi pressure–121°C.

$CaCl_2 \cdot 2H_2O$ Solution:
Composition per 10.0mL:

$CaCl_2 \cdot 2H_2O$	0.06g

Preparation of $CaCl_2 \cdot 2H_2O$ Solution: Add 0.06g of $CaCl_2 \cdot 2H_2O$ to distilled/deionized water and bring volume to 10.0mL. Mix thoroughly. Sparge with 100% N_2. Autoclave for 15 min at 15 psi pressure–121°C.

L-Cysteine·HCl·H_2O Solution:
Composition per 10.0mL:

L-Cysteine·HCl·H_2O	0.3g

Preparation of L-Cysteine·HCl·H_2O Solution: Add L-cysteine·HCl·H_2O to distilled/deionized water and bring volume to 10.0mL. Mix thoroughly. Sparge with 100% N_2. Autoclave for 15 min at 15 psi pressure–121°C.

Preparation of Medium: Prepare and dispense medium under 100% N_2. Add components, except glucose solution, $CaCl_2 \cdot 2H_2O$ solution, and L-cysteine·HCl·H_2O solution, to distilled/deionized water and bring volume to 970.0mL. Mix thoroughly. Adjust pH to 6.8-7.0. Sparge with 100% N_2. Autoclave for 15 min at 15 psi pressure–121°C. Aseptically and anaerobically add 10.0mL of sterile glucose solution, 10.0mL of sterile $CaCl_2 \cdot 2H_2O$ solution, and 10.0mL of sterile L-cysteine·HCl·H_2O solution. Mix thoroughly. Final pH should be 6.8–7.0.

Use: For the cultivation of *Thermosipho* species.

TF(A) Medium (DSMZ Medium 740)

Composition per liter:

K_2HPO_4	1.6g
Yeast extract	2.0g
Trypticase	2.0g
$Na_2HPO_4 \cdot H_2O$	1.0g
NH_4Cl	0.5g
$MgSO_4 \cdot 7H_2O$	0.16g
Resazurin	0.5mg
Calcium chloride solution	10.0mL
Glucose solution	10.0mL
Cysteine solution	10.0mL
$Na_2S \cdot 9H_2O$ solution	10.0mL
Trace elements solution	10.0mL
Vitamin solution	10.0mL

pH 6.8 ± 0.2 at 25°C

Calcium Chloride Solution:
Composition per 10.0mL:

$CaCl_2 \cdot 2H_2O$	0.06g

Preparation of Calcium Chloride Solution: Add $CaCl_2 \cdot 2H_2O$ to distilled/deionized water and bring volume to 10.0mL. Mix thoroughly. Sparge with 100% N_2. Autoclave for 15 min at 15 psi pressure–121°C. Cool to 25°C.

Cysteine Solution:
Composition per 10.0mL:

L-Cysteine·HCl·H_2O	0.3g

Preparation of Cysteine Solution: Add L-cysteine·HCl·H_2O to distilled/deionized water and bring volume to 10.0mL. Mix thoroughly. Sparge with 100% N_2. Autoclave for 15 min at 15 psi pressure–121°C. Cool to 25°C.

Glucose Solution:
Composition per 10mL:

Glucose	3.0g

Preparation of Glucose Solution: Add glucose to distilled/deionized water and bring volume to 10.0mL. Mix thoroughly. Sparge with 100% N_2. Filter sterilize.

$Na_2S \cdot 9H_2O$ Solution:
Composition per 10.0mL:

$Na_2S \cdot 9H_2O$	0.3g

Preparation of $Na_2S \cdot 9H_2O$ Solution: Add $Na_2S \cdot 9H_2O$ to distilled/deionized water and bring volume to 10.0mL. Sparge with N_2. Autoclave for 15 min at 15 psi pressure–121°C. Cool to 25°C. Store anaerobically.

Trace Elements Solution:
Composition per liter:

$MgSO_4 \cdot 7H_2O$	3.0g
Nitrilotriacetic acid	1.5g
NaCl	1.0g
$MnSO_4 \cdot 2H_2O$	0.5g
$CoSO_4 \cdot 7H_2O$	0.18g
$ZnSO_4 \cdot 7H_2O$	0.18g
$CaCl_2 \cdot 2H_2O$	0.1g
$FeSO_4 \cdot 7H_2O$	0.1g
$NiCl_2 \cdot 6H_2O$	0.025g
$KAl(SO_4)_2 \cdot 12H_2O$	0.02g
H_3BO_3	0.01g

$Na_2MoO_4 \cdot 4H_2O$ 0.01g
$CuSO_4 \cdot 5H_2O$ 0.01g
$Na_2SeO_3 \cdot 5H_2O$ 0.3mg

Preparation of Trace Elements Solution: Add nitrilotriacetic acid to 500.0mL of distilled/deionized water. Dissolve by adjusting pH to 6.5 with KOH. Add remaining components. Add distilled/deionized water to 1.0L. Mix thoroughly. Filter sterilize.

Vitamin Solution:
Composition per liter:
Pyridoxine-HCl 10.0mg
Thiamine-HCl·$2H_2O$ 5.0mg
Riboflavin 5.0mg
Nicotinic acid 5.0mg
D-Ca-pantothenate 5.0mg
p-Aminobenzoic acid 5.0mg
Lipoic acid 5.0mg
Biotin 2.0mg
Folic acid 2.0mg
Vitamin B_{12} 0.1mg

Preparation of Vitamin Solution: Add components to distilled/deionized water and bring volume to 1.0L. Mix thoroughly. Sparge with 100% N_2. Filter sterilize.

Preparation of Medium: Prepare and dispense medium under an oxygen-free 100% N_2. Add components, except vitamin solution, cysteine solution, calcium chloride solution, glucose solution, trace elements solution, and $Na_2S \cdot 9H_2O$ solution, to distilled/deionized water and bring volume to 940.0mL. Mix thoroughly. Sparge with 100% N_2. Autoclave for 15 min at 15 psi pressure–121°C. Cool to 25°C. Aseptically and anaerobically add 10.0mL sterile vitamin solution, 10.0mL of sterile cysteine solution, 10.0mL sterile glucose solution, 10.0mL sterile calcium chloride solution, 10.0mL sterile trace elements solution, and 10.0mL of sterile $Na_2S \cdot 9H_2O$ solution. Mix thoroughly. Adjust pH to 6.8. Aseptically and anaerobically distribute into sterile tubes or flasks.

Use: For the cultivation of *Fervidobacterium pennivorans, Fervidobacterium gondwanense,* and *Fervidobacterium* sp.

TGE Broth

Composition per liter:
Pancreatic digest of casein 10.0g
Beef extract 6.0g
Glucose 2.0g
pH 7.0 ± 0.2 at 25°C

Source: This medium is available as a premixed powder from BD Diagnostic Systems.

Preparation of Medium: Add components to distilled/deionized water and bring volume to 1.0L. Mix thoroughly. Distribute into tubes or flasks. Autoclave for 15 min at 15 psi pressure–121°C.

Use: For the enumeration of bacteria by the membrane filter method.

Thauera aromatica AR-1 Medium (DSMZ Medium 855)

Composition per 992.0mL:
Solution A 870.0mL
Solution C 100.0mL
Solution D 10.0mL
Solution E (Vitamin solution) 10.0mL
Solution B (Trace elements solution SL-10) 1.0mL
Selenite-tungstate solution 1.0mL
pH 7.2 ± 0.2 at 25°C

Solution A:
Composition per 870.0mL:
Na_2SO_4 .... 3.0g
NaCl 1.0g
KNO_3 .... 0.6g
KCl 0.5g
$MgCl_2 \cdot 6H_2O$ 0.4g
NH_4Cl 0.3g
KH_2PO_4 .... 0.2g
$CaCl_2 \cdot 2H_2O$ 0.15g
Resazurin 1.0mg

Preparation of Solution A: Add components to distilled/deionized water and bring volume to 870.0mL. Mix thoroughly.

Solution B (Trace Elements Solution SL-10):
Composition per liter:
$FeCl_2 \cdot 4H_2O$ 1.5g
$CoCl_2 \cdot 6H_2O$ 190.0mg
$MnCl_2 \cdot 4H_2O$ 100.0mg
$ZnCl_2$ 70.0mg
$Na_2MoO_4 \cdot 2H_2O$ 36.0mg
$NiCl_2 \cdot 6H_2O$ 24.0mg
H_3BO_3 .... 6.0mg
$CuCl_2 \cdot 2H_2O$ 2.0mg
HCl (25% solution) 10.0mL

Preparation of Solution B (Trace Elements Solution SL-10): Add $FeCl_2 \cdot 4H_2O$ to 10.0mL of HCl solution. Mix thoroughly. Add distilled/deionized water and bring volume to 1.0L. Add remaining components. Mix thoroughly. Sparge with 100% N_2. Autoclave for 15 min at 15 psi pressure–121°C.

Solution C:
Composition per 100.0mL:
$NaHCO_3$ 5.0g

Preparation of Solution C: Add $NaHCO_3$ to distilled/deionized water and bring volume to 100.0mL Mix thoroughly. Filter sterilize. Flush with 80% N_2 + 20% CO_2 to remove dissolved oxygen.

Solution D:
Composition per 10.0mL:
Na-benzoate 0.7g

Preparation of Solution D: Add Na-benzoate to distilled/deionized water and bring volume to 10.0mL. Mix thoroughly. Sparge with 100% N_2. Autoclave for 15 min at 15 psi pressure–121°C.

Solution E (Vitamin Solution):
Composition per liter:
Pyridoxine-HCl 10.0mg
Thiamine-HCl·$2H_2O$ 5.0mg
Riboflavin 5.0mg
Nicotinic acid 5.0mg
D-Ca-pantothenate 5.0mg
p-Aminobenzoic acid 5.0mg
Lipoic acid 5.0mg
Biotin 2.0mg
Folic acid 2.0mg
Vitamin B_{12} 0.10mg

Solution E (Vitamin Solution): Add components to distilled/deionized water and bring volume to 1.0L. Mix thoroughly. Sparge with 100% N_2. Autoclave for 15 min at 15 psi pressure–121°C.

Selenite-Tungstate Solution:
Composition per liter:

NaOH	0.5g
$Na_2WO_4 \cdot 2H_2O$	4.0mg
$Na_2SeO_3 \cdot 5H_2O$	3.0mg

Preparation of Selenite-Tungstate Solution: Add components to distilled/deionized water and bring volume to 1.0L. Mix thoroughly. Sparge with 100% N_2. Filter sterilize.

Preparation of Medium: Gently heat solution A and bring to boiling. Boil solution A for a few minutes. Cool to room temperature. Gas with 80% N_2 + 20% CO_2 gas mixture to reach a pH below 6. Autoclave for 15 min at 15 psi pressure–121°C. Cool to room temperature. Sequentially add 1.0mL solution B, 100.0mL solution C, 10.0mL solution D, 10.0mL solution E, and 1.0mL sterile selenite-tungstate solution. Distribute anaerobically under 80% N_2 + 20% CO_2 into appropriate vessels. Addition of 10–20mg sodium dithionite per liter from a 5% (w/v) solution, freshly prepared under N_2 and filter-sterilized, may stimulate growth.

Use: For the anaerobic cultivation of *Thauera aromatica.*

Thauera mechernichi Medium (DSMZ Medium 918)

Composition per liter:

Na_2HPO_4	4.20g
Na-acetate	2.93g
KH_2PO_4	1.5g
NH_4Cl	0.3g
$MgSO_4 \cdot 7H_2O$	0.1g
Trace elements solution	2.0mL

pH 7.1 ± 0.2 at 25°C

Trace Elements Solution:
Composition per liter:

Na_2-EDTA	50.0g
$CaCl_2 \cdot 2H_2O$	5.5g
$MnCl_2 \cdot 4H_2O$	5.06g
$FeSO_4 \cdot 7H_2O$	5.0g
$ZnSO_4 \cdot 7H_2O$	2.0g
$CoCl_2 \cdot 6H_2O$	1.61g
$CuSO_4 \cdot 5H_2O$	1.57g
$(NH_4)_6Mo_7O_{24} \cdot 4H_2O$	1.1g

Preparation of Trace Elements Solution: Add components to distilled/deionized water and bring volume to 1.0L. Mix thoroughly. Adjust pH to 6.0.

Preparation of Medium: Add components to distilled/deionized water and bring volume to 1.0L. Mix thoroughly. Distribute into tubes or flasks. Autoclave for 15 min at 15 psi pressure–121°C.

Use: For the cultivation of *Thauera mechernichensis.*

Thermoacetogenium phaeum Medium (DSMZ Medium 880)

Composition per liter:

$KHCO_3$	3.5g
NH_4Cl	1.0g
NaCl	0.6g
KH_2PO_4	0.3g
$MgCl_2 \cdot 6H_2O$	0.1g
$CaCl_2 \cdot 2H_2O$	0.08g
Resazurin	0.5mg
Sodium pyruvate solution	50.0mL
Vitamin solution	10.0mL
$Na_2S \cdot 9H_2O$ solution	10.0mL
Cysteine solution	10.0mL
Trace elements solution	1.0mL
Selenite-tungstate solution	1.0mL

pH 7.0-7.1 at 25°C

Sodium Pyruvate Solution:
Composition per 50.0mL:

Sodium pyruvate	5.0g

Preparation of Sodium Pyruvate Solution: Add sodium pyruvate to distilled/deionized water and bring volume to 50.0mL. Mix thoroughly. Sparge with 100% N_2. Filter sterilize.

Selenite-Tungstate Solution:
Composition per liter:

NaOH	0.5g
$Na_2WO_4 \cdot 2H_2O$	4.0mg
$Na_2SeO_3 \cdot 5H_2O$	3.0mg

Preparation of Selenite-Tungstate Solution: Add components to distilled/deionized water and bring volume to 1.0L. Mix thoroughly. Sparge with 100% N_2. Filter sterilize.

$Na_2S \cdot 9H_2O$ Solution:
Composition per 10.0mL:

$Na_2S \cdot 9H_2O$	0.3g

Preparation of $Na_2S \cdot 9H_2O$ Solution: Add $Na_2S \cdot 9H_2O$ to distilled/deionized water and bring volume to 10.0mL. Mix thoroughly. Sparge with 100% N_2. Autoclave for 15 min at 15 psi pressure–121°C.

Cysteine Solution:
Composition per 10.0mL:

L-Cysteine·HCl·H_2O	0.3g

Preparation of Cysteine Solution: Add L-cysteine·HCl·H_2O to distilled/deionized water and bring volume to 10.0mL. Mix thoroughly. Sparge with 100% N_2. Autoclave for 15 min at 15 psi pressure–121°C.

Trace Elements Solution SL-10:
Composition per liter:

$FeCl_2 \cdot 4H_2O$	1.5g
Na_2-EDTA	0.5g
$CoCl_2 \cdot 6H_2O$	190.0mg
$MnCl_2 \cdot 4H_2O$	100.0mg
$ZnCl_2$	70.0mg
$Na_2MoO_4 \cdot 2H_2O$	36.0mg
$NiCl_2 \cdot 6H_2O$	24.0mg
H_3BO_3	6.0mg
$CuCl_2 \cdot 2H_2O$	2.0mg
HCl (25% solution)	10.0mL

Preparation of Trace Elements Solution SL-10: Add $FeCl_2 \cdot 4H_2O$ to 10.0mL of HCl solution. Mix thoroughly. Add distilled/deionized water and bring volume to 1.0L. Add remaining components. Mix thoroughly. Adjust pH to 6.5. Sparge with 100% N_2. Autoclave for 15 min at 15 psi pressure–121°C.

Vitamin Solution:
Composition per liter:

Pyridoxine-HCl	10.0mg
Thiamine-HCl·$2H_2O$	5.0mg
Riboflavin	5.0mg
Nicotinic acid	5.0mg
D-Ca-pantothenate	5.0mg
p-Aminobenzoic acid	5.0mg

Lipoic acid 5.0mg
Biotin 2.0mg
Folic acid 2.0mg
Vitamin B_{12} 0.1mg

Preparation of Vitamin Solution: Add components to distilled/deionized water and bring volume to 1.0L. Mix thoroughly. Sparge with 80% H_2 + 20% CO_2. Filter sterilize.

Preparation of Medium: Prepare and dispense medium under 80% N_2 + 20% CO_2 gas atmosphere. Add components, except $KHCO_3$, sodium pyruvate solution, cysteine solution, and $Na_2S·9H_2O$ solution, to distilled/deionized water and bring volume to 930.0mL. Mix thoroughly. Gently heat and bring to boiling. Boil for 10 min. Cool to room temperature while sparging with 80% N_2 + 20% CO_2. Add 3.5g $KHCO_3$. Mix thoroughly while sparging with 80% N_2 + 20% CO_2 gas atmosphere. Autoclave for 15 min at 15 psi pressure–121°C. Aseptically and anaerobically add 50.0mL sodium pyruvate solution, 10.0mL cysteine solution, and 10.0mL $Na_2S·9H_2O$ solution. Mix thoroughly. Final pH is 7.0-7.1. Aseptically and anaerobically distribute into sterile tubes or bottles.

Use: For the cultivation of *Thermoacetogenium phaeum.*

Thermoactinopolyspora Medium

Composition per liter:

Maltose 20.0g
Agar 15.0g
Papaic digest of soybean meal 15.0g
Yeast extract 2.0g
pH 7.2 ± 0.2 at 25°C

Preparation of Medium: Add components to tap water and bring volume to 1.0L. Mix thoroughly. Gently heat and bring to boiling. Distribute into tubes or flasks. Autoclave for 15 min at 15 psi pressure–121°C. Pour into sterile Petri dishes or leave in tubes.

Use: For the cultivation and maintenance of *Thermoactinomyces* and *Thermoactinopolyspora* species.

Thermoanaerobacter ethanolicus Medium

Composition per liter:

Glucose 8.0g
$Na_2HPO_4·12H_2O$ 4.2g
Yeast extract 2.0g
KH_2PO_4 .... 1.5g
NH_4Cl 0.5g
$MgCl_2·6H_2O$ 0.18g
Reducing solution 40.0mL
Wolfe's modified mineral solution 5.0mL
Resazurin (0.1% solution) 1.0mL
Vitamin solution 0.5mL

Caution: This medium contains Na_2S, and H_2S production will occur, especially upon prolonged boiling. H_2S is hazardous and preparation of this medium should be done in a chemical fume hood.

Reducing Solution:
Composition per 200.0mL:

Cysteine·HCl·H_2O 2.5g
$Na_2S·9H_2O$ 2.5g
NaOH (0.2*N* solution) 200.0mL

Preparation of Reducing Solution: Gently heat the NaOH solution and bring to boiling. Gas with 95% N_2 + 5% H_2. Cool to room temperature. Add the cysteine·HCl·H_2O and $Na_2S·9H_2O$. Anaerobically distribute into tubes. Cap with rubber stoppers. Autoclave for 15 min at 15 psi pressure–121°C.

Vitamin Solution:
Composition per 500.0mL:

Pyridoxine·HCl 0.1g
p-Aminobenzoic acid 0.05g
Calcium pantothenate 0.05g
Nicotinic acid 0.05g
Thioctic acid 0.05g
Biotin 0.02g
Folic acid 0.02g
Riboflavin 5.0mg
Thiamine·HCl 5.0mg
Vitamin B_{12} 1.0mg

Preparation of Vitamin Solution: Add components to distilled/deionized water and bring volume to 500.0mL. Mix thoroughly. Store solution in the dark at –10°C.

Wolfe's Modified Mineral Solution:
Composition per liter:

$MgSO_4·7H_2O$ 3.0g
Nitrilotriacetic acid 1.5g
NaCl 1.0g
$MnSO_4·H_2O$ 0.5g
$CaCl_2$ (anhydrous) 0.1g
$Co(NO_3)_2·6H_2O$ 0.1g
$FeSO_4·7H_2O$ 0.1g
$ZnSO_4·7H_2O$ 0.1g
$AlK(SO_4)_2$ (anhydrous) 0.01g
$CuSO_4·5H_2O$ 0.01g
H_3BO_3 .... 0.01g
$Na_2MoO_4·2H_2O$ 0.01g
Na_2SeO_3 (anhydrous) 1.0mg

Preparation of Wolfe's Modified Mineral Solution: Add nitrilotriacetic acid to 500.0mL of distilled/deionized water. Dissolve by adjusting pH to 6.5 with KOH. Add remaining components. Add distilled/deionized water to 1.0L.

Preparation of Medium: Add components, except reducing solution, to distilled/deionized water and bring volume to 1.0L. Gently heat and bring to boiling under 95% N_2 + 5% H_2. Continue boiling until color changes from blue to pink. Add the reducing solution. The pink color will disappear, indicating that the solution has been reduced. Distribute into tubes or flasks under 95% N_2 + 5% H_2 using anerobic techniques. Cap tubes with rubber stoppers. Autoclave for 15 min at 15 psi pressure–121°C.

Use: For the cultivation and maintenance of thermophilic anaerobes such as *Thermoanaerobacter* species and some *Clostridium* species.

Thermoanaerobacter subterraneus Medium (DSMZ Medium 899)

Composition per liter:

Yeast extract 2.0g
$MgCl_2·6H_2O$ 1.0g
NH_4Cl 1.0g
NaCl 0.6g
Cysteine-HCl·H_2O 0.5g
K_2HPO_4 .... 0.3g
KH_2PO_4 .... 0.3g
KCl 0.2g
$CaCl_2·2H_2O$ 0.1g
Resazurin 0.5mg
D-Glucose solution 30.0mL

$NaHCO_3$ solution 20.0mL
Trace mineral solution 10.0mL
$Na_2S_2O_3$ solution 10.0mL
$Na_2S\cdot9H_2O$ solution 10.0mL
pH 7.0 ± 0.2 at 25°C

$Na_2S\cdot9H_2O$ Solution:
Composition per 10mL:
$Na_2S\cdot9H_2O$ 0.45g

Preparation of $Na_2S\cdot9H_2O$ Solution: Add $Na_2S\cdot9H_2O$ to distilled/deionized water and bring volume to 10.0mL. Mix thoroughly. Autoclave under 100% N_2 for 15 min at 15 psi pressure–121°C. Cool to room temperature.

$NaHCO_3$ Solution:
Composition per 20.0mL:
$NaHCO_3$ 4.0g

Preparation of $NaHCO_3$ Solution: Add $NaHCO_3$ to distilled/deionized water and bring volume to 20.0mL. Mix thoroughly. Sparge with 80% N_2 + 20% CO_2. Filter sterilize.

Glucose Solution:
Composition per 30.0mL:
Glucose 4.0g

Preparation of Glucose Solution: Add glucose to distilled/deionized water and bring volume to 30.0mL. Mix thoroughly. Sparge with 100% N_2. Filter sterilize.

$Na_2S_2O_3$ Solution:
Composition per 10mL:
$Na_2S_2O_3\cdot5H_2O$ 2.5g

Preparation of $Na_2S_2O_3$ Solution: Add $Na_2S_2O_3\cdot5H_2O$ to distilled/deionized water and bring volume to 10.0mL. Mix thoroughly. Autoclave under 100% N_2 for 15 min at 15 psi pressure–121°C. Cool to room temperature.

Trace Elements Solution:
Composition per liter:
$MgSO_4\cdot7H_2O$ 3.0g
Nitrilotriacetic acid 1.5g
NaCl 1.0g
$MnSO_4\cdot2H_2O$ 0.5g
$CoSO_4\cdot7H_2O$ 0.18g
$ZnSO_4\cdot7H_2O$ 0.18g
$CaCl_2\cdot2H_2O$ 0.1g
$FeSO_4\cdot7H_2O$ 0.1g
$NiCl_2\cdot6H_2O$ 0.025g
$KAl(SO_4)_2\cdot12H_2O$ 0.02g
H_3BO_3 .... 0.01g
$Na_2MoO_4\cdot4H_2O$ 0.01g
$CuSO_4\cdot5H_2O$ 0.01g
$Na_2SeO_3\cdot5H_2O$ 0.3mg

Preparation of Trace Elements Solution: Add nitrilotriacetic acid to 500.0mL of distilled/deionized water. Dissolve by adjusting pH to 6.5 with KOH. Add remaining components. Add distilled/deionized water to 1.0L. Mix thoroughly.

Preparation of Medium: Prepare and dispense medium under 80% N_2 + 20% CO_2 gas atmosphere. Add components, except $NaHCO_3$ solution, glucose solution, $Na_2S\cdot9H_2O$ solution, and $Na_2S_2O_3$ solution, to distilled/deionized water and bring volume to 930.0mL. Mix thoroughly. Sparge with 80% N_2 + 20% CO_2. Distribute into sterile tubes or bottles. Autoclave for 15 min at 15 psi pressure–121°C. Aseptically and anaerobically per 1.0L of medium add 20.0mL $NaHCO_3$ solution, 30.0mL glucose solution, 10.0mL $Na_2S\cdot9H_2O$ solution, and 10.0mL $Na_2S_2O_3$ solution. Mix thoroughly. The final pH should be 7.0.

Use: For the cultivation of *Thermoanaerobacter subterraneus.*

Thermoanaerobacter sulfurophilus Medium (DSMZ Medium 827)

Composition per 1055mL:
Sulfur, powdered 10.0g
NH_4Cl 0.33g
KCl 0.33g
KH_2PO_4 .... 0.33g
$MgCl_2\cdot6H_2O$ 0.33g
$CaCl_2\cdot2H_2O$ 0.33g
Resazurin 0.5mg
Glucose solution 25.0mL
$Na_2S\cdot9H_2O$ solution 15.0mL
Vitamin solution 10.0mL
Yeast extract solution 5.0mL
Trace elements solution SL10 .... 1.0mL
pH 7.0 ± 0.2 at 25°C

Vitamin Solution:
Composition per liter:
Pyridoxine-HCl 10.0mg
Thiamine-HCl·$2H_2O$ 5.0mg
Riboflavin 5.0mg
Nicotinic acid 5.0mg
D-Ca-pantothenate 5.0mg
p-Aminobenzoic acid 5.0mg
Lipoic acid 5.0mg
Biotin 2.0mg
Folic acid 2.0mg
Vitamin B_{12} 0.1mg

Preparation of Vitamin Solution: Add components to distilled/deionized water and bring volume to 1.0L. Mix thoroughly. Sparge with 80% H_2 + 20% CO_2. Filter sterilize.

Glucose Solution:
Composition per 50mL:
Glucose 10.0g

Preparation of Glucose Solution: Add glucose to distilled/deionized water and bring volume to 50.0mL. Mix thoroughly. Filter sterilize.

Yeast Extract Solution:
Composition per 10mL:
Yeast extract 1.0g

Preparation of Yeast Extract Solution: Add yeast extract to distilled/deionized water and bring volume to 10.0mL. Mix thoroughly. Autoclave under 100% N_2 for 15 min at 15 psi pressure–121°C. Cool to room temperature.

Trace Elements Solution SL-10:
Composition per liter:
$FeCl_2\cdot4H_2O$ 1.5g
$CoCl_2\cdot6H_2O$ 190.0mg
$MnCl_2\cdot4H_2O$ 100.0mg
$ZnCl_2$ 70.0mg
$Na_2MoO_4\cdot2H_2O$ 36.0mg
$NiCl_2\cdot6H_2O$ 24.0mg
H_3BO_3 .... 6.0mg
$CuCl_2\cdot2H_2O$ 2.0mg
HCl (25% solution) 10.0mL

Preparation of Trace Elements Solution SL-10: Add $FeCl_2\cdot4H_2O$ to 10.0mL of HCl solution. Mix thoroughly. Add distilled/deionized water and bring volume to 1.0L.

Add remaining components. Mix thoroughly. Sparge with 80% N_2 + 20% CO_2. Autoclave for 15 min at 15 psi pressure–121°C.

$Na_2S·9H_2O$ Solution:
Composition per 20mL:
$Na_2S·9H_2O$ 0.6g

Preparation of $Na_2S·9H_2O$ Solution: Add $Na_2S·9H_2O$ to distilled/deionized water and bring volume to 20.0mL. Mix thoroughly. Autoclave under 100% N_2 for 15 min at 15 psi pressure–121°C. Cool to room temperature.

Preparation of Medium: Prepare and dispense medium under 80% N_2 + 20% CO_2 gas mixture. Add components, except $Na_2S·9H_2O$ solution, glucose solution, vitamin solution, and yeast extract solution, to distilled/deionized water and bring volume to 1.0L. Mix thoroughly. Heat to 90°C on each of 3 successive days. Aseptically and anaerobically add 25.0mL sterile glucose solution, 15.0mL sterile $Na_2S·9H_2O$ solution, 10.0mL sterile vitamin solution, and 5.0mL sterile yeast extract solution. Mix thoroughly. Aseptically and anaerobically distribute into tubes or bottles. The pH should be 7.0.

Use: For the cultivation of *Thermoanaerobacter sulfurophilus.*

Thermoanaerobacter tengcongensis Medium (DSMZ Medium 965)

Composition per liter:
Soluble Starch 10.0g
NaCl 2.0g
Tryptone 2.0g
NH_4Cl 1.0g
Yeast extract 1.0g
$MgCl_2·6H_2O$ 0.5g
Cysteine-HCl·H_2O 0.5g
K_2HPO_4 .... 0.3g
KH_2PO_4 .... 0.3g
KCl 0.2g
$CaCl_2·2H_2O$ 0.05g
Resazurin 0.5mg
Thiosulfate solution 50.0mL
Trace elements solution 10.0mL
pH 7.5 ± 0.2 at 25°C

Thiosulfate Solution:
Composition per 50mL:
$Na_2S_2O_3·5H_2O$ 5.0g

Preparation of Thiosulfate Solution: Add $Na_2S_2O_3·5H_2O$ to distilled/deionized water and bring volume to 50.0mL. Mix thoroughly. Autoclave for 15 min at 15 psi pressure–121°C. Cool to room temperature.

Trace Elements Solution:
Composition per liter:
$MgSO_4·7H_2O$ 3.0g
Nitrilotriacetic acid 1.5g
NaCl 1.0g
$MnSO_4·2H_2O$ 0.5g
$CoSO_4·7H_2O$ 0.18g
$ZnSO_4·7H_2O$ 0.18g
$CaCl_2·2H_2O$ 0.1g
$FeSO_4·7H_2O$ 0.1g
$NiCl_2·6H_2O$ 0.025g
$KAl(SO_4)_2·12H_2O$ 0.02g
H_3BO_3 .... 0.01g
$Na_2MoO_4·4H_2O$ 0.01g
$CuSO_4·5H_2O$ 0.01g
$Na_2SeO_3·5H_2O$ 0.3mg

Preparation of Trace Elements Solution: Add nitrilotriacetic acid to 500.0mL of distilled/deionized water. Dissolve by adjusting pH to 6.5 with KOH. Add remaining components. Add distilled/deionized water to 1.0L. Mix thoroughly.

Preparation of Medium: Add components, except thiosulfate solution and cysteine-HCl·H_2O, to distilled/deionized water and bring volume to 950.0mL. Mix thoroughly. Gently heat and bring to boiling. Boil for 3 min. Cool to room temperature while sparging with 100% N_2 gas. Adjust pH to 7.5. Distribute into tubes or bottles under 100% N_2 gas. Autoclave for 15 min at 15 psi pressure–121°C. Cool to room temperature. Aseptically and anaerobically under 100% N_2 gas, add 50.0mL sterile thiosulfate solution per liter of medium.

Use: For the cultivation of *Thermoanaerobacter tengcongensis.*

Thermoanaerobacterium Medium (DSMZ Medium 903)

Composition per 1030mL:
KH_2PO_4 .... 0.5g
NaCl 0.4g
$MgCl_2·6H_2O$ 0.33g
Trypticase 0.25g
$CaCl_2·2H_2O$ 0.05g
Resazurin 0.5mg
Sucrose solution 50.0mL
$NaHCO_3$ solution 20.0mL
$Na_2S·9H_2O$ solution 10.0mL
L-Cysteine solution 10.0mL
Selenite solution 1.0mL
Seven vitamin solution 1.0mL
Trace elements solution SL-10 .... 1.0mL
pH 7.0 ± 0.2 at 25°C

Sucrose Solution:
Composition per 50.0mL:
Sucrose 5.0g

Preparation of Sucrose Solution: Add sucrose to distilled/deionized water and bring volume to 50.0mL. Mix thoroughly. Sparge with 100% N_2. Autoclave for 15 min at 15 psi pressure–121°C.

L-Cysteine Solution:
Composition per 10.0mL:
L-Cysteine·HCl·H_2O 0.3g

Preparation of L-Cysteine Solution: Add L-cysteine·HCl·H_2O to distilled/deionized water and bring volume to 10.0mL. Mix thoroughly. Sparge with 100% N_2. Autoclave for 15 min at 15 psi pressure–121°C.

$Na_2S·9H_2O$ Solution:
Composition per 10.0mL:
$Na_2S·9H_2O$ 0.3g

Preparation of $Na_2S·9H_2O$ Solution: Add $Na_2S·9H_2O$ to distilled/deionized water and bring volume to 10.0mL. Mix thoroughly. Sparge with 100% N_2. Autoclave for 15 min at 15 psi pressure–121°C.

$NaHCO_3$ Solution:
Composition per 20.0mL:
$NaHCO_3$ 2.5g

Preparation of $NaHCO_3$ Solution: Add $NaHCO_3$ to distilled/deionized water and bring volume to 20.0mL. Mix thoroughly. Sparge with 80% N_2 + 20% CO_2. Autoclave for 15 min at 15 psi pressure–121°C. Cool to 25°C. Must be prepared freshly.

Selenite Solution:
Composition per liter:

NaOH 0.5g
$Na_2SeO_3 \cdot 5H_2O$ 3.0mg

Preparation of Selenite Solution: Add components to distilled/deionized water and bring volume to 1.0L. Mix thoroughly. Sparge with 100% N_2. Filter sterilize.

Seven Vitamin Solution:
Composition per liter:

Pyridoxine hydrochloride 300.0mg
Thiamine-HCl·$2H_2O$ 200.0mg
Nicotinic acid 200.0mg
Vitamin B_{12} 100.0mg
Calcium pantothenate 100.0mg
p-Aminobenzoic acid 80.0mg
D(+)-Biotin 20.0mg

Preparation of Seven Vitamin Solution: Add components to distilled/deionized water and bring volume to 1.0L. Sparge with 100% N_2. Mix thoroughly. Filter sterilize.

Trace Elements Solution SL-10:
Composition per liter:

$FeCl_2 \cdot 4H_2O$ 1.5g
$CoCl_2 \cdot 6H_2O$ 190.0mg
$MnCl_2 \cdot 4H_2O$ 100.0mg
$ZnCl_2$ 70.0mg
$Na_2MoO_4 \cdot 2H_2O$ 36.0mg
$NiCl_2 \cdot 6H_2O$ 24.0mg
H_3BO_3 .... 6.0mg
$CuCl_2 \cdot 2H_2O$ 2.0mg
HCl (25% solution) 10.0mL

Preparation of Trace Elements Solution SL-10: Add $FeCl_2 \cdot 4H_2O$ to 10.0mL of HCl solution. Mix thoroughly. Add distilled/deionized water and bring volume to 1.0L. Add remaining components. Mix thoroughly. Sparge with 80% N_2 + 20% CO_2. Autoclave for 15 min at 15 psi pressure–121°C.

Preparation of Medium: Prepare and dispense medium under sparge with 80% N_2 + 20% CO_2. Add components, except seven vitamin solution, $NaHCO_3$ solution, sucrose solution, L-cysteine-HCl·H_2O solution, and $Na_2S \cdot 9H_2O$ solution, to distilled/deionized water and bring volume to 940.0mL. Mix thoroughly. Adjust pH to 7.0. Distribute into anaerobe tubes or bottles. Autoclave for 15 min at 15 psi pressure–121°C. Aseptically and anaerobically add per liter, 10.0mL seven vitamin solution, 20.0mL $NaHCO_3$ solution, 50.0mL sucrose solution, 10.0mL L-cysteine-HCl·H_2O solution, and 10.0mL $Na_2S \cdot 9H_2O$. Mix thoroughly. The final pH should be 7.0.

Use: For the cultivation of *Thermoanaerobacterium polysaccharolyticum* and *Thermoanaerobacterium zeae.*

Thermoanaerobium brockii Medium

Composition per liter:

Pancreatic digest of casein 10.0g
Yeast extract 3.0g
K_2HPO_4 .... 1.5g
NH_4Cl 0.9g
NaCl 0.9g
KH_2PO_4 .... 0.75g
$MgCl_2 \cdot 6H_2O$ 0.2g
Glucose solution 25.0mL
$Na_2S \cdot 9H_2O$ (10% solution) 10.0mL
Trace elements solution 9.0mL
Vitamin solution 5.0mL
Resazurin (0.025% solution) 4.0mL
$FeSO_4 \cdot 7H_2O$ (10% solution) 0.03mL
pH 7.3 ± 0.2 at 25°C

Glucose Solution:
Composition per 100.0mL:

Glucose 20.0g

Preparation of Glucose Solution: Add glucose to distilled/deionized water and bring volume to 100.0mL. Mix thoroughly. Filter sterilize.

Trace Elements Solution:
Composition per liter:

Nitrilotriacetic acid 12.5g
NaCl 1.0g
$FeCl_3 \cdot 4H_2O$ 0.2g
$MnCl_2 \cdot 4H_2O$ 0.1g
$CaCl_2 \cdot 2H_2O$ 0.1g
$ZnCl_2$ 0.1g
$CuCl_2$ 0.02g
Na_2SeO_3 .... 0.02g
$CoCl_2 \cdot 6H_2O$ 0.017g
H_3BO_3 .... 0.01g
$Na_2MoO_4 \cdot 2H_2O$ 0.01g

Preparation of Trace Elements Solution: Add nitrilotriacetic acid to 500.0mL of distilled/deionized water. Adjusting pH to 6.5 with KOH. Add remaining components. Add distilled/deionized water to 1.0L.

Wolfe's Vitamin Solution:
Composition per liter:

Pyridoxine·HCl 10.0mg
Thiamine·HCl 5.0mg
Riboflavin 5.0mg
Nicotinic acid 5.0mg
Calcium pantothenate 5.0mg
p-Aminobenzoic acid 5.0mg
Thioctic acid 5.0mg
Biotin 2.0mg
Folic acid 2.0mg
Cyanocobalamin 100.0μg

Preparation of Wolfe's Vitamin Solution: Add components to distilled/deionized water and bring volume to 1.0L. Mix thoroughly.

Preparation of Medium: Add components, except glucose solution, to distilled/deionized water and bring volume to 975.0mL. Mix thoroughly. Autoclave for 15 min at 15 psi pressure–121°C. While still hot, aseptically add 25.0mL of the sterile glucose solution under 97% N_2 + 3% H_2. Adjust pH to 7.3 if necessary. Aseptically and anaerobically distribute into tubes. Cap with rubber stoppers.

Use: For the cultivation and maintenance of *Thermoanaerobium brockii.*

Thermoanaeromonas Medium (DSMZ Medium 963)

Composition per liter:

$NaHCO_3$ 5.0g
K_2HPO_4 .... 0.78g
KH_2PO_4 .... 0.75g
NH_4Cl 0.5g
Glucose 0.5g
Yeast extract 0.5g
$MgSO_4 \cdot 7H_2O$ 0.25g

NaCl 0.2g
Na_3EDTA 0.04g
$CaCl_2 \cdot 2H_2O$ 0.03g
$FeSO_4 \cdot 7H_2O$ 0.01g
Resazurin 0.5mg
Thiosulfate solution 20.0mL
Vitamin solution 10.0mL
Trace elements solution 10.0mL
Cysteine solution 10.0mL
pH 6.5 ± 0.2 at 25°C

Vitamin Solution:
Composition per liter:
Pyridoxine-HCl 10.0mg
Thiamine-$HCl \cdot 2H_2O$ 5.0mg
Riboflavin 5.0mg
Nicotinic acid 5.0mg
D-Ca-pantothenate 5.0mg
p-Aminobenzoic acid 5.0mg
Lipoic acid 5.0mg
Biotin 2.0mg
Folic acid 2.0mg
Vitamin B_{12} 0.1mg

Preparation of Vitamin Solution: Add components to distilled/deionized water and bring volume to 1.0L. Mix thoroughly. Sparge with 80% H_2 + 20% CO_2. Filter sterilize.

Trace Elements Solution:
Composition per liter:
$MgSO_4 \cdot 7H_2O$ 3.0g
Nitrilotriacetic acid 1.5g
NaCl 1.0g
$MnSO_4 \cdot 2H_2O$ 0.5g
$CoSO_4 \cdot 7H_2O$ 0.18g
$ZnSO_4 \cdot 7H_2O$ 0.18g
$CaCl_2 \cdot 2H_2O$ 0.1g
$FeSO_4 \cdot 7H_2O$ 0.1g
$NiCl_2 \cdot 6H_2O$ 0.025g
$KAl(SO_4)_2 \cdot 12H_2O$ 0.02g
H_3BO_3 .... 0.01g
$Na_2MoO_4 \cdot 4H_2O$ 0.01g
$CuSO_4 \cdot 5H_2O$ 0.01g
$Na_2SeO_3 \cdot 5H_2O$ 0.3mg

Preparation of Trace Elements Solution: Add nitrilotriacetic acid to 500.0mL of distilled/deionized water. Dissolve by adjusting pH to 6.5 with KOH. Add remaining components. Add distilled/deionized water to 1.0L. Mix thoroughly.

Cysteine Solution:
Composition per 10.0mL:
L-Cysteine·$HCl \cdot H_2O$ 0.25g

Preparation of Cysteine Solution: Add L-cysteine·$HCl \cdot H_2O$ to distilled/deionized water and bring volume to 10.0mL. Mix thoroughly. Sparge with 100% N_2. Autoclave for 15 min at 15 psi pressure–121°C.

Thiosulfate Solution:
Composition per 20mL:
$Na_2S_2O_3 \cdot 5H_2O$ 1.24g

Preparation of Thiosulfate Solution: Add 1.24g of $Na_2S_2O_3 \cdot 5H_2O$ to distilled/deionized water and bring volume to 20.0mL. Mix thoroughly. Autoclave for 15 min at 15 psi pressure–121°C. Cool to room temperature.

Preparation of Medium: Add components, except $NaHCO_3$, vitamin solution, thiosulfate solution, and cysteine solution, to distilled/deionized water and bring volume to 960.0mL. Mix thoroughly. Gently heat and bring to boiling. Boil for 3 min. Cool to 25°C while sparging with 80% N_2 + 20% CO_2. Add solid $NaHCO_3$. Adjust pH to 6.8-7.0. Distribute into anaerobe tubes or bottles under 80% N_2 + 20% CO_2. Autoclave for 15 min at 15 psi pressure–121°C. Aseptically and anaerobically add per liter of medium, 10.0mL sterile vitamin solution, 10.0mL sterile cysteine solution, and 20.0mL sterile thiosulfate solution. Mix thoroughly. The final pH should be 6.5.

Use: For the cultivation of *Thermoanaeromonas toyohensis.*

Thermoanaerovibrio Medium (DSMZ Medium 873)

Composition per liter:
$NaHCO_3$ 0.8g
NH_4Cl 0.33g
KH_2PO_4 .... 0.33g
$MgCl_2 \cdot 6H_2O$ 0.33g
$CaCl_2 \cdot 2H_2O$ 0.22g
KCl 0.33g
Yeast extract 0.25g
Peptone 0.25g
Resazurin 0.5mg
$NaHCO_3$ solution 20.0mL
$Na_2S \cdot 9H_2O$ solution 10.0mL
Vitamin solution 10.0mL
Glucose solution 10.0mL
Calcium chloride solution 10.0mL
Magnesium chloride solution 10.0mL
Trace elements solution SL-10 .... 1.0mL
pH 7.0-7.3 at 25°C

$NaHCO_3$ Solution:
Composition per 20.0mL:
$NaHCO_3$ 2.0g

Preparation of $NaHCO_3$ Solution: Add $NaHCO_3$ to distilled/deionized water and bring volume to 20.0mL. Mix thoroughly. Sparge with 80% N_2 + 20% CO_2. Filter sterilize.

Magnesium Chloride Solution:
Composition per 10.0mL:
$MgCl_2 \cdot 6H_2O$ 0.33g

Preparation of Magnesium Chloride Solution: Add $MgCl_2 \cdot 6H_2O$ to distilled/deionized water and bring volume to 10.0mL. Mix thoroughly. Sparge with 100% N_2. Filter sterilize.

Calcium Chloride Solution:
Composition per 10.0mL:
$CaCl_2 \cdot 2H_2O$ 0.22g

Preparation of Calcium Chloride Solution: Add $CaCl_2 \cdot 2H_2O$ to distilled/deionized water and bring volume to 10.0mL. Mix thoroughly. Sparge with 100% N_2. Filter sterilize.

Glucose Solution:
Composition per 10.0mL:
Glucose 3.0g

Preparation of Glucose Solution: Add glucose to distilled/deionized water and bring volume to 10.0mL. Mix thoroughly. Sparge with 100% N_2. Filter sterilize.

$Na_2S \cdot 9H_2O$ Solution:
Composition per 10.0mL:
$Na_2S \cdot 9H_2O$ 0.5g

Preparation of $Na_2S{\cdot}9H_2O$ Solution: Add $Na_2S{\cdot}9H_2O$ to distilled/deionized water and bring volume to 10.0mL. Mix thoroughly. Sparge with 100% N_2. Autoclave for 15 min at 15 psi pressure–121°C. Before use, neutralize to pH 7.0 with sterile HCl.

Vitamin Solution:
Composition per liter:

Pyridoxine-HCl	10.0mg
Thiamine-HCl·$2H_2O$	5.0mg
Riboflavin	5.0mg
Nicotinic acid	5.0mg
D-Ca-pantothenate	5.0mg
p-Aminobenzoic acid	5.0mg
Lipoic acid	5.0mg
Biotin	2.0mg
Folic acid	2.0mg
Vitamin B_{12}	0.1mg

Preparation of Vitamin Solution: Add components to distilled/deionized water and bring volume to 1.0L. Mix thoroughly. Sparge with 80% H_2 + 20% CO_2. Filter sterilize.

Trace Elements Solution SL-10:
Composition per liter:

$FeCl_2{\cdot}4H_2O$	1.5g
$CoCl_2{\cdot}6H_2O$	190.0mg
$MnCl_2{\cdot}4H_2O$	100.0mg
$ZnCl_2$	70.0mg
$Na_2MoO_4{\cdot}2H_2O$	36.0mg
$NiCl_2{\cdot}6H_2O$	24.0mg
H_3BO_3	6.0mg
$CuCl_2{\cdot}2H_2O$	2.0mg
HCl (25% solution)	10.0mL

Preparation of Trace Elements Solution SL-10: Add $FeCl_2{\cdot}4H_2O$ to 10.0mL of HCl solution. Mix thoroughly. Add distilled/deionized water and bring volume to 1.0L. Add remaining components. Mix thoroughly. Sparge with 80% N_2 + 20% CO_2. Autoclave for 15 min at 15 psi pressure–121°C.

Preparation of Medium: Prepare and dispense medium under 80% N_2 + 20% CO_2 gas atmosphere. Add components, except $NaHCO_3$ solution, glucose solution, calcium chloride solution, magnesium chloride solution, $Na_2S{\cdot}9H_2O$ solution, vitamin solution, and trace elements solution SL-10, to distilled/deionized water and bring volume to 929.0mL. Mix thoroughly. Sparge with 80% N_2 + 20% CO_2. Autoclave for 15 min at 15 psi pressure–121°C. Aseptically and anaerobically add 10.0mL glucose solution, 10.0mL $Na_2S{\cdot}9H_2O$ solution, 10.0mL magnesium chloride solution, 10.0ml calcium chloride solution, 10.0mL vitamin solution, and 1.0mL trace elements solution SL-10. Mix thoroughly. Adjust pH to 7.0-7.3 with 10.0mL $NaHCO_3$ solution. Aseptically and anaerobically distribute into sterile tubes or bottles.

Use: For the cultivation of *Thermoanaerovibrio velox (Thermosinus velox).*

Thermobacterium Medium

Composition per liter:

Agar	20.0g
$(NH_4)_2SO_4$	1.3g
Yeast extract	1.0g
Pancreatic digest of casein	1.0g
KH_2PO_4	0.28g
$MgSO_4{\cdot}7H_2O$	0.247g
$CaCl_2{\cdot}2H_2O$	0.074g
$FeCl_3.6H_2O$	0.019g
Salt solution	1.0mL

pH 8.5 + 0.2 at 25°C

Salt Solution:
Composition per liter:

$Na_2B_4O_7{\cdot}10H_2O$	4.4g
$MnCl_2{\cdot}4H_2O$	1.8g
$ZnSO_4{\cdot}7H_2O$	0.22g
$CuCl_2{\cdot}H_2O$	0.05g
$Na_2MoO_4.2H_2O$	0.03g
$VOSO_4{\cdot}2H_2O$	0.03g

Preparation of Salt Solution: Add components to distilled/deionized water and bring volume to 1.0L. Mix thoroughly. Adjust pH to 2.0 with H_2SO_4.

Preparation of Medium: Add components to distilled/deionized water and bring volume to 1.0L. Mix thoroughly. Gently heat and bring to boiling. Distribute into tubes in 11.0–12.0mL volumes. Autoclave for 15 min at 15 psi pressure–121°C. Allow tubes to solidify in a slanted position.

Use: For the cultivation and maintenance of *Thermomicrobium roseum.*

Thermobacteroides leptospartum Medium

Composition per 1168.1mL:

Yeast extract	2.0g
Trypticase	2.0g
NaOH solution	1.0L
Glucose solution	113.0mL
Na_2S solution	22.6mL
Solution A	10.0mL
Mineral salts solution	10.0mL
Solution B	2.0mL
Resazurin solution	0.5mL

NaOH Solution:
Composition per liter:

NaOH	4.0g

Preparation of NaOH Solution: Add NaOH to distilled/deionized water and bring volume to 1.0L. Mix thoroughly.

Glucose Solution:
Composition per 100.0mL:

D-Glucose	5.0g

Preparation of Glucose Solution: Add glucose to distilled/deionized water and bring volume to 100.0mL. Mix thoroughly. Sparge with 100% N_2 for 15 min. Autoclave for 15 min at 15 psi pressure–121°C.

Na_2S Solution:

Na_2S	2.5g

Preparation of Na_2S Solution: Gently heat 100.0mL of distilled/deionized water to 100°C. Boil for 5 min. Sparge with 100% N_2 for 15 min. Add the Na_2S. Mix thoroughly. Sparge with 100% N_2 for 10 min. Autoclave for 15 min at 15 psi pressure–121°C.

Solution A:
Composition per liter:

NH_4Cl	100.0g
$MgCl_2{\cdot}H_2O$	100.0g
$CaCl_2{\cdot}2H_2O$	40.0g

Preparation of Solution A: Add components to distilled/deionized water and bring volume to 1.0L. Mix thoroughly. Adjust pH to 4 with HCl.

Mineral Salts Solution:
Composition per liter:

$EDTA \cdot 2H_2O$	0.5g
$CoCl_2 \cdot H_2O$	0.15g
$MnCl_2 \cdot 4H_2O$	0.1g
$FeSO_4 \cdot 7H_2O$	0.1g
$ZnCl_2$	0.1g
$AlCl_3 \cdot H_2O$	40.0mg
$Na_2WO_4 \cdot 2H_2O$	30.0mg
$CuCl_2 \cdot 2H_2O$	20.0mg
$NiSO_4 \cdot H_2O$	20.0mg
H_2SeO_3	10.0mg
H_3BO_4	10.0mg
$NaMoO_4 \cdot 2H_2O$	10.0mg

Preparation of Mineral Salts Solution: Add components to distilled/deionized water and bring volume to 1.0L. Mix thoroughly. Adjust pH to 3 with HCl.

Solution B:
Composition per liter:

$K_2HPO_4 \cdot 3H_2O$	200.0g

Preparation of Solution B: Add $K_2HPO_4 \cdot 3H_2O$ to distilled/deionized water and bring volume to 1.0L. Mix thoroughly.

Resazurin Solution:
Composition per 100.0mL:

Resazurin	0.2g

Preparation of Resazurin Solution: Add resazurin to distilled/deionized water and bring volume to 100.0mL. Mix thoroughly.

Preparation of Medium: Sparge 1.0L of NaOH solution with 100% CO_2 for 30 min. Add 2.0g of yeast extract and 2.0g of Trypticase. Mix thoroughly. Add 10.0mL of solution A, 2.0mL of solution B, 0.5mL of resazurin solution, and 10.0mL of mineral salts solution with pipets which have been flushed a few times with 100% N_2. Mix thoroughly. Anaerobically distribute 9.0mL volumes into anaerobic tubes fitted with butyl rubber stoppers. Autoclave for 15 min at 15 psi pressure–121°C. One hr prior to inoculation, add 1.0mL of sterile glucose solution and 0.2mL of sterile Na_2S solution to each 9.0mL of medium.

Use: For the cultivation of *Thermobacteroides leptospartum.*

Thermobacteroides proteolyticus Medium

Composition per 1010.0mL:

$NaHCO_3$	5.0g
Pancreatic digest of casein	2.0g
Yeast extract	2.0g
$MgCl_2 \cdot 6H_2O$	1.0g
NH_4Cl	1.0g
$CaCl_2 \cdot 2H_2O$	0.4g
K_2HPO_4	0.4g
$Na_2S \cdot 9H_2O$	0.3g
Resazurin	1.0mg
Trace elements solution	10.0mL

pH 7.0 ± 0.2 at 25°C

Trace Elements Solution:
Composition per liter:

$MgSO_4 \cdot 7H_2O$	3.0g
Nitrilotriacetic acid	1.5g
NaCl	1.0g
$MnSO_4 \cdot 2H_2O$	0.5g
$CoSO_4 \cdot 7H_2O$	0.18g
$ZnSO_4 \cdot 7H_2O$	0.18g
$CaCl_2 \cdot 2H_2O$	0.1g
$FeSO_4 \cdot 7H_2O$	0.1g
$KAl(SO_4)_2 \cdot 12H_2O$	0.02g
$CuSO_4 \cdot 5H_2O$	0.01g
H_3BO_3	0.01g
$Na_2MoO_4 \cdot 2H_2O$	0.01g
$NiCl_2 \cdot 6H_2O$	0.025g
$Na_2SeO_3 \cdot 5H_2O$	0.3mg

Preparation of Trace Elements Solution: Add nitrilotriacetic acid to 500.0mL of distilled/deionized water. Adjust pH to 6.5 with KOH. Add remaining components. Add distilled/deionized water to 1.0L.

Preparation of Medium: Prepare and dispense medium under 80% N_2 + 20% CO_2. Add components to distilled/deionized water and bring volume to 1.0L. Mix thoroughly. Distribute into tubes or flasks. Autoclave for 15 min at 15 psi pressure–121°C.

Use: For the cultivation and maintenance of *Thermobacteroides proteolyticus.*

Thermococcus celer Medium

Composition per liter:

NaCl	40.0g
Sulfur	10.0g
Yeast extract	2.0g
$(NH_4)_2SO_4$	1.3g
KH_2PO_4	0.28g
$MgSO_4 \cdot 7H_2O$	0.25g
$CaCl_2 \cdot 2H_2O$	0.07g
$FeCl_2 \cdot 2H_2O$	0.02g
$NaB_4O \cdot 10H_2O$	4.5mg
$MnCl_2 \cdot 4H_2O$	1.8mg
Resazurin	1.0mg
$ZnSO_4 \cdot 7H_2O$	0.22mg
$CuCl_2 \cdot 2H_2O$	0.05mg
$NaMoO_4 \cdot 2H_2O$	0.03mg
$VOSO_4 \cdot 2H_2O$	0.03mg
$CoSO_4$	0.01mg
$Na_2S \cdot 9H_2O$ solution	10.0mL

pH 5.8 ± 0.2 at 25°C

$Na_2S \cdot 9H_2O$ Solution:
Composition per 10.0mL:

$Na_2S \cdot 9H_2O$	0.3g

Preparation of $Na_2S \cdot 9H_2O$ Solution: Add $Na_2S \cdot 9H_2O$ to distilled/deionized water and bring volume to 10.0mL. Mix thoroughly. Sparge with 100% N_2. Autoclave for 15 min at 15 psi pressure–121°C.

Preparation of Medium: Add components, except $Na_2S \cdot 9H_2O$ solution, to distilled/deionized water and bring volume to 1.0L. Mix thoroughly. Adjust pH to 5.8. Do not autoclave. Sterilize by steaming at 100°C for 30 min on 3 consecutive days. Before inoculation, add 10.0mL of sterile $Na_2S \cdot 9H_2O$ solution. Mix thoroughly.

Use: For the cultivation of *Thermococcus celer.*

Thermococcus chitinophagus Medium (DSMZ Medium 766)

Composition per liter:

Chitin, purified	4.0g
$(NH_4)_2SO_4$	0.5g
KH_2PO_4	0.5g
Resazurin	1.0mg
$(NH_4)_2Ni(SO_4)_2$	0.3mg
$Na_2WO_4 \cdot 2H_2O$	0.15mg
Na_2SeO_4	0.15mg

Synthetic seawater 485.0mL
Trace elements SL-6 .... 15.0mL
$Na_2S·9H_2O$ solution 10.0mL
$NaHCO_3$ solution 10.0mL
pH 6.7 ± 0.2 at 25°C

$Na_2S·9H_2O$ Solution:
Composition per 10mL:
$Na_2S·9H_2O$ 0.5g

Preparation of $Na_2S·9H_2O$ Solution: Add $Na_2S·9H_2O$ to distilled/deionized water and bring volume to 10.0mL. Mix thoroughly. Autoclave under 100% N_2 for 15 min at 15 psi pressure–121°C. Cool to room temperature.

$NaHCO_3$ Solution:
Composition per 10.0mL:
$NaHCO_3$ 0.2g

Preparation of $NaHCO_3$ Solution: Add $NaHCO_3$ to distilled/deionized water and bring volume to 10.0mL. Mix thoroughly. Sparge with 80% N_2 + 20% CO_2. Filter sterilize.

Trace Elements Solution SL-6:
Composition per liter:
$MnCl_2·4H_2O$ 0.5g
H_3BO_3 .... 0.3g
$CoCl_2·6H_2O$ 0.2g
$ZnSO_4·7H_2O$ 0.1g
$Na_2MoO_4·2H_2O$ 0.03g
$NiCl_2·6H_2O$ 0.02g
$CuCl_2·2H_2O$ 0.01g

Preparation of Trace Elements Solution SL-6: Add components to distilled/deionized water and bring volume to 1.0L. Mix thoroughly. Autoclave for 15 min at 15 psi pressure–121°C.

Synthetic Seawater:
Composition per liter:
NaCl 23.477g
$MgCl_2·6H_2O$ 4.981g
Na_2SO_4 .... 3.917g
$CaCl_2$ 1.12g
KCl 664.0mg
$NaHCO_3$ 192.0mg
H_3BO_3 .... 26.0mg
$SrCl_2$ 24.0mg
KBr 6.0mg
NaF 3.0mg

Preparation of Synthetic Seawater: Add components to distilled/deionized water and bring volume to 1.0L. Mix thoroughly. Filter sterilize.

Preparation of Purified Chitin: Cool 200.0mL of 37% HCl to 4°C. Add 20.0g chitin (practical grade from crab shells) to the cooled HCl. Mix thoroughly. Stir for 60 min at 4°C. Pour the suspension into 1 liter of distilled water, precooled to 4°C. Filter through filter paper. Wash the residue 5 times with 500mL distilled water. Resuspend in 1L of distilled water. Neutralize the suspension with 10mL of 5*M* KOH to achieve a final pH 6.5. Filter and wash with 3L of distilled water to remove KCl. Allow to air dry.

Preparation of Medium: Add components, except $NaHCO_3$ solution and $Na_2S·9H_2O$ solution, to distilled/deionized water and bring volume to 980.0mL. Genly heat and bring to boiling. Boil for 5 min. Cool to 25°C while sparging with 100% N_2. Add 10.0mL $NaHCO_3$ solution. Adjust pH to 6.7. Distribute the medium into Hungate tubes under an atmosphere of 100% N_2. Autoclave for 15 min at 15 psi pressure–121°C. Cool to room temperature. Reduce the medium by adding 10.0mL $Na_2S·9H_2O$ solution. Aseptically and anaerobically distribute into sterile tubes or bottles.

Use: For the cultivation of *Thermococcus chitinophagus.*

Thermococcus litoralis Medium

Composition per liter:
NaCl 25.0g
Sulfur 10.0g
$(NH_4)_2SO_4$ 1.3g
Yeast extract 1.0g
Peptone 0.5g
KH_2PO_4 .... 0.28g
$MgSO_4·7H_2O$ 0.25g
$CaCl_2·2H_2O$ 0.07g
$FeCl_2·2H_2O$ 0.02g
$NaB_4O·10H_2O$ 4.5mg
$MnCl_2·4H_2O$ 1.8mg
Resazurin 1.0mg
$ZnSO_4·7H_2O$ 0.22mg
$CuCl_2·2H_2O$ 0.05mg
$NaMoO_4·2H_2O$ 0.03mg
$VOSO_4·2H_2O$ 0.03mg
$CoSO_4$ 0.01mg
$Na_2S·9H_2O$ solution 10.0mL
pH 7.2 ± 0.2 at 25°C

$Na_2S·9H_2O$ Solution:
Composition per 10.0mL:
$Na_2S·9H_2O$ 0.3g

Preparation of $Na_2S·9H_2O$ Solution: Add $Na_2S·9H_2O$ to distilled/deionized water and bring volume to 10.0mL. Mix thoroughly. Sparge with 100% N_2. Autoclave for 15 min at 15 psi pressure–121°C.

Preparation of Medium: Add components, except $Na_2S·9H_2O$ solution, to distilled/deionized water and bring volume to 1.0L. Mix thoroughly. Adjust pH to 7.2. Do not autoclave. Sterilize by steaming at 100°C for 30 min on 3 consecutive days. Before inoculation, add 10.0mL of sterile $Na_2S·9H_2O$ solution. Mix thoroughly.

Use: For the cultivation of *Thermococcus litoralis.*

Thermococcus Medium (DSMZ Medium 806)

Composition per liter:
NaCl 18.0g
Sulfur 5.0g
$MgSO_4·7H_2O$ 3.4g
$MgCl_2·2H_2O$ 2.7g
Yeast extract 1.0g
Trypticase 1.0g
$NaHCO_3$ 1.0g
KCl 0.33g
NH_4Cl 0.25g
$CaCl_2·2H_2O$ 0.14g
K_2HPO_4 .... 0.14g
Resazurin 0.001g
Na_2SeO_3 .... 0.001mg
$NiCl_2·6H_2O$ 0.001mg
Trace elements solution 10.0mL
Vitamin solution 10.0mL
Cysteine solution 10.0mL
$Na_2S·9H_2O$ solution 10.0mL
pH 7.2 ± 0.2 at 25°C

$Na_2S \cdot 9H_2O$ Solution:
Composition per 10mL:
$Na_2S \cdot 9H_2O$ 0.5g

Preparation of $Na_2S \cdot 9H_2O$ Solution: Add $Na_2S \cdot 9H_2O$ to distilled/deionized water and bring volume to 10.0mL. Mix thoroughly. Autoclave under 100% N_2 for 15 min at 15 psi pressure–121°C. Cool to room temperature. Adjust pH to 7.0.

Cysteine Solution:
Composition per 10.0mL:
L-Cysteine·HCl·H_2O 0.3g

Preparation of Cysteine Solution: Add L-cysteine·HCl·H_2O to distilled/deionized water and bring volume to 10.0mL. Mix thoroughly. Sparge with 100% N_2. Autoclave for 15 min at 15 psi pressure–121°C. Cool to room temperature.

Trace Elements Solution:
Composition per liter:
$MgSO_4 \cdot 7H_2O$ 3.0g
Nitrilotriacetic acid 1.5g
NaCl 1.0g
$MnSO_4 \cdot 2H_2O$ 0.5g
$CoSO_4 \cdot 7H_2O$ 0.18g
$ZnSO_4 \cdot 7H_2O$ 0.18g
$CaCl_2 \cdot 2H_2O$ 0.1g
$FeSO_4 \cdot 7H_2O$ 0.1g
$NiCl_2 \cdot 6H_2O$ 0.025g
$KAl(SO_4)_2 \cdot 12H_2O$ 0.02g
H_3BO_3 .... 0.01g
$Na_2MoO_4 \cdot 4H_2O$ 0.01g
$CuSO_4 \cdot 5H_2O$ 0.01g
$Na_2SeO_3 \cdot 5H_2O$ 0.3mg

Preparation of Trace Elements Solution: Add nitrilotriacetic acid to 500.0mL of distilled/deionized water. Dissolve by adjusting pH to 6.5 with KOH. Add remaining components. Add distilled/deionized water to 1.0L. Mix thoroughly.

Vitamin Solution:
Composition per liter:
Pyridoxine-HCl 10.0mg
Thiamine-HCl·$2H_2O$ 5.0mg
Riboflavin 5.0mg
Nicotinic acid 5.0mg
D-Ca-pantothenate 5.0mg
p-Aminobenzoic acid 5.0mg
Lipoic acid 5.0mg
Biotin 2.0mg
Folic acid 2.0mg
Vitamin B_{12} 0.1mg

Preparation of Vitamin Solution: Add components to distilled/deionized water and bring volume to 1.0L. Mix thoroughly. Sparge with 100% N_2. Filter sterilize.

Preparation of Medium: Add components, except Fildes enrichment solution, $NaHCO_3$, vitamin solution, cysteine solution, and $Na_2S \cdot 9H_2O$ solution, to distilled/deionized water and bring volume to 970.0mL. Mix thoroughly. Gently heat and bring to boiling. Boil for 5 min. Cool to room temperature under 100% N_2. Add 1.0g solid $NaHCO_3$. Adjust pH to 7.2. Sterilize at 100°C for 3 h on 3 consecutive days. Aseptically and anaerobically add 10.0mL vitamin solution, 10.0mL cysteine solution, and 10.0mL $Na_2S \cdot 9H_2O$ solution. Mix thoroughly. Adjust pH to 7.2. Aseptically and anaerobically distribute into sterile tubes or bottles.

Use: For the cultivation of *Thermococcus* spp.

Thermococcus profundus Medium

Composition per 1010.0mL:
NaCl 25.0g
Sulfur 10.0g
Peptone 5.0g
Yeast extract 1.0g
Resazurin 1.0mg
Salt base solution 1.0L
$Na_2S \cdot 9H_2O$ solution 10.0mL
pH 7.2 ± 0.2 at 25°C

Preparation of Sulfur: Sterilize powdered elemental sulfur by steaming for 3 hr at 0 psi pressure–100°C on 3 successive days.

$Na_2S \cdot 9H_2O$ Solution:
Composition per 10.0mL:
$Na_2S \cdot 9H_2O$ 0.5g

Preparation of $Na_2S \cdot 9H_2O$ Solution: Add $Na_2S \cdot 9H_2O$ to distilled/deionized water and bring volume to 10.0mL. Mix thoroughly. Sparge with 100% N_2. Autoclave for 15 min at 15 psi pressure–121°C. Before use, neutralize to pH 7.0 with sterile HCl.

Salt Base Solution:
Composition per liter:
$(NH_4)_2SO_4$ 1.3g
KH_2PO_4 .... 0.28g
$MgSO_4 \cdot 7H_2O$ 0.25g
$CaCl_2 \cdot 2H_2O$ 0.07g
$FeCl_3 \cdot 6H_2O$ 0.02g
$Na_2B_4O_7 \cdot 10H_2O$ 4.5mg
$MnCl_2 \cdot 4H_2O$ 1.8mg
$ZnSO_4 \cdot 7H_2O$ 0.22mg
$CuCl_2 \cdot 2H_2O$ 0.05mg
$Na_2MoO_4 \cdot 2H_2O$ 0.03mg
$VOSO_4$ 0.03mg
$CoSO_4 \cdot 7H_2O$ 0.02mg

Preparation of Salt Base Solution: Add components to distilled/deionized water and bring volume to 1.0L. Mix thoroughly. Sparge with 100% N_2.

Preparation of Medium: Prepare and dispense medium under 100% N_2. Add components, except $Na_2S \cdot 9H_2O$ solution, to salt base solution and bring volume to 1.0L. Mix thoroughly. Adjust medium pH to 7.2 with H_2SO_4. Autoclave for 15 min at 15 psi pressure–121°C. Immediately prior to use, aseptically and anaerobically add 10.0mL of sterile $Na_2S \cdot 9H_2O$ solution. Mix thoroughly. Aseptically and anaerobically distribute into sterile tubes or bottles.

Use: For the cultivation of *Thermococcus profundus.*

Thermococcus stetteri Medium

Composition per liter:
NaCl 25.0g
Sulfur 10.0g
Yeast extract 3.0g
$(NH_4)_2SO_4$ 1.3g
Peptone 0.5g
KH_2PO_4 .... 0.28g
$MgSO_4 \cdot 7H_2O$ 0.25g
$CaCl_2 \cdot 2H_2O$ 0.07g
$FeCl_2 \cdot 2H_2O$ 0.02g
$NaB_4O \cdot 10H_2O$ 4.5mg
$MnCl_2 \cdot 4H_2O$ 1.8mg
Resazurin 1.0mg
$ZnSO_4 \cdot 7H_2O$ 0.22mg

$CuCl_2 \cdot 2H_2O$ 0.05mg
$NaMoO_4 \cdot 2H_2O$ 0.03mg
$VOSO_4 \cdot 2H_2O$ 0.03mg
$CoSO_4$ 0.01mg
$Na_2S \cdot 9H_2O$ solution 10.0mL

pH 6.5 ± 0.2 at 25°C

$Na_2S \cdot 9H_2O$ Solution:
Composition per 10.0mL:
$Na_2S \cdot 9H_2O$ 0.3g

Preparation of $Na_2S \cdot 9H_2O$ Solution: Add $Na_2S \cdot 9H_2O$ to distilled/deionized water and bring volume to 10.0mL. Mix thoroughly. Sparge with 100% N_2. Autoclave for 15 min at 15 psi pressure–121°C.

Preparation of Medium: Add components, except $Na_2S \cdot 9H_2O$ solution, to distilled/deionized water and bring volume to 1.0L. Mix thoroughly. Adjust pH to 6.5. Do not autoclave. Sterilize by steaming at 100°C for 30 min on 3 consecutive days. Before inoculation, add 10.0mL of sterile $Na_2S \cdot 9H_2O$ solution. Mix thoroughly.

Use: For the cultivation of *Thermococcus stetteri.*

Thermodesulfobacterium Medium

Composition per liter:
Na_2SO_4 .... 3.0g
$Na_2HPO_4 \cdot 12H_2O$ 2.0g
NH_4Cl 1.0g
KH_2PO_4 .... 0.3g
$MgCl_2 \cdot 6H_2O$ 0.2g
$FeSO_4 \cdot 7H_2O$ 1.5mg
Resazurin 1.0mg
Sodium lactate solution 20.0mL
Trace elements solution 10.0mL
Yeast extract solution 10.0mL
$Na_2S \cdot 9H_2O$ solution 10.0mL
Vitamin solution 5.0mL

pH 6.8 ± 0.2 at 25°C

Sodium Lactate Solution:
Composition per 20.0mL:
Sodium lactate 4.0g

Preparation of Sodium Lactate Solution: Add sodium lactate to distilled/deionized water and bring volume to 20.0mL. Mix thoroughly. Sparge with 100% N_2. Autoclave for 15 min at 15 psi pressure–121°C.

Trace Elements Solution:
Composition per liter:
Nitrilotriacetic acid 12.8g
NaCl 1.0g
$FeCl \cdot 4H_2O$ 0.20
$CoCl_2 \cdot 6H_2O$ 0.17g
$CaCl_2 \cdot 2H_2O$ 0.1g
$MnCl_2 \cdot 4H_2O$ 0.1g
$ZnCl_2$ 0.1g
$CuCl_2$ 0.02g
H_3BO_3 .... 0.01g
$Na_2MoO_4 \cdot 2H_2O$ 0.01g
$NiCl_2 \cdot 6H_2O$ 0.026g
$Na_2SeO_3 \cdot 5H_2O$ 0.02g

Preparation of Trace Elements Solution: Add nitrilotriacetic acid to 500.0mL of distilled/deionized water. Adjust pH to 6.5 with KOH. Add remaining components. Add distilled/deionized water to 1.0L.

Yeast Extract Solution:
Composition per 10.0mL:
Yeast extract 1.0g

Preparation of Yeast Extract Solution: Add yeast extract to distilled/deionized water and bring volume to 10.0mL. Mix thoroughly. Sparge with 100% N_2. Autoclave for 15 min at 15 psi pressure–121°C.

$Na_2S \cdot 9H_2O$ Solution:
Composition per 10.0mL:
$Na_2S \cdot 9H_2O$ 0.5g

Preparation of $Na_2S \cdot 9H_2O$ Solution: Add $Na_2S \cdot 9H_2O$ to distilled/deionized water and bring volume to 10.0mL. Mix thoroughly. Sparge with 100% N_2. Autoclave for 15 min at 15 psi pressure–121°C. Prior to use, neutralize solution by dropwise addition of sterile 1*N* HCl.

Vitamin Solution:
Composition per liter:
Pyridoxine·HCl 10.0mg
Calcium DL-pantothenate 5.0mg
Lipoic acid 5.0mg
Nicotinic acid 5.0mg
p-Aminobenzoic acid 5.0mg
Riboflavin 5.0mg
Thiamine·HCl 5.0mg
Biotin 2.0mg
Folic acid 2.0mg
Vitamin B_{12} 0.1mg

Preparation of Vitamin Solution: Add components to distilled/deionized water and bring volume to 1.0L. Mix thoroughly.

Preparation of Medium: Prepare and dispense medium under 100% N_2. Add components, except sodium lactate solution, yeast extract solution, and $Na_2S \cdot 9H_2O$ solution, to distilled/deionized water and bring volume to 960.0mL. Mix thoroughly. Sparge with 100% N_2. Anaerobically distribute into tubes or bottles. Autoclave for 15 min at 15 psi pressure–121°C. Aseptically and anaerobically add to 1.0L of medium 20.0mL of sterile sodium lactate solution, 10.0mL of sterile yeast extract solution, and 10.0mL of sterile $Na_2S \cdot 9H_2O$ solution. Mix thoroughly.

Use: For the cultivation and maintenance of *Thermodesulfobacterium commune.*

Thermodesulforhabdus Medium

Composition per liter:
NaCl 10.0g
Na_2SO_4 .... 7.0g
Sodium acetate·$3H_2O$ 6.2g
$MgCl_2 \cdot 6H_2O$ 3.0g
KH_2PO_4 .... 1.0g
NH_4Cl 0.25g
$CaCl_2 \cdot 2H_2O$ 0.15g
Resazurin 0.5mg
$Na_2S \cdot 9H_2O$ solution 10.0mL
$Na_2S_2O_4$ solution 10.0mL
Trace elements solution SL-10 .... 1.0mL
$NaHCO_3$ solution .... variable

pH 6.8 ± 0.2 at 25°C

$NaHCO_3$ Solution:
Composition per 20.0mL:
$NaHCO_3$ 1.0g

Preparation of $NaHCO_3$ Solution: Add $NaHCO_3$ to distilled/deionized water and bring volume to 20.0mL. Mix thoroughly. Sparge with 80% N_2 + 20% CO_2. Autoclave for 15 min at 15 psi pressure–121°C.

$Na_2S \cdot 9H_2O$ Solution:
Composition per 10.0mL:
$Na_2S \cdot 9H_2O$ 0.15g

Preparation of $Na_2S \cdot 9H_2O$ Solution: Add $Na_2S \cdot 9H_2O$ to distilled/deionized water and bring volume to 10.0mL. Mix thoroughly. Sparge with 100% N_2. Autoclave for 15 min at 15 psi pressure–121°C.

$Na_2S_2O_3$ Solution:
Composition per 10.0mL:
$Na_2S_2O_3 \cdot 5H_2O$ 2.0g

Preparation of $Na_2S_2O_3$ Solution: Add $Na_2S_2O_3 \cdot 5H_2O$ to distilled/deionized water and bring volume to 10.0mL. Mix thoroughly. Sparge with 100% N_2. Autoclave for 15 min at 15 psi pressure–121°C.

Trace Elements Solution SL-10:
Composition per liter:

$FeCl_2 \cdot 4H_2O$	1.5g
$CoCl_2 \cdot 6H_2O$	190.0mg
$MnCl_2 \cdot 4H_2O$	100.0mg
$ZnCl_2$	70.0mg
$Na_2MoO_4 \cdot 2H_2O$	36.0mg
$NiCl_2 \cdot 6H_2O$	24.0mg
H_3BO_3	6.0mg
$CuCl_2 \cdot 2H_2O$	2.0mg
HCl (25% solution)	10.0mL

Preparation of Trace Elements Solution SL-10: Add $FeCl_2 \cdot 4H_2O$ to 10.0mL of HCl solution. Mix thoroughly. Add distilled/deionized water and bring volume to 1.0L. Add remaining components. Mix thoroughly.

Preparation of Medium: Prepare and dispense medium under 100% N_2. Add components, except $Na_2S \cdot 9H_2O$ solution, $Na_2S_2O_3$ solution, and $NaHCO_3$ solution, to distilled/deionized water and bring volume to 980.0mL. Mix thoroughly. Gently heat and bring to boiling. Cool to room temperature while sparging with 100% N_2. Anaerobically distribute 9.8mL volumes into tubes. Autoclave for 15 min at 15 psi pressure–121°C. Aseptically and anaerobically add 0.1mL of sterile $Na_2S \cdot 9H_2O$ solution and 0.1mL of sterile $Na_2S_2O_4$ solution to each tube. Aseptically and anaerobically add a sufficient volume of sterile $NaHCO_3$ solution to each tube to bring the pH to 6.8.

Use: For the cultivation of *Thermodesulforhabdus norvegicus.*

Thermodesulfotobacterium Agar

Composition per liter:

Na_2SO_4	30.0g
Agar	20.0g
Sodium lactate	4.0g
Yeast extract	1.0g
Mineral solution 2	50.0mL
Na_2CO_3 solution	50.0mL
Mineral solution 1	25.0mL
Cysteine-sulfide reducing agent	20.0mL
Wolfe's mineral solution	10.0mL
Wolfe's vitamin solution	10.0mL
Resazurin (0.025% solution)	4.0mL

pH 7.2 ± 0.2 at 25°C

Mineral Solution 1:
Composition per liter:
K_2HPO_4 .... 6.0g

Preparation of Mineral Solution 1: Add K_2HPO_4 to distilled/deionized water and bring volume to 1.0L. Mix thoroughly.

Mineral Solution 2:
Composition per liter:

NaCl	12.0g
KH_2PO_4	6.0g
$(NH_4)_2SO_4$	6.0g
$MgSO_4 \cdot 7H_2O$	2.4g
$CaCl_2 \cdot 2H_2O$	1.6g

Preparation of Mineral Solution 2: Add components to distilled/deionized water and bring volume to 1.0L. Mix thoroughly.

Na_2CO_3 Solution:
Composition per 100.0mL:
Na_2CO_3 .... 8.0g

Preparation of Na_2CO_3 Solution: Add Na_2CO_3 to distilled/deionized water and bring volume to 100.0mL. Mix thoroughly.

Cysteine-Sulfide Reducing Agent:
Composition per 20.0mL:

L-Cysteine·HCl·H_2O	300.0mg
$Na_2S \cdot 9H_2O$	300.0mg

Preparation of Cysteine-Sulfide Reducing Agent: Add L-cysteine·HCl·H_2O to 10.0mL of distilled/deionized water. Mix thoroughly. In a separate tube, add $Na_2S \cdot 9H_2O$ to 10.0mL of distilled/deionized water. Mix thoroughly. Gas both solutions with 100% N_2 and cap tubes. Autoclave both solutions for 15 min at 15 psi pressure–121°C using fast exhaust. Cool to 50°C. Aseptically combine the two solutions under 100% N_2.

Wolfe's Mineral Solution:
Composition per liter

$MgSO_4 \cdot 7H_2O$	3.0g
Nitriloacetic acid	1.5g
NaCl	1.0g
$MnSO_4 \cdot H_2O$	0.5g
$FeSO_4 \cdot 7H_2O$	0.1g
$CoCl_2 \cdot 6H_2O$	0.1g
$CaCl_2$	0.1g
$ZnSO_4 \cdot 7H_2O$	0.1g
$CuSO_4 \cdot 5H_2O$	0.01g
$AlK(SO_4)_2 \cdot 12H_2O$	0.01g
H_3BO_3	0.01g
$Na_2MoO_4 \cdot 2H_2O$	0.01g

Preparation of Wolfe's Mineral Solution: Add nitrilotriacetic acid to 500.0mL of distilled/deionized water. Dissolve by adjusting pH to 6.5 with KOH. Add remaining components. Add distilled/deionized water to 1.0L.

Wolfe's Vitamin Solution:
Composition per liter:

Pyridoxine·HCl	10.0mg
Thiamine·HCl	5.0mg
Riboflavin	5.0mg
Nicotinic acid	5.0mg
Calcium pantothenate	5.0mg
p-Aminobenzoic acid	5.0mg
Thioctic acid	5.0mg
Biotin	2.0mg
Folic acid	2.0mg
Cyanocobalamin	100.0μg

Preparation of Wolfe's Vitamin Solution: Add components to distilled/deionized water and bring volume to 1.0L. Mix thoroughly. Filter sterilize.

Preparation of Medium: Add components, except vitamin solution and cysteine-sulfide reducing agent, to distilled/deionized water and bring volume to 970.0mL. Mix thoroughly. Gently heat and bring to boiling. Autoclave for 15 min at 15 psi pressure–121°C. Cool to 50°–55°C under 80% N_2 + 20% CO_2. Aseptically add the sterile vitamin solution and then the sterile cysteine-sulfide reducing agent. Adjust the pH to 7.2. Distribute aseptically and anaerobically into sterile tubes.

Use: For the cultivation and maintenance of *Thermodesulfobacterium commune* and other *Thermodesulfobacterium* species.

Thermodesulfovibrio yellowstonii Medium (DSMZ Medium 749)

Composition per liter:

Na_2SO_4	4.0g
Na-lactate	2.5g
$NaHCO_3$	1.3g
KCl	0.5g
Yeast extract	0.5g
$MgCl_2·6H_2O$	0.4g
NH_4Cl	0.25g
Na_2HPO_4	0.2g
Na-thioglycolate	0.2g
L-Ascorbic acid	0.2g
$CaCl_2·2H_2O$	0.15g
Resazurin	0.5mg
Trace elements solution	10.0mL
Vitamin solution	10.0mL

pH 7.5 ± 0.2 at 25°C

Trace Elements Solution:

Composition per liter:

$MgSO_4·7H_2O$	3.0g
Nitrilotriacetic acid	1.5g
NaCl	1.0g
$MnSO_4·2H_2O$	0.5g
$CoSO_4·7H_2O$	0.18g
$ZnSO_4·7H_2O$	0.18g
$CaCl_2·2H_2O$	0.1g
$FeSO_4·7H_2O$	0.1g
$NiCl_2·6H_2O$	0.025g
$KAl(SO_4)_2·12H_2O$	0.02g
H_3BO_3	0.01g
$Na_2MoO_4·4H_2O$	0.01g
$CuSO_4·5H_2O$	0.01g
$Na_2SeO_3·5H_2O$	0.3mg

Preparation of Trace Elements Solution: Add nitrilotriacetic acid to 500.0mL of distilled/deionized water. Dissolve by adjusting pH to 6.5 with KOH. Add remaining components. Add distilled/deionized water to 1.0L. Mix thoroughly.

Vitamin Solution:

Composition per liter:

Pyridoxine-HCl	10.0mg
Thiamine-HCl·$2H_2O$	5.0mg
Riboflavin	5.0mg
Nicotinic acid	5.0mg
D-Ca-pantothenate	5.0mg
p-Aminobenzoic acid	5.0mg
Lipoic acid	5.0mg
Biotin	2.0mg
Folic acid	2.0mg
Vitamin B_{12}	0.1mg

Preparation of Vitamin Solution: Add components to distilled/deionized water and bring volume to 1.0L. Mix thoroughly. Sparge with 80% H_2 + 20% CO_2. Filter sterilize.

Preparation of Medium: Prepare and dispense medium under an oxygen-free 100% N_2. Add components, except vitamin solution, Na-thioglycolate, $NaHCO_3$, and L-ascorbic acid, to distilled/deionized water and bring volume to 990.0L. Mix thoroughly. Gently heat and bring to boiling. Cool while sparging with 100% N_2. Add 0.2g Na-thioglycolate, 1.3g $NaHCO_3$, and 0.2g L-ascorbic acid. Mix thoroughly. Adjust pH to 7.5. Autoclave for 15 min at 15 psi pressure–121°C. Cool to 25°C. Aseptically and anaerobically add 10.0mL sterile vitamin solution. Mix thoroughly. Aseptically and anaerobically distribute into sterile tubes or flasks.

Use: For the cultivation of *Thermodesulfovibrio yellowstonii.*

Thermofilum pendens Medium

Composition per liter:

Sulfur, powdered	10.0g
$(NH_4)_2SO_4$	1.3g
KH_2PO_4	0.28g
$MgSO_4·7H_2O$	0.25g
$CaCl_2·2H_2O$	0.07g
$Na_2S·9H_2O$	0.3g
$FeCl_3·6H_2O$	0.02g
$Na_2B_4O_7·10H_2O$	4.5mg
$MnCl_2·4H_2O$	1.8mg
$ZnSO_4·7H_2O$	0.22mg
$CuCl_2·2H_2O$	0.05mg
$Na_2MoO_4·2H_2O$	0.03mg
$VOSO_4·2H_2O$	0.03mg
$CoSO_4·7H_2O$	0.01mg
Yeast extract solution	20.0mL
Sucrose solution	20.0mL
Polar lipid fraction	6.0–12.0mL

pH 5.2 ± 0.2 at 25°C

Yeast Extract Solution:

Composition per 20.0mL:

Yeast extract	2.0g

Preparation of Yeast Extract Solution: Add yeast extract to distilled/deionized water and bring volume to 20.0mL. Mix thoroughly. Gently heat and bring to boiling. Boil for a few minutes. Sparge with 100% N_2. Do not autoclave.

Sucrose Solution:

Composition per 20.0mL:

Sucrose	2.0g

Preparation of Sucrose Solution: Add sucrose to distilled/deionized water and bring volume to 20.0mL. Mix thoroughly. Filter sterilize. Sparge with 100% N_2.

Preparation of Sulfur: Add 10.0g of powdered sulfur to a flask and sterilize by steaming for 3 hr on 3 consecutive days.

Polar Lipid Fraction:

Composition per 20.0mL:

Thermoproteus tenax cells (wet weight)	10.0g
Chloroform	500.0mL
Acetone	500.0mL
Methanol	500.0mL
TA buffer solution	80.0mL
Chloroform/methanol 1:1 (v/v)	20.0mL

TA Buffer Solution:

Composition per 100.0mL:

Tris·HCl	0.79g
β-mercaptoethanol	0.78g

NH_4Cl 0.118g
EDTA 0.029g

Preparation of TA Buffer Solution: Add components to distilled/deionized water and bring volume to 100.0mL. Mix thoroughly.

Preparation of Polar Lipid Fraction: Add 10.0g (wet weight) of *Thermoproteus tenax* cells to 20.0mL of TA buffer solution. Mix thoroughly. Sonicate for 2 min. Centrifuge at 20,000 rpm for 20 min. Resuspend pellet in 20.0mL of fresh TA buffer solution. Recentrifuge at 20,000 rpm for 20 min. Again resuspend pellet in 20.0mL of fresh TA buffer solution. Recentrifuge at 20,000 rpm for 20 min. Resuspend pellet in 20.0mL of fresh TA buffer solution. Centrifuge at 5,000 rpm for 5 min. Decant the supernatant solution and discard the pellet. Extract the supernatant solution twice with 20.0mL of chloroform/methanol (1:1) each time. Chromatograph the extract on a SIL-LC (325 mesh) silicic acid column (20 cm × 2 cm) using 500.0mL of chloroform, followed by 500.0mL of acetone, followed by 500.0mL of methanol. The methanol fraction is further purified by DEAE chromatography using chloroform/methanol 1:1 and methanol.

Preparation of Medium: Prepare and dispense medium under 100% N_2. Add components, except yeast extract solution, sucrose solution, sulfur, and polar lipid fraction, to distilled/deionized water and bring volume to 950.0mL. Mix thoroughly. Sparge with 100% N_2. Anaerobically distribute into tubes or bottles. Autoclave for 15 min at 15 psi pressure–121°C. Aseptically and anaerobically add to 1.0L of medium, 20.0mL of sterile yeast extract solution, 20.0mL of sterile sucrose solution, 10.0g of sterile sulfur, and 6.0–12.0mL of sterile polar lipid fraction. Mix thoroughly.

Use: For the cultivation and maintenance of *Thermofilum pendens*.

Thermoleophilum Medium

Composition per liter:

$NaNO_2$ 2.0g
Na_2HPO_4 .... 0.21g
$MgSO_4 \cdot 7H_2O$ 0.2g
KCl 0.04g
NaH_2PO_4 .... 90.0mg
$CaCl_2$ 15.0mg
$FeSO_4 \cdot 7H_2O$ 1.0mg
$ZnSO_4 \cdot 7H_2O$ 70.0μg
H_3BO_3 .... 10.0μg
$MnSO_4 \cdot 5H_2O$ 10.0μg
MoO_3 .... 10.0μg
$CuSO_4 \cdot 5H_2O$ 5.0μg
n-Heptadecane 1.0mL

pH 7.0 ± 0.2 at 25°C

Preparation of Medium: Add components, except *n*-heptadecane, to distilled/deionized water and bring volume to 1.0L. Mix thoroughly. Autoclave for 15 min at 15 psi pressure–121°C. Aseptically add 1.0mL of *n*-heptadecane. Mix thoroughly. Aseptically distribute into sterile tubes or flasks.

Use: For the cultivation of *Thermoleophilum album* and *Thermoleophilum minutum*.

Thermomicrobium fosteri Agar

Composition per liter:

Agar 20.0g
NH_4Cl 2.0g
Na_2HPO_4 .... 0.21g
$MgSO_4 \cdot 7H_2O$ 0.2g
NaH2PO4 .... 0.09g
KCl 0.04g
$CaCl_2$ 0.015g
$ZnSO_4 \cdot 7H_2O$ 70.0μg
H_3BO_3 .... 10.0μg
$MnSO_4 \cdot 5H_2O$ 10.0μg
MoO_3 .... 10.0μg
$CuSO_4 \cdot 5H_2O$ 5.0μg
$FeSO_4 \cdot 7H_2O$ 1.0mg
Heptadecane, filter sterilized 20.0mL

pH 7.2 ± 0.2 at 25°C

Preparation of Medium: Add components, except heptadecane, to distilled/deionized water and bring volume to 980.0mL. Mix thoroughly. Gently heat and bring to boiling. Autoclave for 15 min at 15 psi pressure–121°C. Cool to 50°–55°C. Aseptically add 20.0mL of sterile heptadecane. Mix thoroughly. Aseptically distribute into sterile tubes. Cool tubes rapidly in a slanted position.

Use: For the cultivation and maintenance of *Thermomicrobium fosteri*.

Thermomicrobium roseum Agar (DSMZ Medium 592)

Composition per liter:

Agar 20.0g
$(NH_4)_2SO_4$ (sublimed) 1.3g
Yeast extract 1.0g
Tryptone 1.0g
$MgSO_4 \cdot 7H_2O$ 0.247g
KH_2PO_4 .... 0.280g
$CaCl_2 \cdot 2H_2O$ 0.074g
$FeCl_3 \cdot 6H_2O$ 0.019g
Salt solution 1.0mL

pH 8.5 ± 0.2 at 25°C

Salt Solution:

Composition per liter:

$Na_2B_4O_7 \cdot 10H_2O$ 4.4g
$MnCl_2 \cdot 4H_2O$ 1.8g
$ZnSO_4 \cdot 7H_2O$ 0.22g
$CuCl_2 \cdot H_2O$ 0.05g
$Na_2MoO_4 \cdot 4H_2O$ 0.03g
$VOSO_4 \cdot 2H_2O$ 0.03g

Preparation of Salt Solution: Add components to distilled/deionized water and bring volume to 100.0mL. Adjust pH to 2.0. Mix thoroughly.

Preparation of Medium: Add components to distilled/deionized water and bring volume to 1.0L. Mix thoroughly. Adjust pH to 8.5. Gently heat and bring to boiling. Autoclave for 15 min at 15 psi pressure–121°C. Pour into Petri dishes or pour short slants with a long butt in screw-capped tubes.

Use: For the cultivation and maintenance of *Thermomicrobium roseum*.

Thermomonospora Medium

Composition per liter:

Sucrose 30.0g
Agar 15.0g
Casamino acids 6.0g
$NaNO_3$ 3.0g
Yeast extract 2.0g
K_2HPO_4 .... 1.0g
$MgSO_4 \cdot 7H_2O$ 0.5g

KCl 0.5g
$FeSO_4 \cdot 7H_2O$ 0.01g
pH 8.0 ± 0.2 at 25°C

Preparation of Medium: Add components to distilled/deionized water and bring volume to 1.0L. Mix thoroughly. Distribute into tubes or flasks. Autoclave for 15 min at 15 psi pressure–121°C. Pour into sterile Petri dishes or leave in tubes.

Use: For the cultivation of *Thermomonospora alba* and *Thermomonospora mesophila.*

Thermophilic *Bacillus* Medium

Composition per liter:

Peptone 8.0g
Yeast extract 4.0g
NaCl 3.0g
pH 7.5 ± 0.2 at 25°C

Preparation of Medium: Add components to distilled/deionized water and bring volume to 1.0L. Mix thoroughly. Distribute into tubes or flasks. Autoclave for 15 min at 15 psi pressure–121°C.

Use: For the cultivation and maintenance of a variety of thermophilic *Bacillus* species.

Thermophilic Hydrogen-Bacteria Medium

Composition per 1000.5mL:

$Na_2HPO_4 \cdot 12H_2O$ 4.5g
KH_2PO_4 .... 1.5g
NaCl 1.0g
NH_4NO_3 .... 1.0g
$MgSO_4 \cdot 7H_2O$ 0.2g
$CaCl_2 \cdot 2H_2O$ 10.0mg
$FeSO_4 \cdot 7H_2O$ 10.0mg
Trace elements solution 0.5mL
pH 7.0 ± 0.2 at 25°C

Trace Elements Solution:
Composition per liter:

$ZnSO_4 \cdot 7H_2O$ 28.0mg
$CoCl_2 \cdot 6H_2O$ 4.0mg
H_3BO_3 .... 4.0mg
$MnSO_4 \cdot 5H_2O$ 4.0mg
MoO_3 .... 4.0mg
$CuSO_4 \cdot 5H_2O$ 2.0mg

Preparation of Trace Elements Solution: Add components to distilled/deionized water and bring volume to 1.0L. Mix thoroughly.

Preparation of Medium: Add components to distilled/deionized water and bring volume to 1.0L. Mix thoroughly. Distribute into tubes or flasks. Autoclave for 15 min at 15 psi pressure–121°C. Incubate cultures in 5% O_2 + 80% H_2 + 10% CO_2.

Use: For the cultivation and maintenance of *Hydrogenobacter thermophilus* and *Pseudomonas* species.

Thermophilic Maintenance Medium

Composition per liter:

$NaHCO_3$ 3.0g
Yeast extract 1.0g
NH_4Cl 1.0g
KH_2PO_4 .... 0.4g
K_2HPO_4 .... 0.4g
$MgSO_4 \cdot 7H_2O$ 0.1g
Cysteine-sulfide reducing solution 40.0mL
Fructose solution 25.0mL
Wolfe's vitamin solution 10.0mL
Wolfe's mineral solution 10.0mL
Resazurin (0.01% solution) 1.0mL
pH 5.6 ± 0.2 at 25°C

Cysteine-Sulfide Reducing Solution:
Composition per 100.0mL:

$Cysteine \cdot HCl \cdot H_2O$ 1.25g
$Na_2S \cdot 9H_2O$ 1.25g

Preparation of L-Cysteine-Sulfide Reducing Solution: Add L-cysteine·HCl·H_2O and $Na_2S \cdot 9H_2O$ to distilled/deionized water and bring volume to 100.0mL. Mix thoroughly.

Fructose Solution:
Composition per 100.0mL:

Fructose 20.0g

Preparation of Fructose Solution: Add fructose to distilled/deionized water and bring volume to 100.0mL. Mix thoroughly. Filter sterilize.

Wolfe's Vitamin Solution:
Composition per liter:

Pyridoxine·HCl 0.01g
Thiamine·HCl 5.0mg
Riboflavin 5.0mg
Nicotinic acid 5.0mg
Calcium pantothenate 5.0mg
p-Aminobenzoic acid 5.0mg
Thioctic acid 5.0mg
Biotin 2.0mg
Folic acid 2.0mg
Cyanocobalamin 100.0μg

Preparation of Wolfe's Vitamin Solution: Add components to distilled/deionized water and bring volume to 1.0L. Mix thoroughly.

Wolfe's Mineral Solution:
Composition per liter

$MgSO_4 \cdot 7H_2O$ 3.0g
Nitriloacetic acid 1.5g
NaCl 1.0g
$MnSO_4 \cdot H_2O$ 0.5g
$FeSO_4 \cdot 7H_2O$ 0.1g
$CoCl_2 \cdot 6H_2O$ 0.1g
$CaCl_2$ 0.1g
$ZnSO_4 \cdot 7H_2O$ 0.1g
$CuSO_4 \cdot 5H_2O$ 0.01g
$AlK(SO_4)_2 \cdot 12H_2O$ 0.01g
H_3BO_3 .... 0.01g
$Na_2MoO_4 \cdot 2H_2O$ 0.01g

Preparation of Wolfe's Mineral Solution: Add nitrilotriacetic acid to 500.0mL of distilled/deionized water. Dissolve by adjusting pH to 6.5 with KOH. Add remaining components. Add distilled/deionized water to 1.0L.

Preparation of Medium: Add components, except fructose solution, to distilled/deionized water and bring volume to 935.0mL. Mix thoroughly. Gently heat and bring to boiling. Continue boiling until resazurin turns colorless, indicating reduction. Add 40.0mL of the cysteine-sulfide reducing solution. Autoclave for 15 min at 15 psi pressure–121°C. Cool to 50°C under O_2-free 90% N_2 + 10% CO_2. Add 25.0mL of the sterile fructose solution. Adjust the pH to 5.6 if necessary. Aseptically and anaerobically distribute into sterile tubes. Cap with rubber stoppers.

Use: For the cultivation and maintenance of a variety of thermophilic anaerobes, including *Clostridium thermoautotrophicum*.

Thermophilic *Methanosarcina* Medium

Composition per 1021.0mL:

NaCl	2.25g
Pancreatic digest of casein	2.0g
Yeast extract	2.0g
$NaHCO_3$	0.85g
$MgSO_4 \cdot 7H_2O$	0.5g
NH_4Cl	0.5g
K_2HPO_4	0.348g
$CaCl_2 \cdot 2H_2O$	0.25g
KH_2PO_4	0.227g
$FeSO_4 \cdot 7H_2O$	2.0mg
Resazurin	1.0mg
Rumen fluid, clarified	50.0mL
Methanol solution	10.0mL
Vitamin solution	10.0mL
L-Cysteine·HCl·H_2O solution	10.0mL
$Na_2S \cdot 9H_2O$ solution	10.0mL
Trace elements solution SL-10	1.0mL

pH 6.5–6.8 at 25°C

Methanol Solution:

Composition per 10.0mL:

Methanol	5.0mL

Preparation of Methanol Solution: Add methanol to distilled/deionized water and bring volume to 10.0mL. Sparge with 100% N_2. Autoclave for 15 min at 15 psi pressure–121°C.

Vitamin Solution:

Composition per liter:

Pyridoxine·HCl	10.0mg
Calcium DL-pantothenate	5.0mg
Lipoic acid	5.0mg
Nicotinic acid	5.0mg
p-Aminobenzoic acid	5.0mg
Riboflavin	5.0mg
Thiamine·HCl	5.0mg
Biotin	2.0mg
Folic acid	2.0mg
Vitamin B_{12}	0.1mg

Preparation of Vitamin Solution: Add components to distilled/deionized water and bring volume to 1.0L. Mix thoroughly. Filter sterilize. Sparge with 80% N_2 + 20% CO_2.

L-Cysteine·HCl·H_2O Solution:

Composition per 10.0mL:

L-Cysteine·HCl	0.3g

Preparation of L-Cysteine·HCl·H_2O Solution: Add L-cysteine·HCl·H_2O to distilled/deionized water and bring volume to 10.0mL. Mix thoroughly. Autoclave under 100% N_2 for 15 min at 15 psi pressure–121°C.

$Na_2S \cdot 9H_2O$ Solution:

Composition per 10.0mL:

$Na_2S \cdot 9H_2O$	0.3g

Preparation of $Na_2S \cdot 9H_2O$ Solution: Add $Na_2S \cdot 9H_2O$ to distilled/deionized water and bring volume to 10.0mL. Mix thoroughly. Sparge with 100% N_2. Autoclave for 15 min at 15 psi pressure–121°C.

Trace Elements Solution SL-10:

Composition per liter:

$FeCl_2 \cdot 4H_2O$	1.5g
$CoCl_2 \cdot 6H_2O$	190.0mg
$MnCl_2 \cdot 4H_2O$	100.0mg
$ZnCl_2$	70.0mg
$Na_2MoO_4 \cdot 2H_2O$	36.0mg
$NiCl_2 \cdot 6H_2O$	24.0mg
H_3BO_3	6.0mg
$CuCl_2 \cdot 2H_2O$	2.0mg
HCl (25% solution)	10.0mL

Preparation of Trace Elements Solution SL-10: Add $FeCl_2 \cdot 4H_2O$ to 10.0mL of HCl solution. Mix thoroughly. Add distilled/deionized water and bring volume to 1.0L. Add remaining components. Mix thoroughly. Sparge with 100% N_2.

Preparation of Medium: Add components, except methanol solution, L-cysteine·HCl·H_2O solution, and $Na_2S \cdot 9H_2O$ solution, to distilled/deionized water and bring volume to 960.0mL. Mix thoroughly. Sparge under 80% N_2 + 20% CO_2 for 3–4 min. Autoclave for 15 min at 15 psi pressure–121°C. Aseptically and anaerobically add 20.0mL of sterile vitamin solution, 10.0mL of sterile L-cysteine·HCl·H_2O solution, and 10.0mL of sterile $Na_2S \cdot 9H_2O$ solution. Mix thoroughly. Aseptically and anaerobically distribute into sterile screw-capped bottles under 80% N_2 + 20% CO_2.

Use: For the cultivation and maintenance of *Methanosarcina thermophila*.

Thermophilic *Methanothrix* Medium

Composition per liter:

NH_4Cl	0.5g
K_2HPO_4	0.4g
$MgCl_2 \cdot 6H_2O$	0.1g
Resazurin	1.0mg
$NaHCO_3$ solution	20.0mL
Trace elements	10.0mL
$CaCl_2 \cdot 2H_2O$ solution	10.0mL
Sodium acetate solution	10.0mL
Vitamin solution	10.0mL
Coenzyme M solution	10.0mL
$Na_2S \cdot 9H_2O$ solution	5.0mL

pH 6.5 ± 0.2 at 25°C

$NaHCO_3$ Solution:

Composition per 20.0mL:

$NaHCO_3$	1.0g

Preparation of $NaHCO_3$ Solution: Add $NaHCO_3$ to distilled/deionized water and bring volume to 20.0mL. Mix thoroughly. Sparge with 80% N_2 + 20% CO_2 for 15 min. Autoclave for 15 min at 15 psi pressure–121°C.

Trace Elements Solution:

Composition per liter:

$MgSO_4 \cdot 7H_2O$	3.0g
Nitrilotriacetic acid	1.5g
NaCl	1.0g
$MnSO_4 \cdot 2H_2O$	0.5g
$CoSO_4 \cdot 7H_2O$	0.18g
$ZnSO_4 \cdot 7H_2O$	0.18g
$CaCl_2 \cdot 2H_2O$	0.1g
$FeSO_4 \cdot 7H_2O$	0.1g
$NiCl_2 \cdot 6H_2O$	0.025g
$KAl(SO_4)_2 \cdot 12H_2O$	0.02g
$CuSO_4 \cdot 5H_2O$	0.01g
H_3BO_3	0.01g
$Na_2MoO_4 \cdot 2H_2O$	0.01g
$Na_2SeO_3 \cdot 5H_2O$	0.3mg

Preparation of Trace Elements Solution: Add nitrilotriacetic acid to approximately 500.0mL of distilled/deionized water. Dissolve by adding KOH and adjust pH to 6.5. Add remaining components. Bring volume to 1.0L with additional distilled/deionized water. Adjust pH to 7.0 with KOH.

$CaCl_2 \cdot 2H_2O$ Solution:
Composition per 10.0mL:
$CaCl_2 \cdot 2H_2O$ 0.1g

Preparation of $CaCl_2 \cdot 2H_2O$ Solution: Add $CaCl_2 \cdot 2H_2O$ to distilled/deionized water and bring volume to 10.0mL. Mix thoroughly. Sparge with 100% N_2. Autoclave for 15 min at 15 psi pressure–121°C.

Sodium Acetate Solution:
Composition per 10.0mL:
Sodium acetate 3.3g

Preparation of Sodium Acetate Solution: Add sodium acetate to distilled/deionized water and bring volume to 10.0mL. Mix thoroughly. Sparge with 100% N_2. Autoclave for 15 min at 15 psi pressure–121°C.

Vitamin Solution:
Composition per liter:
Pyridoxine·HCl 10.0mg
Calcium DL-pantothenate 5.0mg
Lipoic acid 5.0mg
Nicotinic acid 5.0mg
p-Aminobenzoic acid 5.0mg
Riboflavin 5.0mg
Thiamine·HCl 5.0mg
Biotin 2.0mg
Folic acid 2.0mg
Vitamin B_{12} 0.1mg

Preparation of Vitamin Solution: Add components to distilled/deionized water and bring volume to 1.0L. Mix thoroughly. Sparge with 80% N_2 + 20% CO_2.

Coenzyme M Solution:
Composition per 10.0mL:
Coenzyme M 0.142g

Preparation of Coenzyme M Solution: Add coenzyme M to distilled/deionized water and bring volume to 10.0mL. Mix thoroughly. Sparge with 100% N_2. Autoclave for 15 min at 15 psi pressure–121°C.

$Na_2S \cdot 9H_2O$ Solution:
Composition per 10.0mL:
$Na_2S \cdot 9H_2O$ 0.5g

Preparation of $Na_2S \cdot 9H_2O$ Solution: Add $Na_2S \cdot 9H_2O$ to distilled/deionized water and bring volume to 10.0mL. Mix thoroughly. Sparge with 80% N_2 + 20% CO_2. Autoclave for 15 min at 15 psi pressure–121°C.

Preparation of Medium: Add components, except $NaHCO_3$ solution, $CaCl_2 \cdot 2H_2O$ solution, sodium acetate solution, vitamin solution, Coenzyme M solution, and $Na_2S \cdot 9H_2O$ solution, to distilled/deionized water and bring volume to 935.0mL. Gently heat and bring to boiling. Continue boiling for 10 min. Cool to room temperature while sparging with 80% N_2 + 20% CO_2. Gas the medium until pH reaches 5.8. Anaerobically distribute the medium into serum bottles. Autoclave for 15 min at 15 psi pressure–121°C. Aseptically and anaerobically add 20.0mL of sterile $NaHCO_3$ solution, 10.0mL of sterile $CaCl_2 \cdot 2H_2O$ solution, 10.0mL of sterile sodium acetate solution, 10.0mL of sterile vitamin solution, 10.0mL of sterile Coenzyme M solution, and 5.0mL of sterile $Na_2S \cdot 9H_2O$ solution. Bring gas atmosphere in each bottle to 30% CO_2.

Use: For the cultivation and maintenance of *Methanothrix thermophila*.

Thermophilic Spirochete Medium

Composition per 1012.0mL:
Solution A 920.0mL
Solution D 50.0mL
Solution E (Vitamin solution) 20.0mL
Solution F 10.0mL
Solution G 10.0mL
Solution B (Trace elements solution SL-10) 1.0mL
Solution C (Selenite-tungstate solution) 1.0mL
pH 6.9 ± 0.2 at 25°C

Solution A:
Composition per 920.0mL:
NaCl 4.0g
$MgCl_2 \cdot 6H_2O$ 0.8g
KCl 0.5g
NH_4Cl 0.3g
KH_2PO_4 .... 0.2g
$CaCl_2 \cdot 2H_2O$ 0.03g
Resazurin 1.0mg

Preparation of Solution A: Prepare and dispense under 80% N_2 + 20% CO_2. Add components to distilled/deionized water and bring volume to 920.0mL. Mix thoroughly. Gently heat and bring to boiling. Continue boiling for a few minutes. Cool to room temperature while sparging with 80% N_2 + 20% CO_2. Anaerobically distribute into tubes or bottles. Autoclave for 15 min at 15 psi pressure–121°C.

Solution B (Trace Elements Solution SL-10):
Composition per liter:
$FeCl_2 \cdot 4H_2O$ 1.5g
$CoCl_2 \cdot 6H_2O$ 190.0mg
$MnCl_2 \cdot 4H_2O$ 100.0mg
$ZnCl_2$ 70.0mg
$Na_2MoO_4 \cdot 2H_2O$ 36.0mg
$NiCl_2 \cdot 6H_2O$ 24.0mg
H_3BO_3 .... 6.0mg
$CuCl_2 \cdot 2H_2O$ 2.0mg
HCl (25% solution) 10.0mL

Preparation of Solution B (Trace Elements Solution SL-10): Add $FeCl_2 \cdot 4H_2O$ to 10.0mL of HCl solution. Mix thoroughly. Add distilled/deionized water and bring volume to 1.0L. Add remaining components. Mix thoroughly. Sparge with 100% N_2. Autoclave for 15 min at 15 psi pressure–121°C.

Solution C (Selenite-Tungstate Solution):
Composition per liter:
NaOH 0.5g
$Na_2SeO_3 \cdot 5H_2O$ 3.0mg
$Na_2WO_4 \cdot 2H_2O$ 4.0mg

Preparation of Solution C (Selenite-Tungstate Solution): Add components to distilled/deionized water and bring volume to 1.0L. Mix thoroughly. Sparge with 100% N_2. Autoclave for 15 min at 15 psi pressure–121°C.

Solution D:
Composition per 50.0mL:
$NaHCO_3$ 3.0mg

Preparation of Solution D: Add $NaHCO_3$ to distilled/deionized water and bring volume to 50.0mL. Mix thoroughly. Sparge with 80% N_2 + 20% CO_2. Autoclave for 15 min at 15 psi pressure–121°C.

Solution E (Vitamin Solution):

Composition per liter:

Pyridoxine·HCl 10.0mg
Calcium DL-pantothenate 5.0mg
Lipoic acid 5.0mg
Nicotinic acid 5.0mg
p-Aminobenzoic acid 5.0mg
Riboflavin 5.0mg
Thiamine·HCl 5.0mg
Biotin 2.0mg
Folic acid 2.0mg
Vitamin B_{12} 0.1mg

Preparation of Solution E (Vitamin Solution): Add components to distilled/deionized water and bring volume to 1.0L. Mix thoroughly. Sparge with 100% N_2. Autoclave for 15 min at 15 psi pressure–121°C.

Solution F:

Composition per 10.0mL:

Starch, soluble 1.0g

Preparation of Solution F: Add starch to distilled/deionized water and bring volume to 10.0mL. Mix thoroughly. Sparge with 100% N_2. Autoclave for 15 min at 15 psi pressure–121°C.

Solution G:

Composition per 10.0mL:

$Na_2S·9H_2O$ 0.3g

Preparation of Solution G: Add $Na_2S·9H_2O$ to distilled/deionized water and bring volume to 10.0mL. Mix thoroughly. Sparge with 100% N_2. Autoclave for 15 min at 15 psi pressure–121°C.

Preparation of Medium: Prepare and dispense medium under 80% N_2 + 20% CO_2. To 920.0mL of sterile solution A, aseptically and anaerobically add 1.0mL of sterile solution B, 1.0mL of sterile solution C, 50.0mL of sterile solution D, 20.0mL of sterile solution E, 10.0mL of sterile solution F, and 10.0mL of sterile solution G in that order. Mix thoroughly.

Use: For the cultivation and maintenance of *Spirochaeta thermophila.*

Thermophilic Streptomycete Medium

Composition per liter:

Agar 20.0g
Maltose 20.0g
Soybean meal 5.0g
Yeast extract 2.0g

pH 6.5 ± 0.2 at 25°C

Preparation of Medium: Add components to tap water and bring volume to 1.0L. Mix thoroughly. Gently heat and bring to boiling. Distribute into tubes or flasks. Autoclave for 15 min at 15 psi pressure–121°C. Pour into sterile Petri dishes or leave in tubes.

Use: For the isolation and cultivation of thermophilic streptomycetes.

Thermophilic Streptomycete Medium Ia

Composition per liter:

Agar 20.0g
Sucrose 5.0g
Pancreatic digest of casein 5.0g
Yeast extract 3.0g
$MgSO_4·7H_2O$ 0.5g
$FeSO_4·7H_2O$ 0.01g
Dung extract 5.0mL
Molasses 5.0mL
Trace elements solution 1.0mL

pH 7.2 ± 0.2 at 25°C

Dung Extract:

Composition per 100.0mL:

Sheep manure, dried 25.0g

Preparation of Dung Extract: Add dried sheep manure to 100.0mL of tap water. Mix thoroughly. Autoclave for 30 min at 15 psi pressure–121°C. Filter through Whatman #1 filter paper. Store at 4°C under toluene.

Trace Elements Solution:

Composition per 100.0mL:

$Fe(NH_4)_2SO_4$ 0.1g
$ZnSO_4$ 0.1g
$MnSO_4$ 0.05g
$CoSO_4$ 0.01g
H_3BO_3 .... 0.01g
$CuSO_4$ 8.0mg

Preparation of Trace Elements Solution: Add components to distilled/deionized water and bring volume to 100.0mL. Mix thoroughly.

Preparation of Medium: Add components to distilled/deionized water and bring volume to 1.0L. Mix thoroughly. Gently heat and bring to boiling. Distribute into tubes or flasks. Autoclave for 15 min at 15 psi pressure–121°C. Pour into sterile Petri dishes or leave in tubes.

Use: For the isolation and cultivation of thermophilic streptomycetes.

Thermoplasma acidophilum Growth Medium 7B

Composition per liter:

Sucrose 17.0g
$(NH_4)_2SO_4$ 6.8g
KOH 1.22g
Yeast extract 1.0g
$MgSO_4$ 0.5g
$CaCl_2·2H_2O$ 0.25g
H_3PO_4 solution 1.5mL
Antifoam A 10.0μL

pH 1.60 ± 0.2 at 25°C

H_3PO_4 Solution:

Composition per 100.0mL:

H_3PO_4 .... 85.0g

Preparation of H_3PO_4 Solution: Add H_3PO_4 to distilled/deionized water and bring volume to 100.0mL. Mix thoroughly.

Preparation of Medium: Add components, except Antifoam A, to distilled/deionized water and bring volume to 1.0L. Mix thoroughly. Adjust pH to 1.60 with 50% H_2SO_4. Distribute into tubes or flasks. Sterilize by heating to 100°C for 30 min. Allow to stand at room temperature for 24 hr. Add 10.0μL of Antifoam A per liter.

Use: For the cutivation of *Thermococcus acidophilum.*

Thermoplasma acidophilum Medium

Composition per liter:

$(NH_4)_2SO_4$ 1.32g
Yeast extract solution 1.0g
KH_2PO_4 .... 0.372g
$MgSO_4·7H_2O$ 0.247g
$CaCl_2·2H_2O$ 0.074g
Glucose solution 20.0mL

Yeast extract solution 10.0mL
Trace elements solution 10.0mL
pH 1.0–2.0 at 25°C

Glucose Solution:
Composition per 20.0mL:
Glucose 10.0g

Preparation of Glucose Solution: Add glucose to distilled/deionized water and bring volume to 20.0mL. Mix thoroughly. Autoclave for 15 min at 15 psi pressure–121°C.

Yeast Extract Solution:
Composition per 10.0mL:
Yeast extract 1.0g

Preparation of Yeast Extract Solution: Add yeast extract to distilled/deionized water and bring volume to 10.0mL. Mix thoroughly. Autoclave for 15 min at 15 psi pressure–121°C.

Trace Elements Solution:
Composition per liter:
$FeCl_3 \cdot 6H_2O$ 1.93g
$Na_2B_4O_7 \cdot 10H_2O$ 0.45g
$MnCl_2 \cdot 4H_2O$ 0.18g
$ZnSO_4 \cdot 7H_2O$ 22.0mg
$CuCl_2 \cdot 2H_2O$ 5.0mg
$VOSO_4 \cdot 5H_2O$ 3.8mg
$Na_2MoO_4 \cdot 2H_2O$ 3.0mg
$CoSO_4 \cdot 7H_2O$ 2.0mg

Preparation of Trace Elements Solution: Add components to distilled/deionized water and bring volume to 1.0L. Mix thoroughly.

Preparation of Medium: Add components, except glucose solution and yeast extract solution, to distilled/deionized water and bring volume to 970.0mL. Mix thoroughly. Adjust pH to 1.0-2.0 with 10*N* H_2SO_4. Autoclave for 15 min at 15 psi pressure–121°C. Aseptically add 20.0mL of sterile glucose solution and 10.0mL of sterile yeast extract solution. Mix thoroughly. Aseptically distribute into sterile tubes or flasks.

Use: For the cultivation and maintenance of *Thermoplasma acidophilum.*

Thermoplasma acidophilum Medium 7A

Composition per liter:
Glucose 10.0g
$(NH_4)_2SO_4$ 6.8g
KH_2PO_4 .... 3.0g
Yeast extract 1.0g
$MgSO_4$ 0.5g
$CaCl_2 \cdot 2H_2O$ 0.25g
pH 1.65 ± 0.2 at 25°C

Preparation of Medium: Add components to distilled/deionized water and bring volume to 1.0L. Mix thoroughly. Adjust pH to 1.65 with 50% H_2SO_4. Distribute into tubes or flasks. Sterilize by heating to 100°C for 30 min.

Use: For the cutivation of *Thermococcus acidophilum.*

Thermoplasma Agar

Composition per liter:
Basal solution 450.0mL
Solution B 450.0mL
Solution C 100.0mL
pH 2.0 ± 0.2 at 25°C

Basal Solution:
Composition per 500.0mL:
KH_2PO_4 .... 3.0g
Yeast extract 1.0g
$MgSO_4 \cdot 7H_2O$ 0.5g
$CaCl_2 \cdot 2H_2O$ 0.25g
$(NH_4)_2SO_4$ 0.2g

Preparation of Basal Solution: Add components to distilled/deionized water and bring volume to 500.0mL. Mix thoroughly. Adjust pH to 2.0 with 10*N* H_2SO_4. Autoclave for 15 min at 15 psi pressure–121°C. Cool to 55°C.

Solution B:
Composition per 450.0mL:
Noble agar 12.0g

Preparation of Solution B: Add agar to distilled/deionized water and bring volume to 450.0mL. Mix thoroughly. Gently heat and bring to boiling. Autoclave for 15 min at 15 psi pressure–121°C. Cool to 55°C.

Solution C:
Composition per 100.0mL:
Glucose 10.0g

Preparation of Solution C: Add glucose to distilled/deionized water and bring volume to 100.0mL. Mix thoroughly. Filter sterilize.

Preparation of Medium: Aseptically combine the cooled, sterile basal medium with sterile solution B and sterile solution C. Mix thoroughly. Pour into sterile Petri dishes or distribute into sterile tubes.

Use: For the cultivation and maintenance of *Thermoplasma acidophilum* and other *Thermoplasma* species.

Thermoplasma Broth

Composition per liter:
Basal solution 500.0mL
Solution C 100.0mL
pH 2.0 ± 0.2 at 25°C

Basal Solution:
Composition per 500.0mL:
KH_2PO_4 .... 3.0g
Yeast extract 1.0g
$MgSO_4 \cdot 7H_2O$ 0.5g
$CaCl_2 \cdot 2H_2O$ 0.25g
$(NH_4)_2SO_4$ 0.2g

Preparation of Basal Solution: Add components to distilled/deionized water and bring volume to 500.0mL. Mix thoroughly. Adjust pH to 2.0 with 10*N* H_2SO_4.

Solution C:
Composition per 100.0mL:
Glucose 10.0g

Preparation of Solution C: Add glucose to distilled/deionized water and bring volume to 100.0mL. Mix thoroughly. Filter sterilize.

Preparation of Medium: Add 500.0mL of basal solution to 400.0mL of distilled/deionized water. Autoclave for 15 min at 15 psi pressure–121°C. Cool to 55°C. Aseptically add 100.0mL of sterile glucose solution. Mix thoroughly. Aseptically distribute into sterile tubes.

Use: For the cultivation and maintenance of *Thermoplasma acidophilum* and other *Thermoplasma* species.

Thermoplasma volcanium Medium

Composition per liter:

KH_2PO_4	3.0g
$MgSO_4·7H_2O$	1.0g
$CaCl_2·2H_2O$	0.25g
$(NH_4)_2SO_4$	0.2g
Glucose solution	10.0mL
Yeast extract solution	10.0mL

pH 6.5 ± 0.2 at 25°C

Glucose Solution:

Composition per 10.0mL:

Glucose	5.0g

Preparation of Glucose Solution: Add glucose to distilled/deionized water and bring volume to 20.0mL. Mix thoroughly. Filter sterilize.

Yeast Extract Solution:

Composition per 10.0mL:

Yeast extract	1.0g

Preparation of Yeast Extract Solution: Add yeast extract to distilled/deionized water and bring volume to 10.0mL. Mix thoroughly. Filter sterilize.

Preparation of Medium: Add components, except glucose solution and yeast extract solution, to distilled/deionized water and bring volume to 980.0mL. Mix thoroughly. Adjust pH to 2.0 with 10*N* H_2SO_4. Autoclave for 15 min at 15 psi pressure–121°C. Aseptically add 10.0mL of sterile glucose solution and 10.0mL of sterile yeast extract solution. Mix thoroughly. Aseptically distribute into sterile tubes or flasks. Final pH should be 2.0–3.0.

Use: For the aerobic cultivation and maintenance of *Thermoplasma volcanium*.

Thermoproteus Medium

Composition per liter:

Solution A	500.0mL
Solution B	450.0mL
Solution C	50.0mL

pH 4.8–5.6 at 25°C

Solution A:

Composition per 500.0mL:

Glucose	10.0g
$FeSO_4·7H_2O$	0.556g
$MgSO_4·7H_2O$	0.492g
$CaSO_4·2H_2O$	0.344g
$(NH_4)_2SO_4$	0.264g
Yeast extract	0.2g
KH_2PO_4	0.014g
Resazurin	1.0mg
Trace elements	10.0mL

Preparation of Solution A: Add components to distilled/deionized water and bring volume to 500.0mL. Mix thoroughly. Immediately filter sterilize.

Trace Elements:

Composition per liter:

$Na_2B_4O_7·10H_2O$	0.45g
$MnCl_2·4H_2O$	0.18g
$ZnSO_4·7H_2O$	0.022g
$CuCl_2·2H_2O$	5.0mg
$Na_2MoO_4·2H_2O$	3.6mg
$VOSO_4·5H_2O$	3.6mg
$CoSO_4·7H_2O$	1.2mg

Preparation of Trace Elements: Add components to distilled/deionized water and bring volume to 1.0L. Mix thoroughly. Adjust pH to 3.0 with H_2SO_4 to retard precipitation.

Solution B:

Composition per 450.0mL:

Sulfur	10.0g

Preparation of Solution B: Add sulfur to 450.0mL of distilled/deionized water. Autoclave for 30 min at 0 psi pressure–100°C on 3 consecutive days.

Solution C:

Composition per 50.0mL:

$Na_2S·9H_2O$	0.85g

Preparation of Solution C: Add $Na_2S·9H_2O$ to distilled/deionized water and bring volume to 50.0mL. Mix thoroughly. Autoclave for 15 min at 15 psi pressure–121°C.

Preparation of Medium: Aseptically combine solutions A, B, and C under 97% N_2 + 3% H_2. Adjust pH to 4.8–5.6 with H_2SO_4. Aseptically and anaerobically distribute into sterile tubes or flasks under 97% N_2 + 3% H_2.

Use: For the cultivation and maintenance of *Thermoproteus tenax* and other *Thermoproteus* species.

Thermoproteus neutrophilus Medium

Composition per liter:

Sulfur, powdered	8.0g
$(NH_4)_2SO_4$	1.3g
$NaHCO_3$	0.85g
KH_2PO_4	0.28g
$MgSO_4·7H_2O$	0.25g
$CaCl_2·2H_2O$	0.07g
$FeCl_3·6H_2O$	0.02g
$Na_2B_4·10H_2O$	4.5mg
$MnCl_2·4H_2O$	1.8mg
Resazurin	0.4mg
$ZnSO_4·7H_2O$	0.22mg
$CuCl_2·2H_2O$	0.05mg
$Na_2MoO_4·2H_2O$	0.03mg
$VOSO_4·2H_2O$	0.03mg
$CoSO_4$	0.01mg
Dithionite solution	1.0mL

pH 6.5 ± 0.2 at 25°C

Dithionite Solution:

Composition per 10.0mL:

$Na_2S_2O_4$	0.25g

Preparation of Dithionite Solution: Add $Na_2S_2O_4$ to distilled/deionized water and bring volume to 10.0mL. Mix thoroughly. Filter sterilize.

Preparation of Medium: Add components, except sulfur, $NaHCO_3$, resazurin, and dithionite solution, to distilled/deionized water and bring volume to 1.0L. Mix thoroughly. Add sulfur and resazurin. Adjust pH to 6.5 with NaOH. Gently heat and bring to boiling. Continue boiling for 5 min. Cool to room temperature while sparging with 80% H_2 + 20% CO_2. Add $NaHCO_3$. Mix thoroughly. Continue to sparge with 80% H_2 + 20% CO_2 until pH reaches 6.5. Distribute anaerobically 10.0mL of medium into 30.0mL serum bottles. Sterilize medium by heating at 85°C for 1 hr on 3 consecutive days. Prior to inoculation, add 10.0μL of sterile dithionite solution to each bottle.

Use: For the aerobic cultivation and maintenance of *Thermoproteus neutrophilus*.

Thermosipho africanus Medium

Composition per liter:

$NaHCO_3$	4.0g
Sodium acetate	4.0g
Sodium formate	2.0g
Yeast extract	1.0g
L-Cysteine·HCl	0.5g
KH_2PO_4	0.5g
$Na_2S·9H_2O$	0.5g
$MgSO_4·7H_2O$	0.4g
NaCl	0.4g
NH_4Cl	0.4g
$CaCl_2·2H_2O$	0.05g
$NiCl_2·6H_2O$	24.0mg
$FeSO_4·7H_2O$	2.0mg
Resazurin	1.0mg
Sludge fluid	50.0mL
Fatty acid mixture	20.0mL
Trace elements solution SL-10	1.0mL

pH 6.7 ± 0.2 at 25°C

Sludge Fluid:

Composition per 100.0mL:

Sludge	100.0mL
Yeast extract	0.4g

Preparation of Sludge Fluid: To 100.0mL of sludge from an anaerobic digester, add 0.4g of yeast extract. Sparge with 100% N_2 for a few minutes. Incubate at 37°C for 24 hr. Centrifuge the sludge at 13,000 × g for 15 min. Decant the clear supernatant solution. Sparge with 100% N_2 for a few minutes. Store in screw-capped bottles at room temperature in the dark.

Fatty Acid Mixture:

Composition per 20.0mL:

α-Methylbutyric acid	0.5g
Isobutyric acid	0.5g
Isovaleric acid	0.5g
Valeric acid	0.5g

Preparation of Fatty Acid Mixture: Add components to distilled/deionized water and bring volume to 20.0mL. Mix thoroughly. Adjust pH to 7.5 with concentrated NaOH.

Trace Elements Solution SL-10:

Composition per liter:

$FeCl_2·4H_2O$	1.5g
$CoCl_2·6H_2O$	190.0mg
$MnCl_2·4H_2O$	100.0mg
$ZnCl_2$	70.0mg
$Na_2MoO_4·2H_2O$	36.0mg
$NiCl_2·6H_2O$	24.0mg
H_3BO_3	6.0mg
$CuCl_2·2H_2O$	2.0mg
HCl (25% solution)	10.0mL

Preparation of Trace Elements Solution SL-10: Add $FeCl_2·4H_2O$ to 10.0mL of HCl solution. Mix thoroughly. Add distilled/deionized water and bring volume to 1.0L. Add remaining components. Mix thoroughly. Sparge with 100% N_2. Autoclave for 15 min at 15 psi pressure–121°C.

Preparation of Medium: Prepare and dispense medium anaerobically under 80% N_2 + 20% CO_2. Add components to distilled/deionized water and bring volume to 1.0L. Mix thoroughly. Sparge with 80% N_2 + 20% CO_2. Autoclave for 15 min at 15 psi pressure–121°C.

Use: For the cultivation and maintenance of *Thermosipho africanus*.

Thermosphaera Medium (DSMZ Medium 817)

Composition per liter:

$MgCl_2·6H_2O$	2.2g
Yeast extract	1.0g
Peptone	1.0g
NaCl	0.9g
KCl	17.0mg
NH_4Cl	12.5mg
$CaCl_2·2H_2O$	7.0mg
$K_2HPO_4·3H_2O$	7.0mg
Resazurin	0.4mg
$FeCl_3$	0.05mg
Vitamin solution	10.0mL
$Na_2S·9H_2O$ solution	10.0mL
$NaHCO_3$ solution	10.0mL

pH 6.5 ± 0.2 at 25°C

$NaHCO_3$ Solution:

Composition per 10.0mL:

$NaHCO_3$	1.0g

Preparation of $NaHCO_3$ Solution: Add $NaHCO_3$ to distilled/deionized water and bring volume to 10.0mL. Mix thoroughly. Sparge with 100% N_2. Autoclave for 15 min at 15 psi pressure–121°C. Cool to 25°C. Must be prepared freshly.

$Na_2S·9H_2O$ Solution:

Composition per 10.0mL:

$Na_2S·9H_2O$	0.5g

Preparation of $Na_2S·9H_2O$ Solution: Add $Na_2S·9H_2O$ to distilled/deionized water and bring volume to 10.0mL. Sparge with N_2. Autoclave for 15 min at 15 psi pressure–121°C. Cool to 25°C. Store anaerobically.

Vitamin Solution:

Composition per liter:

Pyridoxine-HCl	10.0mg
Thiamine-HCl·$2H_2O$	5.0mg
Riboflavin	5.0mg
Nicotinic acid	5.0mg
D-Ca-pantothenate	5.0mg
p-Aminobenzoic acid	5.0mg
Lipoic acid	5.0mg
Biotin	2.0mg
Folic acid	2.0mg
Vitamin B_{12}	0.1mg

Preparation of Vitamin Solution: Add components to distilled/deionized water and bring volume to 1.0L. Mix thoroughly. Sparge with 80% H_2 + 20% CO_2. Filter sterilize.

Preparation of Medium: Prepare and dispense medium under 80% N_2 + 20% CO_2. Add components, except $Na_2S·9H_2O$ solution, vitamin solution, and $NaHCO_3$ solution, to distilled/deionized water and bring volume to 970.0mL. Mix thoroughly. Flush medium with 80% N_2 + 20% CO_2 for 5 min. Adjust medium pH to 6.53. Autoclave for 15 min at 15 psi pressure–121°C. Cool to 25°C. Aseptically and anaerobically add 10.0mL sterile $Na_2S·9H_2O$ solution, 10.0mL sterile vitamin solution, and 10.0mL sterile $NaHCO_3$ solution. Mix thoroughly. Aseptically and anaerobically distribute into sterile thick walled tubes or thick walled bottles. Pressurize with 2 bar atmosphere of 80% N_2 + 20% CO_2.

Use: For the cultivation of *Thermosphaera aggregans*.

Thermosyntropha Medium (DSMZ Medium 731)

Composition per liter:

Yeast extract	10.0g
Na_2CO_3	3.0g
NH_4Cl	1.0g
NaCl	0.5g
K_2HPO_4	0.3g
KCl	0.3g
$MgCl_2·6H_2O$	0.3g
$CaCl_2·2H_2O$	0.05g
Resazurin	0.5mg
Vitamin solution	40.0mL
Trace elements solution	10.0mL
Cysteine solution	10.0mL
$Na_2S·9H_2O$ solution	10.0mL
$NaHCO_3$ solution	10.0mL
Selenite-tungstate solution	1.0mL

pH 8.5 ± 0.2 at 25°C

Selenite-Tungstate Solution:

Composition per liter:

NaOH	0.5g
$Na_2WO_4·2H_2O$	4.0mg
$Na_2SeO_3·5H_2O$	3.0mg

Preparation of Selenite-Tungstate Solution: Add components to distilled/deionized water and bring volume to 1.0L. Mix thoroughly. Sparge with 100% N_2. Filter sterilize.

$NaHCO_3$ Solution:

Composition per 10.0mL:

$NaHCO_3$	3.0g

Preparation of $NaHCO_3$ Solution: Add $NaHCO_3$ to distilled/deionized water and bring volume to 10.0mL. Mix thoroughly. Autoclave for 15 min at 15 psi pressure–121°C. Cool to 25°C. Must be prepared freshly.

Cysteine Solution:

Composition per 10.0mL:

L-Cysteine-HCl·H_2O	0.15g

Preparation of Cysteine Solution: Add L-cysteine·HCl·H_2O to distilled/deionized water and bring volume to 10.0mL. Mix thoroughly. Sparge with 100% N_2. Autoclave for 15 min at 15 psi pressure–121°C.

Trace Elements Solution:

Composition per liter:

$MgSO_4·7H_2O$	3.0g
Nitrilotriacetic acid	1.5g
NaCl	1.0g
$MnSO_4·2H_2O$	0.5g
$CoSO_4·7H_2O$	0.18g
$ZnSO_4·7H_2O$	0.18g
$CaCl_2·2H_2O$	0.1g
$FeSO_4·7H_2O$	0.1g
$NiCl_2·6H_2O$	0.025g
$KAl(SO_4)_2·12H_2O$	0.02g
H_3BO_3	0.01g
$Na_2MoO_4·4H_2O$	0.01g
$CuSO_4·5H_2O$	0.01g
$Na_2SeO_3·5H_2O$	0.3mg

Preparation of Trace Elements Solution: Add nitrilotriacetic acid to 500.0mL of distilled/deionized water. Dissolve by adjusting pH to 6.5 with KOH. Add remaining components. Add distilled/deionized water to 1.0L. Mix thoroughly.

Vitamin Solution:

Composition per liter:

Pyridoxine-HCl	10.0mg
Thiamine-HCl·$2H_2O$	5.0mg
Riboflavin	5.0mg
Nicotinic acid	5.0mg
D-Ca-pantothenate	5.0mg
p-Aminobenzoic acid	5.0mg
Lipoic acid	5.0mg
Biotin	2.0mg
Folic acid	2.0mg
Vitamin B_{12}	0.1mg

Preparation of Vitamin Solution: Add components to distilled/deionized water and bring volume to 1.0L. Mix thoroughly. Sparge with 80% H_2 + 20% CO_2. Filter sterilize.

$Na_2S·9H_2O$ Solution:

Composition per 10.0mL:

$Na_2S·9H_2O$	0.5g

Preparation of $Na_2S·9H_2O$ Solution: Add $Na_2S·9H_2O$ to distilled/deionized water and bring volume to 10.0mL. Sparge with N_2. Autoclave for 15 min at 15 psi pressure–121°C. Cool to 25°C. Store anaerobically.

Preparation of Medium: Prepare and dispense medium under 100% N_2. Add components, except $NaHCO_3$ solution, $Na_2S·9H_2O$ solution, cysteine solution, vitamin solution, selenite-tungstate solution, and trace elements solution, to distilled/deionized water and bring volume to 919.0mL. Mix thoroughly. Sparge with 100% N_2. Adjust pH to 8.5 with HCl. Autoclave for 15 min at 15 psi pressure–121°C. Aseptically and anaerobically add 10.0mL $NaHCO_3$ solution, 10.0mL $Na_2S·9H_2O$ solution, 10.0mL cysteine solution, 40.0mL vitamin solution, 1.0mL selenite-tungstate solution, and 10.0mL trace elements solution. Mix thoroughly. Aseptically and anaerobically distribute into sterile tubes or bottles.

Use: For the cultivation of *Thermosyntropha lipolytica.*

Thermoterrabacterium Medium (DSMZ Medium 778)

Composition per liter:

$NaHCO_3$	10.0g
Na_2-9,10-anthraquinone-2,6-disulfonate	8.25g
Yeast extract	1.0g
KH_2PO_4	0.33g
NH_4Cl	0.33g
KCl	0.33g
$MgCl_2·6H_2O$	0.33g
$NiCl_2·6H_2O$	200.0μg
$Na_2SeO_3·5H_2O$	120.0μg
$Na_2WO_4·2H_2O$	30.0μg
Calcium chloride solution	10.0mL
Vitamin solution	10.0mL
Glycerol (87%)	3.0mL
Trace elements solution SL-10	1.0mL

pH 6.8 ± 0.2 at 25°C

Calcium Chloride Solution:

Composition per 10mL:

$CaCl_2·2H_2O$	0.33g

Preparation of Calcium Chloride: Add $CaCl_2·2H_2O$ to 10.0mL of distilled/deionized water. Mix thoroughly. Sparge with 100% N_2. Autoclave for 15 min at 15 psi pressure–121°C. Cool to room tempertuature.

Trace Elements Solution SL-10:
Composition per liter:

$FeCl_2 \cdot 4H_2O$	1.5g
$CoCl_2 \cdot 6H_2O$	190.0mg
$MnCl_2 \cdot 4H_2O$	100.0mg
$ZnCl_2$	70.0mg
$Na_2MoO_4 \cdot 2H_2O$	36.0mg
$NiCl_2 \cdot 6H_2O$	24.0mg
H_3BO_3	6.0mg
$CuCl_2 \cdot 2H_2O$	2.0mg
HCl (25% solution)	10.0mL

Preparation of Trace Elements Solution SL-10: Add $FeCl_2 \cdot 4H_2O$ to 10.0mL of HCl solution. Mix thoroughly. Add distilled/deionized water and bring volume to 1.0L. Add remaining components. Mix thoroughly. Sparge with 80% N_2 + 20% CO_2. Autoclave for 15 min at 15 psi pressure–121°C.

Vitamin Solution:
Composition per liter:

Pyridoxine-HCl	10.0mg
Thiamine-HCl·$2H_2O$	5.0mg
Riboflavin	5.0mg
Nicotinic acid	5.0mg
D-Ca-pantothenate	5.0mg
p-Aminobenzoic acid	5.0mg
Lipoic acid	5.0mg
Biotin	2.0mg
Folic acid	2.0mg
Vitamin B_{12}	0.1mg

Preparation of Vitamin Solution: Add components to distilled/deionized water and bring volume to 1.0L. Mix thoroughly. Sparge with 80% H_2 + 20% CO_2. Filter sterilize.

Preparation of Medium: Add components, except $NaHCO_3$, calcium chloride solution, and vitamin solution, to distilled/deionized water and bring volume to 980.0mL. Mix thoroughly. Gently heat and bring to boiling. Boil for several minutes to dissolve the antraquinone. Cool to room temperature under atmosphere of 100% CO_2. Add solid $NaHCO_3$. Adjust pH to 6.8 wtih NaOH. Dispense medium under CO_2 into tubes or bottles. Autoclave for 15 min at 15 psi pressure–121°C. Cool to room temperature. Aseptically and anaerobically distribute into sterile tubes or bottles. Before use, add $CaCl_2$ and vitamins from anaerobic, sterile stock solution.

Use: For the cultivation of *Thermoterrabacterium ferrireducens.*

Thermotoga elfii Medium

Composition per liter:

NaCl	10.0g
Pancreatic digest of casein	2.0g
Yeast extract	2.0g
NH_4Cl	1.0g
K_2HPO_4	0.3g
KH_2PO_4	0.3g
$MgCl_2 \cdot 6H_2O$	0.2g
$CaCl_2 \cdot 2H_2O$	0.1g
KCl	0.1g
L-Cysteine·HCl·H_2O	0.5g
Sodium acetate	0.5g
Resazurin	1.0mg
Na_2CO_3 solution	20.0mL
$Na_2S \cdot 9H_2O$ solution	20.0mL
Wolfe's mineral solution	10.0mL

pH 8.0 ± 0.2 at 25°C

$NaHCO_3$ Solution:
Composition per 20.0mL:

$NaHCO_3$	2.0g

Preparation of $NaHCO_3$ Solution: Add $NaHCO_3$ to distilled/deionized water and bring volume to 20.0mL. Mix thoroughly. Sparge with 80% N_2 + 20% CO_2. Autoclave for 15 min at 15 psi pressure–121°C.

$Na_2S \cdot 9H_2O$ Solution:
Composition per 20.0mL:

$Na_2S \cdot 9H_2O$	0.4g

Preparation of $Na_2S \cdot 9H_2O$ Solution: Add $Na_2S \cdot 9H_2O$ to distilled/deionized water and bring volume to 20.0mL. Mix thoroughly. Sparge with 100% N_2. Autoclave for 15 min at 15 psi pressure–121°C. Before use, neutralize to pH 7.0 with sterile HCl.

Modified Wolfe's Mineral Solution:
Composition per liter:

$MgSO_4 \cdot 7H_2O$	3.0g
Nitrilotriacetic acid	1.5g
NaCl	1.0g
$MnSO_4 \cdot H_2O$	0.5g
$CaCl_2$	0.1g
$CoCl_2 \cdot 6H_2O$	0.1g
$FeSO_4 \cdot 7H_2O$	0.1g
$ZnSO_4 \cdot 7H_2O$	0.1g
$AlK(SO_4)_2 \cdot 12H_2O$	0.01g
$CuSO_4 \cdot 5H_2O$	0.01g
H_3BO_3	0.01g
$Na_2MoO_4 \cdot 2H_2O$	0.01g
Na_2SeO_3	0.01g
$NaWO_4 \cdot 2H_2O$	0.01g
$NiCl_2 \cdot 6H_2O$	0.01g

Preparation of Modified Wolfe's Mineral Solution: Add nitrilotriacetic acid to 500.0mL of distilled/deionized water. Adjust pH to 6.5 with KOH. Add remaining components one at a time. Add distilled/deionized water to 1.0L. Adjust pH to 6.8.

Preparation of Medium: Prepare and dispense medium under 100% N_2. Add components, except Na_2CO_3 solution and $Na_2S \cdot 9H_2O$ solution, to distilled/deionized water and bring volume to 960.0mL. Mix thoroughly. Adjust pH to 8.0 with 10*M* KOH. Gently heat and bring to boiling. Cool to room temperature while sparging with 100% N_2. Anaerobically dispense into tubes in 5.0mL aliquots under an atmosphere of 80% N_2 + 20% CO_2. Autoclave for 45 min at 6 psi pressure–110°C. Just prior to use, aseptically and anaerobically add 0.1mL of sterile Na_2CO_3 solution and 0.1mL of sterile $Na_2S \cdot 9H_2O$ solution to each tube.

Use: For the cultivation of *Thermotoga elfii.*

Thermotoga hypogea Medium (DSMZ Medium 794)

Composition per liter:

NaCl	10.0g
Yeast extract	2.0g
Trypticase	2.0g
NH_4Cl	1.0g
Na-Acetate	0.5g
L-Cysteine	0.5g
K_2HPO_4	0.3g
KH_2PO_4	0.3g
$MgCl_2 \cdot 6H_2O$	0.2g
$CaCl_2 \cdot 2H_2O$	0.1g
KCl	0.1g
Resazurin	0.5mg

Na-thiosulfate solution 20.0mL
$NaHCO_3$ solution 20.0mL
Xylose solution 20.0mL
$Na_2S \cdot 9H_2O$ solution 14.0mL
Trace elements solution 10.0mL

pH 7.3 ± 0.2 at 25°C

Xylose Solution:

Composition per 20.0mL:

Xylose 3.0g

Preparation of Xylose Solution: Add xylose to distilled/deionized water and bring volume to 20.0mL. Mix thoroughly. Sparge with 100% N_2. Filter sterilize.

$NaHCO_3$ Solution:

Composition per 20.0mL:

$NaHCO_3$ 2.0g

Preparation of $NaHCO_3$ Solution: Add $NaHCO_3$ to distilled/deionized water and bring volume to 20.0mL. Mix thoroughly. Autoclave for 15 min at 15 psi pressure–121°C. Cool to 25°C. Must be prepared freshly.

Na-thiosulfate Solution:

Composition per 20.0mL:

$Na_2S_2O_3 \cdot 5H_2O$ 5.0g

Preparation of Na-thiosulfate Solution: Add $Na_2S_2O_3 \cdot 5H_2O$ to distilled/deionized water and bring volume to 20.0mL. Mix thoroughly. Sparge with 100% N_2. Filter sterilize.

$Na_2S \cdot 9H_2O$ Solution:

Composition per 20.0mL:

$Na_2S \cdot 9H_2O$ 0.6g

Preparation of $Na_2S \cdot 9H_2O$ Solution: Add $Na_2S \cdot 9H_2O$ to distilled/deionized water and bring volume to 20.0mL. Sparge with N_2. Autoclave for 15 min at 15 psi pressure–121°C. Cool to 25°C. Store anaerobically.

Trace Elements Solution:

Composition per liter:

$MgSO_4 \cdot 7H_2O$ 3.0g
Nitrilotriacetic acid 1.5g
NaCl 1.0g
$MnSO_4 \cdot 2H_2O$ 0.5g
$CoSO_4 \cdot 7H_2O$ 0.18g
$ZnSO_4 \cdot 7H_2O$ 0.18g
$CaCl_2 \cdot 2H_2O$ 0.1g
$FeSO_4 \cdot 7H_2O$ 0.1g
$NiCl_2 \cdot 6H_2O$ 0.025g
$KAl(SO_4)_2 \cdot 12H_2O$ 0.02g
H_3BO_3 .... 0.01g
$Na_2MoO_4 \cdot 4H_2O$ 0.01g
$CuSO_4 \cdot 5H_2O$ 0.01g
$Na_2SeO_3 \cdot 5H_2O$ 0.3mg

Preparation of Trace Elements Solution: Add nitrilotriacetic acid to 500.0mL of distilled/deionized water. Dissolve by adjusting pH to 6.5 with KOH. Add remaining components. Add distilled/deionized water to 1.0L. Mix thoroughly.

Preparation of Medium: Prepare and dispense medium under 80% N_2 + 20% CO_2 gas mixture. Add components, except xylose solution, $Na_2S \cdot 9H_2O$ solution, Na-thiosulfate solution, and $NaHCO_3$ solution, to 926.0mL distilled/deionized water. Mix thoroughly. Sparge with 80% N_2 + 20% CO_2. Adjust pH to 7.2-7.4. Autoclave for 15 min at 15 psi pressure–121°C. Cool to 25°C while sparging with 80% N_2 + 20% CO_2. Aseptically and anaerobically add 20.0mL sterile xylose solution, 14.0mL $Na_2S \cdot 9H_2O$ solution, 20.0mL Na-thiosulfate solution, and 20.0mL $NaHCO_3$ solution. Mix thoroughly. Aseptically and anaerobically distribute into sterile tubes or flasks.

Use: For the cultivation of *Thermotoga hypogea.*

Thermotoga Medium

Composition per 1017.0mL:

NaCl 20.0g
Starch, soluble 5.0g
$NiCl_2 \cdot 6H_2O$ 2.0g
KH_2PO_4 .... 0.5g
$Na_2S \cdot 9H_2O$ 0.5g
Yeast extract 0.5g
Resazurin 1.0mg
Artificial seawater 250.0mL
Trace elements solution 15.0mL

pH 6.5 ± 0.2 at 25°C

Artificial Seawater:

Composition per liter:

NaCl 27.7g
$MgSO_4 \cdot 7H_2O$ 7.0g
$MgCl_2 \cdot 6H_2O$ 5.5g
$CaCl_2 \cdot 2H_2O$ 2.25g
KCl 0.65g
NaBr 0.1g
H_3BO_3 .... 30.0mg
$SrCl_2 \cdot 6H_2O$ 15.0mg
Citric acid 10.0mg
KI 0.05mg

Preparation of Artificial Seawater: Add components to distilled/deionized water and bring volume to 1.0L. Mix thoroughly.

Trace Elements Solution:

Composition per liter:

$MgSO_4 \cdot 7H_2O$ 3.0g
Nitrilotriacetic acid 1.5g
NaCl 1.0g
$MnSO_4 \cdot 2H_2O$ 0.5g
$CoSO_4 \cdot 7H_2O$ 0.18g
$ZnSO_4 \cdot 7H_2O$ 0.18g
$CaCl_2 \cdot 2H_2O$ 0.1g
$FeSO_4 \cdot 7H_2O$ 0.1g
$KAl(SO_4)_2 \cdot 12H_2O$ 0.02g
$CuSO_4 \cdot 5H_2O$ 0.01g
H_3BO_3 .... 0.01g
$Na_2MoO_4 \cdot 2H_2O$ 0.01g
$NiCl_2 \cdot 6H_2O$ 0.025g
$Na_2SeO_3 \cdot 5H_2O$ 0.3mg

Preparation of Trace Elements Solution: Add nitrilotriacetic acid to 500.0mL of distilled/deionized water. Adjust pH to 6.5 with KOH. Add remaining components. Add distilled/deionized water to 1.0L.

Preparation of Medium: Prepare and dispense medium under 100% N_2. Add components to distilled/deionized water and bring volume to 1.0L. Mix thoroughly. Sparge with 100% N_2. Distribute into tubes or flasks. Autoclave for 15 min at 15 psi pressure–121°C.

Use: For the cultivation and maintenance of *Thermotoga maritima* and *Thermotoga neapolitana.*

Thermotoga 2 Medium

Composition per 1015.0mL:

Starch, soluble 5.0g
NaCl 3.46g
$MgSO_4 \cdot 7H_2O$ 0.88g

EDTA·Na_2	0.768g
$MgCl_2 \cdot 6H_2O$	0.69g
KH_2PO_4	0.5g
$Na_2S \cdot 9H_2O$	0.5g
Yeast extract	0.5g
$CaCl_2 \cdot 2H_2O$	0.14g
KCl	0.08g
NaBr	12.5mg
H_3BO_3	3.75mg
$(NH_4)_2Ni(SO_4)_2$	3.0mg
$SrCl_2 \cdot 6H_2O$	1.9mg
Resazurin	1.0mg
KI	0.006mg
Trace elements solution	15.0mL

pH 7.0 ± 0.2 at 25°C

Trace Elements Solution:
Composition per liter:

$MgSO_4 \cdot 7H_2O$	3.0g
Nitrilotriacetic acid	1.5g
NaCl	1.0g
$MnSO_4 \cdot 2H_2O$	0.5g
$CoSO_4 \cdot 7H_2O$	0.18g
$ZnSO_4 \cdot 7H_2O$	0.18g
$CaCl_2 \cdot 2H_2O$	0.1g
$FeSO_4 \cdot 7H_2O$	0.1g
$KAl(SO_4)_2 \cdot 12H_2O$	0.02g
$CuSO_4 \cdot 5H_2O$	0.01g
H_3BO_3	0.01g
$Na_2MoO_4 \cdot 2H_2O$	0.01g
$NiCl_2 \cdot 6H_2O$	0.025g
$Na_2SeO_3 \cdot 5H_2O$	0.3mg

Preparation of Trace Elements Solution: Add nitrilotriacetic acid to 500.0mL of distilled/deionized water. Adjust pH to 6.5 with KOH. Add remaining components. Add distilled/deionized water to 1.0L.

Preparation of Medium: Prepare and dispense medium under 100% N_2. Add components to distilled/deionized water and bring volume to 1.0L. Mix thoroughly. Sparge with 100% N_2. Distribute into tubes or flasks. Autoclave for 15 min at 15 psi pressure–121°C.

Use: For the cultivation and maintenance of *Thermotoga thermarum.*

Thermotoga petrophila Medium (DSMZ Medium 913)

Composition per liter:

MOPS	5.0g
Yeast extract	2.0g
$(NH_4)_2SO_4$	1.0g
KH_2PO_4	0.5g
Resazurin	0.5mg
Artificial seawater	750.0mL
Trace elements solution	10.0mL
Vitamin solution	10.0mL
$Na_2S \cdot 9H_2O$ solution	10.0mL

pH 7.0 ± 0.2 at 25°C

Artificial Seawater:
Composition per liter:

NaCl	27.7g
$MgSO_4 \cdot 7H_2O$	7.0g
$MgCl_2 \cdot 6H_2O$	5.5g
KCl	0.65g
NaBr	0.1g
H_3BO_3	30.0mg
$SrCl_2 \cdot 6H_2O$	15.0mg
Citric acid	10.0mg
KI	0.05mg
$CaCl_2 \cdot 2H_2O$	2.25 g

Preparation of Artificial Seawater: Add components to distilled/deionized water and bring volume to 1.0L. Mix thoroughly.

Trace Elements Solution:
Composition per liter:

$MgSO_4 \cdot 7H_2O$	3.0g
Nitrilotriacetic acid	1.5g
NaCl	1.0g
$MnSO_4 \cdot 2H_2O$	0.5g
$CoSO_4 \cdot 7H_2O$	0.18g
$ZnSO_4 \cdot 7H_2O$	0.18g
$CaCl_2 \cdot 2H_2O$	0.1g
$FeSO_4 \cdot 7H_2O$	0.1g
$NiCl_2 \cdot 6H_2O$	0.025g
$KAl(SO_4)_2 \cdot 12H_2O$	0.02g
H_3BO_3	0.01g
$Na_2MoO_4 \cdot 4H_2O$	0.01g
$CuSO_4 \cdot 5H_2O$	0.01g
$Na_2SeO_3 \cdot 5H_2O$	0.3mg

Preparation of Trace Elements Solution: Add nitrilotriacetic acid to 500.0mL of distilled/deionized water. Dissolve by adjusting pH to 6.5 with KOH. Add remaining components. Add distilled/deionized water to 1.0L. Mix thoroughly.

Vitamin Solution:
Composition per liter:

Pyridoxine-HCl	10.0mg
Thiamine-HCl·$2H_2O$	5.0mg
Riboflavin	5.0mg
Nicotinic acid	5.0mg
D-Ca-pantothenate	5.0mg
p-Aminobenzoic acid	5.0mg
Lipoic acid	5.0mg
Biotin	2.0mg
Folic acid	2.0mg
Vitamin B_{12}	0.1mg

Preparation of Vitamin Solution: Add components to distilled/deionized water and bring volume to 1.0L. Mix thoroughly. Sparge with 80% H_2 + 20% CO_2. Filter sterilize.

$Na_2S \cdot 9H_2O$ Solution:
Composition per 10.0mL:

$Na_2S \cdot 9H_2O$	0.3g

Preparation of $Na_2S \cdot 9H_2O$ Solution: Add $Na_2S \cdot 9H_2O$ to distilled/deionized water and bring volume to 10.0mL. Sparge with N_2. Autoclave for 15 min at 15 psi pressure–121°C. Cool to 25°C. Store anaerobically.

Preparation of Medium: Add components, except vitamin solution and $Na_2S \cdot 9H_2O$ solution, to 750.0mL artificial seawater. Bring volume to 980.0mL with distilled/deionized water. Mix thoroughly. Gently heat and bring to boiling. Boil for 3 min. Cool to room temperature while sparging with 100% N_2. Distribute under 100% N_2 into anaerobe tubes or bottles. Autoclave for 15 min at 15 psi pressure–121°C. Aseptically and anaerobically add per liter of medium, 10.0mL vitamin solution and 10.0mL $Na_2S \cdot 9H_2O$ solution. Mix thoroughly. The final pH of the medium should be 7.0.

Use: For the cultivation of *Thermotoga petrophila* and *Thermotoga naphthophila.*

Thermotoga subterranea Medium

Composition per liter:

NaCl	12.0g
$MgSO_4 \cdot 7H_2O$	0.5g
PIPES	3.4g
KCl	2.0g
Peptone	1.0g
$Na_2S \cdot 9H_2O$	0.5g
Yeast extract	0.5g
NH_4Cl	0.1g
$CaCl_2 \cdot 2H_2O$	0.025g
K_2HPO_4	0.02g
Resazurin	0.5mg
Wolfe's mineral solution	10.0mL
Wolfe's vitamin solution	10.0mL

pH 7.0 ± 0.2 at 25°C

Wolfe's Vitamin Solution:

Composition per liter:

Pyridoxine·HCl	10.0mg
p-Aminobenzoic acid	5.0mg
Lipoic acid	5.0mg
Nicotinic acid	5.0mg
Riboflavin	5.0mg
Thiamine·HCl	5.0mg
Calcium DL-pantothenate	5.0mg
Biotin	2.0mg
Folic acid	2.0mg
Vitamin B_{12}	0.1mg

Preparation of Wolfe's Vitamin Solution: Add components to distilled/deionized water and bring volume to 1.0L. Mix thoroughly. Filter sterilize.

Wolfe's Mineral Solution:

Composition per liter:

$MgSO_4 \cdot 7H_2O$	3.0g
Nitrilotriacetic acid	1.5g
NaCl	1.0g
$MnSO_4 \cdot 2H_2O$	0.5g
$CoCl_2 \cdot 6H_2O$	0.1g
$ZnSO_4 \cdot 7H_2O$	0.1g
$CaCl_2 \cdot 2H_2O$	0.1g
$FeSO_4 \cdot 7H_2O$	0.1g
$NiCl_2 \cdot 6H_2O$	0.025g
$KAl(SO_4)_2 \cdot 12H_2O$	0.02g
$CuSO_4 \cdot 5H_2O$	0.01g
H_3BO_3	0.01g
$Na_2MoO_4 \cdot 2H_2O$	0.01g
$Na_2SeO_3 \cdot 5H_2O$	0.3mg

Preparation of Wolfe's Mineral Solution: Add nitrilotriacetic acid to 500.0mL of distilled/deionized water. Adjust pH to 6.5 with KOH. Add remaining components. Add distilled/deionized water to 1.0L. Adjust pH to 6.8.

Preparation of Medium: Prepare and dispense medium under 100% N_2. Add components, except Wolfe's vitamin solution, to distilled/deionized water and bring volume to 990.0mL. Mix thoroughly. Adjust pH to 7.0. Sparge with 100% N_2. Autoclave for 15 min at 15 psi pressure–121°C. Aseptically and anaerobically add 10.0mL of sterile Wolfe's vitamin solution. Mix thoroughly. Adjust medium pH to 7.5 by adding sterile anaerobic 1*N* NaOH. Aseptically and anaerobically distribute into sterile tubes or bottles.

Use: For the cultivation of *Thermotoga subterranea.*

Thermovenabulum Medium (DSMZ Medium 962)

Composition per liter:

$NaHCO_3$	0.7g
NH_4Cl	0.33g
KH_2PO_4	0.33g
$MgCl_2 \cdot 6H_2O$	0.33g
$CaCl_2 \cdot 2H_2O$	0.33g
KCl	0.33g
Yeast extract	0.05g
Meat extract solution	50.0mL
Sodium fumarate solution	50.0mL
Vitamin solution	10.0mL
Trace elements solution	10.0mL
Selenite-tungstate solution	1.0mL

pH 7.0 ± 0.2 at 25°C

Selenite-Tungstate Solution:

Composition per liter:

NaOH	0.5g
$Na_2WO_4 \cdot 2H_2O$	4.0mg
$Na_2SeO_3 \cdot 5H_2O$	3.0mg

Preparation of Selenite-Tungstate Solution: Add components to distilled/deionized water and bring volume to 1.0L. Mix thoroughly. Sparge with 100% N_2. Filter sterilize.

Trace Elements Solution:

Composition per liter:

$MgSO_4 \cdot 7H_2O$	3.0g
Nitrilotriacetic acid	1.5g
NaCl	1.0g
$MnSO_4 \cdot 2H_2O$	0.5g
$CoSO_4 \cdot 7H_2O$	0.18g
$ZnSO_4 \cdot 7H_2O$	0.18g
$CaCl_2 \cdot 2H_2O$	0.1g
$FeSO_4 \cdot 7H_2O$	0.1g
$NiCl_2 \cdot 6H_2O$	0.025g
$KAl(SO_4)_2 \cdot 12H_2O$	0.02g
H_3BO_3	0.01g
$Na_2MoO_4 \cdot 4H_2O$	0.01g
$CuSO_4 \cdot 5H_2O$	0.01g
$Na_2SeO_3 \cdot 5H_2O$	0.3mg

Preparation of Trace Elements Solution: Add nitrilotriacetic acid to 500.0mL of distilled/deionized water. Dissolve by adjusting pH to 6.5 with KOH. Add remaining components. Add distilled/deionized water to 1.0L. Mix thoroughly.

Vitamin Solution:

Composition per liter:

Pyridoxine-HCl	10.0mg
Thiamine-HCl·$2H_2O$	5.0mg
Riboflavin	5.0mg
Nicotinic acid	5.0mg
D-Ca-pantothenate	5.0mg
p-Aminobenzoic acid	5.0mg
Lipoic acid	5.0mg
Biotin	2.0mg
Folic acid	2.0mg
Vitamin B_{12}	0.1mg

Preparation of Vitamin Solution: Add components to distilled/deionized water and bring volume to 1.0L. Mix thoroughly. Sparge with 80% H_2 + 20% CO_2. Filter sterilize.

Sodium Fumarate Solution:

Composition per 50.0mL:

Na_2-fumarate	3.20g

Preparation of Sodium Fumarate Solution: Add Na_2-fumarate to distilled/deionized water and bring volume to 50.0mL. Sparge with N_2. Autoclave for 15 min at 15 psi pressure–121°C. Cool to 25°C.

Meat Extract Solution:
Composition per 50.0mL:

Meat extract	3.0g

Preparation of Meat Extract Solution: Add meat extract to distilled/deionized water and bring volume to 50.0mL. Sparge with N_2. Autoclave for 15 min at 15 psi pressure–121°C. Cool to 25°C.

Preparation of Medium: Add components, except $NaHCO_3$, vitamin solution, sodium fumarate solution, and meat extract solution, to distilled/deionized water and bring volume to 890.0mL. Mix thoroughly. Gently heat and bring to boiling. Boil for 3 min. Cool to 25°C while sparging with 80% N_2 + 20% CO_2. Add solid $NaHCO_3$. Adjust pH to 6.8-7.0. Distribute into anaerobe tubes or bottles under 80% N_2 + 20% CO_2. Autoclave for 15 min at 15 psi pressure–121°C. Aseptically and anaerobically add per liter of medium, 10.0mL sterile vitamin solution, 50.0mL sterile meat extract solution, and 50.0mL sterile sodium fumarate solution. Mix thoroughly. The final pH should be 6.5.

Use: For the cultivation of *Thermovenabulum ferriorganovorum.*

Thermovibrio Medium (DSMZ Medium 961)

Composition per liter:

Synthetic seawater	970.0mL
Potassium nitrate solution	10.0mL
Yeast extract solution	10.0mL
$Na_2S{\cdot}9H_2O$ solution	10.0mL

pH 6.5 ± 0.2 at 25°C

Synthetic Seawater:
Composition per liter:

NaCl	27.7g
$MgSO_4{\cdot}7H_2O$	7.0g
$MgCl_2{\cdot}6H_2O$	5.5g
$CaCl_2{\cdot}2H_2O$	0.75g
KCl	0.65g
KH_2PO_4	0.5g
$(NH_4)_2SO_4$	0.25g
NaBr	0.1g
H_3BO_3	0.03g
$SrCl_2{\cdot}6H_2O$	15.0mg
Resazurin	1.0mg
KI	0.05mg

Preparation of Synthetic Seawater: Add components to distilled water and bring volume to 1.0L. Mix thoroughly.

$Na_2S{\cdot}9H_2O$ Solution:
Composition per 10mL:

$Na_2S{\cdot}9H_2O$	0.5g

Preparation of $Na_2S{\cdot}9H_2O$ Solution: Add $Na_2S{\cdot}9H_2O$ to distilled/deionized water and bring volume to 10.0mL. Mix thoroughly. Autoclave under 100% N_2 for 15 min at 15 psi pressure–121°C. Cool to room temperature.

Yeast Extract Solution:
Composition per 10.0mL:

Yeast extract	1.0g

Preparation of Yeast Extract Solution: Add yeast extract to distilled/deionized water and bring volume to 10.0mL. Mix thoroughly. Autoclave under 100% N_2 for 15 min at 15 psi pressure–121°C. Cool to room temperature.

Potassium Nitrate Solution:
Composition per 10.0mL:

KNO_3	0.33g

Preparation of Potassium Nitrate Solution: Add KNO_3 to distilled/deionized water and bring volume to 10.0mL. Mix thoroughly. Autoclave for 15 min at 15 psi pressure–121°C. Cool to room temperature.

Preparation of Medium: Gently heat and bring 1.0L synthetic seawater to boiling. Boil for 3 min. Cool to 25°C while sparging with 80% H_2 + 20% CO_2. Adjust pH to 6.8-7.0. Distribute 20.0mL aliquots into 100mL serum bottles under 80% H_2 + 20% CO_2. Autoclave for 15 min at 15 psi pressure–121°C. Aseptically and anaerobically add per 20.0mL of medium, 0.2mL sterile potassium nitrate solution, 0.2mL sterile yeast extract solution, and 0.2mL sterile $Na_2S{\cdot}9H_2O$ solution. Mix thoroughly. Adjusted pH should be 6.5.

Use: For the cultivation of *Thermovibrio ruber.*

Thermus Agar

Composition per liter:

Agar	28.0g
Pancreatic digest of casein	2.5g
Yeast extract	2.5g
$Na_2HPO_4{\cdot}12H_2O$	0.43g
$MgCl_2{\cdot}6H_2O$	0.2g
Nitrilotriacetic acid	0.1g
KH_2PO_4	54.0mg
$CaSO_4{\cdot}2H_2O$	40.0mg
Micronutrient solution	1.0mL
Fe-citrate solution	0.5mL

pH 7.2 ± 0.2 at 25°C

Micronutrient Solution:
Composition per liter:

$MnSO_4{\cdot}H_2O$	2.28g
H_3BO_3	0.5g
$ZnSO_4{\cdot}7H_2O$	0.5g
$CoCl_2{\cdot}6H_2O$	45.0mg
$CuSO_4{\cdot}5H_2O$	25.0mg
$Na_2MoO_4{\cdot}2H_2O$	25.0mg
Concentrated H_2SO_4	0.5mL

Preparation of Micronutrient Solution: Add components to distilled/deionized water and bring volume to 1.0L. Mix thoroughly.

Fe-Citrate Solution:
Composition per liter:

Fe-citrate	24.5mg

Preparation of Fe-Citrate Solution: Add Fe-citrate to distilled/deionized water and bring volume to 1.0L. Mix thoroughly.

Preparation of Medium: Add nitrilotriacetic acid to 100.0mL of distilled/deionized water. Mix thoroughly. Adjust pH to 6.5 with KOH. Add remaining components and bring volume to 1.0L with distilled/deionized water. Mix thoroughly. Adjust pH to 7.2. Gently heat and bring to boiling. Distribute into tubes or flasks. Autoclave for 15 min at 15 psi pressure–121°C. Pour into sterile Petri dishes or leave in tubes.

Use: For the cultivation and maintenance of *Thermus* species.

Thermus BP Medium (*Thermus* Beef Extract Polypeptone™ Medium)

Composition per liter:

Agar	25.0g
Beef extract	4.0g
Polypeptone	4.0g
K_2HPO_4	3.0g
KH_2PO_4	1.0g

pH 7.0 ± 0.2 at 25°C

Preparation of Medium: Add components to distilled/deionized water and bring volume to 1.0L. Mix thoroughly. Gently heat and bring to boiling. Distribute into tubes or flasks. Autoclave for 15 min at 15 psi pressure–121°C. Pour into sterile Petri dishes or leave in tubes.

Use: For the cultivation and maintenance of *Thermus aquaticus* and other *Thermus* species.

Thermus brockii Medium

Composition per liter:

Pancreatic digest of casein	1.0g
Yeast extract	1.0g
Salts solution	100.0mL

pH 7.6 ± 0.2 at 25°C

Salts Solution:

Composition per liter:

$NaNO_3$	6.89g
KNO_3	1.03g
$MgSO_4 \cdot 7H_2O$	1.0g
Nitrilotriacetic acid	1.0g
$CaSO_4 \cdot 2H_2O$	0.6g
NaCl	80.0mg
$FeCl_3$ solution	10.0mL
Trace elements solution	10.0mL

Preparation of Salts Solution: Add nitrilotriacetic acid to 500.0mL of distilled/deionized water. Adjust pH to 8.2 with 1*M* NaOH. Add remaining components. Add distilled/deionized water to 1.0L.

$FeCl_3$ Solution:

Composition per 100.0mL:

$FeCl_3$	28.0mg

Preparation of $FeCl_3$ Solution: Add $FeCl_3$ to distilled/deionized water and bring volume to 100.0mL. Mix thoroughly.

Trace Elements Solution:

Composition per liter:

$MnSO_4 \cdot H_2O$	2.2g
H_3BO_3	0.5g
$ZnSO_4 \cdot 7H_2O$	0.5g
$CoCl_2 \cdot 6H_2O$	46.0mg
$Na_2MoO_4 \cdot 2H_2O$	25.0mg
$CuSO_4$	16.0mg
H_2SO_4	0.5mL

Preparation of Trace Elements Solution: Add components to distilled/deionized water and bring volume to 1.0L. Mix thoroughly.

Preparation of Medium: Add components to distilled/deionized water and bring volume to 1.0L. Mix thoroughly. Distribute into tubes or flasks. Autoclave for 15 min at 15 psi pressure–121°C.

Use: For the cultivation of *Thermus brockii.*

Thermus Enhanced Agar

Composition per liter:

Agar	28.0g
Pancreatic digest of casein	2.5g
Yeast extract	2.5g
$MgCl_2 \cdot 6H_2O$	0.2g
Nitrilotriacetic acid	0.1g
$CaSO_4 \cdot 2H_2O$	40.0mg
Phosphate buffer solution	100.0mL
Ferric citrate solution	0.5mL
Trace elements solution	0.5mL

Phosphate Buffer Solution:

Composition per liter:

$Na_2HPO_4 \cdot 12H_2O$	43.0g
KH_2PO_4	5.44g

Preparation of Phosphate Buffer: Add components to distilled/deionized water and bring volume to 1.0L. Mix thoroughly. Adjust pH to 7.2. Autoclave for 15 min at 15 psi pressure–121°C. Cool to 50°–55°C.

Ferric Citrate Solution:

Composition per 10.0mL:

Ferric citrate	24.5mg

Preparation of Ferric Citrate Solution: Add ferric citrate to distilled/deionized water and bring volume to 10.0mL. Mix thoroughly.

Trace Elements Solution:

Composition per liter:

Nitrilotriacetic acid	12.8g
$FeCl_2 \cdot 4H_2O$	1.0g
$MnCl_2 \cdot 4H_2O$	0.5g
$CoCl_2 \cdot 6H_2O$	0.3g
$CuCl_2 \cdot 2H_2O$	50.0mg
$Na_2MoO_4 \cdot 2H_2O$	50.0mg
H_3BO_3	20.0mg
$NiCl_2 \cdot 6H_2O$	20.0mg

Preparation of Trace Elements Solution: Add nitrilotriacetic acid to 500.0mL of distilled/deionized water. Adjust pH to 6.5 with KOH. Add remaining components. Add distilled/deionized water to 1.0L.

Preparation of Medium: Add components, except phosphate buffer solution, to distilled/deionized water and bring volume to 900.0mL. Mix thoroughly. Gently heat and bring to boiling. Adjust pH to 7.2 with NaOH. Autoclave for 15 min at 15 psi pressure–121°C. Cool to 50°–55°C. Aseptically add 100.0mL of sterile phosphate buffer solution. Mix thoroughly. Pour into sterile Petri dishes or distribute into sterile tubes.

Use: For the cultivation of *Thermus* species.

Thermus Enhanced Agar with 1% NaCl

Composition per liter:

Agar	28.0g
NaCl	10.0g
Pancreatic digest of casein	2.5g
Yeast extract	2.5g
$MgCl_2 \cdot 6H_2O$	0.2g
Nitrilotriacetic acid	0.1g
$CaSO_4 \cdot 2H_2O$	40.0mg
Phosphate buffer solution	100.0mL
Ferric citrate solution	0.5mL
Trace elements solution	0.5mL

Phosphate Buffer Solution:

Composition per liter:

$Na_2HPO_4 \cdot 12H_2O$	43.0g
KH_2PO_4	5.44g

Preparation of Phosphate Buffer: Add components to distilled/deionized water and bring volume to 1.0L. Mix thoroughly. Adjust pH to 7.2. Autoclave for 15 min at 15 psi pressure–121°C. Cool to 50–55°C.

Ferric Citrate Solution:
Composition per 10.0mL:
Ferric citrate 24.5mg

Preparation of Ferric Citrate Solution: Add ferric citrate to distilled/deionized water and bring volume to 10.0mL. Mix thoroughly.

Trace Elements Solution:
Composition per liter:

Nitrilotriacetic acid	12.8g
$FeCl_2 \cdot 4H_2O$	1.0g
$MnCl_2 \cdot 4H_2O$	0.5g
$CoCl_2 \cdot 6H_2O$	0.3g
$CuCl_2 \cdot 2H_2O$	50.0mg
$Na_2MoO_4 \cdot 2H_2O$	50.0mg
H_3BO_3	20.0mg
$NiCl_2 \cdot 6H_2O$	20.0mg

Preparation of Trace Elements Solution: Add nitrilotriacetic acid to 500.0mL of distilled/deionized water. Adjust pH to 6.5 with KOH. Add remaining components. Add distilled/deionized water to 1.0L.

Preparation of Medium: Add components, except phosphate buffer solution, to distilled/deionized water and bring volume to 900.0mL. Mix thoroughly. Gently heat and bring to boiling. Adjust pH to 7.2 with NaOH. Autoclave for 15 min at 15 psi pressure–121°C. Cool to 50°–55°C. Aseptically add 100.0mL of sterile phosphate buffer solution. Mix thoroughly. Pour into sterile Petri dishes or distribute into sterile tubes.

Use: For the cultivation of *Thermus* species.

Thermus Medium

Composition per liter:

Agar	30.0g
Polypeptone	8.0g
Yeast extract	4.0g
NaCl	2.0g

pH 7.5 ± 0.2 at 25°C

Preparation of Medium: Add components to distilled/deionized water and bring volume to 1.0L. Mix thoroughly. Gently heat and bring to boiling. Distribute into tubes or flasks. Autoclave for 15 min at 15 psi pressure–121°C. Pour into sterile Petri dishes or leave in tubes.

Use: For the cultivation and maintenance of *Thermus aquaticus* and other *Thermus* species.

Thermus 162 Medium (DSMZ Medium 878)

Composition per liter:

Agar	28.0g
Yeast extract	1.0g
Tryptone	1.0g
$MgCl_2 \cdot 6H_2O$	200.0mg
Nitrilotriacetic acid	100.0mg
$CaSO_4 \cdot 2H_2O$	40.0mg
Phosphate buffer	100.0mL
Ferric citrate solution	0.5mL
Trace elements solution	0.5mL

pH 7.2 ± 0.2 at 25°C

Ferric Citrate Solution:
Composition per 10.0mL:
Ferric citrate 24.5mg

Preparation of Ferric Citrate Solution: Add ferric citrate to distilled/deionized water and bring volume to 10.0mL. Mix thoroughly.

Trace Elements Solution:

$CoCl_2 \cdot 6H_2O$	45.0g
$CuSO_4 \cdot 5H_2O$	25.0g
$Na_2MoO_4 \cdot 4H_2O$	25.0g
$MnSO_4 \cdot 2H_2O$	2.28g
$ZnSO_4 \cdot 7H_2O$	0.5g
H_3BO_3	0.5g
H_2SO_4	0.5mL

Preparation of Trace Elements Solution: Add components to distilled/deionized water and bring volume to 1.0L. Mix thoroughly.

Phosphate Buffer:
Composition per liter:

$Na_2HPO_4 \cdot 12H_2O$	43.0g
KH_2PO_4	5.44g

Preparation of Phosphate Buffer: Add components to distilled/deionized water and bring volume to 1.0L. Mix thoroughly. Adjust pH to 7.2. Autoclave for 15 min at 15 psi pressure–121°C. Cool to 50°C.

Preparation of Medium: Add components, except phosphate buffer, to distilled/deionized water and bring volume to 900.0mL. Mix thoroughly. Gently heat and bring to boiling. Autoclave for 15 min at 15 psi pressure–121°C. Cool to 50°C. Add 100.0mL warm phosphate buffer. Mix thoroughly. Pour into Petri dishes or distribute into sterile tubes.

Use: For the cultivation and maintenance of *Thermus* spp., *Rubrobacter xylanophilus, Thermonema rossianum, Deinococcus geothermalis, Deinococcus murrayi,* and *Tepidimonas ignava.*

Thermus 162 Medium

Composition per 1010.0mL:

Agar	28.0g
Tryptone	2.5g
Yeast extract	2.5g
$MgCl_2 \cdot 6H_2O$	0.2g
Nitrilotriacetic acid	0.1g
$CaSO_4 \cdot 2H_2O$	40.0mg
Phosphate buffer solution	100.0mL
Ferric citrate solution	0.5mL
Trace elements solution	0.5mL

pH 7.2 ± 0.2 at 25°C

Phosphate Buffer Solution:
Composition per liter:

$Na_2HPO_4 \cdot 12H_2O$	43.0g
KH_2PO_4	5.44g

Preparation of Phosphate Buffer Solution: Add components to distilled/deionized water and bring volume to 1.0L. Mix thoroughly. Adjust pH to 7.2. Autoclave for 15 min at 15 psi pressure–121°C.

Ferric Citrate Solution:
Composition per 10.0mL:
Ferric citrate 24.5mg

Preparation of Ferric Citrate Solution: Add ferric citrate to distilled/deionized water and bring volume to 10.0mL. Mix thoroughly.

Trace Elements Solution:
Composition per liter:

Nitrilotriacetic acid	12.8g
$FeCl_2 \cdot 4H_2O$	1.0g
$MnCl_2 \cdot 4H_2O$	0.5g
$CoCl_2 \cdot 4H_2O$	0.3g
$CuCl_2 \cdot 2H_2O$	50.0mg
$Na_2MoO_4 \cdot 2H_2O$	50.0mg
H_3BO_3	20.0mg
$NiCl_2 \cdot 6H_2O$	20.0mg

Preparation of Trace Elements Solution: Add nitrilotriacetic acid to 500.0mL of distilled/deionized water. Adjust pH to 6.5 with KOH. Add remaining components. Add distilled/deionized water to 1.0L. Adjust pH to 7.0.

Preparation of Medium: Add components, except phosphate buffer solution, to distilled/deionized water and bring volume to 900.0mL. Mix thoroughly. Gently heat and bring to boiling. Adjust pH to 7.2 with NaOH. Autoclave for 15 min at 15 psi pressure–121°C. Cool to 50°–55°C. Aseptically add 100.0mL of sterile phosphate buffer solution. Mix thoroughly. Pour into sterile Petri dishes or distribute into sterile tubes.

Use: For the cultivation of *Thermus* species.

Thermus Medium Enhanced

Composition per liter:

Agar	28.0g
Yeast extract	2.5g
Tryptone	2.5g
$MgCl_2 \cdot 6H_2O$	0.2g
Nitrilotriacetic acid	0.1g
$CaSO_4 \cdot 2H_2O$	0.04g
Phosphate buffer solution	100.0mL
Ferric citrate	0.5mL
Trace elements solution	0.5mL

Ferric Citrate Solution:
Composition per 100.0mL:

Ferric citrate	0.24g

Preparation of Ferric Citrate Solution: Add ferric citrate to distilled/deionized water and bring volume to 100.0mL. Mix thoroughly.

Phosphate Buffer Solution:
Composition per liter:

$Na_2HPO_4 \cdot 12H_2O$	43.0g
KH_2PO_4	5.44g

Preparation of Phosphate Buffer Solution: Add components to distilled/deionized water and bring volume to 1.0L. Mix thoroughly. Adjust pH to 7.2. Autoclave for 15 min at 15 psi pressure–121°C.

Trace Elements Solution:
Composition per liter:

Nitrilotriacetic acid	12.8g
$FeCl_3 \cdot 4H_2O$	1.0g
$MnCl_2 \cdot 4H_2O$	0.5g
$CoCl_2 \cdot 6H_2O$	0.3g
$CuCl_2 \cdot 2H_2O$	0.05g
$Na_2MoO_4 \cdot 2H_2O$	0.05g
H_3BO_3	0.02g
$NiCl_2 \cdot 6H_2O$	0.02g

Preparation of Trace Elements Solution: Add components to distilled/deionized water and bring volume to 1.0L. Mix thoroughly.

Preparation of Medium: Add components, except phosphate buffer solution, to distilled/deionized water and bring volume to 990.0mL. Mix thoroughly. Gently heat and bring to boiling. Autoclave for 15 min at 15 psi pressure–121°C. Aseptically add 10.0mL of sterile phosphate buffer solution. Mix thoroughly. Pour into sterile Petri dishes or distribute into sterile tubes.

Use: For the cultivation and maintenance of *Thermus* species.

Thermus Medium Enhanced with 1% NaCl

Composition per liter:

Agar	28.0g
NaCl	10.0g
Yeast extract	2.5g
Tryptone	2.5g
$MgCl_2 \cdot 6H_2O$	0.2g
Nitrilotriacetic acid	0.1g
$CaSO_4 \cdot 2H_2O$	0.04g
Phosphate buffer solution	100.0mL
Ferric citrate	0.5mL
Trace elements solution	0.5mL

Ferric Citrate Solution:
Composition per 100.0mL:

Ferric citrate	0.24g

Preparation of Ferric Citrate Solution: Add ferric citrate to distilled/deionized water and bring volume to 100.0mL. Mix thoroughly.

Phosphate Buffer Solution:
Composition per liter:

$Na_2HPO_4 \cdot 12H_2O$	43.0g
KH_2PO_4	5.44g

Preparation of Phosphate Buffer Solution: Add components to distilled/deionized water and bring volume to 1.0L. Mix thoroughly. Adjust pH to 7.2. Autoclave for 15 min at 15 psi pressure–121°C.

Trace Elements Solution:
Composition per liter:

Nitrilotriacetic acid	12.8g
$FeCl_3 \cdot 4H_2O$	1.0g
$MnCl_2 \cdot 4H_2O$	0.5g
$CoCl_2 \cdot 6H_2O$	0.3g
$CuCl_2 \cdot 2H_2O$	0.05g
$Na_2MoO_4 \cdot 2H_2O$	0.05g
H_3BO_3	0.02g
$NiCl_2 \cdot 6H_2O$	0.02g

Preparation of Trace Elements Solution: Add components to distilled/deionized water and bring volume to 1.0L. Mix thoroughly.

Preparation of Medium: Add components, except phosphate buffer solution, to distilled/deionized water and bring volume to 990.0mL. Mix thoroughly. Gently heat and bring to boiling. Autoclave for 15 min at 15 psi pressure–121°C. Aseptically add 10.0mL of sterile phosphate buffer solution. Mix thoroughly. Pour into sterile Petri dishes or distribute into sterile tubes.

Use: For the cultivation and maintenance of *Rhodothermus marinus.*

Thermus PMY Agar (*Thermus* Peptone Meat Extract Yeast Extract Agar)

Composition per liter:

Agar	15.0g
Peptone	5.0g

Meat extract 3.5g
Yeast extract 1.5g
NaCl 1.5g

pH 7.0 ± 0.2 at 25°C

Preparation of Medium: Add components to distilled/deionized water and bring volume to 1.0L. Mix thoroughly. Gently heat and bring to boiling. Distribute into tubes or flasks. Autoclave for 15 min at 15 psi pressure–121°C. Pour into sterile Petri dishes or leave in tubes.

Use: For the cultivation and maintenance of *Thermus aquaticus* and other *Thermus* species.

Thermus ruber Medium

Composition per liter:

Agar 12.0g
Universal peptone 5.0g
Starch, soluble 1.0g
Yeast extract 1.0g

pH 8.0 ± 0.2 at 25°C

Source: Universal peptone is available from Merck, Sharpe, and Dohme.

Preparation of Medium: Add components to distilled/deionized water and bring volume to 1.0L. Mix thoroughly. Gently heat and bring to boiling. Distribute into tubes or flasks. Autoclave for 15 min at 15 psi pressure–121°C. Pour into sterile Petri dishes or leave in tubes.

Use: For the cultivation and maintenance of *Thermus ruber.*

Thermus thermophilus Medium (DSMZ Medium 74)

Polypeptone 8.0g
Yeast extract 4.0g
NaCl 2.0g

pH 7.0 ± 0.2 at 25°C

Preparation of Medium: Add components, to distilled/deionized water and bring volume to 1.0L. Mix thoroughly. Adjust pH to 7.0. Distribute into tubes or flasks. Autoclave for 15 min at 15 psi pressure–121°C.

Use: For the cultivation and maintenance of *Thermus thermophilus.*

Thiobacillus A2 Agar (T3 Agar)

Composition per 1100.0mL:

Solution B 1.0L
Solution A 100.0mL

pH 8.5 ± 0.2 at 25°C

Solution A:

Composition per 100.0mL:

$Na_2S_2O_3 \cdot 5H_2O$ 5.0g
Na_2HPO_4 .... 4.2g
KH_2PO_4 .... 1.5g
NH_4Cl 1.0g
Phenol Red (0.2% solution) 1.0mL

Preparation of Solution A: Add components to distilled/deionized water and bring volume to 100.0mL. Mix thoroughly. Adjust pH to 9.0. Autoclave for 15 min at 15 psi pressure–121°C. Cool to 45°–50°C.

Solution B:

Composition per liter:

Agar 15.0g
$MgSO_4 \cdot 7H_2O$ 0.1g
Trace metals solution 5.0mL

Preparation of Solution B: Add components to distilled/deionized water and bring volume to 1.0mL. Mix thoroughly. Autoclave for 15 min at 15 psi pressure–121°C. Cool to 45°–50°C.

Trace Metal Solution:

Composition per liter:

EDTA 50.0g
$ZnSO_4$ 22.0g
$CaCl_2$ 5.54g
$MnCl_2$ 5.06g
$FeSO_4 \cdot 7H_2O$ 4.99g
$CoCl_2$ 1.61g
$CuSO_4$ 1.57g
$(NH_4)_2MoO_4 \cdot 4H_2O$ 1.1g

Preparation of Trace Metal Solution: Add components to distilled/deionized water and bring volume to 1.0L. Mix thoroughly. Adjust pH to 6.0 with KOH.

Preparation of Medium: Aseptically add 100.0mL of sterile solution A to 1.0L of sterile solution B. Mix thoroughly. Adjust pH to 8.5 if necessary. Pour into sterile Petri dishes or distribute into sterile tubes.

Use: For the cultivation and maintenance of *Thiobacillus versutus* and other *Thiobacillus* species.

Thiobacillus acidophilus Agar

Composition per liter:

Agar 15.0g
$(NH_4)_2SO_4$ 3.0g
$MgSO_4 \cdot 7H_2O$ 1.0g
KH_2PO_4 .... 0.5g
KCl 0.1g
$Ca(NO_3)_2 \cdot 4H_2O$ 18.0mg
$FeSO_4 \cdot 7H_2O$ 0.01mg
Glucose solution 20.0mL

pH 4.5 ± 0.2 at 25°C

Glucose Solution:

Composition per 20.0mL:

Glucose 10.0g

Preparation of Glucose Solution: Add glucose to distilled/deionized water and bring volume to 20.0mL. Mix thoroughly. Autoclave for 15 min at 15 psi pressure–121°C.

Preparation of Medium: Add components, except glucose solution, to distilled/deionized water and bring volume to 980.0mL. Mix thoroughly. Gently heat and bring to boiling. Autoclave for 15 min at 15 psi pressure–121°C. Cool to 50°–55°C. Adjust pH to 4.5 with H_2SO_4. Aseptically add 20.0mL of sterile glucose solution. Mix thoroughly. Pour into sterile Petri dishes or distribute into sterile tubes.

Use: For the cultivation and maintenance of *Thiobacillus acidophilus.*

Thiobacillus acidophilus Medium (DSMZ Medium 108)

Composition per liter:

Agar 15.0g
$(NH_4)_2SO_4$ 3.0g
$MgSO_4 \cdot 7H_2O$ 1.0g
KH_2PO_4 .... 0.5g
KCl 0.1g
$Ca(NO_3)_2 \cdot 4H_2O$ 18.0mg
$FeSO_4 \cdot 7H_2O$ 0.01mg
Glucose solution 50.0mL

pH 4.5 ± 0.2 at 25°C

Glucose Solution:
Composition per 100.0mL:
D-Glucose 10.0g

Preparation of Glucose Solution: Add glucose to distilled/deionized water and bring volume to 100.0mL. Mix thoroughly. Filter sterilize.

Preparation of Medium: Add components, except glucose solution, to distilled/deionized water and bring volume to 950.0mL. Mix thoroughly. Gently heat and bring to boiling. Autoclave for 15 min at 15 psi pressure–121°C. Cool to 45°–50°C. Aseptically add 50.0mL sterile glucose solution. Mix thoroughly. Adjust pH to 4.5 with H_2SO_4. Pour into sterile Petri dishes or distribute into sterile tubes.

Use: For the cultivation and maintenance of *Acidiphilium acidophilum.*

Thiobacillus Agar

Composition per liter:
Ionagar No. 2 .... 12.0g
$Na_2S_2O_3 \cdot 5H_2O$ 10.0g
K_2HPO_4 .... 4.0g
KH_2PO_4 .... 4.0g
$(NH_4)_2 \cdot SO_4$ 0.1g
$MgSO_4 \cdot 7H_2O$ 0.1g
$CaCl_2$ 0.1g
$FeCl_3 \cdot 6H_2O$ 2.0mg
$MnSO_4 \cdot 4H_2O$ 2.0mg
pH 6.6 ± 0.2 at 25°C

Source: Ionagar No. 2 is available from Oxoid Unipath.

Preparation of Medium: Add components, except the agar, to distilled/deionized water and bring volume to 950.0mL. Mix thoroughly. Adjust the pH to 6.6. Add the agar. Bring volume to 1.0L with distilled/deionized water. Gently heat and bring to boiling. Distribute into tubes or flasks. Autoclave for 20 min at 15 psi pressure–121°C. Pour into sterile Petri dishes or leave in tubes.

Use: For the cultivation of nonaciduric *Thiobacillus* species such as *Thiobacillus neapolitanus, Thiobacillus novellus,* and *Thiobacillus thioparus.*

Thiobacillus Agar I for Acidophilic *Thiobacillus*

Composition per liter:
Solution A 500.0mL
Solution B 500.0mL
pH 4.2 ± 0.2 at 25°C

Solution A:
Composition per 500.0mL:
$Na_2S_2O_3$ 5.0g
KH_2PO_4 .... 3.0g
$CaCl_2$ 0.1g
$MgCl_2 \cdot 6H_2O$ 0.1g
NH_4Cl 0.1g

Preparation of Medium: Add components to distilled/deionized water and bring volume to 500.0mL. Adjust pH to 4.2 with 1*N* HCl. Mix thoroughly. Autoclave for 15 min at 15 psi pressure–121°C.

Solution B:
Composition per 500.0mL:
Agar, noble 20.0g

Preparation of Medium: Aseptically combine 500.0mL of sterile solution A and 500.0mL of sterile solution B. Combine the two solutions while still hot. Mix thoroughly. Pour into sterile Petri dishes or distribute into sterile tubes.

Use: For the cultivation of *Thiobacillus thiooxidans.*

Thiobacillus albertis Agar

Composition per 500.0mL:
Solution A 250.0mL
Solution B 250.0mL
pH 4.0 ± 0.2 at 25°C

Solution A:
Composition per 250.0mL:
$Na_2S_2O_3 \cdot 5H_2O$ 5.0g
KH_2PO_4 .... 3.0g
$MgSO_4 \cdot 7H_2O$ 0.5g
$(NH_4)_2SO_4$ 0.4g
$CaCl_2 \cdot 2H_2O$ 0.25g
$FeSO_4 \cdot 7H_2O$ 10.0mg

Preparation of Solution A: Add components to distilled/deionized water and bring volume to 250.0mL. Mix thoroughly. Adjust pH to 4.0. Autoclave for 15 min at 15 psi pressure–121°C. Cool to 50°–55°C.

Solution B:
Composition per 250.0mL:
Agar 15.0g

Preparation of Solution B: Add agar to distilled/deionized water and bring volume to 250.0mL. Mix thoroughly. Gently heat and bring to boiling. Autoclave for 15 min at 15 psi pressure–121°C. Cool to 50°–55°C.

Preparation of Medium: Aseptically combine 250.0mL of sterile solution A with 250.0mL of sterile solution B. Mix thoroughly. Pour into sterile Petri dishes or distribute into sterile tubes.

Use: For the cultivation of *Thiobacillus albertis.*

Thiobacillus aquaesulis Agar

Composition per liter:
Solution B 900.0mL
Solution A 100.0mL
pH 7.4 ± 0.2 at 25°C

Solution A:
Composition per 100.0mL:
$Na_2HPO_4 2H_2O$ 7.9g
KH_2PO_4 .... 1.5g

Preparation of Solution A: Add components to distilled/deionized water and bring volume to 100.0mL. Mix thoroughly. Adjust pH to 7.4. Autoclave for 15 min at 15 psi pressure–121°C. Cool to 50°–55°C.

Solution B:
Composition per 900.0mL:
Agar 15.0g
$Na_2S_2O_3$ 5.0g
NH_4Cl 0.4g
$MgSO_4 \cdot 7H_2O$ 0.1g
Phenol Red 3.0mg
Trace elements solution 10.0mL

Trace Elements Solution:
Composition per liter:
Disodium EDTA 50.0g
NaOH 11.0g
$ZnSO_4 \cdot 7H_2O$ 11.0g
$CaCl_2 \cdot 2H_2O$ 7.34g
$FeSO_4 \cdot 7H_2O$ 5.0g
$MnCl_2 \cdot 4H_2O$ 2.5g

$CoCl_2 \cdot 6H_2O$ 0.5g
$(NH_4)6MoO_{24} \cdot 4H_2O$ 0.5g
$CuSO_4 \cdot 5H_2O$ 0.2g

Preparation of Trace Elements Solution: Add components to distilled/deionized water and bring volume to 1.0L. Mix thoroughly. Adjust pH to 6.0.

Preparation of Solution B: Add components to distilled/deionized water and bring volume to 900.0mL. Mix thoroughly. Gently heat and bring to boiling. Autoclave for 15 min at 15 psi pressure–121°C. Cool to 50°–55°C.

Preparation of Medium: Aseptically combine 100.0mL of cooled, sterile solution A with 900.0mL of cooled, sterile solution B. Mix thoroughly. Pour into sterile Petri dishes or distribute into sterile tubes.

Use: For the cultivation and maintenance of *Thiobacillus aquaesulis*.

Thiobacillus Broth I for Acidophilic *Thiobacillus*

Composition per liter:
$Na_2S_2O_3$ 5.0g
KH_2PO_4 .... 3.0g
$CaCl_2$ 0.1g
$MgCl_2 \cdot 6H_2O$ 0.1g
NH_4Cl 0.1g
pH 4.2 ± 0.2 at 25°C

Preparation of Medium: Add components to distilled/deionized water and bring volume to 1.0L. Adjust pH to 4.2 with 1*N* HCl. Mix thoroughly. Distribute into screw-capped tubes or flasks. Autoclave for 15 min at 15 psi pressure–121°C.

Use: For the cultivation of *Thiobacillus thiooxidans*.

Thiobacillus caldus Agar

Composition per liter:
Solution E 500.0mL
Solution A 460.0mL
Solution D 20.0mL
Solution B 10.0mL
Solution C 10.0mL
pH 2.5 ± 0.2 at 25°C

Solution A:
Composition per 460.0mL:
$Na_2SO_4 \cdot 10H_2O$ 3.2g
$(NH_4)_2SO_4$ 3.0g
$MgSO_4 \cdot 7H_2O$ 0.5g
KCl 0.1g
K_2HPO_4 .... 50.0mg

Preparation of Solution A: Add components to distilled/deionized water and bring volume to 460.0mL. Mix throughly. Adjust pH to 1.75 with H_2SO_4. Autoclave for 15 min at 15 psi pressure–121°C. Cool and maintain above 60°C.

Solution B:
Composition per 10.0mL:
$FeCl_3 \cdot 6H_2O$ 11.0mg
$Ca(NO_3)_2 \cdot 4H_2O$ 10.0mg
H_3BO_3 .... 2.0mg
$MnSO_4 \cdot H_2O$ 2.0mg
$ZnSO_4 \cdot 7H_2O$ 0.9mg
$Na_2MoO_4 \cdot 2H_2O$ 0.8mg
$CoCl_2 \cdot 6H_2O$ 0.6mg
$CuSO_4 \cdot 5H_2O$ 0.5mg

Preparation of Solution B: Add components to distilled/deionized water and bring volume to 10.0mL. Mix thoroughly. Filter sterilize.

Solution C:
Composition per 10.0mL:
Glucose 0.45g

Preparation of Solution C: Add glucose to distilled/deionized water and bring volume to 10.0mL. Mix thoroughly. Filter sterilize.

Solution D:
Composition per 20.0mL:
$Na_2S_4O_6$ 0.77g

Preparation of Solution D: Add $Na_2S_4O_6$ to distilled/deionized water and bring volume to 20.0mL. Mix thoroughly. Filter sterilize.

Solution E:
Composition per 500.0mL:
Phytagel™ (Gellan gum;
available from Sigma Chemical Co.) 15.0g

Preparation of Solution E: Add Phytagel to distilled/deionized water and bring volume to 500.0mL. Mix thoroughly. Autoclave for 15 min at 15 psi pressure–121°C. Cool and maintain above 60°C.

Preparation of Medium: Maintain solutions A and E above 60°C to prevent rapid gelling of medium. Aseptically combine 460.0mL of sterile solution A with 10.0mL of sterile solution B, 10.0mL of sterile solution C, 20.0mL of sterile solution D, and 500.0mL of sterile solution E. Mix throughly. Pour into sterile Petri dishes or distribute into sterile tubes.

Use: For the heterotrophic cultivation and maintenance of *Thiobacillus caldus*.

Thiobacillus caldus Broth

Composition per liter:
Solution A 970.0mL
Solution C 20.0mL
Solution B 10.0mL
pH 2.5 ± 0.2 at 25°C

Solution A:
Composition per 970.0mL:
$Na_2SO_4 \cdot 10H_2O$ 3.2g
$(NH_4)_2SO_4$ 3.0g
$MgSO_4 \cdot 7H_2O$ 0.5g
KCl 0.1g
K_2HPO_4 .... 50.0mg

Preparation of Solution A: Add components to distilled/deionized water and bring volume to 970.0mL. Mix thoroughly. Adjust pH to 1.75 with H_2SO_4. Autoclave for 15 min at 15 psi pressure–121°C.

Solution B:
Composition per 10.0mL:
$FeCl_3 \cdot 6H_2O$ 11.0mg
$Ca(NO_3)_2 \cdot 4H_2O$ 10.0mg
H_3BO_3 .... 2.0mg
$MnSO_4 \cdot H_2O$ 2.0mg
$ZnSO_4 \cdot 7H_2O$ 0.9mg
$Na_2MoO_4 \cdot 2H_2O$ 0.8mg
$CoCl_2 \cdot 6H_2O$ 0.6mg
$CuSO_4 \cdot 5H_2O$ 0.5mg

Preparation of Solution B: Add components to distilled/deionized water and bring volume to 10.0mL. Mix thoroughly. Filter sterilize.

Solution C:
Composition per 20.0mL:

$Na_2S_4O_6$	0.77g

Preparation of Solution C: Add $Na_2S_4O_6$ to distilled/deionized water and bring volume to 20.0mL. Mix thoroughly. Filter sterilize.

Preparation of Medium: Aseptically combine 970.0mL of sterile solution A with 10.0mL of sterile solution B and 20.0mL of sterile solution C. Mix thoroughly. Aseptically distribute into sterile tubes or flasks.

Use: For the chemolithotrophic cultivation of *Thiobacillus caldus.*

Thiobacillus caldus Medium

Composition per liter:

Solution E	500.0mL
Solution A	460.0mL
Solution D	20.0mL
Solution B	10.0mL
Solution C	10.0mL

pH 2.5 ± 0.2 at 25°C

Solution A:
Composition per 460.0mL:

$Na_2SO_4 \cdot 10H_2O$	3.2g
$(NH_4)_2SO_4$	3.0g
$MgSO_4 \cdot 7H_2O$	0.5g
KCl	0.1g
K_2HPO_4	50.0mg

Preparation of Solution A: Add components to distilled/deionized water and bring volume to 460.0mL. Mix thoroughly. Adjust pH to 1.75 with H_2SO_4. Autoclave for 15 min at 15 psi pressure–121°C. Cool to 60°–65°C.

Solution B:
Composition per 10.0mL:

$FeCl_3 \cdot 6H_2O$	11.0mg
$Ca(NO_3)_2 \cdot 4H_2O$	10.0mg
H_3BO_3	2.0mg
$MnSO_4 \cdot H_2O$	2.0mg
$ZnSO_4 \cdot 7H_2O$	0.9mg
$Na_2MoO_4 \cdot 2H_2O$	0.8mg
$CoCl_2 \cdot 6H_2O$	0.6mg
$CuSO_4 \cdot 5H_2O$	0.5mg

Preparation of Solution B: Add components to distilled/deionized water and bring volume to 10.0mL. Mix thoroughly. Filter sterilize. Warm to 60°C.

Solution C:
Composition per 10.0mL:

Glucose	0.45g

Preparation of Solution C: Add glucose to distilled/deionized water and bring volume to 10.0mL. Mix thoroughly. Filter sterilize. Warm to 60°C.

Solution D:
Composition per 20.0mL:

Sodium tetrathionate	0.77g

Preparation of Solution D: Add sodium tetrathionate to distilled/deionized water and bring volume to 60.0mL. Mix thoroughly. Filter sterilize. Warm to 60°C.

Solution E:
Composition per 500.0mL:

Phytagel	15.0g

Preparation of Solution E: Add Phytagel to distilled/deionized water and bring volume to 500.0mL. Mix thoroughly. Autoclave for 15 min at 15 psi pressure–121°C. Cool to 60°–65°C.

Preparation of Medium: Aseptically combine 460.0mL of sterile solution A, 10.0mL of sterile solution B, 10.0mL of sterile solution C, 20.0mL of sterile solution D, and 500.0mL of sterile solution E. Mix thoroughly. Aseptically pour into sterile Petri dishes or distribute into sterile tubes.

Use: For the cultivation of *Thiobacillus caldus.*

Thiobacillus cuprinus Medium

Composition per 1010.0mL:

$MgSO_4 \cdot 7H_2O$	3.45g
$MgCl_2 \cdot 6H_2O$	2.75g
NH_4Cl	1.25g
NaCl	0.5g
Sulfur, powdered	0.5g
KCl	0.33g
$CaCl_2 \cdot 2H_2O$	0.14g
K_2HPO_4	0.14g
$NiCl_2 \cdot 6H_2O$	2.0mg
Trace elements solution	10.0mL

pH 3.5 ± 0.2 at 25°C

Preparation of Sulfur: Sterilize 1.0g of powdered sulfur by steaming for 1 hr on 3 consecutive days.

Trace Elements Solution:
Composition per liter:

$MgSO_4 \cdot 7H_2O$	3.0g
Nitrilotriacetic acid	1.5g
NaCl	1.0g
$MnSO_4 \cdot 2H_2O$	0.5g
$CoSO_4 \cdot 7H_2O$	0.18g
$ZnSO_4 \cdot 7H_2O$	0.18g
$CaCl_2 \cdot 2H_2O$	0.1g
$FeSO_4 \cdot 7H_2O$	0.1g
$KAl(SO_4)_2 \cdot 12H_2O$	0.02g
$CuSO_4 \cdot 5H_2O$	0.01g
H_3BO_3	0.01g
$Na_2MoO_4 \cdot 2H_2O$	0.01g
$NiCl_2 \cdot 6H_2O$	0.025g
$Na_2SeO_3 \cdot 5H_2O$	0.3mg

Preparation of Trace Elements Solution: Add nitrilotriacetic acid to 500.0mL of distilled/deionized water. Adjust pH to 6.5 with KOH. Add remaining components. Add distilled/deionized water to 1.0L.

Preparation of Medium: Add components, except sulfur, to distilled/deionized water and bring volume to 1.0L. Mix thoroughly. Adjust pH to 3.5 with H_2SO_4. Autoclave for 15 min at 15 psi pressure–121°C. Aseptically add 0.5g of sterile sulfur. Mix thoroughly. Aseptically distribute into sterile tubes or flasks.

Use: For the cultivation and maintenance of *Thiobacillus cuprinus.*

Thiobacillus denitrificans Medium

Composition per liter:

KNO_3	5.0g
$Na_2S_2O_3 \cdot 5H_2O$	5.0g
$NaHCO_3$	1.0g

K_2HPO_4 .... 0.2g
$MgCl_2$ 0.1g
pH 7.0 ± 0.2 at 25°C

Preparation of Medium: Add components to distilled/deionized water and bring volume to 1.0L. Mix thoroughly. Distribute into tubes or flasks. Autoclave for 15 min at 15 psi pressure–121°C.

Use: For the cultivation of *Thiobacillus denitrificans.*

Thiobacillus ferrooxidans Medium

Composition per liter:
$Al_2(SO_4)_3·12H_2O$ 1.4g
NaCl 1.0g
KH_2PO_4 .... 0.4g
$MgSO_4·7H_2O$ 0.1g
$(NH_4)_2SO_4$ 0.1g
$CaCl_2$ 0.03g
$MnSO_4·4H_2O$ 0.02g
$FeSO_4·7H_2O$ solution 100.0mL

$FeSO_4·7H_2O$ Solution:
Composition per 100.0mL:
$FeSO_4·7H_2O$ 10.0g
H_2SO_4, concentrated 0.09mL

Preparation of $FeSO_4·7H_2O$ Solution: Add 10.0g of $FeSO_4·7H_2O$ and 0.09mL of H_2SO_4 to distilled/deionized water and bring volume to 100.0mL. Mix thoroughly. Autoclave for 15 min at 15 psi pressure–121°C.

Preparation of Medium: Add components, except $FeSO_4·7H_2O$ solution, to distilled/deionized water and bring volume to 900.0mL. Mix thoroughly. Gently heat and bring to boiling. Distribute into flasks in 90.0mL volumes. Autoclave for 15 min at 15 psi pressure–121°C. Cool to 25°C. Aseptically add 10.0mL of sterile $FeSO_4·7H_2O$ solution to each flask. Mix thoroughly.

Use: For the cultivation of *Thiobacillus ferrooxidans.*

Thiobacillus ferrooxidans Medium

Composition per liter:
Solution I 400.0mL
Solution III 400.0mL
Solution II 200.0mL

Solution I:
Composition per 500.0mL:
K_2HPO_4 .... 0.5g
$MgSO_4·7H_2O$ 0.5g
$(NH_4)_2SO_4$ 0.5g
H_2SO_4 (1*N* solution) 5.0mL

Preparation of Solution I: Add components to distilled/deionized water and bring volume to 500.0mL. Mix thoroughly. Autoclave for 15 min at 15 psi pressure–121°C. Cool to 45°–50°C.

Solution II:
Composition per liter:
$FeSO_4·7H_2O$ 167.0g
1*N* H_2SO_4 .... 50.0mL

Preparation of Solution II: Add components to distilled/deionized water and bring volume to 1.0L. Mix thoroughly. Filter sterilize. Warm to 45°–50°C.

Solution III:
Composition per liter:
Agar 10.0g

Preparation of Solution III: Add agar to distilled/deionized water and bring volume to 1.0L. Mix thoroughly.

Preparation of Medium: Aseptically combine 400.0mL of sterile solution I, 200.0mL of sterile solution II, and 400.0mL of sterile solution III. Mix thoroughly. Aseptically distribute into sterile tubes or flasks.

Use: For the isolation and cultivation of *Thiobacillus ferrooxidans.*

Thiobacillus ferrooxidans Medium, APH

Composition per liter:
$FeSO_4·7H_2O$ 40.0g
$(NH_4)_2SO_4$ 2.0g
K_2HPO_4 .... 0.5g
$MgSO_4·7H_2O$ 0.5g
KCl 0.1g
$Ca(NO_3)_2$ 0.01g
pH 3.0 ± 0.2 at 25°C

Preparation of Medium: Add components to distilled/deionized water and bring volume to 1.0L. Mix thoroughly. Adjust pH to 3.0 with dilute H_2SO_4. Distribute into tubes or flasks. Filter sterilize. Aseptically distribute into sterile tubes or flasks.

Use: For the cultivation and maintenance of *Leptospirillum ferrooxidans* and *Thiobacillus ferrooxidans.*

Thiobacillus ferrooxidans Medium with Ferrous Sulfate

Composition per liter:
$FeSO_4·7H_2O$ 33.3g
KH_2PO_4 .... 0.4g
$MgSO_4·7H_2O$ 0.4g
$(NH_4)_2SO_4$ 0.4g
0.1 *N* H_2SO_4 .... 1000.0mL
pH 1.4 ± 0.2 at 25°C

Preparation of Medium: Add components to distilled/deionized water and bring volume to 1.0L. Mix thoroughly. Adjust pH to 1.4 with H_2SO_4. Distribute into tubes or flasks. Autoclave for 15 min at 15 psi pressure–121°C.

Use: For the cultivation and maintenance of *Clostridium acetobutylicum.*

Thiobacillus ferrooxidans Medium with Tetrathionate

Composition per liter:
$K_2S_4O_6$ 5.0g
KH_2PO_4 .... 3.0g
$(NH_4)_2SO_4$ 3.0g
$MgSO_4·7H_2O$ 0.5g
$CaCl_2·2H_2O$ 0.25g
pH 4.4 ± 0.2 at 25°C

Preparation of Medium: Add components to distilled/deionized water and bring volume to 1.0L. Mix thoroughly. Adjust pH to 4.4 with H_2SO_4. Distribute into tubes or flasks. Autoclave for 15 min at 15 psi pressure–121°C.

Use: For the cultivation and maintenance of *Thiobacillus ferrooxidans.*

Thiobacillus ferrooxidans Medium with Thiosulfate

Composition per liter:
$Na_2S_2O_3·5H_2O$ 5.0g
KH_2PO_4 .... 3.0g
$(NH_4)_2SO_4$ 3.0g
$MgSO_4·7H_2O$ 0.5g
$CaCl_2·2H_2O$ 0.25g
pH 4.4 ± 0.2 at 25°C

Preparation of Medium: Add components to distilled/deionized water and bring volume to 1.0L. Mix thoroughly. Adjust pH to 4.4 with H_2SO_4. Distribute into tubes or flasks. Autoclave for 15 min at 15 psi pressure–121°C.

Use: For the cultivation and maintenance of *Thiobacillus ferrooxidans* and *Thiobacillus thiooxidans*.

Thiobacillus halophilus Agar

Composition per liter:

NaCl	50.0g
Agar	15.0g
$Na_2HPO_4 \cdot 2H_2O$	7.9g
$Na_2S_2O_3 \cdot 5H_2O$	5.0g
KH_2PO_4	1.5g
NH_4Cl	0.4g
$MgSO_4 \cdot 7H_2O$	0.1g
Phenol Red (saturated aqueous solution)	12.5mL
Trace metals solution	10.0mL

pH 7.3 ± 0.2 at 25°C

Trace Metals Solution:

Composition per liter:

Disodium EDTA	50.0g
NaOH	11.0g
$ZnSO_4 \cdot 7H_2O$	11.0g
$CaCl_2 \cdot 2H_2O$	7.34g
$FeSO_4 \cdot 7H_2O$	5.0g
$MnCl_2 \cdot 4H_2O$	2.5g
$CoCl_2 \cdot 6H_2O$	0.5g
$(NH_4)_6Mo_7O_{24} \cdot 4H_2O$	0.5g
$CuSO_4 \cdot 5H_2O$	0.2g

Preparation of Trace Metals Solution: Add components to distilled/deionized water and bring volume to 1.0L. Adjust pH to 6.0. Mix thoroughly.

Preparation of Medium: Add components to distilled/deionized water and bring volume to 1.0L. Mix thoroughly. Gently heat and bring to boiling. Distribute into tubes or flasks. Autoclave for 15 min at 15 psi pressure–121°C. Pour into sterile Petri dishes or leave in tubes.

Use: For the cultivation and maintenance of *Thiobacillus halophilus.*

Thiobacillus halophilus Medium

Composition per liter:

Solution B	900.0mL
Solution A	100.0mL

pH 7.3 ± 0.2 at 25°C

Solution B:

Composition per 900.0mL:

NaCl	50.0g
Agar, noble	15.0g
$Na_2S_2O_3 \cdot 5H_2O$	5.0g
NH_4Cl	0.4g
Phenol Red	10.0mg
Hutner's basal salts solution	20.0mL

Preparation of Solution B: Add components to distilled/deionized water and bring volume to 900.0mL. Mix thoroughly. Gently heat and bring to boiling. Autoclave for 15 min at 15 psi pressure–121°C. Cool to 50°–55°C.

Hutner's Basal Salts Solution:

Composition per liter:

$MgSO_4 \cdot 7H_2O$	29.7g
Nitrilotriacetic acid	10.0g
$CaCl_2 \cdot 2H_2O$	3.335g
$FeSO_4 \cdot 7H_2O$	99.0mg
$(NH_4)_6MoO_7O_{24} \cdot 4H_2O$	9.25mg
"Metals 44"	50.0mL

"Metals 44":

Composition per 100.0mL:

$ZnSO_4 \cdot 7H_2O$	1.095g
$FeSO_4 \cdot 7H_2O$	0.5g
Sodium EDTA	0.25g
$MnSO_4 \cdot H2O$	0.154g
$CuSO_4 \cdot 5H_2O$	39.2mg
$Co(NO_3)_2 \cdot 6H_2O$	24.8mg
$Na_2B_4O_7 \cdot 10H_2O$	17.7mg

Preparation of "Metals 44": Add sodium EDTA to distilled/deionized water and bring volume to 90.0mL. Mix thoroughly. Add a few drops of concentrated H_2SO_4 to retard precipitation of heavy metal ions. Add remaining components. Mix thoroughly. Bring volume to 100.0mL with distilled/deionized water.

Preparation of Hutner's Basal Salts Solution: Add nitrilotriacetic acid to 500.0mL of distilled/deionized water. Adjust pH to 6.5 with KOH. Add remaining components. Add distilled/deionized water to 1.0L. Adjust pH to 6.8.

Solution A:

Composition per 100.0mL:

Na_2HPO_4	6.3g
KH_2PO_4	1.5g

Preparation of Solution A: Add components to distilled/deionized water and bring volume to 100.0mL. Mix thoroughly. Autoclave for 15 min at 15 psi pressure–121°C. Cool to 50°–55°C.

Preparation of Medium: Aseptically combine 900.0mL of sterile solution B with 100.0mL of sterile solution A. Mix thoroughly. Pour into sterile Petri dishes or distribute into sterile tubes.

Use: For the cultivation of *Thiobacillus halophilus.*

Thiobacillus Heterotrophic Medium

Composition per liter:

Glucose	5.0g
$MgSO_4 \cdot 7H_2O$	0.5g
$(NH_4)_2SO_4$	0.15g
KH_2PO_4	0.1g
KCl	0.05g
$Ca(NO_3)_2$	0.01g

pH 3.0 ± 0.2 at 25°C

Preparation of Medium: Add components to distilled/deionized water and bring volume to 1.0L. Mix thoroughly. Filter sterilize.

Use: For the cultivation and maintenance of *Thiobacillus organoparus* and other heterotrophic *Thiobacillus* species.

Thiobacillus intermedius Medium

Composition per 1010.0mL:

$Na_2S_2O_3 \cdot 5H_2O$	10.0g
Solution I	1.0L
Solution II	10.0mL

Solution I:

Composition per liter:

NH_4Cl	1.0g
K_2HPO_4	0.6g
$MgCl_2 \cdot 6H_2O$	0.5g
KH_2PO_4	0.4g
$MgSO_4$	0.3g

$CaCl_2 \cdot 2H_2O$ 0.2g
$FeCl_3 \cdot 6H_2O$ 0.02g

Preparation of Solution I: Add components to distilled/deionized water and bring volume to 1.0L. Mix thoroughly.

Solution II:
Composition per liter:
$CaCl_2 \cdot 2H_2O$ 0.1g
$ZnSO_4 \cdot 7H_2O$ 0.09g
$CuSO_4 \cdot 5H_2O$ 0.04g
$MnSO_4$ 0.02g
$Na_2B_4O_7$ 0.01g
$(NH_4)_6Mo_7O_{24} \cdot 4H_2O$ 5.0mg

Preparation of Solution II: Add components to distilled/deionized water and bring volume to 1.0L. Mix thoroughly.

Preparation of Medium: Combine solution I, 10.0mL of solution II, and 10.0g of $Na_2S_2O_3 \cdot 5H_2O$. Mix thoroughly. Filter sterilize. Aseptically distribute into sterile tubes or flasks.

Use: For the isolation and autotrophic cultivation of *Thiobacillus intermedius*.

Thiobacillus intermedius Medium

Composition per 1010.0mL:
Glucose 10.0g
$Na_2S_2O_3 \cdot 5H_2O$ 10.0g
Solution I 1.0L
Solution II 10.0mL

Solution I:
Composition per liter:
NH_4Cl 1.0g
K_2HPO_4 .... 0.6g
$MgCl_2 \cdot 6H_2O$ 0.5g
KH_2PO_4 .... 0.4g
$MgSO_4$ 0.3g
$CaCl_2 \cdot 2H_2O$ 0.2g
$FeCl_3 \cdot 6H_2O$ 0.02g

Preparation of Solution I: Add components to distilled/deionized water and bring volume to 1.0L. Mix thoroughly.

Solution II:
Composition per liter:
$CaCl_2 \cdot 2H_2O$ 0.1g
$ZnSO_4 \cdot 7H_2O$ 0.09g
$CuSO_4 \cdot 5H_2O$ 0.04g
$MnSO_4$ 0.02g
$Na_2B_4O_7$ 0.01g
$(NH_4)_6Mo_7O_{24} \cdot 4H_2O$ 5.0mg

Preparation of Solution II: Add components to distilled/deionized water and bring volume to 1.0L. Mix thoroughly.

Preparation of Medium: Combine 1.0L of solution I, 10.0mL of solution II, 10.0g of glucose, and 10.0g of $Na_2S_2O_3 \cdot 5H_2O$. Mix thoroughly. Filter sterilize. Aseptically distribute into sterile tubes or flasks.

Use: For the isolation and mixotrophic cultivation of *Thiobacillus intermedius*.

Thiobacillus Medium

Composition per 100.0mL:
$Na_2S_2O_3 \cdot 5H_2O$ 1.0g
KH_2PO_4 .... 0.1g
NH_4Cl 0.1g
$MgCl_2 \cdot 7H_2O$ 0.05g
pH 6.8 ± 0.2 at 25°C

Preparation of Medium: Add components to distilled/deionized water and bring volume to 1.0L. Mix thoroughly. Distribute into tubes or flasks. Autoclave for 15 min at 15 psi pressure–121°C.

Use: For the cultivation of *Thiobacillus thioparus* and *Thiobacillus thiooxidans*.

Thiobacillus Medium

Composition per liter:
$Na_2S_2O_3 \cdot 5H_2O$ 10.0g
K_2HPO_4 .... 4.0g
KH_2PO_4 .... 4.0g
$CaCl_2$ 0.1g
$MgSO_4 \cdot 7H_2O$ 0.1g
$(NH_4)_2SO_4$ 0.1g
$FeCl_3 \cdot 6H_2O$ 0.02g
$MnSO_4 \cdot 4H_2O$ 0.02g
pH 6.6 ± 0.2 at 25°C

Preparation of Medium: Add components to distilled/deionized water and bring volume to 1.0L. Mix thoroughly. Distribute into flasks in 100.0mL volumes. Autoclave for 60 min at 0 psi pressure–100°C on 3 consecutive days.

Use: For the cultivation of nonaciduric *Thiobacillus* species.

Thiobacillus Medium

Composition per liter:
$Na_2S_2O_3 \cdot 5H_2O$ 10.0g
$Na_2HPO_4 \cdot 7H_2O$ 7.9g
Sodium formate 6.8g
Glucose 3.6g
KNO_3 .... 2.0g
KH_2PO_4 .... 1.5g
NH_4Cl 0.3g
$MgSO_4 \cdot 7H_2O$ 0.1g
Trace metals solution 5.0mL
pH 7.6–8.5 at 25°C

Trace Metals Solution:
Composition per liter:
Disodium EDTA 50.0g
NaOH 11.0g
$CaCl_2 \cdot 2H_2O$ 7.34g
$FeSO_4 \cdot 7H_2O$ 5.0g
$MnCl_2 \cdot 2H_2O$ 2.5g
$ZnSO_4 \cdot 7H_2O$ 2.2g
$CoCl_2 \cdot 6H_2O$ 0.5g
$(NH_4)_6Mo_7O_{24} \cdot 4H_2O$ 0.5g
$CuSO_4 \cdot 5H_2O$ 0.2g

Preparation of Trace Metals Solution: Add EDTA to distilled/deionized water and bring volume to 500.0mL. Mix thoroughly. Adjust pH to 6.0 with NaOH. Add remaining components, one by one. Maintain the pH at 6.0. After dissolution of all the salts, adjust the pH to 4.0 with HCl. Store at 4°C.

Preparation of Medium: Add components to distilled/deionized water and bring volume to 1.0L. Mix thoroughly. Adjust pH to 7.6–8.5. Filter sterilize. Aseptically distribute into sterile tubes or flasks.

Use: For the isolation and anaerobic cultivation of *Thiobacillus* species.

Thiobacillus Medium

Composition per liter:

$Na_2S_2O_3 \cdot 5H_2O$	10.0g
$Na_2HPO_4 \cdot 7H_2O$	7.9g
Sodium formate	6.8g
Glucose	3.6g
KH_2PO_4	1.5g
NH_4Cl	0.3g
$MgSO_4 \cdot 7H_2O$	0.1g
Trace metals solution	5.0mL

pH 7.6–8.5 at 25°C

Trace Metals Solution:

Composition per liter:

Disodium EDTA	50.0g
NaOH	11.0g
$CaCl_2 \cdot 2H_2O$	7.34g
$FeSO_4 \cdot 7H_2O$	5.0g
$MnCl_2 \cdot 2H_2O$	2.5g
$ZnSO_4 \cdot 7H_2O$	2.2g
$CoCl_2 \cdot 6H_2O$	0.5g
$(NH_4)_6Mo_7O_{24} \cdot 4H_2O$	0.5g
$CuSO_4 \cdot 5H_2O$	0.2g

Preparation of Trace Metals Solution: Add EDTA to distilled/deionized water and bring volume to 500.0mL. Mix thoroughly. Adjust pH to 6.0 with NaOH. Add remaining components, one by one. Maintain the pH at 6.0. After dissolution of all the salts, adjust the pH to 4.0 with HCl. Store at 4°C.

Preparation of Medium: Add components to distilled/deionized water and bring volume to 1.0L. Mix thoroughly. Adjust pH to 7.6–8.5. Filter sterilize. Aseptically distribute into sterile tubes or flasks.

Use: For the isolation and aerobic cultivation of *Thiobacillus* species.

Thiobacillus Medium B

Composition per liter:

Noble agar	15.0g
$Na_2S_2O_3 \cdot 5H_2O$	5.0g
KH_2PO_4	3.0g
NH_4Cl	0.1g
$MgCl_2$	0.1g
$CaCl_2$	0.1g

pH 4.2 ± 0.2 at 25°C

Preparation of Medium: Add components to distilled/deionized water and bring volume to 1.0L. Mix thoroughly. Gently heat and bring to boiling. Distribute into tubes or flasks. Autoclave for 15 min at 15 psi pressure–121°C. Pour into sterile Petri dishes or leave in tubes.

Use: For the cultivation and maintenance of *Thiobacillus thiooxidans* and *Streptomyces scabies*.

Thiobacillus neapolitanus Medium

Composition per 1002.0mL:

Solution I	1.0L
Solution II	2.0mL

pH 6.2–7.0 at 25°C

Solution I:

Composition per liter:

$Na_2S_2O_3 \cdot 5H_2O$	10.0g
KH_2PO_4	4.0g
K_2HPO_4	4.0g
$MgSO_4 \cdot 7H_2O$	0.8g
$KHCO_3$	0.7g
NH_4Cl	0.4g

Preparation of Solution I: Add components to distilled/deionized water and bring volume to 1.0L. Mix thoroughly.

Solution II:

Composition per liter:

Disodium EDTA	50.0g
NaOH	11.0g
$CaCl_2 \cdot 2H_2O$	7.34g
$FeSO_4 \cdot 7H_2O$	5.0g
$MnCl_2 \cdot 2H_2O$	2.5g
$ZnSO_4 \cdot 7H_2O$	2.2g
$CoCl_2 \cdot 6H_2O$	0.5g
$(NH_4)_6Mo_7O_{24} \cdot 4H_2O$	0.5g
$CuSO_4 \cdot 5H_2O$	0.2g

Preparation of Solution II: Add EDTA to distilled/deionized water and bring volume to 500.0mL. Mix thoroughly. Adjust pH to 6.0 with NaOH. Add remaining components, one by one. Maintain the pH at 6.0. After dissolution of all the salts, adjust the pH to 4.0 with HCl. Store at 4°C.

Preparation of Medium: Aseptically combine 1.0L of solution I and 2.0mL of solution II. Mix thoroughly. Adjust pH to 6.2–7.0. Distribute into tubes or flasks. Autoclave for 15 min at 15 psi pressure–121°C.

Use: For the isolation and cultivation of *Thiobacillus neapolitanus*.

Thiobacillus novellus Medium

Composition per liter:

$Na_2S_2O_3 \cdot 5H_2O$	10.0g
K_2HPO	4.0g
KH_2PO_4	1.5g
$MgSO_4 \cdot 7H_2O$	0.5g
$(NH_4)_2SO_4$	0.3g
Yeast extract	0.3g
Trace metals solution	10.0mL

pH 6.8–7.2 at 25°C

Trace Metals Solution:

Composition per liter:

Disodium EDTA	50.0g
NaOH	11.0g
$CaCl_2 \cdot 2H_2O$	7.34g
$FeSO_4 \cdot 7H_2O$	5.0g
$MnCl_2 \cdot 2H_2O$	2.5g
$ZnSO_4 \cdot 7H_2O$	2.2g
$CoCl_2 \cdot 6H_2O$	0.5g
$(NH_4)_6Mo_7O_{24} \cdot 4H_2O$	0.5g
$CuSO_4 \cdot 5H_2O$	0.2g

Preparation of Trace Metals Solution: Add EDTA to distilled/deionized water and bring volume to 500.0mL. Mix thoroughly. Adjust pH to 6.0 with NaOH. Add remaining components, one by one. Maintain the pH at 6.0. After dissolution of all the salts, adjust the pH to 4.0 with HCl. Store at 4°C.

Preparation of Medium: Add components to distilled/deionized water and bring volume to 1.0L. Mix thoroughly. Distribute into tubes or flasks. Autoclave for 15 min at 15 psi pressure–121°C.

Use: For the isolation and cultivation of *Thiobacillus novellus*.

Thiobacillus perometabolis Medium

Composition per liter:

$Na_2S_2O_3$	10.0g
Yeast extract	5.0g

NH_4Cl 1.0g
K_2HPO_4 .... 0.6g
$MgCl_2$ 0.5g
KH_2PO_4 .... 0.4g
$MgSO_4$ 0.3g
Bromothymol Blue 0.03g
$FeCl_3$ 0.02g
Heavy metal solution 30.0mL
pH 3.0–4.0 at 25°C

Heavy Metal Solution:
Composition per liter:
Ethylenediamine tetraacetate 1.5g
$FeSO_4 \cdot 7H_2O$ 0.2g
$ZnSO_4 \cdot H_2O$ 0.1g
$MnCl_2 \cdot 4H_2O$ 0.02g
Modified Hoagland trace elements solution 6.0mL

Preparation of Heavy Metal Solution: Add ethylenediamine tetraacetate to distilled/deionized water and bring volume to 500.0mL. Mix thoroughly. Adjust pH to 6.8 to dissolve EDTA. Add remaining components. Bring volume to 1.0L with distilled/deionized water. Mix thoroughly. Adjust pH to 6.8.

Preparation of Heavy Metal Solution: Add ethylenediamine tetraacetate to distilled/deionized water and bring volume to 500.0mL. Mix thoroughly. Adjust pH to 6.8 to dissolve EDTA. Add remaining components. Bring volume to 1.0L with distilled/deionized water. Mix thoroughly. Adjust pH to 6.8.

Modified Hoagland Trace Elements Solution:
Composition per liter:
H_3BO_3 .... 11.0g
$MnCl_2 \cdot 4H_2O$ 7.0g
$AlCl_3$ 1.0g
$CoCl_2$ 1.0g
$CuCl_2$ 1.0g
KI 1.0g
$NiCl_2$ 1.0g
$ZnCl_2$ 1.0g
$BaCl_2$ 0.5g
KBr 0.5g
LiCl 0.5g
Na_2MoO_4 .... 0.5g
SeS_2 .... 0.5g
$SnCl_2 \cdot 2H_2O$ 0.5g
$NaVO_3 \cdot H_2O$ 0.1g

Preparation of Modified Hoagland Trace Elements Solution: Add components seqentially to distilled/deionized water and bring volume to 1.0L. Mix thoroughly. Adjust pH to 6.8.

Preparation of Medium: Add components to distilled/deionized water and bring volume to 1.0L. Mix thoroughly. Adjust pH to 3–4. Distribute into tubes or flasks. Autoclave for 15 min at 15 psi pressure–121°C. A flaky yellow precipitate forms after mixing but will change in a few days to a fine, white precipitate. Mix the medium thoroughly before use.

Use: For the cultivation of *Thiobacillus perometabolis* and other bacteria that can utilize thiosulfate as an energy source.

Thiobacillus plumbophilus Medium

Composition per 1010.0mL:
$NiCl_2 \cdot 6H_2O$ 5.0g
$MgSO_4 \cdot 7H_2O$ 3.45g
$MgCl_2 \cdot 6H_2O$ 2.75g
NH_4Cl 1.25g
NaCl 0.5g
KCl 0.3g
$CaCl_2 \cdot 2H_2O$ 0.14g
K_2HPO_4 .... 0.14g
$Na_2WO_4 \cdot 2H_2O$ 0.5mg
Trace elements solution 10.0mL
pH 6.0 ± 0.2 at 25°C

Trace Elements Solution:
Composition per liter:
$MgSO_4 \cdot 7H_2O$ 3.0g
Nitrilotriacetic acid 1.5g
NaCl 1.0g
$MnSO_4 \cdot 2H_2O$ 0.5g
$CoSO_4 \cdot 7H_2O$ 0.18g
$ZnSO_4 \cdot 7H_2O$ 0.18g
$CaCl_2 \cdot 2H_2O$ 0.1g
$FeSO_4 \cdot 7H_2O$ 0.1g
$KAl(SO_4)_2 \cdot 12H_2O$ 0.02g
$CuSO_4 \cdot 5H_2O$ 0.01g
H_3BO_3 .... 0.01g
$Na_2MoO_4 \cdot 2H_2O$ 0.01g
$NiCl_2 \cdot 6H_2O$ 0.025g
$Na_2SeO_3 \cdot 5H_2O$ 0.3mg

Preparation of Trace Elements Solution: Add nitrilotriacetic acid to 500.0mL of distilled/deionized water. Adjust pH to 6.5 with KOH. Add remaining components. Add distilled/deionized water to 1.0L.

Preparation of Medium: Add components to distilled/deionized water and bring volume to 1.0L. Mix thoroughly. Adjust pH to 6.0 with H_2SO_4. Distribute 20.0mL volumes into 100.0mL serum bottles. Sparge with 80% H_2 + 20% CO_2. Autoclave for 15 min at 15 psi pressure–121°C. Check pH of medium after autoclaving and bring to 6.0 if necessary. After inoculation add 3% sterile air (9.0mL per serum bottle) with a syringe. Pressurize bottles to 2 bar with 80% H_2 + 20% CO_2.

Use: For the cultivation and maintenance of *Thiobacillus plumbophilus*.

Thiobacillus prosperus Medium

Composition per 1010.0mL:
$MgSO_4 \cdot 7H_2O$ 3.45g
$MgCl_2 \cdot 6H_2O$ 2.75g
NH_4Cl 1.25g
NaCl 0.5g
Sulfur, powdered 0.5g
KCl 0.33g
$CaCl_2 \cdot 2H_2O$ 0.14g
K_2HPO_4 .... 0.14g
KH_2PO_4 .... 0.14g
$NiCl_2 \cdot 6H_2O$ 2.0mg
Trace elements solution 10.0mL
pH 2.5 ± 0.2 at 25°C

Preparation of Sulfur: Add 10.0g of powdered sulfur to a flask and sterilize by steaming for 3 hr on 3 consecutive days.

Trace Elements Solution:
Composition per liter:
$MgSO_4 \cdot 7H_2O$ 3.0g
Nitrilotriacetic acid 1.5g
NaCl 1.0g
$MnSO_4 \cdot 2H_2O$ 0.5g
$CoSO_4 \cdot 7H_2O$ 0.18g
$ZnSO_4 \cdot 7H_2O$ 0.18g

$CaCl_2 \cdot 2H_2O$ 0.1g
$FeSO_4 \cdot 7H_2O$ 0.1g
$NiCl_2 \cdot 6H_2O$ 0.025g
$KAl(SO_4)_2 \cdot 12H_2O$ 0.02g
$CuSO_4 \cdot 5H_2O$ 0.01g
H_3BO_3 .... 0.01g
$Na_2MoO_4 \cdot 2H_2O$ 0.01g
$Na_2SeO_3 \cdot 5H_2O$ 0.3mg

Preparation of Trace Elements Solution: Add nitrilotriacetic acid to 500.0mL of distilled/deionized water. Adjust pH to 6.5 with KOH. Add remaining components. Add distilled/deionized water to 1.0L.

Preparation of Medium: Add components, except sulfur, to distilled/deionized water and bring volume to 1.0L. Mix thoroughly. Adjust pH to 2.5 with H_2SO_4. Autoclave for 15 min at 15 psi pressure–121°C. Aseptically add 0.5g of sterile sulfur. Mix thoroughly. Aseptically distribute into sterile tubes or flasks.

Use: For the cultivation and maintenance of *Thiobacillus prosperus.*

Thiobacillus tepidarius Medium

Composition per liter:
Agar 10.0g
$Na_2S_2O_3 \cdot 5H_2O$ 4.96g
$MgSO_4 \cdot 7H_2O$ 0.8g
NH_4Cl 0.4g
Phosphate solution 100.0mL
Bromcresol Purple (saturated solution) 2.0mL
Trace metals A-5 .... 1.0mL

Phosphate Solution:
Composition per 100.0mL:
KH_2PO_4 .... 4.0g
K_2HPO_4 .... 4.0g

Preparation of Phosphate Solution: Add components to distilled/deionized water and bring volume to 100.0mL. Mix thoroughly. Autoclave for 15 min at 15 psi pressure–121°C.

Trace Metals A-5:
Composition per liter:
H_3BO_3 .... 2.86g
$MnCl_2 \cdot 4H_2O$ 1.81g
$Na_2MoO_4 \cdot 2H_2O$ 0.39g
$ZnSO_4 \cdot 7H_2O$ 0.222g
$CuSO_4 \cdot 5H_2O$ 0.079g
$Co(NO_3)_2 \cdot 6H_2O$ 49.4mg

Preparation of Trace Metals A-5: Add components to distilled/deionized water and bring volume to 1.0L. Mix thoroughly.

Preparation of Medium: Add components, except phosphate solution, to distilled/deionized water and bring volume to 900.0mL. Autoclave for 15 min at 15 psi pressure–121°C. Aseptically add 100.0mL of the sterile phosphate solution. Mix thoroughly. Aseptically distribute into sterile tubes or flasks.

Use: For the cultivation and maintenance of *Thiobacillus tepidarius.*

Thiobacillus/Thermophilus Medium

Composition per liter:
$Na_2S_2O_3 \cdot 5H_2O$ 5.0g
$NaHCO_3$ 1.0g
Na_2HPO_4 .... 0.2g
$MgCl_2$ 0.1g
$NH_4 \cdot Cl$ 0.1g
pH 7.0 ± 0.2 at 25°C

Preparation of Medium: Add components to distilled/deionized water and bring volume to 1.0L. Mix thoroughly. Distribute into tubes or flasks. Autoclave for 20 min at 6 psi pressure–109°C or filter sterilize.

Use: For the cultivation of *Thiobacillus* species and *Thermophilus* species.

Thiobacillus thiooxidans Medium

Composition per liter:
Sulfur, powdered 10.0g
KH_2PO_4 .... 5.0g
$MgSO_4 \cdot 7H_2O$ 0.5g
$CaCl_2$ 0.25g
$(NH_4)_2SO_4$ 0.2g
$FeSO_4$ 0.01g
pH 7.0 ± 0.2 at 25°C

Preparation of Medium: Add components, except sulfur, to distilled/deionized water and bring volume to 1.0L. Mix thoroughly. Distribute into flasks in 100.0mL volumes. Add 1.0g of sulfur to each flask. Autoclave for 30 min at 0 psi pressure–100°C on 3 consecutive days.

Use: For the cultivation of *Thiobacillus thiooxidans.*

Thiobacillus thioparus Agar

Composition per liter:
Agar, purified 12.0g
$Na_2S_2O_3 \cdot 5H_2O$ 10.0g
K_2HPO_4 .... 4.0g
KH_2PO_4 .... 4.0g
$MgSO_4 \cdot 7H_2O$ 0.1g
$(NH_4)_2SO_4$ 0.1g
$CaCl_2$ 0.1g
$FeCl_3 \cdot 6H_2O$ 0.02g
$MnSO_4 \cdot H_2O$ 0.02g
pH 6.6 ± 0.2 at 25°C

Preparation of Medium: Add components, except agar, to distilled/deionized water and bring volume to 1.0L. Mix thoroughly. Adjust pH to 6.6. Add agar. Gently heat and bring to boiling. Distribute into tubes or flasks. Autoclave for 20 min at 10 psi pressure–115°C. Pour into sterile Petri dishes or leave in tubes.

Use: For the cultivation and maintenance of *Thiobacillus thioparus.*

Thiobacillus thioparus II Medium (DSMZ Medium 486)

Composition per 1003mL:
Solution A 900.0mL
Solution B 100.0mL
Vitamin solution 3.0mL
pH 7.4 ± 0.2 at 25°C

Solution A:
Composition per 900mL:
$Na_2S_2O_3 \cdot 5H_2O$ 5.0g
NH_4Cl 0.4g
Na_2CO_3 .... 0.4g
$MgCl_2 \cdot 6H_2O$ 0.2g
Bromocresol Purple (saturated solution) 2.0mL
Trace metals solution 1.0mL

Preparation of Solution A: Add components to distilled/deionized water and bring volume to 900.0mL. Mix thoroughly. Autoclave for 15 min at 15 psi pressure–121°C. Cool to room temperature.

Solution B:

Composition per 100mL:

KH_2PO_4 .... 2.0g
K_2HPO_4 .... 2.0g

Preparation of Solution B: Add components to distilled/deionized water and bring volume to 100.0mL. Mix thoroughly. Adjust pH to 7.6. Autoclave for 15 min at 15 psi pressure–121°C. Cool to room temperature.

Trace Elements Solution:

Composition per liter:

Na_2-EDTA 50.0g
NaOH 11.0g
$ZnSO_4·7H_2O$ 11.0g
$CaCl_2·2H_2O$ 7.34g
$FeSO_4·7H_2O$ 5.0g
$MnCl_2·4H_2O$ 2.5g
$CoCl_2·6H_2O$ 0.5g
$(NH_4)_6Mo_7O_{24}·4H_2O$ 0.5g
$CuSO_4·5H_2O$ 0.2g

Preparation of Trace Elements Solution: Add components to distilled/deionized water and bring volume to 1.0L. Mix thoroughly.

Vitamin Solution:

Composition per liter:

Riboflavin 20.0mg
Ca-pantothenate 20.0mg
Nicotinic acid 20.0mg
Pyridoxine-HCl 20.0mg
p-Aminobenzoic acid 10.0mg
Thiamine-HCl·$2H_2O$ 10.0mg
Biotin 1.0mg
Vitamin B_{12} 1.0mg

Preparation of Vitamin Solution: Add components to distilled/deionized water and bring volume to 1.0L. Mix thoroughly. Filter sterilize.

Preparation of Medium: Aseptically combine 100.0mL solution A, 900.0mL solution B, and 3.0mL vitamin solution. Mix thoroughly. Aseptically distribute into sterile tubes or flasks.

Use: For the cultivation and maintenance of *Thiobacillus thioparus.*

Thiobacillus thioparusii Agar

Composition per 1003.0mL:

Solution B 900.0mL
Solution A 100.0mL
Vitamin solution 3.0mL

pH 7.1 ± 0.2 at 25°C

Solution A:

Composition per 100.0mL:

K_2HPO_4 .... 2.0g
KH_2PO_4 .... 2.0g

Preparation of Solution A: Add components to distilled/deionized water and bring volume to 100.0mL. Mix thoroughly. Autoclave for 15 min at 15 psi pressure–121°C. Cool to 50°–55°C.

Solution B:

Composition per 900.0mL:

Agar 15.0g
$Na_2S_2O_3·5H_2O$ 5.0g
Na_2CO_3 .... 0.4g
NH_4Cl 0.4g
$MgCl_2·6H_2O$ 0.2g
Bromcresol Purple
(saturated aqueous solution) 2.0mL
Trace elements solution 1.0mL

Trace Elements Solution:

Composition per liter:

Disodium EDTA 50.0g
NaOH 11.0g
$ZnSO_4·7H_2O$ 11.0g
$CaCl_2·2H_2O$ 7.34g
$FeSO_4·7H_2O$ 5.0g
$MnCl_2·4H_2O$ 2.5g
$CoCl_2·6H_2O$ 0.5g
$(NH_4)6MoO_{24}·4H_2O$ 0.5g
$CuSO_4·5H_2O$ 0.2g

Preparation of Trace Elements Solution: Add components to distilled/deionized water and bring volume to 1.0L. Mix thoroughly. Adjust pH to 6.0.

Preparation of Solution B: Add components to distilled/deionized water and bring volume to 900.0mL. Mix thoroughly. Gently heat and bring to boiling. Autoclave for 15 min at 15 psi pressure–121°C. Cool to 50°–55°C.

Vitamin Solution:

Composition per liter:

Calcium DL-pantothenate 20.0mg
Nicotinic acid 20.0mg
Pyridoxine·HCl 20.0mg
Riboflavin 20.0mg
p-Aminobenzoic acid 10.0mg
Thiamine·HCl 10.0mg
Biotin 1.0mg
Vitamin B_{12} 1.0mg

Preparation of Vitamin Solution: Add components to distilled/deionized water and bring volume to 1.0L. Mix thoroughly. Adjust pH to 7.0. Filter sterilize.

Preparation of Medium: Aseptically combine 100.0mL of sterile solution A with 900.0mL of sterile solution B and 10.0mL of sterile vitamin solution. Mix thoroughly. Pour into sterile Petri dishes or distribute into sterile tubes.

Use: For the cultivation and maintenance of *Thiobacillus thioparus.*

Thiobacillus thyasiris Agar

Composition per 1010.0mL:

Solution A 100.0mL
Solution B 900.0mL
Vitamin solution 10.0mL

pH 7.3–7.6 at 25°C

Solution A:

Composition per 100.0mL:

$Na_2HPO_42H_2O$ 7.9g
KH_2PO_4 .... 1.5g

Preparation of Solution A: Add components to distilled/deionized water and bring volume to 100.0mL. Mix thoroughly. Adjust pH to 7.4. Autoclave for 15 min at 15 psi pressure–121°C. Cool to 50°–55°C.

Solution B:

Composition per 900.0mL:

NaCl 25.1g
Agar 15.0g
$Na_2S_2O_3$ 5.0g

NH_4Cl 0.4g
$MgSO_4 \cdot 7H_2O$ 0.1g
Phenol Red 3.0mg
Trace elements solution 10.0mL

Trace Elements Solution:
Composition per liter:

Disodium EDTA 50.0g
NaOH 11.0g
$ZnSO_4 \cdot 7H_2O$ 11.0g
$CaCl_2 \cdot 2H_2O$ 7.34g
$FeSO_4 \cdot 7H_2O$ 5.0g
$MnCl_2 \cdot 4H_2O$ 2.5g
$CoCl_2 \cdot 6H_2O$ 0.5g
$(NH_4)6MoO_{24} \cdot 4H_2O$ 0.5g
$CuSO_4 \cdot 5H_2O$ 0.2g

Preparation of Trace Elements Solution: Add components to distilled/deionized water and bring volume to 1.0L. Mix thoroughly. Adjust pH to 6.0.

Preparation of Solution B: Add components to distilled/deionized water and bring volume to 900.0mL. Mix thoroughly. Gently heat and bring to boiling. Autoclave for 15 min at 15 psi pressure–121°C. Cool to 50°–55°C.

Vitamin Solution:
Composition per liter:

Calcium DL-pantothenate 20.0mg
Nicotinic acid 20.0mg
Pyridoxine·HCl 20.0mg
Riboflavin 20.0mg
p-Aminobenzoic acid 10.0mg
Thiamine·HCl 10.0mg
Biotin 1.0mg
Vitamin B_{12} 1.0mg

Preparation of Vitamin Solution: Add components to distilled/deionized water and bring volume to 1.0L. Mix thoroughly. Adjust pH to 7.0. Filter sterilize.

Preparation of Medium: Aseptically combine 100.0mL of sterile solution A with 900.0mL of sterile solution B and 10.0mL of sterile vitamin solution. Mix thoroughly. Pour into sterile Petri dishes or distribute into sterile tubes.

Use: For the cultivation and maintenance of *Thiobacillus thyasiris*.

Thiobacillus thyasiris/ Thiobacillus halophilus Medium (DSMZ Medium 484)

Composition per 1010mL:

Solution B 900.0mL
Solution A 100.0mL
Vitamin solution 10.0mL

pH 7.3-7.6 at 25°C

Solution A:
Composition per 100mL:

$Na_2HPO_4 \cdot 2H_2O$ 7.9g
KH_2PO_4 .... 1.5g

Preparation of Solution A: Add components to distilled/deionized water and bring volume to 100.0mL. Mix thoroughly. Adjust pH to 7.6. Autoclave for 15 min at 15 psi pressure–121°C. Cool to room temperature.

Solution B:
Composition per 900mL:

NaCl 25.1g
$Na_2S_2O_3$ 5.0g
NH_4Cl 0.4g
$MgSO_4 \cdot 7H_2O$ 0.1g
Phenol Red 3.0mg
Trace elements solution 10.0mL

Preparation of Solution B: Add components to distilled/deionized water and bring volume to 900.0mL. Mix thoroughly. Autoclave for 15 min at 15 psi pressure–121°C. Cool to room temperature.

Trace Elements Solution:
Composition per liter:

Na_2-EDTA 50.0g
NaOH 11.0g
$ZnSO_4 \cdot 7H_2O$ 11.0g
$CaCl_2 \cdot 2H_2O$ 7.34g
$FeSO_4 \cdot 7H_2O$ 5.0g
$MnCl_2 \cdot 4H_2O$ 2.5g
$CoCl_2 \cdot 6H_2O$ 0.5g
$(NH_4)_6Mo_7O_{24} \cdot 4H_2O$ 0.5g
$CuSO_4 \cdot 5H_2O$ 0.2g

Preparation of Trace Elements Solution: Add components to distilled/deionized water and bring volume to 1.0L. Mix thoroughly.

Vitamin Solution:
Composition per liter:

Riboflavin 20.0mg
Ca-pantothenate 20.0mg
Nicotinic acid 20.0mg
Pyridoxine-HCl 20.0mg
p-Aminobenzoic acid 10.0mg
Thiamine-HCl·$2H_2O$ 10.0mg
Biotin 1.0mg
Vitamin B_{12} 1.0mg

Preparation of Vitamin Solution: Add components to distilled/deionized water and bring volume to 1.0L. Mix thoroughly. Filter sterilize.

Preparation of Medium: Aseptically combine 100.0mL solution A, 900.0mL solution B, and 10.0mL vitamin solution. Mix thoroughly. Aseptically distribute into sterile tubes or flasks.

Use: For the cultivation and maintenance of *Thiobacillus thyasiris, Thiobacillus halophilus, Thiomicrospira thyasirae*, and *Halothiobacillus halophilus*.

Thiocapsa halophila Medium

Composition per 1061.0mL:

NaCl 70.0g
$MgCl_2 \cdot 6H_2O$ 4.5g
$MgSO_4 \cdot 7H_2O$ 3.0g
NH_4Cl 0.5g
KH_2PO_4 .... 0.3g
$CaCl_2 \cdot 2H_2O$ 0.05g
$NaHCO_3$ solution 40.0mL
$Na_2S \cdot 9H_2O$ solution 10.0mL
$Na_2S_2O_3$ solution 10.0mL
Vitamin B_{12} solution 1.0mL
Trace elements solution 1.0mL

pH 7.2 ± 0.2 at 25°C

$NaHCO_3$ Solution:
Composition per 100.0mL:

$NaHCO_3$ 5.0g

Preparation of $NaHCO_3$ Solution: Add $NaHCO_3$ to distilled/deionized water and bring volume to 100.0mL. Mix thoroughly. Filter sterilize.

$Na_2S{\cdot}9H_2O$ Solution:
Composition per 10.0mL:
$Na_2S{\cdot}9H_2O$ 0.6g

Preparation of $Na_2S{\cdot}9H_2O$ Solution: Add $Na_2S{\cdot}9H_2O$ to distilled/deionized water and bring volume to 10.0mL. Mix thoroughly. Filter sterilize.

$Na_2S_2O_3$ Solution :
Composition per 10.0mL:
$Na_2S_2O_3{\cdot}5H_2O$ 0.5g

Preparation of $Na_2S_2O_3$ Solution: Add $Na_2S_2O_3{\cdot}5H_2O$ to distilled/deionized water and bring volume to 10.0mL. Mix thoroughly. Filter sterilize.

Vitamin B_{12} Solution:
Composition per 10.0mL:
Vitamin B_{12} 0.2mg

Preparation of Vitamin B_{12} Solution: Add Vitamin B_{12} to distilled/deionized water and bring volume to 10.0mL. Mix thoroughly. Filter sterilize.

Trace Elements Solution SL-12B:
Composition per liter:
$MnCl_2{\cdot}2H_2O$ 50.0g
Disodium EDTA 3.0g
$FeSO_4{\cdot}7H_2O$ 1.1g
H_3BO_3 .... 0.3g
$CoCl_2{\cdot}6H_2O$ 0.19g
$ZnCl_2$ 42.0mg
$NiCl_2{\cdot}6H_2O$ 24.0mg
$Na_2MoO_4{\cdot}2H_2O$ 18.0mg
$CuCl_2{\cdot}2H_2O$ 2.0mg

Preparation of Trace Elements Solution SL-12B: Add components to distilled/deionized water and bring volume to 1.0L. Mix thoroughly.

Preparation of Medium: Add components, except $NaHCO_3$ solution, $Na_2S{\cdot}9H_2O$ solution, $Na_2S_2O_3$ solution, and Vitamin B_{12} solution, to distilled/deionized water and bring volume to 1.0L. Mix thoroughly. Autoclave for 15 min at 15 psi pressure–121°C. Cool to room temperature while sparging with 90% N_2 + 10% CO_2. Aseptically and anaerobically add 40.0mL of sterile $NaHCO_3$ solution, 10.0mL of sterile $Na_2S{\cdot}9H_2O$ solution, 10.0mL of sterile $Na_2S_2O_3$ solution, and 1.0mL of sterile Vitamin B_{12} solution. Mix thoroughly. Aseptically and anaerobically distribute into sterile tubes or flasks.

Use: For the cultivation and maintenance of *Thiocapsa halophila.*

Thiocapsa Medium

Composition per 127.0mL:
Solution 1 .... 76.2mL
Solution 2 + Solution 3 .... 44.8mL
Solution 4 .... 6.0mL

Solution 1:
Composition per 2.5L:
NaCl 39.68g
$CaCl_2$ 2.0g

Preparation of Solution 1: Add components to distilled/deionized water and bring volume to 2.5L. Distribute in 80.0mL volumes into 127.0mL screw-capped bottles. Autoclave for 15 min at 15 psi pressure–121°C.

Solution 2:
Composition per 100.0mL:
Sodium ascorbate 2.4g
KCl 1.0g
KH_2PO_4 .... 1.0g
$MgCl_2{\cdot}6H_2O$ 0.8g
NH_4Cl 0.8g
Heavy metal solution 50.0mL
Vitamin solution 15.0mL
Vitamin B_{12} solution 3.0mL

Preparation of Solution 2: Add components to distilled/deionized water and bring volume to 100.0mL. Mix thoroughly.

Heavy Metal Solution:
Composition per liter:
Ethylenediamine tetraacetate (EDTA) 1.5g
$FeSO_4{\cdot}7H_2O$ 0.2g
$ZnSO_4{\cdot}7H_2O$ 0.1g
$MnCl_2{\cdot}4H_2O$ 0.02g
Modified Hoagland trace elements solution 6.0mL

Preparation of Heavy Metal Solution: Dissolve EDTA in approximately 800.0mL of distilled/deionized water. Add remaining components. Bring volume to 1.0L with distilled/deionized water. Mix thoroughly.

Modified Hoagland Trace Elements Solution:
Composition per 3.6L:
H_3BO_3 .... 11.0g
$MnCl_2{\cdot}4H_2O$ 7.0g
$AlCl_3$ 1.0g
$CoCl_2$ 1.0g
$CuCl_2$ 1.0g
KI 1.0g
$NiCl_2$ 1.0g
$ZnCl_2$ 1.0g
$BaCl_2$ 0.5g
KBr 0.5g
LiCl 0.5g
Na_2MoO_4 .... 0.5g
$SeCl_4$ 0.5g
$SnCl_2{\cdot}2H_2O$ 0.5g
$NaVO_3{\cdot}H_2O$ 0.1g

Preparation of Modified Hoagland Trace Elements Solution: Prepare each component as a separate solution. Dissolve each salt in approximately 100.0mL of distilled/deionized water. Adjust the pH of each solution to below 7.0. Combine all the salt solutions and bring the volume to 3.6L with distilled/deionized water. Adjust the pH to 3–4. A yellow precipitate may form after mixing. After a few days, it will turn into a fine white precipitate. Mix the solution thoroughly before using.

Vitamin Solution:
Composition per 100.0mL:
Pyridoxamine·2HCl 5.0mg
Nicotinic acid 2.0mg
Thiamine 1.0mg
Pantothenic acid 0.5mg
Biotin 0.2mg
p-Aminobenzoic acid 0.1mg

Preparation of Vitamin Solution: Add components to distilled/deionized water and bring volume to 100.0mL. Mix thoroughly.

Vitamin B_{12} Solution:
Composition per 100.0mL:
Vitamin B_{12} (cyanocobalamin) 2.0mg

Preparation of Vitamin B_{12} Solution: Add Vitamin B_{12} to distilled/deionized water and bring volume to 100.0mL. Mix thoroughly.

Solution 3:

Composition per 900.0mL:

$NaHCO_3$ 4.5g

Preparation of Solution 3: Add $NaHCO_3$ to distilled/deionized water and bring volume to 900.0mL. Mix thoroughly. Bubble 100% CO_2 through the solution for 30 min. After CO_2 saturation of solution 3, add solution 2 and immediately filter the mixture through a Seitz filter (or a Millipore) using positive CO_2 pressure to push the liquid through.

Solution 4:

Composition per 200.0mL:

$Na_2S·9H_2O$ 3.0g

Preparation of Solution 4: Add $Na_2S·9H_2O$ to distilled/deionized water and bring volume to 200.0mL. Add a magnetic stir bar to the flask. Autoclave for 15 min at 15 psi pressure–121°C. On a magnetic stirrer, slowly add 2.0mL of sterile 2*M* H_2SO_4. This partially neutralizes the solution. The solution should turn yellow. H_2S gas will be liberated; neutralization and distribution of the solution should be done as rapidly as possible under adequate ventilation.

Preparation of Medium: To the 80.0mL of sterile solution 1 in screw-capped bottles, add combined solutions 2 and 3 immediately after filtration and fill bottles to capacity. Mix thoroughly. Aseptically remove 6.0mL of the medium from the bottles and replace it with 6.0mL of neutralized solution 4. Let stand for 24 hr. The medium should form a fine white precipitate before using. To inoculate, remove 6.0mL of the completed medium from the bottles and replace it with 6.0mL of inoculum.

Use: For the cultivation and maintenance of a variety of *Thiocapsa* species.

Thiocyanate Agar

Composition per liter:

Solution A 800.0mL
Solution B 100.0mL
Solution C 100.0mL

Solution A:

Composition per 800.0mL:

Agar, noble 30.0g
K_2HPO_4 .... 1.0g
KH_2PO_4 .... 1.0g
$MgSO_4·7H_2O$ 0.2g
$CaCl_2$ 20.0mg
$FeCl_3·6H_2O$ (60%) 0.1mL

Preparation of Solution A: Add components to distilled/deionized water and bring volume to 1.0L. Mix thoroughly. Gently heat and bring to boiling. Autoclave for 15 min at 15 psi pressure–121°C. Cool to 50°–55°C.

Solution B:

Composition per 100.0mL:

KCNS 3.6g

Preparation of Solution B: Add KCNS to distilled/deionized water and bring volume to 100.0mL. Mix thoroughly. Autoclave for 15 min at 15 psi pressure–121°C. Cool to 50°–55°C.

Solution C:

Composition per 100.0mL:

Disodium succinate 1.5g

Preparation of Solution C: Add disodium succinate to distilled/deionized water and bring volume to 100.0mL. Mix thoroughly. Autoclave for 15 min at 15 psi pressure–121°C. Cool to 50°–55°C.

Preparation of Medium: Aseptically combine 800.0mL of solution A with 100.0mL of solution B and 100.0mL of solution C. Mix thoroughly. Pour into sterile Petri dishes or aseptically distribute into sterile tubes.

Use: For the cultivation and maintenance of a variety of microorganisms that can utilize thiocyanate as the sole source of nitrogen and sulfur.

Thiocyanate Utilization Medium

Composition per 1225.0mL:

Basal solution 1.0L
Solution C 200.0mL
Solution B 20.0mL
Solution A 5.0mL

Basal Solution:

Composition per liter:

Na_2HPO_4 .... 4.8g
KH_2PO_4 .... 4.4g
$MgSO_4·7H_2O$ 0.5g

Preparation of Basal Solution: Add components to distilled/deionized water and bring volume to 1.0L. Mix thoroughly. Autoclave for 15 min at 15 psi pressure–121°C. Cool to 45°–50°C.

Solution A:

Composition per 100.0mL:

$FeCl_3·6H_2O$ 1.0g
$CaCl_2$ 0.1g

Preparation of Solution A: Add components to distilled/deionized water and bring volume to 100.0mL. Mix thoroughly. Filter sterilize.

Solution B:

Composition per 100.0mL:

D-Glucose 10.0g

Preparation of Solution B: Add glucose to distilled/deionized water and bring volume to 100.0mL. Mix thoroughly. Filter sterilize.

Solution C:

Composition per 200.0mL:

NaSCN 1.0g

Preparation of Solution C: Add NaSCN to distilled/deionized water and bring volume to 200.0mL. Mix thoroughly. Filter sterilize.

Preparation of Medium: To 1.0L of cooled, sterile basal solution, aseptically add 5.0mL of sterile solution A, 20.0mL of sterile solution B, and 200.0mL of sterile solution C. Mix thoroughly. Aseptically distribute into sterile tubes or flasks.

Use: For the cultivation and maintenance of a variety of microorganisms that can utilize thiocyanate as the sole source of nitrogen and sulfur.

Thioglycolate Medium (DSMZ Medium 530)

Composition per liter:

Agar 15.0g
Peptone from casein 5.0g
Meat extract 3.0g
Na-thioglycolate solution 100.0mL

pH 5.5 ± 0.2 at 25°C

Thioglycolate Solution :
Composition per 100mL:

Na-Thioglycolate	0.5g

Preparation of Thioglycolate Solution: Add thioglycolate to distilled/deionized water and bring volume to 100.0mL. Mix thoroughly. Filter sterilize. Warm to 50°C.

Preparation of Medium: Add components, except thioglycolate solution to distilled/deionized water and bring volume to 900.0mL. Mix thoroughly. Gently heat and bring to boiling. Autoclave for 15 min at 15 psi pressure–121°C. Cool to 50°C. Aseptically add 100.0mL warm thioglycolate solution. Adjust pH to 5.5. Pour into sterile Petri dishes or distribute into sterile tubes.

Use: For the cultivation and maintenance of various anaerobic bacteria.

Thiomicrospira denitrificans Agar

Composition per 1001.0mL:

Solution A	940.0mL
Solution B	40.0mL
Solution C	20.0mL
Solution D	1.0mL

pH 7.0 ± 0.2 at 25°C

Solution A:
Composition per 940.0mL:

Agar	15.0g
KH_2PO_4	2.0g
KNO_3	2.0g
NH_4Cl	1.0g
$MgSO_4 \cdot 7H_2O$	0.8g
Trace elements solution SL-4	2.0mL

Preparation of Solution A: Add components to distilled/deionized water and bring volume to 940.0mL. Mix thoroughly. Gently heat and bring to boiling. Adjust pH to 7.0 with NaOH. Autoclave for 15 min at 15 psi pressure–121°C. Cool to 45°–50°C.

Trace Elements Solution SL-4:
Composition per liter:

EDTA	0.5g
$FeSO_4 \cdot 7H_2O$	0.2g
Trace elements solution SL-6	100.0

Preparation of Trace Elements Solution SL-4: Add components to distilled/deionized water and bring volume to 1.0L. Mix thoroughly.

Trace Elements Solution SL-6:
Composition per liter:

H_3BO_3	0.3g
$CoCl_2 \cdot 6H_2O$	0.2g
$ZnSO_4 \cdot 7H_2O$	0.1g
$MnCl_2 \cdot 4H_2O$	0.03g
$Na_2MoO_4 \cdot H_2O$	0.03g
$NiCl_2 \cdot 6H_2O$	0.02g
$CuCl_2 \cdot 2H_2O$	0.01g

Preparation of Trace Elements Solution SL-6: Add components to distilled/deionized water and bring volume to 1.0L. Mix thoroughly. Adjust pH to 3.4.

Solution B:
Composition per 40.0mL:

$Na_2S_2O_3 \cdot 5H_2O$	5.0g

Preparation of Solution B: Add $Na_2S_2O_3 \cdot 5H_2O$ to distilled/deionized water and bring volume to 40.0mL. Mix thoroughly. Autoclave for 15 min at 15 psi pressure–121°C. Cool to 45°–50°C.

Solution C:
Composition per 20.0mL:

$NaHCO_3$	1.0g

Preparation of Solution C: Add $NaHCO_3$ to distilled/deionized water and bring volume to 20.0mL. Mix thoroughly. Filter sterilize.

Solution D:
Composition per liter:

$FeSO_4 \cdot 7H_2O$	2.0mg
H_2SO_4 (0.1*N* solution)	1.0mL

Preparation of Solution D: Add $FeSO_4 \cdot 7H_2O$ to 1.0mL of 0.1*N* H_2SO_4 solution. Mix thoroughly. Autoclave for 15 min at 15 psi pressure–121°C. Cool to 45°–50°C.

Preparation of Medium: Aseptically add 40.0mL of sterile solution B, 20.0mL of sterile solution C, and 1.0mL of sterile solution D to 940.0mL of sterile solution A. Mix thoroughly. Aseptically and anaerobically distribute into sterile tubes under 100% N_2.

Use: For the cultivation and maintenance of *Thiomicrospira denitrificans.*

Thiomicrospira Medium (ATCC Medium 1036)

Composition per liter:

NaCl	25.0g
$Na_2S_2O_3 \cdot 5H_2O$	8.0g
$MgSO_4 \cdot 7H_2O$	1.5g
$(NH_4)_2SO_4$	1.0g
K_2HPO_4	0.5g
$CaCl_2$	0.3g
Vitamin B_{12}	15.0μg
Vishniac and Santer trace metals	0.2mL
Bromcresol Purple (0.05% solution)	0.1mL

pH 7.2 ± 0.2 at 25°C

Vishniac and Santer Trace Metals:
Composition per liter:

Ethylenediamine tetraacetic acid (EDTA)	50.0g
$ZnSO_4 \cdot 7H_2O$	22.0g
$CaCl_2$	5.54g
$MnCl_2 \cdot 4H_2O$	5.06g
$FeSO_4 \cdot 7H_2O$	4.99g
$CoCl_2 \cdot 6H_2O$	1.61g
$CuSO_4 \cdot 5H_2O$	1.57g
$(NH_4)_6Mo_7O_{24} \cdot 4H_2O$	1.1g

Preparation of Vishniac and Santer Trace Metals: Add components to distilled/deionized water and bring volume to 1.0L. Adjust pH to 6.0 with KOH. Mix thoroughly.

Preparation of Medium: Add components to distilled/deionized water and bring volume to 1.0L. Mix thoroughly. Adjust pH to 7.2. Filter sterilize. Aseptically distribute into sterile tubes or flasks.

Use: For the cultivation and maintenance of *Thiomicrospira* species.

Thiomicrospira pelophila Medium

Composition per 1000.2mL:

NaCl	25.0g
$MgSO_4 \cdot 7H_2O$	1.5g
$(NH_4)_2SO_4$	1.0g
$CaCl_2 \cdot 2H_2O$	0.42g
Bromothymol Blue	4.0mg
K_2HPO_4 solution	100.0mL
$Na_2S_2O_3 \cdot 5H_2O$ solution	100.0mL
Vitamin B_{12} solution	10.0mL

Trace elements solution 0.2mL
Na_2CO_3 solution .... variable

pH 7.2 ± 0.2 at 25°C

K_2HPO_4 Solution:
Composition per 100.0mL:
K_2HPO_4..0.5g

Preparation of K_2HPO_4 Solution: Add K_2HPO_4 to distilled/deionized water and bring volume to 100.0mL. Mix thoroughly. Autoclave for 15 min at 15 psi pressure–121°C.

$Na_2S_2O_3$ Solution :
Composition per 10.0mL:
$Na_2S_2O_3 \cdot 5H_2O$..5.0g

Preparation of $Na_2S_2O_3$ Solution: Add $Na_2S_2O_3 \cdot 5H_2O$ to distilled/deionized water and bring volume to 10.0mL. Mix thoroughly. Autoclave for 15 min at 15 psi pressure–121°C.

Vitamin B_{12} Solution:
Composition per 10.0mL:
Vitamin B_{12} 15.0mg

Preparation of Vitamin B_{12} Solution: Add Vitamin B_{12} to distilled/deionized water and bring volume to 10.0mL. Mix thoroughly. Filter sterilize.

Trace Elements Solution:
Composition per liter:
Disodium EDTA 50.0g
$ZnSO_4 \cdot 7H_2O$ 22.0g
$CaCl_2 \cdot 2H_2O$..5.54g
$MnCl_2 \cdot 4H_2O$...5.06g
$FeSO_4 \cdot 7H_2O$...5.0g
$CoCl_2 \cdot 6H_2O$ 1.61g
$CuSO_4 \cdot 5H_2O$...1.57g
$(NH_4)_6Mo_7O_{24} \cdot 4H_2O$ 1.1g

Preparation of Trace Elements Solution: Add components to distilled/deionized water and bring volume to 1.0L. Mix thoroughly.

Na_2CO_3 Solution:
Composition per 100.0mL:
Na_2CO_3 ....0.4g

Preparation of Na_2CO_3 Solution: Add Na_2CO_3 to distilled/deionized water and bring volume to 100.0mL. Mix thoroughly. Autoclave for 15 min at 15 psi pressure–121°C.

Preparation of Medium: Add components, except K_2HPO_4 solution, $Na_2S_2O_3 \cdot 5H_2O$ solution, Vitamin B_{12} solution, and Na_2CO_3 solution, to distilled/deionized water and bring volume to 790.0mL. Mix thoroughly. Adjust pH to 7.2. Autoclave for 15 min at 15 psi pressure–121°C. Aseptically add 100.0mL of sterile K_2HPO_4 solution, 100.0mL of sterile $Na_2S_2O_3 \cdot 5H_2O$ solution, and 10.0mL of sterile Vitamin B_{12} solution. Mix thoroughly. Aseptically adjust pH to 7.2 with the appropriate volume of sterile Na_2CO_3 solution. Aseptically distribute into sterile tubes or flasks.

Use: For the cultivation and maintenance of *Thiomicrospira pelophila*.

Thiorhodococcus Medium (DSMZ Medium 28a)

Composition per 5.0L:
Solution A 4.0L
Solution B 860.0mL
Solution E 100.0mL
Solution F 20.0mL
Solution C...5.0mL
Solution D 5.0mL

pH7.3 at 25°C

Solution A:
Composition per 4.0L:
NaCl 100.0g
$MgCl_2 \cdot 6H_2O$ 5.0g
$MgSO_4$...2.5g
$Na_2S_2O_3 \cdot 5H_2O$..2.5g
KH_2PO_4 ....1.7g
NH_4Cl..1.7g
KCl 1.7g
$CaCl_2 \cdot 2H_2O$ 1.25g

Preparation of Solution A: Add components to 4.0L distilled water. Mix thoroughly. Autoclave for 45 min at 15 psi pressure–121°C in 5-liter special bottle or flask with 4 openings at the top, together with a teflon-coated magnetic bar. In this 5-liter bottle, 2 openings for tubes are in the central, silicon rubber stopper. One is a short, gas-inlet tube with a sterile cotton filter; and the other is an outlet tube for medium, which reaches the bottom of the vessel at one end and has, at the other end, a silicon rubber tube with a pinch cock and a bell for aseptic dispensing of the medium into bottles. The other two openings have gas-tight screw caps. One of these openings is for the addition of sterile solutions and the other serves as a gas outlet. After autoclaving cool solution A to room temperature under a 100% N_2 atmosphere with a positive pressure of 0.05–0.1 atm (a manometer for low pressure will be required). Saturate the cold medium with CO_2 by magnetic stirring for 30 min under a CO_2 atmosphere of 0.05–0.1 atm.

Solution B:
Distilled water 860.0mL

Preparation of Solution B: Autoclave distilled water for 15 min at 15 psi pressure–121°C in a cotton-stoppered Erlenmeyer flask. Cool to room temperature under an atmosphere of N_2 in an anaerobic jar.

Solution C:
Composition per 100.0mL:
Vitamin B_{12}..2.0mg

Preparation of Solution C: Add Vitamin B_{12} to distilled/deionized water and bring volume to 100.0mL. Mix thoroughly. Sparge under 100% N_2 gas for 3 min. Filter sterilize. Store under N_2 gas.

Solution D:
Composition per liter:
Disodium ethylendiamine-tetraacetate
(Disodium EDTA)..3.0g
$FeSO_4 \cdot 7H_2O$ 1.1g
H_3BO_3...0.3g
$CoCl_2 \cdot 6H_2O$...0.19g
$MnCl_2 \cdot 2H_2O$...50.0mg
$ZnCl_2$ 42.0mg
$NiCl_2 \cdot 6H_2O$..24.0mg
$Na_2MoO_4 \cdot 2H_2O$...18.0mg
$CuCl_2 \cdot 2H_2O$..2.0mg

Preparation of Solution D: Add components to distilled/deionized water and bring volume to 1.0L. Mix thoroughly. Autoclave for 15 min at 15 psi pressure–121°C.

Solution E:
Composition per 100.0mL:
$NaHCO_3$ 7.5g

Preparation of Solution E: Add $NaHCO_3$ to distilled/deionized water and bring volume to 100.0mL. Mix thoroughly. Sparge with 100% CO_2 until saturated. Filter sterilize under 100% CO_2 into a sterile, gas-tight 100.0mL screw-capped bottle.

Solution F:
Composition per 100.0mL:

$Na_2S·9H_2O$ 10.0g

Preparation of Solution F: Add $Na_2S·9H_2O$ to distilled/deionized water in a 250.0mL screw-capped bottle fitted with a butyl rubber septum and bring volume to 100.0mL. Mix thoroughly. Sparge under 100% N_2 gas for 3 min. Autoclave for 15 min at 15 psi pressure–121°C. Cool to room temperature.

Neutralized Sulfide Solution:
Composition per 100.0mL:

$Na_2S·9H_2O$ 1.5g

Preparation of Neutralized Sulfide Solution: Add $Na_2S·9H_2O$ to distilled/deionized water in a 250.0mL screw-capped bottle fitted with a butyl rubber septum and bring volume to 100.0mL. Add a magnetic stir bar. Mix thoroughly. Sparge under 100% N_2 gas for 3 min. Autoclave for 15 min at 15 psi pressure–121°C. Cool to room temperature. Adjust pH to about 7.3 with sterile 2*M* H_2SO_4. Do not open the bottle to add H_2SO_4; use a sterile syringe. Stir the solution continuously to avoid precipitation of elemental sulfur. The final solution should be clear and yellow.

Preparation of Medium: Add solutions B, C, D, E, and F to solution A through one of the screw-capped openings against a stream of either N_2 gas or better, a mixture of 95% N_2 and 5% CO_2 while the medium is magnetically stirred. Adjust the pH of the medium with sterile HCl or Na_2CO_3 solution (2 *M* solutions) to pH 7.3. Distribute the medium aseptically through the medium outlet tube into sterile, 100mL bottles (with metal caps and autoclavable rubber seals) using the positive gas pressure (0.05–0.1 atm) of the N_2/CO_2 gas mixture. Leave a small air bubble in each bottle to meet possible pressure changes. The tightly sealed, screw-capped bottles can be stored for several weeks or months in the dark. During the first 24 hr, the iron of the medium precipitates in the form of black flocs. No other sediment should arise in the otherwise clear medium. Incubate in the light at 500–1000 lux intensity. Feed periodically with neutralized solution of sodium sulfide to replenish sulfide and with other supplement solutions.

Use: For the cultivation of *Thiorhodococcus minor.*

Thiosphaera Agar

Composition per liter:

Agar 15.0g
Na_2HPO_4 ..... 4.2g
KH_2PO_4 ..... 1.5g
NH_4Cl 0.3g
$MgSO_4·7H_2O$ 0.1g
KNO_3 ..... 0.1g
Vishniac and Santer trace metals 2.0mL
pH 8.0–8.2 at 25°C

Vishniac and Santer Trace Metals:
Composition per liter:

Ethylenediamine tetraacetic acid (EDTA) 50.0g
$ZnSO_4·7H_2O$ 22.0g
$CaCl_2$ 5.54g
$MnCl_2·4H_2O$ 5.06g
$FeSO_4·7H_2O$ 4.99g
$CoCl_2·6H_2O$ 1.61g
$CuSO_4·5H_2O$ 1.57g
$(NH_4)_6Mo_7O_{24}·4H_2O$ 1.1g

Preparation of Vishniac and Santer Trace Metals: Add components to distilled/deionized water and bring volume to 1.0L. Adjust pH to 6.0 with KOH. Mix thoroughly.

Preparation of Medium: Add components, except agar, to distilled/deionized water and bring volume to 500.0mL. Mix thoroughly. Adjust pH to 8.0–8.2. Filter sterilize. Warm to 45°–50°C. Add agar to distilled/deionized water and bring volume to 500.0mL. Mix thoroughly. Gently heat and bring to boiling. Autoclave for 15 min at 15 psi pressure–121°C. Cool to 45°–50°C. Aseptically combine the two sterile solutions. Mix thoroughly. Pour into sterile Petri dishes or distribute into sterile tubes.

Use: For the cultivation and maintenance of *Thiosphaera pantotropha.*

Thiosphaera pantotropha Medium

Composition per 1001.0mL:

Agar 20.0g
$Na_2HPO_4·2H_2O$ 7.9g
KH_2PO_4 ..... 1.5g
NH_4Cl 0.3g
$MgSO_4·7H_2O$ 0.1g
Yeast extract solution 10.0mL
Trace elements solution SL-10 ..... 1.0mL
pH 7.5 ± 0.2 at 25°C

Yeast Extract Solution:
Composition per 10.0mL:

Yeast extract 1.0g

Preparation of Yeast Extract Solution: Add yeast extract to distilled/deionized water and bring volume to 10.0mL. Mix thoroughly. Autoclave for 15 min at 15 psi pressure–121°C.

Trace Elements Solution SL-10:
Composition per liter:

$FeCl_2·4H_2O$ 1.5g
$CoCl_2·6H_2O$ 190.0mg
$MnCl_2·4H_2O$ 100.0mg
$ZnCl_2$ 70.0mg
$Na_2MoO_4·2H_2O$ 36.0mg
$NiCl_2·6H_2O$ 24.0mg
H_3BO_3 ..... 6.0mg
$CuCl_2·2H_2O$ 2.0mg
HCl (25% solution) 10.0mL

Preparation of Trace Elements Solution SL-10: Add $FeCl_2·4H_2O$ to 10.0mL of HCl solution. Mix thoroughly. Add distilled/deionized water and bring volume to 1.0L. Add remaining components. Mix thoroughly. Sparge with 100% N_2. Autoclave for 15 min at 15 psi pressure–121°C.

Preparation of Medium: Add components, except yeast extract solution, to distilled/deionized water and bring volume to 990.0mL. Mix thoroughly. Gently heat and bring to boiling. Autoclave for 15 min at 15 psi pressure–121°C. Cool to 50°–55°C. Aseptically add 10.0mL of sterile yeast extract solution. Mix thoroughly. Pour into sterile Petri dishes or distribute into sterile tubes.

Use: For the cultivation and maintenance of *Paracoccus denitrificans.*

Thiosulfate-Oxidizing Medium

Composition per liter:

K_2HPO_4 ..... 2.0g
$MgSO_4·7H_2O$ 0.1g

$CaCl_2 \cdot 2H_2O$ 0.1g
$FeCl_3 \cdot 6H_2O$ 0.02g
$(NH_4)_2SO_4$ solution 100.0mL
Thiosulfate solution 100.0mL

pH 7.8 ± 0.2 at 25°C

$(NH_4)_2SO_4$ Solution:
Composition per 100.0mL:
$(NH_4)_2SO_4$ 0.1g

Preparation of $(NH_4)_2SO_4$ Solution: Add the $(NH_4)_2SO_4$ to distilled/deionized water and bring volume to 100.0mL. Mix thoroughly. Autoclave for 15 min at 15 psi pressure–121°C. Cool to 45°–50°C

Thiosulfate Solution:
Composition per 100.0mL:
$Na_2S_2O_3 \cdot 5H_2O$ 10.0g

Preparation of Thiosulfate Solution: Add the $Na_2S_2O_3 \cdot 5H_2O$ to distilled/deionized water and bring volume to 100.0mL. Mix thoroughly. Autoclave for 15 min at 15 psi pressure–121°C. Cool to 45°–50°C.

Preparation of Medium: Add components, except $(NH_4)_2SO_4$ solution and thiosulfate solution, to distilled/deionized water and bring volume to 800.0mL. Mix thoroughly. Autoclave for 15 min at 15 psi pressure–121°C. Cool to 45°–50°C. Aseptically add the sterile $(NH_4)_2SO_4$ solution and the sterile thiosulfate solution. Mix thoroughly. Adjust the pH to 7.8 if necessary. Aseptically distribute into sterile tubes or flasks.

Use: For the isolation and cultivation of iron and sulfur bacteria.

Thiosulfate Salts Broth

Composition per liter:
$Na_2S_2O_3 \cdot 5H_2O$ 24.81g
$NH_4 \cdot Cl$ 2.2g
KH_2PO_4 .... 2.0g
Artificial seawater 500.0mL

Artificial Seawater:
Composition per liter:
NaCl 23.476g
$MgCl_2$ 4.981g
Na_2SO_4 .... 3.917g
$CaCl_2$ 1.102g
KCl 0.664g
$NaHCO_3$ 0.192g
KBr 0.096g
H_3BO_3 .... 0.026g
$SrCl_3$ 0.024g
NaF 3.0mg

pH 5.0 ± 0.2 at 25°C

Preparation of Artificial Seawater: Add components to distilled/deionized water and bring volume to 1.0L. Mix thoroughly.

Preparation of Medium: Add components to distilled/deionized water and bring volume to 1.0L. Mix thoroughly. Adjust pH to 5.0. Distribute into tubes or flasks. Autoclave for 15 min at 15 psi pressure–121°C.

Use: For the cultivation of *Thiobacillus* species.

Thiothrix Agar (DSMZ Medium 573)

Composition per liter:
Agar 12.0g
NH_4Cl 0.2g
Na-acetate 0.1g
K_2HPO_4 .... 0.01g
$MgSO_4 \cdot 7H_2O$ 0.01g
$CaSO_4$ (saturated solution) 20.0mL
$Na_2S \cdot 9H_2O$ solution 10.0mL
Trace elements solution 5.0mL

pH 7.5 ± 0.2 at 25°C

$Na_2S \cdot 9H_2O$ Solution:
Composition per 10.0mL:
$Na_2S \cdot 9H_2O$ 0.3g

Preparation of $Na_2S \cdot 9H_2O$ Solution: Add $Na_2S \cdot 9H_2O$ to distilled/deionized water and bring volume to 10.0mL. Sparge with N_2. Autoclave for 15 min at 15 psi pressure–121°C. Cool to 25°C. Store anaerobically.

Trace Elements Solution:
Composition per liter:
$FeSO_4 \cdot 7H_2O$ 0.7g
EDTA 0.2g
$ZnSO_4 \cdot 7H_2O$ 10.0mg
H_3BO_3 .... 10.0mg
$MnSO_4 \cdot 4H_2O$ 2.0mg
$Co(NO_3)_2$ 1.0mg
$Na_2MoO_4 \cdot 4H_2O$ 1.0mg
$CuSO_4 \cdot 5H_2O$ 5.0μg

Preparation of Trace Elements Solution: Add components to distilled/deionized water and bring volume to 1.0L. Mix thoroughly.

Preparation of Medium: Add components, except $Na_2S \cdot 9H_2O$ solution, to distilled/deionized water and bring volume to 990.0mL. Mix thoroughly. Gently heat and bring to boiling. Autoclave for 15 min at 15 psi pressure–121°C. Cool to 50°C. Aseptically add 10.0mL sterile $Na_2S \cdot 9H_2O$ solution. Mix thoroughly. Pour into Petri dishes or aseptically distribute into sterile tubes.

Use: For the cultivation and maintenance of *Thiothrix nivea.*

Thiothrix Medium (DSMZ Medium 573)

Composition per liter:
NH_4Cl 0.2g
Na-acetate 0.1g
K_2HPO_4 .... 0.01g
$MgSO_4 \cdot 7H_2O$ 0.01g
$CaSO_4$ (saturated solution) 20.0mL
$Na_2S \cdot 9H_2O$ solution 10.0mL
Trace elements solution 5.0mL

pH 7.5 ± 0.2 at 25°C

$Na_2S \cdot 9H_2O$ Solution:
Composition per 10.0mL:
$Na_2S \cdot 9H_2O$ 0.3g

Preparation of $Na_2S \cdot 9H_2O$ Solution: Add $Na_2S \cdot 9H_2O$ to distilled/deionized water and bring volume to 10.0mL. Sparge with N_2. Autoclave for 15 min at 15 psi pressure–121°C. Cool to 25°C. Store anaerobically.

Trace Elements Solution:
Composition per liter:
$FeSO_4 \cdot 7H_2O$ 0.7g
EDTA 0.2g
$ZnSO_4 \cdot 7H_2O$ 10.0mg
H_3BO_3 .... 10.0mg
$MnSO_4 \cdot 4H_2O$ 2.0mg
$Co(NO_3)_2$ 1.0mg
$Na_2MoO_4 \cdot 4H_2O$ 1.0mg
$CuSO_4 \cdot 5H_2O$ 5.0μg

Preparation of Trace Elements Solution: Add components to distilled/deionized water and bring volume to 1.0L. Mix thoroughly.

Preparation of Medium: Add components, except $Na_2S{\cdot}9H_2O$ solution, to distilled/deionized water and bring volume to 990.0mL. Mix thoroughly. Autoclave for 15 min at 15 psi pressure–121°C. Aseptically add 10.0mL sterile $Na_2S{\cdot}9H_2O$ solution. Mix thoroughly. Aseptically distribute into sterile tubes or flasks.

Use: For the cultivation and maintenance of *Thiothrix nivea.*

Tibi Medium

Composition per liter:

Sucrose	100.0–150.0g
Fig, dried, quartered	1
Lemon wedge (0.5cm segment)	1

Preparation of Medium: Add components to tap water and bring volume to 1.0L in a 1.0L Erlenmeyer flask fitted with a cotton stopper. Autoclave for 15 min at 15 psi pressure–121°C. Cool to room temperature. Inoculate with about 50.0mL of Tibi grains.

Use: For the cultivation of osmophilic bacteria and fungi from tibi grains.

TMA Mineral Medium

Composition per liter:

Agar, noble	20.0g
KH_2PO_4	2.78g
Na_2HPO_4	2.78g
$(NH_4)_2SO_4$	1.0g
Tetramethylammonium perchlorate solution	20.0mL
Hutner's basal salts solution	20.0mL

pH 6.8 ± 0.2 at 25°C

Tetramethylammonium Perchlorate Solution:
Composition per 20.0mL:

Tetramethylammonium perchlorate	1.0g

Preparation of Tetramethylammonium Perchlorate Solution: Add tetramethylammonium perchlorate to distilled/deionized water and bring volume to 20.0mL. Mix thoroughly. Filter sterilize.

Hutner's Basal Salts Solution:
Composition per liter:

$MgSO_4{\cdot}7H_2O$	29.7g
Nitrilotriacetic acid	10.0g
$CaCl_2{\cdot}2H_2O$	3.335g
$FeSO_4{\cdot}7H_2O$	99.0mg
$(NH_4)_6MoO_7O_{24}{\cdot}4H_2O$	9.25mg
"Metals 44"	50.0mL

"Metals 44":
Composition per 100.0mL:

$ZnSO_4{\cdot}7H_2O$	1.095g
$FeSO_4{\cdot}7H_2O$	0.5g
Sodium EDTA	0.25g
$MnSO_4{\cdot}H2O$	0.154g
$CuSO_4{\cdot}5H_2O$	39.2mg
$Co(NO_3)_2{\cdot}6H_2O$	24.8mg
$Na_2B_4O_7{\cdot}10H_2O$	17.7mg

Preparation of "Metals 44": Add sodium EDTA to distilled/deionized water and bring volume to 90.0mL. Mix thoroughly. Add a few drops of concentrated H_2SO_4 to retard precipitation of heavy metal ions. Add remaining components. Mix thoroughly. Bring volume to 100.0mL with distilled/deionized water.

Preparation of Hutner's Basal Salts Solution: Add nitrilotriacetic acid to 500.0mL of distilled/deionized water. Adjust pH to 6.5 with KOH. Add remaining components. Add distilled/deionized water to 1.0L. Adjust pH to 6.8.

Preparation of Medium: Add components, except tetramethylammonium perchlorate solution, to distilled/deionized water and bring volume to 980.0mL. Mix thoroughly. Gently heat and bring to boiling. Adjust pH to 6.8. Autoclave for 15 min at 15 psi pressure–121°C. Cool to 50°–55°C. Aseptically add 20.0mL of sterile tetramethylammonium perchlorate solution. Mix thoroughly. Pour into sterile Petri dishes or distribute into sterile tubes.

Use: For the cultivation of *Paracoccus kocurii.*

TMBS4 Medium (DSMZ Medium 559)

Composition per 1002.0mL:

Solution A	870.0mL
Solution C	100.0mL
Solution D	10.0mL
Solution E (Vitamin solution)	10.0mL
Solution F	10.0mL
Solution B (Trace elements solution SL-10)	1.0mL
Solution G	1.0mL
Solution H	1.0mL

pH 7.1-7.4 at 25°C

Solution A:
Composition per 870.0mL:

NaCl	1.0g
$MgCl_2{\cdot}6H_2O$	0.4g
Na_2SO_4	3.0g
KCl	0.5g
NH_4Cl	0.3g
KH_2PO_4	0.2g
$CaCl_2{\cdot}2H_2O$	0.15g
Resazurin	1.0mg

Preparation of Solution A: Add components to distilled/deionized water and bring volume to 870.0mL. Mix thoroughly.

Solution B (Trace Elements Solution SL-10):
Composition per liter:

$FeCl_2{\cdot}4H_2O$	1.5g
$CoCl_2{\cdot}6H_2O$	190.0mg
$MnCl_2{\cdot}4H_2O$	100.0mg
$ZnCl_2$	70.0mg
$Na_2MoO_4{\cdot}2H_2O$	36.0mg
$NiCl_2{\cdot}6H_2O$	24.0mg
H_3BO_3	6.0mg
$CuCl_2{\cdot}2H_2O$	2.0mg
HCl (25% solution)	10.0mL

Preparation of Solution B (Trace Elements Solution SL-10): Add $FeCl_2{\cdot}4H_2O$ to 10.0mL of HCl solution. Mix thoroughly. Add distilled/deionized water and bring volume to 1.0L. Add remaining components. Mix thoroughly. Sparge with 100% N_2. Autoclave for 15 min at 15 psi pressure–121°C.

Solution C:
Composition per 100.0mL:

$NaHCO_3$	5.0g

Preparation of Solution C: Add $NaHCO_3$ to distilled/deionized water and bring volume to 100.0mL Mix thoroughly. Filter sterilize. Flush with 80% N_2 + 20% CO_2 to remove dissolved oxygen.

Solution D:
Composition per 10.0mL:
Syringic acid 0.6g

Preparation of Solution D: Add syringic acid to distilled/deionized water and bring volume to 10.0mL. Adjust pH to 8.0 with NaOH. Mix thoroughly. Sparge with 100% N_2. Autoclave for 15 min at 15 psi pressure–121°C.

Solution E (Vitamin Solution):
Composition per liter:
Pyridoxine-HCl 10.0mg
Thiamine-HCl·$2H_2O$ 5.0mg
Riboflavin 5.0mg
Nicotinic acid 5.0mg
D-Ca-pantothenate 5.0mg
p-Aminobenzoic acid 5.0mg
Lipoic acid 5.0mg
Biotin 2.0mg
Folic acid 2.0mg
Vitamin B_{12} 0.10mg

Solution E (Vitamin Solution): Add components to distilled/deionized water and bring volume to 1.0L. Mix thoroughly. Sparge with 100% N_2. Autoclave for 15 min at 15 psi pressure–121°C.

Solution F:
Composition per 10.0mL:
$Na_2S·9H_2O$ 0.4g

Preparation of Solution F: Add $Na_2S·9H_2O$ to distilled/deionized water and bring volume to 10.0mL. Mix thoroughly. Sparge with 100% N_2. Autoclave for 15 min at 15 psi pressure–121°C.

Solution G:
Composition per 10.0mL:
Na-dithionite 0.25g

Preparation of Solution G: Add Na-dithionite to distilled/deionized water and bring volume to 10.0mL. Mix thoroughly. Sparge with 100% N_2. Autoclave for 15 min at 15 psi pressure–121°C.

Solution H
Composition per liter:
NaOH 0.5g
$Na_2WO_4·2H_2O$ 4.0mg
$Na_2SeO_3·5H_2O$ 3.0mg

Preparation of Solution H: Add components to distilled/deionized water and bring volume to 1.0L. Mix thoroughly. Sparge with 100% N_2. Filter sterilize.

Preparation of Medium: Gently heat solution A and bring to boiling. Boil solution A for a few minutes. Cool to room temperature. Gas with 80% N_2 + 20% CO_2 gas mixture to reach a pH below 6. Autoclave for 15 min at 15 psi pressure–121°C. Cool to room temperature. Sequentially add 1.0mL solution B, 100.0mL solution C, 10.0mL solution D, 10.0mL solution E, 10.0mL solution F, 1.0mL solution G, and 1.0mL solution H. Distribute anaerobically under 80% N_2 + 20% CO_2 into appropriate vessels.

Use: For the cultivation of *Holophaga foetida.*

Treponema bryantii Medium

Composition per liter:
L-Cysteine·HCl 1.0g
NaCl 0.9g
$(NH_4)_2SO_4$ 0.9g
K_2HPO_4 .... 0.45g
KH_2PO_4 .... 0.45g
$MgSO_4·7H_2O$ 0.18g
$CaCl_2·2H_2O$ 0.12g
Resazurin 1.0mg
Rumen fluid, clarified 300.0mL
$NaHCO_3$ solution 100.0mL
Glucose solution (10%w/v) 20.0mL
pH 7.0 ± 0.2 at 25°C

$NaHCO_3$ Solution:
Composition per 100.0mL:
$NaHCO_3$ 5.0g

Preparation of $NaHCO_3$ Solution: Add $NaHCO_3$ to distilled/deionized water and bring volume to 100.0mL. Mix thoroughly. Filter sterilize. Sparge with 100% CO_2.

Glucose Solution:
Composition per 20.0mL:
Glucose 2.0g

Preparation of Glucose Solution: Add glucose to distilled/deionized water and bring volume to 20.0mL. Mix thoroughly. Filter sterilize. Sparge with 100% CO_2.

Preparation of Medium: Prepare and dispense medium under 100% CO_2. Add components, except rumen fluid, $NaHCO_3$ solution, and glucose solution, to distilled/deionized water and bring volume to 580.0mL. Mix thoroughly. Adjust pH to 7.0 with KOH. Sparge with 100% CO_2. Autoclave for 15 min at 15 psi pressure–121°C. Cool to room temperature. Aseptically add 300.0mL of sterile rumen fluid, 100.0mL of sterile $NaHCO_3$ solution, and 20.0mL of sterile glucose solution. Mix thoroughly. Aseptically distribute into sterile tubes or flasks.

Use: For the cultivation and maintenance of *Treponema bryantii.*

Triple Sugar Iron Agar (TSI Agar)

Composition per liter:
Peptone 20.0g
Agar 12.0g
Lactose 10.0g
Sucrose 10.0g
NaCl 5.0g
Beef extract 3.0g
Yeast extract 3.0g
Glucose 1.0g
Ferric citrate 0.3g
$Na_2S_2O_3$ 0.3g
Phenol Red 0.025g
pH 7.4 ± 0.2 at 25°C

Source: This medium is available as a premixed powder from BD Diagnostic Systems and Oxoid Unipath.

Preparation of Medium: Add components to distilled/deionized water and bring volume to 1.0L. Mix thoroughly. Gently heat and bring to boiling. Distribute into tubes or flasks. Autoclave for 15 min at 15 psi pressure–121°C. Allow tubes to cool in a slanted position to form a 1.0-inch butt.

Use: For the differentiation of members of the Enterobacteriaceae based on their fermentation of lactose, sucrose, and glucose and the production of H_2S.

Triple Sugar Iron Agar (TSI Agar)

Composition per liter:

Agar	13.0g
Pancreatic digest of casein	10.0g
Peptic digest of animal tissue	10.0g
Lactose	10.0g
Sucrose	10.0g
NaCl	5.0g
Glucose	1.0g
$Fe(NH_4)_2(SO_4)_2 \cdot 6H_2O$	0.2g
$Na_2S_2O_3$	0.2g
Phenol Red	0.025g

pH 7.3 ± 0.2 at 25°C

Source: This medium is available as a premixed powder from BD Diagnostic Systems.

Preparation of Medium: Add components to distilled/deionized water and bring volume to 1.0L. Mix thoroughly. Gently heat and bring to boiling. Distribute into tubes or flasks. Autoclave for 15 min at 15 psi pressure–121°C. Allow tubes to cool in a slanted position to form a 1.0 inch butt.

Use: For the differentiation of members of the Enterobacteriaceae based on their fermentation of lactose, sucrose, and glucose and the production of H_2S.

Tris YP Agar (Tris Yeast Extract Peptone Agar)

Composition per liter:

Agar	19.0g
Yeast extract	3.0g
Glucose	1.0g
Peptone	0.6g
Tris-buffer (0.05*M*, pH 7.5)	1.0L

pH 7.5 ± 0.2 at 25°C

Preparation of Medium: Add components to distilled/deionized water and bring volume to 1.0L. For top layer agar, add 6.0g of agar instead of 19.0g. Mix thoroughly. Gently heat and bring to boiling. Distribute into tubes or flasks. Autoclave for 15 min at 15 psi pressure–121°C. Pour into sterile Petri dishes.

Use: For the cultivation and maintenance of *Bdellovibrio* species.

Trypticase Serum Seawater Agar (ATCC Medium 1359)

Composition per liter:

Agar	12.0g
Pancreatic digest of casein	1.0g
Seawater	990.0mL
Horse serum	10.0mL

pH 7.3 ± 0.2 at 25°C

Preparation of Medium: Add components, except horse serum, to 990.0mL seawater. Mix thoroughly. Gently heat and bring to boiling. Autoclave for 15 min at 15 psi pressure–121°C. Cool to 45°–50°C. Aseptically add sterile horse serum. Mix thoroughly. Pour into sterile Petri dishes or distribute into sterile tubes.

Use: For the cultivation and maintenance of fastidious marine bacterial species.

Trypticase Soy Agar

Composition per liter:

Pancreatic digest of casein	17.0g
Agar	15.0g
NaCl	5.0g
Papaic digest of soybean meal	3.0g
K_2HPO_4	2.5g
Glucose	2.5g

pH 7.3 ± 0.2 at 25°C

Preparation of Medium: Add components to distilled/deionized water and bring volume to 1.0L. Mix thoroughly. Gently heat and bring to boiling. Distribute into tubes or flasks. Autoclave for 15 min at 15 psi pressure–121°C. Pour into sterile Petri dishes or leave in tubes.

Use: For the cultivation and maintenance of a wide variety of heterotrophic microorganisms.

Trypticase Soy Agar (ATCC Medium 18)

Composition per liter:

Pancreatic digest of casein	17.0g
Agar	15.0g
NaCl	5.0g
Papaic digest of soybean meal	3.0g
K_2HPO_4	2.5g
Glucose	2.5g

pH 7.3 ± 0.2 at 25°C

Preparation of Medium: Add components to distilled/deionized water and bring volume to 1.0L. Mix thoroughly. Distribute into tubes or flasks. Autoclave for 15 min at 15 psi pressure–121°C.

Use: For the cultivation of a wide variety of fastidious and nonfastidious microorganisms from clinical and nonclinical specimens. Also used for the rapid estimation of the bacteriological quality of water.

Trypticase Soy Agar (Tryptic Soy Agar) (Soybean Casein Digest Agar) (ATCC Medium 77)

Composition per liter:

Pancreatic digest of casein	15.0g
Agar	15.0g
Papaic digest of soybean meal	5.0g
NaCl	5.0g

pH 7.3 ± 0.2 at 25°C

Source: This medium is available as a premixed powder from BD Diagnostic Systems.

Preparation of Medium: Add components to distilled/deionized water and bring volume to 1.0L. Mix thoroughly. Gently heat and bring to boiling. Distribute into tubes or flasks. Autoclave for 15 min at 15 psi pressure–121°C. Do not overheat. Pour into sterile Petri dishes or leave in tubes.

Use: For the isolation and cultivation of a wide variety of fastidious as well as nonfastidious microorganisms.

Trypticase Soy Agar with Lecithin and Polysorbate 80 (Microbial Content Test Agar)

Composition per liter:

Pancreatic digest of casein	15.0g
Agar	15.0g
Papaic digest of soybean meal	5.0g
NaCl	5.0g
Polysorbate 80 (Tween 80)	5.0g
Lecithin	0.7g

pH 7.3 ± 0.2 at 25°C

Source: This medium is available as a premixed powder from BD Diagnostic Systems.

Preparation of Medium: Add components to distilled/deionized water and bring volume to 1.0L. Mix thoroughly. Gently heat and bring to boiling. Distribute into tubes or flasks. Autoclave for 15 min at 13 psi pressure–118°C. Cool to 45°–50°C. Pour into sterile Petri dishes in 17.0mL volumes or leave in tubes.

Use: For the detection and enumeration of microorganisms in replicate plating techniques. Also used for the detection and enumeration of microorganisms present on surfaces of sanitary importance.

Trypticase Soy Agar with NaCl (ATCC Medium 176)

Composition per liter:

NaCl	30.0g
Pancreatic digest of casein	15.0g
Agar	15.0g
Papaic digest of soybean meal	5.0g
Bile salts No. 3	1.0g

pH 7.3 ± 0.2 at 25°C

Preparation of Medium: Add components to distilled/deionized water and bring volume to 1.0L. Mix thoroughly. Gently heat and bring to boiling. Distribute into tubes or flasks. Autoclave for 15 min at 15 psi pressure–121°C. Pour into sterile Petri dishes or leave in tubes.

Use: For the cultivation and maintenance of *Vibrio alginolyticus.*

Trypticase Soy Broth (Soybean Casein Digest Broth, USP) (LMG Medium 185)

Composition per liter:

Pancreatic digest of casein	17.0g
NaCl	5.0g
Papaic digest of soybean meal	3.0g
K_2HPO_4	2.5g
Glucose	2.5g

pH 7.3 ± 0.2 at 25°C

Source: This medium is available as a premixed powder from BD Diagnostic Systems.

Preparation of Medium: Add components to distilled/deionized water and bring volume to 1.0L. Mix thoroughly. Distribute into tubes or flasks. Autoclave for 15 min at 15 psi pressure–121°C.

Use: For the cultivation of a wide variety of fastidious and nonfastidious microorganisms from nonclinical specimens. Also used for the rapid estimation of the bacteriological quality of water.

Trypticase Soy Broth with Sea Salts (LMG Medium 249)

Composition per liter:

Sea salts	20.0g
Pancreatic digest of casein	17.0g
NaCl	5.0g
Papaic digest of soybean meal	3.0g
K_2HPO_4	2.5g
Glucose	2.5g
Yeast extract	1.0

pH 7.2 ± 0.2 at 25°C

Preparation of Medium: Add components,to distilled/deionized water and bring volume to 1.0L. Mix thoroughly. Distribute into tubes or flasks. Autoclave for 15 min at 15 psi pressure–121°C.

Use: For cultivation and maintenance of *Cellulophaga baltica* and *Cellulophaga fucicola.*

Tryptone Soya Agar

Composition per liter:

Agar	15.0g
Pancreatic digest of casein	15.0g
NaCl	5.0g
Pancreatic digest of soybean meal	5.0g

pH 7.3 ± 0.2 at 25°C

Source: This medium is available as a premixed powder from Oxoid Unipath.

Preparation of Medium: Add components to distilled/deionized water and bring volume to 1.0L. Mix thoroughly. Gently heat and bring to boiling. Distribute into tubes or flasks. Autoclave for 15 min at 15 psi pressure–121°C. Pour into sterile Petri dishes or leave in tubes.

Use: For the cultivation and maintenance of a wide variety of microorganisms.

Tryptone Thioglycolate Medium (DSMZ Medium 48)

Composition per liter:

Yeast extract	6.0g
K_2HPO_4	5.45g
Peptone	2.0g
Tryptone	2.0g
KH_2PO_4	1.2g
Na-thioglycolate	0.5g
$MgSO_4 \cdot 7H_2O$	0.025g
$CaCl_2 \cdot 2H_2O$	0.015g
$FeSO_4 \cdot 7H_2O$	0.01g
$CoCl_2 \cdot 6H_2O$	2.5mg
$Na_2MoO_4 \cdot 2H_2O$	2.5mg
$MnCl_2 \cdot 4H_2O$	2.0mg
Glucose solution	50.0mL

pH 7.5 ± 0.2 at 25°C

Glucose Solution:

Composition per 50.0mL:

D-Glucose	20.0g

Preparation of Glucose Solution: Add glucose to distilled/deionized water and bring volume to 50.0mL. Mix thoroughly. Filter sterilize.

Preparation of Medium: Add components, except glucose solution to distilled/deionized water and bring volume to 950.0mL. Mix thoroughly. Gently heat and bring to boiling. Autoclave for 15 min at 15 psi pressure–121°C. Cool to 25°C. Aseptically add 50.0mL sterile glucose solution. Mix thoroughly. Aseptically distribute into sterile tubes or flasks.

Use: For the cultivation of *Clostridium beijerinckii.*

Tryptone Yeast Extract Salt Medium

Composition per liter:

Solution A	500.0mL
Solution B	500.0mL

pH 6.8 ± 0.2 at 25°C

Solution A:

Composition per 500.0mL:

NaCl	125.0g
$MgCl_2 \cdot 6H_2O$	50.0g

K_2SO_4 .... 5.0g
$CaCl_2 \cdot 6H_2O$ 0.2g

Preparation of Solution A: Add components to distilled/deionized water and bring volume to 500.0mL. Mix thoroughly. Adjust pH to 6.8. Autoclave for 15 min at 15 psi pressure–121°C.

Solution B:
Composition per 500.0mL:
Pancreatic digest of casein 5.0g
Yeast extract 5.0g

Preparation of Solution B: Add components to distilled/deionized water and bring volume to 500.0mL. Mix thoroughly. Adjust pH to 6.8. Autoclave for 15 min at 15 psi pressure–121°C.

Preparation of Medium: Aseptically combine 500.0mL of solution A with 500.0mL of solution B. Mix thoroughly. Aseptically distribute into sterile tubes or flasks.

Use: For the cultivation of *Haloferax volcanii.*

TS Agar (DSMZ Medium 893)

Composition per liter:
Agar 15.0g
Sucrose 5.0g
Tryptone 5.0g
Beef extract 3.0g
Glucose 1.0g
DpH 7.0 ± 0.2 at 25°C

Preparation of Medium: Add components to distilled/deionized water and bring volume to 1.0L. Mix thoroughly. Gently heat and bring to boiling. Distribute into tubes or flasks. Autoclave for 15 min at 15 psi pressure–121°C. Pour into sterile Petri dishes or leave in tubes.

Use: For the cultivation and maintenance of *Saccharococcus thermophilus.*

TS Medium for *Spirochaeta caldaria*

Composition per liter:
Pancreatic digest of casein 2.0g
Cellobiose 1.0g
Maltose 1.0g
Yeast extract 1.0g
Dithiothreitol 0.15g
Resazurin 1.0mg
pH 7.0 ± 0.2 at 25°C

Preparation of Medium: Prepare and dispense medium under 100% N_2. Add components to distilled/deionized water and bring volume to 1.0L. Mix thoroughly. Gently heat and bring to boiling. Continue boiling for 3 min. Cool to room temperature while sparging with 100% N_2. Anaerobically distribute into anaerobic tubes. Autoclave for 15 min at 15 psi pressure–121°C. Adjust pH to 7.0

Use: For the cultivation of *Spirochaeta caldaria.*

TS Soil Extract (Trypticase Soy Soil Extract)

Composition per liter:
Pancreatic digest of casein 17.0g
Agar 15.0g
NaCl 5.0g
Papaic digest of soybean meal 3.0g
K_2HPO_4 .... 2.5g
Glucose 2.5g
Soil extract 250.0mL

Soil Extract:
Composition per 400.0mL:
African Violet soil 154.0g
Na_2CO_3 .... 0.4g

Preparation of Soil Extract: Add components to tap water and bring volume to 400.0mL. Autoclave for 60 min at 15 psi pressure–121°C. Filter through Whatman filter paper.

Preparation of Medium: Add components to tap water and bring volume to 1.0L. Mix thoroughly. Gently heat and bring to boiling. Distribute into tubes or flasks. Autoclave for 15 min at 15 psi pressure–121°C. Pour into sterile Petri dishes or leave in tubes.

Use: For the cultivation and maintenance of *Bacillus xerothermodurans*.

TTD-Medium (DSMZ Medium 480b)

Composition per liter:
NaCl (marine salts) 25.0g
Sulfur, powder 10.0g
Casitone 5.0g
NH_4Cl 0.33g
$CaCl_2 \cdot 2H_2O$ 0.33g
$MgCl_2 \cdot 6H_2O$ 0.33g
KCl 0.33g
KH_2PO_4 .... 0.33g
Resazurin 1.0mg
$Na_2S \cdot 9H_2O$ solution 10.0mL
Vitamin solution 10.0mL
Trace elements solution SL-10 .... 1.0mL
pH 6.9 ± 0.2 at 25°C

Trace Elements Solution SL-10:
Composition per liter:
$FeCl_2 \cdot 4H_2O$ 1.5g
$CoCl_2 \cdot 6H_2O$ 190.0mg
$MnCl_2 \cdot 4H_2O$ 100.0mg
$ZnCl_2$ 70.0mg
$Na_2MoO_4 \cdot 2H_2O$ 36.0mg
$NiCl_2 \cdot 6H_2O$ 24.0mg
H_3BO_3 .... 6.0mg
$CuCl_2 \cdot 2H_2O$ 2.0mg
HCl (25% solution) 10.0mL

Preparation of Trace Elements Solution SL-10: Add $FeCl_2 \cdot 4H_2O$ to 10.0mL of HCl solution. Mix thoroughly. Add distilled/deionized water and bring volume to 1.0L. Add remaining components. Mix thoroughly. Sparge with 100% N_2. Autoclave for 15 min at 15 psi pressure–121°C.

Vitamin Solution:
Composition per liter:
Pyridoxine-HCl 10.0mg
Thiamine-HCl·$2H_2O$ 5.0mg
Riboflavin 5.0mg
Nicotinic acid 5.0mg
D-Ca-pantothenate 5.0mg
p-Aminobenzoic acid 5.0mg
Lipoic acid 5.0mg
Biotin 2.0mg
Folic acid 2.0mg
Vitamin B_{12} 0.1mg

Preparation of Vitamin Solution: Add components to distilled/deionized water and bring volume to 1.0L. Mix thoroughly. Sparge with 80% H_2 + 20% CO_2. Filter sterilize.

$Na_2S{\cdot}9H_2O$ Solution:
Composition per 10.0mL:
$Na_2S{\cdot}9H_2O$ 0.5g

Preparation of $Na_2S{\cdot}9H_2O$ Solution: Add $Na_2S{\cdot}9H_2O$ to distilled/deionized water and bring volume to 10.0mL. Sparge with N_2. Autoclave for 15 min at 15 psi pressure–121°C. Cool to 25°C. Store anaerobically.

Preparation of Medium: Prepare and dispense medium under an oxygen-free 80% N_2 + 20% CO_2 gas mixture. Add components, except vitamin solution, and $Na_2S{\cdot}9H_2O$ solution to 980.0mL distilled/deionized water. Mix thoroughly. Sparge with 80% N_2 + 20% CO_2. Adjust pH to 5.9 with concentrated NaOH. Sterilize medium by heating for 1 hr at 90-100°C on 3 subsequent days. Sparge with 80% N_2 + 20% CO_2. Before use, aseptically and anaerobically add 10.0mL sterile vitamin solution and 10.0mL sterile $Na_2S{\cdot}9H_2O$ solution. Mix thoroughly. Aseptically and anaerobically distribute into sterile tubes or flasks.

Use: For the cultivation of *Thermococcus stetteri.*

TY Salts Medium

Composition per liter:
Pancreatic digest of casein 1.0g
Yeast extract 1.0g
Salts solution 100.0mL
pH 8.2 ± 0.2 at 25°C

Salts Solution:
Composition per liter:
$NaNO_3$ 6.89g
KNO_3 .... 1.03g
$MgSO_4{\cdot}7H_2O$ 1.0g
Nitrilotriacetic acid 1.0g
$CaSO_4{\cdot}2H_2O$ 0.6g
NaCl 80.0mg
$FeCl_3$ solution 10.0mL
Trace elements solution 10.0mL

Preparation of Salts Solution: Add components to distilled/deionized water and bring volume to 1.0L. Mix thoroughly. Adjust pH to 8.2 with 1*M* NaOH. Autoclave for 15 min at 15 psi pressure–121°C.

$FeCl_3$ Solution:
Composition per 100.0mL:
$FeCl_3$ 28.0mg

Preparation of $FeCl_3$ Solution: Add $FeCl_3$ to distilled/deionized water and bring volume to 100.0mL. Mix thoroughly.

Trace Elements Solution:
Composition per liter:
$MnSO_4{\cdot}H_2O$ 2.2g
H_3BO_3 .... 0.5g
$ZnSO_4{\cdot}7H_2O$ 0.5g
$CoCl_2{\cdot}6H_2O$ 46.0mg
$Na_2MoO_4{\cdot}2H_2O$ 25.0mg
$CuSO_4$ 16.0mg
H_2SO_4 .... 0.5mL

Preparation of Trace Elements Solution: Add components to distilled/deionized water and bring volume to 1.0L. Mix thoroughly.

Preparation of Medium: Add components, except salts solution, to distilled/deionized water and bring volume to 900.0mL. Mix thoroughly. Autoclave for 15 min at 15 psi pressure–121°C. Aseptically add 100.0mL of sterile salts solution. Mix thoroughly. Aseptically distribute into sterile tubes or flasks.

Use: For the cultivation of *Thermomonospora aquaticus, Thermus filiformis, Thermus flavus, Thermus ruber,* and other *Thermus* species.

TYN Medium

Composition per liter:
$Na_2S_2O_3{\cdot}5H_2O$ 10.0g
Pancreatic digest of casein 1.0g
Yeast extract 1.0g
Na_2SO_4 .... 1.0g

Preparation of Medium: Add components to distilled/deionized water and bring volume to 1.0L. Mix thoroughly. Distribute into tubes or flasks. Autoclave for 15 min at 15 psi pressure–121°C.

Use: For the cultivation and maintenance of *Thiobacillus* species.

Tyrosine Casein Nitrate Medium (TCN Medium)

Composition per liter:
Sodium caseinate 25.0g
Agar 15.0g
$NaNO_3$ 10.0g
L-Tyrosine 1.0g

Preparation of Medium: Add components to tap water and bring volume to 1.0L. Mix thoroughly. Gently heat and bring to boiling. Distribute into tubes or flasks. Autoclave for 15 min at 15 psi pressure–121°C. Pour into sterile Petri dishes or leave in tubes.

Use: For the isolation and cultivation of streptomycetes from infected plants.

TZC Selective Medium

Composition per liter:
Agar 17.0g
Peptone 1.0g
Glucose 0.5g
Pancreatic digest of casein 0.1g
2,3,5-Triphenyltetrazolium·HCl solution 10.0mL

2,3,5-Triphenyltetrazolium·HCl Solution:
Composition per 10.0mL:
2,3,5-Triphenyltetrazolium·HCl 0.05g

Preparation of 2,3,5-Triphenyltetrazolium·HCl Solution: Add 2,3,5-triphenyltetrazolium·HCl to distilled/deionized water and bring volume to 10.0mL. Mix thoroughly. Autoclave for 15 min at 15 psi pressure–121°C.

Preparation of Medium: Add components, except 2,3,5-triphenyltetrazolium·HCl solution, to distilled/deionized water and bring volume to 990.0mL. Mix thoroughly. Gently heat and bring to boiling. Autoclave for 15 min at 15 psi pressure–121°C. Cool to 45°–50°C. Aseptically add 10.0mL of sterile 2,3,5-triphenyltetrazolium·HCl solution. Mix thoroughly. Pour into sterile Petri dishes.

Use: For the isolation, cultivation, and differentiation of *Pseudomonas solanacearum.* The virulent, wild-type strains appear as irregular to round white colonies with a pink center. Avirulent mutants, which readily occur in nature, appear as round, deep red colonies with a narrow blue border.

Urea Agar
(Urease Test Agar)
(Urea Agar Base, Christensen)

Composition per liter:

Urea	20.0g
Agar	15.0g
NaCl	5.0g
KH_2PO_4	2.0g
Peptone	1.0g
Glucose	1.0g
Phenol Red	0.012g

pH 6.8 ± 0.2 at 25°C

Source: This medium is available as a premixed powder from BD Diagnostic Systems.

Preparation of Medium: Add components, except agar, to distilled/deionized water and bring volume to 100.0mL. Mix thoroughly. Filter sterilize. Add agar to distilled/deionized water and bring volume to 900.0mL. Mix thoroughly. Gently heat and bring to boiling. Autoclave for 15 min at 15 psi pressure–121°C. Cool to 50°C. Aseptically add the 100.0mL of sterile basal medium. Mix thoroughly. Distribute into sterile tubes. Allow tubes to solidify in a slanted position.

Use: For the differentiation of a variety of microorganisms, especially members of the Enterobacteriaceae, aerobic actinomycetes, streptococci, and nonfermenting Gram-negative bacteria, on the basis of urease production.

Urea R Broth
(Urea Rapid Broth)

Composition per liter:

Urea	20.0g
Yeast extract	0.1g
Na_2HPO_4	0.095g
KH_2PO_4	0.091g
Phenol Red	0.01g

pH 6.9 ± 0.2 at 25°C

Source: This medium is available as a prepared medium from BD Diagnostic Systems.

Preparation of Medium: Add components to distilled/deionized water and bring volume to 1.0L. Mix thoroughly. Filter sterilize. Aseptically distribute into sterile tubes or flasks.

Use: For the differentiation of members of the Enterobacteriaceae based on the rapid detection of urease activity. Urease-positive bacteria turn the medium cerise.

Urea Semisolid Medium

Composition per liter:

Solution A	400.0mL
Solution B	50.0mL

Solution A:

Composition per 400.0mL:

Pancreatic digest of casein	6.0g
Yeast extract	2.0g
NaCl	1.0g
Yeast extract	0.8g
Agar	0.3g
L-Cystine	0.1g
Thioglycollic acid	0.12mL

pH 7.2 ± 0.2 at 25°C

Preparation of Solution A: Add components to distilled/deionized water and bring volume to 400.0mL. Mix thoroughly. Gently heat and bring to boiling. Autoclave for 15 min at 15 psi pressure–121°C. Cool to 60°C.

Solution B:

Composition per 50.0mL:

Urea	8.0g
Na_2HPO_4	3.8g
KH_2PO_4	3.64g
Yeast extract	0.04g
Phenol Red	4.0mg

Preparation of Solution B: Add components to distilled/deionized water and bring volume to 50.0mL. Mix thoroughly. Filter sterilize.

Preparation of Medium: Aseptically combine 400.0mL of sterile solution A and 50.0mL of sterile solution B. Mix thoroughly. Aseptically distribute into sterile screw-capped tubes in 7.0mL volumes. Pass the tubes into an anaerobic chamber containing 85% N_2 + 10% H_2 + 5% CO_2 for 60 min. Close screw caps tightly.

Use: For the cultivation and differentiation of anaerobic bacteria based on their production of urease. Bacteria that produce urease turn the medium bright red.

Urea Test Broth

Composition per liter:

Urea	20.0g
Na_2HPO_4	9.5g
KH_2PO_4	9.1g
Yeast extract	0.1g
Phenol Red	0.01g
Urea solution	100.0mL

Urea Solution:

Composition per 100.0mL:

Urea	20.0g

Preparation of Urea Solution: Add urea to distilled/deionized water and bring volume to 100.0mL. Mix thoroughly. Filter sterilize.

Preparation of Medium: Add components, except urea solution, to distilled/deionized water and bring volume to 900.0mL. Mix thoroughly. Autoclave for 15 min at 15 psi pressure–121°C. Cool to 45°–50°C. Aseptically add sterile urea solution. Mix thoroughly. Aseptically distribute into sterile tubes in 3.0mL volumes.

Use: For the cultivation and differentiation of members of the Enterobacteriaceae and aerobic actinomycetes based on their production of urease. Bacteria that produce urease turn the medium bright red.

Urease Test Broth
(Urea Broth)

Composition per liter:

Urea	20.0g
Na_2HPO_4	9.5g
KH_2PO_4	9.1g
Yeast extract	0.1g
Phenol Red	0.01g

pH 6.8 ± 0.2 at 25°C

Source: This medium is available as a premixed powder from BD Diagnostic Systems.

Preparation of Medium: Add components to distilled/deionized water and bring volume to 1.0L. Mix thoroughly. Filter sterilize. Aseptically distribute into sterile tubes or flasks.

Use: For the differentiation of organisms, especially the Enterobacteriaceae, on the basis of urease production. Urease-positive bacteria turn the medium pink.

Ustilago Complete Agar II

Composition per liter:

Agar	20.0g
Glucose	10.0g
Hydrolyzed casein	2.5g
NH_4NO_3	1.5g
Yeast extract	1.0g
Salt solution	62.5mL
Vitamin solution	10.0mL
Hydrolyzed nucleic acids solution	5.0mL

pH 7.0 ± 0.2 at 25°C

Salt Solution:

Composition per liter:

KH_2PO_4	16.0g
KCl	8.0g
Na_2SO_4	4.0g
$MgSO_4$	2.0g
$CaCl_2$	1.0g
Trace elements solution	8.0mL

Trace Elements Solution:

Composition per liter:

$CuSO_4 \cdot 5H_2O$	0.4g
$ZnCl_2$	0.4g
$MnCl_2 \cdot 4H_2O$	0.14g
$FeCl_3 \cdot 6H_2O$	100.0mg
H_3BO_3	60.0mg
$Na_2MoO_4 \cdot 2H_2O$	40.0mg

Preparation of Trace Elements Solution: Add components to distilled/deionized water and bring volume to 1.0L. Mix thoroughly.

Preparation of Salt Solution: Add components to distilled/deionized water and bring volume to 1.0L. Mix thoroughly.

Vitamin Solution:

Composition per liter:

Inositol	1.0g
Calcium pantothenate	0.2g
Choline chloride	0.2g
Nicotinic acid	0.2g
Thiamine	100.0mg
p-Aminobenzoic acid	50.0mg
Pyridoxin	50.0mg
Riboflavin	50.0mg

Preparation of Vitamin Solution: Add components to distilled/deionized water and bring volume to 1.0L. Mix thoroughly.

Hydrolyzed Nucleic Acids Solution:

Composition per 80.0mL:

DNA, calf thymus	2.0g
RNA	2.0g
HCl (1*M* solution)	30.0mL
NaOH (1*M* solution)	30.0mL

Preparation of Hydrolyzed Nucleic Acids Solution: Add DNA to 30.0mL of 1*M* NaOH solution. Add RNA to 30.0mL of 1*M* HCl solution. Autoclave the 2 solutions separately for 10 min at 15 psi pressure–121°C. Mix the two solutions. Adjust the pH to 6.0. Centrifuge at 5000 × g for 10 min. Decant supernatant solution and filter. Bring volume to 80.0mL with distilled/deionized water. Store at –20°C.

Preparation of Medium: Add components to distilled/deionized water and bring volume to 1.0L. Mix thoroughly. Gently heat and bring to boiling. Adjust pH to 7.0. Distribute into tubes or flasks. Autoclave for 15 min at 15 psi pressure–121°C. Pour into sterile Petri dishes or leave in tubes.

Use: For the cultivation and maintenance of *Ustilago* species.

Ustilago Complete Broth II

Composition per liter:

Glucose	10.0g
Hydrolyzed casein	2.5g
NH_4NO_3	1.5g
Yeast extract	1.0g
Salt solution	62.5mL
Vitamin solution	10.0mL
Hydrolyzed nucleic acids solution	5.0mL

pH 7.0 ± 0.2 at 25°C

Salt Solution:

Composition per liter:

KH_2PO_4	16.0g
KCl	8.0g
Na_2SO_4	4.0g
$MgSO_4$	2.0g
$CaCl_2$	1.0g
Trace elements solution	8.0mL

Trace Elements Solution:

Composition per liter:

$CuSO_4 \cdot 5H_2O$	0.4g
$ZnCl_2$	0.4g
$MnCl_2 \cdot 4H_2O$	0.14g
$FeCl_3 \cdot 6H_2O$	100.0mg
H_3BO_3	60.0mg
$Na_2MoO_4 \cdot 2H_2O$	40.0mg

Preparation of Trace Elements Solution: Add components to distilled/deionized water and bring volume to 1.0L. Mix thoroughly.

Preparation of Salt Solution: Add components to distilled/deionized water and bring volume to 1.0L. Mix thoroughly.

Vitamin Solution:

Composition per liter:

Inositol	1.0g
Calcium pantothenate	0.2g
Choline chloride	0.2g
Nicotinic acid	0.2g
Thiamine	100.0mg
p-Aminobenzoic acid	50.0mg
Pyridoxin	50.0mg
Riboflavin	50.0mg

Preparation of Vitamin Solution: Add components to distilled/deionized water and bring volume to 1.0L. Mix thoroughly.

Hydrolyzed Nucleic Acids Solution:

Composition per 80.0mL:

DNA, calf thymus	2.0g
RNA	2.0g
HCl (1*M* solution)	30.0mL
NaOH (1*M* solution)	30.0mL

Preparation of Hydrolyzed Nucleic Acids Solution: Add DNA to 30.0mL of 1*M* NaOH solution. Add RNA to 30.0mL of 1*M* HCl solution. Autoclave the 2 solutions separately for 10 min at 15 psi pressure–121°C. Mix the two solutions. Adjust the pH to 6.0. Centrifuge at 5000 × g for 10 min. Decant supernatant solution and filter. Bring volume to 80.0mL with distilled/deionized water. Store at −20°C.

Preparation of Medium: Add components to distilled/deionized water and bring volume to 1.0L. Mix thoroughly. Adjust pH to 7.0. Distribute into tubes or flasks. Autoclave for 15 min at 15 psi pressure–121°C.

Use: For the cultivation of *Ustilago* species.

Ustilago Medium

Composition per liter:

Yeast extract	11.0g
Glucose	10.0g
NH_4NO_3	1.5g
Salt solution	62.5mL
Vitamin solution	10.0mL

Salt Solution:

Composition per liter:

KH_2PO_4	16.0g
KCl	8.0g
Na_2SO_4	4.0g
$MgSO_4·7H_2O$	2.0g
$CaCl_2$	1.0g
Trace elements solution	8.0mL

Preparation of Salt Solution: Add components to distilled/deionized water and bring volume to 1.0L. Mix thoroughly.

Trace Elements Solution:

Composition per 500.0mL:

$CuSO_4·5H_2O$	0.2g
$ZnCl_2$	0.2g
$MnCl_2·4H_2O$	0.07g
$FeCl_3·6H_2O$	0.05g
H_3BO_3	0.03g
$Na_2MoO_4·2H_2O$	0.02g

Preparation of Trace Elements Solution: Add components to distilled/deionized water and bring volume to 500.0mL. Mix thoroughly.

Vitamin Solution:

Composition per liter:

Inositol	0.4g
Calcium pantothenate	0.2g
Choline chloride	0.2g
Nicotinic acid	0.2g
Thiamine	0.1g
Pyridoxine	0.05g
Riboflavin	0.05g

Preparation of Vitamin Solution: Add components to distilled/deionized water and bring volume to 1.0L. Mix thoroughly. Filter sterilize.

Preparation of Medium: Add components, except vitamin solution, to distilled/deionized water and bring volume to 990.0mL. Mix thoroughly. Gently heat and bring to boiling. Autoclave for 15 min at 15 psi pressure–121°C. Cool to 45°–50°C. Aseptically add 10.0mL of sterile vitamin solution. Mix thoroughly. Aseptically distribute into sterile tubes or flasks.

Use: For the cultivation of *Ustilago* species.

Ustilago Minimal Medium

Composition per liter:

Glucose	10.0g
KNO_3	3.0g
Salt solution	62.5mL

Salt Solution:

Composition per liter:

KH_2PO_4	16.0g
KCl	8.0g
Na_2SO_4	4.0g
$MgSO_4·7H_2O$	2.0g
$CaCl_2$	1.0g
Trace elements solution	8.0mL

Preparation of Salt Solution: Add components to distilled/deionized water and bring volume to 1.0L. Mix thoroughly.

Trace Elements Solution:

Composition per 500.0mL:

$CuSO_4·5H_2O$	0.2g
$ZnCl_2$	0.2g
$MnCl_2·4H_2O$	0.07g
$FeCl_3·6H_2O$	0.05g
H_3BO_3	0.03g
$Na_2MoO_4·2H_2O$	0.02g

Preparation of Trace Elements Solution: Add components to distilled/deionized water and bring volume to 500.0mL. Mix thoroughly.

Preparation of Medium: Add components to distilled/deionized water and bring volume to 1.0L. Mix thoroughly. Distribute into tubes or flasks. Autoclave for 15 min at 15 psi pressure–121°C.

Use: For the cultivation of *Ustilago* species.

V-8™ Agar

Composition per liter:

Agar	15.0g
$CaCO_3$	2.0g
V-8 canned vegetable juice	200.0mL

Preparation of Medium: Add components to distilled/deionized water and bring volume to 1.0L. Mix thoroughly. Gently heat and bring to boiling. Distribute into tubes or flasks. Autoclave for 15 min at 15 psi pressure–121°C. Pour into sterile Petri dishes or leave in tubes.

Use: For the cultivation of numerous yeasts and filamentous fungi.

V-8-0 Agar

Composition per liter:

Agar	15.0 g
$CaCO_3$	5.0 g
$CaCl_3$	100.0mg
β-Sitosterol	30.0mg
Tryptophan	20.0mg
Thiamine	1.0mg
V-8 canned vegetable juice	354.0mL

Preparation of Medium: Add $CaCO_3$ to V-8. Centrifuge for 20 min at 4000rpm. Decant the supernatant. Add the supernatant to distilled/deionized water and bring volume to 1.0L. Mix thoroughly. Add remaining components. Gently heat and bring to boiling. Distribute into tubes or flasks. Autoclave for 15 min at 15 psi pressure–121°C. Pour into sterile Petri dishes or leave in tubes.

Use: For the cultivation of *Phytophthora syringae, Phytophthora palmivora, Phytophthora nicotianae, Phytophthora erythroseptica, Phytophthora drechsleri,* and *Phytophthora citrophthora.*

V-8 Juice Seawater Agar

Composition per liter:

Agar .... 15.0 g
$CaCO_3$ .... 3.0 g
Artificial seawater .... 800.0 mL
V-8 canned vegetable juice, unsalted .... 200.0 mL
pH 7.0 ± 0.2 at 25°C

Artificial Seawater:

NaCl .... 20.0 g
$MgCl_2 \cdot 6H_2O$ .... 5.38 g
$MgSO_4 \cdot 7H_2O$ .... 6.78 g
KCl .... 0.72 g
$NaHCO_3$ .... 0.2 g
$CaCl_2 \cdot 2H_2O$ .... 1.4 g

Preparation of Artificial Seawater: Add components to distilled/deionized water and bring volume to 1.0L. Mix thoroughly.

Preparation of Medium: Combine components. Mix thoroughly. Adjust pH to 7.0. Gently heat and bring to boiling. Distribute into tubes or flasks. Autoclave for 15 min at 15 psi pressure–121°C. Pour into sterile Petri dishes or leave in tubes.

Use: For the cultivation of *Halophytophthora masteri* and *Halophytophthora tartarea.*

V-8 Rye Agar

Composition per 1050.0mL:

Agar .... 20.0 g
$CaCO_3$ .... 0.2 g
Rye broth .... 1.0 L
V-8 canned vegetable juice .... 50.0 mL

Rye broth:

Composition per liter:

Whole rye grains, ... 50.0 g

Preparation of Rye Broth: Soak the rye grains in 1.1L of distilled/deionized water for 24–36 hr at 24°C. Autoclave at 15 psi pressure–121°C for 30 min. Filter through four layers of cheesecloth. Bring the final filtrate volume to 1.0L with distilled/deionized water.

Preparation of Medium: Add components to distilled/deionized water and bring volume to 1.0L. Mix thoroughly. Gently heat and bring to boiling. Distribute into tubes or flasks. Autoclave for 15 min at 15 psi pressure–121°C. Pour into sterile Petri dishes or leave in tubes.

Use: For the cultivation and maintenance of *Phytophthora infestans.*

V24N Medium (DSMZ Medium 88b)

Starch, soluble 2.0g
$(NH_4)_2SO_4$ 1.3g
Sulfur, powdered 1.0g
$Na_2S \cdot 9H_2O$ 0.5g
KH_2PO_4 .... 0.28g
$MgSO_4 \cdot 7H_2O$ 0.25g
Yeast extract 0.2g
$CaCl_2 \cdot 2H_2O$ 0.07g
$FeCl_3 \cdot 6H_2O$ 0.02g
$Na_2B_4O_7 \cdot 10H_2O$ 4.5mg
$MnCl_2 \cdot 4H_2O$ 1.8mg
Resazurin 0.4mg
$ZnSO_4 \cdot 7H_2O$ 0.22mg
$CuCl_2 \cdot 2H_2O$ 0.05mg
$Na_2MoO_4 \cdot 2H_2O$ 0.03mg
$VOSO_4 \cdot 2H_2O$ 0.03mg
$CoSO_4$ 0.01mg
pH 6.0± 0.2 at 25°C

Preparation of Medium: Add components to distilled/deionized water and bring volume to 1.0L Mix thoroughly. Adjust the pH to 6.0 with H_2SO_4 (25% v/v). Sparge medium with 100% N_2 gas. Distribute anaerobically into rubber stoppered tubes or bottles that have been presterilized by autoclaving. On 2 successive days heat the medium to 85 °C for 2 hr in order to sterilize.

Use: For the growth and maintenance of *Thermofilum librum.*

Van Niel's Agar

Composition per liter:

Agar 20.0g
Yeast extract 10.0g
K_2HPO_4 .... 1.0g
$MgSO_4$ 0.5g
pH 7.0 ± 0.2 at 25°C

Preparation of Medium: Add components to distilled/deionized water and bring volume to 1.0L. Mix thoroughly. Gently heat and bring to boiling. Distribute into tubes or flasks. Autoclave for 15 min at 15 psi pressure–121°C. Pour into sterile Petri dishes or leave in tubes.

Use: For the cultivation of *Rhodomicrobium vannielii.*

Van Niel's Medium, Modified

Composition per liter:

Yeast extract 10.0g
$MgSO_4$ 0.1g
EDTA 2.0mg
Trace elements solution 10.0mL
K_2HPO_4 solution 2.5mL
pH 7.1 ± 0.2 at 25°C

Trace Elements Solution:

Composition per 100.0mL:

$CaCl_2 \cdot 2H_2O$ 0.3g
Ferric ammonium citrate 0.2g

Preparation of Trace Elements Solution: Add components to distilled/deionized water and bring volume to 100.0mL. Mix thoroughly. Filter sterilize.

K_2HPO_4 Solution:

Composition per 100.0mL:

K_2HPO_4 .... 4.0g

Preparation of K_2HPO_4 Solution: Add K_2HPO_4 to distilled/deionized water and bring volume to 100.0mL. Mix thoroughly. Filter sterilize.

Preparation of Medium: Add components, except K_2HPO_4 and trace elements solution, to distilled/deionized water and bring volume to 987.5mL. Mix thoroughly. Autoclave for 15 min at 15 psi pressure–121°C. Cool to 50°C. Aseptically add trace elements and K_2HPO_4 solutions. Mix thoroughly. Distribute into sterile tubes or flasks.

Use: For the cultivation and maintenance of *Rhodobacter sphaeroides.*

Van Niel's Yeast Agar with Glutamate (ATCC Medium 1370)

Composition per liter:

Agar 20.0g
Yeast extract 10.0g
K_2HPO_4 .... 1.0g

Glutamate 0.7g
$MgSO_4$ 0.5g
pH 7.1 ± 0.2 at 25°C

Preparation of Medium: Add components to tap water and bring volume to 1.0L. Mix thoroughly. Gently heat and bring to boiling. Distribute into tubes or flasks. Autoclave for 15 min at 15 psi pressure–121°C. Pour into sterile Petri dishes or leave in tubes.

Use: For the cultivation of *Rhodomicrobium spp.*

Van Niel's Yeast Agar with NaCl

Composition per liter:
NaCl 25.0g
Agar 20.0g
Yeast extract 10.0g
K_2HPO_4 .... 1.0g
$MgSO_4$ 0.5g
pH 7.1 ± 0.2 at 25°C

Preparation of Medium: Add components to tap water and bring volume to 1.0L. Mix thoroughly. Gently heat and bring to boiling. Distribute into tubes or flasks. Autoclave for 15 min at 15 psi pressure–121°C. Pour into sterile Petri dishes or leave in tubes.

Use: For the cultivation of halophilic *Rhodomicrobium spp.*

Van Niel's Yeast Agar with 25% NaCl (ATCC Medium 217)

Composition per liter:
NaCl 250.0g
Agar 20.0g
Yeast extract 10.0g
K_2HPO_4 .... 1.0g
$MgSO_4$ 0.5g
pH 7.1 ± 0.2 at 25°C

Preparation of Medium: Add components to tap water and bring volume to 1.0L. Mix thoroughly. Gently heat and bring to boiling. Distribute into tubes or flasks. Autoclave for 15 min at 15 psi pressure–121°C. Pour into sterile Petri dishes or leave in tubes.

Use: For the cultivation of halophilic *Rhodomicrobium* spp.

Van Niel's Yeast Agar with Succinate (ATCC Medium 1243)

Composition per liter:
Agar 20.0g
Yeast extract 10.0g
K_2HPO_4 .... 1.0g
Succinate 0.6g
$MgSO_4$ 0.5g
pH 7.1 ± 0.2 at 25°C

Preparation of Medium: Add components to tap water and bring volume to 1.0L. Mix thoroughly. Gently heat and bring to boiling. Distribute into tubes or flasks. Autoclave for 15 min at 15 psi pressure–121°C. Pour into sterile Petri dishes or leave in tubes.

Use: For the cultivation of *Rhodomicrobium spp.*

Van Niel's Yeast Medium with Pyruvate, Modified

Composition per liter:
Yeast extract 10.0g
$MgSO_4$ 0.1g
EDTA 2.0mg
Sodium pyruvate solution 100.0mL
Trace elements solution 10.0mL
K_2HPO_4 solution 5.0mL
Trace metal A-5 solution 1.0mL
pH 7.1± 0.1 at 25°C

Sodium Pyruvate Solution:
Composition per 100.0mL:
Sodium pyruvate 1.1g

Preparation of K_2HPO_4 Solution: Add sodium pyruvate to distilled/deionized water and bring volume to 100.0mL. Mix thoroughly. Filter sterilize.

Trace Elements Solution:
Composition per 100.0mL:
$CaCl_2 \cdot 2H_2O$ 0.3g
Ferric ammonium citrate 0.2g

Preparation of Trace Elements Solution: Add components to distilled/deionized water and bring volume to 100.0mL. Mix thoroughly. Filter sterilize.

K_2HPO_4 Solution:
Composition per 100.0mL:
K_2HPO_4 .... 4.0g

Preparation of K_2HPO_4 Solution: Add K_2HPO_4 to distilled/deionized water and bring volume to 100.0mL. Mix thoroughly. Filter sterilize.

Trace Metal A-5 Solution:
Composition per liter:
H_3BO_3 .... 2.86g
$MnCl_2 \cdot 4H_2O$ 1.81g
$Na_2MoO_4 \cdot 2H_2O$ 0.39g
$ZnSO_4 \cdot 7H_2O$ 0.222g
$CuSO_4 \cdot 5H_2O$ 0.079g
$Co(NO_3)_2 \cdot 6H_2O$ 0.049.g

Preparation of Trace Metal A-5 Solution: Add components to distilled/deionized water and bring volume to 1.0L. Mix thoroughly. Filter sterilize.

Preparation of Medium: Add components, except sodium pyruvate, trace elements, K_2HPO_4, and trace metal A-5 solutions, to distilled/deionized water and bring volume to 884.0mL. Mix thoroughly. Autoclave for 15 min at 15 psi pressure–121°C. Cool to 25°C. Aseptically add 100.0mL of sterile sodium pyruvate solution, 10.0mL of sterile trace elements soultion, 5.0mL of sterile K_2HPO_4 solution, and 1.0mL of sterile trace metal A-5 solution. Mix thoroughly. Adjust pH to 7.1± 0.1. Aseptically distribute into sterile tubes or flasks.

Use: For the cultivation and maintenance of photosynthetic bacteria, such as *Heliobacillus mobilis* and *Rhodopseudomonas palustris*.

Vanillate Medium

Composition per liter:
Agar 20.0g
$(NH_4)_2SO_4$ 1.0g
KH_2PO_4 .... 0.4g
Yeast extract 0.1g
$MgSO_4 \cdot 7H_2O$ 0.01g
Trace elements solution 10.0mL
Vanillic acid solution 10.0mL

Trace Elements Solution:
Composition per liter:
$MnSO_4 \cdot 4H_2O$ 0.4g
H_3BO_3 .... 0.5mg
$ZnSO_4 \cdot 7H_2O$ 0.4mg

$FeCl_3$ 0.2mg
$(NH_4)_6Mo_7O_{24}·4H_2O$ 0.2mg
KI 0.1mg
$CuSO_4·5H_2O$ 0.04mg

Preparation of Trace Elements Solution: Add components to distilled/deionized water and bring volume to 1.0L. Mix thoroughly. Filter sterilize.

Vanillic Acid Solution:
Composition per 10.0mL:
Vanillic acid, sodium salt 1.5g

Preparation of Vanillic Acid Solution: Add vanillic acid to distilled/deionized water and bring volume to 10.0mL. Mix thoroughly. Filter sterilize.

Preparation of Medium: Add components, except vanillic acid solution and trace elements solution, to distilled/deionized water and bring volume to 980.0mL. Mix thoroughly. Gently heat and bring to boiling. Autoclave for 15 min at 15 psi pressure–121°C. Cool to 50°–55°C. Warm the vanillic acid solution and the trace elements solution to 50°–55°C. Aseptically add 10.0mL of sterile vanillic acid solution and 10.0mL of sterile trace elements solution. Mix thoroughly. Pour into sterile Petri dishes or distribute into sterile tubes.

Use: For the cultivation of *Pseudomonas fluorescens*.

VCR Medium

Composition per 1001.0mL:
Agar, noble 15.0g
$NaNO_3$ 2.0g
KH_2PO_4 .... 1.5g
K_2HPO_4 .... 1.2g
NH_4Cl 0.5g
$MgSO_4·7H_2O$ 0.2g
$FeCl_3$ 0.01g
$CaCl_2·2H_2O$ 15.0mg
$CuSO_4·5H_2O$ 1.0mg
Vitamin B_{12} solution 1.0mL
pH 7.2 ± 0.2 at 25°C

Vitamin B_{12} Solution:
Composition per 10.0mL:
Vitamin B_{12} 10.0μg

Preparation of Vitamin B_{12} Solution: Add Vitamin B_{12} to distilled/deionized water and bring volume to 10.0mL. Mix thoroughly. Filter sterilize.

Preparation of Medium: Add components, except Vitamin B_{12} solution, to distilled/deionized water and bring volume to 1.0L. Mix thoroughly. Gently heat and bring to boiling. Autoclave for 15 min at 15 psi pressure–121°C. Cool to 50°–55°C. Aseptically add 1.0mL of sterile Vitamin B_{12} solution. Mix thoroughly. Pour into sterile Petri dishes or distribute into sterile tubes.

Use: For the cultivation of *Sphaerobacter thermomethanica*.

VEN CHI2 Medium (DSMZ Medium 293a)

Composition per liter:
NaCl 20.0g
$MgCl_2·6H_2O$ 3.0g
KCl 0.5g
NH_4Cl 0.25g
KH_2PO_4 .... 0.2g
$CaCl_2·2H_2O$ 0.15g
Resazurin 1.0mg
$NaHCO_3$ solution 10.0mL
Na_2-succinate solution 10.0mL
$Na_2S·9H_2O$ solution 10.0mL
Vitamin solution 10.0mL
Quinic acid solution 10.0mL
Trace elements solution SL-10 .... 1.0mL
pH 7.2 ± 0.2 at 25°C

$Na_2S·9H_2O$ Solution:
Composition per 10mL:
$Na_2S·9H_2O$ 0.36g

Preparation of $Na_2S·9H_2O$ Solution: Add $Na_2S·9H_2O$ to distilled/deionized water and bring volume to 10.0mL. Mix thoroughly. Autoclave under 100% N_2 for 15 min at 15 psi pressure–121°C. Cool to room temperature.

$NaHCO_3$ Solution:
Composition per 10.0mL:
$NaHCO_3$ 2.5g

Preparation of $NaHCO_3$ Solution: Add $NaHCO_3$ to distilled/deionized water and bring volume to 10.0mL. Mix thoroughly. Sparge with 80% N_2 + 20% CO_2. Filter sterilize.

Na_2-Succinate Solution:
Composition per 10.0mL:
Na_2-succinate 3.25g

Preparation of Na_2-Succinate Solution: Add Na_2-succinate to distilled/deionized water and bring volume to 10.0mL. Mix thoroughly. Sparge with 100% N_2. Filter sterilize.

Vitamin Solution:
Composition per liter:
Pyridoxine-HCl 10.0mg
Thiamine-HCl·$2H_2O$ 5.0mg
Riboflavin 5.0mg
Nicotinic acid 5.0mg
D-Ca-pantothenate 5.0mg
p-Aminobenzoic acid 5.0mg
Lipoic acid 5.0mg
Biotin 2.0mg
Folic acid 2.0mg
Vitamin B_{12} 0.1mg

Preparation of Vitamin Solution: Add components to distilled/deionized water and bring volume to 1.0L. Mix thoroughly. Sparge with 80% H_2 + 20% CO_2. Filter sterilize.

Quinic Acid Solution:
Composition per 10.0mL:
Quinic acid 1.0g

Preparation of Quinic Acid Solution: Add quinic acid to distilled/deionized water and bring volume to 10.0mL. Mix thoroughly. Adjust pH to 7.0. Sparge with 80% N_2 + 20% CO_2. Filter sterilize.

Trace Elements Solution SL-10:
Composition per liter:
$FeCl_2·4H_2O$ 1.5g
$CoCl_2·6H_2O$ 190.0mg
$MnCl_2·4H_2O$ 100.0mg
$ZnCl_2$ 70.0mg
$Na_2MoO_4·2H_2O$ 36.0mg
$NiCl_2·6H_2O$ 24.0mg
H_3BO_3 .... 6.0mg
$CuCl_2·2H_2O$ 2.0mg
HCl (25% solution) 10.0mL

Preparation of Trace Elements Solution SL-10: Add $FeCl_2 \cdot 4H_2O$ to 10.0mL of HCl solution. Mix thoroughly. Add distilled/deionized water and bring volume to 1.0L. Add remaining components. Mix thoroughly. Sparge with 80% N_2 + 20% CO_2. Filter sterilize.

Preparation of Medium: Prepare and dispense medium under 80% N_2 + 20% CO_2 gas atmosphere. Add components, except $NaHCO_3$ solution, Na_2-succinate solution, $Na_2S \cdot 9H_2O$ solution, vitamin solution, quinic acid solution, and trace elements solution SL-10, to distilled/deionized water and bring volume to 949.0mL. Mix thoroughly. Adjust pH to 7.2. Sparge with 80% N_2 + 20% CO_2. Autoclave for 15 min at 15 psi pressure–121°C. Aseptically and anaerobically add 10.0mL $NaHCO_3$ solution, 10.0mL Na_2-succinate solution, 10.0mL $Na_2S \cdot 9H_2O$ solution, 10.0mL vitamin solution, 10.0mL quinic acid solution, and 1.0mL trace elements solution SL-10. Mix thoroughly. Aseptically and anaerobically distribute into sterile tubes or bottles. After inoculation, flush and repressurize the gas head space of culture bottles with sterile 80% N_2 + 20% CO_2 to 1 bar overpressure.

Use: For the cultivation of *Ilyobacter insuetus.*

Vibrio Agar

Composition per liter:

Sucrose	20.0g
Agar	15.0g
NaCl	10.0g
Sodium citrate·$2H_2O$	10.0g
$Na_2S_2O_3 \cdot 5H_2O$	6.5g
Oxgall	5.0g
Yeast extract	5.0g
Pancreatic digest of casein	4.0g
Proteose peptone	3.0g
Sodium deoxycholate	1.0g
Sodium lauryl sulfate	0.2g
Water Blue	0.2g
Cresol Red	0.02g

pH 8.5 ± 0.2 at 25°C

Preparation of Medium: Add components to distilled/deionized water and bring volume to 1.0L. Mix thoroughly. Adjust pH to 8.5. Gently heat and bring to boiling. Do not autoclave. Pour into sterile Petri dishes or distribute into sterile tubes.

Use: For the isolation and cultivation of the *Vibrio cholerae.*

Vibrio costicola Medium

Composition per liter:

NaCl	81.0g
$MgSO_4 \cdot 7H_2O$	19.7g
Agar	15.0g
$MgCl \cdot H_2O$	15.0g
Yeast extract	10.0g
Proteose peptone	5.0g
KCl	2.0g
Glucose	1.0g
$CaCl_2 \cdot 2H_2O$	0.48g
$NaHCO_3$	60.0mg
NaBr	26.0mg

Preparation of Medium: Add components to distilled/deionized water and bring volume to 1.0L. Mix thoroughly. Gently heat and bring to boiling. Distribute into tubes or flasks. Autoclave for 15 min at 15 psi pressure–121°C. Pour into sterile Petri dishes or leave in tubes.

Use: For the cultivation and maintenance of *Vibrio costicola.*

Vibrio Medium

Composition per liter:

NaCl	10.0g
Pancreatic digest of casein	10.0g
$MgCl_2 \cdot 6H_2O$	4.0g
KCl	1.0g

pH 7.5 ± 0.2 at 25°C

Preparation of Medium: Add components to distilled/deionized water and bring volume to 1.0L. Mix thoroughly. Distribute into tubes or flasks. Autoclave for 15 min at 15 psi pressure–121°C.

Use: For the cultivation of *Vibrio diazotrophicus.*

Vibrio natriegens Medium

Composition per liter:

Urea	20.0g
NaCl	15.0g
Agar	15.0g
Peptone	5.0g
Meat extract	3.0g

pH 7.0 ± 0.2 at 25°C

Preparation of Medium: Add components to distilled/deionized water and bring volume to 1.0L. Mix thoroughly. Gently heat and bring to boiling. Adjust pH to 7.0. Distribute into tubes or flasks. Autoclave for 15 min at 15 psi pressure–121°C. Pour into sterile Petri dishes or leave in tubes.

Use: For the cultivation of *Vibrio natriegens.*

Vibrio parahaemolyticus Agar (VP Agar)

Composition per liter:

Agar	20.0g
NaCl	20.0g
Sucrose	20.0g
Sodium citrate	10.0g
$Na_2S_2O_3 \cdot 5H_2O$	10.0g
Peptone	10.0g
Sodium taurocholate	5.0g
Yeast extract	5.0g
Sodium lauryl sulfate	0.2g
Bromothymol Blue	0.04g
Thymol Blue	0.04g

pH 8.6 ± 0.2 at 25°C

Preparation of Medium: Add components to distilled/deionized water and bring volume to 1.0L. Mix thoroughly. Gently heat and bring to boiling. Do not autoclave. Pour into sterile Petri dishes.

Use: For the isolation, cultivation, enumeration, and presumptive identification of coliforms in specimens of sanitary significance. Sucrose-fermenting bacteria appear as yellow colonies with pale yellow peripheries. Sucrose-nonfermenting bacteria appear as mucoid, green colonies with a dark green center.

Vibrio vallismortis Medium

Composition per liter:

NaCl	25.0g
$MgSO_4 \cdot 7H_2O$	9.6g
$MgCl_2 \cdot 6H_2O$	7.0g
Glucose	5.0g
KCl	3.8g
Yeast extract	1.0g

$CaCl_2 \cdot 2H_2O$ 0.5g
$K_2HPO_4 \cdot 3H_2O$ 0.4g
$NaHCO_3$ solution 20.0mL
pH 7.0 ± 0.2 at 25°C

$NaHCO_3$ Solution:
Composition per 20.0mL:
$NaHCO_3$ 3.0g

Preparation of $NaHCO_3$ Solution: Add $NaHCO_3$ to distilled/deionized water and bring volume to 20.0mL. Mix thoroughly. Filter sterilize.

Preparation of Medium: Add components, except $NaHCO_3$ solution, to distilled/deionized water and bring volume to 980.0mL. Mix thoroughly. Autoclave for 15 min at 15 psi pressure–121°C. Cool to room temperature. Aseptically add 20.0mL of sterile $NaHCO_3$ solution. Mix thoroughly. Aseptically distribute into sterile tubes or flasks.

Use: For the cultivation of *Vibrio vallismortis.*

Violet Peptone Bile Lactose Broth

Composition per liter:
Lactose 10.0g
Peptone 10.0g
Bile salts 5.0g
Gentian Violet 0.04g
pH 7.6 ± 0.2 at 25°C

Preparation of Medium: Add components to distilled/deionized water and bring volume to 1.0L. Mix thoroughly. Gently heat and bring to boiling. Distribute into tubes or flasks. Autoclave for 15 min at 15 psi pressure–121°C. Pour into sterile Petri dishes or leave in tubes.

Use: For the selective cultivation of members of the Enterobacteriaceae.

Violet Red Bile Agar (VRB Agar)

Composition per liter:
Agar 15.0g
Lactose 10.0g
Pancreatic digest of gelatin 7.0g
NaCl 5.0g
Yeast extract 3.0g
Bile salts 1.5g
Neutral Red 0.03g
Crystal Violet 2.0mg
pH 7.4 ± 0.2 at 25°C

Source: This medium is available as a premixed powder from BD Diagnostic Systems and Oxoid Unipath.

Preparation of Medium: Add components to distilled/deionized water and bring volume to 1.0L. Mix thoroughly. Gently heat while stirring and bring to boiling. Distribute into tubes or flasks. Autoclave for 15 min at 15 psi pressure–121°C. Pour immediately into sterile Petri dishes or leave in tubes.

Use: For the detection of coliform bacteria in water.

VM1 Medium (DSMZ Medium 890)

Composition per liter:
NaCl 20.0g
$MgCl_2 \cdot 6H_2O$ 12.6g
Na_2SO_4 .... 3.24g
$CaCl_2 \cdot 2H_2O$ 2.38g
KCl 0.56g
Sulfur, powdered 0.5g
NH_4Cl 0.3g
K_2HPO_4 .... 0.2g
$NaHCO_3$ 0.16g
Trace elements solution 10.0mL
pH 7.0 ± 0.2 at 25°C

Trace Elements Solution:
Composition per liter:
$MgSO_4 \cdot 7H_2O$ 3.0g
Nitrilotriacetic acid 1.5g
NaCl 1.0g
$MnSO_4 \cdot 2H_2O$ 0.5g
$CoSO_4 \cdot 7H_2O$ 0.18g
$ZnSO_4 \cdot 7H_2O$ 0.18g
$CaCl_2 \cdot 2H_2O$ 0.1g
$FeSO_4 \cdot 7H_2O$ 0.1g
$NiCl_2 \cdot 6H_2O$ 0.025g
$KAl(SO_4)_2 \cdot 12H_2O$ 0.02g
H_3BO_3 .... 0.01g
$Na_2MoO_4 \cdot 4H_2O$ 0.01g
$CuSO_4 \cdot 5H_2O$ 0.01g
$Na_2SeO_3 \cdot 5H_2O$ 0.3mg

Preparation of Trace Elements Solution: Add nitrilotriacetic acid to 500.0mL of distilled/deionized water. Dissolve by adjusting pH to 6.5 with KOH. Add remaining components. Add distilled/deionized water to 1.0L. Mix thoroughly.

Preparation of Medium: Add components to distilled/deionized water and bring volume to 1.0L. Mix thoroughly. Adjust pH to 7.0 using H_2SO_4. Fill 20.0mL medium into 100mL serum bottles and seal with a rubber stopper. Add atmosphere of 78% H_2 + 20% CO_2 + 2% O_2 with an overpressure. Autoclave for 15 min at 15 psi pressure–121°C.

Use: For the autotrophic cultivation of *Hydrogenothermus marinus.*

Von Hofsten & Malmquist Medium B

Composition per liter:
Agar 15.0g
$NaNO_3$ 2.0g
K_2HPO_4 .... 0.5g
$CaCl_2 \cdot H_2O$ 0.02g
$FeSO_4 \cdot 7H_2O$ 0.02g
$MgSO_4 \cdot 7H_2O$ 0.02g
$MnSO_4 \cdot H_2O$ 0.02g
pH 7.5 ± 0.2 at 25°C

Preparation of Medium: Add components to distilled/deionized water and bring volume to 1.0L. Mix thoroughly. Gently heat and bring to boiling. Distribute into tubes or flasks. Autoclave for 15 min at 15 psi pressure–121°C. Pour into sterile Petri dishes or leave in tubes.

Use: For the cultivation of *Alteromonas* species and *Cytophaga saccharophila.*

VP Medium (Voges-Proskauer Medium)

Composition per liter:
Peptone 7.0g
K_2HPO_4 .... 5.0g
Glucose 5.0g
pH 6.9 ± 0.2 at 25°C

Preparation of Medium: Add components to distilled/deionized water and bring volume to 1.0L. Mix thoroughly. Adjust pH to 6.9. Distribute into tubes in 3.0mL volumes. Autoclave for 15 min at 15 psi pressure–121°C.

Use: For the cultivation and differentiation of bacteria based on their ability to produce acetoin.

VY Agar

Composition per liter:

Agar 15.0g
Baker's yeast 10.0g
$CaCl_2 \cdot 2H_2O$ 1.0g
Cyanocobalamin 5.0mg
pH 7.2 ± 0.2 at 25°C

Preparation of Medium: Add components to distilled/deionized water and bring volume to 1.0L. Mix thoroughly. Gently heat and bring to boiling. Distribute into tubes or flasks. Autoclave for 15 min at 15 psi pressure–121°C. Pour into sterile Petri dishes or leave in tubes.

Use: For the cultivation and maintenance of myxobacteria.

VY2 Agar

Composition per liter:

Agar 15.0g
Baker's yeast 5.0g
$CaCl_2 \cdot 2H_2O$ 1.0g
Cyanocobalamin 5.0mg
pH 7.2 ± 0.2 at 25°C

Preparation of Medium: Add components to distilled/deionized water and bring volume to 1.0L. Mix thoroughly. Gently heat and bring to boiling. Distribute into tubes or flasks. Autoclave for 15 min at 15 psi pressure–121°C. Pour into sterile Petri dishes or leave in tubes.

Use: For the cultivation and maintenance of *Myxococcus amylovorans.*

VY5 Agar

Composition per liter:

Agar 15.0g
Baker's yeast 2.0g
$CaCl_2 \cdot 2H_2O$ 1.0g
Cyanocobalamin 5.0mg
pH 7.2 ± 0.2 at 25°C

Preparation of Medium: Add components to distilled/deionized water and bring volume to 1.0L. Mix thoroughly. Gently heat and bring to boiling. Distribute into tubes or flasks. Autoclave for 15 min at 15 psi pressure–121°C. Pour into sterile Petri dishes or leave in tubes.

Use: For the cultivation and maintenance of myxobacteria.

VY/4-SWS Agar (DSMZ Medium 958)

Composition per 1001mL:

NaCl 20.0g
Agar 15.0g
Yeast cell paste (Baker's yeast, washed in deionized water) wet weight 2.5g
Seawater salts solution 1000.0mL
Vitamin B_{12} solution 1.0mL
pH 7.5 ± 0.2 at 25°C

Seawater Salts Solution:
Composition per liter:

$MgSO_4 \cdot 7H_2O$ 8.0g
$CaCl_2 \cdot 2H_2O$ 1.0g
KCl 0.5g
$NaHCO_3$ 0.16g
KBr 0.08g
$SrCl_2 \cdot 6H_2O$ 0.03g
H_3BO_3 .... 0.02g
Ferric citrate 0.01g
di-Na-ß-glycerophosphate 0.01g
Trace elements solution SL-4 .... 1.0mL

Trace Elements Solution SL-4:
Composition per liter:

EDTA 0.5g
$FeSO_4 \cdot 7H_2O$ 0.2g
Trace elements solution SL-6 .... 100.0mL

Preparation of Trace Elements Solution SL-4: Add components to distilled/deionized water and bring volume to 1.0L. Mix thoroughly.

Trace Elements Solution SL-6:
Composition per liter:

H_3BO_3 .... 0.3g
$CoCl_2 \cdot 6H_2O$ 0.2g
$ZnSO_4 \cdot 7H_2O$ 0.1g
$MnCl_2 \cdot 4H_2O$ 0.03g
$Na_2MoO_4 \cdot H_2O$ 0.03g
$NiCl_2 \cdot 6H_2O$ 0.02g
$CuCl_2 \cdot 2H_2O$ 0.01g

Preparation of Trace Elements Solution SL-6: Add components to distilled/deionized water and bring volume to 1.0L. Mix thoroughly. Adjust pH to 3.4.

Preparation of Seawater Salts Solution: Add components to distilled/deionized water and bring volume to 1.0L. Mix thoroughly.

Vitamin B_{12} Solution:
Composition per 10.0mL:

Cyanocobalamine 5.0mg

Preparation of Vitamin B_{12} Solution: Add cyanocobalamine to distilled/deionized water and bring volume to 10.0mL. Mix thoroughly. Sparge with 100% N_2. Filter sterilize.

Preparation of Medium: Add NaCl, agar, and yeast cell paste to 1.0L seawater salts solution. Mix thoroughly. Adjust pH to 7.5 with 1*M* NaOH. Gently heat and bring to boiling. Autoclave for 20 min at 15 psi pressure–121°C. Cool to 50°C. Aseptically add 1.0mL sterile Vitamin B_{12} solution. Mix thoroughly. Pour into sterile Petri dishes or leave in tubes.

Use: For the cultivation and maintenance of *Haliangium ochraceum, Haliangium tepidum, Plesiocystis pacifica,* and *Enhygromyxa salina (Thaxtera salina).*

W Medium

Composition per liter:

Sulfur 10.0g
KH_2PO_4 .... 3.0g
$MgSO_4 \cdot 7H_2O$ 0.5g
$CaCl_2 \cdot 2H_2O$ 0.25g
$(NH_4)_2SO_4$ 0.2g
$FeSO_4 \cdot 7H_2O$ 10.0mg
pH 3.0 ± 0.2 at 25°C

Preparation of Sulfur: Sterilize by steaming at 100°C for 60 min on 3 consecutive days.

Preparation of Medium: Add components, except sulfur, to distilled/deionized water and bring volume to 990.0mL. Mix thoroughly. Autoclave for 15 min at 15 psi pressure–121°C. Aseptically add 10.0g of sterile sulfur. Mix thoroughly. Aseptically distribute into sterile tubes or flasks.

Use: For the cultivation of *Thiobacillus thiooxidans.*

Waksman's Glucose Agar

Composition per liter:

Agar	12.5g
Glucose	10.0g
Peptone	5.0g
Beef extract	5.0g
NaCl	5.0g

pH 7.4-7.6 at 25°C

Preparation of Medium: Add components to distilled/deionized water and bring volume to 1.0L. Mix thoroughly. Gently heat and bring to boiling. Distribute into tubes or flasks. Autoclave for 15 min at 15 psi pressure–121°C. Pour into sterile Petri dishes or leave in tubes.

Use: For the cultivation and maintenance of *Streptomyces* species.

Waksman's Sulfur Medium

Composition per liter:

KH_2PO_4	3.0g
$MgSO_4 \cdot 7H_2O$	0.5g
$(NH_4)_2SO_4$	0.2g
$CaCl_2 \cdot 2H_2O$	0.2g
$Fe_2(SO_4)_3$	0.1mg

Preparation of Medium: Add components to distilled/deionized water and bring volume to 1.0L. Mix thoroughly. It is not necessary to sterilize this medium. Distribute into sterile tubes or flasks.

Use: For the cultivation of sulfate-reducing microorganisms from soil.

Walsby Medium

Composition per liter:

$MgSO_4 \cdot 7H_2O$	0.075g
K_2HPO_4	0.039g
Na_2CO_3	0.02g
$CaCl_2 \cdot 2H_2O$	0.018g
H_3BO_3	2.8mg
$MnSO_4 \cdot 4H_2O$	2.0mg
$ZnSO_4$	0.22mg
MoO_3	0.18mg
$CuSO_4 \cdot 5H_2O$	0.08mg
$Co(NO_3)_2 \cdot 6H_2O$	0.05mg
Iron-EDTA solution	1.0mL

pH 8.5 ± 0.2 at 25°C

Iron-EDTA Solution:

Composition per liter:

EDTA	12.7g
$FeSO_4 \cdot 7H_2O$	4.98g

Preparation of Iron-EDTA Solution: Add components to distilled/deionized water and bring volume to 1.0L. Mix thoroughly.

Preparation of Medium: Add components to distilled/deionized water and bring volume to 1.0L. Mix thoroughly. Distribute into tubes or flasks. Autoclave for 15 min at 15 psi pressure–121°C.

Use: For the isolation and cultivation of planktonic gas-vacuolate cyanobacteria.

Water Agar

Composition per liter:

Agar	15.0g
$CaCl_2 \cdot 2H_2O$	1.0g

pH 7.2 ± 0.2 at 25°C

Preparation of Medium: Add components to distilled/deionized water and bring volume to 1.0L. Mix thoroughly. Gently heat and bring to boiling. Distribute into tubes or flasks. Autoclave for 15 min at 15 psi pressure–121°C. Pour into sterile Petri dishes or leave in tubes.

Use: For the cultivation of myxobacteria.

Waxy Maize Starch Medium

Composition per liter:

Agar	20.0g
Waxy maize starch	5.0g
Pancreatic digest of casein	5.0g
Yeast extract	5.0g
$CoCl_2 \cdot 6H_2O$	0.1g
$CaCl_2 \cdot 2H_2O$	0.1g
Maltose solution	100.0mL

pH 6.7 ± 0.2 at 25°C

Maltose Solution:

Composition per 100.0mL:

Maltose	10.0g

Preparation of Maltose Solution: Add maltose to distilled/deionized water and bring volume to 100.0mL. Mix thoroughly. Filter sterilize.

Preparation of Medium: Add components, except maltose solution, to distilled/deionized water and bring volume to 900.0mL. Mix thoroughly. Gently heat and bring to boiling. Adjust pH to 6.7. Distribute into tubes or flasks. Autoclave for 15 min at 15 psi pressure–121°C. Aseptically add maltose solution. Pour into sterile Petri dishes or leave in tubes.

Use: For the cultivation and maintenance of *Bacillus* species.

WCX Agar

Composition per liter:

Agar	15.0g
$CaCl_2 \cdot 2H_2O$	1.0g
Cycloheximide solution	100.0mL

pH 7.2 ± 0.2 at 25°C

Cycloheximide Solution

Composition per 100.0mL:

Cycloheximide	2.5mg

Preparation of Cycloheximide Solution: Add components to distilled/deionized water and bring volume to 100.0mL. Mix thoroughly. Filter sterilize.

Preparation of Medium: Add components, except cycloheximide solution, to distilled/deionized water and bring volume to 900.0mL. Mix thoroughly. Gently heat and bring to boiling. Autoclave for 15 min at 15 psi pressure–121°C. Cool to 45°–50°C. Aseptically add sterile cycloheximide solution. Mix thoroughly. Pour into sterile Petri dishes or distribute into sterile tubes.

Use: For the cultivation of myxobacteria.

Wilbrinck Agar for *Xanthomonas*

Composition per liter:

Sucrose	20.0g
Agar	12.0g
Peptone	5.0g
K_2HPO_4	0.5g
$MgSO_4 \cdot 7H_2O$	0.25g

pH 7.2 ± 0.2 at 25°C

Preparation of Medium: Add components to distilled/deionized water and bring volume to 1.0L. Mix thoroughly.

Adjust pH to 7.2. Gently heat and bring to boiling. Distribute into tubes or flasks. Autoclave for 15 min at 15 psi pressure–121°C. Pour into sterile Petri dishes or leave in tubes.

Use: For the cultivation and maintenance of *Pseudomonas caryophylli, Xanthomonas albilineans,* and *Xanthomonas axonopodis.*

Wilbrinck Agar for *Xanthomonas albilineans*

Composition per liter:

Agar	20.0g
Sucrose	10.0g
Peptone	5.0g
K_2HPO_4	0.5g
$MgSO_4 \cdot 7H_2O$	0.25g
Na_2SO_3, anhydrous	0.05g

pH 7.2 ± 0.2 at 25°C

Preparation of Medium: Add components to distilled/deionized water and bring volume to 1.0L. Mix thoroughly. Gently heat and bring to boiling. Distribute into tubes or flasks. Autoclave for 15 min at 15 psi pressure–121°C. Pour into sterile Petri dishes or leave in tubes.

Use: For the cultivation and maintenance of *Xanthomonas albilineans* and other *Xanthomonas* species.

Winogradsky's Medium, Modified

Composition per liter:

$CaCO_3$	5.0g
$(NH_4)_2SO_4$	1.0g
K_2HPO_4	1.0g
NaCl	1.0g
$MgSO_4 \cdot 7H_2O$	0.5g
$FeSO_4$	0.4g

Preparation of Medium: Add components to distilled/deionized water and bring volume to 1.0L. Mix thoroughly. Gently heat until dissolved. Do not autoclave. Distribute into tubes or flasks. Swirl flask while dispensing to suspend precipitate.

Use: For the cultivation of nitrifying bacteria.

Winogradsky's N-Free Medium

Composition per liter:

Agar	20.0g
$CaCO_3$	5.0mg
Sugar solution	100.0mL
Concentrated salt solution	5.0mL

pH 7.2 ± 0.2 at 25°C

Sugar Solution:

Composition per 100.0mL:

Sucrose or glucose	10.0g

Preparation of Sugar Solution: Add sucrose or glucose to 100.0mL of distilled/deionized water. Mix thoroughly. Autoclave for 10 min at 10 psi pressure–115°C. Cool to 50°C.

Concentrated Salt Solution:

Composition per liter:

KH_2PO_4	50.0g
$MgSO_4 \cdot 7H_2O$	25.0g
NaCl	25.0g
$FeSO_4 \cdot 7H_2O$	1.0g
$MnSO_4 \cdot 4H_2O$	1.0g
$Na_2MoO_4 \cdot 4H_2O$	1.0g

Preparation of Concentrated Salt Solution: Add components to tap water and bring volume to 1.0L. Mix thoroughly. Filter sterilize.

Preparation of Medium: Add components, except sugar solution, to distilled/deionized water and bring volume to 900.0mL. Mix thoroughly. Gently heat and bring to boiling. Distribute into tubes or flasks. Autoclave for 15 min at 15 psi pressure–121°C. Cool to 50°C. Aseptically add sugar solution. Adjust pH to 7.2. Mix thoroughly. Pour into sterile Petri dishes or leave in tubes.

Use: For the cultivation and maintenance of *Azomonas insignis.*

Winogradsky's Nitrite Medium

Composition per liter:

Agar	15.0g
$NaNO_2$	2.0g
Na_2CO_3, anhydrous	1.0g
K_2HPO_4	0.5g

Preparation of Medium: Add components to distilled/deionized water and bring volume to 1.0L. Mix thoroughly. Gently heat and bring to boiling. Distribute into tubes. Autoclave for 15 min at 15 psi pressure–121°C.

Use: For the selective isolation and cultivation of *Nocardia* species and *Rhodococcus* species.

WMC Medium

Composition per 1010.0mL:

$NaHCO_3$	5.0g
Sodium acetate	1.0g
L-Cysteine	0.5g
Resazurin	1.0mg
Mineral salt solution 2xW	500.0mL
LIP solution	50.0mL
TYC solution	50.0mL
Trace elements solution	10.0mL
$Na_2S \cdot 9H_2O$ solution	10.0mL

pH 7.2 ± 0.2 at 25°C

Mineral Solution 2xW:

Composition per liter:

NaCl	40.0g
$MgCl_2 \cdot 6H_2O$	5.6g
$MgSO_4 \cdot 7H_2O$	0.7g
KCl	0.68g
NH_4Cl	0.5g
$CaCl_2 \cdot 2H_2O$	0.28g
K_2HPO_4	0.28g

Preparation of Mineral Solution 2xW: Add components to distilled/deionized water and bring volume to 1.0L. Mix thoroughly. Filter sterilize. Store at room temperature in the dark.

LIP Solution:

Composition per liter:

L-Isoleucine	10.0g
L-Leucine	5.0g
Pantothenate	0.1g

Preparation of LIP Solution: Add components to distilled/deionized water and bring volume to 1.0L. Mix thoroughly. May be stored unsterilized at -20°C.

TYC Solution:

Composition per liter:

Casamino acids	100.0g
Yeast extract	50.0g
L-Tryptophan	1.0g

Preparation of TYC Solution: Add components to distilled/deionized water and bring volume to 1.0L. Mix thoroughly. Sparge with 100% N_2. Autoclave for 15 min at 15 psi pressure–121°C.

Trace Elements Solution:

Composition per liter:

$MgSO_4 \cdot 7H_2O$	3.0g
Nitrilotriacetic acid	1.5g
NaCl	1.0g
$MnSO_4 \cdot 2H_2O$	0.5g
$CoSO_4 \cdot 7H_2O$	0.18g
$ZnSO_4 \cdot 7H_2O$	0.18g
$FeSO_4 \cdot 7H_2O$	0.1g
$CaCl_2 \cdot 2H_2O$	0.1g
$KAl(SO_4)_2 \cdot 12H_2O$	0.02g
$CuSO_4 \cdot 5H_2O$	0.01g
H_3BO_3	0.01g
$Na_2MoO_4 \cdot 2H_2O$	0.01g
$NiCl_2 \cdot 6H_2O$	0.025g
$Na_2SeO_3 \cdot 5H_2O$	0.3mg

Preparation of Trace Elements Solution: Add nitrilotriacetic acid to 500.0mL of distilled/deionized water. Adjusting pH to 6.5 with KOH. Add remaining components. Add distilled/deionized water to 1.0L.

$Na_2S \cdot 9H_2O$ Solution:

Composition per 10.0mL:

$Na_2S \cdot 9H_2O$	0.5g

Preparation of $Na_2S \cdot 9H_2O$ Solution: Add $Na_2S \cdot 9H_2O$ to distilled/deionized water and bring volume to 10.0mL. Mix thoroughly. Sparge with 100% N_2. Autoclave for 15 min at 15 psi pressure–121°C.

Preparation of Medium: Prepare and dispense medium under 80% H_2 + 20% CO_2. Add components, except TYC solution and $Na_2S \cdot 9H_2O$ solution, to distilled/deionized water and bring volume to 940.0mL. Mix thoroughly. Sparge with 80% H_2 + 20% CO_2. Autoclave for 15 min at 15 psi pressure–121°C. Aseptically and anaerobically add 50.0mL of sterile TYC solution and 10.0mL of sterile $Na_2S \cdot 9H_2O$ solution. Mix thoroughly.

Use: For the cultivation and maintenance of *Methanococcus voltae*.

Woods and Welton Agar

Composition per liter:

NaCl	23.4g
Casein hydrolysate	17.0g
Agar	15.0g
Glycerol	10.0g
Pancreatic digest of gelatin	5.0g
Glucose	5.0g
Papaic digest of soybean meal	3.0g
Beef extract	3.0g
Yeast extract	2.0g
Casamino acids, vitamin free	0.5g
Pancreatic digest of casein	0.5g
Na_2SO_3	0.1g

pH 7.6 ± 0.2 at 25°C

Preparation of Medium: Add components to distilled/deionized water and bring volume to 1.0L. Mix thoroughly. Gently heat and bring to boiling. Adjust pH to 7.6. Distribute into tubes or flasks. Autoclave for 15 min at 15 psi pressure–121°C. Pour into sterile Petri dishes or leave in tubes.

Use: For the cultivation of *Vibrio alginolyticus*.

Xanthine Agar

Composition per liter:

Solution 1	900.0mL
Solution 2	100.0mL

pH 7.0 ± 0.2 at 25°C

Solution 1:

Composition per 900.0mL:

Agar	15.0g
Pancreatic digest of gelatin	5.0g
Beef extract	3.0g

Preparation of Solution 1: Add components to distilled/deionized water and bring volume to 900.0mL. Mix thoroughly. Gently heat and bring to boiling.

Solution 2:

Composition per 100.0mL:

Xanthine	4.0g

Preparation of Solution 2: Add xanthine to distilled/deionized water and bring volume to 100.0mL. Mix thoroughly. Gently heat and bring to boiling.

Preparation of Medium: Combine solutions 1 and 2. Mix thoroughly. Distribute into tubes or flasks. Autoclave for 15 min at 15 psi pressure–121°C. Pour into sterile Petri dishes or leave in tubes.

Use: For the differentiation of aerobic *Actinomycete* species. Clearing around a colony indicates utilization of xanthine. *Streptomyces* species utilize xanthine; most *Nocardia* and *Actinomadura* species do not utilize xanthine.

Xanthobacter agilis Agar

Composition per 1100.0mL:

Solution A	1.0L
Solution B	100.0mL

Solution A:

Composition per liter:

Agar	15.0g
$NaH_2PO_4 \cdot 12H_2O$	9.0g
KH_2PO_4	1.5g
$NH_4 \cdot Cl$	1.0g
Sodium propionate or 3-hydroxybutyrate	1.0g
$MgSO_4 \cdot 7H_2O$	0.2g
Trace elements solution	1.0mL

pH 7.0 ± 0.2 at 25°C

Trace Elements Solution:

Composition per 2.0mL:

H_3BO_4	560.0μg
$ZnSO_4 \cdot 7H_2O$	350.0μg
$NiCl_2 \cdot H_2O$	160.0μg
$Na_2MoO_4 \cdot 2H_2O$	100.0μg
$CuSO_4 \cdot 5H_2O$	16.0μg
$MnCl_2 \cdot 4H_2O$	16.0μg

Preparation of Trace Elements Solution: Add components to distilled/deionized water and bring volume to 2.0mL. Mix thoroughly.

Preparation of Solution A: Add components to distilled/deionized water and bring volume to 1.0L. Mix thoroughly. Gently heat and bring to boiling. Autoclave for 15 min at 15 psi pressure–121°C. Cool to 50°–55°C.

Solution B:

Composition per 100.0mL:

Ferric ammonium citrate	50.0mg
$CaCl_2 \cdot 2H_2O$	100.0mg

Preparation of Solution B: Add components to distilled/deionized water and bring volume to 100.0mL. Mix thoroughly. Autoclave for 15 min at 15 psi pressure–121°C. Cool to 50°–55°C.

Preparation of Medium: Aseptically combine 1.0L of sterile solution A with 100.0mL of sterile solution B. Mix thoroughly. Pour into sterile Petri dishes or distribute into sterile tubes.

Use: For the cultivation of *Xanthobacter agilis,* a dinitrogen-fixing, hydrogen-oxidizing bacterium.

Xanthomonas Agar

Composition per liter:

Agar	15.0g
Pancreatic digest of gelatin	10.0g
Sucrose	10.0g
Beef extract	6.0g

pH 6.8 ± 0.2 at 25°C

Preparation of Medium: Add components to distilled/deionized water and bring volume to 1.0L. Mix thoroughly. Gently heat and bring to boiling. Distribute into tubes or flasks. Autoclave for 15 min at 15 psi pressure–121°C. Pour into sterile Petri dishes or leave in tubes.

Use: For the cultivation and maintenance of *Xanthomonas* species.

Xanthomonas albilineans Agar

Composition per liter:

Sucrose	20.0g
Agar	15.0g
Peptone	10.0g
Yeast extract	5.0g

pH 7.0 ± 0.2 at 25°C

Preparation of Medium: Add components to distilled/deionized water and bring volume to 1.0L. Mix thoroughly. Adjust pH to 7.0. Gently heat and bring to boiling. Distribute into tubes or flasks. Autoclave for 15 min at 15 psi pressure–121°C. Pour into sterile Petri dishes or leave in tubes.

Use: For the cultivation of *Xanthomonas alblineans*.

Xanthomonas maltophilia Medium

Composition per liter:

Trizma® (tris[Hydroxymethyl] aminomethane) base	6.04g
Glucose	5.0g
KCl	1.0g
NaCl	1.0g
L-Phenylalanine	0.9g
$MgSO_4$	0.2g
L-Arginine	0.1g
L-Methionine	0.1g
$(NH_4)_2SO_4$	0.1g
NH_4Cl	0.1g
Glycerol	0.68g
L-Serine	0.22g
L-Alanine	0.18g

pH 7.2 ± 0.2 at 25°C

Preparation of Medium: Add components to distilled/deionized water and bring volume to 1.0L. Mix thoroughly. Adjust pH to 7.2. Filter sterilize. Aseptically distribute into sterile tubes or flasks.

Use: For the cultivation of *Stenotrophomonas maltophilia.*

Xanthomonas Medium

Composition per liter:

Pancreatic digest of gelatin	10.0g
Sucrose	10.0g
Beef extract	6.0g

pH 6.8 ± 0.2 at 25°C

Preparation of Medium: Add components to distilled/deionized water and bring volume to 1.0L. Mix thoroughly. Gently heat with mixing. Distribute into screw-capped test tubes. Autoclave for 15 min at 15 psi pressure–121°C.

Use: For the cultivation and maintenance of *Xanthomonas* species.

Xanthomonas TYG Agar (*Xanthomonas* Tryptone Yeast Extract Glucose Agar)

Composition per liter:

Agar	20.0g
Pancreatic digest of casein	5.0g
Glucose	5.0g
Yeast extract	3.0g
K_2HPO_4	0.7g
$MgSO_4 \cdot 7H_2O$	0.25g

Preparation of Medium: Add components to distilled/deionized water and bring volume to 1.0L. Mix thoroughly. Gently heat and bring to boiling. Distribute into tubes or flasks. Autoclave for 15 min at 15 psi pressure–121°C. Pour into sterile Petri dishes or leave in tubes.

Use: For the cultivation and maintenance of *Xanthomonas* species.

Xenorhabdus Agar

Composition per liter:

Agar	15.0g
Peptone	10.0g
NaCl	5.0g
Yeast extract	5.0g

pH 7.2 ± 0.2 at 25°C

Preparation of Medium: Add components to distilled/deionized water and bring volume to 1.0L. Mix thoroughly. Gently heat and bring to boiling. Adjust pH to 7.2. Distribute into tubes or flasks. Autoclave for 20 min at 15 psi pressure–121°C. Pour into sterile Petri dishes or leave in tubes.

Use: For the cultivation and maintenance of *Bacteroides galacturonicus* and *Xenorhabdus nematophilus*.

XLD Agar (Xylose Lysine Deoxycholate Agar)

Composition per liter:

Agar	13.5g
Lactose	7.5g
Sucrose	7.5g
$Na_2S_2O_3$	6.8g
L-Lysine	5.0g
NaCl	5.0g
Xylose	3.5g
Yeast extract	3.0g
Sodium deoxycholate	2.5g
Ferric ammonium citrate	0.8g
Phenol Red	0.08g

pH 7.5 ± 0.2 at 25°C

Source: This medium is available as a premixed powder from BD Diagnostic Systems and Oxoid Unipath.

Preparation of Medium: Add components to distilled/deionized water and bring volume to 1.0L. Mix thoroughly. Gently heat and bring to boiling. Do not overheat. Distribute into tubes or flasks. Autoclave for 15 min at 15 psi pressure–121°C. Pour into sterile Petri dishes or leave in tubes.

Plates should be poured as soon as possible to avoid precipitation.

Use: For the isolation and differentiation of enteric pathogens, especially *Shigella* and *Providencia* species. Nonfermenting xylose/lactose/sucrose bacteria appear as red colonies. Xylose-fermenting, lysine-decarboxylating bacteria appear as red colonies. Xylose-fermenting, lysine-nondecarboxylating bacteria appear as opaque yellow colonies. Lactose or sucrose-fermenting bacteria appear as yellow colonies.

XPS Agar

Composition per liter:

Solution A 500.0mL
Solution B 500.0mL

pH5.1 ± 0.2 at 25°C

Solution A:
Composition per 500mL:

Potatoes, infusion from 40.0g
Sucrose 15.0g
Peptone 5.0g
Glucose 4.0g
Casamino acids 1.0g
Na_2HPO_4 .... 0.79g
$Ca(NO_3)_2 \cdot 4H_2O$ solution 10.0mL

Potatoes, Infusion From:
Composition per 500.0mL:

Potatoes 4.0g

Preparation of Potatoes, Infusion From: Peel and dice potatoes. Add 400.0mL of distilled/deionized water. Gently heat and bring to boiling. Continue boiling for 30 min. Filter through cheesecloth.

Ca(NO3)$_2$·4H$_2$O Solution:
Composition per 10.0mL:

$Ca(NO_3)_2 \cdot 4H_2O$ 0.5g

Preparation of $Ca(NO_3)_2 \cdot 4H_2O$ Solution: Add $Ca(NO_3)_2 \cdot 4H_2O$ to distilled/deionized water and bring volume to 10.0mL. Mix thoroughly. Filter sterilize.

Preparation of Solution A: Add components, except $Ca(NO_3)_2 \cdot 4H_2O$ solution, to distilled/deionized water and bring volume to 490.0mL. Mix thoroughly. Autoclave for 15 min at 15 psi pressure–121°C. Aseptically add 10.0mL of sterile $Ca(NO_3)_2 \cdot 4H_2O$ solution. Mix thoroughly. Cool to 50°–55°C.

Solution B:
Composition per 500mL:

Agar 20.0g

Preparation of Solution A: Add agar to distilled/deionized water and bring volume to 500.0mL. Mix thoroughly. Gently heat and bring to boiling. Autoclave for 15 min at 15 psi pressure–121°C. Cool to 50°–55°C.

Preparation of Medium: Aseptically combine 500.0mL of solution A with 500.0mL of solution B. Mix thoroughly. Aseptically pour into sterile Petri dishes or distribute into sterile tubes.

Use: For the cultivation of *Xanthomonas campestris*.

XPS Broth with Thymidine (Thymidine Auxotroph XPS Medium)

Composition per liter:

Potatoes, infusion from 40.0g
Sucrose 15.0g
Peptone 5.0g
Glucose 4.0g
Casamino acids 1.0g
Na_2HPO_4 .... 0.79g
Thymidine 10.0mg
$Ca(NO_3)_2 \cdot 4H_2O$ solution 10.0mL

pH 5.1 ± 0.2 at 25°C

Potatoes, Infusion From:
Composition per 500.0mL:

Potatoes 4.0g

Preparation of Potatoes, Infusion From: Peel and dice potatoes. Add 500.0mL of distilled/deionized water. Gently heat and bring to boiling. Continue boiling for 30 min. Filter through cheesecloth.

$Ca(NO_3)_2 \cdot 4H_2O$ Solution:
Composition per 10.0mL:

$Ca(NO_3)_2 \cdot 4H_2O$ 0.5g

Preparation of $Ca(NO_3)_2 \cdot 4H_2O$ Solution: Add $Ca(NO_3)_2 \cdot 4H_2O$ to distilled/deionized water and bring volume to 10.0mL. Mix thoroughly. Filter sterilize.

Preparation of Medium: Add components, except $Ca(NO_3)_2 \cdot 4H_2O$ solution, to distilled/deionized water and bring volume to 990.0mL. Mix thoroughly. Autoclave for 15 min at 15 psi pressure–121°C. Aseptically add 10.0mL of sterile $Ca(NO_3)_2 \cdot 4H_2O$ solution. Mix thoroughly. Aseptically distribute into sterile tubes or flasks.

Use: For the cultivation of *Xanthomonas oryzae*.

Xylella Agar (LMG Medium 115)

Composition per liter:

Agar 17.0g
Yeast extract 10.0g
ACES 10.0g
Activated charcoal 2.0g
α-ketoglutarate 1.0g
KOH, 1*N* 40mL
Cysteine iron solution 20.0mL

pH 6.9 ± 0.2 at 25°C

Cysteine Iron Solution:
Composition per 20.0.0mL:

L-Cysteine·HCl 0.4g
$Fe_4(P_2O_7)_3$ 0.25g

Preparation of Cysteine Iron Solution: Add L-cysteine·HCl and $Fe_4(P_2O_7)_3$ to 20.0mL distilled/deionized water. Mix thoroughly. Filter sterilize.

Preparation of Medium: Add ACES to 500.0mL distilled water at 50° C. Combine with a solution containing 40.0mL of 1 *N* KOH in 440.0mL distilled water. Add other components except cysteine iron solution. Mix thoroughly. Gently heat and bring to boiling. Autoclave for 15 min at 15 psi pressure–121°C. Cool to 50°C. Aseptically add 20.0mL sterile cysteine iron solution. Pour into sterile Petri dishes or distribute into sterile tubes.

Use: For the cultivation of *Xylella* spp.

Xylella fastidiosa Medium (LMG 115)

Composition per liter:

Agar 17.0g
Yeast extract 10.0g
ACES buffer 10.0g
Activated charcoal 2.0g
L-Cysteine-iron solution 20.0mL

pH 6.9 ± 0.2 at 25°C

L-Cysteine-Iron Solution:
Composition per 20.0mL:

L-Cysteine·HCl	0.4g
$Fe_4(P_2O_7)_3$	0.25g

Preparation of L-Cysteine-Iron Solution: Add components to distilled/deionized water and bring volume to 20.0mL. Mix thoroughly. Filter sterilize.

Preparation of Medium: Add ACES to 500.0mL of distilled/deionized water at 50°C. Add a solution containing 40.0mL of 1*N* KOH in 440.0mL of distilled water. Mix thoroughly. Add the remaining components, except L-cysteine-iron solution. Mix thoroughly. Gently heat and bring to boiling. Adjust pH to 6.9 with KOH. Autoclave for 15 min at 15 psi pressure–121°C. Cool to 50°C. Aseptically add 20.0mL of sterile L-cysteine-iron solution. Mix thoroughly. Pour into sterile Petri dishes or distribute into sterile tubes.

Use: For the cultivation of *Xylella fastidiosa.*

Xylophilus Medium

Composition per liter:

$CaCO_3$	20.0g
Agar	15.0g
D-Galactose	10.0g
Yeast extract	10.0g
Ferric ammonium citrate solution	10.0mL

Ferric Ammonium Citrate Solution:
Composition per 10.0mL:

Ferric ammonium citrate	0.25g

Preparation of Ferric Ammonium Citrate Solution: Add ferric ammonium citrate to distilled/deionized water and bring volume to 10.0mL. Mix thoroughly. Autoclave for 15 min at 15 psi pressure–121°C.

Preparation of Medium: Add components, except ferric ammonium citrate solution, to distilled/deionized water and bring volume to 990.0mL. Mix thoroughly. Gently heat and bring to boiling. Autoclave for 15 min at 15 psi pressure–121°C. Cool to 50°–55°C. Aseptically add 10.0mL of sterile ferric ammonium citrate solution. Mix thoroughly. Pour into sterile Petri dishes or distribute into sterile tubes.

Use: For the cultivation and maintenance of *Xylophilus ampelina.*

Xylose Sodium Deoxycholate Citrate Agar

Composition per liter:

Agar	12.0g
Xylose	10.0g
Sodium citrate	5.0g
$Na_2S_2O_3·5H_2O$	5.0g
Beef extract	5.0g
Peptone	5.0g
NaCl	2.5g
Sodium deoxycholate	2.5g
Ferric ammonium citrate	1.0g
Neutral Red (1% solution)	2.5mL

pH 7.5 ± 0.2 at 25°C

Preparation of Medium: Add components to distilled/deionized water and bring volume to 1.0L. Mix thoroughly. Gently heat and bring to boiling for 20 sec. Do not autoclave. Cool to 45°–50°C. Pour into sterile Petri dishes.

Use: For the cultivation of *Salmonella* species and some *Shigella* species.

YA Halophile Medium

Composition per liter:

NaCl	100.0g
Agar	15.0g
Sodium acetate·$3H_2O$	10.0g
Na_2HPO_4	3.8g
KH_2PO_4	1.3g
$Mg(NO_3)_2·6H_2O$	1.0g
$(NH_4)_2SO_4$	1.0g
Yeast extract	1.0g

pH 7.2 ± 0.2 at 25°C

Preparation of Medium: Add components, except magnesium nitrate, to tap water and bring volume to 1.0L. Mix thoroughly. Distribute into tubes or flasks. Autoclave for 15 min at 15 psi pressure–121°C. Aseptically add magnesium nitrate. Adjust pH 7.2 with sterile KOH. Pour into sterile Petri dishes or leave in tubes.

Use: For the cultivation and maintenance of halophilic microorganisms, including *Bacillus halodenitrificans.*

YB Medium (Yeast Extract Beef Extract Medium)

Composition per liter:

Agar	20.0g
Peptone	10.0g
Beef extract	7.0g
Yeast extract	5.0g
NaCl	3.0g
Thiourea	0.1g
Methanol	20.0mL

pH 7.2 ± 0.2 at 25°C

Preparation of Medium: Add components, except methanol, to distilled/deionized water and bring volume to 1.0L. Mix thoroughly. Distribute into tubes or flasks. Autoclave for 15 min at 15 psi pressure–121°C. Aseptically add filter-sterilized methanol. Pour into sterile Petri dishes or leave in tubes.

Use: For the cultivation and maintenance of bacteria that can utilize methanol as a carbon source, including *Achromobacter methanolophila, Methanomonas methylovora, Methylobacterium* species, and *Pseudomonas methanolica.*

YDC Medium

Composition per liter:

$CaCO_3$	20.0g
Glucose	20.0g
Agar	15.0g
Yeast extract	10.0g

pH 7.2 ± 0.2 at 25°C

Preparation of Medium: Add components to distilled/deionized water and bring volume to 1.0L. Mix thoroughly. Gently heat and bring to boiling. Distribute into tubes or flasks. Autoclave for 15 min at 15 psi pressure–121°C. Pour into sterile Petri dishes or leave in tubes.

Use: For the cultivation of *Bdellovibrio* species.

Yeast Agar, Van Niel's

Composition per liter:

Agar	20.0g
Yeast extract	10.0g
K_2HPO_4	1.0g
$MgSO_4·7H_2O$	0.5g

pH 7.0–7.2 at 25°C

Preparation of Medium: Add components to tap water and bring volume to 1.0L. Mix thoroughly. Gently heat and bring to boiling. Distribute into tubes or flasks. Autoclave for 15 min at 15 psi pressure–121°C. Pour into sterile Petri dishes or leave in tubes.

Use: For the cultivation and maintenance of a variety of microorganisms, including *Cytophaga* species, *Heliobacterium chlorum, Rhodomicrobium vannielii, Lysobacter enzymogenes, Rhodobacter* species, *Rhodocyclus gelatinosus, Rhodopseudomonas palustris,* and *Rhodospirillum rubrum.*

Yeast Agar, Van Niel's with Glutamate

Composition per liter:

Agar	20.0g
Yeast extract	10.0g
K_2HPO_4	1.0g
$MgSO_4$	0.5g
Monosodium glutamate	0.85g

pH 7.0–7.2 at 25°C

Preparation of Medium: Add components to tap water and bring volume to 1.0L. Mix thoroughly. Gently heat and bring to boiling. Distribute into tubes or flasks. Autoclave for 15 min at 15 psi pressure–121°C. Pour into sterile Petri dishes or leave in tubes.

Use: For the cultivation and maintenance of a variety of bacteria, including *Bacillus firmus, Cytophaga johnsonae, Heliobacterium chlorum, Lysobacter enzymogenes, Rhodobacter capsulatus, Rhodomicrobium vannielii, Rhodobacter sphaeroides, Rhodocyclus gelatinosus, Rhodocyclus gelatinosus, Rhodopseudomonas palustris,* and *Rhodospirillum rubrum.*

Yeast Agar, Van Niel's with 2.5% NaCl (ATCC Medium 1370)

Composition per liter:

NaCl	25.0g
Agar	20.0g
Yeast extract	10.0g
K_2HPO_4	1.0g
$MgSO_4$	0.5g

pH 7.0–7.2 at 25°C

Preparation of Medium: Add components to tap water and bring volume to 1.0L. Mix thoroughly. Gently heat and bring to boiling. Distribute into tubes or flasks. Autoclave for 15 min at 15 psi pressure–121°C. Pour into sterile Petri dishes or leave in tubes.

Use: For the cultivation and maintenance of *Chromatium vinosum* and *Rhodopseudomonas* species.

Yeast Agar, Van Niel's with 25% NaCl

Composition per liter:

NaCl	250.0g
Agar	20.0g
Yeast extract	10.0g
K_2HPO_4	1.0g
$MgSO_4$	0.5g

pH 7.0–7.2 at 25°C

Preparation of Medium: Add components to tap water and bring volume to 1.0L. Mix thoroughly. Gently heat and bring to boiling. Distribute into tubes or flasks. Autoclave for 15 min at 15 psi pressure–121°C. Pour into sterile Petri dishes or leave in tubes.

Use: For the cultivation and maintenance of halophilic bacteria, including *Haloarcula vallismortis, Halococcus morrhuae,* and *Halobacterium* species.

Yeast Agar, Van Niel's with Succinate

Composition per liter:

Agar	20.0g
Yeast extract	10.0g
Sodium succinate	1.35g
K_2HPO_4	1.0g
$MgSO_4 \cdot 7H_2O$	0.5g

pH 7.0–7.2 at 25°C

Preparation of Medium: Add components to tap water and bring volume to 1.0L. Mix thoroughly. Gently heat and bring to boiling. Distribute into tubes or flasks. Autoclave for 15 min at 15 psi pressure–121°C. Pour into sterile Petri dishes or leave in tubes.

Use: For the cultivation and maintenance of *Rhodobacter capsulatus.*

Yeast Extract Agar

Composition per liter:

Agar	15.0g
Peptone	5.0g
Yeast extract	3.0g

pH 7.2 ± 0.2 at 25°C

Source: This medium is available as a premixed powder from Oxoid Unipath.

Preparation of Medium: Add components to distilled/deionized water and bring volume to 1.0L. Mix thoroughly. Gently heat and bring to boiling. Distribute into tubes or flasks. Autoclave for 15 min at 15 psi pressure–121°C. Pour into sterile Petri dishes or leave in tubes.

Use: For the enumeration of microorganisms in potable and freshwater samples.

Yeast Extract Agar

Composition per liter:

Agar	20.0g
Yeast extract	1.0g
Buffer solution	2.0mL

pH 6.0 ± 0.2 at 25°C

Buffer Solution:

Composition per 400.0mL:

KH_2PO_4	60.0g
Na_2HPO_4	40.0g

Preparation of Buffer Solution: Add 40.0g of Na_2HPO_4 to 300.0mL of distilled/deionized water. Mix thoroughly. Add 60.0g of KH_2PO_4. Mix thoroughly. Adjust pH to 6.0.

Preparation of Medium: Add components to distilled/deionized water and bring volume to 1.0L. Mix thoroughly. Autoclave for 15 min at 15 psi pressure–121°C. Pour into sterile Petri dishes.

Use: For the identification of *Histoplasma capsulatum, Blastomyces dermatitidis,* and *Coccidioides immitis.*

Yeast Extract Glucose Calcium Carbonate Agar

Composition per liter:

$CaCO_3$	20.0g
Glucose	20.0g
Agar	15.0g
Yeast extract	10.0g

Preparation of Medium: Add components to distilled/deionized water and bring volume to 1.0L. Mix thoroughly. Gently heat and bring to boiling. Distribute into tubes or

flasks. Autoclave for 15 min at 15 psi pressure–121°C. Pour into sterile Petri dishes or leave in tubes.

Use: For the isolation and cultivation of *Erwinia* species.

Yeast Extract Glucose Medium

Composition per liter:

Agar	15.0g
Yeast extract	10.0g
Glucose	10.0g

Preparation of Medium: Add components to tap water and bring volume to 1.0L. Mix thoroughly. Gently heat and bring to boiling. Distribute into tubes or flasks. Autoclave for 15 min at 15 psi pressure–121°C. Pour into sterile Petri dishes or leave in tubes.

Use: For the cultivation of a variety of bacteria, including *Streptomyces* species, *Rhodococcus* species, and others.

Yeast Extract Mannitol Agar

Composition per liter:

Agar	15.0g
Mannitol	10.0g
$CaCO_3$	4.0g
K_2HPO_4	0.5g
Yeast extract	0.4g
$MgSO_4 \cdot 7H_2O$	0.2g
NaCl	0.1g

pH 6.8–7.0 ± 0.2 at 25°C

Preparation of Medium: Add components to distilled/deionized water and bring volume to 1.0L. Omit $CaCO_3$ if a clear solution is needed. Mix thoroughly. Gently heat and bring to boiling. Distribute into tubes or flasks. Autoclave for 15 min at 15 psi pressure–121°C. Pour into sterile Petri dishes or leave in tubes.

Use: For the cultivation of members of the Rhizobiaceae.

Yeast Extract Medium

Composition per liter:

Yeast extract	10.0g

Preparation of Medium: Add yeast extract to distilled/deionized water and bring volume to 1.0L. Mix thoroughly. Distribute into tubes or flasks. Autoclave for 15 min at 15 psi pressure–121°C.

Use: For the cultivation of *Pseudomonas cepacia.*

Yeast Extract Medium with Sodium Sulfide

Composition per liter:

Yeast extract	10.0g
Na_2S	0.15g

Preparation of Medium: Add yeast extract to distilled/deionized water and bring volume to 1.0L. Mix thoroughly. Autoclave for 15 min at 15 psi pressure–121°C. Immediately prior to inoculation, add 0.15g of Na_2S. Mix thoroughly. Aseptically distribute into sterile tubes or flasks.

Use: For the cultivation of *Rhodospirillum molischianum.*

Yeast Extract Mineral Agar

Composition per liter:

Agar	15.0g
Yeast extract	4.0g
$NaHPO_4 \cdot 12H_2O$	3.5g
K_2HPO_4	1.0g
NH_4Cl	0.5g
$MgSO_4 \cdot 7H_2O$	0.03g

pH 7.0 ± 0.2 at 25°C

Preparation of Medium: Add components to distilled/deionized water and bring volume to 1.0L. Mix thoroughly. Gently heat and bring to boiling. Distribute into tubes or flasks. Autoclave for 15 min at 15 psi pressure–121°C. Pour into sterile Petri dishes or leave in tubes.

Use: For the cultivation of *Bacillus azotoformans.*

Yeast Extract Mineral Medium (DSMZ Medium 259)

Composition per liter:

Agar	15.0g
Yeast extract	4.0g
$Na_2HPO_4 \cdot 12H_2O$	3.5g
K_2HPO_4	1.0g
NH_4Cl	0.5g
$MgSO_4 \cdot 7H_2O$	0.03g

pH 7.1 ± 0.2 at 25°C

Preparation of Medium: Add components to distilled/deionized water and bring volume to 1.0L. Mix thoroughly. Gently heat and bring to boiling. Distribute into tubes or flasks. Autoclave for 15 min at 15 psi pressure–121°C. Pour into sterile Petri dishes or leave in tubes.

Use: For the cultivation and maintenance of *Bacillus azotoformans.*

Yeast Glucose Broth (YGB)

Composition per liter:

Beef extract	10.0g
Peptone	10.0g
NaCl	5.0g
Glucose	5.0g
Yeast extract	3.0g

pH 6.8 ± 0.2 at 25°C

Preparation of Medium: Add components to distilled/deionized water and bring volume to 1.0L. Mix thoroughly. Adjust ph to 6.8. Distribute into tubes or flasks. Autoclave for 15 min at 15 psi pressure–121°C.

Use: For the cultivation of *Enterococcus faecalis, Streptococcus anginosus,* and *Rhodobacter sphaeroides.*

Yeast Malate Medium

Composition per liter:

Yeast extract	5.0g
Sodium malate	1.0g

pH 7.0 ± 0.2 at 25°C

Preparation of Medium: Add components to distilled/deionized water and bring volume to 1.0L. Mix thoroughly. Distribute into tubes or flasks. Autoclave for 15 min at 15 psi pressure–121°C.

Use: For the cultivation of *Rhodopseudomonas viridis.*

Yeast Malate Medium

Composition per liter:

$KH_2PO_4 \cdot 12H_2O$	1.0g
$NaHCO_3$	1.0g
Sodium malate	1.0g
$(NH_4)_2SO_4$	0.5g
Trace elements solution	1.0mL

pH 6.8 ± 0.2 at 25°C

Trace Elements Solution:
Composition per liter:
H_3BO_3 .... 0.3g
$CoCl_2 \cdot 6H_2O$ 0.2g
$ZnSO_4 \cdot 7H_2O$ 0.1g
$MnCl_2 \cdot 4H_2O$ 0.03g
$Na_2MoO_4 \cdot 2H_2O$ 0.03g
$NiCl_2 \cdot 6H_2O$ 0.02g
$CuCl_2 \cdot 2H_2O$ 0.01g

Preparation of Trace Elements Solution: Add components to distilled/deionized water and bring volume to 1.0L. Mix thoroughly. Filter sterilize.

Preparation of Medium: Add components, except trace elements solution, to distilled/deionized water and bring volume to 999.0mL. Mix thoroughly. Autoclave for 15 min at 15 psi pressure–121°C. Aseptically add 1.0mL of sterile trace elements solution. Mix thoroughly. Aseptically distribute into sterile tubes or flasks.

Use: For the cultivation of *Rhodopseudomonas viridis.*

Yeast Malate Medium (LMG 176)

Composition per 1001.0mL:
$KH_2PO_4 \cdot 12H_2O$ 1.0g
$NaHCO_3$ 1.0g
Sodium malate 1.0g
Yeast extract 1.0g
$(NH_4)_2SO_4$ 0.5g
Trace elements solution 1.0mL
pH 6.8–7.0 at 25°C

Trace Elements Solution:
Composition per liter:
H_3BO_3 .... 0.3g
$CoCl_2 \cdot 6H_2O$ 0.2g
$ZnSO_4 \cdot 7H_2O$ 0.1g
$MnCl_2 \cdot 4H_2O$ 0.03g
$Na_2MoO_4 \cdot 2H_2O$ 0.03g
$NiCl_2 \cdot 6H_2O$ 0.02g
$CuCl_2 \cdot 2H_2O$ 0.01g

Preparation of Trace Elements Solution: Add components to distilled/deionized water and bring volume to 1.0L. Mix thoroughly.

Preparation of Medium: Add components to distilled/deionized water and bring volume to 1.0L. Mix thoroughly. Adjust pH to 6.8–7.0. Distribute into tubes or flasks. Autoclave for 15 min at 15 psi pressure–121°C.

Use: For the cultivation of *Rhodopseudomonas viridis.*

Yeast Mannitol Agar

Composition per liter:
Agar 15.0g
Mannitol 10.0g
K_2HPO_4 .... 0.5g
Yeast extract 0.4g
$MgSO_4 \cdot 7H_2O$ 0.2g
NaCl 0.1g
pH 6.8 ± 0.2 at 25°C

Preparation of Medium: Add components to distilled/deionized water and bring volume to 1.0L. Mix thoroughly. Gently heat and bring to boiling. Distribute into tubes or flasks. Autoclave for 15 min at 15 psi pressure–121°C. Pour into sterile Petri dishes or leave in tubes.

Use: For the cultivation of *Rhizobium* and *Azorhizobium* species.

Yeast Mannitol Agar

Composition per liter:
Agar 20.0g
Mannitol 10.0g
$Na_2HPO_4 \cdot 12H_2O$ 1.2g
KH_2PO_4 .... 0.55g
$MgSO_4 \cdot 7H_2O$ 0.25g
NaCl 0.25g
Yeast extract 0.25g
$CaSO_4 \cdot 2H_2O$ 30.0mg
$FeSO_4 \cdot 7H_2O$ 3.5mg
H_3BO_3 .... 500.0μg
$MnSO_4 \cdot 4H_2O$ 400.0μg
$ZnSO_4 \cdot 7H_2O$ 160.0μg
$CuSO_4 \cdot 5H_2O$ 80.0μg
pH 7.0 ± 0.2 at 25°C

Preparation of Medium: Add components to distilled/deionized water and bring volume to 1.0L. Mix thoroughly. Gently heat and bring to boiling. Distribute into tubes or flasks. Autoclave for 15 min at 15 psi pressure–121°C. Pour into sterile Petri dishes or leave in tubes.

Use: For the cultivation of *Bradyrhizobium japonicum* and *Rhizobium leguminosarum.*

Yeast Mannitol Agar, Modified

Composition per liter:
Agar 15.0g
Mannitol 10.0g
$CaCO_3$ 4.0g
K_2HPO_4 .... 0.5g
Yeast extract 0.4g
$MgSO_4 \cdot 7H_2O$ 0.2g
NaCl 0.1g
pH 6.8 ± 0.2 at 25°C

Preparation of Medium: Add components to distilled/deionized water and bring volume to 1.0L. Mix thoroughly. Gently heat and bring to boiling. Distribute into tubes or flasks. Autoclave for 15 min at 15 psi pressure–121°C. Pour into sterile Petri dishes or leave in tubes.

Use: For the cultivation of *Amycolata autotrophica, Amycolatopsis orientalis, Nocardioides albus, Promicromonospora citrea, Rhizobium fredii, Rhizobium galegae, Rhizobium leguminosarum, Rhizobium loti, Rhizobium meliloti,* and *Rhizobium tropici.*

Yeast Nitrogen Base Glucose Broth

Composition per liter:
Yeast nitrogen base 25.0mL
Glucose solution 25.0mL
pH 5.6 ± 0.2 at 25°C

Yeast Nitrogen Base:
Composition per 500.0mL:
$(NH_4)_2SO_4$ 5.0g
KH_2PO_4 .... 1.0g
$MgSO_4 \cdot 7H_2O$ 0.5g
NaCl 0.1g
$CaCl_2 \cdot 2H_2O$ 0.1g
DL-Methionine 0.02g
DL-Tryptophan 0.02g
L-Histidine·HCl 0.01g
Inositol 2.0mg
H_3BO_3 .... 0.5mg

$ZnSO_4 \cdot 7H_2O$	0.4mg
$MnSO_4 \cdot 4H_2O$	0.4mg
Thiamine·HCl	0.4mg
Pyroxidine·HCl	0.4mg
Niacin	0.4mg
Calcium pantothenate	0.4mg
p-Aminobenzoic acid	0.2mg
Riboflavin	0.2mg
$FeCl_3$	0.2mg
$Na_2MoO_4 \cdot 4H_2O$	0.2mg
KI	0.1mg
$CuSO_4 \cdot 5H_2O$	0.04mg
Folic acid	2.0μg
Biotin	2.0μg

Source: Yeast nitrogen base is available as a premixed powder from BD Diagnostic Systems.

Preparation of Yeast Nitrogen Base: Add components to distilled/deionized water and bring volume to 500.0mL. Mix thoroughly. Filter sterilize.

Glucose Solution:
Composition per 500.0mL:

Glucose	10.0g

Preparation of Glucose Solution: Add glucose to distilled/deionized water and bring volume to 500.0mL. Mix thoroughly. Filter sterilize.

Preparation of Medium: Aseptically combine 25.0mL of sterile yeast nitrogen base and 25.0mL of sterile glucose solution. Mix thoroughly.

Use: For the cultivation and enrichment of yeast from sewage and polluted waters.

Yeast Peptone Agar (ATCC 1858)

Composition per liter:

Agar	15.0g
Yeast extract	2.5g
Peptone	2.5g

pH 7.0 ± 0.4 at 25°C

Preparation of Medium: Add components to distilled/deionized water and bring volume to 1.0L. Mix thoroughly. Gently heat and bring to boiling. Distribute into tubes or flasks. Autoclave for 15 min at 15 psi pressure–121°C. Pour into sterile Petri dishes or leave in tubes.

Use: For the cultivation of *Cytophaga lytica, Pseudomonas lanceolata, Rhodobacter capsulatus, Rhodobacter sphaeroides, Rhodopseudomonas blastica, Rhodopseudomonas palustris, Rhodospirillum rubrum, Rubrivivax gelatinosus, Sphaerotilus natans,* and *Zoogloea ramigera.*

Yeast Peptone Broth

Composition per liter:

Yeast extract	2.5g
Peptone	2.5g

pH 7.0 ± 0.2 at 25°C

Preparation of Medium: Add components to distilled/deionized water and bring volume to 1.0L. Mix thoroughly. Distribute into tubes or flasks. Autoclave for 15 min at 15 psi pressure–121°C.

Use: For the cultivation of *Rhodopseudomonas* species.

Yeast Peptone Salt Medium

Composition per liter:

Agar	15.0g
Peptone	2.5g
Yeast extract	2.5g
NaCl	1.25g

pH 7.0 ± 0.4 at 25°C

Preparation of Medium: Add components to distilled/deionized water and bring volume to 1.0L. Mix thoroughly. Gently heat and bring to boiling. Distribute into tubes or flasks. Autoclave for 15 min at 15 psi pressure–121°C. Pour into sterile Petri dishes or leave in tubes.

Use: For the cultivation of *Deinobacter grandis.*

Yeast Water Agar

Composition per liter:

Glucose	20.0g
Agar	15.0g
Casein hydrolysate	5.0g
Yeast extract	4.0g
KH_2PO_4	0.55g
KCl	0.4g
$CaCl_2$	0.13g
$MgCl_2 \cdot 7H_2O$	0.13g
$FeCl_3 \cdot 6H_2O$	2.5mg
$MnSO_4 \cdot 4H_2O$	2.5mg
Bromcresol Green solution	1.0mL

Bromcresol Green Solution:
Composition per 10.0mL:

Bromcresol Green	0.22g
Ethanol	10.0mL

Preparation of Bromcresol Green Solution: Add Bromcresol Green to 10.0mL of ethanol. Mix thoroughly. Filter sterilize.

Preparation of Medium: Add components to distilled/deionized water and bring volume to 1.0L. Mix thoroughly. Gently heat and bring to boiling. Distribute into tubes or flasks. Autoclave for 15 min at 15 psi pressure–121°C. Pour into sterile Petri dishes or leave in tubes.

Use: For the cultivation of *Zymomonas* species.

Yeastrel Agar

Composition per liter:

Agar	15.0g
Peptone	9.5g
Yeastrel	7.0g
Lab-Lemco (meat extract)	5.0g
NaCl	5.0g

pH 7.0 ± 0.2 at 25°C

Source: Lab-lemco is available from Oxoid Unipath. Yeastrel is produced by Mapleton's Foods Ltd., Moss Street, Liverpool and is available from health food shops.

Preparation of Medium: Add components to distilled/deionized water and bring volume to 1.0L. Mix thoroughly. Gently heat and bring to boiling. Distribute into tubes or flasks. Autoclave for 15 min at 15 psi pressure–121°C. Pour into sterile Petri dishes or leave in tubes.

Use: For the cultivation of *Aeromonas salmonicida.*

YEPG Medium

Composition per liter:

Glucose	1.0g
Polypeptone	2.0g

NH_4NO_3 .... 0.2g
Yeast extract 0.2g
pH 7.0 ± 0.2 at 25°C

Preparation of Medium: Add components to distilled/deionized water and bring volume to 1.0L. Mix thoroughly. Adjust pH to 7.0. Distribute into tubes or flasks. Autoclave for 15 min at 15 psi pressure–121°C.

Use: For the cultivation of *Pseudomonas putida* and *Pseudomonas fluorescens.*

YEPP Medium (Yeast Extract Proteose Peptone Medium)

Composition per liter:
Agar 15.0g
Proteose peptone 10.0g
NaCl 5.0g
Yeast extract 3.0g
pH 7.2-7.4 ± 0.2 at 25°C

Preparation of Medium: Add components to distilled/deionized water and bring volume to 1.0L. Mix thoroughly. Gently heat and bring to boiling. Distribute into tubes or flasks. Autoclave for 15 min at 15 psi pressure–121°C. Pour into sterile Petri dishes or leave in tubes.

Use: For the cultivation and maintenance of *Pseudomonas* species.

YGC Medium (Yeast Extract Glucose Carbonate Medium) (ATCC Medium 73)

Composition per liter:
Agar 20.0g
Glucose 20.0g
$CaCO_3$ 20.0g
Yeast extract 10.0g

Preparation of Medium: Add components to distilled/deionized water and bring volume to 1.0L. Mix thoroughly. Gently heat and bring to boiling. Distribute into tubes or flasks. Autoclave for 30 min at 10 psi pressure–115°C. Cool to 48°C. Mix thoroughly. Pour into sterile Petri dishes or leave in tubes.

Use: For the cultivation of *Xanthomonas* species, *Erwinia* species, *Kluyvera* species, *Rhodococcus* species, *Streptomyces* species, *Pseudomonas psueudoalcaligenes*, and *Xylophilus ampelinus*.

YGC Medium with Glutamic Acid (Yeast Extract Glucose Carbonate Medium with Glutamic Acid)

Composition per liter:
Agar 20.0g
Glucose 20.0g
$CaCO_3$ 20.0g
Agar 20.0g
Yeast extract 10.0g
Glutamic acid 0.1g

Preparation of Medium: Add components to distilled/deionized water and bring volume to 1.0L. Mix thoroughly. Gently heat and bring to boiling. Distribute into tubes or flasks. Autoclave for 30 min at 10 psi pressure–115°C. Cool to 48°C. Mix thoroughly. Pour into sterile Petri dishes or leave in tubes.

Use: For the cultivation and maintenance of *Xanthomonas campestris.*

YGCB Salt Medium

Composition per liter:
NaCl 50.0g
Beef broth 10.0g
Glucose 10.0g
Peptone 10,0g
Triammonium citrate 5.0g
Yeast extract 5.0g
Sodium acetate 2.0g
$MgSO_4 \cdot 7H_2O$ 0.2g
$MnSO_4 \cdot 4H_2O$ 50.0mg
Tween 80 .... 1.0mL
pH 6.7 ± 0.2 at 25°C

Preparation of Medium: Add components to distilled/deionized water and bring volume to 1.0L. Mix thoroughly. Distribute into tubes or flasks. Autoclave for 15 min at 15 psi pressure–121°C.

Use: For the cultivation of *Tetragenococcus halophila.*

YGCP Medium (Yeast Extract Glucose Carbonate Peptone Medium)

Composition per liter:
Glucose 20.0g
Agar 17.5g
$CaCO_3$ 10.0g
Yeast extract 2.5g
Peptone 2.5g
NaCl 1.0g
K_2HPO_4 .... 1.0g
$MgSO_4$ 0.5g

Preparation of Medium: Add components, except calcium carbonate, to distilled/deionized water and bring volume to 1.0L. Mix thoroughly. Gently heat and bring to boiling. Adjust pH to 7.2. Add calcium carbonate. Mix thoroughly. Distribute into tubes or flasks. Autoclave for 15 min at 15 psi pressure–121°C. Pour into sterile Petri dishes or leave in tubes.

Use: For the cultivation and maintenance of *Xanthomonas campestris* and *Xanthomonas oryzae.*

YI-S Medium

Composition per liter:
YI broth 880.0mL
Bovine serum, heat inactivated 100.0mL
Vitamin mixture 18 .... 20.0mL

Source: Vitamin mixture 18 is available from Bio-fluids, Inc., Rockville, MD.

YI Broth:
Composition per liter:
YI base stock 780.0mL
10× Glucose buffer stock 100.0mL

YI Base Stock:
Composition per 780.0mL:
Yeast extract 30.0g
L-Cysteine·HCl 1.0g
NaCl 1.0g
Ascorbic acid 0.2g
Ferric ammonium citrate 228.0mg

10× Glucose Buffer Stock:
Composition per 100.0.0mL:
Glucose 10.0g

K_2HPO_4	1.0g
KH_2PO_4	0.6g

Preparation of 10× Glucose Buffer Stock: Add components to distilled/deionized water and bring volume to 100.0mL. Mix thoroughly. Filter-sterilize.

Preparation of YI Base Stock: Add components to 600.0mL of distilled/deionized water. Mix thoroughly. Bring volume to 780.0mL with distilled/deionized water. Adjust pH to 6.8 with 1*N* NaOH. Distribute in 78.0mL aliquots to 100.0mL screw-capped bottles. Autoclave for 15 min at 15 psi pressure–121°C. Cool to room temperature.

Preparation of YI Broth: Aseptically add 10.0mL of 10× glucose buffer stock to 78.0mL of cooled YI base stock. Adjust osmolarity with NaCl to 380.0milliosmols/kg

Preparation of Medium: Aseptically add 2.0mL of vitamin mixture 18 and 10.0mL of heat-inactivated bovine serum to 88.0mL of YI broth. Distribute in 13.0mL aliquots to 16 x 125mm screw-capped test tubes. Store at 4°C in the dark with the caps screwed on tightly. Use within 96 hr.

Use: For the cultivation of *Entamoeba* species.

YMA Agar

Composition per liter:

Agar	15.0g
Mannitol	10.0g
$CaCO_3$	4.0g
KH_2PO_4	0.5g
Yeast extract	0.4g
$MgSO_4 \cdot 7H_2O$	0.2g
NaCl	0.1g

pH 6.8 ± 0.2 at 25°C

Preparation of Medium: Add components to distilled/deionized water and bring volume to 1.0L. Mix thoroughly. Gently heat and bring to boiling. Distribute into tubes or flasks. Autoclave for 15 min at 15 psi pressure–121°C. Pour into sterile Petri dishes or leave in tubes.

Use: For the cultivation of *Rhizobium fredii, Rhizobium galegae, Rhizobium huakuii, Rhizobium leguminosarum, Rhizobium loti, Rhizobium meliloti, Rhizobium phaseoli,* and *Rhizobium trifolii.*

Yopp's Medium

Composition per liter:

NaCl	116.88g
$MgCl_2 \cdot 6H_2O$	10.68g
$MgSO_4 \cdot 7H_2O$	10.0g
KCl	2.0g
$CaNO_3 \cdot 4H_2O$	1.0g
Glycyl-glycine buffer	0.5g
$K_2HPO_4 \cdot 3H_2O$	0.065g
Ferric EDTA	5.0mg
Trace metal solution	1.0mL

pH 7.8 ± 0.2 at 25°C

Trace Metal Solution:

Composition per liter:

$MnCl_2 \cdot 4H_2O$	2.0g
H_3BO_3	0.5g
$ZnNO_3 \cdot 6H_2O$	0.5g
$Co(NO_3)_2 \cdot 6H_2O$	0.025g
$CuCl_2 \cdot 2H_2O$	0.025g
$Na_2MoO_4 \cdot 2H_2O$	0.025g
$VOSO_4 \cdot 6H_2O$	0.025g
HCl	3.0mL

Preparation of Trace Metal Solution: Add components to distilled/deionized water and bring volume to 1.0L. Mix thoroughly.

Preparation of Medium: Add components to distilled/deionized water and bring volume to 1.0L. Mix thoroughly. Gently heat and bring to boiling. Distribute into tubes or flasks. Autoclave for 15 min at 15 psi pressure–121°C.

Use: For the isolation and cultivation of halophilic cyanobacteria.

YP87 Medium

Composition per liter:

Na_2SO_4	4.0g
$NaHCO_3$	1.3g
KCl	0.5g
Yeast extract	0.5g
$MgCl_2 \cdot 6H_2O$	0.4g
NH_4Cl	0.25g
L-Ascorbic acid	0.2g
Na_2HPO4	0.2g
Sodium thioglycolate	0.2g
$CaCl_2 \cdot 2H_2O$	0.15g
Resazurin	1.0mg
Modified Wolfe's mineral solution	10.0mL
Wolfe's vitamin solution	10.0mL
Sodium lactate, 60% syrup	3.0mL

pH 7.5 ± 0.2 at 25°C

Modified Wolfe's Mineral Solution:

Composition per liter:

$MgSO_4 \cdot 7H_2O$	3.0g
Nitrilotriacetic acid	1.5g
NaCl	1.0g
$MnSO_4 \cdot H_2O$	0.5g
$CaCl_2$	0.1g
$CoCl_2 \cdot 6H_2O$	0.1g
$FeSO_4 \cdot 7H_2O$	0.1g
$ZnSO_4 \cdot 7H_2O$	0.1g
$AlK(SO_4)_2 \cdot 12H_2O$	0.01g
$CuSO_4 \cdot 5H_2O$	0.01g
H_3BO_3	0.01g
$Na_2MoO_4 \cdot 2H_2O$	0.01g
Na_2SeO_3	0.01g
$NaWO_4 \cdot 2H_2O$	0.01g
$NiCl_2 \cdot 6H_2O$	0.01g

Preparation of Modified Wolfe's Mineral Solution: Add nitrilotriacetic acid to 500.0mL of distilled/deionized water. Adjust pH to 6.5 with KOH. Add remaining components one at a time. Add distilled/deionized water to 1.0L. Adjust pH to 6.8.

Wolfe's Vitamin Solution:

Composition per liter:

Pyridoxine·HCl	10.0mg
p-Aminobenzoic acid	5.0mg
Lipoic acid	5.0mg
Nicotinic acid	5.0mg
Riboflavin	5.0mg
Thiamine·HCl	5.0mg
Calcium DL-pantothenate	5.0mg
Biotin	2.0mg
Folic acid	2.0mg
Vitamin B_{12}	0.1mg

Preparation of Wolfe's Vitamin Solution: Add components to distilled/deionized water and bring volume to 1.0L. Mix thoroughly.

Preparation of Medium: Prepare and dispense medium under 100% N_2. Add components, except L-ascorbic acid, $NaHCO_3$, and sodium thioglycolate, to distilled/deionized water and bring volume to 1.0L. Mix thoroughly. Adjust pH to 7.5. Gently heat and bring to boiling. Cool while sparging with 100% N_2. Add L-ascorbic acid, $NaHCO_3$, and sodium thioglycolate. Mix thoroughly. Sparge with 100% N_2. Distribute into tubes or bottles. Autoclave for 15 min at 15 psi pressure–121°C.

Use: For the cultivation of *Thermodesulfovibrio yellowstonii.*

YPSC Agar (Yeast Extract Peptone Sulfate Cysteine Agar)

Composition per liter:

Agar 15.0g
Yeast extract 1.0g
Peptone 1.0g
Sodium acetate·$3H_2O$ 0.5g
$MgSO_4 \cdot 7H_2O$ 0.25g
$CaCl_2 \cdot 2H_2O$ 0.25g
L-Cysteine·HCl·H_2O 0.05g
pH 7.5 ± 0.2 at 25°C

Preparation of Medium: Add components to distilled/deionized water and bring volume to 1.0L. Mix thoroughly. Gently heat and bring to boiling. Distribute into tubes or flasks. Autoclave for 15 min at 15 psi pressure–121°C. Adjust pH to 7.5 with sterile 10*M* NaOH. Pour into sterile Petri dishes or leave in tubes.

Use: For the cultivation and maintenance of *Bdellovibrio* species.

YPSC Agar, Cation-Supplemented

Composition per liter:

Sodium acetate·$3H_2O$ 50.0g
Agar 15.0g
Peptone 10.0g
Yeast extract 10.0g
$MgSO_4 \cdot 7H_2O$ 0.74g
$CaCl_2 \cdot 2H_2O$ 0.29g
L-Cysteine·HCl·H_2O 0.05g
Bacitracin solution 10.0mL

Bacitracin Solution:
Composition per 10.0mL:

Bacitracin 6,000U

Preparation of Bacitracin Solution: Add bacitracin to distilled/deionized water and bring volume to 10.0mL. Mix thoroughly. Filter sterilize.

Preparation of Medium: Add components, except bacitracin solution, to distilled/deionized water and bring volume to 990.0mL. Mix thoroughly. Gently heat and bring to boiling. Autoclave for 15 min at 15 psi pressure–121°C. Cool to 45°–50°C. Aseptically add sterile bacitracin solution. Mix thoroughly. Pour into sterile Petri dishes or distribute into sterile tubes.

Use: For the cultivation and enumeration of *Bdellovibrio* species.

YPSC Medium (Yeast Extract Peptone Sulfate Cysteine Medium)

Composition per liter:

Yeast extract 1.0g
Peptone 1.0g
Sodium acetate·$3H_2O$ 0.5g
$MgSO_4 \cdot 7H_2O$ 0.25g
$CaCl_2 \cdot 2H_2O$ 0.25g
L-Cysteine·HCl·H_2O 0.05g
pH 7.5 ± 0.2 at 25°C

Preparation of Medium: Add components to distilled/deionized water and bring volume to 1.0L. Mix thoroughly. Distribute into tubes or flasks. Autoclave for 15 min at 15 psi pressure–121°C. Adjust pH to 7.5 with sterile 10*M* NaOH.

Use: For the cultivation and maintenance of *Bdellovibrio* species.

YSP Agar

Composition per liter:

Sucrose 20.0g
Agar 12.0g
Peptone 10.0g
Yeast extract 5.0g
pH 7.0 ± 0.2 at 25°C

Preparation of Medium: Add components to distilled/deionized water and bring volume to 1.0L. Mix thoroughly. Gently heat and bring to boiling. Distribute into tubes or flasks. Autoclave for 15 min at 15 psi pressure–121°C. Pour into sterile Petri dishes or leave in tubes.

Use: For the cultivation of *Xanthomonas albilineans.*

YTG Medium

Composition per liter:

Tryptone 10.0g
Na_2CO_3 .... 5.3g
Yeast extract 5.0g
$Na_2HPO_4 \cdot 2H_2O$ 0.356g
L-Cysteine·HCl 0.2g
$Na_2S \cdot 9H_2O$ 0.2g
KCl 0.075g
Resazurin 1.0mg
Glucose solution 20.0mL
pH 10.1 ± 0.2 at 25°C

Glucose Solution:
Composition per 20.0mL:

D-Glucose 3.0g

Preparation of Glucose Solution: Add glucose to distilled/deionized water and bring volume to 20.0mL. Mix thoroughly. Sparge with 100% N_2. Autoclave for 15 min at 15 psi pressure–121°C.

Preparation of Medium: Prepare and dispense medium under 100% N_2. Add components, except glucose solution, to distilled/deionized water and bring volume to 980.0L. Mix thoroughly. Sparge with 100% N_2 for 30 min. Anaerobically distribute 9.8mL volumes into tubes. Autoclave for 15 min at 15 psi pressure–121°C. Aseptically and anaerobically add 0.2mL of sterile glucose solution to each tube. Adjust pH to 10.1 with sterile anaerobic 3*N* NaOH solution.

Use: For the cultivation of *Clostridium paradoxum* and *Clostridium thermoalcaliphilum.*

YTSS Medium, Half-Strength (DSMZ Medium 974)

Composition per liter:

Sea salts 20.0g
Yeast extract 2.0g
Tryptone 1.25g
pH 7.0 ± 0.2 at 25°C

Preparation of Medium: Add components to distilled/deionized water and bring volume to 1.0L. Mix thoroughly. Distribute into tubes or flasks. Autoclave for 15 min at 15 psi pressure–121°C.

Use: For the cultivation of *Roseovarius nubinhibens,* a budding bacterium from a hypersaline lake, and *Silicibacter pomeroyi,* a marine bacterium.

Z Medium

Composition per liter:

Casein hydrolysate	10.0g
NaCl	10.0g
Yeast extract	5.0g
Glucose	1.0g
$CaCl_2 \cdot 2H_2O$	0.367g

Preparation of Medium: Add components to distilled/deionized water and bring volume to 1.0L. Mix thoroughly. Distribute into tubes or flasks. Autoclave for 15 min at 15 psi pressure–121°C.

Use: For the cultivation of *Alcaligenes eutrophus (Ralsonia eutropha),* a hydrogen-producing phytopathogen.

ZF2 Medium (DSMZ Medium 943)

Composition per liter:

$NaHCO_3$	3.8g
Yeast extract	3.0g
$(NH_4)HCO_3$	0.45g
$MgSO_4 \cdot 6H_2O$	0.13g
$CaCl_2 \cdot 2H_2O$	0.12g
Resazurin	0.5mg
Phosphate buffer	10.0mL
Glycine solution	5.0mL
Arginine solution	5.0mL
$Na_2S \cdot 9H_2O$ solution	5.0mL
Dithionite solution	1.0mL
Seven vitamin solution	1.0mL
Trace elements solution SL-10	1.0mL
Selenite-tungstate solution	1.0mL

pH 7.3 ± 0.2 at 25°C

Arginine Solution:
Composition per 10.0mL:

Arginine-HCl	3.5g

Preparation of Arginine Solution: Add arginine-HCl to distilled/deionized water and bring volume to 10.0mL. Mix thoroughly. Filter sterilize.

Glycine Solution:
Composition per 100.0mL:

Glycine	15.0g

Preparation of Glycine Solution: Add glycine to distilled/deionized water and bring volume to 100.0mL. Mix thoroughly. Sparge with 100% N_2. Autoclave for 15 min at 15 psi pressure–121°C. Cool to room temperature.

$Na_2S \cdot 9H_2O$ Solution:
Composition per 10mL:

$Na_2S \cdot 9H_2O$	0.3g

Preparation of $Na_2S \cdot 9H_2O$ Solution: Add $Na_2S \cdot 9H_2O$ to distilled/deionized water and bring volume to 10.0mL. Mix thoroughly. Autoclave under 100% N_2 for 15 min at 15 psi pressure–121°C. Cool to room temperature.

Dithionite Solution:
Composition per 10mL:

Na_2-dithionite	0.25g

Preparation of Dithionite Solution: Add Na_2-dithionite to distilled/deionized water and bring volume to 10.0mL. Mix thoroughly. Autoclave under 100% N_2 for 15 min at 15 psi pressure–121°C. Cool to 25°C.

Seven Vitamin Solution:
Composition per liter:

Pyridoxine hydrochloride	300.0mg
Thiamine-HCl·$2H_2O$	200.0mg
Nicotinic acid	200.0mg
Vitamin B_{12}	100.0mg
Calcium pantothenate	100.0mg
p-Aminobenzoic acid	80.0mg
D(+)-Biotin	20.0mg

Preparation of Seven Vitamin Solution: Add components to distilled/deionized water and bring volume to 1.0L. Sparge with 100% N_2. Mix thoroughly. Filter sterilize.

Phosphate Buffer:
Composition per liter:

$Na_2HPO_4 \cdot 12H_2O$	43.0g
KH_2PO_4	5.44g

Preparation of Phosphate Buffer: Add components to distilled/deionized water and bring volume to 1.0L. Mix thoroughly. Adjust pH to 7.3. Autoclave for 15 min at 15 psi pressure–121°C. Cool to 50°C.

Trace Elements Solution SL-10:
Composition per liter:

$FeCl_2 \cdot 4H_2O$	1.5g
$CoCl_2 \cdot 6H_2O$	190.0mg
$MnCl_2 \cdot 4H_2O$	100.0mg
$ZnCl_2$	70.0mg
$Na_2MoO_4 \cdot 2H_2O$	36.0mg
$NiCl_2 \cdot 6H_2O$	24.0mg
H_3BO_3	6.0mg
$CuCl_2 \cdot 2H_2O$	2.0mg
HCl (25% solution)	10.0mL

Preparation of Trace Elements Solution SL-10: Add $FeCl_2 \cdot 4H_2O$ to 10.0mL of HCl solution. Mix thoroughly. Add distilled/deionized water and bring volume to 1.0L. Add remaining components. Mix thoroughly. Sparge with 80% N_2 + 20% CO_2. Autoclave for 15 min at 15 psi pressure–121°C.

Selenite-Tungstate Solution:
Composition per liter:

NaOH	0.5g
$Na_2WO_4 \cdot 2H_2O$	4.0mg
$Na_2SeO_3 \cdot 5H_2O$	3.0mg

Preparation of Selenite-Tungstate Solution: Add components to distilled/deionized water and bring volume to 1.0L. Mix thoroughly. Sparge with 100% N_2. Filter sterilize.

Preparation of Medium: Prepare and dispense medium under 80% N_2 + 20% CO_2 gas mixture. Add components, except phosphate buffer, glycine solution, arginine solution, dithionite solution, seven vitamin solution, and $Na_2S \cdot 9H_2O$ solution, to distilled/deionized water and bring volume to 973.0mL. Mix thoroughly. Equilibrate with 80% N_2 + 20% CO_2 to reach pH 7.3. Autoclave for 15 min at 15 psi pressure–121°C. Cool to 25°C. Aseptically and anaerobically add 10.0mL phosphate buffer, 5.0mL glycine solution, 5.0mL arginine solution, 1.0mL dithionite solution, 1.0mL seven vitamin solution, and 5.0mL $Na_2S \cdot 9H_2O$ solution. Mix thoroughly. Aseptically and anaerobically distribute into sterile tubes or flasks.

Use: For the cultivation of *Sedimentibacter saalensis* from freshwater sediment.

Zoogloea Medium

Composition per liter:

Agar	15.0g
Pancreatic digest of casein	5.0g
Glycerol	5.0g
Yeast autolysate	1.0g
Sodium lactate	0.5g

Preparation of Medium: Add components to distilled/deionized water and bring volume to 1.0L. Mix thoroughly. Gently heat and bring to boiling. Distribute into screw capped test tubes. Autoclave for 15 min at 15 psi pressure–121°C. Pour into sterile Petri dishes or leave in tubes.

Use: For the cultivation and maintenance of *Zoogloea ramigera* and other *Zoogloea* species.

Zymomonas Agar

Composition per liter:

Glucose	20.0g
Agar	15.0g
Yeast extract	5.0g

Preparation of Medium: Add components to distilled/deionized water and bring volume to 1.0L. Mix thoroughly. Gently heat and bring to boiling. Distribute into tubes or flasks. Autoclave for 15 min at 15 psi pressure–121°C. Pour into sterile Petri dishes or leave in tubes.

Use: For the cultivation and maintenance of *Zymomonas mobilis* and other *Zymomonas* species.

Printed and bound by CPI Group (UK) Ltd, Croydon, CR0 4YY
23/10/2024
01778250-0017